Handbook of Bird Biology

Second Edition

Sandy Podulka, Ronald W. Rohrbaugh, Jr.,
and Rick Bonney, Editors

Published by the Cornell Lab of Ornithology
in association with Princeton University Press

CORNELL LAB *of* ORNITHOLOGY
159 Sapsucker Woods Road • Ithaca, New York 14850 USA
www.birds.cornell.edu

CORNELL LAB *of* ORNITHOLOGY

159 Sapsucker Woods Road • Ithaca, NY 14850 • (607) 254–BIRD (2473) • www.birds.cornell.edu

A membership institution working to interpret and conserve the earth's biological diversity through research, education, and citizen science focused on birds.

The Cornell Lab of Ornithology's Handbook of Bird Biology, Second Edition

Published in association with Princeton University Press, 41 William Street, Princeton, New Jersey 08540 and, in the United Kingdom, 3 Market Place, Woodstock, Oxfordshire, OX20 1SY www.nathist.princeton.edu

Ronald W. Rohrbaugh, Jr., Project Manager

Sandy Podulka, Ronald W. Rohrbaugh, Jr., and Rick Bonney, Editors
Marie Read, Photo and Illustration Editor
N. John Schmitt, Illustrator
Diane L. Tessaglia-Hymes, Graphic Designer

Editorial Assistance: Daniel R. Otis, Marie Eckhardt, J. B. Heiser, and Terry Mingle
Design Assistance: Patty Porupski, Kathy Schaufler, Terry Mingle, and Richanna Patrick
Additional Illustrations: Christi Sobel
CD Narrator: Margaret A. Barker

Acknowledgments

In addition to the authors and other major contributors, we thank the following individuals for their help with the enormous task of completing this project:

Ken Able, Carl D. Barrantine, Louis B. Best, John Bower, Jack Bradbury, Greg Budney, Russ Charif, Natalie Demong, Kenneth P. Dial, Tim Dillon, Robert J. Dooling, Lang Elliott, Steve Emlen, Bill Evans, John Fitzpatrick, Adam Frankel, Kurt Fristrup, Melissa Fowler, Tim Gallagher, Harry W. Greene, John Greenly, Bob Grotke, Michael J. Hamas, Mark E. Hauber, John Hermanson, Leslie Intemann, Paul Kerlinger, George Kloppel, Stephen W. Kress, Donald E. Kroodsma, Kevin McGowan, Jason Mobley, Frank K. Moore, Eugene S. Morton, Drew M. Noden, Stephen Nowicki, Steve Pantle, Irene Pepperberg, Bill Podulka, Suzann Regetz, Ken Rosenberg, Karel A. Schat, Roger Slothower, Laura Stenzler, Peter Stettenheim, Sandra Vehrencamp, Peter Wrege

Cover Photographs

Front, left to right: Whooping Crane preening by Arthur Morris/Birds as Art; American Robin nest in crab apple by Marie Read; Lilac-breasted Roller feeding on grasshopper by Eberhard Brunner.
Spine: Prothonotary Warbler by Lang Elliott.
Back, left to right: Birders in Sapsucker Woods by Tim Gallagher; Berlin specimen of *Archaeopteryx*, courtesy of Humboldt Museum für Naturkunde, Berlin; Great-blue Herons nest building by Arthur Morris/Birds as Art.

Course production has been underwritten in part by an anonymous foundation and the National Science Foundation (ESI9618945).

Library of Congress Control Number 2004105098
ISBN 0-938-02762-X
British Library Cataloging-in-Publication Data is available

When citing this publication, please use the following format as a template. This example cites information presented in Chapter 9:

Temple, S. A. 2001. Individuals, populations, and communities: The ecology of birds. *In* Handbook of Bird Biology (S. Podulka, R. Rohrbaugh, Jr., and R. Bonney, *eds.*) The Cornell Lab of Ornithology. Ithaca, NY.

Preface

Welcome to the second edition of the Cornell Lab of Ornithology's Handbook of Bird Biology! Some readers may recognize the content of this book as the backbone for the second edition of the Lab's popular and long running Home Study Course in Bird Biology. While this book will continue to satisfy that role, thanks to the modified format and an innovative partnership with Princeton University Press, it will also serve as the most useful general ornithology reference currently available.

The continued expansion of ornithology as a premier scientific discipline and the rapid growth of bird watching as a recreational pastime indicate that the need for a modern, top-quality textbook and distance-learning course in ornithology never has been greater. The Cornell Lab of Ornithology is proud to be filling this niche. Our Home Study Course originally was conceived to provide comprehensive, college-level information on birds and their environments in a manner accessible to nonscientists and teachers the world over, and we hope that this book will be useful to everyone with an interest in birds.

The *Handbook of Bird Biology* and the Home Study Course build upon the highly successful first edition released in 1972, edited by well-known ornithologist and former director of the Cornell Lab of Ornithology, Olin Sewall Pettingill, Jr. Dr. Pettingill's groundbreaking course included nine "seminars," each focused on a different ornithological subject, and each was followed by an exam that students completed and returned to the Lab for grading. From 1972 to 1998, more than 10,000 students completed the course, a testament to its popularity and to Pettingill's skills as a writer, editor, and teacher.

Throughout the 1970s and '80s the course underwent several revisions to reflect new findings in ornithology and related disciplines. By the mid-1990s, however, minor revisions no longer were sufficient to reflect new and relevant ornithological findings in vibrant research areas such as animal communication, conservation biology, animal behavior, and evolutionary biology. Therefore, in 1998 the Lab of Ornithology temporarily stopped enrolling new students in the Home Study Course so that science education staff at the Lab could begin the challenge of overhauling the course from top to bottom and expanding its coverage by engaging some of the most knowledgeable professional ornithologists to bring the Home Study Course into the modern era of ornithology.

The second edition contains new text, photographs, illustrations, graphs, and tables. It covers all of the major topics addressed in the first edition, from anatomy and physiology to ecology and behavior. In addition, we've added a complete chapter on bird identification and another on conservation to make it more useful for birders and to introduce the relatively new science of conservation biology. We've also added a chapter on vocal communication, which, in keeping with the Lab's tradition of recording and producing media using high-quality animal sounds for education and research, is accompanied by an audio CD of bird vocalizations that are used to illustrate the many elements of bioacoustics.

We are confident that you will find the *Handbook of Bird Biology* to be an essential resource for all of your bird-related questions.

Home Study Course and Certificate of Completion

In addition to serving as a general ornithology reference, this book is meant to accompany the Home Study Course in Bird Biology (HSC) administered by the Lab of Ornithology Education staff. To successfully complete the course you must read each chapter at your own pace and complete an open-book exam (paper or online) for each of the 10 chapters. Exams are graded and returned so that students can review their answers and keep track of their performance. Questions and comments can be submitted to the course instructor, who will help guide students through the most challenging material. After completing all 10 chapters and exams with passing grades, you will receive a certificate of completion signed by the Louis Agassiz Fuertes Director of the Lab of Ornithology. We are constantly working to incorporate new information, supplemental materials, and distance learning tools on our web site to keep the HSC content current, provide additional resources, and create an environment that will keep students engaged and motivate them through completion and certification. Please visit our web site <**www. birds.cornell.edu/homestudy**> for the latest HSC information. To enroll call us at 800-843-BIRD (outside the United States at 607-254-2452), sign up on our web site, or mail the card inserted into this book.

Whether you acquired the *Handbook of Bird Biology* to use as a general ornithology reference or received it as a part of your enrollment in the Home Study Course, we wish you the very best in your quest for more knowledge and awareness about the many interesting facets that characterize the science of ornithology. We hope that this book provides you with an accessible gateway to the information you seek, and that it functions as a top-notch reference for years to come.

Table of Contents

CHAPTER 1 — INTRODUCTION: THE WORLD OF BIRDS

Sidebars

Chapter 2—A Guide to Bird Watching

CHAPTER 3 — FORM AND FUNCTION: THE EXTERNAL BIRD

SIDEBARS

CHAPTER 4 — WHAT'S INSIDE: ANATOMY AND PHYSIOLOGY

Sidebars

Chapter 5 — Birds on the Move: Flight and Migration

Sidebars

EVOLUTION OF BIRDS AND AVIAN FLIGHT

CHAPTER 6 — UNDERSTANDING BIRD BEHAVIOR

Sidebars

Chapter 7 — Vocal Behavior

Sidebars

CHAPTER 8 — NESTS, EGGS, AND YOUNG:
BREEDING BIOLOGY OF BIRDS

Sidebars

CHAPTER 9 — INDIVIDUALS, POPULATIONS, AND COMMUNITIES: THE ECOLOGY OF BIRDS

CHAPTER 10 — BIRD CONSERVATION

Sidebars

Birds and Humans:
A Historical Perspective

Sandy Podulka, Marie Eckhardt, and Daniel Otis

 The relationship between birds and humans undoubtedly began as soon as people appeared on the scene. By the time the first ancestral human beings appeared, some 14 million years ago, birds had been flying and running about for 136 million years. Modern humans with skeletons much like ours first appeared 125,000 years ago, and the countrysides where they lived were also home to birds. Humans evolved in a world saturated day and night, summer and winter, with birds.

Knowing and appreciating birds as we do today, we can readily surmise how strongly they impressed themselves on the minds of our ancestors—how their forms, colors, and sounds appealed to the senses, how their flight spurred imagination, how their periodic absences, timed to the seasons, aroused curiosity. They accordingly became a pervasive part of our heritage—a prehistoric and historic influence on our language and literature, our religion and mythology, our art and our music.

More recently, birds have been a focus of scientific inquiry, and it is the *science* of birds that is the focus of this course. Science, however, is a recent innovation. Only in the last few hundred years, and particularly in the last century, has it evolved as one of our most effective tools for understanding the world. Before this great burst of scientific discovery, birds and humans had coexisted for eons. Hunters, foragers, and farmers, observing the birds that passed through their lives, asked some of the same questions as modern scientists. Based on their own understanding of how the world works, they also gave their own answers. Often these explanations are charming and ingenious, reflecting an odd mix of careful observation and astute deductions with wild rumor, conjecture, and supernaturalism.

Typical of prescientific interpretations are many of the accounts in Aristotle's *Historia Animalium.* Aristotle (384–322 B.C.) knew, for instance, that Common Cuckoos of Europe lay their eggs in the nests of other birds, and that the consequences for the other nestlings were usually dire; in this he is in accord with modern science. He departs from the modern view, however, by attributing this practice to the cuckoo's wisdom in recognizing its own character defects: "This bird is pre-eminent among birds in the way of cowardice; it allows itself to be pecked at by little birds, and flies away from their attacks." (A modern scientist might suggest that the cuckoos were being driven off by parent birds trying to protect their own genetic inheritance.) "The fact is, the mother-cuckoo is quite conscious of her own cowardice and of the fact that she could never help her young one in an emergency, and so, for the security of the young one, she makes of him a… child in an alien nest." In other words, he believed that the bird behaves like any other responsible person who is unable to care for a child.

More accurately, Aristotle unambiguously asserts that certain birds migrate with the seasons, something European naturalists were still arguing about 2,000 years later. Some creatures stay put in winter, he says, but "others migrate, quitting…the cold countries after the autumnal equinox to avoid the approaching winter, and after the spring equinox migrating from warm lands to cool lands to avoid the coming heat. In some cases they migrate from places near at hand, in others they may be said to come from the ends of the world."

Although the myriad observations on bird life woven into every aspect of human culture rarely meet today's standards as objective "scientific" facts, they are fascinating in themselves. If they don't reveal reliable truths about how and why birds function as they do, they tell us about something equally real—the intimate, ancient, and continuing relationship between people and birds.

In the following chapters, we present the current scientific knowledge on bird anatomy, inside and out. We consider bird evolution and the geographic distribution of birds over the face of the earth. We explain bird movements during migration, their use of song, and many other aspects of their behavior. We look at their life cycles, nesting practices, and the threats birds face in a world utterly dominated by our species. But before we begin our science-based examination of birds, we want to briefly note the influence of birds in other realms of human culture. We also consider the origins and early development of the field that has become ornithology. The account, roughly chronological and roughly organized by theme, might be considered a brief history of ornithology's predecessors and offshoots. We can only cover a few of the major highlights, but even a scanty knowledge of the influence of birds on the diverse facets of human life can't help but enrich our scientific appreciation of birds.

Birds as Food

For early humans, the greatest attraction of birds was probably gustatory. Beside the recently discovered bones of a Neanderthal man, who lived some 50 to 90 thousand years ago, lay the bones of a Great Auk, possibly the remains of his last meal.

Although our common sense tells us that humans feasted on birds in prehistoric times, we have little tangible proof. Fragile bird bones do not fossilize well, especially on land, and the few scraps of bird bones that turn up in digs often rest in storage sheds, sometimes for years, while the paleontologists study larger, more dramatic finds.

In spite of this neglect, we know that early humans hunted birds and that their weapons became increasingly sophisticated, progressing from stones and clubs that they held in their hands, to sticks, darts, and spears that they threw, and on to snares, bolas, traps, and bows and arrows that required ingenuity to construct and skill to use **(Fig. H–1)**. At the same time, they must have relied, as primitive people do even today, on the eggs and nestlings of birds. Until recently, the Bushmen of the Kalahari Desert in South Africa gathered the eggs of weavers as part of their food. Flightless, molting waterfowl may have been another seasonal staple; they are easily captured if one knows their hiding places.

Humans eventually learned that taming and rearing birds ensured a reliable food supply, and several species were domesticated. In the Old World, specifically Asia, the most notable triumphs were the domestic chicken, duck, and goose. The chicken is probably a form of the Red Junglefowl (see Fig. 1–80), which still lives in the wild in southeastern Asia. Already domesticated in India as early as 3200 B.C., chickens appeared in Egypt about 1500 B.C. and somewhat later in Europe. The domestic duck, a form of Mallard, and the domestic goose, a form of the Greylag Goose of Eurasia, were domesticated at least as early as the chicken.

Figure H–1. Trapping Waterfowl for Food in Ancient Egypt: This wall painting from the tomb of the Egyptian Khnum-Hotpe (about 1900 B.C.) shows the operation of a trap designed to capture various species of geese, ducks, and grebes. Additional waterfowl, including recognizable pintails, flock nearby. In the surrounding shrubbery are a hoopoe, a redstart, a dove, and several shrikes. Photo: All rights reserved, Metropolitan Museum of Art (33.8.18).

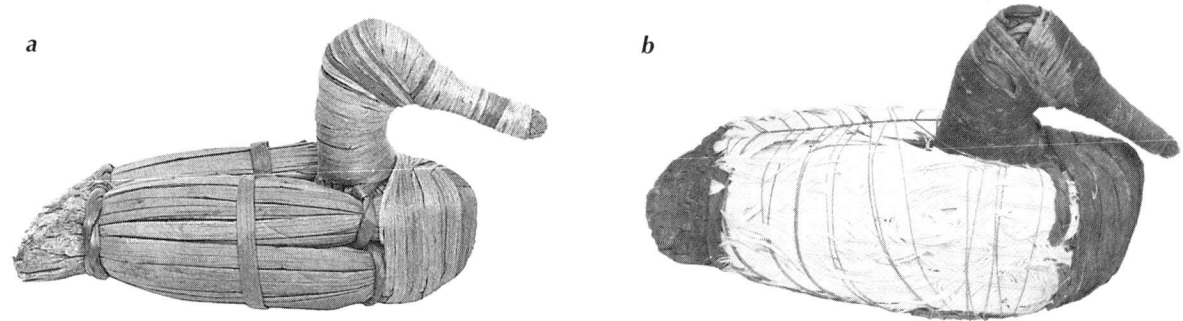

a   b

Figure H–2. Ancient Duck Decoys: *These waterfowl decoys, dating from 1000 A.D., were discovered in 1924 during an archaeological excavation of Lovelock Cave, Nevada. The decoys were formed of bulrush stems and either painted or stuffed.* **a.** *The body of the decoy was formed by binding together bundles of bulrush stems. Note the careful and accurate shaping of the head, which was constructed separately and sewn to the body.* **b.** *A painted decoy representing a Canvasback drake. Black and reddish-brown pigments were used to color the head, neck, and tail; then white feathers were tied to the body using fine hemp cord. "Stuffed" decoys (not shown) were topped with the stuffed head of an actual duck, and sometimes part of the duck's body skin was stretched over the bulrush body. From* Lovelock Cave, *by Llewellyn L. Loud and M. R. Harrington, 1929, published by the University of California Press.*

It is noteworthy that throughout much of history, most people were intimately acquainted with the habits of birds, albeit domestic ones, because most people lived in the countryside. Only in the past half-century has a chicken dinner meant a trip to the grocery store rather than a trip to the chicken coop. As our scientific understanding of birds proliferates as never before, people outside the ranks of bird watchers have less personal experience with birds than ever before.

In the New World, the Native Americans in southern North America domesticated the resident Wild Turkey, though no one knows how early. The Cortés expedition in 1519 found the domestic turkey already widely distributed in Mexico and Central America. Soon thereafter the explorers took these domesticated turkeys to Europe, and later the colonists brought them back to North America. Thus, the domesticated turkey that we eat at Thanksgiving came to us from southern North America by way of Europe; it is not a direct descendant of the Wild Turkeys the pilgrims found in the woods near Plymouth, Massachusetts.

In the New World outside of Mexico and Central America, people depended on wild birds much longer than in the Old World and became very skilled at capturing them. By 1000 A.D., the Native Americans were using decoys to attract ducks. Explorations of Lovelock Cave, Nevada, yielded two types of decoys ingeniously wrought of rushes, one of which had a stuffed skin for a head (**Fig. H–2**). From sites on St. Lawrence Island, Alaska, occupied from about 900 A.D. to the late 19th century, have come decoys of 45 identifiable bird species.

Figure H–3. Sitting Bull Wearing Feathered War Bonnet: *Photographed around 1885, the famous Sioux chief, Sitting Bull, is shown here wearing a war bonnet of eagle feathers. In traditional Native American society, the number of feathers in a chief's headdress symbolized his importance. Photo copyright Corbis.*

Use of Skins and Feathers

■ The people of many societies used the feathers and skins of wild birds for warmth, and some still do so today. Eagle feathers were units of currency in some North American Indian tribes. The Eskimos used waterfowl feathers for pillows and coverlets and skins for coats. In Greenland, the Eskimos made blankets from the skins of eiders. After

a

b

plucking the contour feathers to leave only the down, they sewed the skins together, edging them with the colorful feathers from the head of the male birds. Each blanket required more than 100 skins.

Bird skins and feathers have served as ornaments in many societies, including our own. The royalty in a number of New World Indian tribes wore capes, cloaks, and headdresses fashioned from feathers **(Fig. H–3)**. We are all familiar with the flowing headdress of eagle feathers, the glory of the Native American chief of the plains, at least in the movies. To the Crees, Cherokees, Natchez, Zunis, and many of the Great Plains tribes, the eagle was sacred. On the Missouri River in 1742, Pierre de la Vérendrye found Assiniboine Indians trading with the Mandans for deer skins "carefully dressed with fur and feathers" and for "painted feathers;" he noticed also that the Mandans "worked very delicately in hair and feathers."

Parrot feathers were highly esteemed by the prehistoric pueblo dwellers of New Mexico, who traded for them with peoples from the south. Macaw feathers, especially, seem to have been a status symbol. The aborigines of Australia wore feathers of cockatoos and Emus in their hair until recently. The Alaskan Eskimos and the Aleuts decorated their rain garments with tufts of brightly colored feathers sewn in parallel seams; the Pomo Indians decorated their baskets with feathers.

In the Pacific, the Hawaiians made exquisite cloaks from the feathers of honeycreepers; samples can be seen in the Bernice P. Bishop Museum in Honolulu **(Fig. H–4)**. The natives of Borneo made similar cloaks from the feathers of the Rhinoceros Hornbill (see Fig. 1–82). And to the north, the Chinese fashioned jewelry from feathers.

In the United States in the late 19th century, plumes were a fashionable appendage to women's hats, which led to the decline of egret

Figure H–4. Hawaiian Feather Cloaks: *Elaborate feather work was already being used in ceremonial garments for high-ranking Hawaiian chiefs when Captain Cook visited the Hawaiian Islands in the late 18th century. The red and yellow feathers of Hawaiian honeycreepers and honeyeaters—small, endemic forest birds—were especially revered. Most important were the completely red Iiwi and Apapane, and two black species, the Hawaii Oo and the Hawaii Mamo, which have yellow feathers under their wings, thighs, or tail. During harvesting, the latter two species were freed after removing their sparse yellow feathers. a. These full-length cloaks of red and yellow feathers, made in the early 1800s, once belonged to the chiefs Kiwalao, Kalaniopuu, and Kamehameha. The cloak on the far right is constructed from approximately half a million Hawaii Mamo feathers, representing 80,000 to 90,000 birds. b. A tattooed officer wears a feathered cloak and headdress in this engraving from 1817. Both photos courtesy of the Bishop Museum, Honolulu, Hawaii.*

Figure H–5. Woman Wearing Hat Decorated with Egret Plumes: *Photographed in Manhattan around 1886, a woman wears a hat typical of the fashions of late 19th century America and Europe. Obtaining plumes for the millinery trades of London, Paris, and New York became a lucrative occupation. Plume harvesting ended in North America when the public was made aware of the numbers of herons, egrets, and other wading birds being killed for their feathers in Florida rookeries. The resulting outcry led to the beginnings of the conservation movement with the establishment in 1896 of the first Audubon Society in Massachusetts (see Chapter 10). Photo courtesy of the National Audubon Society.*

and other wading bird populations **(Fig. H–5)**. One of the strands of early conservation movements coalesced around opposition to this capricious waste of bird life.

Birds in Literature, Culture, and Religion

Art

If birds have been important materially, as tangible creatures, in our diets and as sources of materials for warmth and decoration, the *idea* of the bird has perhaps been just as important from our prehistoric beginnings. The paintings made by the Cro-Magnon people on the walls of Lascaux Cave near Montignac, France, include an occasional bird among the other animals. In one of these paintings made about 17,000 years ago, one of the few humans depicted in cave art has a mask or birdlike face; nearby is a long-legged bird, or perhaps a bird at the top of a stick or spear-throwing device **(Fig. H–6)**. Most authorities agree that the paintings, because of their secluded location, probably had a sacred or ritualistic purpose rather than being merely decorative. The same may be true of the drawings of human beings, birds, and other animals done about 6,000 years ago on the walls of Tajo Segura in Spain. The Lascaux and other paintings suggest that even at this early date, birds may have assumed some kind of symbolic meaning in the human mind, and the pictures represent more than they actually portray.

In the historic period, birds are present in the visual art of virtually every culture in every part of the world. In the Arctic, the Eskimos carved birds in walrus ivory. For the Senufo of the western Sudan, a bird was a tribal

Figure H–6. The Bird-man of Lascaux: *In this detail of a well-known Paleolithic cave painting from Lascaux Cave in southern France, a bird-headed man (perhaps wearing a bird-shaped mask) is depicted with a charging bison. Nearby is a bird, variously interpreted as being long-legged or perching on a pole or spear-throwing device. Photo by Charles and Josette Lenars / Corbis.*

emblem used to decorate masks and other objects. The court painters of Mughal India painted birds exquisitely, both as incidental elements in court and landscape scenes and as the main subjects in natural history paintings. The Aborigines of Australia painted Emus on rocks. The standards of Roman armies bore the image of eagles. Birds decorated the gold cups of kings in Mycenae in ancient Greece and the musical instruments of the Anatolian kingdoms of what is now Turkey. Stone-carved birds sit in stone-carved trees in Angkor Wat in Thailand. The image of the eaglelike "thunderbird" was often portrayed by the Native Americans of North America, to whom it was an important element of mythology, capable of producing rain, thunder, and lightning **(Fig. H–7)**. The Navaho could make an owl playing the cat's cradle game with string; the Pomo of California could construct a hummingbird. Bird motifs appear often in the art of the Inca and Aztec. From China, an ancient ceramic pot almost 5,000 years old depicts a stork holding a fish at the tip of its bill. By 3000 B.C., the Egyptians began

Figure H–7. Thunderbird Totem Pole: *The eaglelike thunderbird is an important mythological figure to native cultures throughout North America. They believe it has the power to produce rain, thunder, and lightning. Certain Native American groups of northwestern North America believe the thunderbird created the world, and its carved image appears frequently on totem poles and masks. Here, a thunderbird sits atop a totem pole in Stanley Park, Vancouver, British Columbia. Photo by The Purcell Team / Corbis.*

Figure H–8. Shrike and Bamboo: *Painted by Li An-chung, an artist at the Hsuan-ho Painting Academy during the reign of Emperor Hui-Tzung of the North Sung Dynasty, "Shrike and Bamboo" is typical of the "bird-and-flower" genre of Chinese Painting. The bird is re-created in exquisite detail in a simple setting of bamboo leaves and branches. The medium is color and ink on silk. Painting courtesy of The National Palace Museum, Taipei, Taiwan, Republic of China.*

using paintings and sculptures of birds to adorn the friezes of buildings and the tombs of royalty. One picture from a tomb depicts five species of geese, three so faithfully drawn as to be identifiable.

Noteworthy among the world's ornithological art are the bird paintings of China. Even a thousand years ago they were so popular that they constituted an entire genre known as "bird-and-flower painting" **(Fig. H–8)**. Although the artists took pains to observe birds with great care and paint them with extraordinary accuracy, the paintings often had a symbolic dimension as well, different birds representing certain Buddhist or Confucian qualities and ideals. Far from being stiff, analytical case studies, these paintings present lively, graceful, harmonious scenes of perhaps slightly idealized birds in natural surroundings. The serene portrayal of birds in a carefully selected context of a few flowers, vines, grasses, and rocks continues today as a motif in Chinese painting.

The array of uses to which birds have been put in Western art is much broader. Of the legions of Western artists who have portrayed birds in some fashion, we can only mention a few. The depictions reflect an extraordinary range of human ingenuity. Some painters have been intrigued by the birds themselves and portray them with varying degrees of realism, and sometimes birds are incidental elements in landscapes, as in Breughal's "The Return of the Hunters." Albrecht Dürer had clearly studied bird wings carefully; the wings of some of the angels and cherubim floating about in his paintings are strikingly realistic. In the United States, among those who have painted birds is Andrew Wyeth, who painted two dead crows dangling against a white woodshed wall in bright sunlight; he seems equally interested in the abstract pattern of the corpses against the wall and in the crows as dead creatures.

Other painters use birds more as a means to an end, to express the multitude of qualities we associate with them. Brancusi's highly abstract "Bird in Space" has an aerodynamic grace. In a few Georgia O'Keefe paintings, crows appear as black blurs slicing through the sky. The birds in Van Gogh's last painting, "Wheat Field with Crows," seem rather ominous. Birds may also appear as stylized, iconic elements, as in Paul Klee's "Landscape with Yellow Birds" and "Twittering Machine." Finally, in some paintings the fact that it is a bird that is being portrayed may be almost incidental—it is the form that interests the painter. Or, the use of a bird may be deliberately incongruous, as in the work of some of the surrealists.

Figure H–9. The Phoenix: The phoenix, a fabulous mythical bird that undergoes regular rebirth by fire, appears in the myths of numerous human cultures throughout history. Its physical appearance and story vary greatly among cultures. According to one current version of the myth, every 500 years the phoenix, sensing its own death approaching, builds its own funeral pyre and sets fire to itself, being reborn out of its own ashes. Its first appearance is in ancient Egypt in the guise of the sacred, heron-like "Benu" bird, which re-creates itself daily in the rays of the rising sun. The Chinese version of the phoenix, often seen in paintings, is a fanciful creature called the "Feng-Huang," whose elaborate plumage resembles that of a peacock. In western cultures, the phoenix is an eaglelike bird, often depicted in medieval illustrations or on coats of arms as arising out of flames. Christians adopted the phoenix as a symbol of resurrection and immortality. The symbolism of rebirth is important even in modern times: the illustration shown here is adapted from the logo of the Phoenix Assurance Company, symbolizing the need for fire insurance. From Symbols, Signs and Their Meaning and Uses in Design, Second Edition, by Arnold Whittick, 1971. Published by L. Hill, London.

Religion

From about 1300 B.C., the polytheistic religion of the Egyptians embraced both real and imaginary birds. To them the embodiment of the sun was the fanciful phoenix, a beautiful bird that burned itself up every five hundred years and rose from its own ashes youthful again, a symbol of immortality **(Fig. H–9)**. The Egyptians held the very real ibis so sacred that in many a royal tomb they buried an embalmed ibis wrapped in cloth and painted.

Modern historians regard the first events of the Bible as taking place in about the 20th century B.C. Almost 40 species of birds are mentioned—the doves most frequently—in ways that show a knowledge of their habits and an appreciation for their grace and beauty. Indeed, it doesn't seem farfetched to suggest that the idea of the angel owes something to observations of birds, which often appear as symbols of transcendence and as messengers from the gods.

Like Aristotle in our earlier anecdote, the writer in Job 39:13–17 sees accurately but reports somewhat anthropomorphically: "The wings of the ostrich wave proudly; but are they the pinions and plumage of love? For she leaves her eggs to the earth, and lets them be warmed on the ground, forgetting that a foot may crush them, and that the wild beast may trample them. She deals cruelly with her young, as if they were not hers; though her labor be in vain, yet she has no fear; because God has made her forget wisdom, and given her no share in understanding."

In the religious festivals of many peoples throughout the ages, feathers were used as symbols, and participants often wore feather gar-

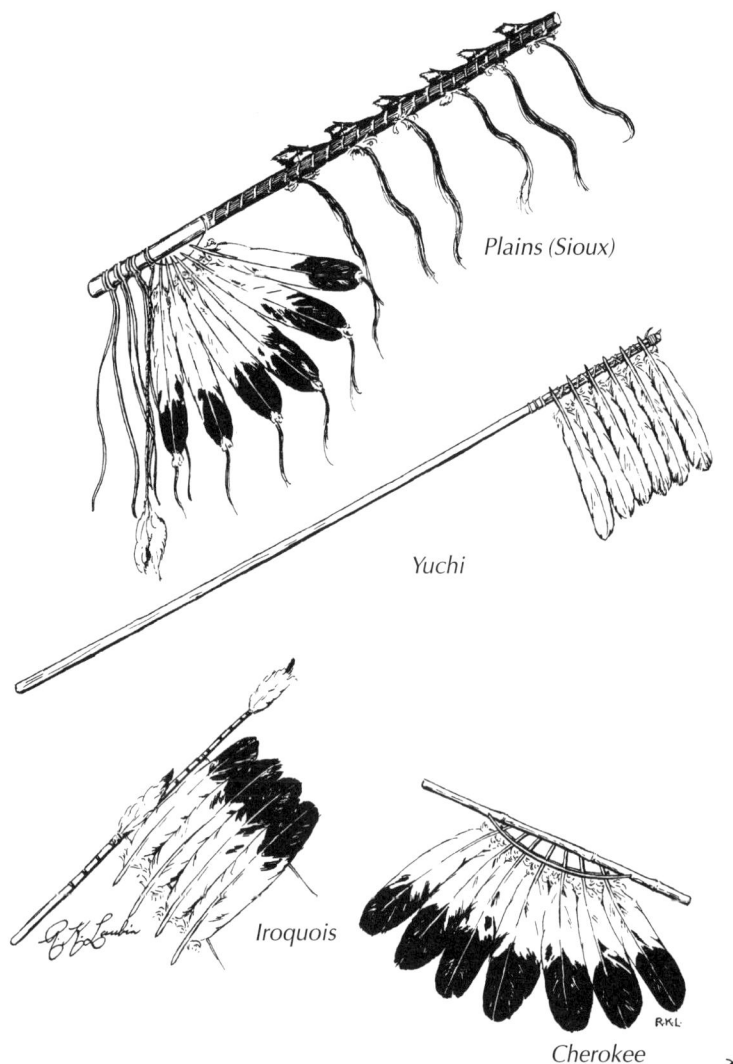

Plains (Sioux)

Yuchi

Iroquois

Cherokee

Figure H–10. Types of Wands used in Native American Eagle, Calumet, and Feather Dances: *Calumet (ceremonial pipe) and Eagle Dances were used by many Native American tribes to greet strangers, to create ceremonial friendships, to bring success in hunting or war, to bring good luck or counter bad luck, to cure sickness, or to make peace between warring tribes. The original calumet was a pipe decorated with a fan of eagle feathers. Over time, it became merely a wand with feather decorations, as shown by the four examples here. The Plains (Sioux) wand is also decorated with Pileated Woodpecker scalps.* From Indian Dances of North America *by Reginald and Gladys Laubin. Copyright 1977 by the University of Oklahoma Press. Reprinted with permission.*

ments on these special occasions **(Fig. H–10)**. The Plains Indians carried dried bird skins in their medicine bundles and decorated the calumet, the sacred "peace pipe," with feathers, heads, or skins —red for war, white for peace. The Pueblo Indians fastened feathers to their wands and prayer sticks. Several North American tribes still use feathers of eagles and other raptors in ceremonial displays.

Folklore

Peoples whose cultures seem quite unrelated often have similar legends and folklore motifs. The Old Testament story of Noah sending forth a dove to find land, for instance, has many parallels, including the Delaware Indian belief that a loon led the survivors of a flood to dry land. On a similar theme, according to the Crow Indians of Montana, diving ducks sent by the Creator dove down under primeval waters and brought up mud to make the Earth.

Symbols, too, are often the same in different cultures. Large raptors are almost always symbols of power and strength. Thus, the eagle feather was fitting for the headdress of a Native American chief, and the bird itself was thought fitting to represent Zeus. Owls were symbols of mystery, appropriate companions of witches **(Fig. H–11)**. Because of the owl's ghostly habit of hooting at night, they portended disaster. On the other hand, the eye of an owl, worn about the neck, warded off evil spirits; a cure for failing eyesight was to eat the eye of an owl; owl soup was considered helpful with whooping cough. Owls also symbolized wisdom. Because of the binocular vision that gives the owl a knowing expression, a number of stories feature the "wise old owl."

All the nightjars, the family to which the Whip-poor-will, Chuck-will's-widow, and Common Poorwill (see Fig. 4–128) belong, are nocturnal and have very large eyes and low calls that in some societies spell doom. The call of the Whip-poor-will foretold a death to a New England settler. The calls of its relatives foretold the length of life to a South American Indian. Even today, the natives of certain parts of South America and the West Indies believe that the calls of the nightjars are the voices of lost souls or ghosts.

Like the owl, both the raven and crow have dual characters—good and evil. For the crow, there was a chant: "One for sorrow, two for mirth, three for marriage, four for birth." In most folklores, the black-plumaged birds are threatening, their once-white feathers blackened by the gods for their evil deeds. According to one legend, the gods caught the pure white raven stealing water and punished it by turning its feathers black. To some people, a raven's harsh call foretells death; to others it means that the bird is carrying a message to the gods. One has only to see a raven fly croaking into the distance to sympathize with early humans fearing that the raven had seen their misdeeds and was on its way to report them to the gods.

Throughout Europe, the noisy, conspicuous cuckoo forecast coming events—rain, the arrival of spring, or portents in the affairs of men. In Japan, outings to hear the first notes of the cuckoo were once a popular spring entertainment.

Other species of birds figuring prominently in folklore include the loon for madness, the dove for peace, the swan for pride, the goose for confusion, and the swallow for travel. As with the crow, a single magpie meant sorrow and two portended joy. Even the behavior of the bird was sometimes important—in Germany, a woodpecker flying to the right was a sign of good luck. Indeed, a bird's behavior could even influence the destiny of nations. Pliny the Elder (23–79 A.D.) tells of special Roman fighting cocks whose manner of eating grain was thought to foretell the likelihood of success of state endeavors and to influence whether they were actually undertaken.

In India, among the Jains, a sect noted for its extreme reverence for life in all its forms, it is considered an act of virtue to free a caged bird. Paradoxically, a trade has grown up of capturing birds and bringing them to the market so the Jains can free them. The Jains also operate a hospital for sick and injured birds; carnivorous birds such as hawks and owls are treated, but only as outpatients.

Folklore also recounts how certain birds got some of their characteristics. According to legend (see Turner 1985, p. 233, in Suggested Readings), the Red Crossbill (see Fig. 4–119) "twisted its beak on the nails of the Cross," and the European Robin (see Fig. 21b) "scorched its breast bright red while stealing fire from the sun." The Yellow-bellied Sapsucker is said to have "acquired its many hues as consolatory gifts from other birds because it had been too tipsy on birch juice to attend the official distribution of colors."

Figure H–11. Eastern Screech-Owl: *With their ghostly screeching and hooting calls given under cover of darkness, it is not surprising that owls have featured prominently in folklore through the centuries. They are often considered omens of bad fortune or symbols of mystery. Yet, because of their large eyes and intelligent expression, owls also symbolize wisdom. Photo courtesy of Mike Hopiak/CLO.*

Literature

At the foundation of many societies' cultural heritage are epic poems and stories based partly on ancient fact and passed on by word of mouth and later in writing. Some tales tell of the origin of the people, their early way of life, and their beliefs; others are amusing. Many include birds in one way or another.

By the 16th century B.C., the Greeks had a written language, partly hieroglyphic and partly alphabet. The *Illiad* and *The Odyssey*, the two

great works attributed to Homer, were written in the 9th century B.C. These epic poems poetically immortalized the Greek gods, many of whom were associated with birds—an owl with Athena, goddess of wisdom; a dove with Aphrodite, goddess of beauty; a falcon with Apollo, god of the sun; and an eagle with Zeus, king of the gods.

Aesop's fables, originating orally about the 6th century B.C. and appearing in early English in the 9th century A.D., contained many bird characters, including the cock, crane, swallow, stork, jay, peacock, goose, nightingale, eagle, and crow. Each fable had a moral to it, such as "Use your Wits" or "Little by Little Does the Trick" from "The Crow and the Pitcher." The story involves a crow, dying of thirst, who finds a pitcher with a little water in the bottom. He cannot reach far enough into the pitcher to get the water, so he drops pebbles into the pitcher, one by one, until the water is at the top. He then drinks the water and is saved (**Fig. H–12**).

Birds as a literary device to hold a group of stories together originated as early as the 11th century A.D., when "Suka Saptati," 70 tales of a parrot, appeared. In this Indian collection, translated as "The Enchanted Parrot," a parrot tells 69 tales poking fun at women to keep her philandering mistress from taking several lovers while her husband is away from home.

In a number of ancient tales, the authors used birds to explain natural phenomena. In the Middle East, for example, the Garuda, a giant bird of prey, carried the sun from east to west each day. Vikar, another bird of prey, stirred up the winds by beating its wings.

William Shakespeare (1564–1616) enlivened passage after passage in his plays with vivid bird analogies and metaphors. No one knows how much of his knowledge of birds stemmed from his reading and how much from boyhood memories of woodlands along the Avon. He did, however, draw from the Greek and Roman classics and medieval bestiaries, and from scriptures, fables, and epic poems. His birds, it would seem, are symbolic rather than real: the raven and crow meant blackness and evil; the dove and swan, whiteness and purity; the nightingale, evening; the lark, the dawn.

In later centuries birds remained an occasional focus of poets and novelists. Japanese haiku master Issa (1763–1827) felt a special affinity for sparrows. In Europe, we have, among many others, Shelley's "To a Skylark," Keats' "Ode to a Nightingale," Wordsworth's "The Sparrow's Nest," and Yeats' "Leda and the Swan." Poets of the New World were not slow to immortalize native birds—the Bobolink in Bryant's "Robert of Lincoln," the Black-capped Chickadee in Emerson's "The Titmouse," and the mockingbird in Whitman's "Out of the Cradle Endlessly Rocking." More recently, Wallace Stevens, in what may be his most famous poem, told of 13 ways of looking at a blackbird. Robert Frost's bird poems include "The Oven Bird," "A

Minor Bird," and "On a Bird Singing in Its Sleep." In "Dust of Snow," he wrote

The way a crow
shook down on me
a dust of snow
from a hemlock tree
has given my heart
a change of mood
and saved some part
of a day I had rued.

Many writings for children deal with birds, including Edward Lear's "The Owl and the Pussycat," Robert McClusky's *Make Way for Ducklings*, and E. B. White's classics *The Trumpet of the Swan* and *Stewart Little*. (Stewart himself is a mouse, but his romantic interest is a small songbird named Margalo.)

If we venture yet further into the realm of popular culture we again find swarms of birds—literally, in the case of Hitchcock's famous movie *The Birds*. And we mustn't forget Daffy and Donald, Foghorn Leghorn, Heckle and Jeckle, Roadrunner, and Tweetie-Bird, although we'll admit that they probably don't often attract the attention of serious ornithologists.

Music and Dance

Birds, with their tremendous diversity of plumage, displays, habits, songs, and calls, have inspired the imagination and wonder of people throughout time—and have thus been included in some form in many ceremonies, dances, and musical compositions.

Music and Dance of Indigenous Cultures

Music and dance are today considered forms of entertainment by most people, but to traditional native peoples, both past and present, they may be powerful forces in the human experience. Their power comes from the fact that music and dance are often important components of rituals and ceremonies that have great spiritual significance. Birds, because they were often seen as intermediaries between the physical and spiritual worlds, were often represented in many ceremonies. But the way many indigenous cultures think about their relationship with birds and the spiritual world is so different from the way western cultures view these things that it is extremely difficult to adequately explain their beliefs in this brief passage; indeed, most westerners may find the concepts nearly impossible to grasp without long-term study.

To many indigenous peoples, the land is a source of life. But it is more than just a source of food and livelihood, for there is a core perception that all aspects of existence—including spirituality, culture, and social life—are inseparably connected to the inanimate world of mountains, rivers, skies, and rocks and the living world of people, other animals, and plants. Spiritual meaning, including a connection

Figure H–13. Feathered Ceremonial Costumes of Indigenous Peoples: a. A Native American dance costume featuring an elaborate bustle is worn by a dancer performing at the Yakima Indian Nation Powwow in White Swan, Washington. Photo by Jay Syverson/Corbis. b. A New Guinean tribesman at Mount Hagen, Papua New Guinea wearing an ornate ceremonial headdress composed of many feathers of birds-of-paradise and other species. Photo by Quadrillion/Corbis.

a   b

to the past (ancestors), the present, and the future is found in many of the connections within this complex web. Ceremonies, rites, songs, and dances were created to maintain harmony and balance within this web—and these often incorporated birds because of their spiritual or ecological significance. Sometimes just the physical appearance of birds was imitated or represented, but at other times their displays, vocalizations, and other behaviors were incorporated as well. The specific beliefs, legends, and ceremonies are as varied as the cultures. Bird spirits were invoked to render strength, to provide guidance and wisdom, to heal the sick, to communicate with ancestors, to enhance fertility, and to provide food and rain. In the following paragraphs are just a few examples of the many ways birds were, and often still are, incorporated into the music, dance, and other types of ceremonies of indigenous peoples.

Birds frequently appeared in ceremonies in the form of feathers or physical images (real or mythical). Familiar to most people are the feathered headdresses and bustles that mimic the spread tail feathers of strutting birds **(Fig. H–13)**. Feathers are also used to adorn ceremonial objects such as the prayer sticks of the Zuni of Arizona and New Mexico. Each prayer stick is decorated with a specific type and number of feathers, each feather having a special meaning (Bol 1998). Bird images are re-created in ceremonial masks, totem poles, and other objects. The Tlingit people of Alaska, for example, often used the image of the raven because of its significance in their traditions. Many birds are represented as kachinas (dolls representing ancestral spirits) in Hopi and Pueblo tradition.

One bird that is represented in traditional native dances all across the North American continent is the eagle. To many North American tribes, it is the greatest and most powerful of birds, ruler of the air and the creatures in it; powerful, fierce, and fearless. The Iroquois Eagle Dance, still performed today, was used throughout time to bring luck

Figure H–14. The Seneca Eagle Dance: In this painting from around 1900 by Ernest Smith of the Tonawanda Reservation, New York, four dancers perform the Seneca Eagle Dance. Each wears an Iroquois-style costume, including leggings, a decorated breechcloth or kilt, armbands, and a close-fitting cap with a single eagle feather. Notice, also, the feathered wands. Members of the Seneca branch of the Iroquois Nation still perform an Eagle Dance to this day. Originally it was associated with peace and war, but in modern times it is performed to celebrate friendships, cure disease, bring rain, uplift the downhearted, and as a thanksgiving. Recent performances have had two dancers, but accounts from the early 1900s, such as this painting, show four dancers. Smithsonian Institution, Bureau of American Ethnology.

in hunting. Although it evolved from ceremonies about peace and war, it now serves to celebrate existing friendships, to heal illness, and as a giving of thanks **(Fig. H–14)**.

Because they noticed birds performing certain behaviors at the time of year when rain falls, many traditional peoples assumed that the birds were able to bring rain. Thus they developed dances to imitate these behaviors so that they, too, could bring rain. For example, the Tarahumara Indians of Mexico perform two springtime dances that imitate the strutting courtship ritual of the male Wild Turkey—a display that occurs during the rainy spring season. Similarly, the Zuni and Hopi of the southwestern United States try to bring rain by invoking the hummingbird as the mediator between humans and the gods (because hummingbirds migrate back from their wintering grounds and appear in the southwest during the rainy springtime).

Bird activities were also associated with hunting and food. In one ceremony in the Torres Strait, between New Guinea and Australia, dancers mimic the Torresian Imperial-Pigeon, which swings its head up and down during one of its displays **(Fig. H–15)**. The ceremony serves as a young man's initiation and recognition, as well as an appeal for an abundant food supply (the pigeon is an important source of food in the region).

Figure H–15. Torresian Imperial-Pigeon: In a ceremony in the Torres Strait between New Guinea and Australia, dancers imitate the head-bobbing displays of the Torresian Imperial-Pigeon. Detail of plate by Lilian Medland, from Birds of New Guinea, by Tom Iredale, 1956. Melbourne, Australia: Georgian House.

Birds with elaborate courtship displays were often imitated by traditional cultures. Of particular interest were birds that displayed in leks (see Ch. 6, Reproductive Behavior: Lek Polygyny): in these species, males gather at traditional sites and perform showy displays, competing with each other for the attentions of the females, who visit

a

Figure H–16. The Prairie-Chicken Dance: a. Ethnographers Reginald and Gladys Laubin studied and performed the dances of American Plains Indians from the 1930s through the 1950s. Here, Mr. Laubin performs the Prairie-Chicken Dance. Many tribes, including the La-kota, Oglala Sioux, Blackfeet, and Plains Cree, originally had dances to honor and imitate the springtime displays of prairie-chickens and other grouse. The Prairie-Chicken Dance, while entertaining, also celebrated the tribe's survival through the winter and the coming of spring. The dancer struts and jumps, stamping his feet, shaking his shoulders, and quivering his whole body, while uttering gut-tural, clucking calls. Photo from Indian Dances of North America *by Reginald and Gladys Laubin. Copyright 1977 by the University of Oklahoma Press. Re-printed with permission. b. During its courtship display the male Greater Prai-rie-Chicken performs moves similar to those of its human imitator. Photo cour-tesy of Mary Tremaine / CLO.*

the leks only to choose a male and mate. One group that mimics a lekking species is the Chukchi of Si-beria, who imitate the display of the Ruff. Male Ruffs gather in leks, take up stances about a foot apart, and silently confront each other in simulated attacks (see Fig. 6–40). In the Bavarian highlands, performers imi-tate the lekking display of the "Black Cock" (probably a local common name for either the Black Grouse or the Eurasian Capercaillie), in which the birds posture, dance, and engage in mock fights. In the traditional Bavarian dance called the *Schuhplattler*, a man jumps along behind his partner, clicking his tongue and clap-ping. Eventually he strikes the ground with one or both hands and bounds toward her with arms outspread or hanging down. Although the dance is now performed for recreation, its similarity to other bird dances in-dicates that it originated long ago from beliefs asso-ciating birds and the spiritual world (Armstrong 1975). The Blackfoot Indians and many other tribes from the plains of North America imitate the dance of the Sage Grouse—another lekking species—which engages in spectacular displays involving hundreds of birds. Many tribes from the North American plains also had a Prairie-Chicken Dance in which men imitated the energetic lekking display of male prairie-chickens **(Fig. H–16)**. The Jivaro Indians of Peru and Ecuador imitate the lekking antics of the Andean Cock-of-the-Rock (see Fig. 7–28). In addition, the Es-kimos had a Ruffed Grouse Dance in which the prin-cipal dancer would dance on his knees and imitate a male grouse drumming (see Ch. 7, Sidebar 1, Fig. A). Although Ruffed Grouse do not assemble in leks, their displays are quite impressive.

The courtship displays of cranes have attracted the attention of many cultures around the globe. Although they do not display in leks, male and female cranes of many species perform elaborate dances in spring either to strengthen an existing pair bond or to establish a new one (see Fig. 6–19). The circular movements of the dances of some spe-cies were associated with the seasonal movements of the sun, and in many cultures these movements came to represent fertility and death. For example, the Oskrits of Siberia perform a funeral dance wearing crane skins.

Bird vocalizations are of particular importance to people of the rain forests, and have been incorporated into many aspects of their cul-

Figure H–17. Temiar Trancing Ceremony: During trancing ceremonies of the Temiar people of Malaysia, percussive sounds are used to alter participants' emotional states. The pulsing call of the Golden-throated Barbet alternates between high and low notes, and is reminiscent of a beating heart. Hearing the bird calling in the forest invokes a feeling of longing in the Temiar. The ceremonial bamboo percussion instrument imitates this sound: two bamboo tubes of different lengths are struck alternately against a log: the shorter tube produces a higher frequency sound, the longer tube, a lower sound. The constant pulsing rhythm brings about a trance-like state in the participants. Here, female percussionists (right) play bamboo-tube stampers during a trancing ceremony behind a male medium (left). Inset shows a Golden-throated Barbet. Photo courtesy of Dr. Marina Roseman.

tures, including music and dance. For the Temiars of Malaysia, knowing a path through the jungle is essential to survival. To lose one's path can be fatal. This importance of "a path" is a metaphor for how they perceive themselves relative to their environment, their society, and the cosmos. Consequently, they believe that illness results when a person's soul gets waylaid. Treatment involves singing what's called a "way" (a song that represents a pathway), finding the soul, and leading it home. During ceremonies, musical sounds are used to modify the feelings of the participants in many different ways. One of the sounds invoked is the call of the Golden-throated Barbet. Its two-toned, pulsing sound brings about a sense of longing. The Temiar believe that it beats in rhythm with one's heart, moving the heart to long for a loved one or for a deceased relative. The rhythm is re-created during ceremonies by bamboo percussion and drums, and is used to bring about a trance state and connect with a spirit guide (Roseman 1991) **(Fig. H–17)**.

The Kaluli people of the tropical rain forest in Papua, New Guinea integrate birds into every aspect of their traditions, including their music and dance. Birds are considered spirit reflections of their dead; mediators between the living world and Kaluli who have died and reappeared in the form of birds. Birds are thus not seen simply as creatures separate from humans, and their sounds have profound effects on living people. The different types of birds represent spirits from different social categories (men, women, old, young) and from different temperaments (angry, docile, hostile, cranky, and so on). One example of the integration of birds and their sounds with Kaluli music and dance is the "weeping ceremony" of sorrow or loss. In these ceremonies the performer takes on characteristics of a fruit-dove, making

H·18
            Sandy Podulka, Marie Eckhardt, and Daniel Otis

H

melodic sounds based on the sorrowful call of these birds. Which of the many species of fruit-doves the weeper becomes depends on the degree of sadness (Feld 1990).

Music and Dance of Western Cultures

  Western cultures also incorporated birds into their music and dance in a variety of different ways, but to them birds were not as connected to the spiritual world, and not as intertwined with their very existence. In western music and dance, especially that of Europeans and Americans as discussed here, birds served more as the subject of the music, or as part of the scenery, rather than being the very essence of it.

  Western music most often involves birds by using instruments or voices to imitate their sounds—the sound of the European cuckoo (see below) is a favorite subject, as in the earliest known English secular piece, the 13th century canon "Sumer is icumen in." But sometimes a musical passage is designed to create a general impression of a certain bird, as in "The Peacock" from Ravel's "Histoires Naturelles." In this piece, the peacock's majestic yet arrogant manner is portrayed, as well as the spreading of the great tail (with opposing black key glissandi on the piano) and the harsh cry (with dissonance). Contemporary composers may use actual recorded bird sounds in their music; American composer James Fassett constructed his "Symphony of Birds" entirely from recorded sounds. Respighi (1879–1936) was among the first to use recorded sound: his tone poem "The Pines of Rome" included a gramophone recording of nightingale song.

  In many compositions, birds are the main focus, but in others, birds are used to set the scene or indicate time of day, particularly dawn or dusk. The familiar two-note call of the Eurasian Common Cuckoo (the model for the "cuckoo clock") is often used to set a pastoral scene (as in Beethoven's Sixth Symphony), since it is widely recognized and easy to imitate.

  Many birds besides the cuckoo have inspired composers through their sounds. Even though their song is complex and difficult to reproduce faithfully, nightingales are often imitated in music, as in Rameau's "Air du Rossignol" from his early 18th century opera "Hippolyte et Aricie." Doves are another favorite subject. Several species whose songs are less-often portrayed include the Scarlet Tanager, in Dvorak's "American String Quartet" and the Great Tit, in the 1st movement of Bruckner's Fourth Symphony. Bruckner strings together several Great Tit songs that sound very realistic, but yet are well-integrated with the music. Twentieth century French composer Olivier Messiaen used the calls of 260 species throughout his works. In "Reveille des Oiseaux" he used 21 species to represent a dawn chorus, and in "Des Canyons aux Etoiles" (1974) he used 60 species from Hawaii and Utah.

  Other birds have impressed composers with their form or movement. Flight, an inspiration to poets and writers as well as composers, lends itself well to musical interpretation. Benjamin Britten (1913–1976) used flutes to create swallow flight in one of his airs, and piccolos for the flight of the skylark in "A Midsummer Night's Dream." Vaughan Williams also portrayed the skylark's flight in "The Lark Ascending."

The rapid movement and twittering of goldfinches is portrayed by Vivaldi in his flute concerto "Il Gardellino." The graceful movement of swans is another common subject of musical compositions, most notably in Tchaikovsky's ballet "Swan Lake" **(Fig. H–18)**. The court-ship display of the lyrebird (see Fig. 1–88) even figures in at least one one-act ballet, "The Display," by Australian composer Malcolm Williamson (b. 1931).

Figure H–18. London City Ballet Production of Swan Lake: With their elegant forms and graceful movements, birds have often inspired western composers and choreographers. The elegant motions of swans were the inspiration for Tchaikovsky's well-known ballet Swan Lake. Photo by Robbie Jack / Corbis.

The Evolution of North American Ornithology

The Early Years: From Aristotle to the 17th Century

The first known scientific list of birds was compiled by Aristotle (384–322 B.C.). His descriptions of the habits of about 74 of the 170 species he recorded were good enough that we can recognize these birds today. Pliny the Elder (23–79 A.D.), a naturalist and army officer, carried on the tradition in the 1st century A.D. with his *Historia Naturalis*, an encyclopedia of natural science that included birds.

Archaeologists have uncovered masterful paintings of birds in the ruins of the ancient Roman city of Pompeii, destroyed in 79 A.D. **(Fig. H–19)**. The Roman Empire fell in 400 A.D., beginning the period of intellectual stagnation we now know as the Dark Ages—a time when the threat of the stake hung over all who dared to question, reason, or explore. Although moralizing distorted much of the human knowledge existing at this time, birds were not forgotten. Amid the bestiaries and allegories of this era, one very important book stands out: *The Art of Falconry*, by Frederick II of Hohenstaufen (1194–1250), who had the

Figure H–19. Birds in a Garden Mural from Pompeii: *Excavations at the city of Pompeii, which was buried and yet preserved by the volcanic eruption of Vesuvius in 79 A.D., have provided a window into ancient Roman life. Murals from the many walled gardens show how important nature was to the inhabitants. Shown are three panels (as a composite image) of one such garden mural. In the left panel, a jay is shown among plants such as lilies, poppies, morning glories, and a palm. The small center panel shows what has been described as a bustard. The right panel shows (from bottom to top) a Rock Partridge and a male and female Eurasian Golden Oriole. Photos courtesy of Foto Foglia, from The Gardens of Pompeii, by Wilhelmina Feemster A. Jashemski, 1993. Published by Aristide D. Cavatzas, New Rochelle, New York.*

courage to carefully analyze and experiment with the natural world. The book, in addition to details and directions for falconry, gives much factual information about birds and is the first scientific work known to contain bird illustrations **(Fig. H–20)**.

Until nearly the end of the Dark Ages, substantive material about birds had been scarce. With the coming of the Renaissance in the 14th century and the invention of the printing press in 1448, a surge of interest in birds swept western Europe. In the next two centuries, with new opportunities for inquiry and knowledge and for scientific and literary expression, birds received their share of attention. Three natural history encyclopedias, published in Zurich, Paris, and Bologna, recorded current information about birds, including the fanciful as well as the factual.

With the intellectual expansion came the urge to broaden physical horizons. Explorers set out from all parts of Europe. Many were bird watchers, albeit mostly for practical reasons. Navigators followed migrating birds to land, or they took land birds to sea and released them, hoping that the birds would lead them to new territories and new riches.

The use of birds in navigation was not new. The Norse tale of the discovery of Iceland in 874 A.D. described the practice of sending a bird from shipboard to find land. Earlier, Pliny had written of the Ceylonese

Figure H–20. Falconers with Falcons:
Shown here is a drawing from The Art
of Falconry *by Emperor Frederick II of
Hohenstaufen. Written in 1248, the
book provided detailed information on
the husbandry of falcons, and the techniques and trappings of falconry. This
important early book provided much
information about the natural history of
game birds and is the first scientific work
known to contain bird illustrations.*

doing the same in the 5th century B.C. The "Dialogues of the Buddha"
mention it as being a habit of the seagoing merchants. The Polynesians
may have been following the migration routes of birds, possibly the
Long-tailed Koel, when they journeyed to the Hawaiian Islands and
New Zealand.

Christopher Columbus crossed one migratory pathway of North
American birds at just the right time and followed the ever-moving
stream south to the Bahamas. In the journal of his first voyage in 1492,
he wrote:

> *October 8—There were many small land-birds and*
> *[the sailors] took one which was flying to the*
> *south-west. There were jays, ducks, and a pelican.*
> *October 9—All night [the sailors] heard birds passing.*

For over 300 years after Columbus' notation in his diary—probably the first written comment on North American birds—the knowledge of the birds of this continent came piecemeal. The colonists and
explorers, most of them untrained observers, spent more time in providing for their own survival than in noticing birds. When they did
observe birds, they thought of them in terms of their native European
species. The common name of the familiar American Robin is one example. The early settlers named it after the smaller, brighter, European
Robin, which is a very different bird **(Fig. H–21)**. The same is true for
New World cuckoos, named after the Common Cuckoo of Europe. The
American Redstart is also named after a similar European species.

Nevertheless, there are some early records: Cabeza de Vaca, in
Florida in 1528, probably in the state's northwestern lake country,
writes in his *La Relacion* (1542): "Geese in great numbers. Ducks,
mallards, royal ducks, flycatchers, night herons and partridges abound.
We saw many falcons, gerfalcons, sparrowhawks, merlins." On the

a   b

Figure H–21. European and American Robins: *Early European settlers in North America, not well-versed in avian taxonomy, tended to name the birds they encountered after Old World species that looked similar. For instance, the American Robin (**a**) was so called because its chestnut-red breast reminded them of the familiar European Robin from back home. The European Robin (**b**) is a compact, chatlike bird, which systematists currently place in the family Muscicapidae (Old World flycatchers). The American Robin is a larger and more strongly built bird in the family Turdidae (thrushes). Photo a courtesy of J. R. Woodward / CLO; photo b by John Cancalosi / VIREO.*

Pecos River in 1535, apparently near the Texas/New Mexico border, he feasted on quail brought in by native archers for the evening meal. Castenada, writing of the Coronado expedition of 1541 to 1542, observed turkeys and tame eagles in Arizona and cranes, wild geese, crows, and blackbirds in New Mexico, reports that must surely be among the first ornithological observations made in the continental United States. As early as 1585, John White, a member of Sir Walter Raleigh's second expedition to Roanoke Island, Virginia, painted a series of watercolors of American birds. The paintings, well drawn, accurately tinted, and more important, the results of first-hand observations, were first published in 1590 in de Bry's *Virginia*; they may be enjoyed today in all their glowing colors in Stefan Lorant's *The New World* (1946). Between 1630 and 1646, William Bradford, governor of the Plymouth colony, wrote that wild turkeys and waterfowl had been abundant in 1621, but that waterfowl decreased thereafter. In 1674, John Josselyn noted that young turkeys were once abundant in the woods, "But…the English and the Indians [have] now destroyed the breed, so that tis very rare to meet with a wild turkie in the woods."

Although the explorers and colonists sent "home" living and dead specimens of birds during this period, the investigation of American bird life had to wait for the further development of science in Europe. In 1678 appeared *The Ornithology of Francis Willughby*, by Willughby and his friend John Ray. This publication was a scientific landmark, the first major work based on careful observation of both the structures and habits of birds. The authors adopted the most valid classification and species concepts of the day, and the illustrations were far more accurate than any previously printed. In preparing the illustrations, the artists used engraved plates that permitted far more delicacy of detail than the woodcuts used previously in the early encyclopedias (**Fig. H–22**).

The 18th Century

With the 18th century came important American scientific events. From 1712 to 1719 and 1722 to 1726, Mark Catesby, a young Englishman, roamed the shores and woods of the southeastern colonies and islands. His ambitious plan was to paint, describe, and name every bird, and other vertebrates as well, that occurred "between the 30th and 45th degrees of latitude."

Short of funds, he published his book in sections from 1731 to 1743. Physically, the complete volume, *The Natural History of Carolina, Florida, and the Bahama Islands*, is luxurious, nearly 14 by 20 inches (35 by 50 cm) in size, with 220 hand-painted engravings, all but two of which Catesby did himself **(Fig. H–23)**. Scientifically, the illustrations and the accompanying discussions earned for the author the well-deserved title "Founder of American Ornithology." Catesby, for his time, was remarkably unbiased and insisted on verifying hearsay with personal observation. His drawings and descriptions were the basis for over one-third of the American birds that Linnaeus later described. Artistically, Catesby provided several innovations: he painted living birds, naturally posed in their native habitats. He also considered shading and composition, using bold patterns of light and dark in some of his paintings. The best were unsurpassed for over 100 years.

In 1758, Linnaeus published his *Systema Naturae*, which laid the groundwork for modern binomial nomenclature and kindled in travelers the desire to find and name different kinds of organisms, including birds.

Specimens of plants and animals flooded into Europe from the New World. Aviaries with pet birds from the colonies were much in vogue. In England, George Edwards, Thomas Pennant, and John Latham, important "armchair" ornithologists, compiled great volumes on birds from specimens of species they never saw alive. Their comments, however, based on second- or third-hand information, were not always reliable. Edwards' *A Natural History of Birds* (1743–1751) was worldwide in scope; Pennant's *Arctic Zoology* (1784–1785) dealt

Figure H–22. Waterfowl from The Ornithology of Francis Willughby: *Willughby was an English naturalist known especially for his early classification work on birds and fishes. He toured Europe with John Ray, collecting material for his book* Ornithologia *(1676, in Latin), translated into English by Ray as* The Ornithology of Francis Willughby *(1678). The illustrations, more accurate than any previously printed, were reproduced using engraved plates, permitting fine detail. Shown here is a plate of waterfowl that includes Canada Goose, Common Shelduck, and Black Scoter. Note that both the scientific and most frequently used common names of the latter two species have changed since Willughby's time. Courtesy of Blacker-Wood Library, McGill University.*

Figure H–23. Mockingbird by Mark Catesby: In the early 1700s, English naturalist and artist Mark Catesby set out to paint, describe, and name every bird and other vertebrate in the southeastern American colonies and islands, eventually to be published in his book The Natural History of Carolina, Florida, and the Bahama Islands. *His paintings, based on personal observation as well as specimens, presented birds naturally posed in native habitats, as exemplified by this lively mockingbird perched amidst dogwood. This illustration is Plate 27 in Volume I of Catesby's book, published in 1731.*

A

HISTORY

OF

BRITISH BIRDS.

THE FIGURES ENGRAVED ON WOOD BY T. BEWICK.

PART I.

CONTAINING THE

HISTORY AND DESCRIPTION OF LAND BIRDS.

NEWCASTLE:

PRINTED BY EDWARD WALKER, FOR T. BEWICK: SOLD BY HIM, AND
LONGMAN AND CO. LONDON.

1809.

Figure H–24. Title Page and Illustrations from A History of British Birds, by Thomas Bewick: *In the 1790s, Bewick developed a new printing method for illustrations. He used end-grain woodcuts, eliminating the need for expensive copper plates yet allowing fine detail. His 1809 A History of British Birds was the first bird book readily available to the general public. The illustrations show, from top to bottom, Common Cuckoo (the model voice for the popular cuckoo clock), European Goldfinch, and Northern Lapwing (called in the original by its colloquial name "Pee-Wit").*

primarily with North American birds; and Latham's *A General History of Birds* (1821–1828) included the first attempt to systematically treat all the birds of the world.

In the 1790s, Thomas Bewick devised a new method for producing illustrations by using end-grain woodcuts that increased detail to a degree formerly impossible in wood and eliminated the need for costly copper engravings. His *A History of British Birds* (1809) was the first well-illustrated bird book available to everyone (**Fig. H–24**).

In 1791 appeared *Travels Through North and South Carolina*, by William Bartram, a professional bird watcher and son of a famous Philadelphia naturalist. Although Bartram's notes on birds and their habits were, at times, more poetic than scientific, he stands out as one of the first Americans to contribute to ornithological knowledge. Louis Vieillot, a Frenchman, was the first to point out variations in plumage with season and sex in his *L'Histoire Naturelle des Oiseaux de L'Amerique Septentrionale* (1807). John Abbot, an Englishman, worked from 1804 to 1827 on several drafts of a manuscript that contained notes on life histories and illustrative paintings of the birds of Georgia. Unfortunately, none of his work has ever been published.

In the New World through the 18th and into the 19th centuries, most people, still recovering from the American Revolution and intent on establishing a new nation, were concerned with more practical affairs than studies of birds. Almost from the beginning, however, the eagle in one form or another was a symbol of the United States. The two-headed eagle, first proposed as the national symbol of the United States in 1776, resembled the imperial German eagle and was not accepted. Congress considered a number of other designs and finally, in 1782, chose a crested, stylized eagle for the official seal. Since that time, minor alterations in the seal have included the elimination of the bird's crest and modifications in form and color so that the bird on the seal today more closely resembles the Bald Eagle than did the original (**Fig. H–25**).

The selection of the Bald Eagle as the American symbol did not please everyone. Benjamin Franklin thought that the eagle "does not get his living honestly…too lazy to fish for himself, he watches the labor of the fishing-hawk, and when that diligent bird has at length taken a fish…the bald eagle pursues him and takes it from him." Franklin much preferred the Wild Turkey, which, though a little "vain and silly," was "a bird of courage." Today ornithologists try to avoid judging birds by human standards of morality.

Figure H–25. Great Seal of the United States: The eagle has been the official symbol of the United States since the nation's beginning. Its form, however, has varied over the years. In 1776, congress rejected the two-headed bird proposed as the national symbol, finally accepting a crested, stylized eagle in 1782. Since then, the crest has been omitted, and the shape and color modified so that today's symbol more closely represents an actual Bald Eagle. The eagle clutches arrows and an olive branch—symbols, respectively, of war and peace. The style of eagle currently used in the Great Seal, as shown here, was first adopted around 1885. The Great Seal is used in numerous official ways, such as to seal certain government documents, on certain letters and envelopes, and on the one-dollar bill. The lithograph shown here was created by Andrew B. Graham sometime around the turn of the 20th century. Courtesy of Library of Congress.

The 19th Century

In a century that included many gifted "literary scientists," Henry David Thoreau was perhaps the finest. Although his works were often

Figure H–26. John Burroughs and John Muir: As an essayist and naturalist, John Burroughs inspired a generation of American nature-lovers and conservationists in the late 19th and early 20th centuries. Nicknamed "John O' Birds," Burroughs (right) imbued his bird writings with emotional and poetic descriptions, while his excellent birding skills assured the accuracy of his observations. John Muir (left), shown here visiting Burroughs for the latter's 75th birthday celebration, was a naturalist, conservationist, and bird enthusiast from the western United States. Nicknamed "John O' Mountains," Muir became well known as a proponent of creating national parks and conserving forests. Photo courtesy of Library of Congress.

emotional and philosophical, he was an exceedingly acute and accurate observer, and he is considered by many to be the first truly great American nature writer. Thoreau had little interest in what we would call laboratory studies of birds. In his words, ornithology at its best was a "window opened wide to nature."

In the last half of the 19th century, John Burroughs **(Fig. H–26)**, a serious student of birds with a critical and inquiring mind, was perhaps the most widely read and loved of the nature writers in spite of the fact that he did not hesitate to personalize birds, as in his "Wake Robin." The John Burroughs Society, established in his honor, continues today and each year honors the author of the most outstanding nature book published in the preceding year.

At the end of the 19th century three authors of books on birds stand out. John Muir (see Fig. H–26), naturalist, conservationist, and also bird enthusiast, wrote about the Water-Ouzel in his book *The Mountains of California.* The approach of Bradford Torrey, writer and ornithologist, varied from the anthropomorphic in *Birds in the Bush* to a detailed study of species in *The Foot-Path Way.* Frank Bolles had a brief career but showed great promise in his last book, *At the North of Bearcamp Water.*

The nature writers of the 19th century wrote charming, pleasant books for a general audience, and they contributed much information about birds, often mixing the information with philosophy and humor. Popular magazines also did their part. *The Atlantic Monthly, Harper's,* and *Scribner's Monthly* all carried nature essays and articles about birds, as did *Appleton's Journal, Harper's Bazaar, The Independent,* and *MacMillan's Magazine.* Several children's periodicals, *Harper's*

Young People, St. Nicholas, and *The Youth's Companion*, entreated their young readers to protect birds. *Birds*, a magazine started in 1896, served as a guide to nature education in the schools, as did four books by women: *The First Book of Birds* by Olive Thorne Miller, *Birds of Village and Field* by Florence A. Merriam, *Citizen Bird* by Mabel Osgood Wright, and *How to Attract Birds* by Neltje Blanchan.

Meanwhile, the ornithologists were accumulating, assimilating, and beginning to publish vast amounts of information. Lacking modern photography, the artists among them portrayed the species as they saw them.

Alexander Wilson, poet, artist, and weaver, arrived in America from Scotland in 1794 and within 14 years had completed the first volume in a projected 10-volume work on American birds that would earn for him the title, "Father of American Ornithology." Short of funds and forced to depend on an inadequate number of subscribers for the entire set of *American Ornithology*, Wilson walked through New England and south along the coast to Georgia, carrying Volume I (1808) and collecting, painting, and studying birds as well as seeking subscribers. He took a second walk westward, down the Ohio and Mississippi Rivers to Natchez, and then east across the country to Philadelphia.

He found support for this project in Philadelphia: William Bartram and Charles Wilson Peale encouraged and helped him; Peale's museum supplied him with specimens. He remained near Philadelphia for the rest of his life, writing, painting, and collecting. When he died in 1813 at the age of 47, he had completed eight volumes. The ninth, edited by George Ord, was published in 1814.

The nine volumes that appeared in just over six years contained illustrations of 320 species—39 that had never been illustrated before—on 76 full-page engravings **(Fig. H–27)**. Looking at them today, we admire the skillful drawing and forget, if we can, the awkwardness caused by crowding too many birds on one plate as an economy measure. The text raised the level of ornithology by including accurate first-hand observations, measurements of specimens, and notes on the fresh colors of the bill, irises, and other soft parts that fade quickly once the bird is dead.

The paintings of John James Audubon, however, were soon to surpass those of Wilson even though Wilson may have inspired Audubon to publish his work. The two men met in Louisville, Kentucky, and reports of this meeting are numerous and conflicting. Later there was ill feeling between the two artists.

Figure H–27. Plate from American Ornithology *by Alexander Wilson: This plate, from Volume I of the nine volume set, demonstrates the crowding of several species on a single plate as a means of reducing publishing costs. Shown here are Pine Siskin, Rose-breasted Grosbeak, Black-throated Green Warbler, Yellow-rumped Warbler, Cerulean Warbler, and Blue-headed Vireo. Note that some of these species' names have changed since this book was published.*

Figure H–28. Osprey by John James Audubon: Born in the Caribbean in 1785 and raised in France, Audubon began painting birds as a child. His dramatic paintings of birds from the New World were first published by an English engraver, Havells of London. Between 1827 and 1838 Havells produced the four elephant-folio volumes of The Birds of America, *containing a total of 435 plates, using the technique of aquatinted copper engraving. Typical of this artist's portrayal of action is this Osprey carrying its catch.*

Audubon, born in 1785, grew up in France and painted birds from childhood. His early inclination became an obsession as he hunted and studied birds in the Ohio and Mississippi Valleys of the New World. In 1824, he journeyed to Philadelphia in search of a publisher. But Wilson was dead and those of his friends who were still alive refused to support this young challenger to Wilson's fame. Audubon found a publisher in England—Havells of London. From 1827 to 1838, Havells published *The Birds of America* in four elephant-folio volumes with 435 aquatinted copper engravings by Robert Havell, engravings that retain a remarkable fidelity to the original paintings **(Fig. H–28)**. Robert Havell, Jr., and others, including Audubon's son John, occasionally completed the background details of the paintings—the vegetation or landscape. Three of Audubon's birds are line copies of Wilson's drawings. Nevertheless, Audubon was the dominant spirit behind the spectacular accomplishment and deserves credit for a remarkable achievement.

Audubon painted from fresh specimens arranged in natural poses, just as Catesby had done 100 years before. Some earlier artists had painted a single species on a plate, life-size against a natural background; others had considered the artist's composition of the paintings. Audubon absorbed all their ideas and created something uniquely his. His *Ornithological Biography*, however, shows that he was not as acute an observer of birds as Wilson.

Meanwhile, in England, painter John Gould had a far more ambitious plan—to illustrate all the birds of the world. By painting from bird skins rather than living birds and by organizing others to help him, Gould produced nearly 50 sumptuous volumes, illustrated with over 3,000 plates by a relatively new method, lithography **(Fig. H–29)**. The volumes were monographs on taxonomic groups of birds and attempted to show the relationships among the species covered.

In the 19th century, knowledge of the North American avifauna increased rapidly with the exploration of the West. The members of government expeditions, though mostly untrained in science, did their best to record the natural history of the areas they visited. Such

Figure H–29. Spotted Forktails from Birds of Asia, by John Gould: Author of many books and scientific articles, English ornithologist and painter John Gould (1804–1881) aspired to illustrate all the birds of the world. He eventually published 41 folio volumes, in which the illustrations were reproduced by the relatively new method of lithography. His artistic wife contributed greatly by handling the technical aspects of the lithographic process, and several other artists contributed plates. Gould's monographic works included the birds of Australia, Asia, Europe, Great Britain, and New Guinea, and his Monograph of the Trochilidae *(hummingbird family) (1849–1861) is considered his masterpiece.*

was the case with the Lewis and Clark Expedition. Without a trained zoologist to guide them, the captains relied on designations in common use in the East, naming new birds in terms of species with which they were familiar. As a result, they called the California Quail "a bird of the quail kind" and McCown's Longspur a "small bird resembling a lark," but the vivid descriptions of the two species leave little doubt about their identity. So well and in such detail did Lewis and Clark record their observations of new species that later ornithologists were able to place the proper technical names with these descriptions. Nevertheless, very little ornithological renown has been bestowed on the captains, even though their expedition returned with observations of about 130 birds, many, such as the Clark's Grebe, Tundra Swan, Lewis' Woodpecker, Sage Grouse, Broad-tailed Hummingbird, and Western Tanager, new to science. True, "Clark's Grebe," "Clark's Nutcracker," and "Lewis' Woodpecker" memorialize their names, but these ascriptions reveal little of their achievement. One searches the lists of alternative common names in vain for mention of Lewis and Clark. The American Ornithologists' Union Check-list, at least, refers to two names the captains first used, the "Whistling Swan" (Tundra Swan) and the "Prairie Hen" (Sharp-tailed Grouse). Later naturalists generously availed themselves of the captains' patient labors, almost always without a shred of acknowledgment.

　　Some expeditions were fortunate in having ornithologists whose careful records contributed a great deal to the expedition's data. Titian

Peale accompanied the Long Expedition of 1819–1820, painted several birds, and no doubt laid the groundwork for his later illustrations in Charles Lucien Bonaparte's extension of Wilson's *American Ornithology*, published in four volumes in 1825. From Peale's brush came all but one of the illustrations in Volume One, including the plate of Say's Phoebe, first described by his fellow traveler on the Long Expedition, Dr. Thomas Say. Peale also went on the Wilkes Expedition of 1838–1842 from Virginia to Rio de Janeiro, around Cape Horn to Peru, Antarctica, New Zealand, the northwestern United States, the Philippines, and back to New York by way of the Cape of Good Hope. The data from this voyage, compiled and edited by ornithologists who had not made the journey, became the first major publication on birds sponsored by the government. John Cassin compiled several of the volumes, among them *Illustrations of the Birds of California, Texas, Oregon, British and Russian America ... and a General Synopsis of North American Ornithology* (1853–1856).

Several naturalists joined the railroad explorations that moved west from the Mississippi River to the Pacific between 1853 and 1856. Their data, added to information from private collections, were edited chiefly by Spencer Fullerton Baird, assistant secretary of the Smithsonian Institution, and became the second part of the ninth volume of *Reports of Explorations and Surveys to Ascertain the Most Practicable and Economical Route for a Railroad from the Mississippi River to the Pacific Ocean* (1858), one of the most important federal publications of the century. This purely ornithological section, known as Baird's *General Report* for political reasons, included descriptions of 738 species of birds and was the first book on the birds of the entire continental United States. Later, the section was reissued privately as *The Birds of North America* (1860). Although the information was more detailed and accurate than any yet published, the book lacked the charm and spontaneity of the works by Wilson and Audubon.

Natural history inventories and the collection of natural history objects (including birds and their eggs) led to the creation of many natural history museums in the United States in the 1800s. Among these were the American Museum of Natural History in New York City; the United States National Museum (Smithsonian) in Washington, D. C. **(Fig. H–30)**; the Academy of Natural Sciences in Philadelphia; the Museum of Comparative Zoology in Cambridge, Massachusetts; the Field Museum of Chicago; and the Carnegie Museum in Pittsburgh. Although there were many private collectors, these early museums soon became centers for the study of birds.

In the 19th century, ornithology was still a descriptive science. No formal academic training was available and the few fortunate men em-

Figure H–30. Smithsonian Institution Bird Gallery: This 1885 photograph depicts taxidermic mounts and study skins atop a long case of specimen drawers at the Smithsonian Institution in Washington, D. C. Many natural history museums centered on bird collections were established in the United States during the 1800s. Negative number 96–3532. Smithsonian Institution Bird Gallery, 1885. Smithsonian Institution Archives, Record Unit 7006, Alexander Wetmore Papers, Box 195.

ployed as "professional" ornithologists were among the directors and curators of the early natural history museums. They held no degrees in ornithology, and in fact, many had no college education at all. Instead, they were either self-taught or apprenticed under someone knowledgeable in the field. Expertise was gained by firsthand experience in the field, collecting and observing birds. A personal collection of bird specimens was considered essential for those serious about bird study. Along with the "professionals," there was a well-to-do upper class who enjoyed learning about birds and had the financial means to spend time in the field creating their own personal collections. They also relied heavily on the work of collectors, taxidermists, and sport hunters to expand their collections of bird specimens.

The professional ornithologists affiliated with the early American museums played a critical role in advancing ornithology. Notable among this group were Robert Ridgway, curator of birds from about 1881 until his death in 1929 at the United States National Museum in Washington, D. C., and Frank Chapman, curator of birds at the American Museum of Natural History in New York City. Robert Ridgway (1850–1929), a founder and president of the American Ornithologists' Union, served as curator of birds at the United States National Museum (USNM) for over 50 years and was the author of two important works in ornithology: *A Manual of North American Birds* (1887) and a series of volumes known as *Birds of North and Middle America* (1901–1919). Frank Chapman (1864–1945), was considered the Dean of American Ornithology **(Fig. H–31)**. He was a pioneer in bird photography, a lecturer, and a prolific writer with over 225 articles and 17 books. His *Handbook of Birds of Eastern North America*, published in 1895, was the most widely used guide of its time.

By the late 1800s, regional bird clubs, societies, and organizations were sprouting up all over the country, and some remain active to this day. The Nuttall Ornithological Club, created in the early 1870s, was the first ornithological club in the United States. The requirements for joining this group were minimal—you had to be male and have a special interest in birds. The club soon included a diverse group of men,

Figure H–31. Frank Chapman: Although he had no formal ornithological training, Chapman was a gifted writer, lecturer, and scientist. He served as the curator of birds at the American Museum of Natural History and bridged the gap between amateur and professional ornithologists with hundreds of articles, 17 books, and a popular magazine, Bird-Lore, *which eventually became* Audubon *magazine. Negative number 12930, courtesy of the Department of Library Services, American Museum of Natural History.*

Figure H–32. Theodore Roosevelt and John Muir: *This photo, circa 1906, shows two conservation pioneers in California's Yosemite Valley. While at Harvard University, Theodore Roosevelt (left) was a member of the nation's first birding club, the Nuttall Ornithological Club. As president of the United States (1901–1909), Roosevelt championed many wildlife conservation issues. John Muir, like Roosevelt, was a strong proponent of creating national parks and conserving forest lands. Copyright Bettman/Corbis.*

including Theodore Roosevelt, who was a freshman at nearby Harvard University **(Fig. H–32)**. The Nuttall Ornithological Club—still in existence (but now accepting women)—soon became embroiled in the controversy surrounding the introduction of House Sparrows into the United States, a program begun in the early 1850s. Known as the "Sparrow Wars," the controversy ultimately led to negative publicity and a temporary decline of the club.

Another organization of this era was The Linnaean Society of New York. It formed in 1878 as a local (New York City area) natural history society whose main interest was birds. An offshoot of this group was the Bronx County Bird Club, perhaps the most enthusiastic group of bird watchers anywhere. Although these organizations contributed to the growing interest in bird studies, the establishment of the first national organization—the American Ornithologists' Union—really laid the foundation for 20th century ornithology.

The American Ornithologists' Union and the U. S. Biological Survey

Twenty-three men were present at the founding of the American Ornithologists' Union (AOU) in 1883. Although there were many regional natural history organizations and bird clubs at the time, this was the first national organization and its creation was a turning point for ornithology. Two important committees were created at the first meeting: the Committee on Bird Migration and the Committee on Classification and Nomenclature.

C. Hart Merriam, the first chair of the Committee on Bird Migration, immediately began an ambitious program to study bird migration. He was interested in learning arrival and departure times, the influence of weather, and many other aspects of this poorly understood behavior. Merriam accomplished his goal by setting up an information network. He sent a circular to 800 newspapers soliciting the support of all interested people—sportsmen, ornithologists, field collectors, and nature enthusiasts—to help gather information on these topics. The response was overwhelming. Merriam and the AOU petitioned congress to create a Division of Economic Ornithology within the Department of Agriculture **(Fig. H–33)**. Because the public was eager to categorize the effects of different types of birds on agriculture as either beneficial or harmful, Merriam and the AOU were able to gain federal support by emphasizing the economic value of bird migration data. Of necessity, the division's early publications, such as *The Hawks and Owls of the United States in their Relation to Agriculture,* and numerous bulletins on the food of different bird species, were oriented toward farmers. In 1885, the Division of Economic Ornithology was authorized as a unit within the Division of Entomology. By 1886, it was separate from the Division of Entomology, and in 1896 it became the U. S. Biological Survey, whose focus was to study the geographic distribution of both birds and mammals. This later became the U. S. Fish and Wildlife Service within the Department of the Interior. From the work of one committee, the foundation for a federal agency with a mandate at least partly in ornithology was created, and the first study of birds using data collected from a large number of volunteer participants—Merriam's bird migration study—was carried out.

The other original AOU committee, the Committee on Classification and Nomenclature, developed the first "Code of Nomenclature"—a standardized list of common and scientific names for North American birds. This was a significant step because before the "Code" there were two competing lists of North American birds. Standardizing bird names was essential for any accurate inventory of bird distributions. The Committee on Classification and Nomenclature has also been responsible for determining which groups of birds constitute species: whether certain groups traditionally considered separate species should be put together as one (lumped) or whether certain species should be divided into several (split). These issues are often fiercely debated. Each year at the annual AOU meeting, this committee reviews new research concerning the taxonomy of North American birds, determining which species to split and which to lump, as well as deciding on any name changes for species. The AOU regularly publishes updates of the *Check-list of North American Birds*. Anyone who thinks taxonomy is a stagnant science, should consider that there were over 150 changes between the sixth edition published in 1983 and the seventh edition published in 1998.

By the late 1800s, there was growing concern over declines of many bird species. Habitat destruction and market hunting for plumes and meat had taken their toll on a number of species, including the now extinct Passenger Pigeon and Carolina Parakeet. The AOU Committee on Protection of North American Birds, created in 1884, played an important role in fostering bird protection in America. Their first priority was to alert the public to the decline of birds. They also proposed legislation, which was passed in some states, making it illegal for anyone to kill, purchase, or sell non-game birds or their nests and eggs.

The First Audubon Movement

At the same time the AOU was publishing its first bulletin on bird protection, George Bird Grinnell was creating the first Audubon Society. Editor-in-chief of *Forest and Stream*, and at one time a student of Lucy Audubon, the widow of John James Audubon, Grinnell was a leading voice in condemning the commercial exploitation of wildlife. He was also one of the founding members of the AOU. Grinnell's long-term goal was to create a national society organized into local chapters and he named his new organization after the legendary bird artist John James Audubon. Grinnell's Audubon Society enjoyed early success, but unfortunately it wasn't sustained. By 1888, the work of both the AOU bird protection committee and Grinnell's Audubon Society had stalled. Although short-lived, Grinnell's Audubon Society strengthened the public's growing concern about bird destruction and established a public arm to the bird protection movement.

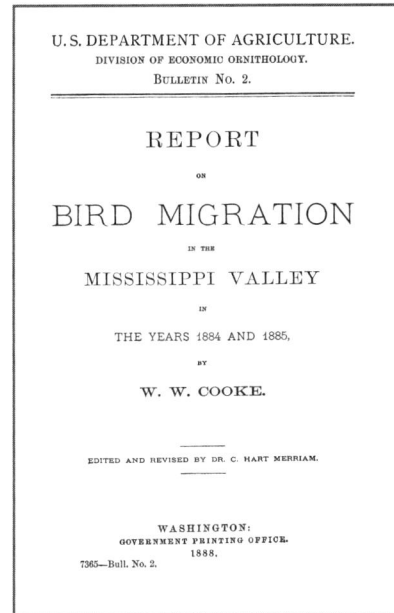

Figure H–33. Bird Migration Report: *Shown here is a report on bird migration from the U. S. Department of Agriculture's Division of Economic Orni-thology. Through the urging of C. Hart Merriam and the American Orni-thologists' Union, congress authorized the establishment of the Division of Economic Ornithology in 1885. Although the original intent of the division was to study the effects of birds on agricultural production, over the years the scope broadened to include the general study of birds. By 1896 the Division of Economic Ornithology had become the U. S. Biological Survey, which eventually became the U. S. Fish and Wildlife Service within the Department of the Interior.*

Women, initially excluded from the male-dominated "professional organizations," were instrumental in popularizing the study of birds and bird protection. In 1886 Florence Merriam (the sister of C. Hart Merriam) created a local Audubon chapter while she was a student at Smith College. In 1887 she began writing a series of articles ("Hints to Audubon Workers: Fifty Birds and How to Know Them") for Grinnell's *Audubon Magazine*. At the conclusion of the series in 1888, she included a field key incorporating color, bill shape, song type, behavior, habitat preference, and nesting habits as characters to help identify the fifty species. In 1889 she published her first book, *Birds Through an Opera Glass*, in which she expanded her field key to seventy species. It became a bestseller, inspiring thousands to take up bird watching. Many similar publications by other authors soon followed.

The Second Audubon Movement

The bird protection movement was revitalized in 1896 by two Boston women, Harriet Hemenway and her cousin Minna Hall, who were instrumental in creating the Massachusetts Audubon Society. Inspired by horrific accounts of birds slaughtered only to adorn women's apparel, the mission of the organization was to discourage the use of feathers for ornamentation and to promote the protection of birds. Concerned people in other states soon established similar organizations.

At about the same time, the AOU bird protection committee was strengthened as William Dutcher became its chairman. Dutcher was dedicated to establishing state Audubon societies and promoting bird protection through public education. In 1905, Dutcher created the National Association of Audubon Societies. Having a national association relieved the AOU of much of its role in bird protection—a role that was increasingly divisive among its members, as many were actively collecting birds. The AOU did not revive its bird protection activities until 1930.

Meanwhile, the American Museum of Natural History was profoundly affecting ornithology. Frank Chapman, first hired by the museum in 1888, eventually became curator and served in that capacity until 1945. Although he had no formal education, Chapman was a gifted writer, lecturer, and scientist (see Fig. H–31). In a field that was becoming increasingly divided, Chapman formed a bridge between the professional and amateur. He created the popular bird magazine *Bird-Lore* in 1898, in part to supplement his income, but also to assist the AOU bird protection committee in distributing information on bird protection to the public. **(Fig. H–34)**. *Bird-Lore* introduced a new generation of people to the study of birds and it served as the voice of the emerging National Association of Audubon Societies. The National Audubon Society assumed the publication of *Bird-Lore* in 1941, renaming it *Audubon Magazine*. In 1953, the name was shortened to *Audubon*, the publication familiar to us today.

During the last quarter of the 19th century, ornithological publications began to proliferate. Some were sponsored by state governments,

Figure H–34. **Bird-Lore:** *One of the first popular bird magazines, Bird-Lore, begun in 1898 by Frank Chapman, introduced a new generation of people to the study of birds and served as the voice of the National Association of Audubon Societies.* Bird-Lore *was renamed* Audubon Magazine *in 1941, and the name was shortened to* Audubon *in 1953. Today,* Audubon *serves as the primary publication of the National Audubon Society. Courtesy of the National Audubon Society.*

but many others were produced by private publishers. The literature included state and regional works on birds and, most importantly, Elliot Coues' *Key to North American Birds*, the classic that first introduced ornithology to many thousands of people **(Fig. H–35)**.

The 20th Century and the Expanding Role of the Bird Watcher

The growing number of bird watchers in the late 19th century presented both challenges and opportunities to professional ornithologists. As bird collecting became more restricted, many ornithological societies focused more on field studies and cooperative research, especially in the areas of geographical distribution, migration, and life history studies. Lynds Jones, a founding member of the Wilson Ornithological Society, and his student, William Dawson, advocated systematically keeping detailed checklists of birds observed at different locations and times of the year. The value of such lists, they lobbied, was that they could be used to estimate bird abundance. Dawson and Jones soon had hundreds of observers competing to compile the longest lists. Getting out into the field to count birds replaced collecting in numerous ways, providing many of the same benefits without killing birds. The competitive aspect also attracted the interest of a whole new breed of bird watchers. Observers were encouraged to submit their lists for publication in the *Wilson Bulletin*, the journal of the Wilson Ornithological Society.

The idea of counting birds inspired Frank Chapman to propose a Christmas Bird Count in 1900. Chapman encouraged people to count birds instead of participating in the traditional Christmas Day

Figure H–35. Elliot Coues: Coues' Key to North American Birds *was the most influential of the numerous American ornithological publications that proliferated during the late 19th century. Its two illustrated volumes contained information about the structure and classification of living and fossil birds as well as a manual on collecting, preparing, and preserving bird specimens.*

Figure H–36. Arthur Cleveland Bent: During the early 1900s, A. C. Bent produced a series of texts on the life histories of many groups of birds, such as the Life Histories of North American Birds of Prey *and the* Life Histories of North American Shorebirds. *Bent obtained much of the information for his books from the newly founded U. S. Biological Survey and by corresponding with other keen bird watchers. To this day, Bent's texts are used by birders and ornithologists as a source of life history information. Photo courtesy of D. L. Garrison.*

Hunt, promising to publish the results in *Bird-Lore*. From this modest beginning, the Christmas Bird Count has become a major ornithological tradition, collecting valuable information on bird abundance throughout the United States to this day.

By the early 1900s, the U. S. Biological Survey had amassed an enormous amount of data on bird migration in addition to data on bird abundance from the surveys. This information was used by A. C. Bent to compile his monumental *Life Histories* series **(Fig. H–36)** and by the AOU to prepare their Check-lists of North American Birds. While the U. S. Biological Survey's census programs waxed and waned, the National Audubon Society began its own spring count in 1937, the "Breeding Bird Census." Still held each June throughout the United States, the Breeding Bird Census is a major effort to count birds in the breeding season.

Bird banding was quickly gaining momentum as people realized it was useful for studying bird movements. At the 1909 AOU meeting, banding enthusiasts organized the National Bird Banding Association (NBBA), which was administered through several different organizations before its demise. In its wake, regional organizations were formed to fill the void: the New England Bird Banding Association in 1922 (renamed the North American Bird Banding Association in 1924), the Inland Bird Banding Association (1922), the Eastern Bird Banding Association (1923), and the Western Bird Banding Association (1925). As a result of the growth of bird banding in the 1920s and '30s, the U. S. Biological Survey (now the Biological Resources Division of the U. S. Geological Survey) began to coordinate banding activities. This organization and its Canadian counterpart have overseen the activities of dedicated North American bird banders ever since.

The Development of the Field Guide

While bird watchers were amassing all kinds of information about birds, the scientific community was becoming increasingly skeptical about the accuracy of many reports. With no specimens to back them up, reports of questionable birds were hard to verify and thus were of little scientific value. Yet, as collecting became more restricted by law, scientists had to rely increasingly on observational reports. One scientist who believed that just about anyone with an interest in birds could learn to accurately identify them was Ludlow Griscom, an ornithologist at the American Museum of Natural History.

Griscom, an early Ph.D. student of Arthur Allen at Cornell University, was instrumental in promoting the field identification of birds based on characteristics that could be easily observed in wild birds. His book *Birds of the New York City Region*, published in 1923, was an inspiration to many young bird watchers. It was especially important to the young enthusiasts of the Bronx County Bird Club, a local group that regularly watched birds in the New York City region during the 1920s and '30s. One member of this group was the now-legendary Roger Tory Peterson. An artist and teacher, Peterson had an uncanny ability to recognize birds in the field, and he took Griscom's ideas to the next level. By organizing groups of birds in similar poses, using a system of arrows to indicate key field marks, and shortening the text to the minimum needed to identify a bird, Peterson produced a reference book that revolutionized bird watching. His first field guide, *A Field Guide to the Birds: Giving Field Marks of all Species Found in Eastern North America*, was published in 1934 **(Fig. H–37)**. In 1980, range maps reflecting the seasonal movements of each species were added to the guide. *A Field Guide to the Birds* was the first of a series of field guides that would make publishing history. More than anyone else in the world, Peterson introduced an enormous number of people to birds—an achievement that far surpasses the commercial success of his series. As species recognition is the first step toward preservation, his work also has been pivotal to the 20th century conservation movement.

Academic Training in Ornithology

Ornithology became increasingly professional in the early part of the 20th century. Although no formal degree programs were yet available, a number of academic institutions offered classes in ornithology as part of zoology or nature study programs. A period of rapid expansion soon followed, with more students interested in ornithology, and more in-depth, scientifically-based studies of birds being carried out. As the field grew, graduate degrees became standard requirements for serious researchers. Biology itself was rapidly transforming from a descriptive discipline to an experimental one, with a number of specialized fields such as genetics, embryology, and the field-based studies of ecology and animal behavior.

During the early 20th century, Cornell University was the leading institution for graduate training in ornithology. From its beginning, Cornell had a strong program in zoology, but once Dr. Arthur Allen arrived it earned a reputation as an important center for bird studies. Allen wasn't the first person to earn a Ph.D. by studying birds, but his research was the first to be widely publicized. His Ph.D. dissertation, *The Red-winged Blackbird: A Study in the Ecology of the Cattail Marsh*, published in 1914, set a new standard for documenting the ecology and life history of a bird species. A year after he graduated, Allen was offered a position as an instructor of zoology at Cornell. In 1915 he was promoted to assistant professor of ornithology and developed the first graduate program in ornithology in America.

Many of Allen's students, including Ludlow Griscom, John Emlen, Peter Paul Kellogg, Olin S. Pettingill, Jr., and George M. Sutton, went on to become leaders in emerging biological disciplines. In addition

Figure H–37. Plate from an Early Peterson Field Guide: *Roger Tory Peterson revolutionized bird watching in 1934 when he published* A Field Guide to the Birds: Giving Field Marks of all Species Found in Eastern North America. *This plate of terns and skimmers from a 1939 edition of the book shows the unique system of arrows Peterson used to indicate important field marks on each bird. Today, Peterson guides are the standard by which all other field guides are judged. "Terns and Skimmers" from* A Field Guide to the Birds: Giving Field Marks of All Species found East of the Rockies. *Copyright 1939 by Roger Tory Peterson. Reprinted by permission of Houghton Mifflin Company. All rights reserved.*

Figure H–38. Peter Paul Kellogg, Arthur A. Allen, and James Tanner: *Seen here with photographic and sound recording equipment, Peter Paul Kellogg (left) and Arthur Allen (center) were leaders in developing the technology to record bird sounds. The success of this technology, along with the belief that the public could contribute to the professional study of birds, led Allen to found the Cornell Lab of Ornithology and its Library of Natural Sounds. Photo courtesy of Cornell Lab of Ornithology Archives.*

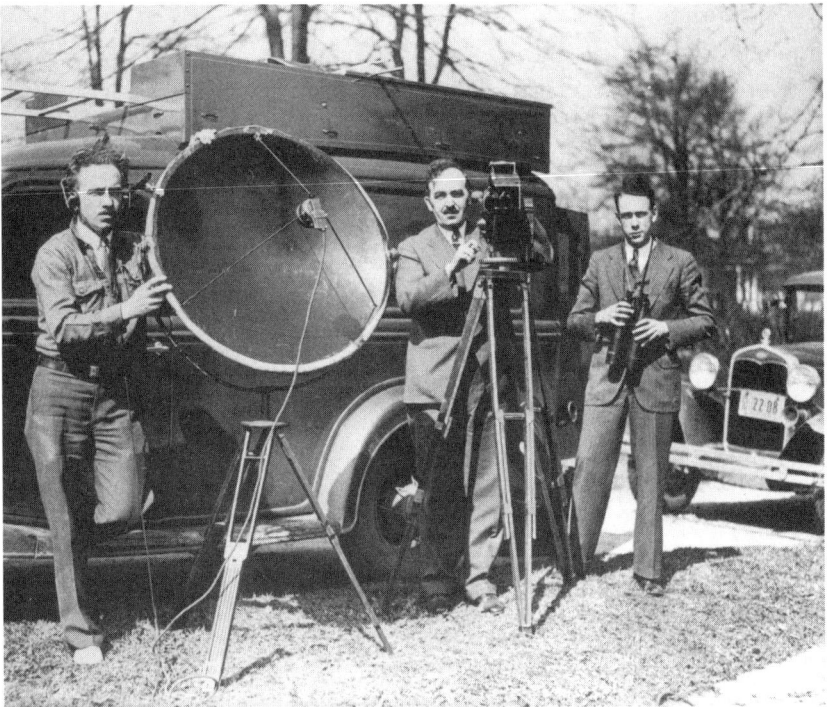

to his impressive list of academic progeny, Allen was one of the first professors to teach courses in wildlife conservation and management. He also was a leader in popularizing birds—offering many courses to the public on bird identification and photography, and writing many articles for *Bird-Lore* and *National Geographic*. Through the work of Arthur Allen and Peter Paul Kellogg, along with the financial support of Albert Brand, Cornell became a leader in developing the technology to record bird sounds **(Fig. H–38)**. This tradition continues to grow today with the Library of Natural Sounds and Bioacoustics Program at the Cornell Lab of Ornithology.

Close on Allen's heels, academically, was Joseph P. Grinnell of the Museum of Vertebrate Zoology at the University of California at Berkeley. Annie Alexander, a wealthy naturalist and collector, founded the museum in 1909 and appointed Grinnell its first director. Grinnell earned his Ph.D. in 1913 at Stanford University by studying the mammals and birds of the Lower Colorado Valley; a study now considered a classic in biogeography. After he earned his degree, Grinnell's appointment was expanded to include a professorship in the Department of Zoology at U.C. Berkeley. Grinnell raised ornithology to new scientific levels by setting new standards for collecting expeditions—emphasizing careful note-taking and record keeping, and accurate labeling of specimens. He also placed considerable emphasis on accumulating and carefully documenting data on each species. Like Allen at Cornell, Grinnell produced an impressive academic line, including students such as Alden Miller, Frank Pitelka, Ned Johnson, Charles Sibley, and Robert Storer.

Although Cornell University and the University of California at Berkeley were the most prolific academic institutions in the 1920s and '30s, several other universities also established active ornithology programs. Among these were Case Western Reserve in Ohio, the University of Kansas, and the University of Michigan. Today over thirty universities in North America offer graduate degrees with the opportunity to pursue advanced studies of birds.

As academic research has expanded, so has the body of knowledge concerning birds. Beyond classification and life history studies, birds have served as models for studies in a diversity of biological fields, including biogeography, evolution, mating systems, population dynamics, ecology, animal communication and learning, neurobiology, and conservation. In this way, birds have helped us develop a better understanding of the natural world.

Bird Conservation, Bird Watching, and the Age of Technology

Birds and bird watchers have been critical to the field of wildlife conservation. Because birds are relatively accessible to observers, they are important warning flags for the wide-scale environmental changes caused by people. Growing public concern about the fate of birds has been key to passing federal legislation protecting birds and other wildlife. The Lacey Act of 1900 was the first federal legislation offering some protection for birds, and it came about largely through the efforts of the AOU bird protection committee, the U. S. Biological Survey, and Audubon supporters. These groups also helped to create the first national wildlife refuge—Pelican Island in Florida—home to a colony of nesting Brown Pelicans. The Audubon Society also played a key role in passing the Federal Migratory Bird Treaty Act of 1918, a law that for the first time protected all migratory birds in America. The Endangered Species Act of 1973, re-authorized in 1988, gave even more weight to securing the future of endangered and threatened birds. The Wild Bird Conservation Law, passed in 1992, banned the import of all birds listed by CITES (the Convention on International Trade in Endangered Species of Wild Fauna and Flora). For more information on legislation affecting birds, see Chapter 10.

Figure H–39. Aldo Leopold: Considered the father of modern wildlife management, Aldo Leopold is shown here preparing to weigh an American Woodcock as part of his field research. Leopold's 1949 book, A Sand County Almanac, focused on environmental stewardship and the ethical use of land. Photo by Robert Ockting, courtesy of The Aldo Leopold Foundation Archives.

Twentieth century nature writers and artists have done their part to instill an environmental ethic in the American public. Among the classics are Aldo Leopold's *A Sand County Almanac* (1949), which focused on the ethical use of land **(Fig. H–39)**, and Rachel Carson's *Silent Spring* (1962), which indicted the use of pesticides. There are many more—each book nudging the reader forward to greater awareness and appreciation of the natural world. Among the 20th century artists, Louis Agassiz Fuertes stands out for his ability to capture, with pen and brush, the essence of

Figure H–40. American Kestrel by Louis Agassiz Fuertes: Fuertes' ability to capture the essence of a bird's character is exemplified by this American Kestrel feeding on a grasshopper. This painting hangs in the halls of the Cornell Lab of Ornithology in Ithaca, New York, and is one of the most popular among the general public. Courtesy of Cornell Lab of Ornithology.

the living bird **(Fig. H–40)**. Although he died in 1927, his work is still unsurpassed and his influence is apparent in the paintings of many artists working today. Other great bird artists of the 20th century include George Miksch Sutton, Roger Tory Peterson, Larry McQueen, Robert Bateman, and Don Eckleberry.

Today, bird watching has become big business. Bird feeders, bird foods, bird baths, binoculars, telescopes, field guides, books, magazines, audio and video tapes, computer software, guided birding tours, checklists, and birding festivals all contribute to a multi-billion dollar industry. Binoculars and telescopes, once engineered for military use, are now tailored to the needs of bird watchers. Beyond fueling the growth of this lucrative industry, however, volunteer bird watchers have become vital to the science of ornithology—counting, atlassing, monitoring, studying, and conserving North America's bird populations.

Today, computer and Internet technologies are poised to revolutionize bird watching and the role of the volunteer bird observer once again. An unprecedented amount of information is available on the Internet to help people identify birds by sight and sound, and understand more about their life histories. For example, BirdSource—an interactive web site administered by the Cornell Lab of Ornithology and the National Audubon Society—allows users to submit and retrieve massive amounts of data collected through several different programs, such as Project FeederWatch and the Christmas Bird Count. Using a complex database and associated computer software, BirdSource can generate geographical distribution maps and other sophisticated graphics and tables almost immediately. Such rapid data processing

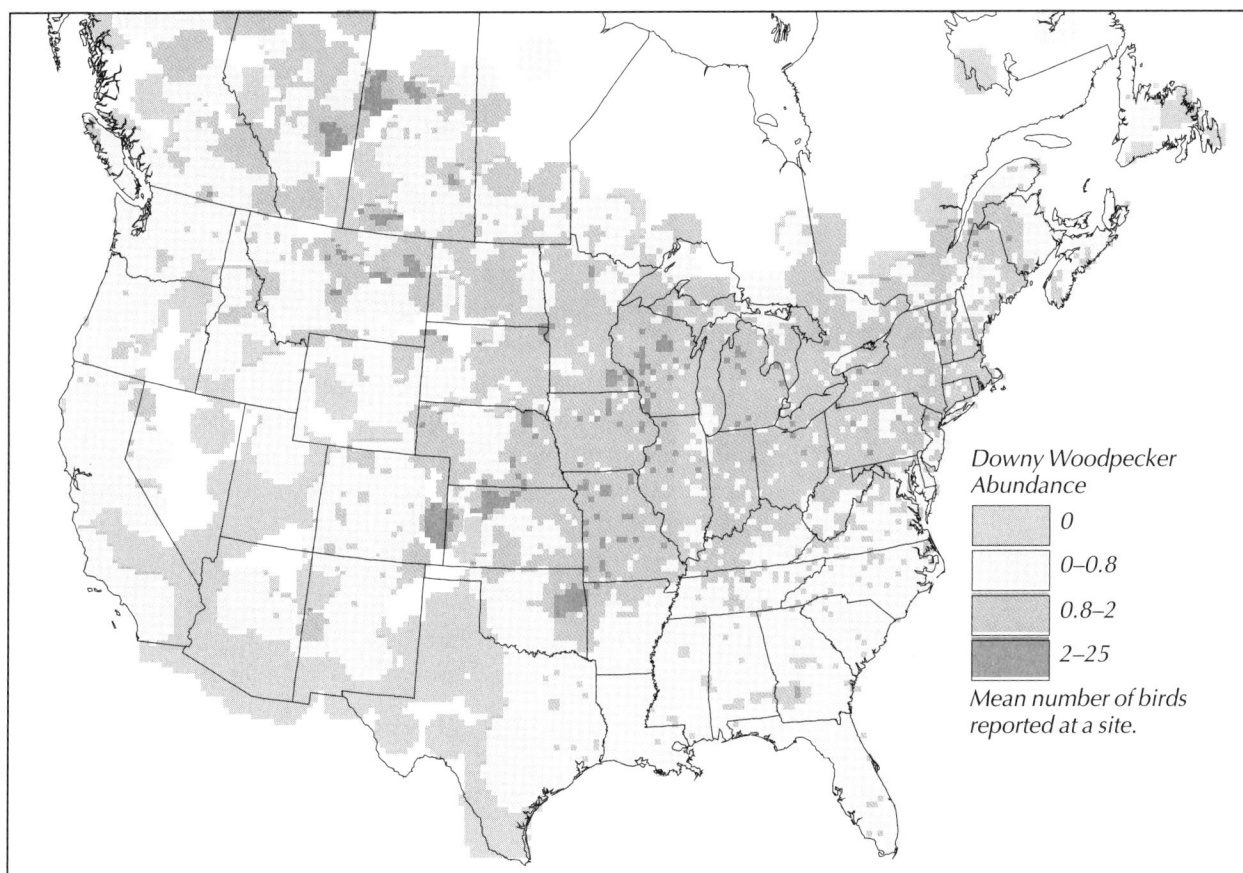

Downy Woodpecker
Abundance

0

0–0.8

0.8–2

2–25

Mean number of birds
reported at a site.

gives birders and researchers all over the world a chance to view and interpret the results instantaneously. Other web sites, such as one operated by the United States Geological Survey's Patuxent Wildlife Research Center, also function as a clearinghouse for information on bird populations. At this one site, birders and researchers can access data and other information from a variety of sources, including the North American Breeding Bird Survey, hawk migration counts, night bird monitoring, marsh bird monitoring, and the Colonial Waterbirds Inventory and Monitoring Program.

Because the Internet can handle bird distribution and abundance data from huge numbers of observers so rapidly—analyzing and displaying in minutes data sets that previously might have taken years to input and process—it is a powerful conservation tool **(Fig. H–41).** Having instant access to large amounts of data permits scientists to evaluate changes in bird populations over time, allowing conservation biologists to take action while species are still relatively common. When bird watchers take part in these on-line counts and other bird watching programs, their observations are making valuable contributions to bird conservation.

Figure H–41. Abundance and Distribution of Downy Woodpeckers as Recorded by Great Backyard Bird Count Participants in February 2000: The power of the Internet is shown in this example from the Great Backyard Bird Count (GBBC), an annual continentwide survey that occurs during a four-day period each February. This map, which depicts the late-winter distribution of Downy Woodpeckers throughout North America, is based on 22,500 individual reports submitted over the Internet during the GBBC of February 2000. The GBBC is a joint project of the National Audubon Society and the Cornell Lab of Ornithology. Copyright BirdSource.

Suggested Readings

Barrow, Mark V. Jr. 1998. *A Passion for Birds, American Ornithology after Audubon.* Princeton, New Jersey: Princeton University Press.

Buxton, E. J. M. 1985. "Birds in Poetry." pp. 475–478 in *A Dictionary of Birds,* ed. by B. Campbell and E. Lack. South Dakota: Buteo Books. 670 pp.

Feld, Steven. 1990. *Sound and Sentiment; Birds, Weeping, Poetics, and Song in Kaluli Expression, Second Edition.* Philadelphia, PA: University of Pennsylvania Press.

Hall-Craggs, J. M. and R. E. Jellis. 1985. "Birds in Music." pp. 369–372 in *A Dictionary of Birds,* ed. by B. Campbell and E. Lack. South Dakota: Buteo Books. 670 pp.

Lambourne, L. 1985. "Birds in Art." pp. 23–25 in *A Dictionary of Birds,* ed. by B. Campbell and E. Lack. South Dakota: Buteo Books. 670 pp.

Laubin, Reginald and Gladys. 1976. *Indian Dances of North America; Their Importance to Indian Life.* OK: University of Oklahoma Press.

Roseman, Marina. 1991. *Healing Sounds from the Malaysian Rainforest, Temiar Music and Medicine.* Berkeley and Los Angeles: University of California Press.

Streseman, Erwin. 1975. *Ornithology, From Aristotle to the Present.* Cambridge, MA: Harvard University Press.

Turner, G. E. S. 1985. "Birds in Folklore." pp. 233–234 in *A Dictionary of Birds,* ed. by B. Campbell and E. Lack. South Dakota: Buteo Books. 670 pp.

Introduction:
The World of Birds

Kevin J. McGowan

 Every child knows what a bird is. Birds exist on every continent, in every climate, and over every body of water. Everyone has seen them and recognized them. But what exactly makes a bird a bird?

Like fish, amphibians, reptiles, and mammals, birds are **vertebrates**. That is, they are supported along the back by a series of small bones, the vertebral column. Like many vertebrates, birds lay eggs. Only reptiles and birds lay eggs with a tough shell that allows the embryo within to remain alive out of water. Like some reptiles (crocodiles and alligators) and all mammals, birds have four chambers in their hearts; like mammals, they are **endothermic** ("warm-blooded"). Endothermic animals can regulate their body temperatures physiologically, and thus are less dependent on ambient temperatures for survival than **ectothermic**, or "cold-blooded," animals. Using the heat energy they produce by burning (digesting) their food, endothermic animals keep their bodies relatively warm—at the optimal temperature for the chemical reactions necessary for life. Being endothermic allows birds and mammals to be active on days so cold that ectothermic animals such as snakes and insects could not move. Endothermy has its costs, however; most significant is the large amount of energy, and therefore the great volume of food, that is required to produce heat. An ectothermic snake needs to eat only once every few weeks, whereas an endothermic bird would starve in just one day without food.

Figure 1–1. Barn Swallow: *Feathers are unique to birds. Here, a Barn Swallow stretches its right wing, providing an excellent view of the individual feathers. Photo by Marie Read.*

Unique to birds are **feathers**, outgrowths of the skin that cover and streamline the body **(Fig. 1–1)**. Another special feature of birds, perhaps the most obvious, is their ability to fly **(Fig. 1–2)**. Although flight is not exclusive to birds—bats, for example, have also evolved true flapping flight—no other vertebrate is so thoroughly modified for proficiency in the air. Most birds can fly, and those that cannot, such as penguins, evolved from ancestors that were capable of flight.

Figure 1–2. Northern Harrier in Flight: *Although not alone in their ability to fly, birds are thoroughly modified for proficiency in the air. Northern Harriers—slender, long-tailed hawks with long wings and a conspicuous white rump patch—typically fly low over open fields and marshes, gliding and tilting from side to side with the wings held in a shallow V, as they search for meadow voles and other prey. Drawing by Orville O. Rice.*

Ornithological Terms

■ Ornithology, the scientific study of birds, embraces a vast store of knowledge acquired by many thousands of researchers. Over the years, these scientists have developed a terminology for certain structures, processes, and concepts that may strike nonscientists as being unnecessarily technical. For some terms this may be true, but many other technical terms are used because no adequate or concise substitutes exist in everyday language. Technical terms also help scientists to convey very specific information with a minimum of confusion, in the same way that legal wording allows members of the legal profession to avoid ambiguity in contracts. This *Handbook of Bird Biology* minimizes the use of technical terms, using them only when necessary to clearly explain the material. To get you started, a few basic directional terms are described in **Sidebar 1: Which Way is Up?**

The Form of a Bird

■ Undoubtedly, you know what "a bird" looks like. But how closely and carefully have you really looked at a bird? Have you ever noticed the scales on the legs? Have you seen the small, whisker-like feathers around the bills of insect-eating birds? Do you know where a bird's knee is, or its ankle? One goal of the first section of this chapter is to help you understand the basic form of a bird. This knowledge provides a framework within which you will be able to identify, describe, compare, and eventually classify the different kinds of birds.

A live bird in the hand is an ideal aid for studying bird form. Any pet bird such as a parakeet, finch, domestic fowl, or Rock Dove (pigeon) will do, but a fresh road or window casualty also can be used. In the United States and Canada, however, the only casualties that can be legally removed from the road or window sill for personal study are non-native species: Rock Dove, House Sparrow, and European Starling. All other North American birds are protected by the Migratory Bird Treaty Act (see Chapter 10); it is illegal to possess a carcass, a living specimen, a nest or egg, or even a feather of protected species without the appropriate permits. To study some avian features, even a whole chicken or turkey from the grocery store can be helpful. If you can get one with the head and feet still on it, great, but you can learn a few things even from one that is ready to cook. If you prefer to learn from live, wild birds, you might go to the park and study the pigeons, House Sparrows, or ducks that come to the bread you throw out, or watch the birds at a feeder next to your window. In any way you can, get up close to birds and observe the details.

For descriptive purposes, the bird's body is arbitrarily divided into seven major regions; each of these, and their major parts, are considered in turn. On your subject bird, first note the shape of the body—how it tapers at both ends, perfectly streamlined for flight. Then identify the seven main parts: beak, head, neck, trunk, wings, tail, and hind limbs **(Fig. 1–3)**.

(Continued on p. 1·6)

Sidebar 1: WHICH WAY IS UP?
Kevin J. McGowan

Left or right? Whose left, yours or mine? Behind or in front? Of you or me? Directions can be confusing when they use arbitrary reference points. When face to face, my left is your right, and vice versa. When discussing anatomy, keeping directions straight is imperative. Unfortunately, many directional terms used in everyday speech are not specific enough when dealing with parts of the body. Such words as "lower" or "upper" and "back" or "front" can be confusing when trying to discuss parts of an animal. For example, stand up straight with your hands hanging at your sides. Are your fingers above or below your elbows? Hold your hands over your head. Now where are your fingers in relation to your elbows? What if you were standing on your head? It is useful to have terms that can be understood no matter what the orientation of an animal is, just as it is useful to have similar terms for a boat (port and starboard, fore and aft). Whether your bird subject is upside down, on its back, or on its belly, you need one term to indicate the direction "toward the head" and another to indicate "toward the tail;" you also need a way to indicate whether the position of one structure compared to another is closer to, or more distant from, the middle of the body. Such terms exist, and the following list defines those used throughout the *Handbook of Bird Biology* to describe the positions of anatomical structures (**Fig. A**).

Dorsal: Toward the back (the vertebral column).
Ventral: Toward the belly.
These terms do not depend on the orientation of the body. The breast of a penguin is ventral to the back whether the bird is tobogganing along the snow or standing upright.

Lateral: To the side of the body; away from the midline.
Medial: Toward the midline of the body.
Median: On the midline of the body.
The beak of a bird is median. A bird's outer tail feathers are lateral to the inner tail feathers.

Distal: Away from the center of the body or from the origin of the structure.
Proximal: Toward the center of the body or toward the origin of the structure.
The elbow is distal to the shoulder, but the elbow is also proximal to the wrist. The tip of a feather is distal to its base (the base of a feather is where it attaches to the bird). On a tree, the ends of the branches are distal to the inner branches.

Left: Always refers to the animal's left side, not that of the observer.

Right: Always refers to the animal's right side, not that of the observer.

Caudal: Toward the tail.
Cranial: Toward the head.
The body is always caudal to the head; the neck is always cranial to the tail.

When you reach the head or neck and still want to describe forward positions, what do you do? Use the following:

Rostral: Toward the beak. For positions on the head and neck.
The nasal opening is rostral to the ear opening. Although rostral is the official term for directions toward the beak on the head and neck, in much of the literature and in everyday language, cranial and anterior are often used. This course uses the terms interchangeably, when their meaning is clear.

Anterior: Toward the front.
Posterior: Toward the back.
Anterior and posterior can be confusing, because they use an outside frame of reference—the earth. For a standing penguin or person, the belly is anterior and the back is posterior. On a swimming penguin, however, the head is anterior and the tail, posterior. Officially, the terms anterior and posterior should be used only within the eye and inner ear of a bird. In much of the literature and in everyday language, however, anterior and posterior are used interchangeably with cranial and caudal. In places where anterior and posterior are not confusing, this course uses them interchangeably with cranial and caudal.

Transverse Plane: A vertical plane through a bird, dividing the body into cranial and caudal portions.
Frontal Plane: A (usually) horizontal plane through a bird, dividing the body into dorsal and ventral portions.
Sagittal Plane: A vertical plane through the long axis of a bird, extending from head to tail. It divides the body into left and right portions.
These terms refer to planes of sectioning through an animal, and are used in the Handbook of Bird Biology *to describe the perspective from which various sectional views of a bird's internal anatomy are shown.*

Now, if you stand straight up with your hands hanging at your sides, your fingers are distal to your elbows. With your hands held over your head, they are still distal. Even when you stand on your head, your fingers remain distal to your elbows. ∎

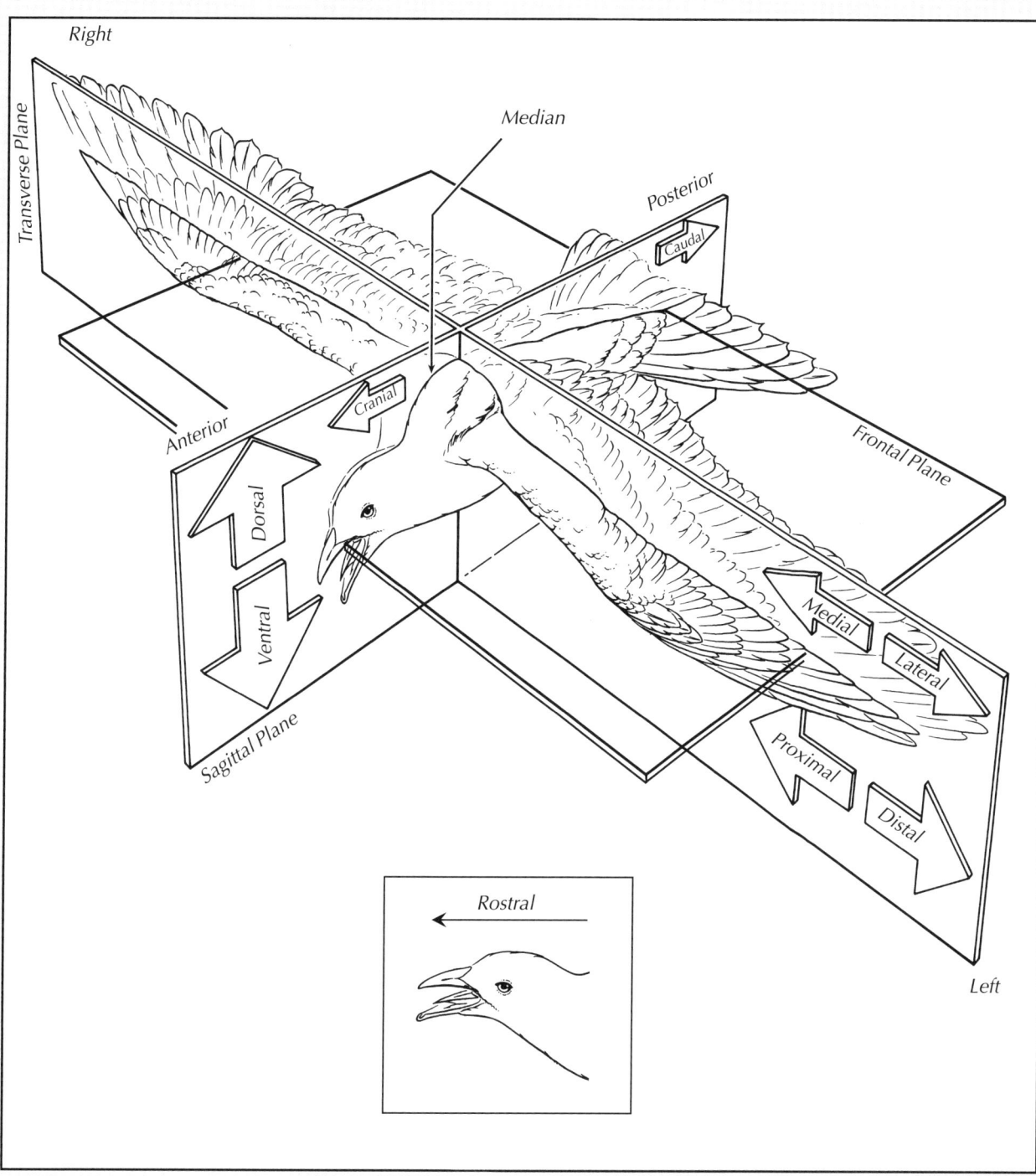

Figure A. Anatomical Directions: *Researchers use specific terms to refer unambiguously to directions and relative locations on the bodies of animals, as illustrated here. Toward the back is termed* **dorsal**, *and toward the belly is* **ventral**; *toward the midline of the body is* **medial**, *toward the side is* **lateral**, *and something positioned on the midline of the body is* **median**; *toward the center of the body is* **proximal**, *and away from the center is* **distal**. *To describe directions toward the head or tail, cranial and caudal are generally used, respectively; but anterior and posterior may be used as well, although their use is sometimes limited to sites within the inner ear and eye. To refer to something in the direction of the tip of the beak, from a point of reference on the head (see inset), the term* **rostral** *is used. The terms* **left** *and* **right** *refer to the animal's own left and right, not those of the observer. It is also useful, for certain anatomical sectional views, to refer to planes cut through an animal. For a bird in the position illustrated, a* **sagittal plane** *extends vertically from head to tail, a* **transverse plane** *extends vertically from side to side, and a* **frontal plane** *extends horizontally. Reprinted and adapted from* Manual of Ornithology, *by Noble S. Proctor and Patrick J. Lynch, with permission of the publisher. Copyright 1993, Yale University Press.*

Figure 1–3. Bird Topography: *Birds have seven main topographic regions: beak, head, neck, trunk, wings, tail, and hind limbs. To facilitate identification and description, ornithologists divide many of these regions further. Drawing by Charles L. Ripper.*

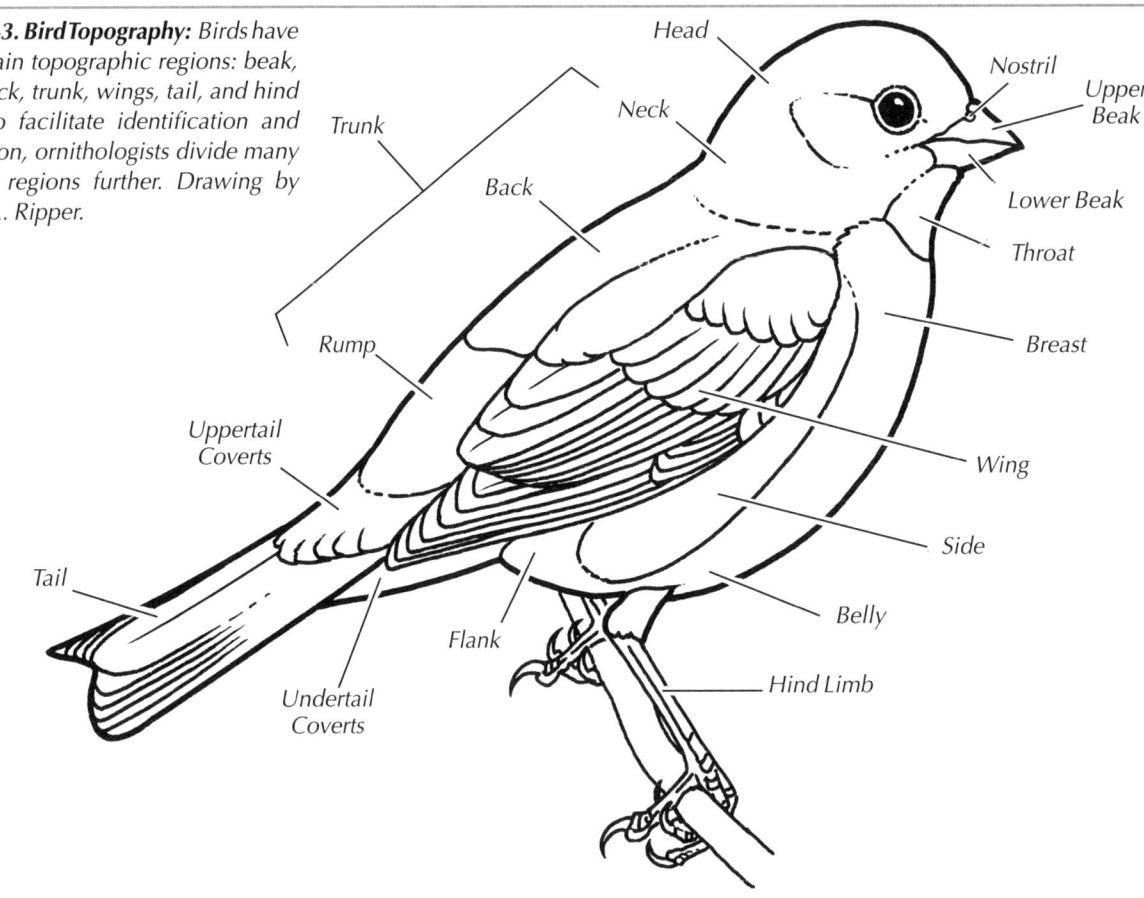

Bill

The bill has two parts, the **upper beak** and the **lower beak**; the upper beak slightly overlaps the lower beak when the bill is closed. As you study birds, you undoubtedly will come across other terms for the upper and lower beak, such as the "maxilla" (upper) and "mandible" (lower), or the "upper and lower mandible." Because "maxilla" technically refers to a specific bone, and "mandible" can have various meanings, this course follows the use by Lucas and Stettenheim (1972) of "upper and lower beak" to refer to the bird's jaws. The bones that make up the beak are the **premaxilla** (upper beak) and **dentary** (lower beak) (see Figs. 3–33 and 4–8).

Although today the terms "bill" and "beak" are used interchangeably, the term "beak" was originally used to describe the decurved (downwardly curved) bills of birds of prey. Each half of the jaws has a bony core that is attached to (and generally considered a part of) the skull, and a horny outer covering, or sheath, called the **rhamphotheca** (see Fig. 3–33). The rhamphotheca—which also makes up the tip and sharp, biting edges of the beak—grows throughout the life of a bird, like claws and fingernails, but continual abrasion during feeding normally keeps its length constant. Birds in captivity often do not get enough to chew on to wear the rhamphotheca down properly, so the bill grows too long and must be trimmed. Toward the base of the upper beak, on each side, is a **nostril** or **naris**. Its shape varies among different species of birds; in some it is partially concealed by a tuft of feathers, and in others it may be absent.

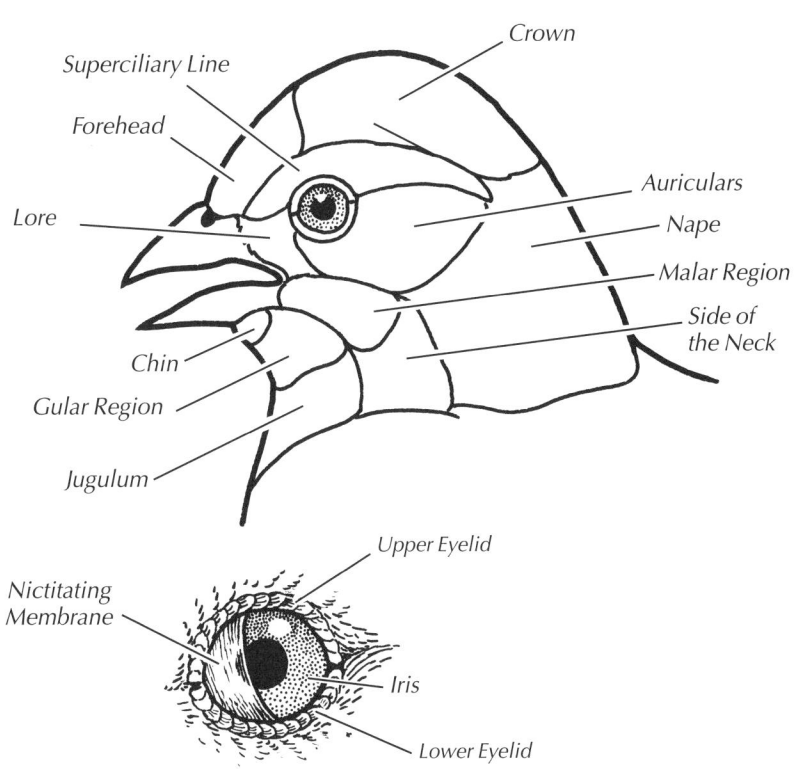

Superciliary Line

Forehead

Lore

Crown

Auriculars

Nape

Malar Region

Side of the Neck

Chin

Gular Region

Jugulum

Nictitating Membrane

Upper Eyelid

Iris

Lower Eyelid

*Figure 1–4. Regions of the Head and Neck: The head and neck regions are subdivided into a number of smaller areas. These subdivisions allow researchers and birders to give precise descriptions of a bird. Note that the **superciliary line** is also called the **eyebrow stripe**, or simply the "eyebrow"; and a stripe in the malar region may be called a **malar stripe**, a **mustache**, or a **whisker stripe**. The close-up view of the eye shows the upper and lower eyelids fully open. A third eyelid, termed the **nictitating membrane**, is partially covering the eye, as it sweeps to close from left to right. This translucent, inner eyelid protects the eye while still permitting birds to see. Drawing by Charles L. Ripper.*

Head and Neck

As you read this section, refer to the profile of the head and neck in **Fig. 1–4**, as well as Fig. 2–8. Caudal to the upper beak are the **forehead**, **crown**, and **nape**. The nape is actually part of the neck. Running back from the upper beak and below (ventral to) the boundary of the forehead and crown is the **eyebrow stripe** or **superciliary line** (also called simply the "eyebrow"), which is distinctively colored in some birds.

The **eye** is very large. Only the dark **pupil** and surrounding colored **iris** show; much more of the eyeball lies under the skin. In most birds, the combined size of the eyes is larger than the brain! In some species, the eye changes color as the bird ages. American Crows have grayish blue eyes while in the nest, but their eyes quickly turn the same dark brown as adults' eyes within a month or so. (Oddly, Australian crows and ravens start out with dark eyes that turn to pale white as adults.) Birds have three eyelids: the **upper eyelid**, the **lower eyelid**, and a "third eyelid," the **nictitating membrane**—a thin, translucent fold that sweeps across the eye sideways from front to back. The nictitating membrane moistens and cleans the eye and protects its surface. All birds blink from time to time, but most blink regularly only with the nictitating membrane. In raptors and other predatory birds, the nictitating membrane protects the eyes as the bird pursues prey through heavy cover, such as a blackberry thicket. Birds close their eyes—usually by raising the lower lid—when they sleep or when their eyes are threatened by something. If you have a live bird, gently move your finger toward the eye and watch the response. Does the

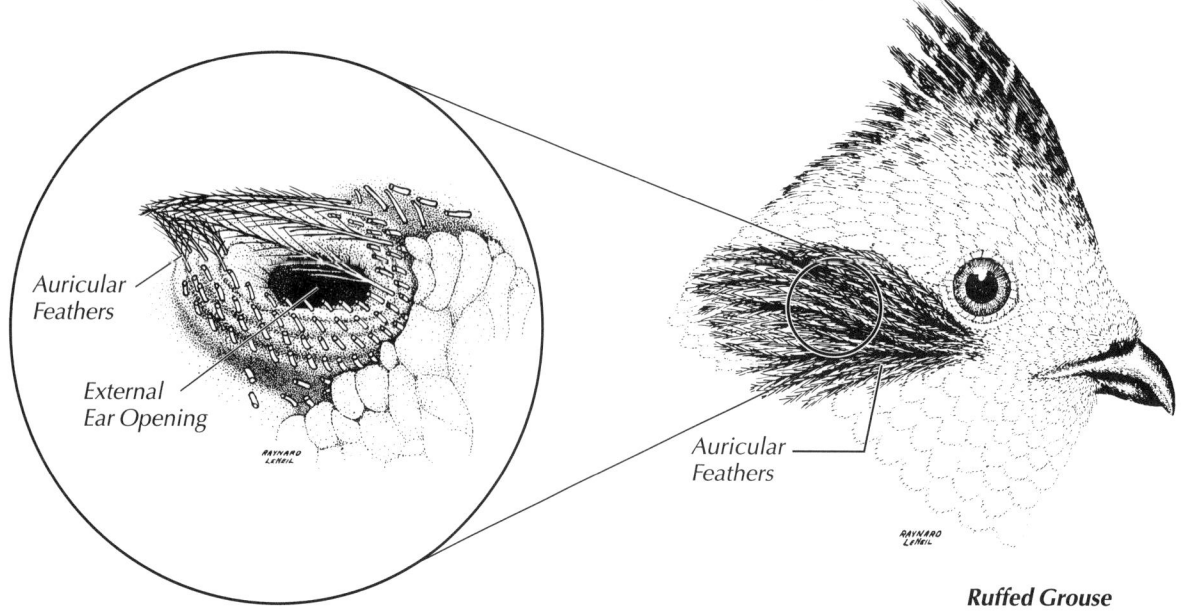

Ruffed Grouse

Figure 1–5. Auricular Feathers of a Ruffed Grouse: The auricular feathers cover the external opening of the ear. Unlike most feathers, the auriculars have an open texture, providing a protective screen from wind noise and debris, while at the same time helping to channel sounds into the ear—much like the external ear flaps of mammals. In many birds, the auriculars are only visible upon close inspection. Inset shows detail for a domestic chicken, with many auricular feathers cropped near the base to reveal the external ear opening. From Lucas and Stettenheim (1972, pp. 99 and 100).

nictitating membrane react first, or one of the eyelids? Around the eye many birds have a circle of differently colored feathers or skin referred to as the **eye ring**.

The small space between the eye and the base of the upper beak is the **lore**. In some birds the lores are distinctively colored; in a few birds they are unfeathered. The small space caudal to the base of the lower beak is the cheek or **malar region**. A **malar stripe** is sometimes referred to as a **mustache** or **whisker stripe**.

Birds, obviously, have no external ear flaps. Below and behind the eye lies a patch of feathers, the **auriculars**, which conceals the ear opening **(Fig. 1–5)**. The feathers are specially formed, having an open texture that provides a protective screen, yet helps to channel sounds into the ear, much like the ear flaps of mammals. The auricular feathers also act like a wind screen on a microphone, reducing the noise of the wind in the ear opening.

Under the lower beak is a very small area, the **chin**, followed backward by the **gular region**, and finally the **jugulum**, which is the lower part of the neck. When describing plumage, people often combine the gular region and jugulum into the "throat." Lying on each side between the jugulum and the nape is the **side of the neck**.

All birds have long necks. Although some birds, such as swans and herons, obviously have long necks, even birds that appear to have no necks at all, such as quail or chickadees, in fact have long necks. These birds typically keep their neck folded in an S shape, and the covering of feathers hides the details **(Fig. 1–6)**. If birds were mammals, none would have necks relatively shorter than a deer or horse. When a bird dies, the muscles relax and the long neck becomes apparent. The long, floppy neck of dead birds has given rise to the misconception that birds commonly die by breaking their necks. (In fact, most birds that strike windows die of head trauma or other internal bleeding.)

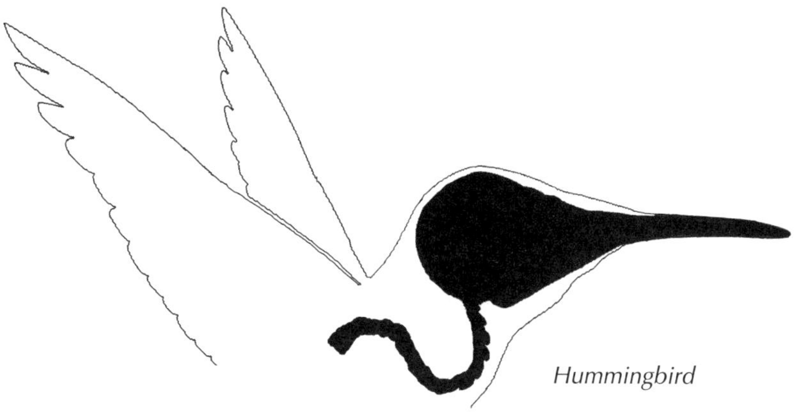

Hummingbird

Figure 1–6. Bird Neck: *All birds have long necks. Their necks do not appear long because they are folded in an S shape and concealed beneath feathers. Even birds such as quail, chickadees, and the hummingbird pictured here, which appear to have little or no neck, actually have relatively long necks. The large dark area in this diagram is the skull, and the long, thin, dark structure curving down from it indicates the chain of vertebrae that make up the neck. Adapted from Gill (1995, p. 94).*

Trunk

The rather compact trunk of birds is divided into the **back, rump, breast, belly, sides,** and **flanks** (see Fig. 1–3). Often the sides and flanks are partly concealed by the wings and visible only when the bird is in flight.

Wings

The wing of a bird **(Fig. 1–7)** is structured like the forelimb of any amphibian, reptile, or mammal, with the same three divisions: the upper arm, or **brachium** (pronounced BRAKE-e-um); the forearm, or **antebrachium;** and the hand, or **manus.** To understand the wing, compare it to the human forelimb as you go along **(Fig. 1–8).** If you have a chicken wing, boil it until the meat comes off and look at the bones.

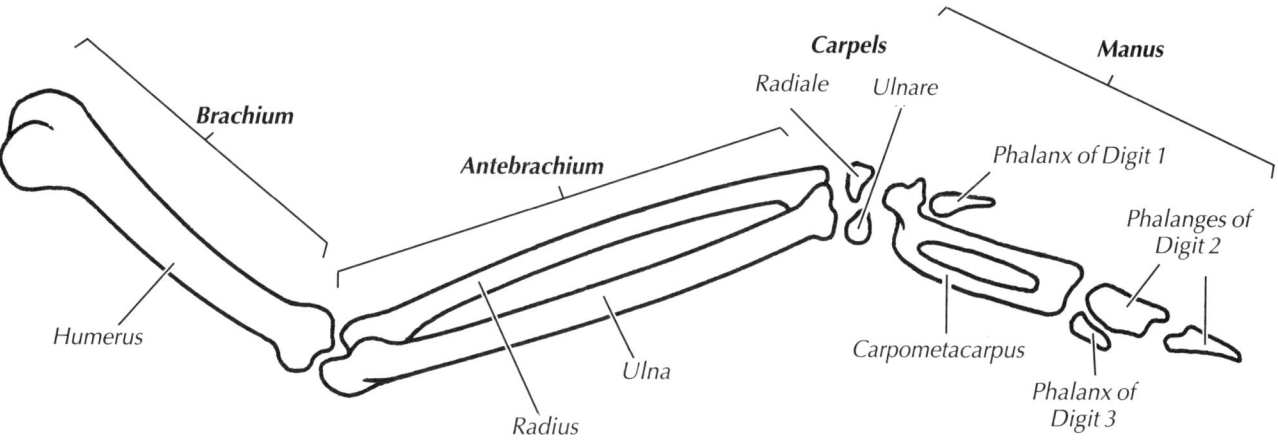

Figure 1–7. Bones of the Wing: *The wing of a bird, like the forelimb of any amphibian, reptile, or mammal, is divided into three sections: the* **brachium** *(upper arm), the* **antebrachium** *(forearm), and the* **manus** *(hand). The brachium is supported by a single long bone, the* **humerus***; the antebrachium, by two long bones of unequal size, the thicker* **ulna** *and the smaller* **radius***. The manus consists of a series of bones that vary in number and size. Notice that the bird "hand" has fewer bones than the highly dexterous human hand (see Fig. 1–8) because many bones are absent or are fused to form a rigid structure important for flight. In birds, the carpals (wrist bones) are reduced to two bones, the* **radiale** *and* **ulnare***, and the metacarpals (palm bones) are fused with some of the carpals to form a single, large* **carpometacarpus***. The finger bones (***phalanges***) of only the first three digits remain, with only digit 2 having more than one bone. Drawing by Charles L. Ripper.*

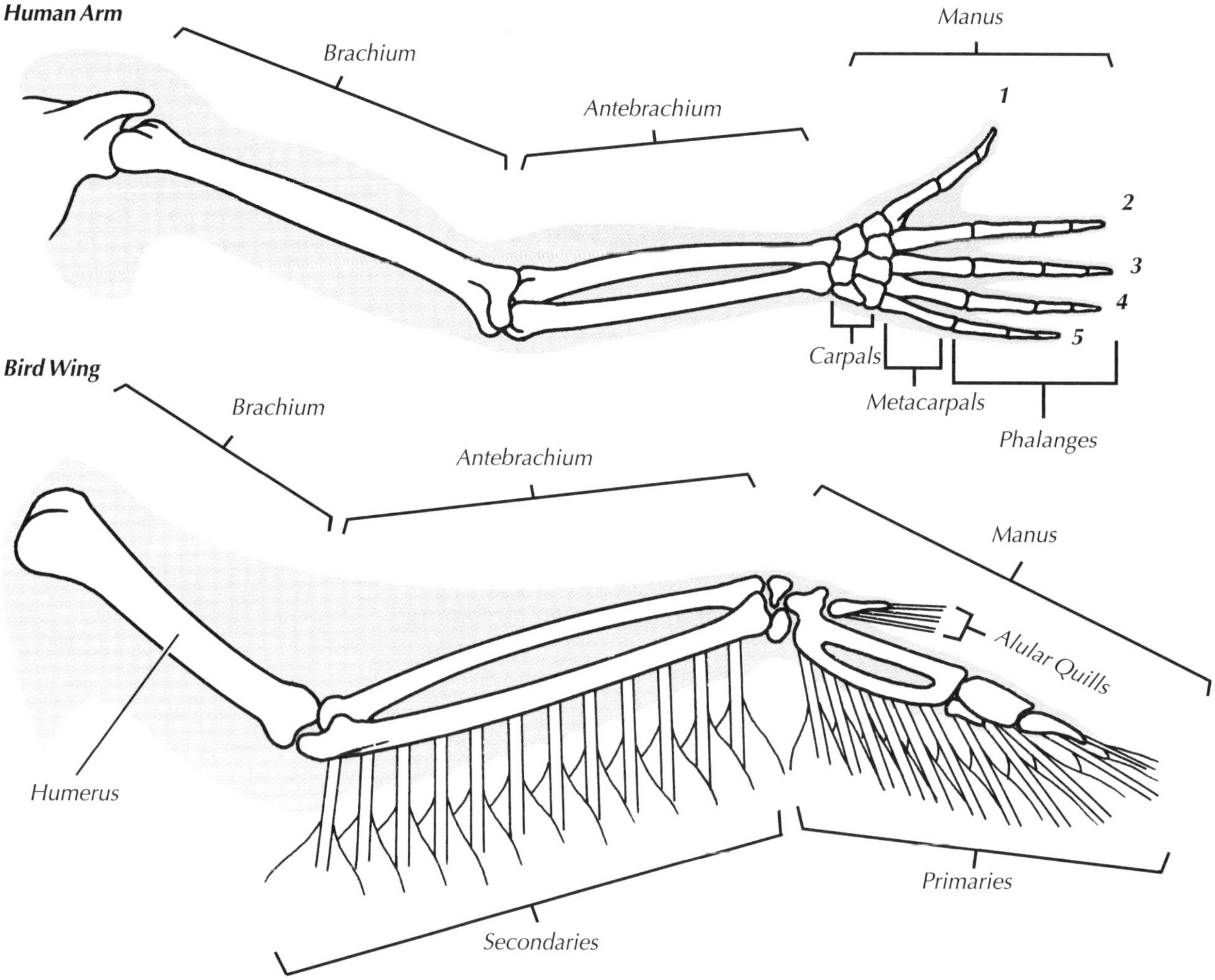

Human Arm

Bird Wing

Figure 1–8. Bird Wing Compared to Human Arm: *Although the wing and arm have the same three main sections, the manus and antebrachium make up a larger portion of the bird wing, providing a long attachment site for the primary and secondary flight feathers. These two sections are what you normally see as the wing when you watch a bird in flight, as the brachium is short and close to the body—the division between it and the antebrachium is generally obscured by feathers. The large muscles that move the wings attach to the humerus, moving the entire wing by moving the humerus. The* **alular quills**—*a small group of feathers attached to the first digit—form the* **alula**, *which helps to keep air flowing smoothly over the upper surface of the wing. Drawing by Charles L. Ripper.*

In both the wing and human arm, note that the brachium is supported by one long bone, the **humerus**; the antebrachium, by two long bones of unequal size, the **radius** and the larger and thicker **ulna**; and the manus, by a series of bones that vary in number, size, and thickness. These are wrist bones, or **carpals**; palm bones, or **metacarpals**; and finger bones, or **phalanges** (singular **phalanx**). Observe that the wing has fewer bones than the human arm because many are fused or absent, leaving two carpals, the **radiale** and **ulnare**; one big, fused palm bone, the **carpometacarpus**; and four phalanges, all the bones that remain of the first three fingers. Only the middle finger, with two of the four phalanges, is still large.

The skeleton of the wing is significantly lighter than that of the forelimb of any terrestrial vertebrate. The long bones are actually hollow; the humerus is even invaded by an air sac from the respiratory system (see Fig. 4–82). Furthermore, compared to your arm, most wings are additionally lightened by having no large muscles. Their principal movements are controlled by tendons coming from huge muscles on the breast. This arrangement takes weight away from the wing and brings it nearer to the bird's center of gravity—a more stable arrangement for a creature that must fly.

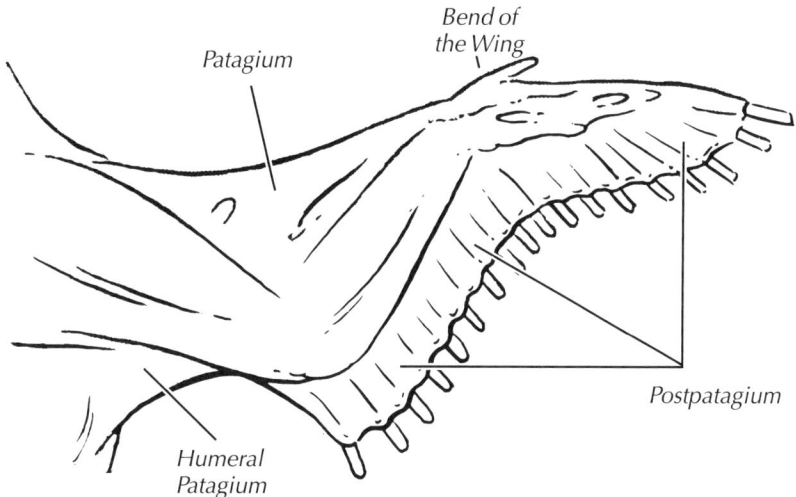

Figure 1–9. Patagia of the Wing: A fold of tough skin, the **patagium**, extends from the brachium to the antebrachium, connecting the shoulder to the wrist. Covered with feathers, it forms the leading edge of the inner wing during flight. A smaller fold of skin, the **humeral patagium**, extends from the brachium to the trunk. During flight, the long wing feathers must be able to withstand the force of the air moving by them. A band of tough, tendinous tissue, the **postpatagium**, surrounds the bases of the flight feathers, supporting and holding them in place. Adapted from Proctor and Lynch (1993, p. 101).

Spread out the wing of your specimen and, by feeling underneath, identify the big bones under the skin. Observe the fold of skin, the **patagium**, that extends from the brachium to the antebrachium, essentially connecting the shoulder to the wrist **(Fig. 1–9)**. The patagium is covered with feathers, and forms the leading edge of the inner wing in flight. A smaller **humeral patagium** extends from the brachium to the trunk. The prominent angle at the wrist is commonly called the **bend of the wing**. Gently open and close the wing, observing the action at each main joint. The wing of a bird is not as mobile as your arm. Note that the wing cannot rotate in a full circle at the shoulder as your arm can, nor can the manus rotate at the wrist as yours can. The joints of the wing are formed only for specific movements in flight, whereas those in your arm are designed for a multitude of different functions. Chapter 5 describes precisely how a bird uses its wings in flight.

Refer to **Fig. 1–10** as you work through the following information on the feather groups of the wing. The longest wing feathers are the flight feathers, or **remiges** (singular, **remex**), the long, stiff quills that extend distally from the bones. The remiges are stiff feathers that form the predominant air-catching portion of the wing. Above these feathers are the **coverts**, the smaller feathers that overlap the flight feathers at their bases like evenly spaced shingles on a roof. From two to six (usually three) **alular quills** (pronounced AL-you-lar; a quill is a feather) project from the phalanx of the first finger (the bird's "thumb"; see Fig. 1–8) at the bend of the wing. The alular quills make up the **alula**, sometimes known as the "bastard wing," which can be spread apart from the rest of the wing and is used for fine control of airflow over the wing.

The remiges emerging from the manus are the **primaries** and those from the antebrachium are the **secondaries**. By feeling their bases under the wing, you will find (in flying birds) that all are firmly attached to the skeleton by ligaments—the primaries to the bones of the manus and the secondaries to the ulna. A tough band of tendinous tissue, the **postpatagium**, also holds the remiges firmly in place and supports each quill (see Fig. 1–9). Each individual within a species normally has the same number of remiges. Overlying each remex on the upper surface of the wing is one **greater primary covert** or **greater secondary covert**. Overlying the greater coverts are the **median coverts**, and overlying them are the **lesser coverts**. All are in distinct rows. The remaining feathers that complete the forward roofing of the wing are the **marginal coverts**. Finally, note again the alular quills overlying the bases of the greater primary coverts.

Other features of the feathering that you should note are the **scapulars**, a group of feathers emerging from the upper surface of the brachium and shoulder, but not attached to the bone; the **underwing coverts**, often collectively called the "lining of the wing"; and the **axillaries**, a cluster of feathers in the "armpit" that are recognizably longer than those lining the wing.

Tail

The tail of a bird is technically a small bony and fleshy structure marking the end of the vertebral column, but most people, including ornithologists, use the term "tail" to mean the feathers arising from the "official" tail. The long, stiff flight feathers of the tail are the **rectrices** (singular, **rectrix**); the shorter feathers overlying their bases, above and below, are the **coverts** (called the **uppertail coverts** and **undertail coverts**, respectively). The rectrices are paired, one member of each pair on each side of the tail, with no feather in the middle (see Fig. 3–7). Most birds have five or six pairs, but some, like the Ruffed Grouse, have more. Like the flight feathers of the wing, the number is the same in each species (with only the occasional mutant). Note that when the tail is not spread, each rectrix is overlapped by the one next to it; the exception is one of the middle pair, which lies on top. Because the middle (or "deck") feathers protect the other feathers, they often are the most badly worn of the rectrices.

Hind Limbs

As in the wing, the structure of the bird's hind limb is similar to that of a human, with three divisions: the upper leg, or **thigh**; the lower leg, or **crus**; and the **foot**. Unlike the hind limb of a human, however, the foot has been elongated and modified into two functional sections, more like the hind limb of a horse or dog.

In most birds, the thigh bone is rather short and often hidden by body feathers. Because of the elongated lower leg and foot, a bird has three elements to the leg, not just two as in the human leg. Note that birds' knees bend in the same direction that yours do, but the actual location of the knee can be confusing. Just as the length of the neck is hidden by posture and feathers, so, too, is the thigh hidden from sight, making the ankle appear to be the knee.

The hind limb of a bird and human are compared in **Figure 1–11**. First, notice that in both, the thigh is supported by one long bone, the **femur**. Next, note that in humans the crus (lower leg) is supported by two long bones, the **tibia** and **fibula**, and the foot is composed of a series of bones: the ankle bones (**tarsals**), the instep bones (**metatarsals**),

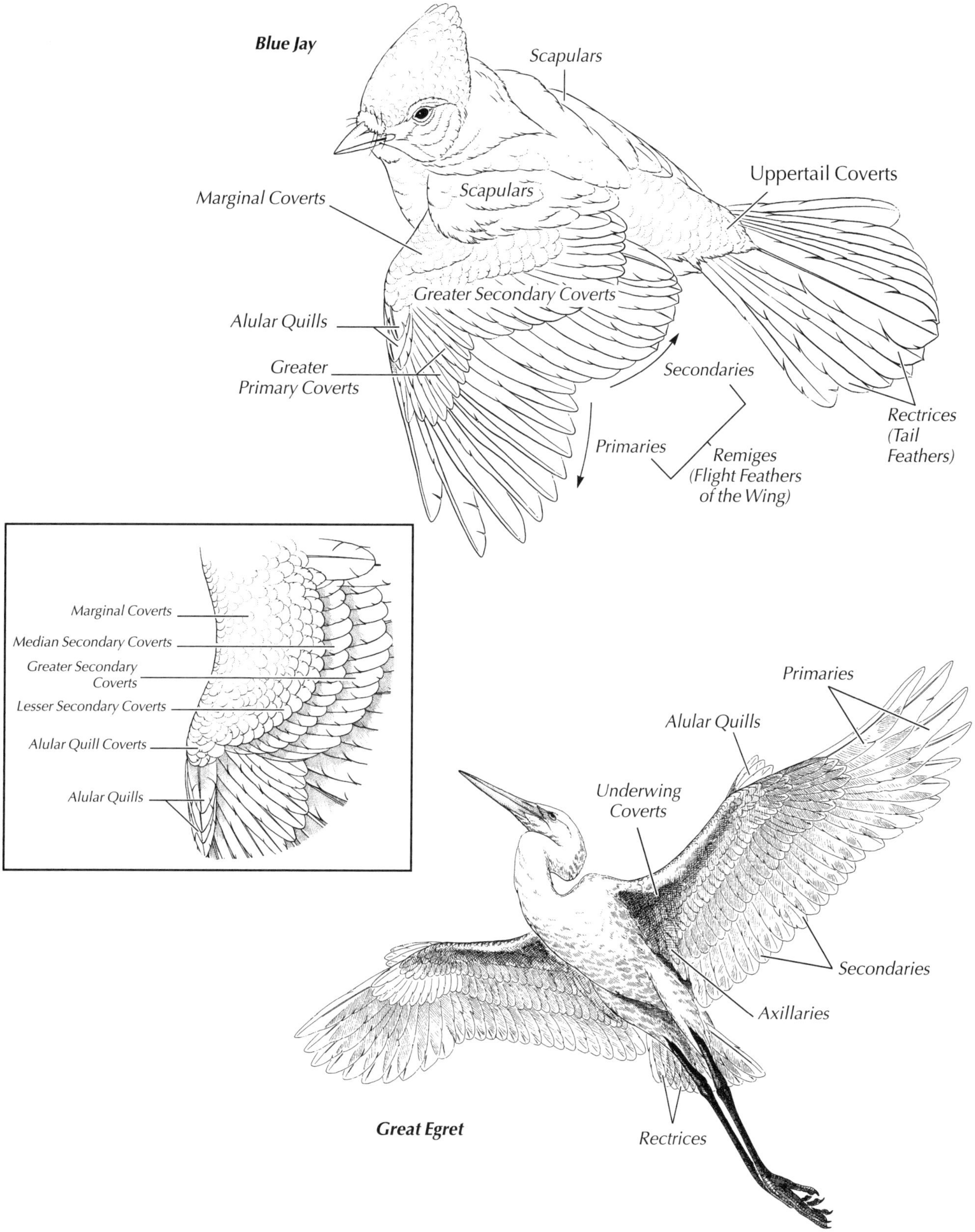

Blue Jay

Scapulars

Scapulars

Uppertail Coverts

Marginal Coverts

Greater Secondary Coverts

Alular Quills

Greater
Primary Coverts

Secondaries

Rectrices
(Tail
Feathers)

Primaries

Remiges
(Flight Feathers
of the Wing)

Marginal Coverts

Median Secondary Coverts

Greater Secondary
Coverts

Lesser Secondary Coverts

Alular Quill Coverts

Alular Quills

Primaries

Alular Quills

Underwing
Coverts

Secondaries

Axillaries

Rectrices

Great Egret

Figure 1–10. Feathers of the Wings and Tail: *The major groups of feathers are illustrated in dorsal view on a Blue Jay, and in ventral view on a Great Egret. Note that the details of feather arrangement—such as number, size, and shape—vary dramatically among different species, but that the same main groups of feathers are present in most birds. The inset, a dorsal view of a Rock Dove wing, shows a detailed view of the secondary coverts. See text for a description of each feather group. Blue Jay and Great Egret reprinted from* Manual of Ornithology, *by Noble S. Proctor and Patrick J. Lynch, with permission of the publisher. Copyright 1993, Yale University Press. Inset adapted from Proctor and Lynch (1993, p. 59).*

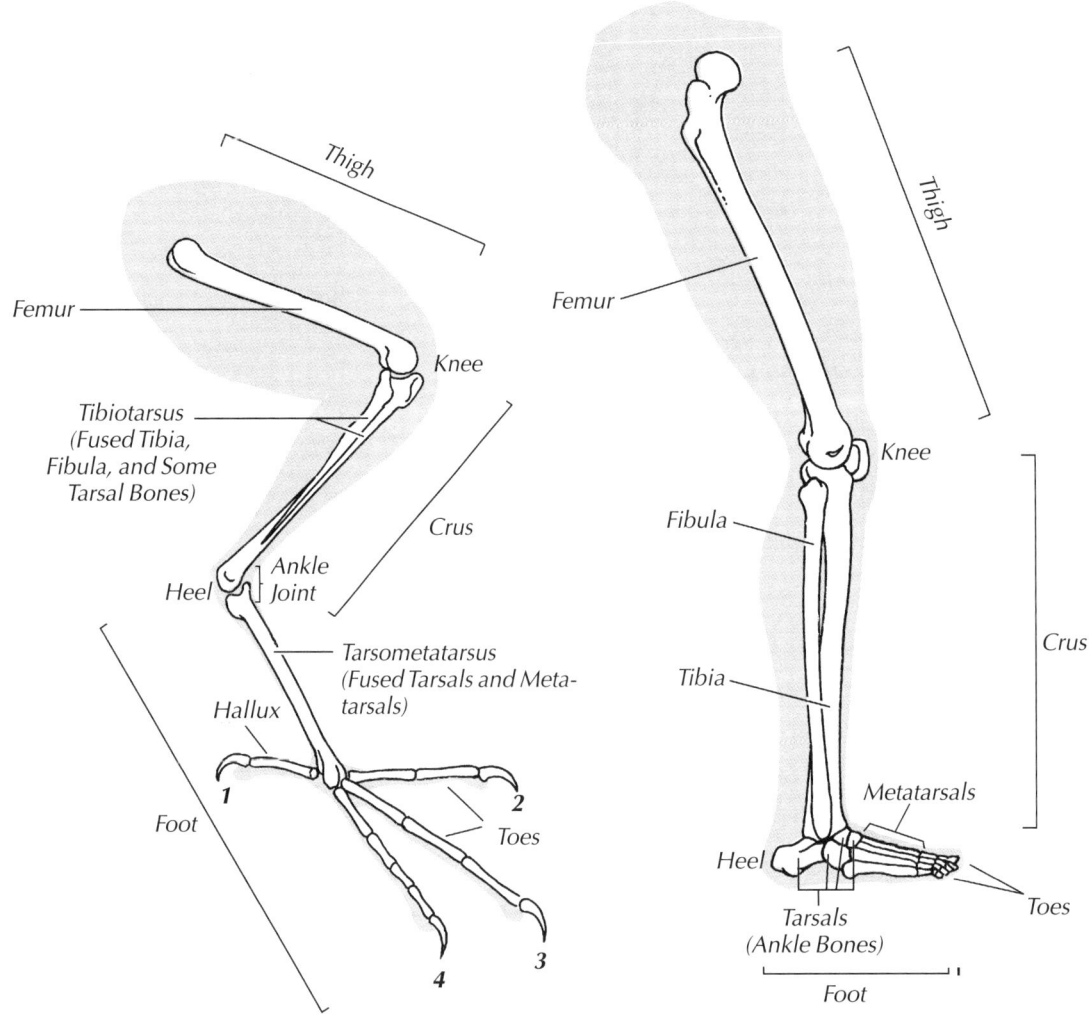

Figure 1–11. Comparison of Human Leg and Bird Leg: *The leg of both birds and humans has three main divisions: the* **thigh** *(upper leg), the* **crus** *(lower leg), and the* **foot**. *In both birds and humans, the thigh is supported by a single bone, the* **femur**. *Two bones, the* **tibia** *and* **fibula**, *support the crus in humans. In birds, however, the crus is supported by a single long bone, the* **tibiotarsus**—*the familiar "drumstick" of a cooked turkey. The tibiotarsus is formed by the fusion of the tibia with two or more of the* **tarsal** *(ankle) bones. The fibula is present, but reduced to a fine, needlelike bone running about two-thirds of the way down the side of the tibiotarsus. In addition to the tarsal (ankle) bones, the human foot consists of* **metatarsals** *(instep bones) and* **phalanges** *(toe bones). The bird foot, however, is quite different. It has been elongated and modified into two functional sections. The upper section consists of a single long bone, the* **tarsometatarsus** *(also called simply the* **metatarsus**), *formed by the fusion of the remaining tarsals and metatarsals. The lower section consists of the* **phalanges**. *Birds have four toes, at most, although their arrangement differs from species to species. One common arrangement, shown here for a bird's right foot, is with the first toe, termed the* **hallux**, *pointing back, and three toes in front, numbered from medial to lateral. At first glance, the sections of a bird's leg can be confusing. The thigh is short and, along with the knee, usually hidden by the body feathers. Thus, the joint that appears to be a knee bending backward is actually the ankle. A bird's knee bends in the same direction as a human's. Birds walk up on their toes, with their heel in the air, as do horses and dogs. Drawing by Charles L. Ripper.*

and the toe bones (**phalanges**). Compare these bones with those in the bird and you will see numerous differences.

In the crus of a bird there is only one long bone, the **tibiotarsus**, extending between the knee and heel (this is the chicken's "drumstick"). The tibiotarsus is a fusion of the tibia with two or more of the tarsal bones. What there is of the fibula is a needlelike bone (perhaps you have noticed it when gnawing on a drumstick) running two-thirds of the way down the side of the tibiotarsus. In some birds, such as the turkey, the tendons that connect the leg muscles with the toes are calcified (stiffened by deposited calcium salts, like those in bone) and may appear to be small, thin bones.

The foot of a bird contains one long bone, the **tarsometatarsus**, and the phalanges of the toes. The tarsometatarsus—sometimes called simply the **metatarsus**—represents a fusion of tarsals (ankle bones) and metatarsals (instep bones) and forms the skeleton of the **tarsus**, the general name for the section of the foot between the heel and toes. Birds have, at most, four toes, and each one bears a claw. The toes are given numbers to correspond with those of birds' ancestors. The earliest vertebrates had five toes, all of them pointed forward (**Fig. 1–12**), numbered (by researchers) from the inside out (medial to lateral). In birds, the first toe (if present) projects backward and is called the **hallux**. Thus, the second toe of early vertebrates is the innermost toe of birds. Many species of birds have no hallux, with only three forward-facing toes. Ostriches have only two toes (**Fig. 1–13**). Because the tarsus is elevated, with the toes attached to the distal end, a bird walks on its toes only, keeping its heel off the ground.

A bird's leg and foot, like the wing, have been structurally lightened. The femur, like the humerus, receives an air sac from the respiratory system. The toes and most, if not all, of the tarsus have only a thin, scaly covering. Movements of the toes are controlled by tendons extending from muscles in the crus.

If you have a live bird or a freshly killed specimen, slowly extend its hind limb to full length, then flex it close to the body, while watching the movement of the toes. You will observe that when you extend the limb the toes open, and when you flex it, the toes close into a position for grasping. The toes close because of tension placed on the tendons as the heel bends (**Fig. 1–14**). When a bird squats on a perch to sleep, the toes thus automatically grip the perch and stay locked onto it until the bird awakens and stands up.

Diversity in Bird Form

■ Now that you are familiar with the form of a bird and have become acquainted with the names for the different parts, you are prepared to take a broad look at the ways birds differ from one another in form and appearance. This step is essential before learning how birds are classified and named.

Although the ability to fly imposes certain restrictions on size and weight, the range in birds is astonishing—from the male Bee Hum-

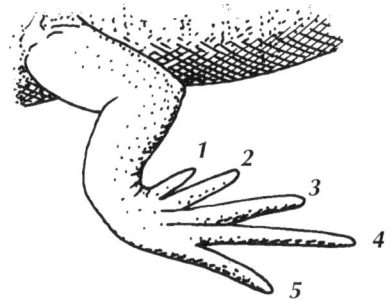

Figure 1–12. Primitive Foot: The feet of the earliest terrestrial vertebrates had five forward-pointing toes, numbered from medial to lateral. The four (at most) toes of birds are thought to correspond to toes one through four of these vertebrates.

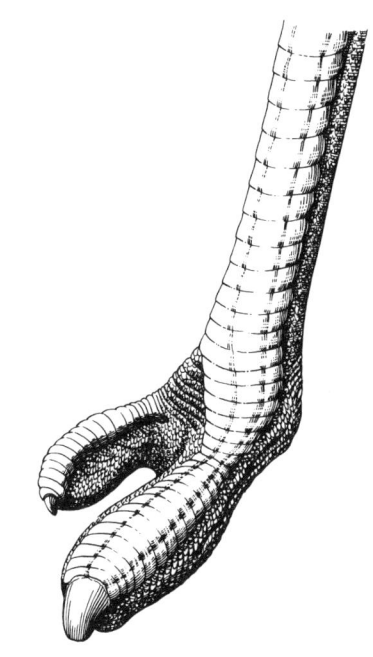

Figure 1–13. Ostrich Foot: In many fast-running animals, natural selection has reduced the surface area contacting the ground. Just as the feet of horses are reduced to hooves, the feet of Ostriches are reduced to two short, stout toes, one poorly developed. On the bottoms of both toes are soft, elastic pads, which prevent the feet from sinking into soft sand. The flightless Ostriches rely on their ability to maintain a steady running speed of about 31 miles (50 km) per hour—to escape predators in the open savannas and deserts of Africa. From Birds: Readings from Scientific American, edited by Barry W. Wilson. Copyright 1980 by W. H. Freeman and Company. Used with permission.

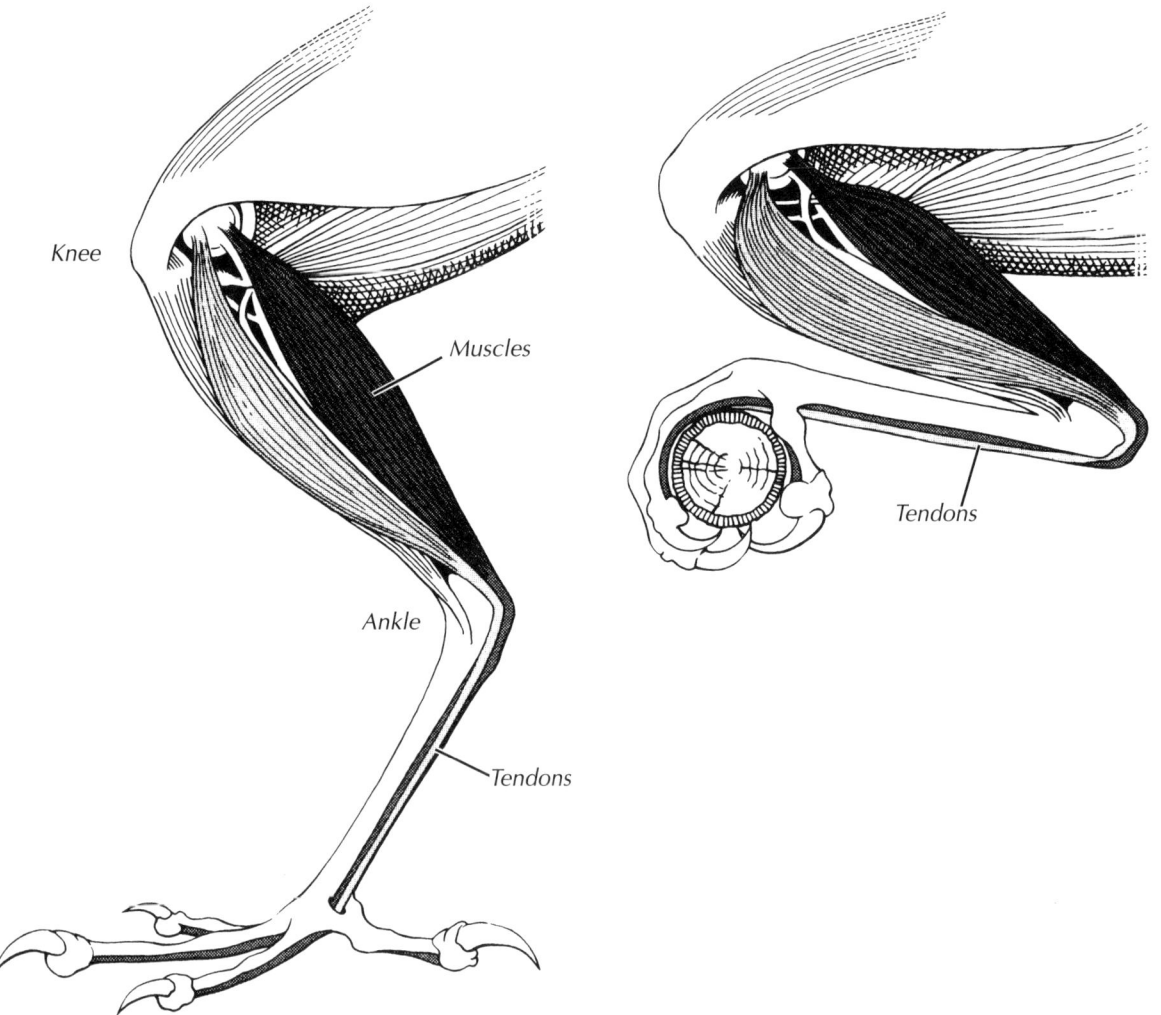

Knee

Muscles

Tendons

Ankle

Tendons

Figure 1–14. The Mechanics of Perching: In the legs of tree-dwelling birds, the tendons from certain muscles extend down the leg behind the ankle to attach to the tips of the toes. When a perched bird bends its ankle to lower itself onto a branch, the bending automatically creates tension on these tendons, which pulls on the toes, forcing them to close around the branch. Thus, when the bird is at rest, its toes tightly grasp the perch. *From* Birds: Readings from Scientific American, *edited by Barry W. Wilson. Copyright 1980 by W. H. Freeman and Company. Used with permission.*

mingbird with a wingspan of 2.6 inches (65 mm) and a weight of less than 1/14 ounce (2 gm) to the Wandering Albatross with a wingspan of over 11.5 feet (3.5 m) and a weight of as much as 19 pounds (8.6 kg) **(Fig. 1–15).** The heaviest birds cannot fly; the flightless Ostrich may weigh 345 pounds (156.5 kg); the Emperor Penguin, 100 pounds (45.3 kg). The heaviest of flying birds—the Mute Swan, Wild Turkey, Kori Bustard, and Great Bustard—weigh up to 20 to 30 pounds (11 to 15 kg), with extreme individuals of the large bustards reported up to 44 pounds (20 kg).

Perhaps because of their adaptations for flight, all birds are similar in general configuration, and all are immediately identifiable as birds. Within these general limits the external form of birds varies widely, reflecting their specific adaptations both to the habitats in which they live and to their methods of acquiring food. The following sections describe some of the more common variations in the bill, wings, tail, and feet.

The Bill

The bills of birds have different shapes for reaching, picking up, and manipulating different types of food **(Fig. 1–16).** Bills can be short,

Figure 1–15. Bird Sizes: *Birds range in size from the tiny male Bee Hummingbird with a wingspan of 2.6 inches (65 mm) to the Wandering Albatross with a wingspan of over 11.5 feet (3.5 m). Here, a hummingbird perches on a flight feather from the wing of a large bird. Drawing by Charles L. Ripper.*

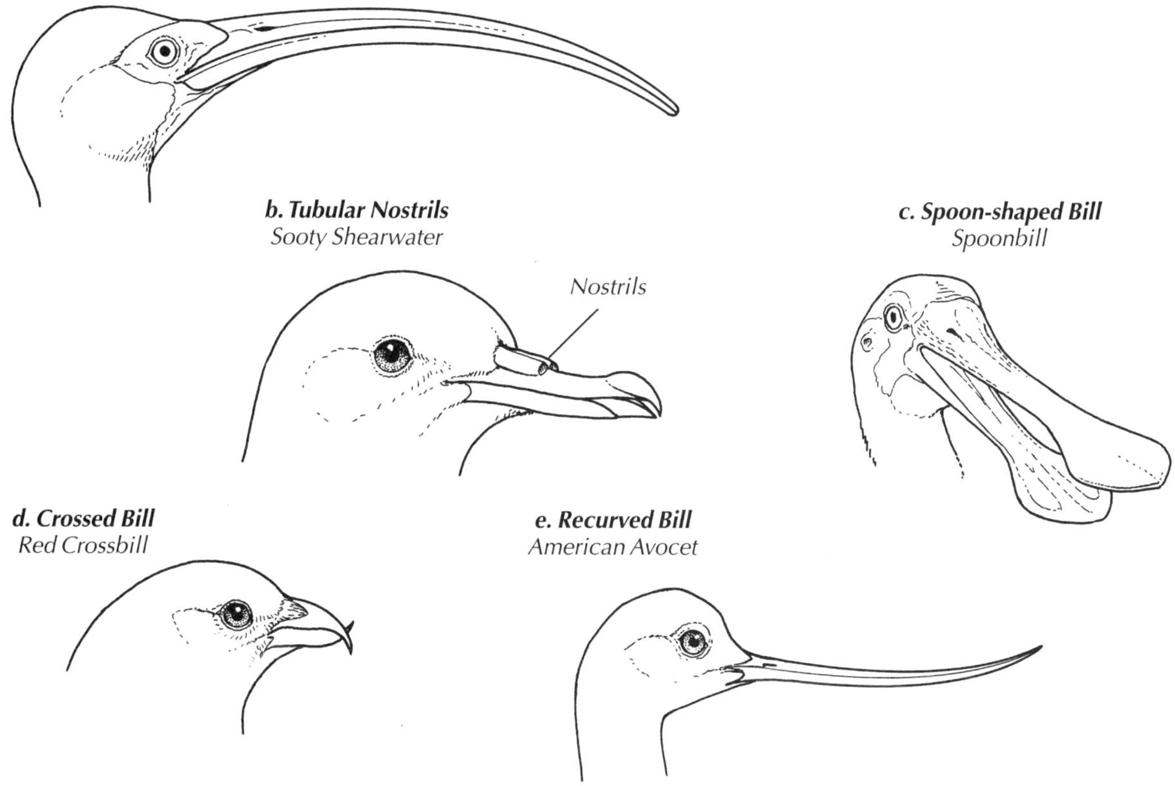

a. Decurved Bill
White Ibis

b. Tubular Nostrils
Sooty Shearwater

Nostrils

c. Spoon-shaped Bill
Spoonbill

d. Crossed Bill
Red Crossbill

e. Recurved Bill
American Avocet

Figure 1–16. Bill Diversity: *The bills of birds have been molded by natural selection into a great variety of shapes and sizes, each suited to a particular foraging strategy. In addition to gathering food, birds use their bills for fending off predators, courting, nest building, and preening.* **a. Decurved Bill of White Ibis:** *The White Ibis uses its long,* **decurved** *(downward curved) bill to probe mud and sand for invertebrates. It also sweeps its bill from side to side in the manner of the spoonbill (see* **c**). **b. Tubular Nostrils of Sooty Shearwater:** *Many birds that spend much of their time at sea, and thus must drink salt water, have their nostrils at the ends of long tubes on top of the bill. These* **tubular nostrils** *help with the elimination of excess salt (see Fig. 3–34b).* **c. Spoon-shaped Bill of Spoonbill:** *The spoonbill sweeps its* broad, flat bill through shallow water from side to side to search for small aquatic animals such as crustacea and mollusks. In murky water, however, the spoonbill sweeps its bill just above the bottom, the curved dorsal and flat ventral surfaces creating swirling currents that pull prey up off the bottom and into the water, where they can be captured more easily. **d. Crossed Bill of Red Crossbill:** *The crossed tips of crossbill bills allow them to efficiently pry seeds from deep within open or closed cones of pines or other conifers.* **e. Recurved Bill of American Avocet:** *The American Avocet sweeps its long, sensitive,* **recurved** *(upward curved) bill from side to side in water, in a similar manner to that of the spoonbill. Drawings by Charles L. Ripper.*

long, stout, thin, pointed, or blunt. Bills may curve up or down, or may be faintly or conspicuously notched, spoon-shaped, or crossed. Bills can be specialized for cutting flesh, filtering small organisms from water, stabbing, grasping, hammering, picking up small insects, or opening large, hard fruits (see Ch. 4, Oral Cavity, Bill).

The nostrils (nares) also may have various shapes: oval, circular, or slit-like; sometimes they have bony tubercles in their centers or are located at the ends of elongated tubes; sometimes they lack a septum between the two sides (you can look in one and out the other!); and sometimes they are surrounded by a fleshy **cere** or overarched by a fleshy **operculum** (see Fig. 3–34). Some birds, such as boobies and gannets, lack external nostrils altogether. In some cases the function of the peculiarities is known; in many cases, it is not. Nostrils and their various forms are discussed further in Ch. 3.

The Wings

The wings are considered **long** when the distance from the bend of the wing to the tip is longer than the trunk of the bird, or **short** when this distance is the same or less. Wings are **rounded** when the middle primaries are longest, or **pointed** when the outermost primaries are longest. They are considered **narrow** when all the remiges are short, or **broad** when all the remiges are long; **concave** when the curvature of the wing's underside is extreme, or **flat** when the curvature is unusually slight (**Fig. 1–17**).

Figure 1–17. Wing Diversity: Wings have evolved a tremendous diversity of shapes, each suited to a different flight style. The four examples here are viewed from above. In **a** and **b** the curves labeled "wing curvature" are cross sections through the outstretched wing from the leading, cranial edge (thicker part of curve, at left) to the trailing, caudal edge (thinner part of curve, at right). They represent the curvature of the top and bottom surfaces of the wing. **a. Flat Wings of a Swallow:** Swallow wings have little curvature and are considered **flat**. **b. Concave Wings of a Grouse:** Grouse wings are curved below, and are considered **concave**. **c. Round Wings of a Hawk:** In some hawks, the middle primary feathers are the longest, creating a **rounded** wing tip. **d. Pointed Wings of a Gull:** In gulls, the outermost primary feathers are the longest, creating a **pointed** wing. Drawings by Charles L. Ripper.

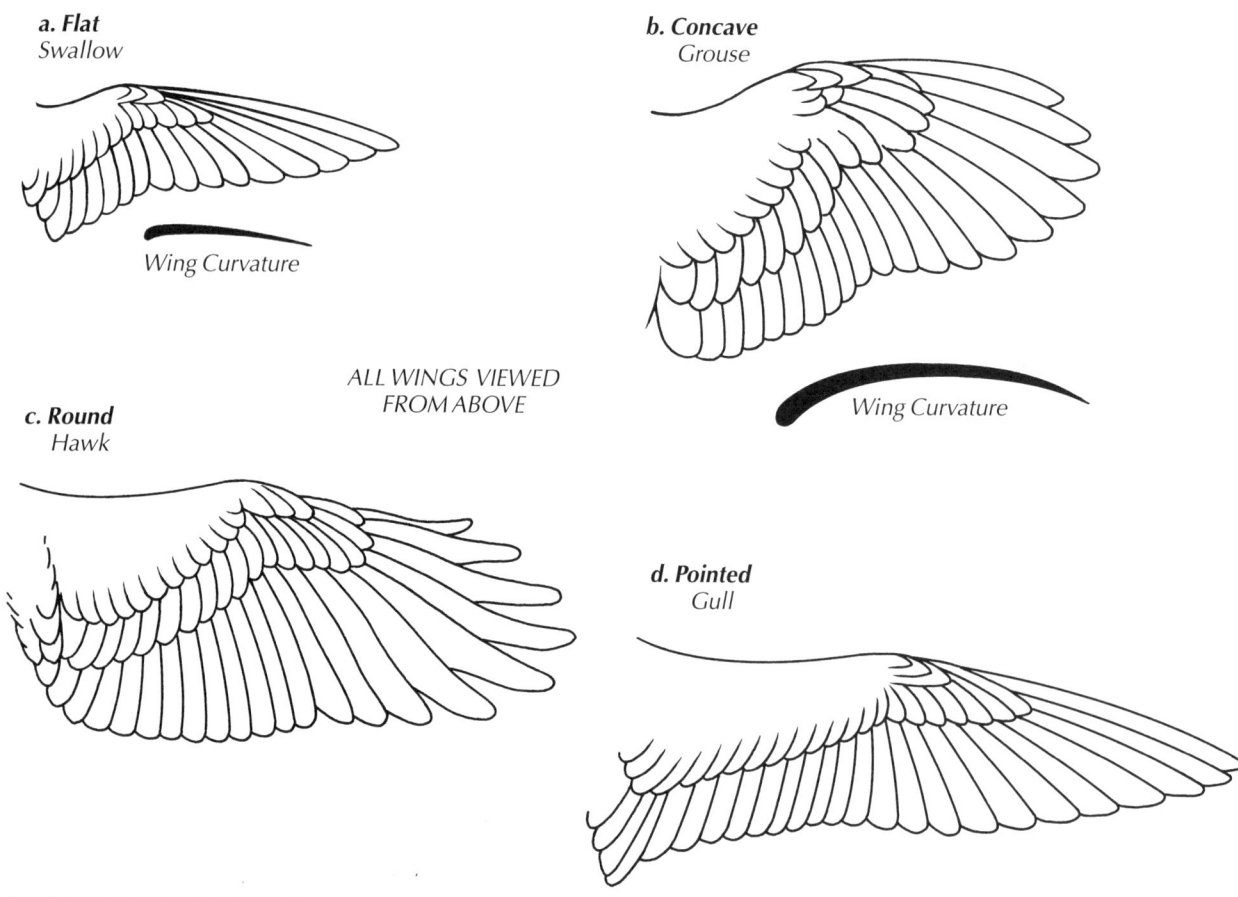

a. Flat
Swallow

b. Concave
Grouse

Wing Curvature

ALL WINGS VIEWED FROM ABOVE

Wing Curvature

c. Round
Hawk

d. Pointed
Gull

The shape of birds' wings is not arbitrary—each is suited to its own specific purpose. The shape of a wing directly affects the way a bird can fly, affecting such things as the amount of lift and drag that the wing creates when moving through the air. Long, narrow, pointed wings are best suited to long-distance flying and soaring over the seas; long, broad, rounded wings to long-distance flying or soaring over land; short, rounded wings to short flights in forests and fields; rounded, concave wings for quick take-off and rapid escape over short distances; and pointed, flat wings for quick wing action and swift flight. Wing shape and its effect on flight is discussed in more detail in Chapter 5.

The Tail

The tail is considered **long** if it is obviously longer than the trunk, or **short** if it is the same length or shorter than the trunk. It is practically absent in a few birds, such as grebes. Usually the tail owes its shape to the relative lengths of the rectrices and the way they terminate at the trailing margin **(Fig. 1–18)**. Thus, the tail is **square** at the end if the rectrices are all about the same length (for example, Clark's Nutcracker); **rounded**, if they become slightly longer from the outside in (for example, American Crow); **graduated**, if they become abruptly longer from the outside in (for example, Black-billed Magpie); **pointed**

Figure 1–18. Tail Shapes: Like beaks and wings, bird tails come in a variety of shapes and lengths. Although the functions of different tail shapes are not well understood, the tails of birds are important in flight and various types of displays. Tail shape is determined by the relative lengths and shapes of the rectrices (tail feathers). All tails shown here are viewed from below. **a. Round (American Crow):** *The rectrices become slightly longer from the outside in (lateral to medial).* **b. Graduated (Black-billed Cuckoo):** *The rectrices become abruptly longer from the outside in.* **c. Forked (Common Tern):** *The rectrices become abruptly longer from the inside out.* **d. Pointed or Acute (Ring-necked Pheasant):** *The middle rectrices are much longer than the others.* **e. Emarginate or Notched (Pine Siskin):** *The rectrices become slightly longer from the inside out.* **f. Square (Sharp-shinned Hawk):** *The rectrices are all about the same length. Drawings by Charles L. Ripper.*

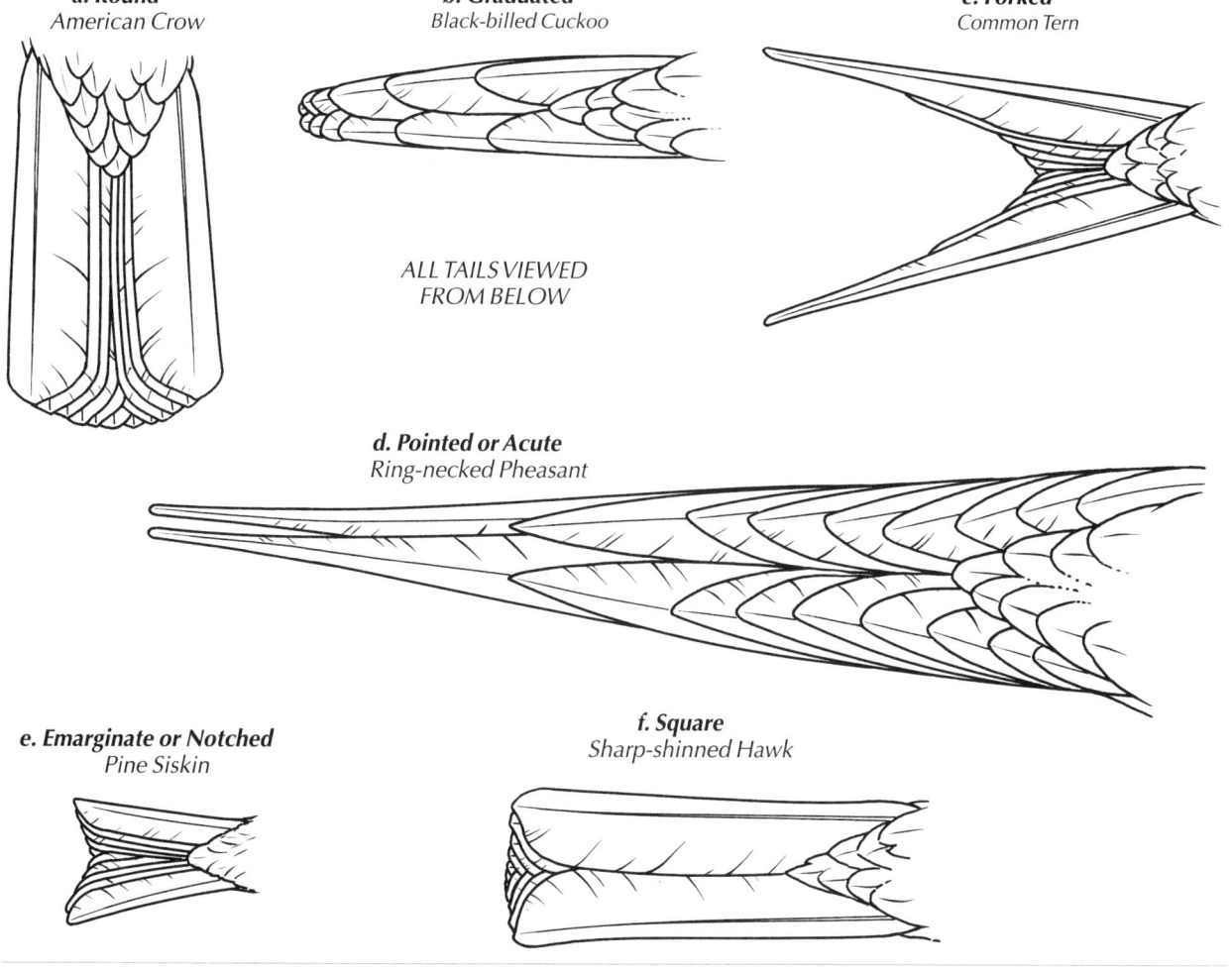

a. Round
American Crow

b. Graduated
Black-billed Cuckoo

c. Forked
Common Tern

ALL TAILS VIEWED
FROM BELOW

d. Pointed or Acute
Ring-necked Pheasant

e. Emarginate or Notched
Pine Siskin

f. Square
Sharp-shinned Hawk

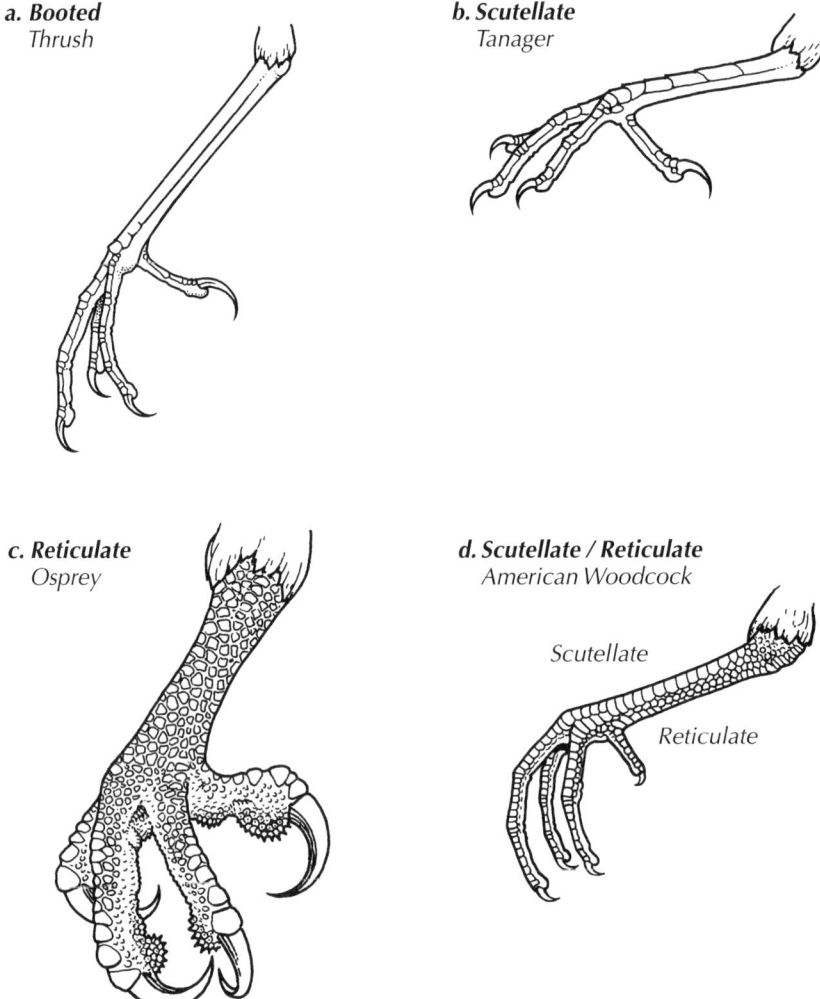

Figure 1–19. Forms of the Podotheca: *The tough skin covering the tarsus, termed the **podotheca**, has different forms in different types of birds. The functions of these forms, if any, remain unknown. **a. Booted (Thrush):** In thrushes the podotheca is smooth, divided into long, continuous, nonoverlapping scales. **b. Scutellate (Tanager):** In most birds with bare legs, such as the songbirds, the podotheca is broken up into overlapping scales. **c. Reticulate (Osprey):** In some birds, such as falcons, plovers, and the Osprey, the podotheca is divided into a network of small, irregular, nonoverlapping plates. **d. Scutellate/Reticulate (American Woodcock):** Different sides of the feet and legs may have different forms of podotheca in some species. In the American Woodcock, for example, the front of the leg and top of the foot is scutellate, and the back and bottom surfaces are reticulate. Drawings by Charles L. Ripper.*

a. Booted
Thrush

b. Scutellate
Tanager

c. Reticulate
Osprey

d. Scutellate / Reticulate
American Woodcock

Scutellate

Reticulate

or **acute**, if the middle rectrices are much longer than the others (for example, Mourning Dove); **emarginate** or **notched**, if the rectrices become slightly longer from the inside out (for example, Violet-green Swallow); or **forked**, if they become abruptly longer from the inside out (for example, Swallow-tailed Kite).

The Feet

The tough skin covering the tarsus, the **podotheca**, has different forms in different types of birds **(Fig. 1–19)**. It may be smooth (**booted**), broken up into overlapping scales (**scutellate**), or divided into numerous small, irregular plates (**reticulate**). Sometimes the tarsus is booted only in the back and scutellate in the front, or scutellate in front and reticulate in back. The functions, if any, of these different forms are not known.

Most birds have four toes with one toe (the **hallux**) directed backward, the others, forward. This arrangement is known as **anisodactyl** **(Fig. 1–20)**. In a few birds, however, either the second or fourth toe joins the hallux in its backward direction. In birds with **zygodactyl** feet, such as woodpeckers, cuckoos, parrots, and others, the reversed toe is the fourth. (In some, such as the Osprey and owls, the outer toe

TYPE OF FOOT

TOE CONFIGURATION
(Right Foot)

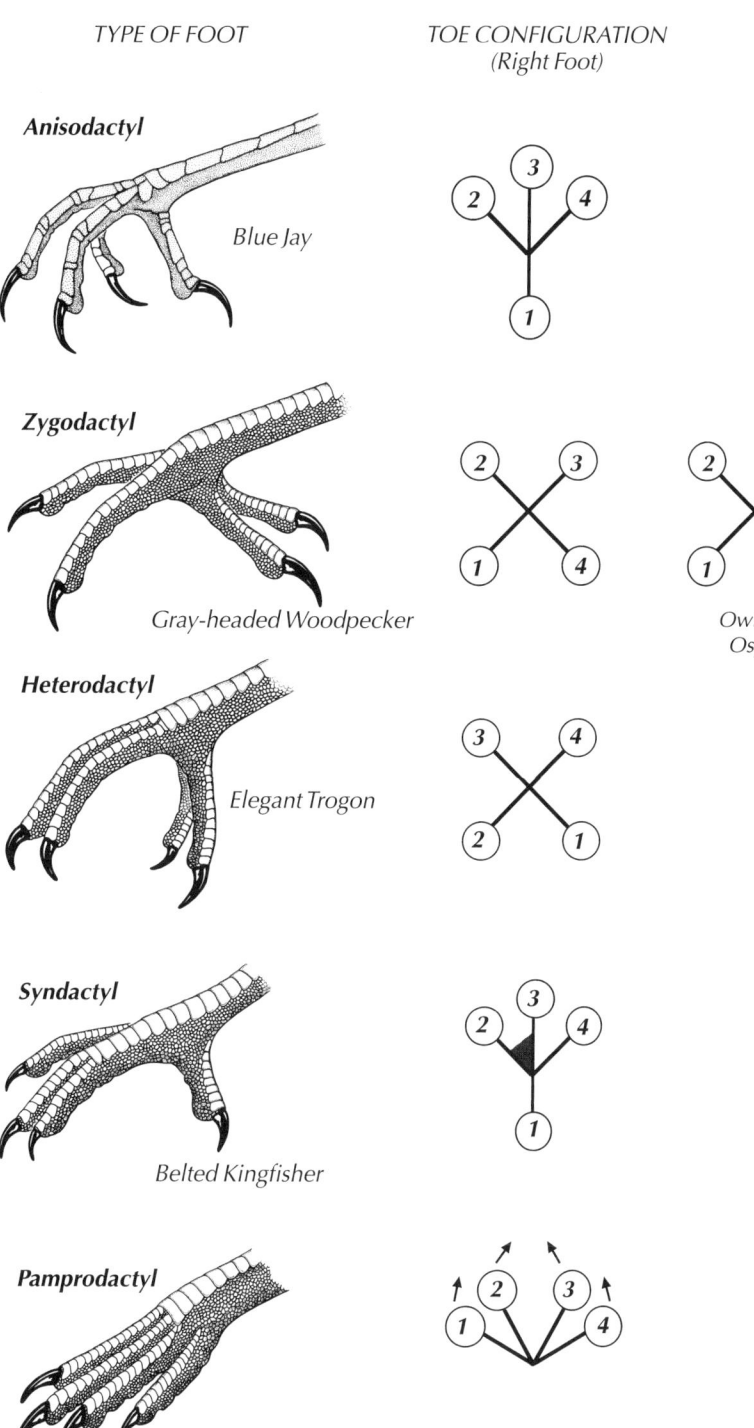

Anisodactyl

Blue Jay

Zygodactyl

Gray-headed Woodpecker

Owls and
Osprey

Heterodactyl

Elegant Trogon

Syndactyl

Belted Kingfisher

Pamprodactyl

Chimney Swift

Figure 1–20. Toe Arrangements: Although most birds have four toes, their arrangement differs among different groups of birds. Most birds, including nearly all perching (passerine) birds such as the Blue Jay, have an **anisodactyl** foot, with three toes directed forward and a hallux directed backward. Woodpeckers, cuckoos, toucans, owls, Osprey, turacos, most parrots, and some other birds have a **zygodactyl** foot, in which two toes point forward and two point backward. (In each schematic, which represents the right foot, the backward-directed digits are at the bottom; thus in the zygodactyl foot, toes 1 and 4 point backward.) Note that the fourth toe of some birds with zygodactyl feet, such as owls and Osprey, is reversible, allowing the bird to use it in either a forward or backward position. The zygodactyl feet of many of the different bird groups appear to have evolved independently, and thus are considered convergent structures. Trogons also have two toes reversed, but in their feet, termed **heterodactyl**, toes 1 and 2 point backward. Anisodactyl, zygodactyl, and heterodactyl feet are all well-adapted to grasp branches. In the **syndactyl** feet of many kingfishers and hornbills, the toes are arranged as in the anisodactyl foot, but toes 2 and 3 are fused for part of their length. In the **pamprodactyl** feet of some swifts, such as the Chimney Swift, all four toes point forward, allowing them to cling to vertical surfaces such as the inside walls of chimneys and hollow trees (see Fig. 1–28). They can also use toes 1 and 2 against toes 3 and 4 in a pincerlike fashion, to grasp plants or nest material. Reprinted and modified from Manual of Ornithology, by Noble S. Proctor and Patrick J. Lynch, with permission of the publisher. Copyright 1993, Yale University Press.

is reversible—the bird can use it pointing forward or backward. Only trogons have **heterodactyl** feet, with the second toe reversed. In the **pamprodactyl** feet of some swifts, all four toes, including the hallux, point forward. Swifts can use their small feet as hooks to hang from the inside walls of chimneys, caves, and hollow trees. In addition, they can grasp plant material or nest material with a lateral, pincerlike motion, using toes one and two against toes three and four. Kingfishers have three toes forward and one behind, but their feet are termed **syndactyl** because the inner and middle toes are united for much of their length—the ecological advantages, if any, of this arrangement are not known. Mousebirds (Coliidae)—small, long-tailed, African birds—have highly versatile feet, which, although often classified as pamprodactyl, can actually be used in a variety of configurations: all four toes pointed forward, toes one and four pointed back, or any combination in between these positions.

The form of the feet of many birds clearly suits their habits and environment (**Fig. 1–21**). Having one or two toes reversed helps a bird to grasp perches, but anisodactyl, zygodactyl, and heterodactyl feet

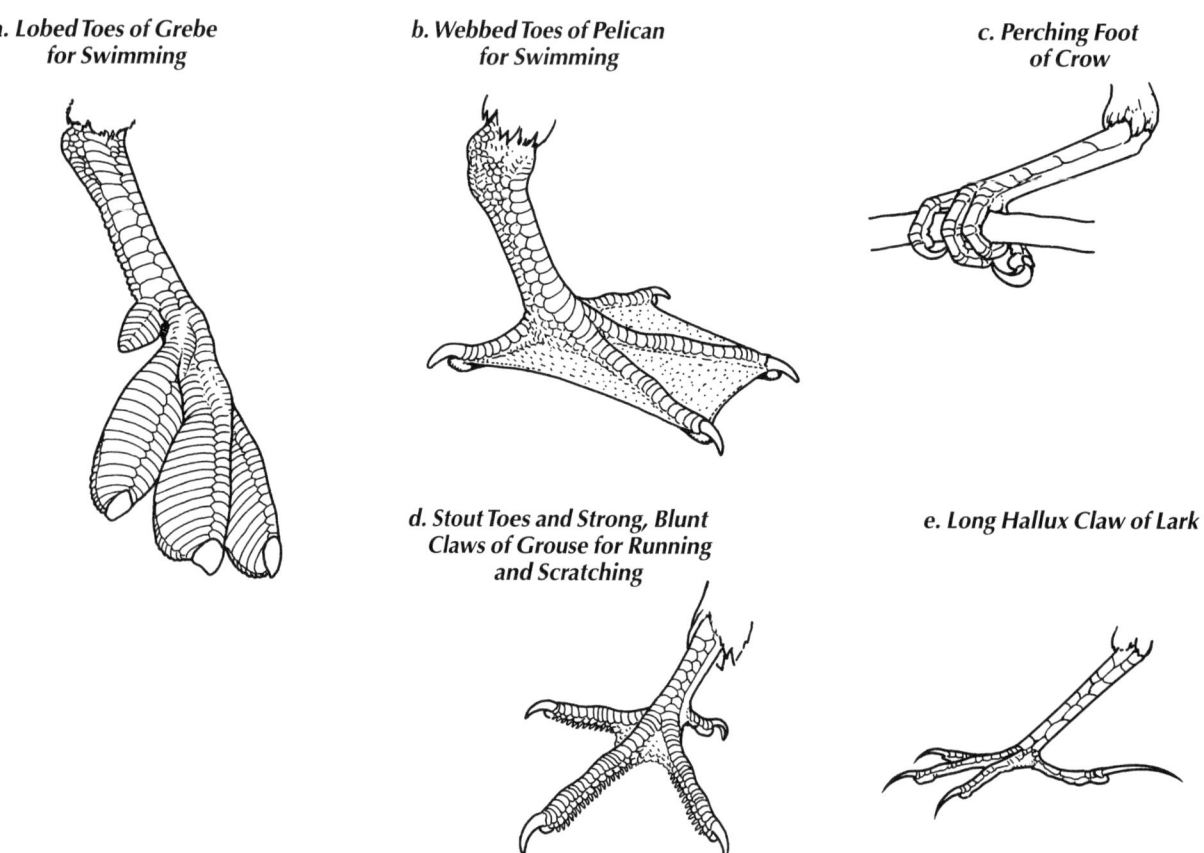

a. Lobed Toes of Grebe for Swimming

b. Webbed Toes of Pelican for Swimming

c. Perching Foot of Crow

d. Stout Toes and Strong, Blunt Claws of Grouse for Running and Scratching

e. Long Hallux Claw of Lark

Figure 1–21. Foot Diversity: The form of a bird's foot reflects its behavior and habitat. Feet modified for swimming have a large surface area to push against the water. This may be accomplished by having lobed toes, as in coots and grebes (a), or by having webbed toes. In birds such as ducks and gulls, only the three forward toes are webbed, but in others, such as cormorants and pelicans (b), all four toes are webbed. Feet for perching in trees, climbing, capturing prey, and carrying and manipulating food commonly have toes with sharp, curved claws, as in crows (c). Birds such as grouse (d) that spend a great deal of time on the ground, often running and scratching in the dirt for food, generally have stout toes with strong, blunt claws. Some foot adaptations are not well understood, such as the long claw on the hallux of larks and some other open-country birds (e). Drawing by Charles L. Ripper.

all seem to do this equally well—so any ecological advantages that the different toe arrangements might confer remain a mystery. Toes for perching in trees, climbing, capturing prey, and carrying and manipulating food are equipped with sharply curved and pointed claws. Toes for running and scratching are robust with strong, rather blunt claws (as in turkeys, quail, grouse, and others). Toes for swimming may be webbed, either involving just the three forward toes (ducks and gulls) or including the back one as well (pelicans and cormorants); or all the toes may be lobed, as in grebes and coots. In some birds, the toes bear peculiar features whose functions, if any, are unknown. For example, the hallux claw of larks and some other open-country birds is very long, and the forward toes of some shorebirds (such as the Semipalmated Sandpiper) may be partly webbed, even though the birds do not normally swim.

Feathering

In addition to variations in the feathering of the wings and tail that may distinguish different birds, some of the remiges may be notched or greatly narrowed and stiffened (as in the American Woodcock), whereas the rectrices may be noticeably stiffened and pointed (woodpeckers) **(Fig. 1–22)**, or may feature partly bare shafts (motmots). Elsewhere on different birds, the feathering may be just as distinctive, if not more so. On the head, the feathers may form crests (cardinals, titmice), discs about the face (owls) **(Fig. 1–23)**, and bristles at the base of the bill (many flycatchers), or they may be absent on parts or all of the head (vultures). Feathers on the head, neck, back, rump, and even the flanks

Figure 1–22. Female Northern Flicker: *The stiff, pointed tail feathers of woodpeckers, such as the Northern Flicker pictured here, act as a prop when the bird is perched vertically or ascending a tree trunk. They also brace the woodpecker's body while drumming, excavating a nest cavity, or pecking for insects—allowing the head and neck to pound forcefully against a tree trunk. Although it drums on trees and excavates nest cavities, the Northern Flicker forages primarily on the ground, consuming ants and beetles. Photo courtesy of Isidor Jeklin/CLO.*

Figure 1–23. Barn Owl Facial Disc: *As nocturnal predators, owls rely heavily on sounds to help them locate prey. The facial disc of feathers on many owls—heart-shaped in the Barn Owl illustrated here—enhances the owl's sensitive hearing by focusing sounds into the ear openings, which are located posterior to the eyes. Photo courtesy of J. Robert Woodward/CLO.*

may be greatly lengthened and modified to form conspicuous plumes or adornments (birds-of-paradise, peafowl).

The colors of feathers not only enhance adornments but quite obviously provide color patterns that characterize different birds. Thus, some birds have spotted plumage, others, streaked. Some birds have color patterns that make them inconspicuous, whereas others have gaudy patterns. The same species may have colors that conceal it on its nest and other colors that it uses to attract attention when displaying.

Although birds may seem to have feathers growing from all parts of the body, feathers in most birds grow from tracts separated by bare areas. The bare areas are usually concealed, because the feathers from the neighboring tracts overlap and hide them. Feathers and other external features of birds are discussed in detail in Chapter 3.

Internal Anatomy

The diversity of birds is no less apparent in their internal systems—skeletal, muscular, nervous, circulatory, respiratory, digestive, and urogenital. Indeed, some birds differ more sharply internally than they do superficially, thus providing a sound basis for distinguishing them. The internal anatomical differences among species are discussed in detail in Chapter 4.

Diversity in Bird Movement

■ Birds are masters of the air, but many of them also move quite well on land: walking, hopping, or running on the ground; or climbing in trees. Many species have become proficient around water: swimming, diving, or plunging for food. Here, the different methods that birds use to move on land and in water are briefly discussed. Bird flight is explored at length in Chapter 5.

Movement on Land

The first birds were probably tree-dwellers; when not flying from tree to tree, they moved from branch to branch by hopping—that is, moving by a series of jumps with feet together. Later, some of the tree-dwellers, such as woodpeckers and creepers, achieved an ability to climb by hitching themselves up, using their tail feathers as supports, or, like the nuthatches, by jerking themselves up at an angle or climbing down headfirst, keeping one foot high for suspension and the other low for support **(Fig. 1–24)**.

When many birds left the trees for life on the ground, they began moving by walking and running—moving the legs alternately. The changeover from hopping to walking and running was gradual, and gradations still can be seen. Among the cuckoos, for example, are the awkwardly hopping, arboreal Yellow-billed and Black-billed cuckoos, and the fast-paced, terrestrial Greater Roadrunner. Although most passerine species still hop much of the time, crows and the European Starling ordinarily walk; adult larks, pipits, and wagtails walk and run

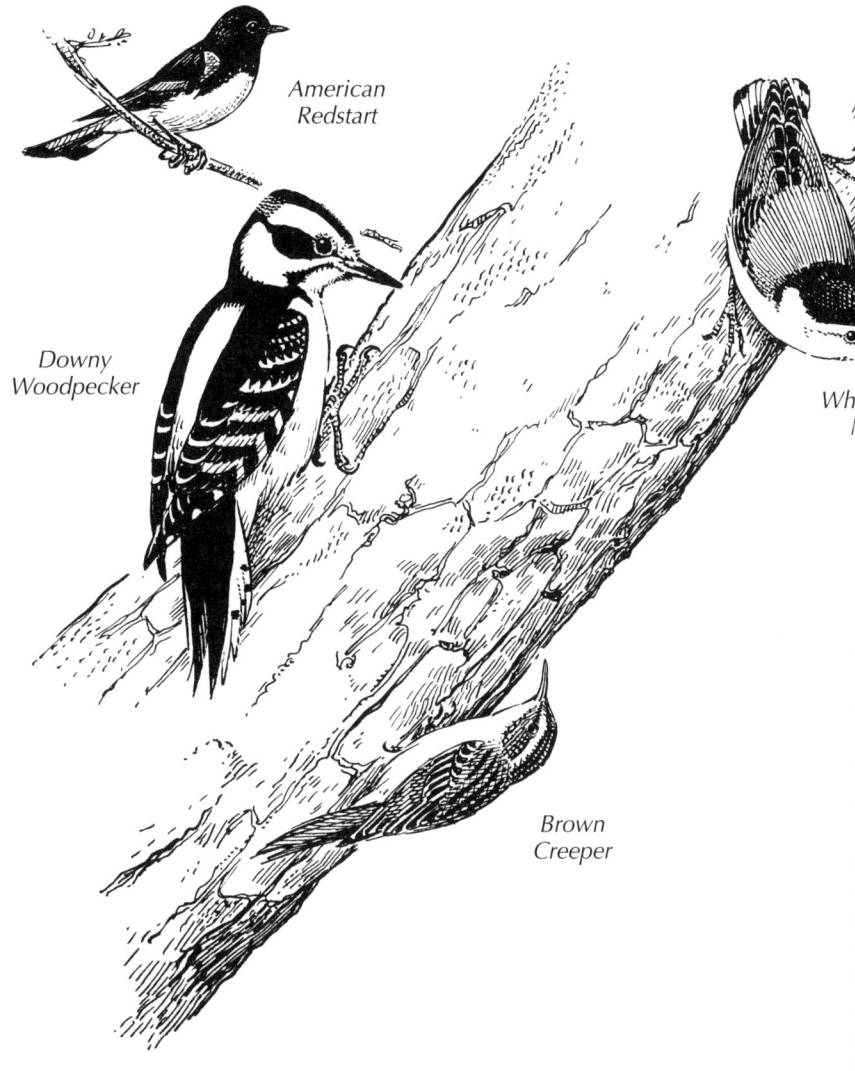

American
Redstart

Downy
Woodpecker

White-breasted
Nuthatch

Brown
Creeper

Figure 1–24. Movement in Trees: *Many birds are well-adapted to perch and move about in trees. Most perching (passerine) birds, like the American Redstart, have anisodactyl feet, allowing them to easily grasp a branch. Woodpeckers, creepers, and nuthatches have long, curved claws that act as climbing hooks, helping them to creep along tree trunks. Woodpeckers and creepers hitch themselves up tree trunks, using their toes and tail for support as they move. Nuthatches either work their way up at an angle, or descend headfirst, often keeping one foot high for suspension and the other low for support, since their tails are not modified to act as props. The claw of their hallux is larger and more strongly curved than the claws of their forward-facing toes, providing the strong grip that allows them to descend headfirst. Drawings by Robert Gillmor.*

but their fledglings hop; and the American Robin hops, walks, and runs between pauses when foraging for earthworms on a lawn **(Fig. 1–25)**.

For walking in water, some birds—herons, cranes, Limpkins, and many shorebirds—have long legs with the accompanying requirement of long necks for reaching far down to feed, and for counterbalancing the long legs during flight. For walking in mud and on aquatic vegetation, some species have exceptionally long toes (gallinules) (see Fig. 3–39e) and claws (jacanas), which prevent them from sinking down **(Fig. 1–26)**.

Penguins, because of their torpedo-shaped bodies and stubby legs set far back from their center of gravity, must walk upright with a short-stepping gait **(Fig. 1–27)**. If hard pressed to move fast, penguins resort to "tobogganing" on their bellies, propelling themselves vigorously with both their flipperlike wings and their feet.

Extremes in the ability to move on land are best represented by the powerful Ostrich, which can run as fast as 43 miles (70 km) per hour, and by the swifts, the most aerial of all birds, which cannot hop or run, and can use their weak feet only for clinging to vertical surfaces **(Fig. 1–28)**.

Figure 1–25. Movement on the Ground: The first birds probably lived in trees, flying or hopping from branch to branch. As some birds adapted to life on the ground, they gradually evolved from hopping (moving both legs at the same time) to walking or running (moving the legs alternately). For example, the Greater Roadrunner runs quickly along the ground, but when its arboreal relatives—Yellow-billed and Black-billed cuckoos—venture to the ground, they hop. Although most passerines hop, the European Starling walks when on the ground. Most birds either hop or walk and run, but watch an American Robin for any length of time and you will no doubt see that it uses a combination of movements: walking, hopping, and running as it searches a lawn for earthworms. Drawings by Robert Gillmor.

Greater Roadrunner

American Robin

European Starling

Figure 1–26. Walking in Water: Some birds—especially herons, cranes, Limpkins, and a number of shorebirds—have long legs for wading in water, the length reflecting the water depth in which they may feed. The Great Blue Heron, with its long legs and neck, usually stalks fish in shallow water, but sometimes wades belly deep along the shorelines of oceans, as well as freshwater lakes, streams, and marshes. The Black-necked Stilt, with somewhat shorter legs, favors shallower water and flooded fields, where it searches mainly for aquatic insects. Other birds, notably jacanas and some rails, have long toes that distribute their weight over a large surface area, like snowshoes, allowing them to walk across muddy areas and floating vegetation such as lily pads. The Northern Jacana, like other jacanas, has long claws as well as long toes. Drawing by Robert Gillmor.

Great Blue Heron

Northern Jacana

Black-necked Stilt

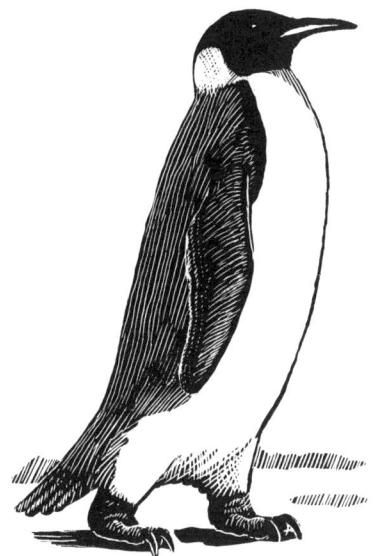

a. Walking

b. Tobogganing

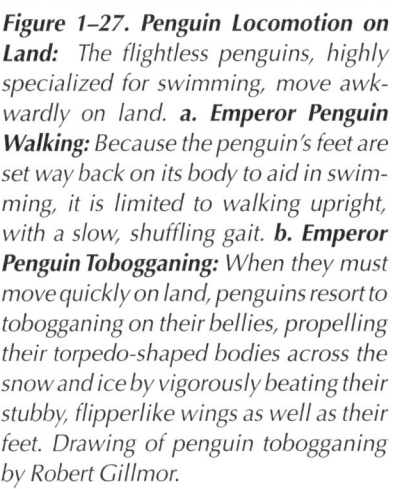

Figure 1–27. Penguin Locomotion on Land: *The flightless penguins, highly specialized for swimming, move awkwardly on land.* **a. Emperor Penguin Walking:** *Because the penguin's feet are set way back on its body to aid in swimming, it is limited to walking upright, with a slow, shuffling gait.* **b. Emperor Penguin Tobogganing:** *When they must move quickly on land, penguins resort to tobogganing on their bellies, propelling their torpedo-shaped bodies across the snow and ice by vigorously beating their stubby, flipperlike wings as well as their feet. Drawing of penguin tobogganing by Robert Gillmor.*

Figure 1–28. Chimney Swift Clinging to a Vertical Surface: *Chimney Swifts spend most of their time in the air—foraging, courting, and even copulating in flight. These streamlined "flying cigars" are highly specialized for an aerial life, with feet so weak that they cannot walk, run, or hop. Their feet, with all four toes forward, can be used like small hooks to cling to vertical surfaces of chimneys, trees, barns, and other structures. Swifts can also move toes 1 and 2 against toes 3 and 4, pincerlike (see Fig. 1–20), to grasp plants and nest material. Although they once nested primarily in hollow trees, an increasing number of Chimney Swifts now nest and roost in chimneys. At dusk on a late summer evening, hundreds of Chimney Swifts may gather to roost in a single chimney. Photo courtesy of Michael Hopiak/CLO.*

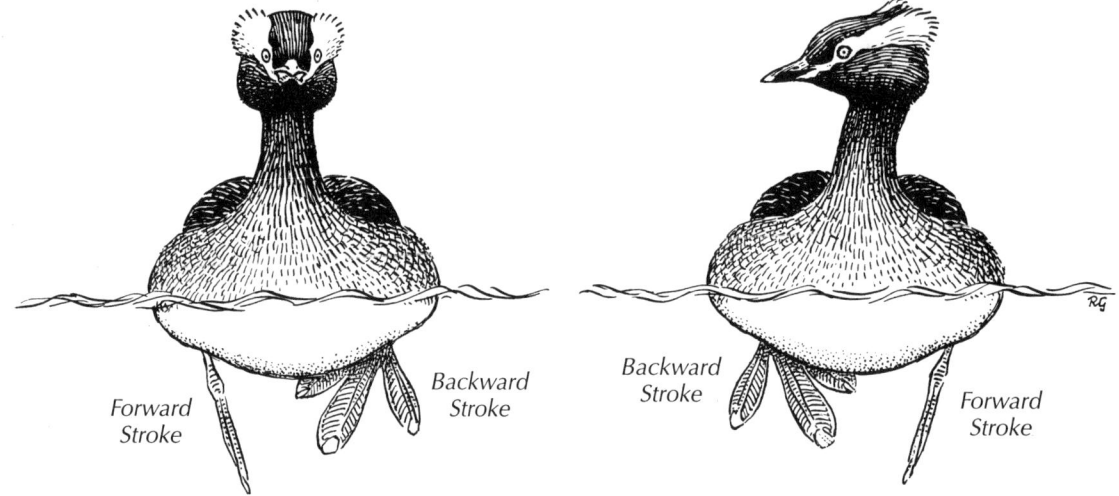

Forward Stroke   Backward Stroke   Backward Stroke   Forward Stroke

Figure 1–29. Horned Grebe Swimming:
To increase the surface area that pushes against the water, many swimming birds have either webbed or lobed toes. Coots and grebes, such as the Horned Grebe pictured here, have lobed toes. To swim, they move their legs alternately, as in walking. On the backward stroke, which generates power, the lobes push the bird forward like oars or paddles. On the forward recovery stroke the lobes of coots are folded and the foot of grebes is turned, both actions presenting the least possible amount of resistance to the water. Drawing by Robert Gillmor.

Movement in Water

Although the ancestors of birds were terrestrial and most birds still depend on land for nesting, many species have adapted so strongly to an aquatic existence that they spend much of their time swimming. A great number also dive.

Most birds swim by alternately moving their feet in the water. Swimming birds commonly are aided by either webbed or lobed toes although some birds, such as moorhens and phalaropes, are excellent swimmers despite having no special toe modifications. Both webs and lobes act as paddles, pushing against the water on the backward (power) stroke, then folding or turning to resist as little water as possible on the forward (recovery) stroke **(Fig. 1–29)**.

The size of the feet in proportion to body size is commensurate with the importance of swimming in a bird's life. Storm-petrels have small feet because they swim infrequently; conversely, pelicans have enormous feet because they spend much of their time swimming. In most birds, swimming is further aided by broad, almost raftlike bodies that give stability, and by a dense feather coat that holds air for buoyancy.

Swimming is the principal movement in the water for albatrosses, shearwaters, petrels in the genus *Pterodroma,* gulls and other aquatic scavengers, and phalaropes—all of which gather food on or just below the water's surface. But more than swimming is required of birds that obtain food under water. Swans, geese, and the dabbling ducks—teal, Mallard, Northern Pintail, Wood Duck, and so on—tip up in shallow water, reaching down to forage on the bottom **(Fig. 1–30)**.

Osprey, Brown Pelicans, boobies, gannets, and terns plunge into the water from the air, gaining momentum as they descend. The height of the dive depends on their size and the depths to be attained. The heavy-bodied gannets may plummet from 100 feet (30.5 m) to reach depths of 10 to 12 feet (3 to 3.7 m). An adaptation consisting of many minute air sacs under the skin acts as a shock absorber when they strike the water. The lighter-bodied terns plunge from only a few feet, because they go no deeper than a few inches **(Fig. 1–31)**.

Mallard

Tundra Swan

Lesser Scaup

Horned Grebe

Figure 1–30. Foraging in Water: *Birds search for food in water in a number of different ways. Swans, geese, and the so-called "dabbling ducks" may dip just the head and neck below the surface to nibble on aquatic plants, or may turn "bottom-up" (as in the Mallard pictured here) to reach deeper vegetation. Other birds dive from the water's surface, using the wings, feet, or both to propel themselves through water in search of plants, fish, or other small aquatic animals. Surface divers include penguins, shearwaters, diving-petrels, loons, grebes, cormorants, anhingas, mergansers, alcids, coots, and the "diving ducks"—such as the Lesser Scaup illustrated here. Drawing by Robert Gillmor.*

Figure 1–31. Diving from Air: *Many birds "plunge dive" into water from the air, the height of the dive depending on the bird's size and the depth to be attained. The heavy-bodied Northern Gannet scans for schools of fish and then may plummet from as high as 100 feet (30.5 m) to reach underwater depths of 10 to 12 feet (3 to 3.7 m). The lighter-bodied tern dives from much lower, barely going below the water's surface. Terns hover in place before diving, watching for fish near the water's surface. Osprey, Brown Pelicans, boobies, and kingfishers also dive into water from air. Drawing by Robert Gillmor.*

Northern Gannet

Tern

Arctic Loon

Gentoo Penguin

Long-tailed Duck

Figure 1–32. Swimming Underwater: *Birds specialized for underwater swimming have streamlined, torpedo-shaped bodies with the feet located far back on the body. In this position, the feet can generate much power and also act as effective rudders. Loons, grebes, cormorants, anhingas, and coots use primarily their feet to propel themselves through water. Mergansers and diving ducks, including the Long-tailed Duck, may use just the feet while foraging along the bottom, keeping their wings at their sides, but use both wings and feet to swim to great depths. Although they usually feed within 30 feet (9 m) of the surface, Long-tailed Ducks have been observed as deep as 200 feet—deeper than any other duck! Penguins, however, are the champion avian divers. Like alcids, shearwaters, and diving-petrels, they use only their short, paddle-like wings to "fly" through the water. They regularly forage at depths of 65 feet (20 m) or more, the large Emperor Penguins going as deep as 1,752 feet (534 m). Drawing by Robert Gillmor.*

Penguins, shearwaters, diving-petrels, loons, grebes, cormorants, anhingas, diving ducks, mergansers, alcids, and coots dive from the water's surface (**Fig. 1–32;** see Fig. 1–30). Lacking the momentum of an aerial plunge, they power their dives with their feet or wings, or sometimes both. Their feet are suitably located near the rear of the body, just as a ship's propeller is located at the stern. They are heavier than land birds of similar size because their bones are less pneumatic (air-filled). Furthermore, to reduce their buoyancy during a dive, they compress their feather coat and internal air sacs to reduce the air content. Penguins, shearwaters, diving-petrels, and alcids use only their short, bladelike wings to propel themselves through water. Loons, grebes, cormorants, anhingas, and coots mainly use their feet, which are large and powerful. The diving ducks and mergansers can use both their feet and wings. This versatility is particularly characteristic of the diving ducks that frequent the sea. For example, a Long-tailed Duck foraging on the bottom propels itself only with its feet, keeping its wings at its sides, but when diving to reach great depths, it powers itself by both wings and feet.

Twenty feet (6 m) is probably the limit to which most birds dive, although there are numerous records of birds going much deeper. The Common Loon can descend to 90 feet (27 m) and the Long-tailed Duck to about 200 feet (61 m). Although diving birds usually stay under water for a minute or less, the Long-tailed Duck has been timed under water for three minutes or more. Most diving birds can extend submergence—if frightened, for instance—by relying on some residual air in the air sacs and large amounts of oxygen stored in the muscles. The length of a bird's endurance, if forcibly held under water, is about 15 minutes.

Not surprisingly, the master divers of the bird world are the penguins. Many species regularly dive deeper than 65 feet (20 m). The two largest species, the Emperor and King penguins, are the undisputed champions. The large King Penguin has been recorded (with depth gauges) to dive up to 1,059 feet (323 m), and the even larger Emperor Penguin has been recorded to a fantastic 1,752 feet (534 m). Even the smallest penguin, the Little Penguin of Australia, which normally dives to less than 6 feet (2 m), has been recorded diving to 88 feet (27 m). To dive this deeply, penguins can hold their breath for long periods, up to 15.8 minutes for the Emperor Penguin, which regularly stays under water for 2 to 4 minutes. Birds diving for such long periods actually stay under water longer than would be predicted by the amount of oxygen in their bodies. Penguins have a number of complex physiological adaptations, not all completely understood, that allow them to accomplish these long and deep dives.

A few other birds, not water birds per se, exploit the food resources of inland waterways by expertly diving or swimming. The Belted Kingfisher is skilled in plunging for fish, although it cannot swim; the American Dipper, residing along swift mountain streams in western North America, swims—actually flies—under water and walks on the stream bottom to take insect larvae attached to rocks. Neither species has specializations for maneuvering in water.

One more unique, water-related behavior is often observed and is worth noting. When mildly alarmed, some grebes, particularly Pied-billed Grebes, may squeeze air from their feathers and air sacs, thus increasing their density, and slowly sink straight down into the water **(Fig. 1–33)**. They may remain completely submerged, or with the eyes just peering out across the water, until all danger has passed.

Often, birds can capably perform certain movements to which they are not primarily adapted. Ducks and gulls have webbed feet, and coots lobed feet, for swimming, yet these birds can forage successfully on dry land far from water. Under compelling circumstances, practically all swimmers can dive to some extent, and many species can swim even though they lack webbed or lobed feet. This is true of all gallinules, rails, and shorebirds. Watch a Spotted Sandpiper attacked by a predator and do not be surprised to see it take to the water and dive repeatedly!

Figure 1–33. Pied-billed Grebe Sinking: *When mildly alarmed, the Pied-billed Grebe may squeeze air from its feathers and air sacs, decreasing its buoyancy and thus slowly sinking straight down into the water. It may submerge to various degrees, sometimes remaining completely underwater, sometimes leaving the neck and body partly visible, and sometimes leaving just the eyes and nostrils above the surface, to check for danger. Photos by Tom Vezo.*

Naming and Classification of Birds

■ Have you ever looked in a field guide and wondered why the brightly colored orioles and meadowlarks are all lumped together in the same family (Icteridae) with the mostly black grackles and Red-winged Blackbirds, whereas the soot-black starlings and crows are not? Why are the coots put in with the rails and cranes and not with the ducks, which they resemble so much when swimming around in a pond? Why are the swifts and swallows that look and act so similar not listed together? And why are the cranes on different pages than the similar-looking herons? What these simple questions really translate to is "How, and why, do people classify birds?"

History

At least since Aristotle (384 to 322 B.C.), and probably from time immemorial, people have been giving names to birds and trying to put them into logical groups, that is, making classifications **(Fig. 1–34)**. Humans love to put things into pigeonholes, whether they go easily or not. Exactly *how* people arrange birds into categories however, depends on what the categories are supposed to represent. For example, in some primitive societies, birds might be divided into two groups: edible and inedible. (Even today some people tend to think of "useful" and "harmful" birds.) Aristotle and other early Western thinkers trying to group "similar" birds based their classifications on such things as habitat (for example, water versus land), locomotion, food habits, and obvious physical characters (for example, bill size and shape). Aristotle divided birds into three categories: those that live on land, those that live in the water, and those that live at the edge of water. In the mid 16th century, Pierre Belon, following the ideas of Aristotle, classified birds into six groups: raptors, waterfowl with webbed feet, marsh birds without webbed feet, terrestrial birds, large arboreal birds, and small arboreal birds. (His classification lumped Ostriches, chickens, and larks into the same terrestrial category!) Morphological characters (the actual form of body structures) were first used as a basis for classification by Francis Willughby and John Ray in 1676, and their work was one of the primary sources used in the mid 1700s by a Swedish naturalist, Carolus Linnaeus, in his classification of animals, the *Systema Naturae*, discussed in more detail later in this chapter.

Various systems of classification based on differing ideas of habitat, behavior, and physical similarities floated around for the next hundred years or so, with no consensus on what to represent. With the understanding of evolution by natural selection as expounded by Darwin in 1859 **(Sidebar 2: The Evolution of an Idea: Darwin's Theory)**, the classifications based on some vague "natural" or "logical" order were replaced with ones that grouped organisms because of common ancestry. In other words, organisms thought to be related to one another were put in the same group. The classifications, therefore, were an attempt to represent the **phylogeny**, or evolutionary history, of birds. (Note that **taxonomy** is the classification—assigning names and

Figure 1–34. Significant Steps in the History of Bird Classification

384–322 B.C.	Aristotle groups birds into three categories: those that live on land, those that live in water, and those that live at the edge of water.
Mid 1500s	Pierre Belon separates birds into six groups: raptors, waterfowl with webbed feet, marsh birds without webbed feet, terrestrial birds, large arboreal birds, and small arboreal birds.
1676	Francis Willughby and John Ray are the first to use morphological characters to classify birds.
Mid 1700s	Carolus Linnaeus devises system of binomial nomenclature for classifying organisms.
1859	Charles Darwin publishes his theory of evolution in his book *On the Origin of Species by Means of Natural Selection.* From this time on, researchers try to classify birds based on common ancestry.
1888 & 1892	Maximilian Fürbinger (1888) and Hans Gadow (1892) each devise modern classifications of birds based on a host of anatomical characters. For the first time, all passerine songbirds are grouped together, swifts are separated from swallows, and cuckoos are separated from woodpeckers.
1930–1960	Alexander Wetmore uses new information to adjust the classifications of Gadow, including organizing the songbirds. Wetmore's publications are responsible for the ordering of groups found in most American field guides.
1980s	Charles Sibley and Jon Ahlquist propose a major rearrangement of the orders and families

relationships—of organisms, whereas **phylogeny** refers to the actual evolutionary relationships that the taxonomist hopes to represent.) Just what characters to use and where to draw the boundaries have been, and continue to be, subjects of great debate.

The first modern, evolutionary classifications of birds were proposed by Maximilian Fürbinger in 1888 and Hans Gadow in 1892, based on a host of anatomical characters. Finally, all the songbirds were grouped together, and the swifts (because of such features as their unusual wing structure) were separated from swallows, and cuckoos from woodpeckers. Subsequent bird classifications were heavily based on these works, and few major changes were made in the first half of the 20th century.

Especially relevant is the work of Smithsonian researcher Alexander Wetmore. Although his classifications published between 1930 and 1960 differed little from Gadow's, he did incorporate some new information. He also organized the songbirds, a group that Gadow viewed as uniform, and had left largely undifferentiated. Wetmore's

(Continued on p. 1·38)

Sidebar 2: THE EVOLUTION OF AN IDEA: DARWIN'S THEORY
Sandy Podulka

In 1859, Charles Darwin rocked both the scientific and nonscientific world with his radical theory on how species change over time. His book, *On the Origin of Species by Means of Natural Selection*, set forth the idea of evolution by natural selection, which remains a cornerstone of biology to this day. It addressed many puzzling aspects of living things, such as their tremendous diversity, their origins and relationships to one another, their similarities and differences, their bewildering geographical distribution, and their startlingly appropriate adaptations to their own environments. It also unified the field of biology by beautifully fitting together the array of scattered observations and facts that previously had composed the discipline. It is, indeed, the one grand theory through which much of the natural world can be understood.

Darwin's theory had two main parts—both revolutionary ideas for the time:

(1) Organisms Change through Time (Evolution)

By definition, **evolution** is any change over time. Just as an idea or fashion trend can evolve, so can living things. But when biologists speak of evolution, they generally mean a change in the most common characteristics of a species or population over many generations, not a change in an individual over its lifetime. The idea that organisms change had been around for a while, but was not accepted by most people. Darwin provided a great deal of scientific evidence to support it.

(2) Natural Selection is the Mechanism by which Organisms Change

The process of natural selection, detailed below, was a completely new concept.

The Evidence for Evolution

Although Darwin lived at a time when most people believed in Divine Creation, a number of different observations, including the following, convinced him that organisms did indeed change. First, Darwin recognized that the fossil record held many species that no longer existed, and different species were found in different rock layers. In some places, organisms preserved in adjacent layers showed only slight differences, indicating a pattern of gradual change through time. Second, Darwin noted the strong resemblances between certain aspects of living things, such as the bones of the forearm of a number of different vertebrates (**Fig. A**). He reasoned that they must be so similar because they all were modified from a common ancestor—the same basic design could not

possibly be the one best starting point from which to build appendages for flying, swimming, walking, running, *and* handling objects. Third, during his voyage on the *H. M. S. Beagle*, Darwin observed oddities in the distribution of species around the globe. In particular, he noted that the flora and fauna of temperate South America more closely resembled the species in nearby tropical South America than they did the species of temperate areas of Europe. Thus, he reasoned, the similarities must result from common ancestry, and not because species were created to fit certain habitats. Fourth, Darwin saw that plant and animal breeders were able to produce changes in domestic organisms—such as chickens, pigeons, or roses—through the generations, even though no one understood how these changes came about. He assumed that if humans could cause changes, then changes could occur naturally as well.

Natural Selection: The Mechanism of Change

Figuring out the mechanism of change was more involved, but again, Darwin got a hint from domestic breeders. From a group of offspring with a variety of traits (say, different feather colors in pigeons), they would choose those individuals with the desired traits and breed them again, repeating this process through the generations until all or nearly all of the offspring had the desired coloration. Through these and other observations, Darwin noted two important facts about natural populations, which formed a basis for natural selection:

(1) **In any generation of a species, many more young are born than will survive to reproduce.** (Recently, someone calculated that if all the offspring of all the house flies on earth survived, in just six months the entire planet would be 47 feet deep in house flies.)

(2) **Variation exists among the individuals of a species: they are not identical in all of their characteristics.**

From these facts, Darwin inferred the following:

(1) **Individuals with certain traits have a better chance of surviving and reproducing than individuals with other traits, because they compete better.**

Because essential resources such as food, water, and suitable habitat are not usually superabundant, individuals must compete for them—whether directly or indirectly. They also "compete" in their ability to avoid predators, to obtain mates, to locate appropriate breeding sites, to produce and rear offspring, and so on. This "survival of the fittest" is the basis for natural selection: the individuals with the advantageous traits survive; in essence, they've been "selected by nature."

(2) Some of the variation between individuals is hereditary.

Although the idea of receiving traits through our genes is common knowledge today (indeed, it is now "fact," rather than "inference"), Darwin lived at a time when nothing was known about inheritance. He proposed his theory long before chromosomes and genes were discovered, and just before the work of Gregor Mendel, an Austrian monk working with pea plants, provided the basis for modern genetics.

(3) When more individuals with certain (advantageous) traits survive and reproduce, more of the advantageous traits appear in the next generation.

Over many generations, the characteristics of a species may change. Through this process, evolution occurs. For example, if flying faster allows a Mourning Dove to more easily avoid becoming a meal for a Cooper's Hawk, then in any generation of Mourning Doves, more of the faster individuals should survive and reproduce. If flight speed is hereditary, the next generation should have a greater percent of faster individuals, and so on. (In evolutionary jargon, the faster birds are said to have a greater **fitness**—the likelihood of producing offspring that will survive to reproduce—relative to the others; being fast is considered an **adaptation**—a genetically controlled trait that increases an individual's relative fitness.)

This simple example assumes that flying fast carries no disadvantages, which is undoubtedly false—high speed brings energetic, aerodynamic, and maneuverability concerns. Thus in reality, the flight speed of Mourning Doves results from a compromise: it has been honed by natural selection to meet an array of concerns. Natural selection, a truly simple process, is intriguingly complex in its application.

One further assumption was essential to Darwin's theory: that enormous spans of time are available for slow, gradual change. In Darwin's time, people believed that the earth was created a few thousand years before the birth of Christ. But Darwin's friend, geologist Charles

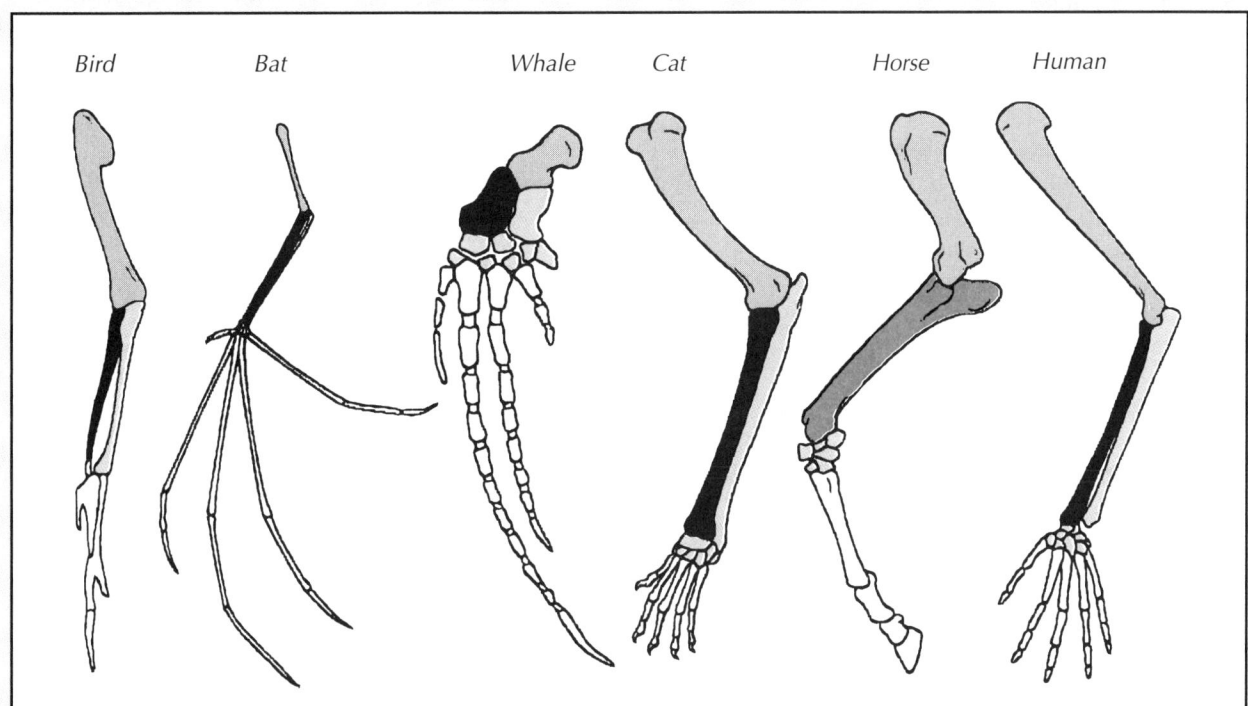

Figure A. Vertebrate Forelimbs: *The forelimbs of all vertebrates are constructed from the same basic bones, although they have been modified through natural selection to carry out widely different functions. In this drawing, corresponding bones are shaded similarly for each animal. Darwin noticed the similarity among vertebrate forearms, and reasoned that it must result from each being modified from the same common ancestor. Otherwise, if the structures were unrelated, each "designed" from scratch, one would have to accept the unlikely theory that the same basic set of bones was the best starting point for constructing appendages for flying, swimming, running, and handling delicate objects. Note that the bat's palm and finger bones are greatly elongated to support the membranous wing, and the whale's bones are shortened and thickened to form a strong flipper. Both the cat and horse walk up on their toes, with the palm bones elevated—the horse, on one toe, the hoof being modified from the claw of just one toe. Reprinted and adapted from* Inquiry into Life, *5th Edition, by Sylvia S. Mader. Copyright 1988, William C. Brown, publisher. Reproduced with permission of the McGraw-Hill Companies.*

Lyell, provided him with evidence that the earth was much older. Today we know that the planet has existed for at least 4.5 billion years, and life of some form has been present for at least 3.5 billion. Because natural selection often acts on minute differences between individuals, producing slight changes from one generation to the next, long time spans are necessary to explain the kinds of changes evident in the fossil record.

How Natural Selection Acts

Although species or populations evolve, natural selection does not act directly on them as a whole. Natural selection acts *only* on individuals—each individual lives or dies accordingly—but it results in the *evolution* of populations and species. Furthermore, natural selection acts on the *whole individual.* Because each individual is the sum of many different genetic traits, the relative advantage of each trait depends on its genetic context. For example, a trait increasing the growth rate of a tree's trunk and branches is useless, possibly even detrimental, unless accompanied by a trait to increase the growth of the roots.

The Source of Genetic Variation

The variability between individuals of the same species ultimately arises from **mutations:** actual changes in the basic structure of DNA. Mutations can arise spontaneously, but also are induced by radiation and certain chemicals. Although mutations can occur in any cell in the body, only those in the cells that produce sperm and eggs can be passed on to an individual's offspring. Over hundreds and thousands of generations, many mutations arise. Neutral and advantageous ones may persist, while disadvantageous ones may die out. Over time, much variability may be produced in a population (for example, green, blue, and brown eyes; blond, black, and red hair) from neutral or near-neutral mutations.

A key point, misunderstood by many people, is that *mutations are random.* They arise by chance, not for a purpose. In the previous hypothetical example, the variation in the flight speed of Mourning Doves did not develop because a need for faster birds arose, but because either (1) by chance, an array of flight speeds existed in the population before pre-

dation by Cooper's Hawks became a survival factor, or (2) a mutation increasing flight speed arose randomly in a Mourning Dove, allowing it to more readily escape hawk predation. Once the variability exists, natural selection proceeds in a very nonrandom fashion, but the type of variability that arises in the first place is random.

The random nature of mutations explains why evolution is often so slow. Nearly all mutations are detrimental. As Cornell astronomer Carl Sagan explained to his students: "Think of any living thing as a finely tuned watch. What is the chance that dropping your watch will improve its function?" Natural selection proceeds slowly because it has no predetermined purpose: it is not goal-oriented. Natural selection works on the random efforts of a "blind watchmaker" (Dawkins 1987), but natural selection itself is anything but random. With time, it can produce a masterpiece of adaptation: from the colorful, elaborate feathers in a peacock's train to the sharp, fish-catching beak of a Black Skimmer **(Fig. B)**. Natural selection did not, however, have to produce anything like birds or humans—both are the result of natural selection, at each step along the way, merely responding to the conditions at hand.

Figure B. Black Skimmer Foraging: *The beak of the ternlike skimmers is highly specialized for their unique, fish-catching technique. With its beak open, a Black Skimmer (as pictured here) flies low over a river or ocean, its lower beak slicing through the water. When it contacts a fish, it snaps the upper and lower beak together, grasping its prey tightly. Natural selection has modified the lower beak to a vertically narrow, knife-like blade, which creates minimal friction as it moves through the water. It also contains many sensory receptors, allowing it to detect precisely when a fish is contacted. Furthermore, the lower jaw can open widely to allow the lower beak to penetrate below the water's surface in flight, while the upper beak also opens widely to avoid touching the water. Also see Figures 4–87 and 4–88. Photo by Tom Vezo.*

The Effects of Natural Selection

Natural selection does not always result in change. Some organisms, such as cockroaches and horseshoe crabs, have changed only slightly over vast stretches of geologic time, whereas others, such as mammals, have changed dramatically. If a species is well adapted to its environment, natural selection may act *against* any mutation that would change it. For example, if Mourning Doves already fly at the optimum speed, selection will act against mutations that would make that speed either faster or slower. This type of selection, termed **stabilizing selection**, works to keep the species at the optimal middle point. Only when conditions change (for example, Cooper's Hawks become less abundant) and the former traits are no longer optimal, does natural selection produce changes.

When studying living things, it is easy to forget that what you see is not a finished product. Natural selection is an ongoing process that will continue into the future, further modifying the organisms alive today as conditions change.

Evolution can occur on several different levels. The evolution of new species over thousands of years is often referred to as **macroevolution**. In contrast, the frequency of certain characteristics in a population can change relatively quickly—a type of evolution termed **microevolution**. Some changes are so rapid that humans can observe them in their lifetimes—the evolution of drug-resistance in certain strains of bacteria, for example.

The words "evolution" or "natural selection" often bring to mind the expression "survival of the fittest," but this phrase can be very misleading. Although the fittest do indeed survive, and the natural world can be a pretty gory place, evolution is responsible for a number of more pleasant aspects of life as well. In some circumstances, the fittest individuals are those who cooperate with others of their own species. Many social birds, including American Crows, Florida Scrub-Jays **(Fig. C)**, and White-fronted Bee-eaters, as well as humans, fit this model. In these species, individuals benefit by cooperating with others, and very complex social systems have evolved. Such admired human traits as motherly love, courage, compassion, and honesty all can, in various social situations, act to increase an individual's relative fitness. ■

Figure C. Cooperative Breeding in Florida Scrub-Jays: Found only in patches of stunted scrub oak in central Florida, the threatened Florida Scrub-Jay has evolved a cooperative breeding system. These jays live year round in extended family groups of up to eight birds, consisting of a permanently mated pair plus offspring produced in previous years. Young Florida Scrub-Jays delay their own dispersal for up to six years, remaining with their parents as "helpers." In this photo, a three-year-old daughter (on nest, at right) helps to defend the nest of the breeding pair (her four-year-old stepfather, at top; and her eight-year-old mother, at left). Two nestlings are visible beneath the daughter. All three adults have drawn near the nest, joining forces against the photographer. Helpers assist significantly with territory defense, nest defense, and nestling care, increasing the chance of nesting success.

Although helpers delay their own breeding to help, these long-lived birds may actually increase their lifetime genetic contributions to successive generations by doing so. The Florida Scrub-Jay's habitat is rare and patchy—the jays can occupy only those areas that recently have been burned by wildfire. As a result, nearly all suitable areas are constantly being defended vigorously by breeding birds. Thus, a newly fledged male seeking a territory on his own stands little chance of success. Because bigger groups can defend larger territories, a male that remains with his parents can increase the size of their territory, which he may inherit when they die, or from which he may "bud off" a territory of his own. Similarly, female helpers benefit from having a "home base" from which to search a wide area for an unpaired territorial male—a rare and ephemeral commod-

ity. Young scrub-jays that disperse in their first year face great difficulty competing for breeding opportunities, and often die before succeeding. Besides providing a safe "home base," delaying their dispersal allows helpers to contribute some of their own genes to future generations by rearing their own siblings (who have some of the same genes as the helper because they share parents), while they wait for a chance to become breeders themselves.

As remarkable as it may seem, natural selection can produce cooperative breeding systems even more complex than the Florida Scrub-Jay's, simply because helpers benefit more by helping their parents than they would by attempting to breed early on their own. See Ch. 6, Parental Behavior: Why are there "Helpers at the Nest" that Care for Someone Else's Offspring. Photo courtesy of John W. Fitzpatrick.

publications are responsible for the ordering of groups found in most American field guides. For example, as a result of Wetmore's work, the lobe-footed, short-billed coots are shown with the rails and not with the web-footed, spatulate-billed ducks, which they most superficially resemble when swimming **(Fig. 1–35)**.

Figure 1–35. Wetmore's Classification of Coots, Rails, and Ducks: *For many years, coots were grouped with the ducks because, like ducks, they swim on the water's surface and dive for food. Alexander Wetmore was the first to recognize that the coots, with their long, lobed toes, medium-long legs, and short, thick bills should be grouped with the rails (see Sora photo)—who have similar characteristics (but whose long toes are unlobed). Wetmore published his classifications between 1930 and 1960. Coot and Mallard photos by Marie Read. Sora photo courtesy of Lang Elliott/CLO.*

American Coot

Sora

Mallard

Although a few odd reorganizations occasionally have been proposed, none has gained much support in the ornithological community. In fact, inertia has played a great part in maintaining the classification of birds throughout the past century. Unless strong evidence to the contrary has been available, bird relationships have stayed just where Gadow put them. Species get moved around regularly, and some families have emerged while others have been discarded, but very little movement of families among orders (the next category of classification above families—see Fig. 1–41) has taken place.

In the 1980s, Charles Sibley and Jon Ahlquist proposed an entirely new and somewhat radical phylogeny of birds. Sibley had long been interested in the higher groupings of birds and was a pioneer in using chemical (instead of physical) methods to explore bird relationships. The studies of Sibley and Ahlquist resulted in the first major rearrangement of the orders and families of the world's birds in nearly half a century. Many of the relationships they proposed were familiar (ducks and geese still went together, as did hawks and falcons), many corroborated previous "heretical" revisions proposed by others (New World vultures and condors are really modified storks and not hawks), and still others were surprisingly novel (frigatebirds, penguins, loons, and tubenoses formed a group). Perhaps one of the most unexpected findings was that the numerous passerine land birds native to Australia were not related to the warblers and thrushes they resembled, but were related to each other in a remarkably diverse array of forms at least as spectacular as the better-known marsupial mammals.

Unfortunately, this work was not without controversy. Some of the results were tainted by suspect methods and unexplained correction factors. As a result, acceptance of the new ordering of birds has been slow and guarded. Gradually other researchers are testing the proposed novel relationships, and generally finding corroboration for most. It will take a while, however, for the dust to settle and for any arrangement of birds of the world to become universally accepted again. The American Ornithologists' Union, the largest body of professional ornithologists in North America, in its 7th edition of the *Check-list of North American Birds* (commonly called the "*AOU Check-list*"), used the following philosophy regarding taxonomy: "Because of wide acceptance of the *Check-list* as an authoritative standard, the Committee responsible for its preparation feels it necessary to avoid hasty decisions that risk quick reversal, thereby fostering instability.... Our general stance has been conservative and cautious when judging recently published proposals for novel classifications, schemes of relationship, and species limits.... The Committee established a policy for this edition whereby changes in classification of major groups require concordant evidence from two or more independent data sets." In the future, as more of the relationships are examined and more information is gathered, the placement of some species may change. Taxonomy is not a static endeavor, but rather, like much of science, ideas change as more information becomes available.

Methods Used to Classify Birds

Through the years, a large number of physical characters have been used to classify birds. Some of the obvious external characters, such as color, bill shape, and foot shape, have turned out to be of little use because of the problem of **convergence**, one of the most troubling problems for biologists. Researchers now realize that some animals (and plants) resemble each other not because they are related, but because they evolved to do the same thing; that is, they independently developed a similar solution to a similar problem. This pattern of evolution is called **convergent evolution**. For example, horses and antelope both have long legs not because they came from the same ancestor, but because they both use the strategy of running fast to escape predators, and long legs are good for that. Both swifts and swallows have long, pointed wings and short bills because they both chase flying insects and capture them in their mouths, and their wing and mouth shapes help them to do that. On examination of other characters, such as their toes, wing bones, and voice box structure, it becomes apparent that swifts and swallows look alike only on the outside; inside they are very different **(Fig. 1–36)**. They have independently converged on the same overall body plan because they each have evolved to do the same thing. Because they did not start from the same point (in other words, they had different ancestors), their exact "solutions" are somewhat different: antelope have two toes at the end of their long, running legs, whereas horses have only one; swallows have long wings owing to a very long radius and ulna (an elongated antebrachium), whereas in swifts the long wings result from an elongated manus.

Sometimes separating similarity owing to lifestyle from similarity due to common descent is relatively easy, but often it is difficult. Certain physical characters are subject to very strong selection, especially characters that help birds find food or escape enemies. A bird with a stronger bill can open larger seeds; a bird with an efficient wing shape for its flight style uses less energy. All these differences affect the

Figure 1–36. Convergence of Swifts and Swallows: At first glance, swifts and swallows are very similar in appearance. Specialized for life in the air, they both have streamlined bodies; long, pointed wings; and short bills with which they capture insects in midair. A closer examination, however, reveals many significant anatomical differences. Swallows have a more complex syrinx (voice box) than swifts, a different toe arrangement (anisodactyl, compared to the pamprodactyl feet of swifts [see Fig. 1–20]), and different wing proportions. In swallows, the wings are long as a result of the long radius and ulna (in the forearm, or antebrachium), whereas the swift's long wings result from an elongated manus (hand). These anatomical differences indicate that the similarities between the two bird groups are due to convergence—the evolution of similar features in response to a similar environment and lifestyle—rather than to common ancestry.

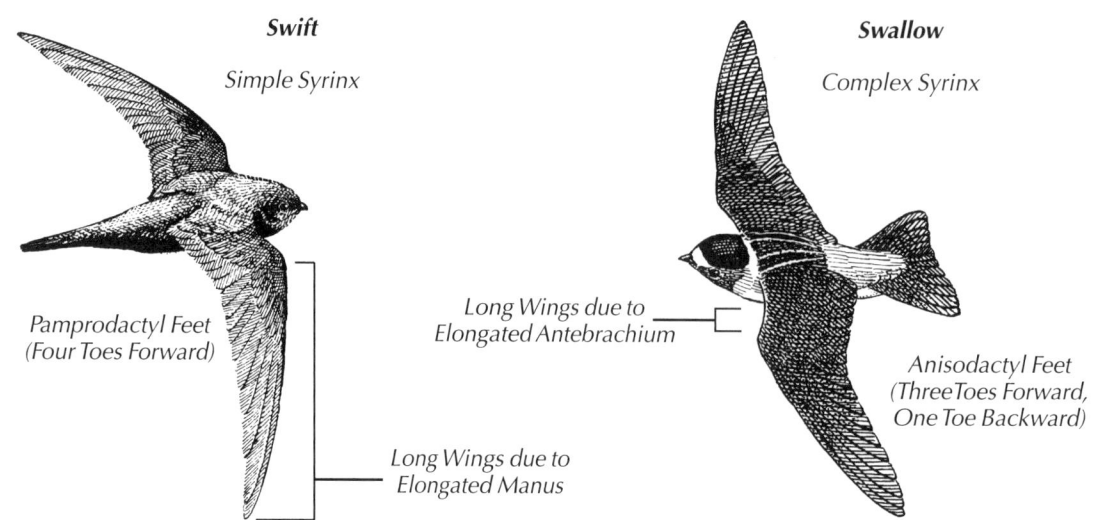

Swift
Simple Syrinx

*Pamprodactyl Feet
(Four Toes Forward)*

*Long Wings due to
Elongated Manus*

Swallow
Complex Syrinx

*Long Wings due to
Elongated Antebrachium*

*Anisodactyl Feet
(Three Toes Forward,
One Toe Backward)*

survival and reproductive success of the individual bird, and consequently can be rather easily modified by natural selection. As a result, the bill and overall shape of birds, which are often striking, may be of little help in discovering true phylogenetic relationships.

More useful in determining relationships are characters that vary among groups but that are not so easily changed by natural selection—using these often minimizes the confusion resulting from convergence. Internal characters often have been used because they are considered (whether rightly or wrongly) to be more evolutionarily conservative than external characters. Examples of such characters are the shapes of feather tracts, the shape and construction of the oil gland at the base of the tail, the shapes of many different bones in the skull, the shapes of bones of the inner ear, the arrangement of muscle and tendon attachments, the construction of the **syrinx** (voice box), the arrangement of the blood vessels, the construction of the intestines, the presence of specialized feathers, the number of feathers in the wing, and various aspects of behavior (especially reproductive behavior). All of these have been used to classify birds.

The discovery of the structure of **DNA** in 1953 by James Watson and Francis Crick helped researchers to understand the "genetic code"—how the genes provide instructions for making **proteins.** Proteins are complex molecules composed of strings of amino acids. They act both as structural material in the body and as **enzymes**, chemicals that **catalyze** (assist) chemical reactions. Understanding the process of protein synthesis from DNA codes allowed researchers to see how information is passed from parent to offspring, and how it is modified in the process of natural selection. An organism's traits (both internal and external) are determined by the sequence of merely four different chemical bases in the DNA molecule **(Fig. 1–37)**. These bases are the "letters" and, taken three at a time, the "words" in the instructions for building an organism. Each "word" of three bases codes for a specific amino acid, and the sequences of amino acids produced determine the types of proteins created. The proteins direct the production of all the other molecules needed, thereby determining the characteristics of the organism. DNA is truly a blueprint for making an organism.

Most organisms, including humans and birds, have two copies of each **gene**. (Precisely defined, a gene is the sequence of base pairs that codes for one specific protein.) You received one copy from your mother, and the other from your father. Because you are using some of the same blueprints to construct yourself that your parents used, some of your constructed parts resemble those of your parents. (Note that you don't have your mother's *eyes*, you have her *recipe* for eyes.) Sometimes, when copies are being made to put in the sperm or egg, the instructions get garbled; such a change in the base sequence of DNA is called a **mutation**. This change will then be passed on to the offspring. Physical modifications of the body made during the lifetime of an organism do not get passed on—only changes in DNA. For example, for hundreds of years, farmers cut the horns off domestic cattle to make them safer to be around. Each of those de-horned cows, though, gave birth to offspring that would grow horns. At some point,

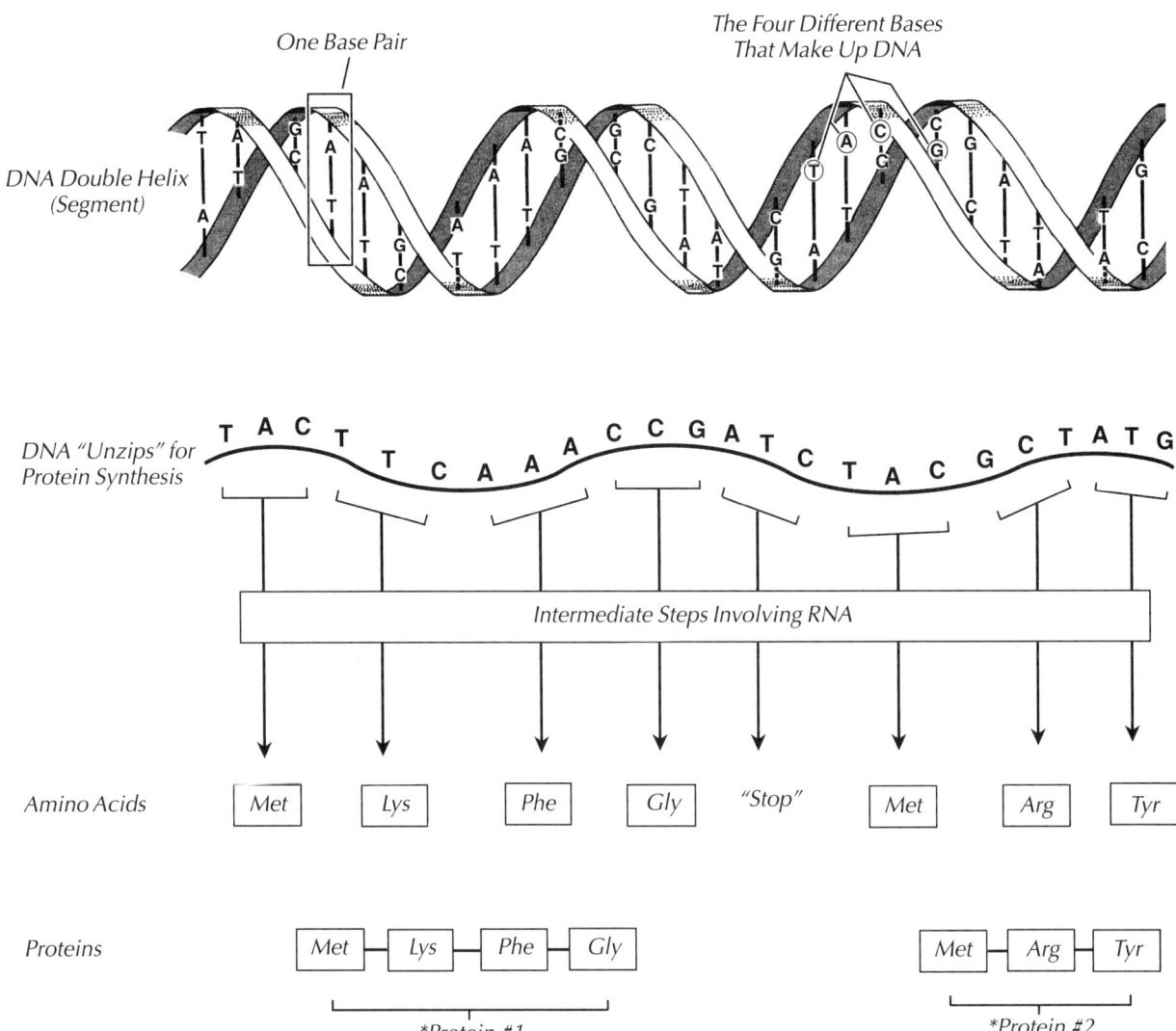

In reality, proteins are much longer chains of amino acids.

Figure 1–37. The Genetic Code and Protein Synthesis: *The main genetic material, DNA, contains within its structure the instructions for making proteins, which in turn determine every characteristic of an organism. DNA is composed of subunits called nucleotides, which differ only in the type of nitrogenous (nitrogen-containing) base they contain. (A base is a type of chemical, such as ammonia, which acts in an opposite manner to an acid in solution.) The nucleotides of DNA each contain only one of four types of bases, adenine (designated by the letter A), guanine (G), thymine (T), and cytosine (C). The DNA molecule is composed of two long chains of nucleotides, with their attached bases. Each base in a chain is paired with one base in the other chain, A only with T, and C only with G; this double chain is twisted to form a double spiral known as a "double helix." Only a short section of a DNA double helix is shown here, since an entire DNA molecule is many millions of base pairs long.*

During protein synthesis, the bonds between some of the paired bases break temporarily, so that a portion of the DNA molecule, in effect, "unzips" to create a single chain. The key to the genetic code is the sequence of base pairs of DNA. Through a multitude of complex biochemical steps involving another type of genetic material, RNA, each set of three adjacent bases is translated to indicate a particular amino acid. The cell then bonds the amino acids together into long chains, in the sequence indicated, to form proteins. The sequence of amino acids determines the type of protein. Since there are only 20 types of amino acids, and 64 possible sequences of three bases, there is some redundancy: several different codes may indicate the same amino acid, and some codes indicate "stop" (the end of one protein) and "start" (the beginning of another protein). For simplicity, proteins are indicated here as short chains of amino acids, but in reality they are much longer. Amino acid abbreviations used here are met (methionine), lys (lysine), phe (phenylalanine), gly (glycine), arg (arginine), and tyr (tyrosine).

just by chance, a mutation cropped up that changed the DNA code for making horns, and a cow grew up hornless. It passed this mutation on to its offspring, and those cattle with the new hornless gene were able to produce hornless young.

With the understanding of DNA and the development of molecular techniques, systematists (scientists who study how organisms are related to one another) began to look in different places for useful characters to construct classifications. They began to look at the chemical makeup of birds, not just their outward (or internal) appearance. Because DNA molecules are very small, direct observation of the sequences of bases is not possible. Researchers using chemical methods to classify birds started by looking at things one step removed—not at the blueprints, but at the building blocks of organisms, the proteins. Certain proteins can be chemically isolated from an organism. Different types of animals make slightly different forms of the same protein (the forms all have the same function), and these differences can be detected. One important tool used to compare protein forms is **electrophoresis**. By watching how far each form of a protein migrates up a gel in an electric current (somewhat the way ink "bleeds" up a strip of paper), different forms of the same protein can be separated **(Fig. 1–38)**. Smaller or differently charged forms travel at different speeds than do others. This method was one of the first ways to look at chemical differences quite apart from (and thus unbiased by) external morphology.

Although proteins are the main building blocks of animals, they are still a step away from DNA. Ideally, what should be examined to determine relationships are not the products of protein synthesis as translated from DNA, but the actual sequences of bases in the DNA molecules themselves.

DNA molecules, however, are too small for their structure to be "seen" directly, so various indirect methods are used to determine their base sequences. A number of different techniques have been developed to look at DNA, and each is useful for a different level of com-

Figure 1–38. Gel Electrophoresis of Proteins: Because protein molecules are too small to see, even under a powerful microscope, scientists must use indirect means to determine their form. Gel electrophoresis separates large molecules, such as protein molecules with different lengths, by their rate of movement through a thin slab of gel in an electric field. To create the electric field, electrodes (one negative and one positive) are attached to opposite ends of the gel, and the electric current is turned on. Researchers treat the protein molecules so that they carry a negative charge, and thus the proteins are attracted to the positive electrode, migrating slowly toward it when the current is on. Because larger molecules are slowed down more by the gel, they will not migrate as far in a given amount of time. Thus, if a sample of protein molecules of different lengths is applied to the gel, they will appear as separate "locations" on the gel after migration. Originally, electrophoresis was used only to separate proteins, but more recently it also has been used to separate DNA molecules of differing lengths. Adapted from Campbell (1990, p. 403).

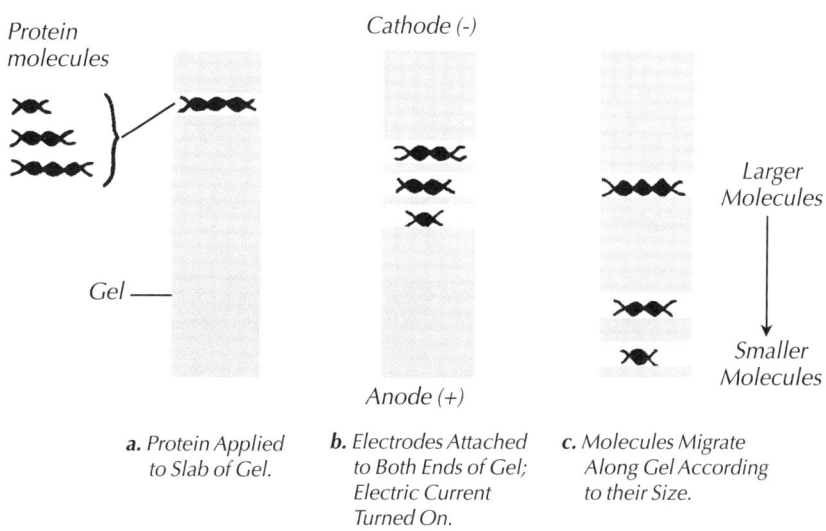

Protein molecules

Gel —

a. Protein Applied to Slab of Gel.

Cathode (-)

Anode (+)

b. Electrodes Attached to Both Ends of Gel; Electric Current Turned On.

Larger Molecules

Smaller Molecules

c. Molecules Migrate Along Gel According to their Size.

parison. For example, the information needed to distinguish the more closely related two out of three chickadee species is different from the information needed to determine which two out of three groups, say, ducks, hawks, and sparrows, are more closely related. Here, a brief overview of the techniques that are currently most important for classifying birds is presented.

Through a variety of techniques, past researchers were able to actually read the sequence of bases in DNA, a process termed "sequencing." One major problem was that a single DNA molecule for a relatively simple organism, such as a nematode worm, contains *millions* of base pairs in its DNA. Sequencing takes time, and no one could possibly read the entire sequence and make sense of it. So, the first techniques used to read DNA did not concentrate on the DNA contained in the **nucleus** (command center) of a cell, where the plans for the whole organism lie. Instead, they took advantage of the fact that little structures within the **cytoplasm** (material outside the nucleus) of each cell, the **mitochondria** (singular: mitochondrion) have their own DNA, which is much smaller and simpler than that found in the nucleus **(Fig. 1–39)**. The mitochondria are the "powerhouses of the cell," where most of the energy generation takes place. They are probably ancient bacteria that somehow got incorporated into cells at a very early stage of evolution. Part of the evidence supporting the bacterial origin of mitochondria is that their DNA is very similar to that of bacteria: relatively short and arranged in a circle—unlike the long, complex strands of DNA, termed **chromosomes**, found in the nuclei of most cells. Researchers were able to isolate the mitochondrial DNA (mDNA) and sequence specific regions of it.

Figure 1–39. Nuclear Versus Mitochondrial DNA: *DNA is found in both the nucleus and mitochondria of animal cells. In the nucleus (command center), individual DNA molecules exist as extremely long, thin, coiled strands known as **chromosomes**. Bird chromosomes vary widely in size and their lengths are not fully known, but they are on the order of tens of millions of base pairs (see Fig. 1–37) long. In the close-up view, the chromosome is shown as two identical DNA strands held together in the middle. This is the form DNA takes just before cell division (in which the two strands pull apart and one goes to each new cell), and is the usual way chromosomes are illustrated. Mitochondria are small structures in the **cytoplasm**, the cellular material found outside the nucleus. Their single chromosome is much shorter (an average of 16,000 base pairs long in vertebrates) and in the form of a ring.*

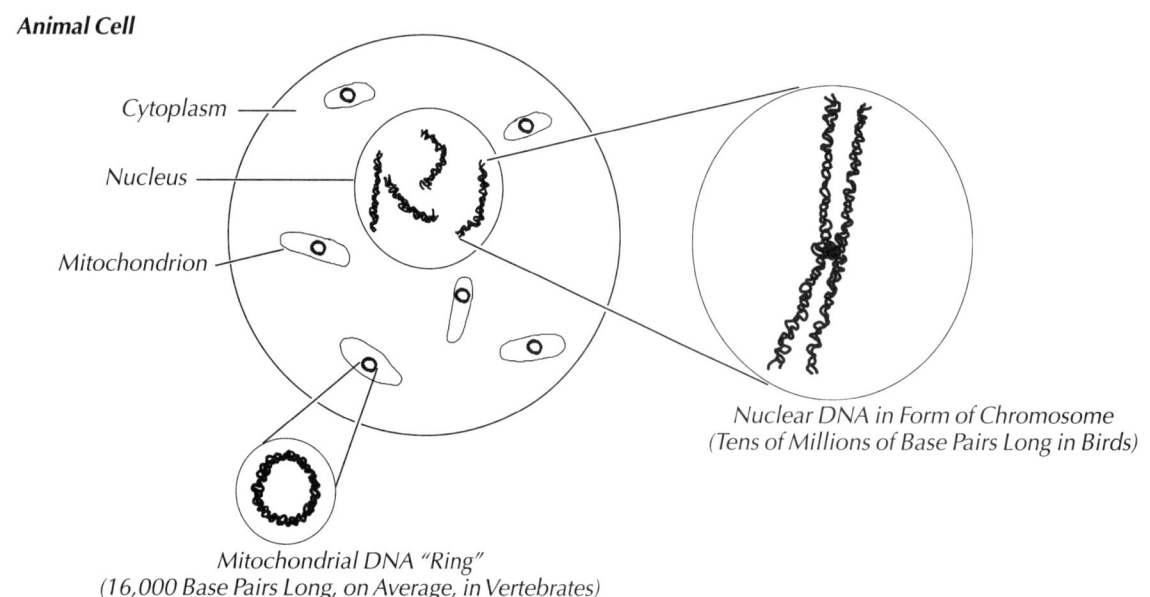

Animal Cell

Cytoplasm

Nucleus

Mitochondrion

Nuclear DNA in Form of Chromosome
(Tens of Millions of Base Pairs Long in Birds)

Mitochondrial DNA "Ring"
(16,000 Base Pairs Long, on Average, in Vertebrates)

Again, problems were encountered. Mitochondria are not passed from both parents to their offspring, but instead are transmitted intact from mother to offspring. Therefore, all your mitochondria came from your mother, and all of hers came from her mother, and so on back through time. Although useful for many studies, maternal-only inheritance presents some problems for understanding evolution. Nevertheless, examining mDNA remains an important tool in exploring evolution and the relationships of organisms, especially at the population level (a **population** is a group of individuals of the same species that live in the same general area).

Systematists recently have turned to sequencing nuclear DNA, but still are limited by how little of the whole DNA molecule they can sequence. Most researchers, therefore, are restricting their work to specific regions of the DNA molecule, sequencing single genes for comparisons between organisms. This technique creates the same problems that faced earlier systematists who used morphological traits to classify birds, namely that a single character is used to determine a relationship. To get around this problem, many systematists use a number of known DNA sequences together with morphological characters to investigate the relationships among groups of birds.

Another technique using DNA for detecting relationships, especially at the higher levels of comparison, is **DNA-DNA hybridization**. This technique was used by Sibley and Ahlquist to construct their new phylogeny. They collected and analyzed tissue samples from an astonishing one-sixth of the bird species of the world—the largest set of DNA comparisons made for any group studied to date.

In the process of DNA-DNA hybridization, the double-stranded molecules of purified DNA (which look roughly like incredibly small but very long, twisted zippers) from two species are "unzipped" **(Fig. 1–40)**. That is, the molecules are chemically broken apart down the middle, between the millions of base pair combinations. Because a DNA molecule is more stable double-stranded (zipped) than single-stranded (unzipped), the doubled-stranded molecules will tend to reform. If unzipped DNA from one species is put in contact with unzipped DNA from another species, the molecules will try to fuse with one another. Of course, some of the "words" in the genetic code of one species will be different from those in the other species (some of the zipper teeth will not fit together), so the base pairs of the hybrid DNA molecules will not be perfectly matched. The key is that the fit of those DNA strands that are most similar will be most stable, and that the stability can be measured. The more stable the hybrid DNA molecules, the more closely the two species are related.

Binomial Nomenclature and Classification System

Early in the 18th century, the time of great explorations to all parts of the globe, expeditions returned to Europe bearing new plants and animals by the score. Naturalists in different countries gave them different names; there was no uniformity, no standard followed by all naturalists in all countries. How long this chaotic situation might have

Figure 1–40. DNA-DNA Hybridization:
DNA-DNA hybridization is used to determine the degree of similarity between
two different samples of DNA. At the top
of this schematic, a short section of DNA
from two different species is illustrated;
in the actual process, however, the entire
molecules would be used. The different
shapes along the line down the middle of
each represent the different nitrogenous
bases. For example, the square "peg"
might be adenine, and the square
"hole," thymine (see Fig. 1–37); and, the
triangular "peg" might be cytosine, and
the triangular "hole," guanine. First, the
DNA samples are heated to break the
bonds between the base pairs, disassoc-
iating (separating) each double-stranded
molecule into two separate strands
(middle drawing)—a process termed
"melting." Because DNA is more stable
in the double-stranded form, single DNA
strands in the same sample will bond
together to form double strands, unless
kept apart by high temperatures. Therefore, "melted" samples from each species are combined, and held at a lower
temperature, allowing double-stranded
molecules to form. Although many of the
single strands will bond with another
strand from the same species, some of
the strands will "hybridize," bonding
with the DNA of the other species (bottom drawing). Because hybrid double
strands do not have perfectly matched
base pairs, they are less stable, and thus
separate more readily when reheated.
Furthermore, the greater the difference
between the base sequences of the two
species' DNA, the lower the melting
temperature of the mixture of reformed,
double-stranded DNA molecules—giving researchers a quantitative measure
of the degree of similarity between the
DNA molecules from the two species.
Adapted from Proctor and Lynch (1993,
p. 23).

DNA Samples from Two Species are
Collected and Purified

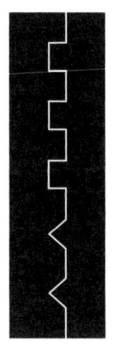 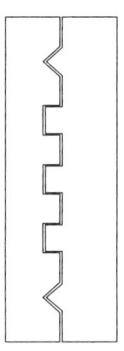

Species 1
DNA Strands

Species 2
DNA Strands

The DNA is Heated to Disassociate the
Molecules into Individual Strands. The
Samples are then Combined to Form
Hybrid DNA.

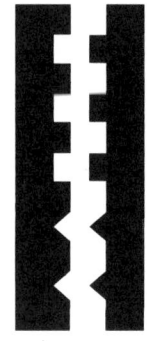 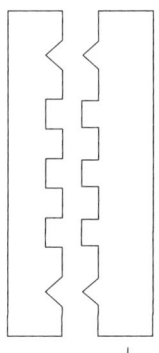

The Hybrid DNA is
then Reheated and
the Melting Tempera-
ture is Analyzed

Mismatched
Areas are
Weak and
More Readily
Disassociate
when Heated

Hybrid DNA
from Both Species

continued is anybody's guess, had it not been for Swedish naturalist Carolus Linnaeus (1707 to 1778). A genius at organization, he set up a system, **binomial nomenclature**, for all plants and animals. This method was so useful that it soon became the standard system and biologists have used it ever since.

Linnaeus designated each plant or animal by two Latin names, the first denoting the **genus** (plural **genera**; always capitalized) and the second, the **species** (plural also **species**; never capitalized). He chose Latin because it was the universal language of scholars, and he used two names because there were not enough single names in the language for all the species. Thus, although the second word is officially called the "species" designation, *both* the genus and species must be written together to indicate a unique organism, because there may be more than one unique organism with the same "species" designation (for example, the Least Flycatcher, *Empidonax minimus,* and the Bushtit, *Psaltriparus minimus*). To allow easy recognition of these two-word Latin (or "scientific") names for species, they are always underlined or placed in italics.

The indisputable value of Linnaeus' system can be demonstrated today by examining all the common names for a certain large bird of prey that ranges widely over much of the world. The "official" English name is Osprey, while in some parts of both the United States and Canada it is known as the Fish Hawk. The Swedes call it Fiskgjuse; the Germans, Fischadler; the Dutch, Visarend; the South Africans, Visvalk; the Burmese, Wun-let; and the Argentines, Sangual. But to all ornithologists, regardless of the language they speak, the bird is *Pandion haliaetus.* This illustrates why guides to identification and other authoritative treatises give the scientific names after the common names—to eliminate any doubt in anyone's mind about the identities of the birds mentioned.

Every scientific name means something **(Sidebar 3: Latin and Greek Roots of Biological Terms)**. The names are concocted from Latin or Greek roots, or at least made into the form of Latin words. Although some names may have been created to honor a person, many of the translated roots are descriptive, and usually provide some information about the organism. For example, the familiar American Crow is known as *Corvus brachyrhynchos. Corvus* comes from the Latin for "crow," and *brachyrhynchos* means "short beak," which a crow has, compared to a raven.

Linnaeus used very few higher categories to express the organizational affinities of genera. But as more organisms were described and fitted into the classification scheme, new categories had to be added to supplement and organize the Linnaean categories of genus and species. Each type or form of animal or plant is described by placing it in a hierarchy of categories. As mentioned previously, since the work of Darwin, classifications generally have been attempts to reflect evolutionary relationships. Individuals of a recognizable type are considered members of the same **species**. Different species that are

(Continued on p. 1·52)

Sidebar 3: LATIN and GREEK ROOTS of BIOLOGICAL TERMS
Marie Eckhardt

Students new to the field of biology may find the terminology daunting. Many biological terms and names, however, are derived from Latin or Greek roots. Once you become familiar with these roots, you will find that biological terms are more comprehensible. You may already be acquainted with some of these terms. Many are present in our everyday language as prefixes or suffixes embedded in familiar words.

The Latin and Greek roots presented here are each spelled in a form that is commonly encountered in biological words, but these are not necessarily the original forms.

Additional information may be found in Jaeger (1955) and Pough et al. (1996).

Root	Meaning of Root	Example and Definition
a, ab	away from	**ab**ductor muscles: muscles that draw bones away from the center of the body
a, an	without, lacking, not	**a**biotic: not living; **a**pteria: areas of bird skin that lack feather tracts
ad	to, toward, attached to	**ad**ductor muscles: muscles that draw bones toward the center of the body
aeros	the air	**aero**dynamics: branch of science dealing with the motion of air and objects in air
al, alula	a wing	**alula**: feathers attached to the first digit in a bird's wing
alb	white	**alb**ino: lacking color pigmentation
all, allo	other	**allo**preening: the preening of one bird by another
aqua	water	**aqua**tic: living primarily in or on water
amphi	both, double	**amphi**bian: a class of vertebrates that spend part of their lives in water and part on the land
andro	male	poly**andr**y: mating system in which a single female mates with several males in a breeding season
ante	before	**ante**brachium: the forearm of a bird—the part before the brachium
anthro	human	**anthro**pomorphic: ascribing human qualities to nonhumans
arch	beginning, first time	**Arch**aeopteryx: the earliest known fossil bird
audi	to hear	**audi**tory: related to hearing
av	a bird	**av**ian: relating to birds
bi, bis	two	**bi**pedal locomotion: walking on two feet
bio	related to life	**bio**logy: the study of life
blast	bud, sprout	**blast**odisc: flattened sphere of cells that is the first stage of development in the bird embryo
brachi	arm	**brachi**al artery: main artery of the human arm and bird wing
carn	flesh	**carn**ivorous: meat-eating
caud	tail	**caud**al vertebrae: bones of the tail
ceno	new, recent	**Ceno**zoic: the most recent geological era
cord	guts, a string	spinal **cord**: cable or thick "string" of nerves running the length of the vertebral column
chorio	skin, membrane	**chorio**n: membrane surrounding the embryo, yolk sac, amnion, and allantois in a bird egg

Root	Meaning of Root	Example and Definition
chrom	color	**chrom**atophore: a pigment-bearing cell
cide	killing	pesti**cide**: an agent used to destroy pests
circa	about, approximately	**circa**dian rhythm: a biological rhythm of about one day
circum	around	**circum**navigate: to go or travel completely around something
cloac	a sewer	**cloac**a: in birds, the chamber in which fecal material, urine, and eggs or sperm collect before being discharged from the body
cran	skull	**cran**ial: belonging to the skull
dactyl	finger, toe	aniso**dactyl**: the toe arrangement in songbirds and many other perching birds
de	removal of, off	**de**hydration: an abnormal loss of body fluids
derm	skin, covering	epi**derm**is: the outer skin of vertebrates
di, diplo	two, double	**di**morphic: occurring in two distinct forms
e, ex	out of, from, without	**ex**cretion: elimination of metabolic waste products from the body
eco, oikos	house, home	**eco**logy: the study of the relationship between organisms and their environment
ect	outside	**ect**oparasite: a parasite living on the outside of the body
end	within, internal	**end**oparasite: a parasite living inside the body
epi	upon, above	**epi**dermis: the outermost layer of skin in vertebrates
extra	beyond, outside	**extra**pair copulation: copulation with someone other than your mate
frug	fruit	**frug**ivorous: fruit-eating
gallus	poultry	**gall**inaceous birds: chicken-like birds in the order Galliformes, including quail, pheasants, turkeys, domestic chickens, grouse, and guineafowl
gastr	belly, stomach	**gastr**ointestinal tract: digestive tract
gen	origin	**gen**etics: the study of genes, which are essentially an organism's "origin"
geo	earth	**geo**logy: the study of the earth
gloss	tongue	ento**gloss**al process: a bone in the tongue of birds
gon	seed	**gon**ads: sexual reproductive organs
gul	throat	**gul**ar fluttering: rapid vibration of muscles and bones in the throat, which helps birds cool down in hot weather
gyn	female	poly**gyn**ous: mating system in which a single male mates with multiple females within a single breeding season
hemi	half	**hemi**sphere: half of a spherical body
hepat	the liver	**hepat**ic veins: large veins that carry blood from the liver to the caudal vena cava
hetero	different, other	**hetero**geneous: consisting of ingredients or constituents that differ from one another
homo	alike	**homo**logous: having the same relative value, position, structure, or origin
hydr, hydro	pertaining to water	**hydro**philic: having a strong affinity for water
hyper	above, excessive	**hyper**sensitive: abnormally sensitive

Root	Meaning of Root	Example and Definition
hypo	below, insufficient	**hypo**thermia: subnormal temperature of the body
inter	between	**inter**specific: occurring or existing between different species
intra	within, inside	**intra**cellular: occurring or functioning within a cell
leuc, leuk	white	**leuk**ocyte: white blood cell
kin	movement	cranial **kin**esis: the ability of birds to raise the upper beak while simultaneously depressing the lower beak
klepto	to steal	**klepto**parasitism: robbing another bird species of its prey
mela	black	**mela**nin: a dark brown or black pigment
meso	middle	**meso**bronchus: main air tube running down the middle of each bird lung
meta	to change	**meta**bolism: all the physical and chemical processes of an organism, including those by which food is converted to energy or body structures
micro	small	**micro**climate: the local climate of a small site or habitat
milli	a thousand	**mill**ennium: a period of 1,000 years
mono	one, single	**mono**gamous: having just one mate in a given period of time
morph	shape	**morph**ology: the study of the form and structure of something
nephro	kidney	**nephro**n: the excretory unit of the vertebrate kidney
neuro	nerve	**neuro**toxin: a poison that acts on the nervous system
nidi	nest	**nidi**colous: reared in a nest
nomen	a name	**nomen**clature: a system of naming
not	the back	**not**ochord: a flexible, rod-like support running along the back of all chordate embryos
omni	all	**omni**vorous: feeding on both animal and vegetable materials
oo	pertaining to an egg	**oo**cyte: an immature egg cell
ology	study of	ornith**ology**: the study of birds
opercul	a cover, lid	**opercul**um: a protective flap partly covering the nostrils of some birds such as starlings, pigeons, and chickens
optic	eye	**optic** lobes: the portions of the brain that process visual information
ornith	bird	**ornith**ologist: one who studies birds
os	bone	**os**sicle: a small bone or bony structure
ov, ovi	pertaining to an egg	**ov**ary: female reproductive organ that produces eggs
paleo	ancient	**paleo**ntology: the study of past geological periods as known from fossils
par, para	beside	**para**site: an organism that lives at the expense of another species, often in or upon it
parous	bearing, giving birth to	ovi**parous**: producing eggs that develop and hatch outside the female's body
path	disease	**path**ogen: disease-causing agent
peri	around	**peri**pheral vision: the outer part of the field of view
pheno	visible	**pheno**type: the visible properties of an organism

Root	Meaning of Root	Example and Definition
phyl	tribe, race	**phyl**um: one of the primary divisions of the animal kingdom
plasm	formative substance	cyto**plasm**: formative substance in cells
pod	foot	mega**pod**e: a ground-dwelling bird that uses its large, strong feet to construct mound nests
poly	many	**poly**gyny: having more than one female mate at one time
post	after	**post**nuptial: occurring after breeding
pre	before, in advance of	**pre**cocial: young bird that hatches in an advanced state of maturity—capable of much independent activity before (at an earlier age than) altricial young
prim	first	**prim**ary feathers: outermost feathers of a bird's wing
pro	before, favoring	**pro**ventriculus: the glandular or true stomach of birds that is situated before the gizzard
pter	wing, feather	**pter**osaurs: extinct flying reptiles
pyg	the rump	**pyg**ostyle: the last bone in the tail of a bird, formed by the fusion of several tail vertebrae
soma	body	**soma**tic cells: the body cells, in contrast to the sex cells (the eggs and sperm)
sonus	sound	**son**agram: an image produced by sound
sub	below, under, smaller	**sub**ordinate: occupying a lower rank, class, or position
supra	above, over	**supra**coracoideus: a principle flight muscle that is attached to the humerus, a bone located above the coracoid
sym, syn	together, with	**sym**patric: occurring in the same area
tax	arrangement of	**tax**onomy: the classification of organisms
tele	far off, distant	**tele**scope: a tubular optical instrument for viewing distant objects
therm	heat	**therm**ometer: an instrument that measures temperature
tri	three	Foramen **tri**osseum: a hole formed by the junction of three bones in the shoulder joint of birds
troph	nourishment	**troph**ic level: level in the food chain at which an organism seeks its nourishment
ultra	beyond	**ultra**violet: a portion of the light spectrum beyond the blue and violet wavelengths
uni	one, single	**uni**form: consistent, having a single form, manner, or degree
ura, oura	tail or urine	**ur**opygial gland: oil gland at the base of the tail in birds
vas	a vessel	**vas** deferens: duct that carries sperm in vertebrates
ventr	the belly	**ventr**al: anatomically, the lower or abdominal side of the body
verm	worm	**verm**ivorous: worm-eating
viv	live	**viv**iparous: giving birth to live young
vor	to devour	herbi**vor**ous: plant-eating
xer	dry	**xer**ic habitat: a habitat with very little moisture
zoo	animal	**zoo**logy: the study of animals
zyg	union, coupling	**zyg**ote: the cell formed by the union of the sperm and egg cells of animals

somewhat similar are grouped within the same **genus**. Similar genera are placed within a **family**, similar families within an **order**, similar orders within a **class**, similar classes within a **phylum**, and finally, similar phyla within a **kingdom (Fig. 1–41)**.

To help understand the classification system, look again at the crow. The species American Crow, *Corvus brachyrhynchos*, lives over most (but not all) of North America, from British Columbia to Mexico and Florida. It can be separated from other species of crows and ravens by its calls, the size of its body parts, its tail shape, and the shape of the back and throat feathers. All crow and raven species are placed in the genus *Corvus*. Although not all of the 50 species of crows and ravens found throughout the world are entirely black, most would instantly be

CATEGORY	Classification of the American Crow	Other Examples of Category
KINGDOM	**Animalia**	All Other Kingdoms: Plantae, Fungi, Monera (Bacteria), Protista (Misc. Simple Organisms)
PHYLUM	**Chordata (Chordates)**	Examples of Other Animal Phyla: Porifera (Sponges), Nematoda (Roundworms), Arthropoda (Arthropods), Mollusca (Mollusks)
CLASS	**Aves (Birds)**	All Other Chordate Classes: Agnatha (Jawless Fish), Chondrichthyes (Cartilagenous Fish), Amphibia, Reptilia, Mammalia
ORDER	**Passeriformes (Perching Birds)**	Examples of Other Avian Orders: Anseriformes (Waterfowl and Screamers), Columbiformes (Pigeons and Doves), Apodiformes (Swifts and Hummingbirds)
FAMILY	**Corvidae (Crows, Ravens, Jays, Magpies)**	Examples of Other Passerine Families: Tyrannidae (New World Flycatchers), Vireonidae (Vireos), Turdidae (Thrushes), Parulidae (Wood-Warblers)
GENUS	***Corvus* (Crows and Ravens)**	Examples of Other Corvid Genera: *Nucifraga* (Nutcrackers), *Cyanocorax* (Certain Neotropical Jays), *Cyanocitta* (Blue Jay and Steller's Jay)
SPECIES	***Corvus brachyrhynchos* American Crow**	Examples of Other Species in the Genus *Corvus*: *Corvus monedula* (Eurasian Jackdaw), *C. corax* (Common Raven), *C. frugilegus* (Rook), *C. caurinus* (Northwestern Crow)

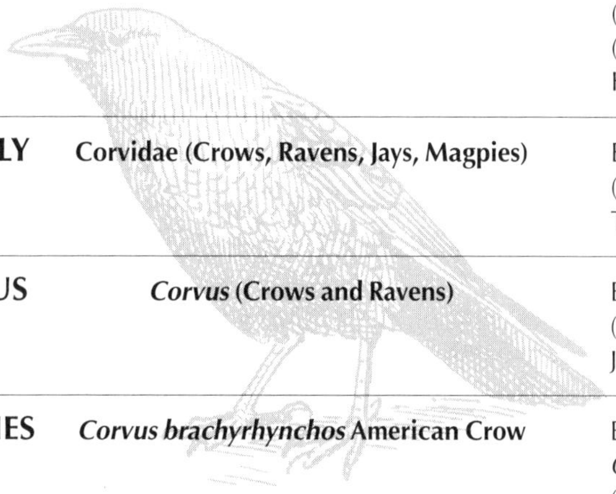

*Figure 1–41. Classification System for Living Organisms: Scientists classify all living organisms using the hierarchical system pictured here. Each species is designated by a unique, two-word scientific or "Latin" name, according to the system developed by Carolus Linnaeus in the 18th century. The first word of the name is the **genus**, and the second, the **species**. Both are underlined or italicized, and the genus (but never the species) is capitalized. For example, the scientific name of the bird species commonly called the American Crow is* Corvus brachyrhynchos. *There are other species within the genus* Corvus, *and each has a different word for the "species" designation. As shown for the American Crow, related genera are placed within the same **family**, related families within the same **order**, related orders within the same **class**, and so on. Note that for the kingdoms and chordate classes, all possibilities are shown, but for the other categories, only a selection of examples is given. People use various mnemonics to remember the levels in the classification scheme, but one effective one is: "King Philip Came Over For Good Spaghetti," in which the first letter of each word gives the first letter of each category, in the proper sequence.*

recognizable as similar to the American Crow: they are large, mostly black, omnivorous, intelligent birds. Crows and ravens, together with jays and magpies, make up the family Corvidae. Most corvids are large, omnivorous birds with feathers covering the nostrils on their stout beaks. Corvidae is just one family in the order Passeriformes, a group containing all of the familiar songbirds. Passeriformes are recognized by their overall perching physique (three toes forward, with a well-developed hallux), and especially by the uniquely complex structure of their syrinx (voice box). The 31 orders of living birds make up the class Aves, or birds. Among living organisms, birds are instantly recognizable by their feathers. Together with the various other animals (fishes, mammals, reptiles, amphibians, and the strange lancelets and tunicates) whose embryos share certain characteristics such as a notochord (a rod-like support along the back) and gill slits, birds make up the phylum Chordata, within the kingdom Animalia.

Linnaeus' basic system of classification is still in use today, albeit with many changes in names and positions of the organisms. It is, of course, his *system* that is important, not necessarily the classifications he made. For example, he lumped whales in with fish, even though two thousand years earlier Aristotle already had realized that they were different. Linnaeus' classification of birds was based heavily on the structure of bills and feet, and consequently grouped together such odd bedmates as parrots, woodpeckers, crows, and orioles.

The Species

The species is the basic unit of classification of living organisms. Species serve as the basis for describing and analyzing biological diversity. By the prevailing Biological Species Concept (BSC), a species is a group of potentially interbreeding individuals that share distinctive characteristics and are unlikely to breed with individuals of other species. Although most species are recognizable by obvious structural and behavioral peculiarities—how else could they be identified as different species in the field?—the ultimate criterion for defining them is whether they are reproductively separate. To retain its distinctiveness and thus its "species" status, a group of organisms must not breed with other species; that is, there should be no **gene flow** (movement of genetic material, as between generations) between species. There are exceptions to this general criterion for defining species, however. Hybrids between different but closely related species do occur in nature. Notable examples include the Blue-winged and Golden-winged warblers and the Mallard and Black Duck. In such cases, ornithologists have concluded that the frequency of hybridization is insufficient to consider calling the birds the same species.

When just one region is considered, species seem easy to identify, but when the entire range of organisms across the globe is considered, the boundaries between species become murky. Life is complex, and organisms do not always fall into the neat little categories that people desire. The main criterion for the BSC, that individuals of two different species will not interbreed, is often useless in practice. Similar birds may live in quite separate localities and thus have no chance of interbreeding (for example, American Crows and the similar looking Carrion Crows of Eurasia). Are they different species? What would happen if they came into contact? How could anyone begin to guess? And does it really matter if a few individuals interbreed? Would that mean that the two forms share most of their genes, or just that they do not recognize each other as different forms? Because of dissatisfaction with problems such as these, a number of people dislike the BSC, and have proposed other definitions of species.

One species concept that has gained much support is the Phylogenetic Species Concept (PSC). This idea is more concerned with

Figure 1–42. Song Sparrow Cline: *Song Sparrows found on the Pacific coast of North America show a gradual change (termed a* **cline***) in body size, plumage coloration, and song characteristics, when considered from north to south. The large, dark Aleutian subspecies appears dramatically different from the small, pale subspecies of the Southwestern desert region, with the Pacific Northwest and California coast subspecies showing intermediate characteristics. Reprinted from* Manual of Ornithology, *by Noble S. Proctor and Patrick J. Lynch, with permission of the publisher. Copyright 1993, Yale University Press.*

Aleutian Subspecies
Melospiza melodia maxima

Pacific Northwest Subspecies
Melospiza melodia morphna

California Coastal Subspecies
Melospiza melodia heermanni

Desert Southwest Subspecies
Melospiza melodia saltonis

An Example of a Cline:
Pacific Coast Subspecies of the Song Sparrow
(Melospiza melodia)

separate evolutionary histories than interbreeding individuals. If two forms became different (either morphologically or chemically) because they diverged genetically at some point in their evolutionary history, the PSC considers them separate species. The PSC is not based on hybridization or guesses about whether species will interbreed, but is concerned only with discrete, recognizably different forms with separate evolutionary histories. Under this concept, the number of bird species recognized in the world would be approximately double the number recognized by the BSC. For example, consider two forms of the Northern Flicker—the Red-shafted and Yellow-shafted flickers—named for the brightly colored feather shafts of the wing and tail. In addition to shaft color, a number of differences in the color patterns of the plumage distinguish the two groups. The red-shafted form is found throughout western North America and the yellow-shafted form, throughout eastern North America, but they interbreed within a narrow north-south zone through the Great Plains, extending northwest to southern Alaska. The BSC would proclaim these forms a single species (their current official designation), because of the extensive interbreeding. The PSC, however, would declare them separate species, considering the distinct plumage differences to indicate separate lines of evolution.

Like the human species, all bird species show individual variation. If a species ranges widely over geographical areas that encompass shifts in environmental conditions, such as from warm to cold or humid to arid climates, it is likely to show a gradual change (called a **cline**) in certain characters from one population to the next. The variations between populations may be in one or more of a number of characters, including coloration, body size, bill size, song, or number of eggs laid; these variations are roughly correlated with changes in geographical areas **(Fig. 1–42)**.

Sometimes the changes in characters are abrupt, especially if physical barriers such as a large body of water, a mountain range, or a desert separate the populations. Whether gradual or abrupt, populations in certain geographical areas may be sufficiently distinct to warrant the designation **subspecies**, sometimes called **race (Fig. 1–43)**. When a subspecies is formally described in the scientific literature, it receives a third name after the binomial. For example, the common American Crow in New York is *Corvus brachyrhynchos brachyrhynchos,* whereas the smaller form common in California is known as *Corvus brachyrhynchos hesperis.*

A subspecies is defined as a population of a species that has some unique characters and some that are shared with other populations, and which can interbreed with other populations when they meet. Subspecies, then, are separated only geographically, not reproductively. In practice, reproductive isolation is difficult to detect, so under the PSC many geographically isolated forms would be considered full species.

The Formation of Species

As discussed above, widespread species are composed of populations that, through geographical separation, may become sufficiently distinct to warrant their designation as subspecies. If the isolating fac-

Figure 1–43. Races of the Dark-eyed Junco: *Familiar for its pale beak and white outer tail feathers, flashed conspicuously in flight or during displays, the Dark-eyed Junco has a number of different subspecies or races. Males of three of the races are pictured here. All breed mainly in coniferous or mixed coniferous/deciduous forests. The* **Slate-colored** *race is widespread in North America, but much more common in the east. Males are dark gray, with a contrasting white belly that gives them a "hooded" appearance. Females are brownish gray, with a similar pattern. The* **Pink-sided** *race, found in the central Rocky Mountains, has pinkish-brown sides, a blue-gray hood, black lores, and a white belly. The* **Oregon** *race of western North America is quite variable, with paler colors found in the more southerly populations. In all Oregon forms, the black hood of the male contrasts sharply with the brown back, pinkish to buff sides, and white belly. The hoods of females are gray. Some researchers consider the Pink-sided race a pale version of the Oregon race. Not shown, are Gray-headed and White-winged races. Where their ranges overlap, the Dark-eyed Junco races interbreed frequently, producing offspring with intermediate coloration. Hybrids between the Slate-colored and Oregon races are particularly common, especially in the Great Plains. The fact that the Dark-eyed Junco races are, in general, separated geographically but not reproductively is evidence that the different forms, once each considered separate species, are actually races of a single species. Slate-colored race photo by Marie Read; Pink-sided and Oregon race photos courtesy of Ted Willcox/CLO.*

Slate-colored

Pink-sided

Oregon

tors continue operating, the subspecies may eventually develop still stronger distinctions until their populations, unable to interbreed, become species **(Fig. 1–44)**. Evolution of this sort (termed **divergent evolution** because one group "diverges" into two or more) results in the generation of new species—a process known as **speciation.** The geographical separation that leads to speciation is frequently caused

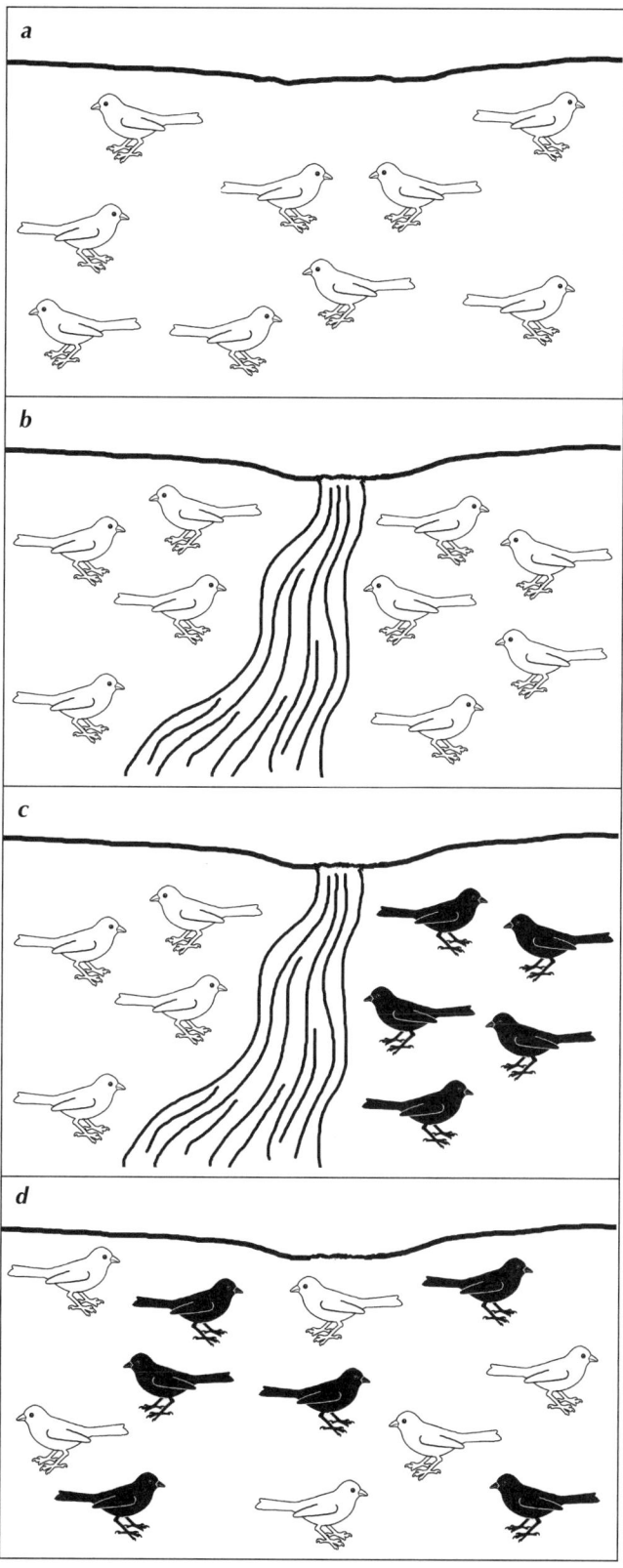

Figure 1–44. Speciation through Geographic Isolation: *When populations of a species become geographically isolated from one another,* **speciation** *(the formation of new species) may occur. As an example, consider the population of "white" birds shown in* **(a)**. *It becomes divided into two populations that do not interbreed because they are geographically isolated by the formation of a large river* **(b)**. *Over time, each population is exposed to different ecological factors, which cause natural selection to favor different traits in each—represented by the black versus white plumage in* **(c)**, *and the two populations "diverge" in a process known as* **divergent evolution.** *Such ecological factors might include slightly different food supplies, predators, habitats, nesting materials, and competitors. If the two populations become so different that they can no longer interbreed, they are considered separate species. In* **(d)**, *the two populations have reestablished contact with one another (due to range expansion or removal of the geographic barrier), yet they have retained their separate identities—demonstrating that they are probably not interbreeding and are, indeed, separate species. Adapted from Mader (1988, p. 522).*

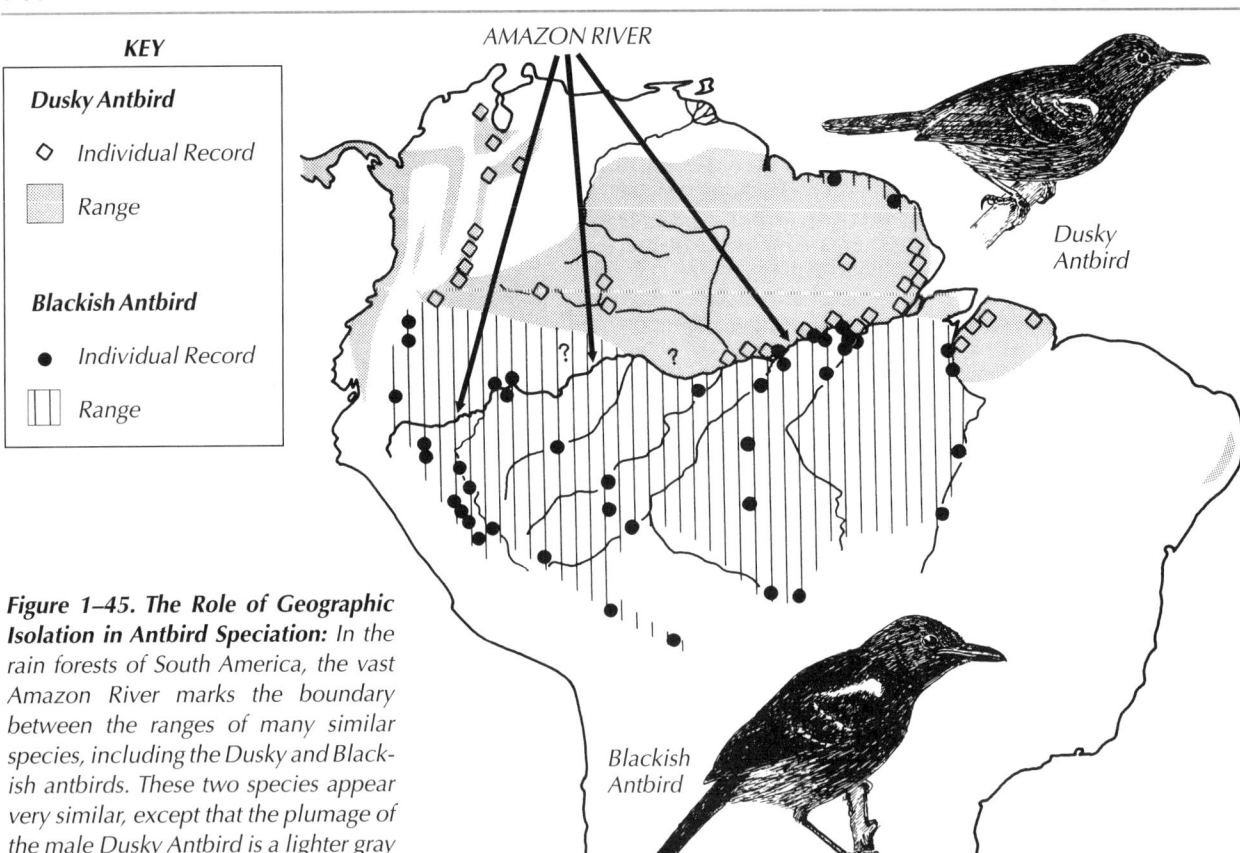

KEY

Dusky Antbird

◇ Individual Record

▨ Range

Blackish Antbird

● Individual Record

▥ Range

AMAZON RIVER

Dusky Antbird

Blackish Antbird

Figure 1–45. The Role of Geographic Isolation in Antbird Speciation: *In the rain forests of South America, the vast Amazon River marks the boundary between the ranges of many similar species, including the Dusky and Blackish antbirds. These two species appear very similar, except that the plumage of the male Dusky Antbird is a lighter gray than that of the male Blackish Antbird, as illustrated here. What role the Amazon River played in the divergence of these species is not clearly understood. Perhaps the Amazon, instead of serving as an absolute barrier to the intermingling of the two groups, maintained slightly different ecological conditions or sets of competing species on its two banks, which, together with the partial water barrier, discouraged a mixing of the individuals from the two sides. Where the two species occur together, on the northern bank of the Amazon and in the coastal lowlands of the Guianas, they occupy slightly different habitats.* From Neotropical Ornithology, *edited by P. A. Buckley, Mercedes S. Foster, Eugene S. Morton, Robert S. Ridgely, and Francine G. Buckley. Ornithological Monographs No. 36, published by the American Ornithologists' Union, 1985. Reprinted by permission of the American Ornithologists' Union.*

by physical barriers. The formation of new mountain ranges, the formation of new large bodies of water, or changes in climate and the consequent changes in the distribution of specific habitats can be responsible for separation of bird populations and the formation of new species **(Fig. 1–45)**.

Because populations exist at every conceivable intermediate stage between recently isolated populations and completely separate species, decisions on how to draw the artificial "species/subspecies" line are usually controversial. Often, there simply is not enough information to determine exactly where on this continuum the forms lie. In some cases, future work may shed light on how different the forms are and how much gene flow exists. In other cases, the distinctions are more a matter of perspective: different people would categorize the populations differently. In some cases "superspecies"—groups of closely related species—can be described, such as the Great Blue Heron of North America, the Gray Heron of the Old World, and the White-necked Heron of South America. Differentiating between superspecies and groups of subspecies always will be contentious.

New species also frequently evolve when a small population of birds colonizes a distant island, and further gene flow with the parent population is minimal or nonexistent. Over time the founding population evolves to be better adapted to the local environment: some traits disappear, while others become more widespread. Mutations occur, and if advantageous, spread across the island. When a whole *group* of isolated islands is colonized, natural selection may proceed differently on the different islands, eventually producing an array of new species that all evolved from the same founding species **(Fig. 1–46)**. The formation, from a common ancestor, of a variety of different species adapted to different niches and behaviors, usually showing different morphologies (sometimes drastically different), is known as **adaptive radiation** (or simply "radiation").

Although adaptive radiation is not restricted to islands, very isolated island chains present some of the most striking examples. The Galapagos, a cluster of islands 600 miles (965 km) off the west coast

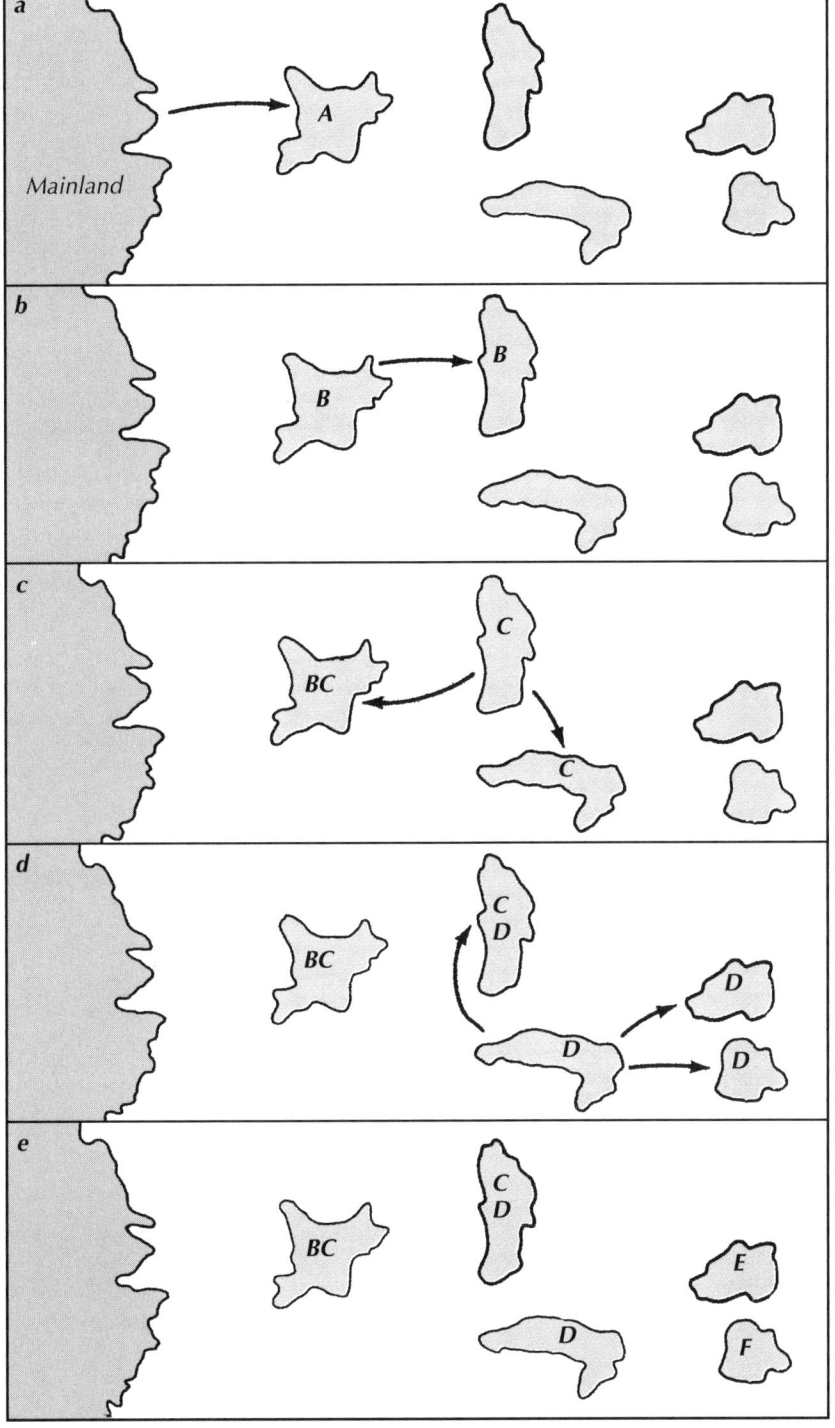

Figure 1–46. Adaptive Radiation on Islands: New species frequently evolve when a small population of birds colonizes a distant island and gene flow with the parent population becomes minimal or nonexistent. When a group of islands is colonized, natural selection may proceed differently on the different islands, eventually producing an array of new species in a process termed **adaptive radiation. a.** In this hypothetical example, species A from the mainland colonizes the closest island. **b.** Over time, different selection pressures (factors that favor one trait over another) on the island cause the founding population to evolve enough differences to become a new species, indicated as B. Species B spreads to the next closest island. **c.** Over more time, the population of B on the second island adapts to its new environment, evolving into new species C, which spreads to the next island as well as to the first, where it is unable to interbreed with species B and remains as a separate species. **d.** Eventually, species C on the third island evolves to be a new species, D, which spreads to the last two islands and back to the second island, where it remains separate from C. **e.** Finally, the populations of species D on the last two islands each adapt to the conditions on their own island, forming two new species, E and F. Adapted from Campbell (1990, p. 468).

Figure 1–47. Adaptive Radiation in Darwin's Finches: *The Galapagos Islands are home to a group of finches known as "Galapagos Finches" or "Darwin's Finches." Now placed within the family Emberizidae, the 14 species are a classic example of adaptive radiation, as they are all thought to have evolved from a common ancestor that reached the islands from South America. Although they are similar in appearance—gray, brown, black, or greenish, sparrow-sized birds with short wings and tails—the shape and relative size of their beaks differ from species to species. Beak shape is not an infallible identification aid, though, as it varies widely among individuals within some species, especially among populations of the same species on different islands. Based on feeding habits, the finches fall roughly into three groups, the ground-finches, the warbler-like finches, and the tree-finches, but there is considerable diversity within the groups. See text for details. From Pough, F. H., J. B. Heiser, and W. N. McFarland,* Vertebrate Life, *4th Edition. Copyright 1996 Prentice-Hall. Reproduced by permission of Prentice-Hall, Inc.*

of Ecuador, are a veritable laboratory of recent speciation. Never connected to one another or to South America, the small volcanic islands were at first without life. Gradually, they acquired habitats that could support the animals that reached them accidentally from overseas. The separate forms of mockingbirds and tortoises found on each island provided an important stimulus to Charles Darwin as he developed his theory of evolution.

The Galapagos also are host to an array of finches. These "Galapagos finches" (order Passeriformes, family Emberizidae) are a classic example of divergent forms evolved from a common ancestor. These birds are commonly known as Darwin's Finches, although Darwin did not notice their similarities until the eminent ornithologist John Gould pointed them out while viewing Darwin's collected specimens. The 14 species found today probably radiated from a single ancestral finch species that arrived on the Galapagos relatively recently from South America. Limited dispersal, isolation, and adaptation to different types of foods found on the islands eventually produced an array of species with notably different beak types—although most are still clearly variations on the typical seed-eating finch beak **(Fig. 1–47)**. Based on feeding habits, the finches fall roughly into three groups:

Darwin's Finches

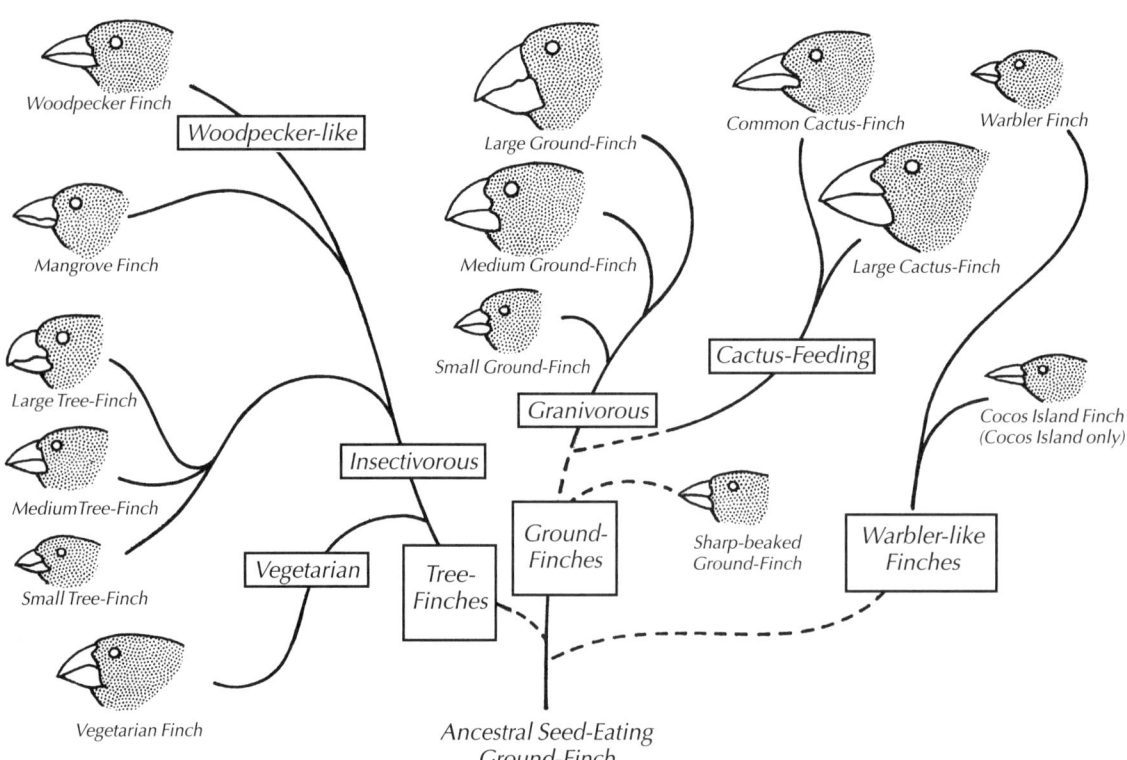

Ground-Finches: The Small, Medium, and Large ground-finches use their strong, conical beaks to crack and eat seeds—their names reflecting the relative sizes of their beaks. Smaller species are restricted to eating smaller and softer seeds. The Sharp-beaked Ground-Finch has a longer, more pointed beak, which it uses to probe flowers for nectar and leaf-litter for insects and seeds. On some islands, birds of this species have even longer beaks, which they use to draw and drink the blood of large seabirds (see Fig. 6–30m) and to break open seabird eggs and drink the contents. The Small, Medium, and Sharp-beaked ground-finches also remove ticks from iguanas and tortoises. The Common Cactus-Finch has a longer, thinner beak that it uses in a variety of ways, most notably to extract nectar, pollen, and pulp from prickly pear cacti. The Large Cactus-Finch has a large, heavy beak that allows it to feed on larger seeds and insects than most other ground-finches.

Warbler-like Finches: These two species have distinctively thin, pointed, warbler-like beaks used for picking insects and spiders from flowers, leaves, and twigs, as well as for probing flowers for nectar. The Cocos Island Finch is found only on Cocos Island, 390 miles (630 km) northeast of the Galapagos, where being the only Darwin's Finch, its foraging methods and diet have diversified tremendously. But, interestingly, instead of individual birds each using a variety of foraging techniques, different individuals within the population have adopted different specific foraging methods.

Tree-Finches: These birds spend significantly more time foraging in trees than do the ground-finches. The Vegetarian Finch has a short, broad beak for crushing and eating fruits, leaves, and buds. The Small, Medium, and Large tree-finches have conical, somewhat parrot-like beaks, which are able to apply force at the tip, and are used primarily to probe into decaying wood to extract insects. The Woodpecker Finch has a long, stout, tanager-like beak for prying large insects from bark and soft wood. The Mangrove Finch, found in dense mangrove swamps, has a similar, but smaller, beak, used to capture insects and spiders. Woodpecker and Mangrove finches are famous for their ability to use "tools"—using cactus spines or twigs to pry insect larvae and pupae from holes in dead branches (see Fig. 6–10a).

Even more spectacular, however, are the Hawaiian honeycreepers. The Hawaiian Islands, the most isolated group of islands in the world, once was home to an extensive bird fauna, now mostly extinct. From a goldfinch-like ancestor arose several dozen species of small birds that show a remarkable array of bill sizes and shapes, all of which evolved to exploit different niches on different islands **(Fig. 1–48)**.

Orders and Families of World Birds

The living species of birds in the world total about 9,600. All are named in accordance with the binomial system established by

Linnaeus and now controlled by the International Commission of Zoological Nomenclature. Through its code, the Commission stipulates how the names shall be written, makes decisions on accepting or rejecting names proposed, and sometimes authorizes changing or otherwise altering names already in use. **Appendix A** lists the 31 orders and all families of living birds. **Figure 1–49** depicts the 31 orders in their primary habitats.

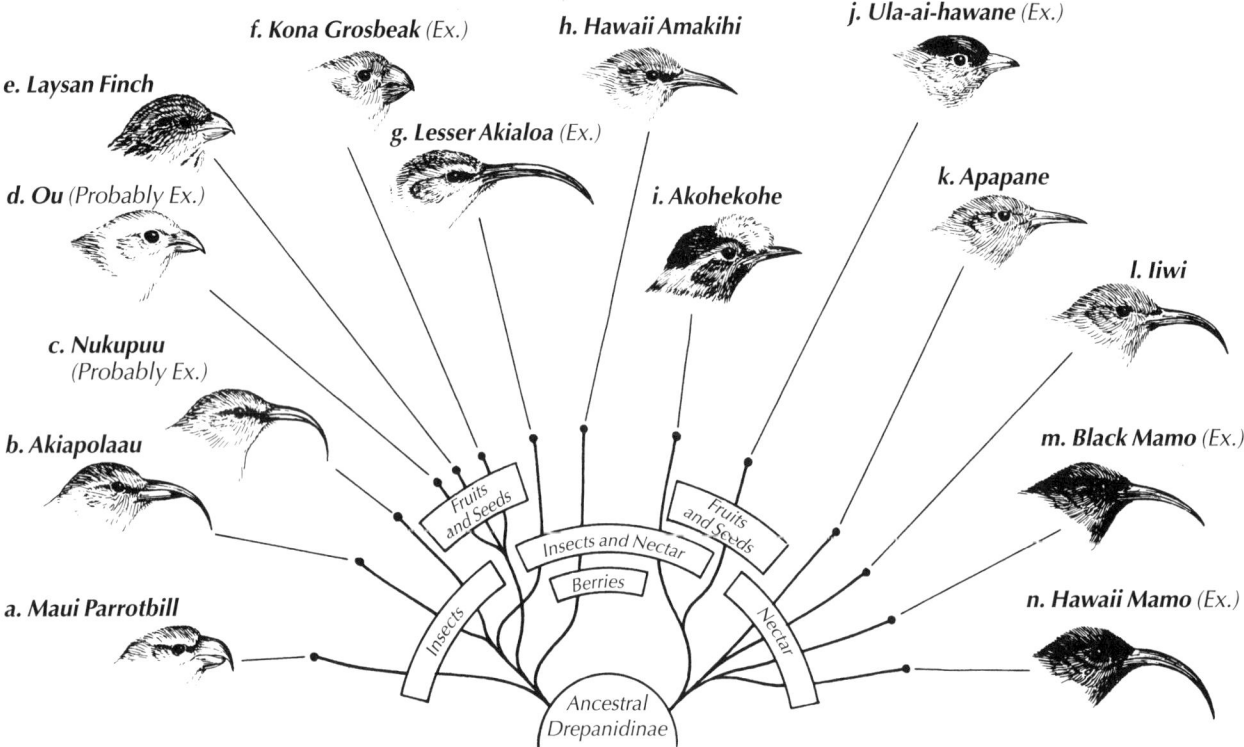

Figure 1–48. Adaptive Radiation of Hawaiian Honeycreepers: *Illustrated here are 14 of the 32 known species of Hawaiian Honeycreepers, a subfamily (Drepanidinae, within the family Fringillidae) of small, often colorful birds with a bewildering variety of beak shapes. Extinct species are noted with (Ex.) after the name. The ancestral honeycreepers were probably a flock of cardueline finches from North America that strayed out over the Pacific Ocean and landed on one of the Hawaiian Islands. They subsequently spread to the other islands, radiating into numerous species with a dramatic diversity of beak shapes and feeding habits.*

*Three members of the same genus (**b, c,** and **g**) illustrate the correlation between beak shape and specific function. The Lesser Akialoa (**g**) had a long, slender, decurved beak for picking insects from bark crevices as the bird hopped along tree limbs. The Nukupuu (**c**) has a long, slender, strongly decurved upper beak and a shorter, thicker lower beak; the lower beak is used alone to chip and pry away loose bark as the bird searches for insects on tree trunks. In the bizarre beak of the Akiapolaau (**b**), the lower beak is straight and stout, whereas the upper beak is slender, sickle-shaped, and nearly twice as long. Holding its beak open to keep the upper beak out of the*

way, the Akiapolaau chisels, woodpecker-like, into soft wood with the lower beak, then uses the long upper beak as a probe to reach insects.

*The beaks of some Hawaiian Honeycreepers are short and stout. The thick, powerful beak of the Kona Grosbeak (**f**) was used for cracking the small, extremely hard fruits of naio trees; and the thick, parrot-like beak of the Maui Parrotbill (**a**) is used to search for insect larvae and pupae by ripping into decaying wood, small twigs, ripe fruits, and plant stems. Other beaks are slender and decurved, adapted for gathering nectar from flowers, as in the Iiwi (**l**), the Black Mamo (**m**), and the Hawaii Mamo (**n**). Still others, not shown, are slender and warblerlike for gleaning insects and probing flowers for nectar.*

*The most abundant and widespread Hawaiian Honeycreeper, the Apapane (**k**), may resemble the ancestral form. It uses its long, slightly decurved beak to glean insects and probe ohia flowers for nectar. It is well-known for its long-distance flights in search of flowering ohia trees.*

Adapted from Ecology and Field Biology, *Second Edition, by Robert Leo Smith (1974, p. 476). Copyright 1966 and 1974, by Robert Leo Smith. Reprinted by permission of Addison-Wesley Education Publishers, Inc.*

Figure 1–49. Living Orders of World Birds: *The world's 31 orders of living birds are shown here in their primary habitats—water, shore, open ground, trees, or air. This schematic only approximates the real world, however, because some groups actually contain members in several different habitats, and because subtle variations among different types of each major habitat are not shown. The evolutionary relationships among the different orders are not well established and remain highly controversial among ornithologists. Note that the bird silhouettes are not drawn to scale.*

Orders and Families of North American Birds

Appendix B lists the living orders and families of birds that occur in North America north of Mexico, including Hawaii.

How Naming and Classification Can Help You

Now that you have read the information on the naming and classification of birds, peruse the latest edition of your field guide. In most guides the names and sequence of taxa are roughly in accordance with the lists in Appendices A and B. In your guide, be sure to note the following:

(1) **The sequence of orders and families.** Try to learn as much of the sequencing as you can. Besides helping you to crack the book quickly at the right place for a particular group of birds, the sequence will give you recent scientific thinking on evolutionary relationships among the groups.

(2) **The sequence of species within a family.** Although purely conjectural, the sequence is an attempt to put related species close together.

(3) **The scientific names of species in a large family or subfamily, such as the wood-warblers, Parulidae.** The words, when translated, may be helpful by describing the bird's appearance, behavior, range, or habitat. They also may be useful to you by expressing relationships. For example, the Black-and-white Warbler is so different from all the other warblers that it warrants its own genus, *Mniotilta*; the Tennessee, Orange-crowned, Nashville, Virginia's, Colima, and Lucy's warblers are so much more like one another than like the rest of the warblers that they share one genus, *Vermivora* (**Fig. 1–50**). Knowing this, you can better understand why these birds are so similar in color patterns, in the way they nest, and in many of their songs and calls.

The Use of Common Names

Unlike so many groups of organisms—plants, worms, insects, and so on—all bird species have common names in English. Unfortunately, not everyone agrees on just which common names should be "official," so reference to the scientific names is still important. In North America, the common names are so well established and familiar, thanks to consistent following of the *AOU Check-list*, that both professionals and amateurs speak of species by their common names, confident that everyone will know which species they are referring to.

In this course, therefore, birds are referred to by their common names. The common names given in the *42nd Supplement to the AOU Check-list* (AOU 2000) are used for all species covered by that list. All species not covered by the *AOU Check-list* are referred to by the English names provided by James Clements in his *Birds of the World: A Checklist* (2000). A list of the scientific names of all species mentioned

Black-and-white Warbler

Tennessee Warbler

Orange-crowned Warbler

Figure 1–50. Vermivora *and* Mniotilta Wood-Warblers: *By noting the scientific names of bird species, and which ones are placed in the same genera, you may gain useful information on the relatedness among similar species. For example, among the numerous species of New World wood-warblers (family Parulidae), the Black-and-white Warbler is distinct enough to warrant its own genus, Mniotilta, whereas the Tennessee, Orange-crowned, Nashville, Virginia's, Colima, and Lucy's warblers are more similar to one another than to other warblers, and thus are placed in the same genus, Vermivora. The Vermivora warblers have slender, finely pointed bills for picking insects from leaves and branches; muted colors—usually olive, green, brown, gray, white, and/or yellow; and songs that include trilled notes. The Black-and-white Warbler is most distinctive in its foraging behavior: it creeps like a nuthatch along tree trunks and large branches, probing into bark for insects and spiders with its slightly decurved bill. Its placement in a separate genus is based primarily on anatomical adaptations for this foraging style: it has a much longer hallux, shorter tarsus, larger feet, and stronger legs than other warblers. Its genus name, Mniotilta (literally, "moss-plucking"), also reflects this habit and its species designation, varia ("variegated"), refers to its bold, black-and-white striped plumage. The Black-and-white Warbler also commonly forages among foliage, gleaning insects like other wood-warblers. Photos of Tennessee and Black-and-white warblers courtesy of Bill Dyer/CLO. Photo of Orange-crowned warbler courtesy of Donald Waite/CLO.*

in the *Handbook of Bird Biology*, alphabetical by common name, is located after the references. For any species that occurs both in North America and in other countries where it has a different common name, the North American name is used. For example, Common Loon is used when speaking of the same species in Europe, where the English name is Great Northern Diver.

So that you will recognize when the common name of a particular species is mentioned, it is capitalized—for example, "Yellow Warbler." Therefore, a Yellow Warbler is a specific species, *Dendroica petechia*, and a yellow warbler is any warbler that is yellow.

Throughout the *Handbook of Bird Biology*, the following terms are used to cover particular groups of birds:

Ratites: All birds lacking a keel on the sternum. Includes the flightless Ostrich, rheas, Emu, cassowaries, and kiwis, as well as the tinamous, which are fully capable of flight.

Waterfowl: Ducks, geese, and swans. Thus, refers only to the family Anatidae.

Water Birds or Aquatic Birds: All species with webbed feet that commonly swim, including the Anatidae; also, all deep-water waders belonging to the order Ciconiiformes, such as herons and storks.

Seabirds or Marine Birds: All species directly associated with the open seas and consistently dependent on the seas for food.

Shorebirds: Oystercatchers, plovers, snipes, sandpipers, curlews, phalaropes, and sheathbills. Ornithologists in Britain and the British Commonwealth, except Canada, speak of shorebirds as "waders."

Land Birds or Terrestrial Birds: All species not included in the aquatic groups above.

Gallinaceous Birds: Grouse, quails, turkeys, pheasants, and all other Galliformes; includes domestic chickens.

Raptors: All diurnal and nocturnal birds of prey, namely Falconiformes and Strigiformes.

Perching Birds, Passerine Birds, or Passerines: All species of Passeriformes.

Songbirds: All passerines in the suborder Passeri. These birds have particularly complex voice boxes.

Occasionally, for lack of a suitably inclusive name for a group of birds, the ordinal or familial name will be shortened; for example, "charadriiform birds" for the order of shorebirds, gulls, and auks; "alcids" for the family of auks, dovekies, guillemots, murres, and puffins. If the names are unfamiliar to you, refer to the list of orders and families in Appendices A and B.

Evolution of Birds and Avian Flight

■ Did birds evolve 230 million years ago from small, arboreal reptiles called thecodonts, or 150 million years ago from terrestrial theropod dinosaurs? Did feathers, which probably first evolved as insulation, then enlarge to promote gliding from trees, leaping on the ground, or insect-catching? And what were the few birds like that survived the massive extinctions at the end of the Cretaceous period—to become the wellspring of all modern birds? The answers to all of these questions remain elusive, but the debates rage on.

These complex and controversial topics are beyond the scope of this course, but we have included an in-depth discussion of them in an optional section located at the beginning of Part 2. This is an exciting area of research because so many fossil birds continue to be discovered—as you may hear through the news media. These new findings compel researchers to alter their theories, so our view of the relationship between birds and reptiles is continually evolving.

Bird Distribution

■ Although birds are found nearly everywhere on earth, their distribution is quite uneven. Some habitats, such as tropical rain forests, have numerous species, whereas others, such as deserts and alpine zones, have few. Even among similar habitats the distribution is uneven—the rain forests of the Amazon Basin in South America host many more species per square mile than those of the Congo Basin in Africa, for example. Furthermore, species' ranges vary tremendously in size: some species, such as the Northern Harrier and Tundra Swan, are found throughout the northern portions of North America, Europe, and Asia, but most species have somewhat limited distributions. A few, such as the Kirtland's Warbler and Guadalcanal Honeyeater, are restricted to very small areas. Similarly, some bird families, such as the swallows (Hirundinidae) and larks (Alaudidae), have representatives on most continents, whereas others are found on just one or two. The earth has not always looked as it does today, and appreciating why birds are distributed in these ways requires an understanding of historical changes in climate and the movements of the continents, as well as past speciation, dispersal, and extinction patterns in different lineages of birds. For a chart of the major geological time periods, and a summary of the changes in the global climate, the arrangement of the continents, and the diversity of living things, see **Appendix C: Geological Time Scale**.

Two hundred forty-five million years ago, a single land mass known as Pangea formed. It then broke apart into two large masses: Laurasia, in the north, and Gondwanaland, in the south. These eventually fragmented into separate continents **(Fig. 1–51)**. Attached to great plates that float on the molten rock of the earth's mantle, the continents slowly drifted into different arrangements—a process that continues to this day. Over time, continental collisions and volcanic activity have created mountains, and erosion has lowered them; huge basins have filled with water, and others have drained; and the output of energy from the sun has varied—causing dramatic changes in the global climate.

The early diversification of birds took place on a very different earth; neither the arrangement of the continents nor the distribution of climates resembled those of today. During the time that modern orders of birds were evolving, in the early Tertiary about 50 to 60 million years ago, Gondwanaland already was breaking apart: South America had separated from Africa and India had split from Antarctica and was moving north to collide with Asia. North America, Europe, and Asia remained joined as Laurasia, however.

During much of the Tertiary period the world's climates were warm from pole to pole, and birds moved through the tropical-subtropical or warm temperate forests that covered Eurasia and North America. At first, movement was easy across a broad North Atlantic land bridge. After the separation of North America from Europe during the Eocene, however, the Bering land bridge between Siberia and Alaska became the main corridor for faunal exchange between Eurasia and North America.

Similarly, the warm climate of Gondwanaland allowed much exchange of fauna among Africa, South America, and Australia. Dispersal between Australia and South America was possible via the temperate climates and forests of Antarctica. The early flightless ratites may have taken advantage of this dispersal route, resulting in Ostriches in Africa, Emus in Australia, and rheas in South America. Penguins, too, may have moved among all three continents via Antarctica.

At the close of the Tertiary, the earth's climates cooled, especially in the polar regions, and became more strongly seasonal. Tropical birds became restricted to equatorial latitudes. During the Pleistocene, cli-

*Figure 1–51. Continental Drift: Attached to plates that float on the molten rock of the earth's mantle, the continents slowly have drifted into different arrangements over time—a process that continues today. **a. Triassic:** Around 245 million years ago, at the end of the Paleozoic era, all the continents came together to form one large supercontinent called **Pangea**, which persisted throughout the Triassic period. **b. Jurassic:** Around 200 million years ago, during the Jurassic period, Pangea split into two large masses, a northern **Laurasia**, consisting of present-day North America, Europe, and Asia; and a southern **Gondwanaland**, consisting of South America, Africa, Madagascar, India, Australia, New Zealand, and Antarctica. During this time, the earliest birds evolved. **c. Cretaceous:** During the late Cretaceous period, Gondwanaland fragmented—South America split from Africa and India split from Antarctica and drifted north toward Asia. Laurasia remained more or less intact. This is the approximate configuration of the continents during the time period when most of the modern orders of birds evolved, in the very early Tertiary. **d. Eocene Epoch of Tertiary:** At this time, North America separated from Eurasia, but a land bridge between Alaska and Asia remained, allowing the exchange of fauna. The components of the former Gondwanaland continued to drift apart, heading toward their current positions. Just 10 million years ago, India collided with Asia, forming the towering Himalayas. Adapted from Gill (1990, p. 470).*

a. Triassic

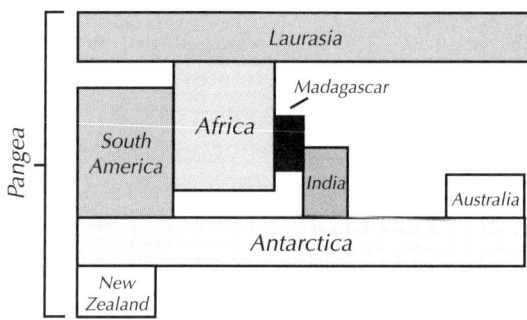

b. Jurassic

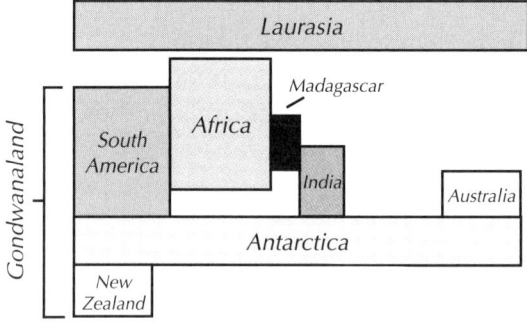

c. Cretaceous

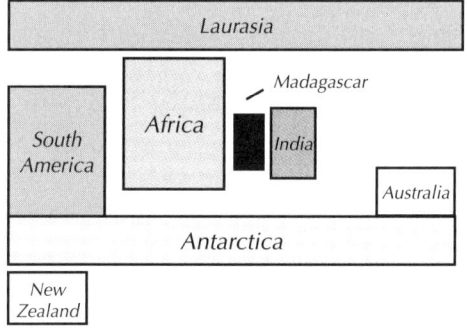

d. Eocene Epoch of Tertiary

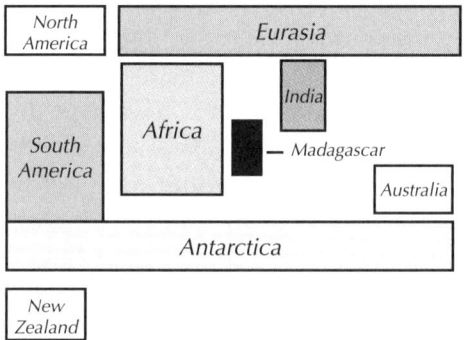

matic changes and glaciers drastically altered the distributions of birds throughout the world. Dry, cool climates alternating with wet, warm ones split the geographic ranges of birds and promoted both speciation and extinction. Remnant or relict populations are one major consequence of historical habitat changes. For example, Ostriches, now restricted to Africa, once roamed throughout Asia, and todies—tiny, colorful kingfisher relatives found only on the Greater Antilles of the West Indies—once lived in Wyoming and France.

Distribution of Land Birds

Birds can be a window on the world in many ways. Knowing that some North American birds migrate south for the winter, you can use them to tell the seasons. For instance, if you saw a picture or a film from Ithaca, New York, and it included a Rose-breasted Grosbeak, you would know that the picture must have been taken between May and October. Similarly, you can observe geography using birds as a lens. If you were blindfolded and whisked away to some distant part of the globe, and if you knew your birds of the world well enough, you could tell where you were, within perhaps a thousand miles or so, simply by observing birds.

Scientists looking at the fauna of the world, including birds, have divided the globe into six general regions, with boundaries where the distributions of many different types of animals all seem to change. These regions roughly mirror the continents, but many boundaries are climatic rather than physical **(Fig. 1–52)**. The following sections describe the **avifauna** (set of bird species) of each region.

Palearctic Region

The Palearctic region encompasses most of the large landmass of Eurasia, as well as northern Africa and most of the Sahara desert. It stretches from the Atlantic to the Pacific and from the Arctic to the Himalayas, and is by far the largest region. The major physical features and habitats run roughly in east-west belts and are, from north to south: the arctic tundra, the boreal forests (also known as coniferous forests, or taiga), a chain of deserts stretching from the Sahara in Africa east to the Gobi in Mongolia, and a nearly continuous chain of mountains (from the Pyrenees and Alps east to the Himalayas). The climate ranges from the high arctic to the subtropical, stopping short of the true tropics.

Because so much of this region is either **temperate** (free from extreme heat and cold, but experiencing some of both) or arctic, a large proportion of the avifauna migrates south to winter in the tropics or beyond. In winter, insect food is scarce, but a small number of birds, such as the titmice (chickadee relatives), woodpeckers, and nuthatches, have adapted to remain and continue their insect diet throughout the year. When winter is over and the northern lands warm again, insect populations increase rapidly, and the food available for insectivorous birds becomes tremendous. Migrants return in the spring and take advantage of the abundant insects to raise their young.

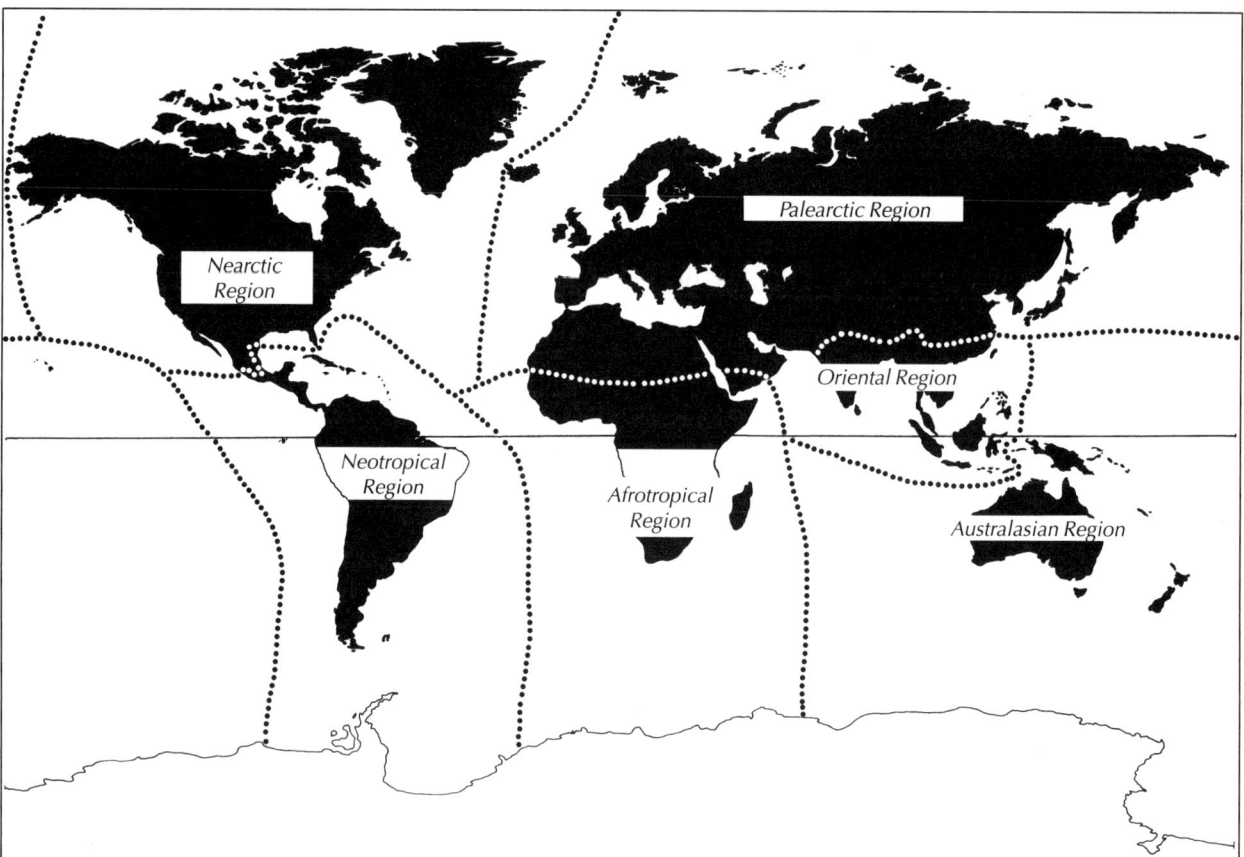

Figure 1–52. The Six Major Zoogeographic Regions: *Scientists studying the world's fauna have divided the land areas of the globe into six major regions. At the boundaries of the regions, distributions of many different types of animals, including birds, change. Many of the boundaries are climatic rather than physical, but the regions generally mirror the continents. In some cases, different researchers divide the globe differently. For example, Greenland may be included with the Palearctic region instead of with the Nearctic; Madagascar and New Zealand may each be placed in separate regions; and the islands of Oceania in the mid and southern Pacific Ocean, including Hawaii, may constitute a separate region. In some cases, the Afrotropical region is referred to as the "Ethiopian" region, and the Australasian region as the "Australian." Drawing by Charles L. Ripper.*

The land bridge across the Bering Strait once allowed the exchange of species between Eurasia and North America, and the similar avifaunas of the two continents reflect this connection—although the Palearctic also shares many species and higher taxa with the other regions it borders. Many Palearctic birds would be familiar to a visiting North American, as approximately 13 percent of Palearctic species and 35 percent of the genera also are found in the Nearctic. In fact, so many groups of birds (for example, loons and alcids) occur in these two regions and nowhere else on earth that some authorities prefer to group the Palearctic and the Nearctic regions into one, the **Holarctic**. The Holarctic has the fewest bird species per land area of any of the regions, presumably owing to the harsh climatic conditions found in many areas. Birds characteristic of, but not limited to, the Holarctic include hawks, owls, grouse, woodpeckers, swallows, thrushes, kinglets, tit-mice, creepers, crows, jays, and many shorebirds and water birds that breed in the Arctic.

Perhaps surprisingly, despite its size, the Palearctic region has but one family that is **endemic** (found only in that region), the **Prunellidae** (accentors and Dunnock)—sparrow-like birds with slender, pointed bills. Twelve of the thirteen species of this family inhabit high mountains—ranging from northwest Africa and western Europe as far east as Japan. The other species, the Dunnock **(Fig. 1–53)**, breeds at somewhat lower elevations in Europe, and is a familiar garden bird in Great Britain.

Bird groups with a great array of species in the Palearctic include the buntings (genus *Emberiza*), Old World warblers (family Sylviidae), and cardueline finches (family Fringillidae, subfamily Carduelinae).

Other typical Palearctic birds include larks, Old World fly-catchers, pipits, and wagtails **(Fig 1–54)**.

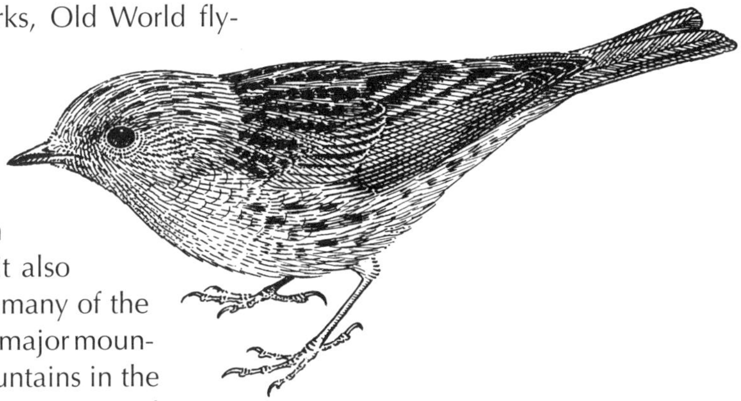

Nearctic Region

The Nearctic region includes arctic, temperate, and subtropical North America, reaching south to the northern border of tropical rain forest in Mexico; it also includes Greenland. Although it contains many of the same physical features as the Palearctic, the major mountain ranges of the Nearctic (the Rocky Mountains in the west and the smaller Appalachians to the east) run north-south, adding a layer of complexity to latitudinal climate belts. As in the Palearctic, east-west belts of arctic tundra and boreal forest exist, with patches of deciduous forest extending southward wherever there is sufficient rainfall to support it (as in the southeastern United States). The center of the continent, with lower rainfall, consists of prairie-covered plains; areas further west, with even less rainfall, are semi-arid and desert. The southern tip of Florida and the extreme southwestern United States and northern Mexico are subtropical.

As in the Palearctic, a large proportion of the bird species are migratory, taking advantage of the abundance of spring insects to raise their young, but leaving northern Nearctic areas after breeding—many to winter in the Neotropics. These birds, termed **Neotropical migrants**, include wood-warblers, tanagers, and orioles. The springtime return of multitudes of these colorful birds is one of the most spectacular birding events of the region.

Despite the diversity of Nearctic habitats, no avian families are endemic to the region. Instead, the Nearctic is more distinctive for its blend of Palearctic and New World groups, which reflects the continent's varied history: periods of isolation as well as periods when land bridges connected with South America or Eurasia (either via Greenland or the Bering Strait).

Figure 1–53. Dunnock: Formerly called the Hedge Sparrow, the Dunnock is the best-known representative of the only bird family endemic to the Palearctic region, the Prunellidae. Although the Dunnock is common in gardens, woodlands, scrublands, and cultivated areas throughout western Europe, in central and eastern Europe, it breeds mainly in mountainous areas. It is brownish above and grayish below, and generally sparrow-like in appearance, but has a more slender, pointed bill than most sparrows. Drawing by Robert Gillmor.

Figure 1–54. White Wagtail: Inhabiting farmlands, grasslands, and other open areas throughout Eurasia, often near water, wagtails (family Motacillidae) are slender, titmouse-sized birds with long tails. Their habit of frequently wagging their tails gives them their name. Aided by their strong legs and long toes, wagtails typically run across the ground in pursuit of insect prey, and occasionally snatch insects from the air. The White Wagtail shown here (formerly called the Pied Wagtail) is common and well-known because it often lives near humans, running along roads and rooftops after insects. It breeds in northwestern Alaska as well as Eurasia. Photo by T. J. Ulrich/VIREO.

Figure 1–55. Great Tit: *A member of the same family, Paridae, as the familiar North American chickadees and titmice, the Great Tit, with its yellow breast, is a bit more colorful than its New World relatives. Found throughout much of the Palearctic and Oriental regions, the Great Tit readily breeds in nest boxes, and thus has been the subject of many detailed, long-term studies. It is very similar in behavior to the chickadees of North America. Drawing by Robert Gillmor.*

Figure 1–56. Winter Wren: *The small, spunky wrens are almost entirely a New World family, but the tiny, stubby-tailed Winter Wren of North America is also found throughout much of the Palearctic region—where it is known as the "European Wren" or simply as the "Wren." In North America, it breeds mainly in moist, coniferous woods, often near streams, but it frequents a wide variety of Palearctic habitats. Drawing by Charles L. Ripper.*

Bird groups shared with the Palearctic, in addition to those already listed under the Palearctic region, include larks, pipits, nuthatches, cranes, pigeons and doves, shrikes, and kingfishers. Although the particular species represented may differ in the two regions, they often are similar in ecology and basic appearance. For example, a Great Tit **(Fig. 1–55)** or a Blue Tit might come to a bird feeder in Europe, instead of a Black-capped Chickadee or a Tufted Titmouse, but any North American birder would notice the similarities in behavior and appearance immediately.

The New World families—those occurring in both the Nearctic and Neotropical regions but nowhere else—include the New World vultures, New World quail, Limpkin, hummingbirds, tyrant flycatchers, mockingbirds and thrashers, vireos, blackbirds and orioles, and wood-warblers. Only one of the 78 species of wrens (Troglodytidae) occurs outside the New World—the Winter Wren, known simply as the "Wren" in Great Britain **(Fig. 1–56)**. The distinctive turkeys (family Phasianidae, subfamily Meleagridinae) also are restricted to the New World **(Fig. 1–57)**.

Neotropical Region

The Neotropical region includes South America, Central America north to the northern edge of tropical forests in Mexico, and the West Indies. It has by far the greatest diversity and the greatest number of bird species per area of any of the faunal regions, hosting approximately one-third of all bird species found on earth. About one-third of the families found here are endemic: Tinamidae (tinamous), Rheidae (rheas), Anhimidae (screamers), Cracidae (curassows, guans, and chachalacas), Eurypygidae (Sunbittern), Psophiidae (trumpeters), Cariamidae (seriemas), Pluvianellidae (Magellanic Plover), Thinocoridae (seedsnipes), Opisthocomidae (Hoatzin), Steatornithidae (Oilbird), Nyctibiidae (potoos), Todidae (todies), Momotidae (motmots), Bucconidae (puffbirds), Galbulidae (jacamars), Ramphastidae (toucans and New

World barbets), Furnariidae (ovenbirds), Dendrocolaptidae (wood-creepers), Thamnophilidae (antbirds), Formicariidae (antthrushes and antpittas), Conopophagidae (gnateaters), Rhinocryptidae (tapaculos), Cotingidae (cotingas), Phytotomidae (plantcutters), Pipridae (manakins), Oxyruncidae (Sharpbill), Dulidae (Palmchat), and Coerebidae (Bananaquit).

The tremendous avian diversity of the region results partly from the types of environments found there—particularly the extensive tropical rain forest of the Amazon Basin, which covers about one-third of South America. Rain forests, with their complex structure, warm climate, and abundant resources, contain more species per land area than any other habitat on earth. There are also vast grasslands—some north of the Amazon River, but even more extensive ones to the south, the pampas, stretching from Mato Grosso in Brazil south to Patagonia. These grasslands host numerous species of birds and other wildlife. Within the southern grasslands, extending across parts of Brazil, Paraguay, and Bolivia, is an expansive region, the Pantanal, which is transformed each year into a huge marsh by torrential rains. The Pantanal has one of the greatest concentrations of wildlife on the continent, including numerous waders and other water birds, as well as the few remaining Hyacinth Macaws. South America narrows so much toward the south that not much land area is found within the temperate zone—a climate that typically contains fewer species. To the west, the towering Andes, adding an array of habitats at different altitudes, run the length of the continent. Along the Pacific coast south of the equator, the cold and fish-laden Peru (Humboldt) current flows northward from the Antarctic, supporting a rich diversity of oceanic birds (see Fig. 1–102). This cold current sets up a temperature inversion by cooling a layer of air directly above the ocean, which is overtopped by the warm, tropical air typical of most of the area. The layering creates the Atacama Desert along the Pacific coasts of Ecuador, Peru, and northern Chile, because fog and clouds often form, but rarely rain. Although the Atacama is one of the driest deserts on earth and thus has few bird species, it covers a relatively small portion of the Neotropical region.

The historical isolation of the Neotropics also contributes to its extraordinary avian diversity. Over much of the last 60 million years, the Central American land bridge between North and South America was transformed into a series of islands by high water levels. As a result, South America was effectively a huge island, its only connection with another faunal region reduced to the island "stepping-stones" across Central America. During this time, a huge array of new birds evolved in South America, resulting in the numerous endemic groups found today.

A further factor contributing to the plethora of Neotropical species is the fluctuating climatic conditions (the series of "Ice Ages") during the Pleistocene, which created habitat "islands" within both the Andes mountains and the lowland areas of the Amazon Basin. In the Andes, many of the high forests have cool temperatures year round,

Figure 1–57. Wild Turkey: Native to North America, the Wild Turkey inhabits mature forests, roosting high in trees at night, and by day scratching the ground for nuts with its large, strong feet. Turkeys also eat grass seeds, buds, bulbs, berries, and some small animals. Wild Turkeys are the only native New World birds to be widely domesticated. Explorers in the 1500s found them already domesticated in Mexico and Central America, and took these birds back to Europe. Colonists eventually brought them back to North America. When displaying to females, the males raise and fan their tails, assuming a posture familiar to every schoolchild. They also lower and spread their wings, raise their back feathers, and give loud gobbles, while the bare skin of their head turns red, blue, and white. Drawing by Charles L. Ripper.

and many bird species are restricted to these forests and elevations. During the previous ages when the world was cooler, the forests on the main chain of the Andes were connected by low-lying cool forest to the forests on peaks lying out away from the main chain. Each time the world heated up, the cool forests retreated to the higher elevations, and those on the outlying peaks became isolated, separated from the others by a vast expanse of hot lowland forest, as they are today. Although no obvious physical barriers existed, cool forest specialists found their desired habitats isolated from similar ones on "sky islands." These mountain islands acted as genuine islands, and many new species evolved in these very restricted areas. New species continue to be discovered in these areas even today, as scientists explore the most remote of these mountaintop islands. Similarly, periods of glaciation caused areas of the high Northern Andes above the tree line that are covered by glaciers and a humid grassland habitat termed **paramo** to become connected, allowing a greater exchange of species **(Fig. 1–58)**. In the Amazon Basin, the cool periods (which were also dry) caused the extensive rain forests to shrink to isolated patches known as **refugia**. These refugia were widely separated by grasslands. Existing species thus were fragmented into a number of isolated populations, many of which experienced different selection pressures and evolved into new species. When warmer conditions returned, these new species expanded their ranges with the expanding forests. In this way, formerly widespread species evolved into a number of different species that had either more limited ranges or overlapping ranges, or some combination. The current distribution of certain toucan species reflects this pattern of speciation **(Fig. 1–59)**.

Of the many endemic bird families in the Neotropics, some contain just a few species, but others have radiated into numerous species. Some of the smaller, but more extraordinary, families include the following:

rheas (2 species): large, flightless ratites of the temperate open country

tinamous (46 species): primitive, grouse-like birds whose eerie calls haunt both forests and pampas (see Fig. 7–21); tinamous are the only ratites capable of flight

screamers (3 species): heavy-bodied, goose-like birds with far-reaching calls (see Fig. 4–18a)

Hoatzin (Fig. 1–60) (1 species): an odd-looking, leaf-eating bird whose young clamber about in trees on all fours, aided by small claws on their wings (see Fig. 3–41)

seedsnipes (4 species): ptarmigan-like birds that nest both in southern lowlands and in high mountains

Oilbird (1 species): a large, nocturnal frugivore, related to nighthawks, that uses echolocation to reach its nest deep within caves (see Fig. 4–53)

trumpeters (3 species): large, hump-backed, chicken-like birds that roam the rain forest floor in flocks, sometimes following army ants for the arthropod prey they disturb

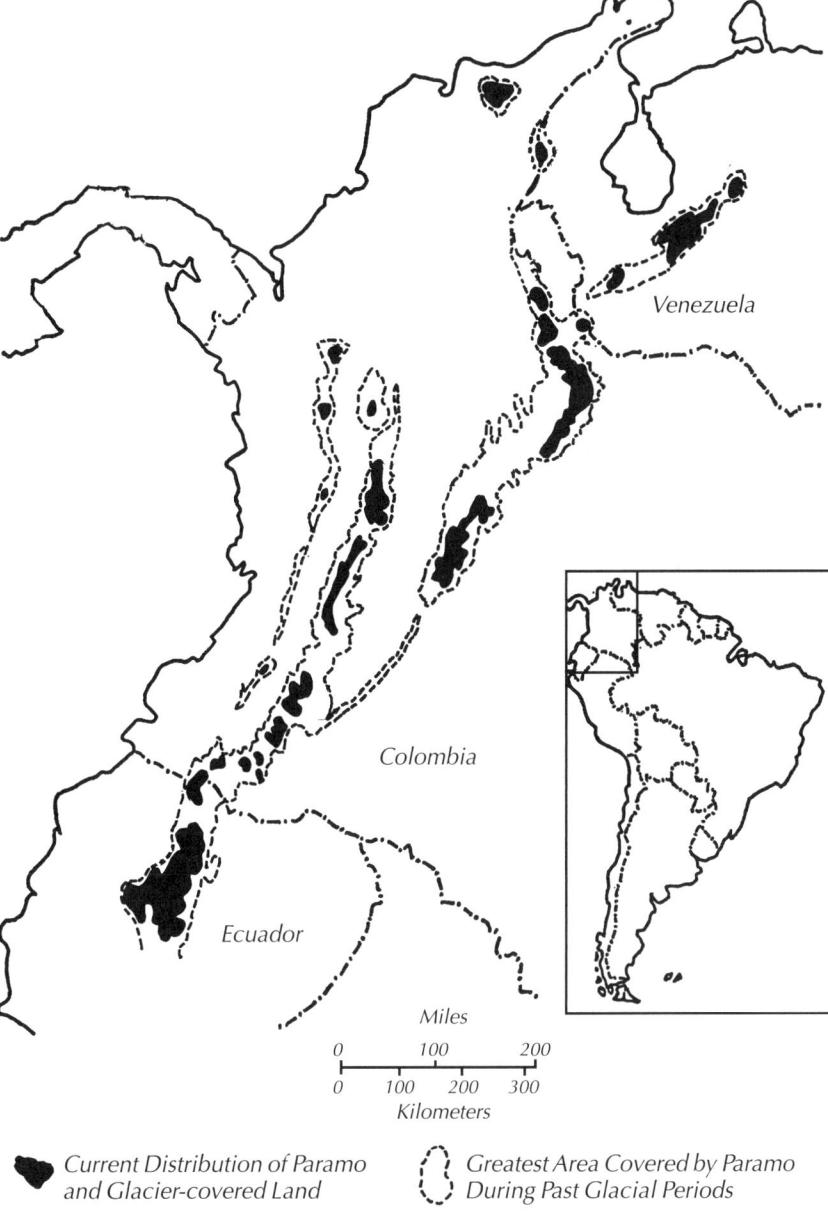

Venezuela

Colombia

Ecuador

Miles

0   100   200

0   100   200   300

Kilometers

🖤 Current Distribution of Paramo   ⬭ Greatest Area Covered by Paramo
and Glacier-covered Land   During Past Glacial Periods

*Figure 1–58. Paramo "Islands" of the High Andes: In the high northern Andes of South America, areas above the tree line but below the zone of permanent snow (glaciers) are covered by a habitat called **paramo**—humid grasslands with some shrubs, dotted with lakes and bogs. Currently, paramo and glacier-covered land areas (shown in black) exist as chains of isolated islands on the higher peaks, separated by areas of lowland or mountain forest. During the Pleistocene, however, the cool, dry climate accompanying each period of glaciation allowed the glaciers and paramo to expand, creating long, continuous regions of paramo connecting the islands. The greatest extent of this continuous paramo habitat is shown by dashed lines. The alternating warm and cool climates during the series of Ice Ages caused the glaciers and paramo islands to shrink and expand repeatedly—creating conditions highly favorable for speciation. Many new forms or species could evolve in the isolated islands during warm periods, and then come in contact with one another during cooler periods. When similar species or forms come together, natural selection often causes them to rapidly diverge to reduce competition for similar resources. The result of these alternating conditions, then, is a great number of endemic bird species, many with quite restricted ranges. In a similar way, the distribution of high forests in the Andes alternated between islands and larger, connected areas, also favoring large-scale speciation and the evolution of numerous endemic species with restricted ranges. Reprinted with permission from Vuilleumier, Beryl Simpson, 1971. Pleistocene Changes in the Fauna and Flora of South America. Science, Vol. 173, Number 3999. Copyright 1971, American Association for the Advancement of Science.*

seriemas (2 species): fast, long-legged, grassland birds that chase down their reptile prey on foot

⌇⌇⌇

One of the most important features of the Neotropical avifauna is the extensive radiation of the suboscines (Tyranni). This suborder of the huge order Passeriformes is distinguished from the other suborder, Passeri (oscines or songbirds), by the relatively simple syrinx. No region on earth has as many suboscines as the Neotropics; in fact, most regions have none or only a few. Fully a third of the Neotropical bird species are suboscines. In general, the Neotropical families with the most species tend to be suboscine, including the following:

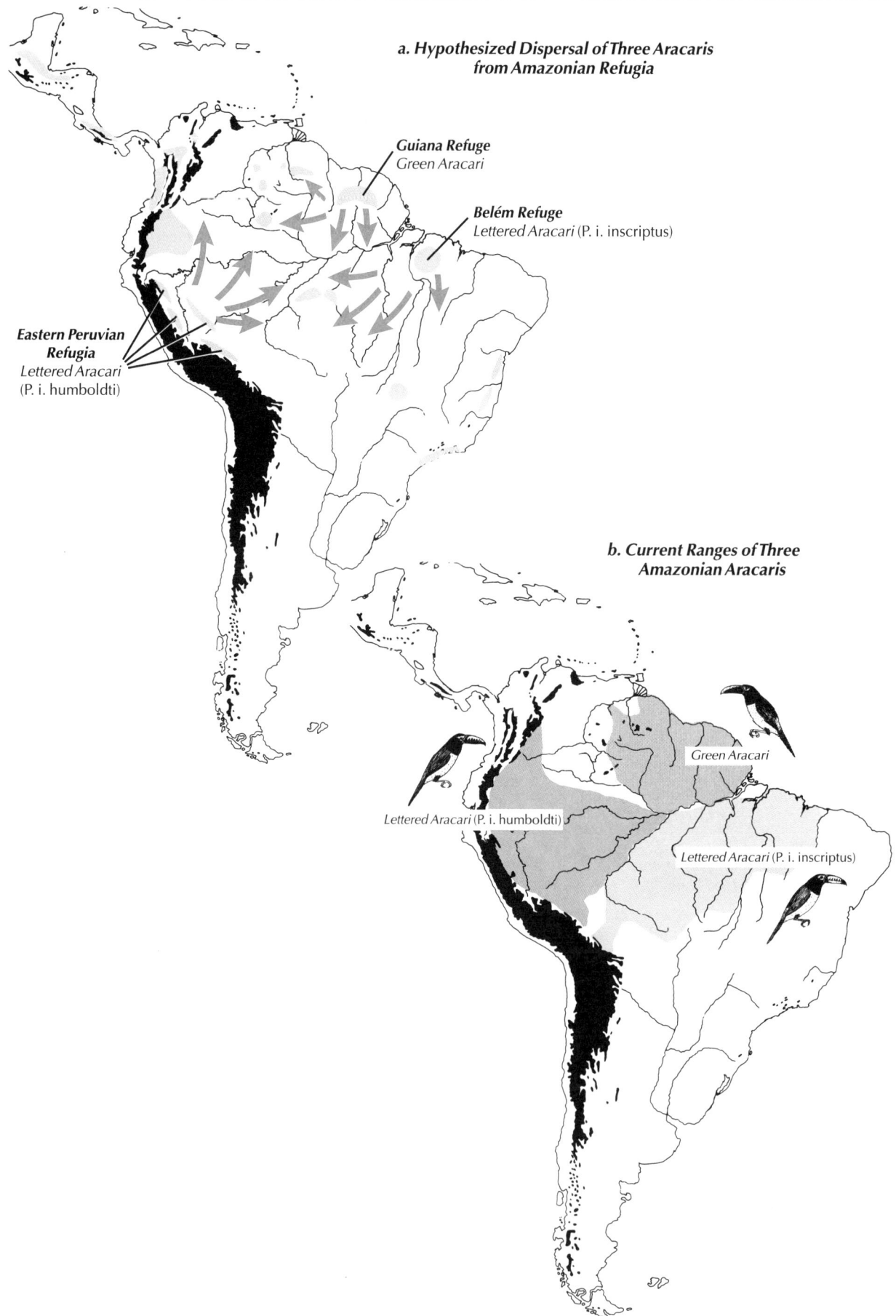

a. Hypothesized Dispersal of Three Aracaris from Amazonian Refugia

Guiana Refuge
Green Aracari

Belém Refuge
Lettered Aracari (P. i. inscriptus)

Eastern Peruvian Refugia
Lettered Aracari
(P. i. humboldti)

b. Current Ranges of Three Amazonian Aracaris

Green Aracari

Lettered Aracari (P. i. humboldti)

Lettered Aracari (P. i. inscriptus)

Figure 1–59. Amazonian Rain Forest Refugia and the Distribution of Aracaris: *In the Amazon Basin, the cool, dry glacial periods of the Pleistocene caused the extensive rain forests to shrink to isolated patches separated by grasslands. These isolated patches are known as* **refugia**. *Existing species thus were fragmented into a number of isolated populations, many of which experienced different selection pressures and evolved into new species. When warmer conditions returned—between the Ice Ages, and in the time since the last glacial period—the new species expanded their ranges with the expanding forests. In some cases, such as the aracaris illustrated here, a widespread, ancestral species apparently evolved into a number of species with more limited ranges. Aracaris are slender, small to medium-sized toucans. The three forms discussed here, the Green Aracari and two subspecies of the Lettered Aracari (considered subspecies because they interbreed along their common boundary), differ mostly in the color of the bill. These three aracaris are so similar that their taxonomic relationships remain unclear. Clements (2000) considers Pteroglossus inscriptus humboldti a subspecies of the Green Aracari, instead of a subspecies of the Lettered Aracari, as shown here.* **a. Hypothesized Dispersal of Three Aracaris from Amazonian Forest Refugia:** *Colored areas indicate the presumed locations of large forest refugia in tropical South and Central America during the cool, dry glacial periods of the Pleistocene. Dark areas indicate mountains above 6,600 feet (2,000 m). Researchers hypothesize that a single aracari species was fragmented into populations in the Guiana Refuge, the Belém Refuge, and a third site in the upper Amazon (possibly a group of refuges known as the Eastern Peruvian Refugia). In isolation, they each evolved into new forms. As the forests expanded during warmer times, the new forms expanded their ranges as indicated by the arrows.* **b. Current Ranges of Three Amazonian Aracaris:** *The current ranges of these three forms, indicated by the different types of shading, overlap very little. After Haffer (1974). Copyright Nuttall Ornithological Club.*

Figure 1–60. Hoatzin: *The bizarre, chunky, pheasant-sized Hoatzins, with their blue, bare facial skin, bright red eyes, and ragged crest, appear to be caught in a perpetual bad-hair day. Found in noisy groups crashing around the foliage along slow-moving streams and oxbow lakes of the Amazon and Orinoco Basins, Hoatzins place their crude, communal nests low in branches overhanging streams. Nonbreeding adults assist the breeding pair in tending the eggs and young. To escape danger, the nestlings may drop into the water. They climb back to the nest using small claws on their wings (see Fig. 3–41). Hoatzins are among the few bird species that eat mostly leaves, including many containing toxic compounds; the leaves are digested and detoxified by bacteria in their gut. Drawing by Robert Gillmor.*

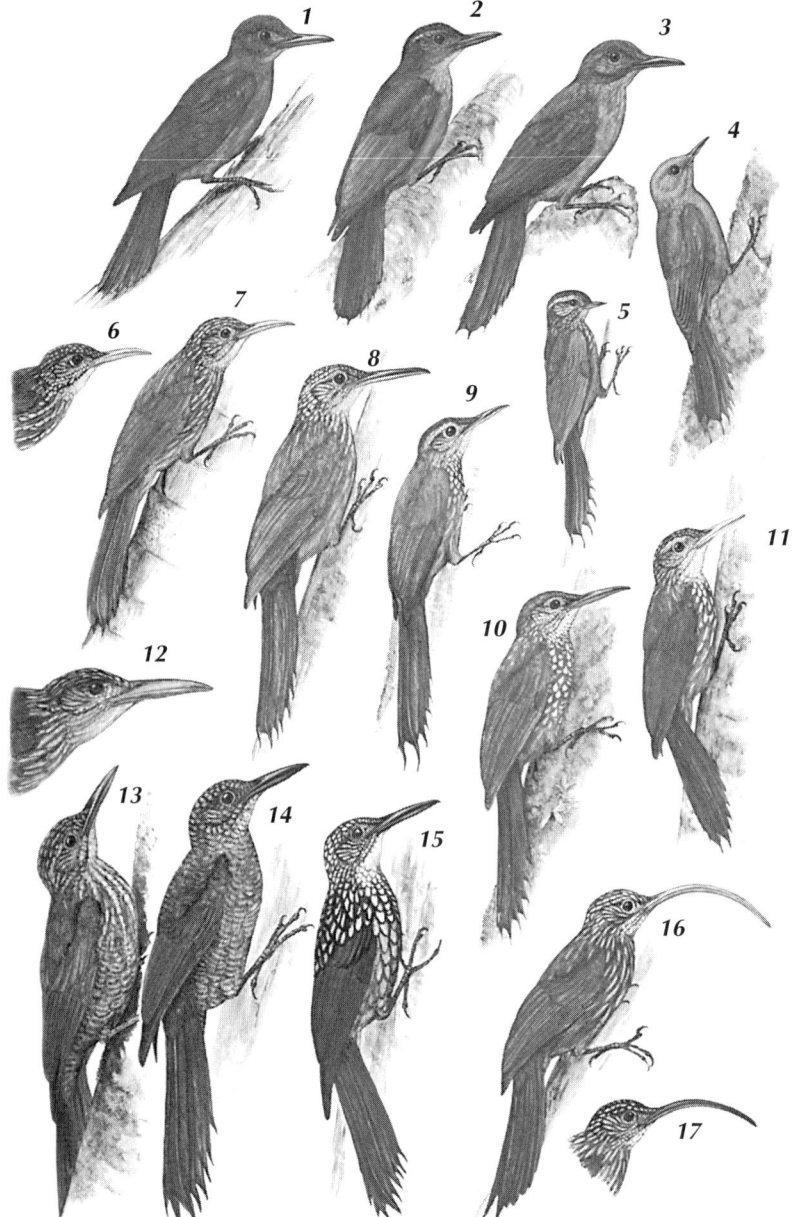

Figure 1-61. Woodcreepers: *Found in forests, forest borders, and mangroves throughout Central and South America, the sub-oscine woodcreepers (family Dendrocolaptidae) resemble the unrelated oscine creepers (family Certhiidae: Brown Creeper of North America and treecreepers of Eurasia) in both appearance and habits. Using its stiff tail as a brace, a woodcreeper hitches its way along branches and up tree trunks, peering and probing for insects and other small animals under bark, in tree crevices, or among mosses and epiphytic plants (plants, such as orchids and bromeliads, that grow on other plants). The tail is well-adapted to wedge into irregularities in the bark as the bird climbs, because the feather shafts extend beyond the vanes and curve downward. The 51 species are remarkably similar to one another in appearance, their rufous or olive plumage usually barred or spotted with lighter colors in the head and breast regions. However, as can be seen from this plate of 17 species from Panama, bill shape varies widely, from short and pointed (bird #5); to long, strong, and woodpecker-like with varying amounts of curvature (for instance, 7, 12, 13, and 14);*
and to slender and strongly downcurved, as in the scythebills (16 and 17). These bill variations undoubtedly reflect subtle differences in foraging methods. A few species, notably the Plain-brown Woodcreeper, are regular followers of army ants, capturing the insects flushed as an ant swarm passes by (see Ch. 9, Sidebar 3: Ant Followers). Species shown here are **(1)** *Ruddy Woodcreeper,* **(2)** *Tawny-winged Woodcreeper,* **(3)** *Plain-brown Woodcreeper,* **(4)** *Olivaceous Woodcreeper,* **(5)** *Wedge-billed Woodcreeper,* **(6)** *Spot-crowned Wood-creeper,* **(7)** *Streak-headed Woodcreeper,* **(8)** *Buff-throated Woodcreeper,* **(9)** *Long-tailed Woodcreeper,* **(10)** *Spotted Woodcreeper,* **(11)** *Straight-billed Woodcreeper,* **(12)** *Strong-billed Woodcreeper,* **(13)** *Black-banded Woodcreeper,* **(14)** *Amazonian Barred-Woodcreeper,* **(15)** *Black-striped Woodcreeper,* **(16)** *Red-billed Scythebill, and* **(17)** *Brown-billed Scythebill. Painting by John A. Gwynne (adapted), from* A Guide to the Birds of Panama *by Robert S. Ridgely and John A. Gwynne, 1989, Plate 19. Published by Princeton University Press. Used with permission.*

woodcreepers (51 species): similar-looking, rust-colored birds with a wide array of beak shapes (**Fig. 1–61**); they forage in bark much like the unrelated Brown Creeper of North America

antbirds (197 species): small, insectivorous, forest birds; some specialize on following columns of army ants to prey on the insects and other arthropods stirred up by the numerous moving ants (**Fig. 1–62**; also see Fig. 7–28 and Ch. 9, Sidebar 3: Ant Followers)

antthrushes and antpittas (60 species): small, drab birds with loud, ringing songs; many haunt the rain forest floor and may follow army ant swarms

ovenbirds (240 species): a diverse group, especially numerous in temperate South America, named for the oven-shaped, clay nests of some species (see Fig. 4–118)

tapaculos (52 species): antbird relatives with cocked tails, frequenting more southerly, open, and dry areas than antbirds

Figure 1–62. White-plumed Antbird: Antbirds are small, often boldly patterned birds that frequently move through the rain forest in mixed-species foraging flocks. Most species pick insects and other arthropods from foliage, with different species feeding at different levels in the forest. The group gets its name from a few species, informally termed the "professional antbirds," which specialize on following columns of army ants as they march across the forest floor. These antbirds prey on arthropods stirred up by the moving ants. The White-plumed Antbird, striking with its white face and crest, black head, blue-gray back and wings, and chestnut breast and tail, is among the most common of the professional antbirds in northern Amazonia. Photo by Doug Wechsler/ VIREO.

The most numerous of all suboscines are the flycatchers. Although many "fly-catching" birds around the world are called "flycatchers," the huge family **Tyrannidae** (tyrant flycatchers) is entirely restricted to the New World. Only a handful of the nearly 400 tyrannid species make it to the Nearctic, and they are all rather similar, generalized, aerial insectivores (see Figs. 2–15 and 7–43). In the Neotropics, the tyrant flycatchers have evolved to fill a large number of ecological roles, specializing into a diverse assemblage (**Fig. 1–63**).

The Neotropical avifauna also includes many bird groups that are shared with the Nearctic or Holarctic regions. In the near future, DNA comparisons may provide useful clues on the origins of these groups, but for now scientists can only speculate based on the distribution of species. The overwhelming number of species of both hummingbirds and tyrant flycatchers in the Neotropics compared to the Nearctic suggests that both these groups originated in the Neotropics and later reached the Nearctic. Other shared groups include New World vultures, waterfowl, hawks, New World quail, pigeons, owls, woodpeckers, vireos, jays, wrens, thrushes (Turdidae), New World warblers (Parulidae), tanagers, cardinals (Cardinalidae), and blackbirds and New World orioles (Icteridae). Regardless of where these groups originated, most of them radiated in the Neotropics to such an extent that they currently contain many more Neotropical than Nearctic species. Two important families, the fruit-eating tanagers and the closely related, insect-eating New World warblers, contain numerous species. A few members of each migrate to the Nearctic to breed, but these are essentially Neotropical groups. Although both titmice (Paridae) and nuthatches (Sittidae) are widespread throughout the rest of the world, they have no Neotropical members.

Most of the non-endemic bird families of the Neotropics are shared only with the Nearctic or Holarctic, but several groups, such as the parrots and trogons (**Fig. 1–64**), are found in tropical regions

a. Vermilion
Flycatcher

b. Cliff Flycatcher

c. Cock-tailed
Tyrant

d. Strange-
tailed Tyrant

g. Long-tailed
Tyrant

e. Spectacled
Tyrant

f. Short-tailed
Pygmy-Tyrant

k. White-crested Spadebill

l. Boat-billed Flycatcher

h. Sharp-
tailed Tyrant

i. Many-
colored
Rush-Tyrant

j. Royal
Flycatcher

n. Short-tailed
Field-Tyrant

m. Ringed
Antpipit

fitzpatrick
'81

Figure 1–63. Tyrant Flycatcher Diversity: *The suboscine family Tyrannidae (tyrant flycatchers), found only in the New World, has more species than any other family of birds. Of the nearly 400 species, just over three dozen regularly breed in the Ne- arctic. Most of these—including most* Empidonax *flycatchers, some kingbirds, the Great Crested Flycatcher, wood-pewees, and phoebes—are similar in general shape and aerial insect- catching behavior. In the Neotropics, however, the tyrant flycatchers have radiated into a wide array of ecological roles, making up roughly 10 percent of all bird species, and 20 to 25 percent of the passerines. Some of the more spectacular tyrant flycatchers pictured here are: the brightly colored Ver- milion Flycatcher (a) and Many-colored Rush-Tyrant (i), the long-tailed Strange-tailed Tyrant (d) and Long-tailed Tyrant (g); the bizarre Royal Flycatcher (j), which flashes its fan-shaped orange crest only during alarm or aggression; the Cock-tailed (c) and Sharp-tailed (h) tyrants, with modified tail feathers; the miniature Short-tailed Pygmy-Tyrant (f); and the Spectacled Tyrant (e), with its conspicuous yellow eye wattle.*

Tyrant flycatchers live in nearly every habitat—from rain for- est and savanna to the high, treeless paramo; and forage at every forest level—from the ground, understory, and forest edge, to the canopy and above. Like the Nearctic-breeding tyrannids, many Neotropical tyrannids sally out to grab flying insects from an exposed perch, but numerous variations exist. Some have even evolved to eat fruit, fish, young birds, or frogs. Some fly- catchers reach for insects while perched or hovering, whereas others fly out to grab insects from branches, foliage, or even the surface of water. The Cliff Flycatcher (b) looks and behaves like a swallow, perching on cliffs to search for aerial insects. The Boat-billed Flycatcher (l) has a huge, wide beak with which it may capture large insects, and the tiny White-crested Spadebill (k) uses its wide, flat beak like a shovel, scooping insects off the undersides of leaves. The long-legged Ringed Antpipit (m) walks on the rain forest floor, looking up at the undersides of leaves and then hopping up to grab any insects it finds. The Short-tailed Field-Tyrant (n) is also terrestrial. Original paint- ing by John W. Fitzpatrick, from Traylor and Fitzpatrick (1982). Reprinted with permission of John W. Fitzpatrick.

throughout the world. Birds distributed in this way are said to be **pantropical**.

Because so much of the Neotropical region is covered by lowland tropical forest, fruits and flowers are particularly abundant, and **frugivory** (specializing on eating fruit) is particularly common. Entire families of frugivorous birds have evolved, such as the cotingas, manakins, most of the tangers, and the toucans **(Fig. 1–65)**. Abundant flowers feed the large number of hummingbirds, a successful group containing more than 300 species, all but a dozen of which are found only in the Neotropics. Easy food, such as abundant fruits, allowed many Neotropical birds to evolve a mating system in which the males take no part in raising young, but instead display elaborately for mates—often congregating in traditional display areas that females visit. The colorful Neotropical manakins (see Figs. 6–42 and 6–47) and cotingas—including the fantastically plumaged umbrellabirds **(Fig. 1–66)** and the gaudy cocks-of-the-rock **(Fig. 1–67)**—are challenged only by the New Guinean birds-of-paradise for having the most elaborate plumages and displays in the avian world.

Afrotropical Region

The Afrotropical region includes Madagascar, southern Arabia, and all of Africa south of the Sahara. Isolated by water as well as by sand—the vast Sahara and Arabian Deserts—the region has many endemic families and is second only to the Neotropical region in the number of species. Although much of the region is relatively warm, it is also dry, receiving only about half as much rain as South America. As a result, a greater proportion of the region is covered by desert, scrub, grassland, and savanna, and the areas of tropical rain forest are somewhat more restricted than those of the Neotropics. Although mountainous regions are present, there is no

Figure 1–64. Slaty-tailed Trogon: *Trogons are chunky, colorful birds that tend to perch on branches in a characteristic, upright posture with their long, square tail pointing down. They often remain perfectly still for long periods of time. Thus, in spite of their brilliant colors, they can be difficult to see, but their frequent "cow, cow, cow" call may give away their location. These fruit- and insect-eaters dig their nest cavities in dead trees or in active termite or wasp nests. Although most trogon species live in the Neotropics, they are also found in tropical Africa and Asia. The Slaty-tailed Trogon is found in rain forests from Mexico to Ecuador. Photo by Marie Read.*

Figure 1–65. Choco Toucan: *Toucans, instantly recognizable with their enormous, gaudy bills, live in Neotropical forests, nesting and roosting in tree cavities. In body shape and some aspects of their ecology, they are convergent with the Old World hornbills (see Fig. 1–82). Toucans use the tips of their large, lightweight bill to reach out and grab ripe fruits. They then flip back their head and toss the fruit down their throat. Although primarily frugivorous, toucans also eat small animals, including the eggs and nestlings of other birds. The Choco Toucan—striking with its yellow and chestnut beak, yellow face and front, black back, white rump, and red undertail coverts—lives in wet forests on the Pacific slopes of western Colombia and Ecuador. Photo by S. Holt/VIREO.*

Figure 1–66. Long-wattled Umbrellabird: Like other members of the suboscine family Cotingidae, umbrellabirds are large, stocky fruit-eaters of the Neotropical forests. Many cotingids are beautiful and interesting, with bizarre courtship rituals. The Long-wattled Umbrellabird, a bit larger than an American Crow, is black with a bluish gloss. Males have a large, umbrella-shaped crest on the head and a long, fleshy, inflatable wattle—covered with feathers—hanging down from the throat. Females are similar with a smaller wattle. Males display in the rain forest canopy by spreading their crests, inflating their wattles, and jumping from branch to branch while giving a long, low grunt. Drawing by Robert Gillmor.

Figure 1–67. Guianan Cock-of-the-rock: Another cotingid (see Fig. 1–66), the flicker-sized, male Guianan Cock-of-the-rock is among the gaudiest of Neotropical rain forest birds. It is almost entirely vivid, day-glo orange, with some white and black markings. A large, disk-like crest nearly hides the beak. Females are dull, olive brown, with a smaller crest. Males display in groups of up to 40 or 50 birds at traditional sites, usually near rocky outcrops. Each male clears leaves and debris from a small area of ground to form his own display "court." When a female arrives to choose a mate, the males erupt into a cacophony of calls and displays, but much of their "showing off" consists of static posturing directed toward the female. Typically just a few males do most of the mating. The female performs all of the nesting activities—sticking her cup nest of mud to a rock wall near the entrance to a cave, incubating for approximately 40 days, and then feeding the young exclusively on fruit.

major chain comparable to the Andes, with its tremendous diversity of altitudinal zones. All of these factors undoubtedly contribute to the lower species diversity of the region compared to the Neotropics (1,560 versus 3,300 species). The drier climate hosts relatively few water birds, but a great diversity of terrestrial and seed-eating birds.

The different climate zones of Africa exist roughly as a series of latitudinal belts that become more open and dry farther from the equator. The eastern and southern thirds of the continent, however, are raised into wide plateaus 3,000 feet (1,000 m) or more above sea level, too dry to support forests. Approximately 37 percent of the Afrotropical region is above 3,000 feet, compared to less than 17 percent of the Neotropical region. Around the equator, in the western two-thirds of the continent, is the lowland tropical rain forest of the Congo Basin. As in the Amazon Basin, the dry, cool glacial periods of the Pleistocene created a series of refugia here, which are thought to have served as centers for species diversification across the east-west axis of today's equatorial tropical forests.

Farther from the equator, and in parts of the plateau regions, are vast areas of open woodlands, savannas, and grasslands. Characteristic birds of the grassy areas are **cisticolas** (small, drab, insectivorous warblers), **weavers** (a diverse group of colorful seed-eaters, many of which weave large, complex nests), and **waxbills** (another group of colorful seed-eaters; see Australasian region for more information). Deserts and semi-arid scrublands, with sparse vegetation and scattered trees such as acacia (and in the east, baobab), are found in three main areas: (1) in the Sahel region, just south of the Sahara, (2) in the northeast, between the Gulf of Aden and northern Tanzania, and (3) in the southwest—the Kalahari Desert region. Typical birds of these open areas are larks, bustards, and coursers. **Bustards** are large, heavy-bodied, flat-headed birds with long legs and necks **(Fig.1–68)**. They often assume bizarre postures during their elaborate courtship displays. **Coursers** are slender, plover-like ground nesters that sometimes cool their eggs or young by partially burying them in sand or by bringing water to the nest in their breast feathers. In addition, the trees support a great diversity of other species, including many Palearctic migrants.

Families endemic to the Afrotropical region include the following:

Struthionidae (Ostrich): the largest living bird, this familiar ratite of deserts and savannas is almost entirely herbivorous; Ostriches breed communally, with several females laying eggs in the same nest

Balaenicipitidae (Shoebill): a large, stork-like water bird with a huge, shoe-shaped, hooked bill for seizing big fish **(Fig. 1–69)**

Figure 1–68. Kori Bustard: Stout birds with long legs and necks, the ground-dwelling bustards frequent a variety of open habitats, from semi-desert and grassland to dense scrub. They eat a variety of plant and animal foods ranging from shoots, flowers, and berries to insects, frogs, and small mammals. Although most species live in Africa, a few are found in the Oriental, Australasian, and southern Palearctic regions. The long-lived Bustards range in size from as small as a Ring-billed Gull to as large as a Tundra Swan. They have strong legs and feet designed for running, with three forward-facing toes and no hallux. Their plumage is cryptically colored in buff or brown overall. Some of the smaller species perform dramatic aerial courtship displays: males of the Buff-crested Bustard fluff their feathers and fly straight up into the air, sometimes as high as 100 feet (30.5 m), then parachute back down to earth. The large Great Bustards and Kori Bustards (pictured here), display on the ground. These courting males cock their outspread tails forward, inflate their gular (throat) sacs, retract their heads, and elevate the light-colored neck and undertail feathers—creating a feathered entity that only barely resembles a bird! Drawing by C. J. F. Coombs, from A Dictionary of Birds, edited by Bruce Campbell and Elizabeth Lack, p. 75. Copyright 1985, The British Ornithologists' Union. Reproduced with the kind permission of the British Ornithologists' Union.

Figure 1–69. Shoebill: The only member of its family, the Shoebill (also known as the Whale-headed Stork) has a huge, shoe-shaped bill with a prominent, hooked nail at the tip. Standing nearly 4 feet (1.2 m) tall, the Shoebill hunts freshwater swamps of central and eastern Africa by waiting motionless for long periods, watching the water for unwary fish, amphibians, aquatic snakes, and small birds. Its nest is a large, flattened mound of grass on floating plants, hidden among dense vegetation. Nesting Shoebills regularly bring water to the nest in their large bills, pouring it onto their eggs and downy young to cool them. Photo by A. Morris/VIREO.

Figure 1–70. Secretary-bird: The sole member of its family, the gray and black Secretary-bird is distinctive with its elongated central tail feathers and long, loose crest of quills. Over 3 feet (1 m) tall, this long-legged bird-of-prey hunts by striding through the savannas and grasslands, stamping its feet to scare potential prey—large insects, reptiles, rodents, and other small animals—out of hiding. It is known for capturing snakes and then crushing them against the ground with its blunt talons. Despite being ground hunters, Secretary-birds fly and soar well, performing spectacular aerial courtship displays with undulating dives and croaking calls. Their nest, used year after year, is a large platform of sticks in the top of a shrub or tree, often in the flat-topped acacia trees common throughout the African savannas. Drawing by Robert Gillmor.

Scopidae (Hamerkop): a stork-like water bird whose shaggy, crested nape and stout, tapering bill make the head appear hammer-shaped (see Fig. 8–38)

Sagittariidae (Secretary-bird): a long-legged bird of prey that stalks the savannas on foot for snakes, other small vertebrates, and large insects **(Fig. 1–70)**

Coliidae (mousebirds or colies): long-tailed birds about the size of a Mourning Dove that climb through vegetation using their stout, hooked beaks; they frequently perch in "chin-up position," hanging vertically down from a branch **(Fig. 1–71)**

Musophagidae (turacos): noisy, arboreal relatives of cuckoos that run squirrel-like along branches; the plumage is soft green (one of the few green pigments known from birds), blue, or gray, often with red on the wings **(Fig. 1–72)**

Phoeniculidae (woodhoopoes): sociable birds with glossy, dark plumage and long tails; woodhoopoes nest in tree cavities and breed cooperatively—with additional adult birds helping the parents to tend the nest

Lybiidae (African barbets): small, colorful, stocky birds with large, sometimes serrated, beaks; they dig their nest cavities in trees, earthen banks, or termite nests

Picathartidae (rockfowl): dull-colored birds with colorful, bare heads; they hop along the rain forest floor in rocky areas and nest colonially in caves

Promeropidae (sugarbirds): two long-billed, long-tailed, nectar-feeders that specialize on *Protea* plants on the mountainous slopes of South Africa

Also endemic to the Afrotropical region are the **guineafowl**, six species that form subfamily **Numidinae** within the pheasant family (Phasianidae). These gregarious, chicken-like birds have distinctively spotted and striped plumage that often lands them in zoos **(Fig. 1–73)**. They have loud, harsh calls and are often domesticated.

The Afrotropical region shares many families with the Oriental region, as well as about 30 percent of the genera. But, only about two percent of the *species* are shared, probably reflecting how long the birds of the two regions have evolved independently. A significant annual event is the huge influx of migrants that occurs as roughly one-third of the Palearctic species, mostly insect-eaters, move south to overwinter throughout Africa.

Many Afrotropical birds are passerines, including larks, weavers, shrikes, and **sunbirds (Fig. 1–74)**. All these groups have numerous species in the Afrotropical region. Other passerine groups that are well-represented include thrushes, cisticolas, starlings, bulbuls, bushshrikes, and waxbills.

Important nonpasserine components of the avifauna include hawks, hornbills (see Oriental region), bee-eaters, and rollers. **Bee-eaters** are brightly colored birds with long, slender beaks. They catch stinging bees, wasps, and ants flycatcher-like, beating them to remove

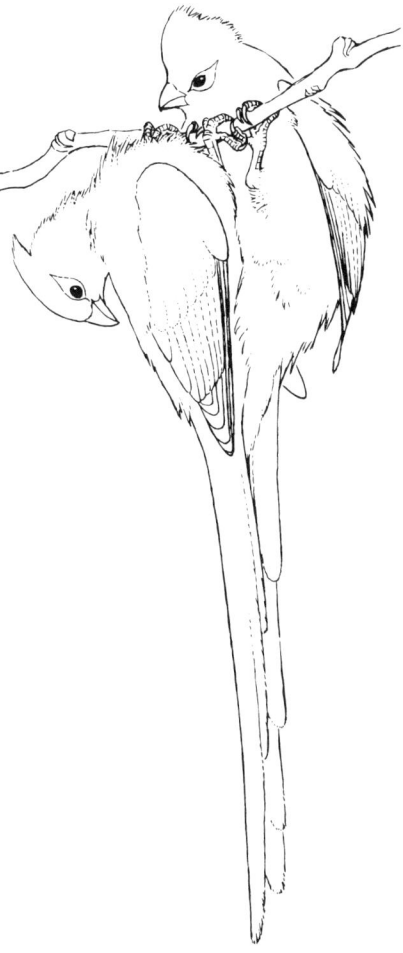

Figure 1–71. Speckled Mousebirds: The grayish or brownish mousebirds (also known as colies) get their common name from their long, drooping tails and habit of scurrying along the tops of branches, mouse-like. They clamber through vegetation with great agility, and often cling upside down, like parrots, to feed on fruits, foliage, and nectar. Strong claws and reversible first and fourth toes aid their acrobatic abilities. Highly gregarious, mousebirds live year round in groups of a dozen or more, breeding cooperatively, and constantly chattering and whistling as they forage in the wooded savannas and scrublands of sub-Saharan Africa. They roost in tight clusters, often dust-bathing and playing together, preening each other, or perching breast-to-breast for warmth—as pictured here. The Speckled Mousebird is the most widespread species, ranging throughout eastern, central, and southeastern Africa. Drawing by Ian Willis, from The Birds of Africa, Vol. 3, p. 252. C. H. Fry, S. Keith, and E. Urban, editors. Copyright 1988 Academic Press Limited. Used with permission.

Figure 1–72. Schalow's Turaco: *Together with plantain-eaters and go-away-birds, turacos make up the family Musophagidae. These crow-sized, long-tailed, arboreal birds are weak fliers, preferring to run along branches much like squirrels. Their outer toe can be rotated forward or back, but is often used at right angles to the main axis of the foot—allowing them to nimbly traverse branches in search of fruit, foliage, and buds. Their harsh, barking calls are familiar sounds in the African forests. The forest-dwelling turacos and plantain-eaters are usually soft green, violet, blue, or some combination of these colors, with crimson on the wings. They also have colorful crests, eye rings, and bills. The gray and white go-away-birds inhabit more open, dry woodlands. This family of birds is famous for its unique pigments, found in no other animals. The reds are produced by a copper-containing pigment called turacin, and the greens, by turacoverdin. Schalow's Turaco, found in the humid forests of south-central Africa, is mostly green, with crimson primaries, a scarlet eye ring, and a green crest tipped with white. Photo by P. Davey/VIREO.*

Figure 1–73. Vulturine Guineafowl: *The large, chicken-like guineafowl have dark plumage distinctively spotted and striped with white. The bare skin of the head and neck is often brightly colored and may be adorned with wattles. Some species have bony casques or feathered crests on the head. Guineafowl are gregarious when not breeding, often gathering into flocks of 50 or more birds. Their raucous voices can be heard from nearly every African habitat—from dense forest to semi-desert. The Vulturine Guineafowl has particularly beautiful plumage, including brilliant blue on its breast and back. It occurs in dry, scrubby regions of eastern Kenya, Somalia, and Ethiopia. Photo by Marie Read.*

Figure 1–74. Mount Apo Sunbird: *The tiny sunbirds—some smaller than a Ruby-throated Hummingbird—have long, slender, downcurved bills, and the males of many species are brilliantly colored, often iridescent. Most of the 124 species inhabit Africa and Madagascar, frequenting a variety of habitats. Some, however, are found in the Oriental, Australasian, and southern Palearctic regions. Although the family name (Nectariniidae) hints that sunbirds feed on nectar, they also eat insects. Short-billed species tend to eat more insects than nectar, whereas long-billed species primarily feed on the nectar of tubular flowers—much like the unrelated hummingbirds. Like hummingbirds, sunbirds act as pollinators of flowers, and their bills have evolved to match the shape of the flowers they visit. Unlike hummingbirds, however, sunbirds cannot hover to feed on the wing. Instead they cling, often upside down, to a flower as they probe for nectar. Shown here is a Mount Apo Sunbird from the Philippines. Photo by Doug Wechsler/VIREO.*

Figure 1–75. Red-billed Oxpeckers on a Black Rhinoceros: The bluebird-sized oxpeckers inhabit savannas throughout much of eastern and southern Africa, feeding on ectoparasites that infest the skin of large mammals. They forage in small flocks, hopping and climbing over the bodies of grazing zebras, giraffes, buffalo, rhinos, warthogs, antelopes, and even domestic cattle. The mammals are usually indifferent to, and indeed benefit from, the birds' gleanings. Oxpeckers have short legs; sharp, curved claws; and stiff tails that act as braces—all adaptations helping them to cling to their hosts. Their bills are laterally flattened, and the birds open and close them rapidly in a scissor-like action as they push them through the fur or over the naked skin of a mammal, removing ticks and other skin parasites. They sometimes capture flies that land upon the host's skin and drink fluid from around the host's eyes. Oxpeckers breed cooperatively and, like their starling relatives, roost communally at traditional sites. Photo by Marie Read.

the venom before eating them. Some species are cooperative breeders that live in complex societies. **Rollers**, named for their rolling or rocking dives during courtship flights, appear somewhat like bee-eaters, but with more muted plumage colors mainly in shades of blue, pink, olive, or chestnut. They often forage like bluebirds, perching and then flying to the ground to grab large arthropods.

Particularly interesting in their habits are the oxpeckers (members of the starling family, Sturnidae) and the nonpasserine honeyguides. The sociable **oxpeckers**, also called "tickbirds," climb upon large African mammals as they graze, removing ticks, insects, and the scabs of skin wounds—a relationship benefitting both the birds and their hosts **(Fig. 1–75)**. The **honeyguides** are peculiar in their ability to digest wax, especially beeswax, in addition to their insect prey. At least two species use distinctive calls and behaviors to lead humans, baboons, or honey-badgers to bees' nests—feeding on the wax and larvae once the mammalian helper has opened the nest for its honey.

The island of Madagascar, off the east coast of Africa, deserves special mention. Although it shares most of its fauna with mainland Africa at the family level, it has been isolated long enough to have evolved a number of unique families and a host of endemic species. Its habitats, although much denuded by humans, vary from dry woodlands and scrub in the west, through central highlands, to rain forest in the east. Families unique to Madagascar and the surrounding islands include:

mesites (Mesitornithidae): rail-like, ground-dwelling birds, about the size of a Mourning Dove

ground-rollers (Brachypteraciidae): solitary, terrestrial insect-eaters with stout bills, short wings, and moderately long legs and tails **(Fig. 1–76)**

Cuckoo-Roller (Leptosomatidae): a single, crow-sized, arboreal species with a stout, broad bill

asities and false sunbirds (Philepittidae): suboscines, the asities feed on fruits, and the false sunbirds, on insects and nectar

vangas (Vangidae): a diverse passerine group of 14 shrike-like species; most are gregarious and noisy, gleaning insects and other small animals as they move through the trees **(Fig. 1–77)**

Madagascar was the home of the Giant Elephantbird (Aepyornithidae), a huge, flightless creature long extinct, which weighed half a ton and laid two-gallon (eight-liter) eggs (see Fig. 5–48). The Dodo (Raphidae), a flightless bird the size of a turkey, lived only on the island of Mauritius, east of Madagascar; it has the unhappy distinction of being one of the first birds known to be eliminated by "civilized" humans, in the 17th century (see Fig. 9–75).

Figure 1–76. Long-tailed Ground-Roller: The five species of ground-rollers make up the family Brachypteraciidae, which is endemic to Madagascar. These stocky, ground-dwelling birds have large heads and eyes, stout bills, short wings, and relatively long tails and legs. All are insectivorous, occasionally preying on small lizards, snakes, or snails. They nest in tunnels dug into the side of a bank. Like most other birds of Madagascar, their populations are in severe decline due to human-induced habitat destruction. The Long-tailed Ground-Roller, shown here, is about the size of a magpie and is mottled brown and black above, with blue on the wings and outer tail feathers and a black band across the pale breast. It inhabits desert scrub, whereas the other four species live in dense evergreen forests. Drawing by N. A. Arlott, from A Dictionary of Birds, *edited by Bruce Campbell and Elizabeth Lack, p. 257. Copyright 1985, The British Ornithologists' Union. Reproduced with the kind permission of the British Ornithologists' Union.*

Oriental Region

The Oriental region includes all of Asia south and east of the Himalayan Mountains (India and Southeast Asia), as well as southern China and the islands of Indonesia and the Philippines. Although the Himalayas form a clear boundary with much of the Palearctic region, the distinction is less clear in China, where some mixing of the Oriental and Palearctic faunas occurs, with thrushes, accentors, dippers, and tits being shared between the two regions. Similarly, the Australasian and Oriental faunas grade into each other in Indonesia. The boundary has been chosen somewhat arbitrarily, just to the east of the islands of Timor and Sulawesi, where the proportions of the two faunas are roughly equal. The Oriental region is mostly tropical and subtropical, much of it in rain forest, although there are smaller areas of drier habitats such as dry forest, scrub, savanna, and desert.

Less isolated than the other avifaunal regions, the Oriental hosts only three endemic families, the Irenidae (arboreal songbirds called leafbirds and fairy-bluebirds), the Megalaimidae (Asian Barbets), and the Aegithinidae (ioras). The oriole-sized **leafbirds**, mostly green and yellow, feed mainly on insects and fruit. The two **fairy-bluebirds**, named for the brilliant blue and black plumage of the males, are slightly

Figure 1–77. Helmet Vanga: The vanga family (Vangidae) consists of 14 species, all endemic to Madagascar. Gregarious, shrike-like birds, vangas move through the trees in noisy flocks, capturing insects. Most species have heavy bills, which are hooked in some. Their plumage is boldly patterned and sometimes glossy, and varies in color—being black and white, blue and white, or a combination of black, rufous, and gray. The Helmet Vanga, shown here, has bold black-and-rufous plumage and a greatly enlarged bill, which it uses to capture insects as well as tree frogs and small reptiles. Drawing by C. E. Talbot-Kelly, from A Dictionary of Birds, *edited by Bruce Campbell and Elizabeth Lack, p. 619. Copyright 1985, The British Ornithologists' Union. Reproduced with the kind permission of the British Ornithologists' Union.*

larger, feeding primarily on figs and other fruits **(Fig. 1–78)**. The **Asian barbets** are chunky birds slightly smaller than a Belted Kingfisher, with thick bills and gaudy, clashing colors; many are green with red, blue, or yellow markings. They feed mostly on fruits and insects **(Fig. 1–79)**. The four species of **ioras** are small, arboreal songbirds that search leaves, often in dense foliage, for insects. Males of some species perform elaborate courtship displays with vertical leaps and parachuting flights.

Birds of the Oriental region—especially in and near India—resemble those of tropical Africa more than they resemble any other faunal group, reflecting the long period in geologic history that India spent attached to Africa via Madagascar. The fact that many families (for example, hornbills, honeyguides, broadbills, bulbuls, sunbirds, and weavers), but few species, are shared between the Oriental and Afrotropical avifaunas indicates the long period of time—about 85 million years—that they have been evolving separately.

Particularly well-represented in the Oriental region are members of the family **Phasianidae**: Old World quail, pheasants, partridges, and grouse—as well as the spectacular peafowl (see Fig. 3–6). The domestic chicken has descended from the Red Junglefowl of this family **(Fig. 1–80)**. Additional families with numerous species include pigeons, corvids, sunbirds, several families of finches, and the following:

Figure 1–78. Asian Fairy-bluebird: Together with the leafbirds, fairy-bluebirds make up the Irenidae, a family of songbirds endemic to the Oriental region. Male fairy-bluebirds have brilliant, metallic blue plumage with contrasting areas of velvety black; females are similar, with somewhat duller colors. Slightly larger than orioles, fairy-bluebirds inhabit the canopy of tall, semi-deciduous or evergreen forests, where they roam widely in search of fruiting trees, especially figs. Drawing by C. E. Talbot-Kelly, from A Dictionary of Birds, *by Bruce Campbell and Elizabeth Lack, p. 201. Copyright 1985, The British Ornithologists' Union. Reproduced with the kind permission of the British Ornithologists' Union.*

Figure 1–79. Red-crowned Barbet: Stocky and colorful, Asian barbets (family Megalaimidae) are one of just three families endemic to the Oriental region. They have stout, sharp beaks with which they pluck fruits and insects, and excavate their nest holes. Asian barbets were once placed in the same family with African barbets and South American barbets, but the three groups are now considered separate families (with the South American barbets in the same family as toucans). The Red-crowned Barbet, pictured here, is mostly glossy green with a red crown, and has red, yellow, and blue facial markings. Its prominent rictal bristles—the large "hairs" visible projecting from the base of the bill—are characteristic of all barbets, and give them their common name. Photo by Tim Laman/VIREO.

Figure 1–80. Red Junglefowl: The family Phasianidae—pheasants, partridges, grouse, turkeys, peafowl, and Old World quail—is particularly well-represented in the Oriental region. Most species are ground-dwelling seed-eaters, and many—particularly the pheasants and peafowl—have elaborately patterned, colorful plumage. Many of the species are important game birds, and some have been domesticated. The Red Junglefowl of Southeast Asia is thought to have given rise to the domestic chicken. Although the exact date and location of domestication are not known, domestic chickens were recorded in India as early as 3200 B.C. The male Red Junglefowl, shown here, is glossy red and orange above, with dark, iridescent underparts and an elaborate tail. Drawing by Robert Gillmor, from Lack (1968).

babblers: a diverse group of gregarious, insectivorous birds, many of whom have complex social systems and breed cooperatively

pittas: secretive, stocky birds of the tropical forest floor with long legs and short tails; many are brightly colored below and cryptic above; pittas use their heavy bills to catch a variety of insects and other small animals, especially snails (see Fig. 7–28)

flowerpeckers: small, often-colorful songbirds, much like sunbirds in their busy, noisy behavior as they forage high in the trees on berries, nectar, and insects

broadbills: chunky, brightly colored forest birds most closely related to cotingas and tyrant flycatchers, they use their wide, flat, colorful bills to snatch up large insects; 10 of the 15 species of broadbills are found in the Oriental region (**Fig. 1–81**)

Figure 1–81. Black-and-red Broadbill: Found throughout the Old World tropics, the broadbills are stocky, brightly colored, suboscine birds with large, wide, colorful bills. They occur in a variety of forest and scrub habitats, and feed by grabbing insects from foliage or out of the air. Some species are large enough to take large grasshoppers and small lizards. Broadbill nests are elaborate, pear-shaped bags, camouflaged with many tendrils, plant fibers, spider webs, and lichens hanging below them like a beard. They are often hung from an inaccessible vine or branch across an open space or above water. The Black-and-red Broadbill of Southeast Asia, Sumatra, and Borneo is black above and crimson below, with a blue and yellow bill. Photo by Doug Wechsler/VIREO.

Figure 1–82. Rhinoceros Hornbill at Nest Cavity: *Found throughout the Oriental and Afrotropical regions, hornbills have long tails and strong, decurved bills with a distinctive casque on the top. Many species have black-and-white plumage, but the bill is often colorful, and the head may have brightly colored areas of bare skin or feathers. Hornbill species are found in a variety of habitats, and vary from 2 to 4 feet (38 to 126 cm) in length. They eat large insects and other arthropods, small vertebrates, and fruit. Most interesting is their breeding behavior: the female (sometimes with help from the male) seals herself into the nest cavity, where she remains throughout incubation and much of the nestling period, receiving all of her food from her mate through a narrow slit in the mud-sealed entrance. During this time, females of many species undergo a rapid molt and are flightless. The Rhinoceros Hornbill, one of the largest species, inhabits Malaysia and Indonesia. In this photo, a male brings food to a female in the nest cavity. Photo by M. Strange/VIREO.*

Other birds characteristic of the region include owls, parrots, woodpeckers, shrikes, thrushes, warblers, mynas, and Old World flycatchers. In addition, many Palearctic migrants overwinter in the Oriental region.

Two particularly bizarre bird families found in the Oriental region are the hornbills (Bucerotidae) and the frogmouths (Podargidae). The **hornbills** have huge, down-curved bills topped by a peculiar casque **(Fig. 1–82)**, and are famous for their unique nesting behavior, in which the female seals herself inside the nest cavity with the eggs and young nestlings, and depends on the male to bring food (see Figs. 3–32, 8–44 and their associated text sections). The **frogmouths** resemble their nightjar relatives in both cryptic appearance and behavior **(Fig. 1–83)**. These nocturnal forest birds sally out from a branch and use their wide, "frog-like" beaks to snatch up small animal prey—including mice, frogs, and small birds—from the ground or branches.

Australasian Region

The Australasian region encompasses the islands south and east of the Oriental region, including the Moluccas, New Guinea, Australia, Tasmania,

Figure 1–83. Papuan Frogmouth: *Found in forests of the Oriental and Australasian regions, the frogmouths resemble their smaller nightjar relatives in both cryptic appearance and behavior. They have mottled gray or brown plumage and heavy bills with an enormously wide gape. From an exposed perch, they wait for suitable prey—large insects or other arthropods, frogs, or lizards—to appear, and then swoop down to snatch them from the ground or branches with their wide bills. If disturbed during the day, the nocturnal frogmouths assume a cryptic pose, freezing with the neck outstretched and bill angled upward, resembling a broken branch. Here, a Papuan Frogmouth shows its large, night-adapted eyes and its wide gape. It inhabits northeastern Australia and New Guinea, and islands in between. Photo by W. Peckover/VIREO.*

Figure 1–84. Varied Sittellas: *Because the Australasian region has been isolated from other land masses for so long, many of its birds are unrelated to those in other regions. Through convergent evolution, however, many have come to resemble other bird groups. One example is the sittellas, which—as their common name reflects—resemble nuthatches (Sittidae) in appearance and behavior. The two species of sittellas, however, differ from nuthatches in subtle ways (such as leg musculature and bill form), as well as in breeding habits, and are placed in the family Pachycephalidae, with whistlers. Also known as "treerunners," the forest-dwelling sittellas explore tree branches in tight spirals, working from tip to trunk, and move headfirst down trunks, prying under bark and probing into crevices for insects and other arthropods. They occasionally use a twig or bark chip as a probing tool. The Varied Sittella shown here is common in open* Eucalyptus *and* Acacia *woodlands of Australia, and is uncommon in New Guinea, where it inhabits moist mountain forests. Unlike nuthatches, which nest in cavities, the Varied Sittella builds a deep cup nest in a tree fork. Composed primarily of spider webs and cocoons, it is camouflaged on the outside with bark and lichens. The Varied Sittella is highly social—living, roosting, and breeding in groups. All group members help to build the nest and feed the young, although normally only one female lays and incubates the eggs. Drawing by N. W. Cusa, from* A Dictionary of Birds, *edited by Bruce Campbell and Elizabeth Lack, p. 539. Copyright 1985, The British Ornithologists' Union. Reproduced with the kind permission of the British Ornithologists' Union.*

New Zealand, and the numerous smaller islands in the mid-Pacific Ocean (Micronesia, Melanesia, and Polynesia—including Hawaii). The region, about half tropical and half temperate, is relatively flat, with mountainous areas mainly in New Guinea (tropical) and New Zealand (temperate).

The region is somewhat restricted in total bird species, probably as a result of at least three factors: (1) The large number of islands. As discussed earlier, islands tend to have fewer species than continental regions, although the species they have may be unique. (2) The relatively small land mass compared to the other avifaunal regions. (3) The relatively arid climate. Although the region has some species-rich tropical rain forest (mostly in New Guinea and northeastern Australia), compared to other regions a greater percentage of the land is occupied by drier habitats—woodland, savanna, grassland, scrub, and desert. Fully 70 percent of Australia is desert, the harsh conditions hosting fewer than two dozen species of birds, including the nomadic Budgerigars (see Fig. 4–132).

The region is rich, however, in endemic forms—second only to the Neotropics in the number of endemic families. Australia has been isolated from other land masses for so long that its fauna include many lineages that originated and radiated there. From their DNA-DNA hybridization work, Sibley and Ahlquist (1985) suggested that many Australasian songbird families, although resembling birds from other regions, actually result from a huge radiation of a lineage of crow relatives—often referred to as parvorder Corvida (see Fig. 7–22). This group diversified in the region in the early Tertiary, taking on the ecological roles of unrelated birds in other parts of the world. Thus, the Australasian avifauna provides many examples of convergent evolution, such as the Australo-Papuan warblers of the genus *Gerygone*, which closely resemble and behave like the unrelated Sylviid and Parulid warblers of Eurasia and the Americas, respectively. Similarly, the sittellas (Pachycephalidae) have a striking resemblance to nuthatches (**Fig. 1–84**). When Australia drifted closer to Southeast Asia during the late Tertiary, some of the crow relatives—including the crows and jays of family Corvidae—presumably dispersed to Asia and eventually to other regions, radiating further in these areas.

The Australasian region has a number of endemic bird families: Casuariidae (cassowaries), Dromiceidae (Emu), Apterygidae (kiwis), Rhynochetidae (Kagu), Aegothelidae (owlet-nightjars), Acanthisittidae (New Zealand wrens), Climacteridae (Australasian treecreepers), Menuridae (lyrebirds), Atrichornithidae (scrub-birds), Ptilonorhynchidae (bowerbirds), Maluridae (fairywrens), Eopsaltriidae (Australasian robins), Orthonychidae

Female

Male

(logrunners), Pomatostomidae (pseudo-babblers), Corcoracidae (White-winged Chough and Apostle-bird), Paradisaeidae (birds-of-paradise), Cracticidae (butcherbirds, Australasian Magpie, and currawongs), Callaeidae (wattlebirds), Grallinidae (mudnest builders), Melanocharitidae (berrypeckers and longbills), Paramythiidae (Tit Berrypecker and Crested Berrypecker).

Prominent nonpasserines in the Australasian avifauna include parrots, pigeons and doves, and kingfishers (especially in New Guinea). In addition, many songbird families have had extensive radiations here, including the birds-of-paradise, the whistlers and allies, the monarch flycatchers, the honeyeaters, and the Australo-Papuan warblers. Compared to temperate areas of Europe and North America, the temperate Australasian habitats have fewer small seed-eaters (some parrots fill this role) and many more nectar-feeders, many of which act as pollinators. Notably absent from the avifauna are Old World vultures, pheasants, woodpeckers, trogons, cardueline finches, and emberizids. Relatively few Palearctic migrants overwinter in the region—mostly waders and a scattering of other birds—although many of the Australasian species have complex migratory patterns within the region.

Some of the more spectacular or peculiar birds of the region include the ratite **Emu (Fig. 1–85)** and **cassowaries** (see Fig. 3–44, and Fig. 17 in Evolution of Birds and Avian Flight); the chicken-like **megapodes**, in which the males care for eggs from several females in huge, warm mounds of decaying vegetation (see Fig. 6–36); the strange, flightless **Kagu (Fig. 1–86)**—a single species, presumably related to the cranes and rails, which lives on the verge of extinction in the forest underbrush of New Caledonia, tapping the ground to locate its earthworm prey; the diverse, forest-dwelling **birds-of-paradise**—unsurpassed for male

Figure 1–85. Emu: At roughly six feet (two meters) tall, the flightless Emu is the second largest living bird, next to the Ostrich. The only member of its family, it has powerful legs with three toes; coarse, loose, drooping feathers; and bare blue skin on the face and upper neck. It is common in open areas throughout Australia, occupying habitats ranging from near desert conditions to scrub, woodlands, and alpine pastures. It eats a wide variety of foods, including insects such as grasshoppers and caterpillars, fruits, seeds, grains, flowers, and grasses. It has adapted well to living in agricultural areas, where it may come in conflict with humans by damaging fences and crops. The male Emu, alone, incubates the eggs and raises the young (for four to six months), while the female either remains nearby or leaves to mate with another male. Drawing by Robert Gillmor.

Figure 1–86. Kagu: The few remaining Kagu live on the verge of extinction in the dense, forest undergrowth on New Caledonia, a large island northeast of Australia. The only member of its family, the Kagu can run quickly on its strong legs and feet, and although flightless, can glide down slopes. It taps the ground to locate its primary food, earthworms, and digs up any discovered prey with its strong beak. Drawing by Robert Gillmor.

Figure 1–87. King Bird-of-paradise: *The birds-of-paradise, found mostly in New Guinea, are famous for their spectacular, colorful plumages and displays. In the past, they were hunted intensively to provide feathers for European hats. They inhabit wet forests at a variety of altitudes, and the numerous species differ significantly in diet, size, and mating system. Males of the King Bird-of-paradise display solitarily from trees. Photo by W. Peckover/VIREO.*

ornamentation (**Fig. 1–87**; also see Fig. 3–10); and the **bowerbirds**—builders of remarkably complex bowers, which they decorate to attract females (see Fig. 6–41). Rivaling the plumage and constructions of the previous two groups are the songs and calls of the two species of **lyrebirds**. Named for their elaborate, harp-shaped tails (**Fig. 1–88**), these large passerines are fantastic mimics, including in their songs mechanical sounds as well as the songs of other birds. They are reported to give convincing imitations of a passing flock of cockatoos, complete with wing noise and doppler effect (the drop in pitch when calls arrive from a receding bird), and also may mimic logging trucks and chain saws.

Some bird families are represented by numerous species in the region. The 181 species of **honeyeaters**, for example, have diversified into nearly every habitat (**Fig. 1–89**). These arboreal, relatively dull-colored birds have medium-length, curved bills, and busily feed on nectar, insects, and fruits, often congregating at flowering trees. They are important pollinators, and have a distinctive brush-tipped tongue that they move in and out up to 10 times per second, lapping up nectar.

Figure 1–88. Superb Lyrebird: *Among the largest of songbirds, the two pheasant-sized lyrebird species live in the rain forests of Australia. Although they can run quickly, their flight is limited to short distances and gliding. They roost high in trees that they ascend by jumping from branch to branch. Lyrebirds forage for insects and other arthropods by digging and scratching in soil and rotting logs with their strong feet and claws. Their voices, often used to mimic various sounds, may be the most powerful of any songbird. Male lyrebirds were killed in great numbers in the 19th century for their stunning tail plumes. The family is named for the two outer tail feathers of the Superb Lyrebird, which, when raised and spread as shown here, resemble the frame of a lyre. Male Superb Lyrebirds display to females from mounds on the ground by arching the tail forward and turning slowing, all the while vocalizing dramatically. Drawing by Robert Gillmor.*

Superb Lyrebird

Figure 1–89. Red-collared Myzomela—a Honeyeater: *The 181 species of honeyeaters form the diverse Australasian family Meliphagidae. These active, gregarious nectar feeders vary widely in color, size (from as small as a hummingbird to nearly as large as a magpie), shape, and lifestyle. In an adaptive radiation nearly as impressive as that of marsupial mammals in the same region, they have spread to every habitat that contains flowering trees and shrubs, evolving into nectar-feeding species whose form and behavior resemble birds as diverse as flycatchers, titmice, nuthatches, woodpeckers, hummingbirds, and jays. Despite their variety, all honeyeaters have a specialized tongue that differs from the tongues of other nectar feeders. The tip of this "brush-tongue" is divided lengthwise into four sections, each with frayed edges, that together form a brush-like structure with which the bird rapidly licks up nectar. They also have a specialized digestive system, which allows the easily-digestible nectar to bypass the stomach and continue straight to the intestines, where it can be absorbed quickly. Although honeyeaters also may eat insects and fruit, their primary food is nectar. They are extremely important pollinators of flowers, and many species are nomadic, congregating in large numbers around flowering trees. The male Red-collared Myzomela shown here, from mountainous regions of New Guinea, is black with a crimson collar. Photo by W. Peckover/VIREO.*

Members of another large family, the **whistlers**, are known for their explosive, often-beautiful songs. These large, round-headed birds with stout, slightly hooked bills have evolved a variety of foraging behaviors—searching trunks, branches, leaves, or the ground for insects. The numerous **Australasian robins** are actually more like flycatchers—although they usually snatch food from the ground—but the orange, pink, or yellow breasts and dull-colored backs of many of them are reminiscent of both New and Old World robins. Other species-rich families include the **fairywrens**—cooperatively breeding wrenlike birds with long, cocked tails (see Fig. 6–43)—and **waxbills**, estrildid finches including the well-known cage bird, the Zebra Finch.

Other characteristic Australasian birds include the following:

pseudo-babblers: noisy, busy, social ground-feeders with a long, curved beak and long, towhee-like tail

cracticids: a distinctive family with stout, straight beaks; loud, melodic calls; and a generalist, crow-like diet of small vertebrates, eggs, insects, and fruits. It includes the voracious, shrike-like butcherbirds named for their habit of wedging prey into a tree fork and dismembering it; the large, crow-like currawongs; and the Australasian Magpie—a crow-sized, black-and-white bird that is common, widespread, and familiar because it has adapted well to open areas with trees, such as orchards, golf courses, gardens, and other suburban areas.

mudnest builders: two striking, black-and-white, robin-sized birds named for their large, cup-shaped mud nests. The Magpie-lark is widespread, abundant, and well-known throughout open areas in much of Australia **(Fig. 1–90)**; and the Torrent-lark inhabits fast-flowing streams in the mountainous areas of New Guinea.

corcoracids: the large, blackbird-like White-winged Chough and smaller, seed-eating Apostlebird. Both are cooperative breeders that range over agricultural fields in huge flocks when not breeding.

Figure 1–90. Magpie-larks: The Magpie-lark of Australia, together with the Torrent-lark of New Guinea, form the family Grallinidae, commonly known as "mudnest builders" or "mudlarks." Magpie-larks are common in a wide range of open areas throughout much of Australia, including farmlands and city suburbs. Their striking black-and-white markings, loud calls and songs (including duetting between a mated male and female), and conspicuous behavior render them one of the best-known birds in Australia. Magpie-larks forage on the ground for insects, spiders, snails, frogs, and seeds. Pairs mate for life, both members actively and aggressively defending the territory year round.

They are particularly common near water sources where they can obtain mud to build their large, bowl-shaped nests, which are placed far out on a horizontal branch, often over water. In dry years, when mud is in short supply, breeding may be curtailed, but in wet years, two or occasionally three broods may be raised.

The black-and-white head patterns of Magpie-larks differ by sex and age. Note how the head pattern of the male (bottom) differs from that of the female (center). The juvenile (top) has a pattern intermediate between those of its parents. Drawing by N. W. Cusa, from A Dictionary of Birds, edited by Bruce Campbell and Elizabeth Lack, p. 335. Copyright 1985, The British Ornithologists' Union. Reproduced with the kind permission of the British Ornithologists' Union.

Figure 1–91. Rifleman: The Rifleman is one of four known species of the family of tiny, suboscine New Zealand wrens. One of these species, the Stephen Island Wren, became extinct at the end of the 19th century. New Zealand wrens build domed nests inside tree cavities or rock crevices, and augment their insectivorous diet with some fruit. Just three inches (eight centimeters) long, the Rifleman is three-quarters of the size of a kinglet. The female, illustrated here, is slightly larger than the male, and differs from him in appearance. The Rifleman is most abundant in mountainous forests, where, creeper-like, it extracts insects from bark crevices and from mosses and lichens growing on trees and shrubs.

cuckoo-shrikes: a diverse group of arboreal songbirds that are slender like cuckoos, with a shrike-like bill, but related to neither

woodswallows: small, chunky birds with a graceful, swallow-like flight, known for huddling together on branches or in tree cavities in groups of up to 50 or more birds

New Zealand separated from Australia so long ago (in the late Cretaceous) that its fauna is quite distinctive. Indeed, some researchers believe that New Zealand should be considered a separate zoogeographic region. Currently, three bird families are endemic there: the flightless, and almost wingless, ratite **kiwis**—grouse-sized, nocturnal birds that probe the soil with their long beaks, using their keen sense of smell to locate earthworms (see Figs. 4–54c and 5–48); the tiny, nearly tailless **New Zealand wrens (Fig. 1–91)**; and the odd, forest-dwelling **wattlebirds**, aptly named for the fleshy wattles at the corners of their mouths **(Fig. 1–92)**. Not so long ago at least 22 species of **moas** (ratites in the family Dinornithidae) roamed the open foothills and tussock lands of interior New Zealand. The smallest was turkey-sized, the largest, *Dinornis*

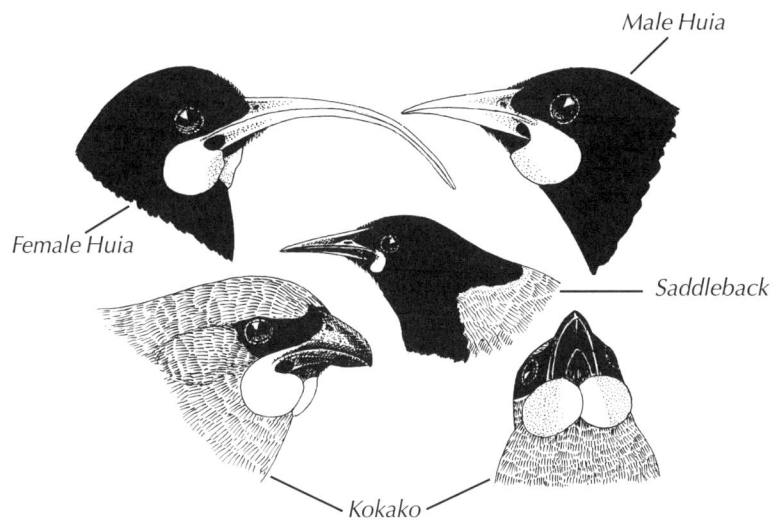

Male Huia

Female Huia

Saddleback

Kokako

Figure 1–92. Wattlebirds: *Three species of wattlebirds, one of which has recently become extinct, form the family Callaeidae, endemic to New Zealand. Named for the pair of colorful, fleshy wattles, which develop from a fold of skin at the base of their bill, wattlebirds are dark-colored forest birds somewhat resembling slender grackles. The mostly black Huia, eliminated by 1910 as a result of overhunting, feather collecting, and habitat destruction, had orange or yellow wattles. The bills of male and female Huias differed greatly: the female had a long, thin downcurved bill with which she probed insect tunnels for larvae and adults, whereas the male had a straighter, shorter bill, used to chisel out insect prey (also see Fig. 9–28). The stocky, black Saddleback has a chestnut brown back and red wattles, and feeds on fruits and insects—the latter, by probing and tearing apart rotten wood. The large, gray Kokako (shown in both side and ventral views) has blue or yellow wattles, and eats mostly young leaves and fruit. Like other species that were once widespread in New Zealand, both the Kokako and the Saddleback are severely declining as a result of predation by introduced mammals such as rats. Conservationists have been successful in establishing viable populations of both species on small, nearby islands that contain no predatory mammals. Drawing by Sir Charles Alexander Fleming, from* A Dictionary of Birds, *edited by Bruce Campbell and Elizabeth Lack, p. 645. Copyright 1985, The British Ornithologists' Union. Reproduced with the kind permission of the British Ornithologists' Union.*

maximus, stood 14 feet tall. Being flightless, the moas were easy marks for predatory humans, and all have been extinct for more than 200 years (see Fig. 20 in Evolution of Birds and Avian Flight).

Island Distribution

Islands, as long as they are adequate in size and have suitable habitats and food, almost invariably support breeding populations of land and freshwater birds from the adjacent continents. The more remote the islands, the fewer kinds of birds they support, because the chance of stragglers, or "pioneers," reaching them in sufficient numbers to found colonies decreases with increased distance from the mainland. In general, the species that colonize the more remote islands are not necessarily the best flyers, but the ones that can best adapt to restricted cover and food resources. Some of the weakest flyers—for example, rails and gallinules—have been the most successful colonizers, whereas strong flyers such as swallows have rarely colonized distant islands.

The kinds of birds on islands close to a continent are little different from those on the mainland, because they have not been sufficiently isolated to develop distinctions as separate species. They clearly belong to the avifauna of the continent and its zoogeographic region. However, birds on remote islands and archipelagos have often, through long isolation, developed distinctions so peculiar that they not only have the status of endemic species, but their origins are blurred as well. A classic example is the avifauna of the Hawaiian Islands, which includes 52 endemic species: one goose, two ducks, one hawk, two rails, one coot, five honeyeaters, one crow, one Old World flycatcher, one Old World warbler, five thrushes, and the endemic subfamily of Hawaiian honeycreepers (Drepanidinae) containing 32 species. Of

Figure 1–93. Weka: *Dark, tame, flightless rails a bit larger than a crow, Wekas inhabit the scrublands and forest edges of New Zealand, foraging on a variety of foods as well as scavenging among kelp on beaches and from garbage bins. Wekas have strong beaks and feet and can run very fast, but typically walk slowly, flicking their tails. Photo by D. Hadden/VIREO.*

these birds, at least four honeyeaters, one thrush, and nine honeycreepers recently have become extinct. Some Hawaiian bird families appear to have originated in the Nearctic, and some in the Australasian region, but the origin of others remains unknown. Hawaiian birds are generally placed in the Australasian zoogeographic region based on their location, but some researchers place them in other regions, or in a separate region with other islands of the middle and southern Pacific Ocean.

Bird species native to islands normally have small populations. Sedentary and without predators, many have become flightless and tame **(Fig. 1–93)**. They are consequently vulnerable to extinction from human-imposed causes: direct killing, destruction of habitat, and the introduction of dogs, cats, rats, goats, and other human symbionts. Until humans began eliminating them, dozens (perhaps even hundreds) of species of flightless rails existed on different islands around the world. Many of these are known only from fossil or skeletal specimens, and disappeared as soon as humans colonized their islands. Nine flightless rails have gone extinct over the past 150 years, and at least six more are in danger of extinction—all as a result of humans!

Distribution of Marine Birds

Marine birds, or seabirds, are directly associated with the ocean, consistently depending upon it for food. They are generally divided into pelagic and coastal species. **Pelagic** birds roam the open ocean, feeding primarily on small animals such as fish, squid, crustacea, and carrion at the surface or just below it; they come to land only to nest. Many pelagic birds are in the large order Procellariiformes (tubenosed seabirds), which includes albatrosses; shearwaters, fulmars, and typical petrels; storm-petrels; and diving-petrels. The remaining pelagic groups are the tropicbirds, some penguins, the boobies and gannets, most of the alcids, the skuas and jaegers, the noddies, the kittiwakes, Sabine's Gull, and some terns—notably the Sooty Tern. **Coastal** species primarily occupy the shallower waters around oceanic islands or above the continental shelf, feeding mainly on fish, crustacea, and mollusks, which they find on or near beaches and other shorelines. They frequent land—usually coastal areas—in the nonbreeding season as well as for breeding. These include most penguins, cormorants, pelicans, frigatebirds, gulls, terns, and skimmers. Shorebirds, waders, and waterfowl, many of which depend more heavily on land for feeding and rarely venture far from coastal areas, are not considered marine birds. The exceptions are eiders, scoters, and some phalaropes, which are usually grouped with the coastal seabirds.

Many marine birds, especially the pelagic species, are quite long-lived. The Northern Fulmar, for example, once it has survived to breeding age, lives an average of 44 years. Marine birds also tend to nest colonially, often on islands; lay only one or sometimes two, eggs;

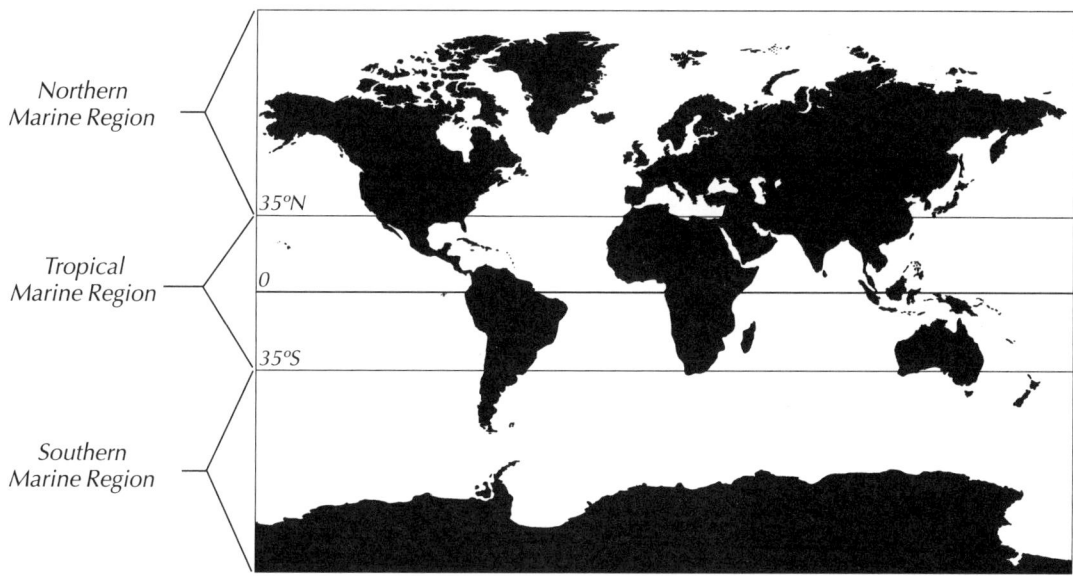

Northern
Marine Region

35°N

0

35°S

Tropical
Marine Region

Southern
Marine Region

and have long incubation, fledging, and "adolescent" periods, many species taking 5 to 10 years to reach reproductive age.

Like the land areas, the seas have been divided by biologists into major faunal regions, as follows **(Fig. 1–94)**:

(1) **Northern Marine Region:** The frigid waters of the Arctic south to about 35 degrees north latitude. The water temperature rises rapidly between 35 and 40 degrees north latitude, an area termed the Subtropical Convergence of the Northern Hemisphere. This change in water temperature has important ecological consequences that affect bird distribution, as discussed below.

(2) **Southern Marine Region:** The frigid waters around Antarctica north to about 35 degrees south latitude. In the Subtropical Convergence of the Southern Hemisphere, between 35 and 40 degrees south latitude, the water temperature rises rapidly, as in the north.

(3) **Tropical Marine Region:** The warm equatorial waters between the Subtropical Convergences of both hemispheres.

Northern Marine Region

This region is characterized by the family Alcidae, which includes 24 species of auks, auklets, murres, murrelets, guillemots, and puffins, most of which live in the colder waters of the higher latitudes **(Fig. 1–95)**. The medium-sized, black-and-white alcids are excellent swimmers, pursuing their fish prey by flapping their wings under water, and are well-known for their upright, penguin-like stance. The family is almost completely restricted to the Northern Marine region, but several species venture south along the coast of southern California—the Craveri's Murrelet all the way to Mexico, breeding on islands in the Gulf of California. Because many more alcid species are found in the North Pacific than in the North Atlantic, the family may have originated there. Other birds of the Northern Marine region include numerous gulls and terns, both kittiwakes, three species of albatross, and an

*Figure 1–94. Marine Faunal Regions: Like land areas, the seas have been divided by biologists into major faunal regions. The **Northern Marine region** extends from the cold Arctic waters south to the Subtropical Convergence of the Northern Hemisphere at about 35 to 40 degrees north latitude, where the water temperature rises rapidly. The **Southern Marine region** extends from the cold Antarctic waters north to the Subtropical Convergence of the Southern Hemisphere, at about 35 to 40 degrees south latitude. The **Tropical Marine region** includes the warm equatorial seas between the two Subtropical Convergences. Adapted from a drawing by Charles L. Ripper.*

Figure 1–95. Common Murres: *Like other members of the family Alcidae, the crow-sized, black-and-white Common Murres are designed for swimming. Their streamlined bodies and feet set back on the body give them an upright, penguin-like stance. These adaptations allow them to pursue fish expertly under water. As shown here, Common Murres nest in dense colonies on ledges of rocky cliffs throughout much of the northern Holarctic. Drawing by Robert Gillmor, from Lack (1968).*

assortment of typical petrels, shearwaters, cormorants, and gannets.

Southern Marine Region

This region is by far the richest of the three regions, not only in the array of species, but also in the numbers of individuals. Even the most disinterested sailors will notice the steady increase in birds as their ship heads southward into the higher southern latitudes. The birds that they see are mainly albatrosses, shearwaters, and petrels. The high density of plankton, the base of the ocean food chain, accounts for the presence of so many of these birds, as described below. The almost constant winds of the region are favorable to the albatross's style of long-distance, soaring flight (see Fig. 5–42), and 10 of the 14 albatross species are found here (see Fig. 5–1). Other bird families characteristic of the region include penguins (only 3 of the 17 species are found outside the region) **(Fig. 1–96)**; the auk-like diving-petrels (3 of 4 species are found here); and the two sheathbills (Chionididae), aberrant, all-white shorebirds **(Fig. 1–97)**—the southernmost birds in the world without webbed feet! The region also hosts numerous species of shearwaters and typical petrels, as well as gannets, terns, gulls, and cormorants. The large number of shearwater and albatross species suggests that these groups originated here.

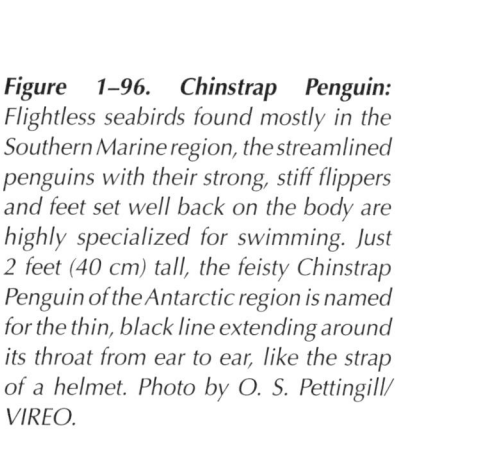

Figure 1–96. Chinstrap Penguin: *Flightless seabirds found mostly in the Southern Marine region, the streamlined penguins with their strong, stiff flippers and feet set well back on the body are highly specialized for swimming. Just 2 feet (40 cm) tall, the feisty Chinstrap Penguin of the Antarctic region is named for the thin, black line extending around its throat from ear to ear, like the strap of a helmet. Photo by O. S. Pettingill/ VIREO.*

Tropical Marine Region

The warm, nutrient-poor, tropical waters of this region are low in plankton, and thus cannot support a rich array of the marine animals, such as fish, squid, and crustacea, which ultimately depend on plankton. Most fish are found in schools close to land, and consequently most distinctive birds of this region—tropicbirds, boobies, frigatebirds, and several species of terns—keep to the inshore waters. Only three albatross species, several typical petrels and shearwaters, and a number of storm-petrels **(Fig. 1–98)** find the plankton supply sufficient for them to occupy the open tropical seas. The tropicbirds, boobies, and frigatebirds **(Fig. 1–99)** are found only in this region. Most Tropical Marine birds feed by hovering and plunging quickly into the water after prey, or skimming prey from the water's surface while hovering. The region hosts notably few gulls.

Plankton and Bird Distribution

The abundance of marine birds in different oceanic regions is directly related to the supply of plankton. As the base of the ocean food chain, plankton are consumed by numerous small marine animals, which are, in turn, prey for larger animals, in a chain of consumers that often ends in seabirds. Where plankton are plentiful, so, too, are fish and seabirds. **Plankton** consists of microscopic plants and animals termed phytoplankton and zooplankton, respectively. Just like terrestrial plants, phytoplankton use the sun's energy to convert simple inorganic molecules into carbohydrates; and, like terrestrial plants, they require a supply of nutrients to survive. Although sunlight is plentiful in the open ocean, nutrients are usually very limited: when organisms die they tend to sink to the ocean bottom, decomposing into their component nutrients very slowly. There they remain for centuries, unless something happens to bring them to the water's surface. Coastal areas, including the relatively shallow waters over the continental

Figure 1–97. Snowy Sheathbill: *Named for the horny, yellow-green sheath covering the base of the stout bill and nostrils, the two species of sheathbills are white, pigeon-like shorebirds with bare faces and fleshy wattles around the eyes and base of the beak. Sheathbills live on or near shores in the Antarctic region of the Atlantic and Indian Oceans, often in association with penguin or seal colonies. They scavenge any type of food they can find and are quite tame, frequenting human campsites in the region. Reluctant to fly, they escape predators by running quickly. The Snowy Sheathbill is found in Southern Argentina and Chile, the Falkland Islands, and the Antarctic Peninsula—a long, thin finger of Antarctica extending toward South America. Photo by Tom Vezo/VIREO.*

Figure 1–98. Wilson's Storm-Petrel: *Found throughout most of the world's oceans, the starling-sized, pelagic storm-petrels are among the few bird species that find the low number of plankton and other tiny organisms of the open oceans of the Tropical Marine region sufficient to live on. They fly low over the water's surface, treading their feet in the water and fluttering their wings somewhat like butterflies, as they search for prey. Storm-petrels are tubenosed seabirds of the order Procellariiformes, and breed colonially on islands, digging their own nest burrows. The tiny Wilson's Storm-Petrel, thought to be one of the most abundant seabirds on the globe, breeds in the Antarctic and ranges the northern oceans, except the northern Pacific, in winter. Drawing by Robert Gillmor, from Lack (1968).*

Figure 1–99. Birds of the Tropical Marine Region: *Tropicbirds, boobies, and frigatebirds, found only in the Tropical Marine region, are all pelican relatives in the diverse order Pelecaniformes, whose members are distinguished by having all four toes joined by webbing.* **a. Red-billed Tropicbird:** *The tropicbirds have two long, central tail feathers, which render them unmistakable in flight. With short legs set far back on the body, tropicbirds cannot support the weight of their body on land, and usually nest colonially on steep cliffs—from which they can take off without walking. The flicker-sized (excluding the long tail) Red-billed Tropicbird, found in most tropical seas around the world, plunge-dives into the water from air in search of small fish or squid. Photo by T. J. Ulrich/VIREO.*
b. Masked Booby: *Large birds with long, narrow wings and torpedo-shaped bodies, the boobies catch fish and squid by either plunge-diving or diving from the water's surface. Boobies breed in huge colonies on islands, and incubate their eggs by placing them under the webbing of their feet. The pantropical Masked Booby eats mainly flying fish.*
c. Magnificent Frigatebird: *The dark, eagle-sized frigatebirds, with their long, narrow wings and deeply forked tails, can both soar and maneuver skillfully in flight. They rarely settle on water, as their plumage wets quickly. They frequently harass other seabirds, forcing them to drop or disgorge recently eaten fish. The frigatebirds then catch the dropped item in midair. They also follow schools of tuna or dolphin to grab the flying fish that they scare into the air, swoop over nesting colonies and snatch untended chicks (even of their own species), and scavenge items such as jellyfish, eggs, and baby turtles from on or near the water. Frigatebirds nest colonially on islands, building flimsy stick nests in dense vegetation such as mangroves, where they can perch, as their small, sharply-clawed feet are ineffective for walking or swimming. In courtship, males inflate their bright red throat pouch, vibrate their wings, and rattle their huge, hooked bills. Booby and frigatebird photos by Marie Read.*

a

b

c

shelf, tend to be rich in nutrients, as some are carried from land to the ocean by rivers, and others are brought to the surface by the action of tides, waves, winds, and coastal currents **(Fig. 1–100)**.

In the Tropical Marine region, the sun heats the upper layers of water so much that a significant temperature stratification develops, with cold, dense water below the warmer, lighter surface layers. This layering, much like a temperature inversion in the atmosphere, is an effective barrier to mixing between the layers: the cold, dense water stays below, as do the associated nutrients. The dramatic layering explains why the open oceans of tropical latitudes are so low in plankton and other marine life.

In the Northern and Southern Marine regions, layering is less marked, and thus nutrients, plankton, and other marine life are more plentiful. The growth of phytoplankton is seasonal, however, with a burst of growth in spring—a time when sunlight is more available and just after winter storms have disturbed the water sufficiently to bring up nutrients from deep in the seas. A number of bird species in these regions are migratory, and some shearwaters and storm-petrels move back and forth across the equator to take advantage of spring and summer in both hemispheres. (Birds with this habit are termed **transequatorial migrants**.)

In certain areas where ocean currents meet or where deep currents are forced upward by the topography of the sea floor, large-scale upwelling of deep water occurs, bringing abundant nutrients to the surface. One example is the Antarctic Convergence at roughly 60 de-

*Figure 1–100. Coastal Upwelling: When marine organisms die, they usually sink and decay very slowly in the cold environment of the ocean floor. There, vast amounts of nutrients, released from the decaying organisms, build up over time. In coastal areas, winds blowing from land to sea (a common nighttime occurrence) move the surface water away from shore, causing cold water from the ocean bottom to replace it by moving upward and toward shore, carrying dissolved nutrients. This process—cold, nutrient-rich water from the ocean bottom rising to the surface—is called **upwelling**. The nutrient-rich coastal waters support numerous **phytoplankton** (microscopic plants), which are eaten by **zooplankton** (tiny animals), which in turn support an abundance of fish and other marine life—all components of the highly productive coastal ecosystem. Upwelling also occurs in certain areas where ocean currents meet. From Owens/Chiras/Regangold,* Natural Resource Conservation *7/e, copyright 1998. Reprinted by permission of Prentice-Hall, Upper Saddle River, NJ.*

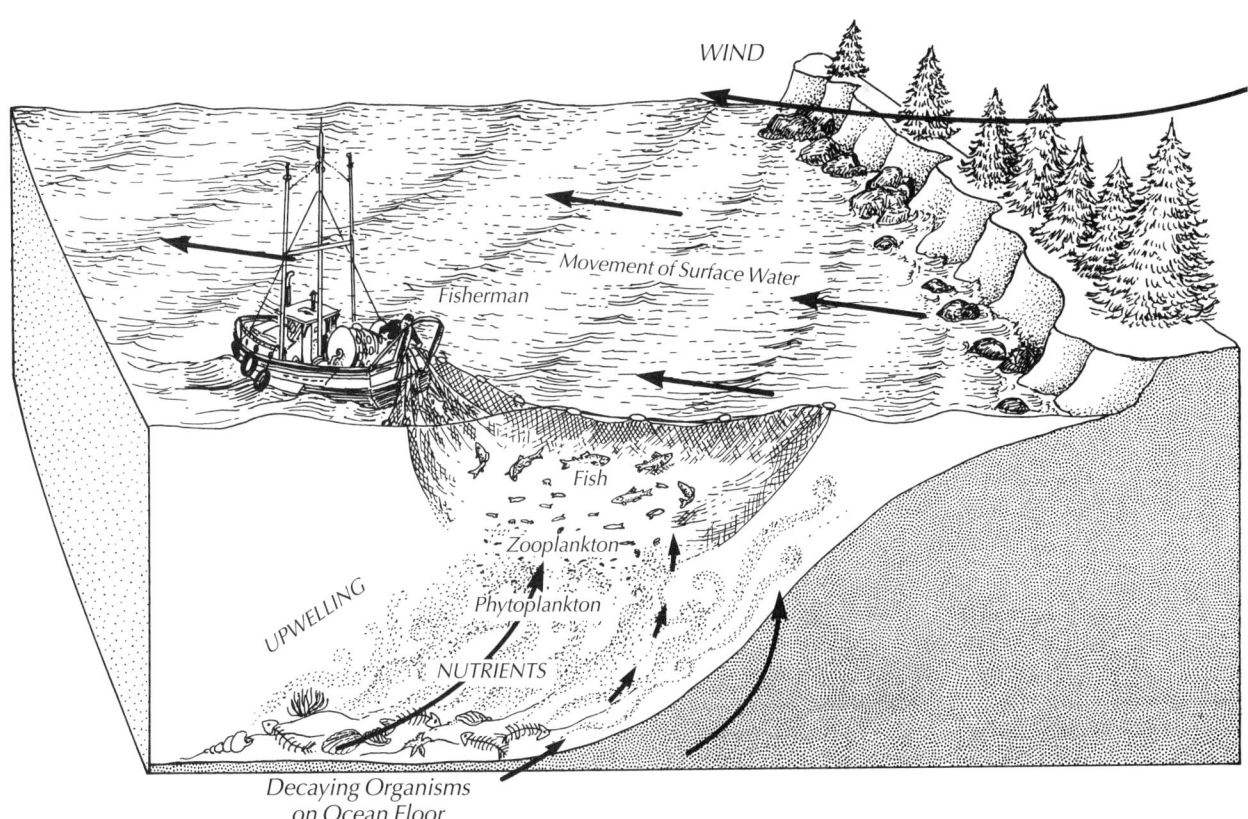

WIND

Movement of Surface Water

Fisherman

Fish

Zooplankton

Phytoplankton

UPWELLING

NUTRIENTS

Decaying Organisms
on Ocean Floor

Figure 1–101. Plankton and Seabird Abundance in the South Atlantic: *The graph on the left shows the average plankton abundance in the upper 55 yards (50 m) of water, in thousands of organisms per liter, at different latitudes in the Atlantic Ocean. On the right, the relative abundance of certain species of pelagic seabirds is illustrated for three different latitudes. Between about 50 and 60 degrees south latitude, in a region known as the* **Antarctic Convergence,** *cold north-flowing and warmer south-flowing currents meet, resulting in large-scale upwelling of nutrient-rich water from the ocean's depths. The plethora of nutrients supports tremendous numbers of plankton, which support higher levels of the food chain, including a great diversity and abundance of seabirds. The nutrient-poor water near the equator supports few plankton, and thus few seabirds, but the numerous, small storm-petrels find the plankton supply adequate. Plankton data originally from Hentschel (1933) and seabird data originally from Spiess (1928). Figure from* The Life of Birds, 4th Edition, *by Joel Carl Welty and Luis Baptista, copyright 1988 by Saunders College Publishing, reproduced by permission of the publisher.*

grees south latitude, an area extremely rich in seabirds **(Fig. 1–101)**. Other notable upwellings occur along the western coasts of several continents, where currents flowing toward the equator from north and south tend to bend away from shore because of the earth's rotation (a phenomenon called the Coriolis effect), drawing deep, nutrient-rich water upward near the coastline **(Fig. 1–102)**. The south-flowing California and north-flowing Humboldt (or Peru) currents, along the west coasts of North and South America, respectively, create such upwellings; along the west coast of Africa, the Canary and Benguela currents act similarly. Although many tropical, pelagic seabirds are worldwide in distribution, the areas of upwelling in tropical seas create isolated patches of abundant food. Many bird species that specialize on these resources have very limited ranges—for example, the Peruvian Booby and Humboldt Penguin of the Peruvian coast; and the Jackass Penguin and several cormorant species of western Africa. Upwelling also occurs in the Arabian Sea from May to September, owing to the action of monsoon winds on the water, which brings a seasonal abundance of plankton and marine birds to an area that hosts few birds during the rest of the year.

Undoubtedly the most famous upwelling site is the southwest coast of Peru. Here the rich water supports a tremendous population of anchovies, a virtual banquet for millions of seabirds, as well as the basis for an extensive fishing industry (the 1969 to 1970 catch was over 10 trillion fish). Many of the seabirds nest on offshore islands, where, because of the arid conditions, their droppings (known as **guano**) accumulate in massive quantities **(Fig. 1–103)**. Droppings from these

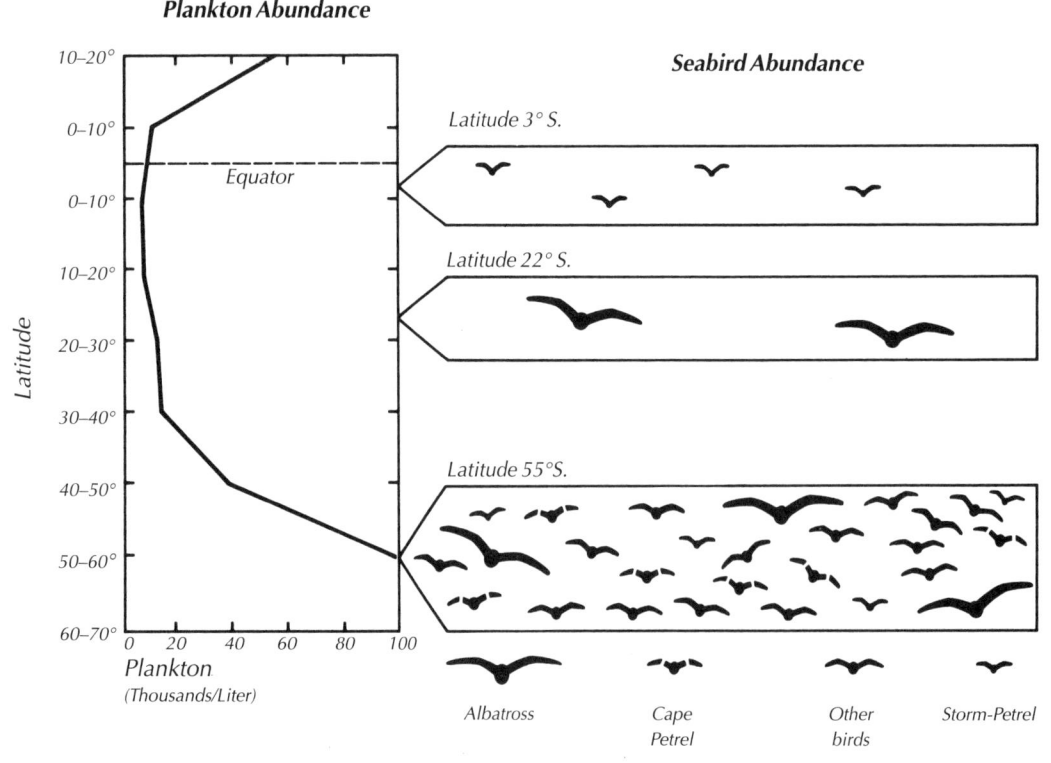

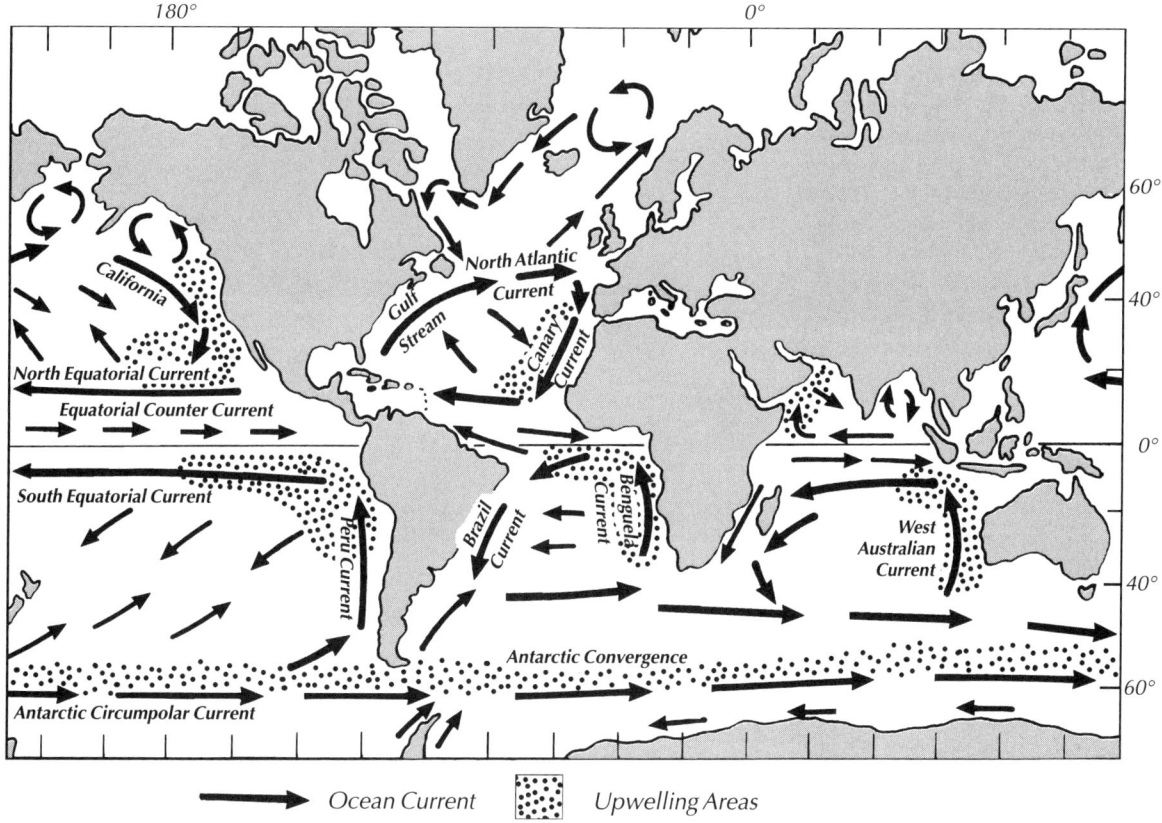

Ocean Current   Upwelling Areas

"guano birds"—especially the Guanay Cormorant, Peruvian Booby, and Peruvian Pelican—historically reached depths of more than 300 feet (90 meters), but have been mined extensively for use as fertilizer since at least the days of the Incas. In 1972, a combination of particularly strong effects of El Niño (triggered by atypical atmospheric conditions, this warm, south-flowing surface current of Ecuador extends southward to flow along the coast of Peru every two to ten years, warming the upper water layers and preventing upwelling) and overfishing caused anchovy and thus seabird populations to plummet to around one million birds, down from a high of 27 million in the late 1950s. Although fish and seabird populations have increased since their decline, they have not returned to their former levels.

Now that you have a feel for the great diversity of birds in the world, as well as the remarkable similarities from one region to another, you may find that paying attention to birds increases your enjoyment of travel—whether around the world or across the country. If you live in Vermont and travel to Utah, for example, most birds you encounter will be quite different. The common, large, blue bird at the feeder will no longer be the Blue Jay, but the Steller's Jay. The juncos, instead of being all gray on top, will have black caps and rusty brown sides, being the Oregon rather than the Slate-colored form. Open any field guide that covers all of North America and examine the range maps. You will find that many species are restricted to the western part of the continent, some are restricted to the eastern, and others can be found all over.

Figure 1–102. Major Ocean Currents and Areas of Upwelling: The water in the world's oceans circulates continually, the patterns resulting from differential heating and cooling of water masses in different parts of the globe, combined with the rotation of the earth and the action of persistent winds. In certain areas, the patterns of currents cause large-scale upwelling of deep water. Upwellings bring abundant nutrients to the surface, and support large, but localized, seabird populations. In addition to massive upwelling just north of Antarctica, between 50 and 60 degrees south latitude (see Fig. 1–101), upwellings occur along the western coasts of Africa, the Americas, and Australia. There, currents flowing toward the equator from north and south tend to bend away from shore due to the Coriolis effect, drawing deep, nutrient-rich water upward along the coasts. The California and Peru (Humboldt) currents off the Americas, the Canary and Benguela currents off Africa, and the West Australian current off Australia create such upwellings. Upwelling also occurs in the Arabian Sea from May to September, due to the action of monsoon winds on the water. Adapted from Grahame (1987).

Figure 1–103. Peruvian Booby Nest Colony on Ballestas Islands, Peru: *Along the southwest coast of Peru, the north-flowing Peru (Humboldt) current bends away from shore as it approaches the equator, pulling nutrient-rich water to the surface in a massive upwelling. This upwelling supports a huge population of anchovies—a virtual banquet for millions of seabirds. The seabirds, including the Peruvian Boobies shown here, form huge nesting colonies on islands in the region, such as the Ballestas. Their droppings accumulate to great depths in the arid environment, and are mined by Peruvians for use as fertilizer. Photo by Sandy Podulka.*

The Importance of Biodiversity

■ Much of this chapter has described the great diversity of birds—diversity in bills, wings, tails, feet, and feathers—as well as the diversity of taxonomic groups that use an array of behaviors to exploit nearly every habitat on earth. Birds are just one component of **biodiversity**—the great wealth of living organisms that occur on earth. As scientists continue to study natural systems, they are becoming increasingly aware of the importance of each component, and of the complex ways these components are related to one another. The study and preservation of biodiversity are major focuses of contemporary scientists around the world.

The earth's biodiversity, including birds, provides direct benefits to humans in such forms as food, clothing, recreation, and aesthetic experiences; but biodiversity is more than that. Maintaining biodiversity is crucial to sustaining the healthy, functioning ecosystems on which all life depends. These ecosystems maintain a balance among organisms, purify and cycle water, recycle nutrients, and ensure adequate reproduction of living things.

As you continue through this course, try to interpret each topic and example in the context of biodiversity. For example, ask yourself, "In terms of conserving biodiversity, why is it important to identify and count birds? Why study the shapes of hummingbird bills? Why determine the exact migration routes of birds?"

Birds are beautiful, fascinating, and intriguing. Having chosen to pursue this course, you are undoubtedly already drawn to birds. But birds in all their diverse forms are also vital components of the world around us, and as you read the following chapters, your respect and appreciation for them will likely grow.

Appendix A:
ORDERS AND FAMILIES OF WORLD BIRDS

Below are listed the 31 orders and all families of living birds. The list was compiled using Clements (2000), Sibley and Monroe (1990), and Sibley and Ahlquist (1990), but for North American species follows the 7th edition of the *Check-list of North American Birds* (1998), prepared by the Committee on Classification and Nomenclature of the American Ornithologists' Union.

Ornithological names are pronounced following the rules listed in any standard English dictionary. Note that all order names end in "-iformes," all family names in "-idae," and all subfamily names in "-inae." The names of orders and families are shown in boldface type. Those in color are represented by species occurring in North America north of Mexico, or near its coasts, including Hawaii. The number of living species in each family is given in parentheses; it includes some species that have gone extinct very recently. Several orders and families of birds that recently went extinct are included in square brackets.

With the 7th edition of the *Check-list*, the AOU took the bold step of making several groups *incertae sedis*, that is, having "no certain affinity." Instead of putting a confusing bird somewhere just to give it a home, as has been the general practice in the past, the committee, in effect, admitted that they don't know where these birds fit in. Future studies may shed more light on the real relationships of these "orphaned" taxa.

Tinamiformes
    Tinamidae—Tinamous (46)
Rheiformes
    Rheidae—-Rheas (2)
Struthioniformes
    Struthionidae—Ostrich (1)
Casuariiformes—Cassowaries and Emu
    Dromiceidae—Emu (1)
    Casuariidae—Cassowaries (3)
[**Aepyornithiformes** (extinct)]
    [**Aepyornithidae**—Elephantbirds (extinct)]
Dinornithiformes
    [**Dinornithidae**—Moas (extinct)]
    Apterygidae—Kiwis (3)
Gaviiformes
    Gaviidae—Loons (5)
Podicipediformes
    Podicipedidae—Grebes (22)
Sphenisciformes
    Spheniscidae—Penguins (17)
Procellariiformes—Albatrosses, Shearwaters, and Petrels
    Diomedeidae—Albatrosses (14)
    Procellariidae—Shearwaters, Fulmars, and Typical Petrels (76)
    Hydrobatidae—Storm-Petrels (21)
    Pelecanoididae—Diving-Petrels (4)
Pelecaniformes—Pelicans, Cormorants, and Allies
    Phaethontidae—Tropicbirds (3)
    Sulidae—Boobies and Gannets (9)
    Pelecanidae—Pelicans (8)
    Phalacrocoracidae—Cormorants (38)
    Anhingidae—Anhingas (4)
    Fregatidae—Frigatebirds (5)
Ciconiiformes—Herons, Ibises, Storks, New World Vultures, and Allies
    Ardeidae—Herons, Egrets, and Bitterns (65)
    Balaenicipitidae—Shoebill (1)
    Scopidae—Hamerkop (1)
    Threskiornithidae—Ibises and Spoonbills (33)
    Ciconiidae—Storks (19)
    Cathartidae—New World Vultures (7)
Phoenicopteriformes
    Phoenicopteridae—Flamingos (5)
Anseriformes—Screamers and Waterfowl
    Anhimidae—Screamers (3)
    Anatidae—Ducks, Geese, and Swans (161)
Falconiformes—Vultures and Diurnal Birds of Prey
    Sagittariidae—Secretary-bird (1)
    Accipitridae—Kites, Eagles, Hawks, Old World Vultures, and Allies (237)
    Falconidae—Caracaras and Falcons (64)
Galliformes—Gallinaceous Birds
    Megapodiidae—Megapodes (21)
    Cracidae—Curassows, Guans, and Chachalacas (50)
    Phasianidae—Pheasants, Partridges, Grouse, Turkeys, Old World Quail, and Guineafowl (177)
    Odontophoridae—New World Quail (30)
Gruiformes—Cranes, Rails, and Allies
    Turnicidae—Buttonquail (17)
    Rallidae—Rails, Gallinules, and Coots (144)
    Heliornithidae—Finfoots (3)
    Rhynochetidae—Kagu (1)
    Eurypygidae—Sunbittern (1)
    Mesitornithidae—Mesites (3)
    Aramidae—Limpkin (1)
    Gruidae—Cranes (15)
    Psophiidae—Trumpeters (3)

Cariamidae—Seriemas (2)
Otididae—Bustards (25)
Charadriiformes—Shorebirds, Gulls, and Auks
  Burhinidae—Thick-knees (9)
  Charadriidae—Plovers and Allies (66)
  Haematopodidae—Oystercatchers (11)
  Recurvirostridae—Avocets and Stilts (10)
  Jacanidae—Jacanas (8)
  Rostratulidae—Painted-Snipes (2)
  Scolopacidae—Sandpipers and Allies (88)
  Dromadidae—Crab Plover (1)
  Glareolidae—Coursers and Pratincoles (17)
  Chionididae—Sheathbills (2)
  Pluvianellidae—Magellanic Plover (1)
  Pedionomidae—Plains-wanderer (1)
  Thinocoridae—Seedsnipes (4)
  Ibidorhynchidae—Ibisbill (1)
  Laridae—Gulls, Terns, Skuas, and Skimmers (106)
  Alcidae—Auks, Murres, and Puffins (24)
Family INCERTAE SEDIS = "Family with No Order" [as
  by AOU; or Pteroclidiformes? or Charadriiformes?]
  Pteroclidae—Sandgrouse (16)
Columbiformes—Pigeons and Doves
  [**Raphidae**—Dodo and Solitaires, (3) extinct]
  Columbidae—Pigeons and Doves (314)
Psittaciformes
  Psittacidae—Parrots, Parakeets, Lories, and
    Macaws (359)
Coliiformes
  Coliidae—Mousebirds (6)
Musophagiformes
  Musophagidae—Turacos (23)
Cuculiformes
  Cuculidae—Cuckoos, Roadrunners, and Anis (142)
Opisthocomiformes
  Opisthocomidae—Hoatzin (1)
Strigiformes—Owls
  Tytonidae—Barn Owls (17)
  Strigidae—Typical Owls (170)
Caprimulgiformes—Oilbird and Goatsuckers
  Steatornithidae—Oilbird (1)
  Podargidae—Frogmouths (14)
  Aegothelidae—Owlet-Nightjars (8)
  Nyctibiidae—Potoos (7)
  Caprimulgidae—Goatsuckers (82)
Apodiformes—Swifts and Hummingbirds
  Hemiprocnidae—Treeswifts (4)
  Apodidae—Swifts (101)
  Trochilidae—Hummingbirds (329)
Trogoniformes
  Trogonidae—Trogons (39)
Upupiformes—Hoopoes and Woodhoopoes
  Upupidae—Hoopoes (2)
  Phoeniculidae—Woodhoopoes (8)
Coraciiformes—Kingfishers and Allies

Todidae—Todies (5)
Momotidae—Motmots (9)
Alcedinidae—Kingfishers (95)
Meropidae—Bee-eaters (26)
Coraciidae—Rollers (12)
Brachypteraciidae—Ground-Rollers (5)
Leptosomatidae—Cuckoo-Roller (1)
Bucerotidae—Hornbills (55)
Bucorvidae—Ground-Hornbills (2)
Piciformes—Woodpeckers and Allies
  Bucconidae—Puffbirds (33)
  Galbulidae—Jacamars (18)
  Indicatoridae—Honeyguides (17)
  Megalaimidae—Asian Barbets (26)
  Lybiidae—African Barbets (42)
  Ramphastidae—Toucans and New World Barbets (55)
  Picidae—Woodpeckers and Allies (217)
Passeriformes—Perching or Passerine Birds
Suborder Tyranni, Suboscines
  Acanthisittidae—New Zealand Wrens (4)
  Pittidae—Pittas (31)
  Eurylaimidae—Broadbills (15)
  Philepittidae—Asities and False Sunbirds (4)
  Furnariidae—Ovenbirds (240)
  Dendrocolaptidae—Woodcreepers (51)
  Thamnophilidae—Antbirds (197)
  Formicariidae—Antthrushes and Antpittas (60)
  Conopophagidae—Gnateaters (8)
  Rhinocryptidae—Tapaculos (52)
  Tyrannidae—New World Flycatchers or Tyrant
    Flycatchers (398)
  Cotingidae—Cotingas (61)
  Phytotomidae—Plantcutters (3)
  Pipridae—Manakins (44)
  Oxyruncidae—Sharpbill (1)
Suborder Passeri, Oscines
  Climacteridae—Australasian Treecreepers (7)
  Menuridae—Lyrebirds (2)
  Atrichornithidae—Scrub-birds (2)
  Ptilonorhynchidae—Bowerbirds (20)
  Maluridae—Fairywrens (26)
  Meliphagidae—Honeyeaters (181)
  Pardalotidae—Australo-Papuan Warblers (68)
  Eopsaltriidae—Australasian Robins (44)
  Irenidae—Leafbirds and Fairy-bluebirds (10)
  Aegithinidae—Ioras (4)
  Orthonychidae—Logrunners (2)
  Pomatostomidae—Pseudo-Babblers (5)
  Laniidae—Shrikes (30)
  Vireonidae—Vireos (52)
  Cinclosomatidae—Whipbirds and Quail-thrushes (15)
  Corcoracidae—White-winged Chough and
    Apostlebird (2)
  Pachycephalidae—Whistlers and Allies (58)
  Corvidae—Crows, Magpies, and Jays (115)

Paradisaeidae—Birds-of-Paradise (46)
Artamidae—Woodswallows (14)
Cracticidae—Butcherbirds, Australasian Magpie, and Currawongs (10)
Oriolidae—Old World Orioles (29)
Campephagidae—Cuckoo-shrikes and Minivets (82)
Dicruridae—Drongos (24)
Monarchidae—Monarch Flycatchers (139)
Malaconotidae—Bushshrikes and Allies (93)
Vangidae—Vangas (14)
Callaeidae—Wattlebirds (3)
Grallinidae—Mudnest Builders (2)
Picathartidae—Rockfowl (2)
Alaudidae—Larks (91)
Hirundinidae—Swallows (91)
Paridae—Chickadees and Titmice (58)
Remizidae—Verdins and Penduline Tits (14)
Aegithalidae—Bushtit and Long-tailed Tits (8)
Sittidae—Nuthatches and Allies (25)
Certhiidae—Creepers (7)
Troglodytidae—Wrens (78)
Cinclidae—Dippers (5)
Pycnonotidae—Bulbuls (130)
Regulidae—Kinglets (6)
Sylviidae—Old World Warblers and Gnatcatchers (291)
Muscicapidae—Old World Flycatchers (118)
Turdidae—Thrushes (182)
Timaliidae—Babblers (267)
Panuridae—Parrotbills (20)
Rhabdornithidae—Rhabdornis (3)
Cisticolidae—Cisticolas (117)
Zosteropidae—White-eyes (95)

Mimidae—Mockingbirds and Thrashers (35)
Sturnidae—Starlings (114)
Prunellidae—Accentors (13)
Motacillidae—Wagtails and Pipits (63)
Hypocolidae—Hypocolius (1)
Bombycillidae—Waxwings (3)
Ptilogonatidae—Silky-flycatchers (4)
Dulidae—Palmchat (1)
Promeropidae—Sugarbirds (2)
Dicaeidae—Flowerpeckers (43)
Nectariniidae—Sunbirds (124)
Melanocharitidae—Berrypeckers and Longbills (10)
Paramythiidae—Tit Berrypecker and Crested Berrypecker (2)
Peucedramidae—Olive Warbler (1)
Parulidae—New World Warblers or Wood-Warblers (115)
Coerebidae—Bananaquit (1)
Thraupidae—Tanagers (252)
Emberizidae—Seedeaters, New World Sparrows, and Buntings (323)
Cardinalidae—Cardinals and Allies (44)
Icteridae—Blackbirds and New World Orioles (99)
Fringillidae—Finches and Hawaiian Honeycreepers
    Fringillinae, Chaffinches (3)
    Carduelinae, Finches (135)
    Drepanidinae, Hawaiian Honeycreepers (32)
Catamblyrhynchidae—Plush-capped Finch (1)
Passeridae—Old World Sparrows (35)
Ploceidae—Weavers (117)
Estrildidae—Waxbills and Whydahs (159)

Appendix B:
ORDERS AND FAMILIES OF NORTH AMERICAN BIRDS

Below are the living orders and families of birds that occur in North America north of Mexico, or near its coasts, including Hawaii, as listed in the 7th edition of the *Check-list of North American Birds* (1998), prepared by the Committee on Classification and Nomenclature of the American Ornithologists' Union. Included here are some bird groups that are accidental to this area (marked with an A) and some that have been introduced (marked with an I). Bird groups that, within this region, are found only on Hawaii, are marked with an H. For each family (and some subfamilies), the number of species known to occur in this region is given in parentheses; it includes some species that have gone extinct very recently. The names of orders are shown in boldface type. Endings and pronunciation guidelines are the same as for the world list.

Gaviiformes
Gaviidae—Loons (5)
Podicipediformes
Podicipedidae—Grebes (7)
Procellariiformes
Diomedeidae—Albatrosses (8)
Procellariidae—Shearwaters, Fulmars, and
Petrels (27)
Hydrobatidae—Storm-Petrels (12)
Pelecaniformes
Phaethontidae—Tropicbirds (3)
Sulidae—Boobies and Gannets (5)
Pelecanidae—Pelicans (2)
Phalacrocoracidae—Cormorants (6)
Anhingidae—Anhingas (1)
Fregatidae—Frigatebirds (3)
Ciconiiformes
Ardeidae—Herons, Egrets, and Bitterns (16)
Threskiornithidae—Ibises and Spoonbills (5)
Threskiornithinae—Ibises (4)
Plataleinae—Spoonbills (1)
Ciconiidae—Storks (2)
Cathartidae—New World Vultures (3)
Phoenicopteriformes
Phoenicopteridae—Flamingos (1) A
Anseriformes
Anatidae—Ducks, Geese, and Swans (62)
Dendrocygninae—Whistling-Ducks and Allies
Anserinae—Geese and Swans
Tadorninae—Shelducks and Allies
Anatinae—True Ducks
Falconiformes
Accipitridae—Hawks, Kites, Eagles, and Allies (30)
Pandioninae—Osprey (1)
Accipitrinae—Kites, Eagles, and Hawks (29)
Falconidae—Caracaras and Falcons (10)
Micrasturinae—Forest-Falcons (1) A
Caracarinae—Caracaras (1)
Falconinae—True Falcons and Laughing Falcons (8)

Galliformes
Cracidae—Curassows, Guans, and Chachalacas (1)
Phasianidae—Pheasants, Partridges, Grouse, Turkeys, Old World Quail, and Guineafowl (15)
Phasianinae—Partridges and Pheasants (11)
Tetraoninae—Grouse (10)
Meleagridinae—Turkeys (1)
Numidinae—Guineafowl (1) I, H
Odontophoridae—New World Quail (6)
Gruiformes
Rallidae—Rails, Gallinules, and Coots (17)
Aramidae—Limpkin (1)
Gruidae—Cranes (3)
Gruinae—Typical Cranes
Charadriiformes
Burhinidae—Thick-knees (1) A
Charadriidae—Lapwings and Plovers (16)
Vanellinae—Lapwings A
Charadriinae—Plovers
Haematopodidae—Oystercatchers (3)
Recurvirostridae—Stilts and Avocets (3)
Jacanidae—Jacanas (1) A
Scolopacidae—Sandpipers, Phalaropes, and Allies (64)
Scolopacinae—Sandpipers and Allies
Phalaropodinae—Phalaropes
Glariolidae—Coursers and Pratincoles (1) A
Laridae—Gulls, Terns, Skuas, and Skimmers (57)
Stercorariinae—Skuas and Jaegers
Larinae—Gulls
Sterninae—Terns
Rynchopinae—Skimmers
Alcidae—Auks, Murres, and Puffins (22)
Family INCERTAE SEDIS = "Family with No Order"
Pteroclididae—Sandgrouse (1) I, H
Columbiformes
Columbidae—Pigeons and Doves (20)
Psittaciformes
Psittacidae—Lories, Parakeets, Macaws, and Parrots (6)
Platycercinae—Australian Parakeets and
Rosellas I

mya	Period		Era
—208	TRIASSIC	**MASS EXTINCTIONS** Pangea is elevated and shallow seas drain. Extensive deserts. The land animals that survived the Permian extinction diversify and spread to vacant niches. **Thecodonts replace therapsids as the dominant vertebrates. By end of period, dinosaurs, pterosaurs, marine reptiles, crocodiles, lepidosaurs (ancestral snakes and lizards), froglike amphibians, bony fish, and true mammals appear.**	
—245	PERMIAN	**PERMIAN EXTINCTION: MOST EXTENSIVE EXTINCTIONS IN EARTH'S HISTORY** Cold climate warms throughout period; widespread aridity. A single world continent, Pangea, forms at end of period. Cone-bearing plants (gymnosperms) replace spore-producing plants. **Early reptiles and mammal-like reptiles (therapsids) diversify and dominate the land.** Most extensive extinction event in earth's history occurs on both land and in the sea at the period's end, defining the end of the Paleozoic Era.	*PALEOZOIC*
—290	CARBONIFEROUS	**Age of Amphibians.** Major glaciation in second half of period. Coal swamps prevalent in tropical areas, dominated by spore-producing plants such as mosses and ferns. Amphibians and fishes diversify and spread. **Insects diversify greatly on land, providing a food source that spurred the radiation of terrestrial vertebrates.** A major evolutionary advance is the **amniotic (terrestrial) egg**, which frees ancestral reptiles from a dependence on water. **First reptiles** appear; mammal-like reptiles are present by end of period.	
—363	DEVONIAN	**MASS EXTINCTIONS** **Age of Fish.** Major mountain building in N. America and Europe. Climate cooler. Oceans dominated by reef-building organisms (corals). **First forests and amphibians.** Fishes diversify but many go extinct at end of period.	
—409	SILURIAN	Shallow seas still extensive. **Vascular plants and arthropods first appear on land.** Jawless fish radiate; first jawed fish appear.	
—439	ORDOVICIAN	**MASS EXTINCTIONS** Shallow seas widespread over continents. Moderate climate at end of period. First complex land plants. **Marine animals, including first jawless fish, diversify.**	
—490	CAMBRIAN	**Age of Marine Invertebrates.** Continental masses break up and are covered by shallow seas. Chordates and invertebrates with shells first appear. **First vertebrates** may appear at end of period.	
—545		**Large land masses form. Oxygen first appears in atmosphere. Multicellular organisms** (algae, fungi, invertebrates) appear and diversify.	*PRECAMBRIAN*
—2,500		**Formation of the earth.** Variable climate. **First fossils** known from 3.5 billion years ago.	
—4,600			

*mya = million years ago

Appendix C:
GEOLOGICAL TIME SCALE

To understand the significance of the fossil record of birds and other organisms, it is essential to place them in geological time—time that begins with the formation of the earth and continues to the present. Geological time is measured in millions, even billions, of years—a magnitude difficult for humans to comprehend. This enormous span of time is subdivided into progressively shorter units termed **eras**, **periods**, and **epochs**. Within the rock layers laid down sequentially throughout geological time, fossilized plants and animals succeed one another in a recognizable order, which reflects the evolution of living organisms through time. The ages of rock layers and fossils are determined by radioactive dating techniques. To read this table in chronological order, begin at the bottom of the table with the formation of the earth and proceed upward. Modern humans first appear in this sequence approximately 200,000 years ago.

Large-scale extinctions of many species, termed **extinction events**, have occurred throughout earth's history. Although the cause of each event is not well understood, scientists attribute the extinctions to either catastrophic events, such as meteorite or asteroid showers, or a major shift in earth's environmental conditions due to volcanism, glaciation, global climate change, or changes in the salinity or oxygen level of the ocean. Five major and many lesser extinction events have occurred in the last 600 million years. Best known is the K-T event at the Cretaceous-Tertiary boundary 65 million years ago, which claimed dinosaurs, marine reptiles, pterosaurs, and many marine invertebrates. However, the most extensive mass extinction in earth's history, the Permian Extinction, occurred at the Permian-Triassic boundary 245 million years ago. At this time, 90 to 95 percent of marine species and many terrestrial species were eliminated. Adapted from Pough et al., (1999).

(Open)

ERA	PERIOD	EPOCH	TIME mya*	MAJOR EVENTS
CENOZOIC	QUATERNARY	Holocene (Recent)		**Rise and expansion of human civilizations.** Passerines are the most diverse and abundant group of modern birds. **Human activities cause many birds, particularly island birds, to go extinct.**
		Pleistocene	0.01	Time of repeated glaciation, changing sea levels, and widespread extinctions claiming many birds and large mammals. Hominids spread globally. **Modern humans first appear in the fossil record approximately 200,000 years ago. Most modern species of birds evolve.**
		Pliocene	1.7	Lower latitudes remain warm, while higher latitudes cool further. Mountain building in western North and South America; Panamanian land bridge forms. Sea level drops. Extensive grasslands and deserts develop as forests contract. **Hominids (early humans) first appear. Birds reach their maximum species diversity.**
		Miocene	5.2	Climate continues to dry and cool. The Alps and Himalayas form, and grasslands dominate the plains of Asia and N. America. **Passerines diversify explosively and spread to new niches. Modern bird genera begin to appear.**
	TERTIARY	Oligocene	23	Climate begins to dry and cool, especially at poles; great forests spread to cover most land masses. **Order Passeriformes evolves late Eocene to early Oligocene. Nearly all families of nonpasserines are present.**
		Eocene	36.6	Global climate mild and humid. N. America and Europe separate. Mammals diversify. **Epoch of greatest diversification of birds: most modern orders are present by 50 mya,** and most modern families are present by beginning of next epoch. Shorebirds, flamingo-like birds, rail-like birds, and cranelike birds are diverse and abundant.
		Paleocene	55	Mild climate worldwide. Shallow continental seas disappear. **First primitive primates.** Lithornithids (medium-sized, flying birds possibly ancestral to ratites) appear in N. Hemisphere. Giant, flightless, predatory birds appear: *Diatryma* in N. Hemisphere and phorusrhacids in S. America. **Many modern orders of birds begin to appear.**
			65	**K-T EVENT: MASSIVE WORLDWIDE EXTINCTIONS**
MESOZOIC	CRETACEOUS			Warm tropical and subtropical climate; slight cooling at end of period. Gondwanaland fragments and most of world is covered by shallow seas. **Flowering plants (angiosperms) appear and soon become the dominant land plants.** Marine reptiles such as plesiosaurs flourish in the shallow seas. **Dinosaurs dominate the land, while small mammals, pterosaurs, and medium-sized birds diversify. The dominant birds are the enantiornithines (opposite birds).** Dromaeosaurs appear. Birds present include the toothed birds *Hesperornis* and *Ichthyornis*; and the earliest toothless bird, *Confuciusornis* (125 mya). **A mass extinction at the end of the period claims dinosaurs, pterosaurs, marine reptiles, and many marine invertebrates.**
	JURASSIC		145.5	Pangea splits into Laurasia and Gondwanaland. Atlantic Ocean forms from rifts in the continental crust. Climate warms worldwide. Lush vegetation dominated by gymnosperms. Dinosaurs diversify while mammals remain small. The first birds, lizards, and salamanders ...teryx fossils appear 150 mya.

1

A Guide to Bird Watching

Stephen W. Kress

 When you see something new for the first time—a bright green beetle, a startling yellow flower, or a warbler with a golden head and breast—your first question is usually "What is it?" To make sense of the bewildering variety of living things, and to think about how they relate to each other, we need names. Using names helps us to separate the jumble of bird voices and fleeing brownish blurs into distinct species with habits of their own, and opens doors to the wonders of courtship, nesting, migration, and other aspects of bird biology.

The ability to recognize birds also can help efforts in bird conservation. Much of what ornithologists know about bird numbers and distributions, and about how they change in response to alterations in our environment, comes from bird watchers who report sightings in their local areas. In fact, there are many organized conservation projects in which birders can participate (see Ch. 10). But to help scientists follow trends in wild bird populations and develop conservation plans, you must know how to identify bird species. So, one of the first steps in becoming a better earth steward is to learn the cast of bird characters inhabiting our planet.

Figure 2–1. Using Silhouette to Identify Bird Groups: *Birds in the same taxonomic group typically have the same body shape and proportions, although they may vary in size. Silhouette alone offers many clues to a bird's identity, and may allow a birder to assign a bird to the correct group or even the exact species. European Starling and Belted Kingfisher, for example, both can be identified by silhouette.*

1: Belted Kingfisher, **2:** Mallard, **3:** woodpecker, **4:** quail, **5:** mockingbird, **6:** kingbird, **7:** nuthatch, **8:** screech-owl, **9:** jay, **10:** vireo, **11:** cardinal, **12:** European Starling, **13:** grackle, **14:** warbler, **15:** dove, **16:** finch, **17:** swallow, **18:** kestrel, **19:** crow, **20:** wren, **21:** shrike, **22:** Killdeer, **23:** meadowlark.

How to Identify Birds

■ Novices invariably are awed by how quickly an experienced birder can differentiate similar-looking birds zipping overhead or dashing through dense tangles. Learning to identify birds is actually similar to getting to know your human neighbors. When you move into a new community, at first everyone is a stranger. But soon you learn to distinguish your neighbors as you unconsciously build a catalog of pertinent details. One neighbor wears Bermuda shorts all summer, another emerges only for church on Sunday, a third is always surrounded by a gaggle of children. One shuffles along slowly, another is always in a hurry. Some jabber constantly, others never utter a peep. As their habits, silhouettes, styles of walking, and "habitats" become more familiar, you learn to recognize your neighbors in a flash—even at a distance.

In a similar manner, paying attention to differences will help you identify your bird neighbors. You can recognize many birds simply by observing their shapes and postures. Rapid assessments of this sort are based on **jizz**, a birding term that harkens back to the "general impression of size and shape" (G. I. S. S.) that British observers used during World War II to distinguish between enemy and friendly aircraft. As an example, in their book *Hawks in Flight*, Pete Dunne, David Sibley, and Clay Sutton share a tip for using jizz to separate two similar-looking falcons. They write, "A Merlin is to a kestrel what a Harley-Davidson is to a scooter." Of course, to get to the point at which you can identify falcons from general impressions, you must look closely at many falcons of each kind, so a mastery of jizz can take years to develop.

Beginners should start by learning to identify the general groups of birds. These groups, such as warblers and flycatchers, contain birds whose members all share certain similarities. Warblers, for instance, are generally small, brightly colored birds that glean insects from leaves and twigs; flycatchers usually perch upright on exposed branches, making frequent sorties to capture flying insects. Examples of other groups include woodpeckers, which extract insects from tree trunks and large branches, using their tails as props; kinglets, which hover near branches while picking off tiny insects; and wrens, small, energetic, brown birds that dart through underbrush with upright tails. The differences among groups, and among species within groups, can be daunting at first but will become clear with experience.

During the initial phase of birding, you will see many birds that you don't recognize. As you begin learning what they are, focus on the features described in the following sections. Remember, though, that in most cases several features must be considered together to make a final identification.

Remember also that the following section presents only a general overview of identification features. To explain all there is to know about identifying birds would take an entire book, in fact, several books. Fortunately, such books exist, and many of these "field guides" are excellent. An annotated list appears at the end of this chapter.

Shape

Although birds in the same general group can vary dramatically in size, they usually have the same body shape and proportions. For example, doves have chunky bodies, whereas blackbirds are more slender. Having a knowledge of bird shapes and silhouettes allows you to assign birds to the correct group **(Fig. 2–1)**. Sometimes shape can even reveal the exact species. For instance, the European Starling and Belted Kingfisher can be identified by their silhouettes alone.

Shape is not infallible as an identification clue, though. Once I spotted a "fat-bodied" bird with a long tail sitting at the top of a smoking chimney. Thoughts of rare exotics floated past as I waited for the bird to move. When it flushed from its warm perch, it flashed white patches on the wing—a Northern Mockingbird fluffed up against the frigid winter.

Postures and Flight Patterns

Similarities of postures and flight patterns can also help to place birds in their proper groups. If you watch one common member of the thrush family, the American Robin, strut across a yard, you'll see that the bird takes several steps, then adopts an alert, upright stance with its breast held forward. Hermit and Wood thrushes have similar postures. Other birds that strike vertical poses include hawks, flycatchers, and larks, whereas birds that usually perch horizontally include shrikes, crows, and vireos **(Fig. 2–2)**.

Many bird groups also have characteristic flight patterns **(Fig. 2–3)**. Finches exhibit a steep, roller-coaster flight; woodpeckers generally fly in a pattern of moderate rises and falls. Accipiters such as Sharp-shinned Hawks, Cooper's Hawks, and Northern Goshawks typically make several wing flaps followed by a glide, unlike buteos such as Red-tailed Hawks, which are usually seen soaring.

Posture and flight pattern can sometimes help to identify a bird's species. The American Crow flies with regular, flapping wingbeats, whereas the similar-looking Common Raven flaps occasionally and frequently soars like a hawk. Soaring Turkey Vultures look a lot like hawks, but they typically hold their wings in a shallow V shape over their backs, whereas most hawks and eagles hold their wings flat.

Least Flycatcher

Red-eyed Vireo

American Robin

Loggerhead Shrike

Red-tailed Hawk

American Crow

Figure 2–2. Using Posture to Identify Bird Groups: *Posture can be a clue to placing a bird in its correct group. Flycatchers, thrushes, and hawks usually stand or perch with an upright stance. Vireos, shrikes, and crows usually perch horizontally. Distant perched crows and hawks may look similar, but noting their different postures may help to distinguish them.*

a

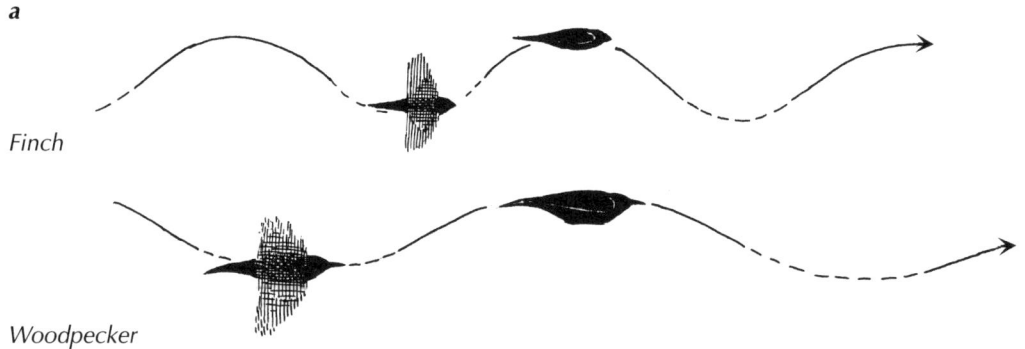

Finch

Woodpecker

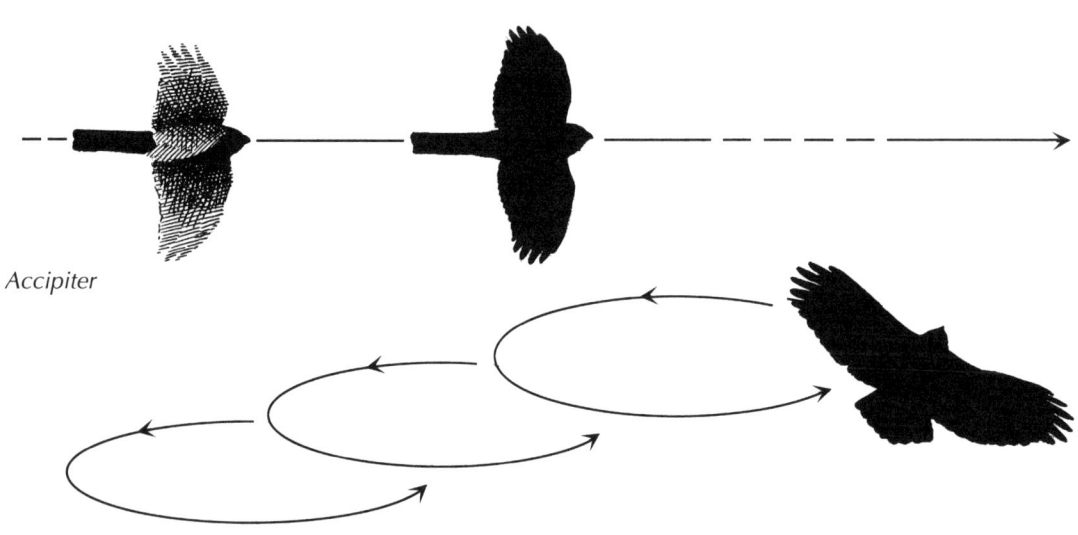

Accipiter

Buteo

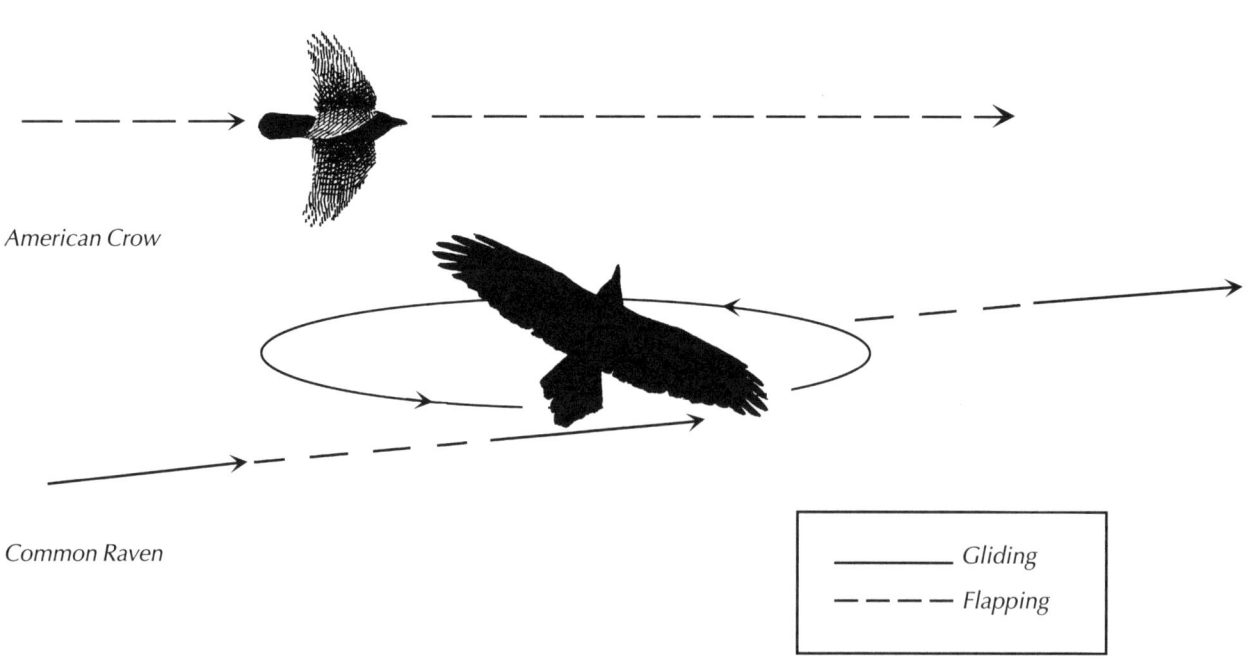

American Crow

Common Raven

Gliding
Flapping

b

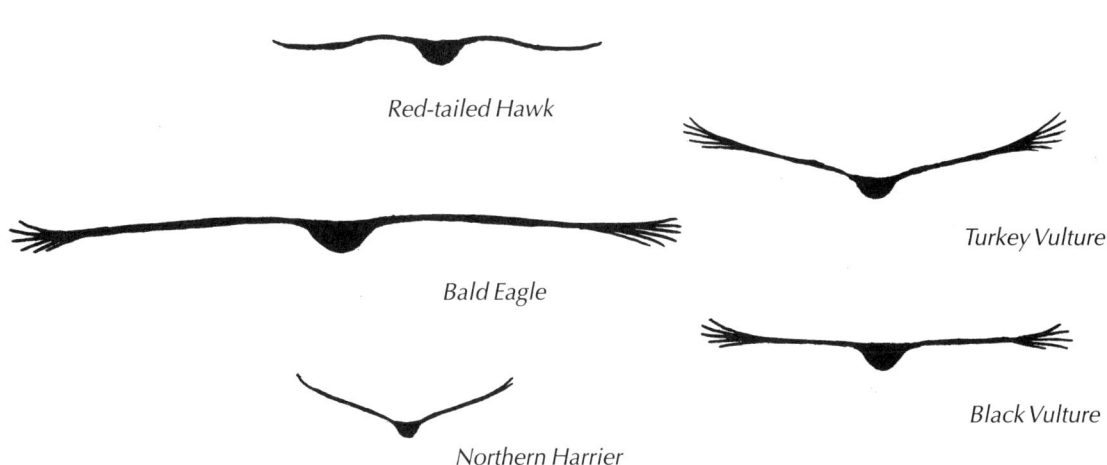

Red-tailed Hawk

Turkey Vulture

Bald Eagle

Black Vulture

Northern Harrier

Behaviors

Sometimes a bird's behavior is your best clue for determining its group. For instance, warblers and vireos have similar postures and shapes, but the two groups are readily differentiated by behavior. Watch them feed and you'll see that warblers are quick, energetic birds that constantly dart from place to place as they pick tiny insects from leaves and branches. In contrast, vireos often perch for several minutes in one place, waiting until they see a large insect, and then they dash forward to snatch up their prey. A close view of warbler and vireo beaks helps to explain this behavioral difference. Warblers have trim, pointed beaks—ideal tools for eating insects such as mosquitoes or aphids. Because these prey are so small, it takes a lot of them to make a meal for a bird, so warblers are always on the move. Vireos have stouter beaks with a distinct hook at the tip, so they can subdue and hold much larger prey—hence their wait-and-attack behavior.

You can also differentiate the various groups of ducks by observing their behavior. Dabbling ducks such as Mallards and Gadwalls tip their tails up to feed in shallow water, whereas Canvasbacks, Redheads, and other diving ducks completely disappear in search of bottom-dwelling fish and plants.

Behaviors can also help distinguish individual species. For instance, Fox Sparrows scratch the leaf litter looking for spiders and insects; Song Sparrows pump their tails in flight as they dash from one shrub to the next. Or, consider Mourning Doves and American Kestrels. These two birds are about the same size, have similar silhouettes and postures, and perch on telephone lines in open farm country. However, kestrels tend to pump their tails frequently while perched, and doves do not **(Fig. 2–4)**.

Size

Once you have assigned a bird to the correct group by observing its shape, posture, flight pattern, and behavior, you can use several

*Figure 2–3. Using Flight Patterns as Identification Clues: Many bird groups have diagnostic flight patterns. **a:** Finch flight is steeply undulating, whereas woodpecker flight has more moderate rises and falls. Accipiters typically fly with several wingbeats followed by a glide, unlike buteos, which usually soar. Flight pattern also can help to distinguish similar species: the American Crow has deliberate, flapping wingbeats, whereas the Common Raven often alternates flapping with hawklike soaring. **b:** Head-on flight profiles also may give identity clues: Soaring Turkey Vultures resemble hawks, but hold their wings in a shallow V-shape, whereas most hawks and eagles hold their wings out flat. Black Vultures also have a flatter, more hawk-like profile. Northern Harriers hold their wings in more of a V shape, but their hovering behavior generally gives away their identity. Note how the Bald Eagle's profile is even more flat than that of a typical hawk, such as the Red-tailed Hawk.*

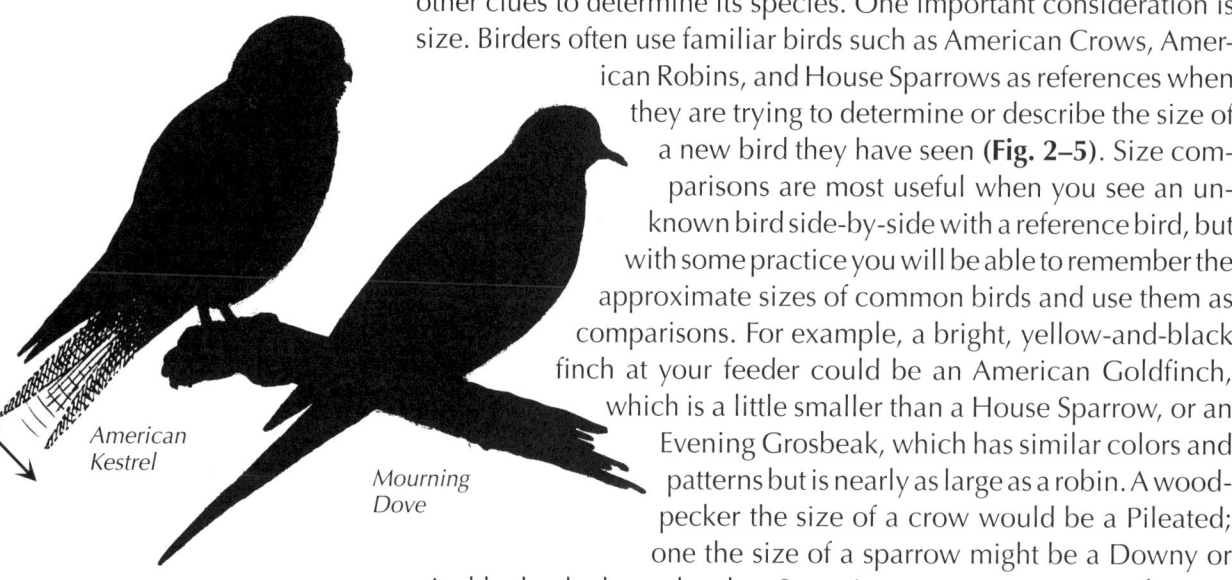

other clues to determine its species. One important consideration is size. Birders often use familiar birds such as American Crows, American Robins, and House Sparrows as references when they are trying to determine or describe the size of a new bird they have seen **(Fig. 2–5)**. Size comparisons are most useful when you see an unknown bird side-by-side with a reference bird, but with some practice you will be able to remember the approximate sizes of common birds and use them as comparisons. For example, a bright, yellow-and-black finch at your feeder could be an American Goldfinch, which is a little smaller than a House Sparrow, or an Evening Grosbeak, which has similar colors and patterns but is nearly as large as a robin. A woodpecker the size of a crow would be a Pileated; one the size of a sparrow might be a Downy or a Ladder-backed woodpecker. Sometimes you can use two reference birds for comparison. For example, waxwings are larger than sparrows but slightly smaller than robins. Jays are larger than robins, but smaller than crows.

Like shape, size is also fallible as a bird-identification clue. Apparent size can be affected by lighting conditions and distance, and size may be especially hard to judge in rain or fog and at dusk or dawn when, in silhouette, perched blackbirds can look like crows and crows can look like hawks. Birds also can change their apparent size. Dur-

Figure 2–4. Distinguishing Birds by Behavior: *Similar-looking species may be distinguishable by behavior. For instance, American Kestrels and Mourning Doves are about the same size, have similar silhouettes and postures, and choose similar perches in their open-country habitats. Perched kestrels, however, frequently pump their tails up and down, whereas doves do not.*

Figure 2–5. Using Familiar Birds as Size References: *Use the sizes of well-known birds, such as the American Crow, American Robin, and House Sparrow, as references when trying to identify an unfamiliar bird. For instance, a crow-sized woodpecker would be a Pileated, but one the size of a sparrow might be a Downy or a Ladder-backed woodpecker. A yellow-and-black finch smaller than a sparrow is probably an American or a Lesser Goldfinch; Evening Grosbeaks have similar colors and patterns but are almost robin-sized. Sometimes you need two reference birds for comparison. For instance, a waxwing is bigger than a sparrow but smaller than a robin. A Blue Jay is larger than a robin but smaller than a crow.*

ing hot weather birds may hold their feathers tightly to their bodies, which makes them look smaller. Conversely, on frigid days they may fluff themselves up to provide better insulation, so a chickadee might look like a mockingbird.

Also, birds of the same species sometimes differ in size. Male gulls, turkeys, and pheasants are larger than females, but female hawks, eagles, and owls are larger than males. And when young birds leave the nest they can be bulkier than their parents. Exercise soon trims them to adult size.

Comparing Body Features

You can sometimes identify species of birds by taking a careful look at their body features, especially extremities such as beaks, heads, and tails. As an example, consider the Hairy and Downy woodpeckers, two similar-looking species that often live in the same woodlot and frequently appear at feeders. At first glance, the two species—which have nearly identical plumage patterns—look hopelessly similar. The Hairy Woodpecker is noticeably larger than the Downy Woodpecker, but what if you see one of these birds alone? The key is to note the proportions of the bird's beak relative to its head (**Fig. 2–6**). On careful inspection, you can see that the Hairy Woodpecker's wood-drilling beak is nearly as long as its head. However, the beak of the smaller, bark-picking Downy Woodpecker is only as long as the distance from the base of its beak to the back of its eye, or about half as long as its head. This may seem like a subtle distinction, but it works!

Once you are tuned in to this type of proportional difference in size and shape, similar species become much easier to sort out. Other examples abound: the nearly identical Cooper's and Sharp-shinned hawks can be distinguished when flying overhead because the Cooper's Hawk head protrudes far ahead of its wings, whereas the head of the smaller Sharp-shinned Hawk barely extends beyond its wings (**Fig. 2–7**). Greater and Lesser scaups look similar—the Greater Scaup is bigger, but size is hard to judge when you're looking at birds halfway across a lake—so look at the head: the Greater Scaup has a rounded head whereas the Lesser Scaup's head is more peaked.

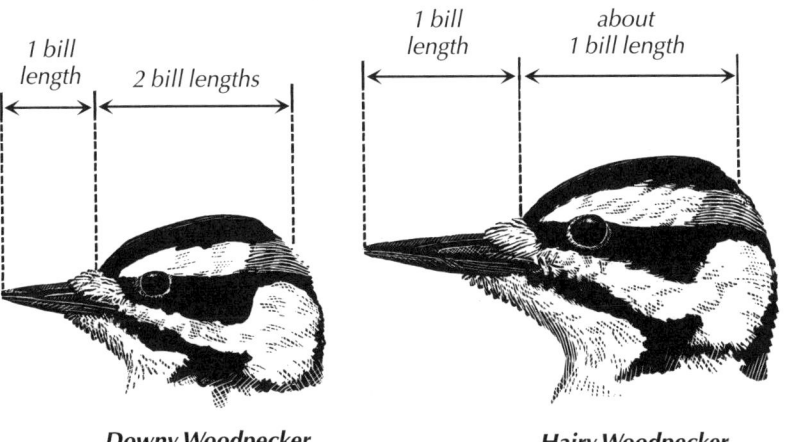

1 bill length   *2 bill lengths*

1 bill length   *about 1 bill length*

Downy Woodpecker

Hairy Woodpecker

Figure 2–6. Using Body Proportions to Distinguish Similar Species: Pay attention to body features, especially beaks, heads, and tails, to tell similar species apart. The Hairy and Downy woodpecker, for instance, have almost identical plumage patterns and habitats. The best way to tell them apart is to compare the length of each bird's beak to the length of its head. The Downy Woodpecker's beak is only about half as long as its head, whereas the Hairy Woodpecker's beak is proportionately much longer, almost the same length as its head.

Field Marks

Birds display a huge variety of patterns and colors, which they have evolved in part to recognize other members of their own species for mating. Fortunately, bird watchers can use the same features to help distinguish species. For instance, the three different but similar-looking species of sea ducks known as scoters are easy to differentiate by the pattern of white patches on the tops of their heads. Among adult males, the White-winged Scoter has one white patch under the eye, the Surf Scoter has white patches on the forehead and nape, and the Black Scoter has an all-black head.

When identifying an unknown bird, the following features are particularly important. You may find it useful to review the parts of a bird illustrated in Fig. 1–3.

Head

Check whether the bird's head has a **crest** (tuft), which will narrow the list of possible species dramatically. Also check for a stripe over the eye (**eyebrow stripe**), a line through the eye (**eyeline**), or a ring of color around the eye (**eye ring**) (**Fig. 2–8**). These field marks can be very

Figure 2–7. Sharp-shinned Hawk Versus Cooper's Hawk Flying Overhead: The Sharp-shinned Hawk and the Cooper's Hawk are almost identical in appearance and their size ranges partially overlap; thus, if you see a lone bird flying overhead, it can be difficult to identify. A good distinguishing characteristic is the length of the head. The Cooper's head protrudes far ahead of its wings, whereas the Sharp-shinned Hawk's head barely extends beyond its wings.

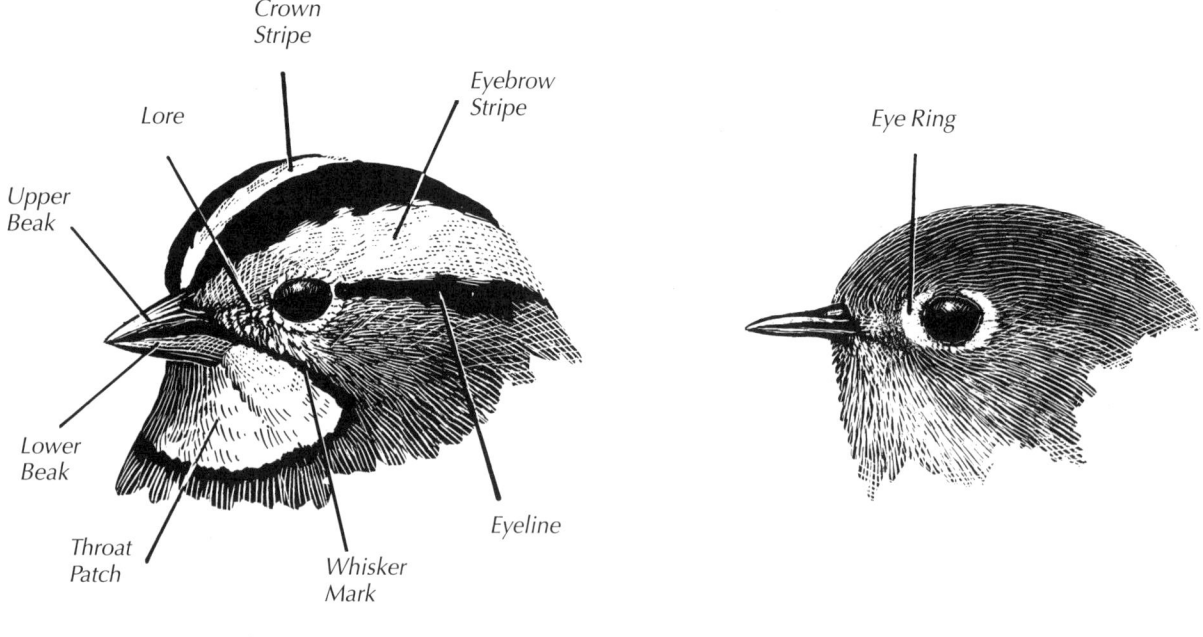

Figure 2–8. Field Marks on the Head: The following features of the head, if present, serve as good field marks: A stripe over the eye (eyebrow stripe), a line through the eye (eyeline), a stripe in the midline of the head (crown stripe), a ring of color around the eye (eye ring), and a throat patch. Pay attention to the colors of the upper and lower beak, and the area between the base of the bill and the eye, known as the lore.

useful. For example, the black eyeline of a Red-breasted Nuthatch separates it from the White-breasted Nuthatch, which completely lacks an eyeline **(Fig. 2–9)**. This field mark can actually be more diagnostic than color—many female Red-breasted Nuthatches have very light breasts, and in dim light or backlighting the colors can be difficult to see. As another example, the Ruby-crowned Kinglet has a white eye ring, whereas the similar Golden-crowned Kinglet has a white eyebrow stripe. And as one more example, Field and Chipping sparrows both have rufous caps and plain gray breasts. But the Chipping Sparrow has a crisp, black eyeline and white eyebrow stripe, whereas the Field Sparrow shows at most a hint of a brown eyeline with an indistinct grayish eyebrow area.

Bill Shape and Color

Bill shape can help to identify both general bird groups and individual species. Most members of the family that includes blackbirds, orioles, and meadowlarks, for example, have long, pointed beaks. Flycatchers have beaks that are flattened with a hook on the end, which improves their ability to grip large insects; warblers generally have pointed beaks that lack a hook; and vireos have beaks intermediate between warblers and flycatchers, thickened from the sides with a hook for holding large, squirming insects. Considering individual species, study the beaks of the similar Greater and Lesser yellowlegs and you'll see that the beak of the greater tilts slightly upward whereas the lesser has a shorter, straight beak.

Beak color is most helpful for identifying individual species. The yellow lower beak of the Eastern Wood-Pewee distinguishes it from the

Figure 2–9. Use of Field Marks in Bird Identification: Prominent field marks often facilitate bird identification. For instance, the black eyeline of the Red-breasted Nuthatch (left) readily distinguishes it from the White-breasted Nuthatch (right), which completely lacks an eyeline. Under difficult lighting conditions this field mark can be more diagnostic than the birds' coloration. Photographs by Marie Read.

Red-breasted Nuthatch   White-breasted Nuthatch

Eastern Phoebe. Field Sparrows have pink beaks, whereas the similar American Tree and Chipping sparrows have black ones. Snowy Egrets have black beaks, whereas Great Egrets have bright yellow beaks.

Wings

Check for wing bars or wing patches **(Fig. 2–10).** If the warbler you see has a large white wing patch, you can be sure it is either a Cape May, Blackburnian, or Magnolia warbler; or a Painted Redstart. Just a glimpse of a small white dot on the wing announces a Black-throated Blue Warbler, even in fall. In fact, wing bars are so useful in identifying warblers and vireos that some field guides separate each of these groups into birds with and without wing bars. The orange-brown wing bars of the Blue Grosbeak distinguish it from the similar all-blue Indigo Bunting.

When viewing a perched or standing bird, note the length of the wings compared with the tip of the tail. With this information you will be able to distinguish certain terns, gulls, and sandpipers more readily. For example, the wings of the less common Baird's and White-rumped sandpipers extend a bit beyond the tail, whereas those of the similar but more common Least, Western, and Semipalmated sandpipers are about the same length as the tail. Birders determined to find an unusual species may spend hours scanning a distant flock of these "peeps" to pick out the one or two different birds. Comparing wing-to-tail lengths also will help you differentiate perched Red-tailed Hawks from Swainson's Hawks **(Fig. 2–11).**

Tail

Check the length and shape of the tail. Note if the end is notched, rounded, or straight, and look for any white spots or white outer feath-

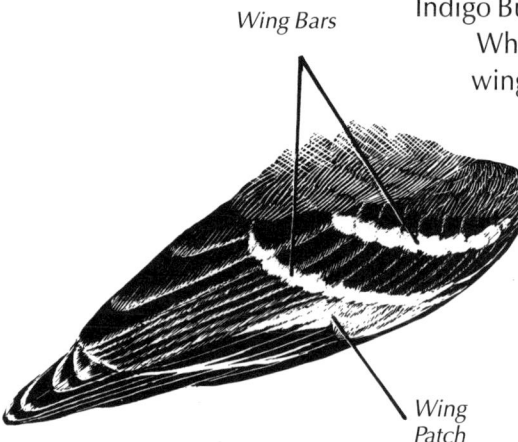

Wing Bars

Wing Patch

Figure 2–10. Field Marks on the Wing: *Check for wing patches and wing bars. In a few groups, most notably the warblers and vireos, wing markings can provide positive identification even if the bird is in nonbreeding plumage.*

Red-tailed Hawk   Swainson's Hawk

Figure 2–11. Wing Versus Tail Length: *In perched or standing birds, the length of the folded wing compared to the tip of the tail is a useful distinguishing character, particularly for challenging groups such as terns, gulls, and sandpipers. Comparing wing-to-tail length also may help differentiate perched buteos: notice how the wings of the Swainson's Hawk (right) extend beyond the tip of its tail, unlike those of the Red-tailed Hawk (left).*

ers. In towhees, juncos, meadowlarks, mockingbirds, and Vesper Sparrows, for example, the feathers on the outside of both sides of the tail are white—an especially prominent field mark when the bird flies. House and Winter wrens are quite similar in both appearance and behavior, skulking in dense vegetation with their tails pointed upward. But the short, stubby tail of the Winter Wren immediately distinguishes it from the House Wren. At first glance, flying crows and ravens look very similar—both are big, black birds. But pay careful attention to the shape of their tails, and you will be able to distinguish them. Crows have tails that are smoothly rounded or straight across the end, whereas the raven's tail is "wedge-shaped," coming to a broad point in the middle **(Fig. 2–12).**

American Crow

Common Raven

Figure 2–12. Raven Versus Crow Tail Shape: *In flight, an American Crow appears similar to a Common Raven, but the two species can be distinguished by tail shape: the crow's tail is smoothly rounded or straight at the end, whereas the raven's tail is wedge-shaped.*

Legs

Note the length and color of the legs, two features especially useful in identifying shorebirds, gulls, and egrets (see Ch. 3, Nonfeathered Areas, Legs and Feet, for examples). Keep in mind, however, that no matter how bright yellow a Least Sandpiper's legs are, they will still look brown after it walks in mud!

Colors and Plumage Patterns

Although color is often useful in bird identification, it also can be misleading. A bird's apparent color often varies with the angle of view, time of day, and lighting direction. The Indigo Bunting presents a dramatic example: in direct light this bird flashes a brilliant indigo glow, but if it is lit from behind, it appears jet black. Color is especially difficult to see over water. Most ducks have spectacular colors, but they can be difficult to see because of glare from surrounding water. Instead, look for the distinctive dark and light patterns of their plumage. The first edition of Roger Tory Peterson's *Field Guide to the Birds* didn't even show duck colors, and it worked fine for identification.

As another example, puffin beaks display a rainbow of brilliant colors that you can sometimes use to identify the birds, but you must see the beak in just the right light to enjoy the display. Rather than counting on seeing a colorful beak, first learn to recognize puffins as members of the auk family, a group of plump seabirds with bold black-and-white patterning. Each auk species has a slightly different black-and-white pattern that helps birders (and probably birds, too) recognize it at a distance **(Fig. 2–13)**. The shape of a puffin's beak—vertically broad—is the second clue to its identity. The beak's flashy colors,

Figure 2–13. Plumage Patterns of Alcids: *Because of glare, the true coloration of birds is especially difficult to see accurately over water, hindering the identification of distant, swimming birds such as ducks and members of the auk family (the alcids). When identifying these birds, distinctive light and dark plumage patterns are often more useful than plumage color. Notice the different black-and-white patterns of the Black Guillemot (top left), Atlantic Puffin (top right), Razorbill (bottom left), and Ancient Murrelet (bottom right).*

Northern Flicker

White Rump

Figure 2–14. Conspicuous Plumage Patterns of Land Birds: *Bold plumage patterns such as breast stripes, dark caps, and white rumps—as in the Northern Flicker—are visible under most lighting conditions and are often conspicuous, even at a distance.*

like the body colors of ducks, are less useful for field identification than the black-and-white body pattern.

Land birds, too, have distinctive contrasting patterns that are more useful than colors for identification. Patterns such as striped breasts (Ovenbird, waterthrushes), dark caps (Eastern and Say's phoebe, Blackpoll Warbler), and light-colored rumps (Northern Flicker) are visible under most lighting conditions, and are often conspicuous even at a considerable distance **(Fig. 2–14)**.

Trying to identify birds by color can present another problem: the amount of color an individual bird displays often varies with age, sex, diet, and time of year. And each fall, many birds such as shorebirds, tanagers, and warblers undergo complete body molts—they lose all their bright, breeding-color feathers as they change to more somber winter plumages (see Ch. 3).

Comparing living birds with color plates in field guides will show a final problem: color reproductions can only approximate bird colors, and the quality of the color can vary from one printing to the next. To get a feel for the extent of color variability in pictures, look up the same bird in several field guides and compare the colors of the different paintings and photos.

Despite these limitations, color can be helpful when you are observing birds nearby and in direct light (not backlighting). The similar Least and Semipalmated sandpipers, for instance, are very much alike in size and behavior, but Least Sandpipers have brown backs and straw-colored legs, whereas the semipalmated has black legs and a gray back. Male Orchard and Baltimore orioles can also be distinguished through color; the orchard's chestnut body is strikingly different from the vivid orange Baltimore. The females differ, too, the Orchard Oriole being more yellow, the Baltimore Oriole more orange.

Songs

Experienced birders can identify most birds by their songs or calls alone, and knowing bird vocalizations can help you find birds you would otherwise overlook, because they are far away, skulking in dense vegetation, or otherwise hard to see. Knowing bird songs also helps in bird-monitoring projects, when it's not practical to observe all birds in an area by sight. Hearing a bird tells you as clearly as seeing it that the species is present, and population estimates based on counting calls and songs are vital in assessing the health of wild bird populations.

Becoming familiar with bird songs usually comes after learning to recognize species by sight, and it takes a great deal of practice in the field. The best way to begin is to focus on one or two species at a time. Whenever possible, track down singing birds to discover their identities. Watching a bird sing its song will improve your memory of that song. Another good approach is to accompany a knowledgeable birder who can identify the singers—but you still should try to watch the birds sing, to establish the visual connection. Listening to recordings of bird songs can also be useful and, unlike listening to wild birds, you can replay the songs as often as you want. Recordings are most useful, however, as reviews for songs you have already heard outdoors. For novice birders, videos of singing birds may be more helpful than audio tapes, as they allow you to see and hear a bird at the same time. In addition, CD-ROM field guides and birding games are now widely available, giving you the added advantage of interactive learning while seeing and hearing birds vocalize.

Bird songs come in a huge variety. Winter Wrens, for example, have the longest song of any North American bird—each song includes about 40 notes and can last more than 10 seconds. The high-pitched, musical notes ring through the forest in a rapid, ever-changing, piccolo-like run. The singer of a song fitting this description, heard in a forest, is easy to identify—and it's a good thing, because these mouse-like birds are seldom seen. At the other extreme is the Henslow's Sparrow, whose song once was described by Roger Tory Peterson as "a feeble hiccup."

Many bird watchers learn new songs most easily by associating them with songs they already know. So, mastering a few basic songs will give you a framework for comparison. When you hear a new song, think whether it is similar to one with which you are already familiar. Is it faster, slower, higher, lower? How do the rhythm and tone differ? The American Robin, for example, has a clear, musical song consisting of a string of short phrases delivered as one song, with very short pauses between the phrases. The Scarlet Tanager, though not related to the robin, has a similar song—the rhythm is almost exactly the same—but it has a raspier quality. And the Rose-breasted Grosbeak—again no relation—also sounds much like a robin, though it strings its phrases together more closely, and has a whistled tone, like "a robin with a cold in a hurry." Rose-breasted Grosbeaks also may utter a very sharp, high-pitched *peek*—it could be mistaken only for a tree squeaking in the wind—in the middle of the song or between songs.

Not only do unrelated birds often sound alike, but related birds can sound very different. In fact, members of a single family often produce a diverse array of songs. The 57 species of North American warblers present a good example. Although none of them actually warble, some produce insect-like buzzes (Blue-winged Warbler), some trill (Pine and Worm-eating warblers), and others sing loud, rollicking songs (Hooded and Kentucky warblers). The different song characteristics may have a lot more to do with the kind of habitat in which a bird lives than with its particular family (see Fig. 7–20).

Certain terms are useful for describing and mentally organizing songs. The Chipping Sparrow, Swamp Sparrow, Dark-eyed Junco, Pine Warbler, and Worm-eating Warbler, for instance, have similar songs that are best described as **trills**, in which a single note or note cluster is repeated again and again. To tell these similar-sounding birds apart takes a very practiced ear. A few birds make **buzzy** sounds, somewhat like a very loud bee. By counting buzzes, you can distinguish the slow, lazy *bee-buzz* of the Blue-winged Warbler from the *bee-buzz-buzz-buzz* of the Golden-winged Warbler. Some of our most melodic songsters **whistle**, producing clear tones one pitch at a time. Northern Cardinals, Black-capped and Carolina chickadees, Tufted Titmice, Eastern Meadowlarks, Baltimore Orioles, and Broad-winged Hawks all whistle their songs. These are some of the easiest songs for us to mimic—a whistled two-note imitation of the Black-capped Chickadee song in spring will nearly always prompt a response if a male is within earshot. Other songs are not as beautiful to our ears. Kingfishers produce a harsh trill we term a **rattle**, Common Grackles **squeak**, parrots **squawk**, and herons, gallinules, coots, grebes, and mergansers give hoarse **croaks**.

Some birders use **mnemonic devices**, such as associating phrases from human speech with the songs of particular birds, to help them remember bird songs more easily. It doesn't take too much imagination to picture a Barred Owl chanting *Who cooks for you, who cooks for you-all?* or an Eastern Wood-Pewee whistling *pee-a-wee* (although it helps if someone slowly mouths the words for you while you listen to the bird's song). But other commonly used mnemonics can be misleading. Olive-sided Flycatchers supposedly say *quick-three-beers*, and White-eyed Vireos reportedly say *pick-up-the-beer-check-quick*, but making these connections takes some creativity (and perhaps a cold beer on a hot day). Muddling matters further is the tendency for different authors to produce strikingly different mnemonics for the same call. In his journals, for instance, Henry David Thoreau transcribed the call of the Olive-sided Flycatcher as *whip-ter-phe-ee* and cited an authority who paraphrased the call of the White-eyed Vireo as *tshippewee-wa-say*. However, even when mnemonics don't sound exactly like the song, they may still help you remember a song's general features.

Although different mnemonics usually result from a listener's idiosyncrasies in hearing, sometimes the calls or songs of one species can vary between regions (see Ch. 7, Song Dialects). Donald Borror, an important figure in the field of bioacoustics, became so familiar with the local dialects of White-throated Sparrows migrating through central Ohio that he could tell his students where individual birds were heading! Songs can also vary among individuals, sometimes even in the same bird. For this reason, learning song patterns can be important. That is, rather than learning precisely what the bird says, learn to recognize the bird's *voice*—the speed, duration, pitch, and tonal quality of notes (buzzy, whistled, raspy, screechy, harsh, musical, chirpy). Song Sparrows, for example, sing from 5 to 15 different songs, and many individuals sing their own unique songs. Even so, you can recognize the voice of these common songsters quickly because they start each type of song with the same two or three clear introductory notes. Thoreau noted that the people of New Bedford, Massachusetts described the bird's song as *maids, maids, maids—hang on your tea-kettle-ettle-ettle-ettle-ettle.* Whereas the *tea-kettle-ettle* part is debatable, the two or three *maids* followed by a variable jumble of notes is nearly always diagnostic.

Recognizing patterns can help birders distinguish many other songs as well. Brown Thrashers hold the record as the most varied singers in eastern North America: an individual bird can sing 2,000 different songs, many of which mimic phrases from other birds. Still, Brown Thrasher songs are easy to recognize once you focus on the pattern—they usually repeat each phrase twice, in couplet fashion, going on and on in a harsh voice, and changing phrases continually. But even though patterns are helpful, you still must learn tones. For example, other birds also sing in couplets, including the Indigo Bunting—whose song is shorter, sweeter, and more emphatic than the thrasher's and is exactly the same in every rendition that a particular bird sings—and the Yellow-throated Warbler, whose song has a much sweeter, fluid quality. All of these birds could possibly be found singing in the same location.

As another example of pattern, consider the vireos. Most of these treetop birds sing similar songs consisting of whistled notes delivered in short phrases—but the quality and pacing of the songs vary. Red-eyed Vireos have a chanting, repetitive song often paraphrased as *Here I am, way up high, over here, look at me,* the phrases continuing on and on, in no particular order, with no long pauses. Yellow-throated Vireos sing a similar song in a similar phrasing, often paraphrased as *three-eight, eight-three, three-eight, eight-three* and so on, but their song has a raspy quality akin to the burry voice of a Scarlet Tanager. Blue-headed Vireos also follow the general pattern, but compared with the Red-eyed Vireo's, their song has a more pure, whistled quality, is often slower, and includes a pause after every two or three phrases—all characteristics that give away their identity as they perch hidden in the leafy canopy of a deciduous forest.

Some birds are actually easier to identify by song than by sight. For instance, song is one of the best ways to distinguish the Alder, Willow, Least, and Acadian flycatchers. These small birds look so much alike that it's hard to tell them apart even in the hand (**Fig. 2–15**). Indeed, the birds themselves probably recognize members of their own species by song rather than appearance. Although all four have short, harsh, emphatic songs, they differ enough that a little practice will allow you to tell them apart. The song of the Willow Flycatcher has a sharp, abrupt beginning: *FITZ-bew.* The alder's song has a softer beginning: *free-BEE-er.* The acadian's song is more shrill than the others: *PI-zza.* And the Least Flycatcher—unlike the other birds who sing their song

che-BECK,
che-BECK....

Least Flycatcher

free-BEE-er

Alder Flycatcher

FITZ-bew

Willow Flycatcher

PI-zza

Acadian Flycatcher

Figure 2–15. Empidonax Flycatcher Identification by Song: Some birds are easier to identify by song than by sight. Least, Alder, Willow, and Acadian flycatchers (all of the genus Empidonax) *are mostly drab olive in color, with faint eye rings and wing bars, and are notoriously difficult to distinguish by plumage. However, each has a short, distinctive song that is easy for birders to paraphrase and recognize in the field.*

just once—repeats his song over and over in a long series: *che-BECK, che-BECK, che-BECK.*

Habitat

Each species of bird is predictably found in a particular habitat, and each plant community—a spruce-fir forest, a meadow, a freshwater marsh—contains a predictable assortment of birds. For example, Swamp Sparrows and King Rails are usually found in freshwater marshes. In a salt marsh, the Swamp Sparrow would be replaced by a sharp-tailed or Seaside sparrow, and the King Rail by a Clapper Rail. Learn which birds to expect in each habitat and, faced with an unfamiliar bird, you'll be able to eliminate from consideration species that usually live in other habitats **(Fig. 2–16)**.

As an example, knowledge of breeding habitat would help you identify three birds mentioned earlier that sing similar trilled songs: the Dark-eyed Junco, Chipping Sparrow, and Swamp Sparrow. Their ranges overlap, and—especially in the Northeast—all three may appear in the same region. But their habitat preferences differ, and taking habitat into account when you hear a trill can help you decide which bird you are hearing. A slow trill emanating from a mass of cattails or a wet, shrubby area is almost certainly a Swamp Sparrow. Chipping Sparrows are more likely to be suburbanites, favoring lawns, parks, grassy fields, and forest edges. And juncos are most common in the interior of coniferous or mixed woods. But be careful: Chipping Sparrows and juncos do overlap in habitat, so you'll need to catch a glimpse to be certain which bird you are hearing.

Breeding habitat can also help distinguish the similar-looking Northern and Louisiana waterthrushes. Both species nest on the ground, have similar, harsh chip notes, and true to their names, are usually found near water. (Untrue to their names, they are warblers, not thrushes.) Nevertheless, they are rarely found together except during migration. The Northern Waterthrush sticks to the quiet, slow-moving, or stagnant waters of woodland bogs or swamps, whereas the Louisiana Waterthrush lives in wooded ravines or gorges with streams.

Although having a knowledge of bird habitat preferences is one of the best ways to sort out which birds you are most likely to encounter, surprises do occur. During spring and fall migrations, for example, birds often settle down when they get tired, regardless of habitat. And loons sometimes land on wet highways during rainstorms, presumably mistaking the broad, winding, slick surfaces for rivers. (This is a serious mistake—with their small wings and heavy bodies, these birds need a water runway and can't take off from land, so they're stuck.) Tired bitterns sometimes land in grassy backyards; so accustomed are they to hiding among cattails that they still hold their heads in their characteristic vertical posture **(Fig. 2–17)**.

Range and Abundance

Although birds can travel fast and often show up in out-of-the-way places, each species usually stays within a certain geographic area called its **range**. You can find information about bird ranges in any North American field guide; those with maps are easiest to use in the field. Some maps show both breeding and wintering ranges and give dates for the arrival of migratory species.

Range maps are invaluable for determining which of several similar species might appear in your region. For instance, suppose you are sure that you've seen a titmouse, but you're not sure which of the four North American species it may be. Range maps will show you that in most parts of the continent you can figure this out by range alone, because the species scarcely overlap **(Fig. 2–18)**.

Knowing the relative abundance of the different species in your area is also helpful. Some species are common, others are rare. Learn the common birds first, then you'll be more likely to spot unusual birds that look different. Checklists showing the relative abundance of birds are available in many regions.

A warning: when using range and abundance as guides, remember that the birds have not seen the maps or read the books. Ranges change, and wandering individuals occur in most species. In fact, your alert observations can help to document these events, improving our understanding of regional, national, or global changes in climate and habitat.

Time of Year

Some birds, such as nuthatches, chickadees, titmice, and most woodpeckers, are year-round residents, staying in the same general area throughout the seasons. By contrast, most flycatchers, thrushes, tanagers, warblers, and vireos spend the breeding season in northern latitudes but winter in the southern United States, Mexico, or Central or South America. During migration, they pass through areas in between. Knowing which birds live in or visit your locale at different times of year can help distinguish similar-looking species.

For example, some closely related and similar-appearing species, such as American Tree and Field sparrows, neatly divide the year with only a little overlap **(Fig. 2–19)**. During late spring and summer, American Tree Sparrows breed in the northern tundra, while Field Sparrows are breeding throughout southern Canada and the northern United States. In late fall and winter, American Tree Sparrows migrate south

a. Abandoned Field

b. Mixed Deciduous/Coniferous Forest

Figure 2–16. Common Birds of Common Plant Communities: *Each bird species requires a certain combination of habitat components, and each plant community supports a predictable assortment of species. By knowing which birds to expect in each habitat, you may be able to identify an unfamiliar bird by the process of elimination.* **a: Abandoned Field Inhabitants:** *Field Sparrow, House Wren, Red-tailed Hawk, Blue-winged Warbler.* **b: Mixed Deciduous/Coniferous Forest Inhabitants:** *White-throated*

c. Sonoran Desert

d. Cattail Marsh

Sparrow, Winter Wren, Northern Goshawk, Black-throated Green Warbler. **c: Sonoran Desert Inhabitants:** Black-throated Sparrow, Cactus Wren, Harris' Hawk, Lucy's Warbler. **d: Cattail Marsh Inhabitants:** Swamp Sparrow, Marsh Wren, Northern Harrier, Common Yellowthroat.

Figure 2–17. Habitat Surprises: *During spring or fall migration, exhausted birds may land anywhere, regardless of habitat. A tired American Bittern, landing in a backyard, will automatically assume its typical vertical posture if approached; this behavior renders it nearly invisible when performed in its normal habitat of dense marsh vegetation, but provides little concealment in a backyard.*

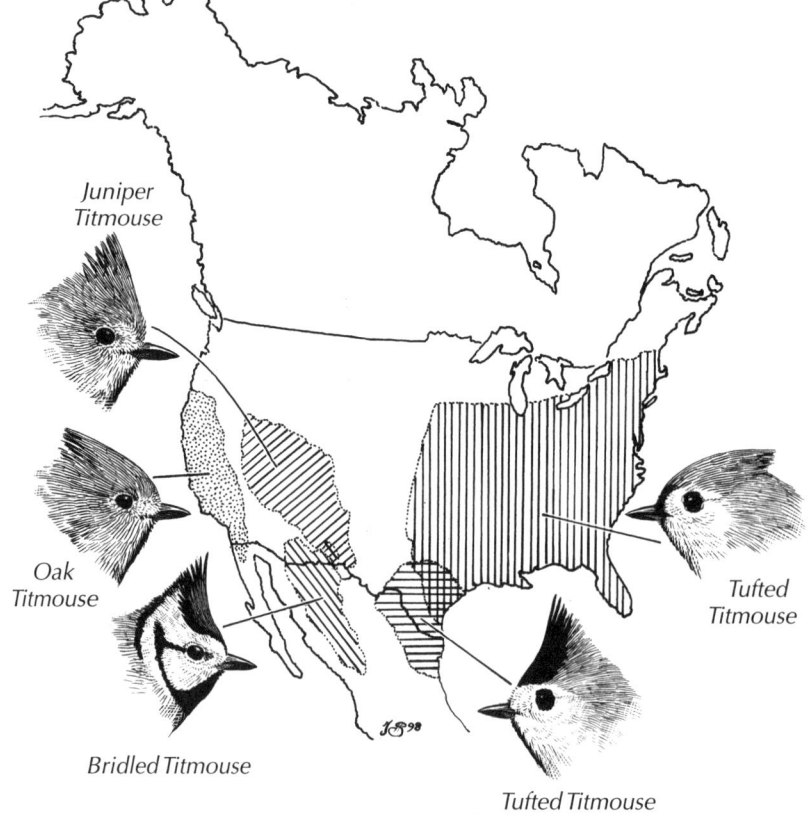

Figure 2–18. Titmouse Range Maps: *In most parts of the United States, titmice can be identified to species by range alone, because the ranges of the four North American titmice barely overlap. Shown are three western species—the Juniper, Oak, and Bridled titmouse— and the eastern Tufted Titmouse with its "black-crested" form (the latter two were considered separate species until recently).*

Juniper Titmouse

Oak Titmouse

Bridled Titmouse

Tufted Titmouse

Tufted Titmouse "Black-crested" form

into the breeding range of Field Sparrows, but Field Sparrows head for the southern United States. Therefore,
these potentially confusing species are seldom seen
together. Field marks, such as the American Tree
Sparrow's black chest spot and the Field Sparrow's
pink bill, will confirm an identification.

American Tree Sparrow

   You can profit from the birding experience of others
by seeking out local bird checklists produced by Audubon
Society chapters and other bird clubs, or the staffs of county,
state, and national parks and forests, national wildlife refuges, nature centers, and similar organizations (see Fig. 2–34). Such
checklists tell approximately when each species occurs in an area and
describe its relative abundance; sometimes they provide arrival and
departure dates. It's worth studying the list for any area in which you
are birding, whether you're a traveler, a new resident, or a longtime resident but beginning bird watcher.

Sorting Out Birds

   The process of bird identification begins when you
note the features—shape, posture, and behavior—that
permit placing a bird in the correct group. Identifying the group greatly reduces the number of possibilities; you need only consider which members
of the group are likely to be in that habitat at that

Field Sparrow

time of year. Then, look for field marks and listen to the bird's song to
make your final identification. With attention to these details, you'll
soon be able to identify your bird neighbors as quickly as you do your
human neighbors.

*Figure 2–19. Time of Year as an Identi-
fication Clue: Knowing which species
are found in a given location at different
times of year can help sort them out.
The American Tree Sparrow breeds in
the northern tundra, whereas the similar-looking and closely related Field
Sparrow breeds in southern Canada
and the northern United States. In late
fall and winter, American Tree Sparrows
migrate south into the breeding range of
the Field Sparrow, which in turn heads
for the southern United States. So even
though the two potentially confusing
species overlap in range, they do not
overlap much in time, and are seldom
seen together.*

Closing the Distance

■ Most birds are wary of approaching people; usually they fly off or
retreat into dense vegetation. Nevertheless, you can close the distance
between you and the birds with a few tricks (also see **Sidebar 1: Attracting Birds to Your Yard**).

Sitting Quietly

   One of the best ways to observe birds up close is also one of the
simplest: just sit quietly in a likely location until the birds no longer
notice you—try it for at least 10 minutes, but preferably half an hour.
You'll be pleasantly surprised by the birds and other animals you will
notice in almost any location—forest, field, or even a yard—but you
will see even more if you choose a site that is especially attractive to
birds, such as a marsh, pond, or streamside, or the edge where several
different habitats meet.

   On a typical nature walk, many of the behaviors you see are those
of birds alarmed by your intrusions, not birds going about the ordinary

(Continued on p. 2·26)

Sidebar 1: ATTRACTING BIRDS TO YOUR YARD
Sandy Podulka

If you already enjoy watching and identifying birds, you can add a new dimension to your relationship with them by enticing them to visit and nest in your yard. Birds, like humans, have a few basic needs that, when missing, limit their ability to live in an area. At a minimum, all must have food, water, cover, and nest sites; some also need song perches, foraging perches, dust-bathing sites, and specialized nest materials.

The simplest way to attract birds is to put out bird feeders. The key is to provide a variety of food types (suet, black oil sunflower seed, cracked corn, niger seed, and so on) and feeding situations (ground, hanging, and platform feeders) throughout the year.

Bird feeding is a good way to lure birds close enough to see them well and to study certain types of behavior, and it is invaluable as a way to share your love of birds with friends or to interest children in nature. Whether or not feeding actually benefits birds significantly is extremely difficult to determine, and there is very little clear data on this subject. Bird feeding does not appear to harm bird populations, and it undoubtedly increases the survival of some individual birds in times of food shortage or severe weather. But detailed, long-term studies incorporating the many other factors that affect bird survival and mortality will be needed before we can fully understand the impact of feeding on bird populations.

If you want to be sure to assist wild bird populations, landscaping to improve the quality of your yard for them is one of the best actions you can take. The trees, shrubs, flowers, and groundcovers you plant can provide cover, food, nest sites, and perches all year long, attracting a greater diversity of birds than will feeders alone. The best strategy is to select an assortment of plants with a variety of food types (tasty buds, seeds, nuts, cones, and fruits), fruiting seasons, heights, and cover types (evergreen and deciduous trees and shrubs, thorny bushes, unmowed grasses, brush piles, hedgerows, and so on).

Water also will entice birds to your yard. A simple birdbath can attract a surprising number of birds, especially in areas where there is little suitable natural water, as in many suburban and urban areas. If you have more space and ambition, you can construct a small built-in pool or even a pond. For birds, an essential component of any of these water resources is a shallow area one to two inches deep for bathing.

Nest boxes are effective in attracting hole-nesting birds, because natural cavities are usually in short supply around modern-day yards and gardens. For other species, you can build nest shelves or platforms. One more way to assist birds is, whenever possible, to leave old dead trees (called "snags") standing, as they supply food, cover, perches, and nest sites for birds.

Improving your yard habitat is important for birds and other wildlife because the biggest obstacle they face today is the loss of natural habitat. Fields, forests, and swamps are being rapidly converted to housing, highways, airports, and parking lots. Large farms with few crop types have replaced brushy fencerows, farm woodlots, and wetlands—varied habitats that are essential in maintaining abundant and diverse bird populations. Trackless horizons of single crop plantings (corn or wheat "deserts") or timber plantations result in a monotony of birdlife. Similarly, in suburban areas natural habitats are often replaced with sterile lawns and pavement. Wild habitats that are not obliterated are often degraded by pollution or chopped into isolated fragments too small to meet the needs of many species.

Although public nature preserves, wildlife refuges, and parks are superb ways to provide high-quality habitats for birds, much of the good and potentially good bird habitat in the United States is on private land. It is therefore up to each of us, whether our "yard" is 100 square feet or 100 acres, to improve or preserve the land under our stewardship. For example, a single small yard may not be able to provide all the space or resources a breeding Northern Cardinal needs. But together, several neighboring yards may form the ideal habitat and may "share" a cardinal or an American Robin. Adjacent private lands, or a combination of private and public lands, can form "corridors" of suitable habitat critical for the successful migration and dispersal of birds and other wildlife. So, when you landscape your yard for birds, your effect may reach well beyond your own boundaries. ■

Selected References on Attracting Birds

Feeding Birds

Henderson, Carrol L. 1995. *Wild About Birds: The DNR Bird Feeding Guide. Minnesota Department of Natural Resources*, 278 pages. If not at bookstores, order from Minnesota's Bookstore: 117 University Ave, St. Paul, MN 55155. (800) 657–3757.

An excellent, thorough guide to feeding birds. Includes sections on specific birds and how to attract them, as well as sections that focus on each different type of bird food, on bird feeder types and how to build them, and on troubleshooting. The best all-round reference to bird feeding.

Dennis, John V. 1983. *A Complete Guide to Bird Feeding.* NY: Alfred A. Knopf, 288 pages.

Discusses different types of feeders, nontraditional foods to offer birds, and problems at feeders. Gives information on the behavior and identification of feeder birds, and the food preferences of specific feeder birds.

Stokes, Donald and Lillian. 1987. *The Bird Feeder Book.* Boston: Little, Brown and Company, 90 pages.

Gives basic information on feeder types, feeder maintenance, problems at bird feeders, and bird behavior. Then discusses each common feeder bird in detail, including information on how to identify and attract it, and its behavior at feeders.

Landscaping

Henderson, Carrol L. 1987. *Landscaping for Wildlife.* Minnesota Department of Natural Resources, 144 pages. If not at bookstores, order from Minnesota's Bookstore: 117 University Ave, St. Paul, MN 55155. (800) 657–3757.

A comprehensive guide to bringing wildlife to your yard or a larger piece of land. Contains everything from landscape plans to tips on building brush piles, bird feeders, and frog ponds. Detailed charts of plants and their use by wildlife. Most applicable to the Midwest and Northeastern United States.

Kress, Stephen W. 1995. *The Bird Garden.* London: Dorling Kindersley Limited, 176 pages.

Produced by the National Audubon Society. Discusses many ways to attract birds to your backyard, from bird feeders and nest structures, to ponds and gardens. For each region of the United States, contains a guide to the plants that are most effective in attracting birds.

Providing Water

Low, Jim. 1992. "Wetting Your Whistlers." *Birder's World,* June 1992, pp. 50–54.

A thorough discussion on how to provide water for birds, from bird baths to garden pools and ponds.

Putting Up Nest Boxes

Henderson, Carrol L. 1992. *Woodworking for Wildlife.* Minnesota Department of Natural Resources, 111 pages. If not at bookstores, order from Minnesota's Bookstore: 117 University Ave, St. Paul, MN 55155. (800) 657–3757.

Detailed plans for constructing nest boxes, shelves, platforms, and roost boxes, as well as information on how to locate and maintain the structures.

Stokes, Donald and Lillian. 1990. The Complete Birdhouse Book. Boston: Little, Brown and Company, 95 pages.

Gives several basic nest-box plans, with dimensions and modifications for most cavity nesters in North America. Also includes information on buying, locating, and maintaining nest boxes, as well as on the nesting behavior of the birds.

tasks of their daily lives. Once you have settled down for quiet observation, you may see behaviors you've never noticed before: a Scarlet Tanager gleaning insects from leaves, an Ovenbird poking about in the forest debris, a Barn Swallow gathering mud for its nest. If you have the patience to be completely still, birds may move within a few feet of you. One day after photographing a Black Tern nest from her canoe, a friend of mine retreated about three feet to watch the adults return. In less than 10 minutes, one bird came back and sat on the nest. The other member of the pair flew about for a little longer, then perched on the bow of the canoe. After that, each time my friend returned to track the progress of the nest, the birds sat on her canoe, sometimes even preening. One bird even perched on her head—the most magical moment of all.

Becoming part of the environment for even a few minutes will leave you exhilarated, feeling privileged to have witnessed a few special moments in the natural world as an insider. In spring and summer, be sure to carry insect repellent—swatting bugs will not help you blend into your surroundings. A cushion to sit on also may increase your comfort.

Pishing and Squeaking

When alarmed, many land birds give a call to rally nearby birds, who may collectively chase away a predator such as a snake, owl, or cat. These alarm calls tend to have a similar sound, a sort of "psh." You can sometimes bring in birds for closer observation by imitating this sound with a technique that birders call "pishing." Repeat a syllable like "psh" or "spsh" in a drawn-out, hissing exhalation. While pishing, try to be very inconspicuous by standing against the trunk of a large tree to break up your silhouette, or crouching down to hide the typical human upright shape. Although pishing is most effective during the breeding season when birds are protecting their nesting territories and fledglings, it may work at any time (but not on windy days, because the birds can't hear it). "Squeaking" is a similar attracting noise produced by kissing your clenched fist to make a prolonged squeaking sound.

Not all birds are equally attracted by pishing and squeaking. Skulkers, such as sparrows and Common Yellowthroats, fall for it frequently; treetop birds respond less often. Success generally depends on a bird's breeding condition and level of excitement when you first attempt to attract it. When you do succeed in attracting a bird, it usually darts into view, takes one quick look, then disappears again. Chickadees and some other birds may give their own alarm calls if they get excited by your pishing and squeaking efforts; these calls will help to attract additional birds (**Fig. 2–20**).

Mobbing

Small birds often **mob** potential predators, swooping and dashing at the intruders to chase them out of their territories. Owls, hawks, snakes, and even mammalian predators—foxes, cats, or squirrels—

may be mobbed if they are discovered near active nests. Common mobbers include chickadees, titmice, jays, blackbirds, grackles, and crows, and sometimes these birds, which may have overlapping territories during the nesting season, will combine their efforts to mob a common threat (see the more detailed discussion of mobbing in Ch. 6, Antipredator Behavior: Why do Some Birds Mob Predators?).

Owls spark the most intense mobbing behaviors **(Fig. 2–21)**. Screech-Owls and Great Horned Owls often prey upon sleeping birds, so small birds that chase owls out of their territories during the day may be safer at night. You can use this fact to lure birds in for close view: simply play a recorded screech-owl call and watch the reaction. The owl's trembling whistle usually attracts small birds, which flock to the sound, ready to mob. Even a whistled imitation of a screech-owl call may rally a local songbird congregation.

Figure 2–20. Attracting Birds Using Pishing and Squeaking: Sometimes you can bring in birds for close observation by imitating their alarm calls, with a technique that birders call "pishing": repeating a syllable like "psh" or "spsh" in a drawn-out, hissing exhalation. The technique works best for skulking species, and is most effective during the breeding season. A squeaking sound, produced by kissing your clenched fist or the back of your hand, has a similar attractive effect on certain birds. Avid birders quickly overcome their self-consciousness about producing these sounds in public!

You can trigger an even stronger mobbing reaction by playing a screech-owl recording in the presence of an owl model. Make an owl from papier-mâché strips over a balloon (be sure to give it big yellow eyes), or buy one at a hardware store (they are often sold to scare birds from gardens, though they are usually ineffective at this). Mount the model in a conspicuous place, turn on the tape recorder, and hide nearby to watch the reaction. Note the calls given by the mobbing birds and how long the reaction lasts. Don't overdo this activity—remove the owl after about 15 minutes and let the birds sense a victory over the owl invader. Otherwise, you may cause birds to squander important energy reserves.

Playback Songs

One function of bird song is to alert males of the same species that a breeding territory is occupied. A newcomer who begins to sing within an established territory will soon be confronted by the resident male (some females, such as cardinals and orioles, also sing to repel other females). Persistent singing by the "challenger" male is usually met with a chase from the territory's first "owner" (see Ch. 7 for more on this topic).

Figure 2–21. American Crows Mobbing a Great Horned Owl: *Birds often mob potential predators, especially if their nests or young are threatened. Owls trigger the most intense mobbing activity. Birders can take advantage of this behavior; try playing a recorded screech-owl call to lure birds in for a close view, but don't overdo this activity, because it may cause birds to use up energy reserves they could put to better use. Drawing by Anne Senechal Faust, from* The National Audubon Society Handbook for Birders, *by Stephen W. Kress, 1981.*

You can use this chase response to lure seldom-seen birds into view. Play a tape recording of a species' song within its territory and the territorial male will quickly appear. Even hard-to-see birds that live in treetops or dense tangles may come into view to challenge a newcomer.

Commercial recordings serve nicely for playback, especially those on CDs, which permit you to quickly retrieve the species you want by simply punching in the location code. You can also obtain calls for playback by making your own recordings of the territorial male you are trying to observe. Birds do not recognize their own voices, so a singing male will come to defend his territory against any rival, even his own recorded voice!

Although an occasional confrontation with a tape recorder probably has little effect on a breeding bird, you should take care to avoid excess use. Once the bird you are seeking has appeared, turn off the recorder and let the male sing his song without competition. *Never* use tape recordings to attract rare or endangered birds or any bird nesting outside its normal range. Such disturbance to a bird already in precarious circumstances may threaten its breeding success. These same precautions apply to using pishing and squeaking, as well as owl calls and models, to attract birds.

Bird Blinds

Bird blinds are generally used for close observation and photography of nests, but they also are great to set up near watering places and special feeding spots such as the edges of marshes and wetlands **(Fig. 2–22)**.

Most birds apparently recognize the human shape by our distinctive two-leggedness, so anything that hides the human form, especially the legs, can function as a blind. Even a burlap bag or loose poncho draped over your back will hide your waist and legs, but find a comfortable tree for a back prop. Such simple blinds, however, are usually un-satisfactory for long waiting periods; your "blind" moves when you do, so you must keep completely still, which gets downright painful before too long.

You can construct a simple blind that allows some freedom of movement by attaching a skirt to a large umbrella bound to the top of a sharpened stake. Or use a lightweight card table as the roof and frame for your blind—just toss a prestitched cover over the top and you're ready to hide. A number of commercial blinds are also available through catalogs and advertisements in nature photography and bird-watching magazines.

When setting up a blind near a nest, always keep the best interests of the birds in mind. If you choose a commercial blind, look for one that you can erect quickly, because prolonged commotion near a nest can cause birds to abandon it. Stay in the blind for long stints (several hours), and never disturb the vegetation near a nest. Overhanging leaves and branches, which hide the nest from predators and offer shade to keep the young from overheating, are especially important.

Figure 2–22. A Homemade Bird Blind: To disguise your human shape, sit in a blind to observe birds near a watering area or feeding spot, such as the edge of a wetland; birds will usually forget their wariness and approach, allowing close observation. Photograph by Marie Read.

Viewing Birds

Using Binoculars

Binoculars are a virtual necessity for locating birds. If you don't yet own a pair, you'll find information on selecting and caring for binoculars in the next section. For information on adjusting a pair of binoculars to work best with your eyes, see **Sidebar 2: How to Calibrate Binoculars For Your Eyes**. Meanwhile, just a note on using them in the field: Don't get discouraged. Birds are moving targets, and both skill and practice are needed to find a bird in a binocular's narrow field.

The most important tip is: first spot the bird with your unaided eyes and then, *holding your head still and keeping your eyes on the bird,* lift the binoculars to your eyes and look through them. Avoid

(Continued on p. 2·32)

Sidebar 2: HOW TO CALIBRATE BINOCULARS FOR YOUR EYES
Stephen W. Kress

Most binoculars have a center focusing wheel that adjusts the focus of both eyepieces simultaneously and a separate diopter adjustment that allows you to focus one eyepiece independently, to make up for the differences in vision between your left and right eye. To determine the correct diopter adjustment on your binoculars, stand about 30 feet away from a sign with clear lettering—make sure that it is in the middle of the focal range of your binoculars—and follow these steps:

Step 1

Notice that the two binocular barrels pivot on a **hinge post**, allowing the eyepieces to fit your eyes comfortably. Facing the sign, spread the barrels as wide as you can. Then, put the binoculars to your eyes and press the barrels together until the two images converge into one. (If you cannot push the eyepieces close enough together to see through comfortably, reject those binoculars; the "interpupillary distance" of that model may be too wide to accommodate the narrowness of the space between your eyes.) The number (angle) indicated on the hinge post will always be the same for your eyes, on any pair of binoculars.

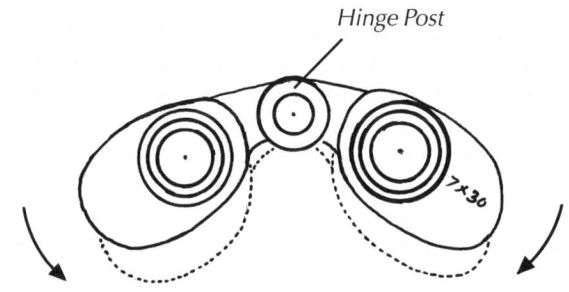

Hinge Post

Step 1: Form one image

Step 2

Turn the center focusing wheel counterclockwise as far as it will go. On most binoculars, one of the eyepieces (usually the right one) is marked with calibrations and can move independently. This is called the diopter adjustment ring. Turn this ring in a counterclockwise direction until it stops. Now both eyepieces should be out of focus. (Please note: some binoculars have a separate knob in the center or another mechanism for diopter adjustments; if so, consult the manufacturer's instructions.)

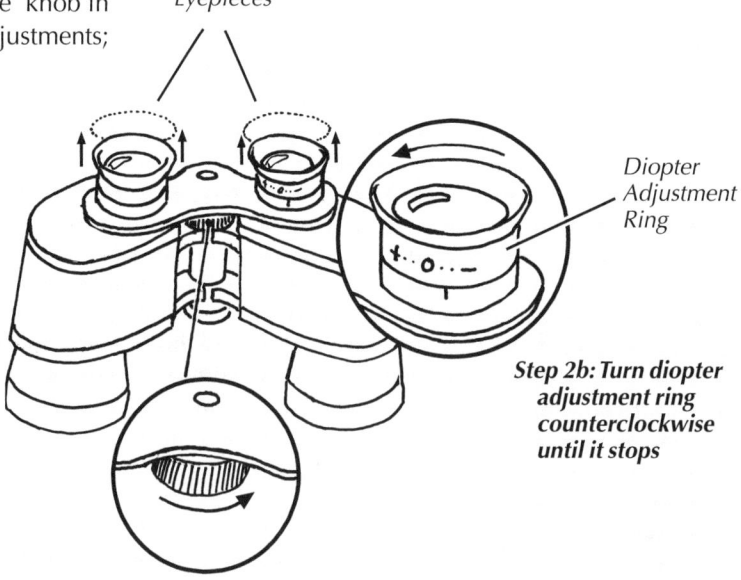

Eyepieces

Diopter Adjustment Ring

Step 2b: Turn diopter adjustment ring counterclockwise until it stops

Step 2a: Turn center focusing wheel counterclockwise until it stops

Step 3

Facing the sign, lift the binoculars into position and cover the end of the right binocular barrel. With both eyes open, turn the center focusing wheel until the lettering comes into sharp focus. To be sure you have the sharpest possible focus, pass the sharpest point and then back up to find it again.

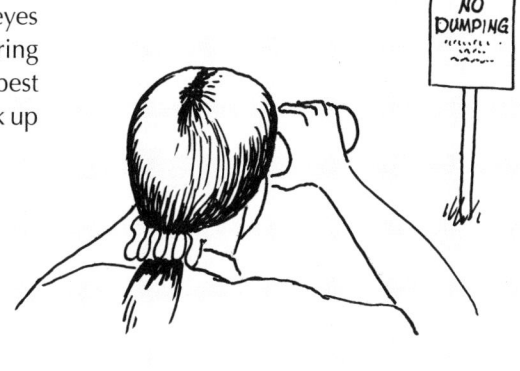

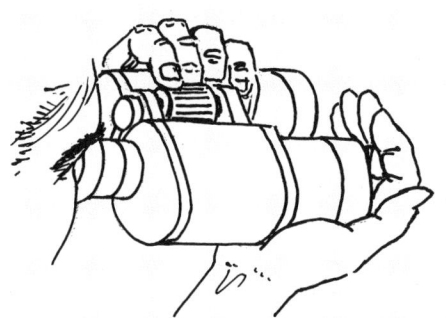

Step 3: Focus left eyepiece with central focusing wheel

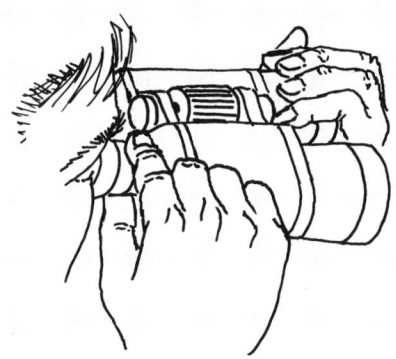

Step 4

Now cover the left barrel (keep both eyes open) and turn the diopter adjustment ring clockwise to bring the lettering into focus. Be sure to leave the center focus in exactly the same position as before. Pass the point of sharp focus and then back up to where the lettering is sharpest. Uncover the left barrel and the binoculars should be in perfect focus and calibrated for your eyes.

Step 4: Focus right eyepiece by turning the diopter adjustment ring

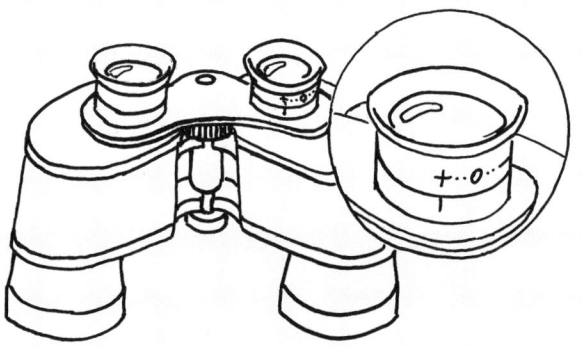

Step 5

Note the diopter setting because it is now adjusted to your eyes. That setting should remain constant, unless your vision changes. Some people put a piece of tape over the diopter adjustment to prevent it from shifting accidentally. Once this adjustment is set, you need only adjust the center wheel to focus both eyepieces.

Step 5: Note your diopter adjustment ring setting

2

Figure 2–23. Pointing Out the Location of a Bird to Other Observers: *Use precise descriptions rather than vague directions. Start with close landmarks that all the observers can see, then narrow the field until you come to the bird. For example, to describe the location of the screech-owl pictured here, you might say: "See the birch tree that has been chewed by a beaver? Beyond it and to the right there's a broken snag with fungi on it. To the right of that is a large maple with a double trunk. Follow the rightmost trunk up to the second branch on the right. The screech-owl is about halfway out from the trunk on that branch."*

scanning wildly through the trees. Practice locating stationary objects first—birdhouses, feeders, flowers, tree branches. Start with large objects, then try to find progressively smaller ones.

Pointing Out Birds to Others

If you are with a group of birders and someone cannot find a bird that you see, describe the location precisely. Vague directions, such as "It's in that tree," "It's over there," and "Look where I'm pointing," are no help and only increase the chance that the bird will fly away before others see it. Here are a few tips for describing a bird's location:

• Refer to the most obvious landmark near the bird, then narrow the field until you come to the bird. For example, if you spot a hawk in a farm field, you might describe its location this way: "See that large red barn with the white silo? Look over the top of the silo to the fence on the hillside behind it. Count eight fence posts to the right and there's the hawk sitting on top of the post. Do you see it?"

In wooded areas, try referring to an unusual-looking tree trunk or other natural landmark in the foreground to make sure everyone is looking at the same place. Using the reference point, successively indicate trees closer to the bird until you lead others to your discovery (**Fig. 2–23**). For maximum success, check often to make certain your directions are clear. At moments of excitement, calmly sharing a bird discovery takes as much skill as locating the bird in the first place.

• For a bird in a tree, use the "clock" technique to describe its position (**Fig. 2–24**). Mentally superimpose an hour hand onto the tree and use it to point to the bird. This system works especially well for birds near the edge of the tree. A bird in the top of the tree is at twelve o'clock; a bird halfway down the right side is at three o'clock. If a bird is not at the edge, then the hour designation is only the first step in describing its position. You must give additional pointers, such as "Find two o'clock in the largest sycamore, then move in halfway to the center of the tree. The bird is in front of the largest branch near a large woodpecker hole. See it?" Avoid using distance measurements, such as "20 feet from the top of the tree." Most people find

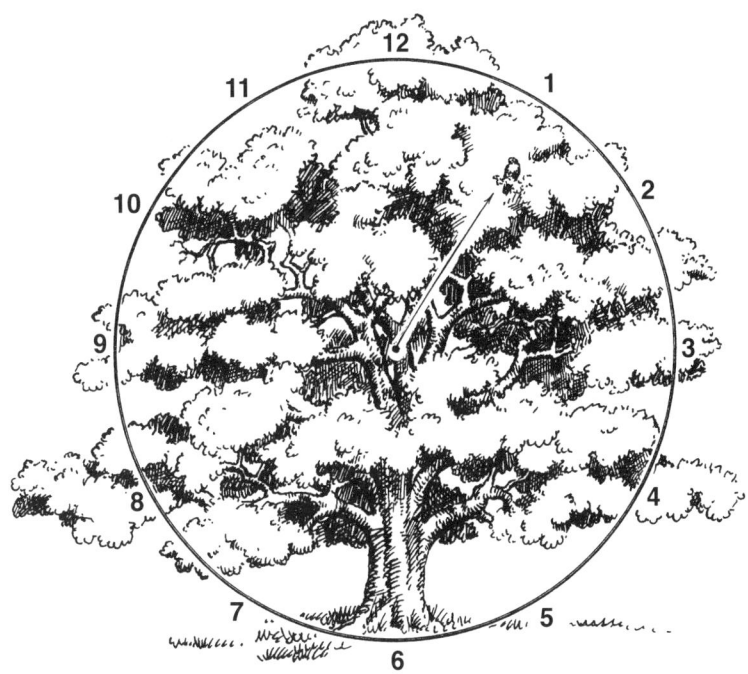

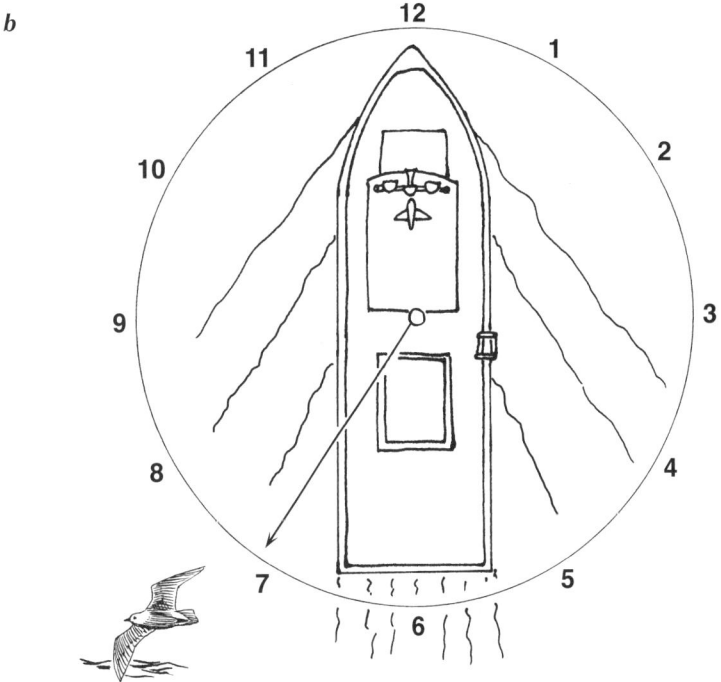

Figure 2–24. The Clock Method for Describing the Location of a Bird: a: For a bird in a tree, mentally superimpose a clock face onto the tree, with twelve o'clock at the top and six o'clock at the bottom. Then use an imaginary hour hand to point to the bird; in this case the bird is at one o'clock. This system works well for a bird at the edge of the tree. If the bird is not at the edge, the hour designation must be supplemented with additional directions. b: To use the clock system when birding from a boat, the imaginary clock face is oriented with twelve o'clock at the bow, and six o'clock at the stern. In this case the Ross's Gull is at seven o'clock.

it difficult to agree on exact distances. The best check on your success in giving directions is to ask if people see the bird.

- The clock system also works for spotting birds from a moving vehicle, such as a bus or a boat. For nautical birding, the clock is oriented with the twelve at the bow and the six at the stern (see Fig. 2–24). Calling out "Ross's Gull at seven o'clock" would send people rushing back to search for the bird just to the left of the stern. The clock also can be superimposed on land in a horizontal position: twelve o'clock is usually north or toward some predetermined landmark. This system is sometimes used to point out migratory hawks at hawk-watching locations.

Selecting Binoculars

Binoculars are probably the most important tool for watching birds, but choosing the best type, brand, and model for your needs can be bewildering. Magnification power, field of view, brightness, lens coating, size, weight—all are important. So is price: binoculars range in cost from less than $100 to well over $1,000. You must decide which features are most important to you and how much you're willing to spend. Here are some tips to help you make your selection. Remember, a wise choice will give you much pleasure and will last for years.

Magnification Power

Examine the flat upper surface of a binocular housing and you'll find two numbers—for example, 7x35 (pronounced "seven by thirty-five") or 10x40. The first number always designates the power of the binoculars; 7x (pronounced "seven ex") means the binoculars make subjects appear seven times closer than they would without magnification. (The second number is the diameter of the binoculars' **objective lenses**—those farthest from the eye—in millimeters; see next section.) Some birders prefer binoculars as powerful as 10x for viewing birds such as hawks, waterfowl, and shorebirds, which are likely to be seen in relatively open areas. However, the majority of bird watchers prefer 7x or 8x binoculars, for a couple of reasons. First, the more powerful your binoculars, the more difficult they are to hold steady for comfortable viewing—the effects of "hand shake" are greatly increased in binoculars with a magnification power greater than 8x. Also, lower-power binoculars tend to have greater light-gathering ability and a wider field of view than more powerful models and generally can be focused on closer objects.

Although some "zoom" binoculars offer the ability to quickly increase magnification power from 7x to 15x, the convenience is a poor trade-off for the bulk and weight: at the higher magnifications the binoculars are so difficult to hold steady and the image is so dark that it's almost impossible to see important field marks.

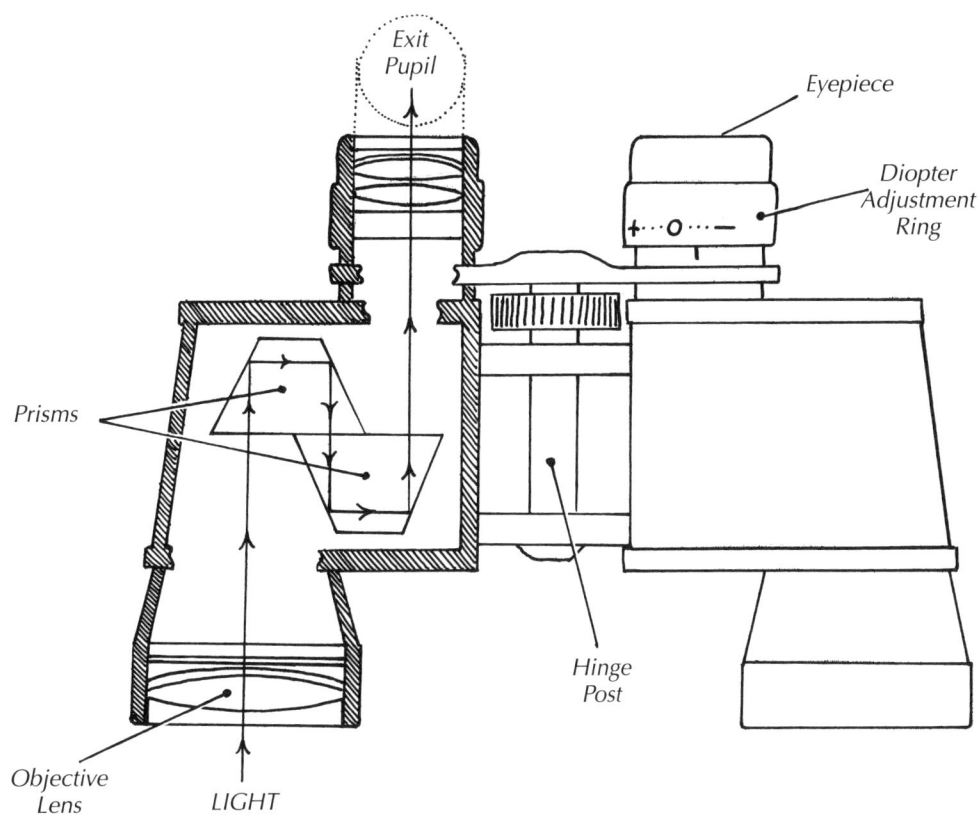

Exit
Pupil

Eyepiece

Diopter
Adjustment
Ring

+ ··· o ··· —

Prisms

Hinge
Post

Objective
Lens

LIGHT

Light-gathering Capacity

To a birder, the light-gathering capacity of binoculars is nearly as important as image sharpness. Only a bright image reveals the subtle nuances of field marks and the full beauty of bird colors.

Light enters binoculars through the objective lenses **(Fig. 2–25)**. As mentioned above, the diameter of these lenses in millimeters is the second number in the binoculars' designation—so 7x35 binoculars have 35 mm objective lenses. The bigger the objective lens, the more light that can be gathered and the brighter the image. Therefore, 7x50 binoculars have the same magnification power as a pair of 7x35, but the 7x50, with their 50 mm objective lenses, have a significantly greater light-gathering ability. Just as an owl's large eyes gather sufficient light to permit nocturnal vision, binoculars with large objective lenses provide an advantage for bird watching in low light, such as at dawn or dusk, or in dark, forested habitats.

The best measure of a binocular's brightness is the size of the **exit pupil**, the hole that the observer is looking through. You can see the exit pupil by holding your binoculars at arm's length and looking into the eyepieces **(Fig. 2–26)**. Depending on the binoculars, the exit pupil may vary in appearance from a dark hole to a brilliant, clear circle. To determine the exit pupil size, divide the size of the objective lens by the magnification number. Thus, 7x35 binoculars have an exit pupil of 5 mm, whereas 7x50 binoculars have an exit pupil of 7.1 mm, which

Figure 2–25. Porro Prism Binoculars: *Light enters the binoculars through the objective lens, and passes through a series of prisms before leaving through the exit pupil in the eyepiece and entering the observer's eye. The binocular barrels pivot around the hinge post. Binocular optics can be adjusted to your eyes by means of the diopter adjustment ring (see Sidebar 2: How to Calibrate Binoculars for Your Eyes).*

7 X 35

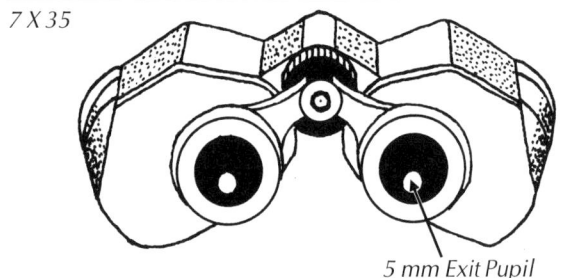

5 mm Exit Pupil

7 X 50

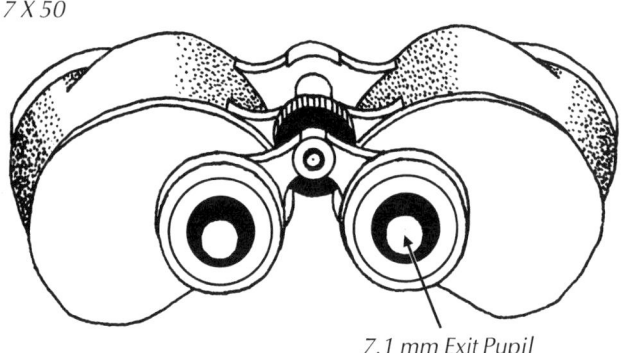

7.1 mm Exit Pupil

Figure 2–26. Exit Pupil Comparison: *A binocular's brightness can be judged by the size of its exit pupil: the larger the exit pupil, the brighter the image. Hold the binoculars at arm's length and look into the eyepieces to see the exit pupil. To calculate the size of the exit pupil, divide the size of the objective lens by the magnification number. For example, 7 X 35 binoculars have a 5 mm exit pupil, whereas 7 X 50 binoculars have an exit pupil of 7.1 mm, providing a brighter image. Drawing by Anne Senechal Faust, from* The National Audubon Society Handbook for Birders, *by Stephen W. Kress, 1981.*

Figure 2–27. Exit Pupil and Optical Quality: *Binoculars with poor quality optical components can have poor light-gathering abilities despite large objective lenses. Holding the binoculars at arm's length, examine the edge of the exit pupil: it should form a complete, bright circle, as in* **(a)**. *If only the center of the exit pupil is bright, as in* **(b)**, *inferior optics are blocking some of the light, counteracting the advantages of large objective lenses. Drawing by Anne Senechal Faust, from* The National Audubon Society Handbook for Birders, *by Stephen W. Kress, 1981.*

provides a much brighter image.

Birders using binoculars on boats will find that an exit pupil of at least 5 mm offers a distinct advantage. When motion causes your binoculars to move in all directions around your eyes, you may experience image blackouts as the exit pupil moves away from your eye's pupil. In bright daylight, when your eye has a pupil opening of about 2 mm, binoculars with a 5 mm exit pupil provide 3 mm of leeway to adjust to the movement.

Although binoculars with larger exit pupils are better for boating and generally offer brighter images, they do have drawbacks—principally the additional size and weight of the objective lenses and the larger housing necessary to support them. Fortunately, the best binoculars made today offer remarkable brightness with moderate weight by using high-quality optical glass and incorporating design improvements.

Check the following table to determine which exit pupil size meets your needs:

Exit Pupil Size	Appropriate Situations
2–4 mm	Bright-light situations (such as open farmland, mountains, shorelines)
4–5 mm	Shaded situations (such as forests)
Over 5 mm	Dusk and dawn, boating

Binoculars with large objective lenses can have poor light-gathering abilities if the optics are poor. As one test, carefully examine the edge of the exit pupil to see if it forms a complete, bright circle or if it is shaded in gray, resulting in a bright central area **(Fig. 2–27)**. If only the center of the exit pupil is bright, then inferior optics are blocking some of the light, and the advantages of the large objective lenses are not being realized.

Light entering the objective lens must pass through as many as eight pieces of optical glass in each barrel. At each glass surface some light is reflected backward rather than passing through the prisms and lenses. The optics of well-made binoculars are coated with a nonreflec-

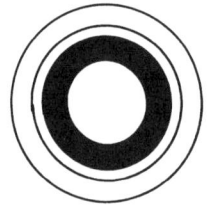 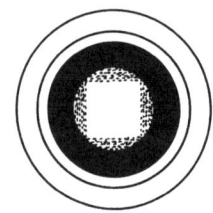

a. High Quality Optics   *b. Low Quality Optics*

tive film that helps deliver more than 90 percent of the light gathered by the objective lenses. Without this nonreflective coating, binoculars may lose up to 60 percent of the light that enters the objective lenses. Coated optics also are a great aid when you're looking at backlit subjects. Light reflects within uncoated binoculars, causing annoying glare. (But even with coated optics, never look directly at the sun; it could cause permanent eye damage.) Make sure the binoculars you purchase have "fully coated" optics. Although most manufacturers coat the exterior lenses, some inexpensive binoculars may have uncoated internal optics, which will cause a significant loss of light.

Field of View

The term **field of view** refers to the width of the area you see while looking through your binoculars. It is usually described as the width of the area visible at 1,000 yards from the observer—for example, some binoculars show an area 400 feet wide at 1,000 yards. If all else is equal, binoculars with a higher magnification power will have a smaller field of view than those with a lower magnification power. Sometimes the manufacturer of a particular binocular model expresses the field of view in degrees. If you wish, you can convert degrees to feet simply by multiplying the number of degrees by 52.5, the number of feet in 1 degree at 1,000 yards. Thus, a 6-degree field of view would show an area 315 feet wide at 1,000 yards (6 degrees x 52.5 feet/degree = 315 feet) **(Fig. 2–28)**.

The wider the field of view, the easier it is to locate birds with your binoculars. Wide-angle binoculars are especially useful for beginning bird watchers, because the larger field of view they provide makes it easier to find birds—especially if they are flying or skulking in dense vegetation. Manufacturers of extra-wide-angle binoculars expand the field of view by increasing the size and number of lenses in the binoculars' ocular system. The additional optics increase the cost of the binoculars and make them heavy and bulky. Because producing binoculars that have sharp images across their entire field of view is difficult and expensive, beware of low-cost, extra-wide-angle binoculars. They are probably only sharp in the center of the field. Most experienced bird watchers find that a standard field of view is adequate for most situations and that investing in extra-wide-angle binoculars is unnecessary.

Resolution

Resolution is a function of the quality of the optical glass used in the manufacture of binoculars. High-

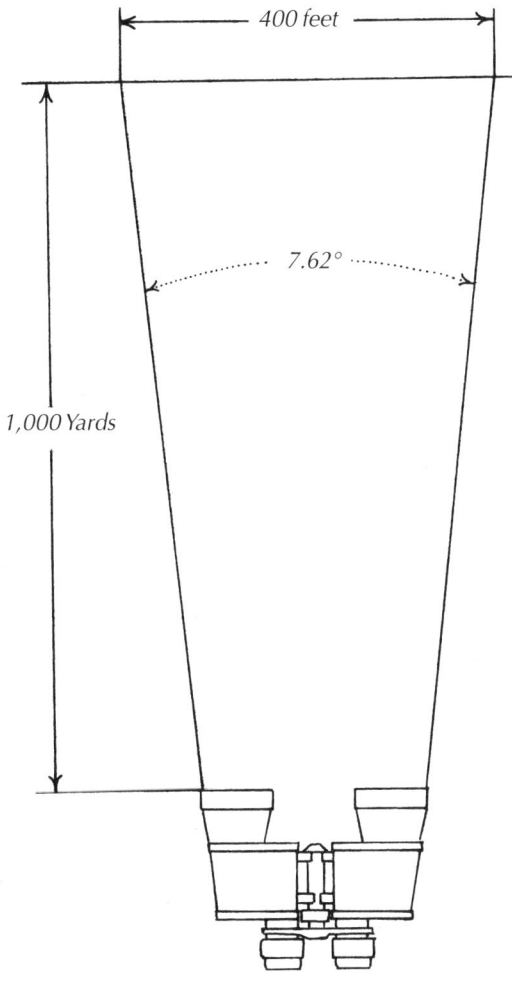

Figure 2–28. Field of View: *The field of view is the width of the area you see while looking through your binoculars. It usually is expressed as the width of the area visible at 1,000 yards (which, in this example, is 400 feet). Sometimes a manufacturer gives the field of view in degrees (in this example it is 7.62 degrees). Convert degrees to feet by multiplying the number of degrees by 52.5 (the number of feet in 1 degree at 1,000 yards). Binoculars with higher magnification usually have a narrower field of view than those with lower magnification.*

quality optical glass is extremely expensive, and each lens and prism must be professionally ground and mounted with expert precision. Top-of-the-line binoculars are finely crafted instruments. Manufacturers of lesser products cut corners throughout production, often by using less expensive glass and looser quality control.

High-priced binoculars usually have excellent optics, producing tack-sharp, crisp images from the center to the edge of the field of view. You can check the center-to-edge resolution of a pair of binoculars by focusing them on a map or newspaper tacked to a wall. Stand back about 25 feet and see if you can read the print at both the center and edge of the field of view.

Alignment

Because binoculars consist of two separate optical instruments—basically, an individual telescope for each eye—it is vitally important that they stay in proper alignment. When binoculars are functioning properly, both sides focus on the same field of view, but a sharp jolt can easily throw them out of alignment so that the two fields no longer overlap. Looking through misaligned binoculars, your eyes attempt to bring the two views together. If the binoculars are severely misaligned, you will see a double image and the subject will look blurry (when both your eyes are open). In some ways, binoculars that are only slightly out of alignment may be more of a problem, because your eyes strain to bring the two images together; this quickly results in eye fatigue and a headache.

Inexpensive binoculars are more likely to go out of alignment than higher-priced models. Prisms and lenses in cut-rate models may be glued in place rather than securely strapped by metal brackets. Temperature changes or slight jars can easily throw inexpensive binoculars out of alignment. And realigning binoculars is not a simple task. They must be taken apart by an experienced technician and recalibrated using special equipment. It makes far more sense to invest in good binoculars in the first place than to repeatedly replace or repair inexpensive binoculars each time they get bumped in the field. (And birding can be very tough on optical equipment.) Top-quality binoculars are more likely to withstand the stress of constant field use and, if you treat them with reasonable care, should last a lifetime. To check the alignment of your binoculars, try the simple test shown in **Figure 2–29**.

Binocular Designs

You'll find three basic designs in modern binoculars—Porro prism, reverse Porro prism, and roof prism **(Fig. 2–30)**. You can easily recognize standard Porro prism binoculars, the most common, because their eyepieces are closer together than their objective lenses. Reverse Porro prism binoculars have an inverted design, with the objective lenses placed closer together than the eyepieces. (Several compact binocular models employ this design.) Roof prism binoculars have straight barrels, with the eyepieces and objective lenses directly

a. Binoculars In Alignment
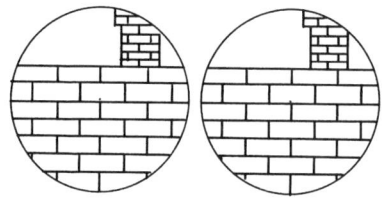

b. Binoculars Out of Alignment
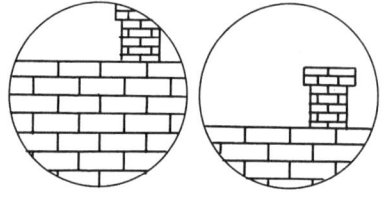

Figure 2–29. Binocular Alignment: To check the alignment of your binoculars, try this simple test. Look at the roof of a house through them, then, continuing to look through the eyepieces, move the binoculars about eight inches away from your eyes. If the binoculars are in alignment, the horizontal line of the roof should be at the same level in both fields (a). If the roofline appears offset (b) the binoculars are out of alignment. Drawing by Anne Senechal Faust, from The National Audubon Society Handbook for Birders, by Stephen W. Kress, 1981.

in line—a feature achieved by placing the two prisms in each barrel close together. Roof prism binoculars offer several advantages over Porro prism binoculars. Most roof prism models are compact and lightweight, and they provide excellent image resolution without sacrificing brightness or field of view. Many of them focus internally by moving lens elements back and forth inside the casing to achieve focus, rather than moving the eyepiece assembly back and forth externally as do most Porro prism binoculars. Internal-focus binoculars can be sealed more effectively and tend to be more resistant to moisture and dirt. On the other hand, roof prism binoculars are usually much more expensive than Porro prism binoculars, their depth perception is not as good, and they don't focus as well on nearby objects unless they've been specially designed or retrofitted to improve their close-focusing ability.

Mini Binoculars

Palm-sized binoculars are becoming increasingly popular among birders. More than 40 models are currently available, ranging in price from about $50 to more than $600. They generally use a reverse Porro prism or roof prism design, and some models deliver quite sharp images. They appeal to many people because they are small and lightweight, but birders with large hands and long fingers may find them uncomfortable to hold.

Beware of lower-priced mini binoculars, which often have poor light-gathering capacity. Mini binoculars in the upper price range, however, are usually finely crafted instruments with excellent optics. Tested to withstand the rigors of temperature extremes and sudden jolts, they are a good option for bird watchers who are already encumbered by bulky camera gear, tape-recording equipment, and field guides.

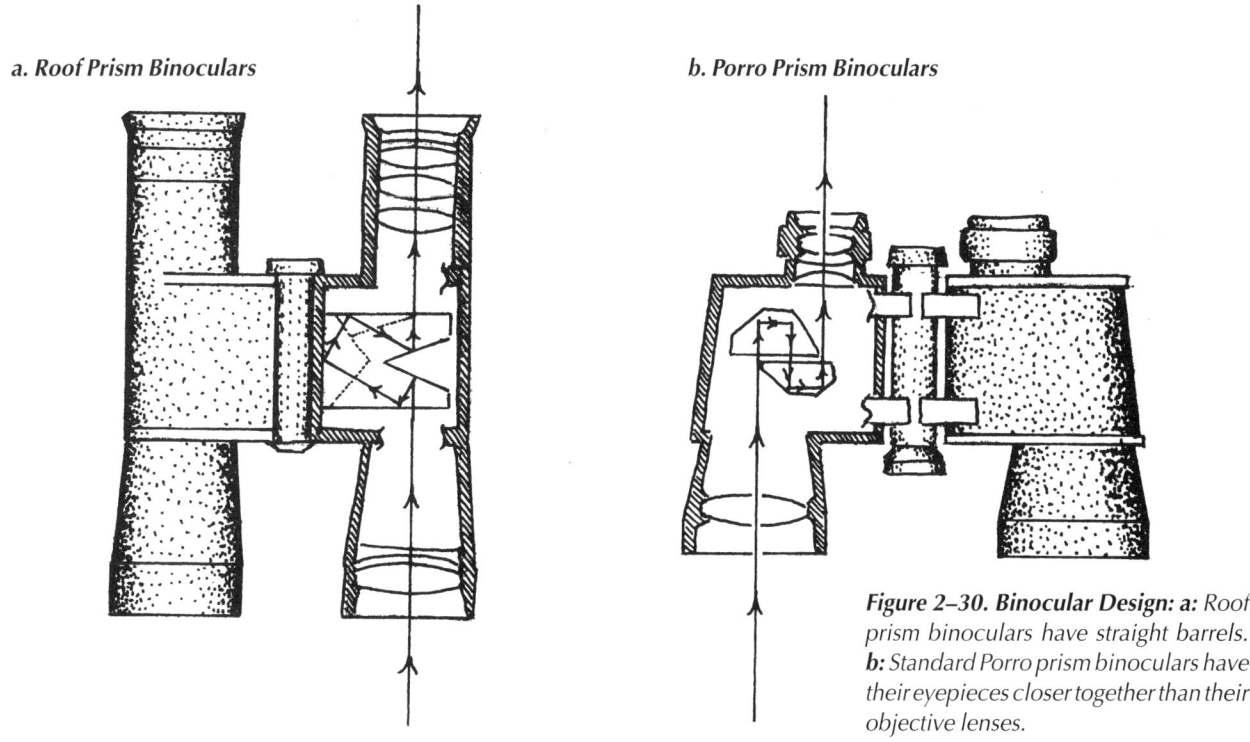

a. Roof Prism Binoculars

b. Porro Prism Binoculars

Figure 2–30. Binocular Design: a: Roof prism binoculars have straight barrels. b: Standard Porro prism binoculars have their eyepieces closer together than their objective lenses.

Binoculars for Eyeglass Wearers

People who wear eyeglasses should always leave their glasses in place when using binoculars. (You'll never be able to find and focus quickly on a bird if you always have to remove your eyeglasses before looking through your binoculars.) Of course, eyeglasses do get in the way: they prevent your eyes from getting as close to the eyepieces as they should to obtain the full field of view. Most binocular manufacturers now have rubber eyecups that you can either roll or pop down to minimize this problem, but some work better than others (**Fig. 2–31**). If you wear eyeglasses, look through several models and see which work best for you.

How to Shop for Binoculars

Once you've narrowed down your choice of magnification power, objective lens size, and field of view, try the following tests on the array of suitable binoculars behind the store counter. Save your final decision regarding price until you've examined what's available.

- Compare binoculars of the same magnification power by holding one above the other. Alternately look through each binocular, comparing them for brightness and clarity. Then compare the best binoculars from your first selection with a third group—each time choosing the binoculars with the best characteristics. Continue this process of elimination until you have thoroughly examined everything that's available.

- Holding the binoculars at arm's length, check the exit pupils to see if they are blocked at the edges by gray shadows. Nearly all binoculars under $100 have a gray border obstructing the exit pupil.

- Look into the objective lenses to make sure that all optical surfaces are coated with an even purple-violet or amber hue. Carefully examine the objective and ocular lenses for scratches.

- Be sure that all the mechanical parts move smoothly and that the bridge supporting the barrels does not wobble.

- Outside the store, check alignment by looking at a rooftop or horizontal power line. Carefully examine the print on a billboard or sign to see if you can read the lettering at the edge of the field as well as at the center.

- Look at the edge of a backlit sign or building to see if it is fringed with a band of bright color. This fringing indicates an inferior optical system that cannot focus light of different wavelengths to the same point.

After narrowing the field to a few choices, select the highest-priced binoculars you can afford. Price is often a good measure of craftsmanship and materials. To produce lower-priced binoculars, manufacturers have to make compromises with the quality of their products. But even inexpensive binoculars can be good enough to launch your enjoyment of bird watching. You can always retire your

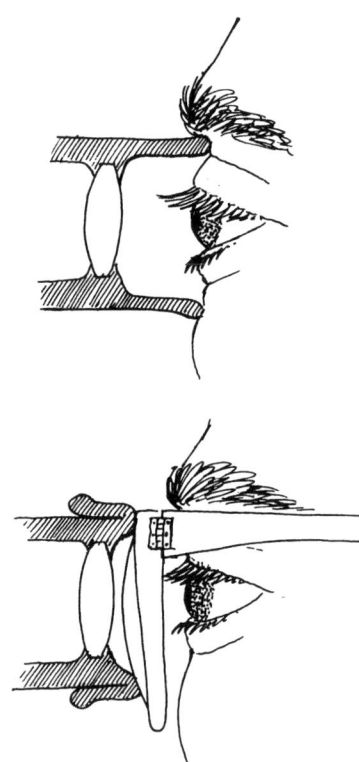

Figure 2–31. Binoculars for Eyeglass Wearers: To locate a bird and focus binoculars quickly on it, birders who wear eyeglasses should always keep their eyeglasses in place. Eyeglasses may prohibit the full field of view, though, by preventing the eyes from being close to the eyepieces. Many binoculars have rubber eyecups that can be rolled or folded down to minimize this problem.

first pair to backup status, or better yet, donate them to someone else who will appreciate them. Many bird observatories, clubs, nature centers, and schools are happy to receive donated binoculars.

How to Clean Binoculars

■ Binoculars should be cleaned frequently, following these suggestions:

- Thoroughly wipe off metal parts and lightly brush all lenses with a wad of lens-cleaning tissue or a soft camel's-hair brush to dislodge particles of sand and grit. Removing this debris keeps you from scratching the lens and its coating during the cleaning process. Hold binoculars upside down so that dirt will fall away from the lens surface.

- Fold a piece of lens-cleaning tissue so that it is at least four layers thick. This prevents oil from your fingers from soaking through the lens tissue and onto the lens surface. Use a circular movement to gently wipe all lens surfaces.

- If there is a film of oil on the lens, put a drop of lens cleaner on the tissue and repeat the circular wiping movement.

- Look for dirt on all the internal optics by holding the binoculars up to the light and looking into the objective lenses. Never attempt to open the binoculars; you can easily disrupt their alignment.

Although it's expensive, leave internal cleaning to the professionals.

Protecting Binoculars

- Never stroll through the woods swinging binoculars by the strap; banging them on a tree could throw them completely out of alignment. Always keep your binoculars around your neck in the field.

- When you have to jump across a ditch, climb a rocky slope, get into a boat, or do any other active maneuver, always tuck your binoculars inside your jacket or secure them under your arm.

- Never leave your binoculars on your car seat—a quick stop will send them flying—a sure way to knock them out of alignment. And never leave your binoculars out in the open in your car, especially on a hot summer day. If thieves don't find them, the sun may soften the lens coatings, causing them to crack and separate from the lenses.

- Keep binoculars under cover as much as possible if it starts raining. Water can leak into the housing, causing internal fogging and carrying in dirt, which can stain the internal optics. Rain guards offer some protection during light rain and drizzle, but they are not adequate protection for heavy rain. If your binoculars do fog up on the inside, set them in a warm, dry place, and they will probably dry out in a couple of days. Otherwise, fungus may start growing on

Figure 2–32. Binoculars Face Many Hazards: *Intrepid birders expose their binoculars to many perils. Binoculars must be rugged enough to withstand precipitation, salt spray, impact damage from scrambling over rocky shores and in and out of boats, abrasion from dust and sand, and exposure to extreme temperatures.*

the lens coating. Alternatively, leave them overnight in a sealed bag with some desiccant (purchased at a camera store) that will absorb the excess moisture. It is prudent to bring desiccant on birding vacations to humid climates, where your binoculars may not dry out on their own.

• If your binoculars fall into fresh water, have them professionally cleaned as soon as possible to avoid rusting. If you drop them in salt water, rinse them thoroughly in fresh water, seal them in a plastic bag, and rush them to a professional service department immediately. Salt water is amazingly corrosive and can turn fine binoculars into junk in just a few days.

In bird watching, binoculars face many hazards that they would never be subjected to at opera houses and football stadiums. They must be able to withstand precipitation and highly corrosive salt spray, and must be rugged enough to accompany birders as they scramble up rocky slopes, climb in and out of boats, lie down on sandy beaches, and hike through both wet and arid bird habitats **(Fig. 2–32)**. External-focus binoculars are particularly vulnerable to water and dirt, which may enter through the focusing apparatus. Dirty binoculars provide neither sharp detail nor crisp colors.

Selecting a Spotting Scope and Tripod

■ Spotting scopes are medium-range telescopes, usually with a magnification power between 15x and 60x. Most of them use interchangeable fixed-focal-length eyepieces or a zoom eyepiece to change magnification power or field of view. Telescopes designed for stargazing tend to be much more powerful, but they usually don't have sufficient light-gathering ability for effective bird watching. Spotting scopes provide the magnification necessary to see distant birds and to admire the detail at closer ranges.

The problems of less light and more vibration that accompany greater magnification power in binoculars also apply to spotting scopes. High powers magnify the air as well as the subject, often producing hazy images or distracting shimmering from heat vibrations over water and other flat expanses.

With good observation conditions and a steady tripod, the extra magnification power of a good scope will help you to spot birds and distinguish field marks that may be impossible to see with binoculars. When you're scanning an area with a spotting scope, however, it's best to start with a low-power eyepiece (or the lowest setting on a zoom eyepiece), and then switch to a higher power once you've located the birds you want to examine closely.

Zoom lenses offer the convenience of being able to change magnification power from 20x to 45x or even 60x with a single, simple adjustment. But viewing conditions are seldom good enough to go beyond 45x—the image generally becomes too dark to see much detail as you move toward 60x. The best all-purpose magnification power is 25x. The top spotting scopes are made with "ED" (Extra-low Dispersion) glass or have fluorite-coated lenses. The difference in brightness and image clarity between these special scopes and identical non-ED or nonfluorite scopes made by the same manufacturers is very noticeable, particularly in difficult, low-light viewing conditions.

For overall stability when you're using a spotting scope, a tripod can't be beat. If you don't like the weight or bulk of a tripod, however, you can mount a scope on a modified rifle stock, but don't use a more powerful eyepiece than 15x or 20x—you won't be able to hold it steady enough. Commercially built stocks for spotting scopes and telephoto lenses are available.

A rifle-stock-mounted scope is difficult to share with a group of bird watchers. If you want to give everyone in your group a good look at a bird, you should use a tripod. But buying a good tripod for birding can sometimes be more difficult than buying a scope. Tripods come in numerous heights and weights and with a wide assortment of heads. Some tripods are clumsy to use—they may have as many as nine different locks and clamps to control the extension of their legs. Not infrequently, you get the last leg secured just as the bird leaves. Some tripods are too heavy to carry around in the field, so they end up getting left at home. Other tripods are just too flimsy.

For birding, look for a moderate-weight tripod with a minimum of clamps and twisting parts. The most efficient for birding have "flip locks" to adjust leg length. They're easy to operate because once the locks are released, the legs fall to their own level and you can fasten them in place with the locks conveniently located at the top of each leg.

"Window mounts" are also available that allow you to temporarily attach your spotting scope to a partially open car window. These are particularly useful on long birding trips at locations where you must stay in your car, as at some National Wildlife Refuges, or in situations in which getting out of your car would scare or disturb the birds you are viewing. A car can be an excellent bird blind.

Although birders use scopes most often for long-distance viewing of birds that live in expansive, open habitats, spotting scopes also can provide intimate views of small land birds perched at close range. Such scope-aided views frequently reveal the intricate beauty of a bird's plumage and allow you to observe behavior that might otherwise go unseen. Even tiny, secretive warblers sometimes sit still long enough to view with a spotting scope, especially when they're singing on their territories rather than flitting through the forest canopy in search of insects.

How to Shop for a Spotting Scope

- The best all-around eyepiece for a birding spotting scope is 25x. Because of the effects of heat distortion and loss of light, eyepieces larger than 45x usually are useless for birding.

- Ideally, the objective lens (the one farthest from your eye) should be at least 60 mm in diameter to provide adequate light.

- Zoom lenses that vary in power from 20x to 45x are ideal for most bird watching. They permit convenient scanning at low power and then a quick shift to higher power for looking at details. But many of the less expensive zoom eyepieces are optically poor. The only good zoom eyepieces I've seen are the ones made by the top optical companies for their high-quality scopes.

- Don't buy a cheap spotting scope. Inexpensive scopes deliver fuzzy, distorted images. The shortcuts the manufacturer took to deliver a low-cost product will only give you disappointing field performance and splitting headaches.

- Select a rigid tripod with as few leg adjustments as possible. The flip-lock design provides a secure mount for your scope and a quick way to set the legs on uneven terrain.

Binoculars are best for close-up birding, but for distant birds such as waterfowl and hawks, spotting scopes can expand your vision at least three times beyond that of binoculars. You'll be amazed what a difference that makes.

Recording Observations

As I discussed at the beginning of this chapter, giving a name to a creature you've encountered in nature opens a door to exploring the many facets of its life. To keep the door open, to turn your ephemeral memories of your daily experiences into a durable record of the natural world as it existed in a certain place at a certain time, you must record your observations.

This can be a source of personal pleasure; the notes will help you relive your field experiences. In an article about keeping field notes, for instance, the Lab of Ornithology's Director of Education, Rick Bonney, wrote, "My field notes from a trip to the Everglades in 1986 tell me that on March 15, at 11:05 A.M., I saw a Snail Kite hunting over the marsh

behind the Miccosukee Restaurant—an entry that immediately recalls the smell of coffee and refried beans." But even more important, your written records make the benefits of your time in the field available to all who have reason to care about the abundance of birds. Rick Bonney (1991) also wrote this account of how documenting what you see can have scientific value:

Figure 2–33. Male Blue Grosbeak: *Always keep accurate field notes–you can never predict when they might have scientific value, such as documenting the occurrence of a species outside its normal range. Photograph by Tom Vezo.*

> *I remember the morning well. It was early spring 1981, with lifting fog and smell of earth. The season's first bird songs had lured me out of bed, and I was tramping through the state forest behind my home looking for early migrants. Around 8:00 A.M. I had just started back toward the house when I heard a strange song, one I knew I'd heard before but couldn't place. Slowly I crept toward the bird until I could see it silhouetted against the sky, perched atop a large shrub. At first I thought it was an Indigo Bunting singing a weird song, but after a harder look I realized that the bird was a Blue Grosbeak, a species I knew well from the South but had never before seen in Upstate New York* **(Fig. 2–33)**.

> *As it turned out, very few people had seen it, perhaps only one. For no good reason I didn't think to look up the bird's status right away, but several months later I was perusing* Birds of New York State *by John Bull and read this: "As far as I am aware, the only Upstate report of a Blue Grosbeak with details is that of a male observed near Lake Champlain on June 17, 1964."*

Wow! I thought. My Blue Grosbeak had been a really good bird. I should have reported it. I should still report it. When had I seen it? Let's see, it was late April—or was it early May? Already I couldn't remember. And not only had I neglected to record the date, I hadn't recorded any information about the bird at all—so even if I could reconstruct the timing, I had no documentation that would prove to anyone other than myself that a Blue Grosbeak had decided to visit Willseyville, New York, in the spring of 1981.

The moral is quite simple: take field notes. . . . Such observations have scientific value. My sighting, for example, would have been useful to ornithologist Janet Carroll when she compiled information on the spread of the Blue Grosbeak into New York for the state's breeding bird atlas, published in 1988—that is, it would have been useful had it been properly documented and recorded.

Although we all enjoy seeing rare birds, it's not just unusual sightings that warrant documentation. Even lists of birds common to a certain place are valuable if they include numbers of birds seen and are carefully made. Bird populations change constantly, and their ups and downs often reflect changes in the environment. Only birders meticulously recording the numbers of birds seen at various localities can properly document these changes so that scientists can look for patterns and try to explain what the changing populations mean.

What should you actually write when describing your forays into the field? There's a broad range of options. Your choice will depend on the nature of your field excursions, your goals in keeping notes, and on how much time you can devote to your record keeping.

Checklists

The kind of field record used most often is the **checklist**, a printed list of the birds found in a particular area. Unfortunately, most lists contribute little to our understanding of bird distributions and abundance because observers check off the species they saw on a particular trip without indicating how many they saw or exactly where they saw them. And, in their search for new and unusual species, many birders ignore the most common birds. But it is precisely these birds, and not the solitary wanderer far from his usual haunts, that can be sensitive indicators of environmental changes. Checklists are much more valuable when they include an actual count or an estimate of the numbers for *all* species **(Fig. 2–34)**.

Most checklists provide space for some other crucial information: the exact location (compass direction and distance in air miles to the nearest town), the number of hours you spent in the field, and the weather conditions. These basics ensure that your checklist will have meaning for anyone who wants to use the data. Most observers, especially beginners, find that the best way to keep their checklists up to date is to fill them out in the field; the complications of life thwart our

The Edwin B. Forsythe National Wildlife Refuge's Brigantine and Barnegat Divisions contain more than 42,000 acres of southern New Jersey coastal habitat. Refuge headquarters and public use facilities, including an eight-mile Wildlife Drive, observation towers and two short nature trails, are at the Brigantine Division. Best birdwatching opportunities occur during spring and fall migrations. A "Guide to Seasonal Wildlife Activity" is available in the refuge's general brochure.

This folder identifies 293 species that have been observed at the Brigantine and Barnegat Divisions. Names and order of listing are in accordance with the Sixth American Ornithologists' Union Checklist.

Most birds are migratory. Their seasonal occurrence is coded as follows:

SEASON

s	Spring	March – May
S	Summer	June – August
F	Fall	September – November
W	Winter	December – February

• Birds known to nest on or near the refuge
Italics indicate threatened/endangered species

RELATIVE ABUNDANCE

a	abundant	a species which is very numerous
c	common	likely to be seen or heard in suitable habitat
u	uncommon	present, but not certain to be seen
o	occasional	seen only a few times during a season
r	rare	may be present but not every year
x	accidental	seen only once or twice on refuge

NOTES

Location FORSYTHE / BRIGANTINE DIV.
Date MAY 9, 1998 Time 1:30 – 5:00 pm
Observers BRUCE TRACEY + CLASS
Weather MAX. TEMP 55°F. FREQUENT RAIN.
NW WIND 10-15 mph, WITH GUSTS UP
TO 25-30 mph.

LOONS – GREBES	s	S	F	W
___ Red-throated Loon	o		o	u
150+ Common Loon	o	r	o	o
___ Pied-billed Grebe	u	o	u	o
___ Horned Grebe	u		u	u

SHEARWATERS – PELICANS – CORMORANTS	s	S	F	W
___ Sooty Shearwater	r	r		
___ Northern Gannet	r	r	u	
___ American White Pelican	r			r
___ Brown Pelican		u		
___ Great Cormorant				r
12 Double-crested Cormorant	u	o	c	u

BITTERNS – HERONS – IBISES	s	S	F	W
___ • American Bittern	u	o	u	u
___ • Least Bittern	u	u	u	
2 • Great Blue Heron	c	c	c	u
25 • Great Egret	c	c	c	o
50+ • Snowy Egret	c	c	c	o
___ • Little Blue Heron	u	u	u	o
___ • Tricolored Heron	u	u	u	o
11 • Cattle Egret	u	u	u	
___ • Green-backed Heron	u	u	u	
1 • Black-crowned Night-Heron	u	u	u	u
___ • Yellow-crowned Night-Heron	o	o	o	
___ White Ibis		r	r	
12 • Glossy Ibis	a	a	o	o
___ • White-faced Ibis		x		

SWANS – GEESE – DUCKS	s	S	F	W
___ Tundra Swan	o		u	o
___ • Mute Swan	c	c	c	u
___ Greater White-fronted Goose			o	o
1 Snow Goose	a	o	a	a
___ Ross' Goose			r	r
200+ Brant	a	o	c	c
100+ Canada Goose	c	c	c	c
___ • Wood Duck	u	o	u	r
___ • Green-winged Teal	c	o	a	c
6 • American Black Duck	a	c	a	a
4 • Mallard	c	u	c	o
___ • Northern Pintail	c	o	a	u
___ • Blue-winged Teal	c	u	a	o
___ • Northern Shoveler	c	o	c	c
___ • Gadwall	c	c	c	u
___ Eurasian Wigeon			r	r

resolve to fill out our records "as soon as we get home," and days of observations are likely to be lost.

Your complete daily checklists may be useful in preparing and updating local and regional checklists, and their value increases over time because they are an important source of baseline data for detecting population changes. Information from a number of observers over a broad area can provide an early warning signal that populations of a formerly abundant species are declining. Your daily records will become historic accounts that someday could help restore species to habitats where they once flourished.

Journals

Recording field observations in a **journal** suits the purposes of many bird watchers willing to spend just a little extra time **(Fig. 2–35)**.

Figure 2–34. Filling Out a Bird Checklist: A checklist is a printed list of the birds found in a particular area. Filled out correctly and completely, checklists can be meaningful sources of data on bird distribution and changing abundance. Don't just check off the species; include an actual count or an estimate of the numbers of birds of all species seen. Fill out the exact location, number of hours spent in the field, and the weather conditions. For accuracy's sake, the information should be filled out in the field. Checklist reproduced with permission, Edwin B. Forsythe National Wildlife Refuge, United States Fish and Wildlife Service.

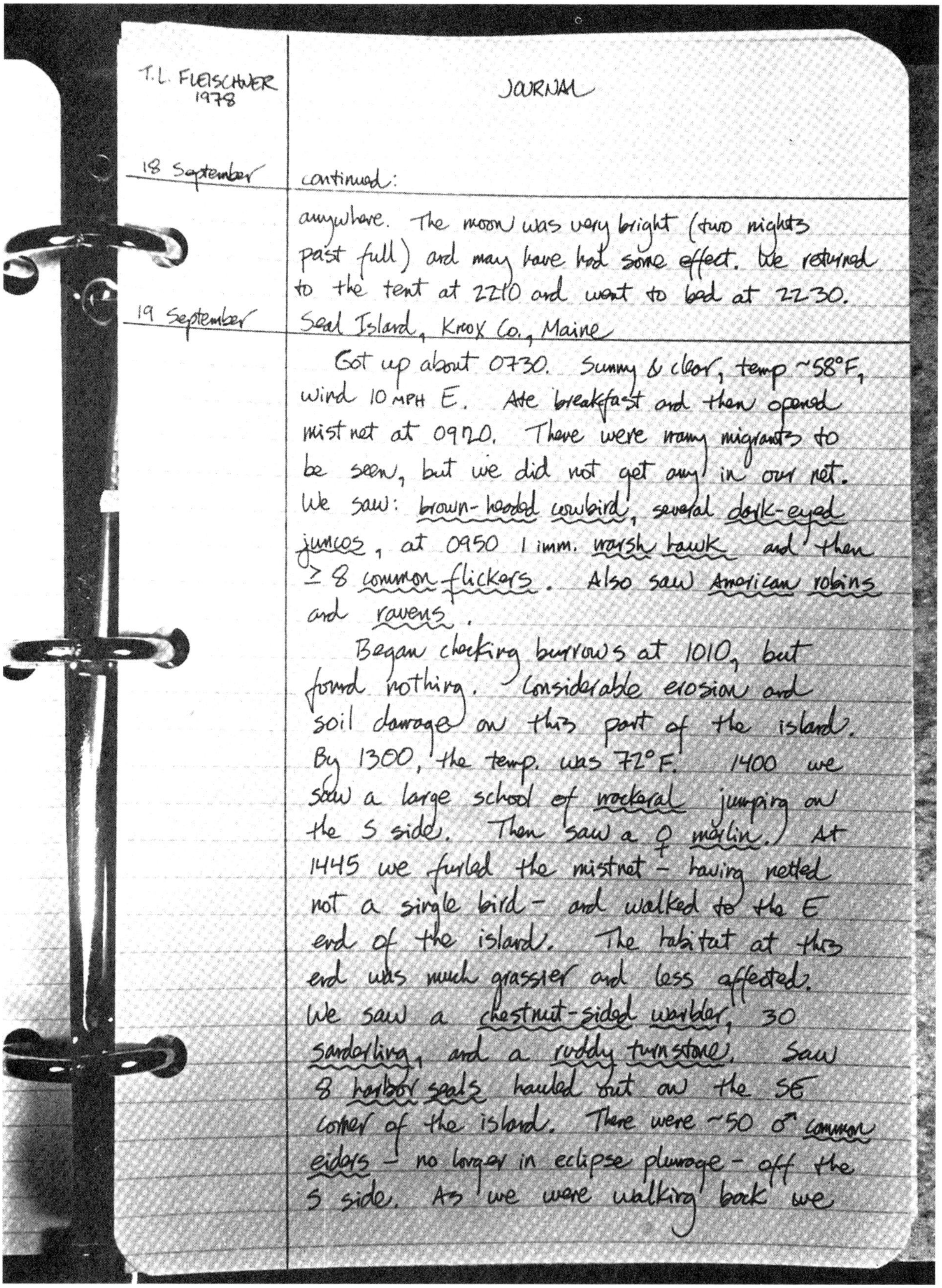

T.L. FLEISCHNER
1978

JOURNAL

18 September | continued:

anywhere. The moon was very bright (two nights past full) and may have had some effect. We returned to the tent at 2210 and went to bed at 2230.

19 September | Seal Island, Knox Co., Maine

Got up about 0730. Sunny & clear, temp ~58°F, wind 10 MPH E. Ate breakfast and then opened mist net at 0920. There were many migrants to be seen, but we did not get any in our net. We saw: brown-headed cowbird, several dark-eyed juncos, at 0950 1 imm. marsh hawk and then ≥ 8 common flickers. Also saw American robins and ravens.

Began checking burrows at 1010, but found nothing. Considerable erosion and soil damage on this part of the island. By 1300, the temp. was 72°F. 1400 we saw a large school of mackeral jumping on the S side. Then saw a ♀ marlin.) At 1445 we furled the mistnet — having netted not a single bird — and walked to the E end of the island. The habitat at this end was much grassier and less affected. We saw a chestnut-sided warbler, 30 sanderling, and a ruddy turnstone. Saw 8 harbor seals hauled out on the SE corner of the island. There were ~50 ♂ common eiders — no longer in eclipse plumage — off the S side. As we were walking back we

Figure 2–35. Keeping a Field Journal: *Recording observations in a field journal is worth the extra time and effort. As with a checklist, include exact location, date, and weather conditions, along with detailed information about the birds seen. Underline species names with wavy lines so they will be easy to locate at a glance. Supplement written material with sketches, where needed. Courtesy of Stephen W. Kress, from* The National Audubon Society Handbook for Birders, *1981.*

As with a checklist, certain information is crucial to note: date, time of day, location (distance and compass direction to nearest town, name of county, state, and country if you're abroad), weather conditions, and persons with you, if any. Then, simply write down what you see—species, numbers, ages, sexes, and other identifying characteristics. Underline species names with wavy or double lines, so they will be easy to locate at a glance. A good technique for recording field marks is to start by looking at the bird's head and work your way back to the tail. With practice, you can use quick sketches to map field marks and capture behaviors **(Sidebar 3: Sketching Birds in the Field)**.

One benefit of taking notes is that it improves your powers of observation and memory. Think of a bird you have seen many times, perhaps a Black-capped, Carolina, or Mountain chickadee. Can you sketch its black-and-white pattern from memory? Exactly where do the black cap and bib begin and end? Once you have looked closely at a bird and tried to write a description or sketch it, you will notice and remember more about it.

Besides recording the species you see and their descriptions, you may want to describe the birds' behaviors. If you find a nest, observe it for a period of time and record the birds' activities. Accurate field notes about behavior are just as critical as detailed descriptions of bird sightings. For the field biologist, behavioral notes are a source of data on how a bird relates to its environment, just as laboratory experiments provide the database for a physical scientist.

Take complete notes. You never can tell what seemingly unimportant facts may later become decisive. Greg Butcher, a former biologist at the Lab of Ornithology and editor of *Birder's World* magazine, tells about watching what he thought was a Hooded Oriole (a resident of the southwestern United States) singing in Niantic, Connecticut. The song sounded funny, though, so he wrote it down, syllable by syllable. Later he discovered the exact same description of an oriole's song in his field guide—and learned that his bird had actually been a first-year male Orchard Oriole, a common bird in his location.

A few nuts-and-bolts considerations: you can keep your notes in a loose-leaf notebook or a bound notebook. Waterproof ink is best, but you can use pencil. If you do use ink, select paper with a high rag content; it will hold the ink better and will not yellow with age. You'll also need a technical pen with a tip of approximately 0.35 mm; pens that draw a narrower line are too likely to clog. Such pens are available at art supply stores. Keep them in plastic bags, especially when traveling, to contain leaks.

The cardinal rule, and one that requires a good bit of self-discipline, is to record everything in the field as it occurs. You can't possibly remember, at the end of a long day, everything that happened. Write things down at once, before your memories slip away. If you hate writing in the field, take along a cassette recorder, record your stream-of-consciousness account, then transcribe your notes after you get home. Microcassette players that fit in a shirt pocket are great for this purpose. Be

(Continued on p. 2·52)

Sidebar 3: SKETCHING BIRDS IN THE FIELD
Stephen W. Kress

A quick field sketch, with pertinent field marks noted, can be invaluable when you are trying to recall exactly how a bird looked or behaved. Because a sketch can replace part of your written description, it also can save you some writing time in the field, although it should be accompanied by thorough notes. In addition, a sketch will allow you to convey your memories to others much more clearly than you could with just a written record.

A few basic techniques will allow anyone, no matter how little drawing experience he or she has had, to make a useful sketch. Start your field sketch with an oval that approximates the general proportions of the bird. Regardless of whether you intend to sketch an owl, heron, or robin, they all have oval (egg-shaped) bodies (**Fig. A**). It is the differences in wings, tails, and legs that give each species a distinctive form. Watch carefully to see at what angle the bird holds its body, then begin to assemble body parts, outlining

Figure A. Use Ovals to Begin your Bird Sketch: *The bodies and heads of most bird species are roughly oval in shape, so you can begin drawing by placing ovals of the right size in the orientation you wish to portray. Then add the distinctive legs, wings, tail, and beak for your species. Drawing by Anne Senechal Faust, from* The National Audubon Society Handbook for Birders, *by Stephen W. Kress, 1981.*

head and neck, wings, tail, and legs. Concern yourself with the proportions of the different parts to one another and the position of attachment, always referring to the living bird.

Draw with smooth, flowing lines to achieve an outline sketch of the bird. Don't worry about erasing mistakes or lines you don't like. The goal should be to capture shape and posture with as few lines as possible before the bird flies away.

Portray most behaviors by changing the posture of the bird's body (position of the oval) and the position of its appendages. If you see an unusual bird or one you can't identify, quickly draw a standard perching posture and then add details to illustrate distinctive field marks, carefully noting these in the margins of the page **(Fig. B)**.

Practice by sketching tame birds, such as captive parakeets, pigeons, or feeder birds. Perched postures are easiest, but it won't take long before a few pencil lines will also capture the movement of birds in flight. ■

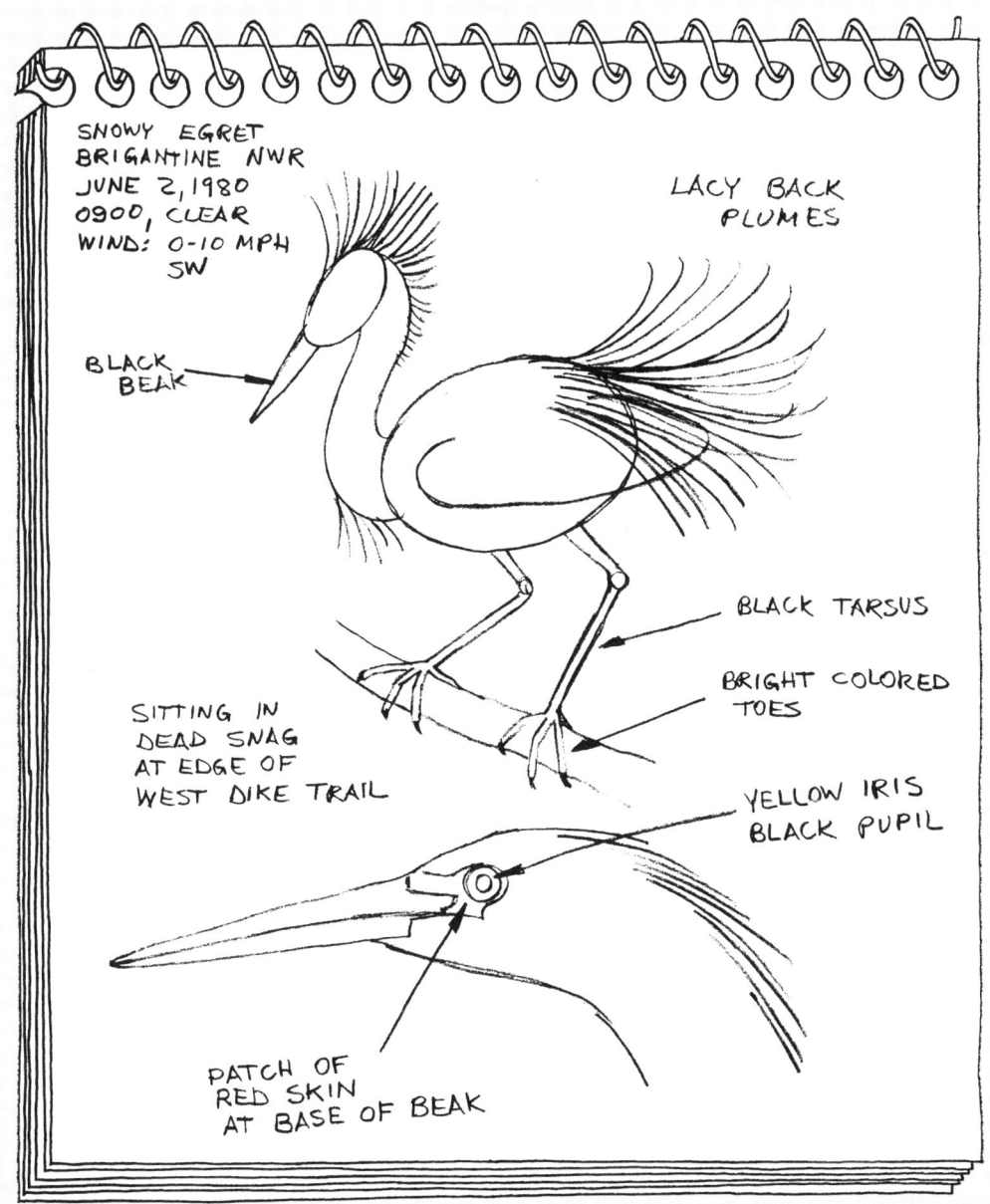

Figure B. Sketching for Identification: *When you encounter a bird you do not recognize, first make a quick sketch of the general shape. Then add notes in the margins to detail all the special field marks and features you observe. Drawing by Anne Senechal Faust, from* The National Audubon Society Handbook for Birders, *by Stephen W. Kress, 1981.*

aware, though, that it's easy to end up with a shelf full of untranscribed cassettes. If you need to refer back to a specific incident, finding the right tape can take a while. Still, tapes are better than no record at all. Keep in mind, too, that cassette recorders break and batteries fail. Pencil points can always be sharpened with a pocket knife or even with your teeth in the field.

Finally, regardless of the technology you use to record your observations, follow these three rules at all times: record your observations as soon as possible after making them; don't consult references before writing up your field notes—your impression of what you *actually* saw may be influenced by what you think you *should* have seen, rendering your notes less accurate and useful; and never change your notes—these are the records of your observations and should remain as you first made them.

Once you've completed your initial field notes on your excursion into the field, you have several options. Some observers consider their recordkeeping done at this point. Others enjoy using their field notes to compose a more structured set of records, organizing the material in various ways for easy access. The most devoted observers may use the straightforward, standardized note-taking system established by Joseph Grinnell in the early 1900s, which is described in Herman (1986). This system in its entirety is too demanding for most recreational birders to use routinely, but aspects of it can be adapted to the needs of the weekend naturalist.

One enjoyable way to learn a lot about individual species is to keep **species accounts**, transferring observations from your field notebook into another notebook that you've organized by species **(Fig. 2–36)**. Then all your observations on, say, the Black-capped Chickadee will be grouped together. A loose-leaf notebook is the best choice for this effort, because it allows you to add pages to your existing notes on each species. If you follow this procedure for a few species that are of particular interest to you, you'll soon become an expert on them.

Now, what are you going to do with all this dutifully recorded information? At first, it's not crucial that the data you've gathered be published. As you learn to observe birds closely, you are learning far more about the specific bird you are watching than you could ever learn from a book. But even in the beginning you should get in the habit of reporting your sightings to your local bird club, which will probably use them to update local and regional checklists. You might even want to file your daily checklists with the club. The value of your lists increases with time, providing important baseline data for detecting population changes. Information from a number of observers—you and your fellow club members—over a broad area can serve as an early warning signal that a formerly abundant species is experiencing a serious population decline.

Also, many journals produced by state bird organizations welcome carefully documented reports for publication. Your local bird club can probably put you in touch with the regional editor for your locality. And, if you live near a city, county, state, or national park or any other sanctuary or refuge, check with officials there to see if they

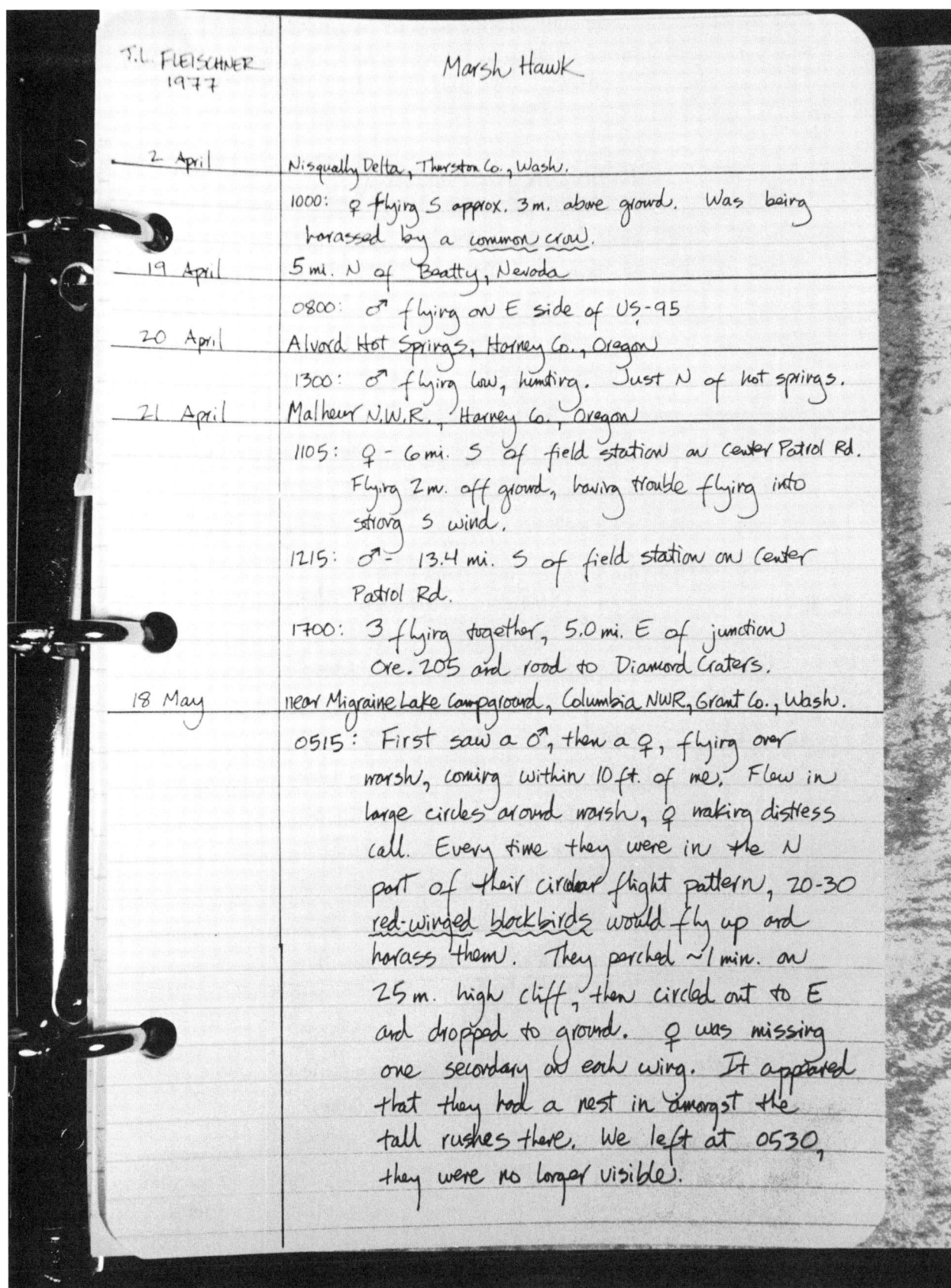

Figure 2–36. Sample Page from a Species Account: *An enjoyable way to learn about particular birds is to organize your observations by species, accumulating them over time into species accounts. Courtesy of Stephen W. Kress, from* The National Audubon Society Handbook for Birders, *1981.*

would like reports of your sightings to maintain their checklists.

Finally, if you begin keeping detailed species accounts, you will probably accumulate enough information to write short reports on those species for publications, particularly if you pick species that have not been well studied. You could become the regional expert.

Reporting Rare Birds

If you see an unusual bird, make your notes as detailed as possible, because validation of such sightings may well depend on your notes. Record details of color, plumage, and behavior, as well as the conditions under which you are making the sighting, such as the lighting and your approximate distance from the bird. You might even make a quick sketch. If possible, photograph the bird and tape record its songs and calls.

When you report a rare bird or even a common bird in an unexpected season (a Blackburnian Warbler in Michigan in winter, for instance), your observation may need to be verified before it is accepted as part of the official local, state, or national record. Procedures vary from state to state, but as an example, the New York State Avian Records Committee (NYSARC), a group of experienced birders, reviews the accuracy and completeness of the field description in each report, then decides whether to accept or reject the sighting.

Listing Birds

Many birders enjoy keeping a variety of separate bird lists, not necessarily as checklists. The possibilities are countless—lists of birds at the feeder, in the garden, or on a field trip; daily lists, weekly lists, yearly lists, or a "yard list." Listing is especially fascinating during the migration periods. A daily list in the spring, for example, follows the changes from the seed-eaters, such as finches, to insect-eaters, such as warblers and flycatchers. In species with sexes of different colors, you can keep separate records for males and females. You can compare the lists for different seasons and note which birds migrate and which do not. Remember, the cardinal that nests near your back door may not be the same one you fed all winter (although it often will be). A yearly list may reveal important population trends in your area corresponding to habitat changes.

Many bird watchers keep a **life list**—a record of every species they have ever seen with the date and place of the first sighting. The life list, like solitaire, is your own game in which you make your own rules concerning which species to count. Few birders would consider the colorless blur that flashed past just as someone called out "juvenile Lincoln's Sparrow" an honest candidate for their life list, however. The constant challenge of the life list widens the scope of your bird-watching activities and leads to exciting new experiences with birds. Standing as constant goals are the North American life lists of some expert bird watchers, a few of whom topped the 700 mark some time ago and are now aiming for 800 North American species. Few of us will ever attain such lists, although it is fun to try.

People vary in their attitudes toward life lists. Some derive their greatest pleasure from just watching familiar birds in their own habitat; others find great satisfaction in traveling from place to place, checking off the "lifers." But it is sad when the life list becomes a birder's only goal, because so many opportunities for enjoyment and contributions to bird conservation are overlooked. Horace's golden mean—"Moderation in all things"—applies to bird watching as well as it does to other human activities.

Counting Birds

■ Counting birds accurately requires lots of practice. It's not hard to count a few chickadees or a small flock of crows, of course, but when the birds fly past you in multitudes rather than dozens, or when several species flock together, counting accurately can be very challenging. How can you judge the size of a mixed flock of Canada and Snow geese that stretches from horizon to horizon or a huge, roiling mass of foraging blackbirds? But even difficult, laborious counts are more useful than "ballpark estimates." The challenge is to give as accurate a count as possible.

If the lighting is poor or the birds are far away, making an exact count impossible, presenting round numbers is best. A flock containing 100 or fewer individuals should be rounded off to the nearest 5 or 10 birds; a flock with more than 100 birds may be rounded to the nearest 25 or 50, depending on your viewing opportunities. Trained observers usually can give a good estimate of large numbers, and such estimates are better than ranges, which are more difficult to compare than specific numbers.

Flocks of flying birds, such as waterfowl, shorebirds, and blackbirds, are among the most difficult to count. Their speed, movement, and habit of flying in dense, three-dimensional flocks contribute to the problem of making reasonable estimates. Continued practice and a few additional techniques will help. To determine the number of birds congregating at a certain point (herons or blackbirds returning to a roost, for example), count how many pass a tree, house, or other fixed point for one-minute periods throughout the time during which the birds are returning. Average your one-minute counts and multiply the average by 60 to find the number of birds passing the reference point in one hour. To compute a grand total, multiply the average number of birds-per-minute by the total duration of the procession in minutes.

Exact counts are usually possible if a flock contains fewer than 30 birds. For larger flocks, try a technique called **blocking (Fig. 2–37)**. This approach entails counting the birds in a "block" of typical density from the trailing end of the flock (so that birds are not flying into your projection) and then visually superimposing this block onto the rest of the flock to see how many times it will fit. If a flock contains about 100 birds, count the trailing 20 and fit this onto the remainder of the flock; it should "fit" about five more times. For huge flocks, start by choosing

Figure 2–37. The "Blocking" Method of Counting Birds: First count the birds in an imaginary block of typical density from the trailing end of the flock (this avoids the distraction of birds flying into the block). Then visually superimpose the block onto the entire flock and estimate how many times it fits. Finally, multiply this number by the number of birds in the original block. In this example , the block contains 17 birds and fits into the flock about 3 times, giving an estimate of 51 birds. There are actually 60 birds in the flock.

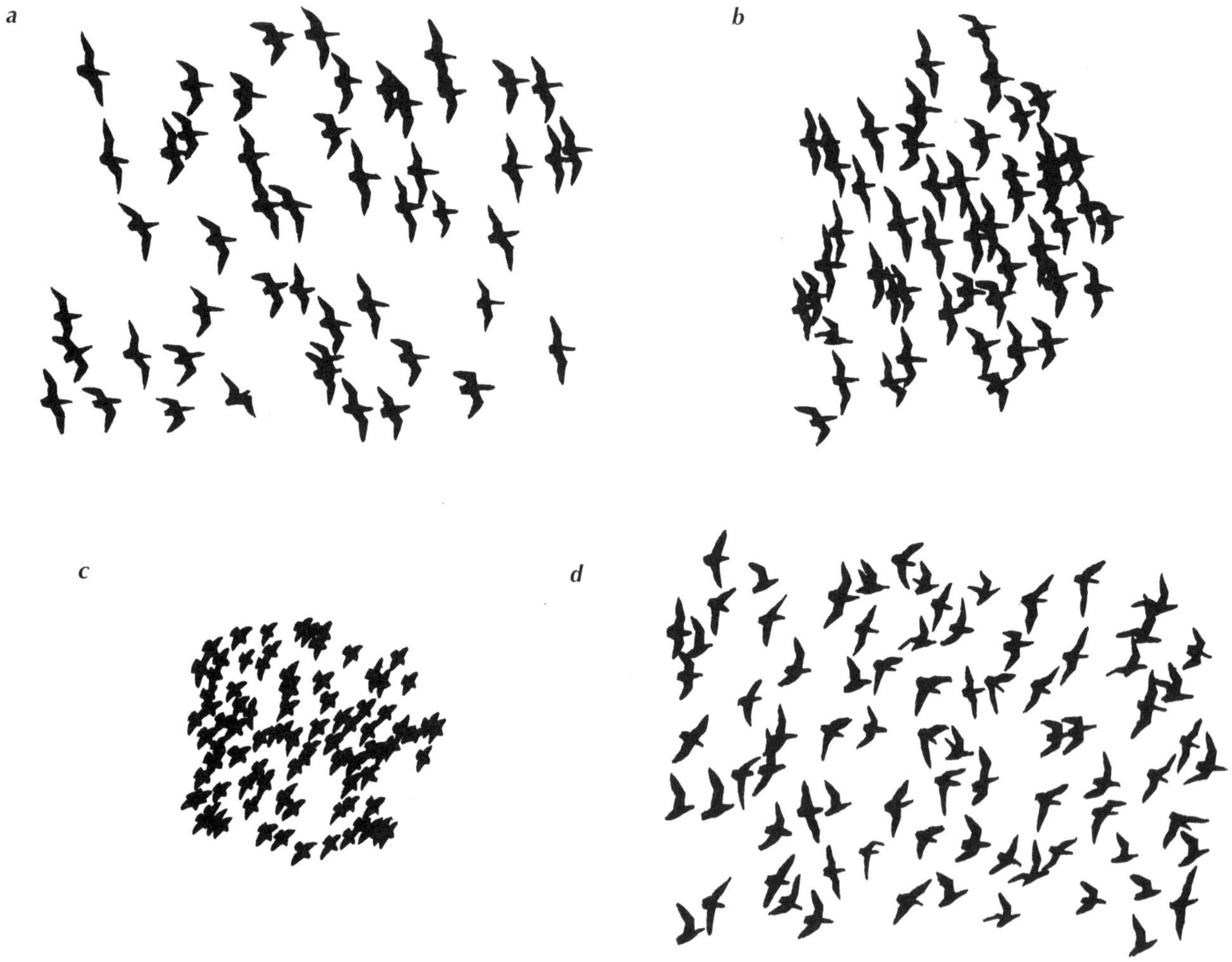

a

b

c

d

Figure 2–38. Practice Flocks: Bird flocks vary widely in size, shape, and density. Use these examples to practice your counting technique. The actual number of birds in each flock is given at the end of the chapter.

a group that represents 50 or 100 birds and see how many times this block fits on the flock as a whole.

Concentrate on memorizing impressions of what flocks of different sizes look like. Practice by throwing rice grains out on a table-top, making a quick estimate, and then checking your success. With practice you can develop mental images of different-sized flocks of various shapes **(Fig. 2–38)**.

The estimates of different observers looking at the same birds vary to a surprising degree, but accuracy is important. Many North American bird censuses and surveys rely on amateur participants, and the success of these important studies depends largely upon the counting skills of participants.

Figure 2–39. Great Blue Heron: The magical experience of watching a Great Blue Heron take flight from a misty wetland may be all that a newcomer needs to begin a lifetime of fascination with birds, and with it a concern for the natural environment. Photomontage by Marie Read.

Conclusion

■ We are all teachers. Our fascination with birds is one of the greatest gifts we can pass on to others—friends, neighbors, family, and especially children. More than 60 million Americans already watch birds, so it appears that birds themselves may be the ultimate environmental educators. All we need to do is direct more of the uninitiated into the realm of birds, closing the critical gap of distance between bird and human, and let the birds do the rest with their magic. Initiate a newcomer with a close encounter with a Great Blue Heron lifting out of a misty wetland, and the conversion starts **(Fig. 2–39)**. You can continue by sharing intimate views of elegant waxwings plucking apple blossoms and bluebirds resplendent in glowing color. "How could I have missed all of this?" they ask. "No longer," they insist.

Perhaps it is envy of their flight, colors, and stamina, or delight in their enchanting songs and remarkable behaviors. Perhaps it is awe at their mysterious migration and boundless vitality. For whatever reason, birds capture our attention and our imagination, whether we stay at home and let the migrant flocks flow into our lives and out again, or we pursue them by foot, bike, plane, or ship. Once people notice birds, commitment to their well-being usually follows. This connection is at the soul of birding: the birds' future is intimately tied to our own.

The Birder's Essential Resource Guide

■ There is a mountain of information available these days to birders with various levels of interest and experience in the form of field and audio guides, checklists and travel guides, textbooks, popular magazines, scientific journals, web sites, general and leisure reading about birds. Due to the volume of information available and the continual release of new publications of interest to bird watchers and other enthusiasts, to list just a few of these resources here would be vastly incomplete. Please visit the Home Study Course website <**www.birds.cornell.edu/homestudy**> to view the Birder's Essential Resource Guide online and see what resources, old and new, that we believe might be of interest to you.

Form and Function:
The External Bird

George A. Clark, Jr.

 One reason people love birds is that they are so beautiful and so varied. Consider color alone, ignoring for the moment the many variations in features such as bill length and feather shape. Birds come in every conceivable hue. Any experienced birder can easily think of a bird for every color of the rainbow, even among the birds of North America: Northern Cardinal, Baltimore Oriole, Yellow Warbler, Green Jay, Eastern Bluebird, Indigo Bunting, and Purple Martin. In fact, the blues alone vary from the subtle hues of the Blue-gray Gnatcatcher and Cerulean Warbler to the deeper tones of the Blue Grosbeak and Steller's Jay. Furthermore, some species, such as the Painted Bunting, simultaneously display an amazing array of these colors. Nearly every color we can imagine can be found on at least one of the nearly 10,000 species of birds.

Most of the remarkable colors and other aspects of a bird's exterior are specializations of its skin—not only the coverings of the face, beak, and legs, but even the feathers. Feathers occur on no other living animal: if it has feathers, it is a bird. Some years ago a mutant breed of nearly featherless chickens was studied with the idea that these birds could be sold cheaply because they would need little or no plucking. It turned out that the extra heating required to keep these birds warm more than offset the savings on plucking; featherless birds, it would seem, don't have much of a future.

This chapter introduces feathers—their locations, structure, function, care, and development, including their replacement through molting. Then we move to other parts of the skin, such as the bill and legs, where a feathered covering would in many cases be a liability. Each part of a bird's exterior contributes to the unified whole, and we give special attention to the color and pattern of the entire bird, considering, for example, why some birds have vivid colors but others are dull.

Feather Tracts

Examine a plucked chicken or turkey ready for the oven. You can see the sites where feathers were attached, known as **follicles**, as small indentations in the skin. In most birds, the feathers are not attached uniformly over the body, but are grouped into **feather tracts** called **pterylae** (singular: **pteryla**). Between the tracts are regions of bare or less-feathered skin called **apteria** (singular: **apterium**) **(Fig. 3–1)**.

Pterylosis, the arrangement of feather tracts and bare patches, varies from one taxonomic group to another, and some groups have unique patterns. For instance, the corvid family, which includes ravens, crows, jays, nutcrackers, and magpies, has a characteristic apterium in the midline of the pteryla located on the back. Historically, such distinctive patterns have been important in classifying the main groups of birds. The feather patterns are symmetrical from one side of the body to the other and are shared by all members of a species.

Although we know little about the possible functions of the many taxonomic variations in pterylosis, we can deduce that, for any given group, strategically locating feathers in tracts may allow birds to get away with fewer feathers overall, thus reducing the baggage they must carry in flight. Much as balding men comb their remaining hair to cover a bald spot, feathers from the feather tracts cover the apteria, forming flight surfaces and providing insulation without requiring a solid mass of feathers. Grouping feathers in tracts also may allow the muscles that move them to be smaller and more localized—lightening the bird even more. As discussed in Chapter 5, birds have evolved many such adaptations to reduce their weight and create an aerodynamic shape. In addition, apteria may aid heat loss, as many birds raise their feathers to expose bare skin when becoming overheated.

A few kinds of birds, most notably penguins, have a continuous pterylosis with no apteria. This arrangement helps to prevent water from penetrating to the penguin's skin and chilling it. Penguin feathers are so good at trapping an insulating layer of air near the body that the skin stays dry even while the bird is swimming and diving. Adult Ostriches from Africa also lack apteria, but their embryos have them. Because many researchers believe the stages that developing embryos pass through indicate the stages their ancestors evolved through, the lack of apteria in adult Ostriches is probably a secondary condition. The ancestors of Ostriches undoubtedly had a more typical feathering and perhaps those of penguins did as well.

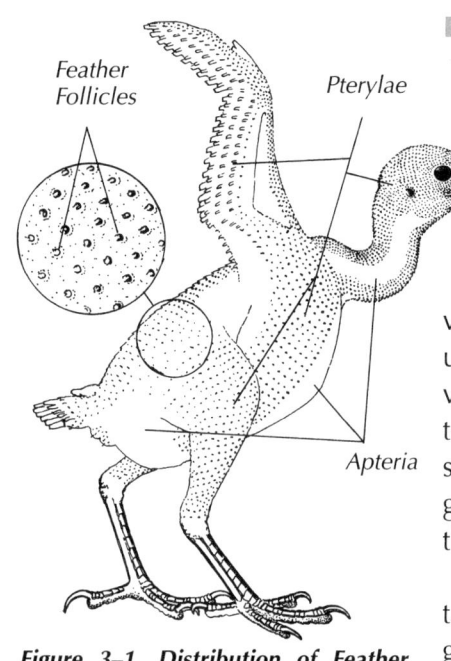

Feather Follicles

Pterylae

Apteria

Figure 3–1. Distribution of Feather Tracts on Plucked Bird: *Dots represent feather* **follicles**, *sites of attachment of feathers. Groups of dots, often in a linear pattern, are feather tracts, termed* **pterylae**. *Featherless areas between tracts are* **apteria**.

Feathers lie on the body and wings with impressive neatness. Such orderliness helps to produce the tidy, streamlined cover of feathers that is crucial to a bird's survival.

Feather Form and Function

■ From an engineering viewpoint, feathers are magnificent—they accomplish so much with so little material. Of their many functions, the most important are (1) insulating and protecting the skin and body, (2) providing the smooth, streamlined surface area required for efficient flight, and (3) providing pattern and color, which are important in social behavior, as discussed later in this chapter and in Chapter 6.

Feather Structure

Examine a feather from the wing or tail of a bird. Note the rather stiff central **shaft** and the two broad **vanes** extending from opposite sides of the shaft **(Fig. 3–2)**. In birds capable of flight, the outer wing and tail feathers are typically asymmetrical, with one vane narrower than the other. This asymmetry produces greater rigidity on the leading (narrower-vaned) edge of the wing, which is needed to maintain the streamlined shape of feathers during flight. It also causes individual wing and tail feathers to twist as they move through the air, which is essential for flight (see Figs. 5–17 and 5–19).

In general, vane width becomes less symmetrical as you move farther from the center of the body. Thus, a feather from the middle of the tail will be more symmetrical than one from the edge of the tail. In the same way, the vanes of the innermost flight feathers of the wing may be almost symmetrical, while the vanes from the outermost feathers are quite different in width **(Fig. 3–3)**.

When flight has been lost secondarily through evolution, as in certain rails on isolated oceanic islands, the primaries have lost their asymmetry. The asymmetry in the fossilized primary feathers of *Archaeopteryx* indicates that this earliest known bird could indeed fly, but whether it was a strong flier remains controversial.

The central shaft is divisible into two sections. The lower portion, part of which lies beneath the skin, is the **calamus**; it is hollow and has no vanes. Above the calamus lies the **rachis,** which is essentially solid **(Fig. 3–4)**. The vanes, extending from the rachis, are made up of a series of parallel branches called **barbs.** At right angles to the barbs, and in the same plane, are branchlets called **barbules.** The barbules, by hooking together (the distal barbule of one barb catching upon the adjacent proximal barbule of the next), hold the vane intact. The whole effect is somewhat like a series of tiny strips of Velcro.

Run your fingers down a feather from tip to base, and notice how you can separate the barbs by pulling them apart. Now press the barbs

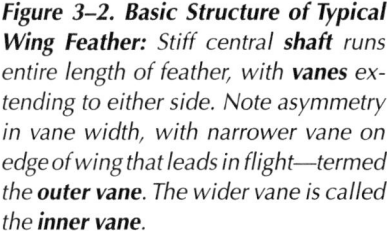

*Figure 3–2. Basic Structure of Typical Wing Feather: Stiff central **shaft** runs entire length of feather, with **vanes** extending to either side. Note asymmetry in vane width, with narrower vane on edge of wing that leads in flight—termed the **outer vane**. The wider vane is called the **inner vane**.*

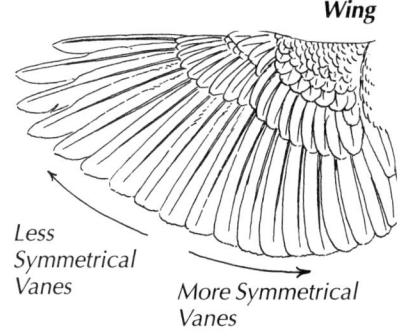

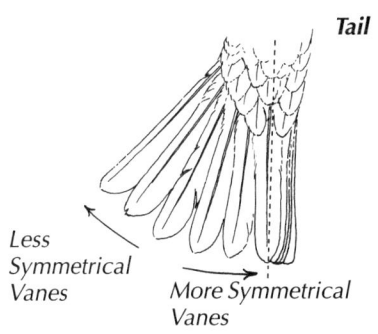

Figure 3–3. Asymmetry of Vane Width in Wing and Tail Feathers: The narrower (outer) vane of each feather is located toward the leading edge of the wing, and toward the outside of the tail. Note that asymmetry in vane width increases farther from the center of the body.

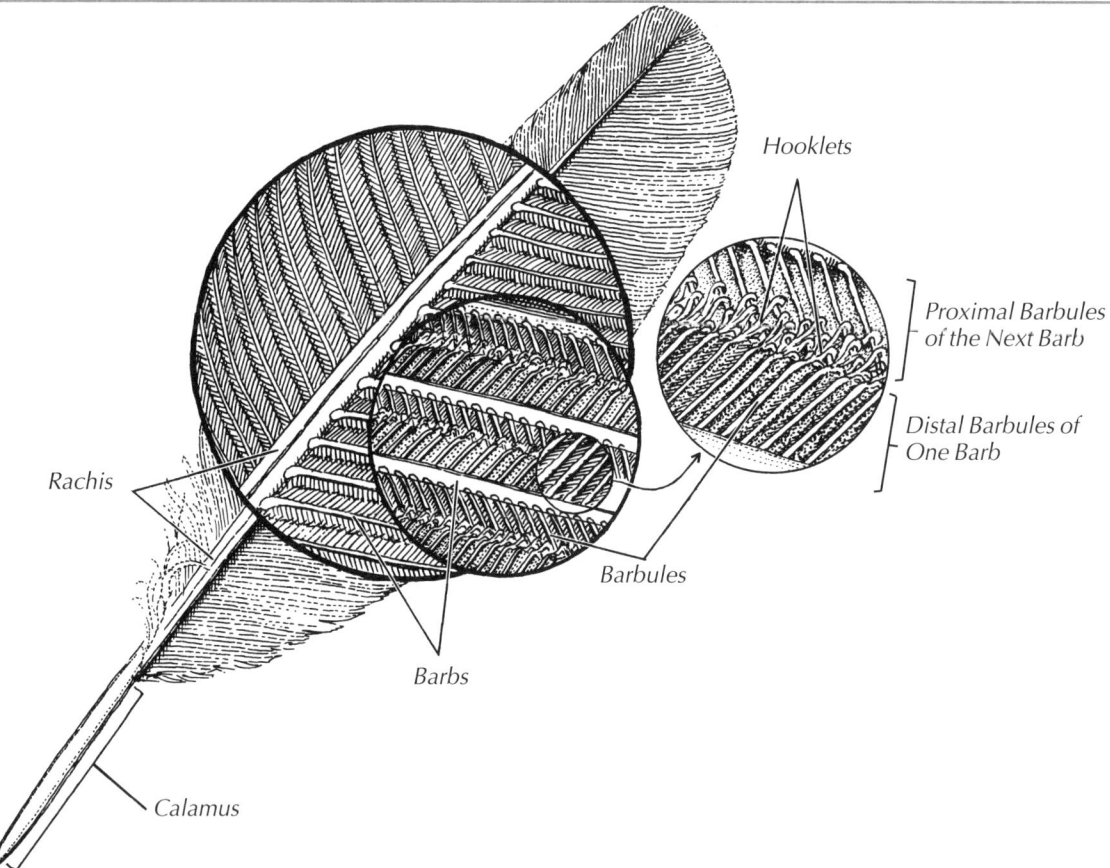

Figure 3–4. Structural Detail of a Typical Contour Feather: *The feather shaft is divided into two segments: the lower, hollow, vaneless **calamus**; and the upper, solid **rachis**, to which the vanes attach. The vanes consist of parallel branches called **barbs**, which are likewise flanked by parallel branchlets called **barbules**. Tiny **hooklets** on each distal barbule catch onto each adjacent proximal barbule, holding the barbs together lightly, somewhat like strips of Velcro. Adapted from Faaborg and Chaplin (1988, p. 15, Fig. 2–1).*

together by stroking from the base toward the tip, and see how neatly they fit against one another, forming a smooth, continuous surface. This motion is exactly what birds do while they preen, to smooth and adjust their feathers, "zipping" them together. The large surface area of wing feathers is particularly important in providing lift and thrust for flight. The barbules not only hold the barbs together but allow them to slide relative to one another, contributing to the feather's extraordinary flexibility.

Types of Feathers

When we think of a feather, we often visualize a flight feather from the wing; however, this is just one of the many types of feathers that birds possess. A single bird may have delicate and fragile down feathers, largely vaneless bristles, and strong feathers used in flight. Furthermore, each type of feather differs from species to species, and within a species, by sex and age. Differences occur in size, shape, pattern, color, and microscopic structure.

Given a single feather, an experienced person with a microscope—and access to a large collection of bird specimens for comparisons—usually can determine the species, sex, age category, and the feather's original position on the body **(Sidebar 1: Feather Detective)**. Information of this sort can be valuable. Researchers studying the diet of nesting Sharp-shinned Hawks, for example, cannot kill these protected species to examine their stomach contents. But by studying feathers of prey that are dropped from the hawk's nest, they can learn the species, sex, and age of the prey.

Contour Feathers

The **contour feathers** are what you see when you look at a bird. Forming the outer shell of the bird's feather coat, they are termed contour feathers because they give the bird its characteristic shape or outline. Unlike other major types of feathers, most contours have tightly knit vanes that form a relatively impenetrable surface. These feathers streamline and shape the wings to provide propulsion and lift in flight. The tightly knit vanes deflect air currents, minimizing the effects of wind and allowing the bird to slice through the air. When beaten against the wind, they also provide strong resistance to air; the bird uses this resistance in aerial maneuvers, and in stopping. Try to blow through the intact vane of a large wing feather and you'll see that only an extremely strong wind can penetrate such a surface.

Contour feathers also provide insulation against extremes of temperature. Because these feathers are aligned one atop another like shingles on a roof, water tends to run off. A bird in the rain, therefore, remains dry at the skin thanks to its feathers. If this protection becomes disrupted (for example, if the feathers become oil-soaked), the bird can die within seconds from hypothermia.

Healthy birds, you may have noticed, always look trim and neat. Each contour feather has a set of specialized muscles, located beneath the surface of the skin around the follicle, which helps to position the feather. Without such muscles to hold feathers in the right positions, the feather coat would be disheveled. In some cases, muscle sheets immediately beneath the skin act together with the individual feather muscles to coordinate movement of the feathers within a tract. Individual feathers usually can't move very much, but their movement can be highly conspicuous when, for example, a displaying male grackle elevates the feathers over most of his body **(Fig. 3–5)** or when an adult male Red-winged Blackbird displays his bright red **epaulettes** (shoulder patches) (see Fig. 6–38). Even more spectacular is a male peafowl spreading his train of feathers **(Fig. 3–6)**.

The largest contour feathers in many birds are the large flight feathers on the wing, called **remiges,** and the tail feathers, called **rectrices.** Because the remiges of all flying birds provide lift and propulsion, it is important that they not be displaced by strong air currents. Thus, they are attached firmly to the bones of the wing either directly or indirectly via ligaments, unlike most other feathers, which attach only to the skin. The rectrices are connected to one another by ligaments, with only the innermost attaching directly to the tail bone (pygostyle) via ligaments.

Figure 3–5. Common Grackle Display with Feathers Erect: Each contour feather has a set of specialized muscles that helps to position the feather, keeping the feather coat smooth and neat. Although birds have only limited control over feather placement, some, like this Common Grackle, take advantage of that control in their displays. Photo by Tom Vezo.

(Continued on p. 3·10)

Sidebar 1: FEATHER DETECTIVE
Mike Lipske

(Excerpted from "To Down Detective Roxie Laybourne, a Feather's the Clue," by Mike Lipske, in Smithsonian, *March 1982. Reprinted with permission from the author and Smithsonian.)*

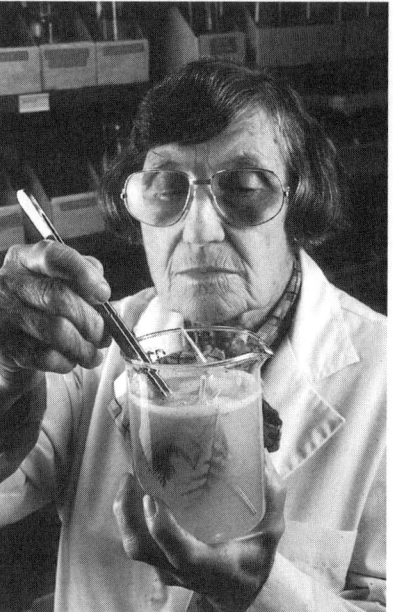

Figure A. Roxie Laybourne Washes Soon-to-be-identified Feathers. *Photo courtesy of Chip Clark.*

On her way to becoming this country's absolute last word on the identification of feathers, Roxie C. Laybourne has not hewn to the tidy path. The clutter on her desk at the Smithsonian Institution stands a solid six inches deep at the left rear corner and slopes to a depth of two inches near the right front. From mountain to plain spreads a layered mass of routing envelopes, official letters, scientific texts, seminar proceedings and scattered scraps of paper. One small brown-and-white feather, pennant on a sinking ship, pokes up from the pile. "It's from Thailand, but I haven't worked it out," she says of the feather.

On an adjacent table are more papers, more books, her black microscope, a plastic bread bag filled with eagle feathers, as well as several small glass bottles containing charred bits of down floating in Ivory Snow and water.

Feathers, ones in need of a name, flow steadily to Mrs. Laybourne's tiny office in the National Museum of Natural History. Aircraft engineers, designing planes to withstand collisions with birds, send her tiny shreds of feathers scraped from jet engines. An archaeologist in New Mexico drills a hole in a 500-year-old sealed pot, plucks bright yellow and orange feathers from inside, and mails them to Mrs. Laybourne in Washington. Feathers come routinely from field agents of the U. S. Fish and Wildlife Service (FWS) and the Federal Bureau of Investigation.

A systematic ornithologist and research associate in the Smithsonian's Division of Birds, Mrs. Laybourne is also a part-time zoologist with the Division of Law Enforcement of the FWS. Short, square-faced, she speaks in a thick Carolina drawl, and punctuates her conversation with rolling hand gestures.

Combing quickly through the crammed drawers of her desk, she pulls out plastic bags holding bird remains. From one, she dumps a few bits of mangled fluff recovered from an airplane engine following a bird strike in Texas last year. Reading the scant evidence, she has conjured up a female American Kestrel.

"This strike occurred in Turkey," she says of the next one, a grayish wing and one severed foot, sent by the U. S. Air Force. "Golden Plover."

On a microscope slide are mounted a few tiny, white strands of down: "I knew for certain it was a Buteo (a genus of soaring hawks). It turned out to be a Red-tailed Hawk. *Probably* a Red-tailed Hawk. You have to be careful not to overstep your bounds."

"Remarkable woman," says Douglas Sutton, at the time a Smithsonian post-doctoral fellow for whom she was identifying feathers found at archaeological sites in Labrador. "That work she does with feathers, I've never seen anyone do. Bright as hell. Feathers, for God's sake! That's a very difficult thing."

"I grew up as an ornithologist," says Mrs. Laybourne. "I've been interested in natural history and birds ever since I could walk. You get used to thinking of birds in certain terms. People who band birds know the bird from handling it, or seeing it in the field. I know birds better when they're dead."

She was born in Fayetteville, North Carolina. The year, she says, is not "pertinent." Her parents moved soon after to Farmville, in Pitt County. "It was a small town. You had plenty of fields and woods to roam in. Most people don't. But I did. I liked to hunt and fish."

She was graduated in 1932 from Meredith College in Raleigh, with a degree in science and mathematics, and went on to work at the North Carolina State Museum of Natural History, where she learned taxidermy and spent summers at the federal fisheries station in Beaufort, collecting birds, fish and other specimens for the museum and puzzling out problems in shark identification. Graduate study in zoology at North Carolina State University was followed by a master's degree in botany from George Washington University in Washington, D.C.

She came to work at the Smithsonian in June 1944, intending to leave Washington after a year. But by November 1946, she had transferred to a job at the Fish and Wildlife Service, where she has worked since, for many years with the Bird and Mammal Laboratories, now with the Division of Law Enforcement. As a Smithsonian research associate, she has an office and access to the vast collection of bird skins and skeletons in the Museum of Natural History. "You have all the privileges of an employee but no salary," she explains.

It was not until 1960 that Roxie Laybourne's peculiar knack for identifying feathers began to reveal itself. On October 4 of that year a turboprop Lockheed Electra took off from Boston's Logan Airport, burst into flames and plunged into Winthrop Bay with 72 aboard. Eight passengers and two crew members survived the 47-second flight—cut short by a flock of birds sucked into the Electra's engines.

The Federal Aviation Administration turned to Mrs. Laybourne and John W. Aldrich, her supervisor at the time, for help in identifying fragments taken from the engines. "That was our first case," remembers Aldrich, also a research associate in the Division of Birds. "And it was a tough one. They were all mangled. Finally we did find one whole starling feather."

The Boston crash caused an immediate scurrying in the industry, and a feeling that the effects of birds on airplanes was something that had to be understood. "It was one of the first that ever happened with birds," said one industry employee. "Lives were lost and a plane came down out of the sky."

The aircraft industry prefers to remain quiet about its bird strike research. "People get hysterical when we take dead pigeons and throw them into running engines," says a Pratt & Whitney public relations man. "And we do."

Al Weaver, a Pratt & Whitney engineer who has been sending Mrs. Laybourne feather fragments for ten years, says he is not especially curious about the species or sex of birds that fly into engines. But he does want to know the size and if it is a flocking bird, like a gull, "or a loner like a hawk." Which is where Roxie Laybourne becomes a good person to know.

"Feathers," she points out, "will take a lot of beating. They're pretty tough. That's one reason you can identify them after they go through something, like a jet aircraft engine, and they're all chewed up."

The first thing she does with the chewed-up feathers that Weaver and other engineers send to her is wash them **(Fig. A)**. "You cannot identify dirty feathers." After the feather fragments have been soaked in soapy liquid, rinsed and dried, she puts one on a slide and under her microscope.

She is a student of the fine, diagnostic structures on the barbules of down **(Fig. B)**. These barbules are microscopic parts of a feather that extend from an individual barb. The barbs, in turn, are tiny featherlets, some small as an eyelash, that grow from the quill and make up the soft vane of a feather. In a six-inch feather taken from the wing of a city pigeon, about 1,200 barbs and about 990,000 barbules make up the vane. Here, in that 990,000th part of a feather, is where the scientist goes to work.

Feathers Stand Up to Jets

She uses down for her investigations because the microscopic structures on downy feathers vary less than do similar structures on the barbs of flight feathers. Also, the downy feathers, which are more flexible, stand up better to the sort of wear and tear that comes of being inhaled by a jet engine.

Why birds have evolved microscopic trademarks in their feathers is something Mrs. Laybourne cannot say: "All I know is they have it." The microscopic process pins identification down to family in most cases, but Mrs. Laybourne does not attempt to identify the particular species through her microscope. Instead, once she has determined the family of birds that a fluff of down has come from, she takes accompanying feather fragments from the tidbits that have been sent to her and compares them with feathers from specimens in the Smithsonian's skin collection.

Mrs. Laybourne says the only thing special about her method is that she

has worked at it so hard, examining thousands upon thousands of feathers and committing their trademarks to memory during the 21 years she has studied the structure of down. But even her doggedness, and her memory for minute variation, would be insufficient if she did not have access to the reference library that is the Smithsonian's collection of bird skins—"somewhere approaching half a million specimens," according to George Watson, curator of birds.

Stored in drawers in a complex of cases on the sixth floor of the Natural History Museum's East Wing **(Fig. C),** the collection ranks first for North American species of birds, but trails the American Museum of Natural History, in New York City, for birds of the world. Possibly a third of the Smithsonian's skins were collected for the old Biological Survey of the Department of Agriculture, which became today's Fish and Wildlife Service. The Museum's array of bird skeletons and preserved specimens is preeminent.

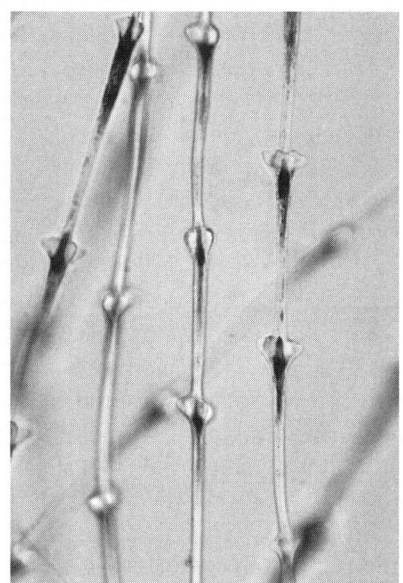

Figure B. Microscopic Detail of Barn Owl Down Feather: Magnified 400 times, the microscopic structure of a downy feather recovered from the engine of a Boeing 747 reveals it to be from a Barn Owl. The engine maker, Pratt & Whitney, sought identification. Photo courtesy of Douglas W. Deedrick.

Feather Identification and the FBI

The resource is often tapped by "outsiders" who, in turn, add their own knowledge. It was, for example, an FBI microscopist who helped Mrs. Laybourne come up with a permanent slide mounting medium for use in her feather studies and, more important, eased her burden of routine identifications. Douglas Deedrick, a 32-year-old agent working with hair and fibers in the microscopic analysis unit of the FBI laboratory in Washington, met Mrs. Laybourne in early 1978 **(Fig. D)**. Deedrick showed an interest in her work and she, who had been identifying feathers for the FBI for 18 years, saw in the young agent the answer to her wish that someone at the Bureau would "learn feathers." She had taken on as many as a dozen cases a year for the FBI, handling feather evidence in crimes such as robbery, kidnapping and murder. Criminal convictions seldom hinge on feather identifications. But having a scientist testify that a feather fragment adhering to a suspect's knife matches the filling in the victim's down-lined jacket cannot hurt the government's case.

The carefully dressed-and-pressed FBI man and the white-coated research scientist are an unlikely pair of crime fighters. Yet she speaks of Deedrick as a favored nephew, and the agent is effusive in his praise of her: "She's a stand-up person. She never sits down. She can walk twice as fast as I can. Her memory is sharp. She can associate all the different areas she's worked in—botany and birds and animals—and put it all together. She can take something most people wouldn't find that interesting and put herself into it. She's a true scientist."

Thirty-seven years into her career working at the Museum of Natural History, and 21 years into her research on the structure of down, she has no intention of retiring to Virginia. "I could go out to my farm and pull up weeds. But who wants to do that all the time? The weeds will come back. You never win that battle. This field, I'm just beginning to get to where I can go somewhere with it. I've just got enough background now to begin learning. You have to know so much to begin asking yourself questions. I figure in 20 years I might learn something."

15 YEARS LATER
Daniel Otis

In 1997, 15 years after publication of the above Smithsonian article, Roxie Laybourne was still at work full time, still in love with her job, still the ultimate authority on bird IDs for the Air Force and commercial airlines, and still eager to surpass herself at her craft. She had the advantage of the long view, and in a phone conversation I asked her how her job had changed over the years.

Not much, she told me, when it comes to bird IDs. Mrs. Laybourne still received chewed-up bits of feathers from around the world, cleaned them off, and tracked down the species using the Smithsonian's bird collections as a reference. When I asked if she always used the microscope, she mentioned that she could sometimes tell a bird's family by "how its feather behaves in the cleaning solution." So the fact that her approach had not changed may be unsurprising. Expertise of this caliber evolves over decades and is not easily rendered obsolete by technological developments.

But the world from which she obtained her birds had changed. "There's more awareness now—people are more interested," she said. "Back when I started, people weren't paying much attention to feathers. At the FBI, they didn't think feathers could be used as evidence. Now they realize that a tiny bit of feather can include a lot of information. There are more planes flying, too, and they're faster and quieter. Birds can't evade them the way they could when they were slower and noisier."

So there were more birds being struck, and since the Air Force organized its Bird Aircraft Strike Hazard (BASH) Team, personnel at air bases became more aware of air strikes and better at collecting feather samples. The consequence is that she got to do more birds.

That's how she put it: "I get to do more birds." Sixty-five years after she graduated from college, and nearly 40 years after she began research on feather identification as part of her duties at the U. S. Fish and Wildlife Service, her pleasure in her work was undiminished. Her persistence and dedication earned her some notoriety; a brochure advertising her 1996–97 exhibit at the Smithsonian, "Feather Focus," referred to her as "world-famous feather detective Roxie Laybourne." And although you got the impression that bragging was alien to her character, she couldn't quite conceal her pride in being an acknowledged master. She laughed easily and often. Her enjoyment of her craft was evident in her tone of voice—eager, careful, kind, like that of a favorite aunt, still rich with the cadence of North Carolina. It was also evident from her comments.

"To me, the whole process is like

Figure C. Smithsonian Institution's Bird Skin Collection: *Roxie Laybourne displays just a few of the Smithsonian's numerous drawers of bird skins. The collection, used for reference, research, and teaching, contains specimens from all over the world. Photo courtesy of Douglas Deedrick.*

a puzzle—every time you get bird remains, you don't know which species you have. With every bird you're being put to a test. Even if it is the same species, the feathers could well be from different parts of the bird. And the date and locality of the strike are important. You see, you're putting a puzzle together, trying to figure out where the pieces go.

"I warn people, don't come to work with me—you may become addicted. When people say to me 'you don't make much money in museum work,' I always say 'you don't need as much—you don't need to go on vacations to enjoy yourself. You've enjoyed your day. Your work is your recreation.'"

When I spoke to her in mid-September, her current recreation was identifying birds for the Air Force. Planes from a base in North Carolina were hitting birds at night—"So it's migrating birds," she said, fitting a piece of the puzzle. "They're anxious to know what they're hitting. So far, I've identified Mourning Doves and Killdeer—birds that like the grassland habitat of airfields. Also three bats." The Air Force had compiled information on the birds that live at all the different bases. If they have bird problems, they may change flight patterns, discourage birds by changing habitat, or even move flights to different airfields.

"Overall, the bird hit most often is probably the Horned Lark," she told me. "In the United States and Europe, commercial airlines are most likely to hit gulls. In Africa and Asia, it's Black Kites. Air Force planes often hit Turkey Vultures."

The Horned Larks weigh about one and one-half ounces (42.5g), so like the vast majority of bird strikes, they do no damage and endanger no lives. Nevertheless, she said, "A strike is always a serious problem, because you never know when it will cause damage." Mrs. Laybourne's years of work have helped aircraft manu-

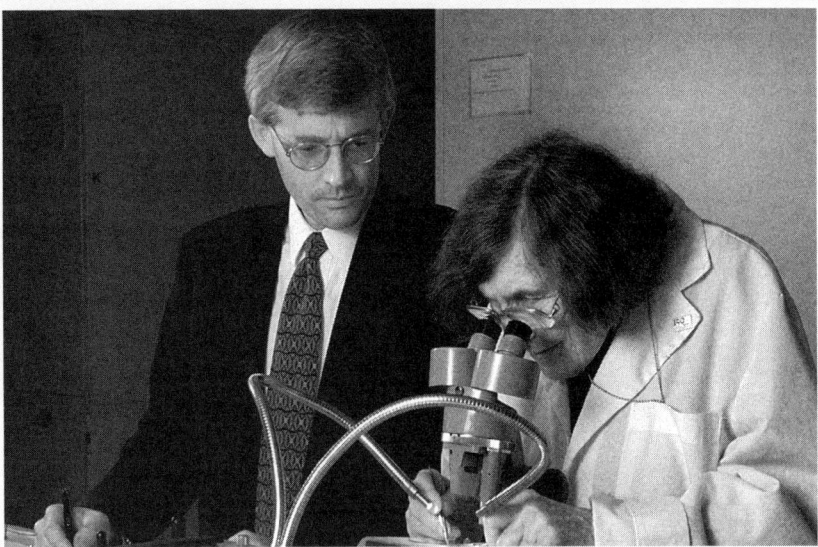

Figure D. Douglas Deedrick and Roxie Laybourne: *Photo courtesy of Chip Clark.*

facturers improve the safety of their aircraft. One use to which her identifications are put is developing aircraft specifications. When designing a windshield for a jet, for instance, it helps to know whether a strike by a sparrow or a goose is more likely.

Having worked so long to accumulate her skills, she was determined to pass them on to others. "When you have some knowledge," she said, "it's your responsibly to share it. No sense in other people having to start at the beginning the way I did. They should be able to build on what you've learned, and then they'll be able to go on that much further."

That's another change: she had more students. One of her first, Douglas Deedrick, has taken over her work for the FBI. In classes she taught at the Smithsonian she had students as young as six years old. "It's good to get them young," she said. "Because to do this, to specialize, you've got to know birds in the field, have a background in bird watching. You've got to know systematics. Otherwise you don't know where your piece of the puzzle fits." More typical, though, are undergraduate and graduate students, postdocs, and professionals. She had forensics students from California and other students of bird

strikes from Britain, Slovakia, and elsewhere.

If her own enthusiasm testified to the work's addictiveness, her comments on the demands of the work emphasized its daunting aspects. She took her responsibility very seriously. "I try to instill in my students how careful you have to be in this work, the importance of being absolutely accurate," she says. "This is not glamorous, and some people wouldn't like it. The first qualification is perseverance. You have to work hard for a long time without results. You need a good memory. You're thinking all the time. You've got to know where you're going. You can't just go hunting through the collections trying to match a feather. It's like heading out on a cross-country drive—you can't just take off, you've got to have at least some idea of your destination."

It seems not too presumptuous to suggest that her attitude toward her work was an important source of her achievement. Despite the difficulties, she told me, "I am happy with it, I enjoy it every day, and I still learn something new every day—it's just as true now as it ever was." ∎

On August 7, 2003, Roxie died peacefully at her cabin in Manassas, Virginia.

Figure 3–6. Display of the Male Indian Peafowl: *In a spectacular display of feather control, the peacock faces one or more females and lifts and fans his "train" while rustling his wing feathers and stamping his feet. The "train" consists of nearly 150 uppertail coverts with as many "eyes" and is supported behind by the unadorned rectrices. Photo courtesy of Isidor Jeklin/CLO.*

The outermost remiges are the **primaries**, recognized by their attachment to the skeleton of the "hand" (see Fig. 1–8). To meet the demands of powered flight, the primaries must be strong yet flexible. Depending on the species, birds commonly have between 9 and 12 well-developed primaries.

Projecting rearward from the forearm of the bird wing is another row of remiges; these are known as **secondaries**. In both flapping and soaring flight, the secondaries provide lift and thus function like the fixed wing of an airplane. Among birds as a whole, the number of secondaries varies more than the number of primaries, ranging from 8 to 32 (Van Tyne and Berger 1959). The higher numbers occur among birds with long wings, such as albatrosses, which need a large wing surface for soaring flight over water. In most flying birds, the secondaries attach directly to the ulna bone in the forearm via ligaments. On many isolated ulnas, such as might be washed up on a beach, you can see the bony bumps that were points of attachment for the secondaries.

The other major set of flight feathers in most birds, the rectrices (tail feathers), are particularly important for stability and control in flight, somewhat like the tail of a child's kite (Thomas and Balmford 1995). For instance, forest-dwelling hawks, such as the Northern Goshawk, use their tails as rudders to help guide them through an obstacle course of trees. Among flying birds the number of rectrices ranges from 6 to 32, with the higher numbers occurring in larger birds **(Sidebar 2: Feather Facts)**. To make accurate references to individual feathers, ornithologists have developed a standardized numbering system **(Fig. 3–7)**.

Bordering and overlying the remiges and rectrices on both the upper and lower sides are rows of feathers called **coverts,** which provide insulation, color, and pattern on the wing surface and, perhaps most important, contribute to the streamlined shape of the wing and

(Continued on p. 3·12)

Figure 3–7. Numbering System for Rectrices and Remiges: *The rectrices are numbered from the central two feathers toward the left and right sides of the bird. The primaries are numbered from the innermost toward the tip of the wing, and the secondaries are numbered from the innermost toward the center of the bird. The number of each type of feather is constant within a species, but varies among species. There are typically between 6 and 32 rectrices, between 9 and 12 primaries, and between 8 and 32 secondaries.*

Sidebar 2: FEATHER FACTS
Sandy Podulka

Number of Feathers

How many feathers does a bird have? As you might expect, big birds tend to have more feathers than small birds. The record low count (contour feathers only) is 940—from a Ruby-throated Hummingbird (Wetmore 1936). The highest number was reported by Ammann (1937), who (patiently) counted 25,216 contour feathers on a Tundra Swan. Of these, 80% (20,177) were on the head and neck. This distribution is not surprising, since swans are so clearly long-necked, but even short-necked birds have a high percentage of their feathers on the head and neck, where many small feathers are packed closely together.

Feather number also varies with a bird's need for insulation. For example, birds living in cold climates have more feathers in winter than in summer. Wetmore (1936) found that in winter White-throated Sparrows have about 2,500 contour feathers, but in summer they have only 1,500—a 40% decrease. In addition, aquatic birds tend to have more feathers than terrestrial species of a similar size—undoubtedly due to the rapid loss of heat experienced by birds in water.

Heat retention is probably also the reason smaller birds tend to have more feathers per unit body weight than do large birds. Like a human, a bird generates heat through the metabolic reactions that take place throughout its body, and it loses much of this heat across its body surface. Therefore, the more surface area a bird has relative to its heat-generating volume—a comparison known as the **surface-to-volume ratio**—the more quickly it will lose heat. Because small birds tend to have a higher surface-to-volume ratio, they lose heat rapidly, and thus need more feathers for each unit of body weight than does a larger bird.

Size of Feathers

Bigger birds not only have more feathers, they have bigger feathers. Thus, if you find a feather on the ground, its size is a useful clue to the size of the bird it came from.

Not surprisingly, the smallest contour feathers known are from the world's smallest bird: the feathers from the eyelids of the Bee Hummingbird are merely 1/63 of an inch (0.4 mm) long! In contrast, the contour feathers forming the peacock's tail are nearly 4,000 times longer—up to 5 feet (1.5 m) in length.

Weight of Feathers

Feathers are, indeed, "light as a feather." Although the feathers of some birds, such as owls, seem to make up half their volume, the feather coat commonly accounts for only 5 to 10% of a bird's weight. Thus, a chickadee weighing four-tenths of an ounce (10 g)—about as much as two quarters—would have a feather coat weighing less than a dollar bill (four-hundredths of an ounce [1 g]). And yet, as light as they are, a bird's feathers are still usually 2 to 3 times heavier than its skeleton. ∎

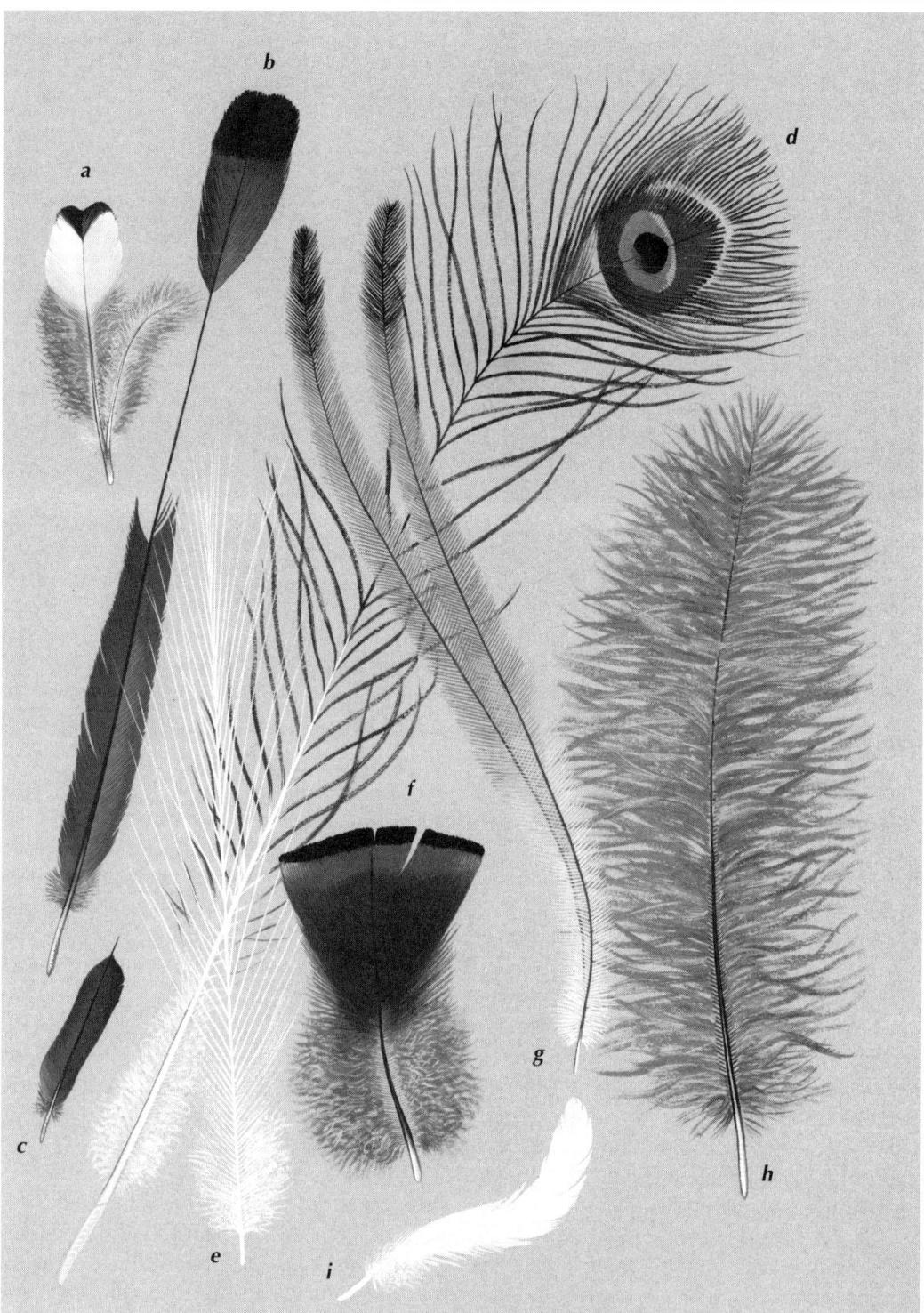

Figure 3–8. Examples of Modified Contour Feathers: a: *Body feather of a pheasant. It has a smaller afterfeather with all of the barbs free and soft.* **b:** *One of the two long, central rectrices of a motmot (Momotidae) from the New World tropics. Along the terminal half of the shaft the barbs have fallen away, leaving a racketlike tip.* **c:** *Rectrix of a Chimney Swift. The tip end is devoid of barbs and spinelike.* **d:** *One of the shorter feathers in the "train" of the male Indian Peafowl. All of the barbs are free except near the tip, where they are hooked together to form a continuous surface for the colorful "eye."* **e:** *The plume from the back of an egret. The few barbs are little more than willowy filaments.* **f:** *Breast feather of a turkey. The barbs in the vanes toward the tip gradually shorten, producing the truncated or squared-off effect.* **g:** *Body feather of an Emu. Its afterfeather is so similar in size and structure to the main feather that it is indistinguishable, thus forming a double feather. All of the barbs are free.* **h:** *Body feather of an Ostrich. The many long, soft barbs with their countless barbules, unhooked and fuzzy, give the feather its characteristic fluffiness.* **i:** *Crest feather from the Sulphur-Crested Cockatoo of Australia and New Guinea. The curliness results partly from one side of the shaft being exceedingly flat. Drawing by Charles L. Ripper.*

tail. Just like aircraft, flying animals need streamlining to reduce friction with the air; streamlining greatly reduces the energy needed to fly.

Some contour feathers have an **afterfeather**, resembling the main feather but in miniature, growing from the lower shaft. Well-developed afterfeathers contain a shaft and two vanes **(Fig. 3–8a)**. Afterfeathers provide extra insulation and are especially well developed in grouse, many of which live in seasonally cold or arctic regions. In the flightless Australian Emus, the afterfeathers are as large as the main feathers **(Fig. 3–8g)**. These big, fluffy contour feathers help to form a thick protective coat that is useful in the dense brush of the Australian

scrublands. The feather coat is so tough that these birds can cross the Australian outback's sturdy four-foot-high sheep fences topped with barbed wire simply by running fast, crashing into them, and somersaulting over. In the process, they leave behind a pile of feathers hanging on the barbed wire **(Fig. 3–9)**. It's quite a testimony to the protective value of the feather coat—one hates to think what would happen to a human who tried to cross such a fence by smashing into it!

Figure 3–9. Holy cow! I hope that new hair replacement stuff works on feathers.

Some contour feathers have become highly modified for special functions (see Fig. 3–8). Among the most bizarre are those used in courtship displays, such as the elaborate ornaments of the male birds-of-paradise **(Fig. 3–10)**. Motmots, nightjars, and other species have odd feathers with a wirelike rachis for their displays **(Figs. 3–8b, 3–11)**. Other modifications are more subtle. In most owls, for instance, the leading edge of the first several primary feathers has a loose fringe, and the dorsal surface of the inner vanes of most flight feathers has a soft "pile" **(Fig. 3–12)**. These modifications render the feather coat very soft to the touch. More important, they allow owls to fly very quietly, creating little noise even in the high-frequency range that their rodent prey hear so well.

Other special contour feathers include the waxlike tips of some wing and tail feathers of waxwings (see Ch. 7, Sidebar 5, Fig. A); the curly feathers in the erectile crests of the Australian cockatoos **(Fig. 3–8i)**; and the stiff, strawlike crown feathers of the Black Crowned-Crane of Africa **(Fig. 3–13)**. Some birds, such as the American Woodcock and Common Snipe, have contour feathers modified to produce sounds (see Ch. 7, Sidebar 1, Figs. C and E).

Bristles are highly specialized contour feathers in which the rachis is stiffened and lacks barbs along its outermost parts (Stettenheim 1974). Among the best-known bristles are **rictal bristles**, which project from the base of the beak in birds that catch insects, such as flycatchers, nightjars, and some New World Warblers **(Fig. 3–14a)**. Some ornithologists have suggested that rictal bristles might funnel insects into the mouth,

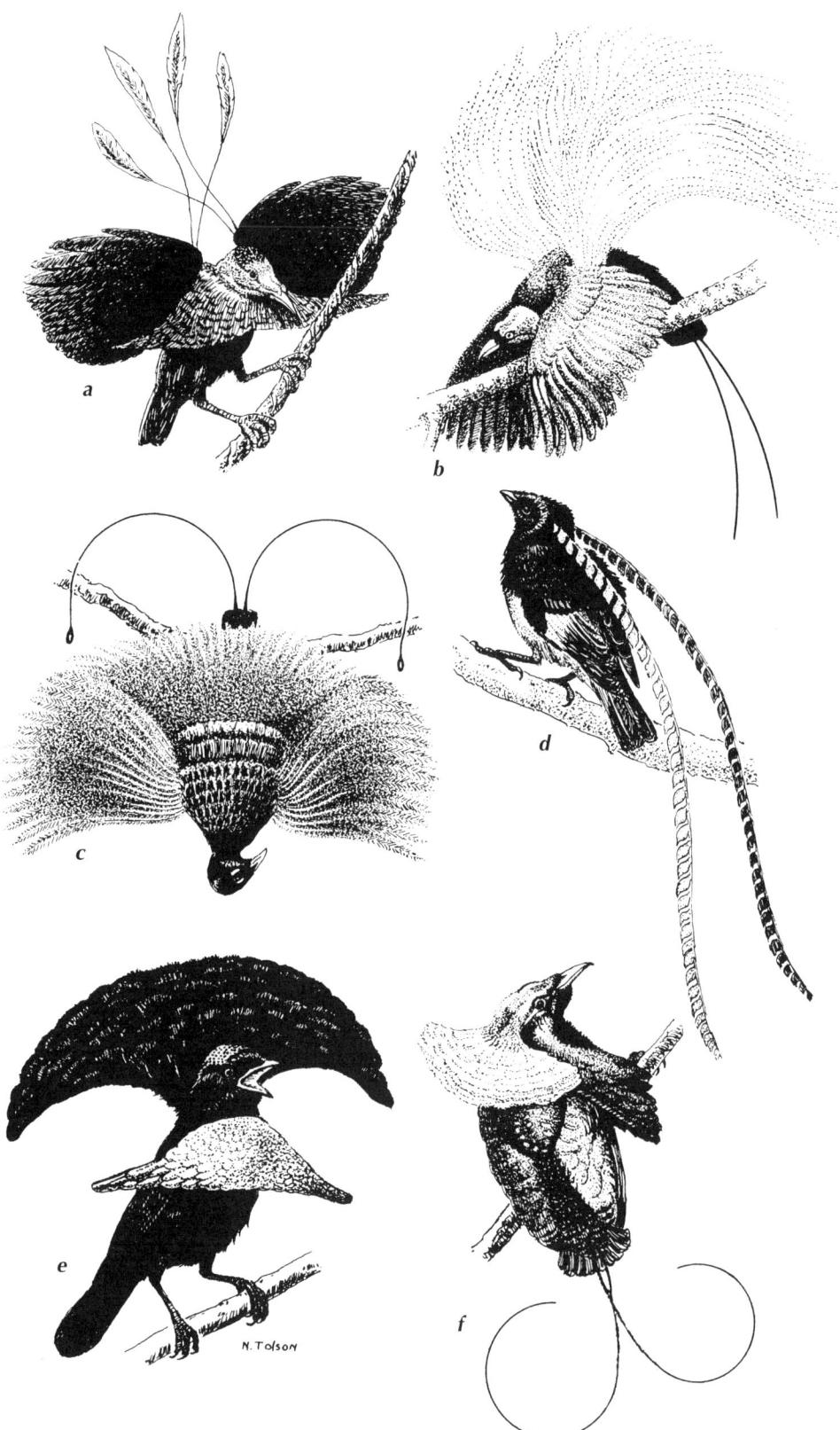

Figure 3–10. Display Feathers of Male Birds-of-paradise from New Guinea: a: *Wallace's Standardwing.* **b:** *Greater Bird-of-paradise.* **c:** *Blue Bird-of-paradise.* **d:** *King-of-Saxony Bird-of-paradise.* **e:** *Superb Bird-of-paradise.* **f:** *Magnificent Bird-of-paradise. Note in b and c, the smoky effect of the long filamentous feathers; in c and f, the long, barbless "wires;" in e, the erectile cape and bib; and in d, the two extraordinarily long crown plumes, which have vanes on only one side, and which are so deeply scalloped that they suggest a series of tiny pennants. Drawing by N. Tolson.*

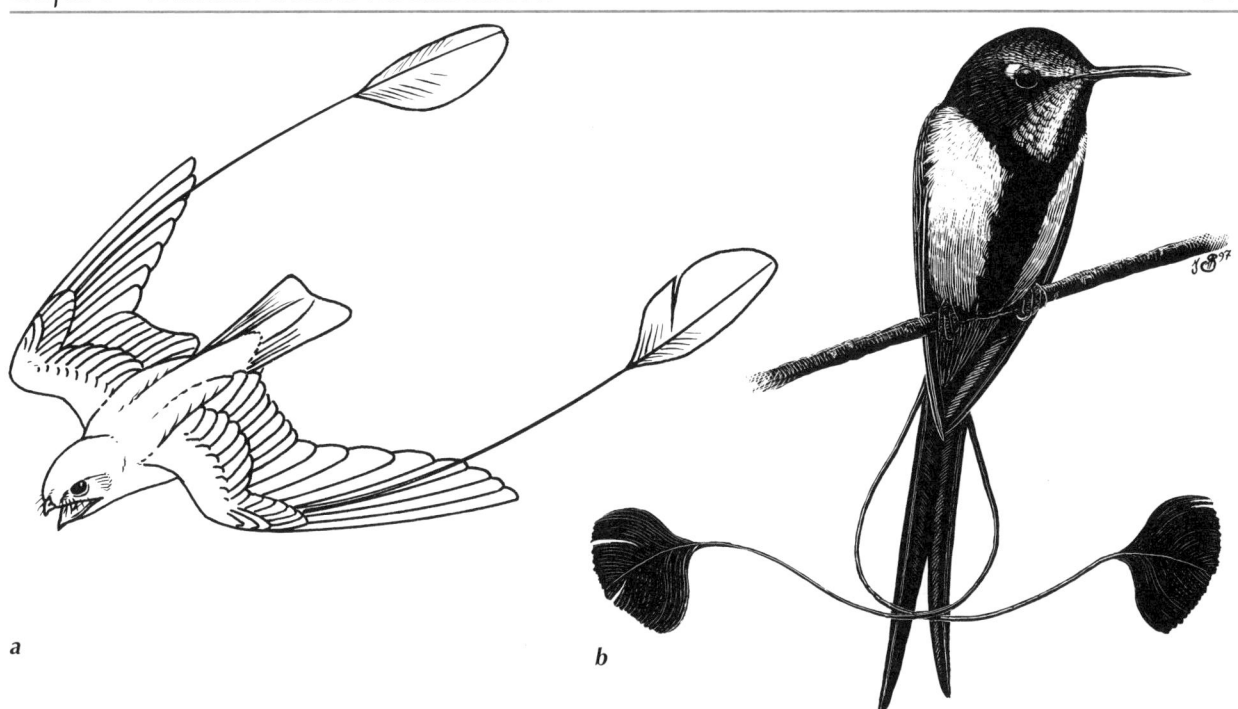

a   b

Figure 3–11. Modified Flight Feathers: a: *Two extremely modified remiges adorn the male Standard-winged Nightjar, a small, nighthawk-like bird from central Africa. The second primary on each wing features a broad vane toward the tip of a very slender shaft. It is approximately 11 inches (28 cm) long, and extends well behind the other wing feathers. When flying in the twilight with the shafts of the two feathers practically invisible and their terminal vanes flapping, the starling-sized bird looks as though it were being pursued by two little bats. Drawing by Charles L. Ripper.* **b:** *Two extremely modified rectrices grace the unmistakable male Marvelous Spatuletail, a hummingbird from the Andes of northern Peru.*

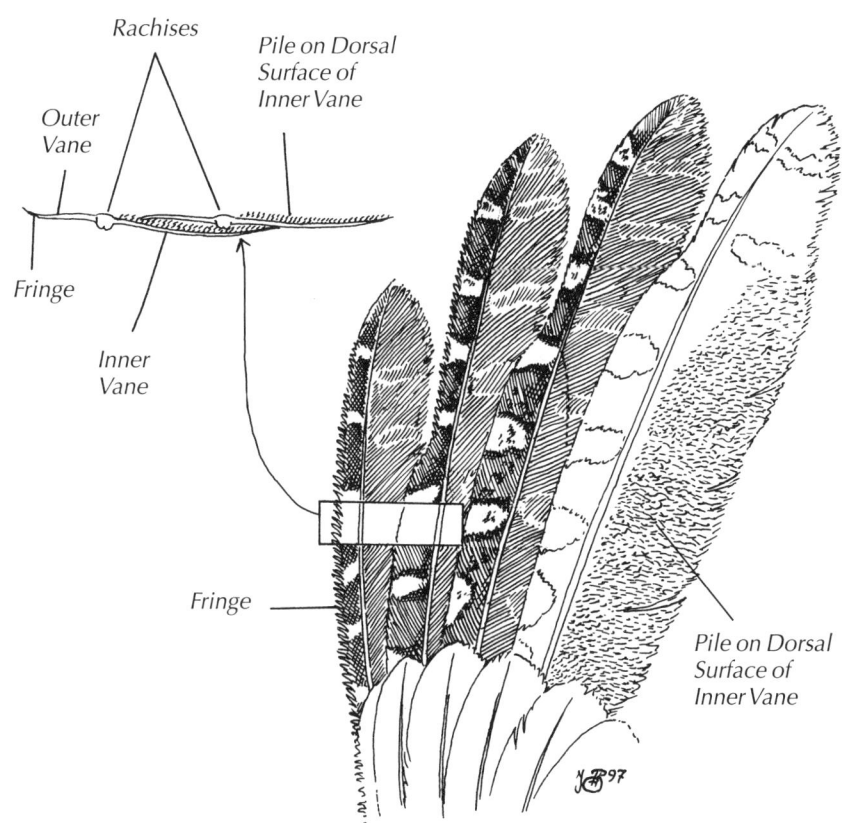

Figure 3–12. Features of Owl Feathers that Produce Silent Flight: *Owls achieve very quiet flight with two specializations of their feathers. First, a soft* **fringe** *on the margin of the outer vane of the first two or three primary feathers softens the contact between the air and the leading edge of the wing. Second, a velvety pile is located on the dorsal surface of the inner vanes of all remiges and, to a lesser extent, the rectrices. This pile, a fur-like, filamentous component of the barbs and barbules, is very efficient at deadening any scraping sound that may be made as the remiges slide against each other during flight. In addition to owls, Northern Harriers exhibit this pile on their feathers and probably benefit greatly from silent flight as they course or hover over meadows in pursuit of rodents.*

Rachises

Pile on Dorsal Surface of Inner Vane

Outer Vane

Fringe

Inner Vane

Fringe

Pile on Dorsal Surface of Inner Vane

but little supporting data exist. In Willow Flycatchers, Conover and Miller (1980) found that experimentally taping down or removing the bristles did not affect the birds' success in capturing insects. By using a wind tunnel, however, these researchers demonstrated that bristles protect the eyes from flying insects and other debris. They released particles in front of the mouths of Willow Flycatchers and found that more particles struck the eyes of birds whose rictal bristles had been removed. Rictal bristles are highly developed among certain species (such as puffbirds—Neotropical relatives of kingfishers) that regularly capture scaly moths, butterflies, and other large, noxious insects; apparently the bristles protect the face and eyes of these birds from their prey (John W. Fitzpatrick, personal communication). Rictal bristles also may help birds detect movements of prey held in the beak, functioning like the whiskers of some mammals.

Down Feathers

Down feathers are fluffy and soft, typically lacking a rachis. If present, the rachis is always shorter than the longest barbs. Because the barbules lack hooks, each flexible barb can wave about independently, giving the feathers their characteristic "downy" appearance **(Fig. 3–14b)**. Down feathers are excellent, lightweight insulators because the barbs and barbules form a loose tangle of air pockets.

Figure 3–13. Modified Contour Feathers of Black Crowned-Crane: The distinctive crown of this large African bird is formed from stiffened, strawlike contour feathers. Photo courtesy of Michelle Burgess/CLO.

Down feathers include the **body downs** and the **natal downs.** Body downs, the down feathers of adults, are most common in water birds such as penguins, loons, petrels, auks, geese, and ducks, as well as in hawks—all of whom undoubtedly benefit from the extra thermal insulation. Some birds, notably the woodpeckers, generally lack body downs.

Many female waterfowl pluck body downs from their bellies to line their nests. Of these, the most famous is the Common Eider (see Fig. 4–124a). On islands off coastal Norway, people collect the down from nests during incubation and sell it. The harvesters avoid harming the birds and leave enough down so that the birds will come back to nest each summer. Some eiders nest under shelters set up by the harvesters, and a few even nest under the residents' houses. Eider down clothing and bedding is notoriously expensive because accumulating just one pound requires down from 35 to 40 nests, the down must be gathered by hand, and the supply is limited. Nevertheless, eiders provide the highest quality down used in jackets and sleeping bags.

Natal downs are present only around the time of hatching. They are prominent on newly hatched chickens and ducklings, giving these birds their soft, fluffy look. These first feathers characteristically lack barbules on the outer portions of the outermost barbs and are probably important for insulation, camouflage, and social behavior, as discussed below. Natal downs arise from the same follicles that later produce contour feathers, and are typically carried on the tips of the growing contour feathers. In contrast, body downs of adults arise from follicles that produce only down feathers.

The old saying "naked as a jay bird" is quite correct for young jays, such as the Blue Jay and Steller's Jay, which have no natal downs at hatching. The newly hatched young of many other birds, including most passerines, have a much-reduced natal down. Clues from embryonic development, however, indicate that the ancestors of these birds had a complete coat of natal down. The less down a new nestling has, the more easily it is warmed by its parent. Therefore, scientists hypothesize that naked chicks save energy both by not producing down and by quickly absorbing their parents' body heat, which may allow them to develop more quickly. Chicks that leave the nest promptly after hatching, such as ducks and pheasants, must carry their own insulation from the moment of hatching, and so have kept their down coats (see Fig. 8–113).

Many kinds of young that cannot fly for weeks after hatching have an insulating cover of downy feathers that lasts for much of the flightless period—during which the young grow enormously. Presumably, it is most efficient to postpone development of contour feathers until the bird requires them for flight. Hawks, for example, have a prolonged downy cover in early life. Their coat of natal down is followed by a second downy plumage, consisting of body downs, which transforms the appearance of the young **(Fig. 3–15)**. The gradual replacement of natal downs with darker body downs, and eventually with juvenal contour feathers, makes these hawk nestlings look like clean, white snowmen, growing increasingly dirty and angular in the winter's afternoon sun.

Semiplumes

The **semiplumes (Fig. 3–14c)** occur in a continuum of forms between down and contour feathers. Unlike contour feathers, the barbules lack hooks and thus the barbs do not cling together as a vane, but unlike down feathers, the rachis is longer than the longest barb. They lie at the edges of the contour feather tracts, sometimes visible, sometimes hidden beneath the contour feathers. Semiplumes provide insulation and help maintain the streamlined form in the overlying feathers, functioning much like tissue paper stuffed into a pair of good shoes that are not being worn.

Filoplumes

Scattered among, and usually hidden by, the contour feathers of most kinds of birds are feathers of a third major type known as **filoplumes**. Hairlike but relatively stiff, filoplumes are simple structures—a

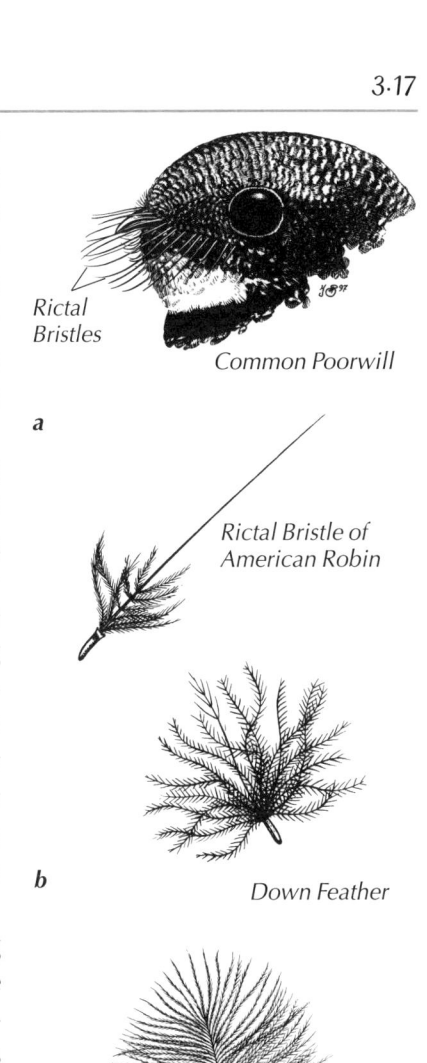

Rictal Bristles

Common Poorwill

a

Rictal Bristle of American Robin

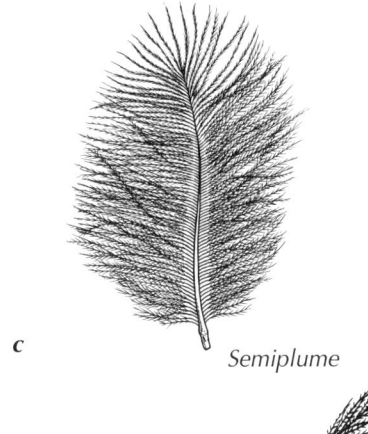

b   *Down Feather*

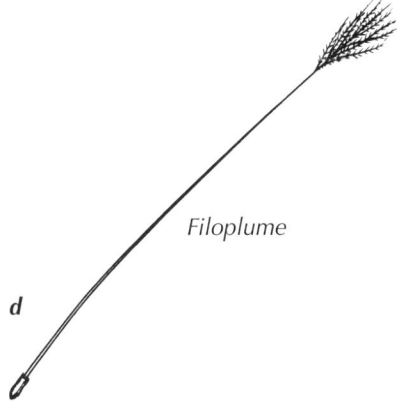

c   *Semiplume*

Filoplume

d

Figure 3–14. Specialized Feathers: a: *Rictal bristles on a Common Poorwill and from an American Robin.* **b:** *Down Feather.* **c:** *Semiplume.* **d:** *Filoplume. Drawings b, c, and d by Charles L. Ripper.*

rachis, usually bare, with barbs, if any, only on the tip **(Fig. 3–14d)**. These are the "hairs" that you see on a plucked supermarket chicken. Unlike contour feathers and body downs, filoplumes lack feather muscles. They do, however, have sensory receptors in the skin next to their follicles, which monitor movement within the feather coat (Necker 1985). Why are such receptors necessary? Fully grown feathers are dead structures analogous to human hair and fingernails; therefore, much of a bird's body is covered by a flexible shell of lifeless feathers. To monitor goings-on in this feather coat, the bird depends partly on filoplume movements. Birds also have sensory receptors in the skin away from the filoplumes, so they receive several kinds of tactile information on events inside the feather shell. Birds presumably use this information to monitor feather positions (Brown and Fedde 1993) and to detect changes of the type that might be caused by wind or body movements. Much remains to be learned about the sensory and neural processes involved in monitoring of this type.

Figure 3–15. Nestling Red-tailed Hawk in Second Coat of Down: In hawks, the natal down is followed by a coat of body down that insulates them during much of their long nestling stage. Photo courtesy of Jim Weaver/CLO.

Powder Downs

Perhaps the strangest of all feathers are the **powder downs.** They are never molted, but grow continually, disintegrating at their tips to produce a fine powder something like talcum powder. The powder permeates the plumage, possibly helping to waterproof and prevent staining of the feathers, although its functions are not yet clearly understood. In form, powder downs may appear like somewhat fluffy contour feathers, or more like body downs **(Fig. 3–16)**. They occur only in certain taxonomic groups, such as herons and pigeons, and may be scattered throughout the body downs, or clustered in patches (Lucas and Stettenheim 1972).

Figure 3–16. Typical Powder Down Feather from a Pigeon: Adapted from Lucas and Stettenheim (1972, p. 337, Fig. 227 C).

Care of Feathers

■ Feathers are essential to the health and survival of birds, so birds spend a good deal of time caring for them. Because feathers are dead structures, they have no active circulatory system to maintain them from the inside. Therefore, a bird provides all their care from the outside, by preening, water-bathing, dust-bathing, sunning, or anting.

Preening

In **preening**, a bird grasps a feather near its base, then nibbles along the shaft toward the tip with a quivering motion **(Fig. 3–17)**, removing stale oil and dirt. The bird may also draw the feather through its bill, smoothing the barbs so that they will lock together. The bird fluffs the feathers in the section of the body it is preening and turns and twists in a variety of movements which—if one follows them closely—are all quite stereotyped, about the same for each part of the body and each feather every time. Preening keeps the feathers neat, preserving their streamlining and insulating effects as well as their color pattern. Preening also removes external parasites (**ectoparasites**), some of which are described below.

Birds preen even without sensory stimulation from the region being preened, as shown by experiments in which severing sensory nerves from certain skin areas did not eliminate preening of those areas (Delius 1988). Thus, preening and perhaps other grooming behaviors are practiced even in the absence of a "tickle" or other stimulation of the skin. The advantage of this kind of unstimulated preening may be that the bird routinely cleans and sorts through its feathers, removing ectoparasites and other disturbances in the feather coat before they become problematic, thereby improving the coat's overall sanitation and health.

Figure 3–17. American Oystercatcher Preening: *When preening, a bird grasps a feather near its base, then nibbles down the shaft toward the tip, removing stale oil and dirt. Photo by Brian Kenney.*

Although researchers have not examined closely the seasonal frequency of grooming behaviors for many birds, temperate-zone birds do appear to groom less in the colder months. During short winter days, small birds must spend much of their time foraging to get enough to eat, so they have less time to groom. Furthermore, because ectoparasitic insects and mites are cold-blooded (ectothermic), they are presumably less likely to move onto or between hosts when it's cold. We would expect birds in tropical regions with little seasonal change to have more uniform grooming patterns, but this apparently has not been investigated.

As you might expect, birds groom more often when they are molting. Also, birds that have spent the night incubating eggs (for example, female passerines) often engage in a prolonged bout of preening just after leaving the nest for the first time each morning. If you observe this behavior, you know you're watching an incubating bird.

Certain kinds of birds with strong social ties, such as parrots and crows, are well known for their **allopreening**, in which one bird preens another. Birds often direct their allopreening to the back of the neck and other areas that the recipient cannot reach with its own bill. Allopreening presumably helps to remove ectoparasites, to keep the plumage in order, and to establish and enhance social bonds between birds. Allopreening also serves as an indication of dominance-subordinance relationships, as subordinate individuals offer themselves to be preened by dominant individuals. Allopreening may even occur between different species. For example, Red-winged Blackbirds occasionally allopreen Brown-headed Cowbirds. Why these birds allopreen is not clearly understood, but the behavior appears to function more in aggression and dominance interactions between these species than in feather care (Post and Wiley 1992).

Oiling

For many kinds of birds, preening includes using the bill to spread fatty secretions from the **oil gland** (also called the **uropygial gland** or **preen gland**) over their feathers. The oil gland, one of the few glands in bird skin, is located on the rump immediately in front of the rectrices. On a hand-held songbird you can see the small pimple-like gland by blowing gently in this region to part the feathers **(Fig. 3–18)**.

Researchers have suggested that oil gland secretions might waterproof the feathers, but there is no evidence for this. When the oil glands of ducks were removed, their feathers did not lose their waterproofing ability. Instead, the results indicated that the secretion is a conditioner that keeps the outer surface of the skin supple and prevents feathers and scales from becoming brittle and breaking prematurely (Jacob and Ziswiler 1982). Actual waterproofing of feathers seems to result primarily from the microstructural arrangement of the barbs and barbules, which provide an evenly spaced surface of ridges with narrow gaps between them that sheds water very effectively. Thus, spreading oil gland secretion seems to facilitate waterproofing only indirectly; feathers provide the waterproofing, but oil gland secretions make feathers last longer.

Researchers also suspect that the secretions help both to control the growth of undesirable fungi on the feathers and to promote the growth of favorable fungi that chemically inhibit lice from living among the feathers. Another suggestion is that, for some species, the secretion's strong odor might help deter mammalian predators. The cavity-nesting hoopoes and woodhoopoes have particularly noxious uropygial odors, making them the bird equivalents of skunks. The birds cannot spray their secretions, however, so they are repulsive only at close range. It also has been proposed that the oil gland secretions, when exposed to sunlight and then consumed in preening, might be a source of vitamin D for birds, but recent experiments failed to confirm any role for these secretions in vitamin D production.

Head-Scratching

To oil the feathers that it cannot reach with its bill—those on the head, for instance—the bird rubs oil on its feet with its bill and then scratches the head and other areas. Birds also scratch their heads when not applying secretions. They use two techniques, bringing the foot forward either under the wing, as in doves, or over the wing, as in many tyrannid flycatchers **(Fig. 3–19)**. Virtually all species use only

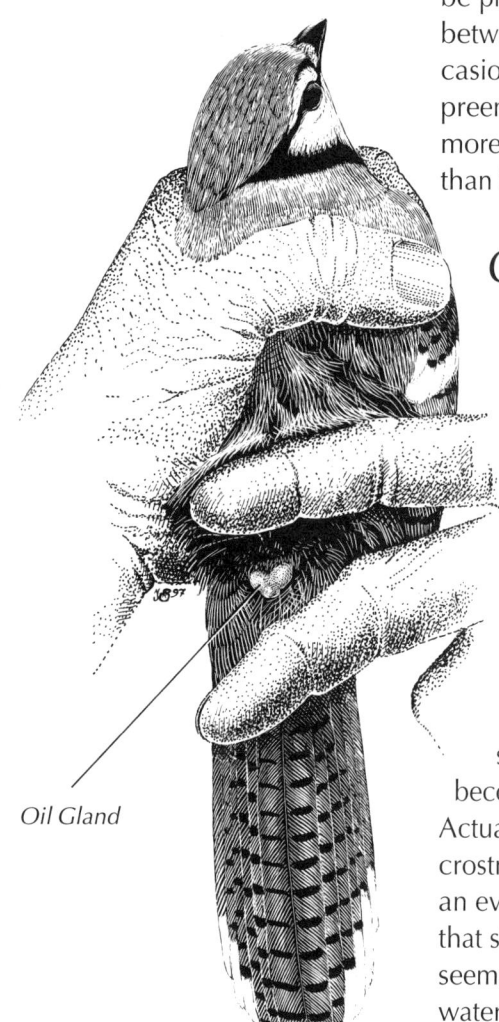

Oil Gland

Figure 3–18. Oil Gland of Hand-Held Blue Jay: *By carefully spreading the feathers, and gently blowing on the rump area of a hand-held bird, you can see the small pimple-like* **oil gland** *(also known as the* **preen gland** *or* **uropygial gland**)*. Secretions from the gland are used in preening to keep the skin supple and to prevent the feathers and scales from becoming brittle.*

one of the two techniques, but some birds, such as Blue Jays, use both methods. Why different birds use different head-scratching methods is not known, so a good research opportunity for a dedicated birder exists here. Birds that can't scratch their heads, either through accidental injuries or experimental use of neck collars, tend to have more ectoparasites living on their heads (Clayton 1991).

Bathing

Another feather maintenance behavior is **bathing**. Birds may bathe in water, snow, or even dust.

Figure 3–19. Head-Scratching Methods: *Birds scratch their heads either by (a) bringing the leg under the wing, as in doves, or (b) over the wing, as in many tyrannid flycatchers. Most species use only one of the two techniques.*

In water, the manner of bathing tends to be quite consistent within various families. Swallows and swifts drop down repeatedly and skim the surface of a lake, pond, or stream to wet their feathers. Terns plunge into the water from the air, and kingfishers, from an overhanging branch. Many forest birds bathe in rainwater or dew that has collected on leaves. Most songbirds such as American Robins and European Starlings stand or squat in shallow puddles, vigorously ruffling the feathers and shaking both wings, creating a veritable shower. Hummingbirds sometimes dart in and out and swing back and forth in the spray from a garden hose on a hot day. Common Nighthawks may bathe on the wing during intense downpours, alternately flapping rapidly and ruffling the body feathers amid the deluge.

A thoroughly wet bird can be quite helpless, and many are unable to fly efficiently, making them particularly vulnerable to cats and other predators. Because birds must dry as soon as possible, all have special drying movements. After bathing, a bird typically shakes its body to throw off extra water, whirs the wings, fluffs its feathers to promote evaporation, and begins to preen and scratch its head. Similar body motions are used in snow. During hot weather, bathing in water might help a bird's body temperature to stay down.

Dust-bathing is most often reported for species that spend much time on the ground, such as House Sparrows and Wild Turkeys. This behavior is not a substitute for water bathing; several families of birds do both. In dust-bathing **(Fig. 3–20)**, a bird usually squats or lies down in a dusty area, often a dirt road, and drives fine particles through its plumage by rolling its body, fluffing its feathers, wiping with its head and bill, or even, in the manner of the Greater Rhea of South America, picking up dust and throwing it over its body. The bird removes some of the dust by violent shaking, but often much dust remains, staining the bird's plumage the color of the soil.

Figure 3–20. Wild Turkey Dust-Bathing: *Lying in the dust, a dust-bathing bird rolls and wiggles its body to work the dirt into its feathers, tossing dust up over its back using its bill or wings. Photo by Sandy Podulka.*

Presumably all kinds of bathing can help to control ectoparasites, but this has not been studied experimentally. Ectoparasites breathe through small holes in their outer skeletons, which are vulnerable to clogging with dust. Comparing birds with and without opportunities for dust-bathing has shown that dust-bathing helps to remove substances coating the feathers, such as old oil gland secretions.

Sunning

Sunning may also help maintain the feathers (Simmons 1986). Sunning birds adopt quite varied and extremely unusual postures. Commonly the bird's feathers are fluffed, the tail is spread against the ground, and a wing is extended on at least one side, sometimes both. Sunning birds frequently lie on the ground in a warm place, with the head lowered and tipped to the side, remaining nearly motionless for many seconds or minutes. Although sunning has been described for numerous species of birds, few experimental studies have tested its functions. Suggested purposes include conditioning the feathers by keeping them supple through limited heating; harming or repositioning ectoparasites; and saving energy by taking up solar heat through the feather coat. The possibility that birds might sun simply because it feels good is difficult to test.

A number of large birds, including cormorants, anhingas, pelicans, storks, and New World vultures, stand for many moments with their wings extended to the side in a pose known as the **spread-wing posture (Fig. 3–21)**. Although cormorants and anhingas are closely related, they use the spread-wing postures in slightly different ways. Both types of birds prey on fish, but the feathers of anhingas are more permeable to water. This reduces the birds' buoyancy and allows them to swim for long periods with just their necks and heads above the surface. Both cormorants and anhingas use the spread-wing posture to dry their feathers after swimming, but anhingas take longer to dry because their permeable feathers take up extra water. Furthermore, anhingas generate less heat internally, and in cooler areas, they may use the spread-wing posture to absorb sunlight. The difference in anhinga and cormorant energy production is reflected in their geographic distributions—cormorants can live in much cooler regions than can anhingas.

Anting

Like bathing and sunning, **anting** is believed to help control ectoparasites. In **passive anting** a bird simply stations itself among a swarm of ants, permitting them to run all over its body and move in and out among the feathers **(Fig. 3–22)**. In **active anting** a bird picks

Figure 3–21. Anhinga Drying Wings: See text for explanation. Photo courtesy of Isidor Jeklin/CLO.

up an ant or other chemically potent object, such as a millipede, and deliberately rubs it in the feathers. Objects rubbed in the feathers during "anting" (in its broad sense) include insects, plant material, and even cigarette butts. Ornithologists presume that the rubbed materials contain chemicals that are noxious to ectoparasites. Because birds sometimes eat the arthropods used in anting, it's possible that the process also helps rid a potential food item of its most noxious chemical protection.

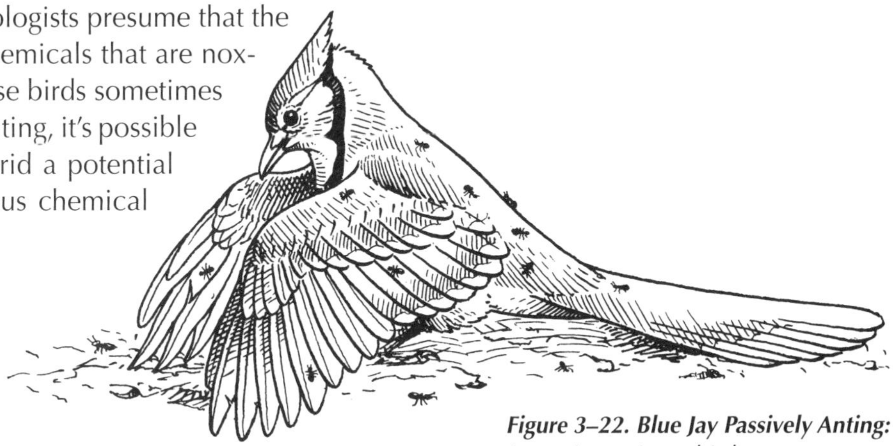

Ectoparasites

Controlling ecto-parasites is important to a bird's fitness. Although carrying a few ectoparasites may have little effect on a bird, a heavy ectoparasite load can impair a bird's health and reproduction. Scientists believe that many feather-maintenance behaviors evolved to control these parasites, either physically or chemically. Many birds, including hawks, House Sparrows, and European Starlings, add pieces of fresh plants to their nests after they complete

Figure 3–22. Blue Jay Passively Anting: In passive anting, a bird squats among a group of ants with its wing (and sometimes tail) feathers spread, and lets the ants run in and out of the feathers and over its body. Anting is thought to help in the control of ectoparasites, but is not well understood. Drawing by Charles L. Ripper.

construction. Some of these plants either produce chemicals noxious to ectoparasites, or have antibacterial effects. Clark and Mason (1985; 1988) found that at least one bird, the European Starling, appears to choose plants specifically for these properties.

Birds are host to many kinds of bloodsucking external parasites, especially flies, ticks, fleas, lice, and mites. The habits and life histories of these parasites are fascinating, although in many cases they remain poorly studied.

Many ground-nesting birds carry the same species of ticks found on mammals in the same habitat; the Ruffed Grouse, for example, is often heavily infested with rabbit ticks. In North America certain birds, especially those that frequent the ground, may carry the ticks that sometimes contain bacteria that produce Lyme disease, a serious human health problem.

Another group of ectoparasites that feeds on the blood of mature birds is the hippoboscid flies, which superficially resemble house flies but are vertically flattened so they can slip readily between the contour feathers **(Fig. 3–23a)**. Nestlings, especially those of Tree Swallows, bluebirds, and other cavity nesters, often are infested with the large larvae of another type of fly called a blowfly, which feed on their blood **(Fig. 3–23b)**.

Bird lice of the order Mallophaga feed on feathers and the surface layers of the skin **(Fig. 3–23c)**. Quite harmless to humans, a few species do live on mammals, feeding on hair. Many are highly host-specific; that is, a given species of louse lives only on members of a single family or genus of birds. Like the hippoboscid flies, most bird lice are flat, and live among the feathers. Most species are adapted for life on specific portions of a bird's plumage, rarely venturing away from the head, neck, body plumage, flight feathers, or whatever area the louse calls home. One very specialized genus of lice lives in the throat pouches of pelicans and cormorants.

Several kinds of mites occur on birds, including itch-mites, nasal mites, and red mites. The most specialized are tiny feather mites that live among, on, or even within the feathers, feeding on the feather itself or on the skin **(Fig. 3–23d)**. Some are so specialized that they live only in certain areas of plumage on certain species; one mite species, for example, lives only on the white areas of the remiges of the Eurasian Nightjar.

Feathers provide some protection against nonparasitic biting insects, such as mosquitoes. Mosquitoes can often get a blood meal, however, by biting where the feathers are short, such as around the eyelid, base of the beak, or on the legs. Mosquitoes can transmit encephalitis viruses to birds, and they sometimes carry these viruses from birds to people. In humans, the disease can be fatal; fortunately, it is uncommon. Different kinds of birds differ in their resistance to biting mosquitoes. Active birds such as White Ibises move constantly and so are less likely to be bitten than relatively sluggish birds such as Black-crowned Night Herons. Most insect bites, however, occur at night while birds are roosting.

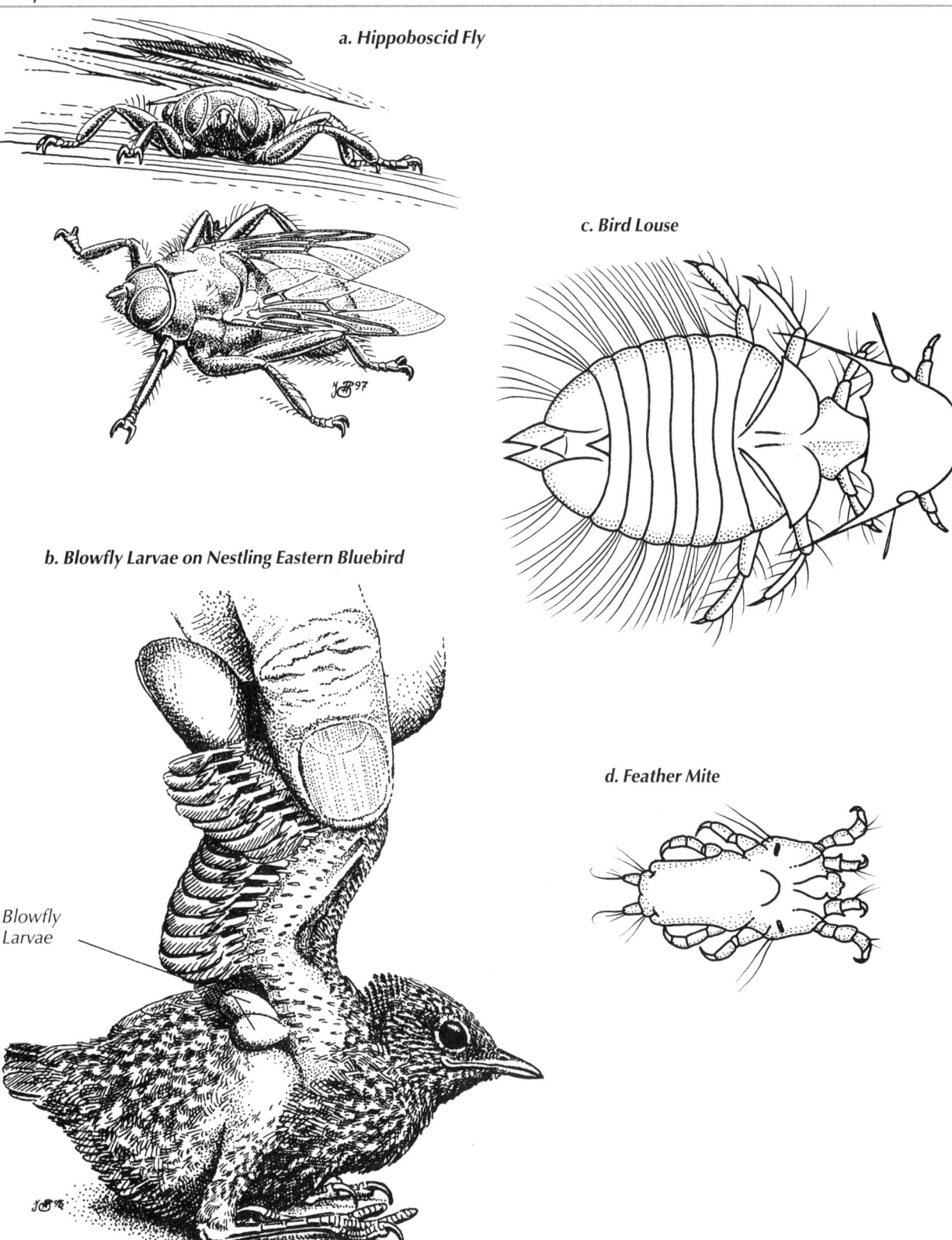

a. Hippoboscid Fly

c. Bird Louse

b. Blowfly Larvae on Nestling Eastern Bluebird

d. Feather Mite

Blowfly
Larvae

Figure 3–23. Avian Ectoparasites: a: *Hippoboscid fly. Note extremely flattened body that allows it to crawl easily between contour feathers. Hippoboscids, also called "feather-flies," are bloodsuckers that live on many different species of birds, especially birds of prey.* **b:** *Nestling Eastern Bluebird with two attached blowfly larvae. These are particularly common bloodsuckers on nestlings of cavity-nesting birds.* **c:** *Bird louse (Order Mallophaga) from a turkey. These ectoparasites are also flattened, and they feed on bird feathers and skin.* **d:** *Feather mite from a passerine bird. Mites live among, on, or within the feathers, feeding on feathers and skin. Illustrations are not drawn to scale. Drawings c and d by Charles L. Ripper.*

Development of Feathers

■ The skin of a bird, like our own, consists of two major parts, the inner **dermis** and the outer **epidermis**. Throughout life, the epidermis continually renews itself by the growth of new cells in its lower layers and the hardening, drying, and sloughing off of the outer layers. During embryonic development, the surface of the skin is covered with little bumps known as **papillae (Fig. 3–24a)** arranged in the eventual pattern of the bird's feathers. Each papilla is an outgrowth of dermis forming a core beneath a rapidly multiplying layer of epidermal cells. As the proliferation of epidermal cells continues, outstripping the dermis, the epidermis doubles inward around the papilla, forming an epidermis-lined pit, the **follicle (Fig. 3–24b)**. From each follicle a succession of feathers is produced by the bird during its lifetime.

As the epidermal cells multiply, the papilla elongates into a cone. The outer cells harden, fuse, and form an **epidermal collar** that surrounds the dermal portion of the original papilla **(Fig. 3–24c)**. Most conspicuous structures in the fully formed feather originate in this epidermal collar. The inner layers of the collar continue to grow toward the center, forming a tubular series of ridges that eventually solidify to become the rachis and barbs of the feather. Because the feather grows from its base outward, the outer part is always the oldest, as in a human fingernail.

The dermal portion of the papilla remains in the follicle for the life of the bird. While each feather is developing, blood vessels extend outward from the papilla into the feather shaft, providing a temporary source of nourishment for the growing feather. When the feather is fully formed, the blood supply is cut off as the vessels are resorbed into the papilla through a hole at the very end of the calamus. On large feathers, you can see this hole with the naked eye, although it appears more as a depression than a hole.

A growing feather is surrounded by a thin **feather sheath**, which acts somewhat like a mailing tube **(Fig. 3–24d)**. When the sheath breaks open, the feather vanes unfurl from the tubular packing into the broad, mature feather. The unfurling occurs slowly over many days. Packaging feathers in sheaths permits them to grow much more densely than they could if each feather somehow emerged fully unfolded directly from the skin.

Growing feathers can be recognized by the intact sheaths, which are typically gray or bluish, and look like the fat tips of knitting needles sticking out of a bird's skin **(Fig. 3–25)**. Before the sheaths open, these growing feathers are termed **pin feathers**. Look for a pin feather on a supermarket chicken, and break it open. Notice that the material inside is soft and moist. This is the developing feather; on a live bird, the cells would be alive and multiplying. As the feather grows, its cells eventually die and become hardened as the sheath splits and falls away. In many cases the outer (distal) portion of the sheath dies and breaks open, while growth continues in the parts of the feather still in the sheath.

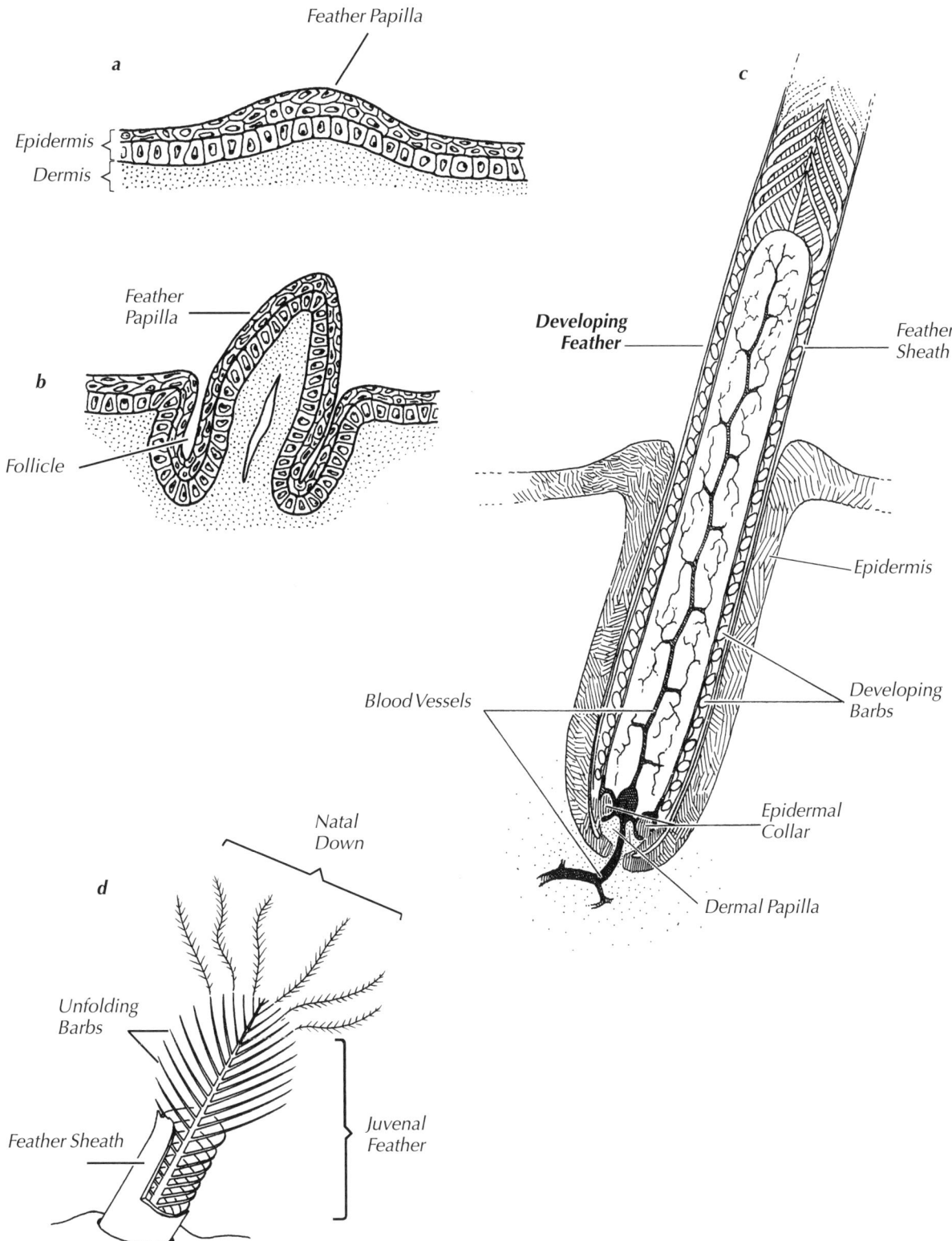

Figure 3–24. Development of a Feather: a: *The skin with a* **feather papilla** *beginning to form.* **b:** *As epidermal cells multiply more quickly than dermal cells, the epidermis folds inward around the papilla, forming an epidermis-lined pit, the* **follicle. c:** *As epidermal cells continue to multiply, the papilla elongates to form a cone, or "pin feather." The dermis remains as the* **dermal papilla** *at the base of the developing feather, and is surrounded by a* **collar** *of epidermal cells that multiply to produce most structures in the developing feather, including the thin* **feather sheath,** *and a series of doughnut-shaped ridges that eventually form the barbs and rachis. The growing feather is nourished via blood vessels that extend into the shaft.* **d:** *The feather sheath gradually splits open from the tip and falls away, allowing the feather vanes to begin unfurling. In this case a juvenal feather has developed and carries out the old natal down on its tip. Drawings a, b, and d by Charles L. Ripper.*

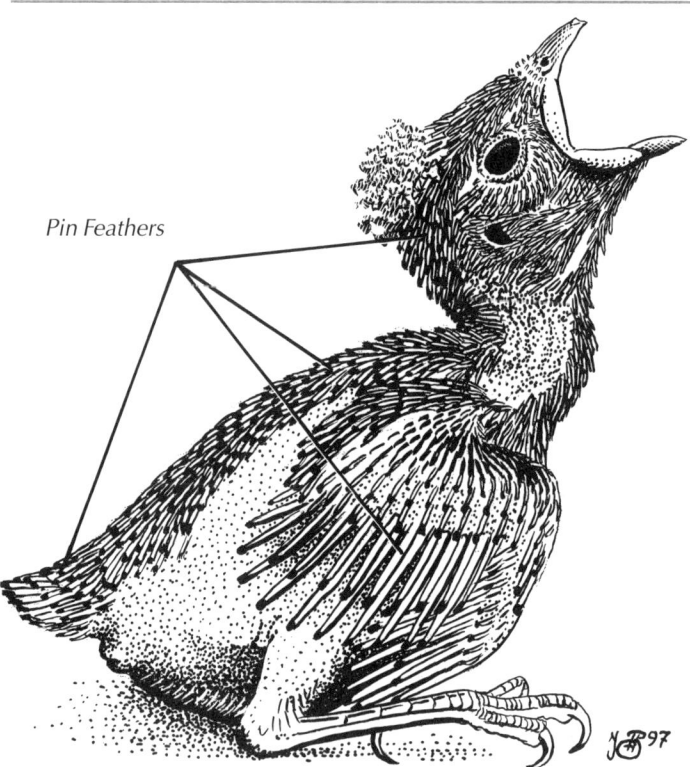

Pin Feathers

Figure 3–25. Nestling in Pin-Feather Stage: *With each developing feather still completely enclosed in its sheath, a young bird in the pin-feather stage looks something like a small porcupine or pin-cushion.*

When feather development is completed, the cells cease to multiply. Yet the original papilla, with its dermal core and epidermal covering, waits deep within the follicle, ready to restart the process when it is time for a new feather. Then a new collar, growing upward and outward, will push out the old feather.

The hardening of a maturing feather is caused by formation of the protein **keratin**, a metabolic product of the dying cells. Indeed, keratin is the primary structural component of mature feathers. Originally produced inside cells, the keratin from one cell may bond strongly with the keratin in adjoining cells, developing over time into such familiar structures as scales, claws, and feathers. Avian keratins differ in their amino acid sequences from all other known keratins, including those in the scales of living reptiles such as lizards and crocodiles, and in the nails and hair of humans. Therefore, the evolution of feathers must have involved major evolutionary innovations (Brush 1993). Keratins are not readily digested by any of the chemicals ordinarily found in the digestive tracts of animals, so feathers last longer than most other discarded animal parts. But we are not chest-deep in loose feathers, even though birds drop billions of feathers annually during molts, so their keratin is broken down somehow. Researchers assume that bacteria and fungi are responsible, but almost nothing is known about the specific organisms that digest feathers in nature.

Many environmental factors can influence the development of feather structure. For example, nutritional deficiencies can lead to minor changes in barbule structure (Murphy et al. 1989). This information is useful to researchers in the field. For species that periodically can be recaptured, researchers can pluck feathers and then, when the bird is captured again, examine the structure of the regrown feather to determine the bird's nutritional state (Grubb 1995).

Molts and Plumages

◼ A feather is a dead structure; it cannot be repaired if it becomes worn or broken. Yet the feathers of most birds, especially those living in open, sandy, or grassy areas, become brittle, faded, and frayed in just a few months. To restore their feather coats and produce feathers appropriate to their age and sex, birds regularly must grow new sets of feathers. Replacing all or part of the feather coat is called **molting**. In a **complete molt**, all feathers are replaced; in a **partial molt**, only some feathers are replaced. If an entire feather is lost between molts, it grows back right away. Damaged, broken, or worn feathers are replaced only during the regular molt cycle, however.

During a typical molt, growth of a new feather pushes the old one from its follicle. In juvenile birds, each natal down feather may be physically attached to the tip of its successor, the **juvenal feather**, which is the first true contour feather; thus the new juvenal feather carries the old natal down out as it grows (see Fig. 3–24d). (Note that the term "juvenal" is applied to feathers and plumages, whereas "juvenile" designates a young bird; the two words often are confused.) The strength of the connection between the downs and the contour feathers varies greatly. In some birds, the down is so delicate that it breaks at the merest touch, and finding museum specimens that show the connection can be difficult—the down has broken off either in the nest or during the preparation of the specimen. In other birds, such as the European Starling and American Robin, the connection is very strong, and you can often see the fluffy natal downs still attached to the heads of juveniles who have recently left their nests **(Fig. 3–26)**.

Figure 3–26. Fledgling Eastern Bluebird: *Note wispy down feathers still attached to the head of this newly fledged bird. Photo courtesy of the North American Bluebird Society.*

Because producing new feathers takes a lot of energy, birds generally molt when they are not engaged in other energetically demanding activities, such as feeding young or migrating. The molting process also reduces flight efficiency, lowering a bird's ability to evade predators and procure food. Therefore, long-distance migrants may molt before they migrate, after they migrate, or partly before and partly after. In the temperate zone, adults of many species of North American songbirds molt during late summer, when they are no longer caring for young but before fall migration. Because temperatures are often still warm and energy demands are relatively low, this is a good time to molt. We humans notice this molt when some of our more colorful birds acquire their subdued winter tones: male Scarlet Tanagers become yellow-green; the bright red head of male Western Tanagers becomes yellow; and many striking male warblers, such as the Chestnut-sided Warbler, Yellow-rumped Warbler, Magnolia Warbler, and Common Yellowthroat, lose their bold spring patterns.

Annual Molt and Wear Cycles

Birds vary greatly from species to species in the number, type, and timing of their molts. Some birds, such as American Crows, undergo only one complete molt per year. Mature crows look much the same before and after this annual molt; the new feathers are less worn than the old ones, but this is evident only at close quarters. Many other North American passerines, such as the American Goldfinch, undergo a complete molt following the breeding season, and then a partial molt in winter or spring. This partial molt includes all the contour feathers except those of the wings and tail, and provides brighter colors for the breeding season. Only a few species are known to have two complete molts per year. Most of these live in harsh habitats that quickly wear out feathers. Examples include Marsh Wrens and Bobolinks, which move within abrasive vegetation, and African

Figure 3–27. The Effect of Wear on European Starling Feathers: a: After the fall molt, newly grown European Starling body feathers have pointed, buff-colored tips. b: The buffy tips of the European Starling's fall feathers give it a mottled appearance. c: By spring, the pointed, buffy feather tips have worn off, and the starling is a glossy black. Photos a and b courtesy of Carrol Henderson. Photo c by Marie Read.

larks that dwell in windy, sandy deserts.

Some passerine species have only one complete molt per year, but nevertheless change their appearance seasonally as a result of wear on the feathers. For example, after they molt in mid to late summer, European Starlings have prominent buff-colored tips on their belly and breast feathers. By the time they are ready to breed the following spring, the buff-colored tips have worn off and the underparts are dark **(Fig. 3–27)**. Similarly, adult male Purple Finches have dull red heads after their annual molt in late summer, but feather wear slowly removes the outer barbules, revealing the bright, pink-red color below. And male Northern Cardinals in fresh fall plumage have a grayish cast, especially on the back, which wears away as winter progresses. By late spring the cardinal is a photogenic bright red, yet has undergone no intervening molt.

Subadult and Definitive Plumages

Young birds also molt; they must replace their feather coats as they mature. As young birds grow, the small feathers produced early in life, such as natal downs, are no longer large enough or able to carry out the functions necessary for an older bird. So, young birds pass through one or more **subadult** (immature) plumages, eventually reaching the **definitive** plumages, those of a mature bird.

The ages at which birds reach definitive plumage vary. Many songbirds take less than a year; thus, a one-year-old Song or Chipping sparrow may be indistinguishable from an older one. One-year-old Red-winged Blackbirds, on the other hand, do not have the brilliant, crisply defined, red-and-yellow epaulettes of older birds; their "shoulders" (actually located at the wrist, or bend of the wing) are duller, usually more of an orange-red and buff, with the buff portion often blurred and marked with black.

In general, long-lived species such as large raptors, gulls, and pelagic seabirds retain their subadult plumages for a relatively longer period of time. For example, the familiar definitive plumage of the Bald Eagle is not reached until a bird's fourth or fifth year. Before acquiring a clear white head and tail, young birds go through several subadult plumages that are much less tidy, with a blotchy or mottled brown-and-white appearance. Many gulls also take several years to reach definitive plumage; in a "three-year gull" such as the Ring-billed Gull, birders can distinguish first-winter, second-winter, and adult birds. Four-year gulls, such as the Herring, Great Black-backed, California, and Western gulls, sport at least four different plumages on their way to maturity **(Fig. 3–28)**. The male Steller's Eider acquires his definitive plumage in his third fall; the males of the larger Common Eider and the King Eider, in their fourth fall. Slowly maturing birds such as albatrosses may take as

Adult Breeding
(Alternate)

Third Winter
(Third Basic)

Second Winter
(Second Basic)

First Winter
(First Basic)

Natal Down

Juvenal

Figure 3–28. Complete Plumage Series of the Herring Gull: *Some birds take more than one year to reach their definitive adult plumage. After shedding its natal down, the Herring Gull passes through four different subadult plumages before reaching the familiar white and gray definitive plumage of the adult—at four years of age!*

long as seven or eight years to reach their definitive plumage. Given a choice, many birds such as gulls select mates in definitive plumage. Sometimes, however, birds in subadult plumage do get a chance to breed. Breeding by subadults occurs principally among species with relatively short life spans, or when birds in definitive plumage are in short supply.

In some species, one sex remains in subadult plumage for longer than the other, a situation termed **delayed plumage maturation**. The male American Redstart, for instance, does not wear his orange-and-black definitive breeding plumage until his second breeding season; his first breeding plumage is nearly identical to that of the adult female, who has yellowish patches, and who acquires her definitive plumage by her first breeding season. However, at least some male redstarts breed at one year of age, despite their subadult plumage. Why plumage maturation is delayed in this species is not known (Morse 1989); the trait is not shared by most other New World warblers.

How delayed plumage maturation evolved, and the advantages it brings, is a hot topic of research among ornithologists. A number of hypotheses have been proposed to explain why one sex (usually the male) might remain in subadult plumage longer than the other. The "female mimicry" hypothesis (Rohwer et al. 1980) suggests that during the breeding season, subadult males may be able to fool adult males by mimicking females. Thus disguised, the subadults may be able to defend their own territories or may gain access to adult territories, thereby procuring more food resources, or even mating with the adult female while her mate is away. In contrast, the "status signalling" hypothesis (McDonald 1989) suggests that subadult plumages accurately convey a male's younger age to adults, thus decreasing the amount of aggression it receives from older, more dominant, males. This hypothesis is most relevant to species, such as Long-tailed Manakins (see Fig. 6–47), with well-developed age hierarchies in which younger males rarely mate, and in which adult males have elaborate plumages or behaviors that are energetically costly to maintain. Other hypotheses focus on potential advantages conveyed by subadult plumages in winter. For instance, Butcher and Rohwer (1988) proposed that subadult plumages may either decrease a bird's conspicuousness to predators (the "winter crypsis" hypothesis) or may decrease male-male aggression (the "winter status signalling" hypothesis). In either case, subadults might enjoy higher winter survival rates. Butcher and Rohwer suggested that although the advantages are gained in winter, birds might retain their subadult plumage through the breeding season to avoid the high energetic costs of spring molt. (Most birds undergo their most extensive molt in fall.) Undoubtedly, different species gain different advantages from delayed plumage maturation, but researchers attempting to determine which hypotheses apply to which bird species have met with mixed success.

The willingness of female birds-of-paradise to mate with males in subadult plumage when males in definitive plumage are not available may have saved some species of these beautiful birds. During the early 20th century, feathers of wild birds were in great demand for decorating women's hats—a practice no longer fashionable and now banned in many places. In New Guinea, for example, killing male birds-of-paradise in their spectacular definitive plumages (see Fig. 3–10) was so extensive that these males were nearly extirpated from accessible areas. Ordinarily, females do not mate with males in the duller subadult plumages, but they had to when males in definitive plumage became in short supply. The delayed plumage maturation of young males therefore proved fortuitous; otherwise, the feather trade might have completely wiped out some of the showiest bird species in the world.

Plumage Naming Systems

In its broadest sense, a **plumage** is a bird's entire feather coat; thus, you might say, "The plumage of the American Robin is orange and gray." Some authorities, however, use the word to designate the set of feathers produced in a particular molt. In this usage, after a partial molt, a bird simultaneously wears parts of two different plumages. In this chapter we will use the former definition and refer to the bird's entire feather coat as its plumage.

Historically, the common practice was to name plumages for the presumed presence or absence of breeding activity at the time the bird wore the particular plumage. Thus, in spring and summer birds were said to be in nuptial (breeding) plumage, and in fall and winter, in postnuptial or winter plumage. This system was somewhat misleading, however, because different plumages are not necessarily linked with reproductive activities or seasons of the year. Among birds in the tropics, where there is no real winter, "winter" plumages may be correlated with the wet season, a time when few birds breed. Also, some seabirds breed either less or more often than once a year, foregoing an annual cycle.

For these reasons, the **Humphrey-Parkes** nomenclature has come into use. Under this system, an adult's main plumage each year, usually produced by a complete molt, is termed the **basic plumage**. Although bird enthusiasts in the temperate zone generally think of a bird's "breeding plumage" as the main plumage—because it is usually the most noticeable and colorful—it is rarely worn for as long as the "nonbreeding" plumage. Thus, the nonbreeding plumage is termed the basic plumage of most birds. The majority of birds, including most jays, chickadees, woodpeckers, flycatchers, thrushes, vireos, swallows, hummingbirds, hawks, and owls, molt only once a year and thus have only a basic plumage as adults.

If a partial molt occurs before breeding, it produces an additional plumage, the **alternate plumage.** This is the plumage that temperate zone birders usually think of as the "breeding plumage." It is this molt that produces the brilliant red of the male Scarlet Tanager **(Fig. 3–29)**.

If, in addition to the basic and alternate plumages, the bird produces another, it is termed **supplemental**; supplemental plumages occur in ptarmigans and certain buntings.

Equivalents for the traditional and Humphrey-Parkes nomenclatures for plumages and molts are shown in **Table 3–1**. Note that molts are named for the feathers they *produce*, not for the feathers they shed.

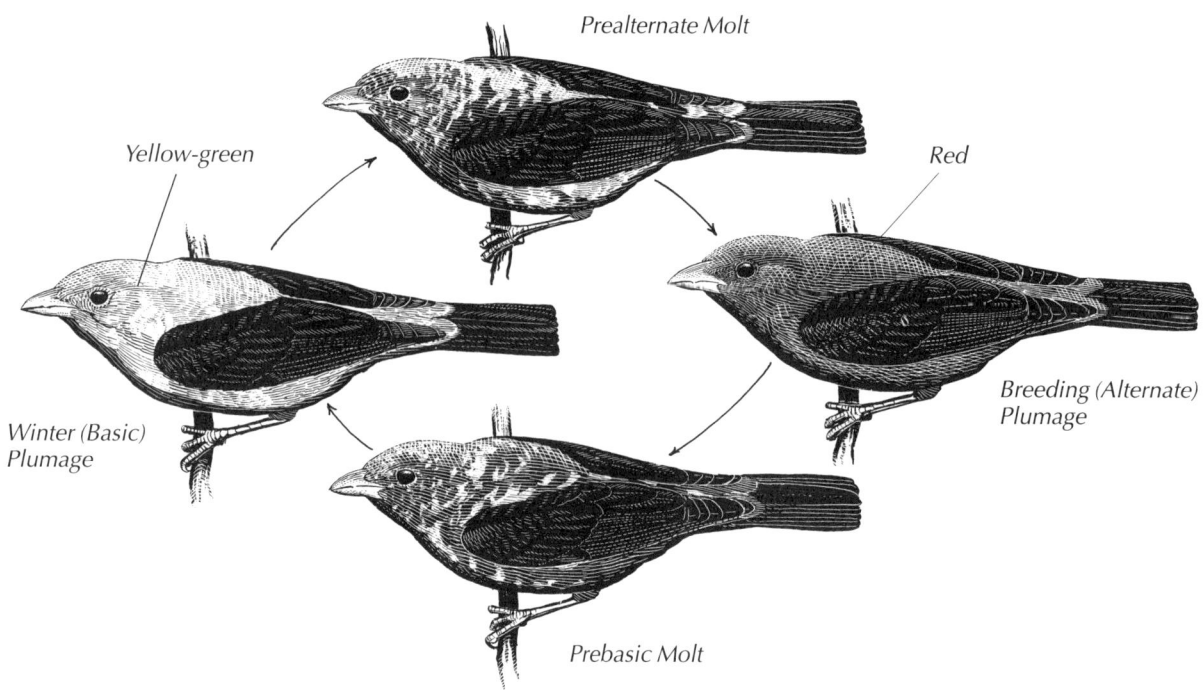

Prealternate Molt

Yellow-green

Red

Winter (Basic)
Plumage

Breeding (Alternate)
Plumage

Prebasic Molt

Figure 3–29. Scarlet Tanager: *The adult male Scarlet Tanager alternates between a brilliant red breeding plumage and a dull yellow-green plumage during the nonbreeding season. The wings and tail remain black in all plumages. After the breeding season, the birds undergo a prebasic molt of all the feathers, during which the red body feathers are replaced by yellow-green feathers. In the prealternate molt just before breeding, only the body feathers are shed, being replaced by the familiar scarlet feathers of the male.*

This naming system was developed because the energy required at molting and other physiological changes at this time are more related to producing the incoming feathers than to discarding the old ones. Note also that subadult plumages are indicated by numbering them as "1st," "2nd," "3rd," and so forth. Once definitive plumage is reached, these numbers are dropped.

The Humphrey-Parkes system also works well for molt and plumage sequences that are highly specialized, such as that of the Mallard. In early summer after mating, the male Mallard has a complete (prebasic) molt, producing a dull-colored basic plumage aptly termed the **eclipse plumage**. Most male ducks in eclipse plumage resemble females **(Fig. 3–30)**. Female Mallards undergo a prebasic molt a bit later in summer, after their young are independent, although their appearance does not change at all. The basic plumage of both sexes is soon lost in a molt of the body feathers—a partial prealternate molt—which produces the brightly colored head and other distinctive features of the male. The occurrence of the prealternate molt immediately after the prebasic molt (much sooner than in most species) is related to the timing of courtship in Mallards, which begins in the fall. Thus, male Mallards are in basic plumage for only a few weeks of the year. For field observers, male ducks in alternate plumage are the easiest to identify to species. Females or males in the basic (eclipse) plumage are generally much less distinctive, so duck identification is often more difficult in mid to late summer.

The Progression of a Molt

In most birds, especially land birds, molt occurs through an orderly, sequential replacement of feathers along a feather tract so that

Table 3–1. Plumage Naming Sytems

TRADITIONAL SYSTEM	HUMPHREY-PARKES SYSTEM
NATAL DOWN (IF ANY)	**NATAL DOWN (IF ANY)**
Postnatal Molt	Prejuvenal Molt
JUVENAL PLUMAGE	**JUVENAL PLUMAGE**
Postjuvenal Molt	1st Prebasic Molt
1ST WINTER PLUMAGE	**1ST BASIC PLUMAGE**
1st Prenuptial Molt	1st Prealternate Molt
1ST NUPTIAL PLUMAGE	**1ST ALTERNATE PLUMAGE**
1st Postnuptial Molt	2nd Prebasic Molt
2ND WINTER PLUMAGE	**2ND BASIC PLUMAGE**
2nd Prenuptial Molt	2nd Prealternate Molt
2ND NUPTIAL PLUMAGE	**2ND ALTERNATE PLUMAGE**
2nd Postnuptial Molt	3rd Prebasic Molt
3RD WINTER PLUMAGE	**3RD BASIC PLUMAGE**

When a bird reaches its definitive plumage, the numerical designations are dropped. For this example, we assume definitive plumage is reached after the 3rd winter or 3rd basic plumage:

Prenuptial Molt	Prealternate Molt
NUPTIAL PLUMAGE	**ALTERNATE PLUMAGE**
Postnuptial Molt	Prebasic Molt
WINTER PLUMAGE	**BASIC PLUMAGE**
ETC.	**ETC.**

Female—Breeding (Alternate)

Male—Breeding (Alternate)

Male—Eclipse (Basic)

Figure 3–30. Mallard Plumages: *The male Mallard in eclipse plumage looks remarkably like the female, but his bill is light olive green, whereas the female's bill is orange marked with black. The basic plumage of the female is identical to her alternate (breeding) plumage.*

in any particular tract, only a minority of feathers are replaced at one time **(Fig. 3–31)**. Thus, you will rarely see a land bird missing more than a few wing or tail feathers. This gradual replacement of feathers spreads out the energetic cost of molting over time, while minimally impairing the flight, insulation, and other functions of the feathers. In woodpeckers, for example, the stiffened rectrices act as a prop to support the bird while it searches up and down tree trunks. Molting of these feathers starts with the second innermost rectrices and progresses to the outside. The long central pair of rectrices is replaced only after the other tail feathers have grown in, so the tail can serve as a sturdy brace throughout the molt cycle.

Many water birds, including loons, grebes, anhingas, swans, geese, ducks, and even dippers, drop all the remiges simultaneously, leaving the birds unable to fly during their prebasic molt. These water birds typically have a high ratio of body mass to wing area; thus the loss of a small amount of wing surface area would seriously impair flight performance. By replacing all the major wing feathers at once, they avoid the prolonged period of impaired flight that would result if the remiges were replaced gradually. Being completely flightless for

Wing

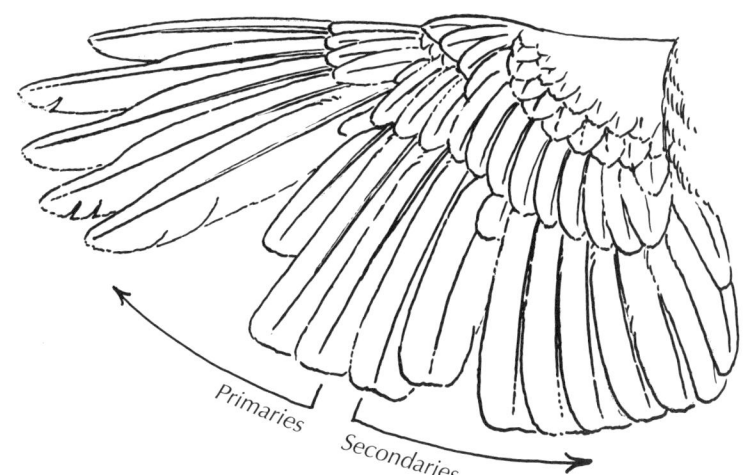

Primaries

Secondaries

Figure 3–31. Typical Progression of Molt in the Flight Feathers: Most birds molt only a few flight feathers at one time, in an orderly progression. In the tail, the central feathers usually are lost first, and as they grow back, successive feathers toward each side are molted. In this example, the central tail feathers have completely regrown, while numbers 2 and 3 have only partially regrown. Tail feathers 4 and 5 have not yet been molted. In the wing, molt begins simultaneously in the inner primaries and the outer secondaries and proceeds in opposite directions, as indicated by the arrows.

Tail

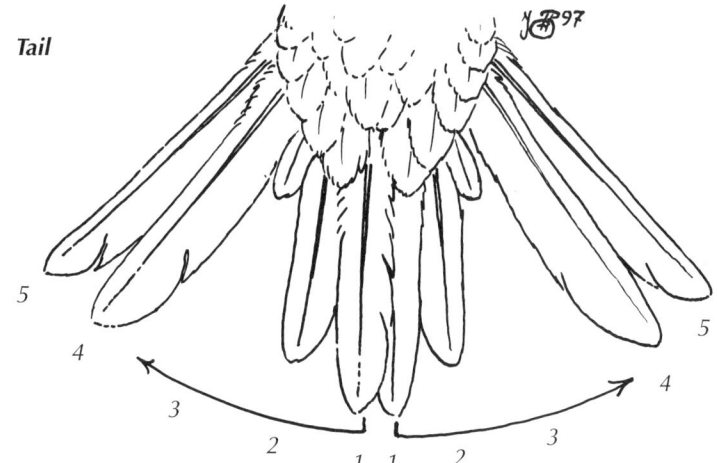

a time probably makes these birds more vulnerable to predators, but they often molt in secluded places, and if they are attacked, most can escape by swimming underwater. For many male ducks, recall that the prebasic molt in which all the flight feathers are lost produces a dull eclipse plumage; this helps them remain inconspicuous. Furthermore, the molt in ducks occurs in several stages, such that the flight feathers are retained until *after* the brightly colored body feathers have been replaced.

A few groups, including pigeons, parrots, and cuckoos, replace their tail feathers irregularly. In contrast, some of the smaller owls, such as the Northern Pygmy-Owl, the Little Owl, and the Burrowing Owl, molt all the rectrices at once. This is also true of some of the other short-tailed birds—the alcids and rails—which use their tails very little in flying. Simultaneous tail molt is most conspicuous in the Boat-tailed Grackle; the males of this long-tailed species appear ludicrous during their stub-tailed period.

Females of a number of hornbill species also molt abruptly. In these Old World cavity nesters, the females stay in the tree cavity throughout incubation and the early part of nestling life; while sequestered they undergo a full molt, including an extended flightless period.

Figure 3–32. Male Rhinoceros Hornbill Bringing Food to Imprisoned Female: *During the incubation and early nestling stages, the female hornbill remains sealed in her nest cavity, depending solely on food brought to her by her mate. Photo by M. Strange/VIREO.*

The birds plaster a mixture of mud and saliva around the entrance hole of the nest, creating a barrier to exclude predators such as climbing snakes and monkeys. The male passes all food to the imprisoned female and young through a small opening; the female and young depend on his survival for their own **(Fig. 3–32)**.

Occasionally, a bird will shed many feathers all at once. This **shock molt** or **fright molt** has been reported for dozens of species in numerous taxonomic families, but occurs only in unusual circumstances. For instance, a bird may shed its tail feathers when the tail is grabbed by a predator—clearly an adaptive strategy. It occasionally occurs during handling by people, and reports say that it can be induced by violent natural events such as earthquakes and tornados. The mechanism of shock molt has not been determined, but clearly some relaxation of the muscles holding each feather shaft must be involved: under normal circumstances, a force of roughly one to two pounds is required to pull out a single feather!

The speed of molt varies to some extent with latitude and the extent of seasonal change in climate. Molt is often more rapid, for example, in Arctic-nesting gulls than in closely related gulls of the temperate zone. Birds in the Arctic simply do not have time for a slow-paced molt after breeding, as they must soon head south for the winter. The abundance of food and continual daylight for foraging in the Arctic undoubtedly help these birds in gathering the resources they need to molt rapidly. In some tropical species, on the other hand, molts may be prolonged. The Rufous-collared Sparrow in northern South America takes two months to molt. Breeding seasons in the tropics also may last a long time, and the energy demands of some species are low enough that they are able to overlap molt and breeding activity.

Once in a while, a bird turns up with very few feathers on its head. The head may be bare, with the ear opening clearly visible, or may have a short layer of barely emerging feathers. Generally the entire head from the neck up is bare, but sometimes just patches of feathers are missing. These "bald" birds (often Blue Jays or Northern Cardinals) are seen most often in the fall, although they occasionally appear at other times of the year. Most bald birds seen in fall are probably juveniles born earlier in the year undergoing their first prebasic molt. For unknown reasons, some of these birds drop all of the head feathers at one time and quickly regrow them, rather than losing and replacing them a few at a time. Why birds occasionally become bald at other times of year, and why adult birds are sometimes affected, remain a mystery. It's possible that some type of atypical molt causes baldness in these birds, or that an infestation of feather lice or feather mites is responsible. Whatever the cause, bald birds certainly command the attention of birders—their prehistoric, vulture-like appearance reminds us of their ancestral ties to reptiles.

Nonfeathered Areas

■ We now consider parts of the bird normally not covered by feathers, such as the eyes, bill, legs, and feet. Frequently, their peculiarities in color or form play a significant role in the bird's life, and are useful in identifying particular species.

Eyes

No thorough description of a bird is complete without mention of the color of its **iris** (see Fig. 3–34c). Although most birds have a dark, commonly brown, iris around a black pupil, many species have more conspicuously colored irises—white, yellow, orange, blue, red, or even, as in some grebes, concentric circles of two or more colors. The Yellow-eyed Junco and the Red-eyed Vireo are named for the color of the iris in adults, as is the White-backed Fire-eye of the Neotropics. Iris color may help some birds in species recognition. In Malaysia, mixed flocks may include several species of bulbuls almost identical in size and plumage color, but quite different in eye color—the differences may help both birds and bird watchers to distinguish among the species!

Sometimes the sexes differ in iris color. Mature male Brewer's Blackbirds have a yellow iris, but females have a brown iris. The effect of iris color on social behavior has not been studied extensively, but when iris color differs by age or sex, it presumably reflects different social positions and, like differences in plumage color, plays a role in social interactions. Commonly, if the adult has a bright iris color, the juvenile's iris is duller, so iris color can indicate the age of a bird. For example, juvenile Red-eyed Vireos have brown irises during their first fall. This difference may help the birds recognize juveniles, especially when there are no obvious plumage differences.

In some species, eye color continues to change as the bird ages. A Cooper's Hawk begins life with a bright yellow iris, but as the bird grows older the eye changes to light orange, then medium orange, then dark orange, and finally to a deep, dark red. This pattern of color change is so predictable that scientists can determine the bird's approximate age by eye color alone (Rosenfield et al. 1992).

Bill

The terms **beak** and **bill** are synonymous. The visible portion of the bill consists of a sheath of skin, the **rhamphotheca**, which covers the projecting portion of the bony jaws (**Fig. 3–33**). In most birds, the rhamphotheca is hard and hornlike; it is softer and more leathery in most waterfowl, sandpipers, plovers, and pigeons.

Although the outer layers of the rhamphotheca consist of dead skin worn away by bill-wiping and abrasion during feeding, the inner layers are alive, and they renew the outer cover of skin. At the tip,

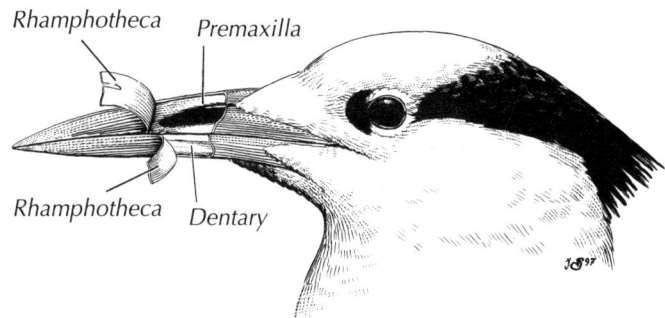

Rhamphotheca   *Premaxilla*

Rhamphotheca   *Dentary*

*Figure 3–33. Bill Anatomy: The **premaxilla** and **dentary** bones, which make up most of the hard, internal structure of the bill, are covered by a thin sheath of skin called the **rhamphotheca**. In most birds, the rhamphotheca is hard and hornlike, but it is softer and leathery in most waterfowl, sandpipers, plovers, and pigeons.*

growth is almost continuous, but it normally is not noticeable because of a continual wearing away. If an injury breaks one tip or puts the two jaws out of line, such that one is not worn away by contact with the other, the bill may grow extremely long until the bird cannot eat and starves to death. Species that feed on a soft diet of insects in summer and then switch to hard seeds during the cooler months have seasonal differences in bill length. In House Sparrows, for example, bills are longer in the summer months and are worn down by the harder foods and grit taken during winter; these differences in bill length are small, and can be detected only by detailed measurements. Captive birds may not eat hard enough foods to wear down their bills properly, so their owners often give them cuttlefish bone, which provides an abrasive surface to chew on, as well as minerals.

In certain kinds of birds, including ratites, albatrosses, petrels, pelicans, and cormorants, grooves extend along the rhamphotheca. In many cases, the function of these grooves is unknown. In albatrosses and pelicans, however, the grooves carry salty fluids emitted from salt-excreting glands near the eyes (see Fig 4–131), guiding them from the nostrils to the bill tip, where they drip away **(Fig. 3–34a)**. (These are *not* the types of grooves for which the Groove-billed Ani is named.)

The nostrils or **nares** (singular: **naris**) are on the upper part of the bill, usually near its base. The shape of the nares differs along taxonomic lines. The most highly pelagic order of birds, which includes albatrosses, shearwaters, petrels, storm-petrels, and diving-petrels, is characterized by tubular nares **(Fig. 3–34b)**, whose function is uncertain. These birds, often called "tubenoses," excrete large amounts of excess salts in a fluid that is released into the nasal cavities. Some biologists believe that the protection afforded by the tube reduces heat and airflow at the point where salty fluid is excreted from the salt gland. Thus, evaporation of the salty fluid occurs at a point farther down the bill, reducing the possibility that salts left behind after evaporation will clog the gland. Another specialization occurs in some ground-feeding birds such as starlings, pigeons, and domestic chickens. These birds have a fixed protective flap, the **operculum**, partially covering the nares; this flap may help keep out debris **(Fig. 3–34c)**. Perhaps most bizarre are the kiwis of New Zealand, whose nares are uniquely situated at the tip of their lengthy bills (see Fig. 4–54c). Kiwis use their well-developed sense of smell to probe for earthworms at night. The nares of hawks, pigeons, and some parrots are located in a leathery band of skin known as the **cere,** which extends across the base of the upper part of the bill and presumably protects the nostril openings **(Fig. 3–34d)**.

Many land birds, including hummingbirds, woodpeckers, and passerines, regularly clean their bills by **bill-wiping** on tree branches, the ground, or other surfaces, especially after eating messy foods such as oily insects or suet. Other land birds, such as doves, rub off debris by scratching with their toes and never have been reported to bill-wipe. Ducks and many other water birds clean their bills by bathing and rubbing them against their feathers.

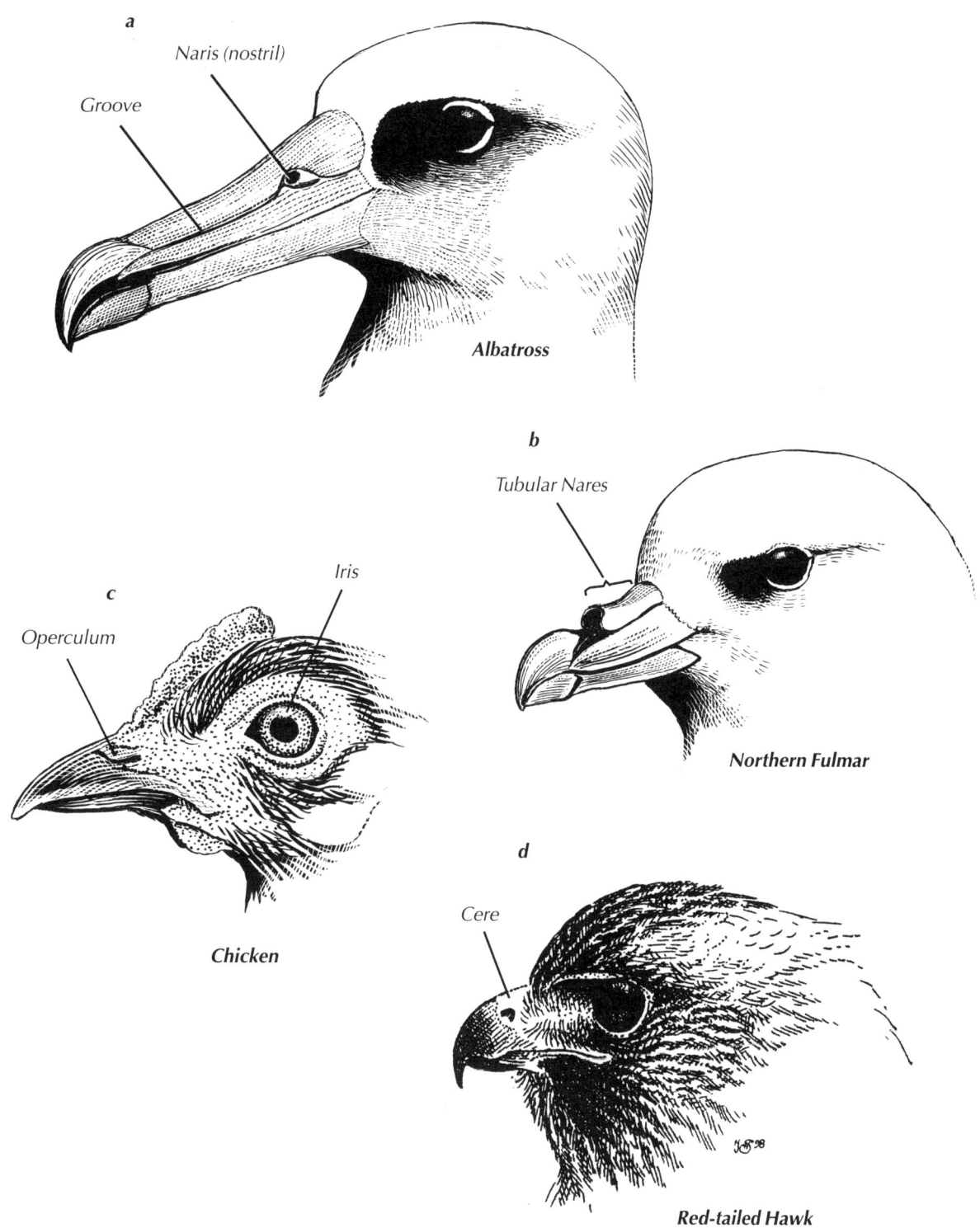

a
Naris (nostril)
Groove
Albatross

b
Tubular Nares
Northern Fulmar

c
Iris
Operculum
Chicken

d
Cere
Red-tailed Hawk

Figure 3–34. Adaptations of the Bill: a: *A groove on each side of the albatross bill carries excreted salty fluid from the bird's naris to the tip of the bill, where it drips away.* **b:** *In the Northern Fulmar, like other pelagic birds termed "tubenoses," a tubular structure covers the site where salty fluid is excreted from the salt gland. The fluid flows through the tube and out the opening in the end. These* **tubular nares** *may reduce heat and airflow at the salt gland so that evaporation occurs farther down the bill, reducing the possibility that salts left behind will clog the gland.* **c:** *In some ground-feeding birds, such as the domestic chicken, a flap termed the* **operculum** *partially covers the nares, probably helping to keep out debris.* **d:** *Many parrots, pigeons, and birds of prey, such as this Red-tailed Hawk, have a leathery band of skin termed the* **cere** *at the base of the bill, into which the nares open. The cere is thought to protect the nares.*

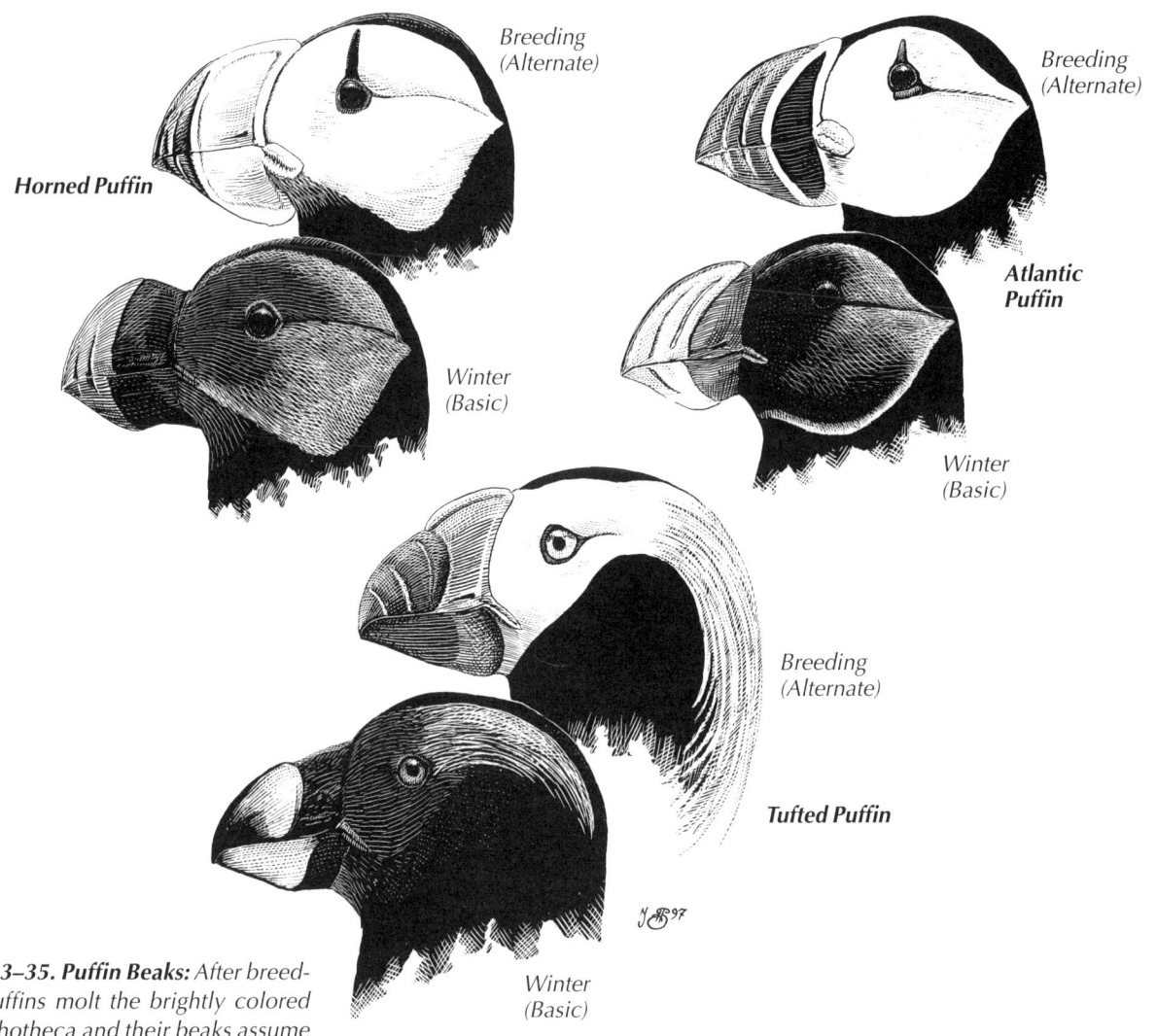

Figure 3–35. Puffin Beaks: After breeding, puffins molt the brightly colored rhamphotheca and their beaks assume the more subdued tones of the new rhamphotheca.

In most birds the bill is black, but bills do occur in virtually every color, from the stunning red bill of a Common Merganser to the bold sky-blue beak of a breeding male Ruddy Duck. In some birds, including puffins and many toucans, the bill is the most colorful area of the body.

The colors of some bills change with the seasons. Atlantic, Horned, and Tufted puffins molt the colorful outer sheath of their bills after breeding **(Fig. 3–35)**. More typically, though, color changes occur by a gradual wearing away of the rhamphotheca, exposing the new, brighter color of the growing skin layers below. The European Starling and the American Robin have dark brown bills in the fall, but they turn yellow in the breeding season; indeed, the increased yellow of the starling's bill in February is one of a birder's signs of spring. The reverse occurs in the male Bobolink and House Sparrow, whose bills are black in the breeding season and yellowish or light brown in the fall, and in the Evening Grosbeak, whose bill turns from bright apple-green in the breeding season to olive-yellow in the fall. Most brightly colored bills do not attain their full coloration until the bird matures sexually.

Bill color also may differ by sex. In the European Starling, during breeding season the basal portion of the lower beak becomes bluish in males, and pinkish-brown in females. The bill of the Mallard is olive-green to olive-yellow in the male (varying with the season), and bright orange mottled with black in the female.

Bill color markings serve several functions. The red mark on the lower beak of breeding adult Herring Gulls is a target for begging nestlings. When the chick pecks at this spot, the adult opens its mouth and disgorges food. In the nonbreeding season this region of the bill becomes dark. Nestling cuckoos, woodpeckers, and passerines have brightly colored temporary enlargements called **oral flanges (Figs. 3–36** and **3–37)** at the base of the bill, which are targets for adults feeding the young (Clark 1969). Juveniles of cavity-nesting songbirds such as starlings typically have bright yellow oral flanges; these remain visible in the dim light resulting when a parent blocks the nest entrance on returning with food. Colorful bill markings also may serve a social function, especially in some highly gregarious birds such as toucans, which have bold, colorful patterns on oversized bills.

Figure 3–36. Begging Black-and-white Warbler Nestlings: At the slightest hint that a parent may have returned with food, hungry young birds "open wide," fully displaying the bold **oral flanges** (see also Fig. 3–37). These temporary swellings at the side of the mouth act as targets for adults feeding young. Photo courtesy of Isidor Jeklin/CLO.

Legs and Feet

Typically, the unfeathered part of the leg is the lower portion of the **tarsometatarsus,** simply called the **tarsus** by most ornithologists. One exception is the Ostrich, whose entire leg, including the thigh, is featherless. Many birds have tarsi and feet of gray, black, or another dark color, but others show a wide variety of colors and patterns. Bird watchers often use leg and foot color to recognize particular kinds of birds: the bright yellow legs of the Greater and Lesser yellowlegs are a beacon to the identity of these birds, which otherwise look a great deal like many other members of the sandpiper family. Leg color also is helpful in distinguishing among the white egrets. These all-white waders look similar at a distance, and size comparisons may be difficult to judge, but their feet and beak colors quickly give away their identities. Great Egrets have yellow beaks and black legs; Snowy Egrets, black beaks and legs with bright yellow feet; Cattle Egrets, yellow beaks and legs; and white-phase Reddish Egrets, pink beaks with a black tip and dark blue legs. In many herons and egrets, leg colors become especially bright

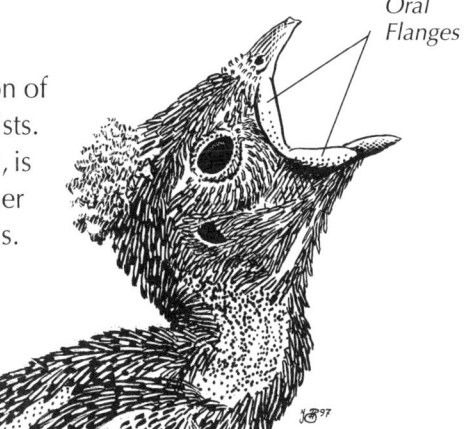

Figure 3–37. Oral Flanges of a Nestling: Begging nestling displays its prominent **oral flanges** at the sides of its mouth (see also Fig. 3–36).

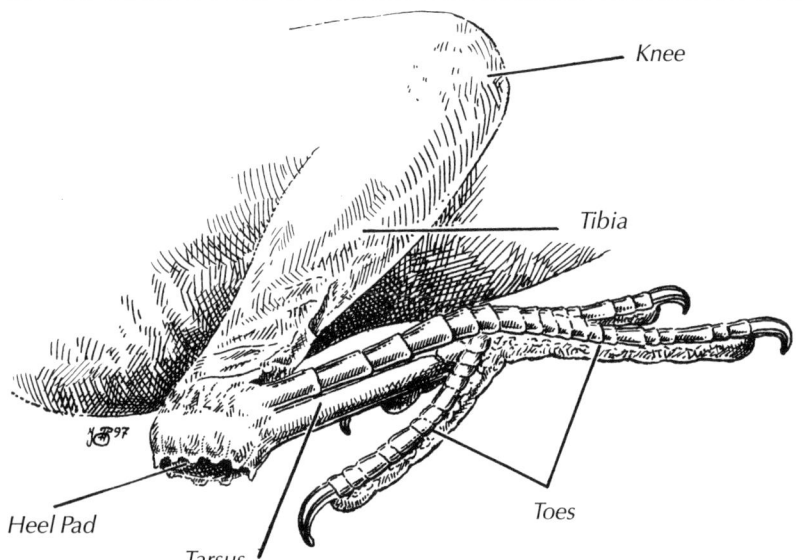

Figure 3–38. Heel Pad of Nestling Toucan: Many nestling cavity nesters, such as woodpeckers and toucans, have an enlargement at the upper end of the tarsus termed the **heel pad**. It is thought to reduce the abrasion of the tarsus from the rough lining of the nest cavity.

for the breeding season. The bright orange feet of a breeding Green Heron are unmistakable as it walks in mud or along dark logs. Like the bill, legs change color through the sloughing off of dead outer skin layers to expose the replacement layer produced underneath.

In certain cavity-nesting birds such as woodpeckers, nestlings have a special enlargement of the upper end of the tarsus, the **heel pad (Fig. 3–38)**. Heel pads probably reduce abrasion of the tarsus caused by the rough lining of the nest, in the way that elbow pads protect a hockey player. Heel pads are shed at about the time the young leave the nest.

A few kinds of birds, such as ptarmigans and most owls, have feathers covering the legs and feet **(Fig. 3–39a)**. Ptarmigans have extra feathers on their feet during winter. These provide insulation and offer a larger surface for support on snow, like a snowshoe. Owls apparently have feathered feet to help suppress flight sounds that might alert prey. Unlike the silently flying nocturnal owls, daytime fishing owls lack feathered feet, and their flight is relatively noisy.

In most birds the outer end of the leg has some type of skin covering that resists abrasion, and depending on the type of bird, may include scales, papillae, or leathery skin. Patterns of papillae and folds on the undersides of the toes often reflect the functions the feet must perform. The feet of Osprey, for example, have spiny-tipped papillae on the underside to firmly grip slippery fish **(Fig. 3–39b)**. In Ruffed Grouse, an enlargement of scales along the sides of the toes creates supporting winter "snowshoes" **(Fig. 3–39c)**.

In many birds, papillae patterns on the feet vary considerably and can be used to identify specific individuals (Clark 1972; Smith et al. 1993), much as human fingerprints are used to identify individual people. In the case of rare birds worth tens of thousands of dollars, such as certain falcons and parrots, knowledge of papillae patterns could help identify individuals stolen or illegally taken from the wild.

The claws at the ends of toes often reflect a bird's habits. Species that climb tree trunks, such as nuthatches, Brown Creepers, and Black-and-White Warblers, have claws more curved than those of nonclimbing species **(Fig. 3–39d)**. These claws help them grasp irregularities on bark without noticeably impairing their ability to perch. It is remarkable how well nuthatches can climb on vertical trunks without slipping or falling off. The curved claws of *Archaeopteryx* are among the characteristics indicating that it was arboreal in its habits (Feduccia 1993).

Ground-dwelling songbirds such as larks and pipits are noted for their long hind claws. These conceivably could help them to avoid sinking into mud or other soft surfaces, although this idea remains

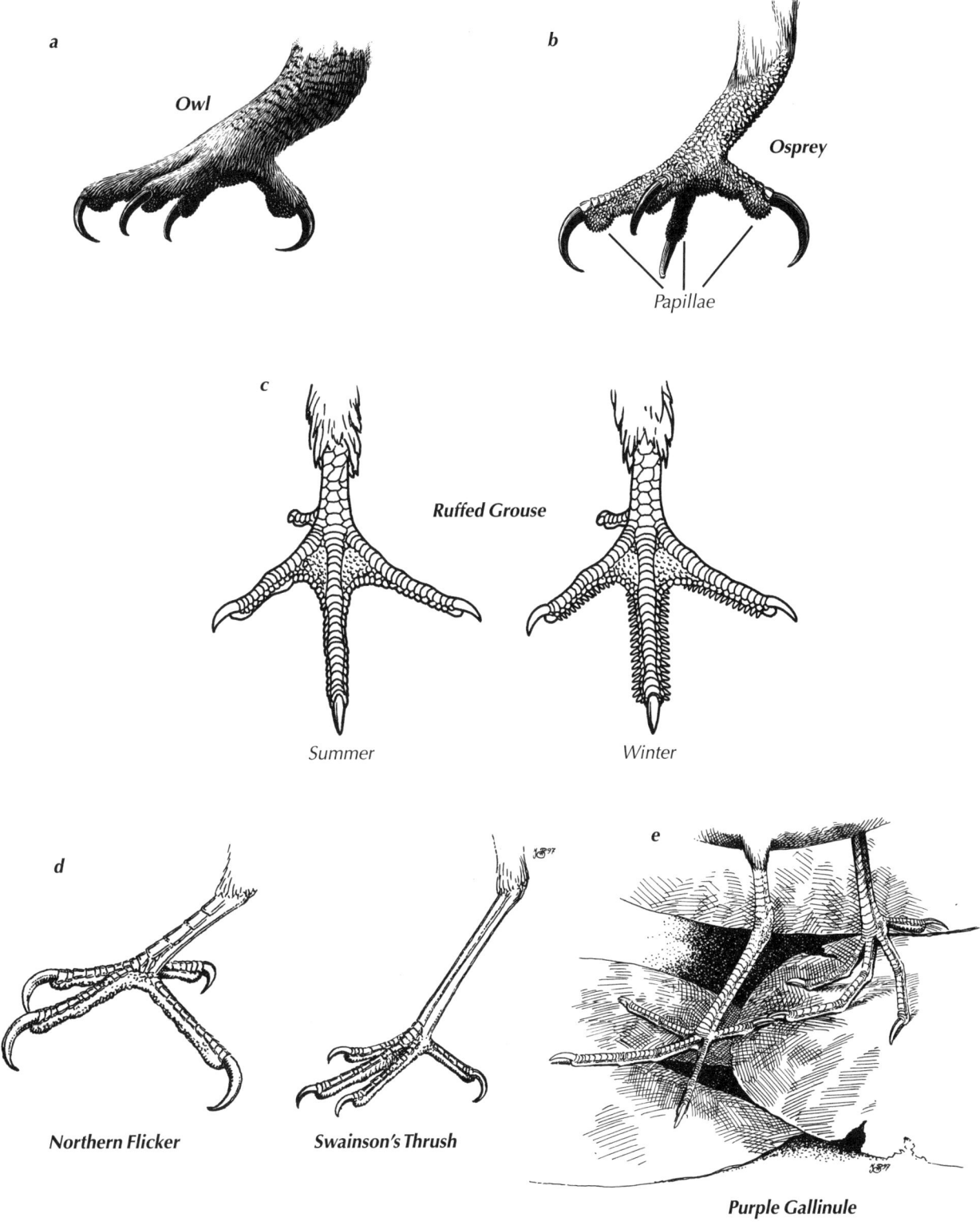

Figure 3–39. Adaptations of Bird Feet: a: *The feet and legs of most owls are covered with feathers, which apparently suppress flight sounds that might alert prey.* **b:** *The undersides of Osprey feet are covered with spiny-tipped* **papillae** *to help them firmly grasp slippery fish.* **c:** *In winter, the scales on the toes of Ruffed Grouse enlarge, forming "snowshoes" that provide a larger surface area to help support their weight on snow. Drawing by Charles L. Ripper.* **d:** *Climbing birds, such as the Northern Flicker, have claws curved more than those of nonclimbing birds, such as the Swainson's Thrush.* **e:** *Purple Gallinules have long toes to help distribute their weight so they can walk on floating lily pads.*

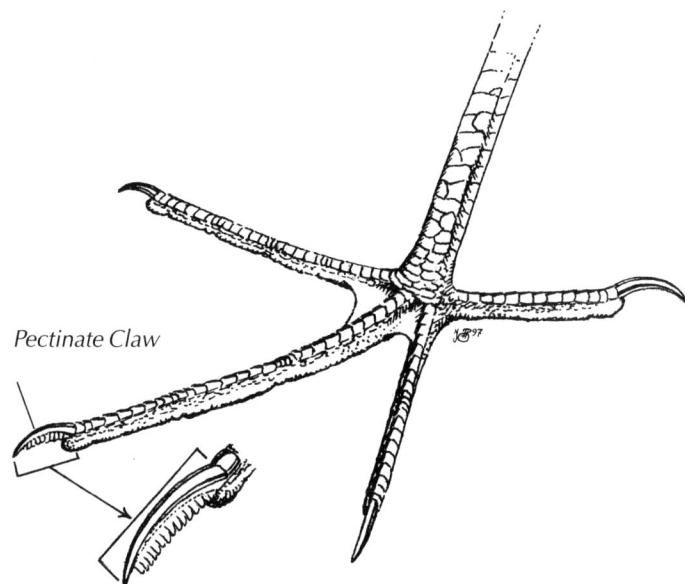

Pectinate Claw

Figure 3–40. Pectinate Claw of Heron: *The middle claw in some birds, such as this heron, has a comb-like, serrated edge used as a preening tool. The structure is termed a **pectinate claw** or a **feather comb**.*

speculative. The Purple Gallinule uses its long toes in a similar way—they distribute the bird's weight so it can walk on floating lily pads **(Fig. 3–39e)**.

In a few birds, including Barn Owls, nightjars, bitterns, and herons, the side of the middle claw has a comb-like, serrated edge **(Fig. 3–40)**. This **pectinate claw** (sometimes called a "feather comb") is used as a preening tool.

Claws, like beaks, are subject to constant wear, which helps to maintain the normal length. When wear is reduced, however, as in captive birds, the skin covers of both claws and beaks may grow exceptionally long and must be trimmed.

Embryos and juveniles of many species bear claws on the wings as well as on the feet. Most remarkable are those of the Hoatzin of Amazonian South America, which builds its nest a few yards above quiet water. When disturbed, nestlings leap into the water beneath and swim away. When danger is gone, they use their wing claws to climb back up into the vegetation **(Fig. 3–41)**. Wing claws in the embryos or young of most living species, however, are **vestigial**: nonfunctional holdovers of an ancestral structure. *Archaeopteryx* had well-developed wing claws, which it presumably used to climb in vegetation.

Other Unfeathered Areas

Areas of unfeathered or sparsely feathered skin occur in birds of almost every order. These vary from the relatively inconspicuous to the bold and beautiful. In many species, areas of bare skin function in display, but they have other uses as well. Carrion-feeding birds, including the New and Old World vultures and some of the storks, have heads that are largely bare except for bristles. This is reasonable because face and head feathers—if they had them—would become badly soiled when the bird reached inside messy carcasses to feed. Among birds that feed regurgitated, partly digested fish to their nestlings, such as pelicans, the young tend to have few or no feathers on their faces.

Some vultures also use their unfeathered, highly vascularized head and neck to help regulate body temperature (Arad, Midtgard, and Bernstein 1989) **(Fig. 3–42)**. For example, when a Turkey Vulture is too hot, it stretches out its neck and increases the flow of blood through that region. Excess body heat carried by the blood is then given off through the skin, from blood vessels near the skin's surface. When cold, the bird retracts its head and reduces blood flow through the superficial blood vessels, thus conserving body heat.

The unfeathered areas of most birds are on the head and neck. Some of these areas are normally inconspicuous, but become very prominent during displays. Certain North American grouse have

Wing Claw

Figure 3–41. Hoatzin Adults and Young:
a: The bizarre, pheasant-sized Hoatzins of Amazonian South America build their nest in low branches overhanging quiet water. b: Young Hoatzins bear claws on the leading edges of their small wings. When disturbed, they leap from the nest into nearby vegetation or water, then use the claws of all four appendages to climb back up.

Figure 3–42. Turkey Vulture: Turkey Vultures and many other bare-headed birds can eliminate excess heat by stretching out their necks to expand the area of bare skin and increasing blood flow through the skin of that region. Photo courtesy of Ray Martorelli/CLO.

Figure 3–43. Bearded Bellbird: The Bearded Bellbird, a jay-sized cotinga of northern South America, features a mass of slender, fleshy wattles, black in color, which suggests a beard. Photo by C. H. Greenewalt/VIREO.

highly colored **booming sacs** on the sides of their neck (see Fig. 4–97). In Sage Grouse, the sacs lie on the breast and are inflated by the expansion of the walls of the esophagus. The bare throats of male frigatebirds inflate like huge red balloons during courtship displays (see Fig 6–44b). Other bare areas change seasonally: the enormous orange knob at the base of the bill of the male King Eider shrinks after the breeding season, as do the frontal shields of some coots and other members of the rail family.

Bare or sparsely feathered areas of the head often have peculiar outgrowths of the skin, such as the **comb** and throat **wattles** of the Domestic Fowl, the warty bumps on the face of the Muscovy Duck, or the **snood** of the Wild Turkey—a limp, red, fingerlike projection from the forehead. Pendulous wattles also hang from the heads of New Zealand wattlebirds (see Figs. 1–92 and 9–28) and the spectacular Neotropical bellbirds **(Fig. 3–43)**. A number of species have feather-free **casques** on top of the head, including the cassowaries of Australia and New Guinea **(Fig. 3–44)** and the Maleo, a brush-turkey (megapode) native to the Indonesian island of Sulawesi.

Brilliantly colored eyelids occur in birds of many families, including plovers, pigeons, cuckoos, trogons, thrushes, and Old World flycatchers. In some, such as the Wattled Broadbill and the Yellow-wattled Bulbul—both of the Philippine Islands—the brightly colored eyelid is also enlarged and fleshy. In some birds, such as toucans and some parrots and honeyeaters, an area around the eye is bare and brightly colored. In many herons the bare area is between the eye and the base of the bill, and the color varies with season.

Colors

The tremendous range of bird colors, on both feathered and unfeathered parts, suggests immediately that birds have well-developed color vision. Indeed, experiments have verified that many birds do see colors, and the structure of their eyes indicates that they may be able to discriminate a greater variety of colors than can humans. In addition, many birds can see certain types of ultraviolet (UV) light, which is invisible to the unaided human eye **(Fig. 3–45)** (Bennett et al. 1994). What advantages might this give them? The answer remains a mystery, but for a discussion of this topic, see Sidebar 1: The Amazing World of Avian ESP, in Chapter 4. Altogether, the bird-human differences in visual perception are so great that it is hard for us to understand exactly how birds see the world (Bennett et al. 1994).

To get a sense of the full palette of colors and array of patterns found in the more than 9,000 species of birds, you need only skim through a work such as *Birds of the World,* by James F. Clements. Strikingly named birds appear on every page, and the names suggest the profusion of colors: Claret-breasted Fruit-Dove, Silvery-throated Spinetail, Saffron-breasted Redstart, Purple-bearded Bee-eater, Scarlet-hooded Barbet, Fire-maned Bowerbird, Pink-throated Brilliant, Ruby-topaz Hummingbird, Violet-necked Lory, Opal-crowned Manakin, Cobalt-winged Parakeet, Pearly-breasted Cuckoo, Buff-throated Purpletuft, Lilac-tailed Parrotlet, Sapphire-rumped Parrotlet, Malachite Sunbird, Gilt-edged Tanager, Glistening-green Tanager, Citron-headed Yellow-Finch. As these names imply, the assort-

Figure 3–44. Featherless Adornments of the Southern Cassowary: This Australian bird displays a huge bony helmet—possibly a protection against thorny vegetation and vines in the bird's rain forest habitat; grotesque, floppy wattles tipped bright pinkish red; and cobalt blue bare skin on the face and neck. These features, combined with the bird's coat of glossy black body feathers and five-foot stature, give this largest of the cassowaries a truly awesome aspect. Photo by C. Volpe/VIREO.

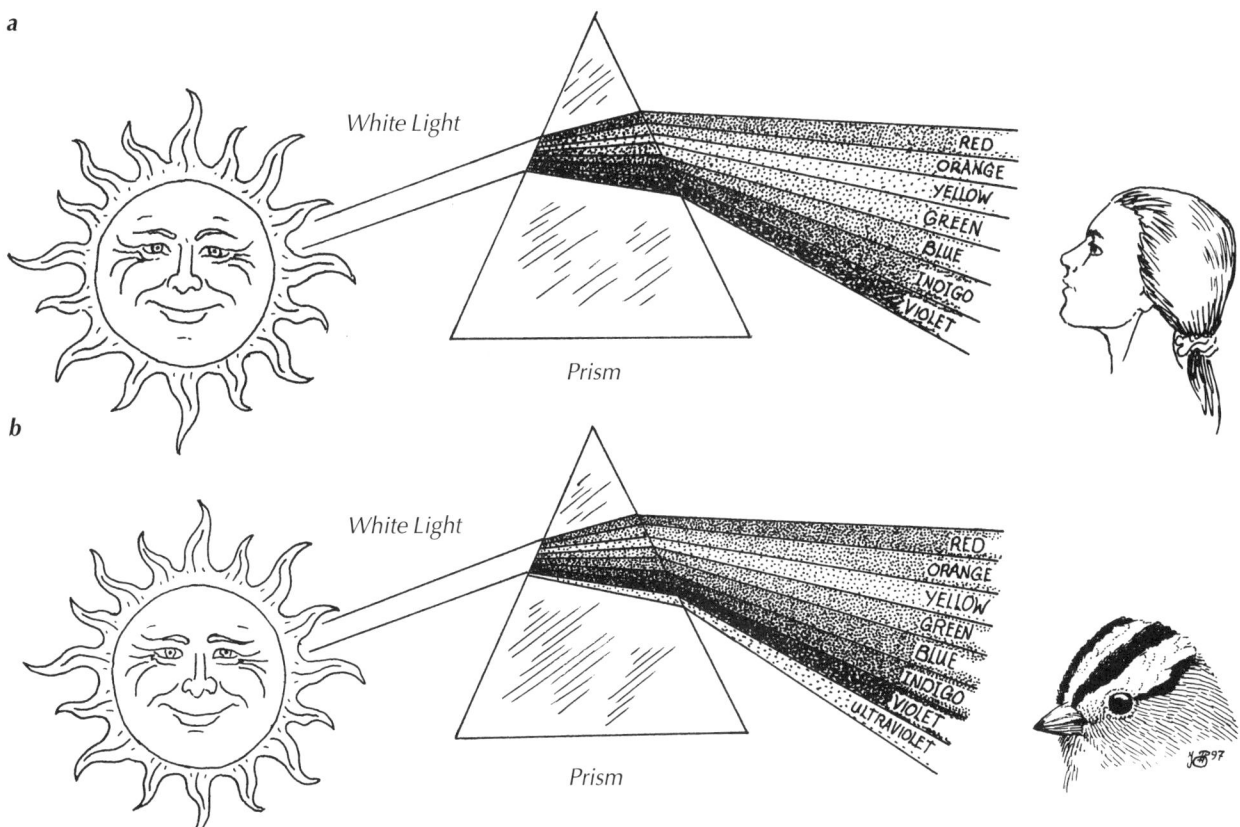

Figure 3–45. Human Versus Avian Visual Spectrum: When white light strikes a prism, the different wavelengths it contains are each bent to a different degree, forming a spectrum. Longer wavelengths such as red are bent less than shorter wavelengths such as violet. In the atmosphere, water droplets can act as prisms to produce a rainbow. *a:* The portion of the sun's spectrum visible to humans. This is normally termed "visible light." *b:* The portion of the spectrum visible to birds that can see UV light in addition to "visible light."

ment of patterns is just as spectacular: birds are freckled, chevroned, naped, collared, capped, crowned, hooded, shouldered, browed, spectacled, mustached, necklaced, speckled, striped, blotched, bordered, banded, vented, chinned, backed, cheeked, and crested with this amazing array of colors.

How are all these beautiful colors produced? To understand this, you first need to realize that sunlight is a mixture of many different wavelengths of light. Seen all together they look white, but when a prism or water droplet in the air bends each wavelength to a different degree, the wavelengths separate, and we see each as a different color, as in a rainbow (see Fig. 3–45). When white light strikes an object, some wavelengths are absorbed by the object while others are reflected back to our eyes. The reflected wavelengths are what we see as the object's color. A red ball, a cardinal, or red paint appears red to us because it reflects only red light; the other colors are absorbed and never get to our eyes.

Birds employ two different systems to break light apart into its component wavelengths—pigments and the microscopic structure of feathers. **Pigments** are colored substances that can, in principle, be extracted from feathers or other parts of the skin. (In practice, their actual extraction may be very difficult.) The color of pigments can be seen even after the physical structure of the material in which they are embodied—a green leaf, a red bowl, or a brown feather—has been destroyed. In contrast, **structural colors** depend on the actual physical structure of the feather to reflect certain wavelengths of light. Destroy that structure and you lose the color, just as you lose the rainbow if you destroy the prism.

Pigments

Three main types of pigments are found in birds: melanins, carotenoids, and porphyrins. **Melanins**, which usually occur as tiny granules in the skin and feathers, are the most common. Depending on their concentration and distribution, melanins can produce any shade from the darkest blacks through browns, red-browns, yellow-browns, and pale yellows. Melanin colors the all-black Common Raven and American Crow, the reddish phase of the Eastern Screech-Owl, and the yellowish down of a young chicken. Birds synthesize their own melanins from amino acids, which they obtain from the proteins in their diet.

Melanin provides more than color for feathers, though. Feathers containing melanin are stronger and more resistant to wear than feathers with other pigments. Some researchers suggest that the melanin granules themselves provide the added strength, but others attribute it to the higher levels of keratin found in feathers with melanin; indeed, both factors may be involved. White feathers, which lack any type of pigment, are the flimsiest. It is no surprise, then, that most birds, especially white or lightly colored ones, have dark wings or wing tips—extra protection for the feathers most vulnerable to abrasion during flight (Burtt 1986). Dark wing tips are especially common in birds that fly at high speeds or spend a lot of time flying, such as gannets, terns, and gulls, including Common and Arctic terns and Ring-billed and Herring gulls **(Fig. 3–46)**.

Carotenoids produce reds as in Northern Cardinals, oranges as in male Blackburnian Warblers, and many yellows, especially the bright yellows of birds such as Yellow Warblers, goldfinches, and canaries. Unlike melanins, carotenoids are synthesized only by plants. Birds must therefore acquire them preformed in their diet, either by eating plants or by eating something that has eaten plants (Brush 1990; Hudon 1991). In some bird species, differences in the specific types of carotenoids in the diets of individuals inhabiting different locations produce regional color variations. This occurs in the European Great Tit, whose young are more yellow underneath when fed on caterpillars from deciduous rather than coniferous woodlands (Slagsvold and Lifjeld 1985). Some colors, such as the olive-green of the female Scarlet Tanager, result from the interaction of both melanins and carotenoids in the same feathers.

The third group of pigments, **porphyrins**, produce reds, browns, greens, or pinks in a number of different avian orders. They interact with melanins to produce the browns of many owls, and are also found in pigeons and gallinaceous birds. In eggshells, they may appear pink or red, or they may be masked by melanins. Porphyrins are most striking, however, in the brilliant reds and greens of the turacos, a family of

Figure 3–46. Light Birds with Dark Wing Tips: *Many white or light-colored birds that spend a lot of time flying have dark wing tips colored with the pigment melanin. Feathers with much melanin are stronger and more resistant to wear than feathers with no pigments, or other types of pigments. Species from left to right: Northern Gannet, Common Tern, Herring Gull, American White Pelican, Laysan Albatross.*

colorful African birds related to New World cuckoos. The porphyrin pigment turacoverdin, which produces the green body plumage of many turacos, is one of the few green pigments known from birds. Most greens on birds, as in the parrots, are produced as structural colors that are modified by overlying carotenoid pigments.

Porphyrins are complex, nitrogen-containing molecules related to hemoglobin, which birds (and other living things) make by modifying amino acids. Although the exact chemical structure varies from pigment to pigment, all porphyrins share one significant feature: they fluoresce bright red under ultraviolet light. If you have ever seen an exhibit of minerals at a natural history museum, you may recall the dazzling colors of the fluorescent rocks displayed under ultraviolet (or "black") light. Feathers with porphyrins glow in the same manner.

Abnormalities and Variations in Pigment Colors

If you take a careful look at the birds coming to your feeder, you will see that not all individuals of the same species look alike. For example, House Finches vary in the amount of streaking on their breasts and in the extent of the red areas, as well as in the exact shade of red. Color variations like these are common and normal. But sometimes a truly unusual individual shows up: perhaps a House Finch with a white head, or an American Crow with white patches under its wings. These types of color abnormalities are discussed here.

Some odd colorations are caused by abnormalities in a bird's pigmentation. The most common conditions involve the reduction or absence of melanin. Birds with reduced melanin are significantly paler than normal, whereas birds with increased levels are darker—a relatively common example is the dark phase of the Rough-legged Hawk, illustrated in most field guides. Sometimes melanin is completely absent from the entire plumage, or from certain parts, as in the crow example above. Colors produced by other pigments remain, sometimes resulting in odd-looking birds. For example, male Downy Woodpeckers lacking melanin are pure white with a red spot on the back of the head, orioles are white and yellow instead of black and yellow, and Cedar Waxwings are white with yellow abdomens and tail tips **(Fig. 3–47)**.

A bird that lacks all *melanin* is termed an **albino**. Albinism results from a genetic mutation that interferes with the production of tyrosinase, an enzyme that helps to produce melanin. Birds that lack not just melanin but all types of pigments in the plumage, eyes, and skin are rare and are called **complete albinos**. These birds have white feathers, but their unfeathered areas, including the eyes, legs, and feet, usually appear pink due to the hemoglobin within the blood vessels near the surface of the skin.

Abnormal pigmentation is usually under genetic control, but disease, injury, and diet also can be factors. Wild flamingos, Roseate Spoonbills, and Scarlet Ibis, for example, derive their pink color from the carotenoids produced by certain crustacea that they eat. When held in captivity they eventually become white unless zookeepers add carotenoids to their diet, often in the form of chopped shrimp.

Downy Woodpecker

Baltimore Oriole

Figure 3–47. Downy Woodpecker and Baltimore Oriole with No Melanin: *When a genetic mutation prevents a bird from producing melanin, an albino results. Because other pigments may still be produced, the outcome may be very odd indeed. A Downy Woodpecker may be pure white with a red cap (produced by carotenoids), and a Baltimore Oriole may retain its bright orange or yellow color but lose its distinctive black markings.*

Plumage colors may be altered by all sorts of environmental factors. Exposure to sunlight, for example, may lighten feather colors. Birds in open areas that receive lots of direct sunlight, such as deserts, are especially susceptible. Consider American Kestrels, found in a wide range of habitats. Birds from Arizona deserts look very much like birds from Michigan when their feathers are fresh in the fall. After a few months, however, Arizona birds are so much paler than Michigan birds that the two look like different species. Before such color changes were understood, museum scientists sometimes classified the sun-bleached and unbleached specimens as separate races or species.

In industrial areas, air pollution may affect feathers, giving birds a coating of soot that may mask their true colors. When the use of soft coal as fuel was restricted in Pittsburgh, Pennsylvania in 1940, the natural history museum began getting telephone calls from bird watchers trying to identify "a handsome little bird in chestnut, gray, and white, with a black bib." Many were disappointed to learn that the "new" bird was only the male House Sparrow without a coating of soot. Even in relatively clean areas, tree-climbing birds such as woodpeckers and creepers may dirty their plumage by contact with bark.

Colorful soils and foods also may stain feathers. Birds that dust-bathe in soils of certain consistencies cannot always remove all the dust by shaking, and may thus acquire the general color of the soil. In southern Brazil, for example, certain House Sparrows may acquire a pinkish tinge from the reddish soil of the region—a bit of a change

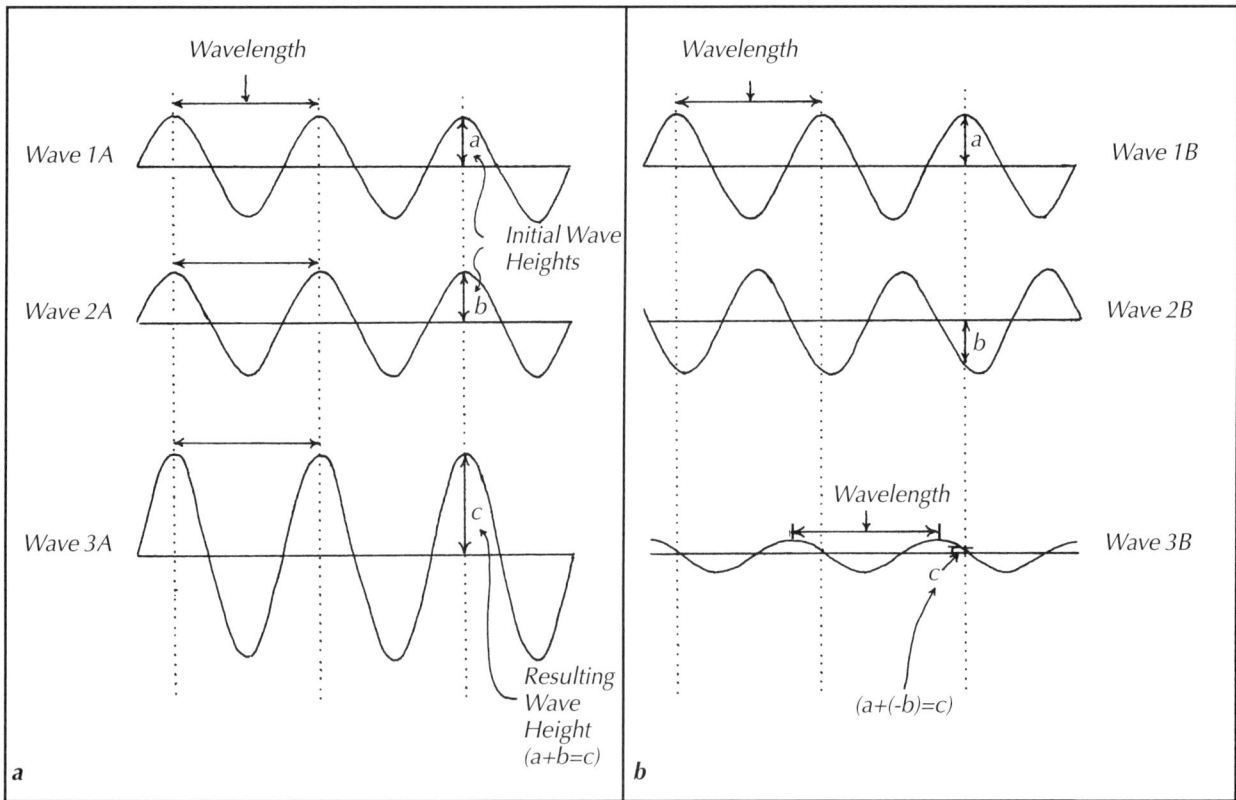

Figure 3–48. Patterns of Interference from Two Light Waves: *a* and *b* = initial wave heights, which may be positive (+) or negative (-); *c* = wave height resulting from interference (determines perceived brightness). **a:** The combination (or "interference") of two light waves (1A and 2A) in which crests and troughs line up perfectly results in a new wave (3A) with crests twice as high. The wavelength, which determines the color, remains the same, but the color will appear twice as bright. **b:** Waves 1B and 2B interfere, resulting in wave 3B. In this case, however, the wave crests and troughs nearly cancel each other out, producing very low waves, which will be seen as a dull, nearly black, version of the original color. Again, the actual color remains the same because the wavelength is unaltered. If wave crests and troughs overlapped perfectly, the result would be a straight line, viewed as black.

from the House Sparrow's usual browns and grays. Some geese, ducks, swans, and cranes that feed in water containing iron oxide (rust) turn reddish as the iron precipitates out on their feathers; this can be especially noticeable on the heads of Snow Geese. Birds even acquire a few of their colors while eating. Berries may stain the feathers around the face and vent of berry-eating birds; pollen may discolor the faces of birds probing flowers for nectar or insects; and half-digested shrimps may turn the pure white fronts of penguins a splotchy pink.

Structural Colors

Two kinds of structural colors occur: iridescent and noniridescent. Both result from the microscopic structure of the feathers, which causes only certain wavelengths of light to be reflected. **Iridescent colors**, as on soap bubbles, the "eyes" of a Peacock's plumes, and the throats of many hummingbirds, are very bright and change with the angle of view—so they appear to glisten and shimmer as the bird moves. Iridescence is produced when light waves reflected off certain structural layers within flattened barbules "interfere" with one another, somewhat the way that widening circles from pebbles tossed into a pond interact when they contact each other. Depending on whether the contacts are wave crests or troughs, the result can be taller waves, shorter waves, or a complete cancellation of the waves **(Fig. 3–48)**. The eye interprets these different wave heights as different brightnesses of color. Only interference patterns can produce the "super bright" colors seen on iridescent parts of feathers.

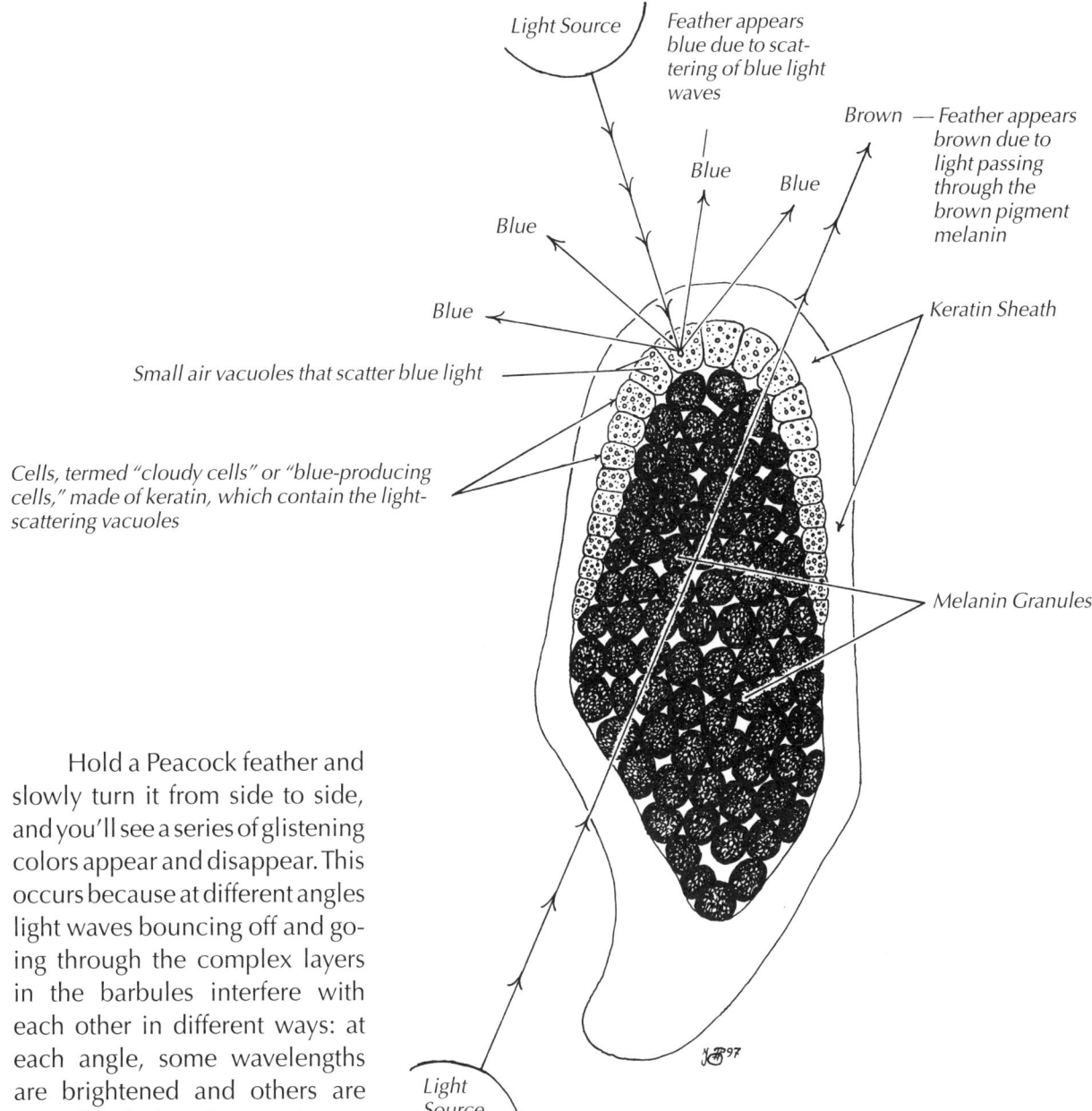

Light Source — Feather appears blue due to scattering of blue light waves

Brown — Feather appears brown due to light passing through the brown pigment melanin

Blue

Blue

Blue

Blue

Blue

Keratin Sheath

Small air vacuoles that scatter blue light

Cells, termed "cloudy cells" or "blue-producing cells," made of keratin, which contain the light-scattering vacuoles

Melanin Granules

Light Source

Hold a Peacock feather and slowly turn it from side to side, and you'll see a series of glistening colors appear and disappear. This occurs because at different angles light waves bouncing off and going through the complex layers in the barbules interfere with each other in different ways: at each angle, some wavelengths are brightened and others are cancelled before they reach your eye. At each position, you see only the brightened colors **(Sidebar 3: Iridescence)**.

Iridescent colors rarely occur on a bird's flight feathers, probably because the flattened barbules required to reflect light would not be sturdy enough for flight, and because any wear would change the feather structure, destroying its color (Greenewalt et al. 1960). Furthermore, flashing iridescent colors on moving wings might hinder a bird's ability to hide from predators.

Noniridescent structural colors arise when tiny vacuoles (pockets) of air within cells in the barbs scatter incoming light **(Fig. 3–49)**. According to the laws of physics, whenever particles smaller than a particular wavelength of light are separated by distances greater than that wavelength, all incoming wavelengths of roughly that size or larger will be scattered: absorbed by the particles and re-emitted in a new

(Continued on p. 3·58)

Figure 3–49. Cross Section through a Blue Jay Feather Barb: When light is scattered by air vacuoles in the blue-producing cells, the feather appears blue. When light is transmitted through the feather, the feather appears brown. In some birds the keratin layer contains pigment. For example, many green birds, such as parrots, have a yellow pigment here that adds to the scattered blue light to produce green. Adapted from Welty and Baptista (1988, p. 49, Fig. 3–19 A).

Sidebar 3: IRIDESCENCE
Sandy and Bill Podulka

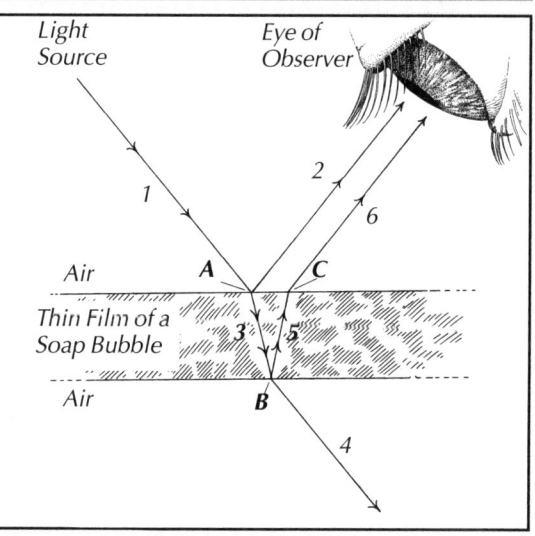

To understand the details of how feather structure can produce iridescent colors, you must first understand the physics of how a thin film affects light. For this discussion, please refer to Fig. A. When a beam of light (1) hits a thin, translucent film, such as a coating of oil on water, or the membrane around a soap bubble, some of it is reflected (2) and some enters the film (3). Whenever light moves between media with different densities, it is bent (refracted)—as occurs at point A—and the degree of refraction depends on the difference between the two densities. Some of the refracted light may travel on and exit the film (4), but some is reflected off the bottom surface (5). Since the eye of any observer is much larger than the distance between the paths of light waves 2 and 6, they will both be seen at the same time. When these two light waves meet inside the eye, the combination produces interference, and the type of interference depends on whether wave crests or troughs arrive at the eye at the same time (see Fig. 3–48). Whether crests or troughs arrive together depends on the extra distance in the film (A to B to C) that light leaving in beam 6 has traveled, compared to light leaving in beam 2. That extra distance, in turn, depends on three things: the angle at which the incoming light strikes the film, the degree of refraction, and the thickness of the film.

In the simplest case, in which light of only one color, such as green, is involved, the result seen by the observer (light waves 2+6) will be either bright green (when wave crests meet), dull green (when partial crests and/or troughs meet), or black (when wave crests and troughs meet and cancel each other out). For a stationary observer watching a moving film (such as a drifting soap bubble), under green light the color will appear to change from bright green to dull green to black, as the film moves, changing the angle at which the incoming light hits the film.

When sunlight, a mix of all colors, strikes a film, at each angle between the observer and the film, just one color will be brightened—the one for which wave crests meet—and it will be seen to the exclusion of the others. This is why you see a series of colors as you turn a peacock feather in your hand, changing the angle between your eye and the feather.

The iridescent colors of feathers, like those of a soap bubble, result from light interference created by thin films. The structural complexity of iridescent feather barbules, however, is much greater than that of a single thin film. Although the details vary from one group of birds to another, all use the same basic plan: layers of melanin granules, running parallel to the surface of the barbule, embedded in keratin. In most hummingbirds, as pictured in **Fig. B**, there are 7 to 15 layers of pancake-shaped granules, and each granule contains a layer of air-filled vacuoles (a vacuole is a parcel of air or some other substance, contained by a membrane). Light waves reflect off the upper and lower surfaces of the vacuoles. The vacuoles thus serve the same function as the top and bottom (or outer and inner) surfaces of the skin of a soap bubble.

The key to the barbules' exquisite design, however, is that each layer of melanin granules (and thus vacuoles) is the same thickness. As a result,

Figure A. Reflection and Refraction from a Thin Film: Light beam striking a thin film—the skin of a soap bubble. See text for explanation. Adapted from Simon (1971, p. 71).

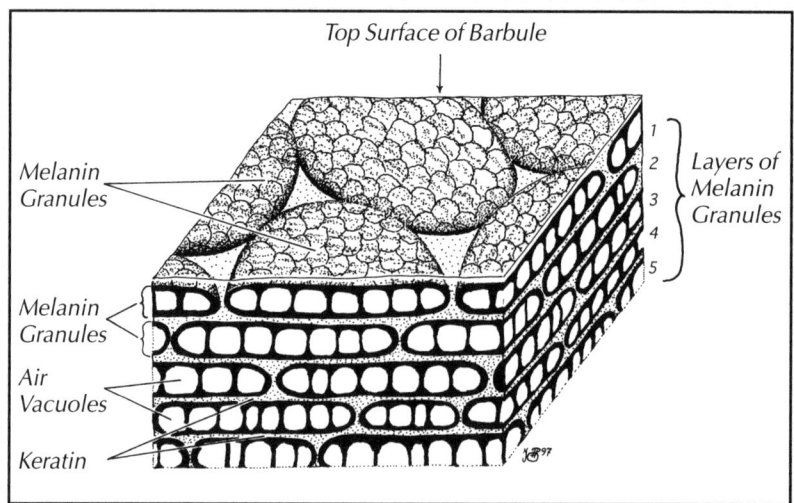

Figure B. Detail of a Hummingbird Feather Barbule: Three-dimensional cross section of approximately one-half of a barbule from a typical hummingbird feather. See text for explanation. Adapted from Simon (1971, p. 160).

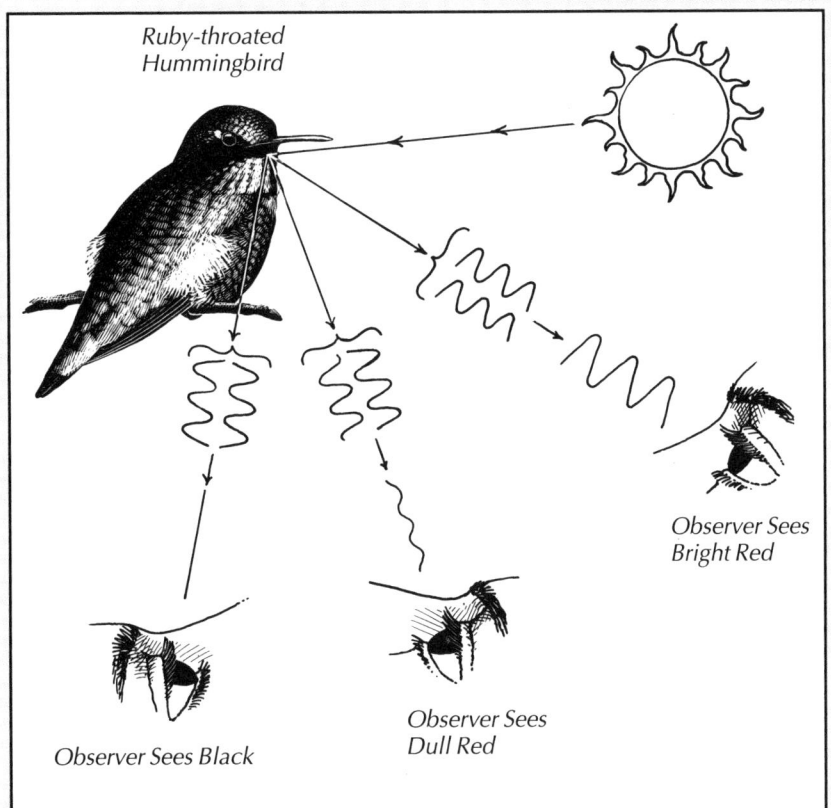

Ruby-throated
Hummingbird

Observer Sees
Bright Red

Observer Sees
Dull Red

Observer Sees Black

Figure C. Angles, Interference, and Iridescence: *Schematic representation of how different angles between the sun, Ruby-throated Hummingbird throat, and observer produce different interference patterns and thus different brightnesses of red. Angles and resulting interference patterns are not necessarily accurate.*

light reflecting off the top surface of the vacuoles in layer 1 interferes with light reflecting off the bottom surface of the vacuoles in layer 1 in exactly the same way as it interferes with light reflecting off the bottom of the vacuoles in layer 2, or 3, or any other layer. Therefore, for a given incoming light angle, the color that is brightened is the same not only for each layer individually, but between different layers, as well. The total brightness you see is the sum of all these layers acting together, and the more layers involved, the brighter the color.

If you have ever watched a male Ruby-throated Hummingbird probing a flower, you may remember his flashing, changing colors, his throat varying from brilliant red to near black. As he moved, he altered the angle formed by your eye, the reflecting surface, and the light source. His throat barbules are structured so that interference patterns can only

brighten the red wavelengths, and the brightness depends on the angle of view. At some angles all wavelengths cancel each other out, and you see the throat as black (**Fig. C**).

Variations on this structural theme are as endless as the array of iridescent colors they produce. The melanin may be solid or air-filled, and may be in flat disks, or in rounded, hexagonal, or flattened rods. Sometimes the melanin granules are tightly packed, and at other times they have spaces between them, roughly forming a grid. The colors produced may be as pure and brilliant as the red throat of a Ruby-throated Hummingbird or may contain a duller mix of wavelengths—as in the greens of some trogons. Some colors result from a combination of iridescent colors—as in the copper (a mix of red and yellow-green iridescence) on some African sunbirds.

The precise arrangement of the layers of melanin granules restricts

the range of colors produced. Hummingbirds (as mentioned above) are notorious for their quickly disappearing colors—changing from glittering brilliance to black with just a slight change in angle. Some, such as the Ruby-throated Hummingbird, can only produce red or black on the throat feathers, but others may display several colors in addition to blackness. The colors of a peacock feather are numerous—ranging from bronze through blue-green, but they never completely disappear to form black. One of the most beautiful birds in the world, the Resplendent Quetzal of Central American cloud forests, varies between gold-green and blue-green, with a touch of violet at times. The intensity and beauty of its colors has placed it as the national symbol of Guatemala; in the past, it was worshiped by the Mayans and Aztecs.

Although some of the most spectacular iridescent colors occur on tropical species, many North American birds also display iridescence—and some of these can be seen at your feeder. European Starlings, many grackles and cowbirds, Wild Turkeys, Ring-necked Pheasants, and Wood Ducks all glitter with iridescent hues. But iridescence is not limited to birds. Some snakes (such as boas), lizards, fish (rainbow trout and many popular aquarium species such as neon tetras), and insects (especially butterflies, moths, and beetles) produce iridescence. Although they use different substances and structures to form the colors, they still use the principle of interference from thin films.

Perhaps the most awe-inspiring aspect of iridescence—even beyond the beauty of the colors themselves—is that the purity, range, and brilliance of the colors produced by each species are tightly controlled by minute structures whose size, density, and shape must be highly accurate—sometimes to within 4 ten-millionths of an inch (0.00001 mm). ∎

Blue

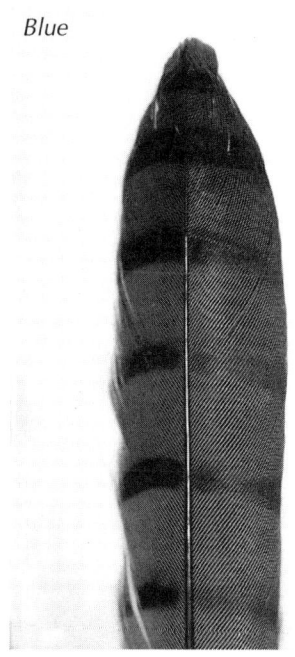

a

Brown

b

Figure 3–50. Blue Jay Feather Under Different Light Conditions: a: *When light reflects off the top surface of a Blue Jay feather, the feather appears bright blue due to the detailed structure of the feather.* **b:** *When light is transmitted through a Blue Jay feather, the feather appears brown due to the pigment melanin. (See text for further explanation.) Photos courtesy of Carrol Henderson.*

direction. The shorter the wavelength, the more strongly it is scattered. In a feather, the air vacuoles act as scattering particles, and are so tiny—smaller in diameter than the wavelength of blue light (about 19 millionths of an inch [0.0005 mm])—that all wavelengths of visible light are affected. But the shades of blue light (blue, indigo, and violet) are affected the most, because they have the shortest wavelengths. The blue can be seen from any angle, because the wavelengths are scattered in all directions. Scattering of this sort produces the blues of bluebirds, Indigo Buntings, and Steller's Jays. It also produces the blue of the sky as dust particles scatter sunlight.

Melanin is also present in blue feathers, just below the layer of cells containing the light-scattering air vacuoles. It absorbs most of the longer wavelengths of light, such as red, that are not scattered as much by the overlying air vacuoles, creating a dark background that intensifies the blue you see. (These longer wavelengths tend to move straight through the cell layer that contains the scattering particles, reaching the melanin below.)

If you hold a blue feather up with a light behind it, or look at a backlit Blue Jay, it looks brown. In these situations the light you see has passed through the feathers, rather than reflecting off them, so the blue disappears and you see only the brown of the melanin **(Fig. 3–50)**. In contrast, if you hold a yellow feather up to light it still looks yellow, because the color is produced by a pigment, not the structure of the feather. Blue pigments are rare in birds, occurring in only a few species such as the Blue-capped Fruit-Dove of New Guinea; most blues are structural colors.

One more way to demonstrate to yourself the difference between structural and pigment colors is to crush or grind up a blue feather. When the blue-producing structure is destroyed, the feather will appear dark due to the melanin. In contrast, when you crush a pigmented feather—try this with a yellow or red one—it will remain the same color when its structure is destroyed (Simon 1971).

In addition to blue, a few other noniridescent colors may be produced structurally. When the light-scattering air vacuoles are a bit bigger (bigger than the wavelength of blue light), the result is green (since blue is no longer scattered, and green wavelengths are now scattered the most), as in some parrots. With even larger air vacuoles, no wavelengths are scattered, but all are reflected, producing white light and thus plumage that we perceive as white; white does not exist as a pigment in birds. The feather structure of many birds also reflects ultraviolet light (Burkhardt 1989), which birds, but not humans, can see. Thus, birds may appear very different to each other than they do to us (Andersson 1996) **(Fig. 3–51)**.

A few colors result from a combination of pigment and structure. The greens of some parrots are caused by yellow pigment overlying the blue-reflecting structure of the barbs. When the yellow pigment fails to develop, a bird such as the Yellow-headed Parrot, which is normally green with a yellow head, is blue with a white head—a rare genetic mutation. When the overlying pigment is red instead of yellow, purples and violets result, as on the heads of some Indian parrots.

a

b

Figure 3–51. Thrushes Under Different Lighting Conditions: *Black-and-white photographs of Old World thrushes.* **a:** *Birds il-luminated by light conditions visible to humans.* **b:** *Birds illuminated only by ultraviolet light (wavelengths 320 to 400 nm). Note that b does* not *show the birds as they would appear to a bird whose vision is sensitive to both visible and ultraviolet light, but does* show the *additional ultraviolet plumage features that might be visible to a bird, but not to a human. The way these plumages might actually appear to a bird that sees in both the visible and ultraviolet ranges is probably somewhere in between the images shown in a and b. Species from top to bottom are: male Eurasian Blackbird, Song Thrush, Taiwan Whistling-Thrush, and male Blue Whistling-Thrush. Photos courtesy of Staffan Andersson.*

For in-depth explanations and clear illustrations of how structural colors are produced in birds, see Simon (1971) and Greenewalt (1960).

Functions of Color and Color Patterns

■ Feather structure and pigments combine to make birds among the most colorful animals on earth. Their bright colors and striking patterns are rivaled only by those of coral reef fishes and butterflies. For all animals, including birds, coloration is an evolutionary compromise between hiding from predators and being conspicuous for social interactions such as territorial defense, courtship, and mate choice. For most animals, however, concealment is the more critical need. Just as the dull, cryptic colors of most mammals, reptiles, amphibians, and fish help them to remain inconspicuous, so, too, do most birds sport earth tones that render them hard to see. Nevertheless, a large number are brightly colored—perhaps because flight allows them to escape most predators, perhaps because bright colors are concealing in some habitats (sunlit, fruit-and-flower-laden forest canopies, for instance), or perhaps because colors are so crucial for birds' social interactions.

We probably will never know exactly why each bird looks the way it does, but we do know that color and pattern, like most other attributes, result from generations of natural selection. Individuals whose colors best meet their needs raise more young and pass to those offspring their favorable colors and patterns. Over time, this "hand of evolution" continues to fine-tune each bird's appearance. Thus, when we view a bird, we must remember that it is not a finished product, but one under continuous construction. Some patterns remain a mystery to us. Why, for example, is the Bobolink light above and dark below, not the reverse, like most birds? With careful observation and interpretation, however, we often can deduce which selective forces produced the colors and patterns we see in birds today.

Throughout the following discussion of bird colors and their functions, keep in mind that most observations and experiments with birds have assumed that their color vision, as well as that of their predators, is similar to ours. Only in the early 1970s did researchers begin to realize that birds see some types of UV light, and that some birds may be able to discriminate a greater number of colors than we do. Our understanding has thus been hampered by our own perception of the colors around us. As we struggle to grasp the way the world really looks to birds and their predators, some of our long-held beliefs about the roles of colors and patterns on birds may be challenged.

Cryptic Coloration and Patterns

Predation is the chief cause of death in many birds, especially smaller ones. Avoiding predation, therefore, appears to be the main reason that so many birds have evolved cryptic markings. Not only is concealment important to prey—we assume, for example, that ground-nesting meadowlarks are more likely to avoid the sharp eyes of hungry hawks if they look like their grassy surroundings—but the predators themselves may be more successful if they are hard to detect. For example, we assume that even a bird as large as a Snowy Owl can go unnoticed by its lemming prey if it blends into the white arctic landscape. The only data to support this idea, however, come from experiments on captive Black-headed Gulls, which became much less effective at capturing fish in a tank from the air when the undersides of their ordinarily white wings were dyed black (Götmark 1987).

Figure 3–52. American Woodcock: The mottled brown-and-white feathers on the back of an American Woodcock render it virtually invisible against the leaf litter of the forest floor. Photo courtesy of John Trott/CLO.

Different predators, however, have different visual capabilities, so we must not assume that all prey appear as conspicuous or as cryptic to predators as they do to us. For instance, most mammalian predators such as cats and raccoons, as well as nocturnal avian predators such as owls, apparently lack color vision. Many bright yellow or red birds may thus be inconspicuous to these predators, which see the colors as shades of gray. On the other hand, hawks, eagles, and other avian predators that are active by day probably have full color vision including some ultraviolet wavelengths. Thus, markings on a "prey" bird such as a goldfinch may be cryptic for one kind of predator, but not for another.

Blending In

Birds that spend much time on the ground generally are colored more cryptically than arboreal species; this camouflage probably results from heavy predation. The mottled patterns on many birds of the forest floor, such as the American Woodcock, Ruffed Grouse, and many of the nightjars including the Whip-poor-will, resemble or mimic the leaf litter beneath them **(Fig. 3–52)**. Nearly all shorebirds, from Least Sandpipers to Long-billed Curlews, have mottled brownish backs that conceal these ground nesters during incubation as well as while they are foraging on beaches, mud flats, or in grasses. Birds of grasses and reeds—the bitterns, snipes, and grass-loving sparrows such as the Savannah Sparrow—often have patterns with longitudinal streaks. Among sparrows and thrushes that feed on the ground, such as Song Sparrows and Wood Thrushes, brownish-patterned plumages are common. The tundra-dwelling ptarmigans molt from their brown summer plumage to a white garb that blends with heavy snow cover in winter **(Fig. 3–53)**.

Perhaps the most striking adaptations of color and pattern to match their environment occur among some of the Old World larks. Even within a single species, the birds of different populations resemble the different types of ground on which they live—black lava, reddish brown earth, or white sand. Individuals rarely stray into an environment with the "wrong" background, for if they do, they become more conspicuous to their sight-oriented predators. Natural selection favors those individuals that best resemble their background and that

a

b

c

Figure 3–53. Cryptic Coloration of the Willow Ptarmigan: a: In winter, the white plumage of both males and females disappears against the snow-covered tundra. Photo by Dan Stosits/CLO. b: In spring and fall, as the male molts, his patchy brown-and-white plumage matches the patchy snow conditions of the tundra. Photo by Johnny Johnson/Bruce Coleman Ltd. c: In summer, the female's mottled brown feathers make her nearly impossible to see as she incubates her clutch. Photo courtesy of Al Cornell/CLO.

"stay put" in it. The pale Piping Plover, for example, spends much time on white sandy beaches, where it is difficult for even the most persistent bird watcher to spot.

Many bird species with **precocial** young—those, such as ducklings, that hatch with feathers, ready to leave the nest and forage with their parents—are most cryptically colored in the period before they can fly. Then, their best defense against predators is to be "invisible." Young terns and skimmers on beaches, for instance, blend so well into their background of sand and gravel that they are in danger of people stepping on them.

Birds that we think of as gaudy in pictures or captivity may actually blend into their natural surroundings. A visitor to the tropics may spend a long time staring at a huge squawking tree before finally glimpsing one of the many large, bright green parrots within. Even birds as brilliantly colored as Keel-billed Toucans and Paradise Tanagers virtually disappear against the sun-dappled tropical canopy overhead.

Disruptive Coloration

In addition to resembling their backgrounds, some birds have **disruptive** coloration. Like the "camouflage" of military gear, disruptive plumage has patches, streaks, or other bold patterns of color that break up the shape of the bird, catching the eye and distracting the observer from recognizing the whole bird as one form. For example, the two bold, black neck bands of the Killdeer combined with its brown back make it virtually impossible to distinguish against a pebbly field or beach **(Fig. 3–54)**. Similarly, the dark or light eyelines found in so many birds are thought to help disguise the otherwise conspicuous eye. Disruptive coloration has the added advantage of providing camouflage on a variety of backgrounds. Downy young shorebirds in the arctic tundra have variegated plumages that break up their outline so much that predators (and humans, as well) have trouble seeing them in a range of different vegetation types (see Fig. 8–125).

Countershading

Many birds are hard to see because they are darker on top than below, a pattern known as **countershading**. Eastern King-birds, Northern Mockingbirds, most gulls, and many of the smaller plovers, sandpipers, and songbirds are familiar examples **(Fig. 3–55)**. With the darker back lit by the bright sun overhead and the lighter underparts in the shade of the bird's own body, the bird appears as nearly one color, and until it moves, predators have difficulty seeing it as a three-dimensional object. In addition, white underparts may reflect the color of the ground, producing even more effective camouflage. Countershading is best developed in open-country birds that are vulnerable to distant predators hunting by sight. It also is common in birds that spend much time on the water, such as loons, grebes, ducks, auks, and many pelagic birds.

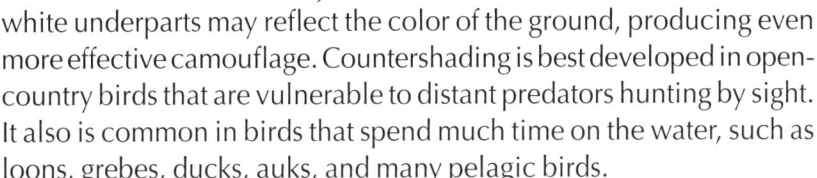

Figure 3–54. Disruptive Coloration in Killdeer: Note how the black neck bands of the adult and two chicks break up each bird's outline, making it hard to distinguish against the busy, pebbly background. Photo courtesy of Mary M. Tremaine/CLO.

Behaviors that Aid Concealment

Birds may increase the benefits of cryptic plumage with certain behaviors. Simply holding still is one good way to avoid attracting predators. By freezing into immobility and flattening against the beach, young terns and gulls eliminate their shadows and look just like bumps on the sand. The nocturnal potoos, odd Neotropical relatives of night-hawks, also complement their mottled brown-and-gray plumage with behavior. They spend most of the day just sitting, often perched on the stub of a dead tree limb, with neck stretched up and large eyes closed,

Figure 3–55. Countershading in Black Turnstones: The shadow cast on the light-colored underparts of a standing Black Turnstone helps to offset the contrast between the dark upper and light lower portions, making the bird appear flatter against the background, and more difficult to see. This effect would not work if the bird were light above and dark below, as demonstrated by the upside down individual, which is much easier to spot in its dark, rocky habitat. This pattern of dark above and light below is known as **countershading**, and is found in many species of birds.

Figure 3–56. Bird or Stick? *The Common Potoo of Central and South America, a relative of North American nighthawks, spends most of its day motionless, often perched at the end of a snag. Photo by J. V. Remsen/VIREO.*

looking for all the world like an extension of the stub **(Fig. 3–56)**. Many owls, such as Long-eared Owls and screech-owls, adopt similar "stick postures." The freezing strategy also works for many predators, such as herons prospecting for fish or frogs along a shoreline. In some cases, however, movement enhances concealment. American Bitterns blend their streaked pattern with reeds by standing with their bills pointed skyward. If wind blows the reeds, the bird, too, may sway slightly from side to side (see Fig. 2–17).

Conspicuous Markings and Predation

Many cryptically colored birds have conspicuous markings that become obvious only when the bird flies off. The white outer tail feathers of juncos, meadowlarks, and Hooded Warblers; the black-and-white flight feathers of Willets; the white "windows" in the primary wing feathers of nighthawks; the bright orange-and-white rump patch of Killdeer; and the white rump of Northern Flickers are examples **(Fig. 3–57)**. When such markings are suddenly revealed they may startle an approaching predator, leading it to hesitate or to strike in the wrong place and miss its target. Because many of these conspicuous markings disappear when the bird lands, suddenly hiding the bird's position, they also may confuse predators that had focused primarily on the bold markings during pursuit.

The evidence that this **deflective coloration** serves an antipredator function is scant, however. Birds sometimes reveal these hidden markings even without flight. A junco or Hooded Warbler hopping on the ground, for instance, sometimes spreads and contracts its tail feathers slightly, flashing its white markings prominently. The display of hidden conspicuous markings may actually function more in social interactions between members of the same species than in interactions between species, but more research is needed to evaluate the various possibilities.

Prominent markings in other animals, such as insects and frogs, may warn predators that the individual is distasteful. However, no example of a conspicuous, distasteful bird has yet been found. Moreover, the few known poisonous birds, such as certain members of the genus *Pitohui* from New Guinea, show no obvious warning coloration. These poisonous birds are reddish brown—no more striking in appearance than the nonpoisonous species of the same genus in other parts of New Guinea.

Reduction of Glare for Foraging

Birds foraging in bright sunlight may experience glare as light is reflected into their eyes from light-colored areas around the eyes, lores, and upper beak. The glare may interfere with their vision, reducing their foraging efficiency. Dark colors near the eye may reduce such glare, as in the face mask of the Common Yellowthroat, the dark eye stripe of the Cedar Waxwing, and the dark head and upper beak of many flycatchers, including the wood-pewees and phoebes. Some

Dark-eyed Junco

Northern Flicker

Willet

Figure 3–57. Deflective Coloration in Birds: *Many otherwise cryptically colored birds have conspicuous markings that become visible only when the bird flies or spreads its wings or tail. Such markings, termed **deflective coloration**, include the white outer tail feathers of Dark-eyed Juncos, the white rump of Northern Flickers, and the black-and-white patches in the flight feathers of Willets. Although researchers have long assumed these markings evolved as antipredator defenses—the birds revealing them suddenly to startle or confuse approaching predators, there is little evidence for this. Alternatively, they may function in social interactions between conspecifics. Drawings by Charles L. Ripper.*

evidence for this hypothesis exists. When Burtt (1984) painted the normally dark upper beaks of Willow Flycatchers white, the birds foraged in the shade more often than those with the normal beak color. When the birds with whitened upper beaks did forage in the sun, they had lower foraging success. In addition, Burtt (1986) noted that species of North American wood warblers with dark upper beaks forage in sunlight significantly more often than those with light upper beaks.

The Role of Color and Pattern in Social Behavior

Birds, like humans, may use color and pattern to distinguish among species, sexes, ages, or individuals. The more that researchers study coloration, however, the more they realize its importance in male-male (or female-female) competition for mates, and in mate choice. Many ways that coloration functions in interactions between individuals are discussed below.

Species Recognition

A bird must recognize other members of its species (termed **conspecifics**) to find a mate. Matings between species (**hybridizations**) often are evolutionary dead ends, either because no young are produced or because the hybrid offspring do not survive or reproduce as well as offspring from two parents of the same species. Hybridization sometimes occurs when conspecific mates are hard to find. For example, hybrids between the Blue-winged Warbler and Golden-winged Warbler, known as Brewster's and Lawrence's warblers, are selected as mates primarily when pure blue-winged or golden-winged individuals are not available (Ficken and Ficken 1968). Hybrid pairing between a Blue Jay and a Florida Scrub-Jay occurred when the scrub-jay was the last female of its species remaining in the coastal scrubs of northeastern Florida.

Hybrids occur frequently in captivity, where the choice of potential mates is limited. This is especially common among ducks. For example, the Wood Duck of North America and the Mandarin Duck of Asia do not hybridize in nature, partly because they never meet. They have hybridized in captivity, however, when their mate choices were restricted.

Because individuals that mate with their own species usually produce more young, this tendency is strongly favored by natural selection. As a result, birds have evolved many different mechanisms to ensure that they mate with the correct species. Appearance, vocalizations, courtship displays (see Ch. 6, Courtship Displays), anatomical and physiological differences (size, sperm-egg incompatibility), and other behaviors probably all play a role.

Although researchers have long assumed that color and pattern both function in species recognition, devising experiments that clearly distinguish their importance is difficult. Therefore, little supporting data exist for this idea. Moreover, when obvious species-specific markings, such as the red epaulettes of the male Red-winged Blackbird, are experimentally covered up, the sexes do continue to establish pairs (Searcy and Yasukawa 1983); these experimentally altered males are recognized as conspecifics by the other males. Interestingly, the altered males were less able to defend their territories. Clearly, species recognition involves a variety of cues acting together, behavior plays a key role, and species-specific color patterns also influence how conspecifics treat one another.

Age Recognition

As discussed earlier, many different birds (gulls, American Redstarts, Red-winged Blackbirds, and others) have distinctive subadult plumages before they become sexually mature. We know that birds recognize age differences because, given a choice, they usually select mates in definitive plumage. But we do not know for certain if they choose mates on the basis of plumage or some other indicator of age. Subadult plumages clearly serve other functions besides mate selection.

Sex Recognition

To a bird looking for a mate, knowing the sex of another individual is key, but sex recognition also is critical in numerous other social interactions, including birds defending territories or mates. In species in which the sexes differ dramatically in appearance, such as Northern Cardinals, American Kestrels, and Red-winged Blackbirds, males and females are easy to distinguish. Differences between sexes are more subtle in other species. In many woodpeckers, for example, the amount of red on the head reveals the sex. In the Downy, Hairy, and Ladder-backed woodpeckers, *only* males have red on the back of the head. In others, such as the Red-bellied, Pileated, and Acorn woodpeckers, males have *more* red on the head. Birds undoubtedly use the same plumage cues that we do. In Northern Flickers, for example, only males have the black malar (whisker) stripe **(Fig. 3–58)**. One researcher (who probably also drew mustaches on posters) painted malar stripes on females, and their mates chased them away (Noble, 1936)! When students working with a flock of California Quail dyed the plumage of the females to resemble that of the males, the males treated their former mates as males. In species with similar-looking sexes, males and females seem to recognize each other by behavior.

Note that birds do not *require* different plumages to recognize the opposite sex. About half of all songbirds, for example, have similar plumages in both sexes. In these species, males and females recognize each other by behavior. This fact reminds us that sexual differences in plumage may arise for a variety of reasons besides mate selection; reasons that relate to the different roles played by each sex—in courtship, nest attendance, and territorial defense, for example.

Individual Recognition

To us, one Black-capped Chickadee or Yellow Warbler looks much like the next. Occasionally we might recognize a bird with a distinctive color abnormality or a missing feather, and with much work we can learn to recognize individuals of some species, such as Downy or Pileated woodpeckers, by differences in the patterns on the backs of their heads. Among wintering Tundra Swans in western England, researchers can distinguish hundreds of individuals, mostly by variations in their bill markings. But individual recognition in most species remains a mystery to us—unless, of course, we have colorbanded or tagged birds for personal identification. Many birds, however, clearly recognize each other, especially their mates. How? Birds probably identify familiar individuals the same way we recognize other people, by subtle differences in color, head and body shape, facial features, posture, and voice. And some are very good at this—a Northern Pintail can recognize its mate 300 yards away!

Female

Male

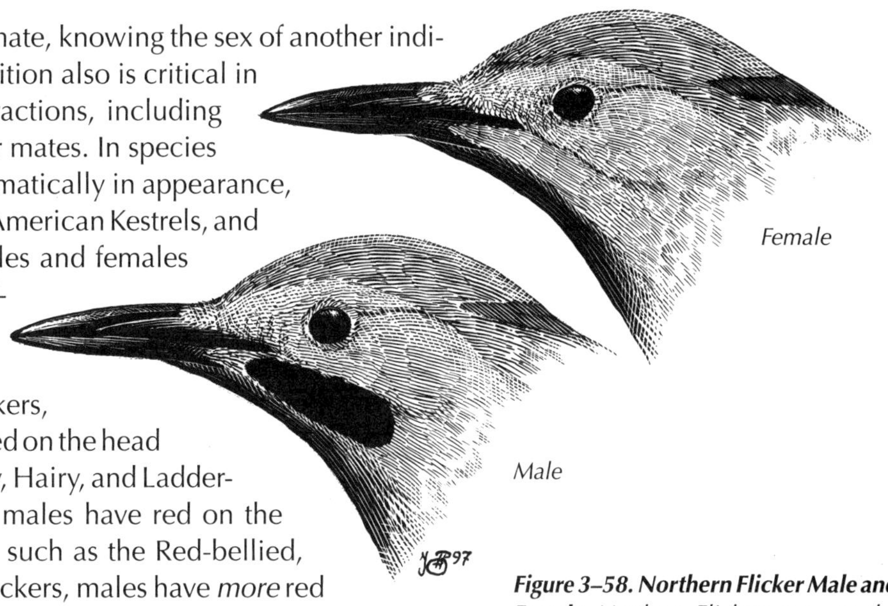

Figure 3–58. Northern Flicker Male and Female: *Northern Flickers may use the black malar (whisker) stripe, found only on the male, to determine the sex of a bird. When a researcher painted malar stripes on females, their mates chased them away! (Noble 1936)*

Several studies have demonstrated that birds use color and pattern to recognize individuals. When Ruddy Turnstones—shorebirds with highly variable plumage—were shown models that mimicked their neighbors or strangers, they reacted aggressively only to the strangers (Whitfield 1986), strong evidence that they could distinguish them by appearance alone. In addition, several colonially nesting species, including Ring-billed Gulls, can pick their own young from a "crowd" of young by their facial markings.

Flock Attraction

Individuals of gregarious species such as gulls, gannets, swans, egrets, vultures, crows, or blackbirds often locate other individuals or feeding flocks by sight. Most of these species are black, white, or black-and-white—conspicuous colors that show up well from a distance and may aid flocking (Savalli 1995).

Sexual Selection

Throughout this section we have discussed how birds may use differences in appearance to avoid mating with other species and to recognize each other in various ways. But the more we learn about birds, the more evidence we find that successfully competing for a mate *within* the species may be one of the most important functions of distinctive colors and patterns (Butcher and Rohwer 1989; Savalli 1995).

Well over a century ago, Charles Darwin recognized that certain inherited traits, such as the gaudy plumes of a peacock or the red epaulettes of a Red-winged Blackbird, do not necessarily promote individual *survival.* Instead, these features improve a bird's chance of *acquiring a mate,* and therefore spread in a population because more young are produced that bear the advantageous traits. Darwin gave this process its own name, **sexual selection**, although it is, in fact, a special type of natural selection.

Sexual selection can operate in two related but distinct ways. First, it can promote traits that increase a bird's ability to compete with other members of its sex, thus gaining favor with or access to mates. As an example, consider the black bib of the male House Sparrow, thought to have evolved this way **(Fig. 3–59)**. Researchers have demonstrated that male House Sparrows with larger bibs are more dominant, hold better quality territories, and chase and copulate with females other than their mates more often than males with smaller bibs. The black bib, by itself, does not bring these advantages, but serves as an accurate way for a male to visually signal his quality to other males, without fighting. Presumably, males that can avoid the risk of injury associated with fighting have an evolutionary advantage over those who must fight. Similarly, the red epaulettes of Red-winged Blackbirds probably evolved because they give males an advantage in male-male competition for territories, and females choose their mates on the basis

Figure 3–59. Male and Female House Sparrow: The black bib of the male House Sparrow, thought to have evolved through sexual selection, appears to act as an indicator of male quality. Males with larger black bibs are more dominant, hold better quality territories, and copulate with females other than their mates more often than do males with smaller bibs. Drawing by Charles L. Ripper.

of territory quality. A second way that sexual selection can operate is by directly increasing an individual's attractiveness to the opposite sex. Female House Finches choose mates with the brightest red coloration, and female Great Snipes (an Old World species) choose males at least partly by the amount of white on their tails. We assume, therefore, that these features—bright red House Finch feathers, and white in Great Snipe tails—have evolved through sexual selection.

A trait may be attractive to members of the opposite sex for a number of reasons, most of which are discussed in Chapter 6. However, directly related to plumage color is the suggestion by Hamilton and Zuk (1982) that birds in good health, or with fewer ectoparasites, may be better able to sport brightly colored feathers than are less healthy birds. Indeed, bright colors actually might have evolved as an honest indicator of health. Therefore, in choosing their mates, males and females should select the more brightly colored individuals. A number of studies have supported the relationship between health and bright colors. For example, in House Finches, males with brighter feathers are apparently in better nutritional condition (Hill and Montgomerie 1994) and the colorful combs of chickens and wattles of turkeys fade in brightness when the birds become sick (Stephen T. Emlen, personal communication).

Why don't birds "cheat" and produce colors or markings that advertise a higher quality than the bird actually is? In most cases, producing the attractive colors bears a cost, which only birds of high quality can afford. For example, House Sparrows with large bibs may be challenged by others with large bibs, and if they don't have the fighting ability they claim, they may be badly beaten in a fight. And unhealthy or otherwise poor-quality individuals may be unable to spare the energy needed to produce and maintain brightly colored plumage.

Sexual selection is most apparent when it produces extreme and colorful plumages, which usually occurs in species in which a few males mate with many different females, as in peacocks and the gaudy, bright-orange South American Cock-of-the Rock. But it also can operate in species with the most common type of mating system—monogamy—in which birds appear to form a pair bond with only a single mate. As researchers look more closely at these species, even sampling the DNA of young and adults to determine the paternity of the young, they are finding that males and females of many species thought to be monogamous commonly copulate with individuals in addition to their own mates. In some cases, birds may use plumage coloration as a quick way to assess the quality of their potential consorts. Sexual selection can even operate in species in which a single male and female truly do pair, as birds with more "attractive" plumages may acquire a better-quality mate or may be chosen earlier in the breeding season. Sexual selection is discussed in greater depth in Chapter 6.

We have now scrutinized the "outsides" of birds from beak to foot. This close look has revealed how exquisitely natural selection has fine-tuned birds to meet the demands of their physical and social environments, from the way that barbules interlock on a hummingbird feather and sparrows sputter about during dust baths, to the "snow-shoes" of a Ruffed Grouse in winter and the cryptic brown streaking on an American Bittern. We do not know the significance of every structure or color pattern we've considered, and some may remain a mystery forever. But the more we find out about each pattern, the better we understand how evolution works and why birds behave as they do. These are just some of the reasons why birds are such a rich and important group to study and enjoy endlessly.

3

4

What's Inside: Anatomy and Physiology

Howard E. Evans and J. B. Heiser

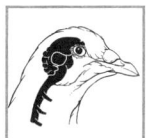 Much of the thrill we experience watching birds is in seeing them fly, listening to their songs, discovering them at their nest, or simply watching them go about their daily rounds caring for their plumage and feeding. Rarely do we consider just how the bird's body actually accomplishes these wonders. Exactly what are the various parts and pieces that go together to make a bird—and how do they perform, in concert with one another, to produce the marvelous lives that give us so much pleasure?

That is the subject of this chapter—the anatomy and physiology of birds. The subject is so vast that we can barely scratch its surface. The topic is somewhat simplified because much of the form and function of birds is very much like that of our own, thanks to a shared four-legged terrestrial ancestor. However, birds do have senses and capabilities that we can only imagine, so different are they from ours.

Most of the distinctive features of avian anatomy and physiology are adaptations to flight: not only are birds light in weight, but the demands of flight make birds some of the highest energy users for their size in the animal kingdom. This energy powers flight; heats a small, easily cooled body; and supports the highly active life birds live even when not flying. While exploring the anatomy and physiology of a group of organisms, one often finds that the study of one specialized characteristic leads to another and then to another. This is especially true with birds, however, so intimately are their adaptations interconnected; flight affects everything about them.

Throughout this chapter we will take a functional approach, concentrating in each section on anatomical structures with similar functions; these have been grouped together into functional systems. As in all living things, these systems are ultimately composed of **cells**, the basic units of life. Cells with related and often very similar characteristics are aggregated to form **tissues**. Tissues, often of quite distinctive character and function, combine to form discrete **organs**. A group of organs whose various functions are coordinated to accomplish one or more of the basic functions of life is recognized as an **organ system**. For example, cardiac muscle cells together form the smooth muscle tissue of the heart, which, together with heart valves and the inner and outer covering of the heart, form most of the heart organ. The heart, blood vessels, and the blood together constitute the circulatory system **(Fig. 4–1)**.

In this chapter we discuss the important features that are "inside" a bird, considering the internal organ systems one by one to understand the component parts. Interaction and integration of these systems are of vital importance and are repeatedly pointed out. The function of each system (its physiology) is then investigated with special attention to flight, reproduction, and metabolism (energy use). The systems dis-

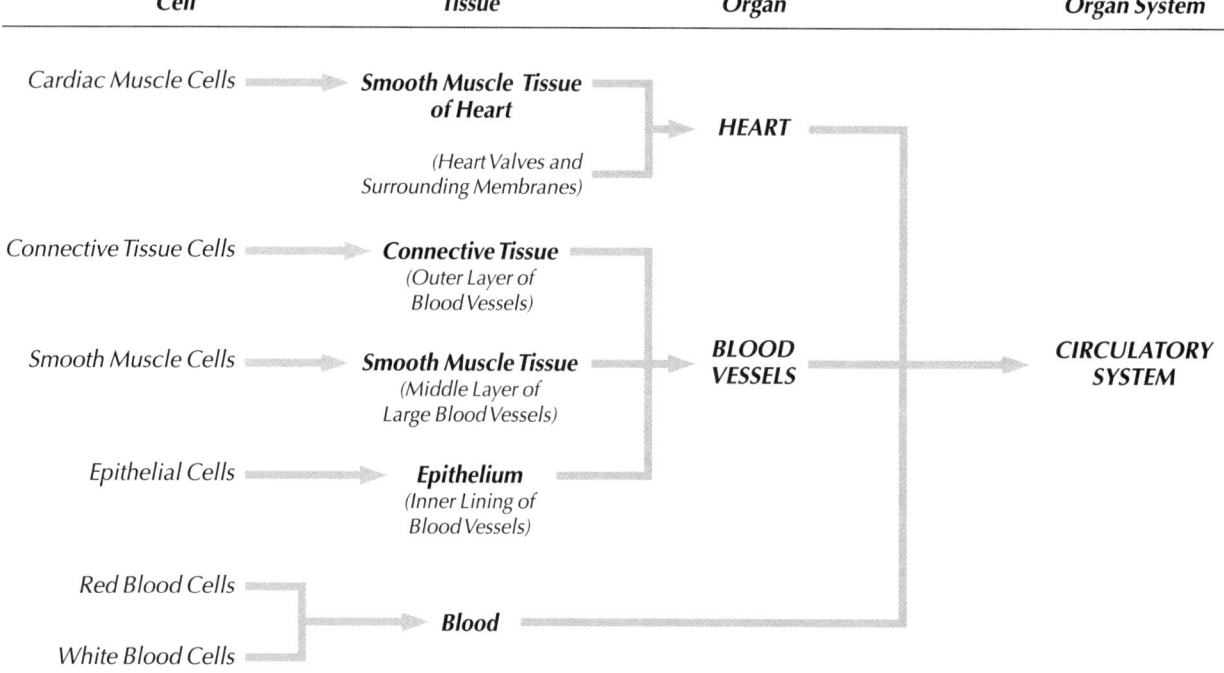

Figure 4–1. Functional Organization of the Vertebrate Body: Cells are the basic units of living things. Cells with similar characteristics group together to form tissues, and different types of tissues combine to form organs. A group of organs whose function is coordinated to carry out one or more basic life processes is considered an organ system. As an example, consider the components of the organ system called the circulatory system, pictured here. Some of the basic units are connective tissue cells, smooth muscle cells, and epithelial cells. These cell types aggregate with their own kind to form, respectively, connective tissue, smooth muscle tissue, and epithelium. Together, these three tissue types make up the organs known as blood vessels. Another circulatory organ, the heart, is composed of smooth muscle tissue as well as heart valves and membranes. (The latter two components are really not tissues or organs, but organ parts.) The heart, blood vessels, and blood (a liquid tissue composed of white and red blood cells) together make up the circulatory system.

cussed are: the **skeletal system**; **muscular system** (excluding the skin muscles that move the feathers); **nervous system** including the **sense organs**; **endocrine system**; **circulatory system**; **respiratory system**; **digestive system**; and **urogenital system**. The skin and structures that are produced by the skin such as feathers, color pigments, scales, claws, beak, wattles, and comb make up the **integumentary system**, which was considered in Chapter 3. This chapter concludes with a look at bird metabolism.

At the end of the chapter is a table summarizing the major anatomical differences between birds and mammals, arranged by organ system. You may wish to refer to the table from time to time as a quick review of each system.

The Skeletal System

■ The **skeletal system**, composed of bone and cartilage as well as associated joints, tendons, and ligaments, supports and protects the soft structures of the body; provides for the attachment of muscles that move the skeleton; and serves as a storehouse for calcium, phosphorus, and other elements. You might think that adult bones, once formed, were "dead" structures and always remained the same, similar to the steel girders of a building. But this is an entirely false perspective. Bones are very much alive: they are sheathed and riddled with living cells. Bones are constantly changing in shape and composition in response to physical stress such as that produced by exercise, or by variations in the vitamins and minerals in the diet, or in the body's demands for the minerals that bones contain. Certain bones of birds also contain cavities filled with red marrow, the primary blood-forming tissue in the adult.

Bone is a tissue composed of living cells in a matrix (the substance between cells in which they are embedded) called osteoid. The matrix contains microscopic fluid spaces and has a blood supply that constantly deposits or moves the mineral components, primarily hydroxyapatite—a calcium phosphate mineral—from one place to another. The calcium residing in one bone in the morning may well be in another by afternoon, or in an eggshell the next day, should the bird be laying a clutch.

Bones have distinctive shapes, and are constantly being remodeled as they grow. Bones act mechanically as levers for the action of muscles, and the continued use of a muscle can result in the formation of a bump or process on the bone where the muscle attaches. This occurs because when a muscle contracts and pulls on a bone it activates deposition of calcium within the matrix of the bone, especially at the site of muscle attachment, which may alter the shape or size of the bone. These features are so characteristic that they can serve as landmarks for the identification of a particular bone.

When a bird is forming an egg prior to egg laying, calcium is taken from the bones, transported by the blood, and deposited as shell on the egg in the uterus. If too little calcium is deposited on the egg, the shell will be thin and might break when the bird sits on it during incubation.

But if too much calcium is taken from the bones during eggshell formation, the bones may become fracture-prone. Much of the delicate balancing act that makes possible both successful reproduction *and* the continued existence of the parent bird is controlled by the endocrine system (see later in this chapter) and its regulation of calcium metabolism, especially through the dynamics of living bones.

Two general features were acquired by the skeleton of birds during the evolution of flight. One is **rigidity** and the other is **lightness (Fig. 4–2)**.

Rigidity of the skeleton is the result of various fusions of neighboring bones, particularly parts of the vertebral column. Consequently, birds (unlike most other vertebrates) are notoriously stiff-backed. As compensation they have long and highly moveable necks. Their body rigidity strengthens the skeleton for the stressful actions of flying and landing as well as running and jumping. Likewise, fusion of the skull bones allows the use of the beak as a lever or as a "hammer and chisel"

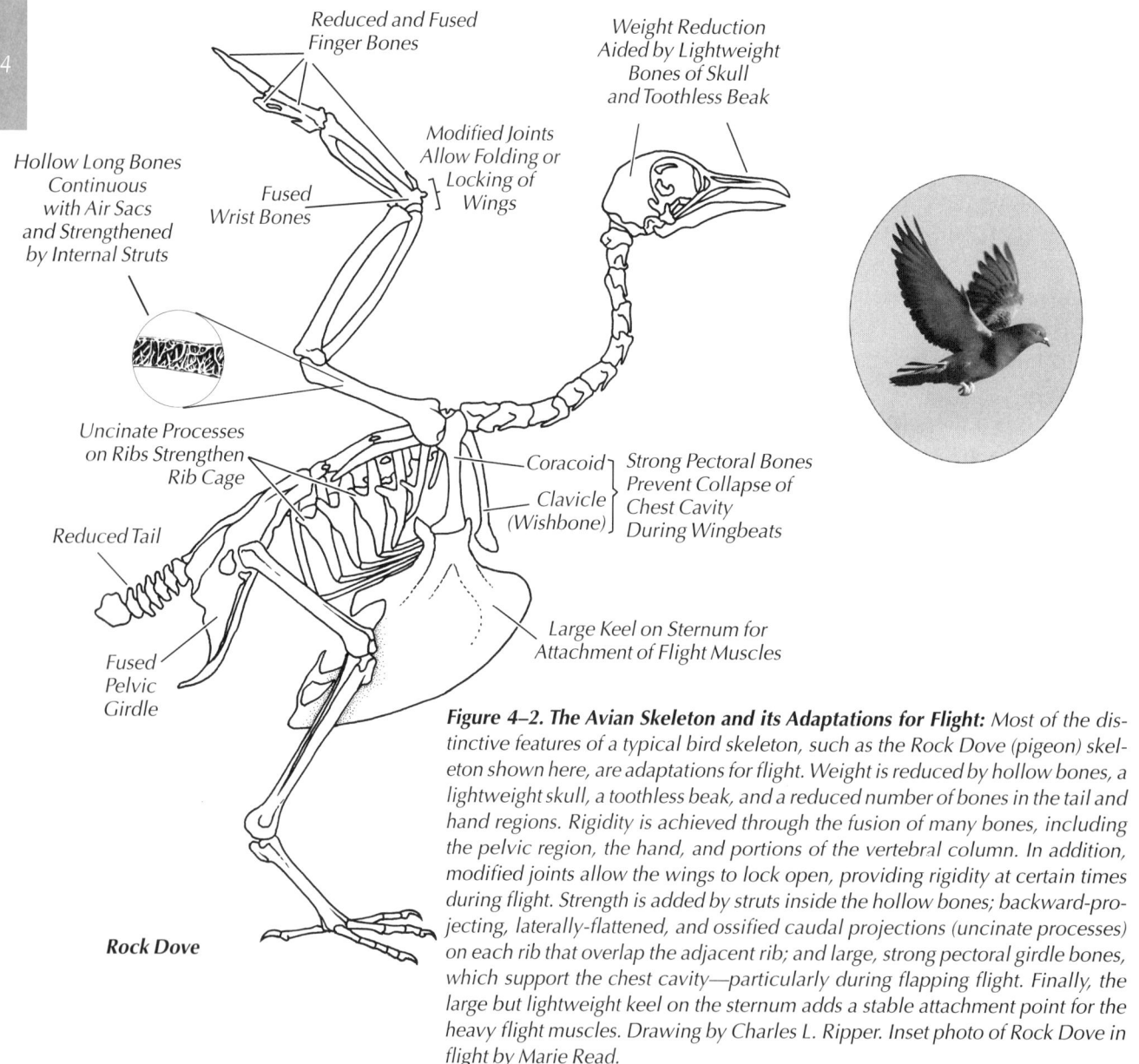

Reduced and Fused Finger Bones

Weight Reduction Aided by Lightweight Bones of Skull and Toothless Beak

Modified Joints Allow Folding or Locking of Wings

Hollow Long Bones Continuous with Air Sacs and Strengthened by Internal Struts

Fused Wrist Bones

Uncinate Processes on Ribs Strengthen Rib Cage

Reduced Tail

Coracoid
Clavicle (Wishbone)
Strong Pectoral Bones Prevent Collapse of Chest Cavity During Wingbeats

Large Keel on Sternum for Attachment of Flight Muscles

Fused Pelvic Girdle

Rock Dove

Figure 4–2. The Avian Skeleton and its Adaptations for Flight: Most of the distinctive features of a typical bird skeleton, such as the Rock Dove (pigeon) skeleton shown here, are adaptations for flight. Weight is reduced by hollow bones, a lightweight skull, a toothless beak, and a reduced number of bones in the tail and hand regions. Rigidity is achieved through the fusion of many bones, including the pelvic region, the hand, and portions of the vertebral column. In addition, modified joints allow the wings to lock open, providing rigidity at certain times during flight. Strength is added by struts inside the hollow bones; backward-projecting, laterally-flattened, and ossified caudal projections (uncinate processes) on each rib that overlap the adjacent rib; and large, strong pectoral girdle bones, which support the chest cavity—particularly during flapping flight. Finally, the large but lightweight keel on the sternum adds a stable attachment point for the heavy flight muscles. Drawing by Charles L. Ripper. Inset photo of Rock Dove in flight by Marie Read.

as in woodpeckers. (How the woodpecker protects its brain from the violent impacts of its hammering is still an open question.)

Lightness comes from cavities or spaces that develop within almost all bones as a bird grows. These spaces connect to the respiratory system and thus contain air. We call such bones **pneumatic** (noo–MAT–ik) and speak of the skeletal system as being **pneumatized**. Because they are filled with air spaces and not bone and marrow, bird bones are lighter in weight than similar-sized bones of other vertebrate groups. For example **(Fig. 4–3)**, a male Mallard and a male mink may both weigh about two and one-half pounds (just over one kilogram) and be nearly the same lengths (not counting the tails), but they appear quite different in "size"—the duck has a much greater volume. The bird's lower density is due in part to the bird's hollow bones. Actually the skeletons of the two animals are very similar in weight (about 2.25 ounces [60 to 65 grams] when clean and dry). But almost every bone of the bird is larger in some or all dimensions, with more surface area for muscle attachment.

The air spaces in skull bones arise from nasal passageways, whereas those in the vertebrae, sternum, ribs, pelvis, humerus, and femur are connected to either the air sacs or the lungs directly (see Fig. 4–82b). The largest and most efficient flying birds, such as the albatrosses and frigatebirds, have interconnected air spaces passing from the humerus to the tips of the digits across all of the joint spaces. In contrast, a penguin, a flightless bird that swims with its short but powerful wings and can dive to considerable depths, has solid bones that lack pneumatic spaces; thus the skeleton is relatively heavy and functions as ballast (weight) for diving. Loons (which make long, deep dives) also have less pneumatization than most nondivers. The degree of pneumatization does not always indicate the flying or diving ability of a bird, however. The air sacs of the skull are actually fewer and smaller in birds such as

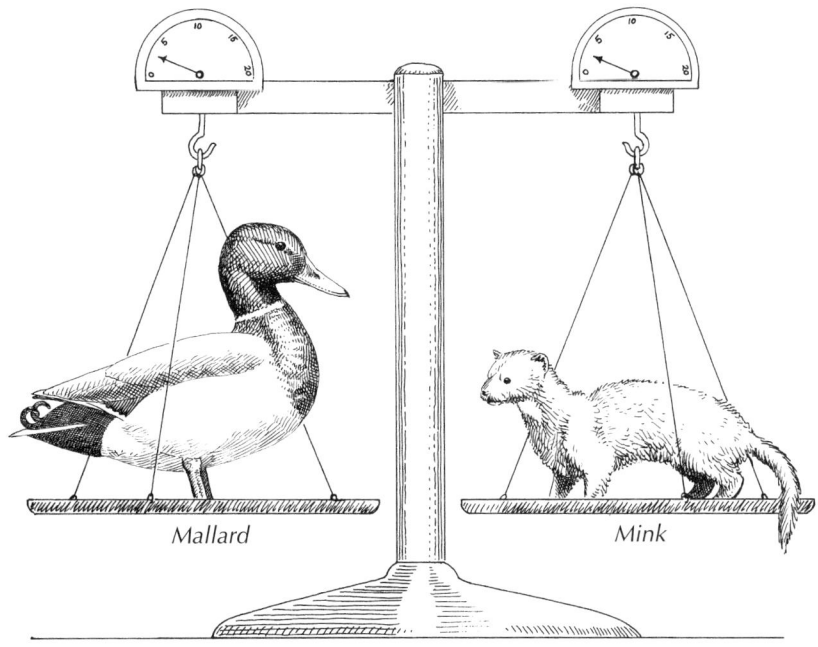

Mallard   Mink

Figure 4–3. Mallard and Mink: A male Mallard and a male mink both weigh about 2.5 pounds (just over 1 kg), and have about the same body length. The Mallard, however, has a much larger volume and thus a lower body density, due in part to its lightweight, air-filled bones.

swifts that fly fast, perhaps to allow the head to be smaller and more compact as an aid to streamlining. Other fast-flying birds, such as shorebirds, also have reduced pneumatization. Extensive skull pneumatization is also decreased in some birds that dive from the air into water, such as terns and kingfishers, and in woodpeckers that "hammer" wood with their bills and need very strong skulls.

Another distinctive and uniquely important tissue of the skeletal system is **cartilage**, the same tissue found in human joints. **Cartilage**, like bone, is a tissue with living cells embedded in a nonmineralized matrix, capable of growth or resorption as well as transformation into bone. In the growing bird embryo, the initial skeleton entirely of cartilage **ossifies** (becomes bone) very rapidly around the time of hatching and continues ossifying at a slower rate throughout life. A similar sequence occurs in other vertebrates, including humans. Bone also can be formed directly in tissues without going through a cartilage stage. Such direct **ossification** is seen in tendons of the hind-limb muscles in most birds, and is familiar to aficionados of the turkey drumstick.

Tendons, which connect muscles to bones, and **ligaments**, which connect one bone to another across a joint, are soft, pliable, and above all elastic tissues when they are not ossified. They have few living cells and a restricted blood supply, which explains why they heal from injury so slowly.

By convention the skeleton is divided into an **axial skeleton** and an **appendicular skeleton**. The axial skeleton consists of the vertebral column of the neck, trunk, and tail, and the exquisitely complicated skull, with its associated hyoid apparatus (the supporting framework of the tongue). The **appendicular skeleton** consists of the sternum or breastbone, the pectoral girdle with wings, and the pelvic girdle with legs.

Shown in Figs. 4–4, 4–5, and 4–6 are the articulated skeletons (that is, the bones are joined together as they would be in life) of a domestic chicken, a Budgerigar (a type of parakeet), and a Golden Eagle. You may wish to remove these pages and keep them readily at hand while reading about both the skeletal and muscular systems. The chicken has its wings extended as if it were about to take off in flight. This is the standard position used in mounting or illustrating a bird for anatomical study. The parakeet has its wings folded in a normal perching position, and the Golden Eagle has its left wing lowered and pulled slightly away from the body. Refer to these drawings frequently as you proceed in your reading; they will orient you to the parts being discussed. You will note differences between the three species of birds illustrated. The skeletons of all bird species differ from one another in ways that are not apparent externally but are often significant to evolutionary and systematic ornithologists. If you have at hand the actual skeleton of a bird, or even some isolated bones, so much the better, for you will be able to see how they appear in three dimensions. (Perhaps it would be useful to plan on a whole roast chicken dinner before reading further!)

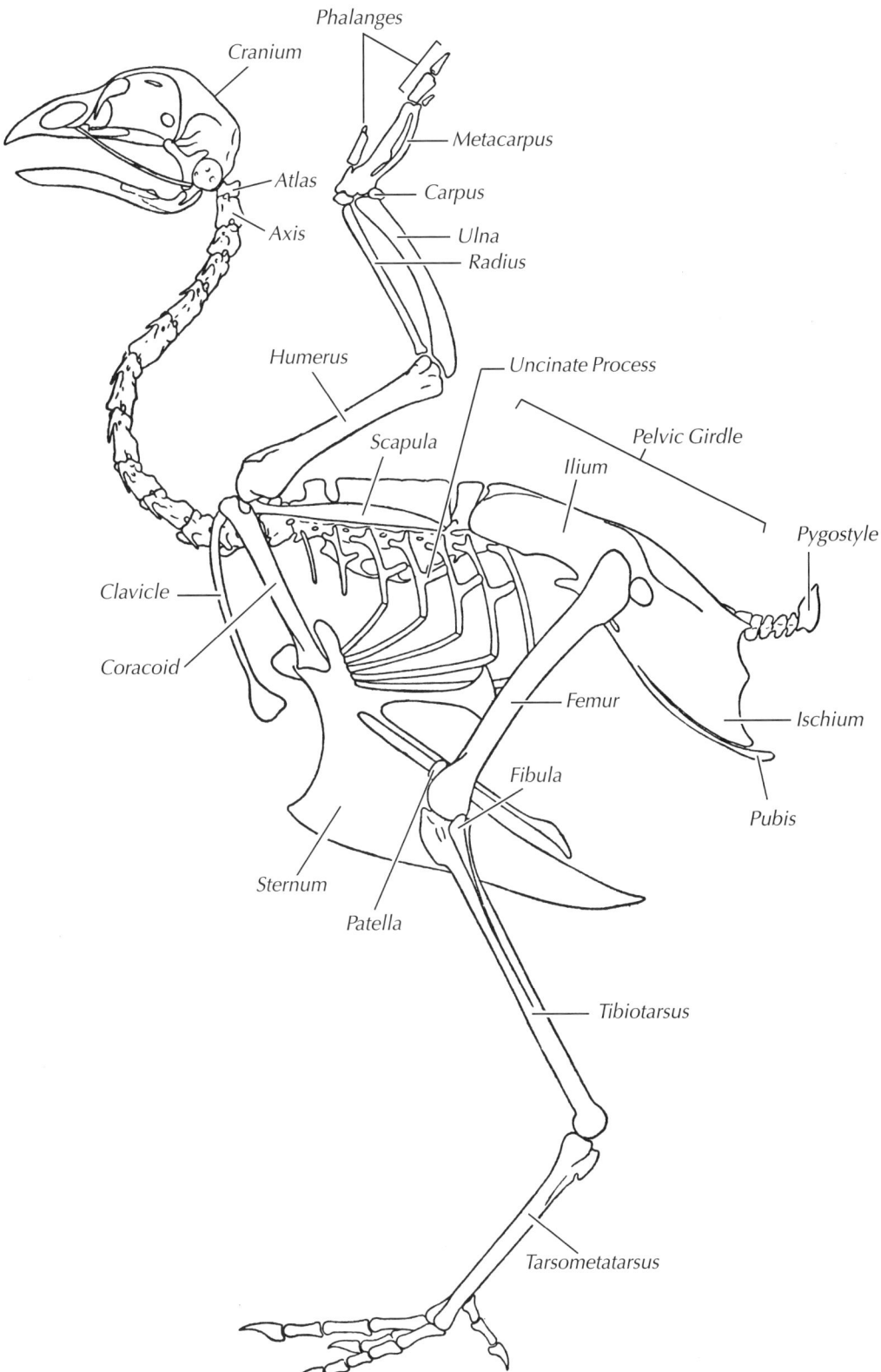

Figure 4–4. Skeleton of the Chicken: *Viewed from the left side, the chicken skeleton is shown with the left wing raised over the body. Compare with Figures 4–5 and 4–6, in which different bird species in different positions provide additional perspectives of the articulated skeletal system. Not all skeletal components are equally developed, visible, or labeled in each example. The adaptations of these three species for their different lifestyles are evident in their bones. On the chicken, notice the robustness of the leg and foot bones, and the relatively large pelvic girdle, indications that this is a ground-dwelling species. Paired structures, such as the wings, legs, and ribs, are only shown for the left side of the bird.*

4

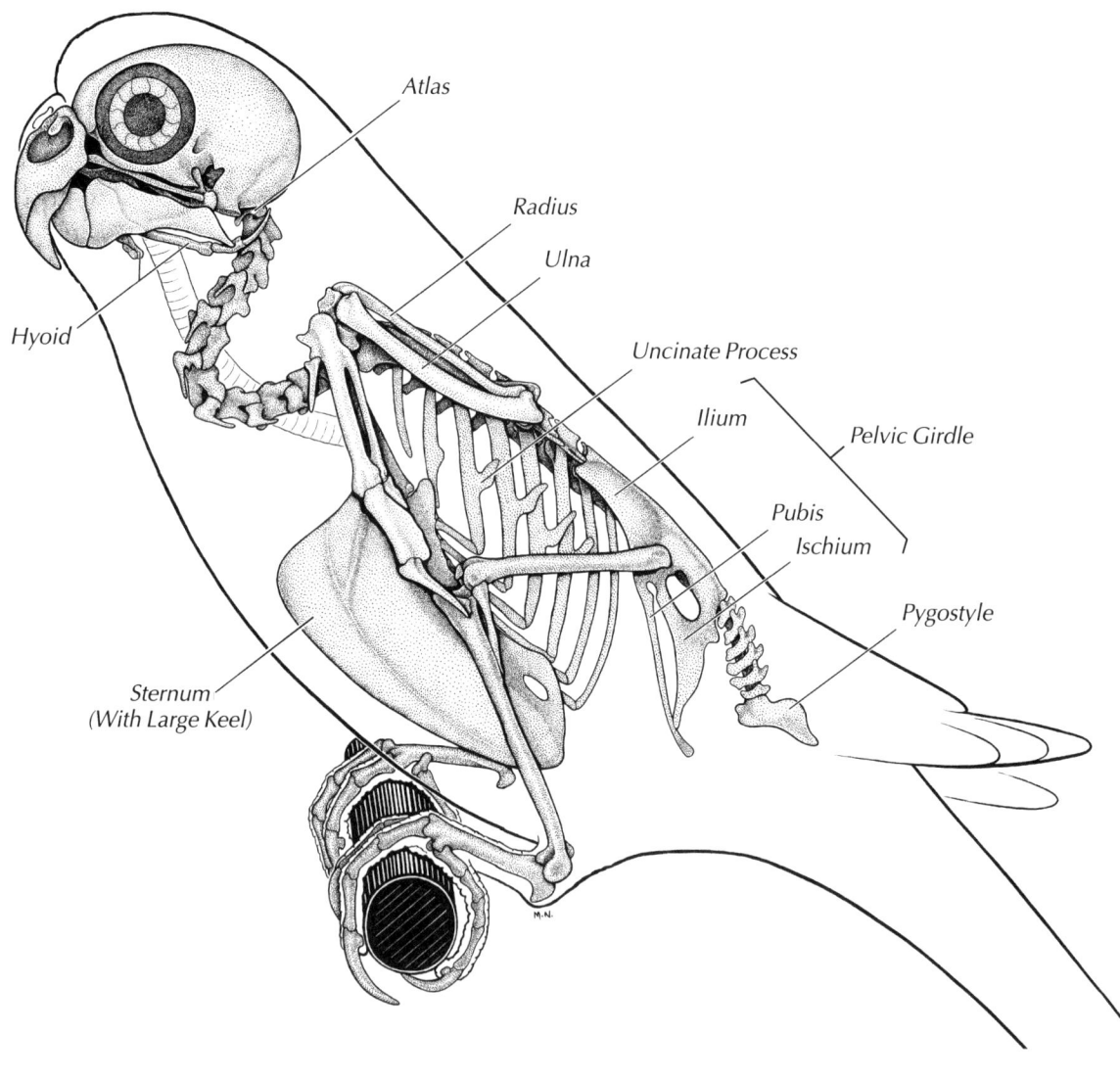

Atlas

Radius

Ulna

Uncinate Process

Ilium

Pelvic Girdle

Pubis
Ischium

Pygostyle

Hyoid

Sternum
(With Large Keel)

Figure 4–5. Skeleton of the Budgerigar: *The skeleton of this Budgerigar (a species of parakeet) is viewed from the left side, in a normal perching position, with the left wing folded against the body. In this view the clavicle and coracoid are hidden by the folded wing, but notice the large keel on the sternum indicating that the bird is a strong flyer. Note, too, the relatively small pelvic girdle and less-developed leg and foot bones—evidence that this bird spends much of its time perching, rather than walking or running on the ground. The differences in skull anatomy between this species and the chicken in Figure 4–4 are conspicuous. Drawing from Evans (1996).*

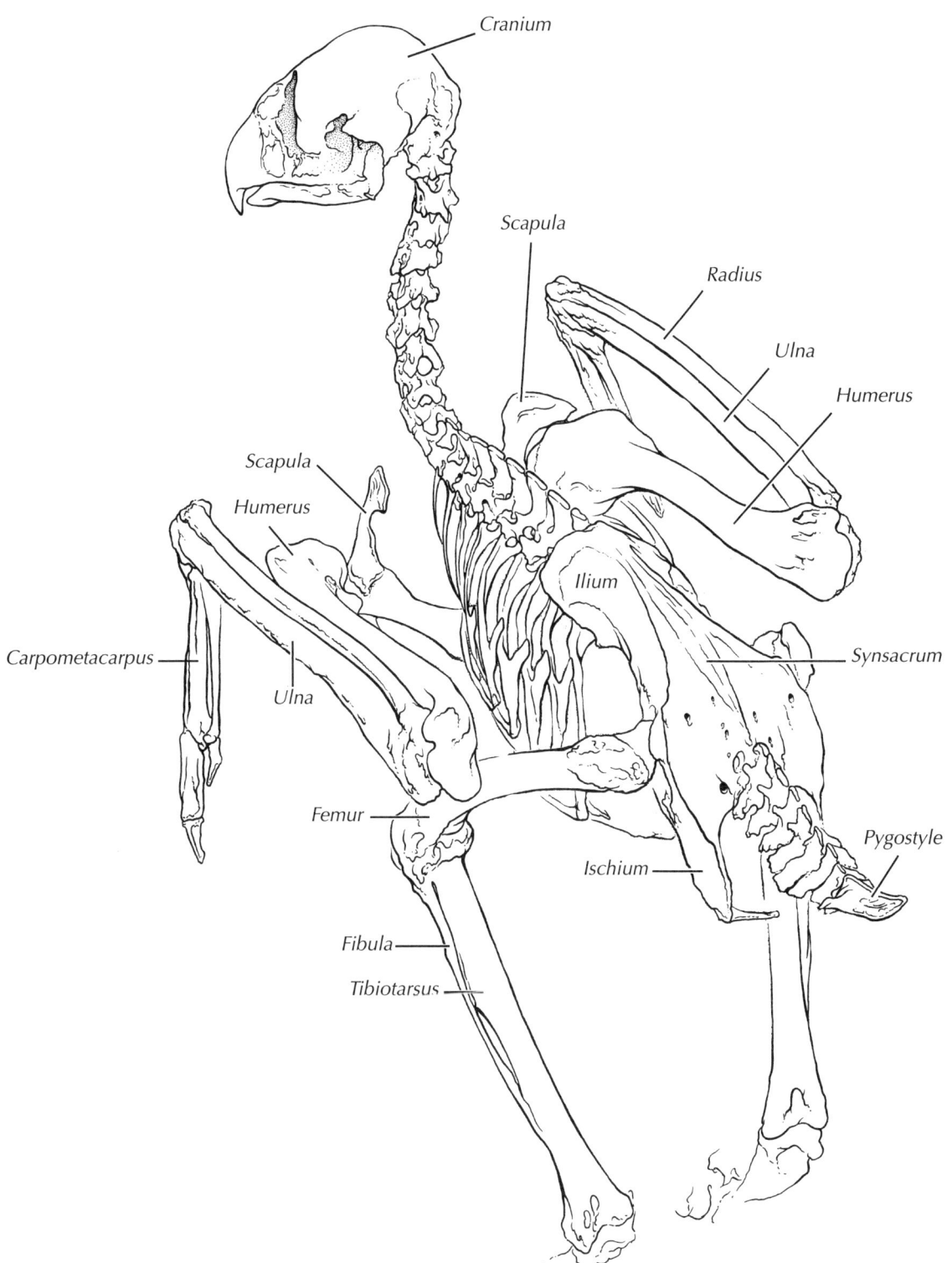

Figure 4–6. Skeleton of the Golden Eagle: *This skeleton is viewed at an angle: from the rear and slightly to the left side, and from somewhat above. The feet are not shown. The left wing is lowered and pulled away from the body slightly, to demonstrate the great mobility of the pectoral girdle (see Fig. 4–16). Notice the eagle's massive wing and leg bones, evidence that it is a powerful flier with equally powerful legs used to capture its prey. The robust skull with its massive hooked beak is yet another adaptation for a predatory lifestyle. Reprinted from* Manual of Ornithology, *by Noble S. Proctor and Patrick J. Lynch, with permission of the publisher. Copyright 1993, Yale University Press.*

Figure 4–7. Age-related Changes in the Ossification and Pneumatization of the Passerine Skull: *The cranium of a newly fledged passerine is composed of a single layer of cartilage and bone. As the young bird ages, a second layer develops under the first, the two layers being slightly separated by air spaces and joined by small columns of ossifying bone. The development of these layers is called skull pneumatization, and is followed by full ossification, usually complete by the time the bird is one year old. Although there is variability among species, the process typically follows one of two patterns. In the peripheral pattern, pneumatization develops from the outer edges of the cranium inward, whereas in the median line pattern, pneumatization starts along a central line and develops on both sides of the cranium simultaneously. Unpneumatized areas of a passerine skull appear pinkish in color, whereas pneumatized areas appear grayish or whitish and develop small white dots as the bony columns ossify. If the head feathers of a bird in the hand are parted to expose the skin overlying the skull, the degree of skull coloration may be visible. Experienced bird banders use skull ossification and pneumatization to reliably determine the age of passerines captured in the fall, a technique known as* **skulling.** *Adapted from Pyle et al. (1987, p.9).*

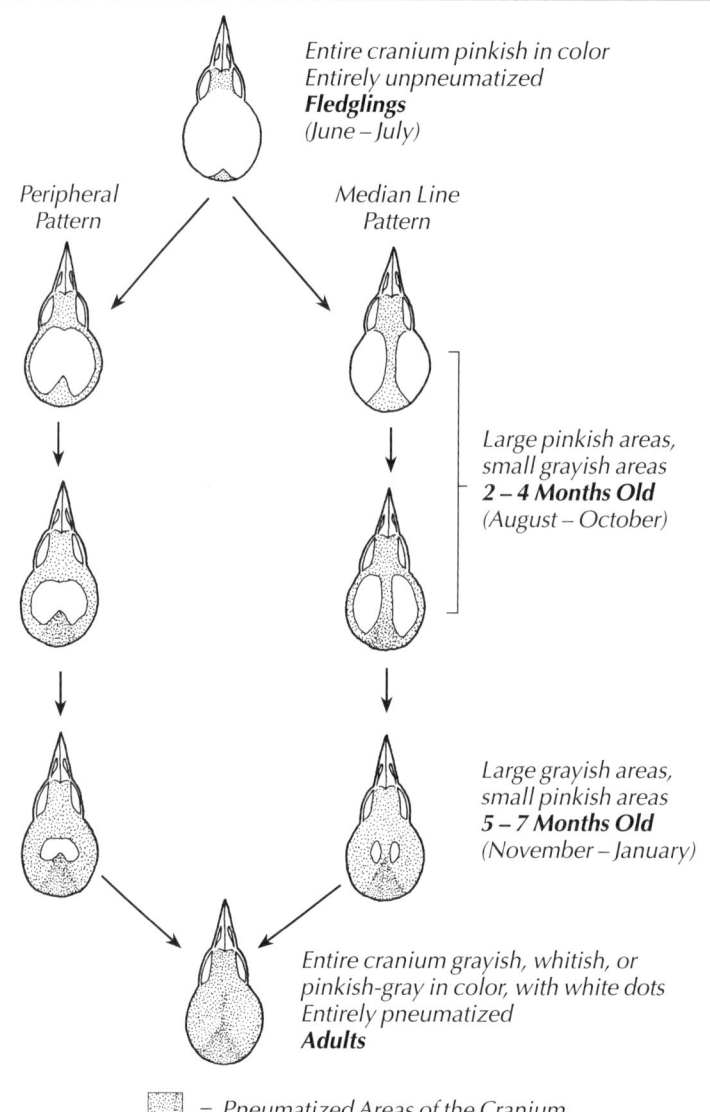

Entire cranium pinkish in color
Entirely unpneumatized
Fledglings
(June – July)

Peripheral Pattern   *Median Line Pattern*

Large pinkish areas, small grayish areas
2 – 4 Months Old
(August – October)

Large grayish areas, small pinkish areas
5 – 7 Months Old
(November – January)

Entire cranium grayish, whitish, or pinkish-gray in color, with white dots
Entirely pneumatized
Adults

▨ = Pneumatized Areas of the Cranium

Axial Skeleton

Skull

The **skull** is the skeleton of the head. Most skull bones are so completely fused in adult birds that we cannot distinguish any **sutures** or boundary lines between them. The presence of visible sutures (which can be seen through the skin of a living bird) is an indication that the skull in question is that of a very young bird. In newly-fledged passerines, the skull is composed of a single layer of cartilage and bone. As the bird ages, however, a second layer develops under the first, the two layers being slightly separated by air spaces and joined by columns of ossifying bone. The development of these layers is called skull pneumatization, and is followed by full ossification of the skull, both of which are usually complete by the time the bird is one year old. The degree of skull ossification and pneumatization are often used by bird banders to age a live passerine bird in the hand, enabling them to distinguish a bird fledged that year from older individuals **(Fig. 4–7).**

a. Dorsal View

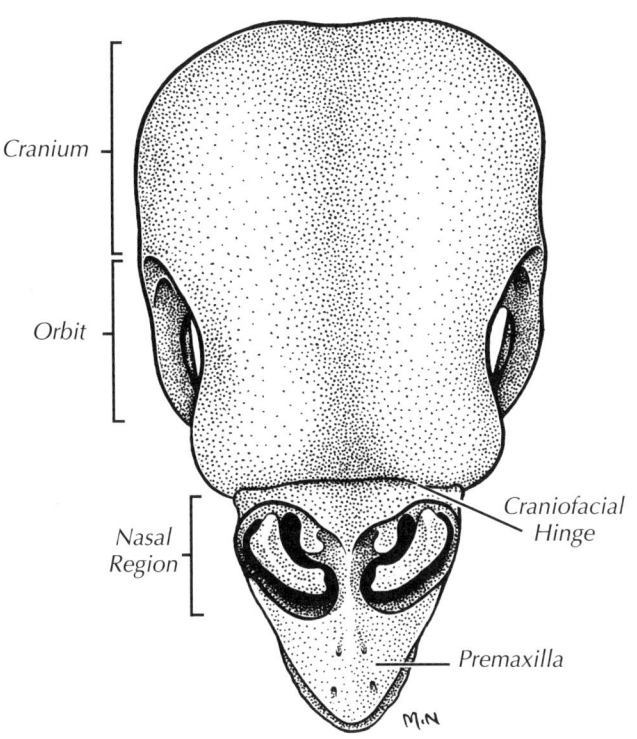

Cranium

Orbit

Nasal Region

Craniofacial Hinge

Premaxilla

b. Lateral View

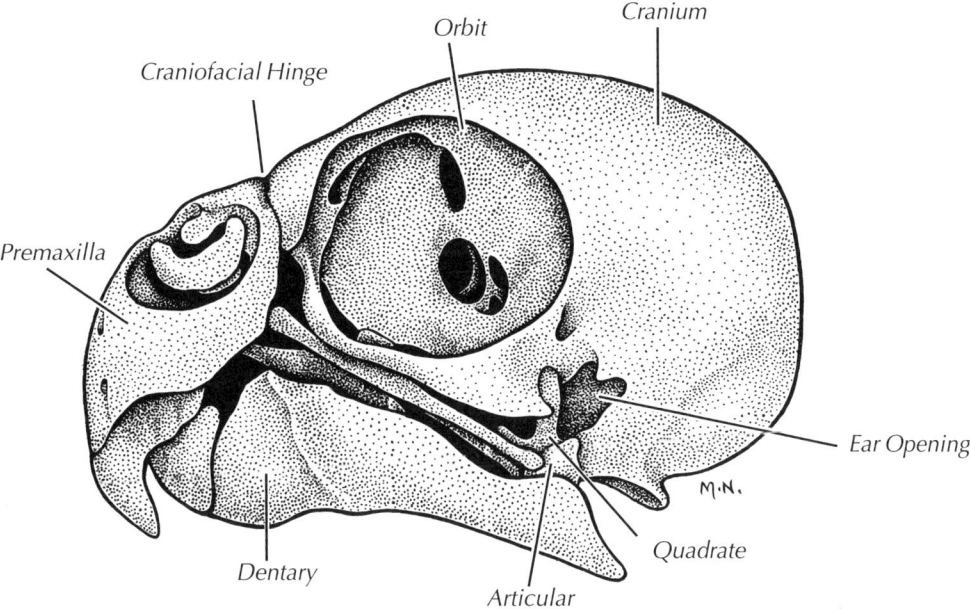

Craniofacial Hinge

Orbit

Cranium

Premaxilla

Dentary

Articular

Quadrate

Ear Opening

Figure 4–8. The Skull of the Budgerigar:
*a. Dorsal View: The **cranium**, or brain-case, is composed of several bones that are so completely fused in the adult bird that the boundaries, or sutures, between them are no longer visible. Notice the large **orbits** in which the eyes reside. The skeleton of the upper beak consists of the **nasal region** containing the nostrils, and the fused **premaxillary** bones. The upper beak connects to the rest of the skull at the flexible **craniofacial hinge** (also called the nasal-frontal hinge).*
b. Lateral View: *This view shows more clearly the cranium's position relative to the bones of the upper and lower jaws, as well as the ear openings, which are slightly caudal and ventral to the orbits. Compare this view to that of the chicken in Figure 4–9. Most obvious is the difference in shape between the premaxilla and dentary bones of the two species. The Budgerigar has a strong, hooked beak typical of parrots and parakeets, used for cracking hard seeds, whereas the chicken has a straight, stout beak for pecking at food on the ground. Drawings from Evans (1996).*

The skull **(Fig. 4–8)** is composed of a braincase or **cranium**, which incorporates the ear or **otic region** on each side. The median **nasal region** and **mouth** support respiratory and digestive openings. The lower jaw or **mandible** of the bird consists of right and left parts (the **dentary bones**) fused at the tip of the beak. The apical region of the upper jaw is formed by right and left **premaxillary bones**. This proper use is still lacking in much ornithological literature in which the potentially confusing terms "upper and lower mandible" are still in common use (see Baumel et al. 1993). Throughout this course we will use "upper

Figure 4–9. The Avian Jaw and Cranial Kinesis: *Shown here is a lateral view of a chicken skull. The premaxillary bones of the upper jaw articulate with the cranium at the flexible craniofacial hinge, allowing the upper jaw to move at the same time as the lower jaw—a process known as* **cranial kinesis.** *The bird's lower jaw consists of left and right dentary bones, fused at their tip. The lower jaw is linked to the upper jaw via the* **articular bone** *and the adjacent* **quadrate bone,** *which play an essential role in cranial kinesis.* **a. Opening the Jaws:** *The process of bill opening begins when a set of muscles acts to drop the lower jaw. As the lower jaw moves downward, the articular bone applies pressure on the quadrate bone, causing the quadrate to rotate such that its lower surface moves forward. As it moves, the quadrate pushes against two sets of bony rods, the* **palatine** *and the* **jugal arch,** *which push against the premaxillary bones, raising the upper jaw while the lower jaw is being depressed.* **b. Closing the Jaws:** *To close the bill, another set of muscles depresses the premaxillary bones while it raises the lower beak. Adapted from King and McLelland (1975, p. 17).*

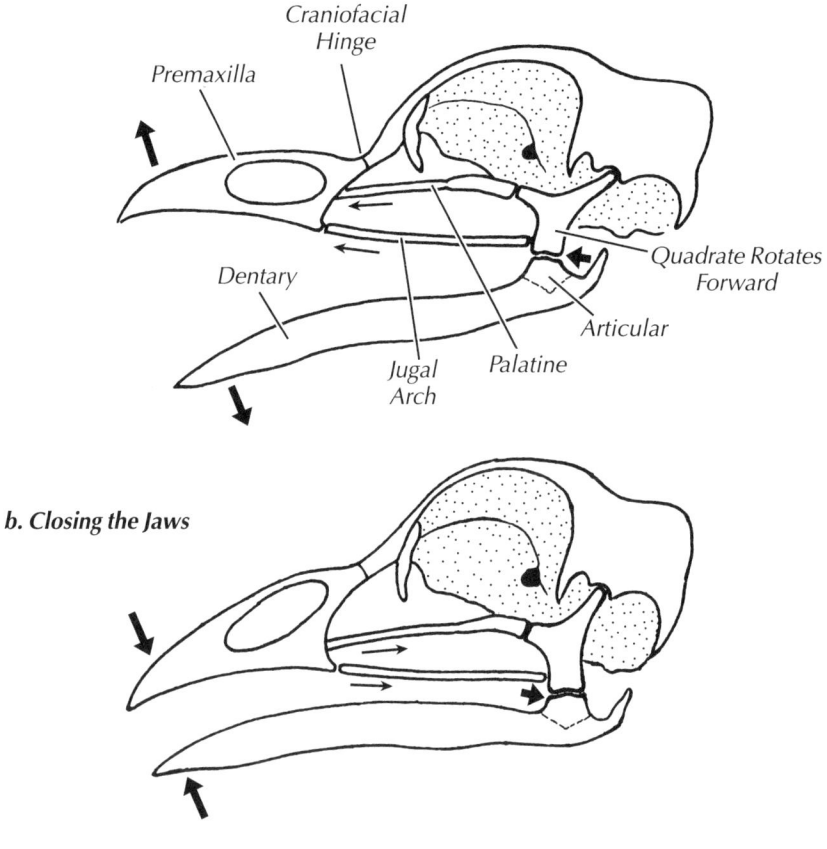

a. Opening the Jaws

b. Closing the Jaws

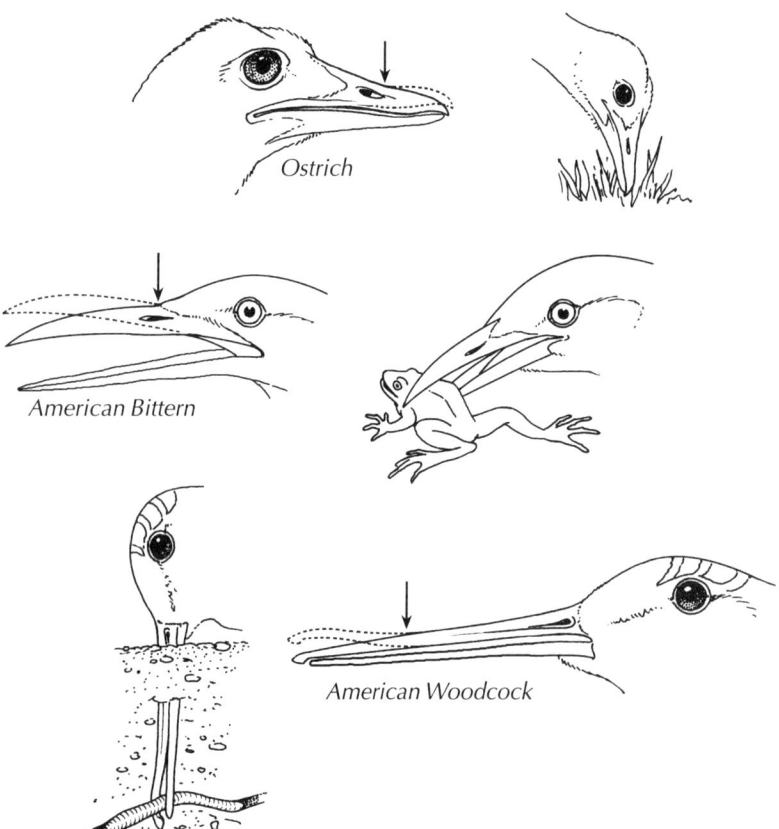

Ostrich

American Bittern

American Woodcock

Figure 4–10. Cranial Kinesis in Action: *The upper jaw of birds, unlike the upper jaw of humans, is capable of movement, giving the beak a pincer-like motion known as cranial kinesis (see Fig. 4–9). The degree to which cranial kinesis occurs in various species is related to their feeding habits. In the Ostrich, a grazing bird, cranial kinesis is poorly developed: to feed on grasses and other vegetation this species does not need to open its beak wide. In contrast, the American Bittern has considerable kinesis, opening its beak wide to seize a frog. In the American Woodcock, movement is greatest at the tip of the beak, allowing the bird to grasp earthworms while the beak is immersed in soft soil. Drawings by Charles L. Ripper.*

beak" and "lower beak" when referring to the intact, living bird.

The lower jaw articulates with the moveable **quadrate bone** on each side of the skull; this articulation allows the mouth to open widely **(Fig. 4–9)**. Because the quadrate bone can also pivot forward at its articulation with the skull, it allows the lower jaw to be protruded while the upper jaw is raised by extreme extension at the craniofacial (nasal-frontal) hinge. One set of muscles attached to the quadrate, cranium, and lower jaw opens the bill by simultaneously lowering the lower beak and rocking the quadrate forward, forcing the upper beak upward. Another set of muscles closes the bill by depressing the premaxillary bones and at the same time raising the lower jaw. This flexibility of the jaw joints, which allows the upper jaw to be raised at the same time that the lower jaw is depressed, is called **cranial kinesis**. Movement of the bird's upper jaw contrasts sharply with the fused immobility of our upper jaw in relation to the rest of our skull, and is associated with the forceps-like functioning of the beak to grasp food **(Fig. 4–10)**.

No living bird has teeth, although several birds, particularly fish-eaters such as the mergansers, have serrated bills. Teeth (structurally like those of other vertebrates) *were* present, however, in the most ancient Jurassic birds of Bavaria, such as *Archaeopteryx,* and continued to be present in Cretaceous birds of North America, such as *Hesperornis* and *Ichthyornis.* These ancient teeth are but one of the many characteristics that birds inherited from their ancestors. Through natural selection, birds subsequently discarded teeth in favor of the lighter bill.

The large **orbit** or cavity for the eye is so deep that the eyes almost meet on the midline of the skull. Only a thin, sometimes incomplete, interorbital septum separates them. Aside from the beak, the orbits are the most prominent feature of the bird skull. In Figure 4–8 notice how the orbits crowd the braincase backward. The **ear opening** lies close to the lower rim of the orbit on each side. In owls, the opening to the ear is at a slightly different level on the right and left sides. This asymmetry allows for a more accurate pinpointing of the direction of a sound source by increasing the disparity in arrival times of a sound at the two ears (see Fig. 4–47).

Hyoid Apparatus

Between the two halves of the lower jaw is the **hyoid apparatus**, a series of articulated bones that support both the **tongue** and the muscles that provide for tongue movement **(Fig. 4–11)**.

a. Ventral View of Rock Dove Hyoid Apparatus

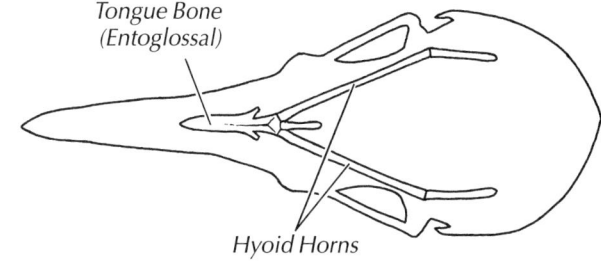

Tongue Bone (Entoglossal)

Hyoid Horns

b. Lateral View of Budgerigar Hyoid Apparatus

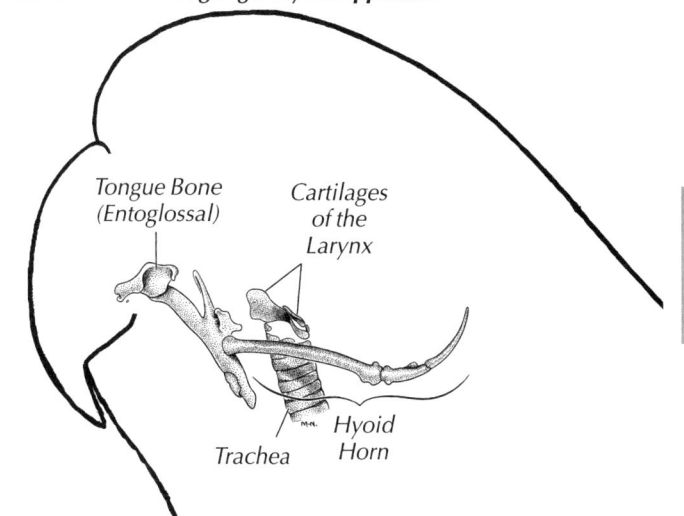

Tongue Bone (Entoglossal)

Cartilages of the Larynx

Trachea

Hyoid Horn

Figure 4–11. The Hyoid Apparatus: The bones and cartilaginous structures forming the skeleton of the tongue are collectively termed the hyoid apparatus. Muscles attached to this compound structure extend and retract the tongue. **a. Ventral View of Rock Dove Hyoid Apparatus:** Note the two **hyoid horns**, each composed of two bones, which extend backward from the **tongue bone** and continue beneath the skull, curving upward around the back of the head in some species. Drawing by Charles L. Ripper. **b. Lateral View of Budgerigar Hyoid Apparatus:** The length of the hyoid horns varies considerably among species. In the Budgerigar, the horns are short, indicating that these birds use their tongues to manipulate food within the mouth rather than extending the tongue to feed. The tongue bone itself may be spear-shaped, as in the Rock Dove; blunt, as in the Budgerigar; or flexible, as in hummingbirds; the differences reflect the different structures and functions of the tongue. The hyoid apparatus surrounds and is attached to the larynx. Adapted from Evans (1996).

The term "apparatus" is a collective name for all the bones and cartilages that compose this V-shaped structure. The muscles attached to the bones of the hyoid extend and retract the tongue. In Figure 4–11b note that each **horn** of the hyoid is composed of two bones that extend backward (caudally) beneath the skull and then curve around the back of the head. The horns are particularly long in woodpeckers and other birds that can extensively project the tongue **(Fig. 4–12)**. The Eurasian Green Woodpecker has the longest hyoid horns for its body size, and presumably can extend its tongue the farthest. Short hyoid horns, as in parrots, indicate that the birds use their tongue to manipulate food within the mouth rather than extending it to collect food. In birds that eat worms and grubs, the bone that supports the tongue—called the tongue bone, or **entoglossal**—may be spear-shaped, and the integumentary portion of the tongue may bear spines; in birds that manipulate seeds or fruits, the entoglossal is blunt and padded by the tongue tissue; and in nectar feeders, it is flexible and covered by a brush-like tongue. More will be said about the tongue in the discussion of the digestive system.

Vertebral Column

The **vertebral column**, commonly called the "backbone," is not a single bone but a series of complicated, uniquely articulating or rigidly fused **vertebrae** (singular, **vertebra**) that vary in number among species. As in other vertebrates, the vertebrae are named by region (and

Figure 4–12. The Hyoid Apparatus and Tongue Protrusion in the Northern Flicker: In woodpeckers, a highly protrusible tongue is an essential food-gathering tool. Many woodpecker species drill holes in dead trees and then probe with their tongues for insects. Others, such as the Northern Flicker illustrated here, feed on the ground, using their tongues to extract ants from subterranean tunnels. Woodpeckers have elongated hyoid horns with elaborate musculature, enabling them to greatly extend their tongues—in some species up to four times the length of the bill. ***a. Tongue Retracted:*** *The long, slender hyoid horns are sheathed by muscle for most of their length. Anchored by muscles in the flicker's lower beak, the horns run separately on either side around the back of the head, outside the skull, then together enter the right nostril (see dorsal view), attaching to the skeleton of the upper beak.* ***b. Tongue Protruded:*** *When the muscles of the sheath contract, the hyoid horns are squeezed into tight contact with the skull and pushed forward, protruding the flicker's tongue from its mouth. Adapted from drawings by Charles L. Ripper.*

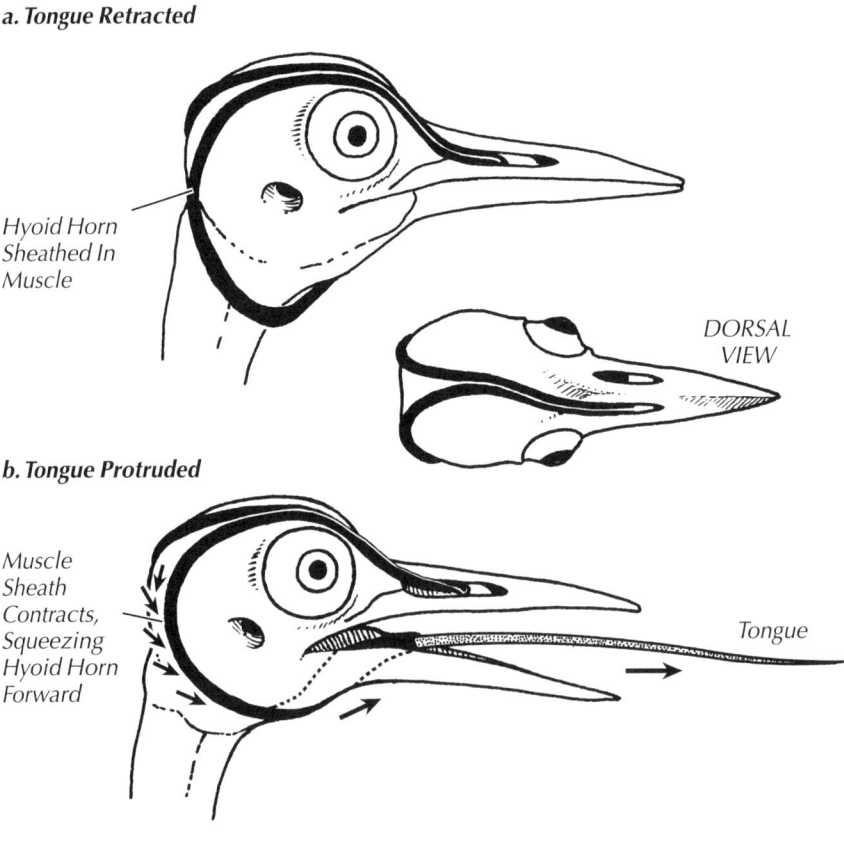

a. Lateral View of Cervical Vertebrae

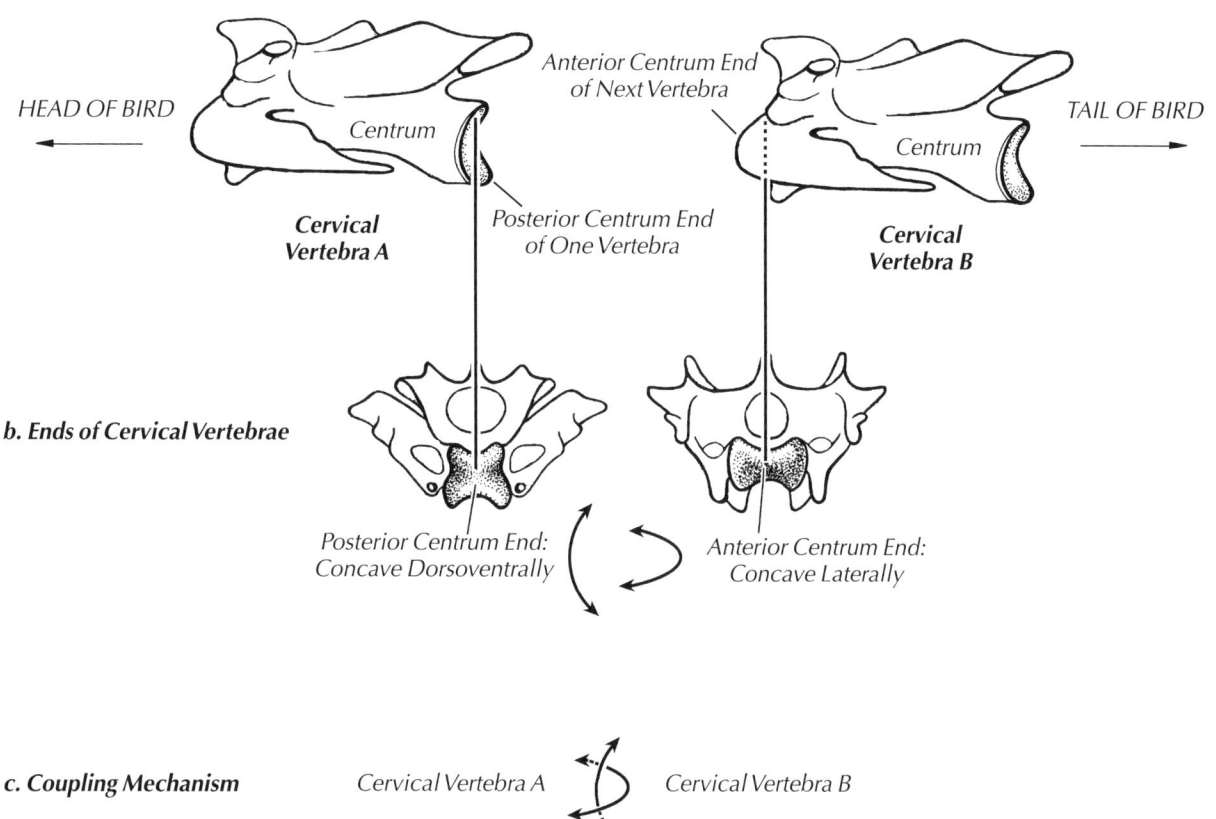

b. Ends of Cervical Vertebrae

c. Coupling Mechanism

Figure 4–13. Cervical Vertebrae and Flexibility of the Bird Neck: *Birds are well known for the flexibility of their necks, facilitated by having a large number of interlocking cervical vertebrae that can rotate against one another freely in all directions. The two ends of the main body (**centrum**, see **a**) of each vertebra are shaped differently. The anterior end of the centrum is concave (saddle-shaped) in a lateral direction, whereas the posterior end of the centrum is concave in a dorso-ventral direction (see **b**). This condition of the centrum ends is termed **heterocoelous**. (Note that the vertical lines between **a** and **b** connect the same points on the centrum end between the two different views.) When the posterior end of Vertebra A contacts the anterior end of Vertebra B (see **c**) the two vertebrae can rotate around each other in all planes of movement. Such flexibility allows many birds to rotate their heads 180 degrees in either direction, an ability particularly well developed in owls (see Fig. 4–45).*

then numbered within each region): **cervical** in the neck; **thoracic** in the rib cage; **lumbar** in the lower back; **sacral** in the pelvic region; and **caudal** in the tail. Nevertheless, fusions within and between regions in different groups of birds blur these distinctions.

Nearly all mammals have seven cervical vertebrae, whether they have short necks as in primates, or long necks as in the giraffe. In birds the number varies from 12 in some cuckoos and passerine birds (or even as few as 11 in the Old World hornbills, in which the first 2 vertebrae are fused and counted as a single unit) to as many as 25 in some swans. Most birds have 14 or 15. The relatively large number of freely articulating cervical vertebrae allows a marked suppleness of the neck and turning ability of the head. The flexibility of the neck, in a sense, compensates for the rigidity of the back. Most birds—not just owls—can turn their heads 180 degrees in either direction, thanks in large part to the saddle-shaped, interlocking ends of the vertebrae **(Fig. 4–13)**. This avian condition of **heterocoelous centrum ends** is unique among living vertebrates, and immediately identifies isolated bird vertebrae.

The first cervical vertebra, termed the **atlas**, is small and articulates with a prominent protrusion, the **occipital condyle**, on the base of the skull **(Fig. 4–14a)**. The atlas is named for the divinity from Greek mythology who supported the sky, in allusion to the vertebra's support of the skull. Uniquely, the atlas has a hole or notch on the ventral surface of its central opening (the vertebral canal) into which the peg-like **dens** of the second cervical vertebra, the **axis**, fits. The dens, which embryologically began as part of the atlas, is attached to the skull by a ligament. Thus, as in all terrestrial vertebrates, the first two cervical vertebrae, the atlas and axis, are highly modified for the articulation of the skull with the rest of the vertebral column. The remaining cervical vertebrae have either short, fused rib remnants **(Fig. 4–14b)** or may

a. Atlas and Axis of Rock Dove

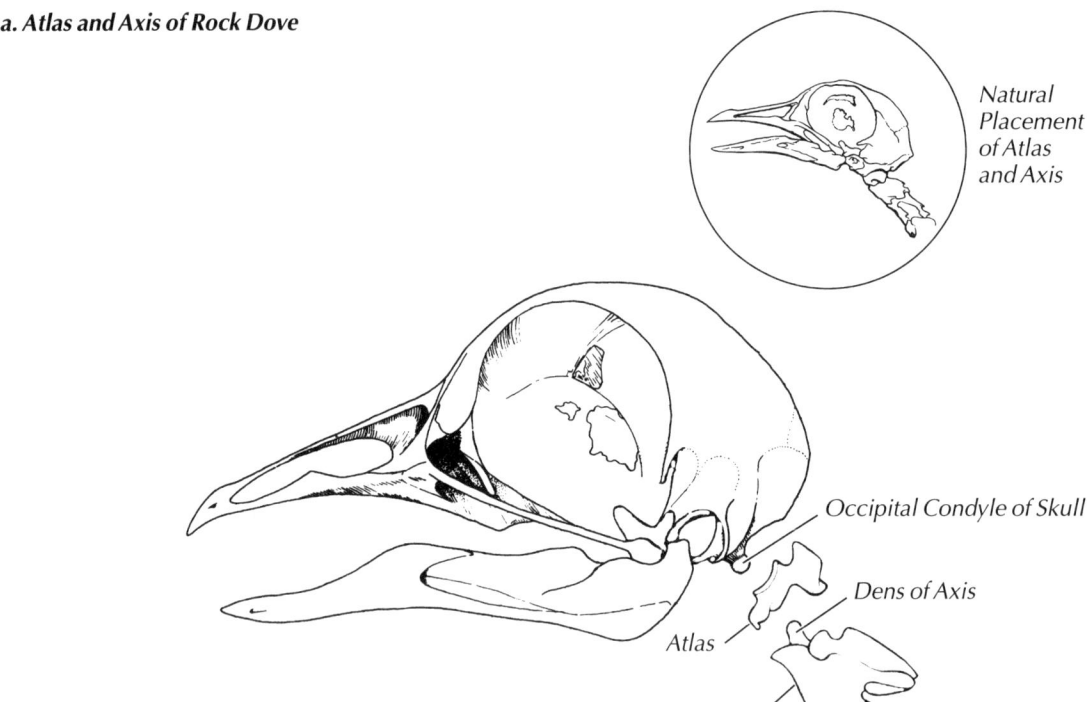

Natural Placement of Atlas and Axis

Occipital Condyle of Skull

Dens of Axis

Atlas

Axis

Figure 4–14. Types of Vertebrae: a. Atlas and Axis of Rock Dove: *The cervical vertebrae, forming the neck, vary in number among bird species. The two cervical vertebrae closest to the head are small and specialized to connect the skull with the rest of the vertebral column. The large drawing shows the skull of a Rock Dove with these two vertebrae separated, to show their articulating surfaces; the inset shows their natural placement. The first cervical vertebra, the* **atlas,** *supports the head by articulating with a prominent peg, the* **occipital condyle,** *on the base of the skull. The atlas has a notch or hole on the ventral surface of its posterior end, which receives the peg-like* **dens** *of the second cervical vertebra, the* **axis.** *These connections permit the bird's head to rotate freely.* **b. Cervical Vertebra of Chicken, Lateral View:** *The remaining cervical vertebrae typically have short, fused rib remnants or small, moveable ribs projecting laterally.* **c. Thoracic Vertebrae and Rib Attachment:** *Thoracic vertebrae make up the thorax or chest region of the vertebral column. As* shown in **c1,** *the lateral view of the first through sixth thoracic vertebrae from the chicken, several thoracic vertebrae are typically fused together, adding rigidity to the vertebral column. The* **spinous processes** *or* **spines,** *ridges of bone projecting from the dorsal surface of all vertebrae, are particularly well developed in the thoracic vertebrae and are fused into a strong, vertical ridge of bone to which the large back muscles attach. On their lateral surfaces, thoracic vertebrae have prominent facets for the attachment of true* **ribs,** *which, unlike those found on the cervical vertebrae, connect with the sternum. Placement of the ribs is shown in* **c2,** *the lateral view of three thoracic vertebrae, and in* **c3,** *the anterior view of a single thoracic vertebra from a Rock Dove. Note that only the upper segment of each rib (termed the vertebral rib) is shown. All drawings adapted from Ede (1964, pp. 30 and 31), except for the lower two in* **c,** *which are adapted from Proctor and Lynch (1993, p. 131).*

b. Cervical Vertebra of the Chicken, Lateral View

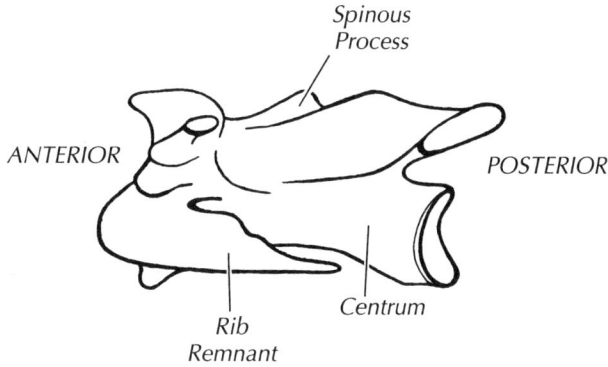

c. Thoracic Vertebrae and Rib Attachment

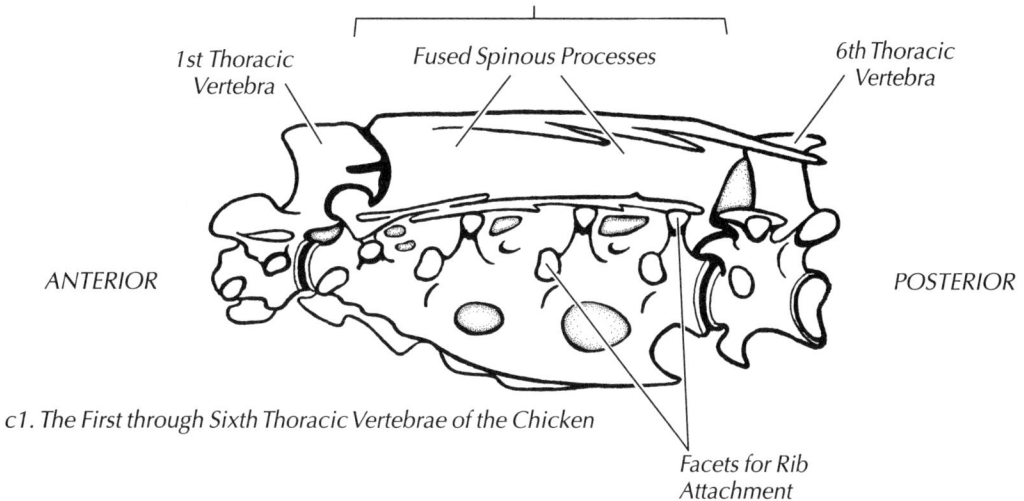

c1. The First through Sixth Thoracic Vertebrae of the Chicken

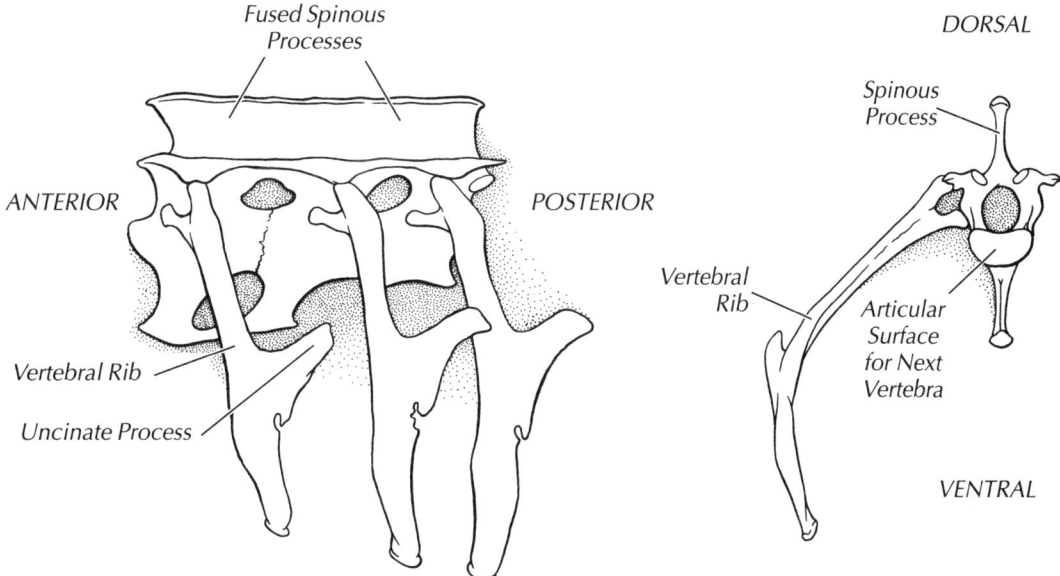

c2. Lateral View of Three Thoracic Vertebrae of
the Rock Dove with Ribs Attached

c3. Anterior View of Thoracic Vertebra of
the Rock Dove with Rib Attached

bear small, moveable ribs. In either case there is an opening between the vertebra and the base of each of the forked ribs. When cervical vertebrae bear moveable ribs, they are difficult to distinguish from thoracic vertebrae, and the term "cervicodorsal" vertebrae may be used. If the rib articulates with the sternum, either directly or by a ligament, it is considered to be a thoracic vertebra.

Thoracic vertebrae (from the thorax or chest region) can be distinguished from other vertebrae by their facets for rib articulation (**Fig. 4–14c**). They also are characterized by well-developed spines (spinous processes) on their dorsal surfaces for the attachment of deep back muscles. Most birds have between four and six thoracic vertebrae; the Rock Dove has five. Some of the thoracic vertebrae may become fused with one another to form a "notarium" for added rigidity of the backbone, a benefit in flying and landing.

In Figures 4–6 and 4–20 note the **synsacrum**, another unique feature of the bird's vertebral column. It consists of a fusion of a variable number of thoracic vertebrae with all of the lumbar, all of the sacral, and the first few caudal vertebrae. This rigid segment is in turn fused on either side with the ilium bones of the pelvis. The number of vertebrae involved in the bird's synsacrum varies among species from 10 to 23.

The **tail** of a bird consists of from four to nine free caudal vertebrae and a terminal bone called the **pygostyle**, formed by several fused vertebrae (see Figs. 4–4 to 4–6). The pygostyle is the shape of a plowshare and provides attachment for the flight feathers of the tail. On top of the pygostyle rests the oil gland.

The **ribs,** together with the thoracic vertebrae above and the sternum below, form a bony "**thoracic cage**" (rib cage) enclosing the heart, liver, and lungs, as well as the thoracic air sacs (**Fig. 4–15**). Each thoracic rib has a dorsal and a ventral part with a hinge between them. The upper or **vertebral rib** articulates with a thoracic vertebra; the lower segment or **sternal rib** articulates with the sternum. This hinged arrangement allows the thorax to be expanded and compressed for breathing, and thus act as a bellows. The sternum moves downward and forward for expansion during inspiration, then upward and backward for compression during expiration. The pattern of inhalation and exhalation while flying may be quite different. During flight, pectoral muscles spread the furcula and thus participate in breathing (see Fig. 5–5). Projecting caudally from the vertebral segment of each rib is an **uncinate process** that overlaps the rib behind it and helps to strengthen the rib cage (see Fig. 4–5). ("Process," in the anatomical sense, means a projection or extension from a bone; *uncinus* is Latin for "hook.")

Appendicular Skeleton

The **appendicular skeleton** consists of the bones of the wings and hind limbs, together with their supporting pectoral and pelvic girdles, and the sternum, which articulates with the pectoral girdle as well as with the axial skeleton.

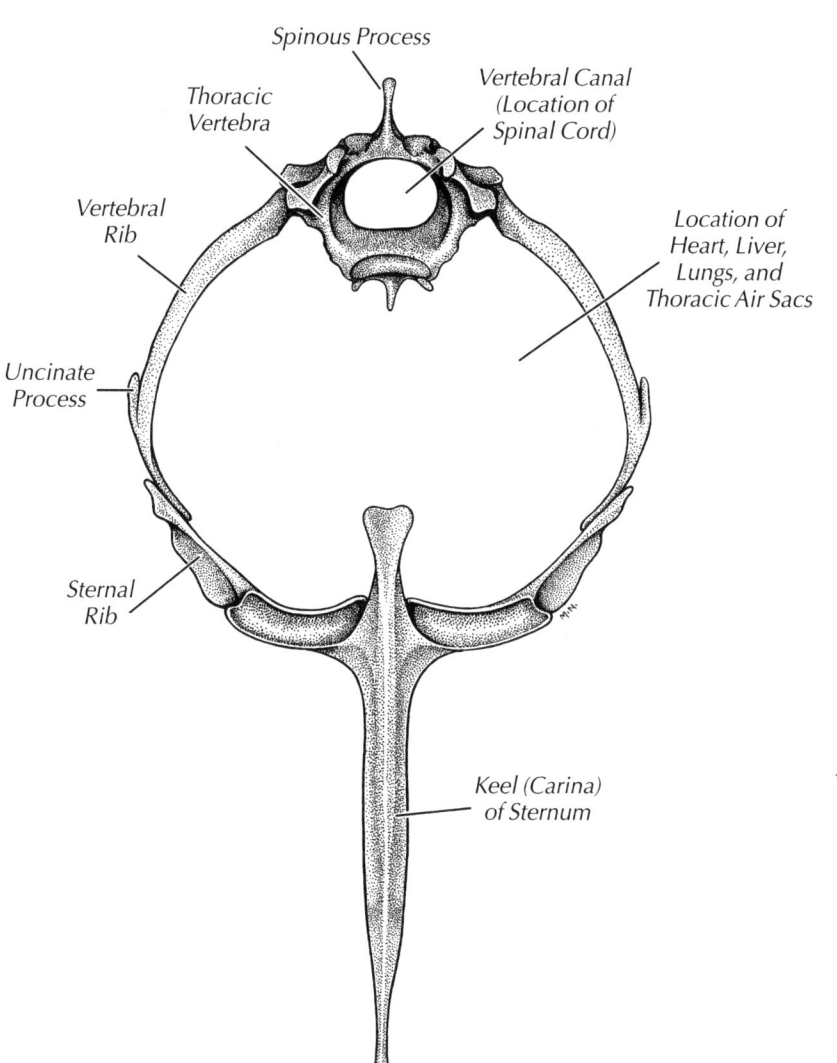

Spinous Process

Thoracic
Vertebra

Vertebral Canal
(Location of
Spinal Cord)

Vertebral
Rib

Location of
Heart, Liver,
Lungs, and
Thoracic Air Sacs

Uncinate
Process

Sternal
Rib

Keel (Carina)
of Sternum

Figure 4–15. The Thoracic Cage: This cranial view shows one "segment" of the thoracic cage of the Budgerigar, at the level of the first thoracic vertebra. The thoracic cage consists of the ribs connected to the thoracic vertebrae above and to the sternum (breastbone) below. It forms a flexible but strong protective enclosure for the bird's heart, liver, lungs, and thoracic air sacs. Each rib consists of two hinged sections, the upper **vertebral rib** and the lower **sternal rib**. The hinge allows the thorax to expand and contract during breathing. From each vertebral rib, an **uncinate process** projects backward, overlapping the rib behind it and thereby strengthening the thoracic cage. Notice the large surface area created by the keel (carina) of the sternum, which is the site of attachment of the powerful flight muscles. The spinal cord is located in the vertebral canal, a "tube" formed by the openings of successive vertebrae. Adapted from Evans (1996).

Pectoral Girdle

The **pectoral girdle** (from the Latin *pectus*, meaning breast) is formed by three bones on each side of the body (**Fig. 4–16**): the **clavicle** (collar bone), the **coracoid,** and the **scapula** (shoulder blade). In nearly all birds the right and left clavicles are fused with a small interclavicle bone to form a single V-shaped bone, the **furcula,** popularly known as the "wishbone." In most birds the ventral end of the furcula is attached to the sternum by a ligament, but in some birds, such as pelicans, this connection ossifies and thus strengthens the support of the shoulder joint, probably an advantage in diving for food. The parakeet has only small remnants of the clavicle on each side, and some parrots have lost the clavicle completely. Although the clavicle provides extra strength, it may also limit shoulder rotation, used in balance. Because parrots often climb tree trunks and branches, maintaining their balance is apparently more important than the extra strength that a clavicle provides.

The **coracoids** are the stoutest and strongest bones of the pectoral girdle. They function as a powerful brace holding the shoulder joint, and thus the wing, away from the body while the pectoral muscles are

Figure 4–16. The Avian Pectoral Girdle:
The pectoral girdle consists of three bones on each side of the body: the clavicle, the scapula, and the coracoid. In most birds, as in the Rock Dove pictured here, the clavicles are fused to form the V-shaped **furcula** *or "wishbone." The scapula and coracoid meet to form a cup-shaped depression, the* **glenoid fossa,** *which receives the rounded end of the humerus, forming a ball-and-socket joint that enables the humerus to rotate freely around the shoulder joint. The upper end of the coracoid articulates with the clavicle as well as with the scapula, and at this three-way joint is an opening termed the* **foramen triosseum** *(also known as the triosseal canal or supracoracoid foramen). Through this opening passes the tendon of the supracoracoideus, a powerful flight muscle that raises the wing (see Fig. 5–6). The lower drawing shows the position of the pectoral girdle (shaded areas) within the skeleton of a Golden Eagle, seen in posteriolateral view, with the left wing lowered. (The furcula is not visible.) This view illustrates that the only bone-to-bone connection between the pectoral girdle (and hence the wing) and the axial skeleton is the junction between the base of each coracoid and the sternum. Further connections are provided by muscles that hold the scapula in place against the rib cage. This arrangement creates a "free-floating" pectoral girdle, allowing the extreme mobility so essential to flight. Main drawing reprinted from* Manual of Ornithology, *by Noble S. Proctor and Patrick J. Lynch, with permission of the publisher. Copyright 1993, Yale University Press. Lower drawing adapted from Proctor and Lynch (1993, p. 139).*

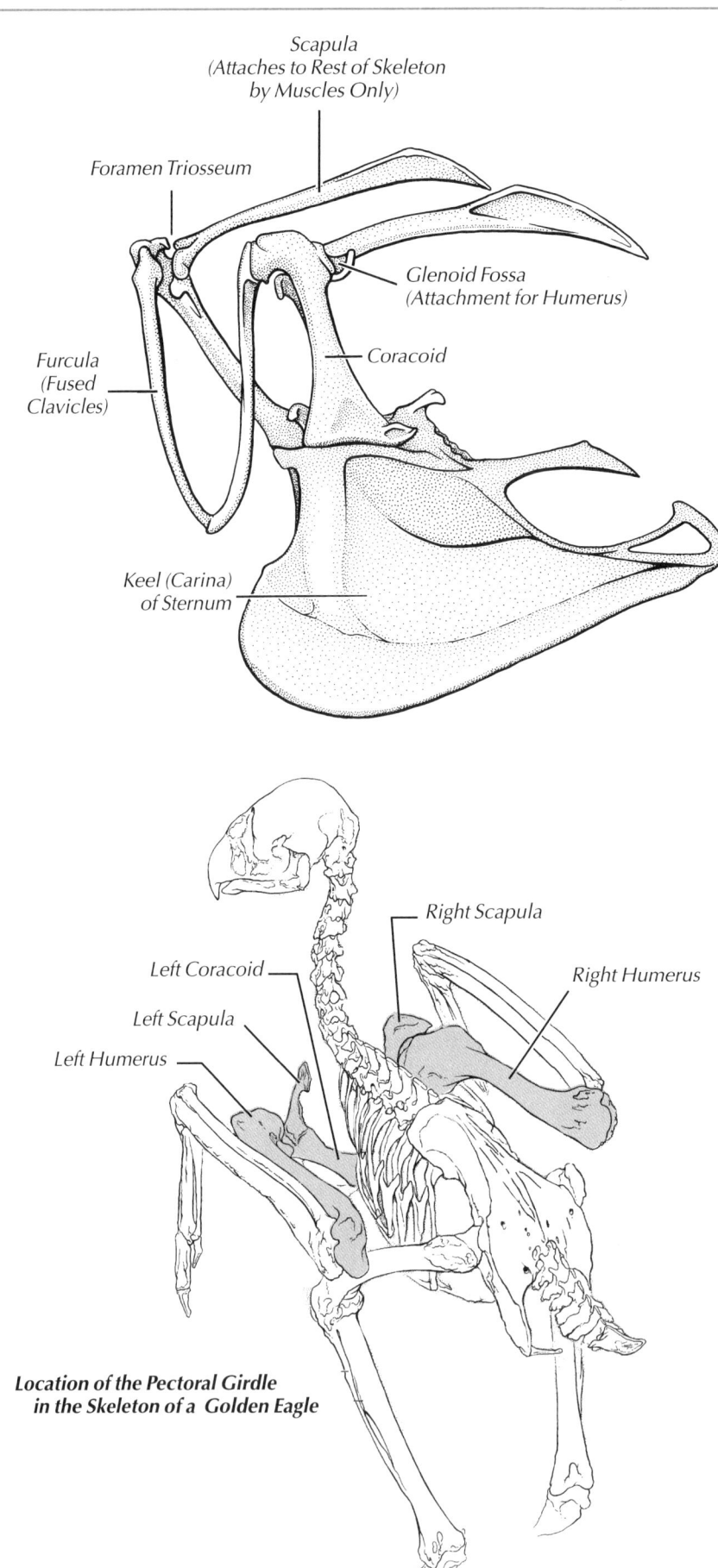

Scapula
(Attaches to Rest of Skeleton
by Muscles Only)

Foramen Triosseum

Glenoid Fossa
(Attachment for Humerus)

Furcula
(Fused
Clavicles)

Coracoid

Keel (Carina)
of Sternum

Left Coracoid

Left Scapula

Left Humerus

Right Scapula

Right Humerus

**Location of the Pectoral Girdle
in the Skeleton of a Golden Eagle**

pulling oppositely on the wing during flight. The broad base of each coracoid bone fits into a groove on the cranial end of the sternum. At its other end, the coracoid meets the scapula at the shoulder joint to form a shallow depression, the glenoid fossa, with which the base of the wing articulates. The upper end of the coracoid bone articulates not only with the scapula, but also with the clavicle, forming an opening at the three-way joint through which the tendon of a powerful flight muscle passes. This **supracoracoid** (meaning "to go above the coracoid") **muscle** raises the wing, and the opening, called the **foramen triosseum** or supracoracoid foramen, serves as part of a pulley system allowing the downward force of the contracting supracoracoid muscle to be redirected to an upward pull on the dorsal surface of the wing (see Fig. 5–6).

Bones of the Wing

Evolution has modified bird forelimbs **(Fig. 4–17)** by reducing the number and length of the bones that correspond to those of our palm (metacarpals) and our fingers (digits—made up of **phalanges**). The wrist has been reduced to two carpals, the radiale and ulnare, by formation of a fused carpometacarpus in which several of the wrist bones have been fused with some of the palm bones. All flying birds

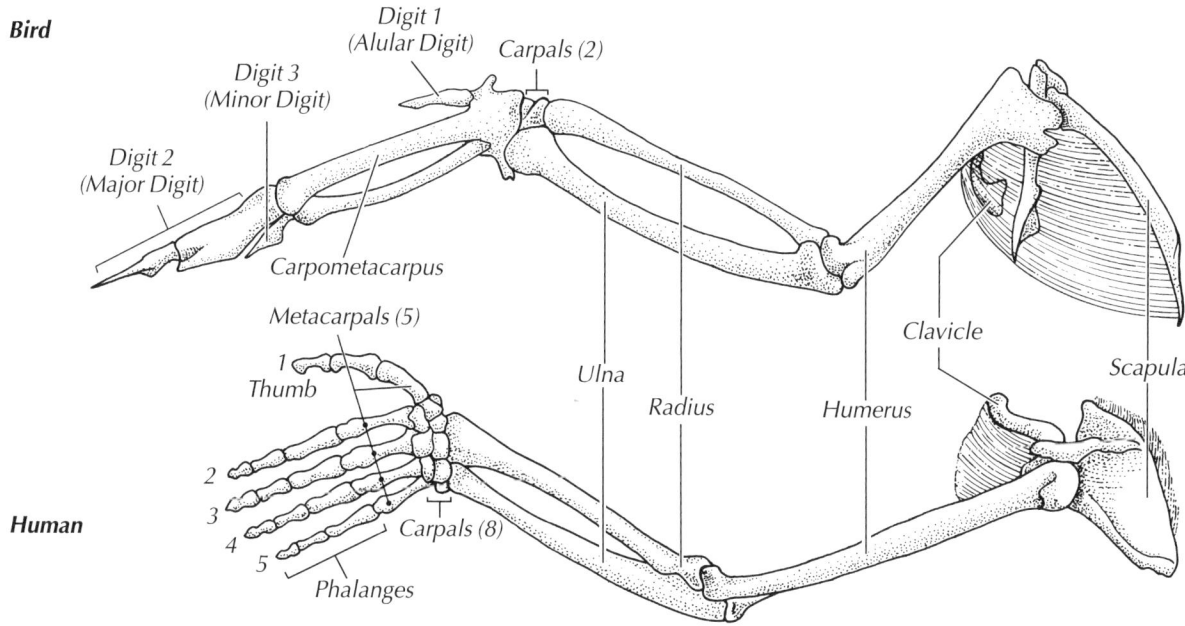

Figure 4–17. The Bones of Human and Bird Forelimb: *The forelimb of a flying bird and a human are compared to show the correspondence between the individual bones. Natural selection has modified the bird's forelimb from that of its reptilian ancestor in several ways. One example is the reduction in the number and length of the bones that correspond to the human palm (metacarpals) and the human fingers (digits, composed of phalanges). Furthermore, the single fused carpometacarpus in birds replaces some of the bones of the human wrist (carpals) and palm (metacarpals). The three digits in birds are thought by some researchers to correspond to the human thumb and first two fingers, but embryological evidence indicates that they correspond to the second, third, and fourth human digits, so their origin is still in debate. Digit 1 in birds (termed the alular digit) carries the alula, and digits 2 and 3 (the major and minor digits, respectively), along with the carpometacarpus, carry the primary wing feathers. Secondary feathers are attached along the ulna, the thicker of the two lower arm bones.* From The Cambridge Encyclopedia of Ornithology, *edited by Michael Brooke and Tom Birkhead, 1991. Copyright Cambridge University Press, reprinted with permission.*

Figure 4–18. Wing Spurs: *Spurs are bony outgrowths that may occur at various sites on the skeleton.* **Wing spurs** *are outgrowths of the carpometacarpus, found in various birds such as cassowaries, plovers, sheathbills, screamers, and jacanas.* **a. Northern Screamer:** *This goose-like, South American bird has paired spurs on each wing, used in aggressive displays and in fights with other Northern Screamers, during which they can inflict considerable damage. Drawing by Charles L. Ripper.* **b. Wattled Jacana:** *The yellow wing spurs of the Wattled Jacana contrast strongly with its black plumage, especially during the raised-wing display shown here, which occurs during alarm or aggression. Photo by Marie Read.*

Wing Spurs

a. Northern Screamer

b. Wattled Jacana

have the same arrangement and number of wing bones. The fingers are invariably three in number with their individual bone segments (phalanges; singular, phalanx) embedded and concealed in a continuous skin covering. Which digits are represented is still a matter of debate. Some researchers say they correspond to our first, second, and third digits. Embryological evidence, however, indicates that they correspond to our second, third, and fourth digits. The avian scientific nomenclature committee, a group of researchers who decide current correct usage, has decided to call the digits: the alular digit (with one phalanx or two), major digit (with two phalanges), and minor digit (with one phalanx). The terminal phalanges (especially of the alular digit) in a variety of birds—swans, ducks, cranes, rails, owls, passerines, and others—may bear a claw. During their first weeks, young South American Hoatzins have temporary claws on one or more of their wing digits. These claws allow them to climb back into the nest, from which they jump at times of danger (see Fig. 3–41).

One should not confuse wing claws with **wing spurs (Fig. 4–18)**. Claws are associated with digits, whereas spurs can be located anywhere on the skeleton. Some birds have both. Wing spurs are bony outgrowths from the carpometacarpus, but they are not digits. Birds with wing spurs include cassowaries, screamers (Anhimidae), plovers, jacanas, and Antarctica's sheathbills. They are used in aggressive display and fighting, especially with other members of the species.

Flightless birds have reduced wing bones. Modern flightless birds show the same basic arrangement of the wing skeleton as flying birds,

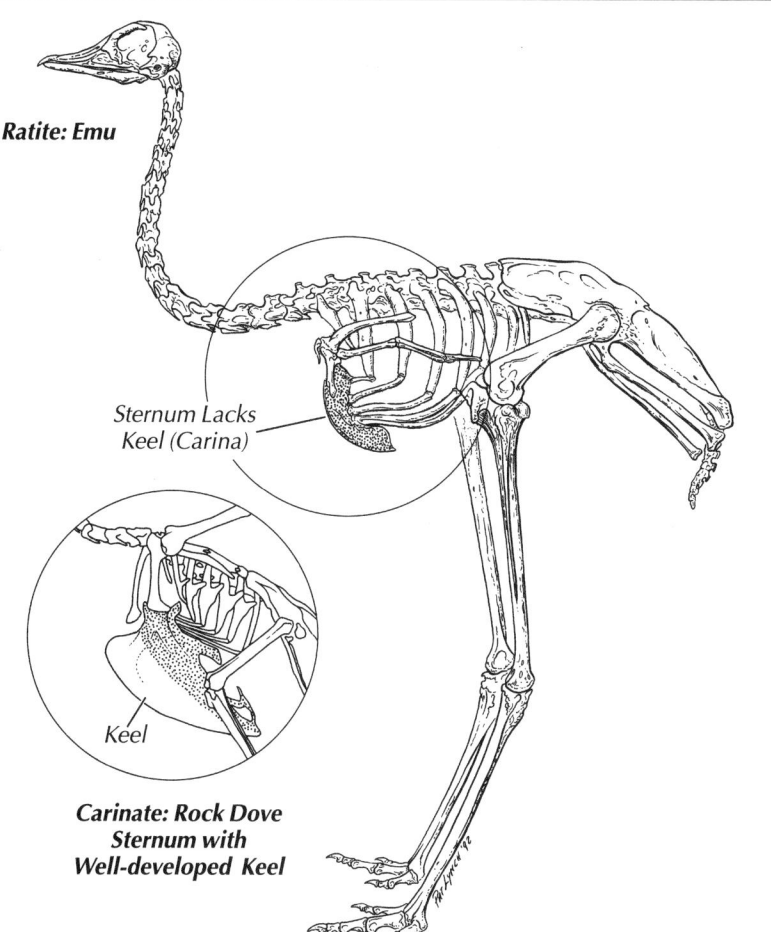

Ratite: Emu

Sternum Lacks
Keel (Carina)

Keel

**Carinate: Rock Dove
Sternum with
Well-developed Keel**

Figure 4–19. The Sternum in a Flying Versus a Flightless Bird: In flying birds, the sternum has a midventral **keel**, or **carina**—a ridge of bone that projects out from the sternum and to which the pectoral flight muscles attach. The size of the keel is closely related to the size of the pectoral muscles and thus reflects the bird's flying ability. Birds with a keel, such as the Rock Dove (inset), are termed **carinates**. In flightless birds, such as the Emu shown here, the sternum is small and shaped like a very shallow bowl. It completely lacks a midventral keel, reflecting the reduced wings and poorly developed pectoral muscles associated with the bird's flightlessness. Birds lacking a keel are termed **ratites**. Note that in both drawings, corresponding parts of the sternum (that is, those not including the keel) are stippled. Emu reprinted from *Manual of Ornithology, by Noble S. Proctor and Patrick J. Lynch,* with permission of the publisher. Copyright 1993, Yale University Press. Inset drawing by Charles L. Ripper.

a fact strongly indicating that their ancestors could fly (see Ch. 5, Loss of Flight). In penguins, the bones are simply shortened, flattened, broadened, and generally strengthened into paddle-like flippers suitable for "flight" under water. Among the ratites (see Fig. 5–48) the wing bones have degenerated in both size and strength, especially in the kiwis of New Zealand, whose wings are mere stubs hidden beneath the feathers.

Sternum

The **sternum** or breastbone of all flying birds has a midventral **keel** or **carina** to which the pectoral breast muscles attach **(Fig. 4–19)**. A general term for birds with a keel is **carinates**. The relative size of the keel is directly correlated with the development of the pectoralis muscle, which provides power for the downward wing stroke, propelling the bird upward and forward in flight. Therefore one can judge, by looking at a bird's sternum, the relative development of its pectoral muscles and consequently its flying ability. A hummingbird has the largest keel of any bird relative to its body size, and is truly the "king" of the carinates. Its flying ability is no surprise. The sternum of large flightless birds such as the Ostrich, Emu, rhea, and cassowary, as well as some others, is flat and plate- or raft-like. Such birds that lack a keel are often collectively spoken of as **ratites** (from the Latin for raft or flat-bottomed boat), but this term may lack taxonomic or evolutionary

Rock Dove

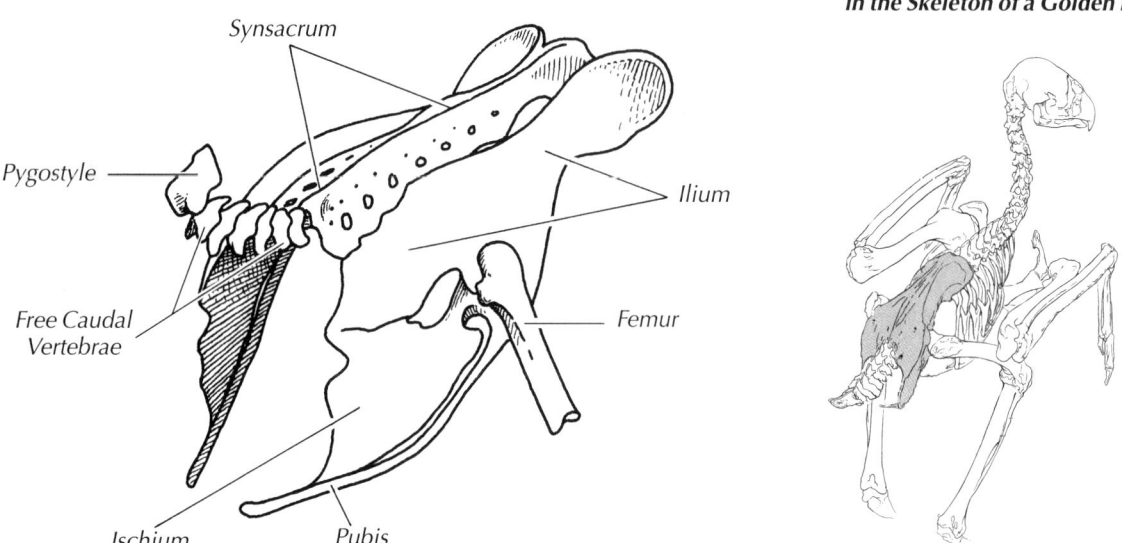

*Location of the Pelvic Girdle and Synsacrum
in the Skeleton of a Golden Eagle*

Synsacrum

Pygostyle

Ilium

Free Caudal
Vertebrae

Femur

Ischium   Pubis

Figure 4–20. The Avian Pelvic Girdle:
The **pelvic girdle**, *shown above in dorsolateral view, consists of three bones on
each side of the body: the* **ilium** *(plural:
ilia), the* **ischium**, *and the* **pubis**. *The ilium forms the cranial and lateral part of
the girdle, and is completely fused with
the ischium, which forms the caudal and
lateral part. The ilium has a cup-shaped
depression for the attachment of the
femur (upper leg bone). The long, thin
pubis runs backward along the outermost edge of the ischium. In the pelvic
region, the vertebrae are fused into the
rigid* **synsacrum**, *consisting of a few
thoracic vertebrae, all the lumbar and
sacral vertebrae, and the first few caudal
vertebrae. The ilia are fused with the synsacrum on either side of the bird's body,
forming a strong but lightweight structure for the attachment of the muscles of
the legs, tail, and abdomen, and providing protection for the abdominal organs.
The right drawing shows the location of
the pelvic girdle and synsacrum within
the skeleton of a Golden Eagle, in dorsolateral view. As a demonstration of how
the form of the pelvis is influenced by
function, compare the size of the pelvic girdle of the ground-dwelling Emu
(Fig. 4–19) with that of a flying bird such
as the Budgerigar (Fig. 4–5). The Emu's
huge pelvic girdle is evidence that legs,
not wings, are its means of transport.
Rock Dove pelvis by Charles L. Ripper.
Right drawing adapted from Proctor and
Lynch (1993, p. 139).*

significance, because it is unclear whether these birds are closely related. The lack of a keel on the sternum of *Archaeopteryx* prompts some
paleontologists to assume that this ancestral bird glided from elevated
perches, and was incapable of flapping flight.

Pelvic Girdle

The **pelvic girdle** or **pelvis** (from the Latin for "basin") is formed by
three bones on each side of the body: the **ilium**, **ischium** (ISK–ee–um),
and **pubis (Fig. 4–20)**. The right and left ilia are fused to the series of
fused vertebrae of that region, the synsacrum, to form a rigid support
for each half of the pelvis. In all birds except rheas, the pelvis is open
below, because the right and left ischia and pubes do not meet. In female birds this open pelvis facilitates the laying of eggs that are large
relative to the size of the parent.

Bones of the Hind Limb

The hind limb, like the forelimb, is composed of a series of bones
articulated end to end **(Fig. 4–21)**. The **femur** or thighbone is relatively
short in all birds, and is usually less than half the length of the next bone
farther down the leg (the **tibiotarsus**) in large wading birds. The small
head of the femur fits deeply into the hip joint's socket or **acetabulum**.
At the lower end of the femur, the **patella** or kneecap (an ossification in
a tendon) glides in a deep groove and adds stability to the knee joint.
When the knee is bent, the patella raises the tendon away from the
knee joint. This increases the tendon's angle of pull on the lower leg,
making the pulling muscle's action more effective.

The **tibiotarsus** (drumstick bone) is usually the bird's longest leg
bone. It is composed of the **tibia** fused at its outer (distal) end with the
proximal tarsal (ankle) bones. The splint-like bone articulating with the
lateral condyle (a process) at the distal end of the femur is the **fibula**,
whose distal end (nearest the ankle) does not ossify in birds.

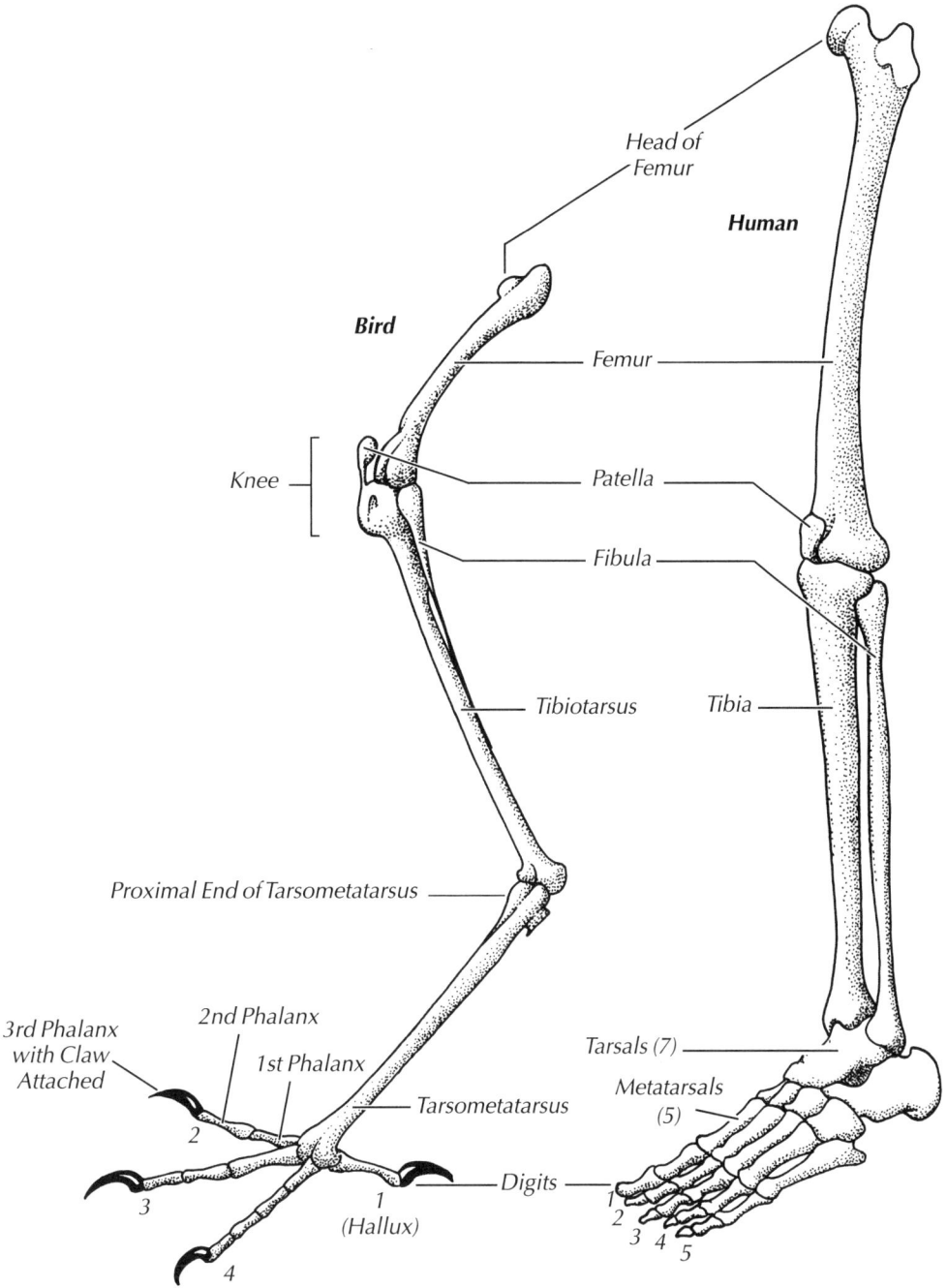

Figure 4–21. The Bones of the Human and Bird (Left) Hind Limb: *The legs of a bird and human are compared to demonstrate the correspondence between the individual bones. The upper leg of the bird resembles that of the human in basic structure, although the lengths of the bones differ. For instance, the* **femur,** *or thighbone, is relatively short in birds, and the* **fibula** *is much reduced. The* **tibiotarsus,** *consisting of the tibia fused to the first few ankle (tarsal) bones, may be very long in some wading birds. The bird's lower leg and foot have been greatly modified through natural selection. The elongated* **tarsometatarsus** *is made up of fused metatarsal bones (the sole of the foot in humans) and the distal tarsal bones (forming the human heel). The proximal end of the tarsometatarsus most closely approximates the heel in humans, yet is elevated such that it may be misidentified as the knee, giving the mistaken impression that a bird's knees point backward! The actual knee is often partly hidden by feathers. All birds lack an outermost fifth toe. The most common arrangement of the four avian toes (digits) is shown here, with one toe facing backward and three facing forward. The rearward-facing toe, termed the* **hallux,** *corresponds to the human big toe (digit 1). As in the fingers, the toes of birds are composed of bones called phalanges. Each toe has one more phalanx than its position number, with the terminal phalanx bearing a claw. The only portions of the terminal phalanges visible in this drawing are the claws. From* The Cambridge Encyclopedia of Ornithology, *edited by Michael Brooke and Tim Birkhead. 1991. Copyright Cambridge University Press, reprinted with permission.*

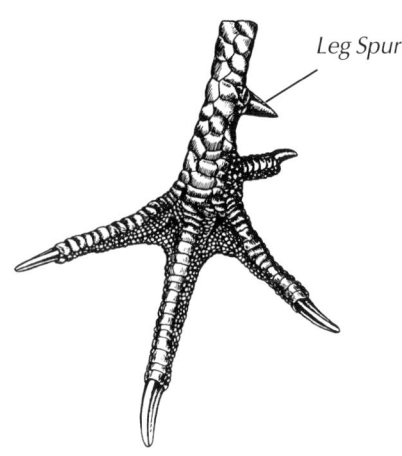

Leg Spur

Figure 4–22. Leg Spur of a Chicken:
Males of certain birds, notably chickens,
peafowl, and other pheasant relatives,
develop bony outgrowths of the lower
*tarsometatarsus known as **leg spurs**,*
which are used as weapons during ag-
gressive interactions with rival males
of the same species. From Lucas and
Stettenheim (1972).

The **tarsometatarsus** represents the fusion of the second, third, and fourth metatarsals (which form the bones of the foot's sole in humans) with the distal tarsal bones. Thus separate tarsal (ankle) bones do not exist in birds, and the joint of the ankle is known as an intratarsal joint.

Leg spurs have developed as weapons of offense or defense in male chickens and other pheasantlike birds **(Fig. 4–22)**. They grow from a spur papilla of the skin that stimulates bone development on the lower caudal surface of the **tarsometatarsus**. Peafowl use their spurs quite aggressively, as do game cocks.

All birds lack the outermost or fifth toe (see Fig. 4–21). The number of bones or phalanges in the four remaining toes is nearly constant, each toe having one more phalanx than its ordinal (position) number. Thus toe number one (corresponding to the innermost, "big toe" of humans), also known as the **hallux**, has two phalanges, toe number two has three, and so on. A number of birds with widely different habits have only three toes, the hallux having been lost evolutionarily. These include such flightless running birds as rheas, Emus, and cassowaries; shorebirds such as the Lesser Golden Plover and Sanderling; and two tree-climbing birds, the Black-backed and Three-toed woodpeckers. In addition, some birds such as the diving petrels and the auks, murres, and puffins—all notable for their ability to swim underwater—also have lost the hallux. The Ostrich is the sole bird with only two toes, as both the outermost and the innermost toes have been lost.

The Muscular System

■ There are three general types of muscles—skeletal, smooth, and cardiac—distinguished by their function, shape, and microscopic structure.

Skeletal Muscle

Skeletal muscles move the bones and constitute what we call the "meat" of an animal, whether it be red as in steak, white as in chicken breast, or any intermediate color. Because skeletal muscle action is under conscious control, skeletal muscles are often called **voluntary muscles**.

Skeletal muscle is a tissue composed of contractile cells (cells that can contract) with nuclei near the cell surface. Skeletal muscle cells have characteristic alternating dark and light striations (stripes) at right angles to the length of the cell, which are visible under a microscope (after the cells are put through a standard laboratory procedure that stains them using dyes that make different substances turn different colors).

The long, cylindrical cells are bound together as **muscle fibers** that can shorten when stimulated by a nerve impulse. Once a muscle fiber contracts it must be forcibly stretched to regain its "resting" length; in other words, muscles cannot "push," they can only "pull." Thus, muscles are generally arranged in opposing pairs; one to stretch the

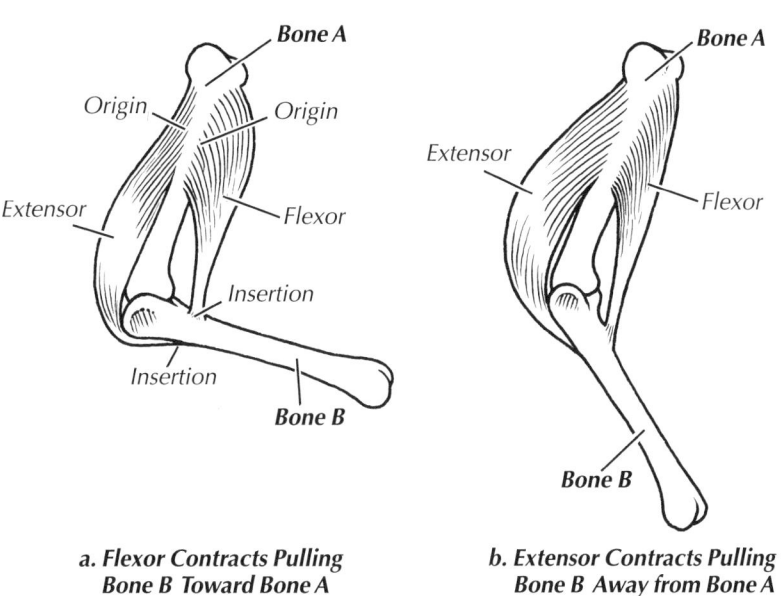

a. Flexor Contracts Pulling Bone B Toward Bone A

b. Extensor Contracts Pulling Bone B Away from Bone A

*Figure 4–23. How Muscles Act to Move Body Parts: Skeletal muscles, through their attachments to bones, are responsible for moving the body parts. Muscles, however, can only pull—they cannot push—so skeletal muscles are arranged in pairs, one opposing the pulling action of the other, to produce smooth movements of the bones. Each muscle usually bridges one or more joints between bones, moving the joint when it contracts. Each muscle has two points of attachment to the skeleton, its **origin** and its **insertion**. The origin is defined as the end of the muscle whose point of attachment moves the least during the muscle's contraction. Muscles are termed **flexors** or **extensors** depending on the actions they exert on the body parts. This figure shows the actions of two hypothetical bones and muscles. a. Flexor Contracts: Bone B is pulled toward Bone A (a movement termed flexion) when the flexor muscle contracts, the flexor becoming shorter and thicker while the extensor muscle relaxes. b. Extensor Contracts: Bone B is pulled away from Bone A (a movement termed extension) when the extensor muscle contracts. At the same time, the flexor muscle relaxes, becoming longer and thinner. Drawing by Charles L. Ripper.*

other and perform the opposite action on their portion of the body **(Fig. 4–23)**. When muscles contract they produce both movement and heat—actually more heat than work! In effect, muscles are the furnaces of the body, and the heat they produce is distributed by the blood moving through the circulatory system. **Shivering**—uncoordinated muscle fiber contraction—is a way to produce heat without directed movement by muscle contraction. To keep the heat generated by exercise or shivering from escaping the body, birds may fluff their feathers, trapping a layer of insulating air.

Each skeletal muscle consists of several hundred to several thousand muscle fibers bound together by connective tissue called **fascia**. These bundles of fibers have two sites of attachment to the skeleton (and occasionally to other structures); one is called the **origin**, the other, the **insertion**. Usually a muscle bridges one or more joints, producing movement at the joint when it contracts. By convention, the end of the muscle whose point of attachment moves least during contraction is designated as the origin. The connecting fascia may be in the form of a **tendon** (which in bird limbs may ossify) or in the form of shiny, broad sheets called **aponeuroses**. Every muscle is innervated (supplied with nerves) and kept alive by these nerves. If the nerve to a muscle is cut, the muscle will eventually shrivel and **atrophy** (die) unless new nerve fibers grow into it, a process that usually takes several weeks.

Skeletal muscles are often given names that indicate their function, location, shape, or derivation. Most anatomical terms were first created to describe human anatomy, then uncritically applied to birds at a later date. Thus, many inappropriate names for bird muscles have had to be changed to more properly reflect their evolutionary origin and avian function (see Baumel et al. [1993]—the *Handbook of Avian Anatomy*—listed with references).

Skeletal muscles that move the wings and limbs **(Figs. 4–24 and 4–25)** act antagonistically, so that when one contracts, the other relaxes in a continuous fashion, producing smooth rather than jerky movement. (Handy examples are the muscles of our upper arm. When we raise a cup to our lips, we bend (flex or close) the elbow joint by contracting the biceps muscle that crosses the inner or flexor surface of the elbow. In the process, the antagonistic triceps muscle—which crosses the outer or extensor surface of the elbow joint—gradually relaxes, allowing the arm's extension to be continuous and smooth. The reverse is true as well. When we lower the cup and open the elbow joint, the triceps muscle contracts, while the biceps relaxes.)

The large breast muscles of birds, the **pectoralis** and **supracoracoideus,** are good examples of antagonistic muscles (see Fig. 5–6). Although they lie in similar positions on the keel of the sternum, they have different actions. This is because the tendon of the supracoracoideus passes through the **foramen triosseum** ("foramen" means "hole"), an opening at the shoulder insertion, forming a pulley system that redirects the force of the supracoracoideus. The pectoralis attaches to the ventral surface of the humerus and draws it and the wing downward. The tendon of the supracoracoideus passes through the foramen triosseum and inserts on the *dorsal* surface of the humerus so it can *raise* (elevate) the wing. Thus the two primary opposing muscles for flight, the pectoralis for the downstroke and the supracoracoideus for the upstroke, both originate on the sternum, one on top of the other.

Even in the chicken and turkey, not known for their flying prowess, the pectoral muscles account for about one-fifth of the bird's weight. Because of the pulley function of the foramen triosseum, the large flight-powering muscle mass can be carried entirely below the supporting wings during flight—a considerably more stable weight configuration than if the wing elevator muscle were situated on the back, above the wings. Because it takes the most force to get lift (and thrust) from the wing's downstroke, the pectoralis muscle is the largest muscle in flying birds. It is placed superficially, thus its bulging during contraction is not hampered by overlying muscle. For more information on the flight muscles of birds, see Ch. 5, Functions of the Flight Muscles.

Refer to **Figure 4–26** for more information on the major skeletal muscles of birds.

Smooth Muscle

Smooth muscle, also known as **involuntary muscle** because it is not controlled consciously, is composed of spindle-shaped cells with a centrally located nucleus and no evident pattern of striations along the cell. Smooth muscles are found in the walls of hollow organs such as the stomach and blood vessels larger than capillaries. Smooth muscle cells are especially characteristic of the vessels of the arterial system, but also are found in the venous system. Smooth muscles are innervated by a separate, non-voluntary (**autonomic**) portion of the nervous system and also are under direct chemical control from

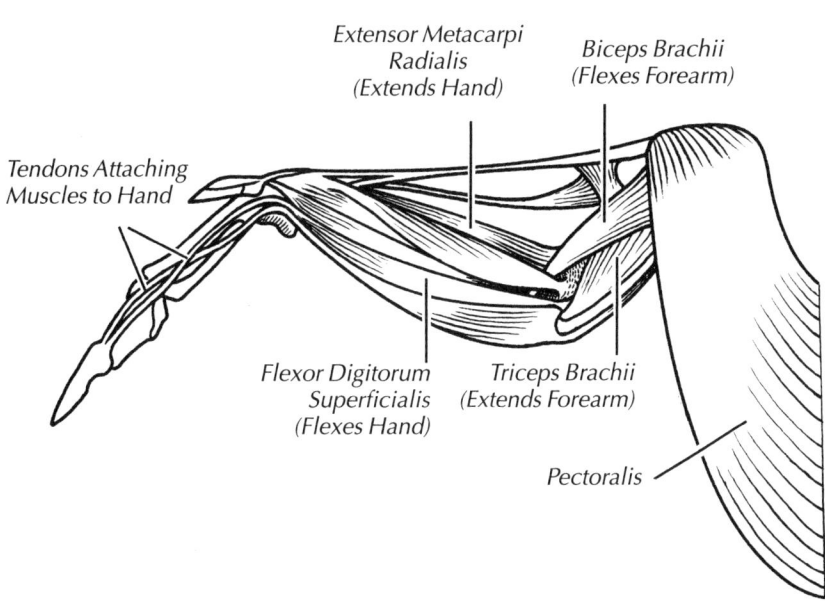

Tendons Attaching
Muscles to Hand

Extensor Metacarpi
Radialis
(Extends Hand)

Biceps Brachii
(Flexes Forearm)

Flexor Digitorum
Superficialis
(Flexes Hand)

Triceps Brachii
(Extends Forearm)

Pectoralis

Figure 4–24. Selected Muscles of the Avian Wing: In this drawing, only some of the muscles are labeled. As is the convention in anatomical texts, the Latin names, which reflect a muscle's function and points of attachment, are used. The **biceps brachii** originates from the shoulder region and inserts on the upper surface of the radius and ulna near the elbow, and flexes the forearm. Its antagonist (the muscle that opposes its action), the **triceps brachii**, also originates from the shoulder region, but inserts on the under surface of the ulna near the elbow, and extends the forearm. The **extensor metacarpi radialis** originates above the elbow, inserts on the carpometacarpus by a long tendon, and extends the hand. Its antagonist, the **flexor digitorum superficialis**, originates below the elbow, inserts on the carpometacarpus, and flexes the hand. Drawing by Charles L. Ripper.

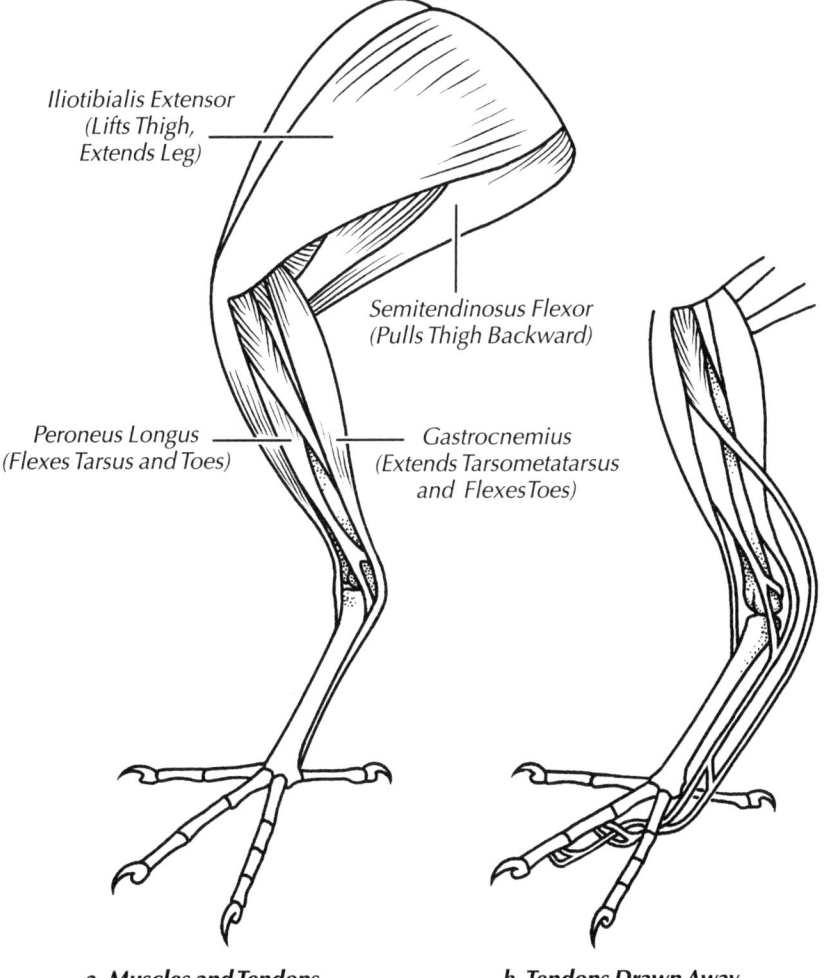

Iliotibialis Extensor
(Lifts Thigh,
Extends Leg)

Semitendinosus Flexor
(Pulls Thigh Backward)

Peroneus Longus
(Flexes Tarsus and Toes)

Gastrocnemius
(Extends Tarsometatarsus
and Flexes Toes)

**a. Muscles and Tendons
in Natural Position**

**b. Tendons Drawn Away
from the Skeleton
to Show Their Insertions**

Figure 4–25. Selected Muscles of the Avian Leg: Shown here are some of the bird's leg muscles and their arrangements for moving the legs and feet. **a. Muscles and Tendons in Natural Position:** The **iliotibialis extensor**, originating from the pelvic girdle, inserts at the knee where it lifts the thigh away from the midline of the body and extends the leg. The **semitendinosus flexor**, also arising from the pelvic girdle, inserts on the distal end of the femur and pulls the thigh backward. The **peroneus longus** originates at the proximal end of the tibiotarsus, and the **gastrocnemius**, from the distal end of the femur. These and other muscles coming from the vicinity of the knee insert by long tendons on the tarsometatarsus and phalanges, moving the tarsus and toes. **b. Tendons Drawn Away from the Skeleton to Show Their Insertions:** The extremities of the hind limb, like those of the wing, are controlled by these tendons, like the strings of a puppet, from muscles located closer to the bird's center of gravity. Drawing by Charles L. Ripper.

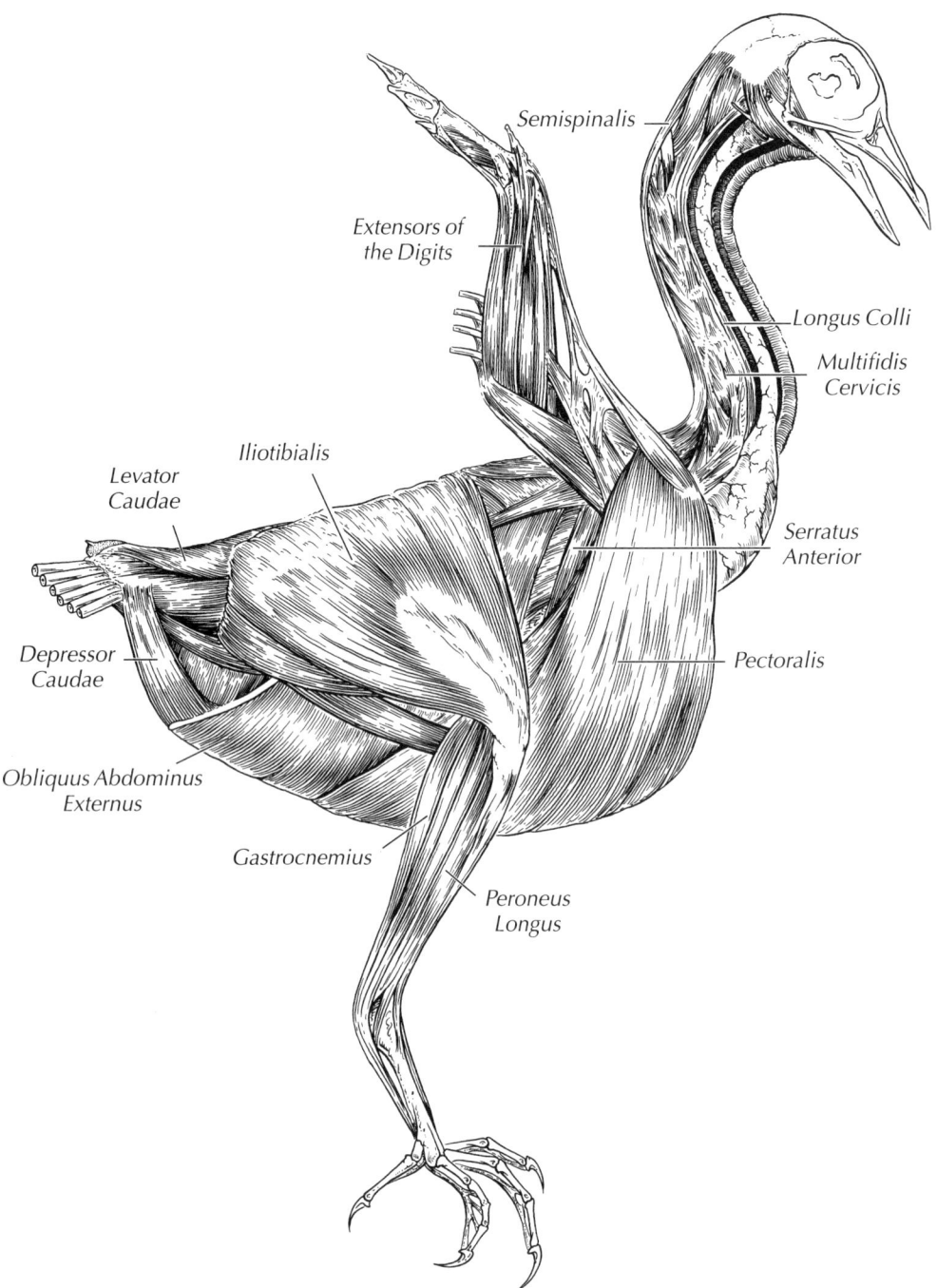

Semispinalis

Extensors of
the Digits

Longus Colli

Multifidis
Cervicis

Iliotibialis

Levator
Caudae

Serratus
Anterior

Pectoralis

Depressor
Caudae

Obliquus Abdominus
Externus

Gastrocnemius

Peroneus
Longus

Figure 4–26. Selected Muscles of the Rock Dove: *This drawing shows many of the superficial muscles of the right side in lateral view. Notice the complex network of small muscles along the neck, the* **multifidis cervicis**, *each surrounding and controlling a cervical vertebra and thus contributing to the neck's flexibility. Long muscles, the* **semispinalis** *and the* **longus colli**, *move the neck up and back, and down and forward, respectively. In flying birds, such as this Rock Dove, the* **pectoralis** *and the underlying* **supracoracoideus** *muscles, which provide the power for flapping the wings, make up between 20 and 30 percent of the body weight, and their importance can clearly be seen here. The muscles of the thorax, for instance the* **serratus anterior** *and others obscured by the uplifted wing in this drawing, support the rib cage, provide the power for breathing, and help to attach the pectoral girdle to the body. The muscles of the abdomen, such as the* **obliquus abdominus externus**, *lie in sheets at right angles to each other, providing strength and protecting the underlying viscera. The pelvis and synsacrum provide a strong base for the attachment of the thigh muscles, such as the* **iliotibialis**. *They also firmly anchor the tail muscles, such as the* **levator** *and* **depressor caudae**, *permitting the tail to function as a powerful rudder and brake during flight. Notice that the large muscle masses are located primarily ventrally, below the wings and near the bird's center of gravity, providing a stable arrangement for flight. The muscles controlling wing and leg movement also are concentrated near the center of the body, moving the limbs via a system of long tendons. Reprinted from* Manual of Ornithology, *by Noble S. Proctor and Patrick J. Lynch, with permission of the publisher. Copyright 1993, Yale University Press.*

substances circulating in the blood. Smooth muscle is also found in the respiratory and urogenital systems, in addition to the digestive and circulatory systems already mentioned—all systems and organs that are concerned with vital life processes and over which the bird has little or no control. Smooth muscle occurs in the skin for the movement of feathers in birds, and for the movement of hair in humans—sometimes causing "goose flesh." Smooth muscle is also essential in the eye for changing focus.

Cardiac Muscle

Cardiac muscle is a special type of smooth muscle that forms the bulk of the heart. The muscle fibers are arranged in a fused network and have cross-striations but centrally located nuclei.

Cardiac muscle has an **innate rhythmicity**—the ability to contract without being stimulated by nerves. Actually, the heart of an embryo begins to beat rhythmically before any nerves have grown to reach it. The nerves that do reach the heart are part of the autonomic nervous system (see later in this chapter), but they do not start the contractions of the heart. Instead, they regulate and modify the rate of the beat.

The Nervous System

■ The nervous system is responsible for all the bird sees, hears, smells, tastes, feels, thinks, and does. Thus the nervous system transmits sensory stimuli, evokes appropriate motor responses, and regulates all internal body functions. The structural parts of the nervous system are similar in all vertebrates, but they differ in their degree of complexity. Mammals have the most complex brains of all vertebrates, whereas the avian brain has traditionally been considered less complex. Calling someone a "bird brain" is generally not intended to be complimentary! However, rather than being less complex overall, the bird's brain is differently organized than that of a mammal. In other words, the brain of birds is composed of the same basic "components" as that of mammals—due to inheritance of the basic structure from our common ancestor—but the avian brain is "wired" differently.

To appreciate how any animal perceives the outside world, we must consider its simple sensory nerve endings and complex sense organs, which are constantly gathering and transmitting information about internal and external conditions. Birds have some sensory capabilities for species recognition and orientation, especially in migration and homing, which we do not yet fully understand (see Ch. 5, Orientation and Navigation). These include the bird's ability to see ultraviolet light, and the probability that they can hear infrasound and ultrasound.

The **central nervous system** consists of the **brain** and **spinal cord**. The **peripheral nervous system** consists of bundles of nerve cell fibers, called **nerves**, and collections of nerve cell bodies clustered in aggregations called **ganglia**. The **cranial** and **spinal nerves** serve various, very specific parts of the body. The **autonomic nervous system** consists of

Figure 4–27. Neuron Structure: *The* **neuron** *or nerve cell is the basic component of the nervous system. The two fundamental types of neurons, sensory and motor, differ in their functions, but have the same basic structure. Each consists of a* **cell body**, *many rootlet-like extensions called* **dendrites**, *and a very long, cable-like extension called the* **axon**, *which is surrounded along its entire length by a fatty, insulating layer called the* **myelin sheath**. **Motor neurons** *carry nerve impulses away from the central nervous system to the muscles and organs, resulting in muscle movement or organ activity.* **Sensory neurons** *carry nerve impulses to the central nervous system from muscles, sensory organs, and receptors scattered throughout the body. Drawing by Charles L. Ripper.*

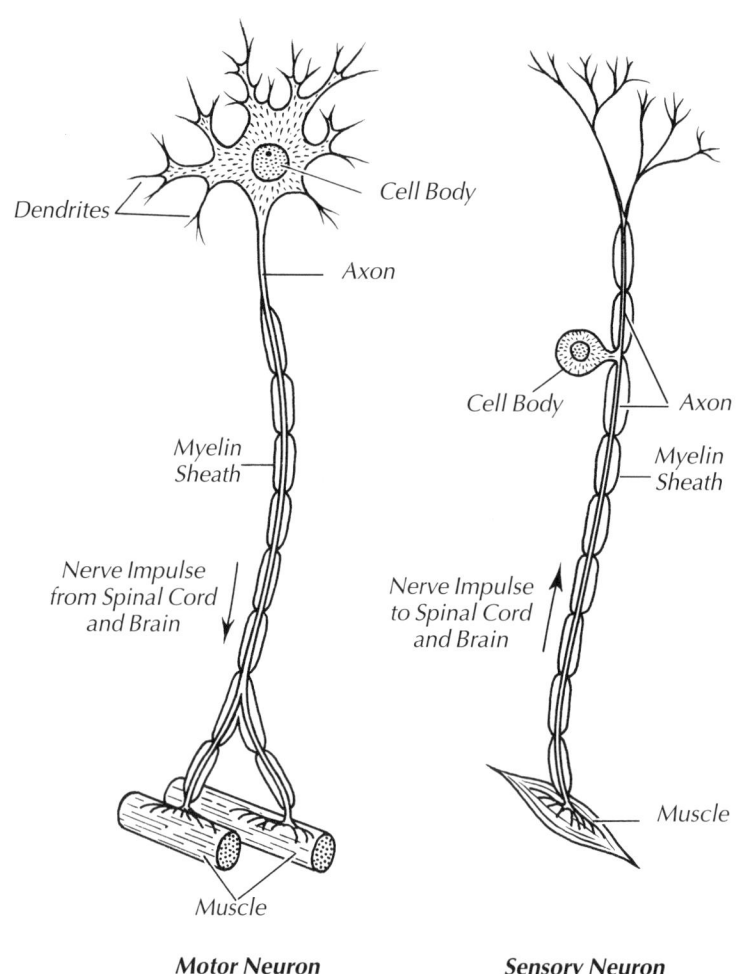

those nerves that regulate the smooth muscle of the viscera, glands, and blood vessels. One cannot tell the difference between cranial, spinal, and autonomic nerves by any means other than their origin, destination, and function. Physically, all nerves look alike.

The Neuron

The basic unit of the nervous system is the **neuron** or **nerve cell** (**Fig. 4–27**). What we call a **nerve** is a collection of specialized portions of many nerve cells, surrounded and bound together by connective tissue (cells that give support and protection), and large enough to be seen by the naked eye. The simplest neurons have a cell body with long, rootlet-like extensions (called **dendrites**) at one end, and a cable-like extension (called the **axon**) issuing from the other end. Bioelectric impulses (nerve impulses) travel along neurons, from one end to the other. A bioelectric impulse consists of a wave of change in electrical charge sweeping along the surface of the neuron due to ion movement across the cell membrane. Transmission of an impulse from one neuron to another takes place across a small gap, or **synapse**, usually from the axon of one neuron to one of the dendrites of another. The cell body of a neuron is microscopic in size, yet its axon, also microscopic in diameter, may be several feet in length. A wrapping of cells containing fatty material called **myelin** surrounds many axons. These myelin sheaths function, in part, like the insulation around household wiring. Because the fatty coats appear pale or white, nerves stand out visually from most other tissues.

The deposition of myelin (**myelination**) in the nerve sheath is important in the functioning of the nerve fibers. In humans, myelination begins in the embryo, generally in the more primitive nerve paths, reaching the more advanced paths after birth. For example, myelination of the most important motor pathway—the nerves that govern walking—begins after birth and proceeds most rapidly between the ages of 12 and 16 months. The human infant begins walking when myelination of that motor pathway develops to a certain point, and not before; so a child cannot be taught or forced to walk until his nerves and muscles are capable. We assume the process is similar in birds. The parent bird never "teaches" the nestling to fly. The young bird flaps its wings, stretches its legs, stands in the nest, and flies only when the nerves and muscles have developed to a certain point. After myelination of the nerves governing flight is complete, nothing, except confinement or injury, will keep the bird from flying.

Sensory and Motor Neurons

Two types of neurons are distinguished, not on the basis of structural differences, but on the function and direction of the impulses they carry (see Fig. 4–27).

A **sensory neuron** conveys impulses *to* the spinal cord and brain. These impulses are interpreted as sensations, which may be conscious—such as visual images of food items, sounds of a predator, or pain from an injured wing—or they may be subconscious impulses from muscles, tendons, and joints, informing control centers of the position of the limbs and muscles. This gives the bird a "body-parts-position sense," and allows it to stand on one leg or fly without visual cues. These latter impulses are called **proprioceptive**, meaning "muscle sense" or "tendon sense." Position sense is very important for proper functioning of skeletal muscles. Because of the proprioceptive neurons, you can move your arms and legs with your eyes closed and still know their exact position. Imagine how important these neurons and the knowledge they provide must be to a bird flying at night, or to species that live deep in caves.

Motor neurons convey impulses *from* the brain and spinal cord to stimulate a muscle to contract or permit it to relax, or to cause a gland to secrete. For example, upon seeing a banana split, most of us feel a sensation of delight resulting from sensory neurons sending an image message to the brain. The brain will likely send a command by motor neurons to the salivary glands, making our mouth water even before the first bite.

The motor nerves stimulating skeletal muscles are called voluntary motor nerves because they are under conscious control. They are of vital importance to the well-being of the muscle. If a motor nerve to a skeletal muscle is cut or dies of polio virus, the muscle is paralyzed and no longer can be moved voluntarily. With the passage of time, such denervated muscle wastes away, and its fibers degenerate unless they are kept alive by regrowth of nerve or artificial electrical stimulation. Destruction of nerves within the brain and spinal cord is usually permanent; overall, little or no regeneration occurs, although some clusters of neurons deep in the brain show some small regeneration potential. If an axon outside the brain and spinal cord is severed, however, it will regenerate. The growth is slow, about 0.04 inches (one millimeter) per day in humans. The length of time between the loss of innervation and the re-establishment of nerve contact determines whether the muscle will function again. A wild bird is so dependent on flight that it will die from an accidentally denervated wing before any significant regeneration of nerves can occur.

The existence of sensory and motor neurons provides a basis for the simplest kinds of behavioral actions, called **reflexes (Fig. 4–28)**, which may be either automatic or learned. Sensory neurons carry impulses from all parts of the body to the central nervous system. There, neural activity is transferred by synapse to a motor neuron that stimulates instant action (the reflex) in the muscle or gland it governs. One example is the speed with which we withdraw our hand from a hot stove; the arm muscles automatically contract to withdraw the hand before our brain tells us the stove is hot. Identical automatic reflexes protect the bird from injury; "learned" reflexes provide greater scope for other, more complex activities. For example, flying is difficult for a young bird just out of the nest; it has trouble taking off and landing properly. Gradually, with continued practice, the many muscular con-

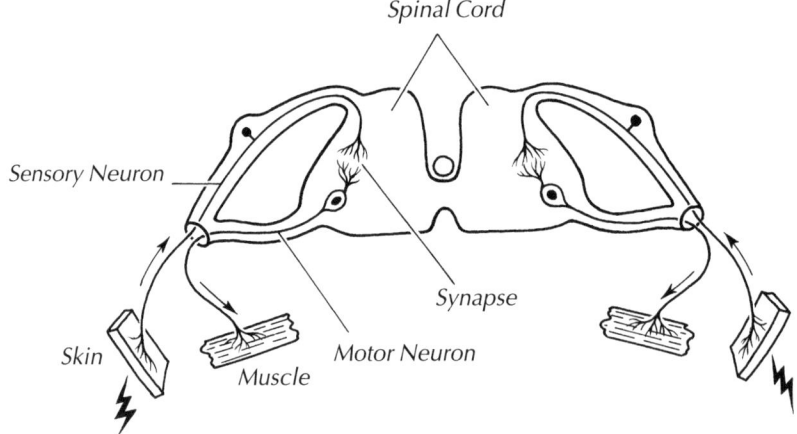

Figure 4–28: A Simple Reflex: A nerve impulse from the skin is carried by a sensory neuron to the spinal cord. There it is transmitted across a synapse between the sensory dendrites and the dendrites of a motor neuron, through which it travels to a muscle where it stimulates instant, automatic action. An example of a simple reflex is the unconscious way we pull our hand away from a hot surface; our arm muscles contract automatically to withdraw our hand before our brain tells us the surface is hot. Drawing by Charles L. Ripper.

tractions necessary for flight are coordinated. Then the bird can turn its attention to other things—to catching food, watching for predators, and even to "navigating," a complex process that we still do not fully understand.

Reflexes are usually much more complex than we have described, because the motor and sensory neurons involved have interneurons within the spinal cord interposed between them **(Fig. 4–29)**. Interneurons allow for transfers up and down the cord to other regions, as well as between the left and right sides of the body. Furthermore, the

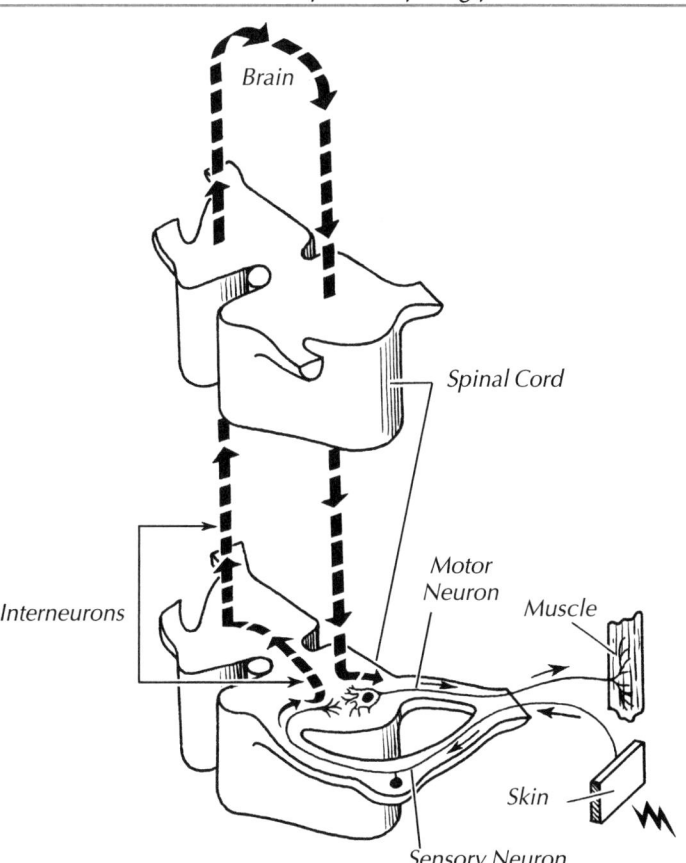

Figure 4–29: A Complex Reflex: *A nerve impulse from the skin is carried by a sensory neuron into the spinal cord. Through synapses with interneurons within the spinal cord, the impulse may be transmitted from one side of the body to the other, up and down the cord, or to and from the brain. Interpretive input from the central nervous system, for instance in the form of experience and memory, may modify the reflex action. Drawing by Charles L. Ripper.*

neurons involved in reflexes connect in the spinal cord with additional neurons that carry sensations up to the brain. They also synapse with still other neurons that bring impulses directly from the brain. Brain input often modifies a reflex action. For example, when a professional chef grabs a hot pan, he may not release it instantly in a reflex action as most people would. Instead, his brain may tell him to hold on even though it is hot, because it is important to the success of his job.

"Wiring diagrams" of the nervous system from its inputs to its outputs thus become exceedingly complex, even at these relatively simple reflex levels of neural integration and behavior. Imagine what occurs in the nervous system of a Clark's Nutcracker as it recalls the location of a pine nut stashed months earlier, and then flies directly to the site and retrieves it! (See Figs. 6–12 and 6–13.)

Central Nervous System

The **central nervous system** (CNS) consists of the **brain** and the **spinal cord**. Both are composed of millions of neurons receiving sensory information, relaying it to many other CNS centers, and sending out motor impulses. Within the CNS, clusters of nerve cell bodies (equivalent to the ganglia of the peripheral nervous system [PNS]) are called **nuclei**, and bundles of axons and their myelin sheaths (equivalent to PNS nerves) are called **tracts (Fig. 4–30)**. The many adjacent tracts within the CNS are known as "white matter" because of the color

Figure 4–30. Organization of the Vertebrate Nervous System: The brain and spinal cord together are termed the **central nervous system** (CNS). The **peripheral nervous system** (PNS) consists of all other nervous system structures—the cranial and spinal nerves and their associated ganglia. Within the CNS, the axons (fibers) of neurons are gathered into **tracts**, whereas clusters of neuron cell bodies are termed **nuclei**. The collective name for CNS tracts is "white matter," from the white color of the axons' myelin sheaths. The collective name for CNS nuclei is "gray matter," from the darker color of cell bodies. Within the PNS, the terminology is different. Neuronal axons are gathered into **nerves**, bundles surrounded by a protective layer of connective tissue that are often large enough to be visible to the naked eye. Neuron cell bodies are collected into rounded aggregations called **ganglia** (singular, ganglion).

CENTRAL NERVOUS SYSTEM (Brain and Spinal Cord)	PERIPHERAL NERVOUS SYSTEM (All Other Nervous System Structures)
Neuron Axons form *Tracts* *Tracts* collectively form *White Matter*	**Neuron Axons** form *Nerves*
Neuron Cell Bodies form *Nuclei* *Nuclei* collectively form *Gray Matter*	**Neuron Cell Bodies** form *Ganglia*

of the myelin sheaths. Areas without myelin, where nerve cell bodies are concentrated, are darker and called "gray matter." In addition, millions of non-neural cells (**neuroglia**) form a supporting and protective felt-like bed for the neurons. Surrounding the brain and spinal cord are vascularized membranes called **meninges**. They consist of an outer fibrous **dura** and inner **arachnoid** and **pia** layers, tightly applied to the surface of the brain and cord (see Fig. 4–33). This complex of non-neural tissues provide sustenance and waste removal for the cells of the brain and spinal cord—vital functions because no blood vessels penetrate these organs to perform those duties. At the same time the meninges establish a **blood/brain barrier** that protects the delicate CNS from many potentially toxic substances circulating in the body.

Brain

The brain of a bird is short, bulbous, and very large in relation to the size of the skull **(Fig. 4–31)**. Most of the brain lies caudal to the orbits, and it fills the cranial cavity completely. The skull is thin and the neck vertebrae exceptionally flexible, therefore the brain and cervical spinal cord can be injured easily, as illustrated by the number of birds that die after flying headlong into windows. The main features of the bird brain are similar to those of the mammal brain: the two large, smooth **cerebral hemispheres** of the **forebrain**; the two large **optic lobes** of the **midbrain**; and a single large, median **cerebellum** of the **hindbrain,** with typical transverse folds. Between the forebrain and the midbrain is a region that serves as the brain's central switchboard for all incoming and outgoing nerve impulses. The ventral portion, the **hypothalamus** (which contains the **pituitary gland)**, plays a major role in the hormonal control of body processes (the endocrine system).

The forebrain, specifically the forwardmost portion known as the **telencephalon,** primitively associated with the sense of smell (olfaction), is narrow rostrally and wide caudally where it overlaps and partially hides the more caudally positioned optic lobes. The forebrain of modern birds consists of two smooth cerebral hemispheres, pointed at their rostral ends where the olfactory nerves from the nasal cavity enter. The cerebral hemispheres are coordinating and control centers for most of the bird's complex behaviors, including memories and learning. Of all birds, the largest olfactory area is that of kiwis—primitive,

a. Location of the Brain Within the Skull

Rock Dove

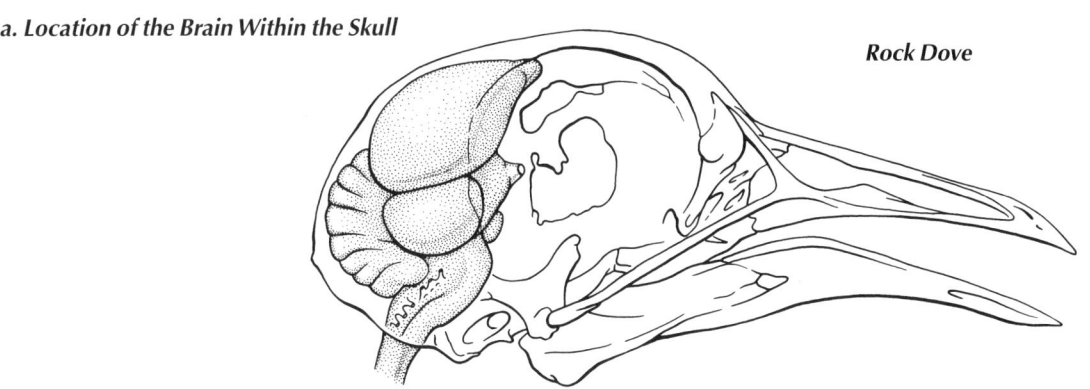

b. Structure of the Brain

Cerebral Hemisphere

Rock Dove

4

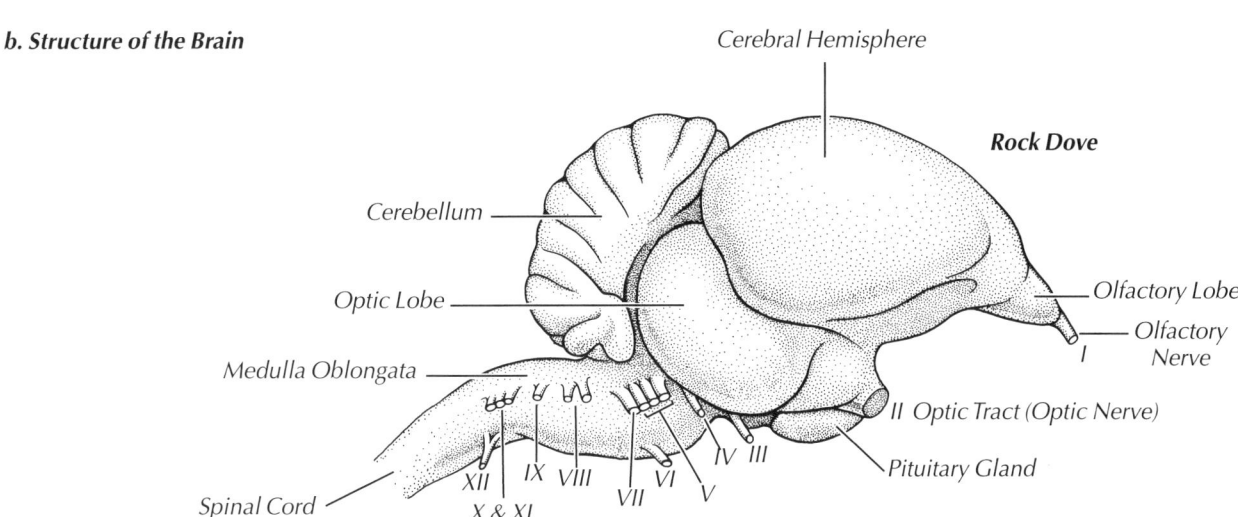

Cerebellum

Optic Lobe

Medulla Oblongata

Olfactory Lobe

Olfactory Nerve

I

II Optic Tract (Optic Nerve)

Pituitary Gland

XII

IX *VIII*

VII

VI

V

IV *III*

Spinal Cord

X & XI

I Through XII = Cranial Nerve Endings

Figure 4–31. Location and External Structure of the Brain: a. Location of the Brain Within the Skull: *The bird's brain is positioned toward the rear of the skull, behind the large eye sockets. The orientation of the brain in the skull varies considerably, but in many species it is oriented nearly vertically, as in the Rock Dove shown here, in contrast to the classic mammalian orientation in which the brain lies horizontally.* **b. Structure of the Brain:** *The Rock Dove brain is shown here in lateral view. Dominating the forebrain are the large, smooth* **cerebral hemispheres,** *which coordinate and control complex behaviors. At the anterior end of the hemispheres are the small* **olfactory lobes,** *concerned with the sense of smell. The midbrain is dominated by the large, paired* **optic lobes,** *which receive the optic tracts from the eyes. Upon exiting the eyes, the optic tracts partially cross each other within the* **optic chiasma** *(not visible) before entering the optic lobes. Thus part of what the right eye sees goes to the left side of the brain, and vice versa. The* **pituitary gland** *is attached to the ventral side of the brain by a stalk, which is not visible here. On the dorsal surface of the hindbrain is the large, deeply folded* **cerebellum,** *which controls muscular coordination, and has important roles in balance, posture, and proprioception (the sense of the position and activity of the limbs). Ventral to the cerebellum is the* **medulla oblongata,** *where the nuclei of most of the* **cranial nerves** *are located, and from which these nerves extend out to the head, neck, and thorax. The cranial nerves are numbered I through XII; only their ends are illustrated. The medulla oblongata (also called the brain stem) extends caudally to become the spinal cord. Modified from Manual of Ornithology, by Noble S. Proctor and Patrick J. Lynch, with permission of the publisher. Copyright 1993, Yale University Press.*

Lateral View, Right Side

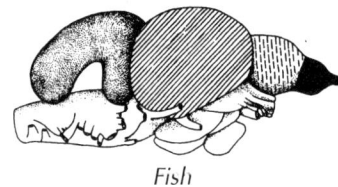

Fish

Amphibian

Reptile

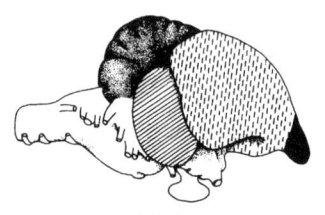

Bird

Mammal

- Cerebral Hemisphere
- Cerebellum
- Optic Lobe (barely visible in mammal)
- Olfactory Bulb

flightless birds of New Zealand. They are the only birds that have their nostrils at the tip of the beak rather than nearer the base of the beak. This position facilitates efficient olfaction, as they feed by probing into the forest leaf litter. The tubenosed seabirds and some New World vultures also have special nasal passages and/or relatively large olfactory areas and excellent senses of smell. Most birds, however, appear to have few olfactory talents, depending much more on vision.

The bird's midbrain or **mesencephalon** is the region where visual inputs are regulated and sent to other parts of the brain for integration into a response to what has been seen. It consists primarily of two greatly enlarged optic lobes, which receive the **optic tracts** from the eyes. These connections between the eyes and the brain are called tracts rather than nerves, because they (and the sensitive layers of the eyes) are outgrowths and modified extensions of the brain rather than bundles of peripheral sensory axons. Exiting the eyes, the tracts cross each other before entering the brain. Thus, what the right eye sees goes to the left side of the brain and vice versa, as in humans. The large optic lobes of birds are relatively much larger than their corresponding part in mammals **(Fig. 4–32)**. On the other hand, the auditory (hearing) and vestibular (balance) components of the avian midbrain are not as conspicuous as they are in mammals. At its caudal end, the mesencephalon is joined to the cerebellum.

The **cerebellum** is attached to the dorsal side of the brain stem (**medulla oblongata**) by two pairs of stout neural tracts. The cerebellum controls posture and the movements of the legs and wings; it thus regulates the highly complex muscular actions necessary for flight. The bird's cerebellum is large relative to other portions of the brain, when compared to mammals, doubtless because of the demands of flight. Much of the medulla oblongata is hidden from dorsal view by the large cerebellum. The medulla extends caudally through the foramen magnum, the large opening at the rear of the skull, where the medulla bends sharply and narrows to become the spinal cord. Important nerves have their nuclei (recall that these are aggregations of cell bodies) within the medulla, and extend from it out to the head, neck, and thorax.

Spinal Cord

The **spinal cord** passes through the neck and trunk in the protective, but in some regions highly flexible, **vertebral canal** formed by the vertebral arches of successive vertebrae (**Fig. 4–33;** also see Fig. 4–15). Essentially, the spinal cord is a cable of neurons conducting impulses to and from the brain. Independently of the brain, however, reflexes within the cord control many of the bird's muscular functions

Figure 4–32. Comparison of Bird and Other Vertebrate Brains: *Shown here are the brains of a bony fish, an amphibian, a reptile, a bird, and a mammal, in lateral view, with the parts of the brain shaded identically in each. Differences in the proportions of various brain parts reflect the varying abilities, needs, and habits of the different vertebrate groups. For example, the optic lobes of birds are relatively much larger than those of mammals, whose optic lobes are obscured by the large, complex cerebral hemispheres. This shows the importance of sight to birds. The bird's cerebellum is also relatively large compared to that of mammals, reflecting the importance of balance during flight. Note the relative differences in the sizes of the olfactory lobes: while small in birds, they are well-developed in fish, reptiles, and mammals, evidence that these latter three have a sense of smell superior to that of birds. Adapted from Pough, Janis, and Heiser (1999, p. 102).*

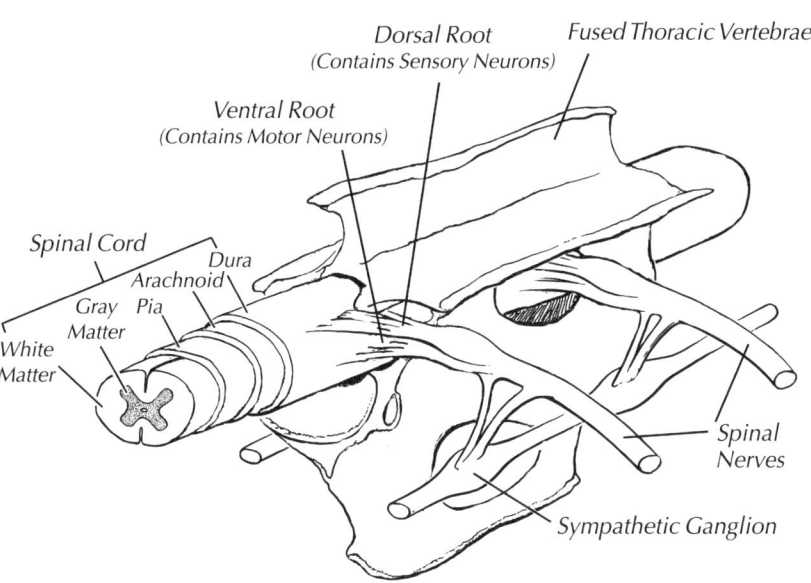

Dorsal Root
(Contains Sensory Neurons)

Fused Thoracic Vertebrae

Ventral Root
(Contains Motor Neurons)

Spinal Cord
Arachnoid
Gray   Pia
Matter
White
Matter

Dura

Spinal
Nerves

Sympathetic Ganglion

Figure 4–33. The Spinal Cord in Relation to the Vertebrae: *The spinal cord lies within the protective confines of the vertebral canal, a bony tube formed by the interconnecting vertebrae. This drawing shows the section of spinal cord within part of the thoracic region of the vertebral column. Surrounding the spinal cord are the meninges, vascularized membranes that nourish the nervous tissue. They consist of a fibrous outer layer, the* **dura,** *and two inner layers, the* **arachnoid** *and* **pia.** *The spinal cord itself can be seen in cross section, showing the* **white matter** *made up of nerve axons, and the* **gray matter,** *made up of nerve cell bodies. At regular intervals along its length, the spinal cord gives off bundles of sensory and motor neurons that exit from the cord in pairs between the vertebrae; these are termed the dorsal and ventral nerve roots, respectively. These bundles occur symmetrically on both the left and right sides of the spinal cord. The members of each dorsal-ventral pair of roots immediately combine to form a single* **spinal nerve,** *which innervates nearby muscles or organs. Associated with the spinal nerves is a chain of sympathetic ganglia running parallel to the spinal cord on each side. These ganglia contain the cell bodies of sympathetic neurons, part of the autonomic nervous system (see Fig. 4–35). Adapted from Proctor and Lynch (1993, p. 247).*

involved in flying and walking. Because of this independent spinal cord function, when a bird is decapitated, its legs and wings continue to function for a brief period (the proverbial "running around like a chicken with its head cut off").

Along its course, the spinal cord gives off spinal nerves between the vertebrae. Each spinal nerve has separate sensory and motor segments that exit from the cord, but they are combined into one spinal nerve outside the vertebral canal. At both the level of the wings and the hind limbs, the spinal nerves are large and join each other to form complex web- or net-like structures, called **plexuses**, outside the vertebral canal **(Fig. 4–34)**. The **brachial plexus** of the wing is associated with a cervical enlargement of the spinal cord within the vertebral canal. The **lumbosacral plexus** of the hind limb likewise corresponds to a lumbosacral enlargement of the spinal cord. Whether one of these spinal cord swellings is the same size or larger than the other depends on a bird's main type of locomotion. In a bird that relies on walking, such as a kiwi, the number and size of the nerves going to and from the legs is certain to be large, as are the lumbosacral enlargements of the spinal cord. In a bird that seldom walks, such as an albatross, the cervical enlargement of the spinal cord and the nerves going to the wings are larger.

A feature unique to birds is an opening—the **rhomboid sinus**—on the dorsal midline of the lumbosacral enlargement, which contains a gelatinous mass of supporting neuroglial cells. The mass is rich in the nutritive sugar glycogen, and thus is known as the **glycogen body**. The function of this structure is still unknown. Perhaps it is an energy reserve that sustains the reflex activity necessary for roosting or other chronic leg posturing, such as during long flights.

Figure 4–34. The Spinal Cord Showing Nerve Plexuses: *At two locations along the spinal cord, spinal nerves exiting the cord are particularly large and joined into complex web- or net-like structures called* **plexuses***. The* **brachial plexus** *is found at the level of the wings, and is associated with a swelling of the spinal cord called the* **cervical enlargement***. The* **lumbosacral plexus***, at the level of the hind limbs, is associated with a* **lumbosacral enlargement***. Which of these spinal cord swellings is larger depends on the style of locomotion of the particular species. The Rock Dove, shown here, relies on both flight and walking, and the two plexuses are similar in size. In a flying bird that seldom walks, such as an albatross, the nerves going to the wings are larger and more numerous than those going to the legs, and thus the cervical enlargement also is larger. The* **rhomboid sinus** *is an opening in the lumbosacral enlargement containing a mass of neuroglial cells rich in the sugar glycogen, known as the* **glycogen body***. Its function is unknown. Drawing by Charles L. Ripper.*

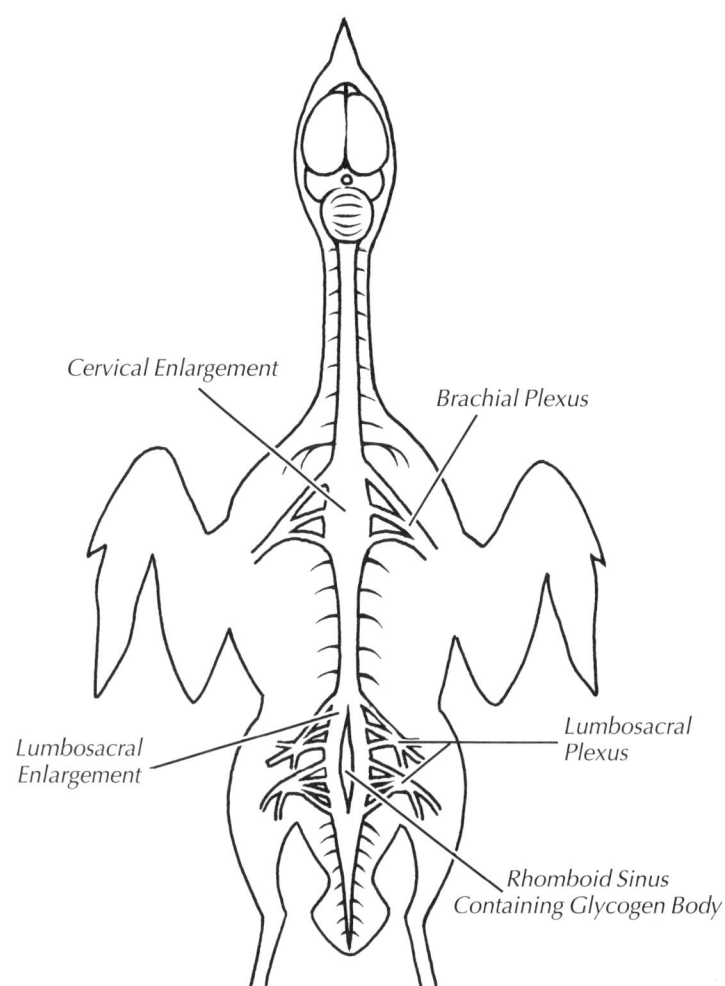

Peripheral Nervous System

The **peripheral nervous system** consists of **cranial nerves** that leave the brain and exit from the skull, **spinal nerves** that leave the spinal cord and exit from the vertebral canal, and the **ganglia** associated with them.

Cranial Nerves

Cranial nerves (see Fig. 4–31) occur as twelve sets of bilaterally paired nerves that vary in function: some are sensory, others motor, and some a mixture of both. They are named according to function and numbered according to the order in which they exit from the brain, from rostral to caudal.

I. Olfactory Nerve

The olfactory nerve carries the "sensations of smell" from the lining of the nasal cavity to the olfactory bulb of the brain. As mentioned previously, olfaction appears to be poor in most birds.

II. Optic Nerve

This large "nerve" of vision is actually a sensory tract from the ganglion cells of the eye's retina, rather than a standard nerve. The optic, nevertheless, is by tradition enumerated as one of the cranial "nerves." All visual sensations from the millions of closely packed visual cells in the retina of a bird are transmitted to the brain over these robust cables. Each optic tract enters via its own optic foramen into the skull. It then crosses to the opposite side of the brain at the **optic chiasma** and enters the brain. The optic tracts are larger than any other cranial "nerve." As would be expected, the optic nerve is particularly large in visual predators such as hawks, and small in many nocturnal birds.

III. Oculomotor Nerve

This motor nerve moves the eye by controlling contractions of some of the muscles that run from the bony orbit to the surface of the eyeball. It also supplies the eyelid muscles and the tear gland of the nictitating membrane. The nerve originates in the midbrain and branches to its target structures after exiting the skull.

IV. Trochlear Nerve

This small motor nerve controls just one eye muscle not innervated by the oculomotor nerve. It originates on the dorsal surface of the brain stem, and exits through its own foramen in the skull.

V. Trigeminal Nerve

This second largest cranial nerve has both sensory and motor components. It divides into **ophthalmic**, **maxillary**, and **mandibular** nerves after exiting from the brain, thus its name, which means "triplet." The ophthalmic nerve is sensory from the nasal cavity (for nonolfactory nasal sensations), eyeball (for nonvisual eye sensations), upper eyelid, forehead, and upper beak. It is very important to ducks and geese, in which it innervates the specialized bill tip organ: a concentration of touch or mechanoreceptors with which the bird seeks submerged aquatic vegetation by feel. The maxillary nerve is sensory from the skin of the face, upper jaw, upper eyelid, and conjunctiva (eye covering tissue). The mandibular nerve is sensory from the lower beak and corner of the mouth, and motor to the muscles of lower beak movement, which correspond to the chewing muscles of mammals.

VI. Abducent Nerve

This motor nerve stimulates the one remaining muscle of the eyeball and the two skeletal muscles that pull the nictitating membrane across the eyeball.

VII. Facial Nerve

This motor nerve stimulates the muscles for protruding the tongue (muscles attached to or associated with the hyoid skeleton), the depressor muscle that lowers the lower beak, the constrictors of the neck, and the muscle that tenses the columella ear bone (see Fig. 4–48). It also may carry some taste fibers from the tongue.

VIII. Vestibulocochlear Nerve

Formerly called the auditory, acoustic, or statoacoustic nerve, this is the sensory nerve for balance and hearing, as the Latin name implies. It is composed of two parts: the **vestibular part** for *balance*, and the **cochlear part** for *hearing*. The vestibular part innervates the region of the inner ear that detects the direction of gravity as well as accelerations in three-dimensional space. The cochlear part contains axons coming from a distinctly different region of the inner ear responsible for the sensation of sound (see Fig. 4–50). No part of this nerve ever leaves the skull.

IX. Glossopharyngeal Nerve

This sensory and motor nerve innervates the tongue, pharynx, esophagus, and throat. It also carries fibers from a specialized ganglion on their way to certain blood vessels, and motor fibers to the salivary glands of the tongue.

X. Vagus Nerve

This is the major visceral sensory and motor nerve to the abdominal organs. It arises from the medulla by a series of segments in close association with the glossopharyngeal and accessory nerves, with which it exchanges fibers. After leaving the skull, the vagus sends fibers to the pharynx and larynx before passing down the neck on the surface of the jugular vein. Within the thorax it branches to the heart and lungs before sending branches to the gizzard, liver, and intestine.

XI. Accessory Nerve

This nerve is of visceral motor function and is distributed in the head region with the vagus to innervate the constrictor muscles of the neck. It exits from the skull with the vagus. Because of its exit site and function it is included in the "cranial" nerves, but it has a very circuitous origin within the upper neck that is thought to reflect an ancient evolutionary "capture" of a spinal nerve for "cranial nerve functions."

XII. Hypoglossal Nerve

The hypoglossal nerve joins with cranial nerves IX and X to form a combined trunk that controls movement of the tongue, larynx, tracheal muscles, and syringeal (of the syrinx) muscles. The tongue in most birds is much less muscular and mobile than that of mammals, and thus requires less hypoglossal control. Parrots, however, have highly manipulative tongues, and thus the brain stem nuclei for this portion of a parrot's hypoglossal nerve are large in size. The syrinx is unique to birds, and in those that have complex song, the hypoglossal branch that supplies it is of major importance (see Syrinx, later in this chapter).

Spinal Nerves

Spinal nerves are paired in all cases—one on each side of, and attached to, the spinal cord. They are sensory and motor to a very specific (and usually closely adjacent) region of skin, muscles, and organs. Their numbers vary directly with the number of vertebrae in the vertebral column, which is generally related to the size of the bird. For example, the Rock Dove has 39 pairs of spinal nerves; the Ostrich has 51. Spinal nerves differ in size according to their particular function. The large spinal nerves governing the muscles of the wings and legs fuse and branch repeatedly to form networks or plexuses. Although we mention the spinal nerves here, in conjunction with the central nervous system, they are actually part of the peripheral nervous system.

Subdividing the nervous system into two parts—the central and peripheral—makes it easier to organize the material for study. But it is important to understand that neither the central nervous system nor the peripheral nervous system can function by itself. Both are simply parts of one elaborately integrated system that also includes the autonomic nervous system.

Autonomic Nervous System

The **autonomic nervous system** is not a series of discrete and distinct structures as are the other systems considered here. Rather, it is a concept designed to facilitate understanding of the visceral nerve network and its physiological responses that operate without conscious control. The system, which controls the "guts" as opposed to the "meat" of the bird, is much more complicated than we describe. To most of the "rules" we present here, there are some exceptions. Even the most basic premise that the system is without conscious control is not totally true. We know, for instance, that some people can voluntarily alter their heart rate and blood pressure, and control gastric secretion. Nevertheless, the concept of an autonomic nervous system is helpful to understanding how birds work. When first described, the autonomic nervous system was thought to be exclusively a motor system to the smooth muscle of various structures. But we now know that some of the nerves involved carry sensory fibers as well.

Each motor component of the autonomic nervous system is a *two-neuron chain*, with a synapse between the neurons **(Fig. 4–35)**. This contrasts with the voluntary motor nerves (such as those of reflexes discussed above), whose single cell bodies lie within the CNS. From the cell bodies of these voluntary motor nerves, axons run directly to the target muscle without any other peripheral neuron interposed. Thus the autonomic system, while potentially a bit slower than a simple voluntary reflex, is capable of a great deal of subtle modification outside the CNS.

As in mammals, the bird's autonomic nervous system can be divided into two segments based on their regions of origin from the central nervous system as well as on their functions. The two divisions are the **parasympathetic system**, with nerves originating in the cranial and sacral regions, and the **sympathetic system**, with nerves originating in the thoracic and lumbar regions.

The **parasympathetic system** has its origin in cranial nerves III, VII, IX, and X as well as from three sacral spinal nerves. These nerves act on smooth muscle to promote feeding, egg laying, and other "peace-

Figure 4–35. The Autonomic Nervous System: *This complex branch of the nervous system controls the automatic functions of the body's internal organs, acting primarily unconsciously. It consists of two subsystems, each of which innervates the same organs, but with opposing effects. The* **sympathetic system** *functions under conditions of stress—for example, speeding up a bird's heartbeat and breathing rate by means of the neurotransmitter epinephrine (adrenaline), to prepare for "fight or flight." The* **parasympathetic system** *calms the bird, promoting feeding, digestion, and other relaxed activities. All autonomic pathways consist of two neurons in series, but the organization of these two neurons differs between the sympathetic and parasympathetic systems. Sympathetic neurons (solid lines) arise from the thoracic and lumbar regions of the spinal cord. The cell bodies of the* **first sympathetic neurons** *have migrated outside the spinal cord, gathering into a* **sympathetic ganglion** *near the ventral surface of each vertebra (see Fig. 4–33). A chain of these ganglia runs the length of the spinal cord. The axons of the first sympathetic neurons run from the sympathetic ganglion to another ganglion (open circles) closer to the organ being innervated, where they synapse with the cell bodies of the* **second sympathetic neurons**, *whose axons provide the final link to the target organ. Parasympathetic neurons (dashed lines) originate in the vagus and other cranial nerves, and in the sacral region of the spinal cord. Cell bodies of the first parasympathetic neurons reside within the brain or spinal cord, and their axons run all the way to the target organ, sometimes a great distance, synapsing with the second parasympathetic neurons (not shown), located in or near the target organ.*

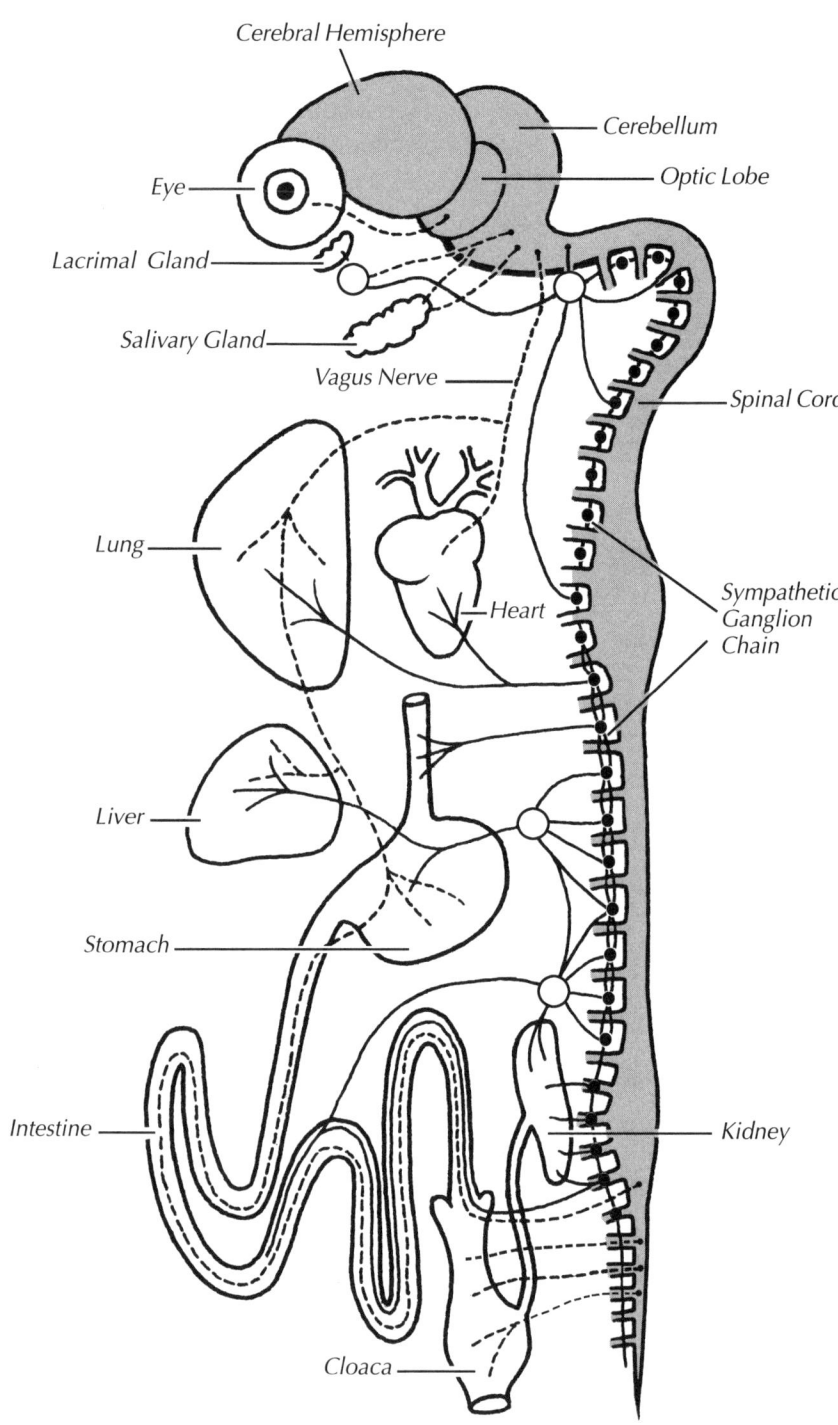

Symptomatic and parasympathetic labels:

- ——— Sympathetic Neurons
- - - - - Parasympathetic Neurons
- Ganglion of Sympathetic Ganglion Chain (Contains Cell Bodies of First Sympathetic Neurons)
- ○ Ganglion Containing Cell Bodies of Second Sympathetic Neurons

ful" activity. Parasympathetic nerve stimulation quiets the bird. Some parasympathetic nerves carry impulses that reduce the heart rate; others promote gastric secretion and peristalsis, thereby facilitating digestion.

In the parasympathetic system, the cell body of the first neuron lies within the brain or spinal cord. Its axon must then pass into the organ being innervated before it meets the cell body of the second neuron. Firing of the second neuron causes contraction of the smooth muscle in the wall of the organ. Thus a vagal neuron affecting the stomach, for example, is a very long neuron, indeed, leaving the brain and reaching all the way to the wall of the stomach, where it synapses with the second neuron of the pair. In the sacral region the distance that the first neuron has to reach is much shorter, because the target organs are closer to the nerve origins in the spinal cord.

The **sympathetic system** consists of nerves that leave the spinal cord from the thoracic and lumbar levels. Sometimes called the "fight or flight" system, it functions in emergencies. Its neurons speed up the heart rate, increase the blood pressure, provide deeper breathing, and allow for greater muscular contraction. The sympathetic system also dilates the pupil of the eye to produce that look of being "wide-eyed with fright." When the sympathetic system is activated, digestion slows or stops, and the bird may vomit or defecate to better prepare its body for fight or flight. Numerous seabirds, frightened at the nest, defecate or disgorge the contents of their crops or stomachs, sometimes most effectively in the direction of the intruder! (See Ch. 6, Sidebar 3, Fig. D.) Vultures do the same from the air.

In the sympathetic system, the cell bodies of the first neurons have migrated outside the spinal cord, where they form visible nodules close to the ventral surface of the vertebrae. These nodules (called sympathetic ganglia) appear as a chain, because their axons may travel up or down, parallel to the vertebral column but just outside it, before passing to their target organs by running along the surface of blood vessels. The cell bodies of the second axons in a sympathetic chain are grouped in visible masses, also called ganglia, on blood vessels very close to the organs being innervated. As an example, a large cranial cervical ganglion is located on each side of the head/neck junction where axons from the first neurons (coming from the thorax) synapse with the second neurons, which innervate arterioles of the head, causing their constriction, and therefore an increase in blood pressure.

The Senses

Vision

Sight is very important to birds, which are thought to have the best vision among vertebrates. To provide wide views and bright images, eyes must be large. Indeed, a bird's eyes are so large that sometimes their weight may equal, or even exceed, the weight of the brain. The largest eyes of any land animal are those of the Ostrich **(Fig. 4–36)**, nearly two inches (50 mm) in diameter! Some birds have the most acute

Figure 4–36. The Large Eyes of an Ostrich: Among vertebrates, birds have the best vision, reflected by the fact that birds' eyes are large, sometimes equaling or exceeding the weight of the brain. The eyes of the Ostrich, nearly two inches (50 mm) in diameter, are the largest of any land animal alive today. Photo by H. Cruickshank/VIREO.

(sharpest or best resolving) vision in the animal kingdom. The Golden Eagle exceeds the visual acuity of humans by two or three times and is capable of spotting movements of small prey such as rabbits from more than a mile away!

Unfortunately, the desire to know how our vision (or any of our other senses) compares with that of birds is fraught with obstacles. First of all, there are approximately 10,000 different species of living birds worldwide, and no blanket statement can encompass the visual abilities of them all. Second, we cannot experience the perceptions of other organisms. We must use indirect means—comparative anatomy, physiology, and behavior—to compare a bird's capabilities to our own. This is a tricky business, but worth the try. Just how do avian eyes stack up to our own? Let's begin by examining eye structures.

The Structure of the Eye

In addition to the familiar upper and lower eyelids, birds have a **nictitating membrane** or third eyelid. It moves sideways across the eye, at right angles to the regular eyelids, cleaning the eye's surface and keeping it moist (**Fig. 4–37**; see also Fig. 1–4). In some aquatic birds, such as loons, cormorants, diving ducks, and alcids, the nictitating membrane has a special window-like area in the center. These birds presumably "wear" their nictitating membranes as swim goggles to improve their underwater vision. Dippers, such as the American Dipper, have an opaque, white third eyelid which, curiously, can be of no use in searching for prey underwater.

On each side of the eye are two tear glands, which moisten the eye and nourish the cornea. One is the **lacrimal gland,** which lies in the lower part of the orbit and has many ducts that enter the space between the lower lid and the cornea. The other is the gland of the third eyelid, which secretes into the space between the third eyelid and the cornea. Tears from both glands drain into lacrimal canals in the corner of the eye near the beak and nostrils to enter the nasal cavity.

The **eyeball (Fig. 4–38)** has a tough outer layer of connective tissue, the **sclera**, which is stiffened by a ring of bony **scleral ossicles (Fig. 4–39)** near the front of the eye. Scleral ossicles are present in the eyeballs of all birds, lizards, turtles, and fishes. The anterior surface of the opaque white sclera is specialized as the transparent **cornea**, which allows light into the eye. The layer deep to the sclera and cornea is the vascular **choroid**, which forms the **iris** (Latin for "rainbow"), the colored part we see when we look at an eye (see Fig. 3–34c). The iris, which contains smooth muscle fibers, encircles the opening or **pupil** (see Fig. 4–40), regulating its size and thus the amount of light entering the eye.

Within the eyeball, the largest cavity, the **vitreous chamber**, is filled with a clear, jelly-like material, the **vitreous body**, which "inflates" the eye and maintains its shape. Projecting into the vitreous body from the site where the optic nerve exits the eyeball is a non-sensory, vascular structure of the choroid called the **pecten**. The pecten is believed to nourish the retina and to control the pH (acidity) of the vitreous body. The pecten takes many forms, but it is present in all birds and known elsewhere only from some reptiles.

The **lens**, a crystalline-like structure composed of regularly oriented layers of collagen fibers, is the primary modifier of focus. It is spherical to ovoid in shape, depending on the species, and is unusually soft and pliable in comparison with that of other vertebrates. The lens is especially soft and pliable in many diving birds, allowing it to be squeezed and stretched to a variety of shapes, so it can achieve the best possible vision under a variety of circumstances. The lens is held in place between the vitreous chamber and the iris by **ciliary processes**. Only in birds do these processes attach directly to the lens. Ciliary muscles attach to the processes, moving them when they contract. The lens-distorting power of these muscles makes possible the variable-focusing powers of the lens. A small chamber in front of the lens is partially divided in two by the iris: a narrow **posterior chamber** between the lens and iris; and a larger **anterior chamber** between the iris and the cornea. Both chambers are filled with **aqueous fluid**, a cell-free fluid similar to blood plasma (see Blood, later in this chapter), which is constantly produced from the blood and secreted into the posterior chamber. The aqueous fluid from the posterior chamber passes through the pupil into the anterior chamber, nourishing and removing wastes as it flows. It then drains into sinus spaces and veins at the base of the iris to re-enter the blood.

Figure 4–37. Great Horned Owl Showing Nictitating Membrane: *The nictitating membrane, or third eyelid, lies between the bird's regular eyelids and the surface of the eye. It moves across the eye at right angles to the eyelids, cleaning and lubricating the eye's outer surface. In this photo it is visible halfway across the eye of an immature Great Horned Owl. Photo by Lang Elliott/CLO.*

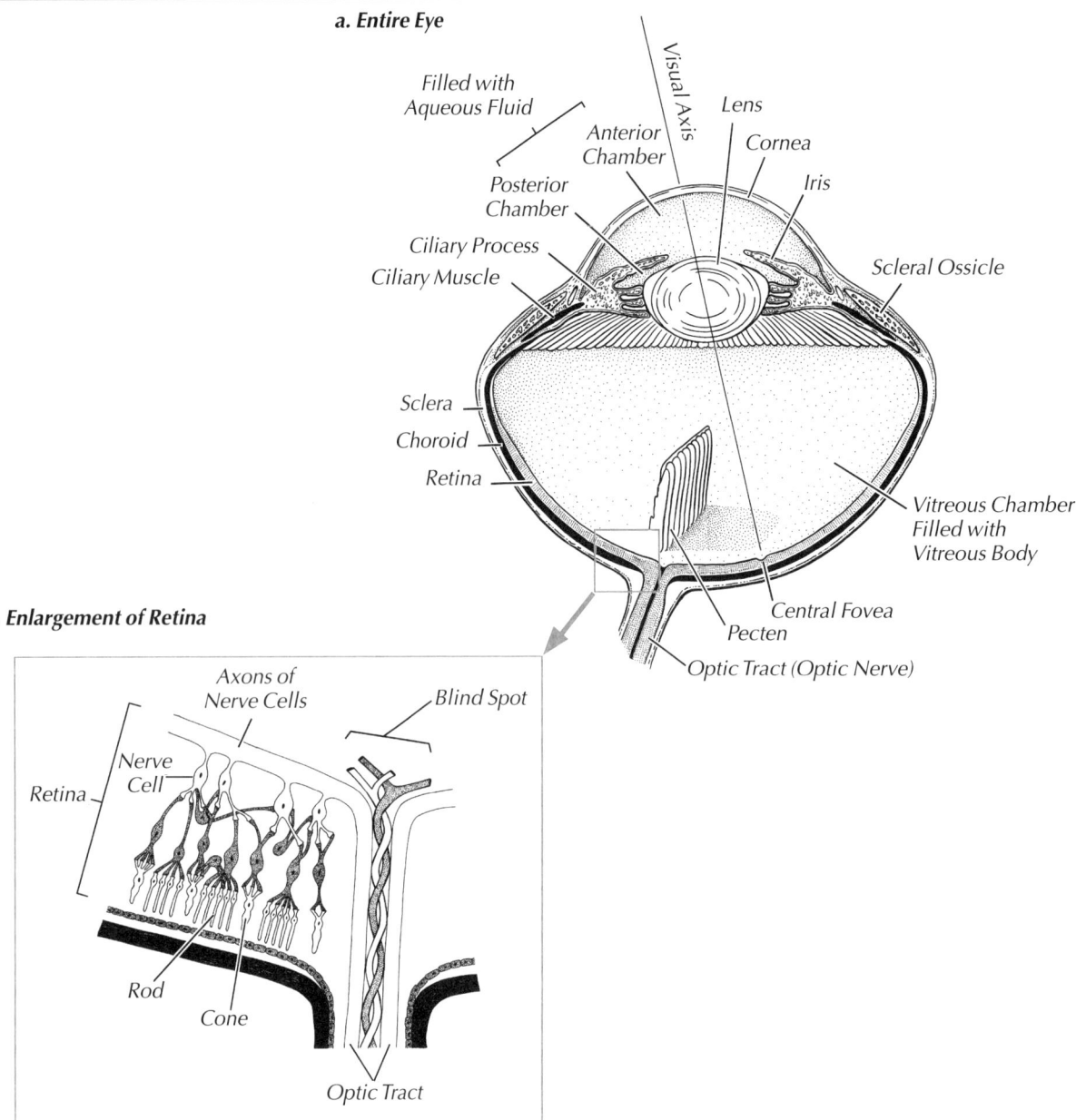

a. Entire Eye

b. Enlargement of Retina

Figure 4–38. The Internal Structure of the Eye: *The eye is positioned with the anterior surface up in both of these cross-sectional views.* **a. Entire Eye:** *The eye consists of three main layers, the sclera, choroid, and retina, which specialize to form various other structures. The tough, whitish outer layer of connective tissue is the* **sclera**. *Toward the anterior end of the eye, the sclera is stiffened by a bony ring of* **scleral ossicles**, *and across the anterior surface of the eye, it is modified to become the transparent* **cornea**. *The middle layer is the* **choroid**, *pigmented and richly supplied with blood vessels. It forms the* **iris**, *the colored part of the eye surrounding the* **pupil** *(see Fig. 4–40). The choroid also forms the* **ciliary processes**, *which attach to the lens and are moved by the* **ciliary muscles**, *thus altering the shape of the crystalline lens to focus images. The innermost layer is the* **retina**, *consisting of light-sensitive cells (see* **b**). *The large central cavity within the eyeball, the* **vitreous chamber**, *is filled with a clear, jelly-like material, the* **vitreous body**, *giving the eyeball rigidity. The* **pecten**, *found in all birds and a few reptiles, projects into*

the vitreous body, and is thought to nourish the retina and control the acidity of the vitreous body. In front of the lens is a small chamber filled with **aqueous fluid** *and divided into the* **anterior chamber** *and the* **posterior chamber**. *The aqueous fluid, a cell-free fluid similar to blood plasma, provides nourishment and removes wastes.* **b. Enlargement of Retina:** *The retina is made up of light-sensitive cells, the* **rods** *and* **cones**, *which contain the visual pigments. Cones are responsible for visual acuity and the sensing of color information. Rods are relatively insensitive to color, but they detect low levels of light and thus are more important than cones in nocturnal birds. Neurons synapsing with the rods and cones form the* **optic tract** *(also called the optic nerve), which forms a* **blind spot** *(where images cannot be detected) on the retina as it exits. The point in the retina where the cones are concentrated and vision is sharpest is known as the* **central fovea**. *Main drawing adapted from Proctor and Lynch (1993, p. 251). Inset by Christi Sobel.*

The innermost layer of the eyeball is the pigmented **retina**, which consists of the light-sensitive cells, the rods and cones. A bird sees an object when light from that object is refracted as it passes through the cornea, lens, and fluid-filled chambers, focusing an image on the retina. There, the rods and cones absorb some of the light's energy, are stimulated, and transmit the resulting signals to the brain via nerve impulses. Acuity (resolving power) depends on close packing of the bright-light-sensitive cone cells. The packing of cones in some regions of the retina of raptors may be as much as 650 million cells per square inch (one million per square millimeter)! This is five times the packing in the human retina.

In addition to being responsible for acuity, cone cells encode information about the spectrum of colors contained in the light focused on the retina. Birds have four to five distinctive light-sensitive cone pigments plus specialized oil droplets in some cones that may function as filters, altering color sensitivity in the same manner as yellow, pink, or some other color of sunglasses. Humans have but three light-sensitive cone pigments, and nothing comparable to the avian oil droplets. The rods of birds are very sensitive to light energy, but are not capable of differentiating much in the way of color information. Because they are good at detecting low light levels, rods are more important than cones in owls and other nocturnal birds. We humans essentially lose our color vision from dusk to dawn, seeing little color differentiation even under bright moonlight, because the light is too dim to stimulate our cones.

Each rod or cone cell has synapses with a complex of nerve axons, many of which pass to the brain. As in all vertebrates, the nerves from the sensory cells lie between the rods and cones and the pupil, blocking some of the light that would otherwise reach the sensitive cells. This appears to be an inefficient way to construct a light-sensitive organ, as the sensory cell layer actually interferes with vision. This arrangement, however, results from the evolutionary history of the vertebrate eye—evolution generally works by modifying existing structures, not by creating new ones that are perfect for the job at hand. Where the nerve layer is thinnest, vision is best. The axons of the neurons leading from the rods and cones pass over the surface of the retina and join to form the **optic nerve**. This nerve leaves the eyeball by penetrating the retina, choroid, and sclera and thus forms a "**blind spot**" in the midst of the retina, where no rods or cones are present to capture light falling on that spot.

In contrast, most birds, like most mammals, have in the central part of their retinas an area, the **central fovea**, where the cones are most concentrated and the neural layer thinned for the sharpest vision. Hawks and other fast-flying diurnal predators have, in addition, another such area, the **temporal fovea**, in the posterior quadrant of the retina.

a. Cross-Sectional Views of Eagle and Owl Eyes

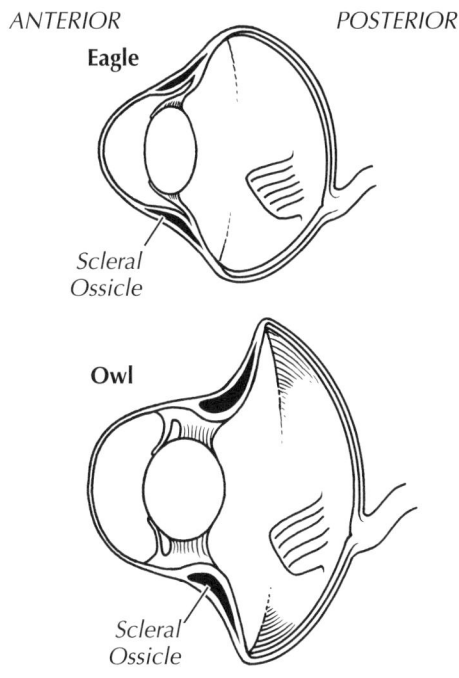

b. Owl Skull

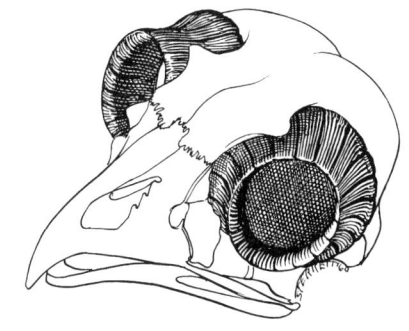

*Figure 4–39. Scleral Ossicles: The eye's outermost layer, the sclera, is stiffened by bony rings called **scleral ossicles**. **a. Cross-Sectional Views of Eagle and Owl Eyes:** Here, sections through the eye of an eagle and an owl show how the large ossicles give each eye its distinctive tubular shape. Drawing by Charles L. Ripper. **b. Owl Skull:** The large scleral ossicles (dark areas) are particularly conspicuous in owls. Drawing from* Bird Study *by Andrew J. Berger (John Wiley & Sons).*

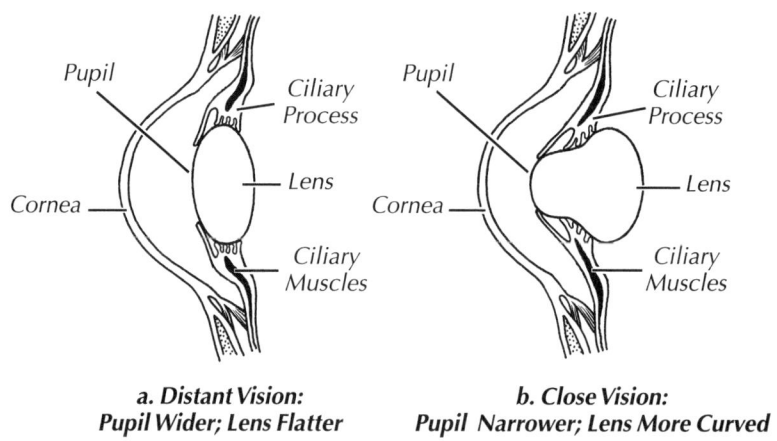

Figure 4–40. Accommodation—The Mechanics of Focusing: To sharply focus images from varying distances on the retina, the eye changes the curvature of the lens by contracting and relaxing the ciliary muscles, which move the ciliary processes, which change the shape of the lens. These focusing adjustments are referred to as **accommodation**. *a. Distant Vision:* To focus distant objects the lens flattens; at the same time, the pupil becomes wider, allowing more light into the eye. *b. Close Vision:* For close objects the lens becomes more rounded and the pupil narrows. Drawings by Charles L. Ripper.

How Birds See

How does a bird use the structural features of its eye to see? What information can help us judge how well birds see? Let's begin by examining the process of focusing.

To allow the retina to obtain sharp images at varying distances, the eye changes the curvature of the lens **(Fig. 4–40)**. This is accomplished by the action of the ciliary muscles, which move the ciliary processes, which exert pressure on the lens. The lens flattens to focus on objects far away, and becomes more rounded for objects close at hand. The curvature of the cornea changes as well. These focusing adjustments are termed **accommodation**. At the same time, the amount of light entering the eye is regulated by changing the size of the pupil—opening it for more light and closing it when there is too much light. This is achieved through the action of muscles in the iris, which open and close the pupil just as the diaphragm in a camera controls the size of the aperture. The next time you have the opportunity to examine a bird closely (try a parrot, with its large, easily observed eyes), watch the rapid change in its pupil diameter. The pupils first narrow to better see you in the foreground and then widen to keep track of conditions farther away. If you watch a friend doing the same visual task, the much slower response of the human pupillary action will become evident.

Many birds show a remarkable range and speed of accommodation, keeping objects in focus at rapidly changing distances. A stooping falcon, for example, can keep prey in focus until it is in its talons. A warbler flying swiftly through a forest sees well enough ahead to avoid the trunks and branches of trees in its path. Some birds, on the other hand, are nearsighted (**myopic**). For example, penguins are notably myopic on land because the structure of their eyes is better suited to seeing objects, such as food, in the water. (Water's refraction of light is so similar to that of the eye tissues themselves that, when a bird is submerged, accommodation requires large, spherical, inflexible lenses and corneas.)

Based on the presence and complexity of the cone cells, most diurnal bird species are believed to have very good color vision. Experimental results confirm this assumption, as color vision has been demonstrated in a diverse set of birds including penguins, pigeons, ducks, owls, hummingbirds, and a number of passerines. Nocturnal species such as owls probably have little, if any, color vision. In addition, many birds apparently can see certain types of ultraviolet light, which is discussed in greater detail later in this chapter.

Many birds—indeed, many other vertebrates—see the world very differently than we humans do, not because of differences in the internal anatomy of the eye, but because of the placement of the eyes in the head **(Fig. 4–41)**. Many animals have primarily **monocular vision**. This occurs when the eyes are situated on the sides of the head such that an object in the external environment can be seen only by one eye or the other but not by both eyes at the same time. In contrast, when the eyes are located toward the front of the head, objects are seen with both eyes simultaneously, resulting in **binocular vision**. The differences between the two are rather like the difference between going to a regular movie versus putting on those red and blue glasses and watching a 3-D movie. Why, then, isn't all vision binocular like ours? As with many biological alternatives, trade-offs are involved.

Monocular vision can be advantageous because it results in a wide field of view, sometimes as much as 340 degrees, which allows an animal to see both in front and in back at the same time. In contrast, animals with binocular vision, including humans, have a much narrower field of view because the visual fields from each eye overlap extensively. Owls, with their binocular vision, can only see through a field of view of up to 70 degrees.

Figure 4–41. Monocular Versus Binocular Vision: The placement of the eyes on a bird's head affects the size and degree of overlap of the left and right visual fields, which determines the extent of binocular vision and thus depth perception. The circles shown here are the visual fields of an American Woodcock (left) and an Eastern Screech-Owl (right), at the horizontal plane corresponding to the line between points A and B marked on each bird. The woodcock's eyes are set on the sides of its head, giving it a field of view slightly greater than 180 degrees for each eye, but only a small area of overlap to the front and the rear. Because binocular vision occurs only where the fields of view overlap, the woodcock has primarily monocular vision. Although its binocular vision is limited, this species has complete monocular coverage of the hemisphere above and behind the head, an adaptation for spotting a predator approaching from any direction. The owl, in contrast, has eyes located toward the front of the head. The field of view for each eye is therefore smaller than the woodcock's, but the area where the visual fields overlap in front is far greater, allowing considerable binocular vision. This is crucial to the owl for the successful capture of prey; it is gained at a cost, however, as it leaves a blind area behind the owl's head.

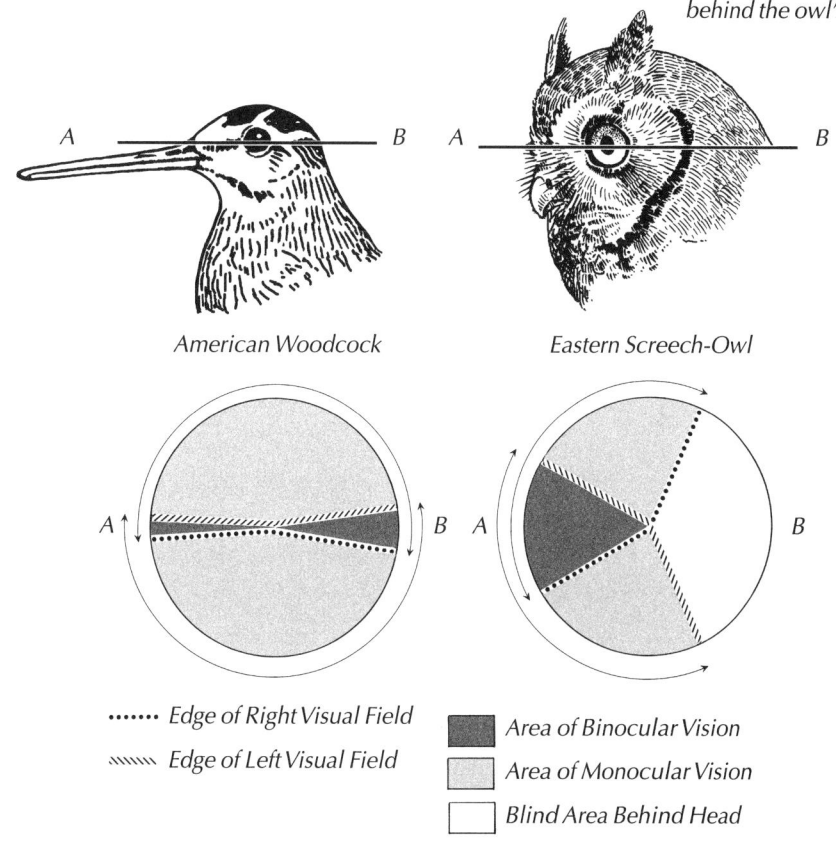

American Woodcock   Eastern Screech-Owl

······ *Edge of Right Visual Field*   ▨ *Area of Binocular Vision*

〰〰〰 *Edge of Left Visual Field*   ▨ *Area of Monocular Vision*

□ *Blind Area Behind Head*

Figure 4–42. A Common Snipe: *A female Common Snipe is viewed from behind as it incubates on its nest. The snipe's eyes are set so far back on its head that its visual fields overlap more behind the head than in front. This produces a greater area of binocular vision to the rear than to the front, providing better protection against predators. Photo by Marie Read.*

Figure 4–43. American Bittern: *The bittern's eyes are set low on the sides of its head. Thus, with the head held horizontally, it can search the water below for food while also seeing ahead. When alarmed, the bittern stretches its head and neck high and points its bill directly upward, as pictured here, blending in with the surrounding grasses (see Fig. 2–17). Even in this defensive posture the position of its eyes permits good vision in front as well as of the sky overhead. Photo by Tom Vezo.*

The eyes of the American Woodcock and the Common Snipe **(Fig. 4–42)** are so far back on the head that these birds can see better behind than in front. This may be a protective feature for watching overhead for enemies while probing in the mud with its long bill. In fact, many animals that are frequently the targets of predators have monocular vision. Another bird with monocular vision, the American Bittern, has eyes set so low on the sides of the head that it can look for food below and see ahead at the same time. When alarmed and in a defensive posture, the bittern stretches its head and neck high and points the bill straight up. In this position it has good vision directly forward as well as upward into the sky **(Fig. 4–43).**

Binocular vision also has an important advantage, however: it enables a bird to have good depth perception and thus to determine distances better. Depth perception results from forward-facing eyes because each eye gets a slightly different view of an object; the closer the object, the more different the two views. The difference allows the brain to distinguish distance. Many predators, such as hawks, eagles, and owls, have binocular vision because it aids them in capturing prey.

A number of birds with restricted binocularity but wide monocularity obtain the benefits of depth perception by head actions. American Robins, peering at prospective food, cock their heads at different angles before picking it up. They are making sure of its identity and capture by determining its form and its precise three-dimensional position. The head-bobbing and teetering movements of some plovers and sandpipers may have the same function. Watch a Rock Dove or a chicken as it walks, alternately jerking the head back and forth with pauses in between movements. In doing this, the bird is getting a series of different views for determining spatial relationships of objects in its surroundings. When the bird's head moves backward relative to its moving body but with little movement relative to the visual scene, acuity is greatest. When the head moves rapidly forward, differences in the apparent movement of objects in the visual field give the bird strong cues about the three-dimensional positions of objects in its field of view: during these head movements, objects closer to the bird appear to move across the visual field faster than those at a distance. We experience a similar visual effect when watching the countryside fly by out the side window of an automobile or train.

Birds that have both central and temporal foveae in each eye (usually predators), tend to have forward-directed eyes for good binocular vision **(Fig. 4–44)**. Because images from each side of their field of view (the regions they are passing through) tend to fall on the central fovea, it provides acute monocular views of these areas. Images from the front, the part of the world they are about to enter, tend to fall on the temporal fovea, providing acute binocular vision for that area. Thus these predators see well on both sides while also seeing ahead—an obvious advantage when pursuing prey. These multiple foveae also

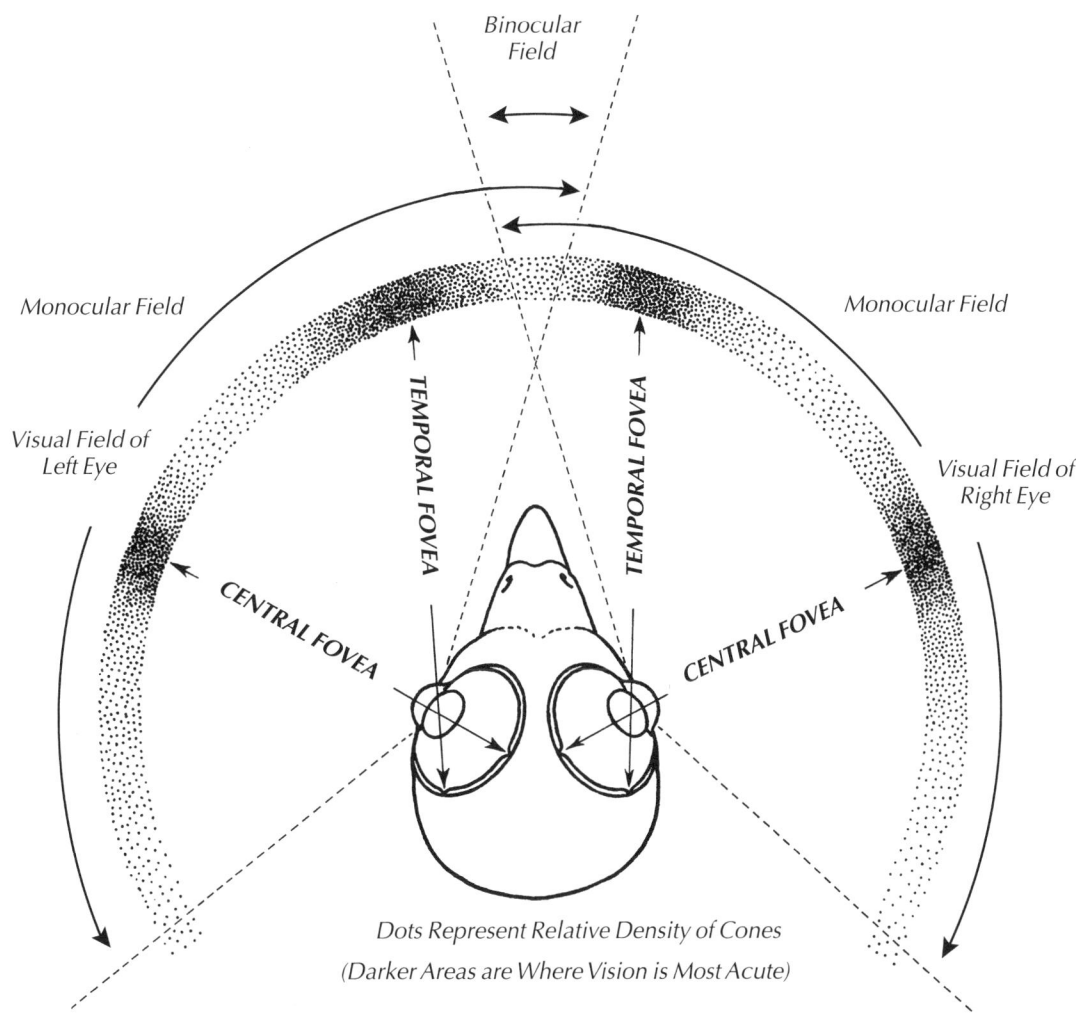

Figure 4–44. Visual Fields and Multiple Foveae of a Hawk: *Foveae are areas of the retina where the cones are densely packed, providing especially sharp vision. Although most birds have only one central fovea in each eye, many hawks and other avian predators have two foveae, central and temporal, in each eye. These birds also tend to have forward-facing eyes that allow good binocular vision. Visual information from the left or right side falls on the central fovea of the corresponding eye, giving sharp monocular views of these areas, whereas information from in front of the bird falls on the temporal fovea, giving a sharp binocular view of the area ahead. Thus these birds enjoy three well-focused views simultaneously—two side views and one forward—a great advantage for locating and pursuing prey. Drawing by Charles L. Ripper.*

create incredible complexity for instantaneous neural processing and analysis! Some shorebirds whose normal habitat is flat, open, and relatively featureless, have a foveal stripe that runs horizontally across the retina. It probably is specialized for detecting conditions, particularly movement, along the horizon.

The eyes of most birds move very little in their sockets. Because of the eyeball's large size, there simply is not much room in the compact, streamlined head for eye muscles. The four rectus and two oblique muscles that move the eyeball in all vertebrates, from fish to humans, are present but much reduced in size. Birds compensate for the lack of eye movement by having very mobile necks **(Fig. 4–45)**. Owl eyes are immovably fixed in the skull, and they have only temporal foveae, which are used for forward binocular vision. For views to the sides, they turn their heads, sometimes halfway around.

So, do birds see better than humans? As should now be evident, merely deciding what we mean by "better" is a problem. Do we mean better acuity or better focus? Better light sensitivity or better color differentiation? To each of these individual components the answer seems to be "Some birds, yes, and some birds, no. Under some conditions some birds, yes, but under other conditions the same birds, no." That is a part of the magic of birds. No matter how hard we try, their variety thwarts any attempt to generalize. The best we may ever be able to do is conclude that some birds "see better" than we do and that vision is the predominant sense of birds as a group.

The Ear and Hearing

Sound is also an important characteristic of the environment, and its detection allows a bird to become aware of predator and prey, to flee impending danger or approach potential mates, and to advertise or warn. The challenge of hearing is to convert pressure waves in the air (see Fig. 7–4) into neural impulses that can be interpreted by the brain. All animals do this by first changing the *varying pressure waves* that make up sounds into *vibrations,* which are then altered further to become *nerve impulses.* Animals with the best hearing can distinguish the direction of origin, and the complexity of frequency variations as well.

A bird's sense of hearing is keen and very important for survival and reproduction. The ear had its evolutionary origin in fishes as membranous, fluid-filled chambers called the **inner ear**, which function in maintaining balance and sensing vibrations. For hearing in air, amphibians, reptiles, and birds all evolved additions to the fluid-filled chambers. These included an air-filled **middle ear** chamber with a small bone, the **columella** or **stapes**, which transmits

Figure 4–45. A Burrowing Owl Looking at the World Upside Down: Unlike humans, birds can move their eyes very little in their sockets. Owls have especially immovable eyes, but like other birds they compensate by having very flexible necks (see Fig. 4–13). Owls regularly turn their heads halfway around to get a better view behind them, or even view the world upside down, as this bird has chosen to do. Photo by Bryan S. Munn.

sound vibrations to the inner ear from an **eardrum**, which stretches across the external opening of the middle ear. These structures are discussed in more detail below. Mammals carried the process even further by adding two more bones on each side, creating a chain of three bones (malleus, incus, and stapes), and a sound-collecting auricle or external ear—the flap we see when we look at a mammal. Do not, however, confuse the "ears" of birds such as the Great Horned Owl and Long-eared Owl with the birds' true ears **(Fig. 4–46)**. The "ears" of these owls are merely tufts of feathers growing from the top of the head, and have no relation to the true ears or to hearing. They are thought to be used in species recognition.

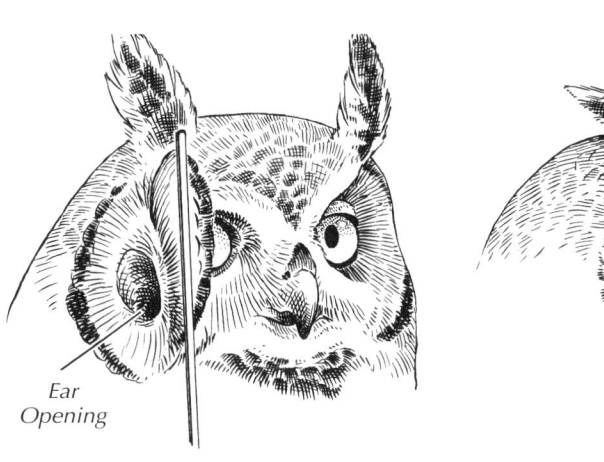

Ear
Opening

Figure 4–46. Location of Great Horned Owl Ear Openings: The owl's facial disk of feathers is shown parted to reveal the large ear opening hidden beneath. Although the tufts of feathers growing from the top of the head of the Great Horned Owl, the Long-eared Owl, and other species are often referred to as "ears," they have no relation to true ears or hearing. Drawing by Charles L. Ripper.

Nevertheless, many birds do have fleshy or feathery specializations (often modified **auricular feathers** [see Fig 1–5] or ear coverts) around their external ear openings. Like the external ears of mammals, these help to concentrate sound waves and enhance the bird's ability to determine the direction from which a sound originated. Because sound travels at a fixed speed through air of a given temperature and humidity, a sound originating at one point will arrive at other points at times that depend on their distance from the origin. Even small differences in distances from the sound's origin, such as the distance between a bird's ears, can create a detectable difference between arrival times of the sound at the two ears. The greater the difference in arrival times, the more precisely the position of the sound's origin can be determined. Also, because a sound weakens as it travels, it will be louder in the ear closest to the sound source. Thus, additional directional cues may be gleaned from differences in a sound's volume (loudness) between the two ears.

Some species of owls that are strictly nocturnal and consequently depend heavily on hearing while foraging have their right and left external ear canals at slightly different levels on the head **(Fig. 4–47)**. This improves the owls' abilities to determine the three-dimensional position of a sound source. Some owls can pull a fold of skin located in front of the ear over the external ear canal to close it off, protecting

Figure 4–47. Asymmetrical Placement of the External Ear Canals of a Boreal Owl: Many species of owls have considerable asymmetry in the placement of the external ear openings, although the structures of the middle and inner ears are located symmetrically. The anterior view shows that the right external canal is located higher than the left. The dorsal view shows further asymmetry, revealing that the right ear canal is located more posteriorly than the left. Asymmetrical ear openings help owls to localize the sources of sounds in three-dimensional space with great accuracy, a feat they achieve by detecting tiny differences between the arrival times of a given sound at each ear. Owls have the most sensitive acoustic systems of all birds, enabling some species, such as the Barn Owl, to capture their prey in complete darkness. From Norberg (1978).

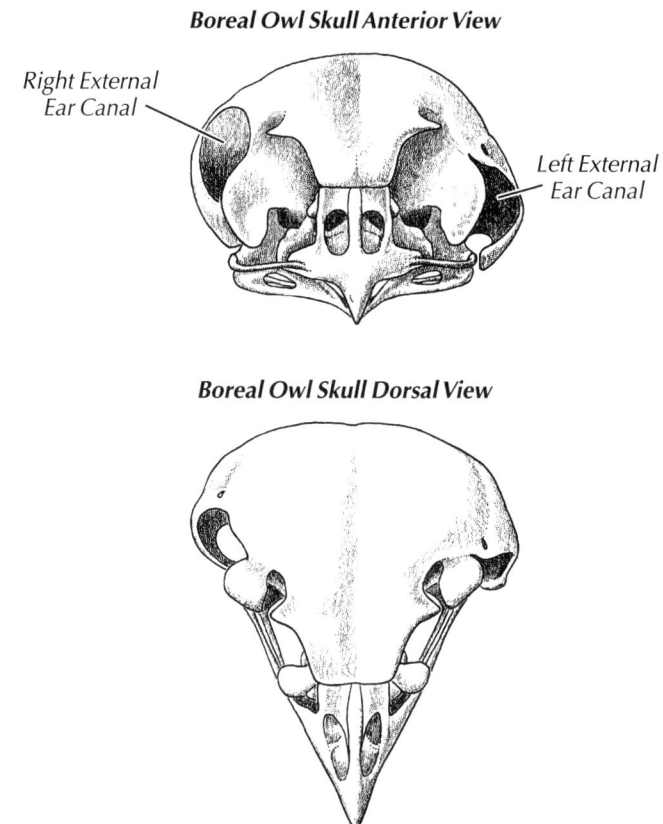

Boreal Owl Skull Anterior View

Right External Ear Canal

Left External Ear Canal

Boreal Owl Skull Dorsal View

sensitive ears from loud sounds; alternatively, the fold can be erected to enhance the detection of sounds coming from behind.

Structure and Function of the Ear

Let's now examine the three parts of the avian ear—the **external ear**, the **middle ear**, and the **inner ear**—in more detail **(Fig. 4–48)**. The external ear canal leads to the **tympanic membrane** or **eardrum**. The eardrum, stretched taut over the ear canal, vibrates when struck by the pressure waves of a sound. The eardrum's movement is then transferred to a piston-like movement of the slender **columella** bone, which is at-

Figure 4–48. Location and Structure of Rock Dove Ear: The avian ear consists of three parts: the external ear, the middle ear, and the inner ear, all embedded in bone. As in humans and many other animals, the **external ear canal** (not shown) leads from outside the body to the **tympanic membrane** or **eardrum**. Attached to the inner surface of the eardrum is a slender bone, the **columella**, which extends across the small, air-filled chamber of the middle ear. At a soft, pliable spot known as the vestibular window, the columella contacts the bony, fluid-filled **inner ear** (see Fig. 4–49), where the organs of hearing and balance reside. Reprinted from Manual of Ornithology, by Noble S. Proctor and Patrick J. Lynch, with permission of the publisher. Copyright 1993, Yale University Press.

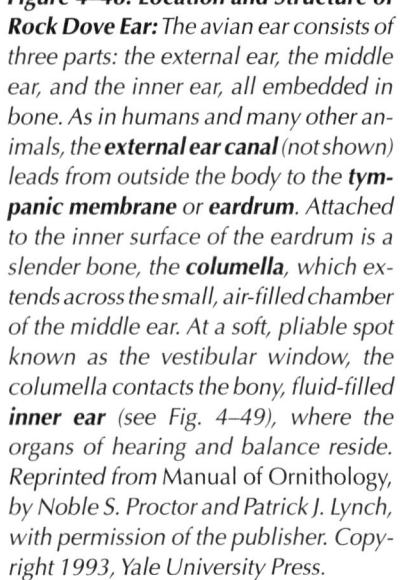

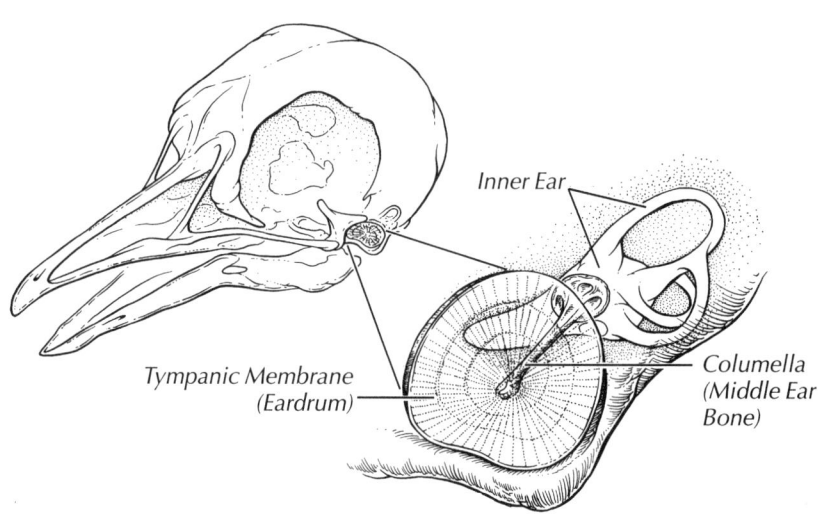

Inner Ear

Tympanic Membrane (Eardrum)

Columella (Middle Ear Bone)

tached to the interior surface of the eardrum. The area internal to the eardrum, which includes the columella, is termed the middle ear; it is a small, air-filled cavity, bounded by bone and open to the throat via the **auditory (Eustachian) tube**. In birds, the auditory tubes join and enter the caudal roof of the mouth through a common opening. Perhaps you have experienced pressure on your eardrums when climbing a mountain or being pressurized in an airplane. You may have relieved the pressure by either blowing your nose, yawning, or swallowing. For a bird constantly changing its altitude in flight, the need to equalize the pressure in the middle ear with that of the external ear must be almost constant. The auditory tubes permit this equalization, allowing air taken in the mouth (and thus at the same pressure as ambient air and the air in the external ear canal) to move in or out of the middle ear until equal background pressure is established on both sides of the eardrum.

The columella extends across the middle ear cavity to contact a soft, pliable spot, the **vestibular window** (formerly called oval window), on the bony **inner ear (Fig. 4–49)**. Unlike the external and middle ear, which are filled with air, the inner ear is filled with fluid. Its entire structure is that of two fluid-filled systems, called labyrinths, one within the other. The inner "sac" is called the **membranous labyrinth,** and is filled with a fluid called **endolymph**. A **bony labyrinth**, filled with a fluid called **perilymph**, encases the membranous labyrinth; thus the delicate membranous labyrinth floats in the perilymph and is well protected. This protection is essential because the sensory cells for each part of the inner ear (termed hair cells) are located on the inner surface of the membranous labyrinth. These sensory cells send messages to the brain via the eighth (vestibulocochlear) cranial nerve.

The inner ear has three major regions: (1) the bony **cochlea**, enclosing the membranous **cochlear duct,** (2) the bony **semicircular canals,** surrounding membranous **semicircular ducts**, and (3) the bony **vestibule**, which encases two chambers called the **utriculus** and **sacculus**—part of the membranous labyrinth. The first region is concerned with hearing, and the second and third are concerned with the sense of balance or equilibrium.

The cochlea of birds is an elongated structure containing three fluid-filled canals **(Fig. 4–50)**. (In mammals the cochlea is curled like a snail's shell, thus giving rise to the name, which means "snail" in Latin.) The upper and lower canals (called **vestibular** and **tympanic canals**, respectively), connected at one end, are part of the bony labyrinth and therefore contain perilymph. Between them is the **cochlear duct**, part of the membranous labyrinth, filled with endolymph. The lower membrane of the cochlear duct is called the **basilar papilla**, and lying close above it is the **tectorial membrane**; all along the basilar papilla are sensory hair cells.

Recall that sound waves in the air vibrate the eardrum, which transfers its motion to the thin, toothpick-like columella. The columella moves the vestibular window and sends pressure waves through the perilymph of the cochlea and endolymph of the cochlear duct. The basilar papilla is set into motion by these pressure waves, moving

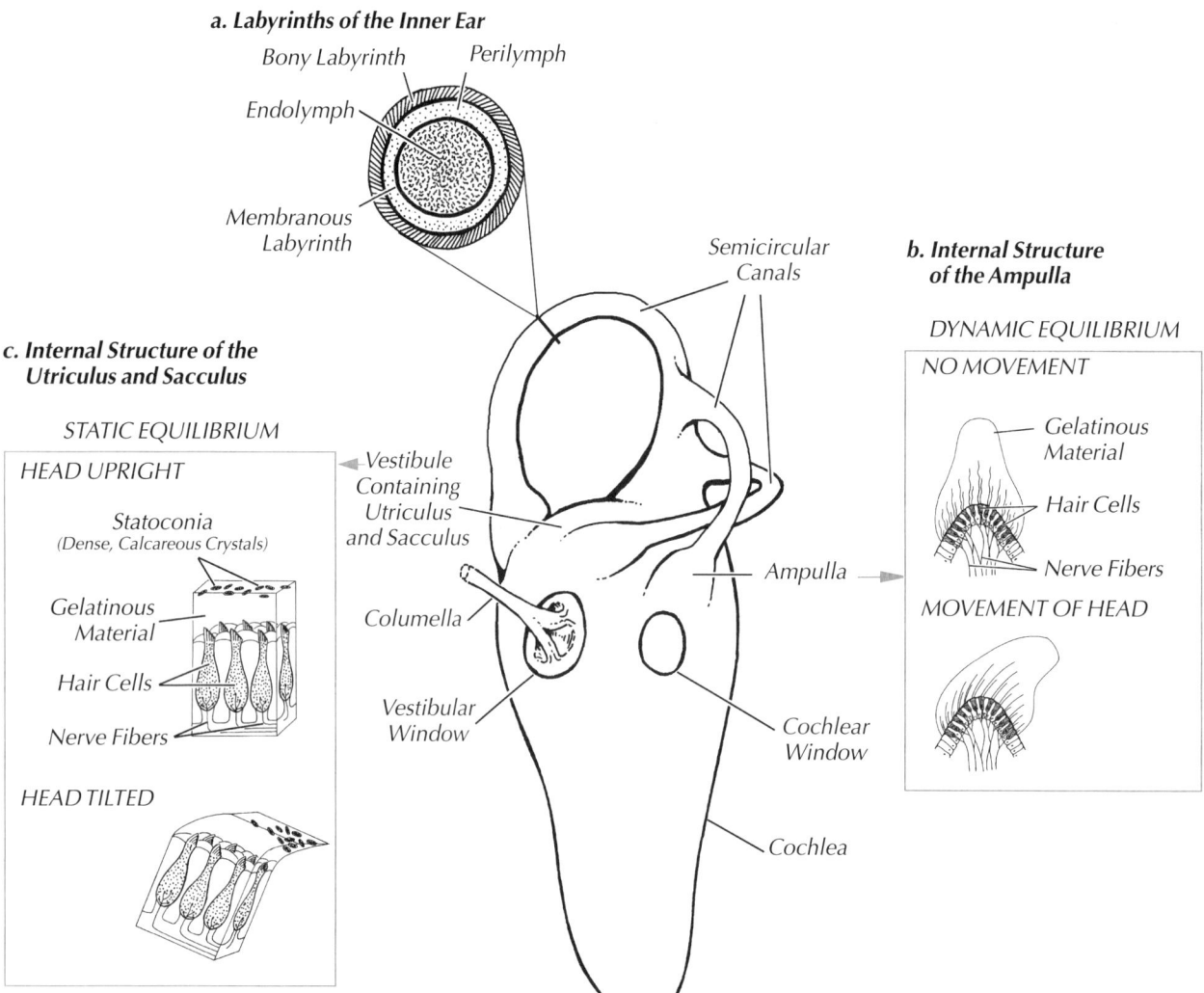

Figure 4–49. Structure and Functions of the Inner Ear: *The bony, fluid-filled inner ear has three major parts: the* **semicircular canals** *and the* **vestibule**, *both of which control balance (equilibrium), and the* **cochlea**, *whose role is in hearing. The* **columella** *of the middle ear contacts the surface of the bony labyrinth (see [a] below) at the* **vestibular window** *(formerly called the oval window), a pliable region marking the base of the cochlea. A second soft spot, the* **cochlear window** *(formerly called the round window), is located nearby. The cochlea and the mechanism of hearing are described in Figure 4–50. The three semicircular canals are arranged approximately at right angles to one another, each lying in a different plane of space.* **a. Labyrinths of the Inner Ear:** *This inset shows a cross-section through one of the semicircular canals, and is representative of the general internal structure of the inner ear: two fluid-filled systems, termed labyrinths, one inside the other. Innermost is the* **membranous labyrinth**, *filled with* **endolymph**, *which floats in the* **perilymph** *inside the* **bony labyrinth**, *the outermost layer of the inner ear.* **b. Internal Structure of the Ampulla:** *The base of each semicircular canal contains a chamber, the* **ampulla**, *which contains sensory hair cells embedded in gelatinous material. Endolymph surrounds the gelatinous material, filling the rest of the ampulla. As the bird changes speed or direction, thus moving the head, the endolymph lags behind, pressing* on the gelatinous material and displacing it, such that it bends the hairs of the hair cells, stimulating them to send impulses to the brain by means of their associated nerve fibers. For any given movement, each of the ampullae and thus its hair cells is stimulated to a different degree, depending on the degree to which each plane of space is involved in the movement. For example, just moving the head horizontally would stimulate only the hair cells in the ampulla oriented in the horizontal plane of space. The brain combines information from each of the three ampullae to interpret the bird's motion in space, a type of balance known as dynamic equilibrium. **c. Internal Structure of the Utriculus and Sacculus:** *The vestibule contains two organs, the* **utriculus** *and the* **sacculus**, *which perceive static equilibrium (the position of the head with respect to gravity). These two chambers each contain hair cells and dense crystals called* **statoconia**, *both embedded in a gelatinous material that is surrounded by endolymph. As the bird changes the position of its head, the statoconia move around and sink in the direction of gravity. This causes the gelatinous material to sag, bending and stimulating certain hair cells. The stimulated hair cells consequently send nervous impulses to the brain, allowing it to determine which direction is down from which particular set of hair cells is stimulated. Drawings by Christi Sobel.*

the hair cells against the tectorial membrane, and triggering a nerve impulse in the affected hair cells. Different frequencies of sound cause different portions of the basilar papilla to vibrate—the highest frequencies causing greater movement in the proximal end of the membrane (near the vestibular and cochlear windows). The brain determines pitch by registering *which region* of hair cells are stimulated **(Fig. 4–51a)**, and is thought to determine tone (the quality of a sound) by the *distribution* of hair cells stimulated. The volume (loudness) of a sound is determined by the amount of pressure of the sound wave. Loud sounds cause more vigorous vibrations of the eardrum and thus of the fluid in the cochlea, and the resulting increased stimulation of hair cells is interpreted by the brain as a loud noise **(Fig. 4–51b)**. At the end of the bony labyrinth farthest from the vestibular window is a second soft spot, the **cochlear window** (formerly called the round window), which abuts the dead space of the middle ear (see Fig. 4–50). The cochlear window acts both as a pressure-release valve and as a damper for the waves in the cochlea. Each wave is dissipated as it expends its remaining energy distending the cochlear window membrane into the middle ear, preparing the inner ear to receive new pressure waves from new sounds.

The delicate structures of the inner ear are protected from very loud sounds by the action of a small muscle that enters the middle ear cavity from the outside and attaches to the columella. When a loud

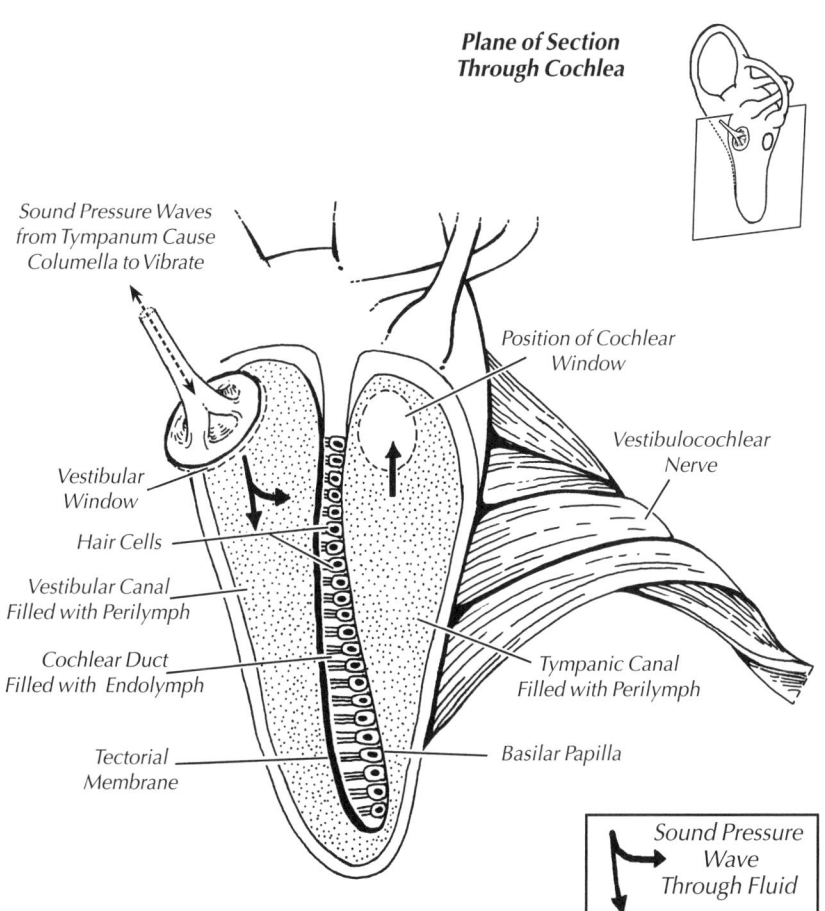

Plane of Section Through Cochlea

Sound Pressure Waves from Tympanum Cause Columella to Vibrate

Position of Cochlear Window

Vestibulocochlear Nerve

Vestibular Window

Hair Cells

Vestibular Canal Filled with Perilymph

Cochlear Duct Filled with Endolymph

Tympanic Canal Filled with Perilymph

Tectorial Membrane

Basilar Papilla

Sound Pressure Wave Through Fluid

*Figure 4–50. Internal Structure and Hearing Mechanism of the Cochlea: The cochlea of birds is an elongated structure containing three fluid-filled canals. Two of them, the **vestibular canal** and the **tympanic canal**, connected to each other at the tip of the cochlea, are part of the bony labyrinth, and therefore contain perilymph. The endolymph-filled canal between them (the **cochlear duct**, part of the membranous labyrinth) consists of a membrane, the **basilar papilla**, along which lie numerous sensory hair cells overlaid by another membrane, the **tectorial membrane**. The hair cells are short at the base of the cochlea, becoming longer toward its tip. The flexible **vestibular window**, at which the columella of the middle ear contacts the cochlea, lies at the base (proximal end) of the vestibular canal. Nearby, at the base of the tympanic canal, is the **cochlear window.** Sounds are perceived in the following way: Sound waves through air vibrate the eardrum, and this motion is transferred to the columella, which in turn vibrates against the vestibular window (dashed arrows), sending sound pressure waves (solid arrows) through the perilymph of the cochlea and the endolymph of the cochlear duct. Different frequencies of waves cause different portions of the basilar papilla to move, moving certain hair cells against the tectorial membrane and bending them, triggering nerve impulses. The impulses are transmitted through the vestibulocochlear (auditory) nerve to the brain, where they are interpreted as sound. The flexible cochlear window acts as a damper in the system, stretching to allow the sound waves to dissipate.*

Figure 4–51. How Pitch and Volume (Loudness) are Perceived: a. Pitch: Sounds of different pitches (frequencies) are distinguished by the cochlea because sound pressure waves of a given frequency travelling through the fluid of the cochlear duct cause a specific region along the basilar papilla to vibrate more strongly than other regions, moving its attached hair cells against the tectorial membrane and thereby triggering nerve impulses. The differential stimulation of hair cells is interpreted by the brain as sound of a certain pitch. The frequency sensitivity of the basilar papilla changes along a gradient, the membrane being most sensitive to high frequencies at the proximal end and becoming more sensitive to low frequencies toward the distal end. b. Volume: Volume (loudness) is a function of the amplitude (height) of the sound wave. The greater the amplitude of a sound, the more vigorous the vibrations of the fluid in the cochlea, and thus the greater the displacement (bending) of the hair cells along the basilar papilla. Stronger stimulation of the hair cells results in more nerve impulses, which the brain interprets as sound of a higher volume.

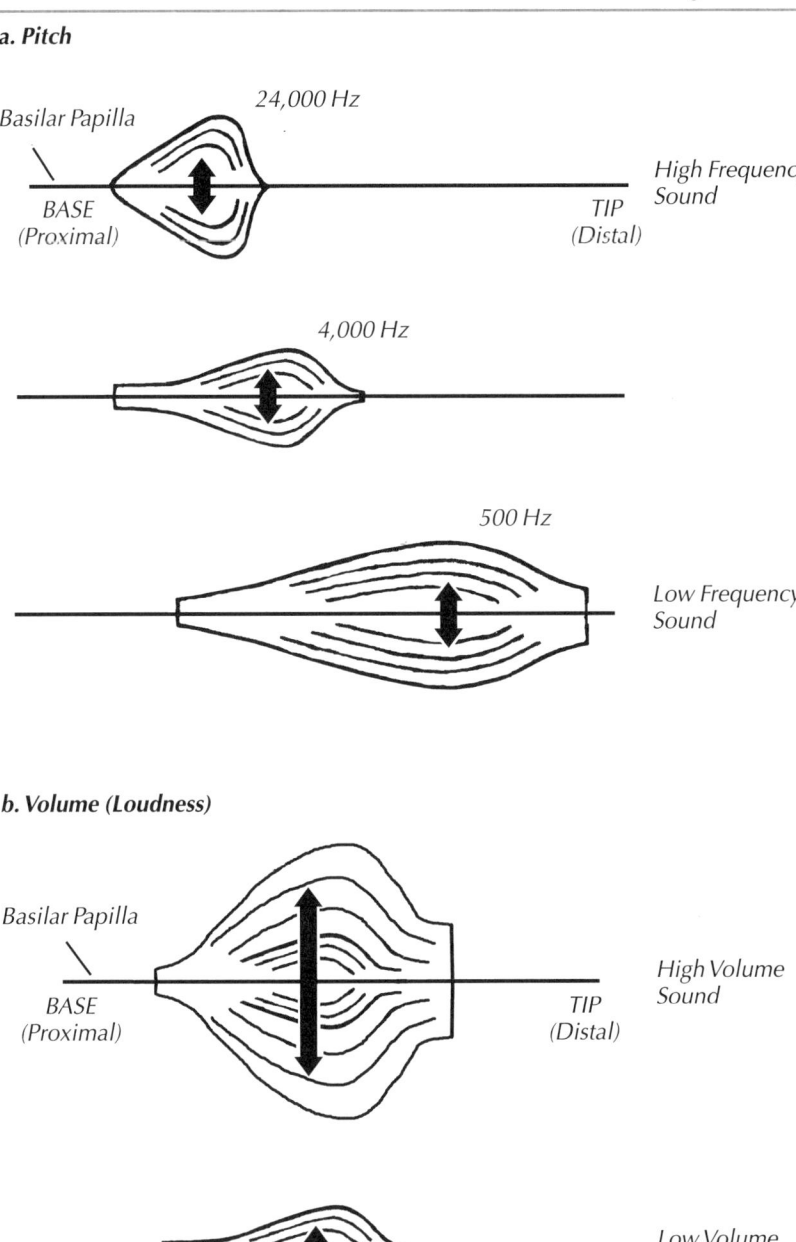

a. Pitch

noise is received, this muscle contracts, restricting transmission of the full force of the eardrum's vibration to the inner ear, thereby protecting it.

In all vertebrates, the inner ear functions in balance as well as in hearing, but the sense of balance is especially important to birds because of the three-dimensional acrobatics of flight. Balance has two components: dynamic, the perception of motion; and static, the perception of gravity. The semicircular canals and ducts function in dynamic balance, and the vestibule, in static balance.

The narrow, ring-like semicircular canals and their enclosed ducts are arranged at approximately right angles to each other, so that one is located in each of the three planes of space (see Fig. 4–49). The base of each semicircular duct (called the ampulla) is widened and contains hair cells embedded in a gelatinous material, surrounded by endolymph. When the bird changes its speed or direction (accelerates), the endolymph in the canals lags behind (because of inertia), pressing on the gelatinous material, which consequently applies greater pressure on the hair cells of whichever planes of space are involved in the acceleration (see Fig. 4–49b). Stimulation of the hair cells causes the vestibular nerve to send impulses to the brain. The brain determines how the bird moved by combining the relative amounts of stimulation it receives from each of the three ampullae.

The vestibule contains two chambers, the **utriculus** and the **sacculus**. Each contains hair cells and dense crystals of calcium carbonate called **statoconia**, both embedded in a gelatinous material (see Fig. 4–49c). The gelatinous material is suspended in the endolymph. The bird's brain determines which direction is down by registering which sensory cells are stimulated by the gravity-induced settling of these dense crystals.

Hearing Ability

We assume that an animal hears, at the very least, the sounds that it and other individuals of the same species produce. We also assume that it hears the relevant sounds of prey and predator. Humans in their prime can usually hear sounds between 16 and 20,000 vibrations, or cycles per second (Hertz). The tested hearing range in a series of birds varied from a low of 40 Hertz in the Budgerigar to a high of 29,000 Hertz in the Chaffinch, a common European bird (see Fig. 7–5). In general, most birds have the greatest sensitivity to sounds in the frequency range of 1,000 to 5,000 Hertz—approximately the top two octaves on a piano. Their ability to detect these sounds appears similar to that of humans within this range of frequencies.

Each species, however, appears to have its own distinct range of frequencies to which it is sensitive. Passerines perceive high-frequency sounds better than most nonpasserines, but nonpasserines perceive low-frequency sounds better than passerines **(Fig. 4–52)**. As discussed later in the chapter, some bird species are apparently able to perceive ultrasound, although whether they actually "hear" it with their ears or detect it in some other way is unknown. In addition, some differences occur between young and adult birds of the same species. Downy young of the chicken respond primarily to the low-pitched clucks of the hen, whereas the hen is especially sensitive to the high-pitched peeps of the chicks. Compared to humans, Great Horned Owls hear low frequency sounds very well, and Barn Owls hear high-frequency sounds

Figure 4–52. Hearing Thresholds of Songbirds Versus Nonsongbirds: This graph shows the hearing thresholds (defined as the lowest intensity, or volume, at which sounds of a given pitch, or frequency, can be perceived) of nine species of songbirds (dark circles) and seven species of nonsongbirds (white circles). The y-axis is volume in decibels (dB), and the x-axis presents frequency on a logarithmic scale in kiloHertz (kHz). Songbirds are able to perceive high frequency sounds at lower volumes than nonsongbirds. For instance, a 10 kHz sound is first heard by songbirds at a volume of about 70 dB, but the same frequency sound must be louder, nearly 90 dB, for a nonsongbird to hear it. Conversely, nonsongbirds are able to perceive low-frequency sounds at lower volumes than songbirds. For example, nonsongbirds first perceive a 0.2 kHz sound at 40 dB, but the sound must be over 50 dB before a songbird can hear it. Songbirds tested were canary, American Crow, European Starling, House Finch, Blue Jay, Brown-headed Cowbird, Red-winged Blackbird, Field Sparrow, and bullfinch. Nonsongbirds tested were Barn Owl, Great Horned Owl, Rock Dove, Budgerigar, Mallard, turkey, and American Kestrel. For more about bird sound and its measurement, see Chapter 7. Adapted from Gill (1995, p. 194).

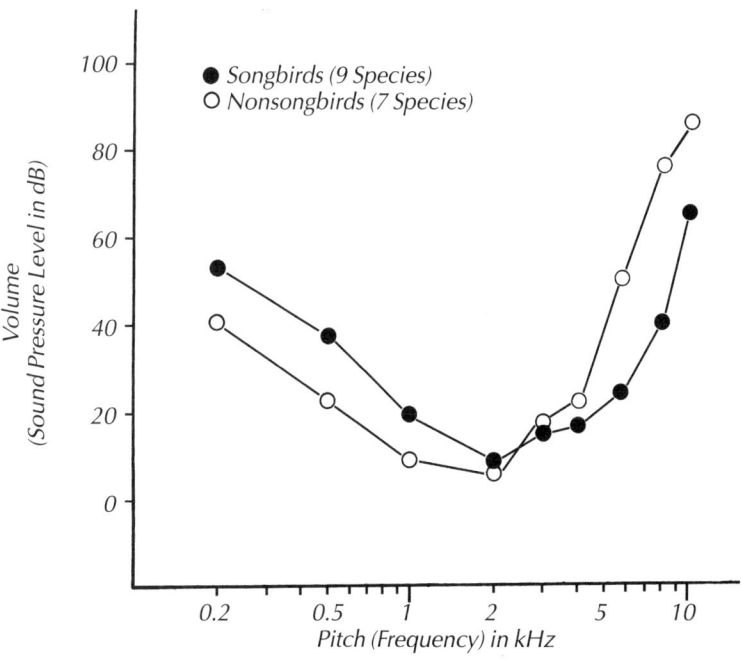

well. In total darkness, many owls (and probably other nocturnal birds) locate prey with surprising accuracy by sound alone.

Experiments show that certain birds use echolocation to avoid obstacles while flying in the dark, just like bats do. This process is well-developed in the Oilbird of Trinidad and northern South America, which nests and roosts in caves **(Fig. 4–53)**. While flying, the Oilbird utters a continuous series of very short, click-like sounds that bounce back from the cave walls, informing the bird of its position in relation to reflecting objects. The response of a flying bird to these sounds is much more rapid than the response of humans to sounds. The Edible-nest Swiftlet, one of several swifts of Southeast Asia whose nests are used for bird's-nest soup, also uses echolocation in flying to and from its nest, which may be as far as 400 yards (366 meters) inside the entrance to its cave.

Olfaction

In all animals, the sense of smell involves the fitting of an airborne (or waterborne) molecule into a highly specifically shaped notch on the surface of an olfactory sensory cell. Much like a lock-and-key system, only the right "key" will turn on a particular sensory cell, which then fires a neural signal recognized as a specific odor in the brain.

As a group, birds have a poor sense of smell. Laboratory tests indicate that birds may generally be only one-third to two-thirds as sensitive to test odors as some fishes and mammals, including humans. The bird's rather weak olfactory sense has been inferred from the relatively small size of the olfactory lobes in most birds' brains. The bird's nasal cavities are discussed later in relation to the respiratory system, but note here that they are relatively small. The lining or surface tissue (epithelium) of each nasal cavity, the **olfactory epithelium**, contains the sensory endings of the olfactory nerves but occupies a very limited area. (The olfactory nerves carry impulses directly to the olfactory lobes in the brain.) Experimental evidence indicates, however, that a variety of birds *without* enlarged olfactory areas may nevertheless have rather keen olfactory abilities.

Most birds with large olfactory lobes and sensitive olfaction are ground-dwelling species, aerially hunting vultures, or marine birds **(Fig. 4–54)**. The ground dwellers include such birds as kiwis and possibly snipes. The kiwis, the only birds whose nostrils are near the tip of the bill, find earthworms and other prey living underground by smelling them. Their olfactory lobes are about 10 times the size of those of other birds **(Fig. 4–55)**. Whether the aerially hunting vultures locate dead animals primarily by smell or by sight has long been debated. Evidence from neuroanatomy, especially olfactory lobe size, indicates that the Turkey Vulture and the King Vulture of the American tropics both depend primarily on olfaction to locate their prey. These scavengers live in forested areas where carrion is usually hidden by the leafy canopy. Other American vultures, such as the Black Vulture, depend primarily on sight, but may follow other species to the carcasses they find by smell. In the Old World vultures, a distinctly different group of

Figure 4–53. Oilbirds in Cave: *The nocturnal Oilbird of Trinidad and northern South America uses echolocation to navigate through deep caves where it nests and roosts colonially. A flying Oilbird produces a series of short clicks, which bounce to and from the cave walls, informing the bird of the presence of stalactites and other obstacles.*

a. Bonin Petrel   **b. King Vulture**   **c. Great Spotted Kiwi**

Figure 4–54. Birds with Sensitive Olfaction: *Most birds do not appear to have a good sense of smell, but there are a few exceptions.* **a. Bonin Petrel:** *Tubenosed marine birds, especially the shearwaters, fulmars, and petrels—such as the Bonin Petrel shown here—are attracted to specific marine odors, helping them to locate the plankton on which they feed. Photo courtesy of Chandler S. Robbins/CLO.* **b. King Vulture:** *Evidence suggests that at least some aerially hunting vultures of the New World, such as the King Vulture shown here and the Turkey Vulture (see Fig. 5–32), locate the carrion on which they feed by its smell. Old World Vultures, however, seem to locate their food by sight. Photo by Steven Holt/VIREO.* **c. Great Spotted Kiwi:** *The nocturnal kiwis, such as the Great Spotted Kiwi shown here, use their highly developed sense of smell to locate their underground prey, primarily earthworms. They probe the ground with their long bill, which has nostrils located at the tip. Photo by B. Chudleigh/VIREO.*

Figure 4–55. Range of Olfactory Lobe Sizes: *Lateral views of the cerebral hemispheres of various birds illustrate the variability in the size of the olfactory lobe (shown in black) at the anterior end of the brain. Birds with large olfactory lobes, and by implication, a well-developed sense of smell, are typically ground-dwelling species, such as snipe, kiwi, and Emus, as well as certain ducks. The kiwi (see Fig. 4–54c), whose olfactory lobes are the largest, with respect to its brain size, of any bird, relies entirely on olfaction to locate the earthworms and other underground prey on which it feeds. Kestrel, parrot, and flycatcher olfactory lobes are very small, and the lobes are barely visible in the magpie, indicating that these species rely on vision rather than smell to find food. From Marshall (1961).*

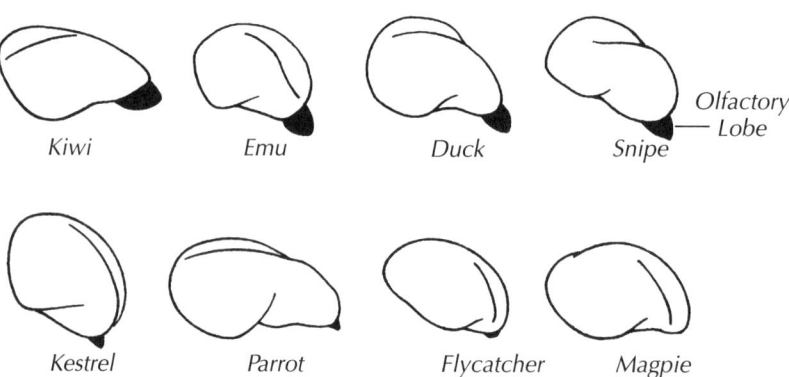

birds that evolved primarily on open savannas and treeless steppes, all species appear to depend entirely on vision to find their food.

Evidence suggests that many of the tubenosed seabirds, especially petrels, shearwaters, and fulmars, are able to smell and home in on specific odors such as the smell of plankton. In these birds, the olfactory lobes are more than one-quarter the size of the brain hemispheres. The attraction of several types of petrels to dimethyl sulfide is particularly interesting. Dimethyl sulfide is an aromatic substance released by phytoplankton (microscopic, drifting marine algae) when zooplankton (tiny, drifting animals such as copepod crustaceans) are feeding upon them (Nevitt 1999). The occasional dense patches of grazing zooplankton—which are the feeding targets of small petrels such as the Wilson's Storm-Petrel—are very difficult to distinguish by sight. Because petrels forage by night as well as by day, these olfactory cues could be critical for locating food. In experiments, artificial slicks of dimethyl sulfide were more attractive to the small, zooplankton-eating petrels than to the large albatrosses that feed on squids and fishes in the same regions **(Sidebar 1: The Amazing World of Avian ESP)**.

Taste

Taste and smell are related; both are types of chemical reception. In humans, both senses determine the flavor of foods. The **taste bud**, a simple structure usually embedded in the epithelium of the oral cavity and the tongue, is the receptor for taste sensations in all vertebrates. Although the sense of taste in birds has been studied very little, it is thought to be poorly developed. Birds have few taste buds when compared to mammals. They number as few as 24 in the chicken, 27 to 59 in the Rock Dove, 62 in the Japanese Quail, and about 300 to 400 in various parrots. By contrast, an adult human—a species not noted for its good sense of taste—has over 10,000 taste buds! Among birds, however, gustatory ability seems not to be correlated closely with the number of taste buds. Birds have few of their taste buds on their tongue and none near the tongue's tip. Most of a bird's taste buds are on the roof of the mouth or deep in the oral cavity **(Fig. 4–56)**. Furthermore,

(Continued on p. 4·69)

*Figure 4–56. Location of Taste Buds in Human and Mallard: Birds' sense of taste seems to be poorly developed, as suggested by the few taste buds present in the oral cavity of most species. An adult human (left) has over 10,000 taste buds, located in circular **vallate papillae** at the back of the tongue and in **fungiform papillae** around the sides and tip of the tongue. In contrast, the Mallard has only about 400 taste buds, none of which are located on the tongue. The schematic view of a Mallard's oral cavity (right), with tongue removed and bill held open wider than possible in life, shows the location of the taste buds. They are scattered mostly in the palate (roof of the mouth), with a few in the floor of the mouth.*

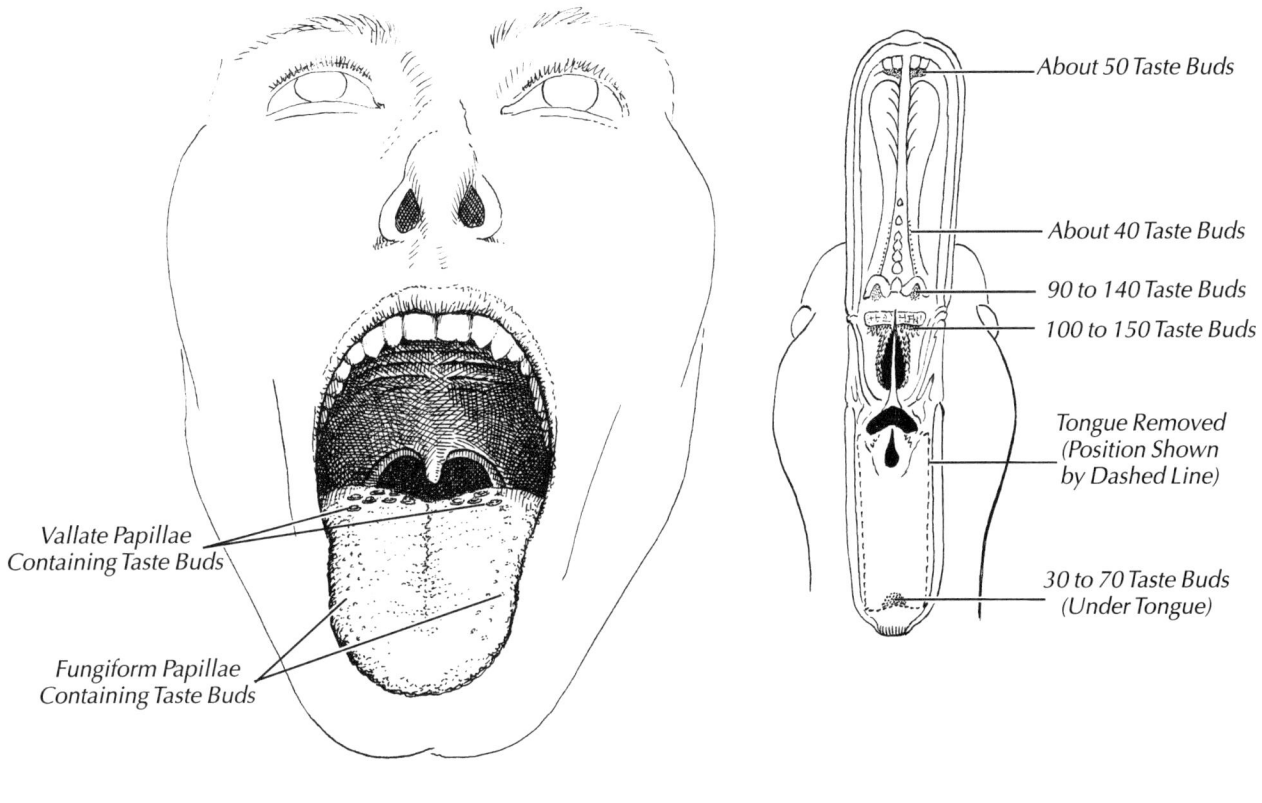

Vallate Papillae
Containing Taste Buds

Fungiform Papillae
Containing Taste Buds

About 50 Taste Buds

About 40 Taste Buds

90 to 140 Taste Buds

100 to 150 Taste Buds

Tongue Removed
(Position Shown
by Dashed Line)

30 to 70 Taste Buds
(Under Tongue)

Human
*About 10,000 Taste Buds,
Mostly on Tongue*

Mallard
*About 400 Taste Buds,
None on Tongue*

Sidebar 1: THE AMAZING WORLD OF AVIAN ESP
J. B. Heiser

Although there is little in the literature about the ability of birds to communicate with their long-deceased ancestors or to see into the future, there is an amazing, and ever-growing literature on the ESP capabilities of birds. The concept of ESP (extrasensory perception) is based on the sensory capacities of humans; whatever is beyond the detection of the human senses is, by definition, "extra sensory." For a long time, our anthropocentric perspective slowed Western science from even considering that the sensory experience of other organisms might be different, let alone that it might be broader and more encompassing, than our own. In the 1820s the observation that younger people could hear high-frequency insect song that older ears could not detect led to questioning whether there might not be sounds of such high frequency that no human ear could hear them. But it was not until 1879 that Sir John Lubbock experimentally proved that ants could see light in the ultraviolet range but were blind to red light. Lubbock thought, but could not provide proof, that ants also make and hear ultrasonic sounds. He did, however, put his finger directly on the essence of animal ESP: "the universe is probably full of music which we cannot perceive...If any apparatus could be devised by which [these sensations could] be brought within the range of our [senses], it is probable that the result would be most interesting." (Pye and Langbauer 1997).

Extrasensory Hearing

This "most interesting" world was entered by Sir Francis Galton when in 1883 he mounted, on the end of his walking stick, a directional, tunable, and high-pitched whistle that he could blow from a distance through a long rubber tube. He struck off "through the whole of the Zoological Gardens" poking his whistle on the end of his walking stick "as near as is safe to the ears of animals" (as quoted in Nowicki and Marler 1988). By observing the reactions of animals to sounds of a higher frequency than humans could hear, Galton was readily able to demonstrate ultrasonic sensitivity in a range of zoo inhabitants, especially members of the cat family. Historically, animal ESP researchers often have had to wait for appropriate technologies to be invented and applied before they could observe phenomena that might be quite common among animals, but had never before been accessible to the human senses.

Unlike Galton's zoo mammals, most birds seem not to be very sensitive to ultrasound (very high sounds ranging from about 15 to 20 kiloHertz up to 200 kiloHertz); only small songbirds make and hear sounds just above our hearing range—to a maximum of 29 kiloHertz. Even those few birds known to echolocate appear to use many frequencies audible to human ears. The South American Oilbird, when flying inside its roosting caves, produces harsh clicks very audible to humans. The Southeast Asian Edible-nest Swiftlets make a sound much like running one's thumb up and down the teeth of a comb as they navigate deep in dark caves.

On the other hand, some birds have been shown to be sensitive to infrasound (very low sounds ranging down from about 20 Hertz to 0.1 Hertz or less). Pigeons have cochlear neurons sensitive to sound frequencies below 20 Hertz and demonstrate behaviorally that they can detect infrasounds as low as 0.05 Hertz! (Schermuly and Klinke 1990). Typically there is little ambient ultra-sound in a natural environment, but quite the opposite is true of infrasound. Weather phenomena such as thunder, wind blowing over and through topographic features, earthquakes, and ocean waves are all sources of infrasound. Infrasounds travel long distances through air with little attenuation (weakening), remaining loud enough to be detected by the kind of sensitivity demonstrated by pigeons, even after propagating hundreds or even thousands of miles from their sources. Although detectable, infrasounds are difficult to *use* in extracting directional information because their wavelengths are measured in yards to tens of yards—longer by far than the distance between the two ears of any but the biggest of animals. Although the idea that animals use Doppler effects (analogous to detecting the change in pitch between the whistle of an approaching versus receding train) to extract directional information from infrasounds has been *proposed*, if and how any animal can get directional information from infrasound remains problematic. A few animals emit infrasounds as communication: some of the great whales—such as blue and fin whales—both species of living elephants, and, among birds, the Eurasian Capercaillie, a giant European grouse.

Extrasensory Olfaction

Although humans have a much better overall sense of smell than do birds, it is clear from experiments done with certain wild birds that in some cases avian sensory perception, even in the area of olfaction, is beyond ours. As discussed in the chapter text, Wilson's Storm-Petrels, one of the smallest of the tubenosed seabirds, can detect and home in on dimethyl sulfide slicks from great distances—demonstrating, for at least this one substance, a sensitivity much greater than that of humans.

Under certain conditions, homing pigeons have been shown to use olfaction in orientation (see Ch. 5, Navigational Maps), depending, it seems, on how and where they were housed and raised; but it remains unlikely that odors play an essential role in bird navigation (Waldvogel 1983; Able 1996). Unfortunately, we currently know little more about avian olfactory ESP.

Extrasensory Vision

Much more is known about the extraordinary visual sensitivities of birds. In addition to their high visual acuity, birds see in a much broader spectrum of "colors" than do humans. Humans are unable (and perhaps unusual in this inability) to perceive ultraviolet (UV) light (wavelengths of 3,000 to 4,000 Angstroms [300 to 400 nanometers]). In contrast, UV vision appears to be a general ability of most birds (Bennett and Cuthill 1994). In fact, birds appear to be more sensitive to UV than to light in the part of the electromagnetic spectrum visible to humans (4,000 to 7,000 Angstroms [400 to 700 nanometers]): the peak sensitivity for the vision of the majority of birds that have been tested is between 3,600 and 3,800 Angstroms (360 and 380 nanometers). Birds can also distinguish *between* wavelengths within the UV portion of the spectrum, perceiving "color" differences where we see no illumination at all. Birds apparently achieve UV vision through having UV-transparent corneas and lenses (ours absorb UV and are thus opaque to it) and by having special cone cells with visual pigments that absorb maximally in the violet or ultraviolet range. Many organic molecules (including DNA) absorb UV wavelengths, but in the process the electromagnetic energy can disrupt the structure of the molecule—reason enough to have the otherwise transparent structures at the front of the human eye absorb UV before it can do damage to retinal pigments. Why, then, should many invertebrates (especially insects), fish, amphibians, reptiles, and a few mammals, as well as most birds, take the risk of UV damage by having eyes sensitive to it? Birds apparently benefit from being sensitive to UV light in three major ways: (1) foraging, (2) species recognition and sexual selection, and (3) orientation and navigation.

Foraging

Urea, the main nitrogenous component of mammalian urine, strongly reflects and fluoresces (re-emits radiation) in the ultraviolet. Voles and other small rodents heavily mark their above-ground trails (termed "runways") with urine and feces so that other individuals will be able to recognize these trails (by their odor). In the wild, Eurasian Kestrels have shown a preference for areas where abandoned runways were freshly treated with vole urine and feces compared to similar areas where runways were untreated. In the laboratory, kestrels spent more time hovering and inspecting urine-treated runways illuminated under UV light than similarly treated runways illuminated by white light (Vitala et al. 1995). It seems that foraging kestrels can judge the probable productivity of a potential hunting area by its ultraviolet "color" caused by rodent urine!

No less exotic is the fact that some flowers and fruits have colors or patterns that are visible only to animals that can see in the ultraviolet—and birds may use these "UV foraging guides" to locate sources of food. It has long been recognized that many insect-pollinated flowers have UV patterns that encourage pollination by guiding insects to the part of the flower where the nectar (as well as the pollen) is located. Future research may show that bird-pollinated flowers have similar UV foraging guides. Because many of the fruits that birds feast upon reflect strongly in the UV, but the leaves surrounding them do not, the fruits stand out brightly—possibly strengthening the ripe fruit's signal that "I'm ripe—so come have a meal and disperse my seeds." In addition, many insects that hide from their bird predators in the leaves match the leaf's reflectivity in both the human visible spectrum and the UV, attaining good camouflage. Some, however, reflect strongly in the UV. Could they be giving off an aposematic (warning) signal?

Species Recognition and Sexual Selection

Many species of birds have plumage and fleshy ornaments that reflect strongly in the UV. Iridescent feathers, brilliant to the human eye, reflect strongly in the UV, but there are also feathers that reflect very little in the human visible spectrum (and thus look black to us) but reflect very strongly in the UV and thus must appear riotously colored to the admiring bird. The Asian whistling-thrushes are blue, purple, violet, and brown birds that have a very different pattern and appearance when seen in the ultraviolet (see Fig. 3–51) (Andersson 1996). Male and female Blue Tits (Eurasian relatives of chickadees) appear nearly identical to the human eye, but in UV their crown patch, which is displayed in courtship, differs between the sexes in the purity of its UV/blue color (Andersson et al. 1998). Blue Tits tend to

4

display in early morning "woodland shade" that is rich in UV light.

Until recently, studies of species recognition and sexual selection have assumed that avian color vision is like our own. All these now seem to be potentially invalidated, as studies that incorporate the ability of birds to see UV often produce unexpected results. For example, when given a choice, Zebra Finches preferred a partner viewed through UV-transparent Plexiglas over a partner viewed through UV-opaque material (Bennett et al. 1996). Hunt et al. (1997) found that even artificial adornment with highly UV-reflective leg bands gave male Zebra Finches a leg up over their less spiffily attired competitors. In other experiments, female starlings ranked males differently when UV light was present than when it was absent. When UV was available, they preferred males with the most UV reflectance, but when UV was absent, they ranked males in a way that was not based on their plumage reflectance in the human visible spectrum (Bennett et al. 1997).

Experiments with the Bluethroat, a Eurasian thrush, have clearly demonstrated the importance of UV-reflectant plumage in reproductive success.

The male Bluethroat actively displays his bright throat pattern during courtship—a colorful, blue signal even to our eyes, but the throat reflects most strongly in the UV. This UV reflection is greater in two-year-old males than in one-year-old males (Andersson and Amundsen 1997). Sunblock (which absorbs very strongly in the UV) applied to the throats of males nearly ruined their sex lives! Females in 13 of 16 trials associated most with males whose UV reflectance had not been eliminated. Sunblock-treated males had greater difficulty attracting mates and thus had a significantly later start to egg laying in their nests compared to the nests of control males. UV-reduced males had lower success in copulating with females other than their mates, and they lost paternity in their own nests, as well, when their mates copulated on the sly more frequently with non-UV-reduced males (Johnsen et al. 1998). Clearly, for a male Bluethroat, losing your UV-reflectance is a serious problem!

Orientation and Navigation

The physical properties of UV light make radiation in this range of wavelengths potentially important for orientation and navigation. Sunlight passing through the atmosphere is scattered by the molecules it strikes and is simultaneously polarized (see Ch. 5, Sidebar 3: Polarized Light). UV light is scattered more and polarized to a higher degree than light of the longer wavelengths of the human visible spectrum (Bennett and Cuthill 1994). Thus, being able to see in the ultraviolet should make it easier for birds to gain significant information about the position of the sun, and thus compass directions, from the patterns of polarized light in the sky. Another consequence of wavelength-dependent scattering is that around dawn and dusk a high proportion of the light available is of short wavelengths, making UV vision particularly adaptive for animals active at these times. Is this how the early bird gets the worm?

Finally, there are senses of birds that are more than just extensions of the senses with which we are so familiar. They are senses for which we personally know no counterpart. An important example is the well-documented magnetic sense of birds. A magnetic compass has been demonstrated in at least 18 species of migrating birds held in the laboratory and exposed to artificial magnetic fields (Wiltschko and Wiltschko 1996). But what are the perception mechanisms used by these birds? Current research is focused on tiny particles of magnetite (an iron-containing, magnetic mineral) that is found in the heads of many birds, and on the possibility that photoreceptive pigments in the eye might provide the sensor—but just how the magnetite and other senses interact to calibrate or set the magnetic compass is not known. The pineal gland is also under consideration as a structure actively responsible for magnetic field detection (Azanza and Del Moral 1994), but little is clearly understood about birds' magnetic ESP. ∎

Suggested Reading

Withgott, Jay. 2000. Taking a bird's-eye view . . . in the UV. *BioScience* 50(10):854–859.

birds do not chew their food; they may manipulate it in their bill or tear off pieces, but then they "bolt it whole" without much processing or time to taste it.

Birds do quickly learn to avoid distasteful substances. Wildlife biologists at Cornell have found that a common flavoring used for grape soda and gum—methyl anthranilate, which naturally occurs in orange blossoms and Concord grapes—is very distasteful to birds. The Environmental Protection Agency has approved the use of this compound at golf courses, airports, and landfills to discourage geese and gulls.

Skin Senses

Birds have many different types of tactile nerve endings in the skin, tongue, palate, muscles, joints, feather follicles, and viscera. The precise function of any one of these sensory endings is debatable. Some sensory organs respond to touch and vibration, others to heat or chemical sensations. Birds' skin sensations originate in many kinds of nerves: in nerve endings wrapped in connective tissue; in dendritic (highly branching) nerve endings; in free simple nerve endings; and in a host of variations with different physiological responses under laboratory conditions. These nerve endings usually have been given names (such as Grandry corpuscles, Herbst corpuscles, Paccinian corpuscles, Ruffini endings, and Merkels complex) based on name of the discoverer.

The **bill tip organ** of ducks and shorebirds is an aggregation of sensory cells best developed in snipe, sandpipers, ducks, and geese **(Fig. 4–57)**. The organ appears to be especially sensitive to rates and changes in rates of vibration stimuli, but its function is unknown. The beaks of parrots have a unique collection of tactile-sensitive cell clusters associated with the shelling and manipulation of seeds and other objects. The brush-turkeys (megapodes) of Australia, New Guinea, and the Indo-Malaysian Archipelago incubate their eggs in unusual ways: in decaying vegetation, in a sun-warmed earth mound, or sometimes in an ash mound heated by volcanic steam! The male, who usually takes sole responsibility for incubation, apparently senses the temperature of the egg site with receptors in his tongue or palate. Males that incubate in rotting vegetation and volcanic ash then uncover or add cover to the mound, as needed, to maintain the optimal incubation temperature (see Fig. 6–36).

The Endocrine System

■ In addition to the fast-acting nervous system, vertebrates have a much slower coordinating system called the **endocrine system**. These two systems interact to produce appropriate responses at a subconscious level. The endocrine system consists of widely separated **endocrine glands**, which secrete specific complex molecules directly into the blood **(Fig. 4–58)**. These endocrine secretions, called **hormones**, are carried by the blood to all parts of the body, where they stimulate or regulate the activities of other glands or organs. Both the nervous and the endocrine systems are concerned with initiating and regulating bodily activities, and the actions of these two systems are very closely integrated. A bird's anatomical development and its specific behaviors, like those of a human, are the result of the interplay between the nervous and endocrine systems. In humans the cerebral cortex of the brain can sometimes modify the basic drives that result from the interaction of hormones and physiological processes. Such brain control appears to be less common in other vertebrates, including birds, which in part accounts for the stereotyped behavior characteristic of birds. The details of the dynamics and interactions of hormones are among the most complex phenomena in the biological sciences. Here, we consider only the most important endocrine glands and some of their principal functions.

Although the endocrine system of the bird and mammal are similar, several differences exist, not

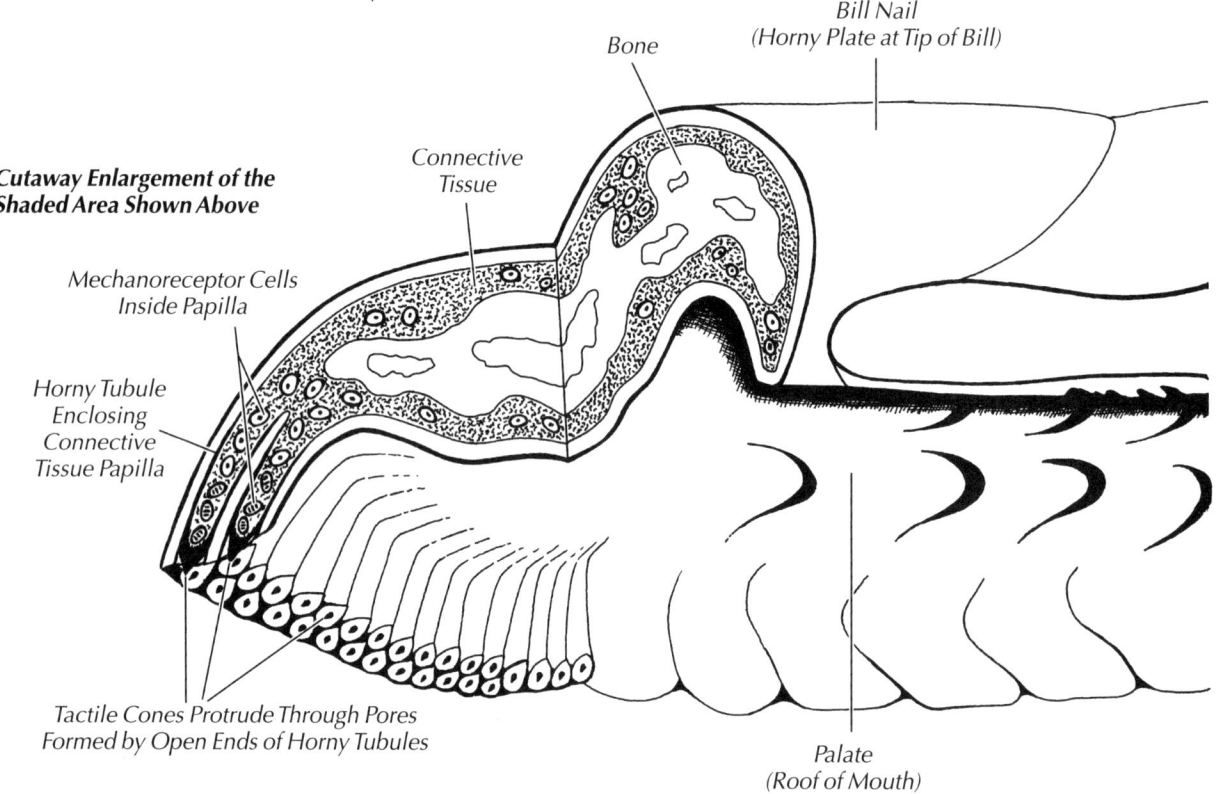

Location of the Bill Tip Organ in the Upper Bill of the Domestic Goose

Cutaway Enlargement of the Shaded Area Shown Above

Mechanoreceptor Cells Inside Papilla

Horny Tubule Enclosing Connective Tissue Papilla

Connective Tissue

Bone

Bill Nail (Horny Plate at Tip of Bill)

Tactile Cones Protrude Through Pores Formed by Open Ends of Horny Tubules

Palate (Roof of Mouth)

Figure 4–57. The Bill Tip Organ of the Domestic Goose: A bill tip organ is found in both the upper and lower bill of geese, ducks, and shorebirds; it is thought to sense tactile stimuli during feeding. In ducks and geese it is located within the bill nail, the hooked structure at the tip of the bill. Externally it appears as double, or sometimes triple, rows of microscopic pores arranged around the inner surface of the tip of the bill. The cutaway view shows the internal structure of the bill tip organ. Note that only the upper bill is shown, and it is slightly rotated away from the viewer. Microscopic examination reveals that each pore is the open end of a **horny tubule** through which protrudes a peg-like structure, the **tactile cone**, made of keratin. This flexible cap of keratin is attached to a long, slender **connective tissue papilla**, originating deep within the dermal tissue of the bill, which fills the horny tubule. The papilla contains numerous **mechanoreceptor cells** of two distinct types—Grandry and Herbst corpuscles—with their associated nerve fibers, which become activated when the tactile cone receives stimulation. Drawing by Christi Sobel, after Gottschaldt and Lausmann (1974).

all of which are fully understood. Most experimental studies of avian endocrinology have involved the Japanese Quail or the chicken, both of which have relatively short life spans. Studies are needed of wild species, from a variety of short-lived passerines to species with longer life spans.

Seabirds have life spans of 20 to 50 years and mature at a late age. Little is known about their reproductive hormones during their long fertile years, which may include 20 or more annual clutches. The Common Tern, for example, may have its first clutch at three to four years of age, but normally reaches full breeding performance only after five years when plasma sex hormones reach adult levels. Some terns have

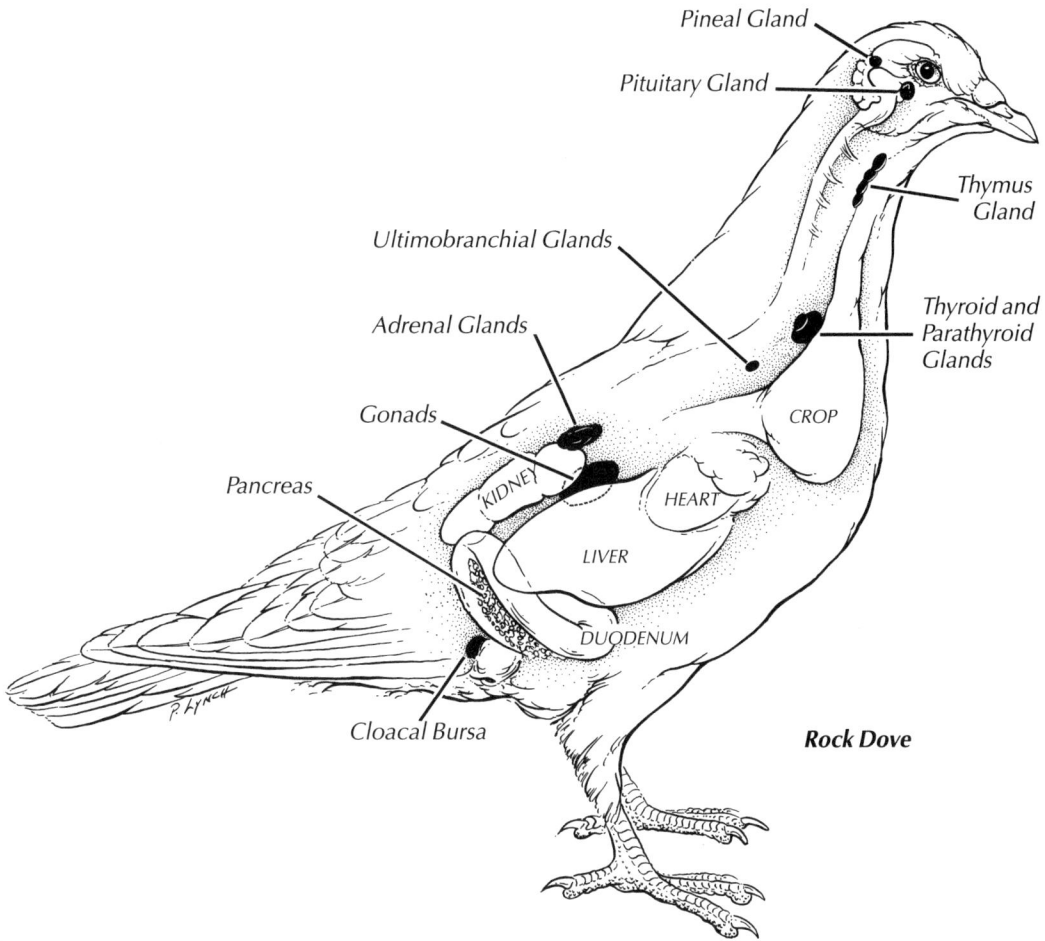

Figure 4–58. Major Endocrine Glands: *The endocrine glands play an essential role in controlling and coordinating body processes by secreting hormones. Transported by the blood-stream to target areas throughout the body, hormones regulate the activities of other glands and organs. Shown here are some of the most important endocrine glands of the Rock Dove. The* **pituitary gland** *is located on the ventral side of the brain, where it is connected by a stalk to the hypothalamus. At the base of the neck are the* **thyroid** *and* **parathyroid glands** *, and nearby are the* **ultimobranchial glands** *. The* **adrenal glands** *are interspersed between the lobes of the cranial end of each kidney. The* **gonads** *(either testes or ovary) secrete hormones in addition to producing sperm or eggs. The* **pancreas** *, in the uppermost loop of the small intestine, carries out nonendocrine functions by secreting digestive juices into the small intestine, but also has hormone-secreting endocrine tissue. The functions of all the above glands and their associated hormones are detailed in the text and in Fig. 4–59. Three endocrine glands shown here are not described in the text. The* **pineal gland** *, located in the dorsal midbrain, secretes the hormone melatonin and plays a role in regulating daily activity cycles, termed circadian rhythms. The* **thymus gland** *, in the upper neck, and the* **cloacal bursa** *(previously known as the bursa of Fabricius) are thought to stimulate tissues of the immune system, and may be important only in young birds. Reprinted (with slight adaptations) from* Manual of Ornithology, *by Noble S. Proctor and Patrick J. Lynch, with permission of the publisher. Copyright 1993, Yale University Press.*

then remained paired for up to seven successive years. After age 12, breeding appears to decline, but some terns continue to lay eggs and raise young until ages 17 to 21 or even older. What hormonal dynamics account for the stability of these breeding cycles over so many years? What accounts for the changes? The basic survey of avian endocrinology presented here will begin to answer these questions. As you read this section, refer to **Figure 4–59**, a schematic of the endocrine system's major glands and hormones and their primary functions.

Figure 4–59. Major Glands of the Endocrine System and their Functions: An overview of the most important endocrine glands (shown as shaded ovals), the hormones they secrete (printed in red), and their effects on target organs. See text for details.

Pituitary Gland

The **pituitary gland** consists of an **anterior** and a **posterior lobe**, connected by a stalk to the **hypothalamus**—a ventral region of the brain. The endocrine system can respond to factors in the external environment, such as day length, temperature, and rainfall, which may

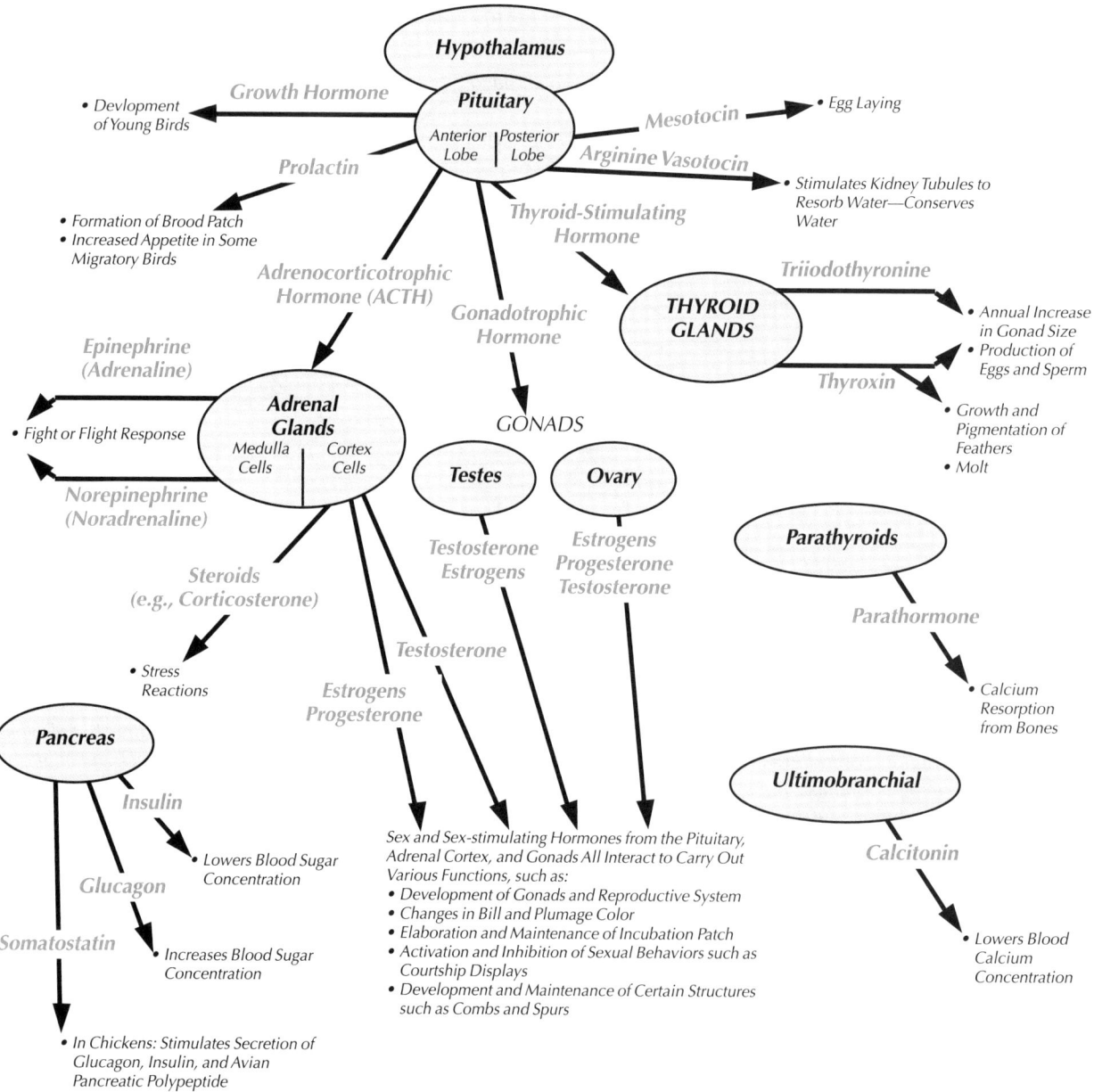

affect seasonal breeders by stimulating the growth of the gonads (ovary and testes). To do this the endocrine system must obtain "messages" from the nervous system and transfer them to other endocrine glands such as the thyroid, adrenals, and gonads. It accomplishes this information transfer via the anterior lobe of the pituitary gland.

This anterior lobe has no nerves carrying stimuli to or away from it, therefore it cannot interact through the "neural" language of cellular electrical activity. Within the hypothalamus, however, are special neurosecretory neurons. These are similar to normal nerve cells, except that they secrete neurohormones into blood vessels leading directly to the pituitary's anterior lobe—allowing it to respond to a "neurosecretory" language. In a sense, the neurosecretory cells are translators from neural to hormonal messages. Depending upon the chemical messages that it receives, the anterior pituitary may secrete any of several hormones into the blood. These hormones perform such functions as stimulating development of the gonads and reproductive system, altering the color of the bill and plumage, and initiating the development of incubation patches. The anterior lobe of the pituitary also plays a role in stimulating or inhibiting certain sexual behaviors, such as courtship displays.

Some hormones secreted by the anterior lobe act directly on other organs, and some act only on other endocrine glands, which then secrete additional hormones that act directly on still other organs and cells. Because it initiates these chains of interaction, the anterior pituitary has been called "the master gland." Examples of anterior pituitary hormones that act directly are: prolactin, which stimulates the formation of brood patches in some species and may stimulate the appetite in some migratory birds; and growth hormone, which acts on many different tissues and organs and is vital in the rapid transformation of a hatchling into a fledgling.

We have noted that the gonads, thyroids, and adrenals are largely under the control of the anterior pituitary's stimulatory hormones. In contrast, the **posterior lobe** of the pituitary does not fabricate hormones, but rather is a storage place for hormones secreted by neurosecretory cells of the hypothalamus. These hormones can then be released by the posterior lobe when it receives the proper signals. One principal hormone released by the posterior lobe is arginine vasotocin, a water-conserving (antidiuretic) hormone. It stimulates resorption of water by the kidney tubules and thus helps to conserve water and change body fluid volume. Another important posterior lobe release is mesotocin, the avian equivalent of mammalian oxytocin, which induces labor. In birds, an egg advances into the vagina under the influence of hormones probably secreted by the most recently ruptured ovarian follicle. Mesotocin then acts on the muscles of the shell gland to cause egg expulsion (laying). Mesotocin injections in birds result in blood vessel "relaxation" to a larger diameter (vasodilation), and a consequential decrease in blood pressure. The roles of both arginine vasotocin and mesotocin in blood pressure regulation of birds need further study.

Thyroid Glands

The **thyroid glands** consist of two pink or red superficial nodules at the base of the neck. They secrete thyroxin and triiodothyronine in response to anterior pituitary secretion of thyroid-stimulating hormone. The thyroid hormones are essential to the annual re-enlargement of the gonads and production of eggs and sperm. Besides its important role in the normal functioning of the gonads, thyroxin is necessary for the normal growth and pigmentation of feathers, and is somehow related to molting. Experiments suggest a possible relationship between thyroxin and migration; injections of thyroxin have induced "migratory restlessness" (see Fig. 5–56) in some passerines.

Parathyroid and Ultimobranchial Glands

One or two small **parathyroid glands** may be located on each thyroid gland as in the Budgerigar, or just caudal to the thyroid glands as in the chicken. The parathyroid glands secrete parathormone, a protein that causes calcium resorption from the bones.

The amount of calcium in the yolk and shell of a chicken egg is about 0.07 oz (2 grams), but the total amount of calcium stored in a laying hen is about 0.7 oz (20 grams), enough for about 10 eggs. A modern laying hen, however, lays about 270 to 300 eggs in a year; thus calcium metabolism in a laying hen must be intense. Not surprisingly, a large number of hormones play a role in fine-tuning the level of calcium in the yolk, shell, bone, and blood of the hen. The interactions among parathyroid hormones and estrogen, androgen, and vitamin D (which is converted in the kidney and liver to a hormone that regulates calcium metabolism) are complex and not well understood. Not surprisingly, they are of intense interest to the egg industry.

The **ultimobranchial glands** are two small, light-colored bodies located near the parathyroid glands, usually on an artery. They secrete a protein hormone, calcitonin, which lowers the blood calcium concentration, as it does in mammals. In birds, however, calcitonin has no known physiological role in bone calcium turnover.

Adrenal Glands

The **adrenal glands** are small yellow or orange bodies located at the cranial end of each kidney. In mammals, the glands have an inner medulla and a surrounding cortex, but in birds the cells characteristic of the medulla (chromaffin cells) are intermixed with the cortical tissue, and both are scattered in pockets among the multiple lobes at the cranial end of the kidneys. Because of this distinction from the mammalian condition, the avian adrenal tissue (and sometimes the gland) is called "interrenal." The interrenal tissue produces secretions that control a variety of physiological processes associated with both circulation and digestion, and it also produces sex hormones like those produced by the gonads.

As in mammals, the avian adrenal medulla cells contain epinephrine (adrenaline), and norepinephrine (noradrenaline). Although the natural functions of these hormones are less understood in birds than in mammals, in some birds they affect blood pressure, carbohydrate metabolism, and perhaps fat metabolism. These effects are related to "fight or flight" responses—how the body responds to urgent or dangerous situations. The cortex tissue secretes a multitude of steroid hormones vital to survival. Among them are corticosteroids (principally corticosterone) associated with "stress" reactions through regulation of carbohydrate and electrolyte (ion) metabolism. The adrenal cortical tissue also secretes androgens (male sex hormones), such as testosterone, as well as estrogens and progesterone (female sex hormones). Note that the adrenal glands in both sexes secrete both male and female sex hormones, and that the effects of these hormones are similar to those produced by the gonads themselves. The adrenal sex hormones, however, may not be as important in the biology of birds as they are in mammals.

Gonads

The **gonads** (testes and ovary) produce sperm and eggs, but they also are important as endocrine glands. The testes secrete the male sex hormone, an androgen called testosterone, and in some birds, they also secrete some female estrogen hormones. The ovaries secrete estrogens, such as estradiol, as well as progesterone; they also secrete some male hormones. An intimate interrelationship thus exists among the sex or sex-stimulating hormones from the pituitary, the cortical interrenal tissue, and the gonads. These secretions influence the activity of the other endocrine glands. In birds, they also have a profound influence on the development and maintenance of anatomical structures such as combs and spurs, and on behavior patterns. In many birds, especially migratory species, the gonads vary tremendously in size and secretory activity on an annual cycle. During the breeding season, for example, the testes of small passerines may grow to several hundred times their nonbreeding volume and weight! They regress again for most of the year, a cycle that probably saves energy during flight.

Pancreas

The **pancreas**, located in the uppermost loop of the small intestine, secretes digestive juices into the small intestine, and it also contains endocrine tissue that produces hormones regulating carbohydrate metabolism and blood sugar levels. The hormones secreted include: insulin, which decreases blood sugar concentrations; glucagon, which increases blood sugar concentrations; and somatostatin, or growth-inhibiting hormone. In chickens, at least, this latter hormone stimulates the secretion of glucagon, insulin, and avian pancreatic polypeptide, a substance of unknown function. Glucagon is apparently more important than insulin in regulating carbohydrate metabolism in birds, whereas in mammals, insulin is the more important hormone. The significance of these hormones may vary considerably among granivorous (seed-eating), herbivorous (foliage-eating), and carnivorous birds, but little comparative research has been done.

Without forming separate glands, the walls of the **stomach** and those of the first part of the small intestine, the **duodenum**, also secrete hormones. These stimulate, respectively, the secretion of other stomach cells and the smooth muscle in the wall of the gall bladder (if present).

The Circulatory System

■ The **circulatory** or **blood vascular system** is a transport system that carries oxygen and nutrients to the cells of the body and removes the waste products of metabolism. It also regulates body temperature by distributing the heat produced by muscles and the gut, and carries antibodies that protect against infection and hormones produced by the endocrine glands. The circulatory system consists of a pump (**heart**)

*Figure 4–60. Internal Structure and Function of the Heart: The muscular heart provides the force that moves blood throughout the body. **a. Longitudinal Section**: The heart consists of two thin-walled receiving chambers, the **atria**, and two thick-walled pumping chambers, the **ventricles**. Between each atrium and its associated ventricle are atrioventricular valves, which prevent the backflow of blood. Deoxygenated blood (dashed arrows) from the body enters the right atrium of the heart, and from there moves into the right ventricle. Contraction of the right ventricle pushes this blood into the pulmonary trunk, which branches into the right and left **pulmonary arteries**, which carry the deoxygenated blood into the lungs. In the lungs the blood picks up oxygen and loses its carbon dioxide. The newly oxygenated blood (solid arrows) re-enters the heart through the **pulmonary veins**, passing into the left atrium and from there into the left ventricle. The highly muscular left ventricle then pumps the blood into the aorta, whose branches distribute it throughout the body. **b. Cross-Section through Points X–X in a:** This view shows the dramatic difference in thickness between the muscular wall of the left ventricle, which must provide adequate force to pump blood throughout the body, and that of the right ventricle, which pumps blood only to the lungs. Adapted from Brooke and Birkhead (1991, p. 32).*

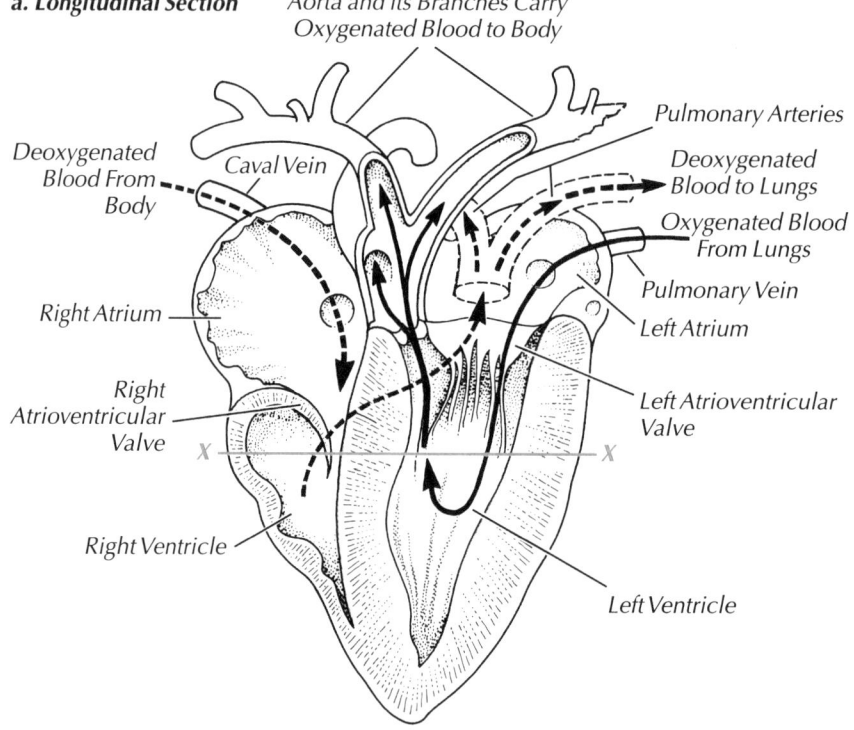

a. Longitudinal Section

Aorta and its Branches Carry Oxygenated Blood to Body

Pulmonary Arteries

Deoxygenated Blood From Body

Caval Vein

Deoxygenated Blood to Lungs

Oxygenated Blood From Lungs

Pulmonary Vein

Right Atrium

Left Atrium

Right Atrioventricular Valve

Left Atrioventricular Valve

X ⸺⸺⸺⸺⸺⸺⸺ X

Right Ventricle

Left Ventricle

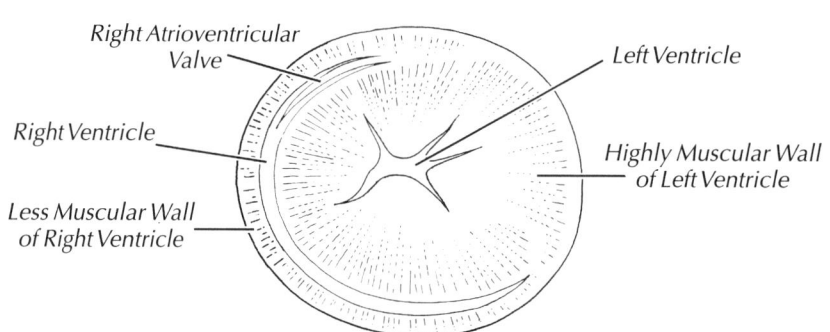

b. Cross-Section View at Line X

Right Atrioventricular Valve

Left Ventricle

Right Ventricle

Highly Muscular Wall of Left Ventricle

Less Muscular Wall of Right Ventricle

and vessels that carry blood away from the heart (**arteries**) and return it to the heart (**veins**). The two systems of vessels are connected by an enormous, fine network of tiny blood vessels, the **capillaries**, the site of dynamic chemical activity.

The **lymphatic system**, an outgrowth from the veins, gathers tissue fluid that has leaked from the capillaries and returns it to the general body circulation. The lymphatic system also releases antibodies and filters out degenerating cells and foreign substances.

The Heart

The heart provides the force that propels blood into the arteries, but venous and lymphatic return of the circulatory fluids is passive, dependent in part on general body movement.

The bird's heart (**Fig. 4–60a**) has two relatively thin-walled receiving chambers called **atria** (singular, **atrium**). The **right atrium** receives blood from the head and body via the **caval veins**. The **left atrium** receives blood from the lungs via the **pulmonary veins**. The two other chambers, the large **ventricles**, are the thick-walled pumping chambers.

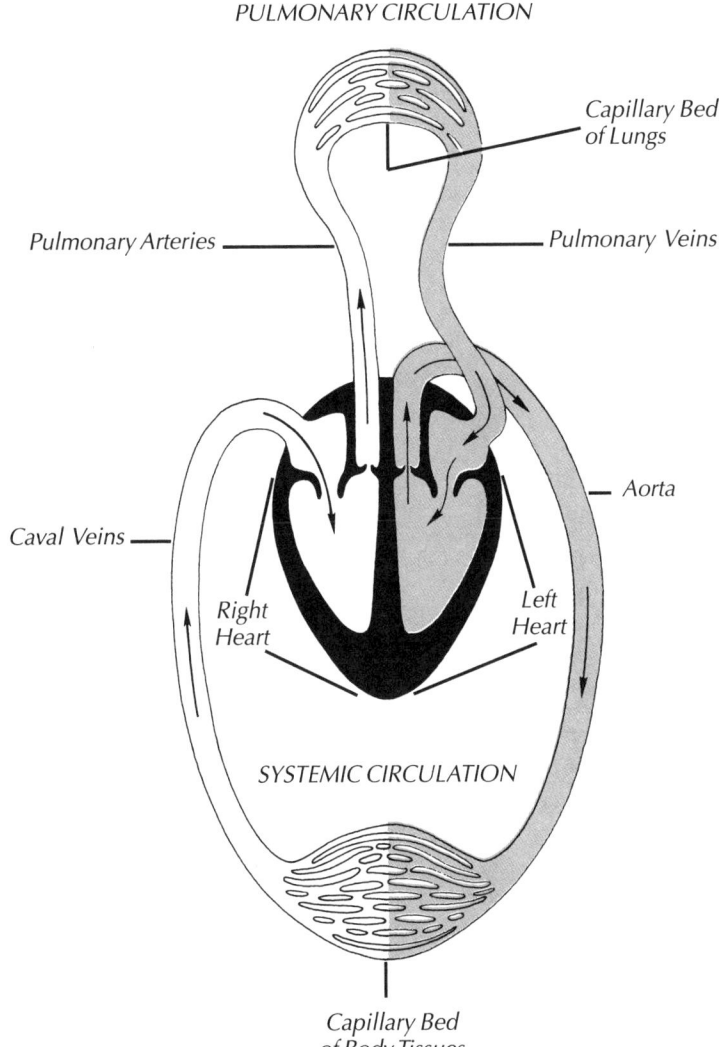

PULMONARY CIRCULATION

Capillary Bed of Lungs

Pulmonary Arteries

Pulmonary Veins

Aorta

Caval Veins

Right Heart

Left Heart

SYSTEMIC CIRCULATION

Capillary Bed of Body Tissues

*Figure 4–61. Schematic of Pulmonary and Systemic Circulation: The heart is functionally divided so that the **pulmonary circulation**, carrying blood to and from the lungs, is separated from the **systemic circulation**, carrying blood to and from the rest of the body. This system ensures that oxygenated blood (colored) and deoxygenated blood (white) do not mix. The oxygen content of arterial blood thus remains high, an essential feature for the optimal function of an animal with a high metabolic rate, such as a bird. Adapted from Keeton (1980, p. 282).*

Partitions divide the heart so that the **pulmonary circulation** of the lungs is separated from the **systemic circulation** of the body (**Fig. 4–61**). Functionally, the right atrium and right ventricle are referred to as the **right heart** (filled with deoxygenated blood). The left atrium and left ventricle are the **left heart** (filled with oxygenated blood).

Venous blood flows into the right atrium and is pumped from there to the right ventricle. Contraction of the right ventricle forces the blood into the **pulmonary trunk**, a large artery that branches to the right and left lungs as the **right** and **left pulmonary arteries**. In the lungs, the blood loses its carbon dioxide and picks up oxygen. **Right and left pulmonary veins** conduct this oxygen-rich blood to the left atrium, from which it is pumped to the left ventricle. Contraction of the left ventricle pumps the blood into the **aorta**, whose branches distribute it to all other parts of the body. Because the left ventricle must pump blood throughout the body, its muscular wall is much thicker than that of the right ventricle, which pumps blood only to the lungs (**Fig. 4–60b**).

Heart Valves

Atrioventricular valves occur, as the name suggests, between each atrium and its corresponding ventricle. They open when the ventricle relaxes, permitting blood to flow into the ventricle, and close when the ventricle contracts, keeping the blood flowing in one direction by preventing backflow into the atrium (**Fig. 4–62**). In nearly

Figure 4–62. Heart Valves and their Role in Blood Circulation: Atrioventricular valves are located between each atrium and its corresponding ventricle, keeping blood flowing in one direction by preventing backflow. Thin arrows indicate the movement of blood. ***a. Blood Flows into Left and Right Atria:*** *This actually occurs as the ventricles contract, as in* ***c,*** *but is shown here as a separate step for simplicity.* ***b. Ventricles Relax:*** *With the relaxation of the ventricles, both atrioventricular valves open (short arrows), allowing blood to flow from each atrium into its ventricle.* ***c. Ventricles Contract:*** *When the ventricles contract, both atrioventricular valves close (short arrows), preventing backflow of blood from the ventricles into the atria. Instead, the blood from the right ventricle flows into the pulmonary trunk and then to the lungs, while blood from the left ventricle flows into the aorta, which carries the blood to the rest of the body. Drawings by Charles L. Ripper.*

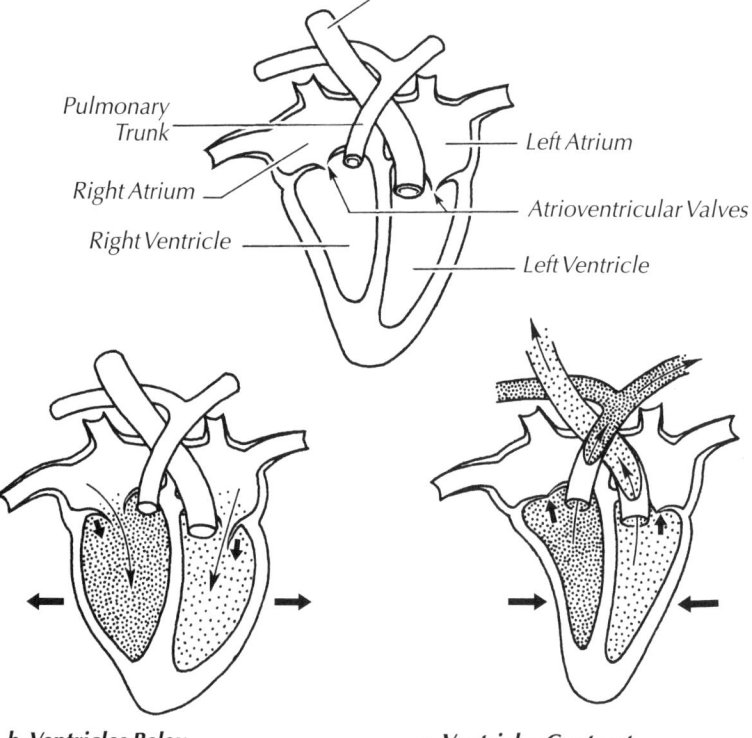

Aorta

Pulmonary Trunk

Left Atrium

Right Atrium

Atrioventricular Valves

Right Ventricle

Left Ventricle

a. Blood Flows into Left and Right Atria

b. Ventricles Relax
- *Atrioventricular Valves Open, Allowing Blood to Flow from Atria into Ventricles*

c. Ventricles Contract
- *Atrioventricular Valves Close, Preventing Backflow of Blood into Atria*
- *Blood Flows into Pulmonary Trunk and Aorta*

all mammals the right atrioventricular valve is composed of three flaps, or pocket-like cusps, whereas the left valve has two. In birds, however, this arrangement is reversed. These equally functional anatomical differences illustrate that evolution, too, has discovered that "there is more than one way to skin a cat"!

Semilunar valves, with three cusps each, are located at the beginning of the pulmonary trunk and the aorta, preventing backflow of blood into the ventricles as the ventricles relax after each beat.

Blood Supply to the Heart Tissue

Even though all blood in the body passes through the heart, only a tiny amount nourishes the heart muscle itself from inside the pumping chambers. Instead, heart tissue is supplied by the coronary arteries, usually two in number, which arise from the first part of the aorta (Fig. 4–63). From there they pass over the surface of the heart, where they send branches into the muscle to nourish its life-long toil. Blockage of a coronary artery starves the heart muscle of oxygen and can lead to heart failure. Coronary veins carry blood and wastes from the heart muscle back to the right atrium.

Conducting System of the Heart

Cardiac muscle has an innate rhythm that causes it to beat before any nerves have grown to it. Two nodes and bundles of conducting nerve fibers govern this innate rhythm. One node, in the wall of the right atrium, initiates the heartbeat and therefore is referred to as the "pacemaker." It stimulates contraction of the right atrium which, in turn, stimulates a second node in the bottom of the septum between the atria. From the second node, a bundle of fibers leads into the septum between the ventricles and causes contraction of the ventricular muscle.

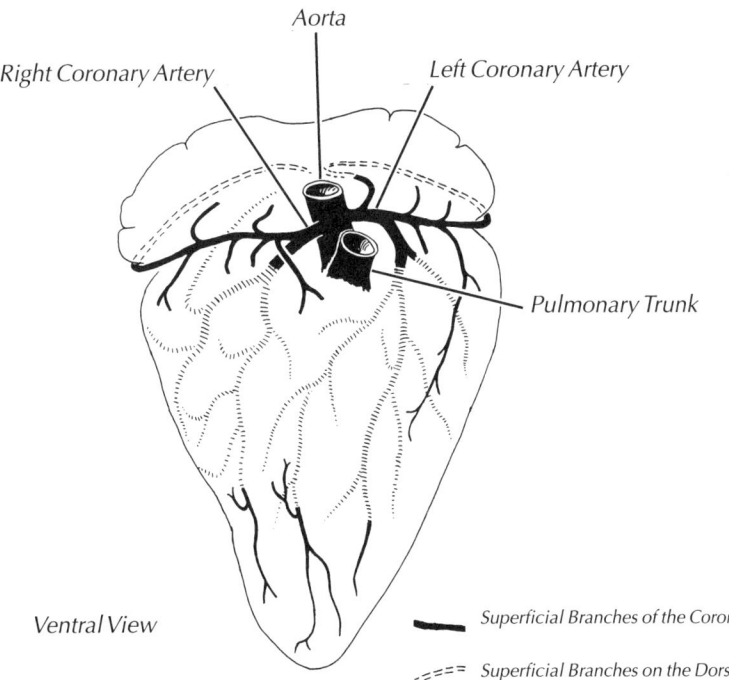

Right Coronary Artery
Aorta
Left Coronary Artery
Pulmonary Trunk
Ventral View

Superficial Branches of the Coronary Artery

Superficial Branches on the Dorsal Side of the Heart

Deeply Embedded Branches of the Coronary Artery

Figure 4–63. Exterior of Chicken Heart Showing the Coronary Arteries: The heart receives little nourishment from the blood being continuously pumped through its chambers. Instead, it has its own blood supply, transported through the right and left coronary arteries, which branch from the aorta soon after it exits the heart. The fine branches of the coronary arteries carry oxygen and nutrients across the surface of the heart and deep into the cardiac muscle. Coronary veins (not shown) remove the waste products of metabolism from the heart muscle, entering the heart's right atrium. Adapted from King and McLelland (1981, Vol. 2, p. 244).

The pacemaker role of the node in the wall of the right atrium has been demonstrated by completely separating the right atrium from the rest of the heart. In this condition, the right atrium continues to beat at the same rate as before; the remainder of the heart will continue to beat but at a slower rate. Even isolated pieces of heart will beat independently for some time.

Although the heart requires no nerve input to *begin* beating, later in development the autonomic nerves begin to regulate the *rate* of the beat. Stimuli received from parasympathetic fibers of the vagus nerve inhibit or decrease the heart rate, whereas stimuli received from sympathetic fibers accelerate the rate in stress or emergency situations.

Location of the Heart

The bird's heart lies on the midline of the thoracic (chest) cavity, below the lungs **(Fig. 4–64)**. In mammals, the thoracic cavity is separated from the abdominal cavity by a muscular diaphragm that functions to fill the lungs during inspiration. Birds, however, lack a diaphragm, inhaling in an entirely different way, and have only a thin, membranous partition that divides the body cavity into a cranial (thoracic) and caudal (abdominal) compartment. The lungs lie around the heart in the cranial compartment just under the vertebral column. The heart itself is surrounded by a thin membrane, the **pericardium**, the inner lining of which secretes pericardial fluid that reduces the friction of the beating heart against the adjacent tissues.

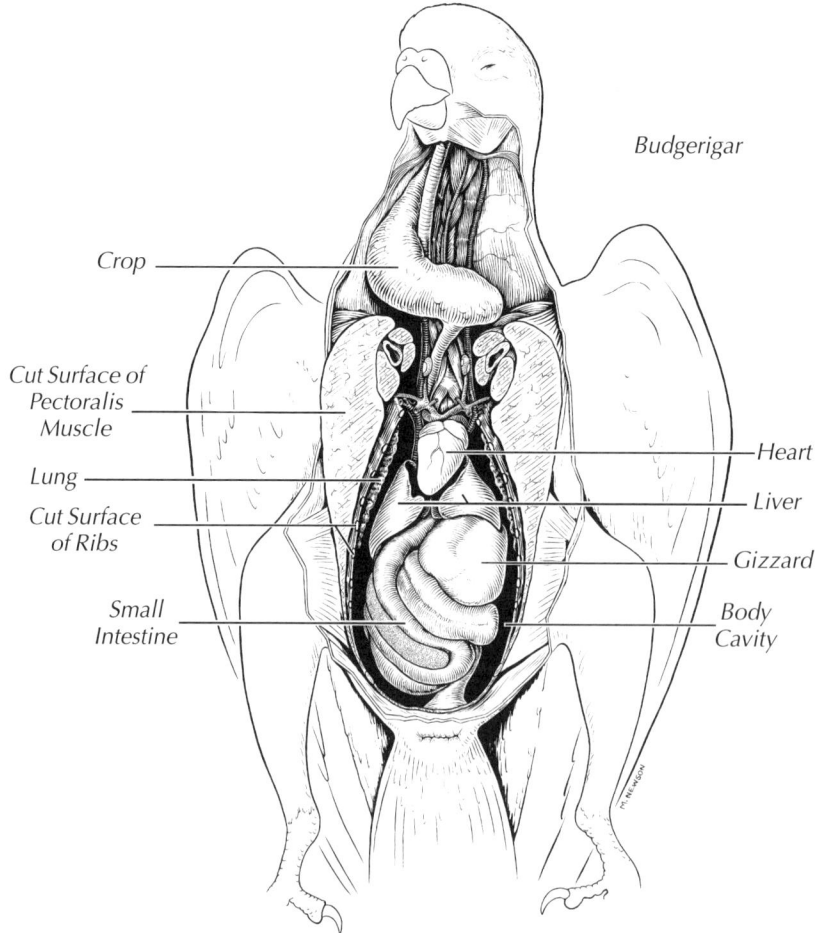

Figure 4–64. Position of the Heart: *In this ventral view, the sternum and the abdominal wall have been removed, and the cut surfaces of the pectoralis muscles and the ribs can be seen. The heart is located on the midline of the thoracic (chest) compartment of the body cavity. In mammals, a muscular diaphragm separates the thoracic compartment from the abdominal compartment, but the diaphragm is absent in birds, replaced by a thin membranous partition (not visible in this diagram). The heart itself is surrounded by the pericardium, a membranous covering too thin to be visible here. The lungs lie around the heart in the thoracic cavity. Caudal to the heart, in the abdominal cavity, are the liver, the intestinal tract, and the remaining abdominal organs. Drawing from Evans (1996).*

Blood Vessels

Blood vessels **(Fig. 4–65)** are named in relation to their course away from or toward the heart, regardless of whether the blood they carry is oxygenated. Arteries carry blood away from the heart, and veins carry blood to the heart. Thus the pulmonary artery carries deoxygenated blood to the lung to be oxygenated, whereas the pulmonary vein carries oxygenated blood to the heart for distribution to the body. In the rest of the body, the arteries carry oxygen-rich blood and the veins, oxygen-poor blood. Arteries and arterioles (small branches of the arteries) have muscle cells in their walls and are capable of constricting, causing a rise in blood pressure.

Capillaries

The smallest vessels, the **capillaries** (Fig. 4–65b), are extremely tiny; some are so small in diameter that only a single red blood cell can pass through at one time. Capillaries are also thin-walled and lack a muscular coat, thus they remain rather constant in diameter. Capillaries connect the arterial and venous systems as a network of anastomosing vessels deeply penetrating almost every body tissue except in the central nervous system and the outer layer of the skin (the

*Figure 4–65. Blood Vessels: a. Arrangement of Blood Vessels: Blood vessels are classified as arteries or veins based on whether the blood they transport is traveling to or from the heart, irrespective of the oxygen content of the blood. **Arteries** transport blood away from the heart, and **veins** transport blood toward the heart. Arteries divide into branches termed **arterioles**. Upon reaching target organs or tissues, arterioles divide into still smaller branches and eventually into the smallest blood vessels, the **capillaries**. Capillaries form a network, or capillary bed, throughout each body tissue except in the central nervous system and the outer layer of the skin. Capillary beds are the interface through which oxygen and nutrients pass into, and wastes pass out of, the tissue cells (see Fig. 4–69). Blood from the capillaries drains into tiny **venules**, which carry it through a series of increasingly larger venules and veins, and eventually enters the heart. **b. Structure of Blood Vessels:** Arteries are thick-walled, consisting of a fibrous outer layer of connective tissue, a thick middle layer of elastic tissue and smooth muscle, and a thin lining termed endothelium. The elasticity of arteries allows them to expand as each heartbeat fills them with blood. Veins may be larger in diameter than their corresponding arteries, but they are thinner-walled and flabbier, with a less muscular middle layer than that of arteries. The venous side of the circulatory system is low-pressure, because less force from the pumping action of the heart is available to keep blood flowing in one direction, so veins have valves that prevent backflow. Capillaries, the smallest blood vessels, are microscopic in diameter, with walls consisting of a single layer of endothelium.*

a. Arrangement of Blood Vessels

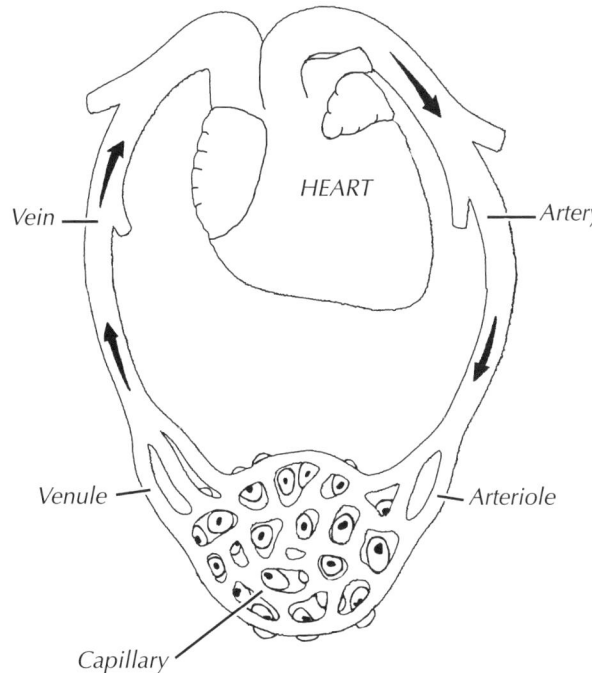

b. Structure of Blood Vessels

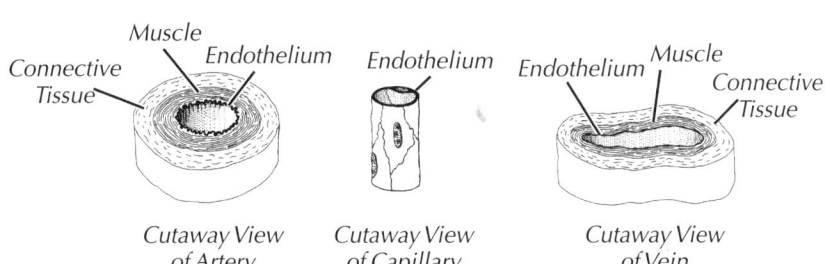

Cutaway View of Artery   Cutaway View of Capillary   Cutaway View of Vein

epidermis). Blood from the arteries to the head, body, and viscera is depleted of oxygen while passing through the capillaries. It is in the capillary beds that the metabolically active cells of the surrounding tissues consume incoming oxygen from the blood and release wastes such as carbon dioxide and ammonia into the blood.

Arterial System

Arteries (Fig. 4–66) conduct blood away from the heart and are thicker-walled than veins. The larger arteries have elastic fibers within their walls, allowing the arteries to expand when the heart pumps blood into them, and to shorten or narrow between heartbeats (Fig. 4–65b). The large arteries branch into distributing arteries whose walls lack elastic tissue.

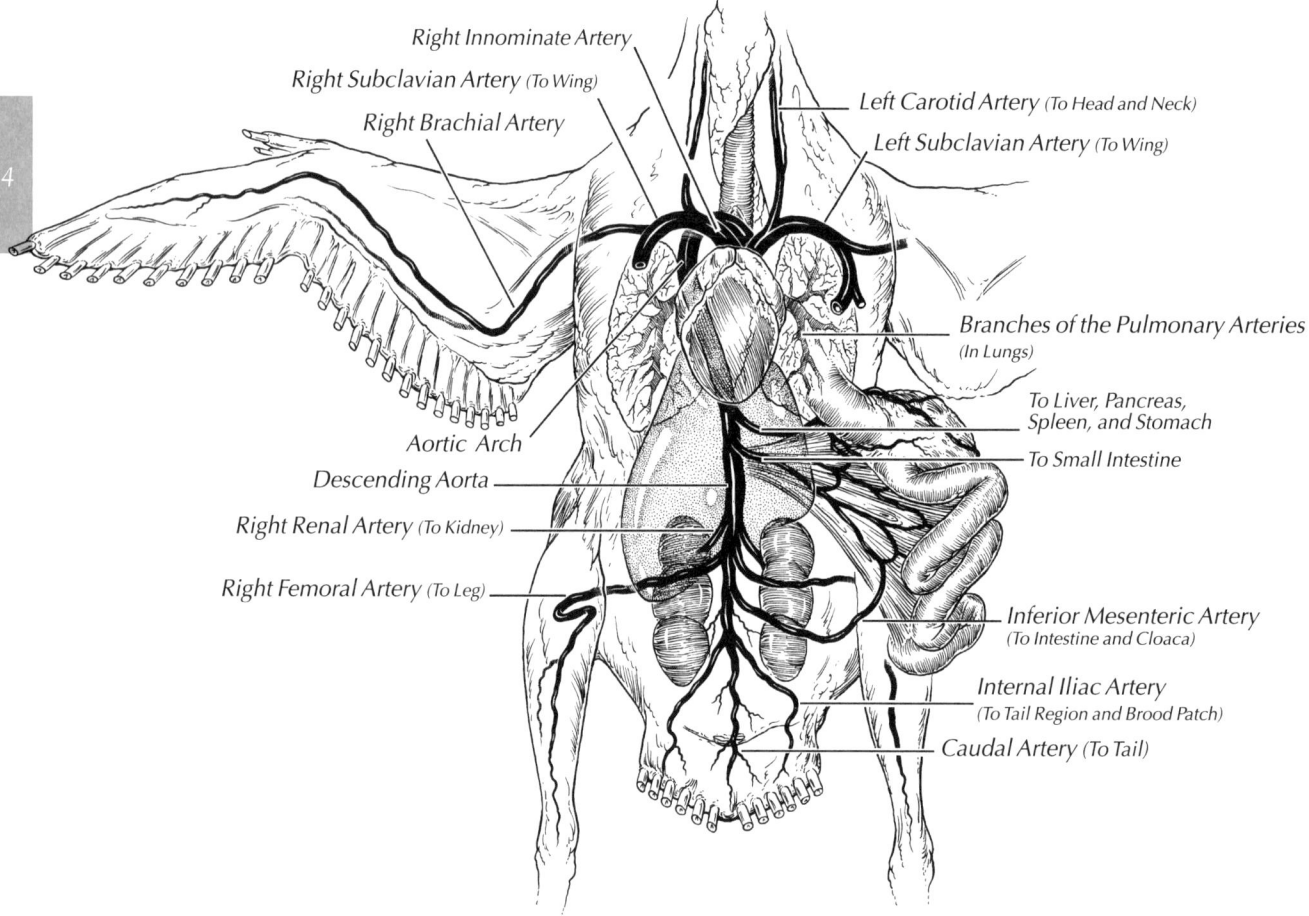

Figure 4–66. Arterial System of the Rock Dove: *The body's largest artery, the aorta, leaves the left ventricle of the heart and curves sharply to the right as the* **aortic arch***, continuing toward the tail as the* **descending aorta***. From the aortic arch arise the right and left innominate arteries, which almost immediately give rise to the paired* **subclavian arteries** *to the wing and the paired* **carotid arteries***, which supply the neck and head. Carotid arteries vary greatly from species to species. Some have a single large carotid on one side or the other, some (as shown here) have two, and some have no carotids at all, blood being carried to the neck by other arteries. Arteries branching from the descending aorta supply blood to the legs, kidneys, intestinal tract, tail, and other caudal regions of the body. Where these arteries are paired, only one side is labeled in this diagram. See text for further details. Reprinted from* Manual of Ornithology, *by Noble S. Proctor and Patrick J. Lynch, with permission of the publisher. Copyright 1993, Yale University Press.*

The **aorta**, the largest artery in the body, leaves the left ventricle and, as the **right aortic arch**, curves to the right as it passes over the heart and toward the backbone. As it approaches the vertebral column, the aorta curves again and passes toward the tail as the **descending aorta**. All vertebrate animals have a descending aorta. Birds are unique, however, in having only a right aortic arch. Reptiles have both a right and left aortic arch; mammals have only a left aortic arch—again a reflection of randomness in evolution. Although no fossil evidence exists, it is believed that having both right and left aortic arches is the ancestral condition for all terrestrial vertebrates. The loss of an arch of

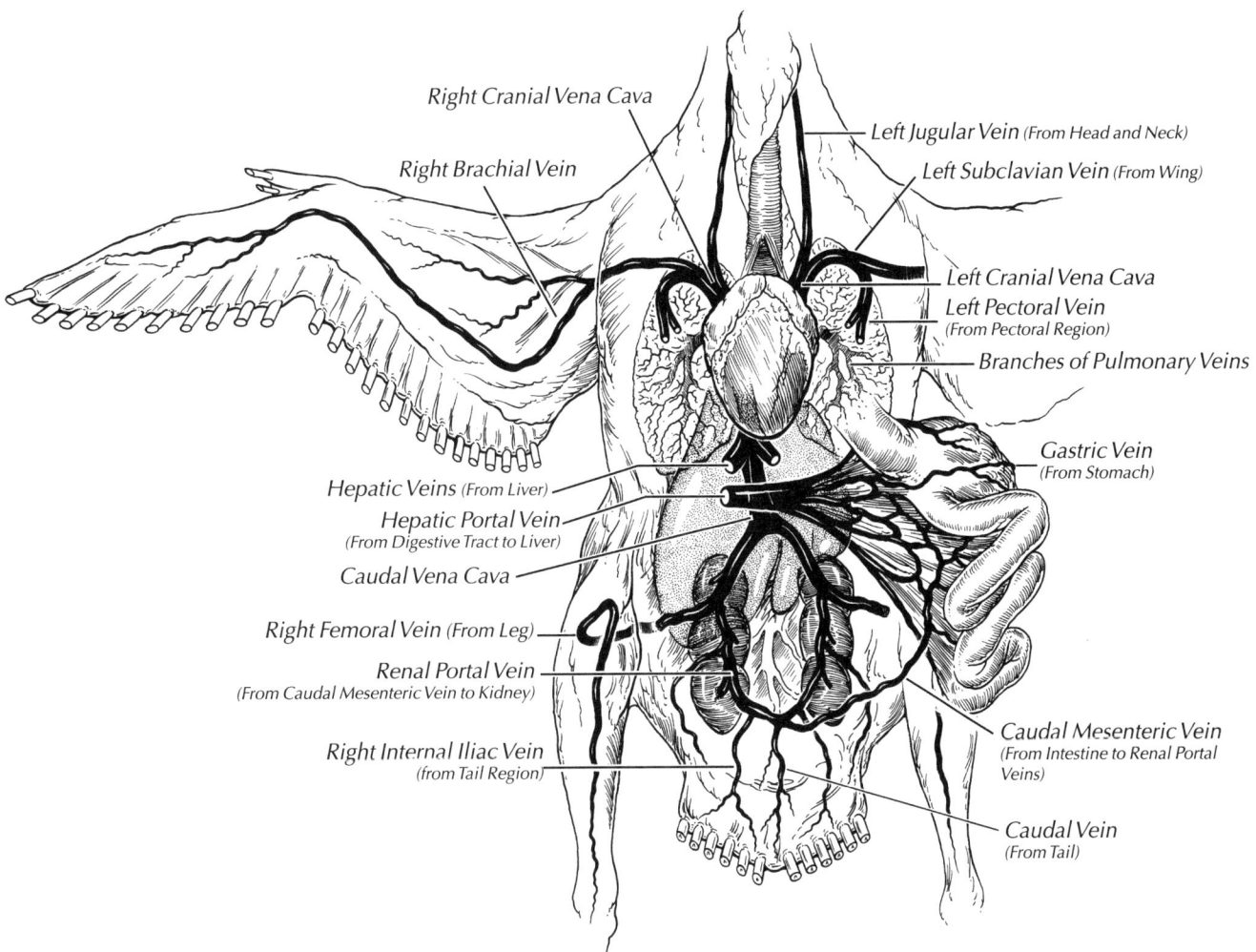

Figure 4–67. Venous System of the Rock Dove: *Venous blood from the wings, head, neck, and pectoral region drains into the* **subclavian**, **jugular**, *and* **pectoral** *veins (only the left of which are labeled here). These three vessels converge on each side to form the* **left** *and* **right cranial vena cavae** *(or caval veins), entering the right atrium of the heart. Blood from the legs, tail, kidneys, and caudal regions of the body drains into the* **caudal vena cava**, *which is mostly obscured by the heart in this diagram. Venous blood from the digestive tract passes through one of two portal systems within the venous circulation. The* **hepatic portal vein** *transports blood from the upper part of the small intestine to the liver, where it undergoes processing before exiting through the* **hepatic veins** *and then flowing into the caudal vena cava and to the heart. In the renal portal system, blood from the lower portion of the small intestine drains into the* **caudal mesenteric vein**, *and is transported to the paired* **renal portal veins** *that form a* **venous ring** *(not labeled) connecting the various lobes of the two kidneys. The blood leaves the kidneys through renal efferent veins (not labeled) and is carried to the heart by the caudal vena cava. Portal systems are described in more detail in Figure 4–68. Reprinted from* Manual of Ornithology, *by Noble S. Proctor and Patrick J. Lynch, with permission of the publisher. Copyright 1993, Yale University Press.*

one side and the strengthening and enlargement of a single conduit occurred in the evolution of increasingly active and environmentally independent forms. The ancestors of birds evolved one way, and mammals, another, without functional significance.

A short distance beyond its origin in the left ventricle, the aortic arch gives rise to large **right** and **left innominate arteries**. These soon branch into the **subclavian artery** to the wing and the **carotid artery** to the neck and head. Presumably because the length of the avian neck varies so greatly in proportion to the size of the body (for instance, in the American Woodcock versus the Great Blue Heron), the carotid arteries of birds exhibit many different patterns. Instead of two equal carotids, one on each side of the neck, some birds have a large carotid artery on one side and a small one on the other. Some have a single carotid artery; and some lack carotid arteries altogether, the blood being carried to the neck by other arteries, usually within canals of the vertebrae.

Branches of the descending aorta carry blood to the rest of the body. A series of paired arteries passes ventrally around the body wall between the ribs to supply the breast and abdominal walls. Three large, single arteries supply: (1) the liver, pancreas, spleen, and stomach, (2) the different parts of the small intestine, and (3) parts of the large and small intestine and the cloaca. As it continues toward the tail, the aorta gives off other paired branches to the kidneys, legs, and organs of the pelvic region. Eventually, the aorta becomes the lone artery supplying the tail.

As the arteries branch and re-branch they become smaller and smaller, becoming **arterioles**, and finally **capillaries**. On the return side the capillaries drain into the tiniest **venules** and then these enter **veins**.

Venous System

Veins (Fig. 4–67) return the blood to the heart. Veins are larger and thinner-walled than arteries, their size depending on the amount of blood they carry. Tiny **venules** receive the blood from even smaller capillaries and conduct it into a series of increasingly larger veins until it reaches the right atrium of the heart (see Fig. 4–65). Correctly then, anatomists speak of the *tributaries* of the veins but the *branches* of the arteries. With a few exceptions, veins run along with arteries and have the same names as the corresponding arteries. One exception is the **jugular vein**, which returns blood from the head and neck to the heart.

Veins possess valves that prevent blood from flowing backward in this low-pressure side of the circulatory system. Contraction of the general body muscles is also important in keeping the venous blood flowing.

Venous blood from the wings, head, and neck drains into subclavian, jugular, and pectoral trunk veins. These veins converge on each side to form the **right** and **left cranial vena cavae**, which connect to the right atrium of the heart. Venous blood from the legs, tail, kidneys, and caudal part of the body wall drains into a single, short **caudal vena cava**, which also empties into the right atrium.

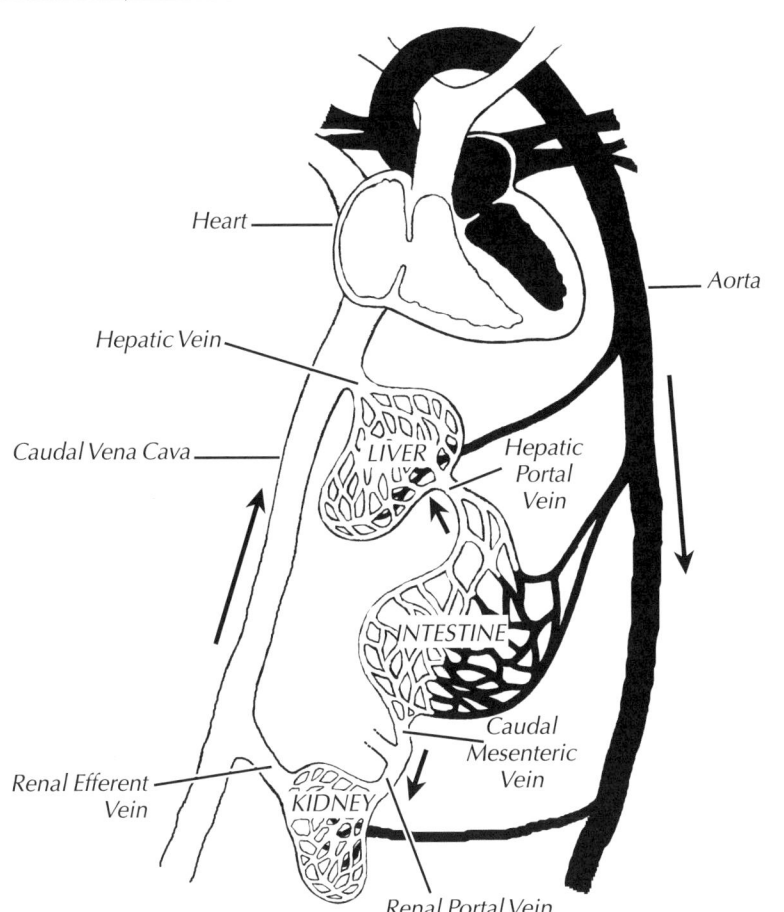

Figure 4–68. Portal Systems: As this schematic of the hepatic and renal portal systems shows, portal veins connect two capillary beds with one another, rather than connecting one capillary bed with the heart as do most veins. Portal systems allow a second organ to process blood before returning the blood to the general circulation via the heart. In the **hepatic portal system**, nutrient-rich blood from the capillary network supplying the microscopic folds of the lining of the small intestine (see Fig. 4–103) is carried to the capillary network of the liver by the **hepatic portal vein**. At the liver, the digestive products originally absorbed from the small intestine undergo further chemical processing. Blood from the liver's capillary bed leaves by way of the **hepatic vein**, returning to the heart via the caudal vena cava. In the **renal portal system**, found in birds and reptiles but not in mammals, blood from the capillary bed surrounding the lower small intestine is collected into the **caudal mesenteric vein**, which joins the **renal portal vein**, through which the blood is conveyed into the capillary bed of each kidney. From here the blood is recollected into **renal efferent veins** and returned to the heart via the caudal vena cava. The function of the renal portal system in birds is unclear. Note that this schematic shows the aorta descending on the bird's left, but in life it actually descends on the right, as shown in Fig. 4–66.

Venous blood from the digestive tract flows through vessels that terminate in the **hepatic portal vein** to the liver **(Fig. 4–68)**. **Portal veins** connect two capillary beds rather than connecting a capillary bed and the heart, as most veins do. The portal veins permit a second organ (in this case, the liver) to process blood chemistry before returning the blood to the general circulation via the heart and lungs. Thus, the simple products of digestion that entered the blood through the capillaries of the small intestine are borne directly to the liver, where the blood passes through the second capillary network. The blood leaves the capillary network of the liver in two large **hepatic veins**, which empty into the caudal vena cava just before it enters the right atrium of the heart.

The venous system of the kidneys of birds, like reptiles but unlike mammals, has a functional **renal portal system** (see Fig. 4–68). This system collects venous blood from the lower portion of the digestive tract via a **caudal mesenteric vein**, and conveys it to the **venous ring** within the kidney. Here some of the blood is passed through a capillary bed in the kidney before being recollected by renal efferent veins and conveyed to the heart via the **caudal vena cava**. In the kidney, venous blood mixes with arterial blood. The function of this peculiar circulatory pattern of birds is unknown. Whatever the function, though, valves present in the veins enable a bird to bypass flow through its renal portal system altogether.

Blood

Blood transports oxygen, nutrients, and hormones to body cells, and carries carbon dioxide, water, and other wastes (especially nitrogenous [nitrogen-containing] wastes), as well as secretions, away from cells. All body cells use oxygen in the "burning" or **oxidation** (breakdown) of digested food, a chemical reaction that releases energy and two waste products, water and carbon dioxide. In the lungs, the blood releases the carbon dioxide and absorbs oxygen. The bird's kidney eliminates excess water and nitrogenous wastes.

Blood is composed of a watery fluid, the **plasma**, containing certain proteins found only in the plasma, as well as inorganic salts and sugars. Blood cells and blood platelets float in the plasma. Blood has nutritive, excretory, regulative, protective, and respiratory functions. The plasma carries: the products of digestion—amino acids, glucose, glycerol, and fatty acids; the nitrogenous waste products of metabolism—ammonium, urea, uric acid, and creatinine; hormones; antibodies to combat infection; and some carbon dioxide.

Some of the fluid portion of the blood, along with nutrients, is forced by the blood pressure through the walls of the capillaries **(Fig. 4–69)**. It enters the minute spaces between the body cells, where it is called tissue fluid. Most of the body cells receive oxygen and nutrients from the tissue fluid and dump the waste products of metabolism into it. Tissue fluid and its dissolved waste materials diffuse back, either into the venous end of the capillary network, or into lymphatic capillaries (see Lymphatic System, later in this chapter), which eventually enter the venous system.

The plasma proteins, including albumins, globulins, and fibrinogen, are too large to diffuse through the membranes in the capillary walls and thus remain in the blood vessels. The presence of plasma proteins in the blood makes it more dense than the tissue fluid outside the circulatory system. This density difference induces water to flow from the tissue fluids into the capillaries by a process termed osmosis, the tendency of water from a less concentrated or "thinner" liquid to flow through a semipermeable membrane toward a more concentrated or "thicker" liquid. The "reconstituted" blood then flows into the venous system.

Two general types of cellular elements (called **blood cells** or **corpuscles**) are present in the blood—the **red blood cells** and the **white blood cells (Fig. 4–70)**. The red blood cells of birds have a nucleus, unlike those of mammals. Hemoglobin, an iron-containing respiratory pigment in the red blood cells, carries large amounts of oxygen and gives the blood its red color. When a vertebrate is in a situation with a high concentration of carbon monoxide in the air, carbon monoxide poisoning can occur because hemoglobin has a special affinity for carbon monoxide. Carbon monoxide binds to hemoglobin, forming such a stable compound that the hemoglobin can no longer carry oxygen to supply the body. Because the brain is especially sensitive to a lack of oxygen, unconsciousness precedes death in carbon monoxide asphyxiation. By taking a canary into the mine with them, coal miners

a. Capillary Bed

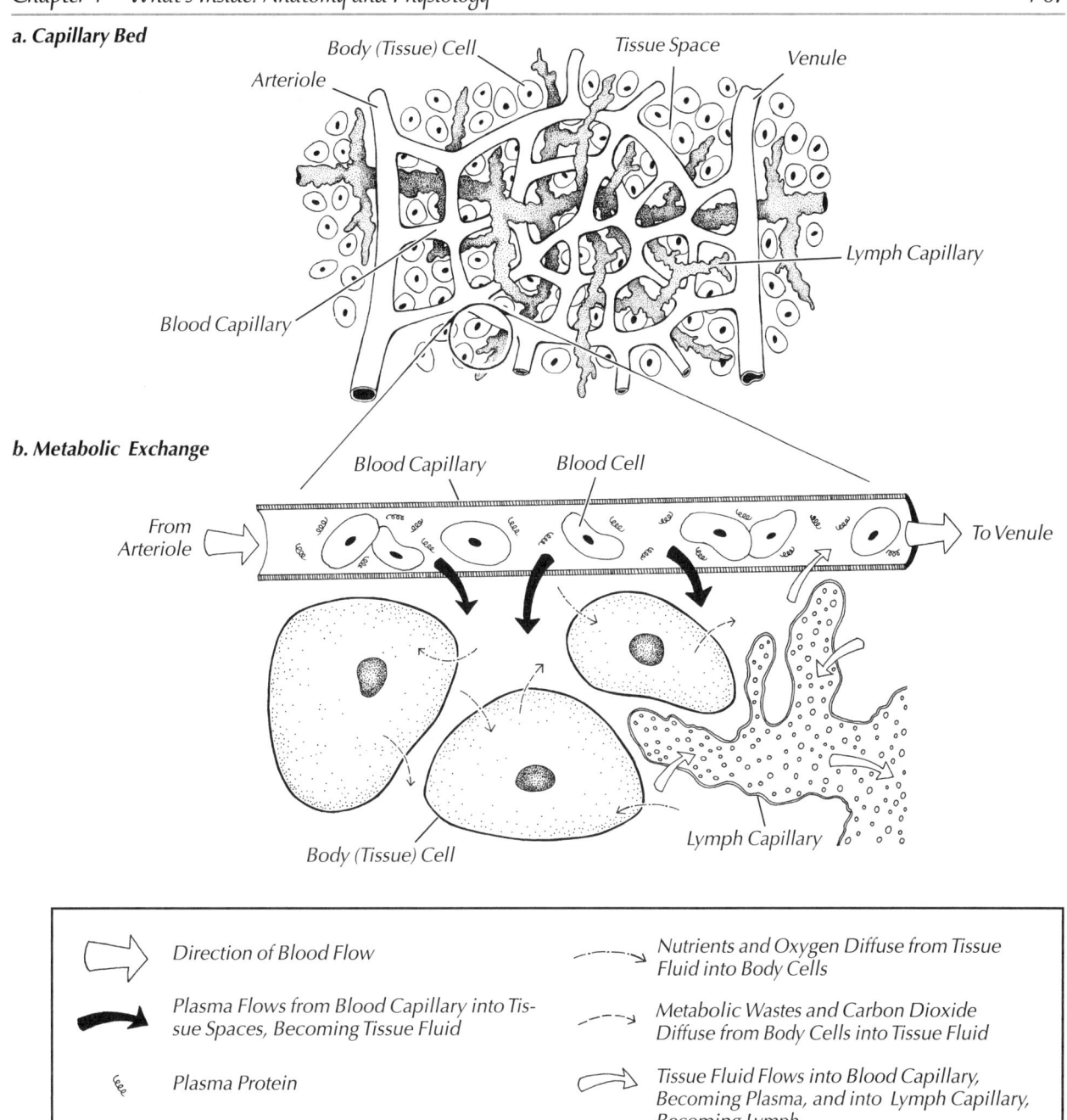

b. Metabolic Exchange

Legend:

➪ Direction of Blood Flow

➡ Plasma Flows from Blood Capillary into Tissue Spaces, Becoming Tissue Fluid

〰 Plasma Protein

- - -→ Nutrients and Oxygen Diffuse from Tissue Fluid into Body Cells

- - → Metabolic Wastes and Carbon Dioxide Diffuse from Body Cells into Tissue Fluid

⇨ Tissue Fluid Flows into Blood Capillary, Becoming Plasma, and into Lymph Capillary, Becoming Lymph

Figure 4–69. Capillaries and Metabolic Exchange: *The circulatory system transports blood to nourish and cleanse the tissues of the body. The actual exchange of materials between the blood and body cells (also called tissue cells) occurs at the blood capillary networks supplying every region of the body. Here, nutrients and oxygen from the blood enter the body cells, and wastes and carbon dioxide from the body cells enter the blood.* **a. Capillary Bed:** *This view shows a network of fine blood capillaries in close association with body cells. Note the presence of a similar network of lymph capillaries, intertwined with the blood capillaries.* **b. Metabolic Exchange:** *An enlargement of a single blood capillary, with surrounding body cells and a lymph capillary, all bathed in tissue fluid, illustrates the process of metabolic exchange. At the arterial end of the blood capillary bed, blood plasma, the fluid portion of the blood, with its dissolved nutrients and oxygen, is forced by blood pressure through the cell membranes of the walls of the capillary, which are only one cell thick. Blood cells and plasma proteins are too large to pass through the cell membranes, so they remain in the blood capillary. The plasma enters the minute, fluid-filled spaces (much exaggerated in size here) between the body cells, and is now called tissue fluid. Oxygen and nutrients diffuse out of the tissue fluid into the body cells. Carbon dioxide and metabolic waste products diffuse out of the body cells and into the tissue fluid. Tissue fluid then diffuses back into the circulatory system, either into the venous end of the blood capillary or into a capillary of the lymphatic system, which leads to progressively larger lymph vessels and eventually enters the venous system (see Fig. 4–71). Once it enters a lymph capillary, the tissue fluid is termed "lymph." Drawings by Christi Sobel.*

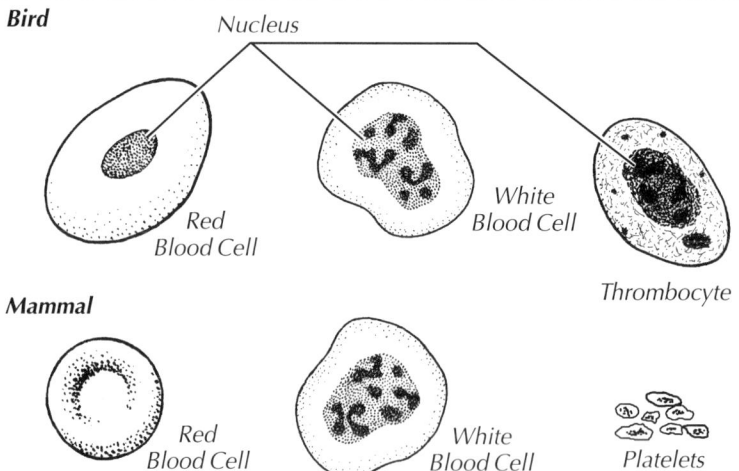

Figure 4–70. Blood Cells of Bird and Mammal: *Blood cells are of two general types whose functions differ. Red blood cells (erythrocytes) are flattened, elliptical cells that play an essential part in cellular respiration. They contain hemoglobin, an iron-containing pigment, which carries oxygen and gives blood its characteristic red color. The red blood cells of birds have nuclei, unlike those of mammals, and therefore tend to be larger than their mammalian counterparts. White blood cells (leukocytes, sometimes spelled leucocytes), which are similar in birds and mammals, are nucleated cells of several types whose shape, size, and internal structure differ. White blood cells function to fight infections. In mammals, small cell fragments known as platelets function in blood clotting. Birds lack platelets; instead their blood clotting is carried out, at least in part, by thrombocytes, nucleated cells that resemble red blood cells but have a more dense and complex internal structure.*

took advantage of this process, which occurs more rapidly in a small bird than in human-sized animals.

There are several types of white blood cells. All contain a nucleus, which varies in size and shape in the different blood cells; none contain hemoglobin. White blood cells can make their way through the walls of the capillaries and move about in the space between the cells. Certain types are important in combating infections because of their ability to engulf bacteria or other foreign substances. **Pus**, the yellowish fluid seen in infections, is a mixture of bacteria, dead white blood cells, and fluid. **Thrombocytes** (see Fig. 4–70), highly specialized blood cells, are important in the life-saving process of blood clotting in birds. Thrombocytes are nucleated, unlike their mammalian counterparts, the blood platelets.

Despite the similarity of its cellular components, bird blood is distinctly different from that of mammals in the proportions of its various chemical constituents. For example, avian blood plasma has a higher sugar and fat content than that of most mammals. Not surprisingly, because it is the main nitrogenous waste product of birds (see Urogenital System, later in this chapter), uric acid also is present at higher concentrations.

Lymphatic System

Recall that the primary function of the lymphatic system is to return to the blood the cell-free tissue fluid that has leaked from the blood capillaries, but that it also releases antibodies, and filters out foreign substances and old or damaged cells. Some of the tissue fluid returns to the circulatory system by way of the blood capillaries, but some enters **lymphatic capillaries**, becoming lymph (see Fig. 4–69). The lymphatic capillaries lead to progressively larger lymphatic vessels, which eventually enter the venous system at one or more sites near the heart. Within the tissues, and on the surface of organs, are also lymphatic capillaries that drain into larger lymph vessels or lymph nodes and eventually enter the great veins leading to the heart. Among birds, only ducks are similar to mammals in having a significant number of lymph nodes that function to filter lymph on its way back to the venous system.

Lymphatic channels in the intestinal wall carry most products of fat digestion to the venous system by way of the **intestinal lymph trunk** and a **thoracic duct (Fig. 4–71)**. The products of the digestion of proteins and carbohydrates, however, are carried by the hepatic portal vein directly to the liver. Although the process is found in (at least) all birds and mammals, exactly *why* digested fats are initially bypassed around the liver, but other digested foods are not, is unclear.

Pressure from body movement and contracting muscles moves the lymph through the lymph vessels. Like veins, lymph vessels contain one-way valves that give a heart-bound direction to the otherwise passive lymph flow. Birds, like reptiles, have pulsating lymph hearts in the pelvic region of the embryo that pump the lymph fluid. In the majority of bird species, most of these lymph hearts disappear after hatching, but some persist in Ostriches, ducks, and some passerines. In ducks and geese, erection of the male copulatory organ is caused by lymphatic pressure, not blood pressure, as is typical in mammals.

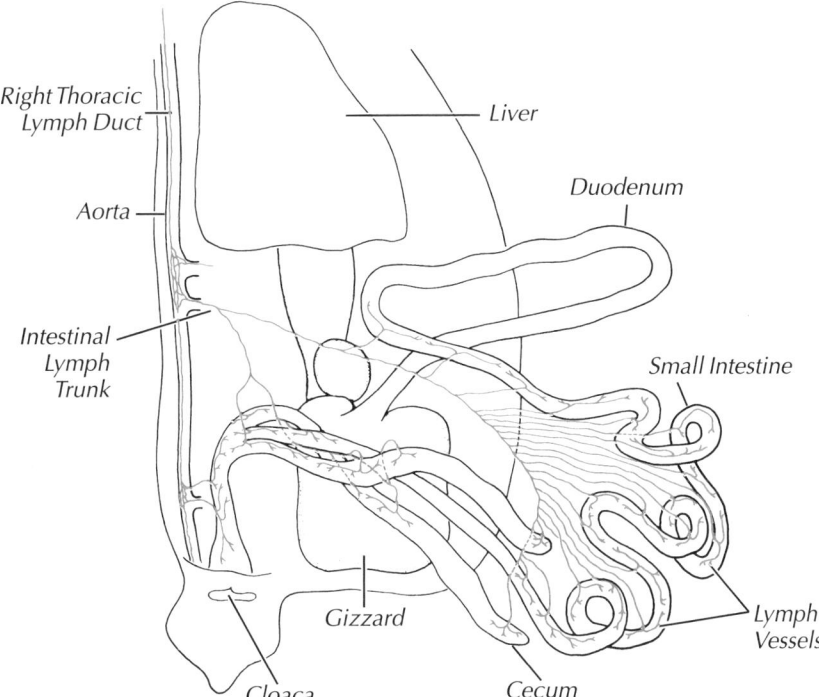

Lymph from Thoracic Duct
Enters Venous System
Through Cranial Vena Cava

Right Thoracic Lymph Duct

Aorta

Intestinal Lymph Trunk

Liver

Duodenum

Small Intestine

Gizzard

Lymph Vessels

Cloaca

Cecum

Figure 4–71. Part of the Chicken Lymphatic System: This diagram is a schematic of one small part of the lymphatic system, with lymph vessels shown in color. The lymphatic system's primary function is to return to the blood the tissue fluid that has leaked from blood capillaries into the spaces between the body cells. This fluid enters lymph capillaries, becoming lymph. Lymph capillaries lead to progressively larger lymphatic vessels that lie along the surface of organs and blood vessels, eventually entering the venous system. This diagram shows part of the lymphatic system that has an additional function, transporting the products of fat digestion from the small intestine to the cranial vena cava, bypassing the liver. Lymph capillaries pick up the products of fat digestion at the small intestine, sending them through larger lymphatic vessels that run along the surface of blood vessels in the intestinal wall. Eventually, they join the **intestinal lymph trunk,** which in turn joins one of the two paired **thoracic lymph ducts** (only the right one is shown here) that travel along the surface of the aorta and eventually enter the venous system through the cranial vena cava just before it enters the heart. The products of protein and carbohydrate digestion are carried from the small intestine directly to the liver, via the blood in the hepatic portal vein. Why the different digested substances are treated differently is not known. Adapted from King and McLelland (1981, Vol. 2, p. 346)

The Respiratory System

We have seen that the blood transports oxygen to the cells of the body and carries away carbon dioxide. Recall, also, that the cells use oxygen in the oxidation of digested food, which releases energy, water, and carbon dioxide. The respiratory system provides for the transfer of oxygen *to* the blood and the expulsion of carbon dioxide and water *from* the blood. This **gas exchange** occurs in the lungs.

The respiratory system consists of external nostrils; nasal cavities and their associated conchae; openings of the nasal cavities into the mouth or throat; air sinuses beneath the orbit; and the pharynx, larynx, trachea, syrinx, bronchi, lungs, air sacs, and pneumatic bones. Recall that birds lack the muscular diaphragm of mammals, relying instead on expansion of the rib cage to draw in air during inspiration. Nevertheless, most components of the avian respiratory system also are found in mammals, but avian "plumbing" and airflow patterns are dramatically different, and are unique among living organisms. The special characteristics of the avian respiratory system are no doubt of great importance in birds' perfection of flight physiology.

Nostrils and Nasal Cavities

In most birds the nostrils open into the nasal cavities at the base of the bill, but several specializations occur. In albatrosses and their relatives, which are known to have a good sense of smell, the nostrils are at the end of hard, horny tubes. These tubes, which appear to function somehow in olfaction, may provide additional benefits during salt excretion (see Figs. 3–34a and b). In nighthawks and their relatives, which have no known olfactory talents, the nostrils are located at the end of soft, flexible tubes. Nostrils are closed over in adult gannets, frigatebirds, cormorants, and anhingas; and are apparently variable in the Red-footed Booby, with some adults having small, slit-like openings and some having none.

A nasal septum usually separates the right nasal cavity from the left **(Fig. 4–72)**. If the septum has an opening or is absent, the condition is called **perforate**. If the septum has no opening, it is **imperforate**. Imperforate septa appear to offer the best arrangement to allow for directional location of olfactory sources, since odors entering one nostril cannot mix with those entering the other. Perforate septa, however, may allow greater sensitivity in detecting an odor. Extending from the lateral wall of each nasal cavity are very thin, scroll-like **conchae**, which are covered with a mucus-secreting membrane in which the endings of neurons of the olfactory nerves are embedded. The mucus

*Figure 4–72. Structure of the Nasal Cavity: This sagittal section of the upper beak gives an internal view of the left nasal cavity, showing two scroll-like **conchae**—the median concha and the anterior concha—extending from its outer wall. The conchae are covered with a mucus-secreting membrane that traps dust, and are embedded with blood vessels that warm the inhaled air, as well as olfactory nerve endings, which sense odors. A **septum** separating the left and right nasal cavities is visible only in the cross-sectional view (taken through the nasal cavity at line X.) The septum shown here completely separates the two nasal cavities, and is thus termed **imperforate** (see text). The two nasal cavities open into the mouth through the palate via a single slit termed the **choana**. Because the choana runs in an anterior-posterior direction, and this sagittal section is taken slightly to the left of the midline, the choana is in front of tissue and not easy to see in the sagittal section; it is visible, however, in the cross-section. Drawing by Charles L. Ripper.*

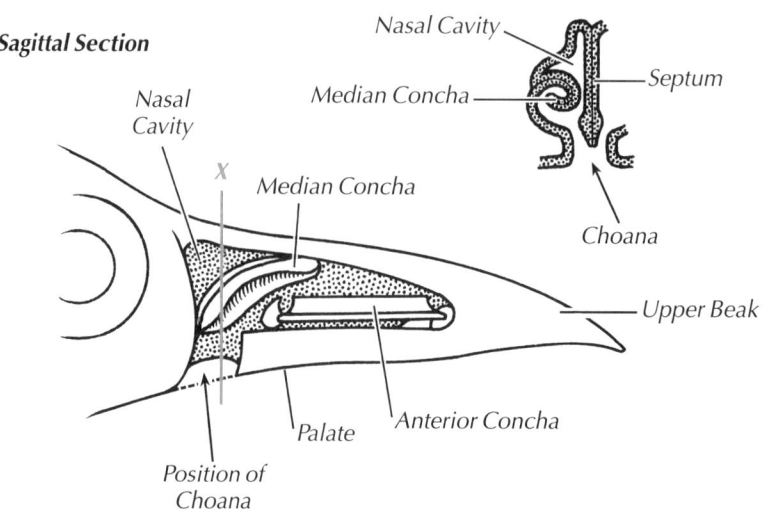

Cross-Section at Line X, Caudal View

Nasal Cavity

Septum

Median Concha

Choana

Upper Beak

Sagittal Section

Nasal Cavity

Median Concha

X

Palate

Anterior Concha

Position of Choana

a. Structure of the Nasal and Oral Cavities of the Budgerigar

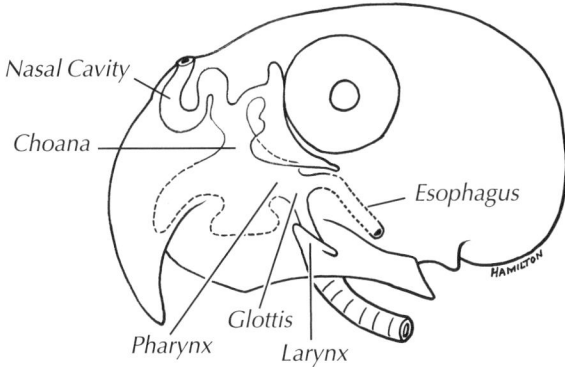

b. Pathways Through the Nasal and Oral Cavities of the Budgerigar

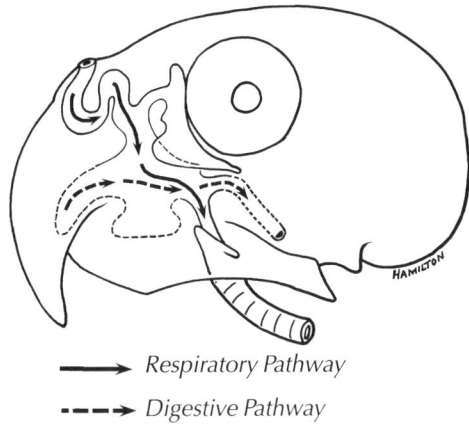

Figure 4–73. Nasal and Oral Cavities of the Budgerigar: a. The structure of the nasal and oral cavities, showing the position of the **pharynx** (throat). **b.** The respiratory and digestive pathways cross in the pharynx. Air in the respiratory pathway (solid arrows) enters through the roof of the mouth at the choana and exits through the larynx en route to the lungs. Food in the digestive pathway (broken arrows) enters through the mouth, crosses the respiratory pathway in the pharynx, and exits through the esophagus en route to the digestive tract. Adapted from Evans (1996).

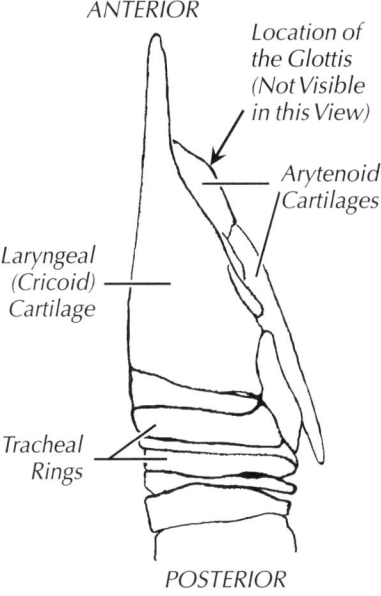

Lateral View of the Larynx of the American Crow

Figure 4–74. The Larynx: The larynx is located beneath the floor of the pharynx, and is surrounded by and attached to the hyoid apparatus (see Fig. 4–11). The laryngeal skeleton consists of several cartilages, the most prominent of which are the paired **arytenoid cartilages** and the large, trough-shaped **laryngeal (cricoid) cartilages**, which lie just below the arytenoids. The arytenoids stiffen and hold the shape of the fleshy folds surrounding the slit-like glottis (not visible in this lateral view), whose muscles regulate the passage of air into the respiratory system. The cricoids form the floor and sides of the larynx. Adapted from King and McLelland (1981, Vol. 4, p. 72).

traps dust, and blood vessels in the membrane warm the inhaled air. The two nasal cavities lead back, opening through the **palate** (roof of the mouth) by a single slit—the **choana**—which is protected between folds of palatal tissue.

Pharynx

The **pharynx** (throat) begins at the back of the tongue and serves as a crossroads for the digestive and respiratory systems **(Fig. 4–73)**. The pathways for food and air cross each other in the pharynx, because the nasal cavities enter the *roof* of the mouth, but the larynx—the next step in the passage of air towards the lungs—lies *ventral* to the esophagus. The auditory tube from the middle ear cavities enters the pharynx on the midline of the palate, and allows air pressure to be balanced on both sides of the tympanic membrane.

Larynx

The **larynx** of birds is not the "voice box" as it is in mammals; thus it is a simpler structure and does not contain vocal cords. It consists of two major cartilages called the **laryngeal (cricoid) cartilages (Fig. 4–74)**. Its primary function is to act as a valve, regulating the flow of air into the trachea, rather than producing sound. The opening to the larynx is the slit-like **glottis** caudal to the tongue, formed by two half-circle-shaped cartilages covered with mucous membrane.

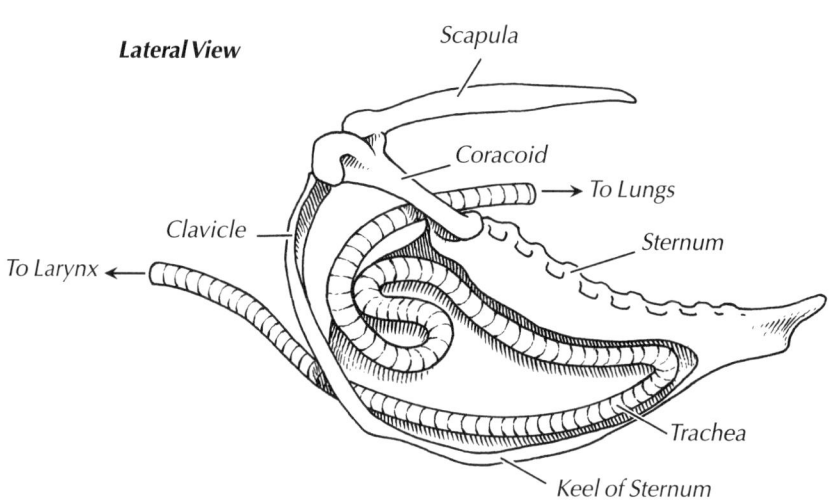

Figure 4–75. Coiled Trachea Within the Sternum of a Whooping Crane: The *trachea*, or windpipe, is a tube supported by cartilaginous rings, which conveys air from the larynx to the lungs. In most birds it follows a straight route, but in some it is elongated and looped. In the Whooping Crane illustrated here, as well as in some swans, its extensive loops lie coiled within the keel (carina) of the sternum. Elongated tracheae are thought to aid vocalization, as well as high-altitude, long-distance flight—warming the inhaled air and allowing more air to be stored within the respiratory system. Drawing by Charles L. Ripper.

Trachea

The **trachea** (see Fig. 4–74), or windpipe, is a tube that conducts air from the larynx to the lungs. A series of cartilaginous rings holds the trachea open for the passage of air. The rings can telescope into each other to shorten the trachea, or can pull apart, exposing soft connective tissue between the rings, to lengthen it. In most birds the trachea follows a straight course from the glottis to the lung bronchi. In some species, however, it is greatly elongated and looped within the neck or even farther afield. For example, in some chachalacas (Cracidae) and the painted snipes (Rostratulidae), the loops lie between the skin and underlying muscles; in some ibises, loops lie within the thorax; in the Plumed Guineafowl, they lie within the furcula or "wishbone"! Perhaps most bizarre, in the Trumpeter Swan, Tundra Swan, and Whooping Crane, the trachea enters the keel of the sternum **(Fig. 4–75)**. Within the keel it passes caudally before returning toward its point of entry, exiting the keel and making another bend to enter the thorax. Some of these looping tracheae are thought to be adaptations for vocalizations; others may be adaptations to high altitude, long-distance flight—functioning to warm the inhaled air and increase the total volume of air that can be stored within the respiratory system.

Other specializations of the trachea include regional expansions along its length that modify the birds' vocalizations. One or two dilations occur between the larynx and the bronchi in South American screamers—odd, gooselike birds whose calls, among the loudest given by any bird, sound something like a creaky but lively series of groans on an old sheep horn. Tracheal expansions also occur in the males of most species of ducks, many of which also have an expanded tracheal bulla (sac) on one side at the lower end of the trachea **(Fig. 4–76)**.

Within the thorax, a short distance before it enters the lungs, the trachea of all birds has an expansion, the **syrinx**, and then divides to form two **bronchi** (singular, **bronchus**). Incomplete rings of cartilage support the walls of each bronchus and continue into the lungs.

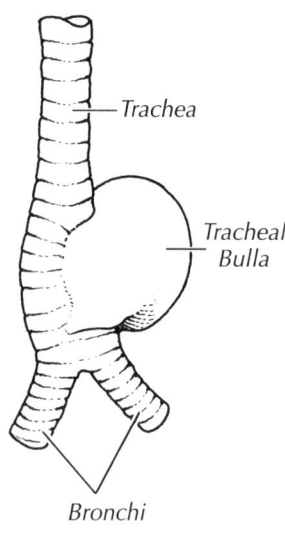

Figure 4–76. Tracheal Bulla of a Duck: Males of many duck species have a **tracheal bulla**—an expanded sac—on one side of the lower end of the trachea. These tracheal bullae and various other tracheal dilations found in other types of birds are thought to modify the sounds produced by the syrinx. Drawing by Charles L. Ripper.

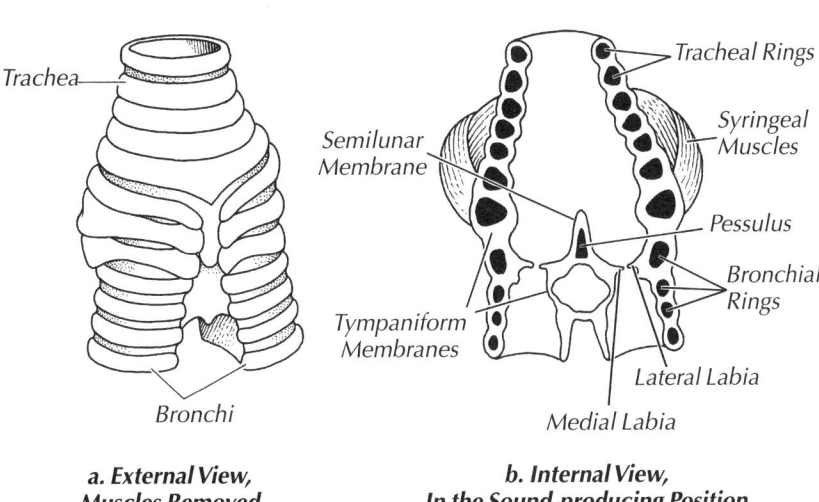

Trachea

Semilunar Membrane

Tympaniform Membranes

Bronchi

Tracheal Rings

Syringeal Muscles

Pessulus

Bronchial Rings

Lateral Labia

Medial Labia

a. External View, Muscles Removed

b. Internal View, In the Sound-producing Position

Figure 4–77. The Syrinx of a Songbird: *The sound-producing organ of birds, the syrinx, is located at the point where the trachea divides to form the two bronchi.* **a. External View:** *The syrinx, shown here with its musculature removed, is formed by an expansion of either the tracheal cartilaginous rings, the bronchial cartilaginous half-rings, or, more commonly, some combination of the two.* **b. Internal View:** *The syrinx here is shown actively engaged in producing sound. Flexible* **tympaniform membranes** *stretch between the cartilaginous rings, permitting the syrinx to change shape as a result of the actions of one or more pairs of* **syringeal muscles***, thereby altering the sounds it can produce. The* **medial and lateral labia** *are particularly important sound-producing membranes (see Sidebar 2: Bird Song: From Oboe and Trombone to Orator and Soprano). The* **semilunar membrane** *and the cartilaginous* **pessulus** *occur only in the syrinx of songbirds, probably contributing to the advanced singing abilities of these birds. Drawings by Charles L. Ripper.*

Syrinx

The **syrinx** is the bird's sound-producing organ (**Fig. 4–77**; and see Fig. 7–34). It is formed by modifications of either the tracheal rings or the bronchial half-rings, or a combination of both. Thus a bird can have a tracheal syrinx, a tracheobronchial syrinx (the most common), or two separate bronchial syrinxes. Internal and external **tympaniform membranes** stretch between the cartilages, allowing the shape of the syrinx to change and altering the sounds it can produce. Additional structures are found within the syrinx of songbirds, the large group of birds with the most complex vocalizations: a median cartilage called the **pessulus** at the bifurcation of the bronchi, and a **semilunar membrane** extending from the pessulus into the cavity of the syrinx. Presumably, these additional structures account, at least in part, for the singing capabilities of these birds. Some birds are able to control the two branches of their syrinx so well that they actually can sing duets with themselves (see Ch. 7, Control of Song, and Fig. 7–35).

One or more pairs of syringeal muscles act on the cartilages of the syrinx and change its shape, thus tensing or relaxing the tympaniform membranes. Air forced past the membranes vibrates them, creating sound waves. The changes in the tension of the membranes and in the diameter of the constricted passage in the syrinx, brought about by the syringeal muscles, result in different types of sounds. Also important in sound modulation is whether one or both bronchial passages are constricted by muscle action and to what degree the right and left bronchial constrictions differ. An increase in the *tension* of the tympaniform membranes increases their frequency of vibration, and thus raises the pitch of the sound; an increase in the *diameter* of the passageway increases the sound's volume (loudness). **(Sidebar 2: Bird Song: From Oboe and Trombone to Orator and Soprano)**

(Continued on p. 4·98)

Sidebar 2: BIRD SONG: FROM OBOE AND TROMBONE TO ORATOR AND SOPRANO

J. B. Heiser

We do not know for how many millennia humans have enjoyed the beauty and musicality of the songs of birds. No less a scientific figure than Charles Darwin felt that birds "On the whole...appear to be the most aesthetic of animals, excepting of course man, and they have nearly the same taste for the beautiful as we have." (The Descent of Man and Selection in Relation to Sex, 1874, p. 697). The comparison of bird song to musical instruments (and vice versa) must have begun with the first whistles and flutes of Paleolithic humans. It may not be by chance alone that one of the oldest surviving musical instruments was made of bird bone: a flute from France estimated to be 10,000 to 15,000 years old!

In spite of our long love of bird song, understanding *how* birds sing has eluded science right up to the present day. How, for instance, are birds with a loud, long, complex, and rapid song able to sustain the airflow needed to perform such extended songs without apparently pausing for a breath? Winter Wrens may continuously vocalize for as long as 41 seconds, and Grasshopper Warblers (an Old World temperate zone species) for 60, and perhaps as much as 117, seconds. Estimates of the air available in their respiratory systems suggest that many birds should run out of breath before completing their feats of song. Careful measurements of tracheal airflow in certain species have shown that they take "minibreaths" of 15 to 50 *milli*seconds in duration between notes or closely spaced groups of notes (Suthers and Goller 1997). These minibreaths are so short, and therefore so shallow, that they probably do little to replenish oxygen—indeed, most of the inhaled air is probably the same air that was just exhaled during sing-

ing, and is thus oxygen-depleted. But the mini-breaths do seem to perform the important function of replacing about as much air as was vented in producing the preceding note(s), thus permitting the singer to emit longer songs. When note repetition rates exceed about 30 per second (what we perceive as fast trills), all the species of songbirds studied so far shift from taking minibreaths between notes to taking breaks between trills; during these "inter-trill" breaks a more substantial inhalation takes place. Apparently there is a threshold at this repetition rate above which the neuromuscular apparatus of breathing cannot function for even the shortest minibreath.

A part of our lack of understanding of how birds sing is due to the unique nature of the avian vocal apparatus, the syrinx. Among living animals, the syrinx is as unique to birds as are feathers and could just as accurately be used as a defining feature of birds. Nevertheless, so much variation exists between the syrinxes of various bird groups that the structure has long been used in constructing evolutionary hypotheses. The structure of some syrinxes is very complex, especially those of the "true" songbirds (Order Passeriformes, Suborder Passeri; also known as the "oscines" from the Latin for "singing bird") (see Fig. 7–22). Although oscines are so numerous that they make up 48 percent of living birds, the syrinxes of all oscines are nearly identical in their complex structure. These consummate singers are almost all small, and the syrinx is so deep in their respiratory tract (at the lower end of the trachea), that direct studies of its activity during song production have been all but impossible.

Simple observations, however, will convince anyone that it is the

complex oscine syrinx that provides for much of the variety and beauty of bird song. Listen at a woodside garden almost anywhere in eastern North America in spring, and you will be treated to a great diversity of passerine songs. Quite striking, however, is the contrast between the aesthetic nature of the songs of the Northern Cardinal, Song Sparrow, and Brown Thrasher (all oscines) and the simple, monotonous, and not very musical quality of the Eastern Phoebe's song. The phoebe, although a passerine, is a "suboscine;" its family and several others worldwide seem to have split off the main passerine lineage early in passerine evolution (see Fig. 7–22). The ancestors of the phoebe became isolated on the South American continent when it was far from any other major land mass and radiated into the tyrannid flycatchers we know today. In spite of the diversity of species that evolved, these flycatchers all retained a relatively simple syrinx **(Fig. A, Eastern Phoebe)**. From the same simple syrinx, the oscines, evolving elsewhere, developed, among other innovations, a complex syrinx with more than three pairs of intrinsic muscles, allowing for a much greater variety of fine control **(Fig. A, American Crow)**. By the time North and South America were united by the formation of Central America, the oscines had populated North America. Thus the Central American land bridge allowed the two long-isolated lineages, the oscines and suboscines, to come together, enriching the avifauna of both continents. The sound of the phoebe presumably has many qualities in common with the songs of early passerines. The songs of true songbirds developed independently, their evolution made possible in part by complex syringeal anatomy.

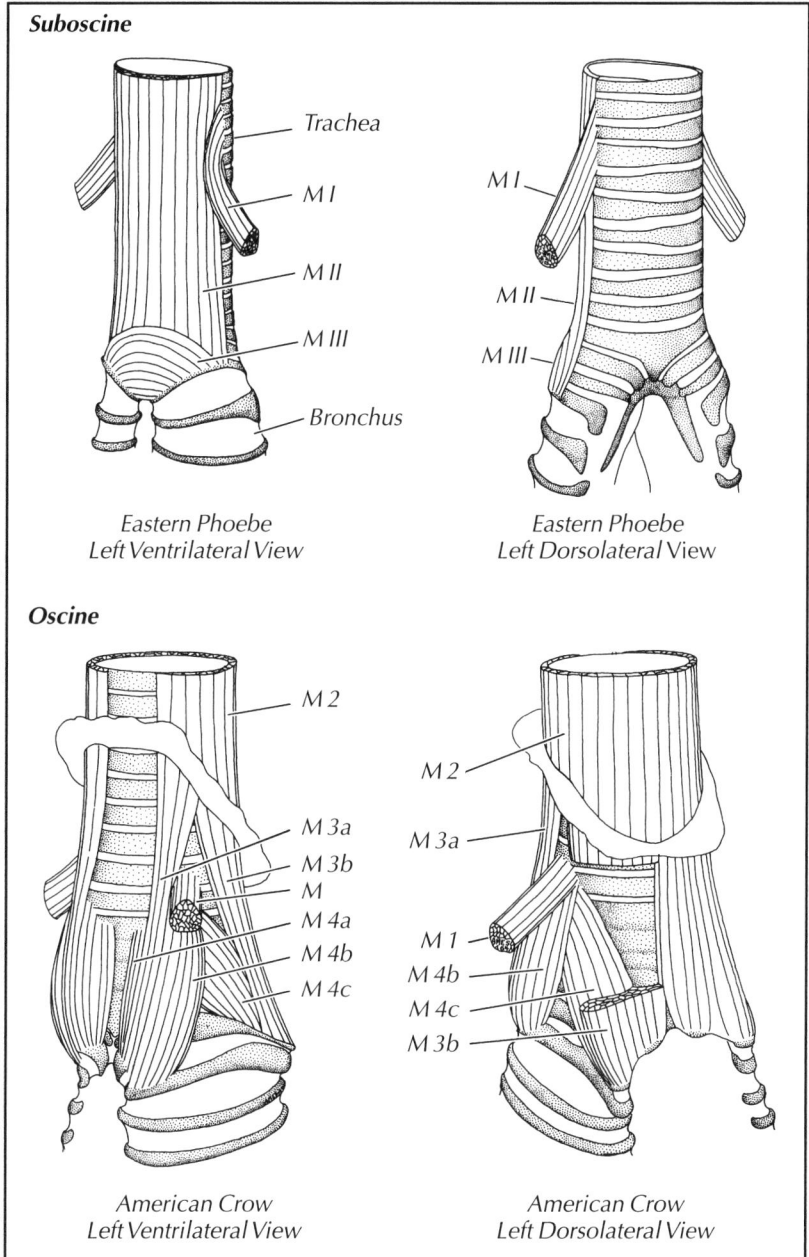

Figure A. Musculature of the Syrinx in Suboscine Versus Oscine Passerines: *The syrinxes of an Eastern Phoebe (a suboscine) and an American Crow (an oscine) are compared in both ventrilateral and dorsolateral views. The muscles are labeled MI through MIII in the phoebe and M1 through M4 in the crow. The phoebe's muscle MI is considered equivalent to the crow's M1, and MII is the same as M2. All the other muscles are unique, however, derived from different developmental origins in each species. Notice the greater complexity of the musculature of the oscine (crow) syrinx, evidence of the superior singing ability of these birds. Adapted from Ames (1971, Plates 5, 6, and 21).*

The majority of birds have a syrinx that contains elements of the right and left bronchi and thus have right and left paired structures. In oscines, and some other birds, these paired structures include, at the cranial end of each bronchus, a distinct set of the membranes that vibrate to produce sound **(Fig. B)**. The possibility thus arises that these birds could produce two harmonically unrelated tones at the same time; in essence, duetting with themselves. This "two voice" phenomenon has been studied in numerous species with surprising results. Brown Thrashers and Gray Catbirds frequently generate different sounds simultaneously, resulting in true two-voice song. At other times, they may switch one side or the other off, with various results. Sometimes they favor the left side, and sometimes each side of the syrinx contributes about equally to the total production of song. Even when the contributions are equal, they are not identical. In all such two-voice cases studied, the left syrinx produces lower-frequency notes than the right in any given series of notes. The left side can produce notes in the range of the right and vice versa, but they seem always to divide the task of singing as do the left and right hands of a keyboard musician. Northern Cardinal songs have sounds that consist of continuous sweeps between 3,000 and 4,000 Hertz. Surprisingly, the lower-toned part of the sweep is performed by the left side of the syrinx and, without any detectable auditory break, the upper tones are seamlessly added by the right side! Some canaries sing about 90 percent of their song elements on the left side of their syrinx, but at the same time the nerves and muscles of the right side are just as active as if they were the source of the sound. Only the muscles that block airflow stay still. It is not entirely clear why this should be the case. With all the variety in syringeal structure and function, in addition to the difficulty of studying the syrinx in a living, singing bird, it is clear that no single functional model is likely to apply to all species; proposing to explain "the function" of "the syrinx" is a fool's task.

The uniqueness of the syrinx as a sound-producing organ and the musicality of the sounds that birds produce has long led people to propose analogies between bird song and musical instruments. Perhaps the earliest scientifically based proposals were those of the French. In 1753 a paper presented to the French Royal Academy claimed that

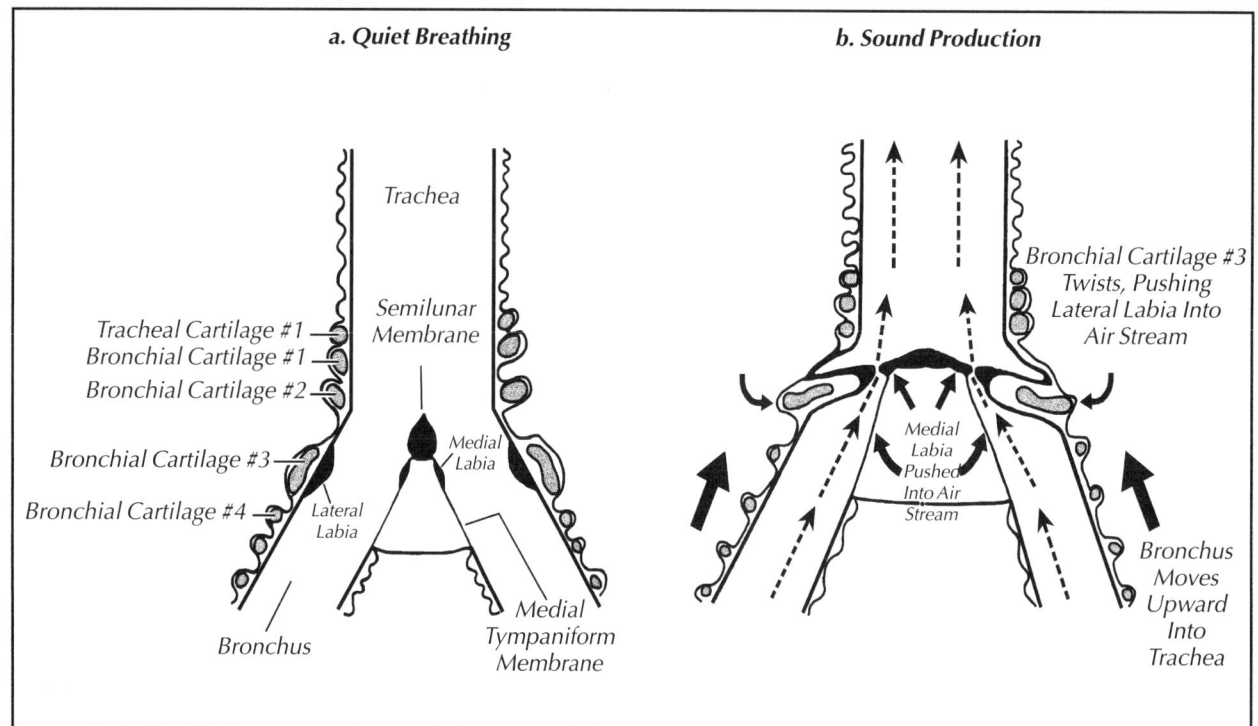

a. Quiet Breathing

b. Sound Production

Figure B. The Oscine Syrinx and Sound Production: *Shown here are schematic, longitudinal sections through the syrinx of an American Crow. **a. Quiet Breathing:** This illustrates the syrinx during quiet breathing, showing the internal structures in relation to one another in the resting position. **b. Sound Production:** This illustrates the syrinx during sound production by one or both sides. Muscles stretch the two bronchi upward into the trachea (large solid arrows), and twist the third bronchial cartilages. The latter push the lateral labia into the path of air flowing out of the respiratory system and into close proximity with the medial labia, which have also been forced into the airway by the movement of the bronchi (medium solid arrows). Sound is produced as air rushes between the lateral and medial labia (dashed arrows), causing these soft tissues to vibrate. Adapted from Goller and*

the syrinx acts like the double reed of an oboe. In the early 1800s Georges Cuvier, famed French anatomist and paleontologist, reckoned that the French horn (naturally!) and the trombone were more accurate comparisons. Subsequently all manner of reed and brass instruments and even organ pipes have been promoted as the best model for how birds sing. None of these analogies has proven close enough to the truth to last for long, but they have some elements in common that may throw light on the mystery of how birds produce their melodious song.

In all of these musical instruments, air rushing through a tube past some flexible structure in its path sets that flexible structure vibrating, producing sound. The features of the tube then modify the original sound in a multitude of ways that result in the characteristic sound qualities by which we can identify the instrument. The avian vocal tract has these same components: abdominal muscles force air from the air sacs and lungs through the syrinx with its several membranous tissues, some of which presumably vibrate, producing sound. The remaining structures in the vocal tract (the trachea, glottis, pharynx, tongue, mouth cavity, and beak) subsequently modify the syrinx's sounds. Although many membranes of the syrinx have been hypothesized to be the ones that vibrate to produce sounds, it appears from tiny fiber-optic endoscopy in the four oscine species studied so far that only the soft tissues known as the lateral and medial labia (lips) (see Fig. B) or thin tissues known as the lateral tympaniform membranes (in the Rock Dove and Cockatiel)

need vibrate to produce natural vocalizations. The actions of the remainder of the vocal tract have been much debated, but as a result of a number of innovative experimental techniques—including recording birds singing in atmospheres composed of gasses of lesser and greater density than air—the contributions of the vocal tract are now generally agreed to be significant. The width of the opening of the beak in White-throated Sparrows and Swamp Sparrows, for example, controls which frequencies produced by the syrinx are muted and which are emitted at full volume (Gaunt and Nowicki 1997).

Significantly, pure, whistle-like tones are apparently not produced in their pure form by the syrinx. Instead, sounds with several simultaneous frequencies (a tone and its harmonics) produced by the syrinx are "cleaned up" or filtered acoustically by the vocal tract, leaving pure tones to be emitted. This "acoustic filter model" for the production of pure tones is, in fact, more like the production of speech by the human voice than the operation of a musical instrument. In our speech the mouth, tongue, and lips are crucial in modifying the sounds produced by the vocal cords, often by suppressing certain frequencies and augmenting others. Thus bird song has recently come to be likened to human speech rather than to musical instruments, whose post-vibration effects are much more limited than those of the human, and apparently the avian, vocal tract.

During the production of the vocalizations of most animals, when the air pressure in the lungs—which drives the vibrating membranes—doubles, the acoustic flow (airflow over the membranes) also doubles (that is, the system behaves linearly), resulting in a predictable shift in the frequency of the sound produced. Experiments with Zebra Finches, however, have demonstrated that, at least in these birds, no such simple relationship exists (Fee et al. 1998; Goller 1998). As air pressure on a Zebra Finch syrinx is slowly increased or decreased, at certain pressures the system spontaneously jumps to a nonlinear output; in other words, the acoustic flow and sound produced by it suddenly are not simply proportional to the driving air pressure. The resulting jumps in the frequency of a vocalization are equivalent to the human voice "breaking" (as in adolescent males) or, in people with some illness or disorder of the vocal cords, having a roughness or "gravelly" quality. These characteristic jumps are not only produced by Zebra Finches during natural song, but also by an isolated Zebra Finch vocal tract when air is passed through it. Such nonlinear behavior must therefore be an intrinsic characteristic of the syrinx. Just how the elastic and oscillatory structure of the syrinx functions to change the type of response it gives to different driving pressures—causing frequency jumps—is not understood. But, the exact functioning of the syrinx is currently a hot topic among researchers.

Another relatively recent discovery leads to a further and final analogy that attempts to explain how bird song is produced: Researchers have found that during a vocalization, parts of the upper vocal tract may directly affect the vibrating syrinx, changing its frequency from what it would emit without this physical feedback (Nowicki and Marler 1988). It is now recognized that, in humans, direct interactions between the vocal cords and the vocal tract are important in determining the quality of the *singing* voice. A trained soprano apparently is able to produce her purest tones by changing the shape of her mouth, the position of her tongue, and the length and shape of her vocal tract, which changes the resonant frequencies within her head and chest; these resonances directly affect the way the vocal cords vibrate, and thus affect the quality of the sound actually produced by the vocal cords. The fact that birds, too, are now known to affect the syrinx's output by controlling the action of the upper vocal tract, leads to speculation that something very similar to a soprano's controlled singing is going on in birds. Perhaps the clear and undeniably beautiful sound of a pure, sung tone is produced in the same way, no matter whether by a bird or a human!

Probably no analogy will ever explain a significant portion of how birds produce their most complicated songs. The syrinx has proved to be not only anatomically unique, but to be marvelously complex in its function and in the scope of its vocal ability as well. ∎

Lungs and Air Tubes

The two surprisingly small **lungs** lie just below (ventral to) the vertebral column and the ribs. As each bronchus enters the lung, it loses the half-rings of cartilage and continues as the primary bronchus or **mesobronchus** through the lung **(Fig. 4–78)**. The mesobronchi gradually decrease in diameter, branching into **secondary bronchi**. Some of these, called **recurrent bronchi**, connect to the air sacs. Other secondary bronchi branch to form **parabronchi**, the major respiratory units of the lung. The longest mesobronchi end at the entrance to the abdominal air sac. Much inhaled air passes directly through the lungs via the mesobronchi to the air sacs without being involved in gas

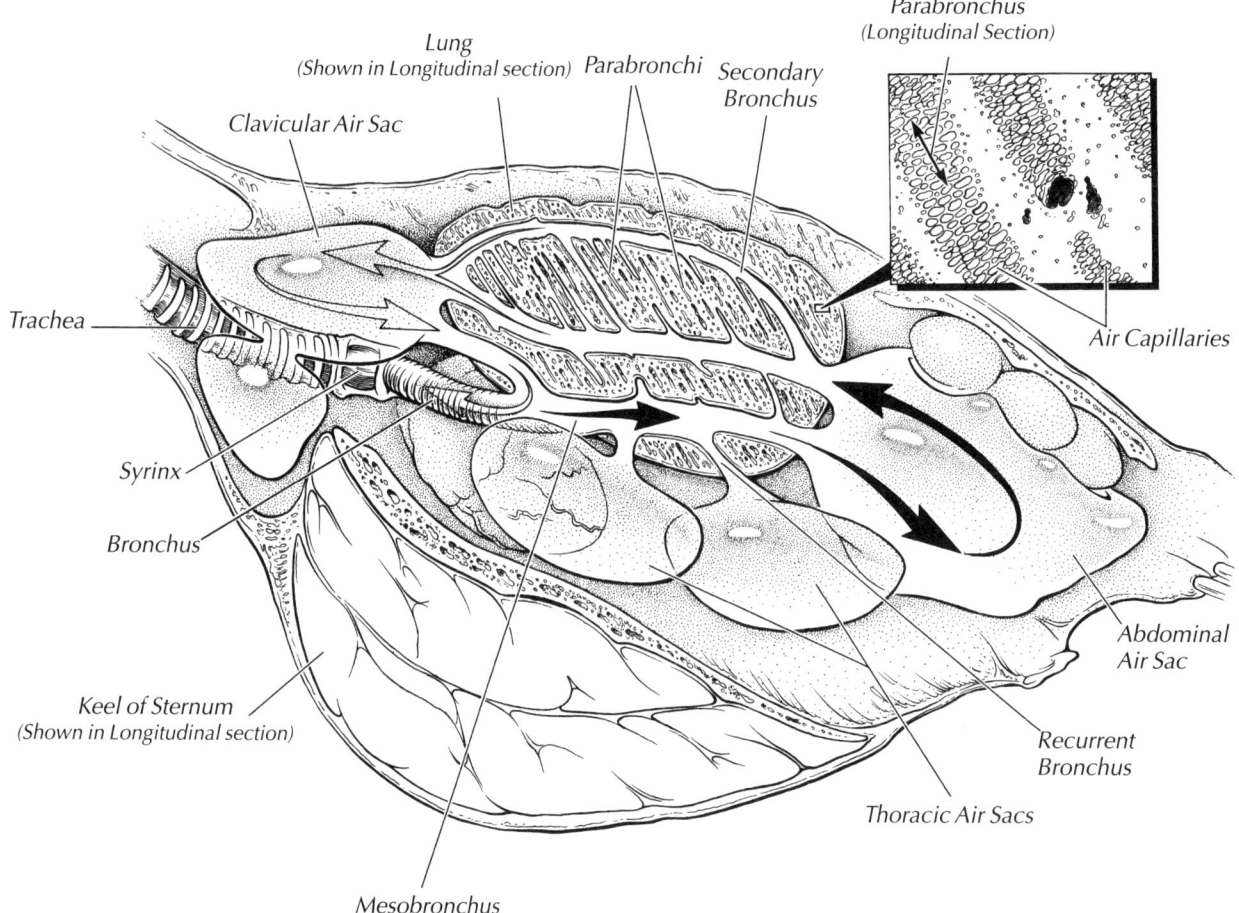

Figure 4–78. The Avian Respiratory System: In this functional diagram, the components of the respiratory system are viewed from the bird's left side. The lung and its internal structures are shown in sagittal section. For paired structures, only one member of the pair is visible. The trachea divides into two **bronchi** (singular **bronchus**), one of which enters each **lung**, losing its cartilaginous half-rings and continuing through the lung as the primary bronchus or **mesobronchus**. The mesobronchi branch into narrower **secondary bronchi**, some of which divide into many fine branches termed **parabronchi**, which are the lung's major respiratory units—the sites of gas exchange. The inset shows several highly magnified parabronchi sectioned longitudinally. Other secondary bronchi, termed **recurrent bronchi**, connect to the **air sacs** outside the lung. Shown here in external view are the **clavicular, thoracic** (two on each side), and **abdominal** air sacs. For clarity, these sacs have been simplified and drawn much smaller than in a live bird; in their natural state, air sacs completely surround the abdominal organs and overlap each other extensively, forming a complex system with connections to air spaces within the bones (see Fig 4–82). Reprinted from Manual of Ornithology, by Noble S. Proctor and Patrick J. Lynch, with permission of the publisher. Copyright 1993, Yale University Press.

exchange in the lungs. Exchange occurs with this air on exhalation, as it passes out of the air sacs, back into the lungs, and through the parabronchi.

The unique internal anatomy of the avian lung is most easily understood by examining the accompanying figures. The bird's lung is a mass of interconnecting air tubes, the parabronchi, too small to be seen with the naked eye (**Fig. 4–79**). Each air tube has openings in its thick wall that allow air to pass into small interconnecting spaces that weave together like a labyrinth through the surrounding blood capillary bed (**Fig. 4–80**). These openings also allow air to pass back from the spaces into the air tubes. The capillary bed surrounding the spaces is supplied by the pulmonary artery and drained by the pulmonary vein. As blood flows through the capillaries (generally at right angles to, or in the opposite direction of, the flow of air—permitting the formation of a countercurrent exchange system), oxygen in the air spaces dissolves into the blood, and carbon dioxide from the blood moves into the air to eventually be exhaled (**Fig. 4–81**). This arrangement of lung tissues is very different from that of mammals, in which respiratory exchange takes place in dead-end pockets (alveoli) at the termination of bronchioles. Birds, however, have no bronchioles or dead-end alveoli. Because the minute air spaces all interconnect in birds, rather than ending blindly, birds can achieve a continuous flow of air across the surface of the capillary bed and thus a continuous, highly efficient oxygen extraction.

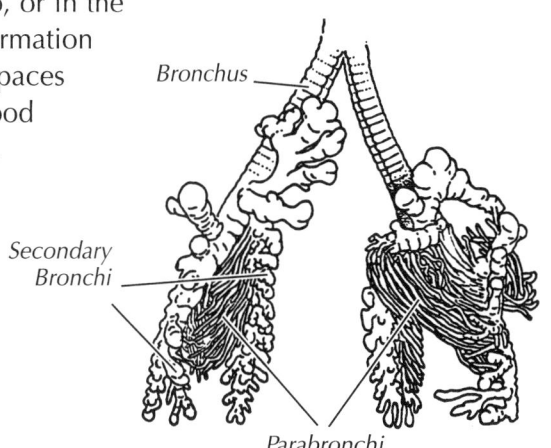

Figure 4–79. Partial Cast of the Lungs of the Chicken: This three-dimensional cast was formed by injecting the lungs with a thick liquid called Wood's Metal, which subsequently hardened. The lung tissue was then chemically stripped away, leaving a solid impression of a portion of the lungs' internal structure. At the top, paired **bronchi** with regularly spaced cartilaginous rings are visible. The thick, short tubes are **secondary bronchi**, and the longer, narrow tubes are the massed interconnecting **parabronchi**, the major respiratory units of the lungs. Drawing courtesy of Howard E. Evans.

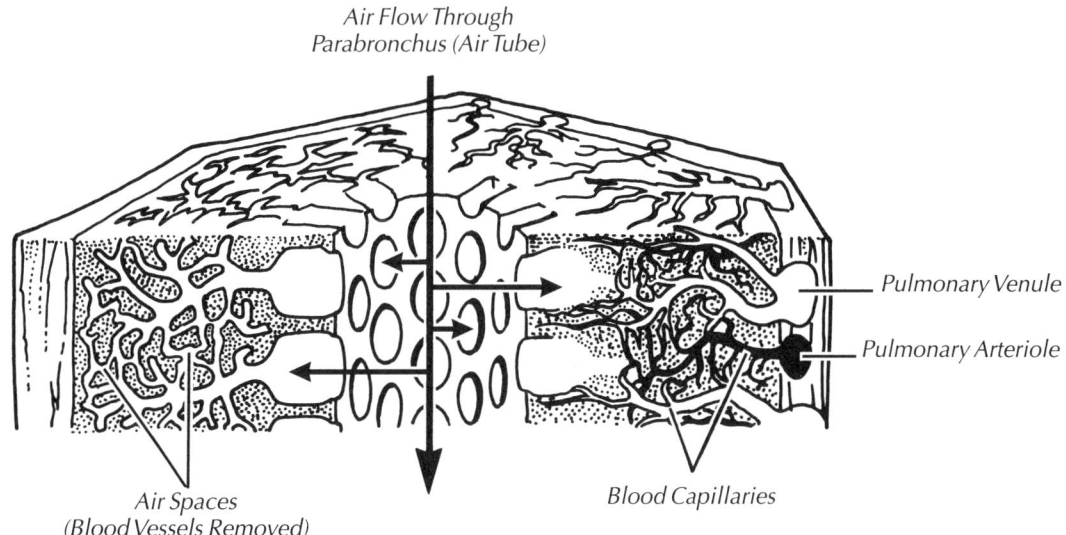

Figure 4–80. Structure of Parabronchus: This cutaway view shows the internal structure of a single **parabronchus** (air tube). Air flows continuously through the parabronchus, moving through openings in its walls into a network of **air spaces**, sometimes referred to as air capillaries. Each of these spaces is surrounded by a network of blood capillaries, and it is between these two structures that oxygen and carbon dioxide are exchanged. For simplicity, the blood capillaries have been removed from the left side of this drawing, and on the right side, the air spaces are mostly obscured by the blood capillaries. **Pulmonary arterioles** supply the system with deoxygenated blood, and **pulmonary venules** transport the oxygenated blood away. Adapted from Brooke and Birkhead (1991, p. 31).

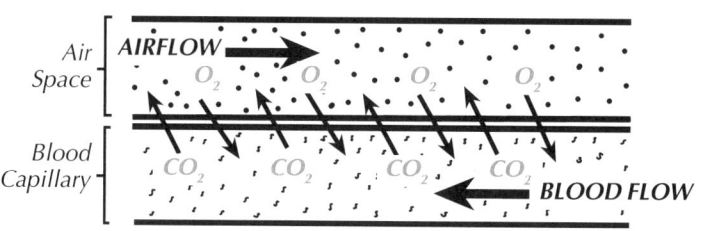

Figure 4–81. Gas Exchange in Parabronchus: *This schematic shows gas exchange between a single air space and adjacent blood capillary in close proximity within a parabronchus. Blood flows in the opposite direction to that of air, establishing a countercurrent system. Oxygen in the air space diffuses into the blood, while carbon dioxide in the blood diffuses into the air space, eventually to be exhaled.*

Air Sacs

Thin-walled **air sacs** (four paired and one median) extend from the mesobronchus or the lung itself to different regions of the body **(Fig. 4–82)**. A **cervical sac** is usually located on each side in the neck region or sometimes a series of cervical sacs are found along the neck, as in geese. A large median **clavicular sac** between the clavicles surrounds the bifurcation of the trachea. The clavicular air sac has extensions on each side that lie in the shoulder joint region and connect through a hole in the humerus, the **pneumatic foramen,** to the pneumatic space within the wing bones. The clavicular sac also has several connections with the pneumatic spaces of the sternum. Two **thoracic air sacs** are fixed in position on each side, but the pair of **abdominal air sacs**, which may have connections into the bones of the pelvis and femur, can shift their position within the abdominal cavity—at different times of the day they may occupy different areas between the organs. This adaptation allows a bird to maintain a streamlined shape for flight even after a large meal or before laying an egg.

The air sacs, so thin-walled that they resemble soap bubbles, have no blood vessels and play no direct part in the exchange of oxygen and carbon dioxide. Instead, they serve as bellows to bring air into the bird and store it until expiration. During expiration, this stored air passes through the **recurrent bronchi** into the parabronchi of the lungs to interact with the respiratory surfaces. Thus birds get fully oxygenated air into the lungs on both inspiration *and* expiration, and consequently have the most efficient respiratory system of any vertebrate.

Breathing and Gas Exchange

In review, the process of breathing and gas exchange occurs as follows **(Fig. 4–83)**: With each inspiration, air passes freely through the interconnecting parabronchi and recurrent bronchi, filling the gas exchange spaces in the lungs, as well as the posterior air sacs, with oxygenated air. With each expiration, air returns from the posterior air sacs and passes through the parabronchi, replacing the now oxygen-depleted air in the gas exchange spaces (which is eventually exhaled) with oxygen-rich air. On the next inspiration, this air (now oxygen-depleted, too) is moved to the anterior air sacs as the next batch of oxygen-rich air is brought into the posterior air sacs and gas exchange spaces. As a result there is little residual, oxygen-poor air in the lungs of birds after exhalation, in contrast to those of mammals. In addition, the bird's arrangement permits oxygen and carbon dioxide to be exchanged

a. Lateral View

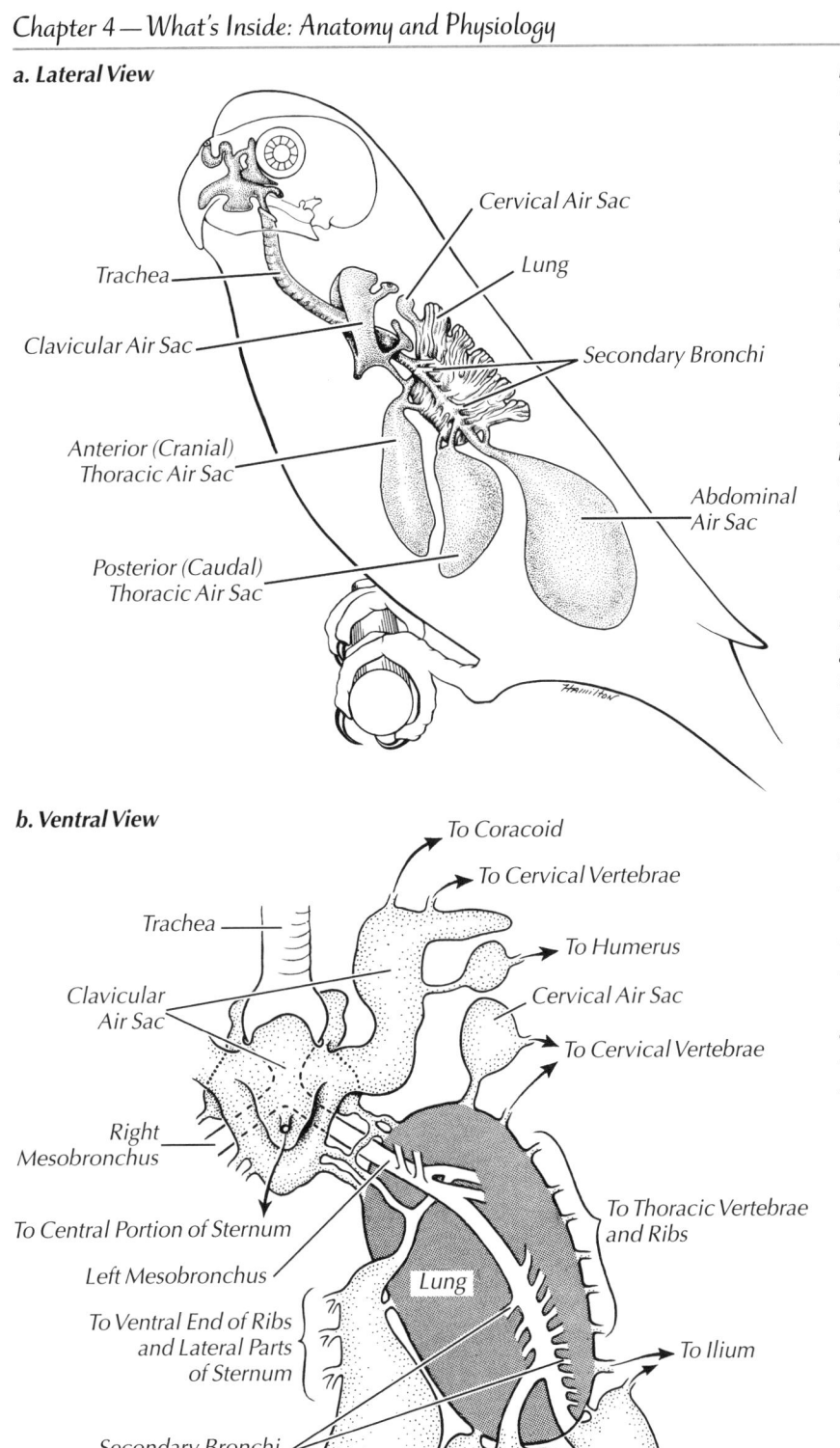

Figure 4–82. Air Sacs: a. Lateral View: This view shows the location of the major air passageways and air sacs of the Budgerigar. A system of thin-walled, transparent air sacs extends from the mesobronchus and the lung to different regions of the body. A single, medial **clavicular air sac** lies around the trachea. A **cervical air sac** extends from each lung. Two **thoracic air sacs**—anterior (cranial) and posterior (caudal)—are fixed in position on each side. One **abdominal air sac** is also located on each side, but its position may shift during the day to allow the bird to maintain a streamlined shape even after a large meal or before laying an egg. No gas exchange occurs in the air sacs themselves; instead, they act as bellows to bring air into the bird and store it until it can pass through the gas exchange spaces of the parabronchi during expiration, maintaining a continuous flow of air through the parabronchi. Contractions of the abdominal muscles and movements of the sternum inflate and deflate the air sacs. **b. Ventral View:** This view of the left side of the air sac system illustrates the many connections of the lung and air sacs with the air spaces within the bones. The continuation of the air sacs inside many bones lightens the bones for flight and allows more air to be stored for continuous gas exchange in the lungs, but leaves the bird more vulnerable to traumatic injury. A bone fracture in a bird is more likely to result in death due to internal bleeding or blood clots than is a similar injury in other types of vertebrates. Drawings from Evans (1996).

b. Ventral View

Figure 4–83. The Mechanics of Breathing in the Bird: *This schematic illustrates the one-way passage of a single, inhaled volume of air (shown in black) through the bird's respiratory system. The precise pathways taken by air varies greatly among species, and the pattern shown here is greatly simplified. In addition, probably no bird maintains discrete parcels of air within its respiratory system: in actuality, some lag and mixing occurs within the air passageways. Note that in contrast to mammals, two inspiratory and expiratory cycles are required for a given volume of air to move through its complete path.* **a. Inspiration 1:** *During the first inspiration, a volume of air enters through the bronchus and into the mesobronchus. At the junction of the mesobronchus and the first secondary bronchus, air is prevented from immediately flowing anteriorly by the presence of an **aerodynamic valve.** This valve is not a physical structure, but a vortex-like movement of air that prevents backflow by forcing the incoming air along the mesobronchus and into the **posterior air sacs** (consisting of the posterior thoracic and the abdominal air sacs), which expand to accommodate it.* **b. Expiration 1:** *With the first expiration, the posterior air sacs contract, pushing the air into secondary (recurrent) bronchi connecting to the lung. An aerodynamic valve here prevents the backflow of air into the mesobronchus. Within the lung the air passes through the parabronchi, gas exchange occurring as it moves.* **c. Inspiration 2:** *With the second inspiration, the air moves out of the parabronchi (further gas exchange occurring as it moves), through secondary bronchi, and into the **anterior air sacs** (consisting of the cervical, clavicular, and anterior thoracic air sacs). (Simultaneously, a new volume of air [not shown], is brought into the posterior air sacs.)* **d. Expiration 2:** *Finally, with the second expiration, the anterior air sacs contract, forcing the air into secondary bronchi and out through the bronchus to be exhaled. (Simultaneously, the newer volume of air [not shown] moves from the posterior air sacs into the parabronchi.) Note that at any given time, two volumes of air are moving through the avian respiratory system. The bird's respiratory system differs from the two-way system of mammals, in which gaseous exchange takes place in dead-end pockets (alveoli). Mammalian airflow to the alveoli is oscillatory: air moves in and out through a common set of air tubes (the bronchioles), with gas exchange occurring only during inspiration. Birds have no bronchioles or dead-end alveoli. Instead, the one-way airflow system allows birds to achieve a nearly continuous flow of air through the gas exchange sites, with the extraction of oxygen occurring during both inspiration (as air leaves the parabronchi) and expiration (as air enters the parabronchi). Adapted from Gill (1995, p. 121).*

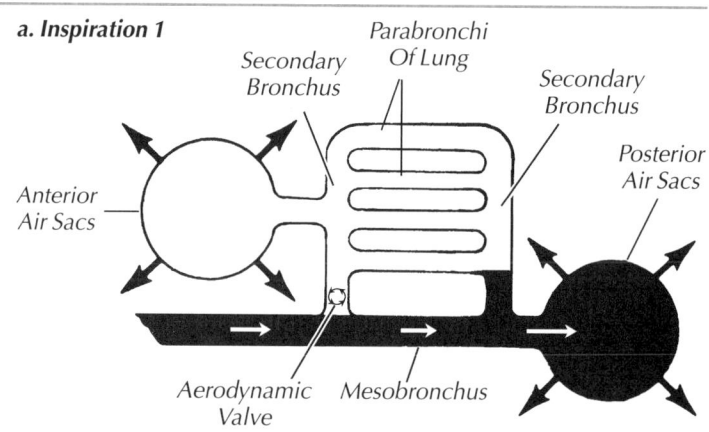

a. Inspiration 1

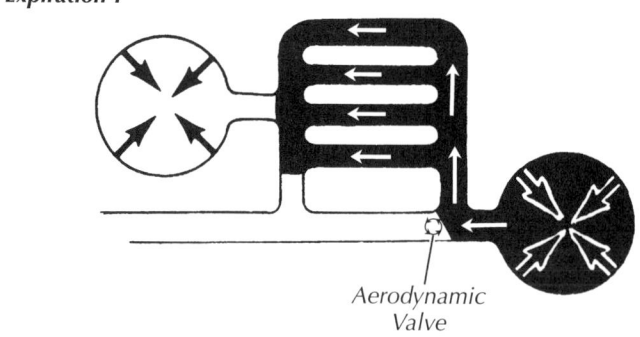

b. Expiration 1

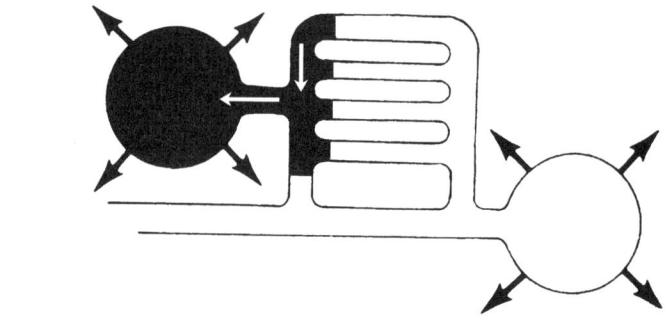

c. Inspiration 2

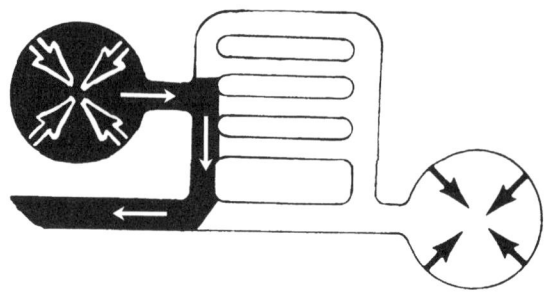

d. Expiration 2

during both inspiration and expiration. Also in contrast to mammals, two inspiratory and expiratory cycles are required for a given volume of air to traverse the complex avian respiratory system. Such a mechanism, obviously more efficient in birds than in mammals—which only exchange oxygen and carbon dioxide on inspiration—is undoubtedly related to the higher rate of metabolism in birds, which is needed to support flight (see Metabolism, later in this chapter for more on this subject). The efficiency of air flushing through the entire respiratory system explains the rapid spread of respiratory illness in domestic birds. The intimate connection of the respiratory system with the skeleton also explains why traumatic injury to a bird so often results in death due to hemorrhage and blood clots.

The Digestive System

■ The digestive system begins with the beak and tongue, which are used for manipulating foods. It continues with an oral cavity (mouth) and pharynx (throat) for swallowing, and an alimentary canal for the passage, digestion, and absorption of food. It ends in the vent, the external opening of the canal from the cloaca, the common single opening for both the alimentary canal and the urogenital system. Along the way, enzymes, hydrochloric acid, and associated gland secretions convert food items into simpler compounds for absorption, storage, or elimination.

The **alimentary canal** includes the esophagus and crop, the two-part stomach, the long small intestine, the ceca, and the short large intestine, which, true to its name, is greater in diameter than the small intestine. The salivary glands, liver, and pancreas are digestive glands that develop in the embryo from its digestive tube and subsequently retain their connection by direct tubular ducts. The secretions from these glands enter the digestive tract through their respective ducts, and aid in digestion.

Oral Cavity

Bill

Because their forelimbs are highly specialized for flight, birds must obtain food with their **bill**, or, in the case of birds of prey, with their feet. Many marvelous variations in the shape and structure of the bill can be seen. In fact, a great deal about a bird's diet can be inferred by examining its bill and tongue, because these structures, and the muscles that operate them, are usually adapted to handle the specific food the bird favors.

The bills of most hummingbirds are a good example, because much variety exists in just this single group of birds **(Fig. 4–84)**. Some hummingbird bills are straight, slender, and rounded in cross section for probing deep into the corolla of flowers for nectar. However, some tropical hummingbirds have very differently shaped bills, from extremely long to extremely short, from sharply recurved (curved

Figure 4–84. Hummingbirds and their Principal Food Flowers: *Nectar-feeding birds, such as the New World hummingbirds and the Old World sunbirds, obtain food by probing their thin beaks into the nectar chambers of flowers. Shown here are four species of hummingbirds from the highlands of Costa Rica, each with its preferred flower. Notice how the shape and length of each bird's beak matches the length and curvature of its principal food flower, which in turn depends on the bird for pollination. The remarkable correspondence between these flowers and birds is an example of coevolution (two or more species evolving with, and in response to, each other). The flowers benefit by having a precise shape that allows just one or a few principal pollinators to reach their nectar, because their pollen is then more likely to be transferred to another flower of the same species. And, the birds benefit because fewer species are competing for the nectar of their principal flowers. Coevolution of this sort is most prevalent in tropical areas, where species diversity is high and a flower's pollen is therefore less likely to reach another member of the same species just by chance. After Wolf et al. (1976).*

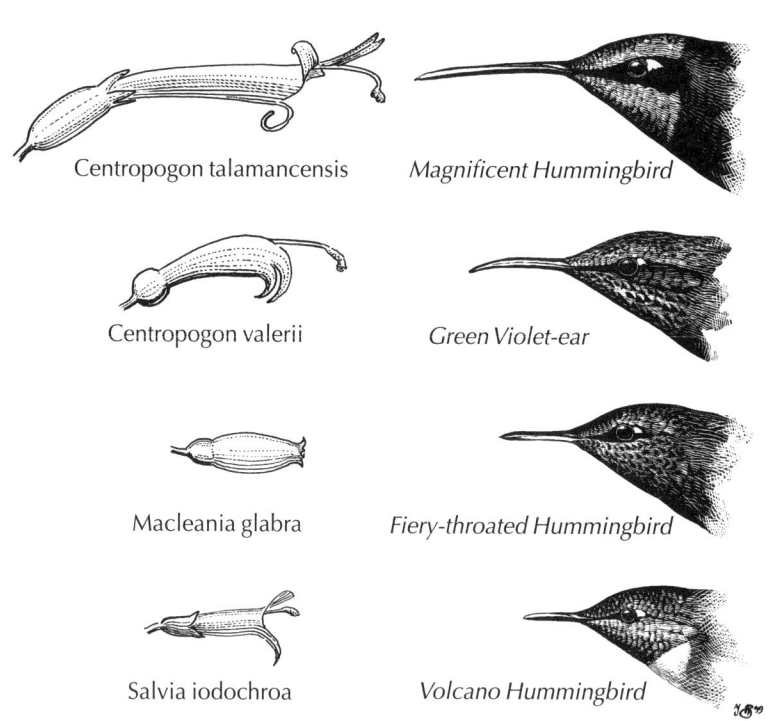

Centropogon talamancensis — Magnificent Hummingbird
Centropogon valerii — Green Violet-ear
Macleania glabra — Fiery-throated Hummingbird
Salvia iodochroa — Volcano Hummingbird

upward) to abruptly downcurved. No doubt such bills give their owners an advantage in obtaining nectar from certain kinds of flowers unreachable by other birds.

Other interesting bills **(Fig. 4–85)** include those of oystercatchers, which are narrow from side to side, functioning as chisels when the birds open oysters, mussels, and other bivalve mollusks. Grackle bills have a sharp keel in the roof of the upper beak; it sticks down into the mouth, and enables them to cut through the hard shells of big seeds. The bill of the Wrybill, a plover from New Zealand, bends to the right in every individual, presumably for prying out food from under pebbles. Researchers have no explanation for the consistency of the bend.

The bills of most fish-eating birds—loons, grebes, tropicbirds, gannets, anhingas, herons, bitterns, terns, and kingfishers—are generally straight and pointed, with sharp edges acting as pincers for grasping, or sometimes spearing, their prey **(Fig. 4–86)**. Bills of mergansers have a further "improvement"—saw-like edges for gripping slippery fish. Pelicans, with hooked upper beaks, big mouths, and distensible throat pouches, can engulf large fish or great numbers of small ones. A puffin, when returning to its young from the sea, may carry as many as a dozen small fish at a time in its parrotlike bill. The bird manages to wedge each one, as soon as it is caught, between ridges in the roof of the upper beak in a very organized manner. Miraculously, it holds each fish firmly in place while it opens its bill to catch another and another!

The Black Skimmer, another fish-eating bird, has a bill in which the lower beak is much longer than the upper beak **(Fig. 4–87)**. With an open bill, the bird skims the surface of the water so that the lower beak cleaves the water **(Fig. 4–88)**. On striking a fish, the bird snaps

a. Oystercatcher

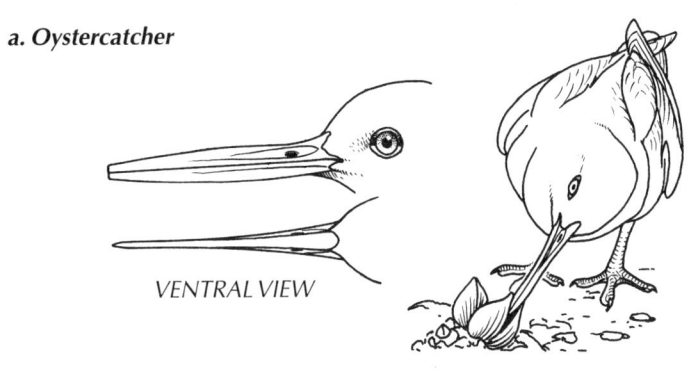

VENTRAL VIEW

b. Grackle

Keel

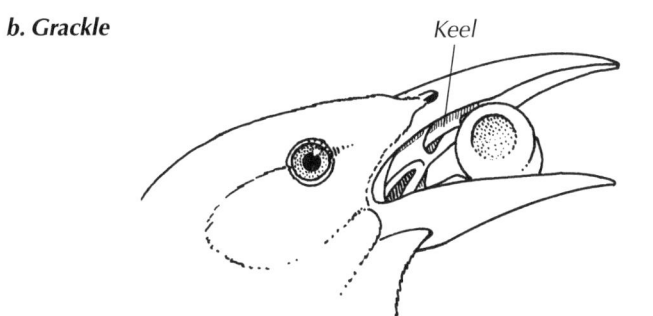

c. Wrybill

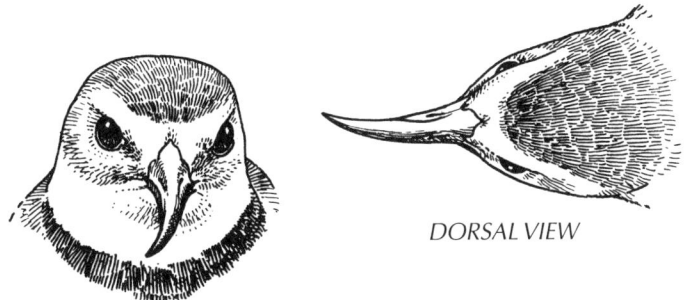

DORSAL VIEW

Figure 4–85. The Diversity of Bird Bills:
The size, shape, and structure of a bird's bill is closely related to its diet, as shown by these examples. ***a. Oystercatcher:*** *The oystercatcher's bill is flattened from side to side (see ventral view), giving it a chisel-like shape with which to pry or hammer open the shells of oysters, mussels, and other bivalve (two-shelled) mollusks.* ***b. Grackle:*** *In the roof of the grackle's bill is a sharp ridge, or keel, sticking down into the mouth, against which the bird can crack open large seeds.* ***c. Wrybill:*** *The bill of the Wrybill, a type of plover from New Zealand, curves to the right side (see dorsal view), allowing it to probe for food under rocks too heavy for it to overturn. Drawings by Charles L. Ripper.*

the two parts of its beak together as its head and neck bend down and underneath the body. The bird then draws the head forward and moves on with the fish in its bill. The evolution of this peculiar feeding behavior involved changes in the bill, head, and neck. The most obvious is the lengthening and narrowing of the lower beak to a blade-like structure that cuts through the water with little friction. The extension of the whole lower jaw and the ability to open the lower beak very wide permit the bird to reach far down while skimming, giving it a better chance of contacting fish. At the same time, the skimmer has the ability to open the shorter, upper beak wider than can its relatives, gulls and terns. This helps to keep the upper beak out of the water where it will not interfere with the fishing. The lower beak has some diagonal ridges with a rich supply of nerve endings, sensitive to the touch of prey. These sensory endings enable skimmers to detect contact with prey, and also allow the birds to feed by night (by touch) as well as by day.

a. Merganser

b. Atlantic Puffin

c. Great Blue Heron

Figure 4–86. Bills of Fish-eating Birds:
a. Merganser: The serrated edges of a
merganser's bill function to grip slippery
fish, giving this bird its nickname, "saw-
bill." Drawing by Charles L. Ripper. **b.
Atlantic Puffin:** Ridges with backward-
facing spines in the palate of the puffin's
upper beak help to secure each fish as it
is captured, allowing this bird to bring
a billful of fish to feed its young. Photo
courtesy of S. Bahrt/CLO. **c. Great Blue
Heron:** The Great Blue Heron has the
characteristic bill of a fish-eating bird—
long, straight, and pointed, for pincerlike grasping. This immature bird has just
captured a fish. Photo by Marie Read. **d.
Brown Pelicans:** Two Brown Pelicans vie
for a fish in the waters off the Galapagos
Islands of Ecuador. The side view of the
open bill of the closer pelican shows
the large, distensible throat pouch. The
second pelican has thrust its bill through
the open bill of the first to try to snatch
the fish. Photo by Tui De Roy.

d. Brown Pelicans

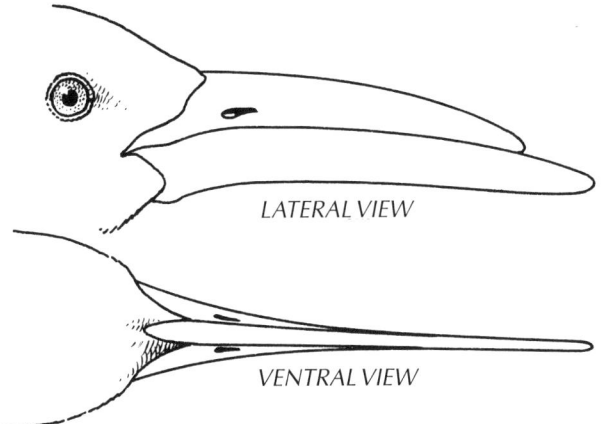

Figure 4–87. Bill of the Black Skimmer: *A fish-eating species related to gulls and terns, the Black Skimmer has a remarkable bill whose lower portion is substantially longer than the upper, as seen in the lateral view. The ventral view reveals the thin, knife-like lower beak for slicing through the water. Drawing by Charles L. Ripper.*

4

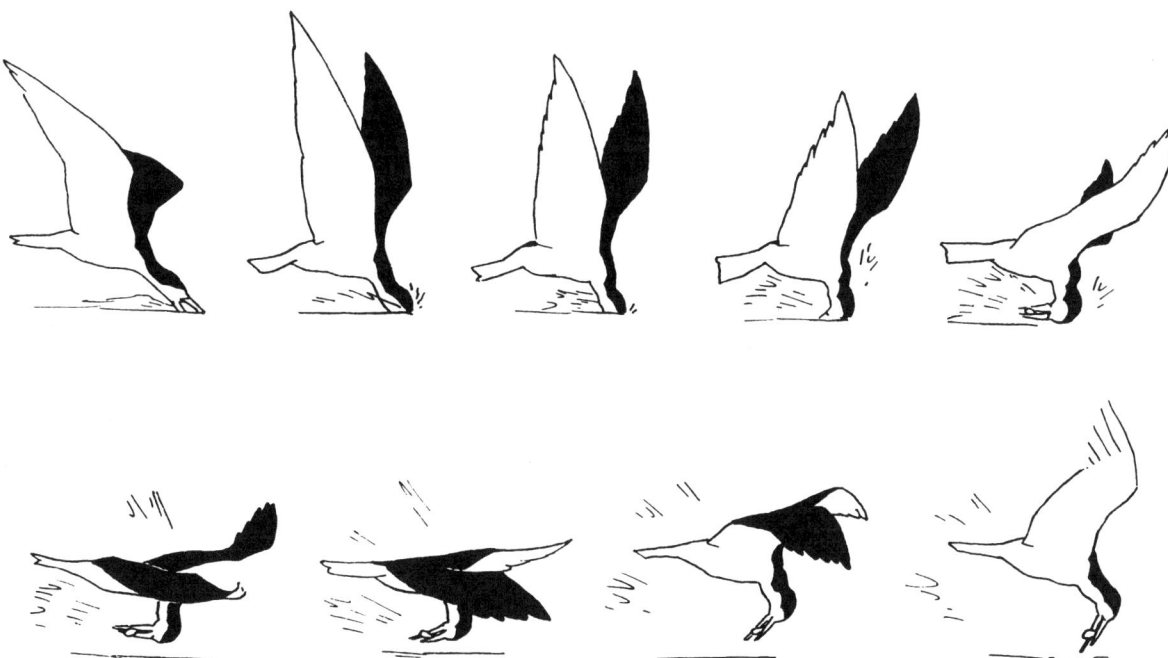

Figure 4–88. Foraging Technique of the Black Skimmer: *This foraging sequence is drawn from a series of movie frames. With its bill open, the skimmer flies low over the surface of the water, its lower beak slicing through it (see also Ch. 1, Sidebar 2, Fig. B). Upon contacting a fish, the bird snaps the upper and lower beak together while bending its head and neck down and underneath its body. The skimmer then draws its head forward and continues flying with the fish in its bill. Several adaptations of the bill, head, and neck permit this unique feeding technique. The lower beak has become narrow and blade-like (see ventral view in Fig. 4–87.) to reduce friction as it moves through the water. The highly extensible lower jaw opens very widely to allow the lower beak to penetrate the water's surface in flight, while the upper beak can also open widely to avoid contact with the water. And finally, the lower beak has numerous sensory nerve endings to allow the bird to detect prey. In addition to these anatomical adaptations, the skimmer flies in a distinctive way during feeding, with its wings held high to avoid accidental contact with the water. Movie frames from Publications of Nuttall Ornithological Club Number 3.*

a. Greater Flamingo

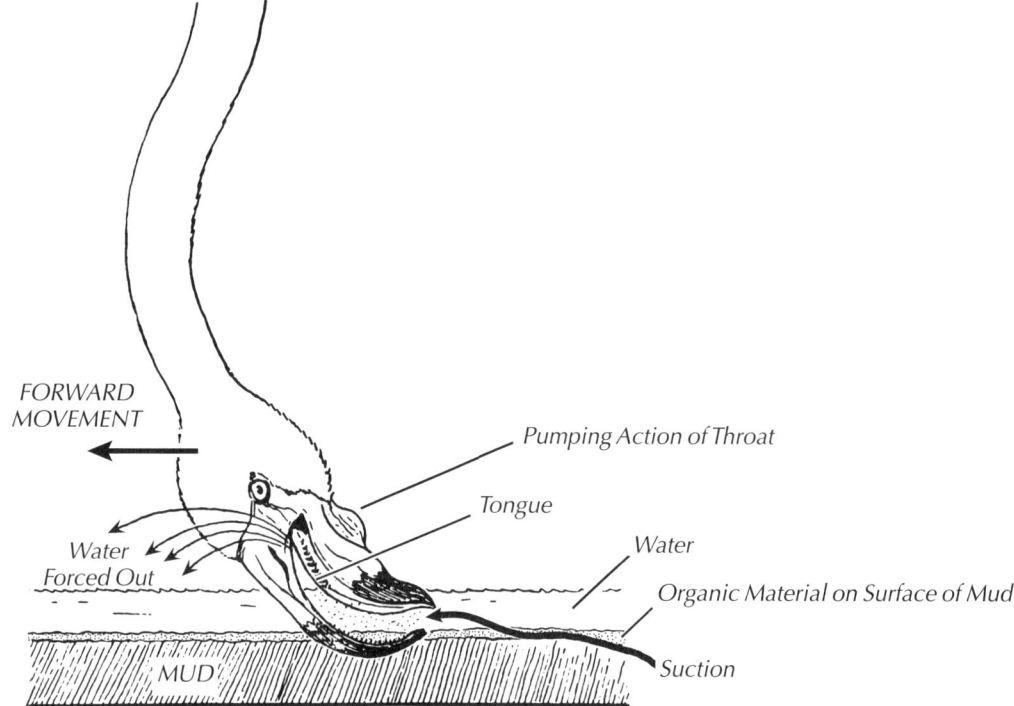

FORWARD
MOVEMENT

Pumping Action of Throat

Tongue

Water
Forced Out

Water

Organic Material on Surface of Mud

Suction

MUD

b. Lesser Flamingos

Figure 4–89. Specialized Bill and Foraging Technique of Flamingos: a. Greater Flamingo: *The flamingo feeds by filtering tiny organisms from shallow water and mud with its specialized bill. In contrast to most birds, the flamingo's lower beak is large and trough-like, whereas the upper beak is thinner and lid-like, and the entire beak is bent strongly downward. While feeding, the bird holds its beak upside down, swinging it from side to side through the water. Pumping action of the throat and rapid piston-like movements of the thick tongue suck a current of water into the mouth, forcing it out past numerous fine, hair-like plates, which fringe the inner edges of both the upper and lower beak and strain out food particles. The Greater Flamingo feeds on small crustaceans, mollusks, and aquatic insects. Its smaller relative, the Lesser Flamingo (**b**), has an even finer food-straining mechanism, allowing it to feed on microscopic algae. Greater Flamingo by Robert P. Allen from* National Audubon Society Research Report Number 5. *Lesser Flamingos by Robert Gillmor.*

Flamingos have a "bent" bill with filtering structures for obtaining their food—tiny organisms—from water and mud **(Fig. 4–89)**. Holding the bill upside down, the birds draw a current of water into the mouth by a pumping action of both the throat and the exceedingly thick, piston-like tongue. They then force the water out, filtering food particles between the many thin, almost hair-like plates fringing the inner edges of the upper and lower beaks. Flamingos can feed only in this highly specialized manner and thus require a very specific set of ecological conditions—food-rich, shallow water—to survive.

Birds that catch insects on the wing have variously shaped bills, reflecting the particular methods they use **(Fig. 4–90)**. Swifts, swallows, and nightjars such as nighthawks have short, broad bills with wide gapes for engulfing insects as they overtake them in the air. On the other hand, tyrannid flycatchers such as kingbirds have strong, flat bills with hooks for snapping up individual insects, which they locate very precisely by vision. If you listen as you watch a kingbird, phoebe, or pewee sally forth from its perch after a passing insect, you will actually hear its jaws close on the prey with a sharp snap. Except for swifts, all birds that catch insects on the wing have bristles at the margins of the gape (see Ch. 3, Types of Feathers, and Fig. 3–14a).

Many orioles and oropendolas (tropical members of the blackbird family) depend on the strength of the gape as they thrust their closed bills into fruit and then open them against the resistance of the skin and pulp. Starlings and meadowlarks use their bills the same way when they probe in the soil for insects.

The size and shape of seed-eaters' bills are an almost legendary reflection of the type of seeds they preferentially select. The Darwin's finches species complex of the Galapagos Islands is the classic example (see Fig. 1–47). In years after El Niño events bring extra rain, there is an abundance of seeds on the islands. The bills of finches show more variation after such years than after years of more meager seed production, when there is presumably stiffer competition for finch survival. Apparently the wide variability in beak structure, which occurs in every new generation of finches, undergoes fierce natural selection during the early life of each generation when seeds are scarce, but survives in rich years.

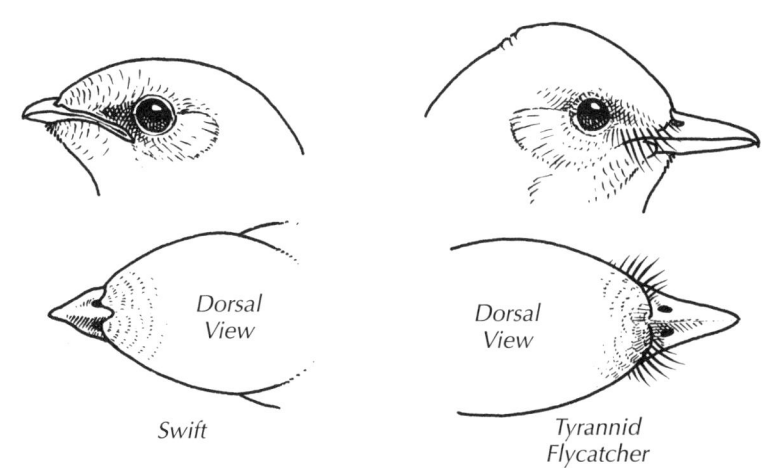

Swift

Dorsal View

Dorsal View

Tyrannid Flycatcher

Figure 4–90. Beaks of Aerial Insectivores: The beaks of birds that catch insects in flight reflect the capture techniques used. The beaks of swifts (left), swallows, and nighthawks are short, broad, and open widely, allowing these birds to engulf insects as they swoop through the air (see also Fig. 6–30a). Tyrannid flycatchers (right) have strong, flat beaks with hooked tips. They scan for flying insects from a perch, then fly out and snatch the prey in midair. All birds that catch insects on the wing, with the exception of swifts, have rictal bristles at the base of their beaks. These are visible in the tyrannid flycatcher. Drawings by Charles L. Ripper.

Figure 4–91. The Diversity of Bird Tongues: *The tongues of birds vary widely in shape, length, and structure depending on the feeding habits of their owners. Shown here (not to scale) are a few of the more elaborate examples. Woodpecker tongues are very long and equipped with backward-projecting barbs at the tip, which help to pry insects or their larvae from crevices or holes that the woodpecker excavates. Sapsuckers, unlike their woodpecker relatives, have moderately short tongues with forward-facing hairs, which they use to draw sap out from the drillings they make in trees. The tongue of the Bananaquit is forked and has fine, brush-like hairs, which it uses to absorb nectar and fruit juices. This small, common, Neotropical bird is familiar to tourists in the West Indies for its other feeding habits—visiting feeders and hotel tables, busily searching for food scraps. The long, rectangular-shaped tongue of the Cinnamon Teal (a dabbling duck, like the well-known Mallard) has many hair-like structures along its outer borders, which it uses to strain food from the water. Drawings after Gill (1995), except sapsucker, by Charles L. Ripper.*

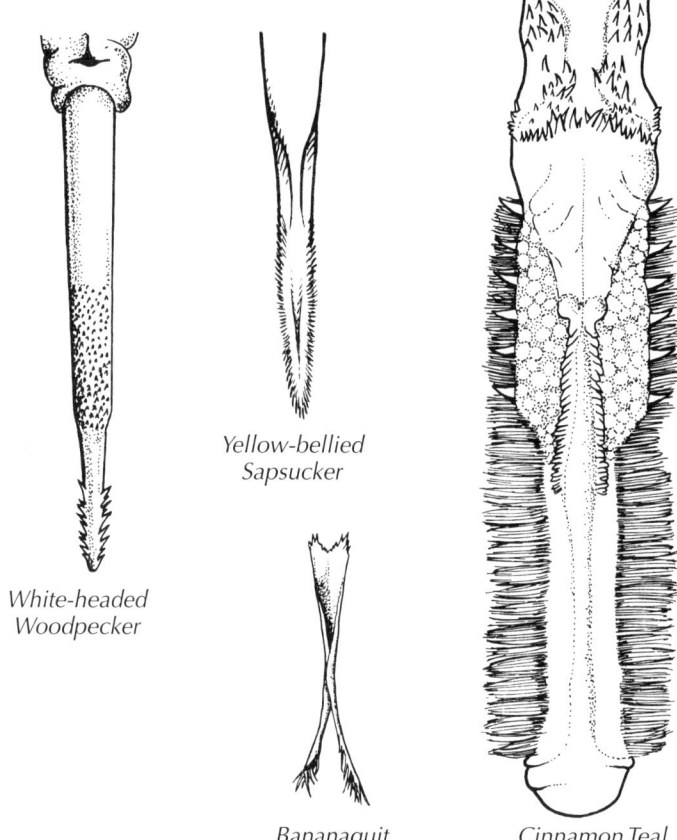

Yellow-bellied
Sapsucker

White-headed
Woodpecker

Bananaquit

Cinnamon Teal

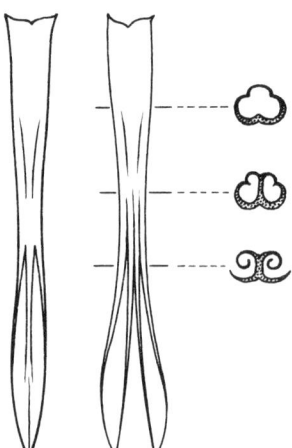

Figure 4–92. Hummingbird Tongues: *Hummingbirds have long, forked tongues whose outer edges become progressively more curled inward from the tip to the base of the tongue. The cross sections taken at three different regions along the tongue illustrate how the curled edges form a hollow trough or channel through which the bird sucks nectar from flowers. The photo shows a hovering female Black-chinned Hummingbird in Arizona extending its long tongue to gather nectar from a cactus flower. Drawing by Charles L. Ripper. Photo by Tom Vezo.*

Tongue

The **tongue**, like the bill, suits the bird's feeding habits. Birds use the tongue, depending on its size and structure, in one or more of the following ways: securing food, manipulating food, swallowing food, and detecting the texture or taste of food. (As discussed in Taste, birds seem to have a poorly refined sense of taste.) The tongue may serve as a probe, a brush, a sieve, a capillary tube, or a rasp. In shape, the tongue may be rectangular, cylindrical, lance-like, flat, cupped, grooved, spoon-shaped, or forked. It may be horny, spiny, fleshy, feathery, or brush-tipped. The tongue is very small and virtually useless in pelicans, gannets, ibises, spoonbills, storks, and some kingfishers.

The long, slender tongues of woodpeckers and hummingbirds protrude for some distance. In these birds, the horns of the hyoid apparatus of the skeleton curve up and over the skull, in some cases reaching as far forward as the level of the nostrils (see Fig. 4–12). These slender, flexible skeletal elements act as sliding anchors for the tongue when it is greatly protruded. In addition to being very long, the tongues of some woodpeckers have backward-projecting barbs at the tips **(Fig. 4–91)**. These help the bird to snag insects and insect larvae when it projects the tongue into a bark crevice or excavation it has made itself. The tongues of sapsuckers, however, are shorter than those of most other woodpeckers and have forward-projecting, hair-like structures. These form a "brush," which helps absorb and draw out tree sap from seeping wounds that the bird previously has made in the trunk and branches.

The forked tongues of hummingbirds **(Fig. 4–92)**, besides being long, tend to be folded at the edges, forming little troughs for bringing nectar from the flowers to the mouth.

Salivary Glands and Saliva

Salivary glands secrete saliva, the primary function of which is to moisten food in the mouth. The presence and development of salivary glands in birds is more or less correlated with the kinds of food they eat. Species eating seeds, plants, and insects have well-developed salivary glands. For example, chickens have several elongate salivary glands beneath the tongue. Birds that normally obtain their food from the water have little need to moisten it, however. Anhingas have no salivary glands, and the same is probably true for related birds.

The Chimney Swift, the European Swift, and certain swiftlets of the Western Pacific and the Orient have salivary glands that secrete an adhesive substance used in nest building. Chimney Swifts use the fluid to attach nesting materials to vertical surfaces. The swiftlets must produce copious saliva, for some species—most notably the Edible-nest Swiftlet **(Fig. 4–93)**—build their nests entirely of saliva! It is the nests of these swiftlets, harvested from the high walls of caves, that are used in "bird's-nest soup." The nests of species that incorporate the fewest

Figure 4–93. The Edible-nest Swiftlet at its Nest: Many swifts and swiftlets have salivary glands that secrete an adhesive substance used to attach nesting material to vertical surfaces. The remarkable nest of the Edible-nest Swiftlet of Indonesia and Malaysia is constructed entirely of saliva. The whitish, bracket-shaped nest is placed high on the vertical wall of a cave, and is surprisingly strong in spite of its translucent appearance. Nests of this species, harvested by skilled, traditional gatherers after the birds have finished nesting, are in great demand as they form the main ingredient of the Asian delicacy "bird's-nest soup."

non-saliva materials in their nests sell for the equivalent of $400 per pound ($880 per kilogram)!

Woodpeckers have a pair of salivary glands that secrete a sticky fluid onto the tongue. When the bird runs its tongue far into an ant burrow, the ants stick to it. A pair of salivary glands in the Gray Jay provides secretions that make a bolus of food sticky enough to hold together and adhere to any surface where the bird attempts to store it, such as under bark or in crevices between branches.

Pharynx

We briefly discussed the pharynx, or throat, in connection with the respiratory system, describing it as a crossroads for the passage of food and air (**Fig. 4–94;** see also Fig. 4–73). The folds of the **palate** on the roof of the pharynx surround the openings of the nasal cavities and the auditory tube. In the floor of the pharynx, the laryngeal folds border the glottis, the entrance to the trachea. Caudal to the glottis is the entrance to the esophagus.

Budgerigar Oral Cavity

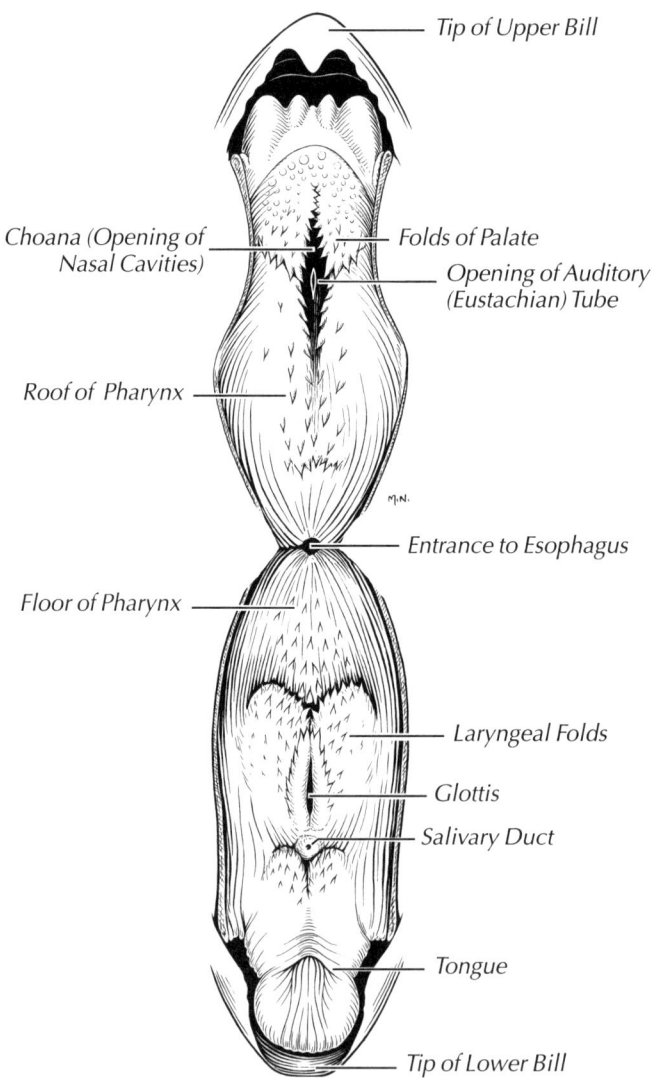

*Figure 4–94. Structures in the Oral Cavity: In this illustration, the upper and lower beak of a Budgerigar have been spread apart widely, in a way never seen in nature, presenting a frontal view of each surface, which allows the structures to be seen more easily. On either side of the entrance to the **esophagus** (the tube carrying food to the stomach) are the upper and lower surfaces of the **pharynx**, or throat. The roof (upper surface) of the pharynx is termed the **palate**. Folds in the palate surround the opening of the nasal cavities, the **choana** (see Fig. 4–72), through which can be seen the opening of the **auditory (eustachian) tube**, which connects the middle ear with the throat, allowing the equalization of pressure in the middle ear with that outside the body. In the floor (lower surface) of the pharynx, **laryngeal folds** surround the **glottis**, the entrance to the respiratory system. A **salivary duct**, which secretes saliva into the mouth, can be see anterior to the glottis in this species; the numbers and locations of such ducts vary considerably among species. Notice the short, rounded, muscular tongue, which the Budgerigar uses to manipulate food; compare this to the other avian tongues in Figs. 4–91 and 4–92. Adapted from Evans (1996).*

Alimentary Canal

Throughout the discussion of the various parts of the alimentary canal, you may wish to refer to **Figure 4–95**.

Esophagus

The **esophagus**, a relatively straight and thin muscular tube, has no digestive glands, but mucus-secreting glands in its walls moisten the food, easing passage. In many species, the esophagus can be greatly expanded to temporarily hold large quantities of food **(Fig. 4–96a)**. Gulls, for example, can hold very large fish in their esophagus while one end is being digested in the stomach.

Among widely different kinds of birds there are outpocketings of the esophagus that play various roles. Some outpocketings are sacs, as in the Emu, grouse, pigeon, and American Bittern, that serve as resonators for sound signals in courtship displays. Grouse such as the Greater and Lesser prairie-chickens **(Fig. 4–97)** have sacs that, when inflated with air, are used to produce audible "booming" signals. While

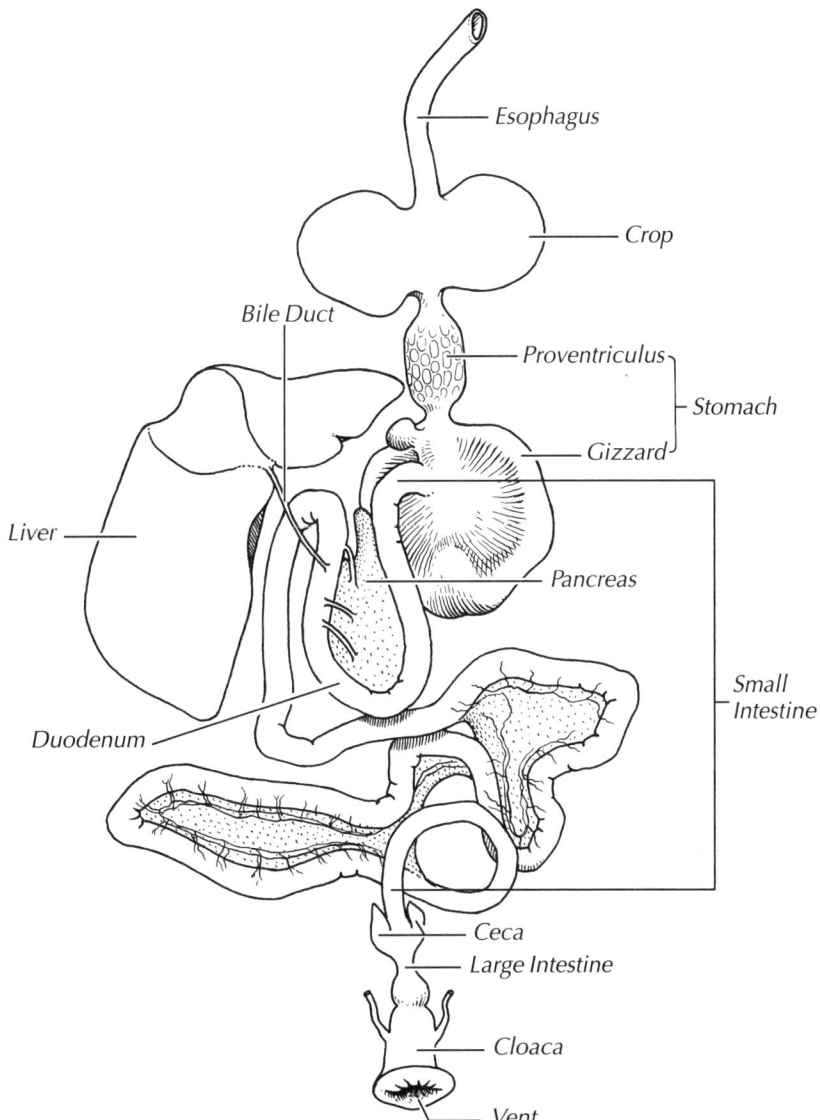

Figure 4–95. The Alimentary Canal: *In this functional view, the components of the **alimentary canal** (**digestive tract**, or **gut**) of a Rock Dove are shown removed from the body and spread apart to illustrate their relationships to each other. The liver has been moved to the side from its usual medial position, and the length of the bile duct has been exaggerated. The various organs are discussed at length in the text and are shown in greater detail in the following figures. For a view of the alimentary canal in a natural position within a bird's body cavity, see Figure 4–64. Drawing by Charles L. Ripper.*

4

Figure 4–96. The Esophagus: *The esophagus is a relatively straight, thin, muscular tube, lacking digestive glands, which conveys food between the mouth and the stomach. A number of different esophageal adaptations for storing food are found among birds.* **a. Cormorant:** *In many birds, such as cormorants, the esophagus can be temporarily enlarged to hold food. Note that part of the cormorant's esophagus has been omitted to shorten the drawing.* **b. Chicken:** *Gallinaceous birds such as the chicken, as well as pigeons, doves, and some passerines, store food in a permanent dilation of the lower esophagus known as the* **crop.** *Crops are generally found in species that eat dry seeds or fruit containing seeds. Drawings by Charles L. Ripper.*

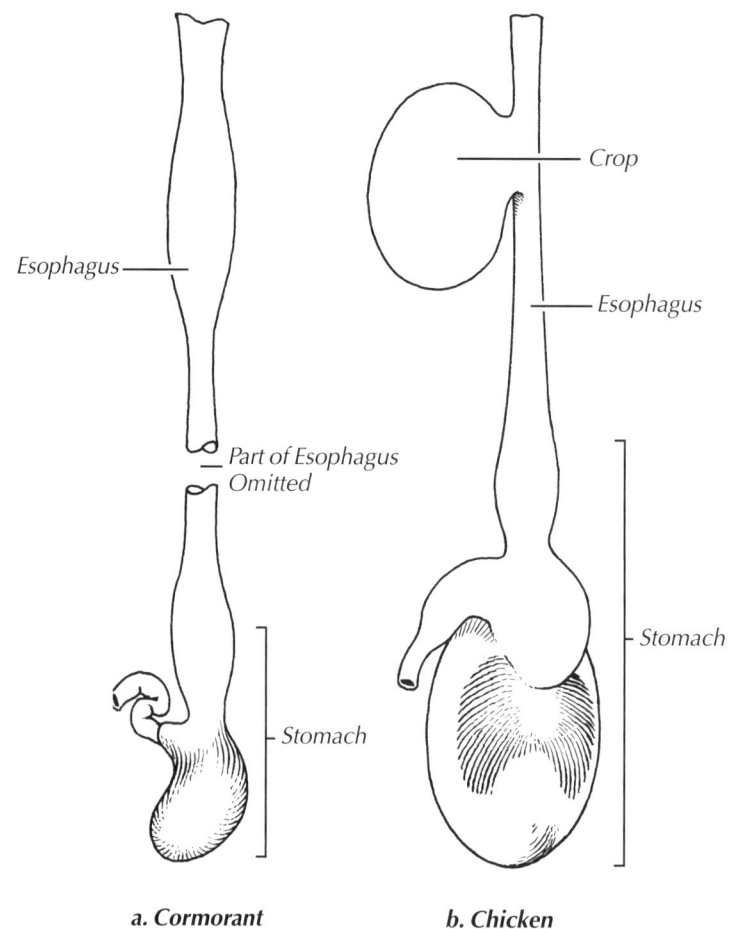

a. Cormorant   b. Chicken

Figure 4–97. Greater Prairie-Chicken with Booming Sacs Expanded: *The esophagus of some birds has out-pocketings that act as resonators for the sounds produced during courtship displays. Grouse, such as the Greater Prairie-Chicken pictured here, have esophageal sacs that inflate with air and are used to create deep "booming" sounds. These sacs are often highly colored, being yellow or orange in prairie-chickens and purple in Rocky Mountain populations of Blue Grouse. Other species with esophageal sacs used in sound production include Emus, bitterns, and pigeons. Photo courtesy of Mary Tremaine/CLO.*

inflated, the sacs protrude dramatically. Being covered with bright yellow or orange skin, they also provide visual signals that enhance the total display.

Other esophageal outpocketings are simple dilations for carrying seeds, as in Common Redpolls and other finches. About 15 families of birds (including the fowl or gallinaceous birds, pigeons and doves, and some passerines), most of which eat dry seeds or fruit containing seeds, have a permanent dilation of the lower esophagus called the **crop (Fig. 4–96b)** for the storage of food. A thin layer of muscle over the surface of the crop squeezes the crop to empty its contents back to the esophagus for passage to the stomach. The crop of the South American Hoatzin (**Fig. 4–98;** see also Fig. 3–41) is thick-walled, muscular, and exceptionally large. Hoatzins feed on the thick leaves of arums, tropical plants related to the popular houseplants often called "Arrowheads," and the crop actually grinds the food and initiates the first stages of digestion through bacterial fermentation—a rather odoriferous process. Consequently, the birds are locally known in several regions as "stinking turkeys"!

Some of the most interesting esophageal specializations have evolved in birds whose adult diet is indigestible or inaccessible to their offspring. Only a small portion of living birds subsists directly on plant material, a feeding specialization that departs from the common, ancestral avian diet of animal tissue. Most plant-feeding specialists initially feed their young on animal matter, usually soft invertebrates such as insects and spiders, further indicating that plant eating prob-

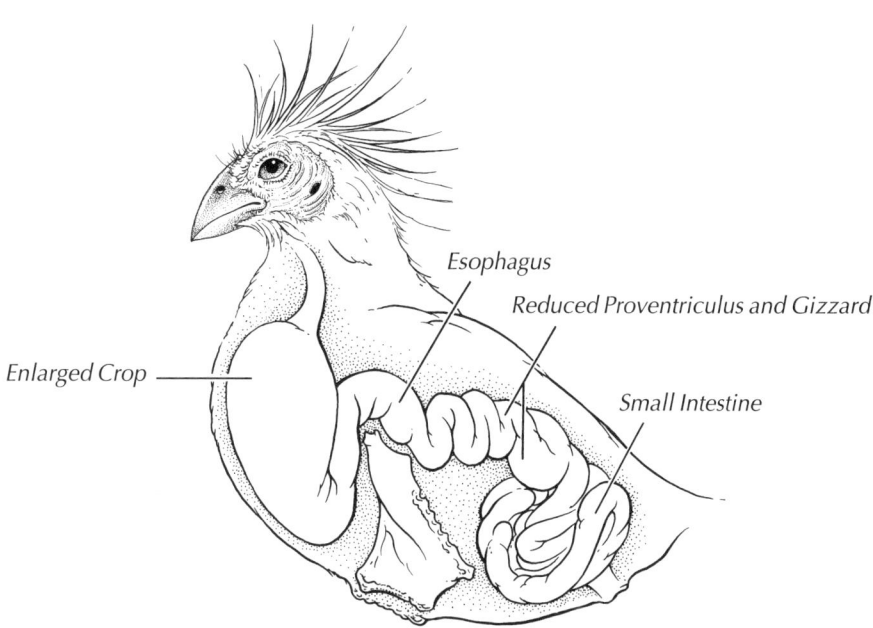

Esophagus

Reduced Proventriculus and Gizzard

Enlarged Crop

Small Intestine

Figure 4–98. The Digestive Tract of The Hoatzin: *The crop of the South American Hoatzin is thick-walled, muscular, and exceptionally large. Hoatzins are unusual birds because they eat mostly leaves, which contain much cellulose and woody materials that are difficult to digest. The Hoatzin's enlarged crop starts the long digestive process by grinding the food and beginning its chemical breakdown through bacterial fermentation. Adapted from Proctor and Lynch (1993, p. 184).*

a. American Goldfinch

b. Mourning Dove

*Figure 4–99. Raising Young Birds on a Diet of Plants: Birds whose adult diet is composed mostly of plant material, which their newly hatched offspring may find indigestible or otherwise inaccessible, have developed special strategies for feeding their young. Many seed-eaters simply feed their young insects and other small invertebrates, increasing the proportion of seeds or fruit as the digestive capabilities of the young birds improve. Other species enlist the aid of their crops. **a. American Goldfinch:** The American Goldfinch feeds almost entirely on seeds as an adult, but also feeds its young exclusively on seeds. Here a female feeds its nestlings a mass of regurgitated seeds, partly broken down and mixed with sticky fluid. Although the finely chopped mass is thought to be regurgitated from the crop, it appears to be partly digested—providing a puzzle for researchers. **b. Mourning Dove:** Pigeons and doves, which eat seeds and fruit as adults, have specialized crops that produce "crop milk" or "pigeon's milk," which they feed to their young. Crop milk consists of a highly nutritious slurry of fluid-filled cells that slough from the lining of the crop. In this photo, an adult Mourning Dove is regurgitating crop milk to a second nestling, mostly hidden by the large nestling in front. As the young doves grow, their diet will include an increasing proportion of seeds and other plant material. Photos courtesy of Mike Hopiak/CLO.*

ably evolved *from* insect eating. The American Goldfinch, however, feeds almost entirely on seeds as an adult, and also feeds its young exclusively on a finely chopped mass of seeds that seems to be partly digested **(Fig. 4–99a)**. The food mass, however, is thought to be regurgitated from the non-digestive crop. The greater availability of seeds later in the plant growing season may explain why goldfinches are such late nesters, not starting nest building until mid to late summer.

More specialized yet is a habit in pigeons and doves, highly specialized seed- or fruit-feeders as adults. Instead of feeding their squabs on insects as most seed-eating birds do, pigeons and doves slough fluid-filled cells from the crop lining to produce "**pigeon's milk**" **(Fig. 4–99b)**. Pituitary prolactin stimulates the crop milk production of both sexes during the last 10 days to one week of incubation. Crop milk alone is fed to hatchlings for the first four or five days, but thereafter is mixed with increasing amounts of seeds through the fledging of the young. Pigeon's crop milk is a lipid-rich material of a "cheesy" consistency, high in vitamins A and B and with a greater protein and fat content than human or cow's milk!

The crop also may come to the aid of the male Emperor Penguin, who somehow survives two months in the dead of the Antarctic winter incubating his single, large egg on the tops of his feet **(Fig. 4–100)**! In the dark and storms of midwinter he may find himself with a hatchling before his mate has found her way across as much as 185 miles (300 kilometers) of ice to bring food to the chick and parental-duty relief to him. In this crisis the male can produce a rich esophagus and crop milk for a few days until his mate arrives. She then takes over, ready to regurgitate the fish and squid she caught at sea, a diet on which the chick will dine thereafter.

Figure 4–100. Male Emperor Penguins Incubating their Eggs: *Huddled together for warmth as snow blows and wind howls during the gloomy Antarctic winter, nesting male Emperor Penguins await the end of their two-month incubation vigil. Soon the single, large egg that each holds on top of his feet, warmed and hidden by a muff of belly skin, will hatch. Immediately after laying the egg, early in the Antarctic winter, the female returned to the sea, leaving her mate to incubate entirely alone, with no possibility of obtaining food. Before the female returns with food, the chick may hatch, getting its first few meals in the form of a curd-like esophageal secretion provided by the male.*

A final case of a specialized esophageal secretion that has evolved in birds whose adult diet is indigestible or inaccessible to their offspring is that of the flamingo. As previously discussed, the adult flamingo's diet (bacteria, algae, aquatic insect larvae and pupae, tiny shrimps, and so on) is captured exclusively by the complex filtering bill and tongue apparatus. For the two months that it takes hatchlings to develop the special filter-feeding apparatus of the adults, young flamingos consume only an esophageal "milk." The secretion is not produced from a crop, but from glands along the esophagus and upper stomach of the parents. Even richer in fat than pigeon crop milk, it contains an abundance of red and white blood cells, and is red in color. Flamingo milk has a liquid consistency—probably the only form of nutrition compatible with the developing filtering apparatus. A thicker secretion, such as that of pigeon's milk, would no doubt clog the chick's growing straining bristles. In addition, flamingo chicks are reared on shadeless islands in high-saline lakes, and thus may require considerable water in their diet.

These cases of upper digestive tract secretions demonstrate convergent evolution in providing the essential energy and especially the

Figure 4–101. Stomachs of Grain-eating Versus Meat-eating Birds: In most birds, the stomach has two parts. The upper part, the **proventriculus**, is elongate and has glands that secrete enzymes that begin the digestion of proteins. The lower part, the **gizzard**, is rounded and has thick, muscular walls, often with hard internal ridges; it functions to grind food. Seed-eating birds, such as the turkey, often eat grit or small stones with their food to aid the gizzard in grinding, but the power for grinding comes from the strong gizzard muscles. In the sections through stomachs shown here, compare the far more muscular gizzard of the turkey to that of the meat-eating hawk. The digestion of seeds requires much more mechanical grinding than the digestion of the proteins in meat. The less muscular gizzards of birds that eat meat or fish mold indigestible material, such as bones, fur, feathers, and the outer skeletons of insects, into compact balls, known as pellets, which are then ejected through the mouth (see Fig. 4–102). The **pylorus**, a muscular sphincter (a circular band of muscle), regulates the passage of food from the stomach into the small intestine. Drawings by Charles L. Ripper.

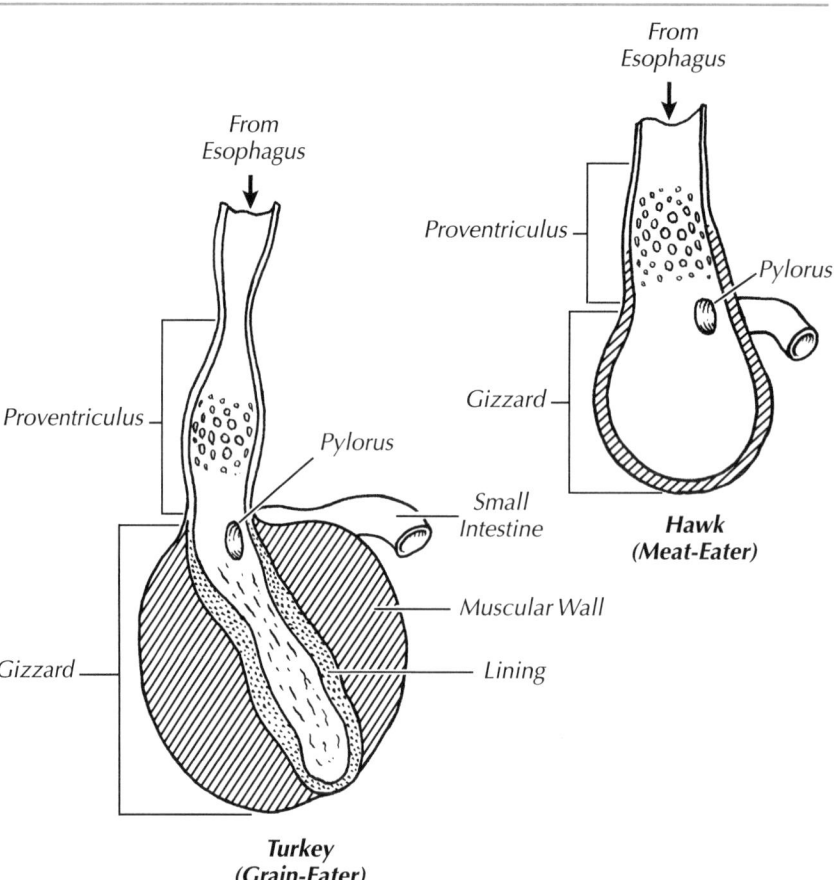

protein needed for hatchling growth in the absence of animal food. They have allowed certain species to reproduce successfully in the face of competition and extreme environmental conditions. As such they parallel, in many ways, the adaptations of mammals for providing the earliest possible nutrition for their offspring.

Stomach

Nearly all birds have a two-part stomach (**Fig. 4–101**). The first portion, the **proventriculus**, is elongate and has gastric glands whose secretions begin the breakdown of proteins in the food. The second part of the stomach is a rather spherical, muscular **gizzard** that functions to grind the food. The stomach ends in a muscular constriction, the **pylorus**.

The **proventriculus** secretes mucus, hydrochloric acid, and an inactive precursor to pepsin, an enzyme that digests protein. The pepsin, activated by the acid in the proventriculus, begins breaking down proteins. In a few birds, for example petrels, cormorants, herons, gulls, terns, some hawks, and some woodpeckers, the proventriculus is expandable and can hold food temporarily. This is an adaptation for feeding the young back at a distant nest or for delayed digestion by the individual. Several birds have enlargements in the pyloric region between the stomach and the duodenum, the first part of the small intestine, probably for storing food. A distinct chamber forms in the Ostrich, some herons (of the genus *Ardea*), cormorants, and anhingas.

The latter, a fish-eater with a long, thin neck, has a diverticulum (out-pocketing) of the proventriculus *and* an enlarged area in the pyloric part of the stomach. These probably store fish near the center of mass, without disrupting neck streamlining, as the food awaits room in the lower gut for digestion.

The **gizzard** or ventriculus has thick, muscular walls and a lining of leathery or sandpaper-like material called **gastric cuticle** or **koilin**. Koilin is a combination of carbohydrate and protein secreted by glands in the wall of the gizzard. When the gizzard is cut open, this lining can be peeled from the wall (as is done in the supermarket before sale). Shedding and resecretion of the lining is probably a more or less regular natural occurrence that maintains the efficiency of food mashing by restoring the toughness of the lining when it becomes worn.

Seed-eating birds with well-developed gizzards, such as the chicken, quail, turkey, and many ducks and swans, eat mineral grit or small stones along with their food to aid the gizzard in grinding. The power for grinding comes from the inner circular muscles of the gizzard, which are enlarged compared to those in the stomachs of gizzardless vertebrates; the outer longitudinal muscles have been lost. The enlarged muscular walls, with their hard internal ridges, contract in alternating directions, grinding the food with the aid of ingested grit. The gizzard, made more efficient by grit, thus performs part of the work done by the teeth in mammals. If you hold a live chicken that has eaten recently against your ear, you can hear the grinding process inside the gizzard. The Ostrich, holder of so many avian big-size records, has been recorded to ingest stones up to one inch (2.5 cm) in diameter.

Unfortunately, some man-made objects are the ideal size for large birds to select as gizzard digestive aids. For example, lead shot from the guns of bird hunters kills enormous numbers of birds that are never hit by gunfire. Stray pellets strewn in the environment have been accumulating for a hundred years or more, for they do not "rust," and thus deteriorate very slowly. When birds ingest the pellets as grit, they are pulverized by the grinding action of the gizzard and lead toxins are absorbed into the bloodstream from the stomach and intestines. A slow death from liver and kidney damage and digestive tract paralysis ensues. In the English Midlands more than half of the dead Mute Swans tested had died from poisoning following ingestion of lead weights lost by anglers!

Fruit-eating pigeons, such as the large, bulky Imperial-Pigeons (genus *Ducula*) of Southeast Asia, have raised bumps like blunt fingers on the inner surface of the gizzard. These interdigitate during contractions and thus act as a toothed "polishing wheel" that functions to peel fruit and remove the flesh from enormous pits such as those of nutmeg; the pits are then voided. Several fish-eating birds, including anhingas, do not have enlarged muscles of the gizzard for grinding and thus the proventriculus and the lower region of the stomach appear similar in structure.

In some carnivorous birds, indigestible or slowly digesting bones, teeth, feathers, fur, and insect parts are formed into a **pellet** in the gizzard that the bird can readily regurgitate after digesting the

Figure 4–102. Pellets of a Barn Owl: In various carnivorous birds, the parts of their prey that are indigestible or difficult to digest, such as bones, teeth, feathers, fur, and the outer skeletons of insects, are compressed into a pellet in the gizzard, and expelled by regurgitation after the bird has digested the flesh. Birds that produce pellets include meat-eaters such as owls, hawks, and shrikes; fish-eaters such as kingfishers and grebes; and insect-eaters such as thrushes, nightjars, and bee-eaters. Owl and hawk pellets may be found beneath nest trees or roost sites. This photo shows two pellets from a Barn Owl. The pellet on the left has been pulled apart to show its contents. Visible within the fur are limb bones, jawbones, and part of the vertebral column of a small rodent. Photo courtesy of Virginia Cutler/CLO.

flesh **(Fig. 4–102)**. This habit is characteristic of owls, hawks, grouse, nightjars, swifts, kingfishers, shrikes, and some thrushes. A pellet from one screech-owl contained a whole tarsal bone of an American Gold-finch, complete with a numbered metal identification band! For the first month after hatching, young kingfishers digest the entire fishes brought to them by their parents, including scales and bones. With the development of flight feathers and full size, however, the demand for calcium and related minerals becomes less and the young king-fishers start producing pellets of fish bones and scales. By the time they actually start flying, the bones and scales of their prey are cast up in such good condition that the species of the catch of the day can be identified!

Grebes, which are primarily fish-eaters, cast pellets consisting of mostly indigestible plant food and their own feathers. Apparently, the strong acid in the grebe's stomach dissolves most of the fish bones. Some ornithologists believe that the feathers (and perhaps the plant matter) the grebe eats act as retainers, keeping the bones in the stomach long enough for them to digest. Also, the indigestible matter may encapsulate sharp spines and bone edges, preventing damage to the stomach while the bones are dissolved, and preventing damage to the esophagus during pellet regurgitation, if any sharp pieces were to remain.

The contents of the proventriculus enter the gizzard very close to the **pylorus** with its muscular sphincter, which marks the beginning of the small intestine. This allows the more liquid portion of the digesting food to bypass the grinding chamber of the gizzard.

Small Intestine

The **small intestine** is the longest part of the digestive tract **(Fig. 4–103)**. This is where the final processes of digestion take place, reducing proteins to amino acids, carbohydrates to simple sugars, and fats to glycerol and fatty acids. The small intestine includes regions, the duodenum, jejunum, and ileum, that do not look any different from each other externally, but whose walls have characteristically distinct cells and functions.

Most of the products of digestion are absorbed by the lining of the intestinal wall and pass into the bloodstream. The products of fat metabolism, however, are absorbed primarily into the channels of the lymphatic system, which eventually dump them into the bloodstream for distribution to the body cells. The inner surface of the small intestine has longitudinal folds and minute, finger-like projections in some birds—hawks and other carnivores—and flattened, leaf-like structures in others—flamingos and many herbivores. Both greatly increase the surface area for absorption, but why there are differences that seem to parallel the birds' primary food types is unknown.

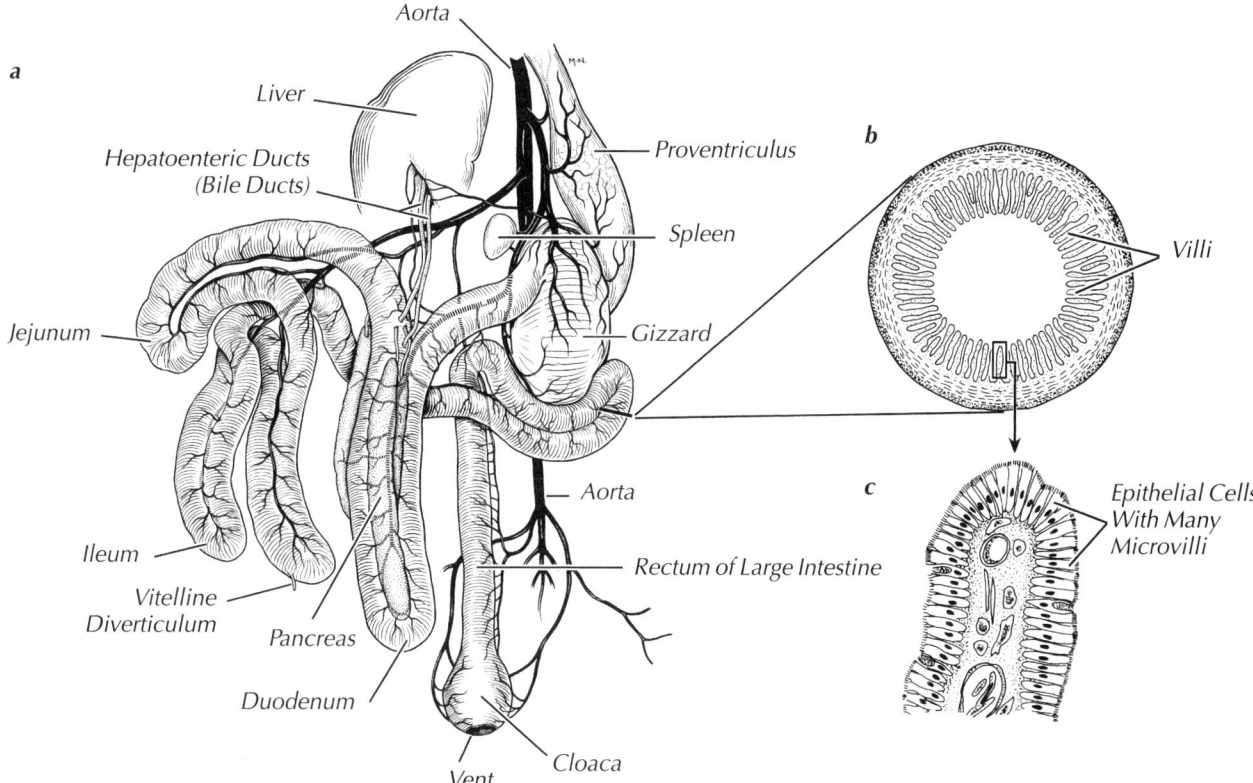

Figure 4–103. The Small Intestine: a. Location within Digestive Tract: *In this functional view of the Budgerigar digestive tract, the small intestine has been spread out so that its parts may be seen more easily. The **small intestine** begins at the pylorus at the lower end of the stomach, and is the longest section of the digestive tract; the site where the final processes of digestion take place. Its successive regions, the **duodenum**, the **jejunum**, and the **ileum**, look similar externally, but their glandular structure and food-processing functions differ. At the junction between the jejunum and ileum there is often a tiny **vitelline diverticulum**, a remnant of the embryo's yolk sac. Secretions into the small intestine from the pancreas and liver carry out digestion. Bile, important in fat digestion, is produced by the liver but stored in the gall bladder. Several small bile ducts carry bile from the gall bladder into the small intestine. Some birds, including the Budgerigar pictured here, lack a gall bladder; in these birds the ducts that transport bile from the liver to the small intestine are* termed **hepatoenteric ducts**. *The small intestine is named for its narrow diameter, not its length, which varies greatly among species, depending on their diet. Birds that feed on foliage or grain have longer small intestines than those that feed on fruit or meat, reflecting the difficulty of digesting the cellulose in plant material. The end of the ileum marks the beginning of the large intestine.* **b. Cross Section through Small Intestine:** *Visible here are numerous folds, known as **villi** (singular **villus**), which greatly enlarge the small intestine's surface area for absorbing nutrients.* **c. Cross Section through a Single Villus:** *Visible in this highly magnified view are the epithelial cells that form the inner lining of each villus. Each cell is edged with microscopic cylindrical processes, termed microvilli, that further increase the surface area inside the small intestine for absorption. Drawing **a** from Evans (1996). Insets from A Dictionary of Birds, by Bruce Campbell and Elizabeth Lack. 1985. Reprinted with permission of the British Ornithologists' Union.*

Secretions from small glands in the wall of the intestine, as well as from the **pancreas**, carry out digestion. Bile, secreted by the **liver**, neutralizes the acid passing into the intestine from the stomach and also emulsifies the fats (divides them into tiny particles that can be more easily digested).

The small intestine is named because of its diameter, not its length. Its length varies, in general, with the diet and size of the bird. Birds that eat grass and other foliage or grain have a small intestine that is relatively longer than that of fruit- and meat-eating birds because cellulose, a main structural component of plant matter that is generally absent from fruits, makes extracting nutrients from leaves and seeds a difficult process. In the grazing Ostrich, for example, the small intestine is 46 feet (14 m) long! Fish-eating birds also tend to have a long small intestine, presumably because fish prey are usually swallowed whole and require much work to digest efficiently.

Among plant-eating birds, the Blue Grouse and Spruce Grouse feed largely on conifer needles—a low-grade food. The small intestines of these grouse are about 28 percent longer than those of similar gallinaceous birds that feed on a higher grade, more nutrient-rich, food. Even in the same species, intestinal length may differ among populations when food habits differ. The race of California Quail living along the humid coast of northern California eats considerably more green food than the race inhabiting the arid interior of the state, where seeds and other richer foods are favored. The greens-eating race has a small intestine that is 11 percent longer than that of the seed-eating race, reflecting the fact that leaves have a higher ratio of cellulose to digestible nutrients than do seeds.

The small intestine is proportionately shorter in birds than in mammals, and the speed with which food passes through the alimentary canal, from mouth to vent, is correspondingly faster. A shrike or a magpie can completely digest a mouse in about three hours; a chicken digests grain in about two and one-half hours. Adult and nestling birds fed berries will pass the seeds within 12 to 30 minutes of first ingesting the berry! In contrast, a small mammal such as a mouse usually takes 24 hours or more to pass food completely through its digestive system (Milo Richmond, personal communication).

The U-shaped **duodenum** is the first loop of the small intestine. The **pylorus**, a circular band of muscle, marks the end of the stomach and the beginning of the duodenum. It serves as a closed gate to the small intestine, its muscles only loosening to let food pass when stomach digestion is complete. Most of the **pancreas** lies within the duodenal loop, and into this loop enter the ducts from the pancreas and several bile ducts from the **liver** and **gall bladder** (when one is present). Many birds lack a gall bladder to collect the bile produced by their livers.

The long **jejunum** and the short **ileum** are the divisions of the remainder of the small intestine. Both are looped and coiled and lack external distinguishing characteristics. At the junction of the jejunum and ileum there is often a **vitelline diverticulum**, a small, blind pouch of tissue, which is a remnant of the yolk sac of the embryo. The ileum becomes the **large intestine** at the point where the ceca attach.

Colic Ceca

The **colic ceca** (also spelled caeca; singular, **cecum** or caecum) are a pair of pouches extending from the junction between the small and large intestines **(Fig. 4–104)**. They function in digestion by holding material long enough for bacterial action to further break it down. They then release the material into the large intestine for absorption of whatever nutrients have been made available by the bacteria. Herons and bitterns have a single cecum rather than a pair, and ceca are absent entirely in some species of woodpeckers, swifts, kingfishers, doves, cuckoos, and parrots. They also are absent or merely buds in anhingas and some hummingbirds.

On the other hand, two pairs of ceca have been reported in the snake-eating Secretary Bird of the African savannahs. The Ostrich and gallinaceous birds, especially the turkey, have very large ceca that often harbor intestinal coccidia, protozoans more commonly associated with parasitic disease. In some species there are outpocketings that enhance the potential fermentation space of the ceca; these are found in the Ostrich, rheas, kiwis, loons, screamers, bustards (at least of the genus *Otis),* and sand grouse (at least of the genus *Pterocles).*

Gallinaceous birds that browse on grass or other foliage have longer ceca than their relatives that eat seeds. For example, Grouse, which browse, have ceca as much as 136 percent longer than the ceca of seed-eating quail. Possibly the bacteria in the ceca break down the cellulose in the tough fibers of the grass. In the two races of California Quail mentioned previously in connection with intestinal length, the race eating more green food has ceca about 19 percent longer than the race eating mostly seeds.

Large Intestine

In most birds, the **large intestine** is a short, straight tube extending from the colic ceca to the cloaca. The large intestine functions to hold intestinal contents while water (and perhaps nutrients made available by the bacteria in the ceca) is being reabsorbed; the indigestible material then passes to the cloaca.

Cloaca

The **cloaca** receives feces from the large intestine, urine from the kidneys, and eggs or sperm from the gonads **(Fig. 4–105)**. Because the intestine ends in the cloaca there is no anus but rather a common **vent** that opens from the cloaca to the exterior of the body. The **cloacal bursa** is a lymphoid organ that opens into the roof of the cloaca in young birds and atrophies in later life. The bursa has no known digestive function. First described by Hieronymus Fabricius in the 17th century (and thus long called the bursa of Fabricius), an understanding of its function has become significant to modern medical research. The structure is important in the immune response because it produces **B cells**, special white blood cells that populate the lymphatic tissues and are a key

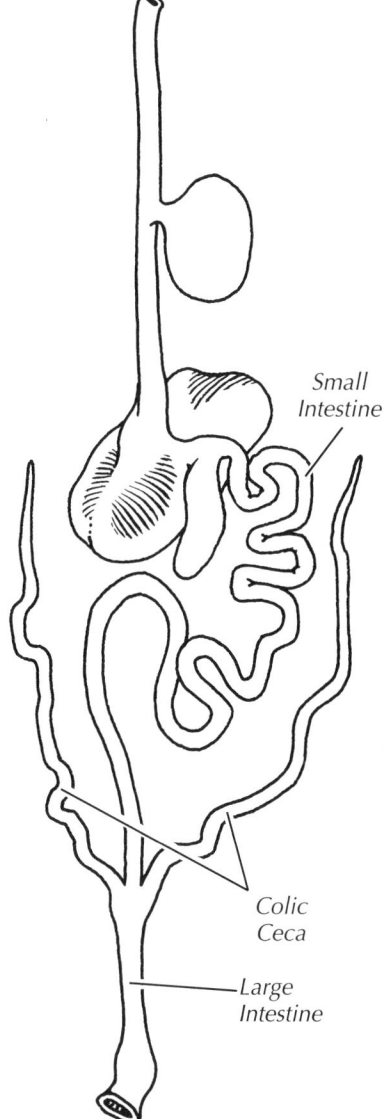

Figure 4–104. Colic Ceca of the Grouse: *Illustrated here is the location of the paired* **colic ceca** *at the junction of the small and large intestines. The ceca aid digestion by holding partly digested material while bacterial action further breaks it down. The ceca then release the material into the large intestine, where any nutrients made available by the bacteria are absorbed. The size and number of ceca vary among species. In the grouse, shown here, they are relatively large and elongated. In the Rock Dove (see Fig. 4–95) they are short and leaf-like. In parrots such as the Budgerigar (see Fig. 4–103), and various other species, ceca are absent entirely, whereas in some species two pairs are present. Drawing by Charles L. Ripper.*

Figure 4–105. The Cloaca: *In contrast to mammals, which have separate exits from the body for the digestive and urogenital systems, birds have a common final chamber, the* **cloaca**, *which is diagrammed here for the Rock Dove in sagittal section. The cloaca receives feces from the large intestine, urine from the kidneys through the paired openings of the ureter (only one is shown here), and eggs or sperm through the paired openings of the oviduct or deferent duct, respectively (only one is shown). The* **cloacal bursa** *(previously known as the bursa of Fabricius) is a lymphatic organ opening into the roof of the cloaca, present only in young birds, which functions in immunity (see text for details). The cloaca opens to the exterior through the muscular vent, through which both the waste materials of digestion and the products of reproduction—eggs or sperm—pass. Drawing by Charles L. Ripper.*

Rock Dove Cloaca, Sagittal Section

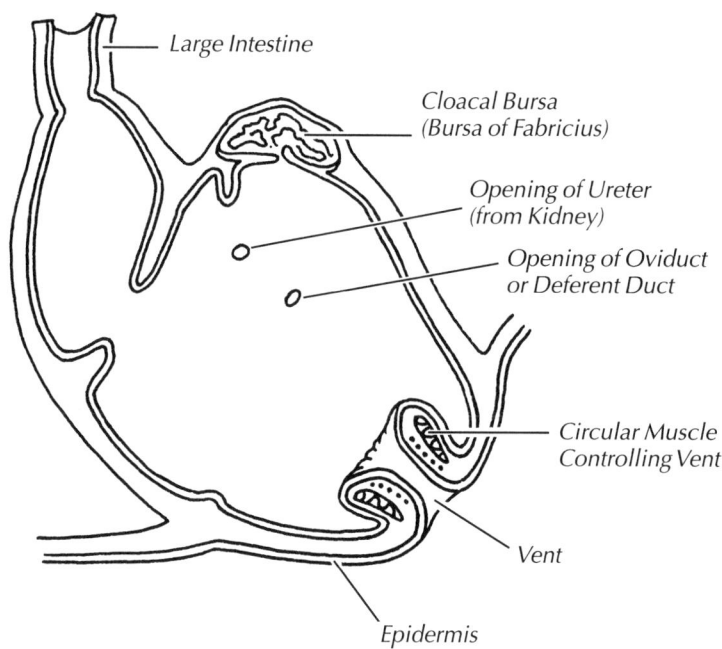

to understanding AIDS development in humans. There are strains of "bursa-less" chickens, bred for research, that are unable to fight infections because they have no B cells.

Liver

The **liver**, the largest internal organ of the body, has two lobes. In most birds, the right lobe is larger than the left. From each lobe several bile ducts lead directly into the duodenum. Many birds have a **gall bladder**, a reservoir for the storage and concentration of bile secreted by the liver. Depending upon the species, the gall bladder may be oval, sac-like, or long and tubular. However, many birds have no gall bladder, including the Ostrich, the Hoatzin, parrots, cuckoos, hummingbirds, many pigeons, and some woodpeckers and passerines, as well as at least the Peregrine Falcon among raptors. Besides producing **bile**, which emulsifies fats for better digestion, the liver has many other functions such as storing sugars and fats, forming uric acid, and removing foreign substances from the blood. Accordingly, the liver is richly supplied by the circulatory system.

The Urogenital System

■ The urogenital system includes the urinary system and the reproductive system. The two systems are considered together because of their intimate developmental and anatomical relationships. The urinary system develops first and is vital in removing toxic nitrogenous wastes from the circulation. Some of its ducts are also used in the adult bird for reproductive functions.

The survival of any species depends upon the reproductive success of its individuals. If its genes are to survive, an individual bird's reproductive rate must be great enough to offset all the losses of offspring from hazards in the environment; thus the sex drive has evolved to be a strong and basic instinct. Natural selection operates through individual reproductive superiority, however slight, over time producing reproductive adaptations that may be exceedingly complex in species that continue to survive. These adaptations provide bird enthusiasts with some of our most exciting and memorable observations.

Unlike other systems in the bird's body, the reproductive system is active only part of the year in many species. This is especially true for birds living in environments that change with the seasons. Following the breeding season, these birds' reproductive organs and ducts (especially the testes, ovary, and oviduct) regress. They remain small until just before the next season, when they enlarge again. This cycle undoubtedly relates to the "cost" of excess baggage in flying animals.

Urinary System

The **urinary system**, sometimes called the **excretory system**, consists of paired **kidneys** and their excretory ducts, the **ureters (Fig. 4–106)**. The kidneys are irregularly shaped structures, with three interconnecting lobes in most birds. Sometimes the left and right kidneys connect caudally to form a horseshoe-shaped kidney. The kidneys lie deep in depressions on the ventral face of the pelvis and the synsacral vertebrae and have obvious blood vessels crossing their ventral surface. A tube from each kidney, the ureter, conducts the thick, white slurry of uric acid to the cloaca. Here it either mixes with the feces, or surrounds it as in the Budgerigar, to form a "bull's-eye" dropping. The only birds with a **urinary bladder** are the South American rheas. This may seem surprising, considering that urinary bladders are a regular feature of vertebrate anatomy. Why should this be? And why is the urine of birds so different from that of mammals?

In all vertebrates, the liver converts the toxic nitrogenous wastes of protein metabolism into less toxic and more efficiently excreted forms. In birds this liver-synthesized substance is **uric acid**, not the urea created by mammals and some other vertebrates **(Fig. 4–107)**. Uric acid is a more complex molecule than urea and thus probably takes more energy to synthesize, but it has two advantages over other excretory compounds. First, it is very concentrated, containing more of the potentially toxic nitrogenous wastes per molecule than does urea—thus ridding the body of more wastes per excreted molecule. Second, it is not nearly so soluble in water as is urea. Thus to excrete nitrogenous wastes, a bird does the following: The kidneys reabsorb most of the water and then excrete a fluid containing a solution of uric acid that is very near to the concentration at which precipitation of the uric acid would begin to occur. When this solution reaches the cloaca, only a small amount of water needs to be withdrawn (an energy-intensive process) through the walls of the cloaca to cause the uric acid to precipitate to a semisolid, white, pasty mass. The water remaining in the cloaca (which the precipitate has left behind) is therefore free of nitrogenous waste and can be reabsorbed by the body relatively easily, and the precipitated uric acid is then excreted. Mammals, on the other hand, must take a solution of urea in water and expend a great deal of cellular energy withdrawing water from it to concentrate it. The urea never precipitates, so in urinating, mammals lose more water per amount of nitrogen waste excreted than do birds. Thus, compared to mammals, birds conserve both water and energy during excretion.

Because uric acid is a compact, low-volume mass, the cloaca is all the storage space most birds require for nitrogenous wastes, explaining why few species have a urinary bladder. The compact nature of bird urine is yet another weight-reduction adaptation for flight—birds could not afford to carry the extra weight of the water that would be required to hold a solution of urea. The great difference

Figure 4–106. Urogenital Systems of the Male and Female: The urogenital system consists of the urinary and reproductive systems, which share developmental and anatomical links. The **urinary system**, like most other body systems, is active year round. In contrast, the **reproductive system** of many birds is active for only part of the year; the reproductive organs, or gonads, and their ducts shrink after breeding, enlarging again just before the next breeding season. This figure shows both the inactive and active states of the urogenital systems of male and female Rock Doves. The urinary system consists of paired, three-lobed **kidneys** with their ducts, the **ureters**, which transport uric acid, the waste product of excretion, to the cloaca. The urinary bladder, which stores urine in mammals, is absent in most birds. The male gonads, the paired **testes**, lie at the cranial end of the kidneys. Sperm produced in each testis is conveyed as semen through the **deferent duct (vas deferens)**, the lower portion of which is enlarged to form a temporary storage receptacle. During copulation, semen from the storage receptacle exits the body through the cloaca. The female gonad, the **ovary**, produces **ova** (eggs) at periodic intervals in a process known as ovulation. In most birds, only the left ovary is functional. The ova pass through a funnel-shaped **infundibulum** and into the **oviduct**, exiting the body at the cloaca. Figure 4–111 details an ovum's journey and the steps that transform it into the familiar, hard-shelled, bird's egg. Drawings by Charles L. Ripper.

Male Urogenital System

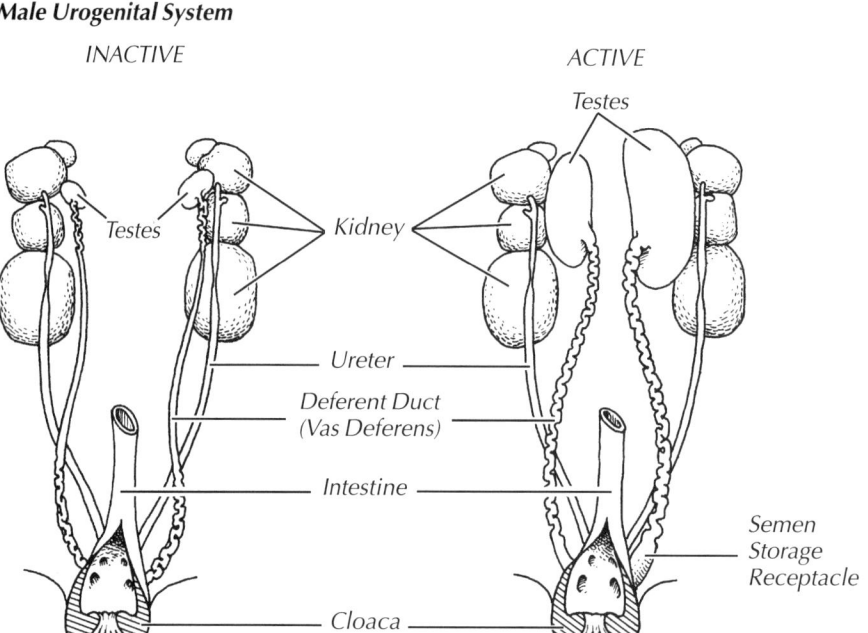

INACTIVE   ACTIVE

Testes

Testes   Kidney

Ureter

Deferent Duct
(Vas Deferens)

Intestine

Semen
Storage
Receptacle

Cloaca

Female Urogenital System

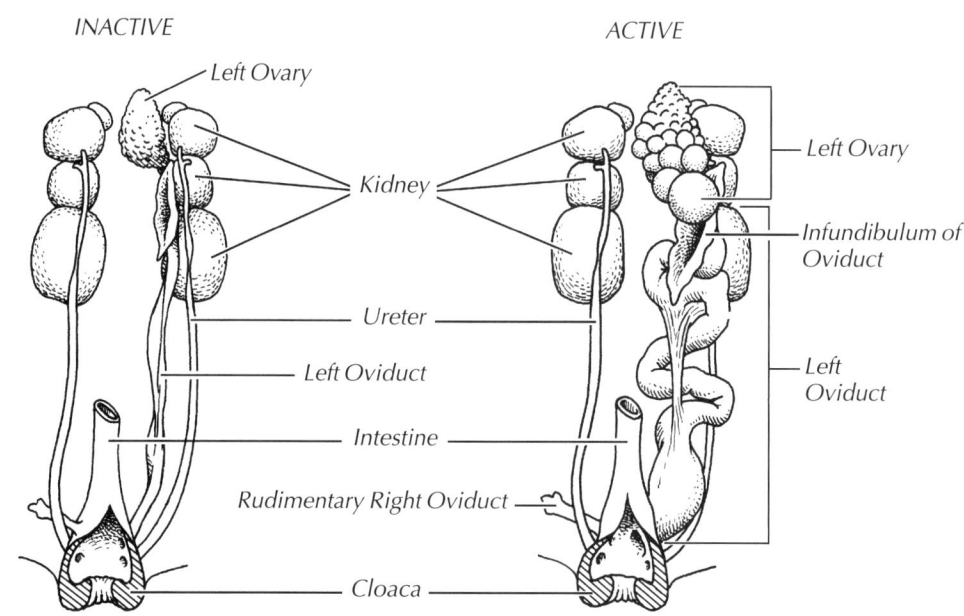

INACTIVE   ACTIVE

Left Ovary

Kidney   Left Ovary

Infundibulum of
Oviduct

Ureter

Left Oviduct   Left
                        Oviduct

Intestine

Rudimentary Right Oviduct

Cloaca

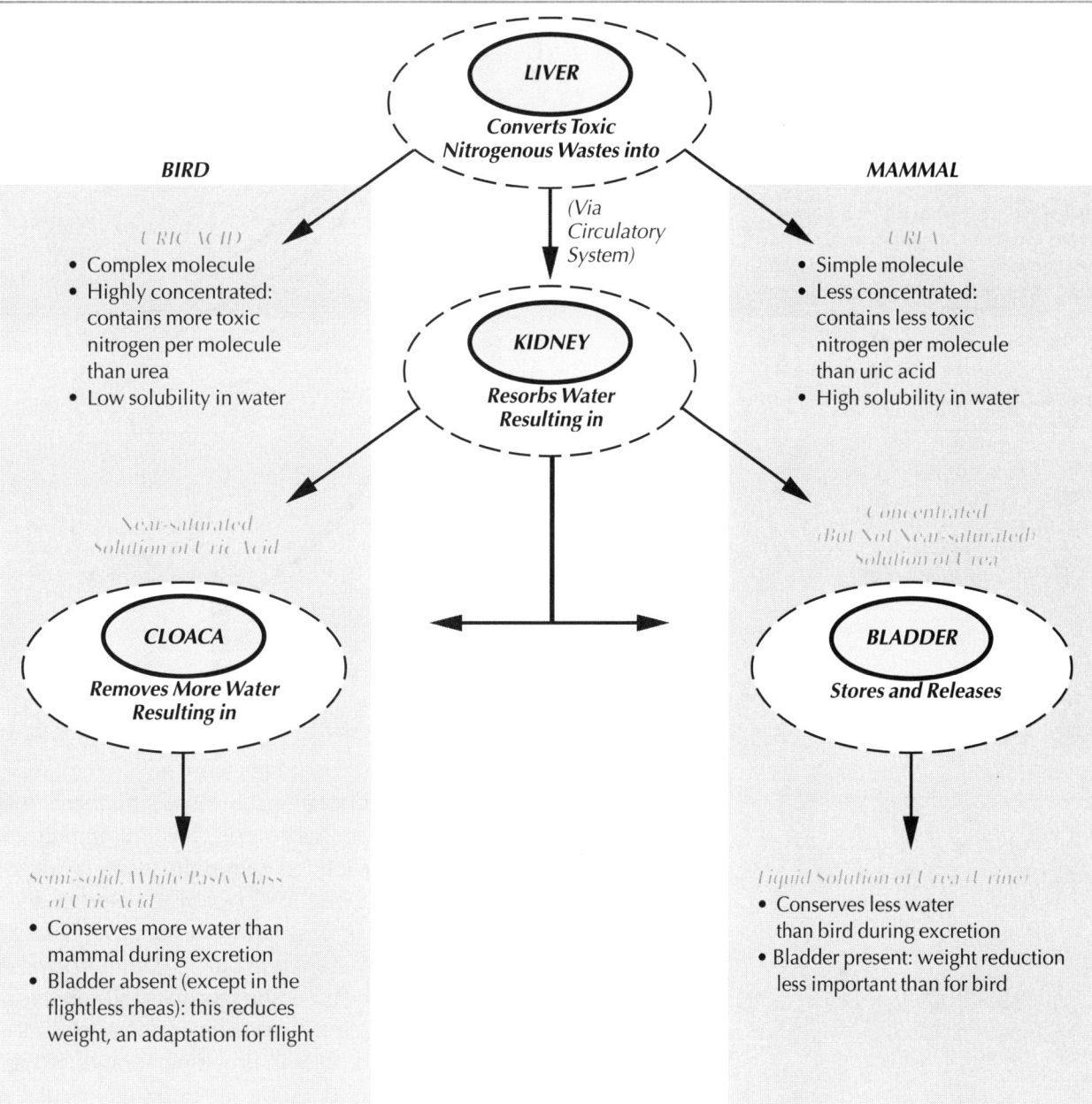

Figure 4–107. The Formation of Uric Acid Versus Urea: *This schematic compares the formation of uric acid in birds with the formation of urea in mammals. See text for details.*

between bird and mammal urine is a clue to how long the two groups of vertebrates have been evolving separately, and to how differently evolution has proceeded in each.

Genital System

Male Genitals (see Fig. 4–106)

The male gonads, **testes** (singular, **testis**), produce the **spermatozoa** (also called **sperm cells** or **sperm**), each of which consists of a single cell composed of a DNA-containing head and a propulsive tail. Birds have two types of sperm cells **(Fig. 4–108):** a short, simple type in nonpasserines, and a longer, spiral-shaped cell in passerines. The testes themselves are oval or elliptical and lie within the body cavity at the cranial end of each kidney. One, usually the left, is larger than

Figure 4–108. Passerine and Nonpasserine Spermatozoa: *The sperm cell, or spermatozoon, consists of a head containing the genetic material DNA, and a long, mobile tail which, during insemination, propels the sperm along the female reproductive tract. This highly magnified view shows the two types of avian sperm. In nonpasserines, such as the chicken, sperm are simple in structure. In passerines, such as the European Greenfinch, the head and much of the tail of the sperm has an elaborate, helical structure, and the entire cell is much longer than that of a nonpasserine. Drawing by Christi Sobel.*

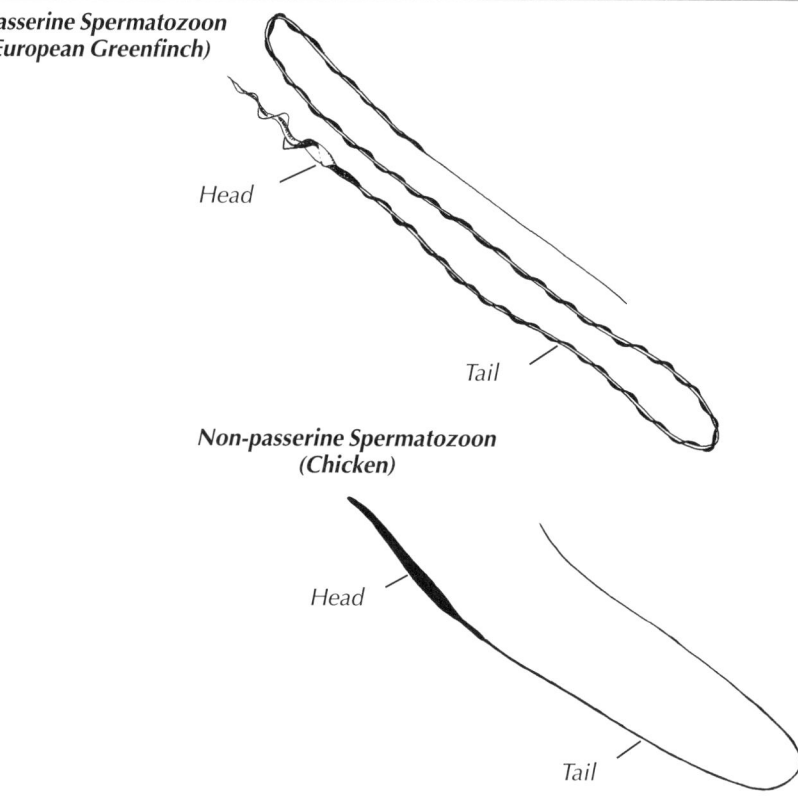

the other. With the approach of the breeding season, the anterior lobe of the pituitary secretes gonad-stimulating hormones that start the growth of the testes, which increase in size from several hundred to nearly a thousand times! Sperm produced within the testes move by the **deferent ducts** (vasa deferentia) to the cloaca. The deferent duct is so convoluted that it appears to have striations across it. A **cloacal protuberance** at the caudal end of the deferent duct is common in passerines. It may enlarge so much in the breeding season that it causes an obvious protrusion of the cloacal region (when seen from the exterior), a positive indication that a bird is a male in breeding condition **(Fig. 4–109)**. In most birds, this protuberance is caused by a swelling at the end of each deferent duct, which opens as a **papilla** into the cloaca. The cloaca is everted during copulation, and the papillae may slightly enter the female's oviduct. In all ratities and waterfowl, there is a better-developed copulatory organ, the cloacal phallus **(Fig. 4–110)**. This structure, often called a penis, is not the same as that of a mammal because it lacks an internal urethra—so sperm must travel on its surface. Also, it erects by lymphatic pressure rather than by the blood vascular network responsible for mammalian erection. Presumably Viagra would not help an aging Leda's swan.

Female Genitals (see Fig. 4–106)

The female gonad (**ovary**) ovulates **eggs** (also called **ova**; singular, **ovum**) at periodic intervals during the breeding season. As in mammals and other vertebrates, all of the eggs that a bird will ovulate in its lifetime are thought to be present in the ovary at birth. Most species of birds have only one functional ovary, usually the left, but in a few spe-

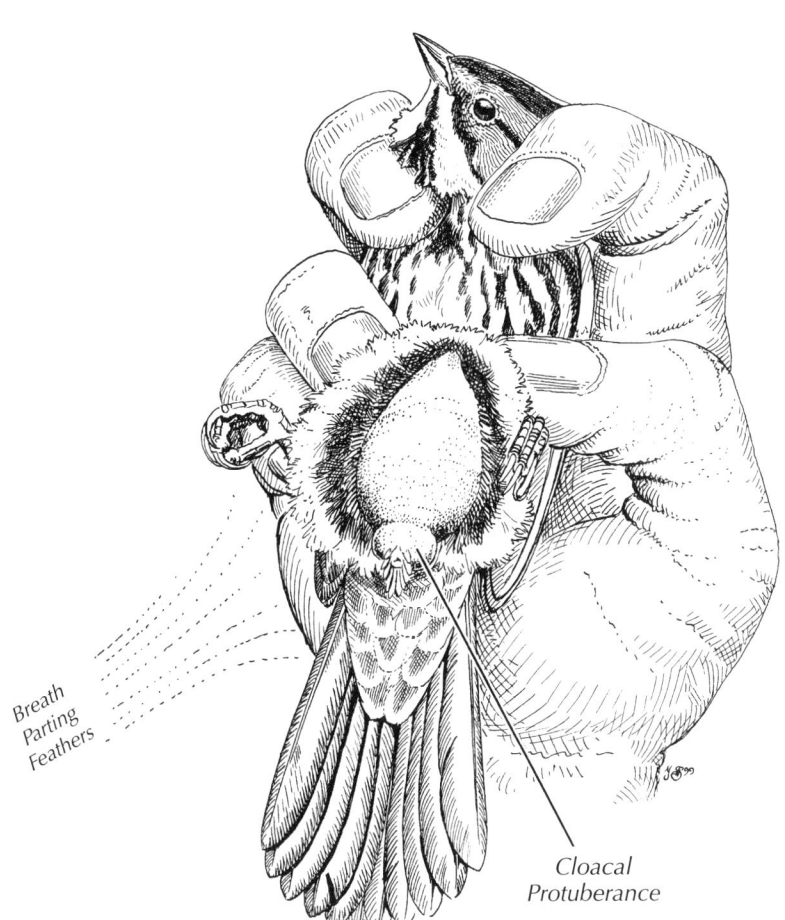

Breath
Parting
Feathers

Cloacal
Protuberance

Figure 4–109. The Cloacal Protuberance of a Breeding Male Song Sparrow: In male passerines, the cloacal region becomes swollen during the breeding season, reaching its peak size when the bird is in full reproductive condition. This cloacal protuberance is visible in a hand-held bird with the vent feathers parted; its presence is used by banders to determine the sex of breeding birds. The protuberance is caused by the seasonal enlargement of structures in the terminal regions of each deferent duct. These structures are the seminal glomus, an elaborately coiled region that develops only in passerines and contains densely packed active sperm, and an adjacent spindle-shaped receptacle, present in all birds, that opens into the cloaca through a flap of tissue called a papilla (see also Fig. 4–110). During copulation the cloaca is everted, and the papillae may actually enter the female's oviduct. The presence of the cloacal protuberance in passerines suggests that maturation of their sperm may be temperature sensitive: the protuberance may help to keep sperm slightly cooler than the body's core temperature, much as the scrotum does in mammals. Why this should be necessary only in passerines remains unclear.

4

Dorsal View

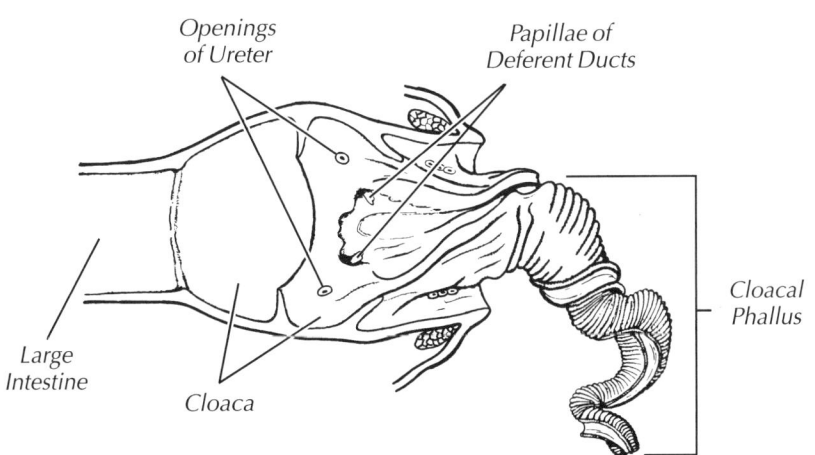

Openings
of Ureter

Papillae of
Deferent Ducts

Cloacal
Phallus

Large
Intestine

Cloaca

Figure 4–110. The Cloacal Phallus of a Male Domestic Duck: In this view the dorsal half of the cloaca has been removed, to show its ventral floor, whereas the spiral-shaped, ridged cloacal phallus, shown in the erect position, is intact. This copulatory organ, present only in waterfowl, and ratites, is erected by means of lymphatic pressure. Note the locations of the papillae of the deferent ducts, through which sperm are transported from the deferent ducts into the cloaca. During copulation, sperm travel along the outer surface of the phallus. Adapted from King and McLelland (1981, Vol. 2, p. 76).

cies, both right and left functional ovaries may be present in over half of the individuals. These species include the Eurasian Great Crested Grebe, Turkey Vulture, Northern Fulmar, kiwis, and a number of birds of prey.

The word "egg" has two meanings: it may refer to the ovum, which is the female reproductive cell (the female equivalent of a sperm cell), or it may refer to the hard-shelled entity with its white albumen and yellow yolk. The yolk, a single cell, is the true ovum. Tremendous in size compared with other cells of the body, it stores highly concentrated food materials to support the developing embryo until hatching time. The single yolk of a chicken egg is 32 percent of the volume of the total hard-shelled egg. The yolk of the much-larger Ostrich egg is an even greater percentage of the total egg volume. In contrast, the ova of mammals are very small and barely visible to the naked eye, having almost no nutritive yolk and thus no concentrated food store.

In most birds, the ovary releases an ovum at daily intervals during the breeding season until a complete set (**clutch**) of eggs is laid (**Fig. 4–111**). The **oviduct**, a tube that transports the egg from the ovary to the cloaca, is suspended from the dorsal body wall by a curtain-like membrane. When the ovary releases an egg, the flattened, funnel-shaped opening of the oviduct—called the **infundibulum**—moves up to the ovary and opens, "swallowing" the egg much like a snake swallows a rat. Thus, an egg being ovulated never crosses a gap to the infundibulum, as it does in mammals. In rare instances the ovum is not swallowed by the infundibulum, but is trapped in the body cavity among the viscera. This condition is a disorder called internal laying and the ovum must be resorbed by the surrounding tissues.

The oviduct is so convoluted that judging its length is difficult. Following the infundibulum is a glandular region or **magnum** that secretes the first of the albumen or "white" of the egg, followed by the **isthmus**, which secretes more fluid albumen and the egg membrane. A line of division can be seen between the magnum and the isthmus. The oviduct's next portion is the well-vascularized **shell gland**, which secretes additional fluid albumen and a calcium-rich shell with or without pigments—whatever is characteristic of the species. (The shell gland has many internal papillae and is sometimes called the "uterus," but it has no nutritive function as does the mammalian uterus, because birds do not bear live young.) Following the shell gland is a short section of oviduct called the **vagina**, which opens into the left side of the cloaca.

Contractions of smooth muscle in the wall of the oviduct move the ovum along its length, where the different glandular areas of its wall add their contributions in succession. An ovum released from the ovary of a chicken requires about 24 hours to become a hard-shelled egg ready for laying. For 18 to 20 hours of this time, the egg rests in the shell gland. Before being laid, most eggs are rotated 180 degrees in the vagina so the blunt end exits first.

Why are there no live-bearing birds? This question has been asked many times, and although several explanations have been given, none

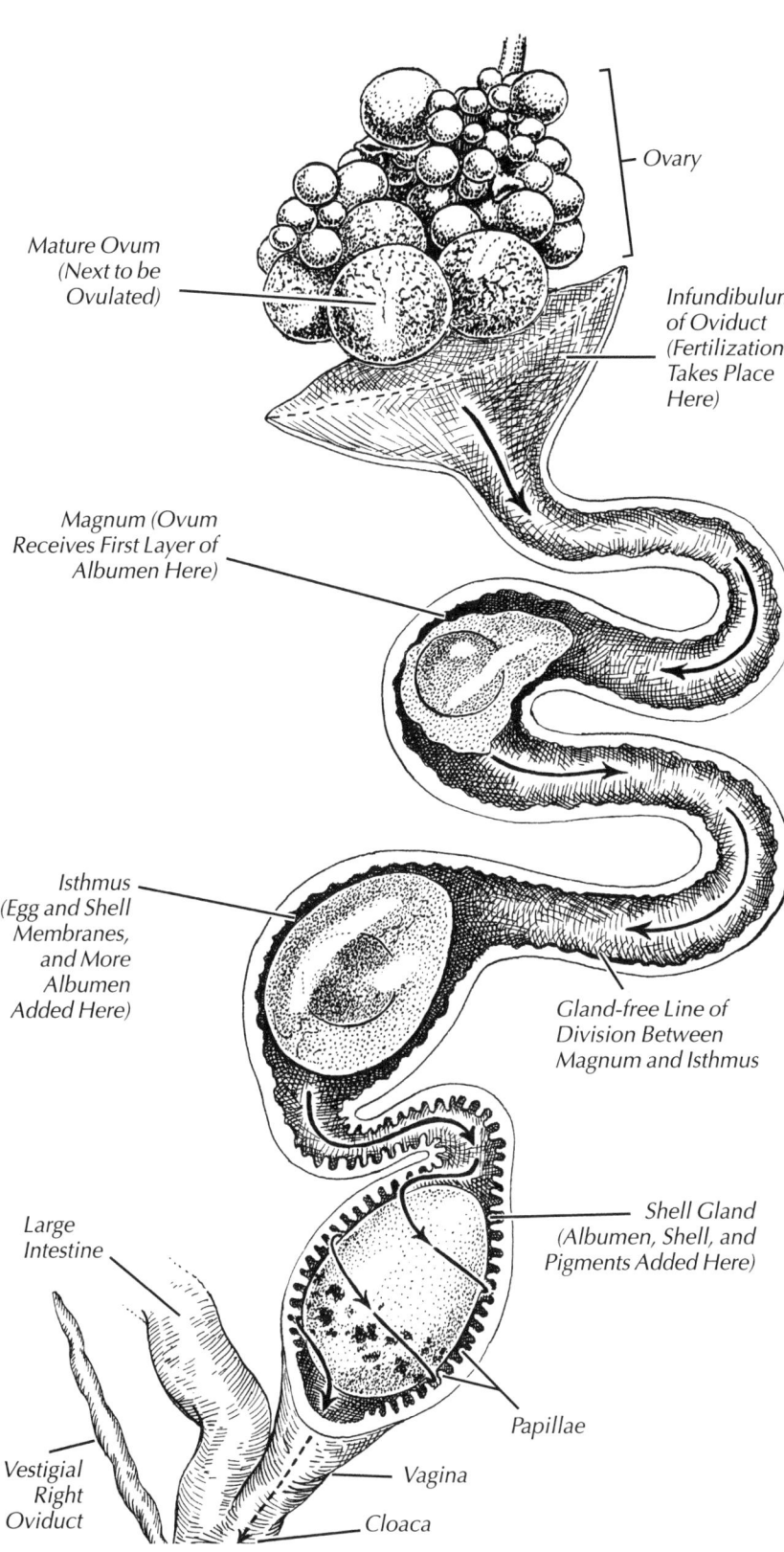

Ovary

Mature Ovum
(Next to be
Ovulated)

Infundibulum
of Oviduct
(Fertilization
Takes Place
Here)

Magnum (Ovum
Receives First Layer of
Albumen Here)

Isthmus
(Egg and Shell
Membranes,
and More
Albumen
Added Here)

Gland-free Line of
Division Between
Magnum and Isthmus

Large
Intestine

Shell Gland
(Albumen, Shell, and
Pigments Added Here)

Papillae

Vestigial
Right
Oviduct

Vagina

Cloaca

Figure 4–111. Stages of Egg Formation in the Oviduct: *This diagram shows the female reproductive tract in cutaway view for most of its length. The **ovary** releases a mature ovum at daily intervals during the breeding season until a complete clutch of eggs has been laid. What we recognize as the "yolk" of a chicken's egg is in fact the ovum, a single cell. As the ovum leaves the ovary, the funnel-shaped **infundibulum** at the cranial end of the **oviduct** (the tube that transports the developing egg to the cloaca) moves up to the ovum, opens, and "swallows" it. Fertilization takes place in the infundibulum, before the ovum has any covering of albumen or membranes. The fertilized ovum passes along the oviduct to the region called the **magnum**, where glands secrete the first layer of albumen (the familiar "white" of an egg) around it, a process that takes about three hours. A short, non-glandular region divides the magnum from the **isthmus**. During the hour or so the ovum spends in the isthmus, egg and shell membranes are deposited around the ovum, and more albumen is added. Next the ovum passes into the **shell gland**, whose many **papillae** secrete still more albumen and the hard, calcium-rich, outer shell. Shell pigmentation also takes place here, the patterns reflecting the speed of the egg's passage and the fact that it rotates as it moves through the shell gland. Rapid movement leads to streaked pigmentation, whereas slow movement leads to a more spotted eggshell. The hard, outer shell takes about 20 hours to complete, after which the egg passes through the lower end of the oviduct, termed the **vagina**, and into the **cloaca**, from which it exits the body. From ovulation to laying, the process of egg production takes about 24 hours.*

4

Figure 4–112. Copulation in Common Terns: *A mated pair of Common Terns copulates near their nest site on a Long Island, New York beach. Copulation in birds is brief, and consists of the coming together of the two birds' cloacas, termed the "cloacal kiss," during which sperm are transferred from the male's cloaca into the cloaca of the female. Photo by Tom Vezo.*

are completely satisfactory. In spite of their great diversity, birds are the only jawed vertebrate class in which all members lay eggs. The reason usually cited is that flying with the "excess baggage" of a growing embryo and fetus would be disadvantageous. This is true for most birds, but for flightless birds it would not be a problem. Yet none are live-bearing. Furthermore, bats seem to manage very successfully, for there are nearly 1,000 species, and all bear live young and can fly.

Another reason cited for the absence of live-bearing birds is the high body temperature of the parent bird (104 to 105.8° F [40 to 41°C]), along with the apparent sensitivity of all terrestrial vertebrate embryos to high temperatures. If birds were to bear live young, once the birth process began and the young were disconnected from the parental blood supply, they would have to be born very quickly because the lack of sufficient oxygen and the parental high temperature could prove lethal. Furthermore, an embryo retained in the oviduct during gestation might not be viable even *with* a parental oxygen supply. Experimental evidence from chickens appears to confirm this possibility. Incubation at temperatures above 104°F results in embryo death, or organ malformation leading to death after hatching. The chicken, however, is not a good example of normal bird reproductive physiology or of adaptability to environmental stress because it has been selectively bred to meet human requirements, sometimes at the expense of traits that would help it to survive in the wild. Consider, instead, birds that bury their eggs in decaying vegetation or volcanic, steam-heated earth mounds, as do the Australasian megapodes (see Fig. 6–36). One wonders what range of temperatures and levels of oxygen are successfully endured by megapode eggs. In ground-nesting birds, bearing live young would reduce the time in the nest—a real advantage in reducing nest predation. And it would, no doubt, be much appreciated by male Emperor Penguins (see Fig. 4–100). Nevertheless, egg laying is the rule in living birds.

Copulation and Fertilization

The transfer of male spermatozoa into the female's cloaca is so brief a contact—sometimes referred to as a "cloacal kiss"—that although it *could* be called copulation, the more general term **insemination** seems preferable **(Fig. 4–112)**. In the process, the male and female cloacas are everted as they are pressed together. This allows the papillae embedded in the male protuberance to contact the lining of the female's cloaca, or even to enter the opening of the left oviduct. As previously mentioned, in waterfowl and ratites a cloacal phallus, erected by lymphatic pressure, accomplishes a more intimate union that could properly be called copulation.

Figure 4–113. Griffon Vulture: Some species of birds are able to store sperm and thereby fertilize an entire clutch of eggs from a single insemination. In contrast, birds that lay only a single egg per season, such as the Griffon Vulture of Europe and Africa, may copulate many times over an extended period before laying. This may be more important for strengthening the pair bond than for insuring fertilization of the egg. Photo by C. H. Greenewalt/VIREO.

The number of sperm ejaculated at one time varies with the species (or the breed in domestic forms) and age of the bird, the time since the last ejaculation, and the time of year. Counts made in the rooster yielded densities of 250,000 to 10,200,000 spermatozoa per cubic millimeter, with an average of 3,200,000. Total volumes of ejaculated rooster semen (from stud roosters) contain about 3 billion sperm! The Rock Dove has about 200 million sperm per ejaculate. Male humans produce about 500 million sperm in one ejaculation.

Sperm live longer in bird oviducts than in those of mammals, with the possible exception of reproductively specialized mammals such as armadillos and bats. Following insemination, female birds can store sperm in "sperm nests" or crypts in the wall of the oviduct at the junction of the vagina and shell gland, or in the region of the infundibulum. Sperm are released from these crypts following the passage of an egg, and make their way to the infundibulum, where they may fertilize subsequent eggs. A female domestic turkey may lay as many as 15 fertile eggs following a single insemination, even up to 30 days after insemination, and 83 percent of the eggs may be fertile. In contrast, birds that lay a single egg each season, for example the Griffon Vulture of Europe and Africa **(Fig. 4–113)**, may copulate frequently for a month before laying. The extended period of copulation may be necessary for strengthening the pair bond, rather than for fertilizing the eggs.

Sperm deposited in the female cloaca enter the left oviduct and make their way to its upper end. Fertilization takes place in the region of the infundibulum as the ovum is being engulfed, before the ovum has any covering of albumen or membranes. The nucleus of the sperm cell and the nucleus of the ovum unite to form a single cell, the fertilized egg or **zygote**. Cell division proceeds as the egg passes down the oviduct, and slows or stops after the egg is laid. Growth does not resume until the egg is warmed by the incubating bird.

Sex Determination

How the sex of a bird is determined at conception is an interesting phenomenon. To understand it, though, one first must understand how the genetic blueprints of plants and animals are stored, sorted, and

Figure 4–114. Chromosomes of Body Cells and Gametes: DNA, the genetic material of nearly all living things, encodes building instructions for all the body's structures and processes. The DNA of an organism is divided up into paired structures called **chromosomes**, which are further divided into **genes**. Each gene contains the instructions for making a particular kind of building block, or information about how or how fast the building block is to be produced. **a. Body Cells:** In the body cells, all chromosomes except the two sex chromosomes are found in pairs whose members appear similar. Every body cell within a given organism contains the same species-specific number of chromosomes. The chicken, for example, has 39 pairs. For simplicity, only four chromosome pairs are shown in this diagram. One member of each chromosome pair was inherited from the individual's mother, and the other was inherited from its father. Although each member of a chromosome pair contains the same sequence of genes, the specific traits indicated may differ. For example, although both might contain a gene directing the pigmentation of the tail feathers of a pigeon, the colors or patterns indicated by each might differ. The sex chromosomes of the individual shown here are different from each other, indicating that it is a female (see Fig. 4–115). **b. Gametes:** The chromosomes of the gametes, or sex cells (eggs or sperm), are not in pairs. This occurs because during its production, each egg or sperm cell receives only one copy of each chromosome from its parent. Furthermore, each egg or sperm cell produced by an individual is different, because it contains a different subset of its parent's chromosomes: some originally came from its parent's mother, and some came from its parent's father.

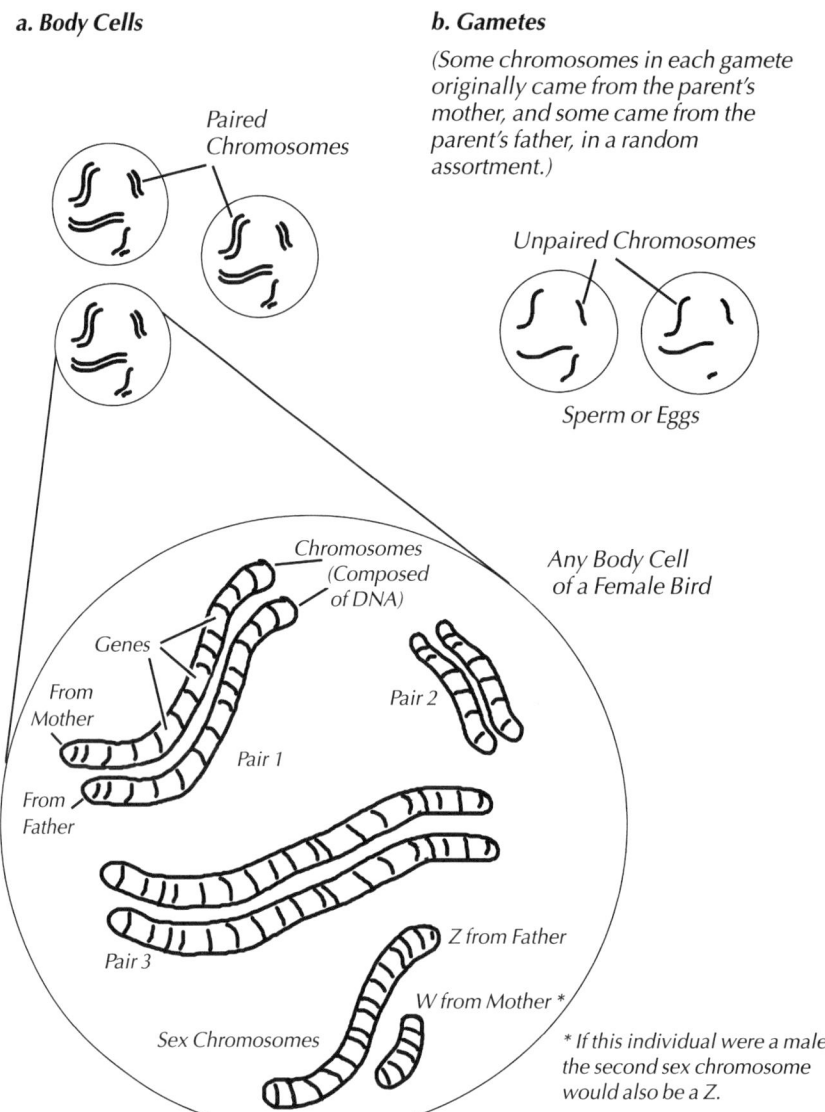

a. Body Cells

Paired Chromosomes

b. Gametes

(Some chromosomes in each gamete originally came from the parent's mother, and some came from the parent's father, in a random assortment.)

Unpaired Chromosomes

Sperm or Eggs

Chromosomes (Composed of DNA)

Genes

From Mother

From Father

Pair 1

Pair 2

Pair 3

Z from Father

W from Mother *

Sex Chromosomes

Any Body Cell of a Female Bird

* If this individual were a male, the second sex chromosome would also be a Z.

passed from one generation to the next. The genetic material DNA encodes each detail of the blueprints necessary for building proteins, including the instructions that result in the anatomical, physiological, and behavioral differences between the sexes. It is the way that DNA instructions are "filed" and "stored," however, that becomes important in understanding sex determination in birds.

Each subset of DNA instructions for making a particular kind of building block of an organism is called a gene. For each building block, other genes code for the conditions under which it is to be produced, and for the rate of its production. In complex organisms, the millions of genes are "stored" within the cell in structures called chromosomes **(Fig. 4–114)**. Each species has a characteristic number of chromosomes, and each body cell within an organism has that same number of chromosomes. The chromosomes can be individually recognized by shape and size, and are found in pairs of similar structure. One member of each pair of chromosomes was inherited from the organism's mother, and the other member, with instructions for all the same materials and processes, was inherited from the father.

The only cells in a complex organism that do not have their chromosomes in pairs are the gametes or sex cells (eggs and sperm). Instead, during their production, eggs and sperm receive only one member of each pair of chromosomes. Interestingly, some of the chromosomes in a single gamete of an individual will be like those of the individual's mother; the others, in a seemingly random fashion, will be like those originally inherited from the individual's father. Furthermore, each different gamete produced by an individual will have a different mix of the chromosomes from the individual's two parents. Each sperm cell of a male, for example, will contain a different mix of the chromosomes that male inherited from his two parents. It is this mixing, along with mutations and a few other types of genetic information sorting, that produces the individual variation in offspring on which natural selection operates to produce evolution.

Somewhat in contrast to most human concepts of a good information storage and retrieval system, the information on any one chromosome is *not* all closely related in "subject matter." Indeed, a chromosome resembles a filing cabinet whose file folders, instead of being arranged in drawers labeled "upper leg," "lower leg," and "foot," have a seemingly random arrangement: a cluster of files having to do with claw design and construction next to another cluster relating to liver cells, followed, perhaps, by some tongue muscle instructions. Any attempt to label such a drawer, much less characterize the entire filing cabinet (chromosome), is doomed to failure! To claim that the genetic filing system is chaotic and random, however, would be incorrect. In a chromosome pair, both chromosomes—whether from father or mother—contain the same files in much the same order, even though the individual genes may differ (reflecting differences in inheritance from mother versus father).

There is an exception to this general picture of chromosomes in birds, mammals, and some other vertebrates, however, and that is found with the positioning of many, perhaps most, genes dealing with sex. The sex genes are concentrated on a single pair of chromosomes. Unlike all other pairs of chromosomes, this pair can be structurally so different, one from the other, that they can easily be told apart in the proper sort of microscopic preparation. In mammals, one of these sex chromosomes always resembles the letter X in a microscopic preparation, but the second member of the pair may be another X or a "Y chromosome." The second chromosome type is designated Y in reference to the next letter in the alphabet, not because it is shaped like the letter Y. Similarly in birds, all individuals have one sex chromosome of a consistent shape and size that has been named the Z chromosome (in keeping with the end-of-the-alphabet mammalian sex chromosome designations, not because of its shape). In parallel with mammals, the second sex chromosome may be the same (another Z) or of a different structure, termed the W chromosome.

When the eggs and sperm are produced, individuals with a pair of similar chromosomes (XX in mammals, ZZ in birds) will produce gametes that all contain the same type of chromosome (X in mammals, Z in birds) **(Fig. 4–115)**. These individuals are thus called **homogametic**. In contrast, in individuals with two different sex chromosomes (XY in mammals, ZW in birds), half of the gametes will be of one type, and half will be of the other type. Individuals that produce two different types of gametes are termed **heterogametic**. When egg and sperm unite during fertilization, the embryo receives either two similar or two different sex chromosomes, one from each parent. This difference is what determines the sex of the individual bird or mammal.

It is here that the similarities in bird and mammal sex determination abruptly end, for the process is exactly opposite in these two lineages of animals. In mammals, heterogametic individuals (XY) develop into males and homogametic ones (XX), into females. In birds, heterogametic individuals (ZW) become females and homogametic ones (ZZ) develop into males. This bird-mammal difference seems to bear no functional significance, but rather reflects the randomness of evolution. It does mean, however, that it is the male's sperm in mammals, but the female's egg in birds that seals the sexual fate of the embryo.

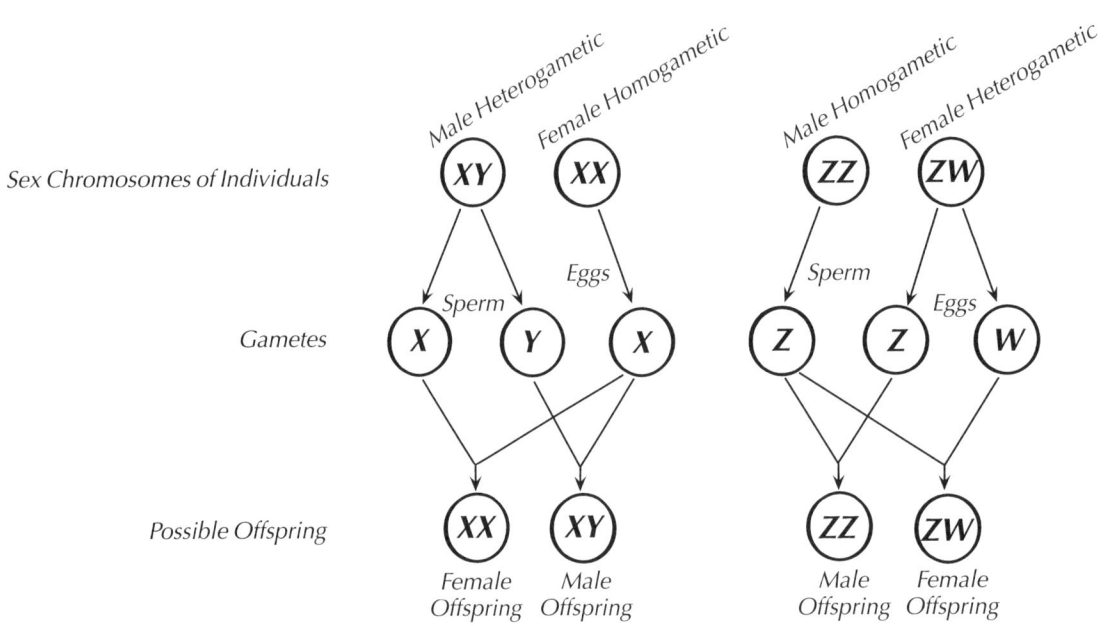

Figure 4–115. Sex Chromosomes and Sex Determination in Birds Versus Mammals: *The sex chromosome "pairs," containing genes determining the gender of an individual, may be dissimilar in appearance, unlike all other chromosome pairs. In mammals, the larger member of the pair is termed the X chromosome, after the letter it resembles, whereas the smaller is called the Y chromosome. In birds, the larger sex chromosome is termed Z, and the smaller is termed W (refer to Fig. 4–114). At conception, the two sex chromosomes an individual receives (one from each of its parents) may be the same, producing a* **homogametic** *individual, or different, producing a* **heterogametic** *individual. Mammals and birds differ in how these differences are translated into an individual's gender. In mammals, homogametic (XX) individuals develop into females, whereas heterogametic (XY) ones become males. In birds the opposite is true: homogametic individuals (ZZ) are males, and heterogametic ones (ZW) are females. During gamete production, male birds can produce sperm with only one type of chromosome (Z), whereas half the eggs a female bird produces contain a Z and half contain a W chromosome. In birds, therefore, it is the type of sex chromosome in the female's egg that determines the gender of the offspring at conception, in contrast to the situation in mammals, in which the sperm determines the sex of the offspring. Note that in any pairing between a male and female bird, approximately half the female's eggs will carry the W chromosome, and half will carry the Z chromosome. Therefore, on average, half of the offspring will be female and*

Sex determination in many other animals is a more complex situation. Most fishes, amphibians, and reptiles, for example, do not have recognizable, differentiated sex chromosomes. The genes encoding sexual characteristics may be widely dispersed across the entire set of chromosomes. What sex an individual becomes, and thus whether it produces eggs or sperm at maturity, may be determined in some of these species by such seemingly strange factors as the temperature at which embryological development takes place (turtles, crocodiles) or the behavior of other animals in an individual's social group (some fishes). In some species of fish and lizards, individuals may be functionally hermaphroditic—producing both eggs and sperm—either at different periods in the life of the individual or even at the same time in its life!

Hormones and Secondary Sex Characters

As discussed in the section on the endocrine system, **sex hormones** are produced by three sources: the anterior lobe of the pituitary gland, the adrenal glands, and, the primary source, the gonads. All sex hormones interact, influencing the development of anatomical structures, physiological processes, and the behaviors essential for successful reproduction. The male sex hormone, an androgen called testosterone, is secreted mostly by cells within the testes, but also by the ovaries (and adrenal glands). The female sex hormones, estrogens, are secreted mainly by the ovary, but the testes of birds of some species also secrete some estrogens, as do the adrenal glands. As you read through this section, refer to **Figure 4–116.**

Seasonal changes in the reproductive organs are controlled by light and other environmental factors, which regulate the activity of the pituitary gland. The pituitary, in turn, governs the secretions from the adrenal glands and the gonads. Hormones from the anterior pituitary activate the growth of the ova in the ovary. Some of the ova mature; most degenerate. Those that develop will produce estrogens that, when secreted into the blood, stimulate the development and preparation of the oviduct to receive the ova. Similarly, under the influence of hormones from the anterior pituitary, the testes begin their cyclic change, culminating in the production of spermatozoa and the production and release of testosterone.

The **secondary sex characters** are the features, besides the sex organs, that distinguish the sexes. Depending on the species of bird, these features occur in the plumage, combs, wattles, color of bill, presence or absence of spurs, size of body, vocalizations, breeding behavior, and so on. Hormones control the secondary sex characters, particularly those that change with the seasons.

The **gonadal hormones**, testosterone and estrogens, besides acting as stimulants for the development of the reproductive system, cause some or all of the following: (1) the appearance of bright plumage in males (and occasionally in females such as phalaropes) just prior to the breeding season; (2) changes in the color of the bill (for example, in the breeding season, the brown bill of the European Starling changes to bright yellow); (3) an increase in singing; (4) aggressive behavior toward other birds, particularly of the same sex and species, and the establishment of territory; (5) courtship displays leading to pair formation and copulation; (6) nest-building behavior; and (7) in most species, the development of an incubation or brood patch.

Late in the reproductive cycle, the anterior pituitary releases a hormone that inhibits further secretion of the other gonadal hormones, retarding the activity of the ovary and the testes. The reduction of these hormones reduces the behaviors initiated by them. By the time the eggs

Figure 4–116. Major Glands and Hormones Involved in the Avian Breeding Cycle: A schematic of the glands involved in reproduction, the hormones they produce, and their effects, as detailed in the text.

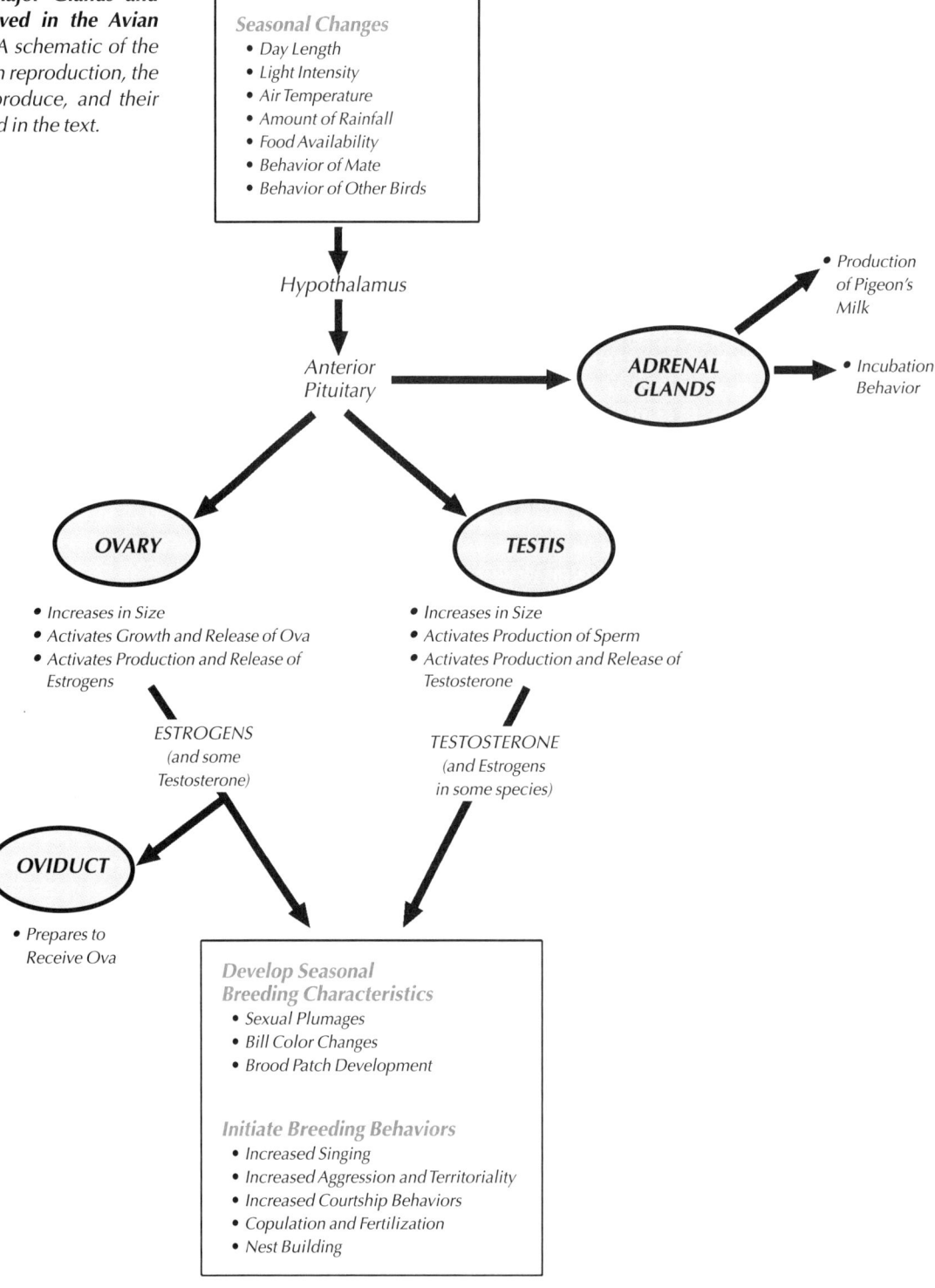

appear, a secretion from the adrenal gland has stimulated incubation behavior and activated the production of "pigeon's milk" in the crop of birds such as pigeons and doves.

The embryo, developing in the egg, requires constant heat. Because feathers are poor conductors of heat, most incubating birds develop one or more incubation or brood patches on the breast that are without feathers (see Fig. 8–89). Heavily suffused with blood vessels, these patches permit direct contact between the warm skin and the eggs. Hormones control development of the incubation patch. Before the female lays her first egg, her patch begins to develop as follows: it loses its feathers, the outer layers of the skin thicken, the blood vessels in the region increase in number and some enlarge, and the spaces between the cells under the skin fill with tissue fluid and remain full during the incubation period. Most male birds that incubate also develop an incubation patch.

The experimental use of hormones on castrated birds (having the testes, and thus the source of testosterone, removed) has yielded some strange results. Female hormones injected into castrated male ducks stop the development of a male-type syrinx and cloacal phallus. This indicates that the female hormones are responsible for sexual dimorphism in the syrinx and cloaca, for without them, male characters develop even in castrates. In contrast, the injection of testosterone into both male and female House Sparrows turns brown bills to black, which is the male breeding condition; injections of estrogens have no effect on the color of the bill. The same is true of European Starlings: testosterone turns brown bills to a bright yellow (the color in breeding adults), but estrogen has no effect on bill color. Thus, while major secondary sex characters appear estrogen-controlled, the seasonal reproductive characters seem to respond to testosterone.

Gonadal hormones also affect behavior. Female canaries injected with testosterone develop a male-type song, exhibit male-like courtship behavior, and become dominant over normal females. Thus testosterone dominates estrogen under these conditions. Surprisingly it is estrogens that have profound effects on the songbird brain. F. Nottebohm and other researchers have shown that major differences exist between brain structures in adult male and female birds of various species. In male brains an enzyme converts androgens to estrogen. For some reason, estrogen, rather than androgens, probably controls the development of the brain areas involved in song acquisition, perception, and production. The brain has at least two neural pathways that contribute to song learning and song production (see Fig. 7–36). One, a motor pathway, controls the muscles of the syrinx for song production; the other neural pathway is required for song learning.

In birds, which sexes sing, how much they sing, and when they sing varies greatly by species. In most birds, only males sing, but in a few species, both sexes sing. Some birds sing year round while others sing only for a short breeding season. In the future, we can expect to hear much more about hormone action in the brain and the neurobiology of bird song, an active and exciting field of research at the present time.

Factors Bringing Birds into Breeding Condition

Many external factors may stimulate the pituitary to signal the other glands involved in reproduction, thus beginning the breeding cycle. They include the amount of light each day, the intensity of light, the temperature, the amount of rainfall, the availability of food, and the actions of other birds, including the behavior of the mate. In this list, the most universally dependable factor is day length and its increase or decrease as the seasons change (at least, in the higher latitudes). The effect of the changing day length on the pituitary gland in birds is an example of **photoperiodism** (any type of response to day length).

Photoperiodism

In most birds of the temperate zone, an increase in day length causes development of the gonads and stimulates migratory behavior. An artificial increase in day length will force some species to come into breeding condition in the dead of winter. Light, therefore, must play a role in regulating the onset of breeding in birds of middle and high latitudes. In some species, an *increase* in the day length starts development of the gonads; in others, exposure to a day length that is *at least* as long as some period seems to be more important, still longer days having no greater effect. Presumably, the retina of the eye sends neural impulses to the brain that eventually stimulate the hypothalamus. Thus stimulated, the hypothalamus secretes hormones that in turn stimulate the pituitary. Much remains to be learned about this process.

In experiments, White-crowned Sparrows and Dark-eyed Juncos respond with a normal increase in gonad size between December and May only if the longest artificial day exceeds 10 hours. Under photoperiods of nine hours, their gonads develop, but at a much slower rate. Just as most avian physiologists were becoming satisfied with the assumption that day length triggered the reproductive cycle, however, an experiment showed that light cues are not necessary for some birds. The testes of domestic ducks that were kept in total darkness for 20 months developed, regressed, and developed again—all without light.

Natural observations also indicate that increasing day length is not *the* single cue for initiating reproduction. The European Robin, for example, begins producing sperm about the first week of January in the foggy midwinter of the English Midlands, when the day length has increased by only a few minutes. In New Zealand, the large mountain parrot called a Kea (**Fig. 4–117**), and in Australia, the Emu and the Superb Lyrebird, all begin producing sperm as the days begin to shorten. All three of these species nest in winter.

Migration is closely connected with the reproductive system in most birds. However, little is understood regarding the role of day length in birds that migrate across the equator from one hemisphere into the other. How can we account for development of the gonads in the Bobolink, which nests in the Northern Hemisphere, migrates to the Southern Hemisphere for the winter, and begins its northward flight when the days there are getting shorter rather than longer? Perhaps a summation of the periods of daylight determines the gonadal response

Figure 4–117. Kea: *A large parrot from the mountainous regions of New Zealand, the Kea begins sperm production as the days begin to shorten, in preparation for its nesting season during the southern hemisphere winter. Photo by T. J. Ulrich/VIREO.*

in transequatorial migrants. However, the birds must become insensitive or reverse their sensitivity as they reach their destinations.

Air Temperature

Every bird species is physiologically adapted to a specific temperature range; any drastic change may affect the beginning of the breeding season. In many birds an unusually cold spell in the early stages of the breeding cycle may delay nesting. This is true for the Great Tit, Blue Tit, Eurasian Blackbird, Pied Flycatcher, European Robin, Chaffinch, and many others. In domestic turkeys and European Starlings, lower temperatures reduce the production of sperm as well as the number of eggs laid. Freezing temperatures can cause frostbite of delicate tissues, which may stop all egg production. Thus farmers may remove the comb or wattles of barnyard birds to prevent such signals from interrupting the harvest.

Temperature changes affect different species in different ways. The Emperor Penguin lays its one egg in the middle of the Antarctic winter when the air temperature is about *minus* 30° F (minus 34.4° C). Thus by hatching time the temperature has become higher (5° F; minus 15° C) for the chick. Possibly, in such species a decreasing day length provides the stimulus for gonadal development.

Rainfall

In arid regions, where rain occurs sporadically between periods of drought that last for months or years, either the rain itself, or the green vegetation resulting from the rain, triggers the breeding cycle. Abert's Towhees in Arizona may begin to nest 10 to 14 days after heavy rains in March or April. Sometimes they nest again following a second period of rain that occurs in July. The beginning of egg-laying by California Quail varies from year to year by about three weeks, depending on the temperature and amount of rainfall.

Figure 4–118. Rufous Hornero at its Nest: *Rainfall may trigger breeding in certain species by providing food, nesting material, or cover. The Rufous Hornero of Argentina, shown here, does not breed until adequate rainfall provides a supply of mud with which it can build its huge oven-like nest. This individual's nest is only partially complete; the finished structure will have a domed roof of mud, and will dry rock-hard in the sun. (See also Fig. 8–34d.) Photo by T. J. Ulrich/VIREO.*

Old World warblers (Family Sylviinae) dwelling in the grasslands of Zimbabwe and Malawi of Africa do not breed until after the seasonal rains—they depend upon rain for the growth of nesting material and cover. The Rufous Hornero in Argentina does not breed until sufficient rain has produced the mud with which it builds huge, oven-like nests **(Fig. 4–118).**

Rainfall is all-important in extreme desert regions because the short period of vegetative growth—possibly little more than a month—must provide nesting material as well as food for the young birds. Birds in the desert regions of Australia are often nomadic, moving about in search of food and water. They nest in any month of the year and, if the rains finally come, may nest twice in six months.

Getting wet from rain, or perhaps even hearing rain, may be a sufficient stimulus to initiate the breeding cycle of some Australian birds. They begin to breed right when the rains begin, before the rain has had a chance to increase the supply of food and nesting material, or to change the general appearance of the environment. The Zebra Finch, Black-faced Woodswallow, Budgerigar, and Australian Tree Swallow are some of the birds that begin nesting at the first drop of rain.

Thus, due to the unpredictability of rain in arid environments, natural selection has favored the evolution of a reproductive system that can spring into action quickly. The reproductive system of birds adapted to such environments can maintain itself without the usual nonbreeding period common in temperate zone birds. In experiments, the Baya Weaver of Asia maintained sexual readiness with a continuous production of sperm for 15 months without any regression of the testes. In contrast, the gonads of the Dark-eyed Junco regress

Figure 4–119. Male Red Crossbill: *No-madic Red Crossbills may breed during any month of the year, initiating nesting when their wanderings take them into areas where bumper crops of their principal foods—pine and other conifer seeds—are available. The crossed tips of their beaks allow them to efficiently pry the conifer seeds from deep within open or closed cones. Food, rather than light, is thought to act as both the proximate and ultimate factors triggering this spe-cies' breeding cycle—because it feeds its young on a regurgitated paste made primarily from seeds. Photo courtesy of Brian Henry/CLO.*

after each breeding period, regardless of the length of light periods in the laboratory.

Food

A bird must time its breeding so that the hatching of young co-incides with the best environmental conditions for their survival. In the Arctic, for example, Pomarine and Long-tailed jaegers and the Snowy Owl depend heavily on small rodents, particularly the brown and collared lemmings, for food. When lemming populations are low, these birds may lay unusually small clutches of eggs, or they may not nest at all.

The Red Crossbill **(Fig. 4–119)** restricts its diet to conifer seeds, particularly those of pine. Its nomadic tendencies and the fact that it may nest during any month of the year suggest that food, rather than light, acts as both the proximate and ultimate factors in initiating the breeding cycle. Artificial light in the laboratory will cause some, but not complete, gonadal development. In the wild, Red Crossbills nest only when the conifers produce a good seed crop, feeding their young a paste of regurgitated seeds, possibly mixed with saliva and some insect matter. They are indifferent to the cold and their young seem to thrive with little or no insect food, a rare phenomenon among most small bird species, regardless of the parents' diet.

Social Interactions

Sometimes light stimulation must be reinforced by other external stimuli before it will trigger the breeding cycle. An important external stimulus for some species is the presence or absence of other individuals of their own species. A captive female Rock Dove lays eggs readily in the presence of a male and less readily when only another female is present. If isolated from all others of her kind, she will not lay at all—unless she has a mirror in her cage. Such social contact is very important in triggering breeding in colonial-nesting birds such as gulls, boobies, and many other marine birds.

The male House Sparrow must be present before the female will lay an egg. European Starlings, normally found in flocks during the nonbreeding season, break up into pairs for breeding. If confined in flocks during the breeding season, female starlings do not lay eggs.

Elaborate courtship behaviors by the male stimulate ovulation in a wide variety of birds, for example the Satin Bowerbird in Australia, and the Chaffinches and robins in Europe. Vocalizations are important for development of the gonads in parakeets, and egg laying is enhanced when three or more birds are present.

In humid equatorial regions, where day length fluctuates very little and a mild climate persists throughout the year, birds have a tendency to prolong the breeding season. Estrildid finches such as the Zebra Finch from Australia breed freely and can raise four or five broods a year in rain forest regions. Some seabird colonies also are active throughout the year. Sooty Terns of Ascension Island breed about every 9.6 months; Audubon's Shearwaters of the Galapagos Islands breed about every 9 months.

Metabolism

The term **metabolism**, in biology, includes all of the chemical changes that take place in the cells and tissues of the body—the use of basic food materials to produce **protoplasm**, living material. Metabolism also includes the conversion of complex substances to simpler ones to produce energy for breaking down the basic foods, for contracting the skeletal muscles, and for producing heat.

The production of new living material is necessary for an animal's growth. It continues throughout life to allow for the repair of cells and tissues, and to permit the continuous turnover of material in the cells. Sugars, other carbohydrates, and fats in the diet provide the energy for this dynamic flux. Enzymes are proteins that act as catalysts in the chemical changes that build up and break down the constituents of protoplasm. Metabolism, then, includes all of the dynamic chemical, physiological activities of cells and tissues. Here, we deal only with the major physiological processes: the maintenance of body temperature, heart rate, and respiratory rate under different environmental conditions; water and salt regulation; and aging.

The contraction of skeletal muscles produces both work, in the form of force applied to the shortening muscle, and heat. Heat is measured in calories; one calorie is the amount of heat required to raise one gram of water one degree Centigrade. The **basal metabolism** of an animal is the number of calories that it uses when completely at rest.

Compared to mammals, BIRDS, as a group, have:

- **HIGHER BASAL METABOLIC RATE**

- **HIGHER BODY TEMPERATURE**

- **FASTER HEART RATE**

- **FASTER (MORE EFFICIENT) RESPIRATORY SYSTEM**

- **HIGHER BLOOD PRESSURE**

- **HIGHER BLOOD SUGAR CONCENTRATION**

Figure 4–120. Bird Versus Mammal Metabolism: In general, birds have a higher basal metabolic rate, body temperature, heart rate, blood pressure, and blood sugar concentration than mammals, and a more efficient respiratory system. But, these differences do not necessarily hold when particular birds are compared to particular mammals, or when birds and mammals of a similar size are compared. For example, the respiratory rate of a bird is actually lower than the respiratory rate of a mammal of the same size. See text for details.

In other words, it is the amount of energy needed to maintain minimal body functions. The basal metabolic rate is, naturally, lower than any of several active metabolic rates that an animal can have.

As a group, birds have a higher basal metabolic rate than mammals—a higher body temperature, faster heart rate, and faster respiratory rate **(Fig. 4–120)**. The amount of food that a bird must eat depends partly on the caloric value of its food, partly on the size of the bird, and very much on the level of activity in which it engages and on the temperature of the environment. Because the higher metabolic rate of birds requires more food calories per unit of time, a bird's blood sugar concentration is much higher than that of mammals.

The size of a bird has important effects on its heat production and loss. The smaller the bird, the larger its body surface area in relation to its volume. The amount of heat produced by an animal is proportional to its volume, because the volume is composed largely of the heat-producing muscle. The amount of heat lost, however, is proportional to the amount of surface area, because most heat loss occurs across the body surface.

The concept is easy to understand if you consider two cubes **(Fig. 4–121)**. The first, 1 x 1 x 1 inches, has a surface area of 6 square inches (6 sides x 1 square inch per side) and a volume (length x width x height) of 1 cubic inch; the surface-to-volume ratio is 6:1. The second cube, 2 x 2 x 2 inches, has a surface area of 24 square inches and a volume of 8 cubic inches; thus the surface-to-volume ratio is only 3:1. Therefore, a small bird must produce more heat in relation to its body size than a large bird, just to offset the high rate of heat loss from the surface of the body.

To determine metabolic rate, researchers may directly measure heat calories produced, but more often they measure the amount of oxygen that an animal consumes. This indirect method works because in order to produce heat, oxygen must be consumed—and this occurs during the continual process of food digestion. Recall that to digest their food, animals combine carbohydrates, fats, and proteins with oxygen, breaking down the food into smaller nutrients that can be absorbed

Figure 4–121. Size and Surface-to-Volume Ratio in Birds: The size of a bird influences its heat production and loss, and therefore has important implications for its metabolic rate. The amount of heat an animal produces is proportional to its volume, because heat-producing muscle makes up much of that volume. The amount of heat lost, however, is proportional to an animal's surface area. Small birds, such as the Black-capped Chickadee, have a greater amount of surface area per unit volume (known as the surface-to-volume ratio) than do large birds, such as the Wild Turkey. This concept is illustrated here using two cubes of different sizes, for simplicity. The cube with 1-inch sides, representing a small bird, has a volume (length x width x height) of 1 cubic inch, and a surface area (length x width x number of sides) of 6 square inches. The cube with 2-inch sides, representing a large bird, has a volume of 8 cubic inches, and a surface area of 24 square inches. The surface-to-volume ratio of the smaller cube is 6:1, whereas that of the larger cube is 3:1. A small bird, with its high surface-to-volume ratio, will lose more heat in relation to its body size than will a large bird, so the small bird must have a higher metabolic rate to compensate

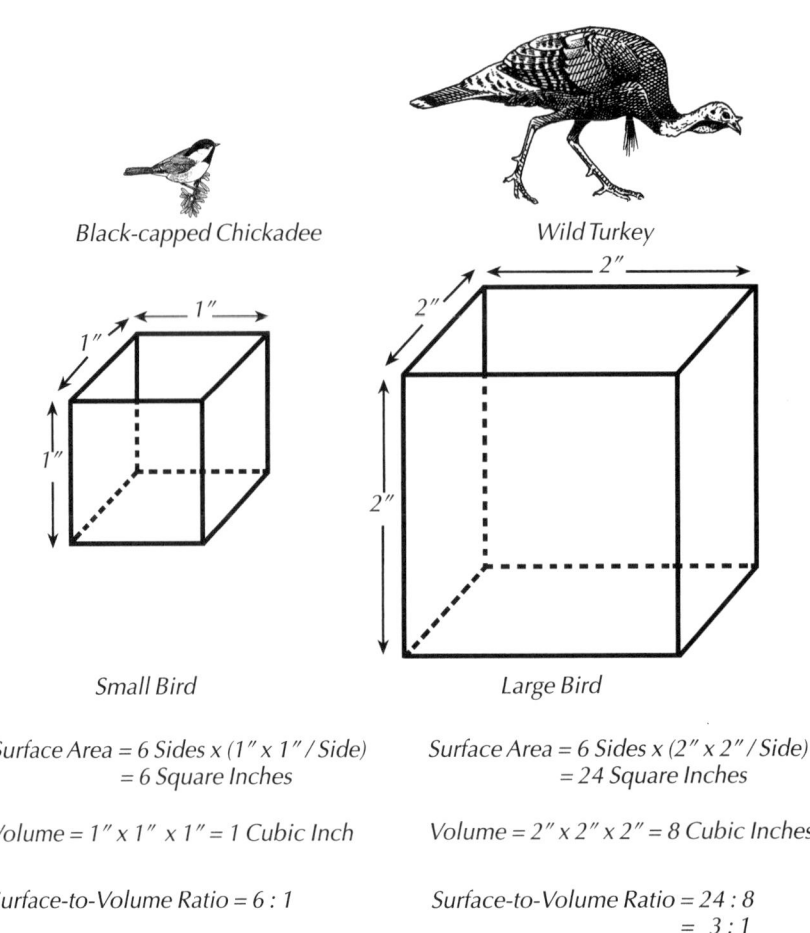

Black-capped Chickadee   Wild Turkey

Small Bird   Large Bird

Surface Area = 6 Sides x (1″ x 1″ / Side)   Surface Area = 6 Sides x (2″ x 2″ / Side)
　　　　　　 = 6 Square Inches　　　　　　　　　　　 = 24 Square Inches

Volume = 1″ x 1″ x 1″ = 1 Cubic Inch   Volume = 2″ x 2″ x 2″ = 8 Cubic Inches

Surface-to-Volume Ratio = 6 : 1   Surface-to-Volume Ratio = 24 : 8
　　　　　　　　　　　　　　　　　　　　　　　　　　　　　　　　　　　　 = 3 : 1

more readily, and producing heat as a by-product. The amount of heat produced is proportional to the amount of oxygen consumed; thus it is possible to use oxygen consumption alone to measure metabolism. For example, when resting, Anna's Hummingbirds and Allen's Hummingbirds **(Fig. 4–122)** use from 10.7 to 16.0 cubic centimeters (cc) of oxygen per gram of body weight per hour (their basal metabolic rate). In hovering flight, Allen's requires 85 cc of oxygen per gram per hour; the slightly larger Anna's requires only 68 cc. Thus basal metabolism is increased nearly sixfold by the activity of flying.

Body Temperature

Birds and mammals are "warm-blooded." This means that they usually maintain their body temperatures within a certain narrow and high range, even when the air temperature changes to far below or considerably above the set body temperature. "Cold-blooded" animals, such as fish, amphibians, and reptiles, do not maintain a constant body temperature in a variable thermal environment. Instead, their body temperatures fluctuate with the surrounding temperature, explaining why they are least active on very warm or very cold days.

Body temperatures in birds range from 99.8 to 112.3° F (37.7 to 44.6° C). The average resting temperature of 311 passerine species has been measured at 105.1° F (40.6° C) and of 90 shorebird species, at

a. Allen's Hummingbird

b. Anna's Hummingbird

about 104.2° F (40.1° C). Thus there do appear to be evolutionary differences in the body temperature set point that large groups of related birds share. The maintenance of body temperature within a normal range depends on the amount of heat the bird produces, the means it has of conserving its heat in cold weather, and the way it gets rid of excess heat in hot weather. The body temperature of nocturnal birds, such as kiwis, owls, and nighthawks, is higher at night, when they are most active. The temperature of the kiwi, generally considered to be a primitive bird, fluctuates more than that of the others.

Although nearly all adult birds maintain the body temperature under varying conditions, most newly hatched young cannot. This is true for both the nearly naked American Robin nestling and the downy Killdeer chick, although the Killdeer chick has some control in air temperatures between 74 and 104° F (23 and 40° C). The temperature control of both nestlings increases rapidly during the first 10 days, and control by the Killdeer chick equals that of the adult by 27 days.

The primary need of newly hatched nestlings that are helpless, almost naked, and asleep most of the time is protection from the elements. In cold weather, the adult broods (covers) them, providing heat from its own body. In hot weather, the adult shades them from the burning sun and sometimes tilts its wings to deflect the slightest breeze onto them **(Fig. 4–123)**.

An adult broods the chicks or nestlings only until they have an efficient temperature-regulating system of their own. To gain this, the young must grow feathers, increase in size (thus decreasing the body surface area in relation to volume), increase neural and glandular controls, and develop air sacs for efficient oxygen delivery to the body cells. In most passerines these changes all take place in about one week. Nestling Field Sparrows and Chipping Sparrows become warm-blooded within about six or seven days.

Most birds conserve heat efficiently. Their thick covering of feathers leaves very little bare skin from which heat may escape. Marine birds, such as penguins and petrels, also have considerable fat under

Figure 4–122. The Metabolic Cost of Hovering Flight: A hovering Allen's Hummingbird (a) consumes about 85 cubic centimeters (cc) of oxygen per gram of body weight per hour, whereas the slightly larger Anna's Hummingbird (b) requires only about 68 cc for the same activity. Hovering is an energy-intensive activity, producing a nearly sixfold increase in the amount of energy used, compared to that of a resting bird. The basal (resting) metabolic rates of these birds range from 10.7 to 16.0 cc of oxygen consumed per gram of body weight per hour. Photos courtesy of Patricia Meacham/CLO.

the skin that helps conserve heat, but all birds depend primarily on their feathers for insulation. The thick down under the contour feathers of the Common Eider constitutes just about the finest insulating material known (Fig. 4–124a). On cold, wintry days, birds commonly "fluff up" their feathers as one fluffs up a down pillow, increasing the air spaces (Fig. 4–124b). The more air spaces, the better the insulation. Frequently during cold weather, birds perch on one leg, drawing the other up under the breast for warmth. Sometimes birds tuck their bills into the feathers of the shoulder for the same reason (Fig. 4–124c). Some birds roost together overnight to conserve heat (see Ch. 6, Sidebar 4, Fig. H). The legs and feet of some birds, such as Herring Gulls, are quite insensitive to cold because a countercurrent exchange of heat takes place between arteries and veins before blood enters the foot.

Figure 4–123. American Robin Shading its Nestlings: Unlike adult birds, most nestlings have little ability to thermoregulate. This is particularly true for altricial young, which hatch with unfeathered bodies and closed eyes. Parent birds, even in species whose young are down-covered at hatching, regularly use their bodies to protect their young from temperature extremes. In cold weather, adults may brood their young, covering them with their fluffed-up belly feathers. In hot weather, adults provide shade from the sun, sometimes spreading their wings to increase the shadow area or to enhance ventilation, as this American Robin is doing. Drawing by Charles L. Ripper.

Countercurrent Heat-Exchange Systems

Countercurrent exchange systems are found in many different regions of animals' bodies and are one of the most important ways in which organisms conserve energy or other vital resources such as water or ions. Thus, taking time to understand the basics of how a countercurrent exchange works not only helps to understand many phenomena in physiology, but heightens appreciation of the problem-solving nature of evolutionary adaptation.

The problem for a gull standing on ice or swimming in frigid water is one of conserving vital body heat while supplying its legs and feet with oxygen and essential nutrients. Fortunately for the gull, its legs and feet are composed primarily of bone, tendon, and scaly skin—all tissues that have low metabolic requirements for oxygen and nutrients. The active movements that they undergo are caused by the contraction of muscles high up on the leg and within the contour of the body, which are transmitted passively to the foot by tendons. Nevertheless, after hours of swimming in the cold, a great deal of body heat could be lost from the trickle of warm blood that must perfuse the legs and feet to supply their minimal needs. Completely shutting off the circulation to the limbs is not an option, because the bird has absolute requirements for a small amount of oxygen, for the removal of accumulating wastes, and for some warming of the extremities—which may be necessary to prevent tissue damage due to freezing.

The anatomical basis for the solution to this dilemma, as with all countercurrent exchanges, is in structural specializations of the blood vessels (Fig. 4–125). In the normal blood supply, arteries become progressively smaller as they branch toward the tissue to be supplied. Eventually the blood flows into a capillary bed in the midst of the target tissue. From the capillary bed the blood recollects into veins of ever-increasing size, making its way back to the heart. In contrast, a second

a. Common Eider Nest

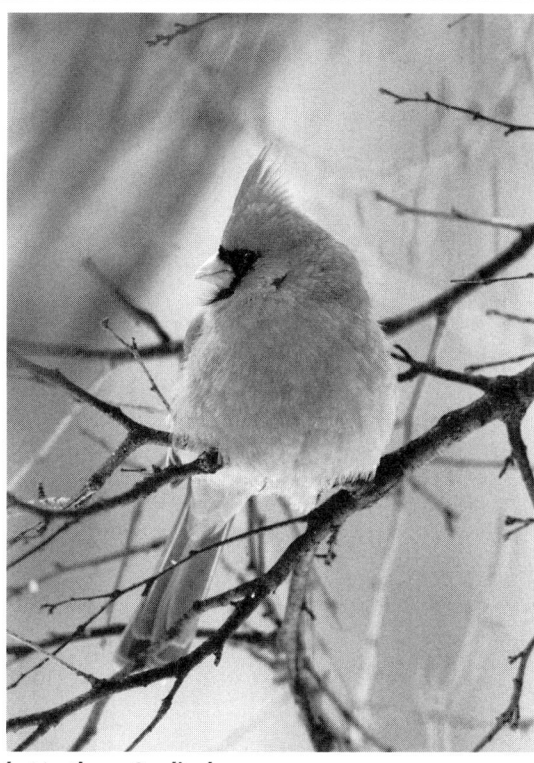

b. Northern Cardinal

Figure 4–124. Conserving Heat: *Birds use a range of methods to conserve heat.* **a. Common Eider Nest:** *A layer of down from the female's breast lines the nest of a Common Eider, lending a thick insulating layer to warm the developing eggs. Photo courtesy of Mary Tremaine/ CLO.* **b. Northern Cardinal:** *A male Northern Cardinal fluffs up its plumage, increasing the number of insulating air spaces among the feathers for protection against the cold of a New York winter. Photo by Marie Read.* **c. Juvenile Black-Crowned Night Heron:** *A juvenile Black-crowned Night Heron tucks one foot up close to its belly and buries its bill in the feathers of its breast and shoulder to keep warm on a chilly New Mexico morning. Photo by Marie Read.*

c. Black-crowned Night Heron

circulatory pattern develops in a countercurrent system. The second system is always characterized by the breakup of the arterial supply vessels into a network or mesh of small vessels *before* the target tissue is reached, *and* a splitting of the venous return vessels into a similar network *after* the veins have collected the blood from tissue capillaries. These arterial and venous networks intertwine, forming the actual "heat exchanger," a distinct structure not embedded among the cells of any nonvascular tissue, and often at some distance from the tissue actually to be nourished and cleansed by the blood they transport. In the case of leg and foot countercurrent heat exchangers, the intertwined vessel networks are stretched out along the upper leg in the region of the lower tibiotarsus bone. In this heat exchanger, the vessels within the network are somewhat larger in diameter and thicker-walled than capillaries, and as in all countercurrent exchangers, they are closely packed together.

The characteristics of the blood entering the two ends of the elongate exchanger are very different. Especially distinctive is the high temperature (near core body temperature) of the arterial blood arriving from above the exchanger and the low temperature (possibly near that of the external environment), of blood returning upward from the tips of the toes. Obviously, heat will flow from the closely packed, small-diameter (and thus large-surface-area-per-volume) arteries to warm the blood in the veins. As it gives up heat energy to the veins, the arterial blood is cooled; thus an exchange of thermal energy occurs from arterial to venous blood.

Note that heat exchange would occur even if blood in the intertwined arteries and veins flowed in the same direction. If this were the situation, however, the blood temperature would quickly come into equilibrium about midway between the arterial and venous temperatures of blood entering the exchanger. The exchange potential of the actual system is of much higher efficiency, however, because the blood flow along the length of the vascular network is always in the opposite direction in the arteries compared to the veins (that is, countercurrent). Because heat energy always will dissipate from an area of higher temperature to an area of lower temperature, chilled arterial blood nearing the end of its flow through the exchanger on its way to the foot is still slightly warmer than the coldest venous blood arriving from the foot, so the little remaining heat energy can flow from the arterial blood to the venous blood (see Fig. 4-125c). Likewise, at the upper end of the exchanger, where venous blood is very much warmed, it still can be warmed further because it is flowing close to arterial blood at body core temperature, the highest in the entire exchanger. The elongate countercurrent nature of the exchanger assures near complete thermal exchange: the arterial blood can be nearly as cold as the foot by the time it exits the exchanger, and the venous blood can be nearly as warm as the core body temperature, instead of each being at some temperature in between.

Although the diameter and wall thickness of the vessels in this type of heat exchanger in the limbs are small and allow heat to flow efficiently from arterial to venous blood, the diameter is too large,

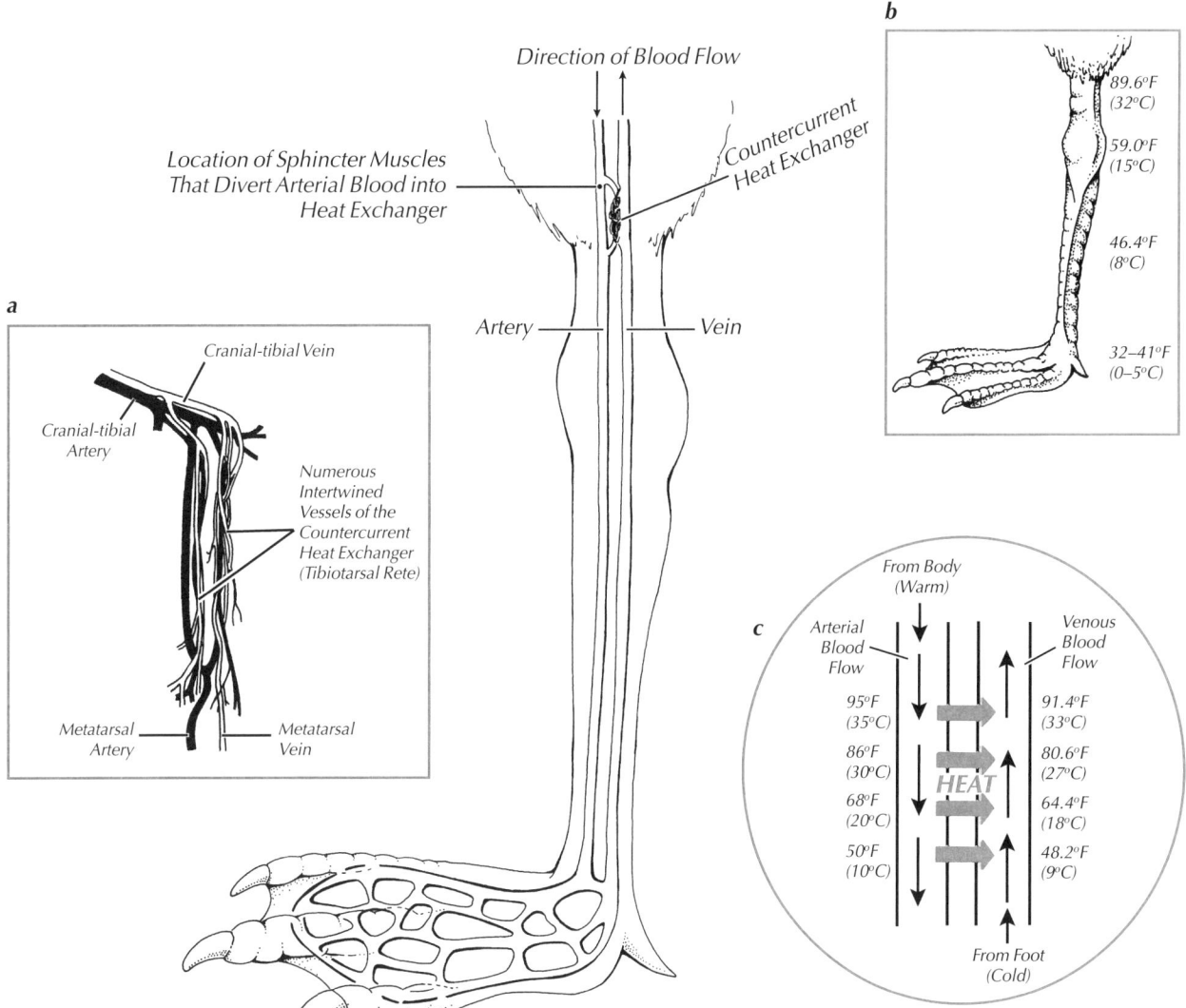

Figure 4–125. Countercurrent Heat-Exchange System in the Leg of a Gull: *Certain birds, such as gulls and waterfowl, have specialized circulatory patterns to reduce the heat lost through their feet when standing on ice or swimming in cold water. In a normal circulatory pattern, arteries become progressively smaller as they approach the tissue being supplied. There, blood passes through a capillary bed in the target tissue, and returns to the heart through a series of increasingly larger veins. In a countercurrent system, the arteries divide into a network or mesh of small vessels before the target tissue is reached, and the veins divide and intertwine with the arterial network after the capillary bed in the target tissue, along the pathway back to the heart. This mesh of intermingling arterioles and venules, which is not embedded among the cells of any target tissue, forms the actual "heat exchanger." In gulls (main drawing) and waterfowl, a countercurrent heat exchanger is located in the upper leg in the region of the lower tibiotarsus, and is technically termed the "tibiotarsal rete." Sphincter muscles located in the artery just below the heat exchanger close to divert blood through the exchanger when external temperatures are sufficiently cold (see Fig. 4–126).* **a. Vessels of the Heat Exchanger:** *A close-up view of the heat exchanger shows the numerous, closely intertwined arterial and venous vessels, formed by the breakup of the cra-* *nial-tibial artery and cranial-tibial vein. The tiny vessels rejoin to form the metatarsal artery and the metatarsal vein, respectively.* **b. Temperature Gradient:** *An example of the temperature gradient of the skin of the leg and foot of a gull standing on ice. The countercurrent heat exchanger allows the bird to keep its body significantly warmer than its feet.* **c. The Mechanism of Heat Exchange:** *Arterial blood arriving from above the heat exchanger is the same high temperature as the core of the body, whereas venous blood arriving from the foot is nearly as cold as the surrounding air. Because of the close proximity of the finely divided blood vessels, heat flows from the arteries into the veins along the length of the heat exchanger, ensuring that the venous blood will be warmed before returning to the rest of the body. The key to the warming power of the system is that the blood flows in opposite directions ("countercurrent flow") in the two vessels: note that even toward the proximal end (top) of the heat exchanger, where the venous blood is fairly well warmed, it continues to receive heat from the arterial blood because it passes near the warmest arterial blood at this point. The numbers illustrate a possible temperature gradient for the blood in each vessel, and are after Campbell (1990). Drawings by Christi Sobel.*

and the walls too thick, to allow oxygen, nutrients, ions, or wastes to diffuse efficiently—so these things are not "exchanged." Diffusion of chemicals requires a very great blood vessel surface area compared to the volume flowing through the vessel, which occurs only with very thin-walled, small-diameter vessels. Thus the limb does not suffer deprivation of oxygen or nutrients.

In addition, both the countercurrent exchanger and the normal arterial system supplying the lower limb are regulated by sphincter muscles in the artery walls. As a result, blood can either be passed through the exchanger to maximize heat conservation or, in times of overheating, can entirely bypass the exchanger, flowing instead through the standard circulatory circuit to the feet to cool the body, using the feet as radiators **(Fig. 4–126)**.

Other types of countercurrent exchangers, with functions other than heat transfer (as in the salt-excreting glands, in the nitrogen-excreting kidneys, and in the parabronchi of the lungs [see Fig. 4–81]), have different characteristics appropriate to their function. The diameter and wall thickness of the vessels may be different, non-blood vessels may compose the network (as in the kidneys and lungs), and so forth. But in all cases, the basic principle of efficiency of exchange through intimate countercurrent flow is the same.

Cooling

Getting rid of excess heat is a problem because birds, unlike mammals, do not have sweat glands. The primary way that animals cool themselves is through evaporation of water from the surface of the

Figure 4–126. How Blood is Diverted into the Heat Exchanger: Blood is diverted into the heat exchanger in a bird's leg (see Fig. 4–125) by sphincter muscles located just beyond the junction between the artery supplying the lower leg and the artery supplying the heat exchanger. In warm weather, when conservation of heat is unnecessary, the sphincter muscles relax, allowing arterial blood from the body to bypass the heat exchanger and proceed to the lower leg, where the heat is lost. In cold weather, the sphincter muscles contract, forcing arterial blood from the body through the heat exchanger, thereby conserving heat.

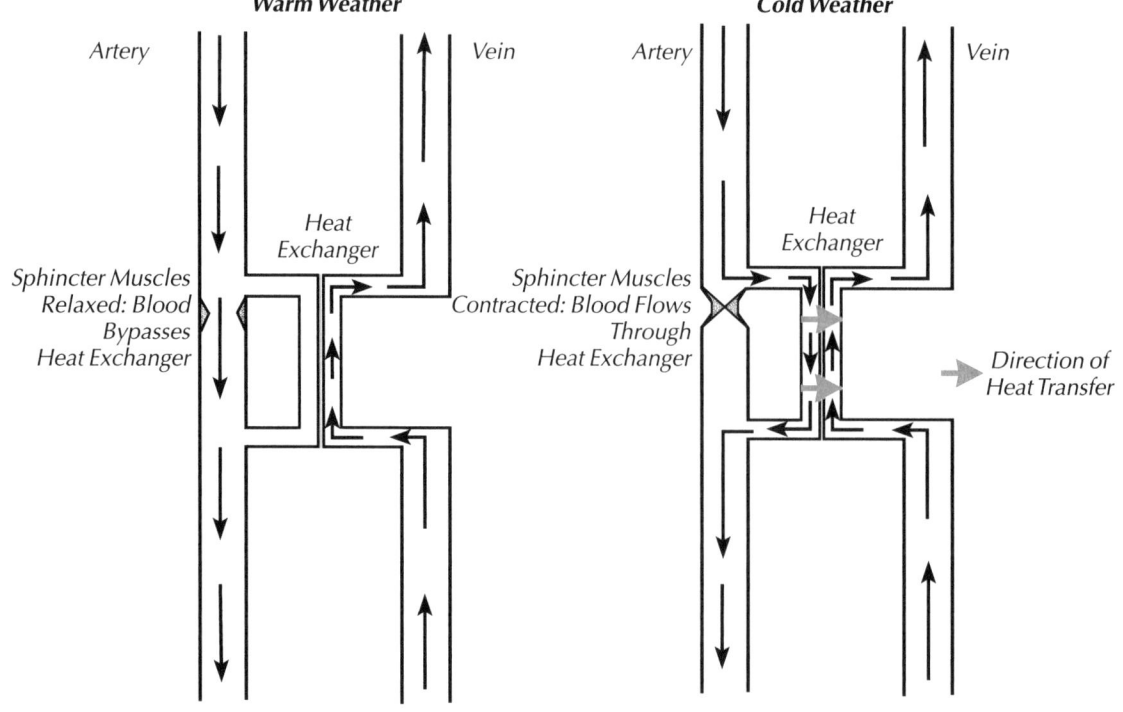

body, which cools the blood flowing just beneath the skin. Some birds open their mouths and **pant**. In this way, the bird increases the area of the body exposed to the air and the amount of air moving across the exposed skin, enhancing the loss of heat. The evaporation of water vapor from the lungs and air sacs also takes away heat. Pelicans, cormorants, herons, owls, and nighthawks have an even more efficient cooling method, called **gular fluttering (Fig. 4–127)**. They open their mouths wide and vibrate the thin, expansive gular membranes of the throat. The movement increases the blood supply in the throat and exposes an even larger featherless area to moving air, thus accelerating heat loss. The blood also loses heat as it flows through any featherless areas on the head, body, or legs. Young pelicans, in addition to gular fluttering, may stand in shallow water during the hottest part of the day, the blood passing through their enormous webbed feet, cooling as it flows.

Torpor

In 1946, a California ornithologist found a Common Poorwill in a rock crevice during the winter **(Fig. 4–128)**. When he picked it up he could detect no heartbeat or respiration, yet the bird was not dead. The bird was torpid (hibernating). Its cloacal temperature was between 64 and 67° F (17.7 and 19.4° C), whereas a Common Poorwill's normal temperature is 106° F (41.1° C). Over an 88-day period, during which the air temperature was around 42° F (5.6° C), the ornithologist handled the bird at two-week intervals, replacing it in the crevice each time. It remained motionless during this entire period, and its cloacal temperature continued to be near 64° F. The weight of the bird remained about the same. After 12 weeks the bird awakened and flew off, just as environmental temperatures began to rise and insects began to appear. When a bird or mammal goes into a profound state of sleep,

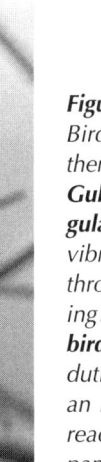

a. Cormorant

Figure 4–127. Avian Cooling Methods: Birds use a variety of methods to cool themselves in hot weather. ***a. Cormorant Gular Fluttering:*** *A cormorant performs* ***gular fluttering****, holding its bill open and vibrating the thin gular membranes of its throat, in order to dissipate heat. Drawing by Charles L. Ripper.* ***b. Eastern Kingbird Cooling Off:*** *Unable to abandon her duties at the nest on a hot summer day, an incubating female Eastern Kingbird reacts to heat stress in several ways: by panting, by raising her body out of the nest, and by elevating her wings slightly, thereby exposing her legs and wings to any breeze. Photo by Marie Read.*

b. Eastern Kingbird

*Figure 4–128. Common Poorwill: Certain birds, such as swifts, hummingbirds, and nighthawks and their relatives—including the Common Poorwill shown here—are known to sometimes enter a state of **torpor** at night or during cold weather. A torpid bird's temperature drops and its metabolic processes and reactivity slow profoundly, allowing it to conserve energy when food is not available. Photo courtesy of Don and Esther Phillips/CLO.*

allowing its body temperature to drop with a consequent slowdown of all metabolic and stimulus-reaction processes, it is said to be in a state of **torpor**.

Poorwills, as determined from later tests, become torpid in temperatures between 35.6 and 66° F (2 and 18.9° C), and in that state use only 0.35 ounces (10 grams) of stored fat in 100 days. We now know that several species of swifts and hummingbirds, and the Lesser Nighthawk, may also regularly enter a torpid state. At night, in a torpid, predator-vulnerable metabolic condition, the body temperatures of both Anna's and Allen's hummingbirds decrease, and both species use less than 3 cc of oxygen per gram of body weight per hour, a two-thirds or more reduction below their basal metabolic rates (10.7 to 16.0 cc oxygen) **(Fig. 4–129)**. By going into torpor, a bird conserves energy when food resources are unavailable. It is not a strategy without risks, however. A coyote finding the above-mentioned poorwill would not have replaced it in its rocky crevice!

Heart Size and Heart Rate

Birds' hearts are larger than those of most mammals of comparable size. In most birds, the greater the body weight, the smaller the heart in proportion to the weight. In the Ostrich and Sandhill Crane, the heart is less than one percent of the body weight; in hummingbirds, the heart may be as much as 2.75 percent of the body weight. A relatively larger heart is necessary for smaller birds, which are generally more active and have a far more rapid metabolism, in part owing to their greater rate of heat loss. In some species, males have larger hearts than females. Heart weight is greater in relation to body weight among species that inhabit higher elevations (with lower temperatures and oxygen concentrations) and higher latitudes (lower temperatures) compared to their lowland or more equatorial kin.

Similarly, the heart rate increases with a decrease in body size **(Table 4–1)**. The resting heart rate of the Ostrich and cassowaries is

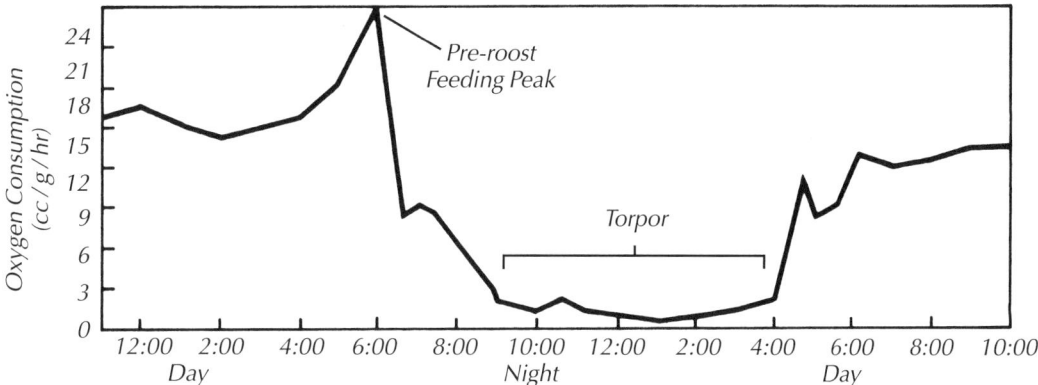

about 70 beats per minute; of hummingbirds, about 615 beats per minute—the fastest in the bird world. In most birds, the heart beat is considerably faster than in mammals of comparable size.

The rate of circulation of the blood and the general metabolic rate increase during flight to meet the oxygen and nutrient demands of an active body. Because the flapping flight of falcons requires more energy than the soaring flight of vultures and hawks, falcons have a faster heart rate and relatively larger hearts than the soaring raptors. In contrast, in the Great Black-backed Gull, the Mallard, and possibly other birds, the volume of blood pumped per heart beat increases during flight, instead of the number of beats per minute increasing. This, of course, also increases the blood pressure.

The heart rate also varies between individuals of a species, depending on the individual bird's activity and body temperature, and the surrounding air temperature. A three-day-old House Wren, essentially cold-blooded at that age, has a heart rate of 121 beats per minute at 70° F (21.1° C), 320 beats at 90° F (32.2° C), and 411 beats per minute at 100° F (37.8° C). Many small birds, when active, double or even triple the heart rate of their resting state.

Figure 4–129. Fluctuations in the Metabolic Rate of a Male Anna's Hummingbird: The metabolic rate, measured as oxygen consumption (cc per gram of body weight per hour), of a single, captive Anna's Hummingbird varies throughout a 24-hour period. During the daytime, the bird alternately rests on a perch and hovers briefly to feed, and its metabolic rate is relatively steady. At around 6:00 P.M. there is a pre-roosting peak of feeding activity, after which the metabolic rate declines precipitously, indicating a period of torpor that lasts through the night. Around 4:00 A.M. metabolic rate again rises as the bird becomes active. By going into torpor, the hummingbird, with its energy-intensive foraging strategy, conserves energy when food resources are temporarily unavailable. Adapted from Welty and Baptista (1988, p. 132); originally from Pearson (1953).

SPECIES	Resting Heart Rate (Beats/Minute)
Mouse	700
Hummingbird	615
Shrew	600
American Robin	570
Black-capped Chickadee	520
White (Laboratory) Rat	350
American Crow	342
Duck	240
Rock Dove	200
Mourning Dove	165
Rabbit	150
Turkey Vulture	132
Dog	100
Domestic Turkey	93
Human	70
Ostrich	60–70
Elephant	25

Table 4–1. Resting Heart Rates of Selected Birds and Mammals: A comparison of the heart rates of various birds shows that as body size decreases the resting heart rate tends to increase. At opposite extremes are the Ostrich at 60 to 70 beats per minute and the hummingbird at 615 beats per minute. A similar relationship between body size and heart rate holds for most mammals. The heart rate of a bird is often higher than that of a mammal of comparable size, however. For example, compare the duck heart rate of 240 beats per minute to that of the similar-sized rabbit at 150 beats per minute.

Blood pressure, like heart rate, also tends to be higher in birds than in mammals. The mean arterial pressure in humans is equal to that of a column of mercury 100 millimeters high. The mean blood pressure of the Rock Dove is 135 mm; the American Robin, 118 mm; and the European Starling, 180 mm. Extreme fright in birds may increase the blood pressure so much that the aorta or atria rupture, resulting in death.

Respiratory Rate

Birds need a great deal of oxygen and food to sustain their high metabolic rate. Although the oxygen requirements of most birds are higher than for most mammals, the respiratory (breathing) rates of birds are slower than for mammals of comparable weight. This results primarily from the more efficient avian respiratory system. The breathing rate of domestic turkeys and chickens at rest is 16 to 38 breaths per minute, about the same as in humans. In smaller birds, however, the rate varies from 45 breaths per minute in the Northern Cardinal to over 80 in the House Wren. The respiratory rates of sleeping Black-capped Chickadees vary from 65 per minute with the air temperature at 50° F (10° C), to 95 per minute with the air temperature at 89° F (31.7° C). In contrast, nestling European Swifts, when in a torpid state during bad weather, may have a respiratory rate as low as eight breaths per minute.

Oxygen requirements and respiratory rate increase in flight. The respiratory rate of the House Sparrow increases from 50 breaths per minute when at rest to 212 when in flight. This increased respiratory rate accompanies an increased rate of heat production from muscular activity, but also an increased rate of heat loss—because much more air is passing more rapidly into and out of the lungs and air sacs, where heat from the body warms the air and eventually is lost to the atmosphere upon exhalation.

Because the lungs are very close to the flexible rib cage, which deforms with each wingbeat (see Fig. 5–5), it seems as though a bird must breathe in and out with each wingbeat. This does occur in pigeons: they inhale on each upstroke and exhale on each downstroke. This is not true, however, for all birds. In flight experiments, some birds with bills encased in rubber balloons demonstrated that breathing during flight is irregular and that they may inhale on either the upstroke or downstroke. In other experiments, a flying Western Gull had an average of 242 wingbeats and 81 breaths per minute; a flying Lesser Scaup had an average of 645 wingbeats and 140 breaths per minute, and a Red-tailed Hawk beat its wings 13 times without taking a breath.

Birds that dive have special respiratory problems. Because they must reduce buoyancy, it seems logical that they exhale upon diving, reducing the air in the lungs and air sacs and thus their buoyancy. Deprived of respiratory gas, the heart and oxygen consumption rates slow. Penguins, in simulated underwater conditions, reduce their consumption of oxygen by 20 or 25 percent. Common Loons are known to survive underwater for at least 15 minutes, and have been underwater long enough to tangle themselves in fish nets more than 180 feet deep.

Both measures are indications of just how dramatically loons can re-duce their short-term need for oxygen **(Fig. 4–130)**.

Figure 4–130. Common Loon Forages Underwater: Special respiratory challenges face birds that dive underwater to find their food. They must balance reducing their buoyancy, which requires exhaling upon diving, with maintaining an oxygen supply for their metabolic needs while underwater. The Common Loon is able to reduce its oxygen requirements dramatically, allowing it to remain underwater for as long as 15 minutes, and to dive as deep as 180 feet below the surface. Photo by Tomas Cajacob/Minnesota Zoo.

Water and Salt Regulation

The correct proportions of salts and water must be maintained within the cells, in the spaces between the cells, and in the blood. If there are too many salts, the body retains water in an attempt to dilute the salt concentration; too few salts, and dehydration occurs. If the wrong proportions of different salts exist, cell membrane function becomes impaired. The correct balance is critical to the survival of the cells and therefore of the animal itself. The system that regulates salt balance, however, is exceptionally complicated. Nevertheless, we must mention one particularly interesting feature of birds with respect to salt balance.

Birds have a challenge in conserving water; they must make up for water lost as vapor from the lungs and air sacs. This conservation becomes especially critical as respiration rate increases. The skin, feathers, and scales all help in preventing the loss of body fluids from the general body surface. The greatest conservers of water within a bird, however, are the kidneys, which resorb most of the water from the urine before it passes from them.

Figure 4–131. Location of the Salt Glands in Marine Birds: Seabirds that drink only salt water have special salt-excreting glands to remove excess salt from their bodies. These paired salt glands (colored areas) are located on top of the head, each in a shallow depression in the skull above or adjacent to the eye, or in some species within the orbit of the eye. Like the kidneys, they remove salt from the bloodstream and concentrate it. The salty fluid produced by the salt glands flows through ducts into the nasal cavity, and then through the nostrils or mouth to the outside. In gulls, the salt solution emerges from the nostrils and drips from the tip of the bill, whereas in cormorants, it flows along the roof of the mouth to the tip of the bill. The pelican has grooves along the upper surface of its bill that channel the fluid to the bill tip, preventing it from entering the bird's pouch and being re-ingested. Each of these birds simply shakes the salty liquid off of its bill tip. In petrels, however, the fluid is forcibly ejected ("sneezed") out of the bird's tubular nostrils. Drawing by Eric Mose, from Schmidt-Nielsen (1959). Used with permission.

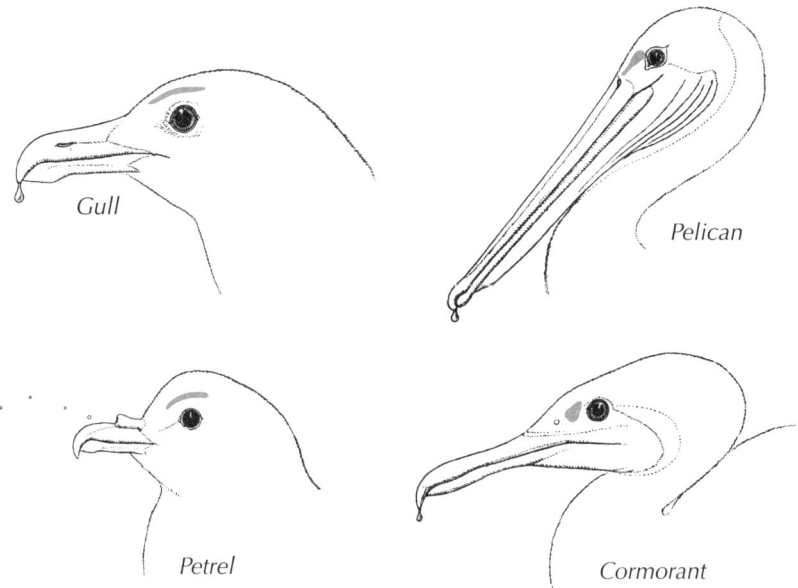

The avian process of excess salt elimination was once puzzling. The kidneys and sweat glands provide this function in mammals, but birds have no sweat glands, and the kidneys eliminate only a portion of the salts. In seabirds that drink only salt water, the kidneys secrete urine that is only half as salty as the seawater they have been drinking. Where does the rest of the salt go?

Scientists have long known that birds eliminate excess salts through salt glands located either in a skull surface depression above the eye, or within the orbit **(Fig. 4–131)**. The excess salts, dissolved in clear fluid, flow from the glands through ducts into the nasal cavities. By way of the nostrils or mouth, the salty fluid then flows to the tip of the bill in droplets that the bird shakes off. This excretion can be seen most readily in seabirds. Several land birds, such as the Australian Budgerigar, have large salt glands that allow them to drink from brine pools in their native habitats without accumulating salts in their bodies **(Fig. 4–132)**.

Figure 4–132. Budgerigars at a Waterhole: The gregarious, nomadic Budgerigar exploits any available water source in its arid native habitat—the deserts of the Australian interior—even drinking from brine pools. It is one of several land birds with large salt glands that prevent salt build-up in its body. Here, a huge flock of Budgerigars descends on a waterhole. Photo by Andrew Henley/Biofotos.

Life Span and Senescence

When compared to mammals, and considering their body weight, birds are generally long-lived **(Table 4–2)**. The Japanese Quail has one of the shortest life spans, at 2 to 5 years. Apparently many birds have evolved mechanisms to protect against rapid aging. Parrots have certainly found the "fountain of youth." As a group they have the longest absolute life spans of any bird, African Grey Parrots living to 60 or 70 years, and macaws of the genus *Ara* living to as much as 90 years! These ages are more than four times as old as would be predicted from their body mass, compared to a mammal. The comparative longevity of birds is surprising, because they have high body temperatures, high blood glucose levels, and high metabolic rates. All of these conditions tend to limit the mammalian life span.

Senescence (aging) occurs as a result of several factors that are not well understood, but it must be related to the breakdown of cellular protective mechanisms. These mechanisms include cells being able to maintain and repair their external and internal membranes, and to successfully divide in a controlled (nonmalignant) manner. Birds may prove very interesting in aging research.

Major Anatomical Differences between Birds and Mammals

■ The table on the following pages contrasts some of the structures seen in present-day birds and mammals to illustrate how they differ. Explanations of the differences cited can be found in the discussions of each organ system in the text or in the figures.

Table 4–2. Maximum Life Span of Selected Wild Birds: *This list presents the maximum known age (in years and months) of a selection of North American birds, determined from banding data from the Bird Banding Laboratory at the Patuxent Wildlife Research Center in Laurel, Maryland. The maximum age is calculated as the difference between the age of the bird at banding and either the date of its subsequent live recapture or the date of the recovery of its band, if the bird died. As of this writing, further longevity information is available from the following Internet web site <**www.pwrc.usgs.gov/bbl/homepage/longvrec.htm**>*

Species	Maximum Age (Years-Months)
Greater Roadrunner	3-09
Blackpoll Warbler	4-03
Golden-crowned Kinglet	5-04
Carolina Wren	6-02
Yellow-bellied Sapsucker	6-09
Western Kingbird	6-11
Roseate Spoonbill	7-09
Western Tanager	7-11
Eastern Bluebird	8-00
Barn Swallow	8-01
Burrowing Owl	8-08
Ruby-throated Hummingbird	9-01
Common Yellowthroat	11-06
Sanderling	12-01
Black-capped Chickadee	12-05
Loggerhead Shrike	12-06
Wild Turkey	12-06
Common Loon	12-11
House Sparrow	13-04
American Kestrel	13-07
Purple Martin	13-09
Red Knot	13-11
American Crow	14-07
Northern Cardinal	15-09
Hairy Woodpecker	15-10
Blue Jay	17-06
Whooping Crane	18-10
American Woodcock	20-11
American Coot	22-04
Bald Eagle	22-09
Great Blue Heron	23-03
Trumpeter Swan	23-10
Red-tailed Hawk	25-09
Mallard	26-04
American White Pelican	26-05
Ring-billed Gull	27-03
Great Horned Owl	27-07
Canada Goose	28-05
Leach's Storm-Petrel	31-01
Mourning Dove	31-04
Atlantic Puffin	31-11
Arctic Tern	34-00
Laysan Albatross	42-05

BIRDS	**MAMMALS**

Skeleton

• light, air-filled (pneumatic) bones	• heavy, marrow-filled bones
• skull has cranio-facial hinge: moveable upper jaw	• no cranio-facial hinge: non-moveable upper jaw
• single occipital condyle for skull	• double occipital condyles
• few sutures visible on skull	• distinct skull sutures
• jaw articulation is quadrate to articular bone	• jaw articulation is dentary to temporal bone, the quadrate having become the incus and the articular having become the malleus of the middle ear
• vertebral regions are variously fused	• vertebrae are distinct
• sternum large and keeled	• sternum small and segmental
• forelimb with three digits	• forelimb usually with five digits
• fusions of limb bones as: a carpometacarpus, tibiotarsus, and tarsometatarsus	• separate carpus, metacarpus, tibia, tarsus, and metatarsus
• pelvic symphysis absent	• pelvic symphysis present
• "wishbone" (fused clavicles and interclavicle) present	• clavicles, if present, not fused as "wishbone"
• coracoid bone acts as a strong brace of the shoulder	• coracoid bone absent
• formula for the number of phalanges in digits 1 to 4 of pelvic limb is 2–3–4–5	• formula for the number of phalanges in digits 1 to 5 is 2–3–3–3–3
• functional ankle joint is *intra*tarsal	• functional ankle joint is *inter*tarsal
• pubis directed caudally	• pubis directed cranially

Muscles

• breast musculature massive	• breast musculature small
• limb tendons often ossified	• limb tendons rarely ossified
• dorsal vertebral muscles reduced and of little in function	• dorsal vertebral musculature of great function locomotion

Nervous System

• cerebral cortex thin; corpus striatum large (serves as major integrative region)	• cerebral cortex thick (serves as major integrative region); corpus striatum relatively small
• corpus callosum (a major connection between the two hemispheres of the brain) lacking	• corpus callosum present
• cerebrum smooth	• cerebrum has folds and grooves
• mesencephalic optic lobes are largest	• telencephalic optic lobes are largest
• small olfactory lobes	• large olfactory lobes
• few taste buds	• many taste buds
• glycogen body in spinal cord	• no glycogen body

BIRDS	**MAMMALS**

Ear

- no external ear
- tympanic membrane convex (curved outward)
- single auditory ossicle: columella (stapes)

- inner ear surrounded by pneumatic bone
- short cochlea

Eye

- shape of eyeball flat to tubular
- bony ossicles in sclera
- ciliary processes attach to lens
- ciliary muscle contraction forces lens to become round by squeezing it

- both lens and cornea change shape to focus object
- pecten present in vitreous chamber

Circulatory System

- aorta derived from right 4th arch
- right and left precavae present
- nucleated red blood cells
- few, if any, lymph nodes
- lymph "hearts" in tail region
- lymphatic erection of phallus
- two portal systems (renal and hepatic)

Respiratory System

- epiglottis absent
- vocal cords absent
- thyroid cartilage absent
- tracheal rings complete
- syrinx present
- small, compact lung
- lung not expansible
- anastomosing parabronchi
- air sacs present
- no diaphragm

MAMMALS

Ear
- large external ear
- tympanic membrane concave (curved inward)
- three auditory ossicles: malleus, incus, and stapes
- inner ear surrounded by dense bone
- long, coiled cochlea

Eye
- eyeball spherical
- no sclerotic bones
- ciliary processes not attached to lens
- ciliary muscle contraction relaxes the lens, allowing it to become round by elastic rebound
- only lens changes shape
- no pecten present

Circulatory System
- aorta derived from left 4th arch
- usually only right precava present
- nonnucleated red blood cells
- many lymph nodes
- no lymph "hearts"
- arterio-venous erection of penis
- only hepatic portal system

Respiratory System
- epiglottis present
- vocal cords usually present
- thyroid cartilage present
- tracheal "rings" open dorsally
- no syrinx
- large, spongy lung
- lung greatly expansible
- dead-end alveoli
- air sacs lacking
- strong diaphragm

(Continued on p. 4·162)

BIRDS

Digestive System
- teeth lacking
- crop usually present
- stomach in two parts: glandular and grinding
- colic ceca usually paired or absent
- cloaca always present

Urogenital System
- kidney recessed in skeleton
- bladder absent
- all egg-laying
- functional ovary only on left side
- left oviduct functional
- mammary glands absent
- testes always internal
- cloacal phallus in some
- female heterogametic (ZW)
- embryo derives nutrition from yolk

MAMMALS

- teeth usually present
- crop absent
- stomach usually single, never a grinding portion
- cecum usually single or absent
- cloaca rarely present

- kidney not recessed in skeleton
- bladder present
- only monotremes lay eggs
- functional ovary on both sides
- both oviducts functional
- mammary glands present
- testes usually external
- external penis except in monotremes
- male heterogametic (XY)
- nutrition from placenta except in monotremes

Suggested Readings

Darling, Lois and Louis. 1962. *Bird*. Boston: Houghton Mifflin Co.

Evans, H. E. 1996. Anatomy of the Budgerigar and other birds. In *Diseases of Cage and Aviary Birds*. W. J. Rosskopf and R. W. Woerpel, editors. Baltimore: Williams & Wilkins, Inc.

King, A. S. and J. McLelland. 1984. *Birds: Their Structure and Function, Second Edition*. London: Bailliere Tindall.

King, A. S. and J. McLelland. 1979–1989. *Form and Function in Birds*. Four Volumes. London: Academic Press.

Leahy, C. 1982. *The Birdwatcher's Companion*. New York: Bramercy Books (Random House Value Publishing, Inc.).

Pough, F. H., C. M. Janis, and J. B. Heiser. 1999. *Vertebrate Life, Fifth Edition*. Upper Saddle River, New Jersey: Prentice Hall.

Proctor, N. S. and P. J. Lynch. 1993 *Manual of Ornithology*. New Haven , CT: Yale University Press. 340 pages.

Van Tyne, J. and A. J. Berger. 1971. *Fundamentals of Ornithology*. New York: Dover Publications, Inc.

Waldvogel, J. A. 1990. The bird's eye view. *American Scientist* 78(4):342–353.

Welty, J. C. and L. Baptista. 1988. *The Life of Birds, Fourth Edition*. Orlando, FL: Saunders College Publishing. 581 pages.

5

Birds on the Move: Flight and Migration

Kenneth P. Able

 Ten miles off the southern coast of New Zealand, I was huddled in the lee of a ship's cabin, my arm wrapped around a railing to prevent being thrown overboard or swept away by the waves crashing over the stern of the boat. On this, my roughest sea voyage, with the ship plunging up and down the waves, I had trouble holding the binoculars steady: the view oscillated between blank walls of water and the sky. The conditions were not conducive to bird watching.

But birds there were. Hundreds of them. Shearwaters and albatrosses for the most part—this world, so alien and difficult for me, was their everyday habitat. Into the gale they flew, effortlessly angling upward, then swooping down among the swells. So perfectly were they adapted that they made it look easy **(Fig. 5–1)**. Shipbound and nauseated, I envied them. Like many before, I wished that I could briefly do what they were doing, feel the world as they felt it, and sense in my muscles the kind of exquisite control of movement they must experience. Impossible, of course, and so, like those of generations before, I simply looked with awe and enjoyed vicariously the splendor of their flight.

The power of flight is the quintessential characteristic of birds, the central adaptation around which many of the most interesting aspects of avian anatomy, physiology, and behavior have been molded.

Figure 5–1. Laysan Albatross: With grace and elegance, an albatross glides low over the water on long, narrow wings. The Laysan Albatross, pictured here, ranges throughout the northern Pacific Ocean, breeding mostly on coral atolls in the Hawaiian Islands. Photo by P. La Tourrette/VIREO.

The birds I was watching from the ship spend most of their lives on the wing. Not only do they fly magnificently in difficult conditions, they travel vast distances, some of them literally circumnavigating the globe. These two characteristics—the power of flight and their incredible migrations—are the main reasons why birds have fascinated people down through the ages.

Although some species of birds have lost the ability to fly during the course of evolution, all modern lineages of birds arose from flying ancestors. In the 450 million years since life first emerged on land, powered flight has evolved in only two phyla, the arthropods (insects, crustaceans, and their relatives) and the chordates (vertebrates, tunicates, and lancelets). Insects, birds, bats, and the extinct pterosaurs (featherless, flying reptiles) **(Fig. 5–2)** all have made flight their major means of moving about. Parachuting and gliding—more limited types of flight—are found in most vertebrate classes (for example, flying fishes, parachuting frogs, gliding agamid lizards, flying squirrels, flying lemurs, and marsupial sugar gliders), and in a mollusc (the flying squid) **(Fig. 5–3)**.

The Flight Syndrome

■ Occupying the fluid medium of air and using powered flight for mobility have required birds to evolve a suite of extraordinary adaptations, which sets birds apart as the most distinctively different class of living vertebrates. These adaptations also constrain their form such that birds, as a group, resemble one another more than do members of most other vertebrate classes. The most notable of these adaptations are briefly described here. Please refer to **Figure 5–4** as you read through adaptations 1 to 7 below.

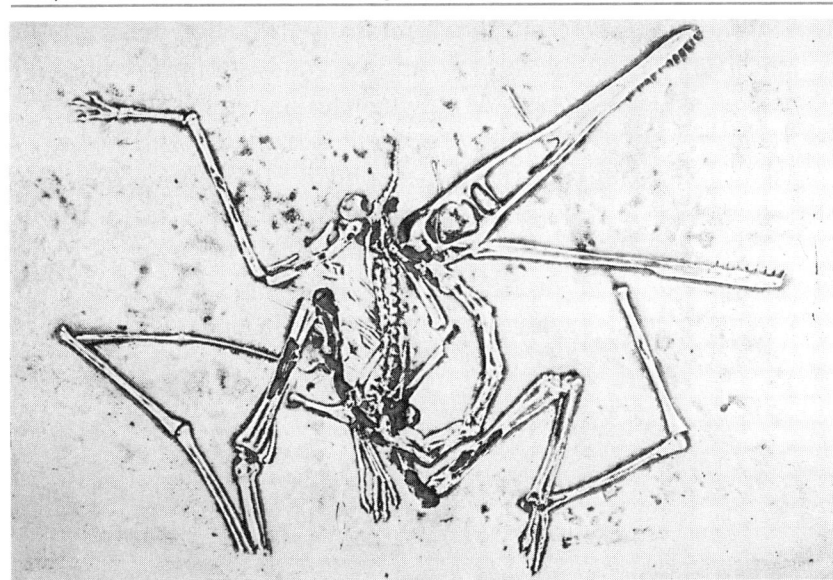

Figure 5–2. Fossil **Pterodactylus:** *Shown here is a fossil of one type of pterosaur (ancient flying reptile), the sparrow-sized* **Pterodactylus.** *The wings of pterosaurs, somewhat like those of bats, consisted of a membrane of skin that stretched from the front limbs to the sides of the body, and possibly to the hind limbs as well. Pterosaurs lived in the Mesozoic era and, like birds, had hollow, long bones; in contrast to flying birds, their sternum had no keel. Note here the long bones of the wings, as well as the flexible neck and long jaws. Photo by GeoScience Features Picture Library.*

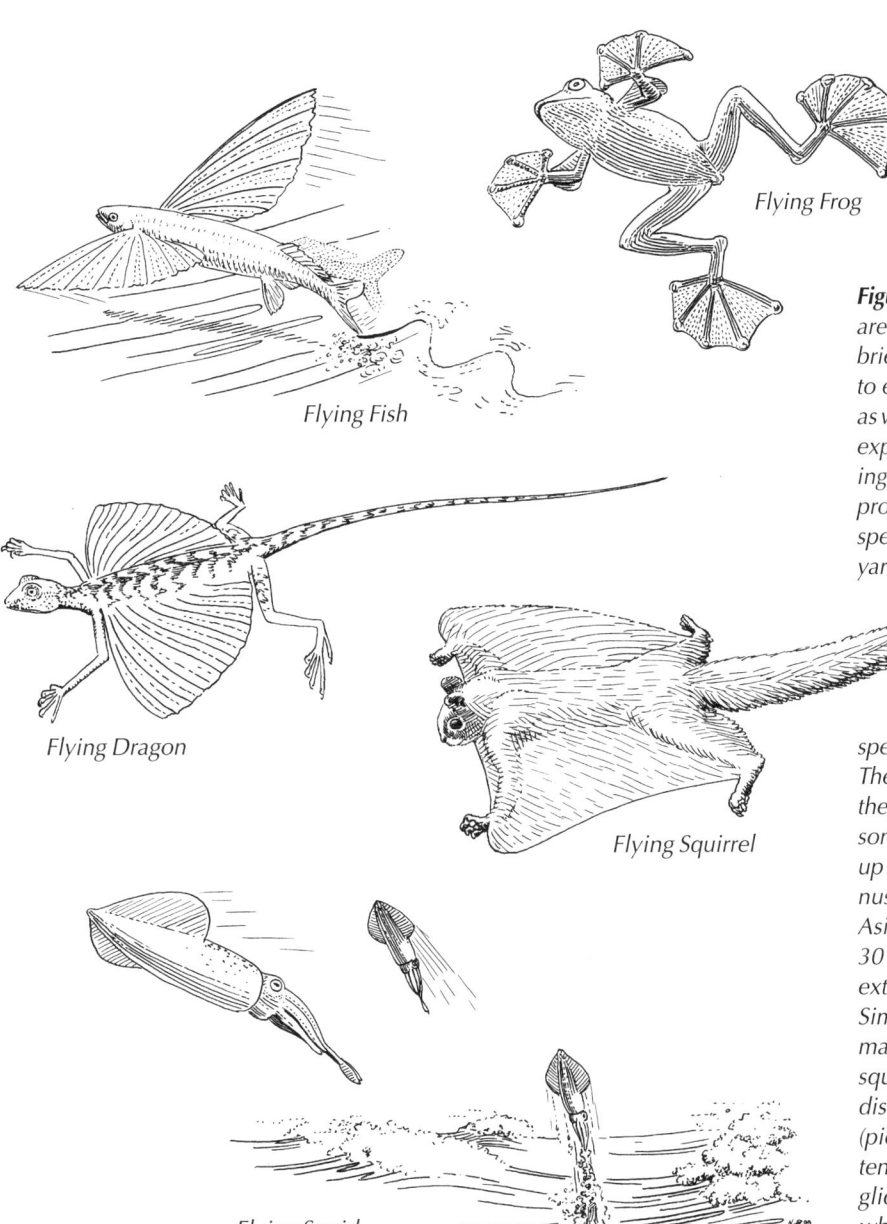

Flying Frog

Flying Fish

Flying Dragon

Flying Squirrel

Flying Squid

Figure 5–3. Gliding Animals: Shown here are a few animals that can glide, at least briefly, through the air. Many use gliding to escape more earth-bound predators, as well as to simply change locations. By expelling a stream of water forward, flying squid (Ommastrephes illecebrosa) project themselves backward with such speed that they often shoot three to four yards out of the water, sometimes even onto the decks of passing ships. The more tropical flying fish (species pictured: Exocoetus volitans) undulate their tails from side to side, first picking up speed underwater, then on the surface. Then, spreading their large pectoral fins, they glide through the air up to 100 yards, sometimes skimming off the surface for up to 400 yards. The flying dragons (genus Draco), forest dwellers of Southeast Asia and the East Indies, can glide for 30 or more yards from the treetops by extending a skin flap between the limbs. Similarly, many small, arboreal mammals, such as the North American flying squirrels (genus Glaucomys), glide great distances between trees. The flying frogs (pictured here, genus Rhacophorus) extend membranes between their toes to glide from treetops, even executing turns while airborne.

Figure 5–4. Avian Adaptations for Flight: *Most of the distinctive features of a typical bird skeleton are adaptations for flight, as demonstrated by this Rock Dove (pigeon) skeleton. Weight is decreased by the hollow bones, light skull, toothless beak, and reduced number of bones in the tail and hand regions. Rigidity is achieved through the fusion of many bones, especially those in the pelvic and hand regions, and in portions of the vertebral column. Drawing by Charles L. Ripper.*

1. A Strong, Light Skeleton: The skeleton of a flying bird must be both light, to enable the animal to fly, and extremely strong, to withstand the stresses placed upon it. As any engineer knows, this combination of characteristics is difficult to reconcile. Compared with the bones of terrestrial vertebrates, bird bones contain much more air. Portions of the hollow long bones of the wings are continuous with the air sacs (part of the bird's respiratory system) and strengthened by series of diagonal internal struts. Both the air sacs and struts are obvious adaptations for flight, and have disappeared in large flightless birds such as Ostriches and Emus.

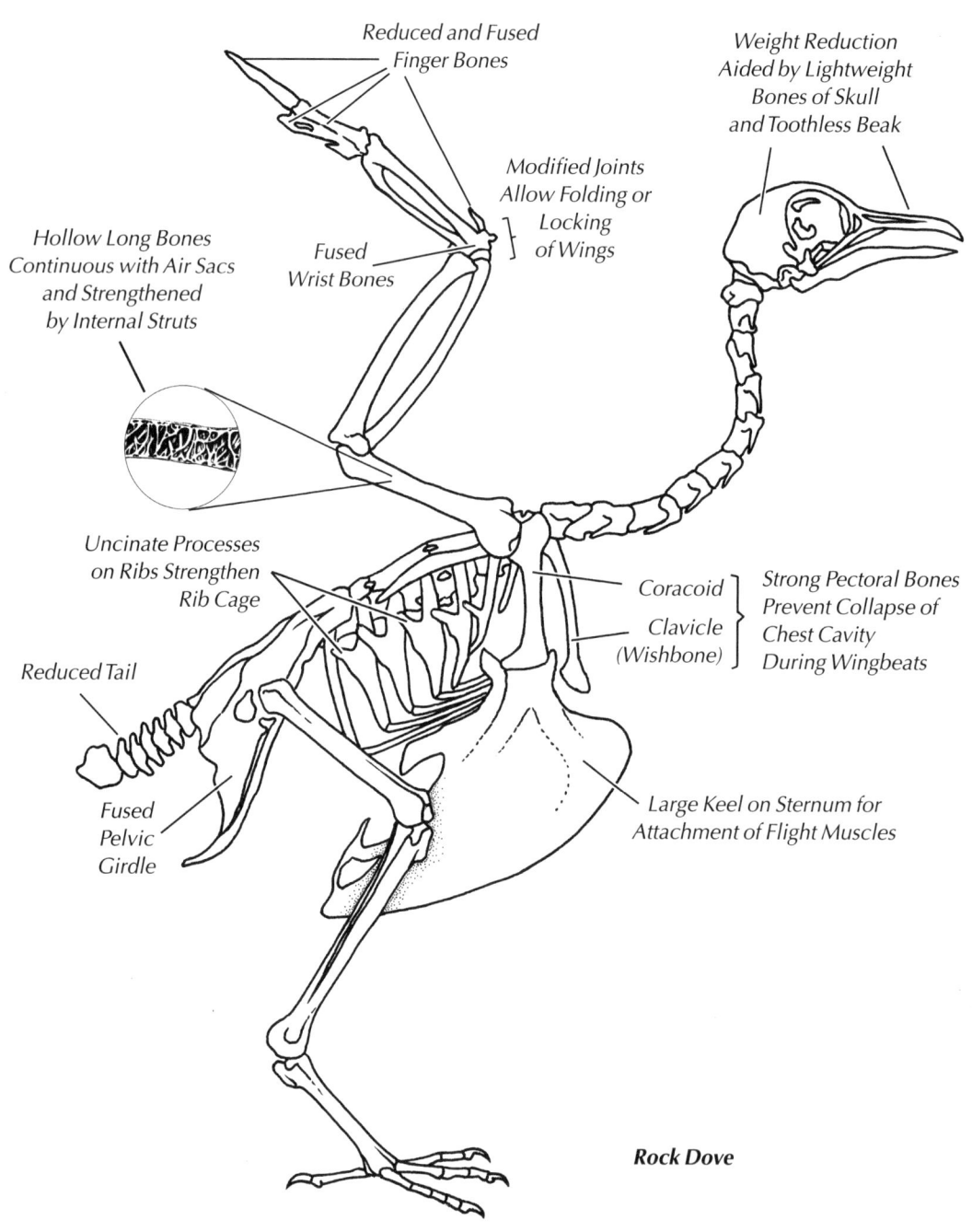

Reduced and Fused Finger Bones

Weight Reduction Aided by Lightweight Bones of Skull and Toothless Beak

Hollow Long Bones Continuous with Air Sacs and Strengthened by Internal Struts

Modified Joints Allow Folding or Locking of Wings

Fused Wrist Bones

Uncinate Processes on Ribs Strengthen Rib Cage

Coracoid

Clavicle (Wishbone)

Strong Pectoral Bones Prevent Collapse of Chest Cavity During Wingbeats

Reduced Tail

Fused Pelvic Girdle

Large Keel on Sternum for Attachment of Flight Muscles

Rock Dove

2. **Reduced Body Weight:** Weight reduction is the theme throughout the avian body. The reptilian teeth have been replaced by a light-weight beak. The small, internal gonads shrink to almost nothing in the nonbreeding season.

3. **A Rigid Skeleton:** Skeletal rigidity is achieved by the fusion of many bones—those of the hand and fingers, most of the wrist bones, and elements in the pectoral and pelvic girdles. Bony **uncinate processes** extend back from the upper part of each rib, overlapping adjacent ribs to reinforce the rib cage.

4. **An Enlarged, Keeled Sternum:** The avian **sternum** (breastbone) is greatly enlarged and has a large **keel** to which the major flight muscles are attached. In poor fliers, the keel and associated flight muscles are smaller, and the sternum of flightless birds lacks a keel altogether (see Fig. 4–19).

5. **Strong Bones in the Pectoral Girdle Prevent Collapse of Chest Cavity During Flight:** The coracoids and the **furcula** (wishbone), formed from a fusion of the clavicles (collar bones), are supporting elements that resist the huge pressures on the chest cavity created by the beating of the wings. High-speed x-ray movies of European Starlings flying in a wind tunnel (Jenkins et al. 1988) show that the furcula bends outward to each side during the downstroke of the wings and recoils like a spring during the upstroke. The sternum moves upward on the downstroke, and downward on the upstroke **(Fig. 5–5)**. This furcular "spring" and sternal "pump" may facilitate the movement of air between the lungs and air sacs independent of breathing. This system is particularly important because birds, un-like mammals, have no muscular diaphragm to drive their breathing apparatus.

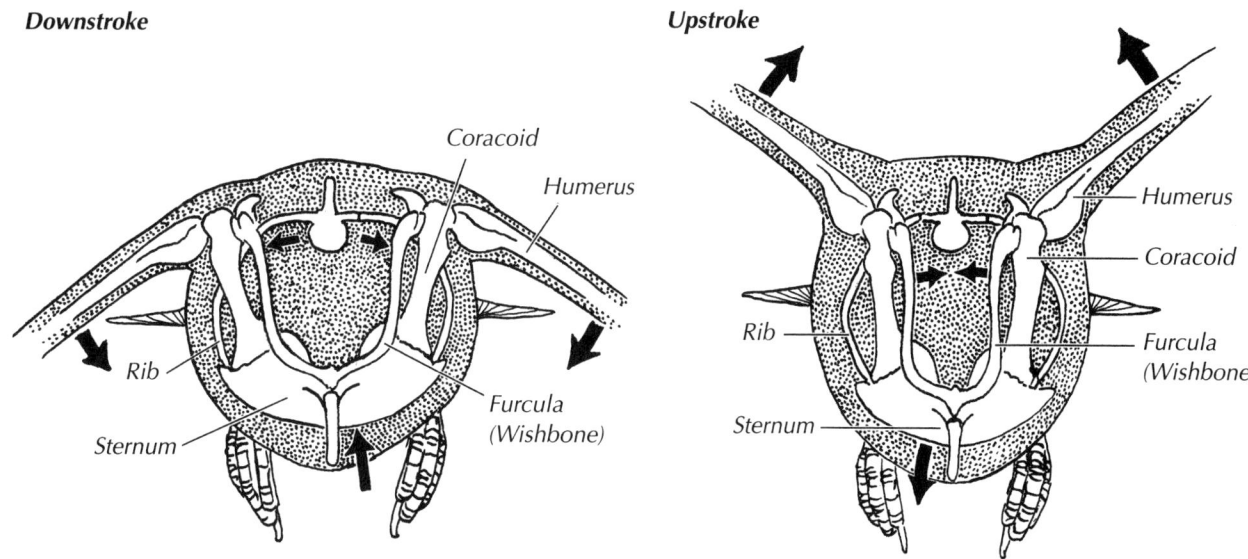

Downstroke   **Upstroke**

Figure 5–5. "Pump and Spring" Mechanism of Furcula and Sternum: *Shown is a cross-sectional view of the pectoral girdle, look-ing head-on toward the tail. Arrows indicate bone movement. During the downstroke, the furcula bends outward to each side and the sternum moves upward; during the upstroke, the furcula recoils inward like a spring, and the sternum moves downward. The action of this furcular "spring" and sternal "pump" probably helps to move air between the lungs and air sacs during flapping flight, supplementing the normal breathing mechanisms.*

a. Downstroke   *b. Upstroke*

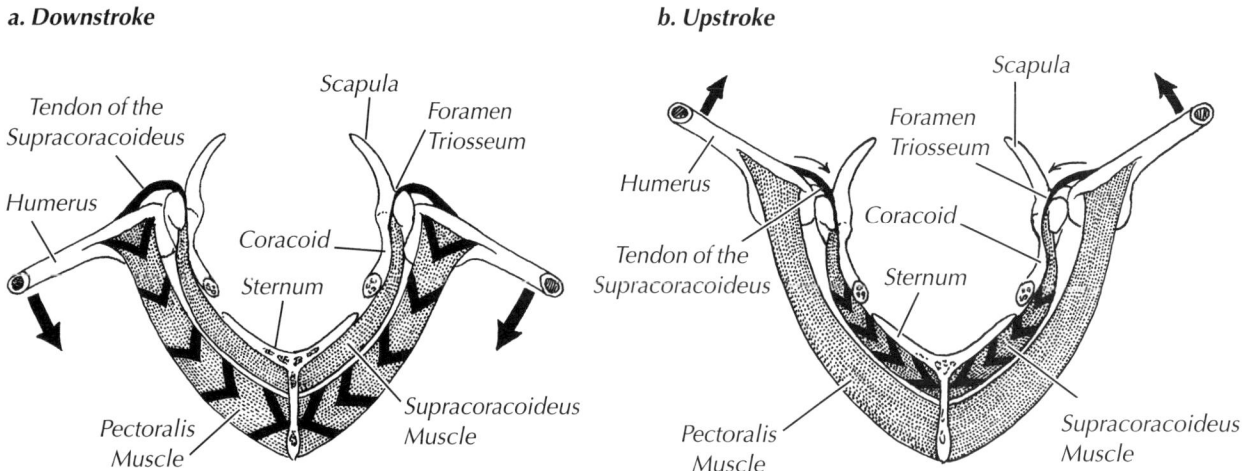

Figure 5–6. Pulley System for Tendon of Supracoracoideus Muscle: *Shown is a cross-sectional view of the large flight muscles and pectoral girdle.* ***a. Downstroke:*** *The pectoralis muscle contracts (as indicated by black Vs), lowering the humerus and thus the wing.* ***b. Upstroke:*** *The supracoracoideus muscle is attached to the dorsal surface of the humerus by a tendon. As a result of its dorsal attachment, when this muscle contracts (black Vs), it raises the humerus and thus the wing. The tendon reaches the top of the humerus by passing through a hole formed by the junction of the coracoid, scapula, and humerus. This hole, termed the* **foramen "triosseum"** *because it is formed by three bones, creates a pulley system that allows the supracoracoideus muscle to remain below the wing and yet function to elevate it. As discussed in Skeletal Muscles, Ch. 4, this weight configuration is much more stable for a flying bird than what would result if the supracoracoideus muscle were located on the back above the wings.*

6. **Modified Joints in Wing Allow Folding or Locking:** The bird wing contains the same bones found in the human forearm, but they are greatly modified. In addition to much fusion of individual bones, which imparts strength to the limb, the joints are modified to permit each wing to fold neatly when the bird is at rest, or to lock rigidly to resist the forces acting upon it during flight.

7. **Large, Powerful Flight Muscles:** A bird's major flight muscles are the **pectoralis** and **supracoracoideus (Fig. 5–6)**. The larger pectoralis is proportionately the most massive paired muscle in any four-limbed animal. It accounts for as much as 15 to 25 percent of a flying bird's total body mass. These large muscles and a host of smaller ones not only provide the power to fly, but permit exquisite control over all aspects of wing movement.

For ease of discussion, each of these adaptations has been treated as a more or less independent entity. Actually, they are intimately related parts of a package that we may call the flight syndrome. Even though natural selection works through details—a small change in the mass of the pectoralis muscle, say, or in the length of a feather—it is the individual as a whole that natural selection works against or favors. In other words, it is how well the bird flies under natural conditions—which House Finch escapes from the Cooper's Hawk and which doesn't—that determines which characteristics natural selection sustains. In this way, natural selection has sculpted most birds into exquisite flying machines. Note, however, that although birds surely look like perfection in the air, natural selection doesn't produce perfection. It selects what works best from among the existing variations.

Functions of the Flight Muscles

■ Both of the major flight muscles, the pectoralis and the supracoracoideus, have their origin on the sternum and their insertion on the humerus of the wing. The traditional view of the action of these muscles during flight was that the pectoralis pulled the wing down on each downstroke, and the supracoracoideus lifted the wing back up during each recovery stroke.

New studies have shown, however, that the process is considerably more complicated. Some of these studies use a technique (electromyography or EMG) in which fine wires are implanted in a muscle, recording the electrical activity associated with muscle contraction. These analyses have revealed that the pectoralis is divided into two portions which have rather different functions. The larger portion serves as the primary depressor of the humerus, but also slows down the wing at the end of the upstroke and pulls the wing forward. The smaller portion also acts as a depressor, but unlike the larger portion, is positioned to pull the wing backward. The supracoracoideus slows down the wing at the end of the downstroke, and accelerates it at the beginning of the upstroke.

Different species have varied flight requirements that are reflected by numerous adaptive differences in flight muscles. For example, in birds such as hummingbirds, which use the upstroke as well as the downstroke to generate power, the supracoracoideus is quite large. Another example of adaptive difference is that the fibers of bird flight muscles are of several types. The most obvious are the **red fibers**, which we know as "dark meat," and the **white fibers**, which constitute "white meat." Red fibers get their color from their massive capillary beds containing blood and **myoglobin**, a substance within muscle cells that carries oxygen, the energy source for most cells. Using this oxygen (in a process called aerobic respiration), red fibers oxidize fats and sugars to power sustained flight. The pectoral muscles of most long-distance fliers are composed largely or entirely of red fibers, explaining why ducks, for example, have so much dark meat. White fibers are of larger diameter than red fibers, have fewer capillaries associated with each muscle fiber, and contain little myoglobin. White fibers derive their energy from anaerobic ("without oxygen") respiration, and because this process releases a byproduct called lactic acid—the same substance that builds up in the leg muscles of long-distance runners—these fibers fatigue quickly. They are ideal for quick bursts of action, but not sustained exercise. For example, the large amount of white fibers in the breast muscles of turkeys enables them to burst into flight from the ground and climb rapidly to clear trees. However, a turkey cannot fly very far. I once watched a turkey try to fly across about a quarter mile (0.4 km) of open water. It ran out of gas, plopped into the water about 20 yards (18 m) from shore, and swam the rest of the way to land.

5

How Do Birds Fly?

■Since the beginnings of recorded history, members of our essentially earthbound species have been fascinated by the ability to fly. The first attempts at human-powered flight took place in machines designed to replicate the flapping action of bird wings. These **ornithopters** depended on humans flapping their arms to lift the crude wings **(Fig. 5–7)**. Of course, they didn't get off the ground because wing flapping is much more complex than a simple up-and-down movement. In addition, human breast muscles are not proportionally as strong as the breast muscles of birds, and the human body is too heavy and lacks essential streamlining for flight. Physicists have calculated that a 150-pound (68kg) human would require a breastbone projecting 6 feet (1.8 m) forward to support ample flight muscles!

Mythological stories relating the unfortunate consequences of Icarus's attempts to fly are familiar to everyone. But some progress has been made: hang gliding and parasailing bear witness to the power of modern, computer-aided engineering coupled with space-age materials. But we still have to jump off high places, or be dropped from or towed behind airplanes, to get off the ground. Even with modern technology, we have reached only the stage of the gigantic pterosaurs and earliest birds, which could apparently glide but perhaps not initiate powered flight from the ground. Compare these simple feats with those of Ruffed Grouse or even Wild Turkeys, which can burst upward from the ground at a very steep angle, and accelerate over the treetops faster than we can aim our binoculars at them.

Although gliding is an important component of bird flight, in most species the flight process is much more complex—so complicated, in fact, that researchers still do not completely understand it. The fol-

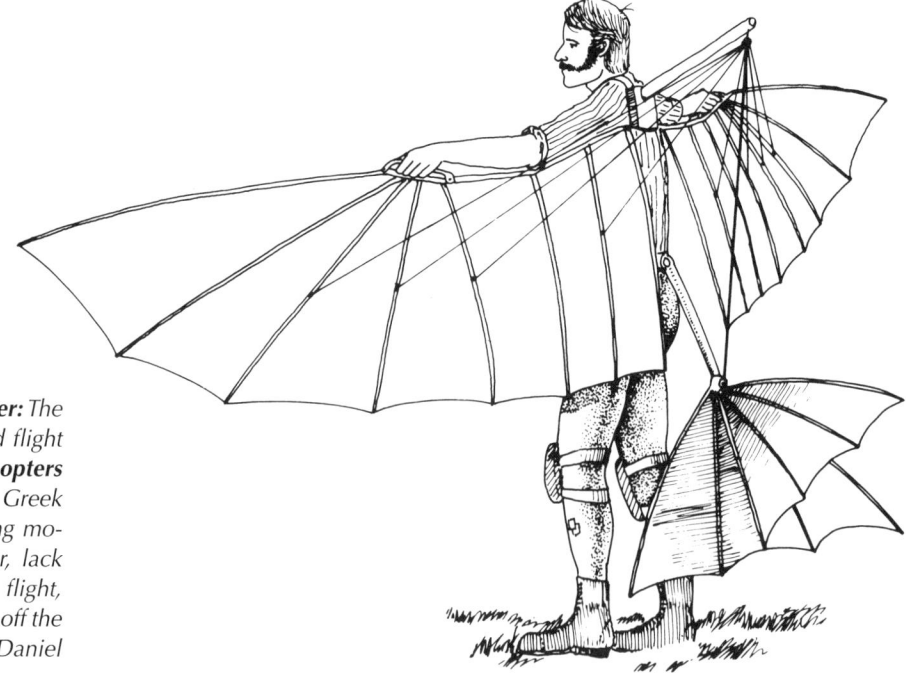

*Figure 5–7. Man with Ornithopter: The first attempts at human-powered flight were in machines termed **ornithopters** (literally, "bird wings," from the Greek roots) that mimicked the flapping motion of birds. Humans, however, lack certain critical adaptations for flight, so these machines never made it off the ground. Drawing courtesy of Daniel*

lowing takes a simplified look at flight, concentrating on the basic physical forces that act upon a bird (or bat, or butterfly, or airplane) moving through the air.

Forces Acting on a Bird in Flight

Most of us have watched some very large bird, such as a Great Blue Heron, in flight. The next time you see one, pay careful attention to exactly what the bird does. I recently saw a Great Blue Heron perched in the top of a large white pine on the edge of a woods. As I watched, the heron leapt into flight, bending its legs and thrusting upward with them as it began to beat its gigantic wings **(Fig. 5–8)**. Through this effort it rose slightly, but once fully airborne it lost altitude for a few wingbeats until it finally achieved stable, level flight, its neck drawn into an S and its legs and feet extended straight back. The bird was headed for a nearby pond to hunt, so it flapped steadily, maintaining its altitude for some distance. As it approached the pond, it set its wings and began to glide, gradually losing altitude, approaching the ground at a rate that would bring it precisely to the shore. In the last seconds of flight, it extended its neck, dropped its legs, adopted a more vertical body alignment, and again began to flap. The wings, previously oriented to slice through the air like knives, now were brought up to beat against the onrushing air to brake the bird's forward and downward speed and avert a crash

Figure 5–8. Great Blue Heron Flight Sequence: As a heron takes off it bends its legs to a crouch, then jumps up into the air as it opens its huge wings and begins flapping. After a few wingbeats, it reaches level flight with legs extended straight back and neck drawn into an S. As it prepares to land, it glides to the desired altitude or location, then extends the neck and lowers the legs, bringing its body into a more vertical position. The wings now flap more front-to-back than up-and-down, thus "catching" the onrushing air and slowing the bird so that it lands gently on its outstretched legs. Lower left photo courtesy of Lee Kuhn/CLO. All other photos by Marie Read.

5

landing. Ultimately the bird was moving too slowly to remain airborne, and it landed gently on its outstretched feet.

This was an everyday observation of a feat that herons and other birds perform thousands of times during their lives. Flying is, after all, as natural to birds as walking is to humans. But such a simple flight can raise a host of interesting questions. Once in flight, why does the heron initially drop in altitude? Why doesn't it continue to drop? When it begins to glide it again loses altitude. Why, and what controls its rate of descent? How does it manage to stop and alight more or less gracefully? No one yet understands all the minute steps that control even this simple flight and descent, but one can begin to grasp the fundamentals by understanding the basic forces that act upon a bird in flight: gravity, lift, drag, and thrust.

Gravity

The first and most familiar force is **gravity**, the attractive force between masses of matter. For our purposes, gravity is the force tending to draw objects toward the center of the Earth. For an object to stay aloft, it must overcome the pull of gravity, a tricky endeavor in the insubstantial, fluid medium of the atmosphere. Balloonists defeat gravity by filling their balloons with enough light, warm air—or lighter-than-air gas—to compensate for the weight of balloon and occupant; rockets burn fuel to create sufficient power to overcome the pull of gravity by brute force.

Birds use a completely different strategy: they remain airborne by manipulating the motion of air past their wings. The key term here is "motion," and it is in this respect that the wing differs from other means of overcoming gravity. A person standing on the ground need not move to keep from sinking; a balloon will float in a completely still atmosphere; a rocket doesn't even require an atmosphere. For a bird's wing to overcome the pull of gravity, however, the air must be flowing over and under it, and it doesn't matter whether this flow results from the wing moving through the air or, on a windy day, from the air moving past the wing.

Gravity so pervasively controls every aspect of the physical world that we tend not to explicitly consider its influence. But a moment's reflection shows that gravity—or, rather, the need to overcome it—is the single most crucial influence on a bird's form. Most obviously, gravity is the force that determines weight, and the need to reduce weight dominates the structure of every part of the bird's body, from the lightweight beak to the hollow feathers and bones.

Lift

Being light in weight, however, is only part of a bird's challenge. To fly, a bird must be able to counteract the force of gravity. The force that serves this need, called lift, is provided by the special shape of a bird's wing. This shape, known as an airfoil, is curved such that it is convex on top, concave below, and tapers at the rear edge **(Fig. 5–9)**. When it moves through the air, the airfoil cleaves the air into two separate airstreams, one on the top and one on the bottom. During the

Figure 5–9. Airfoil: A typical airfoil, such as the wing of a bird or airplane, is convex (rounded) on top and concave (curved inward) below. This shape creates lift as the airfoil moves through the air. (To learn how an airfoil creates lift, see text and Fig. 5–11.)

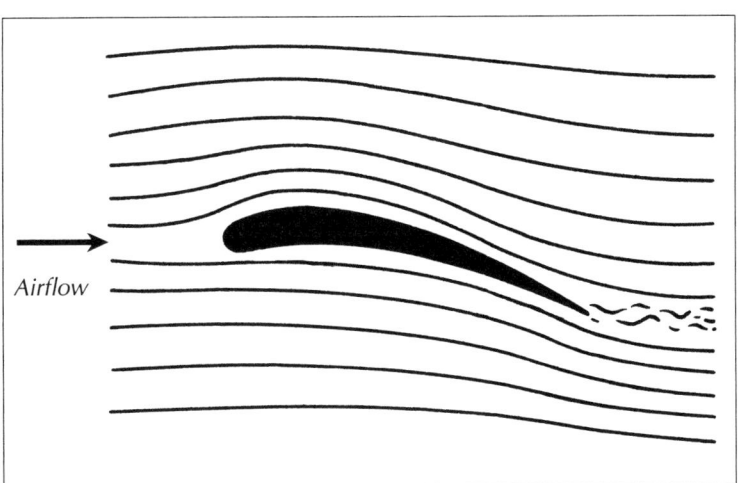

Figure 5–10. Airfoil in Airstream: Shown here is an airfoil in a stream of air. As the airfoil moves through the air, it splits the oncoming airstream. Note that the air layers below the airfoil remain roughly parallel to each other, while the air layers above the airfoil are pushed upward and crowded together. Note also that the forces are the same whether the airfoil moves through stationary air, or the air moves past a stationary airfoil. This airfoil is approximately the same shape as a bird's wing (viewed in cross section), and behaves similarly.

Airflow

time that the air is physically split into two airstreams, each airstream acts as a separate physical system, and they behave differently from one another because of the shape of the wing. This difference in airstream behavior makes flight possible. Because flight is the quintessential characteristic that rules almost every aspect of bird evolution, form, and ultimately behavior, it is worth looking at in some detail. The following explanation is adapted in part from the account in Rüppell's *Bird Flight* (1975), cited in the Suggested Readings. For simplicity, this discussion will be limited to gliding flight, in which a bird moves forward through the air without flapping.

Figure 5–10 shows how the layers of air move around a wing as it passes through an airstream. You get a picture like this when you put an airfoil in a wind tunnel and trace the motion of the air with streams of smoke. Note that the layers of air below the wing remain parallel, but those passing over the top of the wing are crowded together. The crowding above the wing occurs because oncoming air is pushed up and over the convex surface of the airfoil, but the air already above the wing resists this additional input of air by pushing back. Thus the air forced over the wing is "constricted" to an area near the upper wing surface.

Constriction of the airflow increases its speed. You demonstrate this every time you have a birthday: if you try to blow out candles with your mouth wide open, you can't generate much of a puff; if you purse your lips and constrict the airstream, it speeds up and you get them all. The same principle is at work when you make your garden hose spray farther by narrowing the nozzle or putting your finger partially over the end to squirt your brother.

Because of the constriction, the air flows more quickly over the top of the wing than under the bottom. How does this produce lift? To answer this, you must understand two properties of moving air and the way these properties are related.

The first property is **static pressure**: the force, produced by random motion of molecules, that air exerts uniformly in all directions. For example, when you squeeze a balloon, you feel the static pressure from the air inside. The atmosphere outside the balloon also has static pressure. The second property, **dynamic pressure**, is the pressure of move-

Figure 5–11. How an Airfoil Creates Lift: *Horizontal lines represent airstream; lengths of arrows indicate relative magnitudes of the forces they represent.* **a. Airfoil in Still Air:** *Because of the random motion of air molecules,* **static pressure** *is equal above and below the airfoil.* **b. Symmetrical "Non-airfoil" in Moving Air:** *Moving air creates* **dynamic pressure**, *the force you feel when the wind blows against your face. The symmetrical shape in this airstream constricts the air equally above and below, so the air speed is increased the same amount above and below. Therefore, the dynamic pressure (and static pressure) above and below are the same, and no lift is created.* **c. Airfoil in Moving Air:** *Dynamic pressure and static pressure offset each other: if one increases, the other must decrease. An airfoil constricts only the air flowing above it, increasing the speed and thus the dynamic pressure above the airfoil. Because the dynamic pressure increases, static pressure must decrease (Bernoulli's law). Below the airfoil, the air is not constricted and thus air speed, dynamic pressure, and static pressure are unaffected. The result is higher static pressure below the airfoil, creating an upward force known as* **lift,** *which keeps a flying bird aloft.*

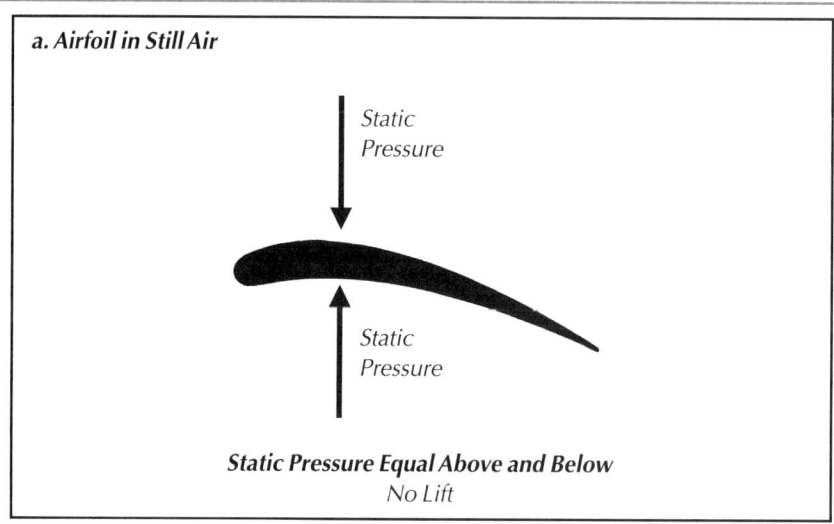

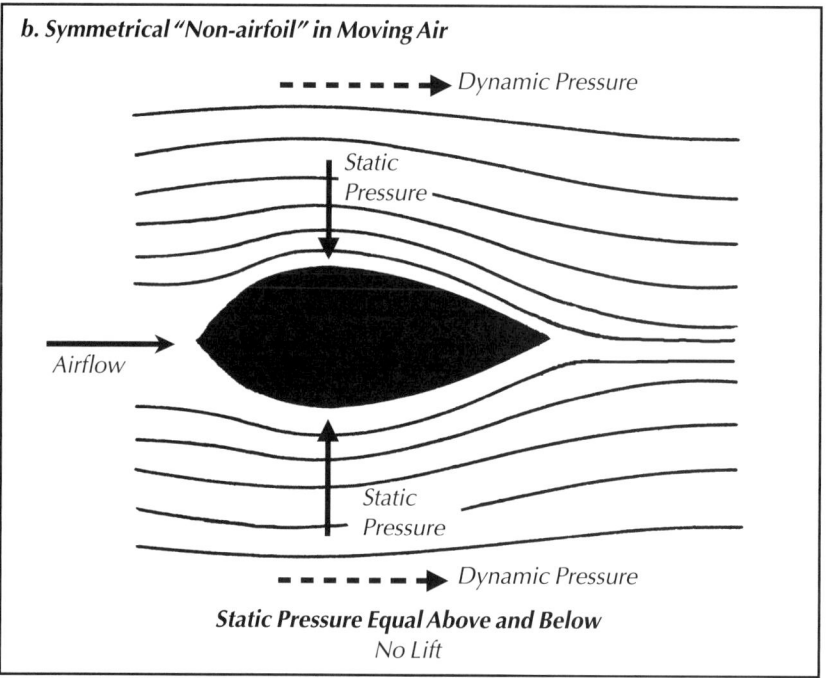

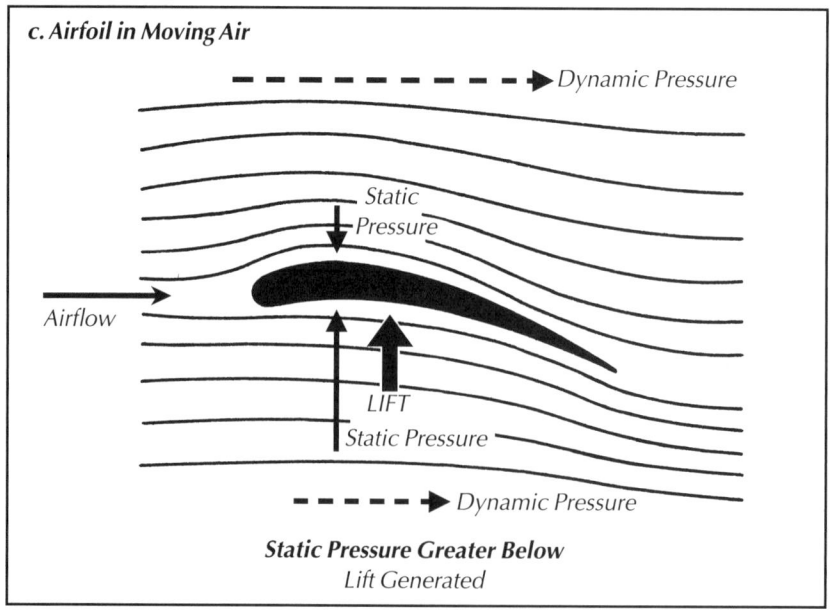

ment. You feel dynamic pressure when wind blows against your face; the harder the wind blows, the more dynamic pressure you feel.

In a given system of airflow, a law of physics termed Bernoulli's law governs the relationship between static and dynamic pressure. It states that these two types of pressure must always add up to a constant. That is, when one increases, the other necessarily decreases; when one decreases, the other must increase. A full explanation of Bernoulli's law is beyond the scope of this course. But it's important to know that this relationship between static and dynamic pressure is a consequence of the law of conservation of energy: the total energy in a given system remains constant regardless of changes within the system.

Consider once more the air flowing over the top of a bird's wing. Because it is constricted, it flows more quickly, and because it flows more quickly, its dynamic pressure increases. Therefore, in accordance with Bernoulli's law, its static pressure decreases. So the static pressure of the air flowing over the top of the wing decreases. But the airflow *below* the wing is not constricted, so the dynamic and static pressures there remain the same. Thus, the static pressure above the wing is lower than the static pressure below the wing, creating an upward (if the wing is horizontal) force known as **lift**. If the pressure difference provides enough lift to compensate for the bird's weight, the bird remains airborne **(Fig. 5–11)**.

"Lift" always operates perpendicular to the flow of air over the wing. It doesn't always "lift" the bird vertically off the ground, however, so the term is not really appropriate. If a bird dives straight down with its wings spread, for instance, "lift" will deflect its descent horizontally in the direction of the top of the wing. In a bird flying upside-down, as some kites and hawks do briefly while diving or during courtship displays, "lift" produces downward movement! Lift operates relative to airflow, not relative to the ground—a concept that may be difficult for terrestrial creatures such as ourselves to appreciate **(Fig. 5–12)**.

A bird can vary the amount of lift that its wings generate by changing the angle between the wing and the oncoming airstream, an angle known as the **angle of attack**. Lowering the front edge of the wing below the horizontal so that airflow strikes the upper wing surface, for instance, generates a net downward force. Elevating the wing's front edge increases lift, up to a point. Too great an angle of attack, however, separates the airflow from the upper surface of the wing and causes **turbulence**—a disorderly flow of air, quite different from the smooth (**laminar**) flow seen at lower attack angles (see Turbulence, later in this chapter). When this occurs, the requirements for lift are no longer met, and the bird stalls **(Fig. 5–13a, b,** and **c)**.

For lift to keep a bird airborne, air must flow over its wings at a certain rate. At some slower speed (which depends on the bird's weight, wing shape, and other factors), the bird will stall and fall vertically. This happens when the lift generated is no longer greater than the bird's weight (gravity). Birds may deliberately stall as part of their landing procedure, but sometimes birds need to fly slowly—or with the wings at a steep angle of attack—without stalling or landing. Some

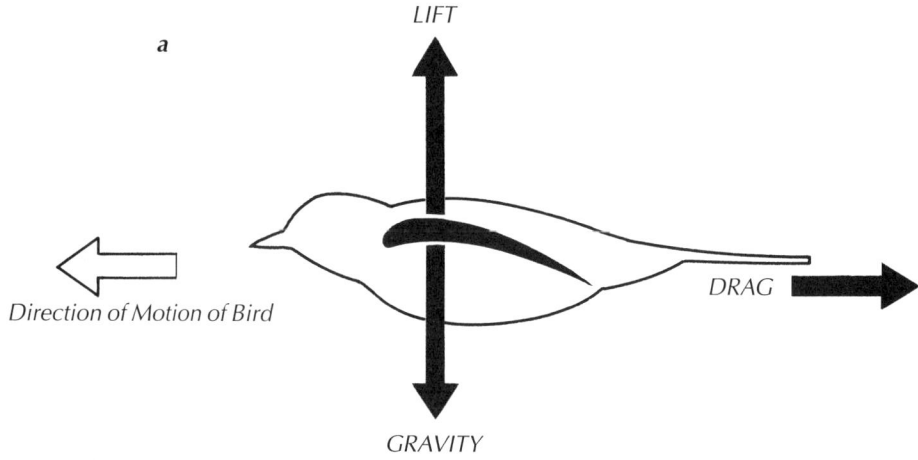

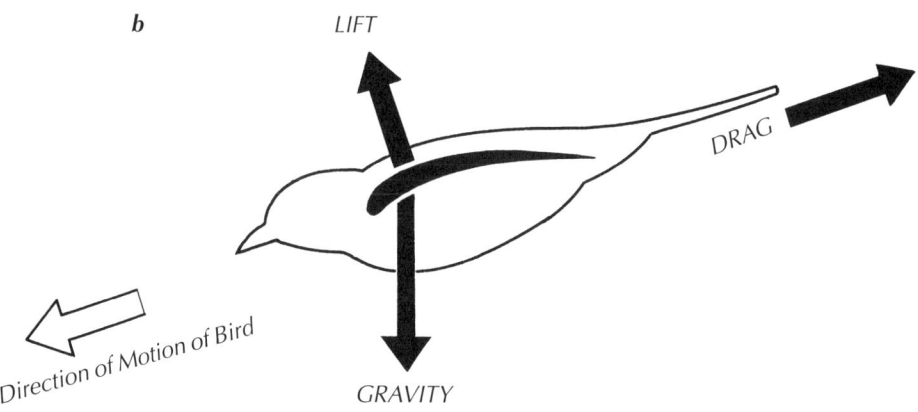

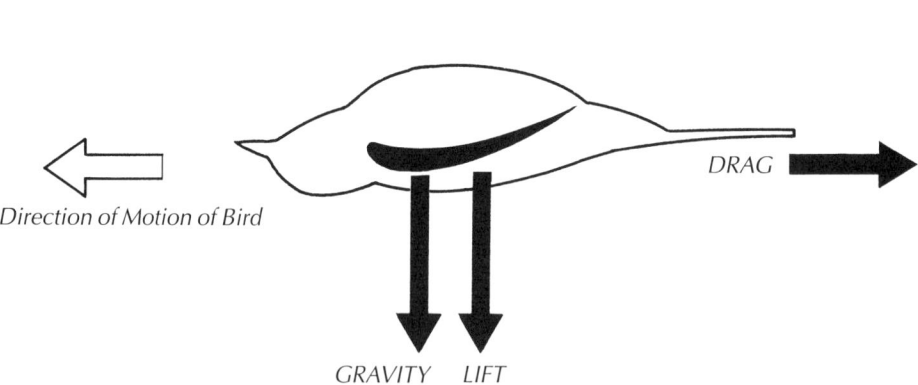

Figure 5–12. Gliding Bird in Airstream: *The force of lift always operates perpendicular (at right angles) to the flow of air over the airfoil—in this case, the bird's wing. Gravity, however, always pulls straight down.* **a. Bird in Level Flight:** *The only time that lift operates straight up is when a bird is flying level. Note that a gliding bird cannot fly level for long, as gravity will bring it to earth.* **b. Bird in Angled Flight:** *When a bird angles toward the ground, lift still operates perpendicular to the airfoil. The upward portion of the lift force keeps the bird aloft, and the small forward component of the force propels the bird forward.* **c. Bird Gliding Upside Down:** *In this hypothetical example, "lift" actually operates to pull the bird toward the ground. Some birds may be briefly oriented upside down during aerial courtship or prey-capture maneuvers, but these generally involve flapping flight to keep the bird aloft.*

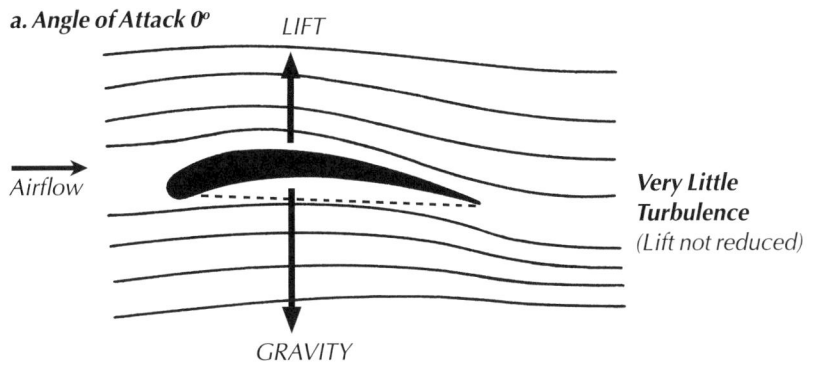

a. Angle of Attack 0°

LIFT

Airflow

GRAVITY

Very Little Turbulence
(Lift not reduced)

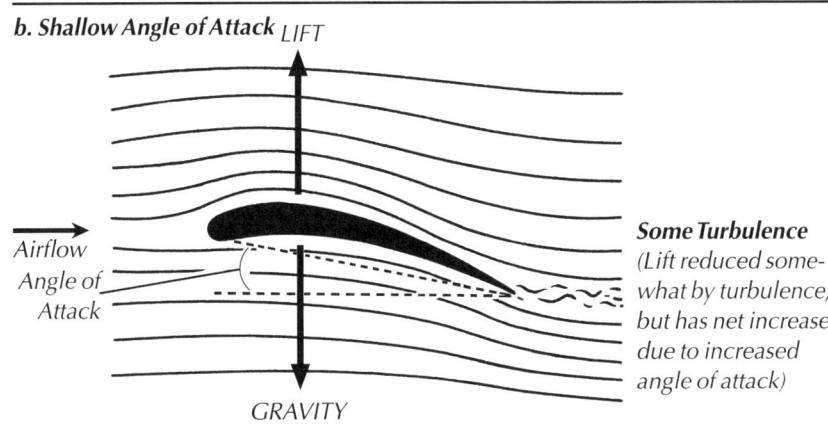

b. Shallow Angle of Attack LIFT

Airflow
Angle of Attack

GRAVITY

Some Turbulence
(Lift reduced some-what by turbulence, but has net increase due to increased angle of attack)

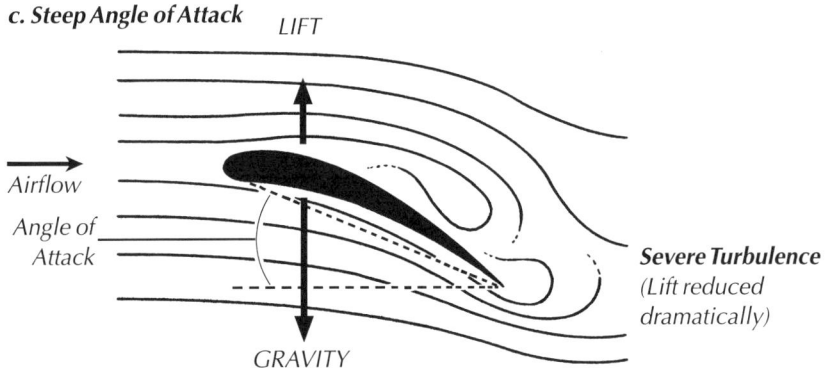

c. Steep Angle of Attack LIFT

Airflow
Angle of Attack

GRAVITY

Severe Turbulence
(Lift reduced dramatically)

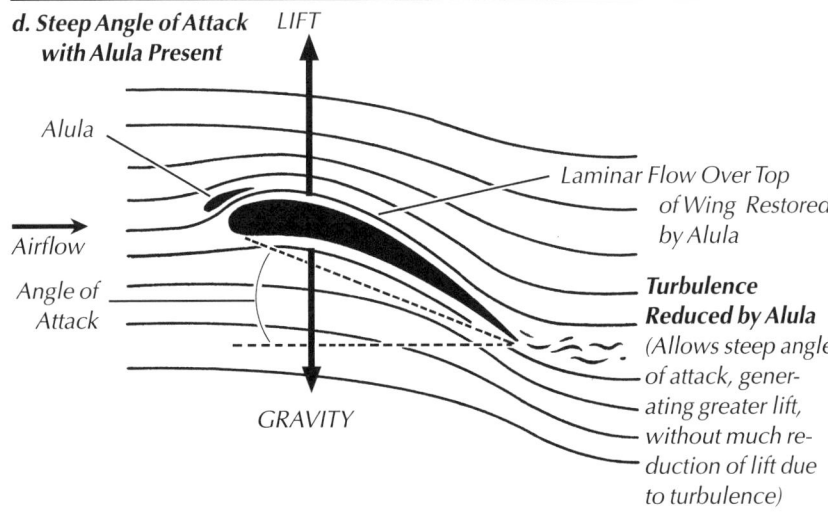

d. Steep Angle of Attack with Alula Present LIFT

Alula

Airflow
Angle of Attack

GRAVITY

Laminar Flow Over Top of Wing Restored by Alula

Turbulence Reduced by Alula
(Allows steep angle of attack, gener-ating greater lift, without much re-duction of lift due to turbulence)

Figure 5–13. The Effect of the Angle of Attack on Turbulence in a Gliding Bird: Horizontal and curved lines represent the airstream; dotted lines represent the **angle of attack** (the angle between the airstream and the horizontal axis of the airfoil); lengths of arrows indicate relative magnitudes of the forces they represent.

a. Airfoil Parallel to Airstream: When the angle of attack is zero, some lift and very little turbulence are produced.

b. Shallow Angle of Attack: As the angle of attack is increased slightly (by tilting the front edge of the wing upward), more lift is created because the airstream is more constricted above the wing. More **turbulence** (shown by curved lines behind the airfoil) is also created.

c. Steep Angle of Attack: When the front edge of the wing is tilted steeply upward, the smooth flow of air over the airfoil (termed **laminar flow**) is disrupted, and the airstream separates from the airfoil in swirls of turbulence. Gains in lift from increased air constriction above are cancelled out by turbulence, resulting in a net decrease in lift, and consequently, stalling.

d. Steep Angle of Attack with Alula Present on Wing: When the **alula** (a group of three small feathers; see Figs. 1–8 and 5–14) is extended, it forms a slot just above the wing that forces air above the wing to flow quickly and close to the top of the wing (in other words, restoring laminar flow). This decreases turbulence and therefore increases lift. Birds that need to fly slowly or at steep angles of attack without stalling (as in hovering or landing) often employ the alula.

5

birds address this problem by creating a slot above the front of the wing **(Fig. 5–14)**. This slot, formed by the **alula** (see Fig. 1–8), forces the air above the wing to flow quickly and down across the top wing surface (laminar flow), preserving the lift that would otherwise be lost as the airstream separated from the top of the wing **(Fig. 5–13d)**.

Because even the most efficient airfoil is useless in still air, a bird must somehow start a flow of air over its wings to become airborne. Our heron did this by launching itself from a high perch. If the wind is strong enough, a bird standing on an exposed bluff can generate enough lift to get off the ground just by extending its wings and facing into the wind, as gulls and albatrosses sometimes do. In the right circumstances, these passive means of achieving lift can be quite effective.

Drag

The force that slows down a gliding bird, or any moving bird, eventually to the point at which it can no longer maintain the lift necessary to overcome gravity, is called **drag**. Essentially, drag is friction between air and a moving body. For instance, when you put your arm out your car window you feel drag; the faster you drive your car, the more drag you feel, because drag increases with increased air speed. The size, shape, and surface area of the object also influence drag. Hold your hand out the window palm down and it slices through the air; turn it so the palm faces the onrushing air and it blows backward.

Drag operates in opposition to the motion of the body; in other words, it has a slowing effect (see Fig. 5–12). If you're driving due east, the drag force on your car is directed due west. If you are a bird flying straight up, the drag force is straight down. If you parachute straight down from a plane, the drag force is straight up.

Recall the example of a heron taking flight. If the bird merely leapt from the tree and extended its wings but did not flap them, its wings would be somewhat analogous to a parachute. By extending its wings, it increases friction with air and generates an upward drag force that slows its fall. Over time, it would descend passively to the ground like a parachutist.

Thrust

So far this chapter has discussed only gliding flight. A gliding bird counteracts gravity with lift, but drag eventually brings the bird to the ground. The heron I was watching that day did not glide down to the ground, however, but sustained level flight for some distance. To do this, it must be able to overcome drag, and it does so by producing another force, **thrust**, the final force to be discussed here.

Thrust propels a bird forward through the air. It is created only in flapping flight, not when a bird is simply gliding. Flapping the wings also creates lift, and allows a bird to start a flow of air over its wings, which is required to begin flight without a high perch or strong winds.

Like lift, thrust is also derived from the airfoil shape moving through an airstream. Are you reading this in the privacy of your home?

Figure 5–14. Hovering American Kestrel with Alulae Spread: *The kestrel has a particularly large alula (small feathers projecting above the bend in the wing), which allows it to hover with a steep angle of attack at very slow speeds. Kestrels typically hunt for prey by hovering over fields and grassy areas, and are often noted by motorists, who see them hovering over roadsides. Illustration by Evan W. Barbour.*

If so, try this: Stand up and hold your arms straight out from your sides, palms down. These are your wings; the palms are the undersides and the backs of your hands are the tops of the wings. Holding your arms out, run around the room. (If you like, you can chirp, coo, peep, or squawk, too.) As you run, the airstreams moving over and under your airfoil-hands create lift.

Now stand still, arms still out to your sides, and turn your hands so the thumbs point down and the palms back. Holding your hands in this position, start to flap your wings vertically, with a stroke down toward the ground. Now you will note that the airstreams, as they move up and around your hand-wing, are passing over and under your "tilted" wing such that the "lift" operates, not vertically, away from the ground, but horizontally, to pull you forward. This is the source of thrust.

In what portion of a wingbeat does a bird generate thrust? A bird's wingbeats are not just up and down. In typical powered flight the wing is pulled down and forward as it approaches the bottom of the downstroke, and backward and upward on the upstroke **(Fig. 5–15)**. In all birds except hummingbirds, thrust is produced primarily on the downstroke. A common misconception is that birds achieve thrust through pushing back against the air, much as a rower pushes a boat

Figure 5–15. Scaup in Flapping Flight: *In the first frame, the downstroke is just beginning, with the primaries overlapped and curved upward from pressure against the air. As the wings continue downward, the primaries act as propellers by pulling the wings forward and the whole bird with them. Meanwhile, the secondaries provide most of the lift. In the second frame, the downstroke is completed, with the wings reaching forward and downward to the maximum extent. In the third frame, the upstroke is under way, with the primaries separated and drawn toward the body. During this recovery stroke, the primaries may push backward slightly against the air to propel the bird forward while the secondaries provide lift. In the fourth frame, the upstroke nears completion as the primaries begin reaching far backward and upward. In the final frame, the upstroke is completed and the wings are about to undertake another downstroke. Drawing by Robert Gillmor.*

Downstroke | Upstroke

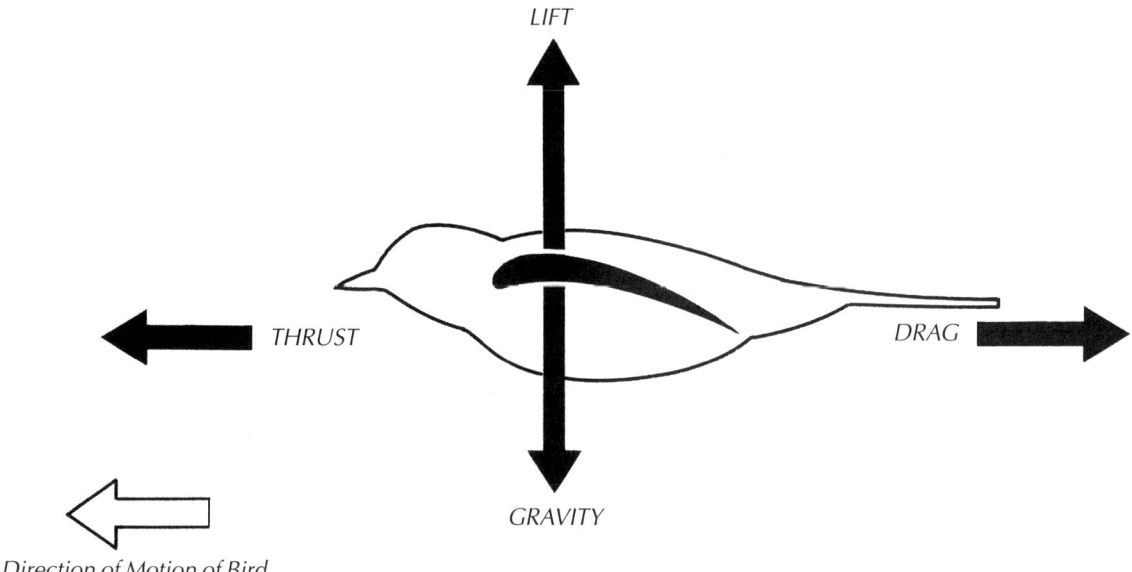

LIFT

THRUST

DRAG

GRAVITY

Direction of Motion of Bird

Figure 5–16. Bird in Flapping Flight:
By flapping its wings, a bird creates a force called **thrust**, *which propels it forward. Thrust is opposed by drag, which always operates in a direction opposite to the bird's motion. As a bird is moved forward by thrust, airflow over the wings creates lift, which keeps the bird aloft. To fly on the level at a steady speed, a bird must create enough thrust to exactly compensate for drag, and enough lift to exactly oppose gravity.*

forward by pressing the oars back against the water. Indeed, the term for the flight feathers of the wings, the "remiges," is based on Latin terms that mean, roughly, "that which rows." What actually happens may look analogous to rowing, but is actually quite different, and is based on "forward lift." The wing motion that produces thrust is complicated and difficult to grasp intuitively, and will be discussed later, in Sidebar 1: Flapping Flight.

How much thrust does a bird need to stay aloft? Consider a sled. To push the sled over snow you must apply sufficient force to overcome friction (drag). The more force you apply, the more quickly you accelerate and the faster you go; decrease the force and friction eventually drags the sled to a stop.

A similar process occurs when a bird is flying. Like a sled, a bird experiences drag only when it is moving. However, a sled that is not being pushed hard enough to overcome friction merely slows down and ultimately stops. A bird producing insufficient thrust to overcome its own drag also slows down, but it also loses lift, and eventually descends.

Therefore, to fly horizontally at steady speed, thrust must completely compensate for drag **(Fig. 5–16)**. When thrust equals drag, the bird flies at a steady speed. When the bird's thrust exceeds drag it flies faster, which produces more lift and causes the bird to ascend. If the bird's thrust falls below the drag, the bird slows, produces less lift, and descends.

Except during take-off, when a bird may use its legs to generate momentum, thrust is produced by the action of the wings. The Great Blue Heron I observed used its legs and wings to generate thrust, overcome gravity, and launch itself into the air. For a second or two, it actually climbed before descending and then leveling off. Could it have continued to climb? Certainly, but at great energetic cost, and in this case there was no need.

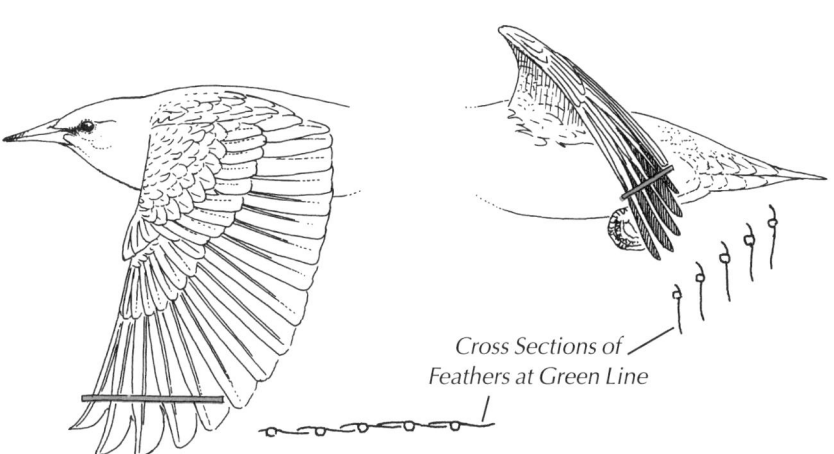

Cross Sections of
Feathers at Green Line

a. Downstroke   *b. Upstroke*

Figure 5–17. Twisting of Primary Feathers During Flapping Flight: *Because the primary feathers have a narrower vane on the edge that leads in flight (outer vane) than on the trailing edge (inner vane), they tend to twist somewhat like venetian blinds as the wing moves through the air during flapping flight.* **a. Downstroke:** *As the wing moves down and forward, air pressure pushes each broad inner vane up against the outer vane of the feather over it, creating an unbroken surface (like closed venetian blinds).* **b. Upstroke:** *As the wing moves upward and back, air pressure from above pushes each broad inner vane down, twisting the primaries open. On the upstroke, air is able to pass through the primaries, thus reducing drag.*

So, the bird's wing generates lift and provides thrust. However, different parts of the wing make different contributions to these two forces. Thrust is produced mainly by the movement of the primary feathers attached to the manus (the outer, "hand" portion) of the wing. In contrast, the proximal portion of the wing, with the secondary feathers attached, provides most of the lift.

Study **Figure 5–17** and note the positions of the primary feathers during flapping flight. The individual primaries are shaped such that each behaves as an individual airfoil, each generating some lift, as does the wing as a whole. In addition, their asymmetrical shape and directional flexibility cause them to twist such that on the downstroke, the air pressure pushes the broader trailing edge (the inner vane) of each primary up against the outer vane of the feather over it. This produces an unbroken surface, much like closed venetian blinds, moving through the air. On the upstroke, air pressure twists the primaries, opening them, like the slats on opened venetian blinds, so that air may pass through **(Fig. 5–18)**. This arrangement of the primaries generates a downstroke having about ten times as much air resistance as the upstroke.

You can demonstrate this twisting action of the primaries by lightly holding two primary feathers (of somewhat similar size, the larger the better) parallel to each

Figure 5–18. Carolina Chickadee During Upstroke: *During the upstroke, air pressure on the broader inner vanes causes the primaries to twist open, resulting in less resistance to the air. Photo by C. H. Greenewalt/VIREO.*

VIEW FROM ABOVE

Figure 5–19. Demonstration of Feather Twisting: To roughly imitate the twisting of the primary feathers during a wing-beat, hold two large primaries of similar size parallel to each other between your index and middle finger, as illustrated. Your fingers should be parallel to the ground, and the feathers perpendicular to the ground. Face the leading edges toward your finger tips, and allow the leading edge vane of one to partially overtop the trailing edge vane of the next, as in a bird's wing. *a. Upstroke:* Move your hand straight upward quickly, and notice how the feathers twist to create a slot between them. *b. Downstroke:* Move your hand straight downward quickly, and the feathers twist and overlap, closing the slot. Photo courtesy of Marie Read/CLO.

a. Upstroke
Hand Moves Upward
(Toward Viewer)
Feathers Twist and Separate

b. Downstroke
Hand Moves Downward
(Away From Viewer)
Feathers Flatten and Overlap

Figure 5–20. Bald Eagle Using the Tail to Steer: A Bald Eagle steers to its left, its tail acting as a rudder to aid in turning. It also tilts and turns the body and lowers one wing in the chosen direction. Photo by Tom Vezo.

other between your index and middle fingers **(Fig. 5–19)**. Hold them so the barbs overlap loosely as they would on a bird, with the feathers at right angles to your fingers, and your fingers parallel to the ground. Then, quickly move your hand straight up and notice that the feathers twist to open a gap between them. Now quickly move your hand down; the feathers twist to close tightly together. Note that on a bird, the feathers stay attached at the base, so the barbs closer to the feather tip twist more than those closer to the base. In your demonstration, however, the entire feather twists in your fingers.

As mentioned earlier, a bird's wing moves down and forward on the downstroke and backward and upward on the upstroke, but this description is greatly simplified. Lift and thrust on different parts of the wing are constantly changing during wing strokes, and the many muscles in the wing (50 or more) integrate these complex dynamics **(Sidebar 1: Flapping Flight)**.

Function of the Tail

When we think about flying birds we naturally focus on the wings, but the tail also plays a role in flight. This role may be limited—most birds can fly without their tails, and males of species such as birds-of-paradise (see Fig. 3–10), with extremely long tails, can probably fly better without them. The tail's main function in flight seems to be to act as a rudder to help the bird steer **(Fig. 5–20)**. Accipiters such as the Cooper's Hawk apparently evolved long tails because they help the birds maneuver as they chase their avian prey through dense vegetation. Birds with particularly short tails, such as loons, grebes, auks, and many ducks, can fly quickly (because a short tail reduces weight and drag), but have a reduced ability to make sharp turns. Some of these birds instead use their webbed feet to steer and brake. The tail is also used during take-off and landing (see next section), and by perched birds to stabilize themselves in a wind. In pigeons, research using EMG has revealed that the tail muscles are active with each wingbeat, and the patterns of activity change with the varying demands of take-off, slow flapping, landing, and other actions. In a walking pigeon, on the other hand, most of the tail muscles are inactive.

Figure 5–21. Red-tailed Hawk Using Tail To Land: During landing, many birds, such as this western morph Red-tailed Hawk, lower and spread the tail. The additional area of the tail acts somewhat like a third wing, providing extra lift and thus allowing the bird to keep flying at a very slow speed without stalling, until just the right moment for a controlled landing. Photo courtesy of Rick Kline/ CLO.

Landing

Controlled landing is in many ways more difficult than taking off or maintaining flight, because the bird must stop its forward momentum and coordinate its movements to bring about a stall at precisely the altitude and speed that will allow the extended legs to make contact gently and avoid a crash.

During landing, the tail is typically lowered and spread, much like a jet that extends and lowers the flaps on the trailing edge of its wings. Some birds spread their rectrices wide on both sides to form a kind of rear wing, as seen in high-speed photographs of birds taking off or landing **(Fig. 5–21)**. This "tail wing" provides extra lift to prevent stalling until the last seconds, permitting the bird to remain airborne at slower speeds that are more amenable to a controlled landing. The action of the tail also helps to suck air downward over the wing, reducing turbulence and again increasing lift to permit slower flight.

(Continued on page 5·26)

Sidebar 1: FLAPPING FLIGHT
edited by Sandy Podulka

Flapping flight is remarkable for its automatic, unlearned performance. A young bird on its maiden flight uses a form of locomotion so complex that it defies precise analysis in physical and aerodynamic terms. The nestlings of some species develop in confined spaces, such as burrows in the ground or cavities in tree trunks, where they cannot spread their wings and practice flapping before they leave the nest. Despite this seeming handicap, many of them can fly considerable distances on their first flights. Diving petrels may fly as far as six miles (10 km) the first time out of their burrows! On the other hand, young birds reared in open nests frequently flap their wings vigorously in the wind for several days before flying—especially large birds such as albatrosses, storks, vultures, and eagles. Such flapping may help to develop muscles, but it is unlikely that these birds are learning to fly; however, a bird's flying abilities do improve with practice for a period after it leaves the nest.

Flapping flight involves so many variables that understanding exactly how it works is difficult. A beating wing is flexible and yields to air pressure, unlike the fixed wing of an airplane. As a wing moves through its cycle of motion, its shape, camber, angle with respect to the body, and the position of the individual feathers all change remarkably. This is a formidable list of variables, and thus it is no wonder that flapping flight has not yet fully yielded to explanation in aerodynamic terms. Nevertheless, the general properties of a flapping wing can be described and analyzed.

Consider, first, the flapping cycle of a small bird (**Fig. A**; also see Fig. 5–15). On the downstroke, the inner section of the wing (from shoulder to wrist), where the secondary feath-

ers attach, simply moves down; whereas the outer wing, where the primaries attach, moves both down and forward. This movement occurs because the flapping wing extends at the wrist joint between the primary and secondary feathers. In addition, the wing twists at the wrist joint dur-

ing the downstroke so that the leading edge of the outer wing is tipped downward relative to the body axis, while the inner section of the wing remains roughly parallel to the body axis. The upstroke is both upward and backward, returning the wing to its original position.

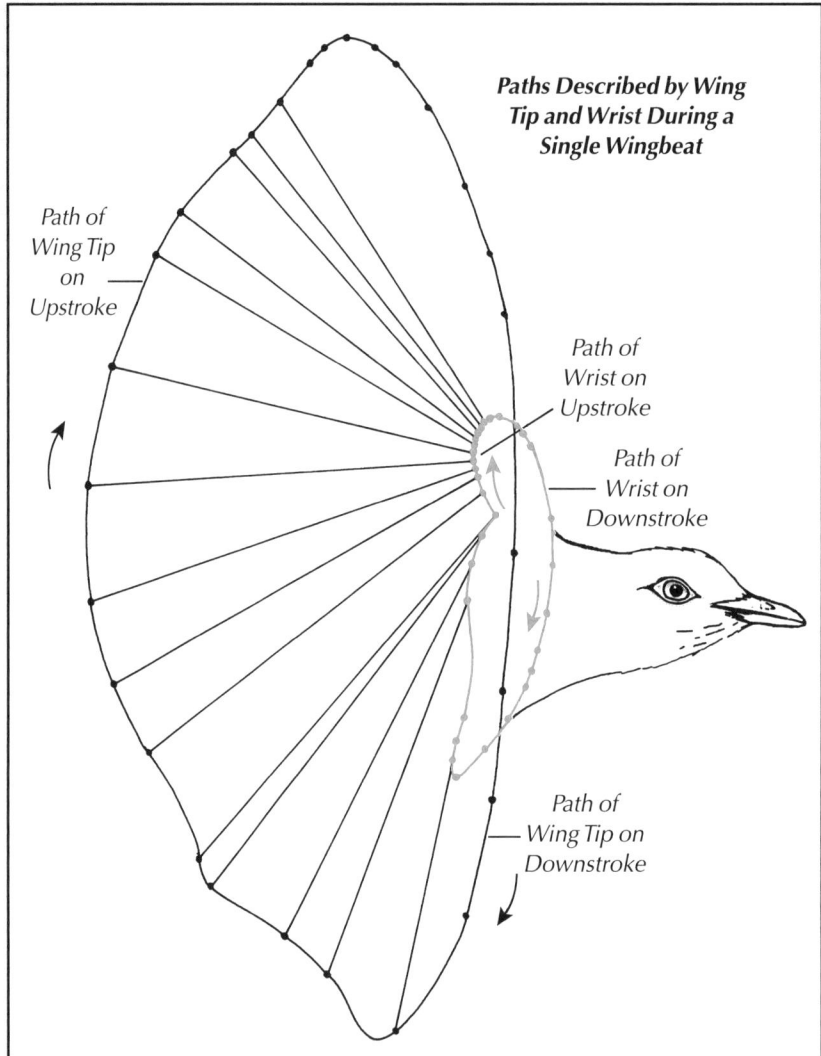

Figure A. Paths Described by Wing Tip and Wrist During Wingbeat: *Colored curves show path of wrist (joint between outer and inner wing) and black curves show path of wing tip for one complete wingbeat. The paths do not show the forward motion of the bird. Dots are wing positions at evenly spaced points in time, and thus indicate the relative speed of the wing: the farther apart, the faster that portion of the wing is moving. Straight lines connecting inner and outer dots, like wheel spokes, link the two parts of the wing at the same point in time. Note that because of twisting at the wrist, the wrist and wing tip take very different paths through the air, and the wing tip covers much more area than the wrist. The exact path of the wing varies a great deal between species. Adapted from Burton (1990, p. 40).*

How do these motions contribute to flight? As discussed earlier in this chapter, a bird maintains level flight by developing a force (thrust) that counteracts the drag operating against forward movement. The flapping of the wings, especially the outer wings, produces this thrust. The inner wings primarily generate lift.

It is easiest to consider the forces operating on the inner wing (secondary feathers) and outer wing (primary feathers) separately (**Fig. B**). Most of the lift, and also most of the drag, results from the forces acting on the inner wing and body of a flying bird (**Fig. C**). The inner wing acts much as if the bird were gliding: since it does not move up and down much during flapping, most of the effective airflow over the inner wing comes from straight ahead, due to the motion of the whole bird through the air. Unlike the outer wing, the inner wing is not tipped forward, so the

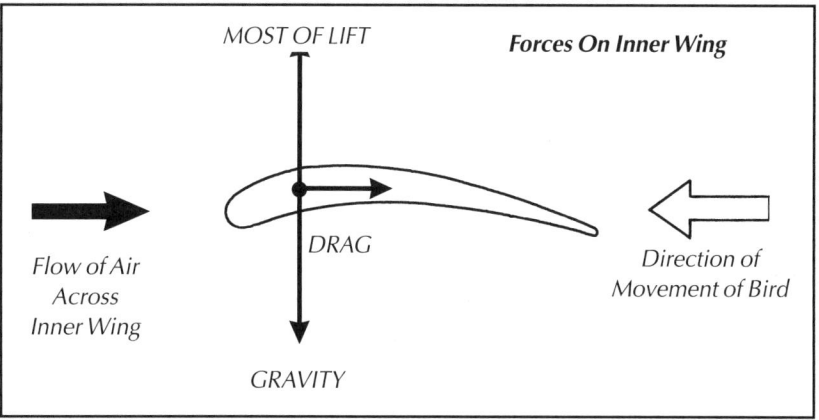

Figure C. The Inner Wing During Flapping Flight: Shown here is a cross section through the inner wing (see Fig. B). During flapping flight, the inner wing acts much like the wing of a gliding bird: because it is relatively stationary, not moving up and down as much as the outer wing, most of the airflow comes from the front–a result of the bird's forward motion. Because the inner wing meets the airstream head-on, instead of being tipped forward like the outer wing, air flowing over the airfoil generates lift nearly straight up during level flight.

movement of air over the airfoil produces lift nearly straight up.

On the outer wing, however, the situation is very different (**Fig. Da**). There, the airflow is mostly upward and slightly backward with respect to the body axis of the bird. The upward flow results from the rapid movement of the outer wing downward, and the backward flow results both from the forward motion of the wing during a downstroke and the forward motion of the bird. Even though the wing is tipped forward, the rapid downward flap produces such a large upward airflow that the effective angle of attack is very steep. The wing would stall if that were all that happened, and no thrust would be generated. What actually happens, however, is that the force of air from below causes the primary feathers to twist such that the leading edge rotates forward and down (recall that the primaries are asymmetric, with a wider vein on the trailing edge than on the leading edge). In this orientation, each primary feather acts like a small, individual airfoil. The twisting brings each primary feather more in line with the airflow, reducing the angle of attack, and preventing stalling (**Fig. Db**). The resulting force on

the outer wing is both upward and forward, because, as you will recall, lift always operates perpendicular to the flow of air over an airfoil. This resulting lift force, directed at an angle with respect to the bird's body, can be divided into two components: one force directed forward, along the body axis of the bird, and another force directed straight upward. The forward-directed component is called thrust; the upward component is an additional source of lift, and adds to the upward lift force generated by the inner wing.

An essential ingredient in the production of lift and thrust is that the wings move at a high speed with respect to the air. For thrust, this is provided by the high rate at which birds flap their wings. Most of the thrust is generated as the wings move rapidly downward and forward on the downstroke, but in some cases thrust is generated on the upstroke as well (see below). Lift, however, is mainly provided by the airflow created by the forward motion of the bird, which results from thrust. Throughout the flapping cycle, the secondaries act much as if the bird were gliding, providing some lift at all times because air continues to

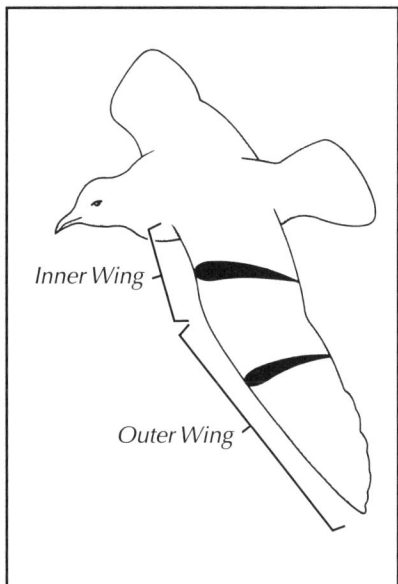

*Figure B. Wing Cross Sections: Cross sections through the wing at two locations are indicated: the **inner wing**, the portion of the wing from the wrist to the shoulder, where the secondary feathers attach; and the **outer wing**, from the wrist to the wing tip, where the primary feathers attach. Note that both wing sections are the shape of an airfoil. Adapted from Burton (1990, p. 32).*

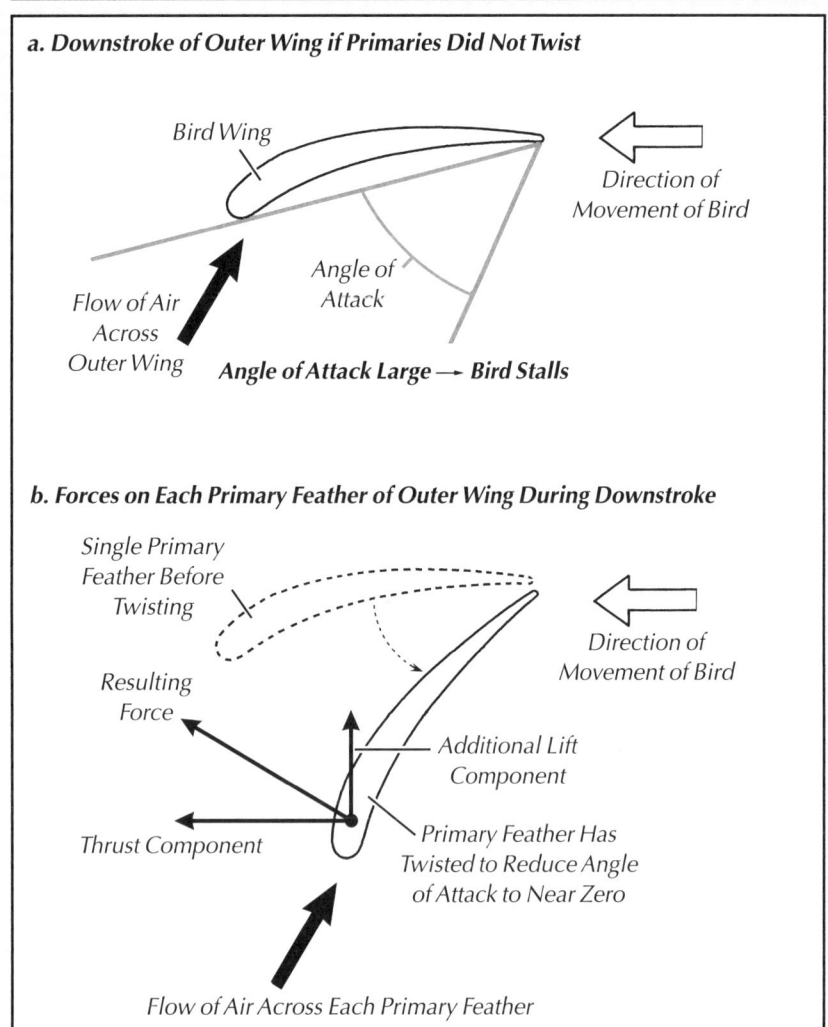

a. Downstroke of Outer Wing if Primaries Did Not Twist

Bird Wing

Direction of Movement of Bird

Angle of Attack

Flow of Air Across Outer Wing   **Angle of Attack Large → Bird Stalls**

b. Forces on Each Primary Feather of Outer Wing During Downstroke

Single Primary Feather Before Twisting

Direction of Movement of Bird

Resulting Force

Additional Lift Component

Thrust Component

Primary Feather Has Twisted to Reduce Angle of Attack to Near Zero

Flow of Air Across Each Primary Feather

*Figure D. Forces and Airflow on Outer Wing During Downstroke: Because the outer wing moves down and forward rapidly on the downstroke, and this movement is much faster than the forward movement of the bird, the airflow over the outer wing is mostly up and somewhat to the rear of the bird. **a. Outer Wing if Primaries Did Not Twist:** Shown here is a cross section through the outer wing (see Fig. B). Note that the leading edge of the wing is tipped slightly down and forward, unlike the inner wing shown in Fig. C. If the primary feathers did not twist during the downstroke, the upward flow of air would produce a very large angle of attack (shown in color). Try turning the page clockwise to orient the airflow horizontally, as in previous figures in this chapter, to more easily visualize the large angle of attack of the wing. Large angles of attack create turbulence and can cause the bird to stall. **b. Primary Feather Showing Twisting:** The airfoil in this diagram represents a cross section through a single primary feather during the downstroke, not an entire wing section, as in the previous drawings. The asymmetry of the vanes causes each primary feather to twist (in actuality, just the tip of the primary feather twists) as the wing moves through the air on the downstroke, such that the leading edge rotates forward and down (dashed lines show position of feather before twisting). In its new position, each feather acts like an individual airfoil, meeting the flow of air head-on, reducing the angle of attack to near zero, and preventing stalling. Moving through the air in this position, the primaries generate a lift force (arrow labeled "resulting force") perpendicular to the direction of airflow. This force, because it has magnitude and direction, can be thought of as a **vector**. Refer to Fig. 5–62, for information on vectors and their addition. This resulting force vector can be divided into two components: (1) a horizontal force directed forward, called **thrust**, which propels the bird forward through the air, and (2) a vertical force directed upward, which, together with the lift generated by the inner wings, keeps the moving bird aloft.*

flow from front to back over the inner part of the wing as long as the bird is moving forward.

A flying bird increases its speed by increasing the depth of its wingbeats, but the frequency of wingbeats remains nearly constant at all speeds during level flight. During take-off, however, some birds increase the speed of their wingbeats briefly, to give them a bit of extra lift until they reach level flight. Large birds have slower wingbeat frequencies than do small birds, and strong fliers usually have slower beat frequencies than do weak fliers.

Not all birds perform flapping flight in exactly the same way. Of particular interest are differences between small and large birds. In both, the thrust produced on the downstroke causes an increase in speed. In small birds, the upstroke is mainly passive: the heavier body of the bird falls slightly with respect to the lighter, more drag-vulnerable, wings; thus the wings rise slightly with respect to the body, contributing to the upstroke. Thus, little or no thrust is produced and the bird slows down during the upstroke. In larger birds with slower wingbeats, the duration of the upstroke is too long to spend in a state of deceleration. A similar situation exists when any bird takes off: to get going, it needs thrust on both the downstroke and the upstroke. Thrust on the upstroke is produced by bending the wings slightly at the wrists and elbow and by rotating the humerus upward and backward (an action powered by the supracoracoideus muscle). This movement causes the *upper surfaces* of the twisted primaries to push against the air and to provide thrust **(Fig. E)**. In this type of flight the wing tip describes a rough figure eight through the air. As speed increases the figure-eight pattern is reduced. In species that rely on a powered upstroke for fast, steep takeoffs, for hovering, or for fast aerial pursuit, the supracoracoideus muscle is relatively large. In fact, the ratio of the weight of the pectoralis

Downstroke	Upstroke

Figure E. One Complete Wingbeat of a Rock Dove Taking Flight: When taking flight, birds need extra force (thrust), which they get from the upstroke. Illustrated here is a sequence of wing positions as a Rock Dove takes flight. The figure "eight" below each picture shows the path of the wing tips during a complete wingbeat; the dark arrow indicates the portion of that cycle illustrated in the frame above. During the downstroke, the wings sweep forward and down, the bases of the primaries flattening against each other to form a large, unbroken surface (see also Figs. 5–17 and 5–19). On the upstroke, the wings bend at the wrist and the primaries twist open as the wings move up and back. During this backward sweep (fourth and fifth frame), the upper surfaces of the twisted primaries push against the air and generate some thrust. At the end of the upstroke, the outer wing twists and flicks upward (last frame), generating some extra lift. Adapted from Burton (1990, p. 62).

(responsible for the downstroke) to the supracoradoideus is a good indication of a bird's reliance on a powered upstroke; such ratios vary from 3:1 to 20:1.

We tend to think only of birds when considering movement that involves creating lift, but this type of locomotion is not restricted to birds. Many fish, mammals, and insects use lift-based mechanisms in their locomotory repertoire. Although we often think that fish swim by beating their tails from side to side, many fishes, such as the surfperches (familiar sport fish along the Pacific coast of North America), flap their long pectoral fins much like birds flap their wings to achieve forward movement. If you watch one swim, you will notice that it holds the tail still while the pectoral fins beat steadily up and down, producing forward thrust via a lift-based technique. In addition to fish, some of the largest mammals on earth, the whales (and many of their relatives), use lift-based strategies to generate thrust by undulating their flukes up and down. These broad flukes function essentially as airfoils, generating thrust to propel themselves forward on both the upstroke and downstroke. Humpback whales have remarkably long pectoral fins that may be used to produce lift, directing the animals either upward or downward in the water.

Birds share the air with another group of vertebrates, the bats; with roughly 900 species, bats are among the most successful mammals. Bats have highly modified wing membranes spanning their digits and extending to the shoulders and legs or ankles. The mechanics of bat flight are similar to those of bird flight: the wing tip describes a figure-eight pattern, and most of the lift and thrust is generated during the downstroke. Like birds, bats have highly developed pectoralis muscles (accounting for up to 10 percent of their total body mass), but also use a few additional muscles on the downstroke. Bats do not have a supracoracoideus muscle to facilitate the upstroke, however, instead recruiting several other muscles of the shoulder region. The other highly successful aerial animals, the insects, use a huge range of lift-based techniques and a variety of wing movements in their mastery of the air.

In summary, note that the only difference between the model of flapping flight discussed here and a simple analysis of a wing being held rigid, as in gliding, is that with flapping flight you need to visualize the complex twists and turns carried out by a bird's wing during the downstroke and upstroke. These twists and turns form a complicated motion that is difficult to analyze, but they are essential in generating the forces that allow a bird to fly. ∎

Portions of this sidebar were reprinted from Vertebrate Life, 4th Edition, by F. Harvey Pough, John B. Heiser, and William N. McFarland, 1996, pp. 528–531. ©1985. Reprinted by permission of Prentice-Hall, Inc., Upper Saddle River, NJ. Other portions of the sidebar were written by Sandy and Bill Podulka, in consultation with Professor John Hermanson of Cornell University.

Bluebird

Scaup

Figure 5–22. Bluebird and Scaup Landing: *When descending to a perch, the ground, or water, a bird slows down by braking against the onrushing air—it positions the body upright (increasing the angle of attack to slow forward momentum); spreads the tail to resist the air; and "back-strokes" with horizontal wingbeats. Meanwhile the bird brings its feet forward, ready for impact. Drawings by Robert Gillmor.*

Birds often reduce their momentum by landing facing into the wind, increasing the angle of attack of the wings, and beating the wings horizontally against the direction of airflow **(Fig. 5–22)**. Many birds swoop upward when landing, using gravity to counter their momentum **(Fig. 5–23)**. Species with webbed feet, as well as Old World Vultures with their unwebbed feet, often extend and spread their feet as an air brake when landing.

Figure 5–23. Downy Woodpecker Landing: *Many birds, including the Downy Woodpecker shown here, have a flight pattern termed **bounding** in which they alternate flapping (during which they rise slightly) with glides on closed wings (during which they descend slightly). To land, they swoop upward to contact a tree trunk, using gravity to counter their momentum. Drawing by Robert Gillmor.*

Hovering

Many species of birds with a variety of wing shapes **hover** at least occasionally. Hovering is an energetically expensive mode of flight achieved by beating the wings more or less horizontally. Forward thrust must be balanced by wind speed, and gravity exactly compensated for by lift **(Figs. 5–24, 5–25)**.

Figure 5–24. Arctic Tern Hovering: *Hovering is an energetically expensive type of flight, but it is the only way for a foraging Arctic Tern to get an aerial view as it hunts for fish. Neither the open water nor the flat, treeless Arctic Tundra where it lives offer good vantage points for searching for prey. The Arctic Tern alternates slow, low-level flight over the water with periods of hovering, similar to the hunting style of the American Kestrel and Northern Harrier—which also hunt in open areas with few good perches. Photo by Tom Vezo.*

Hovering has been carried to its ultimate in the hummingbirds— the only group of birds that can hover for any substantial length of time in still air, beating their wings up to 80 times per second. Using subtle adjustments of the wings that become visible only with high-speed photography (first carried out by Greenewalt [1960]), hummingbirds can fly forward, backward, or hover in one place with apparent ease.

Lift

Wing Motion

Wing Motion

Gravity

Figure 5–25. Belted Kingfisher Hovering: *When hesitating in midair for one reason or another, most birds, such as this Belted Kingfisher, hover by positioning the body more or less vertically and simply flapping the wings forward and backward horizontally to provide lift but not thrust. Hovering uses a great deal of energy because to stay aloft the bird must flap hard enough to generate an upward force equal to its own weight. Drawing by Robert Gillmor.*

Figure 5–26. Hummingbird Hovering:
When hovering with the body motionless, as when taking nectar from a flower, a hummingbird moves its wings in a unique pattern that, unlike the wing motion of other hovering birds, generates lift on both the forward and backward stroke. This is possible because the hummingbird wing differs from that of other birds; the arm bones are reduced such that the hand (outer wing) makes up most of the wing area, and the elbow and wrist joints are locked to form a nearly rigid, unbending wing that moves from the specialized shoulder joint. Pictured, from top to bottom, is the sequence of wing positions involved in one complete wingbeat. Note that on the backward stroke, the wing does not fold. Instead, it rotates such that the lower surface faces up. Adapted from Burton (1990,

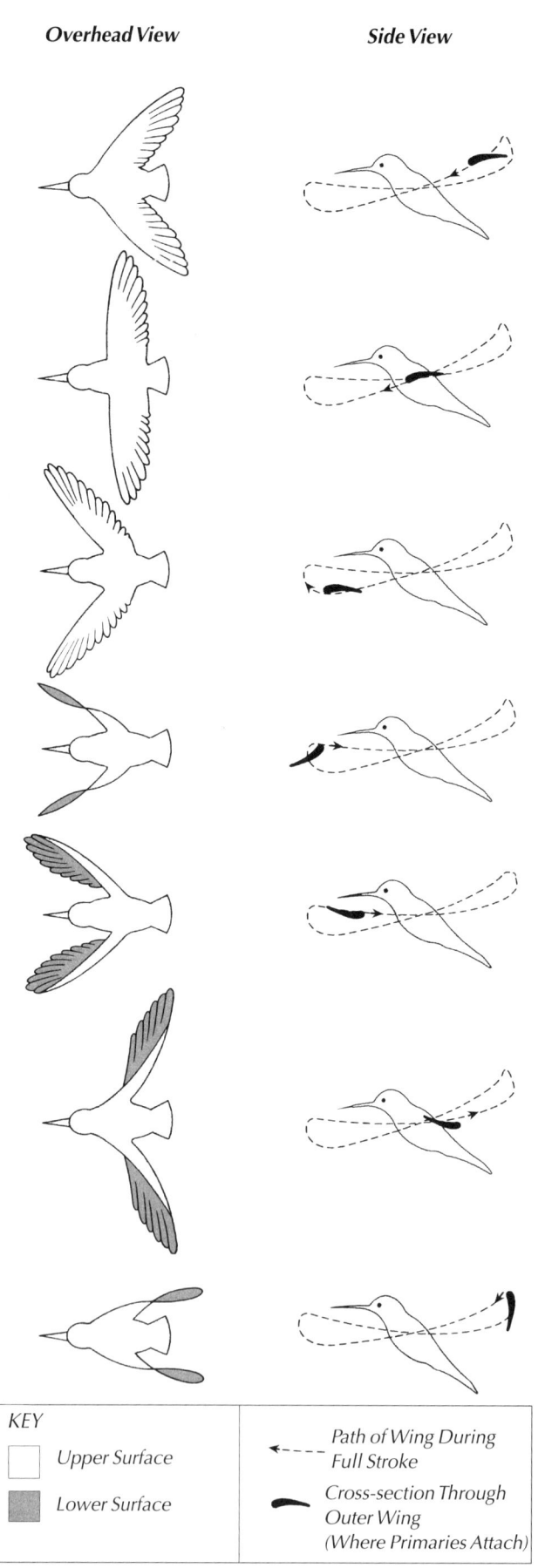

Overhead View   **Side View**

KEY

☐ Upper Surface

▨ Lower Surface

◀--- Path of Wing During Full Stroke

🮡 Cross-section Through Outer Wing (Where Primaries Attach)

Because of the unique structure of the hummingbird's humerus and its articulation with the pectoral girdle, its shoulder can rotate, allowing the wing to push air in almost any direction, depending on the angle **(Fig. 5–26)**.

The small, pelagic storm-petrels use a unique form of hovering to search the ocean for tiny organisms. Storm-petrels hover low over the water with their feet treading ("pattering") on and just below the surface, looking very much as if they were "walking on water" **(Fig. 5–27)**. This unique behavior probably earned them their name, in reference to St. Peter's attempt to "walk on water." Storm-petrels search the surface of the open sea for prey by letting the wind blow them along. Then, when they need to look more closely, they hover or fly very slowly while facing into the wind, pattering their feet. The storm-petrel's flight is typically called "hovering," but it is really more like soaring into the wind. Instead of expending much energy, as in other forms of hovering, the storm-petrel actually uses the wind's energy to stay aloft. In addition, unlike most types of hovering, the wings remain relatively still, held out over the back in a V. (The wings do flip over and back, but how they generate lift in this way is beyond the scope of this course.)

Storm-petrels soaring in this fashion would be blown backward with the wind and would lose lift as the speed of the airflow relative to their wings dropped to zero, if they did not do something to hold themselves back. By dangling their feet in the water, storm-petrels create drag that helps to "anchor" them in place against the wind, like someone holding the string of a kite (Withers 1979). If the feet were motionless in the water, the birds would still be pushed backward by the wind, but much more slowly than if their feet were not in the water. By pattering their feet, storm-petrels can hold themselves steady,

Figure 5–27. Wilson's Storm-Petrel "Walking on Water": A Wilson's Storm-Petrel hovers low over the ocean with its feet treading the water, searching for small marine animal prey. Looking much like large butterflies, storm-petrels hold their wings over their backs in a V and actually "soar" into the wind, dangling their legs in the water to create enough drag to hold them in place against the wind. Wilson's Storm-Petrels are possibly the most numerous seabirds in the world. Photo by Doug Allan/Oxford Scientific Films.

or perhaps even push themselves slowly forward, against the force of the wind. The storm-petrels' particularly light wing loading allows them to generate sufficient lift to stay above the water's surface in winds that are light enough to allow their foot-dragging technique to work.

Complex Control of Flight

Watch a Barn Swallow coursing over a hay field as it chases flying insects, and you witness an unceasing array of swirling, swooping flight maneuvers of consummate grace. To describe such a performance in terms of gravity, drag, lift, thrust, and a little use of the tail, is like explaining the boundless flexibility of the human hand and arm by saying "It all results from the contraction and relaxation of the muscles." True enough, but that doesn't really explain how it works.

Therefore, one must imagine the forces that control flight not as separate elements that the bird controls using a few static techniques at all times—as a pilot controls a plane—but as a constantly changing melange of forces which the

Figure 5–28. Rufous Hummingbird at Columbine: Birds carry out all sorts of aerial maneuvers with such elegance that, to us, their flight appears effortless. To the contrary, flight is precisely controlled by strong muscles acting in complex ways, much like ballet. Photo courtesy of Donald Waite/CLO.

bird manipulates with almost infinite flexibility.

Within rather wide limits, birds can vary their wing area, wing camber (the arch of the wing's airfoil), and angle of attack. They can glide; they can flap their wings slowly or quickly; they can flap continuously, like a duck; or in spurts, producing the bounding flight of woodpeckers and various finches, and the flap-glide flight of ravens, hawks, and other large birds. They can spread the alular or primary feathers at will, or tilt their wings backward. Birds can spread their tail feathers, arc them up and down, and tilt them to either side. Birds also can move each wing independently of the other, allowing them to twist and turn in the air, sometimes quite abruptly. Birds can alter the proportions of lift and thrust in the course of a single stroke of the wing, and many birds are capable of producing lift on the upstroke as well as the downstroke. Gravity is the antagonist of lift, but birds constantly use gravity to gain momentum or to counteract it when landing; drag opposes thrust, but birds use drag to maneuver and stop.

Like ballet, flight appears effortless and free, but only because the muscles are strong and the movements involved are precisely controlled. Indeed, the fine control of the complexities of flight, involving muscular control over not only the wings and tail but also individual feathers, is so fluid and so complicated that it is still not well understood **(Fig. 5–28)**.

Wing Loading

An important factor influencing how a bird flies is its **wing loading**, the ratio of body weight to wing area, or how much "load" each unit area of wing must carry. The wing loadings of birds vary tremendously between species. The Leach's Storm-Petrel spends much of its life in flight over the sea, has very large wings for its weight, and has a low wing loading of 0.0015 lb/in^2 (0.1 g/cm^2). Larger birds with relatively smaller wings (loons, auks, albatrosses) may have wing loadings of up to 0.03 lb/in^2 (2.1 g/cm^2) or more. The American Crow has an average sort of wing loading at 0.0058 lb/in^2 (0.41 g/cm^2). In contrast, the wing loading of a Boeing 747 aircraft is much heavier, at 0.805 lb/in^2 (56 g/cm^2).

In general, smaller birds have lighter wing loadings than larger birds, but flight style is also a factor. For example, large soaring birds, such as eagles and vultures, have lighter wing loadings than other similar-sized birds **(Table 5–1)**.

Table 5–1. Wing Loadings of Various Birds: *Presented here are the wing loadings (how much "load" each unit area of wing must carry) for a variety of bird species, arranged from lightest to heaviest body weight. Note that, in general, larger birds carry a heavier wing loading, but flight style and other aspects of a bird's lifestyle play a role as well. Weights are presented in ounces (oz) and pounds (lb), and areas in square inches (in^2). Metric conversions for each are given in grams (g) and square centimeters (cm^2). Adapted from Welty and Baptista (1988, p. 473). Originally from Poole (1938).*

SPECIES	WEIGHT oz (g)		WING AREA in^2 (cm^2)		WING LOADING lb / in^2 (g / cm^2)
Ruby-throated Hummingbird	0.11	(3.0)	1.92	(12.4)	0.0033 (0.24)
House Wren	0.39	(11.0)	7.50	(48.4)	0.0033 (0.23)
Black-capped Chickadee	0.44	(12.5)	11.78	(76.0)	0.0023 (0.16)
Barn Swallow	0.60	(17.0)	18.37	(118.5)	0.0020 (0.14)
Chimney Swift	0.61	(17.3)	16.12	(104.0)	0.0024 (0.17)
Song Sparrow	0.78	(22.0)	13.41	(86.5)	0.0036 (0.25)
Leach's Storm-Petrel	0.93	(26.5)	38.92	(251.0)	0.0015 (0.11)
Purple Martin	1.52	(43.0)	28.76	(185.5)	0.0033 (0.23)
Red-winged Blackbird	2.47	(70.0)	37.98	(245.0)	0.0041 (0.29)
European Starling	2.96	(84.0)	29.50	(190.3)	0.0063 (0.44)
Mourning Dove	4.59	(130.0)	55.35	(357.0)	0.0052 (0.36)
Pied-billed Grebe	12.12	(343.5)	45.12	(291.0)	0.0168 (1.18)
Barn Owl	17.81	(505.0)	260.93	(1683.0)	0.0043 (0.30)
American Crow	19.47	(552.0)	208.37	(1344.0)	0.0058 (0.41)
Herring Gull	29.98	(850.0)	311.01	(2006.0)	0.0060 (0.42)
Peregrine Falcon	43.12	(1222.5)	208.06	(1342.0)	0.0130 (0.91)
Mallard	49.66	(1408.0)	159.54	(1029.0)	0.0195 (1.37)
Great Blue Heron	67.20	(1905.0)	687.75	(4436.0)	0.0061 (0.43)
Common Loon	85.54	(2425.0)	210.54	(1358.0)	0.0254 (1.79)
Golden Eagle	164.52	(4664.0)	1010.85	(6520.0)	0.0102 (0.72)
Canada Goose	199.72	(5662.0)	437.21	(2820.0)	0.0286 (2.01)
Mute Swan	409.24	(11602.0)	1055.50	(6808.0)	0.0242 (1.70)

Wing loading imposes the ultimate limit on the size of flying animals. One might think there would be an optimum wing loading, so that larger birds would simply have proportionally larger wings. As the size of birds increases, however, their volume (which determines their weight) increases faster than their surface area. To maintain the same wing loading as small birds, then, large birds would require wings that were proportionally much larger with respect to the rest of the body **(Fig. 5–29)**. For example, for a bird the size of a swan to have a wing loading as light as that of a small passerine, its wings would require more than ten times the surface area that swan wings actually have. Such immense wings would be impossible to power or to control. Thus, larger birds tend to have large wing loadings.

Figure 5–29. Wing Loading in Relation to Body Size: *For a large bird to have the same **wing loading** (the ratio of body weight to wing area) as a small bird, the wings must be much larger with respect to the body. This is true because as the overall size of a bird increases, the volume (which is directly related to weight) increases faster than the wing area. Because large birds are generally unable to power the huge wings they would need to have the same wing loading as a typical small bird, large birds tend to have heavier wing loadings.*

Volume is assumed to roughly determine weight:

L= length   **W = width**   **D= depth**

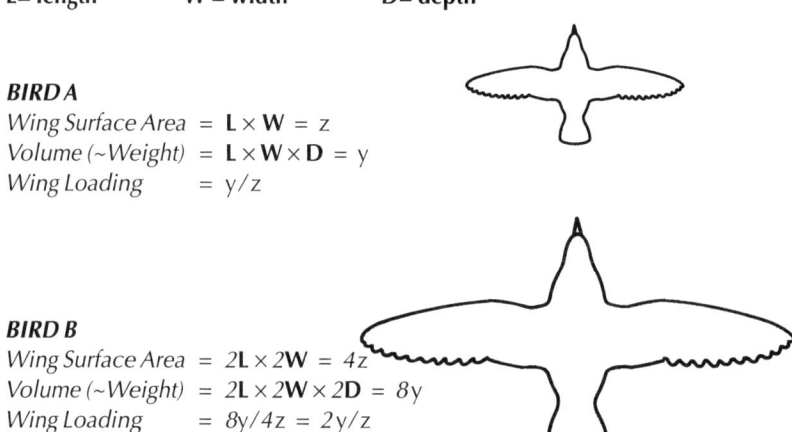

BIRD A
Wing Surface Area = $\mathbf{L} \times \mathbf{W}$ = z
Volume (~Weight) = $\mathbf{L} \times \mathbf{W} \times \mathbf{D}$ = y
Wing Loading = y/z

BIRD B
Wing Surface Area = $2\mathbf{L} \times 2\mathbf{W}$ = 4z
Volume (~Weight) = $2\mathbf{L} \times 2\mathbf{W} \times 2\mathbf{D}$ = 8y
Wing Loading = 8y/4z = 2y/z

If the overall size of the bird doubles, the wing surface area increases by a factor of 4, and the volume increases by a factor of 8. Thus the bird's wing loading is twice as great as that of Bird A.

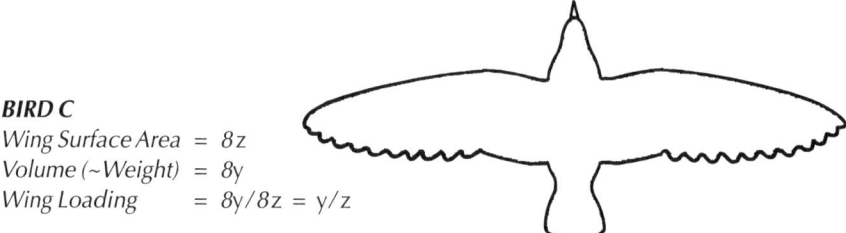

BIRD C
Wing Surface Area = 8z
Volume (~Weight) = 8y
Wing Loading = 8y/8z = y/z

As in Bird B, the overall size of this bird's body has doubled (over Bird A) such that the volume is approximately 8 times that of Bird A, and is equal to that of Bird B. But here the wing surface area is increased even further, so that the wing loading is equal to that of Bird A. Note that for a bird's overall size to increase without an increase in wing loading, the wings must become much larger in proportion to the rest of the body.

To simplify the math here, it is assumed that the large wings do not contribute significantly to the weight. If the increased weight of the wings were taken into account, their surface area would have to be even larger to keep wing loading constant from Bird A to Bird C.

Because higher wing loadings make flight more difficult, larger birds must compensate in various ways. Most solve the problem by increasing speed. Birds with low wing loadings, such as grouse, quail, or dabbling ducks, can leap into the air and start flying. But birds with higher wing loadings must speed along the ground to become airborne, and must maintain higher speeds to stay aloft. Albatrosses get a running start or launch themselves from high ground; diving ducks and loons paddle across the water **(Fig. 5–30)**. Alcids (auks and puffins), which have very high wing loadings so that they can dive under water, often hold their webbed feet, toes spread, on either side of their tails while flying. This apparently increases the available lift-generating surface.

Considerations of wing loading and power requirements suggest that the maximum weight for a flying bird is near 26 lb (11.8 kg).

Several living species from different lineages are near this size: the Kori Bustard, American White Pelican, Trumpeter Swan, and Andean Condor. The largest known flying bird was the Pleistocene condor, *Teratornis incredibilis*, which is estimated to have weighed about 44 lb (20 kg) and to have had a wingspan of 16 feet (4.9 m). Presumably it flew almost entirely by soaring; how it got airborne remains unclear. For a person of average weight to take off under his or her own wing power, he or she would need about 30 square feet (2.8 square meters) of wing area; each wing would weigh about 250 lbs (113.5 kg)—much more than a human could hold out horizontally!

Figure 5–30. Scaup Taking Flight: Birds with high wing loadings, such as this scaup, other diving ducks, loons, swans, geese, and alcids, must reach high ground speeds before they can generate sufficient lift to become airborne. They do this by running and flapping across the water's surface during takeoff. Some heavy birds are unable to take flight from land or from small bodies of water. Drawing by Robert Gillmor.

Turbulence

Although birds are beautifully streamlined, certain aspects of their movement through air still generate turbulence. In particular, turbulence may be created when smooth airflow is disrupted by very high angles of attack, or by friction with surfaces such as a bird's wing feathers. Because turbulence, in turn, increases such friction, it is an important source of drag, and can create unwanted stalling, requiring a bird to use excessive amounts of energy.

Turbulence is often formed at the trailing edge of the wing as the air closest to the top surface of the wing is slowed by friction with the feathers. As mentioned previously (see Fig. 5–13), the airflow may separate from the top of the wing as turbulence in the form of swirls of air that move forward from the trailing edge of the wing. This type of turbulence is greatest at slow air speeds and at high angles of attack with heavy loads.

Another type of turbulence, called **tip vortex**, is created at the tip of the wing. It consists of currents of air spiraling off the wing tip behind the bird **(Fig. 5–31)**. (A bird flying close behind another can take advantage of the lift created by the rising portion of these spirals; see Flocking and Flying in Formation, later in this chapter). Recall that the static air pressure below a moving wing is higher than that above, due to the airfoil shape and Bernoulli's law. Because air tends to flow in a direction that decreases pressure differences, air from the high-pressure area below the wing flows toward the low-pressure area above the wing. This airflow, combined with air flowing from front to back across the top of the wing, creates the spirals of tip vortex. These eddies sometimes may be seen as trails of white behind the wing tips of large airplanes as they take off or land, and are particularly visible on rainy days. These rings of air circulate in a counterclockwise direction on the left side of the bird or aircraft and clockwise on the right, when viewed from the front.

In a flapping bird, the rings of tip vortex are generated as the wings move up and down—a motion accompanied, as you will recall, by twisting and turning of the outer wings. The exact orientation of the vortex ring will depend on the angle of the wing stroke. The size of the vortex ring and its velocity of circulation will depend on both the angle of attack relative to the airflow and the degree of camber of the wing. These latter two variables affect tip vortex because they determine how much pressure difference is created as the air flows across the upper and lower wing surfaces.

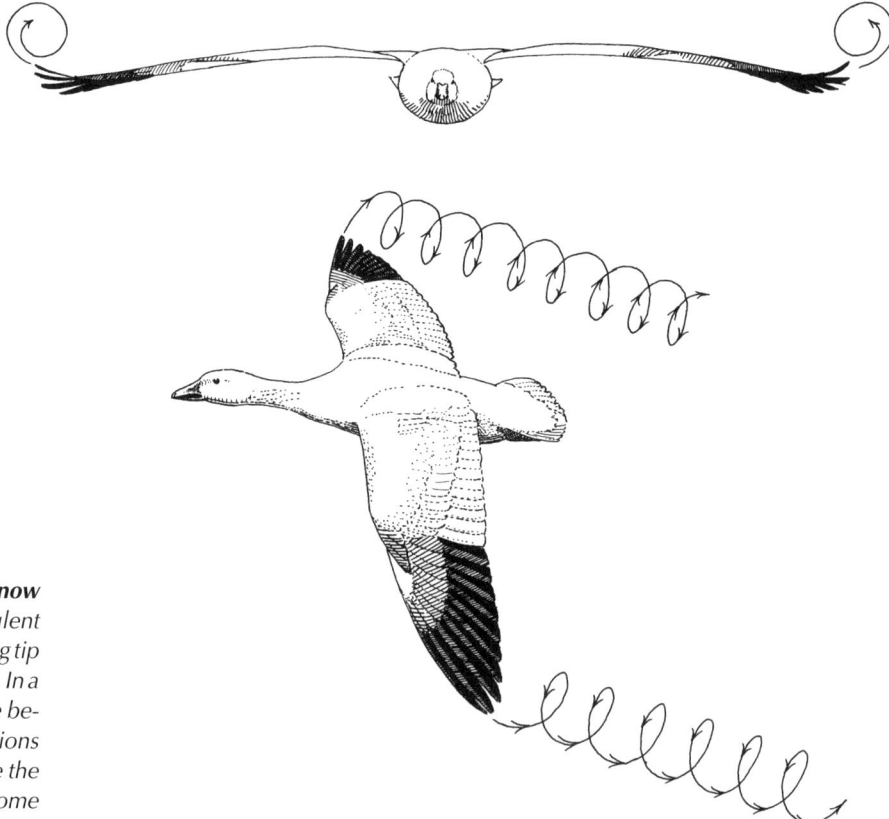

*Figure 5–31. Tip Vortex on a Flying Snow Goose: As a bird flies, swirls of turbulent air called **tip vortex** spiral off the wing tip behind the bird, and trail downward. In a flock, a bird flying at an angle close behind another bird (as in the V formations of Snow and Canada geese) can use the rising portion of the spirals to gain some "free" lift.*

Like other types of turbulence, tip vortex creates drag that interferes with lift, and this drag is greatest at slow air speeds. The drag can be offset in several ways. One method is to elongate the wings (holding width constant), which increases the surface area between the tips that creates lift and is not affected by drag. Elongation improves the ratio of lift to drag (termed the **lift-to-drag ratio**, an important aerodynamic property) and thus compensates for the drag induced at the tip. The effect is most pronounced in long, narrow wings, because the leading edge of the wing produces most of the lift. Long wings also decrease drag because they keep the areas of turbulence at opposite wing tips farther apart. Because gliding birds use high lift-to-drag ratios to stay aloft, one would expect them to have relatively long, thin wings—and most, such as storks and albatrosses, do. Some albatrosses glide so well that they can cover 50 to 60 feet (15 to 18 m) while losing only 3 feet (1 m) of altitude (Kress 1988).

Another way to reduce the drag created by tip vortex is to have narrow, pointed wing tips. The smaller the wing tip area, the less the pressure difference below and above the tip, so the lower the amount of turbulence, and therefore drag. In contrast, broad, rounded wing tips create the most tip vortex.

A final way to decrease tip vortex is to have wing tips with a high degree of **slotting**. Slots are gaps between the feathers of the wing tips, created when a bird having narrow-tipped primaries spreads them during flight. Common in large soaring birds such as eagles, vultures, condors, and certain hawks, slotting makes the wings look as though they were tipped by widespread fingers **(Fig. 5–32)**. Slotting reduces tip vortex by turning each primary feather into an individual, narrow, pointed "wing tip." It also increases lift, because each separated primary feather acts as an individual airfoil (even without a flapping motion), generating its own lift. Slotting thus allows a bird with a broad, rounded wing tip to increase its lift-to-drag ratio.

As the understanding of tip vortices and their effect on bird flight has improved, researchers have been better able to assess the energetic costs of flight and to explain why such a diversity of flight styles exists. For example, the bounding flight of woodpeckers, in which the birds alternate flapping with gliding on closed wings, may be an adaptation that decreases the effect of tip vortices: closing the wings temporarily gets rid of the vortices, and thus allows a gliding phase with less turbulence. In the future, advances in the understanding of vortices will undoubtedly provide even greater insight into the mechanics and energetics of bird flight.

*Figure 5–32. Soaring Turkey Vulture: Large soaring birds, such as this Turkey Vulture, usually counter the lift-reducing effects of tip vortex by having wings with a high degree of **slotting**. Slots are gaps between the feathers of the wing tips that make the wings look as though they were tipped by widespread fingers; they are created by having primary feathers with narrow tips and by spreading the primary feathers during soaring. Photo by Marie Read.*

Variations in Wing Shape and Flight Style

Spend some time watching birds fly, and you quickly will notice the variations: big birds fly differently than little birds, fast birds fly differently than slow ones. Those with short, stubby wings fly differently than those with long wings. From the buzzy flight of the tiniest hummingbird to the sedate movements of giants such as condors and albatrosses, birds fill the air with a diversity of flight styles.

Flying requires a tremendous amount of energy, and birds are under intense selection pressure to reduce energy demands any way they can. One important factor is wing shape. For typical flapping flight, the most aerodynamically efficient wings are large, long, and relatively narrow. But such wings also have costs: they are harder to control and maneuver, they may interfere with takeoff, and they do not necessarily allow the most rapid flight. Other wing shapes, although less energy-efficient while flapping, perform some of these tasks better. Birds also can reduce the energy demands of flight through gliding and soaring, the latter of which is best performed with a wing shape not well suited for flapping flight.

All types of flight—both gliding and flapping—require lift, and flapping flight also requires thrust. Around that basic theme, however, the flight styles of birds are as varied as the habitats they occupy and the lifestyles they lead. Indeed, each lifestyle *requires* different skills and different flight styles. So, the wing shape of each species is a result of evolutionary compromises that allow the bird to meet the total array of life challenges—not just those challenges posed by efficient flapping flight. You may want to refer to **Fig. 5–33** during the following discussion of flight styles and wing shapes.

Bird wings have been classified into four general types based on shape and aerodynamic performance: elliptical, high-speed, slotted high-lift, and high-as-

Figure 5–33. Types of Flight: *The flight styles of birds are as varied as their food habits and plumage colorations, but they can be grouped into a few basic categories. Note that a given species or individual bird may use different types of flight under different circumstances.*

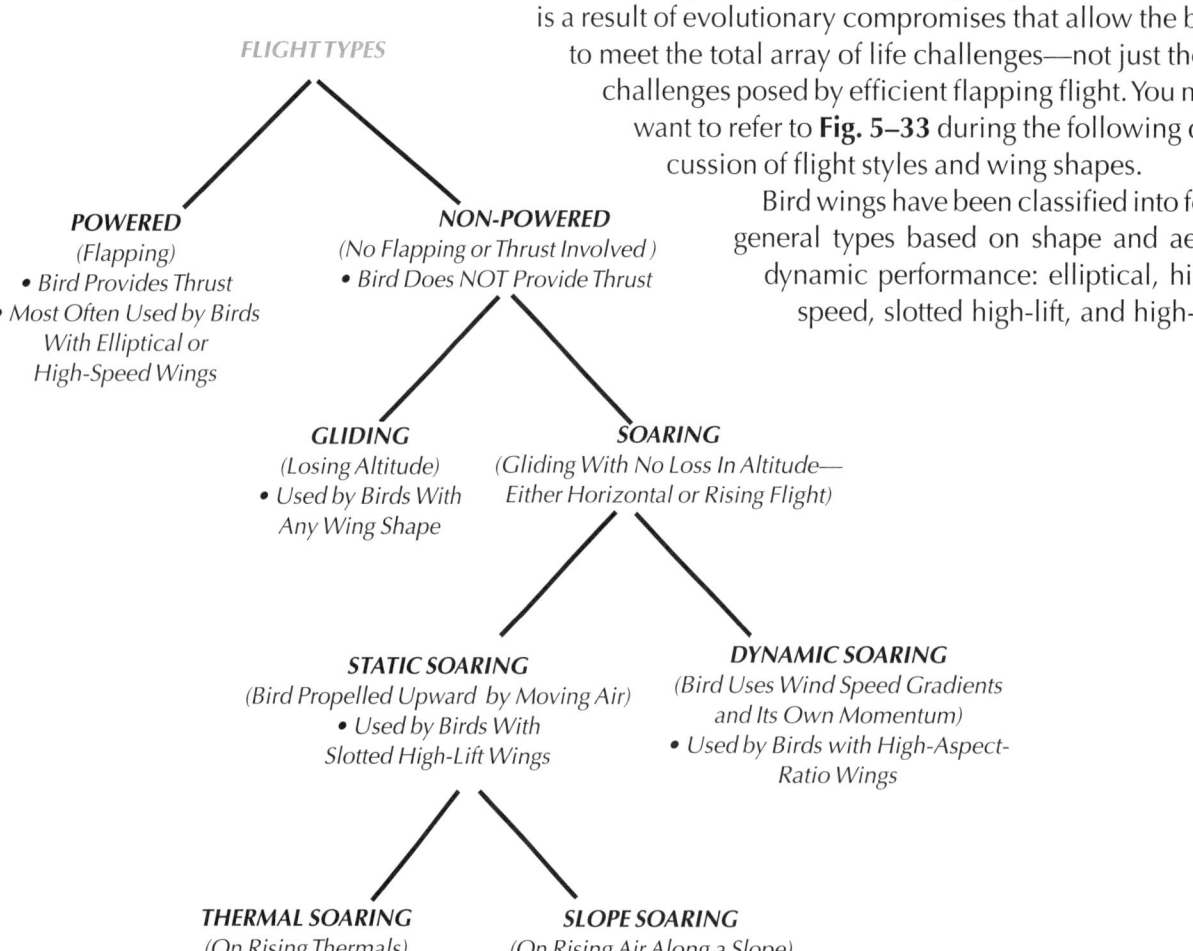

FLIGHT TYPES

POWERED
(Flapping)
• *Bird Provides Thrust*
• *Most Often Used by Birds With Elliptical or High-Speed Wings*

NON-POWERED
(No Flapping or Thrust Involved)
• *Bird Does NOT Provide Thrust*

GLIDING
(Losing Altitude)
• *Used by Birds With Any Wing Shape*

SOARING
(Gliding With No Loss In Altitude— Either Horizontal or Rising Flight)

STATIC SOARING
(Bird Propelled Upward by Moving Air)
• *Used by Birds With Slotted High-Lift Wings*

DYNAMIC SOARING
(Bird Uses Wind Speed Gradients and Its Own Momentum)
• *Used by Birds with High-Aspect-Ratio Wings*

THERMAL SOARING
(On Rising Thermals)

SLOPE SOARING
(On Rising Air Along a Slope)

pect-ratio **(Fig. 5–34)**. Although these categories are arbitrary and falsely group many flight styles, they can help to simplify the extraordinarily wide array of wing shapes found in nature.

Elliptical Wings

Most birds that live in forests, woodlands, or shrubby areas where they must maneuver around and through dense vegetation have **elliptical** wings. Most songbirds, crows, grouse, and quail fall into this category, even though they are not closely related. Exact wing shapes among this group vary a good deal, but in general, elliptical wings are short and broad—that is, they have a low **aspect ratio**. The aspect ratio is the ratio of the length to the width of a wing. Long, narrow wings have a high aspect ratio; and short, broad wings have a low aspect ratio. The turbulence created by the broad tips of elliptical wings is offset somewhat by a high degree of slotting of the primary feathers, which increases lift.

Birds with elliptical wings have traded the aerodynamic advantages of a longer wing for the maneuverability of a shorter wing. The breadth of their wings creates less lift for their size, but helps to reduce wing loading and thus further increases maneuverability. This wing shape also produces a relatively slow flight.

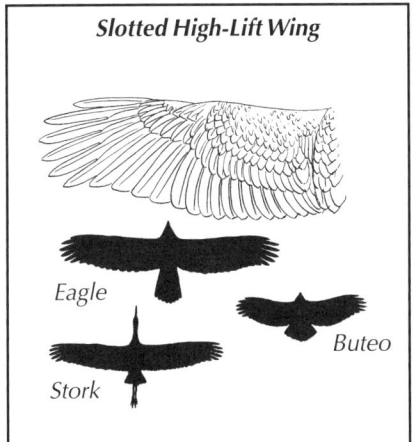

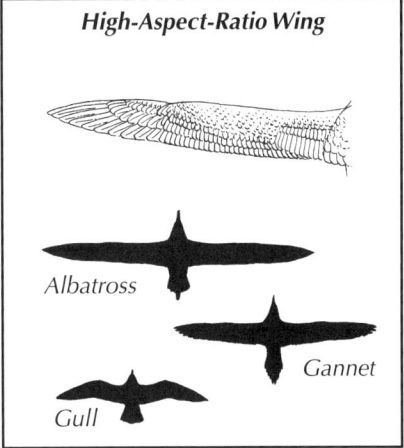

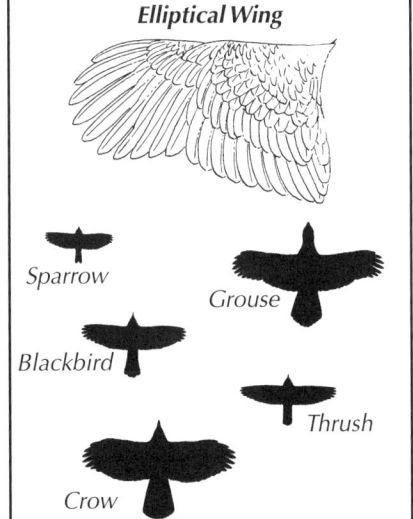

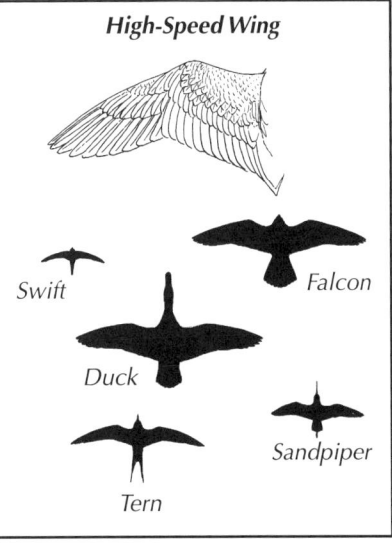

Figure 5–34. Major Wing Types: The tremendous diversity of bird wings have been classified by ornithologists into four major types based on both shape and flight performance. Although these categories are imposed by humans onto a characteristic that actually varies through a continuum, they are helpful in making sense of the overwhelming variety of bird wings. See text for detailed descriptions of each wing type and the flight styles that make use of it.

The importance of habitat in selecting for elliptical wings can be seen by comparing closely related birds of different habitats. Forest hawks and owls, such as the Cooper's Hawk, Sharp-shinned Hawk, and Barred Owl, have short, rounded wings, whereas their open-country counterparts, Northern Harriers and Short-eared Owls, have long, narrow wings **(Fig. 5–35)**.

High-speed Wings

The **high-speed wing** is the tapered, pointed, often swept-back wing characteristic of falcons, swifts, swallows, terns, ducks, and many shorebirds. The primaries have no slotting, and the wings have a high aspect ratio **(Fig. 5–36)**. Flying with this type of wing is energetically expensive because the birds must flap constantly to move fast enough to generate sufficient lift.

Birds with this wing shape generally feed on the wing or migrate very long distances, situations in which high speed and control are crucial. But, they have traded efficient lift generation for these benefits. Birds that need a high wing loading, such as those that dive or

Forest Raptors

Barred Owl

Cooper's Hawk

Open-country Raptors

Figure 5–35. Wing Shapes of Forest versus Open-country Raptors: The Cooper's Hawk and Barred Owl, both forest birds, have short, rounded ("elliptical") wings that allow them to maneuver easily among the trees. Open-country raptors, such as the Northern Harrier and Short-eared Owl, have longer, narrower wings, trading in the maneuverability of short wings for several aerodynamic advantages, such as reducing wing tip turbulence and increasing lift.

Short-eared Owl

Northern Harrier

Cornell Laboratory of Ornithology

swim under water, may also need high-speed wings to remain airborne.

Slotted High-lift Wings

Another group of birds has evolved a type of flight called **static soaring**. They take advantage of the energy in rising air masses to obtain lift with little or no energy expenditure on their part. The most familiar examples are eagles, vultures, storks, and some owls and hawks, such as Broad-winged Hawks and Red-tailed Hawks. These birds seek air masses rising fast enough to propel them upward, and remain in them long enough to rise to great heights, often by soaring in a series of tight spirals. They then use their height to glide effortlessly for a long time, either to cover a territory, to migrate, or to search for food.

Soaring of this sort requires a **slotted high-lift wing**. As the name implies, such a wing tends to be broad with a deep camber and to have very prominent slotting (see Fig. 5–32). The aspect ratio is moderate—between that of elliptical wings and high-aspect-ratio wings. The broad wings help to catch rising air, somewhat like a kite, and help to reduce wing loading. They also allow slow flight speed, which enables the birds to turn in tight spirals. The slotting produces additional lift (while reducing tip vortex), increasing a soaring bird's ability to fly slowly. The extra lift also allows these birds to carry heavy prey. Many soarers can fold or spread their primaries and tail feathers, changing their wingspread and tail shape to increase maneuverability.

*Figure 5–36. **High-speed Wings of a Forster's Tern:** The wings of this winter (basic) plumage Forster's Tern are typical of birds with "high-speed wings;" they are tapered, pointed, swept-back, long and narrow, and not slotted. Like other birds with this wing shape, Forster's Terns must flap constantly to move fast enough to generate sufficient lift to stay aloft. Photo by Marie Read.*

Static Soaring

Static soaring takes several different forms. Perhaps the most familiar is **thermal soaring**. Thermals are columns of rising warm air that result from differential heating of land surfaces by the sun. Dark surfaces such as plowed fields or asphalt parking lots heat up faster than adjacent forests or water bodies, and the south-facing slopes of hills heat up faster than those that face north. It is over these warmer areas that thermals form and "kettles" (large aggregations of rising, spiraling, birds—usually hawks) assemble **(Fig. 5–37)**. Broad-winged Hawks are famous for their kettles containing hundreds of birds during migration (see Ch. 6, Sidebar 4, Fig. Fa). Thermals are most common in warmer parts of the world and in the interior of continents, where lateral winds are less likely to break up the pattern of air movement. Thermal soaring is used by species as diverse as Old and New World vultures, diurnal raptors, cranes, storks, pelicans, and swifts. In many of these species, notably the vultures, hawks, and eagles, thermal soaring is used to search for food. Many of these species are near the maximum weight for flying birds and must cover large areas to find food. Thus, the subsidy they derive from the energy of the atmosphere is crucial to their way of life.

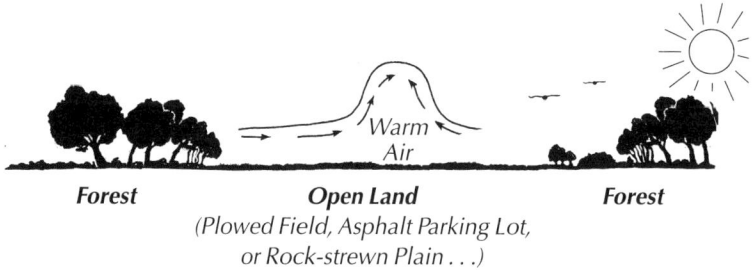

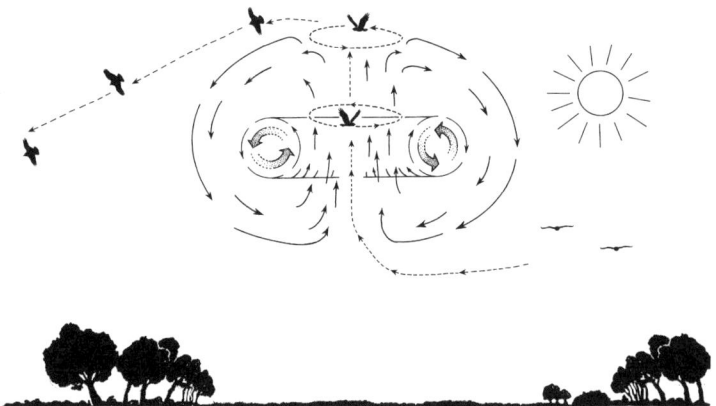

Figure 5–37. Development of a Thermal: *Thermals, rising columns or bubbles of warm air, are created by sunlight. They frequently develop over isolated patches of ground that heat up more quickly than their surroundings; these are often large, dark surfaces such as plowed fields or asphalt parking lots. The late-morning sunlight strikes these areas, causing them to radiate heat, warming the air just over them and forming a warm mass of air surrounded above and to the sides by cool air (top drawing). Note that adjacent forested areas, although they may be dark as well, do not radiate heat as fast, partly because the high moisture content and complex structure of the forest causes it to retain its heat. As the warm air mass heats up further, the air expands and, because it is lighter than the surrounding cool air, rises like the bubbles that form on the bottom of a heated tea kettle (middle drawing). Cool air moves in below the rising bubble; soon it, too, is warmed and forms another rising bubble. Within each bubble the air circulates upward in the center and downward on the outside, producing a revolving ring of warm air somewhat like a smoke ring (bottom drawing).*

Soaring birds use thermals to gain altitude with little input of their own energy. They enter the thermal and circle upward on the rising central column of air. When the speed of the rising air currents is sufficient to offset a bird's weight, it can glide in the circle with little or no flapping, gaining altitude as the bubble continues to rise.

Thermal soaring also can be used to move cross-country **(Fig. 5–38)**. The bird climbs in a thermal (propelled upward by the rising airstream), often reaching a height of 6,000 feet (1,830 m) or more, then glides out in the chosen direction, losing altitude as it goes. When it locates another thermal (possibly by the presence of other soaring birds), the bird repeats the process. This strategy is used by migrating hawks, vultures, storks, and cranes. Energetically, it is very efficient. Some estimates indicate that the total rate of energy expenditure is about one-thirtieth that required to fly the same distance under power.

Turkey Vultures and Black Vultures, although both static soarers possessing slotted high-lift wings, have different wing shapes that are correlated with their differences in behavior and distribution. Turkey Vultures have much longer wings and consequently a lighter wing loading. Thus they can take advantage of weaker thermals, and can

Figure 5–38. Moving Cross-Country with Thermals: In a process similar to that described in Figure 5–37, thermals may develop in temperate areas over south-facing slopes. During the spring and fall, when the sun's rays strike these slopes, they may be warmed more than north-facing slopes, which remain in shadow. Because of this differential heating, series of thermals may develop along south-facing ridges. By entering one thermal, rising in it, and gliding out to the next thermal, birds may use these thermals to move cross-country. Migrating hawks, vultures, storks, and cranes may cover great distances in this manner, using very little of their own energy.

live in more northerly areas, where thermals are not as strong or as common. Turkey Vultures are also active in the earlier and colder part of the day, whereas Black Vultures must wait for stronger thermals to develop before taking flight **(Fig. 5–39)**.

In **slope soaring**, a bird derives lift from the rising air deflected upward when wind strikes a hill or ridge. This kind of soaring is common along seacoasts, where gulls, terns, fulmars, and gannets may soar for hours above the tops of windward sea cliffs **(Figs. 5–40, 5–41)**. In strong winds, birds may remain motionless relative to the ground for long periods. Even birds with high wing loading, such as alcids or cormorants, can slope soar when the wind is strong enough. Ravens and crows often slope soar, and the small American Kestrel frequently uses updrafts from road cuts to hang in the air while it watches for prey. This phenomenon also produces the concentration of migrating hawks along mountain ridges, especially those that parallel the basic migration direction. Hawk Mountain, a high, slender part of the north-south Kittatinny Ridge in Pennsylvania, is one of the many excellent places in eastern North America to view autumn hawk migration. The flow of hawks is particularly strong on clear, windy days following the passage of a cold front (see Fig. 5–64).

High-Aspect-Ratio Wings

Birds that spend most of their lives soaring possess what is called a **high-aspect-ratio wing**. Like the high-speed wing, it is narrow and unslotted; however, the proximal area of the wing that generates lift is greatly elongated. This wing is highly efficient at producing lift at

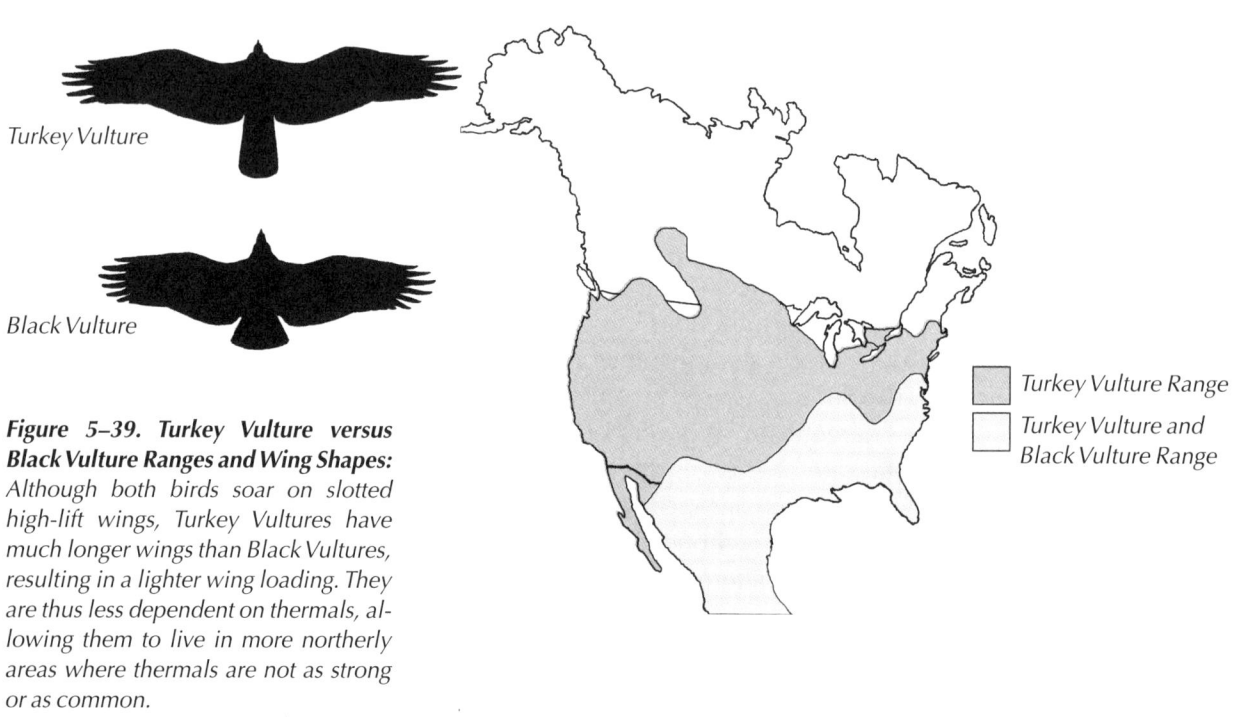

Figure 5–39. Turkey Vulture versus Black Vulture Ranges and Wing Shapes: *Although both birds soar on slotted high-lift wings, Turkey Vultures have much longer wings than Black Vultures, resulting in a lighter wing loading. They are thus less dependent on thermals, allowing them to live in more northerly areas where thermals are not as strong or as common.*

Turkey Vulture

Black Vulture

Turkey Vulture Range

Turkey Vulture and Black Vulture Range

a. Onshore Wind

b. Offshore Wind

Figure 5–40. Slope Soaring: Birds frequently fly along cliffs by simply gliding and rising on the updrafts created by (a) oncoming winds striking the perpendicular surfaces and deflecting upward, or by (b) winds from the opposite direction spilling over the cliffs and eddying upward. Gulls are notorious for riding updrafts along sea cliffs or updrafts caused by ships. Although slope soaring may occur along any windy ridge, it is most common along seacoasts, where winds are steady and strong. Drawing by Charles L. Ripper.

Figure 5–41. Northern Gannets Slope Soaring: Northern Gannets hang in the air over a seaside cliff in Scotland, held aloft by updrafts created as wind striking the cliff face deflects upward. Photo by B. Gadsby/VIREO.

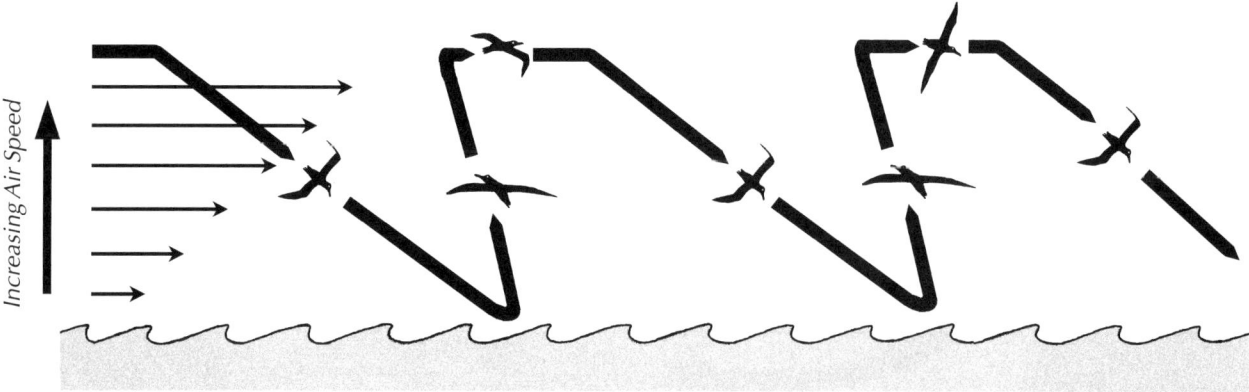

Figure 5–42. Dynamic Soaring: *Albatrosses and other seabirds with high-aspect-ratio wings use the wind gradient over the ocean to travel long distances without spending much of their own energy. Over the southern oceans near 40 degrees latitude, winds are strong and steady, but air moving over the surface of the ocean is slowed by friction with the water, with the air layers closest to the water being affected the most. The result is a vertical gradient in wind speed (arrow length represents relative wind speed). In dynamic soaring, a bird glides down the wind gradient at an angle, then turns and abruptly rises into the wind, using its momentum to gain height quickly. It then turns and glides downwind again, crossing the ocean in large zig-zags or loops, with little effort. See text for details. Adapted from Burton (1990, p. 106).*

relatively high flight speeds, but maneuverability and ease of takeoff are sacrificed as a result of the length. Such wings are found in a few seabirds that are highly specialized for soaring over the ocean, such as albatrosses, shearwaters, petrels, and to a lesser extent, gulls (see Fig. 5–1).

Dynamic Soaring

The high-aspect-ratio wing allows a special type of soaring called **dynamic soaring**, which is usually associated with albatrosses. Dynamic soaring is possible only in regions where winds are strong and constant, such as the so-called "roaring forties"—the belt of water in the southern oceans around 40° latitude. Because air moving over the surface of the ocean is slowed by friction, winds are slowest close to the water's surface, and progressively increase with height up to about 50 feet (15 meters).

An albatross at the top of the gradient glides downwind at an angle, increasing its ground speed **(Fig. 5–42)**. As it nears the water's surface it turns back the way it came and glides upward into the wind, using its momentum to rise, much as a car coasting down a hill can coast part-way up the next hill without using any gasoline. As it rises, the bird encounters progressively faster winds, which increase its lift (recall that greater wind speeds produce more lift on an airfoil). Gliding into the wind decreases the albatross's speed, however, so it does not cover as much ground as it did on its glide with the wind. When it reaches the top of the wind speed gradient, it no longer can use the energy of the wind to rise, so it turns forward and glides downwind once again. Thus, an albatross flies over the ocean in a series of vertical zig-zags or loops. In the roaring forties, albatrosses circumnavigate the globe using the westerlies to cover vast distances while expending remarkably little energy. When not dynamic soaring, an albatross may slope soar on the air currents along the windward face of a wave, until it encounters an upward gust of wind sufficient to initiate dynamic soaring.

Some Flight Facts and Figures

Air Speed

Much information about matters such as the fastest or highest flying bird is based on anecdotal observations, many of which took place only once. The numbers garnered are often suspect because of the means used to measure them. Based on reliable data from tracking and Doppler radar **(Fig. 5–43)**, most small songbirds fly at air speeds of between 20 and 30 mph (32 and 48 km/h). **Air speed** is a bird's speed relative to the air it is moving through; air speed does not include increases of speed caused by being carried along by the wind, so it may or may not reflect a bird's speed relative to the ground. Waterfowl and shorebirds can fly at higher sustained speeds of 55 to 70 mph (89 to 113 km/h); a Red-breasted Merganser pursued by a plane was clocked at 80 mph (129 km/h). House Sparrows are among the slowest species timed, at 15 to 18 mph (24 to 29 km/h). The Peregrine Falcon has a reputation for great speed, although measurements appear to be few. Peregrines apparently cruise at 40 to 62 mph (64 to 100 km/h) and are reported to be capable of reaching 175 mph (282 km/h) in a dive. Dive speed is limited, because if a peregrine reached too high a speed during an "attack dive," it might break apart on impact. In straight, powered flight, the fastest bird is reported to be the Spine-tailed Swift of India at up to 217 mph (349 km/h).

Figure 5–43. Radio-tracking a Gray-cheeked Thrush: One technique used to follow a bird in migration, gaining information on route and speed, is radio-tagging: attaching a tiny radio transmitter to a bird's back and tracking the bird from an airplane equipped with a radio to receive impulses from the transmitter. A Gray-cheeked Thrush radio-tagged in the spring at Champaign, Illinois and tracked by air from the moment it started north in the early evening, is fancifully pictured here as it passes over Chicago and heads east and then north into the darkness over Lake Michigan. Drawing courtesy of Guy Coheleach. Originally from Graber (1965).

Wingbeat Frequency

The wingbeat frequencies of most songbirds are in the range of 10 to 25 beats per second during the flapping portion of their undulating flight. A male Ruby-throated Hummingbird has been recorded at 70 beats per second and a chickadee at 27 beats per second. Larger birds generally have substantially slower wingbeat frequencies. Large vultures, for example, may flap their wings only once per second.

Flocking and Flying in Formation

Some bird species are characteristically encountered in flocks, which form in many different circumstances (see Ch. 6, Sidebar 4: Living in Groups). For example, large numbers of crows assemble in communal nighttime roosts during winter. Shorebirds and gulls typically rest and feed in groups, and in winter many songbirds—especially those that inhabit open country, such as Horned Larks and longspurs—move about in flocks.

Figure 5–44. V-shaped Flock of Snow Geese: *A bird flying close behind another bird, at an angle, may gain some extra lift from the rising portion of swirling air currents trailing behind the wing tips of the bird ahead (termed tip vortex, see Fig. 5–31). By flying in V formations, birds appear to be able to reap this energetic advantage, but evidence so far comes only from theoretical models. Photo by Tom Vezo.*

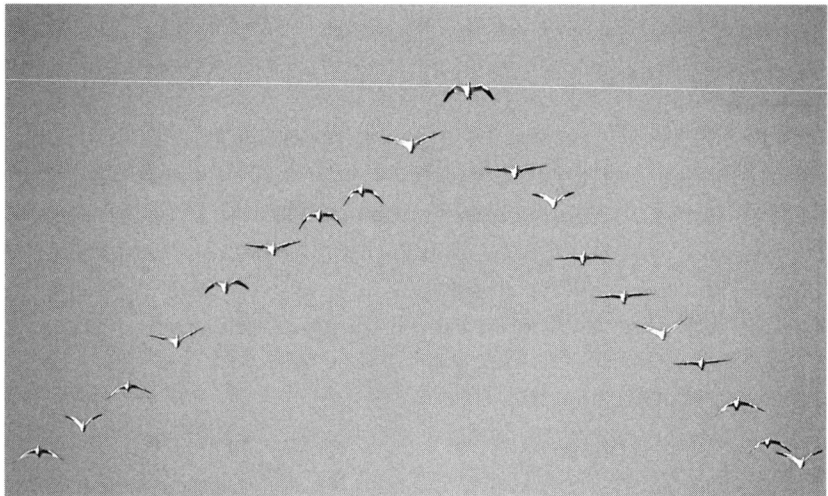

Some species move in flocks during migration. These include most waterfowl, cranes, cormorants, pelicans, the Common Nighthawk, and Broad-winged and Swainson's hawks. Whether flock members pool their navigational expertise in determining which way to fly is not known. For some, such as geese and cranes, the flocks may include family groups of older, experienced birds and youngsters on their first migration. Migrating in family groups provides an opportunity for first-time migrants to learn the migration route and important feeding and resting places along the way. Family groups of Tundra Swans, for instance, migrate together until their arrival back on the breeding grounds the following year.

Most songbirds that migrate at night do not move in flocks, at least not in tight, well-integrated ones. They may move in loose assemblages, however, as is suggested by observations of night migrants that are forced to continue to migrate into daytime because they find themselves over water at dawn—as happens regularly to spring migrants crossing the Gulf of Mexico. As observed on radar, the loosely dispersed migrants flying in darkness ascend and form into flocks at dawn. An observer watching these flocks arrive at the coast will find

Figure 5–45. Single-line Flock of Double-crested Cormorants: *Each bird flying closely behind another bird in a single-line formation may save energy in the same way that a car saves energy by closely following a truck on the highway—a technique termed "drafting." See text and Fig. 5–46 for more information. Photo by Tom Vezo.*

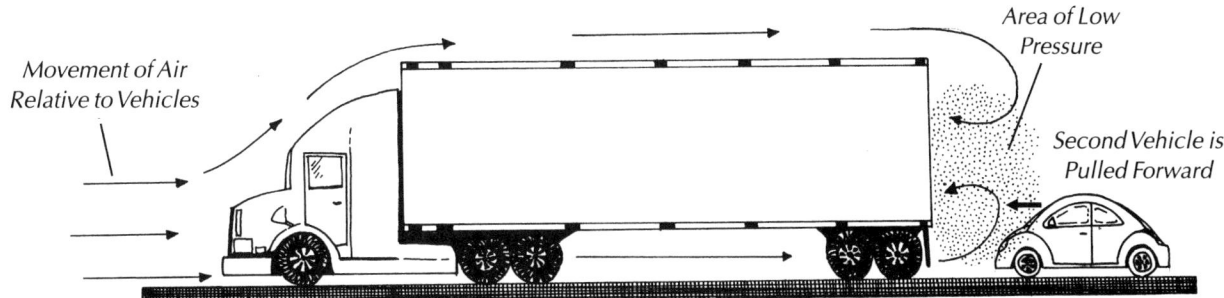

Figure 5–46. Drafting: *A moving vehicle, such as the truck pictured here, continuously displaces air by moving forward, creating an area of low pressure behind itself. Air moving back over the top of the truck swirls down into the low pressure area, and air from below swirls up toward the area, creating a current of air that pulls the second vehicle forward, if it is close enough. A bird flying close behind another in a single-line formation may be pulled forward in a similar manner, reducing the amount of energy that it must expend to produce thrust through flapping flight. Drawing by Christi Sobel.*

that many are made up of a single species (for example, Eastern King-birds, Blue Grosbeaks, or Orchard Orioles). It is hard to imagine how they could sort themselves out in this way unless they already were together in a loose aggregation. Why they form flocks at daybreak is not clear, but the selective advantage of flocking for most birds is certainly the protection it affords against predation.

Many species fly in various kinds of **formations**. Most familiar are the V-shaped formations used by many geese and cranes **(Fig. 5–44)** and the single-file formations of Brown Pelicans and cormorants. Although not all specialists agree, flying in such formations probably confers an energetic advantage to all except the flock leader. In V formations, for example, followers probably take advantage of the rising portion of swirling air currents trailing behind the wing tips of the bird ahead (termed tip vortex, see Fig. 5–31). Theoretical models suggest that geese may save nearly 20 percent of their flight energy by flying in V formation (Berthold 1993), but no one knows whether this energy savings is actually achieved, and what other factors may be involved. The point (lead) bird in such a formation does not receive much energetic benefit and changes frequently during flight. Researchers do not know, however, if the leaders actually change to save energy. In single-file formations, too, each bird that flies directly behind another bird may derive energetic benefits **(Fig. 5–45)**.

Flying straight behind a flockmate is similar to a driving technique known as "drafting," often employed by race car drivers by following each other closely, or by cars closely following a truck on the highway **(Fig. 5–46)**. In each case, the vehicle in front continually displaces air by moving forward, creating a low pressure area behind itself. Air moving back over the top of the first vehicle swirls down toward the low pressure area, and air from below swirls up, creating an air current that pulls the second vehicle forward. In this same way, a bird that flies closely behind another is pulled forward, reducing the amount of energy that it must expend to fly.

One of the most impressive flocking maneuvers is the synchronized wheeling and zigzagging flight of dense flocks of shorebirds or

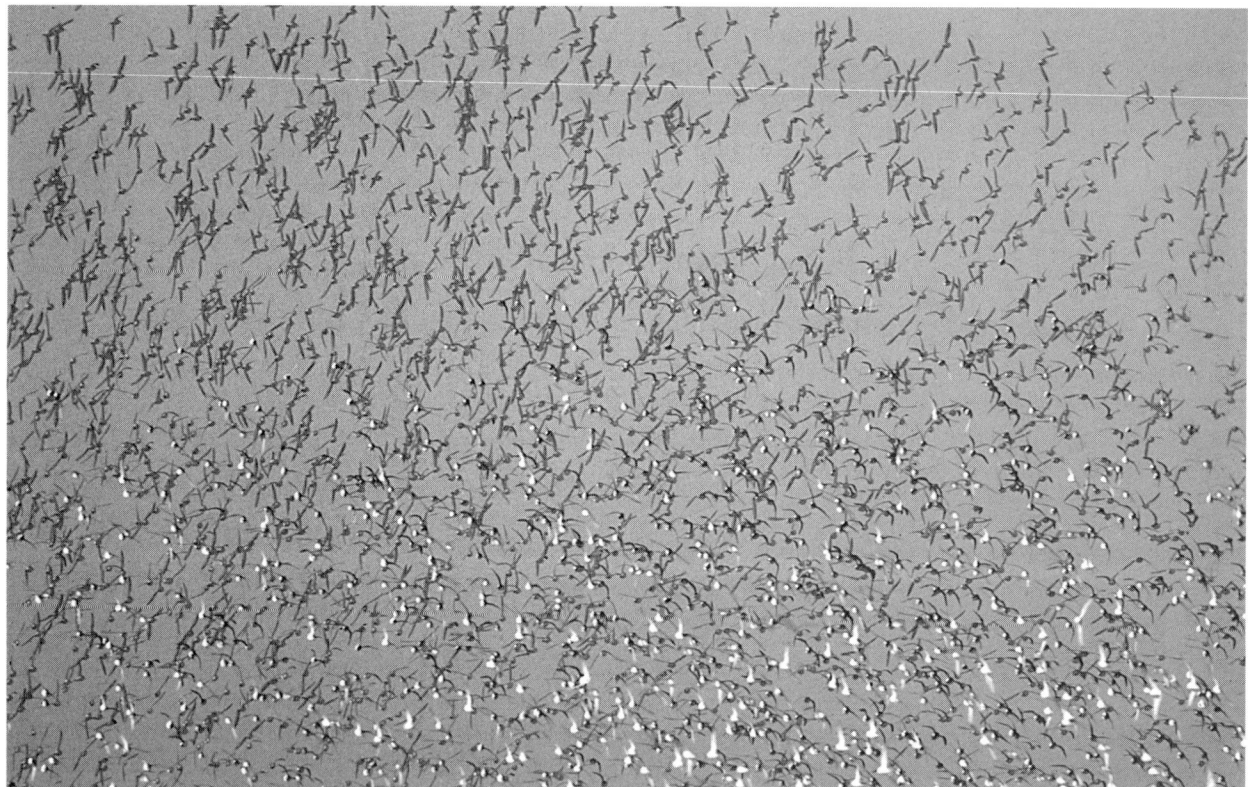

Figure 5–47. Shorebird Flock Maneuvering in Air: The seemingly impossible, split-second twists and turns of a dense flock of birds is one of the most awe-inspiring sights in nature. Recent studies with high-speed photography have revealed that any bird can initiate a flock maneuver, and that once begun, it spreads rapidly much like a "wave" from fans at a sporting event: birds do not wait for their near neighbors to react, but watch the movements of more distant birds so they can anticipate the appropriate time to change direction. Shown here, a winter flock of nearly 4,000 Dunlin (with a few Western Sandpipers), fills the air at Stone Harbor, New Jersey. Photo by Kevin T. Karlson.

starlings in the presence of a flying hawk, especially an accipiter or falcon **(Fig. 5–47)**. In fact, a wheeling flock of starlings often is a hint that a hawk is flying nearby. How such groups manage to move as one organism without individuals colliding into each other remains one of the most intriguing aspects of bird flight.

Recent studies using high-speed photography have revealed some of the mechanisms that permit such precision (Potts 1984). First, there is no consistent flock leader. Birds change position frequently, and any individual can initiate a flock maneuver, which then spreads through the rest of the flock in a wave. Some rules are followed, however. For example, flock members always seem to follow the lead of individuals that bank toward the center of the flock. Such arbitrary rules probably help prevent indecision and allow a flock to respond rapidly during attacks by birds of prey. Once one of these wave maneuvers has begun, it travels through the flock very rapidly—in fact, at a speed much greater than should be possible based on the birds' individual reaction times. Propagation time from neighbor to neighbor is about 15 milliseconds, nearly three times faster than would occur if birds were simply following the action of adjacent flock members. Such rapid reaction apparently is possible because birds pay attention to more distant individuals in the flock, allowing them to anticipate an approaching change in direction in much the same way that fans coordinate "waves" at sporting events.

Loss of Flight

Although flight gives birds a tremendous advantage in exploiting habitats and avoiding predators, it is an energetically expensive way

Ostrich

Kiwi

Elephantbird

to move around. Therefore, the power of flight may disappear when, for many generations, a bird population finds itself in a situation in which the ability to fly provides no strong advantage. In all flightless birds, the keel on the sternum and the mass of the flight muscles are both reduced—thereby saving the energy of maintaining and moving around with these costly structures (see Fig. 4–19).

Flightlessness has evolved in many lineages of birds. Best known, perhaps, are the penguins and ratites (the Ostrich of Africa, rheas of South America, the Emu of Australia, cassowaries of New Guinea and Australia, the kiwis and recently extinct moas of New Zealand, and the extinct 970-lb (440-kg) elephantbirds of Madagascar) **(Fig. 5–48)**.

Penguins apparently evolved flightlessness very early in their history. Many diving birds (for example, loons, auks, and some diving ducks) have legs positioned far to the rear of the body, and the wings are used for swimming as well as flying. But penguins carry these tendencies to an extreme, and literally use their flipperlike wings to fly under water. In the Northern Hemisphere, the extinct Great Auk, the only known flightless member of its family (Alcidae), closely re-sembled the penguins.

Because flying and swimming are very different modes of travel, requiring different body adaptations, a bird that has evolved to do just one or the other will generally do it better than a bird that does both **(Fig. 5–49)**. The Antarctic land mass and the plethora of nearby islands may have provided low, safe nesting places for penguins, thus reducing their need to fly, and allowing them to evolve into swimming specialists. In contrast, many *Arctic* islands are steep-sided, with fewer good nest sites for flightless northern swimmers. Perhaps competition

Figure 5–48. Sample Ratites: One of the best-known groups of flightless birds is the ratites. Most, like the Ostrich, are large, diurnal birds of open country, too heavy to fly, which rely on their long, powerful legs to outrun predators. The three species of kiwi are atypical rat-ites. Duck-sized and nocturnal, they roam the forests of New Zealand using their highly developed sense of smell to search the ground for earthworms and other small arthropods. The elephant-birds, a group of huge (970-lb; 440-kg), extinct, flightless birds that once lived on the island of Madagascar, are among the largest birds ever known. Their egg, measuring up to 1.3 by 9.5 inches (33 by 24 cm) is considered the largest single cell in the animal kingdom. Drawings of Ostrich and kiwi by Charles L. Ripper. Elephantbird redrawn from Pough et al. (1996, p. 543).

5

Emperor Penguin

Pigeon Guillemot

Figure 5–49. Penguin versus Alcid Adaptations for Swimming: *Penguins are the supreme underwater swimmers of the bird world. Their feet, positioned at the far end of the body, are in the best possible location to serve as steering rudders for their streamlined, torpedo-like bodies. Their rigid wings (flippers) lack typical flight feathers, allowing them to use the entire wing to "fly" underwater. By tilting the leading edge of the wing up on the upstroke and down on the downstroke, penguins can produce thrust on both strokes, the wing literally acting as an underwater "airfoil" (hydrofoil). Alcids (auks), such as the Pigeon Guillemot pictured here, still retain the ability to fly and thus have not become completely specialized for swimming. Their feet, like those of penguins, are used for steering, but are not set as far back on the body. Pigeon Guillemots, unlike other alcids, also use their feet for propulsion. The wings of auks are short with short inner secondary feathers, and are kept partly folded at the elbow when underwater, forming a smaller, broader "paddle" than that used in flight.*

for these limited nest sites resulted in only the largest of the alcids, the now-extinct Great Auk, being able to use the low sites and consequently give up flying. The fact that this species restricted its nesting to a couple of low islands in the North Atlantic Ocean undoubtedly hastened its demise.

Most of the flightless ratites are large birds too heavy to fly. Many live in open country where they rely on long, powerful legs to outrun predators. Cassowaries have massive legs and a long, sharp claw on each foot. They defend themselves by jumping off the ground: by kicking forward powerfully, they can disembowel dog-sized mammals. Kiwis also have thick, powerful legs, but they live in forests and are nocturnal. They evolved in New Zealand in the absence of terrestrial predators and apparently relied on the cover of darkness, cryptic plumage, and spending the day in burrows to avoid a now-extinct, huge eagle.

Flightlessness, then, appears to have been able to evolve in ratites because the birds could avoid predators by their large size, or because they originated on an island that had no terrestrial predators. Furthermore, none of the ratites had a need to fly into trees, as they obtained their food on the ground, either by grazing on grasses, or by probing the soil for earthworms (kiwi). In fact, in all ratites except the kiwis, predator avoidance by their Eocene ancestors apparently depended on increased body size and strength of legs and feet at the expense of their wings.

Flightless species occur in many other groups. These include flightless grebes on Lake Atitlan (now apparently extinct) and Lake Titicaca in the Andes, the Flightless Cormorant of the Galapagos, ducks in southern South America and New Zealand, pigeons on several islands in the South Pacific and Indian Oceans (including the extinct Dodo), a flightless parrot in New Zealand, and flightless rails on many islands throughout the world. Only a few flightless passerines are known. The most famous is the tiny Stephens Island Wren (family Acanthisittidae), which lived on an island in Cook Strait between the North and South Islands of New Zealand. It is known only from a few specimens brought in by the lighthouse keeper's cat, which subsequently killed all the rest of the species.

The fossil record reveals an amazing array of flightless birds, from the gigantic predatory *Diatrymas* (see Fig. 22 in Evolution of Birds and Avian Flight), Phorusrhacids, and the bizarre, ducklike, Hawaiian Moa-nalos to a three-foot-tall owl that hunted ground sloths and other mammals in the West Indies.

Flightlessness has evolved most frequently on islands where bird populations found themselves in environments free of predators. When humans arrived on these islands, bringing with them dogs, cats, rats, and other predators, many flightless birds were quickly driven to extinction, and many that remain today are severely threatened.

The anatomy and brain structure of all flightless birds shows conclusively that they evolved from flying ancestors, and some groups seem more prone to the evolution of flightlessness than others. The rails, for example, include many flightless and nearly flightless forms, and in all of them, the sternum does not finish its growth until the bird is nearly fully grown. This pattern of development may predispose rails to lose the power of flight through an evolutionary phenomenon known as **neoteny**, the retention of juvenile traits into adulthood. One could thus speculate that flightlessness in rails may have evolved as the sternum eventually stopped growing before it had reached full size in some species.

Migration

■ Animals that live in strongly seasonal environments where they can meet their life requirements during only part of the year face a serious problem. Many species of mammals, reptiles, amphibians, insects, and other invertebrates have solved this problem by entering an inactive, dormant state known as torpor or hibernation. A very few birds, including some hummingbirds, swifts, and especially nightjars, are also known to enter a torpid state of lowered body temperature and metabolism. The Common Poorwill may remain torpid for weeks or months (see Fig. 4–128). Highly mobile animals have another option: they can leave the inhospitable area entirely. Endowed with strong powers of flight, most birds deal with fluctuating environments by migrating.

When ornithologists speak of migration, they usually mean *seasonal migration*, such as movement between breeding and overwin-

Figure 5–50. Arctic Tern Fall Migration Route: *The Arctic Tern has one of the longest migration routes known, traveling up to 11,000 miles (17,700 km) from its breeding grounds on islands in the far north (dark green) to the outlying pack ice waters of Antarctica, where it overwinters. Drawing by Robert Gillmor.*

 Breeding Grounds

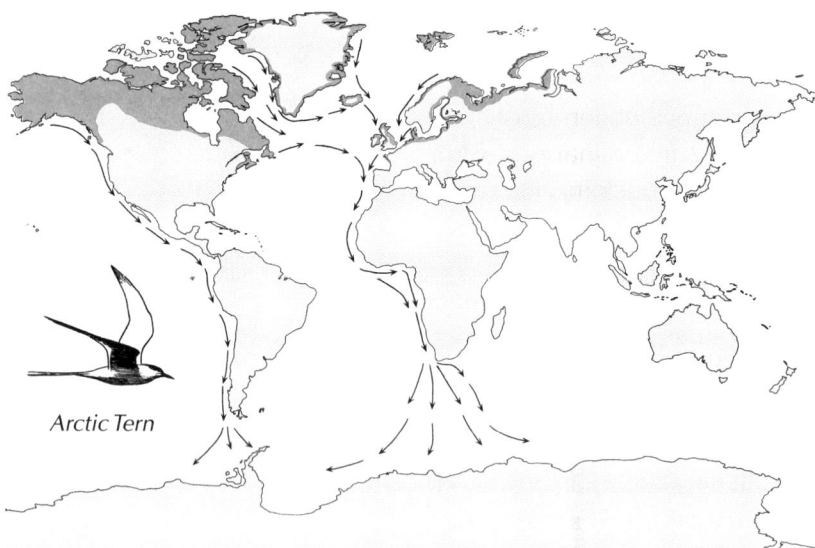

Arctic Tern

tering areas. An extreme case is that of the Arctic Tern. Breeding on islands in the far north, it migrates as far south as the waters around Antarctica in the winter, a flight that may be as much as 11,000 miles (17,700 km) one way **(Fig. 5–50)**. During the course of a year, an individual Arctic Tern may cover a distance equivalent to flying around the world. Equally amazing is the Blackpoll Warbler, which weighs about one ounce (31 grams) before migration, when laden with stored fat. In autumn, blackpolls depart from the coast of the northeastern United States and fly over the waters of the Atlantic Ocean to South America **(Fig. 5–51)**. Unlike the tern, they cannot land on water or feed along the way. They fly nonstop, day and night, for four or more days to make this 2,480-mile (4,000-km) trip.

Patterns of Migration

Migration comprises not only long-distance round trips, but a broad continuum of seasonal movements, ranging from sporadic mass movements over relatively short distances to journeys spanning a hemisphere.

Figure 5–51. Blackpoll Warbler Fall Migration Route: *In the fall, postbreeding family groups of blackpolls gather into larger and larger flocks, then make their way from their breeding grounds in the boreal forests of Canada to the northeastern United States. From this staging area, they set out over the Atlantic Ocean to fly nonstop, day and night, for four or more days to their wintering grounds in South America, a total distance of 2,480 miles (4,000 km). Reprinted from Kenneth P. Able, ed.: Gatherings of Angels: Migrating Birds and Their Ecology. Copyright ©1999, Cornell University. Used by permission of the publisher, Cornell University Press.*

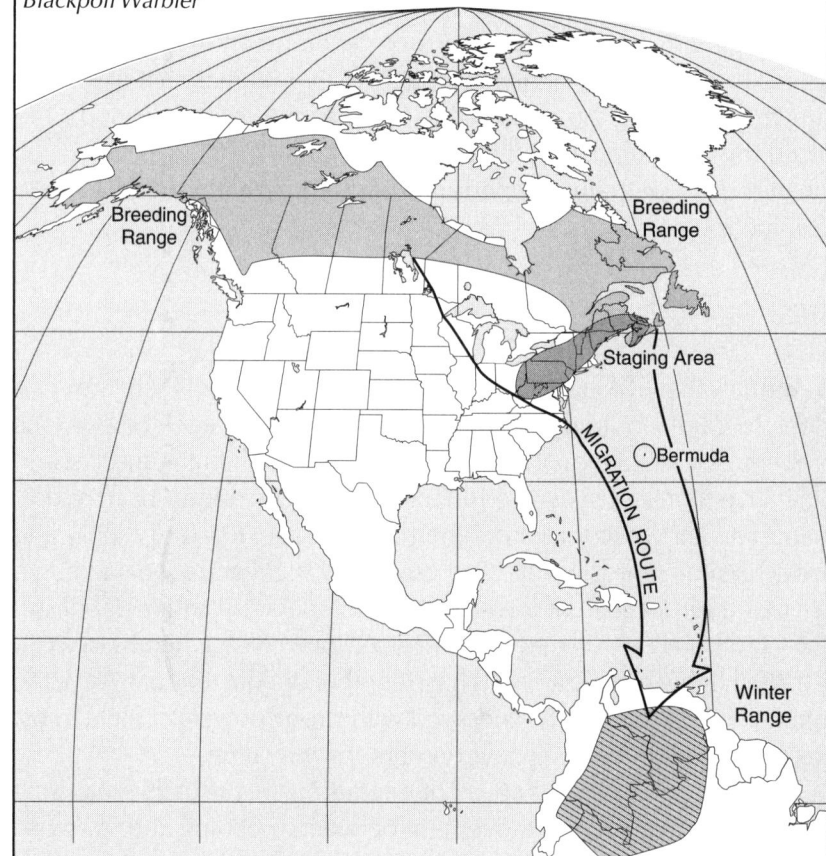

Furthermore, migrations are not always in the north-south direction, and are not always synchronized with spring and fall. Many tropical birds such as certain quetzals, hummingbirds, and parrots, migrate in response to the seasonal alternation of rainfall and drought, which changes the availability of fruits, seeds, insects, and nectar. These migrations are often "vertical"—up and down mountainsides. Blue Grouse in western North America also migrate vertically (**Fig. 5–52**).

The diversity of migration strategies can best be understood in an ecological context: the variability and predictability of the resources—food, water, cover, and other necessities—that the bird depends on determine what type of migratory behavior is likely to evolve. By considering the spectrum of bird movement patterns in terms of seasonal resource variability and predictability, one can bring order to what might otherwise seem like a collection of disparate behaviors. Refer to **Figure 5–53** as you read through the following examples of bird movement patterns.

Figure 5–52. Vertical Migration of the Blue Grouse: *The Blue Grouse, a resident of mountainous regions of western North America, breeds in a wide range of open habitats, from various shrub and grassland communities to open montane forests and edges. Birds from every type of breeding range, however, all seem to overwinter in coniferous forests, feeding mainly on conifer needles. Most populations of Blue Grouse migrate seasonally, but not in the north-south direction typical of most temperate zone birds. Instead, they migrate "vertically," moving up mountainsides in the fall to reach coniferous forests, and moving back down to the breeding grounds in the spring.*

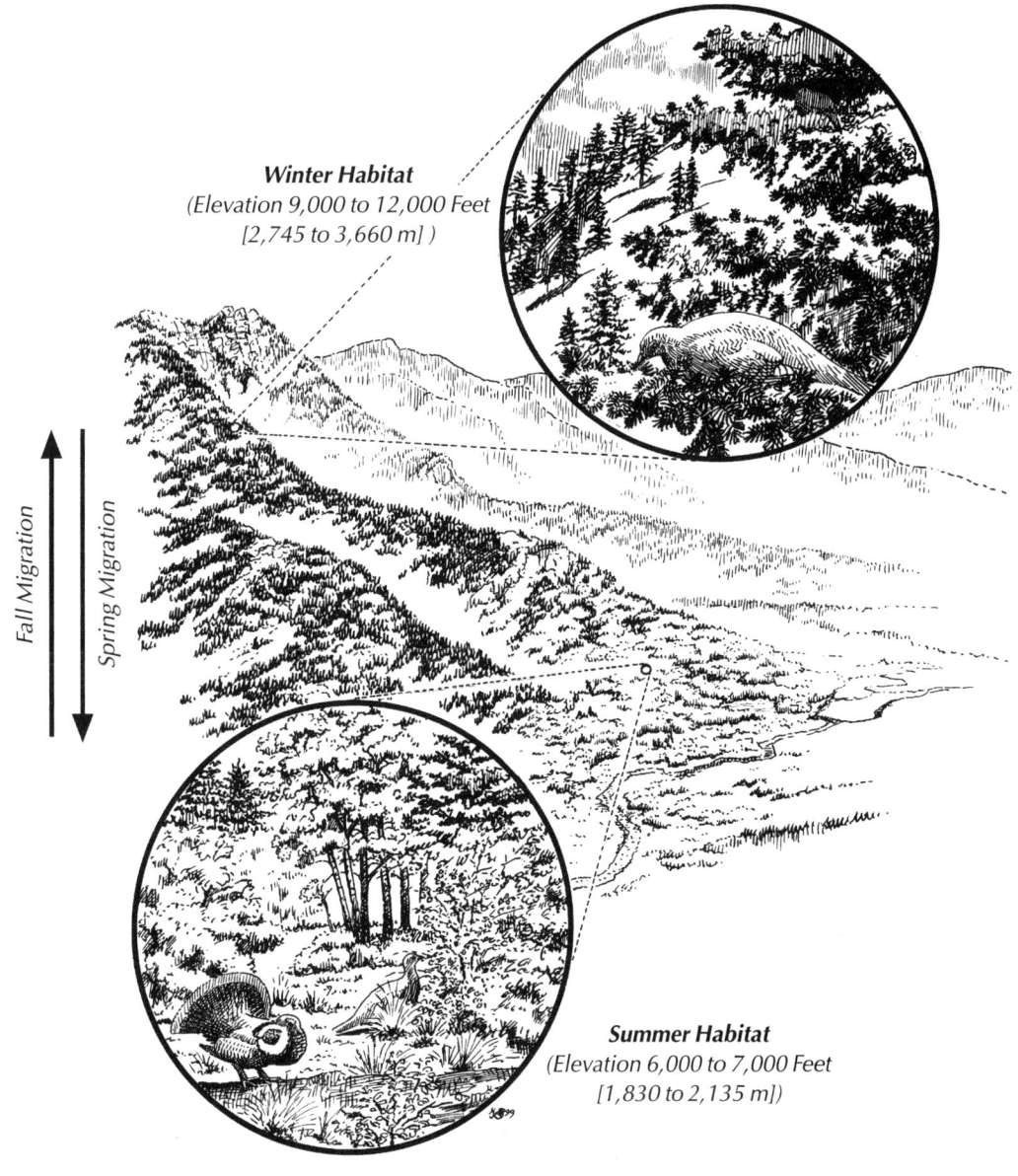

Winter Habitat
(Elevation 9,000 to 12,000 Feet [2,745 to 3,660 m])

Fall Migration

Spring Migration

Summer Habitat
(Elevation 6,000 to 7,000 Feet [1,830 to 2,135 m])

Migration is costly in terms of both energy and risk **(Table 5–2)**. For example, migrants have annual adult survival rates of about 50 percent compared to tropical residents, which have rates on the order of 80 to 90 percent (Gill 1990). However, tropical species have much lower reproductive success than birds that breed in temperate areas (both migrants and temperate zone residents). Therefore migrants have apparently traded in higher adult survival rates for higher productivity in any one year. (Temperate zone residents have even lower annual survival rates, in the range of 20 to 50 percent, since they must face the hardships of winter.) But migration has other disadvantages. Migrants that do return alive may be weakened from their journey. They also must expend energy setting up a new territory each spring—and temperate residents may acquire the best breeding territories before migrants arrive.

In areas where all necessary resources are predictably available year round, natural selection favors individuals who are **resident**. Northern Cardinals, Northern Mockingbirds, and most chickadees, titmice, and woodpeckers are permanent temperate zone residents; they can count on adequate reserves of seeds, berries, and dormant insects to see them through the winter.

When resources in the breeding area differ greatly from season to season, and when this pattern is highly predictable, as it is for birds nesting at high latitudes that eat insects or nectar, the pattern that evolves is **obligatory annual migration**, in which all individuals migrate toward the equator for the winter. This type of pattern is seen in many flycatchers, thrushes, vireos, hummingbirds, and wood warblers. Selection against individuals that make the mistake of trying to stay in the north all year is strong—a flycatcher that decides to try catching flies in the winter will die rather quickly. Evolution equips

Figure 5–53. Relationship between Migration Strategies and Food Resources: *In temperate climates, migration strategies are related to seasonal variation and predictability of food resources in the breeding area. Shown here are six migration strategies that may be defined based on resource predictability (Y axis) and variability (X axis). These strategies exist on a multidimensional continuum, as represented by the central arrow cluster, such that not all individuals within a species or population necessarily conform exactly to the six strategies shown on the graph. For example, Eastern Bluebirds may be facultative partial migrants in some geographic locations, and residents in others. Most species, however, follow the basic behavioral patterns shown here. Birds that are year-round residents, for example, occupy environments with high resource predictability and low seasonal variation in resources. In other words, resident birds can count on a roughly constant supply of food throughout the year. At the other end of the continuum, irruptive migrants contend with low predictability and high variability in food resources. These conditions have led irruptive migrants to adopt a strategy of sporadic movements to areas with plentiful resources when food supplies in the breeding area suddenly plummet. See text for descriptions of migration strategies.*

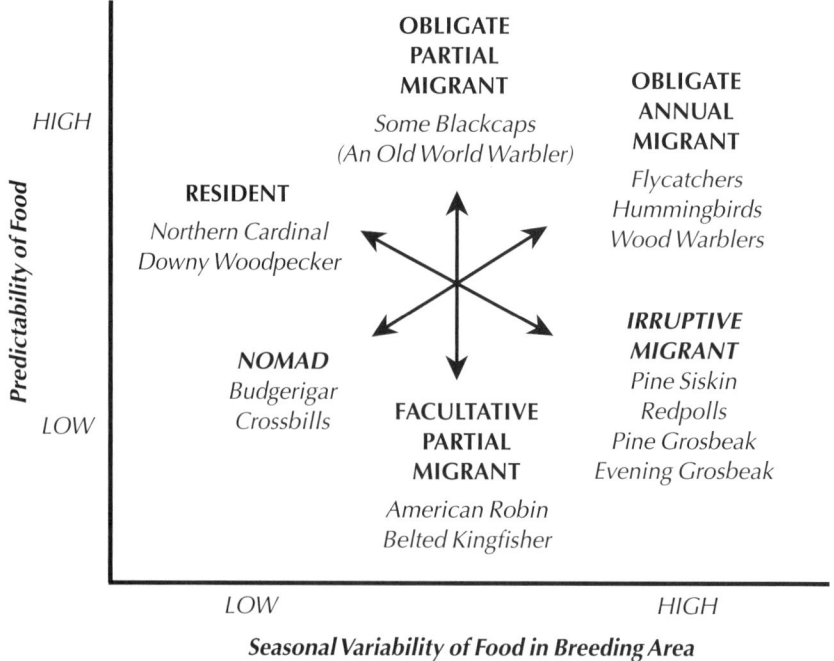

Table 5–2. Survival and Reproductive Success of Residents and Migrants: *Estimates of annual adult survival (percent of adult birds surviving to the next year) and annual reproductive success (number of young fledged per year) are provided for birds that reside year round in the tropics, birds that reside year round in temperate areas, and birds that breed in temperate areas but migrate to the tropics for the winter. Birds in each category appear to have evolved to produce as many young as possible over their lifetimes, but their strategies differ: tropical residents live long, producing few young each year; temperate residents have shorter lives, but produce more young each year. Migrants live moderately long, and raise a moderate number of young. Adapted from Gill (1990, p. 247).*

	TROPICAL RESIDENT	MIGRANT	TEMPERATE RESIDENT
ANNUAL ADULT SURVIVAL	*Tropical residents do not face the hazards of migration or harsh winter weather.* **High (80–90%)**	*Migrants escape winter, but many die during migration.* **Moderate (50%)**	*Temperate residents must endure the harsh weather and restricted food supply in winter.* **Low (20–50%)**
ANNUAL REPRO-DUCTIVE SUCCESS	*Tropical residents have no seasonally abundant source of food for their young.* **Low**	*Migrants feed their young on seasonally abundant insects in the temperate zone, but each year divert time and energy from breeding into migrating and setting up new territories.* **Moderate**	*Temperate residents can take advantage of the huge, seasonal flush of insects to feed their young.* **High**

these species with relatively inflexible "migratory programs," which ensure that virtually all individuals will leave the breeding areas before food resources reach dangerously low levels.

In environments where the variability is less extreme and less predictable, birds adopt more flexible behaviors (these are the nomads, facultative partial migrants, and irruptive migrants of Fig. 5–53, as discussed below). Birds of these persuasions may stop their southward migrations at any of a number of points, depending on conditions. If they can meet all of their needs in a northerly location, they may try to spend the winter there but be prepared to move on if conditions worsen. The Lapland Longspur and many sparrows and water birds seem to fall into these categories, as do some fruit-eaters such as American Robins and Eastern Bluebirds.

Migration schedules do not necessarily follow seasonal changes in climate. The seeds and buds eaten by various finches such as redpolls and Pine Grosbeaks fluctuate dramatically in abundance, not only seasonally but also from year to year and region to region. These fluctuations may be quite unpredictable. Migration in these species must be facultative (flexible), directly responding to local food availability. Such movements have been termed **irruptive** because in some years large numbers of birds move out of the northern forests and in other years they stay put. Similar irruptive migrations are well known among the tundra predators of lemmings such as Snowy Owls, Rough-legged Hawks, and jaegers.

Between the extremes of predictability and variability lie a range of intermediate situations that seem to select for a type of behavior termed **partial migration**—some individuals in a population migrate while others behave like residents. Partial migration may arise from two different types of control mechanisms. In one type, the difference between a migrant and a resident individual might be genetically determined; these birds are called **obligate partial migrants**. The individuals with genes for migration would always migrate, and the nonmigrant genotypes would always behave as residents. This arrangement most likely evolves in environments where resources are always sufficient to enable some, but not all, individuals to overwinter successfully in the breeding area, and when the number of individuals who can stay is relatively stable from year to year. This kind of genetic polymorphism seems to exist in the European Robin and in southern European populations of the sylviid warbler, the Blackcap (Berthold 1996).

When the number of birds that the environment can support varies from year to year, we can expect a more flexible strategy to evolve. In this situation, known as **facultative partial migration**, the number and identity of the individuals migrating varies from year to year in direct response to resource availability. Which birds migrate is not predetermined genetically, although the migrants may be predominantly young or socially subordinate members of the population. The Blue Tit of Europe seems to fit this pattern, and some North American chickadees may too, during their occasional flight years.

It is not clear that any bird species is truly **nomadic** (constantly on the move, showing no tendency to return to previously occupied places). Frequently cited examples are species that occupy the arid interior of Australia, such as the Budgerigar, but researchers need more information about their movement patterns to determine whether these birds are true nomads.

Thus, migration and residence behavior patterns span a continuum from permanent residency to obligate long-distance migration, and many cases may not be easily categorized. Even within one species, different populations may have very different migration strategies. As an example, consider the White-crowned Sparrow populations of coastal western North America **(Fig. 5–54)**. The northernmost breeding

Figure 5–54. White-crowned Sparrow: Along the western coast of North America, different populations of this species have very different patterns of migration, ranging from year-round residency to long-distance migration. Photo by Marie Read.

subspecies, *Zonotrichia leucophrys gambelii*, migrates from Alaska and the Yukon south to the southern plains and into northern Mexico. Farther south, the subspecies *Z. l. pugetensis* breeds from southern British Columbia southward to northern California. It migrates a shorter distance, overwintering in the lowlands of central and southern California. A third subspecies, *Z. l. nuttalli*, is a year-round resident in coastal California, sharing its habitat in winter with migrants from farther north.

The Origin and Evolution of Migration

Today, a diverse range of birds migrates with the seasons, including some penguins (on foot), loons, storks, hawks, owls, hummingbirds, parrots, and songbirds. Migration occurs on all continents, wherever the environment changes periodically, whether the change results from temperature cycles, alternation of rainy and dry seasons, or some other factor. Although the earliest origins of migratory behavior are probably lost forever, it is reasonable to conclude that migratory behavior has evolved (and disappeared) repeatedly within the avian lineage. It may have appeared almost as soon as birds could fly well enough to travel long distances. Although some current migration patterns in the Northern Hemisphere undoubtedly were shaped by events during the Pleistocene, when episodes of glaciation alternated with warmer periods, large-scale migration surely existed long before the glacial epochs.

In the simplest terms, migration begins to evolve when individuals that move from one area to another produce more offspring than those that do not move. When the environment changes, migratory behavior apparently can develop dramatically in just a few generations. In the early 1940s, for instance, House Finches from a sedentary population in California were released on Long Island, New York. The introduction was wildly successful, and House Finches have become one of the most abundant birds in urban and suburban areas of the northeastern United States. Because the environmental conditions of the Northeast change much more with the seasons than do conditions in California, migration appeared within 20 years after the birds were introduced. Although some individuals remain year-round residents in the Northeast, others migrate back and forth to the Gulf States. Therefore, the eastern House Finch has become a partial migrant.

Migratory behavior also can be lost rapidly from a population. It was surely some migrating Dark-eyed Juncos gone astray that colonized Guadeloupe Island, some 155 miles (250 km) off the coast of Baja California. The species is now established there as a sedentary population. Similarly, if less dramatically, populations of White-crowned and Savannah sparrows along the coast of California have lost the migratory habit. Because none of these populations (juncos or sparrows) has evolved enough differences to be considered a new species, researchers assume that their isolation is recent, and that the loss of migratory behavior must have been fairly rapid.

A number of different theories have been proposed for the way in which seasonal migration evolved, including the following:

1. **Climatic Changes:** As conditions changed slowly, due to glacial advances or retreats or shifting continents, birds migrated to return to favorable conditions. For example, temperate zone migrants may have evolved when the region's climate was more tropical, thus they leave in winter to live in the warmer conditions to which they are better adapted.

2. **Lack of Needed Resources:** For example, birds may leave the temperate zone in winter to move to warmer areas where more food, such as fruit, nectar, or insects, is available.

3. **Seasonal Interspecific Dominance Interactions:** In certain seasons, the competition between species (for food, nest sites, or some other resource) is so great that some individuals must leave to find the resource elsewhere.

4. **Seasonal Intraspecific Dominance Interactions:** In certain seasons, the aggression of dominants toward subordinants causes some members of a species to migrate to other areas.

5. **Seasonal Tracking of Fruit or Nectar:** Some birds migrate to follow the fruiting or flowering of their key food plants.

As discussed by Rappole (1995; see Suggested Readings), none of these theories satisfactorily explains the seasonal movements of birds, such as Neotropical migrants, which move both north and south. Migration is a two-way street, and thus arguments for its existence need to explain both why birds migrate *from* an area and why they *return.* If (as outlined by Theories 1 through 4 above) birds leave for a better climate, more resources, to avoid competition with other species, or to avoid aggression from conspecifics, why do they return? Theory 5 above can explain migration in several directions, as birds move back and forth to areas with the most abundant or preferred fruits or flowers, but seems to apply mainly to migration within tropical areas, as in quetzals, hummingbirds, parrots, and others. Tropical species have not needed to migrate as far as the temperate zone to find an abundance of fruit or nectar, and species that rely on these foods do not appear to have evolved originally in the temperate zone (see below).

Theories 1 and 2 above, in order to apply to Neotropical migrants, would have to assume that the birds or their ancestors evolved in the temperate zone, and then developed migration away from the area to avoid adverse situations during winter. In contrast, researchers believe that most land birds that migrate to the Neotropics to overwinter, are, in a sense, returning home. Strong evidence for this tropical origin of migration is that most of "our" vireos, flycatchers, tanagers, warblers, orioles, and swallows have evolved from Neotropical forms. In fact, 78 percent of all Nearctic migrant species have close relatives in the same genus or even populations of conspecifics that reside in the Neotropics year round.

A more acceptable theory for how Neotropical migrants evolved migration is that through many generations, the tropical ancestors of "our" migrants dispersed away from their tropical breeding sites toward the temperate zone, lured by the seasonal flush of insect food and greater day length, which allowed them to raise more young annually (four to six, on average) than their resident tropical relatives (two to three young, on average). As their breeding sites moved to the temperate zone, the birds developed migration to return to hospitable habitats during the colder part of the year. Thus, in the Northern Hemisphere, selection favored a long-distance migration north to breed, and then south to overwinter. In the Southern Hemisphere the directions are reversed.

Many aspects of migratory behavior vary from individual to individual, even within the same species. For a behavior or any other trait to evolve under the influence of natural selection, the individuals of a population must exhibit variation in the trait, and the variation must have a genetic basis (Ch. 1, Sidebar 2: The Evolution of an Idea: Darwin's Theory). Although scientists have assumed that variations in migratory behavior are inherited, the details have only recently begun to emerge. These details are the result of studies performed on a small Eurasian warbler, the Blackcap, by Peter Berthold and his colleagues in Germany. The Blackcap breeds throughout western Europe from Scandinavia southward to areas around the Mediterranean Sea (**Fig. 5–55**). Blackcap populations also exist on the Canary and Cape Verde islands. Conveniently, this species exhibits a wide range of migratory habits, from long-distance obligate migrants in the northern part of the breeding range to partially migratory populations around the Mediterranean, and resident populations on Cape Verde. Blackcaps from all of these populations breed with one another in captivity, making it possible to study the genetic control of their migratory behavior.

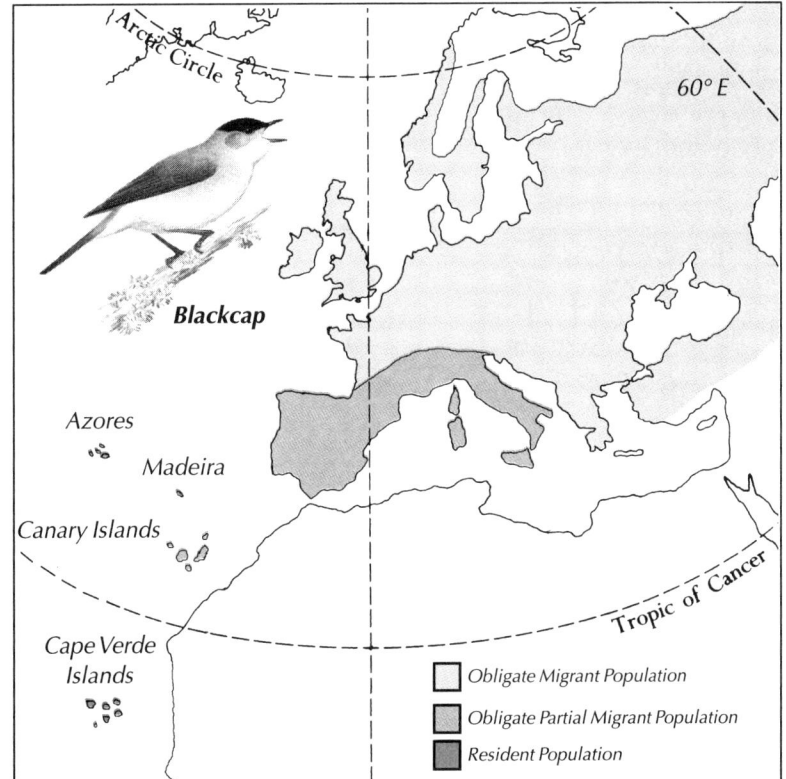

Figure 5–55. Variation in Migratory Habits of Blackcaps: Various populations of Blackcaps (a species of Old World warbler) differ in their migratory habits. Birds that breed in northern Europe and Asia (light red) are obligate annual migrants: all individuals migrate to Africa for the winter. Birds in southwestern Europe and on the Canary Islands (medium red) are partial migrants: some individuals migrate to Africa for the winter, while others remain within the breeding grounds. Because experiments have shown a genetic basis for this difference, these birds may be considered obligate partial migrants. Birds breeding in the Azores, on Madeira, and in the Cape Verde Islands (dark red) are year-round residents.

Berthold's cross-breeding experiments have revealed a remarkable degree of genetic control over a number of facets of migratory behavior. For example, when members of long-distance migratory populations were bred with partially migratory birds from the Canary Islands, the offspring showed intermediate amounts of migratory activity, which was monitored by the amount of hopping and fluttering at night (Berthold 1996) **(Fig. 5–56)**. This **migratory restlessness** was first discovered by German scientists, who termed it *Zugunruhe*, literally "migratory unrest." Because night migrants are inactive and sleep at night except during the migration period, this restlessness provides a good measure of an individual's desire to migrate. Similarly, the direction of orientation can be recorded by using circular "orientation cages" and recording the compass direction of the side of the cage the birds hop and flutter against the most **(Fig. 5–57)**. During their first migration, the compass direction of orientation of the hybrid Blackcap offspring also showed high heritability. Blackcaps from the western part of central Europe migrate southwestward and go around the western end of the Mediterranean and then into Africa. Those from eastern Europe fly around the eastern end of the Mediterranean. Hybrids between individuals from the two sides of this migratory divide showed intermediate orientation—to the south—a direction that apparently has been selected against in wild populations (Berthold 1996) **(Fig. 5–58)**.

Partially migratory populations provide particularly good opportunities to study the genetic control of migration. About 79 percent of Blackcaps from southern France are migrants. The remainder are residents. In captivity, one can selectively mate migrant types with migrants and resident types with residents. In a sense, such an experiment simulates, in an extreme way, what might happen if the environment suddenly began to strongly favor either the migrant type or the resident type. Only three generations of such selective breeding of migrants were necessary to produce a group of birds that were all migrants (again, as measured by the amount of *Zugunruhe*) (Berthold 1996). In six generations of selection for residents (nonmigrants), all evidence

Figure 5–56. Degree of Migratory Restlessness in Captive Blackcaps: Researchers studying nocturnal migrants have long used the degree of **migratory restlessness** (hopping and fluttering in a cage at night) as a measure of a bird's desire to migrate—because these birds are typically inactive at night except during the migration period. The captive breeding experiment with Blackcaps illustrated here determined the number of 30-minute periods each night during which birds displayed migratory restlessness (vertical axis). Each point plotted is the mean for a 10-day period, and the vertical line at each point shows one standard error of the mean (a measure of the variability in the data). The horizontal axis indicates the number of days after migratory restlessness began. Because individual birds may have begun their periods of migratory restlessness on different days, the data is adjusted seasonally to align all the periods. When obligate annual migrants from a population in southern Germany (top line) were bred with partial migrants from the Canary Islands (bottom line), the resulting offspring (hybrids) showed an intermediate amount of migratory restlessness (middle line). This demonstrates, at least for Blackcaps, a genetic basis for their migratory behavior. Adapted from Berthold and Querner

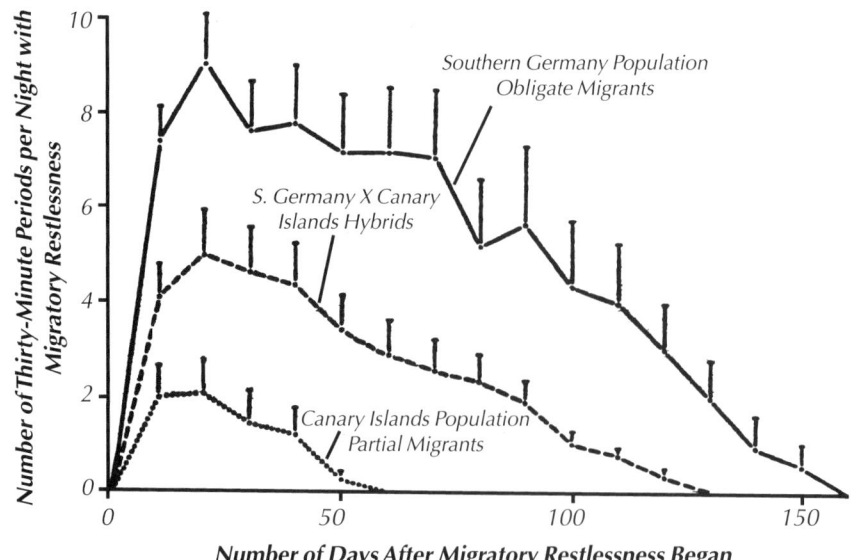

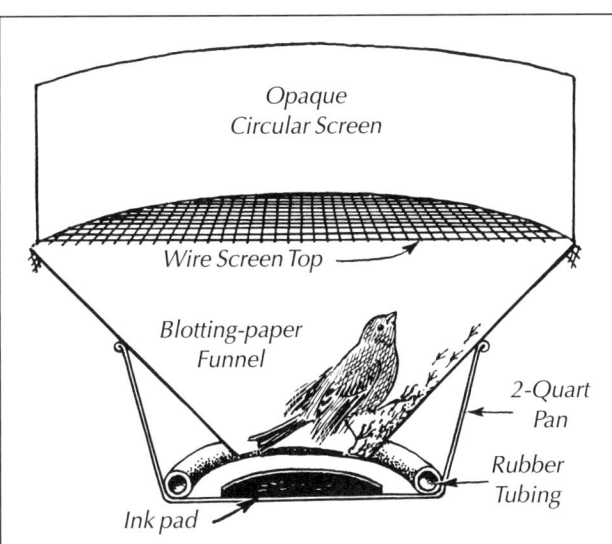

Figure 5–57. Circular Orientation Cage: *Reseachers often determine the direction in which a bird orients its migratory restlessness by placing the bird in a circular orientation cage for the night. A typical cage, as shown here, is screened on top but open to the sky. A piece of white blotting paper makes up the sides of the funnel, and an ink pad, the floor. As the bird hops and flutters against the sides of the funnel, it leaves inky footprints. The compass orientation of the footprints indicates which direction the bird would tend to fly, if it were free to migrate, and the degree to which the footprints are concentrated in one direction indicates the strength of the bird's tendency. Drawing by Robert Gillmor.*

of migratory activity was eliminated from the population. The rapidity with which the behavior responded to selection indicates strong genetic control. This suggests that in nature, given potent selection by the environment, many aspects of migratory behavior have the potential to evolve quite rapidly.

The Blackcap provides yet another illuminating case study. Over the last 25 years or so, central European Blackcaps have developed a new migratory pattern. Instead of migrating to African winter quarters, some individuals have begun to fly northwest, overwintering in the British Isles. The success of this evolutionary adventure has apparently depended in part on the increase in feeding of wild birds in Britain and Ireland in recent years. The Blackcaps use feeders in winter and this, coupled with the lower cost of migrating to Britain rather than to central Africa, may result in higher overwinter survival. The birds wintering in Britain may also return earlier to their breeding areas on the continent and thus enjoy a reproductive advantage.

The key to the evolution of this new migratory pattern lies in the existence, before winter bird feeding increased in the British Isles, of individuals with the genetic trait for migrating in a northwestward direction. A minority of birds captured in Germany orient in that direction, as do most birds from the British winter population brought back to Germany and tested in the fall (Berthold 1996). In earlier times, this genetic variant would presumably have been selected against in the population, because conditions did not favor overwinter survival in Britain. With the increase in bird feeders, this migration direction became a more favorable option, and the genes that code for it increased in the population.

Controlling and Synchronizing the Annual Cycle

The daily lives of most organisms are regulated by internal biological clocks, which produce daily cycles of behavioral and physiological events called **circadian rhythms**. For example, some plants move their leaves **(Fig. 5–59)**, open and close their flowers, or produce

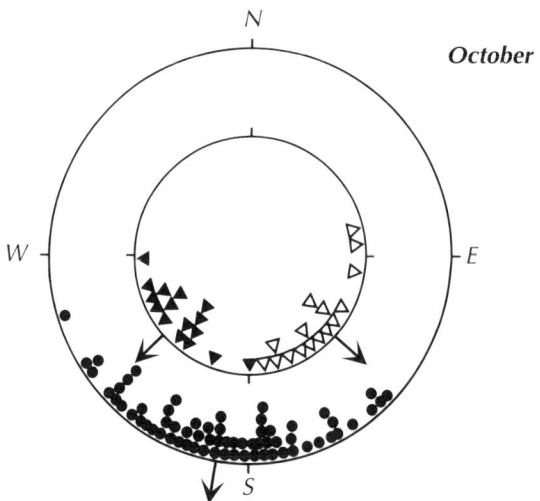

Figure 5–58. Genetic Control of Orientation Direction in Migrant Blackcaps: *In fall, Blackcaps from western central Europe migrate southwestward, flying around the western end of the Mediterranean to reach Africa; those from eastern central Europe migrate southeastward, flying around the eastern end of the Mediterranean. When birds from each of these groups were tested in circular orientation cages, their migratory restlessness was oriented in the same direction as in the wild: birds from Germany (western Europe) oriented to the southwest (solid triangles), and birds from Austria (eastern Europe) oriented to the southeast (open triangles). When birds from these two populations were hybridized, their offspring showed orientation in an intermediate direction: to the south (solid dots). In this diagram the circles each represent a circular orientation cage (see Fig. 5–57); the inner circle shows results for the parents, and the outer circle, the offspring. Each dot or triangle represents the average direction chosen by an individual bird. Each arrow indicates the average direction for all birds from a particular group. Adapted from Berthold (1993, p. 145). Originally from Helbig (1989).*

▼ *Blackcaps From Western Europe*
▽ *Blackcaps From Eastern Europe*
● *Hybrids*

nectar in daily cycles; some crabs are darker in the morning than in the afternoon; flying squirrels have daily activity patterns; birds have daily cycles of activity and body temperature; and humans show daily cycles in body temperature, blood pressure, and sensitivity to drugs, among other things. All these cycles persist even when the organisms are kept under constant conditions. The physiological mechanism by which biological clocks operate is not well understood.

Birds and other creatures also have annual cycles, which are controlled by a clock with a much longer periodicity. Migration is one component of a suite of integrated events that constitutes the annual cycle of a bird. Other important annual events include one or more molts per year and breeding. These cycles are called **circannual rhythms** because they have a periodicity of about ("circa") one year. Examples of nonavian circannual rhythms include yearly cycles of hibernation and weight loss in golden-mantled ground squirrels, and the shedding and growth of antlers in some deer.

The existence of these circannual rhythms is revealed by holding birds captive for very long periods under "constant conditions" in which they have no hint of the changing seasons. For instance, the light may be constantly dim, or the days of constant length. Birds treated in this way continue to go through cycles of migratory activity, breeding

Figure 5–59. Circadian Rhythm of Leaf Movement in Bean Seedlings: *The leaves of some plants, such as these bean seedlings (Phaseolus coccineus), move in a daily cycle. During the day the leaves move outward, perpendicular to the sun's rays; at night, they drop to a more vertical position. Rhythms such as these, which persist even when the plant (or other organism) is placed under constant conditions of light, temperature, humidity, and other variables, are called* **circadian rhythms.** *In these constant conditions, however, the cycles usually become slightly longer or shorter than 24 hours because they lack daily environmental cues.*

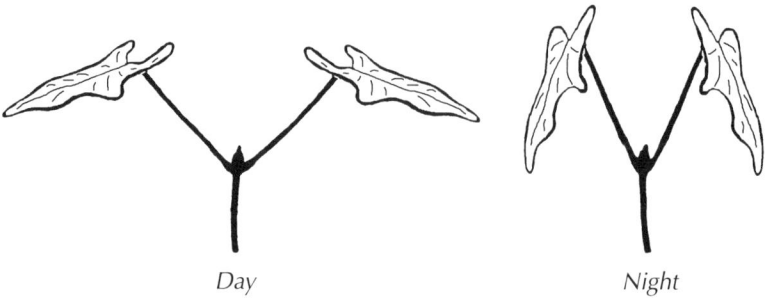

Day   *Night*

condition, molt, and other circannual rhythms in the same sequence as wild birds. In small songbirds, these cycles have persisted for more than ten years—longer than the birds would be expected to live in nature. In obligate migratory species, then, the primary stimulus that triggers migratory behavior arises from the bird's internal physiology rather than from immediate external environmental conditions.

Under constant conditions in captivity, the cycle lengths are not exactly 12 months long, so the birds' rhythms gradually drift out of phase with the actual seasons. Under natural conditions, however, circadian and circannual rhythms are synchronized, or "entrained," by daily cues, such as cycles of daylight and darkness, and by seasonal cues, such as seasonal variation in the **photoperiod** (day length).

A bird's specific response to the internal migratory stimulus may be modified by environmental conditions, which may determine whether a bird begins to migrate on a particular day or how far it will fly. A species that develops a powerful urge to migrate as the days grow shorter, for instance, may decide to sit tight on a particular day if the wind is blowing in the wrong direction, or to take off if the wind is right.

When birds come into migratory condition, a host of physiological changes takes place. Their daily patterns of activity and rest change. Night migrants become active during darkness, and the metabolism of many species changes such that they begin to deposit large quantities of fat, which serves as fuel for flight. The biological clock, fine-tuned by changes in photoperiod, controls all these circannual rhythms. They are further mediated through the nervous and endocrine systems, although the hormonal control of migratory behavior is not well understood.

The life of a migratory bird is played out on an enormous stage, and the success of migration as an evolutionary strategy depends on a host of variables, including conditions on the breeding and wintering grounds and along the path between them. In many ways, both energetically and in terms of risk, migration is the most demanding event on a bird's annual calendar. To cope with the rigors of migrating long distances and thereby increase the odds of survival, birds possess an elaborate suite of behavioral and physiological adaptations, which are described next.

The Physiology of Migration

Flying is strenuous, and even though birds are beautifully suited to their aerial life with such adaptations as extraordinarily efficient hemoglobin, a lung and air-sac system that enables maximum oxygen uptake, and hollow bones, a calorie is still a calorie, and it takes a lot of them to propel a body through the air for hundreds of miles. Fat provides most of the needed fuel.

To produce fat, birds change their feeding behavior as the migration season approaches. They dramatically increase the amount they eat by as much as 25 to 30 percent (a phenomenon termed **hyperphagia**, literally "overeating"). The diet may change, too, especially in autumn when insect populations decline and many plants produce fruit. Fruits are relatively easy to digest and high in carbohydrates, which can be easily converted to fat. Many species of migrants, such as thrushes, warblers, and sparrows, and also some shorebirds, ducks, and gulls, increase the amount of fruit in their diet immediately before and during migration. Laboratory studies have shown that birds can select food items with the highest fat content from an array of choices of equal caloric value. Bird metabolism also appears to change before migration. Production of fat increases, and energy reserves already stored as carbohydrate may be converted to fat. The efficiency of food digestion and absorption also may increase.

Fat is the most energy-rich substance that animals can store in their bodies. When oxidized (burned), it yields about twice as much energy per gram as carbohydrate or protein. Migratory fat is deposited all over a bird's body, even within the muscles and internal organs such as the heart and liver. Most, however, is laid down in **fat bodies** just under the skin; the most conspicuous fat bodies lie over the abdomen and in the depression formed anterior to the breast muscles where the clavicles fuse to form the wishbone. If you part the feathers of a long-distance migrant with a full fat load, you can see that the bird is nearly encased in a layer of yellowish fat **(Fig. 5–60)**. Bird banders and other researchers routinely examine birds in this way, rating the amount of fat deposited as an index of the birds' energetic condition and potential to continue migration.

The amount of fat deposited varies greatly among migratory species. Typically, long-distance migrants or those that must cross large ecological barriers where food is scarce lay down more fat prior to migrating. A typical nonmigratory bird carries 3 to 5 percent of its lean (fat-free) body mass as fat. For passerine migrants, the figure is 60 to 100 percent. In terms of total body mass (including fat), long-distance migrants often carry 30 to 50 percent as fat when they begin a flight, short-distance migrants, 10 to 25 percent. Carrying this much fat is

*Figure 5–60. Nonmigratory versus Pre-migratory Fat Load: During the weeks before migrating, birds eat more than usual, and also may eat more fat and carbohydrates (often as fruits), which are easily converted to fat. Migratory fat is deposited all over the bodies of birds, but most is laid down in **fat bodies** just under the skin. To determine the physical condition of migratory birds, researchers often inspect and "score" the amount of fat deposits in the conspicuous interclavicular and abdominal regions. When present, the clumps of pale yellow fat in these regions are easily observed compared to the surrounding pink, vascularized skin.*

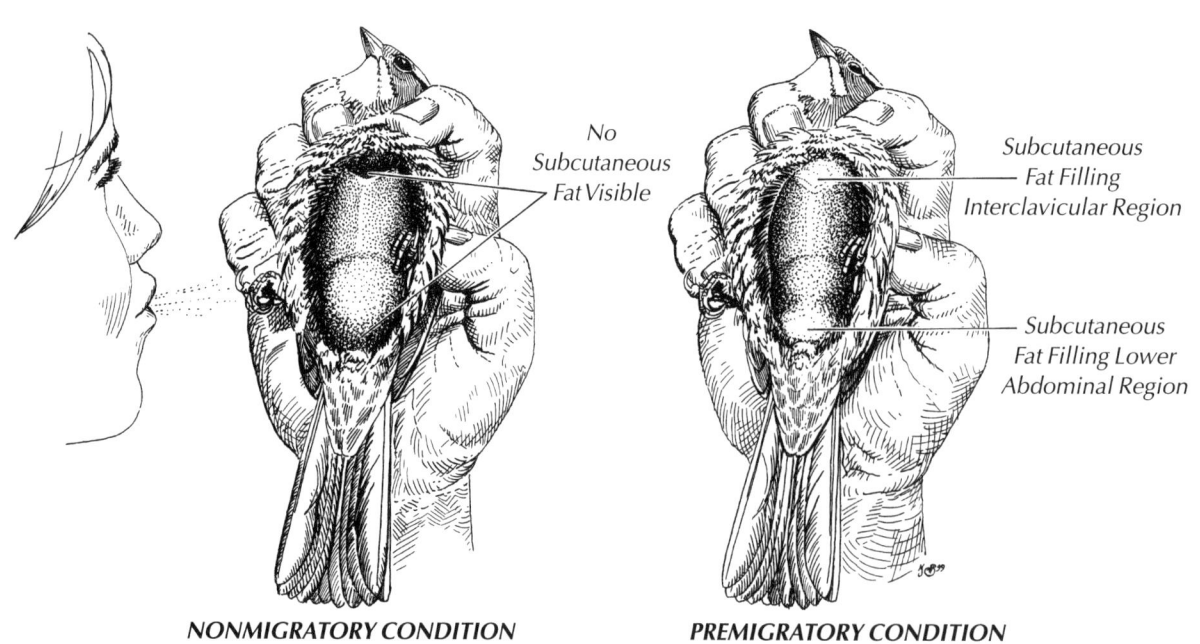

No Subcutaneous Fat Visible

Subcutaneous Fat Filling Interclavicular Region

Subcutaneous Fat Filling Lower Abdominal Region

NONMIGRATORY CONDITION   **PREMIGRATORY CONDITION**

a burden, of course, and it can reduce a bird's maximum range on a single flight by as much as one-half.

Using information on the energy yields of fat and the typical flight speeds of birds, one can construct models to predict how far a bird should be able to fly with a given amount of fat. Several types of models exist and they do not always agree, but they do yield some approximate figures. On a single flight without refueling, some shorebirds can go up to 6,000 miles (9,600 km), many passerines can cover 620 miles (1,000 km), and even the Ruby-throated Hummingbird can fly across the Gulf of Mexico (500 miles [800 km]) in spring. This means that birds operate on the same order of performance as large aircraft!

As a migrant flies it depletes its fat stores, which must be replenished at stopover locations. The quality of the habitat a migrant finds determines how rapidly it can refuel and be on its way. Daily fat deposition for migrating songbirds is typically 2 to 3 percent of their lean body mass, but it may be as high as 10 percent. Depleted fat reserves reduce a bird's propensity to initiate migration, and unless the bird finds itself in a place where it can gain a little weight, it is likely to remain grounded until it stores enough fat for the next leg of the journey. Ironically, some of the best places for bird watchers to see large numbers of grounded migrating birds may be among the worst from a bird's point of view. Offshore islands such as the Dry Tortugas in Florida attract large numbers of exhausted birds, but such sites often lack sufficient food and fresh water. The longer birds remain in such places, the less likely they will be to reach suitable refueling places. Likewise, migrants lured to tiny islands of natural vegetation amid large urban areas might do better by continuing to fly.

A bird in flight generates heat and therefore loses water by evaporation, so birds can become both overheated and dehydrated on long flights. Nevertheless, field and laboratory studies suggest that fuel, not the need to avoid dehydration or overheating, is the main factor that limits migratory flights. The need to avoid heat and water stress may, however, have been a selective factor that favored night migration in many species.

Daily Timing of Migration

Some birds migrate during daylight. Soaring birds such as hawks, cranes, and storks—which rely on thermals for lift—must migrate during the day, when thermals appear. Swifts and swallows, which feed as they fly, also migrate by day. Most waterfowl and shorebirds migrate during either the day or night, depending on weather conditions. Other diurnal migrants include some woodpeckers, kingbirds, crows and jays, larks, pipits, bluebirds, American Robins, blackbirds, and cardueline finches.

The majority of species and individual passerine birds however, migrate almost exclusively at night. (When forced to cross large ecological barriers such as the Gulf of Mexico that cannot be passed in one night, these species of course continue flying in the daytime.) In addition, many species make low-altitude movements during the early morning hours. Why so many normally diurnal birds migrate at night has been the subject of speculation for years. As with so many "why" questions in biology, the hypotheses on this topic are often untestable: you can make up any story you like and no one can prove you wrong. The best guesses are (1) that migrating at night allows birds more time during the day for feeding and replenishing fat stores; (2) that the structure of the atmosphere at night is more stable (there is less turbulence from convection) and more conducive to flight by slow-flying birds; and (3) that the generally cooler air at night reduces stress from heat and dehydration. Of course several different forces could favor nocturnal migration; natural selection takes whatever advantage it can.

Studies using moon-watching (observing the silhouettes of migrants as they pass in front of the moon), radar, and radio-tracking have shown that most nocturnal migration begins 30 to 45 minutes after sunset. The number of birds aloft then increases dramatically, reaching a peak before midnight and then decreasing steadily until dawn. Most night migrants have landed long before daylight. Radar and visual studies show that diurnal migrants begin their flights shortly after dawn; migration peaks around 10:00 A.M. and declines thereafter. Of course, unusual weather may alter these typical patterns.

The Altitude of Migration

Not until the advent of surveillance radar during World War II could we accurately measure how high migrating birds fly. Although there are many old claims that birds migrate at fantastic altitudes, most nocturnal songbird migrants fly at low altitudes over land. At night, in the absence of heat from the sun, there is probably no reason to fly higher in the stronger winds and colder, thinner air, which contains less oxygen for respiration. The majority of songbirds usually migrate below 2,000 feet (610 m), and some 90 percent typically fly below 6,500 feet (1,983 m).

Depending on weather conditions, particularly at the altitudes at which favorable winds may be found, nocturnal passerine migrants may fly up to 15,000 feet (4,575 m) or even higher, but this is unusual. Birds also tend to fly higher when making long crossings over water. Larger, stronger-flying shorebirds sometimes fly much higher. Even over land it is not unusual to find them at 15,000 to 20,000 feet (4,575 to 6,100 m). Daytime land bird migration over land usually occurs at quite low altitudes, but soaring raptors, waterfowl, and gulls may fly at considerable heights (**Fig. 5–61**).

Headwinds or cloud cover usually lower flight altitudes. When cloud layers are not too thick, birds may ascend through them and reach the clear skies above. This occurs most often when favorable winds exist above the clouds, although how the birds know about these winds is a mystery.

Most of the really high altitude records of bird flight come from mountainous regions. In the Himalayas, Lammergeiers (large vultures) and Yellow-billed Choughs (relatives of crows) have been observed flying at 24,600 and 26,900 feet (7,503 and 8,205 m), respectively. Bar-headed Geese regularly migrate over the highest Himalayas and

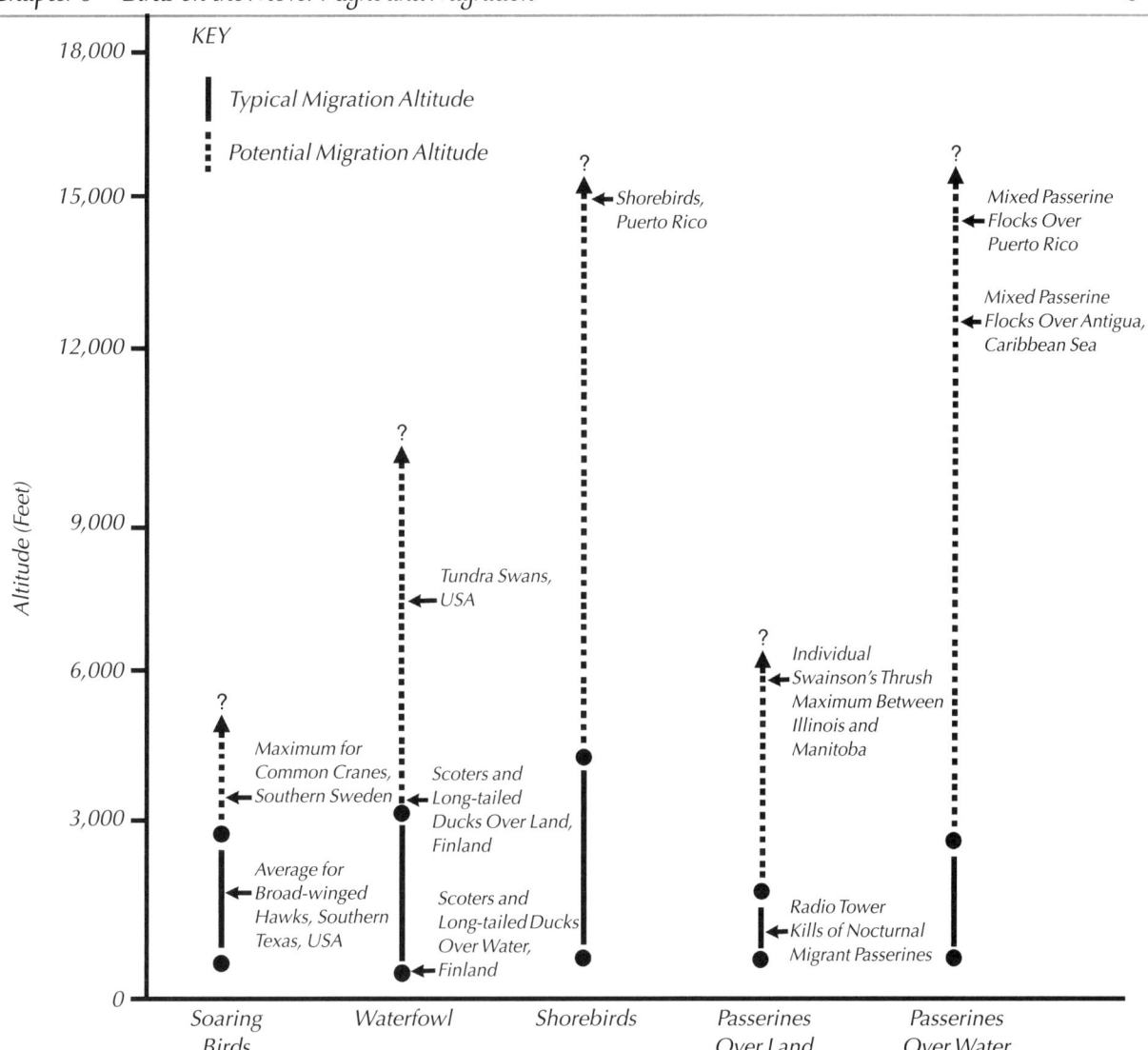

KEY

▮ Typical Migration Altitude

┇ Potential Migration Altitude

Figure 5–61. Migration Altitude of Various Bird Groups: *The height at which birds migrate is influenced by the physical characteristics of the birds themselves, as well as by many environmental variables, including whether the flight is over land or water, the topography, and most importantly, the weather conditions—especially wind speed and wind direction. Consequently variation within groups is high and generalities are difficult to arrive at. This figure presents, for five bird groupings, typical altitude ranges (solid lines) and potential ranges (dotted lines)—altitudes at which migrants have been occasionally observed, but migration may be more common than observations suggest—as well as examples of individual observations for each group. The ranges were determined by an exhaustive search of the literature, but new information from modern satellite-tracking techniques may change our understanding of migration altitude in the future.*

Soaring migrants such as hawks, vultures, storks, and cranes, rely on thermal activity and thus tend to migrate below 3,000 feet (1,000 m), but are sometimes found as high as 4,500 feet (1,500 m). Migrating waterfowl may fly as high as 10,500 feet (3,500 m), but often fly as low as 300 to 1,200 feet (100 to 400 m) during daylight. Migrating shorebirds typically fly higher than most other birds, consistently higher than 3,000 feet (1,000 m), and sometimes considerably higher.

Nocturnal migrant passerines over land usually fly 2,100 to 2,400 feet (700 to 800 m) above ground, but often fly much lower, as demonstrated by the numerous migrants killed by flying into large radio, television, and cellular phone towers. Some passerines, however, do fly higher than this over land. For example, an individual Swainson's Thrush fitted with a radio transmitter was tracked for seven nights from east-central Illinois into Manitoba, Canada, during which its average altitude was 1,065 feet (355 m), with a maximum of 6,000 feet (2,000 m) (Cochran 1987).

Passerines migrating over the water may fly much higher. Using an extensive radar network, Williams et al. (1977) obtained altitude information for a mixed flock of migrating passerines (and possibly small shorebirds) during their 1,875-mile (3,000-km), nonstop fall flight from North America to the coast of South America. From the North American coast to Bermuda the bulk of migrants flew below 6,000 feet (2,000 m) above sea level. Approaching the island of Antigua, however, the birds climbed dramatically, to well above 12,000 feet (4,000 m), gradually descending as they approached South America. Concurrent weather observations from Antigua suggested that the birds increased their altitude to avoid strong winds in the region. Adapted from Kerlinger and Moore (1989, p. 122).

Figure 5–62. Flight Path of a Bird: *The flight speed of a bird with respect to the ground (**ground speed**) is the sum of two other variables: the bird's **air speed** (how fast it moves with respect to the air) and the **wind speed** (the speed of the air pushing on the bird). The direction a bird is actually pointing its beak as it flies is called its **heading**, but any wind not coming from directly behind or in front of the bird will blow it sideways to some degree. The direction the bird actually travels with respect to the ground is called its **track**.*

*These speeds and directions are illustrated here with vectors for a bird heading due south in a wind blowing from the west. A **vector** is a variable with both magnitude (size) and direction, and is typically shown as an arrow whose length represents magnitude and whose orientation represents compass direction. Both ground speed and wind speed are actually vectors, since they have magnitude (how fast the bird is going) and direction (which way the bird is going). To add two vectors, place the beginning of one at the end of the other, maintaining the orientation of each. Thus, to add wind speed to air speed, one could move the wind speed vector to the end of the air speed vector (dotted arrow). The end of the wind speed arrow in its new position gives the end point for the vector sum. For this example, then, the track of a bird heading south in a west wind is to the southeast, and the length of the ground speed arrow represents how fast the bird is moving over the ground.*

have been recorded at least as high as 27,880 feet (8,503 m), and a Mallard struck a plane at 21,000 feet (6,405 m) over Nevada. Andean Condors have been documented at 19,800 feet (6,039 m). A large African vulture, the Rueppell's Griffon, collided with an aircraft at 37,000 feet (11,285 m) over the Ivory Coast!

Flight Speed and the Progress of Migration

The progress of migration is determined by a bird's **ground speed**, or how fast it moves over the earth's surface. This speed is a function of two variables: the forward propulsion of the bird with respect to the air (its **air speed**), and the action of the wind upon its flight path. The direction in which a bird is pointing its beak and propelling itself through the air is called its **heading**. Unless the bird is flying in a direct headwind, direct tailwind, or still air, its actual direction of movement over the ground differs from its heading, because the wind blowing from an angle affects its direction. This actual direction of movement over the ground is called its **track**. For example, a bird heading due south in a wind blowing from the west has a track to the southeast, because the bird is pushed to the east by the wind. Its ground speed will be greater than its air speed because a component of the west wind pushes it to the southeast, adding to its speed like a tailwind **(Fig. 5–62)**.

During migration, most passerines fly at relatively slow air speeds of 20 to 30 mph (32 to 48 km/h). Waterfowl and shorebirds fly considerably faster, at 30 to 50 mph (48 to 80 km/h). By selecting weather conditions that provide tailwinds, birds can achieve ground speeds up to two or more times faster than their air speeds. As explained in the next section, a tailwind is often critical to the success of a migratory flight.

Migrating birds seem very attuned to their air and ground speeds. Radar studies have shown that when flying with a tailwind, birds slightly reduce their flight effort (and therefore their air speed); when flying into a headwind they increase their air speed and work harder.

The progress of migration on the ground depends not only on flight speed, but on the duration and frequency of the migratory flights. Evidence suggests that spring migration proceeds more rapidly than fall migration, at least for songbirds. This seems to result from shorter stopovers and perhaps longer flights; the speed at which individuals fly does not seem to change from spring to fall. Presumably, the shorter spring migration is a response to selection pressure—birds that arrive early to establish a breeding territory may have a reproductive advantage.

Assuming they do not have to cross a large ecological barrier, passerines tend to migrate in a series of relatively short flights of up to 200 miles (320 km) or so, interspersed with one to three days of rest, depending upon weather, the birds' fat loads, and

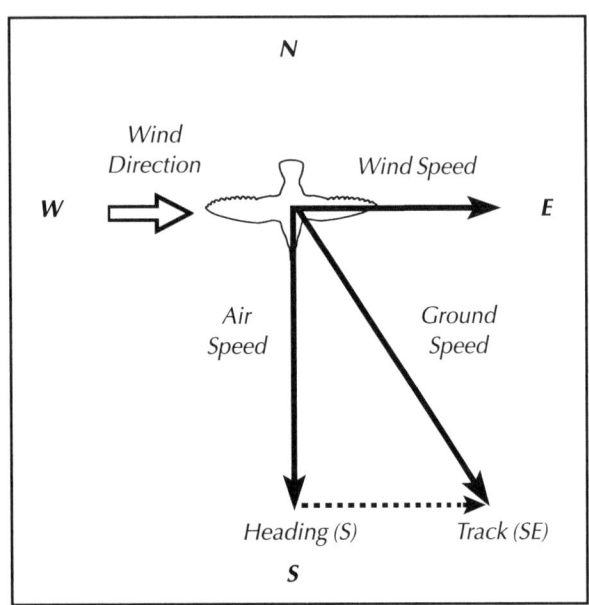

a. Low Pressure System

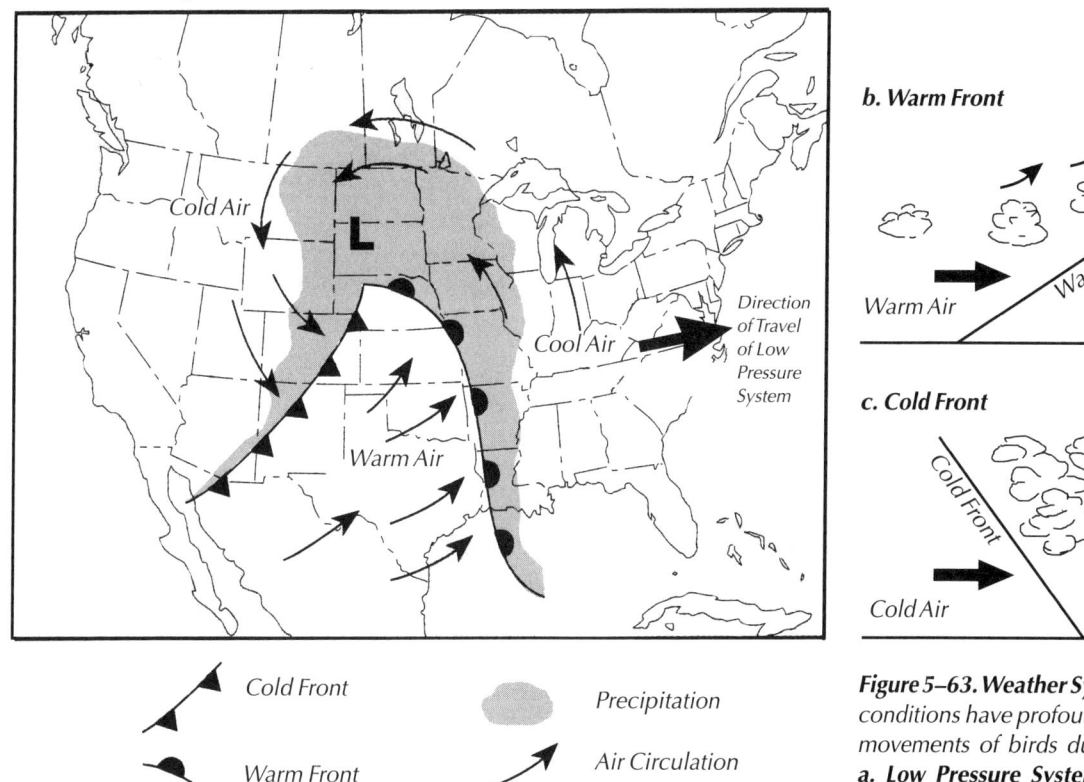

Cold Front

Warm Front

Precipitation

Air Circulation

L Low Pressure Area

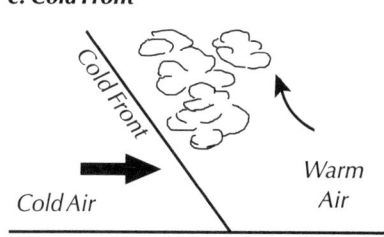

b. Warm Front

Warm Air   Warm Front   Cold Air

c. Cold Front

Cold Front   Cold Air   Warm Air

Figure 5–63. Weather Systems: *Weather conditions have profound effects on the movements of birds during migration.* **a. Low Pressure System:** *The passage of low pressure systems across North America is particularly important to migrating birds in this region. These large systems, which consist of cold and warm air masses circulating around a low pressure area (L), are found only in the middle latitudes, and usually travel from west to east. In the Northern Hemisphere, winds circulate counter-clockwise around the low pressure area (in the Southern Hemisphere, they circulate clockwise).* **b. Warm Front:** *Where the warm air mass overtakes cold air, it pushes up over the denser cold air, is cooled, and thus forms clouds and often precipitation: this interface between the two air masses is called a* **warm front.** **c. Cold Front:** *Where cold air overtakes warm air (at a* **cold front***), the dense, cold air tends to wedge under the warm air mass, forcing the warm air up and cooling it quite abruptly, forming pre-cipitation that is often accompanied by strong winds and lightning, and is heavier and of shorter duration than the precipitation associated with a warm front. Adapted from Lutgens and Tarbuck (1998, p. 211).*

refueling conditions at the stopover sites. Waterfowl and shorebirds tend to make longer nonstop flights, even when flying over land; ducks and geese have been observed flying up to 1,865 miles (3,000 km) from Canada to their Gulf Coast wintering areas in two days.

Weather and Migration

Just as long-term changes in climate may mold the evolution of bird migration, seasonal and day-to-day changes in weather dramati-cally influence the timing and rate of progress of migration. The number of migrants may vary by up to a thousandfold from night to night. In favorable weather, a staggering number of birds may be in flight, prob-ably hundreds of millions over North America alone. I vividly recall a September night in northern Georgia when weather radar showed 200,000 birds crossing a line one mile (1.6 km) long every hour. This passage went on for much of the night!

Many analyses of the relationships between weather and mi-gration volume have been conducted, and they consistently show that temperature and wind direction are the two most important factors. In spring, northbound birds migrate in the warming temperatures and southerly winds that characterize the western sides of high pressure systems. In autumn, migrants favor the falling temperatures and north winds that occur after the passage of a cold front. **Figures 5–63** through **5–65** illustrate these patterns. Birds avoid migrating in rain, clouds, fog,

Figure 5–64. How Birds Use Weather Systems During Migration: *(Circled numbers refer to sections of Fig. 5–65; see that figure for weather conditions at each number.) Low pressure systems can occur at any time of year. In different seasons, migrating birds rely on different sectors of a low pressure system to aid them in their travels.*

In spring, northbound migrants move in great numbers with the southerly winds that accompany a warm front. Precipitation ahead of the front often prevents migration, but once the front has passed, waves of migrants may stream by. These warm fronts usually occur before the passage of a low pressure region and after a high pressure system (winds circulate clockwise around a high in the Northern Hemisphere, so winds are from the south after their passage). Spring migration stops when a cold front, with its north and northwest winds, passes by.

In fall, southbound migrants take flight after the passage of a cold front, when skies clear and northerly winds provide a favorable tailwind. These cold fronts usually occur after a low pressure system and before a high. Migration may continue until the winds change to come from the south, as when a warm front arrives.

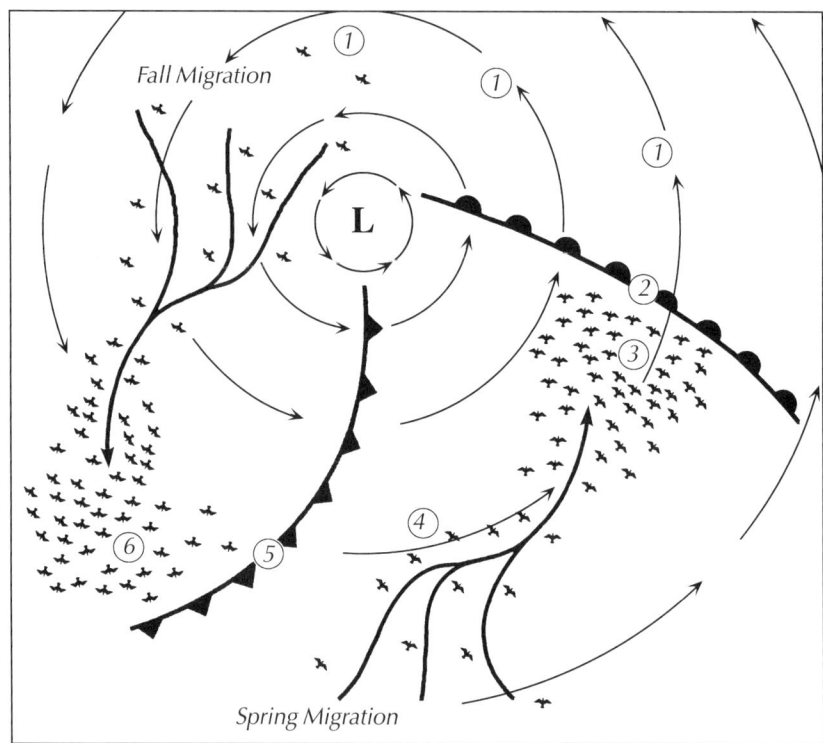

KEY

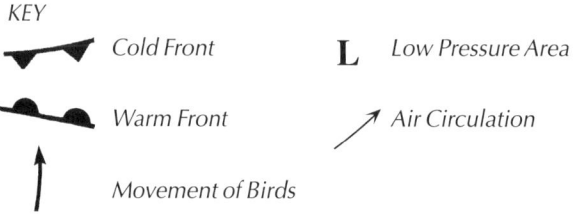

and strong winds, and may even stop their journey if they encounter deteriorating weather while over land. For example, in the spring, land birds crossing the Gulf of Mexico from the south usually come to earth far inland. If a cold front meets them on the coast, however, they descend immediately. At times small birds "flood" the coastal areas of the Gulf states and at other times there seem to be no birds at all.

Like the weather conditions through which the migrant must fly, conditions at the destination also are important. In many species, males tend to migrate earlier in the season than females, and thus arrive on the breeding grounds ahead of their potential mates. As mentioned earlier, a male who arrives early gains a considerable premium in terms of establishing a territory, obtaining a mate, and getting an early start on breeding. But birds that arrive early in spring are more likely to encounter bad weather, and for a bird who has just completed a long, tiring trip, cold, snow, or even heavy rain can be dangerous.

In the fall, when most birds head toward more favorable climates, the risk of arriving early is not so great. Although many water birds linger in the north until freezing temperatures force their departure, in most species, natural selection has favored individuals that anticipate seasonal change and migrate well before conditions turn bad. Thus, most insectivorous songbirds depart breeding areas in late summer when the weather is fine and food is still plentiful. The stimulus that

WARM FRONT

(1) Ahead of Warm Front:
High clouds develop slowly, becoming lower and thicker over 12 to 24 hours. Winds from the **east**. Light precipitation begins as front nears.

Spring Migration: None

Fall Migration: May be in progress

(2) Front Passes:
Overcast skies. Heavier but steady precipitation. Temperature rises. Winds change from **east** to **south**.

None

Stops

(3) Behind Warm Front:
Clear skies. Warm temperatures. **South** winds.

In full force

None

COLD FRONT

(4) Ahead of Cold Front:
Rapidly approaching low, dense clouds. Gusty winds, from the **south** to **southwest**.

Spring Migration: Slowed

Fall Migration: None

(5) Front Passes:
Heavily overcast skies. Severe to heavy precipitation of short duration (including possible thunderstorms and hailstorms). Winds from the **west**.

Stops

None

(6) Behind Cold Front:
Skies clear rapidly. Temperature drops. Winds from the **northwest**.

None

In full force

Figure 5–65. The Effects of Warm and Cold Fronts on Migration: Numbers 1 to 6 refer to the circled numbers on Fig. 5–64. In this chart, detailed information on precipitation, wind direction, cloud cover, and the progress of spring and fall migration is given for the six regions indicated within the low pressure system pictured in Fig. 5–64. Note that there are several locations labeled number 1—ahead of a warm front—on Fig. 5–64 because the figure only shows one low pressure system. In reality, series of low pressure systems move across North America, one following another, and the region "ahead of a warm front" may be right behind a cold front, or quite far behind a cold front, depending on how closely the second low pressure system follows the first. Fall migration, at peak after a cold front, may remain in progress ahead of a warm front until either the winds become unfavorable, or precipitation becomes too heavy.

sends them on their way is not the weather, but internal changes in physiology, as discussed earlier.

Birds whose migration routes take them across large bodies of water or deserts face the ultimate migration challenge. In these cases, selecting the right weather conditions for migrating—for example, tailwinds—can mean the difference between life and death. In autumn the Blackpoll Warbler flies nonstop over the western Atlantic Ocean from the northeastern United States to South America, an amazing feat that may take three to four days and nights (see Fig. 5–51). The birds depart North America under the clear skies and north winds that follow the passage of a cold front. But favorable conditions at the outset do not guarantee an uneventful passage; trouble, such as late-season tropical storms, may arise when the birds are so far along that turning back is impossible. When this happens, some individuals land on Bermuda or other islands; others may be displaced off-route to Florida. Untold numbers undoubtedly perish.

Clearly, a migrant's life depends upon its ability to correctly assess the weather. To what extent can it do so? Researchers are not sure. Pigeons are extremely sensitive to small changes in barometric pressure; it is possible, though not yet demonstrated, that by monitoring pressure changes, birds can anticipate weather changes long before more immediate signs appear.

Even if migrating birds can foretell the weather, they still may be at risk from bad weather. Many birds are blown off course by wind. In extreme cases, hurricanes and other intense cyclonic storms transport birds over hundreds or thousands of miles. Numerous instances of seabirds such as albatrosses, shearwaters, petrels, frigatebirds, and tropicbirds appearing far inland result from violent storms (**Fig. 5–66**).

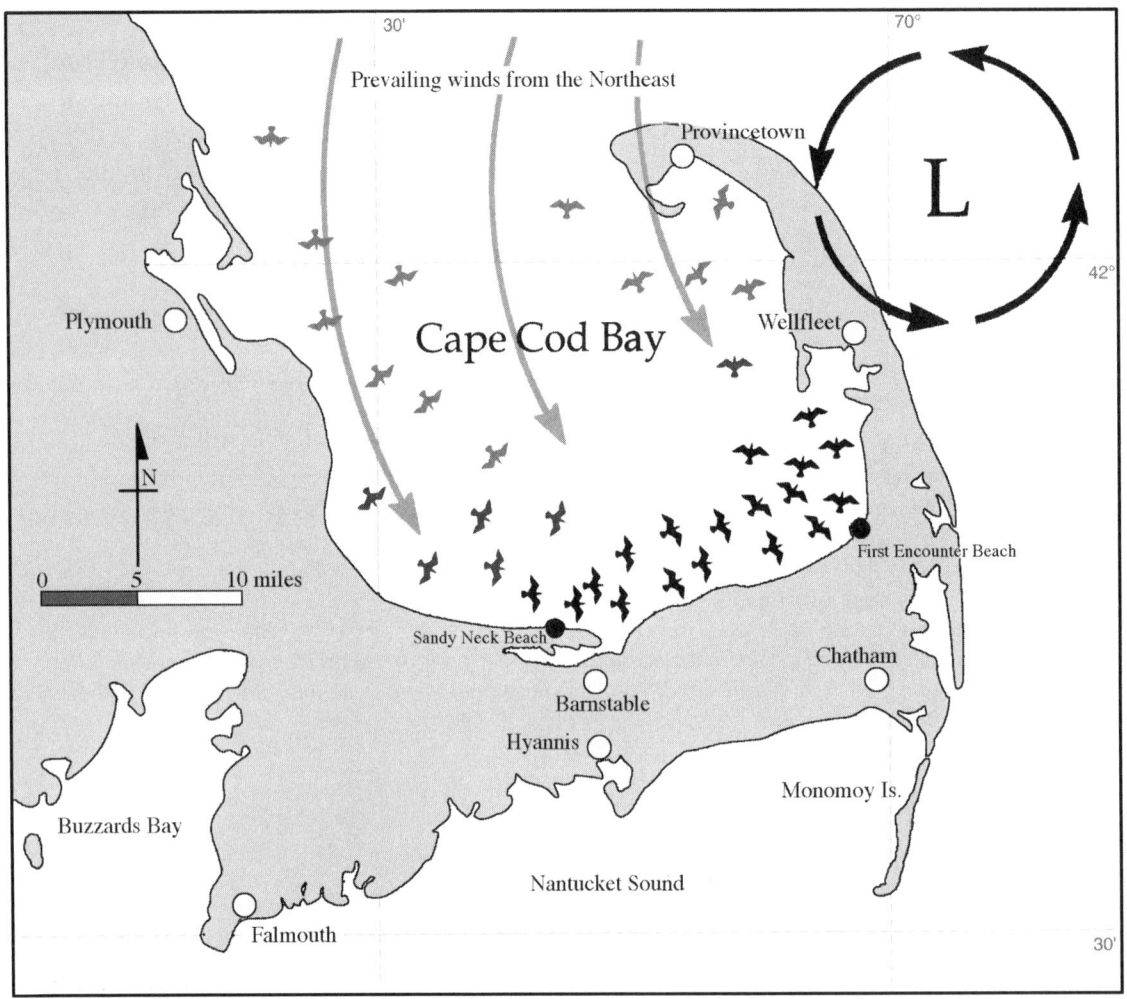

Figure 5–66. Northeaster off New England in Winter: *Each fall and winter, seabirds such as alcids, sea ducks, and phalaropes (pelagic members of the shorebird family) wander the open ocean. These relatively weak fliers are often at the mercy of wind and weather. During New England winters, storms called* **northeasters** *(or Nor'easters) sweep in from the northeast Atlantic, causing spectacular seabird sightings along the coast from Maine to Long Island. Severe northeasters can devastate pelagic bird populations, often pushing birds ashore or even inland in what are known as seabird wrecks, in which many birds die. Seabird wrecks usually occur at capes and along the shores of enclosed bays such as Cape Cod Bay. Northeasters often create excellent birding conditions along the shores of Cape Cod as seabirds swept from the ocean by strong winds are forced into the mouth of the bay. Once trapped there, they parallel the coast in a counterclockwise pattern. Bird watchers along the shore have ringside seats as normally inaccessible pelagic birds pass in review. Northern Gannets may dive for fish mere yards beyond the surf and thousands of shearwaters, alcids, sea ducks, and phalaropes may fly within binocular range. Reprinted from* Manual of Ornithology, *by Noble S. Proctor and Patrick J. Lynch, with permission of the publisher. Copyright © 1993, Yale University Press.*

Along the New England coast in autumn, large numbers of night migrants may be blown offshore in the northwest winds that follow strong cold fronts. Many of these birds undoubtedly fall into the sea, but others land on ships and offshore islands; during the day they can be seen flying back toward the mainland. Large numbers of migrants blown off course sometimes may establish populations in new locations; the colonization of Greenland in 1937 by the Fieldfare, a Eurasian thrush, is an often-cited example.

Migration Routes

Each migratory species has a characteristic general route of travel between its nesting and winter range, but for most species these migration routes are quite broad. Waterfowl tend to follow narrower corridors, which are often determined by the availability of suitable stopover habitat. In fact, biologists once thought that individual waterfowl populations followed distinct, narrow flyways (the Atlantic flyway, the Mississippi flyway, and others). However, leg band recoveries showing that individuals from one nesting area could be found migrating in several different flyways put that idea to rest.

In North America, however, there are some general patterns of migration flow. Songbird migration in the eastern half of the continent tends to move northeastward in spring, and southwestward in fall. In a large number of species, the round-trip migration path forms an ellipse. A classic, extreme example of this elliptical pattern can be seen in the American Golden-Plover, which in spring migrates from its wintering grounds in the pampas of South America to the high Arctic by passing through the interior of North America **(Fig. 5–67)**. In fall, it first flies southeastward to the Canadian Maritimes and then over the western Atlantic to South America. This route takes advantage of prevailing seasonal wind patterns and is used by a number of other shorebird species, such as the Hudsonian Godwit, Buff-breasted Sandpiper, and the nearly extinct Eskimo Curlew.

Radar surveillance indicates that nocturnal migrants move in a dispersed fashion with slight regard to what lies below. But birds migrating by day tend to follow topographical features trending north and south, such as mountain ranges, chains of lakes, river valleys, and peninsulas extending into large bodies of water. As these features narrow, the flight lanes narrow correspondingly, causing migratory movements to be concentrated and conspicuous.

Hawks migrating in the fall pass in large numbers along the Great Lakes, where they skirt around or parallel these wide bodies of water rather than cross them **(Fig. 5–68)**. On days after a cold front when westerly winds blow strongly,

Figure 5–67. Elliptical Migration Route of the American Golden-Plover: The American Golden-Plover migrates from its breeding grounds in the high Arctic tundra to winter on the grasslands of South America, a distance of nearly 8,000 miles (12,800 km) one-way. Most members of this species follow an elliptical migration route, heading south over the Atlantic Ocean in fall and returning by a different, more westerly, route over land in spring. These routes take advantage of the seasonal wind patterns and are used by a number of other shorebird species. Drawing by Robert Gillmor.

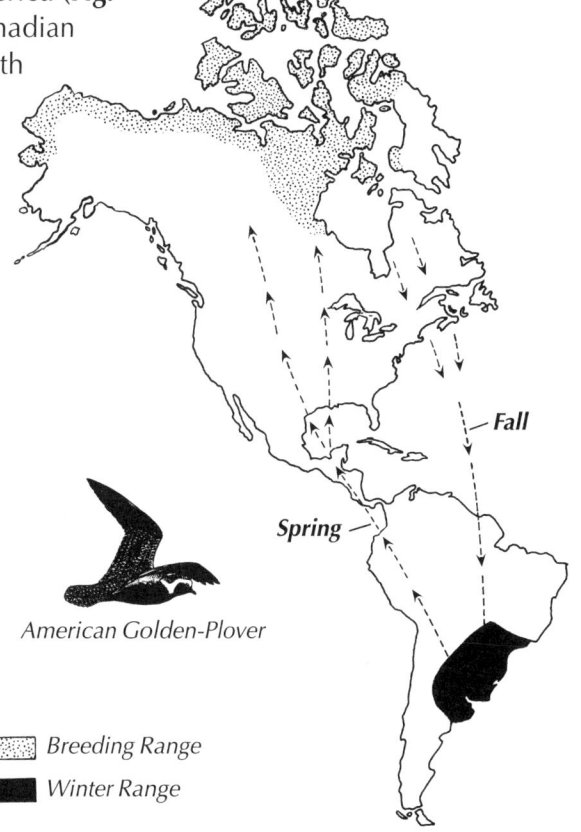

American Golden-Plover

Fall

Spring

▨ Breeding Range

■ Winter Range

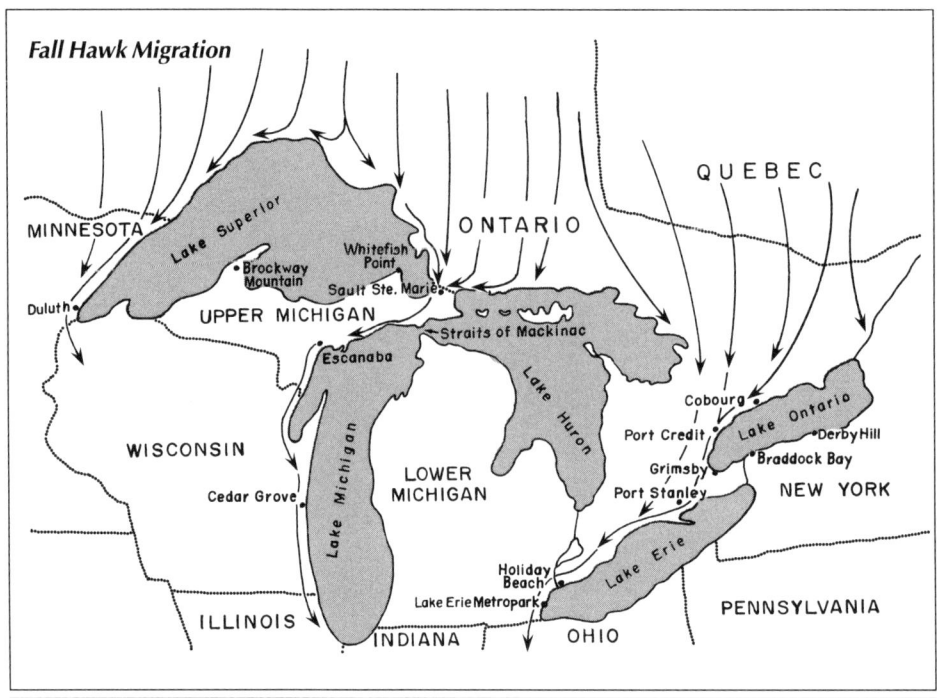

Figure 5–68. Spring and Fall Hawk Migration Routes Around the Great Lakes: *Migrating hawks are generally reluctant to fly over large bodies of water, such as the Great Lakes, and thus tend to follow the shorelines until they find a way around. In fall, southbound birds that reach Lake Superior move either east or west along the lakeshore, large numbers flying by Hawk Ridge in Duluth, Minnesota (55,000 per season, on average), or being channeled into Upper Michigan at Sault Sainte Marie. Farther east, southbound birds are concentrated between Lakes Huron, Ontario, and Erie, passing in abundance by such sites as Grimsby (14,000 per year) and Holiday Beach (81,000 per year), Ontario, as well as two Michigan sites averaging 100,000 hawks per year that are monitored by the Southeastern Michigan Raptor Research—Lake Erie Metropark and the adjacent* Point Mouille State Game Area.

In the spring, northbound hawks moving through Lower Michigan are channeled by Lake Michigan and Lake Huron and come together to cross the water at the Straits of Mackinac. Birds migrating north through Wisconsin and Upper Michigan move along the south shore of Lake Superior, passing in great numbers by such locations as Brockway Mountain on Upper Michigan's Keweenaw Peninsula, Whitefish Point (14,000 per year), and Sault Sainte Marie. Northbound birds that reach Lake Erie or Lake Ontario tend to move northeast along the shorelines, an average of 30,000 birds per season passing by well-known spring hawkwatch sites such as Braddock Bay and Derby Hill, New York. Drawings by Robert Gillmor.

the passage of these daytime predators is impressive in such spots as Port Credit and Amherstburg (Ontario), Cedar Grove (Wisconsin), and Duluth (Minnesota).

In the spring, large numbers of hawks, northbound through Lower Michigan, come together at the northern tip of Lower Michigan on the south side of the Straits of Mackinac. Here, reluctant to fly over the water if the weather is rainy and windless, they settle on trees and other perches until the next clear day with a favorable wind, when they spiral up and head north over the Straits.

Site Fidelity

Individual birds often show amazing loyalty to places they occupied during previous breeding and nonbreeding seasons, and to stopover points between the two, a phenomenon known as **site fidelity**.

For example, banding studies have shown that individual Eastern Phoebes are very likely to return to the same breeding site from one year to the next, even pairing with the same mate. Individual Barn Swallows, too, remain faithful to previously used colonies, often returning to the same nests and mates year after year. And the Bobolink, despite having one of the longest migration routes of any songbird, shows high breeding site fidelity, as individuals regularly return to the hayfields in which they nested the previous year.

Birds also may demonstrate fidelity to wintering areas; the same individual Ovenbirds have been recaptured in successive years from the same locality in southern Mexico (Ely et al. 1977), and individual Northern Waterthrushes return to the same areas in Venezuela, Panama, Trinidad, Belize, and Jamaica. Individual Wood Thrushes inhabit the same nonbreeding territories each year on their wintering areas in Veracruz, Mexico, and also return fairly consistently to the same breeding territories in the United States.

Fidelity to stopover sites during migration is particularly common among large birds—waterfowl, cranes, and storks—that often migrate in flocks composed of family groups of older, experienced birds and youngsters. Sandhill Crane family groups, for example, gather into enormous flocks at traditional stopover sites, with endangered Whooping Cranes sometimes joining them. In spring, 80 to 90 percent of the mid-continent population of Sandhill Cranes stop in the North Platte and Platte River Valleys of Nebraska. Individual Sandhill Cranes also return year after year to the same breeding territories and wintering grounds. Shorebirds, too, exhibit strong loyalty to traditional staging areas along their migration routes **(Sidebar 2: Showdown at Delaware Bay)**.

Most songbirds, especially nocturnal migrants, seem to migrate alone or in loosely defined aggregations. Whether they show high fidelity to particular migration routes or stopover sites is hard to detect, because songbirds are much less likely to be recaptured than waterfowl, cranes, and other large birds.

(Continued on p. 5.79)

Sidebar 2: SHOWDOWN AT DELAWARE BAY
Paul Kerlinger

For Joan Walsh, a biologist at the New Jersey Audubon Society's Cape May Bird Observatory, spring of 1997 was ominously quiet. By mid-May, the shores of Delaware Bay that Walsh frequently visits are usually packed, not with beachgoers but with throngs of mating horseshoe crabs, tens of thousands of migrating shorebirds, and the biologists, birders, and simply curious who come to see them.

For millennia, horseshoe crabs (these ancient animals are more closely related to spiders than to true crabs) have hauled their tanklike bodies onto the shores of Delaware Bay by the millions to mate (**Fig. A**). Some beaches are carpeted black as the paired crabs (the males attach themselves to the larger females) emerge from the surf. As the female deposits her eggs in the sand, the male fertilizes them. Then he may help the female cover the eggs before the couple heads back to sea. The female can lay up to 80,000 eggs a season (**Fig. B**), so the crabs might repeat this mating act several more times before they move into the deeper waters of the bay and ocean for another year. The eggs hatch in about a month, and the young, too, go to sea.

Not all horseshoe crab eggs hatch, but few go to waste. Just as the crabs are laying their eggs, anywhere from 500,000 to 1.5 million shorebirds—the majority of them Red Knots, Sanderlings, Ruddy Turnstones, and Semipalmated Sandpipers—stop along the bay shores of New Jersey and Delaware on their migration from South America to the Canadian Arctic (**Fig. C**). They gorge on the abundant eggs, which provide the fat and protein to fuel the next 1,500 to 2,000 miles (2,400 to 3,200 km) of their trek. Red Knots may gain 50 percent or more of their body

Figure A. Horseshoe Crabs Mating: *A pair of horseshoe crabs, the smaller male attached to the female, comes ashore to lay eggs at Reeds Beach on Delaware Bay. The female digs a nest in the sand six to eight inches (15 to 20 cm) deep and deposits her eggs. As she crawls away, the male passes over the nest, fertilizing the eggs. If horseshoe crab densities on the beach are high enough, laying females will accidentally dig up the nests of females who laid earlier, littering the beach with eggs that are thus made available to migrating shorebirds. With lower numbers of horseshoe crabs, the eggs remain safely buried in the sand. Photo by Sandy Podulka.*

Figure B. Horseshoe Crab Eggs: *Coating the beach like tiny, greenish pebbles, billions of horseshoe crab eggs provide a much-needed meal of fat and protein for the hundreds of thousands of migrating shorebirds who stop at Delaware Bay to refuel, en route from South America to their breeding grounds in the Arctic tundra. Photo by Kevin T. Karlson.*

weight in fat within 10 days to two weeks while feeding furiously at the egg-rich beaches—something like a 160-pound (73-kilogram) human putting on 80 pounds (36 kilograms) over the holidays. Some 80 percent of the North American population of Red Knots stops over each year on the shores of New Jersey and Delaware; horseshoe crab eggs are therefore

Figure C. Shorebirds Refueling at Reeds Beach: *In mid-May, hundreds of thousands of migrating shorebirds, most of them Red Knots, Ruddy Turnstones, and Sanderlings, stop along the shores of Delaware Bay in New Jersey to feast on horseshoe crab eggs before continuing their journey to breeding grounds in the Arctic tundra. Visible with the shorebirds are Laughing Gulls and Herring Gulls, who also take advantage of the banquet. Photo by Kevin T. Karlson.*

critical to the birds' successful migration.

This crossroads of mating horseshoe crabs (locally called king crabs) and migrating shorebirds now draws visitors from around the world. It was brought to public attention in the 1970s by, among others, Cape May Bird Observatory director Pete Dunne and shorebird biologist Brian Harrington. Birding records from the 1930s through the 1960s make no mention of the annual spectacle, presumably because it didn't happen. Between the 1880s and 1920s, about a million horseshoe crabs were harvested each year for fertilizer and hog fodder. Although the market waned, 50 years passed before crab numbers rebounded, shorebirds came back in huge numbers, and the phenomenon was "rediscovered."

Today, horseshoe crabs are again a valuable commodity, but this time as bait. According to fishermen, eels and whelks (locally called conch) are particularly fond of egg-bearing females; the demand is therefore high for those same animals that provide spring fare for shorebirds. Traditionally, the fishermen picked the crabs off the beaches by hand and at no cost; this harvest was relatively small. But with the burgeoning market for eel and whelk both here and abroad, a market for the crabs began to develop along the entire East Coast, and anyone selling them could get 50 cents per crab. This price attracted commercial fishermen, whose incomes had suffered from stringent regulations and the overharvest of other fish, as well as those who saw a fast and easy buck. Some fishermen converted their boats to trawlers, which can drag a net across the sea floor and ensnare tens of thousands of crabs in a single day. In the past 10 years, the harvest has grown from a cottage industry to a regional business complete with middlemen.

By the early 1990s, pickups and even large refrigerated trucks—some coming from as far away as Massachusetts and South Carolina—were lining up on the bay shore in May and early June **(Fig. D)**. Some scientists, and even casual observers, reported seeing fewer horseshoe crabs, especially on the New Jersey side of the bay. In the late 1980s, biologist Mark Botton of Fordham University estimated that the population of this species in the neighboring Atlantic was between 2.3 and 4.1 million crabs. But Botton and others believe that annual harvests since then of more than half a million crabs, combined with the slow maturation of this species (they start reproducing at about 10 years old), have led to a decline.

Fishermen remained unconvinced, even when four separate scientific studies by Botton, Limuli Laboratory (an independent, for-profit company), the New Jersey Bureau of Marine Fisheries, and the Delaware Division of Wildlife all reached similar conclusions.

Figure D. Truckload of Bait: *The harvest of horseshoe crabs, used as bait in eel and whelk fishing, increased dramatically in the 1990s, as the market for eel and whelk flourished. By the early 1990s, pickup trucks and large refrigerated trucks lined the shores of Delaware Bay, waiting to haul loads of crabs. Photo by Kevin T. Karlson.*

Relying either on estimates of adult crab numbers or on population estimates based on surveys of eggs deposited on bay-shore beaches, all four studies concluded that since about 1990, horseshoe crab populations have declined by more than 50 percent. Aerial surveys conducted by Kathy Clark of the New Jersey Endangered and Nongame Species Program (New Jersey Division of Fish, Game, and Wildlife) show that fewer Red Knots are using the New Jersey beaches and that more are now found on the Delaware side, where horseshoe crabs are more abundant. Whether the decline of horseshoe crabs is already having an effect on the hemisphere's population of Red Knots is not known, but without abundant crab eggs to feed on, the future of these birds is in jeopardy.

In January 1995, an unusual and large coalition of environmental groups and businesses (including the New Jersey Audubon Society, National Audubon Society, American Littoral Society, New Jersey Conservation Foundation, DuPont, Mobil Oil, and Atlantic Electric) called for emergency regulations on horseshoe crab fishing, citing not only ecological but also medical and economic reasons for protecting the crabs. Pharmaceutical companies use the crabs' blood to detect bacterial contamination in pharmaceutical products and to screen for diseases such as gonorrhea and spinal meningitis. Once captured and bled, the animals are returned to the water, where their survival rate is about 90 percent. The "harvest" is thus small in comparison with that of the bait business. The May spectacle of horseshoe crabs and shorebirds also helps support a growing ecotourism industry in southern New Jersey. A study by the Cape May Bird Observatory found that birders and other wildlife watchers annually bring millions of dollars to local businesses.

In response to the coalition, the New Jersey Department of Environmental Protection (DEP) restricted hand-harvesting to two nights per week. But trawling continued full force. By May 1997, the issue got the attention of New Jersey Governor Christine Todd Whitman, who issued executive orders temporarily banning all trawling and hand-harvesting. When the New Jersey Marine Fisheries Council failed to adopt new standards, the DEP drew up rules based on Whitman's orders. But in September 1997, the council vetoed the DEP rules. In October, the council reversed its decision in an out-of-court settlement with the New Jersey Audubon Society and the American Littoral Society, and the bans were reinstated. While fishermen aren't happy with the new restrictions, they can take solace in the governor's decision to commit $80,000 to research the population size of the crabs and the migrating shorebirds. The study should help answer the lingering question: How many crabs can be harvested without tipping the delicate balance?

Horseshoe crabs, meanwhile, having already bounced back once after decades of wholesale harvest, are likely to recover again. As a species, these rugged creatures have survived 200 million years of changing climate and habitat. And according to Botton, a synthetic horseshoe-crab "scent" that is being developed would reduce the need for the animals as bait.

The situation for migrating Red Knots and other shorebirds is more delicate. Good stopover sites are few and far between, and no substitute exists for Delaware Bay. Years could pass before horseshoe crab populations are restored and they lay eggs in such abundance that peak numbers of shorebirds are again lured to New Jersey. Biologist Walsh thinks that real conservation will come only when we see horseshoe crabs not as bait but as a national treasure. ∎

Reprinted with permission of Natural History *(May, 1998). Copyright the American Museum of Natural History (1998).*

Orientation and Navigation

■ Birds and many other animals have the ability to return to precisely the same sites they occupied previously, and birds may travel halfway around the globe to do so. This amazing feat raises an obvious question—how do they do it?—and that question provides the impetus for scientists who study bird navigation. Although homing ability has been known and exploited in pigeons since the time of the early Greeks, rigorous studies of the mechanisms of orientation and navigation did not get under way until well into the 20th century.

To understand orientation and navigation, it is important to comprehend how young, inexperienced birds reach the overwintering area of their species or population on their first migration. The problem is especially acute for the many species in which young do not migrate with experienced birds from whom they might learn how to reach the winter quarters. The solution to this problem began to emerge in the 1950s, when Dutch ornithologist A. C. Perdeck conducted an extensive study with banded European Starlings (Perdeck 1958). He captured 11,000 starlings during autumn migration on the Dutch coast, banded them, and transported them by plane to be released in Switzerland. Perdeck advertised his study widely to alert ornithologists, hunters, and others to report recoveries of the birds. He had great success, as 354 were reported. His experiment is diagrammed in **Fig. 5–69**.

The most important finding was that birds captured as adults, who had migrated to the winter grounds at least once before, tended to move from their release site in a northwestward direction toward the correct overwintering area for this population. First-time migrants (birds born that year), on the other hand, moved primarily toward the southwest, in the compass direction their migration along the North Sea coast would have taken had it not been interrupted by capture. The conclusion from this classic study, now supported by much additional evidence, is that a young bird on its first migration has some innate knowledge of the direction (and the approximate distance) that it should fly. It does not, however, seem to know the specific goal of its migration. After arriving within its winter range, the young bird apparently locates an appropriate site, to which it imprints (see Ch. 6, Learned Behavior) over the winter. After this experience, the bird can navigate toward this goal even if displaced while en route.

A bird's first migration is controlled by a circannual rhythm, as discussed earlier in this chapter. The execution of this so-called **migratory program** has been termed **vector navigation** and can in theory take a first-time migrant from its natal area to a point within the winter range of its population on an appropriate schedule, provided it is not waylaid by some researcher or storm. The same phenomenon can be demonstrated with birds exhibiting migratory hopping in orientation cages. The European Garden Warbler, for instance, first migrates southwest from northern Europe toward Spain and Portugal and then turns south into Africa. Hand-raised warblers held for the entire migration season in Frankfurt, Germany, showed the same change in

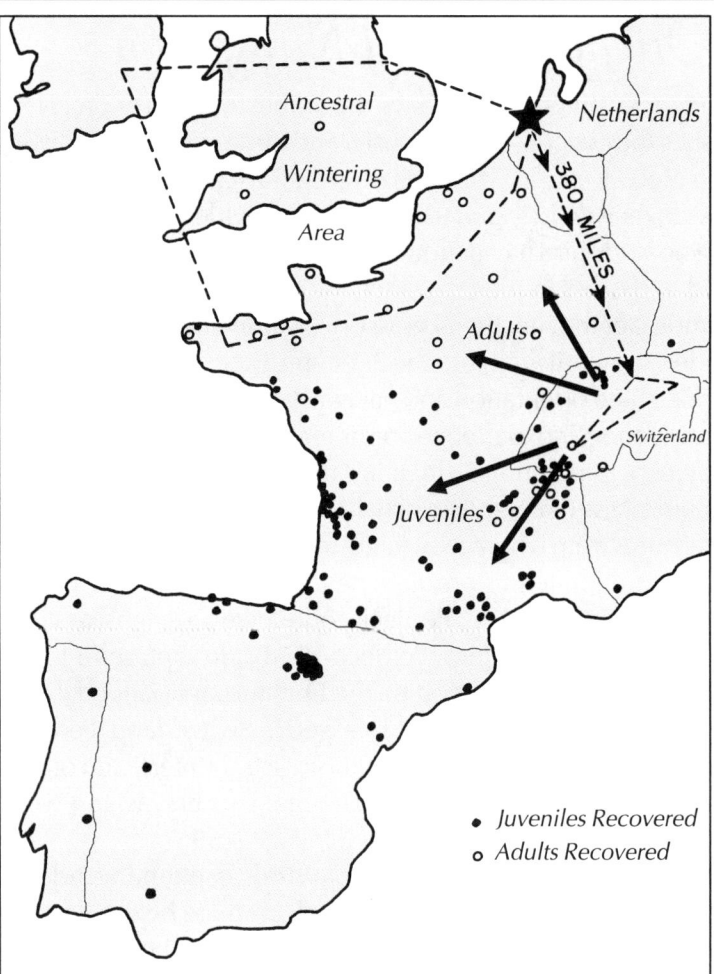

Figure 5–69. Perdeck's Displaced Starling Experiment: In the 1950s, A. C. Perdeck captured 11,000 European Starlings at The Hague in the Netherlands (star) during fall migration, banded them, transported them 380 miles to the southeast (dashed arrows), and released them in Switzerland (triangle). During the winter, 354 of the birds were recaptured. The distribution of recovery sites for adult birds (open circles)—birds that had migrated to the wintering area at least once previously—shows that these birds somehow adjusted for their new location and flew back toward their ancestral wintering area, in some cases even reaching the usual wintering grounds. However, the distribution of recovery sites for juveniles (solid circles)—birds on their first migration—shows that they flew the same direction (southwest) and distance that they should have if they were still in the Netherlands. Thus, first-time migrants appear to have some innate knowledge of the direction and distance they should migrate, but not of the specific goal of that migration. Once birds reach appropriate wintering grounds, they apparently learn more specific information about the area that allows them to navigate toward that goal in the future. Drawing by Robert Gillmor.

direction in their hopping at approximately the right time during the migration season (Able 1995) **(Fig. 5–70)**.

The European Starling experiment of Perdeck described earlier illustrates a fundamental difference in the navigational abilities of experienced and inexperienced birds. In the simplest terms, it shows that for a bird to be able to return to (or **home** to) a specific place, it must have some direct experience with that place. We do not know precisely what birds learn about places that enables them subsequently to home to those spots, but it seems to require at least spending some time moving around in the local "target" area. What we do know is that most (and perhaps all) birds, once experienced with a place, can home to that location from great distances and from places well beyond those that the birds are familiar with.

Most experimental data come from homing pigeons, but extraordinary homing feats are known from other species, both migratory and nonmigratory (Papi and Wallraff 1992). The classic examples involve strong-flying seabirds **(Fig. 5–71)**. For example, Manx Shearwaters once were flown by plane from nesting islands off the coast of Great Britain to Boston, Massachusetts and Venice, Italy and then released. Shearwaters do not fly over land, so they must have taken an overwater route to return to their nest burrows, which they did in 12 to 14 days, covering approximately 250 miles (400 km) per day. To accomplish this, especially from Venice, they must possess some very sophisticated navigational equipment. In other studies, Laysan Albatrosses from Midway Island have been displaced over vast distances in the Pacific basin. At least 82 percent returned, from distances of up to 4,100 miles (6,600 km), and at speeds of up to 317 miles (510 km) per day. The more prosaic White-crowned Sparrow has performed feats of similar magnitude. Several hundred sparrows wintering near San José, California, were captured and flown to Baton Rouge, Louisiana (1,800 miles [2,900 km]), and to Laurel, Maryland (2,400 miles [3,860 km]). Thirty-four of these birds were recaptured the following winter on the same quarter-acre (one-tenth-hectare) plot in California, apparently having visited their northern breeding areas in the interim. How might birds accomplish such homing feats?

In theory, a homing animal might employ one or more of the following mechanisms: (1) the animal might maintain direct or indirect sensory contact with the home area; that is, it might be able to see, hear,

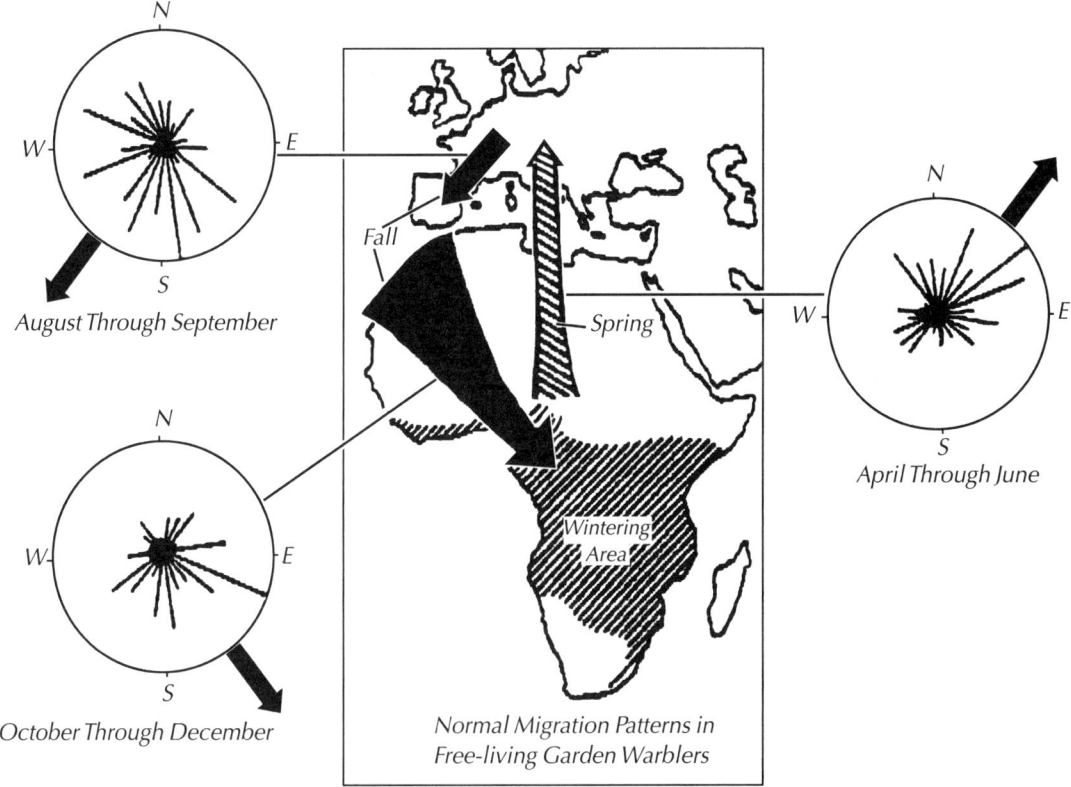

Figure 5–70. Migratory Orientation in Free Versus Caged Garden Warblers: *Garden Warblers breeding in Europe migrate to south-central Africa for the winter. But in fall, their migratory route (solid arrows on map) is not straight to the south. Migrants from northern Europe first head southwest toward Spain and Portugal and then turn southeast across Africa. In the spring, the migrants head more or less straight north (hatched arrow) to their breeding grounds. As demonstrated by experiments with hand-reared Garden Warblers (three circles), the birds apparently have an innate migratory program, controlled by a circannual rhythm, which takes them in the correct direction at the right time of year. When lab-reared birds were tested in circular orientation cages, they oriented in the same direction as the wild migrants in each of the three periods of time tested. The length of each line in the circles represents the relative amount of nocturnal migratory hopping in a given direction. The solid arrow at the periphery of each circle gives the average direction for all the data in that circle. Caged birds were held in a constant cycle of 12 hours light and 12 hours dark, with no view of the sky. They were, however, exposed to the Earth's normal magnetic field. Adapted from Gwinner (1986). Map originally from Gwinner and Wiltschko (1978 and 1980).*

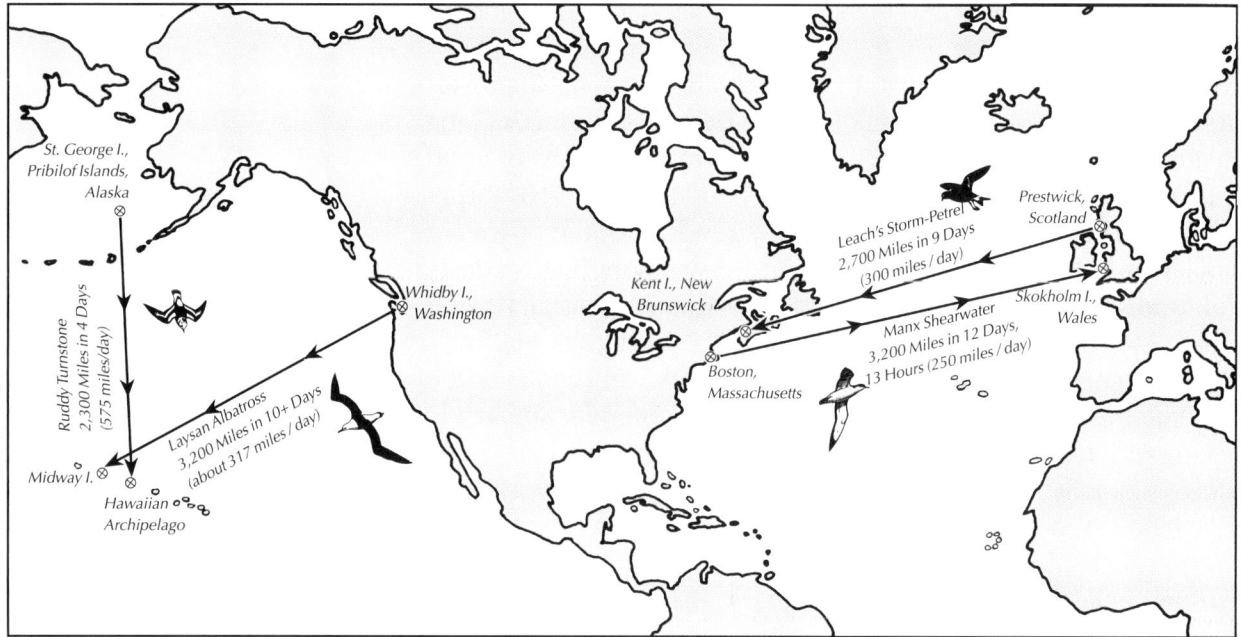

Figure 5–71. Flight Speeds of Four Birds: *Evidence from banding studies (Ruddy Turnstone) and experimentally displaced seabirds (other three species) shows that birds have a remarkable ability to navigate for long distances at relatively high speeds. A Ruddy Turnstone banded and released in late August at St. George Island in the Pribilofs of the Bering Sea was shot four days later in the Hawaiian Islands after migrating 2,300 miles (3,700 km), an average of 575 miles (925 km) per day. A Manx Shearwater transported by plane to Boston returned 3,200 miles (5,150 km) to its nesting burrow on Skokholm, an island off the coast of Wales, in 12 days and about 13 hours, an average of 250 miles (400 km) per day. A Leach's Storm-Petrel returning 2,700 miles (4,340 km) from Prestwick, Scotland to its nest on Kent Island, New Brunswick, in the Bay of Fundy, averaged about 300 miles (480 km) per day for nine days; and a Laysan Albatross flew 3,200 miles (5,150 km) from Whidby Island in Washington State to its nest on Midway Island near Hawaii in just over 10 days, an average of 317 miles (510 km) per day. The flight of the Ruddy Turnstone, a shorebird requiring land to rest, was undoubtedly nonstop, whereas the flights of the others, all pelagic seabirds able to rest on the water and feed on the wing, probably included stops for resting and feeding. Drawing by Robert Gillmor.*

or smell its goal. A male silkworm moth, for instance, follows the odor gradient of the mating pheromone (an odorous substance) emitted by the female and finds her at the end of the trail. During his flight, he is in continuous sensory contact with the female via the odor plume. (2) The animal might use some random or patterned search strategy until it encountered familiar ground. A bird that flew in an ever-increasing spiral, for example, would eventually encounter territory it had seen before. (3) It might perform some sort of **inertial navigation**, logging in its brain all the turns and accelerations of the outward trip and integrating these to compute the route back home. (4) It might refer to a learned, familiar-area "map" (formed in its brain, based on its own experience) to localize its position, relying on familiar landmarks that could be detected by sight, sound, smell, or some other sense. (5) Finally, it might possess a more extensive map that extends well beyond areas of familiarity, presumably based on extensive gradients that could act as analogs of latitude and longitude.

These mechanisms are not mutually exclusive; a bird facing a navigational problem might call on any or all of them. Good evidence indicates that homing pigeons use familiar landmarks when available, and that they also consult information perceived during the outward journey, such as direction of travel and odors encountered (Papi and Wallraff 1992). Evidence from homing pigeons also shows, however, that neither of these types of information is necessary for homing. Schmidt-Koenig and Schlichte (1972), in experiments both in Germany and at Cornell University, fitted pigeons with frosted contact lenses or eye covers that rendered them unable to see objects more than a few yards away, but did let in light **(Fig. 5–72)**. These birds appeared to home just as well as control birds wearing clear lenses, but did not recognize their home loft when they arrived there and so fluttered down to the ground and waited for the researchers to pick them up and carry them inside. In another experiment, Hans Wallraff, a German investigator, transported pigeons to very distant release sites under rigorously controlled conditions that prevented them from perceiving any navigational information—visual, acoustic, magnetic, olfactory, or inertial—during the trip (Wallraff 1980). The birds were carefully transported in closed, airtight cylinders and provided with bottled air. Light in the cylinders was turned on and off at random, and loud white noise was played. The cylinders were enclosed in magnetic coils that provided a changing magnetic field, and were placed on a tilting turntable hooked to a computer that varied both the rotation and tilt at random. Yet when these pigeons were released at very distant, unfamiliar places, they showed no deficit in their ability to fly in the homeward direction (once they recovered from some initial nausea). This experiment, more than any other, demonstrated that homing pigeons could navigate based only on information they gained at the release site.

How pigeons and other birds use all the means available to them to solve a homing problem remains an interesting question; in the end, we must conclude that at least some birds possess a very extensive map that is sufficient to enable them to home from completely unfamiliar sites at vast distances from known areas. Ornithologists call this ability to take the proper course toward a specific goal **true navigation**.

To explain how birds perform such navigation, researchers during the first half of this century suggested a number of sweeping theories, including astronomical navigation systems based on the sun or on the stars, and a map based on both the physical coordinates of the magnetic field (providing latitude) and the Coriolis force (providing longitude). Each of these approaches, however, has been rejected on the basis of experimental tests employing homing pigeons. Nevertheless, the magnetic field alone may form the basis for one type of navigational map, as discussed later.

Gustav Kramer, a German pioneer in bird migration and orientation studies, discovered in 1950 that birds could use the position of the sun as a compass, as is discussed in more detail in the next section. He considered how possessing such a compass might be incorporated into a homing navigation system. But a compass alone is not sufficient to enable homing from an unfamiliar locale. Imagine yourself dropped off in the middle of an unfamiliar forest and given a compass. How would you find your way home? A compass will indicate directions, but that is useless information unless you know where you are relative to your goal—whether your destination is north, south, east, or west. In short, you need a map. Once you know your position on a map, you can tell which direction you must go to get home, and for that a compass is useful. This idea is the basis

Figure 5–72. Homing Pigeon Fitted with Frosted Eye Covers: *Experiments with homing pigeons often use some type of frosted eye covers or frosted contact lenses that let in light, but prevent birds from seeing objects more than a few yards away. Photo courtesy of Steve Johnson.*

5

of Kramer's **map and compass model** of homing navigation, which provides the theoretical basis for nearly all current research on homing (Kramer 1953). Virtually all data from pigeon homing experiments are consistent with the notion that true navigation is a two-step process involving one mechanism to identify spatial position (a map) and another to identify directions (a compass). Let us turn first to what is known about bird compasses.

Compass Mechanisms

Some sort of compass sense is involved in both homing navigation and migratory orientation. Such a sense told Perdeck's young starlings which direction was southwest. To learn how bird compasses work, ornithologists have relied primarily on two sources of information: the initial flight directions of homing pigeons released at sites far from home; and the spontaneous, oriented hopping of migratory birds in several types of circular orientation cages. Researchers now know that birds possess several different compass capabilities, which are described here in the order of their discovery.

Sun Compass

Gustav Kramer discovered the sun compass in 1951 by performing experiments on European Starlings in orientation cages (Kramer 1951). He used mirrors to shift the apparent position of the sun as viewed by starlings in their cage, and found that the birds shifted the direction of their migratory restlessness to match the compass directions indicated by the altered position of the sun **(Fig. 5–73)**. This demonstrated that in choosing directions, the birds compensated for the changing position of the sun as the earth rotated on its axis. At the time of Kramer's discovery, little was known about biological clocks in animals, but researchers soon showed that the sun compass was coupled with the circadian clock, which provided the means for **time compensation** (making allowances for the changes in the sun's position in the sky over the course of a day).

Working with homing pigeons, another German researcher, Klaus Schmidt-Koenig, demonstrated how this time-compensated sun compass operates (Schmidt-Koenig 1960). He placed pigeons in a closed room for several days with an altered cycle of light and dark, thereby resetting their circadian clocks. When he released the birds on a sunny day, they interpreted the position of the sun on the basis of their internal clock, which was now out of phase with real time. They thus inferred that the sun's position indicated a compass direction that it did not, and made a predictable error in choosing a homeward direction (Fig. 5–74). This experiment has been done many times and reveals a peculiarity concerning how pigeons (and other animals using a sun compass) use the sun. When their clock has been shifted in a certain way, they may mistake a noontime sun for a rising or setting sun. This is a mistake that humans would not make, because at noon the sun is high in the sky, whereas at sunrise or sunset it is near the horizon. Animals apparently ignore these differences in the sun's elevation; instead, they

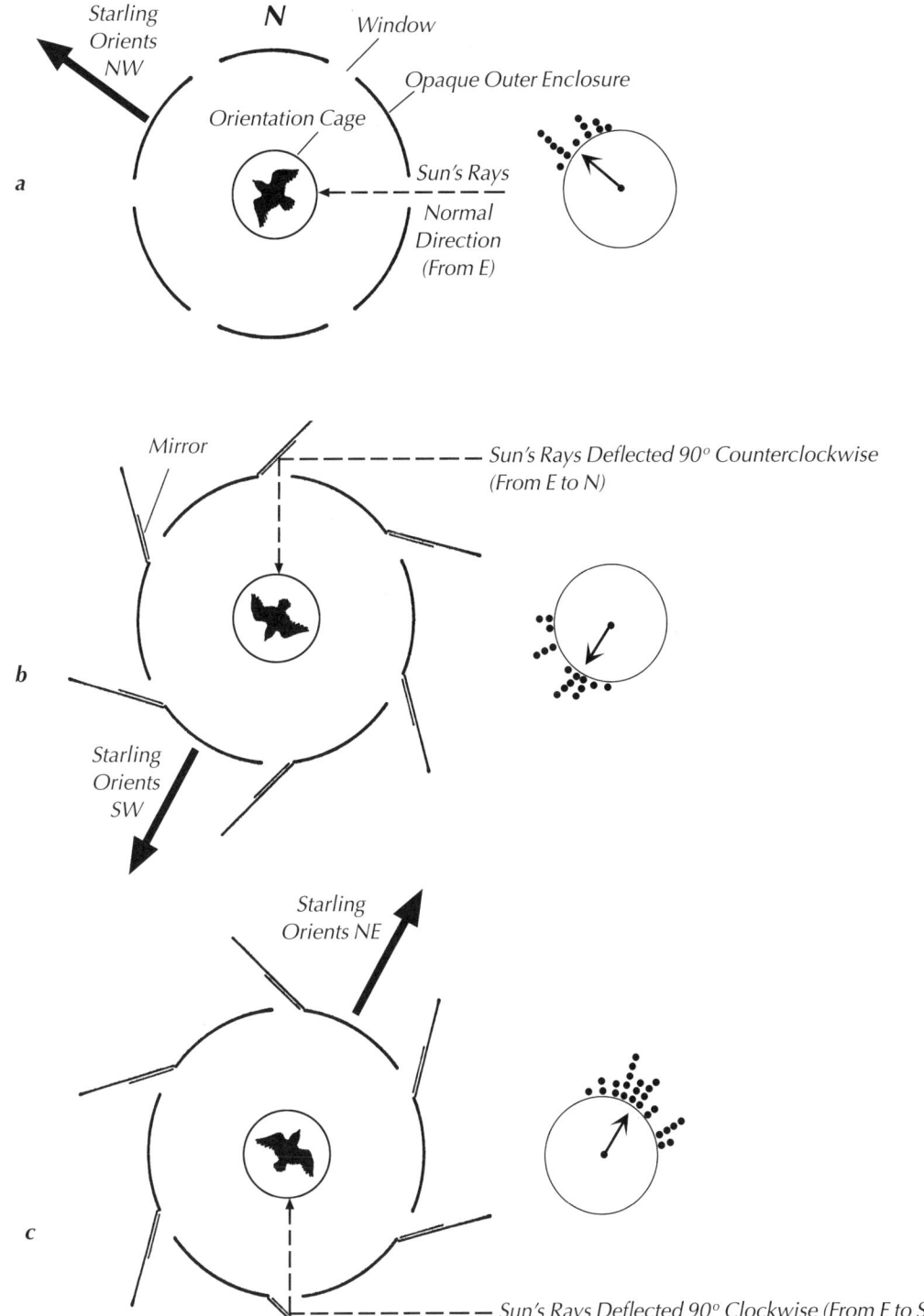

Figure 5–73. Starling Orientation with Respect to the Sun: *Gustav Kramer performed numerous experiments, one of which is shown here, demonstrating that European Starlings use the sun's position to orient themselves.* **a.** *A European Starling was placed in a circular orientation cage (inner circle) within an opaque outer enclosure containing six windows through which the sun's position in the sky could be viewed directly. When the sun was in the east, the starling oriented to the northwest (solid arrow)—135 degrees counterclockwise from the sun—the direction the bird would normally migrate at that time of year (spring). The dots in the circle diagram to the right are the recorded directions in which the starling fluttered during migratory restlessness; the arrow inside the circle is the average direction of the observations.* **b.** *When Kramer added to each window an opaque screen containing a mirror that deflected the sun's rays 90 degrees counterclockwise, so that the sun appeared to be in the north instead of the east, the starling adjusted its orientation accordingly. It now oriented to the southwest, still 135 degrees counterclockwise from the apparent position of the sun.* **c.** *When the screens and mirrors were adjusted so that the sun appeared to be in the south rather than the east, the starling oriented to the northeast, still 135 degrees counterclockwise from the sun. Adapted from Griffin (1974, p. 123).*

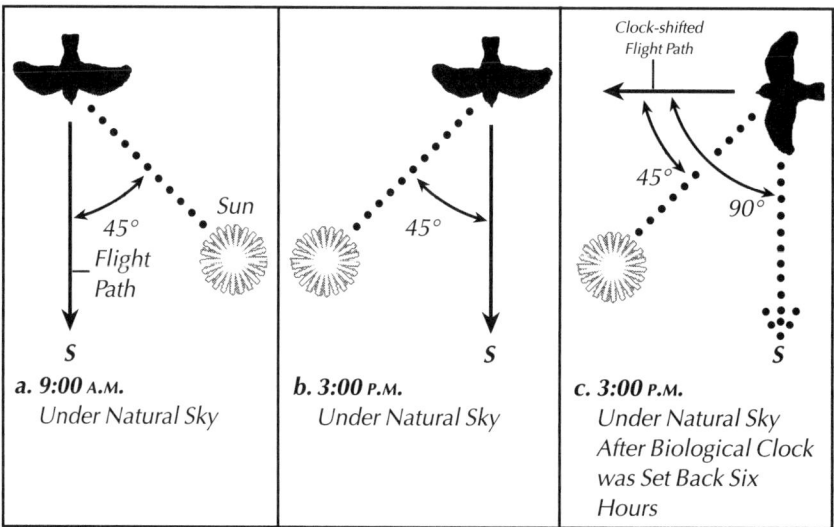

Figure 5–74. Hypothetical Experiment Demonstrating How the Time-compensated Sun Compass Works: a. A bird flying south at 9:00 A.M. might have a flight path 45 degrees to the right (from the bird's perspective) of the direction to the sun (dotted line). **b.** By 3:00 P.M. the sun will have moved approximately 90 degrees in the sky. For the bird to continue to fly to the south, it must be able to adjust its orientation with respect to the sun according to the time of day. In this case, it now flies 45 degrees to the left of the sun. **c.** If birds are held in captivity in a specific, altered cycle of light and dark for a few days, their internal clock can be reset to six hours earlier than the actual time: thus the bird "thinks" it is 9:00 A.M. when it is really 3:00 P.M. A bird clock-shifted in this way, trying to fly south at 3:00 P.M. (when it should orient 45 degrees to the left of the sun) will actually orient 45 degrees to the right of the sun, and head west. The direction it takes is the correct direction according to its own, internal, shifted clock. Adapted from Goodenough et al. (1993). Originally from Palmer (1966).

consider only its azimuth direction, the compass direction at which a vertical line from the sun to the ground intersects the horizon.

The sun compass is the compass of first choice in homing pigeons. Researchers know this because whenever the sun is visible, releasing clock-shifted pigeons results in the predicted deflection of their flight directions relative to control birds whose clocks are running on real time. If pigeons preferred a different compass system, clock-shifting them would affect their orientation only when they were unable to use their preferred system. Whether the sun compass plays any role in *migrating* birds is unclear at present.

To be able to use the sun as a compass, pigeons must *learn* its path. If, for example, young pigeons are allowed to see the sun only in the morning, they will not be able to use it as a compass in the afternoon. How the pigeon knows, for example, that the sun rises in the east and sets in the west is not completely clear, but some evidence suggests that it may use its magnetic compass (see below) to assign compass directions to the azimuths of the sun's path.

Star Compass

Shortly after Kramer's discovery of the sun compass, another German team, Franz and Eleanor Sauer, took up the study of the mechanisms of orientation that might serve nocturnal migrants. In a series of classic experiments under the dome of a planetarium, they showed that night migrants use the stars as a compass (Sauer 1957). In keeping with much of the thinking of the time, the Sauers believed that birds were born with a genetically encoded star map in their heads, but extensive analysis of the stellar orientation system of the Indigo Bunting by an American, Stephen Emlen of Cornell University, showed that this was not the case (Emlen 1967a and b). Emlen showed that young buntings observe the rotation of the night sky that results from the earth's rotation around its axis. By learning the center of this axis of celestial rotation (the North Star, Polaris) they can locate true north. (This is true at any time of night, since Polaris does not appear to move much in the sky.)

Figure 5–75. Star Compass Research in a Planetarium: To determine how birds use the night sky to orient, professor Stephen T. Emlen of Cornell University tested Indigo Buntings in circular orientation cages (funnel-shaped boxes seen here on stepstools) in a small planetarium. He was able to expose the birds to normal and manipulated skies, even changing the pattern of star movement so that the night sky appeared to rotate around Betelgeuse instead of the North Star, Polaris. Photo courtesy of Cornell University.

Figure 5–76. How Indigo Buntings Use the Night Sky to Orient: Experiments by Stephen T. Emlen at Cornell University in the 1960s demonstrated that Indigo Buntings (in the Northern Hemisphere) determine north by the patterns of stars surrounding the North Star, Polaris, which is always nearly due north in the night sky. Although all other stars in the night sky appear to rotate around the North Star, the positions of these stars, relative to each other, *do not change.* **a.** At 9:00 P.M. on a spring night, captive Indigo Buntings in a planetarium under a typical night sky for that time oriented north, as would a spring migrant in the wild. **b.** At 3:00 A.M., with the planetarium sky showing the typical night sky for that time, the caged buntings still oriented to the north. **c.** Caged birds still oriented to the north when they were shown a typical 3:00 A.M. sky at 9:00 P.M. If they were using the rotational positions of the stars to navigate, they would have assumed the stars were in the correct position for 9:00 P.M. and oriented in the wrong direction. Drawing by Adolph E. Brotman, from Emlen (1975).

Through a series of experiments in a planetarium **(Fig. 5–75)**, Emlen showed that the star patterns near Polaris (called **circumpolar constellations:** the Big Dipper, Little Dipper, Draco, Cepheus, and Cassiopeia) are then apparently memorized in this context, so that by the time of migration, the birds can select the proper migratory direction even under a stationary planetarium sky **(Figs. 5–76, 5–77)**. Although many other animals might use a star compass, it has been demonstrated only in birds.

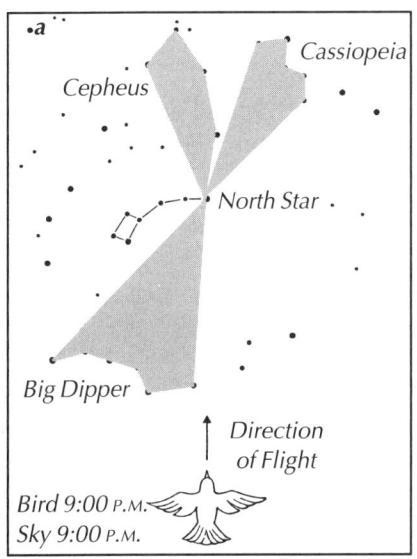

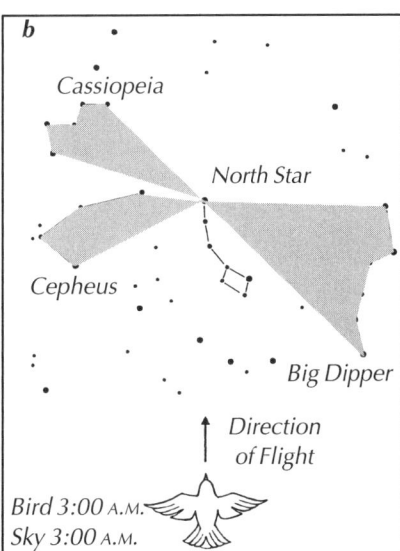

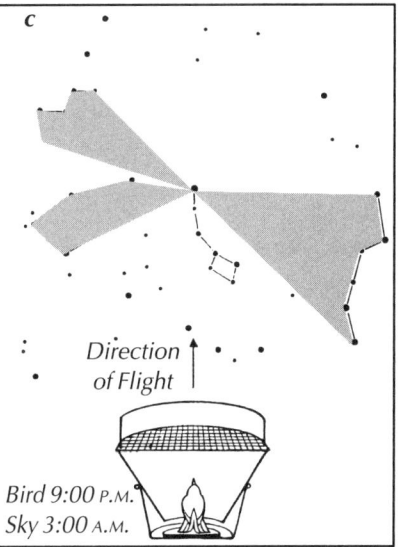

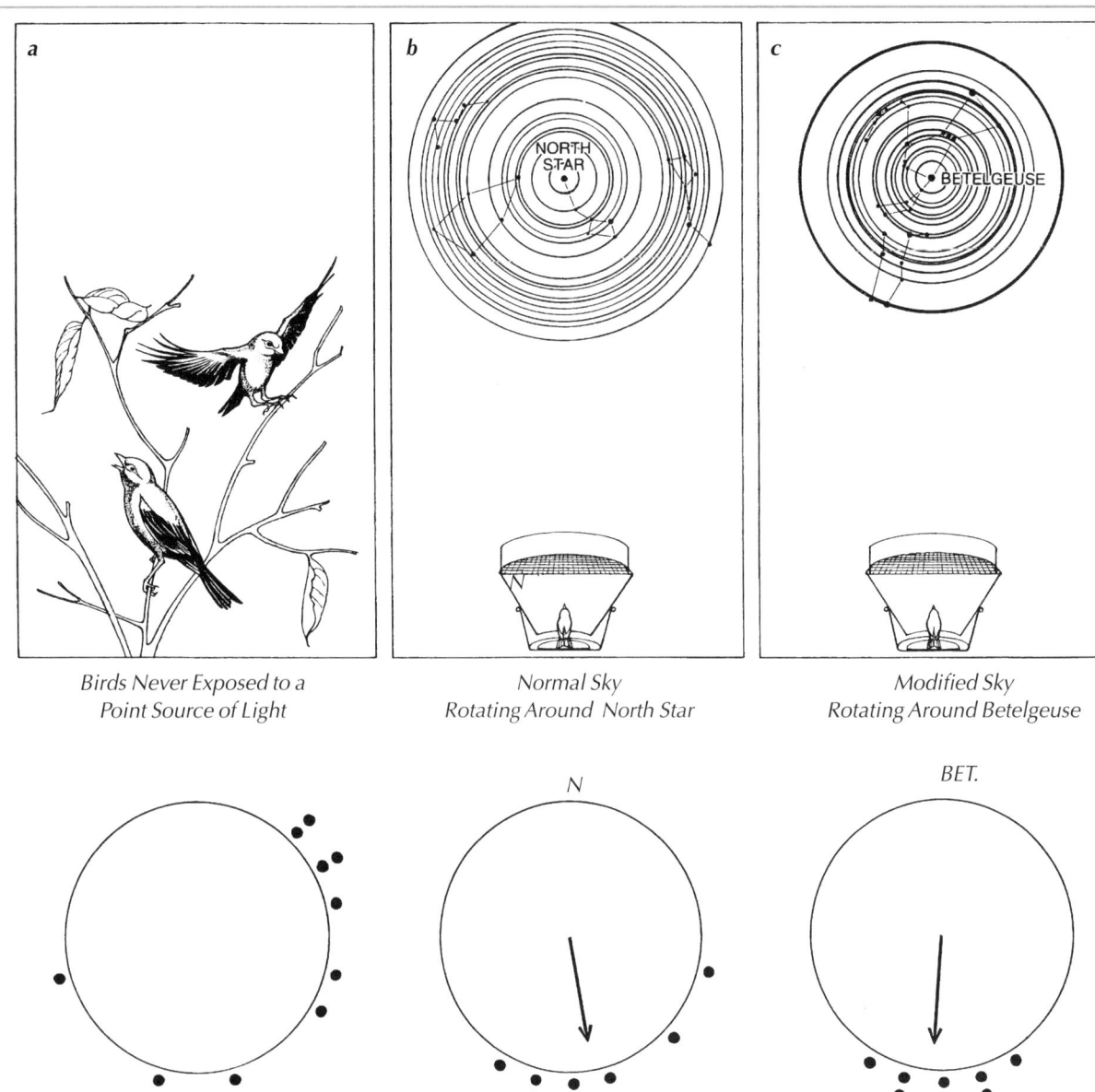

Birds Never Exposed to a
Point Source of Light

Normal Sky
Rotating Around North Star

Modified Sky
Rotating Around Betelgeuse

No Directional Orientation

Migratory Orientation
Away From North Star (South)

Migratory Orientation
Away From Betelgeuse

Figure 5–77. How Indigo Buntings Learn to Use the Night Sky to Orient: *Experiments by Stephen T. Emlen at Cornell University showed that the early visual experience of young Indigo Buntings plays an important role in the development of their celestial-orientation abilities. Three groups of nestlings were captured and hand-reared in the laboratory.* **a.** *The first group lived in a windowless room with diffused lighting and never saw a point source of light. In the fall the birds began to display intense nocturnal activity. When they were tested in circular orientation cages under a stationary night sky in a planetarium, they did not orient in any particular direction.* **b.** *The second group never saw the sun and was exposed to a night sky in a planetarium every other night for two months. Normal celestial rotation (around the North Star) was simulated. When the birds were tested in a planetarium under a normal sky during their fall migratory period, they oriented away from the North Star,*
to the south, the appropriate direction for fall migration. (Arrow indicates the mean direction taken by birds tested.) **c.** *The third group never saw the sun and was exposed to a modified night sky every other night for two months; Betelgeuse, a star in Orion, became the new polestar around which all other stars rotated. When these birds were tested in the fall under a normal night sky, they continued to regard Betelgeuse as the polestar and oriented their activity away from it. Thus young buntings initially learn the north-south axis from the rotation of stars; star patterns by themselves are not useful cues to a naive bunting. The star patterns take on directional meaning only after they have become part of the bird's general orientational framework, the formation of which is influenced, at least in part, by observing the rotation of stars. Drawing by Adolph E. Brotman, from Emlen (1975).*

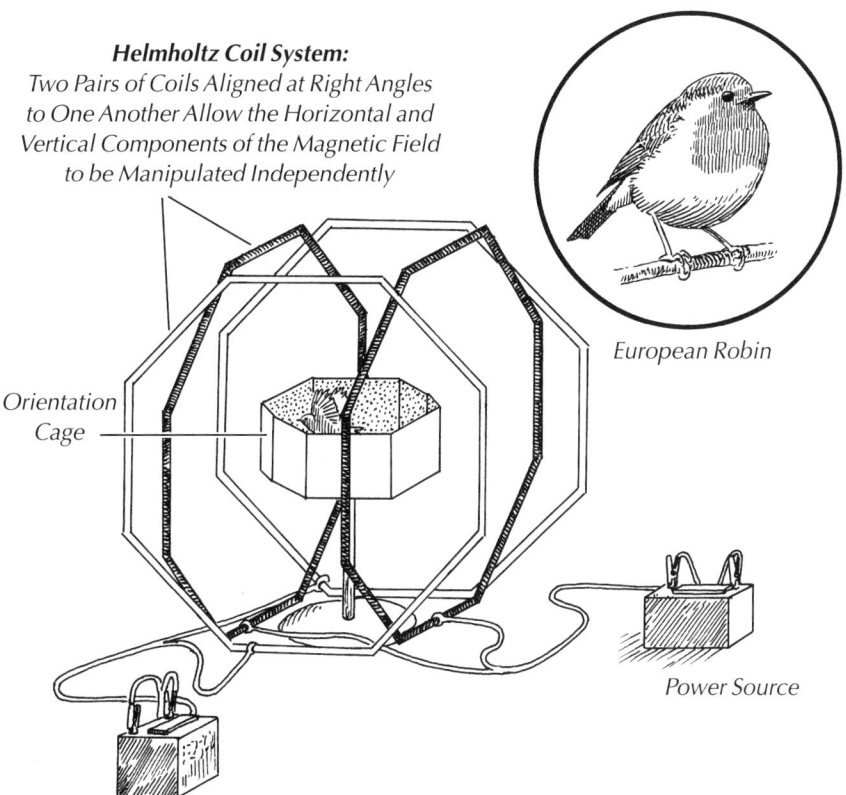

Helmholtz Coil System:
Two Pairs of Coils Aligned at Right Angles to One Another Allow the Horizontal and Vertical Components of the Magnetic Field to be Manipulated Independently

Orientation Cage

European Robin

Power Source

Power Source

Figure 5–78. Helmholtz Coil System for Manipulating the Magnetic Field: By placing two sets (indicated by the solid and open shading) of Helmholtz coils, oriented at right angles to each other, around a bird in a circular orientation cage, researchers can independently manipulate the horizontal component (direction to magnetic north and south) and vertical component (related to dip angle, see Fig. 5–79) of the magnetic field experienced by the test bird. Changing the amount of electric current running through the coils creates magnetic fields of different intensities. The pioneers of this technique, Friedrich Merkel and Wolfgang Wiltschko, were able to change the migratory orientation of caged European Robins (that were prevented from using sun, star, and light cues) in predictable ways, by altering the direction of the magnetic field.

Magnetic Compass

That animals might sense the earth's magnetic field and use it as a compass is a very old idea. The first empirical demonstration of magnetic orientation, however, came in the 1960s from the laboratory of Friedrich Merkel and Wolfgang Wiltschko in Frankfurt, Germany. Working with the European Robin, they showed that birds in migratory condition would hop in appropriate migratory directions when tested in covered cages in a closed laboratory room to eliminate sun, star, and light cues. Most important, by employing Helmholtz coils surrounding the cage to change the direction of the magnetic field experienced by the birds, they were able to change predictably the orientation of *Zugunruhe* in the birds (Wiltschko 1968) **(Fig. 5–78).** Although these results were initially met with skepticism, rigorous demonstrations of magnetic orientation are now available from 18 species of migratory birds as well as the homing pigeon.

Many additional experiments by Wolfgang and Roswitha Wiltschko revealed details concerning how the magnetic compass works. Interestingly, unlike our compass instruments, the avian magnetic compass is not based upon the polarity (the distinction between north and south) of the field. It is as if the bird has a magnetic compass of the sort that you might use on a hiking trip, but the compass needle is identical on both ends. It can detect the north-south axis, but can't tell which of the two directions is north. To determine north, the bird uses the fact that magnetic lines of force not only point toward the poles, but dip toward the earth's surface at an angle **(Fig. 5–79).** Birds can apparently detect that dip angle in some way that we do not yet understand. Our

Figure 5–79. The Magnetic Field of the Earth: *The earth, with its two different magnetic poles, acts like a large magnet. Arrows indicate the direction of the magnetic field of the earth. On the right half of the diagram, complete magnetic lines have not been drawn: the arrows intersecting the earth's surface indicate the direction of the magnetic field at each point, and the length of each arrow is proportional to the strength of the magnetic field at that point on the earth. The magnetic field is stronger at the poles and weaker at the equator. The* **dip angle***, the angle at which the magnetic field lines contact the earth, is 0 degrees at the magnetic equator, and approaches 90 degrees as you approach the magnetic poles. Note that the magnetic field lines point away from the earth in the Southern Hemisphere and toward the earth in the Northern Hemisphere, paralleling the earth at the magnetic equator. Note also that the magnetic poles and equator do not quite coincide with the geographic poles and equator. If you had a compass whose needle was free to dip up and down as well as to swing around, it would align itself at the same angle as the magnetic field lines, with its north end pointing down in the north and up in the south. Adapted from Able (1994). Originally from Waterman (1989) and Wiltschko and Wiltschko (1991).*

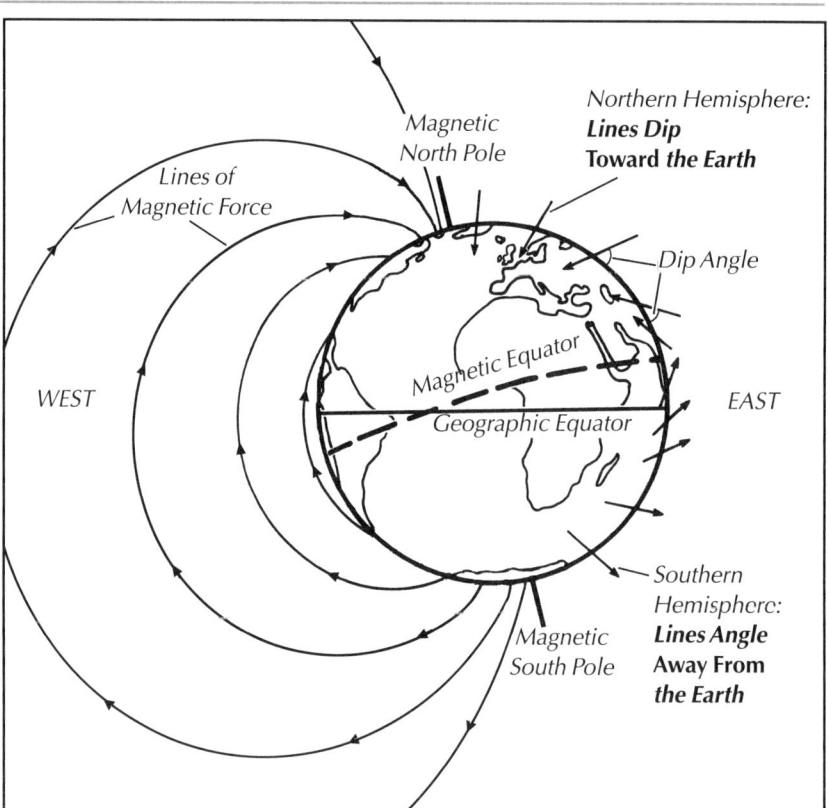

human-made compasses work because the needle aligns itself with the earth's magnetic lines of force; the needle on many compasses does not dip up and down because it is held in one plane. If, however, you have a compass whose needle, instead of just spinning around, is also free to dip up and down, then in the Northern Hemisphere the end of the needle that points north will dip down, and the end that points south will tip up. The opposite will be true in the Southern Hemisphere.

In some very clever experiments that involved changing the dip angle of the magnetic field lines, the Wiltschkos showed that a European Robin in the Northern Hemisphere considers the direction in which the field lines dip downward to be north (Wiltschko and Wiltschko 1995) **(Fig. 5–80)**.

The magnetic compass seems to develop spontaneously in young birds. All that is required is that they grow up in a normal magnetic field; they need not have experience with the sun or stars. Researchers still do not know how birds sense the very weak magnetic field of the earth. Current research is focused on tiny particles of magnetite (an iron-containing, magnetic mineral) that have been found in the heads of many birds, as well as in the abdomens of bees and in certain bacteria, both of which can orient to a magnetic field, and on the possibility that photoreceptive pigments in the eye might provide the sensor (Walker et al. 1997). The magnetic compass is an important component in the orientation equipment of migratory birds, and in homing pigeons it seems to serve as a back-up compass that is used when the sun is not visible.

Figure 5–80. How European Robins Use the Earth's Magnetic Field to Determine North: *Results of a series of classic experiments by Wolfgang and Roswitha Wiltschko in which the orientation of European Robins under different magnetic field conditions was tested in circular orientation cages. The drawings at left are side views, with geologic north to the left and south to the right; a line pointing toward the bottom of the page would represent a line going straight into the earth. Each drawing shows the magnetic field condition tested, and the average response of birds tested. For simplicity, the picture of the bird faces either north or south, although the birds actually oriented either northeast or southwest. The birds were tested in Frankfurt, Germany in the spring, when they would normally be migrating to the northeast.* **a. Normal Magnetic Field:** *When tested in the normal, unmodified magnetic field of Frankfurt, the birds oriented to the northeast.* **b. Magnetic Field Direction Reversed:** *Only the direction of the magnetic field lines were modified, such that magnetic north was now toward the earth's geologic south pole (the "north-seeking end" of a compass needle would point south in this situation). Although it may appear that the direction in which the magnetic lines dip has also been reversed, from the bird's point of view it is still the same, as the lines dip toward magnetic north (the north end of the compass needle would dip down as it pointed toward magnetic north [geologic south]). In this situation, the birds reversed their orientation, and tried to head southwest.* **c. Magnetic Field Di-**

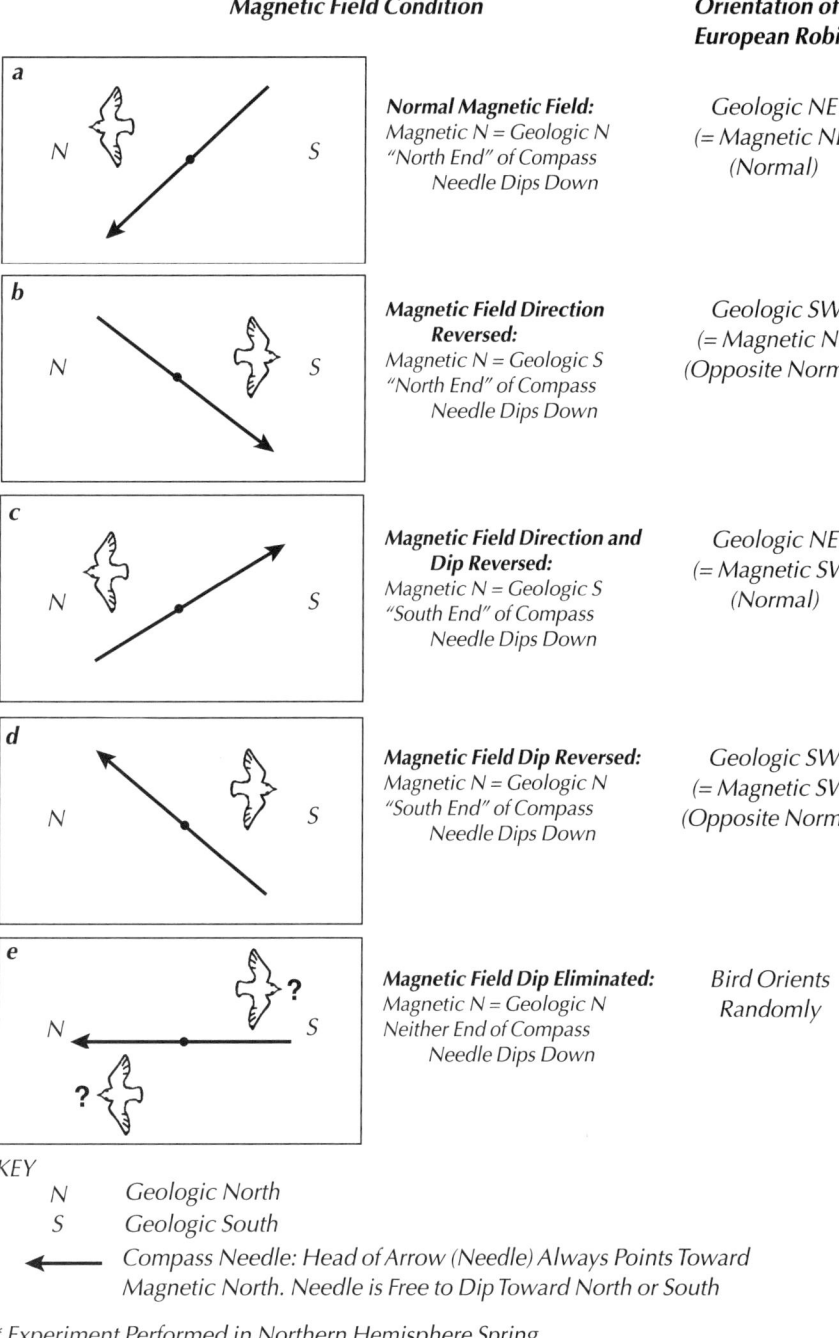

Magnetic Field Condition

Orientation of European Robin *

a **Normal Magnetic Field:** *Magnetic N = Geologic N "North End" of Compass Needle Dips Down* — Geologic NE (= Magnetic NE) (Normal)

b **Magnetic Field Direction Reversed:** *Magnetic N = Geologic S "North End" of Compass Needle Dips Down* — Geologic SW (= Magnetic NE) (Opposite Normal)

c **Magnetic Field Direction and Dip Reversed:** *Magnetic N = Geologic S "South End" of Compass Needle Dips Down* — Geologic NE (= Magnetic SW) (Normal)

d **Magnetic Field Dip Reversed:** *Magnetic N = Geologic N "South End" of Compass Needle Dips Down* — Geologic SW (= Magnetic SW) (Opposite Normal)

e **Magnetic Field Dip Eliminated:** *Magnetic N = Geologic N Neither End of Compass Needle Dips Down* — Bird Orients Randomly

KEY
N   Geologic North
S   Geologic South
⟵   Compass Needle: Head of Arrow (Needle) Always Points Toward Magnetic North. Needle is Free to Dip Toward North or South

* *Experiment Performed in Northern Hemisphere Spring*

rection and Dip Reversed: *As in (b), the direction of the magnetic field lines were reversed, but in addition, the direction in which they dip was reversed so that the north end of the compass needle would now angle up as it pointed to magnetic north (geologic south). In this case, the birds oriented to the northeast, taking the direction in which the field lines dip, rather than the direction to magnetic north, as an indication of geologic north.* **d. Magnetic Field Dip Reversed:** *As in (a), the magnetic field direction was left normal, but as in (c), the direction in which the lines dipped was reversed, such that the north end of the compass would point up as it pointed to magnetic (and geologic) north. Birds under these conditions oriented to the southwest, again taking the direction in which the lines dipped, rather than magnetic north, as an indication of which way was geologic north.* **e. Magnetic Field Dip Eliminated:** *When the magnetic field direction was again left normal, but the dip angle was eliminated (magnetic lines of force were parallel to the earth's surface, as they would be at the equator), the birds did not orient in any particular direction. This series of experiments clearly shows that birds use the dip angle to determine north or south: if they, instead, used magnetic north and south to determine direction, they would always fly toward magnetic northeast (thus birds in groups (c) and (d) should have oriented in a direction opposite to the direction they took, and birds in group (e) should have oriented to the northeast); but if the birds used the dip angle to determine north or south, they would always fly in the direction in which the arrow dips—as was found in every case. Using this type of magnetic-sensing system, one would expect the birds in group (e) to be unable to orient—as was the case. Note that condition (c) is not found on the earth, and condition (d) simulates the Southern Hemisphere. Adapted from Able (1994). Originally from Wiltschko and Wiltschko (1972).*

Sunset Cues

Most nocturnal migrants initiate flights shortly after sunset. Recent orientation cage studies have shown that visual cues around the time of sunset are very important in the decisions these birds make about which way to fly. Work in my own laboratory showed that birds possess yet another compass, this one based on patterns of polarized light in the sky. These patterns are present throughout the day, but nocturnal migrants apparently use them most at sunset, when they need directional information to begin their flights **(Sidebar 3: Polarized Light)**.

One might ask why birds have so many different ways of determining compass directions, and given that they have many, how are they related to one another? The answer to the first question is only conjecture. Presumably it is advantageous to have backup systems to cope with the problems a bird can face during migration, such as places where the magnetic field is severely disturbed, or cloudy weather that obscures visual cues. With regard to how these compass mechanisms are related to one another, there is a good deal of information, but researchers still do not know the whole story. The various compass systems interact in complex ways both during their development in young birds and in older individuals during migration. There seems to be a great deal of calibrating of one system by another and there are comparable data from too few species to draw any very firm generalizations. The best evidence at the moment suggests that the earth's rotation, as indicated by the rotation of the stars at night or changes in polarized skylight patterns in the daytime, is important to young birds in the calibration of orientation mechanisms with each other. In birds on migration, faced with the short-term problem of selecting a migration direction, magnetic cues seem to take precedence over stars, and polarized skylight at dusk seems to override both of those stimuli. Thus, the different compass capabilities are arranged in a hierarchy of subordination. Recall that in homing pigeons, in contrast, the sun compass is the preferred method of orientation.

Navigational Maps

Experimental evidence very strongly suggests that homing pigeons and other birds possess some sort of navigational map, but discovering the physical basis of that map has been one of the most enduring and contentious issues in the study of animal behavior. To some extent, it remains one today. It is important to bear in mind that, as with the compasses of birds, there may be more than one map. At present, there are two possible hypotheses regarding what may constitute the navigational map, and both are derived almost entirely from work with homing pigeons.

The first is the surprising idea, put forth by Floriano Papi and colleagues from the University of Pisa, that atmospheric odors may form the physical basis of the map (Papi et al. 1991). The hypothesis is that pigeons learn an odor map of the vicinity of their loft by associating airborne odors with the directions from which winds carry the odors past the loft area. This local map, much like a topographic map of a familiar region, may gradually be extended through exploratory flights **(Fig. 5–81)**.

Think of a pigeon sitting in its home loft. As winds blow from the east, for example, the pigeon smells the odor of a pine forest. If the pigeon is later taken away from home and released in a place where the odor of pine is very strong, it might conclude that it had been transported eastward from its loft. Papi's group and Hans Wallraff of Germany have performed many experiments to test this **olfactory map hypothesis**. Their evidence that olfactory information is involved in homing by pigeons at distances of 310 miles (500 km) or more from the loft has led Wallraff to propose that odors might form an extensive gradient map. A gradient odor map is based on small but systematic changes in the intensity or composition of odors over a large area. As one moves in a given direction, each particular odor becomes steadily stronger or weaker. An animal could even, in theory, extrapolate its map to areas well beyond those with which it is directly familiar.

(Continued on page 5·96)

Figure 5–81. Odor Gradient Map: Homing pigeons living in a loft southwest of a farm and west of a pine forest might use the odors coming from these sites to develop a map of their home region. The map would be based on the direction and intensity of the odors. When released at unfamiliar sites within the home region, the birds might be able to determine their location with respect to home by sensing the intensities of the two odors and comparing them to their memory of the odor intensities at their home loft. For example, at Release Site 1, a bird would sense that both the farm and forest odors were stronger and determine that it was northeast of home. At Release Site 2, a bird would find the farm odor stronger and the forest odor weaker, thus placing itself northwest of home.

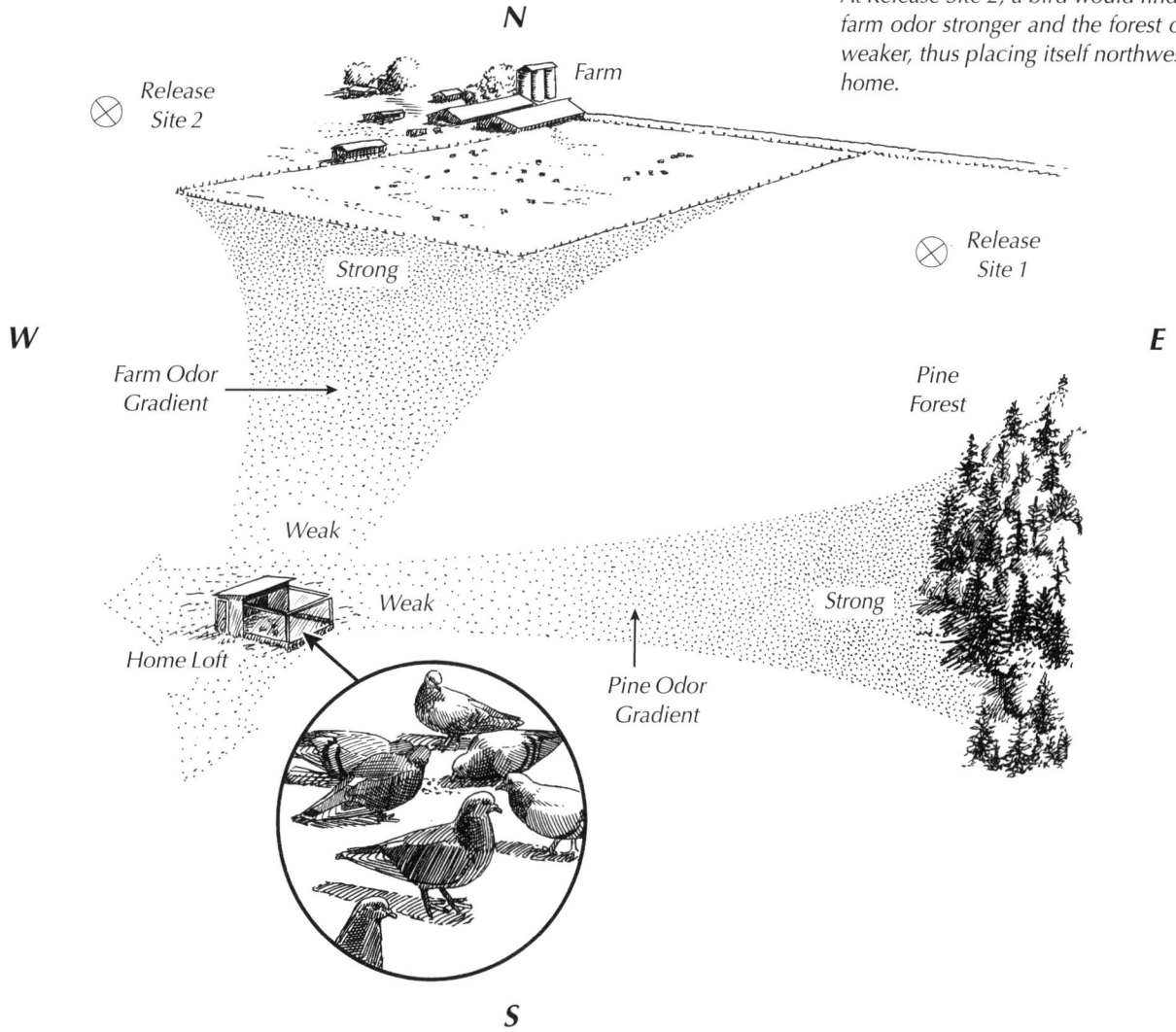

Sidebar 3: POLARIZED LIGHT
Sandy Podulka

Although humans are able (with practice) to see some types of polarized light, we generally do not notice or use it in everyday life—so it remains a bit of a mystery to us. Recall from Chapter 3 that light is in the form of a wave. You can represent this wave with a simple demonstration (Highland 1963). Tie one end of a rope to a doorknob, and shake the other end up and down, creating a vertical wave (an action you probably carried out with a garden hose when you were a child) **(Fig. A).** You can also shake your hand right and left creating a horizontal wave, or between upper right and lower left creating a wave at an angle. In fact, you can shake your hand in any orientation that is perpendicular to the line of rope between you and the door, and still create a wave. An ordinary light beam is a random mixture of all these waves with different orientations.

Now place a cardboard box with a deep vertical slot between you and the door **(Fig. B)**, and run the rope through it. With this setup you can still shake your hand in any orien-tation to create waves, but only the vertical waves will reach the doorknob. All others have been stopped or absorbed by the box. You have just created a **polarized light beam:** one that is made up of a reduced number of different orientations of waves, in this case, only vertical waves. If you, instead, made a horizontal slot, only horizontal waves would pass through. The cardboard box in these situations acts like a **polarizing lens** or **filter,** by selectively transmitting waves with certain orientations.

Polarizing lenses that transmit only vertical waves are used in sunglasses and car windshields to cut out other waves that may create glare from horizontal surfaces such as the road. Polarizing lenses are also used in microscopes and cameras. In contrast, a **depolarizing material** can be used to take a light beam that has been polarized and vibrate it in all directions, creating waves of all orientations, to form an unpolarized beam.

The atmosphere acts like a weak polarizing filter, and in the clear parts of the sky, light from the sun is polarized as it is scattered by passing through the molecules in the atmosphere. Humans and certain species of insects, amphibians, fish, and birds are all able to detect some patterns of polarized light in the sky. The details of how these polarized light patterns are produced and move are beyond the scope of this course, but a few basic points will be mentioned here.

The atmosphere produces a gradient of polarization with the most weakly polarized light areas near the sun and directly opposite the sun, and the most strongly polarized areas 90 degrees from the sun **(Fig. C).** This pattern of polarization maintains its orientation to the sun as the sun's position moves in the sky. In addition, the earth's rotation causes the pattern, like the stars, to rotate around Polaris in the Northern Hemisphere. (Although the polarized light patterns are not visible at night and Polaris is not visible by day, the patterns still rotate around the *location* of Polaris, which does not move in the sky and is always very close to north in the Northern Hemisphere.)

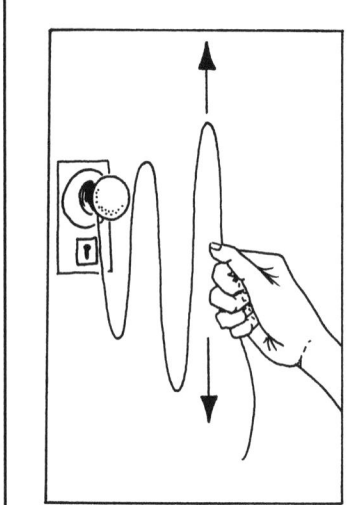 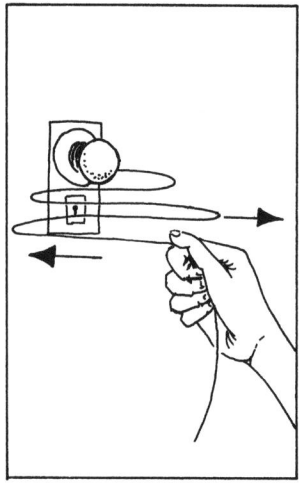 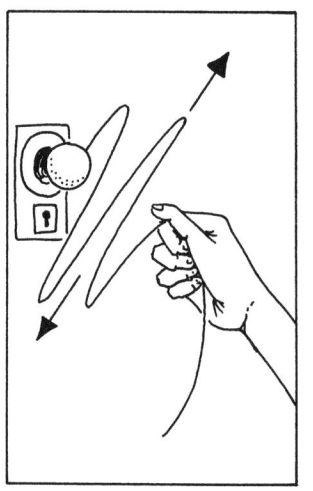

Figure A. Light Wave Simulation: *By tying one end of a rope to a doorknob, and shaking the other end in various directions, you can simulate light waves with various orientations: vertical, horizontal, and angled. Unpolarized light is a random mixture of light waves with these orientations as well as everything in between.*

Thus the polarized light patterns can indicate the sun's position even when it is hidden behind clouds (although the patterns themselves can only be seen in areas of clear blue sky). And, the motion of the light patterns can reveal the direction to true north.

By using Polaroid filters to alter the polarized light patterns seen by birds in orientation cages and by employing depolarizing material to eliminate polarized light, researchers in my laboratory have shown that birds can use polarized light in the sky as a compass (Able 1989). But, exactly how birds extract directional information from the polarized light patterns is not yet understood. ■

Suggested Reading

Able, Kenneth. P. 1982. Skylight polarization patterns at dusk influence migratory orientation in birds. *Nature* 299:550–551.

Figure B. Polarized Light Beam Simulation: *If, to your demonstration in Figure A, you add between you and the door a cardboard box with a deep vertical slot, and you pass the rope through the slot, you can create a beam of "polarized light." The box, like a polarizing filter, allows only light waves with a certain orientation (in this case, vertical) to pass through. See sidebar text for details. Adapted from Highland (1963, p. 42).*

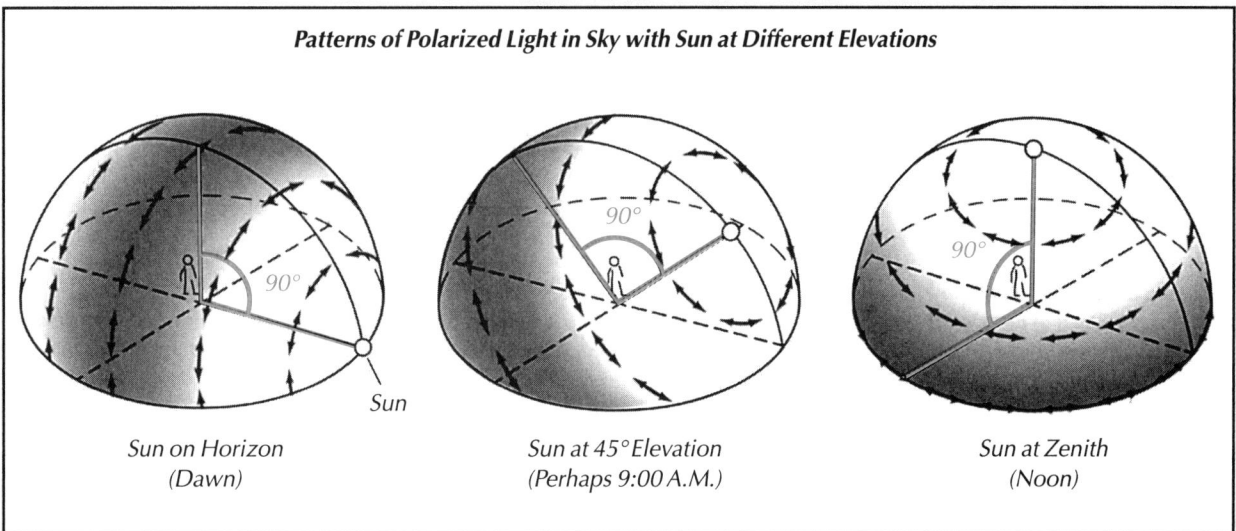

Figure C. Patterns of Polarized Light in the Sky: *The atmosphere, acting as a weak polarizing filter on sunlight, produces a gradient of polarized light in the sky, which moves with the position of the sun. The most strongly polarized area (darkened region on each diagram) is a band across the sky 90 degrees from the sun. Areas of the sky with more weakly polarized light (light areas on each diagram) are close to the sun, and opposite the sun, in the sky. Positions of polarized light bands are shown for the sun (small circle) on the horizon, at 45 degrees (around 9:00 A.M.), and directly overhead. Animals able to perceive these patterns of polarization can use them to orient. Adapted from* Animal Navigation *by T. H. Waterman. Copyright ©1989, Scientific American Library. Used with permission by W. H. Freeman and Company.*

Figure 5–82. Manipulating the Olfactory Environment of Homing Pigeons: *To determine whether homing pigeons use olfactory information to form a navigational map of their home region, Ioale, Nozzolini, and Papi (1990) conducted the experiment illustrated here. The experimental pigeons lived in a loft that was exposed to the natural odors and breezes of the area; they were also exposed, at times, to the odor of benzaldehyde, artificially blown toward them from northwest of the loft. Control birds were exposed only to natural odors. During homing trials, both control and experimental birds were exposed to the odor of benzaldehyde—both during transportation to the release site and at the site. Birds were taken to a site east of the home loft and released singly. Researchers recorded the compass direction each bird was flying when it vanished on the horizon. The circle diagram shows vanishing bearings for 10 experimental and 10 control birds. The arrows in the middle of the circle indicate the average direction taken by birds in each group. Control birds headed west, the correct direction to home; their orientation was unaltered by the odor of benzaldehyde. The experimental birds, however, headed southeast. This would be expected if they had interpreted the odor of benzaldehyde at the release site as an indication that the release site was northwest of the loft. When experimental birds were not exposed to benzaldehyde during a release, they oriented toward home, just like the control birds. Adapted from Goodenough et al. (1993).*

Although hundreds of olfactory experiments have been performed and scores of papers published, the olfactory navigation hypothesis remains controversial (Wallraff 1996; Wiltschko 1996; Able 1996). In general, two types of experiments have been performed. First are those that attempt to *eliminate* the sense of smell—by plugging nostrils, cutting the olfactory nerves (see Ch. 4, Olfaction), applying local anesthetics to the olfactory membrane, transporting the pigeons in sealed containers provided with bottled air, or some combination of these methods. Experiments of this type have been used to eliminate access to environmental odors during transport to release sites (to eliminate outward journey information), at the release sites, or both. Experiments of a second type are designed to *manipulate* the odor environment experienced by the pigeons **(Fig. 5–82)**. This approach has been used to attempt to alter the development of the odor map by birds at the loft using various cage and enclosure designs, fan-produced winds, and artificial odors; it also has been used to predictably change the pigeons' perception of the route of displacement from the loft by performing detour experiments of various types and by exposing pigeons to samples of air from different routes or release sites. However, attempts by other workers to replicate some of the experiments have been unsuccessful, and all have been criticized for one reason or another. The most compelling experiments are those that directly manipulate the odor environment, but we have little evidence as to which atmospheric substances might provide the physical basis for an olfactory map. Wallraff has recently documented spatially stable and directionally distinct gradients in ratios of a number of commonly occurring atmospheric hydrocarbons that are sufficiently reliable to account for the known precision of pigeon homing navigation (Wallraff

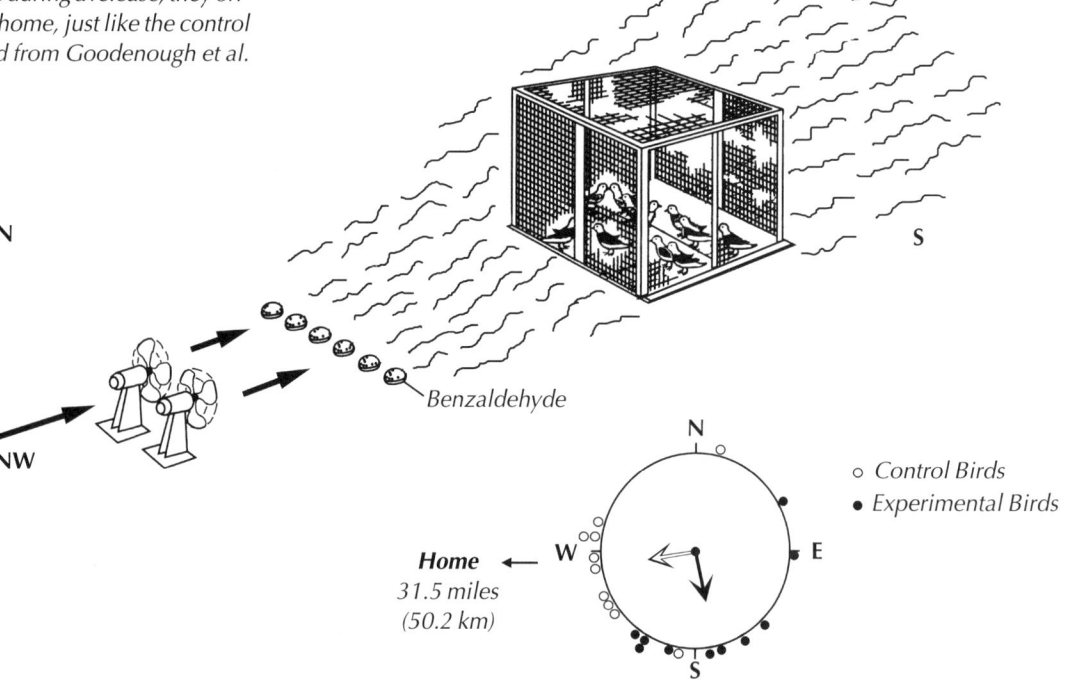

and Andreae 2000; Wallraff 2000). Whether or not pigeons use these gradients, however, remains unknown.

The other current hypothesis on the nature of the navigational map is that birds might possess a **magnetic map**. Using the earth's magnetic field as a map is much more difficult than using it as a compass. For example, because the earth's magnetic field gets stronger as you move from the equator toward either pole, an animal might use the strength of the field at any given point to estimate latitude. But to use the magnetic field in this way requires a level of sensitivity much higher than that needed to use the field as a compass, because the changes in magnetic field strength are very small even over long distances. There is, however, considerable evidence that pigeons are responsive to magnetic field changes of the order necessary to extract the requisite information.

Figure 5–83. Flight Paths of Homing Pigeons in Magnetic Anomalies: Contour lines show the earth's magnetic field over a small region; peaks indicate **magnetic anomalies** where the strength of the magnetic field is elevated by up to 3,000 nanoTesla (nT)—approximately 1 percent above the average. **a. Birds Released in Normal Magnetic Field:** Birds released outside a magnetic anomaly near Worcester, Massachusetts generally flew straight over the anomaly, heading in the correct direction (ENE) toward home. **b. Birds Released at Magnetic Anomaly:** When released at the magnetic anomaly at Iron Mine Hill, Rhode Island, homing pigeons became disoriented and remained so for some time after leaving the vicinity of the anomaly, but they eventually head for home. Each of the two plots shows the flight paths of a different set of released birds. These results suggest that magnetic anomalies affect a pigeon's "map," but not its compass system. Adapted from Baker (1984).

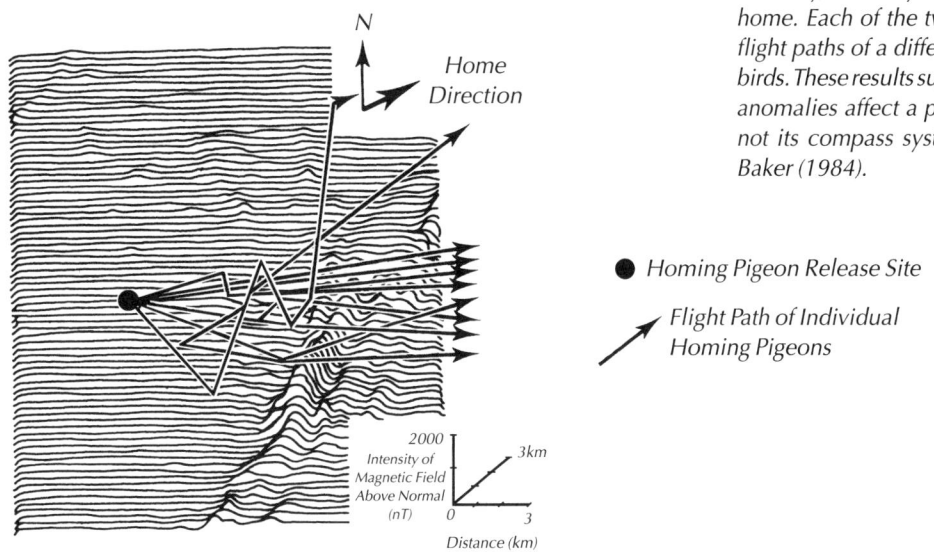

a. Birds Released in Normal Magnetic Field

● *Homing Pigeon Release Site*

↗ *Flight Path of Individual Homing Pigeons*

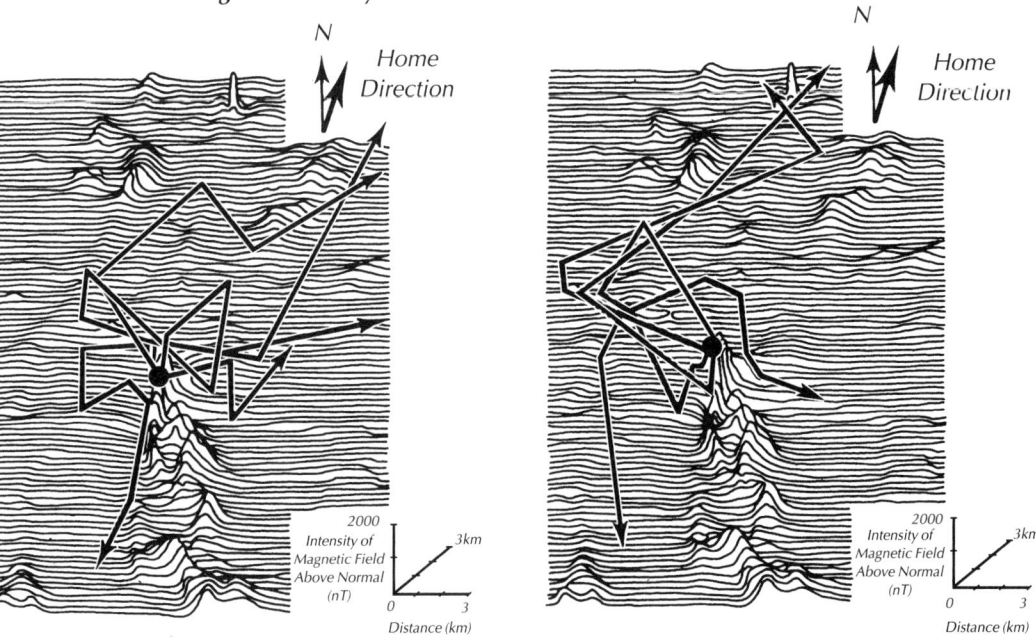

b. Birds Released at Magnetic Anomaly

Empirical evidence supporting the existence of a magnetic map comes almost entirely from studies of homing pigeons released at magnetic anomalies—places where the earth's magnetic field is disturbed, usually by large deposits of iron near the surface **(Fig. 5–83)**. American researcher Charles Walcott discovered that pigeons released within a magnetic anomaly tended to fly off in random directions (Walcott 1996). Once outside the anomaly, however, the birds corrected and headed for home. This effect was found even on sunny days, when pigeons use their sun compass, which suggests that the effect of the anomaly is not on the magnetic compass (which they were not using at that time). In addition, the effect is found only in pigeons released right at the anomaly; pigeons required to fly across a magnetic anomaly on the way home are unaffected. Both of these facts suggest that being released at a magnetic anomaly disrupts only the critical map step of Kramer's map-and-compass model.

Recently, Walcott discovered that pigeons from some lofts are not disoriented by release at magnetic anomalies. This led to an interesting new finding: birds that lived in a loft situated in an area with a rather steep, regular gradient of change in magnetic intensity of the sort likely to provide useful navigational information were disoriented when released at the anomaly; those from a loft where the magnetic field had very little gradient were not affected.

This observation is consistent with the idea that pigeons, and perhaps other birds, possess a flexible navigation system in which the information they learn to rely on depends in part on the availability and reliability of several potential components. Data from experiments performed in the Wiltschko lab suggest that a similar scenario may work with the olfactory map (Wiltschko et al. 1989). Pigeons raised exposed to winds and airflow that would provide good olfactory cues were strongly affected by anesthetizing their olfactory membranes prior to release, whereas those that grew up in a sheltered locale were unaffected and flew homeward immediately.

In the case of migratory birds, we know that many species exhibit remarkable fidelity to breeding and overwintering sites, but we know next to nothing about when and how they navigate to these places. We do not know, for example, whether most of migration is accomplished by compass orientation alone, with navigation to the specific goal occurring only in the final stages of the journey, or whether the birds are oriented toward a specific destination throughout their trip.

Despite the many startling discoveries about bird migration during the last 50 years, we obviously still have much to learn. Animal navigation is a very active area of current research, and important new findings are being made every year. Just 25 years ago, most workers in the field were very skeptical about the reality of magnetic orientation, and almost no one except the original discoverers believed that pigeons used odors in homing navigation. Perhaps the next decade or so will see the discovery of new and important pieces of the navigation puzzle. What is certain is that we cannot yet explain in step-by-step detail how a migratory bird does what we know it does—return with incredible precision to specific spots on the earth after traveling thousands of miles over often unfamiliar terrain.

Suggested Readings

Able, Kenneth P. 1993. Orientation cues used by migratory birds: a review of cue-conflict experiments. *Trends in Ecology and Evolution* 8(10):367–371.

A technical article that reviews the various bird compasses and their interactions.

Able, Kenneth P. 1995. Orientation and navigation: a perspective on fifty years of research. *Condor* 97:592–604.

An historical account and perspective on bird orientation research.

Able, Kenneth P., editor. 1999. *Gatherings of Angels: Migrating Birds and Their Ecology.* Comstock Books, Ithaca, NY. 193 pp.

The world's most knowledgeable migration researchers share their personal and professional experiences regarding some of the most fascinating aspects of bird migration.

Alerstam, Thomas. 1990. *Bird Migration.* Oxford: Oxford University Press. 420 pp.

General reference covering all aspects of migration; uses primarily European examples.

Baker, R. Robin. 1984. *Bird Navigation: The Solution of a Mystery?* New York: Holmes & Meier. 256 pp.

Somewhat slanted overview of bird orientation and navigation mechanisms.

Berthold, Peter, editor. 1991. *Orientation in Birds.* Basel: Birkhauser. 331 pp.

Collection of technical reviews of all aspects of orientation and navigation, each written by an expert in the field.

Berthold, Peter. 1993. *Bird Migration: A General Survey.* Oxford: Oxford University Press. 239 pp.

General review of all aspects of bird migration studies.

Burton, Robert. 1990. *Bird Flight: An Illustrated Study of Birds' Aerial Mastery.* New York: Facts on File. 160 pp.

Beautifully illustrated, readable account of many aspects of bird flight, from ecological to mechanical.

Burton, Robert. 1992. *Bird Migration: An Illustrated Account.* New York: Facts on File. 160 pp.

Beautifully illustrated, readable account of all aspects of bird migration.

Dingle, Hugh. 1996. *Migration: The Biology of Life on the Move.* Oxford: Oxford University Press. 474 pp.

A detailed and technical treatise that presents a unified look at migration in all animal groups.

Gauthreaux, Sidney A., Jr. 1982. The ecology and evolution of avian migration systems. In *Avian Biology*, Vol. 6, edited by D. S. Farner, J. R. King, and K. C. Parkes. New York: Academic Press.

In-depth review of the evolutionary and ecological aspects of bird migration.

Goslow, G. E., Jr., K. P. Dial, and F. A. Jenkins, Jr. 1990. Bird flight: insights and complications. *BioScience* 40:108–115.

Readable account of some of the recent discoveries concerning mechanisms of flight.

Gwinner, Eberhard, editor. 1990. *Bird Migration: Physiology and Ecophysiology.* Berlin: Springer-Verlag. 435 pp.

Collection of technical reviews on all aspects of the physiological control and energetics of migration, each written by an expert in the field.

James, Helen F. and Storrs L. Olson. 1983. *Flightless birds.* Natural History 92:30–40.

Very readable account.

Kerlinger, Paul N. 1989. *Flight Strategies of Migrating Hawks.* Chicago: University of Chicago Press. 375 pp.

Technical treatise on hawk flight, use of thermals, and strategies of cross-country flight.

Kerlinger, Paul N. 1995. *How Birds Migrate.* PA: Stackpole Books. 228 pp.

Covers all aspects of how birds migrate, described clearly and concisely.

Pennycuick, Colin J. 1972. *Animal Flight.* London: Edward Arnold. 68 pp.

Short, readable account of the mechanics of animal flight.

Rappole, John H. 1995. *The Ecology of Migrant Birds: A Neotropical Perspective.* Washington and London: Smithsonian Institution Press. 269 pp.

This book covers the topic of migration in the New World. It includes chapters on the habitats of migrant birds, their resource use, their impact and membership in tropical communities, and comparisons with Old World migration systems. Of particular interest is the book's focus on the evolution of migration from tropical latitudes.

Waterman, Talbot H. 1989. *Animal Navigation.* New York: Scientific American Library, W. H. Freeman. 243 pp.

Readable, well-illustrated account of homing and navigation in all animals.

Evolution of Birds
and Avian Flight

Alan Feduccia

 Most of the *Handbook of Bird Biology* focuses on living birds, but no in-depth discussion of a group of organisms would be complete without exploring its fossils and evolutionary history. The full story of the origin of birds, however, remains elusive. As researchers seek more pieces of the puzzle, they continue to discover new birdlike fossils. These new finds usually generate lively discussion in the scientific community as well as in the news media, often compelling scientists to revise their theories—so our view of the origin and evolution of birds is continually evolving.

Because this topic is so complex, and students may vary widely in their backgrounds and interest in it, this chapter is optional. For students taking the Home Study Course in Bird Biology, exam questions for this chapter are provided to help you review the most important points, but you do not need to submit your answers to Home Study Course staff.

This chapter discusses early birdlike fossils and their similarities with different groups of ancient reptiles; the major theories on the evolution of birds, bird flight, and feathers; the massive extinction of birds, dinosaurs, and most other living things at the end of the Cretaceous period; and how the few birds thought to have survived the Cretaceous extinctions may have given rise to modern birds. Throughout these discussions, you may wish to refer to the two bird evolution diagrams included as Appendices A and B to this chapter, as well as to the Geological Time Scale in Appendix C of Chapter 1. You will note that the first references to the most important fossil organisms discussed in this chapter are printed in color. These indicate organisms whose illustrations and detailed descriptions may be found in Appendix C to this chapter.

Archaeopteryx and Other Urvogels

■ The fossil record allows us to place organisms and their lineages on the geological time scale, and permits us to see changes in diversity and morphology through time. The oldest known bird, *Archaeopteryx*, was preserved in late Jurassic period limestone, dated at about 150 million years old. **Figure E–1** illustrates all the fossil specimens of *Archaeopteryx*, as well as a single feather. You will want to refer to it as you read through this section.

Each of the seven known specimens of *Archaeopteryx* was recovered from the fine-grained Solnhofen limestone, named for the nearby village of Solnhofen in Bavaria. Although fossils are rare in this limestone, the meticulous mining of this highly prized stone, especially for the lithographic printing process, has produced an array of amazingly well-preserved fossils of a variety of plants, invertebrates, fishes, and reptiles. They provide a remarkable window on the past, allowing us to form some idea of the late Jurassic habitat and the conditions that led to its wealth of fossils. The diversity and numbers of fossilized insects in the Solnhofen deposits indicate that they could not have formed far from land. And, many insects found at Solnhofen, including mayflies, caddisflies, and most of the dragonflies, could not have been permanently away from fresh water or brackish coastal water, where they deposit their eggs and undergo development. These and other observations have led researchers to assume that the Solnhofen deposits were formed from marine sediments that were deposited in an arid, tropical climate near the coast. In addition to arthropods, the fossils include numerous fishes, turtles, ichthyosaurs, plesiosaurs, lizards, and crocodiles. Of special interest is a fossil of a single, small theropod dinosaur, *Compsognathus*.

In 1861, Hermann von Meyer reported that an impression of a single feather had been discovered in rock from a quarry near Solnhofen. Even though this first specimen was merely a feather, the unexpected discovery caused a sensation, because it clearly showed that birds dated from the Mesozoic—the Age of Reptiles. This first fossil feather was a secondary wing feather, 2.5 inches (60 mm) long and 0.44 inches (11 mm) wide, with the vane on one side of the shaft roughly half as wide as that on the other side—the same asymmetry as in the flight feathers of modern birds **(Fig. E–2)**. Within a month of his announcement, Meyer reported the discovery of a complete fossil skeleton from the same deposit, but from another quarry. Because this first skeletal specimen of *Archaeopteryx* ended up in the British Museum, it is popularly known as the "London Specimen." It had a long, reptilian tail exhibiting many vertebrae, but attached to each vertebra was what appeared to be a

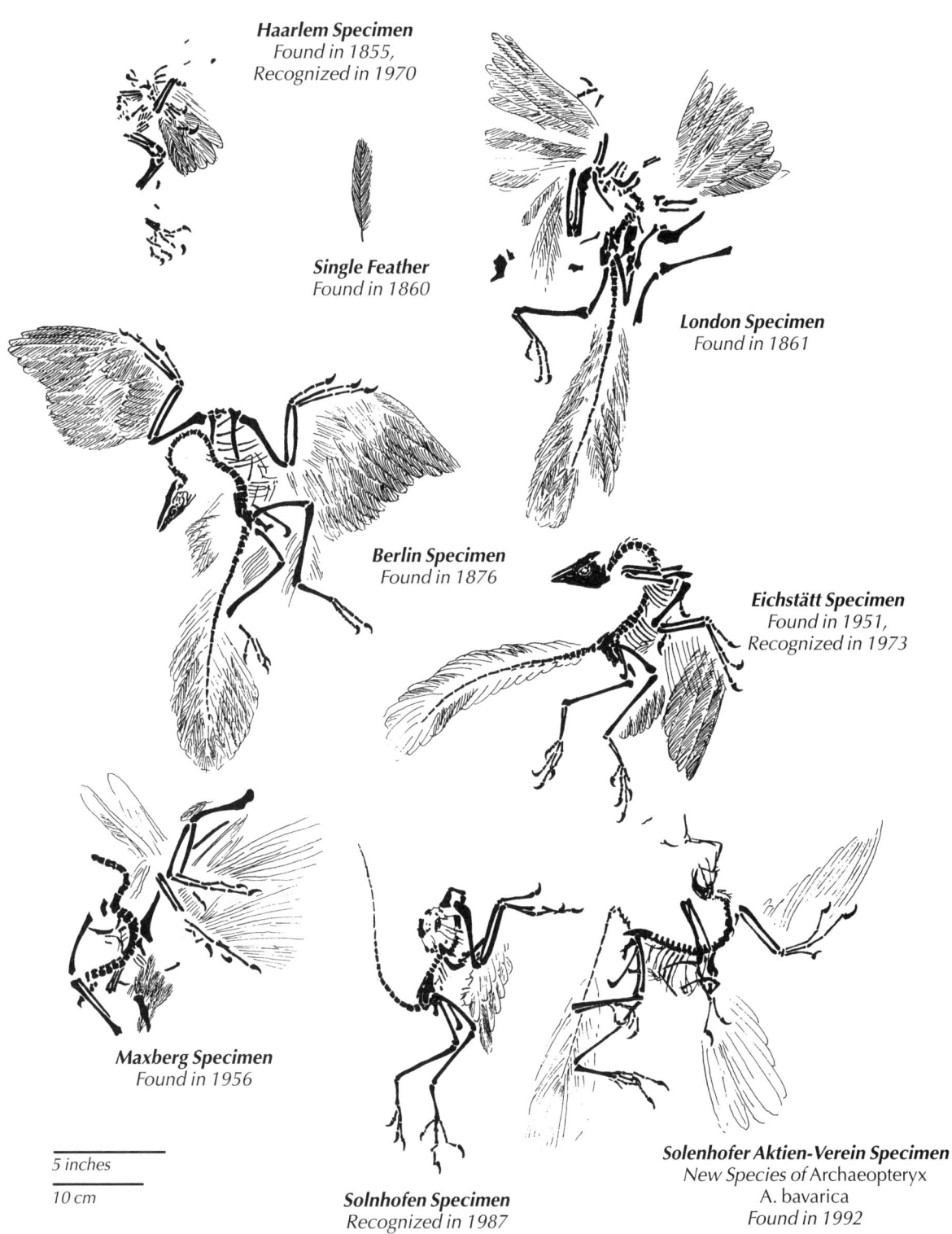

Haarlem Specimen
Found in 1855,
Recognized in 1970

Single Feather
Found in 1860

London Specimen
Found in 1861

Berlin Specimen
Found in 1876

Eichstätt Specimen
Found in 1951,
Recognized in 1973

5 inches

10 cm

Maxberg Specimen
Found in 1956

Solnhofen Specimen
Recognized in 1987

Solenhofer Aktien-Verein Specimen
New Species of Archaeopteryx
A. bavarica
Found in 1992

***Figure E–1. Seven Fossils and a Feather of* Archaeopteryx:** *Shown here are the seven fossil skeletons and a solitary feather of* Archaeopteryx, *including the popular name and the year of discovery or recognition. See text for details on each fossil. From Chatterjee, Sankar.* The Rise of Birds: 225 Million Years of Evolution, *p. 84. Copyright 1997. The Johns Hopkins University Press.*

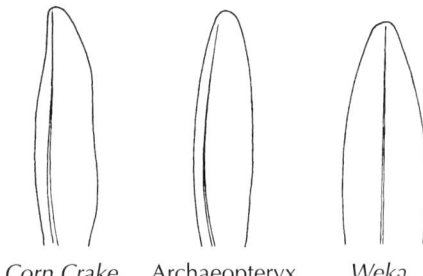

Corn Crake   Archaeopteryx   Weka

Figure E–2. Comparison of Flight Feathers from **Archaeopteryx** *and* **Modern Birds:** Archaeopteryx *had primary and secondary feathers with asymmetric vanes similar to those of modern flying birds such as the Corn Crake. Flightless birds, however, have symmetrical flight feathers, as in the Weka from New Zealand. Because asymmetrical feathers play a key role in the mechanics of flight, their presence in* Archaeopteryx *suggests that the bird may have been capable of flight. Modified from Feduccia and Tordoff (1979); drawing by Ellen Paige. Copyright 1979 AAAS.*

pair of short tail feathers. The limestone had captured these subtle impressions, as well as the startling image of feathered wings. The skull exhibited teeth in the upper and lower jaws. Here, clearly, was a mosaic of avian and reptilian characteristics. The Germans use the general term "**urvogels**" to refer to early, primitive birds such as these.

From its fossilized bones alone, the creature would have been classified as a reptile. But, because the uniquely fine-grained Solnhofen limestone was able to preserve the details of such structures as feathers, this fossil appeared to be more than "just another reptile." Meyer gave it the genus name *Archaeopteryx*, which means "ancient wing" (*archaos*, ancient; *pteryx*, wing), and the species designation *lithographica*, in reference to the lithographic limestone in which it was preserved. In 1877, while scientists all over the world were debating the significance of this Jurassic bird specimen, another discovery was announced from another quarry, near the town of Eichstätt. Known as the "Berlin specimen," this beautifully preserved fossil ended up in the Humboldt Museum in Berlin, and is considered a veritable "Rosetta Stone" of avian evolution **(Fig. E–3)**. Its wings are outstretched and articulated in a natural pose, and attached to the outspread arms and hands are complete impressions of primary and secondary flight feathers, nearly identical in detail to those of modern birds. The perfectly preserved skull, with upper and lower teeth, is arched back over the neck. In addition, each of the three fingers exhibits a sharp, decurved, terminal claw. The feet were those of a perching bird, with three toes directed forward, and a hindward, opposable first toe or hallux. The long tail shows a pair of tail feathers attached symmetrically to each vertebra. This Berlin specimen is no doubt the most widely known and illustrated of all fossil animals.

Remarkably, another specimen of *Archaeopteryx* came to light almost a hundred years later. The third specimen, discovered in a quarry shed, was mainly a torso and was badly articulated; it is known as the "Maxberg specimen." In 1970 still another specimen was recognized. Now known as the "Teyler" or "Haarlem specimen," it was originally recovered in 1855 and described in 1857 as a pterosaur (flying reptile), *Pterodactylus crassipes*. But 120 years later it was shown to be an *Archaeopteryx*! In 1973, another small specimen of *Archaeopteryx* was announced; beautifully preserved, it had very faint feather impressions and for some 20 years had been misidentified as the small theropod dinosaur *Compsognathus*. This "Eichstätt specimen" is roughly one-third smaller than the London specimen. An additional specimen was discovered in 1987—the "Solnhofen specimen"—which was about 10 percent larger than the London specimen.

In 1992, Peter Wellnhofer discovered a new specimen, this one smaller than the London specimen, with longer hind limbs—especially the tibiae. Wellnhofer hypothesized that the strongly curved claws of the foot were adapted for perching **(Fig. E–4)**. This "Solenhofer Aktien-Verein specimen" (named after the quarry company that owns it) is the youngest of the *Archaeopteryx* fossils, and was sufficiently different from the others to be named a new species, *Archaeopteryx bavarica*.

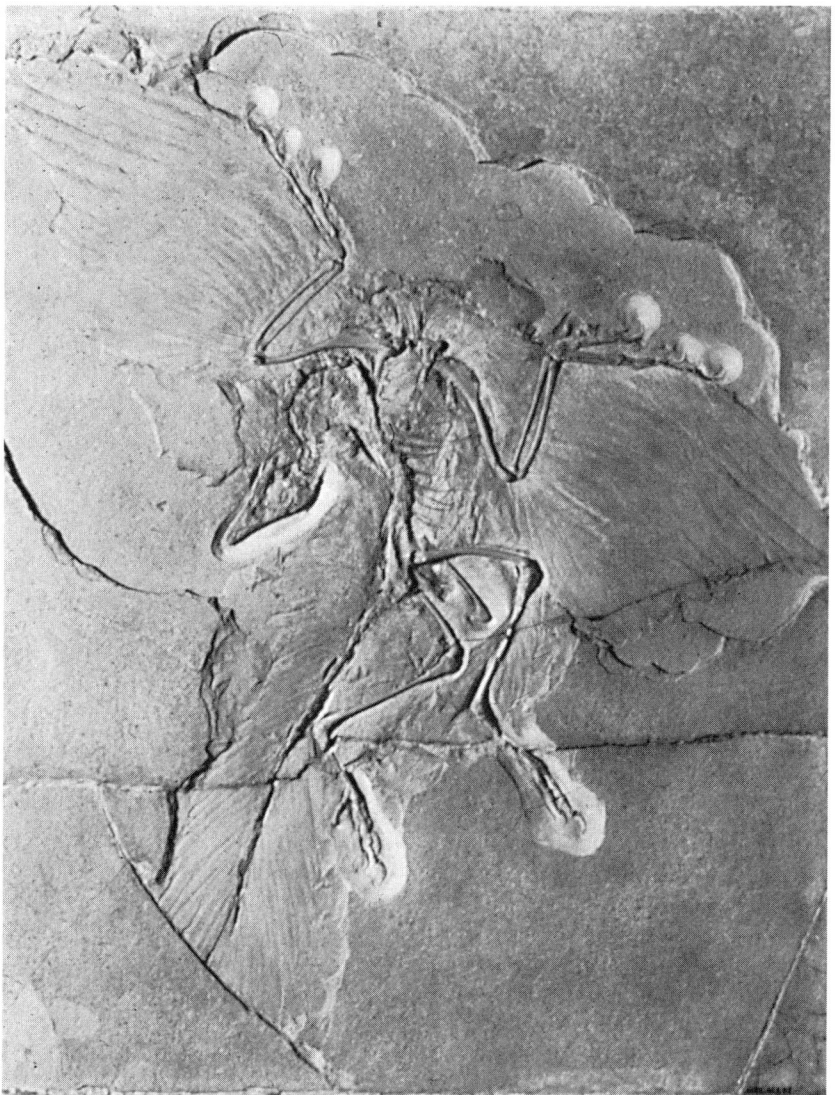

Figure E–3. Berlin Specimen of Archaeopteryx: *Considered the "Rosetta Stone" of avian paleontology, the Berlin specimen of* Archaeopteryx *was discovered in 1876 in a limestone quarry near the Bavarian town of Eichstätt. Courtesy of Humboldt Museum für Naturkunde, Berlin.*

In this specimen, a sternum (breast bone) was preserved for the first time. The presence of an ossified sternum, even though it was flat and without a keel (a downward-projecting ridge of bone to which the major flight muscles of modern birds attach), suggests that *Archaeopteryx* was capable of powered flight.

A rival for *Archaeopteryx*, named *Protoavis*, was announced in August of 1986. Fragmentary bones from two crow-sized creatures—dating from 225-million-year-old (late Triassic) rocks of Texas, and thus alive at the dawn of the age of dinosaurs—were claimed to be those of a fully volant (flying) bird. Evidence that these bones represent an early bird is scant, however, and there is no evidence of feathers. For the time being, *Protoavis* must be considered of uncertain affinities.

In 1995 another urvogel emerged, this time from the early Cretaceous of China (about 25 million years after *Archaeopteryx*). Named *Confuciusornis* and represented by several species, the predominant being *C. sanctus*, this urvogel occurred in incredible numbers around freshwater lakes. *Confuciusornis* resembled *Archaeopteryx* in having three free, clawed fingers and a primitive pelvic region, but the tail was

Figure E–4. Form and Function of Claws: a. Scale Drawings of Claws of Archaeopteryx and Modern Birds: Lateral View: Archaeopteryx *had two types of claws: (1) claws on its toes, and (2) three claws on the first digit of the manus of each wing. Comparing the shape of each claw type to the shapes of the claws of modern birds with known habits can provide hints as to the lifestyle of* Archaeopteryx. *Shown here are the claws of magpies—perching birds that forage on the ground, Long-eared Owls and Eurasian Sparrowhawks—predatory birds that use their feet to grasp prey, Alpine Swifts—cliff dwellers, and White-backed and Lesser Spotted woodpeckers—both tree climbers. Because the toe claws of* Archaeopteryx *appear most similar in shape to those of the perching magpies, and the wing claws appear most similar to those of the tree-climbing woodpeckers, researchers believe that* Archaeopteryx *was probably a perching bird with strong climbing abilities. Adapted from Feduccia (1996, p. 106). Originally from Yalden (1985).*
b. Claw Arc Comparisons: *Claws are very complicated geometrically, but to compare those of different species, researchers often consider them as simple arcs* **(see inset)**, *measuring them in relative degrees of a circle. The angle (y)—from A to B—is a measure of the degrees of arc, and is termed the* **claw arc.** *Compared here are claw arcs, in degrees of curvature, for 30 species of birds. Each vertical column of data points presents a range of values for a single species. Note the almost complete segregation of the claw arc values of ground dwellers, whereas perchers and climbers have some overlap. The mean value for* Archaeopteryx *toe claws (about 120 degrees) is indicated by the line in the center of the perchers, and the mean for wing claws (about 145 degrees), by the line in the middle of the climbers. These data, as in (a) above, suggest that* Archaeopteryx *was a perching bird with strong climbing abilities. Inset and drawing (b) reprinted with permission from Feduccia (1993). Copyright 1993 AAAS.*

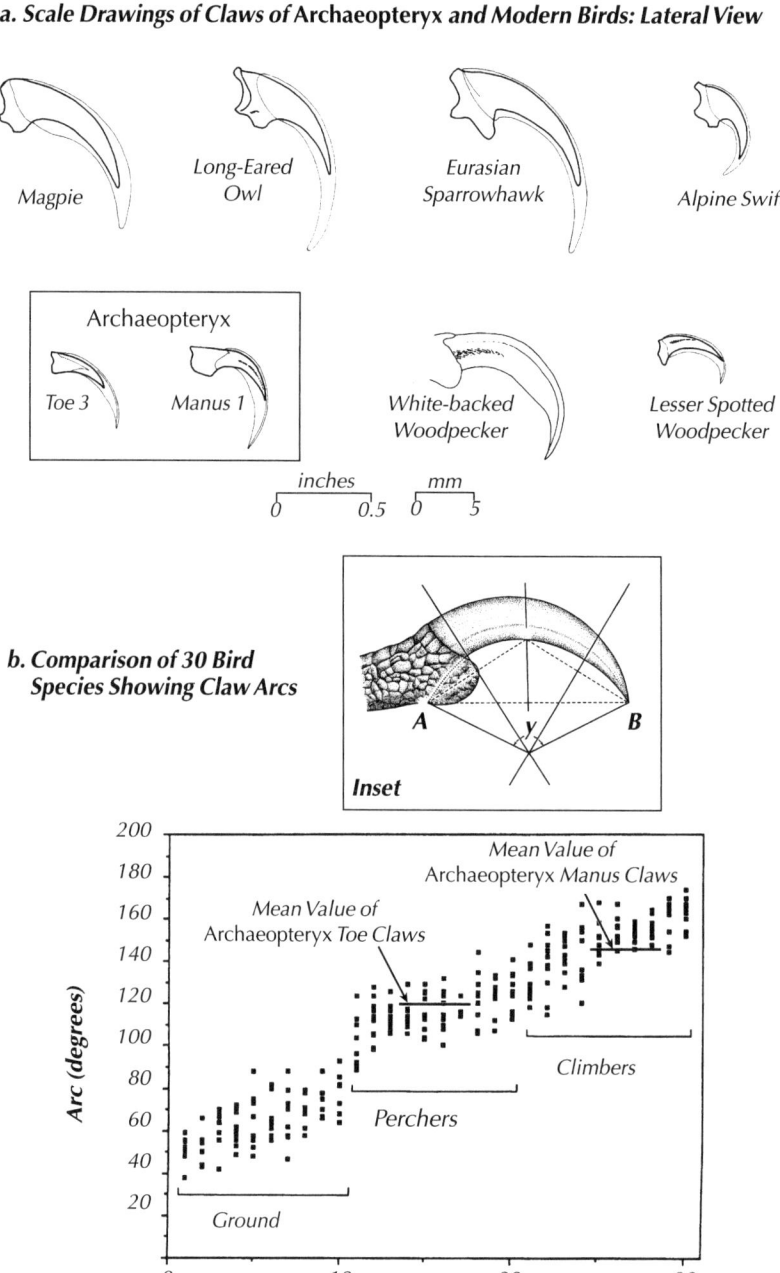

a. Scale Drawings of Claws of Archaeopteryx and Modern Birds: Lateral View

Magpie

Long-Eared Owl

Eurasian Sparrowhawk

Alpine Swift

Archaeopteryx

Toe 3   Manus 1

White-backed Woodpecker

Lesser Spotted Woodpecker

inches   mm
0   0.5   0   5

b. Comparison of 30 Bird Species Showing Claw Arcs

Inset

Arc (degrees)

Mean Value of Archaeopteryx Manus Claws

Mean Value of Archaeopteryx Toe Claws

Climbers

Perchers

Ground

Species

reduced to a long, fleshy pygostyle (the tail bone of modern birds, formed by fusion of the last few vertebrae). The flight architecture was much more advanced than that of *Archaeopteryx*, and the beak was like that of a modern bird. Some of the skull features were more primitive than those of *Archaeopteryx*, however, including the temporal region (the sides of the forehead, or temples), which was fully diapsid (had two openings on each side), like that of early archosaurs **(Fig. E–5)**. Like *Archaeopteryx*, *Confuciusornis* was a perching bird, with a typical perching foot with three anterior toes and a well-developed posterior hallux. Now known from literally hundreds of specimens, *Confuciusornis* must have been a highly colonial species, and one in twenty specimens exhibits long tail plumes resembling those of certain birds-of-paradise. Scientists consider the long tail plumes evidence of some type of complex mating system.

Other more controversial birds to emerge from the same deposits that produced *Confuciusornis* include two feathered, but flightless creatures recently described as *Protarchaeopteryx* and *Caudipteryx*. They were proclaimed to be "feathered dinosaurs," evidence that dinosaurs were ancestral to birds. However, an alternative analysis suggests that they are secondarily flightless Mesozoic birds resembling kiwis (see Fig. 5–48). If this latter interpretation proves correct, they have little, if anything, to do with the evolution of birds. Despite their superficial resemblance to small dinosaurs, they show anatomical distinctions associated with the flightless condition in modern birds. *Caudipteryx* is preserved with a mass of gizzard stones, and must have been able to grind very resistant food. It may have been an herbivore, unlike the carnivorous theropod dinosaurs considered by some to have given rise to birds.

The Descent of Birds

■ Charles Darwin and Thomas Huxley argued that *Archaeopteryx* was really a sideline of early avian evolution and not on the direct line leading to modern birds. In fact, the entire issue of flight origins (see next section) has been sidetracked by *Archaeopteryx*, which, despite some popular belief to the contrary, was already a bird in the modern sense, with fully developed flight adaptations. *Archaeopteryx*, therefore, can tell us very little about the initial stages in the evolution of flight, although the fossil is important in exploring the relationship between birds and reptiles. Recent discoveries in China and elsewhere have shown clearly that early avian evolution was not linear, that there was an early split in avian evolution involving the sauriurine and ornithurine birds (see The Early Fossil Record of Birds, later in this chapter), and that early avian evolution was more bushlike, with many lineages becoming extinct.

What, then, were the reptilian ancestors from which *Archaeopteryx* and other birds evolved? Most of the major groups of Mesozoic reptiles—from lizards to pterosaurs and from crocodiles to dinosaurs—have, at one time or another, been considered the ancestors of birds. After much more than a century of investigation and a fossil record of reptiles that is fairly satisfactory, bird ancestry remains highly controversial. Two major theories are widely debated today, the pseudosuchian thecodont hypothesis and the dinosaur theory. They differ with respect to specific lines of descent, and equally important, they differ as to the time when the first bird appeared. By tracing the genealogy of reptiles, one can see where the two theories diverge. You may want to refer to **Appendix A: Bird Evolution Theories and Early Diapsid Reptiles** as you read through the rest of this section.

All modern reptiles except turtles evolved from diapsid reptiles—the same group that also gave rise to birds. Diapsid reptiles first appeared in the late Carboniferous. By late Permian to early Triassic

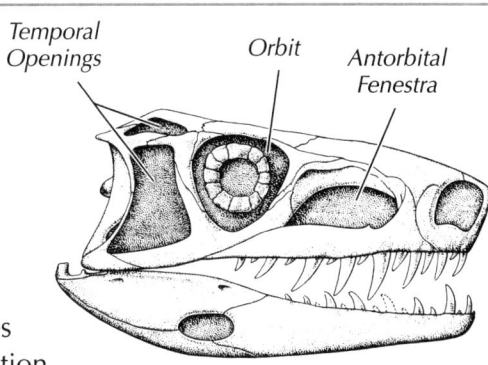

Figure E–5. Diapsid Skull of a Primitive Archosaur: *Diapsid reptiles are those with two openings (fenestrae) on each side in the temporal region of the skull, posterior to the orbit (eye socket). Diapsids include basal archosaurs or thecodonts, as well as snakes and lizards. Illustrated here is a skull of the thecodont* Euparkeria *from early Triassic deposits of South Africa. Thecodonts are a diverse assemblage of early reptiles, united by having teeth set in sockets (the term "thecodont" [literally, a "receptacle for teeth"] indicates this), and an antorbital fenestra, as well as the two temporal fenestrae found in all diapsids. From Romer, A. S. 1966.* Vertebrate Paleontology, *3rd Edition. The University of Chicago Press. Copyright 1966 by The University of Chicago. All rights reserved.*

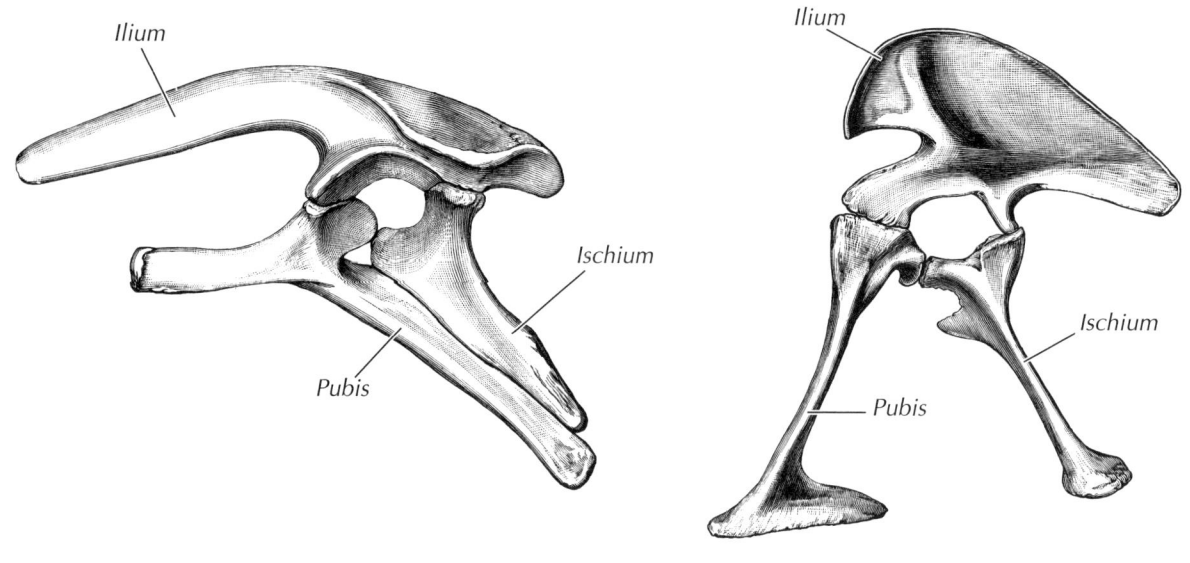

Ornithischian Pelvis   **Saurischian Pelvis**

Figure E–6. Comparison of Ornithischi-
an and Saurischian Hips: *Dinosaurs are*
divided into two main groups, the ornith-
ischian or "bird-hipped" dinosaurs and
the saurischian or "reptile-hipped" dino-
saurs, based on the orientation of the pel-
vic girdle bones. Compared here are the
pelvises of Stegosaurus, *an ornithischian*
(bird-hipped) dinosaur and Allosaurus,
a saurischian (reptile-hipped) dinosaur.
Compare these to the pelvic girdle of a
modern bird, illustrated in Fig. 4–20. Al-
though their pelvic girdles are similarly
oriented, ornithischian dinosaurs are not
ancestral to modern birds. Drawing from
Marsh (1896).

times, some 245 million years ago, two groups of diapsids can be distinguished clearly, one containing the snakes and lizards (Lepido-sauromorpha), and the other containing the archosaurs (Archosauro-morpha). The latter includes thecodonts (basal, or early, archosaurs) and their descendants: crocodiles, ornithischian and saurischian di-nosaurs **(Fig. E–6)**, pterosaurs, and the highly derived (evolutionarily modified) birds. Most of the Triassic archosaurs are included within the Thecodontia, a catchall term for the various early archosaurs, some of which gave rise to the various groups of derived archosaurs, includ-ing the dinosaurs. The name thecodont refers to the fact that the teeth are set in sockets. Archosaurs also are identified by the presence of an opening in front of the orbit (eye socket) called an antorbital fenestra (see Fig. E–5). The thecodonts were ancestral to all the Mesozoic ruling reptiles, including the dinosaurs **(Fig. E–7)**; through one forebear or another, thecodonts gave rise to birds.

Proponents of the pseudosuchian thecodont hypothesis of bird ancestry place the origin of the first bird at this point—the early to middle Triassic—suggesting that birds descended directly from thec-odonts about 230 million years ago. The dinosaur theory, however, postulates the entry of birds into the evolutionary arena much later, after thecodonts had given rise to the saurischian dinosaurs, and after the saurischians had split into distinctive lineages. According to the dinosaur theory, birds descended directly from the theropods, a later lineage of carnivorous dinosaurs that evolved from bipedal thecodonts with shortened forelimbs. More specifically, most recent advocates of the dinosaur theory picture birds evolving directly from a group of theropods called dromaeosaurs, typified by the early Cretaceous *Deinonychus* and the late Cretaceous *Velociraptor.*

Thomas Huxley originated the theory that birds evolved from dinosaurs. Huxley (1868, p. 74) found the resemblance between the small Solnhofen theropod *Compsognathus* and birds very telling: "Surely there is nothing very wild or illegitimate in the hypothesis

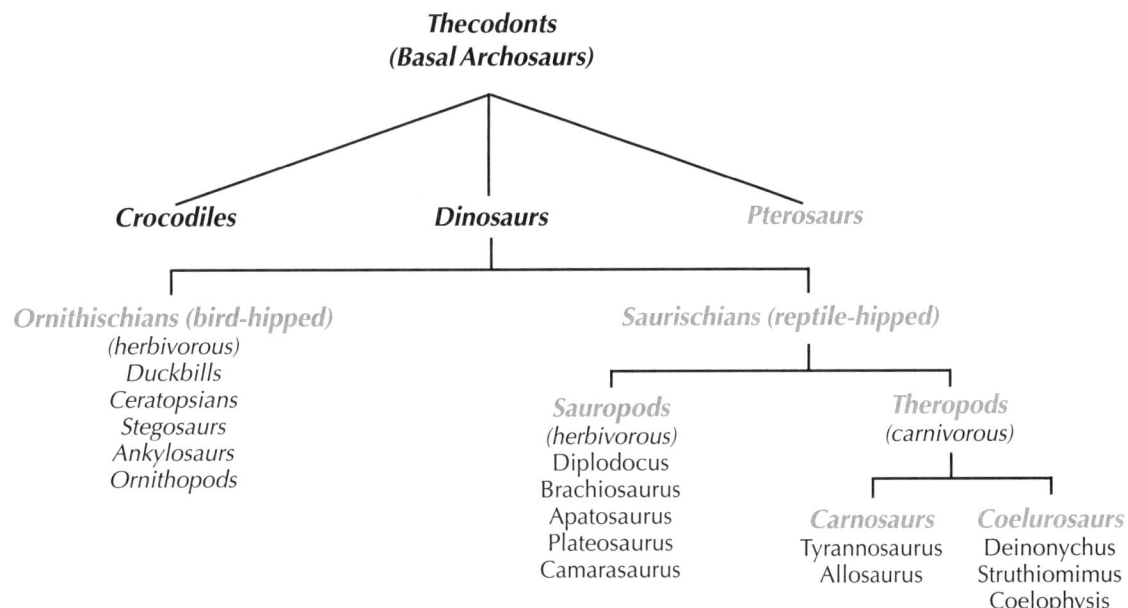

Figure E–7. Descent of Ruling Reptiles from Thecodonts: *This diagram shows the general descent of the Mesozoic ruling reptiles from thecodonts. Most scientists accept the descent of pterosaurs and the branching of the dinosaurs shown here, regardless of their specific views on the origin of birds. Representatives of groups printed in color are illustrated in Appendix C.*

that the phylum of the Class of Aves has its foot in the Dinosaurian Reptiles—that these, passing through a series of such modifications as are exhibited in one of their phases by *Compsognathus*, [have] given rise to [birds]." Most of Huxley's comparisons involved the similarities between the hind limbs of theropod dinosaurs and chickens (both ground-dwelling runners), with little attention paid to the earliest known bird, *Archaeopteryx*.

The argument shifted when Robert Broom, a prominent South African paleontologist, first proposed what has become known as the pseudosuchian thecodont hypothesis for bird evolution. "Pseudo-suchian thecodont" (*pseudosuchian* meaning literally "false crocodile") was used historically to refer to the basal archosaurs, referred to simply as "thecodonts" in this course. In 1913 Broom described from the rich early Triassic deposits of South Africa the pseudosuchian *Euparkeria*, which he believed was ancestral not only to birds, but also to the dinosaurs. *Euparkeria* was a small, 230-million-year-old thecodont, still quadrupedal (moving on four legs) but tending toward bipedality, and it appeared to have all the necessary anatomical qualifications for bird *and* dinosaurian ancestry. No longer was it necessary to deal with the problem that most dinosaurs seemed too specialized to have been ancestral to birds. Broom, then, was arguing that birds and theropod dinosaurs evolved from a common ancestor (a thecodont), not that birds descended *from* dinosaurs.

The publication of Gerhard Heilmann's *The Origin of Birds* in 1926 was a major event in the bird evolution debate. Heilmann considered Broom's *Euparkeria* the key to bird ancestry, and he argued forcefully that birds evolved from thecodonts. He also argued for an arboreal theory for the origin of bird flight: that flight originated among tree-dwelling avian ancestors. Heilmann's book was engaging and well documented; it had all the earmarks of authority. One of his major arguments was the lack of a typically birdlike furcula (wishbone—see Fig. 4–16) in theropod dinosaurs. We now know, however, that a number of theropods have a furcula, although it may not be truly homologous with that of birds. A furcula is now also known in an early archosaur, so the evolutionary importance of this structure remains uncertain. Heilmann's theory of an early thecodontian bird ancestor was repeated in virtually every subsequent textbook and paper on avian evolution until the early 1970s, when the dinosaur theory of bird evolution re-emerged.

In 1973, Professor John H. Ostrom of Yale University published a page-long paper in the British journal *Nature*, outlining the basis of his new version of the dinosaurian origin of birds. Ostrom's hypothesis, simply put, is that birds not only are descended from theropod dinosaurs as proposed by Huxley, but are close to forms such as the Cretaceous coelurosaur *Deinonychus*, in which Ostrom saw many similarities with the earliest bird, *Archaeopteryx*. Ostrom's theory gained considerable momentum during the ensuing years, and now is widely accepted by vertebrate paleontologists. The major concern regarding Ostrom's argument is whether the similarities between theropods and *Archaeopteryx* result from common ancestry or convergent evolution. Many ornithologists remain skeptical of the dinosaur theory because they see many of the anatomical similarities between birds and theropods as superficial, because they think the fossil record shows that the timing is wrong for an ancestor/descendent relationship between birds and dinosaurs, and because they do not support the idea that avian flight originated in ground-dwelling animals—a hypothesis that is often coupled with the dinosaur theory.

For vertebrate paleontologists, *Archaeopteryx* represents little more than a feathered dinosaur, and many consider it a small, ground-dwelling predator. However, recent studies show *Archaeopteryx* to be more birdlike than previously thought **(Fig. E–8)**. It has a birdlike quadrate (a bone of the skull), birdlike occiput (skull base), birdlike brain, and conical teeth that are devoid of the serrations that characterize theropod teeth. It also has a birdlike foot with a fully reversed hallux (found in no dinosaur), and highly decurved claws, which are typical of modern perching birds. On the forelimb it has decurved, flattened wing claws, similar to those of modern trunk-foraging birds and climbing mammals (see Fig. E–4). It has the wings and feathers of modern birds, which have remained essentially unchanged in 150 million years of bird evolution. Its feathers are identical in microstructure to those of modern birds, and the primary and secondary flight feathers have asymmetric vanes—a feature that turns each feather into an individual airfoil during flapping flight, and is correlated with flight capability in modern birds (see Fig. E–2). In birds that become secondarily flightless, such as the Ostrich or flightless rails, the flight feathers tend to lose their asymmetry, the vanes becoming more nearly symmetrical.

Archaeopteryx had an elliptical wing that was similar in profile to that of woodland birds such as woodcock or quail (see Fig. 5–34). Such wings are designed to produce high lift at low speeds, and to provide maneuverability in tight, woodland settings. There can be little question that *Archaeopteryx* was a flying, most probably arboreal, bird.

The reversed first toe of *Archaeopteryx* is further evidence that it was not primarily a ground dweller or a dinosaur. Why would a ground-dwelling dinosaur have a hallux? It only would be a hindrance in running, and no dinosaur is known to have had one. The hallux is a

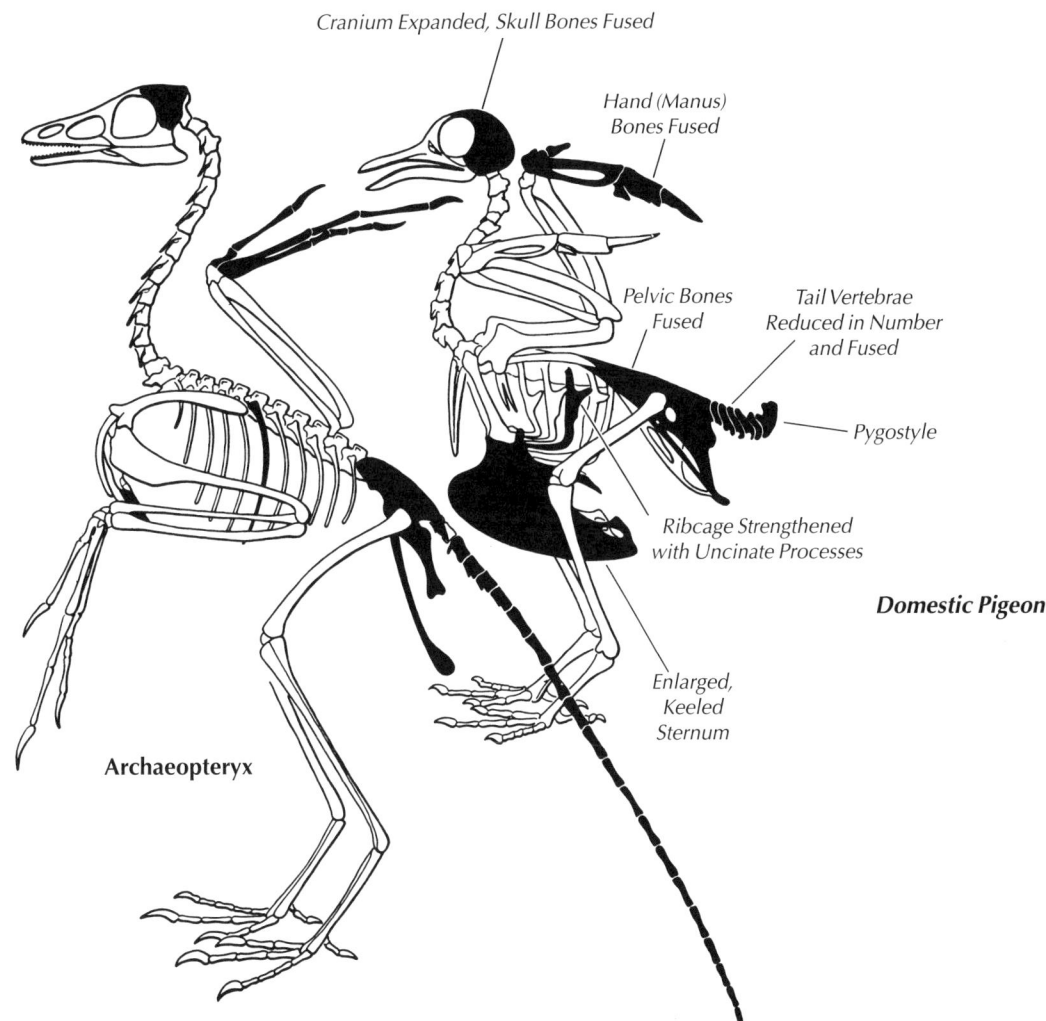

Figure E–8. Comparison of Archaeopteryx and Pigeon Skeletal Features: *The skeleton of* Archaeopteryx *appears quite birdlike, but modern birds have a number of skeletal features not found in* Archaeopteryx. *Most of these are thought to be refinements to improve flying ability. In modern birds (illustrated here by the pigeon), but not in* Archaeopteryx: *(1) the cranium (braincase) is expanded and skull bones are fused; (2) the separate hand (manus) and finger bones are fused, forming rigid wing elements; (3) the separate pelvic bones (ilium, ischium, and pubis) are fused into a single sturdy pelvic girdle; (4) there is a reduction and fusion of tail vertebrae to form one bone, the pygostyle; (5) the ribs are strengthened by horizontal uncinate processes; and (6) the keel of the sternum is expanded for the attachment of flight muscles. From* Evolution of Vertebrates, *by E. H. Colbert. Copyright 1955 John Wiley & Sons, Inc. Reprinted by permission of Wiley-Liss, Inc., a subsidiary of John Wiley & Sons, Inc.*

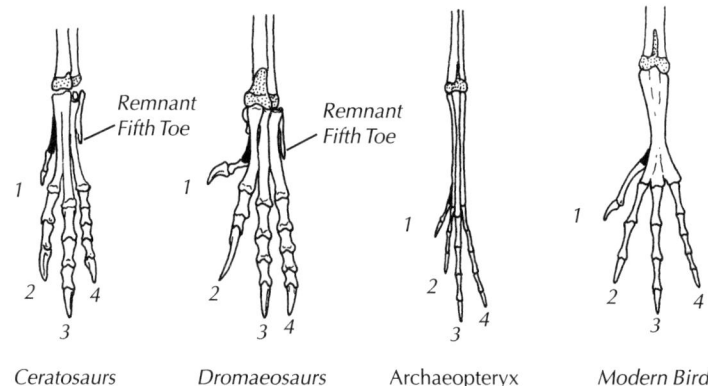

Figure E–9. Feet of Two Theropods, Archaeopteryx, and a Modern Bird: *Compared here are the left feet of two types of theropod dinosaurs (ceratosaurs and dromaeosaurs), Archaeopteryx, and a modern bird—represented by the chicken, Gallus. Note that the theropods have a remnant fifth toe, but Archaeopteryx and modern birds have, at most, four toes. In many modern birds, such as the chicken, the first toe is reversed and points toward the bird's rear. This reversed toe, called the hallux, is an adaptation for grasping branches while perching, and is unique to birds. Unlike theropods, whose first toe is not reversed, Archaeopteryx had a hallux—further evidence of its arboreal lifestyle. Modified from Chatterjee, Sankar.* The Rise of Birds: 225 Million Years of Evolution, *p. 214. Copyright 1997. The Johns Hopkins University Press.*

unique avian adaptation for perching in trees, and many ground-dwelling birds reduce or eliminate it **(Fig. E–9)**.

Birds differ from dinosaurs in many additional ways. Although both birds and dinosaurs have a hand reduced to three fingers, the origins of the digits differ **(Fig. E–10)**. In theropod dinosaurs, the fingers that form the hand are the thumb (digit 1) and the next two, digits 2 and 3; in late Triassic forms, digits 4 and 5 appear as small vestigial structures. In contrast, virtually all embryological evidence indicates that the bird hand consists of digits 2, 3, and 4, the middle three fingers. Theropods have decurved, flattened, serrated teeth, whereas early birds have simple, peglike, conical teeth, constricted at the base and devoid of serrations. The five or so theropod dinosaurs in which the skin is nicely preserved all show typical thick, tuberculated (nodule-covered) reptilian skin, with no hint of anything remotely resembling feathers. Simply put, you have to put a round peg in a square hole to turn a bird into a dinosaur.

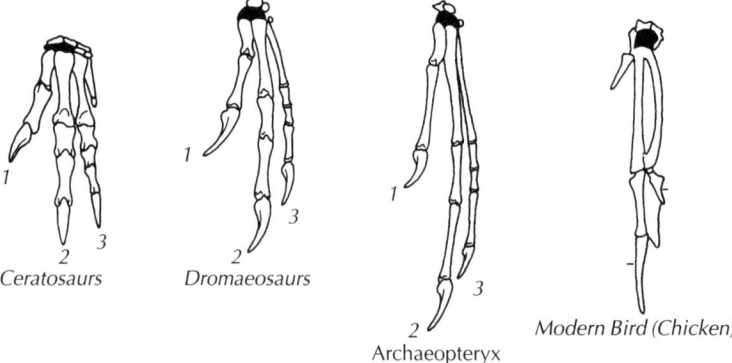

Figure E–10. Left Manus of Two Theropods, Archaeopteryx, and a Modern Bird: *Compared here is the left hand or manus (part of the wing) from two types of theropod dinosaurs (ceratosaurs and dromaeosaurs), Archaeopteryx, and a modern bird—represented by the chicken, Gallus. Both birds and dinosaurs have fewer fingers in the hand than do mammals. In theropods and Archaeopteryx, the digits are the thumb (digit 1), and digits 2 and 3. Three digits are also present in the wing bones of modern birds, but which three is a subject of debate. Embryological evidence suggests that they correspond to digits 2, 3, and 4, but many researchers believe they correspond to ancestral digits 1, 2, and 3 (numbers in parentheses). In either case, there is a reduction and fusion of the hand bones in modern birds. Modified from Chatterjee, Sankar.* The Rise of Birds: 225 Million Years of Evolution, *p. 212. Copyright 1997. The Johns Hopkins University Press.*

There also is the problem of timing. The most superficially birdlike theropods are from the end of the Cretaceous, with forms such as *Velociraptor* appearing almost at the K–T event (the massive, worldwide extinctions at the Cretaceous–Tertiary boundary), some 80 million years after *Archaeopteryx*. The entire group of theropods thought to have given rise to birds, the dromaeosaurs, is restricted to the Cretaceous, the geological period *after* the Jurassic of *Archaeopteryx*. If *Archaeopteryx* is, indeed, an early bird and not merely an evolutionary sideline, this raises a "temporal paradox": you can't be your own grandmother!

Finally, consider the problem of the origin of avian flight. A ground-dwelling animal evolving the ability to fly is a near biophysical impossibility, yet that is the usual flight-origin theory linked with the dinosaur theory of bird ancestry—because early theropods already appeared to be specialized for a terrestrial life. In contrast, the evolution of flight in a tree-dwelling animal is easily explained, and this scenario fits well with the theory that birds evolved from thecodonts—because many thecodonts were small and arboreal. The following section considers the evolution of bird flight in more depth.

Archaeopteryx was well on its way to becoming a true bird, and could well be descended from some form intermediate between reptiles and birds that occurred much earlier than the first dinosaurs. As Columbia University's Walter Bock (1999, p. 568) has noted, until we know more about the actual ancestors of birds, "...it is best to consider birds as part of the great archosaurian radiation without being more specific, as has been agreed by zoologists for more than a century."

Flight Origins

When they began to fly, ancestral birds either lifted themselves up from the ground or glided down from high places; imagining an alternative is difficult. In either case, the anatomical changes required for flight must have evolved in a sequence of very small steps, because nothing we know about evolution suggests that feathered wings could have appeared abruptly as an innovation in avian anatomy. Each new modification of body plan or limbs must have made some contribution to fitness long before the day when a jumping or gliding creature gave the first strong beat of its forelimbs and ceased simply falling back to earth.

To reconstruct the evolution of modern birds, researchers must account for the sequence of changes that replaced reptilian scales with feathers, and, along the way, must answer certain questions: Were the reptilian ancestors of birds runners and then jumpers, or parachuters and then gliders? Was *Archaeopteryx* itself at home on the ground or in trees? Could it only glide, or did it already have the ability to sustain flight by flapping? What was the original advantage of feathers or their epidermal (skin) precursors?

Consider other vertebrates that have taken to the air. Parachuters are known among the frogs, snakes, and geckos; and numerous

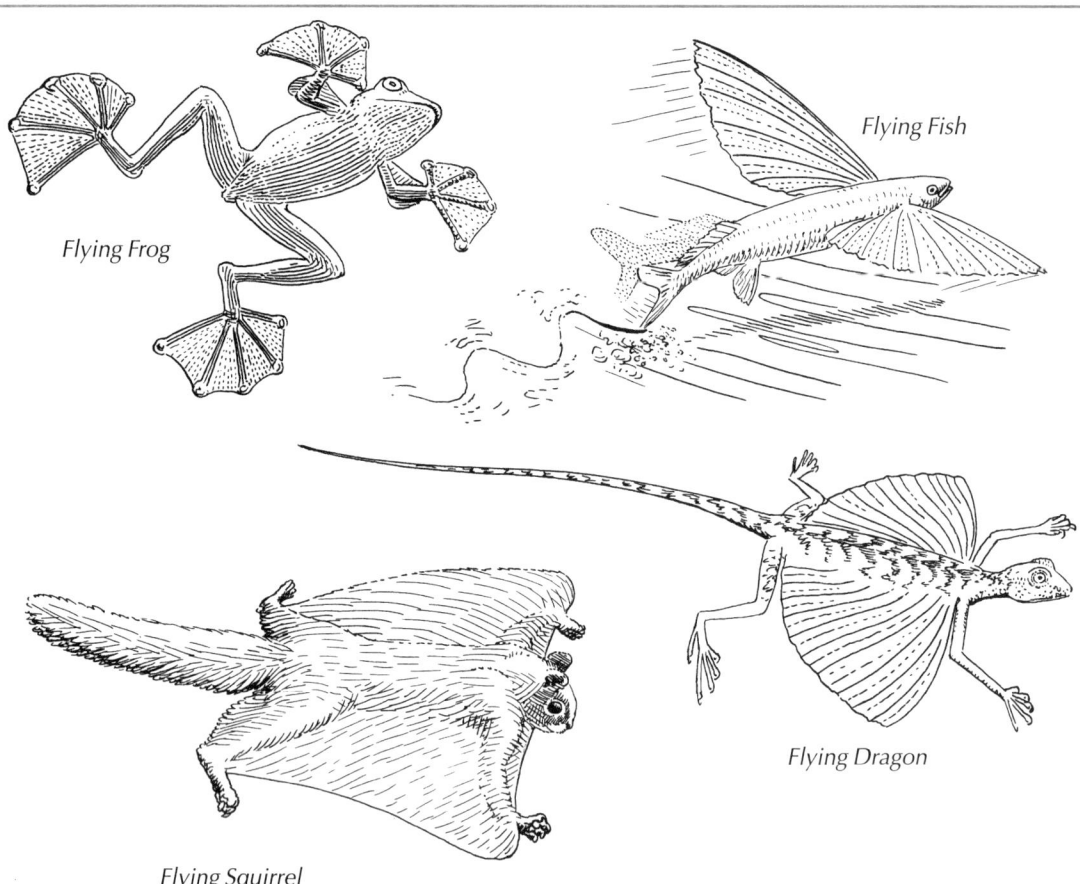

Flying Fish

Flying Frog

Flying Dragon

Flying Squirrel

Figure E–11. A Sample of Gliding and Parachuting Vertebrates: In addition to birds, other vertebrates—including fishes, amphibians, reptiles, and mammals—have evolved a limited ability to "fly" by gliding or parachuting. Shown here are a parachuting frog (genus Rhacophorus), *flying fish* (Exocoetus volitans), *flying dragon (genus* Draco), *and flying squirrel (genus* Glaucomys).

lizards, both living and fossil, have evolved the ability to glide using specialized expansions of their rib cage. Powered flight evolved among the extinct flying reptiles, the pterosaurs. Among mammals, flight now ranges from full-powered flight in bats, to gliding in marsupials (sugar gliders), rodents (flying squirrels), and colugos (flying lemurs; primitive mammals resembling flying squirrels), and modest parachuting in primates **(Fig. E–11)**. All of these flying or gliding vertebrates evolved flight while living in trees, with the possible exception of the pterosaurs, which may have glided out from sea cliffs. None began to fly from a purely ground-dwelling existence. Early in their evolution, fliers use the energy provided by gravity; they climb up and coast down. They do not start their flight with the burst of effort needed to rise directly off the ground. This fact argues strongly against the theory that birds began flight as runners and jumpers, which seems essential to a theory that birds evolved from theropod dinosaurs.

Ground-Up (Cursorial) Theory

Since it was first proposed in 1879, many researchers have advocated a cursorial theory (from the Latin *cursus*, a rapid running motion) to explain the origin of flight **(Fig. E–12)**, probably because they believed that birds descended from cursorial dinosaurs rather than tree dwellers. As late as 1877, paleontologist and dinosaur hunter O. C. Marsh supported a dinosaurian origin of birds. Then in 1880,

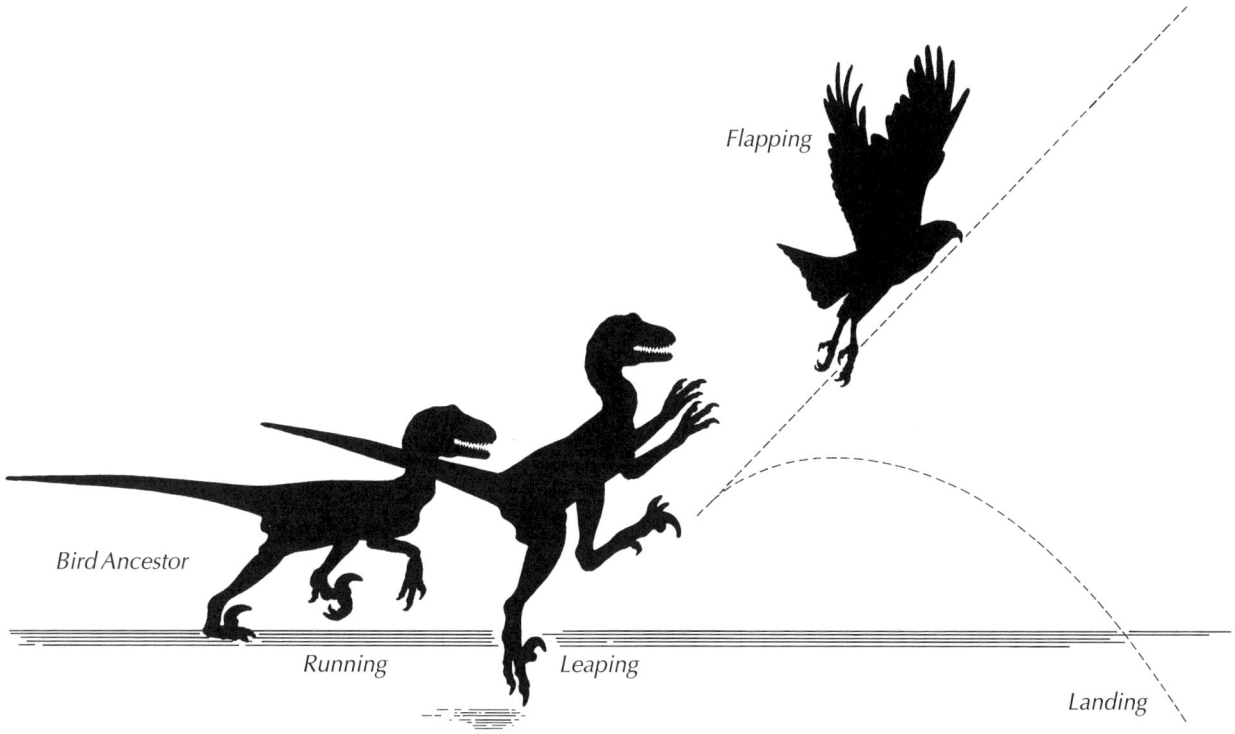

Figure E–12. Cursorial Theory of the Evolution of Avian Flight: *The cursorial theory, first proposed in 1879 by Samuel Williston, an expert on fossil reptiles and a dinosaur collector for paleontologist O. C. Marsh, suggests that ancestors of birds first ran along the ground and eventually began to jump and leap. Wings (and feathers) developed to extend their ability to leap by aiding in propulsion and balance, leading eventually to flight. This theory is closely linked to the belief that birds descended from small, ground-dwelling, theropod dinosaurs. Modified from Chatterjee, Sankar.* The Rise of Birds: 225 Million Years of Evolution, *p. 150. Copyright 1997. The Johns Hopkins University Press.*

Marsh proposed, apparently for the first time, an arboreal theory on the origin of flight. He suggested that small, tree-dwelling, reptilian birds jumping from branch to branch eventually began gliding, aided by rudimentary feathers on the forelimbs. Intuitively facile, this arboreal theory has been favored by many since then. In fact, the same general theory was proposed by Charles Darwin in 1859 to account for the origin of flight in bats.

Nevertheless, the cursorial theory for the origin of avian flight continued to have numerous adherents and experienced a revival early in the 1900s. The revival resulted partly from an examination of modern quadrupedal animals that can rise up to run on their hind legs, as do the living lizards called basilisks *(Basiliscus)* of Central America and the frilled lizards *(Chlamydosaurus)* of Australia. Both the basilisks and the Australian frilled lizard are agile tree climbers, and basilisks can run across water for short distances to escape predation **(Fig. E–13)**! What these lizards illustrate best, though, is the extreme behavioral plasticity of animals, and the near impossibility of ascribing behavioral repertoires to fossil creatures: animals are always capable of at least twice the behavior that their anatomy alone would suggest. In fact, throughout the history of vertebrates, adaptive behavior has probably evolved *before* anatomy in the invasion of virtually every major new habitat or niche.

Figure E–13. Plumed Basilisk Running Across Water: *Basilisks (genus Basiliscus) of Central America can rear up and run on their hind legs in a semi-erect posture. Some paleontologists believe that an animal much like these lizards was the first step in the transition from reptile to bird. The hind feet of basilisks have long toes lined with flaps of skin that spread their weight over a larger area, permitting them to run rapidly over the surface of water for short distances, earning them the nickname "Jesus Christ Lizards." Photo by Joe McDonald.*

Imagine the incredible energy that a cursorial creature would have to expend to become airborne, as the cost of running and leaping into the air is much greater than that of climbing and then gliding. The flailing forelimbs, augmented by the development of feathers, would add to the thrust and speed of the running biped, but once aloft, where could the would-be flier find the power to stay in flight? The main thrust, provided by the traction of the hind feet on the ground, would have disappeared. Proponents of this theory have suggested flapping forearm propellers, but these seem insufficient to prevent the bird from crashing promptly back to earth. The cursorial theory simply ignores the fact that the animal would be fighting gravity all the time.

In 1976, nearly a hundred years after the first cursorial theory was advanced, John Ostrom proposed a very different version, which has been termed the "insect-net" theory **(Fig. E–14)**. Unlike its predecessors, Ostrom's theory was widely accepted, especially by paleontologists and those advocating a dinosaurian origin of birds, and his view of *Archaeopteryx* as a nonflying, reptilian "fly swatter" is found in many textbooks.

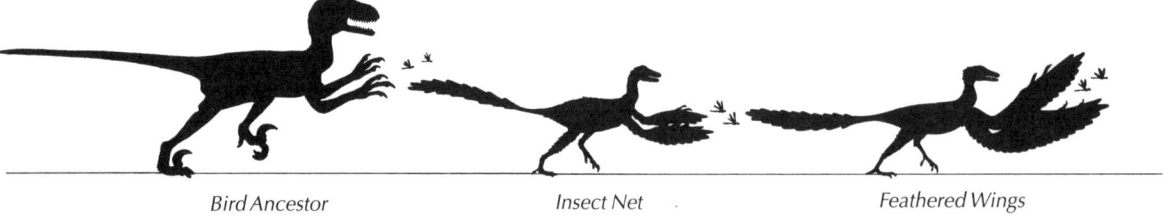

Bird Ancestor   *Insect Net*   *Feathered Wings*

Figure E–14. "Insect-Net" Theory of the Evolution of Avian Flight: *In 1976, John Ostrom, a paleontologist from Yale University, proposed the "insect-net" theory, a variation on the cursorial theory for the evolution of flight. He envisioned* Archaeopteryx *as a small, terrestrial theropod dinosaur using its wings as an insect trap. As the forelimb feathers elongated, they became more efficient for catching prey. Eventually the swatting motion became flapping flight. Ostrom also suggested that* Archaeopteryx *and dinosaurs contemporary with it were warm-blooded (endothermic) and that the first feathers evolved not for flight, but for insulation. Modified from Chatterjee, Sankar.* The Rise of Birds: 225 Million Years of Evolution, *p. 152. Copyright 1997. The Johns Hopkins University Press.*

Ostrom developed his insect-net theory considering features of *Archaeopteryx* anatomy as well as dinosaurs contemporary with it. He argued that "warm-bloodedness" (endothermy) first evolved among dinosaurs. In this scenario, the first feathers served certain groups of dinosaurs as a thermoregulatory pelt. Accordingly, he argued that the coelurosaurs, the small theropod dinosaurs that *Archaeopteryx* anatomically resembled, at least superficially, were warm-blooded animals, and that the first feathers evolved as insulating material, not as aids to flight **(Fig. E–15)**. Ostrom also argued that the head and mouth of *Archaeopteryx* indicate that it preyed on relatively small animals, such as insects, lizards, and small mammals. Running after such creatures on its two hind legs, *Archaeopteryx* used its forelimbs to catch them. In time, elongation of the forelimb feathers made them more efficient for trapping prey, and they became a kind of insect net.

In the 1980s, Jeremy Rayner of the University of Bristol calculated that if a 0.44-pound (0.2-kg), bipedal, cursorial theropod were to jump while running at speeds of up to 6.6 feet (2 meters) per second, its speed would drop by 30 to 40 percent, which would present a serious problem in attaining any type of flight (Rayner 1985). He claimed that the first "flights" of a fluttering proto-flapper would have been at low speeds, at which the energetic demands of flight are at their most extreme, and the wingbeat cycle in living fliers is at its most complex. Rayner suggested that the fluttering model fails because it ignores the extreme morphological, physiological, and behavioral specializations required for flight. He argued that the strategy of running, jumping, and gliding produces air speeds that are apparently too slow to favor flying, and that reaching speeds at which flapping is mechanically straightforward requires more energy than appears possible. The costs of flight at the low speeds attainable by runners are so high that demands on the forelimb musculature become extreme. Given these problems, Rayner hypothesized that a cursorial runner would be unlikely to get off the ground.

Furthermore, Walter Bock (1986) has pointed out that no small, four-legged animals about the size of *Archaeopteryx* that are primarily terrestrial (in other words, not flying-running forms, or secondarily flightless or degenerate flying forms) use their forelimbs for balance during fast running or during leaping. A biophysically convincing

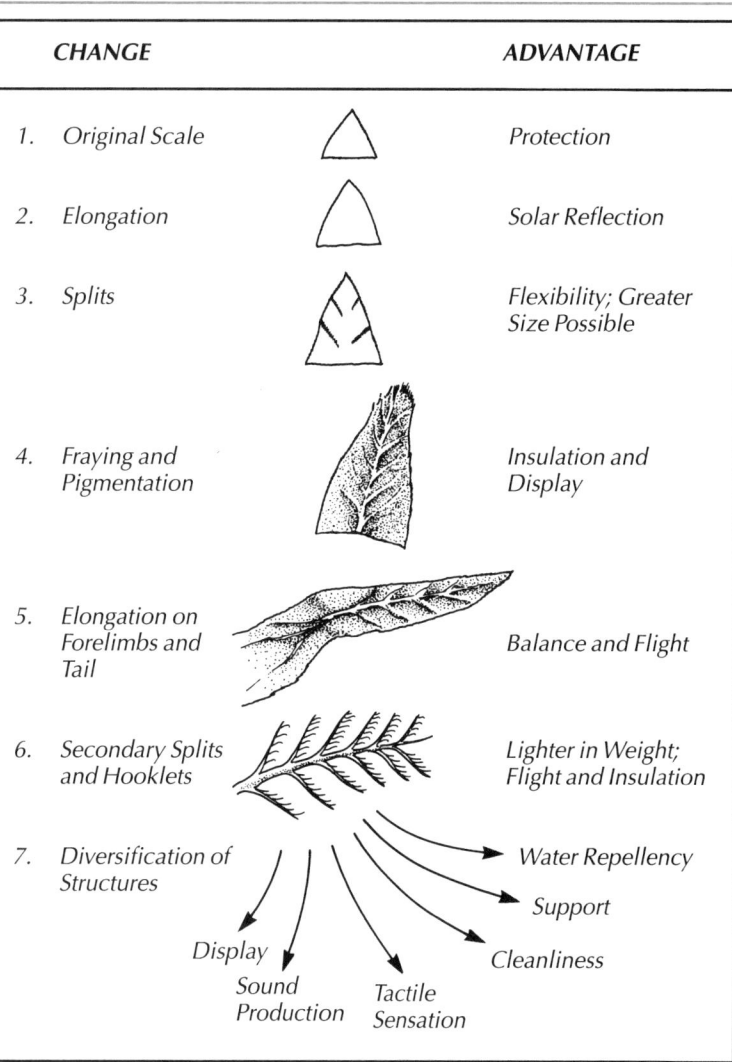

CHANGE		ADVANTAGE
1. Original Scale		Protection
2. Elongation		Solar Reflection
3. Splits		Flexibility; Greater Size Possible
4. Fraying and Pigmentation		Insulation and Display
5. Elongation on Forelimbs and Tail		Balance and Flight
6. Secondary Splits and Hooklets		Lighter in Weight; Flight and Insulation
7. Diversification of Structures		Water Repellency; Support; Cleanliness; Display; Sound Production; Tactile Sensation

Figure E–15. Hypothetical Steps in the Evolution of Feathers from Scales: Elongation and splitting of reptilian scales aided reflection of solar heat and permitted larger, flexible scales. Increased fraying and pigmentation of the larger scales made them more effective in insulation and displays. Elongation of feathers on the forelimbs and tail improved balance on extended leaps and ultimately led to flight. Secondary splitting led to the evolution of branches with interlocking hooklets, and eventually to the modern feather structure that aids flight and insulation. In addition, this versatile structure is easily modified for special purposes, including sound production, tactile sensation, support, and water repellency. Adapted from Ornithology, by Frank B. Gill, Copyright 1990, 1995 by W. H. Freeman and Company. Used with permission.

model for the cursorial origin of avian flight, which would be a unique pattern among flying vertebrates, both living and extinct, has yet to be developed.

Trees-Down (Arboreal) Theory

A theory that contradicts none of the evidence from either *Archaeopteryx* or other fossil finds is the arboreal hypothesis that was first sketched out in 1880 by O. C. Marsh (**Fig. E–16**):

> *The power of flight probably originated among small arboreal forms of reptilian birds. How this may have commenced, we have an indication in the flight of* Galeopithecus (Cynocephalus, *the flying "lemur"), the flying squirrels* (Pteromys), *the flying lizards* (Draco), *and the flying tree frog* (Rhacophorus). *In the early arboreal birds, which jumped from branch to branch, even rudimentary feathers on the fore limbs would be an advantage as they would tend to lengthen a downward leap or break the force of a fall. (1880, p. 189)*

Marsh's views were widely accepted, and were tremendously bolstered in 1926 by Gerhard Heilmann in his *The Origin of Birds*. Heilmann reconstructed, in convincing detail, the hypothetical stages of evolution from terrestrial to tree-dwelling to flying animals. However, he paid little attention to the adaptive advantage of each small step that eventually led to the large-scale change from reptile to bird. More recently, Walter Bock has analyzed the arboreal theory and identified the adaptive purpose of each intermediate stage, showing that evolution could, indeed, account for the changes occurring at each (Bock 1985; 1986). Bock's model depicts an evolutionary pathway following a simple, direct route without elaborate intermediate steps.

Figure E–16. The Arboreal Theory of the Evolution of Avian Flight: *The arboreal theory, first proposed by O. C. Marsh in 1880, suggests that flight probably originated in small, arboreal, reptile-like birds that may have jumped from branch to branch or climbed a tree and glided to the next tree. Gliding is an energy-saving form of locomotion, as it costs an animal less to climb a tree and glide to the next one, than to climb up and down a tree and run to the next. Feathered forelimbs would have allowed the evolution of gliding and thus increased the horizontal distance covered by a downward leap. The arboreal theory has been widely accepted since it was first proposed. Modified from Chatterjee, Sankar. The Rise of Birds: 225 Million Years of Evolution, p. 156. Copyright 1997. The Johns Hopkins University Press.*

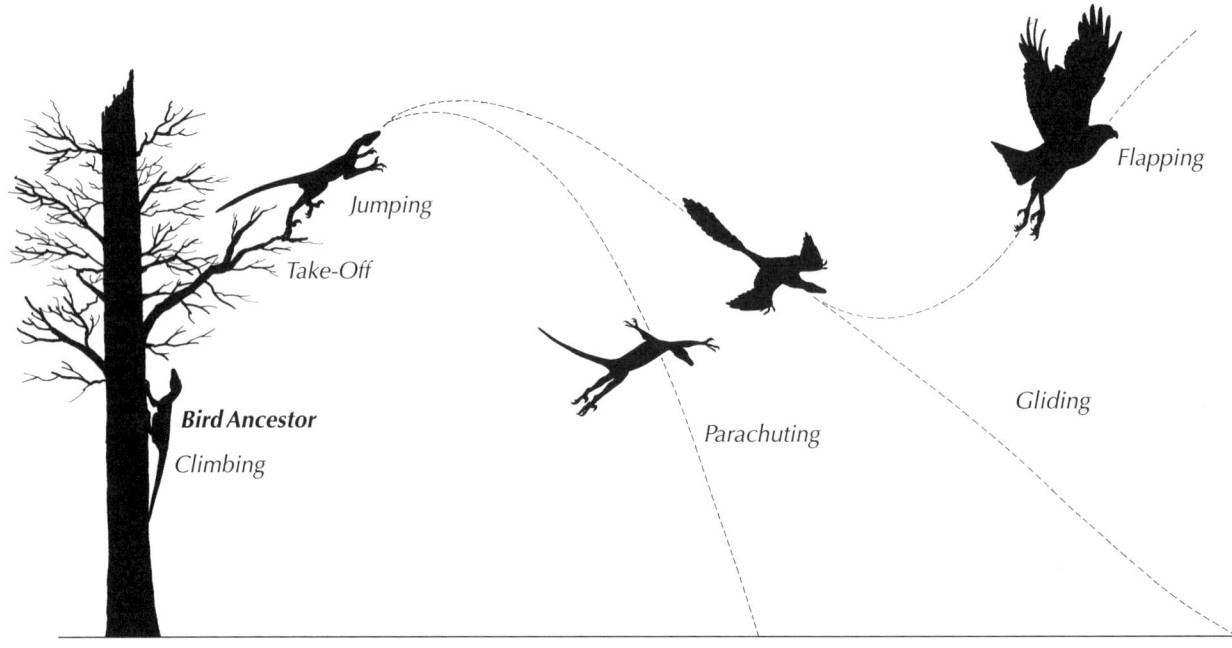

Jumping

Take-Off

Flapping

Bird Ancestor

Climbing

Parachuting

Gliding

The beginning stage involves an ancestral, fairly small, ground-dwelling reptile with either bipedal or quadrupedal locomotion. The animal might have been essentially warm-blooded, but able to use behavioral mechanisms (such as sunbathing or seeking shelter from heat) to supplement the internal regulation of its body temperature. A critical point in Bock's arboreal theory is the invasion of the trees, which he suggests evolved for nesting, hiding, or sleeping in a place safe from predation. He views the invasion of trees, which shifted the animal into a cooler microclimate, as the main reason that natural selection favored the evolution of endothermy and of feathers for insulation. Once this ancestral bird took to climbing and the arboreal life, it presumably began leaping from tree to tree—so selection would favor any adaptation that would decrease the rate of descent or lessen the impact. For example, flattening the body and spreading the limbs horizontally would increase body surface area and lessen impact. Also, increasing the length of feathers would improve parachuting ability. This ancestral bird could then slowly expand its repertoire to include not only parachuting, but gliding, and finally, active, powered flight.

Using an elegant aerodynamic model, Ulla Norberg has shown that the transition to active, powered flight from gliding is both mechanically and aerodynamically feasible. Norberg (1990, p. 260) showed that "...for every step along the hypothetical route from gliding, through stages of incipient flapping, to fully powered flight, there would have been an advantage over previous stages in terms of length and control of the flight path."

Early Bird Flight

A particularly important factor in the evolution of flight, usually given scant attention, is body size. Rayner pointed out that because of mechanical considerations, flight must have evolved in relatively small animals, certainly much smaller than the closest known nonfeathered dinosaur, *Deinonychus,* which was about 10 feet (3 m) long.

Indeed, small size is absolutely essential for both arboreal life and the origin of arboreal flight. Sam Tarsitano (1985, p. 321) wrote "Parachuting requires that the [ancestral bird] be small, lightly built, and able to extend its limbs to present as much surface area as possible to the airflow." Specialization for arboreal life required a decrease in size, which conferred a favorable ratio of mass to surface area, lessening the effect of an impact should the animal jump or fall to the ground. A smaller-sized animal also can move through the air at a lower rate of speed on proportionately smaller wings.

An animal that climbs a tree and then glides to the next tree uses less energy than it would if it climbed up and down a tree and then ran to the next. Ulla Norberg has suggested that maximization of net energy gain during foraging in trees might have augmented selection pressure for increased gliding performance. Once gliding evolved, an animal could dramatically increase its foraging efficiency and drastically reduce its locomotion time during foraging. Parachuting or gliding to escape from enemies also may have been an important aspect of avian evolution. Norberg (1990, p. 259) has pointed out that "Gliding must have been used only for commuting and not for insect catching, which would require high maneuverability, which did not evolve until true flight was well established."

In conclusion, avian flight most logically originated in the trees, taking advantage of the energy-saving locomotion provided by high places and the power of gravity. It follows, then, that the actual ancestors of birds are most likely represented by small, arboreal archosaurs from the Triassic or early Jurassic. Many thecodonts, although structurally close to dinosaur "grade" in form, were small, had not yet become fully erect and bipedal, and were arboreal, with little reduction in their forelimbs. These thecodonts could easily have retained or evolved the elongation of the forelimbs seen in *Archaeopteryx* and presumably present in early birds. In contrast, the early theropods already had attained the erect posture and greatly reduced forelimbs suggestive of terrestrial, rather than arboreal, locomotion. For early theropods to develop lengthened forelimbs from reduced and somewhat vestigial hands would demand a dramatic reversal in evolution.

The assumption that the numerous anatomical similarities between birds and certain dinosaurs are due to common ancestry has been seriously challenged. If the challenges are valid, then late Cretaceous, birdlike dinosaurs were convergent with, rather than related to, birds.

The Early Fossil Record of Birds

*Throughout this section, you may want to refer to **Appendix B: Hypothesized Relationships Among Ancient and Modern Bird Groups.***

■ The primitive birds mentioned previously are placed within the Subclass Sauriurae; most other known birds are placed within the Subclass Ornithurae. The Sauriurae also contains another major group of archaic birds known as the "opposite birds" or enantiornithines. The predominant land birds of the Mesozoic, the enantiornithines are thought to be close allies of *Archaeopteryx*. Now known from Cretaceous deposits throughout the world (Spain, China, South America, Europe, Australia, and Madagascar), these birds had a well-developed flight architecture, but a relatively archaic pelvic region and a long, fleshy pygostyle (tail bone). They are called opposite birds because their three metatarsals (middle foot bones; see Fig. 1–11) fuse in a direction *opposite* (from the proximal to distal end) to that of modern birds. They also have a distinctive formation of the foramen triosseum—the opening among the shoulder bones through which passes an important tendon associated with the flight muscles. Most opposite birds had teeth on both upper and lower jaws that closely resembled those of *Archaeopteryx*. Indeed, the skull of the early Cretaceous forms is very similar to that of *Archaeopteryx*. Particularly well-described genera from the early Cretaceous include the Chinese *Sinornis* and *Cathayornis*, and the Spanish *Iberomesornis* and *Eoalulavis*. The latter have a well-developed, modern type of alula, indicating that they had achieved a fully developed, modern flight apparatus. All opposite birds became extinct at the end of the Cretaceous period, along with the dinosaurs.

Other Mesozoic birds belonged to the Subclass Ornithurae; some of these go back to the earliest Cretaceous and were contemporary with the opposite birds. The Chinese *Lianingornis* and *Chaoyangia* are typical. *Gansus*, known only from a foot, was apparently a shore-dwelling ornithurine. Nevertheless, *Gansus* illustrates a basically modern type of avian tarsometatarsus and toe structure.

A better-known group of Cretaceous ornithurine birds includes the famous Hesperornithiformes (including *Hesperornis* and *Baptornis*), foot-propelled, toothed divers that superficially resembled modern loons through convergent evolution. They were denizens of the great seaways characteristic of the Cretaceous period. Hesperornithiformes became extinct at the close of the Cretaceous, along with their ternlike contemporaries, the Ichthyornithiformes, which included the well-known *Ichthyornis* and *Apatornis*. The early Cretaceous Mongolian *Ambiortus* was a fully volant ornithurine the size of a pigeon. It possessed a well-developed sternal keel and other features of the pectoral region typical of modern birds, confirming that true flying birds existed by about 12 million years after the appearance of *Archaeopteryx*.

The presence of both opposite birds and archaic ornithurine birds shortly after *Archaeopteryx* points to a major dichotomy in the early evolution of birds. The mosaic nature of the anatomy of *Confuciusornis*—more primitive than *Archaeopteryx* in parts of the skull, but more advanced in having a horny beak and a well-developed flight apparatus—shows that, like the early evolution of mammals, the early diversification of birds was probably a complicated bush, not a linear evolutionary pattern. There must have been many extinct lineages that at one time may have been more advanced in some features of their anatomy than their ultimately more successful contemporaries. *Archaeopteryx* was not the ancestor of modern birds, but a sideline of archaic birds that became extinct by the end of the Jurassic, and was replaced by the opposite birds and ornithurines of the Cretaceous.

Palaeognathous Birds

The modern avian radiation consists only of ornithurine birds, and these are divided into two major groups, the Superorders Palaeognathae (ratites) and Neognathae (all other forms). The living ratites consist of flightless birds such as Ostriches, rheas, Emus, cassowaries, and kiwis, as well as their South American, chicken-like relatives, the fully volant tinamous **(Fig. E–17)**. Although unique among ratites in their ability to fly, tinamous, like other ratites, have a "palaeognathous palate," different from that of other modern birds **(Fig. E–18)**. The ratites once were thought to represent a very ancient group of birds that dated well back into the Mesozoic, but no fossils of these birds are known prior to the beginning of the Tertiary period, 65 million years

Northern Cassowary

Emu

Brown Kiwi

Ostrich

Greater Rhea

Red-winged Tinamou

Figure E–17. The Living Palaeognathous Birds (Ratites): *Although all ratites have a palaeognathous palate (see Fig. 18), they are a diverse assemblage of mostly flightless birds whose evolutionary relationships remain controversial. Species representing the modern families of ratites are the Northern Cassowary (Casuariidae), Emu (Dromiceidae), Brown Kiwi (Apterygidae), Ostrich (Struthionidae), Greater Rhea (Rheidae), and Red-winged Tinamou (Tinamidae). Drawing by George Miksch Sutton, used with permission from Alan Feduccia, who was granted use by the late Dr. Dorothy S. Fuller, sister of George Miksch Sutton.*

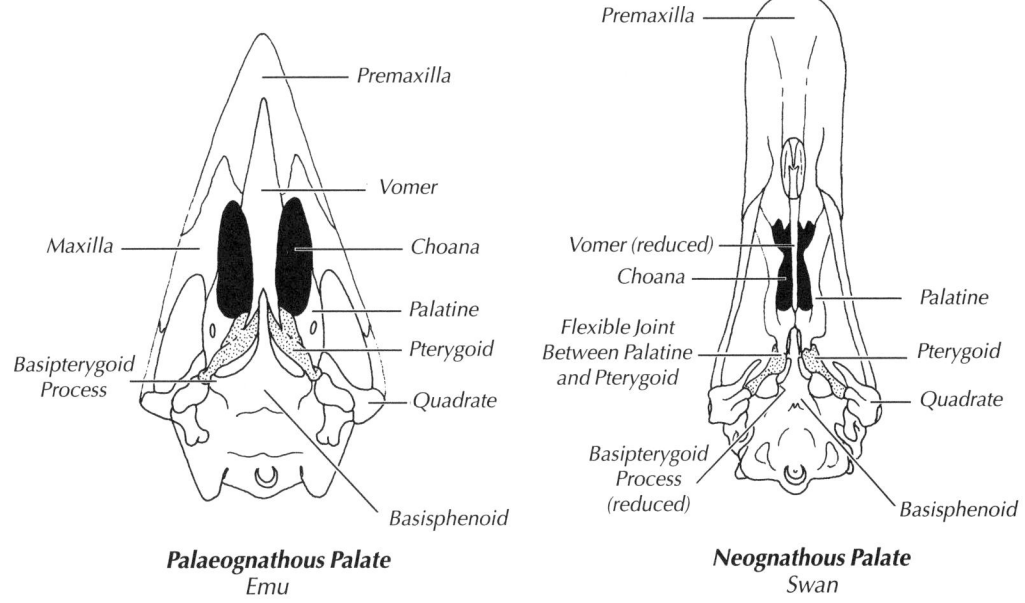

Palaeognathous Palate
Emu

Neognathous Palate
Swan

Figure E–18. Comparison of Palaeognathous and Neognathous Palates: *Modern birds are divided into two major groups, the palaeognathous birds and the neognathous birds. This division is largely based on the structure of the palate, an area of the skull more commonly referred to as the roof of the mouth. In neognathous birds, such as the swan pictured here, the vomer and basipterygoid process are reduced and a flexible joint forms between the pterygoid and palatine bones. In the palaeognathous Emu, the roof of the mouth is formed by larger, more rigid bones. Modified from Chatterjee, Sankar.* The Rise of Birds: 225 Million Years of Evolution, *p. 260. Copyright 1997. The Johns Hopkins University Press.*

ago. More recently, researchers have discovered fully volant, tinamou-like palaeognathous birds known as lithornithids, thought to have been common in North America and Europe during the early Tertiary (Paleocene and Eocene). These chicken-like birds are thought to be the ancestral stock that gave rise to large, flightless forms all over the earth: the lithornithids may have flown to remote parts of the world and given rise to various lineages of flightless birds. Other examples of extinct ratites include huge birds on the islands of Madagascar and New Zealand. The elephantbirds **(Fig. E–19)** of Madagascar were contemporaneous with the native peoples of the island, but probably became extinct in historic times. The same is true in New Zealand, where some dozen species of large moas evolved and survived until the arrival of the Polynesians, who subsequently extirpated them **(Fig. E–20)**. Elephantbirds and moas left no fossils earlier than 10 to 20 million years ago, in the Miocene, perhaps indicating that the ancestors of these giants did not arrive on these islands much earlier. Whether or not the elephantbirds and moas are closely related to each other, or to the ratites, is unclear. Another Tertiary group of giant birds (some as large as the largest elephantbirds), known as the dromornithids, existed

Figure E–19. Elephantbird: *The elephantbirds (Aepyornithidae) are an extinct group of ratites. These large, flightless birds lived on the island of Madagascar, but were exterminated by human activity. The "giant" of the elephantbirds was* Aepyornis maximus, *which stood about 10 feet (3 meters) tall and weighed about 1,000 lbs (450 kg). Fossil elephantbirds are found primarily in Pleistocene and Holocene deposits. Redrawn from Pough et al. (1996, p. 543).*

Figure E–20. Moa: Moas (Euryapteryx) are huge, extinct ratites that flourished in New Zealand until the arrival of the Polynesians. Shown here is a photograph of a museum exhibit of a moa on South Island, New Zealand, during postglacial times (5,000 years ago). Courtesy of Department of Library Services, American Museum of Natural History, Neg. No. 322337.

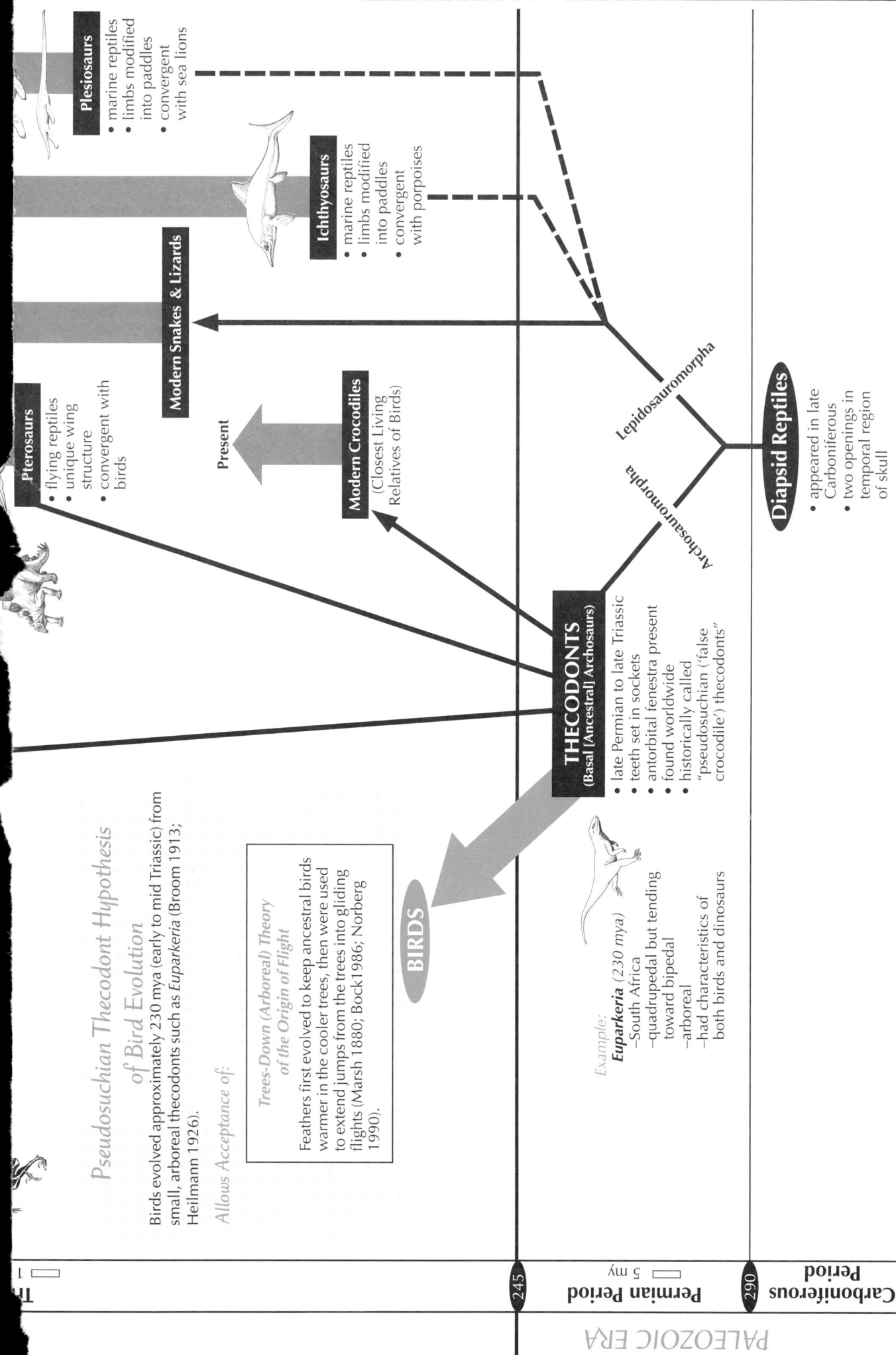

Plesiosaurs
- marine reptiles
- limbs modified into paddles
- convergent with sea lions

Ichthyosaurs
- marine reptiles
- limbs modified into paddles
- convergent with porpoises

Modern Snakes & Lizards

Pterosaurs
- flying reptiles
- unique wing structure
- convergent with birds

Present

Modern Crocodiles
(Closest Living Relatives of Birds)

Lepidosauromorpha

Archosauromorpha

Diapsid Reptiles
- appeared in late Carboniferous
- two openings in temporal region of skull

THECODONTS
(Basal [Ancestral] Archosaurs)
- late Permian to late Triassic
- teeth set in sockets
- antorbital fenestra present
- found worldwide
- historically called "pseudosuchian ('false crocodile') thecodonts"

BIRDS

Pseudosuchian Thecodont Hypothesis of Bird Evolution

Birds evolved approximately 230 mya (early to mid Triassic) from small, arboreal thecodonts such as *Euparkeria* (Broom 1913; Heilmann 1926).

Allows Acceptance of:

Trees-Down (Arboreal) Theory of the Origin of Flight

Feathers first evolved to keep ancestral birds warmer in the cooler trees, then were used to extend jumps from the trees into gliding flights (Marsh 1880; Bock1986; Norberg 1990).

Example:
Euparkeria *(230 mya)*
– South Africa
– quadrupedal but tending toward bipedal
– arboreal
– had characteristics of both birds and dinosaurs

PALEOZOIC ERA

Carboniferous Period
290

Permian Period
5 my
245

Tr
1

Appendix A:
BIRD EVOLUTION THEORIES AND EARLY DIAPSID REPTILES

The two primary theories on the origin of birds both postulate that birds arose from diapsid reptiles. They differ, however, in the time period and specific group from which birds evolved. The pseudosuchian thecodont hypothesis suggests that birds evolved approximately 230 million years ago from thecodonts, whereas the dinosaur theory suggests that birds evolved approximately 150 million years ago from theropods. This diagram shows the major groups of diapsid reptiles and their possible relationships to early birds. Note that the amount of space between one-million-year increments on the time scale is not the same in the different geologic periods.

E

(Open)

K-T EXTINCTIONS

MESOZOIC ERA

KEY

	Fossil Record Time Span
	Relationships Fairly Well Established
	Unknown Relationships
	Time When a Group Went Extinct
mya	Million Years Ago

Present

Dinosaur Theory of Bird Evolution

Birds evolved from theropods such as *Compsognathus* approxi-mately 150 mya (Huxley 1868) or Dromaeosaurs such as *Deinonychus* approximately 110 to 120 mya (Ostrom 1973).

Linked with:

Ground-Up (Cursorial) Theory of the Origin of Flight

Feathers first evolved to extend leaps on the ground, eventually leading to flight (Williston 1879).

Or

Insect-Net Theory of the Origin of Flight
(A variation on the Cursorial Theory)

Feathers first evolved for thermal insulation, then were used for catching prey, and were later used in flapping flight (Ostrom 1976).

DINOSAURS

Saurischia
(Reptile-hipped Dinosaurs)

Ornithischia
(Bird-hipped Dinosaurs)
• herbivorous
Examples:
Hadrosaurus (Duckbill)
Ankylosaurus
Styracosaurus

Sauropods
• herbivorous
• quadrupedal
• some very large
Examples:
Plateosaurus
Camarasaurus
Diplodocus

Carnosaurs
• larger theropods
Examples:
Allosaurus (Late Jurassic)
Tyrannosaurus (Late Cretaceous)

Theropods
• carnivorous
• bipedal with reduced forelimbs

BIRDS

Coelurosaurs
• smaller theropods
Examples:
Compsognathus (Late Jurassic) –chicken sized

Struthiomimus (Late Cretaceous) –convergent with Ostriches

Coelophysis (Late Triassic)

Dromaeosaurs
Examples:
Deinonychus (Early Cretaceous)

Velociraptor (Late Cretaceous)

mya
65
Cretaceous Period

145.5

□ ≈ 20 my
Jurassic Period

208

Triassic Period

million years (my)

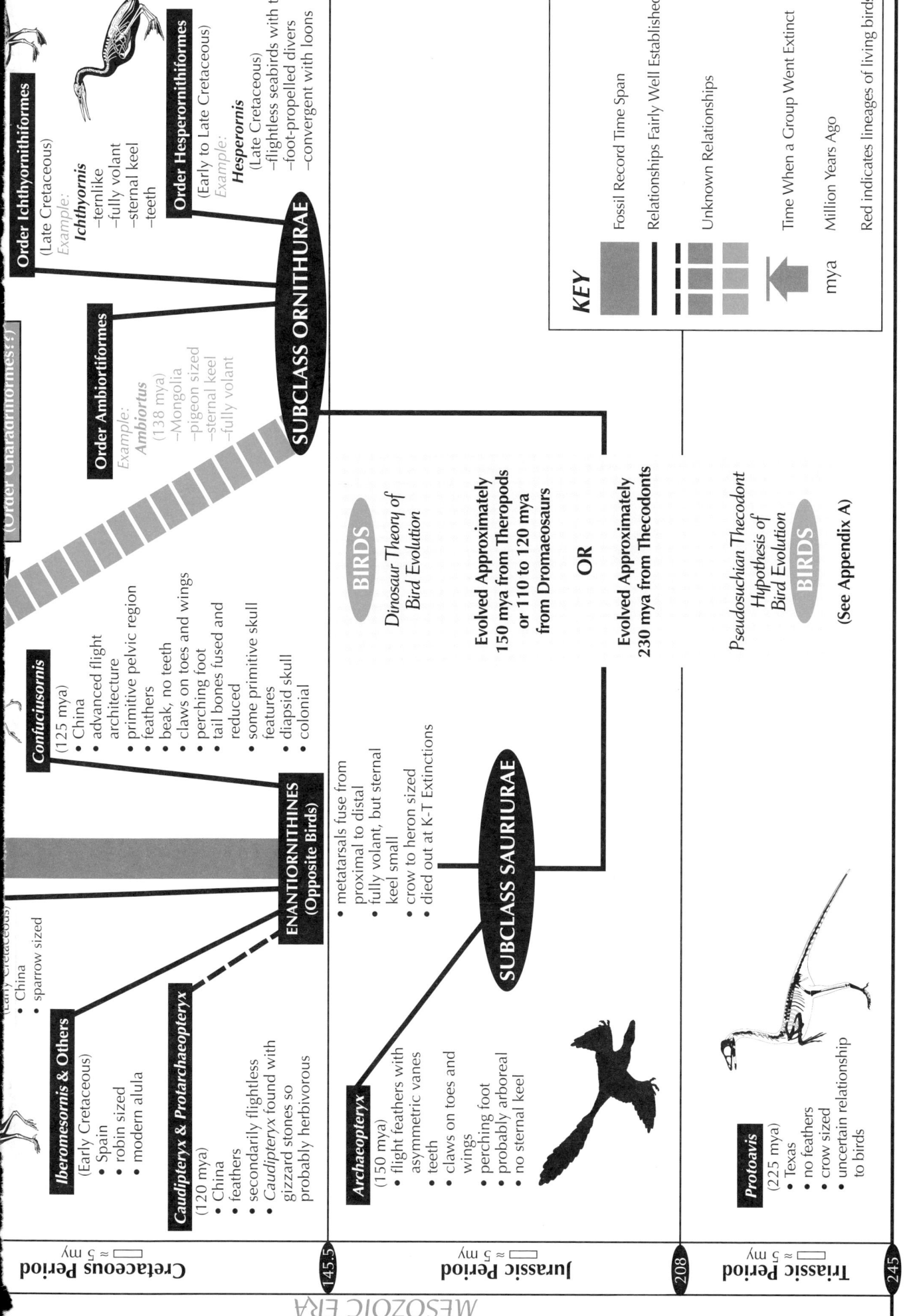

Appendix B:
HYPOTHESIZED RELATIONSHIPS AMONG ANCIENT AND MODERN BIRD GROUPS

The relationships among early fossils resembling birds and the groups of birds alive today are poorly understood. This diagram presents the most widely accepted ideas concerning these relationships, but our view of the origin of birds continues to evolve as more fossils are discovered. Note that the amount of space between one-million-year increments on the time scale is not the same in the different geologic periods, and that time from the Eocene to present is not shown to scale.

E

(Open)

E

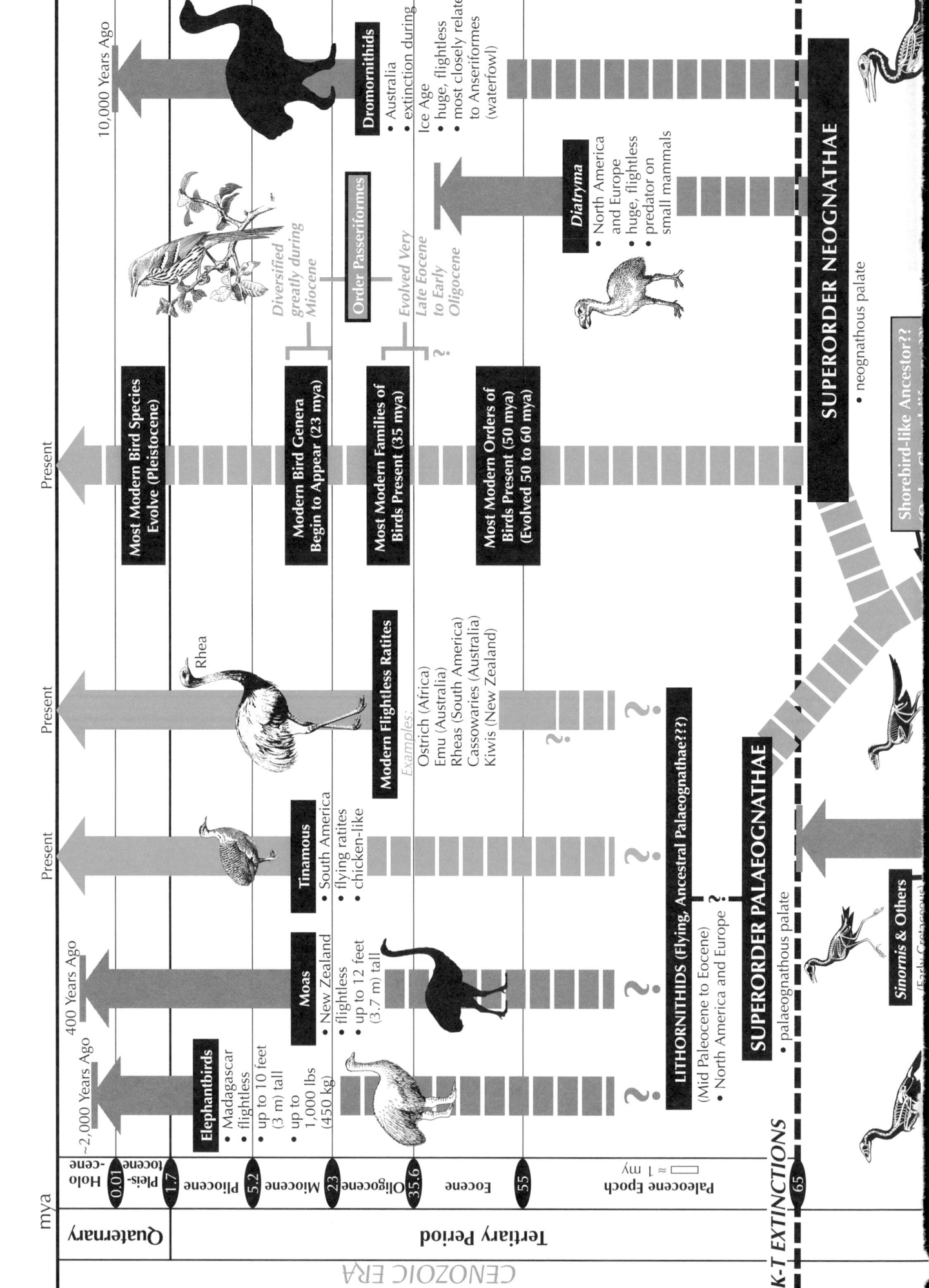

Appendix C:
INDEX TO FOSSIL ORGANISMS

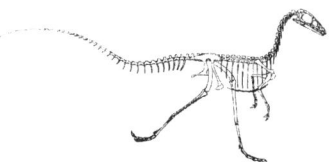

Archaeopteryx: a feathered reptile from 150-million-year-old Jurassic limestone deposits; it possessed a mosaic of bird and reptile characteristics. Since its discovery in the early 1860s, *Archaeopteryx* has been one of the most important and controversial fossils in paleontology, raising many questions about the origin of bird flight and the evolution of birds.

Compsognathus: a chicken-sized theropod dinosaur so closely resembling *Archaeopteryx* that for many years an *Archaeopteryx* fossil was misidentified as *Compsognathus*. Adding to the confusion, fossil remains of *Compsognathus* were recovered from the same German Solnhofen limestone deposits as *Archaeopteryx*.

Confuciusornis: a pigeon-sized fossil bird from 125-million-year-old Jurassic-Cretaceous boundary deposits in China. Currently grouped with the enantiornithine birds ("opposite birds"), *Confuciusornis* is a mosaic of primitive and advanced features. It is the earliest-known toothless bird and possessed contour feathers, indicating that it might have been endothermic.

Deinonychus: early Cretaceous (110-million-year-old) dromaeosaur discovered by John Ostrom in 1964. The name, meaning "terrible-claw," was derived from the large, curved claws on the feet, which were apparently used to rip open prey. Based on *Deinonychus* and similar organisms, Ostrom modified Huxley's dinosaur theory for the origin of birds, suggesting that birds descended not from the 150-million-year-old theropod *Compsognathus*, but from more recent theropods similar to *Deinonychus*.

Diatryma: a giant, flightless bird that inhabited the Northern Hemisphere during the Paleocene and Eocene epochs. *Diatryma* stood over 6.5 feet (2 m) tall and weighed about 385 lbs (175 kg).

Enantiornithine Birds: termed the "opposite" birds, their fossils were first discovered in Argentina, and are now known from around the world. This group of small to medium-sized birds flourished in the Cretaceous between 70 and 140 million years ago. They are called "opposite birds" because their metatarsals (the instep bones of humans) fuse to form part of the tarsometatarsus (see Fig. 1–11) from the proximal end to the distal end, a direction *opposite* to that of modern birds. Genera include ***Iberomesornis*** (upper left), ***Enantiornis*** (upper right), and ***Sinornis*** (bottom).

Euparkeria: a small, Triassic (230-million-year-old) thecodont that was discovered in South Africa by paleontologist Robert Broom and described in 1913. This quadrupedal dinosaur, which was tending toward bipedalism, appeared to be ancestral to both birds and dinosaurs. Robert Broom argued that birds and dinosaurs arose from a common ancestor (the pseudosuchian thecodont hypothesis) based on the characteristics of *Euparkeria*.

Hesperornis: a genus in the extinct order Hesperornithiformes—a group of toothed seabirds that thrived in the late Cretaceous. *Hesperornis* was a flightless, foot-propelled swimmer and diver that superficially resembled modern loons.

Ichthyornis: a genus in the extinct order Ichthyornithiformes. These toothed, ternlike birds were found in the same shallow sea deposits as *Hesperornis*. They were flying birds with a well-developed keel on their sternum and short backs and tails.

Ichthyosaur: the most specialized marine reptiles of the Mesozoic, reaching their peak diversity in the Jurassic. Their limbs were modified into paddles and many aspects of their body resembled those of modern porpoises.

Ornithischian or "Bird-hipped" Dinosaurs: one of the two major groups of dinosaurs. Called ornithischian because of the superficial resemblance of their hips to modern bird hips, ornithischian dinosaurs were highly specialized herbivores that included the duckbills (**Hadrosaurus,** upper left), armored ankylosaurs (**Ankylosaurus,** upper right), plated stegosaurs (**Stegosaurus,** lower left), and horned ceratopsians (**Styracosaurus,** lower right).

Plesiosaur: marine reptiles with their front and hind limbs modified into large paddles used to row the body through water, much like today's sea lions. They appear in the fossil record from the late Triassic to the Cretaceous.

Protoavis: a 225-million-year-old fossil discovered in Texas and described by Sankar Chatterjee in 1991. The fossil pre-dates *Archaeopteryx* by 75 million years and is thought by Chatterjee to be a closer relative to living birds than is *Archaeopteryx*. Because of the fragmentary nature of the fossil and the lack of evidence for feathers, its position in bird evolution remains uncertain.

Pterosaurs: flying reptiles from the Triassic that radiated and diversified in the Jurassic and Cretaceous. Although many pterosaur features, such as hollow bones and a slight keel on the sternum, were convergent with those of birds, their highly developed wings were structurally unique. Each of the large membranous wings was supported by a single, greatly elongated, fourth finger and attached to the side of the body and possibly also to the hind limb.

Saurischian or "Reptile-hipped" Dinosaurs: one of the two major groups of dinosaurs, the saurischian dinosaurs branch into two evolutionary lines, the herbivorous sauropods and the carnivorous theropods.

Sauropods: the herbivorous group of saurischian dinosaurs. It includes such giants as **_Plateosaurus_** (left), **_Diplodocus_** (center), and **_Camarasaurus_** (right).

Theropods: the carnivorous group of saurischian dinosaurs. Theropods are further divided into two groups, the carnosaurs and the coelurosaurs, distinguished by size and other skeletal features.

Carnosaurs: the group of larger theropods, including the familiar **_Tyrannosaurus_** (pictured here) and *Allosaurus*.

Coelurosaurs: the group of smaller theropods, including a bird mimic **_Struthiomimus_** (top)—a dinosaur from the late Cretaceous that is convergent with modern Ostriches (genus *Struthio*); the chicken-sized **_Compsognathus_** (lower left); **_Coelophysis_** (lower center)—a Triassic ceratosaur about 9 feet (3 m) long; and the dromaeosaurs, such as *Deinonychus* and **_Velociraptor_** (lower right).

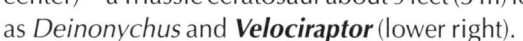

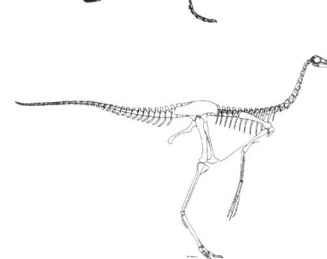

Thecodonts: also known as **basal archosaurs**, thecodonts are a diverse group of reptiles from the early Mesozoic. They are united by three key characteristics: a diapsid skull, teeth set in sockets (termed thecodont teeth), and an antorbital fenestra. Many paleontologists believe they are the common ancestor from which dinosaurs, crocodiles, pterosaurs, and birds evolved—a theory widely promoted by Gerhard Heilmann in the 1920s. The term **"pseudosuchian thecodont"** was used historically to refer to the thecodonts or basal archosaurs, most notably by Robert Broom in his "pseudosuchian thecodont hypothesis" for bird origins. *Euparkeria*, illustrated here, is a thecodont.

Figure Credits for Appendix C

***Confuciusornis, Iberomesornis, Sinornis, Enantiornis, Hesperornis, Ichthy-
ornis, Velociraptor,*** and ***Archaeopteryx*** from: Chatterjee, Sankar. *The Rise
of Birds: 225 Million Years of Evolution*, pp. 9, 96, 103, 110, 114, and 156.
Copyright 1997. The Johns Hopkins University Press.

***Deinonychus, Diatryma, Hadrosaurus, Ankylosaurus, Stegosaurus, Styra-
cosaurus, Plateosaurus, Diplodocus, Camarasaurus, Tyrannosaurus,*** and
Coelophysis from: Pough, F. H., C. M. Janis, and J. B. Heiser. *Vertebrate
Life, 5th Edition.* Copyright 1999 Prentice-Hall. Reproduced by permission
of Prentice-Hall, Inc.

Euparkeria from: Pough, F. H., J. B. Heiser, and W. N. McFarland. *Vertebrate
Life, 4th Edition.* Copyright 1996 Prentice-Hall. Reproduced by permission
of Prentice-Hall, Inc.

Plesiosaurs from: Watson, D. M. S. *Paleontology and Modern Biology.* Copyright 1951 Yale University Press.

Compsognathus from: *Origins of the Higher Groups of Tetrapods: Controversy
and Consensus,* edited by Hans-Peter Schultz and Linda Trueb. Copyright
1991 by Cornell University. Used by permission of the publisher, Cornell
University Press.

Protoavis from: Chatterjee, Sankar. 1991. *Philosophical Transactions of the
Royal Society of London B,* Vol. 332, pp. 227–342. Used by permission of
the Royal Society of London and S. Chatterjee.

Struthiomimus from: Russell, D. A. 1972. Ostrich dinosaurs from the late
Cretaceous of Western Canada. *Canadian Journal of Earth Sciences*
9:375–402. Courtesy of Dale Russell.

6

Understanding Bird Behavior

John Alcock

 Millions of people love to identify birds, but few get the opportunity to study and experience the pleasure of understanding bird behavior. Everything birds do—from selecting nest sites and detecting elusive prey to the squabbles among Black-capped Chickadees at bird feeders and the migrations of massive flocks of shorebirds into and out of the High Arctic—raises fascinating and challenging questions. Does a robin hear the worms it captures, or see them? How can Yellow Warblers be so naive as to devote themselves to the care and nourishment of parasitic cowbird fledglings twice their size? Do Clark's Nutcrackers remember where they stored hundreds of caches of pinyon nuts on an Arizona mountainside? If so, how do they do it? Why does a male Greater Prairie-Chicken spend hours stamping around in circles on a little patch of prairie in the springtime? Why do bands of Mexican Jays defend communal territories, even though only two members of the group may get to breed there in a given year? Do gannets and other seabirds breed in large colonies to protect themselves from predators, or do these groups assemble for other reasons and actually attract more predators?

We can ask endless questions of this sort, and behavioral biologists have begun to explore some of them. This chapter focuses on some of the intriguing puzzles of bird behavior and how ornithologists have tried to solve them.

Questions About Behavior

In organizing our thinking about bird behavior, we can start by recognizing that all questions about behavior can be placed in just two fundamentally different but complementary categories. Let's illustrate this point by thinking about the many questions raised by a robin hunting worms, which it does by hopping across a lawn, stopping to cock its head to one side, and sometimes stabbing its beak downward, coming up with a worm that it pulls steadily out of its burrow (**Fig. 6–1**). How do robins locate worms? Can they see subtle visual cues made by worms moving at or near the surface of the ground? Or do they hear worms sliding through their underground tunnels? How does the organization of the robin's nervous system help it to stab accurately? Is the robin born with the ability to identify worms as a delectable food? Are certain robin genes key in developing worm-hunting behavior? Do robins get better at worm hunting as they grow older, learning from experience where to hunt, how to detect a victim, and how to grab it so that it doesn't get away?

Figure 6–1. American Robin Capturing an Earthworm: A foraging robin hopping across a lawn stops, cocks its head to one side, and stabs downward with its beak, sometimes coming up with a worm, which it pulls steadily up out of its burrow, as shown here. Even an act as simple as this raises many questions about behavior, but they all can be put into two major categories—proximate and ultimate. See text for descriptions of these categories. Photo by A. Carey/ VIREO.

Every one of these questions deals with factors *within* robins, with the design of the birds' physiological systems, which enable them to hunt worms in a particular way. In the jargon of behavioral biology, these questions are labeled **proximate** because they are concerned with the immediate, internal causes of the robin's response to worms. To use an automobile analogy, a proximate question would ask how the car's engine causes the car to move down the highway. With animals, proximate questions require investigations into how the internal living "machinery" of the animal operates to give it certain special abilities.

But let's say that we eventually figured out everything one could know about how the robin's genes, developmental processes, and nervous system all function to help it hunt, kill, and eat worms. Our curiosity would not be completely satisfied because an entire battery of unanswered questions remains, questions about *why* the robin possessed these genes and not others, *why* it had its own particular array of visual and acoustical abilities, and *why* it preferred some foods over others. For example, why are robins worm-hunting experts and not seed-eaters? What reproductive advantage does the worm-eating robin gain from its behavior? In other words, why has it *evolved* to behave like a robin and not like an Eastern Towhee or a cardinal? These questions require a long-term, evolutionary perspective that examines how the history of the robin species has shaped its abilities and attributes. In the lingo of behavioral biologists, these kinds of questions are labelled **evolutionary** or **ultimate**. To return to our car analogy, evolutionary questions would ask about the long history behind a car's design, which would help us to understand the transformation of a horse-drawn carriage to a steam-driven truck, and eventually to a modern sedan with an internal combustion engine.

Because proximate questions ask how things work in the immediate sense, whereas evolutionary questions ask how they got that way historically, the two kinds of questions (and their answers) complement each other. To have a complete picture of any animal's behavior, we must understand its internal operating rules *and* the forces that led species over time to evolve the special proximate mechanisms found in living individuals today.

Figure 6–2. Common Yellowthroat Feeding a Brown-headed Cowbird Fledgling: Why does this male Common Yellowthroat feed a fledgling that is so obviously (to us) bigger than itself, and thus not its own young? Many possible explanations exist, on both the proximate and ultimate levels—and each of these points of view contributes significantly to our understanding of bird behavior. Photo courtesy of John Gavin/CLO.

To reinforce this point, consider some possible proximate reasons why a pair of Yellow Warblers or other small songbirds "permit" themselves to be exploited by a cowbird fledgling **(Fig. 6–2)**. Maybe the young cowbird has a certain appearance or makes certain sounds that stimulate a parental response from the warbler. Or maybe warbler genes somehow guide the development of parental abilities that cowbirds exploit. These are proximate possibilities because they deal with what goes on inside a Yellow Warbler that might cause it to react to nestling cowbirds in a particular way.

On the evolutionary level, we could ask why cowbird "adoption" has persisted from generation to generation in Yellow Warblers. Perhaps today's warblers received the hereditary basis for cowbird adoption from their ancestors. It could be that the behaviors of warblers caring for cowbirds are so helpful for warblers rearing their *own* offspring that they have been maintained over evolutionary time, even though they sometimes lead to victimization by cowbirds.

Knowing that immediate causes for individual actions and long-term evolutionary causes for species' characteristics both exist keeps us from wasting time in mistaken arguments. Imagine, for example, someone contending that a White-throated Sparrow's drive to defend its territory is caused by high levels of the hormone testosterone in its bloodstream. If another person said, "No, no. The sparrow is territorial because it belongs to a species in which territorial individuals reproduced more than nonterritorial ones in the past," the debate would be needless. The hormone hypothesis is proximate because it has to do with the immediate internal causes of the bird's territoriality; the second hypothesis is evolutionary because it seeks to explain territorial drive as the long-term outcome of historical processes within the species. The two explanations are not mutually exclusive; they both could be right, in which case each would contribute to our understanding of the behavior.

Another way to look at this issue is to recognize that evolutionary

6

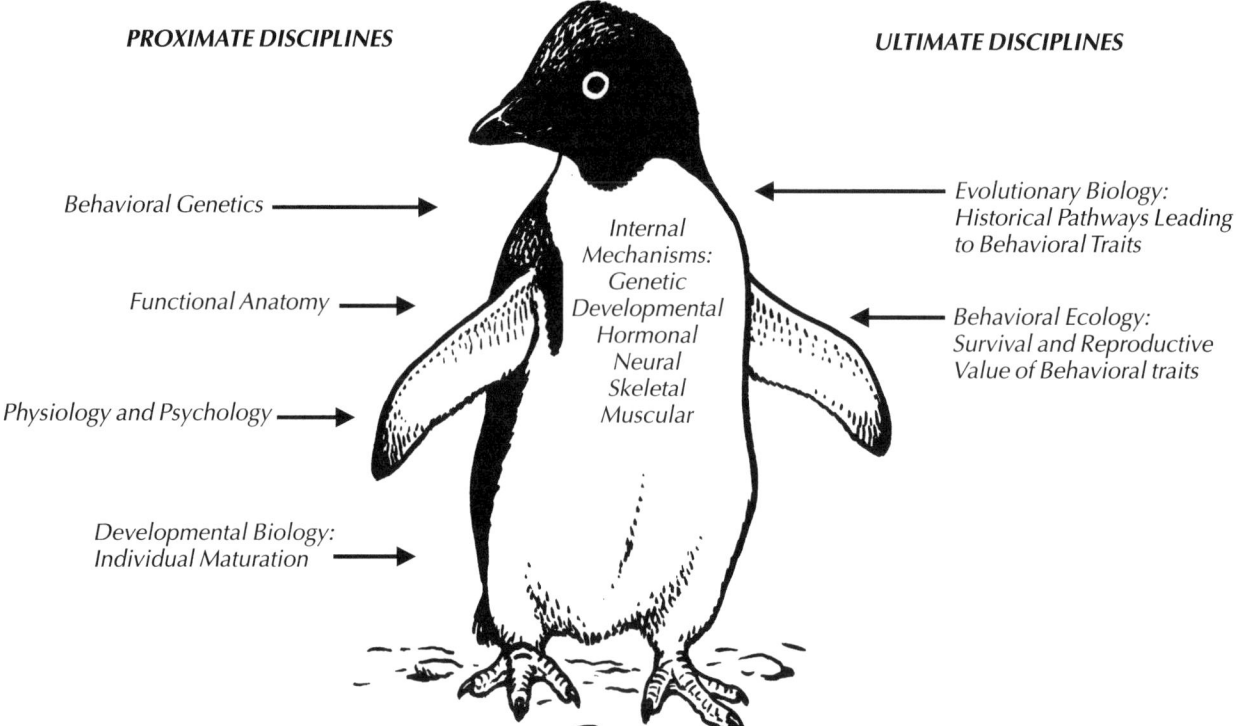

PROXIMATE DISCIPLINES   **ULTIMATE DISCIPLINES**

Behavioral Genetics

Functional Anatomy

Physiology and Psychology

Developmental Biology:
Individual Maturation

Internal
Mechanisms:
Genetic
Developmental
Hormonal
Neural
Skeletal
Muscular

Evolutionary Biology:
Historical Pathways Leading
to Behavioral Traits

Behavioral Ecology:
Survival and Reproductive
Value of Behavioral traits

Figure 6–3. Proximate Versus Ultimate Disciplines: The internal mechanisms underlying behavior (proximate mechanisms) are investigated by different biological disciplines than those that focus on the evolutionary basis of behavioral traits (ultimate mechanisms).

explanations about behavior can tell us why a particular proximate mechanism, such as a hormonal system or neural network, happened to spread through populations in the past. This past history of a species determines its current internal machinery, and thus in this sense, the two levels of explanation in biology are linked together.

The Proximate Basis of Bird Behavior

◼ The fact that proximate and ultimate approaches complement each other is reflected in the different disciplines that explore these two levels of research in behavioral biology **(Fig. 6–3)**. Remember that both approaches are required for a total picture of the causes of any behavioral trait.

To illustrate some interesting proximate puzzles about bird behavior, consider the many seemingly "unintelligent" things that birds do. In Australia, I was amused on several occasions to see a Willie-wagtail, a small black-and-white bird, assaulting his image in the side-view mirror of my van. The bird fluttered in front of the mirror, launching peck after peck at the reflective surface before flying up to land on it. He would remain calm for a moment, but if he bent down far enough to catch a glimpse of himself, he went off in another paroxysm of senseless rage, even though the mirror did not respond like a live rival wagtail.

Similarly, Konrad Lorenz (1952) reports in his wonderful book, *King Solomon's Ring* (see Suggested Readings), that a band of jackdaws, members of the crow family, went berserk when they saw him carrying a pair of black bathing trunks in his hand. Although these birds normally seemed to trust him thoroughly, on this occasion the whole flock attacked him, filling the air with alarm cries. How can jackdaws,

Willie-wagtails, and many other birds be so competent in so many aspects of their behavior, yet make simple mistakes of this sort?

A proximate explanation for these "errors" is that the bird's nervous system is designed to activate specific responses whenever the bird detects certain simple cues in its environment. For a wagtail attacking a car mirror or a cardinal throwing itself at its reflection in a window, the key cues are apparently visual. Give a territorial male wagtail the right set of color cues and he switches into territorial defense mode. Let a jackdaw see a black, dangling, jackdaw-sized object, and it activates the "attack the predator" response, even when it is familiar with the animal holding the object.

Therefore, if we see a pair of Yellow Warblers or some other small passerines feeding a big brood parasite such as a cowbird, one proximate explanation is that the cowbird provides the special visual and acoustical cues that "turn on" the feeding response in parental warblers and the like. These cues are examples of **releasers**, specific objects, physical features, or behaviors that activate (or "release") a specific response in an individual. They also are present in young Yellow Warblers, and usually stimulate parent birds to feed large, healthy young of their own **(Fig. 6–4)**.

This **hypothesis**, or tentative explanation, is based on the assumption that as a Yellow Warbler matures, development of its nervous system is controlled by an interaction between the bird's genes and its environment. (The term "environment" is used here to embrace everything from the foods the bird eats, which provide the molecular building blocks for its growing nervous system, to the bird's various sensory experiences, some of which can influence the development of certain neural circuits.) As a result, a Yellow Warbler's nervous system has particular components designed to detect key stimuli (for example, the begging calls of nestlings) and activate specific responses to those stimuli (for example, placing food in the mouth of the begging bird).

The "releaser-activated response" hypothesis has been tested for some birds in the following way. If the hypothesis is correct, it follows logically that when an experimenter presents the right stimuli, the bird will produce the specialized response. For example, Lars von Haartman (1953) predicted that the begging calls of the European Pied Flycatcher were releasers that caused the parents to feed the young. To test his prediction, he placed several hungry nestling flycatchers out of sight behind the wall of a nest box that held a pair of parent birds and their offspring. The parents fed their own brood until they were full and had stopped calling. But the hungry hidden youngsters continued to beg, and the adults continued to bring food to their silent, gorged offspring. Clearly, the acoustical cue provided by begging youngsters stimulated parental feeding behavior, a conclusion supported by recent additional work with the European Pied Flycatcher (Ottosson et al. 1997).

The next question is whether the flycatcher's response to the begging calls of nestlings is **instinctive.** An instinctive (or "innate") behavior is one that is triggered in full form, without any learning, the first time an individual responds to the releaser. A human infant, for

Figure 6–4. Begging Fledgling and Adult Northern Cardinal: *Begging birds, whether juvenile or adult, provide visual and acoustic cues that stimulate feeding by another individual. These cues are examples of* **releasers**. *a. Begging Fledgling: A Northern Cardinal fledgling (left), begging to be fed by its parent, crouches open-billed, fluttering its wings and calling persistently. Begging postures, releasers for adult feeding of young, are so similar among birds that sometimes adults may be stimulated to feed young of the "wrong" species (see Fig. 8–103). A range of stimuli may trigger begging: calls from the parents, sight of a parent or other object above the nest, vibrations of the nest (as when a parent lands), darkening of the nest hole (in cavity nesters, as when an adult blocks incoming light by appearing at the nest entrance), or even air currents (in hummingbirds and Chimney Swifts, as from the motion of adults arriving at the nest). Photo by Marie Read.* **b. Courtship Feeding:** *An adult female Northern Cardinal begs from her mate during courtship. In many birds the female begs to her mate using postures and calls similar to those of begging young, stimulating him to feed her. The function of this ritualized* **courtship feeding**, *which occurs only during courtship and incubation, remains a matter of some speculation. Traditionally, researchers have thought courtship feeding in many species strengthened the pair bond, rather than provided nutrients to the female—especially because some females beg at bird feeders or when their bills are full of food. But increasing evidence suggests that the food may, indeed, improve the female's condition (see Fig. 8–100). Furthermore, in some species the male's competence in courtship feeding may allow the female to gauge his ability to provide food for the young—if he's not good enough, she may reject him as a mate! In Common Terns, for example, males who bring more food during courtship are better at feeding their young later in the season.*

a. Begging Fledgling

b. Courtship Feeding

example, will smile instinctively at another human face (or even at a releaser as simple as two dark spots drawn on a white circle), and will grasp tightly with its hand when it feels something touching its palm. To determine whether the flycatcher's response is instinctive, it would be helpful to perform von Haartman's experiment on adults nesting for the first time. If inexperienced parents fed their young when they heard begging calls, one could claim with some confidence that these vocalizations were an innate releaser of parental feeding behavior in this species.

If a begging call also prompts instinctive feeding in Yellow Warblers, the feeding response of adults hosting a cowbird can be explained in part, at the proximate level, as an innate reaction to the begging calls of the nestling cowbird, which resemble those of nestling warblers closely enough to release parental feeding behavior. Later we will explore the ultimate bases of the intriguing relationship between cowbirds and their hosts.

Ethology, Ornithology, and Instincts

The study of instincts in birds and other animals was of particular interest to the early **ethologists**, a group of European biologists who, beginning in the 1930s, developed the scientific study of animal behavior into a flourishing discipline. Unlike American psychologists interested in animal behavior, who focused primarily on laboratory studies of domesticated rats and mice, the ethologists emphasized field studies of species ranging from insects to free-ranging birds. Niko Tinbergen and Konrad Lorenz pioneered the field, and described their research in some wonderfully written books for the public (see Suggested Readings). The importance of their work was eventually recognized when they received the Nobel Prize in Medicine in 1973.

Although the study of animal behavior has changed a great deal since the era of Tinbergen and Lorenz, their tests of the instinct hypothesis are still admirably instructive and useful. For example, Tinbergen and his co-workers proposed that the gaping behavior of hungry nestling Eurasian Blackbirds was released by very simple stimuli. Through a series of experiments with eight-day-old birds whose eyes had just opened, Tinbergen established that *any* object that (1) moved, (2) had a vaguely bill-shaped projection at least 0.12 inches (3 mm) wide, and (3) was above the head of the nestling, would elicit the begging response, even if the object had a most imperfect resemblance to a living, three-dimensional adult (Tinbergen 1951) **(Fig. 6–5)**.

Likewise, Tinbergen and Lorenz showed that an incubating adult Greylag Goose would retrieve almost any round object, even one vastly larger than a Greylag egg, that was placed outside its nest. It did so by extending its neck, placing the bill on the outer side of the "egg," and rolling it carefully back into the nest.

Not Moving → No Begging

Moving Above Nestling's Head → Begging

Moving Below Nestling's Head → No Begging

Figure 6–5. Characteristics of Begging Releaser in Nestling Eurasian Blackbirds: Tinbergen and his co-workers presented objects of various shapes to nestling Eurasian Blackbirds whose eyes had just opened, in an experiment to determine which characteristics of the objects were necessary to activate the nestling begging response. They discovered that any object that moved, had an approximately bill-shaped and bill-sized projection, and was above the head of the nestling, would elicit begging (Tinbergen 1951).

In these and other cases in which the instinct hypothesis was confirmed, the ethologists demonstrated that the animal was responding to only a fraction of the available stimuli, and that simple cues (releasers) seemed to turn on a particular response. The ethologists called this response a **fixed action pattern (FAP)**. FAPs are behaviors that are played out in complete form even the first time the animal encounters and reacts to the releasing stimulus. Baby thrushes do not need to learn through experience what objects to beg from. Likewise, right from the start, fledgling Turquoise-browed Motmots do not need to handle coral snakes to learn to avoid this potentially lethal "prey;" motmots have an innate avoidance response to long, thin objects with the coral snake color pattern (Smith 1975) **(Fig. 6–6)**.

Learned Behavior

Although many bird behaviors appear to be programmed reactions to releaser stimuli, many other responses seem to be learned. Learning can be defined as behavior modification resulting from particular experiences. For instance, captive Blue Jays will sample a great

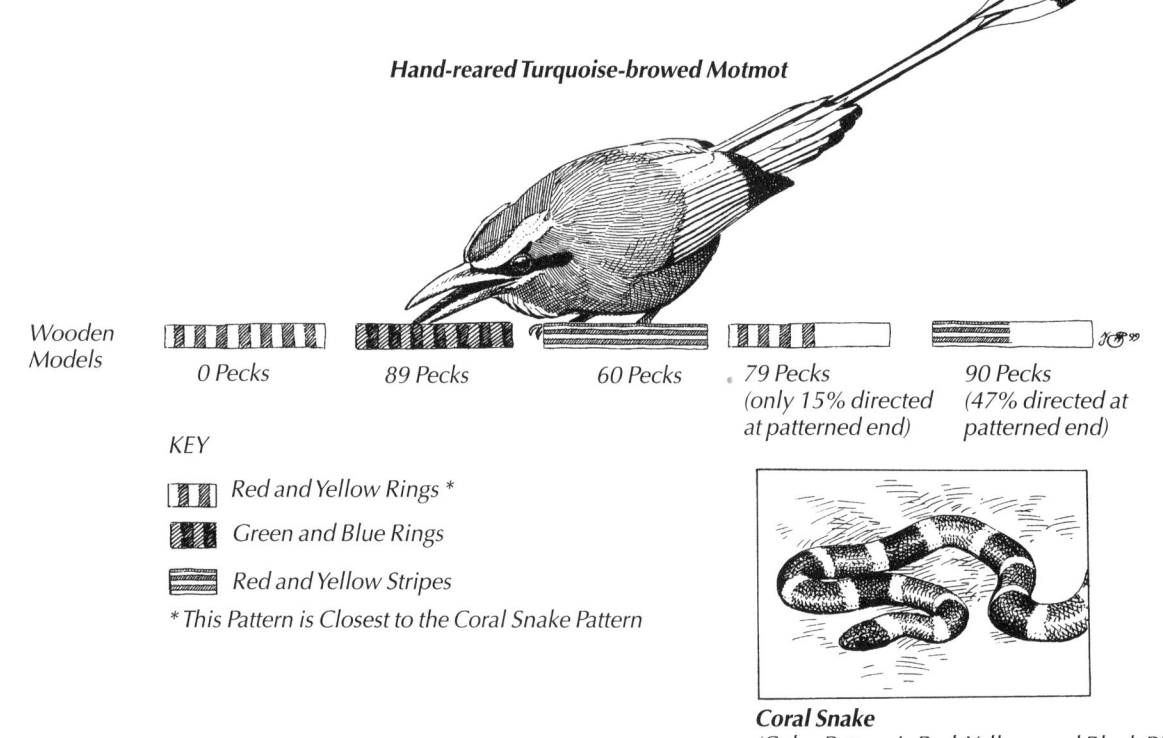

Hand-reared Turquoise-browed Motmot

Wooden Models

0 Pecks   89 Pecks   60 Pecks   79 Pecks   90 Pecks
                                                     (only 15% directed   (47% directed at
                                                      at patterned end)   patterned end)

KEY

Red and Yellow Rings *

Green and Blue Rings

Red and Yellow Stripes

* This Pattern is Closest to the Coral Snake Pattern

Coral Snake
(Color Pattern is Red, Yellow, and Black Rings)

Figure 6–6. Turquoise-browed Motmot with Snake Models: Motmots, spatulate-tailed relatives of kingfishers, inhabit Central and South America where snakes make up part of their natural diet. Hand-reared young Turquoise-browed Motmots in a laboratory setting were presented with thin, wooden models about 3.2 inches (8 cm) long, painted in various colors and patterns. Curious and exploratory birds, they investigated these potential food items by pecking at them. Yet despite having had no previous experience with snakes, the laboratory motmots specifically avoided the model with red and yellow rings, the one that most closely resembled a coral snake (which actually has red, yellow, and black rings), giving it no pecks, although they pecked numerous times at all the other models with different color and pattern combinations. Notice, however, the model with one end ringed red and yellow and the other plain: the motmots pecked much less frequently at the ringed end. These results suggest that the Turquoise-browed Motmot has an innate avoidance response to objects resembling the poisonous coral snake (Smith, 1975).

many potential food items, but after eating a single poisonous monarch butterfly, and vomiting the noxious prey, they learn to touch them no more (Brower 1969) **(Fig. 6–7)**. Likewise, fledgling male White-crowned Sparrows learn their species' full song by listening to adult males. Prevent a young male from hearing his species' song and he never will sing a typical song when he becomes an adult (see Ch. 7, Vocal Development in Songbirds).

Note that to learn *anything*, a Blue Jay or White-crowned Sparrow must have an innate capacity to develop the circuits where the bird stores information from certain experiences, information that it later uses to modify its behavior. Thus, the distinction between instincts and learning is not that one depends on genes and the other does not, or that one requires special features within the nervous system and the other does not. Indeed, they both require genetic information, and they both occur because of special elements within the nervous system. But instincts (FAPs) depend on neural components that can recognize key releaser stimuli (such as the cues provided by a begging nestling) and turn on an "automatic" response. Learning, on the other hand, depends on neural units that store information from certain experiences (such as the nausea associated with eating monarch butterflies), enabling the animal to modify its behavior on the basis of that stored information **(Table 6–1)**.

Figure 6–7. Captive Blue Jay Learns to Avoid Poisonous Prey: Captive Blue Jays will sample any likely food item including the monarch butterfly, which contains toxic and distasteful substances. Eating a monarch causes the Blue Jay to vomit soon after its meal, and from this single negative experience the jay immediately learns to avoid this butterfly species forever. Photos by Lincoln P. Brower.

Table 6–1. Instinctive Versus Learned Behavior

	Instinctive Behavior	Learned Behavior
Development of Response	Responds fully the first time a relevant stimulus is encountered.	Requires previous exposure to stimulus before responding fully.
Requirement for Response	Requires key releaser stimulus.	Requires stimulus and memory of past experience.
Physiology of Response	Ability to recognize stimulus and respond.	Ability to assimilate and store information, and to respond.
Human Example	Baby grasps when pressure is applied to palm of hand.	Baby learns to reach out and grasp an object it wants.
Avian Example	Nestlings beg for food when adults land on edge of nest.	Young birds learn to avoid non-palatable prey, such as monarch butterflies, through trial and error.

The simplest type of learning is habituation, the permanent loss of a response as a result of repeated stimulation without reward or punishment—in other words, learning to ignore unimportant stimuli. Even young birds in the nest learn by habituation. If you tap lightly on the side of a nest when the young still have their eyes closed, they raise their heads and open their mouths wide: in their small world, a light tap means food. If you repeat the tapping without giving the young any food, they eventually stop responding; they have become habituated to the stimulus. Birds that routinely feed along the side of a road habituate to passing cars by learning not to fly up each time a car goes by. Habituation is clearly adaptive because it saves time and energy, and allows an individual to concentrate on more important stimuli. After putting on your clothes, you quickly habituate to the feel of their touch. If you were aware of each brush of fabric against your skin, you would have trouble noticing other, more important stimuli around you **(Fig. 6–8)**.

Another type of learning is trial and error, in which a bird learns to associate its own behaviors with a reward or punishment. Behaviors that result in reward are repeated, and others are abandoned. Just as captive Blue Jays learn to avoid eating monarch butterflies, the diets of recent fledglings are influenced by trial-and-error learning. While a fledgling still depends on its parents for food, it moves about pecking at small objects—pebbles, leaves, anything that contrasts with the background **(Fig. 6–9)**. Eventually it finds an insect and eats it. Once the bird has tasted the insect reward, it pecks more carefully, soon associating a food reward with a certain class of objects. Over time it learns what to peck, what not to peck, and which items to eat. Birds learn to visit bird feeders and eat foods they might never see in the wild through trial and error. Hummingbirds learn that the color red usually provides profuse nectar (since the flowers that have evolved to attract and reward them as pollinators are most often red), so they readily visit hummingbird feeders adorned with red. Tool use **(Fig. 6–10)** probably results from a type of learning similar to trial and error.

Figure 6–8. Turkey Chicks Habituate to Moving Objects Overhead: a. Young turkey chicks crouch in response to all objects moving overhead, from falling leaves to large flying birds. b. Older, more experienced chicks soon habituate to falling leaves and cease crouching whenever one flutters down. Because large flying birds are rare and may represent real danger, young turkeys never habituate to them and continue to respond by crouching. Drawing by Charles L. Ripper.

Figure 6–9. American Robin Fledgling Pecking at a Leaf: Newly fledged robins develop their diet through trial-and-error learning, pecking at any items that stand out from the background, such as leaves and pebbles. When they eventually find something edible, they associate the food reward with the item's characteristics, and over time refine their search for food to certain types of items.

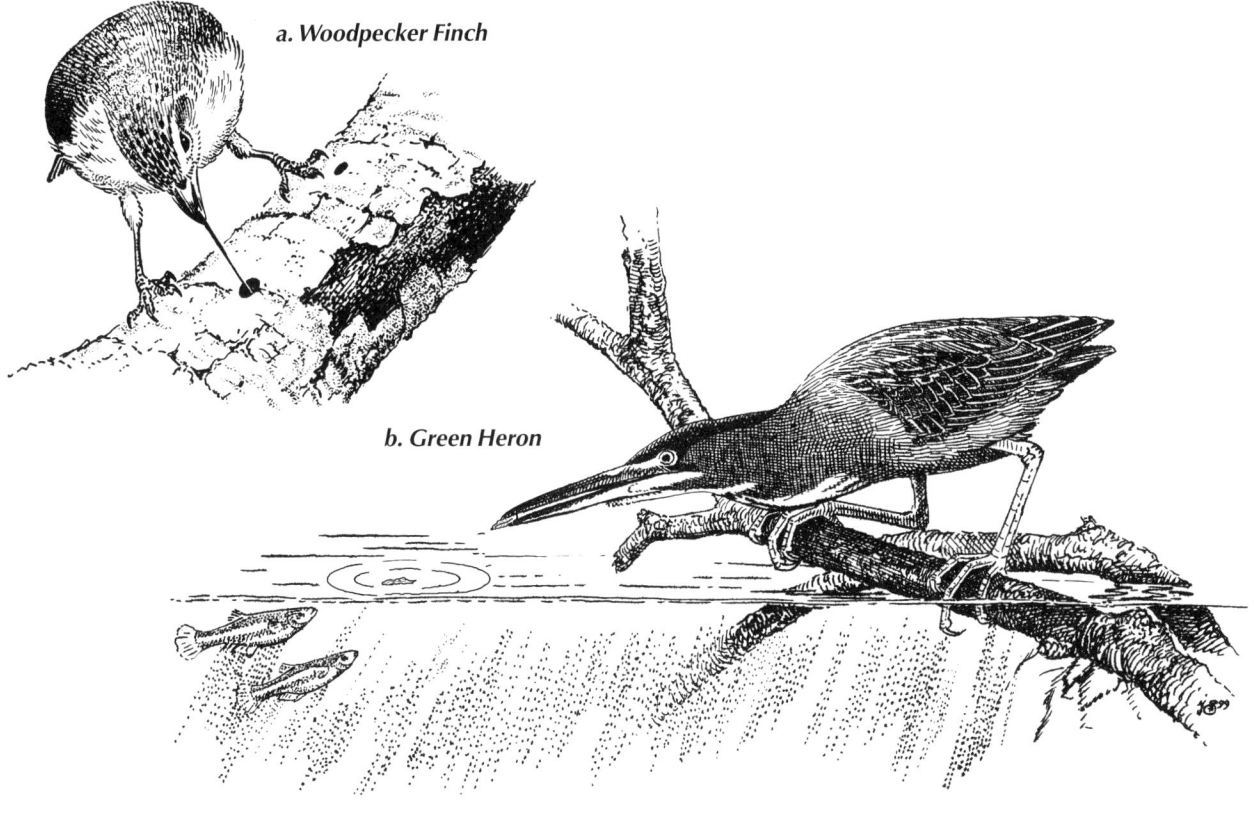

a. Woodpecker Finch

b. Green Heron

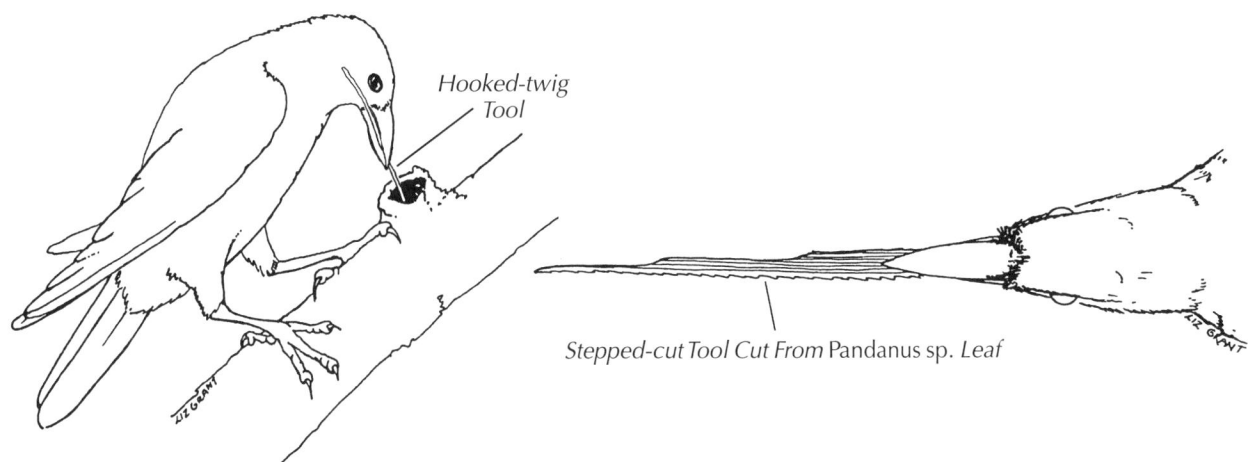

c. New Caledonian Crow

Hooked-twig
Tool

Stepped-cut Tool Cut From Pandanus sp. Leaf

Figure 6–10. Tool Use in Birds: *Using tools was once considered an exclusively human trait, but we now know that over 30 species of birds as well as some chimpanzees use tools. One of the Galapagos finches, the Woodpecker Finch, picks up a cactus spine and uses it as a tool for extracting insects and larvae from holes in trees (a). Similarly, the Brown-headed Nuthatch of the eastern United States uses pieces of bark to probe into holes for food. Some herons, such as the Green Heron (b), have learned to drop insects, leaves, berries, twigs, feathers, or whatever else they can find into the water to lure fish within striking distance (Higuchi 1987). And the Egyptian Vulture actually picks up large stones in its bill and throws them at Ostrich eggs to break the shell and get to the tasty insides.*

More recently, people pointed to the making of standardized tools as a uniquely human skill, but the newly discovered abilities of the New Caledonian Crow are putting even that notion to rest. Hunt (1996) discovered that these crows, from the South Pacific islands of New Caledonia, make and use two distinctly different types of tools (c). One, a **hooked-twig tool**, *is made by stripping the leaves and usually bark from a living twig, leaving a small portion of intersecting twig to form a distinct hook at one end. The other, a* **stepped-cut tool**, *is made by biting a long, jagged piece from the toothed leaves of the Pandanus plant to form a tapered, spined, probe. Both are used to pull insects and worms out of holes in trees and dead wood. The crows carry their tools from site to site as they search for food, and may even stash them in a safe place for a while, returning later to use them again. Drawing a by Robert Gillmor; drawings in c by Liz Grant/Massey University.*

*Figure 6–11. Greylag Goose Goslings Imprinted on Konrad Lorenz: Young birds rapidly learn to follow the first moving object they encounter in their post-hatching experience—an example of the special form of early learning known as **imprinting**. Usually, of course, this object is the parent bird, but if a human takes its place, the young birds will imprint upon the human and follow him or her as they would their parent, as ethologist Konrad Lorenz discovered in his classic experiments with Greylag Geese. Photo by Thomas McAvoy/Life Magazine, copyright Time Inc.*

Learning plays a role in many different aspects of bird behavior, from the trial and error that shapes the diet of some species, to the special effect of acoustical experience on singing, and to the effects of **imprinting**, a special form of early learning in which a young animal quickly acquires specific information for certain experiences. For example, Konrad Lorenz's famous experiments with Greylag Geese demonstrated that recently hatched goslings rapidly learned what to follow on the basis of certain experiences in their first few hours of life outside the egg (Lorenz 1970). Usually, of course, goslings see and hear their mother; when she moves off the nest, the youngsters follow, having imprinted on this moving, calling object. If Lorenz took the place of a parent goose at that time, however, the youngsters walked after him, learning through their special experience with Lorenz that he was the individual to associate with and follow (**Fig. 6–11**).

Even more amazing, when the geese in these experiments reached sexual maturity, the males courted human beings rather than members of their own species. Here the particular experiences that goslings had while following a moving companion led to imprinting on a parental figure *and* to sexual imprinting, with early experience determining their adult mate preferences.

Another form of specialized learning has been explored by Russell Balda and Alan Kamil (1992) in their studies of **spatial learning** (learning the location of objects in space) by Clark's Nutcrackers (**Fig. 6–12**). In nature these birds are unusual in caching supplies of pinyon pine nuts in holes in the ground, to which they may return months later. Do the birds actually remember where they hid food, or do they just poke around in likely places for seeds that they or some other birds might have put away for a hungry day? Testing these alternatives would be difficult in the field, but laboratory experiments can test whether the birds learn precisely where they hid their food caches.

Balda and Kamil built a "food-caching room" with 330 holes in the plywood floor scattered among tree limbs, rocks, and other potential landmarks (**Fig. 6–13**). The holes held sand-filled cups in which

Figure 6–12. Captive Clark's Nutcracker Caches a Pine Seed in a Sand-filled Cup: In the wild this species caches pinyon nuts and other seeds in holes in the ground, depending on these stores as a major source of food in the winter. The fact that individuals seem to remember the location of their sites has led to experiments investigating their spatial memory capabilities. Photo courtesy of Russell P. Balda.

the birds hid pine seeds taken from a feeder in the room. After letting a number of birds cache seeds in about 20 cups, the nutcrackers were ushered out of the room into holding cages elsewhere. Then, between 11 and 285 days (more than nine months!) later, the nutcrackers were allowed to reenter the room when they were hungry. The 20 or so cups in which they had cached food each contained one pine seed; the 300-plus cups in which they had not cached food contained only sand. If the birds remembered where they had stored their caches, they should have been able to find the hidden seeds more quickly than if they hunted randomly through the hundreds of sand-filled cups.

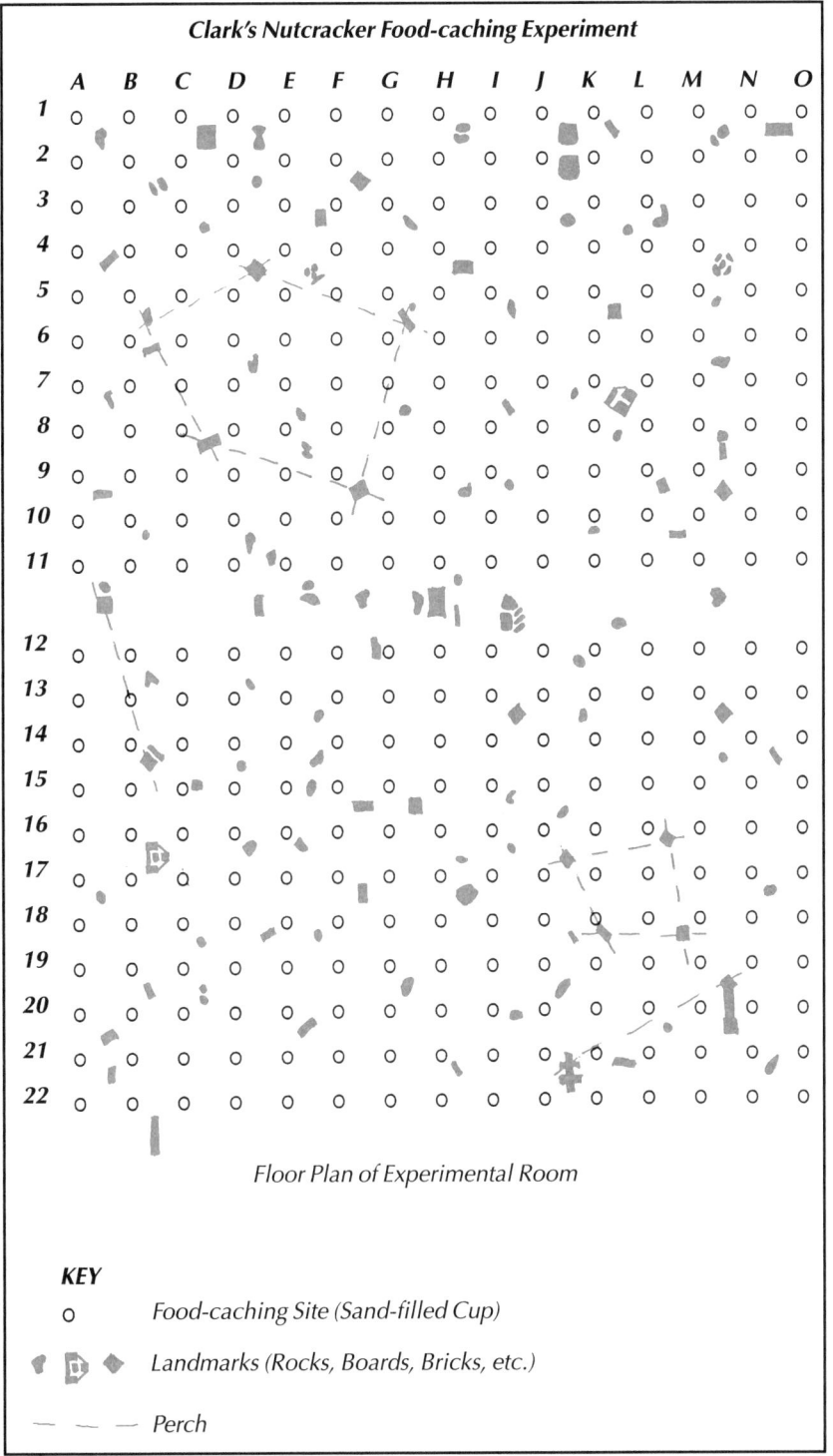

Figure 6–13. Layout of "Food-caching" Room: *Balda and Kamil (1992) built an experimental room to test the extent of long-term spatial memory in the Clark's Nutcracker. 330 holes in the plywood floor held sand-filled cups. The room contained numerous potential land-marks—rocks, boards, and bricks—scattered around the floor, as well as perches, and a feeder filled with pine seeds. Individual nutcrackers were set loose in the room and permitted to cache seeds in the sand-filled cups of their choice. Each bird was then removed for lengths of time varying from 11 to 285 days before being brought back to the room (now feederless) in which the bird's cached seeds remained. The birds found seeds with far fewer probings of the cups than expected by chance alone, suggesting that they remembered where they had hidden their food even after several months had elapsed. Diagram courtesy of Russell P. Balda.*

Figure 6–14. Captive Common Raven Retrieves a Suspended Piece of Meat: When Bernd Heinrich (1995) tied a chunk of meat to the end of a long string and dangled it from a perch in his aviary, some Common Ravens were able to retrieve the piece of meat in the following way: they reached down and grasped the string in their beak, pulled the string up, and then clamped the retrieved section of string to the perch with their foot, while retrieving another section of string. They repeated this sequence until the meat was within reach of their bill. The fact that the ravens were able to solve this complex problem on their first try, with no similar previous experience, suggests that they studied the situation and used insight to solve the problem, rather than trial and error. Photos courtesy of Bernd Heinrich.

When tested, most of the nutcrackers did indeed find seeds with far fewer probing inspections of cups than expected by chance alone. For example, "Karl" investigated 69 cups, relocating 15 of those he had used to make caches six months previously; had he been hunting randomly, he would have been expected to find just four of his caches in 69 tries. The shorter the interval between caching and recovery, the better the birds usually did, but even when many months had elapsed, the nutcrackers used their long-term spatial memory to relocate their cache sites remarkably well.

Finally, we can ask if some birds possess what has been called **insight learning**, modification of behavior that occurs *without* previous experience with a particular problem. Although humans can occasionally think up solutions to novel problems (how to build a certain machine, how to repair a defective device), the possession of analogous abilities by any bird is a matter of debate. Perhaps the most convincing avian example uncovered to date involves the capacity of some, but not all, ravens to reach a chunk of meat suspended at the end of a long string that Bernd Heinrich had tied to a perch in their aviary (Heinrich 1995) **(Fig. 6–14)**. The ravens that appeared to exhibit insight "studied" the situation for a time, then reached down, grasped the string in their beaks, pulled the string up, and clamped it to the perch with their feet, repeating the sequence until they could reach the meat. Some ravens could perform this task on their first try, convincing Heinrich that they used insight to recognize how to get a meal, then carried out the necessary actions without trial-and-error practice, even though they could not have practiced similar behaviors in the wild before they were captured. **(Sidebar 1: Bird Brains)**

A Comparison of Instincts and Learning

In summary, both genes and environment play vital, albeit different, roles in developing nervous systems whose operating rules may permit learned responses to some stimuli and innate responses to others. Learned behavior (with the exception of insight learning) differs from innate behavior in that it requires specific experiences to develop fully. Individuals with the ability to learn do so by storing information from key experiences and using it to modify their behavior accordingly—as when Blue Jays learn to avoid bad-tasting butterflies after eating one. In contrast, an instinctive response (for example, a motmot avoiding a coral snake) appears fully developed the first time that the individual reacts to a particular releaser in its environment.

Many people mistakenly believe that instincts are somehow more "genetic" or "hereditary" than learned behaviors. But whether instinctive or learned, all of a bird's actions are caused by its nervous system, which is in turn the developmental product of an extraordinarily complex *interaction* between the individual's genetic information and the environment in which development occurs. Food from the environment helps build the nervous system. Moreover, the environment is the source of key experiences stored in the bird's memory. To learn a song, to have the capacity to imprint on a moving parent, to remember the appearance of rewarding and punishing foods, to learn anything at all, requires a nervous system with specialized properties. These design features could not develop without genetic input.

By the same token, an innate response to a releaser depends on neural components that "recognize" key stimuli and activate the appropriate response. The neural units in question could not have developed without environmental input. Traditionally, debates have raged among behaviorists and psychologists over dividing behaviors into "genetic" ("nature") versus "environnmental" ("nurture"); but dividing behavior in this way makes no sense. Furthermore, many behavioral actions almost certainly represent a combination of instincts and learning, with animals storing feedback information from innate responses, and learning from experience to modify their original instinctive actions. **(Sidebar 2: Play)**

Ultimate Causes of Bird Behavior

■ Now let us turn from proximate to ultimate causation in behavior. We can explain a young motmot's reluctance to pick up thin objects banded red, yellow, and black as an instinctive response to a particular releaser. This process is mediated by proximate mechanisms in the bird's body—the genetic information that guided the development of a nervous system that could detect the releaser and order the avoidance reaction. But this proximate explanation does not explain why living motmots evolved their special behavioral abilities. All living things have an evolutionary history, so knowing how behavior has changed over many generations provides an avenue for understanding why living species behave the way they do.

In the case of the motmot, which preys on many small snakes, it seems likely that in the past those individuals who lacked an innate avoidance response for thin, black-yellow-and-red-banded objects would have been tempted to pick up coral snakes. Although edible, coral snakes have extremely dangerous venom. If coral snake attackers died young compared to those who had an innate aversion to the coral snake pattern, the genes of attackers would have had less chance of being passed on to offspring than the genes of avoiders. If this process continued generation after generation, eventually only

(Continued on p. 6·22)

Sidebar 1: BIRD BRAINS
Irene Pepperberg

"Bye. You be good. I'm gonna go eat dinner. I'll see you tomorrow." I hear these words most nights as I leave my laboratory. Such utterances would not be surprising were they to come from the lips of my students, but they come from a beak—that of my research subject, a Grey Parrot **(Fig. A)**.

Alex, who is over 20 years old, shares the laboratory with two other Grey Parrots, five-year-old Kyaaro and one-year-old Griffin. For several hours each day these birds interact freely with each other and their human caretakers, requesting toys, food treats, and tickles. But they spend most of their time "in class," as the focus of research to examine their intelligence and communicative abilities.

The birds and researchers communicate vocally, using English speech. Alex in particular understands many labels and how to use them in a limited manner. When you say "What color is the key?" for example, Alex knows to describe the key in a particular way, even though he doesn't understand what keys are for. For

Figure A. Grey Parrot, Alex, with Irene Pepperberg: Dr. Pepperberg shows Alex two items of a similar material (wood) but with different shapes (one is two-cornered [football-shaped], the other, five-cornered). Alex's ability to correctly answer the question "What's the same?" about the objects suggests that he can categorize objects in complex ways. Experiments exploring parrots' conceptual understanding, numerical competence, and communication skills conclude that birds have more advanced mental capacities than formerly thought. Photo courtesy of David Linden.

him, keys are head-scratchers. He can identify over 50 items, including paper, truck, cork, block, chain, gravel, cup, chair, chalk, foods, and water; seven basic colors; and five different shapes, which he designates as two-, three-, four-, five-, or six-corner. (Two-corner items look like minifootballs.) Alex also uses and responds correctly to "no" and "come here," and effectively uses phrases such as "I want [object]" and "wanna go [location]." He learns by watching students demonstrate the meaning of labels or questions—he sees them rewarded for correct responses and scolded for errors. After correctly identifying or responding to a question about an object, Alex can play with it; this procedure helps him firmly associate the object with its label.

Alex's abilities startled the scientific community; traditionally, people assumed that mammals were the smartest creatures. Until the 1970s, however, we knew little about avian intelligence. A few experimenters had shown that parrots and canaries could learn tasks such as distinguishing a three-item set from collections containing other quantities, but most researchers concentrated on pigeons and topics such as "delayed match-to-sample." In a typical delayed match-to-sample experiment, researchers trained a pigeon to peck a red light to start a trial. After a delay of several seconds, the pigeon had to choose between red and green lights; pecking red (a

"match") earned the bird a food reward. Scientists determined how many trials the pigeon needed to learn such a task, and how many additional trials it needed to learn to reverse itself (that is, peck green) if red was no longer rewarded. In such studies, particularly the reversals, pigeons needed many more trials to learn than did mammals. Then in the 1970s the aptly named "cognitive revolution" proposed that levels and types of intelligence in other species formed a continuum with those of humans, and inspired researchers to study a wider range of species and types of learning. Scientists, accustomed to working with primates, adapted their field and laboratory projects for use with various avian species, and data on the advanced capacities of birds began to emerge.

The cognitive revolution, which inaugurated an extraordinary era of study, raised three crucial questions: First, what actually is intelligence? Second, can we judge nonhuman capacities using human tasks and definitions? Third, how do we fairly test creatures with different sensory systems from ours? These questions have not been answered to anyone's complete satisfaction, but ongoing studies provide some preliminary suggestions.

Intelligence probably requires two abilities. The first is simply the ability to use experience to solve current problems. A Cedar Waxwing faced with green and red fruits, for instance, might recall that red indicates ripe and tasty and green indicates unripe and bitter, and therefore choose the red fruit. The second is the more complicated ability to choose, from among many sets of acquired information, the set appropriate to the current problem; that is, to recognize conditions under which the selection of green fruit might be wise (for example, when red indicates spoilage). An organism limited to the first ability has learned some important associations, but lacks the flexibility that is a hallmark of intelligence.

Initially, tasks to determine whether nonhumans have these abilities were based on human intelligence and sensory systems, as in the pigeon study described earlier. But such designs may put nonhumans at a disadvantage: wild pigeons do not peck colored lights for food. To give a reverse analogy, a bird testing human abilities might assess how well singing by men attracted mates and kept intruders out; with perhaps a few exceptions, the bird would conclude that humans were not very competent. Some researchers, therefore, try to design tasks that not only are based on human criteria but that also are relevant to an animal's ecology and physiology. Scientists now test avian spatial memory, for instance, by allowing birds to hide and recover seeds in semi-natural settings. Although researchers can strive to make tasks more species-relevant, these tasks will always have to be evaluated from the standpoint of human sensory systems and perceptions of intelligence: this obstacle is one that humans can't surmount.

Another problem in determining the "intelligence" of birds is that avian abilities differ across species. Migrating birds will probably outperform other species on orientation tasks, whereas songbirds with large repertoires will probably have better auditory discrimination. Although using different tasks for different species precludes exact cross-species comparisons, such an approach demonstrates the range of avian capacities. The following is a small sample of the diverse experiments on avian intelligence.

Birds Categorize Objects

Categorization is how we divide the world into definable bins. Clearly, wild birds must categorize items to survive: food/not-food, predator/not-predator, shelter/not-shelter, mate/not-mate, my species/other species. In the lab, pigeons learn to differentiate slides that show natural stimuli (trees, people) from slides that do not, even when shown slides they haven't seen before.

Some bird species may be limited to dividing the world into "target" and "other" categories, but for my Grey Parrot, at least, categorization is more complex. His appropriate responses to "What color?," "What shape?," and "What matter?" for a green triangle made of wood, for example, show he knows that "green," "three-corner," and "wood" represent instances of different categories; he is not simply sorting "green" from "not-green." Instead, his flexibility in categorizing the same item by different attributes shows that he does not respond rotely to a particular object and, furthermore, that he understands how attribute labels themselves are categorized (such as red and blue under "color"). Alex's competence in this task is comparable to that of a chimpanzee.

Alex also understands more complicated categorical questions. Shown a tray of seven different items, he answers questions such as "What color is item X?" or "What object is shape A?" Moreover, if several items are, say, blue and several are square, but only one has both attributes, he can correctly answer questions such as "What object is blue and square?" On both types of tasks, Alex is as accurate as dolphins and sea lions.

Birds Understand Concepts of "Same" and "Different"

"Same" and "different" are not merely forms of categorization (knowing that A "fits with" A and not with B). To understand these concepts you must understand that two nonidentical blue things are related in the *same* way as are two nonidentical green things—in terms of color—and also know that the blue things are related to each other in a *different* way than are two nonidentical square things. In other words, you must recognize which categories of attributes are the same and different, and more importantly, you must understand how relationships *within* pairs are related to relationships *between* pairs. Such understanding requires complex information processing.

Until my work with Alex, few people believed that birds could understand "same" and "different." Natural behaviors such as song matching and individual recognition (see Chapter 7) require sorting into categories of "same" versus "different," but do not demonstrate (to humans) that a bird truly understands what characteristics are the same or different. Accordingly, researchers must design special tasks to demonstrate such abilities. On these tasks, Alex, like chimpanzees, shows he understands "same" and "different" and their absence. He does not simply state whether objects are identical. Queried "What's same?" or "What's different?" about any two objects, even those he has never seen before, he states which attribute ("color," "material," or "shape") is the same or different, or "none" if no attribute is the same or different. Pigeons do not seem able to accomplish similar tasks.

Birds Have Advanced Numerical Capacities

Birds are sensitive to quantity. Eurasian Blackbirds, wood-pewees, and cardueline finches, for example, apparently respond in different ways depending upon the number of times a neighbor repeats certain vocalizations. In lab experiments, canaries can select a three-item set from a variety of sets containing other quantities; choose the second, third, or fourth object in a group; or eat a specified number of seeds. Pigeons discriminate "more" versus "less." Shown "n" objects, some birds can pick a set of "n" other objects from a choice of several numerical arrays. Grey Parrots, ravens, and jackdaws succeeded for quantities up to eight, pigeons for five or six, and chickens, two or three. Some corvids and parrots also learned to "add" items until they achieved the designated quantity: birds would open boxes containing zero, one, or two seeds until they reached, for example, four if the boxes were red, or six if the boxes were black. To perform this task they also had to understand that colors could represent numbers (for example, "red" meant "four"). Alex can vocally label groups of up to six items, and can do so for novel items, for groups of different items, and for items in no particular pattern. He can also state the number of items in a subset of objects, such as how many corks in a collection of corks and keys.

In all these examples, however, the birds may have determined quantity by using a perceptual strategy akin to pattern recognition, rather than by actually counting: we use such a strategy to determine, without counting, how many spots are on a given domino face; for quantities up to about four, we don't even need a particular pattern to succeed. To eliminate this possibility, we gave Alex a task for which humans cannot use perceptual strategies: quantifying a subset of items distinguished from other subsets by two categories. Thus, Alex saw mixed collections of items such as blue keys, red keys, blue cars, and red cars, and had to answer, for example, "How many blue key?" He performed about as well as humans for subsets up to 6 in collections containing up to 18 objects. Because Alex may be even better than humans at using perceptual strategies, we are currently devising stringent tests to determine whether he can actually count.

Birds Understand "Transitive Inference"

Animals in hierarchical societies may infer linear relationships among group members (for example, A dominates B, B dominates me; therefore A dominates me). Animals lacking such "transitive inference" may engage in fruitless, possibly dangerous, challenges. Yet before the 1970s, few nonhumans were thought capable of transitive inference, which seemed to require complex skills, including language. Interestingly, some humans solve the standard verbal transitive inference task (Jack is taller than Jim, Joe is shorter than Jim; who is tallest?) by representing the relationships in mental images rather than words in their minds. Even without words, transitive inference requires considerable mental effort: one must understand the sentences, remember the items in the sentences, and order the sequences appropriately. Although pigeons respond correctly that B is greater than D after learning A>B>C>D>E, their task may not involve mental representation or transitive inference: pigeons were rewarded for learning pairs of elements in the series (for example, A>B), so their responses might

have been based on remembering patterns of rewards. Further study is necessary to determine how well birds understand transitive inference.

Birds Have Sophisticated Communication Skills

Communication provides insights into avian intelligence. Mockingbirds and Brown Thrashers, for example, sing hundreds of different songs, which suggests extensive learning and memory. Some birds seem sensitive to the importance of serial order. Common Nightingales can acquire strings of over 60 songs by "chunking" the tutors' strings into packages of 3 to 7 songs; they maintain the serial order of the packages, but not necessarily of songs within each package. Marsh Wrens apparently learn the order of their own several hundred songs and those of neighbors and may respond to a singing neighbor by producing the neighbor's next song before he does; how they remember such astonishingly long sequences has inspired much speculation. Studies of certain parrots, crows, and chickens suggest that, like vervet monkeys, some birds use different alarm calls to alert their group to the presence of ground versus aerial predators (for example, raccoons versus hawks). Furthermore, Alex's use of English speech demonstrates striking parallels between avian and primate capacities.

The Future

Researchers are just beginning to examine avian abilities. Current topics include timing capacities (for example, how does a bird learn to time its movements when evading predators, or to maximize food intake over time?) and visual processing (for example, does a flycatcher learn to track the evasive actions of a moth?). Many questions remain. How does Alex *learn* to control his vocal tract to produce human speech, and how do he and other birds, with brains organized so differently from those of mammals, succeed on tasks requiring complex mental capacities? Scientists' answers to these questions will force humans to reevaluate their preconceptions about intelligence and notions about the evolution of mental capacities. At least that is what I reflect upon when Alex greets me each day with "Hiyo…Come here!" ■

Suggested Readings

Pearce, J. M. 1987. *An Introduction to Animal Cognition.* London: Erlbaum Assoc.

Pepperberg, I. M. 1990. Some cognitive capacities of an African Grey Parrot. In P. J. B. Slater, J. S. Rosenblatt, & C. Beer (eds.), *Advances in the Study of Behavior* 19: 357–409. San Diego: Academic Press.

Premack, D. 1976. *Intelligence in Ape and Man.* Hillsdale, NJ: Erlbaum Assoc.

Ristau, C. 1991. *Cognitive Ethology: The Minds of Other Animals.* Hillsdale, NJ: Erlbaum Assoc.

6

Sidebar 2: PLAY
Sandy Podulka

Kittens play. Puppies play. Young otters, foxes, and raccoons play. But do young birds truly play? A kitten pouncing on a ball of yarn is not much different from a young screech-owl pouncing on a wind-blown leaf or a young Peregrine Falcon pouncing on a pine cone. Nevertheless, we are comfortable with the notion of mammal play—we *know* that playful urge—but are not so comfortable with the thought that birds, too, might play. Yet, clearly, they do.

Play in birds, as in mammals, enhances learning of motor and sensory skills and the social behaviors necessary for survival as an adult. The trial-and-error nature of play allows young animals to practice and hone their skills before mistakes become a matter of life and death.

Early researchers tended to label as "play" any behavior that seemed to serve no immediate purpose, but more recent observations have revealed certain features common to both mammal and bird play. For example, play is most frequent in young animals, and often consists of incomplete or rearranged sequences of actions, incomplete movements, exaggerated motions, or highly repetitious actions, often performed out of context (Ficken 1977). But even when we know what kinds of activities play entails, the behavior can be difficult to distinguish because few of its components are restricted exclusively to play. Nevertheless, the context and sequence of actions can provide hints that an animal is playing. For example, corvids (crows, jackdaws, ravens, and related species) use rising air currents to help them travel, but those that soar together on air currents, swoop down to earth, then rise again, over and over, are surely playing.

In birds, play is most prevalent in species with a well-developed forebrain, a portion of the brain important in learning; and indeed, species whose development relies heavily on learning tend to play more. For example, play is more common in the avian orders with altricial species,

Figure A. Young Herring Gull Playing With a Skate Egg Case: *The most common type of play in birds is object play, in which the bird handles an item, often nonedible, repeatedly tossing it into the air and catching it, or dropping it and catching it in midair. Here a young Herring Gull plays with a skate egg case on the seashore, repeatedly shaking it and tossing it into the air—behavior that may help the young bird develop prey-capture and prey-handling skills.*

Figure B. Eiders Play in Tidal Rapids: *In August 1934, B. Roberts watched a group of eiders "playing" by a fjord in Iceland. He described his observation as follows: "At the entrance of the fjord there is a long narrow sand-spit jutting out into the sea, and when the tide is ebbing a very strong current races past the end of this spit into the open sea beyond. I watched a party of Eiders "shooting" these rapids, and was amazed to see them land on the outer side of the spit, walk across it, and immediately "shoot" the rapids again. They were plainly doing it for enjoyment, and repeated the performance over and over again, sometimes even running back across the spit, apparently in great haste to experience the sensation once more." Roberts (1934, p. 252).*

which are born helpless and rely more on learning, than in those that are primarily precocial and are more fully developed at birth (Ortega and Bekoff 1987). These correlations of play behavior with forebrain size and altricial young, although interesting, do not necessarily imply cause and effect.

Although play is more widespread in mammals, avian examples are numerous. The most common is "object play" **(Fig. A)**. Hawks, eagles, gulls, terns, corvids, and others carry items such as twigs, stones, leaves, or dead prey into the air and drop them, only to catch them again in midair. One Hooded Crow repeated this performance dozens of times, catching his "toy" after it had dropped about 36 feet (11 meters) (King 1969). Similarly, cormorants and pelicans play with fish and stones, and swallows

play with feathers. Behaviors such as these undoubtedly help develop the prey-capture and prey-handling skills of predatory birds. Another type of play is loosely termed "locomotor." Common Eiders were observed by Roberts (1934) floating down tidal rapids, then hurrying back to the beginning to ride again, over and over **(Fig. B)**. Four Common Ravens were seen taking turns sliding down a snow bank on their tails, feet first (Bradley 1978), and captive ravens developed a game in which they repeatedly slid down a smooth piece of wood in their cage. The function of these activities is not known, although they may help to improve motor skills.

Social play, relatively uncommon in birds, may help young to develop social ties and to learn sexual, aggressive, and submissive behaviors.

Captive young White-fronted Parrots engaged in various "play" forms of allopreening, bill-nibbling, pseudo-copulation, bill-gaping, and neck-stretching (Skeate 1985). Their most frequent "game," however, was play fighting, in which they slowly and deliberately bit gently at the toes and tarsi of their playmate. In real fighting, the parrots gape and jab rapidly and aggressively at the heads of their opponents. Corvids also engage in extensive social play. Young may develop complicated games similar to "king of the mountain," "follow the leader," or "keep-away," and one Common Raven was seen playing with a dog, the two taking turns chasing each other around a tree (Thorpe 1966a).

Of all birds, corvids have the most complex play behavior. One captive raven repeatedly tossed a rubber ball, pebbles, or snail shells into the air and caught them. At other times, it lay on its back and shifted various playthings between beak and claws (Thorpe 1966a). Other captives observed by Gwinner (1966) learned to fall forward from a perch like an acrobat, in order to hang upside down by their feet, wings outstretched, then let go one foot at a time. While upside down, sometimes they carried pieces of food, or shifted items from beak to feet. One, while holding onto a branch with his feet, learned to propel himself around and around the perch by flapping his wings, like a gymnast on uneven parallel bars in a sort of "loop-the-loop." Gwinner's captive ravens also played balancing games: carefully walking out as far as possible to the end of a tiny branch until it bent downward, turning them upside down; or trying to stand on a stick or bone held in the feet, while balancing it on top of and parallel to a perch made from a thick, round wooden dowel (**Fig. C**). The playful inventiveness of captive corvids demonstrates how well they can modify their behavior to suit their immediate environment—an asset greatly enhanced by the extensive learning period each young corvid experiences. This adaptability may be one major reason corvids can occupy such a diversity of habitats.

Although many possible functions may be hypothesized for these and other examples of play, it is tempting to suggest that they just might be "fun." But then, has the feeling of "fun" evolved because it positively reinforces behaviors that contribute to an individual's survival, or is it an end in itself? Questions such as these remain unanswered—for humans as well as for birds and other animals.∎

Figure C. Raven Balancing a Stick on a Perch: *Corvids, such as crows and ravens, have the most complex play behavior. Here a captive raven stands on a stick, while balancing that stick upon and parallel to a wooden perch—a raven's version of "log rolling," perhaps. This is just one of many games played by the captive ravens observed by Gwinner (1966).*

the genes associated with coral snake avoidance would remain in the motmot population, and all the birds in this species would avoid coral snakes.

The process I just described is one in which evolutionary change occurred by Darwinian natural selection. Charles Darwin realized that if *individuals* within a species had inherited differences that affected their chances of leaving surviving descendants, then those that reproduced more would shape the species in their image. The reproductive failures, on the other hand, would take their hereditary attributes out of the **gene pool**—all the genes existing in a population at any given time.

Territoriality, Dominance Hierarchies, and Ritualized Aggression

Darwinian theory provides a powerful approach to studying the evolution of all living things. The theory suggests that the attributes of species, including their behaviors, have evolved through reproductive competition, and therefore should help *individuals* reproduce successfully. Thus, the question for the evolutionary biologist is how a particular behavioral trait helps an individual leave as many surviving descendants as possible.

Consider the fact that birds are often aggressive toward other members of their species during the breeding season as they stake out breeding territories. When a gannet tries to claim a suitable nest territory in a dense rookery **(Fig. 6–15a)** on Bonaventure Island, its fellow gannets do not assist it in any way. Instead, they regularly assault the newcomer with their formidable beaks, trying to drive it away. The newcomer may be sufficiently persistent and aggressive to force his way into the colony, in which case he has some chance of attracting a mate and rearing a brood of young.

A gannet's territory is a very small patch of bare ground (less than a square yard or meter), but many birds have breeding territories that are substantially larger **(Fig. 6–15b)**, ranging up to 1,000 acres (400 hectares) or so for some larger birds of prey. These territories are essential for successful reproduction, because they contain superior nest sites, food resources, or both. Therefore, it is not suprising that regardless of the size of a territory, its defenders react strongly to intruders. In spring in Arizona, the male Vermilion Flycatchers in the mesquite bosque near my home spend hours calling and watching over the territories they have carved from the open forest. When an intruder approaches a territory holder, the resident immediately chases him to the edge of his territory, calling vigorously all the while. The two birds weave through the mesquite trees in a beautiful aerial ballet with an entirely serious purpose.

If two contestants for a territory come in contact with one another, an all-out fight is possible. Niko Tinbergen's film *Signals for Survival* shows territorial Lesser Black-backed Gulls clubbing each other with their bills and trying to pull feathers from their rivals.

a. Part of a Dense Nesting Colony of Northern Gannets

Figure 6–15. Variation in Size of Breeding Territories: *Bird species vary widely in the size of their breeding territories. Shown here are two extremes* **a. Part of a Dense Nesting Colony of Northern Gannets:** *Notice the regular spacing between the tightly packed individual nests—each gannet's breeding territory is limited to less than a square yard or meter, little more than the area around its nest that it can defend as it incubates. Photo by Tom Vezo.* **b. American Kestrel Territories:** *Some birds of prey defend very large territories. American Kestrel territories average about 50 acres (20 hectares), whereas those of larger raptors may be up to 1,000 acres (400 hectares). These territories typically include food resources as well as a nest site. Inset drawing by Sam J. Norris.*

b. American Kestrel Territories

SCALE 1:24000

1   ½   0   1 MILE

1000   0   1000   2000   3000   4000   5000   6000   7000 FEET

1   .5   0   1 KILOMETER

CONTOUR INTERVAL 20 FEET
NATIONAL GEODETIC VERTICAL DATUM OF 1929

(HAMBURG) 5865 III SW

PENNSYLVANIA

QUADRANGLE LOCATION

NEW RINGGOLD, PA.
40075-F8-TF-024

1992

DMA 5865 III NW–SERIES V831

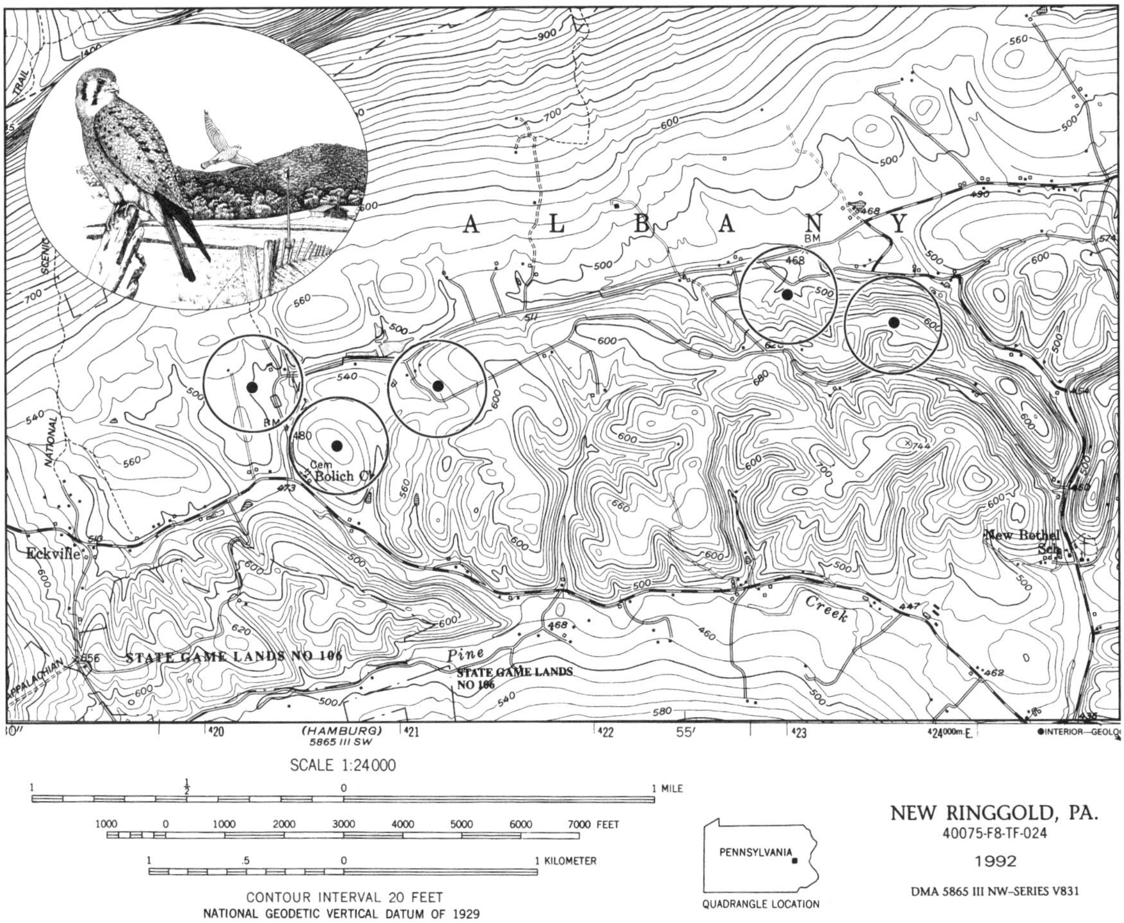

● American Kestrel Nest Site

◯ American Kestrel Breeding Territory
Approximate Size is 50 acres (20 hectares)

Physical aggression, however, is uncommon because most conflicts are resolved without contact between the opponents. The songs of territorial White-throated Sparrows, for instance, are sufficient to deter most other males from invading their domains. When experimenters place tape recorders that play white-throat songs in an area, other white-throats are less likely to enter the site than when the recorder is turned off (Falls 1988). Even when a bird who has no territory invades another's territory, the intruder typically flees at once when challenged by the resident. Moreover, this challenge often takes the form of a ritualized aggressive display, rather than an all-out assault **(Fig. 6–16)**. The puzzle here is that without a breeding territory, which is a scarce commodity, most birds have no chance to reproduce. Yet those who have not acquired a territory appear very reluctant to attempt to take a site from an established rival. Why would a bird so readily give up its chance to raise young and pass on its genes?

A similar evolutionary puzzle is that in winter, some flocking birds with low positions in the pecking order "voluntarily" concede food to others, even though their future reproductive success depends on getting enough to eat. Thus, aggression without combat is the rule in the winter mobs of Dark-eyed Juncos and bands of Black-capped Chickadees gathered about feeders in the northeastern United States. These flocks are highly stratified in a **dominance hierarchy** that determines which birds will have their way and which must step aside **(Fig. 6–17)**. The mere approach of a dominant individual (sometimes supplemented by a noncontact threat display) generally causes a subordinate to give up food or a safe perch to its superior. Why do birds who are low on the totem pole behave this way instead of fighting for their "fair share" **(Fig. 6–18)**?

One explanation of apparent self-sacrifice was proposed in 1962 by V. C. Wynne-Edwards, who argued that territorial behavior and the formation of dominance hierarchies evolved through **group selection**, the differential survival of groups based on variations in group attributes. According to this argument, some birds voluntarily refrain from breeding or from fighting for limited resources when restraint is advantageous for the species as a whole. Thus, if the population of gannets became extremely high, some "excess" individuals could assist their species by not producing young. In this view, too many gannets would put extra pressure on the fish stocks essential for the long-term survival of the gannet species. Wynne-Edwards also argued that dominance hierarchies help species avoid extinction by insuring that when food is in short supply, at least the dominant birds have enough to eat to perpetuate their species.

Wynne-Edwards' theory of group selection had a certain superficial plausibility, but George C. Williams soon pointed out a fatal defect (Williams 1966). Imagine that some gannets really did sacrifice their breeding chances for the good of their group. If the self-sacrificing behavior of these birds had a hereditary basis, what would happen to their genes and the behavior they controlled? Clearly, the genetic basis for stepping aside would become progressively more rare as the

(Continued on p. 6·28)

a. Resting Postures and Threat Displays of Red-winged and Yellow-headed Blackbirds

Red-winged Blackbird   **Yellow-headed Blackbird**

Normal Resting Posture   *Normal Resting Posture*

"Bill-Up" Display   *"Bill-Up Flight" Display*

"Head Forward" Display   *"Wings-Up" Display*

Figure 6–16. Threat Displays: *Threatening birds tend to adopt stylized attack postures: extending the head forward, spreading the tail, raising the wings, or opening the beak. Often these displays make the aggressors appear as large as possible or emphasize their weapons—beaks and wings. Many songbirds, especially thrushes and blackbirds, have a Head-up threat display in which they move the head upward and compress the feathers. Another songbird threat display, particularly among finches and warblers, is Head Forward, in which, holding its body horizontally, the bird points its head at the opponent, often gaping or raising the wings. Threats usually cause the opponent to leave without a fight. The illustrations here and on the following two pages portray threat displays in a range of species.* **a. Resting Postures and Threat Displays of Red-winged and Yellow-headed Blackbirds:** *Territorial male Red-winged Blackbirds respond to intruders with ritualized Bill-up and Head Forward displays. Territorial Yellow-headed Blackbirds employ Bill-up Flight and Wings-up displays. Most intruders flee without fighting with the territorial individuals. Drawing by Gene M. Christman. From Orians, G. H. and G. M. Christman. 1968. A comparative study of the behavior of Red-winged, Tricolored, and Yellow-headed blackbirds. University of California Publications in Zoology, Volume 84.*

(Figure continued on next page)

Figure 6–16. *(Continued)* **b. Great Blue Heron "Forward" Threat Display:** *Seen in a variety of heron species, this display is considered to be a ritualized attack sequence. The bird retracts its neck; erects the plumes of the head, neck, and back; and stabs its bill toward the opponent; sometimes also squawking and snapping its bill. The display most often occurs in the vicinity of the nest as part of territorial defense. From Mock, D. W. 1976.* Pair formation displays of the Great Blue Heron. Wilson Bulletin *88(2):185–376 (p. 206); drawn from movie frames of a filmed display. Drawings reprinted with permission of Douglas W. Mock.* **c. Black-capped Chickadee "Bill-Up" Threat Display:** *The bird tilts back its head until the bill is vertical; in moments of high intensity, the bird also leans its whole body back and points its tail directly downward.* **d. Hairy Woodpecker "Bill-Wave" Threat Display:** *The bird waves its bill from left to right, spreads and flicks its tail from side to side, and may also flick its wings and call. The behavior usually occurs when two woodpeckers of the same sex have a territorial conflict.*

b. Great Blue Heron "Forward" Threat Display

c. Black-capped Chickadee "Bill-Up" Threat Display

d. Hairy Woodpecker "Bill-Wave" Threat Display

(Figure continued on next page)

e. White-breasted Nuthatch "Wing-Spread" Threat Display

g. Common Redpoll "Head Forward" Threat Display

f. Young Great Horned Owl Threat Display

Figure 6–16. (Continued) **e. White-breasted Nuthatch "Wing-Spread" Threat Display:** *The bird raises its body away from its perch, points its bill upward, fully spreads its wings and tail, and sways slowly from side to side for one or two seconds. This display is often given at bird feeders when the nuthatch is competing with birds of its own or other species.* **f. Young Great Horned Owl Threat Display:** *The bird fluffs its feathers and raises its wings, making itself appear larger than it really is. Drawing by Charles L. Ripper.* **g. Common Redpoll "Head Forward" Threat Display:** *The redpoll extends its head forward, and in situations of high intensity may raise its wings and gape at the opponent. Drawing by William C. Dilger.*

	LOSERS								
	RRR	YTT	GC	BGT	RRF	BRF	RC	Total Wins	% Wins
RRR		5	18	8	15	11	7	64	96
YTT	–		4	–	9	6	9	28	85
GC	–	–		2	2	6	1	11	31
BGT	3	–	–		–	3	2	8	44
RRF	–	–	–	–		12	6	18	40
BRF	–	–	2	–	1		6	9	20
RC	–	–	–	–	–	–		0	0

*(Left side label: **WINNERS**)*

Figure 6–17. Dominance Relationships in a Winter Flock of Black-capped Chickadees: *Shown are the results of 138 interactions between the seven members of a Black-capped Chickadee winter flock. For each possible pairing, for example, BRF vs. GC, you can determine the number of times each bird won by locating its "winner" row, and then reading across until it intersects its opponent's "loser" column. Thus, BRF won over GC 2 times, and GC won over BRF 6 times. GC is therefore above BRF in the dominance hierarchy, as indicated by its placement in the row and column headings. This flock had a linear dominance hierarchy with bird RRR at the top as alpha male, followed by YTT, GC, BGT, RRF, BRF, and then RC. Note that in all but 3 of 67 interactions with other flock members, RRR displaced his companions. In 3 of 11 interactions with bird BGT, however, RRR was the loser: BGT is female, and is RRR's mate. Adapted from Hartzler (1970).*

a. Herring Gull "Facing Away" Display

self-sacrificers failed to reproduce and pass on the genes for "good-of-the-group" behavior. In a population composed of both self-sacrificing gannets and reproductively selfish gannets, the genetic basis for successful reproduction should increase, while the genes for species-benefiting self-sacrifice should decrease and then disappear.

Theorists continue to explore whether some other form of group selection might contribute to the evolution of species. For our purposes, however, it is enough to recognize that, unlike Wynne-Edwards' theory of evolution by group selection, the theory of evolution by Darwinian natural selection is based on the defensible premise that hereditary traits which advance the survival and reproductive success of *individuals* will become more and more common. Therefore, if we observe White-throated Sparrows that are unable to acquire territories and are not attempting to reproduce, a Darwinian biologist might propose that by *postponing* breeding attempts when

b. Great Blue Heron "Stretch" Display

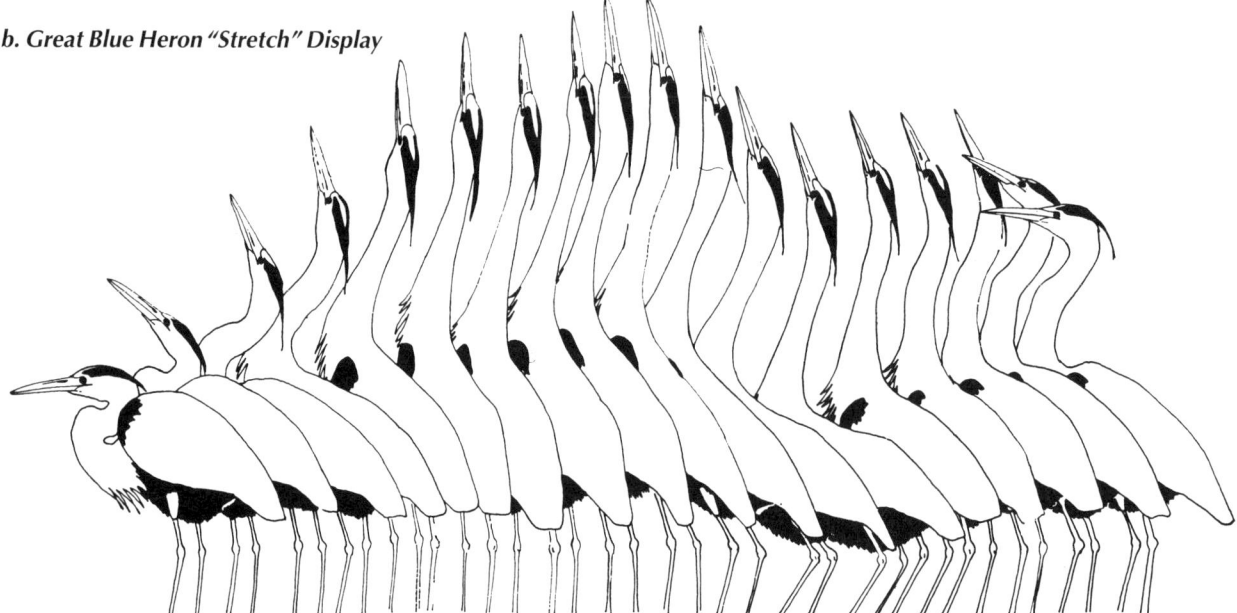

Figure 6–18. Appeasement Displays: *Birds trying to decrease the aggression of a mate or rival may adopt certain postures that have become ritualized to convey submission. These postures are often the opposite of those used in threat displays and tend to de-emphasize the bird's weapons or size and to expose vulnerable parts of the body to the opponent. A submissive bird may point the beak down or away, fold the wings, lower or turn away the head, point the tail down, or adopt some combination of these postures. Many birds use appeasement displays in courtship, presumably to reduce the mate's natural aversion to being so physically close to another individual of the same species.* **a. Herring Gull "Facing Away" Display:** *Many gulls and other birds have a mate-appeasement display in which the members of the pair stand next to each other and turn their faces away. Drawing by Charles L. Ripper.* **b. Great Blue Heron "Stretch" Display:** *The heron smoothly lifts its head and raises*

its bill to the vertical position, with lower neck plumes fully spread. It then gives a long, moaning call that continues as the legs are flexed and the head and bill are lowered to their resting positions. This display is used in a variety of contexts, and in only a couple of them does it seem to have an appeasement function. Mock (1976) has described these as follows: First, when one mate arrives at the nest to relieve the other of incubation duties, the bird on the nest gives a stretch display, which may indicate "I will not attack." And second, during nest building when the female inserts a stick and then sends her mate off to gather another one, she gives this display, presumably to appease him and promote cooperation in the task. From Mock, D. W. 1976. Pair formation displays of the Great Blue Heron. Wilson Bulletin 88(2): 185–376 (p. 188); drawn from movie frames of a filmed display. Drawings reprinted with permission of Douglas W. Mock.

(Figure continued on next page)

Figure 6–18. (Continued) *c. Black-capped Chickadee "General Sleeking" Appeasement Display*

d. Canada Goose "Head-Down" Appeasement Display

e. American Crow "Foot Putting" Appeasement Display

f. European Starling "Submissive Crouch" Display

Fig 6–18. (Continued) *c. Black-capped Chickadee "General Sleeking" Appeasement Display:* This display involves flattening all the head and body feathers and leaning away from a dominant individual. *d. Canada Goose "Head-Down" Appeasement Display:* The head and neck are drawn in toward the breast and the bill is pointed downward. This display is given by an individual who is near a more dominant goose. *e. American Crow "Foot Putting" Appeasement Display:* Subordinate crows near dominant individuals, especially when approaching a breeding pair, may put one foot out to the side, as if trying to keep from being driven away. Yearling crows may even roll over on their backs in submission. *f. European Starling "Submissive Crouch" Display:* Many bird species give a Submissive Crouch display when confronted with a threatening individual.

6

conditions are unfavorable, these birds actually increase their lifetime chances of producing surviving offspring. According to this hypothesis, some white-throats may be weaker than others, and so are unable to fight successfully for the kind of territory essential to rearing young. But by skipping a breeding season, a nonbreeding bird can eventually build up its reserves, and then at some later date win a territory in good habitat where the demanding investment in a reproductive attempt has a reasonable chance of paying off in the currency of surviving offspring.

This Darwinian hypothesis leads to the following testable predictions: (1) nonbreeding individuals should be young, or in relatively poor condition, and therefore at a competitive disadvantage in the struggle to acquire territories, (2) birds that attempt to breed in places which most other members of their species avoid (probably because of poor habitat quality) should tend to have fewer surviving offspring than individuals that secure popular, highly contested territories, and (3) birds that forego reproduction one year should often attempt to reproduce in a subsequent breeding season, without regard to overall population density. These predictions have been found to be true in studies of many bird species. Male Pied Kingfishers, an African species, are much less likely to reproduce successfully in their first adult year than in their second (Reyer 1984). In the European Great Tit, birds forced to settle in "leftover" habitat have fewer surviving offspring on average than those who get into favored areas (Krebs 1971).

The same approach can be applied to acceptance of low status in dominance hierarchies and the use of ritualized aggression to resolve conflicts. Chickadees that cannot defeat a stronger flock member in an all-out fight save their time and energy by not engaging in futile, costly challenges. Instead, they move aside at once when threatened by a higher-status bird, deriving what benefits they can from flock membership (such as safety from predators), and increasing the odds that they will live long enough to move up the dominance hierarchy, at which time they, in turn, will be able to displace rivals from food with just a glance.

Using ritualized signals to settle disputes can advance the reproductive success of both winners and losers. If birds can judge in advance of a fight who would win and who would lose, then both save time and energy and reduce the risk of injury by using threats, rather than a feather-pulling, beak-stabbing battle, to resolve their dispute.

The Evolution of Ritualized Displays

We briefly examined how agonistic (threat and appeasement) displays may enhance the survival of birds, but how, in evolutionary terms, do these and other ritualized displays arise? How, for example, did natural selection produce an action as bizarre as Whooping Crane courtship, described by Slinger (1996, p. 73) as "a couple of NBA centers performing the *Dance of the Sugar Plum Fairies*" **(Fig. 6–19)**?

A close look at displays reveals that some may have arisen from **intention movements**, motions that are either incomplete or that indicate what the "actor" is about to do. For example, an American

Figure 6–19. Whooping Crane Courtship Display: *A displaying pair is shown in two different positions during the spectacular dance, described by Robert Porter Allen (1947, p. 139) in the following way: "Suddenly one bird (the male?) began bowing his head and flapping his wings. At the same time he leaped stiffly into the air, an amazing bounce on stiffened legs that carried him nearly three feet off the ground. In the air he threw his head back so that the bill pointed skyward, neck arched over his back. Throughout this leap the great wings were constantly flapping, their long black flight feathers in striking contrast to the dazzling white of the rest of the plumage. The second bird (the female?) was facing the first individual when he reached the ground after completing the initial bounce. This second bird ran forward a few steps, pumping her head up and down and flapping her wings. Then both birds leaped into the air, wings flapping, necks doubled up over their backs, legs thrust downward stiffly. Again they leaped, bouncing as if on pogo sticks. On the ground they ran towards each other, bowing and spreading their huge wings. Then another leap! The climax was almost frantic, both birds leaping two and three times in succession. Quickly it was all over, after about four minutes, and an extended period of preening followed."*

Figure 6–20. Common Goldeneye "Head-Throw" Display: In this ritualized courtship display, the male repeatedly flicks his head rapidly backward in a smooth arc, pointing his bill upward.

Figure 6–21. Herring Gull Displacement Preening: Displacement activities are actions that seem out of context or inappropriate, as when a Herring Gull suddenly stops to preen in the midst of a territorial conflict or courtship. Ethologists hypothesize that such behaviors may occur because of conflicting motivations or indecision. Drawing by Charles L. Ripper.

Goldfinch about to attack another goldfinch might crouch down and tense its muscles. The better the "observer"—the bird being attacked—is at interpreting the crouched, tensed posture to mean "attack," the better his chance of escaping unharmed, and thus surviving to reproduce more goldfinches with "quick detection" genes. In this way, natural selection would favor observers that could respond to the slightest hint of impending attack. The aggressor, too, benefits by quick, nonviolent resolution of the dispute. Because natural selection favors aggressors who clearly signal their intent to attack, intention movements may become more exaggerated from generation to generation. The evolutionary process by which these and other everyday motions become exaggerated, repeated, and stereotyped into displays presenting a clear message is called **ritualization**. Behaviorists speculate, for example, that the Head-throw courtship display of Common Goldeneyes **(Fig. 6–20)** may have evolved from motions showing intent to leap out of the water (presumably to mate), and that the Forward threat display of many herons (see Fig. 6–16b) evolved from motions indicating intent to attack.

Niko Tinbergen suggested that some displays appear to arise from behaviors performed at times of conflicting motivations or indecision. Often these acts, termed **displacement activities**, seem totally inappropriate or out of context: a Blue Jay in the middle of a fight may suddenly stop to wipe its bill vigorously on a branch; sparring European tits (chickadee relatives) may stop to peck at tree buds; and courting ducks and gulls may interrupt their antics to preen **(Fig. 6–21)**. Other animals, including humans, engage in displacement activities. Think of the person who, while his boss is yelling at him, obsessively straightens up his desk; or the man who, in a heated argument, keeps rolling and unrolling his shirtsleeves or running his hand through his hair.

Behaviorists cannot travel back in time to observe the stages through which a particular display evolved, nor are behaviors preserved in the fossil record. So researchers are limited to comparing similar behaviors in a range of closely related living species for hints on the origin of any particular display. Consider the Head-turning display of many courting ducks, in which the male touches his bill to part of a wing—often the brightly colored speculum, which may have evolved to enhance the effectiveness of the display. In some species, such as the gaudy Mandarin Duck of Asia **(Fig. 6–22a)**, the male merely touches one large, bright orange feather. In other species, such as the Mallard **(Fig. 6–22b)**, the male actually preens. Although certain preening motions were originally displacement activities performed when the male wanted to go after the female yet did not quite dare to, they may have evolved to signal a male's courtship intent. Through time the preening may have become more highly ritualized, to the extent that, in the Mandarin, the bill merely touches the wing. If we saw only the Mandarin's display, we might consider wing-touching an odd and inexplicable element of courtship, but comparing it to the Mallard's preening gives a clue to its possible origin.

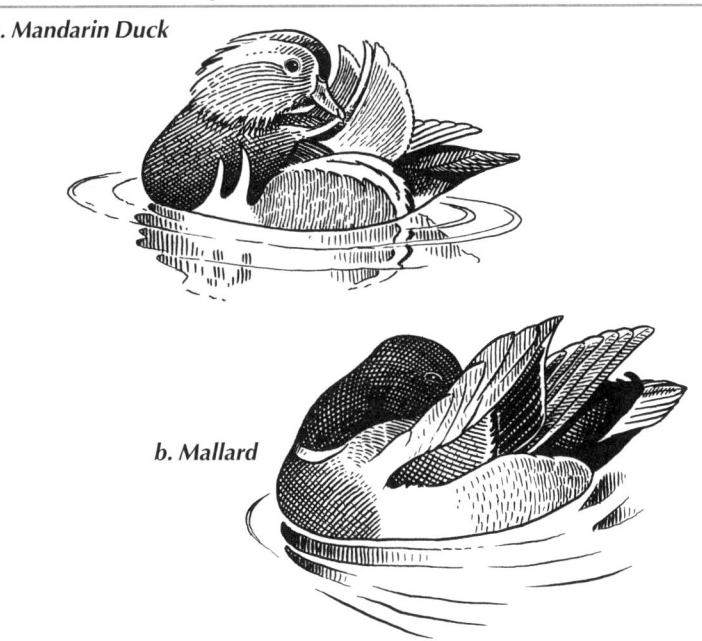

a. Mandarin Duck

b. Mallard

Figure 6–22. Mandarin Duck and Mallard Courtship Displays: To investigate how certain displays might have evolved, behaviorists often compare the actions of related species. The Head-turning display of courting male ducks, which shows off the brightly colored speculum of the wing, is a good example. The male Mandarin Duck (a) merely touches one brightly colored feather with his bill, but the male Mallard (b) actually does displacement preening as part of a similar display, suggesting that the Mandarin's brief Head-turning display may have evolved from displacement preening. Drawings by Charles L. Ripper.

Occasionally a bird directs an appropriate action at an inappropriate subject. These **redirected activities** often appear at times of stress or conflicting motivations. Falcons, irritated yet afraid to attack humans who approach their nest, may attack other birds passing by. And, an Argus Pheasant, perhaps frustrated yet aroused when the object of his attention did not cooperate, courted a stone water trough. Mammals, too, may redirect behaviors: a pet dog or cat, just disciplined by its owner, may take a swipe at a fellow pet, rather than its owner. How often have you come home and yelled at a sibling or spouse when you really wanted to scream at your teacher or boss, but were afraid to? Sometimes redirected behaviors, like displacement behaviors, are ritualized into displays. Male Herring Gulls defending their territory often pull deliberately at nearby grasses **(Fig. 6–23)**. In the past, male Herring Gulls facing an opponent, unsure whether to attack or flee, may have pulled at the grass or pecked at the ground, instead of launching an outright attack. Over time, this behavior may have evolved into a display signaling aggression.

Unfortunately, few things are as simple as they appear. Behaviorists cannot actually talk to birds, so they can never *know* exactly what message is being communicated. They can see what the actor and the observer do before and after a display, and can look for behavioral changes that appear to correlate with the display, but they can only speculate on the exact meaning. To further complicate matters, birds often have several different displays for the same apparent message, and the same display may be used in different contexts.

The traditional hypotheses presented above on the origin and function of displays are appealing, but very difficult to test—so they remain unconfirmed. More recently, researchers have suggested that displays do not simply increase the clarity of a message; rather, they may involve ambiguity, deceit, and bluffing. Perhaps bird displays reveal only what is necessary to appear convincing, and sometimes mis-

*Figure 6–23. Herring Gull Redirected Aggression: A male Herring Gull, defending its territory from another gull, tugs forcefully at a clump of grass. This display may have evolved from **redirected aggression**: in the past, territorial gulls, conflicted as to whether to attack or flee from an opponent, may have directed their aggression toward an inappropriate recipient—such as grass. Over time, this may have evolved into a display signalling aggression. Drawing by Charles L. Ripper.*

lead, in an attempt to manipulate competitors. Other researchers have suggested that displays must honestly advertise a bird's true intents and abilities, or they would not be effective for long. These intriguing ideas are currently the subject of much debate among behaviorists (Dawkins and Krebs 1978; Johnstone 1997). If all the researchers met to discuss their ideas, what types of displays would they use to convince others, and how would an observer from another world interpret the meaning of the gestures and expressions?

Courtship Displays

The formation of a pair bond, whether temporary, seasonal, or lifelong, involves an exchange of signals between a male and female. In most songbirds, the males procure territories through squabbles with other males, and then the females arrive and choose their mates. At first, each male behaves aggressively to all members of his species, both males and females, intruding on his territory. If the sexes differ in color or size, the male may be less aggressive toward the female right away, but in species with identical sexes, he treats her just as he would another male—so the first stages of courtship may be difficult to distinguish. If, on being chased, the female assumes a submissive posture or merely refuses to leave, the male gradually becomes less aggressive and directs his energy toward courtship activities.

Courtship displays are the most complex and intriguing rituals carried out by most species, perhaps because they serve a number of crucial functions. First, the displays have evolved to assure that only members of the same species will copulate—natural selection eliminates most individuals that choose a mate of a different species because hybrid matings usually produce sterile or weak young, or none at all. Second, the displays communicate the sex, breeding status, and sexual readiness of the two potential mates. Finally, courtship helps to stimulate and synchronize the breeding behavior of the partners. It is probably for this reason that species such as White-breasted Nuthatches and Canada Geese, in which mates spend their whole adult lives together, still carry out courtship rituals before each breeding season.

Courtship displays take many forms—from the presentation of elaborate plumes in birds-of-paradise (see Fig. 3–10) to the aerial rolls and loops of some birds of prey—but a few main themes are common. Many displays show off the male's assets—his colorful feathers, his ability to dominate other males, or his parental skills—to his prospective partner. Others include some type of ritualized nesting or breeding activities, such as carrying and presenting nest material by Eastern Bluebirds and Northern Mockingbirds, and feeding of the female by many terns and birds of prey.

On the following pages are examples of the amazing variety of courtship displays used by birds in various stages of mate attraction, pairing, and breeding (**Figs. 6–24 through 6–29**; see also Figs. 7–56 and 7–67).

(Continued on p. 6·42)

a. "Grunt-Whistle" Display *(Male Only)*

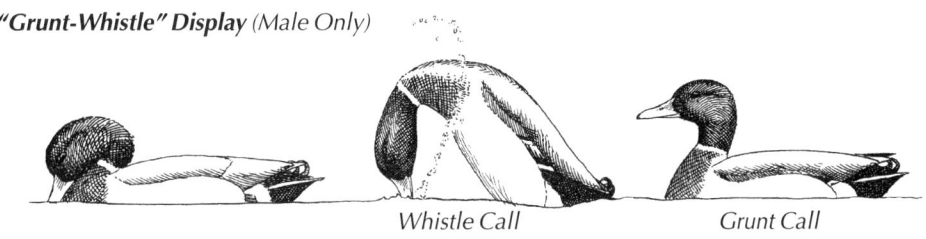

Whistle Call   Grunt Call

b. "Head-Up-Tail-Up" Display *(Male Only)*

Whistle Call

c. "Down-Up" Display *(Male Only)*

Whistle Call

Figure 6–24. Mallard Courtship Displays: *Beginning in the fall, before migration, nearly every body of water on which Mallards and other ducks congregate becomes a stage upon which groups of males "strut their stuff" to females. The courtship displays and pair-bonding rituals continue right through migration and on the wintering grounds. Many individuals arrive on their breeding grounds in spring already paired, but courtship displays are still very common at this time, both because some individuals are not yet paired, and because members of established pairs use displays to strengthen the pair bond and as precursors to copulation. Although copulation also occurs in the fall and winter, it is probably not effective in fertilization until spring, when the testes are no longer regressed. Once females are no longer fertile and begin incubation, the pair bond dissolves and the males leave to join other males and begin molting.*

Mallard courtship behavior includes a variety of displays which, because of the species' familiarity and tameness, provide an excellent opportunity for close observation and study. Some of the displays illustrated are performed by groups of males toward females, one may be done by either sex, and others take place between members of a pair. Because many other duck species, such as Northern Pintail, Gadwall, American Wigeon, American Black Duck, and various teal, have similar displays, learning those of the Mallard will provide insight into the behaviors of a number of different species.

a. "Grunt-Whistle" Display: The Mallard drake (male) lowers his bill into the water, arches his neck, and raises his body upright and almost out of the water, while still keeping his bill in the water. When the neck is most arched, the drake tosses an arc of water droplets into the air with his bill and gives a loud, sharp whistle. As he returns to his normal position he gives a deep grunt. The entire display takes about one second to perform and is typically performed by groups of males.

b. "Head-Up-Tail-Up" Display: With a loud whistle, the Mallard drake quickly draws his head and neck upward and backward, at the same time curving his rump upward and ruffling his rump feathers, making his body short but tall. His folded wings are also briefly raised, showing off his bright blue wing patch and his curly tail feathers. After about a second he returns to his normal position, turning his head toward the female to whom the display was directed. This display is typically performed in groups.

c. "Down-Up" Display: The Mallard drake tips forward and rapidly dips his bill into the water. He then flips his bill up, creating a small arc of water droplets, while giving a whistle followed by a nasal call that sounds like rhaebrhaeb. *The display lasts about two seconds and is typically performed in groups.*

(Figure continued on next page)

d. "Nod-Swimming" Display *(Male or Female)*

e. "Pumping" *(Male and Female)*

f. "Inciting" *(Female)*

Quegeg *Call*

Figure 6–24. *(Continued)*

d. "Nod-Swimming" Display: *The Mallard swims very rapidly for short distances with its neck outstretched and its head flattened down to just graze the surface of the water. This display may be performed by either males or females. When done by females, it is directed toward groups of males who, by their behavior, are expressing an interest in courtship. The female Nod-swims toward each male, swimming in quick arcs around as many of them as possible. This stimulates the males to give their own courtship displays. Drakes perform Nod-swimming during bouts of the Head-up-tail-up display and also immediately after mating.*

e. "Pumping": *The male and female Mallard face each other and begin rhythmically bobbing their heads up and down. The head is thrust upward with the bill held horizontally, then jerked downward. While bobbing, the male and female move their heads in opposing directions: while the male is at the highest point, the female is at the lowest. The pumping display may be repeated many times, and is usually followed by mating.*

f. "Inciting": *The female Mallard follows closely behind her chosen mate, repeatedly flicking her head back over her shoulder while giving a distinctive call that sounds like* quegegegegegegeg. *The display is given when the pair is approached by a strange male.*

Figure 6–25. Great Blue Heron Advertising Displays: Advertising or **mate attraction displays** are performed by the Great Blue Heron at its nesting colony. After selecting a territory—a branch in a tall tree—the male often performs as follows:

    a. He howls, starting this loud call at a fairly high pitch with the head and neck stretched up.
    b and c. He then lowers the pitch as he lowers his head and body. Toward the end of the call, he increases the pitch quickly.
    d. After a bout of howling he may reach out and tug at a twig, or
    e. perform Bill Sharpening.
    f. Eventually, after more howling, a female arrives and proceeds to preen.
    g. When she flies off, he follows. Both birds will return and leave the territory repeatedly until the pair bond is formed.

Drawings by Richard P. Grossenheider.

Northern Gannet

Great Crested Grebe

Adelie Penguin

Eared Grebe

Wandering Albatross

Red-necked Grebe

Figure 6–26. Mutual Displays: *Many long-lived birds that mate for life engage in **mutual displays**—intricate, synchronized dances that appear to stimulate and coordinate breeding behavior between the pair and to reaffirm the pair bond. Typically, male birds do more of the courting, but in these mutual displayers, males and females share courtship equally. Birds that display mutually include gannets, penguins, grebes, albatrosses, cormorants, boobies, herons, and storks. All are large, monogamous birds with sexes of similar appearance. Songbirds, though often monogamous with similar sexes, rarely give mutual displays.*

*   **a. Mutual Displays of Various Species:** Pictured here are Northern Gannet, Great Crested Grebe, Adelie Penguin, Eared Grebe, Wandering Albatross, and Red-necked Grebe pairs engaged in mutual displays. To get an idea of what may be involved, consider in more detail the mutual display of the Northern Gannet. Northern Gannets display mutually during pairing, and at any time throughout the nesting cycle, especially after a temporary separation. When a gannet returns to its mate on the nest from a bout of fishing, both birds stand and face each other, stretch their necks high with bills pointed upward, partially open their wings, and clash their bills together. Then, in a display termed "mutual fencing," they keep their bills in contact while rapidly rolling one bill around the other from side to side, as if in a mock duel. Drawings by C. G. Pritchard.*

(Figure continued on next page)

b. The Weed Ceremony of the Western Grebe

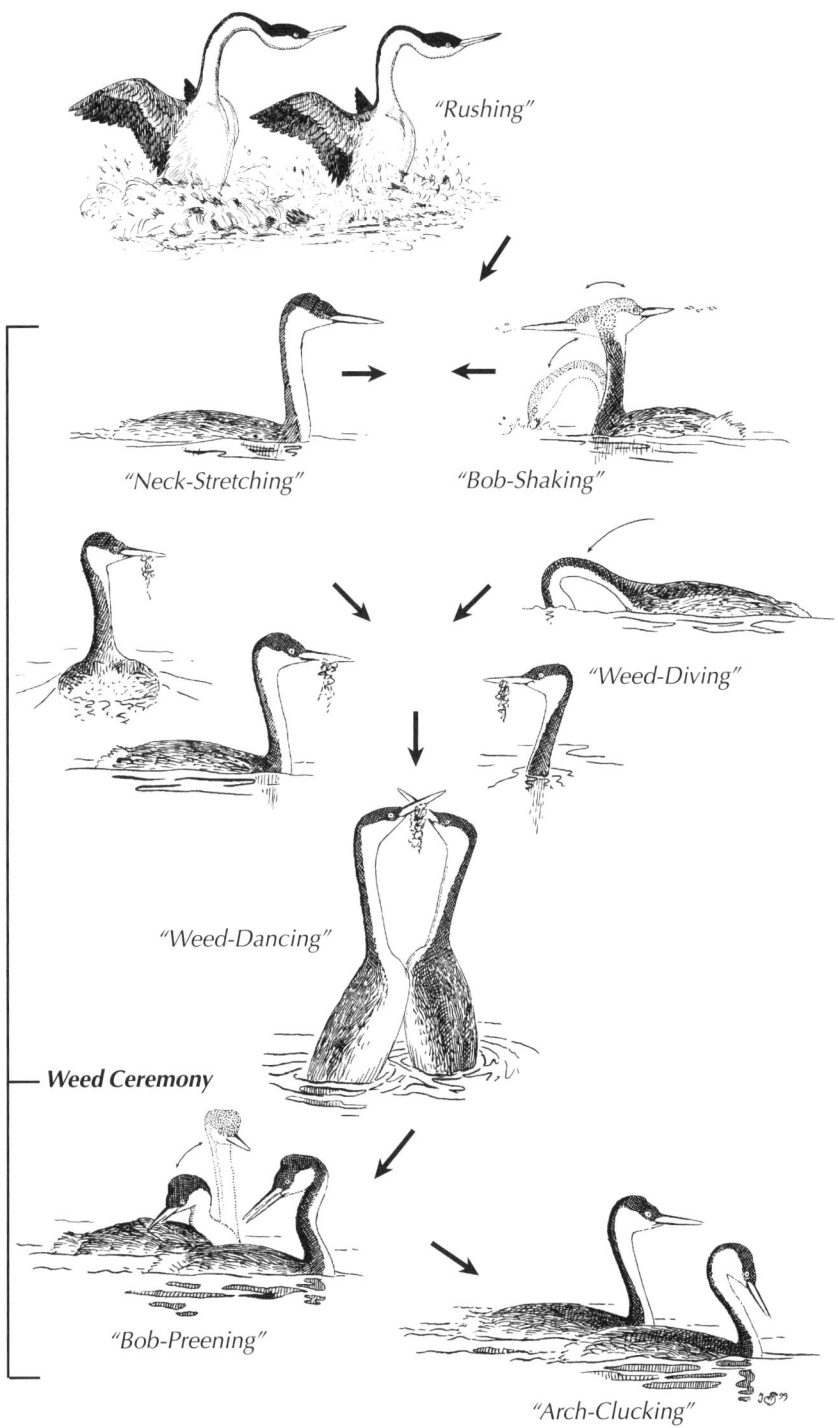

"Rushing"

"Neck-Stretching"

"Bob-Shaking"

"Weed-Diving"

"Weed-Dancing"

Weed Ceremony

"Bob-Preening"

"Arch-Clucking"

Figure 6–26. (Continued) **b. The Weed Ceremony of the Western Grebe:** *The sequence of mutual displays shown here occurs late in the pair-formation process. The conspicuous behavior known as* **Rushing**, *in which two or more Western Grebes run side-by-side across the water's surface, has often been described as a courtship display. It may, however, occur between a male and female, between two males, or between a female and two or more males, so it probably has several different functions, including mate attraction and male-male competition, as well as courtship. When Rushing is performed by a male and a female, it may be followed by the display sequence known as the* **Weed Ceremony:** *Surfacing from a post-Rushing dive, the two courting grebes swim toward each other. They participate in* **Neck-stretching**—*an erect posture with raised crest during which the bird gives a drawn-out trilling call—and/or* **Bob-shaking**, *in which the bill and forehead are dipped into the water, then pulled out and shaken to and fro, all in a ritualized way, creating a slight splash. The birds then begin to* **Weed-dive**, *repeatedly diving from the water's surface to search for submerged plant material. They approach each other closely, each holding its billful of weeds high, then perform* **Weed-dancing**. *Breast to breast, the two grebes rise up vertically, churning the water strongly with their feet, and bring their weeds together over their heads, sometimes spiraling around each other or slowly moving forward in the process, while still maintaining bodily contact. It is thought that Weed-dancing is derived from nest building. After Weed-dancing ends, the birds swim side-by-side, performing mutual* **Bob-preening**—*repeatedly going through preening motions—and mutual* **Arch-clucking**, *in which each bird displays with its neck stretched into a high arch and its crest spread laterally, while giving clucking calls. Adapted from Nuechterlein and Storer (1982).*

6

Figure 6–27. Precopulatory Displays of the Male Red-winged Blackbird: *Immediately before copulation, birds frequently give displays that are primarily invitation or solicitation performances. Shown are some of the postures of the male Red-winged Blackbird that precede copulation.* **a.** *Perched above or near the female, the male arches his body.* **b and c.** *As the intensity of his excitement increases, he leans forward into a crouching position, spreading his wings slightly. He maintains this posture for several seconds.* **d.** *Just prior to copulating he is deeply crouched, with body feathers ruffled, wings drooped, and tail spread.* **e.** *As he approaches the female he flutters his wings in a ritualized way that shows off his red epaulettes. The female, like females of other species, assumes a more submissive posture—flexing her legs, holding her body horizontal or somewhat tipped forward, and elevating her tail (see Figure 7–56). Drawings by Gene M. Christman. From Orians, G. H. and G. M. Christman. 1968. A comparative study of the behavior of Red-winged, Tricolored, and Yellow-headed blackbirds.* University of California Publications in Zoology, *Volume 84.*

Figure 6–28. Copulatory and Postcopulatory Displays of the Great Blue Heron: *These displays take place on the pair's territory. After greeting calls and precopulatory displays, copulation follows, either* **(a)** *on the nest, or* **(b)** *on a branch nearby. This is followed by* **(c)** *shaking and preening, and* **(d)** *Neck-crossing accompanied by Bill-snapping. Drawings by Richard P. Grossenheider.*

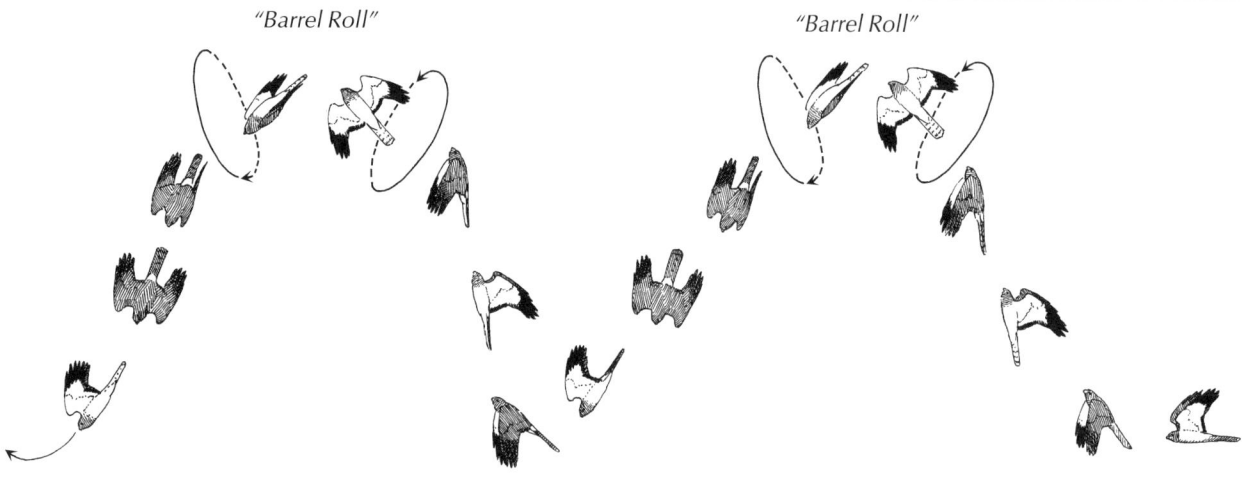

"Barrel Roll"   "Barrel Roll"

Figure 6–29. Courtship Display of the Northern Harrier: *The dramatic aerial courtship display of the male Northern Harrier has been aptly named "sky-dancing." With wings flapping rather slowly and loosely, the bird climbs steeply into the air, on average reaching 60 feet (18 meters), but sometimes ascending much higher. At the peak of the climb he performs a midair "barrel roll" by rolling onto his side, then continuing to rotate in the same direction to right himself. Then he immediately launches into a steep earthward dive, checking just above the ground. Depending on the intensity of the display, this sequence may be repeated many times to create a series of spectacular deep undulations through the sky, covering a distance of up to one-half mile (0.75 km).*

The Use of Darwinian Evolutionary Theory

■ The logic of Darwinian evolutionary theory leads to ultimate hypotheses that focus on a trait's possible reproductive benefit to an individual, not to the species as a whole. The first trick for an evolutionary biologist is to identify how an evolved trait might be **adaptive** (that is, better in promoting individual survival and reproductive success than some alternative form of that characteristic). The next trick is to *test* the adaptationist hypothesis that one thinks might be correct. Many hypotheses that are fully consistent with Darwinian theory have been shown to be incorrect. To pick just one example, in *Ravens in Winter* (see Suggested Readings) Bernd Heinrich describes how he once thought ravens called loudly by dead moose and deer to attract other carrion-eating ravens to the spot. The discoverer would thus not be alone when he went down to the carcass, where a predator might be lurking. According to this argument, the caller gained survival advantages from its behavior, and those that joined him gained a chance for a good meal. Thus, the explanation is a perfectly good Darwinian possibility. But Heinrich found, using dead goats he placed out in the Maine woods, that ravens continued to call long *after* a large group had gathered and begun to feed safely. He therefore rightly concluded that the "dilution-of-risk-of-predation" hypothesis could not be the right explanation for calling by ravens. Actually, Heinrich eliminated a whole spectrum of possible explanations for calling before he found support for the hypothesis that callers were territorial intruders attempting to attract a mob of outsiders to the carcass, the better to overwhelm the defenses of the territory owners. Testing Darwinian ideas is essential.

The really interesting cases for Darwinian biologists to solve through hypothesis testing involve traits whose reproductive *disadvantages* are clear enough, but whose benefits are not so obvious. Adopting cowbirds, for example, surely handicaps a small passerine whose average life expectancy is one year. Cowbird nestlings eject some or all of the host's nestlings from the nest and prevent most survivors from receiving their fair share of food. Therefore, Yellow Warblers that host cowbirds definitely leave fewer descendants on average than those who avoid them.

Nevertheless, as noted above, the warbler's parental drive could be adaptive because it *usually* causes them to direct food to the largest, most vigorously begging of their own offspring. Big, active nestlings and fledglings may be the ones most likely to survive to reproduce and thus propagate the genes of the parents. Parent warblers that refused to feed large, healthy offspring might very well avoid feeding cowbirds, but they also might raise fewer of their own progeny than a less discriminating, more parental member of their species.

Furthermore, Yellow Warblers that abandon parasitized nests altogether could actually rear fewer offspring on average than those that stayed with a cowbird-infested nest. (Yellow Warblers and other small passerines do not have the option of removing a cowbird's egg from their nests because their bills are too small and the cowbird's egg too tough to puncture.) In the north temperate United States and Canada, the breeding season available to small passerines is short, which limits their options. If a pair of Yellow Warblers abandons a nest with a cowbird egg in it (as they sometimes do), they lose valuable time and increase the chance that any fledglings from their second nest will be too small or too young to survive the fall migration.

We can test the hypothesis that when Yellow Warblers accept cowbird eggs, they do so because this option is more adaptive than the alternatives (nest abandonment or burying the parasite egg under a "second-story" new nest [see Fig. 8–144]). This hypothesis predicts that Yellow Warblers further into the nesting season will be more likely to accept the parasite's eggs than those victimized earlier in the season. When the season is advanced, it should be better to stick with one's clutch, even though the nest also has a cowbird egg, because *some* chance of rearing young warblers is better than *none.* In fact, Yellow Warblers *are* more likely to continue to brood a cowbird egg in the second half of the clutch initiation period than in the first half of the season (Sealy 1995).

We now explore the ultimate basis of a spectrum of bird behaviors, showing how an adaptationist approach has illuminated puzzles associated with how birds (1) select food, (2) react to predators, (3) breed colonially, (4) choose mates and reproduce, and (5) behave parentally, five important determinants of an individual's lifetime reproductive success.

Feeding Behavior: Why Do Birds Generally Restrict Their Diets, Ignoring Some Edible Foods in Favor of Others?

The variety of foraging techniques that birds use is as impressively diverse as the range of avian bill sizes and shapes **(Fig. 6–30)**. Ruby-crowned Kinglets hover by pine branches to pluck minute insects from

(Continued on p. 6·48)

Figure 6–30. Foraging Techniques: *Birds use an amazing diversity of techniques to procure their food. Some of the most common methods are shown here and on the following pages.*

Barn Swallow

a. Aerial Insect Capture: *The Barn Swallow pursues insects in flight, capturing them in midair in its mouth, a technique known as* **sweeping**. *Birds that sweep for insects generally have huge mouths and small beaks. Swifts, nighthawks, and many swallows use this feeding style.* **Hawking** *is a very different technique for catching aerial insects, performed by many flycatchers, kingbirds, bee-eaters, and sometimes by waxwings and certain woodpeckers. A hawking bird sits very still on a high or exposed perch with a good view until it sees an insect. Then it flies out and snatches the insect from the air, and returns to the same or a nearby perch to eat its prey.*

b. Insect Gleaning: *Taking insects and other small invertebrates, such as spiders or slugs, from the surface of vegetation or other substrate is known as* **gleaning**. *Like many wood-warblers, the Yellow-rumped Warbler, shown here, practices* **perch gleaning**—*capturing prey without flying from its perch, by searching among the leaves, stems, twigs, and bark for prey. Red-eyed Vireos, chickadees, titmice, and certain small flycatchers such as Least, Acadian, and Willow flycatchers practice* **sally gleaning**. *With this technique, a bird sits still and watches the surrounding vegetation until it sees an insect move. It then flies out to pluck the prey item from the distant surface.* **Hover gleaning**, *hovering while taking food from vegetation surfaces, is performed by kinglets, phoebes, and Great Crested Flycatchers.*

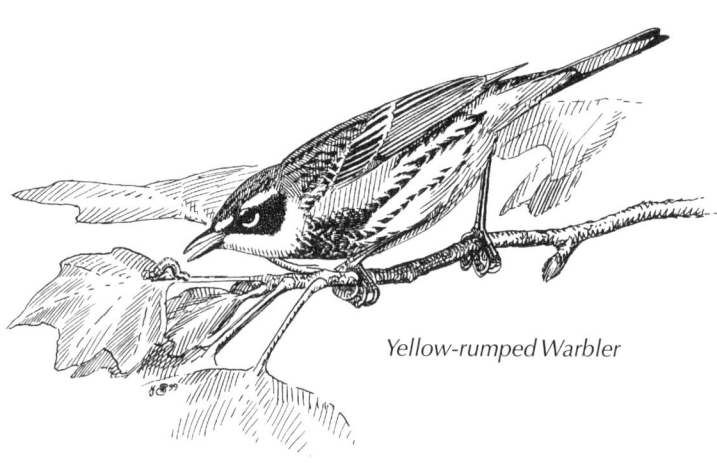

Yellow-rumped Warbler

c. Probing: *A* **probing** *bird plunges its beak into the substrate to search for prey. Some wading birds, such as the White Ibis pictured here, and many shorebirds, including woodcock and snipe, probe in mud for their prey; the specific type of motion used may be characteristic enough to help humans to identify them. Other birds, such as Brown Creepers, Black-and-white Warblers, and nuthatches, probe under tree bark and into crevices for their prey.*

White Ibis

Figure 6–30. *(Continued)* **d. Chiseling and Pounding:** *Most woodpeckers, such as the Pileated Woodpecker shown here, feed by hammering on tree trunks and limbs. The **pounding** may disturb insects in the wood enough that they come to the surface, and the **chiseling** creates holes through which the woodpecker can insert its long, barbed tongue to grab the insects. Sapsuckers also chisel living wood to create sap flows: they return later to these sites to drink the oozing sap and eat the insects attracted to it. In addition, oystercatchers pound open the shells of shellfish to reach the flesh inside.*

Pileated Woodpecker

Great Horned Owl

e. Raptorial Predation: *The Great Horned Owl captures a rabbit by **pouncing** upon it from midair and pinning it against the ground—a method used by many other owls and hawks. Falcons, such as the Peregrine Falcon, employ **stooping**—dropping through the air at great speed in pursuit of a flying bird or insect, stunning and snatching it in midair.*

6

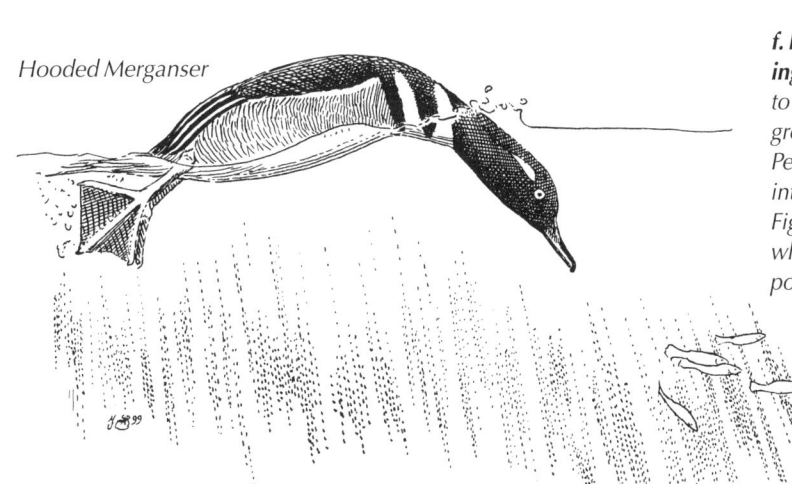

Hooded Merganser

f. Diving: *The Hooded Merganser uses **surface diving** to submerge itself from a swimming position to pursue aquatic prey, as do other diving ducks, grebes, cormorants, and loons. Kingfishers, Brown Pelicans, auks, gannets, and Ospreys **plunge dive** into water from the air to capture their prey (see Fig. 1–31); Ospreys capture fish with their talons, whereas the others catch their prey in their long, pointed bills.*

(Figure continued on next page)

Snowy Egret

Figure 6–30. (Continued) **g. Stalking and Stabbing:** Various herons, such as the Snowy Egret portrayed here, use the **stalk and watch** method to find their prey. They slowly walk or wade, or simply stand still and watch for fish or other aquatic prey to come within reach, then suddenly stab or snatch the creatures from shallow water. Birds such as plovers, robins and other thrushes, and larks use similar methods on land, taking their prey from the soil's surface. A more active approach involves **foot-raking** and **stamping**, sometimes used by herons and egrets to scare prey hiding in the sediments or in submerged vegetation out into the open. **Wing-flashing** is also done by some herons, egrets, and storks such as the Wood Stork—one or both wings are quickly raised or brought forward as the bird moves along through the water. This behavior may frighten prey out of hiding, or alternatively may provide a shady place where unsuspecting prey can try to hide; either way brings the prey within the bird's reach.

h. Sifting: Spoonbills, such as the Roseate Spoonbill, strain small animals and plant material from mud or water by sweeping the partly open bill from side to side while wading. Some ducks, including Northern Shovelers, sift for food as they swim in shallow water. Sifting birds have bills with specialized edges that trap small aquatic creatures as the water drains out of the bill. Flamingos use a similar technique, but actually pump the water through their specialized bills (see Figure 4–89).

Roseate Spoonbill

i. Pecking and Biting: The Ruffed Grouse feeds by **pruning**—biting off and eating plant buds. Grouse and Wild Turkeys also **pluck** ripe fruits from their stems, as do waxwings, robins, and catbirds. Goldfinches **peck** at plant seedheads to remove the seeds, as do grosbeaks feeding on tree seeds. Pheasants, juncos, sparrows, and blackbirds of various species also peck on the ground to pick up fallen seeds. **Grazing** involves biting off clumps of grass or other vegetation with the bill, as done by geese.

Ruffed Grouse

Hummingbird

Figure 6–30. (Continued) *j. Nectar Feeding:* Hummingbirds hover at flowers, plunging in their long bills and using their long tongues to sip the nectar. Other nectar-eaters cling to flowers to feed. These include the Old World sunbirds, the honeyeaters of the Southwestern Pacific region, some Hawaiian honeycreepers, tropical honeycreepers and other tanagers, and orioles.

Northern Shovelers

k: Water Surface Feeding: The pair of Northern Shovelers shown here is **dabbling**—moving the beak rapidly on the surface of shallow water to pick up small aquatic animals and plant material. Northern Shovelers and other dabbling ducks also tip their tails up and reach down under the water's surface to obtain submerged food. Unlike most dabblers, the Northern Shoveler has a specialized beak for straining out food (see Fig. 6–30h). Other surface feeders include storm-petrels, who patter across the ocean's surface picking up floating food (see Figure 5–27). Phalaropes also pick up food from the ocean's surface, spinning around to create a vortex that brings up food from below the water to within the bird's reach.

6

Eastern Towhee

l. Searching through Leaf Litter: Many ground-feeding birds, especially in forested habitats, search for food by **leaf-tossing**—throwing aside the leaf litter. Eastern Towhees keep both feet together as they kick back leaves, looking for insects and other small invertebrate prey. Fox and White-crowned sparrows also leaf-toss with both feet, whereas grouse and turkeys scratch at leaves with one foot at a time, and thrashers "thrash" with their bills to achieve the same end—a habit that has earned them their name.

(Figure continued on next page)

Figure 6–30. (Continued) m. Specialists: A number of species search for food using specialized techniques. Shown here, a Parasitic Jaeger pursues a Forster's Tern, intent on stealing the tern's fish—an act termed **piracy**. Frigatebirds and various gulls also make a living this way. Another specialized foraging technique is **blood-feeding**. One subspecies of the Sharp-beaked Ground-Finch, a Galapagos finch nicknamed the "Vampire Finch" and confined to one island in the Galapagos archipelago, eats the blood of Red-footed and Masked boobies that nest there (see Fig. 1–47). The finches land on the backs of the boobies and bite the skin at the base of the flight feathers in the elbow region, opening a wound and causing blood to ooze from the skin along the shaft of the feather. These same finches also eat insects and cactus blossoms. Another special foraging strategy is that of the **honeyguides** of Africa. They "lead" humans and honey-badgers to beehives, keeping just 15 or 20 feet ahead of the followers, giving a special call, and fanning their tails. After either the person or the honey-badger has opened the hive, the honeyguides feast on the wax of the honeycomb, a unique food for a bird. As discussed earlier in this chapter, some birds **store food** temporarily, mostly in autumn, for use later when the food becomes scarce or difficult to obtain. A few birds in northern latitudes—the nuthatches and the Acorn Woodpecker, for example—place seeds or insects in cracks and crevices in the bark of trees. The Clark's Nutcracker (see Figs. 6–12 and 6–13) buries pinyon nuts in the ground and retrieves them in the winter when the food supply is low. Many other examples of specialized foraging techniques exist, including the following: **scavenging** (see Sidebar 4: Living in Groups, Fig. D), **shell-smashing by dropping** (see Fig. 6–31), **cooperative feeding** (see Sidebar 4: Living in Groups, Fig. E), and **skimming** (see Fig. 4–88).

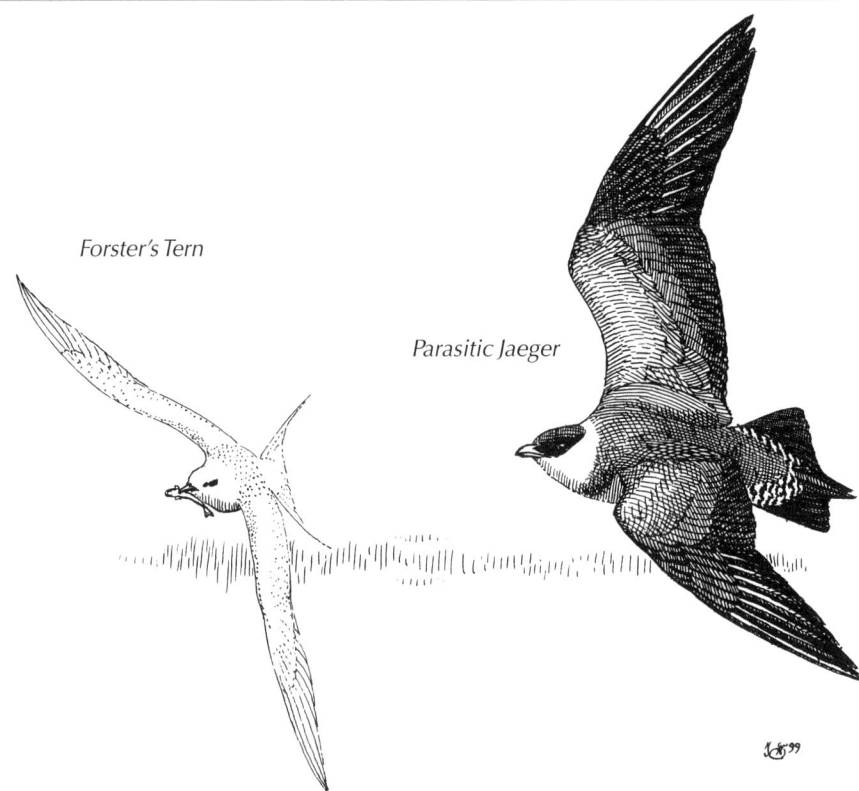

Forster's Tern

Parasitic Jaeger

the foliage. Fish-hunting gannets plunge dive from great heights into the ocean. Olive-sided Flycatchers swoop out from an ambush perch to intercept passing flies and other large insects. Northwestern Crows break open whelks (a type of mollusc) and clams by flying up and dropping them onto a hard surface, such as a breakwater.

The adaptive value of the different foraging methods seems obvious in most cases. To get at the calories and nutrients in clam flesh requires one technique, to feast on fish meat requires another. But a full exploration of the ultimate causes of different foraging methods can go considerably further by once again looking for elements of the behavior that appear costly, disadvantageous, or a handicap to reproductive success.

The Northwestern Crow, for example, is very choosy when it comes to collecting whelks to drop on rocks in the intertidal zone in Washington and British Columbia. The birds bother only with the larger whelks. Why do they limit themselves to a small fraction of the potential prey? Passing over small and medium-sized whelks has clear costs—it takes more time to find a large whelk. Selective birds, one might think, would get fewer calories per hour of hunting than birds that were less fussy about prey size.

In addition to firmly preferring larger whelks, the crows consistently fly up with them for about five meters (16.5 feet) before dropping them. Why not three meters, or four? Lower drop flights would save time and energy that could be spent finding more prey.

On the other hand, if the birds' behavior evolved through natural selection, then their decisions ought to contribute greatly to reproductive success, perhaps by maximizing caloric gain in the time spent foraging. Reto Zach (1979) figured out a way to test this proposition, by accepting the assumption that the unconscious goal of whelk-hunting crows was to maximize their intake of calories during the time they devoted to searching for this food. If true, then (1) large whelks should require fewer drops to break than medium or small specimens, and (2) any improvement in the odds of breaking a whelk from heights greater than five meters should be slight, whereas drops below five meters should yield a markedly lower breakage rate.

Zach tested these predictions by building a simple apparatus for elevating whelks of different sizes to various heights on a small platform; he then pushed the whelks off the platform so that they fell onto a hard surface. He found that as predicted, large whelks were substantially more likely to break open when dropped from any height (**Fig. 6–31**). Thus, they required fewer time- and energy-consuming drop flights than medium or small whelks. Also as predicted, the probability that a large whelk would break open increased sharply up to around five meters in drop height, after which the additional improvement in opening whelks was slight. In other words, crows that dropped their prey from three or four meters would be forced to try and try again, much more than birds using the five-meter drop height. However, crows that took their prey up to six meters or more would be only marginally more likely to break a whelk on any given drop, and therefore they would not be compensated for the greater energetic expense of the higher flights.

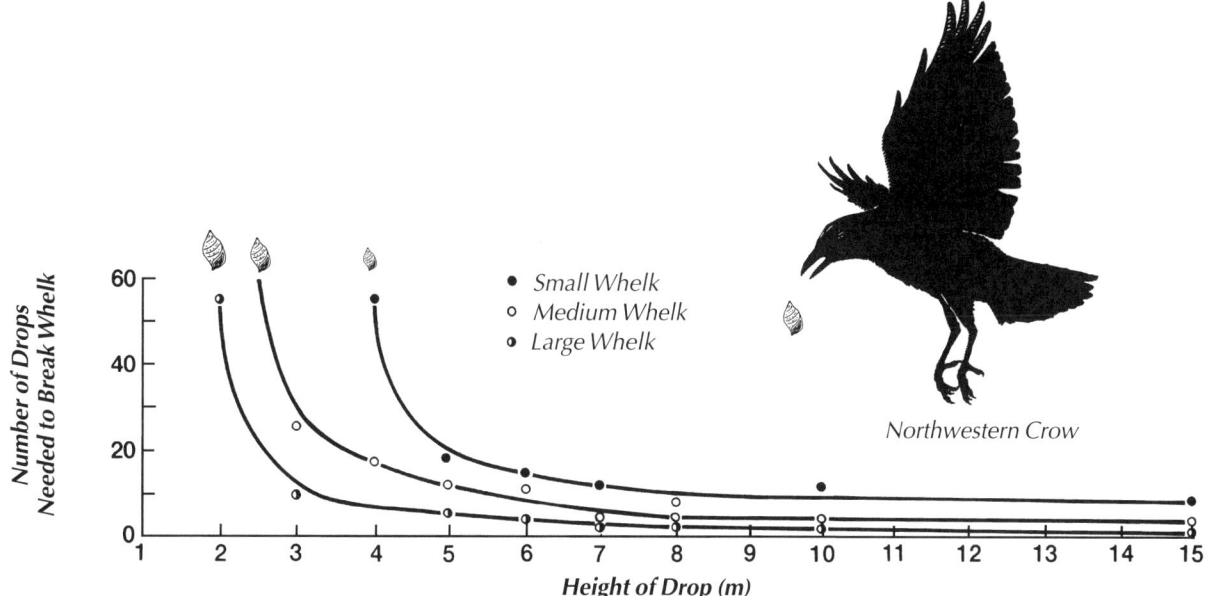

Figure 6–31. How Northwestern Crows Feed On Whelks: *Northwestern Crows break open whelks and other molluscs to obtain their flesh by carrying them up into the air and dropping them onto a hard surface. This graph shows data from Zach (1979) on the number of times a whelk had to be dropped before it broke, for different drop heights (in meters). Note that for any drop height, large whelks break more easily (with fewer drops) than medium or smaller ones; also note that the breakage rate increases sharply with height, but only up to about 5 m (16.5 feet). The crows' behavior seems to match these findings, as they strongly prefer large whelks over small and medium ones, and typically drop the large whelks from a height of about 5 m (16.5 feet). The crows thus maximize their caloric intake through their choices in feeding behavior—an example of optimal foraging. From* The Cambridge Encyclopedia of Ornithology, *edited by Michael Brooke and Tim Birkhead, Cambridge University Press, 1991. Reprinted with permission of Cambridge University Press.*

Because small and medium-sized whelks were actually less likely to open when dropped from five meters, and because large whelks contained more meat, the crows maximized their net caloric gain through their foraging choices. Nevertheless, not every bird is likely to be an energy-maximizer with respect to every item in its diet. Indeed, Northwestern Crows almost certainly make some foraging decisions based on the nutrient value of a food item (as well as its caloric content), the risks required to hunt for it, and other factors. Hypotheses about the possible adaptive value of foraging behavior that take these factors into account will be very different from the energy-maximization explanation illustrated above.

For example, Steven Lima noticed that chickadees often (but not always) leave an exposed bird feeder with a sunflower seed, flying some distance back to dense cover to eat the food before returning for another seed. For a bird attempting to maximize the number of calories eaten per minute of foraging, the sensible approach would be to prepare and eat the food bits at the feeder rather than spending time and energy flying to and from the food source. But Lima recognized that the reproductive success of foraging chickadees may be influenced by more than their efficiency in acquiring calories. Assuming that preparing and eating some foods under concealing shelter is safer than eating them out in the open, he proposed that chickadees compromise energy intake to improve their survival chances (Lima 1985).

Lima tested this hypothesis by predicting that if he could experimentally heighten the chickadees' perception of the risk posed by predators, the birds would take a higher proportion of their food to cover. He arranged for a "predator"—a model of a hawk that "flew" along a wire—to fly over a feeder being visited by a flock of chickadees. As he predicted, after exposure to the "predator," the birds became substantially more likely to take sunflower seeds to cover than they had been before. Thus, chickadees that sensed a predator in their neighborhood modified their foraging behavior, sacrificing short-term energy gain to improve their long-term survival chances.

Antipredator Behavior: Why Do Some Birds Mob Predators?

I still remember vividly an encounter between a Cooper's Hawk and a pigeon that I witnessed more than 30 years ago. The hawk chased the pigeon over an open pasture near my home in southeastern Pennsylvania. The big, gray hawk relentlessly kept pace as the pigeon zigged and zagged, forcing the doomed bird lower and lower until it ran out of room to maneuver. As I watched with racing heart, the Cooper's Hawk slammed into the pigeon, knocking it into the grassy field, where it lay incapacitated with the hawk perched on its body.

Most birds run a gauntlet of predators every day of their lives, and much of their behavior bears the imprint of selection by their enemies: the remarkable aerial agility of pigeons and other prey species, which nevertheless provides no absolute guarantee of escape from all predators; the vigilance of chickadees on a feeder, always ready to dive for cover should they spot danger; and the alarm calls given by Mexican Jays that have detected a snake in a nest tree (**Sidebar 3: Defense Behavior**). Even some seemingly trivial actions, such as parent birds removing eggshells from exposed nests, may have evolved as an antipredator response. Niko Tinbergen showed long ago that egg- and chick-eating crows could use opened eggshells, which are conspicuously white on the inside, as a visual cue to locate nests with vulnerable nestlings or unhatched eggs. Breeding adults have something to gain by removing this cue to their nest's location.

Given that as a general rule birds sensibly seek to minimize contact with their predators, it is surprising that a great many species occasionally seek out deadly enemies (**Fig. 6–32**). Eastern Kingbirds badger Red-tailed Hawks, Northern Mockingbirds swoop over domestic cats, and wood-warblers assemble noisily around Great Horned Owls. The reproductive costs of these activities to a mobber are not hard to discern; they include the chance that the mobber will be captured by the harassed predator (or by another enemy lurking nearby) as well as the time and energy invested in the activity.

(Continued on p. 6·57)

Black-tailed Gnatcatcher

Elf Owl

Figure 6–32. Black-tailed Gnatcatchers Mobbing Elf Owl: Many birds group together to harrass their predators—diving at the predator, flying around it, and making lots of noise (which attracts even more mobbers), but usually keeping just out of the predator's reach. They invest valuable time and energy in this behavior in spite of some risk of capture. Why mobbing occurs has intrigued scientists, who have suggested a variety of benefits that mobbers may reap (see Table 6–2). Drawing by Marilyn Hoff Stewart, courtesy of John Alcock.

6

Sidebar 3: DEFENSE BEHAVIOR
Sandy Podulka

As constant subjects of predation, all birds have numerous behavioral adaptations to promote their survival. Birds are ever alert, ready to respond to danger signals. Sharp sounds or quick movements invariably trigger escape behavior, but if the sounds and movements prove to be harmless and are repeated often enough, the birds become habituated to them and no longer respond. Birds continually watch their surroundings, including the sky. Those typically preyed on by hawks respond with alarm to nearly any shape moving overhead that suggests an avian predator.

How birds respond to danger signals depends on their stage of the breeding cycle, as well as the types of predators they must evade. Some threaten and attack predators most vigorously when they have eggs or young to defend, fleeing or hiding from danger at other times.

Some of the more common reactions of birds to danger signals are discussed below. One group defense, mobbing, was discussed earlier in this chapter and in Ch. 2, Mobbing.

Fleeing and Freezing

Threatened by a predator such as a hawk, many birds flee to the nearest cover, where they "freeze"—stay motionless with feathers sleeked, head in line with body, one eye cocked in the direction of the predator, and legs flexed for a quick take-off. Birds on the ground crouch to eliminate shadows cast by the body. Birds with cryptic coloration commonly respond to danger by freezing in place, a habit that no doubt accompanied the evolution of their concealing coloration (Fig. A). Open-country birds that lack cryptic colors, or cover in which to hide, rely on fast or erratic flight to outdistance or outmaneuver their predators.

Figure A. Piping Plover Chick Defense Behavior: Threatened with danger, some birds, particularly those with cryptic coloration such as young shorebirds, freeze in defense, crouching against the ground to eliminate their shadows—behavior that makes them more difficult to distinguish from the background. Photo by Marie Read.

Figure B. Snake Mimicry: An Antipredator Behavior by a Nesting Black-capped Chickadee: When surprised at its nest hole by a potential predator, such as a squirrel, a Black-capped Chickadee may give a dramatic display that strongly suggests a striking snake. The bird rises up, lifting its head up high, and then lunges forward, bringing its head sharply down while giving a hissing call. At the same time it spreads its wings forcefully, sometimes hitting them audibly against the nest cavity walls.

a. Killdeer

b. Black Skimmer

Figure C. Distraction Displays: a. Killdeer: *Feigning injury by dragging and flapping one wing, a Killdeer gives a broken-wing display accompanied by distress calls, slowly fluttering along to divert the attention of a potential predator away from the bird's nest on the ground. Photo by T. J. Ulrich/VIREO.* ***b. Black Skimmer:*** *Near its nest, a Black Skimmer gives a similar distraction display, involving exaggerated spreading motions of both wings to suggest injury. Photo by A. and E. Morris/VIREO.*

Threatening

Confronted at close quarters with potential danger and unable or disinclined to flee, birds may perform threat displays similar to those they direct toward members of their own species, though usually more exaggerated. These threat displays may include special features to make the performers appear as formidable as possible (see Fig. 6–16).

Snake-mimicry

When a predator draws near the nest, adults of some cavity-nesting birds, such as chickadees, may mimic a snake. From inside the hole the bird hisses, and may even open its mouth and sway back and forth, sometimes thumping its wings against the walls of the hole **(Fig. B)**. Squirrels and raccoons may be startled enough to give up and explore a different hole.

Giving Distraction Displays

Birds with nests and young sometimes feign injury when a human or other predator intrudes upon them. Ground-nesting birds from Ostriches to songbirds fake injury by fanning, beating, and dragging one or both wings; fluffing the back and rump feathers; and spreading the tail—always revealing any bold, attention-grabbing colors or patterns **(Fig. C)**. During the act they give "distress" calls, and alternately move with and away from the intruder. If the intruder follows, the bird suddenly "recovers" just in time to avoid capture. Birds nesting in trees feign injury too, fluttering or parachuting "helplessly" earthward.

Birds that nest on the ground in thick grass or other vegetation may sneak off the nest away from the intruder and then scurry along like a meadow vole, occasionally hopping up into view to attract attention while continuing to run away. This is appropriately termed "rodent-running."

As with other defense behaviors, the intensity of distraction displays varies with the stage in the nesting cycle, being greatest from a few days before the eggs hatch until the young near independence. If the displaying bird is harried for a long time, however, the performance gradually wanes in intensity. Sometimes a species gives a distraction display to one kind of intruder, but attacks another kind. A Killdeer, for instance, feigns injury vigorously when a human discovers its nest, yet attacks any cattle that come near enough to step on the nest. (Whether humans should be flattered or insulted by this difference in treatment is unclear.)

Attacking

Birds with nests and young may attack a predator outright, instead of just threatening it, particularly when they are safe from physical retaliation themselves. A Red-winged Blackbird, for example, will pursue a large hawk or crow and savagely strike it from above as it passes over the home marsh. Nesting Northern Goshawks, Great Horned Owls, and a few of the other large birds of prey may actually strike humans that intrude near their nest, as will some colonial seabirds such as the powerful skuas.

Some birds vomit unpleasant substances in self-defense. Vultures, herons, gulls, and all the tubenoses (albatrosses, petrels, fulmars, storm-petrels, and diving petrels) use this technique to ward off predators (**Fig. D**). Fulmars are the most accom-

Figure D. Northern Fulmar Projectile Vomiting in Self-defense: Fulmars, together with their marine "tubenose" relatives, feed in the open ocean, their stomachs converting fat-rich prey into an oily mix of flesh and fluid, which they regurgitate to feed to their young. They also may use this foul-smelling substance to repel a potential predator approaching the nest: adult and young fulmars alike may vomit this stomach oil forcefully at the intruder, projecting the fluid for several feet.

plished at this art: a bird on the nest can spit foul-smelling oil from the stomach at a predator 2 to 3 feet (0.6 to 0.9 m) away. Even the unhatched young can spit—through small holes in their pipped eggshells.

Massing

Birds with gregarious habits often respond to a predator by massing into a compact formation, making it dif-

ficult for the attacker to isolate and capture just one bird. When threatened, ducks on a pond bunch up on the water, blackbird and starling flocks pull quickly into a dense ball, and shorebirds take flight in a tight flock that swerves and dips while trying to elude the pursuer (**Fig. E**). These dramatic defense maneuvers inspire awe in anyone lucky enough to witness them. For more on the ben-

Figure E. Massing in a Flock of European Starlings: Flocking birds, such as starlings, blackbirds, and shorebirds, often respond to an approaching predator by tightening their flock formation into a compact mass, making it difficult for the attacker to isolate and capture any one individual. **a.** Typical density of an unthreatened starling flock in flight. **b.** The starling flock draws together into a tight ball when a raptor approaches.

efits of flocking, see Sidebar 4: Living in Groups.

Giving Alarm Calls

Nearly all birds have one or more calls that they give in response to nearby danger, usually a predator. Jays, domestic chickens, and many others even distinguish between ground and aerial predators by giving different calls. Some birds, especially the songbirds, give a low, easy-to-locate *chink* to warn of danger on the ground and a harder-to-locate, high *seet* or *zee* when the danger is airborne (see Fig. 7–18 and Track 18 on CD). Having different types of alarm calls allows birds to respond appropriately to the type of danger at hand—taking flight to escape a cat or fox, and freezing or diving into thick cover to avoid a Sharp-shinned Hawk or Merlin. The similarity among the alarm calls of many species is a striking example of convergent evolution and allows birds to avoid predators detected by species besides their own. Even some mammals, notably the ground squirrels, have calls that follow these patterns. By the same token, deer, foxes, and squirrels routinely heed the loud warnings of Blue Jays, and alert birders can discover hawks and owls by tuning in to avian alarm calls.

Florida Scrub-Jays also use different calls for ground and aerial predators. A high-pitched, thin screech warns of a falcon or accipiter, prompting nearby jays to dive for cover, whereas a scolding sound incites others to mob the terrestrial intruder, usually a cat or large snake **(Fig. F)**. But in addition, scrub-jays vary the intensity of their alarm calls to convey the urgency of the situation, giving several rapid screeches when danger is close by, for example (J. W. Fitzpat-

Figure F. Florida Scrub-Jay Giving Alarm Call: *In response to danger, such as the approach of a predator, most birds give alarm calls, both to warn others and to alert the potential attacker to the fact that it has been sighted. Here, a Florida Scrub-Jay serving as a sentinel calls to warn its flockmates of impending danger. Florida Scrub-Jays and some other species have different types of alarm calls for aerial versus terrestrial predators. But scrub-jay vocal complexity goes even further: call intensity may vary with the urgency of the situation. Photo by Brian Kenney.*

rick, personal communication). The most obvious function of alarm calls is to warn others of the approach of a predator in time for them to freeze, flee, hide, or aggregate—whatever it takes to foil an attack. But alarm calls also let the predator itself know it has been sighted. Having lost the advantage of surprise, the predator may decide to abort its attack. If a predator is thwarted each time it stalks a particular prey species or hunts in a particular area, it may eventually hunt elsewhere or seek other types of prey.

Although alarm calls clearly benefit the listening prey animals, the advantages to the caller are not so

apparent. In many cases the caller, by drawing attention to himself, presumably increases his own risk of being preyed on. This assumption is extremely difficult to test in birds, because attacks are so rare and the caller is often hard to locate and follow. In species such as jays and crows that give alarm calls from high, exposed perches, it seems reasonable to assume that the caller incurs an added risk, but no data on this topic currently exist.

Why, then, should any bird give an alarm call, if doing so decreases his chance of survival? Researchers offer several hypotheses to explain this dilemma. First, if the caller can save several relatives at only a small risk to himself, the tendency to sound the alarm can actually be promoted through natural selection, because close relatives share a percentage of genetic traits. In other words, by warning relatives, the caller increases his chances of having at least some of his genetic material passed on to the next generation (see Fig. 6–48 and Mate Choice: Why Cooperate in Courtship Displays?, later in this chapter). Any breeding bird warning a mate or fledglings of impending danger, then, improves the young's chances of survival, an evolutionary benefit. Nonbreeding birds with long-term pair bonds also benefit from warning a mate, both because they will not have to spend energy finding a new mate and because experienced pairs often have greater reproductive success than new pairs. Birds that live in extended family groups or colonies should benefit even more, because their alarm calls are likely to warn more relatives.

We would predict, then, that in colonies with related individuals, those with more relatives nearby would give more alarm calls. This hypothesis has not been tested in birds, but supporting evidence is available for mammals. Belding's Ground Squirrels live in **matrilineal** societies: the females remain in their natal colony, while the males disperse upon reaching adulthood. Thus, in any given colony females are related to each other and to many of the young, but males are only related to their own young. In these societies, females give more alarm calls than males and call the most when close relatives are nearby (Sherman 1977).

An alternative explanation for the existence of alarm calls is that the caller as well as all listeners clearly benefit if, as mentioned before, a predator stops hunting the caller's species or in his area because it is discouraged by repeated run-ins with an efficient alert system.

Some researchers also suggest that alarm calling can benefit the caller if it increases the likelihood that another individual, even an unrelated one, will return the favor at a later date—a very difficult prediction to test.

In at least one species, the White-fronted Bee-eater of East African savannas, females appear to use alarm calls deceitfully (P. H. Wrege and S. T. Emlen, personal communication). During breeding, males frequently chase and attempt to force copulations with females other than their mates. Chases can be tiring for the females, as they are pursued relentlessly both near and away from the nesting colony. Occasionally, a fleeing female gives the alarm call for an aerial predator—causing most birds, including her pursuers, to dive into the bushes, ending the chase. In the open savanna habitat, where bee-eaters may be easy prey for Gabar Goshawks and various harriers, ignoring the aerial alarm call carries a high cost. If females "cried wolf" frequently, the tactic might become less effective, but since they do so

Figure G. Sentinel Behavior in American Crows: *Some birds that spend all or part of their lives in small, stable groups have sentinel systems in which group members take turns watching for predators while their companions forage. American Crows may live in extended family groups encompassing individuals from up to several generations, all of whom help in the nesting attempt. On their breeding territories, American Crow family members take turns acting as sentinels—as pictured here. When crow families gather into larger flocks during the winter, the system of coordinated vigilance appears to break down.*

only occasionally, it is in the best interests of others to heed the warning and take cover.

The questions surrounding who gives alarm calls and when and why they call are exciting to explore, but much more research will be necessary before answers are found.

Posting Sentinels

Some birds that live in small, stable groups have sentinel systems in which group members take turns watching for danger. In Florida Scrub-Jays, which live year round in extended family groups, all individuals take turns sitting on an exposed perch, mechanically scanning the sky for raptors (McGowan and Woolfenden 1989). When the sentinel spots a predator, he or she gives the appropriate type of alarm call.

In some parts of their range, American Crows also live in extended family groups; some contain as many as 15 birds from up to seven generations. On their breeding territories, American Crows take turns acting as sentinels **(Fig. G)**, but when they gather into larger feeding flocks, coordinated vigilance appears to break down. One or more crows often can be seen sitting in trees while others forage on the ground, but they do not appear to take turns, and whether they even serve as "official" lookouts is not clear (K. J. McGowan, personal communication). As these larger flocks are less stable, often changing in composition from day to day, the development of a coordinated sentinel system is probably just not possible. ∎

Table 6–2. Possible Explanations of the Adaptive Value of Mobbing

1. **The "Predator Move On" Hypothesis:** Once the mobbers inform the predator that they are alert to its presence, the predator leaves. The mobbers gain because they don't need to modify their activities to avoid the now-departed enemy, and the predator gains by leaving the area to hunt for unsuspecting prey elsewhere.

2. **The Predator Distraction Hypothesis:** The mobbers protect their offspring by keeping the predator from concentrating on the search for their vulnerable young.

3. **The Alarm Call Hypothesis:** The activities of the mobbers alert others (notably mates and relatives) of the presence of a predator; the alerted birds then can take action to avoid danger.

4. **The "Attract a Predator of the Predator" Hypothesis:** The mobbers' noisy, conspicuous behavior reveals to larger predators the whereabouts of the smaller predator that is being mobbed.

The benefits of mobbing are much more difficult to identify. Kingbirds cannot seriously injure a hawk by pulling at its tail feathers any more than a Yellow-rumped Warbler can physically harm a Great Horned Owl. The puzzle of mobbing is sufficiently intriguing to have attracted a great many behavioral ecologists, who have developed a large number of alternative hypotheses (Curio 1978), a sample of which appear in **Table 6–2**.

Some of these hypotheses have been tested for certain species. The communal mobbing of foxes, hawks, and large gulls by some of the smaller colonial-nesting gulls, for example, has been explained as a form of parental care under the predator-distraction hypothesis. When egg or nestling predators show up, nesting Black-headed Gulls of northern Europe often respond by flying at them in force and trying to strike them with a beak or foot while defecating on them. Although these attacks cannot gravely injure the predators, they may expose them to infection. Perhaps for this reason, a fox or Carrion Crow keeps an eye on mobbing gulls and tries to avoid contact with them, which may compromise its ability to search for eggs or nestlings.

Hans Kruuk (1964) tested this proposition by placing chicken eggs on the ground at intervals, beginning outside a large colony of Black-headed Gulls and continuing into the nesting area itself. The idea was to test whether, as expected, eggs outside the colony were more likely to be eaten by egg-hunting predators (notably Carrion Crows and Herring Gulls) than those inside the colony. If mobbing truly distracts predators, then eggs placed near nesting Black-headed Gulls should be better protected by the colony's aggressive response to intruders.

Kruuk found that "survival" of the experimental eggs was indeed a function of their proximity to the colony **(Fig. 6–33)**. Predators foraging outside the colony were not mobbed by bands of gulls and had little trouble finding and eating Kruuk's donations. But as they approached the nesting area, egg hunters were increasingly likely to be mobbed, and their foraging success declined in proportion to the intensity of

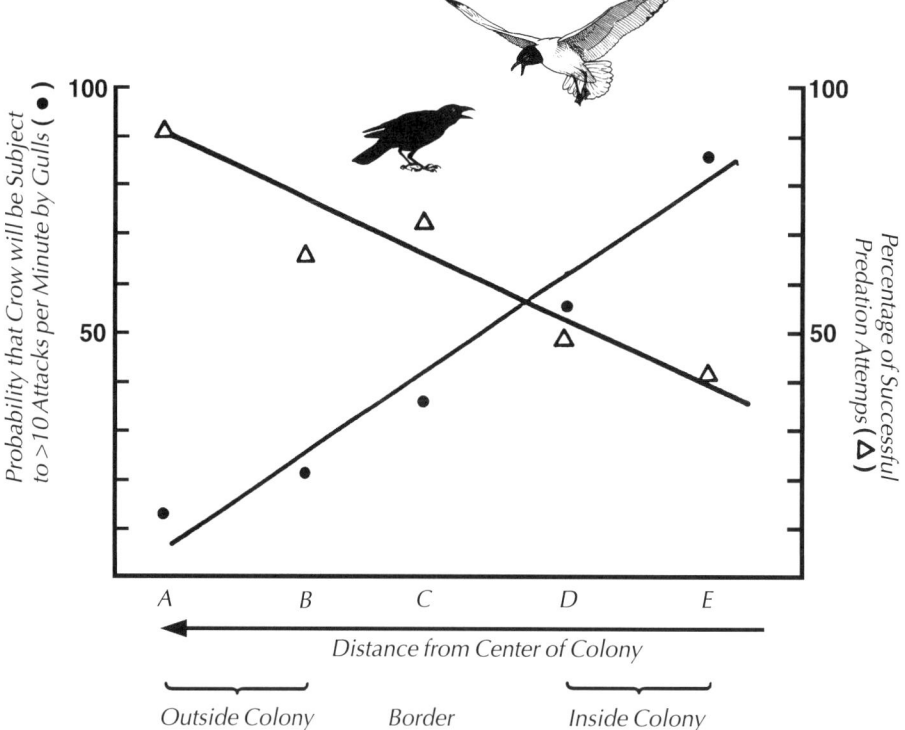

Figure 6–33. Effectiveness of Black-headed Gulls in Mobbing Carrion Crows: *Carrion Crows eat the eggs of Black-headed Gulls—colonial nesters of Europe who communally mob their predators, including Carrion Crows. Kruuk (1964) indirectly tested the effectiveness of this mobbing in deterring Carrion Crow predators by placing chicken eggs on the ground at various distances from the center of the colony (see x-axis of graph). He then gathered two types of data: One, the percentage of eggs eaten by any predator at each distance from the colony's center (open triangles, see y-axis of graph, at right side), and two, the probability of a Carrion Crow being mobbed intensely (more than 10 attacks per minute) by gulls at various distances from the colony (solid dots, see y-axis of graph, at left side). He found that eggs were less likely to be eaten by predators, and that crows were more likely to be mobbed, the closer that each was to the center of the nesting colony. This suggests that mobbing by Black-headed Gulls is effective in deterring the egg-eating crows. Adapted from Kruuk (1964).*

the mobbing. These results support the distraction hypothesis for the evolution of mobbing by nesting Black-headed Gulls.

Similarly, William Shields (1984) found that in nesting colonies of Barn Swallows, the birds that mobbed a potential predator (a stuffed screech-owl in this case) were far from a random sample of the birds in the area. Adult swallows with young in the nest were greatly over-represented in the group that took the initiative in swooping at the stuffed owl. Nonbreeding adults and juvenile birds that happened to be hanging around the colony almost never harassed the owl. Thus, in this case, mobbing occurs because parents try to protect their offspring by distracting predators.

Nest Spacing:
Why Do Some Birds Nest in Large Colonies?

The fact that Black-headed Gulls and Barn Swallows nest colonially enables individuals to gain protection for their eggs through communal mobbing. Perhaps this and other antipredator benefits account for the fact that colonial nesting is a fairly common phenomenon in birds. Clearly, birds that breed in colonies must gain some strong reproductive advantage because, as Richard Alexander (1974) points out, living closely with others has a great many obvious disadvantages in terms of survival and reproductive success. Clumping increases the chances that infectious disease and parasites will spread. For example, the offspring of Cliff Swallows nesting in large, dense colonies are more likely to be infested with blood-sucking parasites than the young of swallows nesting in smaller colonies. Moreover, when individuals live close together they can interfere with each other in many ways, including sneaking copulations and killing each other's offspring. Unmated males of the colonial-nesting Barn Swallow regularly practice infanticide, plucking helpless nestlings by the head from unguarded nests and throwing them to the ground. (In this way, killer males sometimes gain partners who produce a second brood with them rather than with their previous companions.)

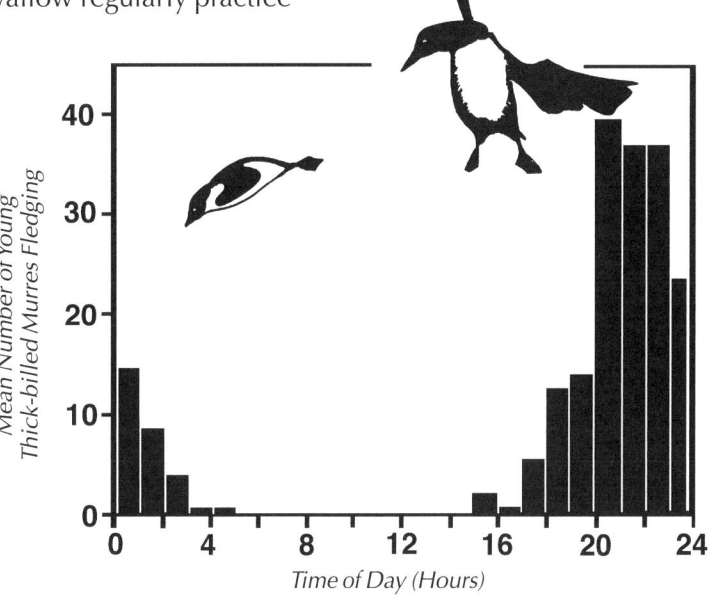

The question becomes, what reproductive advantages does social living offer to outweigh its clear disadvantages? Let us apply this question specifically to seabird colonies, which provide some spectacular examples of aggregated nesting birds (see Fig. 6–15a). For further discussion of the advantages and disadvantages to group living, see **Sidebar 4: Living in Groups**.

Researchers have proposed several hypotheses for colonial breeding, among them the **information center hypothesis** and the **dilution effect hypothesis**. The first hypothesis focuses on the possible foraging benefits that birds living together might enjoy. In colonial species, such as gannets or Cliff Swallows, unsuccessful hunters might find feeding grounds by following previously successful foragers to good hunting spots. In contrast, the dilution effect hypothesis emphasizes the possible antipredator benefits of living together in numbers large enough to overwhelm the consumption capacity of local predators. If local predators kill, say, 100 gannets in a breeding season in a given area, then birds that form a nesting colony of 1,000 members are safer (one chance in ten of being killed) than if the colony were smaller (for example, 200 birds, each with one chance in two of becoming a victim).

Evidence in support of both hypotheses exists for certain species. Thick-billed Murres nest in vast rookeries in the Arctic. Their young fledge before they can fly, jumping off their cliff nest sites in an attempt to reach the ocean far below. Although a parent accompanies each fledgling on its dangerous descent, gulls are waiting to capture them in midair, and foxes hunt for those that reach the ground and are walking to the sea. Most of the young murres leave the colony at about the same time of day during the fledging period **(Fig. 6–34)**. A synchronous exit floods the local gulls and foxes with potential prey, improving the

Figure 6–34. Synchronous Fledging in Thick-billed Murres: This arctic species nests in huge colonies on cliff ledges. The young fledge while still flightless and head immediately for the safety of the ocean. Although always accompanied by a parent, they are extremely vulnerable to predation during their quick descent to the sea. Data from Daan and Tinbergen (1979) show that young Thick-billed Murres (known as Brunnichs' Guillemots in Europe, where the study was carried out) leave the nesting colony primarily during a four-hour period (between 8:00 p.m. and midnight—see graph) each day during the fledging period. The huge numbers of fledgling murres thus overwhelm the local predator population with potential prey, increasing the chance of survival for any individual murre chick. This synchronous fledging is consistent with the **dilution effect** *hypothesis (see text) proposed to explain colonial breeding. Adapted from Daan and Tinbergen (1979).*

(Continued on p. 6·68)

Sidebar 4: LIVING IN GROUPS
Sandy Podulka

Some birds live together in permanent groups, some group only during certain seasons, and a few never group at all. Groups can be small or large, made up of a single or mixed species, and may or may not have a defined social order. The types of groups are as varied as the reasons for joining them. Groups range from winter feeding flocks of Red-winged Blackbirds, communal roosts of Turkey Vultures, migrating flocks of shorebirds or warblers, and nesting colonies of herons or swallows, to permanent complex societies of White-fronted Bee-eaters or jackdaws **(Fig. A)**. But one fact is true for all: life in a group is always a compromise. This sidebar explores some of the reasons that birds live in groups and some of the trade-offs they make to do so. Because colonial nesting is discussed in the text, and complex societies are covered in Sidebar 6: Bird Families as Models for Understanding Ourselves, only the nonbreeding season is considered here.

When they are not breeding, most birds live in groups. Many land birds that are territorial as pairs in the breeding season form flocks at other times. Some, including warblers, tanagers, thrushes, orioles, and robins, migrate south and spend the winter in mixed flocks **(Fig. B)**. Year-round residents such as chickadees and Song Sparrows form small single-species flocks and range over larger areas than their breeding territories, whereas juncos, American Tree Sparrows, White-throated Sparrows, and others form flocks and shift southward, the distance depending on the severity of the winter weather.

During the nonbreeding season, most of a bird's time and energy is devoted to surviving: avoiding predators and getting enough food. Whether a bird lives in a group at

Figure A. Group of Eurasian Jackdaws: *This species is gregarious year round, nesting colonially as well as forming communal roosts in both the breeding and nonbreeding seasons. Drawing by Charles L. Ripper.*

this time depends almost entirely on the types and numbers of predators it must escape and on the amount, type, and distribution of its food.

Groups and Predation

Groups are by nature more conspicuous and easier for predators to locate than individual birds, but group advantages in evading predators may more than compensate for their prominence. Group advantages such as mobbing and the "dilution effect" are discussed in the main text. In addition, a group has many eyes and ears that are alert to danger, allowing earlier predator detection. In experiments with trained goshawks, Kenward (1978) found that they were less successful when their prey, Common Wood-Pigeons, were in larger flocks—mostly because the large flocks spotted the hawk at greater distances and took flight. Group vigilance also leaves

each individual bird in a flock more time to forage. Shorebirds, for example, spend roughly 20% of their time scanning for predators when alone, 5% when in a flock of 5 birds, and 3% when in a flock of 25. On the African savannas, each Ostrich spends a smaller proportion of its time scanning for predators when in a group than when alone, yet the percentage of time that at least one bird is scanning increases with flock size (Bertram 1980). Furthermore, because each Ostrich raises its head to scan for predators independently of the others, it is nearly impossible for a predator such as a lion to sneak up when no bird is looking—the "down times" are unpredictable for groups.

Once an attack is under way, groups, especially aerial flocks, are better than lone birds at avoiding capture. When they spot a hawk, feeding shorebirds take to the air

6

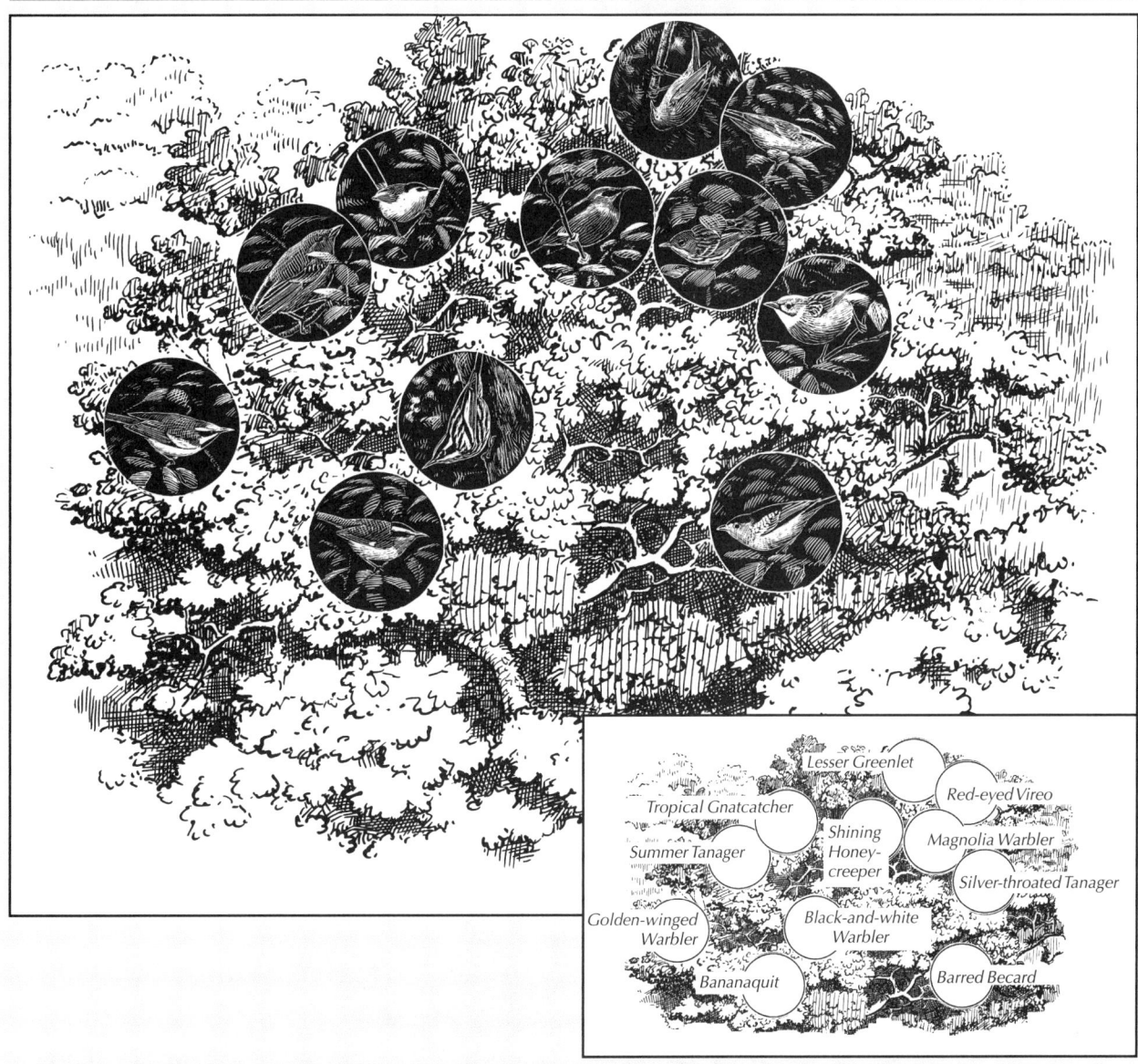

Figure B. Mixed-Species Flock in the Neotropical Forest Canopy: Certain birds, termed Neotropical migrants—including various warblers, tanagers, thrushes, orioles, and robins—migrate from North America to the tropics of Central and South America, often joining mixed feeding flocks of residents and other migrants during the winter. Foraging together for nectar, fruit, and insects in the Neotropical forest canopy may be resident birds such as the Tropical Gnatcatcher, Lesser Greenlet, Shining Honeycreeper, Silver-throated Tanager, Barred Becard, and Bananaquit, right along with migrants such as the Summer Tanager, Golden-winged Warbler, Red-eyed Vireo, Magnolia Warbler, and Black-and-white Warbler. Migrants may join a mixed feeding flock for various reasons: to take advantage of residents' knowledge of good feeding sites, to increase their foraging success as prey are scared up by flockmates, or to obtain the antipredator benefits that come with being in a group.

en masse, and pigeon or blackbird flocks rapidly gather into tight clusters which, as they swoop and turn, confuse the predator, so that isolating and grabbing one individual is difficult (see Sidebar 3, Fig. E). You can simulate the "confusion effect" by throwing tennis balls at a friend. Is it harder or easier for him to catch a ball when three are thrown at once, or when just one is thrown? Gathering into a dense group also may be the result of each flock member attempting to move toward the middle—a safer spot because predators generally attack the edge, where their chance of separating one bird is greatest. Moreover, some researchers hypothesize that a Merlin or Peregrine Falcon might be reluctant to plunge at high speed into a tight flock of starlings or any other prey, since the risk of injury from accidental contact is too great.

Bird flocks, like fish schools, amaze all who watch them. How they move so quickly and gracefully as one "being," how they maintain uniform spacing, and why individuals never crash into each other have long mystified scientists, but frame-by-frame analysis of videos is beginning to reveal some of their secrets. When birds in the front of the flock curve left or right or swoop lower, the birds next to them follow suit, and so on through the flock in a nearly instantaneous chain reaction (see Fig. 5–47). It's hard for us to imagine such quick responses, but birds have much faster reaction times than we do—as becomes clear when you watch two birds twist and maneuver in an aggressive chase through vegetation. Although flock members "follow" whomever is in front at any given time, flocks appear to have no set leaders, since individuals shift positions within the group as they turn one way and another, often changing who holds the lead position. We have certainly just begun exploring how flocks ma-

neuver, and the more we learn, the more impressive they appear; but many mysteries remain.

Groups and Feeding

Groups offer many advantages to foraging birds, especially if their food is found in large patches that provide superabundant food for a brief period, but whose location is unpredictable at any point in time. Large schools of fish, aerial insect swarms, dead animals, and sizable trees or orchards at their peak flowering, fruiting, or seed-setting stages offer far more food than even the most gluttonous gull, tern, swallow, vulture, parrot, or blackbird can eat by itself **(Fig. C)**. Yet these foods are available only for a few hours (fish, insects, carcasses) or days (flowers, fruits, seeds). A bird seeking this type

of food nearly always obtains more in a group, as many eyes can search more effectively for the next ephemeral banquet.

Instead of maintaining close groups, some birds whose food is plentiful yet unpredictable seem to spread out and scan for food, usually keeping an eye or ear tuned to their neighbors. When one vulture finds a dead animal, for example, his neighbors see him and drop in, as do their neighbors, and so on, assembling birds from great distances to the feast **(Fig. D)**. One observer counted 237 vultures at a single horse carcass. On locating a fish school at sea, many petrels, shearwaters, gulls, and terns attract other birds by calls or flight behavior.

Many bird groups use special foraging strategies to exploit their

Figure C. Cormorant Flock Feeding on a School of Fish: *Group foraging may benefit birds whose food is distributed in patches that are unpredictable in their location at any point in time, but are so large that plenty of food is available for many individuals once a patch is located. Examples of this type of food resource include fish schools, insect swarms, and enormous fruiting or flowering trees. By grouping together, foraging birds are more likely to find such food patches than if they searched alone—more eyes and ears are on the lookout. Here, a large flock of cormorants feeds on a school of fish, attracting the attention of terns and gulls who join in the feeding frenzy.*

prey more effectively than a lone bird could. Groups of predators, for example, are much better than single predators at catching prey that is itself in groups: Eleonora's Falcons feeding on flocks of autumnal migrants crossing the Mediterranean are much more successful when they hunt in flocks (Walter 1979). Many water birds use social foraging techniques. For example, upon discovering a fish school, swimming American White Pelicans may form a line, and by beating and splashing their wings and dipping their bills into the water,

drive the fish into shallower water, trapping and concentrating them by circling around them or by forming a semicircle against the shore **(Fig. E)**. Cormorants and mergansers may use similar tactics. Avocets and Black-necked Stilts often band together, forming a wedge of birds wading through shallow water and gleaning aquatic insects, crustaceans, and molluscs along the way. Wintering Surf Scoters, when harassed by Glaucous-winged Gulls attempting to steal their catch, adjust the lengths of their dives so that they all surface

together (Schenkeveld and Ydenberg 1985). Since the gulls cannot hassle all the scoters at once, the chance that any one scoter's prey will be stolen is reduced. Ground-feeding blackbird flocks in fields may advance in a thick line, stirring up insects for each other as they proceed. While the ones in front pause to eat, birds from behind fly up to the front, pause to eat, and in turn are passed. From a distance, the noisy flock appears to roll across the field in a frenzy.

Some birds forage in groups containing many different species. In these "mixed species flocks" the member species usually have different types of prey or different methods of foraging, to avoid competition. To some, the benefits of flocking may be primarily predator evasion, as discussed earlier. But others may actually increase the foraging success of their flockmates by scaring up prey for each other as they move through an area. Shorebirds, waders, waterfowl, and numerous small insectivorous birds in both temperate and tropical forests typically feed in mixed flocks. Migrants passing through an area often join local mixed flocks, gaining some of the advantages discussed above, and possibly benefitting from the residents' knowledge of good feeding sites and ways to avoid local predators. One of the best ways to find migrant warblers in spring or fall is to listen for the chickadee flocks with which they often associate.

Although birds may be better able to find and exploit certain foods in a group, they must share whatever food they can find with other flock members. Competition for food may thus be a major factor in determining whether birds group and how large a group they form. In House Sparrow flocks, the first bird to discover food often gives a *chirrup* call, drawing in others to dine. In experiments by Elgar (1986), House Sparrows given a slice of bread in crumbled form gave the *chirrup* call, but those given a whole slice kept quiet. Although

White-backed Vulture

Rueppell's Griffon

Lappet-faced Vulture

Figure D. Vultures Descend on a Carcass in the African Savanna: *Vultures often search for food in very loose groups—the location of their food is hard to predict, so they spread out and search—but once found, there is plenty for everyone. The soaring vultures pay close attention to their neighbors' behavior, though. If one individual locates a carcass and drops down upon it, others circling nearby notice and follow, attracting the attention of still more distant neighbors who immediately fly in to join the ever-growing crowd of scavengers. Shown are White-backed and Lappet-faced vultures and Rueppell's Griffons descending to feed on a wildebeest carcass.*

Figure E. Group Foraging in American White Pelicans: *Cooperative (or social) foraging enables American White Pelicans to effectively exploit their schools of prey. When they locate a school of fish, groups of foraging pelicans may form a line (a), and by beating and splashing their wings and dipping their bills in the water in a coordinated way, they drive the fish into shallower water, trapping and concentrating them. During this process, the pelicans sometimes form a circle around the prey, and at other times they form a semicircle against the shore (b). American White Pelicans also may forage singly or in small groups, and many variations on the cooperative foraging technique exist. Other species of pelicans, as well as cormorants and mergansers, may use some degree of cooperative foraging at times.*

American White Pelicans

the amount of food was the same each time, a whole slice is not easily shared, so perhaps birds only invite a flock to divisible food. Another drawback to flock foraging is that the presence of many individuals may scare or alert certain types of prey, decreasing the capture rate for each bird. During the day, Common Redshank feed by sight, either solitarily or in widely spaced groups, seeking small shrimp that sit with only their tails protruding from the surface of the mud. At the first patter of redshank feet, the shrimp duck into the mud. At night, redshank feed in flocks by sweeping their beaks through the

mud and grasping any snails they can feel. One possible interpretation is that flocking helps redshank evade predators, but that during the day the cost of disturbing each other's prey outweighs the predator-related advantages (Goss-Custard 1976).

Grouping Versus Territoriality

A very few birds, including hummingbirds, woodpeckers, and birds of prey, do not form groups but remain territorial when not breeding. Nectar-producing flowers, the food of hummingbirds, occur in small patches, and the sites at peak flowering change slowly through the

year. Because nectar is generally not superabundant in any one spot, most hummingbirds defend territories that shift with the seasons. The food of woodpeckers and most hawks is evenly dispersed at low levels, so again, the birds spread out and defend territories. Hawks often appear to migrate in flocks, as thousands pass traditional lookouts such as Hawk Mountain, in Pennsylvania, on peak days. But most of these "flocks" are really concentrations that result from each individual favoring routes with good soaring conditions (for example, along ridges) and avoiding open water; the hawks do not actively seek each other's company **(Fig. F)**. The concentrations break up at the wintering grounds, where hawks once again defend territories. Vultures are the exception among raptors, as they rely on each other to quickly locate perishable carrion. This need to quickly locate isolated, but superabundant, sources of food has led to the development of a more gregarious social system.

We can think of territoriality and grouping as opposite ends of a continuum, with the details of food supply and predation pressure evolutionarily driving species or populations toward one strategy or the other. A stable, defensible food supply leads to territoriality, whereas unpredictable, indefensible, superabundant food promotes flocking. Some species change social systems as their food availability changes. When food becomes patchy, unpredictable, or inadequate, then crows, jays, magpies, and White Wagtails relinquish their territories and join flocks (Davies 1976; Verbeek 1973).

Communal Roosts

A few species spend the night in large aggregations termed communal roosts. Some roost in groups year round, others roost only in the nonbreeding season. Mixed groups of grackles, blackbirds, cowbirds, and robins roosting along the Mississippi River may contain up to 15 million birds. Vultures, ravens, crows, starlings, herons, egrets, and ibis are all notorious for their large, noisy roosts **(Fig. G)**.

There is much speculation on the functions of communal roosts. Some species certainly benefit from each other's warmth: on cold nights, more than 10 Eastern Bluebirds have been found together in one cavity **(Fig. H)**; and Common Bushtits, which normally won't let another bird approach within two inches, can be seen roosting shoulder to shoulder. Chickadees, titmice, Brown Creepers, and Tree Swallows also huddle in groups in cavities on particularly cold nights. Vultures often roost near sites where good early morning thermals develop, allowing them to start the day with less effort. And, as discussed in the chapter text, there is growing evidence that some communal roosts and nest colonies serve as "information centers." Cliff Swallow, Black Vulture, Common Raven, and a few other communal roosts appear to function in this way.

Occasionally roosts come into conflict with human desires, especially when birds gather in urban or suburban sites or areas prized for

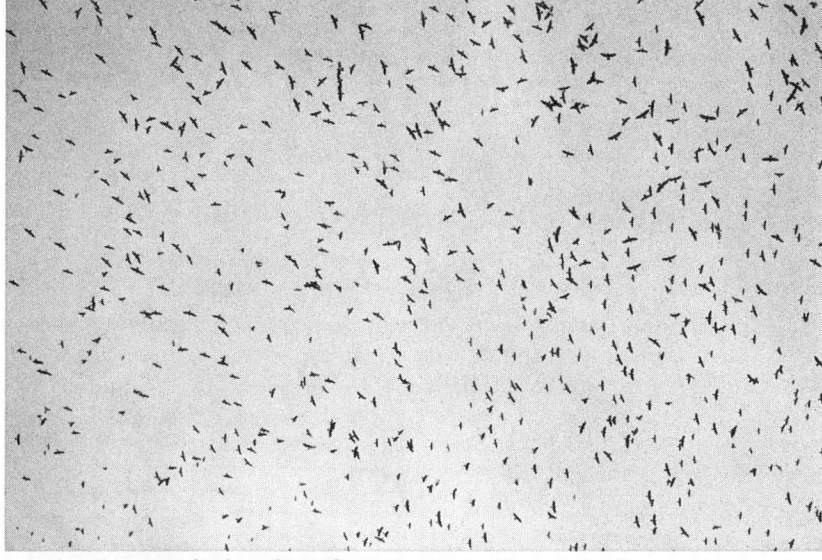

a. Migrating Broad-winged Hawks

b. Foraging Red-tailed Hawk

Figure F. Mass Migration Versus Territoriality in Hawks: a. Huge Kettle of Migrating Broad-winged Hawks: Hawks migrating en masse do not form true flocks—the groups are actually aggregations that result from each individual bird favoring routes with good soaring conditions, such as the thermals along a ridge (see Fig. 5–38), and also from each bird's tendency to avoid flying over open water. Photo by N. G. Smith/VIREO. b. Foraging Red-tailed Hawk Soars Alone Over its Territory: Whether on their breeding grounds or in their wintering areas, most birds of prey (other than vultures) rely on an evenly distributed, defensible food supply, so each individual holds a separate territory. Photo courtesy of Frank Schleicher/CLO.

agriculture or timber. People may be disturbed by the noise, masses of droppings, killing of vegetation due to the toxic effect of the droppings, or even destruction of trees—the extinct Passenger Pigeons roosted in flocks so thick that sometimes trees as large as two feet in diameter toppled under their weight.

But roosts also can provide hours of quiet pleasure for those willing to sit and watch them for a while. Most birds arrive and leave roosts at a particular light intensity, so you can enjoy a stream of birds flying in or out, and can estimate the number of birds at the roost. As sites within the roost may vary in quality—some are too exposed to the elements, some are too vulnerable to predators, some are targets for droppings from above that may soil feathers—you may see aggressive and submissive displays as individuals sort out who may perch where.

One evening on the island of Trinidad off the coast of Venezuela, I was lucky enough to visit a Scarlet Ibis roost in the Caroni Swamp. The engine noise of even our small boat seemed a crass intrusion on the lazy jungle river, so I was relieved when we paused and cut the engine. Ahead lay open water as vivid blue as the sky, cluttered with large, steeply mounded mangrove islands, rising like tortoises against the horizon. We waited, watching the blues and greens darken a bit as dusk approached, and finally, a bright red speck appeared in the sky. As it drew closer I could clearly see the black wing tips against the scarlet body, vibrant in the setting sun. It landed in a nearby mangrove. Another appeared, and another, and soon streams of red were coming in from all directions. As the islands slowly turned from green to scarlet, the clamor increased, as if someone were gradually turning up the volume on a recording of a noisy crowd. Then, in a vision etched forever in my mind, the last direct rays

Figure G. Communal Roost of Turkey Vultures: *Certain bird species spend the night in large communal roosts—some only during the nonbreeding season, others year round. Some roosts consist of a single species such as the Turkey Vultures shown here, whereas others may contain many different species: herons, egrets, and ibis often roost together, as do various blackbirds, grackles, cowbirds, and starlings. Other species that may roost communally at certain times of the year include swallows, crows, ravens, geese, and cranes. Photo by Joe McDonald.*

Figure H. Eastern Bluebirds Roosting Together in Winter: *One of the functions of roosting together may be to keep warm at night in winter. This group of 11 Eastern Bluebirds (females on the right, males on the left) spent chilly nights together in a hollowed-out log. Photo by Michael L. Smith.*

of sun lit the bird-filled mangrove mounds, setting the birds afire, like Christmas lights red as glowing embers.

The experience would have been perfect, except for the actions of other humans. A boat arrived part-way through the show, and motored close to the roost trees, scaring the birds up into a cloud of red that quickly settled down again. Then another boat went by, the people clapping their hands to scare the birds and laughing at their success. This was a typical evening, our guide told us. Although legally protected, the birds were still frequently killed and served as delicacies—often by those responsible for their protection.

Birds that gather in conspicuous groups often have been persecuted by humans. Although the extinction of Passenger Pigeons was partly due to the clearing of their oak-beech forests for agriculture, the birds' huge communal roosts and dense migratory flocks and nest colonies made them easy to kill. People used guns, nets, dynamite, and clubs to slaughter the birds, and burned grass or sulfur below their roosts to suffocate them by the thousands **(Fig. I)**.

Thankfully, most predators do not have such potent destructive forces at their command, so the conspicuous nature of groups is rarely as detrimental as it was for the unlucky Passenger Pigeons. Natural selection continually weighs the benefits and costs of living alone and living in groups. In choice of social systems as in other aspects of behavior, individuals who adopt the best strategy or whose behavior is flexible enough to change as food or predation levels change, leave more of their kind. This process of perpetual fine-tuning allows generation after generation to meet the changing ecological challenges they face.   ∎

Suggested Readings

Krebs, J. R. and N. B. Davies. 1993. Living in Groups. Chapter 6 in: *An Introduction to Behavioural Ecology.* Oxford & Boston: Blackwell Scientific Publications.

Figure I. Shooting Wild Pigeons In Iowa: Birds that gather in conspicuous groups always have been vulnerable to persecution by humans. The Passenger Pigeon, shown here, migrated in dense flocks and roosted in huge aggregations, both of which allowed people to slaughter them by the thousands. People shot multitudes of birds from passing flocks, and used many methods to kill roosting birds en masse. Clearing of oak-beech forests for farming also contributed to the species' extinction. From Leslie's Il-lustrated Newspaper, September 21, 1867. Artwork reprinted with permission of the State Historical Society of Wisconsin (item number WHi(X3)14277).

odds of survival for any one of the departees. These observations are consistent with the predictions of the "dilution effect" hypothesis.

Danielle Clode (1993), however, argues that the clustering of nesting birds actually *attracts* many predators, and thus the "dilution effect" only slightly reduces this negative consequence of colonial life. Instead, she favors the "information center" hypothesis, arguing that the primary benefit to individuals in large seabird colonies lies in foraging gains.

Clode's position is based on an analysis of three categories of birds that feed in or near the ocean during the breeding season: **(1) offshore or pelagic feeders**, which hunt schooling fish far out to sea and far from their nests, **(2) inshore feeders**, which forage close to shore, reasonably near their nesting areas, and **(3) shore feeders**, such as marine waders, which hunt on shore in the intertidal zone, closest of all to their nests. Clode points out that the birds that feed farthest from their nests form the largest and most densely aggregated colonies **(Fig. 6–35)**. Birds such as gannets and murres hunt for patchily distributed prey that is unpredictable in space but extremely abundant when located. Under these circumstances, individuals may well gain by observing their fellow hunters' successes. Moreover, not only can they locate large schools of fish well out at sea by living where they can observe others, they also can join forces with their fellow colony members to dive into the school together. Mass attack defeats the evasive tactics of schooling fish, providing more food for all. In contrast, inshore and shore feeders—for example, the small terns and shorebirds such as sandpipers—pursue prey that is probably more evenly distributed. If so, communal foraging produces fewer benefits because there is no prey "hot spot" to locate, which reduces selection for nesting together.

The fact that pelagic seabirds are highly aggregated, that inshore feeders nest in smaller colonies, and that waders nest solitarily is *consistent* with the information center hypothesis. Nevertheless, Clode's conclusion that information sharing is the *primary* factor in the evolution of large nesting colonies of seabirds is currently under investigation and remains tentative at this time.

Reproductive Behavior: Why Are There Different Kinds of Avian Mating Systems?

Because of Darwinian natural selection, reproduction is the central feature of bird life, around which all other aspects of behavior revolve. One striking aspect of avian reproduction is the extent to which one male cooperates with one female in the attempt to leave surviving descendants. Almost all the birds whose behavior we have mentioned thus far, from Yellow Warblers to gannets, are **monogamous** species in which one male pairs with one female in a given breeding season. In fact, over 90 percent of all birds fall into this category. This observation is puzzling; monogamy is exceptionally rare in all other groups of animals, for which the standard mating system is **polygyny**, in which one male mates with several females in a breeding season.

Figure 6–35. The Relationship Between Seabird Nest Colony Size and Distance From Feeding Area: *Colonial breeders may derive foraging benefits from living together. The* **information center** *hypothesis proposes that individuals living in dense colonies may improve their food-gathering success by following neighbors that are successful foragers to good feeding areas. Evidence supporting this theory comes from analyses of the breeding habits of various seabirds (Clode 1993). Offshore or pelagic feeders, such as Northern Gannets (top), feed on schooling fish far out to sea, at a great distance from their vast nesting colonies. The location of such fish schools at any given time is impossible to predict, but once one is located, there is plenty of food for many predators. The information center theory suggests that individual birds may locate the fish schools more easily if they live communally, where they can observe the success of their fellow colony members. In contrast, inshore feeders (middle), such as Little and other terns, forage closer to shore, reasonably close to their smaller, less-dense breeding colonies. Their prey is less patchily distributed and easier to locate, so an "information center" is less important for their foraging success. Finally, shoreline feeders (bottom), such as the American Oystercatcher and other marine waders, search for their evenly distributed food in the intertidal zone, closest of all to their nests, which are located solitarily. Information sharing with other birds would presumably be of no help to them in locating their prey.*

Monogamy in birds is a Darwinian puzzle for the following reasons. Males, by definition, produce sperm, the smaller of the two kinds of gametes generated within a species. Females, by definition, produce eggs, which are larger and more costly to manufacture. The difference in the sizes of sperm and eggs is enormous in all birds; whereas sperm are microscopic, a single egg constitutes at least 4 percent of the female's body weight in most birds, and may reach 25 percent for certain kiwis and albatrosses. Females are limited in the number of eggs they can produce, but males have the potential to fertilize a huge number of eggs. Therefore, a male's reproductive success would seem to be determined primarily by the number of eggs his sperm fertilize, and the more females inseminated, the more eggs his sperm can reach. The logic of this argument dictates that males should compete among themselves for access to multiple partners, with those winning the competition becoming successful polygynists.

But most birds seem to ignore this adaptive recipe for male behavior. Instead, male birds typically remain with a single female during a breeding season, usually assisting her in nest building, incubation, and feeding the young (**Sidebar 5: Length of the Pair Bond**). In the Northern Mockingbird, males take over exclusive care of young fledglings when their mates start to construct a new nest for a second clutch of eggs. For some days, males are in total charge of "child care," single-handedly feeding and protecting their fledglings against a large array of mockingbird consumers.

Male megapodes are even more impressive examples of paternal birds. Males of the Australian Malleefowl, for example, build a huge nest mound that contains an average of 7,480 pounds (3,400 kg)—nearly 4 tons—of sand, dirt, and decaying vegetation (**Fig. 6–36**). Females, about the size of small hen turkeys, lay their eggs at intervals in the mound. Each time a female visits, the male opens the mound so she can deposit a single egg deep in the pile; then he closes it up again. Moreover, after the eggs are well buried, the male regulates the heat within the "compost heap" by shifting mound material onto or away from the eggs. The average amount of debris a male moves to prepare the mound to receive an egg and then cover it up is about 1,870 pounds (850 kg), about 500 times the bird's mass, or the human equivalent of shoveling about 35.2 tons of dirt (Weathers et al. 1993). Although male Malleefowl are remarkable paragons of paternality, the parental contributions of males of most other bird species are not to be sneezed at.

What factor has selected for the spread of monogamy in bird populations? According to Gordon Orians (1969), the answer is the male bird's unusually high potential for productive parental care. Unlike male mammals, which cannot produce milk and so generally do not feed their young, male birds can do almost everything that a female can—except lay eggs. As a result, the potential to improve the reproductive success of a partner is unusually high in the typical male bird compared to a typical male mammal.

If this argument is correct, experimentally removing a helpful male parent should greatly reduce the production of young by the surviving

(Continued on p. 6·73)

Figure 6–36. Nest Mound of the Australian Malleefowl: a. Male Malleefowl at his Nest Mound. b. Malleefowl Pair at a Nest Mound: *The male, on the left, is scratching material toward the center of the nest mound to regulate the temperature.* **c. Internal Structure of the Nest Mound:** *The male digs a pit and fills it with a mound of leaves, twigs, and other organic matter. This forms the egg chamber, over which a layer of sandy soil is deposited. Fermentation of the organic matter provides the heat necessary for incubation. By poking his bill into the mound, the male tests the temperature, which he then adjusts by moving material to and from the mound. Photos a and b courtesy of John Alcock. Drawing c by Charles L. Ripper.*

a. Male Malleefowl at his Nest Mound

b. Malleefowl Pair at a Nest Mound

c. Internal Structure of the Nest Mound

Egg Chamber

Sand

Organic Matter

Undisturbed Soil

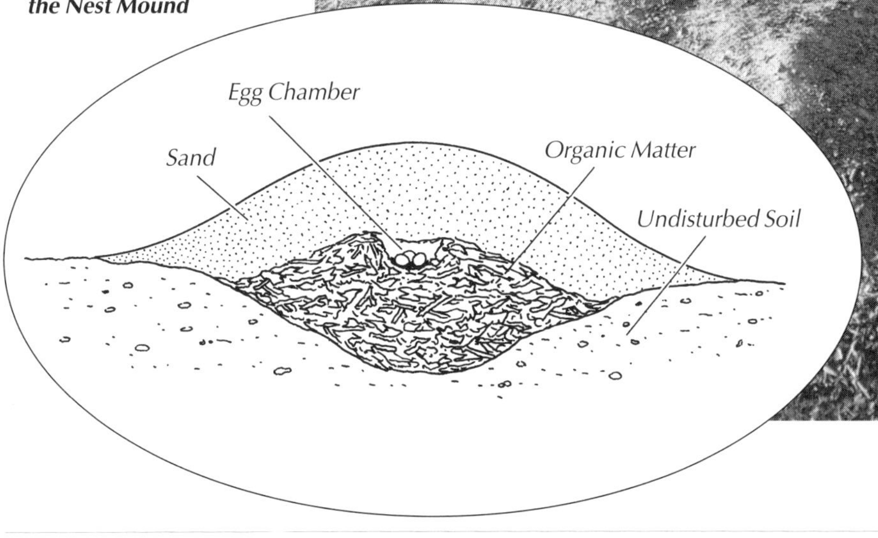

Sidebar 5: LENGTH OF THE PAIR BOND
Sandy Podulka

A few types of birds, such as grouse and manakins (see Fig. 6–42), associate with the opposite sex only to copulate. Others stay together for a few days (Ruby-throated Hummingbirds), and some split after a prolonged courtship and mating period (ducks), but most birds stay together for at least a breeding season. We tend to imagine that birds keep the same mate from year to year—surely the same pair of Eastern Phoebes is nesting under my eaves again?—but data from banding studies show that for most species, this is the exception. Of 200 pairs of Song Sparrows studied by Margaret Nice, for instance, only 8 retained their mates the next breeding season. Between breeding seasons, of course, many birds die, especially migrants. But the rate of mate-changing is much higher than the mortality rate. Occasionally birds do keep the same mates. Certain American Robin, Northern Cardinal, Song Sparrow, and Northern Mockingbird individuals all have been known to pair for three consecutive years, and Downy Woodpeckers, for four.

Some species, especially the long-lived ones, mate for life. Geese, swans, cranes, petrels, albatrosses, oystercatchers, Herring Gulls, some shearwaters and penguins, ravens, some crows, many parrots, roadrunners, Wrentits, Tufted Titmice, White-breasted Nuthatches, Pygmy Nuthatches, Brown Creepers, Cactus Wrens, and some House Sparrows all retain the same mate from one breeding season to the next (**Fig. A**). One Royal Albatross pair was seen together for 15 consecutive years, and a pair of Mute Swans in England lasted 8 years before one mate died. The "widow" waited three years before re-pairing.

Birds that keep their mates from year to year may reap some benefits. Since they begin the breeding season with a mate, they may be able to secure a better territory or start nesting sooner—young fledged earlier in the season have more time to gain strength and experience before winter. Also, experienced couples presumably have already learned how to cooperate in a breeding effort. In Black-legged Kittiwakes, experienced pairs start breeding earlier, lay more eggs, and raise more young than do new pairs. ∎

Figure A. Greater Sandhill Crane Pair: *Many long-lived bird species, such as these Greater Sandhill Cranes (a subspecies of Sandhill Cranes), mate for life. By starting the breeding season already paired (and thus spending less time in mate attraction and courtship), these species may reap some benefits such as obtaining a better territory, or beginning to nest earlier in the season—and young fledged earlier may have a better chance of surviving. Experienced pairs also may benefit by already knowing how to cooperate in a breeding effort. Photo by Marie Read.*

female parent. When the experiment was done with Snow Buntings (Lyon et al. 1987), there was in fact a sharp decrease in the output of young **(Fig. 6–37)**, which supports the hypothesis that monogamous males are compensated for their inability to be polygynous by helping their partner to raise more young.

Reproductive Behavior: Resource-defense and Female-defense Polygyny

Although most birds form monogamous pair bonds, a few do not. Understanding exceptions to the rule is always interesting, so we shall explore the unusual cases at some length to show how evolutionary theory can help to explain the rare mating systems of birds as well as the most common type. Among the small minority of nonmonogamists are those that practice one or another form of polygyny. Male Red-winged Blackbirds caroling in a cattail marsh are familiar polygynous birds. The calling, chasing, and red-epaulette-flashing are part and parcel of the competition among males for territories that may attract more than one mate **(Fig. 6–38)**. Because female redwings choose partners by evaluating the quality of the territories controlled by males, the mating system of this species has been labelled **resource-defense polygyny** by Stephen Emlen and Lewis Oring (1977).

The reproductive benefits of this mating system are clear for the males that achieve polygyny. The puzzle is why female redwings would bond with a male that had already secured one or more mates, if unmated males were available (and they almost always are). Females with polygynist partners must accept less parental assistance for their offspring than females that pair with a previously unclaimed male, who can then provide undiluted care for their progeny. However, when the quality of territories varies greatly, as has been documented in a number of locations, females that share a resource-rich territory with other females can potentially do as well as, or better than, those that pair with a male on a territory that is short of food or safe nesting sites.

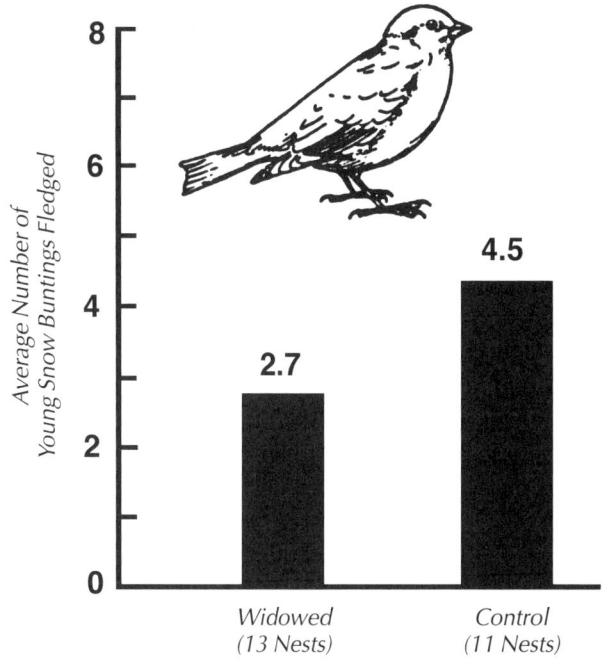

Figure 6–37. Effect of Removing Male on Snow Bunting Reproductive Success: Snow Buntings are monogamous, with both parents bringing food to the young. When Lyon et al. (1987) experimentally removed the male parent from 13 nesting attempts, they found a decrease in the reproductive success of the widowed females compared to that in 11 control nests. Widowed Snow Buntings raised an average of 2.7 young to fledging age, whereas control females who retained their partners raised an average of 4.5. This study provides evidence that males who remain and help to rear young in monogamous situations, instead of leaving to pursue other females, may gain benefits in terms of the number of young they raise. These benefits may offset those potentially lost by not being polygynous. Adapted from Lyon et al. (1987).

Figure 6–38. Displaying Male Red-winged Blackbird: Loud calling and displaying of prominent scarlet epaulettes is part of the competition among male Red-winged Blackbirds for control of prime territories. Females choose mates based on the quality of their territories—an example of resource-defense polygyny. Photo by Marie Read.

Figure 6–39. Montezuma Oropendola Nest Colony: Females of this Central American species weave their long, hanging nests in colonies in certain isolated trees. Males gather at these sites and compete fiercely with each other for control of the clumped females—a mating system known as female-defense polygyny—forming a dominance hierarchy in which the top-ranking male secures the majority of copulations. **Main Photo:** *Montezuma Oropendola nest colony in a tree. Photo by R. and N. Bowers/VIREO.* **Inset Lower Left:** *Close-up view of a male Montezuma Oropendola.* **Inset Lower Right:** *Close-up view of the nests showing their pendulous shape. Inset photos courtesy of Michael Webster.*

A very few birds practice **female-defense polygyny**, in which males fight for control of clusters of nesting females, rather than resource-rich territories. In the Montezuma Oropendolas of Central America studied by Michael Webster (1994), females form nesting colonies in certain trees **(Fig. 6–39)**. Males gather at these sites when females are fertile and compete for access to them. A male dominance hierarchy results, with alpha, beta, gamma, and still more subordinate individuals perching in and around the nesting tree. The top male attacks rivals that approach receptive females and physically prevents them from mating. As a result, the alpha male secures 90 to 100 percent of all copulations at the nesting colony.

Resource-defense and female-defense polygyny may evolve when male parental care is less important to female reproductive success than other factors, such as safe nesting sites or rich food supplies on a territory. Freed from the demands of paternal behavior, males can compete among themselves for key resources that females desire, or can compete directly for clusters of females, depending on the conditions that determine the distribution of females.

Reproductive Behavior: Lek Polygyny

Males of some other polygynous birds defend tiny display sites at a traditional mating location termed a **lek.** The sites do not contain useful resources or nesting colonies of females. Displaying males of these **lek polygynous** species may be visited briefly by sexually receptive females. After mating, the females leave their partners and go to other areas to nest and rear the young entirely without male assistance.

Famous examples of lekking birds include the birds-of-paradise, Ruffs **(Fig. 6–40)**, many of the manakins and bowerbirds **(Fig. 6–41)**, both cock-of-the-rock species, certain pheasants (such as the peacock), and many grouse. Males of the Satin Bowerbird, for example, defend small territories on which they build their bowers, two parallel rows of curved twigs painted with saliva and charcoal (see Fig. 6–41a). In front of the avenue between the twigs, males scatter a variety of generally blue items, especially parrot feathers, which they collect from the forest. Neighboring males drop by from time to time, and if the resident bower builder happens to be absent, they may wreck his bower and steal his decorations, which they take back to ornament their own bowers (Borgia 1986).

Females also occasionally visit a series of bowers; at each stop they are treated to elaborate visual and acoustical displays from the bower owner (male bowerbirds are vocal mimics that incorporate the songs of many other species into their lengthy display routines). Despite the displays, females usually go on their way, but sometimes they enter the bower, crouch down, and permit the male to copulate.

*Figure 6–40. Lek of the Ruff: In lek polygyny, many males perform their courtship displays at a traditional courtship ground termed a **lek**, where each bird defends only a small display site. Females visit the lek only for mating. A well-known example of a lekking species is the Ruff, a shorebird that gathers in the breeding season at established leks in the Eurasian tundra. The male Ruff in breeding plumage has exaggerated ear tufts and a collar of prominent neck feathers, both of which show great individual variability in color and pattern. Males display to females by running excitedly around with fluttering wings and expanded ruffs, and then suddenly freezing in place, bowing low with their spectacular ruffs fully erected, wings outspread and quivering. Shown here are four males displaying to a single female. Drawing by Robert Gillmor.*

a. Satin Bowerbirds at Avenue Bower

b. MacGregor's Bowerbirds at Maypole Bower

Figure 6–41. Bowerbird Courtship: *The bowerbirds of Australia and New Guinea have substituted the building and decorating of complex structures for elaborate displays and plumages. Males of most of the 18 species, true to their name, build bowers on the ground at which they accumulate a variety of brightly colored objects—flowers, butterfly wings, fruit, feathers, bright leaves, and so on. The bower, which may be one of several different types, attracts females who visit for courtship and mating, and then depart to nest alone elsewhere.* ***a. Satin Bowerbirds at Avenue Bower:*** *The Satin Bowerbird builds a type of bower termed an avenue bower. The male constructs two parallel walls of interwoven twigs stuck in the ground, the avenue between the walls giving the bower its name. He aligns the bower along a north-south line, at the north end of which is a display court. This he decorates with numerous colorful objects, which may be natural or human-made; he shows a strong preference for shiny, blue items. A visiting female enters the avenue and watches the displaying male from within it. If the female signifies her willingness, the male enters the avenue and mates with her.* ***b. MacGregor's Bowerbirds at Maypole Bower:*** *The MacGregor's Bowerbird builds a maypole bower, a central pole with a circular display court at its base. The male chooses a thin sapling, which he surrounds for much of its height with horizontal piles of sticks, their ends decorated with hanging items such as regurgitated fruit pulp. At the pole's base he constructs a display court in the form of a circular mat of compressed moss with a rim. Objects such as seeds or woody black fungi are used to decorate this court. During courtship, male and female move back and forth around the court, keeping the maypole between them. When the female stops moving, the male expands his bright orange head plume, shaking it from side to side while remaining behind the maypole; he then moves forward and mates with the female.* ***c. Vogelkop Bowerbirds at Maypole Bower with a Hut:*** *A more complex maypole bower, with an open-sided, hut-like structure over the circular display court, is built by the Vogelkop Bowerbird. Its roof of sticks, several feet high, is supported by a vertical sapling, and the floor beneath it is cleared of litter. The male decorates the floor with numerous neat piles of decorative objects, carefully sorted by color, which in the case of perishable items like flowers and fruit, he replaces daily. The impression is of neatly kept garden beds, and in fact another common name for this species is the Vogelkop Gardener Bowerbird.*

c. Vogelkop Bowerbird with Maypole Bower with a Hut

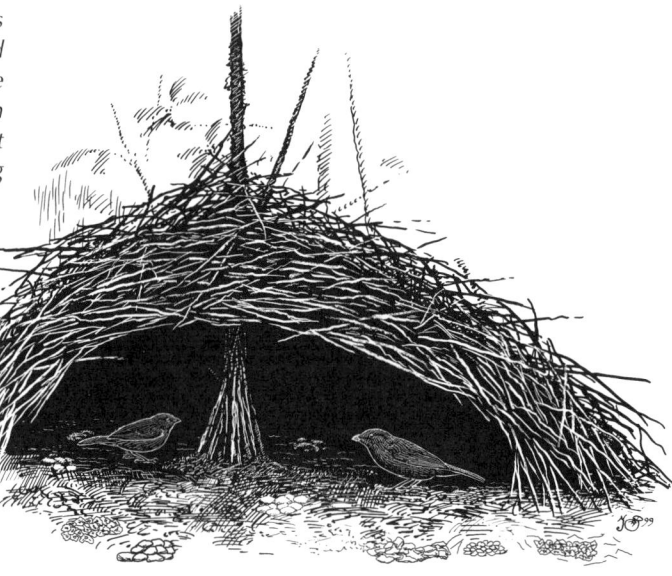

In any population, only one or a very few males are chosen by most of the receptive females over the course of the breeding season. As a result, many male Satin Bowerbirds fail to secure a single partner in a given year, even though they may be inspected by a large number of highly selective females.

The mating behavior of Satin Bowerbirds has many similarities to that of most other lek polygynous birds. For example, males of the White-bearded Manakin of South America defend small saplings within a display court cleared of leaves (Lill 1974) (see Ch. 7, Sidebar 1, Fig. F, as well as Track 16 on CD or cassette). Groups of males have their display courts close together. When a female appears, males rocket from their sapling perches to the ground and back, cracking their clubbed wing feathers together to add an acoustical component to their display. Females appear highly choosey, and most pick the same individual from among the battery of potential mates. After copulating, the female goes off to rear her offspring without male parental assistance; the male resumes his energetic attempts to attract additional mates. Many other manakin species have similarly spectacular displays **(Fig. 6–42)**.

The evolutionary basis of lek polygyny is especially difficult to explain because it would seem more profitable for males to compete for resources that attract mates than to devote their energies to defending territories used solely as a platform for highly bizarre displays. In the environment of some species, however, neither resources nor females may be clumped spatially, greatly raising the energetic cost for males to attempt to defend resources or females. If, in addition, males cannot contribute usefully to the welfare of their young (perhaps because their presence at a nest would make it much more conspicuous to efficient nest predators), parental-assistance monogamy also cannot evolve. Under these special conditions, the absence of any other profitable mating tactics for males makes lek polygyny the default mating system.

Reproductive Behavior: Polyandry

Another kind of mating system that is extremely rare in birds is a form of polyandry in which one female forms pair bonds with two or more males, as seen in the Spotted Sandpiper, most jacanas, and certain phalaropes (Oring 1985). The behavior of female Spotted Sandpipers and Northern Jacanas (see Fig. 4–18b for Wattled Jacana) resembles that of males engaged in resource-defense polygyny. They defend territories that contain safe nesting sites and access to food. Females with adequate territories attract males, who settle there, build nests, accept the clutch of eggs that their partner donates to them, and incubate and care for the young when they hatch—all without female assistance. Instead, females devote themselves to securing sufficient food to produce another clutch of eggs and, in the case of the Spotted Sandpiper, to acquiring another territory in competition with rival females. Females that succeed may attract a second male; when he accepts the clutch, he, too, is abandoned and must provide all the

Figure 6–42. Group Courtship Display of the Blue Manakin: *The manakins, which range from central Mexico to northern Argentina, are noted for their stereotyped courtship behavior. Although the plumages of the males of most of the 59 species are quite colorful, they are nonetheless modest when compared to those of the birds-of-paradise (see Fig. 3–10). Each male manakin performs a species-specific series of precise and intricate courtship maneuvers, which are often accentuated by sounds, but are enhanced only secondarily by plumage displays.*

Male manakins perform in traditional display areas known as courts, many of which have been used for generations. Depending on the species, some courts are on the forest floor, and others are in trees. All the courts meet certain requirements, such as the appropriate number and size of the saplings, twigs, and branches for maneuvers; and must have a clear and unobstructed approach, enabling females to see the performances and reach the courts readily. In some species the males perform on their courts alone, but in others several males perform in interacting groups (see Fig. 6–47).

The Blue Manakin of South America, shown here, is one species that performs a cooperative display. In this species, three, or sometimes more, males display to females from a vine a few feet above the forest floor. In this drawing, birds #1, #2, and #3 are males, and the fourth bird is a female. **a.** *The female perches at one end of the vine and the males crouch along the vine, oriented toward her, quivering and giving rhythmic, twanging calls.* **b.** *The male closest to the female (male #1) jumps up and flies toward her, briefly hovering, then flies toward the far end of the line of males.* **c.** *Male #1 lands on the perch at the far end of the line of males, but faces in a direction opposite to that of the other males. Meanwhile, the other males have all sidestepped along the perch, moving one place forward, male #2 now being the closest to the female.* **d.** *Male #2 now jumps up and hovers before the female, then flies back toward the far end of the line. Male #1 swivels to face in the same direction as male #3.* **e.** *Male #2 lands at the end of the line, facing in a direction opposite to that of the others. This sequence proceeds, each male describing a circular flight path away from the female, attaining such speed that the display eventually resembles a rotating wheel of birds. The joint display ends when the dominant male flies up in front of the other males and gives a shrill call, causing them to disperse. The dominant male then proceeds with a solo display of slow-motion flights to and from the perch, which, if the female is sufficiently impressed, culminates in copulation. Adapted from drawings by William C. Dilger.*

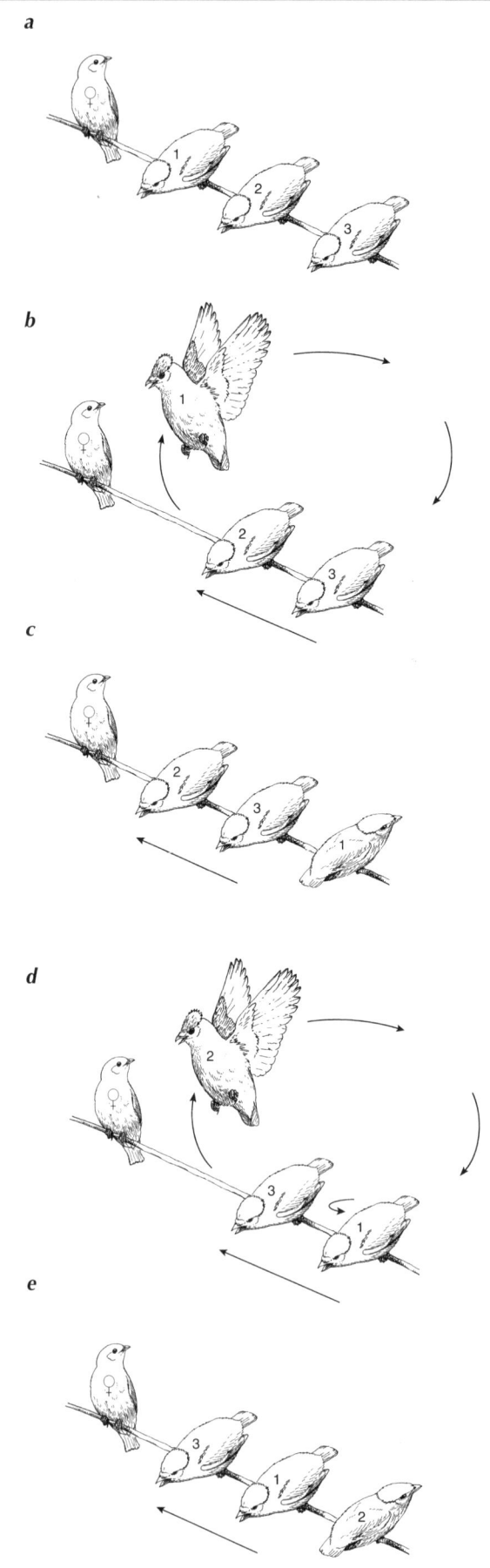

parental care. In the Northern Jacana, a female's territory may be sufficiently large and rich in resources to attract several males, each one of which nests in the successful polyandrist's domain.

In contrast, the polyandrous female phalarope does not defend a useful territory, but instead directly locates a male, defending him against competing females until he accepts her clutch. She then produces a second clutch, if possible, and tries to find and defend another male to incubate her eggs and attend the young for a time after they hatch.

The evolution of polyandrous mating systems of the sort described here poses special problems for evolutionary biologists—puzzles that have not been fully solved. The key puzzle is why males would enter polyandrous pairings, which require that they fertilize only some of the eggs produced by their partner, who has additional mates. But perhaps in cases of this sort one parent is as successful as two would be in caring for a clutch of eggs and the resulting youngsters. Furthermore, there is evidence that, in the Spotted Sandpiper at least, adult males slightly outnumber adult females, providing opportunities for females that desert their first partners. Given these two factors, if females consistently do not care for their eggs, mated males are left with two options: to abandon the clutch, which will then fail, or to stay put, providing the necessary parental care by themselves, and thus making the best of the situation that arises from the absence of an incubating partner.

Mate Choice: Extrapair Copulations in Birds

Although most birds pair off one male to one female, in recent years infidelity has been discovered to be far more common than once believed. Many supposedly monogamous birds are effectively **polygamous**, with individuals of both sexes sometimes copulating with more than one partner in a breeding season (Birkhead and Møller 1992). Copulations with birds other than one's mate are termed **extrapair copulations** (EPCs).

Evidence for extrapair copulations comes from two sources: direct observation of marked individuals showing that some mate with several different partners, and molecular studies of the paternity of clutches of young, which sometimes reveal that the presumptive father cannot possibly have been the actual father for all of the young under his care.

The use of DNA fingerprinting (analysing DNA from different individuals to determine sufficient structural details to identify individuals and their relatives with a high degree of certainty) and other related genetic techniques has established that pair-bonded females sometimes copulate outside the pair bond at times when they are fertile. In one study of wild Zebra Finches, for example, DNA fingerprinting demonstrated that 2 of 82 offspring were definitely not fathered by the male paired to the mother, indicating a relatively low frequency of infidelity in this population. In contrast, about a quarter of all female Tree Swallows in one population were estimated to have engaged in at least one EPC.

Figure 6–43. Superb Fairywren: Male in breeding plumage. These year-round residents of eastern Australia appear monogamous, but close study has revealed that they commonly engage in extrapair copulations. Photo by C. H. Greenewalt/VIREO.

An even higher degree of infidelity has been recorded in the Superb Fairywren (**Fig. 6–43**), which forms breeding pairs that occupy year-round territories in suitable habitat in eastern Australia. Males help their mates defend the territory and feed the young, particularly in the fledgling stage, and so might have been considered more or less typical monogamous birds were it not for careful observation of color-banded individuals by several Australian ornithologists. They found that while a male's partner is building a nest and incubating eggs, he is likely to visit other females on neighboring territories. There the "philandering" male courts his neighbors' mates, often displaying his iridescent blue crown and cheek patches while carrying a yellow flower petal in his beak. Resident males attack intruders to the extent that they can detect them, but "philanderers" are extremely persistent, especially with respect to females in their fertile phase—the week they lay their clutches of eggs.

To determine whether fairywren intruders fertilize some eggs laid by their neighbors' mates, Raoul Mulder (1994, see Suggested Readings) employed DNA fingerprinting to check the paternity of nearly 200 nestlings in his study population in Canberra. He found that more than three-fourths of this sample had been fathered by intruder males—not the mothers' pair-bonded partners. Females in this population clearly engaged in EPCs at times when their eggs could be fertilized.

The benefits of EPCs to outsider males are obvious. They visit nearby territories at times when their own regular partners are not fertile, and if they succeed in fertilizing a neighboring female, their "extra" offspring will be cared for by another male. In this way a successful "philanderer" can raise his lifetime production of surviving offspring at relatively little energetic cost to himself (at least with respect to parental care).

The benefits of EPCs to females are less obvious. Outsider males do not usually assist parentally, eliminating extra parental care as a possible incentive for females. In the European Dunnock, however, females that mate with several males do get help from their multiple partners in taking care of the young.

To account for EPCs in Superb Fairywrens, some researchers have suggested that females gain sperm with unusually "good genes" through extrapair matings with superior males. According to this hypothesis, females evaluate the health, longevity, or competitive ability of neighboring males and then copulate with the best of the lot, endowing their offspring with hereditary traits superior to those of their pair-bonded partner.

If this hypothesis is correct, many females should select the same males for their EPCs. Mulder's DNA fingerprinting work showed that one male, two of his sons, and a grandson were responsible for a highly

a Greater Bird-of-paradise

b. Magnificent Frigatebird

Figure 6–44. Elaborate Male Orna-ments: a. Greater Bird-of-paradise: A male displays his ornate plumage. Drawing by Robert Gillmor. b. Magnificent Frigatebird: A male with its gular sac expanded into a scarlet balloon during display. Photo courtesy of Lee Kuhn/CLO.

disproportionate share of all the extrapair fertilizations in his large study population of Superb Fairywrens.

Mate Choice: Why Do Some Birds Display Elaborate Ornaments?

In a great many species of birds, including any number that are at least superficially monogamous, males possess striking plumage, combs, wattles, inflatable sacs, and the like (**Fig. 6–44**; also see Figs. 3–6, 3–43, and 6–40). Previously, we noted that males of the "monogamous" Superb Fairywren show their iridescent cheek and crown feathers to prospective mates. Males of the familiar Barn Swallow, a similarly "monogamous" species, possess two unusually long outer tail feathers, which females examine when selecting a male with which to breed (**Fig. 6–45**). Anders Møller (1992) has shown that both the length and symmetry of these tail ornaments influence a male's chances of acquiring a mate early in the breeding season. Only males with relatively long and symmetrical outer tail feathers are likely to attract a mate soon enough to be able to father two broods in one breeding season in northern Europe, where Møller did his research.

Although elaborate ornaments and complex displays occur in certain "monogamous" species, lek polygynous species exhibit some of the most extreme examples. The males of one bird-of-paradise have beautiful powder blue plumes, which they can erect and vibrate while hanging upside down from a perch to display to a potential mate. The orange inflatable sacs of Greater Prairie-Chickens come into play

Symmetrical   *Asymmetrical*

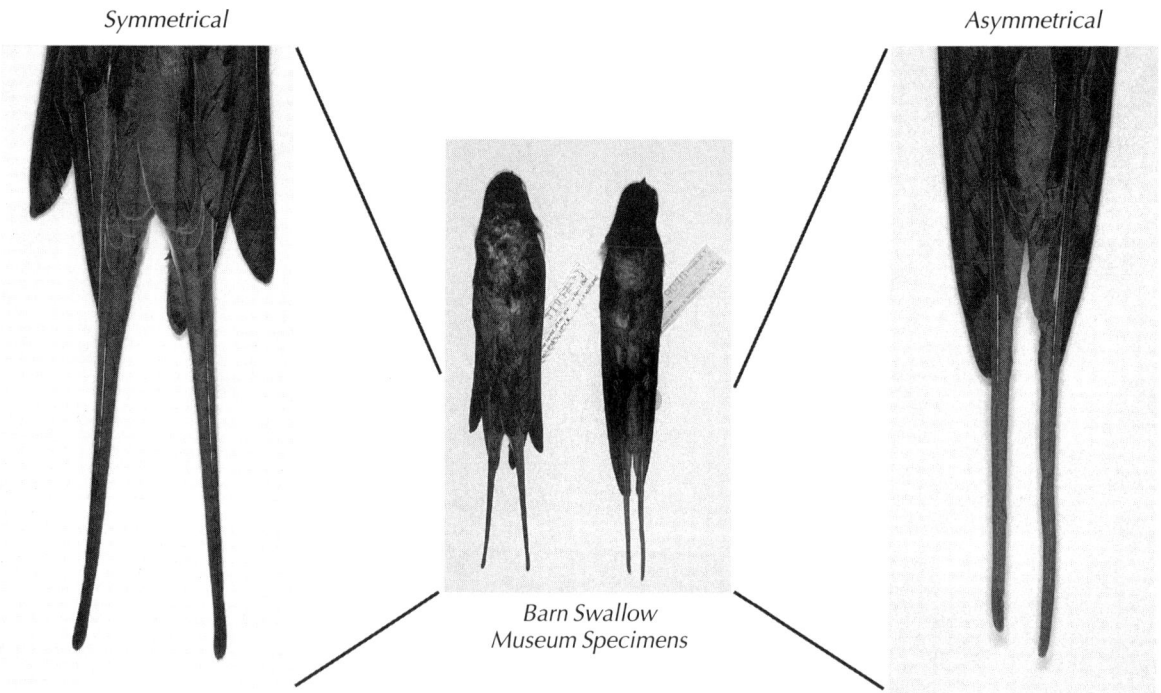

Barn Swallow
Museum Specimens

Figure 6–45. Symmetrical and Asymmetrical Barn Swallow Tails: *According to a study carried out by Møller (1992) in northern Europe, both the length and symmetry of the outer tail feathers of male Barn Swallows appear to influence a male's ability to attract a mate early enough in the breeding season to be able to raise two broods of young. In these photographs of museum specimens, note the similar-length outer tail feathers on the bird on the left, compared to the very different lengths of the two outer tail feathers of the bird on the right. Photo courtesy of Marie Read/CLO.*

when males boom out their mate attraction calls while performing a courtship dance in front of observant females that have come to the lek to select a mate. Wild male peacocks erect, spread, and shiver their overblown, magnificently patterned, blue-and-green "tail" under similar circumstances (see Fig. 3–6).

The peacock's train constitutes the classic example of an elaborate male ornament, but only in recent years have hypotheses on its evolution been rigorously tested, by Marion Petrie and her co-workers. One idea long deemed plausible is that a female preference for males with elaborate trains has produced **sexual selection** via female choice for this male attribute.

Sexual selection is a form of natural selection that occurs if individuals differ in their ability to acquire mates. Such differences typically arise because males differ in their competitive ability or because males differ in their attractiveness to the opposite sex. (In some species, *males* can exert sexual selection on *females* through mate choice, but this is an exception to the rule.) Thus, with respect to the peacock's tail, it is possible that males with elaborate trains are better able to intimidate rival males and so keep them away from females. But it could also be that males with highly decorated tails are especially attractive to receptive females. Petrie chose to examine the second possibility (Petrie and Halliday 1994, see Suggested Readings). She asked whether peahens exerted sexual selection in favor of males whose tails were ornamented with a larger number of eyespots. She predicted that, if they did, the number of eyespots in male trains would be correlated

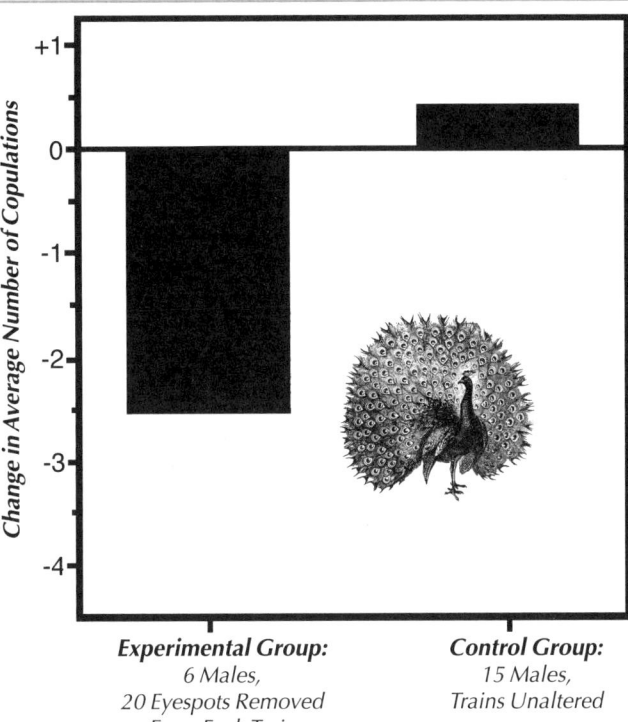

Experimental Group:
6 Males,
20 Eyespots Removed
From Each Train

Control Group:
15 Males,
Trains Unaltered

Figure 6–46. Number of Eyespots Versus Mating Success of the Male Peacock: Petrie and Halliday (1994) captured male Indian Peafowl between breeding seasons and removed 20 eyespots (by snipping them out) from their trains, then returned them to the wild. Males with the number of eyespots reduced averaged 2.5 fewer copulations during the following breeding season than in previous seasons, compared to a control group who showed no such decline in number of copulations. This difference between the experimental and control groups was statistically significant. Previous field observations had shown that the males with more eyespots in their trains obtained more copulations during a breeding season. These data provide evidence that females choose males on the basis of the number of eyespots in their trains, and thus that the elaborate trains of peacocks evolved through female choice. Adapted from Petrie and Halliday (1994).

with the mating success of their owners. She and her colleagues found that 50 percent of the variation in the number of copulations secured by males at one lek could be attributed to variance in eyespot number. In other words, why a few males did very well in the mating game while most did poorly could be explained in large measure by the differences in their numbers of eyespots.

It was possible, however, that females were not focusing on eyespots, but on some other aspect of the male's appearance or behavior that happened by chance to be correlated with eyespot number. Petrie therefore devised an experiment in which she decreased the number of eyespots in some males to test the prediction that these birds would suffer a decline in mating success **(Fig. 6–46)**. She and her co-workers captured six males, and from each snipped out 20 of the outermost eyespots—one from the end of each of 20 tail feathers—and released their subjects. During the following breeding season, they observed the 6 experimental males as well as 15 control males whose tails had not been altered. They found that on average the controls mated about as often as they had in preceding seasons. In contrast, the experimental males (with fewer eyespots) averaged 2.5 fewer matings that year—a significant decline and one that supports the hypothesis that female choice by peahens is responsible for the evolution of elaborate trains with hundreds of eyespot ornaments.

But what do peahens gain by choosing certain males? Let's consider three major alternative explanations: the **good genes**, the **runaway selection**, and the **direct benefits** hypotheses (Ryan 1997; Reynolds and Gross 1990) **(Table 6–3)**. Earlier we noted that the good genes hypothesis had been invoked to explain the readiness of female fairywrens to copulate with males other than their pair-bonded

Table 6–3. The Reproductive Benefits Gained by Females in Choosing Nonpaternal Sexual Partners with Elaborate Ornaments: Three Hypotheses

Hypothesis	Benefit to Choosy Females
Good Genes	Both male and female offspring receive genes that enhance their chances of surviving to reproduce successfully.
Runaway Selection	Male offspring receive genes that produce elaborate traits that enhance their sex appeal but may reduce their longevity.
Direct Benefits	Females gain improved survival or fertility by avoiding diseased males or those likely to provide infertile sperm. The quality and condition of a male's elaborate traits may signal his health status.

partners. However, as discussed below, the evidence cited in favor of this hypothesis (the fact that females tended to prefer a very few males in the same paternal line) actually supports all three competing hypotheses.

Good Genes: The same few males may be chosen because their genes confer survival advantages on their offspring. "Good genes" are those that contain information that translates into superior foraging ability, greater skill in dealing with predators, or better capacity to hold territories—all useful survival aids.

Runaway Selection: Another possibility is that female mate choice is adaptive strictly because the sons of males with elaborate ornaments will inherit these attributes and thereby become irresistible to females. In addition, the daughters of these choosy females will inherit a preference for the extreme ornaments that make males attractive to many other females. As a result, they too will tend to mate with attractive males and produce sons of great sexual attractiveness. Therefore, the evolution of mate choice and sexual ornaments can be driven entirely by the aesthetic preferences of females, whether or not males with the preferred ornaments survive as well as males without them.

Incidentally, the technical term "runaway selection" is used because mathematical models have shown that once female mate preferences and male preferred traits are inherited together, the stage is set for a runaway process in which ever more extreme preferences and bizarre ornaments can spread through a species. The trains of peacocks, the bowers of certain bowerbirds, and the remarkable plumes of some birds-of-paradise are all candidate products of runaway selection, although as noted, these structures conceivably can be produced by good genes or direct benefits selection as well.

Direct Benefits: The third possibility is a straightforward one, namely that females benefit by mating with certain males because they are least likely to infect the females with a disease, mites, or some other affliction. This hypothesis is called the direct benefits hypothesis because the female herself derives reproductive advantages (improved health) through her choosiness in mating.

To discriminate among the three hypotheses is difficult, because they tend to produce many of the same predictions. However, if females benefit by securing sperm with survival-enhancing genes for their offspring from genetically advantaged males (the good genes hypothesis), then the offspring of preferred, highly ornamented males should have lower mortality rates than offspring of less ornamented, less attractive individuals. This prediction has been tested. Petrie discovered that the offspring of peacocks with more elaborate trains grew faster and survived better than youngsters whose fathers had tails with fewer eyespots.

This result is clearly consistent with the good genes hypothesis—offspring apparently inherit genes that improve the viability of both sexes, rather than only making male progeny more attractive to females (as expected from the runaway selection hypothesis). However, to also eliminate the direct benefits hypothesis, we still need to know whether the differences in growth and survival of the young birds occurred because some were infected with contagious disease or parasites that their mother picked up through contact with an infected male. Researchers are still studying peacocks and other species, trying to untangle the possible explanations for why females choose carefully among potential mates.

Mate Choice:

Why Cooperate in Courtship Displays?

A special evolutionary puzzle associated with mate choice is exhibited by males of the Long-tailed Manakin, which cooperate with one another in attracting mates **(Fig. 6–47)**. David McDonald's long-term study (McDonald and Potts 1994, see Suggested Readings) reveals that males of this extraordinarily beautiful Central American bird form courtship display teams of up to 13 members, headed by an alpha and beta male. The alpha and beta males often stay together for two or more years, spending most daylight hours at a perch in the forest where the pair sing their loud and ringing mate-attraction song *to-le-do* in perfect unison over and over again—up to 335 times per hour. Occasionally the alpha or beta male leaves temporarily, in which case another team member may join the remaining bird for duets that last until the higher-ranking individual returns to reclaim his place.

Should a female arrive at the lek arena, the two males begin a joint visual display, leapfrogging over each other on their perch in a stereo-

Figure 6–47. Cooperative Courtship Display of the Long-tailed Manakin: *Males of this beautiful Central American species form long-lasting teams of cooperatively displaying males. Typically two males—an alpha and a beta—call together in a highly coordinated fashion from the forest canopy. Once a female has been attracted, they descend to a display perch to perform an intricately choreographed courtship display that closely resembles that of the Blue Manakin (see Figure 6–42). The top-ranking male invariably secures the majority of matings, leading biologists to speculate on why the beta male should engage in cooperative courtship with little apparent reproductive advantage—especially since DNA analysis has determined that the two males are not related. Nevertheless, by cooperating, the beta male benefits by establishing his right to take over the top rank and monopolize females when the alpha male finally dies (McDonald and Potts 1994). Here, two males, top and bottom left, perform a leapfrogging display, one in midair and the other moving toward the female (bottom right) on the display perch. Photo by Marie Read.*

6

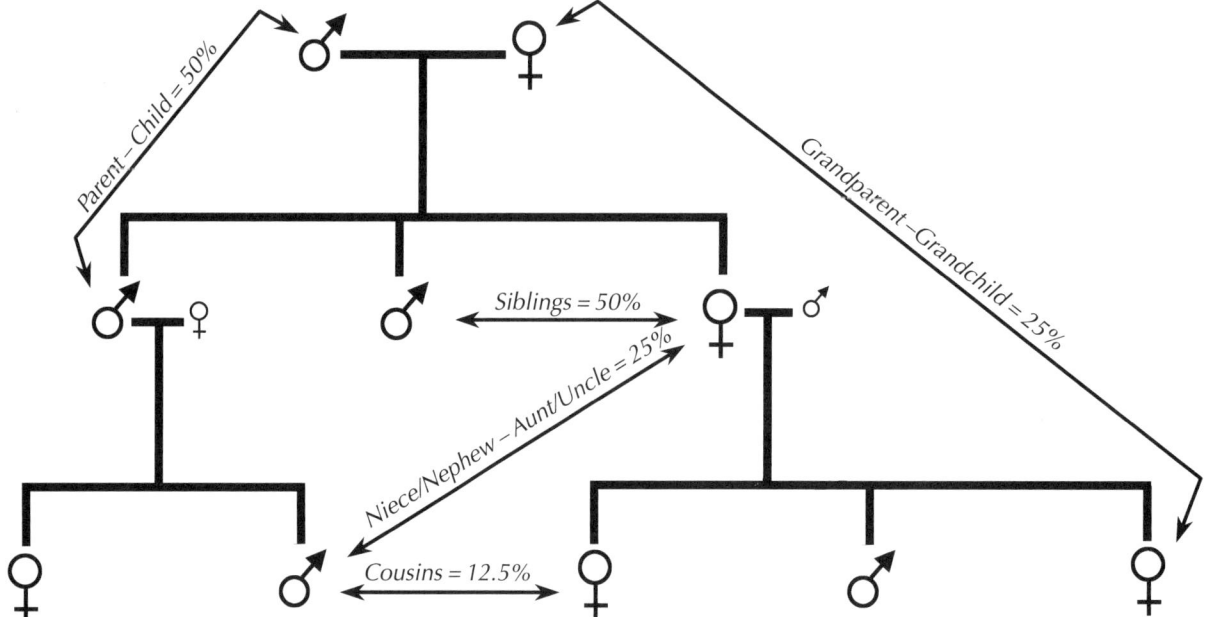

Figure 6–48. How Relatives Share Genes: *Related individuals share some of the same genes, having inherited them from their common ancestors. A parent and his or her offspring share exactly 50% of their genes, because the child receives half of its genetic complement from its father and half from its mother. Full siblings share on average 50% of their genes. This is because, through each parent, there is a 25% chance that any given gene possessed by one sibling will also be in the other (For gene A, for example: there's a 50% chance of Mom giving it to sibling 1, and a 50% chance of Mom giving it to sibling 2. The chance that both those events will occur is 50% x 50%, or 25%). To find the probability that any gene possessed by one sibling is shared by the other, you must add together the probability that it is shared through Mom and the probability that it is shared through Dad (25% through Mom plus 25% through Dad), resulting in the 50% average of shared genes between siblings. Using similar reasoning, grandparents and grandchildren share an average of 25% of their genes, as do aunt/uncle and niece/nephew pairs. Cousins, on average, share 12.5% of their genes.*

typed fashion, or flying together with exaggerated wingbeats back and forth over the arena. On most occasions, females watch and then leave, but once in a while, after a prolonged spell of cooperative display by the two males, females submit to mating. Because McDonald had color-banded 33 alpha-beta pairs, he could determine the distribution of matings among the males in his study. Over 10 years, he recorded 263 copulations, of which 259 were secured by the alpha male, who may retain his top-bird ranking for up to 8 years.

From an evolutionary viewpoint, this result is most surprising. How can it be adaptive for the nonmating male to engage in what may be years of mutual display, if during this time the male has almost no chance to mate? McDonald tested two alternative hypotheses on this phenomenon: (1) the beta male is related to the alpha male, and therefore helps propagate his genes indirectly by raising the reproductive success of his relative with whom he shares some proportion of his genes (the **indirect fitness** hypothesis), and (2) the beta male gains reproductive success himself, albeit long-deferred, by virtue of reaching alpha status when his long-term partner dies (the **direct fitness** hypothesis). The point is that individuals can propagate the genes underlying their behavioral abilities in either way: (1) by helping relatives leave more descendants than they would have without their help, or (2) by reproducing personally and transmitting copies of their genes to their offspring. Helping relatives (Hypothesis 1) can be adaptive because related individuals share some of the same genes as a result of having inherited them from their recent common ancestor. For example, full brothers have an average of 50 percent of their genes in common because they had the same mother and father **(Fig. 6–48)**. A bird that helps his brother reproduce is really reproducing by proxy, because many of his genes will be present in his nephews and nieces.

The indirect fitness hypothesis yields the prediction that alpha and beta males will be fairly closely related to one another. Testing this prediction was difficult, because there seemed to be no way to locate and follow the offspring of a set of females for the long time required before the sons of those females became part of alpha-beta partnerships. (Males do not appear to become betas until they are at least four years old.) Fortunately, McDonald and his co-worker Wayne Potts could use molecular genetics technology to establish that any two alpha-beta partners were no more likely to share genes in common than two individuals drawn at random from the manakin population as a whole. Thus, they could reject the indirect fitness hypothesis.

The direct fitness hypothesis produces testable predictions, too. If beta males cooperate with alpha males to establish their "right" to eventually become the top bird, they should move up as soon as the alphas disappear. McDonald has seen only 11 males become alphas in his 10-year study—a testimony to the longevity of these birds. In every case, however, the new top manakin was the previous beta male, not a lower-ranking individual.

Furthermore, females (who also live for many years) are faithful to a particular mating arena, so when a beta male assumes the top rank, he tends to secure sexual access to the females who came to the arena in previous years to mate with the prior, generally older, alpha. Thus, there is a potential reproductive payoff for the patient cooperator who secures the beta position in a cooperative team.

Figure 6–49. Sibling Rivalry in Nestling Great Egrets: Nestling egrets of various species often fight aggressively with their nestmates. Fierce blows from their strong bills can be lethal, especially if one of the nestlings is smaller and weaker than the others, as happens when egg laying, and thus hatching, is staggered. Here, the nestling on the left pecks at one of its squabbling siblings. Parent egrets ignore sibling aggression, and researchers have suggested that siblicide may, in fact, be adaptive behavior from the perspective of the parent birds. See text for further explanation. Photo courtesy of Douglas Mock.

Parental Behavior: Why Do Some Birds Ignore Lethal Aggression Among Their Nestlings?

In most birds, both males and females help feed and defend their offspring, especially in species with **altricial** young (those that hatch from the egg in a featherless, helpless state). Given the high level of parental care in so many birds, it is surprising to observe parent Cattle Egrets standing calmly on the nest while at their feet one offspring batters a sibling to death. The absence of parental intervention in this and other cases of **siblicide** is yet another major evolutionary puzzle: how can it be adaptive for parents to accept a reduction in the number of their offspring (Mock et al. 1990) **(Fig. 6–49)**?

Researchers have offered several hypotheses on why siblicide is paradoxically adaptive from the parental perspective. One possibility is that parents permit an offspring to eliminate nestlings that would have little chance of surviving to adulthood, due to the scarcity of food for parents to provide to their offspring. Inadequate amounts of food brought to the nest result in increased conflict among nestlings.

This argument is founded on the recognition that "reproductive success" is measured not in terms of the number of eggs laid or nestlings produced, but by the number of offspring that reach the age of reproduction and pass on genes received from their parents. If Cattle Egret parents that divide a limited quantity of food among three nestlings wind up having three small, weak fledglings that all die before reproducing, then they will contribute no genes to the next generation. In contrast, egret parents with siblicidal offspring may wind up feeding just one or two nestlings, so survivors fledge at a heavier weight and may well live long enough to breed.

Parent egrets incubate their eggs in ways that promote siblicide, suggesting that tolerance of siblicide is adaptive. The birds start incubating as soon as the first egg is laid. Because eggs are laid one per day, at one- to two-day intervals, the first egg hatches sooner than the second, which hatches before the third, so the first nestling gets a head start in growth over its siblings. This process is called **asynchronous hatching**. The senior chick's size advantage allows damaging attacks on the smaller siblings, so it is invariably the last-hatched youngster that dies from the sibling's blows, or is forced out of the nest to starve.

Parent egrets could make it much tougher for one offspring to dominate the others if they simply waited (as many other birds do) to begin incubation until the complete clutch had been laid, leading to **synchronous hatching**. Instead, the use of asynchronous hatching suggests that creating unequal, frequently lethal competition within their brood is adaptive for egrets.

This argument has been tested by circumventing the effects of the incubation pattern. Researchers shuffled very young nestling Cattle Egrets among a number of nests. In some nests, the resulting experimental brood consisted of three nestlings all the same age and size; in others the three nestlings were 1.5 days apart in age and of different sizes, as in natural sibling groups. As predicted, adult birds endowed with same-age nestlings actually reared fewer fledglings (1.9 on average) than those that had received nestlings of different ages (2.3 fledglings on average) (Mock and Ploger 1987).

Parental Behavior: Why Are There "Helpers at the Nest" That Care For Someone Else's Offspring?

In a surprising number of bird species, breeding pairs receive "parental" assistance from some other adults that do not reproduce, but instead adopt the role of "helper at the nest." In the Florida Scrub-Jay, one of the best studied of these cooperative breeders, thanks to Glen Woolfenden and his colleagues (Woolfenden and Fitzpatrick 1984), extended family groups of up to eight birds occupy and defend a single territory, communally bringing food to the young and repelling predators such as snakes **(Fig. 6–50)**. Yet only two members of the flock reproduce in any given year.

The behavior of the nonbreeders is surprising from an evolutionary perspective because they appear to reduce their own reproductive success while raising that of others, precisely the sort of self-sacrificing behavior that Darwinian theory cannot accommodate. There are,

Figure 6–50. Helpers at the Nest in the Florida Scrub-Jay: In cooperatively breeding species, such as the Florida Scrub-Jay pictured here, the breeding pair receives help from their non-breeding adult offspring from previous years. These "helpers at the nest" aid their parents in territorial defense, in attacking would-be predators, and in bringing food for the young—who are, in actuality, the helpers' younger siblings.

however, some possible Darwinian solutions to the paradox posed by nest helpers. These hypotheses have historically been placed in two categories by behavioral biologists. Included under the **ecological constraints** category are explanations for helpers that focus on the possible costs to young birds of dispersing from their natal territory. Within the **benefits of philopatry** (staying at home) category are hypotheses that focus on the possible benefits to young adults of remaining with their parents **(Table 6–4)**.

So, according to one of the constraints arguments, young Florida Scrub-Jays may stay at home with their parents because their breeding habitat is already saturated with territory owners (see Ch. 1, Sidebar 2, Fig. C). An attempt to find a suitable site under these conditions is likely to fail, while exposing the young bird to predators and the risk of starvation.

Table 6–4. Two Approaches to Understanding the Puzzle of "Helpers at the Nest"

Disadvantages of Leaving Home: Ecological Constraints

1. Few vacant territories of good quality available
2. Few suitable breeding partners available
3. Little chance of successful reproduction for birds until they gain "parenting" experience

Opportunities for "Stay-at-Homes": Benefits of Philopatry

1. Survival improved via group membership
2. Chance to improve the survival of close relatives
3. Chance to acquire superior territory, either by monitoring vacancies in neighboring sites or by inheriting natal territory

Adapted from Emlen (1994).

In addition, those young scrub-jay adults who stay with their parents are in a position to benefit from their philopatry—they can assist their parents in the rearing of additional siblings. If their helpful behavior results in the production of siblings that otherwise would not have survived, the helpers have in effect advanced the propagation of their own genes. As noted earlier, close relatives share a substantial proportion of their genes. In this sense, the special "self-sacrificing" behavior of helpers has the same genetic consequences as having some offspring of their own (parents are no more closely related to their own offspring than they are to their full siblings; on average, they share 50 percent of their genes with each [see Fig. 6–48]). As a result, helping at the nest can in theory persist over evolutionary time as a consequence of the indirect fitness gains derived by helpers. In the case of the Florida Scrub-Jay, helpers can sometimes more than double the number of fledglings produced by their parents, judging from the fledgling output of adults with and without helpers at the nest.

Stephen Emlen (1982) points out that the two categories of hypotheses, those that focus on the costs of dispersal and those that emphasize the benefits of staying at home, are complementary, not mutually exclusive. If dispersal carries a high cost, the benefits of staying home need not be great for a young bird to gain in the long haul by becoming an adult helper at its parents' nest. Thus, the two factors can work together to create durable families of breeders and helpers.

Furthermore, both types of hypotheses yield the same central prediction: nonbreeding helpers should be the offspring of the adults they assist. The prediction holds in many cases (but not in all). In the Florida Scrub-Jay, for example, long-term studies of marked individuals have revealed that nest helpers are usually adult sons of the breeding pair. The same is true for the Superb Fairywren, whose "monogamous" breeding pairs are sometimes helped by up to three sons from previous breeding attempts. In the White-fronted Bee-eater, both sons and daughters may stay with their families, helping to rear still more siblings.

In species with family groups, a combined cost-benefit approach yields an additional series of predictions about the helpers and their chances for personal reproduction. A central prediction is that if a nonbreeding helper gets a significant chance to reproduce, it will do so, rather than accepting permanent nonbreeding status. In the Florida Scrub-Jay, helpers often become breeders, moving into vacant breeding niches in neighboring territories when reproducing adults disappear, or breeding in their natal territory when a parent or step-parent dies.

Likewise, in the Seychelles Brush-Warbler, ornithologist Jan Komdeur (1992) found nest helpers to be commonplace on one island where the warbler was abundant. If helpers remain where they were born because suitable breeding habitat is in short supply, then we can predict that they should quickly switch to personal reproduction when transplanted into an area with no competitors. To test this prediction, Komdeur moved helpers from their original island to another one where the warblers were previously absent. As predicted, the transplants immediately set up territories, attracted mates, and began nesting. Furthermore, their offspring did not stay at home, but dispersed into the unsaturated habitat on the island, where they, too, attempted to reproduce personally rather than helping at a parental nest. This result provides one more demonstration of the evolved ability of birds to behave in the "best interests" of their genes. **(Sidebar 6: Bird Families as Models for Understanding Ourselves)**

How to Study Bird Behavior Yourself

■ Taking the time to watch how birds behave is something that most amateurs feel is less rewarding than the search for as many species as possible. As a longtime bird-lister myself, I know the thrill of racking up a big total for a day and the even bigger thrill of seeing a species that I have never seen before. But as a behavioral biologist I also have experienced the pleasure of really observing a species—getting to know what it does and why it does it. This experience of discovery and understanding is one I recommend highly. There are plenty of opportunities for you to have this experience as well. The behavior of a huge number of bird species, even common ones, is incompletely described and poorly understood. *The Birder's Handbook* (see Suggested Readings) is a particularly good source for what is and what is not known about the behavior of North American species. With only a few thousand professional ornithologists in the world, dedicated amateurs can make contributions to the field of bird behavior at many different levels.

At one level, the bird watcher who wishes to expand his or her horizons need only slow down and spend some time with the birds encountered on a walk. A little patience, the willingness to sit quietly, a notebook and pencil, and a permanent record book for transcribing one's observations are a starting point. There is real pleasure to be

(Continued on p. 6·96)

Sidebar 6: BIRD FAMILIES AS MODELS
FOR UNDERSTANDING OURSELVES
Stephen T. Emlen and Natalie J. Demong

Cooperatively breeding species such as the Florida Scrub-Jay, the White-fronted Bee-eater, and the Seychelles Brush-Warbler described in the text are of interest for reasons extending far beyond their helping behavior. They are interesting because they live in family groups. Mature offspring frequently delay their dispersal and remain with their parents for significant portions of their adult lives. Parents themselves form long-term pair bonds, and both mother and father contribute to the care of the young. The result is a multigenerational family structure that bears an uncanny resemblance to our own family systems (**Figs. A, B**). The obvious question is: Can we learn anything about ourselves by studying family-dwelling birds?

The study of human behavior (in this case, of human family dynamics) has typically been the research domain of social scientists, particularly sociologists and psychologists. Twenty years ago, most social scientists believed that culture alone determined human behavior, and that heritable influences were of minor, if any, importance. This view has changed radically with the explosion of human genome studies, however, and it now is generally recognized that virtually every human behavior develops through a complex interplay of both genetic and cultural influences.

Let us assume for the moment that genes *do* influence our behavior—that we carry heritable tendencies to behave in certain more or less predictable ways in specific social situations. How can we determine what these predispositions might be, and how might such knowledge be useful?

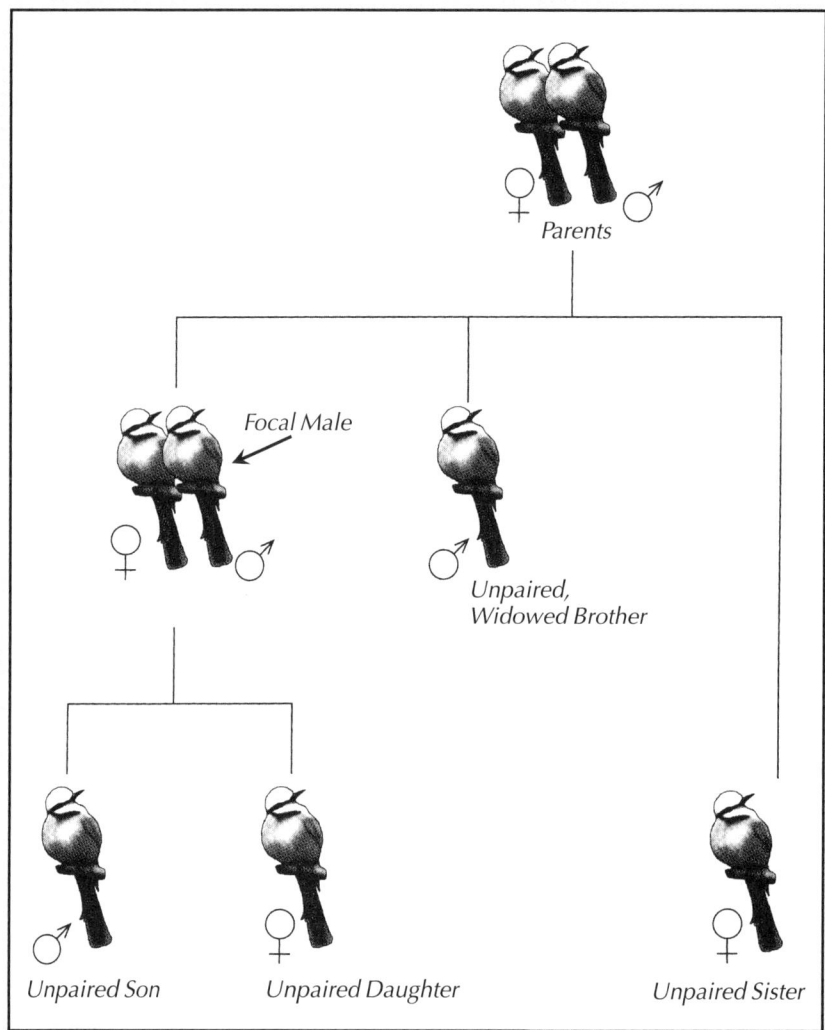

Figure A. Genealogy of a White-fronted Bee-eater Extended Family: *This cooperatively-breeding, East African species lives in extended family groups within large colonies. Fully grown offspring frequently delay their own breeding, instead remaining with their family of origin and helping their parents or other family members raise young. The result is a multigenerational family structure similar to our own. For example, the paired male labeled "focal male" has an unpaired son and daughter, a widowed brother, and an unpaired sister available to help him and his mate in their breeding effort, as well as potential help from his parents, should they fail to breed themselves. From Emlen, S. T., P. H. Wrege, and N. J. Demong. 1995. Making decisions in the family—an evolutionary perspective. American Scientist 83(2): 148–157 (p. 150). Reprinted with permission.*

a. Human Extended Family

b. White-fronted Bee-eater Family

Figure B. All in the Family: a. Human Extended Family: Authors Stephen T. Emlen and Natalie J. Demong are second and third from left, at back. Photo courtesy of Stephen T. Emlen. b. White-fronted Bee-eater Family: Five family members huddle together in the chill of an East African dawn. Photo by Natalie J. Demong.

The evolutionary biologist starts by assuming that whatever heritable behavioral tendencies we possess would have been selected during the tens of thousands of years that we lived as hunter-gatherers. The reasoning is that humans ceased the hunter-gatherer way of life and started living in large, permanent settlements only very recently, after the domestication of food plants some 12,000 years ago. Even though our social way of living has changed immensely since the discovery of agriculture, there has been insufficient time for natural selection to have significantly altered most heritable components of social behavior. Thus, to the degree that we carry heritable behavioral tendencies, they will be for behaviors that were adaptive to our ancestral hunter-gatherer way of life.

What is known about the social life of our pre-agricultural ancestors? Anthropologists agree that the social core of hunter-gatherer societies consisted of multigenerational extended family groups. We thus have had a long history of interacting not only with mates and offspring, but also with replacement mates (stepparents), half- and stepsiblings, in-laws, and other assorted family members. Natural selection has thus had plenty of time to fine-tune our behaviors for interacting optimally with different types of family members.

And this is where the value of avian studies comes in. Birds provide excellent model systems for understanding human family behavior for two reasons. First, over 300 species of birds are known to live in multigenerational family groups very similar to our own. Birds, more than most mammals, and even more than most primates, offer the most comparable family models to humans. Second, avian behavior is largely free of cultural influences. If there are universal rules governing family interactions among birds, then such rules are likely candidates to be heritable predispositions in ourselves. We then can ask whether these same rules are useful in predicting our own behaviors in different family situations today.

Kinship and Cooperation in Intact Avian Families

What might such basic rules be? Let us start by examining intact avian families—those in which the two parents stay together as a breeding pair. There is an important biological reason to expect such families to be more harmonious than other types of social groupings. The reason is that intact families are composed of extremely close genetic relatives—parents and their grown offspring. Such relatives share genes with one another by virtue of common descent (see Fig. 6–48). An individual that helps other family members is thus indirectly helping itself. If such help leads to higher survival of the next brood of young, the result is the production of additional close relatives. If such help reduces the work load of one's parents and allows them to live to breed again, the result is the same. The helper receives indirect genetic benefits in direct proportion to how closely it is related to the birds that it helps. The closer the kinship, the greater the tendency for animals to cooperate. This line of reasoning is known as "kin selection theory."

We should therefore expect to find extensive cooperation and helping within intact families. This expectation is borne out by data. More than 90 percent of bird species that live in multigenerational families practice cooperative breeding, similar to that of the Florida Scrub-Jay, White-fronted Bee-eater, and Seychelles Brush-Warbler, in which nonbreeding adults commonly help to rear offspring that are not their own. In essence, parents living with their grown offspring can count on a built-in "work force" to help with the task of child rearing. In stark contrast, such cooperation is practically unknown in avian species that live in any other type of social grouping.

There is a second reason to expect increased harmony in intact families. Competition over mates is commonplace in most avian species. But there is little or no such sexual competition among members of intact family units. Recall that family members are close genetic relatives. It is well known that incestuous matings (those with close relatives) can have harmful genetic consequences, so natural selection has fostered mechanisms that help birds recognize and avoid inbreeding with close kin. Sons rarely compete with their fathers, and daughters rarely compete with their mothers, for sexual access to a parent's spouse. Likewise, siblings don't seek sexual relations with one another, despite their frequent social interactions. The result of this avoidance of incestuous matings is increased family harmony—especially compared to the courtship disruption, mate guarding, and other sexually-related aggressive behaviors we see in non-family-living species.

Behavior in Avian Stepfamilies

Harmony in family interactions may evaporate if a bird parent dies or divorces. If the remaining parent takes a new mate, the equivalent of a stepfamily is created, and the family dynamics can instantly become contentious on a number of fronts.

Consider the "rule" of avoidance of incestuous matings. Unlike the original biological parent, a stepparent is not genetically related to its new mate's previous offspring. Their relationship thus is not bound by any incest restriction. A stepparent may find a willing partner among its stepoffspring (or vice versa), especially if the stepoffspring has only a minimal chance of finding a mate and a

territory of its own outside of the family. Stepparent-stepoffspring matings have been reported in Stripe-backed Wrens, White-browed Scrubwrens, and White-fronted Bee-eaters. In such cases, the remaining biological parent opposes such matings, resulting in intense aggressive interactions within the family.

Stepparents also have little biological incentive to provide care for their dependent stepchildren. In fact, from a stepparent's point of view, any time or effort spent investing in the mate's previous offspring may be a waste, especially if it delays its own reproduction with the new mate. If food is scarce, competition from the mate's previous offspring may decrease the chances of survival of its own, younger offspring. Thus the stepparent is predicted to offer only minimal, if any, care for its stepoffspring. This prediction is again borne out by data from family-dwelling birds.

Under some circumstances a stepparent may even benefit by forcibly terminating all care from others to its stepoffspring. The extreme example of this behavior, infanticide, is well documented not only in birds, but also in mammals, including rodents, social carnivores such as lions, and many species of primates. The risk of infanticide is greatest when the stepparent is of the dominant sex (in birds, generally the male).

If the new couple succeeds in having offspring of their own, conflicts of interest are even further intensified. This occurs because so many different degrees of relatedness now exist within the family. For example, when the original parent takes a new mate and produces a new brood, it is equally related to all its own offspring, both new and old. But its new mate is related only to the new offspring. The two parents will therefore disagree about how much investment to allot to the various youngsters in the stepfamily unit. And if the stepparent should bring previous offspring of its own into the

new family, the two sets of offspring will be totally unrelated to each other (they will be stepsiblings). The large differences in relatedness will result in a host of competitive interactions within the new family.

Consider the formerly reliable "work force" of grown offspring that routinely helped their parents in intact family situations. In the new stepfamily environment, these older offspring will be related to future young produced by the new breeding pair only as half-siblings (they have only one parent in common). They thus stand to gain only half as much in indirect fitness benefits by helping. Kin selection theory predicts that such offspring will be less willing to help in the rearing of half- or stepsiblings, a prediction borne out by studies of Florida Scrub-Jays, White-fronted Bee-eaters, and Seychelles Brush-Warblers.

When all these data are combined, the unavoidable prediction is that stepfamilies will be less stable than intact families. There are fewer reasons for the offspring to stay home and help, and the reduced degree of kinship among family members will lead to increased conflict, even between the remaining parent and its new mate. We know of no studies analyzing the rate of "divorce" among parents in avian stepfamilies, but we predict it will be high.

Implications for Human Families

The profound disruptions we find in the social dynamics of avian stepfamilies closely mirror those found in human stepfamilies. Large-scale sociological studies consistently show that human stepparents do invest less time and effort in the offspring from their partner's previous marriage than they do in their own children; that stepchildren are at greater risk for sexual and physical abuse than children in intact families; and that children report more conflicts with half-siblings and stepsiblings than with full siblings. Further, stepfamilies are less stable than intact families. The incidence of subsequent divorce is higher in second marriages, and increases with the number of stepchildren present. Children in stepfamilies also leave home significantly earlier than children in intact families.

Of what possible value is knowing that we share with birds a propensity to behave in similar ways in similar family situations? First, it suggests that the same evolutionary logic used to predict avian social interactions also correctly predicts some of our own patterns of family interactions. This justifies the comparative use of avian family systems as models for better understanding ourselves.

Second, such information can help us, as a society, to better anticipate the problems that accompany the changes in family composition occurring today. A century ago, most people lived in extended family situations. Mothers had access to the built-in support system of other extended family members who assisted with the tasks of child rearing. This work force was largely stripped away, however, with the shift from extended to nuclear families that occurred in the early 1900s.

The escalation of divorce rates is leading to a further transformation from nuclear to single-parent (typically mother) families. This further exacerbates the already reduced support system, creating a child-rearing situation that had no antecedent in our ancestral family environments. Whatever our heritable decision-making rules might be, they were never adapted to single-parent child-rearing situations. We have created a culturally novel social situation—one for which our inherited predispositions have prepared us poorly. Incorporating this evolutionary perspective on child care, however, can help us as we seek to build societal alternatives to replace the family-based child support system that operated in our past.

Another lesson to be learned from the study of avian families is how we, as individuals, can better anticipate and deal with family changes in our own lives. Recall the data on increased conflict in avian stepfamilies. Should those of us who are stepparents, stepchildren, or stepgrandparents be discouraged by these findings? We think not. Rather, we should accept what evolutionary theory predicts: conflict and strife are statistically more likely to occur, and to occur more intensely, in stepfamily situations than in intact families. However, as family members (and family counselors) become more aware of this increased potential for conflict, all participants can learn to become more sensitive to it. We then can be better prepared to deal with family problems early on, in ways that promote harmony and stability.

Biological predispositions are just that—predispositions. Unlike birds, we humans can use our intellectual resources to consciously modify behaviors we judge to be undesirable. The study of avian family dynamics contributes to this process by identifying situations where family conflict is most likely to occur, and by helping us to understand why such conflict is sometimes so difficult to eliminate.

So the next time you read about family members in some avian species helping at the nest, stop and think about the potential importance of such discoveries. By seeking and interpreting patterns in bird behavior, we discover that the importance of avian research often extends well beyond the birds themselves. It offers instruction that is directly relevant to the drama of our own lives. ∎

gained from looking at a bird long enough to see what it is eating, what plant it is pollinating, what displays it employs when interacting with conspecifics, and so on.

Perhaps the next step is to join established research projects; many of these are led by university types ranging from professors with teams of observers to graduate students in need of a volunteer. Research projects are also coordinated by nature centers, bird research organizations, government agencies, and various other nonprofit groups. Chapter 10 gives more details on how to get involved in research projects.

The truly dedicated amateur can consider a project of his or her own design. But it is only fair to say that the job is a challenging one. The first task is to identify a behavioral trait that warrants further research from a personal or scientific standpoint. To this end, there is no substitute for curiosity. If you observe a bird doing something puzzling that piques your curiosity, you have a start. As this chapter has emphasized, behavioral biology presents two general kinds of puzzles: the problem of internal or proximate causes, and the problem of evolutionary or ultimate causes. To develop a research project, you must know whether the puzzle requires exploring the underlying proximate causes or the long-term evolutionary causes. So, for example, you may wonder why Blue Jays sometimes stretch on the ground near an ant nest, apparently permitting the ants to "attack" them (see Fig. 3–22). You could focus on the proximate cues for the behavior: Do parasites cause the skin to itch? Do Blue Jays seek ant nests only when the air temperature is high? *Or* you could investigate the reproductive, evolutionary benefit of the odd "anting" behavior: Do the ants remove skin and feather parasites that reduce the birds' vigor? Is the behavior a way to find food, luring the ants close enough to be eaten? The research you do depends entirely on the kind of hypothesis you select to test.

As we have seen, a good way to investigate evolutionary causes of behavior is to ask, "Why would the bird do such-and-such, given that the action carries with it some obvious disadvantages in terms of reproduction?" Anting is of evolutionary interest precisely because anting birds appear to both expose themselves to predators and to waste time that could be spent in other, more productive, behaviors.

Having selected a puzzle and developed one or more possible solutions, the next step is to test the hypothesis, or better still, several alternative hypotheses. This task involves thinking of testable predictions (behaviors or circumstances that you should be able to observe if the hypothesis is correct) **(Table 6–5)**. Testing to see if your expectation matches reality is full of pitfalls and pleasures. Your goal is to collect enough *quantitative* data to determine if the prediction was accurate or not, and thereby to reject or confirm the hypothesis that gave rise to it. This requires, among other things, a great deal of patience, neutral note-taking, precise record-keeping, a clear view of what observations are pertinent, and avoiding overinterpretation or biased analysis of what the birds are doing.

So, for example, to test the hypothesis that anting was food-gathering behavior, you would need to watch carefully for instances of "anting" jays picking up ants and swallowing them. Observations from a single jay would not be adequate. Records that failed to indicate total time of observation, total number of ants picked up, and total number of ants consumed (if any) would be of little value.

At some point early in the process, it would help to have access to a university library carrying ornithological journals such as *The Auk, The Condor, The Wilson Bulletin,* and *The Ibis* as well as behavioral journals such as *Animal Behaviour, Behavioral Ecology and Sociobiology, Behaviour,* and *Ethology.* General ornithology textbooks, the Stokes Nature Guides, and behavioral texts (see Suggested Readings and Ch. 2, The Birder's Essential Resources) also may help you track down what already has been done to address a specific hypothesis or question, so that you do not simply repeat another person's study.

If you can find research papers related to your own interest, you may be able to contact the authors, some of whom may be willing to assist you in various ways. Publishing a paper on a completed research project is a difficult, but highly rewarding, part of the scientific process. It is a great thrill to carry a behavioral project through from start to finish.

Table 6–5. The Scientific Method

(1) Choose a question to study: Questions are usually generated based on prior observations or known problems.

Example: *Is nesting success in American Kestrels related to nest-cavity orientation?*

(2) Restate the question as a hypothesis that can be tested: (Note: not all scientific inquiries involve hypothesis testing.)

Example: *American Kestrels have higher nesting success in southeast-facing cavities compared to cavities that face in other directions.*

(3) Design a systematic study to objectively test the hypothesis: Study designs can be quite complex and should include plans for statistically analyzing the data.

Example: *You will determine the compass azimuth and fate of American Kestrel nesting attempts in your study area. Then you will compare nesting success of kestrels in southeast-facing cavities to that of kestrels nesting in cavities that face in other directions using a chi-square test for independence.*

(4) Collect unbiased data following the methodology set forth by the study design:

Example: *Without biasing nest outcome, monitor each kestrel nest and determine its fate and orientation.*

(5) Analyze the data using appropriate statistical techniques and interpret the results in statistical and biological contexts:

Example: *Statistically compare rates of nesting success for American Kestrels using southeast-facing cavities versus those in cavities that face in all other directions.*

It is found that 52% of southeast-facing nests were successful, whereas 46% of nests facing in other directions were successful.

When a chi-square test for independence is done, the difference in nesting success between southeast-facing cavities and cavities that face in other directions is found to be not statistically significant.

(6) Draw conclusions and confirm or reject the hypothesis:

Example: *Nesting success in American Kestrels is not significantly higher in cavities that face southeast. Reject the hypothesis proposed in step (2).*

(7) Devise new hypotheses for future testing and repeat Steps 1 through 7:

Example: *American Kestrels have higher nesting success in northwest-facing cavities compared to cavities that face in all other directions.*

(8) Report the results of your study to share the information with others

(9) Consider new questions:

Example: *Is nesting success in American Kestrels related to nest-cavity height?*

6

Suggested Readings

Alcock, J. 1993. *Animal Behavior: An Evolutionary Approach.* Fifth Edition. Sunderland, MA: Sinauer Associates.

A general behavior text that covers many of the same topics presented in this chapter in more detail and with a wider range of animal examples.

Brown, J. L. 1987. *Helping and Communal Breeding in Birds.* Princeton, NJ: Princeton University Press.

An advanced text on a highly fascinating evolutionary issue, the occurrence of helpers at the nest and other forms of reproduction by social birds.

Ehrlich, P. R., D. S. Dobkin, and D. Wheye. 1988. *The Birder's Handbook.* New York: Simon & Schuster.

Summarizes much of the knowledge of the behavior of North American birds. Includes a short section on how amateurs can contribute to ornithology.

Heinrich, B. 1989. *Ravens in Winter.* New York: Simon & Schuster.

One of the best books ever on the scientific method applied to a problem in bird behavior.

Lorenz, K. Z. 1952. *King Solomon's Ring: New Light on Animal Ways.* New York: Thomas Y. Crowell Co.

A classic text that presents the ethological perspective, especially on the proximate causes of bird behavior.

Stokes, D. and L. 1979–1989. *A Guide to Bird Behavior, Vols. I to III.* Stokes Nature Guides. Boston: Little, Brown and Company.

For a selected list of common birds, gives detailed life history information; descriptions of displays, songs, and calls; and suggestions on how to observe the behaviors.

Tinbergen, N. 1953. *The Herring Gull's World: A Study in Social Behavior of Birds.* London: Collins.

Another classic account of bird behavior that presents the ethological perspective, especially on the ultimate causes of Herring Gull behavior.

Some representative recent short articles on bird behavior that are exciting and readable:

Heinsohn, R. 1995. Raid of the red-eyed chicknappers. *Natural History* 104(2): 44–51.

McDonald, D. B. and W. K. Potts. 1994. Cooperative display and relatedness among males in a lek-mating bird. *Science* 266:1030–1032.

Mulder, R. A. 1994. Faithful philanderers. *Natural History* 103(11):56–63.

Petrie, M. and T. Halliday. 1994. Experimental and natural changes in the peacock's *(Pavo cristatus)* train can affect mating success. *Behavioral Ecology and Sociobiology* 35:213–217.

Vocal Behavior

Donald E. Kroodsma

Think, every morning when the sun peeps through
The dim, leaf-latticed windows of the grove,
How jubilant the happy birds renew
Their old, melodious madrigals of love!
And when you think of this, remember too
'Tis always morning somewhere, and above
The awakening continents, from shore to shore,
Somewhere the birds are singing evermore.
—Henry Wadsworth Longfellow

 It's four in the morning, Sunday, mid-June, atop my usual fire tower at Mount Lincoln in western Massachusetts, and already I sense the impending frenzy. Off and on all night long the Whip-poor-will has been singing vociferously, earning his alternate name of "nightjar." But others now begin to stir. As the eastern sky hints of the sunrise, Ovenbirds increasingly launch from the ground into the canopy and sometimes beyond, releasing their energetic, jumbled bursts of song, which seem to have been pent up all night long. Minutes later, the American Robins are caroling. The two other thrush songs soon follow, the flutelike *ee-oo-lay* of the Wood Thrush **(Fig. 7–1)** and the ethereal, downward-spiraling *vee-ur, vee-ur, veer, veer* of the Veery. I know the Eastern Towhee will

Figure 7–1. Wood Thrush: *One of the first and last singers of the day, a Wood Thrush serenades residents of eastern forests with its melodic, flutelike phrases. Photo by Lang Elliott.*

Figure 7–2. Black-and-white Warbler: *The Black-and-white Warbler sings his high, thin wee-see, wee-see, wee-see from a high branch in mixed woodlands of eastern North America. Photo by Lang Elliott.*

be next—he almost always is, seeming to flaunt his entire repertoire of four different songs as fast as he can.

As four-thirty approaches, others join in. Three warblers, the Chestnut-sided, Black-and-white (**Fig. 7–2**), and Black-throated Blue, aggressively announce their presence, and the staccato, machine-gun-like fire of the Chipping Sparrow jars the air. Below me, the Eastern Phoebe continues to rapidly alternate his two song forms. I'm sure the distant Black-capped Chickadee has been whistling for some time, too, but he's just now caught my attention. The crows, jays, vireos, and others have all chimed in, and by quarter to five there's no doubt—everyone is present and accounted for.

With my trained ear, I not only can identify who is making each sound, but also can hear a little of what each bird is doing. The Ovenbirds abandon their flight songs and settle low in the canopy to repeat their *tea-CHER, tea-CHER, tea-CHER* song. The towhee gradually eases his pace and increasingly repeats each different song several times before introducing the next. Others slow down, too. The Chipping Sparrow lengthens his familiar, dry trill until he's singing the way the field guides dictate. As the minutes tick away, one of the phoebe's two songs predictably comes to predominate. Near sunrise, the warblers switch from their aggressive dawn song to a daytime song, one used more in the presence of their female friends. During all this time, the chickadee has been whistling his *hey-sweetie* on a series of different frequencies. And so much more is happening, I am sure—my nonavian senses give me just a hint of all that transpires during these dawn events.

Each of these birds, and each of the others whom I've slighted in this all-too-brief glimpse of dawn at Mount Lincoln, is on a mission. The mission continues through the morning, the afternoon, into the next day, the next week, month, year, and throughout life. The mission is to succeed, in evolutionary currency, by surviving and reproducing, and each tiny note of each sound a bird makes during the mission is geared to increasing the likelihood of success. For each bird, the ultimate goal is to perpetuate itself. In all bird species, success for a male requires a female, and success for a female requires a male. Nearly all of what we hear from the birds on their mission is the negotiating, cajoling, persuading, and impressing that accompany the selfish games required for success.

Evolution occurs as each individual attempts to perpetuate itself in whatever selfish way it can. Males interact to establish a hierarchy and to impress potential mates, either directly or indirectly. Females assess possible mates by listening carefully to the male interactions and proclamations. Once paired and raising a family, a male and female further negotiate, on a daily basis, how much effort each contributes. Also, females of ostensibly monogamous species sometimes choose to mate with males other than their social partners; much of the male noise we hear during the dawn chorus may actually be geared toward these extrapair encounters. Just as we would manage a stock portfolio for maximum financial gain, each bird uses its sounds to manage its

Figure 7–3. Blue Jay Tending Nestlings: *The success of any bird is measured in the currency of offspring. Individuals that are better at using their sounds to manage their environments will, on average, produce more young, passing some of their favorable genes for sound production on to those young. Photo courtesy of Isidor Jeklin/CLO.*

social environment for its own selfish gain **(Fig. 7–3)**. Each manager is in turn managed by other individuals, of course, so compromises must be reached and agreements brokered. Sounds play a key role in the entire process (Smith 1977, 1996).

This chapter is all about the sounds birds use to manage their daily lives. The goal is to help you, the student, understand what birds do and why they do it, to the best of our current knowledge. After reviewing what sound actually is, you'll learn about the repertoires of sounds that birds use, how songbirds (and parrots and some hummingbirds) learn many of their sounds just as we learn our speech, and how other species (such as flycatchers) don't seem to learn. You'll also learn how brains control songs in birds, how a bird can sing a duet all by itself with its two voice boxes, how sounds of some birds differ from place to place and from one generation to the next, and why birds duet, mimic, and make the particular sounds they do. In Sidebar 6, Listening on Your Own, I suggest several listening exercises to help you get started.

The objective, in short, is to help you appreciate the extraordinary world of bird sounds that surrounds us. I know that as I revel in the dawn aural feast served to me every summer day, throughout New England and elsewhere as dawn circles the globe, windows are being slammed so that unappreciative humans can slumber another hour or so. My wishes are to open each reader's mind to the world of bird sounds, so that we not only hear but see and feel these sounds; and to hear windows opening wider at dawn, to let the sounds of our natural world permeate our lives.

What is Sound?

■ Consider a musical instrument such as a drum. Pounding on the tightly stretched skin of the drum makes it vibrate, which sets into motion the air around it. As the vibrating skin moves outward, it compresses a band of air; when the skin moves inward, the air that rushes into the space formerly occupied by the skin is thinned. This suc-

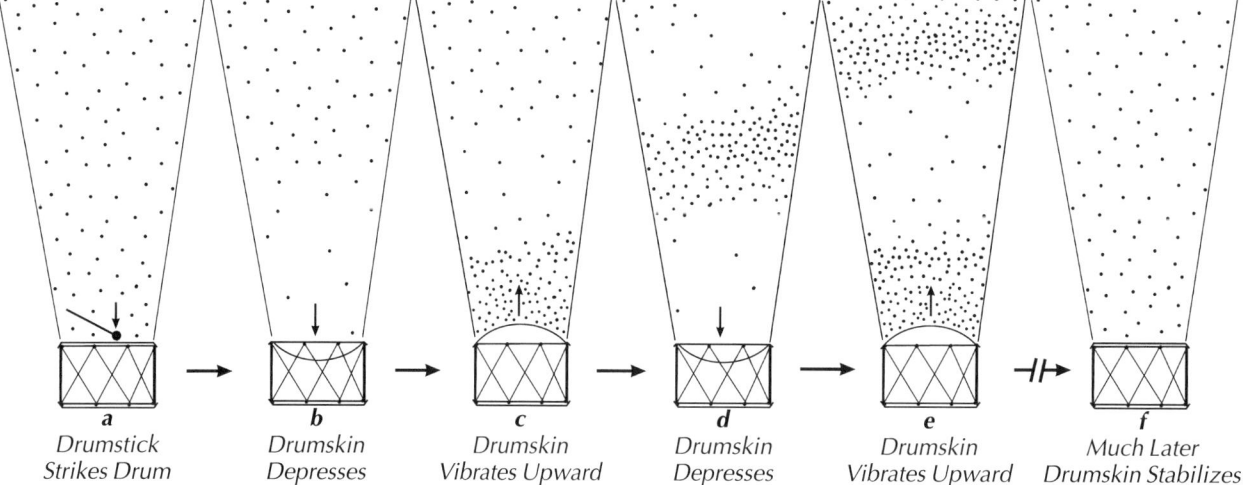

a	**b**	**c**	**d**	**e**	**f**
Drumstick Strikes Drum	*Drumskin Depresses*	*Drumskin Vibrates Upward*	*Drumskin Depresses*	*Drumskin Vibrates Upward*	*Much Later Drumskin Stabilizes*

Dots Represent Molecules of the Gases that Compose Air

a: *Stationary drumskin just before it is struck; air molecules uniformly distributed*
b: *After drumstick hits drumskin, drumskin depresses, thinning nearby air*
c: *Drumskin vibrates upward, compressing nearby air; previous band of thinned air has propagated upward*
d: *Drumskin vibrates downward, thinning another band of air; previous bands of thinned and compressed air have propagated upward*
e: *Drumskin vibrates upward, compressing yet another band of air; previous bands continue to propagate upward*
f: *Much later, after drumskin has ceased to vibrate, air molecules return to uniform distribution*

Figure 7–4. Generation of a Sound Wave Through Air: *A sound wave consists of a series of alternately compressed and thinned bands of air molecules propagating through air. Here, the vibration of a drumskin (shown greatly exaggerated), alternately reduces and increases the density of the air nearby, and this "wave" of sound travels upward and away from the drum. For convenience, the "wave" is shown here in a narrow arc above each drum; in reality, the wave would consist of bands of air moving outward in all directions, like ripples of water from a stone dropped into a pond.*

cessive compression and thinning (or rarefaction) of air molecules is transmitted to air farther away and constitutes a sound wave traveling through space **(Fig. 7–4)**.

On the familiar piano, the heavy wires that produce the lowest C vibrate only 32.7 times per second, and the thin wires that produce the highest C vibrate more than 4,000 times per second. The number of times this cycle of compression and rarefaction occurs per second is the frequency, which is measured in cycles (waves) per second, or Hertz (Hz), named after Heinrich Hertz, a German pioneer in the study of sound. The lowest frequency most humans can hear is about 20 Hz, and the highest is about 20,000 Hz. Fortunately for us, most bird sounds are in the range that we hear well, from about 2,000 Hz (or 2 kilohertz [kHz]; 1 kilohertz = 1,000 Hertz) to about 7,000 Hz (7 kHz) **(Fig. 7–5)**. Some birds, such as pigeons, can hear sounds of a lower frequency than we can, and some owls can hear sounds of lower intensity than we can. For the most part, however, laboratory tests have not shown any truly superior general hearing ability among birds (Dooling 1982).

Seeing Sounds: Sonagrams and Oscillograms

We humans rely heavily on our vision, and our relatively untrained ears often have trouble appreciating sound. One way to enhance our hearing and listening skills is to convert the bird sounds to graphs that we can see and study. These graphs take two forms, called

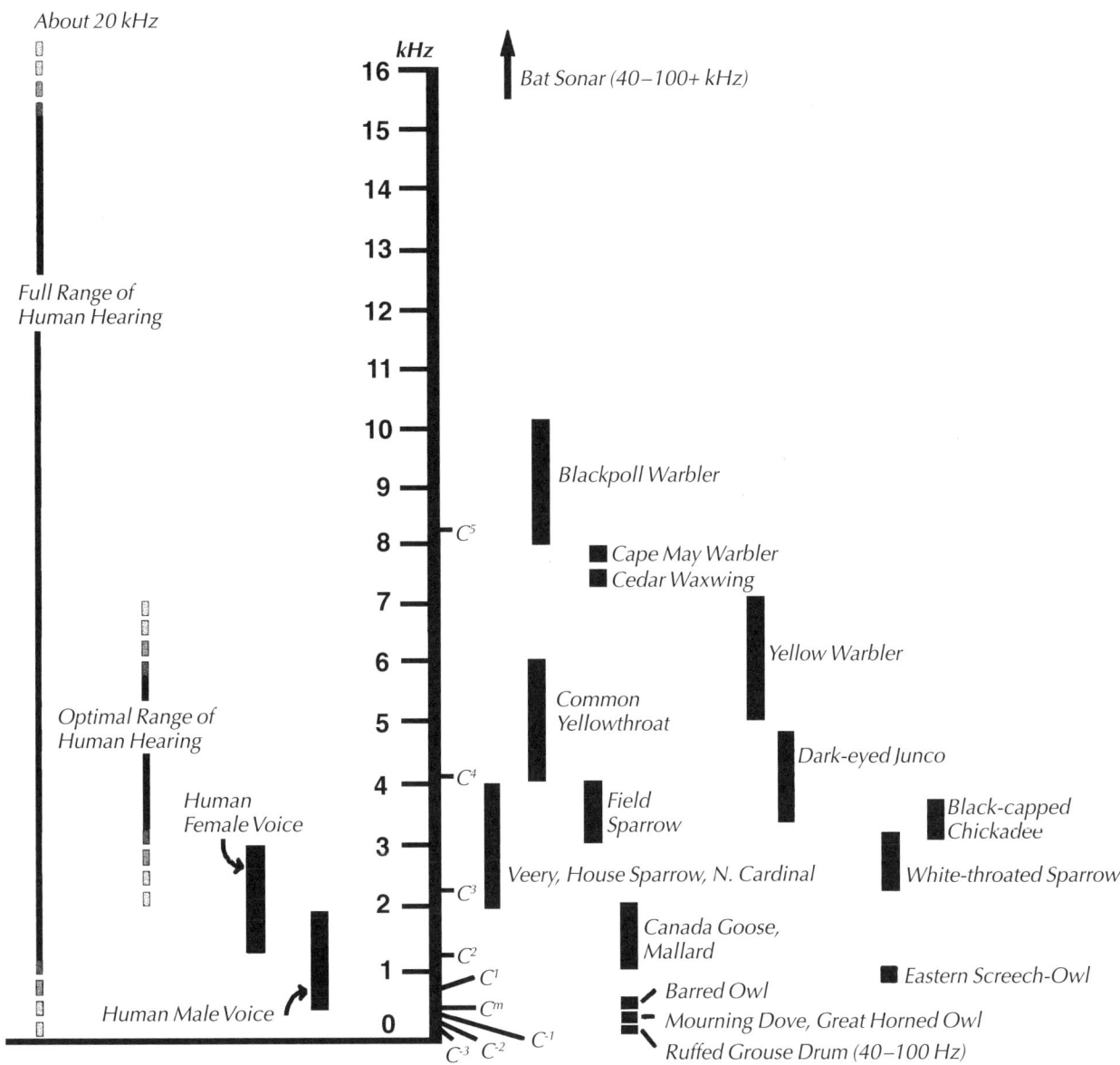

Figure 7–5. Typical Frequency Ranges of Selected Bird and Human Sounds: *Average frequency ranges, shown in kilohertz (kHz; 1 kilohertz = 1,000 Hertz), of various bird songs, bat sonar, and human voices and hearing. Also shown are musical notes. C^m is middle C on a piano, C^1 to C^5 are the first through fifth Cs above middle C (C^4 is the highest C on a piano), and C^{-1} to C^{-3} are the three Cs below middle C. Note that the Cs are not evenly spaced because the relationship between frequency (cycles or waves per second) and pitch (a musical note on a scale) is not linear, but logarithmic. Because human speech is very complex, with many harmonics and different frequency ranges for different types of sounds (for instance, an "s" sound is in a different frequency range than an "ee" sound), the ranges shown here for human voice are only rough approximations. Human hearing ranges vary from person to person, and the exact endpoints are hard to determine, as indicated by the fading bars.*

sonagrams and oscillograms (wave forms). Although studying these graphs seems somewhat daunting at first, a little experience shows that the graphs can be very helpful in understanding how to think about and listen to bird sounds.

First examine the sonagram, using the simple whistled song of the Black-capped Chickadee as an example **(Fig. 7–6a & Track 1; see "Use of CD with Chapter Text," below)**. The vertical axis is frequency, measured in kilohertz. The horizontal axis is time. Field guides describe the chickadee's song as *fee-bee*, or *hey-sweetie,* with the first part slightly higher in frequency than the second. The sonagram of this whistled song clearly shows two whistles, each about 0.4 seconds in duration, with about 0.2 seconds between them. The first whistle is at about 4.0 kHz, roughly the highest C on a piano, and the second one is about 400 Hz lower, at about 3.6 kHz.

One feature of our chickadee's *hey-sweetie* doesn't show very clearly on a sonagram, however, and that is the relative loudness of the sounds. If we listen carefully, we can hear two syllables within the *sweetie,* because a subtle but clear drop in loudness occurs in about the middle of the whistle. Although the relative intensity of a sound is reflected in the darkness of the sonagram, the second graph form, the oscillogram, reveals it far better **(Fig. 7–6b)**. Time in the oscillogram is again shown on the horizontal axis, but relative intensity is on the vertical axis, so the louder a sound is, the greater the height and depth of the wave on the oscillogram. We now see in our two graphs not only patterns in frequency but patterns in amplitude; the whistles are clearly not the same intensity throughout, and the audible break in *sweetie* is evident.

By using both the sonagram and the oscillogram to help us "see" bird sounds, we can begin to appreciate the finer details of what we hear. Consider another example of a song with pure whistles, the

Use of CD with Chapter Text

Please note that there is a CD (compact disc), narrated by Margaret A. Barker, to accompany this chapter. It is attached to the inside back cover of this book. As you read through the chapter and sidebar text, in addition to the usual figure references (for example, Fig. 7–n), you will encounter the word Track followed by a number. This tells you to stop reading and listen to the appropriate selection on your CD; you will be able to directly punch in the number after Track as the track number on the CD. The CD tracks for sidebars are integrated with the tracks for the chapter text, such that if you read the sidebar at the time it is referred to in the text, you can continue your CD tracks in numerical order.

In many instances, a CD track is linked to a sonagram, which appears in the text as a regularly numbered figure. These situations will be indicated by the reference: **(Fig. 7–n & Track n)** in the text. In these cases, the sonagram shows a portion of the CD track you are listening to (for exactly *which* song or portion, refer to the figure caption); you will want to look at the sonagram as you listen to the CD.

Appendix A of this chapter is a list of the tracks and a description of their contents. For many tracks, the information in the CD narration will be sufficient to allow you to understand what you are listening to. For some tracks, however, especially those with no associated sonagram, more detailed information will be helpful. This additional information is included in the track descriptions in Appendix A. ∎

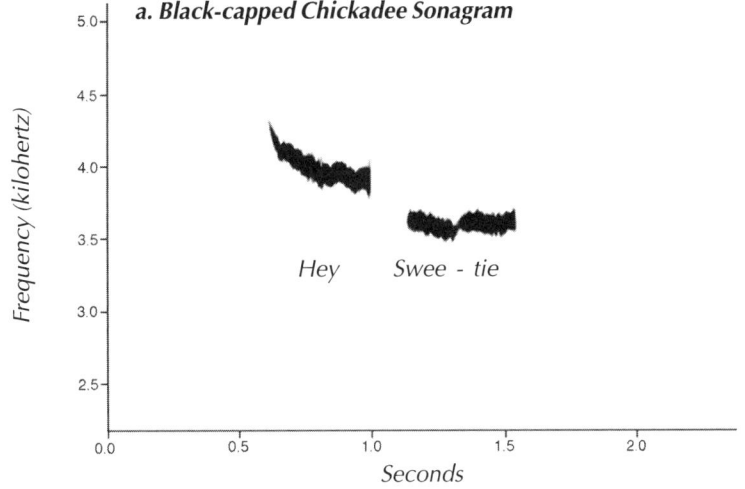

a. Black-capped Chickadee Sonagram

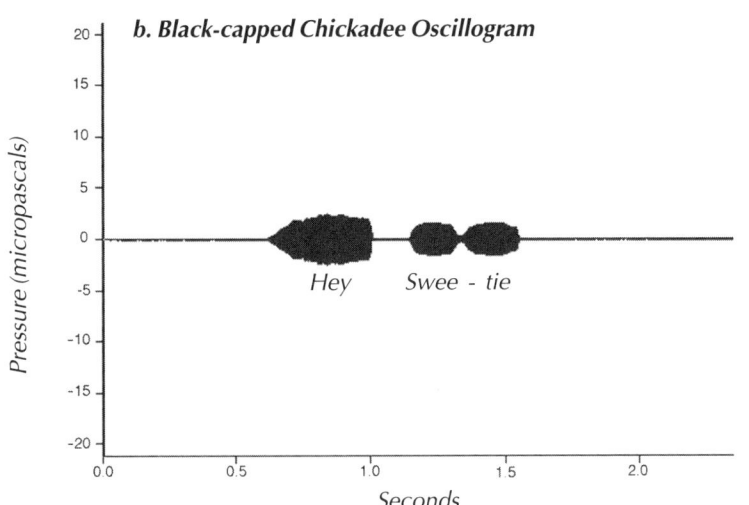

b. Black-capped Chickadee Oscillogram

Figure 7–6. Sonagram and Oscillogram of Black-capped Chickadee Song (Track 1, 2nd Song): a. Sonagram: A sonagram is a representation of a sound plotted as frequency (measured in thousands of cycles per second [kilohertz]) versus time in seconds. The volume (loudness) is roughly indicated by the darkness of the notes. The more pure the tone, the smaller the frequency range of the note at any given point in time. In this sonagram of a Black-capped Chickadee song, the pure, whistled quality of the notes, as well as the rhythm, can be seen easily. *b. Oscillogram:* An oscillogram represents a sound in terms of relative loudness (vertical axis) versus time (horizontal axis). The vertical axis actually gives a measure of the increase and decrease in air pressure (in micropascals) associated with the sound wave (see Fig. 7–4), which determines the loudness. Therefore, the louder the sound, the greater the height and depth on the vertical axis. Note that frequency cannot be determined from an oscillogram. In this oscillogram of the same Black-capped Chickadee song pictured in (*a*), the greater loudness of the first note, and the two-part nature of the second note, the sweetie, can be seen more clearly than in (*a*).

White-throated Sparrow's *old sam peabody peabody peabody* (**Fig. 7–7 & Track 2**). These songs are introduced by two or three relatively long whistles, each on a slightly different frequency; in some individuals successive notes rise in frequency and in some they fall, but the overall pattern is unmistakable. The last part of the song, the *peabody* portion, clearly shows three notes coupled together to form each *peabody*, even though the human ear has to listen carefully to distinguish the three parts. The patterns of frequency are visible in the sonagram, and patterns of amplitude are especially clear in the oscillogram.

In sharp contrast to whistled sounds, such as the chickadee's *hey- sweetie* and the sparrow's *peabody*, are "noisy" sounds, which consist of a broad span of frequencies at an instant in time. Noisy sounds appear on a sonagram as vertical lines, not horizontal ones. On a sonagram, the snap of a finger and the slam of a door appear as noise. When a woodpecker taps on a tree, the sound is noise because different parts of the tree vibrate at many different frequencies.

The drumming sounds of three different woodpecker species nicely illustrate this noise and show how patterns of noise can be highly distinctive (**Fig. 7–8 & Track 3, Track 4**). The first two species, the Downy Woodpecker and Hairy Woodpecker, look a lot alike, but can

Figure 7–7. Sonagram and Oscillogram of White-throated Sparrow Song (Track 2, 1st Song): *Like the song of the Black-capped Chickadee, the song of the White-throated Sparrow consists entirely of pure whistles, each note on nearly a single tone. Although there is some variability among individuals, most songs follow the same general pattern: the first two or three notes are long and solid, and the last three or four notes are divided into three parts each. The first notes may be lower or higher than the last, and occasionally one is divided like the later notes. The song is often described as old sam peabody peabody peabody because it resembles the folk song—but here poor old sam is used at the beginning, because there are three notes.* **a. Sonagram:** *Note how clearly frequency differences are shown.* **b. Oscillogram:** *Rhythm is still evident, but not frequency. Note that the division of the peabody notes into three parts is much more clear in the oscillogram than in the sonagram. In addition, the oscillogram more clearly shows that all the notes are approximately the same loudness.*

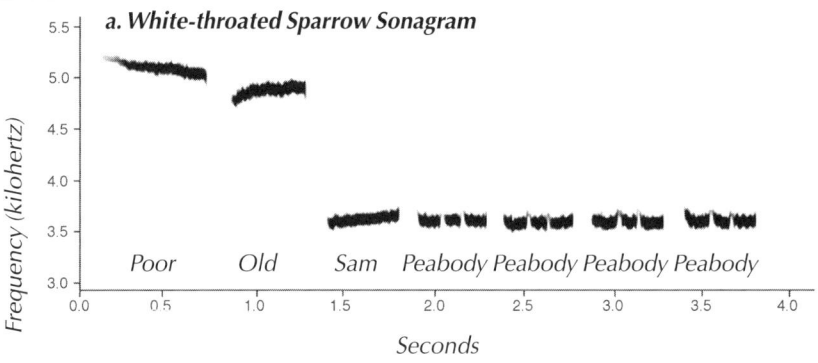

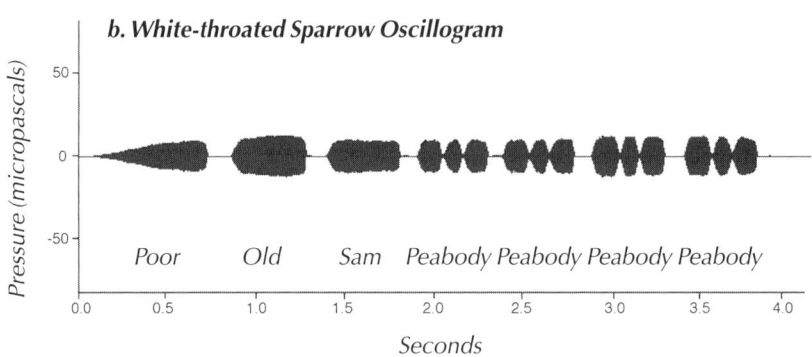

Figure 7–8. Sonagrams of Woodpecker Drums (Track 3, 1st Hairy W. Drum; Track 4, 1st Downy W. Drum; Track 5, 2nd Y-b. Sapsucker Drum): *In contrast to pure tones, harsh, "noisy" sounds appear on a sonagram as tall, vertical bands consisting of many different frequencies at any given point in time. These "broad-band" sounds are typical of woodpecker drums and pecks, as well as many mechanical sounds such as clapping hands or the slamming of a door. The Hairy Woodpecker, in spite of its larger size, drums consistently faster than the Downy Woodpecker. In this example, the Hairy Woodpecker gives 26 drums per second, and the Downy Woodpecker, 15. The Yellow-bellied Sapsucker drum is distinctive, varying in rate and generally slowing down toward the end. Woodpecker drums may be given by both sexes, and function to declare a territory and to attract a mate. They are faster and more regular than the more leisurely pecks used to find food.*

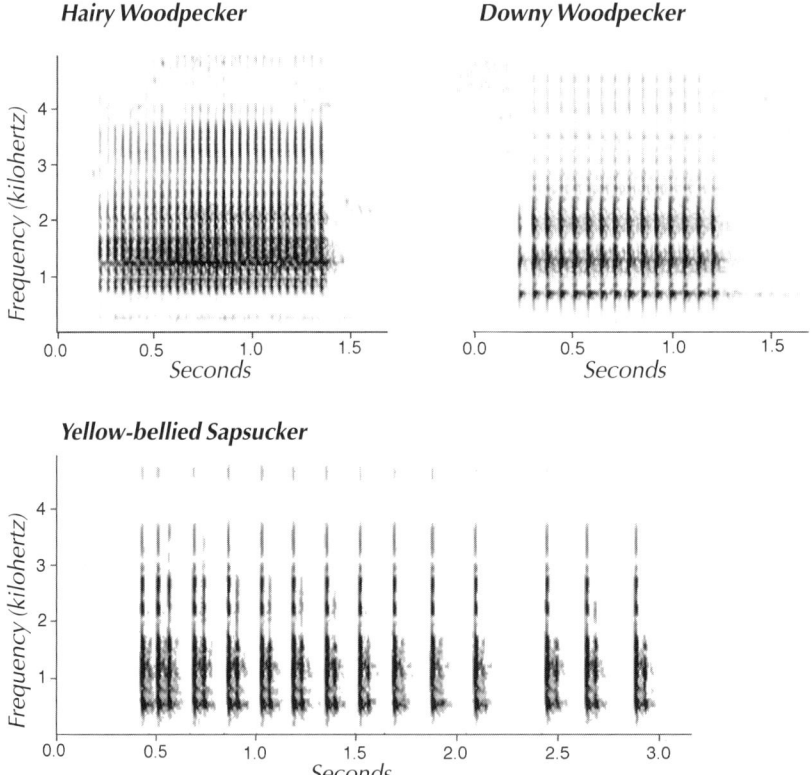

be distinguished by their size or by their unique vocalizations. They differ, too, in the rate at which they drum. Surprisingly, the larger Hairy Woodpecker drums faster than the Downy. Counting the number of pulses in a second shows that, in our example, the Hairy Woodpecker has about 26 drums per second and the Downy only 15. The third species, the Yellow-bellied Sapsucker, after which the Cornell Laboratory of Ornithology's bird sanctuary was named, adds a unique rhythm to its drumming, delivering taps not at the constant rate of the Downy and Hairy, but starting fast and ending at a slower pace **(Fig. 7–8 & Track 5)**.

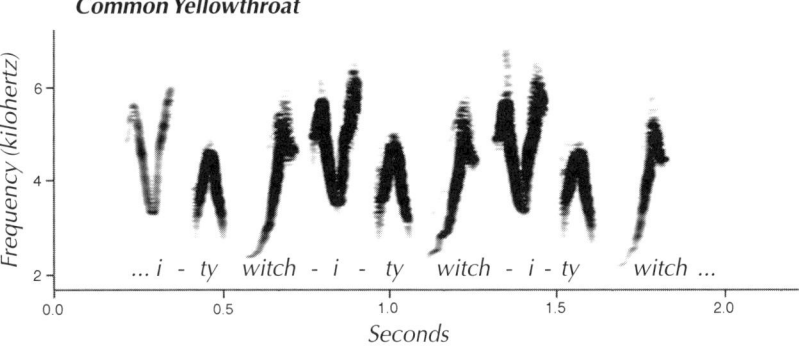

Figure 7–9. Sonagram of Common Yellowthroat Song (Track 6, 3rd Song): Common Yellowthroat songs vary tremendously from bird to bird, but all have in common a pattern of three or four notes (three in this example) usually described by birders as witchity. *This pattern is typically repeated three times in a song.*

Understanding Complex Songs

Sonagrams and oscillograms help us to visualize more complex sounds, too. The song of the Common Yellowthroat can be heard throughout much of North America. The song is rendered as *witchity-witchity-witchity* by field guides, so we know to search on a sonagram for some kind of a complex repeating pattern of individual notes **(Fig. 7–9 & Track 6)**. Each *witchity* consists of three or four dark notes on the sonagram, and successive *witchities* appear identical. This pattern occurs in almost every yellowthroat song. When the tape is slowed to one-half or one-quarter speed, we can more easily follow each note. Halving the tape speed also halves the frequency, of course, so the song sounds much lower.

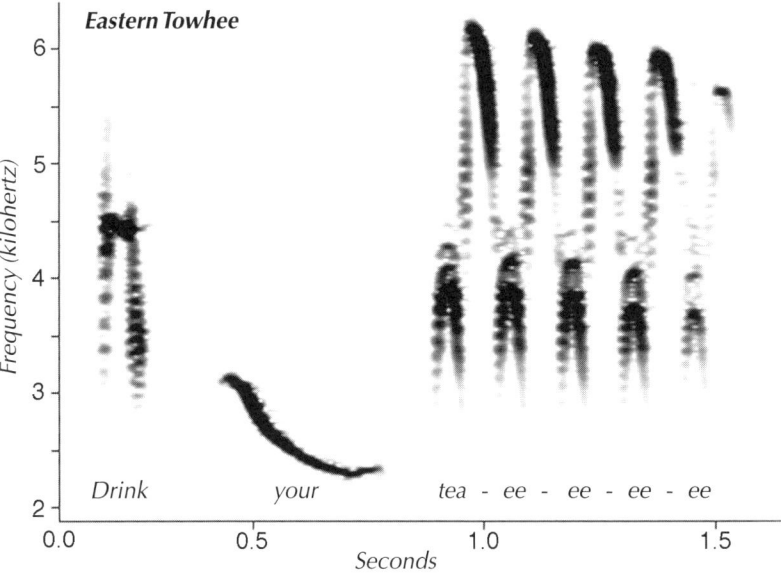

The *drink-your-teaeeee* of the Eastern and Spotted towhees is also familiar throughout North America, although the western Spotted Towhees often seem to omit the *drink-your* in their songs. The classic song has a couple of introductory sounds, starting with a high *drink* followed by a lower-frequency *your*. Then the song typically concludes with a series of identical units, or syllables, often repeated so fast that we cannot count them **(Fig. 7–10 & Track 7)**. Given the repetitive nature of this song ending, perhaps the *teaeeeee* would be more appropriately rendered as *tetetetete*. The chosen example at normal and slowed speeds clearly shows these features of the song.

One final example illustrates especially well how "seeing" the song as a sonagram helps us appreciate what we are hearing. The song

Figure 7–10. Sonagram of Eastern Towhee Song (Track 7, 1st Song): The classic Eastern Towhee song begins with two relatively slow introductory notes—a higher, then a lower one—often described as drink-your. *These are followed by a **trill** (tea-ee-ee-ee-ee)—a series of rapidly repeated units or **syllables**. Songs vary greatly among birds, however, and each individual male may sing up to eight different song types.*

Figure 7–11. Sonagram of Winter Wren Song (Track 8, 1st Song): *The song of the Winter Wren is one of the longest and most complex among North American birds. The high, rapid, piccolo-like sounds race by as a series of different trills, sometimes separated by individual notes.*

Winter Wren

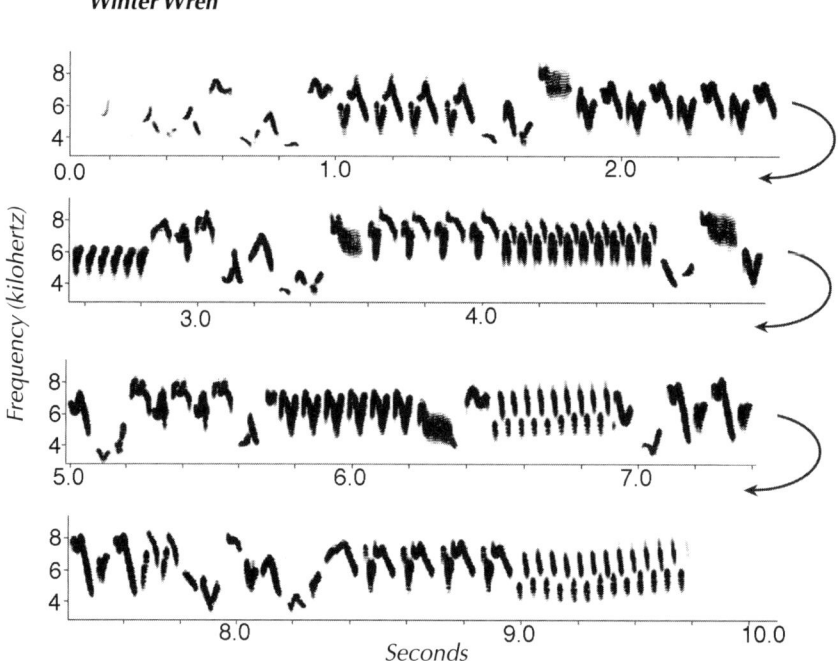

of the Winter Wren, especially from western North America, has been referred to as "the pinnacle of song complexity" (Kroodsma 1981). The songs of these wrens are truly remarkable; they can be 10 seconds or longer, but the wren sings successive notes so rapidly that all we hear is a blur. Once we slow the song down, however, we can appreciate all of the finer details and begin to hear what we see in the sonagram. At one-half and especially at one-quarter normal speed, the listener can clearly identify each note the wren produces **(Fig. 7–11 & Track 8)**.

Dissecting the sounds of the wren and the other species in this fashion gives us a new appreciation of the complexity of bird sounds. By using sonagrams, and to a lesser extent oscillograms, we can also begin to "see" the finer details of what birds do with their sounds. We add a dimension to our listening skills that enhances our overall experience and appreciation for bird sounds. Throughout this chapter we illustrate sonagrams for you to inspect, accompanying them with recordings so that you can improve both your seeing and hearing of bird sounds.

Vocal Repertoires

▮ Our experience tells us that birds produce a tremendous variety of sounds. How can we begin to make sense of this great diversity? The best way is to start with an individual bird. Professionals who study bird sounds often mark a bird, usually with a distinctive combination of colored leg bands, and then follow it, often for days at a time, and sometimes year after year. The researchers record all sounds the bird makes and carefully document the context in which each sound was used. Next, they make sonagrams for each recording so that the anatomy of each sound can be studied in more detail. Finally, they sort all the sonagrams by shape and combination of notes to establish a library of sonagrams representing the vocal repertoire of that individual. In this way, we gradually begin to understand how the bird uses this vocal repertoire in different social contexts.

This approach has revealed that birds have a relatively limited selection of types of **vocal signals** (**signals** are vocalizations that have evolved for a specific function). For now, we'll consider "song" one type of signal; even though a Brown Thrasher, for example, might have thousands of different songs, all seem to serve the same function (see Song Repertoires, later in this chapter). Consider again the Black-capped Chickadee (for good accounts of chickadee behavior, see Hailman 1989; S. M. Smith 1991; and Hailman and Ficken 1996) **(Fig. 7–12)**. Extensive tape recording, creation of sonagrams, and classification of sound structure has shown that chickadees have about 14 different kinds of sounds. Three sounds are frequently used, highly distinctive, and readily recognized by anyone who watches chickadees for even a short time. One of these is the *hey-sweetie*, identified as the "song" because the male usually sings it loudly from an exposed perch, as if he were proclaiming his territory or serenading a female. Another vocalization, the one for which the bird was named, is the *chick-a-dee* call, often heard in winter flocks and near winter bird feeders **(Fig. 7–13a & Track 9)**. The birds give this call in a variety of contexts, including when they seem mildly alarmed, as at humans or other animals. The third common vocalization is the "gargle," a complex sound used mostly by males in aggressive interactions over short distances **(Fig. 7–13b & Track 10)**.

These three sounds are the most common, but careful study has revealed that Black-capped Chickadees make about 11 other vocalizations. These have been given a variety of names: faint *fee-bee*, subsong, broken *dee*, variable *see*, hiss, snarl, twitter, *tseet*, high *zee* (alarm *zee*), scream, and squawk. These vocalizations occur in a wide variety of contexts. Mates in the vicinity of the nest give a "faint *fee-bee*," which is basically a *fee-bee* (or *hey-sweetie*, as I prefer) but much softer. Young male and female birds, beginning at about 20 days of age, give "subsong," a variable whistled performance in which the young birds seem to be practicing their song. Chickadees "hiss" when they are cornered, as in a nest cavity; this hiss may be a kind of mimicry evolved to make predators think they are dealing with a snake at close quarters. Chickadees "snarl" in fights and "twitter" to mates near the nest. They give "high *zees*" when they detect a predator or other alarming object; high-frequency alarm calls can be difficult to locate, warning chickadees (and other birds) of the danger without revealing the caller's location to the predator. Chickadees "scream" when captured, as in a mist net, and "squawk" to the young in the nest.

The Problem of Meaning

The total repertoire or vocabulary has been established for very few species, and is as thoroughly documented for chickadees as for any other birds. Yet those who study such repertoires worry that we humans don't really classify signals the way the birds do. The sonagram

Figure 7–12. Black-capped Chickadee: *Resident in woodlands, clearings, parks, and suburban areas throughout the northern United States and much of Canada, the Black-capped Chickadee is one of the most familiar birds and a common visitor to feeders. The male's distinctive* fee-bee *or* hey-sweetie *song is one of the earliest avian "signs of spring." Sung frequently in late winter as the daylight periods begin to lengthen, the song serves to proclaim a territory and impress females. Photo by Lang Elliott.*

7

Figure 7–13. Sonagrams of Black-capped Chickadee Chick-a-dee and "Gargle" Calls: In addition to its hey-sweetie *song,* the Black-capped Chickadee has about 13 different calls in its vocal repertoire. **a. Chick-a-dee Calls (Track 9, 1st Call):** The chickadee is named for its chick-a-dee call, used in winter flocks and often heard at bird feeders. Each flock has a distinctive chick-a-dee call, and new birds who join a flock change their call to match that of their new flockmates. The call is given frequently when two or more different flocks interact. The call may contain up to four different types of notes, referred to by researchers as A, B, C, and D, in various combinations. Most commonly, however, it consists of a few introductory notes (chosen from A, B, and/or C) followed by a series of D notes—the sonagram clearly shows the broad-band nature of the raspy-sounding D notes. Note that our example contains only three types of notes, A, B, and D. **b. "Gargle" Calls (Track 10, 1st Three Calls):** The "gargle" call is very complex and highly variable—the same individual may give many different versions. It is used most often by males in aggressive interactions over short distances. The sonagram shows three gargles given consecutively by one bird, and even here the variability is apparent.

a. Black-capped Chickadee chick-a-dee Call

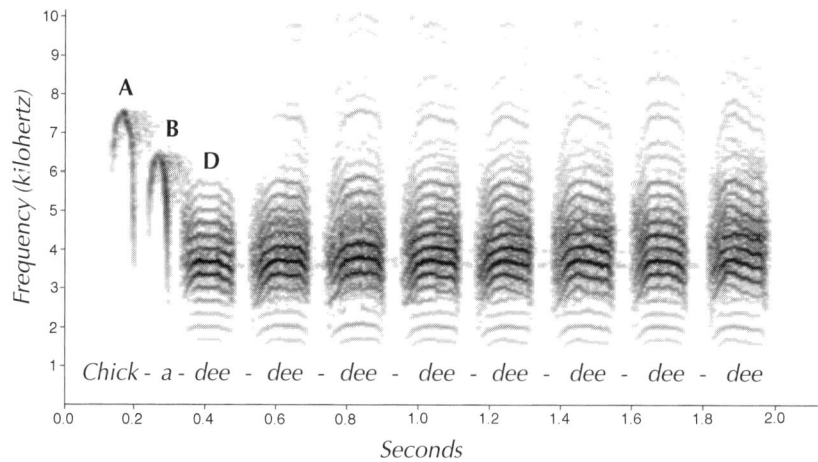

b. Black-capped Chickadee "Gargle" Call

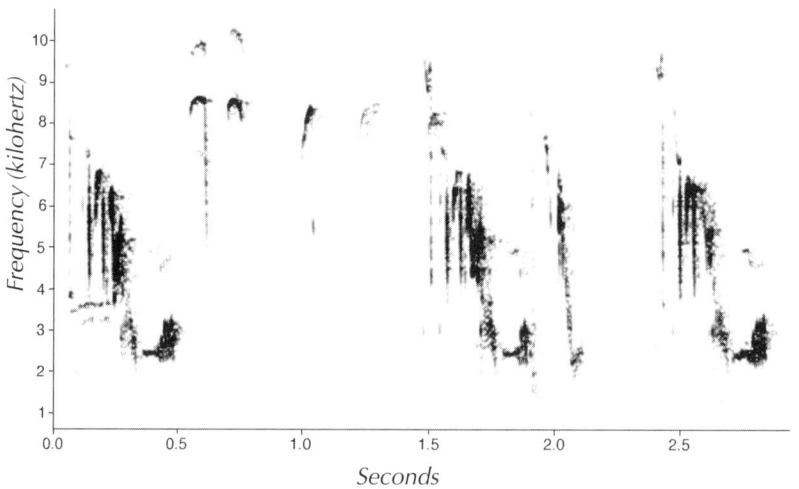

is a crude way to picture a sound, as we appreciated when we compared the oscillogram and the sonagram. The oscillogram nicely shows relative amplitudes, but not frequency; the sonagram shows relative frequency well, but amplitude poorly. The main problem is that we have little idea what features the birds really attend to. We humans can classify sounds by relatively gross features on the sonagrams, and we can correlate these sound categories with general social contexts, but we undoubtedly fail to appreciate the finer details of what birds really hear and respond to.

In fact, the birds themselves may hear and use a far greater variety of sounds than we appreciate. Reconsider that "simple" *chick-a-dee* call (see Fig. 7–13a). This call consists of four different notes, which might be identified as A, B, C, and D. Most *chick-a-dee* calls contain these four notes, but the birds deliver them in different combinations and differing numbers of times, to form hundreds of qualitatively different kinds of calls. Does ABCD convey information different from ABCDD, for example? Does the number of repetitions of a note mean anything, or does absence of a particular note convey something special? Do chickadees send different messages with all of these variations? We simply don't know.

The "gargle" is another "combinable" vocalization, in the sense that the component parts can be rearranged to form different vocalizations. Each gargle consists of two to nine short notes, which are drawn from a pool of about two dozen notes in each local population of birds. The notes used in any one gargle are thus only a sample of the local population, and each bird combines and recombines the notes to produce a tremendous variety of different gargle calls. To simplify our view of chickadee communication, we combine all of those variations into a single category, the gargle. But the birds undoubtedly hear so much more in a given gargle than we humans can begin to appreciate. Although we lump the hundreds of qualitatively different gargles into a single class, the chickadees probably attend not only to different sequences and combinations of elements, but to fine nuances of pronunciation as well. In our attempt to understand a complex situation, we may have unjustly oversimplified what the birds are doing.

Consider an analogy with human speech. The words "I love you" can be said in many ways, and we are primed to hear these subtle nuances in how our language is used. The emphasis, for example, can be on any one of the three words, and the implications for the resulting statement are profound. The statement can stress who loves *(I)*, the type of relationship *(love)*, or who is loved *(you)*. The inflection of the statement can rise, thus questioning the statement, or fall, which can have different meanings depending on the context. An alien unfamiliar with our communication might fail to appreciate these nuances, just as we, aliens to the birds, undoubtedly fail to appreciate the finer points of their expressions.

Even the simple *hey-sweetie* song varies in ways that are only recently being appreciated (Horn et al. 1992). If you listen to a Black-capped Chickadee during the early morning, for example, you hear the typical *hey-sweetie* over and over. The song is simple and doesn't seem to vary. But listen to 10, 20, or 30 songs, and suddenly you'll hear a change that seems, in context, rather dramatic. At some point, the bird transposes this simple *hey-sweetie* by a couple of hundred Hertz, either up or down, and then resumes singing the *hey-sweetie* on this new frequency. A few hundred Hertz may not sound like much, but the human ear is very good at comparing two signals given in succession, and the shift is indeed dramatic (**Track 11**). Careful analysis has now shown that a chickadee can sing its *hey-sweetie* on many different frequencies. So how many different *hey-sweetie* songs does the male chickadee sing? Perhaps an infinite number, if we "split hairs"? Do we vastly oversimplify the chickadee's communication system by saying that it has only one *hey-sweetie* song? Furthermore, even successive *hey-sweetie* songs on the same frequency may vary in subtle ways—in duration or patterns of loudness, for example. We can hear and see these subtle variations when we graph the songs, but we don't yet know whether they are meaningful to the chickadees.

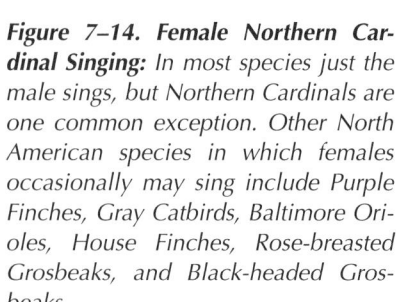

Song

The amazing variation in one type of signal, the songbird's "song," of which the *hey-sweetie* is an example, has received special attention from researchers. "**Song**" has been defined in numerous ways, but it is generally accepted that songs are loud vocalizations, often delivered from an exposed perch, with the presumed function of attracting mates or repelling territorial intruders. It's tempting to let the definition of song go at that, because no entirely adequate definition can be found (Spector 1994). Although the distinction between "song" and "call" is a somewhat arbitrary one devised by humans, our general definition ordinarily works well. It works because most birds do sing—they use special behaviors to try to attract mates or defend territories. Remove a female from a male songbird, for example, and he soon begins singing, announcing with great vigor his bachelorhood and thus his availability. Because songs are often sung loudly and persistently, they are the most noticeable vocalizations that we hear from birds.

In functional terms, woodpeckers "sing" as they drum on a tree, and snipes "sing" during aerial displays as wind rushing over their tail feathers produces winnowing sounds (**Sidebar 1: Winnows, Snaps, and Spring Thunder: Nonvocal Sounds**). Most birds "sing" vocally, however, although the complexity of "song" varies widely among species. In some, such as the Winter Wren, the song is by far the longest and most complex vocalization produced. In others, however, the "song" can be much simpler than "calls"—for example, the *hey-sweetie* song of Black-capped Chickadees is much simpler than their gargle call. Songs can be "musical" or not, according to our ears; what's important to remember is that beauty lies in the ears of the beholder.

In some species both males and females sing (**Fig. 7–14**), but in our northern temperate zone we hear mostly males, because they are the ones who are primarily responsible for defending territories and attracting mates (see Duetting, later in this chapter, for more on female song, especially in the tropics). Hence, in this chapter the singer is usually referred to as "he."

The songbirds are a group of about 4,600 species renowned for their singing ability. Unlike some other bird groups, songbirds learn their songs, which can therefore be especially complex (more on this topic later, in the Vocal Development section). If we accept that sorting based on sonagrams gives a crude indication of how many different songs a bird can sing, we discover vast differences among species. The chickadee is a songbird with a very simple song repertoire, essentially one basic *hey-sweetie* that varies in frequency. Except for these slight differences in frequency, all *hey-sweetie* songs look basically the same on sonagrams. Some other species also have simple repertoires; Chipping Sparrows, Indigo and Lazuli buntings, Common Yellowthroats, Ovenbirds, White-throated Sparrows, and White-crowned Sparrows all use one basic song form in their singing. The sequence of notes in their songs, as revealed by sonagrams, remains essentially unchanged from one song to the next.

Figure 7–14. Female Northern Cardinal Singing: In most species just the male sings, but Northern Cardinals are one common exception. Other North American species in which females occasionally may sing include Purple Finches, Gray Catbirds, Baltimore Orioles, House Finches, Rose-breasted Grosbeaks, and Black-headed Grosbeaks.

(Continued on p. 7·19)

Sidebar 1: WINNOWS, SNAPS, AND SPRING THUNDER— NONVOCAL SOUNDS

Sandy Podulka

The melodies put forth by singing birds appeal to our love of music and dominate the spectrum of natural sounds we hear around us, yet the many nonvocal sounds birds make are equally dramatic and engaging. Discussed here are only a few.

Woodpecker Drumming

You have already listened to the drumming of several different woodpeckers. Most of these short "drum rolls" differ markedly from the more quiet, irregular, leisurely taps woodpeckers make while searching for food or excavating a nest site (Track 12). Drumming, like song, is clearly meant to be noticed by other members of the same species. The performer chooses a dry branch, hollow log, drainpipe, or tin roof—anything that will make a good loud noise—and then lets loose with a roll. One particularly creative Yellow-bellied Sapsucker drummed on a garbage can lid! Both males and females drum, to proclaim territory and as a part of courtship. Although they may have many different "signal posts" within their territory, they usually favor one or two that are particularly well-placed or resonant. Drums are one of the first bird "songs" heard in the late winter woods—a sign that behavior patterns are changing and spring is on its way.

Ruffed Grouse Drumming

In the predawn hours of early May, 1932, Arthur A. Allen hid in Sapsucker Woods near the drumming log of a male Ruffed Grouse, and through slow-motion photography and sound recording, solved the mystery of how "spring thunder," as he called it, was produced. Until then, people had speculated that grouse pounded their wings on a log or beat them together,

Figure A. Ruffed Grouse Drumming: *A male Ruffed Grouse "drums" by cupping his wings and then bringing them upward and forward with such force that he compresses a parcel of air between his chest and wings, creating a sound wave without his wings and chest ever touching. He repeats this "thump" in a series, beating faster and faster until the thumps become a whir. The sound is so low in frequency (40 Hz) that a human may not even notice it, or may write it off as a distant car starting, unless he or she is listening carefully. The male usually drums while standing on a hollow log, which serves as a resonating chamber—amplifying his message that he owns the territory and is available to mate. Drawing by Charles L. Ripper.*

but Doc Allen's frames revealed that the male stood crosswise on his log, braced on his tail, and cupped his wings, bringing them forward and upward with such force that he compressed a parcel of air between his chest and wings, creating a sound wave without wings and chest ever touching (**Fig. A**). Each stroke of his wings produced a muffled thump. He started slowly, stroking faster and faster until the thumps merged to a final whir (**Fig. B & Track 13**). The male proclaiming his territory and availability to any passing female can be heard up to 1/4 mile (0.4 km) away,

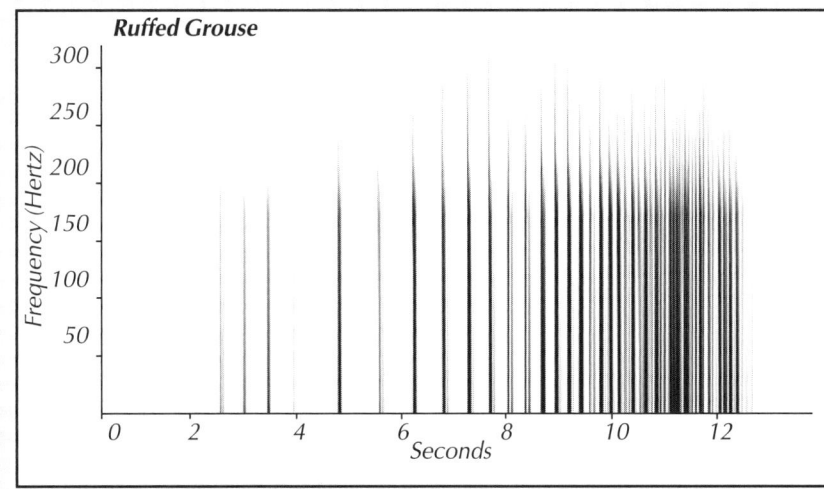

Figure B. Sonagram of Ruffed Grouse Drum (Track 13, 1st Drum): *Each thump produced by a drumming Ruffed Grouse is made up of a broad band of frequencies, but lasts only a short time. Note that the frequency scale here is in Hertz, whereas most other sonagrams in this chapter are scaled in kilohertz (1,000 Hertz).*

but the frequency of 40 Hertz is so low that it sounds soft even from very close. Why so low? No one knows for sure, but the lower limit of hearing of one of the Ruffed Grouse's most powerful enemies, the Great Horned Owl, is 60 Hertz. Perhaps in past generations any grouse drumming at a higher frequency was also proclaiming the availability of dinner to an owl, and was eaten. Through natural selection, then, the sound may have evolved to its current low frequency.

To hear "spring thunder," visit a northern woods in spring at dawn or dusk, when the drumming is most frequent. Males also drum all night under strong moonlight, and occasionally in the fall or during the day. Listen carefully, for the sound is easy to miss. You may even be lucky enough to discover a male's drumming site—usually a big, mossy, hollow log. A pile of droppings and a flattened patch worn in the moss from many years of grouse feet may confirm your find. If a log is not available he may use any elevated site, such as a mossy mound or a stone pile. A grouse may have several other drumming logs in addition to his primary one, so don't be surprised if you hear him drumming from more than one location in his territory.

American Woodcock Display

Another nonvocal performance captivates nature-watchers in eastern North America each spring—the courtship display of the American Woodcock. In open, wet fields, males display briefly at dawn and dusk—the times of day when the light intensity changes most rapidly. The outer three primaries of each wing are narrow and stiff, and when a male flutters his wings rapidly, the air vibrates these specialized feathers to produce a high twittering (**Fig. C).** The performance begins on the ground with a series of nasal, vocalized *peents* similar to the call of a nighthawk. Then, as described by

Specialized Primaries

Figure C. American Woodcock in Flight: *The outer three primary feathers of each wing of the male American Woodcock are unusually narrow and stiff; when he flutters his wings rapidly during the flight portion of his courtship display, the air vibrates these feathers, and they produce a high twittering noise.*

Aldo Leopold in "Skydance" from *A Sand County Almanac,*

> *...suddenly the peenting ceases and the bird flutters skyward in a series of wide spirals, emitting a musical twitter. Up and up he goes . . . the twittering louder and louder, until the performer is only a speck in the sky. Then, without warning, he tumbles like a crippled plane, giving voice in a soft liquid warble that a March bluebird might envy. At a few feet from the ground he levels off and returns to his peenting ground, usually to the exact spot where the performance began, and there resumes his peenting.*

Although the twittering during the upward spirals is nonvocal, at the peak, 300 feet (90 m) in the air, the bird begins a series of vocal chirps—to the careful listener straining to catch a glimpse of the tiny dot against the dusky sky, a clue that he is on his way down **(Fig. D) (Track 14).**

To view the "skydance" of the woodcock, dress in dark clothes and listen in open fields for the peenting call at dusk. When the performer takes to the air, quickly move in a little closer. Before the bird drops out of the sky, crouch down or hide near a shrub, then remain still until he is up again. Displaying males are quite sensitive to disturbance, however, so stay far enough away to let the male continue uninterrupted.

Common Snipe Winnowing

Over wet pastures and meadows throughout northern North America, yet another springtime ritual is carried out each dawn and dusk: the winnowing of the Common Snipe. The displaying bird circles high overhead, 300 to 360 feet (90 to 110

m) up, and then dives to achieve the necessary speed of 24 to 52 mph (39 to 83 km/h). In the midst of the dive, the performer spreads his tail, causing the air to vibrate the stiff, modified outer feather on each side of the tail, which produces a whistle-hum sound that can be heard up to 1/2 mile (0.8 km) away. Because the beating wings send air over the tail in pulses—not in a steady stream—the whistle-hum is a rapidly pulsing *woo-woo-woo-woo,* rather than a continuous sound **(Fig. E) (Track 15).**

Either sex may winnow to proclaim territory, and the male may do so to court a female. That the sound is indeed produced by feathers and not voice was demonstrated by placing tail feathers of a European Snipe in a wind tunnel, producing the whistle-hum sound with no bird present. (Because the airflow over the feathers was steady, this sound did not pulse.)

If you visit a marshy field during twilight, listen carefully, and look for a performer circling overhead. If the sky is light enough, you will be able to watch its dives, and to see it spread its tail feathers as you hear the haunting sound.

Other Nonvocal Noises

Birds seem to use every available method of nonvocal sound production.

A number of birds, including storks, herons, owls, roadrunners, and Tree Swallows, snap their jaws together in various displays, but wing noises are the most common. Grouse and quail wings thunder as the birds burst from cover when disturbed, perhaps startling predators and buying a few precious seconds for the fleeing bird. Mourning Dove and Common Goldeneye wings whistle in flight, and the wings of Rock Doves may hit each other over their backs, making a loud clapping noise in flight. Hairy Woodpeckers, Chuck-will's-widows, and Common Poorwills clap their wings in a similar way in territorial defense, as do

Figure D. American Woodcock Courtship Display: At dusk and dawn, and on moonlit nights, the male American Woodcock performs his spring ritual throughout the eastern United States in small clearings in fields or wet meadows. He begins on the ground with a series of nasal peents during which he often faces different directions—thus the sound may seem to vary in loudness. Then he takes silently to the air and ascends gradually. As he climbs higher he flies in wide spirals, making musical twittering sounds with his wings (see Fig. C). At the peak, as much as 300 feet (approximately 100 m) in the air, he begins vocal chirps as he zig-zags downward. For the last part of the descent he is silent. When he returns to the ground, near where he began, he resumes his "peenting."

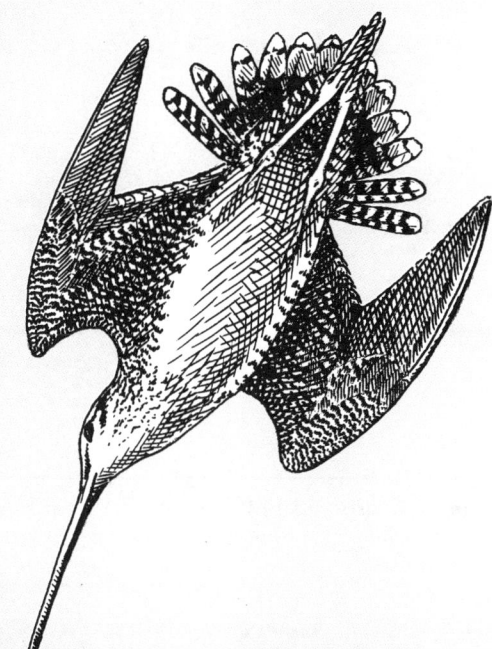

Figure E. Common Snipe Winnowing: Circling more than 300 feet in the air over wet meadows and pastures, Common Snipe perform their territorial and courtship display in the springtime twilight. Termed "winnowing," after the rapid, whistle-hum woo-woo-woo-woo produced during one portion, the display may be given by either males or females. To make the noise, the performer must reach a minimum air speed of around 24 mph, which it achieves by diving. In the middle of the dive it spreads its tail, causing the air to vibrate the outer tail feathers, which produces the haunting sounds that can be heard up to 1/2 mile (0.8 km) away.

Short-eared Owls in their shallow courtship dives. Many hummingbirds, too, make buzzes or shrill whistles with their wings in their territorial and courtship flights. The Common Nighthawk dives deeply in courtship, and at the bottom, his wings make a low roar like someone blowing across the top of an empty bottle.

Manakins, a huge group of Neotropical birds, produce an assortment of nonvocal noises during courtship. White-bearded Manakins form leks of up to 60 males. In dramatic courtship displays, they hop back and forth between saplings, and slide down the trunks, their wings mak-

ing growls and rapid firecracker-like snaps. The sounds are produced by striking together the stiff, narrow, outer primaries, or the thickened secondary feather shafts **(Fig. F)** **(Track 16)**.

This dazzling variety of nonvocal sounds demonstrates the remarkable flexibility of evolution. Natural selection doesn't "design" systems from scratch for a specific purpose; it works instead on existing behaviors and structures, sustaining those that most improve survival and reproduction. Most birds communicate vocally using the syrinx, whose main purpose is sound production. But others use less conventional "in-

struments"—the "jug band crowd" evolved using whatever raw material was available to make their sounds. And they use their feather-whirring, feather-rattling, beak-snapping, and wing-whacking in the same ways other birds use vocalizations—to attract mates and defend territories. ∎

Suggested Readings

Allen, Arthur A. 1987. Spring Thunder in Sapsucker Woods. *Living Bird Quarterly.* 6(4):8–11. (First printed in *The Cornell Plantations,* Spring 1947.)

One of the Lab's founders, Doc Allen, recounts a trip into Sapsucker Woods on the day he and his equipment-laden colleagues captured for the first time the thunder of a drumming Ruffed Grouse.

Leopold, Aldo. "Skydance" in *A Sand County Almanac.* Ballantine Books, NY. 1949.

A beautifully written essay describing the American Woodcock and its aerial display. The book is a classic, containing numerous passionate, yet ecologically informative, essays about wildlife in the American midwest. Aldo Leopold is one of the founders of the field of wildlife biology and the environmental movement in the United States.

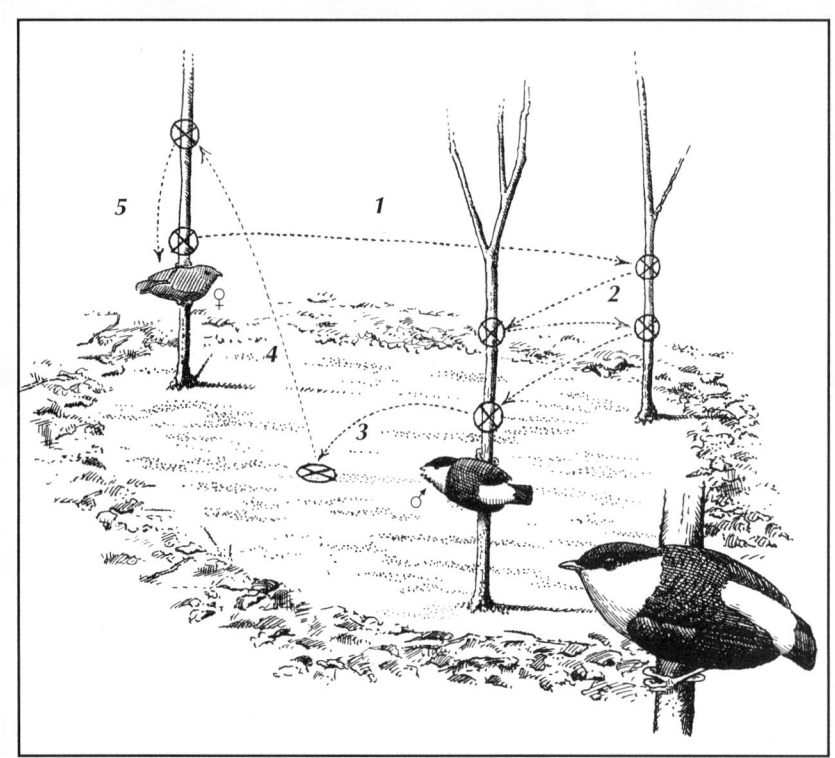

Figure F. Courtship Display of the White-bearded Manakin: *Males of this chickadee-sized lekking species (see Ch. 6, Reproductive Behavior: Lek Polygyny) display in groups of up to 60 individuals, each on his own small arena or "court," often within yards, sometimes feet, of his neighbors on the forest floor. Females actively participate, although males frequently display when no females are present. Each tiny arena is kept clear of leaves by the male, and has several small, thin saplings growing within it. The male leaps rapidly around his arena from sapling to sapling, giving loud, firecracker-like snapping noises with each leap. In intense excitement he may give volleys of snaps, resulting in a loud ripping sound. He may repeatedly travel in one direction (1), suddenly changing to ricochet back and forth between two saplings (2). Each time the male lands, he does so in a horizontal position, flaring out his white beard. When a female joins him, the display becomes still more frenetic, both birds leaping around in the above manner together, sometimes following each other, sometimes moving in opposite directions and passing each other in midair. Often the male jumps to the ground (3), then springs up onto a sapling (4), giving a sound partway between a grunt and a hum, then continues leaping around the arena. If the female accepts the male, she remains immobile on a sapling, and he uses the grunt-jump (4) to land well above her. He then slithers rapidly down the sapling (5) to land on her back, and copulation occurs. The snaps, rips, and grunts are nonvocal sounds, thought to be produced by the male manakin's specialized wing feathers. Females do not make these sounds.*

Figure 7–15. Eastern Towhee: *The male Eastern Towhee sings his song from moderately high, exposed perches within his territory—often a location along a forest edge or in an open woodland. Photo by Lang Elliott.*

Eastern Towhee Repertoire

Song Type 1

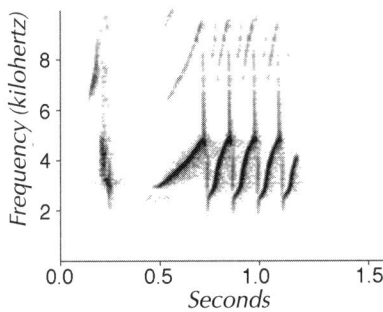

Song Type 2

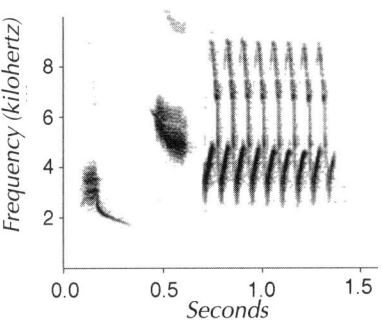

Song Type 3

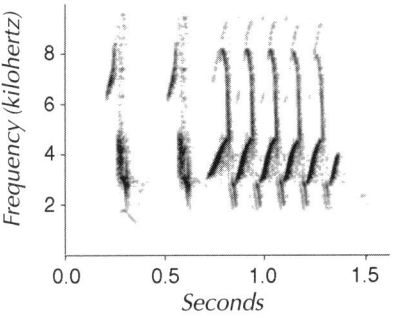

Figure 7–16. Sonagrams of Eastern Towhee Songs (Track 17, 1st, 4th, and 7th Songs): *Three different song types from the same Eastern Towhee. Note that although none of the songs match the "classic" song type shown in Fig. 7–10, they all follow the general pattern of beginning with one or several introductory notes and ending with a trill. The note and trill types differ from song to song. A male may have a repertoire of between 3 and 8 different song types, and he usually sings each one 10 to 30 times before switching to a different type.*

Other species have much larger repertoires (Krebs and Kroodsma 1980; Kroodsma 1986). An Eastern Towhee **(Fig. 7–15)** may sing three to eight different *drink-your-tea* songs; each song form has a distinctive structure and sequence of notes, and the differences are visible on sonagrams and audible to anyone who listens carefully **(Fig. 7–16 & Track 17)**. Over a period of five to ten minutes, a male typically sings ten to thirty renditions of one song form, but he then abruptly switches to another form, and eventually another, until he returns to sing, with great fidelity, the first song you heard. The towhee thus partitions his song performance into three to eight different songs. A male Song Sparrow typically has 8 to 10 different songs. Each one is repeated often, and this repetition is an integral part of his singing activities. Even more impressive are Marsh Wrens; in eastern North America, males have about 50 different songs apiece, and males in the western states have about 150. Northern Mockingbirds typically sing 100 to 200 songs. The largest repertoire found so far, however, is that of the Brown Thrasher. When thousands of sonagrams of a male thrasher were sorted into different categories, he was estimated to sing well over 2,000 different songs **(Fig. 7–17)**.

The Structure and Function of Sounds

The structure of a song or other sound is a combination of all the features that form that sound, as typically detected on a sonagram. These features include the duration of the sound, the overall frequency,

Figure 7–17. Brown Thrasher: The larg-est song repertoire known belongs to the Brown Thrasher, who can sing up to 2,000 different song types, a few of them imitating the songs of other spe-cies. He usually repeats each song type two or three times, quickly moving on to the next.

whether the sound is a relatively pure whistle or reaches across a broad spectrum of frequencies at once, whether individual sound units are repeated, and so on.

To some extent, the functions of vocal signals dictate their struc-ture. When a hawk flies overhead, for example, small birds typically freeze and produce a high, narrow-frequency call that begins softly, increases in amplitude, and then fades away; sounds with these char-acteristics, such as the high *zee* of the chickadee, are difficult to locate **(Fig. 7–18 & Track 18)**. When songbirds "mob" an owl, however, they dive and swoop at the predator, making seemingly daring passes at it; calls in this situation are harsh, broad-band noise, and easily located (Marler 1955) **(Fig. 7–19 & Track 19)**. In general, aggressive sounds are often broad-band noise emphasizing low frequencies, whereas appeasing sounds are higher in frequency (Morton 1982).

The type of habitat in which a bird sings also can influence song structure, because certain songs transmit better in some habitats than in others. Songs of the Great Tit (an Old World chickadee relative) differ consistently between forests and more open woodlands; forest songs are simpler, lower in frequency, and contain more pure tones **(Fig. 7–20)**. Perhaps the most striking effect is the use of low-frequency, fairly pure tones near the ground, especially in tropical forests. Al-though low-frequency sounds travel best in any habitat, in some hab-itats ground-dwelling birds are "forced" to use low-frequency sounds because sounds higher than one or two kilohertz do not transmit well owing to reflections off the ground (Wiley and Richards 1982). The very low-frequency drumming display of the Ruffed Grouse, for example,

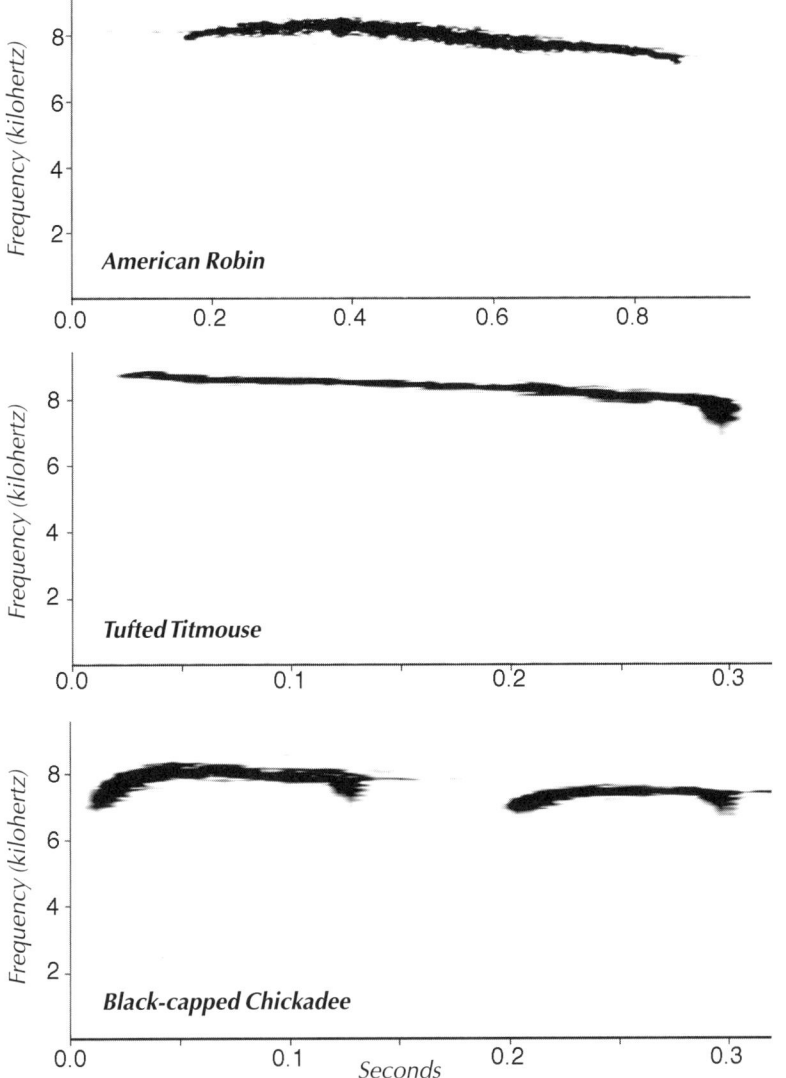

Figure 7–18. Sonagrams of Alarm Calls (Track 18, 1st A. Robin Call, 1st T. Titmouse Call, 1st Two B-c. Chickadee Calls): *The alarm calls of many birds are similar—a high, narrow-frequency note that begins softly, increases in amplitude, and then fades away. Sounds with these characteristics are particularly difficult to locate, thus allowing a bird that spots a predator to warn other individuals (of both its own and other species) without giving away its location to the predator. The examples shown are the high zee calls of the American Robin, Tufted Titmouse, and Black-capped Chickadee. Note that although the high-frequency nature of the calls can be seen from the sonagram, the amplitude changes are not visible—they would require an oscillogram to be distinguished. Note also that the time scale for the American Robin calls (which last longer) is different from that of the other two species.*

carries a long distance through the forest even though it is given from near the ground. The low-frequency, often tremulous whistles associated with tropical forests are in many cases voiced by ground-dwelling birds, such as tinamous **(Fig. 7–21) (Track 20)**.

Our understanding of the meanings of avian vocalizations to other birds is truly in its infancy. Our current classifications may bear little resemblance to how birds actually hear and use their vocal signals, and we must remember that our ultimate goal is to ask the birds themselves how they do this. Experiments in the laboratory and the field in which we attempt to interact with birds by playing sounds to them over loudspeakers are probably the best way to understand birds, and these experiments are just beginning (Dabelsteen and McGregor 1996; McGregor and Dabelsteen 1996). Only when we can actually converse with birds in a meaningful way will we truly begin to appreciate what it is like to be one. Until then, we remain as aliens, pondering the meaning of all that birds say.

Figure 7–19. Sonagrams of Mobbing Calls: In contrast to alarm calls, mobbing calls of many different species tend to be broad-band, raspy sounds. These are relatively easy to locate, allowing an individual bird scolding a predator to quickly attract other individuals, often of several different species, to form a mobbing flock that is more likely to drive the predator away. a. (Track 19, 1st Five Seconds of Blue Jay Calls, 1st Three House Wren Calls, 1st Five Seconds of Tufted Titmouse Calls): Blue Jay, House Wren, and Tufted Titmouse mobbing calls—note the similarity among the different species in the structure of this call. b. (Track 19, 1st Six Seconds of Calls): Mixed-species mobbing flock from Arizona, including Mexican Chickadee, Western Tanager, and Red-breasted Nuthatch, all with similar broad-band calls.

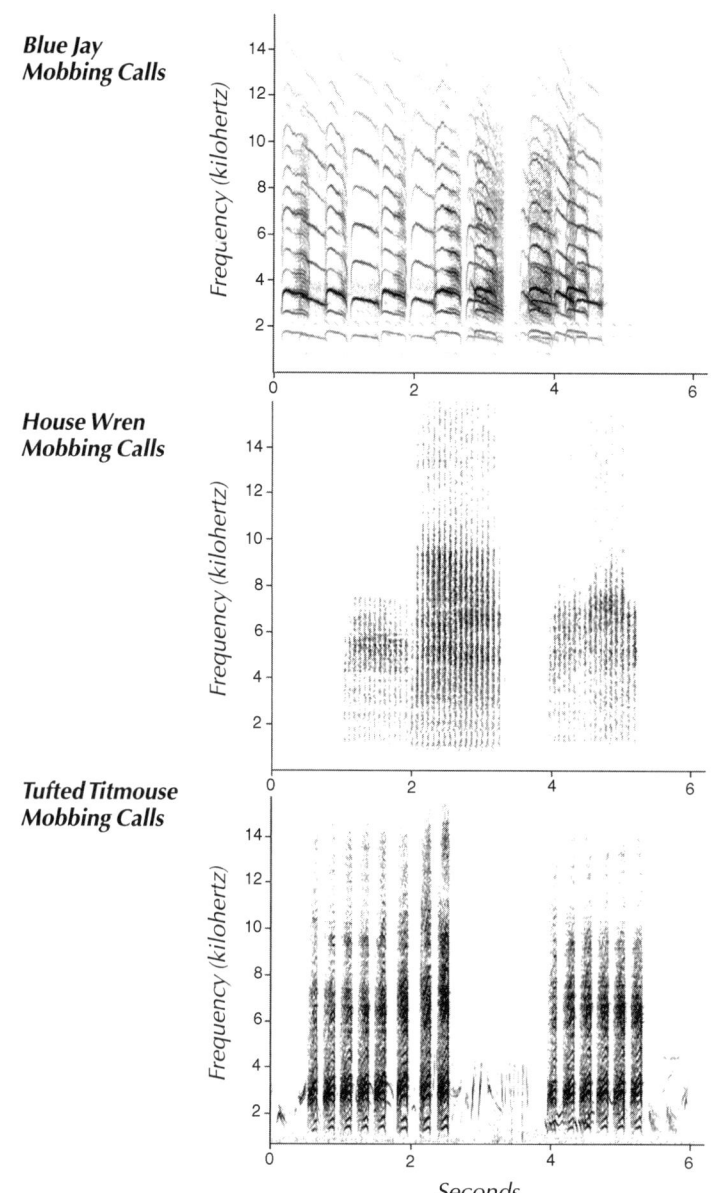

a

Blue Jay Mobbing Calls

House Wren Mobbing Calls

Tufted Titmouse Mobbing Calls

b

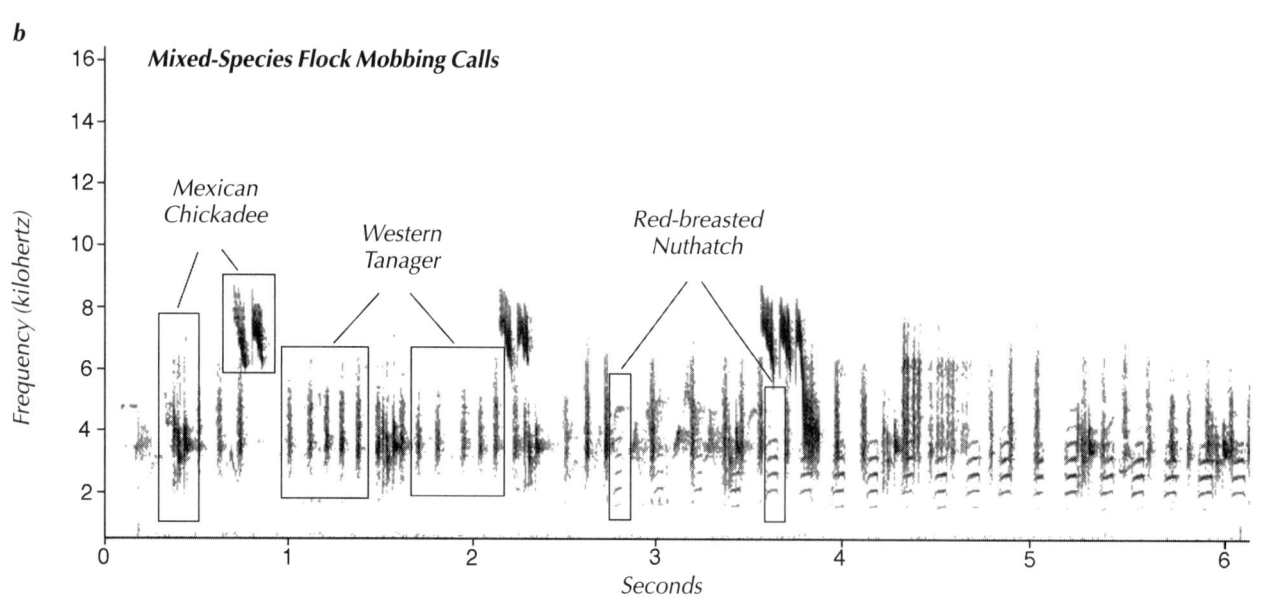

Mixed-Species Flock Mobbing Calls

Mexican Chickadee

Western Tanager

Red-breasted Nuthatch

Figure 7–20. Effect of Habitat on Great Tit Song Structure: *In more forested areas (left), the song of the Great Tit (an Old World chickadee relative) is simpler, of lower frequency, and contains more pure tones than songs from Great Tits that live in more open woodlands (right). The difference is thought to result from the different sound transmission properties of the two habitats. The dense forest vegetation interferes more with the transmission of sound waves, distorting sounds to a greater extent than does the vegetation of open areas.*

Vocal Development

Birds clearly produce a variety of sounds, but how do they know which one to use and when to use it? After a parent incubates an egg for the appropriate length of time, a tiny nestling emerges. Its first sounds are soft peeps, given perhaps in response to siblings or to parents who provide warmth and food. The nestling grows and eventually becomes a fledgling, a yearling, and finally an adult. During this development, various sounds appear in turn to enable the growing bird to manage its social environment. How does the bird know when to use what kind of vocalization? Is it a little robot, with all its vocalizations encoded in its genes, to be uttered automatically in the appropriate circumstances? Or is development more complex, with the young bird learning much of what it knows from other individuals, much as we humans learn our language from adults who were, in evolutionary terms, successful before us?

The reason vocal development has been such an intriguing focus of interest is that certain groups of birds are clearly *not* robots in their development. Most of the interest has been on the song of songbirds, such as wrens, sparrows, thrushes, mockingbirds, warblers, swallows, and crows. In all songbirds that have been studied, researchers have discovered some kind of learning. Just as in humans, this learning involves listening to a model sound, memorizing the model, and practicing until the sound matches with great fidelity the young bird's memory of the original sound.

Figure 7–21. Tinamou in Rain Forest Vegetation: *Tinamous—ground-dwelling, chickenlike birds of Central and South America—give haunting, low-frequency, tremulous whistles that ring through the rain forest in the early and late twilight hours. Low-frequency sounds travel best in any habitat, especially dense jungle vegetation, and may avoid interference from sound waves reflecting off the ground, which may be a problem with sounds higher than one or two kilohertz.*

Vocal Development in Songbirds

Among birds, the "songbirds" are especially successful. Comprising about 4,600 of the world's roughly 10,000 bird species, they are one of two suborders within the Order Passeriformes (the "perching birds") **(Fig. 7–22)**. The songbirds are well known for their com-

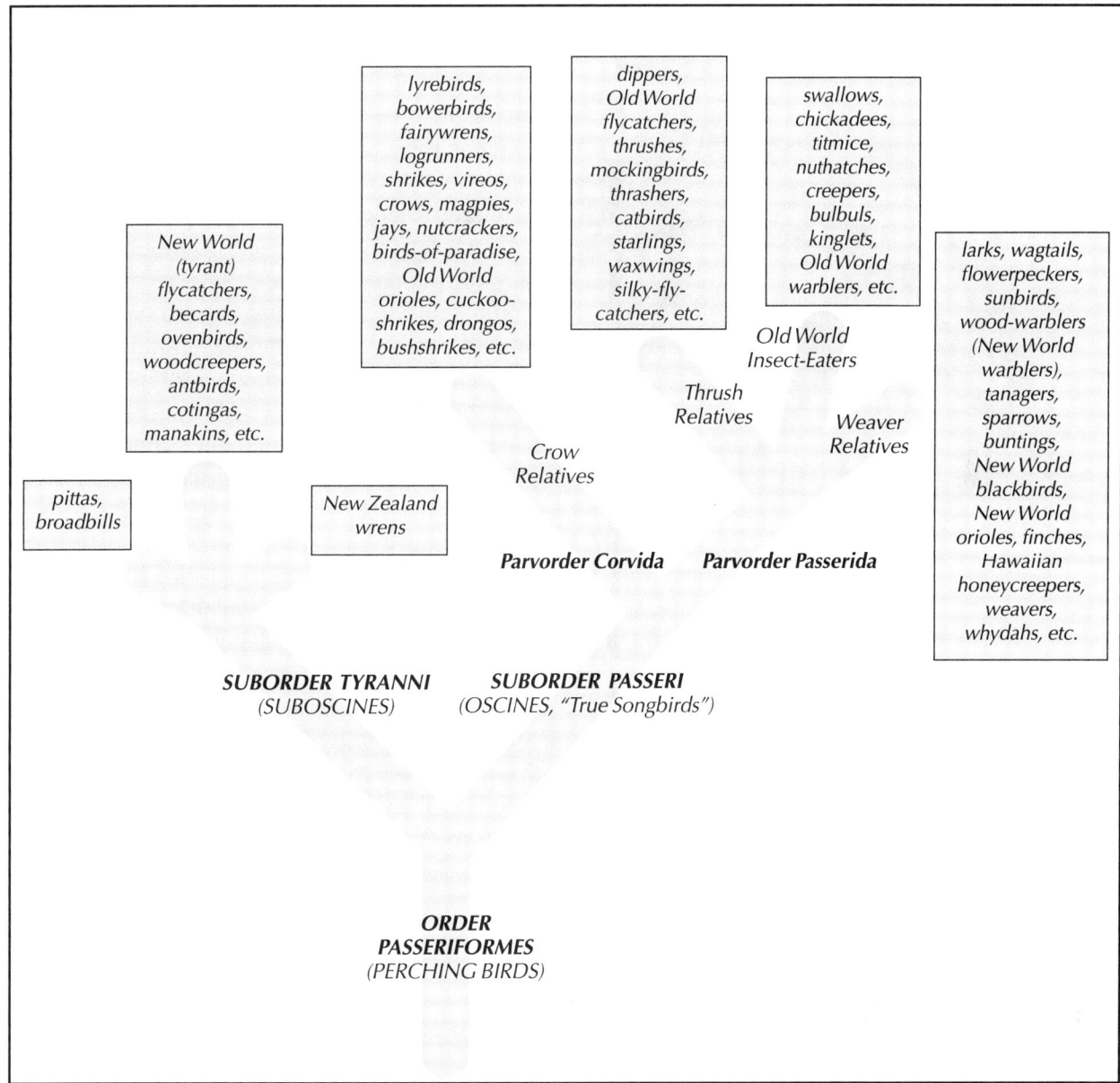

Figure 7–22. Passerine Taxonomic Tree: *Traditional classification of birds was based on anatomical similarities and differences among various species. The order Passeriformes (the passerines or perching birds) has long been recognized to consist of two major divisions, which differ in their syrinx musculature: suborder Tyranni (the suboscines or non-oscine passerines) and suborder Passeri (the oscines or true songbirds). Data from molecular biology, particularly the technique of DNA-DNA hybridization, recently have been used to develop modern theories of how bird groups are interrelated. The much-simplified passerine taxonomic tree presented here is a well-ac- cepted interpretation, based on the work of Charles Sibley and Jon Ahlquist. The suborder Tyranni consists of three groups: one containing the New Zealand wrens, a second containing the pittas and broadbills, and a third composed of the remaining suboscines (entirely restricted to the Americas). The suborder Passeri is divided into two groups: parvorder Corvida contains the crows and their relatives, parvorder Passerida contains the thrush relatives, Old World insect-eaters, and weaver relatives. Source material from* Distribution and Taxonomy of Birds of the World, *by Charles G. Sibley and Burt L. Monroe, Jr., 1990.*

plex songs, which to human ears are often beautiful **(Fig. 7–23)** (Track 21); **(Fig. 7–24)** (Track 22). By definition, most bird species "sing." That is, they use a loud vocalization to attract mates or defend territories. In this way, even nonsongbirds such as shorebirds and owls "sing." Not all songbirds, however, produce "beautiful" songs (consider the song of a crow), but they're songbirds nonetheless, based on their evolutionary history. The songbirds are so named because they typically sing so frequently and so conspicuously—at least from the human perspective.

Songbirds also imitate sounds. Caged birds have been kept for centuries, for example, and trained to sing a variety of sounds often uncharacteristic for their species. Various Island Canary strains have been selected to produce different kinds of songs, so the genes of the different strains now dictate what range of song syllables the birds learn (Mundinger 1995).

In the laboratory, we can readily demonstrate the song-learning potential of birds. In North America, the White-crowned Sparrow has been a favorite study species (Baptista 1996). If researchers take a sparrow from a nest in the wild at eight or nine days of age and keep him in the laboratory where he can't hear any of his species' natural songs, he develops a highly abnormal song **(Fig. 7–25 &** Track 23). Expose the young male to songs of an adult White-crowned Sparrow, however, and he learns to sing the details of the tutor song. Regardless of where the young bird was born, he can learn a wide range of White-crowned Sparrow songs, taken from anywhere in the geographic range of the species. Details of the youngster's song match those of the song with which he was tutored, detail for detail.

Most songbirds seem to have what is called a "**sensitive period**" for song learning. It is during this relatively brief time that birds are best equipped to memorize the details of a tutor song. For the White-crowned Sparrow, model songs from a loudspeaker in the laboratory seem to have the greatest impact from day 15 to about day 50. Songs heard after day 50 seem to be far more difficult to learn. If the young

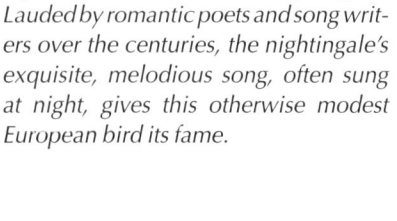

Figure 7–23. Common Nightingale: Lauded by romantic poets and song writers over the centuries, the nightingale's exquisite, melodious song, often sung at night, gives this otherwise modest European bird its fame.

Figure 7–24. Musician Wren: The small, brownish-orange Musician Wren of the Amazonian rain forests is far more often heard than seen. It more than makes up for its small size with its long, loud, complex, melodic song, considered by many people to be among the most beautiful of all bird songs—perhaps because it resembles our notion of music quite closely. Photo by J. Dunning/VIREO.

bird is given a live bird to interact with, learning after day 50 becomes somewhat easier, but never as easy, it seems, as during those early weeks of life. Other songbirds, such as Swamp Sparrows, Marsh Wrens, and Zebra Finches, also learn best during this early period (Baptista 1996).

The extent to which birds can learn later in life varies among species. Northern Mockingbirds can add new sounds to their complex repertoires as adults, so repertoire sizes may increase steadily with age. Other species that mimic also are reported to acquire songs throughout their lives. In most species, however, individuals seem to develop a repertoire of sounds during their first year and then rarely, if ever, modify it. One male Red-winged Blackbird, for example, sang the same six songs over a period of five years in a Tallahassee, Florida, marsh (Kroodsma and James 1994).

Birds are selective in what they learn. A young White-crowned Sparrow does not learn just any bird song it hears. It seems that the components of the song must match, to some extent, characteristics of some innate knowledge of song that the young bird is born with. If he hears both Song Sparrow and White-crowned Sparrow songs, he'll learn the White-crowned Sparrow songs. If he's housed with a singing Song Sparrow, however, his physical and vocal interactions with that bird may override his inborn tendencies, and a young White-crowned Sparrow can then, under some circumstances, learn to sing Song Sparrow songs (Baptista 1996).

How songbirds learn to sing is worth considering in some detail (Kroodsma 1993). Nestlings can hear within a few days of hatching, and they begin their practice singing shortly after leaving the nest, at about three weeks of age. The earliest "**subsong**" is barely perceptible, even to someone listening only a wing's length away. Among birds hand-raised in the laboratory, this early practice typically occurs in a fledgling that recently has been fed. He (and in some species, she) perches, resting comfortably, even appearing to doze with eyes closed and head tilting to one side. All the bird's systems appear to be completely at rest, but soft whispers, accompanied by barely perceptible throat movements, reveal that inwardly the future maestro has begun his work. If this resting state is disturbed, as by a sibling or curious human, the youngster breaks abruptly from his apparent slumber and becomes fully alert, producing no more evidence of practice.

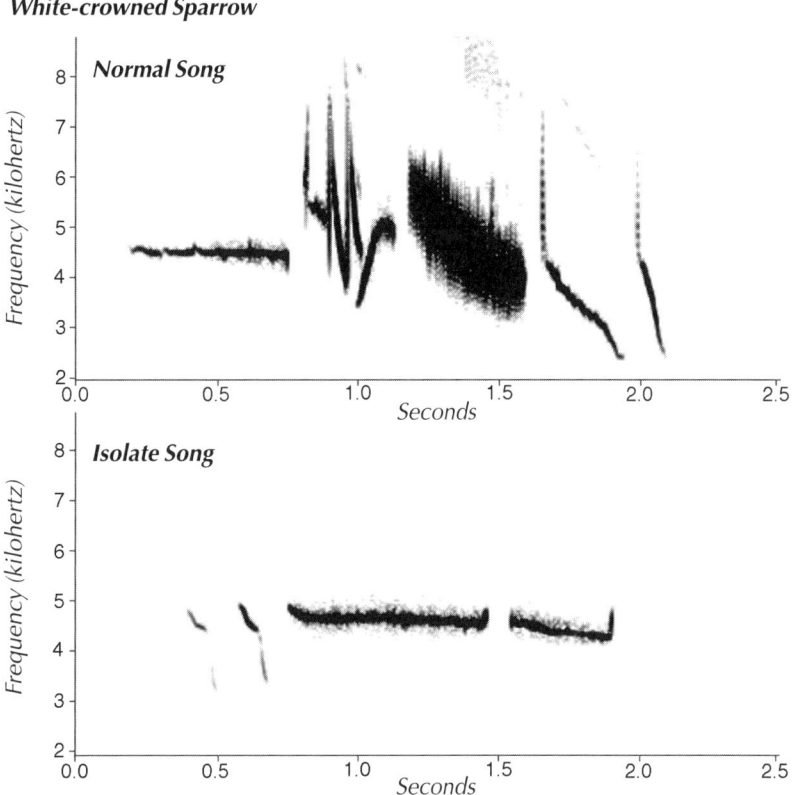

White-crowned Sparrow

Figure 7–25. Sonagrams of White-crowned Sparrow Normal and Isolate Song (Track 23, 1st Normal Song, 1st Isolate Song): The normal song of an adult White-crowned Sparrow (top sonagram) is a complex mix of pure tones, buzzes (the large, dark, broadband note ending at about 1.5 seconds), and other notes. When a young male is taken from a nest in the wild at 8 or 9 days of age and raised in the laboratory without hearing any further songs of his own species (bottom sonagram), he develops an atypical song that is much less complex than normal songs.

This song practice gradually becomes louder, more persistent, and more structured, and the sounds begin to resemble the adult song. In juvenile males of the nonmigratory Bewick's Wrens that I studied in the Willamette Valley of Oregon, this process was especially evident **(Fig. 7–26)**. A typical young male left his parents' territory by four to five weeks of age, and a few weeks later was already defending the territory he would hold for the rest of his life. On that territory, the quality of the practice singing gradually improved, until songs matched, detail for detail, the songs of adult neighbors. Although the young male was capable of learning his father's songs during the first four or five weeks when he was cared for by his parents, he rejected his father's songs in favor of songs from his new home.

The practice singing (subsong) of a young songbird, such as the Bewick's Wren, is remarkably similar to the practice speaking (babbling) through which humans progress as toddlers. Consider my daughter's babbling when she was about a year and a half old. I have numerous recordings of her early speech attempts, but my favorite is when she was sitting in her highchair, looking out the large picture window to the garden, and perhaps to the Oregon Cascades beyond. With no one else in the room, she babbled and babbled. The sounds of one segment (Track 24) were "bow wow . . . wow wow . . . bow wow . . . wee wee wee . . . hi daddy ba ma wow wow wow . . . daddy! da daddy bow wow bow wow . . . dere's da ditty . . . hey ditty . . . hey daddy . . . nyeh . . . no . . . down." Clearly evident in this practice are simple syllables and entire words that she will eventually incorporate into her adult speech. The "bow-wow" refers to a dog, and "ditty" to kitty. "Daddy" is unmistakable, as are "no" and "down." The "wee wee wee" shows good recall of the little piggy's homeward cry. Importantly, no dog or kitty was in the room, nor was Dad, who was hiding around the corner with his microphone. She took all of these elements of the vocabulary out of their appropriate context and strung them together in what was, by adult standards, a nonsensical practice session.

Figure 7–26. Bewick's Wren: The Bewick's Wren, an active, noisy resident of dry, scrubby areas and most common in the western United States and Mexico, is well known for its variable song. Males have a repertoire of 9 to 22 different song types, but repeat each one many times before switching to another. Photo by J. Hoffman/VIREO.

The young Bewick's Wren in the garden just outside my daughter's window was also practicing. His task was to master a vocabulary of about 16 different songs from his male neighbors. His eventual goal was to take one of these 16 songs and sing it 20 to 50 times in succession, giving each rendition crisply, confidently, and consistently, with no mistakes or wavering. Then he would introduce another of the 16 songs, and eventually another, until slowly, methodically, over several hours, he had worked his way through his entire song repertoire. If each song were paraphrased into English, one might imagine an adult Bewick's Wren proclaiming from the treetops a series of 30 "Good morning's," then 30 "How are you's?," 30 "Keep out's," 30 "Be my Valentine's," and so on, in a nice orderly progression, with pauses of five or six seconds between each two-second song. (We're paraphrasing in this way merely to show that the songs are different; to the birds, the songs may all carry the same meaning even though they sound different.)

But the practicing youngster gets it all wrong. Like my daughter, he takes bits of sounds out of context and strings them together in a continuous, nonsensical sequence, such as "BeGoodYouAreKeepIngOut-OrnTimeMy . . ." The sounds lack the crispness and confidence of an adult's, and no two attempts at the same sound are alike. Eventually this young wren will be as competent as his father, just as each young human eventually masters the fine art of his or her spoken language (**Fig. 7–27 & Track 25**).

Bewick's Wren

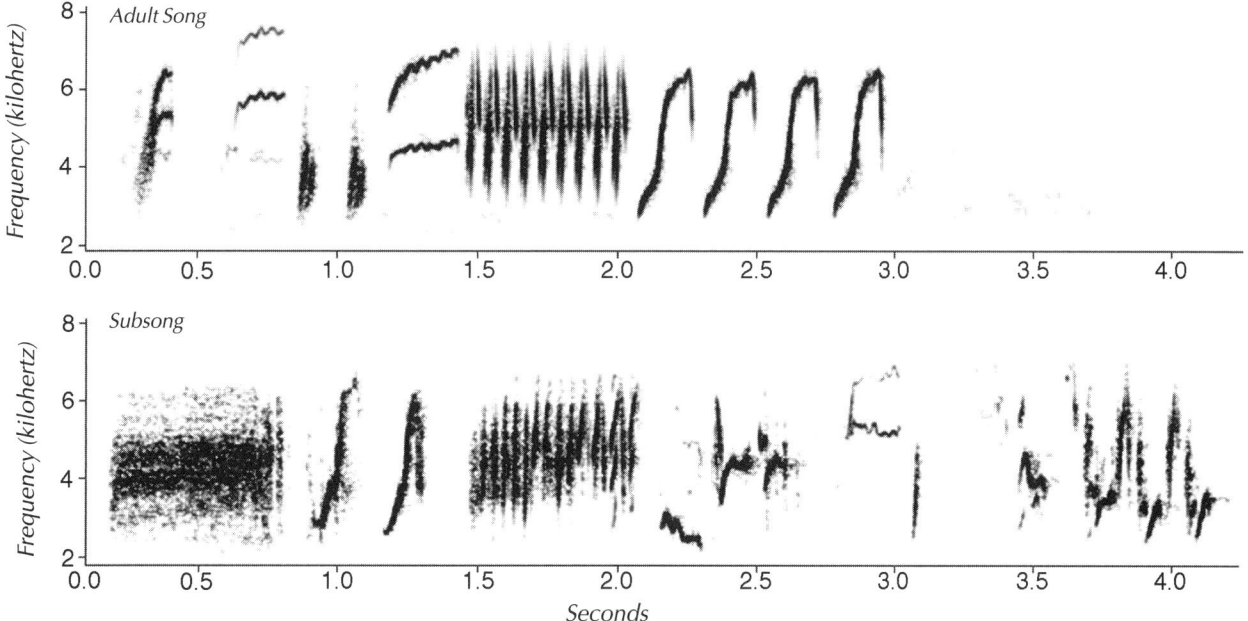

Figure 7–27. Sonagrams of Adult Song and Subsong of Bewick's Wren (Track 25, 4th Adult Song, 1st Subsong Phrase): *The adult song of the Bewick's Wren (top sonagram) is a complex assortment of notes, trills, and buzzes, similar to that of a Song Sparrow. In contrast the **subsong** (practice song) of young Bewick's Wrens (bottom sonagram) is rambly and long, with less distinct notes, and is different each time the bird sings. Note that the young bird's sounds between 1.0 and 2.0 seconds are similar to those produced by the adult between 1.5 and 3.0 seconds, but are in reverse order.*

Memorization, recall, and a good ear all are crucial in this learning process, for both songbirds and humans. The young songbird first memorizes a song, which it can do as early as 15 to 20 days of age. Days, months, or even a year later, when the youngster begins his practice singing, he recalls that memory and tries repeatedly to produce a copy of what he remembers. Successive corrections to the practice sounds eventually result in a perfect copy of the remembered song. Hearing is thus crucial to both the initial memorization and the later production of that sound.

Chances are you'll be able to hear young birds practicing if you listen carefully during late summer and early fall. Practice songs are less structured and more rambling than the adult male song. Bewick's Wrens and Song Sparrows are good subjects, but young of many species practice during the late summer.

Vocal Development in Nonsongbirds

We can better appreciate song development in songbirds by understanding how sounds develop in other bird groups. Consider the close relatives of the songbirds, the suboscines (Kroodsma 1988). The songbirds (also called oscines) and suboscines are the two suborders of the order Passeriformes; they are "sister groups," each other's closest living relatives (see Fig. 7–22). Just a handful of suboscines live in the Old World, but about a thousand species occur in the New World, including antbirds, antwrens, woodcreepers, cotingas, and flycatchers. Of all suboscines, only a few flycatchers reach North America **(Fig. 7–28)**.

Song development by the Alder and Willow flycatchers clearly illustrates the striking difference in song development between suboscines and songbirds. Current field guides describe these two flycatchers as essentially identical in appearance, and only in 1973 did the American Ornithologists' Union officially recognize the then "Traill's Flycatcher" as really consisting of two species (see Fig. 2–15). The only reliable way to identify these two species is by their voices. The Alder Flycatcher sings a *fee-BEE-o* (also described as *free-BEE-er*) song, and the Willow Flycatcher, a *FITZ-bew* song; never does the same bird sing both (**Track 26**).

In an experiment, young Alder and Willow flycatchers were taken from their nests at about 10 days of age. The young Willow Flycatchers were then trained over loudspeakers with Alder Flycatcher songs, and Alder Flycatchers were trained with Willow Flycatcher songs, in an attempt to confuse the birds in their singing. Although the young birds from *fee-BEE-o* nests heard *FITZ-bew* and the young *FITZ-bew* birds heard *fee-BEE-o*, the young birds were not confused. Remarkably, and so unlike songbirds, these young flycatchers developed perfectly normal songs of their own species. The experience with the wrong songs caused no problems (Kroodsma 1996a).

How Alder Flycatchers develop their songs is especially fascinating. This process was first discovered in the field and then confirmed in the laboratory. During the first few days out of the nest, the

Black-billed
Scythebill

Ocellated
Antbird

Scissor-tailed
Flycatcher

Banded
Pitta

Guianan
Cock-of-the-rock

Long-tailed
Manakin

Rifleman

Figure 7–28. Suboscine Diversity: *The suboscines are one of the two suborders of the order Passeriformes. A few species of suboscines live in the Old World, including the secretive* **pittas**, *long-legged ground-dwellers in tropical areas, and the New Zealand wrens (for example, the* **Rifleman***), which may be the most ancient group of passerines. Most suboscines, however, live in the Neotropics, with only one group, the flycatchers, containing species that reach as far north as the United States. Neotropical groups include the woodcreepers, such as the* **Black-billed Scythebill**, *woodpecker-like birds that, like nuthatches, search bark for insects; antbirds, such as the* **Ocellated Antbird**, *which follow army ant swarms to prey on the insects and other small animals flushed up by the ants' passing (see Ch. 9, Sidebar 3: Ant Followers); cotingas, such as the* **Guianan Cock-of-the-rock**, *large, stocky, fruit-eating birds, many of whom have bizarre courtship displays; and the brightly colored manakins (see Fig. 6–42), such as the* **Long-tailed Manakin**, *small, chunky, fruit-eaters, also with unusual courtship behaviors.*

young flycatchers usually remain near each other in a small family flock. When the fledgling brothers and sisters get separated, they use a particular vocalization as a contact call, as if to keep tabs on each other and perhaps to announce their location to their parents, who are still feeding them. This call is an unmistakable, although scratchy and uncertain, rendition of what will eventually become the adult song, the *fee-BEE-o*. Thus, as soon as young Alder Flycatchers leave the nest, they are already uttering what will clearly become their adult song. In the laboratory, the process is the same: a young bird that has just left the nest gives this call repeatedly when separated from its siblings. This call, clearly the precursor of the adult song for these suboscines, is already being used at day 14, the age at which young songbirds are just beginning to memorize the sounds they will later produce.

Follow-up experiments with the Eastern Phoebe, another flycatcher species, reveal that no matter what environment they are raised in, young flycatchers seem to know the proper songs to sing. These three species are only 0.3 percent of the 1,000 or so suboscines, so generalizing to the whole suborder would be unwise. Nevertheless, some clues from other species, such as the contact calls used by young birds and the lack of geographic variation (see Variation in Space and Time, later in this chapter), suggest that the song development of most suboscines may resemble that of these flycatchers.

One additional piece of evidence shows just how different the flycatchers are from the songbirds. Like humans, songbirds must be able to hear themselves vocalize during the developmental process. Deaf humans cannot learn to speak properly, and deaf songbirds cannot learn to sing properly. Their learning is impaired because, to compare their practice sounds to the remembered model sounds, both the young human and the young songbird must hear themselves vocalize. Flycatchers, however, are different. Eastern Phoebes develop normal song even if they can't hear themselves practice. Brain differences between songbirds and suboscines tend to confirm this relationship between learning and hearing, and to further reinforce the differences between the two groups (see Control of Song, later in this chapter).

Vocal development in other orders of birds, collectively called the "nonpasserines" (i.e., "not the order Passeriformes") is less well studied, but what we do know can be compared to the development of songbirds and suboscines. In only two other orders, the parrots (Psittaciformes) and the hummingbirds (Apodiformes), does extensive learning seem to occur. Parrots, of course, are renowned for their ability to imitate human speech in captivity (Pepperberg 1990; see Ch. 6, Sidebar 1: Bird Brains), although we know little about how that ability to imitate is used in the wild. That some hummingbird species learn was demonstrated by raising young Anna's Hummingbirds **(Fig. 7–29)** in the laboratory (Baptista and Schuchmann 1990). Furthermore, the existence of hummingbird dialects, in which neighboring groups of birds sing different songs, also shows that some species, such as the tropical Little Hermit, learn songs. (For more on dialects, see Variation in Space and Time, later in this chapter.)

Figure 7–29. Anna's Hummingbird: *Although most nonpasserines do not appear to learn their songs, hummingbirds are one exception, as was demonstrated by raising young Anna's Hummingbirds, a common species along the Pacific coast of the United States, in the laboratory.*

Vocal development in other groups is more like that of suboscines and seems to involve little, if any, imitation. Doves and pigeons, for example, develop normal vocalizations without learning them from adults; hybrids between dove species produce vocalizations intermediate to those of their parents, again suggesting that the vocalizations are coded in the genes and not learned by imitating other adults. Roosters develop normal crowing calls even if they cannot hear themselves, again suggesting that learning is unnecessary. Scientists have studied vocal development in few other groups, but believe that for most groups, no striking evidence of vocal learning will be found.

The logical question at this point is "Why?" Why do the songbirds learn, whereas most of their close relatives, the suboscines, appear not to? Do the songbirds have some special advantages that the suboscines lack? Although we have a few hints about some subtle aspects of song development (see Songbird Diversity, below, on styles of learning among wrens), we simply do not know the answer. Presumably the ancestor of songbirds acquired the ability to learn, so learning evolved only once in the songbird lineage; in the suboscine lineage, however, learning occurs among bellbirds, but apparently not among most other species, such as flycatchers. But what advantage might a songbird such as a wren have over a suboscine such as a flycatcher? What does learning *accomplish*? Clearly each of the roughly 400 flycatcher species in the world is highly successful at perpetuating itself, as is each of the roughly 75 species of wrens. Once learning evolved, wrens, like many other songbirds, developed large song repertoires and local dialects, but wrens don't appear to be any more successful, in any sense of the word, than flycatchers. We can't say that learning is any better than not learning, even though, as one ranking of the world's best songsters shows, we especially appreciate the music of the songbirds (Hartshorne 1973). We are left with a big mystery on why songbirds are the

way they are—they're special singers because they learn, but we don't know why they evolved the ability to learn in the first place.

Songbird Diversity

With well over 4,000 species of songbirds, trying to generalize about how "the songbird" develops its song is dangerous. Our only valid conclusion may be that some evidence of learning has been found in all songbirds studied to date. But *how* different species learn varies considerably.

Natural selection has presumably molded the developmental process in each species so that individuals develop the best sounds for managing the behavior of other birds of the same species. At one extreme are species such as White-crowned Sparrows, in which young birds must be tutored with adult songs and learn perfect details of songs. At the other extreme are species such as the Gray Catbird (Kroodsma 1996a) **(Fig. 7–30)**. A young catbird in the laboratory does not seem to need exposure to normal catbird song (at least

Figure 7–30. Gray Catbird: The Gray Catbird seems to require less learning to acquire its song than most songbirds, since young taken from nests in the wild at 8 to 10 days and kept in isolation develop normal catbird songs. Like its close relatives, the mockingbirds and thrashers, this mimic is able to imitate the songs of other birds, so some degree of song learning must occur. Photo by Lang Elliott.

not after the age of 8 to 10 days, when the young were taken from the nest in this study). Apparently normal repertoires of hundreds of typical catbird sounds develop in birds isolated from adult songs at this early age. Catbirds are highly atypical in this regard, because the young of most songbird species develop abnormal songs if they are isolated in this way. Nevertheless, studies in both the laboratory and nature show that catbirds can imitate both their own and other species, so vocal learning does occur in catbirds. It's just that imitating other adult catbirds doesn't dominate the song-development process as it does in so many other species.

Why the young sparrow imitates his neighbors so perfectly but the young catbird "improvises" his unique repertoire are mysteries. Before we can understand these differences among species, we need to know more about the daily life of the sparrow and the catbird, more about whom each individual expects to influence with its vocalizations, and how those birds are "persuaded." We must follow them in nature, from hatching to adulthood.

In most studies of this sort, like that of the Bewick's Wren discussed earlier, the young birds have been found to be especially good at learning the songs of other adults in their immediate neighborhood. Consider yearling Indigo Buntings, for example (Payne 1996) (see Fig. 7–50). Work in lower Michigan has shown that "neighborhoods" of different songs occur side by side, and that young males and females typically breed in neighborhoods where they were not born. When a young male settles on a territory during his first breeding attempt, he usually has an odd song, one that is unique to him and that typically bears little resemblance to his father's song. Over time, however, he usually changes his song to match the song details of an immediate territorial neighbor, thereby perpetuating the local neighborhood of songs **(Fig. 7–31 & Track 27)**. In some way, a young male must gain social advantages by matching the songs of his immediate neighbors. A better understanding of the song's function, such as exactly why the male sings and who listens and is influenced by it, may someday help us to understand why young males imitate their neighbors in this way.

Young Song Sparrows develop their songs in much the same fashion, but the process seems more complex because each sparrow has a repertoire of 8 to 10 songs (Beecher 1996) **(Fig. 7–32)**. In a Seattle, Washington population, young males don't settle down immediately, but rather seem to range over the territories of four or five adults late during their hatching year. The young males learn the songs from adults in this small neighborhood, and seem to follow two primary rules in the learning process. First, they learn a complete song from a particular bird and don't cobble together songs by taking pieces from different songs or different males. Second, they learn the most common songs, so that if all five adults in the neighborhood share a song, the youngster is especially likely to learn it; songs unique to a single adult are far less likely to be learned. Over the winter, the young male sparrow either takes over the territory of an adult who has died or simply usurps an area for himself in this neighborhood.

These rules for learning maximize the chance that a male will sing songs of surviving males in the neighborhood where he will breed. Like the Bewick's Wren and the Indigo Bunting, neighboring males then share similar songs. Sharing the same song repertoires must somehow enhance the birds' abilities to interact with and influence other birds in their social environment. Somehow, a bird who imitates the right songs must increase his ability to guard or acquire resources, such as

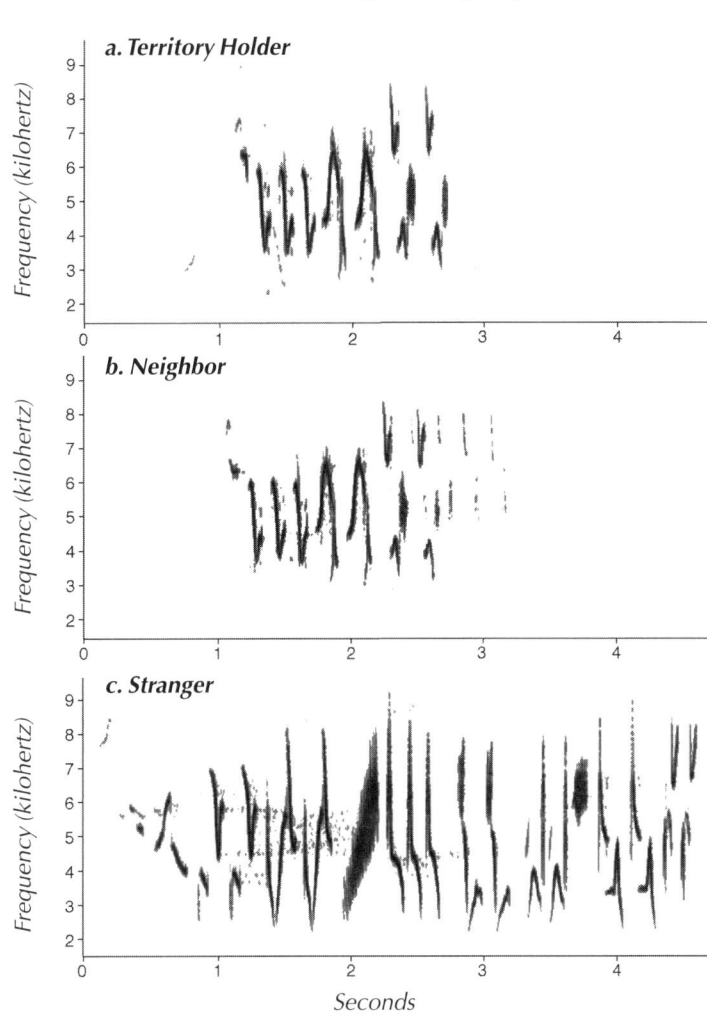

Indigo Bunting Songs

Figure 7–31. Sonagrams of Indigo Bunting Song from Neighbors and Stranger: Songs of Indigo Buntings consist of a few notes or syllables, each generally repeated two or three times. Although each individual bird sings only one song type, songs vary from bird to bird in both length and content. Furthermore, birds often sing shortened versions of their one song type, especially during territorial encounters. Young birds change their songs to resemble those of their neighbors, creating pockets of birds with similar songs. *a. Territory Holder (Track 27, 3rd Song):* Relatively short song of one territorial bird, with three different syllables that are repeated. *b. Neighbor (Track 27, 2nd Neighbor Song):* Song of a bird who holds a territory adjacent to the bird in (**a**). Note how similar the two songs are. *c. Stranger (Track 27, 1st Stranger Song):* Song of a bird who is a stranger to the birds in (**a**) and (**b**). Note how different the song is from that of the other two birds.

Figure 7–32. Song Sparrow: In late summer and fall of their hatching year, the young Song Sparrows studied in Seattle, Washington by Beecher (1996) learned some of their 8 to 10 song types from other males in the area in which they would eventually settle. Young were more likely to learn songs shared among several territorial birds than less common songs. Photo by Marie Read.

a territory or a mate, but the exact process by which the bird gains this advantage is unknown.

One final example illustrates the same phenomenon of neighbors sharing signals and also provides an instructive counterexample. Marsh Wrens in western North America learn each other's songs, so they have nearly identical repertoires (see Fig. 7–51). They often engage in a startling display of what is called **matched countersinging** (for additional discussion, see Song Repertoires, later in this chapter, and Verner 1976). Males have large repertoires of over one hundred song types, and they often hurl identical songs back and forth. One male sings A, and his neighbor counters with A; B from one, B from the neighbor; C, C; D, D; and so on. The songs are sufficiently different that a human listener can readily tell when the males are matching each other in this fashion (Track 28). Many western Marsh Wren populations are resident year round, and males thus come to know each other. They imitate each other's songs, as shown in laboratory studies, and males in a marsh interact in intricate ways with their shared vocabularies. With a premium on neighbors sharing nearly identical vocal repertoires, Marsh Wrens are thus like the Bewick's Wren, the Song Sparrow, and the Indigo Bunting, although the number of signals involved and the complexity of the vocal exchanges seem to have escalated.

You can easily hear this kind of matched countersinging among Marsh Wrens in any marsh in western North America—my favorite is Coyote Hills Regional Park southeast of San Francisco. In the East, the Tufted Titmouse and the Northern Cardinal are good subjects. Listen carefully and you may be able to hear how males within a species interact with each other, hurling the same song form back and forth.

In contrast, the Sedge Wren of North America seems to do little of this kind of intricate countersinging (Kroodsma, Liu, et al. 1999) **(Fig. 7–33)**. In the laboratory, Sedge Wrens do not imitate songs the way Marsh Wrens do. They learn a little, but mostly they seem to "improvise," inventing 100 or more good "Sedge Wren" songs using some kind of inborn song generator. Thus, no two males share identical repertoires, so when they sing to one another, they cannot match each other. This kind of development is apparently ideal for the lifestyle of Sedge Wrens in North America. They are somewhat nomadic birds, unpredictable in their breeding locations within and between seasons, and they seem to go where the habitat is most suitable. As a result, neighboring males are relative strangers to one another, and they communicate not with identical song types, but with generalized songs that declare "Sedge Wren" without getting into matching details.

These examples suggest that careful imitation of neighbors is especially important in stable neighborhoods, where birds get to know each other and can interact with shared identical songs. Nature provides an exciting test of this proposed relationship between learning, improvising, and degree of familiarity with one's neighbors. The Sedge Wren occurs throughout North, Central, and South America, in a variety of habitats. In some places, the Sedge Wren lives in communities that are stable throughout the year, like the Marsh Wren of western North America. In those communities, one would therefore predict that Sedge Wrens should behave more like the Marsh Wrens of western North America than like the Sedge Wrens of North America. A study in Brasilia National Park confirmed just that. There, male Sedge Wrens imitate their neighbors and countersing in the same kinds of intricate matching displays found among western Marsh Wrens. Thus, the stability of the habitat, year-round residency, and familiarity with neighbors seem to influence how songs develop and, consequently, how the songs are used in interactions among birds (Kroodsma 1996a).

Figure 7–33. Sedge Wren: The tiny Sedge Wren breeds in wet, grassy meadows. Unlike its close relative, the Marsh Wren, the Sedge Wren seems to invent most of its 100 or more song types, rather than learning them by copying the songs of other Sedge Wrens. Photo by Lang Elliott.

The focus in this section has been on songbirds and song, because that's where researchers have done the most work. But we have much to learn about songbirds, and even more about other birds and about vocalizations other than songs. Among "nonsong" vocalizations of songbirds, a variety of developmental patterns occur. Both the gargle and the *chick-a-dee* calls of the Black-capped Chickadee are learned, for example. The gargle varies among populations, and birds undoubtedly learn the gargle of the population in which they settle. Also, the *chick-a-dee* call differs among flocks, and when a chickadee joins a winter flock, its *chick-a-dee* call converges on the call structure of its new flock (Nowicki 1989). The other calls of chickadees are less well studied.

A variety of other calls are also learned among songbirds. Some calls of American Goldfinches and related species are learned; the calls of members of a pair may converge on one another, such that pairs within a larger flock can be identified based on call structure. In Europe, the "rain call" of the Chaffinch is also learned, so local dialects of this call can be found (see Song Dialects, later in this chapter).

Learning seems to play no role in the development of many other songbird vocalizations. Calls of the Eastern and Western meadowlarks, for example, seem not to be learned. Where these two species meet in the Great Plains, they can learn each others' *songs,* but the *call notes* enable us (and probably the birds) to identify the birds to the appropriate species (Lanyon 1969). Call notes are discussed in more detail in Sidebar 5: "Call Notes" and Their Functions, near the end of this chapter.

Control of Song

■ Birds have several calls and sometimes hundreds or even thousands of different songs. Some of these are learned, but others are not. How does the brain control these vocalizations? Where in the brain is knowledge of these signals stored? How does the syrinx, or voice box, produce the signals? What seasonal factors influence the endocrine system, and how does the endocrine system in turn control song?

How the brain controls song in songbirds is an exciting research area (Brenowitz and Kroodsma 1996). This brain control system was discovered somewhat by serendipity, during a study of song learning. Researchers had known for some time that the voice box consists of two independent units, located where the trachea splits to form the two bronchi. On top of each bronchus is a voice box (**Fig. 7–34**). One nerve travels down the left side of the neck to innervate the left voice box, another down the right side of the neck to the right voice box. This "dual innervation" gives birds independent control of each voice box. In several species that produce two sounds at once, blocking the signals from one nerve eliminates one of the two sounds. Some bird songs clearly demonstrate this dual innervation and control of the two voice boxes. Wood Thrush songs, for example, often contain two sounds produced simultaneously; one voice box controls the higher-pitched sound and the other, the lower-pitched sound. In essence, the Wood Thrush can sing a duet with itself (**Fig. 7–35 & Track 29**).

*Figure 7–34. The Syrinx: The bird's voice box, the **syrinx**, is really a pair of chambers located along the trachea, where it splits to form the two bronchi heading to the lungs. The chamber of the syrinx, like the rest of the trachea and bronchi, is kept open by rings of cartilage. The muscles of the syrinx control the details of song production (see Fig. 4–77). The more complex syringeal musculature of songbirds allows them to produce more intricate songs than other taxonomic groups of birds (see Ch. 4, Sidebar 2, Fig. A).*

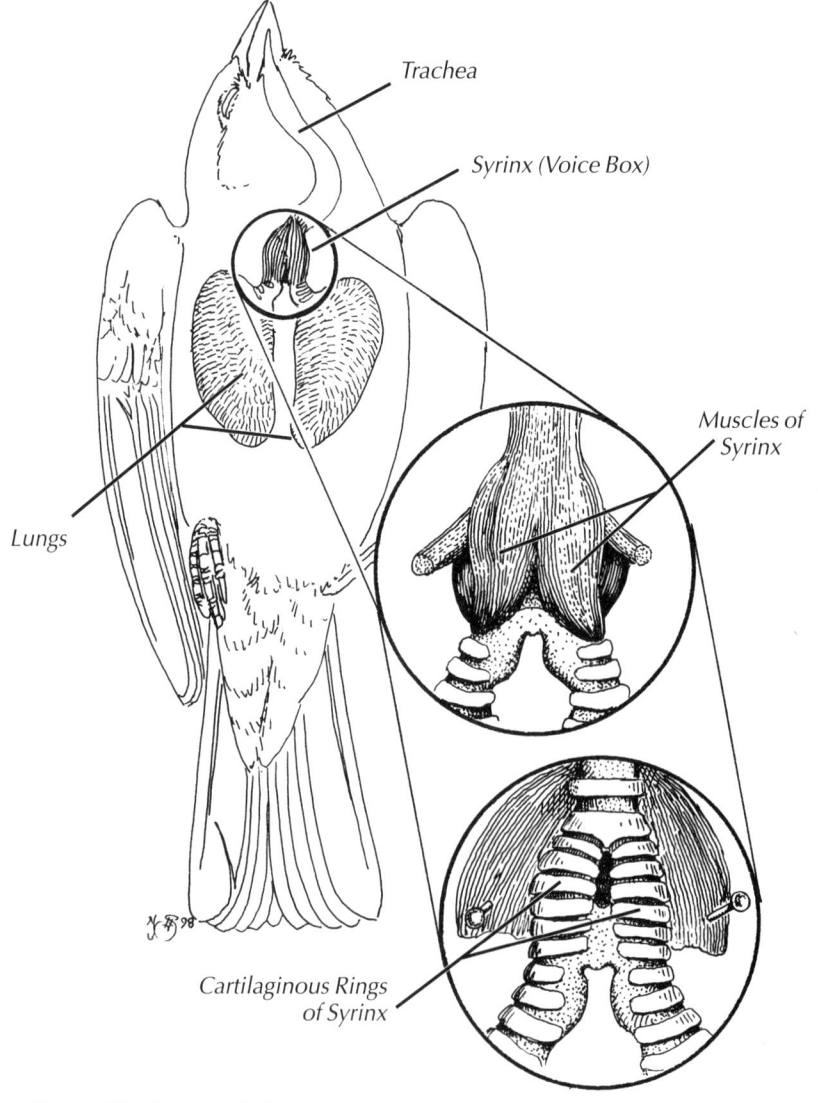

Trachea

Syrinx (Voice Box)

Muscles of Syrinx

Lungs

Cartilaginous Rings of Syrinx

Wood Thrush

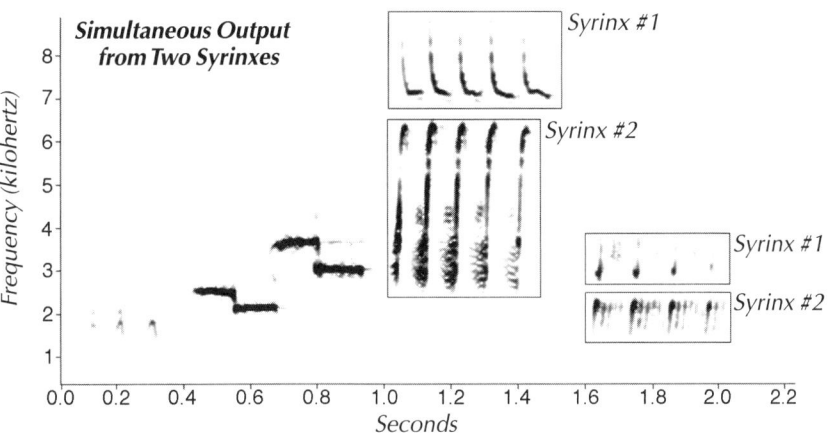

Simultaneous Output from Two Syrinxes

Syrinx #1

Syrinx #2

Syrinx #1

Syrinx #2

Figure 7–35. Sonagram of Wood Thrush Song Showing Contribution from Each Half of the Syrinx (Track 29, 4th Normal Song, 3rd Song at One-Half Speed): The song of the Wood Thrush allows the contributions from each half of the syrinx to be distinguished fairly easily. You can see from the sonagram that the middle and final trills are made up of components that are generated simultaneously—these appear to be produced from different halves of the syrinx. Note that the designations "Syrinx #1" and "Syrinx #2" are purely arbitrary, serving only to differentiate the sounds from the two halves. Which half of the syrinx produces which sounds is not known.

Tracing these nerves back into the brain led to the discovery of a series of cell groups involved in both hearing and producing songs. All cell groups were interconnected in the way one would expect if a young bird had to compare its own songs with a remembered version before trying again (**Fig. 7–36**). The overall size of these control centers in the brain is proportional to the size of the bird's song repertoire. In Zebra Finches, females do not sing, and the song control centers of the female are much smaller than those of the male. In species in which females do sing, these control centers are correspondingly larger. Within species, too, the more songs a male has, the larger his song control centers are likely to be. Thus, male Island Canaries or Marsh Wrens with larger repertoires are likely to have larger song control

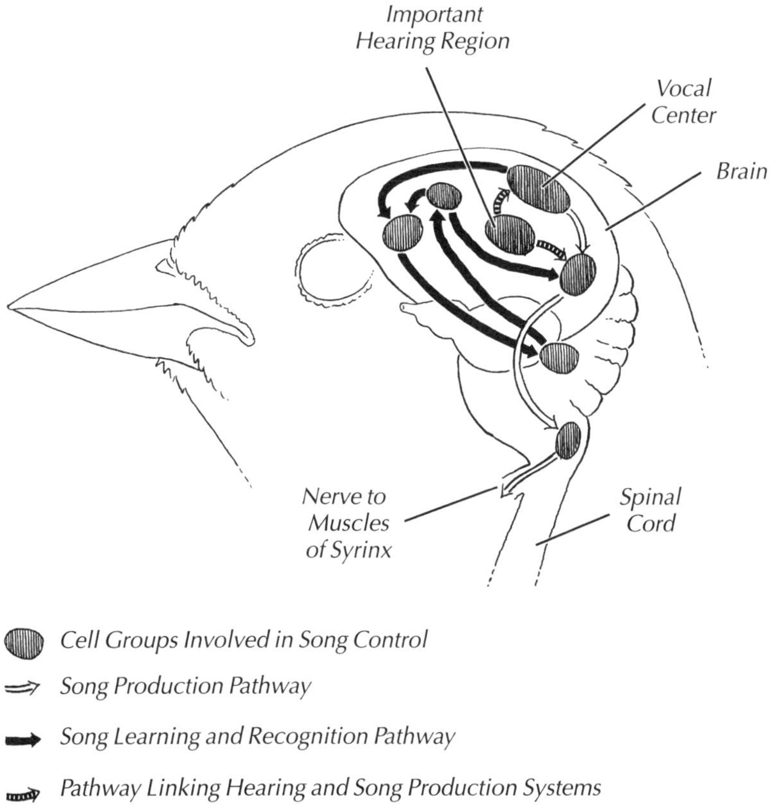

Important Hearing Region

Vocal Center

Brain

Nerve to Muscles of Syrinx

Spinal Cord

Cell Groups Involved in Song Control

Song Production Pathway

Song Learning and Recognition Pathway

Pathway Linking Hearing and Song Production Systems

Figure 7–36. Song Control Centers in Avian Brain (Section through a Songbird's Brain as Seen from the Side): Song development and production are controlled by an elaborate network of nerve cell groups and neural pathways in the avian brain. The system's interconnectedness is essential because, to learn songs correctly, young male songbirds usually must hear and remember the songs of adult male conspecifics, then later compare their own practiced songs with the remembered versions before further refining their efforts. Note that this is a highly simplified diagram of the system. For more information, see Brenowitz and Kroodsma (1996) and DeVoogd and Lauay (2001).

| Day Length Increases | → | Gonad Size Increases | → | Hormone Production Increases | → | Song Control Centers in Brain Stimulated | → | Singing Rate Increases |

→   LATE WINTER   →   EARLY SPRING   →   LATE SPRING   →

Figure 7–37. Pathway by which Day Length Affects Singing Rate: In late winter, the increasing day length stimulates the gonads to grow larger, which in turn increases the output of gonadal hormones. One effect of the gonadal hormones is to stimulate the song control centers in the brain, which causes neural changes that increase the amount of singing.

centers than males with smaller repertoires. Species with huge song repertoires, such as the Brown Thrasher, also have enormous song control centers.

Also intriguing is how the characteristics of these song control centers change with the seasons (Nottebohm 1987). They appear to shrink during the nonbreeding season, when hormone levels are low, and enlarge again during the breeding season, when hormone levels are high. Careful work has also shown that new neurons are being born in some of these control centers—an activity not previously thought to occur in the adult nervous system of any vertebrate! This neural control network is a stimulating model for understanding how brains control behavior, and the songbirds provide an exciting diversity of species in which to explore how nature has shaped control in different ecological circumstances.

Why these brains change with the seasons is somewhat controversial. The first ideas came from studying Island Canaries and Zebra Finches. The canary's brain changes with the seasons and the canary learns new songs each year; the finch brain does not change and the finch does not learn new songs each year. Were the brain changes necessary for the new learning? That idea was quickly refuted when researchers discovered that other species, such as the Eastern Towhee, seasonally change their brains but not their songs. Perhaps, then, the brain changes so that birds can learn to *recognize*, though perhaps not sing, the songs of new neighbors each year. Maybe—and maybe not. As scientists study the brains and songs of more species we may start to see some patterns and better understand why these brains work the way they do.

The endocrine system, too, is intricately involved in bird song. Many of the song control centers in the songbird brain contain receptors for gonadal hormones, so as day length influences the size and hormonal output of the gonads, it indirectly affects the brain. Lengthening days cause the production of more hormones, which activate cells in the song control centers and stimulate neural changes, which in turn cause singing **(Fig. 7–37)**. The combined effect of all these factors first begins to surface in northern climates on those occasional warm sunny days in January. It is then that, at least in the Midwest and northeastern United States, we hear songs from Northern Cardinals, Tufted Titmice, Black-capped Chickadees, and other residents, all of whom are undergoing a physiological transformation, from their gonads up to their brains.

The song control centers of songbirds are clearly involved in the learning and production of songs, and these neural centers have been identified in all major songbird groups. But what about other bird groups (Gahr et al. 1993)? The brains of those suboscines that do not learn their songs differ remarkably from those of songbirds, and searches for a comparable control system among the suboscine groups have failed. To be sure, some neural network in the brain controls vocalizations, but nothing like the one in the songbird brain has been found. Some kinds of control centers have been found, however, in parrots and a hummingbird, the two other bird groups in which vocal learning has been documented. The control centers differ in the three groups that learn, however, suggesting that brain structure and vocal learning arose independently in these three orders.

When I think of how birds sing, I like to recall that Winter Wren song we heard earlier. Study the sonagram and listen to the slowed-down tape once more (see Fig. 7–11 & Track 8). Each song consists of a hundred or more brief sounds, each pronounced with precision in both time and frequency, and all placed in a consistent sequence, so that the wren produces a remarkably complex song with unfailing accuracy. Somehow the young bird memorizes that song from other adults, storing all the bits of information somewhere in his tiny brain. That he recalls the details of his memory and sends the correct neural messages to the voice boxes, and that those tiny muscles contract in such a controlled sequence to produce such eloquence, I find simply astounding. Western Winter Wren males learn not one but dozens of these complex songs, and western Marsh Wrens learn up to two hundred. My mind simply boggles at the thought. For those who know birds, the phrase "bird brain" takes on new meaning!

Variation in Space and Time

What is especially fascinating about signal variation is the great variety of patterns (Kroodsma 1986). Questions abound! Why don't male White-throated Sparrows match the songs of a local neighborhood the way males of the closely related White-crowned Sparrows do? Why does the *hey-sweetie* of the Black-capped Chickadee occur from Maine to British Columbia when songs of other species are so much more local? In short, why do species differ in how their signals vary from one population to the next (over space) and in how their signals vary from one generation to the next (through time)? The birds use each signal in some way to interact with and influence other individuals, and the patterns of variation must be well adapted for communication. The goals in studying this variation are first, to document the patterns of variation, and second, to understand the relationship between the patterns and their functions. In this section, we focus first on how species differ and how we can use these differences to identify each species. Then we study some of the variation that is so important to the birds but is usually less appreciated by humans. We'll learn how signals vary from bird to bird, population to population, and generation to generation. Throughout, our goal is to understand *how* sounds vary and ultimately *why* they vary the way they do, although answers to the "why" are few, so far.

Species Differences

As we listen to the great diversity of bird sounds and study field guides and associated sound guides, we appreciate that each species makes different sounds. We can identify the Chipping Sparrow's dry trill, and the trained, professional ear can (usually!) distinguish the Chipping Sparrow's trill from the similar songs of the Pine Warbler, Dark-eyed Junco, and Swamp Sparrow. During its normal daily activities, each bird typically interacts only with other members of its own species, because they're the ones it must compete or cooperate with to achieve success. Evolution has thus insured that these interacting birds share vocal signals, either by inheritance, for sounds encoded in genes, or by vocal traditions, for sounds transmitted by learning.

Humans use these species differences in a variety of ways. The sounds are indispensable in survey work, for example. In many habitats, especially in dense tropical forests, few birds are seen but many are heard. It is thus the birds' voices that enable us to determine the relative abundance of different species in different habitats. Or consider the work of Bill Evans at the Cornell Laboratory of Ornithology **(Sidebar 2: Listen Up!)**. Most small birds migrate at night, when they are invisible, at least to our eyes. But many of them call, and the calls they use in flight are distinctive enough that sophisticated computer-recognition software can automatically identify many night-recorded sounds to the correct species. By recording sounds raining from the sky, the abundance of some Neotropical migrants can be determined, day after day, year after year, and these kinds of data are extremely valuable in our conservation efforts.

Individual Variation

We humans can easily identify each other by our distinctive voices, and research clearly shows that birds can recognize each other by voice, too (Stoddard 1996). This ability to recognize individuals was first demonstrated with the songs of the Ovenbird (Weeden and Falls 1959). Each male has his own unique rendition of the *tea-CHER* song. Differences in the songs of individuals can be seen clearly in sonagrams, but, more importantly, some ingenious field playback experiments have demonstrated that the birds use those differences to identify each other **(Fig. 7–38)**. Thus, males defending territories get to know their neighbors and the songs they sing. A neighbor singing from the appropriate location may be acknowledged with a few songs in return, but if the neighbor sings from the "wrong" location, or if a strange bird delivers an unfamiliar song anywhere in the area, the territorial male responds more aggressively. Clearly Ovenbirds can identify each other by their distinctive songs, and they take advantage of those differences to help maintain their territories.

This **individual recognition** is especially easy in Ovenbirds, because all males have a single song that is unique. Just as a unique fingerprint identifies each human, a song identifies each Ovenbird.

(Continued on p. 7·47)

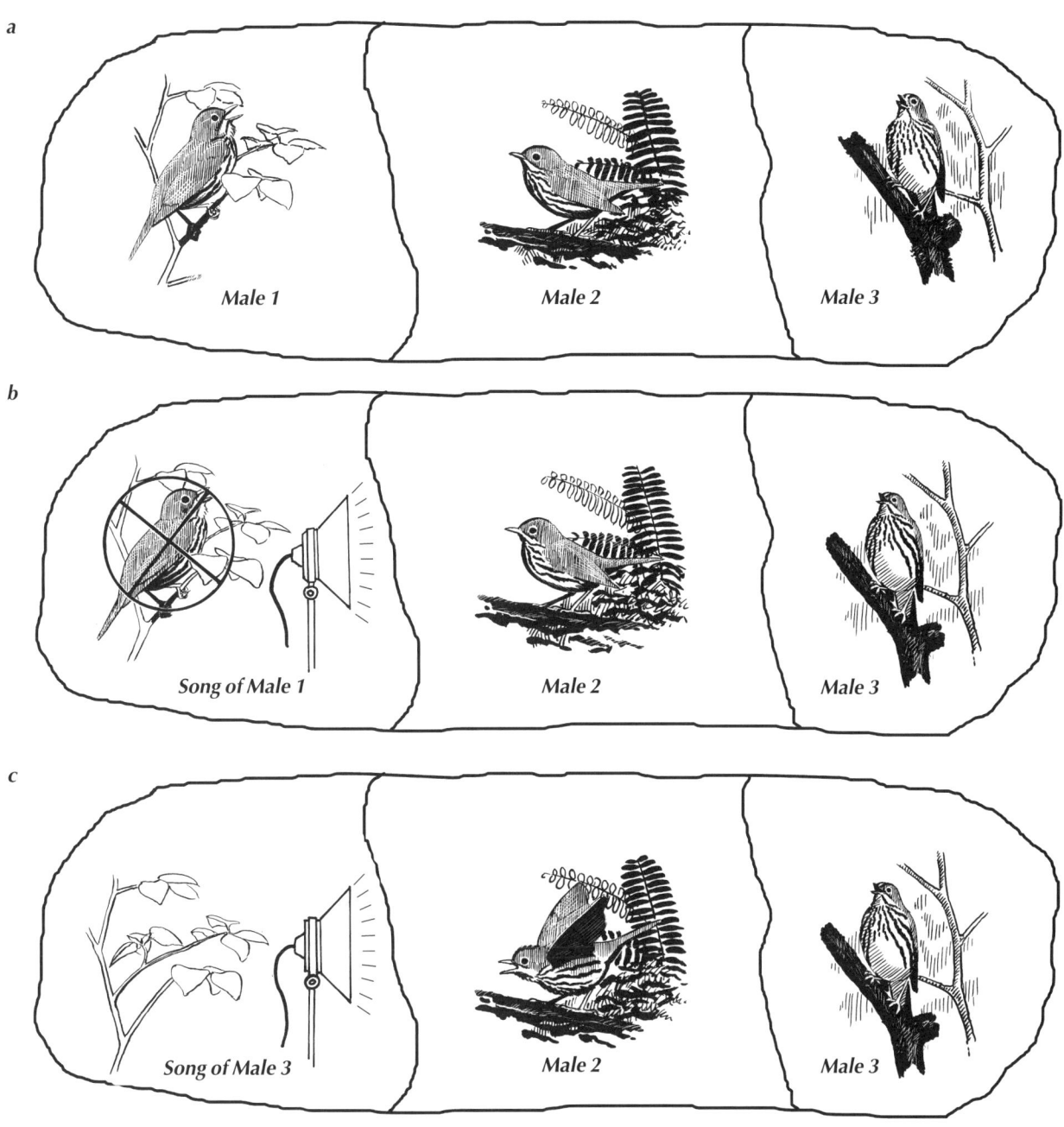

Figure 7–38. Neighbor Recognition in Ovenbirds: *Many territorial birds learn to recognize the songs of their neighbors—a fact established by numerous experiments over the years. A typical experiment might proceed as outlined here:* **a.** *Male Ovenbirds 1, 2, and 3 occupy territories in a line, as if along a stream; males 1 and 2, and males 2 and 3 have shared borders.* **b.** *If male 1 is removed, and his song is played from the common boundary between 1 and 2, male 2 responds little, if at all. Male 2 has come to know that particular song from that location.* **c.** *If, however, the song of male 3 is played on the boundary between males 1 and 2, male 2 responds very aggressively. The wrong song coming from that boundary means that the status quo has been disturbed, and boundaries need to be reestablished. Thus the experiment demonstrates that male 2 can use the differences in the songs of males 1 and 3 to know who is singing from what location.*

Sidebar 2: LISTEN UP!
Bill Evans

One mid-September night my Dad pointed out the calls of night migrating birds—faint *peeps* and *tseeps* passing high above our backyard in southern Minnesota. I was just 15, and 10 more years would pass before I really noticed these voices in transit again. This time I would find my own calling.

Twice a year, millions of birds migrate across the Americas to and from ancestral breeding and wintering grounds. The migrations of most species occur under cover of night, and many vocalize during their flights. The calling has long been thought to help birds keep in contact with one another as they migrate, allowing them to form and maintain in-flight associations in the darkness. Some suggest the calls may serve as air traffic control. But no one knows to what extent they act as a method of information exchange. Whatever their purpose, on a good migration night in eastern

North America, thousands of calls may be heard by listeners on the ground **(Fig. A)**.

In the spring of 1985, I heard an astounding flight while camping on a bluff along the St. Croix river in eastern Minnesota. An avid birder, I lived for the short spring migration and the chance to see colorful migrating flocks refueling before heading to northern breeding grounds. But on the night I heard that incredible flight so clearly, I realized that if I knew the callers' identities, I could sit out at night in a lawn chair and view in my mind the species composition of this clandestine symphonic transit. The thought was overwhelming!

Many people have heard flocks of Canada Geese migrating at night. Even though you can't see them, you know they are Canadas because you've heard the calls during the day. But what most people don't realize is that all North American warblers and sparrows, and many other song-

birds, migrate at night, vocalizing while they fly. Warbler and sparrow calls are typically short, high-pitched notes, not unlike a single field cricket chirp (Track 30) **(Fig. Ba &** Track 31). Thrush calls are beautiful, mellow notes, lower pitched than warbler and sparrow calls. My Dad, after walking home from work early in the morning, told me of a large flight of thrushes he'd heard descending from night migration. He described some of the calls as sounding almost like a cat's meow. I think he may have been hearing Veery night flight calls **(Fig. Bb &** Track 32). Others have likened the night flight notes of the Swainson's Thrush to calls of the spring peeper, a small tree frog common throughout most of eastern North America (Track 33).

Many people, in fact, are unaware of songbird night calls because they mistake them for insects, frogs, or even cats! To hear the calls, find a quiet place away from traffic, loud insect noise, and other environmental noise such as streams or wind. It helps to find the highest place around in order to get as close to the migrants as possible—although you can often hear good flights in low-lying areas as the birds descend from night migration, typically in the hours before dawn. Choose a night during spring or fall migration when the winds are favorable for a flight (see Fig. 5–64). Then sit back, relax, and listen up toward the night sky.

I've taken many people out to listen with me. They often have trouble hearing calls at first. But then, as if some kind of acoustic door opens, they hear a call, and then another, and their ears quickly become focused on what to listen for. It's like magic when, in a matter of moments, they tune in to the abundance of beautiful voices from passing migrants.

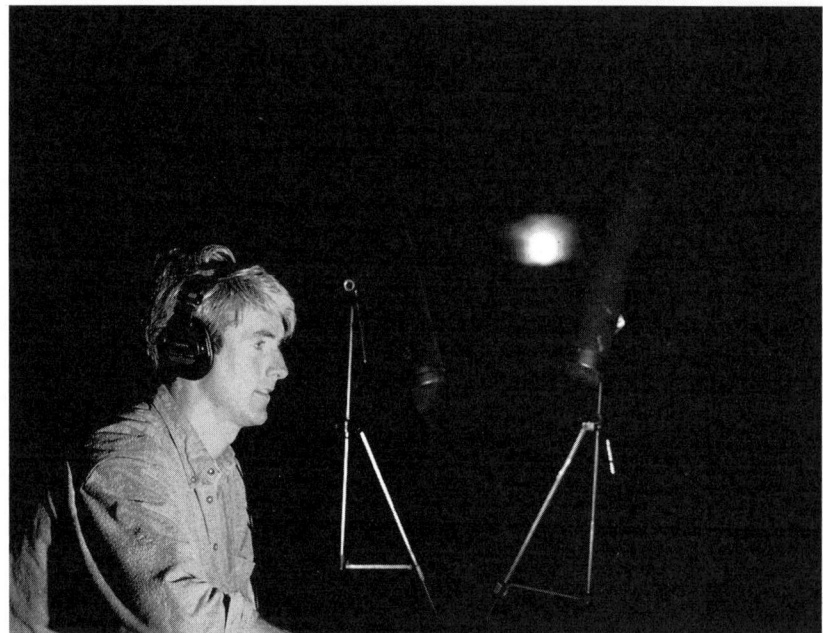

Figure A. Bill Evans Listens to Night Migrants: *Researcher Bill Evans listens to the calls of night migrants flying overhead. On a good migration night in eastern North America, listeners on the ground may hear thousands of calls, even without the help of microphones, from quiet, high places such as hilltops. Photo by Tim Gallagher.*

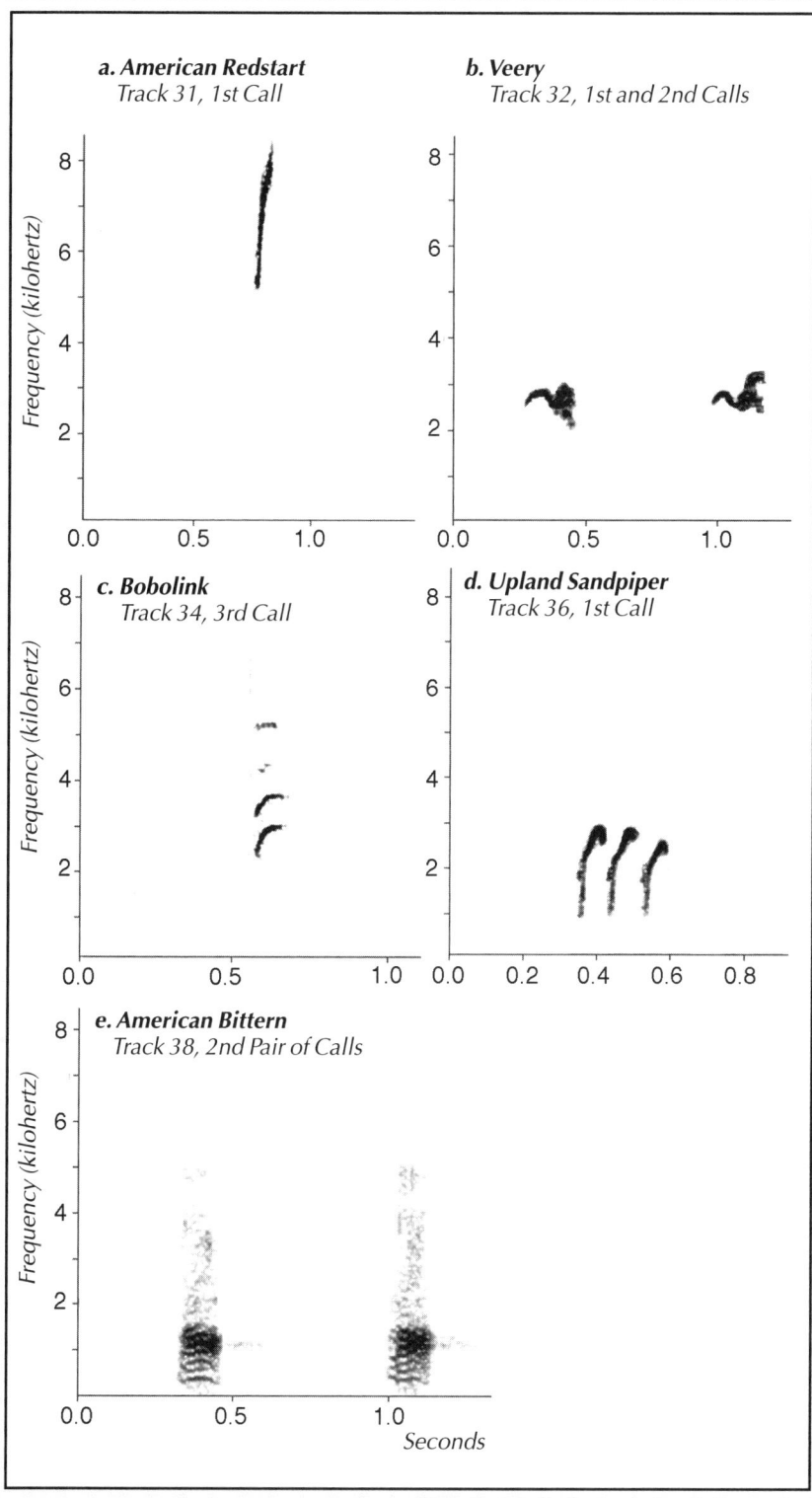

Figure B. Sonagrams of Night Flight Calls of Migrants: a. American Redstart (Track 31, 1st Call): *Most warbler night flight calls are short and high pitched.* **b. Veery (Track 32, 1st and 2nd Calls):** *Thrush night flight calls are lower, longer, more mellow notes than those of warblers. The Veery says* veeree *or* veer *in night flight.* **c. Bobolink (Track 34, 3rd Call):** *The* pink *used during night migration is the same as the common daytime flight call note.* **d. Upland Sandpiper (Track 36, 1st Call):** *The night flight call is a short, upward-slurred whistle—repeated quickly a few times.* **e. American Bittern (Track 38, 2nd Pair of Calls):** *The deep croak of a migrating American Bittern is similar to the croaks of night-herons—note the broad band of frequencies in each note.*

On a good night the myriad voices flow by all night long in ever-varying composition: a wonderful intermingling of soft *phews, quees,* and *wheers* of migrant thrushes and Rose-breasted Grosbeaks; *pinks* of Bobolinks; prehistoric-sounding squawks from herons and bitterns; lisping *tseeps* and *tzeeps* of sparrows and warblers; a plethora of strident shorebird calls; and many other sounds that thrill and perplex the listener (**Fig. Bc &** Track 34) (Track 35) (**Fig. Bd &** Track 36) (Track 37) (**Fig. Be &** Track 38). There is no better way than through hours of such peaceful listening to ponder the mystery of these ancestral migrations: little songbirds traveling thousands of miles twice each year, often to the same wintering and breeding grounds.

One can only imagine what the Native Americans may have heard and thought of the great migratory flights. Night calling has been studied by modern science for the last 100 years or so, although with remarkably little progress. For unlike Canada Geese, whose calls are the same night and day, many migrant land birds use call types during nocturnal migration that are rarely given during the day except in specific behavioral situations. When I began studying night calls, I was able to recognize distinctive call types at night, but even with a decade of birding experience, I could not associate most of them with particular species. I relished the few published articles on night flight calls I could find (see Ball 1952, Graber and Cochran 1959 and 1960, Graber 1968, and Tyler 1916). Then I heard about the Cornell Laboratory of Ornithology's Library of Natural Sounds (LNS), the largest natural sound archive in the world. I came to work at LNS in 1988 with the hope of finding clues to many of my unknown calls.

I gradually realized, however, that no one knew the identity of most night flight calls and only a handful of recordings existed. I was faced with the challenge of identifying nearly all the night flight calls of migrant land birds—a quest I pursued with a three-part strategy. First, because some species' night flight calls are similar to their daytime calls, by intensively listening to diurnal call notes from birds that I could see, I was able to find matches to many of my unknown night flight calls. Second, by extensively recording night flight calls in different geographic regions, I was able to narrow down the possibilities for the unknown calls by correlating the geographic distribution of particular calls with the known geographic regions through which particular species migrate. Similarly, I was able to gain clues by correlating the time of year I recorded certain call types with known migration times of certain species.

In 1994 I began an association with the Cornell Lab of Ornithology's Bioacoustics Research Program (BRP). The computer program "Canary" developed by BRP became my primary tool, allowing me to easily make sonagrams from my recordings of night flight calls. With these detailed "pictures," I could now visually pick out differences between calls that previously had sounded the same to my ear. With sonagrams, it was also much easier to compare my night flight calls to daytime calls from known species—many night flight calls are so brief that it's nearly impossible to hear the differences between them, even if you listen very carefully.

After 10 years of research, I now have a fairly good understanding of the night migration language of birds in eastern North America. I can identify nearly all the night flight calls heard in this region to species or to "species complex"—a group of closely related species. But much work remains. I would like to be able to distinguish the similar calls of each species within the species complexes—to separate the calls

of Tennessee Warbler, Nashville Warbler, Orange-crowned Warbler, and Black-throated Green Warbler, for example. I also would like to be able to document the range of variation of each species' calls. The typical night flight calls of the Ovenbird and the American Redstart, for instance, are clearly distinct, but some call variations of each are quite similar to one another.

Since learning to recognize many night migrants by their calls, one of my major focuses has been to monitor these calls across eastern North America to gather information on their migration routes and changing abundance. Counting the *number of calls* from each species over a particular recording site each night is relatively simple, but translating those totals into the *number of individuals* passing by is a much more difficult task. For example, if I count 35 Great Blue Heron calls in one evening, how many individuals have flown by? One bird might pass over a recording station and call only once, whereas another may call 10 times during its transit. Because of such variations in individual call rates, I cannot simply total the calls each night, but also must look at their pattern of occurrence. There are several ways to do this. One is to carefully examine sequences of calls. As a bird approaches, passes over the recording site, and heads away I often hear a sequence of weak calls, followed by loud ones, then weak ones again. Thus I can acoustically follow the progression of some individual birds over the recording station. Another way to estimate numbers of individuals from a recording of a night's calls is to determine the time it takes for a bird of a particular species to pass overhead—in other words, to appear and disappear on the tape. Luckily, species have fairly consistent flight speeds, so average transit times for

each can be established. In New York State, after listening to thousands of hours of recordings, I have found that thrushes typically pass over one of my recording stations in one to two minutes. Therefore I can assume that thrush calls separated by more than two minutes are very likely to be from different individuals. One additional way to sort a night of calls into individuals is to record in stereo—from two carefully placed microphones. When two birds pass by together, stereo recordings often reveal their separation in space to a careful listener. Although progress so far is encouraging, further research on counting techniques is needed to develop a reliable method of monitoring migration through recording night flight calls.

One exciting possibility is automating the collection and analysis of night flight calls using computers that can detect and identify the various calls from a tape. Harold Mills, in the Lab's Bioacoustics Research Program, recently developed a prototype software program that, when given a tape with a selection of known night flight calls, was able to identify them to species or species complex, and to total the number of calls of each. Once this procedure is perfected, an array of night flight call monitoring stations can be set up across the continent, interconnected via the Internet. Such an array could provide us with real-time bird migration maps analogous to the weather radar loops we watch on television news—we could broadcast bird migration TV into America's living room, follow the progress of our favorite species during their migration, and automatically track changes in bird populations. That's the dream we are currently working toward here at the Lab. Stay tuned! ∎

Thus, the memory needed to recognize all of one's neighbors is limited, and distinguishing neighbors is especially easy. This kind of neighbor recognition has now been shown in a number of species in which each male has essentially a single song form: Indigo Bunting, White-crowned Sparrow, White-throated Sparrow, and Common Yellowthroat (Stoddard 1996).

To easily hear differences among individuals, try listening carefully to males of some species in which the male sings only a single song type. In addition to those mentioned above, Chipping Sparrows, Field Sparrows, and Prothonotary Warblers are good subjects. For many of these species, you can learn to recognize individual birds by their unique songs.

But neighbor recognition also has been demonstrated in species with much larger repertoires. Song Sparrows may have up to 10 different songs apiece, but they can still recognize their neighbors, even though the neighbors may collectively have over 50 songs. This neighbor recognition is based on the birds' remarkable memory for different song forms. Memory of this sort is important because monitoring the behavior of one's competitors is essential to breeding success. A stranger introduces great instability into a neighborhood; the immediate threat is probably to the territory, and later to the female. Songs of strangers thus provoke strong responses. Responses to neighbors are not so strong, because a truce has typically developed, in which each bird has come to accept the other as a neighbor. The neighbor with the familiar song can still pose a great threat, however. Male Red-winged Blackbirds, for example, often inseminate the females on neighboring territories. In Song Sparrows, it is again neighbors who are especially likely to steal real estate. Using sound to monitor these potential threats to one's success is therefore indispensable **(Sidebar 3: Pushing the Limits: New Computer Techniques for Studying Bird Song)**.

Birds also use vocalizations other than territorial song to recognize individuals (Beecher and Stoddard 1990). Bank Swallows live in colonies, and once the young leave the nest, they mix with young from other nests **(Fig. 7–39)**. Parents use the distinctive vocal signatures

(Continued on p. 7·53)

Figure 7–39. Bank Swallow Young: After fledging, young Bank Swallows gather together with young from other nests to form large groups. Parent birds are still able to single out their own young, however, by recognizing their voices. (See Ch. 8, Recognition Between Parents and Young.) Photo courtesy of Mary Tremaine/CLO.

Sidebar 3: PUSHING THE LIMITS: NEW COMPUTER TECHNIQUES FOR STUDYING BIRD SONG

John Bower

My admiration for the pioneers of bird song research, who faced the daunting task of studying bird song without modern technology, never wavers. These turn-of-the-century researchers studied song by translating the songs they heard into standard musical notation, and assigning words to the phrases the birds were singing. Such translations, although the only technique available at the time, were very limiting: at best they were gross approximations of the complex information transmitted when a bird sings (**Fig. Aa & Track 39**; **Fig. Ab & Track 40**). The modern era of bird song studies began in the 1940s when the invention of magnetic tape recorders and the sonagram allowed the preservation and objective analysis of natural song. These two technologies have been the basis for a half century of great advances in our understanding of avian communication. Now, as the 21st century begins, we find ourselves riding a new technological revolution: the development of fast, powerful, and portable computers. Here I describe three new computer-based techniques that hold great promise for furthering our understanding of avian communication.

Interactive Playback

The first new technique, interactive playback, allows researchers to interact more realistically with the birds they are studying. In traditional (non-interactive) playback experiments the researcher plays tape recorded songs through a speaker to a bird, and notes or tape records the subject bird's response (see Fig. 7–55). Traditional playback experiments have played an important role in deciphering the functions of song. Consider the following experiment: after mapping a male Song Sparrow's

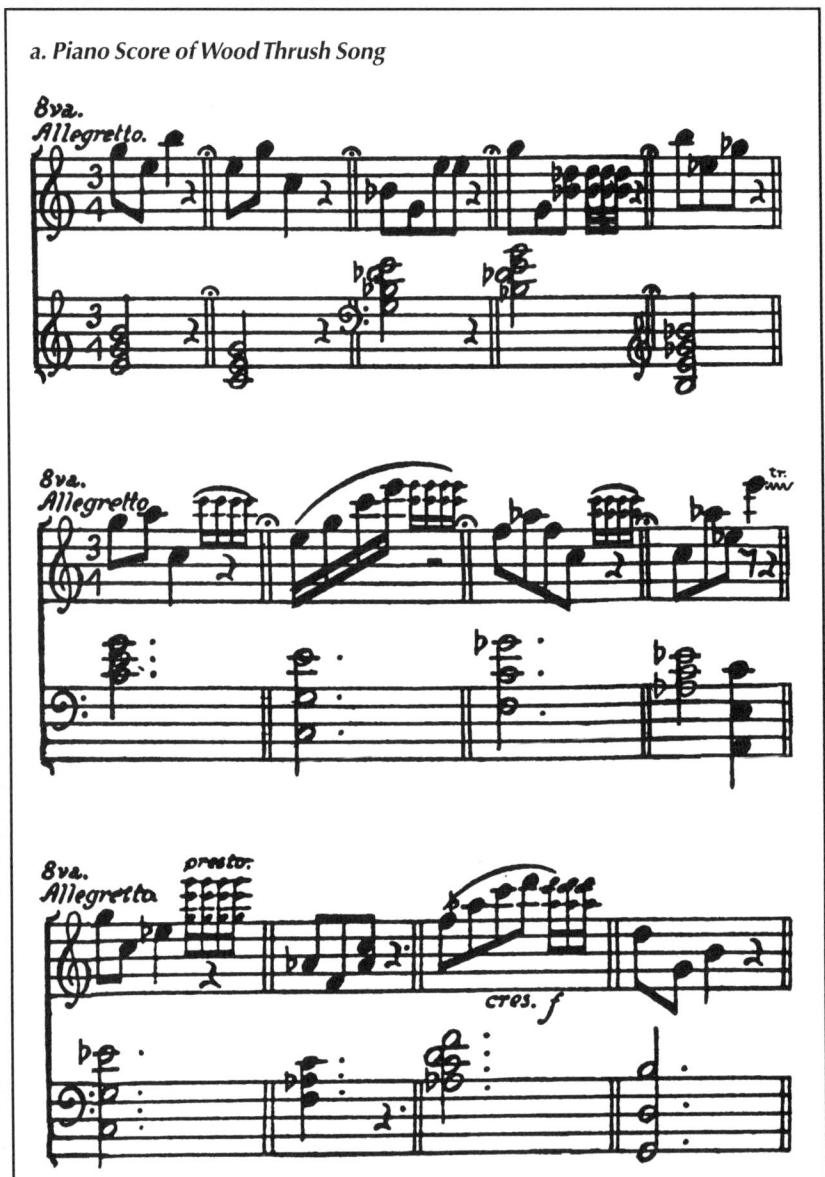

Figure A. Piano Score and Sonagram of Wood Thrush Song: a. Piano Score (Track 39): *Early bird song researchers did not have sophisticated computers, microphones, tape recorders, and other equipment used to study bird song today. They listened carefully, and often used musical notation to describe the songs they heard. In 1921, F. Schuyler Mathews published his* Field Book of Wild Birds and Their Music, *a collection of musical scores representing the songs of birds common in the eastern United States. Shown here is the Wood Thrush score from that book. In Track 39 you will hear this Wood Thrush score played on a piano.* **b. Comparison of Sonagram and Piano Score (Track 40, 3rd Song):** *A comparison of one portion of the Wood Thrush piano score with a sonagram of a single phrase from a Wood Thrush—note that the introductory notes to the song, audible at close range and on Track 40, were removed from the sonagram to better match the piano score.*

Figure A. *(Continued)*
b. Comparison of Piano Score and Sonagram of Wood Thrush Song

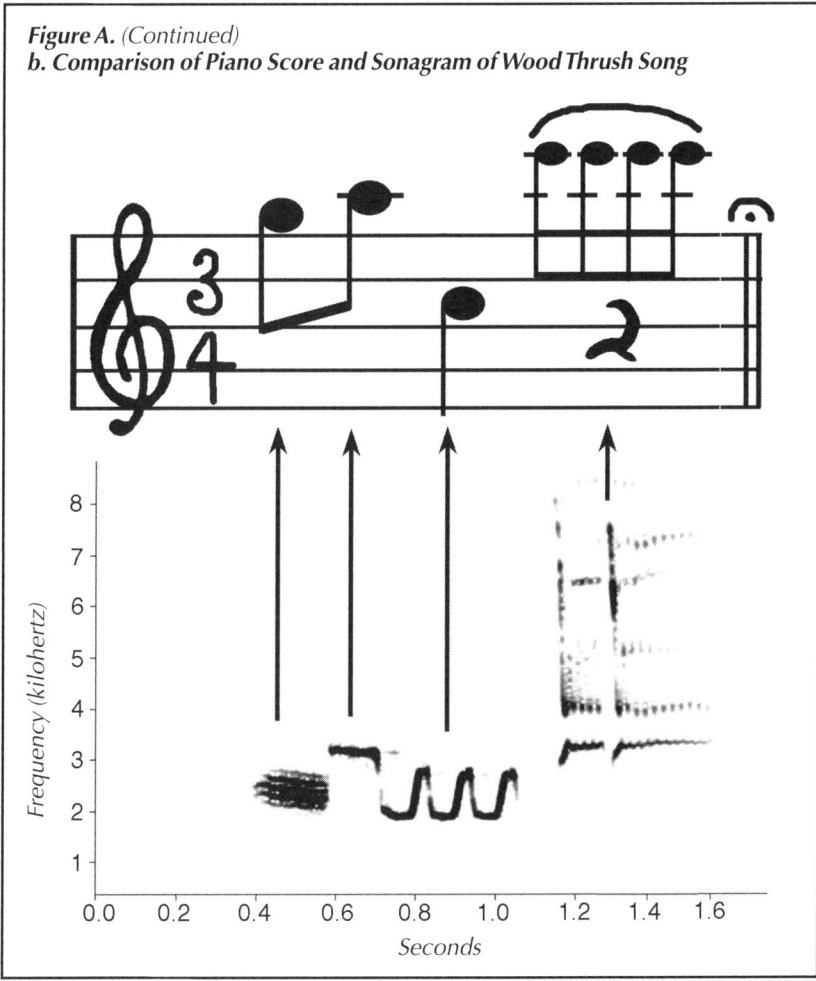

territory a researcher places a speaker just outside the territory boundary and plays a Song Sparrow recording made several miles away (see Fig. 7–32 for Song Sparrow picture). A male Song Sparrow will almost always respond to this experiment by singing and approaching the speaker. Typically, he will continue to sing for long periods if the playback continues. This experiment, repeated for many species, has shown that territorial males consider song sung near their territory by an unfamiliar bird to be a threat, and that the territory holder uses his song to confront intruding males. But *how* do males use their songs during these encounters? What information in the song is important for the communication between the territory holder and his foe? For that matter, how would an intruder use his song in this situation? Traditional playback experiments cannot fully

answer these questions—although they allow the researcher to start an acoustic interaction with a bird, she cannot continue the interaction in any realistic way since she is limited to the series of songs on her playback tape.

In contrast, interactive playback experiments allow researchers to engage the bird in a realistic bout of acoustic interactions. To do this, the researcher first stores many recorded songs on a portable computer. Once she starts an interaction with a bird, she can continue the interaction by choosing from the many songs stored on her computer. Returning to our Song Sparrow playback example, if the researcher is using interactive playback, once she has elicited song from the territorial male she can engage him in different kinds of acoustic interactions. For instance, with a click of a mouse,

the researcher might switch to a song that closely resembles the song that the subject bird is singing, engaging him in "matched countersinging" (see Marsh Wren discussion in Songbird Diversity, earlier in this chapter). Or the researcher can play a longer song than the subject sang, or choose to overlap the subject's song by instructing the computer to play a song while the subject bird is still singing his song.

But do birds really respond differently to the two types of playback experiments? Peter McGregor (Copenhagen University) and co-workers answered this question by measuring 26 different aspects of singing behavior in Great Tits subjected to both traditional playback and interactive playback (in which songs were chosen to match the type and/or structure of the song the subject bird was singing). It turned out that birds approached the speaker in response to either playback experiment, indicating that Great Tits mistook both types of playbacks for intruders. The subjects sang differently in response to the two types of playbacks, however. When interactive playback was used, subjects waited longer before singing, and tended to match aspects of the playback song structure (for example, the number of phrases in the song) more than when traditional playback was used. Another difference was that birds sang shorter songs after the interactive playbacks ended than after traditional playbacks, suggesting that responding to interactive playbacks may be more tiring. Clearly, these birds considered the two types of playbacks to be very different things.

Two other studies have used interactive playback to examine the fine details of how birds interact acoustically. Torben Dabelsteen (Copenhagen University) had observed that territorial male Eurasian Blackbirds matched the "intensity" of their songs (as measured by their singing rate, loudness, and the amount of "twittering" contained in the song) to the level of threat they felt from neighboring males. Using interactive techniques, Dabelsteen then found that his subjects reacted most aggressively when the intensity of the playback song increased in parallel with the intensity of the subject's reaction. Playbacks that presented songs of different intensities in random order elicited less aggression from the subjects. Interactive playback thus enabled Dabelsteen to discover that the order in which an intruder sang songs of different intensities was an important part of the meaning communicated to a threatened territorial male.

Bonnie Nielsen and Sandra Vehrencamp (University of California at San Diego and Cornell University) have used interactive playback to look at how territorial Song Sparrows interact acoustically during intrusions. They hypothesized that male Song Sparrows send aggressive threats to other males by switching song types at the same time as their rival (synchronized song-type switching), and by matching the song being sung by the rival (matched countersinging). They tested both of these hypotheses through a series of interactive playback experiments in which they instructed their computer to engage in synchronized song-type switches, to match the song types sung by the subject, or to do neither. They found that subjects responded with the most aggression when the computer matched their song types, with an intermediate amount of aggression when the computer engaged in coordinated song-type switching without matching, and with less aggression when neither occurred (Nielsen and Vehrencamp 1995). The work of these pioneers in interactive playback already has revealed new layers of complexity in the communication systems of territorial birds. Many more fascinating discoveries await us.

Multichannel Recording

Whether studying naturally occurring bird song, or using traditional or interactive playback, technological limitations have kept scientists from studying the interactions of more than two birds at once. Yet birds do not live in isolation—communication often involves a whole neighborhood of birds. My research, made possible by the engineering feats of computer programmers in the Bioacoustics Research Program at the Cornell Lab of Ornithology, aims to overcome this limitation by recording the vocalizations of an entire neighborhood of birds on multiple, widely spaced microphones. By placing eight microphones along the perimeter of a nine-acre (22-ha) field, I am able to record simultaneously the singing of the 14 territorial Song Sparrows, as well as the songs of many other species. When I return to the lab, I play the eight-channel recording into the computer. During analysis, sonagrams from all eight microphones scroll across the monitor—a visual feast of all the singing from an entire field of birds! **(Fig. B & Track 41)** From these sonagrams, I first document each bird's repertoire of song types so that I can recognize individual singers by their songs alone. With that knowledge, I can examine any tape from my study site and determine which birds are singing, what they are singing, and when they sang. Furthermore, by measuring the differences among the times a given song arrives at each different microphone, the computer can calculate a precise location for each singer. In this way, I am able to preserve and recreate the entire acoustic scene that occurs in my study site.

I am using this technique to examine the acoustic interactions of Song Sparrows when new males challenge established territory boundaries. With my system, I can record how the challenge affects the singing behavior of the terri-

tory holder, as well as monitor the challenger's singing, and the singing of all the other birds in the field. Early results from my study are showing interesting patterns in the singing between the males involved in a territory battle, and among the other birds in the field. Challengers almost always sing at the highest rates in the minutes before they intrude into the established male's territory. Interestingly, the challenged males respond in markedly different ways—ranging from singing in intense bouts at rates that nearly match the challenger, to singing less often and irregularly. These findings raise more questions than they answer. For instance, why would the challenger advertise his attack through a high song rate instead of engaging in a quieter ambush-type attack? Is he using his high song rate to intimidate the bird he is about to challenge? Are challenged territory holders sending specific messages through their differing singing behaviors, and, if so, what are those messages? I hope that questions such as these can be answered by observing more naturally occurring territorial conflicts, and through the use of interactive playback experiments.

If I expand my view to all the birds in the field, I find that the entire neighborhood's song production increases shortly before a territorial battle occurs. Does this mean that the entire neighborhood is aware that a battle is about to take place? Are other males doing the bird equivalent of shouting, "Fight! Fight! Fight!?" A closer look shows that not all birds in the field sing at equal rates. Birds whose territories border the territory of the bird being challenged tend to sing at higher rates than more distant birds. Perhaps these birds are more threatened by the instability that often occurs after a territorial battle, and are sending their own message to the challenger and challenged birds: "Whichever of you loses, don't think about trying to move in on my territory!" Furthermore, neighbors' song rates appear to be closely related to the movements of the challenger. Again, it appears that neighbors may be singing in proportion to their perceived threat from the challenger.

My recordings of an entire field of Song Sparrows are providing some of the first glimpses into the singing dynamics of an entire neighborhood of songbirds.

Computer-Based Song Recognition

A third technique that holds tremendous potential for changing how we study avian communication is computer-based recognition of bird

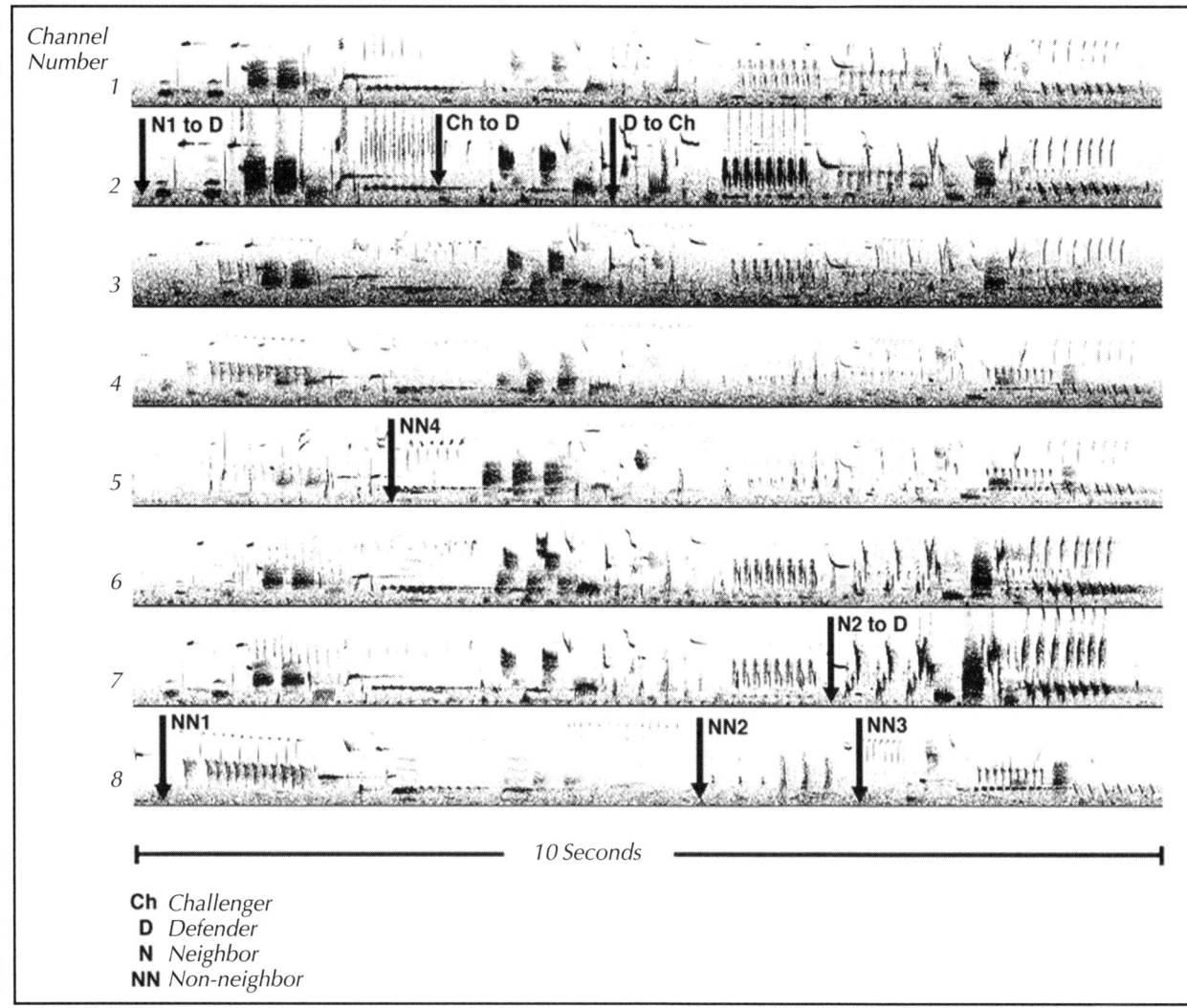

Figure B. Sonagrams of Songs of Several Territorial Song Sparrows Recorded Simultaneously by Eight Microphones (Track 41): This multichannel sonagram is of a 10-second recording made by John Bower at 7:06 A.M. on April 17, 1998 in Ithaca, NY using an array of eight microphones arranged around the edge of a field in which several male Song Sparrows held territories. Simultaneous recordings from the eight microphones are shown stacked from top to bottom, with channel 1 (recorded from microphone 1) on top and channel 8 (from microphone 8) at the bottom. As you read through this visual representation of auditory changes in time and space, you will notice that many of the same songs are recorded at multiple microphones. Computer analysis of the tiny differences in arrival times at different microphones allows the spatial locations of the birds to be calculated.

The recording was made two minutes before a territorial fight between the challenger (Ch) and the defender (D). The challenger was trying to expand his small territory (newly acquired two days earlier) by stealing parts of neighbors' territories. Other Song Sparrows singing in this sonagram were two males (N1 and N2) whose territories were adjacent to those of the combatants, and four non-neighbors (NN1 to NN4) from the other end of the field.

Song Sparrow N1 starts the activity by singing toward the defender (notice how this song is represented strongly in channel 2, less strongly in channels 1 and 3, and weakly in channels 5 through 7). The challenger then sings toward the defender (again strongly picked up in channel 2 but also in other channels) and the defender sings in response. At the same time as this activity, NN1, NN4, and NN2 sing in channels 8 and 5. Neighbor N2 then sings in response to the defender (channels 7 and 6, and represented more faintly in channels 1 through 3 as well). Finally NN3 sings in channel 8, his song appearing in channels 3 and 4 as well.

song. This technique is deceptively simple in concept—the computer compares a recorded song, or some feature of the song, with a stored template to determine which species or even which individual sang the song. In practice, building a template that can distinguish a particular bird's song from natural and man-made background noise is a challenging task. Early development of this technique in the Bioacoustics Research Program is focusing on automated recognition of the calls of nocturnal migrants (see Sidebar 2: Listen Up!), but many other applications are waiting to be tried. For instance, if I want to compare how often each Song Sparrow at my study site sings on days when territorial conflict is high versus days when it is low, I currently need to spend hours and hours wearily "browsing" through tapes stored on the computer. Soon, and much to my relief, the computer may spend the night analyzing my tape while I sleep, greeting me in the morning with a list of who sang what, when, and from where.

Computer-based recognition of bird song also holds great promise for advancing the conservation of Neotropical birds. Ornithologists often warn that the extinction of many bird species is inevitable unless we conserve critical tropical habitat from ever-increasing human destruction. But which habitats are most important to save? A major problem in answering this question is that we know very little about the population distributions of Neotropical birds. Censusing birds in the Neotropics is no easy task. Because visual identification is difficult at best in many tropical habitats, census workers are faced with the daunting task of identifying thousands of Neotropical species by sound alone. Very few ornithologists have the knowledge to do this. This problem was made obvious by the tragic death of Ted Parker, who died in 1993 when the small plane in which he was riding crashed during a recording trip in Ecuador. Ted was arguably the greatest human resource for Neotropical bird song identification who ever lived. His ability to identify bird sounds was legendary. During visits to the Library of Natural Sounds, for example, it was common to sit Ted in a quiet room and play tape after tape of unidentified Neotropical bird sounds recorded by other people. Occasionally Ted would be stumped, but more often he would calmly say something like, "*Phlegopsis nigromaculata,* Black-Spotted Bare-Eye." Ted's knowledge of Neotropical sounds was so complete that he was often able to guess correctly where mystery recordings were made by the complex of species present on the recording. Ted's death was a tremendous loss, but we are fortunate that his legacy of approximately 15,000 carefully cataloged recordings are archived in the Macaulay Library of Natural Sounds. It would be a great contribution to conservation, and an honor to Ted, if we could use those recordings to program a computer with the ability to recognize Neotropical species by their sounds. Such a program would enable tropical bird recordists to record the sounds of a particular habitat and location, and rely on the computer to determine what species were singing at that location. In this way, the knowledge required to make good conservation decisions in the Neotropics would be improved.

These three techniques will be even more powerful when they are used together. For instance, conducting an interactive playback experiment while the microphone array is running will give us a much finer glimpse into how a community of birds responds to the carefully controlled singing of our artificial intruding male. Or, using automated recognition to analyze tapes made with a microphone array in a little-studied tropical habitat may result in a much better look at which animals (birds or otherwise) use that habitat, and where in the habitat they are found. Clearly, exciting times are ahead for the application of new computer technology to the study of avian acoustics. ■

Suggested Readings

Bower, J. L. 2000. *Acoustic Interactions during Naturally Occurring Territorial Conflict in a Song Sparrow Neighborhood.* Dissertation. Cornell University.

Dablesteen, T., and S. B. Pedersen. 1990. Song and information about aggressive responses of blackbirds, *Turdus merula:* evidence from interactive playback experiments with territory owners. *Animal Behaviour* 40: 1158–1168.

McGregor, P. K., T. Dabelsteen, M. Shepard, and S. B. Pedersen. 1992. The signal value of matched singing in Great Tits: evidence from interactive playback experiments. *Animal Behaviour* 42: 987–998.

Nielson, B. M. B., and S. L. Vehrencamp. 1995. Responses of Song Sparrows to song-type matching via interactive playback. *Behavioral Ecology and Sociobiology* 37: 109–117.

Stap, Don. 1994. Remembering Ted Parker. *Living Bird,* Winter 1994, 13(1): 24–25.

of their young to identify them in large flocks of other young Bank Swallows. This kind of parent-offspring recognition seems less well developed in Barn Swallows; young Barn Swallows are less likely to mingle with birds from other nests because their parents don't nest in large colonies like Bank Swallows. Numerous experiments show, too, that in many species, mates recognize each other. Sometimes they do so under conditions that seem overwhelmingly difficult to us, such as in a huge colony of penguins or seabirds. As in humans, recognizing individuals is the foundation for social relationships, and individual recognition by voice can be expected in almost every social situation that birds encounter.

Song Dialects

One feature of variation in the songs of songbirds has attracted special attention: song dialects (Lynch 1996; Payne 1996). We humans have dialects in our speech, of course; Americans all recognize "southern drawls" and r-less Bostonians, and experts can often pinpoint a person's place of origin by the subtleties of these different accents. Songbirds have dialects, too. Dialects in both speech and song are a consequence of vocal learning. Humans learn their speech, and songbirds learn their songs, and if individuals remain at the location where they learned their vocal signals (or if newcomers learn the local dialect), then individuals in a given geographic location come to use the same local dialect.

The term **dialect** is typically used to signify any *clustering* of similar vocalizations that is a consequence of learning. Dialects can thus consist of only a few birds or of thousands, depending on how learning and dispersal affect the distribution of vocalizations in a particular species. Each neighborhood of like-singing Indigo Buntings could be called a dialect, as could the megapopulation of Black-capped Chickadees extending from Maine to British Columbia, whose members sing very similar versions of the *hey-sweetie* song.

Hearing song dialects requires a good ear, but a skillful listener can hear dialect differences in the songs of many songbird species. Dialects have been especially well studied in the White-crowned Sparrows of the coastal chaparral in California (Baker and Cunningham 1985) **(Fig. 7–40)**. There, boundaries between dialects are so sharp that, near Point Reyes Bird Observatory, one can stand facing the Pacific Ocean and hear songs of one dialect to the left and an-

Figure 7–40. White-crowned Sparrow: Many aspects of the White-crowned Sparrow's singing behavior have been well studied, including its **song dialects** (regional differences in song). Because each male sings only one song type, dialects are particularly easy to identify. Photo courtesy of Mike Hopiak/CLO.

other to the right. Distinguishing dialects is especially easy in this species, because each male uses a single song form, and each song thus identifies the dialect (**Fig. 7–41 &** Track 42). Laboratory experiments have shown that males usually learn their songs rather early in life (see Vocal Development in Songbirds, earlier in this chapter), so the dialects clearly result from each male learning his song, then staying within the region where he learned it to breed and defend a territory.

Just how do these sparrow dialects form? As humans who build houses in California chaparral have discovered, sometimes tragically, the chaparral is a fire climax community. Vast stretches of the habitat routinely burn, temporarily destroying good sparrow habitat. The destruction is part of normal renewal, however, and when pockets of suitable habitat again become available, the sparrows reinhabit them. Founding birds can have songs that are incompletely learned or in

Figure 7–41. Sonagrams of White-crowned Sparrow Dialects (Track 42, 2nd Oregon Song, 1st California Song, 2nd Alberta Song): Representative songs from three different White-crowned Sparrow dialects, one in Oregon, one in California, and one in Alberta, Canada. Note that although each song follows the same general pattern of pure tones, buzzes, and other notes, the details of note structure and order vary from song to song.

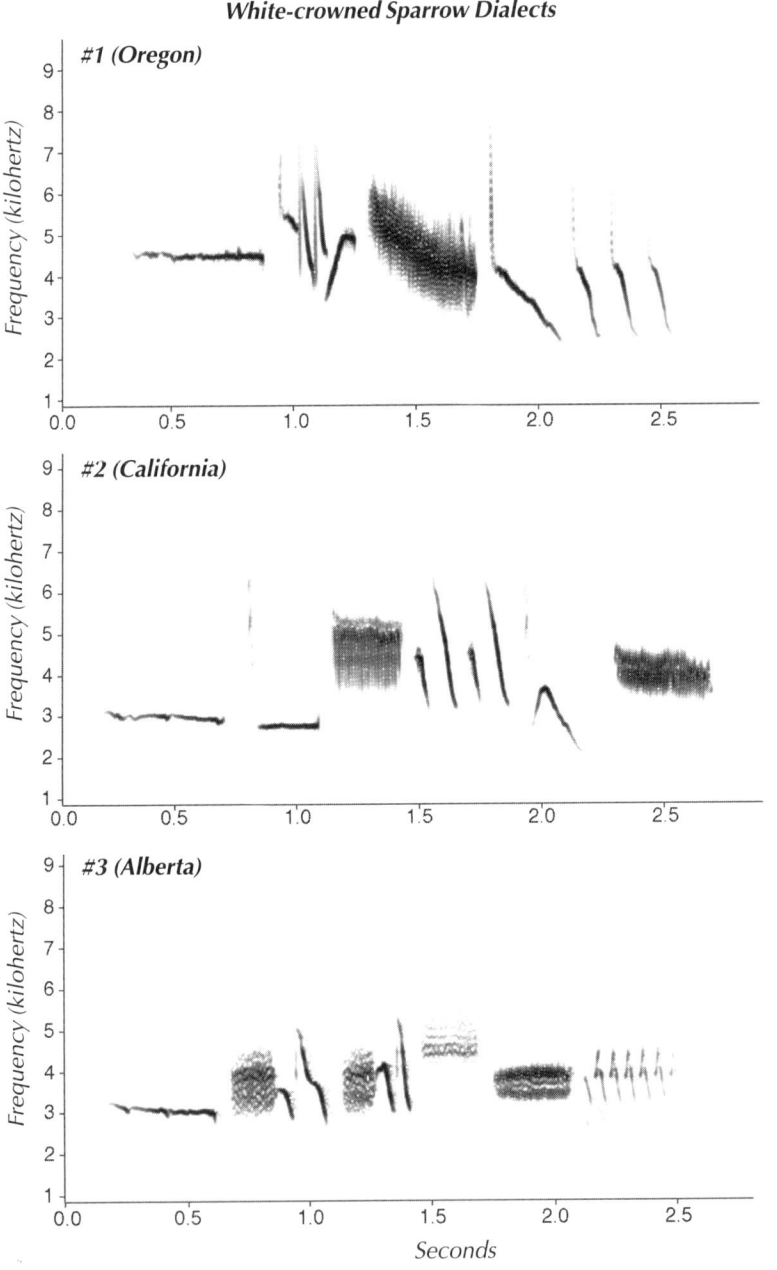

some way different, and these odd songs can then become the basis for a new dialect. As suitable habitat areas expand, they eventually contact other expanding areas of suitable habitat. It is at these locations of secondary contact that boundaries between dialects form **(Fig. 7–42)**.

The formation of dialects in this manner is relatively straightforward. More controversial is what maintains boundaries once they have formed. Dialects and their boundaries remain fairly stable over time, so one can predict that males within a given area will sing a certain dialect. The central question is this: How socially and genetically isolated are the birds in different dialects? The degree of isolation is determined in large part by when and where young sparrows learn their songs and enter the breeding population. Because we know that dialects tend to remain stable, two possibilities exist. First, males may learn their songs before leaving their fathers' territories, keep the same songs in later life, and settle in their fathers' dialect regions. Second, young males may move to new dialect areas and modify their songs to match the new dialect.

Occasionally, females also sing, and like males, young females learn their songs early in life. A female might use her memorized song, whether she sings it or not, to choose a male with the same kind of song. What is especially intriguing is the possibility that most young birds, both males and females, elect to remain in or return to their natal dialect, so that the learned songs promote the isolation of the birds in the different dialects. This kind of isolation could enhance genetic differences among populations, eventually leading to differences so great that the populations become different species. Perhaps, it has been hypothesized, the song-learning ability of songbirds, together with early learning and dispersal more within than between dialect areas, has been instrumental in generating so many songbird species.

Exactly what young White-crowned Sparrows do, however, is largely unknown. Laboratory experiments show that young males certainly are with their fathers long enough to learn their songs, and they learn most readily during that time. But laboratory experiments also show that, under the right social conditions, birds can modify their songs later in life. Banding studies in the field show that some young males remain in their natal dialect area, but some also move across dialect boundaries and learn the songs at the new location. Merely knowing that young males opt for both choices doesn't, however, address the critical question: Does the dialect boundary in any way inhibit dispersal? Do fewer birds cross the dialect boundary than one would expect by chance? If the dialect boundary in any way restricts dispersal, then mating opportunities are also relatively restricted, to birds with a similar singing background. Unfortunately, testing these ideas in the field is extremely difficult, so we still do not know, with confidence, how these dialect boundaries affect dispersal and mating opportunities in the White-crowned Sparrow. Most young songbirds are adept at learning songs while they are with their fathers, but the influence of the father's songs on dispersal and on the range of songs that a young male (or in some cases, female) finds acceptable remains to be determined.

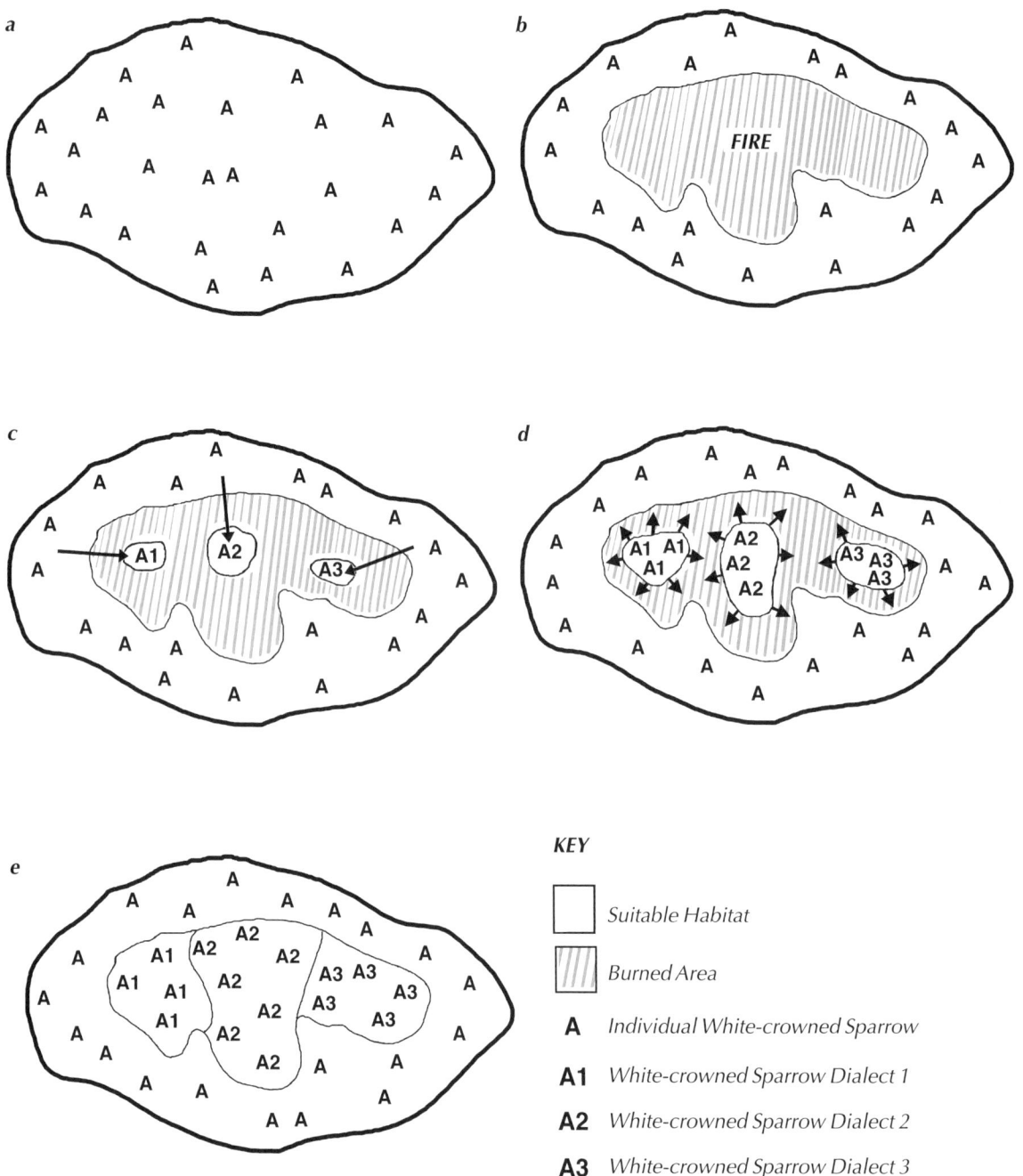

KEY

☐ Suitable Habitat

▨ Burned Area

A Individual White-crowned Sparrow

A1 White-crowned Sparrow Dialect 1

A2 White-crowned Sparrow Dialect 2

A3 White-crowned Sparrow Dialect 3

Figure 7–42. Development of White-crowned Sparrow Dialects: White-crowned Sparrow dialects occur in small patches with sharply defined boundaries in the coastal chaparral (dry areas with low shrubs) of California. These dialect patches develop in concert with the natural cycle of fire and regrowth that maintains the habitat: a. Established population of White-crowned Sparrows. b. Fire temporarily destroys a large part of the suitable sparrow habitat. c. Over time, isolated pockets of suitable habitat develop within the burned area as vegetation regrows. Dispersing White-crowned Sparrows (A1, A2, and A3) colonize these areas of renewal. If a founding sparrow's song happens to be slightly different from that of the original population, it can become the basis for a new dialect. d. As suitable habitat expands, the colonist populations of White-crowned Sparrows increase. Still isolated from the original population, their young learn only the local songs. e. Eventually the subpopulations (A1, A2, and A3) contact other areas of expanding suitable habitat; it is at these locations of secondary contact that boundaries between dialects form.

Pacific-slope Flycatcher

Cordilleran Flycatcher

Geographic Variation in Suboscine Vocalizations

Dialects occur in learned vocalizations, such as the songs of songbirds, but how much do nonlearned vocalizations change over geographic space? Very little, it seems. The songs of the Alder Flycatcher are an unmistakable *fee-BEE-o* from Maine to British Columbia. No detectable local variation occurs in these songs. The same general pattern is found among other suboscines. The songs of these birds are not learned, but instead seem to be a more or less "direct readout" of the genes. Apparently, the genes that dictate the *fee-BEE-o* of the Alder Flycatcher don't vary from coast to coast in North America.

Because songs are a good indication of a bird's genetic background, songs of flycatchers and undoubtedly other suboscines can help us to determine which populations of birds belong to which species—that is, the songs are good "systematic" characters. During the 1960s, for example, investigators discovered that the two groups of the Traill's Flycatcher differed consistently in voice, suggesting that the two groups did not interbreed. The voice difference helped to identify the two populations as two species, the Alder Flycatcher and Willow Flycatcher. More recently, researchers recognized that the Western Flycatcher consisted of two different groups, each with its distinctive songs (Johnson and Marten 1988). Study of genetic characters showed that the songs identified different genetic populations, and the two groups were recognized as distinct species, the Pacific-slope Flycatcher and the Cordilleran Flycatcher (**Fig. 7–43**).

Figure 7–43. Pacific-slope Flycatcher and Cordilleran Flycatcher: Because most suboscines do not learn their songs, little regional and individual variability occurs among groups that interbreed. Differences in songs, then, may indicate genetically separate groups, and can be valuable clues to separating species that appear very similar in other ways. Until 1989, these two birds of western North America were both considered to be Western Flycatchers, but differences in their songs alerted researchers to their other differences. Finding that they were also different genetically allowed researchers to separate them into the coastal Pacific-slope Flycatcher and the more interior Cordilleran Flycatcher. Photos by R. and N. Bowers/VIREO.

The songs of suboscines can be especially useful in charting the unknown avian biodiversity in the tropics (Kroodsma et al. 1996). A remarkable series of studies throughout the Neotropics by Wesley Lanyon used the songs of *Myiarchus* flycatchers (the group containing the Great Crested Flycatcher of North America) to help identify the different species in the group (Lanyon 1978). We know very little, however, about antbirds, woodcreepers, flycatchers, cotingas, manakins, and other suboscines from the New World tropics. Systematists who classify bird species currently differentiate most species on the basis of plumage patterns or other morphological characters, but experience has shown that such characteristics can be misleading **(Fig. 7–44)**. Plumage patterns might not vary much over a vast geographic area simply because the habitat favors a certain type of cryptic pattern. Behaviors such as songs, which are used to identify mates for breeding, could be a more reliable indicator of which birds are capable of mating with one another, which in turn helps us to define species. Tape-recording and analyzing songs from these Neotropical suboscines may thus be the fastest way to chart their diversity, and speed is important, because habitats and the birds themselves are disappearing rapidly.

The Diversity of Geographic Patterns in Songbirds

Some songbird songs appear in geographic patterns other than the sharply defined dialect areas of the White-crowned Sparrow. And, just as the functions and consequences of White-crowned Sparrow dialects remain unknown, so do the reasons for these other patterns (which are described below). In some way, these different patterns must

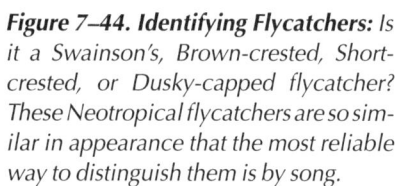

Figure 7–44. Identifying Flycatchers: Is it a Swainson's, Brown-crested, Short-crested, or Dusky-capped flycatcher? These Neotropical flycatchers are so similar in appearance that the most reliable way to distinguish them is by song.

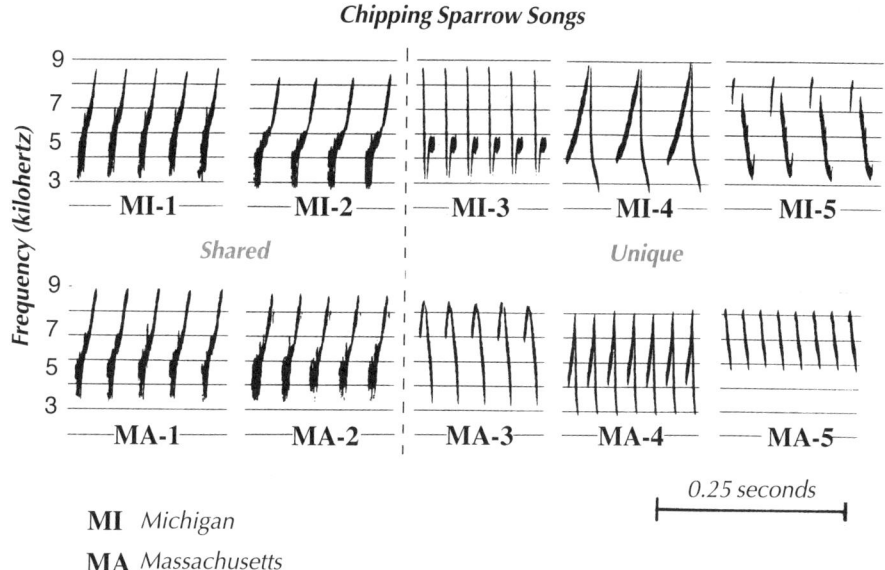

Chipping Sparrow Songs

MI   *Michigan*

MA   *Massachusetts*

make sense—a bird who shares a vocal signal with either immediate neighbors or more distant birds must gain some social advantage that leads to a particular pattern of geographic variation. But learning what these social forces and advantages are awaits further field work.

In some species, the same songs recur throughout the geographic range of the species. Within any population of Chipping Sparrows, for example, many different song types exist with little sharing among males, but songs found in one population can also occur in other populations across North America (Borror 1959). It seems that Chipping Sparrows sing a limited number of song forms, each of which could perhaps occur in any population; these song forms are distributed not in local dialects but throughout the range of the species **(Fig. 7–45)**. Also widely distributed are the hundred or so components from which the Indigo Bunting song is constructed; local dialects in buntings are formed not so much by the types of elements used, but by their particular combinations.

The whistled *hey-sweetie* of the Black-capped Chickadee has an especially puzzling geographic distribution. Like the Alder Flycatcher, the chickadee uses the same song from Maine to British Columbia. Patterns of whistles are the same in both frequency and amplitude, and birds use the same rules to transpose their songs in frequency. Such consistency in a learned song across an entire continent is remarkable. How the chickadees maintain such a stereotyped learned signal over such a vast geographic expanse remains a mystery; the more typical songbird pattern is to show at least some form of local variation.

Some intriguing local populations with song forms other than the *hey-sweetie* do occur, however. One is on Martha's Vineyard, a small island off the coast of southeast Massachusetts (Kroodsma, Byers, et al. 1999). There, males also whistle, but the two whistles are on the same frequency, with no drop between the first and second whistles.

Figure 7–45. Chipping Sparrow Songs from Two Populations: Individual male Chipping Sparrows each sing only one song type—a trill made up of one simple, repeated syllable. The song type comes from an apparently limited pool of song types used by all the populations of this species. The song types are not distributed in dialects, but rather throughout the range of the species, such that in any population, a few songs will be shared among several males, but most will be sung by only one male. Illustrated in the figure are five songs from a Michigan population (MI-1 through MI-5) and five songs from a Massachusetts population (MA-1 through MA-5). Within a given population, some songs are sung by only one bird (songs 3, 4, and 5) and some are shared or at least very similar (in Michigan, MI-1 and MI-2 are the same; in Massachusetts, MA-1 and MA-2 are the same). Between populations, the same is true: songs MI-3, MI-4, MI-5, MA-3, MA-4, and MA-5 are each unique, but some songs are found in both populations: MI-1 and MA-1 are the same, and MI-2 and MA-2 are the same. In this example it is simply by chance that the same two songs shared within populations (songs 1 and 2) are also shared between populations. Note that the "songs" shown are actually just short segments from the entire songs—in actuality, Chipping Sparrow songs consist of much longer trills, repeating the same syllable around 20 or more times. Adapted from Kroodsma (1996a).

Furthermore, song dialects occur on the island, so that males in different areas use different songs **(Fig. 7–46 &** Track 43**)**. On the far western end of the island, at Gay Head, songs typically consist of two whistles, both on the same frequency, but with an amplitude break in the first, not the second, whistle. Instead of *hey-sweetie*, the males sing *sweetie-hey*, and the males sing this song on two frequencies, one noticeably higher than the other. On the eastern end of the island, near Edgartown, however, males have more complex songs, with two amplitude breaks in the first and one in the second whistle *(swesweetie-sweetie).* That song also occurs on a high and low frequency. In other parts of the island, males sing a *sweetie-sweetie*, and sometimes the high frequency song is different from the low frequency song; a typical combination throughout the center of the island is for a male to sing a high frequency *sweetie-sweetie* and a low frequency *sweetie-hey*. A second pocket of song differentiation occurs in Oregon and Washington, where males sing a bewildering variety of whistled songs. The patterns on Martha's Vineyard and in Oregon and Washington thus differ markedly from those elsewhere in North America.

How do these regional differences in song arise? Martha's Vineyard is about four miles off the coast of New England, so birds there may

Figure 7–46. Sonagrams of Black-capped Chickadee Songs from Martha's Vineyard (Track 43, 5th and 6th Western Songs, 3rd Eastern Song): On the island of Martha's Vineyard, Black-capped Chickadee dialects exist that are very different from typical Black-capped Chickadee song. At the far western end of Martha's Vineyard, at Gay Head (upper sonagram), males tend to sing each song all on one frequency, with the amplitude break in the first whistle rather than the second(sweetie-hey). Birds sing the song on two different frequencies, as illustrated in the sonagram. At the eastern end of the island, near Edgartown (lower sonagram), males also sing each song all on one frequency, but have two amplitude breaks in the first whistle, and one break in the second (swesweetie-sweetie). The first swe is not clearly visible in the sonagram, existing as just a small blip on the first "note." It would be easier to distinguish in an oscillogram. Eastern birds, like western birds, sing the song at two different frequencies, but only the higher frequency song is illustrated here.

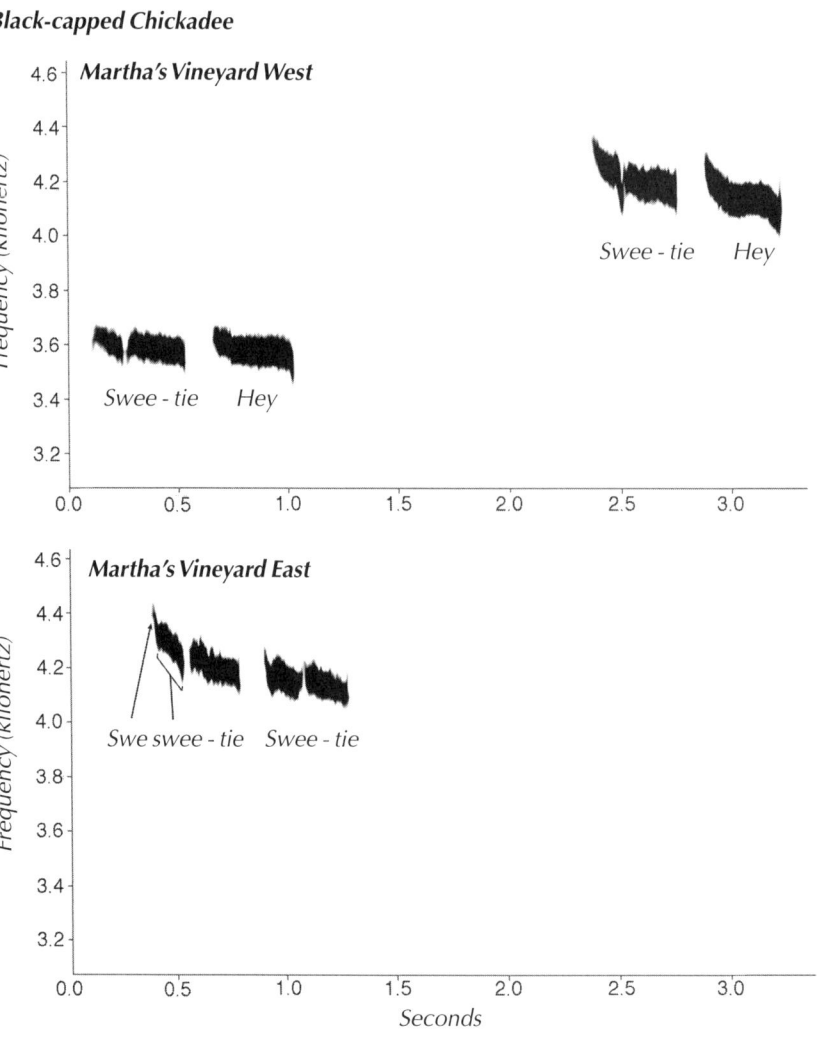

Black-capped Chickadee

be relatively isolated from mainland North America. Perhaps isolated pockets of chickadees also occurred in some western states, so songs in those areas also had a chance to diverge from the standard *hey-sweetie*. We don't know whether these vocally distinct populations are genetically distinct, or if a female of one tradition would accept a male from another.

Work with chickadees in the laboratory has only enhanced the mysterious nature of chickadee song (Kroodsma et al. 1995). If male chickadees from normal *hey-sweetie* populations are tutored with normal *hey-sweetie* songs in the laboratory, they develop highly abnormal songs, unlike anything one would expect from a chickadee in nature. Although the whistled tonal quality remains, the songs consist of from one to seven or more different whistles on a variety of frequencies (Track 44). Furthermore, males develop repertoires of up to three different song forms. Groups of males isolated from one another even develop different "dialects." In the laboratory, the true potential for chickadee singing is unleashed, as it seems to be, at least in part, on Martha's Vineyard. Over most of the North American continent, some social factors must restrict song variation to the simple *hey-sweetie*, transposed in frequency; in the highly artificial laboratory environment, those forces are absent, permitting both song repertoires and dialects. But what are the social forces? Is it the female who sets standards for what songs are to be sung? Among songbirds, the female is believed to select a mate, not vice versa. Female Brown-headed Cowbirds respond differently to the different songs of males, and in that way "instruct" males about which songs to sing to be especially successful in acquiring a mate (West and King 1996). Perhaps the female chickadee, too, has certain inviolate standards for her mate's song. Or do male singing contests in some way limit variation on mainland North America? Future work must, in some way, ask the chickadees to help answer these questions.

The diverse patterns of geographic variation are puzzling, but some hints as to why songs vary in these ways are found in certain warblers, and, perhaps surprisingly, the Black-capped Chickadee again (Kroodsma 1996a). Among certain warbler groups, males have two categories of song forms (Kroodsma 1989; Spector 1992). One category seems to be used especially with females. Examples include the *beee-buzzzz* of the Blue-winged Warbler **(Fig. 7–47)** (Track 45) and the *pleased-pleased-pleased-to-MEETCHA* of the Chestnut-sided Warbler **(Fig. 7–48)** (Track 46). Sonagraphic analyses show that these song forms vary little over the entire geographic range of the species. The birds use a second category of song at dawn and in highly aggressive situations, when males are countersinging with one another, especially near territory boundaries. These songs occur in dialect patterns, with songs often changing over short distances.

Figure 7–47. Blue-winged Warbler: Blue-winged Warblers, Neotropical migrants that breed in brushy fields in the northeastern United States, have two different categories of songs. One, the well-known bee-buzz, is used most in the presence of females, and varies little throughout the range of the species. The other type of song, used at dawn and in very aggressive situations, is more variable and occurs in dialects. Photo by Lang Elliott.

Figure 7–48. Chestnut-sided Warbler: Like the Blue-winged Warbler, male Chestnut-sided Warblers have two different song categories. The familiar pleased-pleased-pleased-to-meet-cha songs, often termed the "accented-ending songs," are used in association with females, and vary little throughout the bird's range. The so-called "unaccented ending song" is used at dawn and in highly aggressive situations, and occurs in dialects. Migrating from the Neotropics to the northeastern United States and Canada, these birds breed in second-growth deciduous woodlands and along forest edges. Photo by Lang Elliott.

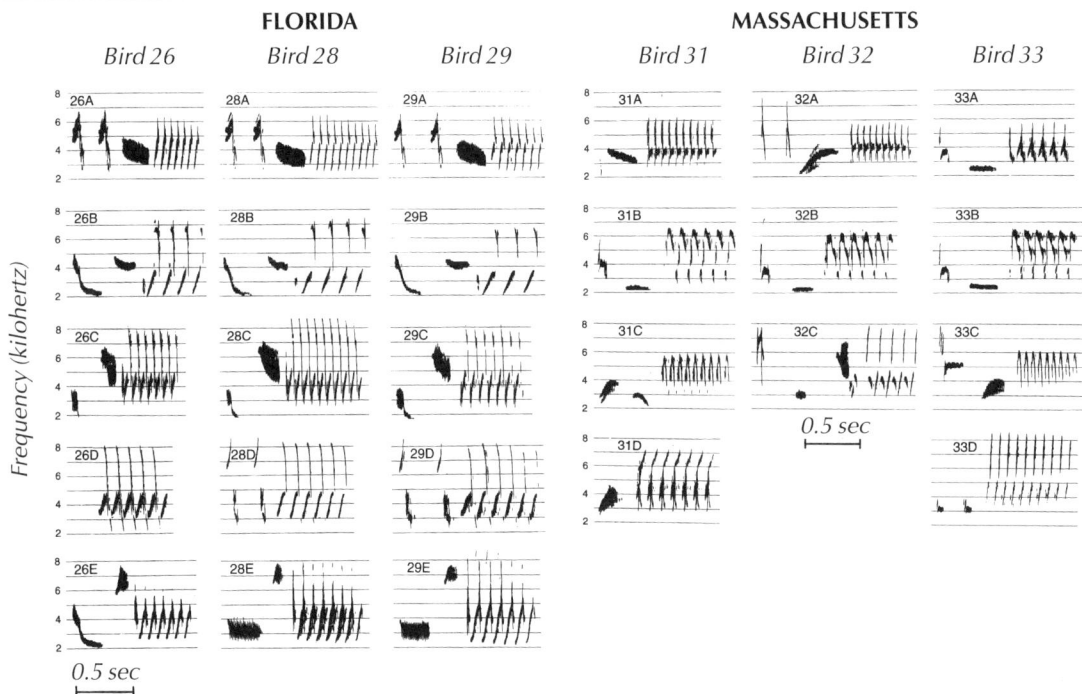

FLORIDA

Bird 26   Bird 28   Bird 29

MASSACHUSETTS

Bird 31   Bird 32   Bird 33

a. Highly Similar Song Types Among Resident Towhees   *b. Less Similar Song Types Among Migratory Towhees*

Figure 7–49. Song Sharing Among Migrant and Resident Eastern Towhees: Within some songbird species with song repertoires, as demonstrated here with the Eastern Towhee, a greater degree of song sharing occurs among populations that are resident than among breeding populations that are migratory. a. Sonagrams of Highly Similar Songs of Neighboring Resident Males from Corkscrew Swamp Sanctuary, Florida: Five songs each are shown from Birds 26, 28, and 29 (columns), although their complete repertoires are slightly larger. Similar song types are compared in rows. Song types A, B, and C are identical in all three birds, and song types D and E are shared between birds 28 and 29. Song types 26D and 26E are unique. b. Sonagrams of Less Similar Song Types of Neighboring Migratory Males from Western Massachusetts: Complete song repertoires of each bird are shown, but only one pair of songs is identical: 32B and 33B. Terminal trills are identical only in the following pairs of songs: 31A and 32A; 32B and 33B; 31C and 33C. Adapted from Ewert and Kroodsma (1994).

Among these warblers, it seems that a male attempting to attract a female succeeds best if he has a highly stereotyped signal that all females can recognize. Females of many songbird species tend to disperse farther than males, so an especially stereotyped male signal will be unmistakably recognizable to all females, no matter where they fledged. When males address other males on neighboring territories, however, some advantage must accrue to having a local signal. Exactly how males sing just a few territories away might not matter, as long as males remain on or return to their own territories year after year, as they often do.

The Black-capped Chickadee helps to solve this puzzle, too. During the breeding season, the whistled song seems to be used (in part) to attract a mate; unpaired males, for example, sing their *hey-sweetie* all day long. After pairing, they use this song less often. Another, more complex vocalization, the gargle, is used in especially aggressive situations. This call varies locally, from one population to the next.

In the warblers and chickadee, then, the vocalization used in more aggressive contexts varies from place to place, in dialects. Far more stereotyped over geographic space are the songs used when a male is unpaired and prospecting for a female. Because these two songbird groups are in different evolutionary lineages, this pattern must have evolved independently at least two times. Together, these groups strongly support the idea that the variability of signals over space relates to their function.

These patterns of variation actually express patterns of familiarity or hoped-for familiarity among individuals. By singing local songs in aggressive contexts, males identify the neighborhood of males in which they hope to have influence. If a male's songs match those of the local neighborhood, he announces that he has a history in that neighborhood and belongs there; others listening know he has survived in the local area long enough to be an acknowledged resident. By advertising stereotyped signals familiar to all birds of their kind, on the other hand, males hope to attract a female, regardless of her locality of origin.

The same patterns of familiarity are undoubtedly expressed in the geographic distribution of songs of other species. The Eastern Towhee provides a good example **(Fig. 7–49)**. At the Archbold Biological Station in Florida, the Eastern Towhee is believed to be a resident species. Males probably stay on their territories throughout their lives and come to know each other well. These males use song repertoires that almost perfectly match the songs of their neighbors. Furthermore, songs change over only short distances, so males within small neighborhoods can match each other with their songs but cannot match males even from nearby areas. In contrast, Eastern Towhees of the northeastern United States are migratory and are on their territories only a few months each year. Consequently, these birds are undoubtedly less familiar with each other, and the songs of neighboring males differ considerably. The tight vocal communities of Florida are nowhere to be found. Residency and familiarity thus seem to be correlated with the tight vocal communities of males who use the same song repertoires (Ewert and Kroodsma 1994).

Song Change Over Time

Because song is learned, it has a high potential to change over both space and time. Our examples have shown that song patterns in space vary considerably from species to species, and even within species. How signals change over time also varies considerably among species.

The most impressive long-term study of how bird song changes over time focused on the Indigo Bunting (Payne 1996) **(Fig. 7–50)**. Throughout their entire geographic range, these buntings use about a hundred different song elements. Each male uses only about six of the hundred to produce his song, however, and the combinations of those elements vary locally, so small neighborhoods occur in which local males come to use the same combination of elements (see Fig. 7–31).

How these dialects form, who copies whom, and how songs change over time are especially fascinating. Consider what you might hear over a summer if you monitored a local bunting population. Buntings migrate, living during the North American winter in the southern United States and on islands in the Caribbean. Older male buntings, those two years and older, typically return to the same territories as in previous years, and their songs usually do not change from year to year. When yearlings (birds hatched the previous year) return to breed, however, their songs are initially unique, typically unlike the songs of any males from the previous year, including their fathers. During their first adult year, many of these yearlings change their songs to match those of the older birds in their immediate neighborhood. In this way, small pockets of birds, all within earshot of one another, come to sing highly similar songs.

Figure 7–50. Indigo Bunting: Breeding along woodland edges and clearings, and in old fields with young trees throughout most of the eastern, central, and southwestern United States, these Neotropical migrants have often been the focus of bird song researchers. Of particular interest is how their dialects develop and change over time. Photo by Lang Elliott.

Figure 7–51. Marsh Wren: *Breeding in cattail and bulrush marshes throughout most of the northern United States and southern Canada, the small, brown Marsh Wren looks much the same from coast to coast. But song differences between eastern and western birds reveal two distinct subgroups that do not appear to interbreed. Males often perch atop a cattail to deliver their bubbly, squeaky songs. Photo by Marie Read.*

Not all copies of bunting song are exactly like the model, however. Perhaps an element is added, or perhaps one is dropped. Thus, annual changes accumulate in a given song as it is copied from male to male. New songs also are introduced to a population when yearling males keep their unique songs, as they sometimes do, or when older males from other locations immigrate into a population. All of these factors, especially the new songs of immigrating males, contribute to a rapid turnover of song forms within local populations. Each year, many songs are introduced and many songs are lost in a local population, and only a few are copied and maintained from one year to the next. The net result of all these processes is that bunting songs in a given neighborhood continually and relatively rapidly evolve over time. Almost a complete turnover of songs occurs in a given neighborhood over a period of 10 or so years.

Songs of most species don't change at such a rapid pace, however. Songs of the White-crowned Sparrows in certain California dialects have remained highly stable over several decades. As more and more species are studied, and as locations are revisited in the future, we will better understand how and why songs change or don't change over time.

Dialects Over Broad Regions

Distributions of nonlearned songs, such as those of flycatchers, provide good clues about evolutionary histories and species relationships, but the learned songs of songbirds also can provide information about evolutionary history (Martens 1996). Several informative patterns of learned song occur in North America, and the Marsh Wren provides a good example (Kroodsma 1983) **(Fig. 7–51)**. Field guides typically identify a single species of Marsh Wren, because the birds look about the same from coast to coast, but careful listening reveals two distinctly different Marsh Wrens in North America. Near Stanton and Norfolk, Nebraska, for example, males typically introduce their songs with a nasal buzz and then produce a relatively musical series of repeated notes. From there to the Atlantic Ocean, that basic song pattern is consistent; even though a male sings fifty or more different song types, each type is of this basic formula **(Fig. 7–52 & Track 47)**. Just a few miles to the west of Stanton and Norfolk lies Erikson, and farther west the Valentine National Wildlife Refuge. Birds at those two locations use songs very different from those of their eastern relatives. These western songs are loud, raucous, buzzy, noisy, and coarse, and some include loud, penetrating whistles modulated rapidly in frequency. Most striking is the tremendous diversity of songs. Never are songs introduced by the faint buzzy note of the east, and often songs include or end with a raspy, noisy note **(Fig. 7–53 & Track 48)**. Further analysis

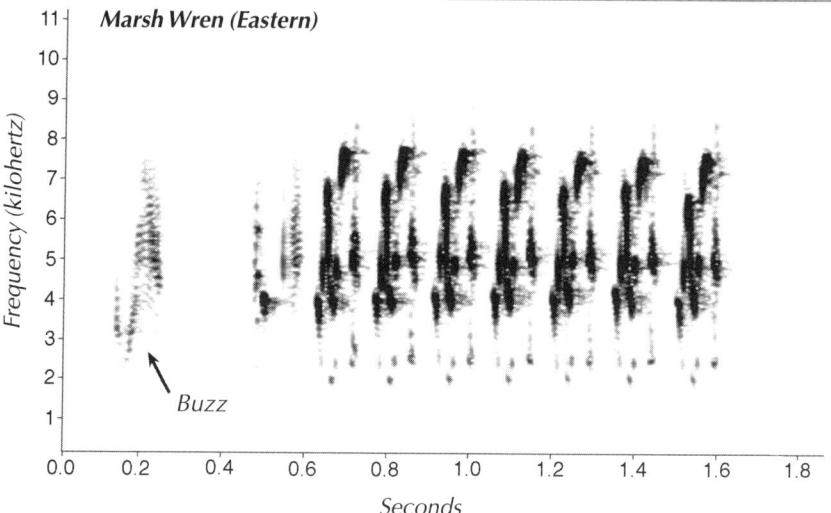

Figure 7–52. Sonagram of Eastern Marsh Wren Song (Track 47, 8th Song): From Stanton and Norfolk, Nebraska east to the Atlantic Ocean, Marsh Wrens typically start their songs with a nasal buzz, then give a musical trill that sounds somewhat like a stick being run quickly along a picket fence. Although each male may sing 50 or more different songs, each follows this basic pattern.

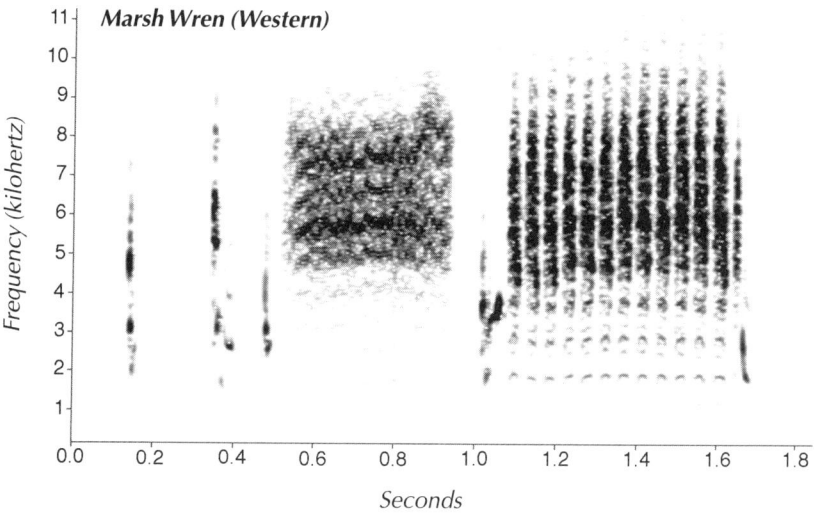

Figure 7–53. Sonagram of Western Marsh Wren Song (Track 48, 4th Song): From Erikson, Nebraska west to the Pacific Coast, Marsh Wrens have loud, raucous, buzzy, and coarse songs, often ending with a raspy, noisy note. Unlike the songs of eastern Marsh Wrens, they do not begin with a faint buzz. Males may sing more than 150 different songs.

reveals another major difference. The western males sing about three times as many songs as the easterners (roughly 150 versus 50). In south-central Saskatchewan, Canada, these eastern and western males occur in the same marshes along the Qu'Appelle River.

The simplest explanation for the origin of these regional differences lies in the distribution of habitats during periods of North American glaciation. When glaciers advanced, they must have isolated two Marsh Wren populations, one in the west and the other in the east. Over thousands of years, the song differences we hear today developed in these two isolated populations. As the glaciers receded, the available habitat increased, and the two populations then met in the central Great Plains. Even though males of the two groups can learn each others' songs, as has been shown in the laboratory, in nature they tend to maintain their distinctiveness, and few hybrid singers occur.

What happens at these zones of secondary contact depends on how much each group has evolved during isolation (Rising 1983). The two Marsh Wrens appear sufficiently distinctive that they breed primarily within their own groups, and future work may confirm that we should recognize two species. The same pattern has occurred in other taxon pairs, such as Eastern and Western meadowlarks, Eastern and Western kingbirds, Indigo and Lazuli buntings, Bullock's and Baltimore orioles, and Rose-breasted and Black-headed grosbeaks. Each of these eastern and western forms is currently recognized as a separate species; presumably the populations evolved sufficient differences during isolation to prevent mating in most cases when they came back in contact. Current field guides lump the eastern and western counterparts of the Northern Flicker into a single species, suggesting that the evolutionary changes were not as great for this bird. Songs of the kingbirds and flickers are not believed to involve learning, and their songs would therefore reflect genetic differences among the populations. Although songs of the songbirds (wrens, meadowlarks, buntings, orioles, grosbeaks, and towhees) are learned, the learned traditions are sufficiently stable that they still reflect past evolutionary history. As illustrated by the laboratory-reared Black-capped Chickadees who sang truly innovative songs and by the eastern or western Marsh Wrens who learned songs of the other wren, these songs have a strong potential for change. Yet some social forces, undoubtedly those that involve individual attempts to manage one another, seem capable of conserving the learned qualities of songs over extraordinarily long periods of time.

Other recently discovered examples occur in North America, too, such as the Winter Wren (Kroodsma and Momose 1991). Males of eastern North America have one to at most three different, relatively complex songs. Males of western North America have even more complex songs and much larger repertoires. We don't know where the dividing line is between the two types or whether they interbreed, but they behave very differently, and future field guides will probably acknowledge these differences and identify them as two different species.

The Functions of Song

Most investigations of the function of bird vocalizations have focused on the loud "songs" of songbirds. These are the most extravagant and noticeable sounds birds make, and it is no wonder that they have attracted so much attention. Just how did such "songs" evolve, and what good are they, anyway? How are they used? What are their functions? **(Sidebar 4: Do Birds Think?)**

Two broad functions for bird song are generally accepted. One is that songs help to defend a territory. If one plays a song of a Chipping Sparrow from within a Chipping Sparrow territory, the local territorial male responds aggressively. Males of many species go so far as to at-

(Continued on p. 7·69)

Sidebar 4: DO BIRDS THINK?
Donald E. Kroodsma

Please Note: In this sidebar, I have used words like "mind," "think," "reason," "saying," or "choose." These are controversial words and are avoided by some scientists who believe these words are too anthropomorphic—attributing human abilities to animals. To be more objective, for example, some would argue that we should substitute the words "brain and behavior" for "mind." I respect those concerns, yet find it convenient to think anthropomorphically. We often can reliably predict the behavior of animals when we attribute human abilities to them. An animal might behave, for example, as if it had a mind that thought or reasoned carefully about the possible choices and their consequences, and then chose an appropriate response. The danger lies in our accepting these descriptions as an explanation of why the animal behaves the way it does, rather than just as a tool to help us predict how animals will behave.

Just what goes on in the "minds" of birds as they communicate with one another? When male Marsh Wrens hurl songs back and forth to one another, or when warblers "choose" to use an appropriate song from their repertoire, are their minds simply acting like programmed robots? When a female hears a male sing, do her mind and body react in some preprogrammed fashion? Do birds respond and counter-respond by using some mechanical, turn-taking rules? Are these interactions just an example of "stimulus and response"? I don't think so.

Scientists are becoming increasingly interested in trying to understand how the minds of animals work. This exciting new area of study in animal behavior is termed "cognitive ethology." We know that, physiologically at least, other animals function much like we do; our bodies all follow certain natural rules in duplicating genes, processing foods, and the like. Perhaps, too, the brains of other animals share many traits with our brains, and perhaps other animals "think," too. We're confident, of course, that our thinking and logic are superior, both qualitatively and quantitatively, and that no other animals are as "self-aware" or "self-conscious" as we are. The goal of much current research is to identify just how different the minds of other animals are.

In certain situations, we know that birds act *as if* they can weigh multiple factors, think about possible

reactions, and then choose an appropriate response. The most impressive example is Alex, an African Grey Parrot, who has been taught to use English to communicate with his trainers; his ability to reason and think is truly impressive (see Ch. 6, Sidebar 1: Bird Brains). He can, for example, examine an object and identify its color, shape, or material. Even more striking, he can survey a tray of objects and identify how they are all similar (for example, color) or all different (for example, shape), but if asked "What's similar," and all the objects are different, he'll respond by saying "nothing." He thus also knows the absence of a quality, an ability that we think of as an abstract concept. Alex clearly begins to threaten some of our unfounded notions about "birdbrains," and at the same time forces us to rethink just how different our own abilities are from a parrot's. Parrots are undoubtedly not unique among birds, and other birds must have some of these same abilities. Ravens, for example, seem remarkably intelligent (Heinrich 1995). Once other birds have been taught to use English or another communication system that we can understand, perhaps we will begin to grasp the differences both among birds and between birds and humans. Eventually, we also hope to improve our understanding of what birds are saying to each other during their daily activities.

As we ponder the mental powers of birds, we begin to realize that we need to know more about how they

behave. Before we can understand how their brains are functioning, we need to describe more carefully just how birds interact with one another. When Marsh Wrens countersing with identical songs, for example, we need to know answers to a whole series of questions. When does this matched countersinging occur? What is the social relationship of the two (or more) males engaged in this behavior? Who leads and who follows in these exchanges? How do these exchanges vary during the day or throughout the season? Does mating status, age, relative dominance, or how well the interactants know each other affect how each male participates? Those of us who study birds are convinced that these wrens must weigh many of these factors and then choose an appropriate way in which to interact with singing neighbors. But, as scientists, we realize that we must go to the field and collect data on how the birds behave. Only when we have fully described these complex interactions will we begin to understand how the wren's mind sees the world.

Consider one final example based on research by Munn (1986). In Peruvian rain forests, flocks of birds move throughout both the understory and the canopy. The composition of each flock is highly predictable, and consists of a dozen or so mated pairs, each of a different species. Foraging behaviors differ among the species within each flock, with the foraging habits of each mated pair

Figure A. White-winged Shrike-Tanager Gives a Deceptive Alarm Call to Gain a Foraging Advantage: White-winged Shrike-Tanagers lead permanent mixed-species flocks in the rain forest canopy of the Peruvian Amazon Basin, maintaining flock cohesion by means of loud vocalizations throughout the day. The shrike-tanager obtains food by capturing insects flushed by the flock's movements. It also acts as flock sentinel, giving alarm calls at the approach of bird-eating hawks and thus warning its flockmates, which react by freezing in place or diving for cover. Ornithologist Charles A. Munn noticed that during aerial chases of insects by one or more of the flock's members, the shrike-tanager sometimes joined in the pursuit and gave its predator alarm call—even though no actual predator was present. He suggests that the shrike-tanager is "crying wolf," giving the alarm call deceptively to distract the other birds and thus enhance its own chance of capturing the insect prize. Here, a White-winged Shrike-Tanager (left), giving deceptive alarm calls, follows a Yellow-crested Tanager (right), which is distracted from chasing a large insect, while other members of the foraging flock react to the false alarm by taking cover.

probably complementing those of other pairs. One pair is of particular interest here—the "sentinel" species. Individuals of this species often perch in the open, sallying out to capture prey items flushed by their flockmates **(Fig. A)**. These sentinels are so named because they are the first to spot predators, and an alarm by the sentinel alerts flock members to impending danger. On occasion the sentinels seem to sound an alarm when no predators are present. As a result, others in the flock freeze or take cover, allowing the sentinel to sally forth and capture a prey item that might have been hotly contested had

others in the flock not been distracted by the false alarm.

If we were the sentinel bird, we might be thinking something like the following: "The dangers from predators are high, and my flockmates must attend to my alarm calls. I'll be honest most of the time, so that I remain credible, but on occasion I can 'fib'. My false alarms will be relatively few, so I won't be found out. An inquiry might be made, but the jury won't be able to distinguish between slight incompetence, a trigger-happy alarm tendency that simply fires falsely on occasion, or a true intent on my part to deceive. It's just a coincidence, I would insist,

that after these false alarms I have uncontested access to some particularly attractive prey item."

Unfortunately, we do not know exactly what transpires in the brain of this sentinel individual. Nor do we know what other birds are thinking as they use their vocalizations during their daily activities. Given how difficult it often is for us fellow humans to understand each other, we may be pessimistic about ever understanding the thoughts, feelings, or motives of these birds. Therein, however, lies the challenge for those of us intrigued with how the minds of our feathered friends work. ∎

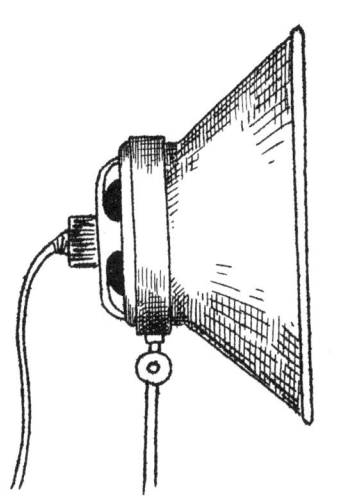

Figure 7–54. House Wren Attacking Loudspeaker: Evidence that one function of bird song is to defend a territory comes from **playback experiments**, in which songs are broadcast from a loudspeaker. If the song of a House Wren is played within the territory of a House Wren, for example, the male usually will respond aggressively, and may even attack the speaker.

tack the loudspeaker **(Fig. 7–54)**. Even better evidence that songs help to keep out other males has been derived from "speaker replacement" experiments (Krebs et al. 1978). In these experiments, researchers remove a male from his territory and replace him with tape players and a series of speakers. Playing songs from different speakers simulates the presence of the territorial male. Intruders are more likely to invade territories from which no song is broadcast, showing that song does say "Keep Out" to other birds **(Fig. 7–55)**.

Speaker replacement experiments remove the male from the territory and leave the songs. In another twist to testing song function, the songs can be removed from the territory, leaving the males. This entails temporarily muting a male by puncturing one of the air sacs in his respiratory system. He then can't maintain enough pressure in his respiratory system to produce songs and typically loses his ability to defend his territory, so intruders are more likely to invade. Both types of experiments show that song is crucial to defending a territory.

Song, however, is not the only vocal keep-out signal that birds use. Birds of many species defend territories during the nonbreeding season, and they typically use a simpler call to broadcast territorial rights at these times. A male or female Northern Mockingbird on a winter territory, for example, uses a loud "chuck" to announce its presence. A warbler on its winter territory in the Caribbean does not sing, but rather uses a simple call note to declare territorial rights.

The other major proposed function for bird song is to attract and stimulate a female (Searcy and Yasukawa 1996). Evidence for this function, although largely circumstantial, is highly believable. Unpaired males of most species typically sing all day long, and paired males sing far less often. If a paired male loses his female, he typically reverts to abundant singing. The amount of song coming from a male is thus a good indication of his pairing status. Experiments with European Starlings and European Pied Flycatchers have shown that nest boxes from which male songs are played attract females. Curiously, the same songs attracted male starlings and male flycatchers, so the situation is a little more complex than it initially seems!

In the laboratory, too, females that have been tested typically prefer songs of their own species. The response used to assay a song is the "precopulatory display," a tail-up display by the female that initiates mating (Fig. 7–56). In these experiments, females typically display more to songs of their own species than to songs of other species, further implicating a mate-stimulation function for song.

All these birds might be reading far more subtle cues than simple presence or absence of the song, however. The amount of song from a territory, for example, could indicate the relative health of a male. A territory rich in food can supply its owner with energy more quickly than a poorer territory, so a male on a good territory can sing more. By experimentally supplementing the territory's food, researchers have documented the relationship between amount of song and quality of

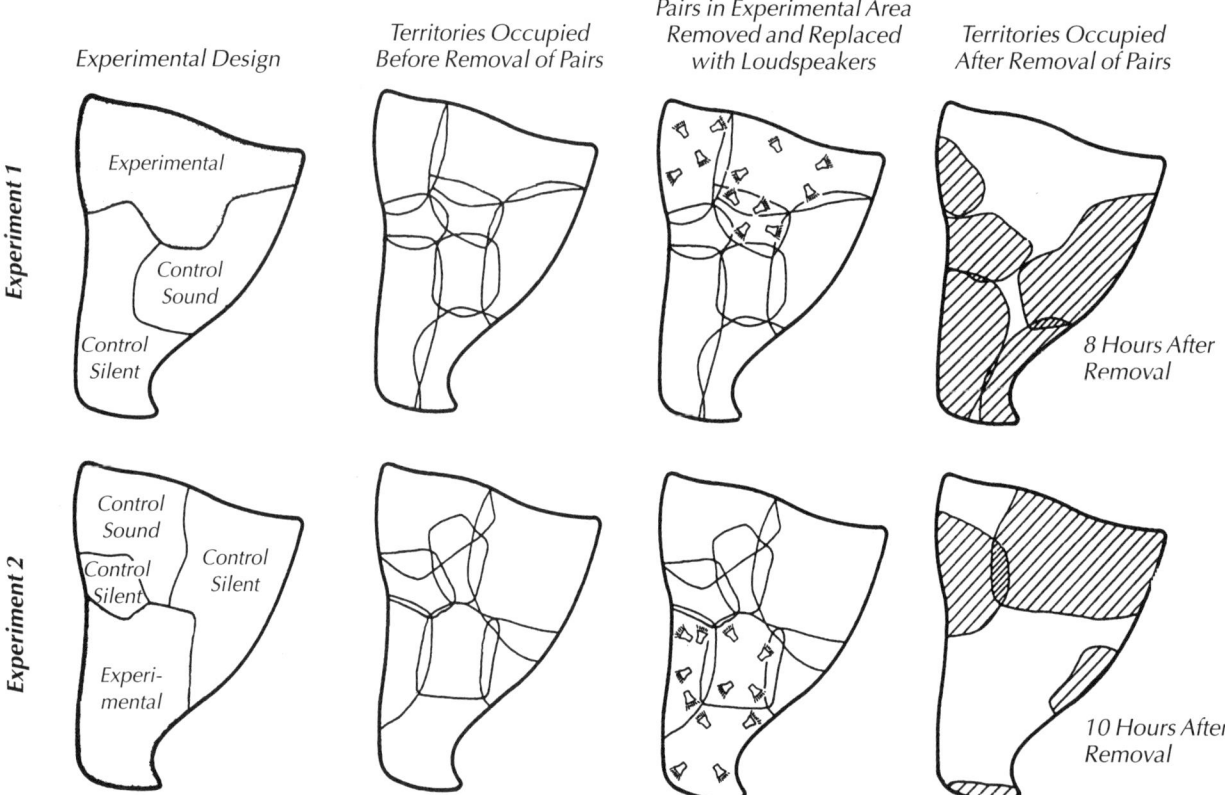

Figure 7–55. Speaker Replacement Experiments: *These experiments with Great Tits (chickadee relatives common in temperate zone woodlands of the Old World) were done to demonstrate that song alone, even without the presence of the territorial pair, keeps intruding birds out of a territory. The experiments were conducted near Oxford, England in early spring, a time when males are prospecting for new territories. In* **Experiment 1,** *the eight territorial pairs in the study area were captured on one morning. Three territories (those in the "experimental" area) each were occupied by four loudspeakers broadcasting the songs of the former residents all day, in a pattern resembling natural singing. Two territories (those in the "control sound" area) were occupied with loudspeakers broadcasting a sound similar to Great Tit song, but played on a tin whistle. The other three territories (in the "control silent" area) were left empty.*

Eight hours later, new males had moved into the five territories in the control areas, but not into the experimental areas. In **Experiment 2,** *conducted a month later, a similar procedure was followed, but the areas receiving control and experimental treatments were changed. Ten hours after birds were removed, four males had moved in, leaving the experimental area empty except for one small portion at the extreme edge. In both experiments, the experimental areas were eventually invaded by new males, but not for about two and a half days. Both experiments clearly demonstrate that Great Tit song functions to defend a territory. Over the longer term, however, intruding birds were able to learn that the loudspeaker was not a real bird, and that the experimental territories were not actually occupied. Adapted from Krebs (1976).*

territory for several species (Searcy and Yasukawa 1996). Males with extra food have been more persistent singers, and in some species the more persistent singers attract the first females in the population. A female could use the amount of song to judge a male's ability to obtain a good territory, or perhaps to indirectly judge the quality of the territory, which might be important as a food supply for herself and her young.

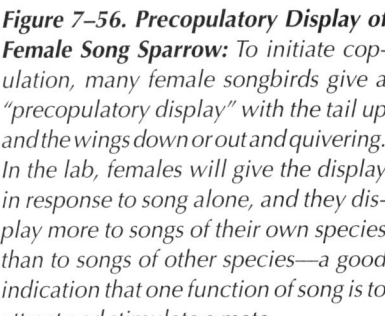

Pause briefly now and think about all of the information available in the singing of a typical songbird in North America. A single song identifies the *species* and (for the vast majority of species) the *sex*. For songs that vary geographically, the song also identifies the *regional location* of the singer. Birds typically don't sing during the nonbreeding season, so a singer is *in reproductive condition, on territory,* and *available for mating* (with either his social partner or any interested female). At any instant, a listener can probably judge the *distance* of the singer, not so much by the loudness but by the degeneration of sound quality over distance. The relative amount that he sings may reveal the *quality of his territory,* because it takes less time to eat on a good territory, which leaves more time to sing. His *general health* can be revealed not only by the amount he sings, but perhaps also by the size of his song repertoire or by the consistency with which he delivers his complex songs. If he sings all day long, he reveals his *mating status*: he's probably a bachelor, or if he already has a mate, he's ready for another (for example, Marsh Wrens are polygynous, with males often having harems of females). If he is of a species that uses different songs in different contexts (for example, a Chestnut-sided Warbler), he immediately reveals his *mood,* because he uses different songs in love and war. His repertoire size may also impart information to a careful listener. In some species, such as Song Sparrows and Great Tits, repertoire size seems to predict *longevity* fairly well. If he's a Northern Mockingbird or another species in which birds add to their repertoires in successive years, he reveals his *age.* Finally, if a male's songs match those of his local neighborhood, he reveals that he probably has *lived in the neighborhood long enough to establish his local identity.*

A singer tries to manage others of his species, and the others try in turn to manage him, all for selfish gain; over evolutionary time, ploy and counterploy have led to a communication system that reveals an extraordinary amount to those who are ready to listen. Birds, of course, are primed to hear all of this information in songs and calls, and to act accordingly. Birds rely far more heavily on their ears than we do (even though, surprisingly, laboratory tests show that our ears are sufficient to hear most of the same details that the birds hear), and birds must be prepared, at any instant, to interpret correctly a sound that warns of a life-threatening emergency or predicts an unexpected opportunity, perhaps for mating. Those who listen and hear succeed reproductively over evolutionary time, so the survivors we listen to today are the summa-cum-laude graduates in the fine art of listening **(Sidebar 5: "Call Notes" and Their Functions).**

Figure 7–56. Precopulatory Display of Female Song Sparrow: *To initiate copulation, many female songbirds give a "precopulatory display" with the tail up and the wings down or out and quivering. In the lab, females will give the display in response to song alone, and they display more to songs of their own species than to songs of other species—a good indication that one function of song is to attract and stimulate a mate.*

(Continued on p. 7·75)

Sidebar 5: "CALL NOTES" AND THEIR FUNCTIONS
Donald E. Kroodsma

Birds use a tremendous diversity of vocalizations. But, of course, they don't classify all of their vocalizations into the two categories of "songs" and "calls," as we so often do, so let's explore exactly what birds do a little more closely.

Perhaps it's most useful to start with the idea that birds use sounds to manage other individuals for selfish gain. For almost all passerine species and many nonpasserines, too, one particular sound seems to stand out. That sound is loud and often repeated persistently from a prominent post. The vocalization can be rather simple, as in a Black-capped Chickadee, or the number of different building blocks within the vocalization can number in the thousands, as in a Brown Thrasher. For temperate zone birds, it's usually the male who "broadcasts" this vocalization, but in more tropical areas, where females and males remain paired year round, females often broadcast, too. It thus seems that most species have a "need" for this particular vocalization, which seems to be to announce ownership of a particular territory or mate, or to try to attract (or maybe impress or stimulate) a mate. We label this particular vocalization a "song."

Cedar Waxwings

But, it seems that not all songbirds sing. Apparently some species don't need a loud, complex vocalization that is broadcast far and wide. The Cedar Waxwing, for example, has no recognizable "song," at least as we have come to know song in many other species (Howell 1973; Witmer et al. 1997). Given that most other songbirds have "songs," it seems likely that waxwings evolved from an ancestor with a typical song, but that the waxwings simply lost that song over evolutionary time.

Waxwings appear to be nonterritorial throughout the year; they nest in loose clusters and forage together at fruit crops when away from the nest. Waxwings thus have no need to broadcast ownership of a territory, because defending the feeding territory at rich, ephemeral food supplies would not be economical (see Ch. 6, Sidebar 4: Living in Groups). But don't males need to "sing" to impress their mates? Apparently not. Intriguingly, the waxy red tips of the secondary feathers on the wing may be an index of a bird's age or health **(Fig. A)** and it is possible that the birds assess and try to impress one another with the quality of their "waxwings" (Mountjoy and Robertson 1988). The "wax" on the wings may be the functional equivalent of intersexual aspects of "song" in other species, and as a feature of waxwing plumage evolved to take on that function, the song may have been lost. That's only a guess, but this line of thinking is instructive, I think, because it focuses on how individuals of a particular species evolve to manage others, and it considers the management game in the context of all of the other attributes of that particular species, such as plumage. Because no two species are alike, we don't expect them to have the same sets of tools to manage one another, either.

If Cedar Waxwings don't sing, what do they do? Their sounds have been broadly classified into two categories, the *bzeee* call and the *seee* call. The difference in these two sounds can be heard easily, because the *bzeee* has a buzzy or rattling quality, and the *seee* is a high-pitched, hissy whistle (Track 49 *[bzeee]*; Track 50 *[seee]*). Careful analysis reveals, however, that the birds use a great diversity of subtle variants (too subtle to be distinguished by the human ear), especially of the *bzeee* call, and

those variants seem to be correlated with particular contexts. Some variants are given during flocking, some when the male and female are nest building, some during courtship feeding, some by nestlings, and so on (Howell 1973). Waxwings undoubtedly use the subtleties of these vocalizations to communicate with each other in ways that we have not yet even imagined.

Red-winged Blackbirds

Our knowledge of vocalizations and their functions in other species is equally rudimentary. Consider the Red-winged Blackbird, for example (see Fig. 7–62). The male clearly has a set of "songs," loud vocalizations that he broadcasts, apparently in an attempt to defend his territory and attract mates. Like a typical songbird, he learns those songs from other male blackbirds (Marler et al. 1972). Females have some loud vocalizations, too. One is the *chit*, which is often delivered immediately after the male's song, so that the two vocalizations overlap; this "duet" between males and females is frequently heard in marshes throughout North America (see Fig. 7–63 and Track 62). The *chit* seems to encourage the male to guard the female's nest against predation and to discourage him from harassing her (Beletsky and Orians 1985). The female has a second loud vocalization, the *teer*, but this one is apparently used in aggressive contexts with other females (Beletsky 1983) (Track 51). These two loud vocalizations apparently do not need to be learned from other females, because they develop normally in females who are isolated from other female vocalizations during and after the nestling stage (Armstrong 1994). If these sounds are not learned, perhaps they are not even controlled by the "song control centers" in

Figure A. Cedar Waxwing: *One songbird without a song is the Cedar Waxwing. These nonterritorial birds nest in loose clusters, feed together on fruit-bearing plants that are often in large clumps, and may use the quality or number of their waxy red wing tips (see inset) to impress potential mates. Because song was not needed to defend a breeding or foraging territory, or to attract a mate, it may have been lost through natural selection. Photo courtesy of John Dunning/CLO. Inset photo courtesy of E. J. Fisk/CLO.*

ent sounds are broadly classified as "alarm and distress calls"; some of these calls seem to be used in specific contexts, such as when hawks are near, but the so-called "hawk alarm call" also is given to raccoons! Both males and females use "feeding calls" when feeding the young, as well as "courtship calls." Males also give "flight calls" when leaving the territory, or just before flying.

Especially intriguing, I think, are a variety of calls that males use as they perch on their territories (Beletsky et al. 1986). Investigators have established seven broad categories of these sounds: *check, chuck, chick, chonk, chink, peet,* and *cheer.* Neighboring males tend to use the same call, such as the *check,* but then they all seem to switch to another call, and eventually another. Waves of call usage spread throughout the marsh, like shock waves spreading out from an epicenter, and anyone patiently sitting and listening beside a red-wing marsh will hear these events. But what are the males doing? Do any of the different sounds that we hear convey a special meaning about a particular circumstance within the marsh? Or perhaps the vocalization itself has no special significance, but rather it is the changing from one vocalization to the next that has particular meaning. Perhaps by simply switching from one vocalization to another the males can monitor changing conditions within the marsh, such as approaches by a predator.

American Goldfinches

Vocalizations of another common bird, the American Goldfinch, illustrate how different one species can be from another. Perhaps the most distinctive sound from goldfinches (and other members of this cardueline subfamily, including siskins and

a female's forebrain; perhaps her nonlearned vocalizations originate from some more primitive part of her brain, as do simpler vocalizations (that is, "calls") in some other species. Notice that we can probably gain a better understanding of the female vocabulary without labeling those vocalizations as either "calls" or "songs." In fact, the more we understand how sounds are used, the

less interesting becomes the question of whether a particular vocalization is a song or a call.

Red-winged Blackbirds use a tremendous variety of other vocalizations, too, as summarized by Yasukawa and Searcy (1995). "Contact calls" are given repeatedly by males on a territory or by both sexes as they forage. "Threat calls" are given in aggressive contexts. Many differ-

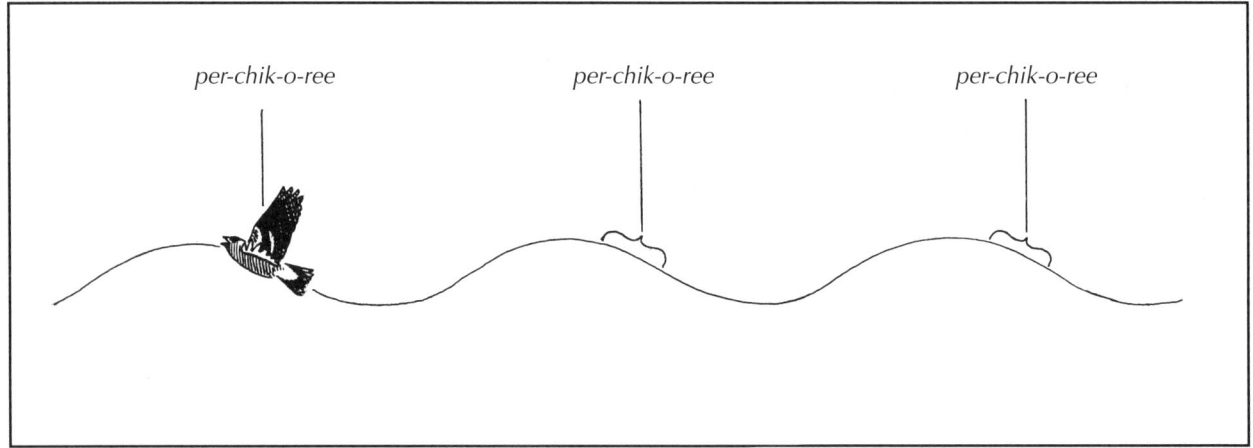

Figure B. American Goldfinch Contact Call: *American Goldfinches give their contact call, a sweet, twittery per-chik-o-ree, during the rising, rapid wingbeat portion of their undulating flight. When a male and female form a pair bond, each changes its contact call to match that of its mate, probably allowing them to distinguish their partners even in a large flock of goldfinches.*

redpolls) is what has been labeled their "contact call." We typically hear this vocalization as goldfinches fly overhead in their characteristic undulating flight pattern, the *per-chik-o-ree* contact call occurring during the rapid wing beats rather than the brief, wings-folded descent **(Fig. B)** (Track 52). A paired male and female change their contact calls to match one another, and this social learning from one's mate undoubtedly enables a bird to identify its partner, even in a large flock of birds. Adult cardueline finches can relearn these calls throughout life, apparently as social partners change. When at least seven years old, for example, one captive adult male Pine Siskin learned the flight call of its new cage-mate, a female Eurasian Siskin. This form of lifelong call learning from social partners probably occurs throughout the cardueline subfamily.

Eastern Phoebes

The vocalizations of the Eastern Phoebe (see Smith 1969) provide a final example illustrating the diversity of vocal behaviors among passerines. The phoebe is a flycatcher and therefore a suboscine, not a songbird. Each male phoebe loudly proclaims his territory and presumed interest in mating with two different "song" forms, the *phoe-be* and the *phoe-bree* (see Sidebar 6, Fig. A; and Track 69), neither of which is learned, but male and female phoebes use a variety of other sounds, too. The most common sound is described by Smith as a "clear, sweet, weak chip," or simply a *tp*, and it is given in a variety of contexts. It typically is used in aggressive encounters, but it also is used by lone birds, such as when foraging during migration or even on the nonbreeding grounds. If a predator is encountered near the nest, this call becomes more emphatic. The same vocalization seems to serve a variety of purposes, and thus the actual meaning of the vocalization may depend on the circumstance in which it is used (Smith 1969). Alternatively, of course, the birds may be more attentive to the details in this chip call than we are. Perhaps they produce and detect subtle differences in these calls, and those subtle differences have meanings to which we are not privy. In a crude first assessment, researchers have lumped all of these different vocalizations together, because they seem to sound alike and, on the sonagrams, look alike. Determining which of these two possibilities is true awaits further study.

The vocabularies of birds are truly rich. In studies of bird sounds, researchers have been initially attracted to study those loud vocalizations produced so persistently ("songs"), and most of the literature on birds sounds is therefore about these "songs." But birds clearly use a variety of other sounds, too. When studying these "call repertoires," our initial approach has been to lump similar sounds into a given category, thereby hoping to see some order in how birds communicate. We thus identify "begging calls" or "alarm and distress calls" or "contact calls." As we explore the details of these vocalizations within and among categories, however, I am convinced that we will be flabbergasted at the richness of the messages that are communicated and at the richness of the details to which birds are attentive. ∎

Dawn Chorus

Some of the most remarkable singing occurs at dawn, especially in temperate zones (Staicer et al. 1996). Most birds are silent the entire night, presumably roosting quietly somewhere on their territory. But beginning an hour or so before sunrise, males begin their dawn chorus. Each species chimes in at a slightly different light level, often beginning with American Robins, and other species join in a rather regular sequence. The quantity and often the quality of singing during this time differ markedly from singing during the rest of the day (**Track 53, Track 54**).

*Figure 7–57. **Chipping Sparrow:** Like many other birds, the Chipping Sparrow sings with more energy at dawn than at other times of the day. At dawn, a male often sings short bursts of song from the ground, quite a contrast from the much longer trills delivered later in the day from high in a tree. Photo by Lang Elliott.*

Species vary considerably in how they behave at dawn. During daytime singing, an Eastern Towhee sings one song form (say, A) over and over, and then introduces another (say, B), then another, and perhaps another, until he has delivered his entire repertoire of three to eight song types over a period of an hour or so. His singing pattern might be illustrated as AAAA . . . BBBB . . . CCCC . . . and so on. The pattern is one of "eventual variety," in which a male repeats one song type many times before "eventually" proceeding to the next type. At dawn, however, the towhee sings with "immediate variety," perhaps delivering all of his song types in 20 to 30 seconds: ABCABCDEDE. The singing is far more energized and dramatic than it is during the daytime.

Chipping Sparrows also sing differently at dawn (**Fig. 7–57**). A typical daytime song is about two seconds long, and consists of perhaps 20 repetitions of a single song element. The male pauses 10 or so seconds between successive songs (**Track 55**). While singing, the male typically sits on an exposed perch high in a tree. At dawn, however, a male often sits on the ground near a male from a neighboring territory and delivers bursts of song as if they were shot from a machine gun. He sings two or three elements of the song, pauses briefly, sings another burst followed by a pause, and so on (**Track 56**). As with the towhee, the vocal display seems highly energized, even frenetic.

Other species use entirely different songs at dawn than during the day. Certain warblers that have two song categories use their aggressive songs for the first 30 to 60 minutes of the morning, after which they lapse into a slower-paced delivery of their other song. A male Yellow Warbler uses about 12 songs delivered with immediate variety during his dawn chorus, and after half an hour or so he switches to his single daytime song type. The male reverts to songs of his aggressive dozen during daytime encounters with other males, too, but the singing is never as energized as at dawn. American Redstarts behave similarly,

Figure 7–58. Eastern Wood-Pewee: *Breeding in woodlands throughout the eastern United States, the Eastern Wood-Pewee gives its plaintive, slurred-whistle song differently at dawn than later in the day. At dawn, the singing rate is faster, and an extra song type, the* ah-di-dee, *may be added to the daytime* pee-ah-wee *and* wee-ur *songs. Photo by Lang Elliott.*

although songs of their aggressive repertoire more typically number only three or four. Some flycatchers, too, such as the Eastern Wood-Pewee **(Fig. 7–58)**, use qualitatively different songs at dawn (Craig 1943) (Track 57; Track 58).

The drama at dawn is truly extraordinary, as if so much pent-up energy is unleashed. But why? Why such a burst of energy at dawn, and again, why does that energy reveal itself in such diverse ways?

Researchers have proposed many ideas for why dawn singing is so dramatic (Staicer et al. 1996). Perhaps males sing at dawn because that is the best time to attract females, especially those who have migrated and arrived overnight. Conditions at dawn are often calm, too, so song at that time carries the maximum distance. Furthermore, during the dawn chorus conditions are often too dark to forage, so singing then is an efficient use of time. Or perhaps singing at dawn is especially important for territory defense; dawn follows the longest period of inactivity, and predation occurs at night, too, so dawn might be an important time for a bird to proclaim "I am still alive" and "This territory is still mine."

The energized displays and often dramatic interactions, such as those of Chipping Sparrows, suggest that during the dawn chorus is when patterns of social dominance are established and daily re-confirmed. Little is known about dominance patterns among males on adjacent territories, but females may attend to such displays and make mating decisions, both intrapair (between mates) and extrapair (with a bird other than the mate), based on male singing exchanges. So many extrapair fertilizations take place in some species, such as the Indigo Bunting and Red-winged Blackbird, that any kind of ritualized display that establishes male hierarchies and broadcasts information about them would benefit listening females. A female Black-capped Chickadee, for example, will mate with males other than her social partner, and her extrapair gambit tends to be with a male higher in the dominance hierarchy than her own mate (S. M. Smith 1988). She could base her decision on her knowledge of dominance hierarchies established during the winter flocks, but that knowledge is probably reinforced by information she gleans from the dawn performances of males in the population. For many species, perhaps singing at dawn is a formal way to establish and monitor social relationships and hier-archies, giving individuals in complex societies the information they need to make wise decisions.

The dawn chorus probably serves multiple purposes, with the emphases undoubtedly varying from species to species, just as social environments and management differ among species. Our failure to understand the full implications of the dawn chorus, however, in no way diminishes the drama played out each morning in these extraor-dinary singing displays **(Fig. 7–59)**.

Figure 7–59. Dawn Chorus: *At dawn, the woods, fields, and marshes come alive with song, as every male bird seems to be proclaiming his territory and existence anew. Although many hypotheses exist to explain why birds sing so much and with such energy and variability at dawn, the phenomenon remains poorly understood—one of the many mysteries surrounding why birds sing as they do.*

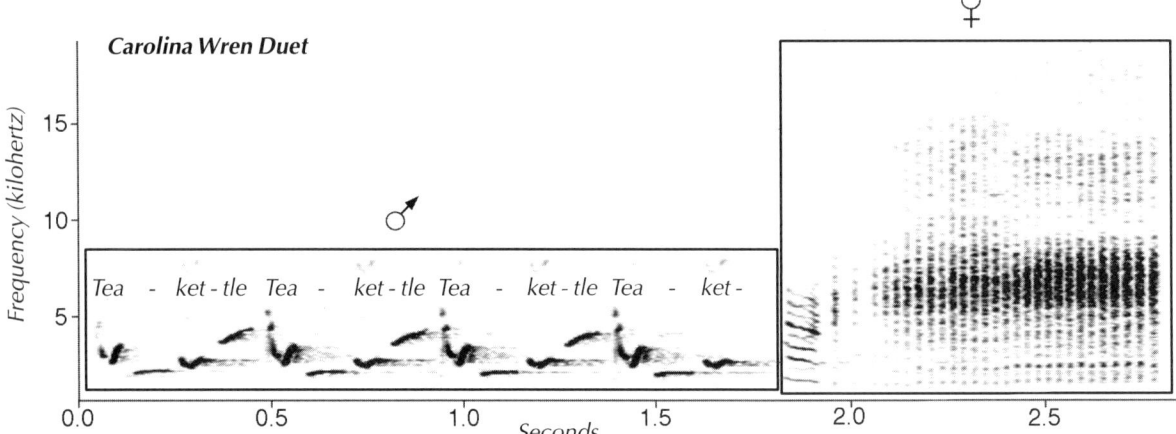

Figure 7–60. Sonagram of Carolina Wren Duet (Track 59, 6th Song): The loud, rolling teakettle-teakettle-teakettle of the male Carolina Wren is a familiar sound in woodlands throughout most of the eastern and midwestern United States. Each male may sing 25 or more different song types, but each consists of a syllable with approximately three parts (tea-ket-tle), usually repeated three or four times. Once in a while, the female may add a coarse rattle to the end of the male's song, creating a simple "duet."

Duetting

A mated male and female communicate frequently, and many species do so vocally. In some species these exchanges are simply overlapping bouts of calling, but in others they are antiphonal—that is, the male and female rapidly alternate their contributions—and are highly coordinated. These "**duets**" of different species differ in a number of ways, including the nature of the overlap between male and female (from synchrony to antiphony), how often mates join in duets, how precisely the vocal exchanges are timed, and the relative quality and complexity of the male and female sounds (Farabaugh 1982).

Wrens of the genus *Thryothorus* provide three good examples of the range of vocal duets. The duet of the Carolina Wren, a common songbird of the eastern United States, is perhaps the least impressive (although we can be confident that it works well for the wrens themselves). Male Carolina Wrens sing a loud, resounding *teakettle-teakettle-teakettle,* each song consisting of a complex syllable *(teakettle)* repeated several times. On occasion, the female appends to this song a coarse rattle or *churr*—unmistakably the female Carolina Wren announcing her relationship with her mate (**Fig. 7–60 & Track 59**). The quality of male and female utterances differs markedly, females are only occasional participants, and the timing of the male and female contributions does not appear to be well coordinated.

The female's performance is more male-like in Rufous-and-white Wrens from Central America. Male and female songs, each a series of pure whistles, appear to be equally complex. In this wren, the female's song is slightly higher in frequency than the song of her mate. The two birds may deliver their songs separately or simultaneously, but if they are delivered together, they simply overlap and are not highly coordinated.

Most extraordinary are the complex, highly coordinated duets some species sing. So coordinated are these singing efforts that the male and female contributions are impossible to differentiate unless the birds are a considerable distance apart. Among the *Thryothorus* wrens, the Buff-breasted Wren, Bay Wren, Riverside Wren, and Plain Wren sing duets of this sort. In these duets, male and female contributions are equally complex, exchanges are highly coordinated and precise, and both sexes participate frequently **(Fig. 7–61 & Track 60)**.

Surveys show that most duetting species live in the tropics. Tropical duetting species include certain types of parrots, woodpeckers, antbirds, flycatchers, shrikes, and wrens, among others. Males and females of many tropical species maintain prolonged monogamous pair bonds and live year round on a given territory. Both males and females thus have a considerable year-round investment in two resources, the mate and the territory. Loss of either resource reduces the chance of surviving and breeding, so it isn't surprising that a female would vocally announce her presence and intentions as consistently and persistently as her mate.

In temperate zones, pair bonds are often markedly different (Morton 1996). In many migratory species, males typically arrive on territory a week or so before females. The male sings to establish the territory and attract a female, and once the female arrives she specializes in egg laying and raising the family. The season is short, and within a few weeks the male and female separate and migrate to nonbreeding quarters again. The many temperate zone females who solicit matings from males who are not their mates don't have the major investment in either the territory or the mate that tropical females do. Accordingly, temperate zone females duet and sing less, and manage their social environment in other ways.

One of the most conspicuous examples of duetting that does occur in North America involves the Red-winged Blackbird, which nests in marshes throughout the continent **(Fig. 7–62)**. Males and females perform a duet similar in complexity to that of the Carolina Wren. In this blackbird, males sing their *konk-a-ree* song over and

Figure 7–61. Sonagram of Bay Wren Duet (Track 60, 1st Song): The Bay Wren, like most duetting species, lives in the tropics. Residents of rain forests from Nicaragua to Ecuador, the male and female sing complex duets so well coordinated and intertwined that their respective contributions to the song are nearly impossible to distinguish without careful study.

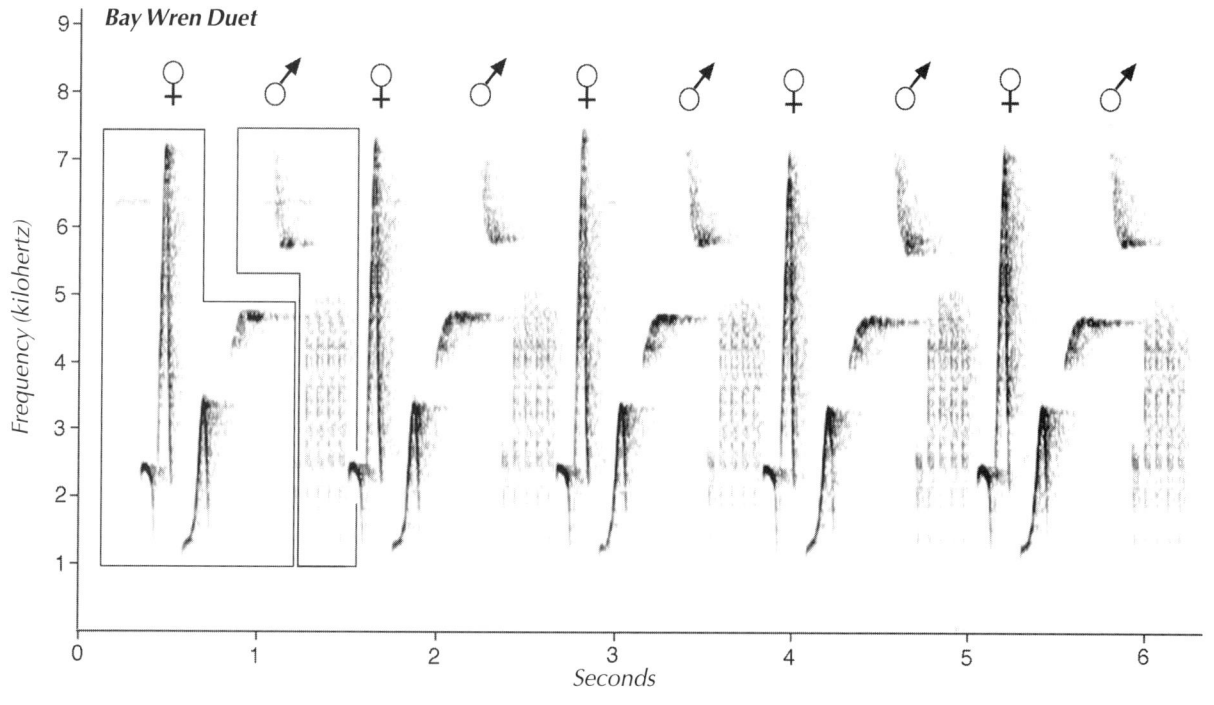

Figure 7–62. Female and Male Red-winged Blackbird Duetting: *Common breeders in fields and marshes throughout North America, male and female Red-winged Blackbirds sometimes perform a simple duet.*

over, eventually switching to another song form. Each male has four to six different *konk-a-ree* songs, and a careful listener can hear when the male switches from one song form to the next (**Track 61**). Males are frequently polygynous, with more than one female nesting on the territory. Females often respond to male songs, producing a loud series of notes during the last half of the male song (**Fig. 7–63 & Track 62**). If you visit any marsh with blackbirds and listen to the male's songs, you can identify how often he changes from one *konk-a-ree* song type to another, and also listen to the patterns of response by the female.

We don't know why males and females interact with each other in exactly the ways they do. We are still learning much about male-female relations; only within the last 10 years, for example, have researchers discovered that a female doesn't always mate with her social partner, and often seeks copulations outside the pair bond (see Ch. 6, Mate

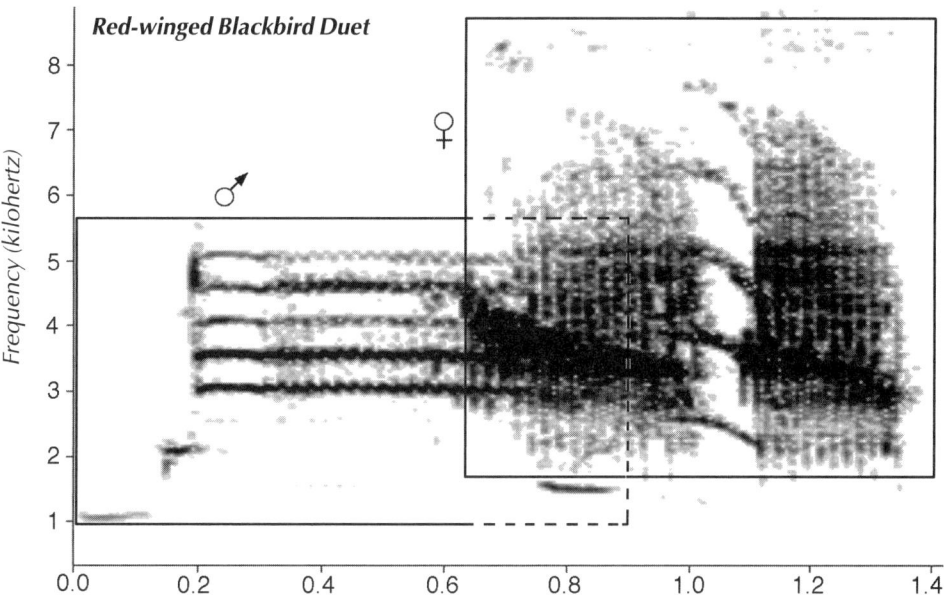

Figure 7–63. Sonagram of Red-winged Blackbird Duet (Track 62, 1st Song): *In the simple duet of the Red-winged Blackbird, the female produces a loud series of notes during the last half of the male's raspy konk-a-ree song. In this sonagram, the dashed lines denote the approximate area in which the male's song is overlapped by the female's; the exact end of the male's song is difficult to distinguish, because it is obscured by the female's song.*

Choice: Extrapair Copulations in Birds). Who mates with whom among these duetting species, how long the pair remains together, how stable local territories are, and other life history characteristics might all affect male and female vocal interactions. Future research must focus on correlations between the types of duets and features of life histories, but as yet the necessary studies simply haven't been done.

Mimicry

One fascinating consequence of song learning is mimicry (Baylis 1982). Some species learn not only their own songs but the songs of other species. In North America, the Northern Mockingbird is perhaps the best example; males sing up to 200 different songs, and a considerable number are clearly derived from other species such as Wood Thrushes, Northern Cardinals, Eastern Phoebes, and Blue Jays (**Fig. 7–64a** & Track 63; Tracks 64–65; **Fig. 7–64b** & Track 66; Track 67). Renowned mimics occur on other continents, too; the lyrebirds in Australia are remarkable, as is the Lawrence's Thrush of South America.

Species not typically thought of as mimics also occasionally learn the vocalizations of other species. Blue Jays imitate the calls of Red-tailed, Red-shouldered, and Broad-winged hawks, for example. European Starlings, closely related to mockingbirds, are excellent mimics. Occasional examples of mimicry in other species are especially intriguing because they show birds' remarkable flexibility and also the risks of song learning. Well-documented examples include a Vesper Sparrow and House Wrens singing songs of the Bewick's Wren, and an Indigo Bunting and a Common Yellowthroat singing a Chestnut-sided Warbler song. It seems that a fairly large number of these occasional mimics are unpaired, suggesting that males who learn the wrong songs often fail to pass their genes to the next generation. Selection against

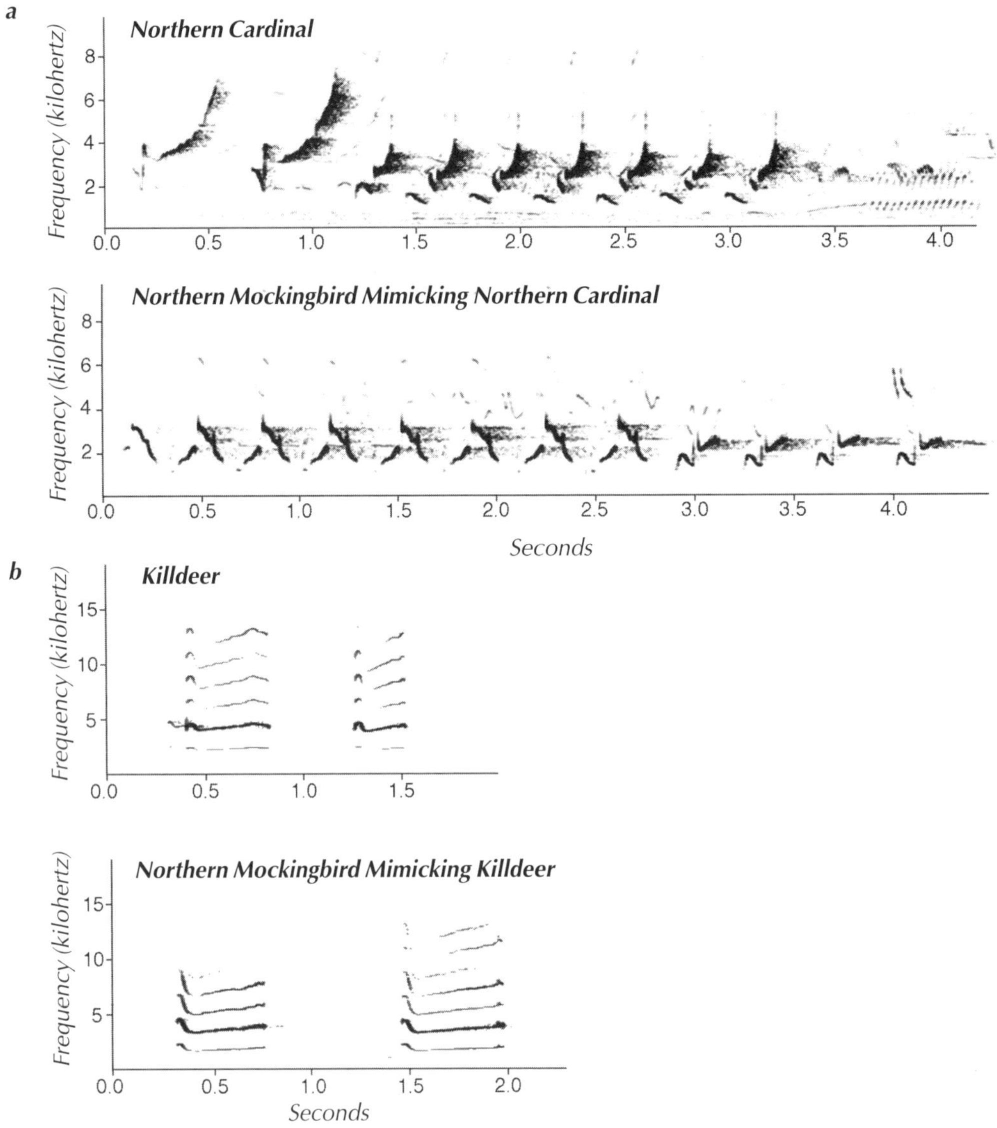

Figure 7–64. Sonagrams of Northern Mockingbird and Model Songs: *Northern Mockingbirds may sing up to 200 different songs, many of them learned by mimicking the songs of other species. Their ability to mimic is so good that not only do the copied songs sound very authentic to our ears, but sonagrams of the mimicked songs closely resemble those of their models. One can readily distinguish mockingbird songs from their models, however, by listening to the pattern of singing: the mockingbird rapidly repeats each song a few times (usually four or more), then moves right on to another song type.* **a. Northern Cardinal (Track 63, 1st N. Cardinal Song, 1st N. Mockingbird Song):** *The* breaker, breaker, breaker *cardinal song.* **b. Killdeer (Not Made from Track 66, but Similar to 1st and 2nd Killdeer Songs; 1st and 2nd N. Mockingbird Songs).**

birds who learn the wrong songs may thus be very strong, so "mistakes" are not perpetuated.

Researchers have proposed a number of hypotheses on why birds mimic. An occasional bird may mimic because it has been deprived of hearing songs of its own species, a condition that can be readily simulated in the laboratory. More interesting, however, are persistent mimics such as the Northern Mockingbird. Mockingbirds are highly aggressive, and they possibly may use other species' songs to threaten them and so keep territories more to themselves. For example, if a Northern Cardinal hears the resident mockingbird utter a cardinal song, does he notice? Is he reminded of aggressive interactions, perhaps back at the winter feeder, or of a confrontation just the day before, all of which might shift the cardinal's use of his territory? Or perhaps the cardinal is oblivious to the mockingbird's crude renditions of his songs. This possibility has never been thoroughly investigated. Perhaps learning the songs of other species is simply an easy way to develop a large repertoire of diverse sounds, which might be used to impress a female. This idea is supported by the Marsh Warbler, which learns an extensive repertoire of African sounds on the nonbreeding grounds during the European winter. The warbler then uses his enhanced repertoire on his European breeding grounds.

Two especially fascinating examples of mimicry suggest very specific functions. Indigobirds of Africa are brood parasites that lay their eggs in the nests of firefinches (Payne 1973) **(Fig. 7–65a)**. Each indigobird species lays its eggs in the nest of one firefinch species, and the firefinch raises the indigobird young. The young of the indigobird have converged in appearance and begging calls on the young of the firefinch, so the firefinch cannot identify the intruders and throw them out of its nest **(Fig. 7–65b)**. Young male indigobirds also mimic the songs of their hosts. Young females undoubtedly imprint on the songs of their host species, too, so male and female indigobirds can use a

a. Adults

HOST

PARASITE

Red-billed Firefinch
(Lagonosticta senegala)

Village Indigobird
(Vidua chalybeata)

b. Nestlings

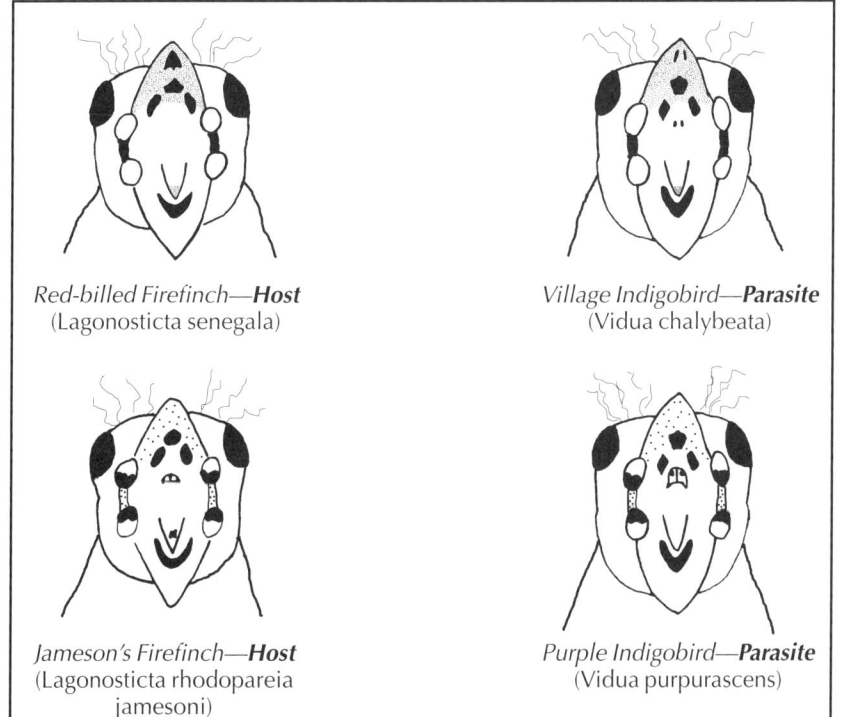

Red-billed Firefinch—**Host**
(Lagonosticta senegala)

Village Indigobird—**Parasite**
(Vidua chalybeata)

Jameson's Firefinch—**Host**
(Lagonosticta rhodopareia
jamesoni)

Purple Indigobird—**Parasite**
(Vidua purpurascens)

Yellow
Pink
Blue-black

Figure 7–65. Firefinch (Host) and Indigobird (Brood Parasite) Adults and Gaping Nestlings: *The indigobirds of Africa parasitize the nests of firefinches; each indigobird species lays its eggs in the nest of one particular species of firefinch, and the firefinch raises the young.* **a. Adult Red-billed Firefinch and its Host, the Village Indigobird:** *Although the adults look fairly different, the eggs and young as well as the songs of the two species are very similar. The same is true for other indigobird-firefinch pairs.* **b. Gaping Nestling Indigobird and Firefinch Pairs:** *The colors and patterns on the gape of each nestling indigobird species have evolved over time to closely resemble those of their host species. Here, note the similarity between the nestling Village Indigobird and its host, the Red-billed Firefinch; and between the nestling Purple Indigobird and its host, the Jameson's Firefinch. Firefinch photo by C. H. Greenewalt/VIREO. Indigobird photo courtesy of Dr. Robert B. Payne. Drawing adapted from Payne (1973).*

7

particular firefinch song to find one another and mate. This complex set of adaptations, involving mimicry not only of songs but of nestlings and their begging calls, occurs throughout the group of indigobirds and firefinches.

The second example involves mimicry of alarm calls (Morton 1976). The female Thick-billed Euphonia of the Neotropics uses the mobbing calls of other species when her nest is threatened. In this way, she attracts birds of the mimicked species to help her ward off predators at her nest. Mimicry thus seems to be a vocal tool that enables euphonias to use other species for selfish gain in predator defense. Those "used" individuals of other species aren't necessarily deceived to their disadvantage, however, because it could be in their best interest, too, to drive off a predator in their general vicinity.

Flight Songs

Although most birds sing while perched, others sing in flight. Many shorebirds deliver their songs high above their breeding grounds. American Woodcock fly above their display grounds, singing not only on the wing but *with* their wings; snipe "sing" with their tails. Other shorebirds produce long, complex songs in flight. Eurasian Sky Larks, the subject of many a romantic poet, impress us by delivering their seemingly unending soliloquy from great heights **(Fig. 7–66)**. Hummingbirds, too, often combine flight and sounds—either vocalized or wing-produced—in their impressive courtship displays **(Fig. 7–67)**.

Singing birds attempt to maximize the influence of their songs. Open-country birds have few high perches from which to broadcast their songs, so taking to the air seems like the obvious choice. Hence, most birds with flight songs are birds of grasslands or the arctic steppe. The extra energy required to deliver the songs must be rewarded by an increase in the area over which the song can be heard.

Figure 7–66. Sky Lark: *The subject of many a romantic poet, a male Sky Lark, a Eurasian species, delivers his long, twittery song from high over his grassy habitat.*

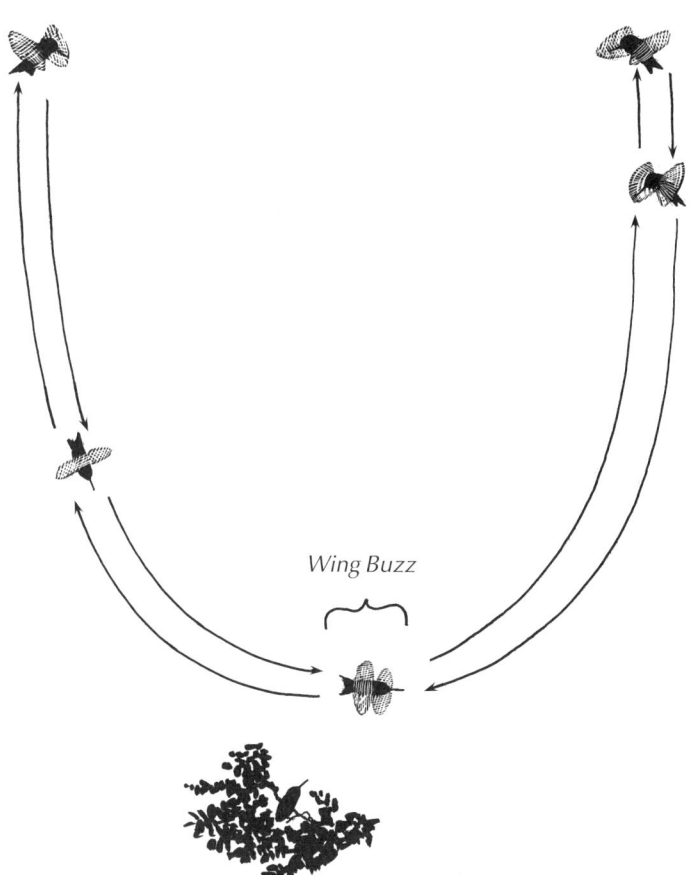

Wing Buzz

Figure 7–67. Ruby-throated Hummingbird Dive Display: *Used in both territorial defense and courtship, the Ruby-throated Hummingbird Dive display combines a stereotyped flight pattern with a wing buzz. The display consists of a U-shaped dive that the bird performs back and forth, making a loud buzzing noise with his wings at the bottom of each dive.*

Some species, however, don't quite fit this comfortable explanation for flight songs. The Ovenbird, a common warbler of forests in northeastern North America, is a prime example. Especially near dawn and dusk, but even in the middle of the night, the Ovenbird launches from near the forest floor up into the canopy and sometimes even beyond. If you listen carefully, you hear the elements of the normal Ovenbird song, the familiar *tea-CHER, tea-CHER, tea-CHER*, embedded in the long, jumbled stream of the flight song, but someone unfamiliar with the Ovenbird would be hard-pressed to identify the source of this odd song (**Track 68**). Why does the Ovenbird perform such "ecstasy flights," as they have been called? We can only guess. To anyone listening, the energetic song makes an emphatic statement, especially in the relative calm of night or predawn. A song delivered high above the forest floor must also carry farther. Who's listening, and who cares? Do females of neighboring territories take note? Are males serving notice of territorial rights? Until we better understand relationships between males and females, both within and between territories, these puzzles will remain unsolved.

Song Repertoires

The style of song development influences song repertoire size (Kroodsma 1988). Species that apparently do not learn songs, such as most suboscines, tend to have small repertoires. The Alder Flycatcher

Table 7–1. Average Repertoire Size and Delivery Style for Selected North American Species: *Average number of songs in the song repertoire of individuals of selected species; even within species, repertoire size may vary widely with geographic location, lifestyle, or age. Delivery style (in what pattern the songs are sung) is given for typical, daytime singing. "Eventual" refers to a bird that sings with "eventual variety," repeating one song type many times before going on to a different song. "Immediate" refers to a bird that sings with "immediate variety," typically singing each song type once, and then moving on to another song type. Several species that sing with eventual variety during the day, such as the the Eastern Towhee and Eastern Phoebe, sing with immediate variety at dawn. Although listed as singing with eventual variety, Eastern Phoebes, when excited, and male Red-winged Blackbirds, when courting a female, may sing with immediate variety. For more information, see Stoddard (1996).*

Species	Average Repertoire Size	Delivery Style
Indigo Bunting	1	-
White-throated Sparrow	1	-
White-crowned Sparrow	1	-
Ovenbird	1	-
Common Yellowthroat	1	-
Alder Flycatcher	1	-
Eastern Phoebe	2	eventual
Eastern Wood-Pewee	3	immediate
Willow Flycatcher	3	immediate
Eastern Towhee	6	eventual
Red-winged Blackbird (male)	6	eventual
Western Meadowlark	8	eventual
Song Sparrow	9	eventual
Northern Cardinal	9	eventual
Tufted Titmouse	10	eventual
Bewick's Wren	16	eventual
Carolina Wren	25	eventual
Red-eyed Vireo	50	immediate
Marsh Wren (Eastern)	50	immediate
Eastern Meadowlark	55	eventual
American Robin	70	immediate
Sedge Wren	100	immediate
Marsh Wren (Western)	150	immediate
Northern Mockingbird	200	immediate
Brown Thrasher	2,000+	immediate

has but one song, the Eastern Phoebe two, the Willow Flycatcher three. In contrast, song learning has enabled the songbirds to evolve huge repertoires. Eastern Meadowlarks have 55 songs, Rock Wrens 200, and Brown Thrashers 2,000 **(Table 7–1)**.

But what do 2,000 songs accomplish for a Brown Thrasher that one song does not do for a Chipping Sparrow? Why is more better in some species? The answer to this question has focused, not surprisingly, on what we presume to be the two main functions of bird song: territory defense, and mate attraction and stimulation. In stylized displays such as the dawn chorus, larger repertoires might help males display their prowess and defend their territories. Common Nightingales in Europe (Todt and Hultsch 1996) and Marsh Wrens in western North America (Verner 1976), for example, seem to duel with their large repertoires; males can match each other with the same song type, refuse to match, jump ahead to the next song in the sequence, overlap a neighbor's songs, or alternate with a neighbor's songs. The complexity of interactions can be dizzying. Verner reported, for example, that Marsh Wrens would match a sequence of as many as 10 songs from a loudspeaker; but because males in the local marsh sang songs in a predictable sequence, the males could sing each song *before* the tape recorder, not after. The male could actually anticipate which song would come next and sing it before the "intruder" did! Recent studies show that this kind of matched countersinging can be important in directing threats to particular individuals and in mediating interactions (Nielsen and Vehrencamp 1995). Females might somehow be more impressed both by larger repertoires and by the way a singer uses his repertoire.

Certain evidence supports these ideas on the functions of repertoires. In the laboratory, females do respond more to a greater variety of song. Measuring a female's responses in nature, however, is more difficult, because the male she seems to pair with isn't necessarily the one from whom she collects sperm to form her family. Recently, by using DNA to identify parentage, Hasselquist et al. (1996) showed that female Great Reed-Warblers in Sweden tend to pick extrapair partners who have especially large repertoires. We need more studies of this sort to determine how females use male song repertoires.

In at least two species, the Great Tit and the Song Sparrow, males with larger repertoires survive longer than males with smaller repertoires (McGregor et al. 1981; Arcese 1989). One is tempted to conclude that having a large song repertoire somehow *causes* a male to live longer, but it is perhaps more likely that good health permits both the longer life and the development of a larger repertoire in youth. Young male Bewick's Wrens, for example, learn more songs if they hatch early in the breeding season than if they hatch late; early-hatching birds have more time to learn songs before their first winter, and they may have a chance to select the best territory, too. In some species, such as the Northern Mockingbird, birds with larger repertoires are older, because males continue to add songs to their repertoires throughout life. For several reasons, then, males with larger repertoires in some species are likely to be healthier and better at surviving, and both males and females are undoubtedly sensitive to these badges of success. How the large repertoires are associated with success and how they might actually help achieve success in some species remain exciting topics for future research **(Sidebar 6: Listening on Your Own)**.

(Continued on p. 7·91)

Sidebar 6: LISTENING ON YOUR OWN
Donald E. Kroodsma

All too often, people stop listening to a singing bird once they have identified it. But that's when the fun can begin! By keeping your ears tuned in a bit longer, you can discover the variety of songs that a particular bird sings and some of the ways he uses his songs. Below I describe several ways to increase your listening enjoyment.

Individual Variety

Follow a male Black-capped Chickadee during the dawn chorus, and hear him shift the frequency of his *hey-sweetie* songs. Or, identify the two song forms of Eastern Phoebes, the *phoe-be* and the *phoe-bree*. Both songs start with essentially the same sound, the *phoe*, but one ends in a raspy *be* and the other ends in a stuttered, higher *bree*. At dawn, males tend to alternate the songs, but later in the day more and more of the songs are of the *phoe-be* type (**Fig. A & Track 69**). Under what circumstances do the proportions of these two songs change in the male's singing?

To hear a song repertoire, listen to a Song Sparrow deliver one song over and over, sometimes varying the ending a bit, and then abruptly move to the next song type. He'll sing eight or so different kinds of songs, and if you have a good ear, you may be able to become familiar with all eight (**Fig. B & Track 70**). The same exercise can be done with Eastern Towhees or Red-winged Blackbirds.

Wood Thrush Song Repertoires

Wood Thrushes provide special opportunities to challenge your listening skills and identify their repertoire size. Each song consists of three parts: a few barely audible notes at the beginning, a loud middle portion, and a softer concluding section. Focus on the loud, highly musical middle portion of the song (**Fig. C & Track 71**). Each male has a repertoire of four to eight phrases he may choose from to use here, and he uses one per song. He often (but not always) sings songs with different "middles" in a regular order: if a male's four types of phrases are identified as A, B, C, and D, for example, the first song will have A, the second B, and so on; every fifth song will have the same sound. By listening carefully, you can learn to identify the different sounds and to determine, by ear, the repertoire size of those middle portions of the song. If your ear is good enough, you should even be able to tell if you have the same Wood Thrush returning to your woodland in successive years. Because our memories are not so good from year to year, consider recording your bird in one year so that you can compare the songs from year to year; even a crude tape recording of the songs will help—don't worry about having expensive equipment.

Northern Mockingbird and Red-eyed Vireo Repertoires

Once your ears are sharpened up, try this special challenge. Northern Mockingbirds and Red-eyed Vireos usually have huge song repertoires, but you may be able to estimate the number of different song types each male sings! These birds tend to sing with immediate variety: cycling through their whole repertoire before repeating any one song type. Listen carefully to an individual singer, and pick out an especially distinctive song. Then count how many other songs occur before you hear the same song again. Repeat this process several times for that unique song, and then calculate the average number of

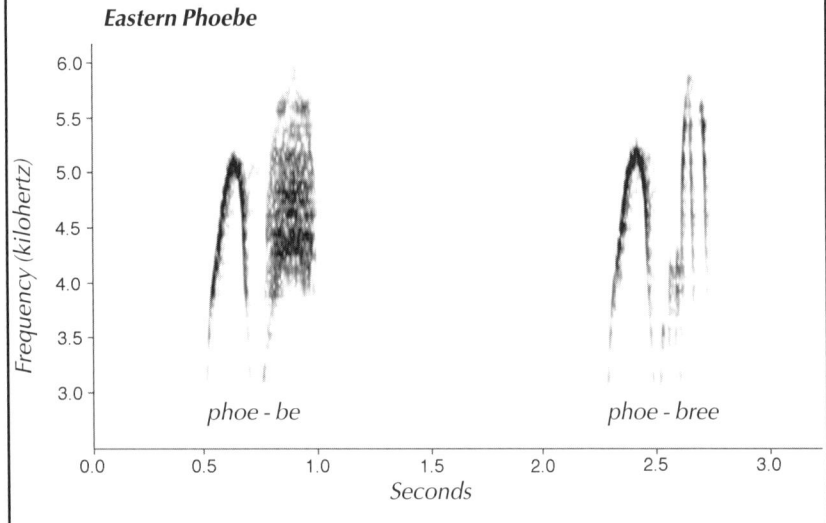

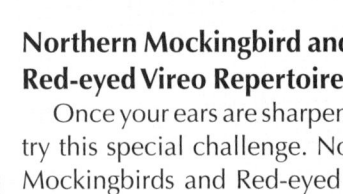

Eastern Phoebe

Figure A. Sonagram of Eastern Phoebe Dawn Song (Track 69, Sonagram not Produced from Track 69, but Similar to 1st and 2nd Songs): *Throughout the eastern United States and much of Canada, the familiar song of this common flycatcher is a welcome sign of spring. The Eastern Phoebe has two song forms,* phoe-be *and* phoe-bree. *Although both begin with a single harsh* phoe *note, the* phoe-be *ends with a raspy* be, *and the* phoe-bree, *with a stuttered, higher* bree. *The phoebe tends to alternate the two songs at dawn—and a portion of dawn singing is shown in this sonagram. Later in the day, mostly* phoe-be *song types are given.*

songs that occur before your song is repeated. Now repeat the entire process for several other songs that you have come to recognize. Calculate the grand average, and you will have a rough estimate of how many different songs the individual sings.

Dawn Chorus

Learn to appreciate how birds of many species sing at dawn, and contrast their dawn behavior with what they do later in the day. For a dramatic difference between dawn and daytime singing, choose the Eastern Towhee, Field Sparrow, or a warbler, such as the Chestnut-sided or Blue-winged. Or, listen to a Song Sparrow, and study how many times he repeats each song type. Pick almost any species, and note how much more energized the singing seems at dawn than later in the morning.

It's also fascinating to document how birds contribute to the dawn chorus of a given location. Each species chimes in at a slightly different light level, for example. In many communities, American Robins seem to be the first to sing, and other species join in a rather regular sequence.

If you are interested in listening more to the sounds of birds, the whole world is your theater. Wherever you travel, birds will be vocalizing, and if you listen carefully you can learn to enjoy eavesdropping on the birds as they go about their daily lives. ∎

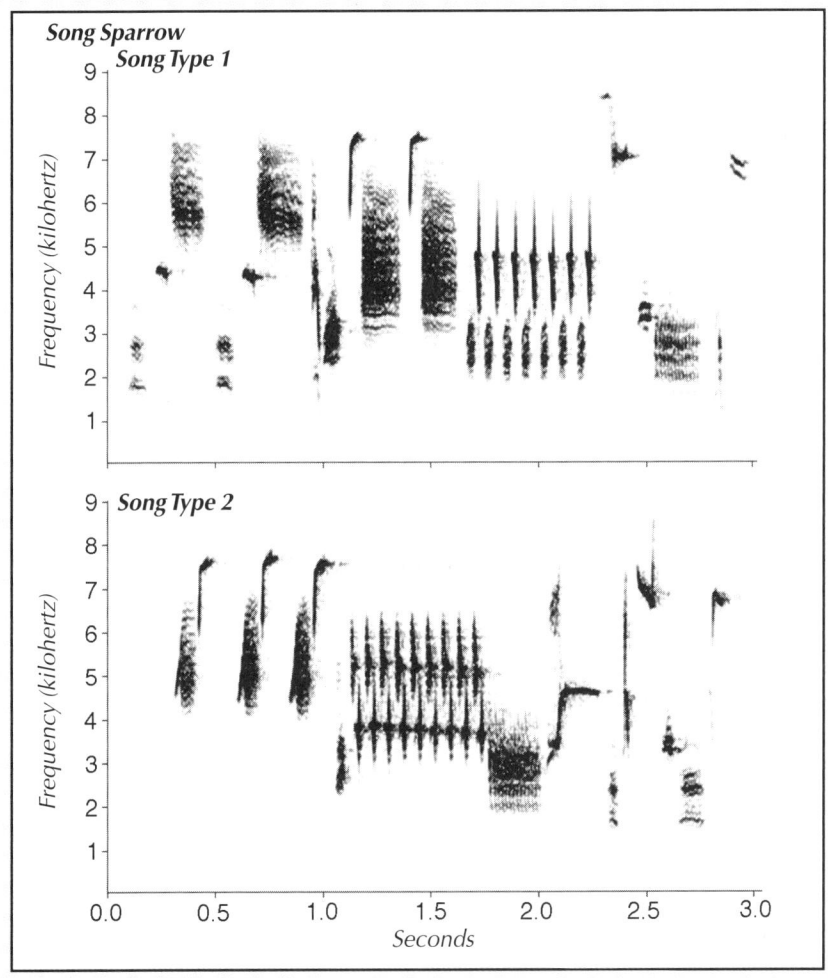

Figure B. Sonagrams of Song Sparrow Song Types (Track 70, 1st Song of Song Type 1, 1st Song of Song Type 2): Common throughout North America, each male Song Sparrow has a repertoire of approximately 8 or 9 different song types, all often clumped under the rough description maids, maids, maids put on the teaeeeee-kettle-ettle-ettle. *Although the details of each song type differ tremendously, many songs stick to the basic pattern of a couple of repeated introductory notes* (maids, maids, maids), *then a jumble of buzzes and notes in the middle* (put on the teaeeeee), *and a final trill* (kettle-ettle-ettle). *If you listen carefully to a singing male, you will hear him repeat a song type many times, perhaps varying the ending a bit; then you will hear him switch to another type, repeating it many times before moving on to another song type. Sometimes a singer delivers most of his repertoire before repeating a song type, and at other times one or more song types may be given much more often than others. The two song types shown here do not fit the general pattern—but they are still typical Song Sparrow songs. Song Type 1 has two different sets of introductory syllables repeated twice each, then a longer trill, and a varied non-trill ending containing a buzz. Song Type 2 has an introductory trill of three syllables, a longer trill, then a jumble of buzzes and notes. In Track 70, the Song Sparrow sings Song Type 1 three times, Song Type 2 three times, and then repeats Song Type 1 three times, varying the endings a bit. This is a more rapid switch in song types than is typical.*

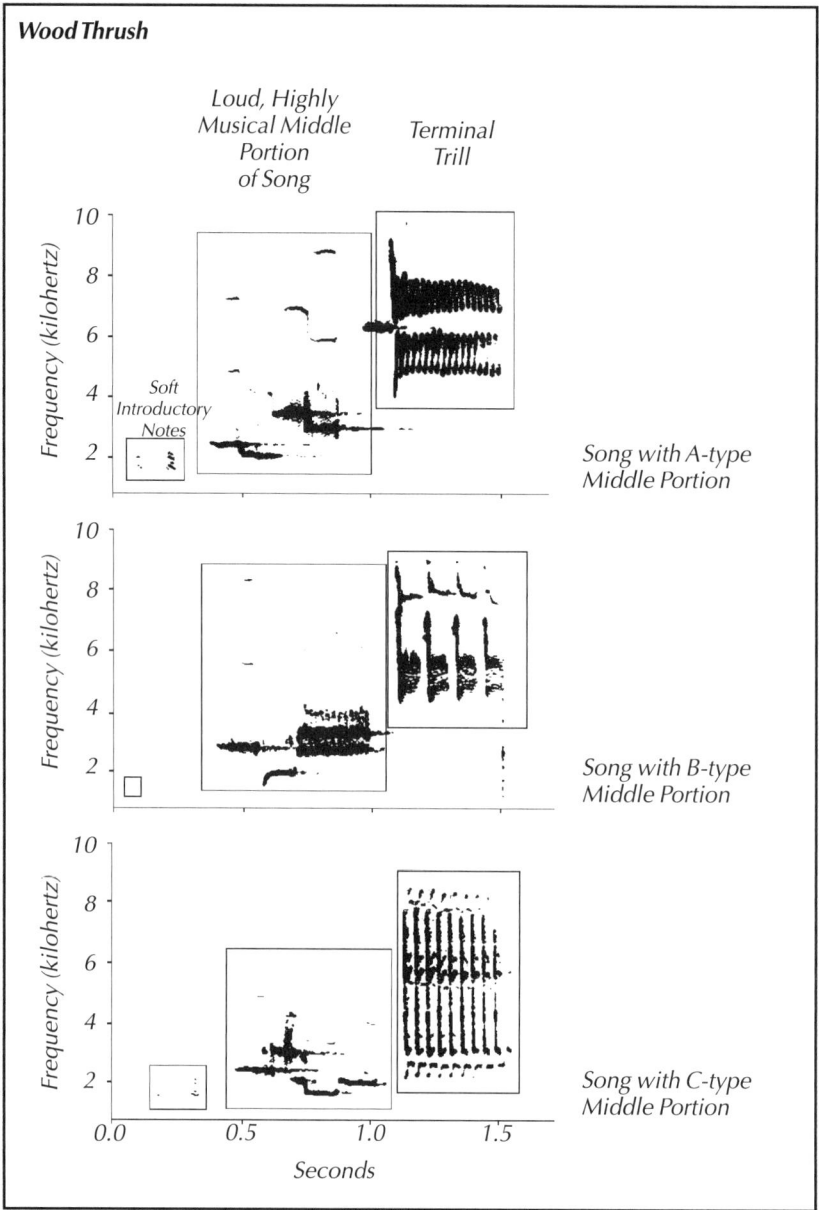

Figure C. Sonagrams of Wood Thrush Song Types (Track 71, 1st Song (A), 2nd Song (B), and 3rd Song (C)): *The beautiful song of the Wood Thrush is made up of three different parts: a few simple, soft introductory notes; a loud, complex, highly musical middle portion; and a final soft trill. The description ee-oo-lay often applied to the song refers only to the middle portion. Although the beginning and end portions may vary, the differences are subtle and often hard for a human to detect by ear. To learn to listen to repertoires, focus, instead, on the middle portion, which also varies. Each male has a repertoire of four to eight different phrases that he may choose from to put here, and he uses one per song. Songs with three different "middles," A, B, and C, are shown here. In Track 71 the Wood Thrush sings songs with their middles in the following order: A, B, C, D, B, C, A, D, A, B.*

The thought of dawn sweeping around the globe every 24 hours is uplifting and powerful for me. Like a giant player piano, the Earth plays on and on, year after year, century after century, from ages past into the indefinite future. Always, somewhere, the birds are waking, and always, somewhere, they are singing. I like to picture humans in all types of dwellings, around the globe, in the northern and southern hemispheres, tracking this dawn chorus by throwing open their windows, letting this marvelous bird music help us greet each new day. The sun's first light, the waking birds, and opening windows thus continually accompany each other around the globe.

And all that goes on as these birds communicate with each other is truly astounding! The goals of each individual, to survive and leave offspring, are rather simple, and the role of sound in achieving this mission is extraordinary. I hear short songs, long songs; simple songs, complex songs; musical songs, harsh songs; duets and solo performances; simple repertoires, huge repertoires; and much, much more. Young songbirds are learning, practicing, babbling as they copy older birds, who pass traditions on to future generations. Ears are poised, brain cells are processing, nerves are firing, and tiny muscles in dual voice boxes respond, creating sounds that we have come to recognize and love.

So many species, each different from the other. And each is successful in its own special way, having survived eons of time to travel, together with us, on this planet Earth. Why each species is the way it is remains one of the great evolutionary puzzles that we face in trying to decipher how these birds communicate with each other. We have so much to learn, and the more we learn, the more we appreciate all of these birds for what they are. May the sounds of birds fill your life. Throw the windows open, and enjoy!

Suggested Readings

Armstrong, E. A. 1963. *A Study of Bird Song.* New York: Dover.

Like the book by Thorpe (see below), this book reviews the study of bird song in its infancy during the 1960s. Armstrong was a British clergyman whose spare-time passion was studying birds, especially the Winter Wren.

Catchpole, C. K., and P. J. B. Slater. 1995. *Bird Song: Biological Themes and Variations.* Cambridge, England: Cambridge University Press.

An excellent review of the study of bird song, written with a wide readership in mind. Chapters cover production and perception, how song develops, getting the message across, when birds sing, recognition and territorial defense, sexual selection and female choice, themes and variations, and variation in time and space.

Greenewalt, C. H. 1968. *Bird Song: Acoustics and Physiology.* Washington, D.C.: Smithsonian Institution Press.

A classic. Written by an "amateur ornithologist" but a superb scientist, winner of the Cornell Lab of Ornithology's Allen Award for his contributions to ornithology. A fine, early treatise on how birds sing.

Hartshorne, C. 1973. *Born to Sing.* Bloomington: Indiana University Press.

One of my favorite books, by a philosopher who loves bird sounds. The author ranks birds by the quality of their music, listing the best singers in the world. Not always accepted as scientifically authoritative, but a delightful tribute to one man's passions.

Jellis, R. 1977. *Bird Sounds and Their Meaning.* London: British Broadcasting Corp.

A good introduction to bird sounds for those who love birds but have little knowledge of bird behavior. Examples in the text are primarily of birds from the British Isles or continental Europe.

Kroodsma, D. E., and E. H. Miller, Editors. 1982. *Acoustic Communication in Birds.* 2 Volumes. New York: Academic Press.

Nineteen authoritative chapters on how birds produce and perceive sounds, on how sounds are designed, and on song learning and its consequences. Includes some of the finest summaries on duetting, mimicry, and other topics that are not a focus of the 1996 volume by the same editors.

Kroodsma, D. E., and E. H. Miller, Editors. 1996. *Ecology and Evolution of Acoustic Communication In Birds.* Ithaca, NY: Cornell University Press.

The most recent compendium, written by 39 experts on bird sounds. Each author wrote about his or her professional interest. The 26 chapters in this book suggest abundant research activities for those wishing to study development, repertoires, geographic variation, signal processing, how birds interact with each other, or a host of other topics.

Smith, W. J. 1977. *The Behavior of Communicating: An Ethological Approach.* Cambridge, MA: Harvard University Press.
An excellent book about communicating in general, not just about birds and not just about vocalizing. Many insights into why animals behave as they do.

Thorpe, W. H. 1961. *Bird Song.* Cambridge, England: Cambridge University Press.

Written by a pioneer in the study of bird song, this book reviews what was known in the early 1960s.

7

Appendix A:
DESCRIPTIONS OF TAPE / CD TRACKS

Track 1: **Black-capped Chickadee *Hey-sweetie* Song (Fig. 7–6):** Recorded February 20, 1988 in New York by Gregory F. Budney.

Track 2: **White-throated Sparrow Song (Fig. 7–7):** Recorded in Ontario, Canada by William W. H. Gunn.

Track 3: **Hairy Woodpecker Drumming (Fig. 7–8):** Recorded in California by Geoffrey A. Keller.

Track 4: **Downy Woodpecker Drumming (Fig. 7–8):** Recorded in Oregon by Geoffrey A. Keller.

Track 5: **Yellow-bellied Sapsucker Drumming (Fig. 7–8):** The third drum, which is quieter than the others, is another bird answering the first. Recorded in Ontario, Canada, by William W. H. Gunn.

Track 6: **Common Yellowthroat Song (Fig. 7–9):** You will hear three songs at normal speed, two at one-half normal speed, and then two at one-quarter normal speed. Recorded in New York by Steven R. Pantle.

Track 7: **Eastern Towhee Song (Fig. 7–10):** You will hear three songs at normal speed, then two at one-half normal speed. Recorded in New York by Arnold van den Berg.

Track 8: **Winter Wren Song (Fig. 7–11):** You will hear two songs at normal speed, one at one-half normal speed, and then one at one-quarter normal speed. Recorded in Oregon by Geoffrey A. Keller.

Track 9: **Black-capped Chickadee *Chick-a-dee* Calls (Fig. 7–13a):** Recorded in New York by Robert C. Stein.

Track 10: **Black-capped Chickadee "Gargle" Calls (Fig. 7–13b):** Recorded on July 4, 1979 in New York by Andrea L. Priori.

Track 11: **Black-capped Chickadee Song Sequence with Frequency Shifts:** You will hear four songs at one frequency, then one at a lower frequency, one at the first frequency, one again at the lower frequency, and then two at the original frequency. All songs from one bird, as sung in the field. Recorded in May 1996 in Massachusetts by Donald E. Kroodsma.

Track 12: **Hairy Woodpecker Foraging Pecks:** Note the irregular rhythm, with pecks paced much more slowly than in drumming. In addition, these foraging pecks are much softer than drumming, as is usually the case, because the bird is not selecting a resonant branch to hit. Recorded in Oregon by David S. Herr.

Track 13: **Ruffed Grouse Drumming (Sidebar 1, Fig. B):** Recorded in Oregon by David S. Herr.

Track 14: **American Woodcock Peenting and Aerial Display:** You will hear three *peents* from the male on the ground, then wing twittering as he takes to the air and spirals upward. At the peak, you will hear the twittering change to a more "chirpy" quality as the male begins to vocalize during his descent. When he reaches the ground, you will hear him give two more *peents*. In an evening or morning of displaying, the male intersperses aerial flights with sequences of peenting that vary in length. Recorded on May 22, 1951 in Ithaca, New York by Arthur A. Allen.

Track 15: **Common Snipe Winnowing:** You will hear three sets of *woo-woo-woo-woo* as the bird dives during its aerial display and spreads its tail. The sound is produced at high flight speeds when air vibrates the spread outer tail feathers. Recorded in Alaska by Leonard J. Peyton.

Track 16: **White-bearded Manakin Wing Sounds at Active Lek:** You will hear occasional growls and various types of wing snaps, all produced by striking together specialized wing feathers. Sounds from three different males displaying simultaneously at lek. Recorded on January 29, 1983 in Cauca, Colombia by Mark Robbins.

Track 17: **Eastern Towhee, Three Song Types (Fig. 7–16):** This track was created so that you will hear three repetitions of Song Type A, three of Song Type B, and then three of Song Type C. In this example, the intervals between songs have been shortened compared to typical Eastern Towhee singing. All songs were recorded in the field from one singing male. Recorded on April 8, 1987 in Florida, by Donald E. Kroodsma.

Track 18: **"High *Zee*" Alarm Calls (Fig. 7–18):** You will hear a series of alarm calls, first from an American Robin, next from a Tufted Titmouse, and finally from a Black-capped Chickadee.

American Robin recorded on August 3, 1990 in Deer Park, in the Upper Peninsula of Michigan, by Lang Elliott. A Broad-winged Hawk was perched nearby, and the robin was clearly responding to the hawk. Tufted Titmouse recorded in Texas by Geoffrey A. Keller. Black-capped Chickadee recorded in Ontario, Canada by William W. H. Gunn.

Track 19: Mobbing Calls (Fig. 7–19): You will hear a series of mobbing calls, first from a Blue Jay, next from a House Wren, and then from a Tufted Titmouse. You will then hear a sequence of mobbing calls from a mixed-species flock that includes Mexican Chickadee, Western Tanager, and Red-breasted Nuthatch. Blue Jay recorded in New York by Hugh McIsaac. House Wren recorded in Alberta, Canada by William W. H. Gunn. Tufted Titmouse recorded in Texas by Geoffrey A. Keller. Mobbing flock recorded in the fall of 1980 in a dry oak woodland in Arizona by Randolph S. Little.

Track 20: Great Tinamou Song: Recorded on December 3, 1992 in Zancudo Cocha, Napo Province, Ecuador, by David L. Ross and Gregory F. Budney.

Track 21: Common Nightingale Song: Recorded April 18, 1997 in Taliouline, Morocco by Arnold van den Berg.

Track 22: Musician Wren Song: Recorded on August 27, 1982 at Cocha Cashu, Manu National Park, Madre de Dios, in Amazonian Peru by Theodore A. Parker, III.

Track 23: White-crowned Sparrow Normal and Isolate Songs (Fig. 7–25): You will first hear typical adult songs, then atypical songs, much less complex, from a bird taken from a nest in the wild at eight or nine days of age and raised in the laboratory without hearing any further songs of his own species. Normal song recorded in Oregon by Geoffrey A. Keller. Isolate song recorded in May 1976 by Mark Konishi.

Track 24: Babbling of 1 ½-year-old Child: A sample of human speech development that parallels the subsong of birds. Donald E. Kroodsma recorded his daughter at home in Oregon in 1972.

Track 25: Bewick's Wren Adult Song and Subsong (Fig. 7–27): You will first hear typical adult songs, then typical practice songs (subsong) from a young bird. Note that the practice songs are long and rambly, with less distinct notes than in the adult songs. Both songs recorded in 1970 in Corvallis, Oregon by Donald E.

Kroodsma. Adult song recorded in May; subsong, in August.

Track 26: Alder and Willow Flycatcher Songs: You will first hear Alder Flycatcher songs, often described as *fee-BEE-o* or *free-BEE-er*. Note the soft beginning to the song, and the accent on the second syllable. Then you will hear Willow Flycatcher songs, usually described as *FITZ-bew*. The latter have a harsh beginning, with the accent on the first syllable. Although the songs sound somewhat similar, they are an easier tool for distinguishing these two species than are their nearly identical appearances. Alder Flycatcher songs recorded in Ontario, Canada by William W. H. Gunn. Willow Flycatcher songs recorded in Oregon by Geoffrey A. Keller.

Track 27: Indigo Bunting Song (Fig. 7–31): You will first hear the song of a territorial male, then the song of a male who holds an adjacent territory. Note how similar the two songs are. Then you will hear the very different song of a male who is a stranger to these two birds. Note that the first song of the territory holder, the second song of the neighbor, and the second and fourth songs of the stranger are noticeably shortened versions of each bird's song. Although each Indigo Bunting has only one song type, males often shorten their songs in various ways, especially during territorial encounters. Territorial male and neighbor recorded at Point Pelee, Ontario by William W. H. Gunn. Stranger recorded in Bloomington, Indiana by Geoffrey A. Keller.

Track 28: Marsh Wrens in Matched Countersinging: The first example you will hear was fabricated using recorded songs from western birds. It illustrates matched countersinging, in which one bird (Bird A) sings a song, then his neighbor (Bird B) sings a very similar song. Then Bird A sings a different song and Bird B again matches the song of Bird A. This pattern continues with a number of different songs, just the way neighboring western Marsh Wrens often sing. The second example is a recording made in the field in California of two Marsh Wrens actually engaged in matched countersinging. The first example was fabricated from songs recorded May 19, 1985 in Coos Bay, Oregon by Geoffrey A. Keller. The second example was recorded March 21, 1996 in Wild Gustin, California by Joe Brazie.

Track 29: Wood Thrush Song (Fig. 7–35): You will hear five songs at normal speed, then three songs at one-half normal speed. Recorded in New York by Arthur A. Allen.

Track 30: Black-and-white Warbler Night Flight Calls: Like most warbler night flight calls, these are short, high, and cricket-like. Recorded in late April 1989 in eastern Florida by Bill Evans.

Track 31: American Redstart Night Flight Calls (Sidebar 2, Fig. Ba): Like the night flight calls of the Black-and-white Warbler, the American Redstart's calls are short, high, and cricket-like. Recorded in mid-May 1989 in eastern Florida by Bill Evans.

Track 32: Veery Night Flight Calls (Sidebar 2, Fig. Bb): Veery night flight calls, like those of other thrushes, are lower, longer, and more mellow than those of warblers. The Veery says *veeree* or *veer* in night flight. Recorded in late August 1988 near Ithaca, New York by Bill Evans.

Track 33: Swainson's Thrush Night Flight Calls: This flock of Swainson's Thrushes migrating overhead sounds a great deal like a pond full of calling spring peepers (small tree frogs). Recorded in mid-September 1990 in Ithaca, New York by Bill Evans.

Track 34: Bobolink Night Flight Calls (Sidebar 2, Fig. Bc): The *pink* note you will hear from a night-migrating Bobolink is the same as the common daytime flight call note. Recorded in mid-May 1989 in eastern Florida by Bill Evans.

Track 35: Rose-breasted Grosbeak Night Flight Calls: The night flight call of the Rose-breasted Grosbeak is a mellow whistle, similar to that of the Swainson's Thrush, but slightly higher in pitch and somewhat off-key. The same call is given by family groups in late summer to maintain contact with one another. Recorded in mid-September 1991 in Oneonta, New York by Bill Evans.

Track 36: Upland Sandpiper Night Flight Calls (Sidebar 2, Fig. Bd): The Upland Sandpiper's night flight call is a short, upward-slurred whistle that is repeated quickly a few times. Recorded in mid-August 1989 in west-central New York by Bill Evans.

Track 37: Barn Owl Night Flight Calls: The Barn Owl's night flight call is a piercing, raspy hiss, much louder than the begging hiss of a juvenile Barred Owl. Recorded in June 1990 in Enfield, New York by Bill Evans.

Track 38: American Bittern Night Flight Calls (Sidebar 2, Fig. Be): The deep croak of a migrating American Bittern is similar to the croaks of

night-herons. Although this example contains pairs of calls, they are more typically given as single notes. Recorded in mid-October 1989 in southern Alabama by Bill Evans.

Track 39: Piano Music Representing Wood Thrush Song (Sidebar 3, Fig. Aa): You will hear the piano music from the score representing Wood Thrush song shown in Sidebar 3, Fig. Aa. This is from *Field Book of Wild Birds and Their Music*, written by F. Schuyler Mathews in 1921.

Track 40: Wood Thrush Song (Sidebar 3, Fig. Ab): You will hear a series of Wood Thrush songs. The third song, chosen for its similarity to one phrase from the piano music played in Track 39, is shown in the sonagram in Fig. Ab. Note that you will hear soft introductory notes in the third song of Track 40, although they were removed from the sonagram in Fig. Ab to better match the piano score. Recorded in New York by Arthur A. Allen.

Track 41: Songs in Song Sparrow Neighborhood (Sidebar 3, Fig. B): This 10-second recording was made using an array of eight microphones arranged around the edge of a field in which several male Song Sparrows held territories. The singing you will hear occurred two minutes before a territorial fight. You will hear the entire 10-second recording twice. Recorded April 17, 1998 at 7:06 a.m. in Ithaca, New York by John Bower.

Track 42: White-crowned Sparrow Song Dialects (Fig. 7–41): You will hear three songs from a dialect in Oregon, three from a dialect in California, and three from a dialect in Alberta, Canada. Oregon dialect recorded by Geoffrey A. Keller, California dialect recorded by William R. Fish, and Alberta dialect recorded by William W. H. Gunn.

Track 43: Black-capped Chickadee Unusual Song from Martha's Vineyard (Fig. 7–46): You will first hear a bird from Gay Head, at the western end of the island, then a bird from Edgartown, at the eastern end. Birds at Gay Head recorded on May 10, 1994 and birds at Edgartown recorded in May 1996, both by Donald E. Kroodsma.

Track 44: Black-capped Chickadee Abnormal Song from Lab-tutored Bird: You will hear highly abnormal songs from a Black-capped Chickadee from a normal population (one that sings the typical *hey-sweetie* song) in Amherst, Massachusetts. This bird was reared in the lab and exposed to typical *hey-sweetie* songs. Under these conditions, male Black-capped Chickadees develop songs with the typical whistled, pure-tone quality, but containing from one to seven or more different notes on a variety of frequencies. Recorded March 19, 1990 by Donald E. Kroodsma.

Track 45: **Blue-winged Warbler Song:** You will first hear the well-known *beee-buzzzz* song, used most by males in the presence of females. This song varies little throughout the range of the species. Then you will hear the male's aggressive song, used at dawn and in very aggressive situations. This song varies more, and occurs in dialects. *Beee-buzzzz* recorded in Ohio by Randolph S. Little. Aggressive song recorded June 11, 1981 in Massachusetts by Donald E. Kroodsma.

Track 46: **Chestnut-sided Warbler Song:** You will first hear the familiar *pleased-pleased-pleased-to-MEETCHA*, often termed the "accented-ending song" because of the emphasis placed on the *meetcha* portion. This is used by males in association with females, and varies little throughout the range of the species. Then you will hear the so-called "unaccented-ending song," used at dawn and in highly aggressive situations. This song occurs in dialects. Accented-ending song recorded in Ontario, Canada by William W. H. Gunn. Unaccented-ending song recorded in June 1990 in the town of Florida in the Berkshire Mountains of western Massachusetts by Donald E. Kroodsma.

Track 47: **Marsh Wren Songs from Eastern North America (Fig. 7–52):** Recorded at Long Lake National Wildlife Refuge, Moffit, North Dakota, by Geoffrey A. Keller.

Track 48: **Marsh Wren Songs from Western North America (Fig. 7–53):** Recorded in Oregon by Geoffrey A. Keller.

Track 49: **Cedar Waxwing *Bzeee* Calls:** Cedar Waxwings have no real "song," in the sense of a loud vocalization that is broadcast repeatedly from a prominent post. Instead they have several different types of calls that researchers have broadly classified into two categories: *bzeee* calls and *seee* calls. The *bzeee* calls, as you will hear, are high-frequency notes with a buzzy or trilled quality, and are quite variable. Although humans cannot determine the subtle differences by ear, they are visible on sonagrams, and the birds appear to use the

different variants in different contexts, such as in flocks, during courtship, during begging, or during nest building. Recorded in Ontario, Canada by William W. H. Gunn.

Track 50: **Cedar Waxwing *See* Calls:** The other category of Cedar Waxwing calls (see Track 49) are the *seee* calls. These are high-frequency, hissy whistles, generally pure tones all on one frequency. *See* calls commonly are given by flock members during take-off or landing, or in flight, and one often can determine when a flock is about to take off by the change from *bzeee* calls to *seee* calls. *See* calls also are given when there is a disturbance, such as a predator, near a bird's nest. Recorded in Oregon by Geoffrey A. Keller.

Track 51: **Red-winged Blackbird Female *Teer*:** Female Red-winged Blackbirds appear to give the loud *teer* vocalization in aggressive encounters with other females. Recorded in Massachusetts by R. Simmers.

Track 52: **American Goldfinch *Per-chik-o-ree* Calls:** The twittery, lilting *per-chik-o-ree* call of the American Goldfinch is typically given by birds in flight. Recorded in Ontario, Canada by William W. H. Gunn.

Track 53: **Dawn Chorus in New York:** This chorus was fabricated from songs recorded by David L. Ross and Gregory F. Budney in three different locations in eastern deciduous forests in New York. Species sing in approximately the following order: Scarlet Tanager, Black-and-white Warbler, Worm-eating Warbler, Mourning Dove, American Crow, Ovenbird, Wood Thrush, Yellow-billed Cuckoo, Red-eyed Vireo, Winter Wren, Yellow Warbler, Indigo Bunting, Tufted Titmouse.

Track 54: **Dawn Chorus in French Glen, Oregon:** Species sing in approximately the following order: Red-winged Blackbird, Yellow-headed Blackbird, Yellow Warbler, Cedar Waxwing. Recorded in habitat dominated by juniper, sage, and bunchgrass on June 13, 1993 by David S. Herr.

Track 55: **Chipping Sparrow Daytime Song:** In typical daytime singing the male Chipping Sparrow perches high in a tree and gives a simple, long trill—approximately 20 repetitions of a single element—usually lasting about two seconds. He then pauses around 10 seconds before singing again. Recorded in Washington by David S. Herr.

Track 56: **Chipping Sparrow Dawn Song:** At dawn the male Chipping Sparrow usually sings from the ground, delivering very short trills separated by short pauses. Note how different this pattern is from that of typical daytime singing (Track 55). Recorded in California by Matthew D. Medler.

Track 57: **Eastern Wood-Pewee Daytime Song:** In typical daytime singing the male Eastern Wood-Pewee irregularly alternates between a whistled *pee-ah-wee* and a *wee-ur*. Recorded in Manitoba, Canada by William W. H. Gunn.

Track 58: **Eastern Wood-Pewee Dawn Song:** At dawn and dusk the Eastern Wood-Pewee gives his "twilight song," which adds a slurred, whistled *ah-di-dee* to the other two daytime song forms (see Track 57). Recorded in Nebraska by Geoffrey A. Keller.

Track 59: **Carolina Wren Duet (Fig. 7–60):** You will hear one song type from a male Carolina Wren; it follows the typical pattern of a syllable with approximately three parts (*tea-ket-tle*) repeated three or four times. But, on the first, fourth, sixth, and eighth songs, the female Carolina Wren adds a coarse rattle to the end, creating the simple duet of this species. This selection also illustrates the variability in this species' duet. Note that in the first song, the female begins her portion well before the end of the male's song; in the second and seventh songs, she inserts just a short buzz in the middle of the male's song. Recorded by Eugene S. Morton.

Track 60: **Bay Wren Duet (Fig. 7–61):** It is nearly impossible to distinguish the male and female contributions to this complex duet from the Bay Wren without carefully following the sonagram of the duet in Figure 7–61. Recorded on March 3, 1991 in the Province of Heredia, Costa Rica by David L. Ross.

Track 61: **Red-winged Blackbird *Konk-a-ree* Song Types:** You will hear one *konk-a-ree* song type from one male (designated Bird A), repeated two times; then you will hear a different *konk-a-ree* song type from a different male (Bird B), repeated two times. First song type recorded in New York by Randolph S. Little; Second song type recorded in Ontario, Canada by William W. H. Gunn.

Track 62: **Red-winged Blackbird Duet (Fig. 7–63):** You will hear the male begin his *konk-a-ree* song, then, toward the end of the song, the female gives several harsh notes that sound very much like the mobbing or scolding calls of many species. (In the first example, the female gives two scolding notes; in the second, she gives four notes.) In this "duet," the female's notes obscure the end of the male's song, at least to our human ears. Recorded in California by William R. Fish.

Track 63: **Northern Cardinal Song; Northern Mockingbird Mimicking Northern Cardinal (Fig. 7–64):** You will first hear the *breaker, breaker, breaker* song of a Northern Cardinal, with several introductory notes. Then you will hear a Northern Mockingbird mimicking the *breaker, breaker, breaker* portion of Northern Cardinal song. Northern Cardinal recorded on May 19, 1989 in Concan, Texas and Northern Mockingbird recorded on May 23, 1988 in Del Rio, Texas, both by Theodore A. Parker, III.

Track 64: **House Sparrow Song; Northern Mockingbird Mimicking House Sparrow:** You will first hear the series of chirps that passes for House Sparrow song. Then you will hear a Northern Mockingbird mimicking House Sparrow chirps. House Sparrow recorded in Massachusetts by C. S. Thomas; Northern Mockingbird recorded in Texas by Theodore A. Parker, III.

Track 65: **Blue Jay *Jaay* Calls; Northern Mockingbird Mimicking Blue Jay:** You will first hear a series of Blue Jay *jaay* calls (used to assemble a flock, or as alarm or mobbing calls). Then you will hear a Northern Mockingbird mimicking the *jaay* calls of Blue Jays. Blue Jays recorded on May 2, 1982 in New York by Andrea L. Priori; Northern Mockingbird recorded in Pennsylvania by J. C. Glase.

Track 66: **Killdeer Songs; Northern Mockingbird Mimicking Killdeer (Fig. 7–64):** You will first hear a series of Killdeer songs. Then you will hear a Northern Mockingbird mimicking Killdeer songs. Killdeer recorded in Montana by Geoffrey A. Keller; Northern Mockingbird recorded in Pennsylvania by J. C. Glase.

Track 67: **Northern Mockingbird Song Sequence:** A delightful sample of singing by a Northern Mockingbird. Note the typical mockingbird singing pattern of rapidly repeating each song type many times, then moving right on to the next song type without any pause. It is this singing pattern that gives away the identity of the singer, not necessarily the specific content of his songs. *Handbook of Bird Biology* editors' best guess as to the sequence of model songs sung is: Red-winged Blackbird, Blue Jay, American Robin *tuck* call note, unknown, American Kestrel *killy killy killy* call, Northern Cardinal, unknown, unknown, unknown, Tufted Titmouse, Wood Thrush, *Empidonax* flycatcher *whit* call (one time), Northern Flicker *kleer* call, Carolina Wren, unknown. Recorded in June 1994 in Groton Plantation in Luray, South Carolina by Gregory F. Budney.

Track 68: **Ovenbird Flight Song:** Note the typical *tea-CHER, tea-CHER, tea-CHER* song of the Ovenbird (a shortened version, with only three *tea-CHERs*) toward the middle of this long jumble of notes given as the singer launches himself high into the air. Recorded at dusk on June 8, 1994 in Seney National Wildlife Refuge in Michigan's Upper Peninsula by Lang Elliott and Ted Mack.

Track 69: **Eastern Phoebe Dawn Song (Sidebar 6, Fig. A):** Note that this recording of Eastern Phoebe dawn song was not used to produce the sonagram in Fig. A, but it is nearly identical to the recording used. In both, the Eastern Phoebe alternates his two song forms, *phoe-be* and *phoe-bree*, beginning with the *phoe-be*. Recorded in Ohio by Randolph S. Little.

Track 70: **Song Sparrow Song Sequence (Sidebar 6, Fig. B):** You will hear one male Song Sparrow sing a song type (designated Song Type 1) three times, then sing another song type (Song Type 2) three times, and finally repeat Song Type 1 three times. This is a natural recording, but the singer switches song types after fewer repetitions than is typical. Recorded in New York by Matthew D. Medler.

Track 71: **Wood Thrush Song Sequence (Sidebar 6, Fig. C):** If you just focus on the musical "middle" portion of the songs, the singer here sings songs with their middles in the following order: A, B, C, D, B, C, A, D, A, B. See caption for Sidebar 6, Figure C for more information. Recorded in New York by Arnold van den Berg.

Nests, Eggs, and Young: Breeding Biology of Birds

David W. Winkler

 *They sit for months at a time, huddled together against winds up to 60 mph (90 km/h) and wind chills of -150 degrees F (-100 degrees C) in the black ever-night of Antarctic winter, each incubating a single egg on top of his feet under a flap of densely feathered skin. When the sun returns to the Antarctic, the mates of these male Emperor Penguins return, after many days' journey over pack ice. They relieve the fathers and feed the recently hatched chick, which until now has been fed only a thick soup from the esophageal lining of the father. As each male departs for the distant ocean, he begins a long series of solitary trips for food to feed the rapidly developing chick—a series of forays, nest duties, and exchanges that, with luck, will culminate over six months from laying with the production of an independent descendant....*

In the dank, steamy forest of eastern Australia, a male brush-turkey (a type of megapode) tends his mound—a pile of rotting vegetation that can be as much as 16 feet (5 m) high and 40 feet (12 m) across, containing many tons of material; the nest site may have been continuously used for over a thousand years. Poking his bill into the mound he methodically monitors the progress of the fermentation within—he is a living thermostat, adding or removing vegetation from the mound to maintain temperatures and humidities near its core that are ideal for

Figure 8–1. Blue Tit at Nest Hole: *Like their North American relatives, the chickadees, Blue Tits of Eurasian woodlands nest in cavities and lay large numbers of eggs. Through natural selection, these short-lived birds have traded a high annual survival rate for the opportunity to rear many offspring each year. Photo by D. J. Saunders/Oxford Scientific Films.*

the development of the eggs of his species. Females are attracted to males according to the quality of the mounds they maintain and defend, and, after laying her egg, a female will never see her young again. The chicks that emerge from these eggs are the most well-developed chicks of any bird, and they flutter away from the mound as soon as they dig their way out, to live lives of independence from the very start....

*Far to the north and months later, in a cool, English oak wood, a female Blue Tit returns to her nest after a short absence (**Fig. 8–1**). She climbs into her nest hole and clears the moss and fur off the eggs—she had covered them just before she emerged to make a quick dash from the nest to defecate and grab a morsel of food. She is sitting on a clutch of 17 eggs, and as soon as these myriad young hatch, she and her mate will engage in a two-week marathon of feeding—gathering food from dawn to dusk and bringing it to the nest at rates of up to 500 visits per day. When the young tits flutter from the nest, they will seek shelter in high, dense vegetation. There they will be fed for another week or so, at ever decreasing rates, until the young become independent and depart their natal territory for a life of perpetual risk....*

At the edge of the same British wood is a most puzzling sight: a small warbler feeding a fledgling that is many times its own size. The hapless parent seems in danger of being swallowed by the offspring it is feeding, as one of ornithology's most intriguing dramas is played out once again. The large young is that of a cuckoo. The bird that inspired the cuckoo clock is actually a so-called "brood parasite"—the original dead-beat parent—relying on other species to incubate and rear its young. This parasitic habit has arisen at least six times in the evolution of birds. It relies on a carefully cultivated deception that prevents host parents from rejecting the parasite's young, a deception that in different parasites has led to the evolution of a suite of mimetic and aggressive ploys unparalleled in the feathered world....

■Despite these different scenarios, however, there is no essential variation in the reproductive challenge that these and all other adult birds face: finding a safe place to lay their eggs; ensuring that the eggs receive the warmth they need for the developing embryos within to hatch; and then ensuring that the young are well fed, increasing their chances of surviving to independence and then perpetuating the parents' heritage. Clearly, birds manage to meet the reproductive challenge in a variety of habitats with a rich diversity of strategies. This chapter explores these strategies, following the breeding cycle from nest building, through egg laying and incubation, to chick rearing and interactions between parents and offspring after the young fledge.

Over the past 40 years biologists have developed a rich set of ideas known as **life history theory** to explain the diversity in breeding strategies. This theory sheds light upon the evolution of **life history traits**, such as number of offspring and age of first reproduction, which are closely connected to the most fundamental processes of life: survival and reproduction. In discussing the life histories of birds, I will try to provide an inkling of the larger theoretical issues as we pass by. Even the most vagabond bird must settle down long enough to rear its young, and ornithologists have thus been able to gather more information about the breeding life histories of birds than any other aspect of avian biology. In writing this chapter, I have taken examples from my own reading and experience; by the chapter's end I hope you will feel empowered to ask questions of the birds in your world, as I have of the birds in mine.

Survival

■ Biologists today view life histories as the ways in which organisms compete with others of their species to leave the greatest number of descendants, and thus genes, in future generations. A key concept in life history theory is that trade-offs exist among different life history traits. The most common trade-off is between increased reproductive output and decreased survival—survival of both the parents and the offspring they produce. These trade-offs are reflected in the considerable variations in the reproductive effort and survival of birds: A Blue Tit may produce more than 15 fledglings in one summer, but has at best a 50 percent chance of surviving to the next year. In contrast, a Royal Albatross produces at most one fledged young every two years, but the adult's annual survival rate is over 95 percent. No matter how prodigious or restrained a bird's reproduction, its lifetime output is usually just adequate, on average, to replace losses to mortality. (You'll hear more about this in Chapter 9.) Variation in mortality is one of the principal ways that natural selection can sift among the variants in a population: the least fit are more likely to die before passing on their genes. Thus, mortality, as unfortunate as it may seem on an individual basis, is one of the wellsprings of the great diversity of reproductive (and all other) traits that we see in birds, and it is a natural place to begin this chapter's exploration of the breeding life histories of birds.

A great deal remains to be learned about the survival rates of birds, but several generalities emerge from the studies available:

(1) **Larger bird species tend to have higher survival rates than smaller ones, especially among closely related species (Table 8–1).** (This rule generally holds among all animals.) For example, adult swans have an 80 to 98 percent chance of surviving to the next year, whereas their small cousins, the ducks, have annual survival rates

Table 8–1. Annual Adult Survival Versus Weight: *Larger adult birds tend to survive better from one year to the next than smaller birds, but some exceptions are evident. For example some seabirds, such as the Sooty Shearwater and Northern Fulmar, are particularly long-lived for their weight. Unless otherwise indicated, all weight data from Dunning (1984) and survival data from Welty and Baptista (1988), after Lack (1954).*

Species	Annual Adult Survival (%)	Average Weight oz (g)	
Blue Tit	30	0.4	(11)*
Barn Swallow	37	0.7	(19)
Song Sparrow	30	0.8	(21)
European Starling	47	2.9	(80) female
		3.0	(85) male
American Robin	52	2.8	(77)
Blue Jay	55	3.1	(87)
California Quail	50	6.1	(170) female
		6.3	(176) male
Sooty Shearwater	91	10.3	(287)
Northern Fulmar	94	17	(479) female
		21.8	(609) male
American Coot	40	20	(560) female
		26	(724) male
Black-crowned Night-Heron	70	31.5	(883)
Herring Gull	70	37.3	(1,044) female
		43.8	(1,226) male
Mallard	52	38.7	(1,082)
White Stork	79	107–125	(3,000–3,500)*
Canada Goose (*canadensis* race)	84	118	(3,314) female
		136	(3,814) male
Yellow-eyed Penguin	90	185	(5,200)*
Tundra Swan	92**	221	(6,200) female
		254	(7,100) male
Royal Albatross	97	296	(8,300)*

*Weight and survival data from Brooke and Birkhead (1991)
**Survival data from Limpert and Earnst (1994).

of only about 40 to 60 percent. In general, most big birds (gull size or larger) have annual adult survival rates between 70 and 95 percent, and most small birds (robin size and smaller) have annual adult survival rates between 30 and 60 percent. A few groups appear to be distinctive, however: penguins have comparatively low survival rates for their size, and swifts and hummingbirds survive longer than one might expect from their small size. The ecological reasons for these distinctions are not known.

(2) Immature birds nearly always have lower survival rates than birds that have reached reproductive age. For example, in their first few years, gulls have average annual survival rates of 30 to 50 percent, whereas older birds survive at about 80 percent per year **(Fig. 8–2).**

(3) Birds in tropical areas tend to have higher survival rates than their relatives at higher latitudes. Studies of color-marked passerines in the tropics indicate that over 80 percent of adults survive each year, whereas only about 50 percent of similar-sized songbirds in the temperate zone survive each year (see Ch. 5, Patterns of Migration, Table 5–2).

(4) Birds have higher survival rates than mammals of a similar size. Because very few birds approach humans in mass, however, most birds have substantially lower survival rates, and consequently shorter life spans, than humans. The mammalian exceptions to this mammal-bird comparison are bats, whose relatively high survival rates approach those of similar-sized birds.

Once the period of low survival early in the life of an immature bird is past, the adult experiences a constant probability of surviving any given year for most of its remaining life. Ecologists define **survivorship** as the proportion of the individuals born at the same time (a **cohort**) that survive to a given age. Plotting survivorship relative to age yields a curve for birds that is distinctly different from our own: like birds, humans have a period of high mortality in the very early years of life, but after that period is past, we have very high survival rates until late in life, at which time our mortality rates climb steeply (see Fig.

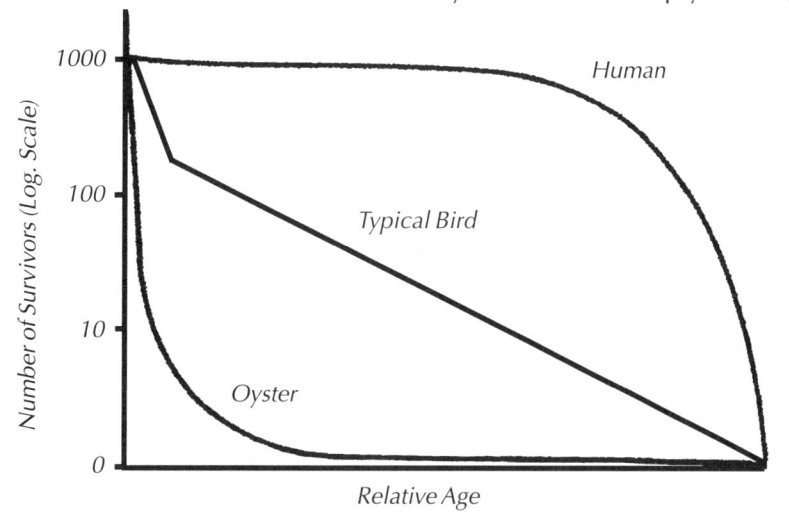

Figure 8–2. Typical Survivorship Curves: Graphs showing the approximate number of individuals of different species remaining alive at different ages, out of a starting group (cohort) of 1,000 individuals of each species. Relative age, rather than absolute age, is shown, because of the very different life spans of the species. Typical invertebrates such as the oyster produce vast numbers of young, but few survive to adulthood (note the sharp reduction in survivors at a young age). Once oysters reach a certain age, however, their chance of surviving each year remains high (as indicated by the nearly horizontal portion of their curve). Humans have a very short period of high mortality just after birth (not shown due to scale), but after that, their chance of surviving each year remains very high until late in life, when mortality rates rise again. Birds also have high mortality rates at a young age. Once they reach about one year of age, however, their survival rate remains quite constant (but lower than that of humans or oysters) for the rest of their lives. Unlike humans, birds rarely live long enough to experience greater mortality due to old age. Bird survivorship curve based on actual Herring Gull data from Paynter (1966), in which only 60 percent of the birds survived their first year, but 90 percent survived each later year.

8–2). Most birds do not live long enough to experience the senescence that causes this climb in mortality rates late in our own lives. This, coupled with the generally lower survival rates of birds, means that they face prospects very different from our own during adult life. So, before reading on, consider for a moment what it would be like to face a constant and substantially higher risk of mortality throughout your adult life; think what this risk would do to the potential for social ties outside the family, and what havoc it would wreak with our insurance premiums!

Some birds put more effort than others into reproduction—generally yielding more offspring, but at a cost of decreased survival for the parents. For example, a Northern Cardinal that puts too much energy into rearing offspring during one breeding attempt may depress its own survival to such an extent that it will end up producing fewer offspring in the long run than another Northern Cardinal that puts less energy into the current breeding attempt and saves more for its own survival, thus elevating its chances of producing more young in the future. Over evolutionary time, the way in which this trade-off between parental survival and offspring production plays itself out may set the optimal balancing of costs and benefits at very different levels of effort for different species **(Fig. 8–3)**. A key factor in this balance is the change in the number of offspring that a bird can expect to produce in the future. For example, if a bird is faced with a future in which the prospects for its own survival are beginning to decline, then it will produce more offspring overall if it puts more into reproduction immediately. In some birds, such as scrub-jays, gulls, and terns, for which much data on banded and recovered individuals has been obtained, researchers can detect a decline in annual survival rates very late in life. Pugesek (1981) has reported that the effort put into reproduction by very old California Gulls increases, possibly as an evolved response to this late senescence. As already noted, however, the great majority of birds never face a

Figure 8–3. Reproductive Strategies: Over their lifetimes, birds produce offspring in a range of different patterns. Shown here are two extremes. The Eurasian Tree Sparrow produces many young each year of its relatively short life, whereas the Yellow-eyed Penguin produces fewer young per year, but lives much longer—sometimes 20 years or more. Producing large numbers of young per year apparently places such an energetic burden on parents that their long-term survival is reduced. Natural selection thus imposes an evolutionary trade-off: species must choose either longer lives or a higher rate of reproduction. Adapted from Ricklefs (1973b, p. 410).

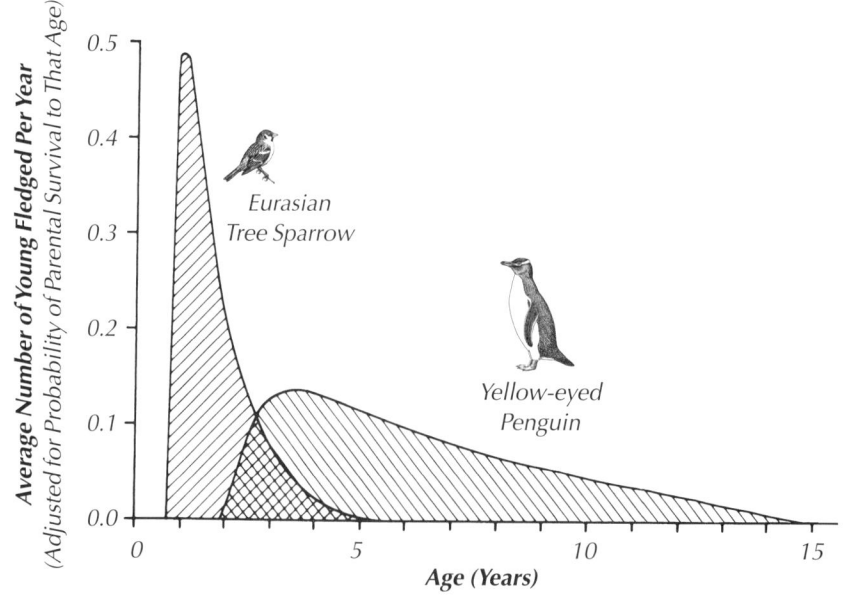

Figure 8–4. Zitting Cisticola: *The diminutive, warblerlike Zitting Cisticola (also known as the Fan-tailed Warbler) holds the record for rapid attainment of reproductive maturity. Formerly classified with Old World Warblers, cisticolas are now placed in their own family. Although most cisticolas are strictly African, this common species ranges from Africa into southern Europe, Asia, and even northern Australia, inhabiting a variety of wet and dry grasslands, marshes, and agricultural fields in both temperate and tropical regions. The polygynous male weaves a number of bottle-shaped nests on his territory, suspending them from grass stems. Individual females choose a nest, add a lining, and raise young with no additional help from the male, while he continues to build more nests and display for more females. The breeding season is long, lasting up to seven months in Mediterranean regions and Japan, and most of the year in tropical Africa. In a Japanese study, Ueda (1985) found that 11 percent of the breeding females were juveniles—nesting in the same breeding season into which they fledged. One female, banded as a nestling, laid her first egg merely 27 days after fledging, and another female, just 46 days after fledging. On average, juvenile females were as successful as adult females in their breeding attempts. The Zitting Cisticola is the only species whose range extends into the temperate zone that is known to complete two generations within the same breeding season. Photo by D. Tipling/VIREO.*

situation in which their annual survival rates are declining, and the curious biologist thinking about the evolution of life history traits in birds can consider a bird that is in the early years of its reproductive life to be essentially equivalent to one that is much farther along. This makes a big difference for the field biologist gathering data on the breeding biology of wild birds: discriminating the ages of breeding birds is much less important than it would be in life-history studies of many other organisms, such as mammals or fish.

Recall that immature birds have lower survival rates than adults, and the transition to a fairly constant survival rate usually occurs when a bird reaches reproductive maturity. Most songbirds reach maturity a little less than a year after fledging. However, females of some small tropical species with very long breeding seasons can initiate their own breeding attempts just a few weeks after fledging **(Fig. 8–4)**—laying eggs and rearing young in the same breeding season into which they hatched! At the other extreme are some large eagles, condors, and albatrosses, which can take up to 12 years to reach reproductive maturity.

Increasing the survival prospects of their young is one of the principal ways that birds can increase their reproductive output, and much of the breeding biology explored here is adapted to achieve this end. Before examining these strategies, however, it is worthwhile to briefly consider the principal threats to young. The most pervasive is predation. Recall that adult birds have higher survival rates for their size than mammals. Many researchers believe that this high survival results from their ability to escape predators through flight. Viewed from this perspective, the act of settling down at a single spot to rear young is a very significant departure from the avian lifestyle. Indeed for many birds, such as swifts and pelagic seabirds, the only time they come down from the sky for a significant period is when they breed. Many nest predators will take sitting adults as well as eggs or young, so incubating birds of all species, not just waterfowl, are

a. California Gull

b. American Crow

Figure 8–5. Predation on Adult Birds at Nest: Although nest predators usually destroy and eat only eggs or nestlings, occasionally they kill adult birds on the nest. a. California Gull: This California Gull incubating in a colony at Honey Lake, California was killed and eaten at night, along with its eggs. Photo courtesy of David W. Winkler. b. American Crow: This female American Crow and her nestling were killed and eaten overnight in their nest by a Great Horned Owl. Photo by Kevin J. McGowan.

literally "sitting ducks" for predators **(Fig. 8–5)**. The loss of an incubating parent is relatively rare, but eggs and young can be consumed by predators at an alarming rate: predators commonly destroy close to half the nests in most forest-nesting passerines **(Table 8–2)**. Some of birds' most engrossing nesting habits are antipredator adaptations to protect both parents and young. From swifts building their nests on cliffs and under waterfalls, Killdeer with their nearly invisible eggs and nests, and lapwings and geese with their pugnacious nest defense, to Southern Penduline-Tits and thornbills whose nests have false openings, much of the breeding biology of birds appears to be an effort to evade or fool predators. Notoriously effective predators include monkeys, raccoons, skunks, weasels, cats, opossums, and arboreal snakes **(Fig. 8–6)**. Among birds, the most effective nest predators in forested habitats appear to be crows and the crowlike, Australasian currawongs and their relatives; in marine habitats, the top predators are skuas, jaegers, gulls, and frigatebirds. Important nest predators sometimes come from surprising groups: mice and chipmunks are significant predators in North America, toucans in the Neotropics, and Great Spotted Woodpeckers in Europe. In Africa, a large bird of prey called the African Harrier-Hawk specializes on bird nests.

Other sources of mortality tend to be much less constant. Weather is probably the most important after predation. Although high rainfall, tides, or winds can cause the wholesale destruction of nests of marsh birds or colonial birds nesting on low islands, weather more often affects the eggs or nestlings indirectly by reducing food supplies. Forced to leave the nest for extended periods in search of food for their own maintenance, parents often lose their eggs or young to exposure or starvation. Birds that feed on flying insects can be especially vulnerable to cold, because their prey may still be present but unable to fly: In three years out of ten, cold fronts lasting two or more days and passing through during the nestling period have halved the fledgling production of Tree Swallows nesting near Ithaca, New York (McCarty and Winkler 1999). In tropical and desert areas, fledgling production can be severely depressed by unpredictable periods of extended rain or drought.

Table 8–2. Nest Predation Rates on North American Passerines: *Presented here for a selection of passerines is the percentage of nests lost to predators each breeding season, on average. Some of the percentages may include cases in which only some of the eggs or young were taken by predators. Note that the sampling methods vary greatly from study to study. In addition, predation rates may vary widely from year to year, and among different regions and habitats. Thus, these predation rates may not be directly comparable to one another, and must be viewed with some caution.*

Species	% Nests Lost to Predation	Location	Reference
Yellow-billed Cuckoo	39%	Arkansas	(Martin 1993a)
Eastern Phoebe	49%	Southern Indiana	(Weeks 1979)
	71%	Central Appalachians	(Hill and Gates 1988)
Greater Pewee	42%	Southeastern Arizona	(Chace and Tweit 1999)
Acadian Flycatcher	39%	Arkansas	(Martin 1993a)
Red-eyed Vireo	49%	Arkansas	(Martin 1993a)
Warbling Vireo	38%	Arizona	(Martin 1993a)
Carolina Wren	64%	Northwestern Alabama	(Haggerty and Morton 1995)
American Robin	39%	Pacific NW of N. America	(Sallabanks and James 1999)
	54%	Arizona	(Martin 1993a)
Wood Thrush	40%	Arkansas	(Martin 1993a)
Hermit Thrush	83%	Arizona	(Martin 1993a)
American Redstart	nearly 52%	New Hampshire	(Sherry and Holmes 1992)
Black-and-white Warbler	26%	Arkansas	(Martin 1993a)
Hooded Warbler	47%	Arkansas	(Martin 1993a)
	44%	Pennsylvania	(Evans Ogden and Stutchbury 1994)
	30%	Ohio	(Evans Ogden and Stutchbury 1994)
	36%	Ontario, Canada	(Evans Ogden and Stutchbury 1994)
Ovenbird	24%	Michigan	(Hann 1937)
	29%	Arkansas	(Martin 1993a)
	70%*	Central Illinois	(Robinson 1992)
Worm-eating Warbler	21%	Arkansas	(Martin 1993a)
Orange-crowned Warbler	33%	Arizona	(Martin 1993a)
Virginia's Warbler	31%	Arizona	(Martin 1993a)
Red-faced Warbler	40%	Arizona	(Martin 1993a)
MacGillivray's Warbler	49%	Arizona	(Martin 1993a)
Yellow-rumped Warbler	38%	Arizona	(Martin 1993a)
Western Tanager	45%	West-Central Idaho	(Hovis et al. 1997)
	30%	Northeastern New Mexico	(Goguen and Mathews 1998)
	46%	Arizona	(Martin 1993a)
Scarlet Tanager	69–78%**	Illinois	(Brawn and Robinson 1996)
Green-tailed Towhee	61%	Arizona	(Martin 1993a)
White-throated Sparrow	35%	Adirondack Park, New York	(Tuttle 1993)
	45%	Algonquin Park, Canada	(Falls and Kopachena 1994)
Dark-eyed Junco	31%	Arizona	(Martin 1993a)
Indigo Bunting	75%	Indiana	(Carey and Nolan 1979)
	43%	Arkansas	(Martin 1993a)
Black-headed Grosbeak	23%	Arizona	(Martin 1993a)

* An unusually high predation rate; data from small forest fragments.

** Sample Size = 4 nests.

a. Eastern Chipmunk

b. Bullsnake

c. Pied Currawong

d. Brown Skuas

Figure 8–6. Nest Predators in Action: *Eggs are vulnerable to attacks from a wide range of predators.* **a. Eastern Chipmunk Raiding American Robin Nest:** *Chipmunks consume significant numbers of eggs in North America, even from nests high in trees. Photo by Bruce D. Thomas.* **b. Bullsnake Eating Long-billed Curlew Egg at Nest:** *Snakes consume numerous eggs and young in a variety of habitats. Photo courtesy of Mary Tremaine/CLO.* **c. Pied Currawong Stealing Egg from Willie-wagtail Nest:** *In forests, crows and the crowlike Australasian currawongs are the most significant nest predators. Photo courtesy of Richard Major, Australian Museum.* **d. Brown Skuas Consume an Egg:** *In marine habitats, the skuas—large, predatory relatives of gulls—are among the most effective nest predators, along with jaegers, gulls, and frigatebirds. Photo by R. and N. Bowers/VIREO.*

Parasites also can have devastating impacts on fledgling production. Seabirds that nest in dense island colonies, such as boobies, pelicans, and gulls (see Fig. 6–15a), can be wiped out by viruses or other pathogens passing swiftly through the colony. For birds nesting in a more dispersed fashion, brood parasites often cause significant mortality. Cowbirds have been found to parasitize up to 100 percent of the Wood Thrush nests in woodlots in Illinois (Robinson and Wilcove 1994) and parasitism by cuckoos in the Old World can foil over 70 percent of the nesting attempts of their hosts (Liversidge 1971).

Many of these high parasitism rates cannot be sustained by the host populations: cowbird parasitism, for example, is threatening the survival of several species of birds in North America, including Kirtland's Warbler and Bell's and Black-capped vireos.

The Timing of Breeding

In his novel, *Ape and Essence*, Aldous Huxley explored how different our social and emotional lives would be if we restricted our breeding to a discrete period in the annual cycle. Imagine how intense those weeks would be! Then consider that most birds have just such a limited

breeding season. In a given year, the length of time over which most temperate zone birds can potentially breed is no more than about three months; in the tropics, the season of potential reproduction may extend to as much as six months.

In an influential early paper, Baker (1938) introduced the concepts of **ultimate** and **proximate factors** (see Ch. 6, Questions About Behavior) to explain variation in the timing of the breeding seasons of birds. One of the most common proximate factors influencing the start of breeding is day length. Through an elaborate chain of physiological steps within the bird's body, day length serves as the cue that tells most birds the season (see Fig. 7–37 for how day length affects one aspect of breeding behavior—singing rate). Ultimate factors, by affecting the survival and reproduction of birds' ancestors, have caused each population to evolve the breeding time that works best for its members, on average. For most species, the food supply for the developing young is the ultimate factor that seems to have most strongly influenced breeding time. In the annual cycle, birds need the greatest food resources when feeding large young, and most species, at least in the temperate zone, appear to time their breeding seasons to match the period of peak demand with the peak in food availability **(Fig. 8–7)**. This synchronization can explain much of the variation in breeding time that we see: Goldfinches are the only common birds in North America that feed their young thistle seeds (most songbirds bring insects to their young), and goldfinches breed one to two months later than other songbirds, coordinating their nestling feeding with thistle availability. Many birds of prey, such as Great Horned Owls, breed very early in the season—laying their eggs as early as November and December in the southeastern United States—because their large young take a long time to develop. By starting early, the owls can synchronize their period of peak energy demand with the peak availability of vulnerable young mammals and birds in early summer. One raptor, Eleonora's Falcon, has switched its predominant prey base for rearing young to the plethora of migrant land birds that pass through the eastern Mediterranean in the fall; its late-summer laying insures that the nestlings will reach the stage at which they require the most energy just when most is available.

One familiar bird, the Rock Dove, comes close to avoiding breeding seasons altogether. Although it breeds more commonly in spring and summer, the Rock Dove can be seen throughout North America incubating or feeding nestlings in the most inclement weather of deepest winter. Its native relative, the Mourning Dove, also has an uncommonly long breeding season—about six months in New York (Bull 1985). Doves and pigeons may be able to breed when other birds cannot because they feed their young nestlings crop milk (see Fig. 4–99b). A parent may be able to secrete crop milk while subsisting on a seed diet that would not meet the nutritional needs of very young nestlings.

Pigeons and doves have uncommonly *long* breeding seasons, but a few birds have uncommonly *variable* breeding seasons. Birds of desert areas, especially in southern Africa and Australia, nest in extremely unpredictable environments. Whether their environments can support

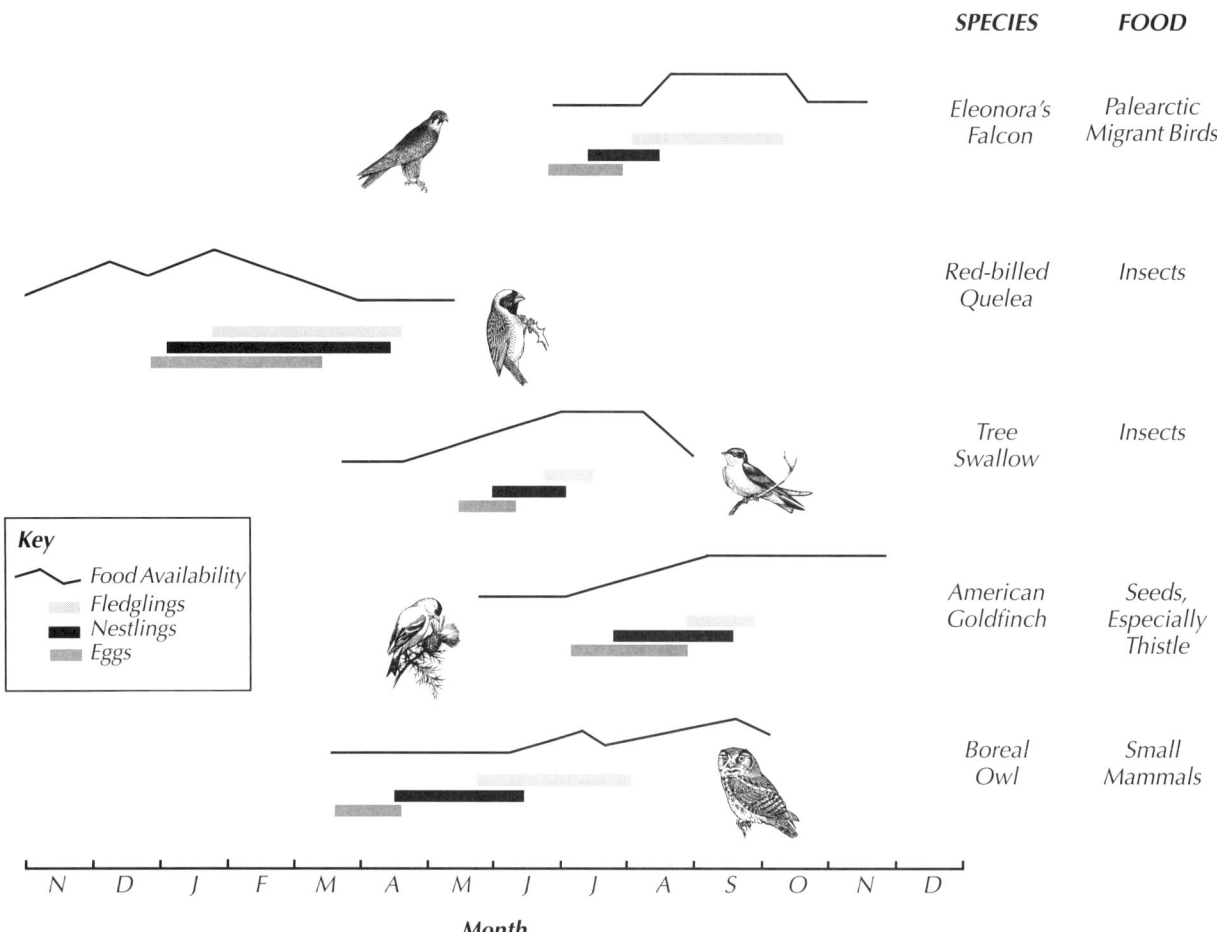

	SPECIES	FOOD
	Eleonora's Falcon	Palearctic Migrant Birds
	Red-billed Quelea	Insects
	Tree Swallow	Insects
	American Goldfinch	Seeds, Especially Thistle
	Boreal Owl	Small Mammals

Key
— Food Availability
Fledglings
Nestlings
Eggs

N   D   J   F   M   A   M   J   J   A   S   O   N   D

Month

Figure 8–7. Breeding Time and Food Abundance: *Most bird species time their breeding so that the period when they need the most food resources—while feeding large young—coincides with the yearly peak in the young's food supply. This chart, indicating the months when eggs, nestlings, and fledglings typically are present, as well as the approximate seasonal patterns in the availability of food for the young, illustrates this correlation for selected species with different types of prey.* **Eleonora's Falcon** *nests relatively late in the year, taking advantage of the abundant avian migrants passing through the eastern Mediterranean region in the fall, which provide ample food for older nestlings and fledglings. Data are based on Walter (1979). Nesting on the African savannas,* **Red-billed Quelea** *feed their nestlings exclusively on insects for their first five days, although the adults are primarily seed-eaters. They breed during the rainy season, when insects are the most abundant. The "food" graph for quelea actually indicates monthly rainfall, which is directly correlated with insect abundance. Data courtesy of Jim Dale.* **Tree Swallows** *and most other temperate zone passerines that feed insects to their young begin breeding in the spring, matching their time feeding older nestlings and fledglings to the high insect abundance of late spring and summer. Data courtesy of David W. Winkler. Unlike most songbirds,* **American Goldfinches** *feed their young on the seeds of thistles and other late summer flowers. They breed one to two months later than other songbirds, apparently to match nestling feeding to thistle abundance. Breeding data based on Middleton (1993). Many birds of prey breed extremely early in the year, temperate zone species frequently laying eggs during winter. Because their young take a long time to develop, starting early allows the parents to have older nestlings and fledglings during early summer, when their prey—vulnerable young mammals and birds—are at peak.* **Boreal Owls** *also breed fairly early, producing large young by summer, but for different reasons. Their primary prey, voles, peak in autumn, well past the fledgling stage, but the vole supply in summer appears to be sufficient for feeding young. Early breeding is probably advantageous because both juveniles and their parents have more time before winter to increase their fat reserves and to complete their postnesting molt. In addition, young fledged earlier have more time to develop their hunting skills before winter, and early breeding adults increase their chances of producing a second brood. Note that the Boreal Owl's small size and consequently faster rate of development allow it to breed slightly later than many larger birds of prey: eggs laid in mid-March produce fledglings by mid-May. Data from Korpimäki (1987) and Korpimäki and Hakkarainen (1991).*

breeding depends entirely on whether rains have fallen recently. The crossbills of the Northern Hemisphere face unpredictability of a similar sort (see Fig. 4–119). Like their relatives the goldfinches, crossbills feed their young on a diet of seeds, but they require the seeds of pine and spruce trees, not thistles. Seed crops in these trees are notoriously variable from year to year, and crossbills range over enormous distances searching for stands of trees with sufficient seed production to support a breeding attempt. As in birds of tropical deserts, the reproductive systems of crossbills are primed to go into action whenever promise of sufficient food for breeding is encountered, and crossbills have been recorded breeding in every month of the year! Note that these birds with highly flexible breeding seasons are responding to food supply as both the proximate cue and the ultimate selective force for timing their breeding.

Breeding Territories

■During the breeding period, nearly all birds defend some kind of territory. Although the size and function may vary widely among different species, most territories include at least the nest site. The majority of species, including warblers, vireos, thrushes, sparrows, most other songbirds, and many woodpeckers, live completely within their territory during the breeding season—mating, nesting, and feeding there.

A number of other species defend smaller territories—usually just the area immediately around the nest—sharing neutral feeding grounds with other conspecifics with little or no aggression. Grebes, swans, harriers, goldfinches, and blackbirds—especially Red-winged and Yellow-headed blackbirds—have this type of territory. Red-wings place their nests rather close together in a marsh, roadside ditch, or wet field and feed together nearby. If they visit your feeder in the spring, you may notice several males eating together peacefully at one time. If a male intrudes near another's nest, however, he is fiercely attacked. In birds that nest colonially, such as penguins, albatrosses, petrels, pelicans, gannets, cormorants, herons, gulls, and terns, the territory may be tiny—restricted to just enough space around the nest to allow mating and nesting activity without physical contact with neighboring pairs. Indeed, some of these territories appear to be separated by the distance an incubating bird can reach with its beak: in exactly the same habitat, nests of the large Peruvian Pelican are placed around two per square yard, whereas those of the smaller Guanay Cormorant are at three per square yard (Welty and Baptista 1988).

A few breeding birds defend territories that do not include the nest site. These include leks, used only for male display and mating (see Ch. 6, Reproductive Behavior: Lek Polgyny); and feeding territories, such as the flower patches defended by many hummingbirds (see Ch. 6, Sidebar 4, Grouping Versus Territoriality). Several species defend separate feeding and nesting territories: Seaside Sparrows defend a feeding area right along the shore in addition to their brushy nest site well above the high-tide line; and in Southern California, Phainopeplas defend their nest site in trees in the canyon bottoms and commute to their hillside foraging territories, which contain stands of buckthorn trees that bear their principal food, a berry (Walsberg 1977).

The territorial system may also vary *within* a species, depending on local ecological factors. Most Song Sparrows, for example, stay within their territories in the breeding season—mating, nesting, and feeding there. But when the density of nesting birds is high, as on some islands off the coast of British Columbia, each male may defend only a small mating and nesting area, joining other males to feed in one area without aggression.

A few birds also defend territories outside the breeding season. For example, Northern Mockingbirds often defend their berry-rich wintering area as vigorously as they do their breeding territory. Snowy Owls, which breed on the arctic tundra, defend winter territories a bit further south that contain fields and edge habitat rich in their chief prey, rodents (Boxall and Lein 1982). Other birds known to defend winter territories include, in North America, Loggerhead Shrikes, American Kestrels, Red-headed Woodpeckers, Sanderlings, and many other shorebirds. On their wintering grounds in Africa, Common Nightingales, European Robins, White Wagtails, and wheatears also defend territories. For the most part, however, nonbreeding birds move in flocks of single or mixed species, or move around an undefended home range either singly or loosely associated with their mate. Even in species that are territorial in winter, individuals can switch to flocking whenever a predator appears or when so many intruders attempt to get at the defended resources that territorial defense is no longer possible.

Functions of Breeding Territories

Although territories are costly to defend—in terms of time, energy, and personal risk—birds benefit from them in many ways. Living in a familiar area allows birds to evade predators more readily, because they know the most protected spots and the best hiding places. Birds also may be able to find food more easily in a familiar area, and defending areas with adequate food secures that food supply for the bird's own use. But why does defending an exclusive area become crucial to most birds in the *breeding season*?

Researchers believe that the primary benefit a bird gains from a breeding territory is that it reduces the chances that other members of the same species will interfere with its breeding effort. Although at first glance most breeding birds appear relatively faithful to their mates and at peace with other members of their species, this scenario rarely survives close scrutiny. Female Warbling Vireos, for example, regularly steal nest material from each other's nests (Howes 1985). Many birds destroy the eggs of conspecifics: Marsh Wrens and House Wrens routinely puncture conspecifics' eggs, and roughly 10 to 20 percent of the nests in Cliff Swallow colonies studied by Brown and Brown (1988) lost at least one egg to conspecifics. In Australia, more than 50 percent of the clutches of Crimson Rosellas (bright red and blue parakeets) observed by Krebs (1998) were destroyed by other Crimson Rosellas during laying. Beissinger et al. (1998) demonstrated how effective nest defense can be: when they set out experimental nest boxes with eggs, 75 percent of the clutches were destroyed by Green-rumped Parrotlets (small Neotropical parrots) within three days, whereas only 4.5 percent of the clutches were destroyed in control nests defended by Green-rumped Parrotlet pairs. Other birds will go so far as to kill the young of conspecifics: unmated male Barn Swallows sometimes kill other Barn Swallows' nestlings by grabbing them and throwing them to the ground.

Breeding interference also may result in birds rearing young that are not their own. Studies with individually marked birds have shown that in most species "monogamous" males will try to sneak copulations with females other than their mates whenever the opportunity arises; these extrapair copulations allow them to father more young, thus further promoting their genes without increasing their parental care duties. Mated females, also, may seek extrapair copulations with mated or unmated males they deem "good genetic material" (see Ch. 6, Mate Choice: Extrapair Copulation in Birds)—and their mates may end up rearing young that are not genetically their own. Very recently, aided by DNA comparisons, researchers have discovered that many species regularly lay eggs in the nests of conspecifics. Over 185 species not previously considered parasitic—especially colonial birds, waterfowl, and other precocial birds—are now known conspecific brood parasites (Rohwer and Freeman 1989; Rothstein and Robinson 1998; Semel and Sherman 2001; Eadie et al. 1998).

By defending an area for breeding, birds isolate themselves from others, thus decreasing their vulnerability to interference with their courtship and nesting activities. Although maintaining any kind of exclusive space is clearly helpful, larger territories provide more isolation than smaller ones—but they also cost more in time and energy to defend.

Nests and Nest Building

■ No branch of ornithology has a greater disparity between our current level of understanding and what our understanding could be than the study of nests. Only insects exceed birds in the diversity and sophistication of the nests they build, and a tremendous amount of literature, much of it from the 19th century, qualitatively describes the nests and nest-building behavior of birds **(Sidebar 1: Neat Nesting Facts)**. However, a quantitative literature that includes a rigorous scientific approach to asking and answering questions about nest function and construction is in its infancy. Across the entire spectrum of birds, from common backyard birds to exotic species in far-off lands, interesting and compelling questions wait to be answered. I will ask a few of these questions in the pages to come, but I wish to emphasize here that many of these questions can be answered with no specialized tools or techniques. I encourage

(Continued on p. 8·18)

Sidebar 1: NEAT NESTING FACTS
Sandy Podulka

Birds can fly and we can't—that alone is enough to inspire our awe and fascination for these small, feathered bundles of energy. But birds also have adapted to almost every habitat on land, and in building their nests and raising their young they regularly carry out tasks that seem nearly impossible to us. Below are some of the feats accomplished by birds—some routine and some quite unusual—as they go about breeding.

Nests

Unusual Nest Sites

Although most birds place their nests on the ground, in vegetation, in cavities, or on rocky ledges, a few individuals have fledged young from more mobile sites. An Ash-throated Flycatcher in Colton, California nested in a crevice in an active steam shovel that moved up to 200 feet (61 m) a day; a Barn Swallow nested on a slow-moving narrow-gauge train in British Columbia, Canada; and both Prothonotary Warblers and Tree Swallows have successfully nested on ferryboats moving back and forth daily across rivers.

Birds also are good at adapting to whatever nest sites are available. On treeless islands, birds that typically nest in trees, such as American Robins, may place their nests right on the ground.

Sizes of Nests

Using primarily their beaks and feet, birds build a bewilderingly diverse array of nests, ranging in size from the tiny cups of the Bee Hummingbird of Cuba, just 0.8 inches (2 cm) across and about 1 inch (2.5 cm) tall (around the size of a large thimble), up to the enormous mounds of megapodes, which weigh many tons. The largest nests in North America, however, are built by Bald Eagles. The record is probably held by one near St. Petersburg, Florida that was 20 feet (6 m) deep and 9.5 feet (3 m) across (Broley 1947). One near Vermilion, Ohio is thought to have weighed about two tons when it came crashing down (Herrick 1934). The mud nest of the Rufous Hornero of South America (see Fig. 8–34d) weighs up to 11 lb (5 kg), nearly 70 times the weight of its 2.6-oz (75-g) builder. We would be busy for quite some time if compelled to construct a 9,000-lb (4,000-kg) mud globe using our lips and toes!

Odd Nest Materials

Birds are equally resourceful when selecting nest materials. Although most species prefer certain materials when they are available, sometimes other substances are more accessible. A Warbling Vireo in California built its nest en-

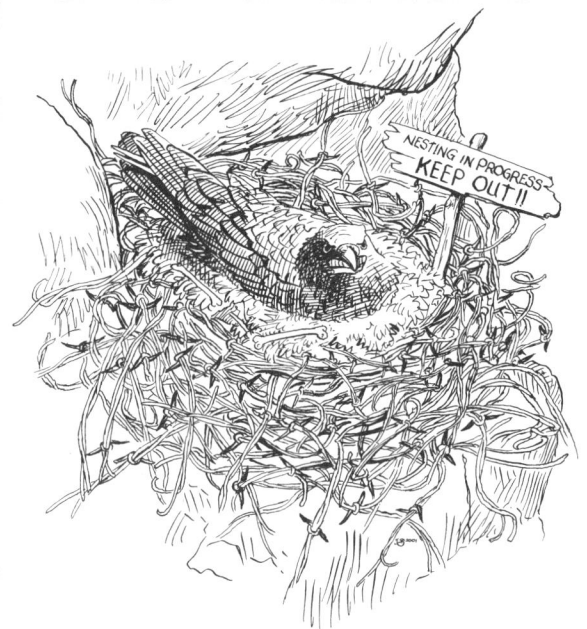

Figure A. White-necked Raven in Barbed Wire Nest: *Birds may make creative use of the materials on hand in constructing their nests. White-necked Ravens sometimes include barbed wire, and occasionally may build a nest entirely from it.*

tirely of facial tissue, White-necked Ravens occasionally use only barbed wire **(Fig. A)**, a Ruby-throated Hummingbird on Long Island, New York relied solely on fiberglass roofing insulation, and a Carolina Wren used mostly hairpins. In Bombay, India, crows included 25 pounds (11 kg) of gold eyeglass frames stolen from an open shop window in the structure of their nest; and, off the coast of Labrador, Double-crested Cormorants—who routinely obtain nest material under water—salvaged pocketknives, men's pipes, hairpins, and combs from a sunken trading vessel to use in their nests (Forbush 1925–1929). In Fresno, California a Canyon Wren placed its nest on the beam of an office building, and keeping with the office theme, constructed it entirely out of office supplies such as paper clips, pins, rubber bands, thumbtacks, shoelaces, needles, wire, matches, and toothpicks—1,791 items in all **(Fig. B)**. The large nest—nearly 8 inches (20 cm) high—weighed a full 2.5 pounds (1.1 kg).

Birds frequently use animal hair in their nests, especially in the lining, but one Loggerhead Shrike in Florida built its nest almost entirely of hair from a nearby dead cow. Birds don't always wait for hair to become available, though. Tufted Titmice may pluck hair from living opossums, woodchucks, squirrels, or people, and Black-capped Chickadees may grab hair from sleeping raccoons (Lynn Leopold, personal communication).

Nest Construction Feats

If you have ever watched a bird build a nest, you probably thought it looked like hard work—and apparently, it is. Barn Swallows may make more than 1,200 trips with a mouthful of mud to build their nest; Black Woodpeckers

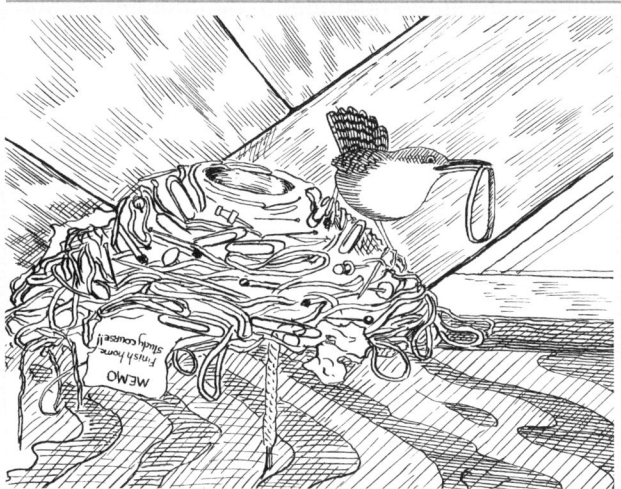

Figure B. Canyon Wren Nest of Office Supplies: *An enterprising Canyon Wren in Fresno, California adds yet another rubber band to its nest. The nest, weighing 2.5 pounds (1.1 kg), was located in an office building and was constructed entirely from office supplies—a total of 1,791 items.*

in Eurasia must hammer on a tree about 100,000 times to excavate their nest cavity (Cuisin 1983); and Hamerkops may make roughly 8,000 trips over four months carrying sticks and grass to build their huge nest mounds (see Fig. 8–38) (Kahl 1967). The pendulous cup nest of one Altamira Oriole in Mexico contained 3,387 pieces of grass, many of which were three to four feet (0.9 to 1.2 m) long. (One wonders whether the bird's nest construction or the researcher's job of counting each piece of nest material was the more difficult task.) The Edible-nest Swiftlet, which constructs its nest almost entirely of saliva (see Fig. 4–93), must work for five or six weeks to build up its nest, even though its salivary glands enlarge greatly during the nesting season (Stresemann 1927–1934).

Nest Sharing and Reuse

Sometimes more than one nesting pair may benefit from a bird's labor in constructing a nest. Large nests, in particular, may be used year after year by the same pair or a succession of different pairs. Bent (1937) reported three different Osprey nests occupied continually for 41, 44, and 45 years. And, the large platform nest of a White Stork in Germany was occupied from 1549 through at least 1930! One nest built in a tall chestnut tree near Saybrook, Connecticut served four species in four years: built by a Cooper's Hawk that used it one year, it was taken over by Great Horned Owls, then Red-tailed Hawks, and finally Barred Owls; each spent a breeding season in the nest (Bent 1938).

Occasionally, two females or pairs of birds may occupy the same nest at the same time. Two Long-billed Curlew females shared a nest containing a large number of eggs, presumably laid by both females. Two Song Sparrow pairs have done likewise, as have two Wood Duck females. In each case, females shared incubation duties—the Wood

Ducks even occasionally incubating simultaneously.

Even more rarely, two different species may share a nest. In Florida, a Great Horned Owl moved into a Bald Eagle's nest, each incubating their own eggs just a few feet apart. (Great Horned Owls often completely take over eagle nests.) In Denver, Colorado, an American Robin and a House Finch shared the robin's nest (Bailey and Niedrach 1936). The female House Finch laid six eggs in the robin's nest, and both pairs fed both species of nestlings. The larger robin young eventually smothered the nestling finches, but the adult House Finches continued to feed the young robins, even after they had left the nest.

Feeding the Young

Young birds, like most babies, are a lot of work right from the start. Experiments with young nestling crows determined that they must eat one-half their body weight each day just to stay alive, but that they could easily consume their full weight daily. In the wild, nestling Belted Kingfishers may devour up to 1.75 times their weight in fish each day (White 1939). In general, older nestlings consume greater amounts of food, but their intake is a smaller fraction of their weight—only about 25 percent. One fledging-age American Robin described by Allen (1961), however, was fed (by researchers) as much as it would eat, and it managed to down a full 14 feet (4.3 m) of earthworms!

To keep the food flowing, most parents make numerous trips to their nest. Skutch (1976) determined that most small songbirds feed each of their nestlings an average of 4 to 12 meals per hour. This can add up quickly: a European Pied Flycatcher, during one nest cycle, may visit the nest 6,200 times with food (Welty and Baptista 1988). One male House Wren in Illinois feeding 12-day-old nestlings by himself may hold the record, however. He brought food to the nest 1,217 times between 4:15 A.M. and 8:00 P.M.—an average of once every 47 seconds! Birds that bring larger prey feed their young much less often. Golden Eagles, for example, bring a rabbit or grouse to their nest about twice a day, and Bald Eagles bring prey four or five times a day.

Some birds travel great distances to find food for their young. Seabirds are famous for this (see Feeding the Young, later in this chapter), but even the Common Swift of Europe may fly a total of 621 miles (1,000 km) each day to gather food for its two or three nestlings (Welty and Baptista 1988). North of the Arctic Circle, parents may gather food over a longer workday (Welty and Baptista 1988): one female Bluethroat (an Old World flycatcher) fed her young from 3:00 A.M. to 11:45 P.M.—a 21-hour day shift! ■

Note: Unless otherwise indicated, the facts presented above are from Terres (1980) and Kress (1988).

Figure 8–8. Lichen-covered Nest of Blue-gray Gnatcatcher: *The small, compact, cup nest of the Blue-gray Gnatcatcher, usually saddled atop a horizontal limb, is neatly bound together and to its supporting branch with spider webs. Pieces of lichen camouflage the outside so well that from a distance the nest may look like just another bump on the branch. Photo courtesy of David Allen/CLO.*

readers to relentlessly ask "why?" Why do goldfinches line their nests with the down from thistle seeds? Why do so many birds (hummingbirds, Blue-gray Gnatcatchers **(Fig. 8–8)**, the nuthatch-like sittellas of Australia and New Guinea, Long-tailed Tits of Eurasia, and others) plaster lichens on the outside of their nest cups? Why does the Cliff Swallow build its nest without any of the vegetation that the Barn Swallow uses to build its mud cup? Why don't all swallows use feathers to line their nests as do Tree, Barn, and Bank swallows? Does the same species nesting in different parts of the continent build a different kind of nest? (Enlist some pen-pals or join the CLO's Birdhouse Network to find out!) These sorts of questions will lead inevitably to questions about the costs (in terms of parental effort and risk) and benefits (in terms of greater nest safety) of various sorts of nest construction and placement, and only when hundreds of people begin asking such questions will a modern science of nest study be possible.

Functions of Nests

Nests are primarily structures to hold and protect the eggs and, in many species, the developing young. In some species, the nest takes on additional functions: some birds roost in their nests even when they are not breeding, and others use them to attract mates. Some nests have secondarily taken on important functions in courtship: many bird species limit their courtship behavior to the vicinity of the nest, and mutual nest building is a common aspect of pair formation and solidification. Nevertheless, providing a safe haven for eggs and young is the main function of nearly all nests.

Clearly, one of the principal requirements for a nest and its site is to thwart predators, and birds have evolved many adaptations to do so. One of the most interesting is placing the nest near insects or other animals that keep predators away. Many tropical birds, for example, nest on or very near the nests of aggressive wasps (Joyce 1993) or in acacia trees tended by aggressive ants (Young et al. 1990), both of which attack mercilessly any of a large suite of potential predators that approach their nest or nest tree **(Fig. 8–9a)**. No one yet knows how (or why) the nesting birds are spared the insects' attacks. Violaceous Trogons of the Neotropics dig their nest cavity right inside a large paper wasp nest, and other trogons, many kingfishers, and various parrots, including the Orange-fronted Parakeet, dig their nest tunnel into an active termite mound **(Fig. 8–9b)**. In the Old World, House Sparrows and European Starlings often nest near Imperial Eagles; and in the tundra, Snow Geese, Brant, and Common Eiders often nest near a Snowy Owl, whose presence discourages attacks by arctic foxes. The ploverlike Water Thick-knee of Africa nests along sandy shorelines near breeding crocodiles (Gill 1990).

a. Nest of Rufous-naped Wren in Ant-Acacia Tree

b. Black-headed Trogon Nesting in Termitary

*Figure 8–9. Birds Nesting in Association with Social Insects: Many tropical birds build their nests near or in the nests of aggressive ants, wasps, or termites. These insects do not bother the birds, but do provide protection from nest predators. **a. Nest of Rufous-naped Wren in Ant-Acacia Tree:** The Rufous-naped Wren—a large, boldly patterned wren of dry tropical forest and thorny scrub in Central America—often places its nest in an **ant-acacia tree**. These trees are covered with large thorns which, formidable predator-deterrents themselves, are inhabited by acacia ants that attack any animal that comes near the tree. In addition, certain aggressive wasps also tend to nest in ant-acacias, for protection against predatory army ants. Rufous-naped Wrens nesting near the nests of these wasps fledge more young than wrens that do not, because they gain the wasps' protection from nest-robbers such as white-faced monkeys (Joyce 1993). Here, two Rufous-naped Wrens approach their nest (center) near a wasp nest (cylindrical structure on the left) in an ant-acacia tree in Santa Rosa National Park, Costa Rica. **b. Black-headed Trogon Nesting in Termitary:** A variety of tropical birds place their nests inside **termitaries** (active termite nests), including this male Black-headed Trogon peering from its nest hole in Santa Rosa National Park, Costa Rica. Termites use digested wood and their own feces to construct their nests, forming a hard, carton-like mound into which birds may dig their nests. As a bird digs into the mound, the termites seal the exposed walls, so that the bird's nest and the termites' living chambers are not in direct contact. Other termitary-nesters include many other New World trogons, several parrots, nearly half the kingfishers, and some jacamars and puffbirds. In Peru, Brightsmith (2000) found that Tui and Cobalt-winged parakeets, as well as Black-tailed Trogons, prefer termitaries that also house colonies of aggressive, biting ants. Brightsmith suggests that the ants, which have a distinctive odor, may protect the birds' nests either directly by attacking potential predators, or indirectly because their odor masks the smell of a bird's nest or in some other way discourages predators from searching for nests. Inset shows male Black-headed Trogon. Photos by Marie Read.*

8

Predation pressure on bird nests is generally thought to be greatest in the tropics, and many species there build nests that are extremely small for the bird's size, apparently to escape detection (Snow 1976) **(Fig. 8–10)**.

Nests also provide shelter from the elements. The microclimates around and in many nests are much more favorable than those in the surrounding environment. For example, temperate-nesting humming-birds choose nest sites under overhanging trees to minimize heat loss at night (Calder 1973); some gulls place their nests in shade to reduce the risk of overheating and dehydration for the nestlings (Winnett-Murray 1979); and the enclosed nests of many desert birds provide shade, preventing the overheating of eggs (and the parents incubating them) during the day (Yom-Tov and Ar 1980 [1981]). Nests also contain the eggs and young to keep them from rolling out: New World orioles and their large, tropical relatives the oropendolas, as well as the Old World weavers (distant relatives of House Sparrows), build their nests at the tips of very long, thin branches, probably as an adaptation against predation by monkeys and other mammals. Their deep, enclosed nests help prevent the eggs from falling out as the nests swing wildly in the wind—a frequent result of their precarious location.

Diversity of Nest Sites

Birds nest in virtually every terrestrial or shallow water habitat on earth—from the surfaces of lakes to rock niches above timberline, and from deep tropical caves to the howling, frozen barrens of Antarctica. Other than the surface of the open ocean and thin air, it is difficult to think of a habitat birds do not use for nesting. Furthermore, the sites birds use within those habitats are similarly diverse. Featureless cliff

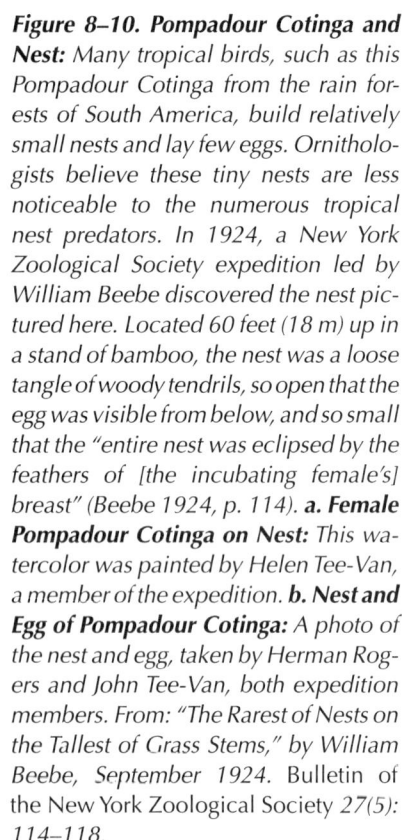

Figure 8–10. Pompadour Cotinga and Nest: Many tropical birds, such as this Pompadour Cotinga from the rain forests of South America, build relatively small nests and lay few eggs. Ornithologists believe these tiny nests are less noticeable to the numerous tropical nest predators. In 1924, a New York Zoological Society expedition led by William Beebe discovered the nest pictured here. Located 60 feet (18 m) up in a stand of bamboo, the nest was a loose tangle of woody tendrils, so open that the egg was visible from below, and so small that the "entire nest was eclipsed by the feathers of [the incubating female's] breast" (Beebe 1924, p. 114). *a. Female Pompadour Cotinga on Nest:* This watercolor was painted by Helen Tee-Van, a member of the expedition. *b. Nest and Egg of Pompadour Cotinga:* A photo of the nest and egg, taken by Herman Rogers and John Tee-Van, both expedition members. From: "The Rarest of Nests on the Tallest of Grass Stems," by William Beebe, September 1924. Bulletin of the New York Zoological Society 27(5): 114–118.

a. Female Pompadour Cotinga on Nest

b. Nest and Egg of Pompadour Cotinga

faces, holes and cracks of every type, branches of all diameters, bare ground, as well as myriad human artifacts (skyscrapers, out-buildings, bridges, telephone poles, signs, oil pumps, old boots and hats): nearly anything that would support a nest has been called into service **(Fig. 8–11)**. Nest sites need not even be stationary: robins, swallows, and phoebes have been reported to nest on active ferries. The variety of nest sites is so diverse as to defy classification. The one factor common to successful sites, however, is that they are protected from predators and the elements, and close to the food supply for hungry nestlings.

Some species seem to have much more specific nest site requirements than others, and the Kirtland's Warbler may be one of the most demanding. This endangered species, which breeds exclusively in a 100-by-60-mile (161-by-97-km) area in northern Lower Michigan, requires a specific combination of plants of particular sizes for successful nesting. It places its nest on the ground among grasses, arched over by the living lower limbs of pines, usually jack pines between 6 and 18 feet (2 and 5.5 m) tall **(Fig. 8–12)**. Smaller pines are too dense near the base; older trees have too many dead limbs at the bottom. The stand of jack pine must be at least 80 acres (32.4 hectares) in size, with numerous openings to admit sunlight and keep the lower limbs of the pines green **(Fig. 8–13)**.

Before humans arrived on the jack pine plains of Lower Michigan, lightning-ignited fires burned sections of the forest at frequent intervals, creating an ideal nesting area for the Kirtland's Warbler. Somewhere in the forest area, newly growing young jack pines of just the right height were always available. When humans began prohibiting forest fires, life became difficult for the Kirtland's Warbler. The pines grew too big and too thick to satisfy its needs. The story has gone full circle, however. Humans, conscious at last of the bird's dilemma, now burn sections of the pine forest regularly on the Kirtland's Warbler Management Area near Lovells, Michigan. This burning, combined with cowbird trapping to reduce brood parasitism, appears to be effective, as the number of Kirtland's Warblers have increased since 1990.

Figure 8-11. Great Kiskadee (Pitangus sulphuratus) and Plain Thornbird (Phacellodomus inornatus) Nests on a Windmill: Birds are known to place their nests in all sorts of seemingly odd places, including various structures built by humans. This Great Kiskadee nest was constructed on top of an existing Plain Thornbird nest in a completely exposed setting within the supporting structure of a windmill on a working ranch in the llanos of central Venezuela. The inset shows a closer view of the nest indicated by the arrow. Photo by Jason A. Mobley.

Figure 8–12. Male Kirtland's Warbler at Nest: The endangered Kirtland's Warbler breeds only amid thick stands of medium-sized jack pines in northern Michigan. It conceals the nest among dense grasses on the ground, usually right near the trunk of a pine (not visible). The lower branches arch over the nest—a cup of dry grasses, sedges, and pine needles lined with finer grasses, hair, and moss stems. Here, a male visits a nest with young nestlings. Photo courtesy of Bill Dyer/CLO.

8

Figure 8–13. Nesting Habitat of Kirtland's Warbler: *The only habitat in which Kirtland's Warblers will nest is jack pine stands of at least 80 acres (32.4 ha), with trees usually 6 to 18 feet (2 to 5.5 m) tall. In such a stand, sufficient sunlight filters through the limbs to keep the lower branches green—a critical nesting requirement for the warbler. This jack pine stand in Crawford County, Michigan, has been burned to provide suitable nesting habitat. The birds first breed in a stand about six years after it has burned. Photo courtesy of J. Surman/CLO.*

Seasonal Changes in Nest Sites

Many species choose different types of nest sites at different points throughout the breeding season. Early in the season, for example, Eastern Towhees, Field Sparrows, and Song Sparrows place most of their nests on the ground. Later, they nest in low shrubs **(Fig. 8–14)**. Because these birds usually begin nest building before the shrubs have leafed out, the ground provides much better cover for their early nests. But once the shrubs have leaves, they apparently provide safer nest sites than the ground. American Robins usually place their first nests of the breeding season low in a protected evergreen tree. Later in the season, however, they usually choose a higher site in a deciduous tree (Sallabanks and James 1999).

Figure 8–14. Seasonal Changes in Nest Height of Field Sparrows: *In a central Illinois study site containing grassland, shrub, and woodland areas, Best (1978) found that the average height of Field Sparrow nests increased as the breeding season progressed. Birds placed most early nests in stands of dead grasses, but placed later ones in shrubs, small trees, and living herbaceous vegetation such as black raspberries and goldenrod, as well as in dead grasses. The latest nests were almost all in shrubs and small trees. These birds begin to nest early in spring before leaves have emerged, when dead grasses provide the best available cover. Increasing nest height as the season progresses may partly reflect additional opportunities for hiding nests in vegetation. Because nest height in trees and shrubs continued to increase even after the leaves were fully out, however, Best suggests that other factors also may influence nest height. Error bars represent one standard deviation.*

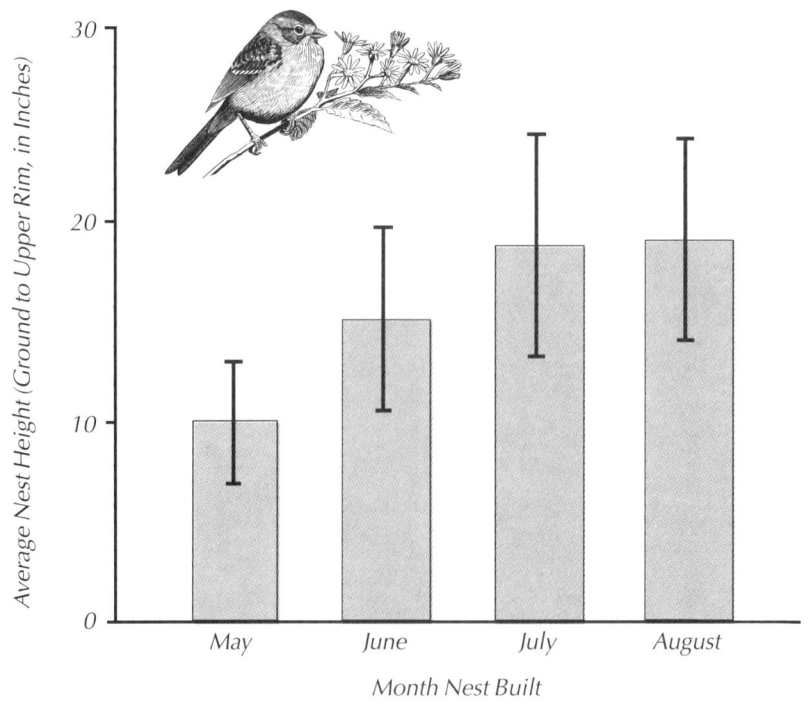

In a California orchard, American Crows nesting early in the season before the trees had leaves chose sites an average of 12.2 rows from the edge. Later in the season, when leaves were out and the foliage was more dense, the crows nested an average of only 8.8 rows from the edge. Thickness of cover was undoubtedly a factor in their choice of nest site (Pettingill 1972).

Nest Site Selection

The decision of precisely where to put the nest is ultimately up to the female (she has veto power!), and in many polygynous species (pheasants, grouse, Ruffs, manakins, cotingas, birds-of-paradise, and others) the female is left with *all* domestic duties. In species in which nest sites are limited, the male often displays for a mate from a potential nest site (as in bluebirds **[Fig. 8–15]**, swallows, and wrens

Figure 8–15. Eastern Bluebird Wing-Wave Display: *In this common visual display of Eastern Bluebirds, the male or female slowly raises one or both wings, sometimes quivering them, while singing or chattering. Wing-Wave may be given during courtship or nest site selection, or as a greeting between the pair in a variety of situations. Nest site selection also involves flight displays in which the male flies with slow, deep, butterflylike wingbeats; hovers in front of the nest hole; or flies oddly, with the wings moving out of synchrony and at varying speeds—all while singing or chipping. Stokes and Stokes (1989, pp. 318–319) describe one common nest site selection scenario:*

When a female first arrives on the [male's] territory, or the pair arrive together, the male will.... [give a flight display], ending up at a prospective nest site. Here he will cling to the entrance or a nearby perch and do Wing-Wave while continuing his singing. After this he may go to the nest hole and, while clinging to the entrance, rock back and forth, putting his head and shoulders in and out of the nest hole and looking around between each rocking motion. He may be carrying a bit of nest material while doing this. The male may also land on the nest box with his back to the female and spread his tail and droop his wings, exposing his vibrant blue back. He may lift and quiver his wings and pivot, appearing to dance.

In this photo, a male raises both wings in a Wing-Wave display. Photo courtesy of Isidor Jecklin/CLO.

Figure 8–16. Male Marsh Wren Displaying at "Courting Nest": Male Marsh Wrens construct numerous (often 20 or more) "courting nests" on their territories. These unlined, globular structures are woven from cattail leaves and grasses and have a side entrance. When a female arrives on a male's territory she visits his nests and then, if she chooses him as a mate, adds a soft lining to one or builds another nest herself. As the female inspects nests the male escorts her, giving a display in which he fluffs his breast feathers and cocks his tail over his head (as pictured here). He also may sing rapidly, quiver his wings, and/or sway from side to side. (Males also give this display to other males during territorial interactions near boundaries.) Researchers hypothesize that the remaining courting nests may help to deceive predators (Leonard and Picman 1987), and they may have other uses as well: the pair may renest in one if their eggs are destroyed; adults may roost in them at night; fledglings may take shelter in them; and they may be used for winter shelter where wrens are year-round residents. Photo by Marie Read.

[**Fig. 8–16**]) or from a territory where the male knows the whereabouts of likely nest sites (as in marsh-nesting blackbirds [**Fig. 8–17**]). In other species—such as gulls, many passerines, and even some polyandrous ones such as jacanas—the choice of a nest site appears to be the result of a negotiation between the male and female, with each considering sites identified by their mate.

In species in which pair members communicate over the choice of a site, that communication is usually mediated by stereotyped displays that often involve the ritualized placement of nest material or pointing at the nest site. This stage of the reproductive cycle is hard to study, because birds tend to be less committed to a breeding attempt when it is just beginning than when they have progressed to the point of laying eggs or hatching chicks: slight disturbances from observers may cause them to abandon the breeding effort. As a result, ornithologists have done less work with individually marked birds at this stage than at later stages. So if you see a bird acting strangely early in the breeding season, pay particular attention, as you may see the details of nest site selection unfold as very few others have.

Diversity of Nests

The nests that birds actually build in their selected sites also are remarkably diverse. I will attempt a provisional classification of nest types, mostly based on shape, but beware that there are many other criteria we could use to classify nests that might be more meaningful to the birds: these groups merely look similar to us.

Figure 8–17. Tricolored Blackbird Nest-Site Demonstration Display: *Tricolored Blackbirds nest in dense colonies in marshes, fields, and blackberry thickets, primarily in central California. Males set up their small territories before pairing, each singing from the highest perch available. **a.** When a male spots a female flying overhead, he gives a Song-Spread display: he lowers and spreads his wings and tail, erects his neck feathers, and sings. **b.** Then he elevates his wings to a V shape and lowers and spreads his tail again. **c.** He moves down into the vegetation, continuing to display, as the female follows. He may point his bill at possible nest sites, move it back and forth rapidly, and pick up nesting material. **d.** At a potential nest site, the female also may pick up nest material and display with the male. This sequence of male-female interactions, described by Orians and Christman (1968), is termed the Nest-Site Demonstration display. Often, when a female flies over a colony of displaying males, many give a Song-Spread and disappear into the vegetation nearly simultaneously. Adapted from a drawing by Gene M. Christman, in Orians and Christman (1968).*

The simplest nest is **no nest**. In a number of species, nesting involves selecting a nest site but no construction of any nest whatsoever. All brood parasites fall within this group, and the final section of this chapter discusses their breeding biology. Some megapodes (chicken-like birds of Australasia) lay their eggs on bare rocks warmed by the sun, murres lay their eggs directly on exposed rock ledges, and New World vultures and condors lay their eggs on the bare floors of shallow caves. White Terns of tropical seas lay their single egg on bare branches **(Fig. 8–18)**, as do the potoos of the Neotropics **(Fig. 8–19**; also see Fig. 3–56). Emperor Penguins transfer their single egg to the top of the male's feet as soon as it is laid, and the "nest" is formed by the male's feet below and a loose bulge of skin above (see Fig. 4–100). Most nightjars, including nighthawks and Whip-poor-wills, build no nest at all **(Fig. 8–20)**, laying their eggs on the forest floor, on sand bars in rivers, or even on flat, graveled roofs. Some species can move their eggs to a different site in the midst of incubation.

The nest of many more species is a simple **scrape (Fig. 8–21)**. For example, the nest of many plovers (including the Killdeer), terns, and skimmers is a very shallow depression, often containing no nest material, or simply lined with a few flat pebbles. These species don't even create a depression if the substrate for nesting is too hard. Many Antarctic-nesting penguins build up their scrape on frozen substrate by moving pebbles around.

a. Egg Perched on Tree Fork

b. Adult and Chick

Figure 8–18. White Tern Nesting: *The pelagic White Tern, nesting on tropical ocean islands, often places its single egg directly in a fork or depression in a horizontal branch, constructing no nest of any kind. Although the egg may be perched precariously, the incubating birds seem to have little trouble settling on or rising from the egg without knocking it down.* ***a. Egg Perched on Tree Fork:*** *The male and female take turns incubating the cryptically colored egg until it hatches at about 34 days. Photo by A. Forbes-Watson/VIREO.* ***b. Adult and Chick:*** *The chick remains on the branch where it hatched, attempting to hold on tightly with its sharp claws, while the parents bring food and provide warmth until it is ready for flight. If the chick falls to the ground, which occasionally happens during severe storms, it is ignored by the parents and soon dies from hunger and exposure. Adults also ignore eggs that are moved just a few inches from where they were laid. The precise location of the egg or young is thus quite important, and birds return year after year to nest in the exact same fork or depression they used the year before. Photo by R. L. Pitman/VIREO.*

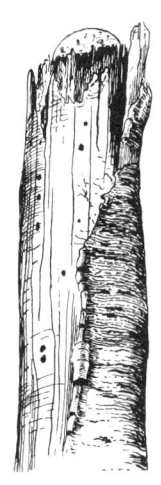

Figure 8–19. Great Potoo Egg on Dead Branch: *Like most of its nightjar relatives, the Great Potoo of South America builds no nest at all. Instead, it may lay its single egg atop a tree stub—the egg fitting into the natural depression so closely that it is difficult to dislodge. Incubating birds sit in an upright posture, looking remarkably like an extension of the tree. Drawing by Dr. C. J. F. Coombs.*

Figure 8–20. Whip-poor-will Eggs in "Nest": *Like most other nightjars, the Whip-poor-will builds no nest. Instead, it lays its two eggs directly on a bed of dead leaves on bare ground in open woodlands, often near the edge of a clearing. The nest site is bathed in a mixture of sunlight and shadows, rendering the incubating bird nearly invisible against its surroundings; it will not flush unless an intruder approaches closely. Photo courtesy of Bill Dyer/CLO.*

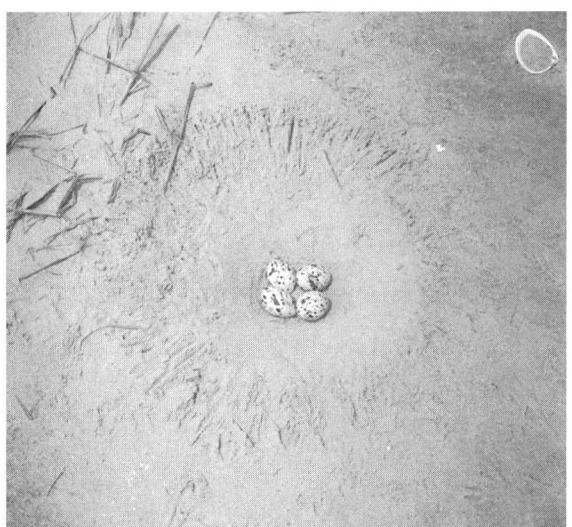

a. Black Skimmer Nest Scrape

b. Gentoo Penguin with Chick at Nest

Figure 8–21. Nest Scrapes: *The nest of many plovers, terns, skimmers, and penguins is a* **scrape***: a shallow depression with little or no nest material, sometimes lined with a few flat pebbles.* **a. Black Skimmer Nest Scrape:** *The Black Skimmer's nest is an unlined, shallow depression 5 to 10 inches (13 to 25 cm) across, within a colony of skimmers on open beaches or sand bars. Although placed above the normal high-tide line, it may be destroyed by unusually high tides during storms. Photo courtesy of Platt/CLO.* **b. Gentoo Penguin with Chick at Nest:** *Gentoo Penguins breed on subantarctic islands and the Antarctic Peninsula, in colonies of 2 to nearly 10,000 pairs. The nest, placed on ice-free ground on a beach, hill, or grassy area, is a platform of stones built up from the ground, 4 to 8 inches (10 to 20 cm) high and 18 inches (45 cm) across. At the center is a small cup lined with smaller stones and, depending on the location, with vegetation. Here, an adult tends its chick in a nest on the Falkland Islands. Photo by Wolfgang Kaehler.*

Figure 8–22. Platform Nest of Turtle Dove: *Like most doves and pigeons, the Turtle Dove builds a small, flimsy platform of twigs for its two white eggs. This nest, lined with finer materials such as roots, grass stems, and leaves, is usually placed about 3 to 8 feet (1 to 2.4 m) off the ground in a tree or shrub, sometimes loosely clustered with other Turtle Dove nests. Breeding in northern Africa and the western Palearctic, Turtle Doves live in mixed habitats with open ground for foraging and trees or shrubs for nesting and roosting. Drawing by Dr. C. J. F. Coombs.*

Figure 8–23. Platform Nest of White Stork: *The beloved White Stork of Eurasia builds a massive platform of sticks and earth up to six feet (two meters) across, topped by a shallow cup of finer materials such as grass and sometimes rags or paper. It often nests on human-built structures such as walls, ruins, churches, roofs, or (as pictured here in Poland) on chimneys. The nests are used year after year, the birds adding new material as necessary: one nest in Germany was occupied from at least 1549 through 1930! Despite centuries of protection in Europe, White Storks have declined dramatically since the early 1900s, much to the dismay of people who believe that a stork nesting on their roof brings good luck. Photo by Liz and Tony Bomford/ Oxford Scientific Films.*

Platform nests consist of a very shallow depression in the top of a mound of nest material. They vary tremendously in complexity, from the flimsy platform of twigs put together by the Mourning Dove and most of its relatives around the world **(Fig. 8–22)**, to the massive structures, sometimes containing thousands of sticks, built by storks or large birds of prey such as Ospreys and eagles **(Fig. 8–23)**. Birds can build platform nests anywhere they find support of sufficient strength, and the type of support used can vary considerably within a species: many herons, cormorants, storks, and raptors build platform nests of twigs in trees, but will nest on the ground on protected islands if no trees are available. Flamingos build short pedestals of mud **(Fig. 8–24)**,

Figure 8–24. Greater Flamingo Nesting Colony: *Found in isolated pockets in certain tropical or high-altitude areas around the globe, flamingos nest in large colonies in the mud flats around shallow, alkaline, soda lakes and salt lagoons. Their nests—circular mounds 14 inches (35 cm) across and up to 12 inches (30 cm) high, wider at the base—are topped by a shallow depression for their single egg. The height protects the egg and nestling from flooding, and also keeps the nest cooler than the surrounding ground—which, in the black mud around Lesser Flamingo nests at Lake Magadi, Kenya, may reach as much as 167 degrees F (75 degrees C) (Welty and Baptista 1988). Photo by E. Bartels/VIREO.*

Figure 8–25. Floating Platform Nest of Black Tern: Nesting in loose groups in shallow freshwater marshes with emergent vegetation such as cattails, the Black Tern builds a flimsy, floating nest that is commonly destroyed by wind, waves, or changes in water level. The nest, just 0.8 to 2.4 inches (2 to 6 cm) high and 0.8 to 2 inches (2 to 5 cm) above water, is often built on an existing floating mat of dead marsh vegetation. The terns pile dead vegetation from the surrounding water upon the mat, then shape it into a shallow cup for the eggs. The eggs are well adapted to their soggy environment, having extra pores to help them tolerate their constantly water-soaked conditions (Davis and Ackerman 1985; Firstencel 1987). Photo by Marie Read.

and grebes and some terns build floating rafts of aquatic vegetation, upon which the shallow nests sit **(Fig. 8–25)**. The Horned Coot of South America builds an interesting variation on this type of platform nest **(Fig. 8–26)**. Nesting on high Andean lakes with little aquatic vegetation, these birds pile stones in the water to make a mound upon which the shallow nest of vegetation is built (Goodfellow 1977). The accumulation of stones, built up by several pairs over a few years, may be 3 feet (1 m) high and up to 13 feet (4 m) in diameter and may weigh over a ton (907 kg).

The Oilbird of northern South America uses quite unusual materials to build its platform-like nest high on a ledge jutting from the wall of a cave. The bulky nest—a short, truncated cone topped by a shallow depression—consists mainly of regurgitated fruits and seeds

Vegetation

Stones

Up to 3 feet (1 meter)

Up to 13 feet (4 meters)

Figure 8–26. Horned Coot Nest on Stone Mound: Breeding on high-altitude Andean lakes with little vegetation, the Horned Coot places its nest on a cone-shaped mound of stones that it creates in shallow water. The stones, which can weigh as much as a pound (0.5 kg) each, are brought to the mound one-by-one. The pile—accumulated by several pairs over several years—may reach up to 3 feet (1 m) high and 13 feet (4 m) in diameter, and may weigh as much as a ton (907 kg). The mound ends below the water's surface, and the birds pile vegetation on top, forming a nest that rises one to two feet (0.3 to 0.6 m) above the water. Although laborious to build, these island nests are protected from terrestrial predators. Drawing by Dr. C. J. F. Coombs.

Figure 8–27. Oilbird at Nest: The nightjar-like Oilbird of northern South America, nearly the size of a small crow, nests high on a ledge inside a cave. The bulky platform nest—a short, truncated cone topped by a shallow depression—consists of regurgitated fruits and seeds mixed with the bird's own excrement. It grows higher as more material is added each year. Adapted from a drawing by Robert Gillmor.

mixed with the bird's own excrement, and grows higher each year as more and more material is added **(Fig. 8–27)**.

The majority of bird species in the world build some sort of **cup nest**. The smallest are those of hummingbirds, and probably the largest are built by large crows and ravens. Cup nests can be composed of a wide variety of materials: most common are small twigs and dried grass, but quite a few species use mud (Rowley 1971). Smaller species, such as hummingbirds and gnatcatchers, often use spider webs in conjunction with lichens and "vegetable down" from seeds **(Fig. 8–28)**.

Ornithologists often distinguish cup nests on the basis of how they are supported. **Statant cups** are built on top of a hard physical support or supports **(Fig. 8–29)**. Some, such as the nests of Horned Larks and Bobolinks, are built directly on the ground. The nests of Horned Larks often are made in open country, commonly placed within the hollow of the hardened hoofprint of a cow or horse—an arrangement that places the eggs slightly below ground surface, thus protecting them from most foot traffic if the large mammals should return. Most nests built in shrubs and trees also are statant, from the loose piles of twigs and grass built by most songbirds, to the massive mud nests built by the enigmatic Magpie-larks, White-winged Choughs, and Apostlebirds of Australia. **Pensile cups** are supported by their rim with the belly of the cup hanging unsupported beneath **(Fig. 8–30)**. Such nests are built by New World blackbirds, vireos, and kinglets; by many songbirds

Figure 8–28. Anna's Hummingbird on Nest on Pine Cone: Hummingbirds, such as this Anna's of western North America, often incorporate spider webs into their tiny cup nests. The sticky webs help attach the nest to a branch, large leaf, or pine cone (as shown here). The webs also may help the female to stick bits of lichen or bark to the outside of the nest so that it blends into nearby branches or tree trunks. Photo taken in the Sonoran Desert, Arizona, by John Cancalosi/Valan Photos.

a. Bobolink

b. American Robin

2 Feet (61 cm)

c. American Crow

d. White-winged Chough

Figure 8–29. Statant Cup Nests: Cup nests built on top of hard physical supports, including the ground, are termed **statant cups**. They vary widely in location, material, and size, and are built by birds from many taxonomic groups. **a. Bobolink:** In hayfields and large, grassy meadows, the Bobolink builds a shallow cup of coarse dead grass and weed stems, lined with finer grasses or sedges. The nest is often placed in a small hollow on the ground—either natural or formed by the female—and well-hidden among dense grass and weeds. Photo by Marie Read. **b. American Robin:** On the fork or horizontal branch of a tree or shrub, or on the ledge of a house, barn, or other building, the American Robin places its familiar cup nest of grasses, weed stems, and mud. See Figure 8–50 for details of structure and nest building. Photo by Hal H. Harrison. **c. American Crow:** In a large tree or shrub, 10 to 70 feet (3 to 21 m) above the ground, the American Crow builds its huge nest on the base of a branch against the trunk or in a vertical fork. A large, bulky basket of sticks, twigs, bark, vines, and sometimes mud, the nest is lined with softer materials such as grasses, plant fibers, moss, feathers, or fur. It usually is about two feet (61 cm) across and about 4.5 inches (11.4 cm) deep. Note one newly hatched chick in this nest. Photo by Kevin McGowan. **d. White-winged Chough:** Inhabiting dry woodlands in eastern Australia, the gregarious, crowlike White-winged Chough lives in an extended family group whose members cooperate to build and tend their mud nest. Saddled across a horizontal branch, this nest is a huge cup of mud lined with shredded bark, grass, and fur. Mixed with the mud are grass stems and bark, which the birds use as wicks to gather mud, and then incorporate into the nest. The structures are 8.5 inches (22 cm) across with walls 3 inches (8 cm) thick, and may weigh 5 pounds (2.3 kg) when dry. They last for many years, and sometimes are reused by the birds. Photo by R. Brown/VIREO. Please note that the photos shown here are not to scale.

a. Red-eyed Vireo   ***b. Golden-crowned Kinglet***

Figure 8–30. Pensile Cup Nests: *Cup nests suspended from branches by their rims are called* **pensile cups. a. Red-eyed Vireo:** *Hanging from a horizontal forked twig in a deciduous or mixed forest, the Red-eyed Vireo's nest is a delicate, compact cup of bark strips, grasses, pine needles, rootlets, and weeds. It usually is held together and bound to its supporting branches with spider webs and cocoons (not visible in this photo), and sometimes is decorated outside with lichen, paper birch bark, or paper from wasps' nests. Photo by Marie Read.* **b. Golden-crowned Kinglet:** *Breeding mainly in old, dense spruce or fir stands in the boreal and subalpine forests of North America, the tiny Golden-crowned Kinglet places its nest high in a conifer, suspended from a twig fork or several small side branches. The deep, thick cup of mosses and lichens is bound together and to its supports with spider webs and hair, and lined with small strips of inner bark, fine black rootlets, hair, and feathers. The well-camouflaged nest is just 2.75 inches (7 cm) across and 3.75 inches (9.5 cm) high, and the rim curves inward at the top, forming a snug hollow below. From* A Guide to the Nests, Eggs, and Nestlings of North American Birds, *2nd Edition, by Paul J. Baicich and Colin J. O. Harrison, 1997. Reprinted by permission of Academic Press.*

in Australia and Asia; and by a host of orioles and flycatchers, and their relatives in the Old World. Exploring whether this type of nest has any intrinsic advantages over cup nests would be interesting. My impression is that most birds would be better off building a cup nest, but that pensile nests are an adaptation for nesting on emergent vegetation in marshes or on thin, distal branches (probably for protection from predators), neither of which offer support from below. Pensile nests with deeper and deeper cups grade into **pendulous cups**, which are entered from the top and have the nest chamber anywhere from around 4 inches (10 cm) to over 1 yard (1 m) below the supports from which they hang **(Fig. 8–31)**. Such nests are usually woven from plant strips or fibers, and they are characteristic of the New World orioles, oropendolas (see Fig. 6–39), and caciques, as well as some weavers in the Old World. **Adherent cups** are made of mud or saliva and rely on chemical forces to hold the nest to a vertical surface. Many swifts use a considerable amount of saliva to hold their nests together and to the nesting surface. The African Palm-Swift goes the extra step of gluing its *eggs* to the padlike vertical nest, incubating them from a vertical position **(Fig. 8–32)**. The use of saliva culminates in the Edible-nest Swiftlets

a. Baltimore Oriole

b. African Masked-Weaver

c. Planalto Hermit

Figure 8–31. Pendulous Cup Nests: *Very deep cup nests suspended from their rims are termed* **pendulous cups**. **a. Baltimore Oriole:** *In open woodlands, forest edges, riverside trees, and shade trees throughout the eastern and central United States, the Baltimore Oriole firmly attaches its nest to a twig fork near the end of a slender, drooping branch of a tall, deciduous tree—usually 25 to 30 feet (7.6 to 9.1 m) above the ground. The deep, hanging pouch, up to 8 inches (20.3 cm) long, is tightly woven from long plant fibers, strips of vine bark, grasses, strings, and sometimes Spanish moss, and lined with fine grass, soft plant down, and hair. Photo by W. Greene/VIREO.* **b. African Masked-Weaver:** *The male African Masked-Weaver weaves an oval nest of grass strips and reeds, suspending it from a drooping branch of a tree or between several reed stems, often over water. The striking, yellow-and-black male, about the size of a sparrow, displays by hanging under the nest (as shown here), fanning his wings and calling. If a female accepts a nest, she lines it with leaves and grass flower heads, then lays eggs and tends the nest herself, while the polygamous male courts other females. Photo by M. P. Kahl/Photo Researchers.* **c. Planalto Hermit:** *This South American hummingbird creates an interesting pendulous cup variation. It suspends its minute nest cup from a twig or leaf tip by a single support, adding a long streamer of vegetation below as a counterweight—to keep the nest upright. The Sooty-capped Hermit builds a similar nest, but for its counterweight attaches tiny pebbles or lumps of dry clay to a spider web streamer. Both birds must build much of their nest on the wing, as their nest sites provide no place to perch. Drawing by Dr. C. J. F. Coombs.*

of Southeast Asia, which often build their nests entirely of saliva (see Fig. 4–93). These nests, the key ingredient of the great Asian delicacy bird's-nest soup, are harvested in the vast nesting caves by workers on rickety skyscraperlike scaffolds of bamboo (Valli and Summers 1990). When the first nest is taken a swiftlet pair will often hurriedly build another, but the replacement nests usually contain bits of vegetation, and thus are much less valuable than those of the first harvest. The Barn Swallow, as well as many other swallows in the Old World, build their cup nests of mud mixed with straw **(Fig. 8–33)**. Similar adherent nests are sometimes made by phoebes.

Some species, especially those that nest on the ground amid some vegetative cover, build **domed nests**. These are cups with a woven dome overhead that probably mainly helps to conceal the eggs or nestlings. In North America, species such as meadowlarks and snipe build domed nests in grassy areas, and Ovenbirds build them in forests **(Fig. 8–34a and b)**. The House Martin, an Old World relative of the Barn Swallow, builds the walls of its mud cup all the way up to an overhanging surface, thus incorporating the overhang as the top of its nest and leaving only a small opening to enter the cavity—presumably a strategy to exclude predators **(Fig. 8–34c)**. This species often nests in dense colonies on favorable sites inaccessible to predators, however, leading some researchers to suggest that the nest closure may also reduce the frequency of unsolicited extrapair copulation attempts from neighboring males in the colony. Note that I have placed the House Martin's nest with domed nests, even though it does not actually build the top of the dome, as do the other species mentioned here. This is a good example of how fuzzy and subjective these nest categories really are. The Rufous Hornero builds a nest that is even more difficult to categorize: a cup of grasses enclosed in hardened mud, complete with an entrance chamber **(Fig. 8–34d)**.

Figure 8–32. African Palm-Swift Eggs Glued to Vertical Nest: Most swifts use saliva to hold their nest material together and to glue their nests to vertical surfaces, but the African Palm-Swift also uses saliva to glue its eggs to the nest! The soft, padlike nest is glued to the underside of a vertical palm leaf, and thus the eggs would fall out if not held in place. The bird incubates in a vertical position, holding onto the nest with its feet. Drawing by Charles L. Ripper.

Figure 8–33. Adherent Cup Nest of Barn Swallow: Inhabiting open areas throughout most of North America and Eurasia, the Barn Swallow originally nested in caves. By the mid-1900s, however, it had shifted to nesting almost exclusively in open buildings such as barns; and under bridges, docks, culverts, and similar structures. To build their nests, which are frequently reused, the birds gather mud—often mixed with grass stems—in their beaks, then plaster it as pellets (up to 1,400 [Møller 1994]) to a vertical wall, slightly below the roof, as in the nest shown here. The shallow, semicircular cup is lined first with fine grass stems, horsehair, or algae strands, and then finally with abundant feathers from poultry. Nests also are placed on a horizontal support such as a beam, and often are in small, loose colonies. Photo courtesy of J. R. Woodward/CLO.

a. Eastern Meadowlark

b. Ovenbird

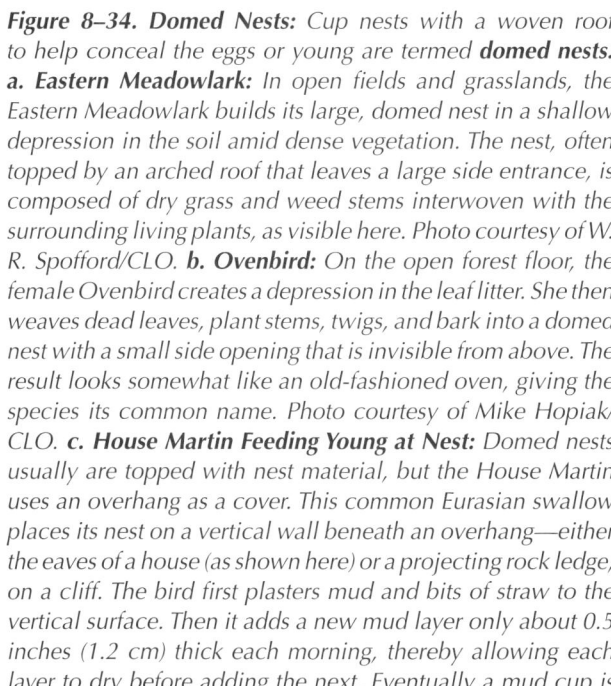

c. House Martin

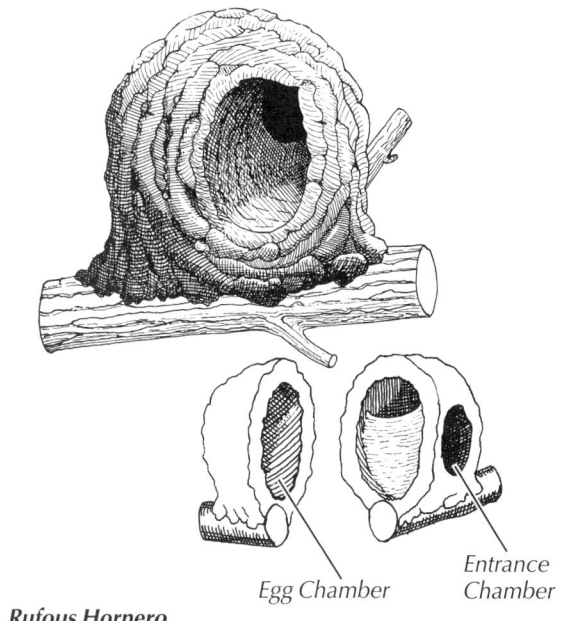

Egg Chamber

Entrance Chamber

d. Rufous Hornero

Figure 8–34. Domed Nests: *Cup nests with a woven roof to help conceal the eggs or young are termed **domed nests.*** *a. Eastern Meadowlark: In open fields and grasslands, the Eastern Meadowlark builds its large, domed nest in a shallow depression in the soil amid dense vegetation. The nest, often topped by an arched roof that leaves a large side entrance, is composed of dry grass and weed stems interwoven with the surrounding living plants, as visible here. Photo courtesy of W. R. Spofford/CLO. **b. Ovenbird:** On the open forest floor, the female Ovenbird creates a depression in the leaf litter. She then weaves dead leaves, plant stems, twigs, and bark into a domed nest with a small side opening that is invisible from above. The result looks somewhat like an old-fashioned oven, giving the species its common name. Photo courtesy of Mike Hopiak/ CLO. **c. House Martin Feeding Young at Nest:** Domed nests usually are topped with nest material, but the House Martin uses an overhang as a cover. This common Eurasian swallow places its nest on a vertical wall beneath an overhang—either the eaves of a house (as shown here) or a projecting rock ledge, on a cliff. The bird first plasters mud and bits of straw to the vertical surface. Then it adds a new mud layer only about 0.5 inches (1.2 cm) thick each morning, thereby allowing each layer to dry before adding the next. Eventually a mud cup is formed, and the martin builds the walls right up to the overhang, leaving only a tiny opening into the nest cavity, which is lined with soft materials. Photo by Mark Hamblin/Oxford Scientific Films. **d. Rufous Hornero:** In the grassy plains of southern South America, the thrush-sized Rufous Hornero (called "El Hornero" or "the ovenbird" in South America, owing to its nest shape) constructs a two-chambered nest of mud (or, in some areas, sand and cow dung), with some grass and hair mixed in. The domed nest, nearly one foot (30 cm) high, usually sits fully exposed atop a stump, post, or limb, and may weigh up to 11 pounds (5 kg). The upper diagram is a front view of the nest, showing the entrance chamber. Below is a vertically sectioned nest: the right half contains the entrance chamber and dividing partition; the left holds the egg chamber, lined with grasses and feathers (not visible here). The nest hardens to the consistency of cement and is so crack-proof that in 1958 Brazilian health workers carefully copied the bird's sand/dung formula and plastered 200,000 huts, reducing infestations of the reduviid bug, which thrives in crevices and carries the fatal Chagas' disease. Nests can last up to eight years, and may be reused by many other species. For photo of bird and partially completed nest, see Fig. 4–118. Adapted from Welty and Baptista (1988), after Pycraft (1910).*

A **globular nest** results when the top of a dome nest is completely enclosed. Usually entered through a hole in the side, these spherical nests are characteristic of most wrens. Anyone who has seen the grassy sphere built by Cactus Wrens, usually within the spiny arms of a cholla cactus, can appreciate how nest construction combined with nest placement can protect the young from both the elements and predators **(Fig. 8–35)**. In a spiny shrub in the same desert habitat, the chickadee-like Verdin builds a smaller globe nest out of spiny twigs, orienting the spines to the outside. Thus spines on both the nest and its supporting shrub discourage predators. The Black-billed Magpie of Eurasia and western North America builds a similar domed nest of thorny twigs, occasionally inserting short strands of barbed wire! Southern Penduline-Tits, African relatives of the Verdin, build globular nests of grass and plant down that can include a false entrance, apparently to foil predators **(Fig. 8–36)**. The false entrance is a prominent hole on the side of the nest that leads to a dead-end chamber, whereas the true entrance is concealed in a slit in the projecting upper lip of the false entrance.

A globular nest with an entrance tunnel is termed a **retort nest**. These are common among the mud-nesting swallows, and in the grass nests of many African weavers and New World swifts in the genus *Panyptila* (for example, Goodfellow 1977) **(Fig. 8–37)**. The lengths of these tunnels can vary from an inch or two (a few cm) to over a yard (meter). Once again, the traditional explanation for the elongation of these tunnels has been defense against predators—but this does not seem to be the whole story. Access to the nests of mud-nesting swallows, especially those of the large species that build very thick mud walls on their tunnels, can be very difficult for all predators except snakes. But predators such as monkeys and large birds can rip into grass nests, no matter how long their tunnels. The length of the tunnel often varies considerably among closely related species, and evaluating their effectiveness against predators of all sizes, as well as their possible role in reducing extrapair copulations and brood parasitism within and between species, would be interesting.

*Figure 8–35. Cactus Wren Nest in Cholla Cactus: Inhabiting open areas of the southwestern United States and northern Mexico, the large Cactus Wren usually places its conspicuous nest in a cholla cactus, as shown here, or in a thorny shrub. The **globular nest** is a bulky sphere of plant stems and grasses one foot (30 cm) or more across with an entrance tunnel about 6 inches (15 cm) long. The nest deters predators through its shape and location, rather than camouflage. Photo by Jim Merli/Valan Photos.*

Figure 8–36. Southern Penduline-Tit Nest with False Entrance: Hanging from a twig several yards off the ground, the globular nest of the Southern Penduline-Tit of southern Africa has two entrance holes on the side. The bottom hole, more prominent, is a false entrance leading to a small, dead-end chamber. Just above this entrance is the true entrance, a narrow slit concealed in the projecting upper lip of the false entrance. Most of the nest is made from fine, woolly plant and animal materials felted into a tough, clothlike material; but the lips, floor, and roof of the true entrance tube are coated with coarse spiderwebs. Thus, when the bird closes the tube the sides cling together, making the true entrance virtually invisible, and presumably thwarting predators. When not breeding, up to 18 of these social birds may roost in one nest. **a. Photos of Nest in Cape Province, South Africa:** In the left photo, only the false entrance is visible. In the right photo, the true entrance also is visible as a wide slit above the false entrance. Photos courtesy of C. J. Skead. Reprinted by permission of Birdlife South Africa/The Ostrich. **b. Cross Section through Nest:** In the left diagram, the true entrance is closed. In the right diagram, the entering bird has pushed the true entrance open.

a

b

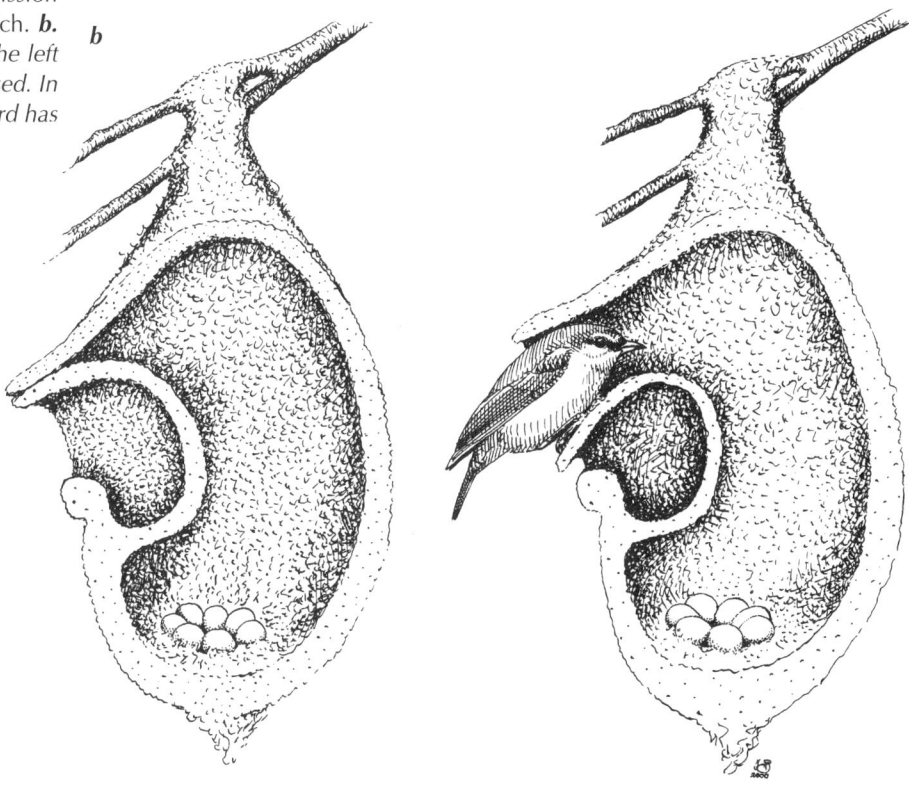

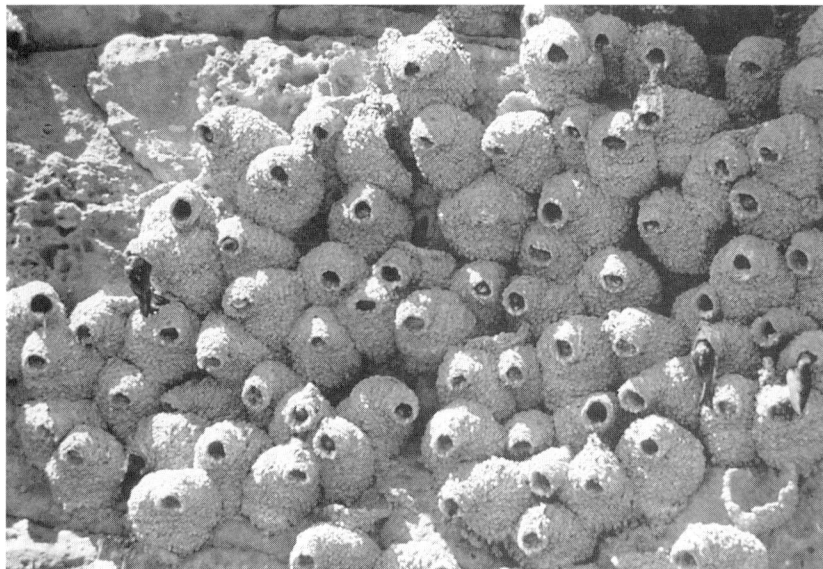

a. Cliff Swallow

Figure 8–37. Retort Nests: Globular nests with entrance tunnels are called retorts. a. Cliff Swallow: Breeding in dense colonies of up to 3,500 active nests under overhangs of bridges, cliffs, or roofs, Cliff Swallows prefer to place their grass-lined, mud nests on a vertical wall just below an overhang. Late arrivals to the colony, finding this prime location used up, usually build their nests right below and connected to the first tier, slightly offset (see top left photo, showing a vertical wall face). The gourd-shaped nests, with entrance tunnels projecting outward 6 to 8 inches (15 to 20 cm) and facing down at the end, are constructed entirely from pellets of mud (each representing one beakful), visible as tiny lumps in the two photos at right. When building nests, Cliff Swallows apparently assess the composition of nearby mud, and choose the types that best adhere to walls (Robidoux and Cyr 1989). The mud walls of the nest protect eggs and young from wind and rain, and also keep the inside of the nest warm at night—up to 13 degrees F (7 degrees C) warmer than the surrounding air. The entrance tunnel traditionally has been thought to deter predators, but it also may reduce interference from neighboring Cliff Swallows, which frequently steal grass (and wet mud) from each other's nests, and may destroy eggs and parasitize nearby nests either by laying directly in them or by transferring their own egg to them in their beaks. Note the pale foreheads of adults, visible in some entrance holes, and the partially completed nest in the lower right corner of the left photo. Top left photo courtesy of Mary Tremaine/CLO; top right, courtesy of Peter Stettenheim; bottom right, courtesy of Isidor Jeklin/CLO. b. Red-vented Malimbe: This African weaver builds a retort nest with a spectacular entrance tunnel up to 2 feet (60 cm) long. Although the walls of the nest chamber are thickly woven, the walls of the tunnel are more open, the fibers forming a criss-cross pattern that acts like a transparent net—allowing a clear view of a bird inside. Crook (1960) described an entering, in-flight bird as dramatically sweeping downward to the tube entrance and then "diving upwards into [it] with closed wings, the momentum carrying [it] to the top of the tube...without [touching] the fabric." Drawing by Dr. C. J. F. Coombs.

b. Red-vented Malimbe

A few species build a **mound nest**. This chapter began with a brief account of a mound nest built by a male brush-turkey, a type of megapode (see Fig. 6–36). Not all megapodes build mounds, but among those that do, considerable variation exists in the type and size of mound built. Megapodes are the only birds that use the nest itself, rather than the parent's body, as the source of heat for the developing embryos. Among passerines, only a few species build true mounds. Social Weavers—small, seed-eating, African relatives of House Sparrows—build a large, colonial nest that is essentially a large haystack in a tree **(Sidebar 2: Social Weavers)**. On a smaller scale, Palmchats build colonial nests up to one yard (one meter) in diameter high in palm trees in the Dominican Republic, and Monk Parakeets build large, colonial stick nests in their native Patagonia (southern South America) and in parts of the United States from Miami to Chicago in which escapees have established colonies. The other notable mound builder is the odd, heronlike Hamerkop of Africa, pairs of which build nest mounds of sticks up to 6.5 feet (2 m) high and wide, usually in a large tree **(Fig. 8–38)**. These nests contain up to 8,000 sticks and weigh up to five hundred pounds (several hundred kg) (Goodfellow 1977). The nest chamber lies in the center of the mound and is connected to the outside by a mud-lined tunnel. For some unknown reason, the birds add a variety of carrion, feces, and scraps to the top of the mound after the eggs have hatched; perhaps they are erecting an olfactory smoke screen to protect their young?

The final set of nest types is **burrows** or **holes**. Some species actually excavate their own holes, whereas others, the **cavity adopters**, obtain a nest cavity created by physical forces, such as decay or erosion, or by other species. Among the excavators, there are two types: those that excavate wood and those that excavate sandy soil. Some wood excavators are familiar to everyone: all woodpeckers in North

(Continued on p. 8·41)

Figure 8–38. Mound Nest of Hamerkop: The heronlike Hamerkop of Africa builds a massive nest mound of sticks, often in a tree fork, measuring up to 6.5 feet (2 m) in diameter and weighing up to 500 pounds (907 kg). The central nest chamber is lined with sticks positioned so that no jagged ends project inward. It connects to a mud-lined entrance tunnel up to 2 feet (0.6 m) long, which opens to the outside near the bottom of one side of the mound (the opening is not visible in this drawing). The roof, reinforced with mud and grass to make it waterproof, may be up to 3 feet (1 m) thick, and has been known to support the weight of a human. The nest may take six or seven weeks to build, and may be reused in later years. Other species sometimes benefit from the Hamerkop's construction: small birds such as weavers, mynas, and pigeons may attach their nests to active Hamerkop mounds; small mammals such as genets may move in; and various raptors may evict the Hamerkop and take over its nest site. In addition, old nests may be occupied by cavity-nesting geese or ducks, or by snakes. Drawing by Dr. C. J. F. Coombs.

Sidebar 2: SOCIAL WEAVERS
Sandy Podulka

Figure A. Social Weaver: *Looking something like small, brown-capped versions of their relatives, the House Sparrows, Social Weavers live in the arid scrublands of southwest Africa, where they forage on the ground for insects and seeds. These highly social birds feed and nest in groups, roosting communally in their nests at night and during the hottest parts of the day. Drawing by Robert Gillmor.*

In the arid scrublands of southwest Africa, Social Weavers—small, seed-eating House Sparrow relatives **(Fig. A)**—build enormous communal nest mounds in sturdy, isolated trees or on telephone poles **(Fig. B)**. The mounds, which are occupied year round, may house more than 100 pairs of birds, and may be more than 25 feet (7.5 m) long, 15 feet (4.5 m) wide, and 6 feet (2 m) high.

The mound is not woven, but thatched from piles of coarse, dry grasses brought by flock members. Numerous fine twigs are also incorporated into the huge domed roof, which provides protection from predators and helps to shed the rain. Once the roof is complete, the birds excavate separate nests from below by pushing in or nipping off grass stems to create an oval chamber and then adding grasses to extend an entrance tunnel downward from the chamber **(Fig. C)**. Eventually, a monogamous pair will occupy each nest chamber. The mounds may be used for many generations, and as new nests are added below, the old ones may be filled in with straws.

In addition to providing a communal breeding site for this highly social species, the huge nest mounds provide thermal mass to help moderate the wide daily temperature fluctuations typical of the arid habitat. By absorbing the sun's heat during the day, the grass mound buffers the actual nest chambers from becoming too hot; during the cool nights the mound radiates back its absorbed heat, adding to the heat of hundreds of bird bodies to keep the nests warm. On cold winter nights, internal nest temperatures have been measured at 32 to 41 degrees F (18 to 23 degrees C) above the external air temperature (White et al. 1975). These "heated" nest chambers reduce the metabolic needs of roosting Social Weavers, which reduces the amount of food they must eat. They also allow the birds to breed at any time the erratic rains provide a sufficient food supply, even in winter when night temperatures routinely fall below freezing. Winter breeding may be advantageous because the birds' major predator, the cape cobra, is inactive then. The greater effectiveness of large nest mounds at buffering nest chamber temperatures is thought to have driven the evolution of such large communal nest structures in this species (Collias and Collias 1984). ∎

Figure B. Communal Nest of Social Weavers: *This large nest mound in an isolated tree in Kalahari Gemsbok National Park, South Africa is typical of the communal breeding mounds built by flocks of Social Weavers. Individual nest tunnels are recognizable as loose clumps of grass hanging down from the bottom. Note that at least two birds are visible—one flying upward to enter a nest tunnel, and another perched on the bare limb extending to the lower right—both dwarfed by the massive nest. Photo by Richard Packwood/Oxford Scientific Films.*

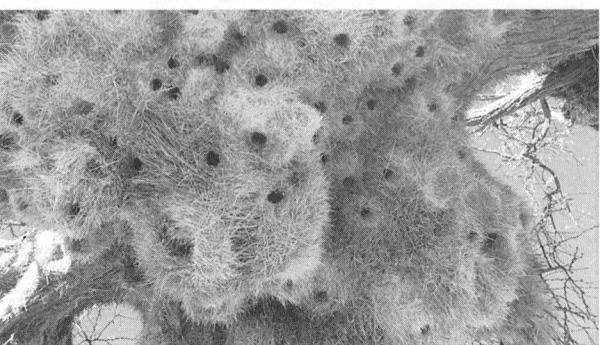

Figure C. Individual Nest Tunnels of Social Weavers: *This photo looks straight up at the bottom of a Social Weaver nest mound in a tree in Namibia. Each hole is the entrance to a tunnel that leads to the oval nest chamber of an individual pair. Photo by Jan Halaska/Photo Researchers.*

Figure 8–39. Woodpecker Nest Cavity: *This longitudinal section shows the narrow entrance and deep, unlined nest chamber in firm wood typical of most woodpecker nests. Photo courtesy of John H. Gerard.*

America are capable of excavating their own nest cavity (although some reuse cavities from year to year) **(Fig. 8–39)**. Worldwide, no other birds are as adept as woodpeckers at excavating cavities, although many species of parrots, barbets, toucans, nuthatches, and chickadees are capable of modifying a pre-existing hole to suit their needs. Many species in many different avian families excavate cavities in sandy soil, and, in contrast to the machinery needed to be able to excavate sound wood (strong, sharp beaks; sturdy skull bones; stiff tail feathers for support), soil excavators often have no specialized adaptations for the task. Good examples in the temperate zone are the Bank Swallow **(Fig. 8–40)** and the Belted Kingfisher **(Fig. 8–41)**, both of which loosen soil with their bills (which are adapted for specialized foraging, not digging) and clear the tunnel of soil with their weak feet. Similar examples of soil excavators abound in the tropics (bee-eaters **[Fig. 8–42]**, motmots, Crab Plovers) and among seabirds (shearwaters, petrels, storm-petrels, puffins).

Many cavity adopters that nest in tree holes rely on other species (usually woodpeckers) to make their holes. The avifauna of Australia, however, contains a particularly high proportion of cavity adopters, yet no woodpeckers live in Australia. How can this be? Apparently, the *Eucalyptus* trees that dominate the Australian landscape are particularly liable to develop cavities through fungal action on scars left by fallen limbs. And indeed, many species throughout the world adopt naturally occurring tree holes or niches in rock. The wheatears of Eurasia are particularly interesting in this regard. These sparrow-sized birds usually build their nests in little caves beneath boulders, and they sometimes

(Continued on p. 8·44)

a. Bank Swallow Colony

b. Two Nestlings Survey their Surroundings

c. Chicks Inside Nest Chamber

Figure 8–40. Bank Swallow Nest Tunnels: *Ranging through much of North America and Eurasia, Bank Swallows breed in dense colonies often having several hundred active nests, but occasionally containing as many as 2,000.* **a. Bank Swallow Colony:** *Bank Swallows locate their colonies in nearly vertical sand or dirt banks, usually near water. Common sites include sand and gravel pits (as shown here), road cuts, and eroded stream banks—places where the substrate is fairly crumbly for digging, but not so loose that tunnels will collapse. Although not evident from this photo, the birds usually dig their tunnels near the top of a bank, where the soil is firm and the nests are farthest from climbing predators. Not all burrows visible in a colony are active—some remain empty from previous years, and others were started by young or unmated birds and never completed. Both sexes dig the burrow, using their beaks and feet, progressing up to 5 inches (12.8 cm) each day. Tunnels average 2 to 3 feet (0.6 to 0.9 m) in length, but occasionally reach 6 feet (1.8 m). Photo courtesy of Lang Elliott/CLO.* **b. Two Nestlings Survey their Surroundings:** *Photo courtesy of Lang Elliott/CLO.* **c. Chicks Inside Nest Chamber:** *At the end of the tunnel, the male and female build a nest of grass, weeds, and rootlets, adding a lining of feathers after the eggs are laid. These three nestlings are almost ready to fledge. Photo by Mike Birkhead/Oxford Scientific Films.*

a. Belted Kingfisher Excavating Nest Tunnel

b. Belted Kingfisher Nest Tunnel

c. Common Kingfisher Feeding Nestlings

Figure 8–41. Kingfisher Nest Tunnels: *All kingfishers nest in holes. Some adopt cavities in trees, others dig nests in active termite mounds, and others, as shown here, dig burrows in earthen banks.* *a. Belted Kingfisher Excavating Nest Tunnel:* *While his mate calls nearby, a male Belted Kingfisher takes his turn digging the nest tunnel with his beak, kicking out loosened dirt. Birds usually dig in the early morning, their tunnels often extending 12 inches (30 cm) after the first day. The entrance hole is about 4 inches (10 cm) in diameter, and the bottom of an active hole has a distinct furrow on each side where an entering or exiting bird's feet scrape the dirt.* *b. Belted Kingfisher Nest Tunnel (Side View):* *Belted Kingfishers usually nest near water, digging their tunnels into steep dirt or sand banks with no vegetation on the face and usually just herbaceous plants on the top, minimizing the chance of hitting roots. The tunnel entrance is high, 14 to 15 inches (36 to 38 cm) below the top of the bank, probably to deter predators and avoid flooding. The tunnel is usually 3.3 to 6.6 feet (1 to 2 m) long, occasionally reaching 15 feet (4.5 m), and it slopes upward to the nest chamber, probably to keep out water. The nest chamber is unlined, but once birds begin to incubate, regurgitated pellets of undigested fish and insect parts may accumulate, somewhat cushioning the eggs (see inset). Nest information from Hamas (1994) and Michael J. Hamas (personal communication).* *c. Common Kingfisher Feeding 14-Day-Old Nestlings:* *After the first few days, parent Common Kingfishers of Eurasia bring small whole fish to their young. The six or seven chicks each may consume 15 or more fish per day, keeping their parents busy. Photo by Angelo Gandolfi/BBC Natural History Unit.*

8

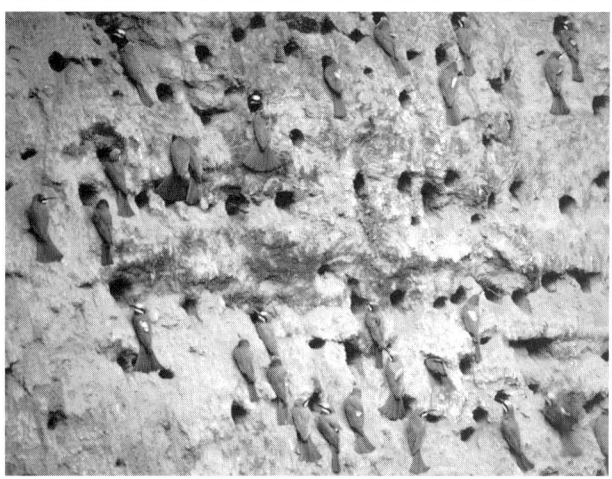

a. White-fronted Bee-eater Colony

b. European Bee-eater Feeding Nestling

*Figure 8–42. Bee-eater Nest Tunnels: Like many of their kingfisher relatives, the colorful bee-eaters of the southern Old World nest in tunnels that they dig into a bank or flat ground. **a. White-fronted Bee-eaters at Nest Colony:** Inhabiting the hot African savannas, White-fronted Bee-eaters breed during the dry season, but dig their tunnels into sand or dirt banks at the end of the previous wet season, before the soil is too hard-baked. White-fronted Bee-eaters nest in colonies, and each tunnel is dug and tended by the members of a complex, extended family group. The tunnels, which may be up to 3 feet (1 m) long, end in an unlined, oval nest chamber, which may acquire regurgitated pellets of hard insect parts as incubation proceeds. In this intensively studied colony in Kenya, note the colored wing tags (appearing mostly as light patches), which researchers have placed on the birds for individual recognition. The birds in this photo are socializing at their nest tunnels, as they do each morning and evening. Photo by Peter H. Wrege. **b. European Bee-eater Feeding Nestling:** Although bee-eaters specialize on venomous, flying insects such as honeybees, they often bring larger prey, such as this dragonfly, to their young. The nestlings have enlarged papillae along the ankle (the upper section of the tarsometatarsus—see Fig. 1–11), and they shuffle around the nest and tunnel on their entire foot, as do humans, rather than on their toes, as do most other birds. The enlarged ankle is visible in this one-week-old nestling, which has crawled out of the nest chamber to meet its parent. Photo by Alain Christof/Oxford Scientific Films.*

line the nest and its entrance with large numbers of pebbles—up to thousands in long-established nests. Debate continues about the function of this behavior, but the pebbles may, in part, signal the quality of a male to his mate before egg laying, and also may aid in thermoregulation of the nest or in protecting the nest or the incubating female from predation (Leader and Yom-Tov 1998). Apparently, all nuthatches are cavity adopters, and many species use mud to narrow the entrance to a hole in a tree branch or under a flap of tree bark. Rock Nuthatches of the Middle East have no trees in which to nest, so they use a niche in a rock face instead. Upon this niche they add a large mud nest with walls over 0.75 inches (2 cm) thick, up to 10 inches (25 cm) across, and weighing up to 77 pounds (35 kg) when dry (Goodfellow 1977) **(Fig. 8–43)**. Many other birds use mud to modify adopted cavities, but most design the entrance to allow the passage of the parents. The spectacular exception is the hornbills of Africa and Asia, named for their large bills, which superficially resemble those of toucans. Most species in this family adopt a tree cavity for their nest. The cavity is enclosed with mud, and the female is left inside, incubating the eggs and later feeding the nestlings with food supplied by the male through a small slit in the mud wall of the nest. The female goes through a rapid molt of her feathers while in the nest, and her old molted feathers, together with the shed feather shafts of her young, combine with many weeks' feces to create a mess without rival in the feathered world **(Fig. 8–44)**. In some species the female emerges part-way through the nestling period to help the male gather food, and in others, mother and young emerge together. It must be a special day for the hornbill family when, at the end of the nestling period, mother and young burst through the mud wall to pursue life outside in fresh air!

The Evolution of Nest Construction

The remarkable diversity in nest construction and placement displayed by birds around the world tempts ornithologists to try to reconstruct the evolution of nest-building behavior. Unfortunately, no simple scenario seems widely applicable. There are certainly birds considered to be relatively primitive that build very simple

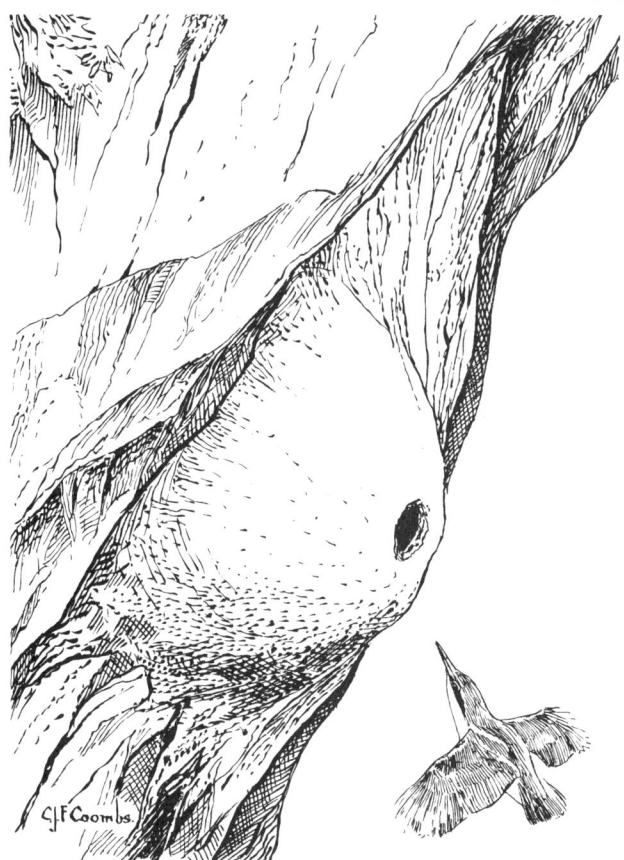

Figure 8–43. Mud "Cavity Nest" of Rock Nuthatch: *In their treeless, rocky habitats in the Middle East, Rock Nuthatches must create their own nest cavities. They plaster mud (or in dry periods, dung) to a rock wall, building up a massive hemisphere with a funnel-shaped entrance tunnel. The thick-walled nests—located under a rocky projection, around a crevice in the rock, or on a steep rock face (as shown here)—may reach 10 inches (25 cm) across and weigh up to 77 pounds (35 kg). Inside, the birds build a large nest cup of moss, hair, and feathers. Drawing by Dr. C. J. F. Coombs.*

Figure 8–44. Red-billed Hornbill Female and Young in Nest: *At the beginning of incubation, Red-billed Hornbills of Africa plaster up most of the entrance hole to their cavity nest with mud, leaving the female captive inside for the 25-day incubation period and part of the 45-day nestling period. There, the female depends on her mate to deliver provisions, both for her and the young, through a tiny slit. During this time, feathers from the molting female combine with feces from both the young and mother to create quite a mess in the nest. Part-way through the nestling stage the female emerges to help the male feed the growing young. One side of this tree cavity was replaced with glass, allowing photographers and researchers to observe nest activity. Photo by Alan Root/Oxford Scientific Films.*

8

nests, but many primitive birds build complicated nests, and many more-recently evolved birds build simple nests. Nest-building behavior has clearly evolved very differently in various bird groups throughout the world, as the groups are molded by quite different selection pressures. As a result, scenarios for the evolution of nest building seem to work best when applied to single groups of birds. For example, Winkler and Sheldon (1993) found an ordered progression of nest types within the swallows of the world **(Fig. 8–45)**: the most primitive swallows today are burrowers, whereas mud nesting and cavity adoption came later in the group's diversification. Among species that build mud nests, it appears that species with open cup nests arose before those with closed cups, which in turn led to species that build retorts. Note, however, that the evolutionary history and relationships among species with these nest types may be very different in a different family of birds.

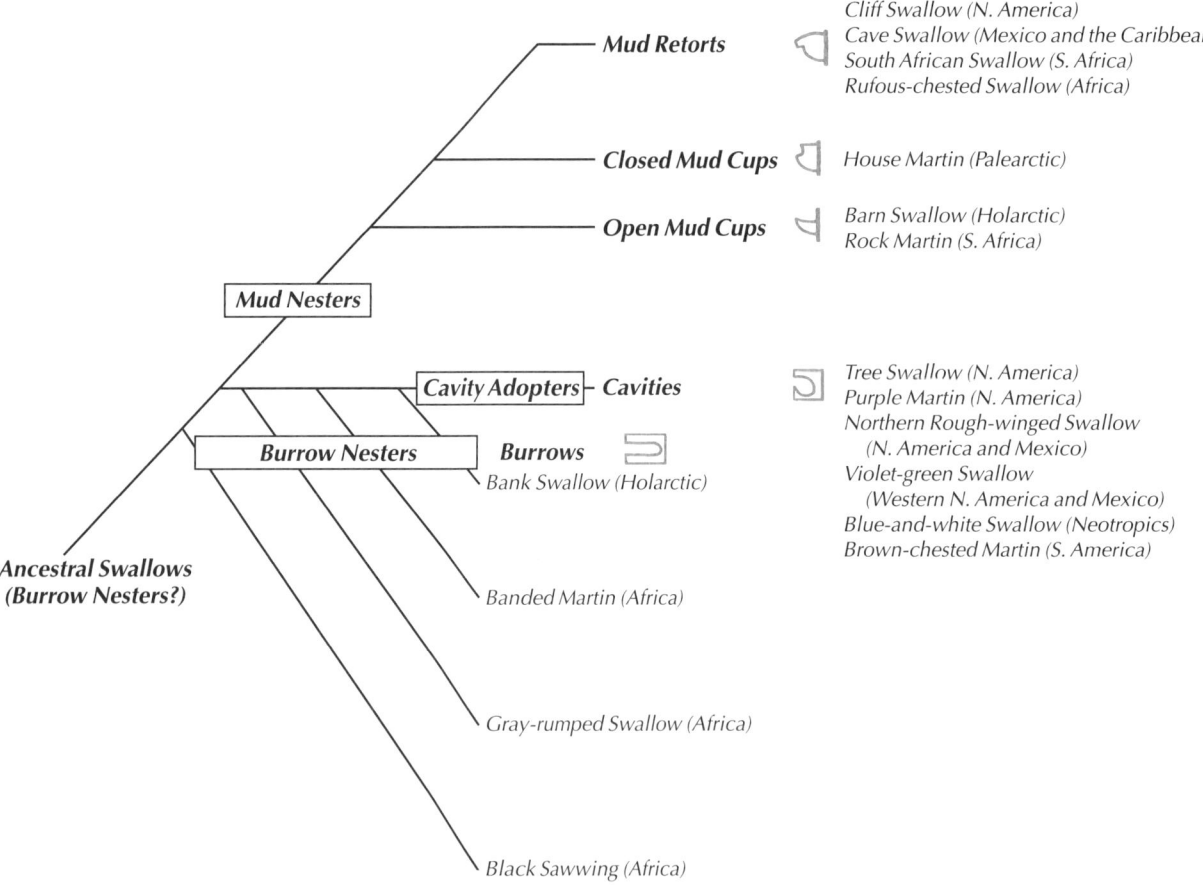

Figure 8–45. The Evolution of Nest-Building Behavior in Swallows: *This diagram, based on the degree of similarity among DNA sequences of the different species, illustrates the probable order in which different types of nest building behaviors evolved in swallows. At each fork, a subset of birds diverged from the ancestral group, following evolutionary paths that led to the representative species alive today. Species listed for each nest type are selected examples, not an inclusive list. The ancestral swallows probably were burrow nesters, from which at least two different burrow-nesting lineages split off. Interestingly, one of these led to cavity adopters—a counter-intuitive progression, as one might expect birds that build their own nests to arise from those that do not. From other burrow nesters arose the mud nesters, first those that build open mud cups, and later those building more closed cups and retorts. See text for more details. Adapted from Winkler and Sheldon (1993).*

Interestingly, in the evolutionary sequence for nest building in swallows, cavity adoption arises *after* at least burrow nesting. Thus, swallows that use no special motor patterns for making a nest evolved from ancestors who already had evolved techniques for digging burrows. This discovery, based on DNA-DNA hybridization studies (Winkler and Sheldon 1993) and confirmed by comparing sequences of mitochondrial DNA (Sheldon et al. 1999) is counter to the intuitively appealing notion (Mayr and Bond 1943) that cavity adopters are more primitive. Not only does this finding remind us how often our intuitions can lead us awry in evolutionary biology, but it highlights the importance of context in understanding the evolution of life history traits. The great majority of cavity-adopting swallows live in the Americas, so this new nesting trait apparently evolved in the New World after an ancestral Old World group colonized there.

Nest Lining

One of the most interesting aspects of the nesting biology of birds is the diversity of materials they use to line their nests **(Fig. 8–46)**. A surprising number of species, ranging from raptors to starlings and swallows, line or adorn their nests with green plant material **(Fig. 8–47)**. Early authors, failing to see a function for this "decoration," wondered if these birds might simply be expressing an aesthetic sense, but recent research indicates that the green plants are carefully selected to bring chemicals that repel or kill insects into the nest (Wimberger 1984; Clark and Mason 1988).

Another habit found in a wide variety of phylogenetic groups that tempts observers to ascribe aesthetic motives to birds is using feathers to line nests. Anyone with a down coat has been comforted by the habit of most female waterfowl to line their nests with down feathers pulled from their breasts (see Fig. 4–124a), but several songbirds line their nests with feathers that are not their own. For instance, many species of swallows collect feathers for their nest lining, and in some species competition for feathers is intense—sometimes even leading to serious fighting **(Fig. 8–48)**. In recent experimental work on Tree Swallows, Winkler (1993) demonstrated that young grow faster in nests lined with more feathers, suggesting that the fights might well be worth the risk in this species. Other birds that gather feathers include kinglets and the Long-tailed Tit of Europe. Long-tailed Tits must be the champions of feather collecting: their oval nests **(Fig. 8–49)**, made of a felt of wool, moss, and spider webs, are lined with between 1,000 and 2,000 feathers (Goodfellow 1977)! One can only wonder how these birds find so many.

Even within a closely related group of birds, interesting differences often exist in the types of nest linings chosen. For example, most thrushes build a nest of twigs, grass, and leaves cemented together with mud and lined with fine grasses. In contrast, the Song Thrush of Eurasia does not add a soft lining, but elaborates the mud portion of the nest into a hard, smooth lining, sometimes in combination with wet rotten

(Continued on p. 8·50)

American Goldfinch Nest Lined With Plant Down

Anhinga Nest Lined With Leaves

Sandhill Crane Nest Lined With Twigs

Cooper's Hawk Nest Lined With Bark

White-breasted Nuthatch Nest Lined With Wool and Fur

Piping Plover Nest Lined With Broken Shells

Figure 8–46. Nest Linings: *Birds line their nests with a remarkable diversity of materials. Many, particularly passerines, use soft plant materials such as grasses, moss, bark fibers, or down—especially cattail and thistle down, as in the American Goldfinch nest pictured here. Anhingas and some waders include leaves in their nest lining. Some raptors and other large birds may use more coarse plant materials such as twigs—as in the Sandhill Crane nest shown here, located in a sphagnum bog—or bark chips, as in the nest of the Cooper's Hawk, which usually is lined with outer bark from oaks or pines. Chickadees, Ovenbirds, White-breasted Nuthatches, and others often construct a soft lining of animal hair. Many beach-nesting shorebirds, such as the Piping Plover, line their simple nest hollow with a few broken shells. Keep in mind that a single species may use widely different nest materials in different habitats, and that these photographs provide only examples of typical nests. American Goldfinch photo by Hal H. Harrison/Photo Researchers; Anhinga and Cooper's Hawk photos by Hal H. Harrison; Sandhill Crane courtesy of L. H. Walkinshaw/CLO; White-breasted Nuthatch courtesy of Bill Duyck/CLO; Piping Plover courtesy of John Gavin/CLO.*

Figure 8–47. Purple Martin Nest with Green Leaves: *Many birds, including Purple Martins, House Sparrows, European Starlings, American Crows, and some raptors, add green leaves to their nests—either placing them around the edges or directly in the nest (as shown here). Many species replenish the leaves daily throughout the incubation and nestling periods. Researchers believe these leaves are added because they repel or kill ectoparasites, but more study of this phenomenon is needed. One species, the European Starling, does appear to choose leaves high in compounds with these properties (Clark and Mason 1988). Because Purple Martins nest colonially in adopted cavities, people often try to attract them by putting up multi-roomed nest boxes or by hanging clusters of hollow gourds with entrance holes. The top of this gourd house has been removed to reveal the nest inside. Photo by Hal H. Harrison.*

a. Tree Swallow Bringing Feather into Nest

b. Tree Swallow Nest Lined with Chicken Feathers

c. Tree Swallows Fighting Over a Feather

Figure 8–48. Tree Swallows and Nest Feathers: *Many swallows, such as Tree Swallows, line their nests with feathers that are not their own.* **a. Tree Swallow Bringing Feather into Nest:** *This female has found a large feather for her nest, which is located in a natural tree cavity. Photo by Marie Read.* **b. Tree Swallow Nest Lined with Chicken Feathers:** *This elegant nest is lavishly lined with chicken feathers. Note the white eggs typical of cavity nesters. Photo by Mark Wilson/ WILDSHOT.* **c. Tree Swallows Fighting Over a Feather:** *Feathers can be hard to find, so competition for them may be intense, sometimes involving serious fights. Because young Tree Swallows grow fastest in nests lined with the most feathers, the fighting among parents may be worth the risk of injury.*

Figure 8–49. Long-tailed Tit at Feather-lined Nest: The 5-inch (13-cm) oval nests of Long-tailed Tits are felted from wool and moss, bound together with spider webs, and covered with lichen. After about nine days of nest building, the pair spends the next one or two weeks gathering feathers for the cozy lining—usually bringing the 1,000 to 2,000 feathers to their nest one at a time, from locations up to several hundred yards away. Small birds in cold regions often line their nests heavily, and these Eurasian tits have found an extremely effective insulating material. Photo by David Hosking/Photo Researchers.

wood, dung, or peat glued together with saliva (Goodfellow 1977). Most warblers line their nests with fine, dried grasses or hair, but Worm-eating Warblers line their nests entirely with the hairlike stalks from the spore capsules of the hair moss (*Polytrichum*) and occasionally with the stalks of maple seeds (Bent 1953). Do the species that make such highly specific and idiosyncratic choices of lining materials reap some advantage? It would be very interesting to find out!

Nest-building Behavior

The behaviors that birds use in building nests vary in complexity with the elaborateness of the nest: a scrape with no lining is clearly a lot less demanding than a hanging retort. Species that build platform nests seldom require any specialized fabrication methods: a foundation of coarse sticks is laid across the supporting branches and other sticks are added until the bulk of the platform is built. Finer sticks and a lining, if any, are then added.

The cup nesters, however, use some specialized motor patterns in construction. To build a cup nest, most species loosely weave together coarser material for the outside, then weave inner layers of finer materials to support the lining, if any **(Fig. 8–50)**. In species with pendulous nests, these weaving motions have been elaborated and perfected through natural selection to produce walls of considerable integrity and tensile strength. But the height of avian weaving technique is reached in the globular nests of the African weavers. These birds use a variety of knots in their constructions, producing nests of considerable durability and unrivaled intricacy **(Fig. 8–51)**. The Common

(Continued on p. 8·54)

a. Gathering Outer
Wall Material

b. Shaping Outer
Wall

c. Gathering Mud

d. Shaping Mud Layer

e. Gathering Lining

f. Shaping Lining

Figure 8–50. Nest Construction in the American Robin: American Robins typically spend five to six days building their cup-shaped nest, placing it on a fork or horizontal branch of a tree, or on the ledge of a building. The nest is about 6 to 7 inches (15 to 17.5 cm) across on the outside, with an inner cup for the eggs about 4 inches (10 cm) wide and 2.5 inches (6.25 cm) deep. Both male and female bring nest material (with the male's contributions being fewer and smaller), but the female actually constructs the nest. It consists of three layers, which are built sequentially. **a. Gathering Outer Wall Material:** For the coarse outer wall, the birds make many trips bringing twigs, rootlets, dead leaves, moss, coarse grasses, and sometimes man-made materials such as paper pieces to the nest site. They often choose wet materials, because they are more pliable. **b. Shaping Outer Wall:** Once the pair has accumulated enough material, the female squats in its midst, rotating left and right (as shown by colored arrow) and pressing down, thereby using her body to shape the material into a rough cup. The birds may then bring more material and repeat the shaping process. **c. Gathering Mud:** The second layer, inside the outer wall, is formed of mud. Visiting banks of streams, edges of puddles, or other muddy places, the female picks up pellets of mud or earthworm castings in her bill, sometimes using a wick of vegetation to carry them (see Fig. 8–53), and plasters them inside the nest. **d. Shaping Mud Layer:** After incorporating numerous mud pellets into the nest walls, the female again squats inside and rotates her body left and right, shaping the mud into a cup by pressing with her breast and the wrists of her wings. **e. Gathering Lining:** While the mud is still damp, the robins bring fine, soft, dead grasses for a nest lining. **f. Shaping Lining:** The female shapes the lining by pressing it into the mud layer, rotating her body as before. Once the nest is complete, she usually begins to lay eggs within a few days.

8

a. Village Weaver Colony

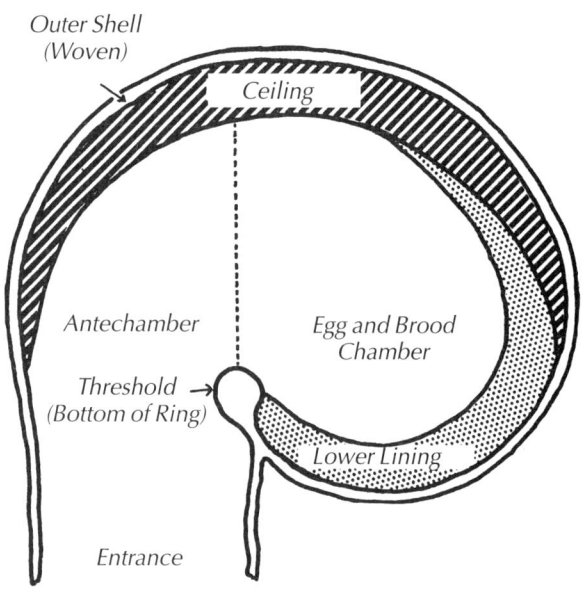

Outer Shell (Woven)

Ceiling

Antechamber

Threshold → (Bottom of Ring)

Egg and Brood Chamber

Lower Lining

Entrance

b. Longitudinal Section Through Village Weaver Nest

c. Male Village Weaver in Nest at Ring Stage

d. Nest Early in Roof-Building Stage

(Figure continued on next page)

8

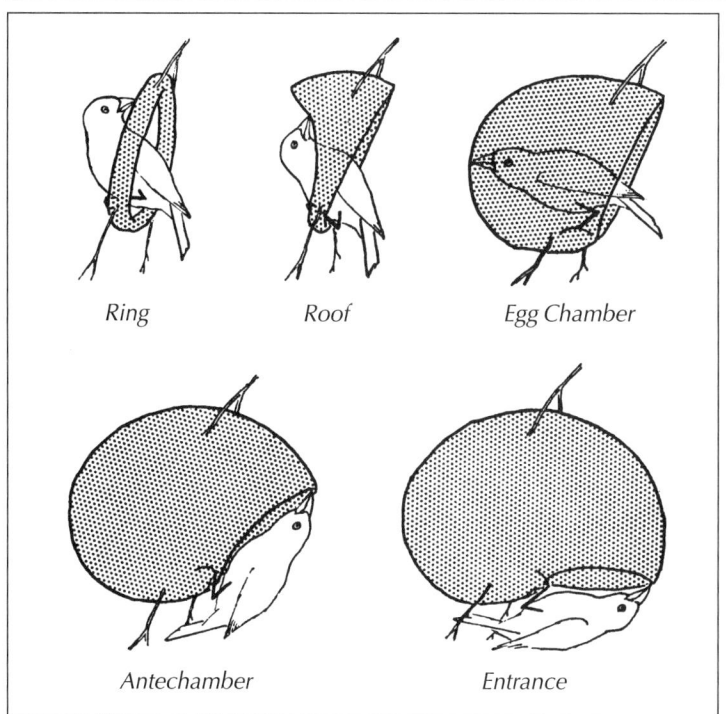

Ring   Roof   Egg Chamber

Antechamber   Entrance

e. Stages in Nest Building

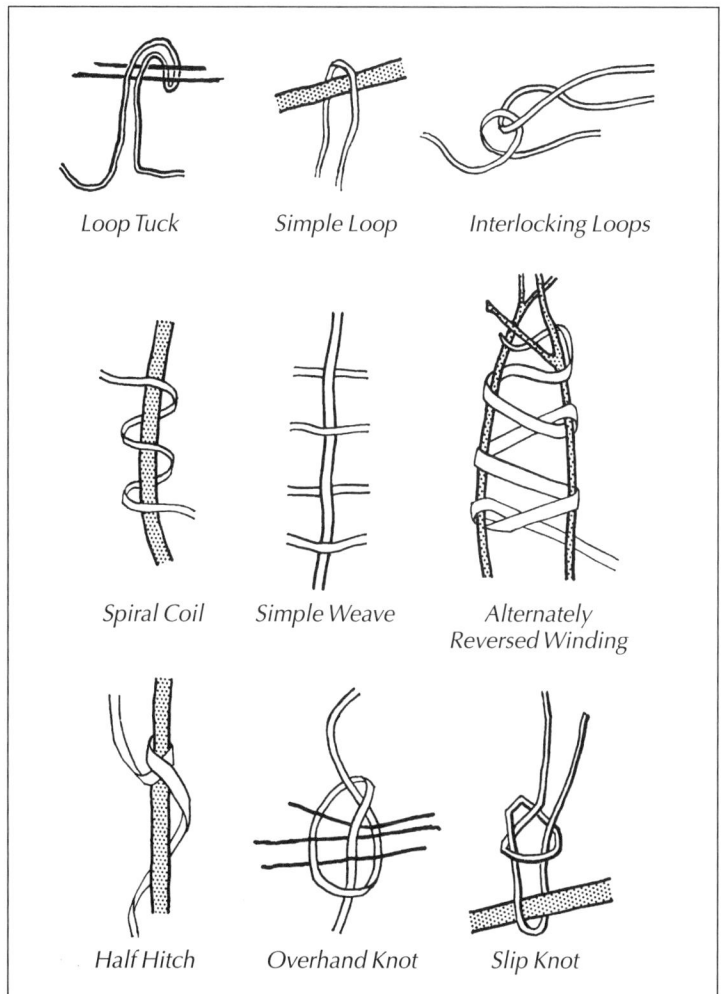

Loop Tuck   Simple Loop   Interlocking Loops

Spiral Coil   Simple Weave   Alternately Reversed Winding

Half Hitch   Overhand Knot   Slip Knot

f. Weaving Techniques

Figure 8–51. Nest Construction in the Village Weaver: Males of the Village Weaver, common in open areas throughout much of Africa, use complex weaving techniques to build their retort nests, which are hung from the ends of long branches. Males learn the details of weaving and nest shaping through experience—the nests of yearlings are usually crude and messy. This species and its nest building have been thoroughly described by Collias and Collias (1984). **a. Nest Colony:** Males place their nests in colonies, most often in acacia trees, but sometimes in other trees such as this palm. **b. Longitudinal Section:** The oval nest is 6 inches (15 cm) wide, with an entrance tunnel 2 to 4 inches (5 to 10 cm) long that may provide protection from predators such as snakes. The thick, thatched ceiling and the outer roof provide protection from rain and sun, and the threshold helps to prevent the eggs and young from rolling out of the nest as it sways in the wind. **c. Male in Nest at Ring Stage:** The male starts his nest by tearing long, flexible strips from living grasses or palm leaves, and winding these securely around a downward-facing, forked twig. He then gathers the dangling ends of the strips and weaves them together, forming a ring that becomes the perch from which he weaves the rest of the nest, beginning with the roof. **d. Nest Early in Roof-Building Stage:** Note the intricate interweaving of the grasses in the roof. **e. Stages in Nest Building:** Perched on the threshold at the bottom of the ring and always facing the same direction, the male weaves the roof, egg chamber, and antechamber. The bright yellow-and-black male then displays from his nest to attract a female, and if one accepts his nest (after careful inspection), he builds an entrance tunnel. The female does no actual weaving, but lines the egg chamber with grass leaves, soft grass heads, and sometimes feathers. The male may then build additional nests to attract more females—sometimes having as many as five active nests at the same time. Although he uses about 300 grass stems in the shell of each nest, he can complete an entire structure in one day, and if any one of his nests is not chosen by a female, he dismantles it and begins another in the same place. **f. Weaving Techniques Used by True Weavers (subfamily Ploceinae) in Nest Building:** The Grosbeak Weaver constructs its primitive nest with simple loops and tucks (top row). The Red-billed Quelea and Red Fody use spiral coils in their crudely woven nests. The Village Weaver, however, uses more sophisticated stitches—attaching the roof to the support twigs with alternately reversed winding, using the simple weave in the egg chamber, and placing half hitches throughout the nest. The more complex overhand and slip knots are only occasionally used by weavers. Photo a by E. Daeschler/VIREO. All other photos and drawings courtesy of Nicholas E. Collias.

Figure 8–52. Common Tailorbird Nests: *The female of this common warbler of India and Southeast Asia creates a funnel-shaped cradle for its fluffy cup nest by piercing a series of small holes near the edges of one or more leaves with its bill, then sewing them together with plant fibers, cobwebs, or silk from cocoons. Nest construction, which lasts about four days, may be hampered by threads breaking and leaves tearing. The bird may form the cradle by curling one large leaf and sewing the edges together (see photo) or by sewing two to four different leaves together (see drawing). The leaf cradle helps to disguise the nest cup and protects it from heavy tropical rains. Although Common Tailorbirds frequently nest near humans, especially in city gardens, their well-camouflaged nests seldom are noticed. Photo by Pat Louis/Valan Photos; drawing by Charles L. Ripper.*

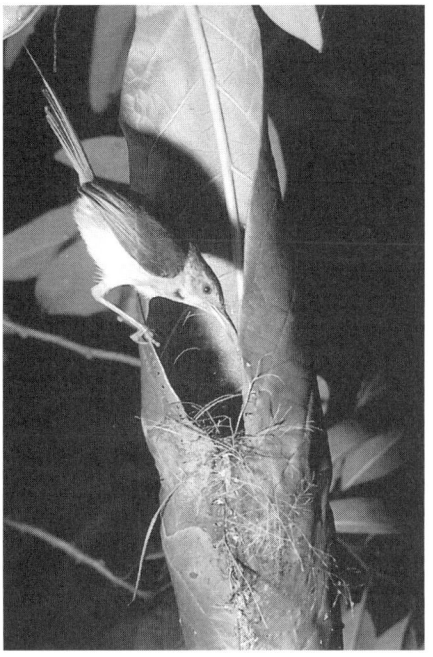
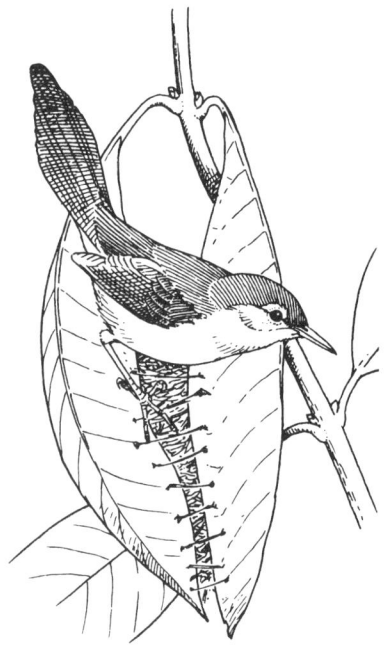

Tailorbird, a well-known warbler (family Sylviidae) in Southeast Asia, creates a funnel-shaped cradle for its soft cup nest by piercing a series of small holes near the edges of two large, living leaves. It then sews the leaves together with plant fibers, cobwebs, or silk from cocoons by placing these fibers in the leaf holes, and drawing the leaves together (**Fig. 8–52**). Much less well understood is how the felted nests of such species as Long-tailed Tits (see Fig. 8–49), Bushtits, and penduline-tits manage to hold together, but certain lichens may provide Velcro™-like adhesion (Hansell 2001).

Within the mud-nesting birds, a great variety of construction methods are used (Rowley 1971). If you have ever watched an American Robin build a nest, you may have seen it take a beakful of vegetation, often dried, dip it into a source of mud, then transfer this wick—and all the mud that adhered to it—back to its nest (**Fig. 8–53**). This is

Figure 8–53. American Robin Using Vegetation Wick to Gather Mud: *Most birds that use mud to construct their nests gather it by taking a beakful of vegetation and dipping it into the mud source. Then they add the entire vegetation "wick" and its mud coating to the nest. This American Robin is adding a mud wick to her nest on a nest shelf. Note that the nest is still unlined at this stage of construction (see Fig. 8–50). Photo courtesy of Hal H. Harrison.*

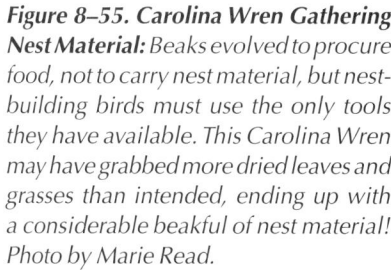

Figure 8–54. Cliff Swallow Gathering a Beakful of Mud: *Cliff Swallows may fly up to several miles from their nests to gather mouthfuls of pure mud from beside streams, lakes, or temporary puddles. Their adherent retort nests are made entirely from mud and then lined with grass. Mud-collecting is carried out by both members of a pair, often in large, synchronized groups of birds from the same colony. Photo courtesy of Cliff Beittel.*

the most common way that birds incorporate mud into a nest, and it is used by thrushes (such as the American Robin), and some flycatchers, such as the Eastern Phoebe. The only birds that build an adherent nest of mud are the mud-nesting swallows. And those species that build a retort (for example, the Cliff Swallow) have evolved the further innovation of transporting mouthfuls of pure mud to the nest **(Fig. 8–54)**; most other mud-nesting swallows use a wick, incorporating it into the nest. Selecting the right mud to meet the extraordinary engineering demands of their nests must require considerable discernment (Kilgore and Knudsen 1977; Robidoux and Cyr 1989); I wonder how well we would do at such a task?

Birds carry out all the digging, weaving, and plastering required to construct their various nest structures with the only tools available to them—their beaks and feet. But as you watch a female oriole weave long strands of grass into an intricate hanging basket, you might forget that beaks and feet evolved primarily for other purposes. Beaks evolved to obtain food efficiently; not to carry sticks, gather mud, pile up vegetation, sew leaves together, and weave grasses **(Fig. 8–55)**: the kingfisher's beak evolved to catch fish, not to dig tunnels in earth. Feet evolved to walk, run, perch, swim, and wade across snow and lily pads; not to dig tunnels, scrape, stamp, and gather and hold nest materials. Woodpeckers are perhaps an exception, as their means of finding food and digging nest cavities are similar: chiseling into wood.

Figure 8–55. Carolina Wren Gathering Nest Material: *Beaks evolved to procure food, not to carry nest material, but nest-building birds must use the only tools they have available. This Carolina Wren may have grabbed more dried leaves and grasses than intended, ending up with a considerable beakful of nest material! Photo by Marie Read.*

Sex Roles in Nest Building

Birds vary considerably in the roles that each sex takes in nest construction, but within certain groups, sexual roles tend to be fairly consistent. For example, in hummingbirds and other species with strong sexual selection on males—such as many cotingas (see Figs. 1–66 and 1–67), birds-of-paradise (see Fig. 3–10), manakins (see Figs. 6–42 and 6–47), and grouse—the female does all the work of nest construction. In polyandrous species, such as phalaropes and jacanas (see Fig. 8–126), the male does all the nest building. In many species in which the male and female are similar in appearance—for example, gulls, corvids, herons, and swallows—the male and female take a near-equal share in nest building. In these species the male often gathers most of the material, and the female actually places it in the nest **(Fig. 8–56)**. In other species whose males and females appear similar, including wrens and raptors, the male takes the lead in nest building as well, with the female often adding only the lining when the nest is finished. Curiously, wrens and raptors also happen to construct many surplus nests on their territories. Although the functions of these extra nests are open to debate, they may allow rapid renesting if a nest is lost to predators or competitors (Newton 1979). Extra nests built by males may also help to attract a female, as in Winter Wrens (Evans and Burn 1996) and some populations of Marsh Wrens (Verner and Engelsen 1970). Is it possible that such an advantage may have "gone wild" in the elaborate courtship structures made by bowerbirds (see Fig. 6–41) (Borgia 1985)? Could the extra nests of some species also serve as decoys to throw off potential nest predators?

*Figure 8–56. **Great Blue Heron Pair Nest Building:** In Great Blue Herons and many other species in which the sexes look alike, the male gathers sticks and twigs and brings them to the female, who stays in the nest and arranges the nest materials. In the left drawing, the male presents a stick to the female. Often this presentation involves much display, especially early in the nest building process, when the female performs a Stretch Display (see Fig. 6–18b) and takes the stick. Then, in another display called Bill Clappering, the male points his bill toward the female, rapidly clicking the bill tips together in the air. In the right drawing, the female inserts a stick into her platform nest while the male looks on. Drawing by Richard P. Grossenheider, from* Miscellaneous Publications—Museum of Zoology, University of Michigan, *Number 102.*

In many sexually dimorphic species with brighter males (for example, tanagers, warblers, and New World blackbirds), the female often takes on the bulk of nest-construction duties, perhaps because she is less likely to be detected by predators. But much is yet to be learned about sex roles in building: why don't male vireos (who look just like their mates) help to build the nest? And why is the gaudy male Magnificent Frigatebird the one who builds the nest with material brought by the female? These and many other questions wait to be answered.

Duration of Nest Building

The length of time that a bird takes to build a nest depends upon the type of nest (simple or elaborate), the season (early or late in the breeding season), and the climate. A passerine building a simple cup nest in northern North America spends about six days: three days on the outer layer and three on the lining. In contrast, a passerine such as an antwren in Panama takes 10 to 15 days (Russell Greenberg, personal communication). In the Neotropics, a Great Kiskadee (a flycatcher) takes 24 days to build its bulky, domed nest; and the Chestnut-headed Oropendola takes about a month to construct its long, pendulous nest. The short breeding season in the higher latitudes apparently forces birds to build faster. In the lower latitudes, where the breeding season is prolonged, there is no such pressure. Birds build leisurely, working on their nests perhaps only a few hours each day.

Some birds take less time to build their second or third nests of the breeding season. American Robins, for example, take an average of five to six days to build their first nest of the year, but may build a new nest in two to three days if the first nest is destroyed (Kendeigh 1952).

Nest Appropriation and Reuse

Most cup-nesting species—even those, such as woodpeckers, that build energy-demanding nests—tend to build a new nest each year. This makes sense because once a nest has been used, especially if it has supported young all the way to fledging, it often has outlived its usefulness. The accumulation of odoriferous nest contents probably makes most used nests more vulnerable to predators than new ones. In addition, old nests often contain large numbers of ectoparasites that could harm the developing young. Nevertheless, eagles, Ospreys, and some other large birds do reuse nests for generations **(Fig. 8–57)**, and recent observations (Davis et al. 1994) suggest that Eastern Bluebirds prefer to nest in boxes that contain an old nest. The ectoparasites in many old nests are part of a complex living community in which some animals feed on feces, uneaten food, and bits of castoff skin and feather sheaths; others prey on living nestlings; and some are actually

Figure 8–57. Huge Platform Nest of Osprey: *Fish-eating Ospreys always nest near water, building their enormous platform nests atop a variety of structures: large dead trees (as pictured here), cliffs, cacti, buoys, utility poles, or nest platforms. On small islands, they may nest on the ground. The nests, built of large sticks—sometimes including debris such as rakes, brooms, shoes, and dolls—are lined with inner bark, grasses, and vines. Nests may be more than 10 feet (3 m) high and often grow bigger each year, as birds reuse them for many generations, adding material with each use. Photo by R. Day/VIREO.*

Figure 8–58. Parasitoid Wasp Laying Eggs in Blowfly Pupa: *One might think that freshly built, insect-free nests would be the most desirable to breeding birds, but some insects that dwell in nests are actually beneficial to birds because they parasitize insects that parasitize the birds. These so-called* **parasitoids**—*such as this gnat-sized female chalcid wasp,* Nasonia vitripennis—*may be more prevalent in used nests, and may render these older nests more desirable to breeding birds. The larvae of this* Nasonia *wasp will hatch and devour the blowfly pupa, thus killing an insect that feeds on the blood of nestlings, especially in cavity nests. Photo courtesy of John H. Werren, Copyright 1980.*

parasites (termed **parasitoids**) on these parasites **(Fig. 8–58)**. Is it possible that in some communities of nest invertebrates, the older nests have a higher preponderance of parasitoids, thus making them more suitable for nesting? Ornithologists are just beginning to investigate this question, and much remains to be learned about the interactions among parasites, parasitoids, and nesting birds. This example illustrates how ornithologists, to really understand the nesting biology of the species they study, need to be broadly trained zoologists!

In some birds, the advantages of using old nests *must* outweigh the disadvantages, because many species appropriate the nests of other species. Solitary Sandpipers and Bonaparte's Gulls, nesting in northern boreal forests (taiga), appropriate the nests of land birds in small spruce trees **(Fig. 8–59)**. House Sparrows use the old nests of a variety of other birds from American Robins to Cliff Swallows, in addition to other niches and cavities of all sorts. Great Horned Owls often use the large nests of other species, such as Red-tailed Hawks or even squirrels. The large, thorny nests of South American thornbirds are used by at least 11 other species, including flycatchers, tanagers, and chachalacas (Lindell 1996). Sometimes the old nest is used simply as physical support for the building of a new nest: Mourning Doves place their flimsy stick nests on almost any elevated support, including the old nests of robins, grackles, or Brown Thrashers. The Little Swift of Africa adopts old mud retorts made by Lesser and Greater striped-swallows, often affixing white feather "decorations" around the entrance.

Some of the most dedicated nest appropriators are species that nest in cavities in trees or in holes in a bank or cliff. Great Crested Flycatchers, chickadees, and Eastern Bluebirds are North American cavity adopters (also called "secondary cavity nesters") who share this habit with close relatives elsewhere in the world. The adoption of tree holes for nesting by American Kestrels and Prothonotary Warblers, however,

Figure 8–59. Bonaparte's Gull Nesting in Spruce Tree: *Bonaparte's Gulls, which nest near lakes, rivers, or bogs in the northern boreal forests of North America, may either use the abandoned nests of land birds or build their own. Their nests are typically located on a horizontal branch of a spruce tree, 4 to 15 feet (1.2 to 4.6 m) above the ground. Photo courtesy of Sam Grimes/CLO.*

Figure 8–60. Prothonotary Warbler in Nest Cavity: Most New World warblers nest on the ground or in vegetation, but the Prothonotary Warbler of the eastern United States, which breeds in swamps with standing dead trees and in wooded bottomlands along streams, places its nest in natural cavities (as shown here), abandoned woodpecker holes, and nest boxes—usually low and over or very close to water. The male places moss (up to 3 inches [8 cm] deep) in a number of potential nest sites, and the female completes the nest cup with more moss, rootlets, bark, and plant down, lining it with finer materials. The only other New World warbler that nests in cavities is the Lucy's Warbler, found in the western United States. Because so few New World warblers nest in cavities, this breeding behavior probably evolved more recently in this group than the building of other types of nests. Photo by G. Bailey/VIREO.

appears to be a more significant innovation, as most other hawks and warblers do not nest in cavities **(Fig. 8–60)**. Most nest appropriators wait until the nest builder is finished with the nest, but European Starlings, House Sparrows, and House Wrens take over active cavities or boxes of other birds; wrens go so far as to puncture the eggs of birds whose nests they fancy, and House Sparrows often kill resident birds to obtain their cavity.

Most New World swallows, including Purple Martins and Tree, Northern Rough-winged, and Violet-green swallows, are totally reliant on cavities made by other species or natural processes, but even in swallow species that build their own nests, such as Barn and Cliff swallows, individuals often will reuse the nest made by another of their species.

Eggs

One of my most pleasant memories from years of studying California Gulls breeding in the Great Basin is the image of hundreds of clutches displayed before me on an expanse of sand as the parents rose into the air, each clutch with its own variation on the typical shape and color of the species, and each glowing softly in the early morning light, like agates or opals being offered by a dealer in gems (Fig. 8–61). Indeed, birds' eggs are one of the most beautiful productions of life on earth.

Egg Structure

The beauty of birds' eggs is testimony to millions of years of elegant engineering by natural selection, allowing them to survive on dry land and thus develop under a variety of environmental conditions. In contrast, the eggs of fish and most amphibians can only survive in water, tying these animals to an existence wholly or partially dependent upon water. Ancestral reptiles, however, evolved a hard-shelled egg with internal membranes—which kept the embryo in the watery medium required for development—freeing these animals from liv-

Figure 8–61. California Gulls at Nest Colony: *The clear blue water of Mono Lake provides a stunning contrast to both the white breasts of nesting California Gulls and the white rocks encrusted with tufa (calcium carbonate) in the background. This breeding colony, located on an island, is one of the largest in North America. The salty lake is high in bicarbonate ions, and where rainwater carrying dissolved calcium flows into the lake, calcium carbonate precipitates out underwater, forming the tufa crust on the volcanic rocks. When the lake water was diverted for use in Los Angeles, the water level dropped dramatically, exposing the tufa-encrusted rocks. Photo courtesy of David W. Winkler.*

ing in or near water. Thus began the long evolution of land animals, including birds.

Bird eggs are large—no bird has an egg as small as that from which each of us sprung. The only mammals with eggs similar to those of birds are the monotremes—echidnas (spiny anteaters) and the platypus of Australasia. Their eggs develop outside the body, unlike those of all other mammals **(Fig. 8–62a)**. Most snakes, lizards, and turtles also lay large eggs **(Fig. 8–62b and c)**. The embryos in the large external eggs of birds, reptiles, and monotremes—unlike most mammalian embryos—must develop entirely independent of their parents for resources. Thus for the long developmental journey, a bird embryo must be packed with all the protein, carbohydrates, fats, and water that it will need to develop into a hatchling. For all the resources it must contain, the egg begins life as a single cell with just one copy of each parent's genes on board: indeed, the egg of an Ostrich is the largest living cell on earth.

Before going into the details of egg structure, step into your kitchen and refresh your memory of the major features of an egg **(Fig. 8–63)**. As you crack the egg, notice how the brittle mineral shell is held together, just like the layers of auto safety glass, by a thin membrane to which it adheres at its inner surface. (There actually are two membranes here: one sticks tightly to the shell; the other, surrounding the egg white or **albumen**, is the one that can be frustrating to peel off a hard-boiled egg.) Once you have broken through these two outer membranes you

a. Short-beaked Echidna Adult (left) and Young Hatching from Egg (right)

Figure 8–62. Egg Diversity: *Eggs that develop outside the parent's body vary widely among different groups of animals, but each must contain all the water and nutrients required by the embryo to develop from a single cell into a hatchling.* **a. Short-beaked Echidna Adult and Young Hatching from Egg:** *The foot-long (1/3-meter), hedgehog-like echidnas (or spiny anteaters) of Australia, Tasmania, and New Guinea use their snouts and long, sticky tongues to feed on ants and termites. Unlike all other mammals, these monotremes—an order of mammals containing only echidnas and the platypus— lay eggs that develop outside the body. After a 21-day gestation period, the female lays a leathery, grape-sized egg and places it in a pouch on her stomach. Ten days later, a semi-transparent hatchling emerges, using its egg tooth and a hard bump on its nose (visible here) to break through the shell. Just one-half inch (13.5 mm) long, it remains with its mother for six to eight months, first nursing while in her pouch, and later in the burrow. Adult photo by Dave Watts/Nature Focus. Hatchling photo by Rismac/Nature Focus.*
b. Female Spotted Salamander Laying Eggs: *Lacking a hard outer shell, the eggs of most amphibians are embedded in a jellylike material, and must remain moist. The female spotted salamander, 6 to 8 inches (15 to 20 cm) long, attaches her single oval egg mass to underwater sticks or plants. About the size of a tennis ball, the mass may contain up to 200 eggs. Photo by Dwight R. Kuhn.*
c. Female Prairie Skink Guarding Eggs: *Reptile eggs, like bird eggs, are adapted to survive away from water, but their shell is less brittle, and more leathery, than that of bird eggs. The female prairie skink—a shiny lizard of the U. S. Midwest—guards her eggs throughout the incubation period. Photo courtesy of Harry W. Greene.*

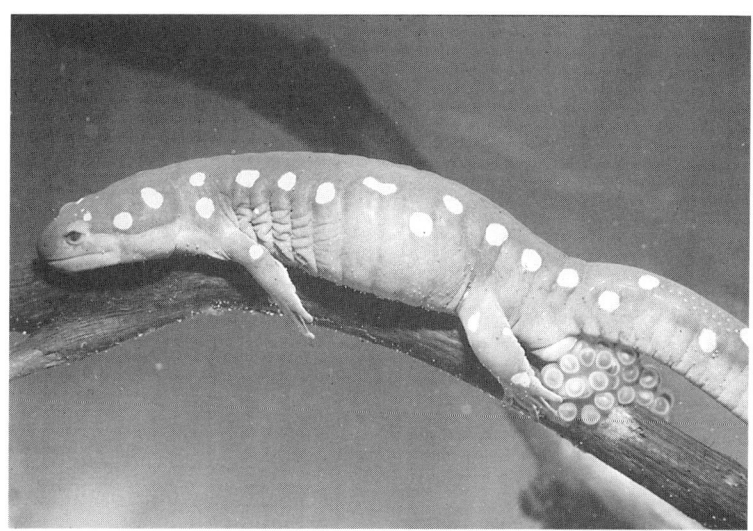

b. Spotted Salamander

c. Prairie Skink

8

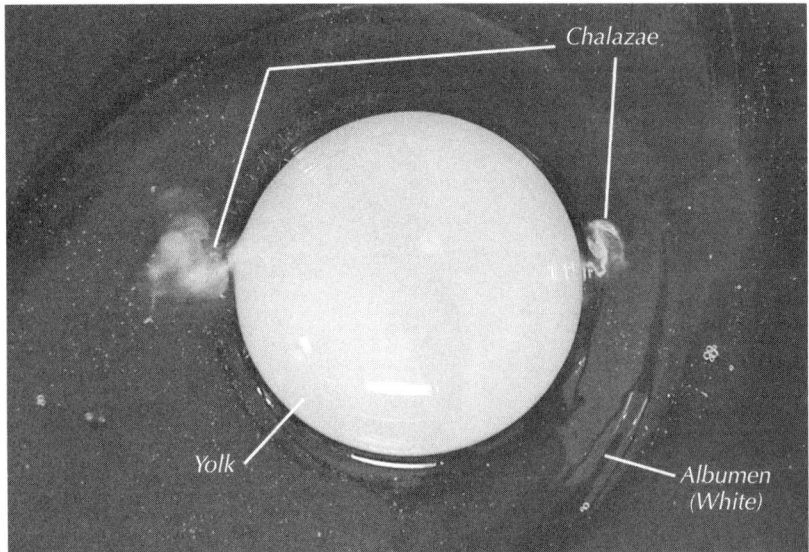
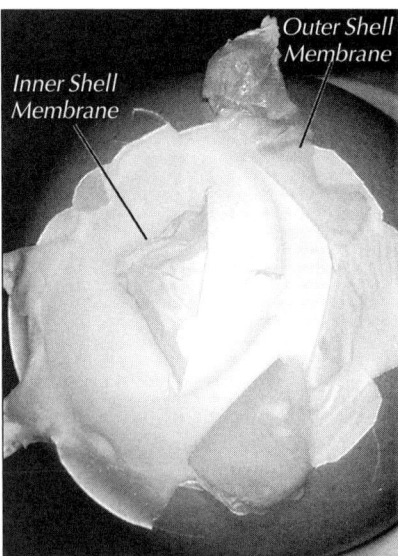

a. Visible Parts of a Raw Egg   **b. Hard-boiled Egg with Membranes**

Figure 8–63. Chicken Egg Photos: a. Visible Parts of a Raw Egg: If you crack a chicken egg into a bowl, you will see the clear, watery **albumen** *(the egg white) and the yellow* **yolk**, *which is held together by the virtually invisible vitelline membrane. The stringy, whitish coils on each side of the yolk are the* **chalazae**. *You also may notice that the yolk rotates such that a circular, white spot on its surface (barely visible in the center of this yolk) faces upward. If the egg were fertilized, this tissue, termed the blastoderm, would make up the tiny developing embryo.* **b. Hard-boiled Egg with Membranes:** *If you gently crack open a hard-boiled egg, you may be able to find the two shell membranes. The* **outer shell membrane** *clings tightly to the inside of the shell, coming off with it. The* **inner shell membrane** *adheres stubbornly to the egg white—usually frustrating anyone who tries to peel it off. Photos courtesy of Marie Read/CLO.*

can spill the contents of the egg into a bowl. There you can see the yolk surrounded by the white, now released from its membrane. The yolk itself is surrounded and held together by the **vitelline membrane**, which is hard to see, but is what you rupture when you "break" a yolk. You may notice that the yolk rotates such that a circular, white spot on its surface is upward in the dish. This tissue, termed the blastoderm, would make up the tiny developing embryo if the egg were fertilized. If you have ever tried to separate a yolk from the white, you have encountered some gelatinous, stringy parts of the albumen, often milky white in color, which are hard to separate from the yolk. These are the **chalazae**, which surround and protect the yolk and may appear twisted.

Now that you know the general layout of an egg, look in greater detail at the structure and function of its parts, beginning with the yolk **(Fig. 8–64)**. The yolk contains essentially all the lipid (fat) and most of the protein for the developing embryo. The relative sizes of the yolk and white vary according to the type of young that will hatch: in species with precocial chicks that are very well developed at hatching (such as ducks, geese, and grouse) the yolks tend to make up a larger proportion of the total egg mass than in species with altricial young (such as passerines).

The yolk is the first part of the egg constructed by the laying female. Before the yolk passes down the oviduct for the addition of albumen and shell (see Fig. 4–111), the yolk of most eggs undergoes rapid growth over a period of about five days to two weeks; but birds, especially songbirds, usually lay an egg a day (and never more). Because the yolk preparation time is longer than the laying interval,

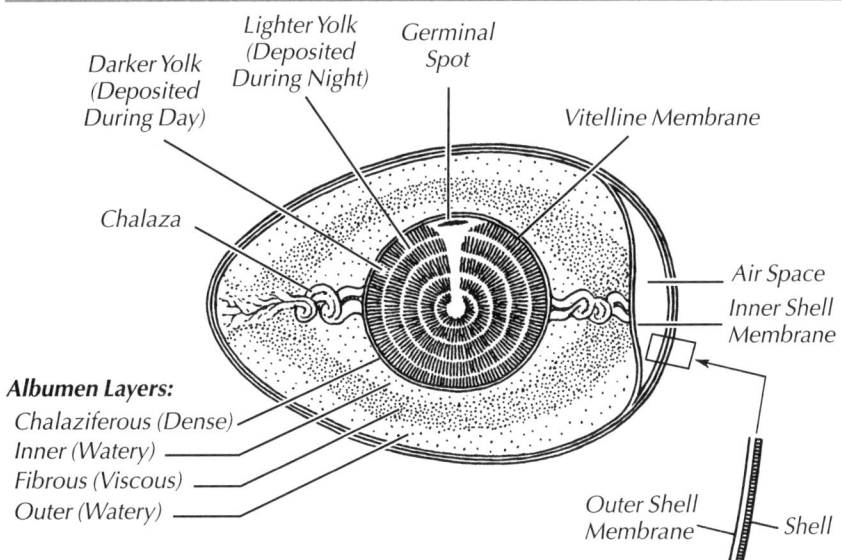

Figure 8–64. Egg Structure: This longitudinal section of a domestic chicken egg shows the major parts of a bird egg. The outer circle around the yolk contains the chalaziferous layer of albumen, which surrounds the yolk and attaches to the far ends of the egg as the chalazae. The next circle in from that is the vitelline membrane, which contains the yolk. Note the five pairs of dark and light yolk bands, which indicate that this egg took five days to "yolk up." See text for a detailed description of each structure. Drawing by Charles L. Ripper.

females often prepare yolks in all the ova they will need for a complete clutch (and often in at least one more ovum) before any eggs are laid. During the preparation period, termed "yolking up," yolk is deposited within the vitelline membrane in alternating bands of darker and lighter yolk, producing a structure resembling the growth rings of a tree in cross section. Because these bands alternate on a daily cycle (light being deposited at night and dark during the day, when the female is ingesting food rich in pigments), the yolk ring structure can be used to determine how long birds take to prepare their yolks for laying (Grau 1976). The one structure interrupting the concentric rings is a cylinder of light yolk that stretches from the yolk's core to its surface, upon which sits the **germinal spot**. This is the site where the embryo will develop, first as a flattened disc called the **blastoderm** on the yolk's surface, and eventually growing to fill the entire egg (**Fig. 8–65**). The lighter-colored yolk of the germinal spot and the column on which it sits also is lighter in weight, and the yolk rotates so that the developing embryo always floats to the top during its development, no matter which way the egg is turned. Unfortunately, the layers of yolk are hard to see without freezing, fixing (adding a stabilizing chemical), and staining the yolks.

The yolk would actually float up against the shell if it were not cradled at the center of the egg by the chalazae, which envelop the yolk and attach to the far ends of the egg. The twisted, cordlike chalazae allow the yolk to rotate (see Fig. 8–64). The chalazae are similar in composition to the remainder of the albumen, which is mostly water and protein. The viscosity of the albumen varies with the proportion of water and protein it contains. Immediately surrounding the yolk and forming the chalazae is a very thin layer of viscous albumen. Around this central complex of yolk and albumen is a thin layer of watery albumen, then a thick layer of more viscous and variably fibrous albumen (the largest component of the albumen), and then finally a thin layer of watery albumen right beneath the shell (for much more on egg

(Continued on p. 8·68)

Figure 8–65. Embryological Development of a Chick: *Selected stages in the development of a domestic chicken embryo. The egg schematics are approximately life-size with the parts drawn to scale. Note that an embryo's stage of development at a particular time varies not only between species, but also within species, owing to changes in incubation temperature, egg freshness at the start of incubation, season, egg size, and other factors. CAM inset for 7-day embryo adapted from Patten (1971). All other photos and drawings courtesy of Drew M. Noden, Department of Biomedical Sciences, Cornell Veterinary College.*

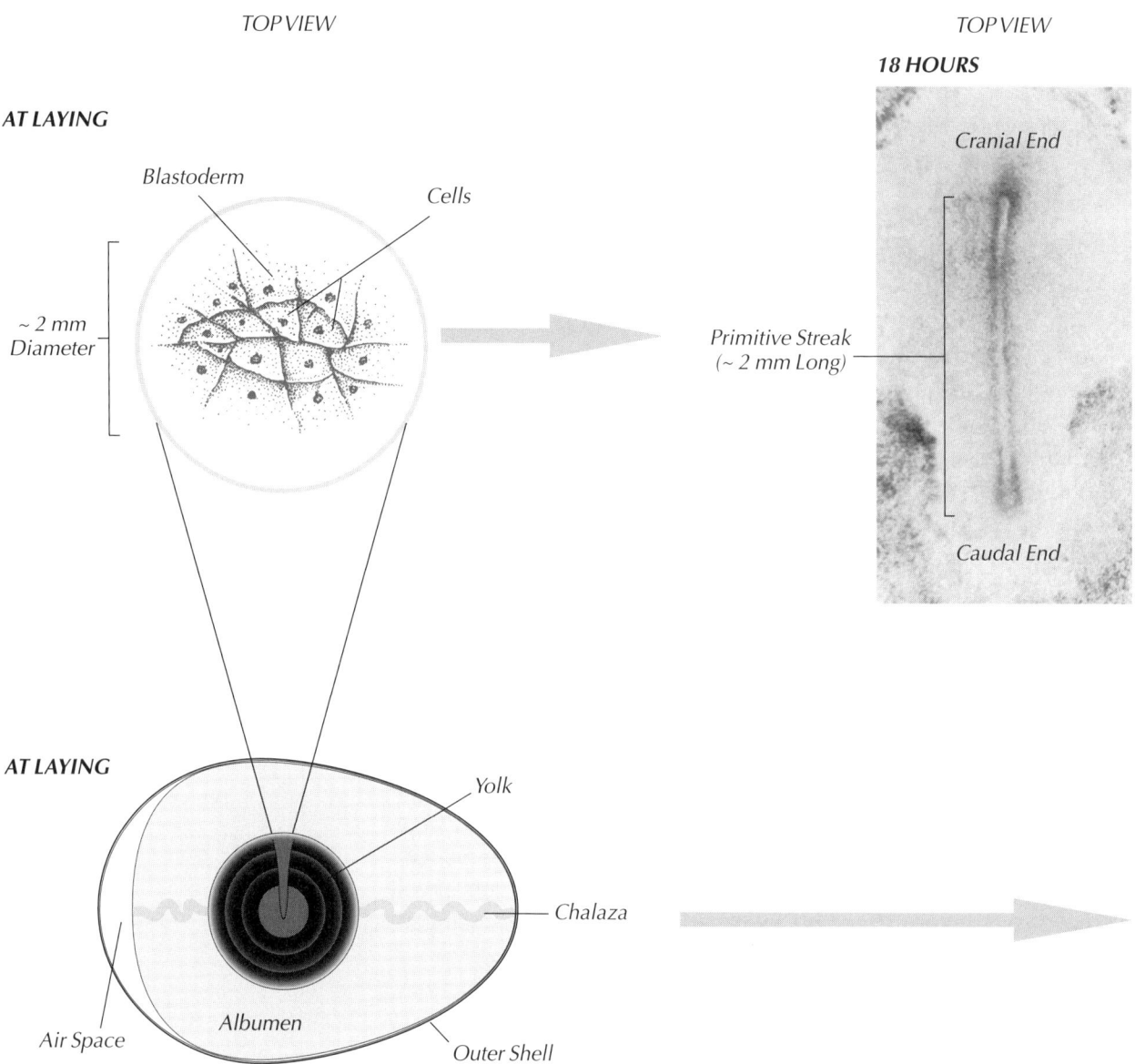

At Laying: *By the time an egg is laid, the egg cell already has begun to divide, producing a flattened disc of cells called the* **blastoderm**, *which lies on the upper surface of the yolk.*

18 Hours: *Some blastoderm cells move to the midline, forming a trough (lighter area) with raised sides (dark lines) called the* **primitive streak**. *Most tissues of the embryo form in or adjacent to this structure, and it establishes the body axes: cranial-caudal, left-right, and dorsal-ventral. Here, the cranial end can be distinguished as the dark area where cells are more closely packed. Eventually, the raised sides fold inward over the groove and meet to form a tube, the neural tube—a precursor to the spinal cord.*

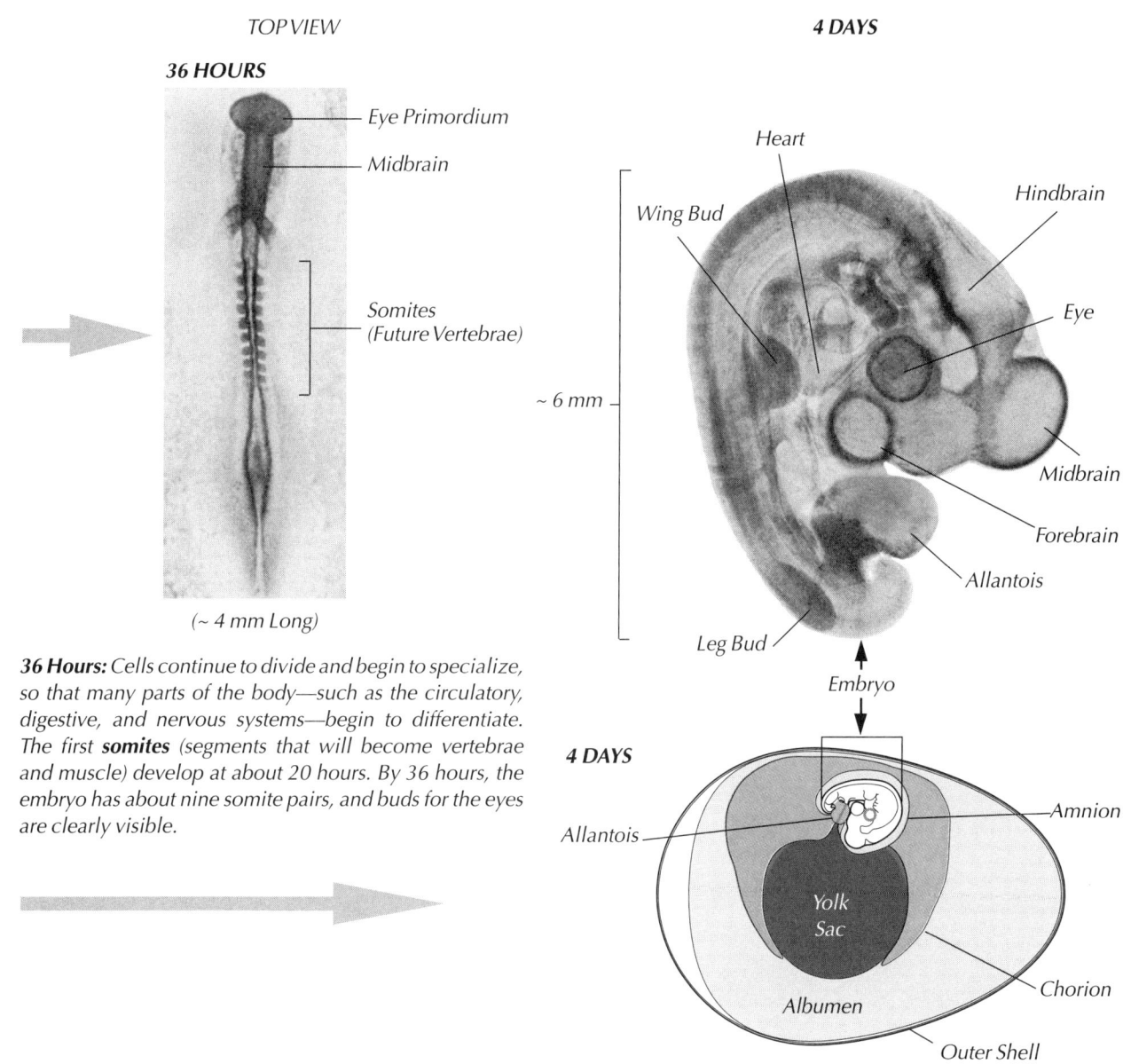

TOP VIEW

36 HOURS

— Eye Primordium

— Midbrain

Somites
(Future Vertebrae)

(~ 4 mm Long)

36 Hours: *Cells continue to divide and begin to specialize, so that many parts of the body—such as the circulatory, digestive, and nervous systems—begin to differentiate. The first* **somites** *(segments that will become vertebrae and muscle) develop at about 20 hours. By 36 hours, the embryo has about nine somite pairs, and buds for the eyes are clearly visible.*

4 DAYS

Heart

Wing Bud

Hindbrain

Eye

~ 6 mm

Midbrain

Forebrain

Allantois

Leg Bud

Embryo

4 DAYS

Allantois

Amnion

Yolk
Sac

Chorion

Albumen

Outer Shell

4 Days: *The four main* **extra-embryonic membranes** *(which protect and nourish the growing embryo, but do not form part of the adult body) have formed as foldings or outpocketings of the embryo. The* **yolk sac** *surrounds the yolk—a store of fat and protein for the embryo; the* **amnion** *becomes filled with fluid and surrounds the embryo, allowing it to move and stay moist, and keeping its various growing parts free from sticking to or blocking one another; the* **allantois** *forms a receptacle for metabolic wastes; and the* **chorion** *surrounds the embryo and other three membranes. The chick's body has bent and twisted so that it lies with its left side on the yolk sac, and numerous body parts, such as the wing and leg buds, are clearly visible. The heart has been beating for a full day.*

(Figure continued on next page)

7 DAYS

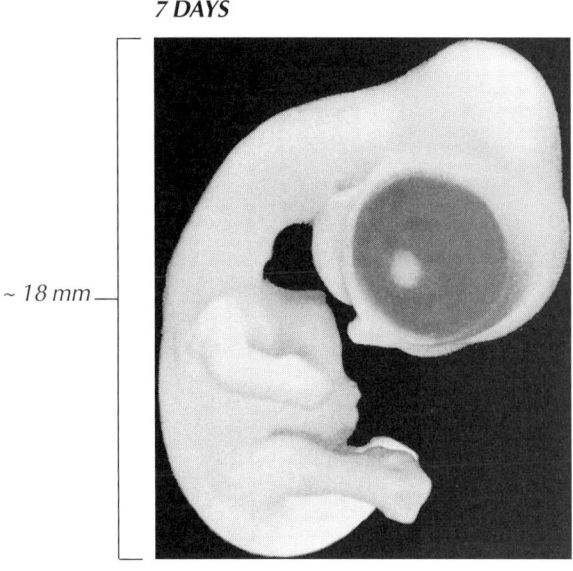

~ 18 mm

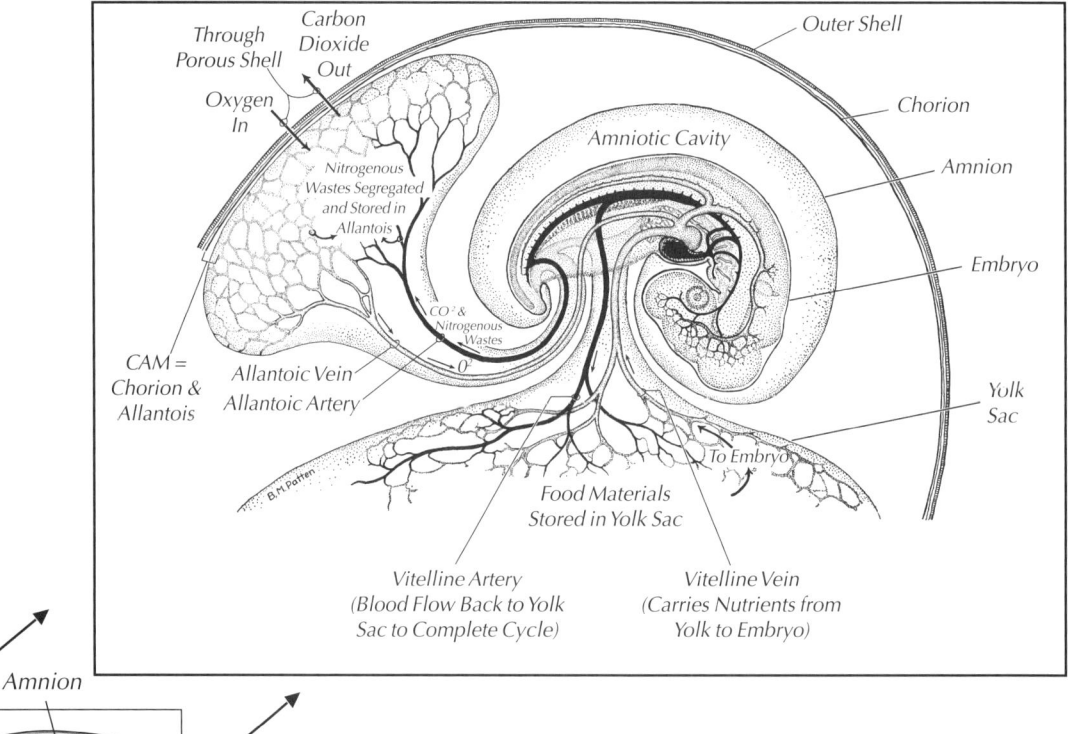

Through Porous Shell · Carbon Dioxide Out · Outer Shell

Oxygen In · Chorion

Amniotic Cavity · Amnion

Nitrogenous Wastes Segregated and Stored in Allantois

CO₂ & Nitrogenous Wastes · Embryo

CAM = Chorion & Allantois · Allantoic Vein · Allantoic Artery · 0²

To Embryo · Yolk Sac

B.M. Patten

Food Materials Stored in Yolk Sac

Vitelline Artery (Blood Flow Back to Yolk Sac to Complete Cycle) · Vitelline Vein (Carries Nutrients from Yolk to Embryo)

7 DAYS

Amnion

Allantois · Chorion

Albumen

Yolk Sac · Outer Shell

7 Days: Most organs are taking shape, notably the huge eyes. Each wing and leg is differentiated into three sections, the digits are separated by grooves or webs, and the beak is prominent, with a tiny egg tooth (see 9-Day photo). The allantois presses against the chorion along a portion of the eggshell, and the two fuse to form the **CAM** (chorioallantoic membrane). Via the numerous blood vessels of the CAM and pores in the eggshell, the embryo receives oxygen from outside the egg and expels carbon dioxide (see inset).

9 DAYS

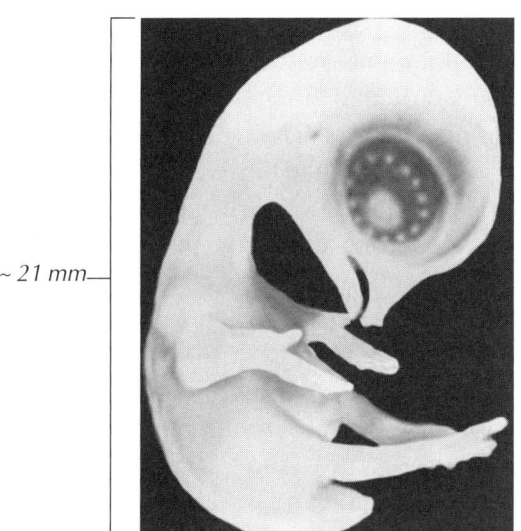

~ 21 mm

15 DAYS

15 Days: *The embryo is much larger and covered with down; it has small claws on the toes and scales on the legs.*

9 DAYS

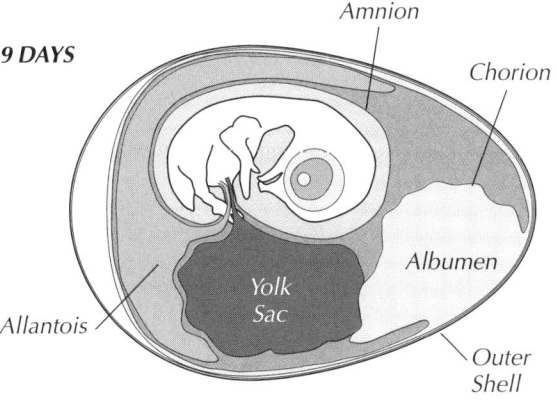

Amnion

Chorion

Albumen

Yolk Sac

Allantois

Outer Shell

9 Days: *The embryo appears much more birdlike, having a longer beak and longer and more distinct digits. It has a skeleton of cartilage, partly formed eyelids, and small bumps on the skin called feather papillae, which are the beginnings of feathers (not visible in photo). The allantois and chorion have fused against more of the eggshell, expanding the CAM. The albumen continues to shrink as the chick uses up water and protein, and water evaporates through the shell.*

20 DAYS

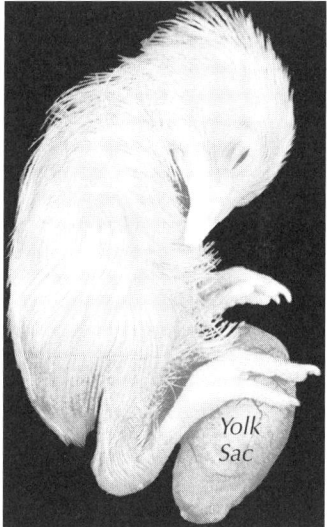

Yolk Sac

20 Days: *The embryo is almost fully developed and will hatch in about one day. It has used up most of the yolk and albumen, but the yolk sac remains attached to its abdomen, and will be brought into the abdomen during hatching. Because the CAM no longer met the embryo's needs, on day 19 the embryo poked its beak into the air space at the blunt end of the egg and began air breathing (see Fig. 8–66b).*

8

structure see the Romanoff's (1949) classic *The Avian Egg*). In addition to providing nearly all the water and much of the protein for the developing embryo, these layers of albumen serve admirably to protect the embryo from physical damage, provided the shell is not broken.

The shell is the developing embryo's first line of defense **(Fig. 8–66a)**, and it is much thicker and stronger than those of other terrestrial egg-laying vertebrates. But increased thickness also has disadvantages, for all the embryo's gas exchange must occur across the shell. The egg contains all the water and other raw materials necessary for embryonic development, but it cannot contain all the oxygen required to fuel metabolism for growth, nor can it hold all the carbon dioxide and waste water produced by the embryo during growth. To get around this problem, the developing embryo must be able to "breathe" through its shell—and breathe it does. Eggs lose, on average, 18 percent of their mass between laying and hatching (Rahn and Ar 1974), mostly from water loss during metabolism of the embryo.

The eggshell's porosity can also put the embryo at risk. Pollutants such as oil from oil spills can enter the egg and poison the embryo, or

a. Outer Layers of the Eggshell

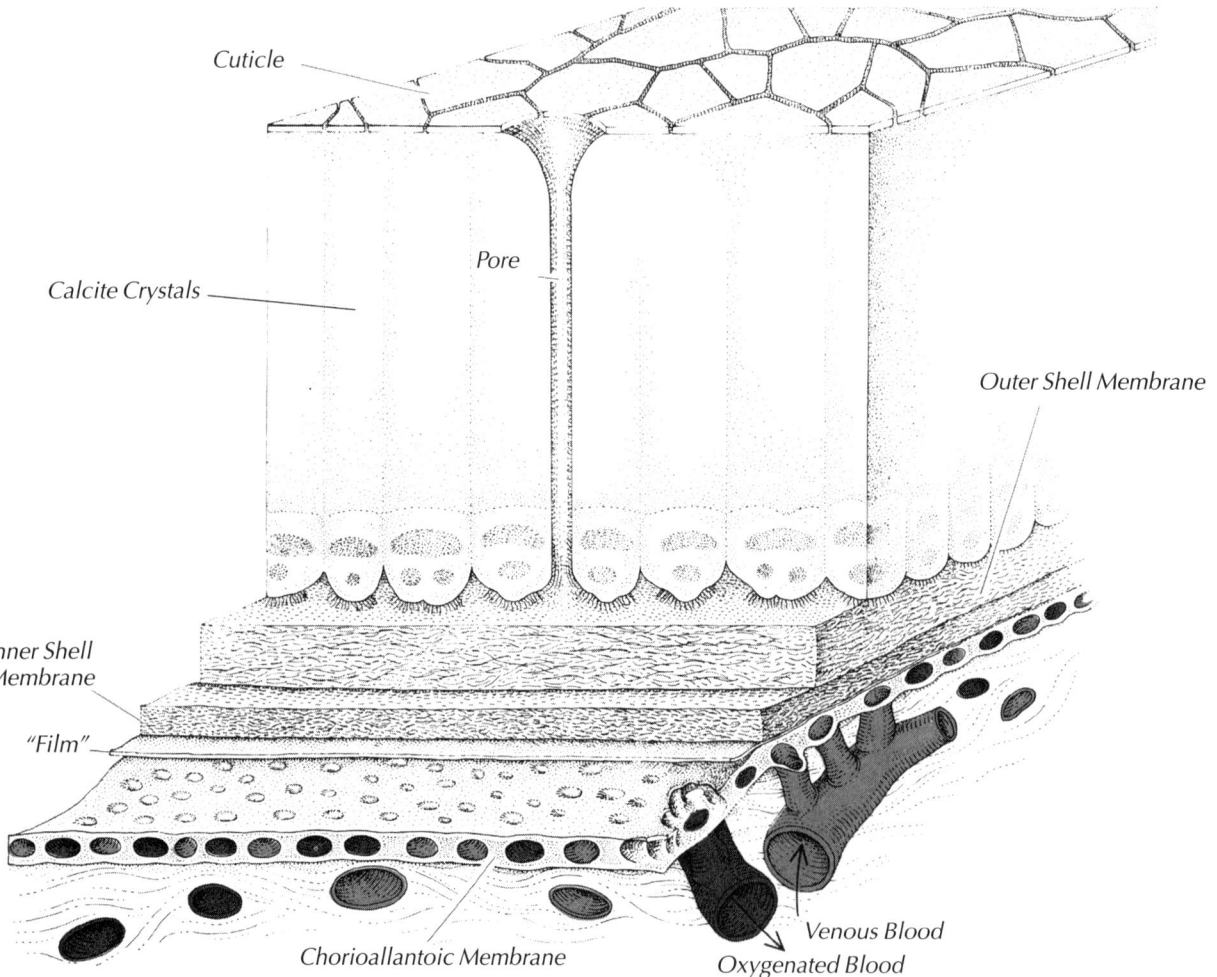

Cuticle

Pore

Calcite Crystals

Outer Shell Membrane

Inner Shell Membrane

"Film"

Chorioallantoic Membrane

Venous Blood

Oxygenated Blood

(Figure continued on next page)

coat the egg's surface, cutting off gas exchange. Researchers have demonstrated that applying just one-half drop of oil per day to a Mallard egg kills the embryo after just a few days. Even one-tenth that amount of oil killed up to 90 percent of the embryos (Graham 1989). Adult birds exposed to low levels of pesticides or oil in water may bring enough back to the nest on their belly feathers to kill their eggs.

How gas exchange takes place is apparent to anyone who opens a fertile chicken egg containing an embryo at an advanced stage of development. The entire inner surface of the shell is covered with a membrane that is richly invested with blood vessels. This **chorioallantoic membrane** ("CAM" for short) is formed from the fusion of two embryonic sacs, the chorion and the allantois (see Fig. 8–65, 7 days). The chorion is the outer membrane surrounding the entire avian embryo, and is homologous (evolutionarily related) to the mammalian membrane, also called the chorion, which forms much of the placenta in most mammals. The allantois is a sac into which the developing bird embryo shunts all metabolic wastes that cannot evaporate through the shell, such as uric acid crystals. The CAM, the remarkable structure

b. Shift to Breathing Air in Chicken

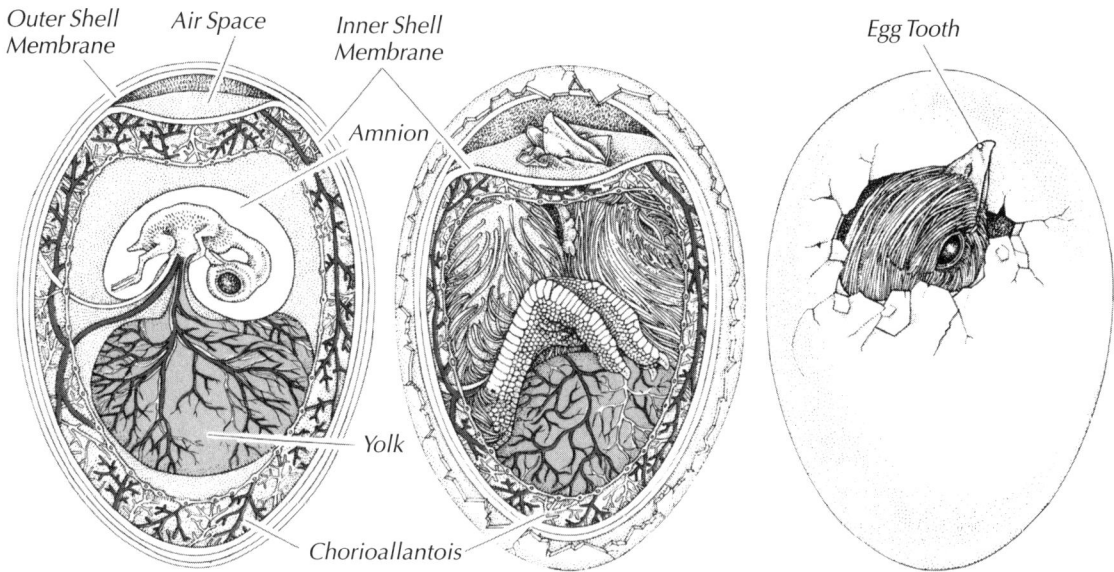

Outer Shell Membrane   Air Space   Inner Shell Membrane   Amnion   Yolk   Chorioallantois   Egg Tooth

Days 5–18: Gas Exchange Through CAM   **Day 19: Air Breathing Begins**   **Day 19 + 6 Hours: Chick Breathes Atmospheric Air**

Figure 8–66. Embryonic Gas Exchange: a. Outer Layers of the Eggshell: *This cross section of the outer layers of a bird egg covers a depth of about 0.016 inches (0.4 mm). The outermost layer is the cuticle, a thin layer of organic material. Underneath is the shell proper, composed of columns of calcite crystals and traversed by pores, which terminate in the loose, fibrous outer shell membrane. Below that is the inner shell membrane, a thinner and less coarse layer whose inner surface, termed the "film," is apparently a continuous sheet. Attached to the film is the chorioallantoic membrane (CAM), the respiratory organ of the embryo, which is homologous to the placenta of mammals. Venous blood (lighter) pumped by the embryonic heart flows to the CAM, where it is replenished with oxygen that has diffused into the egg through the pores. At the same time, carbon dioxide diffuses out of the venous blood. Oxygenated blood (darker) then travels to embryonic tissue. Drawing courtesy of Patricia J. Wynne.* ***b. Shift to Breathing Air in Chicken:*** *On day 5, the CAM begins to cover the inner shell membrane with a network of capillaries that carry out gas exchange for the embryo. Water vapor also diffuses continually from the egg, and the liquid water that evaporates is replaced by gas to form an air space at the blunt end of the egg. On day 19, however, the embryo pokes into the air space with its beak and begins to breathe air, inflating its lungs and air sacs for the first time—although the CAM continues to function. About six hours later (in the chicken) the chick breaks through the eggshell with its egg tooth and begins to breathe atmospheric air, as CAM function begins to wane. Drawing courtesy of Patricia J. Wynne, based on one by Hans-Rainer Duncker of the University of Giessen.*

formed by the fusion of these membranes, serves as the functional equivalent of the embryo's lungs throughout development.

As the embryo develops, the chorioallantoic membrane covers more and more of the inner surface of the shell. At the same time, as water within the egg is used up and evaporated out through the shell, the air space between the two shell membranes at the blunt end of the egg gradually expands **(Fig. 8–66b)**. (This is the air space that you can see at the blunt end when you peel a fresh hard-boiled egg.) The buoyancy imparted by this growing air space is valuable to field ornithologists who wish to check the developmental progress of an egg: they can simply place an egg in water and note the position in which it settles or floats to get a fairly accurate indication of how much longer the embryo within must develop before hatching **(Fig. 8–67)**. (Although water *vapor* moves readily through the shell, the shell membranes do a good job of keeping *liquid* water outside, as well as inside, the egg. So ornithologists needn't worry about drowning the embryo during their brief "float test.")

So, next time you crack open an egg for breakfast, pause for a moment to reflect on this elegant engineering that natural selection hath wrought... And enjoy your omelet!

Egg Size

The largest eggs of living birds are those of the Ostrich, measuring roughly 7 by 5.5 inches (18 by 14 cm) and weighing nearly 3 pounds (1.4 kg) **(Fig. 8–68)**. But Ostrich eggs seem small when compared to those of the extinct elephantbirds of Madagascar (see Fig. 5–48), which measured up to 14.5 by 9.5 inches (37 by 24 cm) and may have weighed as much as 27 pounds (12 kg). One of these eggs could have held the contents of at least 150 chicken eggs. The smallest eggs are laid by hummingbirds. Those of two West Indian species, the smallest of all birds, measure from 0.4 to 0.5 inches (10 to 13 mm) in length and weigh less than 0.04 ounces (1 g): approximately 75 of these would fit inside a large chicken egg **(Fig. 8–69)**.

Figure 8–67. Egg Buoyancy and Embryo Development: As the developing embryo uses water and additional water evaporates from the shell, air moves in to replace it, gradually expanding the air space at the blunt end of the egg. To determine the approximate developmental stage of an egg, ornithologists can place it in a jar of water, noting the position and height at which it floats. A freshly laid egg will remain on the bottom, the air space too small to noticeably affect the egg's buoyancy. By about halfway through development, the air space has grown large enough to raise the blunt end of the egg off the bottom, but will not cause it to float. As the air space enlarges, the egg gradually floats higher in the water, with more protruding above the surface in the later developmental stages. This technique may be used quickly in the field, without harming the egg.

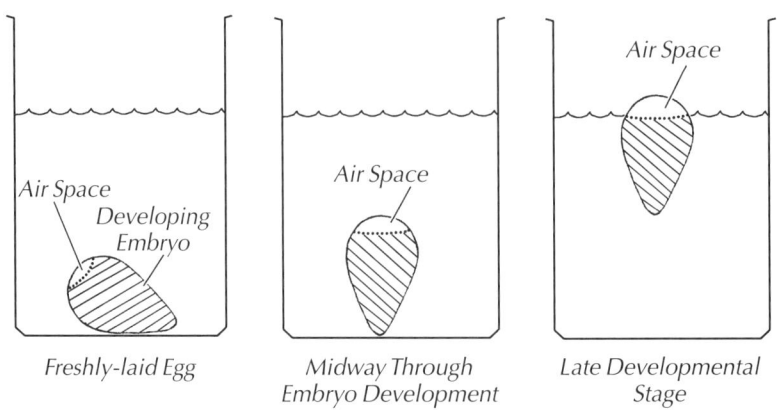

Freshly-laid Egg   Midway Through Embryo Development   Late Developmental Stage

As a general rule, larger birds tend to lay larger eggs, although the eggs of larger species are generally a much smaller proportion of their body size: an Ostrich egg is 1.7 percent of the female's weight, whereas a wren's egg is 13.0 percent of the weight of the female that lays it. Conspicuous and spectacular exceptions to this rule include the kiwis, which lay eggs that are each 25 percent of the female's body weight **(Fig. 8–70)**. Not surprisingly, kiwis lay at most two eggs in one clutch! The ecological pressures that might have selected for such large eggs are not well understood.

Many other interesting variations in egg size occur among birds. In their first nesting attempts, females of many species lay smaller eggs than more experienced breeders, and birds that lay large numbers of eggs have some tendency to lay smaller eggs. The eggs of precocial birds tend to be larger than those of altricial birds of a similar size **(Fig. 8–71)**. For example, a crane (with precocial young) and an eagle (altricial young) of similar body weight lay eggs of about 4.0 and 2.8 percent of the parent's body weight, respectively. Because precocial species spend longer in the egg than altricial species—reaching a more advanced developmental stage before hatching—their eggs must start with a greater supply of nutrients for the embryo. Some brood parasites—the Common Cuckoo of Eurasia, for example—lay small eggs relative to their body size. This could be an adaptation to match the egg sizes of their hosts, or it could allow the parasites to lay more eggs per season, thus placing eggs in more host nests and increasing their chances of hatching and survival. Egg size among closely related species tends to be much less variable than other aspects of birds' breeding biology, such as clutch size and number of broods per season.

Figure 8–68. Ostrich and Hummingbird Egg: Nearly 7 inches (18 cm) long, the egg of the Ostrich is the largest of any living bird. The world's smallest eggs, barely one-half inch (13 mm) long—just the size of a pea—are those of several West Indian hummingbirds. Approximately 5,500 hummingbird eggs would fit inside an Ostrich egg. Photo by Runk/Schoenberger/Grant Heilman Photography.

Figure 8–69. Egg Size Diversity: The eggs of the now-extinct elephantbirds of Madagascar were enormous—14.5 inches (37 cm) long and weighing up to 27 pounds (12 kg). Twice the length of an Ostrich egg, they could have held the contents of more than 11,000 hummingbird eggs! The eggs of most songbirds are sized somewhere between the egg of a chicken and that of a hummingbird.

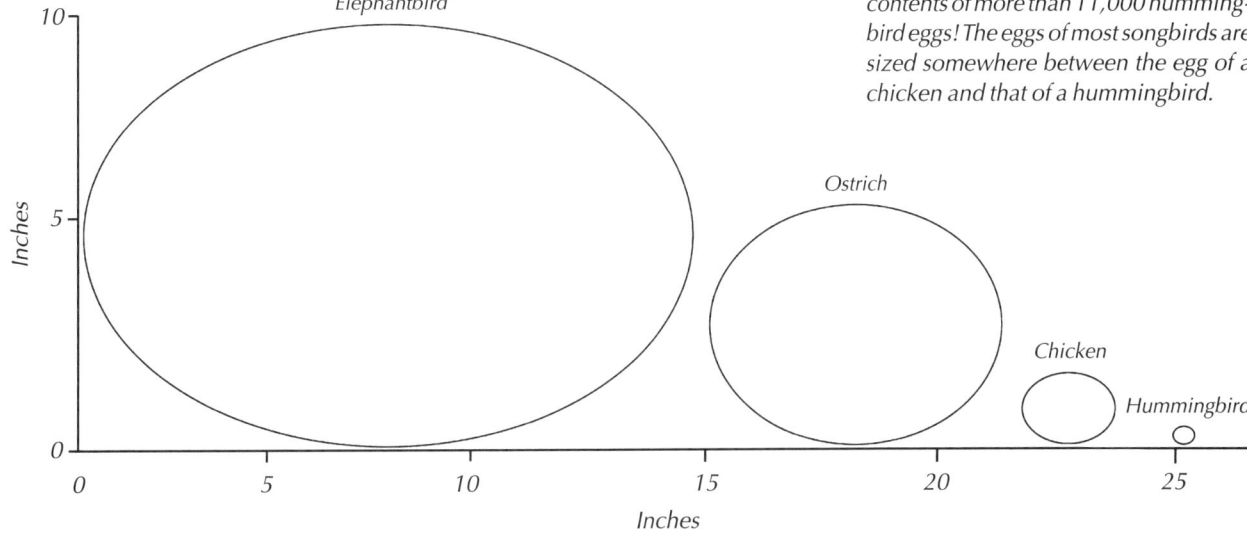

Figure 8–70. X-ray of Brown Kiwi with Egg in Oviduct: *The grouse-sized kiwis of New Zealand lay eggs that are larger with respect to their body size than those of any other bird—typically 18 to 25 percent of the female's body weight. The eggs are incubated for nearly three months, mostly by the male, until the active chicks emerge—fully feathered and with open eyes. This X-ray shows a fully formed egg in the oviduct of a female Brown Kiwi. Photo courtesy of Otorohanga Zoological Society, New Zealand.*

Egg Shape

Egg shape varies from long pyriform, through oval, to nearly spherical, with oval (typified by chicken eggs) being the most common **(Fig. 8–72)**. The diameter and muscular tension of the oviduct during eggshell formation presumably determine the egg's shape, and almost all egg shapes conform to those of a single mathematical family of "path curves" (D. E. Baker, personal communication). But within this family of curves is a broad variety of shapes, and adaptations to the post-laying environment often account for the egg shape of a given species. Murres, colonial seabirds that place their eggs directly on narrow, often slanting, ledges high above the sea, have eggs that are remarkably pointed at one end, and large and rounded at the other (pyriform). These eggs always roll in a tight circle—a distinct advantage for eggs so vulnerable to rolling off a perilous edge **(Fig. 8–73)**. Traditionally, ornithologists have assumed that their shape evolved to prevent them from rolling off their ledge, but the eggs of most other cliff nesters are not pointed and the eggs of all shorebirds have a similar shape despite being laid in flatland nests with no danger of rolling anywhere. Shorebirds usually lay four eggs, pointed at one end, that fit symmetrically

Figure 8–71. Egg Size in Altricial Versus Precocial Birds: *Although adult Killdeer and meadowlarks are approximately the same size, the Killdeer lays larger eggs and incubates them for longer periods—24 to 26 days, compared to the meadowlark's 13 to 15 days. Whereas the **altricial** young meadowlark hatches as a helpless nestling, whose eyes remain closed for the first five days, the Killdeer chick is **precocial**—it emerges from its egg able to see, run around, and pick up its own food. Therefore, the Killdeer requires a larger egg and a longer developmental period inside the egg. Drawing by Charles L. Ripper.*

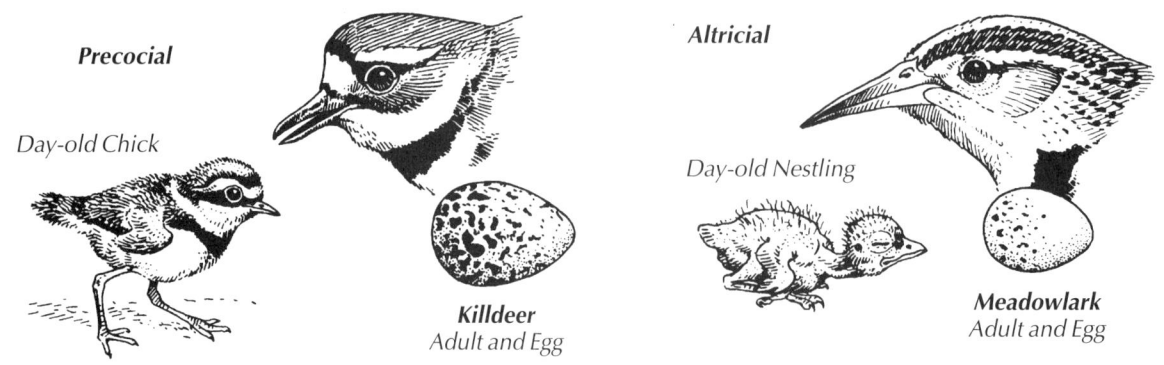

Precocial

Day-old Chick

Killdeer
Adult and Egg

Altricial

Day-old Nestling

Meadowlark
Adult and Egg

a. Egg Shape Chart

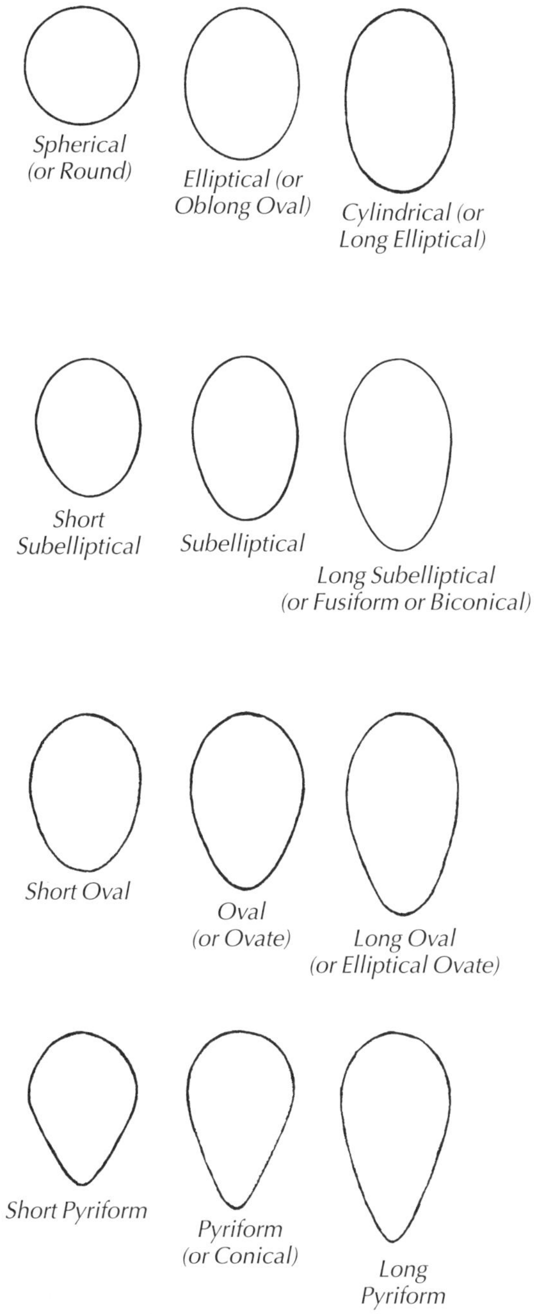

Spherical (or Round)

Elliptical (or Oblong Oval)

Cylindrical (or Long Elliptical)

Short Subelliptical

Subelliptical

Long Subelliptical (or Fusiform or Biconical)

Short Oval

Oval (or Ovate)

Long Oval (or Elliptical Ovate)

Short Pyriform

Pyriform (or Conical)

Long Pyriform

b. Egg Shape Examples

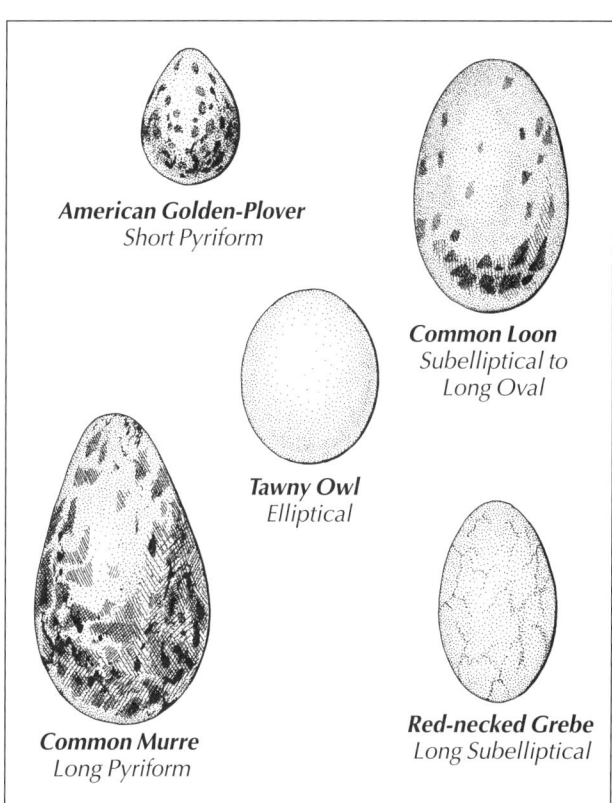

American Golden-Plover
Short Pyriform

Common Loon
Subelliptical to Long Oval

Tawny Owl
Elliptical

Common Murre
Long Pyriform

Red-necked Grebe
Long Subelliptical

Figure 8–72. Egg Shape Diversity: *Bird eggs vary greatly in shape. Not only do they vary from species to species, sometimes they differ within a species or even within a single clutch.* **a. Egg Shape Chart:** *Different references use different terms for egg shapes. Illustrated here are some of the most common terms, based on the system proposed by F. W. Preston (1953). Note that there are four basic shapes (middle column):* **elliptical, subelliptical, oval,** *and* **pyriform,** *with a short and long version of each. A shortened elliptical egg, for example, is spherical.* **b. Egg Shape Examples:** *Some cavity nesters, such as owls and kingfishers, have fairly rounded (elliptical) eggs, whereas all shorebirds (such as the American Golden-Plover) have very pointed (pyriform) eggs, which fit more compactly in the nest (see Fig. 8–74). The large eggs of the cliff-nesting murres also are pointed (long pyriform); several hypotheses have been proposed to explain the function of this shape (see Fig. 8–73). The eggs of loons, grebes, and cormorants are quite elongated. The majority of birds, however, have fairly oval eggs. Reprinted from* Manual of Ornithology, *by Noble S. Proctor and Patrick J. Lynch, with permission of the publisher. Copyright 1993, Yale University Press.*

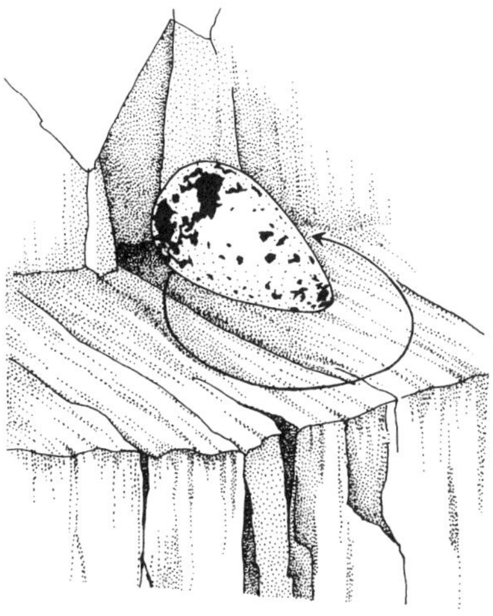

Figure 8–73. Common Murre Egg on Ledge: *The eggs of Common Murres—colonial seabirds that place their eggs directly on narrow, rocky ledges—are pointed at one end and rounded at the other (long pyriform). This shape causes the eggs to roll in a small circle, so ornithologists have traditionally assumed that it evolved to keep the eggs from rolling off their ledges. However, the similar egg shape of non-cliff-nesting shorebirds, and the lack of a similar egg shape in other cliff nesters, both cast some doubt on this explanation (see text for more information). Drawing by Margaret LaFarge, from* The Audubon Society Encyclopedia of North American Birds, *by John K. Terres, 1980. Published by Alfred A. Knopf, New York.*

Figure 8–74. Compact Clutch of Piping Plover: *The eggs of all shorebirds are pointed at one end, allowing the clutch (usually four eggs) to fit together tightly with their pointed ends toward the middle—as in this Piping Plover clutch. Because shorebirds lay large eggs in proportion to their body size and usually nest in cold areas, keeping their eggs warm can be a challenge. Having a compact clutch may help incubating shorebirds to keep their eggs covered. Photo courtesy of Bill Dyer/CLO.*

and compactly in the nest with their pointed ends inward **(Fig. 8–74)**. If you disarrange the eggs in a shorebird nest, the bird will point them all inward again before resuming incubation. The compactness of the clutch may be important because shorebirds lay large eggs in proportion to their body size (see Fig. 8–80), and may need to conserve heat particularly well in their far northern nesting habitats. With the eggs closely packed together, there is no wasted space to interfere with the transfer of heat from the parent's body to the developing offspring.

Egg Surface Texture

Most eggs have a smooth, matte finish like that of a chicken or robin, but the variations are many. The surface may be deeply pitted as in the eggs of Ostriches and storks; chalky as in grebes and flamingos; or glossy, like glazed porcelain, as in tinamous **(Fig. 8–75)**. The egg of the Guira Cuckoo of Asia has a chalky latticework overlaying a deep blue background. The eggs of ducks and geese feature a greasy surface that may be water resistant. The details of how surface texture affects the function of eggshells remain to be worked out.

Emu: Deeply Pitted

Great Blue Heron: Chalky

Figure 8–75. Egg Surface Textures: *Although most eggs, like those of chickens, have a smooth, matte finish, many variations exist. The eggs of Emus, Ostriches, and storks have a rough, deeply pitted surface; those of herons, grebes, flamingos, and cormorants have a chalky surface; and those of kingfishers, woodpeckers, and tinamous have a glossy surface—so glossy in tinamous that the eggs look like glazed porcelain. The ecological significance, if any, of the different surface textures is not known. Photo courtesy of Marie Read/CLO.*

Tinamou: Glossy

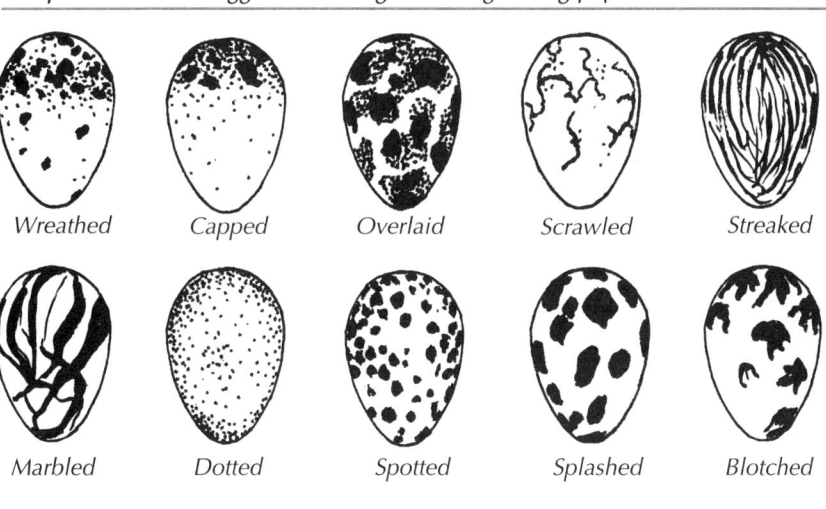

Wreathed   Capped   Overlaid   Scrawled   Streaked

Marbled   Dotted   Spotted   Splashed   Blotched

Figure 8–76. Diversity of Egg Markings: Bird eggs may be decorated with a variety of markings, which they acquire in the female's oviduct as they rotate while pigment is deposited. Rapid rotation and descent through the oviduct results in more streaking, and slower movement leads to more spotting. Because the large end of the egg travels through the oviduct first, it often picks up more pigment, which may be concentrated into a wreath or cap. Eggs of open-nesting birds tend to be pigmented heavily—the markings provide superb camouflage, protecting the eggs from visual predators, and also may prevent the eggs from overheating by shielding them from intense solar radiation (see text). Adapted from Harrison (1975).

Egg Color

Most bird eggs have at least some pigment added to their shells in the oviduct immediately after the shell is formed. Pigment glands in the wall of the oviduct deposit successive layers of color as the egg passes through. The ground color of the egg is usually deposited first, with any spots, streaks, and other darker markings being added later **(Fig. 8–76)**. The colors of eggs often begin to fade soon after the eggs are laid.

In my years of working with the eggs of gulls, I noticed that the last-laid egg in their clutches of three were generally less elongated and had finer markings, with many more narrow squiggles of pigment, often concentrated in the more blunt half of the egg **(Fig. 8–77)**. I cannot explain this pattern, but these differences were not associated with the last egg in two-egg clutches.

The eggs of some species are no less than astonishing in color—for example, the dark, glossy, purplish-chocolate eggs of the Chilean Tinamou and the large, pitted, green, avocado-like eggs of the Emu and cassowaries. As a rule, egg color is fairly constant and therefore

Figure 8–77. Order of Laying in Mew Gull Clutch: In gulls, the last-laid egg in three-egg clutches is usually shorter, with finer markings that often are concentrated at the larger end. In this Mew Gull clutch, the uppermost egg was laid last. Note the smaller spots and narrow squiggles, which form a ring at the broader end. Researchers do not understand how or why this occurs. Photo by A. Morris/VIREO.

Figure 8–78. Variability in Common Murre Eggs: This sample of Common Murre eggs from the Cornell egg collection illustrates their great variability. Markings vary dramatically from none at all to spots, blotches, or lines of light yellow-brown, bright red, deep brown, or black. The background color also varies, from deep blue-green, bright pinkish, and warm ocher to pale bluish, creamy, or white. This variation may help birds to recognize their own egg, which is laid directly on a ledge amid dozens of other single eggs in a dense nesting colony. Photo courtesy of Tim Gallagher.

Figure 8–79. American Crow Eggs: The shape, color, and markings of American Crow eggs vary dramatically, even within a single nest. This variation, whose function is unknown, is demonstrated by the clutch of eggs shown here. In addition to obvious differences in the markings, note that the left, right, and upper eggs have more rounded ends, whereas the middle and lower eggs appear narrower at one end. American Crow eggs range in shape from short to long in both the oval and subelliptical groupings (see Fig. 8–72). Greenish-blue to pale blue, the eggs may have few to many markings, which range from olive to black, with some pale gray or purple. The function of this variability is unknown. Photo by Kevin McGowan.

characteristic of a species, but in some species the egg color varies considerably, even dramatically. For example, the ground color of murre eggs may range from deep blue-green, bright pinkish, and warm ocher to pale bluish, creamy, or white; and the markings may vary from none at all to blotches or lines in light yellow-brown, bright red, rich brown, or black (**Fig. 8–78**). This variation may be an adaptation to enable each bird to recognize its own egg among the dozens on the ledges in densely packed breeding colonies. Experiments show that a murre will accept a strange egg only if it is colored and patterned like its own (Pettingill 1972). Another study (Jackson 1992, 1993) recently documented great variability in egg color among females in the Northern Masked-Weaver of Africa. This strategy may have evolved so that females can detect (and thus remove) eggs laid in their nests by other females of the same species! Surprisingly, other species in which females commonly "dump" eggs in each other's nests have not evolved similar variation in egg color. The clutches of one of the most familiar North American species, the American Crow, display remarkable variability in both color and shape (Kevin J. McGowan, personal communication); further research may reveal some surprising explanations for this variability (**Fig. 8–79**).

Eggs laid in the open, exposed to visual predators such as other birds, are cryptically colored. If you can find the nest of a Killdeer, note how the color and markings of the eggs, together with the small, flat stones used to line the nest, match the surroundings perfectly (**Fig. 8–80**). The heavy pigmentation of eggs in open scrapes, while helping to conceal them, also may shield the embryo from intensive solar radiation. Some dark plumages keep birds relatively cool in hot, windy areas by absorbing solar radiation and holding it in the upper layers, allowing it to

Figure 8–80. Cryptic Killdeer Eggs in Nest: Killdeer lay their four heavily marked eggs in a scrape in an open area, especially where the ground is somewhat textured—often with gravel, stones, or rubble—to help camouflage the eggs and incubating adult. Often the eggs are very difficult to see against the ground or nest lining, which may be pebbles, wood chips, grass, or other debris. Note the large size of these shorebird eggs, compared to the adult. Photo by Jeff Lepore/Photo Researchers.

dissipate into the air before the bird gets too hot (Wolf and Walsberg 2000). Perhaps heavily pigmented eggs dissipate heat similarly, before it penetrates and overheats the egg contents—just as Bedouins in the desert wear black, not white.

Many birds have white eggs. Primarily, these are birds that nest in dark holes—petrels, kingfishers, and woodpeckers (see Fig. 8–48b); begin incubation with the first egg—hawks and owls; or cover their eggs when they leave the nest—grebes and some ducks. Because these eggs are rarely exposed to predators hunting by sight, they do not need colors to lend camouflage, and pigments may be metabolically expensive to produce, or may require dietary precursors that are hard to obtain. For hole-nesting birds, the white color may have an added advantage—individual eggs show up better in the dark!

Egg Laying

Most birds lay their eggs early in the morning. The actual process (**Fig. 8–81**) may take from only a few minutes in most passerines to an hour or more in a larger bird such as a goose. Brood parasites, however, have evolved to lay their eggs quickly, before the host returns to its nest. Common Cuckoos, for instance, can lay their egg and disappear in less than 10 seconds (Seel 1973). This is yet another in the suite of fascinating adaptations that brood parasites have evolved (see Brood Parasite Ploys, later in this chapter).

In all birds, the time required for the oviduct to secrete the various layers around the ovum determines the interval between the laying of one egg and the next. In the smaller shorebirds and grebes, domestic chickens, woodpeckers, rollers, and most passerines, eggs are laid about 24 hours apart. In contrast, many gulls lay their eggs every other day, as do Ostriches, rheas, herons, storks, cranes, bustards, doves, owls, hummingbirds, swifts, kingfishers, some accipiters and cuckoos, and the larger grebes. Larger species tend to take longer between eggs, perhaps because the secretory processes in the oviduct simply take longer to produce a larger egg. For example, within the

Figure 8–81. Greater Rhea Laying an Egg: *The Greater Rhea takes about 10 minutes to lay an egg.* **a.** *The female squats on her tarsi as she experiences abdominal contractions. Here the egg is just visible in her cloaca.* **b.** *Sphincters (rings of muscle) around the cloaca push the egg out.* **c.** *The female sits down or tips her cloaca to the ground as the egg comes out, retracting the cloaca once the egg touches the ground. This photo also shows a male reaching for the egg, which he will roll into his nest with his bill. Many females lay for one male, who incubates the eggs and raises the chicks while the female continues laying for another male. The male is also visible at the left in photo (b). Photos courtesy of Donald F. Bruning.*

waterfowl, many ducks lay eggs every day, whereas large geese and swans lay every other day (Wilmore 1979). Parrots, even the smaller species, have inter-egg intervals of one to three days; penguins, three to six days; and the Masked Booby lays its two eggs as much as a week apart. Common Swifts in Europe, who usually lay 3 eggs at two-day intervals, are more likely to lay two eggs at three-day intervals after a spell of cold weather, when their food—flying insects—is difficult to find (Lack 1973). Species that lay very large eggs for their size also tend to have longer intervals between eggs, for example four to eight days in some megapodes, and 14 to 30 days in kiwis. Most species appear to lay successive eggs at approximately the same time of day, although some species (for example, some herons, bitterns, and parrots) lay at intervals that are not multiples of 24 hours.

Clutch Size

A **clutch** of eggs is the total number laid in an uninterrupted series, for a single nesting, by one female. In the early days of ornithology, many private egg collectors all over the world amassed large collections of bird eggs **(Sidebar 3: Oölogy: From Hobby to Science)**. This practice is illegal today, except by special permit. Although data from collections may give misleading indications of clutch size for many reasons (for example, predators may have removed some of the eggs, two females may have laid in one nest, or egg collectors may have collected only clutches of the larger sizes), for little-known species, clutch size is often the only life history trait for which there is any information. As a result, there has been considerable interest throughout the 20th century in trying to understand the patterns of variation in clutch size. Indeed, no other life history trait of birds has received so much attention. The following paragraphs briefly discuss the patterns observed and some ideas put forth to explain them.

Patterns in Clutch Size Variation

Clutch size varies in many interesting ways **(Table 8–3)**. First, phylogeny significantly affects clutch size: all tubenosed seabirds (albatrosses, petrels, and others) lay a single egg, virtually all hummingbirds and most pigeons and doves lay two eggs, no shorebird normally lays more than four eggs, and

Table 8–3. Clutch Size Trends: Shown here is a list of factors that generally correlate with clutch size. Most researchers believe that the food available to the laying female and/or for feeding young is the most important variable affecting clutch size, and most of these factors are related to the food supply. In nonpasserines that feed their young, brood size is probably limited by the amount of food parents are able to bring to the nest. Latitude, elevation, and the geographic variables all correlate with clutch size in that birds in more harsh areas with high seasonal (short-term) food availability lay larger (but often fewer) clutches. The larger clutches of older females may reflect their greater ability to procure the necessary food resources for egg laying. Laying larger clutches earlier in the season, however, is probably not related to food supply but rather is a female's strategic response to the fact that young fledging earlier in the season usually have higher survival rates. See text for more detailed explanations. Adapted from Gill (1990, p. 419).

	CLUTCH SIZE	
Factors Potentially Affecting Clutch Size	**Small (2–3 Eggs)**	**Large (4–6+ Eggs)**
Food Procurement of Young Nonpasserines	Fed by Adult	Feed Selves
Latitude	Tropics	Temperate/Arctic Regions
Elevation (Weak Trend)	Lowlands	Highlands
Seasonality and Severity of Geographic Area	Lower: Oceanic Islands Coastal Areas	Higher: Mainland Continental Interiors
Availability of Nest Sites	Higher (Open Nests)	Lower (Cavity Adopters)
Age of Female	Younger	Older
Laying Date within Breeding Season	Later	Earlier

so on. Most songbirds lay between 2 eggs (in the tropics) and 6 eggs (in temperate and arctic regions), although extraordinarily large clutches of up to 17 eggs are laid by some Eurasian tits (families Paridae and Aegithalidae). Second, among nonpasserines, those that feed their young (for example, gulls and storks) tend to lay fewer eggs than those whose young feed themselves (for example, waterfowl, grouse, and quail). Third, both within and between species, birds in the tropics tend to lay smaller clutches than those nearer the poles (for example, Klomp 1970; Winkler and Walters 1983; Martin 1987). House Wrens, for example, lay an average of 7 eggs per clutch in Saskatchewan and only 3.5 eggs in Costa Rica (Young 1994). Similarly, birds nesting at higher elevations tend to lay larger clutches than those in the lowlands (for example, Badyaev 1997). Fourth, both within and between species, birds laying on oceanic islands tend to lay fewer eggs than those breeding on the mainland (Blondel 1985); and individuals nesting in continental interiors tend to lay larger clutches than those nesting along the coast (Briggs 1984). Fifth, cavity nesters tend to have larger clutches than open nesters. Sixth, younger females tend to lay smaller clutches. And finally, within most populations, females that lay later in the season tend to lay smaller clutches (for example, see Winkler and Allen 1996; for some exceptions see Crick et al. 1993).

The Evolution of Clutch Size

During the 1940s, David Lack at Oxford University began studies, many of which continue today, of the clutch sizes and life histories

(Continued on p. 8·84)

Sidebar 3: OÖLOGY: FROM HOBBY TO SCIENCE
Lloyd Kiff

Few bird watchers today are aware of the tremendous passion for collecting bird eggs that formerly existed in America. At the turn of the century, most bird enthusiasts collected eggs or had done so during their youth **(Fig. A)**. Indeed, egg collecting was regarded as a natural part of growing up, and the collections were regarded with a reverence both intellectual and aesthetic. Oölogy—the study of eggs—was accepted by persons of learning as a respectable pursuit, and it was touted by its adherents as a legitimate biological discipline.

Just like modern birding, oölogy drew people—mostly men—from every walk of life. Renowned egg collector Fred "Kelly" Truesdale, an itinerant farmhand and cowboy in central California, was welcomed with fanfare into England's prestigious Oölogists' Exchange Club in 1932. One of his frequent field companions was Milton Ray, a wealthy San Francisco industrialist who at one time was the poet laureate of California.

Most of the 19th-century giants of American ornithology, including Robert Ridgway, Elliot Coues, and Charles Bendire, routinely collected eggs as a part of their field work **(Fig. B)**. Some bygone private collectors were leading conservation figures, including longtime National Audubon Society president T. Gilbert Pearson; Audubon biologist and lecturer Alexander Sprunt, Jr.; and William T. Finley, one of the West's best-known conservationists and wildlife photographers.

I was fortunate enough to become acquainted with several of the last of the American oölogists in the 1970s, when most of them were octogenarians or older. Having grown up as an avid life lister and bird bander, I was fascinated to discover the same personality types and drives among the old-time collectors. It seems certain that many of today's prominent birders would have been equally avid egg collectors in a Victorian context. Like hard-core birders, certain egg collectors traveled hundreds of miles over treacherous roads to acquire the eggs of a single species missing from their collection.

Why did people collect eggs? I once speculated that no one would have collected eggs if all eggs were white because the impulse was stimulated by the incredible variety of colors and patterns on eggshells; however, Michael Walters, an ornithologist at the British Museum of Natural History, subsequently showed me a collection made by an English gentleman that contained only white eggs!

Egg collecting reached its zenith in North America at the beginning of the 20th century, but a decline in its popularity soon became apparent. The excesses of some collectors, especially those of commercial dealers in natural history specimens, were viewed with increasing hostility by a new generation of American naturalists armed with binoculars and a conservation conscience. Furthermore, except for Arthur Cleveland Bent's respected *Life Histories of North American Birds*, the oölogical literature rarely rose above the level of anecdotal accounts of collecting trips, and the purported scientific benefits from egg collecting gradually became regarded as little more than justification for a hobby. Such changing attitudes were reflected by the passage of federal and state laws restricting egg collecting by private individuals, and by 1940 the practice had been virtually eliminated in North America.

As a result of the unfavorable image oölogists had acquired, academic interest in egg collections bot-

Figure A. Early Egg Collectors: *The passion for egg collecting that existed in the early 1900s left a valuable legacy of specimens and data for researchers today. Here, brothers Fred L. and Ed J. Court admire Bald Eagle eggs taken from a leafy aerie near Washington, D. C. in 1898. Tree-climber's spikes (foreground) were a valuable aid. Photo courtesy of Raymond J. Quigley, from his personal collection of historical ornithological photographs.*

Figure B. Collecting Golden Eagle Eggs: *Arthur Cleveland Bent (in boots), egg collector and author of the famous* Bent's Life Histories of North American Birds, *and Wright M. Pierce, well-known California collector, at a Golden Eagle nest in the Mohave Desert in San Bernardino County, California in 1929. This nest was still in use as recently as 1989. Photo probably by Lowell Sumner, Jr., courtesy of Raymond J. Quigley, from his personal collection of historical ornithological photographs.*

tomed out between World War II and the early 1960s, and important egg collections languished in museum attics and private closets. The proud heirs of one prominent eastern collector, whose egg display had won a gold medal at a turn-of-the-century world's fair, were dismayed to find that no museum would purchase their father's precious egg collection when they tried to dispose of it in the early 1950s. They could not even find a suitable institution that would accept the collection without a trust fund to ensure its proper care and housing.

The indifferent attitude of the academic community toward egg collections prompted several California naturalists to form a new organization to serve as a depository for unwanted collections, including their own, as well as to continue to field collecting expeditions in unexplored parts of the world. Under the guidance of Los Angeles businessman Ed Harrison, the group originally intended to gather more than bird eggs, so they chose the name Western Foundation of Vertebrate Zoology, to allow such latitude. As the organization took shape, however, birds and their eggs and nests became its focus. Now the foundation holds nearly 300 egg collections from around the world and is the undisputed world center of oölogical research **(Fig. C)**.

It was an ecological tragedy that first suggested the importance of egg collections to ornithologists. In 1961, just a decade after the foundation began collecting collections, the British researcher Derek Ratcliffe documented severe eggshell thinning in English Peregrine Falcons and suggested that the thinning was caused by pesticides in the birds' environment. Soon after, Joseph Hickey and Daniel Anderson of the University of Wisconsin documented a similar phenomenon in several North American species and showed that it resulted from contamination of the birds by high levels of DDE, a metabolite of the ubiquitous pesticide DDT. Ornithologists were quick to perceive the value of long-ignored egg collections as a source of baseline data on eggshell thicknesses that existed before synthetic pesticides came into use in the late 1940s. Egg collections have now been used in hundreds of studies on DDE-induced eggshell thinning throughout the world.

Using eggshells to document environmental change caused by contaminants was only the beginning. Eggs in the foundation's collection are a rich source of information on many topics. Since its formation, foundation collections have been used in more than 3,000 research projects, and the rate of usage continues to grow each year. The collection has been examined by anthropologists to identify eggshells from Indian middens, by historians to establish the travel itineraries of extinct collectors, by physiologists to study the relationship of egg size to the length of incubation periods, by ecologists interested in the evolutionary significance of egg size and color, and by government agency biologists interested in historical and life history information on now-endangered species.

As important as the eggs themselves are the data that accompany them. Although most egg collectors were not professional ornithologists, virtually all had a first-rate knowledge of the birds of their region, and most of them carefully recorded scientific data as they collected their specimens. Those data, preserved on slips that accompany the eggs, contain details on species, date and locality of collection, height and location of nest, and stage of incubation **(Fig. D)**. In aggregate, the slips represent one of the most important

Figure C. Eggs at the Western Foundation of Vertebrate Zoology (WFVZ): *Lloyd Kiff, former director of the WFVZ, shows off part of its huge egg collection. Photo by Frans Lanting.*

bodies of historical data on North American birds, and they are consulted frequently by ornithologists to reconstruct former distributions, to determine the extent and timing of breeding seasons, or to provide information on nesting habits.

In the past five years, the foundation staff has supplied information from data slips to the compilers of at least 25 state breeding bird atlases. It is to their credit that the collectors kept careful and reliable records: the usefulness of these data to modern research efforts supports the oölogists' original claim that they were truly part of the scientific community.

Nevertheless, in the past some ornithologists have been loathe to use egg collections for research, feeling that there was too great a tendency for some collectors to falsify data for commercial or other reasons. It is true that the commercial aspects

of egg collecting did lead to some ethical lapses. I have seen a set of roadrunner eggs painted with bold splotches of dark brown shoe polish and passed off as the eggs of the scarce Snail Kite. Eurasian Sparrow-hawk eggs purchased cheaply and abundantly from English dealers were sometimes sold to unsophisticated American collectors as Sharp-shinned Hawk eggs, just as Common Crane eggs were fraudulently substituted for those of the Sandhill Crane. More than one purported egg of the California Condor, the most desirable egg of all, turned out to be that of a Mute Swan collected from a city park. Critical examinations of egg collections, however, have indicated that only a minority of American collectors made such mistakes, either intentionally or accidentally, and most egg-set data are reliable.

Of greater concern to today's biologists are the biases of legitimate egg collectors, many of whom succumbed to the predictable tendency to place a premium on the unusual—larger clutches, prettier clutches, and eggs of anomalous sizes and shapes. These choices contribute to a collection that may be aesthetically pleasing but is not representative of the average in nature.

The foundation is not the only depository of useful egg collections in North America. Now that the research value of collections has been established, their status has risen considerably. About 100 scientifically important collections, containing in total about 500,000 clutches of eggs, are now housed in other North American institutions. Recently, the foundation compiled a computerized inventory of these collections for the American Orni-

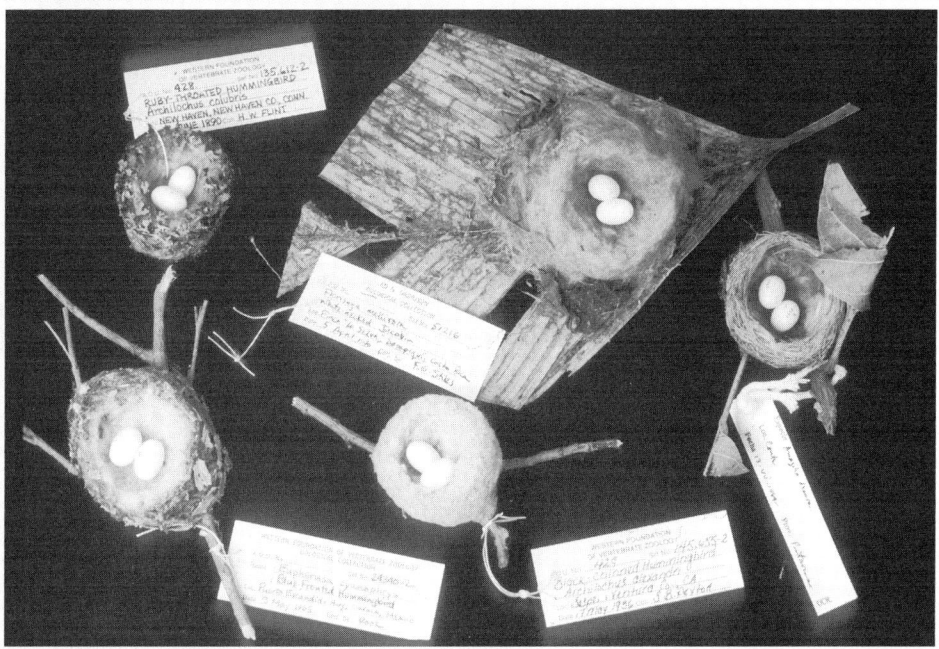

Figure D. Nests, Eggs, and Data: *The data slips that accompany each set of eggs in the Western Foundation of Vertebrate Zoology's collections are an important source of historical information on bird distributions and nesting habits. Photo by Frans Lanting.*

thologists' Union, and researchers were surprised to learn that only about half of the world's bird species are represented; however, the inventory indicates that at least one clutch of eggs exists for every species that breeds on the North American continent.

The fact that oölogists traditionally collected entire clutches or "sets" of eggs rather than removing a single egg from the nest is fortunate for modern researchers. This practice has the advantage of permitting the study of variations in size and color of eggs in the same clutch. It also may have been a better conservation strategy than taking one or two eggs, since the females of most species tend to replace lost whole clutches but not lost individual eggs.

Egg collections are not assemblages of whole eggs but of empty eggshells. The original contents were drained through a single hole drilled in the side of the egg. A stream of air or water was introduced through the hole, and the resulting internal pressure forced the contents of the egg out through the same hole. Because small blowholes were a matter of pride among collectors, recently laid eggs containing only liquid yolk and white were preferred; with patience, however, some collectors managed to coax the contents of even well-incubated eggs through remarkably small holes. The drained shells, which consist mainly of calcium carbonate crystals, seem to last forever if not broken or subjected to excessive light or moisture. Intact eggs of the extinct Giant Elephantbird still turn up occasionally in Madagascar, and one fresh-looking specimen was 14,000 years old. Most large museum collections contain eggs more than 100 years old, which, if they have been protected from exposure to light, cannot be distinguished from recent specimens.

In the United States today, only professional biologists studying environmental contaminants or engaged in other specific research projects may legally collect eggs. No ornithologist would advocate a return to the days of private wildlife collecting, and many present bird populations, already depleted by innumerable causes, could not survive the attentions of a nation of egg collectors. The hobby of egg collecting is an extinct piece of Americana, re-flecting a past generation's different approach to the study and enjoyment of nature.

Nevertheless, thanks in large part to the diligence of the old-time collectors, the serious study of the avian egg is alive and well. The collections are wonderful libraries of information—a public trust. Each specimen is unique and, if properly protected, can be used over and over for future studies, including some not anticipated by past—or present—generations. ■

Currently a biologist with the Peregrine Fund, Lloyd Kiff was director of the Western Foundation of Vertebrate Zoology and Curator of Ornithology at the Los Angeles County Museum of Natural History at the time he wrote this article.

Suggested Reading

Kastner, Joseph. 1986. *A World of Watchers*, Chapter 10. New York: Alfred A. Knopf.

This article originally appeared in Living Bird, *Winter 1989.*

of several local British species. More important, Lack (1954, 1966, and 1968) established the foundation for all subsequent ideas about the evolution of clutch size. Some of his ideas have since been extended—he did not consider a bird's entire reproductive success over its lifetime, for example—but most of what I present here can be found in his original writings.

The Conceptual Framework

Lack's central insight and conviction was that clutch size and other physiological and behavioral traits of birds could be viewed as products of natural selection, just as Darwin and his successors had viewed morphological and anatomical traits. There is good evidence that variations in clutch size are largely heritable. Thus, genes that code for clutch sizes that produce the most surviving young will eventually come to predominate in the population. The optimal clutch size, however, is not always the largest possible. Researchers generally assume that breeding is costly to a bird, in terms of both its own survival and future reproductive prospects. Some evidence supports this assumption. For example, adult birds that do not breed have higher survival rates than those that do (Ekman 1986; Pugesek and Wood 1992), and birds that lay more eggs and rear more young early in a season are less likely to lay again in the same season (Tinbergen 1987) or to survive and breed successfully in the following season (Røskaft 1985). This tug-of-war between current and future reproduction is the central trade-off in life history evolution (Winkler and Wilkinson 1988).

In adjusting clutch sizes, then, we think natural selection accounts for a bird's future reproductive costs—assuming that current breeding does, indeed, come at a cost to future reproduction. Even if this assumption does not hold, however, there are many ways natural selection can act *after* the eggs are laid that will affect a species' clutch size. For example, birds may vary in how effectively they incubate the eggs and protect them from predators, and in the type and amount of food they bring to their young. Young that receive suboptimal kinds or amounts of foods may have a lower chance of surviving and competing for essential resources for reproduction—such as a mate, nest site, or territory—in future years. A female who lays more than the optimal number of eggs in a given environment may actually succeed in *fledging* more offspring, but if those offspring fail to survive and reproduce, her genes coding for the larger clutch size will not be passed, through her, to the next generation. Any step between egg laying and the eventual reproduction of the embryos in those eggs (once they have matured) can affect a female's total reproductive success, causing selection to favor one clutch size over another.

One of my most valuable lessons as an undergraduate came when I made a wisecrack about storm-petrels being way out on an evolutionary limb, unlikely to make it much longer, since they laid only one egg and bred just once (at most) per year. Luckily, my perceptive instructor at the time upbraided me for being too hasty in my judgment, pointing out that natural selection considers more than one or two traits at a time. Evolutionary success has only one proof: the preponderance of one's genes in future generations. A bird that wins that competition by laying a single egg and giving that one offspring extensive parental care is not necessarily less "fit" than one that lays 15 eggs and pays each young scant attention. One of the endless life history marvels is the variety of ways in which birds have succeeded in the struggle for ever-greater representation of their genes in future generations: a penguin (one egg) is no more fit than a partridge (many eggs), and each presents an admirable solution to the complicated trade-offs faced by its ancestors.

Although most people view natural selection as the maximizer of reproductive output (for example, Lack 1966; Wootton et al. 1991), there has been a persistent objection to this view (for example, Wynne-Edwards 1962; Murray and Nolan 1989): How can it be adaptive for some birds to produce large numbers of offspring, have many young die in the ensuing competition, and possibly even degrade the habitat for future generations? Why shouldn't natural selection favor those birds who produce only as many young as they need to replace themselves? Any scenario entailing the possibility that reproductive restraint may have evolved can be unraveled with the simple observation that cheaters will always win: if any bird had genes that caused it to lay more eggs and produce more surviving offspring, it would end up with more representatives in the next generation, and would eventually predominate in the population. Therefore, reproductive restraint, in birds as in humans, cannot evolve by natural selection. Birds, on average, produce just enough young to replace themselves, but this is because of the high mortality rates discussed earlier, which usually fall disproportionately on the youngest birds in the population. This is an unavoidable result of natural selection: the larger the overall production of young, the larger must be their mortality.

Ecological Effects on Clutch Size

We have spent some time considering the evolutionary framework within which clutch sizes (and other life history traits) likely evolve. What specific ecological factors most affect clutch size? Most ornithologists believe that the availability of food for the developing young is the most important influence on the clutch size of birds that feed their young **(Fig. 8–82)**. But what of those species in which the young feed themselves (for example, many waterfowl, pheasants, grouse, and quail; some shorebirds; and megapodes)? In these species, it is pretty clear that the principal limitation on clutch size is not the food supply for raising chicks but the food available for the laying female. The limit on egg production may not be that females are physiologically incapable of laying more eggs, but rather that laying more eggs may so exhaust the female that her survival or subsequent reproduction will be unduly impaired.

If the food available to the laying female can limit the clutch size of species that *do not* feed their young, it also may limit clutch size in species that *do* feed their young (Winkler 1985). Indeed, if food supply for the laying female is the most significant limiting factor (or if the food available for young is more limiting, but the food available at

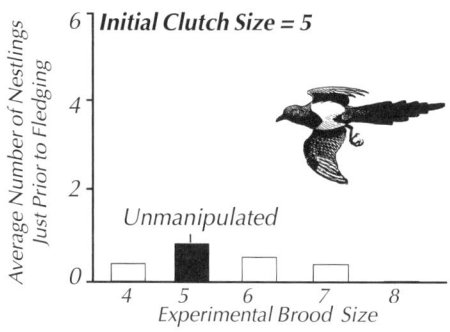

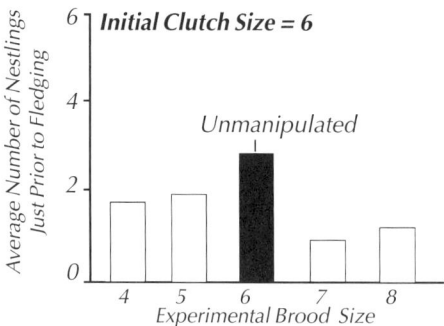

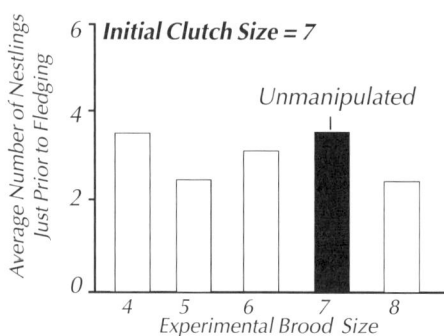

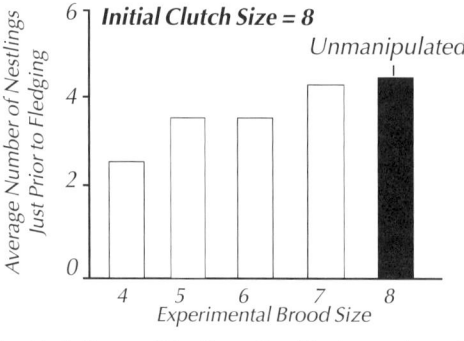

Figure 8–82. Black-billed Magpie Clutch Size Variation and Territory Quality: *In southern Sweden, the clutch size of Black-billed Magpies varies widely. To determine if birds adjust their clutch size to environmental conditions, Högstedt (1980) moved very young nestlings between nests, increasing or decreasing the brood size. Then he observed the number of nestlings in the nest just before fledging, to determine how many the pair was actually able to raise. As illustrated by this series of graphs, at each initial clutch size (5, 6, 7, or 8), birds raised the most young when their brood size was not changed ("unmanipulated"). Enlarging the brood did not increase the number of young raised, because nestlings had higher predation and starvation rates. Therefore, this experiment demonstrated that the female somehow "knew" how many young she could raise, and laid a corresponding number of eggs. Parents with experimentally reduced broods apparently could have raised more young; removing young simply cut their breeding potential. By observing the same set of territories (but not necessarily the same birds) for several years, Högstedt determined that most of the clutch-size variation was linked to which territory a pair occupied—and only slightly linked to an individual female. Thus, each female apparently adjusts her clutch size to the quality of her territory, laying the optimal clutch size for her breeding conditions.*

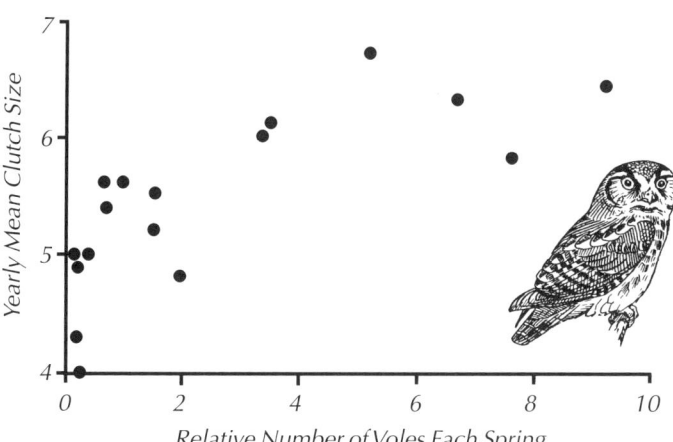

Figure 8–83. Boreal Owl Clutch Size in Relation to Prey Abundance: *In western Finland, the average clutch size of Boreal Owls (known there as "Tengmalm's Owls") varies from year to year according to the availability of voles, their primary prey. In this study area, vole populations typically fluctuate widely in a 3-year cycle. This graph shows the average clutch size for each year from 1973 through 1989 plotted against a relative measure of the density of voles in the spring—determined by extensive trapping. Researchers believe that females base their clutch size on the density of voles during egg formation and laying, which is a fairly good predictor of how abundant food will be during the nestling and post-fledging periods. Adapted from Korpimäki and Hakkarainen (1991).*

laying time is a reliable indicator of the food that will be available for chick rearing), birds can be much more flexible in their clutch sizes, using food supply as a cue to how many eggs they should lay. A prime example of this flexibility is presented by many owls (see for example, Korpimäki 1988; Korpimäki and Hakkarainen 1991; Kaufman 1996) and arctic-nesting raptors (see for example, Newton 1979), in which clutch sizes vary dramatically between years with abundant prey and those with very few **(Fig. 8–83)**. Snowy Owls in a high lemming year may lay and rear 7 to 11 eggs, whereas the same females may lay only 3 to 5 eggs in low lemming years (Parmelee 1992) **(Fig. 8–84)**. This is an interesting case in which the pre-laying food supply could act as both an ultimate and proximate factor, illustrating how fuzzy the borders of our intellectual constructs can be when we subject them to Nature's test.

Why wouldn't all species base their clutch sizes on the amount of food available, laying small clutches in poor food years and large clutches in rich food years? Unfortunately, many species probably do not have that option. When Great Tits are laying eggs in April and May, they have no way of knowing how abundant the winter moth larvae on which they depend to feed the nestlings two weeks later are going to be: there are just no reliable cues available at laying time to tell them what the chick-rearing conditions will be like. For these and a large number

Figure 8–84. Snowy Owl with Lemming Prey: *The Snowy Owl, like the Boreal Owl, adjusts its clutch size to food availability. Breeding in the Arctic tundra and feeding almost exclusively on lemmings (when available), Snowy Owls may lay and rear 7 to 11 eggs when lemmings are abundant. When lemmings are scarce, however, the same females may lay only 3 to 5 eggs, or may not breed at all. The food supply before laying—which in this case is the same type of food that will be used to feed the chicks, apparently is the cue used by females to adjust their clutch size.*

Figure 8–85. Clutch Size Variation with Latitude in the Northern Flicker: *The Northern Flicker, which includes the eastern "Yellow-shafted" and western "Red-shafted" races, breeds throughout much of Canada and the United States. This graph of average clutch sizes at different latitudes, gathered from over 400 nests, shows that clutch size clearly increases toward the North Pole—an increase of approximately one egg for each 10-degree increase in latitude. Error bars represent one standard error around the mean. Adapted from Koenig (1984).*

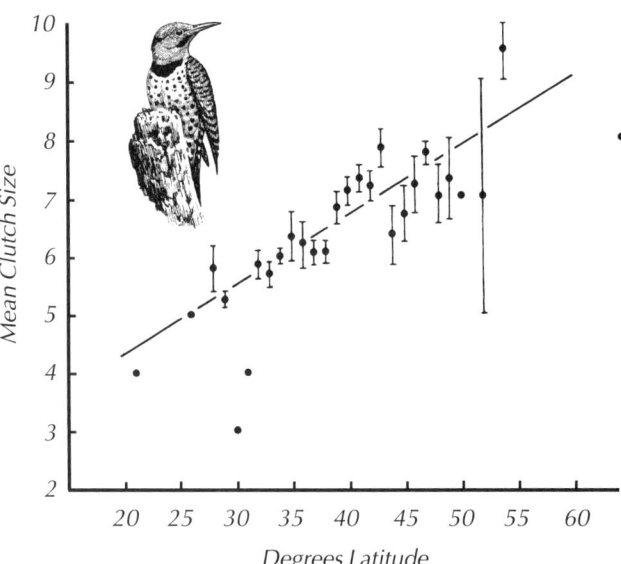

of other birds, the best that they can do to estimate the optimal clutch size is to rely on the experiences of their ancestors, relayed through the genes that each female carries, and to lay a clutch that is the best *on average* at adding the most offspring to the next generation. This average is constantly being adjusted by natural selection and is simply the best possible choice in the absence of any reliable environmental cues about chick-rearing conditions in the future.

Clutch Size and Latitude

Recall that birds tend to lay larger clutches at higher latitudes **(Fig. 8–85)**. Ricklefs (1980a) proposed an interesting hypothesis suggesting how geographic variation in food supply explains the relationship between clutch size and latitude, regardless of whether food is more limiting at laying or chick rearing. This hypothesis, which Ricklefs calls "Ashmole's hypothesis" in honor of the researcher who first proposed it (Ashmole 1963), reasons as follows: Nearer the poles (higher latitudes), the long period of harsh winter weather and limited sunlight severely restricts insect availability and plant growth, forcing most birds to migrate and many nonavian insect and seed predators to go dormant. When spring returns to these areas, there is a huge flush of life as days lengthen and plants kick into high photosynthetic gear. This effect cascades to the seeds and insects that most birds feed upon, resulting in *more* food being available to birds laying eggs and rearing chicks in polar areas than in the tropics. Habitats nearer the equator never experience a large annual variation in day length, and the relatively minor annual variations in plant growth and insect biomass are kept constantly in check, and at a much lower mean level, by complex communities of resident seed-, fruit-, and insect-eaters. Anyone who has visited both arctic and tropical areas during birds' breeding seasons can testify to the substantially higher numbers of insects (remember those mosquitoes?) at the higher latitudes. Thus, factors such as day length and temperature acting at the ecosystem level may determine the amount of food available to birds to form eggs and rear their young, and ultimately, the numbers of eggs that birds lay at different latitudes. Similar effects may help explain higher clutch sizes in interior mainland areas compared to the relatively buffered and less seasonal habitats in coastal and island areas, as well as larger clutches at higher elevations.

Clutch Size in Cavity Nesters

Traditionally, the tendency of cavity nesters to lay larger clutches than open nesters has been attributed to the lower predation rates of cavity nesters (for example, Slagsvold 1982). In most birds, higher predation rates appear to select for fewer young, which are less likely to draw the attention of predators. In addition, increased predation is thought to drive the evolution of faster development of the young, because they spend less time as vulnerable nestlings (see Fig. 8–111). If parents can only bring a given amount of food, they can either use it to rear fewer young faster or more young more slowly. So the longer development times of cavity nesters also allow larger clutches. More recently, however, Martin (1993b) pointed out that only the cavity nesters who *do not* excavate their own holes (such as House Wrens) tend to have large clutches. Since all cavity nesters, regardless of whether or not they excavate, have lower predation rates, the predation hypothesis does not appear to hold. Martin suggested, instead, that the large clutch sizes of nonexcavators results from their uncertainty in obtaining a nest hole: they depend on the availability of either a natural cavity or one made by a different species. When they do manage to obtain a nest hole, they should invest as much as possible in that breeding attempt—hence the large clutch size. Evidence supporting this hypothesis, however, comes mainly from correlations between clutch size and nest type across many species, drawn from studies reported in the literature. More field studies with standardized methods are needed to determine the actual availability of nest sites, the frequency of breeding, and the clutch sizes of various species. These studies would help researchers to see if cavity adopters actually have fewer breeding opportunities than cavity excavators, over their lifetimes.

Clutch Size and Age of Female

Why many young females lay smaller clutches than older ones remains a mystery to ornithologists. In many birds, younger females begin nesting later in the season, and clutch size tends to decline as the season progresses (see the next section). But laying date cannot fully explain the age effect: in many species, younger females still lay smaller clutches than older females laying on the same date. Some researchers have suggested that younger, inexperienced females are not as good at finding the food required to fuel a larger clutch (von Haartman 1971), but this does not hold for all species.

Clutch Size and Season

One of the most interesting and pervasive patterns in avian life histories is that birds that lay later *in a nesting season* tend to lay fewer eggs per clutch (Klomp 1970; for exceptions, see Crick et al. 1993).

Researchers have proposed two general hypotheses to explain this pattern: (1) birds with larger nutrient stores early in the season are somehow forced or constrained by their physiology to lay larger clutches earlier in the season (Price et al. 1988) and (2) birds may lay smaller clutches later because environmental conditions for chick rearing or offspring survival decline later in the season (Perrins 1970; Daan et al. 1988). Whereas the former hypothesis could conceivably explain the seasonal decline in clutch size in birds such as arctic-nesting waterfowl that return to the breeding grounds with all the resources they need for reproduction stored on their bodies (Ankney and MacInnes 1978), my studies of female Tree Swallows (Winkler and Allen 1996) indicate that these and many other passerines, because they are so small and need to retain adequate flight ability, are not able to store up enough resources for egg laying. As a result, they must base their clutch size on the availability of resources during and immediately before laying. Thus, the decline in clutch size appears not to be the result of a physiological constraint linking clutch size and laying date, but rather a strategic adjustment by the laying female to the prospects for the chicks that will hatch from her eggs.

One of the most interesting aspects of this research is how strong the effect of laying date is on clutch size and many other life history traits. Average laying date in a population may shift earlier or later *from year to year* because of changes in weather and food availability, and species respond to these shifts differently: some lay larger clutches when the laying date is earlier, and others do not (von Haartman 1982). In spite of these differences, however, laying date appears to serve as a reliable indicator of individual quality. In most cases, breeding earlier is advantageous because birds may obtain better mates, and because their fledglings have more time to mature and improve their foraging skills before the hazards of winter, migration, or other seasonal changes are upon them. Birds, with their typically high mortality rate, should also have evolved to breed as early as is practical because late breeders are more likely to die before fledging young—on average, earlier breeders pass more genes to the next generation (Goutis and Winkler 1992). In a vast diversity of birds, individuals appear to strive to breed earlier, and the better birds are those that can bring this feat off and live to tell the story....

Egg and Clutch Replacement

Most birds will replace their clutch if it is destroyed, but many single-brooded species (see next section) will not re-lay if the offspring are lost during chick rearing. Laying of a replacement clutch usually requires the female's hormonal system to recycle into laying mode and the yolking up of new ova, and all this usually takes five to eight days in most passerines and longer in larger birds.

Birds respond to the partial removal of a clutch during egg laying in diverse ways. So-called **determinate layers** will not lay another egg if one is removed from the clutch during laying, whereas **indeterminate layers** will lay replacement eggs. Some indeterminate species are capable of prodigious efforts: the removal of one egg every day from the nest of a Northern Flicker induced the bird to lay 71 eggs in 73 days (the normal clutch size is about 6 to 8 eggs); and domestic hens, selected artificially for egg production, can lay as many as 352 eggs in 359 days. Hens, like all other birds, will cease to lay replacement eggs if allowed to sit upon a full clutch and begin incubation. Despite the attractiveness of the determinate-indeterminate dichotomy, the boundary between these types is actually quite fuzzy, as some species, such as gulls, will act as indeterminate layers only if the eggs are removed before the female can sit on two or more (for example, Winkler 1985). There is also variation in how different birds determine that a "complete" clutch has been laid: some seem to count them visually, whereas others seem to rely upon the feel of the eggs under the sitting female (Winkler and Walters 1983).

It is unclear how the existence of various egg-replacement behaviors relates to clutch-size theory. Some might argue that birds able to lay large numbers of replacement eggs are clearly not limited in any way by the availability of food for making eggs; however, apart from domestic hen data, there are no measures of the toll that supplemental egg laying takes on laying females. These artificially manipulated birds may be laying many more eggs than is "good for them." In natural settings, a single egg is rarely taken from a nest (most egg predators take complete clutches), and the birds may not have evolved optimal reactions to specific schedules of egg removal.

No matter what its evolutionary significance, the propensity in raptors and condors to replace eggs taken from the nest during the laying period may have helped to save some of them from extinction. Wildlife biologists were able to "double-clutch" Peregrine Falcons and California Condors, removing some eggs and rearing them in the lab while the birds reared their replacement eggs in the wild. This was an important part of the strategy to increase their critically reduced populations **(Fig. 8–86)**.

Number of Broods per Season

In comparing the clutch sizes of temperate and tropical birds, it is important to note that most tropical species have much longer breeding seasons than do species at higher latitudes. Tropical land birds can often lay eggs over a period of at least four months, whereas species in the temperate zone lay for less than two months, and arctic species may have less than one month. Longer tropical breeding seasons allow most birds at lower latitudes to make many more breeding attempts per season than their counterparts at higher latitudes. Even within North America, species such as the Red-winged Blackbird can raise two broods in New York, but have time for only one in Alberta, Canada. Prothonotary Warblers raise two broods each year in Tennessee, but only one in Michigan. Thus, while the number of eggs in a clutch increases with latitude, the total number of young raised annually by one pair may actually decrease because there is no second brood **(Fig. 8–87)**.

a. Condor Chick Hatching in Lab

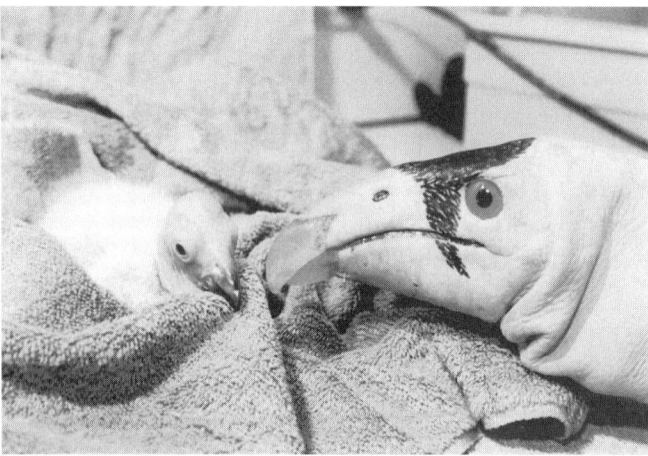

b. Condor Puppet with Chick

Figure 8–86. Double-clutching Aids Captive Breeding of California Condors: In the wild, California Condors have a very low reproductive rate, laying only one egg every other year. Like many birds, however, condors will lay a second clutch (in this case, a single egg) if the first is removed from the nest soon after it is laid. This technique, termed **double-clutching,** *has helped conservationists to keep this endangered bird from becoming extinct. Early in the condor restoration process, biologists removed eggs from wild nests to rear in the lab, while the wild birds re-laid and reared chicks themselves—substantially increasing the number of condors produced each year. Unfortunately, the wild population began to decline more rapidly, because many birds were shot, ate animal carcasses contaminated by lead bullets or poison intended for coyotes, or collided with power lines. Thus, all wild birds were captured and brought to the lab, preserving as much genetic diversity as possible in the group. In 1987, when the last bird was captured, only 27 were left in the world. Currently, the Los Angeles Zoo, the San Diego Wild Animal Park, and the Peregrine Fund's World Center for Birds of Prey in Boise, Idaho are collaborating to breed these birds. To quickly increase the captive population, biologists continue to employ double-clutching. They transfer a pair's egg to an artificial incubator right after laying, inducing the female to lay again within four or five weeks. Sometimes even a third egg may be produced in one breeding season.* **a. Condor Chick Hatching in Lab:** *After nearly two months of incubation, the condor chick hatches. Like wild condor parents, researchers watch closely and provide assistance if necessary.* **b. Condor Puppet with Chick:** *Because the chick eventually will be released to the wild where it will need to fear humans, biologists rear the chick in the company of a lifelike condor hand puppet; this technique prevents the chick from imprinting on humans. The puppet, operated by zookeepers, broods the chick as a live parent would, and interacts with the chick using natural condor postures and sounds. Eventually the chick moves to a larger cage with other chicks, and is released when about six months old. As of August 2001, the world's California Condor population had risen to 183 individuals. Of these, 53 had been released into the wild in California and Arizona. In March 2001, biologists in the Grand Canyon National Park found the first egg laid in the wild by a reintroduced condor. Although the adults inadvertently broke the egg—a common occurrence with first-time condor parents—this is still a major milestone in the condor recovery effort, and it should not be long before some wild California Condors raise a chick of their own. Photographs courtesy of the Los Angeles Zoo.*

Total Eggs Laid at Different Latitudes

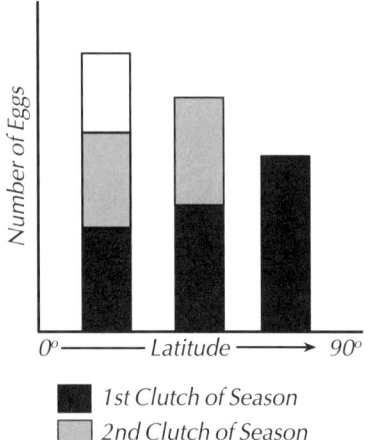

- ■ 1st Clutch of Season
- ▨ 2nd Clutch of Season
- ☐ 3rd Clutch of Season

Figure 8–87. Total Eggs Laid per Breeding Season at Different Latitudes: Although birds at lower latitudes (toward the tropics) may have smaller clutches than birds at higher latitudes, their longer breeding season often allows them to make more breeding attempts and thus to lay more total eggs per year. This graph shows idealized relationships among season, latitude, and clutch size.

Late-season reductions in a species' chief food may limit the number and size of broods. The Great Tit has a second brood only seven percent of the time because it feeds its young on caterpillars, which are usually scarce late in the breeding season. The second clutch is smaller in size, averaging 7 eggs compared to 10 eggs in the first clutch, and fewer young fledge—only 37 percent of those hatched compared to 95 percent in the first brood (Lack 1954).

Incubation

■ Birds are the only major vertebrate group containing no live-bearing species: fish, amphibians, reptiles, and mammals each have at least a few. All birds lay external eggs that must be kept at the proper temperature throughout their development—a process termed **incubation**. Exactly why no live-bearing birds have evolved remains a bit of a mystery, but the most compelling idea (Blackburn and Evans 1986) is that bearing live young would strongly limit either the flight ability of the mother or her brood size. Bats, however, are successful, capable fliers that bear live young, casting some doubt on this idea. Having external eggs does allow both parents (at least potentially) to help in their care, and also allows them to leave the eggs during floods, fires, and so on and consequently live to rear another brood in more favorable conditions.

Incubation is primarily an avian trait: although some insects, fish, amphibians, and reptiles guard their developing eggs until they hatch, none of these animals provide significant temperature regulation. The only two exceptions (Harry W. Greene, personal communication) are crocodiles, who warm their eggs using the heat of decomposing nesting materials (similar to megapodes), and pythons, who use muscle contractions to generate heat to warm their eggs (Slip and Shine 1988). (The only egg-laying mammals, the monotremes, also incubate their eggs.) Depending upon favorable environmental temperatures to help maintain body and egg temperatures has geographically restricted the breeding range of reptiles. But evolving the ability to use their own body heat to regulate their egg temperatures freed birds (and whichever of their ancestors were warm-blooded) from such a dependence, allowing them to breed from the tropics to the polar regions.

The precise temperature needed for bird embryo development varies across species, but virtually all appear to require temperatures around 99 or 100 degrees F (37 or 38 degrees C). Because few environments have a mean air temperature routinely at or above this level, the overwhelming majority of bird species need to provide a source of warmth for their eggs. Most of these species use their own body heat, but two groups have other methods of keeping their eggs warm. Brood parasites leave the incubation of their eggs to their hosts, and megapodes provide heat for incubation with a broad variety of methods, none of which use body heat. The most spectacular megapodes, as mentioned previously, build very large mounds of decaying vegetation to warm their eggs. Other species use geothermal heat—burying

8

their eggs in long tunnels or broad pits where the earth is warmed from nearby hot streams or volcanic cinder fields. A few even rely on solar radiation—burying their eggs in pits or burrows in areas where bare sand or soil is heated by the sun, including forest clearings, beaches along rivers, and coasts (Jones et al. 1995). But these fascinating species are exceptions to the general rule that birds incubate their eggs with heat derived from their own metabolism.

Not all birds need to warm their eggs, however. In very hot environments, the eggs actually need to be cooled: Wilson's Plovers along the Gulf of Mexico and Egyptian "Plovers" (really coursers in the family Glareolidae) of sand bars in African rivers cool their eggs by dipping their breast feathers in water and moistening their eggs **(Fig. 8–88)**: the African bird goes so far as to bury its eggs (and sometimes the chicks) in sand to protect them from the sun. One pair of Black-necked Stilts nesting at the Salton Sea in hot, arid Southern California made 155 trips in one day to soak their belly feathers in water (Grant 1982). Terns and gulls with open nests on desert islands often spend more time standing over their eggs with wings spread, shading their eggs, than they do incubating them (for example, Drent 1975; Howell et al. 1974) (also see Fig. 4–123).

Incubation Patch

A few days before a female lays the first egg of her clutch, an **incubation patch** (also called a **brood patch**) forms: feathers on the breast and belly fall out, and the skin becomes swollen through the retention of large amounts of water in the tissues and the expansion of blood vessels feeding the skin. The incubation patch can be a single large patch, taking up much of the area of the belly and breast (as

Figure 8–88. Egyptian Plover Wetting Eggs: Nesting in a deep scrape on the hot sand bars and sandy shores of large, tropical, African rivers, Egyptian Plovers cool their eggs (and even small chicks) during the day by burying them very shallowly (0.08 to 0.2 inches [2 to 5 mm] deep) in the sand. Then, during the hottest part of the day, the adults continually soak their belly feathers in water and place them over the buried eggs, wetting the sand to lower the egg temperature. The long incubation period (30 days) of this species allows the precocial chicks to be very well developed when they hatch, and the birds' ability to keep the eggs wet probably prevents them from losing moisture during incubation, making the long incubation period possible. Photo courtesy of Thomas R. Howell.

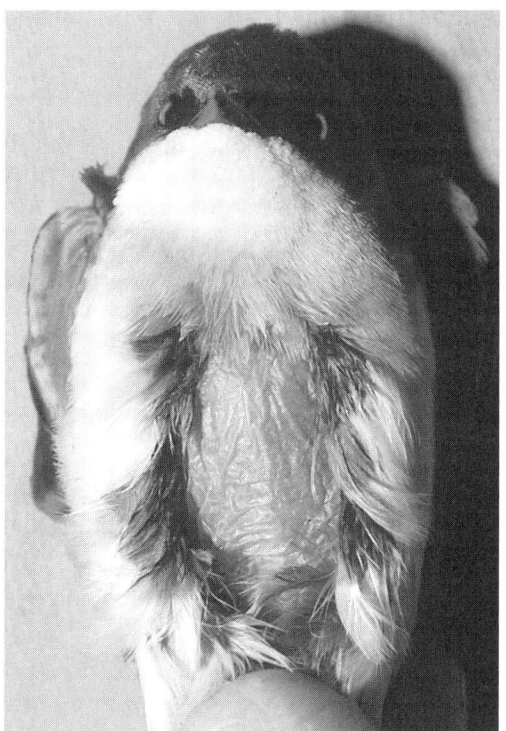
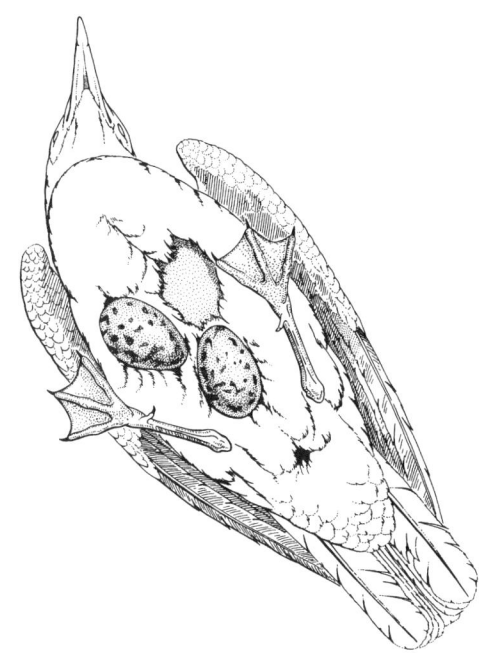

a. Single Incubation Patch of Female Tree Swallow   *b. Multiple Incubation Patches of Herring Gull*

Figure 8–89. Incubation Patches: To facilitate heat transfer to the eggs, most adult birds that incubate develop an **incubation patch***, also known as a* **brood patch***, just before the first egg is laid. Both males and females can develop this bare, swollen patch of skin, which is created as the breast and belly feathers fall out, the underlying tissues retain large amounts of water, and the blood vessels feeding the skin expand. Once parents are no longer brooding their young, the feathers grow back and the skin returns to normal.* **a. Single Incubation Patch of Female Tree Swallow:** *Most passerines, including the Tree Swallow pictured here, develop a single, large brood patch over most of the belly and breast. Photo courtesy of David W. Winkler.* **b. Multiple Incubation Patches of Herring Gull:** *Most shorebirds and gulls have smaller, separate incubation patches—one for each egg in a typical clutch. This Herring Gull has three incubation patches for its three eggs. Reprinted from* Manual of Ornithology, *by Noble S. Proctor and Patrick J. Lynch, with permission of the publisher. Copyright 1993, Yale University Press.*

in passerines) **(Fig. 8–89a)**, or may consist of smaller, discrete areas separated by areas that retain their normal feathering (as in gulls) **(Fig. 8–89b)**. Waterfowl aid the defeathering process by plucking all the down and some contour feathers from the patch and using them to line their nests. To incubate, the bird spreads apart the contour feathers that remain over the patch, so that bare skin rests directly on the eggs **(Fig. 8–90)**. Some albatrosses have an incubation "pouch" into which their single egg fits so tightly that it remains there when the bird stands up (Whittow 1993) **(Fig. 8–91)**. Recent research (Turner 1997) suggests that the high water content of the incubation patch may be an adaptation to increase the efficiency of heat transfer to the egg: heat moves best between materials of similar properties, and the watery nature of the patch may help to match that of the egg's albumen. Most birds continue to provide heat to their young until the developing feathers and metabolism of the young allow them to control their own body temperature. Once this parental **brooding** is over, the feathers grow back—often before the next body molt, the blood vessels shrink to their normal size, and the swollen character of the skin disappears.

In general, the incubation patch develops only in individuals that incubate—for example, only in male phalaropes and only in female

sparrows, but in both sexes of species in which both males and females regularly incubate. Nevertheless, exceptions occur, and little is known about the incubation patches of many species. If the male sometimes incubates or incubates irregularly, as in North American Barn Swallows (Ball 1983), he may or may not have a patch, depending on the importance of his role in incubation.

Incubation Period

The **incubation period** is the time from the start of regular, uninterrupted incubation to hatching. This period ranges from about 11 days in some of the smaller finches to about 80 days in the larger albatrosses. Although larger eggs generally take longer to hatch, there is considerable variation among groups of birds: hummingbird eggs take much longer to hatch than would be expected from their size, as do the eggs of petrels and albatrosses. The rate of development inside the egg is often closely related to that outside the egg: eggs with longer incubation periods generally hatch into young with slower growth rates (Lack 1947–1948). The eggs of precocial species generally take longer to hatch than those of altricial species: although adult Killdeer and meadowlarks are approximately the same body size, the precocial Killdeer's larger egg takes about 24 days to hatch, whereas the altricial meadowlark's egg requires only 13 to 14 days.

Figure 8–90. American Avocet Settling on Eggs: When a bird settles down to incubate, it spreads apart the contour feathers around the incubation patch—as this American Avocet is doing—so that the bare skin of the patch directly contacts the eggs, allowing them to receive as much heat as possible. Photo courtesy of Isidor Jeklin/CLO.

Figure 8–91. Incubation "Pouch" of Laysan Albatross: Like some other albatrosses, the Laysan incubates its single egg in a featherless cavity on its breast that is surrounded by thick feathers. The egg fits into this "pouch" so snugly that it may remain inside even when the bird stands. Photo courtesy of Olin Sewall Pettingill, Jr.

The incubation period may vary a little from nest to nest in a population, and long periods of inclement weather can prolong incubation by a day or two, but incubation periods are actually less variable than the nestling period. If all the eggs in a nest fail to hatch, the stimulus of moving nestlings is not present to change the adult's behavior from "incubation" to "care of the young," and a bird may continue to incubate the eggs for two or even three times its normal period before finally deserting them. Northern Bobwhites have incubated eggs for at least 80 days (considerably beyond the normal period of 23 to 24 days), and Herring Gulls have incubated for at least 103 days beyond their normal period of about 30 days. In our studies of Tree Swallows we have induced females to incubate eggs at least 10 days beyond their normal 14-day period. Although this behavior might seem extreme, there is no reason to believe that these three widely different species are particularly persistent or atypical in their incubation behavior. Incubating well beyond the expected hatch date probably has not been eliminated by natural selection because entire clutches rarely fail to hatch in nature.

Start of Incubation

The time in the laying cycle when birds begin incubation varies greatly among species. Some birds (for example, pelicans, cormorants, herons, storks, eagles, hawks, cranes, parrots, and owls) begin incubation with the laying of the first egg. This protects the eggs from the very beginning, but ties the incubating bird to the nest site for a longer period, extending the adult's exposure to predators. In these species, the embryos of the earliest-laid eggs have already started to develop by the time the later eggs are laid, resulting in staggered hatching, and a brood consisting of young of different ages. Especially in species with large clutches, this can result in a strong hierarchy of sizes and begging abilities among the offspring **(Fig. 8–92)**.

*Figure 8–92. Asynchronously Hatched Brood of Barn Owls: Birds lay their eggs one at a time over a period of several days, but most species delay incubation until the clutch is nearly complete. Because all of the eggs take the same number of days to develop, the young hatch at roughly the same time. Some species, however, begin to incubate immediately after they lay their first egg, so that their young hatch one at a time over a period of days, a situation termed **asynchronous hatching**. Broods of such species contain young of different ages and often quite different sizes, as illustrated by this family of Barn Owls. When food is in short supply, the smaller, later-hatched young may not survive if their older, better-developed siblings outcompete them for the food that their parents are able to bring to the nest. Photo by J. Oakley/VIREO.*

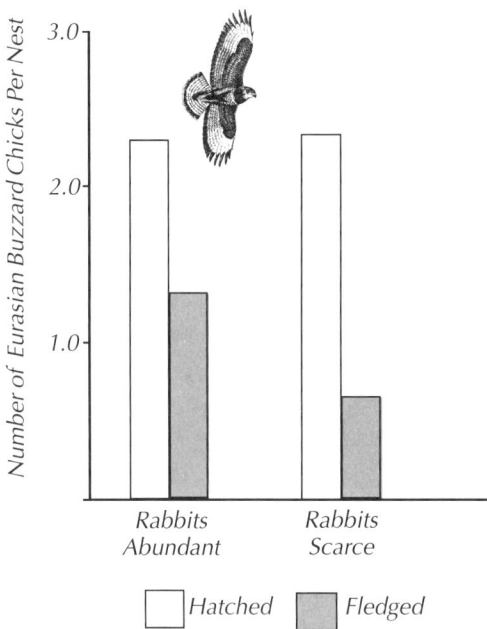

Figure 8–93. Fledging Success Versus Food Supply in Eurasian Buzzards: *Like many raptors, Eurasian Buzzards (related to North American buteos) have young that hatch asynchronously. In Devon, England Eurasian Buzzard pairs hatch approximately the same number of young per nest, whether their territory has many or few rabbits, a primary prey item. Those with many rabbits, however, fledge more than twice as many young (1.33 per nest) as those with few rabbits (0.6 per nest), presumably because when food is abundant, parents can feed their smaller, later-hatched young as well as their larger, first-hatched young. An alternative explanation is possible, however: birds that are able to obtain better territories also may be higher quality parents, capable of rearing more young for reasons other than food supply. Data from Dare (1961).*

Having young of different sizes gives parent birds who are feeding young a systematic way to adjust the number of young to the amount of food available. By first feeding the most vigorously begging bird until it can swallow no more, and then moving on to another, parents ensure that at least some young will survive in years when food is in short supply. At these times the later hatched young, too small to beg competitively, will starve. This parental practice is known as **brood reduction**. When food is abundant, the first-hatched young become satiated, and parents feed the smaller young as well, rearing more young **(Fig. 8–93)**.

A great deal of literature explores the possible evolutionary advantages and disadvantages of this so-called **asynchronous hatching** (for example, Stoleson and Beissinger 1995), but relatively little attention has been paid to the fact that most species (rails, woodpeckers, ducks, geese, gallinaceous birds, and most passerines) avoid extreme hatching asynchrony by waiting to begin incubation until the last or penultimate egg is laid **(Fig. 8–94)**. For precocial species that leave the nest right after hatching, synchronized hatching is essential. These birds leave, at least in part, to avoid the increased predation risk brought on by a nest full of young birds, and having eggs hatch on different days would expose both unhatched eggs and young to greater risks.

Figure 8–94. Synchronously Hatched Brood of Ruffed Grouse: *Most birds delay incubation until their clutch is complete, so their eggs hatch at approximately the same time. This **synchronous hatching** is particularly important for precocial young that leave the nest right after hatching, like these newly hatched Ruffed Grouse chicks. Young birds are more noticeable to predators than are eggs, and if the chicks had to remain in the nest for several days while all the eggs hatched, both unhatched eggs and young would be exposed to greater risks. Photo by Ned Smith/VIREO.*

Role of the Sexes

In many species (for example, gulls, many shorebirds, alcids, and many passerines) both sexes incubate. This is particularly common among species in which both members of the pair look alike. (The Rose-breasted Grosbeak is an exception: the strikingly colored male not only incubates, he often sings while sitting on the nest!) Generally, both sexes in these species share the task about equally, relieving each other at frequent intervals, with the female apparently most often incubating through the night. Members of a pair often perform a species-specific "greeting ceremony" when they exchange duties at the nest. Among some of the larger species, such as albatrosses and boobies, these ceremonies can be spectacular (see Fig. 6–26a), but in most species they are quite subdued. For instance, the male Purple Finch sings an incomplete, scarcely audible song when approaching the nest. An incubating bird usually appears more than ready to give up incubating duties to its mate, but several times I have seen gulls and plovers get into a pushing match, with the relieving bird having to dislodge the previous incubator from the eggs. There must be some very interesting undercurrents in avian family life that we have not yet figured out!

In ducks, geese, hawks, eagles, hummingbirds, most owls, and many passerines, only the female incubates. In only a few species does the male alone incubate. This is the case with the phalaropes. After laying the eggs, the more brilliantly plumaged female leaves care of the eggs, and later the chicks, to the male, and heads off to breed with additional males. In species in which only a single parent incubates, the roles are usually stereotyped, with either the male or female always doing the incubation in a given species. A fascinating exception is the Eurasian Penduline-Tit. In this species, both sexes are capable of incubation, but only one parent takes on the duty. Both parents would like to leave incubation to their mate in order to pursue mating opportunities with other mates, and pair members appear to engage in a serious contest: Within the pendulous nest of vegetable felt (for which the bird is named), the female hides her eggs in the nest lining to keep the male from knowing when she has begun egg laying **(Fig. 8–95)**. The appearance of eggs might provide a clue that the male could use to determine the end of his mate's fertile period, and thus the best time for him to desert. By hiding her eggs, the female has a chance to sneak off first. In the Austrian populations studied by Valera et al. (1997), females appear to end up incubating their first clutch about 50 percent of the time and males about 20 percent, but pairs abandon the nest about 30 percent of the time. Yet another example of how wasteful natural selection can be!

Patterns of Attentiveness

The incubating bird soon adopts a fairly regular rhythm, with periods on the nest (**attentive periods**) alternating with periods off the nest (**inattentive periods**). For most passerine species in northern North America in which the female alone incubates, attentive periods

a. Eurasian Penduline-Tit at its Nest

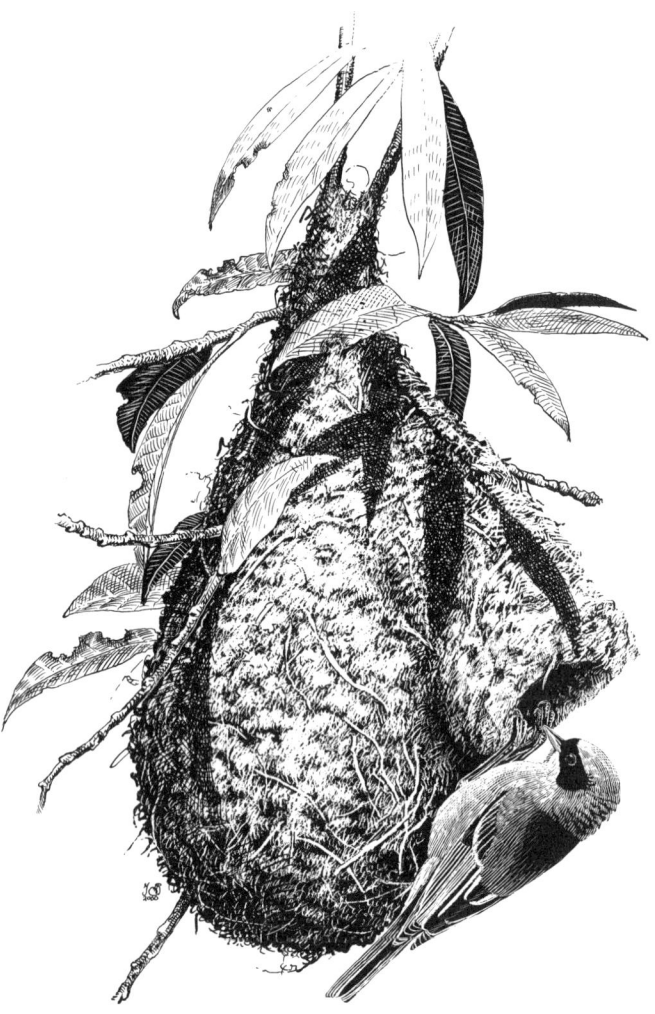

Figure 8–95. Egg Burial and Incubation Avoidance in the Eurasian Penduline-Tit: *The male and female Eurasian Penduline-Tit appear to engage in a contest, each trying to abandon the clutch first and leave its mate to choose between (1) incubating the eggs and rearing the young alone, or (2) abandoning the eggs and allowing them to die. Once free of its nest, the "winner" can find other mates and breed again.* **a. Eurasian Penduline-Tit at its Nest:** *Similar to chickadees but more closely related to the Verdin of the southwestern United States, the Eurasian Penduline-Tit builds a domed nest with an entrance tube. The nest has thick, felted walls of plant down, and is suspended from the tip of a branch that overhangs water—this "pendulous" nest giving the bird its name. Researchers believe that the ability of the nest to keep the eggs and young relatively warm allows a single parent to rear young successfully.* **b. Eggs Buried in Bottom Lining of Nest: Left:** *view of entire nest;* **Middle:** *section through nest with white arrow indicating eggs buried in lining;* **Right:** *schematic showing location of the three buried eggs. During laying, the female hides her eggs in the nest lining, apparently to keep the male from knowing that she has begun laying. Once he sees eggs, the male may be able to determine the end of his mate's fertile period, the time that he can safely stop guarding her and desert. By hiding her eggs, the female has a chance to complete her clutch and sneak off first. Although egg burial originally could have evolved to keep the eggs warmer, to prevent them from falling from the pendulous nest, or to deter predation or parasitism, the fact that females cover eggs only during laying, and uncover them if the male deserts or if he is removed, suggests that females currently bury eggs to hide them from the male. Furthermore, females aggressively try to keep males out of the nest during laying, and those who are better at covering their eggs are more likely to desert first, leaving the male in charge. When researchers uncovered eggs during laying, the males deserted on the same day. Photo and sketch in b, and all data and interpretations from Valera et al. (1997).*

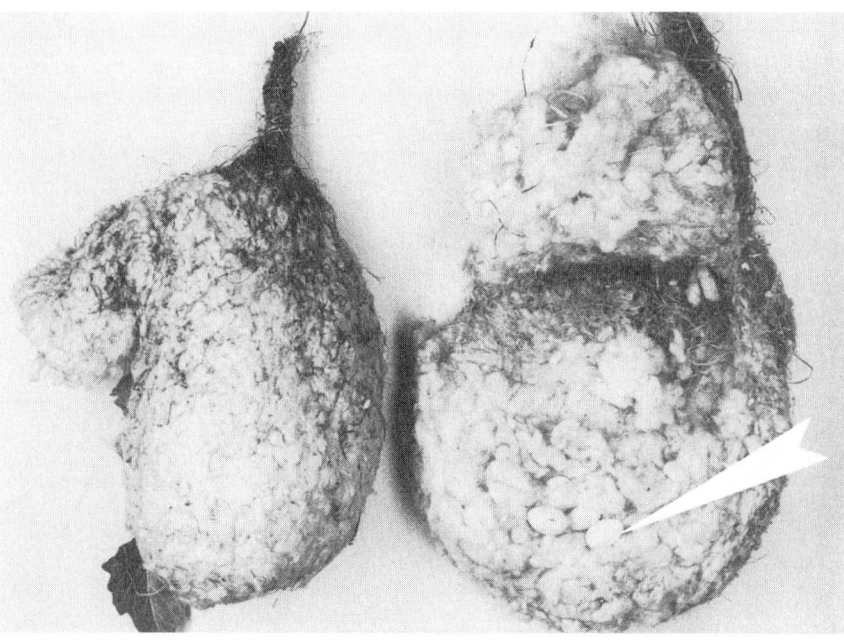

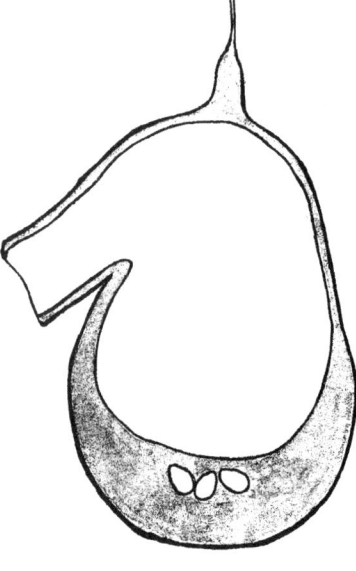

b. Eggs Buried in Bottom Lining of Nest

normally take up about 75 to 80 percent of the nesting bird's time. But many other incubating females, such as arctic-nesting geese, may spend more than 95 percent of their time on the nest.

Birds tend to have longer attentive periods when environmental conditions are harsh, both because the eggs need extra care and because foraging is less productive then; birds tend to leave when they will be able to feed efficiently and return promptly. Thus, most incubating passerines have longer periods on the nest at the beginning and end of the day (the overnight bout is usually uninterrupted), and whenever conditions become cooler and wetter outside the nest. The extreme example is the male Emperor Penguin, who incubates his egg through the latter half of the Antarctic winter in one uninterrupted stretch of 64 days, living off vast stores of body fat.

Species that forage far from their nest may also have longer attentive periods. In many seabirds, such as fulmars, shearwaters, and many petrels, each parent incubates for one to five days while its mate forages far away at sea. Short-tailed Shearwaters may have shifts of 12 to 14 days on the nest before their mates return to relieve them (Serventy 1967).

The duration of attentive bouts is largely determined by the incubating bird's need for food. Species in which the incubating bird is fed by its mate, such as goldfinches and crossbills, tend to have much longer attentive bouts than those in which it is not. And species with high metabolic demands may not have the luxury of long attentive bouts: incubating hummingbirds must leave the nest to feed at least 140 times per day. Predation risk for the parent and egg can also affect the length of attentive bouts (Conway and Martin 2000). At the beginning of any attentive bout, the bird is usually very quiet, often seeming to doze off for long periods, but as the bout wears on, the incubating bird grows more restless, changing its position, turning the eggs, or meddling with the vegetation in or around the nest.

Figure 8–96. Black-necked Stilt Turning its Eggs: *Some species, such as this Black-necked Stilt, turn their eggs with their beaks periodically throughout incubation. Egg turning may help to prevent the embryonic membranes from adhering to the inside of the shell, but embryos of species that do not turn their eggs apparently do not suffer from this problem. Photo by F. Truslow/VIREO.*

Behavior During Incubation

The incubating bird is usually quiet, leaving and returning to the nest secretively and deliberately, especially if it is the only bird incubating. Ground-nesting species leaving a nest often thread their way through the vegetation for several feet before flying up; species leaving elevated nests often drop nearly to the ground and fly level for a short distance before rising to their normal height. The incubating bird will sometimes raise its body slightly, reach down among the eggs with its bill, and gently turn them **(Fig. 8–96)**. Although this turning may help to prevent the premature adhesion of some embryonic membranes, embryos of species that do not turn their eggs, such as megapodes and palm-swifts, seem to suffer no adverse effects.

Figure 8–97. Clapper Rail Returning an Egg to its Nest: *Most ground-nesting birds will retrieve an egg accidentally displaced from the nest by rolling it under the bill. In contrast, the longer-billed rails, such as the Clapper Rail shown here, actually grab the egg in their bill and lift it back into the nest. Eggs beyond reach from the nest are generally ignored by most species. Photos courtesy of Olin Sewall Pettingill, Jr.*

In addition to turning eggs, birds nesting on the ground, such as gulls and waterfowl, will retrieve their eggs if they accidentally roll out of the nest but are still within reach from the nest. Usually, birds roll the eggs back into the nest under their bills, but the longer-billed rails lift the eggs back into the nest in their bills **(Fig. 8–97)**. Murres, whose diversely marked eggs are individually recognizable and laid on the bare rock of a nesting ledge (see Fig. 8–73), will retrieve their eggs from as far away as 25 to 30 feet (7.6 to 9.2 m)! Most other birds pay no attention to eggs that lie out of reach from the nest. Species that build deep cup nests in cavities (for example, Tree Swallows) often ignore eggs outside the nest, even when they are well within sight inside the nest cavity.

Figure 8–98. Female American Woodcock Remaining on Nest: During incubation, many well-camouflaged birds remain on the nest until an intruder is very close—slightly risking their own safety, but relying on their cryptic plumage to render both themselves and their eggs nearly invisible. As shown here, American Woodcocks may not flush until actually touched. A bird's tendency to remain on the nest increases as the incubation period progresses because the bird's investment of time and energy in the nest also increases, making the nest more "valuable" to the bird. (See text for more information.) Photo courtesy of Alfred O. Gross.

Changes in Incubation Behavior

As the incubation period advances, the behavior of the incubating bird changes. Nuthatches, usually noisy, become quiet and secretive. Owls, raptors, gulls, and shorebirds that defend their nest (for example, lapwings) become much more belligerent as hatching approaches. Birds that are cryptically colored remain longer on the nest in the face of intruders: late in incubation, American Woodcocks sometimes do not flush from the nest until touched **(Fig. 8–98)**. Evolutionary biologists have interpreted these changes as expressions of the increasing value of the clutch to the incubating bird as it nears hatching. Not only are eggs near hatching more likely to produce fledglings than eggs just laid, but a replacement clutch becomes less likely to succeed as the breeding season wears on. Conversely, it makes evolutionary sense that birds are much more likely to *abandon* a breeding attempt during nest building, laying, or early in incubation than after the young have hatched (Winkler 1991). However, birds will abandon the clutch or brood at any stage in extreme emergencies—hurricanes, floods, fires, and so on—leaving their young behind, but saving their own lives, and preserving their capacity to breed again. Birders and researchers observing nests need to keep these patterns of nest abandonment in mind, keeping disturbances to an absolute minimum, especially early in the nest-building cycle. But be aware that visiting a nest at any time may leave scents or other cues that may attract predators and lead to the nest's demise.

Feeding the Mate

You may have seen a pair of Northern Harriers, Red-tailed Hawks, or White-tailed Kites perform a food exchange ceremony in which the female flies from the nest to receive food from the male in midair, either taking it directly from his talons or catching it as it falls **(Fig. 8–99)**. These stereotyped displays have long been considered important pri-

Figure 8–99. Aerial Food Exchange in White-tailed Kite Pair: *Males of many bird species feed their mates during the breeding season, with some of the more spectacular food exchanges occurring in raptors. Male White-tailed Kites offer prey to females before egg laying in an impressive aerial maneuver in which the female flies up to meet the male, then turns upside down and grabs the prey from his talons—as shown in this photo. They also may exchange prey while perched. These elegant birds of prey inhabit open areas in the southwestern United States, Mexico, and South America, and feed primarily on small rodents. Photo by Cliff Beittel.*

marily in helping birds establish a pair bond during courtship, and in maintaining it throughout parental care. But increasing evidence suggests that females who receive more food from their mates lay larger clutches (for example, Nisbet 1977) **(Fig. 8–100)**. Mate feeding probably makes it possible for a female to raise more young, by keeping her in good condition and allowing her to put more energy into feeding the young.

Development of Young

■ Whereas all birds hatch from a hard-shelled, terrestrial egg, species vary considerably in how well developed the young are when they hatch. This section explores this variation and its causes and implications relative to the broader nesting ecology of birds.

Hatching

In preparation for hatching, bird embryos develop an **egg tooth**—a short, pointed, calcareous structure on the tip of the upper beak (and sometimes the lower beak as well) (see Fig. 8–112). The avian embryo

Figure 8–100. Male Northern Cardinal Feeding His Mate: *In many bird species, such as the Northern Cardinals pictured here, the male feeds the female during breeding. This male (left) is giving seeds to the female, while both are visiting a feeder. As reflected by the term "courtship feeding" often applied to this behavior, researchers have long considered the primary function to be establishing and maintaining the pair bond during breeding. Increasing evidence suggests, however, that in some species females who receive more food lay larger clutches. Well-fed females also may stay in better condition throughout breeding, which allows them to raise more young by devoting more of their time and energy to feeding them. Photo by David G. Allen.*

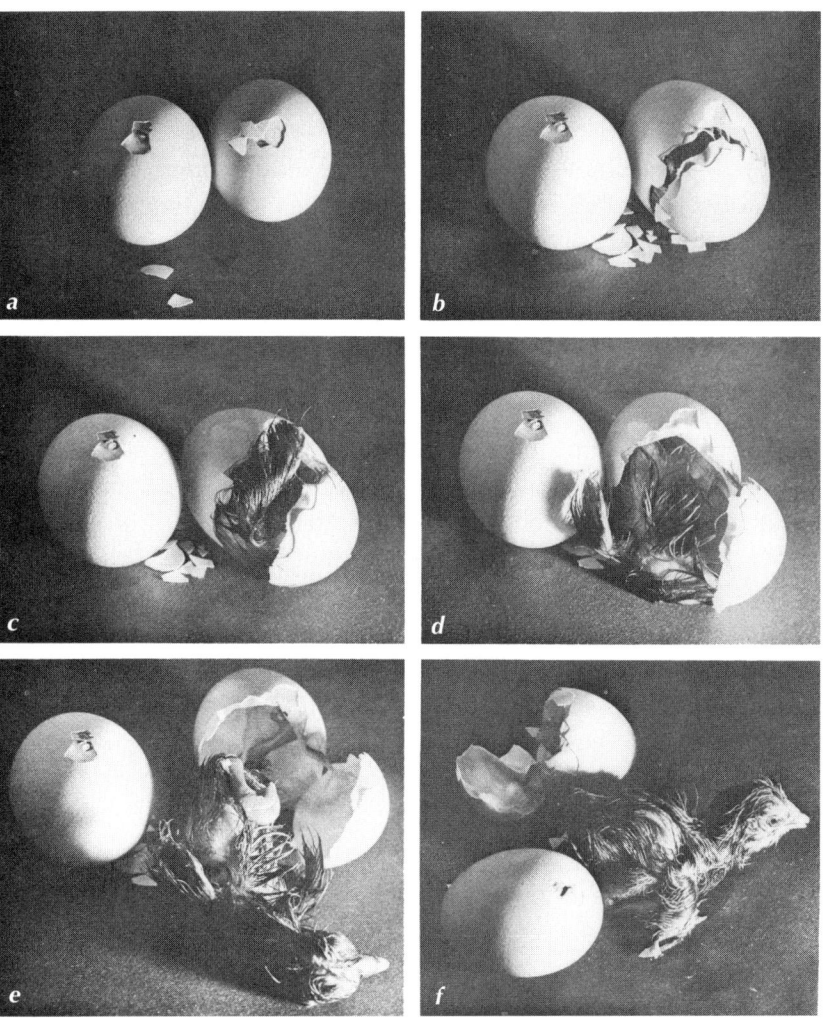

Figure 8–101. Domestic Chicken Hatching from its Egg: *(a) After about 19 days of incubation, the fully developed chicken embryo begins to push and scrape its egg tooth (see Fig. 8–112) against the inner wall of the blunt end of the eggshell. Eventually it punctures a small hole, and the egg is said to be **pipped**. (b) By turning and repeatedly pushing against the shell, the chick eventually punctures a series of holes—either connected or close together—nearly encircling the blunt end of the egg. (c) and (d) The chick's movements eventually push off the end of the eggshell, and (e) the chick struggles free, usually on its 21st day of incubation. (f) After emerging, the exhausted chick lies motionless, drying and apparently gathering its strength. Photos by Richard Nairin.*

is generally in a "fetal position" in the egg, with its head bent forward toward its belly. Near the end of incubation, the fully developed embryo pulls back and up with its neck (often with muscles developed especially for this task), rubbing its egg tooth against the inner wall of the shell, which has been weakened from the absorption of calcium by the developing embryo. Embryos of most species turn in the egg as they repeat this motion many, many times. Eventually, the chick manages to puncture a small hole in the eggshell (at which point the egg is said to be **pipped**), and soon thereafter the chick begins to puncture a series of holes—sometimes connected, sometimes not—which nearly encircle the blunt end of the egg **(Fig. 8–101)**. Once the shell is sufficiently weakened, the chick pushes off the end and struggles free of the egg membranes. This appears to be a time of great exertion for the chick, and most hatchlings lie quite still immediately after they hatch, drying off and to all appearances gathering strength for the long life-struggle ahead! Once dried, most nestlings are about two-thirds the weight of the fresh whole egg from which they started.

The length of time between pipping and hatching varies: many passerines complete the process in five hours, but some of the larger seabirds may take as long as four days. Within a few days after hatching, the egg tooth sloughs off or is resorbed by the growing chick.

In many precocial species, the chicks are able to fine-tune the synchrony of their hatching times quite precisely by vocalizing to one another from inside the egg, beginning a day or two before hatching. Their calls—rapid clicks produced several times per second—vary in speed according to the hatching stage of the young. Researchers have found that physical contact among the eggs in the nest appears to facilitate vocal communication among the unhatched chicks, but there is still much to learn about the exact mechanism by which the embryos affect each other's growth rates.

Development at Hatching

The literature on the stages of development of young at hatching is rich (and confusing). Newly hatched birds vary widely in their extent of feathering, powers of thermoregulation and locomotion, and degree of dependence upon the parents for feeding and protection. This diversity has inspired ornithologists to construct complicated taxonomies to classify chicks into various "types" of development (see recent review in Starck and Ricklefs 1998), but many species do not fit very well into these pigeonholes, no matter how detailed they are made. Every hatchling is a complex mosaic of organ and sensory systems, each of whose degree of development has been molded largely independently by generations of selection to fit the hatchling to the environmental challenges it will face. In general, structures develop when the young will need them, and the behavior of the developing young is the result of a complex interaction between what its structure and function allow it to do and what it is favored to do by conditions in its environment. That said, this section concentrates on the two most common and familiar types of development—altricial and precocial—pointing out some of the interesting variations along the way **(Table 8–4)**.

Table 8–4. Altricial Versus Precocial Features: Characteristics generally found in young classified as either altricial or precocial. Note that these categories are constructed by humans for their convenience, and that in reality nearly every young bird has a unique blend of these features, each developed to a different degree, which creates a continuum of types of young. Adapted from Gill (1990, p. 368)

CHARACTERISTIC	ALTRICIAL YOUNG	PRECOCIAL YOUNG
Eyes at Hatching	Closed	Open
Down at Hatching	Absent or Sparse	Present
Mobility at Hatching	Immobile	Able to Walk/Swim Soon
Parental Care	Very Dependent on Parents	Vary in Parental Care Needs
Egg Size	Smaller (4–10% Female Weight)	Larger (9–25% Female Weight)
Neuromuscular Coordination at Hatching	Poor	Fairly Sophisticated
Muscles for Heat Production at Hatching	Rudimentary	Well Developed

Typical Altricial Young

Altricial Young: Appearance

The newly hatched altricial young appears to be all abdomen and head, with two large eyes bulging against closed lids. Once the bird lifts its head on its wobbly neck and **gapes**, as happens soon after hatching, its open mouth reveals a seemingly bottomless pit between the two eyes **(Fig. 8–102)**. On either side of the mouth, extending from the corner and tapering toward the tip of the bill, are swollen **flanges** supplied with tactile nerve endings (see Figs. 3–36 and 3–37). Touch one of these flanges and the nestling's mouth springs open with the energy of a mechanical toy. The colors in the area of the mouth are often bright and contrasting—"food targets" for the parent birds. The flanges are white or vivid yellow; the lining of the mouth is often a vibrant red, orange, or yellow. Some nestlings also have colored patterns on the roof of the mouth or on the tongue (see Fig. 7–65b). In nestlings confined to nests in cavities, the colors around the mouth tend to be more intense. The combination of the nestling's instant response to a touch on the flanges and its colorful, patterned mouth stimulates and guides the feeding behavior of the parent bird **(Fig. 8–103)**.

No one is likely to describe a young altricial bird as pretty, cute, or even attractive **(Fig. 8–104)**. The nestling's skin, mostly pink from the muscles and blood vessels beneath it, is very thin and quite oily. The viscera and almost empty yolk sac show through the skin of the extended abdomen. If the nestling has any down, it is usually most abundant on the top of the head and on the back, and it springs from the papillae that will eventually produce the juvenal and adult feathers. The color and length of the down varies with the species **(Fig. 8–105)** and may or may not follow the outline of the feather tracts (for information on feather tracts, see Fig. 3–1). The juvenal plumage soon pushes out the down, often retaining comical wisps of the natal down on the tips of juvenal feathers long into development (see Figs. 3–24d and 3–26). The juvenal feathers begin development beneath the skin,

(Continued on p. 8·112)

Figure 8–102. Begging American Crow Nestlings: *Hungry altricial nestlings gape widely, delivering a clear message to their parents, who may view these expectant throats as bottomless pits. Photo by Scott Camazine.*

Figure 8–103. Northern Cardinal Feeding Goldfish: *Parental feeding is often stimulated by the colorful, gaping mouth of the young. Sometimes, however, that mouth may belong to a member of another species, or even another class! This male Northern Cardinal was observed for several days delivering mouthfuls of worms to goldfish gaping at the edge of a garden pool. Perhaps the cardinal first approached the pool for a drink and was met by the open mouths of the fish, accustomed to being fed by people. Photo by Paul Lemmons.*

Adult Cedar Waxwing

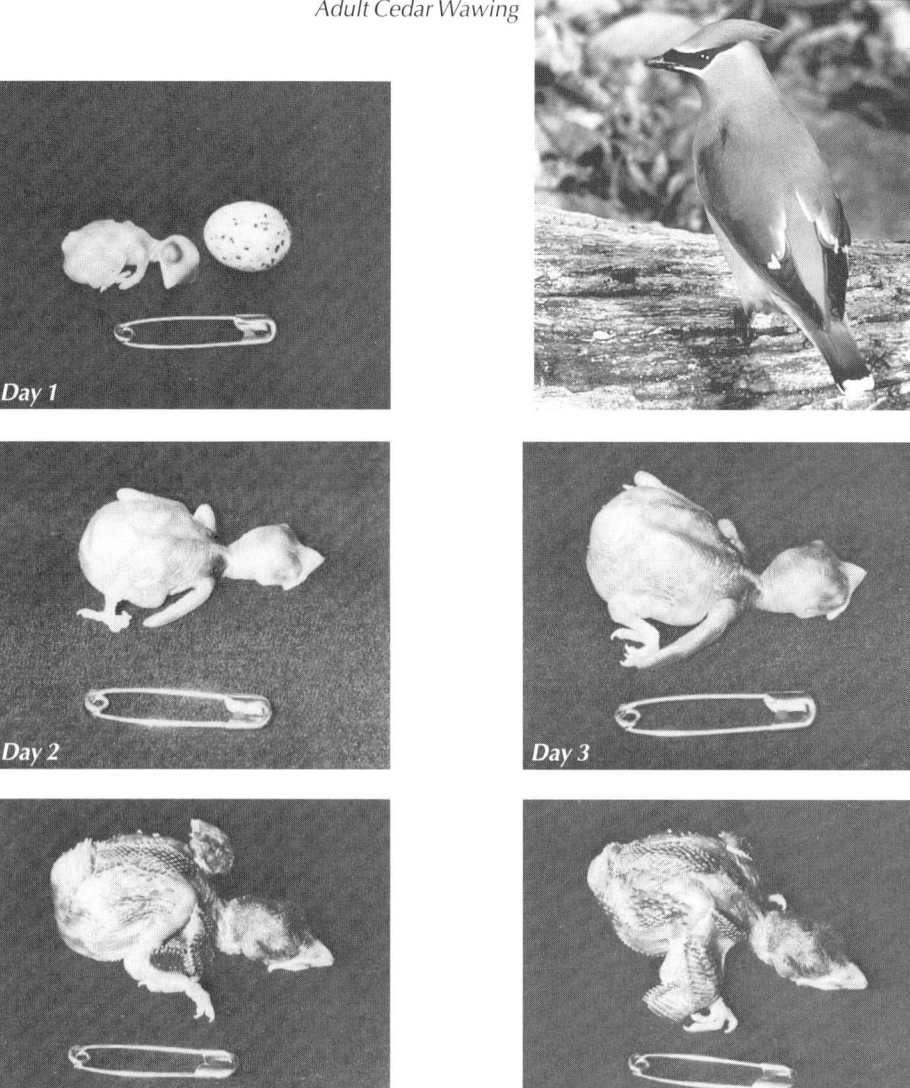

Figure 8–104. Development of an Altricial Nestling (Cedar Waxwing): *Altricial young, such as the Cedar Waxwing pictured in this sequence, develop rapidly from a tiny, weak hatchling into a fully feathered nestling that will continue to be fed by its parents for several weeks after leaving the nest. The safety pin, 1.75 inches (4.45 cm) long, is included as a size reference.* **Day 1:** *The newly hatched chick is all abdomen and head, with closed, bulging eyes. Weighing just 0.1 ounces (3.1 g), it makes no noise and has thin, transparent, pink skin. Although some altricial nestlings have down, the Cedar Waxwing remains naked until its juvenal feathers appear.* **Day 2:** *Although present from hatching, the swollen flanges at the sides of the mouth are more apparent now, and the mouth springs open if the flanges are touched. The nestling can barely raise its head, but it gapes in response to stimuli such as sounds, shading, or vibration of the nest. The skin on some feather tracts appears darker as feathers develop beneath.* **Day 3:** *For the first three days, the female primarily remains with the young, brooding them, while the male brings food. On Day 3, the parents begin a gradual transition in the food type they bring to the young—from mostly insects to mainly fruit, and the female begins feeding the young more*

often. **Day 4:** *The feather sheaths on the back, wings, and tail are now poking through the skin as black nubs termed pin feathers. The chick can grasp with its feet and make uncoordinated motions with its wings.* **Day 5:** *The pin feathers are now visible on most feather tracts.* **Days 6, 7, and 8:** *The ventral pin feathers poke through the skin, and the back and wing feathers continue to grow.* **Day 9:** *The nestling begins to vocalize and can balance on a perch. The eyes open sometime between days 6 and 9.* **Day 10:** *The nestling gives begging cries when a parent visits the nest. It can stretch its neck, preen, and flap its wings. It begins to defecate over the rim of the nest, and the female no longer broods during the day.* **Day 11:** *The feather sheaths disintegrate on day 9 or 10, ending the pin feather stage.* **Days 13 and 14:** *Their juvenal plumage nearly complete, but with short wing and tail feathers, the young begin flapping their wings while perched on the rim of the nest. The female no longer broods them at night.* **Day 15:** *The chick will fledge in about one day.* **Top Right:** *The elegant plumage of adult Cedar Waxwings looks velvety, and the secondaries have waxy, red tips. Developmental information from Witmer et al. (1997). Photo sequence courtesy of Olin Sewall Pettingill, Jr. Inset photo courtesy of J. Robert Woodward/CLO.*

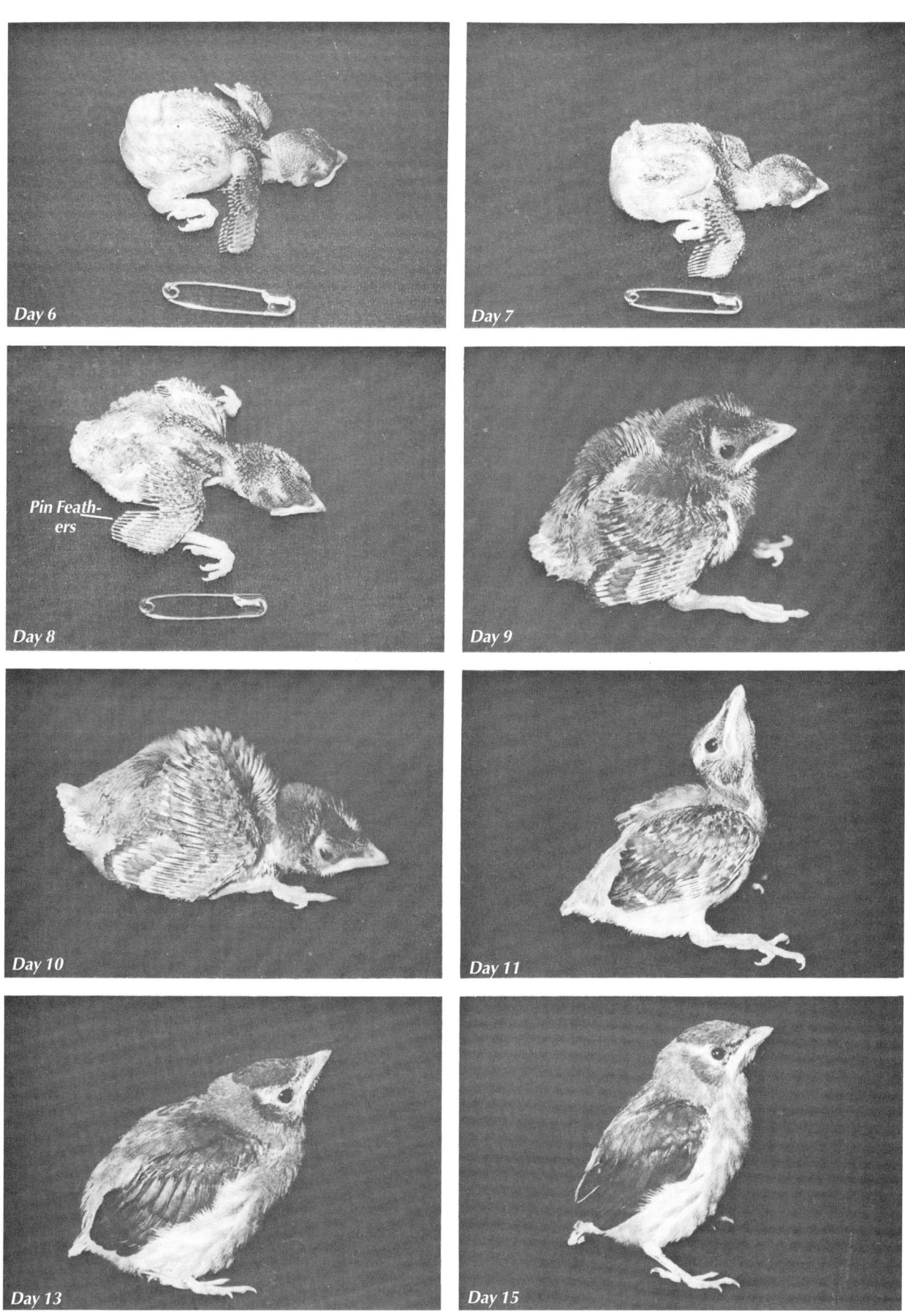

Day 6

Day 7

Pin Feath-
ers

Day 8

Day 9

Day 10

Day 11

Day 13

Day 15

8

Emperor Penguin

Double-crested Cormorant

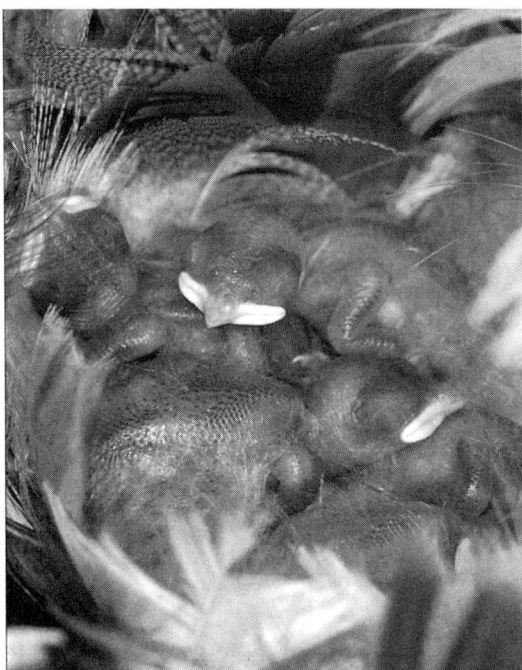

Tree Swallow

Mourning Dove

Great Horned Owl

Figure 8–105. Altricial Nestlings: *Young birds vary tremendously at hatching, each having a unique blend of characteristics that places it somewhere along a continuum between altricial and precocial. Nevertheless, nestlings are generally considered altricial if they hatch with eyes closed, down feathers sparse or lacking, and little ability to walk around or thermoregulate, and are highly dependent on their parents. Within this group, however, nestlings vary widely in their amount of down, rate of development, and degree of dependence on their parents. Some, such as hawks, boobies, herons, and penguins, remain in the nest even though they are physically able to leave at a fairly early age. Penguin offspring also are able to thermoregulate from a very young age. Photo credits: Cedar Waxwing, American Bittern, and Great Horned Owl courtesy of Isidor Jeklin/CLO. Rose-breasted Grosbeak and Red-winged Blackbird courtesy of Michael Hopiak/ CLO. Tree Swallow and Double-crested Cormorant by Marie Read. Mourning Dove courtesy of U. F. Hublit/CLO. Common Kingfisher by David Boag/Oxford Scientific Films. Short-tailed Hawk courtesy of J. C. Ogden/CLO. Blue-footed Booby courtesy of Christopher Crowley/CLO. Emperor Penguin by D. Tipling/VIREO.*

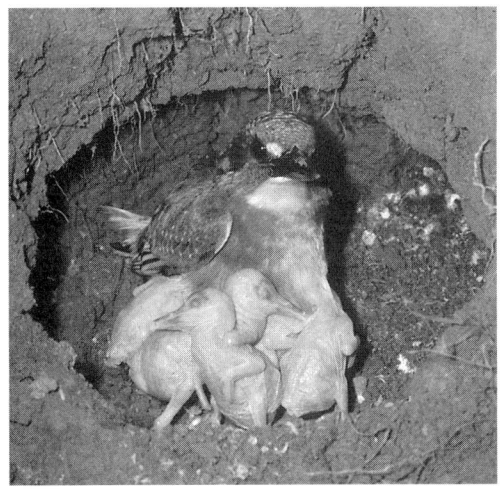

Common Kingfisher

Rose-breasted Grosbeak

Cedar Waxwing

Short-tailed Hawk

Red-winged Blackbird

Blue-footed Booby

American Bittern

each in its own papilla, and as the feathers receive their pigment, the feather tracts appear as dark bands beneath the skin. The papillae enlarge and the feathers, encased in sheaths, break through the skin as stiff, coarse "pin feathers." These appear first in the feather tracts of the head, back, and wings, and finally on the underparts. You may want to review the section Development of Feathers, in Ch. 3. Once the feathers emerge, they grow steadily throughout the nestling period, with the wing feathers, especially, growing continuously until full powers of flight are achieved at or after fledging.

Altricial Young: Temperature Regulation

The newly hatched nestling is unable to regulate its body temperature. It cools to the surrounding temperature when the brooding parent leaves the nest, essentially functioning as though it were "cold-blooded." By brooding it, the parents keep the nestling's body temperature high, near the levels at which digestive and growth enzymes work best. As the nestling gets larger, its surface-to-volume ratio becomes more favorable for heat retention (see the discussion of this phenomenon in Ch. 3, Sidebar 2: Feather Facts), and it begins to grow an insulative feather coat—the juvenal feathers. Eventually, the nestling can thermoregulate itself using the food energy brought by its parents. Most passerine parents begin to reduce brooding around the time the nestling's eyes open, and most altricial young have acquired full powers of thermoregulation by about the time their wings are half grown.

Altricial Young: Weight Gain

After an initial lag of a day or two, during which the digestive system gears up, the nestling develops explosively, often doubling its mass several times in the first ten days. For example, an American Crow weighing only 0.55 ounces (15.6 grams) at hatching weighs 11 ounces (315 grams) just eighteen days later (Ignatiuk and Clark 1991)—a twenty-fold increase. Growth tends to slow later in nestling development **(Fig. 8–106)**, so that by the time most passerine young leave the nest, they are about 70 to 80 percent the weight of their parents. Although they may lose a bit of weight while learning to fly, they generally continue to gain weight slowly until they are completely independent of their parents. (One very important point! Once they reach independence, all birds are essentially the same size as their parents. People often wonder if young birds can be recognized by their size; however, if a young bird flies and looks much like an adult, it is the same size as well **[Fig. 8–107]**.) In some species, the nestlings put on so much weight late in development that they weigh more than their parents. In some cases, this stored energy may help to nourish young birds through the demanding period of learning to forage, which they often endure entirely on their own. In other species, however, the fat stores are mostly depleted by the time the young leave the nest. An extreme example is seen in the altricial young of some pigeons. The extinct Passenger Pigeon is an excellent example: young weighed substantially more than the adults part-way through their development, and they left the nest well before they could fly. Their weight subsequently

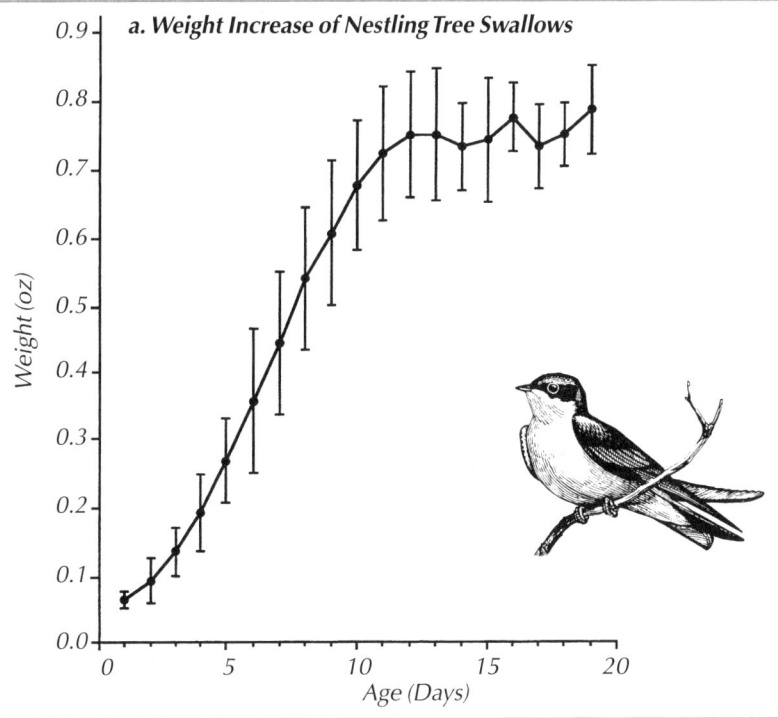

a. Weight Increase of Nestling Tree Swallows

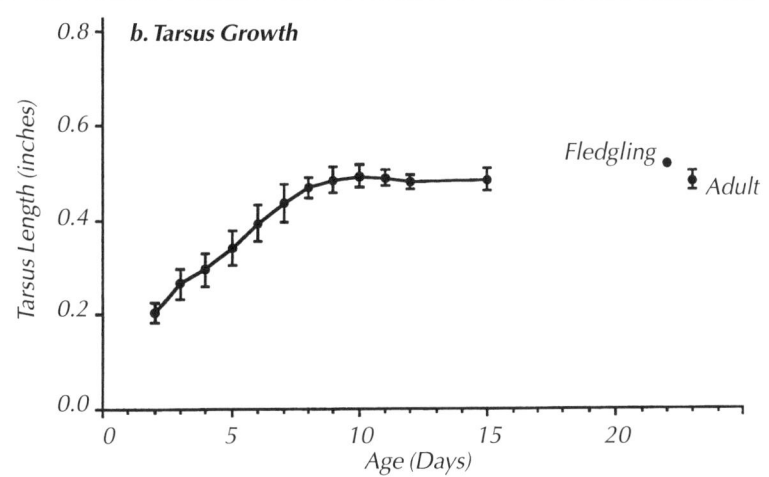

b. Tarsus Growth

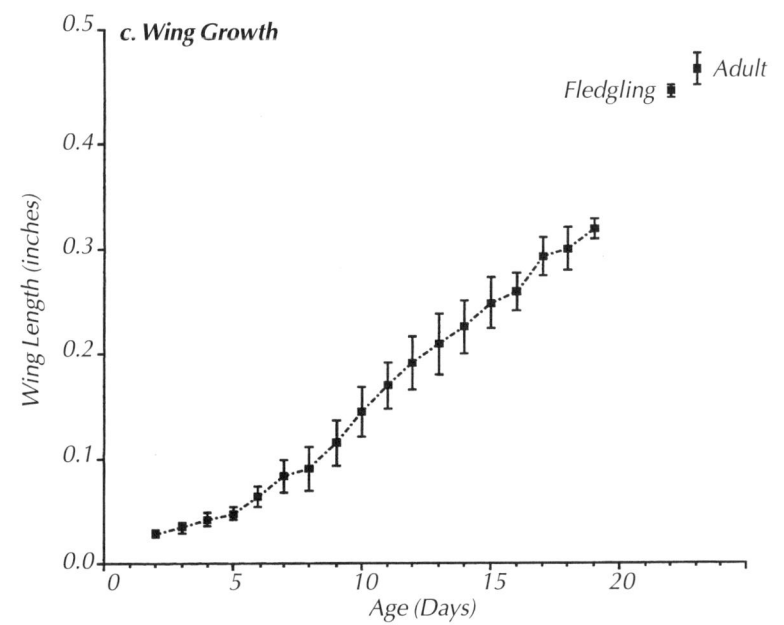

c. Wing Growth

Figure 8–106. Growth of Nestling Tree Swallows: *Most young birds grow rapidly, as illustrated by these measurements of nestling Tree Swallows in Ithaca, New York. Graphs b and c include an average measurement for fledglings and adults, for comparison. Error bars indicate one standard deviation around the mean.* **a. Weight Increase:** *The growth rate is slow for the first one or two days as the digestive system gears up, then it increases rapidly, levelling off when the nestling is about the weight of an adult—well before fledging, which occurs between days 18 and 22. This slightly S-shaped growth curve is typical of all altricial birds. Precocial young may actually lose weight during the first few days, while they digest their remaining egg yolk, but once they learn to forage efficiently, they gain weight in a pattern similar to that of altricial young.* **b. Tarsus Growth:** *The length of the tarsus (the section of the foot between the heel and toes) increases steadily from an early age, growth stopping when the adult size is reached, well before fledging.* **c. Wing Growth:** *The length of the wing section from the wrist to the tip of the longest primary feather, which depends on both bone and feather growth, was measured. During the first five days, the steady increase in wing length results primarily from growth of the wing bones. From the fifth day on, wing growth results almost entirely from feather growth, as the primary feathers begin to emerge from their sheaths and lengthen rapidly. Wing length continues to increase after fledging, even though Tree Swallows leave the nest with better-developed wings than many other altricial birds, probably because of their great reliance on flight and their nearly total independence at fledging. (Note that very old nestlings were not measured because disturbance at the nest causes them to fledge prematurely.) All graphs adapted from McCarty (1995).*

declined as they used their fat stores until, by the time they could fly, their weight approximated that of their parents. What functions, if any, these weight gains and losses play in development is not clear.

Altricial Young: Sensory and Motor Development

As the nestling's body develops, so do its sensory and motor abilities. At hatching, the nestling is all but helpless. It can stay upright only by leaning against the nest. Nevertheless, it gapes, swallows, digests, and defecates: the four behaviors crucial to obtaining and converting food into a rapidly developing young bird. It can also sleep, which it alternates with feeding. Most species cannot even give begging calls their first day or so, but these soon develop—getting louder and more persistent as the nestling grows. Gradually, it develops a repertoire of three or four calls that it uses in the nest. The nestling also seems barely aware of the parent's existence in the first couple of days, and the parent must encourage it to eat, coaxing food into its mouth. But by the time the eyes open, about one quarter of the way to fledging, the nestling gapes at visual stimuli (the adult, a human hand, or even forceps holding food), and it can begin to grasp objects with its feet. Balance is still poor, however. When defecating, the nestling raises its posterior and moves it from side to side; many species also back up to the edge of the nest, making it easier for the parents to grab and dispose of the fecal sacs. A few days after the eyes open, most nestlings develop a crouching response to strange visitors at the nest, and in many species the young soon develop a crouching reaction to parental alarm calls. Whereas the parents initially must place food deep in a young bird's throat to enable it to swallow, by the time most of the feathers are emerging the young is usually able to swallow food placed anywhere in its mouth; at the same time, its begging behavior can become quite aggressive. As the feathers continue to unfurl from their shafts the nestlings begin their first preening movements, and they often begin stretching their wings up, to the side, and back (**Fig. 8–108**).

Figure 8–107. Adult and Juvenile Northern Gannet: Once they are able to fly, most juvenile birds are the same overall size as adults, although their wing and tail feathers may be slightly shorter. Juveniles of many species, however, may be recognized by their distinctive plumages, as illustrated by this juvenile (left) and adult (right) Northern Gannet on Bonaventure Island, Quebec. Photo by Wolfgang Kaehler.

Figure 8–108. Nestling Golden Eagle Stretching its Wings: Once their feathers are partly unfurled from the sheaths, nestlings begin stretching their wings, as demonstrated by this young Golden Eagle. Note the light-colored feather sheaths visible about halfway between each wrist and wing tip. These contain the remaining feather vanes (closer to the bases of the feathers), which will soon emerge. Photo courtesy of Rick Kline/CLO.

Altricial Young: Fledging

Most passerines nesting in the open leave the nest long before they are capable of sustained flight **(Fig. 8–109)**. Many leave their relatively vulnerable nest and disperse into dense foliage, often letting the parents know their location through a series of soft calls. By spreading out in vegetation and exercising some cryptic and evasive movements, the brood is safer than it was in the nest. The fledglings sit on a protected perch, usually begging for food only when a parent gets quite close. While the feathers continue to develop, the fledglings are generally fed less and less often until they are finally independent of parental care and can fly and maneuver unaided through dense vegetation. During this transition to independence, the fledglings develop an enormous number of new skills for finding and catching food. They begin wiping the bill, pecking at objects, picking up food, catching insects, working at grass seedheads, scratching the ground, and probing in bark furrows. They also begin bathing. Parents sometimes help fledglings improve their prey-catching skills—raptors such as Peregrine Falcons may bring back live prey for their young to capture and kill, Northern Harriers may drop mice for their young to catch in midair, and kingfishers may drop recently killed or battered fish into the water as easy targets for their fledglings' first few dives **(Fig. 8–110)**.

In hole-nesting species, the transition to independence can be abrupt, with young learning the needed motor skills without parental feeding visits to keep them going. I have sometimes watched young Tree Swallows, leaving their nest for the first time, fly over the horizon and never look back!

a. Blue-winged Warblers

b. Blue Jay

Figure 8–109. Fledgling Songbirds: *The altricial young of passerines usually leave the nest with short wing and tail feathers, long before they can sustain flight. Most perch in a hidden location, waiting for their parents to bring food, and beg only when a parent approaches closely.* **a. Blue-winged Warbler:** *Leaving the nest just 8 to 11 days after hatching, Blue-winged Warbler young can at first only make short flights with their tiny wings. This photo, probably taken in the first hour after fledging, provides a rare glimpse of siblings still together. The young typically disperse almost immediately after leaving the nest, the parents each caring for a subset of the young. Photo from CLO collection, probably by Arthur A. Allen.* **b. Blue Jay:** *Nestling Blue Jays fledge at 17 to 21 days, and can fly and land well about one week later. The young, fed by their parents for another 1 to 2 months, beg with gaping beak and quivering wings, as shown here. Photo courtesy of Laura Riley/CLO.*

Figure 8–110. Fledgling Belted King-fishers Learning to Dive for Fish: Adult Belted Kingfishers may help their fledglings learn to dive for food by dropping dead or injured fish into the water for them. Here, the parent (top right) has just dropped a dead minnow, which floats on the water—an easy target for one fledgling. As he practices diving, his four siblings look on from a nearby perch. The young, which may be distinguished by their relatively shorter beaks, first capture crayfish and aquatic insects, not mastering the ability to capture live fish for more than a week after fledging. Behavioral information from Michael J. Hamas (personal communication).

Altricial Young: Length of Developmental Periods

Among altricial birds, the time from hatching to independence ranges from approximately 25 days to several months. The majority of passerine species, however, become efficient fliers at about 17 days and are independent of parental care by about 28 days from hatching. Very large altricial birds have the longest periods of nestling life: young Bald Eagles require 10 to 12 weeks in the nest before fledging; young Blue-and-yellow Macaws require more than three months; and young of the California Condor, one of the largest flying land birds, take as long as five months before leaving the nest.

The young of the largest penguins (King and Emperor) are brooded for about six weeks by their parents, then huddle close together with other young in large groups called **creches**. The parents feed them for 7 to 10 months, but the chicks leave before reaching their adult weight (Emperor chicks at only 60 percent adult weight). The parents must stop feeding the young in time to molt and gear up for the next breeding season.

The length of time that eggs and young spend in the nest is directly related to the safety of the nest, and there is a strong correlation between the lengths of the incubation and nestling periods within species **(Fig. 8–111)**. It is as though the growth rate of the developing

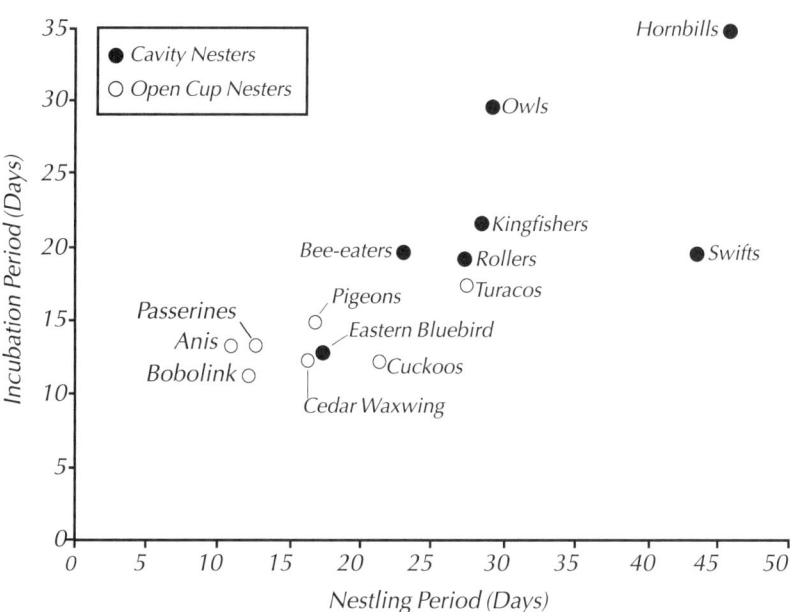

Figure 8–111. Correlation Between Average Incubation and Nestling Periods: The distribution of data points illustrates that birds with short incubation periods also tend to have short nestling periods, and those with long incubation periods also have long nestling periods. This correlation appears to be related to nest safety: birds nesting in secure sites, such as cavities and crevices (dark circles), tend to have longer incubation and nestling periods than those in open, exposed sites (open circles). The shorter developmental periods of open nesters allow the offspring to leave the predation-vulnerable nest in as short a time as possible. Size also affects the rate of development, however: smaller birds tend to have shorter incubation and nestling periods. Nevertheless, note that the relatively small swifts, which nest in various inaccessible sites, have quite long nestling periods. Also note that anis, somewhat larger than most passerines, have relatively short incubation and nestling periods—presumably because they nest low in trees. Data from Brooke and Birkhead (1991).

young has a single setting for both within and outside of the egg. For instance, many ground-nesting birds have both short incubation periods and short nestling periods. The rapid growth that allows the short incubation period continues throughout the nestling period, thus speeding the young bird through its time of greatest vulnerability to predators. In a bluebird-sized bird, for example, a nestling typically spends only 9 to 10 days in an open ground nest, but about 15 days in a less predation-vulnerable cavity nest. The nestling Song Sparrow in its nest near the ground is only slightly smaller than the nestling Cedar Waxwing in its open nest in a tree, yet it develops more quickly than the nestling Cedar Waxwing (which fledges when it is 14 to 18 days old), fledging as early as day 10 (see Fig. 8–104).

Typical Precocial Young

When shorebirds hatch, they are capable of impressive locomotion almost as soon as they dry off. I learned this during a summer of research on Snowy Plovers along the shores of Mono Lake, California. These birds nest at very low density in a vast expanse of basalt sand dunes and salt flats, and their principal defense against predation is to be visually swallowed up by the immense landscape in which they live. To study the relations between plover chicks and their parents we needed to colorband both, and we soon learned that we had to be present at a nest within an hour after hatching to successfully capture and band the chicks (Fig. 8–112). We must have formed an amusing spectacle as we tried to chase down some of the early broods that season, dashing pell-mell across the sand after little balls of fluff just a few hours old. Precocial young can be very precocial indeed....

(Continued on p. 8·120)

Figure 8–112. Newly-Hatched Snowy Plover Chick: Just a tiny ball of fluff weighing less than 0.2 ounces (6 g) at hatching, the Snowy Plover chick can run quickly within a few hours. Although its wings, tail, and bill are much smaller than those of an adult, its legs are almost full size at hatching. Note the prominent egg tooth at the tip of the upper beak; it will drop off within two days. Photo courtesy of Chris Swarth.

Royal Tern

Limpkin

Common Merganser

Common Moorhen

Figure 8–113. Precocial Young: *Although young birds of different species vary tremendously in their developmental stage at hatching, they are generally considered precocial if they hatch covered with down and with their eyes open, and are able to walk or swim and thermoregulate fairly well soon after hatching. Although the wings and tail of these young are small at hatching, the legs and feet are large and muscular—they will be used almost immediately. Note, especially, the large, strong legs of the young Sandhill Crane and Limpkin pictured here. Many precocial young leave the nest soon after hatching, wandering in search of food with their parents. Others, including auks (such as the Razorbill shown here), terns, and gulls, are fed by their parents and remain in the nest until they reach complete independence. Precocial young that feed themselves and thus do not beg to their parents lack the colorful mouth linings and flanges of altricial young. But even those that are fed by their parents have, at most, reduced markings and flanges. Photo credits: American Oystercatcher and Black Skimmer courtesy of L. B. Wales, Jr./CLO. Razorbill courtesy of C. W. Buchneister/CLO. Sandhill Crane courtesy of Marshall Delano/CLO. California Quail courtesy of Eva Schaefer/CLO. American Woodcock courtesy of Michael Hopiak/CLO. Royal Tern courtesy of Michael Costello/CLO. Common Merganser courtesy of Lee Kuhn/CLO. Common Moorhen courtesy of Jack Powell/CLO. Limpkin courtesy of Dana C. Bryan/CLO.*

Black Skimmer

Razorbill

American Oystercatcher

Sandhill Crane

California Quail

American Woodcock

Precocial Young: Appearance

Perhaps the most familiar image of a young bird is a newly hatched chicken. The chick's downy covering, wet from embryonic fluids, dries within two or three hours and then takes on the fluffy appearance that makes the chicks of precocial species so endearing **(Fig. 8–113)**. Precocial embryos spend much more time in the eggs than do those of altricial species, and thus precocial chicks are much better developed at hatching. The motor and sensory capabilities of a newly hatched chicken, in the egg 21 or 22 days, are about as well developed as those of a passerine that has spent only 12 days in the egg but 10 days in the nest. The precocial chick, with eyes wide open and bright, responds immediately to external stimuli, is remarkably adroit in walking and maintaining its balance, and begins exploring its environment with pecks at unfamiliar objects almost as soon as it emerges from the egg **(Fig. 8–114)**. This neuromotor sophistication is perhaps the major hallmark of precocial chicks.

At hatching, the precocial chick's abdomen still contains from one-seventh to one-third of the original contents of the yolk sac, and the digestive tract is essentially fully functional soon after hatching. The egg tooth is conspicuous, no flange (or a reduced one) is present along the margin of the bill, and the mouth lining is usually plain—similar to that of the adult—since most of these young feed themselves and thus have no need to send begging signals to their parents.

Most precocial chicks have large feet and legs at hatching, with muscles far more developed in these body parts than elsewhere in the body (for example, Ricklefs 1980b) **(Fig. 8–115)**. These highly developed muscles probably play a large role in generating heat for the thermoregulating chick (Ricklefs 1980b), and they are crucial to the mobility of precocial chicks, which often leave the nest soon after hatching. Most precocial of all birds, the well-named megapodes ("pod" is Greek for "foot") appear to be the only birds whose embryos lack an egg tooth and break out of the shell using their extremely large and powerful feet (Jones et al. 1995). Megapode chicks also have feathered wings at hatching, and they can fly within a few hours **(Fig. 8–116)**. Chicks so well developed can take care of themselves, and it is thus no surprise that they receive no parental care after hatching. They emerge from their incubation mounds and disperse into the surrounding forest, rarely seeing either of their parents. Except for brood parasites, no other birds have so little exposure to their parents.

Precocial Young: Temperature Regulation

All birds are warm-blooded, regulating their body temperature at about the level of a human fever (see Ch. 4, Body Temperature). Precocial chicks hatch with a thick covering of down—the best natural insulating material known. Precocial chicks are generally larger than the average altricial nestling, such as a sparrow, and thus, with a more favorable surface area-to-volume ratio, a precocial chick loses relatively less heat from its body surface than does an altricial nestling. In fact, it can partly control its body temperature at hatching (well before altricial young can do so), although it still requires some brooding.

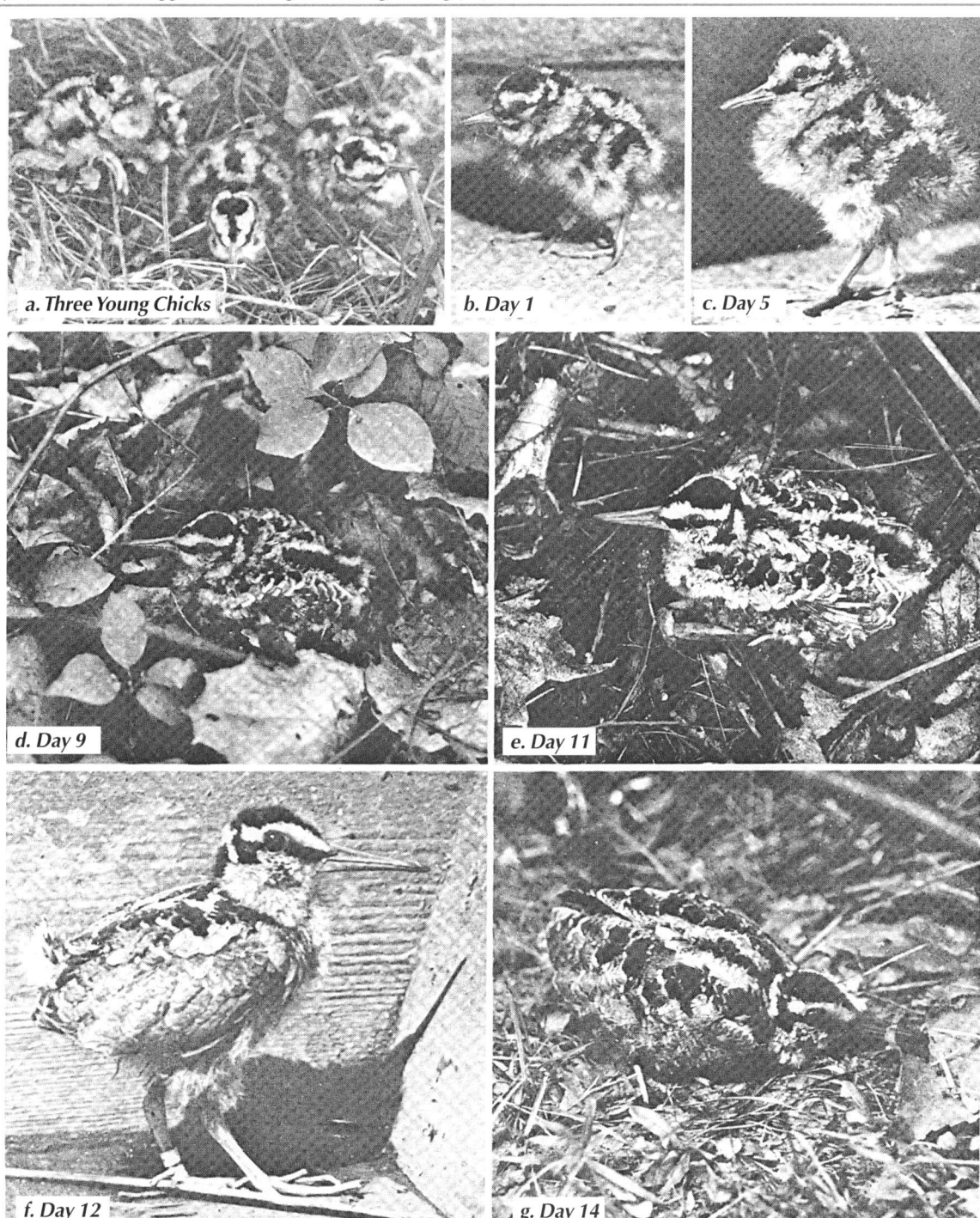

Figure 8–114. Development of Precocial Young (American Woodcock): *Precocial young, such as the American Woodcock pictured in this sequence, hatch downy and well developed. Most are ready to walk and find food within a few hours of hatching, although the degree to which they depend upon their parents varies widely between species. They develop much more slowly than altricial nestlings.* **a. Three Young Chicks:** *When the female brooding them flew off in alarm, these three downy chicks "froze" in position. Even the one (at left) accidentally turned onto its side by its departing mother held its position.* **b. Day 1:** *Weighing just 0.4 to 0.5 ounces (11 to 13 g), the newly hatched woodcock chick is wobbly but active, leaving the nest within a few hours. It does not eat the first day, relying instead upon its yolk supply.* **c. Day 5:** *The chick's down reached full length by day 3. Although it grabs earthworms from its mother's bill for the first week, it is now beginning to probe for food on its own, as well.* **d. Day 9:** *Juvenal feathers are beginning to replace the down, particularly on the wings, and the beak is noticeably longer.* **e. Day 11 and f. Day 12:** *The juvenal feathers on the back are emerging.* **g. Day 14:** *The tail feathers are still very short, but beginning to grow. Many feathers are completely out of their sheaths, but wisps of down, difficult to see in this photo, remain on the tips. The chick can make short flights, and will be able to thermoregulate by itself between days 15 and 20. It will reach full size around day 30 and become independent from its mother between days 31 and 38. Photos courtesy of Olin Sewall Pettingill, Jr.*

a. Crane Chicks   **b. Stork Chicks**

2 Days   32 Days   1.5 Days   29 Days

Figure 8–115. The Timing of Leg and Wing Development in Young Birds: The legs and wings of young birds develop at different ages in different species, depending on when they are needed. **a. Crane Chicks:** *Like many ground-nesting, precocial young that leave the nest shortly after hatching, crane chicks start out with very well-developed legs. At two days they are able to run, and by 32 days the legs have almost finished growing. At this age, however, the wing feathers have scarcely sprouted. The young cranes depend solely on their legs for locomotion until more than two months of age, when they finally are able to fly.* **b. Stork Chicks:** *Although similar to cranes in body shape, storks nest high on roofs or in trees, and the young remain in the nest for several months. In contrast to young cranes, a 1.5-day-old chick is blind and helpless, with neither legs nor wings well developed. By 29 days, however, the wing feathers have matured greatly whereas the legs are still poorly developed. A young nestling stork with functional legs might be in danger of propelling itself to a fall from the nest, but when it finally does leave, its first flight requires fully-functional wings. Adapted from Heinroth (1938).*

a. Malleefowl   **b. Australian Brush-turkey**

Figure 8–116. Megapode Chicks Emerging from Nest Mound: A long incubation period—50 to 70 days—allows megapode chicks to hatch in a more advanced state, in terms of both behavior and physiology, than any other birds. Within a few hours after hatching they can run, find food, regulate their body temperature, and fly. **a. Malleefowl Chick:** Lacking an egg tooth, the hatching megapode cracks open its eggshell with its feet, then spends hours—sometimes several days—digging upward through the mound of rotting vegetation that warmed the egg during incubation. This chick of the Malleefowl, an endangered megapode of southwestern Australia, emerges ready to take care of itself—a necessity because, like other megapode chicks, it never will see its parents. Drawing by Charles L. Ripper. **b. Australian Brush-Turkey:** Common throughout suburbs and national park picnic grounds in eastern Australia, the Australian Brush-Turkey is one of the easiest megapode species to see. This chick is just emerging from its incubation mound. Photo courtesy of Darryl Jones.

From then on, however, temperature control develops slowly, much more so than in altricial nestlings, and thus most precocial chicks do not attain full control until about four weeks of age.

The feather coat of each species develops in a pattern well suited to the life that the newly hatched chick must lead: upland species, such as grouse and shorebirds, probably more vulnerable to radiative heat loss upward, first develop feathers on the back and upper surfaces—and the feathers in these areas are also the longest; aquatic species, such as loons, grebes, and waterfowl, first develop long, thick feathers on the body's undersurface, which help to insulate the chicks while swimming on cold water. The last feathers to grow are the flight feathers and their coverts.

Precocial Young: Weight Gain

Most precocial young, particularly those that feed themselves, are inept at acquiring and ingesting food immediately after they hatch. For the first few days they sustain themselves primarily on the large amount of egg yolk still in their abdomens at hatching. They may actually lose weight during this transitional period, but once they begin to feed themselves efficiently, they gain weight in a pattern similar to that of altricial birds.

As chicks grow bigger, their metabolic rates also increase, and the similarity between precocial and altricial birds in their patterns of weight gain hides the striking differences in the way their metabolic rates increase. The metabolic rate of precocial species, unlike that of altricial species, has two distinct phases of increase: a rapid initial phase shifts quite abruptly into a slower phase about halfway through development (Weathers and Siegel 1995).

Precocial Young: Behavior Out of the Nest

Most precocial chicks leave the nest shortly after their feathers dry, reducing the chance of the entire brood being surprised and eaten by a predator. Thus the adult, rather than the nest, provides security for these chicks, and a strong social bond holds the adult and young together. Location calls keep family members in touch with one another; assembly calls made by the adult bring the young to them; alarm calls made by the adult or young instantly trigger a crouching or freezing response in the young; and distress calls made by the chicks draw the adult to their vicinity. The chicks of some species have additional calls—food calls if the adult still feeds them, and contentment calls if they are comfortable. As the chicks mature, a social hierarchy develops within the group, and siblings often challenge each other with threat calls. In large precocial species, especially those, such as Canada Geese, in which the parents can defend the chicks against most potential predators, a family may stay together all winter (for example, Raveling 1970). In smaller species the family bond is weaker, and large groups of young from several broods may follow first one female and then another. Thus, the flotilla of ducklings paddling along behind a female Common Eider or Red-breasted Merganser may or may not have hatched in her nest.

In addition to helping avoid predation, leaving the nest soon after hatching allows developing young to move to areas of abundant food. Over the course of their development, for example, goslings (young geese) often walk many miles from their nest site in search of rich grass and other nonwoody plants; many species of ducklings gather on particularly productive water bodies; sandpiper and plover chicks congregate with their parents along rich stretches of shore; and groups of young Ostriches and rheas often wander widely through their desert and steppe habitats with their parents, searching for plentiful food **(Fig. 8–117)**.

Figure 8–117. Ostriches: Although resident year round in some of the less arid regions of East Africa, in deserts Ostriches may roam widely looking for water and food—the leaves, flowers, and seeds of various plants. As shown by this group in Kenya, "family" groups often consist of a female (left) and male (right) accompanied by numerous young. Often the adults are unrelated to many of these young, in part because many female Ostriches may lay eggs in the nest of one male, which is tended by the male and his "major hen," and also because hatched broods may merge to form large groups that are attended by just one pair. Photo by P. Davey/VIREO.

Precocial Young: Remaining in the Nest

Despite the significant advantages to leaving the nest, there are quite a few species that are precocial in the degree of their development at hatching, but stay in the nest. Best known are the young of gulls, which are raised on secluded islands with few nest predators. Chicks of most gull species stay in the nest right up until they become independent of their parents and leave the nest on their own. Within the gulls, it is interesting that chicks vary in their tendency to wander from the nest when predators come near. Herring Gulls raised on flat land are often quite willing to abandon the nest site and seek cover in nearby brush or boulders, returning once the predator is gone. Those raised in cliff sites, however, will not abandon the nest site, even when a predator is right nearby (Emlen 1963). Any tendency to wander from a cliffside nest is probably strongly selected against!

Young albatrosses are physically capable of leaving the nest on their first day, but the single nestling remains at the nest with its parents for many months. Although it may wander a few yards away during the day, it generally returns to the nest at night and for food. Young of the Wandering Albatross, one of the largest flying seabirds, exceed all other birds in the length of nestling life—about 9 to 12 months. Since the adults must molt between breedings, they can breed only every other year.

Precocial Young: Length of Flightless Period

The amount of time before precocial young can fly ranges in gallinaceous birds from less than 1 day (megapodes) to two weeks; in shorebirds from 14 to 35 days; and in waterfowl from approximately 1 to 4 months. There is a wide range in these three groups because each includes large species that are slow to attain flight.

Recognition Between Parents and Young

Contact between adults and their young begins before the young hatch. Young near hatching age can be heard peeping while still inside the eggs, and they can apparently hear their parents as well, because many stop peeping at a vocal signal from the adult. Young of some species imprint on their parents' voices at this time. Northern Bobwhite embryos, for example, learn to recognize the call of their own mother while still in the egg (Lickliter and Hellewell 1992).

Adults may or may not recognize young as their own. Cowbird hosts accept the young parasites, and one can interchange the young of most passerine species readily—the adult feeds anything gaping in the nest. Adults of many nonpasserine species—especially colonial-nesting birds—are not as easily fooled. For example Herring Gulls, for two or three days after their young hatch, may accept tiny young from other nests as their own. But after that, when the adults have become accustomed to certain nestlings in their own territory, they refuse any others. In fact, they chase and peck a stray young until it leaves their territory or dies.

When they first hatch, few young birds recognize their parents. Altricial nestlings will beg from nearly any bird that comes to the nest—even from a hawk—unless they are warned by their parent or frightened by the size of the newcomer. As they grow older, however, nestlings learn to recognize their parents.

Young of many colonially-nesting birds, however, recognize the calls of their own parents at an early age—allowing them to quickly prepare for food when their parents arrive, and perhaps in some species, preventing them from begging from the wrong adults, who might react aggressively. Laughing Gull chicks recognize their parents' calls by 6 days of age (Beer 1969); Thick-billed Murre chicks, by 3 days (Lefevre et al. 1998); and Piñon Jay nestlings, by at least 14 days—7 days before fledging. In Piñon Jays, nests are probably most obvious to predators when the young are begging and receiving food, so any reduction in that time probably decreases the likelihood of predation (Balda and Balda 1978). In swallows, the young and parents of colonial species, such as Cliff and Bank swallows, appear to recognize each other's calls by the time the young leave the nest. In these species, young from many nests gather in large groups for the first few days after fledging, but are still fed by their own parents (see Fig. 7–39). In contrast, adults of non-colonial species, such as Barn Swallows (Medvin and Beecher 1986) and Tree Swallows (Leonard et al. 1997), do not recognize the calls of their own young. Individual recognition is less important to these species because the young either remain segregated in family groups or become independent immediately after fledging.

In most species (for example, Australian cockatoos called Galahs [Rowley 1980], Adelie Penguins [Davis 1982], and American White Pelicans [Evans 1984]) whose young gather into creches, the young recognize their parents' calls by the time the creches form **(Sidebar 4: Creches).**

(Continued on p. 8·130)

8

Sidebar 4: CRECHES
Sandy Podulka

Creche Formation

After leaving their nests, the young of some pelicans, flamingos, penguins, parrots, jays, terns, and others participate in an intriguing and somewhat mysterious phenomenon: they assemble into groups called **creches**, attended by one or more adults. In most of these species, parent-young recognition develops before the creche stage, and the adults continue to feed only their own young. The number of young in creches varies widely among species: penguin creches range from small groups with offspring from just a few families (as in Chinstrap, Adelie, Macaroni, and Gentoo penguins) to enormous assemblages containing several thousand young (King and Emperor penguins) **(Fig. A)**; Piñon Jay creches average 20 to 60 young; and Lesser Flamingo creches may contain up to 30,000 young **(Fig. B)**. The age at which young join creches also varies widely: flamingo young cluster at just 3 to 5 days of age, Galahs (Australian cockatoos) at 46 days, Adelie and Gentoo penguins at 2 to 3 weeks, and King and Emperor penguins at 6 weeks.

Functions of Creches

The functions of creches are poorly understood, but undoubtedly vary somewhat among different taxonomic groups. Penguins clearly conserve energy by huddling together in dense groups: the temperature inside a clump of adult male Emperor Penguins averages 18 degrees F (10 degrees C) warmer than the air temperature outside the group (Williams 1995), reducing their metabolic needs by 13 to 37 percent (Le Maho et al. 1976). Further evidence linking penguin creches to heat conservation is that Adelie Penguin chicks are more likely to form creches on days with stronger winds

Figure A. Emperor Penguin Creche: *Beginning at about six weeks of age, Emperor Penguin chicks (foreground) huddle together in creches of up to several thousand young, tended by just a few adults (at back) at any one time. Young in creches are better able to keep warm and avoid predators than single birds, and their parents are free to leave the colony (after the long incubation and brooding periods) to forage at sea, returning every few days to feed their own young. After about 3.5 months in the creche, the chicks leave the colony when their parents stop feeding them. The largest of penguins, Emperors breed only on the ice of the Antarctic Continent. Photo by Wolfgang Kaehler.*

Figure B. Flamingo Creche: *When only three to five days old, young flamingos gather into large creches where they are guarded and shaded by a few adults. Parents continue to feed their own chick, which they apparently locate by its calls. Drawing by Robert Gillmor.*

and colder temperatures (Williams 1995) **(Fig. C)**. There appear, however, to be additional advantages to penguin creches. Clustered young can be guarded by just a few adults at any given time, allowing the parents to forage in the distant seas for longer periods—bringing food back to their young every one to several days and also replenishing their own much-depleted energy stores for the next breeding. In addition, grouped young may be less vulnerable to predators for several reasons. First, the chance of any one young being taken is lower when in a group (the dilution effect). Second, predators may have more trouble singling out any given bird from a group (the confusion effect). Third, there are more eyes and ears to detect the predators.

American White Pelican creches also appear to conserve energy and to allow parents more time away. The creches form at the time the parents

begin leaving the nest to forage for longer periods. At night, when the parents are off foraging, the creches are large and dense, but they break up in the morning when the parents return to feed their young. In Manitoba, Canada, Evans (1984) found that the young huddled closer together in cold weather, and that the birds saved a small amount of energy clustering at 50 degrees F (10 degrees C), but not at 68 degrees F (20 degrees C) or higher. Other functions besides thermoregulation must be important to pelicans, however, because some species that form creches do not live in particularly cold climates (for example, Great White Pelicans in Africa and Dalmation Pelicans in southern Eurasia).

In Spectacled Parrotlets—small, social, Neotropical parrots—several neighboring pairs gather their fledglings into creches in trees during the day, but take them to roost elsewhere at night. Since parents

only return briefly during the day to feed their own young, researchers speculate that the creches allow parents time to forage and prepare for the next breeding, while their young learn socialization skills (Wanker et al. 1996).

The creches of Piñon Jays studied by Balda and Balda (1978) appear to be just one aspect of their highly social lives. Piñon Jays live year round in evergreen woodlands in flocks of 50 to 300 birds, and their first breeding attempt of the year is highly synchronized. Upon leaving the nest at around 22 days of age, the young gather into groups of 20 to 60 fledglings spread out over about 3.7 acres (1.5 hectares). They are guarded by a few adults, some of which may be yearlings without young. The creches allow parents to share sentry and mobbing duties, and free most parents to forage as a flock, as they do in the nonbreeding season. Although the creche of noisy young clearly at-

Figure C. Adelie Penguin Creche: *In contrast to the enormous creches formed by Emperor and King penguins, creches of Adelie Penguins contain only 10 to 20 young, which come from just a few families. The chicks join the creche at about three weeks, and their parents feed them every one to two days until they leave the colony at about two months. At the right, a South Polar Skua—the main predator on both eggs and chicks of Adelie Penguins—feeds on a small chick. Drawing by Robert Gillmor.*

tracts predators, it probably reduces total predation by presenting many more young in one location than the highly territorial major predators (Great Horned Owls, Cooper's Hawks, and Northern Goshawks) can use during one short time period. Parents returning to the creches usually feed only their own young, but the begging calls of their young stimulate other hungry young to beg, too, and about 13 percent of the time parents also feed other young. It is not clear whether these non-offspring feedings are errors, attempts to keep the other young quiet (and thus reduce the chance of attracting predators to the creche), or examples of parents giving extra food to young that may be related.

Brood Amalgamation in Waterfowl

In waterfowl, young from several broods may combine into larger groups by a variety of methods, all of which may be considered forms of **post-hatch brood amalgamation** (Eadie et al. 1988). Pairs may acquire lone chicks peacefully through **adop-** **tion**, or aggressively take over another brood through **kidnapping**; or several broods may combine to form creches with one or more adults attending them. Although these creches may superficially resemble those of other species, there are important differences: waterfowl young are highly precocial, and are thus very mobile and able to feed themselves. Adults tending large groups of young waterfowl, therefore, only provide parental care by guarding the young and guiding them to good foraging areas and safe resting sites.

Brood amalgamation is most common in waterfowl that forage by diving under the water; in these species, the young often feed closer to the surface, and thus have a different food source than the adults. North American diving ducks known to regularly combine broods include Lesser Scaup, Common and King eiders, Harlequin Ducks, Long-tailed Ducks, Surf and White-winged scoters, Common and Barrow's goldeneye, Bufflehead, and Red-breasted and Common mergansers. Brood amalgamation is also common in Canada Geese and Ross's Geese.

The association between foraging styles and creche formation is most notable in Common Eiders, who have different chick-rearing behaviors in different habitats (Gorman and Milne 1972). In Scottish estuaries, the young feed on invertebrates from shallow areas and the adults eat mussels from deeper water. Groups of young are cared for by a succession of females who each arrive at the estuaries with their broods, stay a few days with the group, and then depart to pursue their own foraging needs, leaving their young to the care of later-arriving females **(Fig. D)**. Along rocky coasts, however, young and adults are able to forage side-by-side in the seaweed beds—the young picking out small crustacea and the adults taking mussels. There, the young remain with their own mothers.

Brood amalgamation may be common in many waterfowl because there are few or no costs to rearing extra young, and because young benefit from reduced predation in large groups (as discussed earlier). In contrast, Savard (1987) found that in Bufflehead and Barrow's Gold-

Figure D. Common Eider Creche: *Among waterfowl, species in which the adults dive under water for food are most likely to form creches. In these species, the young often feed closer to the water's surface than do the adults, and because the adults and young have different foraging habits or locations, the parents have more trouble foraging with their young. Thus the family is less likely to stay together and young are more likely to aggregate into creches. Here, Common Eider young from several broods are tended by one female. In some habitats—such as Scottish estuaries—young and adult Common Eiders must forage in different areas, and creches form. See sidebar text for details. Drawing by Robert Gillmor.*

Figure E. Canada Goose Creche: *Brood-rearing strategies of Canada Geese are highly variably and poorly understood, but this cluster of 37 young clearly did not hatch from just one nest. Photo by Stan Osolinski/Oxford Scientific Films.*

eneye, combining broods brought no advantages to either adults or young, and concluded that brood amalgamation was a by-product of aggressive competition among females for brood-rearing areas. In some other species, however, larger groups do appear to survive better. In Snow Geese, at least 13 percent of the broods studied by Williams (1994) were enlarged through adoption. Larger groups competed better for winter food resources, and had greater gosling growth rates and adult survival rates. In Common Eiders (Munro and Bedard 1977) and White-winged Scoters (Kehoe 1989), ducklings from larger groups also survived better. In White-winged Scoters, a single female tends young from one or more broods, none of which may be her own. Females may be closely related to others in their general breeding area, however, so

they possibly may be rearing related young.

Although Canada Geese are widely studied and common throughout most of North America, surprisingly little is clearly understood about their brood-rearing strategies. Canada Geese rear their young in a wide variety of ways, and pairs do not necessarily use the same method every year. Pairs may raise only their own young, adopt one or several young, kidnap young, or join with another pair and raise the young together. Some even combine broods in various ways to form creches attended by one or several pairs **(Fig. E)**. Numerous hypotheses have been proposed for the various brood-rearing strategies—including larger families competing better for winter food, larger groups enjoying decreased predation, accidental mixing, and birds with little expe-

rience or in poor condition giving up their young to pairs with more breeding experience or better health. In some cases, pairs may even give up their goslings to their own parents to rear (Sherwood 1967). Few of these hypotheses have been thoroughly tested, so the reasons Canada Geese may adopt different brood-rearing strategies remain mostly speculative.

Forming creches or combining offspring in other ways are fascinating behaviors to observe and study. Part of their intrigue may arise because they occur in so few species across such a wide variety of taxa. Their similarities and differences from species to species tempt researchers to speculate broadly on their functions, but there is still much to learn about these avian nurseries. ■

Caring for Young

■ Altricial young depend heavily on parental care from the moment they hatch until they become independent—often a considerable length of time after they leave the nest. The parents of altricial young provide care in several ways: they bring copious amounts of food, they keep the nest clean by disposing of the nestlings' feces, and they often defend the nest and young against predators. But even precocial chicks that feed themselves may need considerable care, especially in defense against predators and the elements.

a. Ruby-throated Hummingbird

b. American White Pelican

Figure 8–118. Adult Birds Feeding their Young: *Parents deliver food to their young in a variety of ways.* **a. Ruby-throated Hummingbird:** *To regurgitate food to her tiny nestlings, a female hummingbird pokes her beak deep into the throats of the young, which gape in response to air movement from her wings. Photo courtesy of Isidor Jeklin/CLO.* **b. American White Pelican:** *To feed their older chicks, pelicans open their huge bills and let them dip into the enormous pouch and help themselves to regurgitated fish. Younger chicks may grab regurgitated food from the ground, from the top of the parent's foot webs, or from the tip of the parent's lower beak. Photo courtesy of Allen Cruickshank/CLO.* **c. Double-crested Cormorant:** *Nestling cormorants scramble to be the first to reach a returning parent, then thrust their heads deep into its throat, seizing whole fish directly from the mouth, pouch, or esophagus. Photo by Marie Read.*

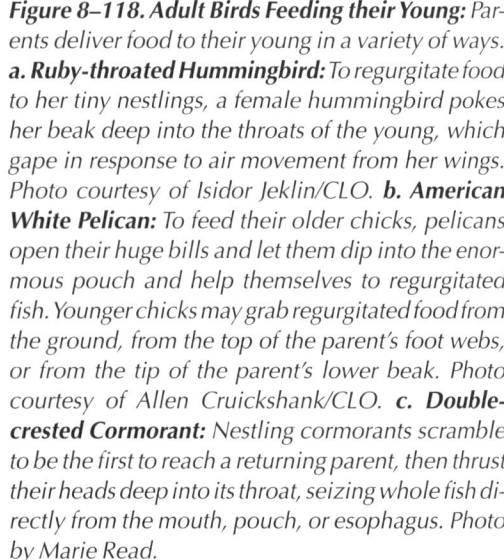

c. Double-crested Cormorant

Feeding the Young

Parents of altricial young must consistently bring food to their growing offspring, and the ways in which the food is delivered can be fascinating **(Fig. 8–118)**. A female hummingbird thrusts her bill down the throats of her young to deliver the nectar they require, often probing so vigorously that it seems she will poke right through her bee-sized offspring! And the young of open-nesting seabirds return the favor: a parent gull, penguin, or cormorant returning home with a load of food is often attacked by its young, as they scramble to get to the parent's harvest. A food supply that was regurgitated to the young when they were smaller gets intercepted by older offspring while it is still in the parent's esophagus—*before* the parent has a chance to deliver it! An observer worries for the safety of the parent in this case, as much as one might for the well-being of the young hummingbirds, in the former.

Some birds, such as raptors, bring large prey items back to the nest. These must be torn up before being fed to the nestlings, at least when they are young **(Fig. 8–119)**. It is no surprise that these birds bring one prey item at a time back to the nest, but it would be highly inefficient for birds with smaller prey to bring just one prey item on each visit to the nest. Puffins, for example, are adept at bringing a well-organized line of up to 50 or more small fish back to their young (see Fig. 4–86b) and I often wonder how they manage to catch the last few fish with their bills already full. Most storm-petrels gather tiny, red crustaceans from the ocean's surface by day and store them in their esophagus. Then they can return to the nest at night, when predatory gulls near the nest colony pose a greatly diminished threat to the adults. Once safe in their burrows, they disgorge a band of crustacean paste—much like pink toothpaste—to their ravenous nestling.

In feeding their young, birds accomplish some impressive feats. Parents of altricial passerines in the north temperate zone commonly bring food to their nests hundreds of times per day. For example, Great Tits in Japan feeding broods of 7 or 8 young bring prey one at a time to their nestlings, at rates of up to 300 to 400 feedings per day (Royama

Figure 8–119. Red-tailed Hawk Feeding Nestlings: Raptors and other birds that bring large prey to the nest must tear it apart for their young. Here, a Red-tailed Hawk pulls pieces from a small rodent to feed its chicks. To get eye level with this nest, which is perched in a saguaro cactus 30 feet above the floor of the Sonoran Desert in Arizona, the photographer erected scaffolding with a blind on top. To avoid scaring the birds, he climbed up the tower in the middle of the night and slept there until there was enough light to take pictures. Photo by John Cancalosi.

1966). And, one Eastern Phoebe feeding four young was reported to make 8,942 nest visits in 17 days! Species that bring more than one prey item per visit average lower visit rates, but they may collect food at an awe-inspiring rate: Tree Swallows bring an average of 19 items per visit (many still alive, and all carried in the mouth!) and feed the typical brood of five an average of 8,000 prey items per day (McCarty 1995). Generally, the more young in the nest, the more food trips the parents make and the greater the amount of food they deliver, but parents seldom increase their efforts in direct proportion to the number of young.

Most temperate zone passerines feed their young a diet dominated by insects. The one exception is finches in the family Fringillidae (for example, House Finches, goldfinches, siskins, crossbills, and canaries), which, like the subtropical Zebra Finches and their relatives (family Estrildidae), feed their young seeds (see Fig. 4–99a). Seeds contain less protein than insects, but even lower in protein are the diets of almost pure fruit that are fed to the young of several groups of frugivorous tropical birds, such as manakins (Foster 1978) and euphonias (Sargent 1993). The most spectacular result of a fruit-only diet is the young of the Oilbird (see Fig. 4–53), a Neotropical relative of nightjars. Young Oilbirds eat only palm fruits, and these are so low in protein that the birds must ingest huge amounts to get the protein necessary for growth **(Fig. 8–120)**. But the fruits are so high in lipid (Bosque and DeParra 1992) that the young grow oily enough to have been harvested by native peoples as a source of fuel! They will literally burn like a candle. At night, Oilbird parents forage over huge home ranges, aided by a unique sonar system based on the echoes of click-like vocalizations, to find palm trees bearing the ripe fruit that their young require.

Most seabird parents explore even larger home ranges. Although many have precocial chicks capable of thermoregulation from an early age, their food is located far from the nest and requires great skill to gather, so the parents must feed their young until they are capable of long-distance flight and effective foraging. Most extraordinary among such seabirds are the albatrosses nesting on tiny islands in the extreme southern oceans. These birds forage for squid and other invertebrates

Figure 8–120. Plump Nestling Oilbirds: *Young of the Oilbird, a Neotropical relative of nightjars, eat only fruits, which are much lower in protein than the insects eaten by most nestlings. The young must consume great quantities of fruit to meet their nutritional needs, and they thus regurgitate numerous seeds, which surround the nest site and chicks, as in this photo. Because the fruits are so high in lipids, the young grow extremely plump and oily; they even have been harvested by native peoples as a fuel source, since they will burn like a candle! Nestlings that are 10 weeks old weigh substantially more than their parents, but they use up their fat in the later stages of development, fledging at about the same weight as their parents. Although they grow fat quickly, the young mature slowly, requiring 3 to 4 months to fledge. Photo by Peter H. Wrege.*

for long periods of time. Although they feed young nestlings almost daily, they may return to the nest to feed older chicks only every few days. Because of the strong and relentless winds at these latitudes, the birds routinely soar downwind around the world between visits to the nest! (See Fig. 5–42.)

As mentioned earlier, the food of some nestlings is not harvested directly from the environment. For up to the first week of the nestlings' lives, pigeons and doves feed their young "crop milk"—a rich soup of epithelial cells from the lining of the esophagus (see Fig. 4–99b). Male Emperor Penguins can produce a thick milk from their crop and esophagus for the first week or so of their chick's development if the female is late in returning to the nest to begin feeding, and flamingos routinely feed their young a liquid esophageal milk.

Some birds also bring water to their young. Common Ravens, anhingas, and others deliver it beak-to-beak (Welty and Baptista 1988). The dovelike sandgrouse, which inhabit deserts and arid grasslands in Africa and Asia, transport water from distant water holes to their young with their spongelike breast feathers (Cade and Maclean 1967) (**Fig. 8–121**; also see Fig. 9–14).

In many precocial species, the parents take an active role in showing the chicks how to find food. Although most can feed themselves soon after hatching, some interesting exceptions exist. American Woodcock chicks, whose bills are at first too short and weak to probe in the soil for their favorite food, earthworms, rely on their female parent to obtain them. As she probes and pulls up the worms, the chicks take them directly from her bill. The young of many diving aquatic species go

Figure 8–121. Namaqua Sandgrouse Chicks Drinking from Father's Belly Feathers: Inhabiting dry grasslands and deserts in Africa, India, and Asia, the dovelike sandgrouse nest far from any source of water. Because the chicks gain little moisture from their seed diet, the male brings them water for several months, until they are able to fly to the closest water holes, which may be as far as 50 miles (80 km) away. The male accomplishes this task using his belly feathers, which are specially adapted to hold water on their inner surfaces, where less will evaporate. When dry, each barbule in the middle of each feather coils into a spiral, lying in the same plane as the vane and holding it together; but when wet, the barbules unfurl and stick inward at right angles, creating a dense bed of "hair" around 0.04 inches (1 mm) deep that retains water like a sponge. To wet his feathers, the male first rubs his belly against dry sand or soil; this action removes any oil, resulting from preening, which might prevent water retention. Then he wades into a water hole up to his belly, raising his wings and tail to keep them dry. For anywhere from a few seconds to 20 minutes, he rocks up and down to saturate his feathers. Then he returns to his chicks, where he stands erect, allowing them to drink from a groove down the middle of his belly feathers. Photo by Ann and Steve Toon/NHPA.

underwater to escape predators, but at first they are so reluctant to dive for food that the adults bring it to the surface. Young grebes, loons, and alcids are fed by their parents for up to 10 weeks, first in the nest and later as the young accompany the parent on the water's surface. Young grebes often ride on their parent's back tucked under the wings, getting some of their first diving experiences as passengers (Fig. 8–122). Young gulls and terns are even more dependent on their parents: they are fed entirely by their parents throughout development, until they become independent and fly away from the nesting colony.

Figure 8–122. Clark's Grebe Carrying Chick on its Back: Young Clark's and Western grebes climb onto a parent's back within minutes of hatching, and spend much of their first four weeks as a passenger. While on board, they are fed by their other parent and brooded. The adult will dive with the chick only to escape imminent danger, and even though it is held firmly beneath the parent's wings under water, the chick often falls off. These birds were photographed from a floating blind in Lower Klamath National Wildlife Refuge in California. Photo courtesy of Jeffrey Rich/CLO.

Defending the Young

Parent birds defend their nests and young to varying degrees, from the deadly effectiveness of the large ratites and raptors to the largely symbolic displays of many small passerines. Few, if any, birds are more dangerous than cassowaries and Ostriches defending their offspring: these birds kick extremely effectively with their powerful legs. And few people attacked near the nest by an aggressive Northern Goshawk or large owl will forget the experience, as these birds have an uncanny ability to sample the scalps of their would-be predators with their talons. Large waterfowl, such as swans and large geese, almost always stand and defend their nests or broods, hissing and flapping their powerful wings at predators (Fig. 8–123). Smaller waterfowl, however, rely on cryptic behavior to evade predators. The same is true for shorebirds: the very large plovers, such as lapwings, defend their young aggressively (Fig. 8–124); whereas the smaller species (from about Killdeer size down) rely on cryptic nests, eggs, and young (Fig. 8–125) (also see Fig. 3–54), coupled with their distinctive distraction

Figure 8–123. Canada Goose Defending Nest Against Striped Skunk: Large waterfowl defend their eggs and young aggressively—hissing, flapping their wings, and sometimes actually hitting intruders with their powerful wings. This Canada Goose is defending its eggs from a Striped Skunk, a significant nest predator. Photo by Thomas Kitchin.

Figure 8–124. Blacksmith Plover Versus Elephant: *Large shorebirds, such as the Blacksmith Plover of Africa, often defend their nests and young aggressively. Normally quiet and shy, when another animal threatens its nest or feeding territory the Blacksmith Plover flies at the intruder giving noisy "klink" calls, which sound like a hammer hitting an anvil, and are the source of the bird's common name. In this particular encounter, the elephant backed down. Photo courtesy of Iain Douglas-Hamilton.*

displays. These involve an extremely effective injury-feigning, usually with hunched back and one wing dragging half-open to the side (see Ch. 6, Sidebar 3: Defense Behavior, Fig. C). As soon as the potential predator moves away from the eggs or young, the seemingly crippled parent undergoes an immediate and complete recovery!

Among passerines, size does not clearly determine defense behavior. Swallows can defend their nests aggressively, with solitary species defending more vigorously than colonial ones (Hoogland and Brown 1986), yet variation within a species can be extreme (Winkler 1992). American Crows in urban areas can be quite vocal and persistent

Figure 8–125. Well-Camouflaged European Golden-Plover Chick: *Although adult golden-plovers are large and may defend their nests and young with fervor, the precocial chicks also escape their many predators through superb camouflage. This recently hatched European Golden-Plover chick crouches, its cryptic markings rendering it virtually invisible amid the tundra vegetation on this Icelandic heath. Photo courtesy of Tim Gallagher.*

Figure 8–126. Male African Jacana Carrying Young Under Wings: *Males of several species of jacanas, including the African Jacana pictured here, carry their young away from danger by sticking them under their wings, with their legs dangling down. Males, who typically provide most or all of the parental care, also may tuck chicks under their wings to shelter them from heavy rains. Jacanas live in tropical freshwater ponds, streams, and marshes. Their long toes and nails help them to walk across aquatic vegetation such as waterlilies (as shown here), earning them the nickname "lily-trotters." Drawing by Charles L. Ripper.*

in nest defense, whereas those nesting in rural areas disappear rather than defend when the nest is threatened (Knight et al. 1987). Presumably, urban crows are more accustomed to people, and are not afraid to stand their ground; but rural crows see few people, and are often shot at.

Some birds take their young away from danger when threatened. Long-billed rails will carry their young to safety in their bills, and jacanas transport their young, with long feet dangling down, under their wings **(Fig. 8–126)**. The male of one Neotropical aquatic bird, the Sungrebe, carries his young to safety, even during flight, by placing them in a pouch under each wing (Alvarez del Toro 1971; Weid 1833) **(Fig. 8–127)**.

Although the young of most species are totally defenseless, some can feint a threat or at least promise a disgusting encounter. A brood full of tits in a cavity nest can do a credible imitation of a snake hissing when a predator comes to investigate their home, and the young of gulls and many other seabirds may release a barrage of feces and regurgitated food when molested by a predator. The young of fulmars are especially revolting in this respect, as their stomachs contain a vile-smelling supply of oily mess—very difficult for a predator to clean from its fur, feathers, or scales (see Ch. 6, Sidebar 3: Defense Behavior, Fig. D).

Nest Sanitation

Because the young of numerous species spend many days to weeks in the nest, the parents try to keep it as clean as possible. As soon as the young hatch, most parents remove the eggshells **(Fig. 8–128)**. In addition, the young of most passerines produce their feces, at least initially, in a tough, flexible bag called a **fecal sac** that is easily removed and disposed of: many parents carry them some distance and drop them **(Fig. 8–129)**. Some species, including swallows and martins, typically drop the sacs over water, and nuthatches and wrens put them on tree branches.

The young of most other species are soon adept at leaving the nest to defecate or backing up and forcibly ejecting the feces over the edge of the nest: hawk and hummingbird young cast their very liquid excreta over the side of the nest, and the area around Osprey and Bald Eagle nests soon becomes whitewashed with splattered feces—so much that one can sometimes estimate the number of young without even seeing them! The feces ejected in pelican, cormorant, and heron colonies—where large numbers of young live close together—soon whitewash and kill the vegetation. Belted Kingfisher young eject their feces against the wall of the nest chamber, then peck at the earth just above the feces, essentially burying the feces as they enlarge the nest chamber slightly (Kilham 1974).

a. Male Sungrebe

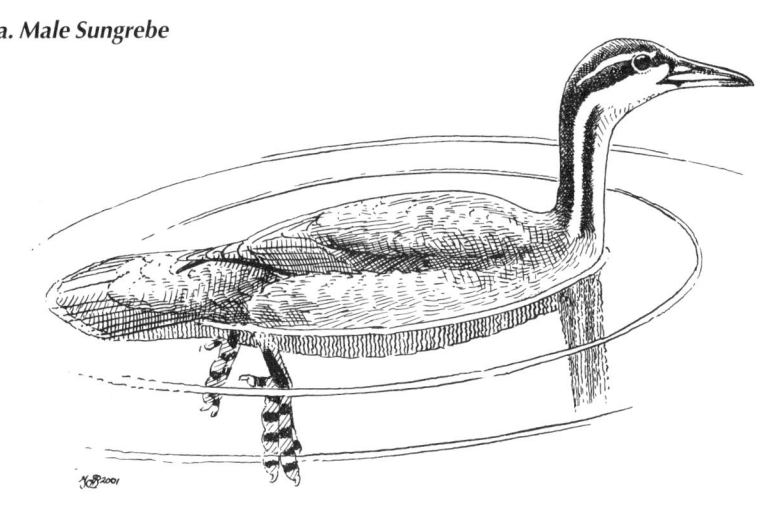

b. Chick in Wing Pouch

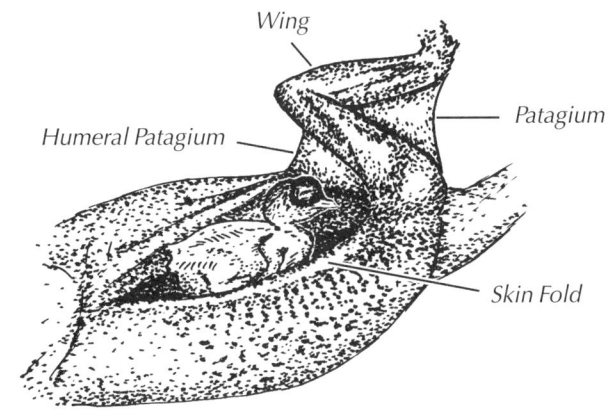

Wing

Humeral Patagium

Patagium

Skin Fold

c. Chick Protected by Feathers in Wing Pouch

Side Feathers

Figure 8–127. Wing Pouch of Male Sungrebe: The male Sungrebe, inhabiting dense, wooded streams in the Neotropics, has a pleat of skin under each wing that forms a pocket for carrying the young. The two young hatch pink, nearly naked, 2 inches (5 cm) long, and with closed eyes and rudimentary feet—after an incubation period of just 10 to 11 days. Although the biology of the species is not well known, ornithologists believe that the male gathers the young into his pouches shortly after they hatch, reaching under his wings to feed them and remove feces, until they are sufficiently developed to swim after their parents in relative safety. This unique transport system—not even present in the female—was first reported in 1833 by Weid, but apparently was not observed again until 1971, when spotted by Alvarez del Toro. *a. Male Sungrebe:* About the size of a Mourning Dove, the Sungrebe acts much like a grebe, but it is more closely related to rails. Its elongated head and neck, lobed toes, and long, stiff tail give it a distinctive appearance. *b. Chick in Wing Pouch:* This sketch, drawn from a specimen, shows a chick tucked into the pocket under the male's wing. For simplicity, only a portion of the raised wing is shown, and all feathers are omitted. (For information on patagia, see Fig. 1–9.) *c. Chick Protected by Feathers in Wing Pouch:* This sketch shows the side feathers of the male curving up and around the chick, keeping it warm and holding it securely in place as the male flies or swims. As a male flies with one chick tucked under each wing, the beaks and heads of the young peak out. Drawings b and c by Miguel Alvarez del Toro, from his article "On the Breeding Biology of the American Finfoot in Southern Mexico," in Living Bird, Vol. 10, 1971, pp. 79–88.

8

Figure 8–128. Least Tern Removing an Eggshell from its Nest: Very soon after a chick hatches, most parents eat the eggshell, feed it to their chicks, or discard it far from the nest. Eggshell removal keeps nests clean and also appears to reduce nest predation, because empty shells in the nest may attract the attention of predators such as crows, drawing them to the remaining unhatched eggs or nestlings. Only precocial species whose chicks vacate the nest soon after hatching leave eggshells in the nest. Photo by Tom Vezo.

Figure 8–129. Male Chestnut-sided Warbler Removing a Fecal Sac: Most passerine nestlings, at least early on, produce their feces in tough, flexible bags called **fecal sacs**. Parents thus can remove the feces easily, often grabbing them right from the cloaca of the nestling—as this male Chestnut-sided Warbler is doing. Parents either eat the sacs or discard them far from the nest. Fecal sac removal keeps nests clean, and, in species that consume them, also may provide a way of recycling scarce nutrients. Photo courtesy of Mike Hopiak/CLO.

Figure 8–130. Used Tree Swallow Nest Box: Nestlings of some species, such as Tree Swallows, stop producing their feces in sacs about halfway through their development. Because the parents can no longer effectively remove the feces, the nests end up a matted mess of feces and nest material. This nest box housed a family of Tree Swallows, who left behind a foul mixture of feces, grasses, and feathers after fledging. Photo courtesy of Marie Read/CLO.

Ascribing these habits to good nest hygiene is tempting, but the costs and benefits of nest hygiene are still poorly understood. Variation in the timing of fecal sac production among various passerines suggests that hygiene may not be everything. The young of many species, such as Eastern Bluebirds, produce their feces in tidy packets right up to the time they leave the nest, such that nests from which young have fledged are nearly absent of feces. In contrast, the nestlings of other species, such as Tree Swallows, quit producing their feces in sacs about midway through nestling life. Therefore, the parents cannot effectively remove feces from the nest, and after the young leave, these nests are a matted mess of feces and nest material **(Fig. 8–130)**. It is amazing that chicks can emerge pristine from a nest so fouled! Perhaps selection for keeping the nest clean has been stronger in bluebirds, which often raise two or three broods in the same nest in one season, than in Tree Swallows or kingfishers, which raise only one brood per summer.

Some ornithologists suggest an interesting alternative interpretation of fecal sac production: perhaps the sacs allow birds to recycle scarce resources. This interpretation springs from the observation that in many species the parents eat the sacs rather than dispose of them. Both the sacs themselves (made of a protein-sugar mixture) and their contents have valuable nutrients that the parents can reuse (Hurd et al. 1991). However the balance between hygiene and nutrient recycling has worked out in various species, fecal sac production is an interesting example of parents and nestlings evolving together over many generations to make parenting more efficient. Exactly what the patterns of fecal sac production are across species, and why this variation exists among species, are interesting, unsolved puzzles in the nestling lives of birds.

Birds carrying fecal sacs are valuable to field ornithologists in several ways. Because the glistening white sacs are easy to see from a distance, birds carrying them are highly visible and can provide clues to the nest location of cryptic species. Many ornithologists consider a bird carrying a fecal sac to be an unequivocal indicator of an active nesting attempt. Thus field observers conducting breeding bird atlases—surveying an area for the most reliable evidence of breeding in each species—can use the sighting of a bird with a fecal sac to confirm breeding. Nest predators may have caught on to this source of information on active nests as well, but I know of no evidence for this.

Brood Parasites

Females of many species (for example, rails, Wood Ducks, Redheads, Cliff Swallows, and Yellow-billed Cuckoos) occasionally lay eggs in the nests of others of their own species (**conspecifics**) while still laying eggs in their own nests. A very few of these species also may lay eggs in the nests of other species—the Redhead, for example, most often parasitizes Canvasbacks (Sorenson 1998) **(Fig. 8–131)**. But one lifestyle stands in stark contrast to the prodigious care that most birds provide to their young: **obligate brood parasitism**. This breeding strategy, in

Figure 8–131. Mixed Brood of Canvas-back and Redhead Ducklings: Female Redheads produce and incubate their own clutch of eggs, but also may lay eggs in the nests of other Redheads, and in those of Canvasbacks. Here, a female Canvasback (third from left) tends a brood of eight Redhead and two Canvasback ducklings (far left, and fourth from left). Photo courtesy of Michael Sorenson.

Figure 8–132. Black-headed Duck Parasitism: The secretive Black-headed Duck, inhabiting wetlands in southern South America, is entirely parasitic, laying its eggs in the nests of coots, rails, ibis, egrets, swans, screamers, Limpkins, gulls, other ducks, and even two raptors—the Chimango Caracara and Snail Kite. Even though females often lay two eggs in each host nest, the young have little effect on their hosts because they leave them to fend for themselves within two days (sometimes just a few hours) of hatching. The young's extra-thick down keeps them warm in the absence of a parent to brood them. Unlike chicks that develop in the company of their parents, these young give no distress calls; they have no one to call for help. a. Black-headed Duck Pair with Red-gartered Coot: One of the Black-headed Duck's most frequent hosts is the Red-gartered Coot. Drawing by Robert Gillmor. b. Black-headed Duck Egg in Red-fronted Coot Nest: The parasite's egg (upper left) does not closely resemble the eggs of this common host. Photo by Milton W. Weller.

which females lay all their eggs in the nests of other species, is very rare; only about one percent of the world's bird species exhibit the trait. It has arisen independently, however, at least seven times during the evolution of birds, in the following groups:

(1) Black-headed Duck

The life history of this duck from southern South America is the logical extension of the Redhead's strategy. The Black-headed Duck is entirely parasitic, choosing as its hosts other species of ducks, as well as coots, ibis, and gulls (**Fig. 8–132**). The ducklings, remarkably precocial, leave their host parents within two days, finding their own food and otherwise caring for themselves without assistance.

(2) Honeyguides (Family Indicatoridae)

This family includes 2 Asian and 15 African species. All are obligate parasites on different species of barbets, Old World warblers, and other hole-nesting species.

Red-gartered Coot (Host)

Black-headed Duck (Parasite)

a. Black-headed Duck Pair with Red-gartered Coot

b. Black-headed Duck Egg in Red-fronted Coot Nest

Figure 8–133. Parasitic Common Cuckoo Fledgling Begging from Winter Wren Host: *Famous for its two-note song widely mimicked by cuckoo clocks, the Common Cuckoo of Eurasia is a formidable brood parasite. Within the first day and a half after hatching, the nestling cuckoo ejects its host's remaining eggs or young from the nest, ensuring that the host will direct all of its energy toward rearing the young cuckoo. A wide range of host species are parasitized, including the tiny Winter Wren pictured here, known just as "The Wren" in Europe. Photo by Maurice Tibbles/Oxford Scientific Films.*

(3) Old World Cuckoos

The nearly 50 species of Old World cuckoos (family Cuculidae, subfamily Cuculinae) are obligate brood parasites, and they use a broad range of host species, from pipits and warblers to honeyeaters and magpies **(Fig. 8–133)**. Old World cuckoos are distinct from the cuckoos in subfamily Coccyzinae, which includes familiar North American species such as the Yellow-billed Cuckoo, Black-billed Cuckoo, and Mangrove Cuckoo. Members of the Coccyzinae are nonparasitic, and although they are not the only cuckoos living in the New World, they are collectively termed "New World cuckoos."

(4) Neomorphine Cuckoos

Only 3 of the 33 species of cuckoos living in the New World (family Cuculidae, subfamilies Coccyzinae and Neomorphinae) are obligate brood parasites, and all of these are in the subfamily Neomorphinae—a small group of tropical species that includes the Pheasant Cuckoo **(Fig. 8–134)**. Very little is known about these New World parasites, but they clearly are more closely related to other nonparasitic New World cuckoos than to the parasitic cuckoos of the Old World. Thus obligate brood parasitism must have arisen separately in New World and Old World cuckoos.

Figure 8–134. Pheasant Cuckoo: *All Old World cuckoos are parasitic, but only three species of New World cuckoos lay their eggs in the nests of other birds. These species—the Pheasant, Striped, and Pavonine cuckoos—are all in the family Cuculidae, subfamily Neomorphinae. The secretive Pheasant Cuckoo pictured here, about the size of a small crow, skulks fairly low in dense thickets of dry Neotropical forests. It primarily parasitizes birds that build cup nests, especially flycatchers, but also some species that construct closed nests. Photo by Marie Read.*

Female

Male

Figure 8–135. Parasitic Weaver: *This yellow, goldfinch-sized bird of African grasslands parasitizes cisticolas in the genera* Cisticola *(often called grass warblers) and* Prinia *(sometimes called wren-warblers). The biology of the Parasitic Weaver—also called the "Cuckoo Finch" and "Cuckoo Weaver"—is poorly known, but birds appear to aggressively defend territories when breeding, and then gather into large flocks after the breeding season, feeding mainly on grass and weed seeds. Ornithologists believe that the host's own young rarely survive in parasitized nests. Painting courtesy of Norman Arlott.*

(5) Parasitic Weaver

Within the large family Ploceidae (weavers), which includes 117 species in Africa and Asia, only this one species—often called the "Cuckoo Weaver"—has evolved to be an obligate brood parasite **(Fig. 8–135)**. Very little is known about the habits of this species, which lives in southern Africa. Recent evidence indicates that the Parasitic Weaver may be more closely related to the indigobirds and whydahs (see below) than to the weavers, with which it is currently classified. If so, obligate brood parasitism may have arisen independently only six times, rather than seven, among birds.

(6) Indigobirds and Whydahs

The 15 species of indigobirds and whydahs **(Fig. 8–136)**—termed "viduine finches" after their genus, *Vidua*—are African members of the large Old World family Estrildidae, which includes the well-known Zebra Finches of Australasia, common in pet shops worldwide. They are particularly interesting because of their close relationship to their hosts—other members of the same family, including firefinches and waxbills. All estrildid nestlings share the distinction of having amazingly detailed and diverse mouth markings, consisting of colorful dots and bars (see Fig. 7–65). Young whydahs and indigobirds have evolved mouth markings that mimic, to a remarkable degree, the markings of the young of their hosts.

(7) Cowbirds

The six species of cowbirds, distinctive members of the family Icteridae (blackbirds and New World orioles), are the most widespread and important brood parasites in the Western Hemisphere **(Fig. 8–137)**. The most recently evolved member, the Brown-headed Cowbird, successfully parasitizes over 140 different host species in North America (Lowther 1993).

Figure 8–136. Pin-tailed Whydah: *Whydahs and indigobirds, all in the African genus* Vidua *of the Old World family Estrildidae, parasitize other members of their family. Young viduines have a unique adaptation: the markings inside their mouths match those of their host's young to a remarkable degree. Most viduines parasitize a single host species, but the Pin-tailed Whydah—named for the narrow, tapering tail feathers of the male, pictured here—parasitizes several species of waxbills with similar mouth markings. Often their hosts are able to raise their own young as well as the young parasites, because the viduines do not evict the host's eggs or young. The finchlike viduines forage by kicking away sand or dust on the ground to uncover grass seeds. Photo by P.F.I.A.O./ VIREO.*

Chestnut-headed
Oropendola

Giant Cowbird

Figure 8–137. Parasitic Giant Cowbird at Nest of Chestnut-headed Oropendola (Host): The Giant Cowbird, about the size of a Common Grackle, parasitizes only oropendolas and caciques, both of which locate their long, pendulous nests in large colonies (see Fig. 6–39). Giant Cowbirds are uncommon but widespread from southeastern Mexico through central South America, inhabiting forest gaps and open country with scattered trees. They either forage on the ground—for insects, seeds, and fruits— or in association with cattle, catching insects flushed by their moving hooves or picking ticks from their skin. Drawing courtesy of Joel Ito, and modified from original.

Evolution and Adaptation Among Obligate Brood Parasites

It is fairly simple to imagine how a duck like the Redhead, which sometimes parasitizes its own and other species, could have evolved into an obligate brood parasite like the Black-headed Duck. But it is much more difficult to envision how some of the more specialized brood parasites, such as Old World cuckoos and African whydahs and indigobirds, evolved. In these groups, the eggs and/or young may mimic those of the host to an astounding degree.

Brood Parasite Ploys

Brood parasites have developed a variety of adaptations to help them deceive and exploit their hosts. Among their most important behavioral attributes is their cryptic movements. Females in search of host nests can be extremely difficult to follow. (Birds that can evade the defenses of their hosts seem to have no problem hiding from humans!) Not only do female brood parasites find host nests without being detected, they often monitor the activity of a large number of potential host nests simultaneously, waiting to lay an egg at the time that it is least likely to be detected (Davies and Brooke 1988). Many cuckoos lay soon after dawn (Arias-de-Reyna 1998), but the Common Cuckoo often lays in the afternoon, when the host is likely to be away. In many Old World cuckoos, the males fly conspicuously up to the nest and

a. Common Cuckoo

b. Brown-headed Cowbird

Figure 8–138. Brood Parasites Removing Host Eggs: *Brood parasites sometimes remove an egg from the nest of their host before they lay their own inside. This behavior, which keeps the number of eggs constant in the host nest, presumably renders the parasite's egg less detectable. It also may prevent the parasitized clutch and brood from becoming too large for the parents to incubate and feed successfully.* ***a. Common Cuckoo:*** *This female Common Cuckoo still holds the egg she removed from this tiny Eurasian Reed-Warbler nest, as she quickly lays her own egg in its place. She may complete the entire process in less than 10 seconds. Photo by Ian Wyllie/Oxford Scientific Films.* ***b. Brown-headed Cowbird:*** *This female Brown-headed Cowbird is removing an egg from a host nest. Drawing by William C. Dilger.*

sing, attracting aggressive chases by the host pair, while the female cuckoo stealthily lays her egg in the host nest (Arias-de-Reyna 1998).

Brood parasites sometimes eat or discard a host's egg, probably making their own egg less detectable (**Fig. 8–138**). They also may destroy entire host clutches without parasitizing them. Some ornithologists have suggested that this destruction may cause host females to produce replacement clutches, creating a longer potential breeding season for the parasite—in effect, the parasites might be farming their hosts! Others have suggested that clutch destruction is essentially blackmail—a punishment for hosts that reject a parasite's egg. Whether brood parasites are most accurately portrayed as farmers or gangsters remains to be seen.

Brood parasites probably suffer much higher losses *per egg* than their hosts, and the parasite's egg-laying abilities have been selected to meet this challenge: a typical passerine might lay at most 10 to 15 eggs in a season, whereas cowbird females are thought to lay up to 50 (Scott and Ankney 1980). Egg laying also can be amazingly swift: a Common Cuckoo can deposit its egg in a host nest in less than ten seconds, and Great Spotted Cuckoo females can lay in two to three seconds (Arias-de-Reyna 1998), whereas a typical *host* would require many minutes to lay an egg. The eggs of brood parasites are often thicker than would be expected from their size, and some species drop their eggs into the host nest, perhaps gaining the extra advantage of cracking some of the host's eggs in the process.

Many other characteristics of brood parasites set them apart from other birds. The young of most parasites have very short incubation periods and rapid growth rates for their size, helping them outcompete the host nestlings (**Fig. 8–139**). Some species, however, leave little to chance. The young of honeyguides and cuckoos employ various means of killing and ejecting the host nestlings with which they share the nest. Young honeyguides, while still blind and naked, use two methods. At hatching, both of their jaws bear a hook on the tip, a modification of

Figure 8–139. Brown-headed Cowbird Chick in Nest of Yellow Warbler: The young of most brood parasites hatch sooner and grow faster than the young of their hosts, allowing them to out-compete their nestmates for food. This Brown-headed Cowbird chick dwarfs its two Yellow Warbler nestmates, grabbing far more than its share of the food brought by the host parent. Photo by Jeff Foott/Valan Photos.

the egg tooth found on the upper beak of most young **(Fig. 8–140)**. In the nest, they either bite the host's young with these vicious hooks until the young die and the host removes them, or they push and shove the host's young, ejecting them one by one from the nest. Young cuckoos, within 10 to 12 hours of hatching, can push away any small object that lies against their sensitive backs between the shoulders, be it an acorn, an egg, or another young **(Fig. 8–141)**. In this way, they quickly eliminate all other eggs or young. Not surprisingly, females of these parasites lay only one egg in each host nest!

A spectacular brood parasite adaptation involves the eggs of the Common Cuckoo, which parasitizes a wide variety of Old World warblers, buntings, wagtails, and pipits. There appear to be several forms of Common Cuckoos, differing only in the coloration of their eggs. The eggs vary dramatically, and in many areas, closely resemble those of the most common host **(Fig. 8–142)**. Females are extraordinarily good at choosing the appropriate host nest in which to lay. How the females know which hosts to choose or how this ability evolved is not yet clear.

Indigobirds and whydahs also have evolved considerable adaptations for the parasitic way of life. Each species has one or a group of preferred hosts, and as mentioned earlier, the plumage and gape of the nestlings closely resemble those of the favored host species. Viduine nestlings also mimic the begging calls of their hosts, and some young whydahs go so far as to copy the begging posture of their

Figure 8–140. Nestling Honeyguide with Hooks on Beak: At hatching, a nestling honeyguide (upper right) has a needle-sharp hook at the tip of both jaws—a modification of the egg tooth. While still blind and naked, it repeatedly bites its nestmates (usually host young) until they die and the parent removes them from the nest. The hooks drop off by about two weeks of age. A young honeyguide also may push the host's young from the nest. In either case, it ensures that it will be the only chick remaining. Drawing by Robert Gillmor.

Figure 8–141. Common Cuckoo Nestling Ejecting Eurasian Reed-Warbler Egg: Common Cuckoo nestlings, while still blind and naked, respond to any object (eggs, nestlings, or even an acorn) that touches the sensitive, shallow depression between their shoulders by hoisting it onto their back, backing up to the rim of the nest, and dumping it out. This behavior develops around eight hours after hatching and disappears after four days. Here, a one-day-old nestling ejects the egg of its host, a Eurasian Reed-Warbler. It will soon eject the remaining egg in the same manner. Photo by Ian Wyllie/Oxford Scientific Films.

host's nestlings—even holding the head upside down! The eggs of both indigobirds and whydahs also resemble those of their hosts in color and size, but no dramatic adaptation appears to have been necessary here, as parasites and hosts all lay fairly generic white eggs.

Host Counterploys and Coevolution

As brood parasites continually evolve more elaborate ways to deceive their hosts, they pressure the hosts to evolve more sophisticated counterstrategies. The hosts with traits that best allow them to avoid or detect parasitism will raise more young with those traits—so the traits will spread in the host population, pressuring the parasites to evolve further tactics. In this way, the interactions between each brood parasite and its hosts have led to a coevolutionary race between the two. Egg mimicry in Old World cuckoos and chick mimicry in viduines, as discussed above, provide two good examples.

Why did the egg mimicry of cuckoos evolve? Why don't female cuckoos just stick with one general type of egg, as do many brood parasites? This seems a clear example of host behavior driving parasite ploys. Because young cuckoos so effectively eliminate their host's nestlings, nearly all parents rearing a young cuckoo fledge only that young and none of their own, in contrast to most Brown-headed Cowbird hosts (see Table 8–5). The cuckoo hosts would gain tremendously by rejecting the parasite's egg—if they could be reasonably sure of the identification! In the past, hosts apparently rejected cuckoo eggs so effectively that cuckoos with eggs more similar to those of their hosts enjoyed a selective advantage over those laying eggs that were more easily rejected. Because hosts of most other parasites often manage to raise some of their own young along with those of the parasite, the selection for egg discrimination and rejection that they experience presumably has not been as strong, and thus has not led to intricate egg-mimicking strategies by their parasites.

a. Cuckoo-Host Egg Pairs

Host   Cuckoo

Garden Warbler
*(Whitish with Gray
and Brown Markings)*

Great Reed-Warbler
*(Greenish with Dark
Markings)*

Common Redstart
(Pale Blue)

White Wagtail
*(Grayish-Blue with Brown
and Gray Markings)*

b. Common Cuckoo Egg in Nest of Great Reed-Warbler

*Figure 8–142. Mimicry of Host Eggs in Common Cuckoos: Throughout Eurasia there are several forms of Common Cuckoos, differing only in the appearance of their eggs. **a. Cuckoo-Host Egg Pairs:** The eggs of different cuckoo forms vary widely in both background color and markings, in most cases closely resembling those of their primary host. In Finland, for example, cuckoos lay plain blue eggs resembling those of Common Redstarts and Winchats, the most common hosts. In Hungary, cuckoo eggs are greenish with large, dark blotches, resembling those of the Great Reed-Warbler, the usual host. In some areas only one form of cuckoo is present, but in other areas up to four may coexist. Not all Common Cuckoos lay eggs that mimic host eggs, however. In southern England, where cuckoos parasitize a range of hosts with dissimilar eggs, they lay more generalized eggs that do not closely match those of any of their hosts. Biologists believe that mimicry of host eggs evolved in Common Cuckoos, and not in most other brood parasites, because cuckoos invariably eliminate all their nestmates, putting strong selection pressure on hosts to detect and remove parasitic eggs. This in turn stimulated Common Cuckoos to evolve eggs that were more difficult for their hosts to detect. Adapted from Rensch (1947). **b. Common Cuckoo Egg in Nest of Great Reed-Warbler:** The cuckoo egg (upper left) closely resembles those of its host, but is slightly larger. Photo by George Reszeter/Oxford Scientific Films.*

The impressive mimicry of host nestlings' mouths by indigobird and whydah nestlings appears to result from a similar coevolutionary battle. All estrildid nestlings have a series of distinctive markings at the corners and roofs of their mouths. These markings vary considerably among the many estrildids that serve as viduine hosts, yet the markings of each parasite nestling match those of its host species remarkably (see Fig. 7–65). Why estrildids evolved these mouth markings in the first place remains a mystery, but once they existed, natural selection on the hosts probably increased their diversity—as a way to better distinguish parasites. Then, selection pressure for reduced host rejection seems to have driven the viduine nestlings to evolve great similarities to their hosts—producing one of the best examples of chick mimicry in the bird world.

What are the limits to this cycle of deception and detection? A recent study of cuckoos in Japan reveals just how complicated the roles of parasite and host can be: Lotem (1992) found that cuckoos were most successful when they parasitized warblers breeding for the first time. By the time the host females began their second year, or even second brood, they had learned to distinguish their own eggs from those of cuckoos. Could an analogous trend exist in young cuckoos, with first-year breeders learning as much about parasitism as their hosts do about rejection? The fact that female Hooded Warblers breeding for the first time attacked Brown-headed Cowbird models (Mark and Stutchbury 1994), however, suggests that there is considerable diversity in the responses of hosts to parasites. These and many other details about brood parasitism remain to be explored.

Evolution and Adaptation in New World Cowbirds

New World cowbirds are not as specialized as parasitic cuckoos, whydahs, and indigobirds. Unlike those of cuckoos, cowbird eggs are neither unusually small in proportion to body size nor always similar in appearance to those of their hosts. And, in only one species (Screaming Cowbirds) have cowbird young evolved to look like their host's young (Fraga 1998). Even though they have no host-mimicry adaptations to interpret, however, the evolution of New World cowbirds is a puzzle.

Scott Lanyon (1992) derived the likely relationships among members of this group by comparing the sequences of mitochondrial DNA (see Fig. 1–39) of different species. His phylogeny shows that the southernmost cowbird species are evolutionarily the oldest, and that cowbirds expanded their range northward as they evolved. Among the six species, a strong trend for the southernmost (and oldest) species to have the fewest hosts is apparent. North America's Brown-headed Cowbird—the northernmost and most recently evolved—is known to parasitize at least 140 hosts, whereas the Screaming Cowbird (inhabiting southern Brazil and nearby countries) parasitizes only a single host.

At least two interpretations of these patterns exist. Lanyon (1992) suggests that, over time, cowbirds have evolved from a specialist into a generalist parasite. (This is counter to the traditional scenario of brood parasites becoming *more* specialized over time.) Thus the northern species—with more general parasitic habits—are able to exploit a greater number of hosts. In contrast, Rothstein et al. (2001) proposed that northern species have more hosts, not because they are generalists, but because their hosts have not yet had time to evolve defenses against these more recently evolved parasites. Southern hosts, plagued by cowbird parasitism for many more generations, have evolved strategies to thwart their parasites. In some cases, the strategies are so effective that the birds no longer are hosts. Given time, Rothstein and his colleagues argue, the hosts of northern species will probably evolve better anti-parasite strategies as well, possibly forcing their parasites to become more and more specialized.

The Brown-headed Cowbird: History and Conservation

As the only common brood parasite in North America north of Mexico, the Brown-headed Cowbird is of particular interest. About 50 species are regular hosts, with another 90 species infrequently rearing cowbird young—most thrushes, vireos, warblers, sparrows, buntings, and a few tyrannid flycatchers are subject, at times, to their

Figure 8–143. Brown-headed Cowbirds and Buffalo: Before the vast forests of eastern North America were cleared in the 1800s, the Brown-headed Cowbird lived only in the grasslands of the Great Plains. Because it often followed herds of buffalo, eating insects stirred up by their hooves, the cowbird was known as the "Buffalo Bird." Drawing by William C. Dilger.

impositions. An additional 80 species have hosted cowbird eggs, but are not known to have successfully raised cowbird young.

Historically, the cowbird inhabited the Great Plains where, because it fed in the grasslands among the feet of the American bison, it was known as the "Buffalo Bird" **(Fig. 8–143)**. Throughout most of the deciduous forest east of the Mississippi River, the Buffalo Bird was almost unknown. On the Great Plains, its regular hosts, especially the Yellow Warbler and the Dickcissel, evolved means of combating its parasitism. The Yellow Warbler builds a new nest atop the old one, eggs and all, whenever a cowbird egg appears **(Fig. 8–144)**. The Dickcissel simply abandons any nest with a foreign egg, rarely succeeding in raising a brood until late summer, when the Buffalo Birds have almost stopped laying.

Figure 8–144. Series of Buried Brown-headed Cowbird Eggs in Yellow Warbler Nest: Most birds that discover a parasitic egg in their nest either toss it out, accept it, or desert the nest, but the Yellow Warbler—a primary host for the Brown-headed Cowbird—sometimes buries the egg, along with any of her own, by adding a new layer of nest material on top. Then she lays a replacement clutch. Here, a Yellow Warbler successively buried four different cowbird eggs before finally completing a clutch of her own. In Ontario, Canada, one Yellow Warbler nest contained 11 cowbird eggs in six layers! A Yellow Warbler is most likely to bury a cowbird egg if it appears when she has just begun laying her own clutch. At other times she may accept it or desert the entire nest. Birds that bury cowbird eggs and lay again raise more young (an average of 0.78 for each of their own eggs laid, including buried ones) than do those that accept the egg (0.53 young per egg) (Clark and Robertson 1981). Photo by W. V. Crich, F. R. P. S.

Recently the Brown-headed Cowbird has expanded its range eastward in North America, parasitizing new hosts along the way. As farmland took over the eastern forests during the 19th century, open pastures with large livestock intermingled with woodlands—creating more of the bird's preferred habitat (open space with large animals)—and the Buffalo Bird became the "Cow Bird" **(Fig. 8–145)**. Today, it haunts both open country and small patches of woodland, hunting for the nests of susceptible hosts. Suburban lawns and other open areas have fragmented large forest areas into a countryside mosaic that allows the open-dwelling cowbirds access to more and more forest birds—because cowbirds are best able to penetrate forests from the edges. Cowbirds are very successful in parasitizing woodland and woodland-edge species in eastern North America because their new hosts, particularly Red-eyed Vireos and Wood Thrushes, have not yet evolved successful anti-parasite defenses.

Because many of the current cowbird hosts did not evolve in association with their parasites, it is no surprise that they are suffering extremely heavy losses as a result of the parasitism **(Table 8–5)**. Kirtland's Warblers and certain populations of Bell's and Black-capped vireos are sustainable at present only because of intensive cowbird-trapping programs in their limited ranges. Many other species are hit hard in certain locations by cowbird predation, but immigration from populations in other parts of their larger ranges buffers local populations against excessive nest losses.

Variation in cowbird parasitism over time and in different locations provides a fascinating glimpse of how changing ecological conditions and animal distributions can have different effects on species with different evolutionary histories. Before us is an instructive example of life history evolution in action—an example made all the more interesting and disturbing by the recent spread into Florida and the southwestern United States of two more cowbird species!

Figure 8–145. Brown-headed Cowbird Males with Cow: As European settlers fragmented North American forests and brought in livestock, the Brown-headed Cowbird expanded its range to include eastern pastures as well as its ancestral area—the grasslands of the Great Plains. Today, the cowbird breeds throughout most of the United States, inhabiting forest/field boundaries and a variety of grassland and forest areas—especially those with livestock nearby, or those that have been fragmented by human activities, such as development and agriculture. Drawing by Orville O. Rice.

Table 8–5. Effect of Brown-headed Cowbird Brood Parasitism on Host Reproductive Success: For the given host species, estimates of either (1) the percent of host eggs laid that survive to produce a fledgling, or (2) the average number of host young fledged per nest, in unparasitized versus parasitized nests. Note that sampling methods vary from study to study, so results are not directly comparable across species. Also, nest parasitism rates may vary widely among different regions and habitats, higher rates often leading to lower host breeding success because each parasitized nest contains more cowbird eggs or because cowbirds remove or damage more host eggs. Adapted from Payne (1977, p. 4).

FLEDGING SUCCESS

Host Species	(% *Eggs Laid that Produce a Fledgling* or *Average #Young Fledged Per Nest*)		Reference
	Unparasitized Nests	**Parasitized Nests**	
Eastern Phoebe	46% eggs	9% eggs	(Klaas 1975)
Acadian Flycatcher	61% eggs	54% eggs	(Walkinshaw 1961)
Red-eyed Vireo	81% eggs	22% eggs	(Southern 1958)
Yellow-throated Vireo (throughout range)	3.4 yg	0.6 yg	(Rodewald and James 1996)
Yellow Warbler (Southern Manitoba)	1.8 yg	0.7 yg	(Goossen and Sealy 1982)
Common Yellowthroat (Michigan)	1.9 yg	0.1 yg	(Stewart 1953)
Kirtland's Warbler (Michigan)	32% eggs	7% eggs	(Mayfield 1960)
Western Tanager (Northeastern New Mexico)	2.44 yg 60.9% eggs	0.9 yg 22.4% eggs	(Hudon 1999) (Hudon 1999)
Indigo Bunting (Michigan)	56.4% nests*	19.5% nests*	(Payne 1989)
Chipping Sparrow (Guelph, Ontario)	3 yg 81.3% eggs	2 yg 28% eggs	(Middleton 1998)
Song Sparrow	3.4 yg**	2.4 yg**	(Nice 1937, 1943)
Lark Sparrow	55% eggs	20% eggs	(Newman 1970)
Orchard Oriole (Kansas)	3 yg	1.5 yg	(Hill 1976)

* Percent of parasitized or unparasitized *nests* fledging host young

** Average number of host young fledged in nests fledging at least one host young

8

Figure 8–146. Black-capped Chickadee Nest: *Nestled in a bed of moss and fur in a rotted stump, these tiny eggs will produce six blind, hungry young in just 12 to 13 days. A mere 16 days and many meals later, the chicks will fly from the nest, never to return. The following spring, many of them will build nests and lay eggs of their own—completing the cycle in which birds go from eggs to parents and back to eggs again. Photo by Hal H. Harrison.*

Conclusion

■ In this chapter I have sketched but a few of the peaks in a rich landscape of ideas and biological facts about breeding biology and behavior. But, I hope you have seen how rewarding exploring the life histories of birds can be. In trying to understand the many variations, we need tools and concepts from both ecology and evolution, and we must put them to work in a mental laboratory that is fully equipped with the myriad details of bird biology. For the majority of species, many details are still poorly known. Yet as we learn more about birds, our understanding of why they reproduce in certain ways will grow richer, and new questions will arise to take our understanding even further.

As important as these intellectual advances can be, I hope you also will share with me an unabashed sense of wonder at the great diversity of ways in which birds go from parents to eggs, and then back to parents again. That, next time you see a chickadee at a bird feeder in the depths of winter, you can marvel at the tiny nest of moss and fur in a rotted stump that will serve as home for its young in the coming spring **(Fig. 8–146)**. And next time you see an Emperor Penguin at the zoo or on TV, you can transport yourself to the most forbidding terrestrial habitat on Earth and marvel that male Emperors will again, this summer, be holding their lonely vigil in the frigid dark while we vacation in the hot pleasure-grounds of the north.

Suggested Readings

Collias, Nicholas E. and Elsie C. 1984. *Nest Building and Bird Behavior.* Princeton, NJ: Princeton University Press. 336 pages.

Davies, N. B. 2000. *Cuckoos, Cowbirds, and Other Cheats.* San Diego, CA: Academic Press. 310 pages.

Goodfellow, Peter. 1977. *Birds as Builders.* New York: Arco Publishing Company, Inc. 168 pages.

Lack, D. 1968. *Ecological Adaptations for Breeding in Birds.* London: Methuen. 409 pages.

O'Connor, Raymond J. 1984. *The Growth and Development of Birds.* New York: Wiley. 315 pages.

Rothstein, S. I. and S. K. Robinson. 1998. *Parasitic Birds and their Hosts.* New York: Oxford University Press. 444 pages.

8

Individuals, Populations, and Communities: The Ecology of Birds

Stanley A. Temple

 As I begin to write this chapter, I am sitting on the veranda of my house on the island of Trinidad in the Caribbean, where I am spending a year as a Visiting Fulbright Professor at the University of the West Indies. Surrounded by lush tropical forest and hundreds of species of birds, my mornings are an ornithologist's dream as a fascinating ecological play with a diverse cast of avian characters unfolds before me.

For beginning ornithologists, the first problem in understanding this play is to identify the actors. Once they are familiar, questions about their roles and why they act as they do become inevitable. The answers require a knowledge of avian ecology: the study of the interrelationships among birds, other wildlife, and their environments.

As a starting point, consider that each bird is struggling to grow, survive, and reproduce. But being successful is not always easy, as birds must select a place to live, cope with weather, obtain food, find a mate, rear offspring, and compete with other individuals. Birds use a variety of tactics and strategies to meet these challenges.

For instance, I watched a male House Wren establish a territory in the backyard and attract a mate **(Fig. 9–1)**. The yard was perfect wren habitat with ample resources, such as food, cover, and nest sites, that the birds require. For days, I watched as the male wren defended his high-quality territory against other males.

Figure 9–1. Life History Strategy of the House Wren: High productivity is the key life history goal of the House Wren. Typical of many small, short-lived birds, the familiar House Wren, found from southern Canada through South America, makes up for its high mortality rate, especially among young birds, by producing many offspring each year. The female lays as many as eight eggs, which hatch in less than two weeks; the nestlings fledge about two weeks later and become independent soon after that. Because of its short nesting cycle, the House Wren may have several broods during a single breeding season. The number of broods depends largely on location: in much of the eastern United States, where House Wrens nest from May through August, they typically attempt two broods each year, but in Trinidad, where their breeding season lasts from March to November, they may attempt five nestings per year. Photo by Marie Read.

The wrens built their first nest in March in a bromeliad (a relative of the pineapple) near the house, where I could easily inspect the nest contents. The first clutch of five eggs disappeared soon after incubation began. I suspect they were taken by the local tree boa, an agile nest predator. Nevertheless, the wrens renested immediately, this time hatching all four eggs of the second clutch. The young grew quickly, leaving the nest after 17 days, and becoming independent just a few days later.

During the next few months the wrens made four additional nesting attempts and fledged a total of 10 offspring, as well as two Shiny Cowbirds, brood parasites that had laid their eggs in the wrens' nest. In late October, nine months after the breeding season started, I banded the young from the wrens' final brood. One of these young nested near the house the next April, having fledged, grown to maturity, and begun reproducing within six months!

In contrast to the wrens, the pair of nesting Channel-billed Toucans using a tree cavity in the nearby forest took nearly 75 days to hatch their eggs and fledge their single surviving nestling **(Fig. 9–2)**. Then the young bird remained with its parents, who provided food and protection for several additional months. Therefore, in the time that the wrens had raised 10 young, the toucans had raised just one, and by the time the young toucan was ready to leave its parents' territory, the first of the young wrens was already breeding. Furthermore, the young toucan would not be ready to breed for another year or two, by which time most of the wrens originally produced in my yard would have bred, raised young, and died!

These contrasting approaches to life are known as life history strategies, and no one strategy is "better" than any other. Rather, each represents a different method of striking a balance among the basic challenges of life: growth, reproduction, and survival. For a wren, the strategy involves rapid growth, early reproduction, a large brood, brief parental care, and a low survival rate. For a toucan, the strategy involves slow growth, delayed sexual maturity, a small brood, prolonged parental care, and a high survival rate. Both strategies result in birds that successfully contribute to future generations. This leads to a fascinating question: How do birds come to evolve different life history strategies?

This chapter is devoted to answering this and many related questions. For example, leaving my veranda and taking a short walk around my tropical neighborhood, I encounter seven pairs of Rufous-browed Peppershrikes, each living in an exclusive territory. I know they are different pairs because the seven males sing individually identifiable songs, which I have been able to learn. Their density seems to be about one pair per 2.5 acres (1 hectare [ha]).

In the evening I also hear Ferruginous Pygmy-Owls. Because they call from specific trees, often at the same time, I can estimate that three pairs are within earshot, an area of about 125 to 150 acres (50 to 60 ha).

Figure 9–2. Life History Strategy of the Channel-billed Toucan: For toucans—such as the Channel-billed Toucan, found in Trinidad, Venezuela, the Guyanas, and Brazil—rearing young is an intensive, long-term process. The huge bills of toucans, adapted for manipulating and eating fruits, render them unable to excavate their own nesting cavities. Instead, they nest in a tree cavity that has been excavated by another species. The eggs—typically four—hatch after just 15 days of incubation, but the nestlings remain in the nest for six to seven weeks. During this time they receive constant care as the adults not only bring food but also remove feces from the nest, keeping it free of waste and insect larvae. Despite this intense parental effort, only one nestling may survive. Even after it fledges, the parents must feed the young bird for several months before it becomes self-sufficient. It will not breed until it is at least two years old. Photo by J. Dunning/VIREO.

Thus, the density of pygmy-owls is about one pair per 50 acres (20 ha), only five percent that of the peppershrikes **(Fig. 9–3)**.

Why do these two common species live at such different population densities? And why do some uncommon species, like the Bat Falcon that hunts over the valley below the house, live at very low densities?

Some birds that frequent the neighborhood have distinctive plumages or colors that reveal their age and sex. For these species, I can unravel some aspects of their population structure, especially the proportions of individuals of each sex and age group. For example, at the end of the nesting season the small flocks of Greater Anis living along the edges of the road have about one immature bird to each adult. I can tell because their eye colors are different. In contrast, the Shiny Cowbirds that congregate in flocks of up to 100 individuals have one adult to every six immatures. In this case, I can tell from their distinctly different plumages. What determines these different ratios?

Next to the veranda, a fruit-filled feeder attracts several species of tanagers, orioles, saltators, and other frugivorous (fruit-eating) birds. The Blue-gray Tanager is the most abundant, and over the span of a few weeks I have trapped 46 individual tanagers, marking each with colored leg bands. During the last six months, this marked population has changed as individuals died or left the area while new individuals were fledged or moved in. My short-term monitoring suggests that 9 birds are no longer in this population, whereas 11 birds have been added, 9 of which are recently fledged young. Despite these changes, my local tanager population has remained stable because losses were counterbalanced by additions. This same dynamic equilibrium is characteristic of most bird populations. What maintains it?

Just as the individuals of a species living in one place make up a **population**, the populations of species living and interacting with one another in the same place make up a **community**. A bird community is complex. Its composition and structure are not established randomly,

a. Rufous-browed Peppershrike

b. Ferruginous Pygmy-Owl

Figure 9–3. What Determines the Density of Bird Populations?: One factor that regulates bird population density is the abundance of food. **a. Rufous-browed Peppershrike:** *Common insectivorous birds of open woodlands and gardens in Central and South America and the Caribbean, Rufous-browed Peppershrikes feed by working their way through the foliage of trees and shrubs, catching insects and eating larvae as they go. In Trinidad, one pair can meet its energy needs in a territory of about 2.5 acres (1 ha), shown by the solid lines on the map.* **b. Ferruginous Pygmy-Owl:** *Birds that eat larger prey, which are more widely dispersed than insects, require large territories in which to forage. In Trinidad, a pair of Ferruginous Pygmy-Owls requires a territory of about 50 acres (20 ha), shown by the dotted line on the map. Therefore the owls live at a population density about five percent that of the peppershrikes. Note that photos are not to scale. Photo of Rufous-browed Peppershrike by J. Dunning/VIREO, and of Ferruginous Pygmy-Owl by Steven Holt/VIREO.*

9

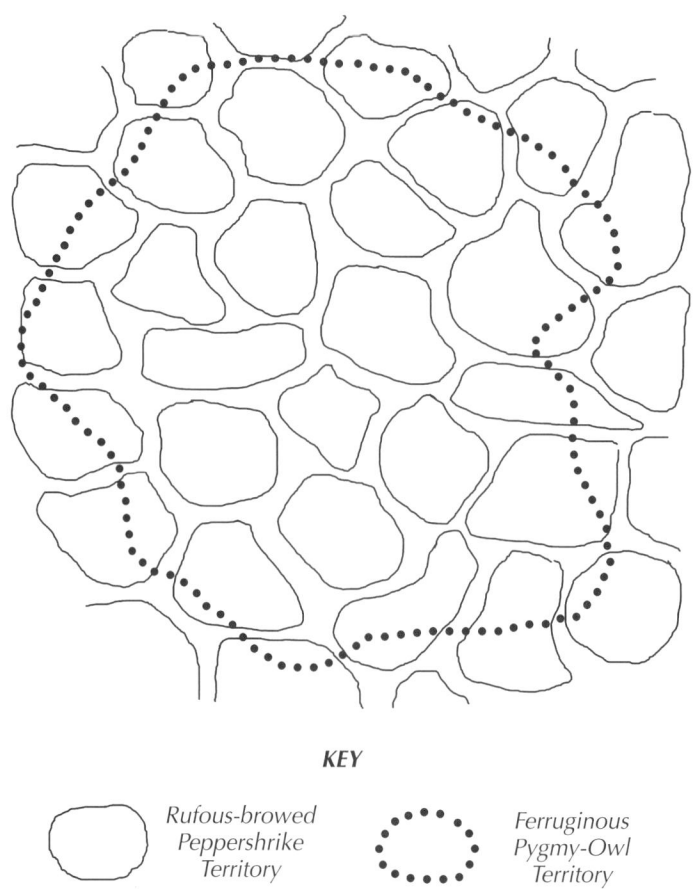

KEY

Rufous-browed
Peppershrike
Territory

Ferruginous
Pygmy-Owl
Territory

but result from interactions among different species and between each species and its environment. Also, bird communities differ greatly from place to place.

Around the Trinidad house I have seen 146 bird species, far more than the 68 species I see regularly from the deck of my home back in the Wisconsin oak savannas. And not only does Trinidad contain more species, it also is home to many more types of birds. Consider the pair of Channel-billed Toucans sunning in the treetops or the raucous flocks of Amazon parrots that awakened me earlier—neither of these birds have Wisconsin counterparts. And even within bird groups, the tropics tend to show more diversity. Back home my flowers attract only the Ruby-throated Hummingbird, whereas my Trinidad flowers are visited by 11 hummingbird species, each having a preference for particular flowers and a specific way of exploiting the nectar they produce.

Why do tropical bird communities contain more species than temperate communities? Why are certain types of birds, like the toucan, found in the tropics but not at temperate latitudes? How can so many similar-appearing hummingbirds coexist in a tropical community?

Many of the birds that appear around the Trinidad house are uncommon, so I don't see them every day or even every week. Why are some birds common while others are rare? Furthermore, some of the less common birds are very specialized. This morning, a pair of Barred Antshrikes is skulking in the understory **(Fig. 9–4)**. These antshrikes and several other insect-eating generalists regularly join company with two species of birds that specialize in following swarms of army ants, the Black-faced Antthrush and the White-bellied Antbird. This mixed foraging flock of ant-following specialists and opportunistic tag-alongs shows up when army ants are moving through the yard and flushing hidden insects that these birds then capture. No doubt the ants will be on the move around the house today. Why are some species so specialized, and often dependent on other species?

Although the interactions among birds in my tropical play have produced a well-organized troupe with specific roles, the cast is actually bigger than just birds, and the stage is bigger than my backyard. Indeed, as they interact with many nonavian species, and also with their nonliving environment, birds are important components of entire **ecosystems**.

For example, at the edge of the forest are patches of *Heliconia*—plants with large banana-like leaves and strange, colorful flowers—that are pollinated by hermits, dull-colored hummingbirds whose long, decurved bills fit perfectly inside. The relationship between hermits and *Heliconia* is known as mutualism, because both plants and birds benefit. In this case, the hermits get nectar and the plants get pollinated **(Fig. 9–5)**.

Or consider the birds that visit the fruit-filled feeders I've placed near the house. Frequently their droppings fall into the garden beds below, which have developed a rich new flora, graphically demonstrating the role these frugivorous birds play in dispersing certain tropical seeds. Some plants actually produce fruits of an ideal size and color to attract birds—insuring that their seeds get dispersed widely.

Figure 9–4. Barred Antshrikes: *Skulking through thick vegetation in the lowland rain forests of Trinidad, a male Barred Antshrike (left) and his mate (right) watch for insects and other arthropods flushed from their hiding places as a swarm of army ants passes through their territory. Barred Antshrikes are one of the approximately 250 species in the family Thamnophilidae, which includes antbirds and antwrens, as well as other antshrikes. Only a few of these species lead the highly specialized lifestyle of obligate, or "professional," army ant followers. Most, including Barred Antshrikes, supplement their typical foraging strategies by taking advantage of army ants as opportunities arise (see also Sidebar 3: Ant Followers).*

Figure 9–5. Mutualism: Certain organisms evolve together in ecological associations called mutualism, in which both organisms benefit from their interactions. Heliconia plants, for example, depend on the pollination services of hermit hummingbirds, such as the White-tipped Sicklebill shown here. The Heliconia flower's large nectar chamber (or nectary) is guarded by a tight passageway at the end of a long, curved, tubular structure, whose shape is matched by the bill of the sicklebill. Instead of hovering, this hummingbird clings to the brightly colored bract that encloses each flower and inserts its bill into the nectary, as shown here. As the sicklebill drinks nectar, the plant's anthers deposit pollen on the sicklebill's forehead, to be carried to the next flower on its rounds. The shape of the flowers discourages foragers less well adapted to transfer the pollen. (Also see Fig. 4–84.) Photo by Doug Wechsler/VIREO.

As one more example of birds' importance in the lives of other species, consider the sphinx moth larvae on the frangipani tree in the corner of my garden. Although the huge caterpillars would seem an attractive meal for any insectivorous bird, they are never eaten by the birds in my yard, probably because the caterpillars acquire a bad taste from the toxic chemicals contained in frangipani leaves. The caterpillars' bright colors—black and yellow stripes and a red head—advertise their chemical defenses, and these physical features apparently evolved to prevent the caterpillars from becoming bird food.

I hope that these observations from my veranda in Trinidad have aroused your curiosity as to how birds interact with each other, with other species, and with their environments. In the rest of this chapter, I discuss how these and many other fascinating aspects of birds' lives can be interpreted ecologically.

Birds as Individuals

To grow, survive, and reproduce, a bird must continually respond to a wide variety of living (**biotic**) and nonliving (**abiotic**) components of its environment. Individuals that survive and reproduce better than other individuals contribute more genes to the next generation, and thus more of their traits are passed on. A bird's degree of success at contributing genes to the next generation is called its **fitness** (see Ch. 1, Sidebar 2: The Evolution of an Idea: Darwin's Theory).

Birds respond to their environment on several different levels. Some responses are in the form of immediate, short-term physiological or behavioral changes. For example, in response to a drop in temperature, a bird can raise its metabolic rate to generate more body heat (a physiological response) or can seek shelter to conserve body heat (a behavioral response).

On another level, birds venturing into a cold climate may, over the long term, "respond" by evolving a thick layer of down feathers.

Through natural selection, of course, birds don't actually develop an attribute because it is needed. But rather, the birds in that climate that *already happened to have* a thicker down layer survive and reproduce better, and pass on more of their genes and thus traits. In this way, over many generations, the population eventually consists of birds with thicker down layers.

Note that both types of responses, and not just long-term ones, have an evolutionary component. Through natural selection, birds develop the *ability* to make different kinds of short-term responses. Both the *range* of possible responses and the *types* of environmental factors that a bird responds to depend on a bird's evolutionary history.

I focus in this section on short-term responses, such as adjustments to body temperature or movements to gather food, because such responses are easy to observe in a bird's daily life. Long-term responses cannot be seen as they develop, although we can view the results: birds that are remarkably well adapted to their environments.

Many short-term responses result from the need that all birds have for energy to maintain their body temperature and to drive activities such as flight. These responses affect their ability to acquire water and nutrients to sustain life processes. Birds usually seek those general habitats in which meeting their needs is easiest, considering their morphology and behavior. Then, they fine-tune themselves to local climate and habitat conditions with short-term adjustments in their physiology and behavior.

Habitat Selection: Choosing a Place to Live

A bird's **habitat**—the place in which it lives—must contain all the resources needed for that individual's growth, survival, and reproduction, including food, water, cover, and roosting and nesting sites. Indeed, few choices that a bird makes during its life will affect its fitness as directly as its choice of habitat. For any bird leaving its birthplace, the process of **habitat selection** must be efficient, so that birds promptly reject places that do not meet their needs, and quickly select a place that does (Cody 1985). The process can be complicated, because each bird is faced with a mosaic of environments from which to choose. Nevertheless, even casual observers can verify that birds are generally precise at selecting habitats. Experienced grouse hunters can tell when they are in good Ruffed Grouse habitat, just as birders learn to recognize the type of forest that will contain Cerulean Warblers or Wood Thrushes **(Fig. 9–6)**. Birds rarely are found in inappropriate places.

Many species are so closely tied to characteristic habitats that ornithologists have named them after those places. A Sedge Wren, for example, lives only in wet meadows and shallow marshes, which often contain sedges. Other examples of species named for their habitats include Greater Prairie-Chickens, Alder and Willow flycatchers, Wood Thrushes, Sage Thrashers, Cedar Waxwings, and Pine Warblers.

Other species are habitat generalists: their requirements can be met in a variety of habitats. Peregrine Falcons are found in many environments from the Arctic to the tropics, and they can adapt to a variety

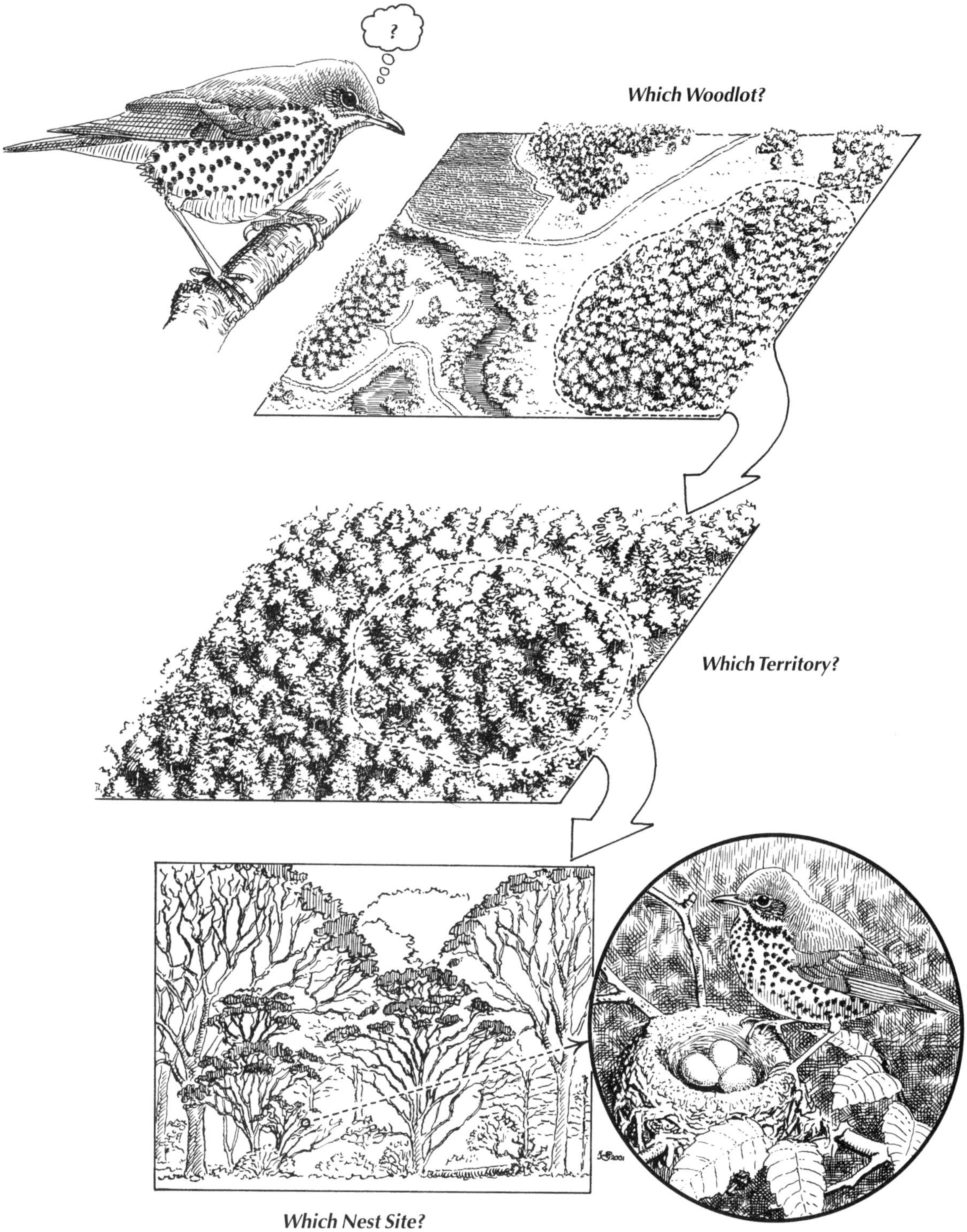

Which Woodlot?

Which Territory?

Which Nest Site?

Figure 9–6. Wood Thrush Habitat Selection: *For many species, including the Wood Thrush, selecting a place to nest involves making choices at several geographic scales. A migrating Wood Thrush returning to its natal breeding area in an agricultural setting is first faced with locating a woodlot suitable for establishing a breeding territory. The woodlot must be of appropriate size, age, shape, and distance to key landscape features, such as water. After selecting a woodlot, the bird must then stake out a territory of about five acres (two ha), usually containing a mix of understory and canopy trees. Finally, the bird must select an appropriate place within the territory to build its nest, which is placed on a forked branch of an understory tree or shrub, such as flowering dogwood.*

Figure 9–7. Urban Cliff Dweller: *This male Peregrine Falcon, known as "Omni," was photographed at its nest ledge on the 28th floor of a Milwaukee, Wisconsin apartment building. Almost extinct in North America by the 1960s, peregrines were reintroduced to the Midwest and eastern United States by hacking, a technique by which young birds, nearing the age of fledging, are placed in cages in their future habitat so that they imprint on their surroundings. As the birds grow older the cages are opened, allowing the birds to fly free. Human attendants continue to place food for the young birds until they are proficient at hunting pigeons and other prey. Later, having imprinted on the site, the hacked birds return to the area to nest. Many peregrines have been hacked in urban areas, including Omni, who was raised in captivity and released in Madison, Wisconsin in 1990. Most urban-nesting peregrines use nest boxes or trays placed for them on bridges and buildings, but some have nested on other structures including cliffs, which are their natural nest sites. Reintroduction coupled with other conservation actions has been very successful. In 1999, after 4,000 captive-raised birds had been released during a 28-year period by The Peregrine Fund, the Peregrine Falcon was officially removed from the U. S. Fish and Wildlife Service Endangered Species List. Photo courtesy of Greg Septon, Milwaukee Public Museum.*

of habitats containing prey and a nesting site, most often a cliff. During the 20th century peregrines even moved into cities, using tall buildings as nesting sites and pigeons for food **(Fig. 9–7)**. This may seem unnatural, but from a peregrine's point of view, cities are ideal habitat, almost as attractive as cliffs and canyons in wilderness areas.

Within a habitat, specific required resources are sometimes missing or rare. This lack of resources can be a **limiting factor** for some species. Examples include tree holes required by cavity-nesting birds, such as bluebirds. Although your neighborhood field with its few scattered trees may appear suitable for bluebirds, if nesting cavities are not available, bluebirds won't live there. Installation of nest boxes would remove this limiting factor, making the area suitable for any bluebirds that discovered it. As another example, Red-tailed Hawks sometimes are limited by the availability of perches from which to hunt. Meadow voles may be abundant in a given habitat, but if red-tails have no place from which to make an attack, they cannot exploit the available prey.

Sometimes limiting factors are things that are present rather than absent, such as potential predators. Blue-winged Teal nest in fields, but they avoid fields containing tall dead trees where predators, such as crows or raptors, can sit and search for females and their broods. As another example, the presence of ice on lakes and streams may preclude Great Blue Herons, Belted Kingfishers, and even Bald Eagles from otherwise excellent winter foraging habitat.

Sometimes two locations may contain all the requisites for success, but may differ in overall habitat quality. For example, one location may contain more food or better nest sites. Such quality differences often are revealed by the order in which migratory birds occupy breeding habitats when they return in the spring. Typically the highest quality habitats are filled first, leaving only lower quality sites for late-arriving prospectors. In species with polygynous mating systems, in which a male often has several mates, females are most attracted to males occupying the highest quality habitats. Studies of Marsh Wrens, Red-winged Blackbirds, Dickcissels, Indigo Buntings, and Bobolinks have shown that males living in the best habitat possess several mates, whereas those occupying poor habitat may have only one mate or none at all (Verner and Willson 1966) **(Fig. 9–8)**.

How do birds choose habitat so accurately? Two separate components of habitat selection seem to be involved: inherited and learned. Apparently, a bird inherits a predisposition to a general range of environments, a fact we acknowledge when we say "she takes to music like a duck takes to water." This predisposition results from generations of natural selection. However, inherited preferences cannot completely explain the precision with which birds select habitat, nor can they explain why individual birds show subtle variations in habitat preference. Clearly, some aspects of habitat selection are learned.

In fact, a bird learns about habitat in several ways and during several phases of its life. First, a young bird learns the characteristics of appropriate habitat while sharing its parents' territory. This early learning, called **habitat imprinting**, seems to occur rapidly and to make an indelible impression, as most adult birds show a strong preference for the type of habitat in which they were reared. This adaptation is appropriate; your parents clearly were successful in the habitat, so mimicking their choice is a good idea. For example, in New York State the American Crow—a habitat generalist that nests in a variety of environments—returns to habitats similar to those in which it was raised. Crows fledged from urban territories tend to return to urban territories to breed, and rural crows return to rural territories (McGowan 2001).

Sometimes learned habitat attributes can take precedence over innate predispositions. For example, the power of habitat imprinting can be demonstrated by **cross-fostering** experiments in which young

Figure 9–8. Habitat Settlement by Breeding Bobolinks: *For returning migrants, habitat quality often determines which territories within a potential breeding area are settled first. In the eastern United States, male Bobolinks prefer to settle in large, mature hayfields, especially those that have not been plowed or reseeded for at least eight years and that are annually mowed (Bollinger and Gavin 1992). These hayfields consist primarily of grasses with scattered broad-leaved plants such as dandelion and clover. Such an area of optimal habitat, where most male Bobolinks hold territories, is shown in the foreground and center of the figure. Less desirable habitats are crop fields (such as alfalfa); younger hayfields (those in which fewer than eight years have passed since reseeding, and which consist mostly of grasses but lack broad-leaved plants); pasture (grazed grassland); and old fields (those no longer mowed, and undergoing invasion by woody vegetation). Female Bobolinks choose males with high quality territories, and in the preferred area males often are polygynous. Lower quality habitats fall to later arriving males, or males unable to claim a territory in the preferred area. Male Bobolinks in these peripheral areas are less likely to obtain mates than males on territories in the preferred area.*

birds are placed with "parents" of another species having different habitat preferences. Genetic predispositions and habitat imprinting are then at odds, and learning typically wins out. In one such experiment, eggs of tree-nesting Mew Gulls were placed in nests of Black-headed Gulls, which nest on the ground in marshes. When the cross-fostered Mew Gulls matured and selected their own habitats, they chose to nest on the ground in marshes rather than in trees **(Fig. 9–9)**.

A second type of learning may occur once a bird has chosen a habitat and settled into an area. Individuals of many species then learn the precise position of their territory, and even if they migrate, they reoccupy the exact location in subsequent years, a phenomenon called **site fidelity** (see Ch. 5, Site Fidelity). Therefore, if a robin nesting in your yard survives both fall and spring migration, it likely will appear in your yard the following year. Site fidelity is strong and efficient, because it brings a bird back to suitable habitat, avoiding the need for annual prospecting. Dilemmas can result, however, if the nesting habitat deteriorates while the bird is away. For example, although Common Terns ideally nest on barren islands, site fidelity induces some long-lived individuals to attempt to nest (unsuccessfully) on islands that formerly were suitable but that have become overgrown with vegetation during their lifetimes. Or, the opposite can happen. In the eastern United States, Northern Goshawks typically nest in mature northern hardwood forests and show a high degree of fidelity to their breeding territories. Occasionally, these territories are logged in winter during the nonbreeding season, forcing the goshawks to abandon site fidelity and use habitat imprinting to identify suitable habitat when they return in the spring.

Most of the time, avian habitat-selection mechanisms ensure that individuals can identify and respond to appropriate habitat. This ability is crucial, especially for young birds striking out on their own. Studies have shown that Northern Spotted Owls (a subspecies of the Spotted Owl) typically die if they don't find a suitable area of vacant, old-growth forest within 12.5 miles (20 km) of their birthplace and within a few months after leaving their parents' territory **(Fig. 9–10)**.

Thermoregulation: Coping with Heat and Cold

You learned in Chapter 1 that birds are **endothermic**, meaning they can use physiological processes to warm their bodies. Birds are also considered **homeothermic**, because they are able to keep their internal body temperature constant even when the outside temperature varies. Therefore, birds can live successfully in environments that are very hot, very cold, or in which temperatures fluctuate widely. To maintain their constant body heat, usually between 104 and 108° F (40 to 42° C), birds employ a variety of physiological and behavioral mechanisms.

In general, bird homeothermy involves a system similar to the one that regulates the temperature of your house. The bird's hypothalamus (part of the brain; see Ch. 4, Brain), acting like a thermostat, measures changes in body temperature. When the temperature drops, the hypo-

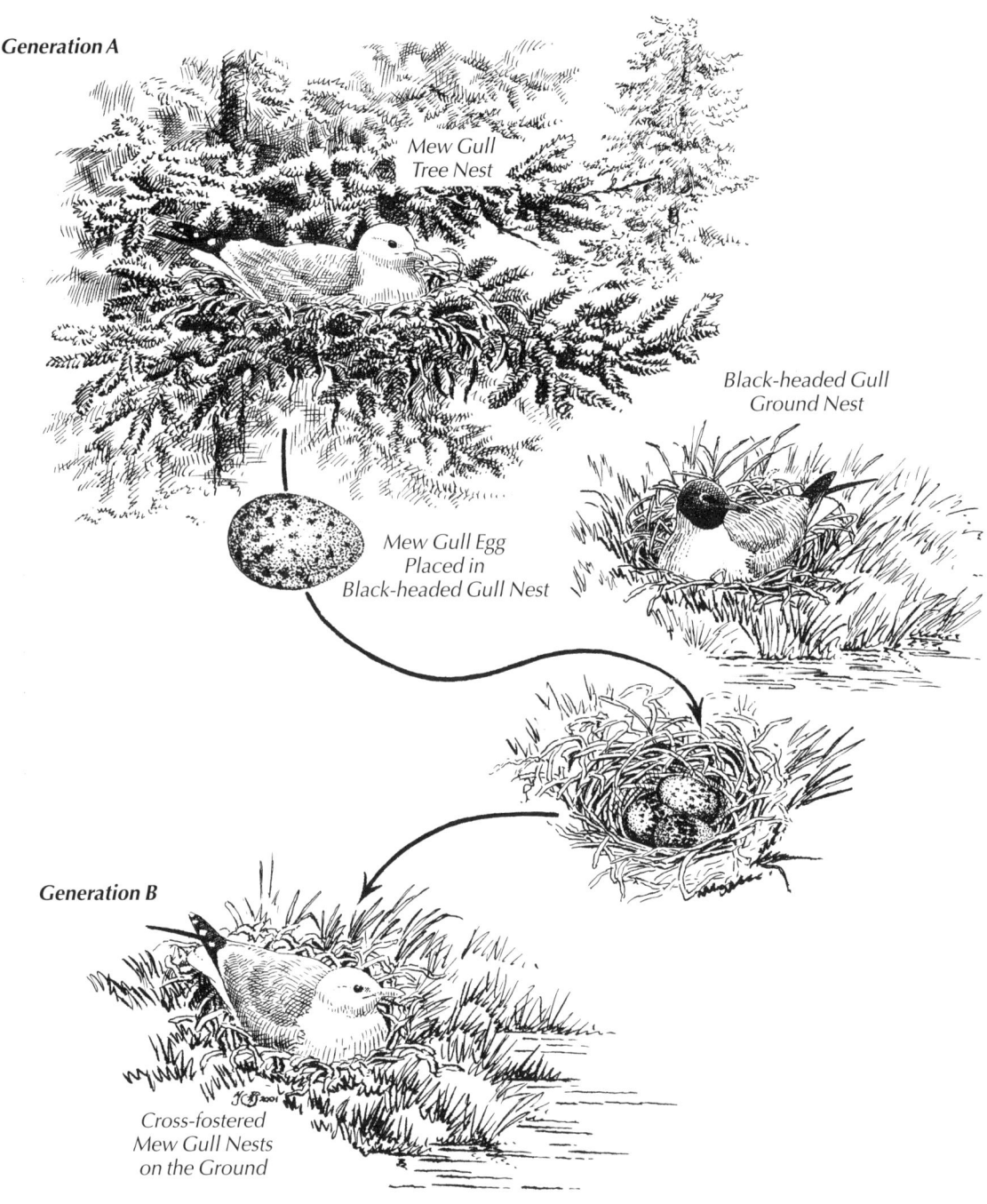

Figure 9–9. Habitat Imprinting in Mew Gulls: *The environment in which a young bird is raised can influence its choice of nesting habitat when it reaches adulthood. Habitat imprinting, which occurs when a bird learns the characteristics of its natal habitat, can be strong enough to supercede innate habitat preference, as shown by cross-fostering experiments in which young birds are raised by different species with different habitat requirements. In the 1930s, German ethologist E. Shuz placed eggs of Mew Gulls, which nest in trees (Generation A), into nests of Black-headed Gulls, which nest on the ground in marshes. When they reached breeding age, some of the cross-fostered Mew Gulls (Generation B) returned to their natal areas in the Black-headed Gull colony. There they made the correct mate choice by pairing with other Mew Gulls. However, they chose to nest on the ground, like their foster parents but unlike their biological parents (Shuz 1940).*

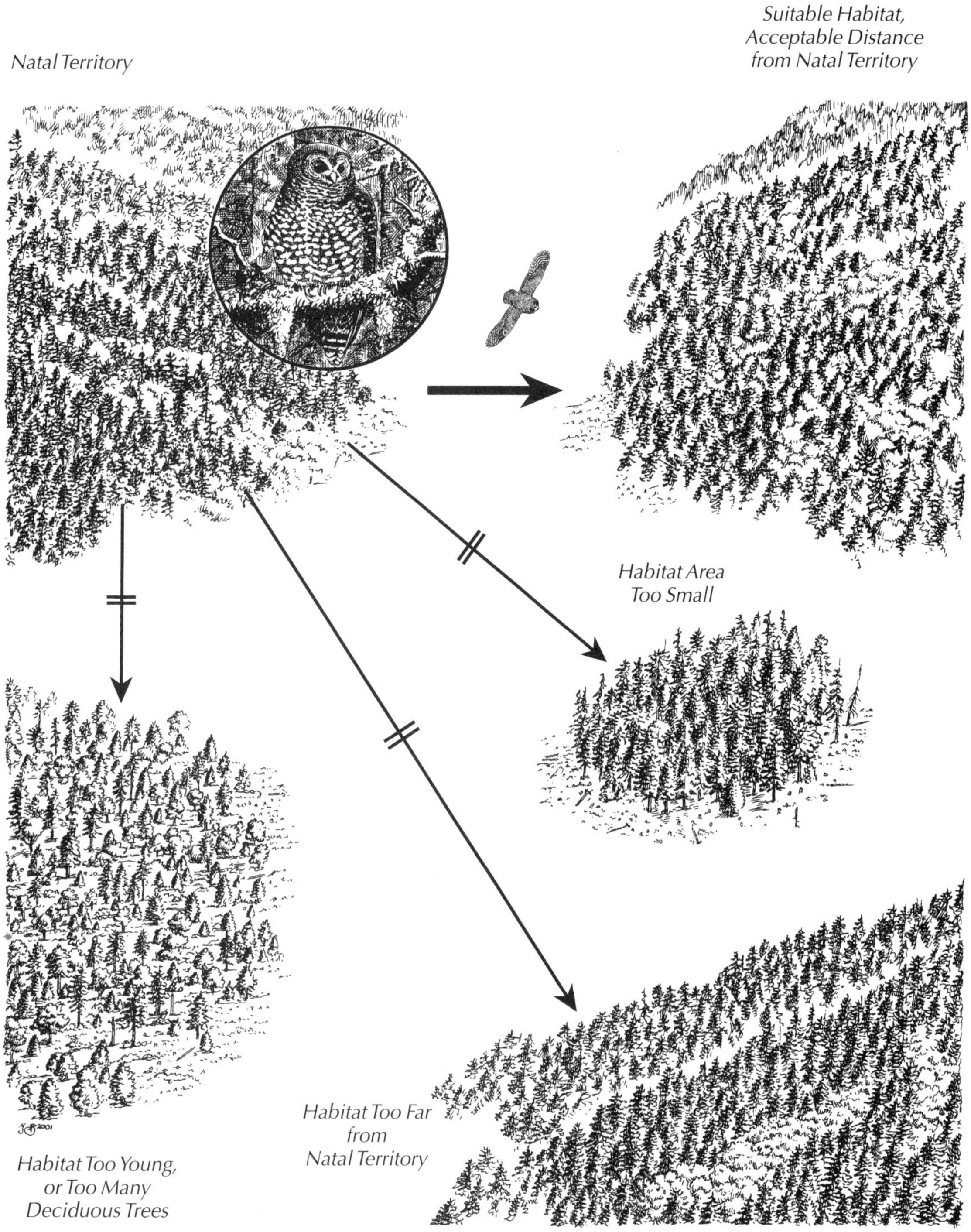

Figure 9–10. Dispersal of Fledgling Northern Spotted Owl: *When young birds disperse from their natal areas, their survival often depends on their ability to find suitable habitat nearby. Dwindling habitat in the Pacific Northwest makes this especially true for the Northern Spotted Owl: loss and fragmentation of the old-growth coniferous forests where the subspecies makes its home leave young owls with few habitat choices and expose them to many risks. Studies indicate that if immature Northern Spotted Owls cannot find suitable habitat within approximately 12 miles (about 20 km) of their natal territory soon after dispersing, they often succumb to starvation or predation by Great Horned Owls. Some old-growth forest patches fail to meet the bird's needs because they are too far away or are too small in area. Other unsuitable forests are too young, or contain a high proportion of deciduous trees.*

thalamus triggers responses that increase heat gain and reduce heat loss. When body temperature rises, it triggers responses that reduce heat gain and increase heat loss.

As an example, consider a Black-capped Chickadee on even a mild, summer day. When the ambient temperature drops below its body temperature, the bird starts to lose heat. Its first responses are behavioral: it fluffs its feathers to increase insulation, tucks its uninsulated feet and legs into its breast feathers, and seeks shelter in a cavity. For a while, these behavioral responses are sufficient to prevent major heat loss.

However, if the ambient temperature drops below a point known as the **lower critical temperature**—for a chickadee, a surprisingly high 79° F (about 26° C)—and behavioral responses become inadequate, physiological responses kick in. The chickadee's metabolic rate increases, producing additional body heat, and if the ambient temperature continues to drop, the metabolic rate continues to rise. Most birds can triple their normal metabolic rate in response to cold, although the process uses a lot of energy **(Fig. 9–11)**.

Eventually, at an ambient temperature known as the **lower lethal temperature**, the chickadee's metabolism reaches its maximum heat production rate. At this point the bird experiences hypothermia, meaning that its body temperature starts to fall. Its metabolic rate then drops uncontrollably, and the bird quickly dies.

Other birds respond to cold in similar ways, but the temperatures that elicit behavioral and physiological responses vary widely. In general, larger birds cope with cold environments more readily than small birds, which reach their lower critical and lower lethal temperatures more quickly. Recall from Chapter 4 (see Fig. 4–121) that small birds have a relatively high surface area-to-volume ratio, meaning they lose heat relatively quickly. Hence, although the 0.35-ounce (10-g) chickadee reaches its lower critical temperature at about 79° F (26° C), a 42-ounce (1,200-g) Common Raven can hold out until 39° F (4° C). Comparing birds of the same size, tropical birds, which generally are poorly insulated, reach their lower critical temperatures sooner than arctic birds, which usually are covered with a thick layer of down. Thus a 0.21-ounce (6-g) Blue-throated Hummingbird living in Mexico reaches its lower critical temperature at a relatively warm 88° F (31° C), whereas a 0.21-ounce (6-g) Golden-crowned Kinglet living in a Vermont forest doesn't reach its lower critical temperature until the ambient temperature drops to 70° F (21° C).

Over evolutionary time, the thermoregulatory advantage of large body size has resulted in a predictable variation in body sizes among members of most species, such that individuals living in colder regions tend to be larger than individuals living in warmer areas. This pattern is called **Bergmann's rule** after its discoverer. For example, Downy Woodpeckers in Florida are about 11 percent smaller than Downy Woodpeckers in Minnesota.

Although we think of very cold temperatures as the most dangerous for birds, body temperatures that *rise* just a few degrees above

a. Fluffing Plumage

b. Tucking Feet and Legs
into Breast Feathers

c. Taking Shelter from Wind

d. Shivering

e. Hypothermia

Figure 9–11. Thermoregulation in Cold Weather in the Black-capped Chickadee: *Faced with falling temperatures on a winter day, a Black-capped Chickadee can prevent its body temperature from dropping in several ways. The first responses are behaviors that conserve heat, including:* **a. Fluffing Plumage:** *The fluffed plumage creates an insulating layer of air that keeps heat trapped against the body surface;* **b. Tucking Feet and Legs into Breast Feathers:** *Covering the unfeathered feet and legs reduces their exposure to the cold; and* **c. Taking Shelter from Wind:** *Finding shelter in a dense shrub or tree cavity helps to reduce convective heat loss. If temperatures continue to drop, the bird must employ physiological methods. At its lower critical temperature the chickadee's metabolic rate starts to rise, producing metabolic heat.* **d. Shivering:** *Involuntary muscle contractions, known as shivering, produce additional metabolic heat.* **e. Hypothermia:** *When a chickadee reaches its lower lethal temperature, and its metabolism can no longer counteract the loss of heat, hypothermia sets in and the bird soon dies.*

normal (**hyperthermia**) can be lethal as well, so birds also employ behavioral and physiological responses to keep their bodies from overheating. For example, the Cactus Wren lives in hot deserts where hyperthermia is a constant threat. Behavioral responses to the deadly heat include seeking shade, sitting in a breeze, and avoiding contact with hot surfaces.

Above the wren's upper critical temperature of about 95° F (35° C), if these behavioral responses become inadequate to stave off the heat, body temperature may start to rise. At this point the bird must employ the physiological response of evaporative cooling, that is, panting and increasing the rate of water evaporation from the body, which dissipates a great deal of heat but, like increased metabolic rate in the cold, uses lots of energy (**Fig. 9–12**). But panting has an additional cost: loss of water.

Strategies for living in hot and cold environments clearly illustrate the prices that birds must pay when dealing with variable environmental conditions. In coping with cold, a major cost is the extra energy required to increase metabolic heat production; a bird must cover this cost by consuming more calories. In coping with heat, a major cost is the water required for evaporative cooling. A bird must replace this water quickly or face dehydration. Yet in hot environments, such as deserts, water often is scarce.

Water: A Matter of Economy

Like all organisms, birds are composed largely of water, and every individual must maintain its water content above a critical level or die of dehydration. Because birds frequently live in dry environments, maintaining water balance can be challenging. A bird can obtain water from three sources: (1) free water; (2) moisture contained in food (preformed water); and (3) metabolic water formed during the process of cellular respiration.

The extent to which a bird living in a challenging environment depends on each of these water sources determines many aspects of its ecology and life history. Its range and habitat can be limited by the need for free water, and its food preferences can determine its independence from free water. Some species, such as the Wild Turkey, are able to obtain water in all three ways, whereas others often rely primarily on one or two methods to obtain water (**Fig. 9–13**).

Free water is simply liquid water that a bird can drink directly, such as water in lakes, rivers, and puddles. Easy access to free water is a requirement for many birds, even some species that live in the desert. In North America these include the House Finch, which can live in the desert only if streams or water holes are located nearby, as well as the Mourning Dove, which is a strong flier and so can venture far from water, but must visit a free water source every day.

Another group of desert birds that visit water holes each day are the sandgrouse of Africa and Asia. To avoid the predators that congregate near water sources, these birds nest far from water on barren ground. Although the adults can fly up to 19 miles (30 km) each day

a. Perching in Shade

b. Sleeking Plumage,
Elevating Wings

c. Perching in Breeze

d. Panting

Figure 9–12. Thermoregulation in Hot Weather in the Cactus Wren: *The desert-living Cactus Wren uses both behavioral and physiological methods to prevent overheating. Behavioral responses include* **(a) perching in the shade**; **(b) sleeking plumage and elevating wings** *to avoid trapping an insulating layer of air next to the skin, at the same time standing tall and elevating the wings to permit heat loss from unfeathered areas; and* **(c) perching in a breeze.** *When behavioral methods are no longer adequate, the physiological response of evaporative cooling takes over and the bird begins* **(d) panting.** *Breathing rapidly with its bill open, the bird exposes the moist mucous membranes of its oral cavity to air, thus increasing the rate of water evaporation from the body. Although a bird can lose much heat through evaporative cooling, it risks becoming dehydrated.*

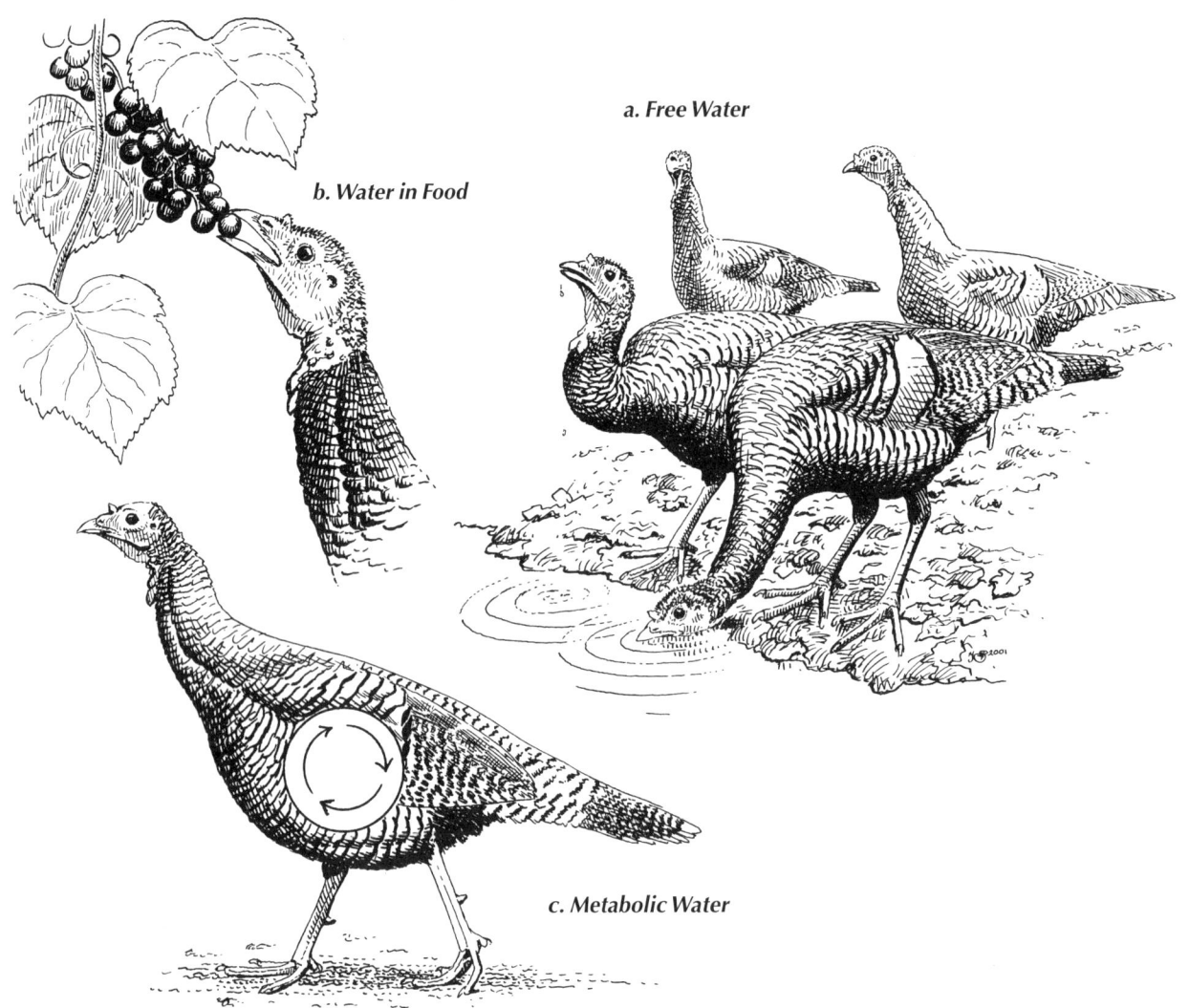

a. Free Water

b. Water in Food

c. Metabolic Water

Figure 9–13. Sources of Water: *Like all living things, birds require water to survive. A bird may satisfy its water needs in three different ways.* **a. Free Water:** *The bird may drink water from lakes, ponds, rain puddles, or streams. Most birds, like these Wild Turkeys, need to drink free water regularly.* **b. Water in Food:** *One important source of water is moisture in food. Soft, watery fruits, such as wild grapes, and juicy insects contain lots of water, but nearly all food contains some water. Many desert birds obtain all their water from food.* **c. Metabolic Water:** *Water also may be obtained as a byproduct of the chemical breakdown of fats, carbohydrates, and proteins within the body. Very few birds rely entirely on metabolic water, however.*

for a drink, the young are confined to the nest, so the parents must bring water back to them. They do this in specially modified breast and belly feathers, which can absorb astonishing quantities—up to 20 mg of water per mg of feather—as the birds stand belly-deep in a pool (Cade and Maclean 1967). After the adults fly back to the nest with their saturated feathers, the nestlings strip the water from the plumage with their bills (**Fig. 9–14**; also see Fig. 8–121).

Other birds survive in dry environments by obtaining water from moisture in the foods they eat. Insectivorous birds such as the Cactus Wren get sufficient water from their insect diet. Similarly, desert predators such as the Greater Roadrunner, or scavengers such as the Turkey Vulture, get plenty of water from their prey (**Fig. 9–15**). Obtaining water from the diet eliminates a bird's need to live near free water, allowing desert survival with few special adaptations.

Finally, a few species satisfy their water needs with metabolic water. When organic compounds, such as fats, carbohydrates, and

a. Namaqua Sandgrouse Soaking Feathers

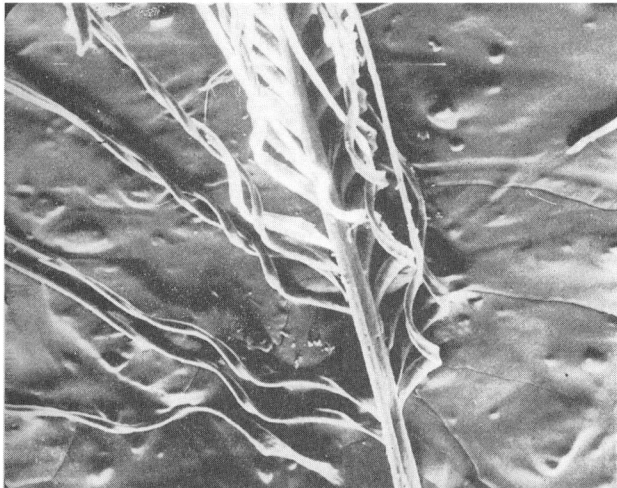

b. Structure of a Sandgrouse Belly Feather

Figure 9–14. Water Transport by Sandgrouse: *Sandgrouse are found in deserts and dry grasslands of Africa and Eurasia. They feed almost entirely on dry seeds, and thus need to obtain free water. Adults fly daily to water holes, typically traveling less than 6 miles (10 km) but sometimes up to 19 miles (30 km). Parent birds, primarily the males, then carry water back to their flightless nestlings in their soaked belly feathers, which are structurally adapted for carrying water.* **a. Namaqua Sandgrouse Soaking Feathers:** *An adult Namaqua Sandgrouse, from southern Africa, soaks its underparts by standing belly-deep in water and rocking back and forth to obtain maximum saturation. When the adult returns to the nest after a flight of 15 minutes or longer, the young birds strip the water from the soaked feathers (see Fig. 8–121).* **b. Structure of a Sandgrouse Belly Feather:** *A scanning electron micrograph reveals the specialized structure of sandgrouse belly feathers (see Fig. 3–4 to review feather structure). The water-carrying feathers differ from typical feathers: their barbules are coiled at the base, and have a straight terminal filament lacking hooklets. When the feather is dry, adjacent barbules intertwine as shown here. When the barbules become wet they uncoil, and their terminal filaments stand up to form a dense mat of hairs, which traps water by capillary action.* **c. Sandgrouse Belly Feather, Dry Versus Wet:** *The difference in appearance between Namaqua Sandgrouse feathers when dry (left) and wet (right) hints at their water-carrying capacity. The average capacity of a fully soaked male, whose water-carrying feathers cover his entire abdomen and lower breast, is estimated to be 0.8 fluid ounces (25 ml) (Maclean 1983), although some of this water may be lost by evaporation during flight. In females the special plumage is limited to the upper belly, so they play a less important role in water transport to their young. Photo of Namaqua Sandgrouse by Charles G. Summers, Jr. Feather photos courtesy of Dr. Gordon L. Maclean.*

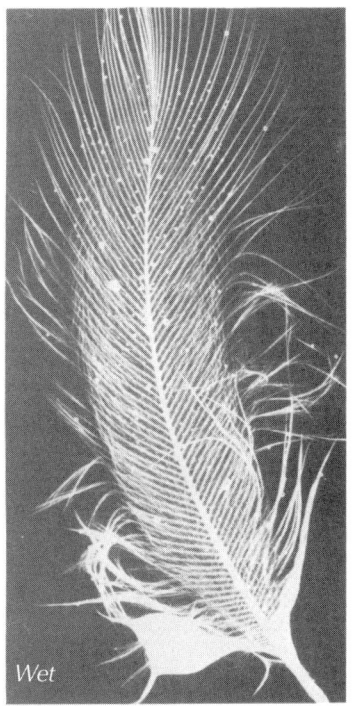

c. Sandgrouse Belly Feather, Dry Versus Wet

proteins, are oxidized (combined with oxygen) during metabolism, they produce water as a by-product. A gram of fat produces 1.07 grams of water, a gram of carbohydrate produces 0.56 grams of water, and a gram of protein produces 0.4 grams of water. Even a diet of dry seeds will yield enough metabolic water for some desert birds, such as the Brewer's Sparrow, to survive for long periods without drinking or eating moist foods. Perhaps you have an Australian Budgerigar or "budgie" parakeet as a pet. Although you surely are conscientious about providing it with water, this desert specialist can get along fine on only metabolic water from its diet of seeds, especially in the moderate environment of your home (Cade and Dybas 1962).

Although obtaining water is important to birds, conserving water also is critical. Birds lose water when they void waste, and keeping this loss to a minimum can be essential for birds in dry climates. Fortunately, birds have some physiological traits that reduce water loss. They eliminate urinary waste as uric acid, which can be concentrated so that little water is required to flush it from the body (see Fig. 4–107). Many birds also can reabsorb water from their feces to excrete very dry droppings. Again, if you have owned a pet Budgerigar, you have seen its dry excrement, common among desert birds.

Evaporation is a special challenge for birds in dry, hot environments because evaporative cooling often is necessary for thermoregulation. The problem is most acute for small birds, whose large surface area-to-volume ratio causes them to evaporate water quickly, requiring them to obtain water often.

Deserts are not the only environments in which birds can have difficulty balancing their water economy. Marine birds must be especially efficient at conserving water because most of their prey comes with a load of salt, the elimination of which requires water. The salt gland (see Fig. 4–131) is a special adaptation allowing marine birds to eliminate salt efficiently, with little water loss.

Figure 9–15. Greater Roadrunner With a Lizard: This New World relative of the cuckoo, native to the southwestern United States and northern Mexico, inhabits arid and semi-arid open country where it often is seen running along the ground. Its omnivorous diet includes insects, spiders and other arthropods, reptiles such as the lizard shown here, small birds and their eggs, rodents, carrion, and occasionally fruits and seeds. Although it will drink copiously if water is available, it usually can meet its water needs entirely from its food. Photo by D. and M. Zimmerman/VIREO.

Foraging Ecology: Meeting Energy and Nutritional Demands

If you watch a bird for a while, you probably will discover that a significant portion of its time and energy is spent searching for food. Birds face tremendous selective pressure to forage efficiently—to get the most food in the least time with the fewest risks—and to select high-quality foods (those that yield the most energy and nutrients). If birds waste time and effort when feeding, or if they select foods that are low in energy and nutrients, they may be too weak to compete successfully for territories or mates, or they may have difficulty rearing young or migrating. As a result of this pressure, most birds are well adapted for foraging.

In Chapter 4 you reviewed the phenomenal variety of anatomical adaptations, especially those of the bill and digestive system, that birds have evolved to improve their food-gathering abilities. From an ecological viewpoint, we are particularly interested in how these specialized foraging tools are used, and the extent to which they limit a bird's foraging options to foods that can be efficiently exploited.

For example, the bills of some hummingbirds, such as the hermits, allow them to take nectar efficiently from only a particular type of flower. This limits the bird's choice of habitats to places where those plants grow. As another example, the mandibles of crossbills are well adapted for extracting seeds from conifer cones, but are clumsy for handling other foods (see Fig. 4–119). This specialization, combined with the high variability in the availability of the birds' food (see Fig. 5–53), forces crossbills to lead a nomadic life. They wander over large areas in search of conifers that are currently bearing cones, which they find only at irregular intervals and in unpredictable geographic areas. These birds have evolved to breed in any month of the year, so they can begin breeding whenever they encounter an area in which cone production is peaking. In contrast to such specialists, birds that have generalized bills, such as crows and jays, can eat a wide range of foods from plants to animals and can live almost anywhere **(Fig. 9–16)**.

How Much Food Does a Bird Need?

All else being equal, large birds need more food than small birds, but small birds need *relatively* more food than large birds. A 10-gram bird (the size of a chickadee) might need 3.5 grams of food per day, which is 35 percent of its body weight. A 100-gram bird (the size of a Blue Jay) might need 10 grams of food a day, which is three times as much food, but only 10 percent of its body weight. And a 1,000-gram bird (the size of a Common Raven) might need to eat 40 grams of food each day—four times that of the Blue Jay, but only 4 percent of its body weight. These disparities relate, in part, to the higher surface area-to-volume ratio of small birds, which causes them to require proportionally more fuel for thermoregulation.

Because smaller birds need relatively more food and generally eat smaller food items, they must spend more time and energy foraging than larger birds. As a result, small birds are more likely to face an

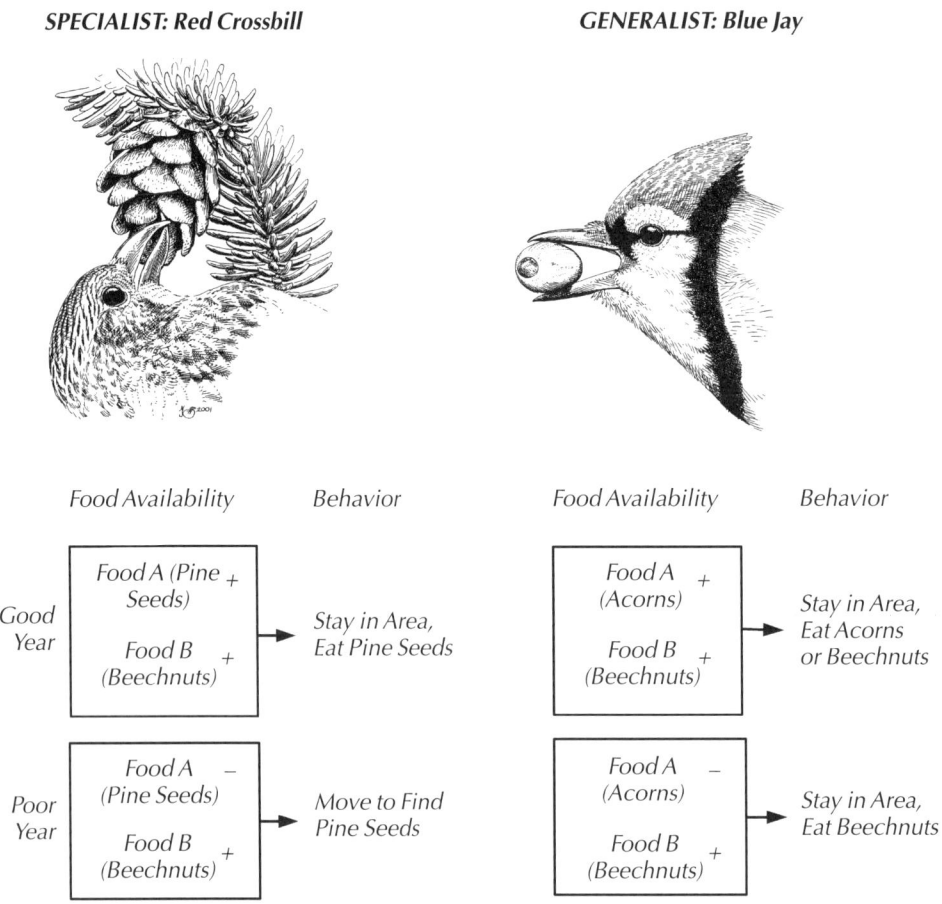

SPECIALIST: Red Crossbill   **GENERALIST: Blue Jay**

Figure 9–16. Feeding Limitations Imposed by Specialized Bills: The bills of some species have become so specialized that they limit a bird to eating only a very few types of foods. In contrast, the bills of generalist species are adapted to handle a variety of foods. The bill of the Red Crossbill, for instance, is adapted for extracting pine seeds from cones, whereas the multipurpose bill of the Blue Jay is capable of handling many food types. The degree of bill specialization has behavioral and ecological consequences for birds. In a year when pine seeds are plentiful in an area (indicated by the + symbol), crossbills can remain in the area to forage and breed. In a year when pine seeds are scarce (– symbol), however, the nomadic crossbill must relocate to find adequate food, because it cannot switch to an alternative food source such as beechnuts. In contrast, if the Blue Jay's preferred food of acorns is scarce, it can easily eat another food such as beechnuts, and thus can remain in one general area to meet its food requirements.

energy crisis if something limits their foraging time, such as inclement weather. If a blizzard keeps a 10-gram Black-capped Chickadee from foraging for one day, its survival may depend on how efficiently it can conserve energy while waiting out the storm **(Fig. 9–17)**. In contrast, a crow could go several days without food before getting into trouble.

A bird will forage until it has satisfied its physiological motivation to continue feeding, which we know as hunger, and which varies considerably with a bird's physiological state. Several conditions, such as heavy exercise, elevate the metabolic rate and cause a bird to eat more. For example, powered flight can increase metabolism to six times the resting rate, with a corresponding increase in energy demand (see Fig. 4–122). Metabolic rates and energy requirements also rise during periods of rapid growth, which occur in young birds or molting adults, as well as when adults are raising young, when birds are in very cold or hot temperatures, and when they are migrating.

In some circumstances, a bird may be physiologically motivated to consume more energy than it requires in the short term (a condition called **hyperphagia**). When such overeating occurs birds convert the extra energy to fat, which they store for use during subsequent periods of extreme energy demand. Many birds become hyperphagic prior to migration, when they accumulate large stores of fat—up to half their body weight—to be used during their lengthy flights (see Fig. 5–60).

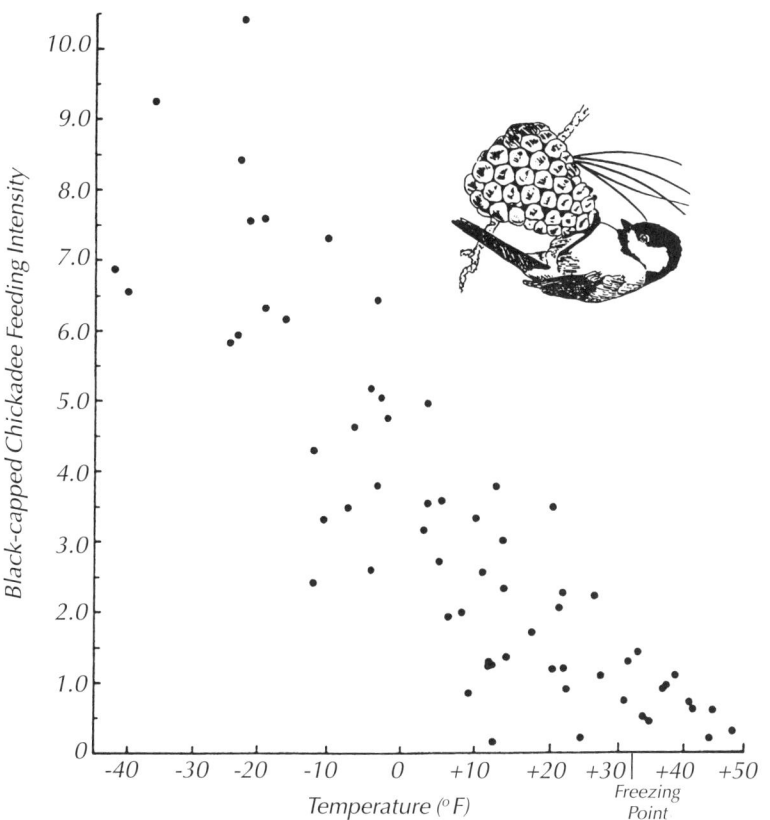

Figure 9–17. Feeding Intensity During Cold Weather: *Like any bird, the tiny Black-capped Chickadee needs more food in cold weather. To obtain a measure of feeding intensity, researchers tracked the amount of time that chickadees spent visiting bird feeders during four winters in Alaska. Then they divided the number of bird-minutes that chickadees spent at the feeders each day (a measure that accounts for the number of chickadees as well as the time that each individual spends feeding) by the total number of minutes the feeder was observed. The measure of feeding intensity climbed as the temperature plunged, because the chickadees increasingly took advantage of the concentrated food source. At warmer temperatures, however, the feeding intensity dropped as the birds spent very little time at the feeders. Adapted from Kessel (1976).*

What Types of Food are Eaten?

Although anatomical specializations determine the range of foods that a species can exploit, the foods that a bird actually consumes are influenced by certain preferences. For example, given a choice of food items that are equal in calories and nutrition, a bird should prefer the food that is most readily available. Often the most available items are those that are most abundant. However, abundance does not necessarily ensure availability. For example, fish may be abundant in a stream, but if the water is murky the fish might not be available to a Belted Kingfisher, which can catch only what it can see. Most birds seem to take the most available items most frequently, a relationship known as a **functional response**.

One might predict a simple linear relationship between the availability of a type of food and the amount of that food in a bird's diet. Such a relationship is rare, however. Instead, birds tend to underuse scarce food items, overuse moderately common foods, and underuse the most available items **(Fig. 9–18)**.

The behavioral explanation for this pattern involves the concept of a **search image**. When a food item is scarce, a bird may not encounter it often enough to learn how to find it efficiently, resulting in "underuse" of that food in relation to its availability. In contrast, a bird can learn to find common items more easily, greatly improving its foraging efficiency, and resulting in "overuse" in relation to its availability. Most of us have experienced a similar phenomenon when we search for hard-to-find items. For example, every Spring when I start searching for morel mushrooms in my woods I have trouble locating

Figure 9–18. Functional Response to Food Availability: As the availability of a bird's preferred food increases in the environment, its percentage in the bird's diet might be expected to increase at the same rate—a simple linear relationship shown by the straight, dashed line on the graph. Instead, however, the bird tends to feed more heavily on that particular food than would be expected (it overuses the item). This probably occurs because the bird develops a search image, allowing it to obtain the food item more efficiently. We see this phenomenon when we observe a flock of Cedar Waxwings stripping a hawthorn tree of every fruit while ignoring other foods nearby. Similarly, birds tend to underuse food items that are particularly scarce. When foods become extremely common, birds tend to use them less often than expected, probably to maintain greater diversity in their diets.

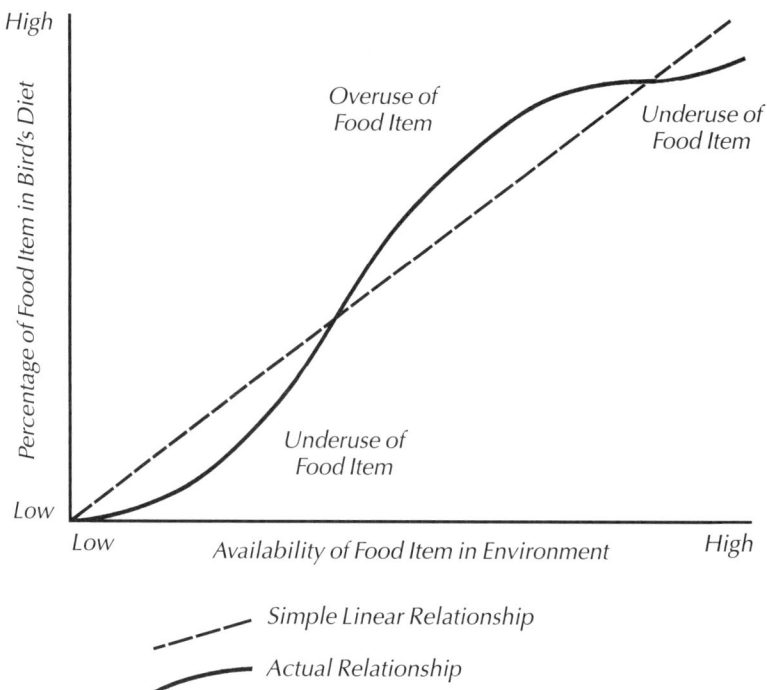

the cryptic fungi, but after I find the first few, I catch on and become increasingly efficient at filling my basket.

In addition, birds may underuse the most available items "on purpose" to keep some diversity in their diet. A diverse diet is more nutritionally balanced than a diet consisting of one type of food. Also, a diverse diet prevents a bird from becoming too dependent on a single food that might unexpectedly fail. Nevertheless, during periodic outbreaks of insects, such as sawflies and spruce budworms in northern forests, or during insect emergences, such as cicadas in deciduous forests, birds often increase their consumption of these insects dramatically.

Birds also learn which foods to avoid and use them less frequently than their availability would predict, or not at all. Many insects have developed chemical defenses against predation, which they advertise by making themselves conspicuous with bold colors and patterns, a strategy called **aposematism**. After a few unpleasant encounters with aposematic insects, birds learn to avoid them, even though they may be abundant. Examples include the previously mentioned sphinx moth larvae on the frangipani tree, as well as the monarch butterfly, which few birds will attempt to eat after sampling a few and suffering the resulting nausea, caused by toxic chemicals that monarch caterpillars acquire from milkweed plants (see Fig. 6–7). The aversion to monarchs is so strong that it provides carry-over protection to the viceroy butterfly, also toxic, which closely mimics the monarch.

When food items are equally available, a bird should prefer those that are most valuable in terms of size, energy, and nutritional content. Large items typically are most valuable because they provide the most energy and nutrients. However, large items may be more difficult for a bird to handle, and the costs of the increased handling effort may cancel the extra value. For example, for seed-eating (**granivorous**) birds, the energetic expense of opening a large, tough seed may not be justified by the energy and nutrients contained inside. As a result, most seed-eaters prefer the largest-sized seed they can handle efficiently, rather than the largest one they can find. Many people with bird feeders can verify that small finches such as American Goldfinches and Pine Siskins prefer small seeds such as those of thistle, whereas larger finches, such as Evening Grosbeaks and Northern Cardinals, prefer sunflower seeds **(Fig. 9–19)**. Similarly, birds of prey generally ignore animals that are too big to subdue and kill without difficulty. Thus by eating foods of appropriate sizes, birds can maximize their net energy and nutritional gains while minimizing wasted time, wasted energy, and risks (see Fig. 6–31).

Size, of course, is not the only measure of food value. Birds also should prefer food items with the highest concentration of energy and nutrients, and evidence suggests that birds can make such choices.

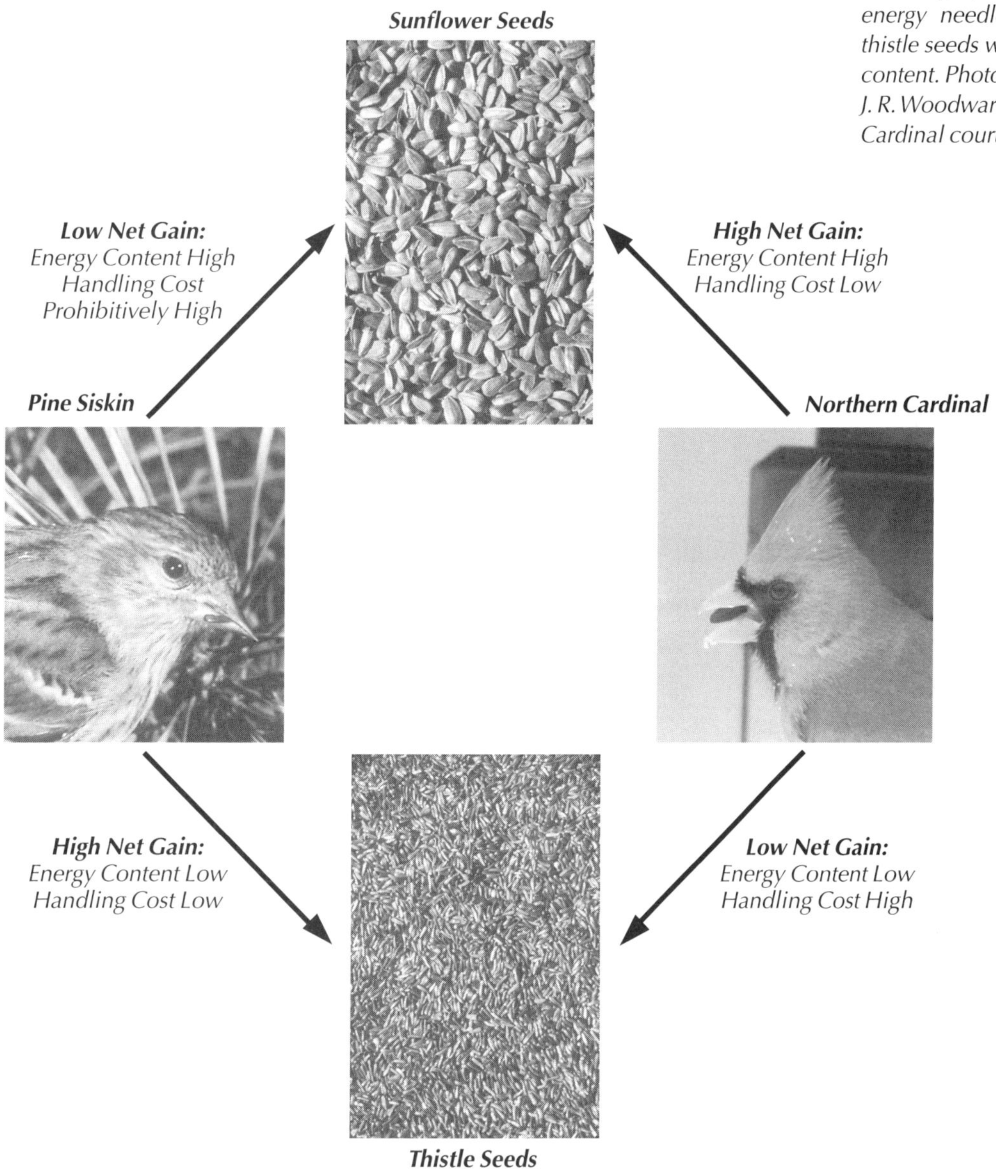

Figure 9–19. Costs and Benefits of Food Handling: Birds typically prefer to eat foods with high nutritional value (energy content). However, a feeding bird also must consider how much energy it expends handling a food item, because net energy gain equals energy content minus handling cost. Although a sunflower seed contains more energy than a thistle seed, its large size makes handling difficult for the small Pine Siskin, increasing the energy expended and decreasing the bird's net energy gain. Similarly, a Northern Cardinal—which can easily manipulate sunflower seeds in its large, powerful bill—would waste energy needlessly by handling small thistle seeds with relatively little energy content. Photo of Pine Siskin courtesy of J. R. Woodward/CLO. Photo of Northern Cardinal courtesy of Marie Read.

Sunflower Seeds

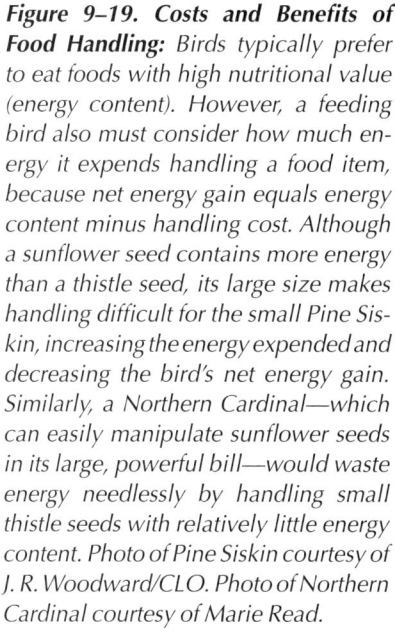

Low Net Gain:
Energy Content High
Handling Cost
Prohibitively High

High Net Gain:
Energy Content High
Handling Cost Low

Pine Siskin

Northern Cardinal

High Net Gain:
Energy Content Low
Handling Cost Low

Low Net Gain:
Energy Content Low
Handling Cost High

Thistle Seeds

Perhaps you have watched American Robins, European Starlings, or Cedar Waxwings ignore the fruits on a backyard tree until they are ripe, then begin feeding on them voraciously. In this case, the preference is related to the peak sugar (and hence, energy) content of ripe fruit. Birds use several cues, such as taste, color, and texture, to determine a food's peak energy content.

Birds sometimes make special efforts to obtain specific nutrients by consuming foods that have little value beyond the nutrients themselves. For example frugivorous birds, such as parrots and macaws, regularly travel considerable distances to visit clay licks where they eat soil containing minerals that are missing from their fruit-laden diets **(Fig. 9–20)**. In a similar way, hummingbirds must supplement their high-energy, high-carbohydrate diet with insects, which provide essential proteins and minerals lacking in nectar.

As another example, vultures feed primarily on the soft tissues of large dead animals (**carrion**), which contain low concentrations of calcium, an essential nutrient. To counter this deficiency, vultures prefer carcasses with small fragments of bone around them. This food requirement presented a problem for Cape Griffons and White-backed Vultures in the savannas of South Africa when most of the large predators were removed to make the areas safe for ranching. Cattle carcasses continued to provide the vultures with food, but because bone-crushing predators such as hyenas were gone, few bone fragments were available. As a result, vulture nestlings developed severe calcium deficiencies, breaking their weak bones and perishing before they could leave the nest. In response, conservationists created "vulture restaurants" near nesting colonies, stocking them with bone fragments that adult vultures eagerly consumed and fed to nestlings **(Fig. 9–21)**.

Food preferences also vary with weather and season. On a cold winter day a Black-capped Chickadee generally selects energy-rich foods containing lots of fats and carbohydrates, such as suet and sunflower seeds. During summer, the same chickadee may forsake sunflower seeds in favor of foods, such as insects, which are rich in proteins and minerals needed for reproduction and growth.

Where and How to Forage

How does a bird decide where to forage? In general, the answer depends on how the bird's food resources are

Figure 9–20. Macaws at a Mineral Lick: Birds often seek out minerals lacking in their diets. Here, Scarlet and Red-and-green macaws, which obtain few minerals in their primarily fruit diet, are using their hooked bills to bite chunks of mineral-rich clay from a steep river bank in Peru. The minerals are released as the clay particles are ground in the gizzard. One mineral often sought in the breeding season is calcium, which females need to form their eggs. Photo by Bruce Lyon.

distributed. Some food items are clumped, so that when a bird discovers one such item, it should search nearby for more. For example, if a Wild Turkey discovers an acorn hidden on the forest floor it stops to scratch in the nearby leaf litter, because acorns usually are concentrated in patches around oak trees. If the bird does not find more acorns quickly it gives up, leaves the site, and searches elsewhere. In contrast, when a Golden Eagle catches a jackrabbit it does not bother to search for more rabbits nearby, because jackrabbits are not clumped on the landscape.

Some food that comes from plants is renewable or continually growing, such as nectar and fruit, and therefore birds often revisit food-bearing plants after allowing them to recover (regrow). Consider a hummingbird that has extracted nectar from a flower. In a few hours the nectar will be replaced, so many hummingbirds visit flowers at regular intervals, a strategy known as "traplining" **(Fig. 9–22)**.

Birds that forage in loose groups rather than defending individual feeding territories frequently locate food by watching other individuals (see Chapter 6, Sidebar 4: Living in Groups). For example, vultures searching for carrion and cormorants looking for fish often converge on a site where other individuals are actively foraging.

Because winter bird feeding has become extremely popular, some birds now have a concentrated, abundant, and reliable source of food, in contrast to natural food sources that often are patchy, scarce, and exhaustible. Do feeder birds switch their foraging tactics to rely primarily or entirely on feeders, reducing the energetic costs of foraging over a large area?

One winter I studied this question at my rural Wisconsin feeder. Each day I observed the number of sunflower seeds taken by dozens of individually marked Black-capped Chickadees **(Fig. 9–23)**. I calculated the total amount of energy contained in all those seeds and compared the result to the estimated daily energy requirements of the birds (corrected for the average daily temperature; remember that energy requirements rise as temperatures fall). I discovered that even though the birds had unlimited access to sunflower seeds, they used them to satisfy only 20 to 25 percent of their energy requirements. Apparently the birds still foraged extensively on natural foods located throughout their large home ranges. In fact, continuing to feed on natural foods makes adaptive sense. Because sufficient food is critical to a chickadee's winter survival, relying on a single food source—which in nature would not be as reliable as a feeder—would be risky. Therefore chickadees show a strong evolutionary response to food unpredictability, maintaining a flexible foraging strategy that prevents them from

Figure 9–21. White-backed Vultures at a Carcass: *The first feeding stations for scavenging birds—dubbed "Vulture Restaurants"—were established in 1966, when conservationists realized that populations of the eight species of vultures in South Africa were declining. Conversion of the African veld to agriculture had reduced populations of wildebeest and other prey animals, resulting in a scarcity of carcasses. Also declining was the hyena, and without this large scavenging mammal to crush the bones of the few available carcasses, White-backed Vulture nestlings were suffering from calcium deficiency owing to a lack of bone fragments in their diet. To prevent vulture bone deformities, volunteers now assume the grisly task of hacking up donated livestock carcasses so that bone fragments will accompany the flesh that parent birds take back to their young. Vulture restaurants also benefit bird watchers, who visit them to observe several different species and their behaviors. Photo by M. P. Kahl/VIREO.*

Figure 9–22. Trapline Foraging in the Long-tailed Hermit: *A Long-tailed Hermit from the lowland rain forests of Central and South America hovers to sip nectar from a flowering Heliconia. The dashed lines show its foraging route, along which it will visit each flower in sequence, returning at regular intervals throughout the day. Hermits—hummingbirds of the subfamily Phaethorninae, which includes the sicklebills (see Fig. 9–5)—use the strategy of "traplining" (circuit foraging) to obtain nectar from flowers scattered through the forest, in contrast to defending flowers within a territory, as do some other hummingbird species. Nectar is a renewable food source, which, once harvested, takes time to replenish. Therefore a hermit optimizing its energy use should not revisit a given flower too soon. The bird, however, must balance the time it waits against the risk that a competitor may reach the flower and harvest its nectar first. By observing uniquely marked individuals, Gill (1988) noted that male Long-tailed Hermits revisited their traplines from three to seven times per morning, at intervals ranging from a few minutes to two hours. Experiments in which artificial hummingbird feeders, refilled at fixed intervals, were included as part of traplines showed that hermits revisited at shorter intervals when competitors also used a feeder than when the birds had exclusive use. Presumably, by visiting more often, the hermits were attempting to prevent losses from competition.*

"putting all their eggs in one basket" (**Sidebar 1: The Winter Banquet**).

Do Birds Always Forage Optimally?

We have discussed how search strategies and preferences for particular foods can allow birds to forage most efficiently. In theory, when a bird uses the smallest amount of energy to gather high-energy, nutritional food, **optimal foraging** is occurring (Krebs and Davies 1981) (**Fig. 9–24**). However, birds do not always forage in this manner, as illustrated by the chickadee example just discussed. Indeed, trade-offs can make an "optimal" strategy less than ideal. If, for example, foraging optimally caused a bird to be more vulnerable to predators, the bird might choose to forage with lower efficiency but greater safety (see Ch. 6, Feeding Behavior: Why do Birds Generally Restrict their Diets, Ignoring Some Edible Foods in Favor of Others?). This trade-off can be observed in many backyards, where birds may be reluctant to visit feeders far from protective cover.

Coping with Environmental Fluctuations

Life would be easier for birds if the environment were constant. Fluctuations are a fact of life, however, and birds must synchronize their activities to biotic and abiotic changes occurring on time scales from hours to years. All of these changes—daily and seasonal cycles of daylight, temperature, moisture, and food availability, as well as competition and predation—can affect birds' lives. To respond appropriately, birds must have a sense of time.

Cycles of light and dark regulate a bird's sense of time and daily activity patterns, and each species has a characteristic daily activity cycle. Most birds are **diurnal** (active during daylight), but some are **nocturnal** (active at night), or **crepuscular** (active at dawn and dusk). Often, obvious environmental factors influence the time that a bird will be most active. For example, desert birds avoid midday activities when the heat is greatest, and hawks often hunt actively around dusk and dawn—when their prey are most active. Furthermore, hawks frequently use thermals created by the midday sun to soar and travel (**Fig. 9–25**).

Differences in activity cycles can affect how birds living in the same habitat interact with each other. Diurnal hawks and nocturnal owls, because they are active at different times of day, can hunt in the same area and avoid competition. The same is true for diurnal swallows and nocturnal nighthawks, which both feed on flying insects.

Seasonal changes also regulate activity patterns of birds. Such changes must be anticipated before they occur, so that birds can begin preparations. A Red-eyed Vireo wintering in the tropics migrates

Figure 9–23. Banded Black-capped Chickadee Takes a Sunflower Seed from a Feeder: Even the most regular visitors to backyard bird feeders are only supplementing their natural diets. Perhaps surprising, this finding is just one of many made possible by the technique of color banding, which allows observers to identify and monitor the activities of individual birds. Research on banded chickadees in Wisconsin found that sunflower seeds provided at feeders account for only 20 to 25 percent of the bird's winter energy requirements. In fall, chickadees cache food in bark crevices, knotholes, and even pine needle clusters; these hidden stores provide an important winter food source. Research indicates that during fall, chickadees can produce new brain cells to handle the additional memory they need to recover the cached food in winter. Photo by Marie Read.

(Continued on p. 9·35)

Sidebar 1: THE WINTER BANQUET
Stephen W. Kress

A few decades ago, most of the birds that fed in North America's backyards ate weed seeds and insects gleaned from crevices in tree bark. Now they're presented with other feeding options. Nearly one-third of the adult human population of North America dispenses about a billion pounds of birdseed each year as well as tons of suet and gourmet "seed cakes." What changes does this largesse impose on our native birds?

Figure A. Black-capped Chickadee at a Bird Feeder: *Because winter bird feeding is a popular pastime throughout much of North America, seed-eating birds often visit backyards to find reliable, concentrated food sources. Nevertheless, under typical winter conditions Black-capped Chickadees still rely mainly on natural foods, obtaining only about 21% of their nutrition from feeders. Photo courtesy of Lang Elliott/CLO.*

There is surprisingly little research on the effects of feeders on individual species, but the limited studies so far suggest that backyard feeders are not creating a population of dependent wintering birds. For example, researchers Margaret Brittingham and Stanley Temple from the University of Wisconsin compared winter flocks of Black-capped Chickadees in two similar woodlands in Wisconsin—one left natural and one equipped with feeders stocked with sunflower seeds. After three years of study, they found that winter survival rates were highest in the woods with the feeders—but only during winters with prolonged periods of extreme cold **(Fig. A)**. This suggests that in milder climates, feeders may have little effect on the winter survival of chickadees. The study also found that nesting populations in both woods were similar the following spring.

Other research by Brittingham and Temple allays the concern that birds may lose their natural talent for finding food and become dependent on the easy life of taking food at feeders. To test the ability of chickadees to switch back to natural food, they removed the feeders from a woodland where birds had been fed for the previous 25 years and compared survival rates with those of chickadees in a nearby woodland where there had been no feeders. They documented that chickadees familiar with feeders were able to switch back immediately to foraging for natural foods and survived the winter as well as chickadees that lived where no feeders had been placed. This was not surprising, since food from feeders had made up only 21% of the birds' daily energy requirement in the previous two years. Clearly, much more work needs to be conducted with many feeder-frequenting species throughout their ranges. These studies, however, suggest that winter feeding does not promote dependency **(Fig. B)**.

Yet anyone who has conducted a Christmas Bird Count in northern latitudes will testify that most wintering land birds are found around bird feeders. That's not surprising, since feeders reduce the time it takes to find food, and the size of the average meal (juicy sunflower kernels and suet) is certainly larger than tiny weed seeds or wintering insects dug out of tree bark. Feeders may also reduce the risk of predation, since feeder birds spend less time foraging and have more time to watch for predators. However, these benefits can be offset by more exposure to disease, increased collisions with windows, and greater vulnerability to house cats that may lurk near feeding stations.

Figure B. Common Redpolls at a Thistle Seed Feeder: As the snow flies on a chilly winter day, Common Redpolls fill the perches on a hanging thistle seed feeder. Although birds certainly benefit from feeding during severe weather, studies suggest that winter feeding does not create populations of dependent birds. Photo by Marie Read.

Feeding birds is certainly not as helpful as improving backyard habitat through landscaping, which provides food for a wider variety of birds, as well as providing shelter, nesting places, and perches. But one thing is certain: feeding helps close the distance between wild birds and people.

Below, a look at some of the most frequently asked feeder questions.

Does Feeding Increase Nest Predation?

Some critics of backyard bird feeding argue that feeders bolster the populations of nest predators such as Blue Jays and grackles, nest competitors such as European Starlings and House Sparrows, and nest parasites such as Brown-headed Cowbirds. But an analysis of the North American Breeding Bird Surveys conducted between 1966 and 1996 found that these species are actually declining—along with many other grassland

and thicket-nesting birds. In contrast, resident forest birds, including Black-capped Chickadees, Hairy Woodpeckers, White-breasted and Red-breasted nuthatches, and Tufted Titmice, are on the increase. Most likely, the regrowth of forest habitats and changing agricultural practices are having far greater effects than backyard feeders.

Does Feeding Birds Affect Migration?

Bird migration is triggered by changes in day length, not the availability of food. Birds that frequent feeders usually will not linger past their normal migration time. Stragglers at feeders are generally birds that are injured or that lack migration talents. Most of them will not survive the winter.

Will Birds Suffer If Feeders Go Empty in the Winter?

In most instances, "your" backyard birds have a feeding tour that includes many neighborhood feeders as well as natural foods. Unless you provide enormous amounts of food and live in an isolated location, it is likely that the birds visiting your feeders will not suffer if you leave your feeders empty for a few weeks of winter vacation.

What Feeders Should I Use?

*Several feeders offering different kinds of foods reduce waste and minimize congestion. Having multiple feeders in different places will also attract a variety of species, since birds typically feed at different levels (**Fig. C**). Woodpeckers, jays, nuthatches, chickadees, and finches readily feed in trees, so they will visit higher-set feeders. Likewise, cardinals, tow-hees, sparrows, and juncos usually feed near the ground. Place feeders close to windows so that you can enjoy the action, but beware of large picture windows that may result in collisions. Avoid ground feeders if there is a risk that house cats will pounce from nearby shrubs.*

Can Birds Catch Diseases at Feeders?

Diseases such as Salmonella can spread at feeders, especially where seeds and droppings mix. Ground-feeding birds, such as doves and finches, are especially vulnerable. To reduce the risk of disease, clean your feeders at least once a year with a 10% bleach solution—one part bleach to nine parts water.

How Can I Keep Squirrels Out of Feeders?

Avoid hanging feeders from trees and eaves. Place them on isolated poles at least five feet off the ground and as far as possible from your house and nearby trees and tall shrubs, keeping in mind that squirrels can leap as far as six feet. Attach to the feeder pole either an in-

Figure C. Red-bellied Woodpecker at a Suet Feeder: *A variety of feeders offering different food types and located in different places attracts a diversity of species, reduces waste, and minimizes overcrowding. Suet feeders regularly attract woodpeckers, jays, nuthatches, chickadees, and titmice, along with occasional wrens, creepers, and warblers. Photo by Marie Read.*

Is it Okay to Leave My Cat Outside as Long as I Put a Bell on Its Collar?

There are about a billion domestic and feral cats in North America, and they kill about 2 million wild birds each day. Cats account for about 30% of birds killed at feeders. Cats are such stealthy hunters that they can stalk and pounce on prey without jingling the bells on their collars. By keeping your cat indoors, you'll not only protect birds but also keep the cat safe from disease, traffic, and fights with neighborhood pets and wildlife.

What Should I Do if a Hawk Starts Killing Birds at My Feeder?

It is important to discourage hawks from dining on feeder birds, because both predator and prey are vulnerable to accidental deaths resulting from window collisions. If a hawk sets up a regular feeding routine at your feeders, stop feeding long enough to let the smaller birds disperse. Soon the hawk will leave to feed elsewhere.

How Can I Stop Birds from Crashing into My Windows?

A recent study found this to be the most common cause of death associated with feeders (about 1 to 10 deaths per building each year). Falcon silhouettes attached to windows seem to have little effect. Fruit-tree netting stretched taut several inches in front of the glass is the best approach.

Will Birds Stick to Metal Feeder Parts During Subfreezing Temperatures?

Our fingers may stick to metal ice cube trays because moisture freezes on contact with frigid metal. However, a bird's feet are covered with dry scales, so there is no surface moisture to freeze to metal perches. Eyes, tongues, and beaks are usually safe from exposure to metal feeder parts. Rapid reflexes prevent the eye from coming in contact with foreign surfaces, and the beak is protected by a horny, dry surface.

This article originally appeared in Audubon, January-February 2000, and is reprinted by permission of Dr. Stephen W. Kress.

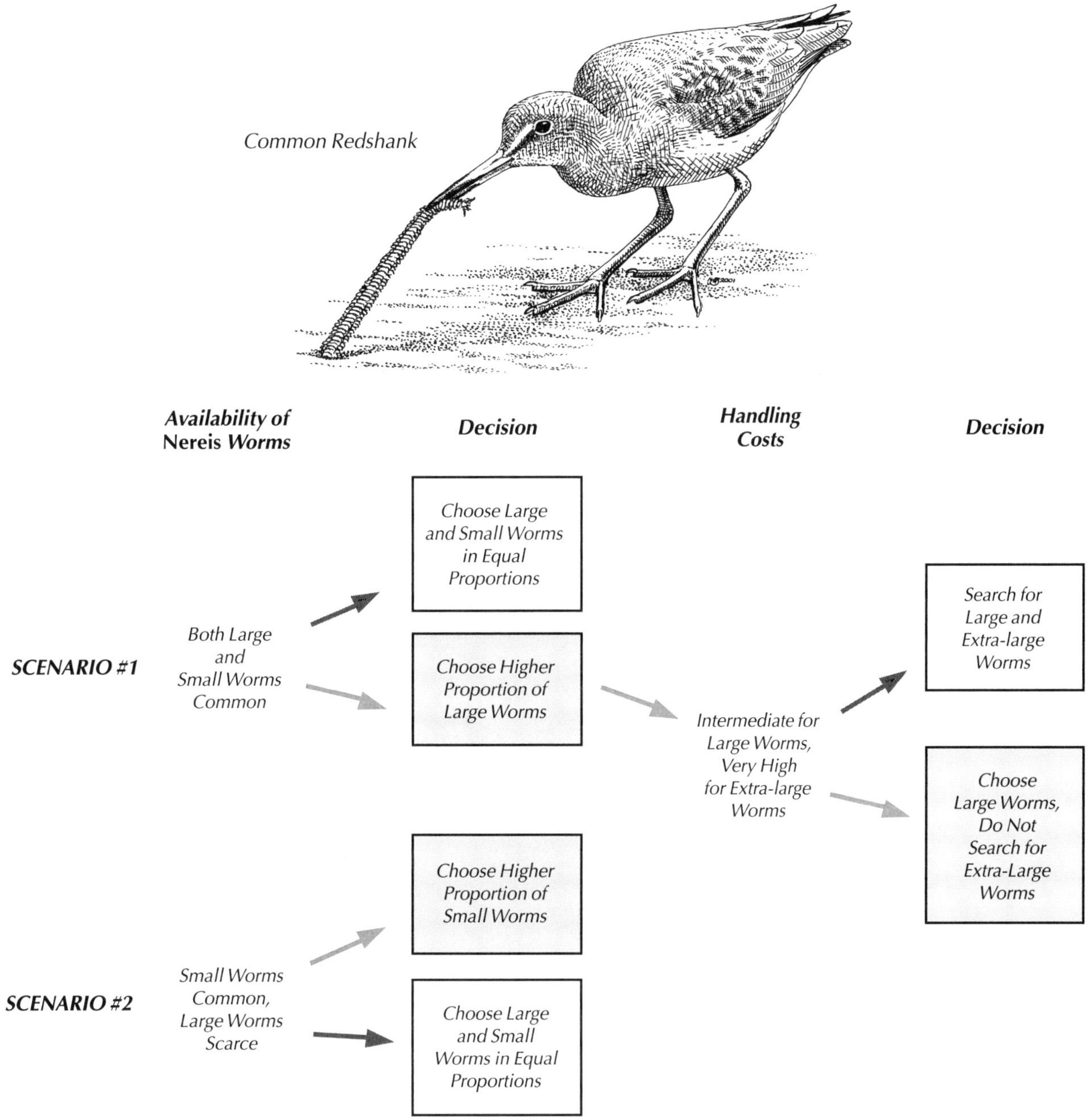

Common Redshank

Availability of Nereis Worms	Decision	Handling Costs	Decision

SCENARIO #1
Both Large and Small Worms Common

Choose Large and Small Worms in Equal Proportions

Choose Higher Proportion of Large Worms

Intermediate for Large Worms, Very High for Extra-large Worms

Search for Large and Extra-large Worms

Choose Large Worms, Do Not Search for Extra-Large Worms

SCENARIO #2
Small Worms Common, Large Worms Scarce

Choose Higher Proportion of Small Worms

Choose Large and Small Worms in Equal Proportions

Figure 9–24. Optimal Foraging in the Common Redshank: *The major assumption underlying the theory of optimal foraging is that an animal should maximize its caloric intake while minimizing the associated costs such as searching, prey capture, and handling. Consider the Common Redshank, a species of shorebird native to Eurasia. In the marine estuaries of the British Isles this bird probes the mud for prey, including* Nereis, *a species of polychaete worm. Studies suggest that, faced with a variety of worm sizes, the redshank does not forage randomly but selects prey sizes that allow it to optimize the weight of prey ingested per unit time (Goss-Custard 1977).*

This diagram illustrates the kinds of decisions that a foraging redshank might make, with the colored boxes showing optimal decisions. In scenario 1, where large and small worms are equally abundant, the redshank chooses a higher proportion of large worms, avoiding small ones. However, the bird also avoids extra-large worms, which would provide even more calories, because they may incur increased handling costs. For instance, increased physical effort may be needed to subdue them. Furthermore, the longer time required to process large prey might expose the redshank to predation by a Peregrine Falcon, or attract the attentions of an aggressive scavenger such as a skua. As shown in scenario 2, only when small worms are common and large ones scarce does the redshank include more small worms in its diet.

Night: Roosting

Early Morning: Hunting

Late Morning: Hunting

Midday: Resting, Preening

Early Afternoon: Move to New Hunting Area

Late Afternoon: Hunting

Late Evening: Return to Roost

Figure 9–25. Daily Activity Cycle of a Red-tailed Hawk: *Each bird species has a typical pattern of activity through each 24-hour period, regulated by the earth's daily cycle of light and darkness. For instance, many diurnal birds are active in the early morning and late afternoon, but are relatively inactive during midday, especially in hot weather. Shown here are the typical daily activities that might be exhibited by a diurnal raptor, the Red-tailed Hawk, during the nonbreeding season. The hawk spends the night roosting in a stand of conifers. At daybreak it moves to one of its habitual perches, then spends the early and late morning hunting for small mammals and reptiles from various perches around its home range. During midday, the red-tail heads for a snag to rest and preen. Early afternoon may find it moving to another hunting area, soaring on thermals if the day is sunny. It continues hunting in the late afternoon. Finally, in late evening the hawk flies back to its nighttime roost.*

thousands of miles to its temperate zone breeding grounds, arriving at the beginning of the flush of summer insects, the perfect time to rear its young (see Fig. 8–7). A Great Horned Owl begins to nest in the January cold; when its young leave the nest three months later, their rodent prey will be at their annual peak. What stimulates these birds to prepare for events that will take place in the distant future?

The answer involves an interplay between long-term and short-term responses. Both vireos and owls have evolved to rear their young at a time when food is abundant. Because it appears to be the main factor underlying the timing of breeding in most birds (see Ch. 8, The Timing of Breeding), food abundance is considered an **ultimate factor** (see Ch. 6, Questions About Behavior). When nesting is properly timed, then nesting success, growth rates of nestlings, and survival of adults and young will all be high, ultimately affecting the birds' fitness.

But short-term responses also are involved, because birds time their activities based on environmental cues called **proximate factors**. These cues do not directly impact a bird's fitness, but they do provide a crucial sense of time. As discussed in Chapters 4 and 5, the proximate factor that determines migration and breeding activity is **photoperiod**. Indeed, the change in day length is the bird's best calendar, because it can be measured precisely and is highly reliable. So in an immediate sense, when birds schedule these critical life events they are responding to the amount of light occurring each day. But over the long-term, they are responding to the seasonal availability of food.

Sometimes birds must cope with environmental cycles lasting more than a year, often longer than the life span of most individuals. In northern forests, snowshoe hares undergo dramatic 10-year cycles in abundance. In the Arctic, lemmings and other rodents experience 4-year abundance cycles, and ptarmigan undergo 10-year cycles. These population cycles of prey animals have serious consequences for avian predators. At regular intervals, Northern Goshawks are faced with shortages of hares; Gyrfalcons with shortages of ptarmigan; and Snowy Owls, Rough-legged Hawks, and Northern Shrikes with a paucity of lemmings. The predators must respond or die. Although a few can stay put and find alternative prey, many disperse to places where prey populations are more abundant. These "irruptions" can take these birds far south of their normal winter ranges **(Fig. 9–26)**.

Figure 9–26. Snowy Owl/Lemming Predator/Prey Cycle: Naturally occurring cycles in prey populations can have significant impacts on the populations of predators, especially in habitats or climatic conditions where alternative prey are limited. In Canada, Snowy Owls feed primarily on lemmings throughout the winter. Because few alternative prey are available in that harsh climate, when lemming populations decline in Canada, Snowy Owls must look elsewhere for food, and they move southward into New England. In this graph, the black line shows the population cycle of lemmings in Canada, and the colored blocks represent the relative abundances of Snowy Owls in the New England states in years in which they move southward in large numbers—when lemming populations begin to decline in Canada. In other years, Snowy Owls are present in New England in much lower numbers. After Shelford (1945).

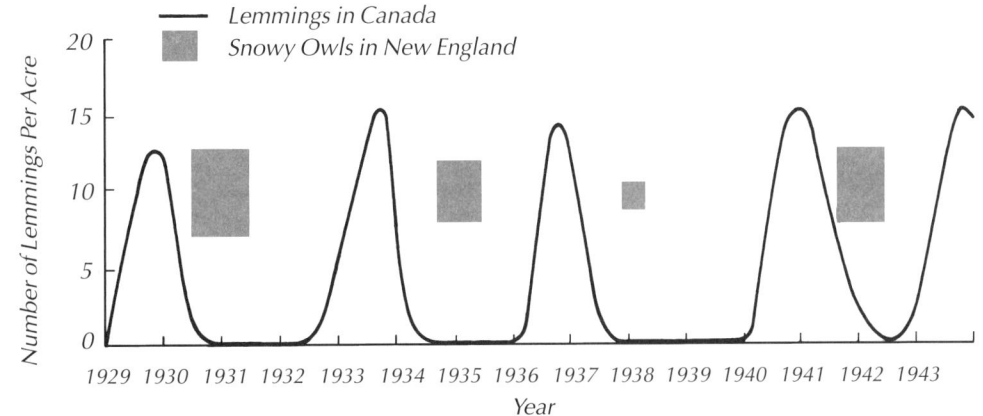

Relationships with Other Individuals

Birds frequently interact with other individuals of the same species, and such **intraspecific** interactions can be cooperative or antagonistic. Cooperative interactions result in reciprocal benefits. Antagonistic interactions yield winners and losers; one individual benefits at the other's expense.

Many examples of cooperation among individuals, usually members of the same pair, family, or flock, were discussed in Chapter 6. However, some cases of cooperation are best understood as adaptations that help an individual cope with its environment. For example, individual nuthatches and creepers sometimes spend cold nights in a shared roosting cavity. By huddling together they become a single large mass with a correspondingly lower surface area-to-volume ratio, so that each bird obtains an advantage by reducing its individual rate of heat loss **(Fig. 9–27)**.

Figure 9–27. Eurasian Treecreepers in a Communal Roost: Small birds that live in harsh climates, such as creepers, nuthatches, and chickadees, can benefit by roosting communally. Here, a group of Eurasian Treecreepers, relatives of the North American Brown Creeper, roost together amid deep crevices in rough bark. By huddling, the birds become a single large mass with a decreased surface-area-to-volume ratio, thereby losing less heat than they would if each bird roosted alone. An even better strategy for keeping warm is roosting communally in a cavity, as do nuthatches, chickadees, and bluebirds (see Ch. 6, Sidebar 4: Living in Groups, Fig. H), because the birds' shared body heat actually warms the air within. By reducing heat loss in cold, windy conditions, birds save energy, which can increase their survival. Illustration by N. John Schmitt, from Birds Asleep, by Alexander F. Skutch. Copyright 1989. Used by permission of the illustrator and the University of Texas Press.

Types of Intraspecific Competition

The majority of intraspecific interactions are antagonistic. Individuals of the same species tend to seek the same resources, which often are in limited supply, so **intraspecific competition**, be it for habitat, food, mates, nest sites, or other essential resources, typically is intense. The resulting antagonism can be a major component of a bird's life, with time, energy, and numerous adaptations devoted to resolving conflicts.

Resolution occurs in two different ways: differential exploitation and interference. **Differential exploitation** allows competing individuals to avoid direct interactions. If one individual can exploit a resource more efficiently than can another, it will receive a disproportionate share, forcing the second individual to use slightly different resources or to use the same resource in different ways. Differential exploitation occurs when male and female American Kestrels use slightly different habitats during winter to avoid competition for the same prey species. **Interference**, on the other hand, involves direct antagonism in which one individual actively prevents another from obtaining a limited resource. Interference occurs when a Northern Mockingbird fiercely defends a fruit-laden shrub against other local mockingbirds.

Differential Exploitation

Differential exploitation may occur in various ways. Often individuals have behavioral differences involving learning that allow them to exploit resources in different ways. For example, older birds typically are more efficient at finding food than younger, inexperienced individuals. As a result, adults and immatures often rely on different resources, avoiding direct competition. Young gulls, for example, often forage in different areas than adults.

Differential exploitation also can occur as a result of physical differences among individuals. The catch, in an evolutionary sense, is that morphological divergence among members of the same species is difficult to achieve in a population of interbreeding birds that share most of their genes. Nonetheless, some species do maintain significant individual variations within a population.

A common example is **sexual dimorphism**, in which males and females of the same species differ in size or form, allowing them to exploit slightly different resources. Because members of a mated pair usually share a home range, having a mate who uses slightly different resources in the shared area is an obvious benefit.

Bills of different sizes and shapes are good examples of sexual dimorphism. In some species, such differences are so dramatic that early naturalists believed males and females to be members of different species. For instance, in the extinct Huia of New Zealand, the bills of males were straight, short, and stout, whereas those of females were decurved, long, and thin (**Fig. 9–28**; also see Fig. 1–92). These bills were obviously suited to very different and perhaps nonoverlapping foraging techniques.

More common sexual dimorphisms are differences in body sizes. Although they are usually the result of **sexual selection** driven by fe-

9

males favoring large males (see Ch. 3, Sexual Selection), some cases of sexual size dimorphism seem to be adaptations that reduce competition. When males and females differ markedly in body size, they often eat different foods, reducing competition between mates.

Sexual size dimorphism is common among predatory birds, including raptors, jaegers, skuas, frigatebirds, and boobies. However, these birds show **reverse sexual size dimorphism**, in which females can be up to twice as large as males. Studies of the diets of strongly dimorphic raptors confirm that males and females do take different sizes and types of prey, but why predatory birds show reversed size dimorphism remains unclear. Bildstein (1992) summarized the many hypotheses that have been proposed to explain reverse sexual size dimorphism among raptors, placing them in the following three categories:

Ecological:
- Reverse sexual size dimorphism reduces food competition between members of a mated pair.

Physiological and Anatomical:
- Large females lay larger (better?) eggs than smaller females.
- Large females better protect developing follicles (eggs) during hunting than smaller females.

Figure 9–28. Sexual Dimorphism in the Huia: *The Huia, an extinct passerine of New Zealand forests, possessed such extreme sexual dimorphism in bill shape that males and females once were thought to be separate species. The shorter, powerful beak of the male (front) was well suited to tearing off bark and chiseling wood in search of insects, whereas the female's more slender, curved beak could probe for prey more deeply and precisely. A 19th-century New Zealand ornithologist once watched a pair feeding from a rotting log. "Sometimes," he wrote, "I observed the male remove the decayed portion without being able to reach the grub, when the female would at once come to his aid, and accomplish with her long slender bill what he had failed to do. I noticed, however, that the female always appropriated to her own use the morsels thus obtained." Their functionally different bills enabled a Huia pair to feed efficiently without competing with each other. Competitive collectors were responsible for the bird's demise, and it was extinct by 1910. Chromolithograph after a painting by J. G. Keulemans, from W. L. Buller's* History of the Birds of New Zealand, *Vol. 1 (London, 1888), plate 2.*

- Large females are better incubators than small females.
- Large females are better able to withstand periods of food shortage during incubation than small females.
- Small males spend less energy providing food for their young than large males.

Behavioral:
- Large females are better protectors of their nests than small females.
- Small males are better protectors of their nests than large females.
- Large females are better preparers of food for their nestlings than small females.
- Large females prevent small males from eating their own young.
- Large females are better able to form and maintain pair bonds than small females.

As you can surmise from the lists above, reverse sexual size dimorphism is likely due to several influences. It is clear, however, that the degree of raptor sexual size dimorphism is closely tied to diet. The most dimorphic species are bird-eaters, including accipiters such as the Sharp-shinned Hawk, and falcons such as the Merlin. Size dimorphism is much less pronounced in species such as Red-tailed Hawks, which eat smaller and slower prey such as mice, and dimorphism is almost nonexistent in carrion-eating raptors such as vultures **(Fig. 9–29)**. This pattern may be related to the fact that a bird-eating raptor frequently has few competitors other than members of its own species. Thus,

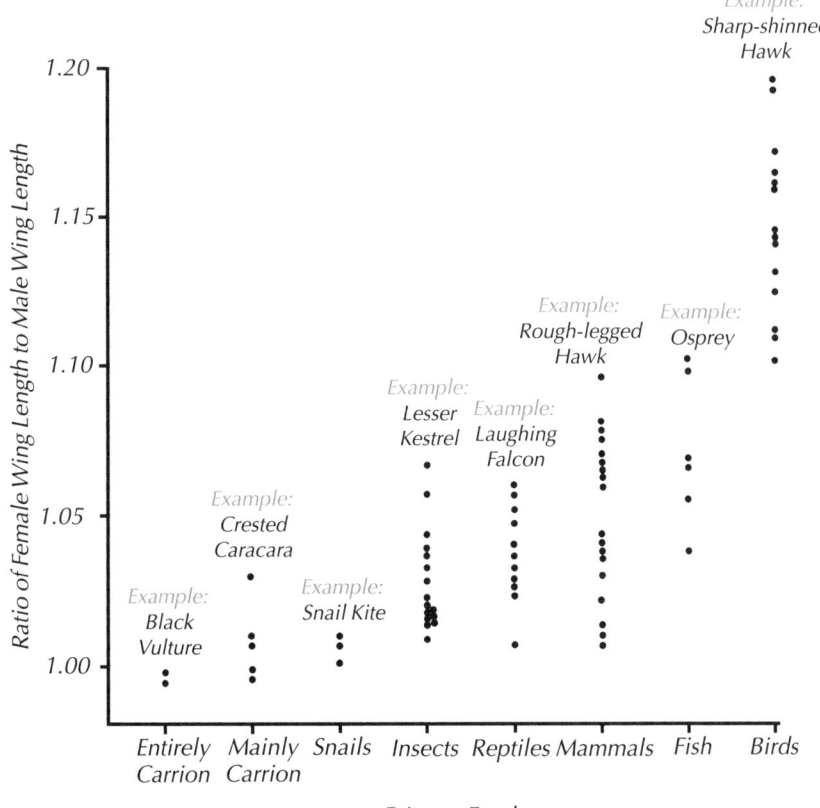

Figure 9–29. Sexual Dimorphism and Diet in Raptors: *Raptors that specialize in capturing fast-moving avian prey tend to have a large selection of prey to choose from because of a lack of competition from other predators. To avoid competition between the sexes, these "bird specialists" have evolved a high degree of sexual size dimorphism, allowing each sex to exploit prey of different sizes. Conversely, raptors feeding on slow-moving prey, such as insects and reptiles, must compete with other generalist predators for the same group of prey, and rarely show much size difference between the sexes. This graph shows the degree of sexual size dimorphism (measured as the ratio of female to male wing length) for raptors with prey of different mobilities. Each dot represents a different raptor species, including the example listed above each prey type. Note that there is a clear relationship between the degree of sexual size dimorphism and the mobility of the raptor's respective prey species. Adapted from Newton (1979).*

bird-eating raptors have a wide array of prey to divide between mates, whereas insect-eating raptors are confined to a narrow subset of their potential prey because they must compete with other bird groups. Because their choice of prey types is so narrow, insectivorous raptors are not able to partition food between the sexes.

In a few species, morphological differences among individuals are unrelated to sex. These **polymorphisms** persist in a population because different types of individuals, called **morphs**, have certain advantages. Ever since I first saw museum specimens of the Hook-billed Kite, a snail-eating Neotropical raptor, I have been fascinated by their strange variations in bill size. When I finally investigated the phenomenon, I discovered that some individuals have bills more than four times as large as others, and that these dramatic differences are unrelated to sex, age, overall body size, or geographic location. Instead, small-billed and large-billed birds coexist in the population **(Fig. 9–30)**. Further investigation (Smith and Temple 1982) revealed that bill dimensions determine the size of snail that a kite can efficiently extract from its shell. Large-billed individuals can extract only large snails, whereas small-billed individuals are best at extracting small snails. Therefore, in areas where snails of different sizes exist (because they belong to either different species or different age classes), polymorphism in bill size appears to be an adaptation allowing different kites to feed on different sizes of prey.

Interference

The other type of competition, more common than differential exploitation, is interference, in which one individual prevents others from obtaining resources by employing various behaviors. These behaviors may include antagonistic interactions that establish dominance hierarchies or territories (see Ch. 6, Territoriality, Dominance Hierarchies, and Ritualized Aggression). In these situations, the competitive struggle between individuals results in clear winners and losers, with the fitness of winners enhanced, and that of losers reduced. For example in dominance hierarchies, the strongest individuals gain access to resources, but subordinates must wait until the higher-ranking individuals have satisfied their needs. In territorial systems, an individual who has successfully defended a territory has nearly exclusive access to its resources.

In a particular species, whether individuals exclude others from resources through dominance hierarchies or territoriality depends largely on the distribution of the resource in demand. When the re-

Figure 9–30. Polymorphism in Bill Size in the Hook-billed Kite: The Hook-billed Kite of Central and South America feeds almost exclusively on tree snails, which it extracts from the shells by enlarging the shell opening and breaking the inner whorls as it drives its bill tip toward the apex of the shell's spiral. The soft body of the snail is then swallowed whole. The bills of individual Hook-billed Kites exist in a number of different sizes. In a single geographic area, some birds may have bills that are two, three, or even four times the size of others. This extreme variation, which is not related to sex or age, appears unique among birds, and affects the sizes of snails that an individual can eat. Small-billed morphs excel at extracting small snails; large-billed morphs specialize on larger snails. Ornithologists believe that all morphs evolved from one ancestral type with a uniformly small bill. Once this ancestral bird had developed its specialized method of snail extraction, different bill sizes evolved to allow the species to exploit a broader range of snails. Such polymorphism probably would not have developed had the kites been competing with other species for large snails. Drawing by John Wiessinger, from Smith and Temple (1982).

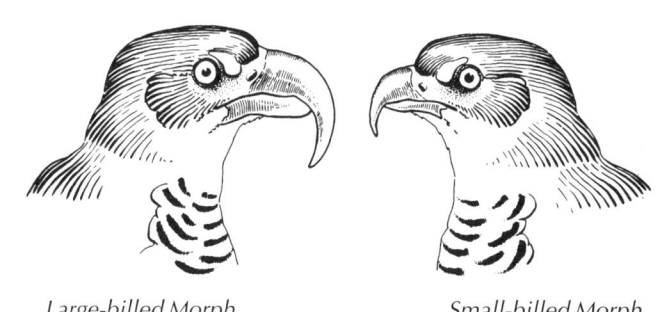

Large-billed Morph   Small-billed Morph

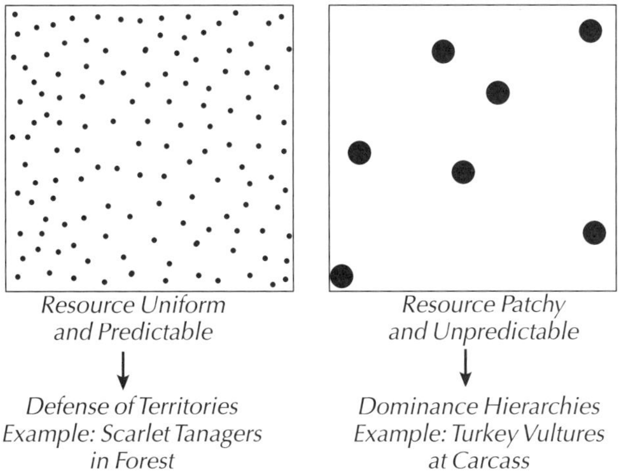

Resource Uniform and Predictable

↓

Defense of Territories
Example: Scarlet Tanagers
in Forest

Resource Patchy and Unpredictable

↓

Dominance Hierarchies
Example: Turkey Vultures
at Carcass

Figure 9–31. Competition for Resources: *The distribution of resources often determines whether the individuals that compete for them will develop a territorial system or a dominance hierarchy. For example, the uniformity of resources in a mature deciduous or mixed forest encourages songbirds such as the Scarlet Tanager to establish evenly spaced breeding territories, which they defend against other tanagers as well as other forest canopy birds, such as Great Crested Flycatchers and Eastern Wood-Pewees. In a territory of just two acres of prime oak habitat, a tanager pair can find all the insects and fruits that it needs. For birds such as the Turkey Vulture that feed on scattered, ephemeral resources such as carcasses, territoriality makes less sense: it would be impossible to defend a large enough area to provide a steady supply of carrion. Therefore, competition for patchy, unpredictable resources among scavengers usually takes the form of dominance hierarchies at a carcass.*

source is moderately abundant, uniformly distributed, and relatively persistent over time, individuals are generally able to defend areas with enough of the resource to meet their needs. In contrast, when a resource is limited and distributed in widely separated patches, none of which are rich enough or sufficiently persistent to support the needs of an individual, competition may be resolved most efficiently through dominance hierarchies **(Fig. 9–31)**.

Sometimes the most appropriate response to intraspecific competition changes over time. For example, during the nesting season, Black-capped Chickadees defend territories containing resources such as nesting sites and abundant insects. During winter, when food resources become sparse and patchy and are not worth defending, territoriality is abandoned. At that time, chickadees form flocks and forage over an area much larger than their breeding-season territories. Within each flock a dominance hierarchy, based largely on sex and age, determines the order in which members have access to contested resources.

Life History Strategies: Putting it All Together

In order to maximize its fitness, each bird, and in fact each living thing, has evolved a set of traits—life span, body size, growth rate, number of young per year, and so on—that collectively make up its **life history strategy**. Recall from Chapter 8 that birds cannot simply maximize fitness by growing fast, living a long time, and producing young at a rapid rate, because various types of trade-offs exist. Producing many young per year, for example, is energetically demanding and appears to decrease a bird's survival.

Because each species may make a different set of trade-offs, a whole range of different life history strategies is possible. For example, a bird may live a short life but raise numerous young by laying many eggs as soon as it becomes reproductively active. Alternatively, it might raise just one or two young per year over a relatively long life. Birds develop different life history strategies through both long-term and short-term responses to their environment.

9

Here I discuss seven life history traits that can influence the fitness of a species: (1) body size, (2) rate of development, (3) age at first breeding, (4) number of eggs produced per nesting attempt, (5) frequency of reproduction, (6) amount of parental care, and (7) longevity.

Among the world's approximately 9,600 species of birds, life history strategies show remarkable diversity. The largest living species is the nearly 350-pound (155-kg) Ostrich (although larger birds have existed, including the extinct 1,100-pound (500-kg) elephantbirds of Madagascar). The smallest bird is the 0.07-ounce (2-g) Bee Hummingbird. The fastest growth rate from egg to fledging occurs in small birds such as sparrows, which develop in just 20 days. In contrast, larger birds such as albatrosses can take more than 200 days to develop. Some small birds are capable of breeding when they are just a few weeks or months old. In some tropical areas with long seasons suitable for nesting, birds may breed less than a year after fledging (see Fig. 8–4). In Trinidad, the House Wren can breed at just five months of age. In contrast, large birds, such as some eagles, condors, and albatrosses, may not breed until they are 12 years old. Clearly, body size, rate of development, and age at first breeding are directly related.

The number of eggs produced each time a bird breeds varies from a clutch of one—as in many penguins, albatrosses, and auks—to clutches of up to 18 in grouse, pheasants, and quail. Brood parasites, such as cuckoos and cowbirds, may lay more than 25 eggs per season because they can invest energy in egg production rather than in parental care. Most birds breed only once each year, but some, such as House Sparrows, can complete up to six breeding cycles annually. A few others, such as some penguins, albatrosses, and condors, can reproduce only once every two or three years. The frequency of breeding often is correlated with the amount of parental care provided to offspring. Some small birds become independent when just a few weeks old, releasing their parents to breed again. In contrast, the offspring of some large species, such as condors, remain dependent on their parents for more than a year, thereby preventing their parents from breeding in the following year.

Within a species, the age individual birds live to varies tremendously. Many individuals die young, with only a few reaching the species' maximum age. The average life span of individuals within a population can be calculated for some well-studied species (Temple 1990b). American Robins live an average of 14 months, whereas Andean Condors live an average of almost nine years. In general, longevity is directly related to body size. The maximum longevity for a bird in the wild is no doubt achieved by the largest birds, such as condors, which probably can live for up to 75 years **(Fig. 9–32)**.

Although theoretically there are an unlimited number of ways that different life history traits could be combined to form a life history strategy, in actuality, only a few of these combinations tend to evolve because most do not result in maximum fitness for any species. In fact, most life history strategies of birds can be considered to fall somewhere along a single continuum. At one end are birds with small body size, rapid development, breeding at an early age, large clutch size, frequent reproduction, brief parental care, and a short life span. This type of strategy is used successfully by many small songbirds. For example, a House Wren weighs 0.35 ounces (10 g), reaches full size in less than three weeks, can begin breeding within a few months, lays clutches of up to nine eggs, produces several broods annually, finishes a reproductive cycle in less than a month, and has an average life span of just over a year. Birds employing this "high output" type of strategy are called **r-selected**. (The variable "r" is often used by population biologists to represent the rate of population growth; this term reflects the fact that these birds are emphasizing an ability to achieve rapid population growth.) These birds produce many young each season, but the young suffer high rates of mortality both in the nest and after fledging.

At the other end of the continuum are birds with a large body size, slow development, delayed onset of breeding, small clutch size, infrequent reproduction, prolonged parental care, and a long life span. At 25 pounds (11.4 kg), an Andean Condor is one of the largest flying birds. It takes nearly five months to reach full size, defers breeding for up to seven years, lays one egg, breeds every other year at most, takes more than a year to complete a breeding cycle, and lives nine years, on average. These birds produce few young per season, but, partly due to the prolonged parental care, the young have high survival rates. Birds using this type of life history strategy, which relies on longevity to maximize reproduction, are called **K-selected (Table 9–1)**. (The variable "K" often represents the carrying capacity of an environment—a population level that can remain stable; use of this term reflects the fact that population levels of these birds typically remain stable and near K.)

None of the various life history strategies employed by birds are "right" or "wrong," and none are better or worse than any other. Instead, they are adaptations that each species has evolved to maximize the number of genes that it can successfully send into future generations.

Most of the life history traits that I have discussed are long-term responses to the environment. For example, the average body size of a species can change, but only by evolving slowly over many generations. In contrast, one important life history trait—clutch size—can sometimes be considered a short-term response, because a female can adjust it to meet changing environmental conditions, in particular the availability of food (see Figs. 8–82 and 8–83).

The optimal clutch size is the number of eggs that will, on average, result in the largest number of offspring surviving and reproducing in the future. If food is in short supply, a large clutch may result in offspring that are in poor condition and do not survive to reproduce. If food is abundant, a small clutch

Figure 9–32. Male Andean Condor: Although the life expectancy of individual birds varies tremendously, longevity is generally related to body size. Small songbirds live an average of one or two years, whereas large species live much longer. The Andean Condor has an average life span of nine years. In captivity, however, a condor may live up to 75 years. Photo courtesy of Alan Lieberman/Zoological Society of San Diego.

Table 9–1. Contrasting Life History Strategies: *Each bird species has evolved a life history strategy—the sum of all aspects of its growth, survival, and reproduction—that maximizes its fitness. This table shows two extremes. Birds characterized by large size, slow growth, and low reproductive rates—such as the Andean Condor—are referred to as K-selected species. Small birds, such as the House Wren, which develop rapidly and reproduce prolifically, are referred to as r-selected. Most species fall somewhere between these two ends of the spectrum of life history strategies.*

	K-selected Species Andean Condor	**r-selected Species** House Wren
Body Size	Large	Small
Rate of Reproduction	Low	High
Rate of Development	Slow	Fast
Age at First Breeding	Old	Young
Parental Care	Long	Short
Survival	High	Low
Life Span	Long	Short

may produce fewer offspring than the parents are capable of rearing **(Fig. 9–33)**.

Species living in generally stable environments, such as deciduous forests, tend to produce clutches with little variation in size from year to year. However, species that live in variable environments, such as the arctic tundra, must adjust their clutch size to current environmental conditions. Snowy Owls, for example, must cope with cycles in the abundance of their lemming prey, as discussed earlier. In years when lemmings are abundant, Snowy Owls lay clutches of up to 11 eggs, whereas in years of lemming scarcity, the owls may lay few eggs or none at all (see Fig. 8–84).

Adjustments of clutch size to local conditions represent a form of "bet-hedging," but they are not the only way that birds fine-tune the production of offspring to unpredictable environmental conditions. As discussed in Chapter 8, some birds use **brood reduction** to systematically reduce the number of young even after eggs have hatched (see Ch. 8, Start of Incubation). In some species with altricial young, such as birds of prey, incubation begins when the first egg is laid, so eggs hatch in the order in which they were laid. Depending on the number of eggs laid and the laying intervals, the first-hatched nestling can be several days old when the last egg hatches **(Fig. 9–34)**. Therefore, when food is in short supply, the first-hatched, and thus larger, nestlings monopolize the limited food while the later-hatched and smaller nestlings starve, thus reducing the brood size. Obviously, the strategy of brood reduction cannot be used to adjust the number of young in birds with synchronously hatched, precocial young.

Within a species, geographic variations in clutch size also can be correlated with local environmental conditions, especially those that affect the bird's food supply. In general, individuals nesting at lower latitudes (closer to the equator) lay smaller clutches than individuals

Figure 9–33. Horned Grebe at Nest: One aspect of a bird's life history that varies depending on environmental conditions is clutch size. The Horned Grebe, which breeds in freshwater wetlands in northern North America and Eurasia, lays from three to eight eggs. In this species clutch size is influenced by date of laying (earlier clutches are larger), weather conditions at the time the first egg is laid, local food abundance, and an individual bird's dominance status in the local population. Each individual strives to lay the optimal clutch size for its environment: that which results in the largest number of young who survive to reproduce in the future. Photo courtesy of Isidor Jeklin/CLO.

nesting at higher latitudes (closer to the poles). Consider House Wrens, which have a large geographic range extending from the boreal forests of North America, through the equatorial tropics, and on to the southern tip of South America. Whereas wrens in the tropics lay clutches of only two or three eggs, those living in the temperate zones of North and South America lay clutches of up to seven (see Fig. 8–85 for additional information on geographic clutch size variation).

In addition, individuals on oceanic islands tend to lay smaller clutches than those nesting on continents. For example, island-nesting Song Sparrows lay smaller clutches than individuals on the mainland. Also, individuals in coastal areas tend to lay smaller clutches than those nesting in continental interiors, and individuals at low elevations tend to lay smaller clutches than those nesting at higher elevations.

Each of these major, well-documented geographic patterns in clutch size (latitudinal, altitudinal, coastal-interior, island-continental) is correlated with seasonal variability in the environment, especially food availability. Locations in which birds lay relatively large clutches

Figure 9–34. Asynchronously Hatched Screech-Owl Young: In some species with altricial young, such as birds of prey, incubation begins when the first egg is laid, resulting in asynchronous hatching. This family of Western Screech-Owls demonstrates the results. At the far right is the adult female. Immediately to her left is the oldest offspring, almost as large as its mother. To the left are four other juveniles of various ages. The leftmost bird is probably the youngest, as shown by its weak grip on the tree branch. Western Screech-Owls lay up to six eggs. The first few may be laid a day apart, with later eggs produced at longer intervals. Thus, the first-hatched nestling may be several days old when the last one hatches, giving it a competitive edge over its younger siblings. If food supplies dwindle, the brood is reduced as the youngest nestlings succumb to starvation, and the oldest nestlings receive enough food to insure their survival. Photo by W. Perry Conway.

9

are characterized by marked seasonal fluctuations in food supply in comparison with places in which smaller clutches are produced. In the tropics, for example, production of many foods remains fairly stable throughout the year, whereas in the temperate zone productivity is much increased during spring. Islands tend to have more stable environments than continental areas at the same latitude because the ocean buffers climatic effects. The greater abundance of food during the breeding season in areas with greater seasonal fluctuations (higher latitudes, higher elevations, interior compared to coastal areas, and continents compared to islands) allows birds that breed in these areas to have larger clutches than their counterparts in less seasonally variable areas.

Another aspect of the relationship between clutch size and the seasonal fluctuation (or "seasonality") of resources can be seen in birds that reside year round in the same region. In areas where seasonal variations in food supply are great, the number of resident birds that survive to breed in the next year is determined by the seasonal low in food availability, which usually occurs during winter. When spring arrives—bringing a flush of food production—the survivors have much food available to them and can produce large clutches.

An important factor in the relationship between clutch size and seasonality is the amount of food available in the breeding season *compared to* the number of breeding birds, and not simply the absolute amount of food available (Ricklefs 1980a): if the amount of food is relatively high or the number of breeding birds is relatively low, both of which tend to hold for highly seasonal areas, there will be more food available per breeding pair, and clutch sizes will be higher. In contrast, because birds living in relatively stable environments do not have their numbers reduced by a lean season, and because there is less of a seasonal flush of food in the breeding season, the amount of food available per breeding pair is lower, and consequently clutch sizes will be smaller **(Fig. 9–35)**.

Birds in Populations

■ In 1962, Rachel Carson's classic book *Silent Spring* alerted the public to the possibility of a future in which pesticide pollution would cause many familiar birds to become scarce. Ornithologists predicted that songbird populations would decline as individuals died from direct pesticide poisoning, whereas numbers of other birds, such as raptors, would drop when adults failed to reproduce after eating contaminated prey. Alarmed by the specter of disappearing birds, many people began to consider how bird populations are related to their environment.

Today, people are much more aware of the ways that bird populations respond to their environments. Christmas Bird Counts and Breeding Bird Surveys conducted by bird watchers provide information about where birds are found and how their numbers are changing. Other programs, such as The Birdhouse Network, monitor reproductive rates, and studies of banded birds, such as MAPS (Monitoring Avian

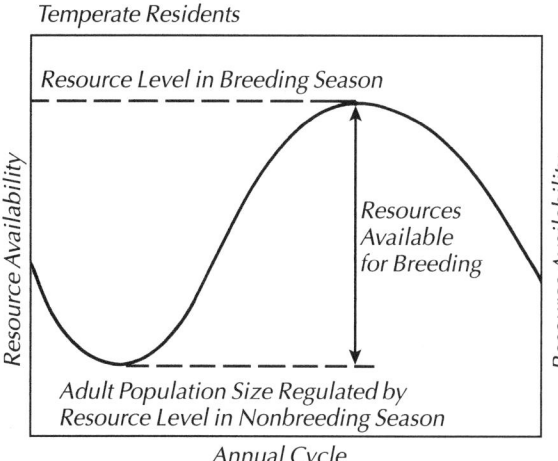

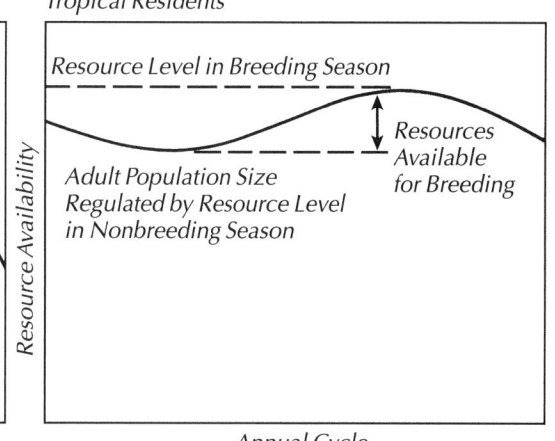

Productivity and Survivorship), track movements and survival rates. Making sense of all this information requires a basic understanding of **avian population ecology**.

Recall that a **population** consists of all individuals of the same species living in a given area. The largest population that one could study would be a **species population**, composed of all living individuals of a single species. However, ornithologists generally study **local** or **regional populations**, such as the population of Blue Jays in one neighborhood or the population of Canada Geese in the central migratory flyway.

Characteristics of Bird Populations

Populations exhibit several characteristics that reflect the cumulative traits of the individuals that compose them. They have a **geographic range** encompassing the home ranges of all individuals. They have a **population size** equivalent to the total number of individuals in the population. Population size is affected by the **birth rate** (or **fecundity**), which is the number of offspring entering the population during a given unit of time, and by the **death rate** (or **mortality rate**), which is the number of individuals dying during a given unit of time. Population size also is affected by **immigration**, the movement of individuals into the population, and **emigration**, the movement of individuals out of the population **(Fig. 9–36)**. Populations are composed of different types of individuals: males and females, young birds and old birds, and birds with different types of genes. The relative proportions of these different individuals result in a **population structure**, which changes continuously depending on what's happening to the individual members of the population. The branch of avian ecology that studies the changes in population size and structure over time is called **avian population dynamics**.

Geographic Distribution Patterns

Every species has a **geographic range**, which is the general area in which it lives, such as the Amazon River Basin or the Sonoran Desert.

Figure 9–35. Food Availability, Reproduction, and Annual Population Cycle of Year-Round Residents: There is a typical relationship between variation in population size and the relative difference in resource availability between the breeding and nonbreeding seasons. This relationship influences the reproductive strategies that birds use in various types of environments. In stable tropical environments where food resources vary little between the breeding and nonbreeding seasons, birds tend to have small clutches of a consistent size and relatively stable populations within each year and from year to year. Conversely, in temperate environments where food supplies and weather are highly variable, birds tend to have large clutches that result in high reproductive output and variable population sizes. High rates of reproduction in the spring permit resident populations to compensate for winter mortality that can drive populations below the carrying capacity of the environment. Adapted from Ricklefs 1980a and Faaborg 1988.

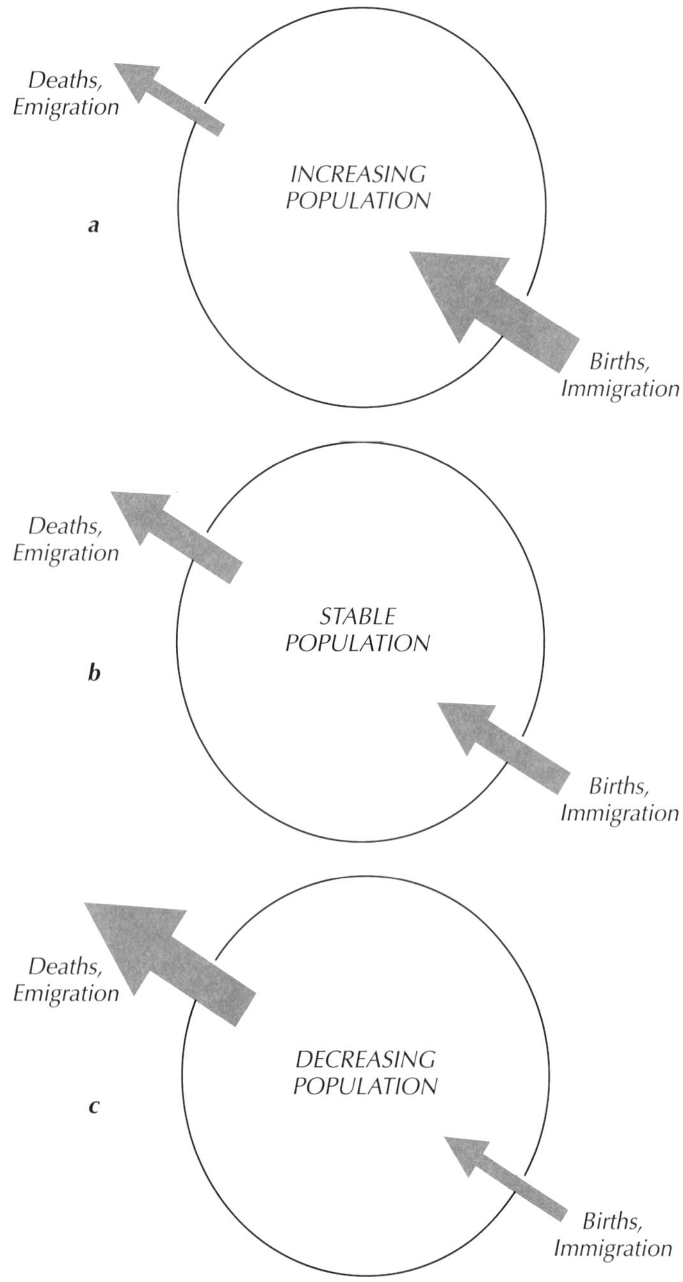

Figure 9–36. Basic Population Dynamics: *Population size depends on four factors—births and immigration, which add to a population, and deaths and emigration, which subtract from the population. If births and immigration exceed deaths and emigration (a), for instance because of improving habitat conditions, a population increases in size. When births and immigration equal deaths and emigration, the population remains stable (b). Finally, if deaths and emigration exceed births and immigration—for instance because of disease—the population will decline (c).*

Birds remain in their ranges except in unusual situations, for example when storms blow them far from home. On the other hand, birds seldom occupy their ranges uniformly. Consider the American Kestrel. Although its range encompasses nearly every state and province in North America, it is absent from heavily forested areas that do not meet its habitat needs. Understanding such distributional patterns, and explaining why birds are found where they are, is one aspect of the field of **avian biogeography**.

Take a few moments to peruse the range maps in your field guide, and you will detect some obvious biogeographic patterns. Some species have wide distributions, yet related species have restricted ranges. Most ranges are continuous but a few are **disjunct**, with a species' distribution divided into geographically distinct areas. For example, compare the ranges of the Yellow Warbler and the Cerulean Warbler. The Yellow Warbler is distributed continuously throughout Canada and the United States, whereas the cerulean has a disjunct distribution in only the eastern United States **(Fig. 9–37)**.

Some species have broadly overlapping ranges, suggesting that common factors may be setting their limits. For example, compare the ranges of the Prothonotary, Prairie, Kentucky, and Hooded warblers **(Fig. 9–38)**.

Some ranges are occupied year round whereas others are occupied seasonally, as birds migrate between breeding and non-breeding regions. For example, compare the range of the Canyon Towhee with that of the Green-tailed Towhee. Although these two species have ranges that overlap in Texas, New Mexico, and Arizona, the Canyon Towhee occupies its range year round, while the Green-tailed Towhee migrates to Mexico during winter.

Finally, geographic ranges are not static; they expand and contract over time. For example, since the 1800s the range of the Northern Cardinal has been expanding, but that of the Red-cockaded Woodpecker has contracted **(Fig. 9–39)**. The explanations for range expansions and contractions involve many factors, both natural and human-caused.

Each species evolved in a particular geographic area, and this **center of origin** strongly influences where it continues to be found. For example, Brown-headed Cowbirds are thought to have evolved on the Great Plains, and this area still marks the center of their distribution, even though they have more than

Deaths, Emigration

a

INCREASING
POPULATION

Births,
Immigration

Deaths,
Emigration

b

STABLE
POPULATION

Births,
Immigration

Deaths,
Emigration

c

DECREASING
POPULATION

Births,
Immigration

(Continued on p. 9·55)

Yellow Warbler

Cerulean Warbler

Continuous Range

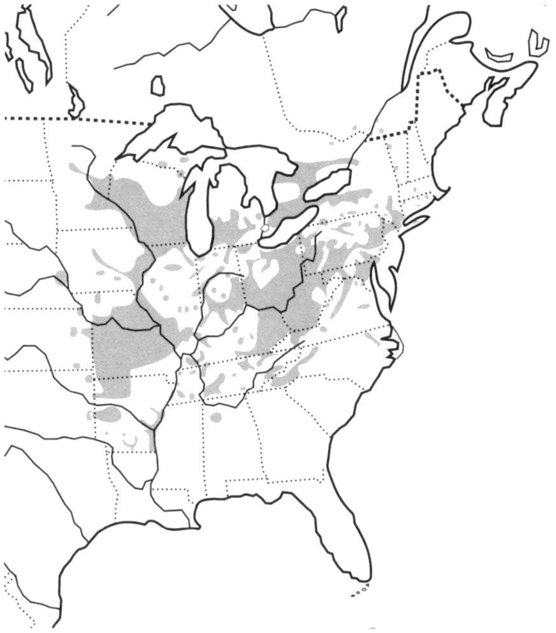

Disjunct Range

Figure 9–37. Continuous and Disjunct Geographic Ranges: *Each species lives in a specific area called its geographic range. Although most species have continuous ranges, others have ranges split into geographically separate regions. The Yellow Warbler (left) has a fairly continuous breeding range. In contrast, the breeding range of the Cerulean Warbler (right) is highly fragmented, or disjunct. In some areas the bird is widespread, for instance in central Missouri, southern Wisconsin, eastern Kentucky, West Virginia, and eastern Ohio. In other areas, such as Illinois, northern Indiana, western Tennessee, and western Kentucky, its distribution is extremely patchy. The difference in ranges probably results from habitat preference. The Yellow Warbler is a bird of wet, deciduous thickets, especially where willows proliferate. It tolerates disturbed and early successional habitats, and hence can thrive near human habitation. In contrast, the Cerulean Warbler is a denizen of mature deciduous forest, where it forages high in the canopy and nests in large, tall trees. Logging in the early 1900s decimated such forests throughout much of the Cerulean Warbler's original range, so that forest remnants and scattered cerulean populations are all that remain. Photo of Yellow Warbler courtesy of Michael Hopiak/CLO, and of Cerulean Warbler courtesy of Bill Dyer/CLO. Maps copyright Birds of North America, Inc.*

Figure 9–38. Overlapping Geographic Ranges of Wood Warblers: *Some species, such as the Prothonotary, Kentucky, Hooded, and Prairie warblers, have broadly overlapping geographic ranges. Their ranges overlap because the habitats, resources, and climatic conditions that these species require occur in roughly the same geographic patterns. The Prothonotary, Kentucky, and Hooded warblers all require large tracts of middle-aged to mature bottomland deciduous forests for successful breeding. These species are often found near streams and rivers or in moist, lush forests. The Prairie Warbler is distributed in a similar geographic pattern, but uses dry, brushy forests on flats, slopes, and sometimes ridge tops. Photo of Prothonotary Warbler by Lang Elliott, Kentucky Warbler courtesy of John S. Dunning/CLO, Hooded Warbler courtesy of William A. Paff/CLO, and Prairie Warbler by Marie Read. Maps copyright Birds of North America, Inc.*

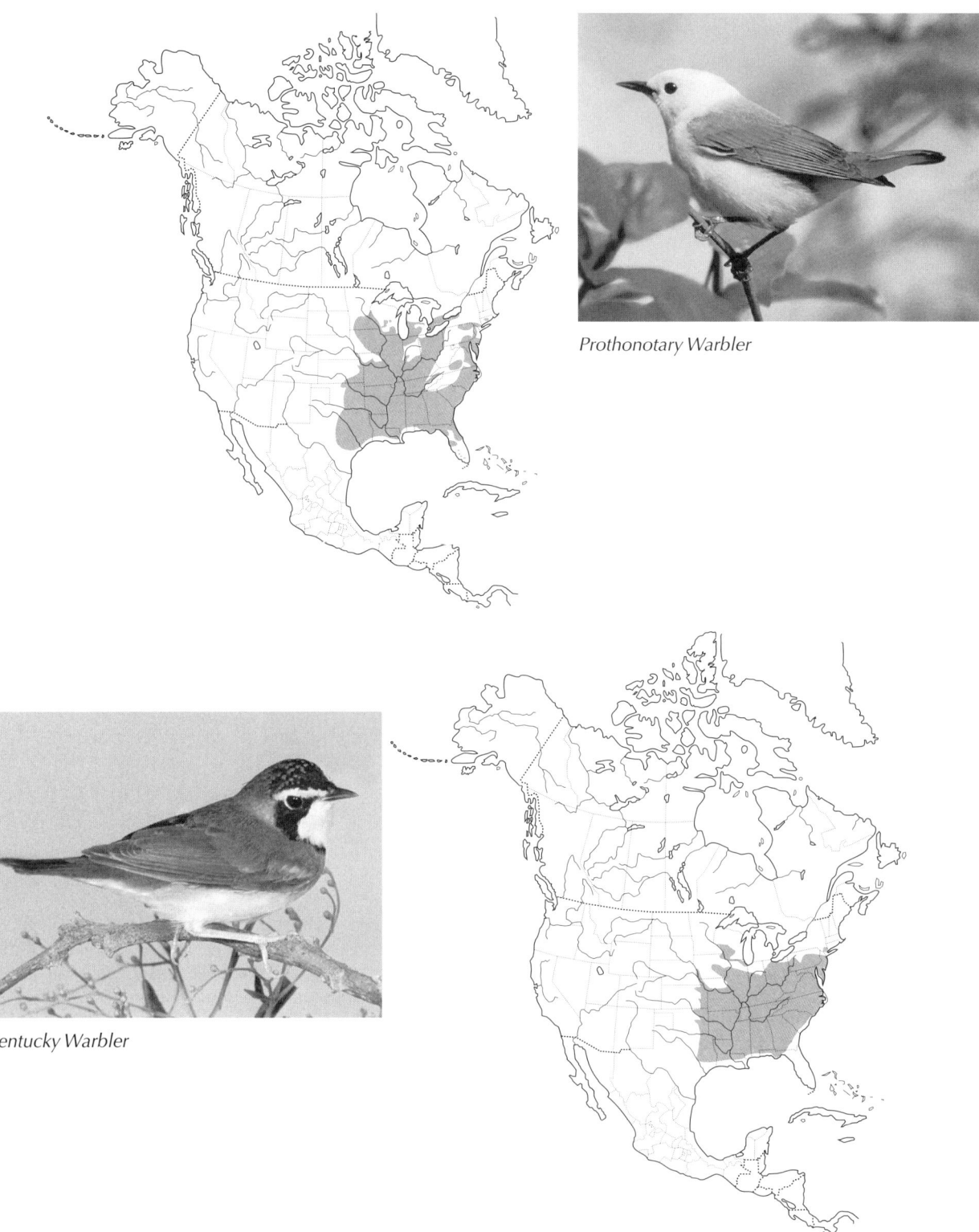

Prothonotary Warbler

Kentucky Warbler

Hooded Warbler

Prairie Warbler

9

Northern Cardinal

Figure 9–39. Expanding and Contracting Geographic Ranges: *Although range maps are static, bird ranges are actually dynamic, changing over time. The familiar Northern Cardinal has been expanding its year-round range northward since the 1800s, taking advantage of moderating winter temperatures, human-induced habitat changes, and the popularity of winter bird feeding. The dark color on the map shows the cardinal's pre-1900 range; the light color shows areas into which it expanded during the 20th century. Approximate dates of first records are given for some locations. In contrast, the range of the Red-cockaded Woodpecker is contracting. Here the dark color depicts the current range, whereas the light color shows the probable extent of the woodpecker's range during much of the 20th century. This species requires pine forests in which to feed and breed. Unfortunately, humans have altered many southern pine forests through development, logging, elimination of the fires that historically maintained healthy longleaf pine stands, and replacement of longleaf pine forests with economically preferable timber species. As a result, the woodpecker's range has contracted, and it has become an endangered species. Maps adapted from* Birds of North America. *Photo of Northern Cardinal courtesy of Craig Larcom/CLO, and of Red-cockaded Woodpecker courtesy of J. H. Dick/CLO.*

Range Expansion of the Northern Cardinal

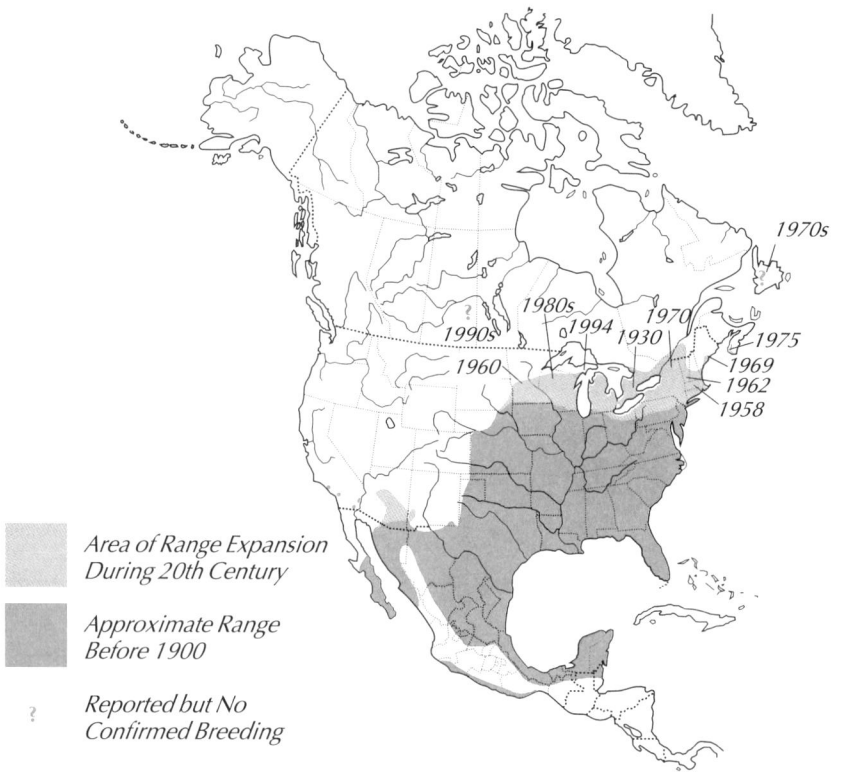

1970s

1980s   1970
1990s   1994 1930
1960   1975
             1969
             1962
             1958

Area of Range Expansion During 20th Century

Approximate Range Before 1900

?   Reported but No Confirmed Breeding

Range Contraction of the Red-cockaded Woodpecker

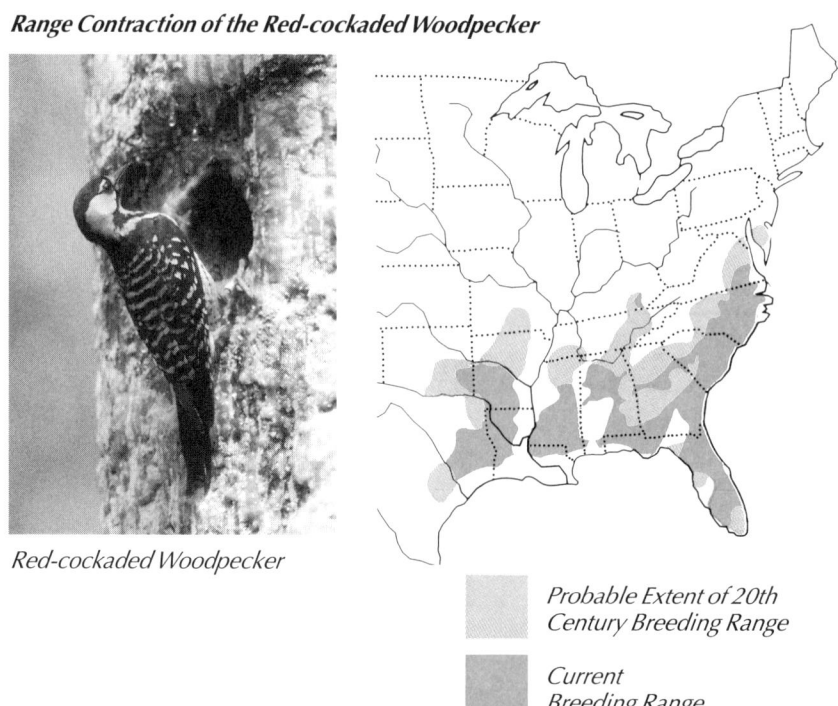

Red-cockaded Woodpecker

Probable Extent of 20th Century Breeding Range

Current Breeding Range

doubled their range in the past century. The most extreme examples are **island endemics**, which are species that originate on an island and never leave. In many cases, however, populations do spread from their centers of origin into new areas. Because birds are so mobile such expansions can proceed rapidly, and ranges can become extensively enlarged.

Sometimes a species is able to expand its range as a result of human actions such as introductions or habitat alterations. When non-native species are introduced to a new environment with the help of humans, they are known as "exotics." Some do very well due to a lack of natural competition or predators, whereas others are unable to adapt to the new conditions. The European Starling and House Sparrow were both introduced from England to the east coast of North America in the late 1800s, and spread across the United States in just a few decades **(Fig. 9–40)**. Another exotic species that has become firmly established throughout North America is the Rock Dove, which was introduced to Nova Scotia from the Mediterranean in the early 1600s.

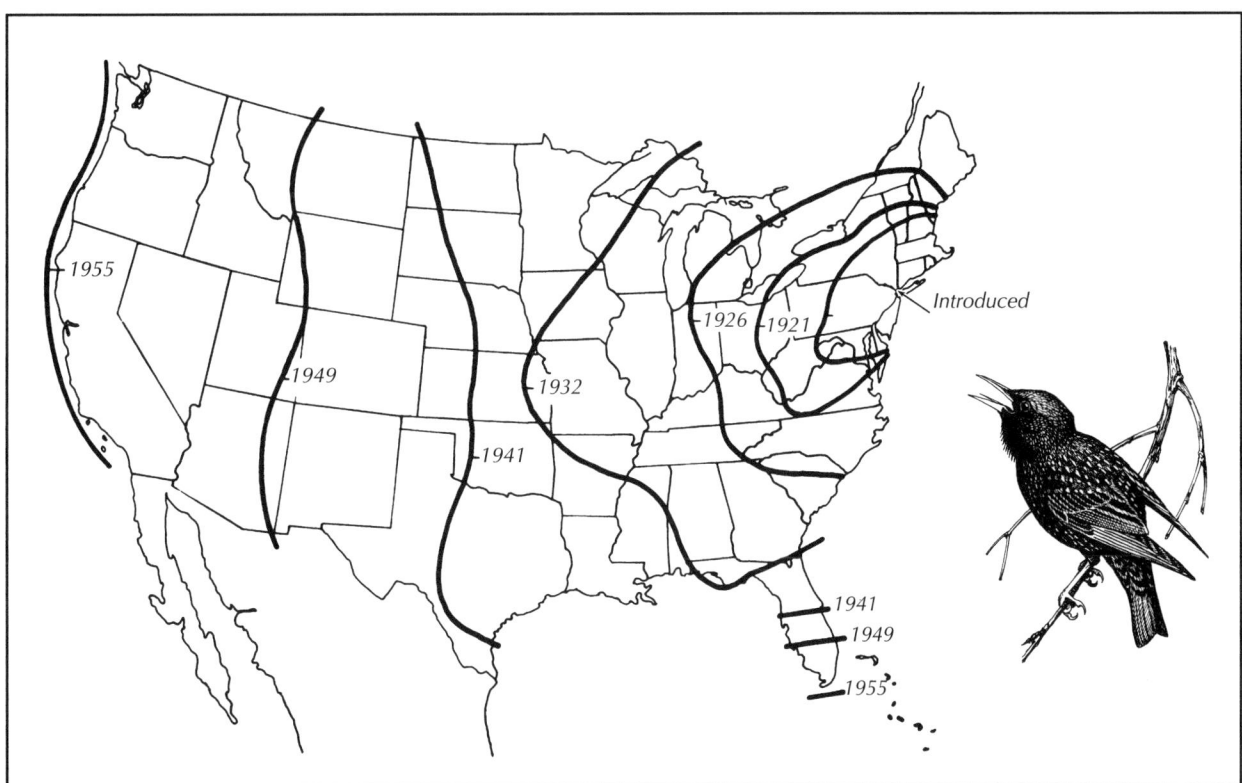

Figure 9–40. Spread of the European Starling: *European Starlings, often disliked by North American bird watchers because of their propensity to displace native species, once held a happier place in history as demonstrated by these lines from a song from Elizabethan England: "See, mine own sweet jewel, what I have for my darling, a robin red breast and a starling." Indeed, the swirling flocks of starlings that we see in North America today were the gift of Eugene Scheifflin, a New York businessman who wanted to bring all the bird species mentioned in Shakespeare to the United States. In 1890 he released 60 starlings into New York City's Central Park, and the next year he loosed another 40. North American habitats proved amenable to the Old World species, and by the mid-20th century the starling had burbled, rattled, and gurgled its way across the continent. The species pursues aggressive breeding strategies that help it colonize new areas. Early in the season, starlings preempt cavity nesters such as kestrels, flycatchers, swallows, wrens, and bluebirds at choice sites; later, when seeking to raise second broods, starlings attack other cavity occupants. Starlings also owe their astonishing success to humans, who have created both an abundance of lawns for foraging and urban structures for communal roosting. Although it amounted to closing the birdcage door after the starlings had flown, Scheifflin's "gift" led to federal laws restricting the introduction of wild exotics. Inset drawing by Orville Rice.*

Eventually, however, most species that expand their ranges encounter some environmental barrier that limits further expansion. Many barriers are physical, such as oceans or mountain ranges. For example, the Pacific Ocean would probably prevent a species that had expanded its range from the Rocky Mountains to the Pacific Coast from leaping to the Hawaiian Islands. Another barrier is climate, which can especially limit northern ranges. During winter, for example, the insectivorous Eastern Phoebe is usually found where the minimum average January temperature is above freezing (Root 1988) **(Fig. 9–41)**.

Range expansion also is limited by biotic factors, particularly habitat. In eastern North America, for example, the ranges of many woodland birds end at the point where forests give way to prairies. Look at a range map for the Eastern Wood-Pewee and you will see that its western limit corresponds to the western forest edge, with fingerlike range projections extending westward along wooded river valleys.

Range limits also may be set by **interspecific competition**, which occurs when one species competes directly with another for limited resources. Generally, closely related species cannot coexist in an area because their needs are too similar. For example, if you compare range maps of Rose-breasted and Black-headed grosbeaks, or Black-capped and Carolina chickadees, you will find little overlap. Or consider the Belted Kingfisher. This species reaches the southern limit of its breeding range near the United States/Mexico border, where it is replaced by both the larger Ringed Kingfisher and the smaller Green Kingfisher. Although Mexico provides suitable habitat for Belted Kingfishers—in fact, most of them spend the winter there—during the breeding season they apparently are excluded from the region by these two species

Eastern Phoebe

Figure 9–41. Winter Distribution of the Eastern Phoebe: *Environmental factors, such as temperature, can limit a species' geographic range. During winter, insectivorous birds are restricted to areas in which average temperatures remain above a certain minimum level. Below this temperature insect food is scarce, and the birds are unable to maintain the metabolic rate necessary for survival. This map shows the winter distribution of the Eastern Phoebe during the 1999–2000 Christmas Bird Count. The two dotted lines are isotherms—lines connecting all locations on a map that have the same average temperature. Notice that few phoebes are reported north of the isotherm along which the average minimum January temperature is 24.8° F (-4° C), and that peak abundance (shown by the darkest color) occurs south of the isotherm along which the average January temperature is 39.2° F (4° C). Drawing by William C. Dilger. Map copyright 2001 BirdSource.*

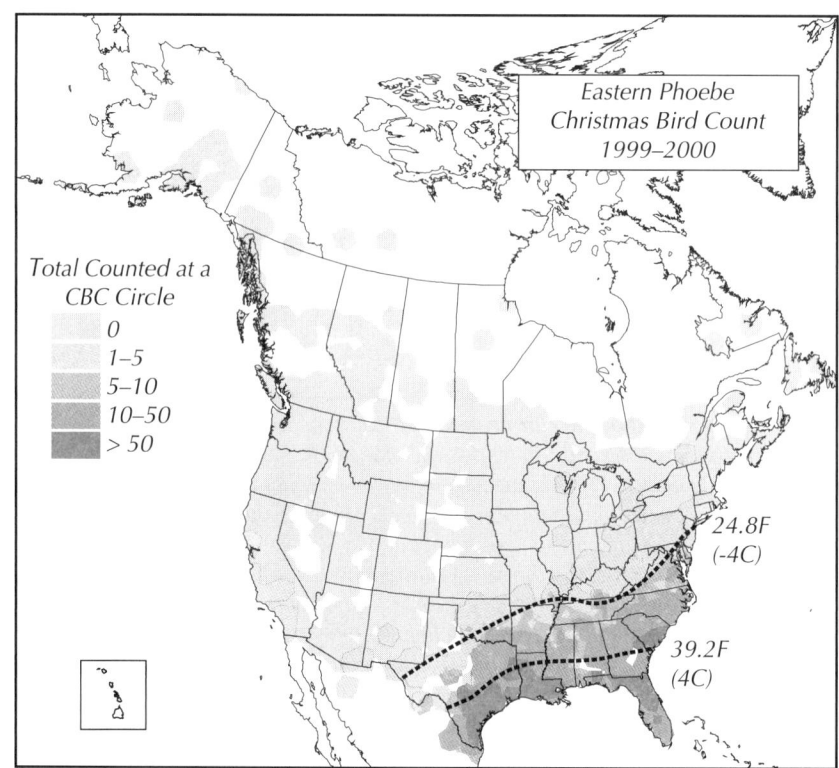

Eastern Phoebe
Christmas Bird Count
1999–2000

Total Counted at a
CBC Circle

0
1–5
5–10
10–50
> 50

24.8F
(-4C)

39.2F
(4C)

that live in the tropics year round and have sandwiched the Belted Kingfisher in a competitive squeeze.

Other dramatic examples of range limits set by competition come from mountainous regions, where closely related species may replace each other at different altitudes. For example, in the Appalachian Mountains of New England, the Veery is found in deciduous forests at elevations of 1,640 feet (500 m) or less, whereas the closely related Hermit, Swainson's, and Gray-cheeked thrushes are found in coniferous forests at higher elevations. In the Great Smoky Mountains, however, where the three latter species are absent, the Veery extends its range upward into coniferous forests **(Fig. 9–42)**.

Some ranges are very small or include several disjunct populations. Often these belong to species that formerly were widespread but that became extinct in certain areas, leaving behind one or more **relict populations**. As recently as 10,000 years ago, the California Condor was found over much of the Southwest, from the Pacific coast to Texas, and from British Columbia to Mexico. Through local extinctions and range contractions, probably owing to human-caused mortality, the condor eventually became restricted to a relict population in central

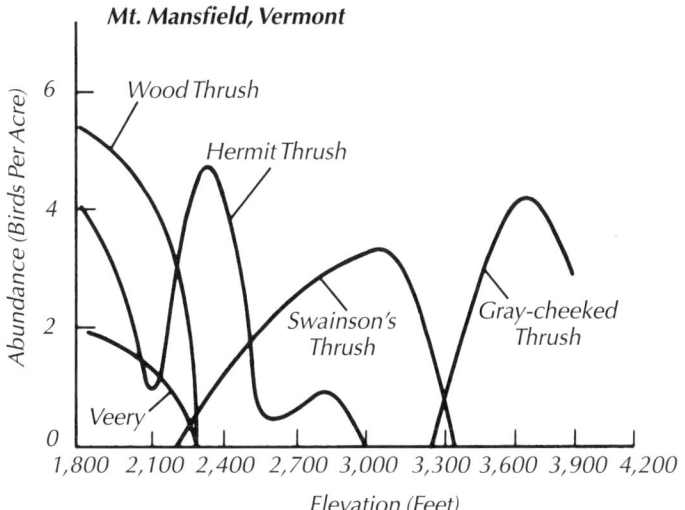

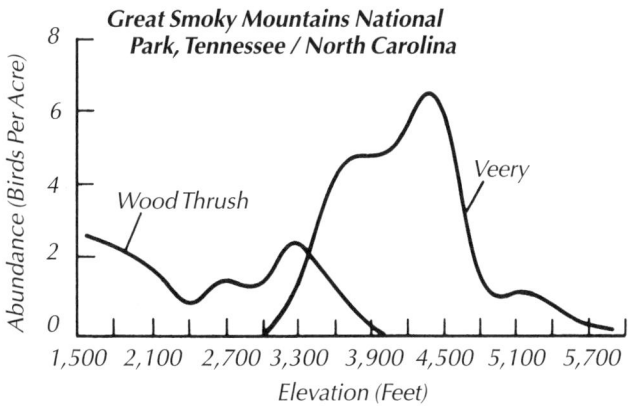

Figure 9–42. The Role of Interspecific Competition in the Distribution of Thrush Species Along an Altitudinal Gradient: Limits for the ranges of some species are set by interspecific competition, with ranges expanding when competing species are absent. The spot-breasted thrushes of eastern North American woodlands—the Veery, Wood Thrush, Hermit Thrush, Swainson's Thrush, and Gray-cheeked Thrush—all breed in the same basic kinds of forests, and all forage for insects on the ground. Nevertheless, the species select slightly different elevations within their common environment. On a study area in Vermont (upper graph), Veeries and Wood Thrushes were most abundant in deciduous forest below about 2,200 feet; Hermit and Swainson's thrushes occupied coniferous forests of middle elevations, up to 3,300 feet; and Gray-cheeked Thrushes were found in the highest areas, up to 4,000 feet. In Tennessee (lower graph) the guild is smaller—only two species—and here, with less competition, the Veery extends its range into coniferous forests up to 5,800 feet and the Wood Thrush ranges up to 4,000 feet. This habitat shift suggests that the Veery and Wood Thrush are excluded from high-altitude zones in New England because of interspecific competition from other thrushes. Adapted from Noon (1981).

California. Similarly, the Greater Prairie-Chicken ranged continuously throughout the prairies until habitat loss and overhunting reduced the species to a few scattered populations.

Current geographic ranges therefore represent a dynamic interplay of several factors: centers of origin, range limitations, range expansions, and range contractions. Furthermore, ranges will continue to fluctuate because of environmental changes that are constantly occurring. Imagine how ranges of North American birds might change over the next century if predictions about global climate change are correct. As the earth warms and rainfall patterns are altered, some birds may be able to expand their ranges northward, whereas the ranges of others could contract. A few species might even become extinct. For instance, the Kirtland's Warbler, already rare within a tiny range, could lose all of its remaining habitat if changing climate causes Jack Pine forests to disappear in the Midwest.

Population Size

Population ecologists seldom know exactly how many individuals of a given species exist. For a few rare species, such as Whooping Cranes or California Condors, counting all individuals is possible. For larger populations, though, the best we usually can do is estimate population size by counting the number of individuals in a small area to determine population **density**, or individuals per acre or hectare. This density estimate then can be multiplied by the total area of suitable habitat within the species' range, yielding an estimate of total population size. In North America, waterfowl biologists who need to set hunting regulations use this technique to estimate the size of fall duck and goose populations. Most ecologists studying avian populations, however, prefer to deal with simple density estimates as opposed to overall population sizes.

The densities at which different species live are determined by a combination of factors, some related to the birds themselves, and some to the environments in which they live. Generally, larger birds live at lower densities than smaller birds, and for birds of the same size, those that feed on plant matter (**herbivores**) live at higher densities than those that feed on other animals (**carnivores**). These patterns reflect the amount of energy available to birds. More energy is available per unit area for small birds feeding on small food items than for large birds feeding on larger items. Similarly, in a given area more energy is available from plants than from animals **(Fig. 9–43)**.

When population density is monitored across a species' entire range, broad patterns of variation often can be detected. For many species, densities are highest near the center of the range. This pattern generally reflects habitat suitability. Near the range's center, environmental conditions typically are ideal, but they tend to deteriorate near the periphery, where both biotic and abiotic factors such as habitat and climate become marginal and eventually unsuitable for the species **(Fig. 9–44)**.

Figure 9–43. The Relationship Between Body Size, Diet, and Population Density: In any given environment, the density of each bird species is inversely related to the size of the home ranges covered by individual birds. Thus species with the largest home ranges have the lowest densities. This relationship is governed by several factors, including body size and diet, as illustrated by this graph. In general, large birds have large home ranges and thus live at low densities. For any given body size, however, herbivorous birds are able to meet their nutritional needs from smaller habitat areas, and so have smaller home ranges and greater population densities. Conversely, carnivorous birds require a greater area to procure adequate food and live at lower population densities. Omnivorous birds (those that eat both plants and animals) typically have home range sizes intermediate to those of herbivores and carnivores.

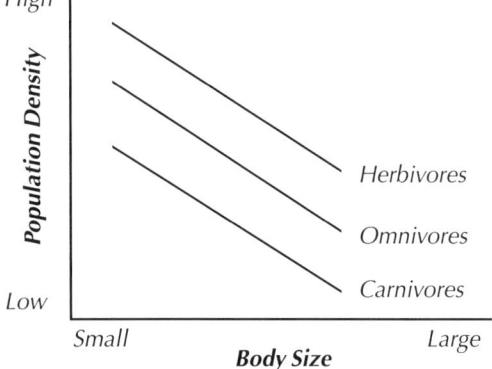

Dickcissel Breeding Range

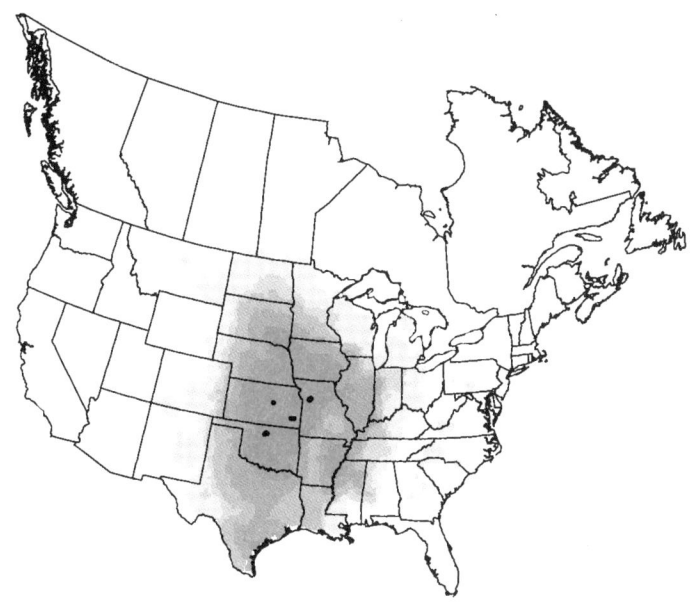

Figure 9–44. Density of the Dickcissel Across its Breeding Range: For any species, population density is heavily influenced by habitat suitability. As in many species, the breeding density of Dickcissels is highest at the center of the range—shown by the darker-colored areas on the map—where the preferred habitat, large open grasslands, is most abundant. Dickcissel density decreases where its grassland habitat is less ideal, at the periphery of its range, shown by the lighter-colored areas on the map. Map adapted from North American Breeding Bird Survey. Photo of Dickcissel by Lang Elliott.

Population density does not always indicate habitat quality, however. Some populations exist at low densities even when the environment is capable of supporting more birds. For example, during some years densities of migratory species may be lower than expected on the breeding grounds, because many individuals died on the wintering grounds or during migration. When I studied breeding populations of Loggerhead Shrikes in the Midwest, I was puzzled that so much suitable habitat hosted so few birds. I discovered that the low densities probably resulted from poor overwinter survival on the Gulf Coast nonbreeding range, where exotic fire ants, introduced accidentally in the 1930s, have decimated populations of ground-dwelling insects and small vertebrates on which shrikes depend for their winter food (**Fig. 9–45**).

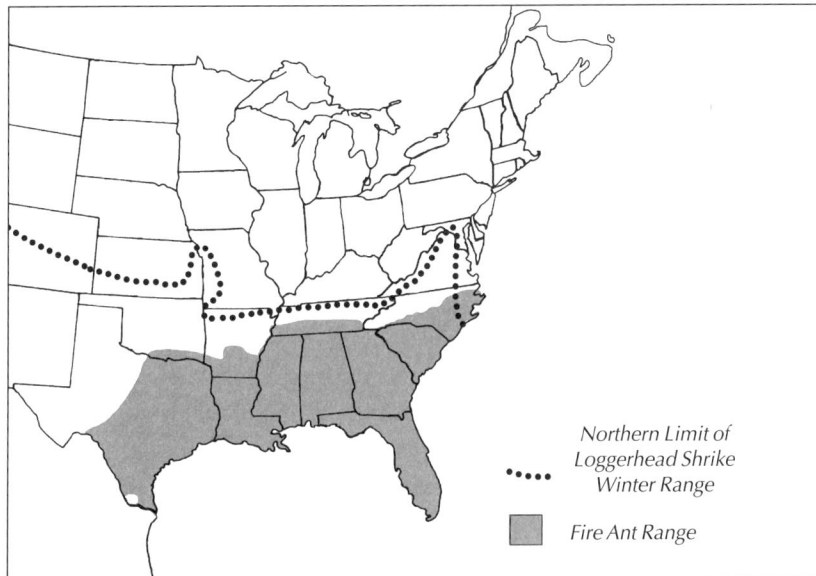

Northern Limit of
Loggerhead Shrike
Winter Range

Fire Ant Range

Figure 9–45. Overlapping Ranges of Fire Ants and Wintering Loggerhead Shrikes: In the Midwest, concern about the low numbers of breeding Loggerhead Shrikes in apparently ideal habitat prompted researchers to look for an explanation outside the birds' breeding range. Finding that introduced fire ants are now found throughout the shrike's winter range, they surmised that ants have decimated populations of the ground-dwelling insects and other small animals upon which the wintering shrikes depend, resulting in low overwinter survival and fewer pairs returning to breed.

9

Population densities also may vary with local changes in food availability, such that densities are highest where food is most concentrated. When local outbreaks of insects occur, such as spruce budworms in coniferous forests, densities of insectivorous birds can increase up to four fold. Wintering raptors, such as Short-eared Owls and Rough-legged Hawks, often concentrate in locations where small rodents are irrupting. And the impressive response of California Gulls to "plagues" of long-horned grasshoppers that threatened Utah crops in 1848 and 1855 is commemorated by a statue of gulls erected by grateful Mormons in Salt Lake City. The gulls moved quickly into the affected areas, saving the crops by eating large numbers of the pests.

How Do We Determine Population Size?

Many birds are secretive or otherwise difficult to see, whereas some are so common that they seem to be everywhere. How, then, does anyone accurately count the birds in a given area? Ornithologists have developed a variety of techniques to cope with the counting challenges posed by birds and their environments. Of course, no technique is completely accurate; each has biases that result in estimates of population size or density that deviate to some extent from the actual number. Some biases are related to the behavior of the birds being counted and the characteristics of the environment in which the count takes place, whereas other biases are introduced by the counters themselves. Usually observers tend to underestimate numbers, because they fail to detect all birds present in an area. Some birds actively avoid the observer, some birds are obscured by their habitat, some birds give few clues to their presence, and some observers are inexperienced or have impaired vision and hearing.

At one international ornithological conference, all participants were required to take a hearing test. A number of ornithologists who routinely censused birds by counting singing individual were surprised to learn that significant hearing impairments (often age-related) were preventing them from detecting high-frequency calls. After training dozens of students to survey birds, I know firsthand that even well-trained observers are rarely comparable to one another, and that even when they survey the same area, they get slightly different results.

For any survey, after each of the possible biases has added a little error, the overall count bias can become great. The best way to avoid unacceptable error is to standardize the surveying technique as much as possible. Problems can be reduced by rigorously training observers, observing strict rules about procedures, and drawing conclusions only from counts that are truly comparable.

The two best-known bird surveys in North America are the Christmas Bird Count and the Breeding Bird Survey. These are both large-scale, long-term monitoring efforts that rely extensively on volunteer observers from among the bird-watching public. Differences between them, however, illustrate some of the problems of collecting and interpreting bird population data.

Christmas Bird Counts (CBCs) have been coordinated by the National Audubon Society since 1900. The counts occur in late December and early January, when tens of thousands of observers go into the field on a single day to count as many birds as possible within hundreds of circular count areas, each of which is 15 miles (24 km) in diameter and usually is centered on a town. Observers are asked to record the number of individuals of each species detected as well as the number of hours that they searched. Results are tallied as the number of individuals of each species within each count circle, and are submitted electronically to the CBC database for analysis.

The results of CBCs are not estimates of population density. Instead, they are an **index**—in this case, birds detected per unit time—which is assumed to be closely correlated with population size. The validity of this assumption can be questioned, however. One reason is that observers often go out of their way to find rare species—after all, that's part of the fun of birding. When this happens, counts of rarities can be biased upward. In addition, the number of observers varies from count to count, and often count areas are not uniformly censused. Finally, although observers are assumed to be competent at identifying birds, skill levels vary widely.

Eastern Bluebird

Despite these biases, CBCs have generated valuable data on continental patterns of winter bird distribution and abundance. They have been particularly useful in tracking long-term trends, because the counts now span almost a century **(Fig. 9–46)**. What CBCs lack in precision they make up for in numbers of observers and birds counted, so that a few unskilled individuals or poorly covered areas have only a minor impact on overall totals.

Figure 9–46. Eastern Bluebird Population Trend from Christmas Bird Count Data: The Christmas Bird Count (CBC), coordinated by the National Audubon Society, has been enjoyed by bird watchers each year since 1900. During one day in late December or early January, thousands of birders across North America count the number of individual birds seen in more than 1,500 count "circles" 15 miles across. CBC results can be calculated as an index of birds detected per unit time, which is assumed to be correlated with bird population density. CBC data are used to show general patterns of bird distribution and abundance; and, because the surveys have been conducted for more than a century, they are particularly useful for showing long-term trends. This graph tracks the Eastern Bluebird from 1959 to 1988. Notice that the curve gradually increases and then becomes steeper. Early in the 20th century, bluebirds and other native cavity nesters declined drastically because of habitat loss and competition for nest sites with aggressive alien species such as the House Sparrow and European Starling. In the late 1970s, the newly formed North American Bluebird Society began advocating nest-box programs to boost bluebird populations. The CBC graph reflects the success of these programs. Graph courtesy of Sauer, J. R., S. Schwartz, and B. Hoover. 1996. The Christmas Bird Count Home Page. Version 95.1. Patuxent Wildlife Research Center, Laurel, MD. Photo by Marie Read.

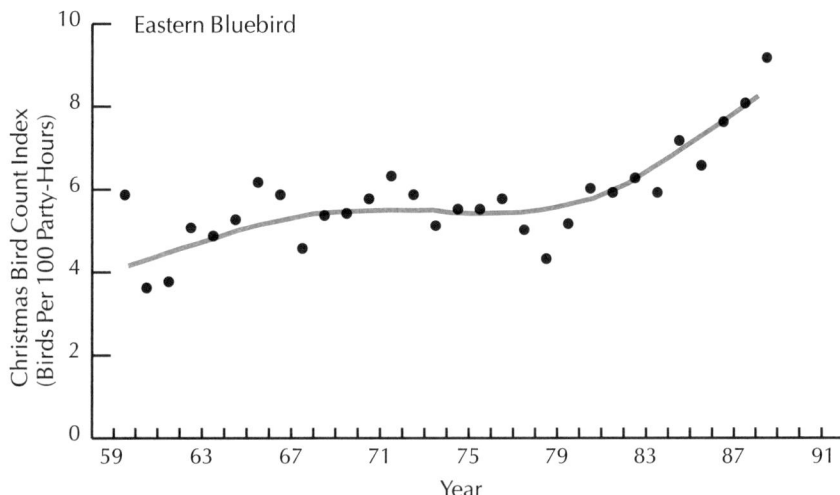

The Breeding Bird Survey (BBS) has been coordinated by the U. S. Fish and Wildlife Service since 1967. This program was carefully designed to provide precise information on bird abundance and distribution, and is based on randomly selected points across the continent. For each point, the nearest road was located, and a 25-mile (40-km) vehicular survey route was established in a randomly selected direction. Because the points and associated routes are randomly located, they sample birds in most North American habitats containing roads.

The BBS works as follows: on one weekend in June, carefully selected observers count birds along each route, beginning one-half hour before sunrise. They stop every half mile (0.8 km), and for three minutes count all birds detected by sight or sound within a quarter mile (0.4 km) of their car (see Fig. 10–23). Then they report their data to the U. S. Fish and Wildlife Service, which tallies the results as an index, in this case individuals of each species per route. Because each stop can be associated with its surrounding habitat, data analysts can correlate changes in bird populations with changes in the local environments. As with CBCs, biases in BBS data do exist, but they have been reduced through randomly selecting routes, using well-trained observers, and observing the many rules about how the surveys are conducted.

Results of the BBS have allowed ornithologists to track population trends and to plot distribution patterns with great precision, and these surveys are now the main source of information used to determine the status of bird populations across North America. They have shown ornithologists that bird populations are much more complex and dynamic than originally thought **(Fig. 9–47)**.

What Affects Population Size?

As discussed earlier, changes in the size and structure of bird populations result from variations in rates of birth, death, and **dispersal** (immigration and emigration). Birth and immigration add individuals to a population, death and emigration remove them, and ultimately population size is determined by the balance between addition and removal rates.

Population ecologists usually measure rates of birth (or **fecundity**) within a population as the average number of young produced per female per unit time (usually one year). Typically, fecundity is determined at the time when young are fully grown and either leave the nest (in the case of **altricial** birds) or leave their brood (in the case of **precocial** birds). Ecologists measure the number of young produced, rather than the number of eggs laid, because many eggs do not result in fledged young.

Whether a given nesting attempt succeeds or fails is influenced by many factors, which are discussed in the next section. Most of these factors are related to habitat quality, and many studies have found the highest success rates in the best habitats. Nest success also varies geographically: a nest at lower latitudes is more likely to fail than one at higher latitudes, primarily because of heavy nest predation in the tropics. Over their lifetimes, however, tropical species are no less successful at producing offspring than are temperate zone species,

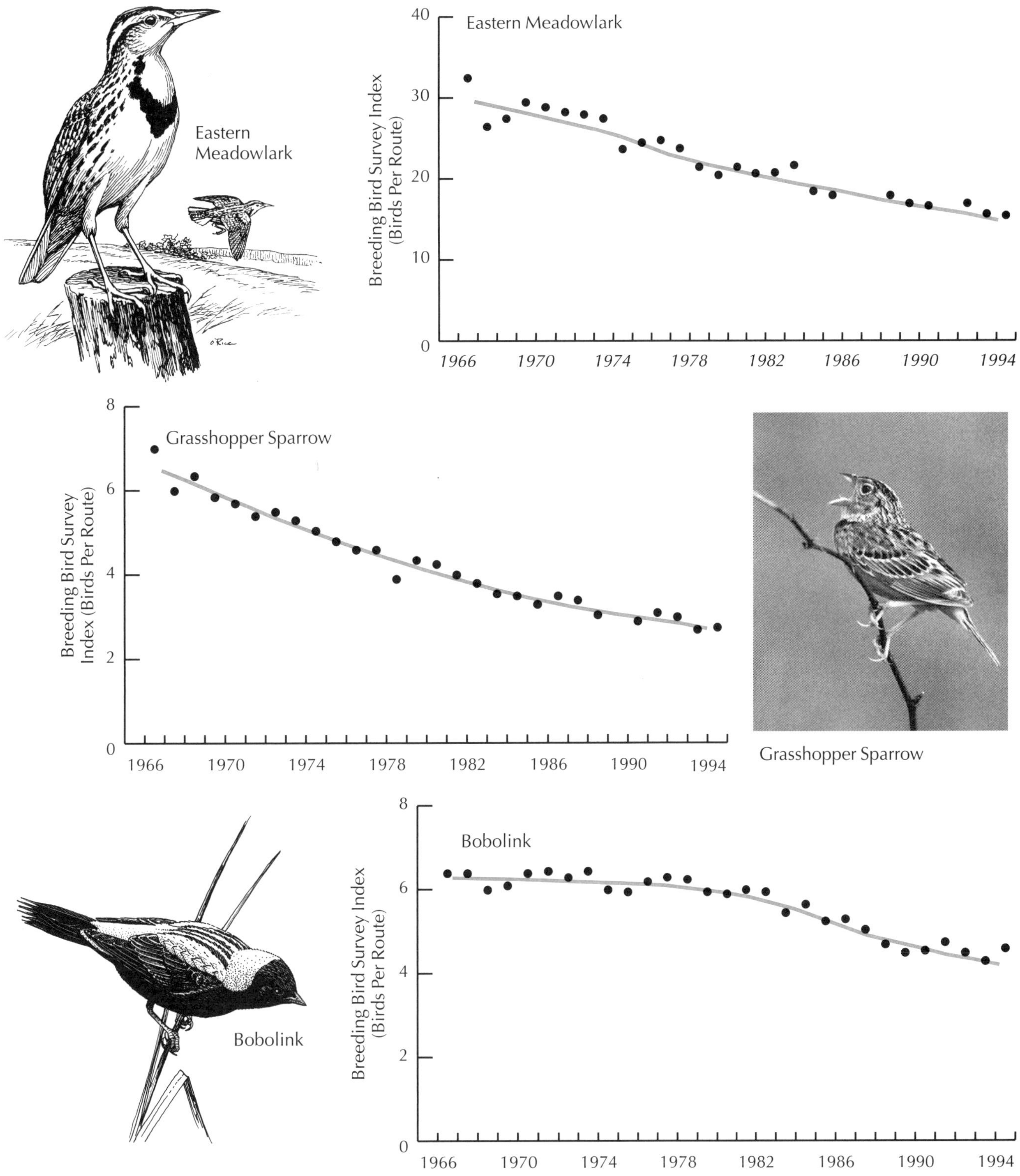

Figure 9–47. Bird Population Trends from Breeding Bird Survey Data: The Breeding Bird Survey (BBS), coordinated by the U. S. Fish and Wildlife Service each June, is designed to provide precise, easily interpreted information about bird abundance and distribution. Trained observers drive specific 25-mile (40-km) survey routes, counting all individuals of each species seen or heard at half-mile (0.8-km) intervals along the way. Data from each stop are associated with the surrounding habitat characteristics, and because the same locations are surveyed each year, changes in bird populations can be linked to environmental changes along the routes. Presented here are graphs of annual BBS indices (measures of individuals counted per route) for three North American grassland species, from 1966 to 1994. They show that populations of Bobolinks, Grasshopper Sparrows, and Eastern Meadowlarks have declined steadily, reflecting the gradual loss and degradation of grassland habitats. Graphs courtesy of Sauer, J. R., B. G. Peterjohn, S. Schwartz, and J. E. Hines. 1995. The Grassland Bird Home Page. Version 95.0. Patuxent Wildlife Research Center, Laurel, MD. Drawings: Bobolink courtesy of William C. Dilger; Eastern Meadowlark courtesy of Orville Rice. Photo of Grasshopper Sparrow courtesy of Lang Elliott/CLO.

because they tend to live longer (see Ch. 5, Patterns of Migration, Table 5–2). The type of nest also influences its success: whereas cavity-nesting birds fledge young from about 66 percent of the eggs they lay, birds that nest in the open succeed with only 45 percent, presumably because cavity nests are better protected from predators. For various reasons, however, open nesters tend to have shorter incubation and nestling periods (see Fig. 8–111), and thus have a greater chance of making several nesting attempts per breeding season, at least partly off-setting their lower success per nest.

The age of a breeding pair also can influence nesting success. First-time breeders frequently produce fewer young than experienced breeders, even when they start with the same number of eggs. Because of this variation, population ecologists calculate **age-specific fecundity** as the average birth rate for females of different ages.

Recall that population size also is affected by the rate at which individuals die, which ecologists measure as either the **survival rate** or the **mortality rate**. The survival rate describes the proportion of individuals in a population that will be alive at some future time (usually one year later). Conversely, the mortality rate describes the proportion that die during the same time interval. Ecologists determine these numbers by marking individuals, usually with leg bands, then recording the rate at which they disappear from a population. Of course such calculations assume that the missing birds all have died, but in fact, a few may simply have left the area.

Annual survival rates vary considerably among species (see Ch. 8, Survival, Table 8–1); and within a species, they vary with age (see Fig. 8 2). Therefore, population ecologists calculate **age-specific survival rates**. Mortality during the first few weeks appears to be particularly high, and among most songbirds, the survival rate of fledglings during their first fall and winter, which may include migration, is typically half that of birds that have reached their first breeding season the following year. Once birds have survived their first winter, however, their annual survival rates increase and tend to remain fairly constant for the rest of their lives **(Table 9–2)**. The highest adult survival rates—95 percent or more—have been recorded in large birds such as penguins, albatrosses, and condors, whereas small songbirds typically have adult rates in the 30 to 60 percent range.

In northern Peru, I once studied a population of 57 Andean Condors, each marked with wing tags and a radio transmitter. Their survival rates revealed the expected pattern. During their first year of life, while still dependent on their parents, 76 percent of young condors survived. Between two and six years of age, the annual survival rate rose to 90 percent. For adult condors more than six years old and potentially capable of breeding, survival rates rose to 94 percent, among the highest recorded for wild birds (Temple and Wallace 1989).

Survival rates also vary geographically. In general, birds at lower latitudes (near the equator) have higher survival rates than birds closer to the poles, and the difference can be significant. For example, adult Blue Tits in the British Isles (about 55 degrees north latitude) have survival rates of only 30 percent, whereas those on

Table 9–2. Life Table for Eurasian Kestrel: Life tables are used to compare annual mortality/survival rates of birds in various age classes. Researchers following a marked population of 245 Eurasian Kestrels in the Netherlands found a typical pattern of survival/mortality over a 10-year period. Mortality rates tend to be very high in young birds, especially in the first few weeks after leaving the nest. Then, as the young birds mature, mortality declines until they begin to succumb to the physical breakdown associated with aging. However, because the risk of death due to predation, starvation, and other factors is so high, wild birds rarely die of old age. Adapted from Newton (1979).

Age in Years	Number Alive at Start of Year	Number Reported Dying During Year	Annual Mortality (%)
0–1	245	126	51
1–2	119	49	41
2–3	70	34	49
3–4	36	14	39
4–5	22	8	36
5–6	14	4	29
6–7	10	5	50
7–8	5	1	20
8–9	4	2	50
9–10	2	2	100

the Canary Islands (about 25 degrees north latitude) have rates of 55 percent, nearly twice as high.

Although birth and death rates significantly influence population size and density, population dynamics also are influenced by movements of individuals into and out of areas. Unfortunately, in most places rates of immigration and emigration are difficult to estimate unless a large percentage of birds are banded. And even then, because birds are more mobile than the ecologists who study them, movements are usually hard to follow. When a new bird moves into a marked population it may be detected, but if a marked bird disappears, knowing whether it emigrated or died is rarely possible.

Nevertheless, studies of banded individuals have shown that dispersal is an important aspect of birds' lives. Often, movements occur between the time of fledging and breeding. This is known as **natal dispersal**. But even though young birds frequently wander, they tend to show fidelity to their natal area—a habit known as **philopatry**—and eventually take up residence near the place where they fledged. How near depends on the species.

Some birds are "homebodies"—establishing breeding territories very near to their parents. In California, 34 banded Song Sparrows established their first breeding territories an average of just 202 yards (185 m) from where they fledged. Many colonial-nesting birds wander extensively before reaching breeding age, then return to their natal colonies to breed. This pattern has been revealed by studies of Herring Gulls banded as nestlings. Whereas immature individuals have been found as far as 2,500 miles (4,000 km) from the colony where they hatched, most adults have been found back at the natal colony.

In some species, the extent of natal dispersal varies with sex, and in general, females tend to disperse farther than males. But in some groups, including waterfowl, females show strong philopatry while males disperse widely **(Fig. 9–48)**. In fact, males often follow the female with whom they paired on the wintering grounds back to her natal wetland. In one well-studied colony of Snow Geese, half the birds breeding for the first time are males that have immigrated from other colonies.

Adult birds are less likely to move than young birds, especially after they have bred successfully. Successful breeding promotes faithfulness to an area. Adults that disperse to a new breeding site even when conditions remain suitable at the old site typically have lower survival and reproductive success than individuals who remain, which demonstrates the importance of site fidelity. However, adults may be forced to move if their habitat is destroyed, if their food supply fails, or if their habitat becomes overcrowded, increasing competition for limited resources.

Dispersal also has genetic costs and benefits. On the plus side, it moves individuals away from close relatives, thus reducing the chances of **inbreeding depression**, the reduction in fitness of offspring produced by closely related parents. But dispersal also results in the mixing of genes with other local populations, increasing **outbreeding**

Female Canvasback

Male Canvasback

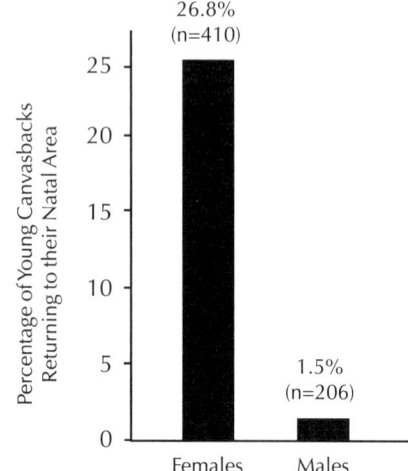

Figure 9–48. Natal Philopatry in Canvasbacks: Female waterfowl are more likely than males to exhibit natal philopatry, that is, to return to their natal areas to breed. This pattern is the opposite of that shown by many other groups of birds, in which females disperse widely while males exhibit strong natal philopatry. The Canvasback is a diving duck that nests in freshwater wetlands in the Midwest, central and western Canada, and Alaska. The species winters on coastal waterways and large lakes. Research shows that, of young birds banded in their natal areas, 26.8 percent of females returned to breed in those same areas in subsequent years, whereas only 1.5 percent of males returned (Anderson 1985). Although both sexes might benefit from returning to a familiar area, the advantages are of particular importance to females, which have the sole responsibility for incubation and subsequent care of the ducklings, and would benefit from being familiar with productive feeding areas and safe resting places. In contrast, once incubation is under way, males of most duck species abandon their mates and leave the breeding areas. Photo of Canvasback female courtesy of Michael Hopiak/CLO; photo of Canvasback male courtesy of L. B. Wales, Jr./CLO.

depression, a disruption of the genetic basis for adaptations that enhance fitness in the local environment. Recall from Chapter 1 that a subspecies is a regional population having a distinctly different genetic structure from other regional populations. Subspecies can retain their identity only as long as little mixing of genes (**gene flow**) occurs as a result of immigration from adjacent populations.

Most bird populations fluctuate throughout the year, reaching an annual peak just after the breeding season when the population is bolstered from all the young hatched that year. From that time the population declines steadily, reaching its annual low just before young begin to fledge in the next year's breeding season. The annual cycle then repeats **(Fig. 9–49)**. The long-term size of a population is regulated more closely, however.

What Regulates Population Size?

Although many factors can affect population size, in natural situations most bird populations remain fairly constant over time. Since Breeding Bird Surveys began tracking North American bird populations in 1966, numbers of most species have shown only minor fluctuations around the now 30-year average **(Fig. 9–50)**. The reason for this stability is that, over time, rates of birth, death, immigration, and emigration counterbalance. More specifically, **homeostatic mechanisms** ensure that when the density of a bird population drops, rates of birth and immigration increase, whereas when density rises, rates of death and emigration increase (Newton 1998).

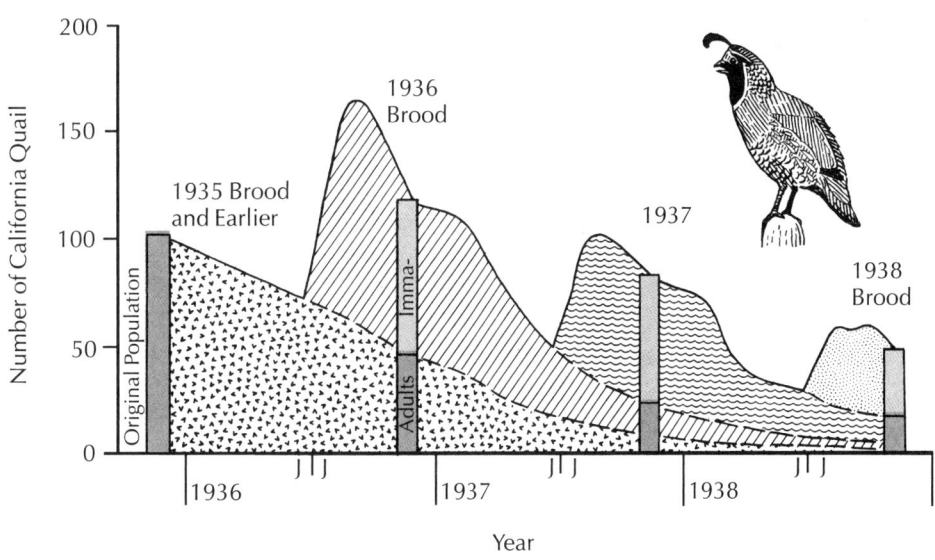

Figure 9–49. Annual Cycle in Population Size: The overall size and age composition of bird populations cycle in a predictable annual pattern. Typically, populations are largest at the end of the breeding season, when the population has been augmented by young birds. This cycle is evident in a California Quail population that was studied at the University of California (Davis) from 1935 to 1938. Notice the large increase in overall population size as immatures are added from each year's broods. Also notice, as indicated by the shaded bars, that the annual addition of many young birds to the population causes the age distribution to be skewed toward immatures following each breeding season (adapted from Emlen 1940).

Gray Catbird

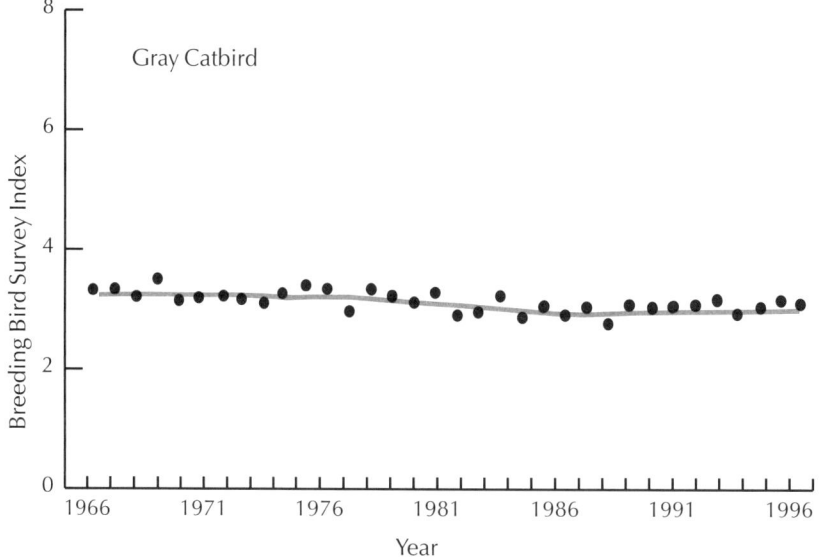

Figure 9–50. Stable Population Trend of the Gray Catbird: Over time, most bird populations remain relatively constant in size unless their environment changes significantly. This graph of Breeding Bird Survey data shows that the Gray Catbird population remained stable from 1966 to 1996, although minor fluctuations occurred. Drawing courtesy of Orville Rice.

Even populations that undergo fairly substantial swings in size and density tend to cycle around a long-term average. Regular fluctuations in some raptors, such as Northern Goshawks (10-year cycle) and Snowy Owls (4-year cycle), occur in response to the cyclic changes in prey populations (see Fig. 9–26). Similarly, Christmas Bird Counts have tracked the cyclic fluctuations of winter finch populations, which also vary in response to food availability, and Project FeederWatch (a winter bird monitoring project administered by the Cornell Lab of Ornithology) has documented cyclic fluctuations in Varied Thrushes **(Fig. 9–51).**

Figure 9–51. Population Cycle of the Varied Thrush: Anecdotal reports of fluctuating Varied Thrush populations in winter led researchers to examine Project FeederWatch and Christmas Bird Count data from the species' winter range in British Columbia, Oregon, and Washington. FeederWatch (solid shapes) and Christmas Bird Count (open circles) data showed a consistent two-year cycle. Researchers surmise that the thrushes rely on a food source that also exhibits biennial abundance. Although in summer these birds consume fruit and insects, during winter they rely primarily on acorns, which are produced abundantly every other year. Adapted from Wells et al. (1996).

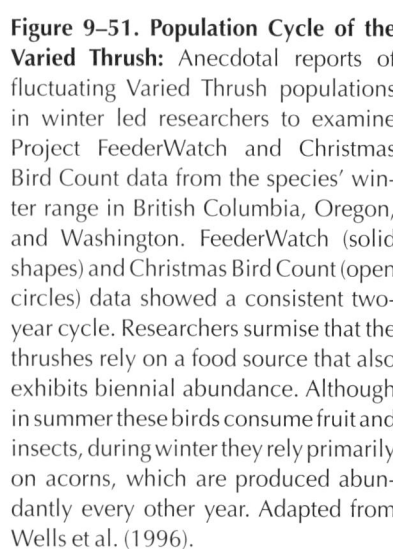

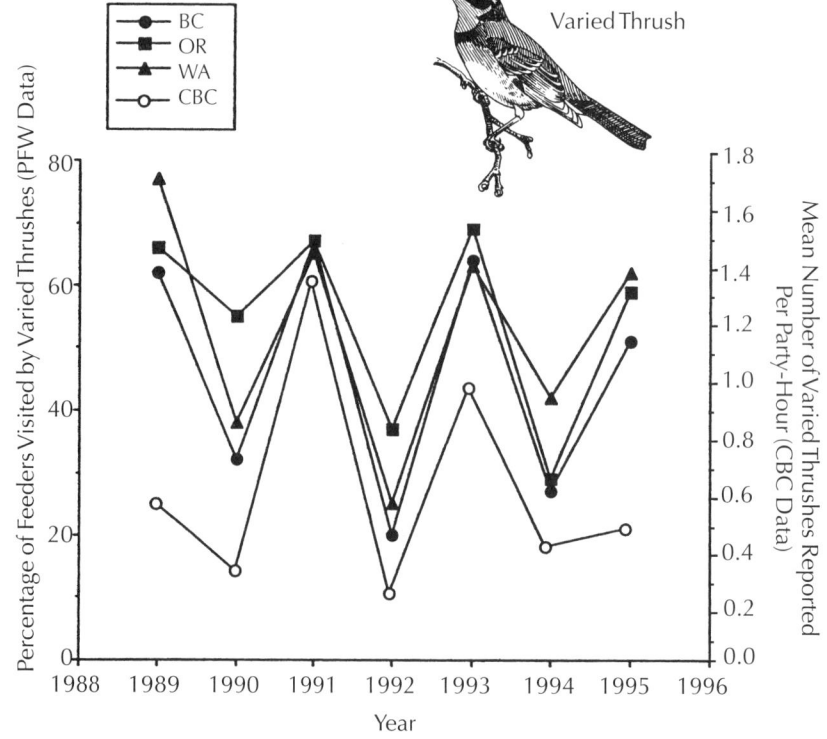

Factors that regulate population size and that exert their influence in relation to a population's density are called **density-dependent factors**. They usually act by changing the rates of birth, death, and dispersal. Such factors include predation, disease, parasites, and competition for food or nest sites **(Fig. 9–52)**.

Consider intraspecific competition. When the amount of a resource such as food is fixed and limited, competition for that resource increases as a population grows. As competition intensifies, it may decrease birth rates and increase mortality, thereby stabilizing the population at a sustainable level.

Predation functions in a similar manner. When prey is scarce in a particular area, predators generally stay away because hunting there is not worth the effort, or if they do forage in the area, their kill rate is low. As prey density increases, however, predators move into the region and take large numbers of prey, raising mortality rates and slowing growth of the prey population so that it remains at equilibrium.

Mortality caused by disease and parasites also can be strongly density dependent because when a population starts to grow, diseases and parasites are transmitted more easily from bird to bird. At high densities, pathogens can spread quickly and cause massive die-offs known as **epizootics**. For example when waterfowl are crowded together during migration or in some urban populations, hundreds or

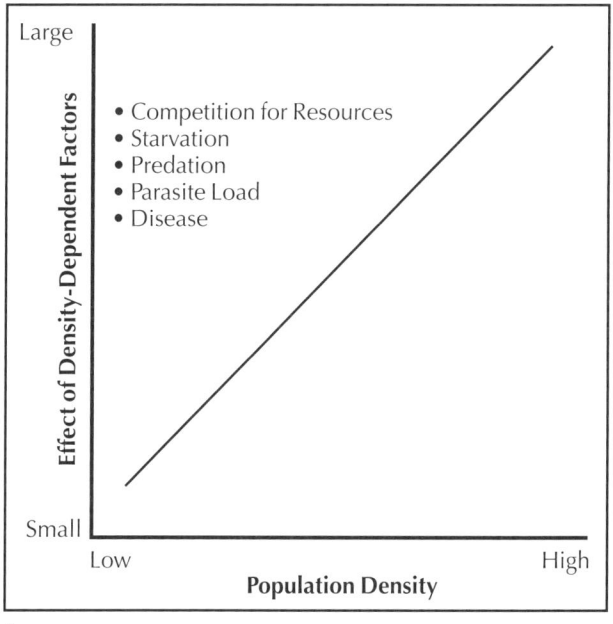

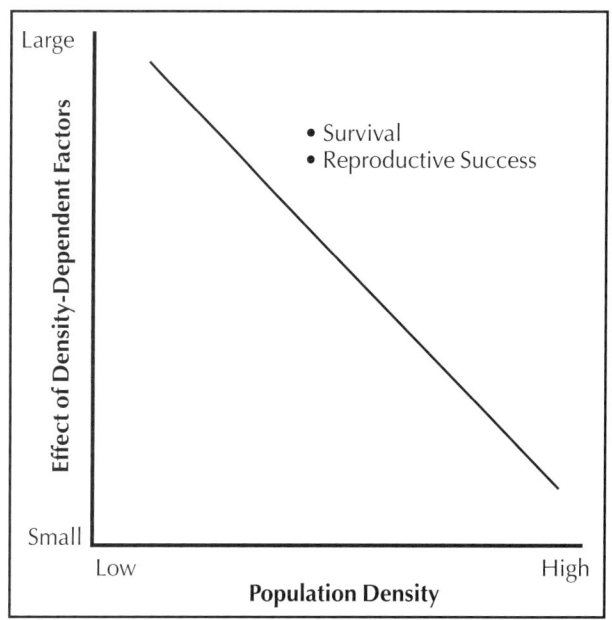

a

b

Figure 9–52. Factors That Regulate Population Density: Population size is regulated by many factors. Those that tend to stabilize populations over time are known as **density-dependent factors**, because the magnitude of their effects on population size is determined by the number of individuals present in the area. An example is intraspecific competition for resources **(a)**. When population densities are low, individuals do not compete with one another for food or nesting sites, and all pairs are likely to breed and raise large broods. But as their numbers grow and prime habitat becomes fully occupied, some birds are forced into poor habitat and fail to rear their broods or do not mate at all. Predation rates, diseases, and parasite infestation likewise increase as population density increases. At high population density, rates of survival and reproductive success are low, but as density declines the chance of survival and successful reproduction increase **(b)**. Eventually these density-dependent factors return the population to its equilibrium density.

even thousands may be killed by a disease known as viral enteritis ("duck plague"). Individuals become infected through direct physical contact with infected individuals, contamination of food and water with pathogens, and the spread of pathogens by intermediary organisms (**vectors**) such as biting insects. Epizootics also can occur in the dense populations of birds that may occur around bird feeders. I once determined that outbreaks of diseases including salmonellosis, trichomoniasis, coccidiosis, aspergillosis, and avian pox had occurred at 16 percent of 624 feeders located in Wisconsin. Mortality was highest among gregarious species, such as House Sparrows, that lived at high densities during winter **(Sidebar 2: The House Finch Hot Zone)**.

Competition for limited space is manifest through territoriality. At low population densities (relative to the equilibrium density), territorial behaviors tend to spread members of a local population around the available habitat, and all individuals find a place to settle in good quality habitat. In fact, these individuals probably will survive and reproduce at rates that will cause the population to increase in size and density. Furthermore, immigration into the population is possible because space is available for new individuals to occupy. Hence, at low population densities, territorial behaviors do not interfere with population growth.

At moderate population densities, territorial individuals completely fill all the best habitat. The remaining individuals in the local population are forced to settle in suboptimal habitat where their fecundity and survival are reduced. Under these conditions, territoriality has set an upper limit to the number of individuals that can live in good habitat, and has forced additional birds to live in places where they will neither survive nor reproduce as well. Also under these conditions, territoriality begins to slow population growth in response to the increased density.

At high population densities, territorial birds completely fill both optimal and suboptimal habitats. The local habitat has reached its **carrying capacity**, the maximum density of individuals that the local resources, in this case space, can accommodate. Additional individuals will be unable to claim a territory and thus have two options: either emigrate in hopes of eventually finding a territory elsewhere, or remain without holding a territory. In both cases, there will be a substantial impact on the population's density. Emigrating individuals, of course, reduce the local density when they leave. Individuals that stay but do not hold a territory (called **floaters**) have very poor survival and are unable to breed (unless they sneak extrapair copulations with mated birds). Floaters, therefore, never add to further population growth. As a result, population growth is slowed as the local density reaches carrying capacity.

Although territoriality may prevent some birds from acquiring territories and breeding at high population densities, even those that have territories may not be as prolific as at lower densities. As population density increases, territory size shrinks, so the territory holders do not have as much food available to them. In response, density-dependent

(Continued on p. 9·75)

Sidebar 2: THE HOUSE FINCH HOT ZONE
André A. Dhondt, Wesley M. Hochachka, Sonia M. Altizer, and Barry K. Hartup

In 1994, a mysterious bird infection emerged seemingly from nowhere. Bird watchers in Maryland suddenly began seeing numerous House Finches with red, crusty, swollen eyes at their backyard feeders. The disease soon spread far beyond the area where it was originally detected, and in just two and a half years, infected birds were being seen across eastern North America. Although people rarely found birds that had been killed by the disease, surely they were numerous. Since the discovery of House Finch eye disease, or mycoplasmal conjunctivitis, House Finch populations in the East have plummeted from an estimated 300 million to 120 million birds—a decrease of 60 percent **(Fig. A)**.

Mycoplasmal conjunctivitis now figures prominently in research conducted at the Cornell Lab of Ornithology. Our research team was one of only 12 initially chosen by the National Institutes of Health (NIH) and other federal agencies to study how large-scale environmental events— such as habitat destruction, biological invasion, and pollution—affect the emergence and spread of disease in both humans and wildlife. How the Lab became involved in studying new diseases shows the serendipity of science and also how valuable the contributions of citizen scientists are in studying the world around us.

A Disease—and a Research Project—Emerge

When rumors of the terrible new bird disease in Maryland first began to circulate in February 1994, no one knew the cause of the infection. But a group from the Southeastern Cooperative Wildlife Disease Study at the University of Georgia and David Ley at North Carolina State University quickly identified the culprit: a novel strain of a well-known and widespread bacterium called Mycoplasma gallisepticum that was common in poultry but had never before been detected in wild birds **(Fig. B)**. Because this bacterium was highly contagious, some scientists feared that the new strain would spread rapidly to many other species of North American birds. Even more frightening, however, was the possibility that Neotropical migrants might carry the new disease to the tropics in the fall—a potentially disastrous scenario.

Because observers were noticing diseased House Finches mostly at feeders, scientists suspected a link between bird feeding and mycoplasmal infection. Would people have to stop feeding birds to slow down the epidemic, or would removing bird feeders cause infected

Figure A. House Finches at a Feeder: The House Finch is a common resident of suburbs and cities and a frequent visitor to backyard bird feeders. Native to the western United States, the House Finch was released on Long Island, New York in the 1940s and since then has spread throughout eastern North America. In 1994, bird watchers noticed the first cases of a deadly, highly contagious eye disease that has spread rapidly through the House Finch population, aided by two aspects of the bird's behavior: juveniles disperse long distances, and the species often forms large, mobile foraging flocks in winter. Since it first appeared, House Finch Eye Disease has claimed the lives of 60 percent of the House Finch population, decreasing it from 300 million to 120 million birds. Photo by Marie Read.

Figure B. House Finch Eye Disease Victim: A male House Finch shows the typical symptom of mycoplasmal conjunctivitis—red, severely swollen (or sometimes crusty-looking) conjunctiva (inner eyelids). Birds sickened by the infection often succumb to starvation or predation, or may simply die from exposure. The cause of the disease is Mycoplasma gallisepticum, a bacterium that causes disease in poultry and other birds. Some strains of the bacterium are known to infect the respiratory tract of chickens and reduce egg production. The "House Finch strain" of Mycoplasma gallisepticum is a previously unknown strain that causes eye infections in wild House Finches, Purple Finches, American Goldfinches, Evening Grosbeaks, and Pine Grosbeaks. It has caused a major epidemic only in House Finches, however. Chickens exposed to the strain do not develop clinical eye disease. Photo courtesy of CLO/House Finch Project.

birds to disperse farther and spread the disease more rapidly? In the middle of this debate, André Dhondt arrived at the Lab as its new director of Bird Population Studies. Our research team soon realized that with the help of citizen scientists, we could answer many critical questions about the new disease.

Participants in the Lab's Project FeederWatch gave us an unprecedented opportunity to track the origin and spread of the disease. They had been collecting information about their winter feeder birds since 1988. With such a network in place, we could mobilize quickly and begin collecting data only months after the first cases of mycoplasmal conjunctivitis had been reported. Because Project FeederWatch participants were collecting data from across the United States and Canada, we could

survey the entire continent and track the spread of the disease. With this revelation, the House Finch Disease Survey was born.

Designing the first survey form was a challenge because we knew so little about the disease. Were all feeder birds susceptible or just House Finches? Were House Finches especially vulnerable because they had been introduced to the East recently and did not have as much genetic variation as native species? Could it be because House Finches lived so closely with humans and gathered in large numbers at feeders? We eventually decided to request information on a number of other species in addition to House Finches—including Purple Finches, House Sparrows, Black-capped Chickadees, and Dark-eyed Juncos—based on how closely related they are to House

Finches and how frequently they use feeders. To leave all options open, we added a category for "other birds."

Another challenge in implementing this survey was to encourage participants to submit negative data—that is, records of healthy finches or even no House Finches at all. We feared that some participants would think their contributions were unimportant if they saw only healthy birds. Thus, rather than ask for a "Yes" or "No" response to the questions "Did you see any House Finches?" and "Did you see sick birds?" we asked observers to fill separate bubbles on the data form for each day they watched their feeders, each day they saw House Finches, and each day they observed diseased birds. Participants who watched their feeder but did not observe any birds, therefore, still reported their efforts.

After we settled on the questions we wanted to ask, our graphic designer drafted a form that could be filled out by observers and scanned into our computers. We ordered 60,000 copies—a big gamble because we had no way to predict whether the disease would fade away or expand into a full-blown epidemic. In September 1994, we mailed six monthly data forms to each of 9,000 eastern FeederWatchers and asked them to begin reporting in November 1994. Then we waited.

Citizen Science On the Trail of an Epidemic

By early December our mailbox was overflowing, as 3,212 participants sent us their observations. We immediately produced a detailed map showing the prevalence of conjunctivitis in eastern House Finches during November 1994. The results were surprising. Within 10 months of its discovery, the disease had invaded an area from eastern Ontario to southern Virginia. The epidemic was well established along the Atlantic coast, reaching western New York, western Pennsylvania, and West Virginia. Isolated reports from as far south as Georgia and as far west as Wisconsin showed that the disease was capable of spreading long distances, carried by migrating or dispersing birds.

As participants continued to send data to the Lab each month, we followed the epidemic across eastern North America **(Fig. C)**. From its origin near Washington, D.C., the pathogen initially spread most rapidly to the north and northeast, probably carried by House Finches that had wintered in Maryland and Virginia before returning to their breeding grounds. By February 1995, just one year after the first infected finches had been reported, FeederWatchers had documented considerable proportions of House Finches with eye infections from Georgia and Kentucky to Vermont and Wisconsin. So many observers reported data that we were able to describe how the disease invaded and became established in state after state as it moved west from Ohio to Wisconsin. Interestingly, as the epidemic moved farther away from the core of the disease's range, multiple local flare-ups of disease appeared before a full-fledged epidemic took off.

Even during the epidemic, however, the vast majority of participants only saw healthy birds. Fortunately, this did not stop them from submitting their observations, because we needed both negative data (no disease seen) and positive data (House Finches with eye infections) to document how the disease spread over space and time. If people had only reported infected birds, we would have severely overestimated the extent of the disease. Moreover, without records of healthy birds in newly invaded sites, we would not have been able to see the geographic expansion of the disease in such detail.

Shortly after we initiated the House Finch Disease Survey, a glaring weakness became apparent. The survey measured the proportion of sites with diseased birds present, but we still needed to establish the

link between the presence of the bacteria and visible symptoms of the disease. Fortunately, in 1996, Barry Hartup, a graduate student at Cornell's College of Veterinary Medicine, decided to address this critical problem. In collaboration with Lab member and bird bander Jean Bickal in New Jersey, Barry trapped hundreds of House Finches and sampled their conjunctiva (inner eyelids) to isolate the pathogen. Barry's work proved that a close link existed between the presence of the Mycoplasma gallisepticum bacteria and symptoms of eye infection. The results from the House Finch Disease Survey, therefore, truly represented the prevalence of mycoplasmal conjunctivitis across time and space.

A Deadly Disease

Because House Finch Disease Survey participants continued to

submit data even when they no longer saw any House Finches at their feeders, we were able to document that in areas with cold winters and a high prevalence of the disease, House Finches disappeared toward the end of winter. This showed that in certain conditions the epidemic caused House Finch populations to decline in numbers, but it still didn't tell us how many House Finches had disappeared.

Geoffrey Hill of Alabama's Auburn University has produced outstanding work on House Finches for more than a decade. When the epidemic swept through his House Finch population on the Auburn campus in late 1996, Hill documented the results in detail. He found that up to 60 percent of his birds contracted eye infections and that some males survived better than others. Hill's team also estimated that the Auburn population crashed

Figure C. The Spread of Mycoplasmal Conjunctivitis Across the Eastern United States: Using data from Project FeederWatch, Cornell Lab of Ornithology Geographic Information System analysts tracked the advance of the mycoplasmal conjunctivitis epidemic. The green dot represents the approximate location in Maryland where diseased House Finches were first observed in February 1994. By November that same year the disease had spread into a region (pale green) from eastern Ontario to southern Virginia. The dotted green line delineates the disease in November 1995, by which time it had reached Georgia in the south, Illinois in the Midwest, and Quebec in the north. By November 1996 the disease had moved westward to invade Missouri, Iowa, and Minnesota (dashed red line). The solid green line represents the extent of the epidemic in March 1997, when it had reached the Dakotas and also had been reported in Texas.

rapidly in the wake of the epidemic. But were House Finch populations crashing across their entire eastern range?

We were able to answer this question only because for more than a century people have been collecting and storing data on the abundance of North American birds during the Christmas Bird Count. By combining data from the Christmas Bird Count with our survey, Wes Hochachka and André Dhondt showed that two to three years after the disease became established, House Finch numbers declined dramatically in regions where populations had previously been stable at high densities. In other regions, the epidemic appeared to slow the growth of the House Finch population. All told, House Finch abundance was about 40 percent of the expected abundance in healthy populations. With this stunning reduction, some 180 million fewer House Finches exist today than would have if this new disease had not emerged.

Perhaps even more surprising than the dramatic decrease in House Finch numbers was the tenacity of the disease. When a disease kills a large part of a population, it tends to fade away rapidly, both because infected individuals die before they can pass on the disease and because pathogens spread less rapidly in less dense populations. A milder disease often persists longer because it can spread effectively for longer periods. House Finch eye disease is unusual because it has persisted for so long, despite having eliminated a large proportion of House Finches in eastern North America.

The Quest Continues

Now in its sixth year, the House Finch Disease Survey has yielded a series of exciting patterns and allowed us to measure seasonal and regional variation in the spread of the disease. We now know that the prevalence of the disease varies markedly through the year, peaking during the fall and winter and declining precipitously during the breeding season. We have documented that the disease has caused declines in the numbers of House Finches. And we know that the disease is not disappearing.

But these results have not quenched our scientific thirst to understand why this is happening. Why House Finches? Why has the disease persisted? Why haven't affected House Finch populations recovered? Answering all of these questions would require the talents of a larger group of scientists and increased funding.

In late 1999, when the National Institutes of Health requested proposals for research on the ecology of infectious diseases, we were in the right place at the right time. We knew a great deal about the disease and all of the questions we wanted to ask. On the strength of our proposal, NIH provided the funding that we needed, not only to describe the patterns of this epidemic in North American birds, but actually to understand the processes underlying seasonal and geographic variation in the spread of disease.

Our current team includes more than 15 collaborators from several institutions: four veterinarians, a mycoplasmologist (who is also a veterinarian), three bird popula-

tion ecologists, two mathematical modelers, two full-time field technicians, and several graduate students and undergraduate volunteers. We collectively offer expertise in a diverse array of fields, including mark-recapture analysis, bacterial genetics, disease ecology, avian health, behavioral ecology, and bird population studies. In our quest to learn how the disease can spread and persist, we are discovering new facts about the winter ecology, behavior, and movements of House Finches in different parts of their range. We are investigating feeder use, flock size, and interactions among flock members. Finally, we are gathering new information on routes of disease transmission and host responses to infection, including remission and recovery.

Although our work might appear to be a routine scientific study, we are relying on an unconventional resource—a powerful and expansive network of citizen scientists—to help answer our questions. Our success depends on having House Finch Disease Survey participants continue to monitor the presence and absence of the disease across the continent. We are also recruiting bird banders to help us follow the movements of House Finches, and possibly the disease, year round. We hope that with such unique collaborations between professional and citizen scientists we can document the emergence and spread of mycoplasmal conjunctivitis in House Finches and contribute to a larger understanding of the dynamics of infectious disease. ∎

This article originally appeared in Living Bird, Autumn 2001.

reductions occur in clutch size, nesting success, and the proportion of birds that might normally make more than one nesting attempt. All of these responses tend to reduce reproductive output and slow population growth.

Some factors that influence the rates of survival, reproduction, and dispersal operate independently of population density and therefore have no potential to regulate population size. These **density-independent factors** include severe weather, failure of food supplies, and other natural disasters, and they can cause catastrophic mortality and massive emigration, often reducing population size dramatically. Many bird populations suffer major declines during extreme winters. For example, in a population of banded Black-capped Chickadees that I studied in Wisconsin, mortality rates increased by 150 percent during a period in which temperatures stayed below 0° F (-18° C) for five days. But even in these situations, numbers usually recover quickly.

Sometimes sustained changes in population size do occur, both upward and downward. Usually these reflect major environmental changes, and as you will learn in Chapter 10, such changes generally have a human origin. For example, populations of Peregrine Falcons, Bald Eagles, and several other raptors declined precipitously after 1945 when their food chain became contaminated with DDT, an insecticide that caused females to lay thin-shelled eggs. Because most eggs broke before hatching, birth rates plummeted. With few individuals being added to the populations, numbers continued to decline for decades until DDT was banned. When reproductive rates returned to normal after the ban, raptor populations began to recover.

In contrast, populations of Red-winged Blackbirds, Common Grackles, and Brown-headed Cowbirds in eastern North America shared a prolonged period of population growth after extensive changes in agricultural practices occurred on their Gulf Coast wintering range. Expanded cultivation of cereal crops, especially rice, provided these birds with unusually abundant food, and their over-winter survival improved dramatically **(Fig. 9–53)**.

Extinction: The Death of the Last Individual in a Population

Although most bird populations maintain their numbers over time, some local populations do fail, first declining to low levels and finally becoming extinct when the last member dies. Usually populations become extinct locally, but some entire species, such as the Passenger Pigeon, also have become extinct. Population ecologists clearly understand the mechanisms that lead to extinction, in part because an alarming number of opportunities to study the process have been available in recent years (see Chapter 10).

Usually, extinction is caused by an extraordinary change in the environment. Although the environments in which bird populations live change naturally over time, the change is usually slow and modest. In contrast, changes caused by human activities can be swift and large scale. Some changes are completely new disturbances with which birds have never had to cope. An example is environmental

Figure 9–53. Increasing Cowbird Abundance from 1900 to 1980: For the first few decades of the 20th century, Brown-headed Cowbirds were rarely seen by birders on Christmas Bird Counts in the Southeast; since then, the percentage of cowbird reports has risen dramatically. Native to the Midwest and Great Plains, cowbirds are brood parasites that now thrive in the fragmented forests and agricultural lands that exist in much of eastern North America (see Ch. 8, The Brown-headed Cowbird: History and Conservation). Farmlands interspersed with small woodlots and residential areas provide cowbirds with much of the edge habitat that makes it easy for them to find host nests. In their wintering habitat, cowbirds feed on waste grain in southern rice fields. The abundance of this food supply has improved their winter survival and thus their overall numbers. Their burgeoning population has had a profound impact on forest songbirds (see also Ch. 10). Adapted from Brittingham and Temple (1983). Inset drawing by Orville Rice.

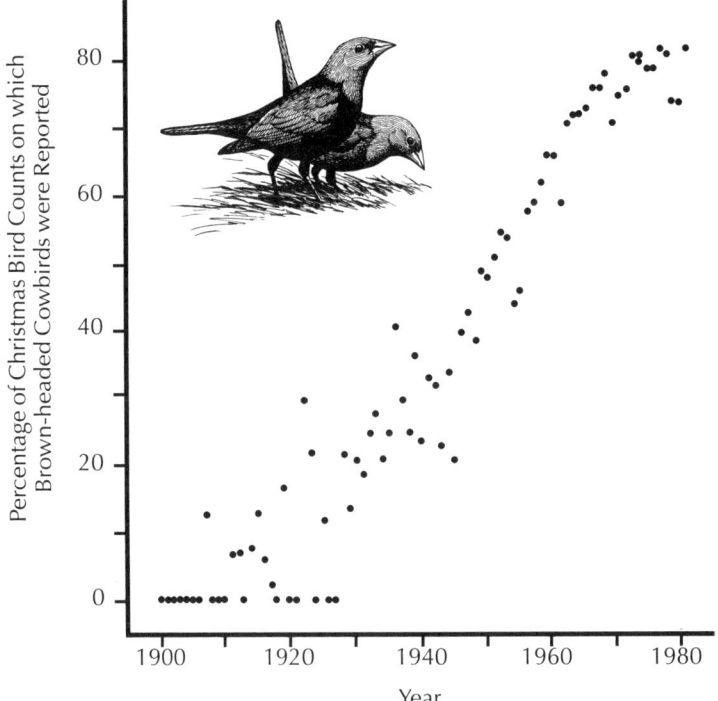

contamination by synthetic chemicals, such as DDT, which exposed birds to toxins having no natural counterparts.

Other disturbances such as fires or droughts may be familiar to birds, but can become extraordinary when they occur more or less often, with greater or lesser intensity, or over larger or smaller areas than they have in the past. Historically, for example, prairie-nesting birds saw frequent and extensive fires, which maintained their nesting habitat by preventing woody plants from invading grasslands. Today, because most fires are controlled, or are small and infrequent, many of the few remaining prairie patches have become overgrown and less than ideal for prairie birds.

As another example, the Northern Spotted Owl is threatened with extinction because of extraordinary changes to its habitat—old-growth forests in the Pacific Northwest. In such a forest under natural conditions, only a few trees fall at the same time, and because they are scattered, the fallen trees change the habitat relatively little. In contrast, clear-cut logging removes trees in large numbers over large areas, completely altering the forest character, and creating a habitat in which the owls no longer can live **(Fig. 9–54)**.

When such extraordinary change occurs and a population begins to decline, mortality rates may rise to levels that cannot be offset by a density-dependent increase in fecundity. Or, fecundity rates may fall to levels that cannot be offset by a density-dependent increase in survival. Either way, extinction of a population becomes a 100 percent certainty. Such an extinction is **deterministic** because the population has no possibility of coping with the changes. For example, the entire regional population of Peregrine Falcons east of the Rocky Mountains became extinct during the 20-year period from 1946 to 1965, when pesticide-induced contamination reduced fecundity to a level so low that density-dependent responses, such as increased survival, could not compensate. Eventually the last individuals in the population died with no young birds to replace them.

Severe declines in population size do not always result in a population's demise, however. Sometimes density-dependent responses do halt the decline, stabilizing the population at a very small size that may be officially recognized as endangered. At this point extinction becomes **stochastic**, meaning that whether the population succumbs is largely up to chance (with very few individuals remaining, the population's chance of going extinct is alarmingly high).

Why does chance play such a major role in determining whether a small population of birds will survive or become extinct? The main reason is that small populations are especially vulnerable to streaks of "bad luck." Consider a small population of birds dealing with a series of density-independent setbacks, such as a severe winter followed by a summer drought followed by another severe winter. Whereas a large population might decline and then rebound, a small population might be pushed over the brink. Such "bad luck" results from **environmental stochasticity**, the fact that random events, such as severe weather, may occur in almost any earthly environment.

Small populations also are threatened disproportionately by random fluctuations in demographics (key population characteristics), such as sex ratios. Consider how the sex ratio of a small population can be severely skewed toward males or females by chance alone. When a bird is conceived, its chance of being either male or female is 50 percent. It's just like flipping a coin: heads it's a male, tails it's a female. However, if you flip a coin only a few times, the resulting ratio of heads to tails will not be exactly 50:50. In fact, you might come up with all heads or all tails. But if you continue to flip the coin, over time the ratio of heads to tails begins to approach 50:50. In this way, a small population with relatively few births each year may end up with an imbalanced sex ratio. If the imbalance persists, the population may not be able to maintain its numbers, and may slide toward extinction.

Finally, small populations are more severely threatened than larger ones by random changes that may occur in a population's genetic characteristics. Because individuals with certain traits may die out just by chance, thus eliminating their traits from the population, the genetic traits existing in small populations tend to be less shaped by natural selection than by chance, so the individuals are less well adapted to their environment. In addition, as traits are lost, small populations become less variable. When environmental conditions change, it is the variability existing in a population that allows the population to adapt—as those individuals with the newly-advantageous traits survive better.

Figure 9–54. Clear-Cut Northern Spotted Owl Habitat: All environments change naturally over time, but natural changes typically occur slowly and have minor impacts on birds. For instance, in the old-growth forests of the Pacific Northwest, random treefalls have little effect on the Northern Spotted Owls that live there. In contrast, habitat changes caused by humans can be rapid and drastic, such as the clear-cut logging seen here in the Olympic National Forest in northwest Washington. Such clear-cutting has led to the owl's near-extinction. Photo by Calvin Larsen/Photo Researchers.

Thus, small populations are not very good at adapting to environmental changes.

Inbreeding depression also is more common in small populations, because mate choices are few and thus birds are more likely to pair with close relatives, such as siblings or parents. Because close relatives share a greater percentage of their genes, they are more likely to share deleterious mutations that, when combined in their offspring, may be expressed and result in reduced rates of growth, survival, and reproduction.

As population size drops, the probability that environmental, demographic, and genetic problems will occur increases. Demographic and genetic problems are particularly common in populations having fewer than 50 breeders, and environmental stochasticity can threaten populations even when they number several hundred birds.

The extinction of the Heath Hen, a subspecies of the Greater Prairie-Chicken formerly found in grasslands along the Atlantic Coast, illustrates stochastic threats to a small population (Halliday 1978). In the face of market hunting and habitat loss, Heath Hens declined during the 1800s, until by 1900 only one small population of about 100 birds survived—on Martha's Vineyard, an island off Cape Cod. In 1907 conservationists created a sanctuary and began managing habitat and controlling predators to help the relict population **(Fig. 9–55)**.

By 1916 the population had rebounded to about 2,000 individuals, but environmental stochasticity took a toll when wildfire swept through the bird's habitat during the nesting season, killing many birds and wiping out the year's reproduction. A second setback occurred the following winter, when snowshoe hare populations in Canada hit the bottom of their regular 10-year cycle, and large numbers of Northern Goshawks and Great Horned Owls dispersed southward looking for alternative prey. Heath Hens fit the bill nicely, and only about 100 birds survived the double setback.

By 1920 the population had climbed to about 200 birds, but then a string of environmental, demographic, and genetic problems began the bird's final decline. First, many birds were killed by a disease brought to Martha's Vineyard by poultry. Those deaths led to a badly imbalanced sex ratio, with males outnumbering females. At that point egg fertility dropped, probably because of inbreeding depression. Finally, in 1932, the last individual died and the population became extinct.

Although none of the events that led to the Heath Hen's demise were particularly unusual, the fact that they occurred in an unfortunate sequence was devastating. Had the population been larger and more widespread, it probably could have survived these natural setbacks. This example illustrates why bird conservationists are so concerned about the many endangered species that currently have populations small enough to be vulnerable to stochastic events. Despite our best efforts to protect them, the future of these small populations will be somewhat dependent on luck.

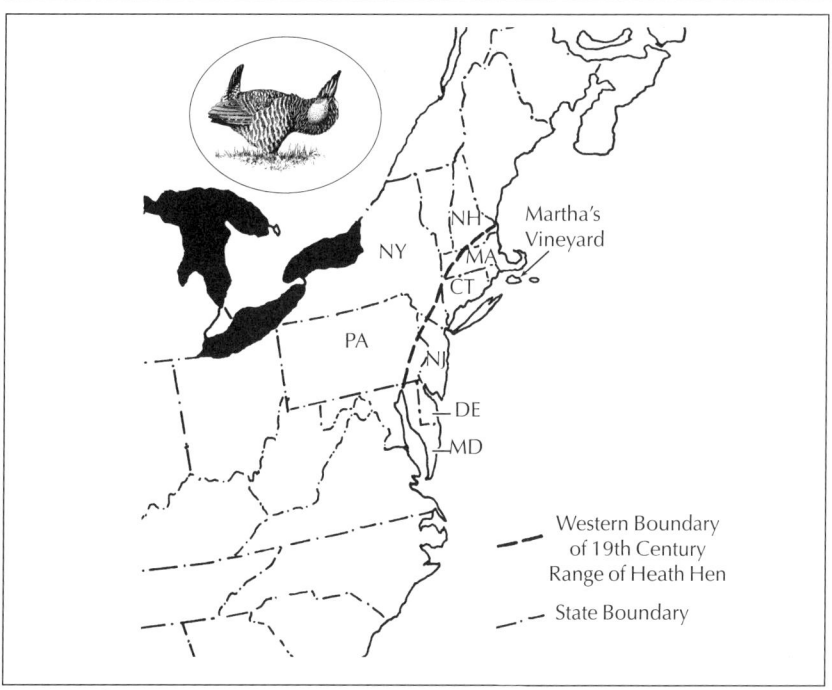

a. Original Range

b. Heath Hen Male and Female

Figure 9–55. The Extinction of the Heath Hen: The demise of this subspecies of the Greater Prairie-Chicken is a classic case of a population that became too small to survive the impacts of random environmental events. **a. Original Range:** Heath Hens once inhabited grasslands and blueberry barrens along the eastern seaboard. "Heathcockes" were often mentioned by early settlers, and in 1840 the governor of Massachusetts told Thomas Nuttall that the birds had once been so common around Boston that "servants stipulated with their employers not to have Heath-Hen brought to table oftener than a few times in the week." The bold dashed line on the map shows the western boundary of the species' 19th-century range. **b. Heath Hen Male and Female:** By the time that the noted artist Louis Agassiz Fuertes painted this pair of Heath Hens, all that remained of the species was a remnant population on the island of Martha's Vineyard, just off the Massachusetts coast. The species had declined so dramatically because market hunters had vied with sportsmen to shoot them, and the birds were vulnerable to domestic cats and diseases from domestic fowl. Once they had been extirpated from their original habitat, efforts to establish a refuge on Martha's Vineyard were too little, too late: a natural fire in 1916 destroyed most of the nests, and irruptions of goshawks and other avian predators, combined with disease and low fertility because of inbreeding, led to the species' extinction in 1932.

Structure of Bird Populations

Recall that any bird population, in addition to having a geographic distribution, size, and density, also has an internal structure based on the different types of individuals that compose it. Because individuals are either male or female, a population has a **sex ratio**. Because individuals are of different ages, a population has an **age structure**. And because individuals in a population have different genes, a population has **genetic structure**. The structure of a bird population is often related to its environment; therefore it is an important issue in avian ecology.

The sex ratio of hatchlings is usually close to 50:50. But soon after hatching, the ratio becomes skewed by differential survival of males and females. The sex ratio of young American Woodcocks during their first fall migration is 103 males to 100 females, but among adults, the ratio in fall is 63 males to 100 females, indicating that females survive better than males. In most species, however, adult males generally outnumber females, reflecting better survival of males. This differential survival usually is attributed to the risky role that females play during breeding, when they become vulnerable to predators and to the stresses of parental care.

This skewing of the adult sex ratio is often revealed by the large numbers of unmated males that are unable to attract mates because too few females are available. In contrast, nearly all breeding-age females are mated. This disparity can be demonstrated by removal experiments. Trap and remove the male of a breeding pair, and he is often replaced promptly. But remove a mated female, and her chances of being replaced are much lower.

In some species, roughly equal numbers of males and females may be present in the population, but the mating system skews the **operational sex ratio**, that is, the ratio of breeding females to breeding males. In species with polygamous mating systems, one individual can monopolize several mates. In polygynous species, the operational sex ratio is skewed toward females, whereas in polyandrous species, the ratio is tipped toward males. In the Sharp-tailed Grouse, a species with male-dominance polygyny, the operational sex ratio can be skewed as much as 14 females to 1 male **(Fig. 9–56)**.

The age structure of a population is the proportion of individuals in each age class, which is determined by the number of young birds produced each year and their subsequent survival. In most bird populations, the proportion of individuals decline in each successively older age class, resulting in a **stable age distribution** in which the proportions of different-aged individuals are the same each year at the start of the breeding season. At the end of the breeding season, the proportion of young birds in the population is temporarily elevated, but the stable age distribution is reestablished by the next breeding season (see Fig. 9–49).

Recall that differing genetic characteristics of individuals in a population can result in physical differences called polymorphisms, which are unrelated to sex, age, locality, or season. Familiar

Figure 9–56. Male Sharp-tailed Grouse Displaying at a Lek: In a population of Sharp-tailed Grouse, a dominance hierarchy determines which males will breed with the available females. As in the Eurasian Ruff (see Fig. 6–40), the males congregate at a lek for several hours each day. When a female arrives, the males inflate their neck sacs, hold their tails erect, droop their wings, and coo. Then they jump and run at each other. Because each female usually chooses the dominant male, whose territory is in the center of the arena, he breeds with most of the hens. Thus, although the number of males and females in the grouse population may be approximately equal, the operational sex ratio —the ratio of breeding females to breeding males—is skewed toward females, being as high as 14 females to one male. Photo courtesy of Dwain Prellwitz/CLO.

polymorphisms include **color phases**, which can be so different that ornithologists can mistake them for separate species. The Snow Goose and "Blue Goose" were thought to be different species for many years before studies showed that they regularly interbreed and are in fact color phases of one species, the Snow Goose. Similarly, the Great White Heron is now considered to be a color phase of the Great Blue Heron.

Color phases are especially common and distinctive among birds of prey and predatory seabirds. In North America, Red-tailed Hawks, Swainson's Hawks, Ferruginous Hawks, Rough-legged Hawks, Short-tailed Hawks, and Gyrfalcons have two or more distinctive color phases. So do Great Skuas, South Polar Skuas, Pomerine Jaegers, Parasitic Jaegers, and Long-tailed Jaegers. All of these predators feed on mammals and birds possessing excellent vision, and thus their color polymorphisms may be adaptations for avoiding detection and recognition by their prey **(Fig. 9–57)**.

Polymorphisms persist in populations because they confer advantages on individuals. What determines the ratios of different morphs? They are not random; instead, they frequently follow a predictable pattern in which one form predominates. This relationship has an underlying genetic basis. Differential survival and reproduction, resulting from environmental influences, can shift morph ratios from the pattern predicted by genetics alone. As a result, morph ratios often vary with environmental conditions. For example, some color phases are known to be related to environmental factors because the ratios of phases within a population vary in predictable ways when the environment changes. Eastern Screech-Owls and Ruffed Grouse have red and gray color phases. In both species, the gray phase predominates in the northern portions of the geographic range, and red predominates in the south. Studies have revealed that gray-phase individuals have better survival rates in cold, snowy weather because they have lower energy demands than red-phase individuals. In Ohio, the proportion of red-phase screech-owls in the population dropped from 23 percent to

Figure 9–57. Color Morphs in Rough-legged Hawks: Color morphs, otherwise known as color phases, are among the more familiar polymorphisms that occur in birds, and are particularly common in birds of prey. The Rough-legged Hawk exhibits two morphs: a light phase and an almost-black dark phase. The light phase seen from below (left) has a darkly blotched belly band across its light underparts, and a dark patch at the wrist of the mostly light underwing. The dark phase seen from below (right) has very dark underparts, with light flight feathers on the mostly dark underwing. The dark phase is more common in the East, and intermediate morphs also may be seen. Just as a bird watcher may be momentarily confused by a species' different plumage morphs, perhaps prey animals may likewise be uncertain of a bird's identity and the threat that it poses. Photos by B. K. Wheeler/VIREO.

Light Phase Rough-legged Hawk   Dark Phase Rough-legged Hawk

15 percent after an unusually cold and snowy winter elevated their mortality rate. However, the color itself does not affect survival, rather, both plumage color and physiological performance are controlled by the same underlying genes.

Bird Communities

■ Nearly every spot on earth except the polar icecaps is home to a **bird community**—two or more bird species that live and interact in one area. Bird communities vary greatly from place to place. Even if you're not an avid birder, you surely have noticed that different environments contain dramatically different kinds and numbers of birds. Major differences are apparent from simple comparisons of birds living in cities with those residing in the nearby countryside, or of birds found in North American forests with those living in South American jungles. Avian ecologists have spent decades attempting to identify patterns in bird communities, and trying to understand the ecological processes that produced them. Some of these patterns and processes are discussed below.

Characteristics of Bird Communities

The composition and structure of a bird community can be largely defined by three characteristics: (1) species richness, (2) the relative abundance of the various species, and (3) the types of ecological niches that the various species fill. **Species richness (Fig. 9–58)** is simply the number of species living in a community. The more species, the greater the richness. **Relative abundance** measures the number of individuals of each species compared with others in the community (or, how

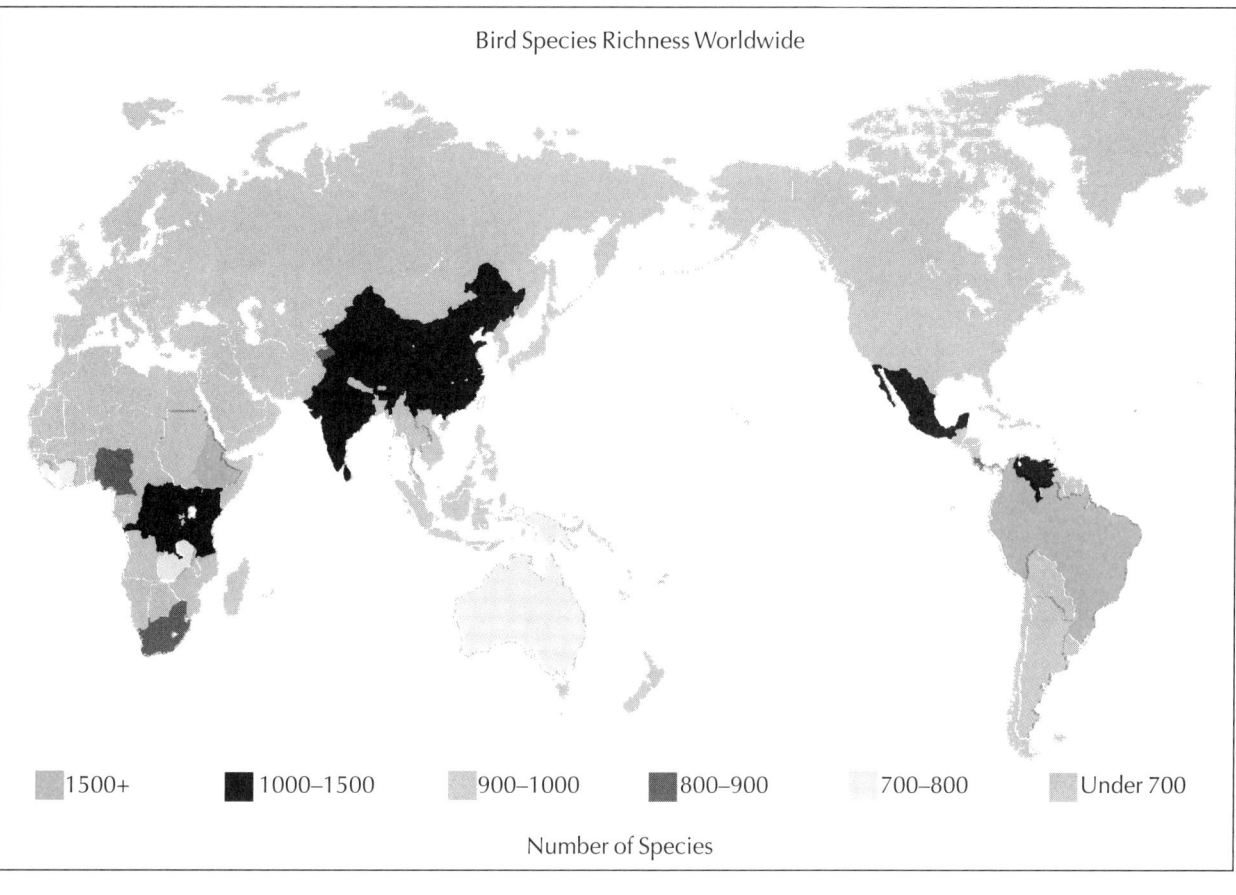

Bird Species Richness Worldwide

1500+   1000–1500   900–1000   800–900   700–800   Under 700

Number of Species

evenly the species are distributed). Finally, the **ecological niches** are the various roles that different species play in the community.

These three characteristics are illustrated by the bird community in a 10-acre (4-ha) oak forest near my home in Wisconsin. When I surveyed this community during one nesting season, I found that it contained 36 individuals of 12 species: two Eastern Screech-Owls, two Downy Woodpeckers, two Northern Flickers, two Great Crested Flycatchers, two Blue Jays, four White-breasted Nuthatches, six House Wrens, four American Robins, two Gray Catbirds, two Indigo Buntings, six Song Sparrows, and two Baltimore Orioles. Because the community contained 12 species, its richness was 12.

Considering relative abundance, the two most abundant species each accounted for 16 percent of the individuals, the two next-most abundant species each accounted for 11 percent, and the eight least-abundant species each accounted for 5 percent. Therefore, the relative abundance of species within this community was uneven.

Finally, the 12 species filled different ecological niches that can be characterized in various ways. Considering activity patterns, one species was nocturnal and 11 were diurnal. Considering diet, one species was a carnivore, six were insectivores, and five were omnivores. Considering foraging habits, two foraged on trunks and limbs of trees, four fed on the ground, five searched for food in foliage, and one took insects in midair. Considering nest site preference, six species nested in cavities, five nested in above-ground cup nests, and one nested on the ground **(Fig. 9–59).**

Figure 9–58. Bird Species Richness Worldwide: The number of bird species within a community is known as species richness, a measure that varies widely between geographic locations. Note that the highest bird species richness is in equatorial South America and Ethiopia, followed by equatorial Africa, Mexico, and Southeast Asia. Contrast this with the low richness in much of the temperate zone of the northern hemisphere. Also notice that most islands, even large ones such as Madagascar and New Zealand, and those of Indonesia, the Philippines, and the West Indies, are relatively impoverished in terms of bird-life. Recognizing patterns such as these is the first step toward understanding the ecological processes that produce them. Climate, latitude, environmental complexity, and island size and isolation are just a few of the factors that contribute to a region's richness, and are explored in the text. Adapted from Clements (2000).

Figure 9–59. Niches of Wisconsin Oak Forest Birds: The bird community of a patch of Wisconsin oak forest (above) contains 12 species. The table (facing page) shows each species' ecological niche, defined in terms of several factors: the time of day that the bird is active, the components of its diet, and its foraging and nesting location. For instance, the Eastern Screech-Owl is a nocturnal, carnivorous bird that nests in cavities and captures its prey on the ground. In contrast, the Gray Catbird is a diurnal omnivore that forages in foliage and builds its cup nest in vegetation.

SPECIES	Eastern Screech-Owl	Downy Woodpecker	Northern Flicker	Great Crested Flycatcher	Blue Jay	White-breasted Nuthatch	House Wren	American Robin	Gray Catbird	Indigo Bunting	Song Sparrow	Baltimore Oriole
ACTIVITY												
Nocturnal	✓											
Diurnal		✓	✓	✓	✓	✓	✓	✓	✓	✓	✓	✓
DIET												
Carnivore	✓											
Insectivore		✓	✓	✓		✓	✓					✓
Omnivore					✓			✓	✓	✓	✓	
FORAGING LOCATION												
Trunks and Limbs		✓	✓			✓						
Ground	✓		✓					✓			✓	
Foliage					✓		✓		✓	✓		✓
Air				✓								
NESTING LOCATION												
Cavity	✓	✓	✓	✓		✓	✓					
Cup in Vegetation					✓			✓	✓	✓		✓
Cup on Ground											✓	

Figure 9–59. (Continued). **Niches of Wisconsin Oak Forest Birds**

Differences in structure and composition among bird communities typically result from variations in habitat type and quality. Evaluating the composition and structure of bird communities can reveal clues to overall ecosystem health. For example, when a niche is left open, it may indicate the presence of a limiting factor. The absence of the nectar-feeding Ruby-throated Hummingbird, for example, was the result of a paucity of flowers.

Patterns of Species Richness

Avian ecologists have identified several patterns of species richness among the bird communities of the world. Four important generalizations are: (1) More species occur at lower latitudes (nearer the equator) than at higher latitudes (nearer the poles); (2) at similar latitudes, more species occur in structurally complex environments, such as forests, than in simple environments, such as grasslands; (3) more species occur in highly productive environments than in less productive environments; and (4) more species occur on large islands than on small islands.

Effects of Latitude

The greater species richness of bird communities located near the equator compared with those located near the poles is dramatic **(Fig. 9–60)**. Across much of North America, a typical 400-mile (644-km)

Figure 9–60. Latitudinal Gradient of Species Richness: Species richness shows many global patterns. For example, richness is higher (more species occur) at lower latitudes (nearer the equator) than at higher latitudes (nearer the poles), as shown by this graph of species richness across the northern hemisphere. A typical square area 400 x 400 miles (644 x 644 km) near the equator in the Amazon Basin contains more than 1,000 species. A similarly sized area in Central America at approximately 10 to 20 degrees north latitude contains 500 to 600 species, whereas in much of North America, between about 30 and 60 degrees north, only 120 to 150 species generally occur. Possible explanations for this pattern are discussed in the text. Note that species richness is shown on a logarithmic scale. After Dobzhansky (1950).

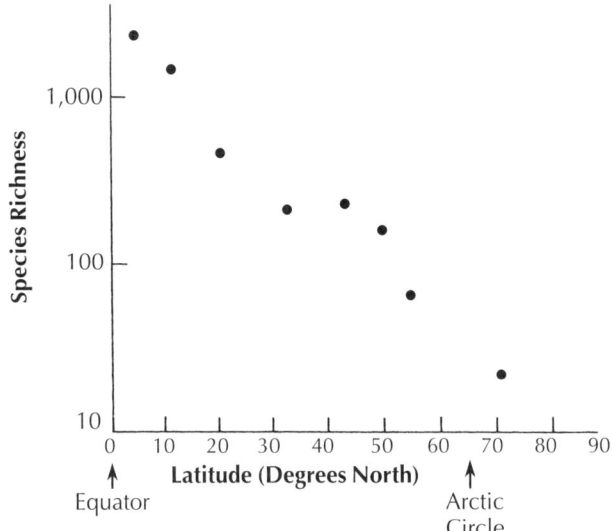

square area contains 120 to 150 species of birds. In Central America, the same size area contains 500 to 600 species, and in the western Amazon Basin, it contains more than 1,000 species. In fact, a single 247-acre (100-ha) plot in the Peruvian Amazon contained over 535 species.

Avian ecologists have proposed many explanations for the huge number of species that live in the tropics, including theories that are based on the mild climate, dependable resources, high productivity, high rates of speciation, and low rates of extinction.

The relatively mild climate of tropical areas compared to higher latitudes has long been considered one factor in the large number of tropical species, because it puts fewer restrictions on the types of species that can evolve. Tropical species do not need to cope with the harshness of winter by adapting to cold or freezing temperatures, or by migrating long distances.

The mild year-round climate in the tropics leads to many food resources being highly dependable year round, which in turn allows animals to specialize on those resources. Unlike temperate areas, tropical areas have, for example, many different kinds of flowers (with nectar), fruits, and insects available all year. This allows many birds—such as hummingbirds, manakins, and flycatchers—to specialize on specific foods. In fact, some tropical birds rely on just one type of food. For example, Snail Kites have specialized, hooked bills that allow them to feed solely on apple snails, and some manakins (see Fig. 6–42) feed almost exclusively on plants in the genus Miconia, which produce nutritious fruits year round.

To illustrate these differences, consider the distribution of hummingbirds: the tiny country of Panama is home to more than 50 hummingbird species and a dozen other nectar feeders, whereas the entire eastern half of North America claims only one obligate nectar feeder, the Ruby-throated Hummingbird. Similarly, the tropics abound with specialized frugivores, such as the manakins and trogons (see Fig. 1–62), whereas few birds that specialize in fruit occur in the temperate zone. Further opportunities for specialization occur in tropical areas because they contain many kinds of resources that rarely (if ever) are present in the temperate zone, such as huge vines, epiphytes (plants that live on other plants), and army ant swarms **(Sidebar 3: Ant Followers)** (also see Figs. 9–4 and 1–62). One comparison between a 40-species bird community in a South Carolina forest and a 206-species community in an Amazonian forest revealed that 56 of the tropical species belonged to ecological groups not represented in the temperate community.

The warm, and in many cases, wet, nature of tropical areas allows a high rate of plant growth (called **primary productivity**), which in turn feeds a huge volume of insects, birds, and other animals. For example, one study found that the total **biomass** (the mass of living organisms) of the breeding birds in a tropical forest in Amazonian Peru was about five times that of temperate forests (Terborgh et al. 1990). The abundance of plant growth, combined with its complex, diverse nature, also promotes specialization.

For several reasons, the rate of speciation (evolution of new species) may be higher in tropical than temperate areas. One reason is the greater reluctance of many tropical forest species to cross rivers or open areas, apparently because they are afraid to leave forest cover. Thus, populations are more likely to become isolated from one another, and evolve into separate species (see Fig. 1–44). Also, the Neotropics have several large geographic features—the Andes Mountains and the Amazon River—that have been particularly effective at isolating populations and promoting speciation (see Fig. 1–45). Another factor, somewhat controversial, is the long period of time (tens of millions of years) that tropical areas have been free of monumental climatic changes. Just 10,000 years ago, much of North America was covered by glaciers, and many temperate zone species became extinct at that time (see Ch. 1, Appendix C: Geological Time Scale). Tropical areas have experienced no such recent, massive extinctions. On the other hand, the series of ice ages did affect tropical areas by alternately isolating and bringing together populations of certain species (by creating habitat islands), which is thought to be a major factor promoting speciation in these areas (see Figs. 1–58 and 1–59).

The mild tropical climate has probably also affected the extinction rate. Recall that species that have small numbers and live in isolated habitats face a high risk of extinction owing to several factors, including periods of unusually severe weather. Because these environmental extremes occur less frequently in the tropics, small populations are more likely to persist there than in temperate areas. Thus, tropical areas have come to contain numerous species with small numbers and restricted geographic ranges. For example, a hummingbird called the Purple-backed Sunbeam is known only from a single valley in western Peru **(Fig. 9–61)**.

All of these theories, then, can help to explain the huge variety of species in the tropics. No doubt other factors are at work as well.

(Continued on p. 9·92)

Sidebar 3: ANT FOLLOWERS
Peter H. Wrege

Slowly the outlines of the forest were becoming visible—huge tree trunks towering up to the canopy, draped in lianas and coated with epiphytes; understory palms and shrubs appearing as vague shapes near the forest floor around me. A friend and I were sitting on a loop of liana (a woody vine) listening to a distant troupe of howler monkeys finishing their roaring statement of territory ownership, while all around us the dawn chorus of forest birds was in full swing. With a cacophony of squawks and screams, several hundred Blue-fronted and Mealy parrots suddenly left their nighttime roost near the town of Gamboa, Panamá and flew out over Soberania National Park, headed for breakfast in fruit-laden trees. The forest seemed almost quiet after the passage of the parrots. Suddenly my colleague perked up and drew my attention to an increasingly noticeable collection of chirrs, whistles, and series of clear, descending notes coming from across the Rio Frijoles. "Army ants!" she said. "Let's go."

We waded the stream and clambered up the opposite bank. When we paused, I could hear a strange sound like soft rain underlying the calls of several different birds that were flitting about close to the forest floor. But it wasn't rain, it was the sound of hundreds of insects, large and small, hopping, flying, and leaping into the air and falling back into the leaf litter. Then I saw them: a seething rope of scurrying ants—Eciton burchelli. Soon I could make out the vague outlines of the swarm front, spread out like a 30-foot-wide umbrella, advancing slowly and inexorably across, and through, and over every structure in its path—insects and small vertebrates fleeing before it. Birds were everywhere. Right in the midst of

the swarm was an Ocellated Antbird **(Fig. A)**, clinging to vertical stems and darting down to snatch a fleeing cockroach or cricket, the bright blue skin around its eyes contrasting with its black throat and black-spotted, rusty chest. There were several Bicolored Antbirds just ahead of the swarm, also with blue eye patches. As one darted for an insect and then hopped up onto a perch near the center of the swarm, it gave a loud, buzzing alarm note as it was displaced by the Ocellated Antbird. All around and higher up were other species: a pair of Spotted Antbirds,

a Gray-headed Tanager, and several small, mostly black birds that I couldn't identify. Just at the edge of the swarm were two Black-crowned Antpittas, hopping on the ground and along fallen branches, gobbling up all manner of insects that were jumping from the army ants' fire into the antpittas' pan! Two woodcreeper species completed the community of ant followers at this swarm—a pair of Plain-brown Woodcreepers and one of the larger woodcreepers, a Northern Barred-Woodcreeper. Both of these foraged from sites that none of the others could use—the trunks of

Figure A. Ocellated Antbird: An Ocellated Antbird—boldly patterned in chestnut and black, with a patch of bare blue skin around its eyes—captures an insect fleeing from an army ant swarm. In Panamá this ant-following bird aggressively dominates the zone directly above the leading edge of ant swarms, a prime location where the highest concentrations of prey are flushed from hiding in the leaf litter on the forest floor.

large trees—where they could move up and down like woodpeckers in pursuit of large prey such as mantises, and even scorpions and spiders.

Most of the birds we observed are considered "professional" ant followers: they obtain more than 50 percent of their food by catching insects scared up by moving army ants. (None of these birds typically eat the ants themselves, which are loaded with formic acid.) Many of the "professionals" are antbirds (family Thamnophilidae), but they come from a wide variety of other groups, including antpittas (family Formicariidae), woodcreepers (family Dendrocolaptidae), and cuckoos (family Cuculidae). The complex communities of birds that regularly follow raiding swarms of army ants are found only in the Neotropics, and here only two species of army ants forage in ways that attract such a following. The birds depend on ants that regularly move en masse through the forest for food, using the tactic of "swarm raiding"—forming an umbrella-like front line that flushes insects and small vertebrates that would otherwise remain hidden in the litter. In Africa, certain species of birds are often observed following two species of driver ants, but these ants often go for days at a time without raiding, or raid at night, and the birds do not depend on them for finding food.

In the Neotropics, colonies of Eciton burchelli often contain 400,000 or more individuals, and raid on at least three out of four days, regardless of weather—usually for the entire day. There are enough colonies of ants in the forest that an active swarm is always within reach of the professional followers. Many of these birds appear to be completely dependent on healthy populations of army ants for their own survival, having lost the ability to forage away from ant swarms efficiently enough to avoid starvation. When forested landscapes are fragmented by logging or for agriculture and ranching,

army ants are among the first species to disappear from the smaller fragments, and the professional antbirds disappear along with them.

As we continued to follow the raiding ants, I was awed by the fact that I was in the midst of a complex and ever-changing web of interactions among numerous species—one of the most fascinating I'd ever seen. Not only were the professional ant-following birds dividing up the foraging space around the front of the swarm, but other animals had been attracted to the raid as well. I could see a small parasitic fly (Stylogaster sp.) darting over the front of the swarm. It was diving at fleeing cockroaches, attempting to single out one on which to lay an egg. From the egg a larva would hatch, then burrow into the cockroach and devour it from inside out! There were other flies in attendance, but more noticeable and elegant were the butterflies—a cluster of swallowtails and a few nymphalids—fluttering just behind the front of the raid and catching flashes of sunlight that made their wings look like stained-glass windows. The butterflies were settling on the fresh droppings of the ant-following birds, where they gathered minerals and amino acids. All of these interactions and interdependencies only begin to describe the dynamic web that unfolds each morning as the ant bivouac (the temporary nighttime resting aggregation of army ants) produces a raiding swarm that heads out across the forest floor.

Various other species of birds (including antthrushes, antshrikes, antwrens, motmots, tanagers, warblers, trumpeters, and small forest raptors), and even a few primates, frequently follow army ant swarms, but only 30 to 35 species—the professionals—get the majority of their food in this way. The location of army ant swarms shifts from day to day and each colony covers many kilometers per year. Because the colonies are unpredictable in their location and are too rich a resource for a single

pair of birds to defend, the professional ant followers are only loosely territorial. They tolerate conspecifics at a swarm within their territory, but are dominant to these individuals and thus can monopolize the best foraging locations at the swarm front. As the swarm moves into the territory of another pair, the dominance hierarchy shifts to favor the new territory owner.

A second hierarchy, this one interspecific, also provides structure within the community of birds feeding with the ants. Although the different bird species broadly overlap in the types and sizes of prey captured, they reduce competition by using different types of perches (thin vertical stems, horizontal twigs, or larger tree trunks) and perches at different heights (**Fig. B**). In addition, within a group of species that prefer the same types of perches, the larger species can dominate and supplant the smaller ones. For example, in Panamá, the Ocellated Antbird controls the highly profitable central region of the swarm front, because it is large and adapted to perch cross-wise on vertical stems right over the most active portions of the swarm. Bicolored Antbirds usually occupy the next region out from the swarm center, with ground-perching birds such as antpittas and the large Rufous-vented Ground-Cuckoo lurking around the edges. The smaller antbirds are also relegated to the periphery, unless one of the more dominant species is absent, in which case these subordinates move up in the hierarchy of preferred foraging locations. Sometimes, when many of the larger, dominant bird species are present, these smaller followers simply give up and abandon the ant swarm altogether. Less specialized species (for example, those without the ability to perch on vertical stems or cling to the bark of trees) perch on overhanging branches and twigs, but the foraging success from these locations is much lower and might explain why none of these are professional ant followers.

Figure B. Niche Division by a Guild of Professional Ant-following Birds: Emerging in a seething column from their overnight bivouac in a rotten log, raiding army ants swarm out to form a broad, umbrella-shaped front as they pursue insects and other arthropods across the forest floor. To reduce interspecific competition, the birds that follow the swarm arrange themselves around the front according to several dominance zones, indicated by dashed lines. These zones are three dimensional: perch height is a factor as well as distance from the swarm, with the smaller, less aggressive, subordinate bird species perching higher than the larger, more aggressive, dominant species. In Panamá, the central zone (A)—directly over the ant swarm front—is dominated by the Ocellated Antbird, shown clinging to a sapling close to the ground, ready to snatch up fleeing insects. Located in the next zone out (B) is the smaller Bicolored Antbird, clinging higher on a sapling, and the Northern Barred-Woodcreeper, perched on a large tree trunk where it is poised to seize any arthropod prey fleeing up the trunk. Even farther away is zone (C), where prey is more scarce. This suboptimal region is the domain of the tiny Spotted Antbird, lowest in the hierarchy of clinging birds. Also found here are the ground-dwelling Black-crowned Antpitta and the large Rufous-vented Ground-Cuckoo, which run in pursuit of small creatures, and the Plain-brown Woodcreeper, shown searching for prey on a tree trunk. Finally, high on an overhanging branch is the Gray-headed Tanager. Less specialized than the other ant followers, it is unable to cling crosswise on saplings. Instead it perches in a more typical passerine fashion, restricting itself to high perches located peripherally around the ant swarm.

Communities of ant-following birds are organized in this manner wherever they occur; only the names and diversity of the interacting species change. As one moves south from southern Mexico and Central America into Colombia and the Amazon Basin, the number of potential participants at an army ant raid increases from 5 in Mexico to about 10 in Panamá and Colombia, and up to 12 species in Amazonia.

When I saw my first army ant swarm and watched the associated birds and insects, I wondered about the relationship between the birds and the ants. Was it **mutualistic** (beneficial to both participants), **commensal** (beneficial only to the birds), or even **parasitic** (beneficial to the birds, but costly to the ants)? Historically, the relationship between ants and ant followers has been considered commensal because it is similar to well-studied cases of commensalism: Cattle Egrets following large herbivorous mammals for the insects they flush; seabirds (for example, boobies, pelicans, and petrels) feeding on schooling fish that have been driven to the surface by pilot whales and orcas; and Double-toothed Kites and Black-fronted Nunbirds following active troupes of monkeys. The ants and birds, however, are both eating insects, so the ant followers actually might be stealing from the ants—removing prey that would otherwise be captured by the swarm. Detailed studies of army ants in Panamá do not support this interpretation, however. They have shown that more than 50 percent of the prey taken by army ants are the eggs, larvae, and pupae of social wasps and other ant species. In contrast, many of the prey taken

by ant-following birds are probably larger than the ants prefer to handle. Unlike the driver ants of Africa, which possess very strong cutting mandibles that can cut apart large prey, the army ants of the Neotropics have weaker mandibles that cannot cut into the cuticle or skin of large insects and small vertebrates.

Edwin Willis and Yoshika Oniki (1978), who have studied the ant-follower bird community in depth, suggest that the ant-bird relationship might even be mutualistic. The many birds and parasitic flies hovering, jumping, and flying near the front of the swarm might actually scare many insects back down into the leaf litter, where they again become available to the ants. These authors have gone so far as to suggest that the benefits from such followers might have led to the evolution of swarm-raiding tactics by the ants in the first place!

━━⌒━━

It was late afternoon in the Panamanian forest when I returned to watch the army ants and their followers. The swarm front had progressed nearly 100 yards through the forest from where we had first encountered it at dawn. The little streams of ants heading back from the raid front were still everywhere, coalescing into larger and larger ropes of ants until they joined the major trunk; and every returning ant seemed to be carrying a tiny egg, cockroach nymph, or larger piece of insect. But now fewer ants were moving in the opposite direction to continue the hunt. The group of ant followers was smaller now: only the Ocellated Antbird, a pair of Spotted Antbirds, and a Plain-brown Woodcreeper still foraged in the area, and they may not have been strongly tied

to the weakening raid. Suddenly the Spotted Antbirds flew up into a small, understory acacia and began singing a duet. As the light faded, more and more birds broke into song, and some frogs along the nearby stream began their nighttime calling. In the dim light I followed the last of the ants back to their bivouac—a huge tangled mass of ants, each with its six legs locked with those of others to form a rough ball in the root mass of a fallen tree. Tomorrow they would stream out again in a different direction, and again the birds would come. But now, the howler monkeys were calling once again from the treetops, and it was time to leave the forest to the creatures of the night. ■

Suggested Readings

Willis, E. O. and Y. Oniki. 1978. Birds and army ants. Annual Review of Ecology and Systematics 9:243–263.

A very readable review by the two authors who have studied ant-following birds most intensively.

Gotwald, W. H., Jr. 1995. Army Ants: The Biology of Social Predation. Ithaca, NY: Cornell University Press. 302 pp.

A complete treatise on all aspects of army ant biology, including a chapter on ant followers that partly updates the 1978 paper by Willis and Oniki.

Kricher, J. 1997. A Neotropical Companion. Princeton, NJ: Princeton University Press. 451 pp.

Includes readable treatments of army ant biology and the community of ant-following birds.

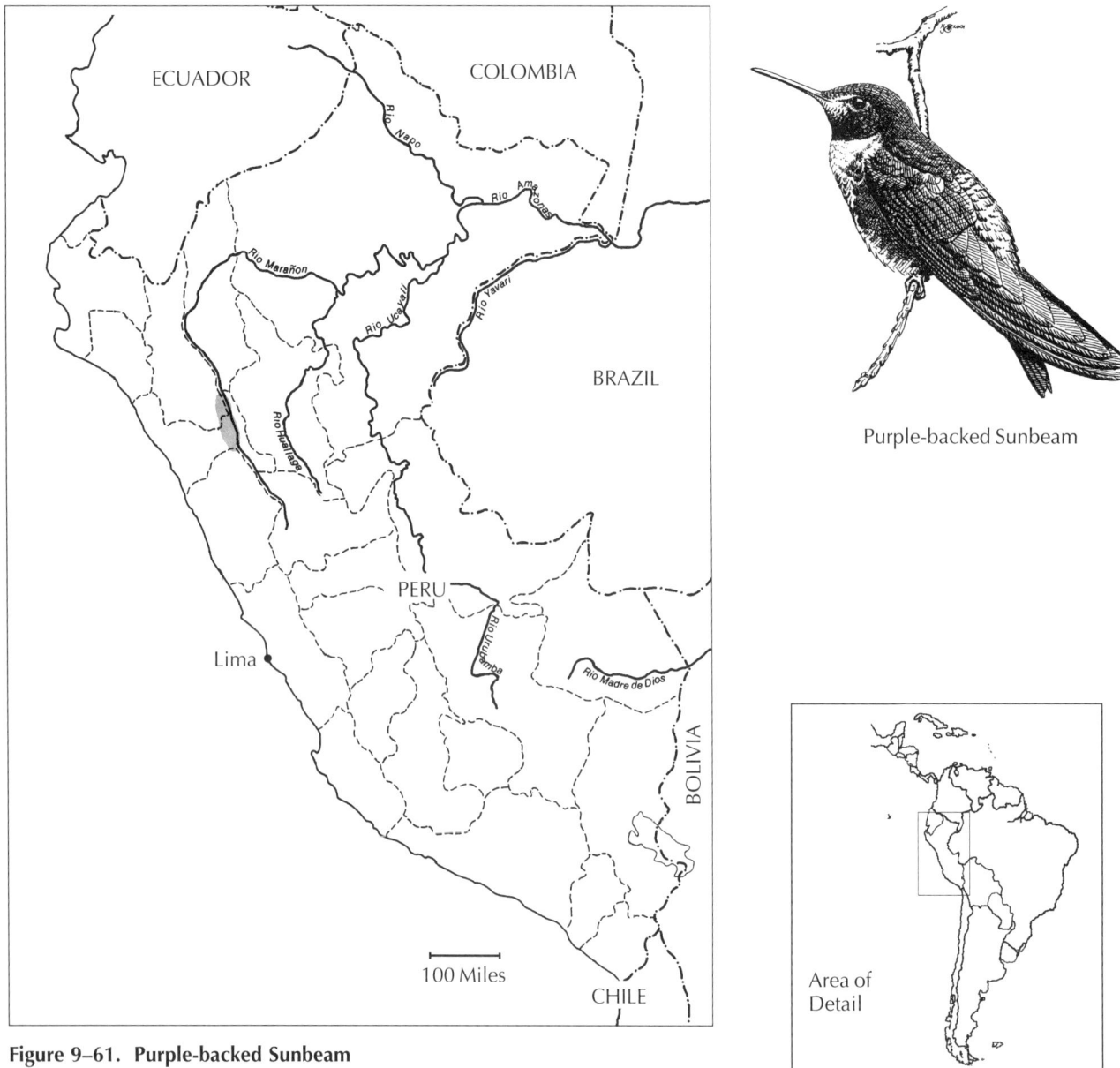

Purple-backed Sunbeam

Area of
Detail

Figure 9–61. Purple-backed Sunbeam Range: The Purple-backed Sunbeam, a tropical hummingbird named for its glittering purple rump and lower back, has a tiny geographic range. As shown by the colored area on the large map, it is found only in a small part of the valley of the Rio Marañon, a river that flows through the Andes mountains of northwestern Peru. Like other species that live in very small populations, the Purple-backed Sunbeam may be able to persist because the climate is relatively stable and predictable.

Effects of Habitat Complexity and Productivity

Although we have been comparing species richness at different latitudes (for example, tropical versus temperate), richness also can vary significantly among bird communities at a single latitude. Generally such variation is explained by both the productivity and complexity of the local environment.

The **productivity** of an area is measured as the number of grams of living material produced per square meter per year ($g/m^2/yr$). Higher productivity allows more species to coexist because each one can specialize on a narrow range of resources. For example, a typical coniferous forest has a productivity of about 800 $g/m^2/yr$ and a bird community comprising 17 species/100 acres (40 ha). Upland deciduous forests (1,000 $g/m^2/yr$) hold about 21 species/100 acres (40 ha), and floodplain deciduous forests (2,000 $g/m^2/yr$) have about 24 species of birds/100 acres (40 ha).

In addition, more species can coexist in an environment with high structural complexity—such as a forest containing several layers of vegetation—than can share a structurally simple environment, such as a grassland. A typical deciduous forest can have several layers, including: ground vegetation, such as ferns; understory vegetation, such as shrubs; mid-story vegetation, such as young trees; and overstory vegetation, such as mature trees, which make up the canopy.

When there is a great diversity of foliage heights, species can partition resources by living in the different layers from ground cover to canopy. For example, in a deciduous forest in temperate North America, the ground and low shrubs are home to Dark-eyed Juncos, Hermit Thrushes, Swainson's Thrushes, Winter Wrens, Ovenbirds, Wood Thrushes, and Veeries. Living in saplings and low canopy are Hairy Woodpeckers, Yellow-bellied Sapsuckers, Downy Woodpeckers, White-breasted Nuthatches, Blue-headed Vireos, Black-capped Chickadees, and Blackburnian Warblers. In the upper layers and high canopy are Black-throated Green Warblers, Red-eyed Vireos, Rose-breasted Grosbeaks, Scarlet Tanagers, Least Flycatchers, and American Redstarts. This subdivision of habitat allows species to avoid competition. In contrast, fewer opportunities for species segregation are available in habitats with a low diversity of foliage heights, such as prairies.

Because habitats in the same general area can differ markedly in both their productivity and structural complexity, bird communities in a single area can vary widely in their number of species. Whereas temperate marshes have a productivity of 2,000 g/m^2/yr—similar to the floodplain deciduous forests that contain 24 bird species/100 acres (40 ha)—marshes are structurally simple, and thus contain an average of just 6 species/100 acres (40 ha). In contrast, whereas deserts are unproductive (70 g/m^2/yr), they hold an average of 14 species/100 acres (40 ha) because they are structurally more complex than marshes.

In areas where the transition between two habitat types is abrupt, for example where a large field meets a mature forest, species richness can be unusually high. This phenomenon, known as the **edge effect**, is caused by the juxtaposition of different groups of birds having different habitat preferences. At the forest edge you can find birds associated specifically with forests, birds associated specifically with fields, birds that use both forests and fields, and birds that prefer the actual edge **(Fig. 9–62)**.

A similar phenomenon occurs at a much larger scale where two ecosystems meet in zones of transition known as **ecotones**, which often have an elevated species richness compared with either of the two ecosystems taken alone. For example, between the eastern deciduous forests and the prairie grasslands is an ecotone known as oak savanna, an area with grassy ground cover and a few scattered trees, usually oaks. Here, many deciduous forest birds reach the western limits of their range, and many prairie birds reach the eastern limits of theirs. The species richness of this ecotone is

Figure 9–62. Local Edge Effects: Species richness may be especially high in the zone where two habitat types meet, a phenomenon known as the **edge effect**. Consider a hayfield adjacent to a patch of eastern deciduous forest. Bird species found in the hayfield include the Savannah Sparrow, Bobolink, and Eastern Meadowlark, whereas the Pileated Woodpecker, Wood Thrush, and Ovenbird are denizens of the forest. Edge dwellers include the Baltimore Oriole, Indigo Bunting, Song Sparrow, and Blue-winged Warbler. However, hayfield birds may visit the edge of the forest, and forest birds may sometimes visit the edge of the hayfield, so avian richness in the edge habitat is higher than in either of the two homogeneous habitats alone.

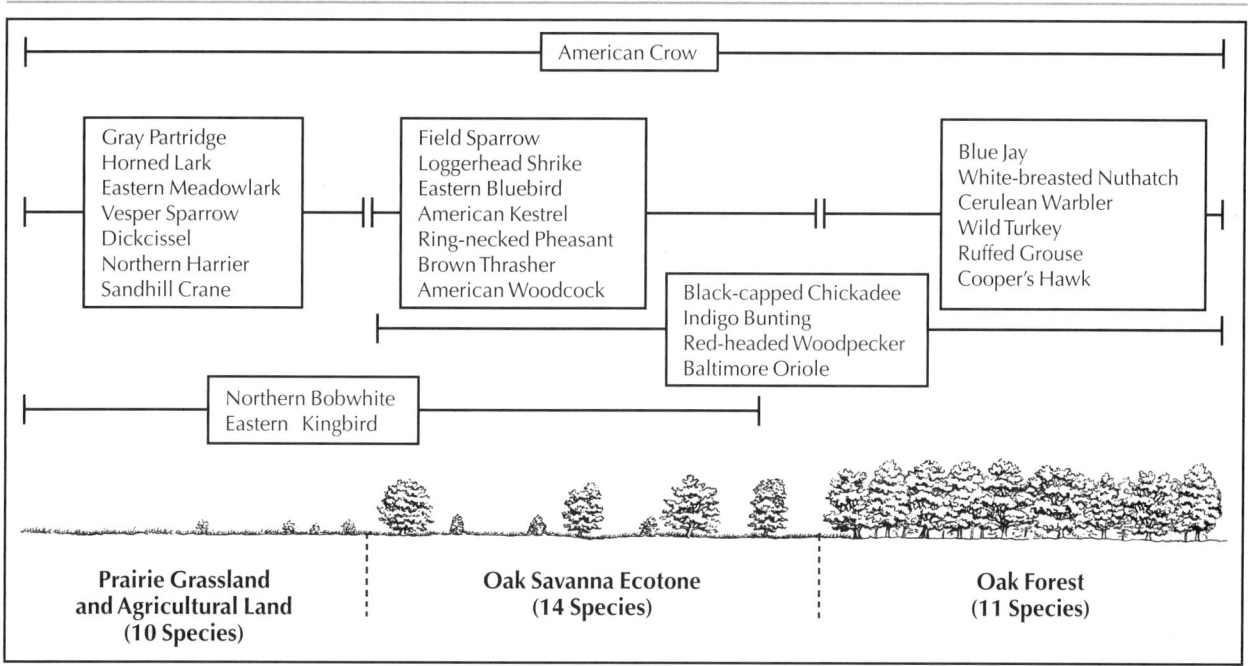

| Prairie Grassland and Agricultural Land (10 Species) | Oak Savanna Ecotone (14 Species) | Oak Forest (11 Species) |

particularly high because, in addition to the species found primarily in the oak savanna, there are prairie species that also live in the oak savanna, and forest species that do the same **(Fig. 9–63)**.

Effects of Habitat Size

In nearly all bird communities in nearly all environments, species richness increases progressively with habitat size. This important pattern, known as the **species-area relationship**, is clearly revealed on islands and island-like environments, such as isolated habitat patches.

Avian ecologists have determined the number of bird species on most of the world's islands, and have found that in general, the number of species doubles with every tenfold increase in island area. For example, the bird community on an island of 39 square miles (100 square km) consists of about 20 species; but on an island of 390 square miles (1,000 square km), about 40 bird species are present (MacArthur and Wilson 1967) **(Fig. 9–64)**. Several theories have been proposed to account for this pattern. The two most popular are the equilibrium theory and the habitat richness theory.

The **equilibrium theory** is based on the observation that the number of species on an island at any one time represents a balance between the number of new species colonizing the island and the number of species becoming extinct. When rates of colonization and extinction are the same, the number of species remains stable. The equilibrium theory proposes that smaller islands contain fewer species because, having smaller populations, species more often become extinct. On larger islands, where each species has a larger population, extinction rates are lower and more species are maintained.

Evidence for this theory comes from the Channel Islands off the California coast, where bird populations have been monitored over several decades. Each year, from 1 to 20 percent of the species on each island have become extinct, with the highest extinction rates on the

Figure 9–63. Species Richness in Oak Savanna Ecotone: The zone of transition between two ecosystems is known as an **ecotone**. Like the edge effect between two habitats, but on a landscape rather than a local level, an ecotone often has higher species richness than either of its adjacent ecosystems. Oak savanna is an ecotone that bridges eastern deciduous forests and prairie grasslands, and is composed of grasses with scattered trees, mostly oaks. Shown here is the typical avian species composition of oak savanna and adjacent ecosystems in southern Wisconsin. Oak forest dwellers include Cerulean Warbler and Cooper's Hawk, whereas the prairie is home to grassland species such as Vesper Sparrows and Horned Larks. Notice that the suite of species occurring within the ecotone, including Loggerhead Shrike and Brown Thrasher, is entirely different from that in the ecosystem on either side. Some birds such as the Northern Bobwhite, however, may be found in both the prairie and the ecotone. Similarly, the Red-headed Woodpecker and the Indigo Bunting may occur in both the oak forest and the ecotone. And, the American Crow is found everywhere. Totaling the numbers, the prairie supports 10 species, and the oak forest has 11, but the oak savanna has higher species richness, being home to 14 species. Information from Wisconsin Birdlife, by Samuel D. Robbins, Jr., 1991.

Figure 9–64. Species-Area Relationship for Islands: As the size of an island increases the number of species living there also increases, as shown by this graph of the number of bird species found on the variously sized Sunda Islands of Indonesia (see map for location). Each numbered dot represents an island, identified by name in the key. As a general rule, the number of species doubles with every tenfold increase in area, a pattern that holds true not only on islands but on island-like patches of habitat—such as remnant woodlots surrounded by agricultural land. Note that both axes are logarithmic scales. Adapted from MacArthur and Wilson (1967).

Location of the Sunda Islands of Indonesia

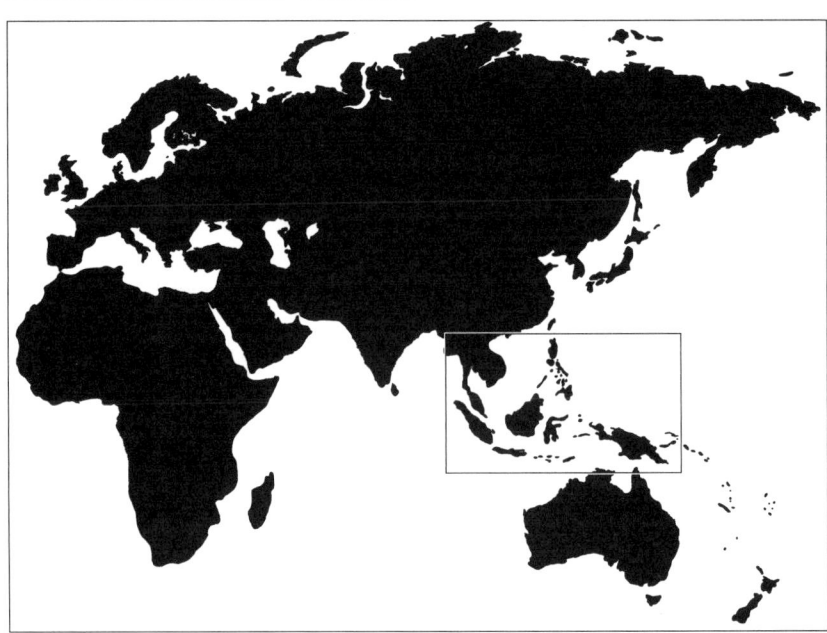

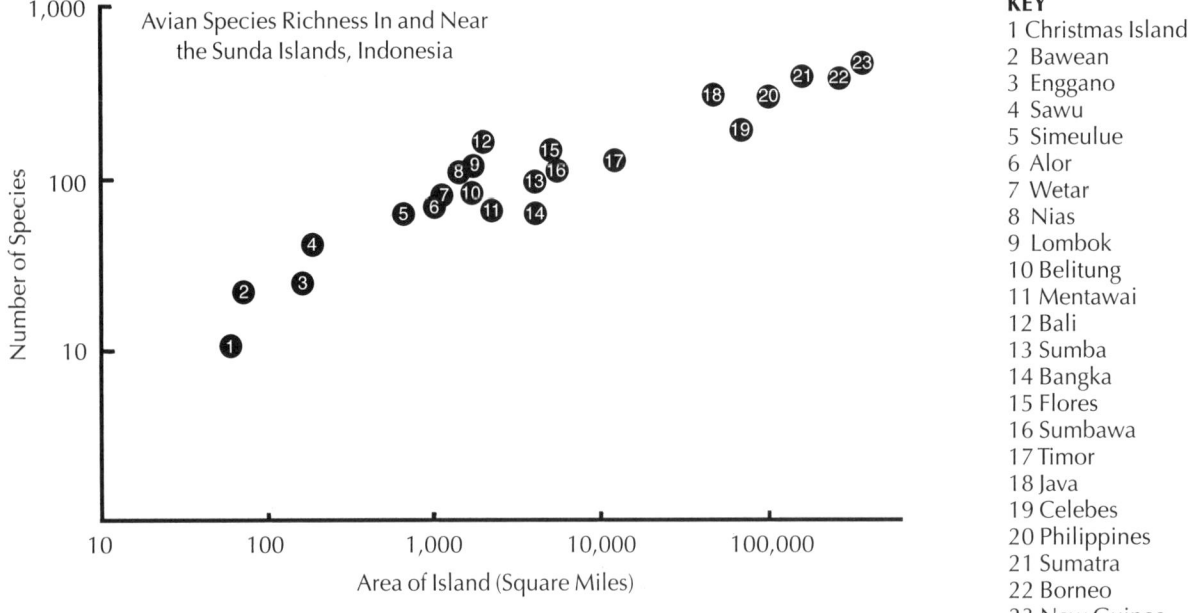

Avian Species Richness In and Near the Sunda Islands, Indonesia

KEY
1 Christmas Island
2 Bawean
3 Enggano
4 Sawu
5 Simeulue
6 Alor
7 Wetar
8 Nias
9 Lombok
10 Belitung
11 Mentawai
12 Bali
13 Sumba
14 Bangka
15 Flores
16 Sumbawa
17 Timor
18 Java
19 Celebes
20 Philippines
21 Sumatra
22 Borneo
23 New Guinea

smallest islands. In all cases, new species have arrived and recolonized each island each year, but the equilibrium number has remained lower on smaller islands than on larger ones.

The **habitat richness theory** is based on the observation that large islands tend to have more habitats than small islands. For example, larger islands frequently have a greater range of elevations, with different environments associated with the different altitudes. Colonizing species have a greater chance of finding suitable habitat on larger islands.

Habitat Patches as "Islands"

As mentioned above, the species-area relationship applies to isolated habitat patches as well as to actual islands. Evidence for this fact comes from a widespread change that humans have introduced to many natural landscapes: **habitat fragmentation**. When a landscape is altered for houses, agriculture, or certain logging practices, large and continuous ecosystems often are subdivided into small, isolated patches or fragments. Through this process landscapes come to resemble a series of islands, and the distributions of birds living in the habitat fragments may come to resemble the distributions of birds on actual islands.

This process is illustrated by the history of the deciduous forest near my home in Wisconsin **(Fig. 9–65).** Since European settlement, the original, heavily forested landscape has been steadily converted to a landscape of small, isolated woodlots, such as the 0.5-acre (0.2-ha) woodlot I have censused near my home. This forest patch now holds just 7 species, in contrast to the 76 species that I have found in a nearby forest of 2,900 acres (1,200 ha). So, the rule of thumb that species richness approximately doubles with every tenfold increase in area applies to these forest fragments in just the same way that it does to islands. The same relationship between species richness and forest fragment size has been documented throughout the eastern deciduous forests **(Fig. 9–66).**

Figure 9–65. Fragmented Wisconsin Landscape: In many areas of the United States, forest fragmentation has drastically altered the amount and quality of the remaining forest habitat. Many forests in the eastern and midwestern United States were fragmented during the 1800s and early 1900s as millions of acres were cleared for housing and agriculture. This time series shows the forest cover (red areas) in Cadiz Township, Wisconsin from 1831 to 1950. During this 119-year span, the forest in Cadiz was converted from one nearly continuous tract to many small and widely separated "islands" of forest. Adapted from Curtis (1956).

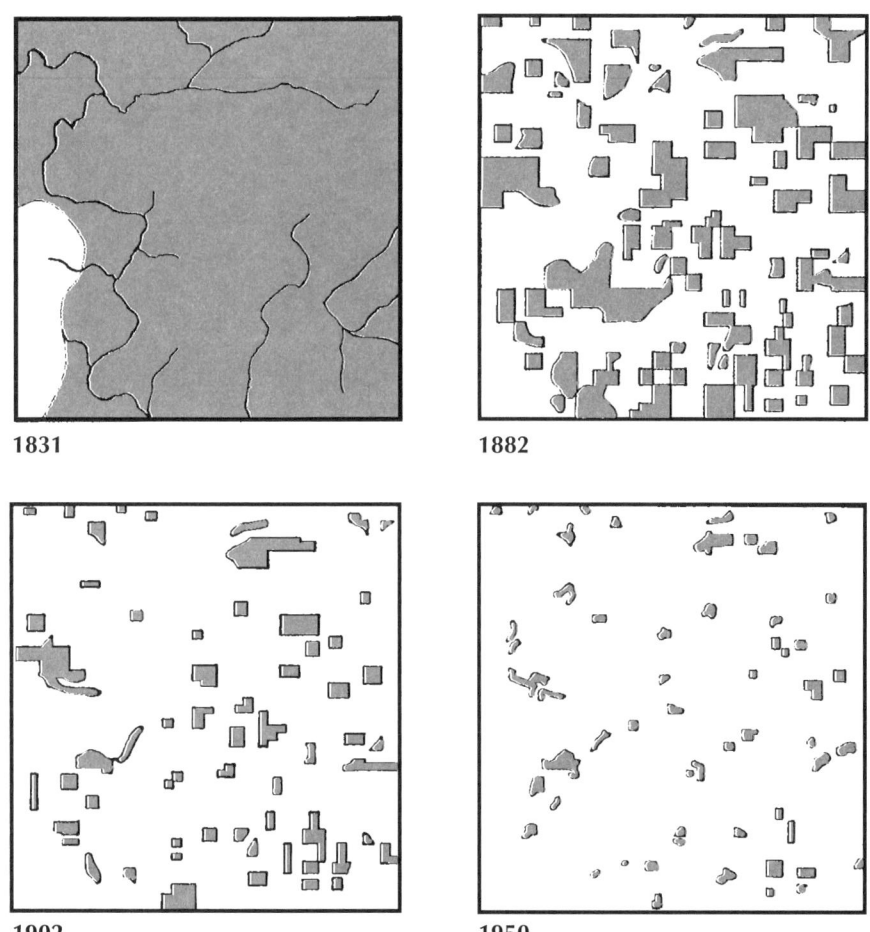

1831

1882

1902

1950

Figure 9–66. Species Richness and Size of Forest Fragments: Forest fragmentation—often caused by human activities such as development or agriculture—causes continuous forested landscapes to be divided into fragments that often have ecological similarities with actual islands. The number of forest-interior species able to successfully survive on these forest fragments is related to the size of the fragment. As the size of a fragment decreases, so does the number of species living there. Smaller fragments lack the habitat diversity and resources needed to support some forest-interior species, such as Ovenbirds, Yellow-throated Vireos, and Cerulean Warblers. This relationship is depicted from three separate studies conducted in Ontario, Illinois, and Missouri. Adapted from Freemark and Collins (1992).

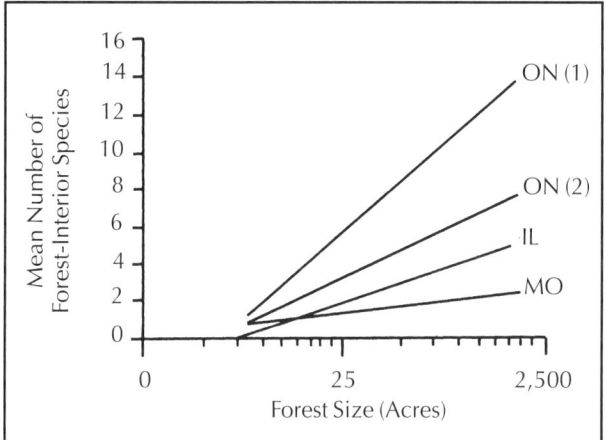

Paradoxically, in heavily fragmented landscapes the local species richness in each of the remnant forest patches may be higher than in a similar-sized area of unfragmented habitat. In the deciduous forest region, for example, the larger number of species in forest fragments results from the addition of edge species to a community previously composed solely of forest birds. But, at a larger scale, regional species richness may decline across a heavily fragmented landscape because of the loss of species that depend on large tracts of unbroken forest **(Fig 9–67)**.

Why do some species become more numerous, yet others disappear, when previously unbroken tracts of forest become heavily fragmented? The species that increase are the ones that can readily take advantage of newly created habitats—those species that live in small forest habitats and near habitat edges. In eastern North America, Brown-headed Cowbirds invaded from the midwestern prairies and savannas after the extensive mixed and deciduous forests became highly fragmented. The species that decline as forests are fragmented are those that, for any of several reasons discussed below, are particularly sensitive to habitat fragmentation.

Some species are sensitive to habitat fragmentation because they have particularly large spatial requirements for a home range or territory. Obviously, when a habitat patch becomes too small to support the home range of a pair or flock of a certain species, that species cannot live in the patch, even if all other habitat conditions are suitable. Not surprisingly, larger species and top predators (both of which tend to have larger home ranges) are affected to the greatest degree. For example, a forested area of 200 acres (81 ha) is too small to support a Pileated Woodpecker, which requires at least 250 acres (100 ha) of forest. It could, however, support a number of Red-bellied Woodpeckers, which generally require 40 acres (16 ha) or less (Shackelford et al. 2000). But, as discussed below, the effects of fragmentation are not only felt by large, space-demanding species.

Other species disappear from fragments much larger than their minimum home range size because they encounter problems with edge-inhabiting species. These forest-interior species must be able to find a home range far from an edge if they are to survive and reproduce

a. Regional-sized Forest (10,000+ acres [4,000+ ha]) with some Natural Fragmentation

- Local species diversity on any 100-acre area is low because few edge species are present.

- Regional species diversity over the entire 10,000+ acres is high because the relatively unfragmented landscape supports many forest-interior species.

Local Area
(Shown Through Time)

b. Regional Forest Becomes Moderately Fragmented

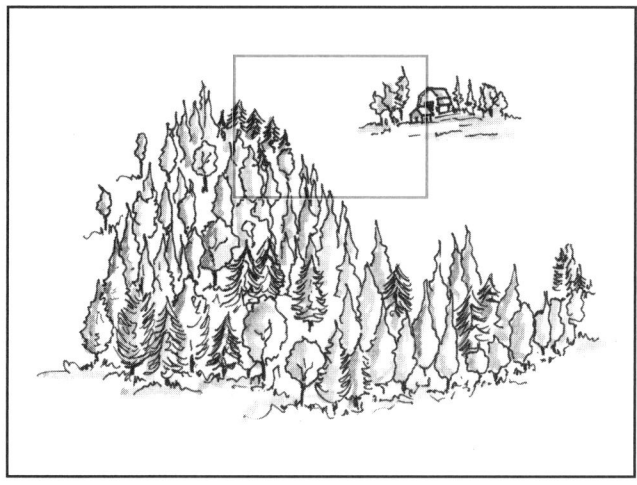

- Local species diversity increases in the newly fragmented areas because edge species are added.

- Regional species diversity remains nearly constant as the gain in the number of edge species offsets the loss of some forest-interior species.

c. Regional Forest Becomes Extensively Fragmented

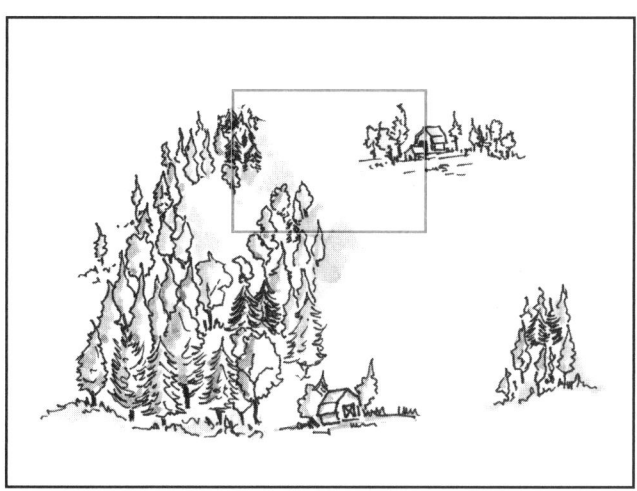

- Local species diversity remains constant as no new edge species are added; however, edge species become distributed over a wider geographical area as more habitat becomes favorable for them.

- Regional species diversity declines because no new edge species are added, but many forest-interior species are lost as habitat becomes increasingly unsuitable.

Figure 9–67. Fragmentation at a Regional Scale: These sketches of a regional-sized forest show how increasing fragmentation leads to more edge habitat and lowered avian diversity at the regional landscape scale. Illustrations by Keila Sydenstricker.

successfully. In eastern forests, for example, these species include many migratory songbirds, such as the Wood Thrush, Ovenbird, and Scarlet Tanager. Rosenberg et al. (1999) showed that in the eastern United States a forest fragment must be at least 172 acres (70 ha) in size to be highly suitable for Scarlet Tanagers, which have a home range size of just a couple of acres. If these sensitive forest-interior species nest near the edge of a forest fragment, they may be subjected to heavy rates of nest predation by edge predators, such as raccoons, feral cats, crows, and jays, which are not usually found in extensive forests. Furthermore, they suffer high levels of brood parasitism from cowbirds, which can penetrate the perimeter of a forest from the surrounding open habitat. For successful nesting, these sensitive forest-interior species require a secure **core area** far from the forest edge. Small fragments are simply not large enough to allow these species to nest far enough away from edges. The minimum distance that the core area must be from the forest edge before a bird will breed in the core varies among species. In addition to size, the shape of a fragment also affects how much core habitat it contains **(Fig 9–68)**.

One of the biggest edge-related problems currently facing forest songbirds is brood parasitism by Brown-headed Cowbirds, which spread into the eastern forest region from the midwestern prairies in the wake of forest clearing for agriculture and development (see Ch. 8, The Brown-headed Cowbird: History and Conservation, and Table 8–5). Taking advantage of newly created open habitat and an abundance of potential hosts with no defenses against parasitism (Robinson et al. 1995), cowbirds exploded in numbers in the East (see Fig 9–53) and are believed to be a significant cause of songbird declines in fragmented forests.

Yet other species suffer in fragmented habitats because they have little ability to disperse through habitats other than their own, and thus

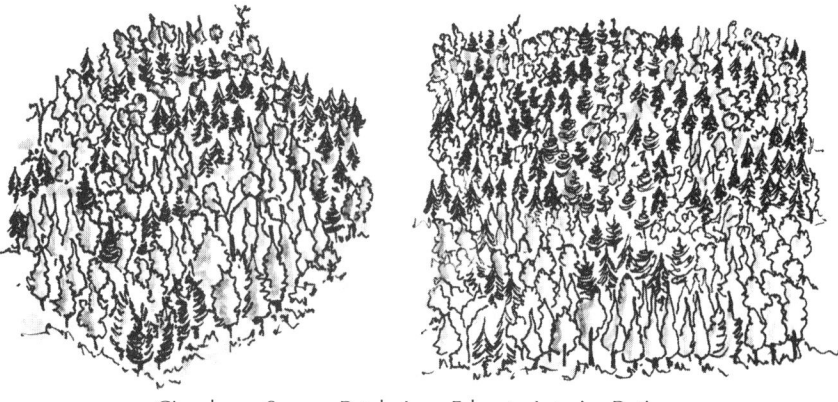

Circular or Square Patch: Low Edge-to-Interior Ratio

Figure 9–68. Edge Habitat and Forest Patch Shape: The shape of a forest patch affects the relative amounts of edge and forest-interior habitat associated with it. Round patches have the least amount of edge relative to the area of forest interior, whereas narrow patches have a relatively high amount of edge compared to forest interior. When managing habitat for forest-interior birds, land managers are encourged to create round or square patches and avoid creating narrow strips of forest. Illustrations by Keila Sydenstricker.

Long, Narrow Patch: High Edge-to-Interior Ratio

become marooned on isolated fragments. Once isolated, the small population on a fragment is subject to local extinction. If the fragment is too far from larger patches of forest, no dispersing individuals will be able to immigrate and recolonize the patch. In tropical forests, for example, many forest-interior birds will not venture out into open habitat, and they easily become marooned on fragments of tropical forest. When fragments of their habitat are too small to support a viable population and too far apart to permit dispersal, these species can become extinct over an entire fragmented forest region.

Some unfortunate species are sensitive to habitat fragmentation in several different ways at once. Consider the problems facing Northern Spotted Owls as logging fragments their old-growth forest habitat. As top predators, they require a large home range, and they are killed by Great Horned Owls living on the edges of old-growth patches. Furthermore, juveniles have trouble dispersing through areas that are not old growth, and often perish if they don't find a suitable patch of old-growth forest near their natal area (see Fig 9–10).

Patterns of Relative Abundance

Most bird communities contain a few species represented by many individuals and a large number of species represented by only a few individuals each. You've probably noticed this pattern at your bird feeder or in your neighborhood, where you may see large numbers of juncos and chickadees, but only a few nuthatches, cardinals, buntings, and woodpeckers. The pattern also is shown by the bird community that I censused in mixed hardwood-coniferous forests on the Apostle Islands of northern Wisconsin. Although the community contained 109 total species, half the individuals belonged to just 5 species: the Red-eyed Vireo, Ovenbird, Black-throated Green Warbler, Black-capped Chickadee, and Veery. The remaining 104 species each accounted for less than five percent of the total individuals, and most accounted for less than one percent **(Fig. 9–69)**.

Several factors interact to create this typical abundance pattern. One factor is that species usually are less common near the edges of their geographic ranges than at the center—where their resources tend to be of the highest quality. In any given community, only a few species are near the center of their range, where they tend to be most abundant. Other species in the community may be limited in various ways by suboptimal resources, and thus are less abundant.

Some birds are limited by the amount of food available to them. As discussed earlier, large predatory birds tend to have the lowest population densities, primarily because they need a lot of food. Thus, large species are almost never common in a bird community. For example,

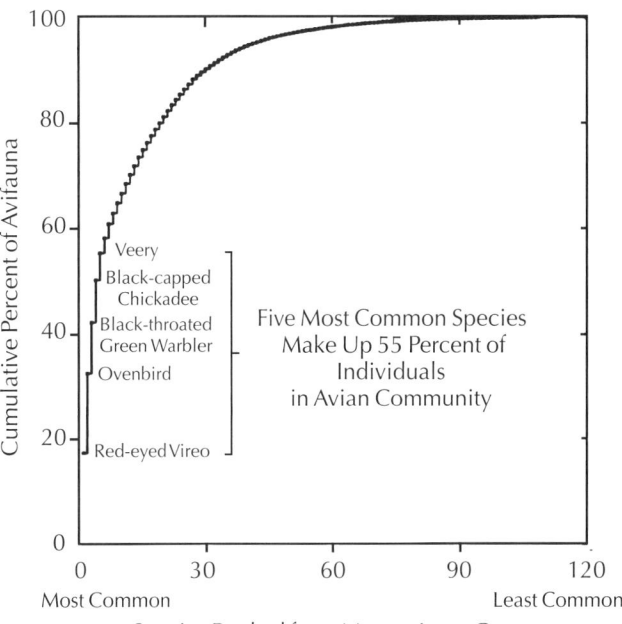

Figure 9–69. Relative Abundance of Birds of the Apostle Islands, Wisconsin: In most bird communities just a few species are common (there are many individuals of that species), whereas many species are rare. For example, on the Apostle Islands in Lake Superior, Wisconsin, the bird community contains 109 species, but just five species account for more than half of the individuals. The graph shows the cumulative percentages of birds ranked from the most to the least common, starting with the Red-eyed Vireo, which makes up about 18 percent of the individuals present. Adding the Ovenbird (14 percent) brings the cumulative percentage up to about 32 percent, adding the Black-throated Green Warbler (10 percent) increases it to about 42 percent, and adding the Black-capped Chickadee (8 percent) brings the total to about 50 percent. Adding the fifth most common species, the Veery (5 percent), increases the cumulative percentage of these five species to about 55 percent of all individuals in the bird community. Adapted from Temple (1990a).

the crow-sized Cooper's Hawk is far less common than the small, insectivorous birds upon which it preys. Smaller bird species also may be limited by food availability if they require specialized food types. Birds of any size may be limited by a lack of specific required resources other than food, such as nest sites, courtship sites, roost sites, foraging perches, or song perches.

As an example of the interaction of these factors, consider the forest-bird community that I censused in northern Wisconsin (see Fig. 9–69). There, the most abundant species were the Red-eyed Vireo and Ovenbird—small, insectivorous birds living near the centers of their ranges, where their specific habitat and food resources are abundant. Two of the rarest species were the Northern Goshawk, a large carnivore whose snowshoe hare prey is scarce in the region, and the Spruce Grouse, which requires a specific type of coniferous habitat uncommon in the area. Wisconsin represents the extreme southern limit of the range for both of these species.

Still other factors may affect relative abundance. For example, predators can help to even out the relative abundances of their prey by feeding on the most abundant species and ignoring the rare ones. Predators thus tend to maintain a community of many prey species with relatively equal (and moderately low) abundances. Indeed, when predators are removed, common species may become overabundant (for example, white-tailed deer in the northeastern United States), and differences in relative abundance within a community can become even more exaggerated.

Ecological Niches

As discussed earlier, the position of each species in a bird community is called its **ecological niche**. The niche encompasses the many ways in which a species interacts with its physical and biotic environment. As simple as this definition sounds, the concept of a niche can be difficult to comprehend and the niche of a species is even harder to measure. In fact, avian ecologists usually try to measure only a few dimensions of niche. For example, a species' foraging niche is defined by what it eats, and when, where, and how it feeds.

A fundamental rule of community organization is the **competitive exclusion principle**, which states that different species within a community cannot completely and simultaneously occupy the same ecological niche. If they do, one of the species will eventually become extinct. Conversely, different species can coexist only if they occupy different niches. For example, Great Horned Owls and Black-capped Chickadees are so different that their niches overlap little, if at all, and they coexist easily within a single community. Other species, such as wood-warblers and vireos, have very similar niches but can coexist because they have evolved **ecological isolating mechanisms** that allow them to finely divide up the resources they share. Groups of ecologically similar birds that use the same resources in similar ways are known as **guilds**. For example, those species, such as the Henslow's

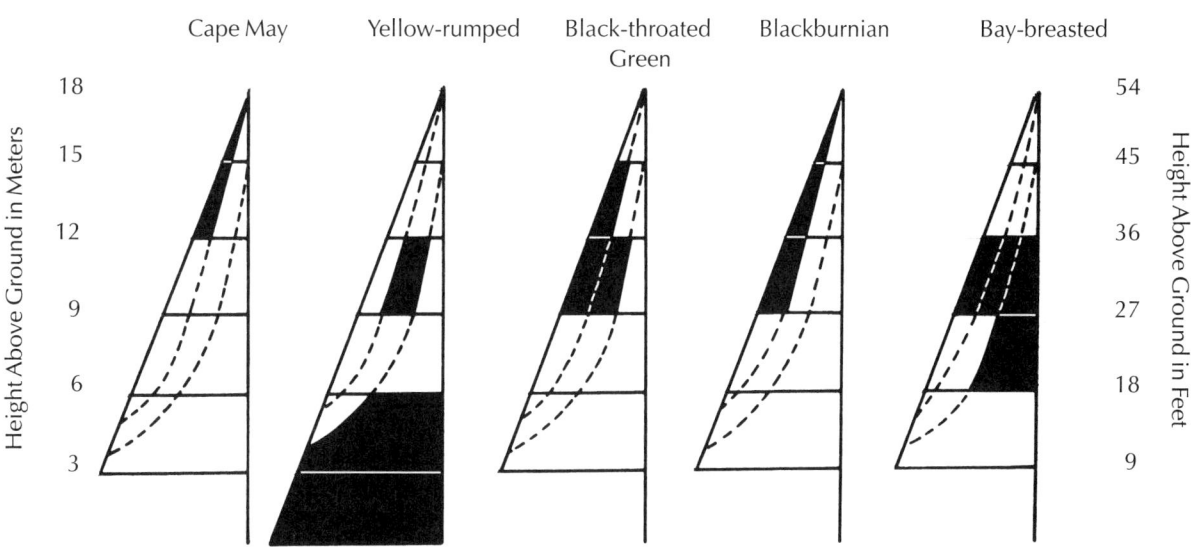

Principal Feeding Positions of Spruce-Woods Warblers in Trees

Sparrow, Grasshopper Sparrow, Eastern Meadowlark, and Bobolink, which nest on the ground in extensive grasslands, are considered to be in the same "grassland nesting guild." Although they are not closely related taxonomically, they use the same habitat in a similar way for nesting, thus placing them in the same nesting guild.

One way that ecologically similar birds segregate their niches is by eating different types and sizes of foods. For example, five species of kingfishers coexist throughout much of the Neotropics. Although each type of kingfisher eats fish, the different kingfishers range in body size, weight, and bill size from the 0.4-ounce (12-g) American Pygmy Kingfisher to the 9.5-ounce (300-g) Ringed Kingfisher. Because of these differences, each species of kingfisher takes a different size of prey, thus avoiding competition with the others.

An even more subtle form of niche segregation was demonstrated by Robert MacArthur in a classic study that examined coexistence mechanisms of five ecologically similar Dendroica warblers in Maine (Morse 1989; MacArthur 1958). All five species—the Cape May, Black-throated Green, Yellow-rumped, Blackburnian, and Bay-breasted warblers—coexist in coniferous forests, and all have similar morphologies, behaviors, and food habits. Close observations, however, revealed that although all the species feed on essentially the same insects in the same patches of forest, they avoid direct competition by foraging in different locations within individual trees **(Fig. 9–70)**.

The existence of ecological isolating mechanisms such as these implies that bird community structure is heavily influenced by competition among the various species, known as **interspecific competition**. Each community member has evolved a niche that is at least slightly different from that of all the others. Further evidence for the importance of interspecific competition is seen in **niche shifting**, in which a bird species occupies a different niche in different communities where it coexists with different groups of competing species.

Figure 9–70. Niche Segregation in Spruce-Woods Warblers: In the spruce forests of northeastern North America, a guild of small, insectivorous birds includes Cape May, Yellow-rumped, Black-throated Green, Blackburnian, and Bay-breasted warblers. The different species coexist by foraging in different parts of the same trees. Shown here are the principal feeding positions of each species, with each triangular diagram representing the "left" half of a spruce tree. Horizontal lines delineate 9-foot (3-m) segments; the lowest segment on each is below most of the foliage. The vertical line on the right of each diagram is the trunk. The area between the trunk and the first set of dashed lines (to its left) represents the inner part of the tree, which has large limbs but sparse foliage. Between the two sets of dashed lines lies the central region of the limbs, characterized by dense foliage. The area between the leftmost dashed line and the solid diagonal line represents the outer regions of the limbs, including the tips of the vegetation. The shaded areas show where each species concentrates its foraging. The Cape May Warbler, for instance, forages primarily in the outermost regions of the treetop. In contrast, the Bay-breasted Warbler forages at mid-height, mostly in the inner and central regions. Adapted from Morse (1989), after MacArthur (1958).

Niche shifting can be illustrated by comparing bird communities on mainlands to those on nearby islands. Because the islands usually contain fewer species, competition is less intense. Therefore, species living on islands may have broader niches than they do on the mainland, a phenomenon known as **ecological release** (the term comes from the idea that with the competitive factors that constrain the niche removed, it is "released" and can expand). As an example, three warblers—the Yellow-rumped, Black-throated Green, and Northern Parula—occur in different combinations on the coast of Maine and nearby islands. On the mainland, where all three species occur together, they segregate their niches by using different foraging zones in trees. On nearby islands where Northern Parulas and Yellow-rumped Warblers live together on islands in the absence of Black-throated Green Warblers, they each expand their niche to forage in the upper portions of the trees, a position normally occupied by Black-throated Green Warblers (Morse 1980) **(Fig. 9–71)**.

Are Bird Communities Organized in Optimal Ways?

Avian ecologists have noticed that bird communities in different parts of the world, although composed of unrelated species, are often very similar. Bird watchers who travel to different countries frequently find unfamiliar birds whose appearance and behavior remind them of familiar species back home. These similarities may not be accidental.

For example, Australian fairywrens are not related to true wrens, but they give similar calls, often cock their tails, and forage in similar ways (see Fig. 6–43). Australian Robins look and behave similarly to American Robins, yet are more closely related to crows. Australian sittelas, although unrelated to nuthatches, resemble them in both appearance and behavior (see Fig. 1–84). Remarkable examples of ecological similarities even occur among bird families: hummingbirds (Trochilidae) and toucans (Rhamphastidae) of South America have ecological counterparts, respectively, in the sunbirds (Nectariniidae) and hornbills (Bucerotidae) of Africa and Asia. As you may recall from Ch. 1, Methods Used to Classify Birds, these similarities result from **convergent evolution**, a process that produces similarities in morphology, behavior, and ecology among unrelated birds that occupy similar environments on different continents **(Fig. 9–72)**.

Although these individual examples are fascinating, avian ecologists have wondered whether convergences also occur among guilds or even entire communities in similar environments. If such similarities exist, the structure of bird communities may be a predictable response to a specific set of environmental conditions, regardless of where the communities occur or the types of birds involved.

Some of the best studies of potential convergences have compared bird communities in the so-called "Mediterranean" ecosystems found in coastal mountains of California, Chile, France, Sardinia, and South Africa. The bird communities of these dry scrub-woodlands

(Continued on p. 9·109)

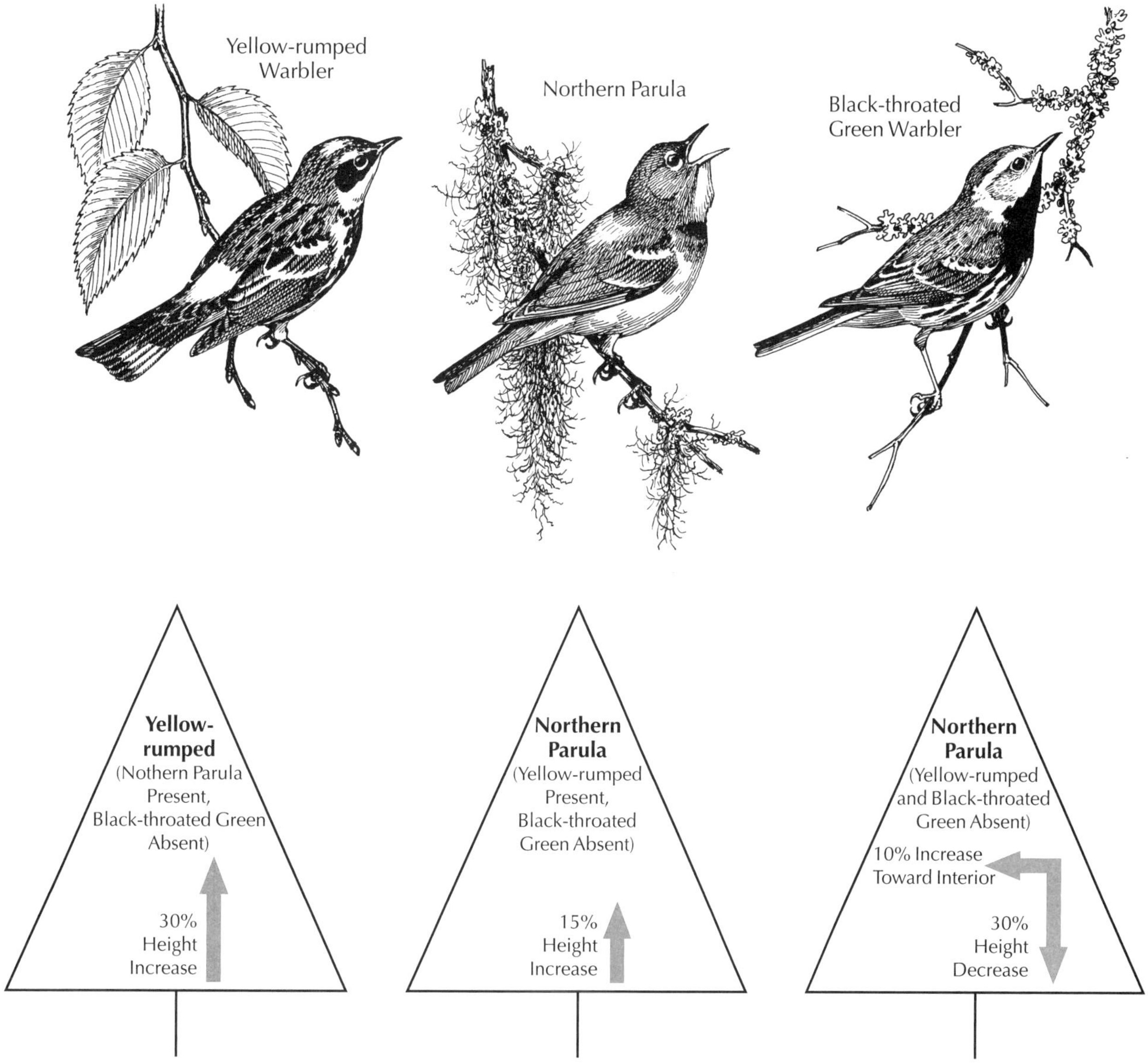

Figure 9–71. Niche Expansion of Northern Parula and Yellow-rumped Warblers When Other Warblers are Absent: Where Yellow-rumped, Northern Parula, and Black-throated Green warblers occur together along the coast of Maine, they partition trees into discrete foraging zones similar to those shown in Fig. 9–70. However, when one or more competitors are absent, as on islands off the Maine coast, the remaining species may expand their niches, a phenomenon known as **ecological release**. The left diagram shows that, when Yellow-rumped Warblers inhabit islands where Northern Parulas are present but Black-throated Green Warblers are absent, the yellow-rumps expand their foraging niche to encompass parts of the tree up to 30 percent higher than those they typically use. The center diagram shows that Northern Parulas on islands where Yellow-rumped Warblers are present but Black-throated Green Warblers are absent forage up to 15 percent higher. And parulas on islands where both Yellow-rumped and Black-throated Green Warblers are absent (right diagram) expand their niche 10 percent toward the center of the tree and 30 percent downward. Because the Black-throated Green Warbler does not expand its foraging area when the other two species are absent it seems to be the dominant species in the triad, constraining the foraging activities of the other two. Warbler drawings courtesy of Orville Rice. Diagrams adapted from Morse (1980).

Figure 9–72. Convergent Evolution Between Species: Unrelated birds from different parts of the world often show striking similarities in appearance or behavior. These similarities are the result of **convergent evolution**, a process through which natural selection produces similar morphology, behavior, and/or ecological characteristics in unrelated birds that occupy similar niches in different world regions.

Eastern Meadowlark
(Family Icteridae)

Yellow-throated Longclaw
(Family Motacillidae)

a. Eastern Meadowlark and Yellow-throated Longclaw: The color patterns of the meadowlarks of North American grasslands and the longclaws of African savannas are a classic example of morphological convergence. Both the Eastern (and its relative, the Western) Meadowlark and the Yellow-throated Longclaw are ground-living birds that inhabit open grasslands, and they are about the same size and shape. Both have streaked brown backs, bright yellow underparts, and a bold, black V-shaped mark on the chest. However, meadowlarks are in the New World blackbird family, Icteridae, whereas longclaws are relatives of the wagtails and pipits in the family Motacillidae. Despite their similarity, a look at their feet tells them apart: the longclaw has extremely long claws on its hind toes. The reason for the similar color patterns remains unclear.

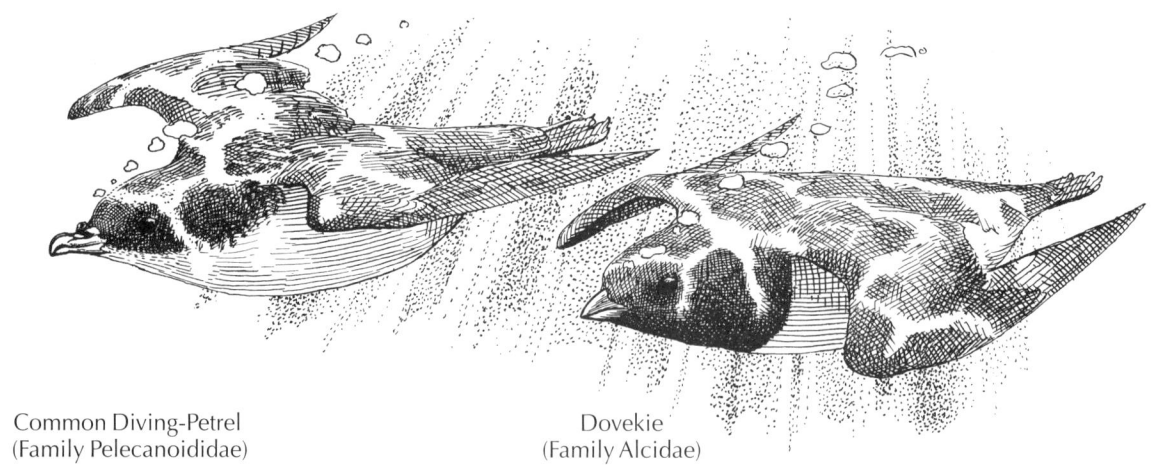

Common Diving-Petrel
(Family Pelecanoididae)

Dovekie
(Family Alcidae)

b. Common Diving-Petrel and Dovekie: Convergence in body shape is shown by the auklets, such as the Dovekie of northern oceans, and the southern ocean diving-petrels, such as the Common Diving-Petrel. Although they live at opposite ends of the globe, both the Dovekie and the diving-petrel are small, black-and-white seabirds with a compact, streamlined body shape. They use the same method of "flying" underwater, using their wings to propel themselves in pursuit of their marine crustacean prey. The Dovekie is in the family Alcidae, related to gulls and shorebirds; the Common Diving-Petrel is in the family Pelecanoididae—its tubular nostrils, lying atop its bill, identifying its affinity to the same order as albatrosses.

(Figure continued on next page)

Violet Sabrewing
(Family Trochilidae)

Scarlet-chested Sunbird
(Family Nectariniidae)

Figure 9–72. (Continued) **c. Scarlet-chested Sunbird and Violet Sabrewing:** Convergence in bill shape is shown by the nectar-feeding sunbirds of the Old World tropics and hummingbirds, which live in the New World. Both groups have long, slender, often down-curved bills ideally suited for probing tubular flowers to reach the enclosed nectar, and both groups play important roles in plant pollination. Also, both groups contain numerous colorful, iridescent species, such as the Scarlet-chested Sunbird and the Violet Sabrewing shown here. Sunbirds are passerines of the family Nectariniidae, and usually perch while feeding. Hummingbirds, well known for their ability to hover while feeding, are more closely related to the swifts, also in the order Apodiformes, and like swifts, they are aerial acrobats.

White-fronted Bee-eater
(Family Meropidae)

Great Jacamar
(Family Galbulidae)

d. Great Jacamar and White-fronted Bee-eater: The jacamars of Central and South America and the bee-eaters of Africa and Asia show remarkable similarities in both appearance and foraging behavior. Entirely insectivorous, these birds sally from an exposed perch to capture flying insects, including butterflies, dragonflies, and numerous bees and wasps. Then, before consuming their prey, they return to their perch and beat the insect against the perch until it is subdued and, in the case of stinging insects, until the stinger is removed. Both jacamars and bee-eaters have long bills suited for insect capture, and nest chiefly in tunnels that they excavate in river banks or cliffs. Jacamars are in the family Galbulidae, order Piciformes, and are related to woodpeckers and toucans. Bee-eaters are in the family Meropidae, order Coraciiformes, and are related to kingfishers and hornbills.

(Figure continued on next page)

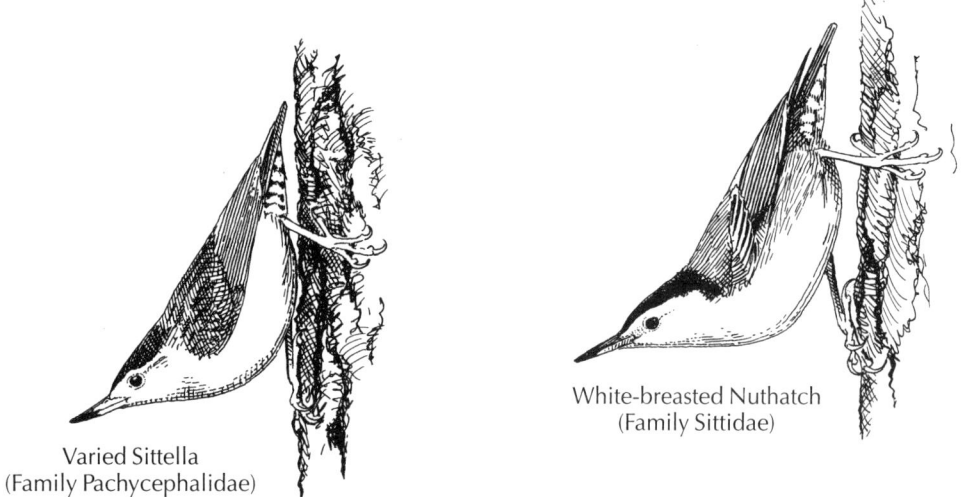

Varied Sittella
(Family Pachycephalidae)

White-breasted Nuthatch
(Family Sittidae)

Figure 9–72. (Continued) **e. Varied Sittella and White-breasted Nuthatch:** The sittellas (otherwise known as "treerunners") of Australia show convergence in behavior with the nuthatches of North America and Eurasia: members of both groups are well known for walking headfirst down tree trunks. Like the familiar White-breasted Nuthatch (family Sittidae) of North American backyards, the Varied Sittella (family Pachycephalidae) of dry Australian woodlands is highly acrobatic. It spirals along branches and hangs upside down from twigs while probing for insect prey by inserting its sharp, wedge-shaped bill into tree crevices and under loose bark.

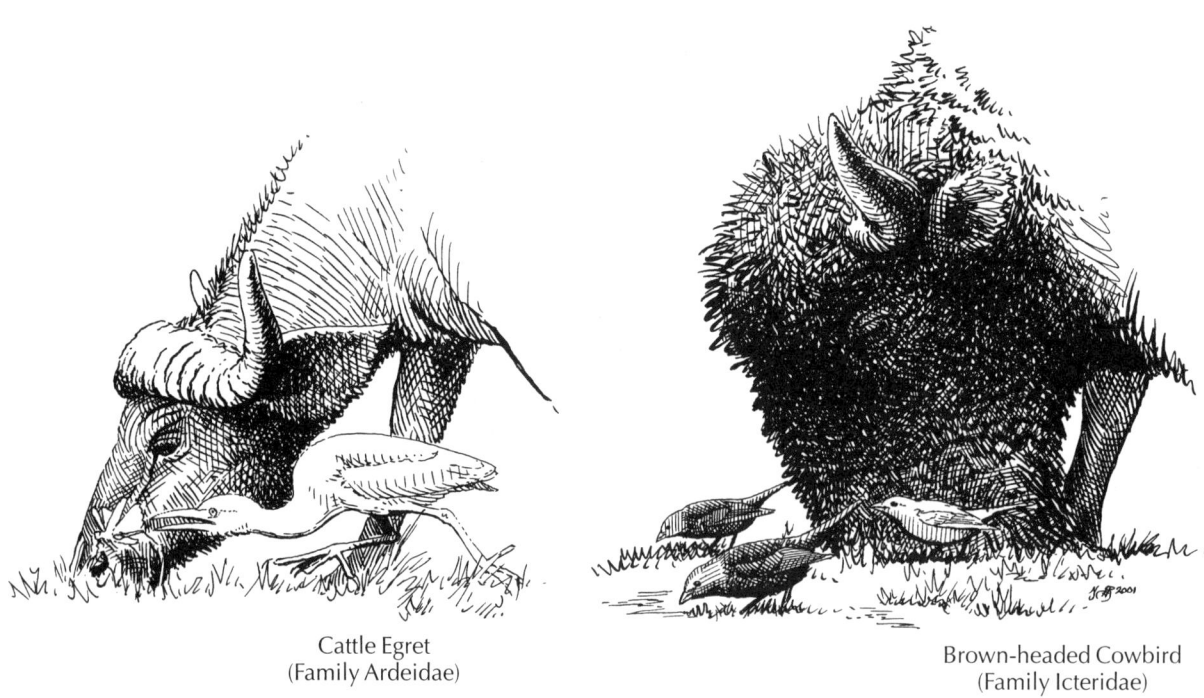

Cattle Egret
(Family Ardeidae)

Brown-headed Cowbird
(Family Icteridae)

f. Cattle Egret and Brown-headed Cowbird: Evolving on two separate continents, the Cattle Egret and the strikingly different Brown-headed Cowbird have converged upon the same foraging strategy: they accompany large mammals whose movements flush prey into view. The Cattle Egret (family Ardeidae) is a common sight on the plains of Africa, where it accompanies feeding rhino or herds of Cape Buffalo, as shown here, stalking under the huge feet of the mammal, ready to snatch up a locust or other insect scared out of hiding. Half a world away on the prairies of North America, the Brown-headed Cowbird, a member of the New World blackbird family (Icteridae), independently developed the same foraging strategy. However, its feeding companion was the American Bison or buffalo. Today both species inhabit the same continent; the enterprising Cattle Egret arrived in the New World in the 1950s, and both birds now forage among North American cattle herds.

show broad similarities, even though they are composed of different, often unrelated, species. The species-area curves for the communities are similar, as are the ways that foraging heights are arranged in the vegetation. Species-for-species matches occur in morphology, behavior, and ecology. Most importantly, these similarities are greater than might be expected by chance alone, and greater than the similarities between these communities and communities associated with other types of environments. Comparisons of grassland bird communities in North and South America, and North American and Australian woodlands, produce similar results. Together these studies suggest that a "best way" of organizing a bird community in a particular environment may exist. All the aspects of avian ecology, such as population dynamics, competition, and niche shifting, may interact to create the "perfect" community of birds.

Birds as Components of Ecosystems

■As discussed above, birds interact with one another, and with other living things, as important members of communities. But each bird also is an important component of its **ecosystem**: the combination of its physical surroundings and all the organisms that live there, and the ecological processes (such as decomposition, soil erosion, and water and nutrient cycling) that bind them all together.

Ecological Distribution of Birds in the Major Terrestrial Ecosystems of North America

In Chapter 1, you examined the avifauna of the six major zoogeographic regions of the world (see Fig. 1–52), and the historical, evolutionary, and biological reasons why bird groups are distributed as they are. This section concentrates on the major terrestrial ecosystems of North America, and describes the types of birds characteristic to each.

In any given area, there is a certain association of plants and animals that, in the absence of large-scale climatic changes, disease, or human disturbance, can perpetuate itself. This community is called the **climax community**. Climax communities develop over long periods of time through a process called **ecological succession**, in which one association of plants and animals is replaced by another, and that one is replaced by yet another, and so on **(Sidebar 4: From Blackberries to Beeches: Ecological Succession in Eastern Deciduous Forests)**. Eventually, an association of plants and animals that remains relatively stable—the climax community—is reached.

In North America north of Mexico, ecologists recognize nine major terrestrial ecosystems (sometimes termed **biomes**). These ecosystems encompass the climax communities for a particular environment, as well as the physical surroundings and ecological processes. Although the types of animals are as characteristic of par-

(Continued on p. 9·114)

Sidebar 4: FROM BLACKBERRIES TO BEECHES: ECOLOGICAL SUCCESSION IN EASTERN DECIDUOUS FORESTS

Ron Rohrbaugh

Birds are integral components of forest ecosystems. Birds use forests for food, cover, and nest sites. They rely on forests to produce foods in the form of seeds, berries, insects, and even other birds—in the case of forest-dwelling hawks and owls. Nonmigratory birds use temperate forests for resting and nesting cover year round, whereas migratory species use forests for nesting during the breeding season and "stopover habitat" while making long migratory journeys. Furthermore, many ecological processes that occur in the forest are dependent upon birds. Seed caching by Blue Jays, for example, is an important mechanism for seed dispersal, and hence forest regeneration.

Not all forests, however, are equally valuable as food, cover, or nest sites for all species of birds. Different species have different habitat requirements based on their ecology and behavior. The age of a forest is a key characteristic that affects its ability to support a given bird species.

The Process of
Ecological Succession

Like forest size, discussed in the chapter text, forest age is often a function of past and present human activity in an area. Timber harvesting, farm abandonment, and strip mining, for example, all create cleared areas upon which forests may grow back. But forests do not grow back randomly. Cleared areas go through an orderly series of changes in vegetation type from field, to old field, to young trees, and eventually to older trees. Although these changes in vegetation type are the most obvious, the entire spectrum of species inhabiting the area also changes.

This series of changes is known as **ecological succession**.

These changes in plant and animal species composition, vegetation structure, and so on are highly predictable. The particular species involved, and sometimes the time required, may vary from one area to another, but the basic process of ecological succession remains the same. At each stage along the way, the plants and animals present change the conditions in such a way that they, themselves, can no longer live there (for example, by changing soil acidity, soil moisture, soil organic content, or the amount of shade). Consequently, they are replaced by species better adapted to the new conditions. Eventually, a community of plants and animals that are able to replace themselves develops. This community, termed the **climax community**, is relatively stable unless climatic conditions change, or disease or human activity interfere. As with the various successional stages, the exact species composition of the climax community (whether it is a deciduous or coniferous forest, for example) is different in different environments.

Types of Ecological Succession

There are two types of ecological succession, and they differ mainly in their starting point. **Primary succession** begins on a substrate—such as rock, sand, or lava—that has never before supported a community. When such a substrate is exposed to the elements, the slow process of soil formation begins, and succession is underway. The process by which a lake gradually fills in to form a bog community is also considered a form of primary succession **(Fig. A)**. Depending on the habitat, primary suc-

cession may take centuries to reach a climax community.

Secondary succession, in contrast, begins with bare soil or an existing community, such as the grasses in your yard (if you were to stop mowing them), a retired farm, or a forest that is not yet at climax. Because soil is already present, secondary succession takes much less time to reach a climax community than does primary succession. Secondary succession is often a recovery process—a reversion back to some pre-existing plant community.

Secondary Succession:
From Old Field to
Eastern Deciduous Forest

To examine secondary succession and its relationship to birds in an eastern deciduous forest, let's take an imaginary walk through space and time.

The Old Field (Scrub/Shrub)

Imagine that we are walking through a 50-acre field in Virginia on which hay production ceased 15 years ago. "Ouch! I just got a thorn in my thigh. I'm exhausted. There must be some way out of these impenetrable briars. Maybe we can push through that opening over there. That's better, now I can see a path." Does this experience sound familiar? If so, you must have visited what is often called an "old field." An old field is essentially the very beginning of a new forest. It is the first successional stage that is not completely dominated by herbaceous plants, such as grasses. It is composed of perennial grasses, forbs such as goldenrod and asters, and woody shrubs and trees that can be scattered or occur in dense clumps. At this stage, the new "forest" has no vertical layering, as most of the plants are the same height **(Fig. B).**

Although the exact species composition will vary from place to place, you can expect to find a fairly predictable community of plants and animals in most old fields in eastern North America. The woody plants tend to be low-growing shrubs and trees that tolerate full sun, such as blackberries, raspberries, hawthorns, honeysuckles, multiflora rose, black locust, and eastern red cedar. A diverse community of birds thrives in old fields. Nesting species include Brown Thrashers, Gray Catbirds, Golden-winged Warblers, Red-winged Blackbirds, Northern Cardinals, Field Sparrows, and Song Sparrows. Many raptors, such as Great Horned Owls, American Kestrels, Red-tailed Hawks, and Cooper's Hawks, use these highly productive old fields for hunting.

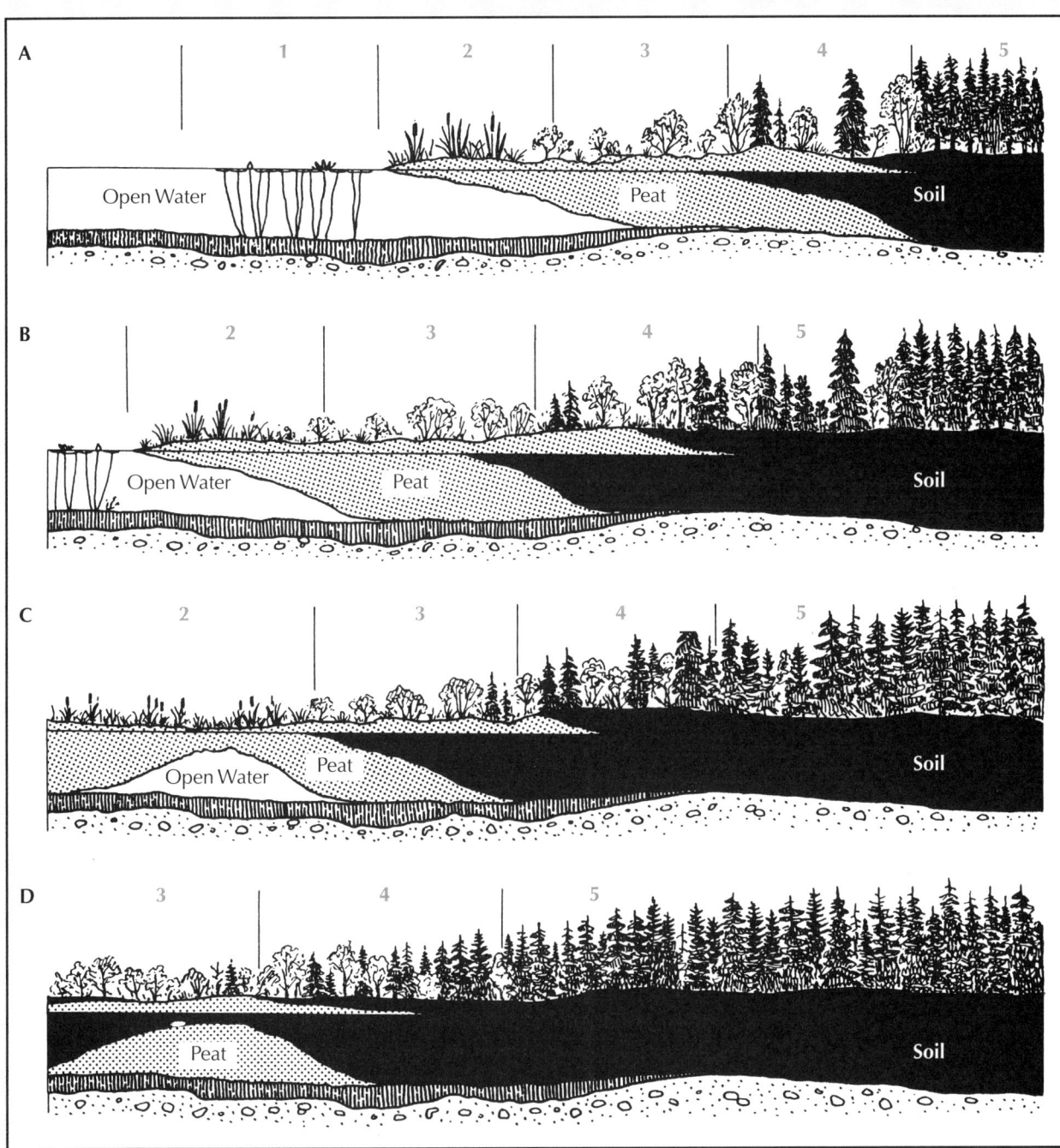

Figure A. Primary Succession in a Bog Community: This is a time series (A to D) showing how primary succession—through four successive stages (1 to 4)—leads to the "take-over" of a circular lake by a climax coniferous forest (5). The floating mat supporting the sedge community (2) gradually moves toward the center of the lake, while the peaty soil that accumulates from the decaying vegetation left behind by the mat enables the low shrub community (3) to follow. The low shrub community "lays the ground" for the tall shrub community (4), and the tall shrub community in turn lays the ground for the climax forest (5). After the mat covers the center of the lake and the peat accumulates sufficiently to fill up the remaining water basin, the low shrub community replaces the sedge community, and so on, until the climax forest occupies the entire site of the former lake.

Figure B. The Old Field: In the eastern United States, shrubby "old field" habitats result when areas, such as pastures and hayfields, previously cleared of trees, are abandoned. Following abandonment, woody plants such as sumac, multiflora rose, honeysuckle, and black locust begin to invade the site, eventually resulting in a uniform distribution of shrubs. Many shrub-nesting bird species, such as Northern Mockingbirds and Song Sparrows, have benefitted from the high rate of farm abandonment during the last half of the 20th century. Photo by David Sieren/Visuals Unlimited.

The Early- and Mid-Successional Forest

Now, imagine that 20 years have passed and you are walking through the same "field" which now looks, smells, and sounds more like a forest. During the last two decades the site has progressed through an early-successional forest stage that was dominated by saplings of roughly uniform age and size. A few species of birds, such as Chestnut-sided Warblers and Eastern Towhees, may have come and gone owing to the changing habitat conditions.

The now 35-year-old mid-successional forest will have a canopy dominated by small and medium-sized trees. You can easily wrap your arms around the trees, and you can get your hands around some of them. An understory of shrubs and young trees probably will be developing below the forest canopy. Thus, the forest will have two vertical layers, the understory and the overstory. Increasing amounts of vertical structure usually increase bird diversity, because a wider selection of foods and nest sites are provided by the

vertical layering **(Fig. C).**

Some of the plants in the understory are tolerant of shade and will continue to grow there even

in later successional stages. These may include mountain laurel, rhododendron, juneberry, witchhazel, and viburnums. Other plants in the understory are young trees such as oaks, maples, hickories, and American beech. These are the offspring of the trees in the canopy. Keep these young trees in mind; they will play an important role in the next successional stage.

What about the birds? We would expect to lose some, if not all, of the breeding species that were present in the old field community. However, these species would be replaced with birds that are adapted for life in the forest. Some residents and migrants that might breed in a typical mid-successional forest include Ruffed Grouse, Great Crested Flycatcher, Blue Jay, Black-capped Chickadee, Tufted Titmouse, American Robin, Black-and-white Warbler, and American Redstart. Although they probably wouldn't breed here, we also would expect to find several species of raptors, such as the Sharp-shinned Hawk, Cooper's Hawk,

Figure C. The Early- and Mid-Successional Forest: Following the old field (shrub) stage, a new forest begins to emerge. The early stages of the forest are often a mixture of shade-intolerant shrubby plants from the old field, and shade-tolerant saplings and "pole-sized" trees that will eventually occupy the mature forest. An important difference between the late stages of the old field and the early-successional forest is the presence of vertical layering. As the forest ages, it begins to develop at least two layers—a shrubby understory and an overstory of maturing trees. Vertical layering provides habitat for a more diverse community of bird species. Photo by Ross Frid/Visuals Unlimited.

Northern Goshawk, Broad-winged Hawk, and Eastern Screech-Owl, hunting in this mid-successional forest.

The Late-Successional Forest

Now let's take a final trek through time. Forty more years have passed, and the oldest trees are now 75 years old—many are too big to fully encircle with your arms. Several important ecological changes have occurred during the past four decades—some trees have grown quite large, some have died, some have fallen, and vertical layering has grown more prominent. These changes have a big impact on the community of birds that inhabit the site.

Large, mature trees provide a consistent source of food by way of fruit production and the insects that live on their leaves and bark. Furthermore, these large trees provide adequate nest sites for canopy-nesting songbirds such as Scarlet Tanagers and forest-nesting raptors such as Red-shouldered Hawks, Broad-winged Hawks, and Northern Goshawks **(Fig. D)**.

Dead and dying trees, known as snags, provide a smorgasbord of food for insectivorous birds. As a tree dies, its bark and decaying wood become home to a host of insects. Birds such as White-breasted Nuthatches and Downy, Hairy, Pileated, and Red-bellied woodpeckers rely on these snags for the ready supply of insect food hidden under the loose bark. These snags also provide nest sites for many cavity-nesting birds, such as woodpeckers, chickadees, and titmice. The largest snags provide nesting cavities for Eastern Screech-Owls and Barred Owls.

Fallen trees provide cover and foraging sites for birds that dwell close to the ground, such as Ruffed Grouse, Ovenbirds, and Dark-eyed Juncos. Fallen trees also create openings or gaps in the forest canopy, thus allowing sunlight to penetrate to the forest floor. These tiny gaps support

herbaceous and sun-loving plants that add habitat diversity to the forest ecosystem.

Late-successional forests usually supply the greatest amount of vertical layering. It's possible to have several layers, which, from the ground up, include the ground cover, an understory of shrubs, an intermediate layer of small trees, a lower canopy of medium-sized trees, and a primary canopy of larger trees. The diverse habitat associated with this vertical layering supports a wide array of birds. For example, you might see Wood Thrushes and Hermit Thrushes in the understory, Red-eyed Vireos and Rose-breasted Grosbeaks in the intermediate layers, and Scarlet Tanagers and Black-throated Green Warblers in the canopy. If late-successional forests are left undisturbed for long enough, a climax community eventually develops.

Interpreting What You See

In your daily travels, or while out birding, it's likely that you see forests of many different successional stages. Examine the structure and composition of these forests and try to interpret what you see. Are all of the trees "even-aged" saplings, suggesting that the site was recently a field and now contains only young trees that established themselves at roughly the same time when the field was abandoned? Does the forest contain a diverse mixture of trees and shrubs with evident vertical layering—characteristic of a mid- to late-successional forest, which contains trees of several age classes?

By combining the answers to these types of questions with your knowledge of the life histories of birds, you can enrich your birding experiences by understanding why birds occur where they do and by predicting which species you are likely to find in a particular location. ∎

Figure D. The Late-Successional Forest: In the late-successional forest, vertical layering becomes even more prominent and other important changes cause the forest to become more complex. Dead and dying trees provide food and nest sites for many birds; gaps in the forest canopy allow small communities of sun-loving plants to thrive, thus increasing habitat diversity; and many trees become large enough to support nesting birds such as hawks, owls, and crows. Photo by Ross Frid/Visuals Unlimited.

ticular ecosystems as the types of plants, each ecosystem is named for certain plant forms or plant associations that are conspicuously dominant. For example, the deciduous forest ecosystem gets its name from its preponderance of deciduous trees, including oaks, maples, beeches, and hickories.

The major ecosystems of North America, each described in detail below, are Tundra, Coniferous Forest, Deciduous Forest, Grassland, Southwestern Oak Woodland, Pinyon-Juniper Woodland, Chaparral, Sagebrush, and Scrub Desert. Each description includes a map indicating its general geographical distribution, but please bear in mind that these large-scale maps do not show the various successional stages because the stages are too numerous, small, and dynamic. Also, these maps do not accurately show the full extent of each major ecosystem in western North America, because some parts of the ecosystems may be scattered, isolated patches that develop wherever the microclimate is just right, such as narrow belts on steep mountain slopes.

Tundra

Extending north from the limit of trees, the **arctic tundra** features swampy flatland interrupted by countless shallow lakes **(Fig. 9–73)**. Only the upper few inches of ground ever thaw; below lies the permanently frozen soil—the permafrost. Winter daylight is brief, and summer daylight is nearly continuous. The growing season is intense but of short duration. All vegetation grows low to the ground. Grasses, sedges, mosses, and lichens grow in hummocky mats; thick shrubs trail over the ground; and many plants produce large, bright flowers.

Few birds are permanent residents other than the Gyrfalcon, Snowy Owl, Common Raven, and two ptarmigan: the Willow and Rock, both of which take on a white plumage in winter (see Fig. 3–53). Many of the migrant birds that breed in the tundra, including waterfowl and shorebirds, arrive already mated and depart for the south as soon as they can complete their accelerated nesting activities. Bird populations are low in the summer, with about 20 to 80 males (of all species, total) per 100 acres (40 ha), and are much lower in the winter. Shorebirds, numbering at least 18 species, are an especially prominent part of the avifauna. Other characteristic nesting species include Red-throated Loon, Tundra Swan, Brant, White-fronted Goose, Snow Goose, Long-tailed Duck, King Eider, Pomarine Jaeger, Parasitic Jaeger, Long-tailed Jaeger, Glaucous Gull, Sabine's Gull, Horned Lark, American Pipit, Lapland Longspur, Smith's Longspur, and Snow Bunting.

The **alpine tundra** occurs above timberline in the high mountains from Alaska south to California and New Mexico, and on isolated peaks in the Appalachians. It features rugged, well-drained terrain interspersed with meadows where, for a few weeks in the early summer, wildflowers bloom in profusion. It resembles the arctic tundra in its low vegetation and short growing season, but differs in its consistently higher winds, colder nights, more intense sunlight, low oxygen content of the air, lack of a permafrost, and lack of long seasonal periods of daylight and darkness.

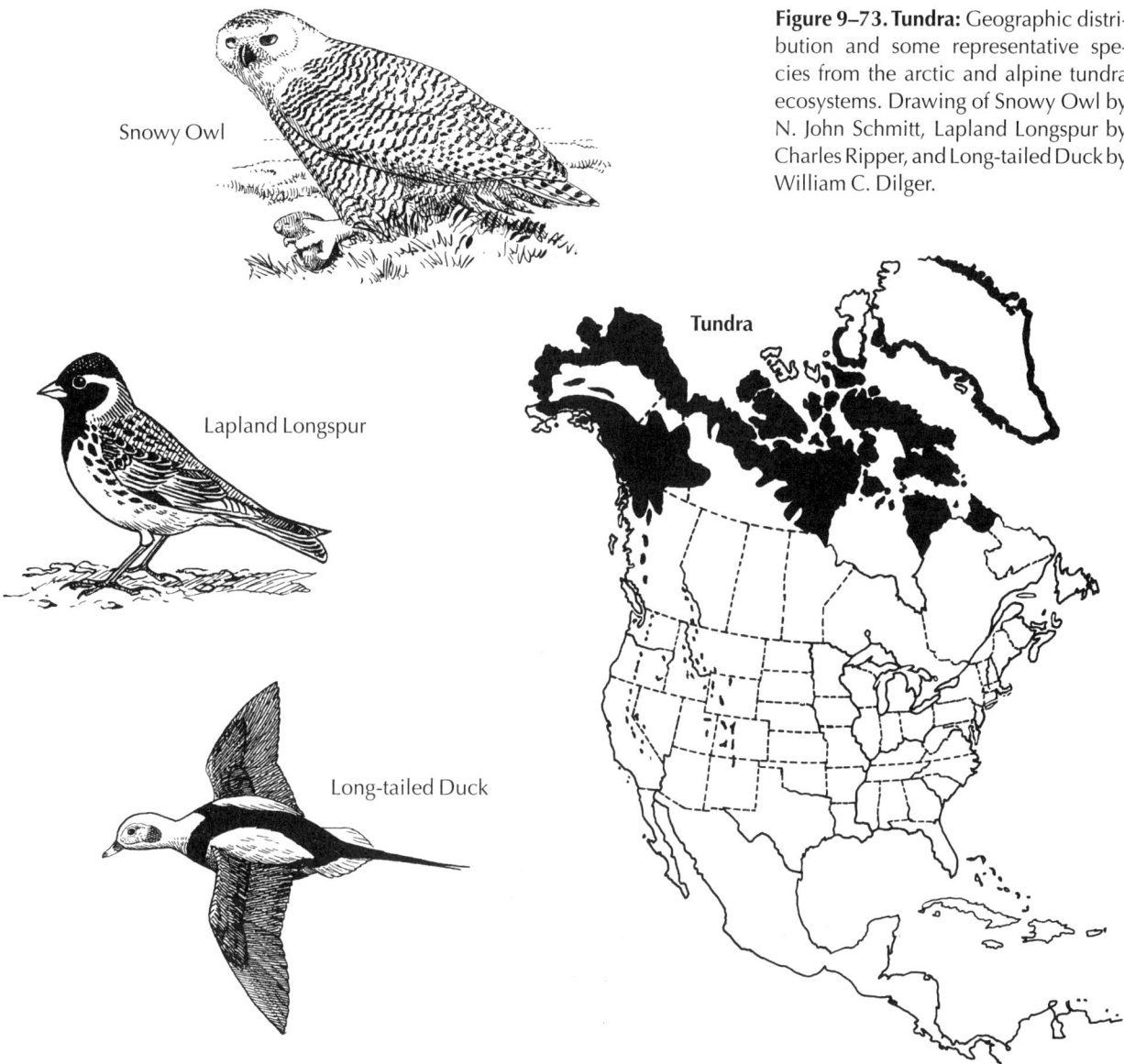

Figure 9–73. Tundra: Geographic distribution and some representative species from the arctic and alpine tundra ecosystems. Drawing of Snowy Owl by N. John Schmitt, Lapland Longspur by Charles Ripper, and Long-tailed Duck by William C. Dilger.

The diversity and density of breeding birds in alpine tundra is very low. Generally the total density is less than 50 males of all species per 100 acres (40 ha). The only species commonly shared with the arctic tundra is the American Pipit. Although the alpine tundra also has a ptarmigan, it is a different species, the White-tailed Ptarmigan. Aside from this species, the only bird unique to the alpine tundra is the Rosy Finch, with several subspecies occupying different mountain ranges. Neither the White-tailed Ptarmigan nor the Rosy Finch, however, occur in the Appalachians. Unlike the arctic tundra, the alpine tundra has no nesting predatory birds, probably because the available foraging areas are too small and unproductive. Golden Eagles and Prairie Falcons appear from lower altitudes in the late summer, to hunt when the fledglings of alpine birds are emerging.

Coniferous Forest

In this ecosystem, where spruce and fir trees form an evergreen shelter and provide cones for winter foraging, some birds such as crossbills have evolved specific foraging adaptations (see Fig. 9–16). Ecologists recognize three common types of coniferous forests **(Fig. 9–74)**.

The **western montane forest** includes the moist coastal environment along the North Pacific slope and the forests of the Cascades, Sierra Nevada, and Rocky Mountains. The species of conifers and the elevations of the forests vary in different areas. On the Pacific slope of the Olympic Mountains in Washington, stands of Douglas fir, western hemlock, and Sitka spruce extend from timberline nearly to sea level. In the Rockies of Colorado, the forests of Engelmann spruce, alpine fir, and Douglas fir reach down only to about 7,500 feet (2,300 m). The **boreal forest**, sometimes called the **taiga**, stretches from Alaska and the Rocky Mountains eastward across Canada to the Atlantic Coast and then branches south on the Appalachians as the **eastern montane forest**. White spruce and balsam fir are the dominant conifers of the boreal forest, giving way to red spruce and Fraser fir farther south in the Appalachians.

Birds that typify the western montane and boreal forests include the Northern Goshawk, Black-backed Woodpecker, Three-toed Woodpecker, Olive-sided Flycatcher, Gray Jay, Red-breasted Nuthatch, Brown Creeper, Winter Wren, Hermit Thrush, Swainson's Thrush, Golden-crowned Kinglet, Ruby-crowned Kinglet, Evening Grosbeak, Pine Grosbeak, Pine Siskin, and Red Crossbill.

The boreal forest supports breeding populations of about 150 to 300 males (of all species, total) per 100 acres (40 ha). Many species are migratory, resulting in winter populations of only about 50 to 100 individuals per 100 acres (40 ha). Some species not generally considered migratory—crossbills, redpolls, and grosbeaks—may move south in large numbers if there is a poor seed crop or an unusually high density of individuals. As you may recall from Ch. 5, this mass movement of typically nonmigratory birds is known as an **irruption**, as birds are "irrupting" from one place to another.

Particularly well represented among the breeding birds are the wood-warblers. Characteristic nesting species include the Spruce Grouse, Yellow-bellied Flycatcher, Boreal Chickadee, Tennessee Warbler, Magnolia Warbler, Cape May Warbler, Yellow-rumped Warbler, Blackburnian Warbler, Bay-breasted Warbler, Palm Warbler, Canada Warbler, Purple Finch, White-winged Crossbill, Dark-eyed Junco, and White-throated Sparrow.

Compared to the boreal forest, the western montane forest has much smaller breeding populations, which tend to be much less migratory. Much of the migration that does exist in the western montane forest is altitudinal rather than latitudinal (see Fig. 5–52). The total number of breeding males of all species averages from 50 to 250 per 100 acres (40 ha), and in winter there may be 25 to 200 individuals per 100 acres (40 ha). Some characteristic nesting species include Blue Grouse, Hammond's Flycatcher, Clark's Nutcracker, Mountain Chickadee,

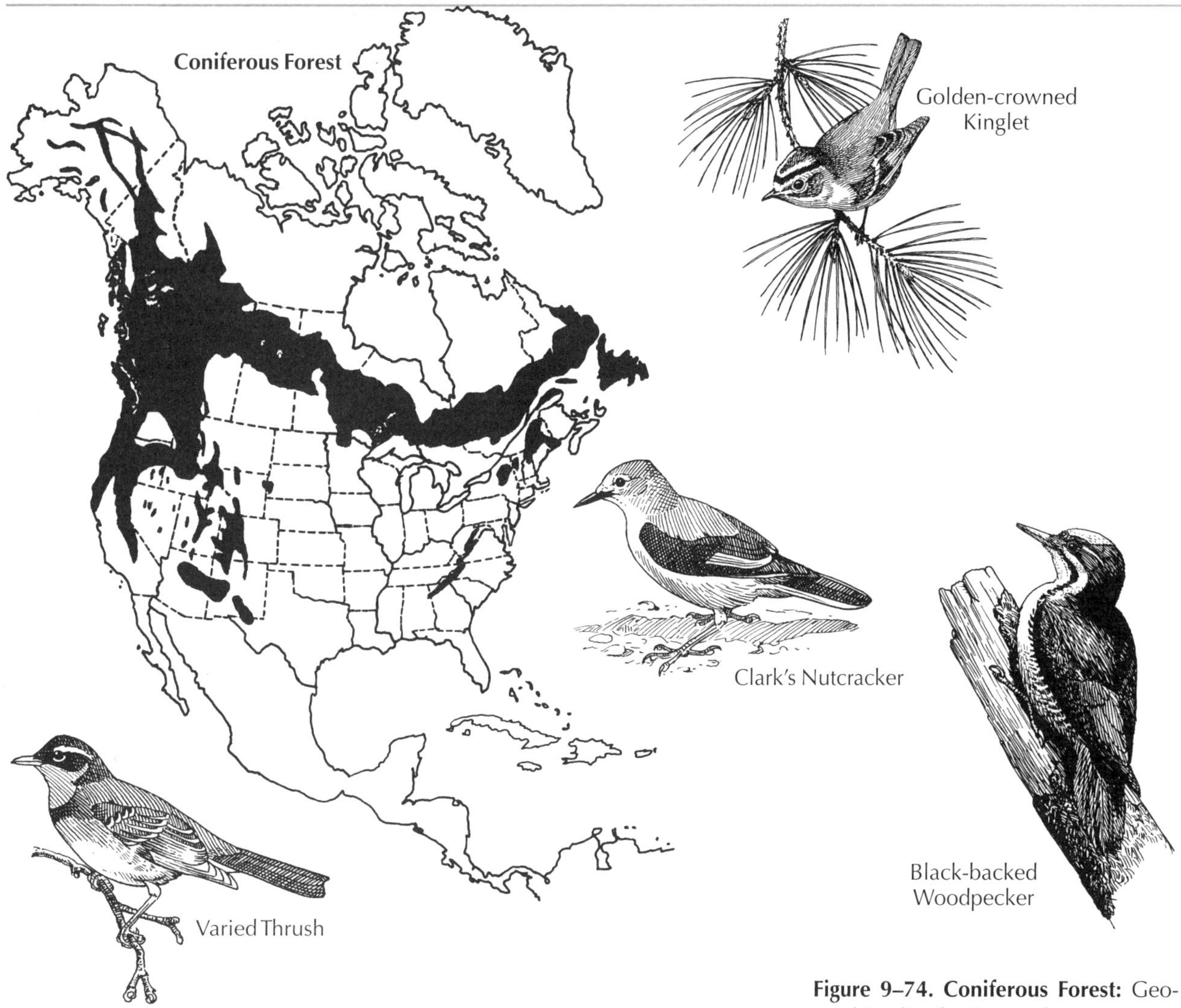

Coniferous Forest

Golden-crowned Kinglet

Clark's Nutcracker

Black-backed Woodpecker

Varied Thrush

Figure 9–74. Coniferous Forest: Geographic distribution and some representative species from the western montane forest, boreal forest, and eastern montane forest ecosystems. Drawing of Black-backed Woodpecker by William C. Dilger, Clark's Nutcracker and Varied Thrush by Charles Ripper, and Golden-crowned Kinglet by Orville Rice.

Varied Thrush (in the northwestern part), Townsend's Warbler, Hermit Warbler, and Cassin's Finch.

The eastern montane forest, except on the few high peaks containing alpine tundra, occupies the higher summits, ridges, and slopes of the Appalachians from northern New England south through Georgia and Alabama. The breeding populations of birds are small, and little altitudinal migration takes place. The few characteristic nesting species, all of which are also found in the boreal forest, include Northern Goshawk, Olive-sided Flycatcher, Red-breasted Nuthatch, Brown Creeper, Winter Wren, Hermit Thrush, Golden-crowned Kinglet, Magnolia Warbler, Blackburnian Warbler, Canada Warbler, Purple Finch, Pine Siskin, Red Crossbill, Dark-eyed Junco, and White-throated Sparrow.

Deciduous Forest

The dominant plants are broad-leaved trees, such as oaks, hickories, maples, and beech, growing close enough together to form a canopy overhead (**Fig. 9–75**). In general, the summer is warm and the winter is cool, with a marked seasonal change. Throughout most of the ecosystem, trees and shrubs shed their leaves in autumn and very little insect food is available for birds during winter.

Many of the typical birds feed, sing, or nest in trees, and have special adaptations for living in an enclosed canopy of trees and shrubs. Total breeding populations range from 150 to 300 males per 100 acres (40 ha). Some of the characteristic nesting species are Red-shouldered Hawk, Broad-winged Hawk, Whip-poor-will, Red-bellied Woodpecker, Hairy Woodpecker, Downy Woodpecker, Great Crested Flycatcher, Acadian Flycatcher, Eastern Wood-Pewee, Carolina Chickadee, Tufted Titmouse, White-breasted Nuthatch, Wood Thrush, Blue-gray Gnatcatcher, Yellow-throated Vireo, Red-eyed Vireo, Worm-eating Warbler, Cerulean Warbler, Kentucky Warbler, and Hooded Warbler.

Grassland

Perennial grasses and forbs, particularly members of the daisy and legume families, are the dominant vegetation. Eastward, where the rainfall is moderate and the terrain flattens, the ecosystem features

Figure 9–75. Deciduous Forest: Geographic distribution and some representative species from the deciduous forest ecosystem. Drawings of Wood Thrush and Red-eyed Vireo by Orville Rice; Red-shouldered Hawk and Hooded Warbler by Charles Ripper.

tallgrass prairie, including big bluestem, little bluestem, Indian grass, switch grass, and many others. Extensive fields of tallgrass prairie are interrupted by groves of trees and streamside woodlands of oaks and hickories. Westward on the higher, more arid Great Plains with typically undulating terrain, the ecosystem features shortgrass prairie chiefly composed of grama and buffalo grasses **(Fig. 9–76)**.

The birds have adapted to life in this open, treeless ecosystem. Many, such as the Bobolink, sing in flight because they have no elevated perches. They nest on the ground, usually under thick grass, for both concealment and shade. Except for some male passerines with conspicuous nuptial plumage, the backs of grassland birds are streaked and mottled to blend with the background. Total breeding densities are rather low, generally less than 100 males per 100 acres (40 ha). Although several species migrate south in winter, many migrants come in from the north so that the winter community is similar in size (but not in species composition) to the summer one. Some of the characteristic nesting species, although in no case uniformly distributed throughout the ecosystem, are Swainson's Hawk, Ferruginous Hawk, Greater Prairie-Chicken, Lesser Prairie-Chicken, Sharp-tailed Grouse, Long-billed Curlew, Burrowing Owl, Horned Lark, Sprague's Pipit, Western Meadowlark, Lark Bunting, Grasshopper Sparrow, Baird's Sparrow, McCown's Longspur, and Chestnut-collared Longspur.

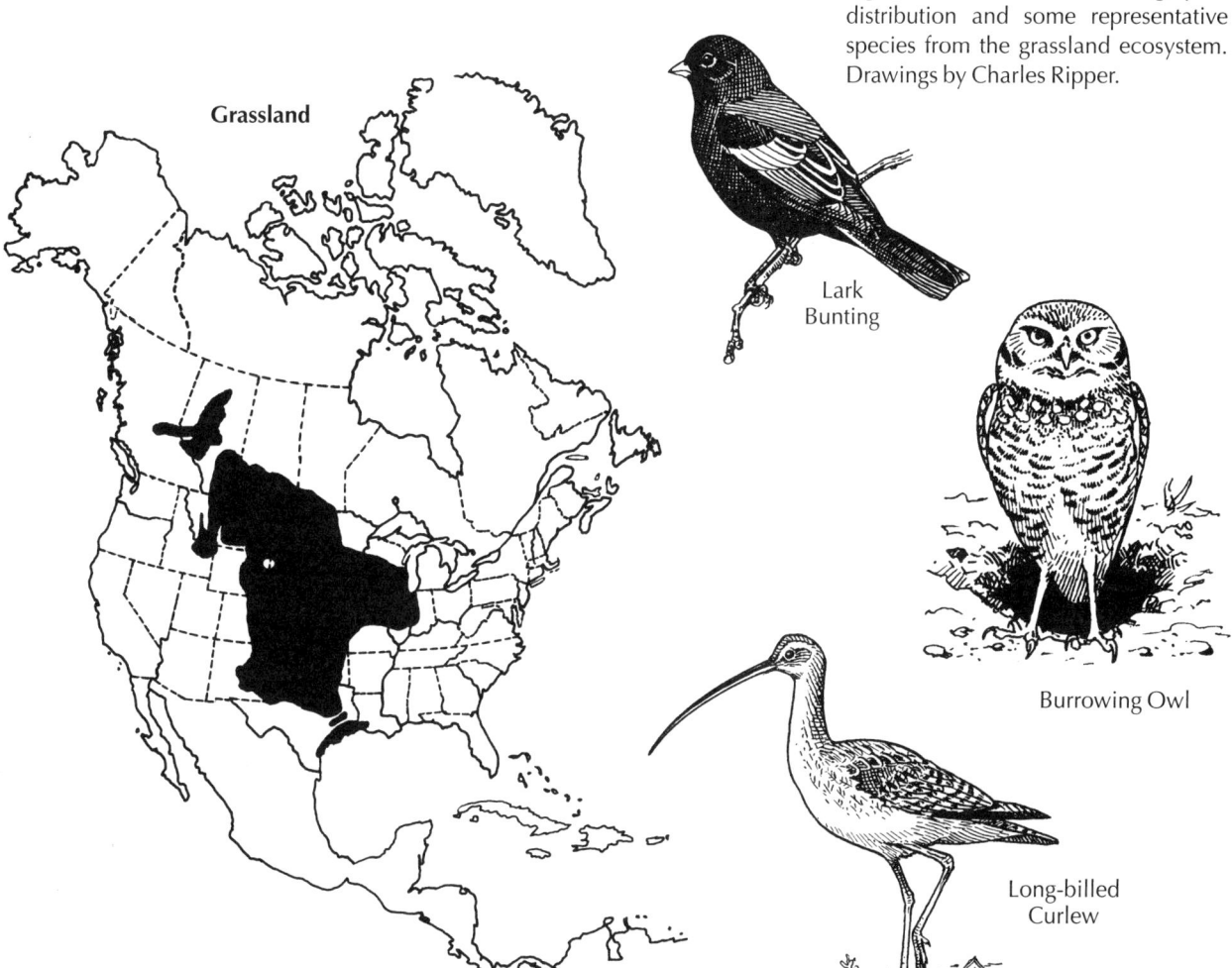

Figure 9–76. Grassland: Geographic distribution and some representative species from the grassland ecosystem. Drawings by Charles Ripper.

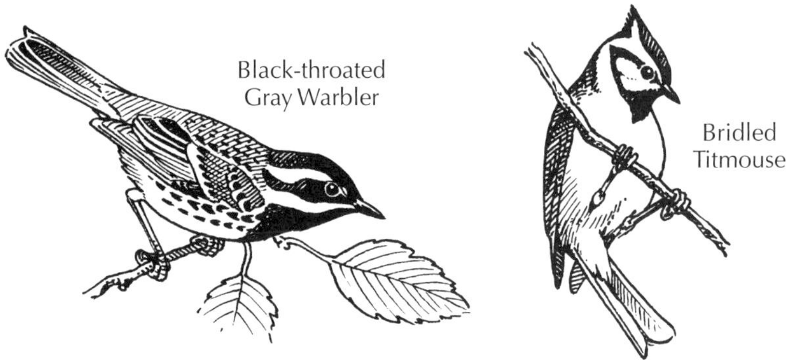

Black-throated
Gray Warbler

Bridled
Titmouse

Figure 9–77. Southwestern Oak Woodland: Representative species from the southwestern oak woodland ecosystem (no map). Drawings by Charles Ripper.

Southwestern Oak Woodland

This ecosystem is scattered on hills and mountain slopes in Utah, Nevada, California, New Mexico, Arizona, and parts of Colorado **(Fig. 9–77)**. Warm summers contrast sharply with cool winters. Rainfall is typically low to moderate. Southwestern oak woodlands are partially open woodlands composed mainly of oaks, but also with scattered, and sometimes very large, ponderosa pines. Some of the characteristic nesting species include Nuttall's Woodpecker, Strickland's Woodpecker, Bridled Titmouse, Hutton's Vireo, Virginia's Warbler, and Black-throated Gray Warbler.

Chaparral

Named for the dense stands of broad-leaved evergreen shrubs, dominated by chamise and manzanita, this ecosystem occupies mainly the low hillsides of southwestern California **(Fig. 9–78)**. Rains fall only in winter. During the summers, which are long, hot, and dry, the dry vegetation together with the accumulation of leaf litter encourages frequent fires. The density of breeding birds is moderate—about 200 males per 100 acres (40 ha). The characteristic nesting birds, in almost

Figure 9–78. Chaparral and Pinyon-Juniper Woodland: Geographic distributions and some representative species from the chaparral and pinyon-juniper woodland ecosystems. Drawings by Charles Ripper.

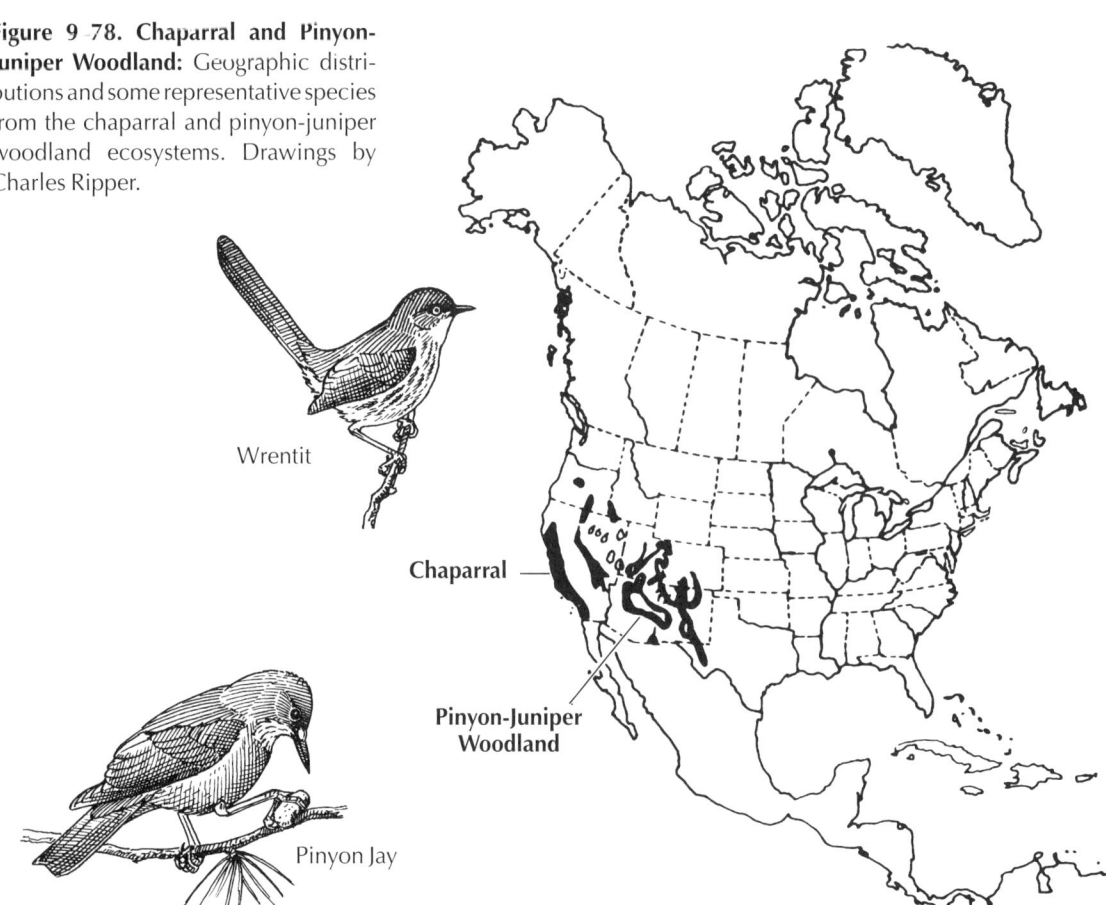

Wrentit

Chaparral

Pinyon-Juniper
Woodland

Pinyon Jay

all cases with long tails, short wings, and dull colors, include Wrentit, California Thrasher, and Gray Vireo.

Pinyon-Juniper Woodland

This ecosystem is clearly defined by two dominant small trees, pinyon pines and junipers, growing in open, park-like stands on hills and mountain slopes above scrub desert or grassland and below the montane forest (see Fig. 9–78). The summers are warm, the winters are cool, and the rainfall is low. Five nesting birds are particularly characteristic: Gray Flycatcher, Pinyon Jay, Juniper Titmouse, Bushtit, and Bewick's Wren.

Sagebrush

An ecosystem of "the cold desert" at elevations above the lower deserts and valley floors of the Great Basin, it has a relatively dry climate with slight rainfall the year round **(Fig. 9–79)**. The high summer temperatures contrast sharply with the low temperatures in the winter. Sagebrush is basically a shrubland consisting of thick sagebrush, shadscale, and other woody growth 2 to 5 feet (0.6 to 1.5 m) high. Populations of breeding birds are low, ranging from zero to somewhat more than 100 males per 100 acres (40 ha) in summer. The four characteristic nesting species are: Sage Grouse, Sage Thrasher, Sage Sparrow, and Brewer's Sparrow.

Figure 9–79. Sagebrush and Scrub Desert: Geographic distributions and some representative species from the sagebrush and scrub desert ecosystems. Drawings of Sage Grouse and Cactus Wren by Charles Ripper; Greater Roadrunner by Robert Gillmor.

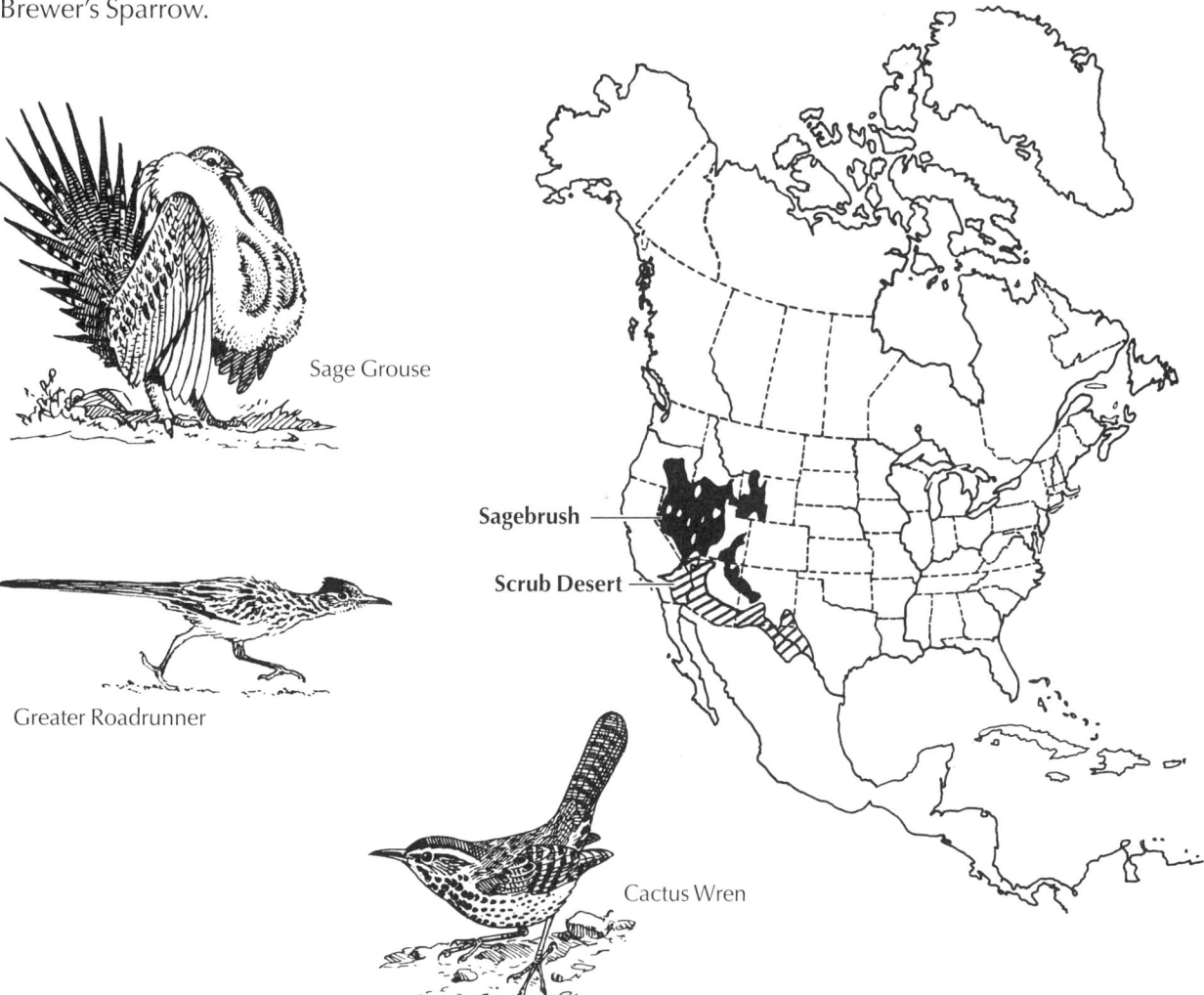

Sage Grouse

Greater Roadrunner

Sagebrush

Scrub Desert

Cactus Wren

Scrub Desert

An ecosystem of "the hot desert" on lowlands and valley floors, it has low rainfall, a high evaporation rate, and extremely high summer temperatures (see Fig. 9–79). Bare ground with widely spaced plants 3 to 6 feet (1 to 3 m) high, such as creosote bush, ocotillo, and bur yuccas, give a true desert appearance. Many of the birds synchronize their nesting with periods of rainfall and their activities with the cooler parts of the day. Characteristic nesting species include Gambel's Quail, Greater Roadrunner, Elf Owl, Lesser Nighthawk, Costa's Hummingbird, Gila Woodpecker, Vermilion Flycatcher, Verdin, Cactus Wren, Bendire's Thrasher, Le Conte's Thrasher, Crissal Thrasher, Black-tailed Gnatcatcher, and Phainopepla.

Two Important Ecotones

Where two major ecosystems overlap there is often a broad ecotone containing not only organisms characteristic of the adjacent ecosystems, but also organisms found almost entirely in the ecotone (see Fig. 9–63). Two particularly extensive ecotones are the arctic tundra/coniferous forest ecotone and the coniferous forest/deciduous forest ecotone **(Fig. 9–80)**.

Figure 9–80. Two Major North American Ecotones: Geographic distributions and some representative species from the arctic tundra/coniferous forest and coniferous forest/deciduous forest ecotones. Drawings by Charles Ripper.

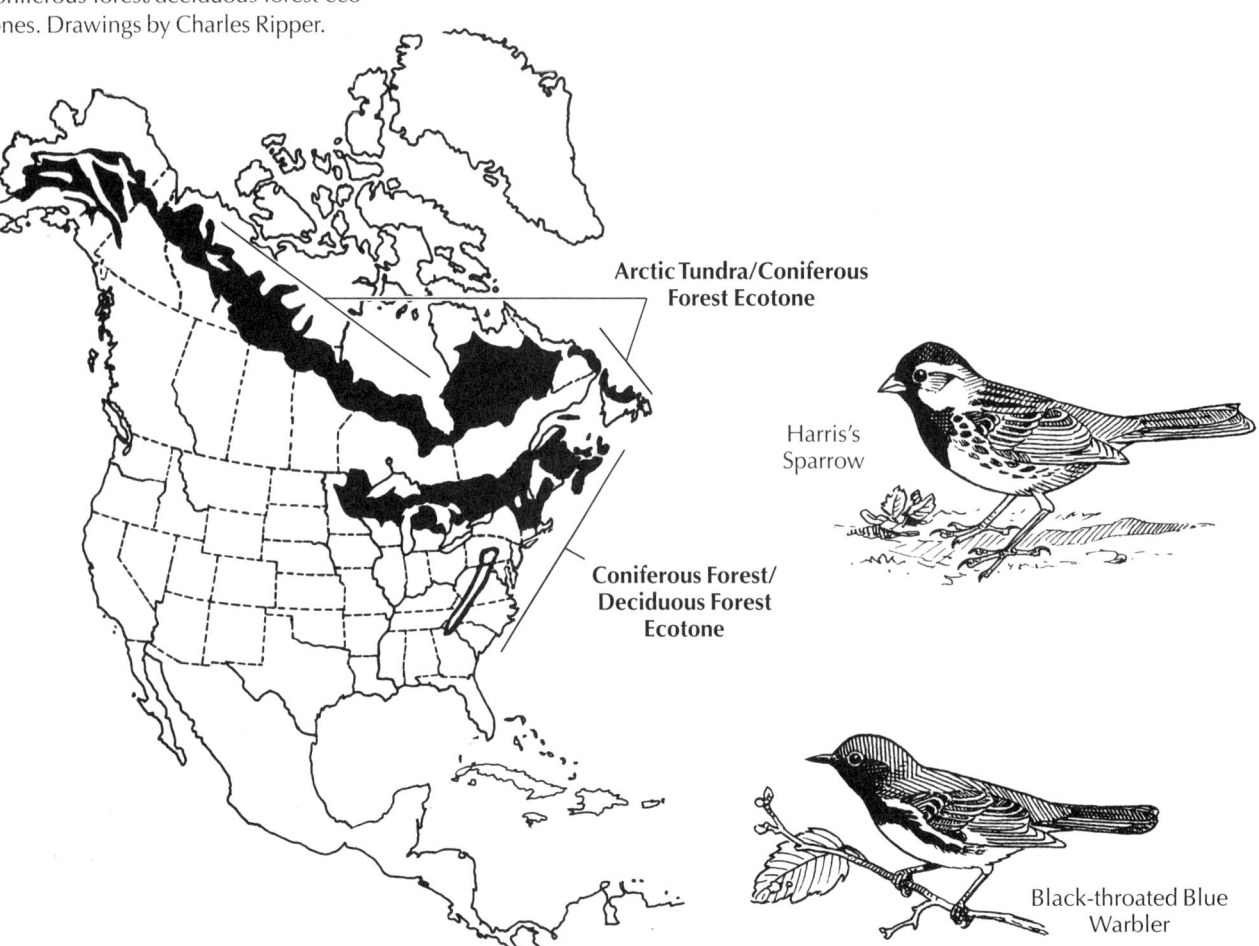

Arctic Tundra/Coniferous
Forest Ecotone

Harris's
Sparrow

Coniferous Forest/
Deciduous Forest
Ecotone

Black-throated Blue
Warbler

The **arctic tundra/coniferous forest ecotone** occurs where the arctic tundra and boreal forest meet and interdigitate. Often referred to as the northern timberline, it is here that fingers of open, stunted spruce forest extend north between fingers of low, shrubby tundra extending south. The Gray-cheeked Thrush, Northern Shrike, Blackpoll Warbler, Common Redpoll, American Tree Sparrow, and Harris's Sparrow are characteristic breeding species. None of them are found nesting regularly in either the uninterrupted tundra or the deep spruce forest.

The **coniferous forest/deciduous forest ecotone** occurs where the two great climax forest ecosystems overlap. Although the general appearance has been radically modified by deforestation, one may still find forests where conifers and deciduous trees naturally intermingle, providing niches suitable to such breeding species as the Blue-headed Vireo and the Black-throated Blue Warbler. Breeding Bird Survey results indicate that this region is the richest in bird species of any in North America.

The Role of Birds in the Food Chain

Understanding entire ecosystems is daunting, so ecologists approach the task by focusing on a few key processes, such as the cycling of nutrients and energy flow. These processes, and the ways in which birds take part in them, are discussed below.

The **trophic**, or "feeding," structure of an ecosystem is based on a **food chain**, the sequence in which organisms feed upon other organisms. A simple food chain in a field, for example, might start with the grass: grasshoppers eat grass, meadowlarks eat grasshoppers, and hawks eat meadowlarks. But even in the simplest ecosystems, the various food chains are so complex and interconnected that the trophic structure is best described as a food web. One can, however, simplify the web somewhat by grouping the organisms into categories based on their position in the food chain **(Fig. 9–81)**.

The ultimate source of energy for all animals, including birds, is the sun. Plants capture energy from the sun and introduce it into the food chain by using it to make foods that can be eaten by animals. Because of this role as food "makers," they are called **producers** . Animals that directly eat plants—herbivores—are called **primary consumers**. Animals that eat primary consumers are called **secondary consumers**, and those that eat secondary consumers are called **tertiary consumers**, and so on. All secondary and tertiary consumers are carnivores (meat-eaters). The bodies of all dead organisms, as well as the waste products of living ones, are consumed by **decomposers**—who break down these complex organic molecules into basic nutrients (such as oxygen, nitrogen, and phosphorus) and return them to the soil for plants to use once again. Many organisms have varied diets and thus occupy more than one position within a food chain. Those that eat both plants and animals are called **omnivores,** and they may act as both primary and secondary consumers.

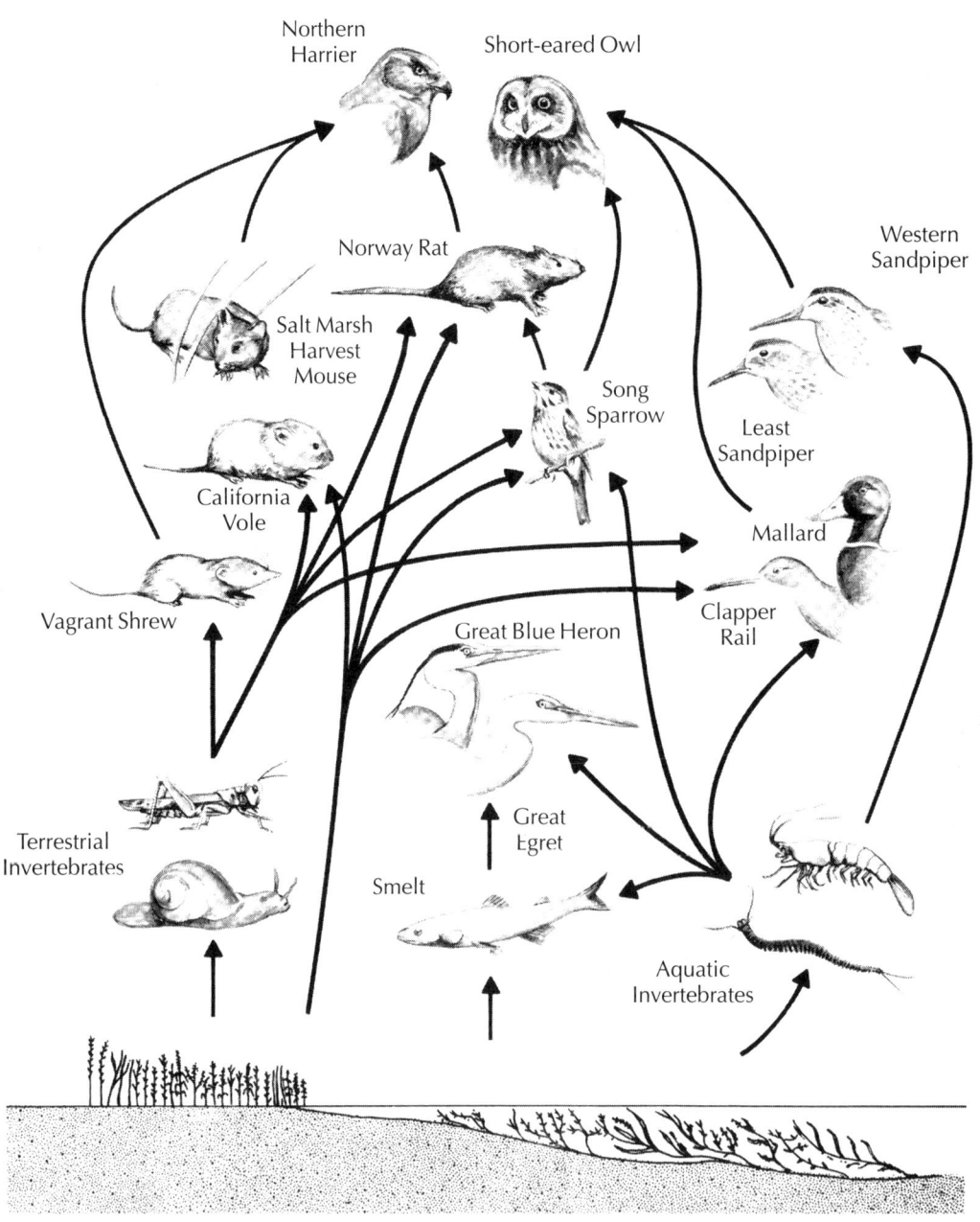

Figure 9–81. Salt Marsh Food Web: Nutrients and energy flow through ecosystems by means of food chains, which interconnect to form complex food webs. Shown here is a food web in a salt marsh in the San Francisco Bay area during winter. **Producers** are the plants, both terrestrial and aquatic, which use the sun's energy to manufacture their own foods. **Primary consumers** are herbivorous animals that eat plants, including terrestrial invertebrates (such as snails and grasshoppers), small fish (such as smelt), aquatic invertebrates (such as worms and crustaceans), and seed-eating birds. **Secondary consumers** are carnivorous animals that eat the primary consumers. Here they include small mammals such as shrews, voles, and rats; fish-eating birds such as the Great Blue Heron and the Great Egret (which also takes a few crustaceans and the occasional frog);

the Clapper Rail (which eats mostly aquatic invertebrates); and sandpipers and other shorebirds that probe in mud for worms and other small animals. Roles are not always discrete, as omnivores can be both primary and secondary consumers. The Song Sparrow, for instance, eats seeds as well as insects, and the Mallard feeds on aquatic plants as well as aquatic insects and other small animals. Topping the food chain are the **tertiary consumers**, also known as top predators, carnivorous animals that feed upon secondary consumers. Here the Northern Harrier, which hunts during the day, and the Short-eared Owl, active both day and night, eat various small mammals and birds. From Environmental Science 3/E, 3rd Edition, by J. Turk and A. Turk, © 1984. Reprinted with permission of Wadsworth, an imprint of the Wadsworth group, a division of Thomson Learning.

Birds occupy all levels of food chains above producer. Most birds that are primary consumers are granivores (seed-eaters), frugivores (fruit-eaters), or nectivores (nectar-feeders), but some birds feed directly on green plant parts such as leaves, shoots, buds, and stems. Fifty-nine bird families, including the Phasianidae (pheasants, turkeys, grouse, and their relatives) and Fringillidae (finches and Hawaiian honeycreepers), contain some species that are primarily granivorous. Seventy-four families, including the Pipridae (manakins) and Trogonidae (trogons), contain some species that are primarily frugivorous, and 16 families, including the Trochilidae (hummingbirds) and Nectariniidae (sunbirds), contain some species that are primarily nectivorous. Although birds from 28 families, including the Anatidae (ducks, geese, and swans), feed on green plant parts, this type of herbivory is less common among birds than among mammals.

Despite the many avian herbivores, most bird species are secondary or tertiary consumers. Bird species that primarily rely on insects and other invertebrates are found in most (175) of the bird families of the world; species whose major food is marine invertebrates come from 29 families; those that eat primarily terrestrial vertebrates, including other birds, come from 50 families; those whose major food is fish come from 41 families; and carrion-eating birds come from 11 families. Birds from 56 families are omnivores, feeding in two or more positions in the food chain.

Although birds occupy all food chain positions and are conspicuous components of many ecosystems, they are not particularly important links in the flow of energy within the environment. Several studies of energy flow through terrestrial food chains show that only about one percent of the energy produced by plants moves upward via birds. This total energy consumption is small in comparison with that of other less-conspicuous animals, such as insects and other invertebrates.

Nonetheless, birds are important conduits of energy flow in some ecosystems. In polar regions, endothermy has allowed birds to become more dominant members of the ecological community. And on remote islands, accessible to birds because they can fly, they have taken over some niches usually filled by other types of animals. In New Zealand, bizarre birds like the extinct moas and the surviving kiwis evolved traits more typical of mammals, which could not reach the islands to colonize. In such ecosystems, the importance of birds in energy flow is greater because they fill more links in the food chain.

Birds, like other animals at the higher levels of the food chain, may be affected seriously by certain toxic chemicals that accumulate in living organisms. The classic example is DDT, a common pesticide that was used in the United States from the mid-1940s to the early 1970s to control Mexican boll weevils, gypsy moths, mosquitoes, and other insect pests. DDT was applied to many of the world's ecosystems in enormous quantities, with disastrous consequences. Two characteristics of DDT allow it to affect animals in ways that no one anticipated originally: (1) it dissolves readily in fat tissue, and (2) it is stable and persists in the environment for many years, breaking down very slowly. When DDT is taken in by a plant or animal, it is quickly metabolized into the persistent toxic compound DDE (see Ch. 10, Chemical Toxins), which is deposited in the organism's fatty tissues, from which DDE is lost at a very slow rate—almost always much more slowly than the rate at which DDT is taken up from the environment. Thus, over an organism's lifetime, the chemical concentration of DDE (and other such toxins) tends to increase in its body, a phenomenon known as **bioaccumulation**. The chemical concentration of these toxins also increases in a step-wise fashion in animals at higher and higher levels in the food chain, a phenomenon termed **bioconcentration** (or **biological magnification**) **(Fig. 9–82)**.

When DDT was applied to environments, primary consumers, such as insects and other invertebrates, ingested small quantities of the pesticide as they fed on contaminated plants or drank contaminated water. Eventually the concentrations of DDE in their body fat rose to levels many times greater than the concentration of DDT in the environment around them.

At the same time, secondary consumers, such as insectivorous birds, fed on the contaminated insects and ingested the heavy doses of bioaccumulated DDE. Because these birds ate many insects over their lifetimes, they accumulated even higher concentrations of DDE in their fatty tissues.

When tertiary consumers such as bird-eating raptors ate the contaminated insectivores, the bioaccumulation of DDE in the prey resulted in dangerously high concentrations in the predators. In estuaries on Long Island, for example, DDE concentrations in the bodies of gulls, terns, and Ospreys became more than 100 times higher than concentrations in the phytoplankton and plants at the base of the local food chains. Such predators then suffered reproductive failures because the high DDE concentrations in their tissues caused them to lay thin-shelled eggs that failed to hatch. DDT is now banned in the United States, but continues to be used in many other countries throughout the world (see Chapter 10, especially Figs. 10–41, 10–42, and 10–43, for more information on DDT and its effects on birds).

What if Birds Disappeared?

Although in most ecosystems birds are not the most important organisms in terms of energy and nutrient flow, they nevertheless play important, often crucial, roles in the lives of other species. Consider the process of ecological succession, discussed earlier and in Sidebar 4.

Some birds play an important role in succession by dispersing plant seeds. After a forest fire, for instance, seed sources for plants that can live in disturbed areas may be too far away for the seeds to reach those areas without help. Birds, being highly mobile, can transport such seeds and deposit them far from the parent plants. In fact, in some ecosytems the ability of birds to disperse seeds has earned them the title of **keystone species**, that is, species that make extraordinary contributions to the ecosystem's functioning **(Sidebar 5: Sapsuckers, Swallows, Willows, Aspen, and Rot)**.

(Continued on p. 9·132)

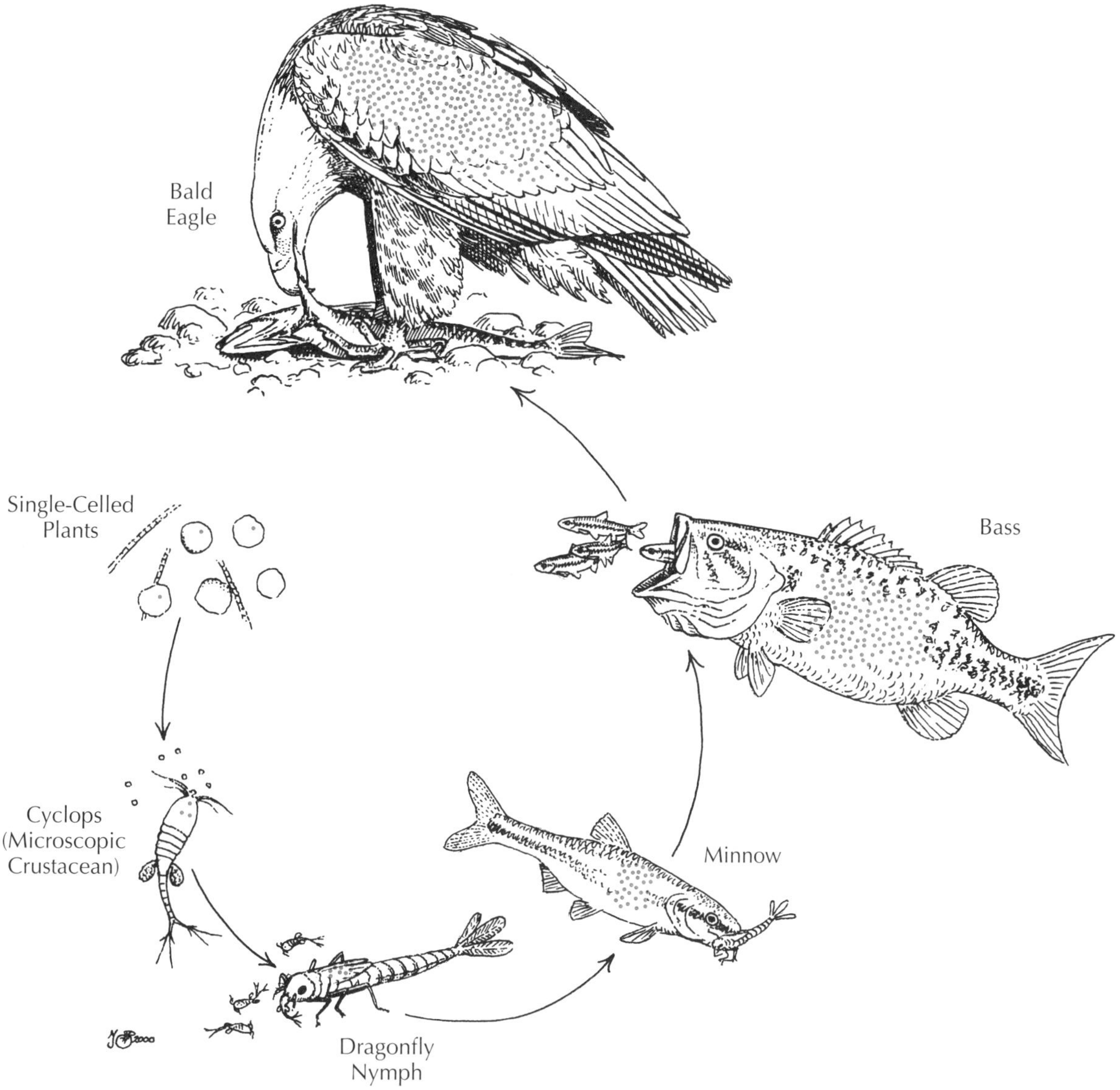

Figure 9–82. Bioaccumulation and Bioconcentration of Toxins: Toxic chemicals moving up a food chain may have drastic effects on birds and other animals. Of particular concern are toxins that are chemically stable and, once ingested, accumulate in body fat (a process termed **bioaccumulation**), where they become non-excretable. In this example, the process begins when single-celled plants absorb tiny amounts of toxin from their aquatic environment, as shown by the colored dots. The chemical does not break down and therefore accumulates in the tissues of Cyclops and other microscopic crustaceans that feed on the plants. Next a dragonfly nymph absorbs large quantities of the chemical when it consumes numerous crustaceans. As it moves up the food chain the toxin becomes more concentrated at each link (a process termed **bioconcentration**), because it takes many dragonfly nymphs to feed a minnow, many minnows to satisfy a predatory fish such as a bass, and many bass to feed a top predator such as a Bald Eagle. Furthermore, the long life span of top predators makes matters worse; an eagle may eat many contaminated fish over its lifetime, so persistent chemicals continue to accumulate in its tissues, often resulting in severe health problems or death. Through bioaccumulation and bioconcentration, a seemingly small amount of environmental pollution can become a heavy toxin load at the top of the chain.

Sidebar 5: SAPSUCKERS, SWALLOWS, WILLOWS, ASPEN, AND ROT

Paul R. Ehrlich and Gretchen C. Daily

At first it seemed nothing could be more fascinating about Red-naped Sapsuckers **(Fig. A)** than their habitual drilling of flowing wells in the bark of shrubby willows and the many other birds, mammals, and insects that crowd into the shrubbery to gorge on the sugary sap. Curiosity drove us to spend months in sapsucker habitat, crawling through and examining the willow thickets for miles around, enduring persistent clouds of mosquitoes and biting flies to film and take data on visitors of the wells, plotting endlessly to outsmart and capture wily sapsuckers and chipmunk visitors, and generally driving our friends crazy with daily sapsucker stories. Instead of satisfying our curiosity, this work further intensified our interest in the sapsucker as a double-duty keystone species and a major architect of willow/aspen groves.

A keystone species is one whose removal from the community would precipitate a reduction of species diversity or produce significant changes in the community structure. One of the most intriguing keystone mysteries was whether or not other species really depended upon the sapsuckers. Orange-crowned Warblers **(Fig. B)** were the most frequent visitors feeding at the sapsucker wells, followed by Rufous and Broad-tailed hummingbirds, chipmunks, and the occasional Red Squirrel. Vespid wasps—some yellow, others black—and a variety of flies rounded out the list.

All these sap-robbers may benefit substantially from exploiting the rich sap resource, supplied when many are breeding and then storing fat for the winter's migration or hibernation. Most of these species are

highly omnivorous, however, making it difficult to determine exactly how much their populations would suffer were the sapsucker, and hence the sap resource, to disappear.

This led us to delve into another feature of the sapsucker's lifestyle that appeared to benefit a second suite of species—the excavating of nest holes. In the vicinity of the Rocky Mountain Biological Laboratory in Gunnison County, Colorado, there are moist meadows largely surrounded by quaking aspen groves and scattered Engelmann

spruce. Approximately one-half the meadow surface is occupied by dense patches of three species of shrubby willow to about six to nine feet (1.8 to 2.7m) in height. A sapsucker pair drills a new nest hole each year in an aspen infected with heartwood fungus **(Fig. C)**. Of 36 active sapsucker nests in the area, only one was situated in an old hole (one that had been excavated the previous year). We tracked the occupancy of old holes and found seven different bird species raising their young in them: Tree Swallows, Violet-green

Figure A. Red-naped Sapsucker at Sap Well: Sapsuckers, such as the Red-naped Sapsucker male shown here, drill wells into the bark of trees and shrubs. The sequence of well construction starts with the sapsucker drilling a series of small, exploratory holes that just penetrate the bark. Holes that exude sap are enlarged into rectangular wells from which the sapsucker subsequently feeds. Sap wells sometimes reach four inches (10 cm) long by over half an inch (1.3 cm) wide and, as can be seen here, penetrate the entire thickness of the bark. As well as forming much of the sapsuckers' diet, the sugary sap is a source of nourishment for many other species of birds, mammals, and insects. Photo courtesy of Gretchen C. Daily and Paul R. Ehrlich.

Swallows, House Wrens, Mountain Bluebirds, Mountain Chickadees, Northern Flickers, and a Williamson's Sapsucker pair.

We found that Red-naped Sapsuckers create at least 10 times as many nest holes as any of the less common woodpeckers near the Rocky Mountain Biological Laboratory. Since a shortage of nest holes may limit the population sizes of species incapable of creating their own cavities, we suspected that the presence of sapsuckers might be vital to these secondary cavity nesters. How could we find out whether the sapsuckers were indeed crucial to the others?

Back in the 1950s, biologists would have simply shot all the sapsuckers near the Biological Laboratory to see whether the populations of other bird species would change as a consequence. Fortunately, the times have

changed, and such brute-force approaches are now rightly condemned. We sought an indirect method that, with luck, would give us the answer. Our strategy involved first identifying a critical feature of habitat required by the sapsuckers themselves. Then, we planned to compare the bird communities in habitat patches with and without that one feature, and thus with and without sapsuckers. This would allow us to infer the effect of removing the sapsucker.

Sensing that all of this would require a lot of work, we enlisted the enthusiastic help of Nick Haddad, then a Stanford honors student with experience in censusing birds. The three of us embarked on this project together during the summer of 1991 at the Rocky Mountain Biological Laboratory.

The most obvious candidate for a critical habitat feature for sup-

porting sapsuckers seemed to be the proximity of suitable willow shrubs (for drilling sap wells) and aspen (for nesting). Our working hypothesis was that the sapsucker would not occur in areas near the Biological Laboratory lacking in either willow or aspen. To test this, we surveyed over 13,000 aspen trees located at varying distances from willow shrubs for signs of sapsucker wells. Sapsuckers drill wells into aspen early in the breeding season, before the willows leaf out; the damage they cause remains distinctive for at least 10 years, providing an indication of habitat occupied by sapsuckers.

Indeed, we found that as many as 35 percent of the aspens in very close proximity (fewer than 50 feet, about 15 m) to willows bore sapsucker damage, whereas fewer than 5 percent that were far (more than 3,280 feet, or 1,000 m) from willows did. Not only was there much more damage on trees close to willows, there were also more nest cavities. In general, we couldn't attribute a nest cavity to any particular primary cavity nester. However, the high prevalence of wells drilled around the nest trees suggested that many were created by sapsuckers.

We also surveyed willow clumps situated close to and far from aspen for signs of damage. Here again, we only found sapsucker damage in willows close to aspen. Willows near large spruce stands or in open, treeless mountain meadows bore no damage at all. These surveys showed us that sapsuckers were only present in areas with both willow and aspen.

This provided an ideal way to test the importance of the sapsucker to other birds. We established census plots of about 13 acres (5.25 ha) in six aspen groves, three near willow (fewer than 65 feet, about 20 m) and three over 0.62 miles (1 km) away from the nearest willow shrub. Then, for the next six weeks, the three of us spent each early morning censusing

Figure B. Orange-crowned Warbler Visits a Sapsucker Well: Many animals other than sapsuckers visit sapsucker wells to feed. In Colorado, Orange-crowned Warblers, such as the immature or molting adult shown here, are the commonest avian visitors, which also include hummingbirds. Small mammals, such as chipmunks and squirrels, and insects also take advantage of the sap. Sapsuckers often are aggressive to these visitors, suggesting that they are competing for the sap resource. Although the non-sapsucker visitors certainly benefit from the availability of sap, many are omnivorous—Orange-crowned Warblers mostly eat insects, for instance—and it is unclear how much they would be affected were the sapsuckers, and the sap they provide, not present. Photo courtesy of Gretchen C. Daily and Paul R. Ehrlich.

the breeding birds in the plots.

Other than the proximity of the willow, the aspen groves were selected to be as similar as possible. We found out, for example, that it was possible to be bitten and sucked dry by voracious mosquitoes in a matter of minutes at all sites! It was certainly encouraging that each site had an insect population high enough to support a rich community of avian insectivores.

As we predicted, sapsuckers were only present in the three sites close to willow. Interestingly, we found both Tree and Violet-green swallows only in the three sites that had sapsuckers

(**Fig. D**). A statistical test showed that the chance of this swallow/sapsucker association being purely coincidental was vanishingly small. Tree Swallows virtually always nest in cavities, and although Violet-green Swallows are known to nest in cliffs, no such opportunity was available at any of our sites.

All of the other secondary cavity nesters were present in each of the six sites. We discovered that they were generally more common in the sapsucker sites, however. Their abundance in the non-sapsucker sites (far from willow) seemed to depend upon the availability of alternative

nesting locations. So, for example, we found many House Wrens in sites littered with fallen, rotting logs, a favorite non-cavity nesting location, and no House Wrens at all in a non-sapsucker site without fallen logs.

It thus seemed that the sapsuckers could be quite important in the persistence of secondary cavity-nesting birds. But how could we be sure that the absence of swallows and lower abundances of other secondary cavity nesters in non-sapsucker sites was not due to some other factor? Perhaps there happened to be less food in the non-sapsucker sites. The swarms of insects present at all sites made that possibility seem unlikely, but such anecdotal evidence is not very admissible in science. That's why we also censused species of insectivorous birds that were not secondary hole nesters, to see whether they too would be much more abundant in the plots with sapsuckers.

We found that open-nesting insectivorous birds occurred in roughly equal abundances on all sites. Most sites had five to six pairs of Dark-eyed Juncos, two to four pairs of American Robins, one or two pairs of Hermit Thrushes, three to six pairs of Warbling Vireos, and a couple of pairs of Yellow-rumped Warblers and Western Wood-Pewees. The Western Wood-Pewee forages aerially upon insects, like swallows do, further making it unlikely that the absence of swallows could be attributed to anything but the absence of nest holes.

This project led to three discoveries about the biological communities around Rocky Mountain Biological Laboratory. First, we found that swallows, and to a lesser extent the other secondary cavity nesters, depend upon the co-occurrence of at least four elements of what we have called a keystone species complex: the Red-naped Sapsucker, aspen trees, certain willow species (in which the sapsuckers can drill wells), and the heartwood fungus. The disappearance of any one element

Figure C. Red-naped Sapsucker at Nest Hole: Each spring a Red-naped Sapsucker pair excavates a new nest cavity, usually in an aspen tree whose wood is rendered easier to chisel because it is infected with heartwood fungus. Common in Colorado, Red-naped Sapsuckers create 10 times as many nest cavities as any of the rarer woodpeckers of the region. They play a keystone role in the community by providing vital nest sites for cavity-nesting birds that cannot excavate their own holes. Photo by Tom J. Ulrich.

Figure D. Secondary Cavity Nesters that Rely on Sapsuckers: A Tree Swallow (left) and a Violet-green Swallow (female, right) cling outside their nest holes, each in an aspen trunk. Both species rely on pre-existing cavities in which to breed. During Ehrlich and Daily's study in Colorado, the two species of swallows were present only at sites that had sapsuckers, further evidence of the sapsuckers' keystone role in the community. Other secondary cavity nesters, such as Mountain Bluebirds and House Wrens, also were more common at sites with sapsuckers. Photos by Marie Read.

could result in the local extinction of the swallows and declines in the populations of the other secondary cavity nesters. Second, the sapsucker has the unusual characteristic of playing two distinct keystone roles: enhancing the persistence of both sap-robbers and cavity nesters. Third, the sapsuckers modify the forest/meadow community by changing the survival and dimensions of the willow clumps and, perhaps, the aspens. The sap wells eventually cause parts of each willow to die and may restructure the willow clumps more

than browsing by mammals. Similar wells drilled in aspen as well as the nest holes chiseled into the aspen trunks may alter their survival and patchiness.

Finally, although the tropics have classically been thought of as supporting species with complex, indirect, and subtle interrelationships, this work suggests that such interdependencies may be common in the temperate zone as well. Saving a species may therefore depend upon the persistence of another species with which it has no obvious interaction.

Or, put another way, the already blinding rate of extinctions may accelerate even more because of the domino effect, in which seemingly independent species all disappear at once from disturbed habitat. Despite the somewhat disturbing conclusions, this work has inspired us to delve deeper into sapsucker biology, seeking solutions to other sapsucker mysteries. ∎

Various versions of this article appeared in American Birds, Spring 1993, and in Whole Earth, summer 1998. Adapted with permission of Paul R. Ehrlich.

9

Clark's Nutcrackers, scrub-jays, Steller's Jays, and Pinyon Jays also play an important role in seed dispersal, by helping pines to colonize openings within coniferous forests in western North America. These corvids disperse pine seeds by carrying them in their bills and crops, and then burying them in "caches" for retrieval during winter. Their effect on the forest may be quite significant: one flock of about 250 Pinyon Jays in New Mexico cached an estimated 30,000 pine seeds per day! Although many of the seeds are recovered, some are forgotten, and because they are buried, they often germinate. The birds' role in seed planting becomes evident when some of the seeds sprout: white-bark pines, the seeds of which the birds often cache in open meadows, frequently grow in tight clusters—the result of seeds in forgotten caches germinating together. Because of their importance as seed dispersers, nutcrackers are keystone species in the western montane forest ecosystem **(Fig. 9–83)**.

Figure 9–83. Clark's Nutcracker: The Clark's Nutcracker lives in the mountainous North American west where, to survive the winter and feed its nestlings in spring, it buries thousands of pine seeds during late summer and fall, usually two seeds per hole. Eight or nine months later, the bird can still find many of its caches, which it apparently recognizes by surrounding landmarks (see Figs. 6–12 and 6–13). The occasionally forgotten caches of pine seeds help western forests regenerate after fire and other disturbances, making Clark's Nutcracker an important keystone species in its ecosystem. Photo courtesy of Mary Clay/CLO.

Many frugivorous birds swallow fruits whole, and the enclosed seeds are dispersed when the bird regurgitates them or passes them in its feces. Having the seeds carried away from the parent plant is important to all plants, because it reduces competition between the seedlings and parent, but it is particularly important to plants that tend to colonize new areas (sometimes called "early successional species"). One genus of plants that colonizes disturbed sites in Neotropical forests is Heliconia (see Fig. 9–22). Its fruits are eaten by many birds, and any seeds that happen to end up in open areas having the appropriate conditions for growth may germinate and begin a new population of Heliconia. When the Heliconias near my house in Trinidad produced fruits—which happened almost continuously throughout the year—they attracted 21 bird species! These birds must disperse the seeds extensively, because I have rarely seen a disturbed site in a tropical forest that was not promptly colonized by Heliconias, even when the site was far from the nearest seed source. The pioneering, sun-loving, Heliconias are gradually replaced by other plants as succession proceeds and the forest canopy closes in.

In addition to dispersing Heliconia seeds, birds also are the primary pollinators of their flowers. Recall that when hummingbirds, especially hermits, feed on nectar in the specially modified flowers, pollen is deposited on their heads (see Fig. 9–5). Then, as the birds move from flower to flower, they cross-pollinate the plants. The birds are very important to Heliconias, which in turn are important plants in the tropical forest ecosystem.

When a species plays an important role in the life of another species, the relationship between the two is referred to as **symbiosis**. Recall that when a symbiotic relationship is beneficial to both interacting species it is referred to as **mutualism**, as in the case of a frugivorous bird obtaining food and its food source getting its seeds dispersed. Sometimes mutualistic relationships become

obligatory, so that the interacting species not only benefit from each other, but they actually become dependent upon each other. Obligatory mutualism is illustrated by the relationships between hummingbirds and Heliconias.

I discovered another example of obligatory bird-plant mutualism, one that demonstrates how catastrophic the loss of a mutualistic partner can be, while working with endangered species on the island of Mauritius in the Indian Ocean. The plant species, a tropical forest tree known locally as the tambalocoque, was in trouble because few if any seeds were germinating, and there were not enough young trees to replace old ones that died. It appeared that this problem had been occurring for several hundred years. The trees produced crops of healthy-looking seeds, but they did not germinate, even when planted in nurseries where they received special attention.

I accidentally got an opportunity to study the large, hard seeds of the tambalocoque one day when I was watching an endangered Mauritius Parakeet feeding on the tree's tasty fruits—it was discarding the seeds, which fell to the ground. While examining the seed it occurred to me that this was an extraordinary seed with a very thick, hard seed coat, and an explanation for the tree's poor germination came to me.

Three hundred years ago there was a frugivorous bird on Mauritius large enough to swallow these seeds. The now-extinct Dodo may have not only swallowed these seeds, but also passed them through its digestive tract, where they would have received a thorough scouring in its powerful gizzard. To protect its seeds from being ground up, the tambalocoque may have evolved a seed coat thick enough to pass through the Dodo unharmed but heavily abraded. Taken to an extreme, the seed coat may have become so thick that it could not germinate without first being eaten and abraded by the Dodo. When the Dodo went extinct, the tambalocoque lost its obligatory partner and may not have been able to reproduce at an adequate rate.

I tested this hypothesis by force feeding some tambalocoque seeds to a reluctant domestic turkey, the best available surrogate for a Dodo. Some of the seeds that passed through the turkey's digestive tract were heavily abraded, and when I planted them, 30 percent germinated. It appears likely that the tambalocoque was disappearing for want of a Dodo. Because its mutualistic partner went extinct, the tambalocoque was itself headed for extinction. Fortunately, the seeds are easy to germinate if they are mechanically abraded, so the tree is safe **(Fig. 9–84)**.

Avian ecologists don't fully understand all of the crucial links that birds have with other species, but we know enough to predict that if there were no birds, the ecosystems of the world would be very different places than they are today. Many species, such as Heliconias and the tambalocoque, that depend directly on birds, would disappear or become much less common. The losses of birds and the species that depended directly on them would set off chain reactions of extinctions among other species indirectly linked to birds throughout the world. Other species that competed with birds or were preyed upon by them would become more abundant. Without birds, the ecosystems of the world would go through a prolonged period of readjustment, just as they did when dinosaurs became extinct 65 million years ago. My own guess is that evolution would eventually "reinvent" organisms much like birds to fill the ecological niches so effectively occupied by today's widespread, diverse, and adaptable avian species.

Figure 9–84. The Dodo: The flightless Dodo once inhabited the island of Mauritius in the Indian Ocean, until overhunting by visiting sailors and settlers, and destruction of the bird's eggs and young by introduced animals, such as pigs and rats, brought about its demise in 1681. Its extinction, although itself a tragedy, also has had severe consequences for the survival of the tree Calvaria major, known locally as tambalocoque, or "Dodo tree," demonstrating how disastrous the loss of one member of a mutualistic relationship can be. In the 1970s, biologists noticed that the tambalocoque was disappearing rapidly. The few remaining individuals produced regular crops of seeds, but very few germinated—a situation that apparently had been occurring for several hundred years. Because the frugivorous Dodo was large enough to have swallowed the tree's large-seeded fruits, one theory suggests that originally the thick-coated tambalocoque seeds needed to be scoured by the Dodo's digestive tract before they could germinate. When the Dodo disappeared, the tree was unable to reproduce successfully. Current conservation efforts use mechanical abrasion to help the seeds germinate. Painting of Dodo from Gleanings of Natural History, by George Edward, London (1758). Courtesy of the History of Science Collections, Cornell University Libraries.

Bird Conservation

John W. Fitzpatrick

 Imagine yourself an American pioneer on a spring morning in the woods of the southern Appalachian Mountains. Entering a clearing, you are startled by a great "whoosh" overhead. Although you've seen this spectacle all your life, you still marvel as a thousand pinkish, long-tailed doves lift up together on powerful wings. Swirling in a dark cloud over the clearing, they trail off to the north. Strangely, this flock does not disappear as it moves northward. Instead, its numbers swell. Suddenly the whole clearing becomes dark as the burgeoning flock blackens the sky overhead. You've never seen this many birds before. Out of curiosity you vow to wait out the flight, so you sit on a log to watch the spectacle. The flock is in steady motion now. The constant sound of thousands of wings whipping overhead resembles a strong wind through pine boughs. Minutes turn to hours, yet the immense flock continues to blot out the sky **(Fig. 10–1)**.

Finally, late for lunch back at your cabin, you point your flintlock upward and fire a load of shot. Eleven plump pigeons hit the ground like stones. Gathering your meal, you head home as the immense flock continues to hurtle northward. They continue all day. Straggler flocks will follow every day for the next week.

The Passenger Pigeon is almost certainly the most abundant bird ever to have existed on earth **(Fig. 10–2)**. Explorers' accounts of its numbers across the eastern United States between 1630 and 1880 read to us like science fiction (Schorger 1955). Audubon wrote of a flock passing for three successive days near Louisville, Kentucky in 1813, stating that "the light of the noonday sun was obscured as by an eclipse." He estimated that this mile-wide flock contained a minimum of 1.1 billion birds. Alexander Wilson, often called the father of American ornithology, considered his own 1806 estimate of a single Kentucky flock—2,230,272,000 birds—to be well below its actual number. How many pigeons existed across the entire range of the species? Nobody knows for sure, but the number may have been 5 billion or more. For perspective, no species of bird alive in the world today even approaches 1 billion, and crude estimates for our most common North American birds do not even come close (for instance , 160 million Red-winged Blackbirds, 80 million Mourning Doves, 62 million American Robins, 11 million Mallards). One thing we do know: as of 1914, just 100 years after Audubon saw his massive flock, the Passenger Pigeon was extinct.

Reading accounts of the Passenger Pigeon spectacle, many written by persons otherwise disinterested in birds, provides a deeply moving glimpse at how different North America is today from what it was like just a few generations ago. The scale and pace of change that humans brought to the world's landscapes and bird populations, especially during the 1800s, are nothing short of astonishing. Many accounts of Passenger Pigeon nesting colonies dwell as much on their frenzied slaughter by greedy hunters as on the natural wonder of the birds themselves (see Ch. 6, Sidebar 4, Fig. I). Even at the last large

Figure 10–1. Passenger Pigeons: When European colonists arrived in eastern North America, vast flocks of Passenger Pigeons—once the most abundant birds on earth—were commonplace. In 1813, John James Audubon observed a flock that he estimated to be one mile (1.6 km) wide and to contain at least one billion individuals. The species ate a specialized diet of acorns, beechnuts, and chestnuts. These nuts—collectively known as "mast"—are produced unpredictably in response to environmental conditions still not well understood. In some years all the trees in a local area produce large quantities of mast; in other years they produce almost none. Huge flocks of Passenger Pigeons, like the one depicted here, roamed widely over the vast eastern forests to take advantage of this rich but locally patchy food source. When they located a mast crop, the birds formed enormous roosts (during winter in the south) or nesting colonies (during late summer in the north). These huge gatherings shifted location from year to year with the mast.

Cornell Laboratory of Ornithology

colony ever documented—in 1878 near Petoskey, Michigan—300 tons of Passenger Pigeons were killed by professional market hunters and shipped by railroad to restaurateurs in Chicago, New York, and Boston.

There is no doubt why the Passenger Pigeon disappeared. Like so many species before it around the world, this gregarious dove was wiped off the earth by a deadly "one-two-three punch" that became the signature of human expansion: (1) uncontrolled exploitation for food, (2) steady advances in technology to aid exploitation as numbers decreased (new methods for trapping, storing, and shipping the carcasses), and (3) large-scale alteration of the landscape for agriculture and human residence (wholesale clearing of the virgin oak and beech forests on which pigeons depended). Last-ditch efforts to avert the Passenger Pigeon's extinction in the 1890s through publicity and legal protection came far too late to make a difference **(Fig. 10–3)**. The world's last individual Passenger Pigeon, known as "Martha," died in the Cincinnati Zoo in 1914. (She remains as a mounted specimen, for us to mourn and reflect upon, at the U. S. National Museum in Washington, D.C.)

Although the conservation efforts begun in the 1890s failed to save the Passenger Pigeon, they were not entirely in vain. Dramatic changes in human values took root at the end of the 19th century in North America. Foremost, perhaps, was the movement to appreciate and protect the earth's natural heritage. This theme would emerge through the 20th century as one of the great new themes in modern culture. Spawned in part by the unique capacity of birds to capture our hearts and passions, the conservation movement would lead the way to some of the most revolutionary actions, organizations, and laws in human history.

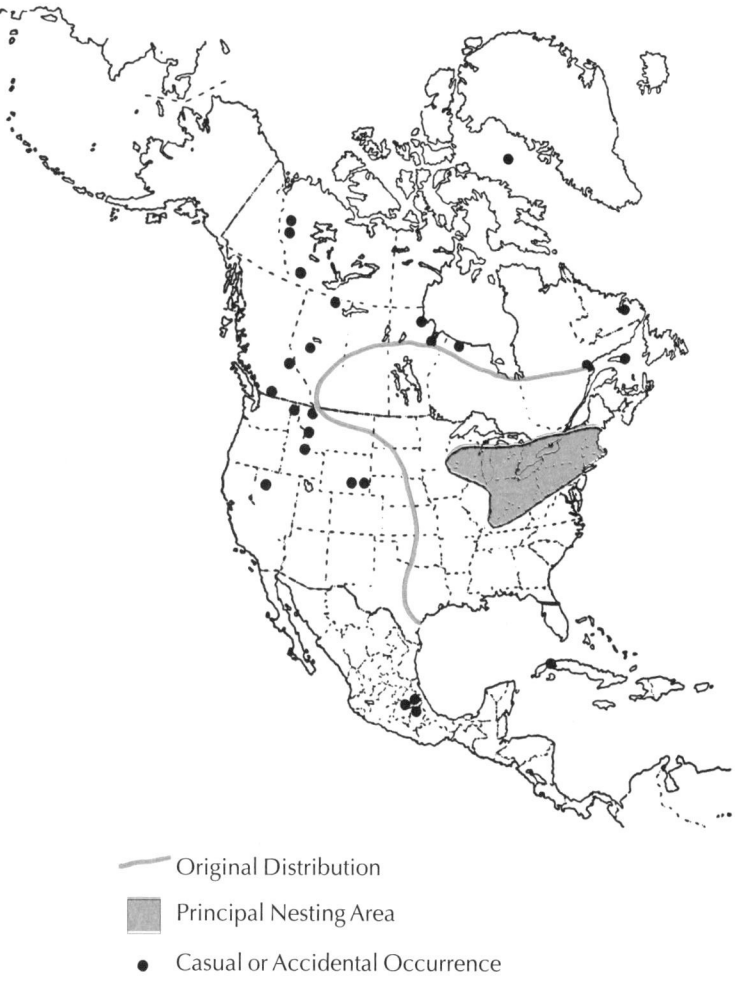

— Original Distribution

▨ Principal Nesting Area

● Casual or Accidental Occurrence

Figure 10–2. Original Distribution of the Passenger Pigeon: The Passenger Pigeon's original distribution in eastern North America (boundary shown by the colored line) mirrors the extent of the deciduous forests that blanketed the continent's eastern side before European settlement. The colored area indicates the principal nesting area, and black dots represent areas of casual or accidental occurrence. Adapted from Schorger (1973, p. 257).

Figure 10–3. Passenger Pigeon Museum Specimen: The Passenger Pigeon's demise resulted from uncontrolled shooting and extensive habitat destruction. Attempts to save the species in the 1890s came too late, and the last individual died in captivity in 1914. Only museum specimens remain, such as this one at the National Academy of Sciences in Philadelphia. Photo by Steven Holt/ VIREO.

In this final chapter, I first explore how and why bird populations declined with the spread of human cultures around the globe, and briefly review the history of bird conservation in the United States. Then I examine the most important causes of population declines among birds; describe the most important methods now used to halt population declines; discuss some reasons why we should care about protecting birds; and outline a few specific steps that individual citizens can take in bird conservation. The chapter emphasizes events in North America from 1850 onward, but also includes a few stories from around the world. Throughout, I stress the profound ways in which humans have changed the ecological settings in which birds originally evolved.

Bird conservation is an accumulation of efforts to reverse population declines and to keep common birds common. These efforts require that we understand ecological balances, deduce how they have been upset, and project how they can be restored. Understanding a population decline requires knowing basic facts about the population's evolutionary history, habitat specialization, and population regulation. As will become clear, declines result from "changes in the rules" affecting either birth rates or death rates. Over the past 2,000 years, most rule changes were created by humans modifying the landscape, thereby altering conditions under which birds evolved. In this sense, then, most concepts discussed in this chapter represent real-world applications of the biological and ecological principles discussed in Chapters 1 through 9.

Historical Context

Birds have been evolving and going extinct since they emerged from the early archosaurian reptiles almost 200 million years ago. Both the origin and the disappearance of bird species have been driven by naturally occurring, long-term processes across the earth's surface. Continents have split apart and slammed together. Volcanic islands have arisen from the sea and then eroded away. Earth's climate has fluctuated from hot to cold and wet to dry. These big events have made the extinction of species very much a natural feature of evolutionary change, as local environments shifted and birds of different regions encountered and competed with one another. In fact, 99 percent of all species that ever lived on earth are now estimated to be extinct. Why, then, should we be concerned about bird species becoming extinct today?

The answer is simple: the wave of mass extinction currently under way is occurring at a faster pace and a larger scale than any other such episode in the earth's history. Furthermore, this catastrophic change is caused by the sudden spread of humans across the globe. Before humans changed the rules, a typical animal species had an average "life span" of about one to two million years. However, the massive worldwide spread of modern Homo sapiens has directly caused the "total loss of 8,000 species or indigenous populations of land birds (including an estimated 2,000 species of flightless rails)," according to recent estimates based on fossil birds (Martin and Steadman 1999). This means that as many as 50 percent of earth's living bird species became extinct within just a few thousand years, which, geologically speaking, is the blink of an eye! Most biologists agree that the current rate of human-caused extinction is at least as great as, and perhaps greater than, the extinction rate during the colossal episode that caused dinosaurs to disappear about 65 million years ago. Such sudden and enormous acceleration of extinction cannot be considered "natural" (except by those who argue that humans represent just another species, and thus eliminating other species for our own use and convenience is perfectly natural).

Fortunately, humans represent a species that can recognize and describe its own mistakes. Unfortunately, not until late in the 20th century did we comprehend both the colossal scale and the long-term implications of the impact we've had on the biological diversity of our planet.

Global Spread of Humans Begins the Extinction Era

Modern humans began actively colonizing the world's continents and islands as long as 50,000 years ago, but until recently we did so in tiny numbers and with relatively little impact. Only in the last few thousand years did we begin modifying places to suit our own needs, but once we started, our pace never faltered. We cleared forests, replaced grasslands with crops, built towns, drained marshes, filled swamplands, irrigated dry areas, suppressed natural fire in some places, and increased fire frequency in others. Everywhere we went, we hunted birds for food, clothing, and ornamentation. In each newly colonized area we encountered birds that had evolved no defenses against our ever-more-advanced hunting techniques and our steadily increasing numbers. On remote islands filled with uniquely adapted species, we introduced new diseases (such as malaria and avian pox) and new predators (for instance cats, rats, and mongooses). On the mainland continents we did the reverse, systematically destroying the largest predators—wolves, bears, and big cats—thereby releasing a host of herbivores, such as white-tailed deer, to overpopulate and destroy native forest understories. To islands and mainlands all over the world we brought domesticated, plant-eating animals (cattle, pigs, goats, horses, and rabbits) and released them in habitats that never had experienced the effects of intensive browsing or grazing **(Fig. 10–4)**. Of course, we made all these changes to create places more accommodating for ourselves. In the process, however, we profoundly altered all the natural systems around us. The effects on birds were devastating.

Humans did not begin to document the world's birds with careful records and preserved specimens until about 1600 A.D. By that time, tropical islands had been colonized and ravaged the world over, resulting in the disappearance of literally thousands of bird species.

Figure 10–4. Goat Browsing: A Saanen goat stands on its hind legs to reach foliage in Tasmania. Introduction of domesticated, herbivorous animals has devastated native foliage in many places throughout the world, profoundly altering the habitat in which local birds evolved, and often leading to their decline. Photo by Antony B. Joyce/Photo Researchers.

Many of these species we may never encounter even as fossils, so we can only guess the extent of our impact based on estimates from places where good fossil evidence exists. Bones and fossils found on various islands of the Pacific Ocean led paleontologist David Steadman (1995) to conclude that prior to recorded history, humans caused the extinction of thousands of bird species in the Pacific alone.

From 1600 A.D. onward, scientists accumulated an increasingly detailed record of the global decline of birds. Today, the **International Union for the Conservation of Nature** (IUCN) identifies 131 bird species that have gone extinct during recorded times. Almost all of these species were confined to islands, and many are known only from a few bones or from descriptions by early explorers. Many species that had colonized predator-free islands had lost the need for flight, resulting in the evolutionary reduction of their wings (see Ch. 5, Loss of Flight). Such flightless creatures included some of the most unusual and spectacular birds ever to have evolved, such as the 1,000-lb (400-kg) elephantbird on Madagascar (the largest known bird; see Fig. 5–48), two dozen species of moas in New Zealand, flightless ibises on Hawaii, flightless owls in the Caribbean, and flightless pigeons in the Indian Ocean.

Oceanic islands were stopping points for human explorers, and the many flightless island birds became easy prey for humans in need of fresh food and water. Across the North Atlantic, sailors feasted on the abundant Great Auk **(Fig. 10–5)** and fishermen even used it for bait. The last one was killed in 1844, and only a few specimens exist today. On Mauritius Island in the Indian Ocean, sailors drove herds of clumsy, flightless Dodos (see Fig. 9–84) aboard their ships to serve as food for many weeks at sea. The species disappeared in the mid-1700s without a single whole specimen being preserved for study. The Dodo's close relative, the Reunion Solitaire, met a similar fate on the island of Reunion. The last of New Zealand's moas were killed during the 1700s, when Passenger Pigeons still flew across North America by the billions. There, colonial settlers were setting the stage for the continental extinctions to follow.

Early Extinctions in North America and the Caribbean

Prior to the arrival of European explorers in 1492, Native American descendants of the first human colonists from Siberia already had populated both North and South America, including the Caribbean Islands. These earliest American people had spread rapidly throughout the Americas around 10,000 years ago, and they represented a wholly new ecological force. Prior to their arrival, life in the Americas had been evolving free from human contact for hundreds of millions of years. Suddenly, just as humans began spreading across the landscape, there occurred a spectacularly swift and catastrophic disappearance of animals. Most dramatic was the nearly complete extinction of the New World "megafauna," a varied assemblage of large mammals that included horses, camels, giant sloths, mammoths and mastodons, glyptodonts, and many others. Exactly why

Figure 10–5. Extinction of the Great Auk: Auk bones in the prehistoric middens and caves of northwestern Europe indicate that this flightless, penguin-like bird was a source of food long before seafarers began exploiting the Grand Banks fishery and the birds that used its rocky coastline for breeding. In the 17th and 18th centuries, the largest colony of Great Auks known to fishermen was on Funk Island, off the Newfoundland coast. There, auks were butchered and their flesh salted down for storage; the fat was rendered for lamp oil. Eggs were taken, too, and live birds were even herded on board to supply sailors with fresh meat during their voyages. One traveler wrote, "I myself do not find them agreeable. They taste of oil. . ." But "there is more meat in one of these than in a goose," observed another, and they were easily driven, "as if God had made the innocency of so poore a creature, to become such an admirable instrument for the sustenation of man." With its low reproductive rate, the Great Auk could not sustain itself in the face of concerted predation by humans. By the early 19th century, the Funk Island colony, once estimated at 100,000 birds, had disappeared. The last individuals that survived elsewhere in the North Atlantic islands were collected for the increasingly popular natural history museums of Europe.

such a large and diverse animal fauna disappeared so suddenly between 12,000 and 8,000 years ago remains a hotly debated question. Some experts point to the major reorganization of habitats and climates associated with the end of the last glaciation as the principal cause. Others argue that Ice Age climates had changed previously without causing major extinctions, and that the tight coincidence in timing between the arrival of humans and the disappearance of the megafauna clearly points to humans as the primary cause of the extinction event. In truth, the cascade of extinctions probably resulted from a combination of factors: rapid habitat changes (Martin and Neuner 1978), direct predation by humans (Martin and Steadman 1999), the spread of human-borne diseases to which the native fauna had no immunity (MacPhee and Marx 1997), and "chain reaction extinctions" caused by the disappearance of certain keystone species (Owen-Smith 1987).

One thing is certain about these extinctions: they included dozens of species of birds as well as mammals. The most spectacular disappearances involved many species of scavengers that lived on the carcasses of large mammals, just as African vultures do today. These extinct vultures, condors, and "teratorns" (giant, soaring scavengers perhaps only distantly related to vultures) **(Fig. 10–6)**, included some of the largest flying birds ever to have lived, and they once occupied nearly every habitat in North America. We can comprehend their sizes by considering them in relation to their largest remaining North American relative, the endangered

Figure 10–6. Skeleton of Merriam's Teratorn: One victim of the major extinction of animal life that occurred in North America about 10,000 years ago was the Merriam's Teratorn (sometimes referred to as Merriam's Giant Condor). Standing 2.5 feet (0.8 m) tall and having an enormous wingspan (see Fig. 10–7), this soaring scavenger fed upon carcasses of prehistoric megafauna that roamed grasslands and open country. When these large mammals disappeared, the teratorn became extinct. Here an articulated skeleton of a Merriam's Teratorn is shown in front of an artist's rendition of the bird's appearance. Notice the huge sternum, an indication of the bird's impressive flying ability. This particular skeleton is one of many fossil bird skeletons, mostly of predators and scavengers, recovered from asphalt pits up to 40,000 years old at Rancho La Brea, Los Angeles, California, one of the world's most famous fossil sites. The birds found at this site were probably trapped while trying to feed on the mammoths, saber-toothed cats, giant ground sloths, and other creatures already mired in the tarry substance. Photo courtesy of the George C. Page Museum.

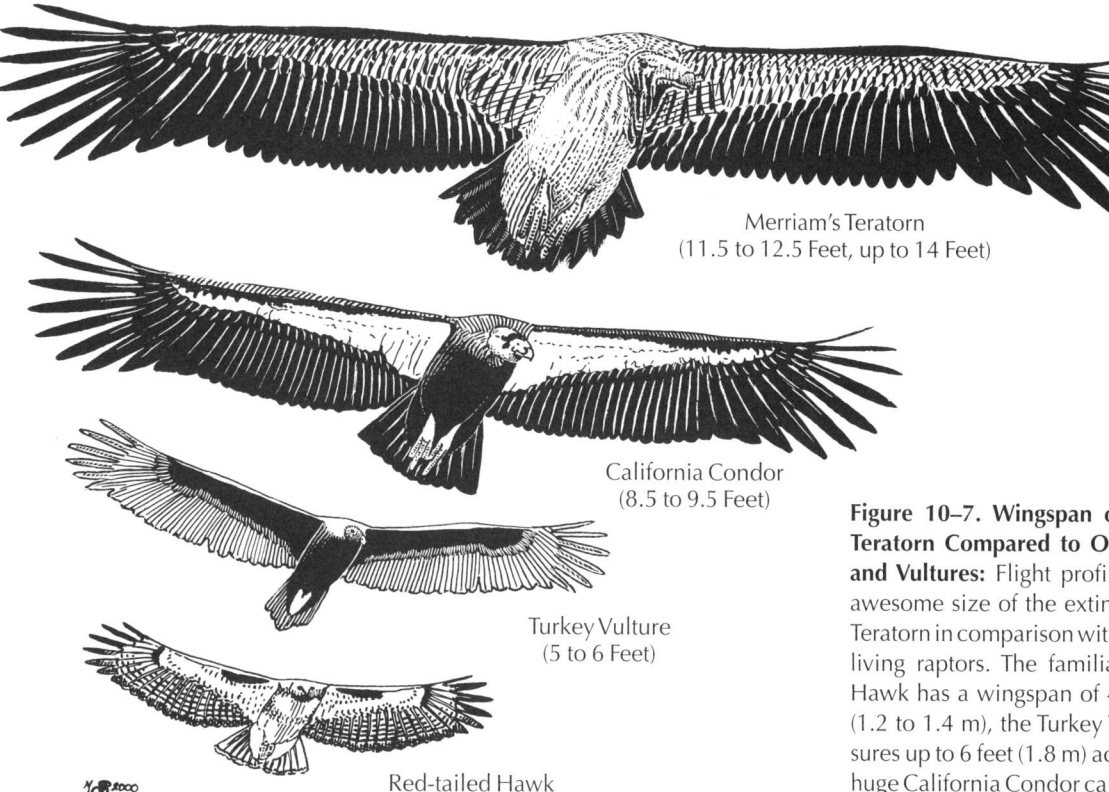

Merriam's Teratorn
(11.5 to 12.5 Feet, up to 14 Feet)

California Condor
(8.5 to 9.5 Feet)

Turkey Vulture
(5 to 6 Feet)

Red-tailed Hawk
(4 to 4.5 Feet)

Figure 10–7. Wingspan of Merriam's Teratorn Compared to Other Raptors and Vultures: Flight profiles show the awesome size of the extinct Merriam's Teratorn in comparison with that of three living raptors. The familiar Red-tailed Hawk has a wingspan of 4 to 4.5 feet (1.2 to 1.4 m), the Turkey Vulture measures up to 6 feet (1.8 m) across, and the huge California Condor can measure up to nearly 10 feet (3 m) across. Merriam's Teratorn dwarfs them all with a wingspan of up to 14 feet (4.3 m).

California Condor **(Fig. 10–7)**, which until a few thousand years ago occurred as far east as Florida and as far north as the Great Lakes. During the era when big mammals roamed the continent, this remarkable bird was just a medium-sized member of the scavenger community!

Over the past 10,000 years a host of extinctions also occurred on the Caribbean Islands, which once harbored remarkably diverse assemblages of now-vanished land birds such as crows, owls, macaws, giant parakeets, and nightjars. Bones of these species are found at human archaeological sites dating back 3,000 to 4,000 years (Pregill et al. 1994), and humans clearly caused a steady disappearance of birds in these fragile tropical islands by cutting down the forests and probably by eating the birds as well. Today, only a small fraction of the presettlement bird fauna persists in the Caribbean, where many species, such as the Puerto Rican Nightjar **(Fig. 10–8)**, are listed among the most seriously threatened in the New World.

Modern Extinctions on Mainland North America

After the great wave of extinctions associated with the first spread of humans in the New World, a smaller wave began as Europeans settled across North America in the 19th century. Continental North America has definitely lost at least three species since 1850 (Labrador Duck, Passenger Pigeon, and Carolina Parakeet), and the

Figure 10–8. Puerto Rican Nightjar: Many species endemic to the Caribbean Islands are listed among the most seriously threatened birds in the New World. The status of the Puerto Rican Nightjar **(a)**, with a current population of only 1,400 to 2,000 individuals, is considered critical. Originally this bird probably inhabited much of Puerto Rico, but now it is restricted to a few small areas in the southwest of the island, shown by the colored areas on the map **(b)**. The nightjar's initial decline, beginning in the late 19th century, may have been caused by introduced mongooses. Now, as humans develop the land, loss of the bird's forest habitat continues to threaten its survival. Painting by Tracy Pederson, from A Guide to the Birds of the West Indies, Princeton University Press, 1998.

a. Puerto Rican Nightjar

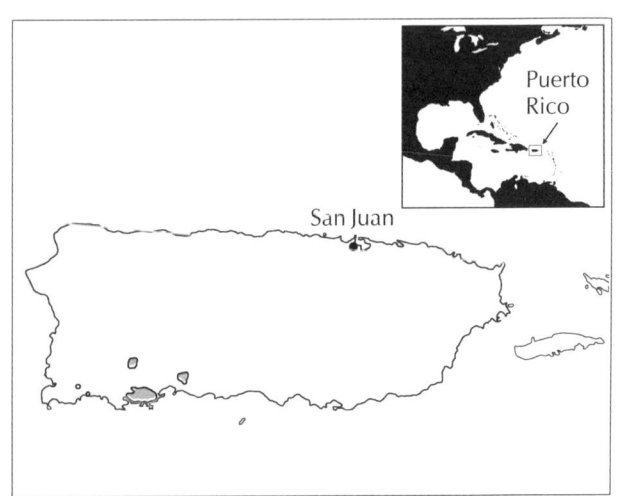

b. Current Distribution of the Puerto Rican Nightjar

continued existence of another three is unlikely (Eskimo Curlew, Ivory-billed Woodpecker, and Bachman's Warbler). These six remarkable species, discussed below, will forever stand as symbols of both rampant exploitation of wildlife and failure to protect native habitats during the late 19th and early 20th centuries.

Labrador Duck—the Mystery Extinction

The Labrador Duck **(Fig. 10–9)** is certainly the most mysterious of the modern North American extinctions. The species never was abundant, and its breeding ecology remains unknown. Audubon never saw a living Labrador Duck, even though they wintered regularly along the eastern North American coast from the Gulf of St. Lawrence south to New Jersey until about 1870. The species apparently bred on rocky islands off the Gulf of St. Lawrence, where it probably was exposed to heavy plume hunting and egg robbing beginning as early as the 18th century. Introduced mammalian predators on the nesting islands probably reduced its nesting success as well. The duck's unique bill was equipped with an extraordinary number of toothlike ridges along very flexible lateral flaps, presumably used for sifting crustacea from silt. This was an ecologically peculiar duck, apparently frequenting sandy shoals very near land and lacking a natural fear of humans. No doubt these features made the bird easy prey for hunters who exploited the New England coastline in winter. The last recorded Labrador Duck was shot off Long Island, New York in 1875.

Figure 10–9. Labrador Duck: This striking sea duck was the first bird species endemic to continental North America to disappear during the wave of extinctions that began soon after European settlement. In breeding plumage, the adult male (left) was patterned black and white, whereas both the immature male (center) and the adult female (right) were brownish gray with a prominent white wing speculum. Probably never abundant, the species disappeared before much could be learned about its biology. Its peculiar bill—with flexible lateral flaps and numerous thin, leaflike plates protruding from the inner surfaces—suggests that the bird sieved shellfish and crustaceans from silt and shallow water. Although the breeding range of the Labrador Duck was never positively identified, it probably bred on rocky islands in the Gulf of St. Lawrence, and possibly on coastal Labrador and points farther north. The bird wintered close to shore along the Atlantic Coast of North America, including the coast of Long Island, New York, where the last specimen was shot in 1875. Its lack of fear of humans, specialized ecological niche, and apparently low population density all may have contributed to its vulnerability. Painting by Louis Agassiz Fuertes. Copyright Cornell Lab of Ornithology.

Passenger Pigeon—Market Hunting at its Worst

The sad story of the Passenger Pigeon surely represents the most spectacular example of avian overexploitation in human history. Perhaps a worthy rival exists today in the massive overfishing of the great cod fisheries in the western North Atlantic, culminating in their closure to commercial fishing in the mid-1990s (Kurlansky 1997). As in the case of these fisheries, intensive market hunting of Passenger Pigeons was aided by ever-improving technology: sophisticated netting techniques allowed the "catch" to be more complete, whole railroad lines were installed to export hundreds of tons of squabs (young pigeons) and adults from the massive colonies in the northern deciduous forests to city markets, and freezer cars facilitated this long-distance shipping. Because the species nested in dense colonies and nestlings were easy and delicious prey, a colony's entire reproductive output could be wiped out during a single season. As a final blow, the Passenger Pigeon's ultimate collapse may have been unusually swift because the remaining birds simply stopped breeding altogether.

Carolina Parakeet—Removal of a Menace

The Carolina Parakeet (**Fig. 10–10**), a bird of the forested bottomlands, once was abundant from Virginia to Florida and west to Louisiana, and occurred at least casually as far north as the Great Lakes. The steady decline of this species throughout the 1800s parallels that of the Passenger Pigeon, as do the causes—large-scale killing and forest clearing. Because of its fruit-eating habits, this species was considered a major agricultural pest, and farmers slaughtered parakeets at every opportunity. The extermination was simplified by the parakeets' fearless behavior toward humans, and their "fatal habit of hovering over their fallen companions" (Bent 1940). Audubon (1842, quoted from Bent) vividly described such a scene:

> "… the Parakeets are destroyed in great numbers, for whilst busily engaged in plucking off the fruits or tearing the grain from the stacks, the husbandman approaches them with perfect ease, and commits great slaughter among them. All the survivors rise, shriek, fly around for a few minutes, and again alight on the very place of most imminent danger. The gun is kept at work; eight or ten, or even twenty, are killed at every discharge. The living birds, as if conscious of the death of their companions, sweep over their bodies, screaming as loud as ever, but still return to the stack to be shot at, until so few remain alive that the farmer does not consider it worth his while to spend more of his ammunition. I have seen hundreds destroyed in this manner in the course of a few hours."

As early as 1860 the Carolina Parakeet had become rare everywhere outside of Florida, and by 1890 it remained only in the most remote forests of that peninsula. Isolated flocks were seen through 1915, and the last reliable reports of Carolina Parakeets ceased in the early 1920s.

Eskimo Curlew—Three Strikes in the Wink of an Eye

The sight of immense flocks of Eskimo Curlews darkening the skies of Nebraska in spring and of Newfoundland in autumn so closely resembled the flights of Passenger Pigeons that the species was sometimes called "Prairie Pigeon." Feeding on grasshoppers and other abundant spring insects in Nebraska, a single spring flock en route north from Argentina was reported to cover 40 to 50 acres (16.2 to 20.3 ha). In the fall, huge flocks descended on blueberry fields and shorelines of the Canadian maritime provinces and New England (**Fig. 10–11**). Despite these numbers, the demise of the Eskimo Curlew occurred over a remarkably short period. The species was common until about 1870, but was rarely seen anywhere after 1890.

Figure 10–10. Carolina Parakeets on Cocklebur: The only parrot native to continental North America north of Mexico, the extinct Carolina Parakeet was a grackle-sized gregarious bird, mostly green with a yellow head and orange cheeks. It inhabited deciduous forests and forest edges in the eastern United States as far north as the Great Lakes region, as well as wooded river bottoms of the Great Plains as far west as Nebraska. It nested in tree cavities and ate the fruits and seeds of many trees and other plants, such as thistles and cockleburs, as shown in this painting. Outside of the breeding season the parakeet formed large, noisy flocks that fed on cultivated fruit, tore apart apples to get at the seeds, and ate corn and other grain crops. It was therefore considered a serious agricultural pest and was slaughtered in huge numbers by wrathful farmers. This killing, combined with forest destruction throughout the bird's range, and hunting for its bright feathers to be used in the millinery trade, caused the Carolina Parakeet to begin declining in the 1800s. The bird was rarely reported outside Florida after 1860, and was considered extinct by the 1920s. Numbered on this painting by John James Audubon are males (1 and 2), female (3), and immature (4). From Birds of North America, Volume 4, 1861.

Figure 10–11. Eskimo Curlews Descend on a Blueberry Field: Until the 1870s, immense flocks of Eskimo Curlews migrating in fall through the Canadian Maritime provinces and New England fattened up on blueberries and fruits of other heathland shrubs before heading south over the Atlantic Ocean to South America. Similarly sized flocks en route north in the spring fed upon grasshoppers and other insects in the Great Plains. Despite its vast numbers, the Eskimo Curlew was decimated by hunters over just a 20-year period, and was rarely seen after 1890. It is now almost certainly extinct.

Such a swift disappearance has spawned numerous theories about its cause, but now can be explained by a lethal combination of three simultaneous events (Gill et al. 1998): (1) After Passenger Pigeons disappeared, the Eskimo Curlew became the target of choice for market hunters in search of new foods to exploit. Hunters in the Great Plains followed the flocks northward in the spring, collecting wagonloads at every opportunity. Fall flocks received similar treatment along the eastern seaboard, especially in Labrador and New England, where, during the 1870s and 1880s, thousands of curlews were shot at single localities over just one or two days. (2) During its passage northward in April and May, the Eskimo Curlew depended almost exclusively on the abundant insect foods of native tallgrass and mixed grass prairies. Curlews descended in huge numbers on areas that had been burned by spring wildfire or trampled by bison herds, where the largest supply of insect food was exposed. In the late 1800s, these critical habitat patches were virtually eliminated by wholesale conversion of prairies to agricultural fields, by widespread suppression of wildfire, and by annihilation of the great bison herds. The specialized Eskimo Curlew was suddenly hard-pressed to find the food resources necessary to sustain itself on the long route to its breeding grounds. (3) Accounts of spring migration in the Great

Plains suggest that Eskimo Curlews relied to a surprising extent on a single prey species, the Rocky Mountain grasshopper (Melonoplus spretus). Around 1900, this swarm-prone grasshopper, a serious agricultural pest in the North American prairies, went mysteriously extinct (Lockwood and DeBrey 1990). Such sudden disappearance of a major food item, combined with major alteration of suitable foraging habitat, compounded the difficulties faced by curlews as they migrated through the killing fields of the Great Plains each spring. Even after the hunting ceased in the late 1890s, population recovery was impossible because of these changes.

Occasional specimens of Eskimo Curlew were taken in South America through the early 1920s, but no confirmed records exist from anywhere within its former range after 1939 **(Fig. 10–12)**. However, between the 1960s and 1990s, experienced birders did report springtime sightings of single birds or small groups along the Texas coast. Thus, we cling to the hope that somewhere a tiny population remains of this delicate bird, once among the most common shorebirds of North America.

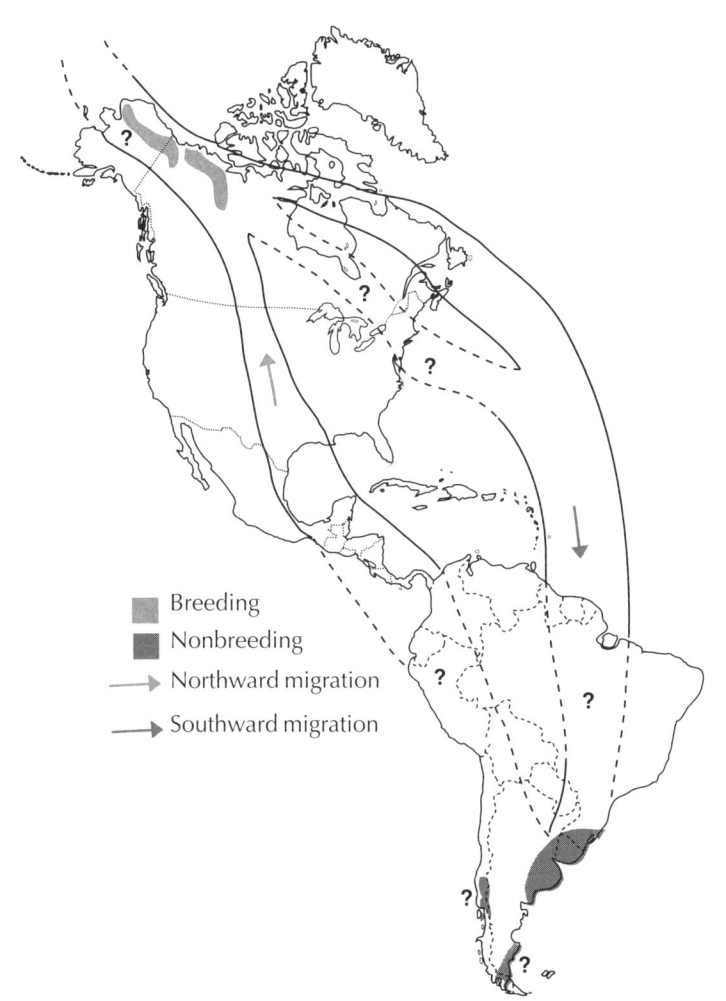

Breeding
Nonbreeding
→ Northward migration
→ Southward migration

Figure 10–12. Breeding and Nonbreeding Ranges and Likely Migration Routes of the Eskimo Curlew: Before its demise in the late 19th century, the Eskimo Curlew nested in the High Arctic (colored areas on map) and wintered in the southern extremes of South America (gray areas on map). Although its complete migration routes are not known, the bird apparently took an elliptical route as shown on the map. The colored arrow indicates the likely northbound leg; the gray arrow shows the probable southbound route. Question marks denote regions where the likely route is unknown. In the mid-to-late 1800s, North American sport and market hunters followed the vast migrating flocks during spring flights in the Great Plains and fall flights in Labrador and New England. Often thousands of birds were killed at a single location within a few days. This uncontrolled shooting—exacerbated by the loss of native prairies, the source of the bird's springtime food—caused the species to disappear. Each spring, birders in coastal Texas still hope for a sighting to confirm that the Eskimo Curlew is not completely extinct, but the prospect is bleak. From Gill, R. E., Jr., P. Canevari, and E. H. Iversen. 1998. Eskimo Curlew (Numenius borealis). In The Birds of North America, No. 347 (A. Poole and F. Gill, eds.). Used with permission. Copyright Birds of North America, Inc., Philadelphia, PA.

Figure 10–13. The Plight of the Imperial Woodpecker: The Imperial Woodpecker—a close relative of the Ivory-billed Woodpecker—was the largest woodpecker in the world. However, with no confirmed sightings since 1956, its future looks grim. This dramatic bird is mostly black with large white wing patches and shoulder lines. The male (left) has a red crest and nape, whereas the female (right) has a curly black crest. The bird originally ranged throughout the Sierra Madre Occidental of northwestern Mexico (see Figure 10–14). There it frequented the high altitude, old-growth, pine-oak forest, where it excavated dead tree trunks searching for the larvae of wood-boring insects and nested in mature pines. Painting courtesy of George Sandström, from Short, 1982, Woodpeckers of the World.

Ivory-billed Woodpecker and Bachman's Warbler—Demise of the Southeastern Forests

Hopes are dim for the survival of these two specialized denizens of the southern forests and bottomlands of North America. Neither species ever was common, and both were clearly associated with extensive stands of old-growth forest. They were found most easily along the Mississippi alluvial plain, but occurred east to the Carolinas and south to Florida. The woodpecker used big trees, and the warbler used dense understory thickets and canebrakes (thickets of bamboo-like grass) associated with forest openings. Today, these habitats remain only in tiny and widely dispersed fragments surrounded by agricultural fields, pine plantations, and towns.

The Ivory-billed Woodpecker is known to have fed principally on large beetle larvae, which it uncovered by chipping the bark from large trees that were dead or dying. To this day, the largest concentrations of these larvae occur after wildfires sweep through a pine forest. Thus, the Ivory-billed Woodpeckers suffered from the dual problem of outright habitat loss (logging of old-growth pines) and suppression of fires within what little was left of the great pine forests. The species was extremely rare by 1900, and no confirmed nesting has occurred since the 1930s. Credible reports do occasionally emerge from Louisiana and Florida, and a tiny population still may persist in the old-growth pine forests of eastern Cuba. Hopeful biologists continue to investigate both areas.

Sadly, the closest relative of the Ivory-billed Woodpecker has met a similar fate, and there is even less hope for its survival. The Imperial Woodpecker **(Fig. 10–13)**—once the largest woodpecker in the world—thrived in the immensely tall pine forests of western Mexico. These great forests have all but disappeared **(Fig. 10–14)**, and the woodpecker has not been seen since 1956. After searching extensively throughout its former range, Mexican ornithologists now believe that this majestic bird has gone extinct.

Bachman's Warblers preferred dense, shrubby forest openings and canebrakes, suggesting that they typically settled in lowland habitats recently battered by hurricanes. The species almost certainly enjoyed a

SANDSTRÖM

brief "population boom" in the middle 1800s as bottomland forests were first penetrated by railroads and then crisscrossed by logging roads (Hamel 1995). These activities created a plethora of openings that quickly grew into impenetrable thickets of raspberry and cane. The boom was short-lived, however, as the logged landscape was converted to farmlands and the rich floodplains were channelized and drained. Worse, the Bachman's Warbler experienced double jeopardy, as its sole wintering grounds (Cuba and the Isle of Pines) also underwent rapid habitat conversion to become sugar cane plantations. The last confirmed Bachman's Warbler sightings were of a singing male that returned to the same location in South Carolina for three successive breeding seasons in 1958, 1959, and 1960 **(Fig. 10–15)**.

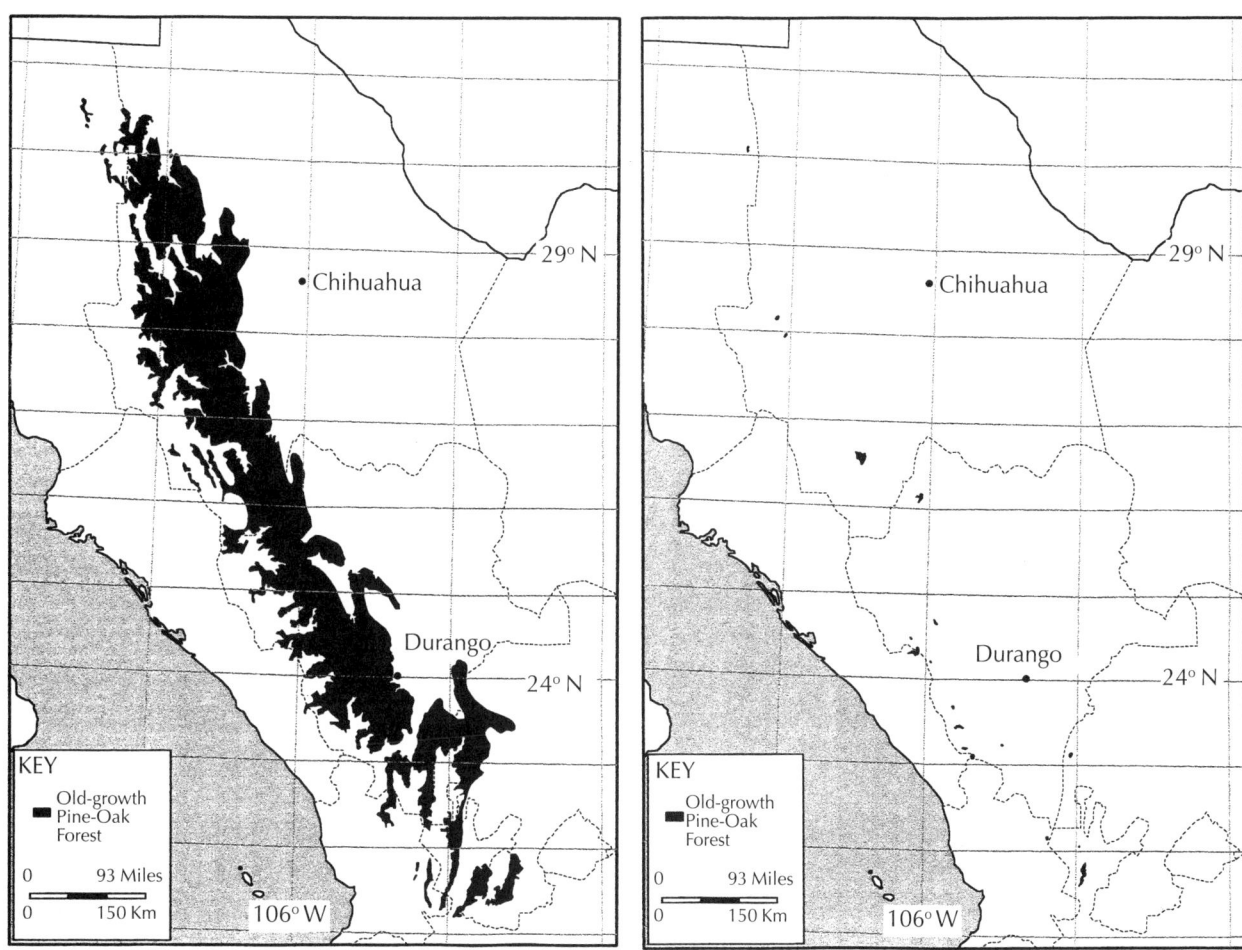

a. Extent of Pine-Oak Forest Above 6,000 ft (2,000 m) Around 1880

b. Extent of Pine-Oak Forest Above 6,000 ft (2,000 m) in October 1995

Figure 10–14. Declining Old-Growth Forests of Northwestern Mexico: Destruction of the large stands of mature pines that once provided food and nest sites for the Imperial Woodpecker, combined with hunting, have led to the bird's precipitous decline, leaving it, if not extinct, the world's rarest woodpecker. Map **a** shows the original extent (black areas) of pine-oak forest in the Sierra Madre Occidental, Mexico around 1880. Map **b** shows the old-growth pine-oak forest (scattered black patches) that remained in October 1995 after many years of logging—a mere 0.61 percent of the original area. Although convincing reports of the woodpecker's existence continue to be made by local people, no Imperial Woodpeckers were found in a survey of remnant forest patches conducted by biologists in 1994 to 1995; thus the species may be extinct. Maps copyright Vanellus Productions, from Lammertink, J. M., et al., 1996, used with permission.

Figure 10–15. Bachman's Warbler: Originally, Bachman's Warbler nested in seasonally flooded forested swamps in the southeastern United States, showing particular affinity for canebrakes (thickets of bamboo-like grass). It wintered in wooded areas of the island of Cuba. Drainage of river bottom swamps and extensive clearing of canebrakes for housing developments and agriculture in the United States, coupled with conversion of native vegetation to sugarcane plantations in Cuba, caused its demise. Although reported sightings of Bachman's Warblers have tantalized birders and ornithologists for decades, the last confirmed observations were of this single male seen in an area near Charleston, South Carolina during the springs of 1958, 1959, and 1960. Each year, the lone bird sang constantly for days on end but never attracted a mate. The site—then a mixed longleaf and loblolly pine forest with heavy undergrowth—is now a heavily developed suburban area. Photo by J. H. Dick/VIREO.

Brief History of Bird Conservation in the United States

As early as the 1840s, John James Audubon already was writing wistfully about the gradual disappearance of birds from the American landscape. By the 1870s, as the last Passenger Pigeon harvests gave way to Eskimo Curlew slaughters, rising voices began to protest the uncontrolled exploitation of birds for food and ornamentation. In 1883 the American Ornithologists' Union (AOU) was founded, and among the most important groups within this fledgling scientific organization was the Bird Protection Committee. In 1886 a charter committee member, George Bird Grinnell, proposed the establishment of an "Audubon Society" dedicated to the protection of birds (Barrow 1998). Grinnell published the first Audubon Magazine in 1887, but the publication folded after just a few years owing to financial difficulties. Nevertheless, as the American public increasingly was blessed with enough leisure time to enjoy watching birds, local Audubon societies were established around the eastern states. By 1887, Grinnell's original Audubon Society boasted over 300 local chapters and had recruited more than 20,000 members.

As the industrial revolution reached its peak in the late 1800s, ladies occupying a wealthy social class in the large eastern cities decorated their hats and fashionable dresses with bird plumes from a wide variety of species (**Fig. 10–16**). Gaudiest were the plumes of several herons, especially Great Egret and Snowy Egret, which bred in large rookeries in the southern swamps. By 1890, the scale and biological effects of commercial plume hunting in the Everglades rivaled the wholesale slaughter of Passenger Pigeons in the 1860s and Eskimo Curlews in the 1870s.

By the end of the 1800s, the poetry and writings of Henry David Thoreau, Ralph Waldo Emerson, and Walt Whitman had opened the eyes of the western world to the philosophical and spiritual importance of nature. Inspired by these authors, there arose a new generation of gifted writer-philosophers who also were expert naturalists. In particular, John Burroughs in the east and John Muir in the west focused on "the little things" in nature and on the importance of nature to the human spirit. American romantic painters were glorifying our spectacular natural places such as Wyoming's Yellowstone National Park and New York's Hudson River. Enthusiastic naturalists across the country were growing increasingly vocal, decrying the wholesale slaughter of birds and lamenting the disappearance of big bird flocks from the skies.

Among the most historic battle cries for bird conservation were those sparked by society ladies in the Boston area. In 1887, Fannie Hardy and Florence Merriam created the Smith College Audubon Society, and ringing renouncements of feather-wearing began sweeping the campus. In 1896, Mrs. Harriet Lawrence Hemenway established the Massachusetts Audubon Society, "to discourage the buying and wearing, for ornamental purposes, of feathers of wild birds, and to otherwise further the protection of native birds" (quoted from Barrow 1998). Massachusetts Audubon's first president was the noted ornithologist William Brewster, whose tenure commenced a pivotal period in which professional scientists and amateur bird-lovers worked together to increase public awareness of bird conservation issues. In 1899 the well-known ornithologist Frank M. Chapman published the first issue of Bird-Lore, precursor of today's Audubon magazine (see Birds and Humans: A Historical Perspective, Fig. 34).

Just prior to 1900 a few states had established laws that controlled or prohibited the commercial harvesting of birds, but substantial trade continued because bird harvesting remained legal in many other states. Then, in 1900—catapulted by the social activism of local Audubon societies—the U. S. government passed the Lacey Act, which banned interstate shipments of birds killed in violation of any state or local law. In 1902, at the urging of Audubon societies, the AOU hired a former plume hunter, Guy Bradley, as the nation's first wildlife warden. For three years Bradley patrolled the great bird rookeries of the western Everglades, until he was murdered by some of the plume hunters he was hired to control. Two more Audubon wardens were murdered at Everglades heron rookeries in 1908.

Despite such setbacks, bird conservation grew steadily in public acceptance and political power. Inspired and staffed by ornithologists, the U. S. Biological Survey was established in 1896 as a branch of the Department of Agriculture. Its charge was to document the economic value of birds and other wildlife. Although the Lacey Act began to slow nationwide traffic in bird plumes, rampant bird trade ended altogether in 1910 with the Audubon Plumage Act,

Figure 10–16. Hat Decorated with Egret Plumes: In this photo taken around 1886 in Manhattan, New York, a woman wears a hat adorned with the white nuptial plumes of egrets. In 19th century America, commercial plume hunting was a lucrative occupation. It catered to the whims of fashionable American and European women until the turn of the century, when public outcry helped to stop the slaughter and began the modern conservation movement. Photo courtesy of the National Audubon Society.

which banned all commercial traffic in plumes from birds native to the United States. In 1911 the Bayne Act prohibited the sale of wild game in markets and restaurants, and in 1913 the United States banned the importation of plumes from foreign countries. In 1918 the movement to protect wild birds reached its peak with passage of the Migratory Bird Treaty Act, a landmark law that to this day prohibits the killing of any native bird not regulated separately as a game species in the United States and Canada.

Besides bringing about regulation of trade in wild birds, bird enthusiasts urged federal officials to begin protecting important pieces of the wild American landscape. In 1903, Theodore Roosevelt established the first National Wildlife Refuge at Pelican Island in eastern Florida, protecting a large coastal breeding area for Brown Pelicans and other seabirds. In 1916 the National Park Service was formed to oversee 14 national parks and 21 national monuments. In 1929, during the depths of the Great Depression, the Norbeck-Audubon Act provided the first annual source of funds for protecting wetlands across the United States. This legislation was augmented in 1937 by the Pittman-Robertson Act, which to this day levies a tax on firearms and ammunition for purposes of purchasing and restoring natural land as wildlife habitat. With this regular revenue at its disposal, in 1939 the U. S. Biological Survey was transformed into the U. S. Fish and Wildlife Service and moved from the Department of Agriculture to the Department of the Interior. Thus, over the first 40 years of the 20th century, conservation of birds and their habitat steadily had become a greater official responsibility of the United States government.

The most influential environmental book ever written in the United States was published in 1949 by Aldo Leopold (**Fig. 10–17**; also see Birds and Humans, Fig. 39), an employee of the U. S. Forest Service. Leopold once had participated in massive, federally sponsored

Figure 10–17. Aldo Leopold: Conservationist Aldo Leopold was trained as a forester and served 19 years in the United States Forest Service before becoming Professor of Game Management at the University of Wisconsin, Madison. Throughout his career Leopold was a respected scientist and conservationist, and was instrumental in developing wildlife, game, and forest conservation policy, and in promoting the protection of wilderness. In 1935 he and his family acquired an abandoned farm in Sand County, Wisconsin, which they dubbed "The Shack." There they spent their weekends repairing the run-down farm building and restoring the ecological integrity of the worked-out land with shovel and axe. In this photograph, Leopold writes in his journal at "The Shack." His ethical attitude toward the land, and his recognition of the need for its respect and protection, is prominent throughout his famous A Sand County Almanac, published posthumously in 1949. This landmark book, a collection of nature sketches and philosophical essays, has inspired generations of conservation-minded readers. Of land conservation, Leopold wrote "A thing is right when it tends to preserve the integrity, stability, and beauty of the biotic community" (from "The Land Ethic," in A Sand County Almanac). Photo courtesy of The Aldo Leopold Foundation Archives.

predator-extermination programs, but had come to see them as folly. In his legendary collection of essays, A Sand County Almanac, Leopold elegantly established both a philosophical and a practical framework for treating the natural landscape as a long-term resource to be used carefully and held in safekeeping for future generations. Sadly, Leopold died just prior to publication of his most influential book, when he suffered a heart attack while fighting a brush fire on a neighbor's farm in Wisconsin. Nevertheless, his name has become synonymous with the concept of the "land ethic," whereby natural resources are seen to warrant the same level of ethical reverence given to humans and human culture.

Leopold's legacy was enhanced by the growing opportunity for the general public to identify the sights and sounds of nature through field guides. Quick guides for identifying birds had been available since the 1890s, and in 1903 Frank M. Chapman had published his Color Key to North American Birds, illustrated by Chester A. Reed. Soon thereafter, Reed began publishing his own books, and the pocket-sized "Reed guides" served as the industry standard until 1934. That year, a young illustrator named Roger Tory Peterson changed the world by publishing A Field Guide to the Birds: Giving Field Marks of All Species Found in Eastern North America. To the publisher's surprise the first printing of 2,000 copies sold out in just a few days. Through successive printings and revised editions, Peterson's remarkable paintings and easy-to-use "field-mark" system played a pivotal role in expanding the popularity of bird watching and nature viewing through the remainder of the 20th century. These hobbies continued to expand as leading outdoor pastimes worldwide, bringing with them ever-increasing commitment to conservation among citizens. The modern field guide, an inexpensive and easy tool by which any person can identify a bird, is the single innovation most responsible for the tremendous growth in bird watching that continues today.

Logging of the great eastern forests (**Figs. 10–18** and **10–19**) had proceeded at a voracious pace from the 1830s onward. Agriculture had spread across the landscape, and the magnificent white pine and mixed hardwood forests had been felled systematically. By 1853, for example, forest cover in the state of Ohio had dropped from nearly 100 percent to 54 percent, and by 1883 the state was only 18 percent forested (Whitney 1994). Midwestern states followed suit through the late 1800s and early 1900s, and by 1950 virgin forest and unplowed prairie had essentially vanished east of the Mississippi River. In 1951, in reaction to this relentless loss of natural places, a group of ecologists in eastern New York founded The Nature Conservancy to protect natural areas possessing singular beauty or rare species of plants and animals. In addition, the federal government continued responding to calls for more action to protect landscapes. In 1964 the United States passed the Wilderness Act, and in 1968 added the National Wild and Scenic River Act, thereby affording federal protection for the country's greatest expanses of wild land and its least spoiled rivers. In 1972 the Clean Water Act was passed, setting into motion important regulations protecting critically important habitats

10

a. 1905   **b. 1958**

Figure 10–18. The Logging of the Great Eastern Forests: A photo of virgin white pine forest in Antrim County in northern Michigan around 1905 **(a)** is a poignant reminder of the scale of North America's original forests. The same scene around 1958 **(b)** shows the area after clear-cut logging. Photos courtesy of Gary Williams. From "One Natural Area—Past and Present," Michigan Botanist, Volume 11, 1972, pp.140–142.

a. Giant Sycamore, Illinois, 1875   **b. Old-Growth Sugar Maples, Michigan, Early 1900s**

Figure 10–19. Old-growth Hardwood Forests—Scenes of a Bygone Era: Logging of the great forests of eastern North America occurred at a rapid pace throughout the late 1800s and early 1900s. By 1950, little virgin forest remained. Historical photographs provide us a glimpse of what we have lost. **a. Giant Sycamore:** The deep fertile soils of the lower Wabash River valley, which forms part of the border between Illinois and Indiana, originally supported huge stands of magnificent hardwoods. Ornithologist Robert Ridgway (left) is shown in this 1875 photograph of a giant sycamore near Mt. Carmel, Illinois. The tree was 15 feet (4.6 m) in diameter, 168 feet (51 m) high, and had a crown 134 feet (41 m) across. Photo used by permission of the Chicago Academy of Science. **b. Old-Growth Sugar Maples:** The sugar maple is characteristic of the northern hardwood forests that once covered the eastern United States. Some of the most impressive specimens were found in the Upper Great Lakes region. Shown here is an old-growth sugar maple forest near Petoskey, Michigan, photographed by S. V. Streator around the beginning of the 20th century, but now long gone. Notice the small human figure dwarfed by the trees. Photo courtesy of The Forest History Society, Durham, N. C.

for birds. In 1985 the Conservation Reserve Program was established, providing financial incentives for farmers and ranchers to set aside portions of their land as fallow grassland.

Despite all the positive steps toward conservation, dozens of bird species across the United States continued sliding toward extinction throughout the mid-20th century. Bald Eagles became so rare that in 1940 special protective legislation was passed to protect them. Programs were initiated in the 1950s for protection and captive rearing of Whooping Cranes and Trumpeter Swans, which had been reduced to tiny numbers by unrestricted hunting and wetland drainage.

With the publication of her landmark book Silent Spring in 1962, Rachel Carson **(Fig. 10–20)** first exposed the potentially devastating environmental effects of pesticides. Within a few years, scientists confirmed the role of these pesticides in causing drastic declines in Peregrine Falcons and other raptors. Such dire cases demanded more formal attention to the plight of disappearing species, leading the United States in 1973 to develop the most far-reaching piece of environmental legislation in history, the Endangered Species Act (ESA). For the first time, individual species were legally recognized as possessing intrinsic value and rights to protection under law, regardless of how trivial or irrelevant any species might seem to human society. Of special importance is the statement of purpose explicitly included in the ESA (Section 2b):

". . . to provide a means whereby the ecosystems upon which endangered species and threatened species depend may be conserved."

Figure 10–20: Rachel Carson: Rachel Carson, writer and scientist, originally was trained in marine biology. Entering government service, she rose to become Director-in-Chief of all publications for the United States Fish and Wildlife Service. A lifelong nature lover who wrote many natural history articles and books about ocean life, Carson is best known for her 1962 classic Silent Spring, a book in which she warned about the long-term and potentially devastating effects that could be expected from widespread and uncontrolled use of synthetic chemical pesticides, especially DDT. Despite attacks by the chemical industry and some government officials, Carson persisted in publicly challenging governmental policy and the practices of agricultural scientists, championing a new view of the role of humankind in the fragile natural world. In this 1945 photo by Shirley A. Briggs, Carson is bird watching at Hawk Mountain, Pennsylvania. Used by permission of the Rachel Carson History Project.

Figure 10–21. Two U. S. Presidents Who Advanced Environmental Protection: In 1973, President Richard M. Nixon **(a)** signed the Endangered Species Act, committing the United States government to take action to prevent the extinction of any species of plant or animal native to the country. Nixon also signed many other landmark environmental laws, and created the Environmental Protection Agency. Photo copyright Bettmann/Corbis. President Jimmy Carter **(b)** created the Department of Energy, led efforts to develop Environmental Super Fund legislation, and signed the landmark Alaska National Interest Lands Conservation Act, which dramatically enlarged the National Park, National Wildlife Refuge, and National Wilderness Area systems. Photo copyright Jimmy Carter Library.

a. President Richard M. Nixon

b. President Jimmy Carter

By this statement, the U. S. Congress acknowledged that no species can persist in the wild without intact natural habitat, and that individual species warrant legal protection because they are indicators for entire ecosystems in need of protection. Key among the ESA's provisions is the formal procedure for using scientific information to establish and maintain a list of threatened and endangered species. Precursors to the endangered species list had been drawn up as early as the 1940s, but now the U. S. government was committed to enforcing a law prohibiting destruction of any such species, and even to devoting resources to the protection and propagation of species on the list. In 1994 the Bald Eagle became the first federally listed species to be "down-listed" from endangered to threatened status, and in 1999 the Peregrine Falcon was formally removed from the list altogether. Both events proved that successful recovery of endangered species could be achieved, given adequate knowledge, effort, and resources.

In many respects, bird conservation in the United States reached maturity during the 1970s. Until that decade, the story of conservation had been largely a litany of losses. Then, through the presidential administrations of Richard Nixon and Jimmy Carter **(Fig. 10–21)** came the banning of DDT and other pesticides, legal protection of endangered species, international agreements banning trade in endangered species, federal regulation of freshwater habitats, accelerated protection of millions of acres of native land by public agencies, major growth of private conservation organizations, and increasing interest in the environment and its protection on the part of millions of citizens. By 1975 conservationists had shown that with commitment of will and resources, bird populations could be protected and even restored. To this day, birds continue to play essential roles both as subjects for the study of conservation strategies, and as flagships for conservation action by humans seeking to protect the earth's natural heritage.

Are North American Birds Disappearing?

Many people today wonder whether the six bird species lost from North America over the past 200 years represent just the beginning of a much larger wave of loss. By the 1980s, bird watchers began asking, "Where have all the birds gone?" well before this question made its way to the title of an influential book (Terborgh 1989). Is it true that even today North American bird populations are still declining overall? If so, which species and which habitats are showing the biggest changes?

Just by looking out the window of an airplane or a tall building and imagining how the landscape once appeared, anyone can tell that the wild habitats in which birds live are less common than they were in the past **(Fig. 10–22)**. Today over much of North America we see cities and towns, highways and shopping malls, or corn and wheat fields in place of the forests, wetlands, and grasslands that existed 100 years ago. Indeed, ornithologists began commenting on declining bird numbers as early as the mid-1800s (examples are listed in Peterjohn et al. 1995). During Civil War days, Coues and Prentiss (1862) noted that songbirds were disappearing around Washington, D.C. as the surrounding human population became more densely settled. Since that time, the explosion of residential and commercial development across North America has dramatically altered which species can live and breed in many areas. Although some species are capable of breeding around densely settled areas, and therefore increase in numbers as the human population grows (for example, American Robin, Northern Mockingbird, Blue Jay, Western Scrub-Jay, and American Crow), far more species are unable to do so. Just how far North American bird populations have fallen since pioneer days probably never will be measured, but we are certain that the changes have been huge. To what extent can we measure changes in bird numbers at all, given that so many different species and habitat requirements exist over such vast areas?

Figure 10–22. Suburban Sprawl in Southern California: Whether lost to sprawling housing tracts and shopping malls, or converted to agricultural fields, native forests and grasslands have rapidly disappeared over the past two centuries. Nowhere is the loss more evident than in California's San Fernando Valley, as shown by this aerial view of typical "suburban sprawl." Photo by Spencer Grant / Photo Researchers.

10

We can thank a modest and far-sighted individual for dreaming up a way to monitor breeding bird populations across the landscape. In 1966 Dr. Chandler S. Robbins of the U. S. Fish and Wildlife Service had just published an extremely popular bird field guide (Robbins et al. 1966). Having led many a bird walk through the Maryland countryside, Robbins recognized that an awesome but largely untapped opportunity for counting birds lay in the energy and skills of a growing population of amateur bird watchers across North America. With several colleagues, Robbins designed a standardized, scientifically rigorous program by which skilled birders counted the birds singing at dawn during prime breeding season along randomly selected pieces of roadway **(Fig. 10–23)**. This program, known as the North American Breeding Bird Survey (BBS), quickly grew into the most important source of long-term information available on North American breeding bird numbers (see Ch. 9, How Do We Determine Population Size?). Every year from 1966 onward, data have been collected, analyzed, and archived at the Migratory Bird Research Lab in Patuxent, Maryland, where they are available to both scientists and the general public <**www.mbr.nbs.gov/bbs/bbs.html**>.

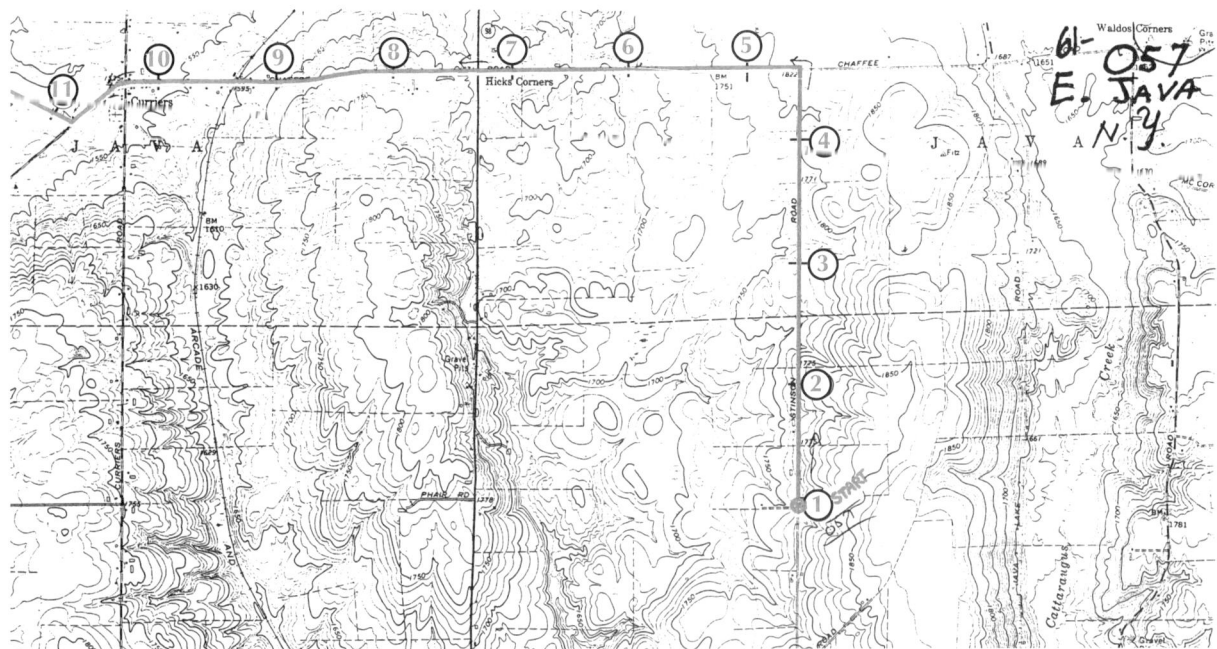

Figure 10–23. The North American Breeding Bird Survey: During the 1960s, Chandler Robbins and his associates at the Migratory Bird Population Station (now the Patuxent Environmental Science Center) in Laurel, Maryland, developed a continentwide monitoring program for all North American breeding birds. The roadside survey methodology was field tested in 1965, and the Breeding Bird Survey (BBS) was formally launched in 1966 when 600 surveys were conducted by amateur birders in the eastern United States and Canada. Today, about 3,700 BBS routes exist throughout North America. Each route is 24.5 miles (39.4 km) long, and is randomly selected to sample representative habitats of the region. Surveys contain a total of 50 stops at half-mile (0.8- km) intervals, and are conducted during the peak of the nesting season, primarily in May and June. This map shows part of a BBS route in East Java, New York. Circled numbers in green show the stop points, at which the birder counts all birds heard or seen within 0.25 miles (0.4 km) during a three-minute period. Results of BBS data analyses are available to scientists and the public at the BBS web site <**www.mbr.nbs.gov/bbs/bbs.html**>.

Figure 10–24. American Redstart: Data from the BBS and other long-term censuses have shown that many forest birds have declined in the eastern United States since the 1960s. Many of these species, such as the American Redstart shown here at its nest, are Neotropical migrants. Research suggests that forest fragmentation on the breeding grounds has led to declines in many Neotropical migrants. Photo courtesy of Betty Darling Cottrille/CLO.

Data from the BBS, along with information from studies focused on individual areas, clearly show that certain bird species indeed have declined from the 1960s onward. Robbins, Sauer, et al. (1989) showed that the individual species with the sharpest declines were certain Neotropical migrants breeding in northeastern North America and migrating each year to the American tropics. Specifically, long-term censuses of small woodlots and parks revealed that populations of many forest birds were declining in many areas of the eastern United States (Askins et al. 1990). Declines were most obvious among forest-interior species (for instance, Red-eyed Vireo, Eastern Wood-Pewee, Wood Thrush, and American Redstart **[Fig. 10–24]**) in landscapes that had lost forest steadily for many decades. Inside the forest fragments that remained in these areas, parasitism by Brown-headed Cowbirds (Robinson et al. 1995) and reduced nesting success because of predation helped to explain the gradual songbird losses (see Fig. 9–66). In addition to outright reduction in numbers, certain bird species were found to be sensitive to fragmentation, meaning that they preferred or even required large, continuous habitat tracts, either shunning the small tracts altogether or failing to breed in them successfully **(Table 10–1)**.

Although many North American bird populations have declined in highly fragmented forest landscapes since the mid-1960s, the total number of birds on the continent has not obviously gone down. In fact, detailed analyses of BBS data have shown that between 1966 and 1979, many more species actually increased than decreased (Peterjohn et al. 1995). This is especially true in large forest tracts, where many populations held steady or increased between 1965 and 1990 (Askins et al. 1990), and many of the species that did decline (such as Least Flycatcher and Philadelphia Vireo) were replaced by other species (such as Black-throated Blue Warbler and Red-eyed Vireo) that increased in total numbers. Long-term changes in bird numbers within forested habitats often result from slight differences in preferred habitats among species. Thus, changes in bird numbers sometimes reflect natural successional changes in forest structure, rather than habitat degradation by humans. Researchers now agree that the question

Table 10–1. Forest and Grassland Bird Species Sensitive to Fragmentation: Species with a high sensitivity are those birds least tolerant of habitat fragmentation. Moderately sensitive species show an intermediate response to habitat fragmentation, and species with a low sensitivity are those most tolerant of fragmentation. Table adapted from Herkert et al. (1993).

FOREST BIRDS *

HIGH SENSITIVITY	MODERATE SENSITIVITY	LOW SENSITIVITY
Broad-winged Hawk	Yellow-billed Cuckoo	Downy Woodpecker
Pileated Woodpecker	Black-billed Cuckoo	Red-headed Woodpecker
Least Flycatcher	Hairy Woodpecker	Red-bellied Woodpecker
Yellow-throated Vireo	Acadian Flycatcher	Great Crested Flycatcher
Black-and-white Warbler	Scarlet Tanager	Eastern Wood-Pewee
Worm-eating Warbler	Summer Tanager	Blue Jay
Cerulean Warbler	Red-eyed Vireo	Brown-headed Cowbird
Ovenbird	Northern Parula	Baltimore Oriole
Mourning Warbler	Yellow-throated Warbler	Common Grackle
Hooded Warbler	Louisiana Waterthrush	Eastern Towhee
American Redstart	Kentucky Warbler	Northern Cardinal
Brown Creeper	White-breasted Nuthatch	Rose-breasted Grosbeak
Veery	Tufted Titmouse	Indigo Bunting
	Blue-gray Gnatcatcher	Gray Catbird
	Wood Thrush	Carolina Wren
		House Wren
		Black-capped Chickadee
		Carolina Chickadee
		American Robin

GRASSLAND BIRDS **

HIGH SENSITIVITY	MODERATE SENSITIVITY	LOW SENSITIVITY
Northern Harrier	Eastern Meadowlark	Northern Bobwhite
Upland Sandpiper	Western Meadowlark	Red-winged Blackbird
Greater Prairie-Chicken	Grasshopper Sparrow	American Goldfinch
Bobolink	Sedge Wren	Vesper Sparrow
Savannah Sparrow		Field Sparrow
Henslow's Sparrow		Song Sparrow
		Dickcissel

* Data compiled primarily from studies of forest fragments in Illinois (Blake and Karr 1984), Missouri (Hayden et al. 1985), and Wisconsin (Temple 1986), with supplemental data from Maryland (Whitcomb et al. 1981), Massachusetts (Tilghman 1987), and the mid-Atlantic United States (Robbins, Dawson, and Dowell 1989).

** Data compiled from studies of grassland fragments in Illinois (Herkert 1991) and Missouri (Hayden 1985).

of whether songbirds are declining cannot be evaluated meaningfully across North America as a whole. The most revealing approach for exposing patterns and causes of population change is to examine individual geographic regions or habitat types.

The Forested Northeast

In the northeastern United States, where intensive agriculture is declining, forest bird populations generally are stable or even increasing. As family farms are abandoned, fields succeed into shrubland and then into forest. As a result, woodland species such as Tufted Titmouse, Red-bellied Woodpecker, Pileated Woodpecker **(Fig. 10–25)**, and Scarlet Tanager are probably more numerous and widespread today than they have been for the preceding 100 years. In contrast, many other familiar species are rapidly declining. Birds such as Eastern Towhee, Brown Thrasher **(Fig. 10–26)**, American Woodcock, and Golden-winged Warbler **(Fig. 10–27)** no doubt exploded during the early and middle

Figure 10–25. Pileated Woodpecker: Many regional changes in North American bird populations are caused by changes in the ways that humans use the land. In the northeastern United States, for instance, widespread abandonment of farms beginning in the 1920s and '30s allowed natural reforestation to proceed. As a result, woodland birds probably were more common in the Northeast in 2001 than they were during the 19th century, when most forested land was logged and converted to agriculture. The Pileated Woodpecker, for instance, was considered rare in the eastern United States prior to 1900. Its scarcity no doubt resulted from a lack of large tracts of mature forest that could provide large-diameter trees in which it could excavate nest cavities and dig for ants and wood-boring beetle larvae. Breeding Bird Surveys carried out between 1966 and 1991, however, show a significant increase in the number of Pileated Woodpeckers across the eastern United States and Canada—evidence that reforestation has helped this species to come back. Photo courtesy of Warren Greene/CLO.

Figure 10–26. Brown Thrasher: Reforestation of the Northeast has led to population declines in species that live in early successional habitats, such as the shrubby "old fields" representing a stage of ecological succession midway between field and forest. The secretive Brown Thrasher, for example, inhabits thickets, hedgerows, forest clearings, and habitat edges in the eastern United States and the Midwest, placing its nest in a low shrub or on the ground. In New York, the species became abundant in the 1950s and '60s as abandoned agricultural land gave way to shrublands. However, as northeastern forests are reappearing, both the shrubby areas and the Brown Thrasher are steadily disappearing. Since 1966, Brown Thrashers have declined at rates of 3 percent to as high as 9 percent per year in the Northeast. Photo courtesy of Michael Hopiak/CLO.

Figure 10–27. Golden-winged Warbler: Thriving in the shrub habitat that followed farm abandonment, logging, and fire, the Golden-winged Warbler expanded its range in the northeastern United States and southeastern Canada for over a century. However, in the 1980s it began to decline, owing primarily to a combination of reforestation and range expansion by the closely related Blue-winged Warbler (with which it both hybridizes and competes). Nest parasitism by Brown-headed Cowbirds and degradation of the warbler's wintering habitat in Latin America may be hastening its decline. Photo courtesy of Isidor Jeklin/CLO.

Figure 10–28. Eastern Meadowlark: Of all North American birds, those inhabiting native grasslands and open country have suffered the most from the agricultural practices of the latter half of the 20ᵗʰ century. North America's early agricultural landscape was dominated by family farms with their diversity of crops and low-technology methods. More recently, however, mechanical and chemical-intensive "big business" farming practices have produced a homogeneous mono-culture habitat that is often unsuitable for grassland birds. For example, BBS data (see graph in Fig. 9–47) show that the Eastern Meadowlark has declined throughout its range in the United States and Canada, with the greatest population drops (up to 10 percent per year) in the urbanized northeast. Photo by Marie Read.

20th century, as abandoned farms became shrubby "old fields." Today, as the taller forests return, most of these shrubland species are declining at rates of 3 to 4 percent per year across northeastern North America.

Grasslands

Since the middle of the 20th century, one entire ecological assemblage of land birds in North America has declined more steadily and more rapidly than any other: species occupying native grasslands and open country, such as Loggerhead Shrike, Eastern Meadowlark **(Fig. 10–28)**, Bobolink, Dickcissel, and a variety of grassland sparrows such as Grasshopper, Henslow's, and Field. These species and many others once occurred in great numbers throughout the extensive open lands from the western sagebrush prairies eastward through the Great Plains, and to the remnant native prairies and heath along North America's east coast.

Diversified agricultural practices during the first half of the 20th century created a patchwork landscape in which crop fields were interspersed with wide grassy margins, hedgerows, and fallow land, resulting in habitat that was excellent for many grassland birds. After about 1950, however, increasingly mechanical and chemical-intensive farming practices transformed this varied agricultural landscape into progressively more homogeneous habitat that was less suitable for grassland birds. Row crops and hay cultivars such as alfalfa were planted as monocultures. Grassy edges and hedgerows suitable for nesting habitat steadily disappeared. What had served as a "pseudo-prairie" (Warner 1994) for a hundred years became an intensively altered landscape of alien plants—habitat that could support few native bird species. As a consequence, the numbers of certain grassland birds dropped as much as 80 to 90 percent during the latter half of the 20th century (Robbins et al. 1986) **(Sidebar 1: A Summer Without Bobolinks)**. Chief among

(Continued on p. 10·36)

Sidebar 1: A SUMMER WITHOUT BOBOLINKS
Les Line

There's no place like a hilltop meadow on a perfect June day. This particular hill is nameless, but we are in the Taconic Range along the New York-Connecticut border, a queue of ancient mountains eroded over millions of years to gentle nubs. I've come to this nameless hill on this perfect June day to check on those harlequins of the hayfields, the Bobolinks, with their natty attire of black, white, and gold **(Fig. A)**.

A farm lane crosses the hill, dividing fields of timothy and alfalfa from an old pasture where oxeye daisies ripple in the breeze. A white-tail doe in ruddy summer coat stares and flicks her flag in mock alarm as I approach. Down the track a bluebird and a goldfinch squabble over some treasure; "riot of color" is a cliché, but how else to describe this unlikely fracas? A pair of Savannah Sparrows bedevil a kestrel. Barn Swallows swoop low, skimming grasshoppers and moths. Turkey Vultures wobble high, sifting the updrafts for the scent of some carcass.

And the birdsong, oh, the birdsong! Here, here, here pipes a Tufted Titmouse in the farm woodlot. Wichity-wichity-wichity answers a Common Yellowthroat in a willow gulch. A mockingbird practices its reper-

Figure A. Male Bobolink: Amid the flowers of a summer meadow, a male Bobolink stands out with its plumage of black, white, and gold. Like other grassland birds, it has suffered steep declines due to dwindling native grasslands, and the conversion of hayfields to crops such as alfalfa and soybeans. Photo by Marie Read.

toire from a weathered stump, while an Eastern Meadowlark whistles a paean to summer. On dried stalks of last year's mullein and burdock, Red-winged Blackbirds balance, heads back, epaulettes flashing, gurgling their konk-a-ree.

Above the glorious clamor floats the medley of a skyful of Bobolinks; for the moment all is right with our hilltop. How does the Bobolink's flight-song sound? "A bubbling delirium of ecstatic music that flows from the gifted throat of the bird like sparkling champagne," ventured the New England birdman Arthur Cleveland Bent. And William Cullen Bryant, in his poem "Robert of Lincoln," rendered it thus: Bob-o-link, bob-o-link; spink, spank, spink.

But words are inadequate to the task. The flood of notes, rising higher in pitch as the vocalist hovers over the grasses and wildflowers **(Fig. B)**, must be heard in person on a day like this, with popcorn clouds drifting in an azure sky. That opportunity, however, is rapidly slipping away in many places across North America, where the dismal prospect of a sum-

Figure B. Male Bobolink Singing in Flight: Singing its bubbling, metallic song, a male Bobolink flutters in slow circles over its hayfield territory. Photo by Marie Read.

10

mer without Bobolinks and other familiar songbirds of our prairies, fields, and pastures is very real.

Northern Illinois, for example, has seen its breeding population of Bobolinks plummet by 93 percent over 25 years. In New York, the count of Eastern Meadowlarks (see Fig. 10–28) is down 76 percent. Western Meadowlarks in Wisconsin, Henslow's Sparrows in Michigan, Lark Sparrows in Texas, Grasshopper Sparrows in Iowa, Dickcissels in Ohio, Lark Buntings in North Dakota, and Loggerhead Shrikes in Oklahoma all are declining at precipitous rates. That's the short list.

Indeed, despite all the attention paid by scientists and conservationists to the plight of North America's forest-nesting songbirds—many of which winter in Central and South America and on Caribbean Islands—grassland species have suffered steeper, more consistent, and more widespread losses over the last quarter-century. Continental populations of almost all grassland-nesting songbirds are declining, and for at least a dozen species that downward spiral is alarming. Two species—the Grasshopper Sparrow (**Fig. C**) and Henslow's Sparrow (see Fig. 10–50)—are losing ground faster than any other native bird; since 1966 their numbers across the United States and Canada have shrunk by nearly 70 percent. (Also see Fig. 9–47.)

"None of these birds is in immediate danger of global extinction," says biologist Jim Herkert of the Illinois Endangered Species Protection Board. "But considering the magnitude of the long-term population declines, local or even regional extinctions are likely unless we can identify and reverse their causes." Cassin's Sparrow, he notes, has all but disappeared from Kansas.

The benchmark for the health of the continent's avifauna is the

North American Breeding Bird Survey (BBS), organized in 1966 by the U. S. Fish and Wildlife Service. Drawing on the expert ears and eyes of amateur ornithologists, the BBS uses annual roadside counts along permanent routes to track changes in bird populations. Those tallies are made at the height of the nesting season, and the reports from thousands of volunteers reflect, for instance, the impact of habitat loss where subdivisions or clearcuts replace fields or forests. "But I never envisioned that the survey would

Figure C. Grasshopper Sparrow: Since 1966, numbers of Grasshopper Sparrows have declined 70 percent across the United States and Canada. Photo courtesy of Lang Elliott/CLO.

tell us we're facing a disaster," says Chandler Robbins, the Fish and Wildlife Service biologist who conceived the BBS.

There's a reason why the spotlight was first turned on birds of the eastern woodlands. Among the dozens of forest migrants are North America's best-known and most-colorful songsters—thrushes, orioles, tanagers, grosbeaks, flycatchers, cuckoos, catbirds, whippoor-wills, vireos, and warblers. We count on their return from the tropics each spring to brighten our

lives with flashing wings and bursts of melody. The first hint of a crisis in the making came from parklands around Washington, D.C., where once-abundant warblers and vireos had disappeared or become rare by the mid-1980s. When an analysis of 20 years of Breeding Bird Survey data revealed that many Neotropical migrants might be in serious trouble, a small army of scientists headed into the forests to find out why. They are still looking. "There could be twenty different reasons why twenty species are declining," says one researcher.

At first there was a lot of finger-pointing at the Third World. Pell-mell deforestation of tropical wintering grounds seemed an obvious cause for the troubling numbers coming from the BBS computer at Patuxent Wildlife Research Center in Maryland. But the focus soon shifted to problems on the breeding grounds. Forest fragmentation from suburban sprawl and rural development was one smoking gun. Isolated islands of habitat, such as well-intended open space preserves of 25 or 50 acres (10 or 20 ha), often are too small to harbor breeding populations of Neotropical migrants. The territory occupied by a pair of warblers may be only an acre in size—but that acre must be part of at least 1,000 acres of continuous habitat. Moreover, songbirds that nest in forest fragments are extra vulnerable to predation and to parasitism by Brown-headed Cowbirds.

The Bobolink and Dickcissel also are Neotropical migrants, and for the Dickcissel (see Fig. 9–44), events in South America almost certainly are contributing to its decline. "In winter the whole world population of Dickcissels settles into the rice-growing area of Venezuela," relates Stan Temple of the University of Wisconsin. To protect their harvest, large rice-growers spray Dickcissel roosts with deadly parathion. "The cropdusters," says Temple, "can kill hundreds of thousands of birds in a matter of minutes."

Most grassland songbirds, however, migrate only a short distance from their nesting grounds to wintering areas in the Sunbelt states and northern Mexico. "Their declines are associated almost entirely with problems in North America," declares Fritz Knopf of the National Ecology Research Center in Colorado. Predators, cowbirds, and the fragmentation of breeding habitat take a toll on the plains as well. And there are unknown factors at work here, too. "The future of grassland birds is clouded by ecological ignorance of wintering habitats and habits," says Knopf. But a smoking cannon has been found at farmlands in the Midwest and Northeast: changing agriculture practices.

Compared to the forest precincts east of the Mississippi, bird life of the open country is impoverished both in number of species and in blockbuster species such as the Scarlet Tanager of deciduous woodlands. "Grasslands cover 17 percent of the North American landscape," explains Knopf, "yet they provide primary habitat for only 5 percent of native bird species." Like the Bobolink, the Western Meadowlark is a world-class chorister, and very handsome with its sunflower vest and black cravat. A male Dickcissel—stuttering its name from the stem of a compass plant—suggests a meadowlark in miniature, and the Lark Bunting is a surprise in its Darth Vader robe. But several of the 20 or so grassland-nesting songbirds are drab, skulking sparrows with weak and decidedly unmusical songs. Bird watchers weaned on showy and boisterous forest inhabitants tend to dismiss them as LBJs—Little Brown Jobs. That's another reason why the predicament of grassland birds has been largely ignored.

"You have to work harder to find them," says Herkert, "but to me sparrows are as appealing as warblers." We had met on a June morning at Goose Lake Prairie, not far from the place where the Kankakee and Des Plains Rivers meet to form the Illinois.

Tallgrass prairie once covered two-thirds of Illinois, but less than 2,500 of the original 21 million acres survived two centuries of plowing. "Prairie birds are resilient," Herkert says, "and they were quite successful in shifting from native habitat to agricultural grasslands. But agricultural grasslands are going the way of the original tallgrass prairie."

Statistics: Between 1964 and 1987, 4,200,000 acres of pastures and hayfields—35 percent of the total—were converted to row crops, mostly soybeans, in Illinois, Iowa, and Indiana. Oats once were important nesting habitat for Bobolinks in Illinois. "You could walk into an oatfield and clouds of Bobolinks would rise up," Herkert recalls. "There would be dozens of males singing in flight at one time." But only 74,000 acres of oats were harvested in Illinois in 1992, down from 1,400,000 acres in the early 1950s. Again, soybeans have taken their place. Illinois farmers planted 9,500,000 acres of soybeans last year. "Now," says Herkert, "the Bobolink is the fastest declining bird in the state."

In the Northeast, meanwhile, millions of acres of hayfields and pastures have disappeared as farms are abandoned and fields slowly return to forest (**Fig. D**). Loss of habitat, then, is one problem weighing heavily on our grassland songbirds. But it gets worse.

a. Native Prairie in North Dakota   **b. Agricultural Grassland in New York**

Figure D. North American Grasslands: a. Native Prairie in North Dakota: Much of the native grassland of North America was converted to hayfields and crops in the 20th century, leaving remnant patches such as this prairie in North Dakota. Most prairie birds shifted successfully from native habitat to agricultural grasslands; however, the recent trend toward converting hayfields and pastures to row crops such as soybeans is rendering even these agricultural landscapes unsuitable as breeding habitat for some species. Photo courtesy of Kenneth V. Rosenberg. **b. Agricultural Grassland in New York:** In the Northeast, agricultural grasslands, such as this one in Montgomery County, New York, are dwindling in number as farms are abandoned and the land reverts to forest. Photo courtesy of Jeffrey Wells.

10

During the first week of May, after a long flight from the South American pampas, Bobolinks return to their shrinking breeding grounds in northern Illinois. The males, splendid in full nuptial dress, are the first to arrive. You can't mistake them. No other songbird is entirely black below and mostly white above, like a tuxedo worn upside-down. Territories will be divvied up by the time the females in their buff plumage drop in a week later, and courtship soon begins, with spectacular aerial chases and strutting dances in the long grass.

By the end of May, every Bobolink nest—a simple hollow on the ground that is carefully hidden by grass and weeds (see Fig. 8–29a)—will be filled with its consignment of five or six spotted eggs. The female will incubate the clutch for about ten days, and she and her mate will care for the nestlings until they take wing two weeks later. If the nest is in a hayfield, however, chances are they won't make it. Like a grim reaper, the farmer's mower probably will destroy the eggs or chicks; the parent birds could also be killed. "Virtually all hay-cutting in Illinois begins before Bobolinks have a chance to raise their young," Herkert relates, "and there's not enough time between mowings for the adults

to raise a second brood." Henslow's Sparrows, Grasshopper Sparrows, and meadowlarks are also hit hard by early and frequent hay-cutting.

It wasn't always so. Hayfields, once planted in timothy and clover, were not mown until late June; the cut hay was left in the fields to dry before being baled. This gave Bobolinks, meadowlarks, and sparrows ample time to raise their broods. "The early 1950s saw a switch to alfalfa, and now it's 90 percent of the Illinois hay crop," says Herkert. "New alfalfa varieties can be mowed earlier and earlier, the first green cut is blown right into the silo, and the fields yield as many as three more cuttings—in a typical year we'll lose up to 95 percent of our hayfield nests."

The story is the same throughout the Midwest as well as in the dairy country of the Northeast, where there were few native grasslands, and Bobolinks arrived only after the original forests had been cleared for farms. In a New York study led by biologist Eric Bollinger, 51 percent of the eggs and nestlings were found to have been destroyed when a hayfield with active nests was mowed (**Fig. E**). "Subsequent nest abandonment, nest predation, raking, and baling increased mortality to 94 percent." The scientists also estimated that hay-cropping killed more than half

of the young Bobolinks that had already fledged. In fields that were undisturbed, however, 80 percent of the Bobolink nestlings survived.

Forget the obvious fix. Farmers concerned with hay production quotas won't delay mowing until the nests are empty. In an article lamenting "a silence of meadowlarks" on Wisconsin's farmlands, economist James Eggert writes: "Despite their sweet song, these birds have no voice economically or politically. They represent a zero within our conventional economic accounting system (we don't even buy birdseed or build birdhouses for meadowlarks). Their disappearance would not create even the tiniest ripple in Commerce Department spreadsheets that supposedly measure our standard of living."

Yet the extinction of meadowlarks, Eggert argues, "would be wrong ethically and also would diminish the quality of our lives." He urges his colleagues to incorporate ecological thinking and values with market thinking and values. "Call it, if you wish, Meadowlark Economics."

In a rare display of vision where national farm policy is concerned, Congress in 1985 passed a Food Security Act that authorized the U. S. Department of Agriculture to lease millions of acres of marginal

Figure E. Hay Mowing: Some changes in farming practice during the past 50 years have negatively impacted breeding success of ground-nesting grassland birds. Today, hayfields are harvested earlier in the season and more frequently than in the past. In many regions of the country, early harvesting times coincide with the incubation period, causing many nests to be destroyed. Frequent harvesting of fields prevents birds from renesting. Photo by Marie Read.

cropland. Landowners are paid up to $50,000 a year for 10 years to plant and maintain perennial vegetation. The objectives of the Conservation Reserve Program (CRP) were to reduce both soil erosion and crop surpluses as well as to restore wildlife habitat. For prairie songbirds, the CRP has been a windfall. Nearly 32 million acres (13 million ha) of grasslands have been enrolled in the program.

A third of those conservation reserve grasslands are spread over eastern Montana, the Dakotas, and western Minnesota, where biologists from the Northern Prairie Wildlife Research Center in North Dakota evaluated their use by breeding songbirds. Lark Buntings and Grasshopper Sparrows, two of the most troubled prairie species, were the most common birds in CRP fields. Population densities of 16 songbirds, including Bobolinks and Western Meadowlarks, were seven times greater in CRP grasslands than in croplands. When CRP leases began to expire in 1996, concerns arose that, unless the program was continued, most farmers would return their grasslands to cultivation whether or not the crops were needed. (See Editor's Note below.)

As with forest tracts, bigger is better where grasslands reserves are concerned. "The number of breeding bird species encountered increases significantly with the size of a grassland fragment," says Herkert. Grasshopper Sparrows, Henslow's Sparrows, Savannah Sparrows, Bobolinks, and Eastern Meadowlarks are more likely to occur on large grasslands, his Illinois studies show. Grasshopper Sparrows were rarely found on an area smaller than 25 acres (10 ha) even though the typical size of their territory is about 2 acres (0.8 ha).

Nest predation also is higher on small grassland fragments, particularly those bordered by a grove of trees or shrubby fields. "Pick your predator," says Herkert. "Snakes,

raccoons, skunks, Blue Jays, crows." Then there is that blackguard of the blackbird tribe, the Brown-headed Cowbird.

We may never unravel the riddle of why the cowbird lost the art of nest building and became what biologists label an "obligate brood parasite." Perhaps it was the birds' choice, in the distant past, of a nomadic lifestyle. They were always on the move following herds of now-extinct ungulates and then bison and finally cattle. A pregnant cowbird would drop her eggs in any convenient nest. Usually she would destroy the clutch of the hapless owners, but if both cowbird and host eggs hatched, the alien nestling, larger and more aggressive, would have the better shot at surviving.

As farms and pastures replaced virgin forests, the cowbird expanded its range eastward from its Great Plains homeland. The eastern forests offered dozens of unsuspecting species to victimize, and biologists have indicted the cowbird in the decline of a number of these Neotropical migrants, including the Wood Thrush and the endangered Kirtland's Warbler. Study after study, moreover, has shown that songbirds nesting in smaller forest fragments are especially vulnerable to parasitism (see Fig. 10–64).

Now, research by Stephen Davis of the University of Manitoba suggests that the same rules apply to fragmented grasslands. Cowbirds parasitized 69 percent of the songbird nests on a 55-acre (22-ha) plot but only 20 percent of the nests on a 158-acre (64-ha) site, Davis reported. It was no surprise that multiple parasitism was prevalent in the Manitoba study, since a single female cowbird can produce more than 40 eggs during the breeding season. Western Meadowlark nests contained more cowbird eggs—as many as eight—than those of any other member of the host community.

A Grasshopper Sparrow was singing—well, buzzing like a certain

long-legged insect—as Jim Herkert and I strolled the tallgrass nature trail at Goose Lake Prairie. It was neat renewing the acquaintance of a Little Brown Job whom I hadn't encountered since my Birdwatching 101 days in Michigan 30 years ago. It was neat still being able to hear it, because the grasshoppery notes are too faint and high-pitched for some aging ears. For a pleasant moment, then, I shoved troublesome questions about habitat loss, fragmentation, hay mowing, predators, cowbirds, and all those dire North American Breeding Bird Survey statistics into the background. For a moment.

The big question, of course, is whether the sweeping declines in grassland songbird populations can be reversed. Scientists will tell us why they are happening, and what might be done. But doing it, except on a modest scale on public lands, seems an impossible dream. There would be some cause for optimism, however, if the Conservation Reserve Program were kept alive and expanded. Call it Dollars for Meadowlarks. That's a fair exchange.

Meantime, the hilltop farm where this grasslands tour began has recently changed hands. The new owners, I'm told, intend to increase the farm's profits. They'll be mowing the hayfields earlier and more often, for sure. And the old pasture could soon be growing corn, not Bobolinks. ∎

This article originally appeared in Wildlife Conservation, July/August 1994, and has been modified with permission.

Editor's Note:
In 1996 (after the original publication of this article), the Conservation Reserve Program (CRP) was renamed the Conservation Reserve Enhancement Program (CREP) and targeted to specific geographic areas. The new program was authorized by Congress through the Federal Agriculture Improvement and Reform Act. For more information, visit <**www.fsa.usda.gov/dafp/cepd/crep/crephome.html**>.

these was the Greater Prairie-Chicken, which once occurred from the Rocky Mountains east to the Atlantic Coast, but whose range today is limited to native tallgrass prairies of the Midwest and Great Plains. Only a few dozen Attwater's Prairie-Chickens (a distinct subspecies of Greater Prairie-Chicken, once widespread in Texas) still display on the native prairies south of Houston **(Fig. 10–29)**. Formerly plentiful, these game birds disappeared as their habitat became fragmented by ranchlands that were ditched, drained, criscrossed with roads, and "improved" for cattle grazing through the introduction of foreign grasses.

Figure 10–29. Attwater's Prairie-Chicken: In 1900, more than one million Attwater's Prairie-Chickens inhabited the gulf coastal prairies of Texas and Louisiana. In 1976, over 2,000 of these unique coastal prairie grouse still survived in the wild. As of June 2001, only an estimated 46 birds were left in the wild in two counties. This distinct race of the Greater Prairie-Chicken is doomed unless protected and restored prairie habitats can be restocked with captive-bred birds. Captive birds currently exist at four conservation and research centers in Texas. Photo copyright 2001 Marc Dantzker.

Southwestern Riparian Habitats

In the arid southwestern regions of North America, an estimated 80 percent of all vertebrate species depend on habitats associated with permanent, seasonal, or subsurface water. Containing cottonwoods, willows, sycamores, and a variety of shrubs and grasses, these "riparian" (river-edge) habitats are far more productive than the dry scrubs, deserts, and grasslands that immediately surround them **(Fig. 10–30a)**. For this reason, riparian habitats are a critically important resource for bird communities throughout lower elevations of the west. However, because these habitats supply shade and direct access to water, they also serve as loafing, feeding, and drinking areas for grazing livestock. Because riparian habitats also supply rich flood plain soils for agriculture, they often are channelized or sacrificed for water impoundments, they are easily invaded by alien plants, and they are used for a wide variety of recreational and commercial purposes by local towns. As a result of these human influences, about 95 percent of all riparian habitat in western North America has been degraded or destroyed over the past 100 years (Krueper 2000). Today, only scattered patches remain for use by the native bird community, and species such as Willow Flycatcher

(Fig. 10–30b), Yellow-billed Cuckoo, Bell's Vireo, and Yellow-breasted Chat have declined precipitously in recent decades (Ohmart 1994). Without significant changes in management strategies, many of these areas stand to lose their native bird diversity altogether.

Shorebirds

As discussed in Ch. 5, Sidebar 2: Showdown at Delaware Bay, many species of sandpipers and plovers are beginning to decline significantly as a consequence of loss or degradation of habitats—natural shorelines, mudflats, seasonal marshes, and flooded agricultural land, for instance—used for refueling during migration. Although firm numbers are still hard to obtain for most of these species, data from

a. San Pedro River

Figure 10–30. Southwestern Riparian Habitat: a. San Pedro River: Riparian, or river-edge, vegetation provides essential habitat for native birds in the otherwise arid environment of the North American southwest. However, birds in riparian habitats are vulnerable to various destructive human activities, such as overgrazing of riverbank vegetation and erosion by livestock, channelization for irrigation, conversion to agriculture, and recreational activities. Studies estimate that 95 percent of the riparian habitat in western North America has been degraded or destroyed over the past century. One of the few remaining areas is the San Pedro River in Arizona. Once ravaged by grazing cattle, this river was protected in 1988 when Congress created the San Pedro National Riparian Conservation Area. Lush cottonwoods, willows, shrubs, and grasses now line its banks, providing habitat for over 400 species of birds. Photo by Marty Cordano. **b. Willow Flycatcher:** The southwestern subspecies of Willow Flycatcher is restricted to dwindling riparian habitats. Although other subspecies inhabit shrubby, wet habitats from Maine to British Columbia, BBS data show that the Willow Flycatcher has declined steeply since 1966, and in 1995 the southwestern subspecies was listed as endangered. Photo courtesy of Leslie Chalmers/CLO.

b. Willow Flycatcher

traditional breeding or stopover locations suggest that species such as Snowy Plover, Red Knot, Red-necked and Wilson's phalaropes, and even the familiar, beach-loving Sanderling have shown stark declines in recent years.

Conservation Problems: The Ecology of Extinction

■ The chief objectives in bird conservation are (1) to identify population declines that signal underlying degradation of resources, habitats, or whole ecosystems, and (2) to derive biological solutions that can be implemented in the real world to repair the underlying cause of the declines. This section and the next treat these two objectives in detail, and show how birds serve as both warning signals and beacons for measuring our ability to maintain healthy natural systems around the world.

Except when a volcano obliterates an ancient island (as on Krakatau in the western Pacific Ocean, in 1883), no bird species becomes extinct all at once. Instead, species decline and go extinct through a combination of forces acting over time. Protecting a species, therefore, requires an understanding of how its population has been regulated naturally, and what forces are disrupting its overall balance of birth and death rates. In short, coming up with conservation solutions first requires identifying the problems.

Describing why the Passenger Pigeon went extinct, for example, seems easy at first glance. Humans ate them in vast numbers while systematically wiping out their breeding habitat. Biologists state this scenario a bit more technically: humans killed the adults and disrupted the production of juveniles, causing net death rate to exceed the rate at which new individuals were added. In fact, however, the final years of the Passenger Pigeon's disappearance still present a biological mystery. The species had become rare over most of its range, but for a decade or more thousands of adults still migrated northward in scattered bands, too few in number to pay the expenses of meat-hungry market hunters. Why, then, couldn't these scattered individuals begin to recover, instead of vanishing completely in just a few years? It is speculated that Passenger Pigeon numbers had fallen below a critical threshold required to stimulate the hormones and behavior patterns necessary for successful reproduction. Isolated individuals and small flocks instinctively searched for large colonies rather than nesting on their own. The species may have succumbed to a specialized feature of its biology known as the **Allee effect** (named after a founder of the study of animal behavior), in which reproductive behavior or social structure become disrupted if the population density falls below a required, threshold level.

The Allee effect is an example of a biological attribute that provides evolutionary advantages under typical circumstances, but also may cause a species to be unusually prone to extinction if circumstances change. Throughout the 20th century we became aware of many such attributes as the scientific disciplines of ecology, population biology, animal behavior, and genetics matured. As it became clear that many species around the world were declining and going extinct, scientists began to apply these separate disciplines together in an effort to suggest solutions. Thus, during the 1970s, the applied science known as **conservation biology** was born **(Fig. 10–31)**. Conservation biology allowed scientists to help predict and shape the future of the natural world, and to this day the field continues to grow in its number of practicing professionals and in the importance of its contributions.

"Applied sciences" combine basic principles of several fields to solve everyday problems and to accomplish specific, real-world tasks. Engineering, for example, is an applied science that uses basic principles of mathematics, physics, and chemistry to build buildings and bridges, to design electronic circuitry, or to construct fuel-efficient automobiles. By the same token, to address the problem of species extinction, scientists employ basic principles of biology to derive solutions that work. However, although engineers can take as much time as necessary to design and build better objects, conservation biologists typically face the grim reality of impending disaster. Moreover, engineers typically work with highly predictable principles and media. In contrast, conservation biologists work with a host of complex principles, many of which remain poorly understood, and they work with ecological systems that differ vastly from one another. Their subjects are as tiny as a minnow and as large as blue whales.

The principles used to interpret changes in natural populations in such varied settings are just now beginning to be understood. The complexity of interactions among natural populations, and the relative youth of the sciences engaged to understand them, help to explain why so many bird species went extinct during the 19th and 20th centuries before humans could save them. Not one of the individual fields of basic science that compose conservation biology existed at the time the Passenger Pigeon and Carolina Parakeet were going extinct.

Figure 10–31. Conservation Biology Premier Issue: With the publication of this first issue of the journal Conservation Biology, in May 1987, the Society for Conservation Biology was born. The first cover artwork was captioned "The Cerulean Paradise-flycatcher of Sangihe Island, Sulawesi, Indonesia, may have become extinct unnoticed. (Painting by Stephen V. Nash.)" Although the suggestion has since proved to be untrue—a tiny population was rediscovered in 1998—the survival of the Cerulean Paradise-flycatcher remains extremely tenuous. Reprinted by permission of Blackwell Science, Inc.

Birth Rates and Death Rates

As discussed in Chapter 9, any bird population that remains stable in nature over the long term must have birth rates and death rates that are in balance. For such a population, numerous life history features (such as age at first breeding, clutch size, period of juvenile dependency, reproductive life span, and migration pattern) have become adapted over thousands of generations to a specific set of environmental conditions. Therefore, virtually any change in these conditions, whether or not the change is caused by humans, will tend to alter the birth rate, the death rate, or both, and thereby will tend to upset the

10

balance. To understand any long-term change in population numbers, then, we must identify what has changed in the environment, and how the change affects birth rate or death rate. Sometimes identifying changes is easy, sometimes not.

Population Increases

Environmental changes that alter birth rate or death rate do not always cause a species to decline, nor do they always affect the entire species. In eastern North America, Northern Cardinals, Tufted Titmice **(Fig. 10–32)**, and Red-bellied Woodpeckers all have spread steadily northward during the 20th century. Such range expansions reflect a shift in the environmental conditions governing the occurrence of these species along their northern border. For cardinals and titmice, death rates during the harsh northern winters probably diminished as a result of additional winter food (sunflower seeds) and shelter (ornamental plantings) provided by humans. For the woodpecker, population expansion probably reflects increased birth rates as 18th and 19th century farmlands revert to forest, providing more nest sites and resources for raising young.

Figure 10–32. Range Expansion of the Tufted Titmouse: A familiar visitor to backyard feeders, the Tufted Titmouse **(a)** spread northward during the 20th century, aided by supplemental food and shelter provided by humans and by the return of forested landscapes. Sequential maps (**b**, facing page) show Christmas Bird Count (CBC) data for 10-year intervals beginning in 1901. Colored dots represent CBC circles where the titmouse was reported; gray dots show circles where it was not reported. From its scattered distribution in 1901, the titmouse has gradually spread into the upper mid-Atlantic states, New England, and southern Canada. In the upper Midwest, the bird's northern boundary has remained stable, presumably because harsh winters in the center of the continent prevent its spread. Photo courtesy of H. Mayfield/CLO. Maps copyright BirdSource/CLO.

a. Tufted Titmouse

Introduced species sometimes undergo major population expansions after long periods of being restricted to a small area, after birth rates and death rates become adjusted to the new environment. House Finches transported from California and released on Long Island in 1940 remained in the New York City area for two decades before undergoing an extraordinary range expansion from 1960 onward. Cattle Egrets **(Fig. 10–33)** appear to have colonized northeastern South America from their native African savanna as early as the late 1800s, but remained confined to the southern Caribbean region until about 1940, when they first colonized Florida. By 1954 they had become established throughout the southeastern United States, and by the 1970s were breeding in most Canadian provinces. The remarkable expansion in the Cattle Egret's range and total numbers continues to this day.

Many wading birds whose numbers were severely depleted during the plume-hunting days (especially Great Egret and Snowy Egret) underwent steady range expansions throughout the 20th century. These rebounds occurred after humans eliminated the factors that had put these populations out of balance: plume hunters had both increased adult death rate (by killing breeding birds for their decorative nuptial plumes) and decreased birth rate (by raiding the breeding colonies and driving nesting success to near zero).

Not all population increases represent good news. During the late 20th century, the number of Snow Geese breeding on the arctic tundra increased steadily as a result of changes in both birth and death rates.

Figure 10–32 (Continued). **Range Expansion of the Tufted Titmouse. b. Christmas Bird Count Data**

Figure 10–33. Cattle Egret: Native to Africa, southern Europe, and tropical Asia, the Cattle Egret underwent explosive, worldwide range expansion during the 20th century. The species colonized northern South America in the late 1800s, apparently by flying across the Atlantic Ocean from West Africa—a feat that might seem unlikely but for the fact that Cattle Egrets are still regularly seen from ships in the mid-Atlantic Ocean. In North America, the Cattle Egret was first sighted in Florida in 1941, was breeding there by 1953, was well established in the eastern United States by 1954, and nested as far north as Minnesota by 1970. It was established in most Canadian provinces by 1974, and was nesting in Saskatchewan by 1981, the same year that it was first reported in southern Alaska. Strongly migratory habits coupled with a tendency of juveniles to wander widely after the breeding season have no doubt promoted the Cattle Egret's range expansion, which continues to this day. Photo courtesy of Lee Kuhn/CLO.

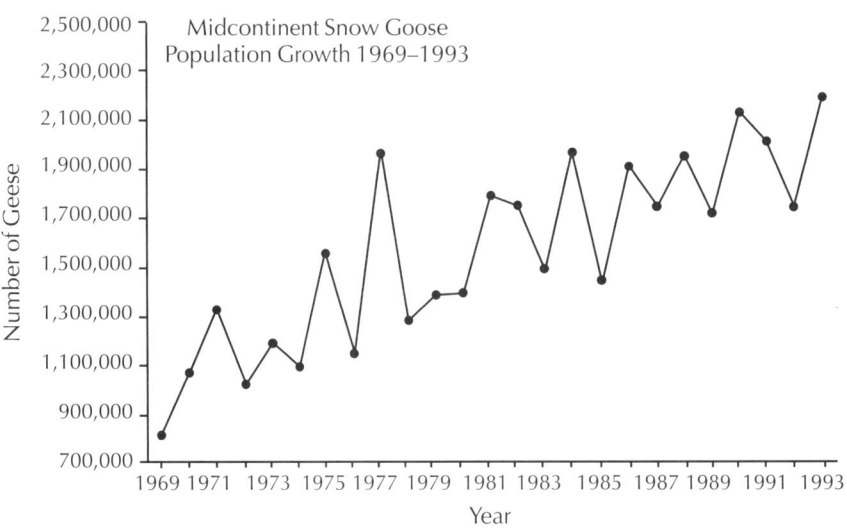

Figure 10–34. Snow Goose Population Growth: The mid-continent Snow Goose population, which breeds in North America's central Arctic, tripled in size between 1969 and 1993, and today numbers well over 3,000,000. The population continues to grow at 5 to 8 percent per year. Several factors are involved. Winter food, inadvertently provided by humans in the form of waste cereal grains, has increased as more land is converted to agriculture, and because farmers often delay post-harvest plowing of crop fields until spring. Many national, state, and private waterfowl refuges have been created to provide sanctuary and to increase waterfowl populations, yet waterfowl hunters have declined in number, so harvest rates have decreased. Adult goose survival rates have increased, probably because of enhanced winter food availability. Finally, recent warmer springs on the breeding grounds have allowed earlier nesting and increased reproductive success. Unfortunately, the burgeoning Snow Goose population is seriously degrading the arctic tundra, especially near Hudson Bay, Canada (see Ch. 10, Sidebar 2). Graph adapted from Rockwell et al. (1997).

The number of geese breeding in central Canada west of Hudson Bay burgeoned from about 800,000 in 1969 to over three million in 1996 **(Fig. 10–34)**, and the number continues to increase. This change in numbers on the breeding ground resulted from changes in agricultural practices on the migration and wintering grounds. As large-scale farms created enormous quantities of rice, corn, and other grains in the Midwest and Gulf states, survival rates of both adult and juvenile geese increased. In addition, adults began to arrive on their breeding grounds in such fattened condition that they could breed earlier and lay larger clutches of eggs. Thus, changes had affected both the birth rate and death rate, causing a genuine population explosion.

The Snow Goose explosion is now causing the reverse effect for many other bird species breeding on the productive arctic tundra. Snow Geese forage by snipping and tugging up the roots of native tundra vegetation, and their numbers in the 1980s and 1990s became so large that vast areas of once-verdant tundra were reduced to sterile, "hypersaline" (extra salty) dirt. Because these areas also were important breeding habitat for many migrant shorebirds (such as Hudsonian Godwit, Lesser Yellowlegs, and American Golden-Plover **[Fig. 10–35]**), these species began to decline. Recovery of devastated tundra vegetation is believed to require hundreds of years, and by the late 1990s many biologists were expressing great concern for the future of the arctic ecosystem **(Sidebar 2: The Best Laid Plans: What Happens When Conservation Efforts Work Too Well?)**.

The question of how to control Snow Goose numbers has become a knotty issue. Many organizations, including the National Audubon Society, favor greatly expanded hunting seasons and even commercial harvesting of Snow Geese to reduce their numbers to a level permitting the tundra ecosystem to recover. To date, no solution has been agreed upon, and the Snow Goose population continues its dramatic increase unabated.

(Continued on p. 10·48)

Figure 10–35. American Golden-Plover: Nesting primarily at arctic and subarctic latitudes, American Golden-Plovers place their simple nests on the ground amid the lichen-covered rocks, mosses, and short grasses typical of tundra habitat. Like many tundra-nesting birds, they and their young rely for a food source on the huge flush of insects, which occurs during the arctic summer. American Golden-Plovers are likely to be profoundly affected by the severe destruction of tundra habitat caused by the burgeoning population of Snow Geese breeding in the Canadian Arctic. Photo courtesy of J. R. Woodward/CLO.

10

Sidebar 2: THE BEST LAID PLANS: WHAT HAPPENS WHEN CONSERVATION EFFORTS WORK TOO WELL?

Robert F. Rockwell, Kenneth F. Abraham, and Robert Jefferies

Each spring and fall vast, undulating skeins of Snow Geese pass overhead, noisily making their way north to arctic and subarctic breeding grounds or south to wintering areas. Who could fail to be awed at seeing thousands of these striking birds speeding powerfully overhead or rising in a massive swarm from a freshly harvested field **(Fig. A)**? For many of us, their migrations herald the arrival of spring and fall each year. It's not surprising that so many biologists, wildlife managers, hunters, naturalists, and other concerned citizens have committed so much time and effort to preserving these birds. Snow Geese are part of our natural heritage and their absence would be unthinkable.

Thanks to the adaptability of these birds and to conservation efforts in the United States and Canada, the Snow Goose population in the central portion of North America has grown nearly 300 percent since the 1960s. Researchers estimate that this population—which breeds primarily along the shores of Hudson Bay and the Foxe Basin and winters from Iowa and Nebraska south to the Gulf Coast—now numbers more than three million birds in midwinter. At a time when many species of birds are declining and others are becoming extinct, you'd think that the Snow Goose's burgeoning population would be cause for celebration. Instead, scientists are concerned that the increasing numbers of geese may soon lead to an ecological catastrophe as these voracious feeders turn their delicate arctic habitat into a barren wasteland.

How did this dire situation come about? And what can be done to alleviate the problem? To understand this dilemma, it is important to trace this species' long-term history and see what a profound impact human actions have had on its numbers.

Most Snow Geese in central North America originally wintered in brackish marshes along the Gulf Coast in Texas and Louisiana. Destructive feeders, they grubbed out and consumed tubers and roots of marsh vegetation with their large, serrated beaks. But the food resources in these natural marshes were finite and acted as a control on the goose population. As the numbers of geese exceeded the available food supply, the birds' mortality rate would increase, reducing their population to a level that their traditional wintering areas could support.

In the two decades following World War II, many coastal marshes were lost or severely degraded through increased commercial development. At the same time, however, farmers increased the production of rice crops on private lands adjacent to the birds' traditional coastal wintering grounds. These agricultural areas provided at least a million acres of wintering habitat with abundant food, not only in the post-harvest rice stubble, but also in adjacent soybean fields and pastures. The winter populations of the geese began to increase and the birds extended their range northward, feeding on rice fields up to 150 miles (241 km) inland from the coast.

Converting inland grasslands to agricultural croplands was an absolute boon for the geese. The crops provided the ever-increasing numbers of geese with a large food subsidy, so that the depletion of the winter food supply in their traditional winter habitats no longer acted as a population control on the birds. Their winter mortality decreased

Figure A. Snow Geese in Flight: A flock of Snow Geese swirls through the sky like a blizzard. In 1996, biologists estimated that the Snow Goose population in central North America had tripled in size since the 1960s. This huge increase has devastated the bird's breeding habitat—the fragile tundra of the shores of Hudson Bay and the Foxe Basin. Photo by Marie Read.

10

Figure B. Experimental Exclosure at La Pérouse Bay Study Area: Voracious feeders, Snow Geese grub out plants by their roots, causing serious damage to the delicate, arctic plant communities. To demonstrate the effects of grazing geese, researchers constructed exclosures surrounded by goose-proof fences. Inside the fence is lush vegetation; outside is a barren wasteland. Photo by Paul Matulonis.

significantly, their body condition in spring improved, and their reproduction rate increased markedly.

Federal and state agencies further expanded the winter range and migration staging areas of the Snow Geese by developing national and state wildlife refuges on the northern prairies. These refuges were intended to restore wetland habitat for breeding and migrating waterfowl. Wildlife managers at the refuges manipulated crops to provide food for the birds—a practice that augmented the large food subsidy the geese were already reaping from private lands where agricultural activity also had increased.

Although hunting pressure was initially intense in areas adjacent to the refuges, political lobbying in the 1970s brought changes, including the establishment of no-hunting zones and restricted goose harvests. The combined effects of less hunting pressure and increased food subsidies contributed to a nearly 50 percent reduction in adult mortality (from 22 percent to 12 percent). This reduction was also influenced by a decline in both the number of hunters and the number of days they hunted.

The size of the winter Snow Goose population increased dramatically as it readjusted to the extensive increase in the quantity and quality of the foraging habitat, both on the wintering grounds and along the migratory flyways. The growth of the goose population in central North America was stimulated further during the late 1960s and 1970s by a temporary warming trend in the Hudson Bay and Foxe Basin nesting region, which resulted in an earlier spring melt, earlier nesting, and increased reproductive success.

Much of our information on the growing Snow Goose population in central North America and the birds' effect on their coastal tundra breeding habitat comes from a long-term study of the goose colony at La Pérouse Bay near Churchill, Manitoba, in Canada. When studies began there in 1968, approximately 2,000 pairs of Snow Geese were nesting in the willow and lyme grass fringes of the coastal salt marsh. By 1990, the colony had grown to 22,500 pairs—an average annual increase of nearly 8 percent. Although much of this account is drawn from that study, research at several other nest-

ing sites indicates that the results are applicable on a broad scale.

Like most arctic-breeding geese, Snow Geese accumulate the nutrient reserves they will need to produce and incubate their eggs during their spring flight north. They also feed at the breeding colony while they search for a nest site and begin laying eggs. Because the geese arrive before the vegetation has begun growing, they initially feed by grubbing below the surface to get the nutrients stored in the roots and rhizomes of plants. By mid-June, grasses and sedges are growing, and the adult geese and their broods of goslings graze on the above-ground portion of their salt-marsh forage plants.

As the population at La Pérouse Bay grew, the overall demand for food throughout the season increased. But grubbing had the most serious impact on the ecosystem **(Fig. B)**. When adults grub for roots and rhizomes early in the season, they destabilize the thin arctic soil so that melting snow and spring rains can cause erosion. In some instances, ponds form and are then enlarged each year as the birds grub along the edges. As the number

Figure C. Degraded Habitat at La Pérouse Bay: Geese aggressively feeding on roots and rhizomes early in spring can destabilize the thin arctic soil, then snowmelt and rain cause erosion. Often ponds form, becoming larger each year as the geese feed around the edges. Photo by Robert F. Rockwell.

and size of these ponds increase, the amount of available forage declines **(Fig. C)**.

The damage caused by the birds' grubbing at La Pérouse Bay was made worse by a series of late spring seasons in the High Arctic. Geese from more northern colonies delayed the last portion of their migration and continued to feed at southern colonies. In 1984, for example, more than 100,000 staging geese destroyed much of the vegetation on one of the main brood-rearing areas on the east side of La Pérouse Bay in less than three days.

Beyond simply removing plants, foraging by Snow Geese—especially grubbing—leads to other changes in the coastal ecosystem. When the vegetation is removed, evaporation from the soil surface increases and inorganic salts from underlying sediments move to the surface, raising soil salinity. As salinity increases, the growth and survival rates of forage plants in the coastal marsh decline. Willows and other vegetation immediately adjacent to the marsh begin dying as the process intensifies. Ponds and bare soil dry out and the surfaces crack. Ultimately, all that remains is a barren forest of dead willows and a few nearly inedible plants

that are capable of surviving in soil with a salinity level that sometimes reaches three times that of sea water **(Fig. D)**.

Thus, the chain of events that began with a single species in a simple food chain consuming the food sources in a habitat too rapidly leads ultimately to the deterioration of the entire ecosystem. The unfortunate consequence of this phenomenon (called a trophic cascade) is that staging, foraging, and nesting habitats are lost—not just to Snow

Geese, but to all the other species sharing the marsh and adjacent areas. These include other species of waterfowl, shorebirds, marsh birds, upland birds, and numerous passerines.

For Snow Geese, the decline in the quantity and quality of foraging habitat at La Pérouse Bay has led to decreases in the size and survival rate of juvenile geese and a reduction in the reproductive success of the adults that continue to use the traditional nesting and foraging areas **(Fig. E)**. You might suppose that this would slow the growth of the goose population and ultimately place a cap on its size, much as the limited winter food resources in the Gulf Coast marshes once did. Unfortunately, this has not been the case. As conditions worsen, increasing numbers of adults are moving to adjacent areas to nest and raise their goslings in coastal marshes that are not yet as degraded as their traditional breeding sites. The success of these dispersers is sufficiently high that the more widely distributed population continues to grow. And with it, the cycle of habitat degradation and dispersal continues to spread across an ever-expanding geographic area.

In addition to La Pérouse Bay,

Figure D. Dead Willows: Grubbing by geese sometimes causes salts to move to the soil surface, killing many plants such as these willows. Photo by Robert F. Rockwell.

Figure E. Snow Goose Family: A pair of Snow Geese and two half-grown goslings search in vain for food in their denuded habitat. Photo by Barbara Pezzanite.

there are at least three other sites on Hudson and James Bays where habitat deterioration and trophic cascades have also begun. This process will continue until adequate forage no longer exists for the members of the dispersing goose families. By that time, miles of coastline and the habitat of numerous other species of birds will have become seriously degraded. And because of the high salinity, cold temperatures, and short growing seasons in these areas, it will take decades for these habitats to recover.

The most hopeful aspect of this saga is that the problem may have been identified in time to take action that can minimize further destruction of the sensitive arctic coastal ecosystems. The United States and Canada have established the Habitat Working Group, a joint study group composed of researchers and land managers set up to examine the ecological impact of Snow Geese and develop a plan to reduce the population of this species to a level that does not threaten the ecological integrity of its breeding habitat. Proposals on the table include increasing traditional egg collecting by native peoples, liberalizing hunting regulations for Snow Geese, and establishing special hunting seasons. The success of these actions—and even whether they can be implemented at all—depends on the willingness of a large and diverse community of scientists, wildlife managers, and naturalists to cooperate in the endeavor. One thing that we must all do is to try to alter our traditional focus on the conservation of single species, whether they are rare or abundant. The dilemma of the Snow Geese is not simply that there are too many gaggles of geese, but that this species is only one member of a diverse and fragile arctic ecosystem. That ecosystem and all its members require and deserve our attention.■

This article originally appeared in Living Bird, Winter 1997.

Figure 10–36. Imperial Parrot: Many of the world's large, tropical parrots are fast declining in the wild because of pressure from the global exotic pet trade. Purple and green in color, the spectacular Imperial Parrot inhabits montane forest on the island of Dominica, in the Lesser Antilles. Its desirability as a cage bird, combined with habitat loss and hunting, has led to a significant population decline, and its wild population reached a low of 80 to 100 birds in 1993. Although poaching still poses a threat, a successful local education program—spearheaded by the nonprofit conservation organization RARE—has helped to reduce local trade. Conservation activities had helped to raise the wild Imperial Parrot population to 250 to 300 birds by the year 2000. Painting by Tracy Pedersen. From A Guide to the Birds of the West Indies, Princeton University Press, 1998.

Direct Exploitation

Most shifts in birth or death rates caused by humans have resulted in population declines. The most obvious cases through history have been the result of direct exploitation, in which the killing of adults was accomplished at such a scale that the birth rate of surviving adults could not compensate for the new death rate imposed by humans. The most familiar examples involved killing of adults for food (Dodo, Passenger Pigeon, Eskimo Curlew), or ornamentation (egrets and herons).

During the late 20th century, global trade in exotic pets began causing an unusual twist to the birth rate-death rate story. Many large, tropical parrots became extremely rare in their native ranges because humans were stealing young parrots from their nests. This did not change the biological birth rate, as surviving adults continued to produce chicks, but it vastly reduced the effective birth rate, as most chicks were smuggled out to foreign countries. In the Lesser Antilles, for example, illegal sale of Imperial Parrots **(Fig. 10–36)** reduced the effective birth rate to near zero in mountaintop forests of Dominica. Without any effective recruitment (the addition of individuals, either through birth or immigration) to compensate for natural adult mortality, the population of this spectacular parrot dropped to several hundred individuals. Today, conservation education throughout the Lesser Antilles (especially through the activities of the RARE Center for Tropical Preservation, a private conservation organization) has greatly reduced the illegal parrot trade on these tiny, forested islands. Consequently, parrot numbers are beginning to stabilize and even increase in this area. Unfortunately, not all stories of exploitation end this happily. In Brazil, the Spix's Macaw **(Fig. 10–37)** appears to have lost its battle as a result of habitat loss and harvesting of chicks for illegal international trade. This species probably has gone extinct in the wild during the year 2000 with the disappearance of the last known individual, a solitary male.

Introduced Predators

Probably the single biggest cause of avian extinctions worldwide has been the introduction of new predators into habitats where the native birds were adapted to environments without them. Rats, cats, mongooses, and ferrets are the most widespread and notorious culprits, but the full list is long and even includes reptiles such as monitor lizards and snakes. The problem is particularly acute on islands where, in the absence of ground predators, native birds evolved flightlessness, ground nesting, and the absence of nest-guarding behavior. The bird fauna of New Zealand today, for example, is a tiny fraction of its size when humans arrived, largely because of predation on eggs, nestlings, and fledglings by the ubiquitous cats, rats, and mongooses that humans brought with them.

The island of Guam provides a spectacular example of the devastating effects of an introduced predator. The island's lush tropical

Figure 10–37. Spix's Macaw: An exquisite blue macaw with an ashy-blue head, the Spix's Macaw **(a)** appears to be extinct in the wild. Much sought after by the pet trade, its last stronghold—where three birds remained—was a small area in northern Bahia, Brazil, as shown on the map **(b)**. These were captured illegally for trade in 1987 and 1988. A single male was rediscovered in 1990 and persisted until late in 2000, after which it disappeared. Although owners of captive Spix's Macaws in zoos and private collections around the world have combined forces to propagate the species in captivity, reintroducing parrots and macaws into the wild is difficult. Young parrots and macaws learn their species' elaborate social system, methods of locating and processing wild foods, and other skills necessary for surviving in the wild from their parents and other members of their species, a process that is difficult to simulate in captivity. Painting by L. Sanz, reproduced with permission, from del Hoyo, J., Elliott, A., and Sargatal, J. eds. (1997), Handbook of the Birds of the World, Vol. 4: Sandgrouse to Cuckoos, Plate 45 (part). Published by Lynx Edicions, Barcelona.

a

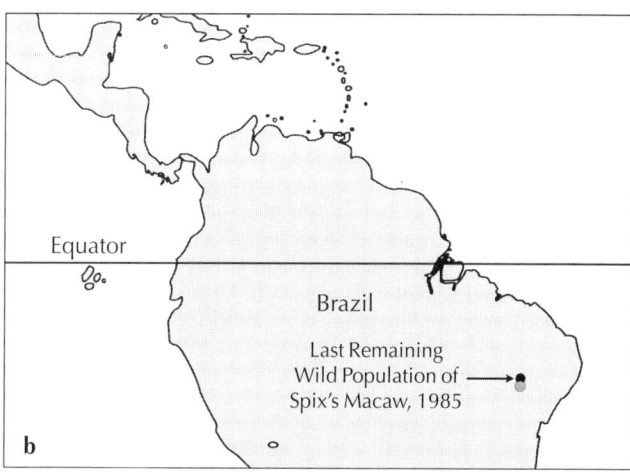

Equator

Brazil

Last Remaining
Wild Population of ⟶
Spix's Macaw, 1985

b

10

Figure 10–38. Guam Rail: Free of terrestrial predators, certain bird species endemic to isolated islands have evolved to be flightless. Introduced predators on these islands have an especially devastating effect on the local avifauna. The Guam Rail is a virtually flightless ground-dwelling bird native to Guam, one of the Mariana Islands in the western Pacific Ocean. Before the 1960s its population numbered over 10,000, but it began a rapid decline in the 1970s, as did many other birds endemic to Guam. By 1983, fewer than 100 Guam Rails remained, and a captive breeding program was begun. The bird became extinct in the wild in 1987. The culprit was the brown tree snake (see Fig. 10–39), introduced from the Solomon Islands. Fortunately, captive breeding of Guam Rails on Guam and in U. S. zoos has been highly successful. Attempts are under way to introduce a self-sustaining rail population on the snake-free island of Rota, and to reintroduce the bird into the wild in snake-free enclosures on Guam. Photo by Sid Lipschutz/VIREO.

forests once contained eleven bird species: a rail, a moorhen, a kingfisher, a fruit-dove, a swiftlet, a crow, and a variety of small songbirds. Until about 1960 all these species remained common and stable, despite extensive use of the island by humans and numerous introduced mammals (including four species of rats). Then, one by one, the bird species began declining dramatically. By the 1980s six species had gone extinct, and the remaining five were perilously close. The last few surviving individuals of the Guam Micronesian Kingfisher (a subspecies of the Micronesian Kingfisher) and the Guam Rail **(Fig. 10–38)** were captured and taken into captivity in the mid-1980s, even before the problem was identified. Biologists then discovered the culprit, a rapidly spreading population of the voracious, bird-eating brown tree snake **(Fig. 10–39)**. This predator is native to the Solomon Islands, and had arrived accidentally on Guam in shipping containers. As the snake spread across the island it killed both adults and juveniles—simultaneously raising the adult death rate and reducing local birth rates to near zero. Even today, the fate of Guam's reduced avifauna still

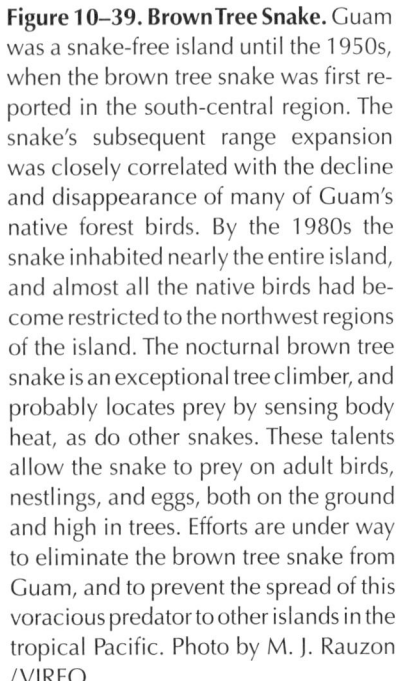

Figure 10–39. Brown Tree Snake. Guam was a snake-free island until the 1950s, when the brown tree snake was first reported in the south-central region. The snake's subsequent range expansion was closely correlated with the decline and disappearance of many of Guam's native forest birds. By the 1980s the snake inhabited nearly the entire island, and almost all the native birds had become restricted to the northwest regions of the island. The nocturnal brown tree snake is an exceptional tree climber, and probably locates prey by sensing body heat, as do other snakes. These talents allow the snake to prey on adult birds, nestlings, and eggs, both on the ground and high in trees. Efforts are under way to eliminate the brown tree snake from Guam, and to prevent the spread of this voracious predator to other islands in the tropical Pacific. Photo by M. J. Rauzon /VIREO.

hangs in the balance. Conservationists are attempting to establish snake-free habitat refuges on nearby, snake-free islands (such as the island of Rota) where some of Guam's nearly extirpated species could be reintroduced. This effort is under way today for both the kingfisher and rail, but it is too early to tell whether either will succeed.

Throughout the islands of the western Pacific, dozens of bird species are poised to join the many that already have gone extinct through centuries of human colonization. Some of the most endangered birds, such as the Guam Rail, Guam Micronesian Kingfisher, and Bali Myna **(Fig. 10–40)**, have so little predator-free native habitat remaining on their original islands that they exist as breeding populations only among the world's zoological parks. Others, such as the Nicobar Pigeon and Rarotonga Monarch, still exist in tiny numbers with uncertain futures, utterly dependent on protection of habitat remnants and the control of rats, cats, and brown tree snakes.

Far out in the Pacific Ocean, Christmas Island once provided a huge, predator-free breeding place for hundreds of thousands of seabirds representing more than a dozen species (especially Sooty Terns, noddy terns, three species of boobies, three frigatebirds, two tropicbirds, and two cormorants). Humans brought cats to the island hundreds of years ago, and in recent years a free-ranging population of these cats has exploded causing precipitous declines in the nesting birds. Seabirds that had been using Christmas Island for hundreds, perhaps thousands of years suddenly dropped to less than half their numbers. Their future remains in doubt unless the cat population can be brought under control. In New Zealand, heroic efforts are under way to restore populations of Chatham Robins, Kakapos, wattlebirds, and many other critically endangered birds by moving breeding adults to offshore islands that have been systematically cleared of introduced predators.

Chemical Toxins

Publication of Rachel Carson's Silent Spring in 1962 alerted the world to the widespread use of harmful chemical herbicides and insecticides throughout the countryside. Carson's worst fears proved well-founded a few years later, when many raptor species began disappearing from long-occupied habitats. Declines were especially

a. Bali Myna   **b. Current Range of the Bali Myna**

Figure 10–40. Bali Myna: a. Bali Myna: Snowy plumage, long drooping crest, black wing tips and tail tips, blue bare skin around the eye, and a yellow bill make the Bali Myna a stunning species, much sought after as a cage bird and commanding black-market prices of up to $2,000 in the 1990s. **b. Current Range:** Illegal trapping has led to the virtual extinction of the wild population, found only on the island of Bali, Indonesia, shown in color on the map. Captive breeding and other conservation programs are in place for this critically endangered species, and around 1,000 individuals are thought to exist in captivity. However, by 1999 the wild population had plummeted to just 12 individuals which, despite being protected in a national park, continue to be highly vulnerable to poaching because of mismanagement and corruption. Photo by S. Lipschutz/VIREO.

Figure 10–41. Crushed Peregrine Falcon Egg: During the 1960s and 1970s, scientists confirmed that numbers of meat-eating predators such as Peregrine Falcons, as well as fish-eaters like the Bald Eagle, Osprey, and Brown Pelican, were sharply declining. After determining that declines were usually caused by reproductive failure, they discovered that modern raptor eggshells were much thinner than the shells of older museum specimens. The problem was eventually traced to contamination of adult birds by the pesticide DDT (see Fig. 10–42). In this photo from the early 1970s, a Cornell Lab of Ornithology researcher holds a Peregrine Falcon egg whose thin shell has been crushed during incubation, killing the developing embryo inside. Photo courtesy of CLO archive.

pronounced for Bald Eagles, Ospreys, and Peregrine Falcons in the United States and for Eurasian Sparrowhawks in Europe. A dedicated ornithologist from Wisconsin named Joseph Hickey and his colleagues discovered the primary cause of reproductive failure: incubating females were crushing the eggs in their nests simply by incubating them (Hickey 1969) **(Fig. 10–41)**. A flurry of studies then produced one of the most famous stories in the history of conservation biology. By comparing the fragments of these crushed eggs with egg specimens stored in museum collections, scientists discovered that the modern eggshells were up to 30 percent thinner than normal (Ratcliffe 1967; Hickey and Anderson 1968; reviews in Cade et al. 1988) **(Fig. 10–42)**. The problem was traced to contamination of the adult birds by chlorinated hydrocarbon pesticides, in particular the common organic pesticide **DDT**, which is quickly metabolized into a stable and highly persistent compound, **DDE** (1,1-dichloro-2,2-bis(p-chlorophenyl)ethylene). Since 1947, DDT had become one of the most widely used pesticides in the world. It was known to be spread around the world on trade winds, and had even appeared in the Antarctic ice sheet! Migratory birds of prey across North America and Europe were eating birds, mammals, and fish that had been accumulating DDE in their tissue. Therefore, these top predators accumulated much higher concentrations of DDE than did species lower down in the food chain (see Fig. 9–82). Physiological effects included higher death rates among adults, but the eggshell thinning was an even bigger problem, as it reduced birth rates throughout the populations. Populations of almost all North American hawks and falcons plummeted during the DDT era (1950 to 1972), and began to recover almost immediately after the banning of the infamous pesticide.

Although DDT has been banned in the United States, vast quantities of organic pesticides including DDT continue to be distributed

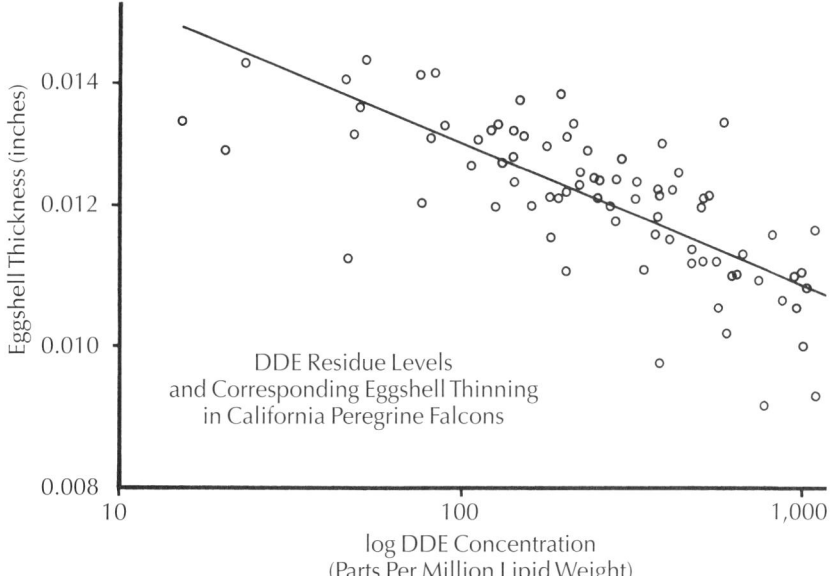

DDE Residue Levels
and Corresponding Eggshell Thinning
in California Peregrine Falcons

log DDE Concentration
(Parts Per Million Lipid Weight)

Figure 10–42. DDE Levels and Eggshell Thinning in California Peregrine Falcons: DDE is a highly stable compound created and stored in the body of birds and mammals when they metabolize the pesticide DDT. As part of the management program for the Peregrine Falcon's endangered California population, eggshell quality and residues of DDE in egg membranes have been monitored since 1975. This famous graph shows that, for a sample of eggs from 1975 to 1983, dramatically thinner eggshells occurred in individuals with the highest levels of DDE contamination. (Note that the x-axis is a logarithmic scale.) Adapted from Peakall and Kiff (1988) in Cade et al. (1988).

Figure 10–43. DDT—A Banned Pesticide: In 1939, scientists in Switzerland first discovered that the chemical dichloro-diphenyl-trichloroethane (DDT), a member of a chemical class known as chlorinated hydrocarbons, was a powerful insecticide. Cheap to produce and relatively nontoxic to mammals, DDT showed great potential for combating mosquitoes that spread malaria, lice that carry typhus, and many agricultural pests. After World War II, use of DDT increased dramatically around the globe. Unfortunately, the pesticide caused numerous environmental problems. Most profound was its effect on predatory birds, which reside at the top of food chains and are thus extremely likely to accumulate environmental toxins. Inside a bird's body tissue DDT is rapidly converted to DDE and stored in fat. There it disrupts calcium metabolism and causes the bird to lay thin-shelled eggs, which break or dent during incubation. As a result of this finding, the sale or use of DDT in the United States was banned in 1972, and raptor populations have since rebounded. Unfortunately, the pesticide is still legal in some other countries, and may even be used inadvertently in the United States. In this photo from the late 1970s, a bottle of DDT lurks in an Idaho garage. Photo by William H. Mullins / Photo Researchers.

around the world **(Fig. 10–43)**. According to the American Bird Conservancy, DDT is still used in six countries in the Western Hemisphere: Argentina, Belize, Ecuador, Guyana, Peru, and Mexico. In the United States, numerous other compounds known to be lethal to birds remain in use. The long-term effects of most pesticides on birth rates and death rates of North American bird populations remain poorly documented **(Sidebar 3: Hawk Deaths Spur Action)**.

The steady decline of the California Condor throughout the 20th century was caused by a combination of factors, such as illegal shooting, eating of carcasses poisoned by humans, and collisions with vehicles and utility wires. Many biologists, however, point to lead poisoning as the chief culprit, as virtually all condors found dead

(Continued on p. 10·56)

Sidebar 3: *HAWK DEATHS SPUR ACTION*

Conservationists from Across the Americas Try to Stem the Tide of Swainson's Hawk Deaths from Pesticide Abuse

In early 1996, researchers discovered that as many as 20,000 Swainson's Hawks had died recently from the ingestion of deadly pesticides. This tragedy resulted in an unprecedented, coordinated, quick response by multiple members of the bird conservation community.

Swainson's Hawks **(Fig. A)** are long-range migrants, traveling from their breeding grounds in western North America to southern South America, especially the "pampas" area of Argentina. In 1994, staff at the Klamath National Forest in California initiated a study of the hawk's migration and wintering ecology after detecting an alarming decline in its population. Hawks were fitted with satellite-capable radio transmitters and monitored during their southbound migration.

In January 1995, Brian Woodbridge of the U. S. Forest Service discovered an entire flock of 700 Swainson's Hawks dead in fields in La Pampa, Argentina. This species spends the winter in large flocks, feeding primarily on grasshoppers. The hawks ingested these grasshoppers, flushed by tractors applying pesticides to their fields, and died. Based on discussions with biologists and ranchers in Argentina, researchers believe such hawk mortality from pesticide ingestion is not unusual, and may be responsible for the decline in the species, whose population is now estimated to be between 350,000 and 400,000 birds.

Woodbridge and other scientists visited Argentina again in January 1996 to locate hawks with transmitters, and thus determine the species' southern range. They counted 4,100 dead Swainson's Hawks and estimated, in the pampas area, as many as 20,000 hawks had been killed **(Fig. B)**. The pesticide responsible was determined to be monocrotophos, an extremely potent organophosphate that is no longer registered for use in the U. S. and Canada.

These frightening statistics have spurred conservation agencies in both North and South America to action. American Bird Conservancy (ABC) has brought the problems (for both humans and birds) of using monocrotophos to the attention of Ciba-Geigy, the major manufacturer. ABC, along with representatives of The Institute for Wildlife and Environmental Toxicology (TIWET) at Clemson University, and other conservationists and toxics experts, hosted a meeting with Ciba-Geigy officials in late summer 1996 to discuss ways to prevent further bird kills.

In February of 1996, the Canadian Wildlife Service provided financial support to Argentina's Instituto Nacional de Tecnologia Agropecuaria (INTA) to conduct surveys and interviews in agricultural areas in the pampas. INTA is working with the Environmental Protection Agency

Figure A. Swainson's Hawk: Swainson's Hawks breed in western North America and winter in southern South America, especially the "pampas" area of Argentina. Concern about the decline of this species prompted biologists to use radio telemetry to study its migration and wintering ecology. Results tied the decline to the deaths of thousands of Swainson's Hawks after they fed on pesticide-contaminated grasshoppers on their wintering grounds. Photo courtesy of Isidor Jeklin/CLO.

Figure B. Poisoned Swainson's Hawks in Argentina: Mike Goldstein, of Clemson University's Institute for Wildlife and Environmental Toxicology, surveys Swainson's Hawks found poisoned by the pesticide monocrotophos at Estancia Chanilao, Argentina. Photo courtesy of Brian Woodbridge.

and TIWET on experimental testing of alternative compounds for grasshopper control. In addition, Argentine agencies including INTA, Instituto Argentino de Sanidad y Calidad Vegetal (IASCV), and the Secretariat of Natural Resources and the Environment have formed a steering committee to address the Swainson's Hawk issue. This effort has resulted in the restriction of monocrotophos use in areas frequented by Swainson's Hawks, and the development of a strategic plan including monitoring, alternative pest management, educational outreach, and toxicological assessment.

Technical assistance and funding for the work is being provided by the U. S. Fish and Wildlife Service, U. S. Department of Agriculture, National Biological Service, Environmental Protection Agency, National Fish and Wildlife Foundation, TIWET, and many others. Conservationists hope this collaborative effort will halt the mass mortality of Swainson's Hawks and prevent the decline of other bird species from pesticide applications.   ∎

This article originally appeared in Bird Conservation, Fall 1996.

10

in the wild had numerous lead pellets in their crops, and the species is unusually sensitive to elevated levels of lead in its blood. Today, although waterfowl hunters have switched from lead shot to steel shot, widespread and unregulated use of lead shot, lead bullets, and lead sinkers in a variety of sporting pastimes continues to cause an enormous environmental problem affecting birds of many different groups. Even the experimental reintroduction of California Condors remains plagued by the prevalence of lead and other toxins in the environment.

Indirect Chemical Pollution

Beginning in the late 19th century and accelerating throughout the 20th, environmental pollution from human sewage, manufacturing plants, oil refineries, coal-powered energy plants, and agricultural fertilizers changed the chemistry of air, water, and soils throughout the industrialized world. These changes reduced prey diversity and altered vegetation structure in lakes, marshes, rivers, and streams, in turn reducing the reproductive success of aquatic birds such as ducks, grebes, bitterns, and loons. Ironically, some of the worst effects resulted from excess nutrients in the water, which allowed one or a few species of algae to explode and take over the entire system, choking out the native food web (a process called **eutrophication**, [Fig. 10–44]).

In the late 20th century, release of sulfur-containing compounds into the atmosphere as a by-product of energy production and steel manufacturing caused a particularly widespread and insidious form of environmental pollution. Acid rain is caused by condensation of water droplets that incorporate atmospheric gases such as hydrogen sulfide in areas downwind of major industrial centers such as northeastern North America and northern and eastern Europe. The environmental effects of acid rain vary geographically, depending largely on the availability of chemical buffers in the soil (especially calcium carbonate from dissolving limestone). In areas lacking buffers, both lakes and soil can become steadily more acidic. In acidified lakes zooplankton and molluscs become scarce, causing an ecological chain reaction culminating in the disappearance of snail-eating and fish-eating birds such

Figure 10–44. Eutrophic Lake: Nutrient-rich runoff containing nitrogen and phosphorus often enters lakes and ponds from agricultural, industrial, and residential sources. Frequently the runoff results in dense growth of algae and aquatic plants such as duckweed, which covers the lake in this photo. Such nutrient overloading, known as **eutrophication**, changes water quality and species composition of the lake, in turn influencing birds and other wildlife. For instance, when the algae and plants die and fall to the bottom of the lake they are decomposed by aerobic bacteria, a process that results in depletion of oxygen in the water, which then leads to the suffocation of fish and other oxygen-breathing aquatic organisms. Fish-eating birds, such as Ospreys and kingfishers, or other birds, such as grebes, that feed on small water creatures, are then deprived of their food source. If nutrients continue to flow into the lake, anaerobic (non-oxygen-requiring) bacteria take over, producing gases such as methane and toxic, odoriferous hydrogen sulfide. At this point the lake is no longer suitable as bird habitat. Although eutrophication can take place gradually via natural erosion and runoff, the process of cultural eutrophication—nutrient input from human sources—is much faster and has led to profound degradation of both freshwater and marine ecosystems worldwide. Photo by Bill Banaszewski/ Visuals Unlimited.

as loons, ducks, and raptors. In acidified soils, both plant composition and animal biomass can become drastically altered. In Europe, this effect appears to be responsible for the disappearance of Eurasian Capercaillie (the world's largest grouse) from large areas that still retain forest cover. Decline or outright disappearance of calcium-bearing molluscs (such as land snails) and arthropods (for example, millipedes) has caused significantly reduced reproductive success among songbirds in European forests, because they cannot obtain enough calcium to produce viable eggs. In the northern Appalachian Mountains of the United States, where acid rain appears to have caused a tenfold reduction in land snails, the population of Bicknell's Thrushes is also declining locally, even within intact forest (Atwood et al. 1996; **Fig. 10–45**). Although acid rain has not been proven to be the cause of the decline, the likelihood of a connection is high because thrushes rely on land snails as a source of calcium during the breeding season.

a. Bicknell's Thrush

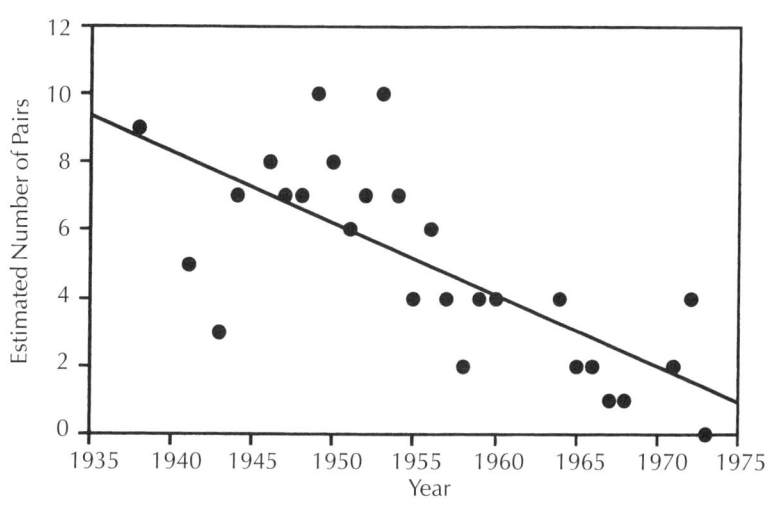

b. Decline and Local Extinction of Bicknell's Thrush on Mt. Greylock, Massachusetts

Figure 10–45. Decline of Bicknell's Thrush: Once considered a subspecies of the Gray-cheeked Thrush, the Bicknell's Thrush **(a)** breeds only in sub-alpine spruce-fir forests atop misty mountains of New England and southeastern Canada. Mt. Greylock, Massachusetts is the only locality where the species has been monitored over an extended period. Estimates suggest that 5 to 10 pairs bred there in the early 1900s. However, records also show that between 1938 and 1973 the species slowly declined on the mountain, finally disappearing in 1973 **(b)**. Intensive surveys on Greylock between 1992 and 1995 failed to relocate the species. Although data are scanty, the Bicknell's Thrush also appears to be declining in other parts of its range. Its patchy distribution and narrow ecological niche place the bird in a precarious position, especially because acid rain may be killing spruce-fir forests in New England and the Adirondack Mountains of New York. Acid rain also reduces populations of land snails, which many species of thrushes eat to obtain calcium necessary for egg production. Graph adapted from Atwood et al. (1996).

Introduced Disease

By the time that Captain Cook first set foot in the Hawaiian Islands in 1778, the Polynesians occupying those islands already had caused the extinction of numerous large bird species. Still, about 70 species of small native birds continued to occupy the forests from sea level to mountaintop. During the second half of the 1800s, however, a baffling wave of disappearances overtook Hawaiian songbirds, island by island. In 1902, ornithologist H. W. Henshaw described the phenomenon: "The author has lived in Hawaii only six years, but within this time large areas of forest, which are yet scarcely touched by the axe save on the edges and except for a few trails, have become almost absolute solitude. One may spend hours in them and not hear the note of a single native bird. Yet a few years ago these areas were abundantly supplied with native birds."

What Henshaw observed was taking place across the entire archipelago except at the highest elevations. Dozens of bird species went extinct almost simultaneously, and throughout the 20th century additional species joined their ranks one by one. Although many factors were responsible for these losses, the introduction of disease-bearing mosquitoes on the island of Maui in 1826 certainly was the single most devastating event for Hawaii's bird fauna since the original arrival of humans one thousand years earlier. Avian malaria and avian pox are the two most virulent agents, and both are carried easily by Culex pipiens, the mosquito brought to Hawaii from Mexico in the water casks of merchant ships. These diseases are common among continental birds throughout the world, in which resistance has evolved through the eons. Indeed, potential sources of both diseases had been present on Hawaii through the millennia in the form of both transient and wintering migratory birds. Before the mosquito's introduction, however, no vector allowing transfer from an infected but resistant migrant to a native bird existed on the islands. Consequently, no native Hawaiian bird had evolved resistance to malaria or pox (**Fig 10–46**).

Figure 10–46. Mosquito Feeds on Apapane: Introduction of disease-bearing mosquitoes to the Hawaiian Islands was the single event most devastating to the native avifauna since the islands' original settlement by humans. Mosquito-borne disease has led to the disappearance of many endemic birds, including several species of Hawaiian honeycreeper, particularly from the lowlands. Because mosquitoes are scarce at high elevations, mountainous areas are the last strongholds of most of Hawaii's endemic birds. The crimson Apapane is the most abundant species of Hawaiian honeycreeper, found on several of the islands. Highest in density in high-elevation forests above the mosquito zone, it searches widely for the flowering trees that provide its main food source—nectar. Ironically, the Apapane's long-distance travel between patches of flowering trees has important implications for disease transmission, because in its travels the bird acts as a carrier of avian malaria and pox. However, there is some evidence that the Apapane is developing resistance to avian malaria, so it may become the subject of important genetic and immunological studies in the future. Photo by Jack Jeffrey.

Once again, humans had changed the rules, this time by gradually transferring mosquitoes among the islands throughout the 1800s. As a result, avian death rates soared. Because mosquitoes are scarce above 4,000 feet (1,300 m), the native avifauna could persist above this elevation, but at the lower elevations the problem is difficult to control because mosquitoes breed in huge numbers in tree-fern trunks hollowed out by introduced pigs. Today, specialists in avian disease have found areas of Hawaii where up to 50 percent of the free-flying mosquitoes below 3,300 feet (1,000 m) are carrying malaria. This infection rate is vastly higher than the percentages known in any continental population. The only long-term hope for the few remaining native forest birds is gradual development of resistance through natural selection. The race is still being lost, however. Through the Hawaiian Islands as a whole, 17 of the 42 native songbird species already have gone extinct, and almost every one of the remainder is poised on the brink **(Table 10–2)**. On Maui, as of the year 2001 only two individual Poo-ulis remained alive along with the parrotbills and Akohekohes. On Kauai, where the last Kauai Oo disappeared in 1989 and the Nukupuu has not been reported reliably for 30 years, the Akikiki and three other native honeycreepers are down to the last few individuals in the wet forests of the Alakai Plateau.

Habitat Loss

The fundamental feature that makes some species vulnerable when the landscape is altered is that each evolved in concert with a particular range of naturally occurring habitats. Physical structure, plumage, foraging habitats, nesting behavior, movement patterns, and life cycle are all tuned to specific and predictable features of the habitat. Certain species ("habitat generalists") successfully occupy many kinds of places , but many species ("habitat specialists") can successfully live and breed only in one kind of place. As humans spread about the planet, and especially over the last several thousand years, we have continually changed the structure, composition, and distribution of habitats. What we could not do was to change the fundamental habitat tolerances of species.

Loss of required habitat is by far the single biggest factor causing population declines. As we converted native prairie to wheat fields or row crops, we reduced the adult population sizes of prairie-inhabiting sparrows or larks. As we drained small wetlands to produce more land area for grazing or cultivation, we reduced the total number of ducks produced across the landscape. As we cut down the tropical forests, we reduced the total number of tropical forest birds.

On the island of Madagascar, where the enormous elephant-bird was hunted to extinction around 1000 A.D., a number of birds are poised at the edge of extinction owing to widespread conversion of the island's habitats into vast grazing lands and rice paddies. The Madagascar Serpent-Eagle **(Fig. 10–47)**, suspected to be extinct since the 1930s, was rediscovered in the early 1990s but exists only as a tiny population in a remote forest tract at the north end of the island. The

10

Table10–2: Status of Hawaiian Songbirds in Jeopardy

Common Name	Recent Island(s)	Status
Hawaiian Crow (Alala)	Hawaii	critically endangered; 2 individuals in wild
Millerbird	Nihoa and Laysan	extinct on Laysan; endangered on Nihoa
Elepaio	Kauai, Oahu, Hawaii	not endangered, but declining
Kamao	Kauai	endangered
Amaui	Oahu	**extinct**
Olomao	Molokai and Lanai	**probably extinct**
Omao	Hawaii	endangered
Puaiohi (Small Kauai Thrush)	Kauai	endangered
Kauai Oo (Ooaa)	Kauai	**extinct**
Oahu Oo	Oahu	**extinct**
Bishop's Oo	Maui, Molokai	**extinct**
Hawaii Oo	Hawaii	**extinct**
Kioea	Hawaii	**extinct**
Laysan Finch	Laysan	endangered
Nihoa Finch	Nihoa	endangered
Ou	All large islands	**probably extinct**
Palila	Hawaii	endangered
Lesser Koa-Finch	Hawaii	**extinct**
Greater Koa-Finch	Hawaii	**extinct**
Kona Grosbeak	Hawaii	**extinct**
Maui Parrotbill	Maui	endangered
Hawaii Amakihi	Molokai, Lanai, Maui, Hawaii	extinct on Lanai, not endangered elsewhere
Oahu Amakihi	Oahu	not endangered
Kauai Amakihi	Kauai	not endangered
Greater Amakihi	Hawaii	**extinct**
Anianiau	Kauai	not endangered
Lesser Akialoa	Hawaii	**extinct**
Greater Akialoa	Kauai	**probably extinct**
Nukupuu	Kauai, Oahu, Maui	**probably extinct**
Akiapolaau	Hawaii	endangered
Akikiki	Kauai	critically endangered
Hawaii Creeper	Hawaii	endangered
Maui Alauahio (Maui Creeper)	Maui, Lanai	extinct on Lanai, not endangered on Maui
Kakawahie (Molokai Creeper)	Molokai	**probably extinct**
Oahu Alauahio	Oahu	**probably extinct**
Akepa	Oahu, Maui, Hawaii	**extinct** on Oahu and Maui, endangered on Hawaii
Ula-ai-hawane	Hawaii	**extinct**
Iiwi	All large islands	extinct on Lanai, sharply declining elsewhere
Hawaii Mamo	Hawaii	**extinct**
Black Mamo	Molokai	**extinct**
Akohekohe (Crested Honeycreeper)	Maui	endangered
Apapane	Laysan, all large islands,	extinct on Laysan, not endangered elsewhere
Poo-uli	Maui	critically endangered; 2 individuals known

a. Madagascar Serpent-Eagle

Figure 10–47. Madagascar Serpent-Eagle: Lack of records of the Madagascar Serpent-Eagle since 1930 led many biologists to believe that the species was extinct until researchers from The Peregrine Fund rediscovered the bird in 1993. Russell Thorstrom found the first known nest of this secretive forest eagle in 1997, and his rare photo **(a)** shows the adult male (right) with its nestling on the nest (left). As the impoverished human inhabitants of Madagascar clear forests for agriculture, the bird's habitat on this large island off the coast of East Africa becomes increasingly fragmented, as shown by the colored areas on the map **(b)**. Although total numbers of the Madagascar Serpent-Eagle remain uncertain, with estimates ranging from as low as 50 to as high as 250, its population is suspected to be in rapid decline. Photo courtesy of Russell Thorstrom.

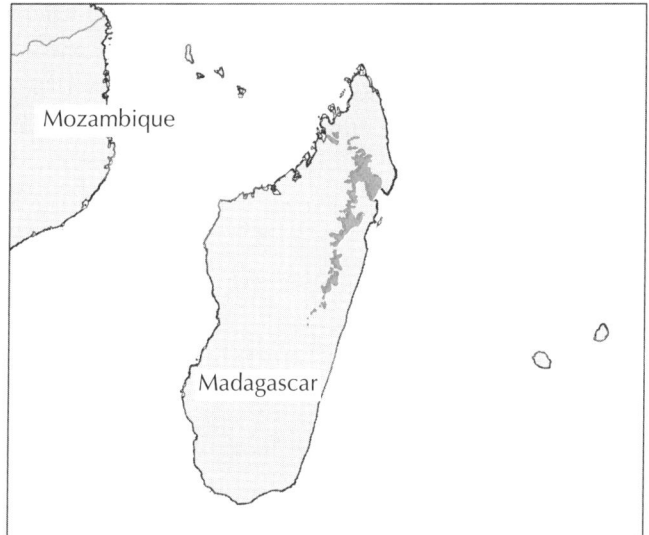

b. Current Range of Madagascar Serpent-Eagle

("habitat specialists") Alaotra Grebe (not seen since 1985) and Madagascar Pochard (not seen alive since 1970, but a male was caught in fishing gear in 1991) appear to have even less hope. Both are restricted to a single marshy lake that is badly degraded by conversion to rice paddies and introduction of exotic fish that are eliminating the native aquatic flora (BirdLife International 2000). Because of the unique flora and fauna of Madagascar, considerable international assistance has been focused on helping its government identify the best remaining patches of native habitat and secure them as ecological preserves.

Habitat Specialization and the "Six Forms of Rarity"

Outright habitat loss causes population declines in any species that depends on that habitat. However, habitat loss does not affect all species equally. Certain species are more vulnerable than others

10

as native habitats disappear. It goes without saying that rare species generally will be more vulnerable than common ones, but what exactly do we mean by the word "rare?" What kinds of species are most vulnerable as the landscape is altered by humans?

Three features of a species' distribution affect its vulnerability once the landscape begins to change. These are overall geographic distribution (range size), overall numbers within its range (population density), and the spectrum of habitats in which the species can live and breed (habitat specificity). In an often-cited paper, Rabinowitz et al. (1986) categorized British plants according to these three variables in order to summarize the different kinds of rarity. That analysis is just as appropriate for birds as it was for plants. Rabinowitz and colleagues identified seven categories of rarity, although one of these actually describes the truly "common" species (that is, species that are widely distributed, with large local population sizes, and broad habitat tolerance). The other six categories provide a broad classification scheme for rare species.

Most species that we typically think of as rare actually are (or once were) quite common in one or a few localized areas. Typically, such species are characterized by having very specific and unusually narrow habitat requirements. The good news is that often we can protect such species simply by protecting those local geographic centers or habitats. The bad news is that if a species' primary centers of occurrence already are destroyed, then protecting such species is extremely challenging. Identifying the rareness category to which a declining species belongs is an important first step in designing its protection or recovery.

Following are the six forms of rarity, and some examples from continental North America:

1. Widely distributed, small local populations, broad habitat tolerance

Bald Eagles **(Fig. 10–48)** occur from Alaska east to the Canadian maritime provinces and south to California and Florida. Although mostly associated with water, the species occurs in habitats as diverse as northern pine forests, southern palm swamplands, and coastal estuaries. Nevertheless, because of its large territory size and low reproductive rate, the Bald Eagle, like other large raptors, is a rare bird everywhere. Because its numbers are so low, this species is vulnerable to ecosystem-wide forces that reduce its ability to produce offspring, such as changes in prey density from overfishing by hu-

Figure 10–48. Bald Eagle: An example of a species having wide distribution and broad habitat tolerance, but living in small numbers locally, the Bald Eagle occurs throughout North America north of Mexico. It lives in many habitats close to water, on coasts as well as inland. However, because the Bald Eagle is a predator at the top of the food chain, with a large territory size and a low reproductive rate, its local populations are always small. Therefore, the species is uncommon throughout nearly all of its range. Photo courtesy of Harold E. Wilson/CLO.

mans, or—as discussed above—metabolic effects of environmental pollution. Today the Bald Eagle is undergoing long-term recovery in the contiguous 48 United States after the banning of DDT. However, it still could be at risk of population declines in northwestern North America as its favored populations of salmon disappear.

2. Widely distributed, large local populations, narrowly specialized habitat requirements

Red-cockaded Woodpeckers **(Fig. 10–49)** once occurred commonly in old-growth pine savannas across the entire southeastern United States from Arkansas and Texas east to the Carolinas and the Florida peninsula. This specialized woodpecker nests only in cavities that it drills into the main trunks of living, mature pines. It protects its nests from predation by snakes and small mammals through an unusual behavior—adults drill tiny "resin-wells" in the bark surrounding the nest cavity, producing a thick coating of sticky pine pitch. The noxious substance repels predators (especially rat snakes) as they attempt to climb toward the hole. Red-cockaded Woodpeckers nest exclusively in mature pine stands in part because the oldest trees are susceptible to a fungus (red heart disease) that renders the center of the trunk soft enough to hollow out. Red-cockaded Woodpecker colonies also require an open grassy understory, and they gradually die out in pine forests that are not regularly burned. Therefore, despite having a wide geographic distribution, local populations of this woodpecker now exist only in the rare places where humans have allowed fully mature pine stands to remain uncut, and where humans permit fires to burn through the forest every two to three years.

Figure 10–49. Red-cockaded Woodpecker: Once common and widespread in the old-growth pine forests that originally covered the southeastern United States, the Red-cockaded Woodpecker now is highly endangered. Although local populations may be relatively large and the species is distributed from Texas to North Carolina and Florida, its range is highly fragmented (see range map, Fig. 9–39). This distribution reflects the remaining extent of the species' narrowly specialized habitat—mature, old-growth pines in which it can drill nest cavities and find food. The species became endangered because southern old-growth pine forests were decimated by logging during the 20th century, and the remnants continue to decline owing to fire suppression. Photo courtesy of Todd Engstrom/CLO.

Figure 10–50. Henslow's Sparrow: Breeding from Minnesota, southern Ontario, and central New York southward to Kansas, Illinois, and North Carolina, the Henslow's Sparrow is widely distributed, although easily overlooked. Where patches of its favored nesting habitat exist—weedy, fallow fields and, in particular, wet, broomsedge meadows—the species can be locally common, forming loose breeding colonies. However, with the decline of grasslands owing to reforestation and conversion to suburbs or monoculture crops, the species has suffered. It has become increasingly rare in the Northeast, and its western range also is contracting. One estimate suggests that the U. S. population declined 68 percent between 1966 and 1991. The Henslow's Sparrow shown in this photo is attending nestlings at its nest, a well-concealed cup just above the ground. Photo courtesy of Betty Darling Cottrille/CLO.

Henslow's Sparrows **(Fig. 10–50)** occur from Minnesota east to Maine, and in appropriate habitat patches they can reach densities up to several pairs per acre (0.4 ha). However, the distribution of their favored habitat (wet broomsedge meadows) changes each year with variable rainfall patterns, and humans have converted most historical breeding grounds into agricultural fields or suburban housing developments. Like many other grassland specialists, Henslow's Sparrow declined precipitously during the latter half of the 20th century, and now it is probably the rarest breeding songbird in the eastern United States. Its persistence depends completely on humans establishing a network of managed grassland preserves across a broad geographic area.

3. Widely distributed, small local populations, narrowly specialized habitat requirements

Piping Plovers breed on barren sandy beaches from central Canada and the Great Lakes east to the Atlantic Coast **(Fig. 10–51)**, but even the largest of these beaches rarely harbor more than two or three

Figure 10–51. Piping Plover: The endangered Piping Plover (inset photo) is widely distributed but has small local populations and narrowly specialized habitat requirements. It nests at low densities on barren sandy beaches where it comes into frequent contact with beach-loving humans, so that its survival now depends on the active protection of nest sites by humans. In the main photo, condominiums encroach upon sand dunes edging a Martha's Vineyard beach, site of a Piping Plover nesting colony. Because the bird is endangered, its colony is fenced and prominently posted with "Keep Out" signs. Each individual nest (a simple ground scrape lined with shell fragments, see Fig. 8–46) is surrounded with a wire mesh enclosure to keep predators out while allowing the birds to move to and from the nest freely. Furthermore, the beach is off limits to unleashed dogs. Nevertheless, the plovers still face some threats—for example, vehicle access to the beach is still permitted, endangering newly fledged young, which tend to crouch in the face of danger rather than to run (see Ch. 6, Sidebar 3, Fig. A). Main photo by Marie Read. Inset photo courtesy of Lang Elliott/CLO.

pairs. Consequently, this species is common nowhere. Furthermore, its extremely narrow habitat tolerance places it directly in competition with humans (who also love sandy beaches). The total number of Piping Plovers could never get very large, simply because of insufficient breeding habitat that can be kept free of human disturbance. The continued existence of this species now depends entirely on active protection of nest sites by humans throughout its range.

4. Small geographic range, large local populations, broad habitat tolerance

This form of rarity describes many bird species that are restricted to islands, and on mainland North America, few species show this combination of characteristics. Perhaps the best example is Nuttall's Woodpecker **(Fig. 10–52)**, a species restricted to western California where it occurs in a variety of oak-dominated habitats of low and middle elevations. The species is not immediately threatened, although its numbers are declining in areas where its density is highest and where it overlaps with the vast suburban sprawl associated with California's major cities.

5. Small geographic range, large local populations, narrowly specialized habitat requirements

This category of rarity probably describes the largest proportion of endangered and threatened bird species in North America and around the world. Florida Scrub-Jays, for example, occur only in peninsular Florida and depend on a stunted, fire-maintained oak scrub habitat. Golden-cheeked Warblers in Texas **(Fig. 10–53)** and Kirtland's Warblers in Michigan (see Fig. 8–12) also have tiny breeding distributions in fire-maintained habitats. A few hundred Kirtland's Warblers return each year from their wintering grounds in the Bahamas to the jack pine plains of Michigan (see Fig. 8–13). Their continued existence depends on periodic fires and on continuation of large-scale control of Brown-headed Cowbirds. California Gnatcatchers occupy only the remnant

Figure 10–52. Nuttall's Woodpecker: Despite having relatively broad habitat tolerance and large local populations, the Nuttall's Woodpecker has a remarkably small geographic range. It is found only in California west of the Sierra Nevada, south to northwestern Baja California. It inhabits oak woodlands from low elevations, such as canyons and river valleys, up to middle elevations in the foothills. Although not currently threatened, Nuttall's Woodpecker is declining in areas of intensive human development. Photo courtesy of Denny Mallory/CLO.

Figure 10–53. Golden-cheeked Warbler: This warbler shares the same category of rarity as most of the world's endangered and threatened birds. It has relatively large local populations, but has a very small geographic range and occupies a narrowly specialized habitat. Restricted to central Texas, Golden-cheeked Warblers prefer to breed in old-growth oak-juniper woodlands at least 250 acres (100 ha) in size. Because urban expansion and clearing of land for agriculture has destroyed much of this habitat, the species was listed as endangered in 1990 despite being locally common. Photo by M. Lockwood/VIREO.

10

Figure 10–54. Tricolored Blackbird: Tricolored Blackbirds can be abundant in certain marshes, but the species has an extremely restricted geographic range and specialized habitat requirements. Its nesting range is limited mostly to California's Central Valley, with scattered additional breeding sites in southern California, Oregon, Nevada, and Washington. Forming the largest breeding colonies of any North American passerine, this species nests only in extensive freshwater marshes, foraging in nearby areas such as grasslands, woodlands, or crop fields. Tricolored Blackbirds once were persecuted as agricultural pests because they ate great quantities of rice and other grains. In land-hungry California today, the Tricolored Blackbird is threatened by habitat loss from development, and its densely colonial nesting behavior makes it vulnerable to mass predation. Currently it is considered a species of concern by the State of California. Photo by Hugh P. Smith/VIREO.

coastal sage scrubs of extreme southern California, where they are declining rapidly as rampant human population growth converts these distinctive scrubs into suburban developments. Tricolored Blackbirds **(Fig. 10–54)** breed only in rich freshwater marshes of central California. All these species once were more widespread than they are today, and have declined dramatically as their specialized habitats were converted for human use. Each one of these species still can be found in large numbers where its specific habitat requirements are met, but persistence of all of them depends entirely on both habitat protection and on proper long-term management of these habitat patches.

6. Small geographic range, small local populations, narrowly specialized habitat requirements

This is the most vulnerable rareness category of all, because virtually any perturbation or catastrophe can extinguish such a species immediately. The category is relatively rare among North American birds, although it is quite common among highly sedentary organisms such as plants, stream fishes, freshwater molluscs, and cave-dwelling arthropods (Stein et al. 2000). Species in this category often represent ancient relics of once formerly widespread forms, now confined to tiny and distinctive habitat patches following major changes in habitat. Among North American birds,

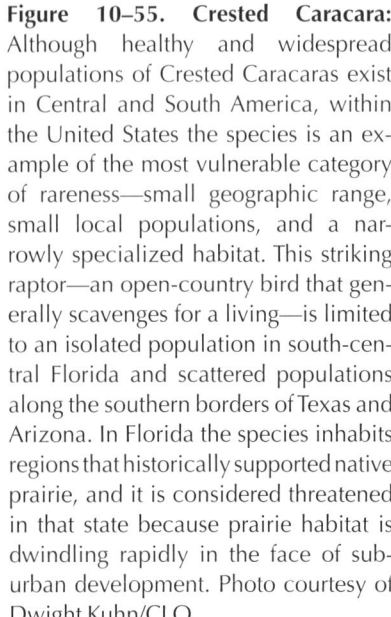

Figure 10–55. Crested Caracara: Although healthy and widespread populations of Crested Caracaras exist in Central and South America, within the United States the species is an example of the most vulnerable category of rareness—small geographic range, small local populations, and a narrowly specialized habitat. This striking raptor—an open-country bird that generally scavenges for a living—is limited to an isolated population in south-central Florida and scattered populations along the southern borders of Texas and Arizona. In Florida the species inhabits regions that historically supported native prairie, and it is considered threatened in that state because prairie habitat is dwindling rapidly in the face of suburban development. Photo courtesy of Dwight Kuhn/CLO.

Figure 10–56. Snail Kite: Like the Crested Caracara, the Snail Kite has extensive populations in Central and South America but only a tiny range within the United States. Formerly known as Everglades Kites, Snail Kites occur in the United States only in large wetlands in peninsular Florida, although a nearby population also exists in Cuba. A highly specialized diet of aquatic apple snails—which the kite extracts from their shells using its perfectly adapted sickle-shaped bill—along with restricted freshwater marsh habitat, small population size, and limited geographic range, place this species in the most vulnerable category of rarity within the United States. Widespread loss and degradation of wetlands in Florida led to its listing as an endangered species in 1967, well before passage of the Endangered Species Act. Shown here is a male Snail Kite at its nest. Photo courtesy of Fred Truslow/CLO.

perhaps the best examples are three Florida raptors, the Crested Caracara **(Fig. 10–55)**, Snail Kite **(Fig. 10–56)**, and Short-tailed Hawk. All three are isolated populations of widespread tropical species that once occurred continuously from Mexico across to Florida. Today only a few hundred of each species remain in remnant subtropical habitats of the southern peninsula (wet prairies, freshwater marshes with giant snails, and subtropical hardwood forests, respectively). The fates of all three species in Florida depend on protection of these narrowly distributed habitats. (Note, however, that all three are widespread in Central and South America, where they are not endangered.)

Unique Problems on Islands

Ever since Charles Darwin's visit to the Galápagos, islands have provided spectacular places for the demonstration of fundamental principles in ecology and evolution. Sadly, for an even longer time they also have demonstrated how vulnerable species can be in the face of environmental change. Most of the avian extinctions that we know about today have been of species restricted to islands. The remaining native avifauna of the Hawaiian and Indo-Pacific Islands, the Caribbean Islands, and the Indian Ocean are dominated by species on the brink of extinction. What makes island species so vulnerable?

Most important, species on islands tend to have small populations. This means that they have less capacity to rebound from any perturbation that greatly reduces their numbers (such as disease, habitat loss, or hurricanes). Even the most stable bird populations in nature always fluctuate through time as a result of year-to-year variations in rainfall, storm tracks, food supply, disease, predation, and other factors. These random ups and downs in population size are referred to as **stochastic (or random) variation (Fig. 10–57)**. Large populations have

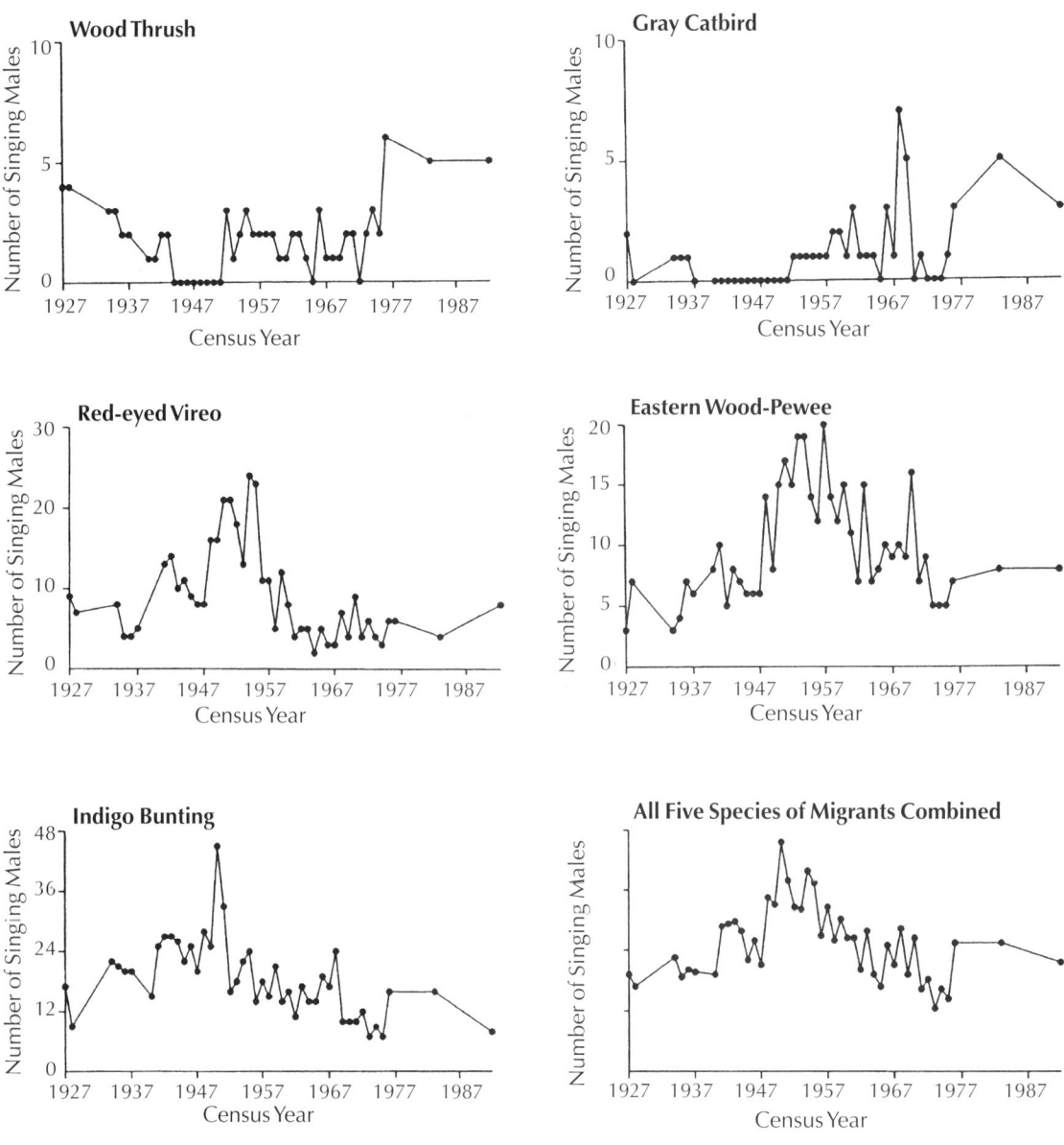

**Estimated Abundance of Neotropical Migrant Birds
in Trelease Woods (Champaign County, Illinois) from 1927 through 1992**

Figure 10–57. Stochastic Variation in Population Density of Neotropical Migrant Birds: Natural fluctuations in population size that result from year-to-year variations in factors such as rainfall, food supply, disease, predation, or habitat alterations from natural causes like violent storms are referred to as **stochastic (random) variation**. These graphs show population fluctuations of five Neotropical migrant species nesting in an Illinois woodlot (Trelease Woods) over the 65-year period from 1927 through 1992. The sixth graph shows numbers of all five species combined. In this study, researchers Brawn and Robinson detected no trends in the overall abundance of any of these species, even though the Illinois landscape was becoming rapidly fragmented. Instead, the populations went up and down over time, with two species (Wood Thrush and Gray Catbird) even disappearing and then recolonizing the woodlot several times. Large populations can withstand stochastic variation, because individual birds can buffer the effects of environmental perturbation by moving from areas of high population density to areas of low density. However, when population size is small—such as on some islands—random fluctuations can cause an entire population to become extinct simply by chance. Adapted from Brawn and Robinson (1996).

Figure 10–58. Dusky Seaside Sparrow: Found along the Atlantic and Gulf coasts of the United States, the Seaside Sparrow inhabits salt and brackish marshes, where it occurs in small, localized populations. Sufficiently distinct from one another to be considered subspecies, several of these isolated populations became vulnerable to chance extinction from random population fluctuations as humans further reduced their available habitat. One such case—a dark-plumaged subspecies known as the Dusky Seaside Sparrow—has lost the battle. Restricted to Merritt Island and the nearby St. John's River marshes of coastal central Florida, as many as 900 pairs of Dusky Seaside Sparrows existed into the late 1960s. By 1980, however, most had succumbed to mosquito-control measures—impoundment and drainage of salt marshes, spraying of insecticides, and uncontrolled fires—and to dramatic habitat changes caused by construction of the Beeline Expressway. Biologists attempted to save this subspecies by taking it into captivity, but the effort came too late. The population had become so small that, by chance, the few survivors all were male, making captive breeding impossible. The last Dusky Seaside Sparrow died in June 1987. Photo by P. W. Sykes, Jr./VIREO.

immense capacity to withstand stochastic variation, because different parts of the population are affected differently at any one time. For instance, population size may be high in one area while low in another. Because birds can disperse to areas of low density, the population is buffered simply by being large.

When population size is small, as it is in many island species (particularly after humans have altered some of the original habitat), even random fluctuations can cause the population to disappear entirely, simply as a matter of chance. "Chance extinctions" are best understood by imagining that each individual of a species flips two coins each year. The first toss determines whether the individual will live or die, and the second determines whether it will reproduce successfully that year. Obviously, in a large population following these rules, approximately half the individuals survive and about half reproduce. In a very small population, however, every individual could lose the coin toss in the same year simply by chance, causing the population to disappear abruptly. Just by chance, for example, the final seven individual Dusky Seaside Sparrows **(Fig. 10–58)** in the salt marshes of eastern Florida were all male, so that even when the remaining birds were brought into captivity as an emergency measure, the species was no longer able to reproduce.

At very small population sizes, even the tiniest natural perturbations can be deadly. On the island of Hawaii, for example, a small population of the endangered, fruit-eating bird called the Ou **(Fig. 10–59)** survived above the mosquito zone until 1984. Unfortunately, that year Mauna Loa erupted and by chance sent a small lava flow directly through the Ou's forested refuge. The species was never seen again. Although the Ou had survived eons of previous lava flows by maintaining a large population on Hawaii, when it was reduced to a small, localized population, it became highly vulnerable and finally succumbed to an isolated, random event.

Figure 10–59. Ou: The Ou, a chunky, green, fruit-eating Hawaiian honeycreeper with a yellow head and finchlike bill, originally was widespread in the Hawaiian Islands. By 1931 it was extirpated from all but Kauai and the island of Hawaii, where small populations persisted until the 1980s. Although habitat destruction, introduced species, and disease led to its initial decline, random natural disasters in the form of volcanic eruptions and hurricanes provided what probably were the final blows. Although Hawaii's native birds originally could withstand volcanic eruptions, such natural habitat perturbations can be lethal to the small populations remaining today. The Ou's last stronghold—the Upper Waiakea Forest Reserve on the island of Hawaii—was partly destroyed by lava flows when Mauna Loa erupted in 1984, and the last confirmed Ou sighting there was in 1987. On Kauai, its habitat has been battered by hurricanes, and the last confirmed Ou sighting was in 1989. Because unconfirmed Ou sightings were made on both islands in the 1990s, some hope for the species remains.

Island populations are especially vulnerable to habitat degradation from grazing and browsing mammals. Introduced goats and sheep on tiny Socorro Island (off the coast of Baja California), for example, have so altered the vegetation throughout the island that its resident specialties, the Socorro Wren, Socorro Dove, and Socorro Mockingbird, are on the verge of extinction. In the 1990s, serious effort began to remove goats and sheep from Socorro with the hope that the endemic birds could be saved through habitat recovery. Many of Hawaii's forest bird species were similarly devastated by large numbers of cattle. Especially hard hit was the Alala, or Hawaiian Crow **(Fig. 10–60)**, which feeds extensively on native fruits once plentiful year round in the forest understory of the Big Island. Because cattle graze preferentially on seedlings and saplings of the many fruit-bearing plants, mature fruit became rare throughout the understory in those few areas where forest remained. As of 2001, the Hawaiian Crow was down to its last free-flying pair, which for the fourth year in a row courted and built a nest but failed to rear any young. These two birds, plus 25 captive individuals, are all that remains of this species. Its recovery will require dramatic habitat restoration, intensive propagation from the captive flock, and "soft-release" in areas where intact habitat has been restored.

The devastating effects of introduced predators and introduced diseases on islands are discussed in several places in this chapter. Feral cats, rats, and mongooses now constitute serious threats to bird popu-

Figure 10–60. Hawaiian Crow (Alala): Many birds have been victimized by the devastation caused when introduced domestic livestock graze on native vegetation. One such victim is the Hawaiian Crow. Because cattle feed upon the seedlings and saplings of fruit-bearing understory plants, the mature native fruits on which the Hawaiian Crow depends have become increasingly rare in Hawaii's dwindling forests. In 1993, when the Hawaiian Crow's wild population plummeted to only 12 individuals, biologists began a captive-rearing program using techniques developed for restoring populations of Peregrine Falcons—removing eggs laid by wild pairs, hatching and rearing young in captivity (sometimes with the help of glove puppets to avoid imprinting the birds onto humans), and finally releasing the young into the wild. Unfortunately, many of the released Hawaiian Crows quickly perished, and the survivors have been brought back into captivity for protection. With only one pair in the wild as of 2001, the Hawaiian Crow was among the first species to receive the attention of the Keauhou Bird Conservation Center. This state-of-the-art facility propagates and prepares for release several endangered Hawaiian species as part of the Hawaiian Forest Bird Recovery Project, a joint partnership between the U. S. Fish and Wildlife Service, the State of Hawaii, the San Diego Zoo, and many foundations, private conservation organizations, and landowners. Photo by Jack Jeffrey.

lations on virtually every tropical island of the world. Because many oceanic islands evolved in the absence of diseases such as malaria, avian pox, and encephalitis, the introduction of these diseases and their vectors can be even more devastating than the introduction of predators.

Habitat Fragmentation: Mainland Habitats as Islands

One of the most far-reaching ecological patterns in nature is the axiom that larger islands are capable of holding more species than smaller islands **(Fig. 10–61)** (see Ch. 9, Effects of Habitat Size). This pattern results in part because population size typically mirrors island size and, as we have just seen, small populations (that is, those on small islands) are more likely to go extinct than large populations. Larger islands also typically contain a greater diversity of habitat types than smaller islands,

Figure 10–61. Number of Bird Species as a Function of Island Area in the West Indies: A long-recognized ecological phenomenon is that the number of species on an island increases as the island's size increases, as demonstrated by the number of bird species on different-sized West Indian islands. On this graph the x-axis (logarithmic scale) shows island area in square kilometers (square miles in parentheses), and the y-axis shows the number of bird species. Note that the scale on the y-axis is not uniform. This species-area relationship also holds for habitat patches on mainlands, which, in ecological terms, behave like islands. Adapted from Ricklefs and Cox (1972).

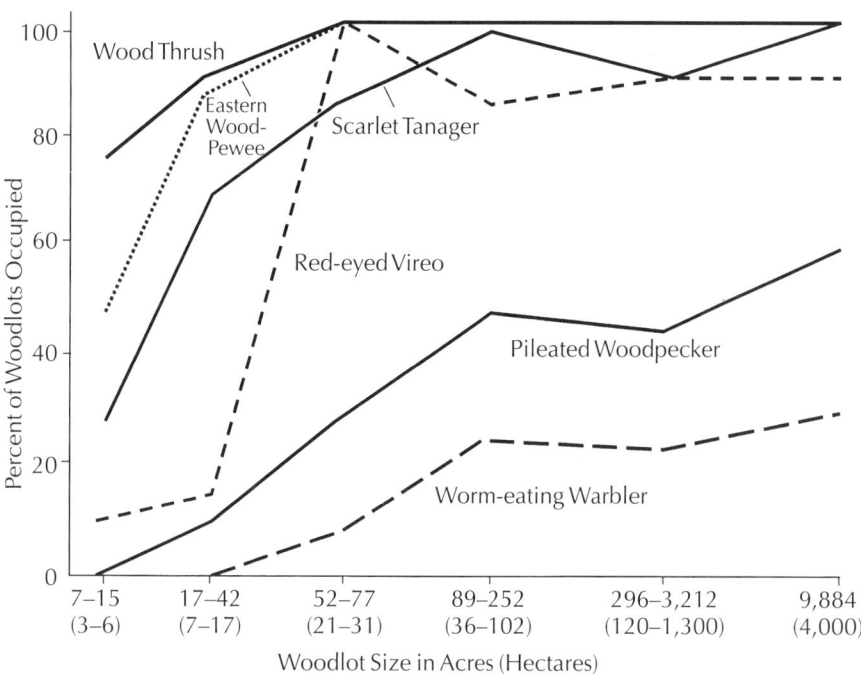

Figure 10–62. Differing Sensitivity to Forest Fragmentation: Species that need large tracts of habitat in which to persist and breed successfully are known as area-sensitive species. Area sensitivity differs greatly among species. Shown here is a graph of area sensitivity—measured by percent occupancy of woodlots of different sizes—for several eastern North American woodland species. Of the species shown here, the Wood Thrush and Eastern Wood-Pewee show the least sensitivity to fragmentation—they occur in most woodlots larger than 50 acres (20 ha), and only in the smallest woodlots do they occur at lower percentages. Scarlet Tanager and Red-eyed Vireo are slightly more sensitive, and species such as Pileated Woodpecker and Worm-eating Warbler are highly area sensitive—they are absent entirely from small woodlots, and their occurrence continues to increase steadily even as woodlot size grows larger than 2,500 acres (1,000 ha).

thereby providing niches for more kinds of species. The importance of this **species-area relationship** lies far beyond its explanatory power for species diversity on islands, however, for most of the continental habitats around the world also act as if they were islands.

In fact, as humans have converted native habitats into patchworks of landscapes designed for our own use, we have fragmented broad expanses of habitat into habitat islands of different sizes, having differing distances from one another, and separated by different intervening habitats. Remnant patches of native habitat are left to persist as islands, often every bit as isolated as if they were surrounded by water. Just as on oceanic islands, these habitat islands inevitably begin to lose their bird species.

Loss of species from habitat patches following their isolation has been demonstrated by hundreds of studies around the world over the past 40 years. One of the most consistent patterns is that species with the smallest populations disappear first. In Java, for example, an arboretum that had been isolated for 50 years (Bogor Botanical Garden) lost 75 percent of the bird species with small populations, while those that were more common largely persisted (Diamond et al. 1987). In Panama, nearly 50 percent of the original bird community disappeared from the island of Barro Colorado after it was formed by the flooding of Lake Gatun during construction of the Panama Canal. The birds that disappeared first were those with the smallest populations, such as ground-cuckoos, raptors, and woodpeckers.

As discussed earlier, species that require the largest tracts of habitat in order to persist and breed successfully are referred to as area-sensitive species, and area-sensitivity differs greatly among species (see Are North American Birds Disappearing?, Table 10–1). In forests of eastern North America, for example, Wood Thrushes, Eastern Wood-Pewees, and Red-eyed Vireos are sensitive only below about 50 acres (20 ha), and occur in most tracts above this size. In contrast, species such as Pileated Woodpeckers and Worm-eating Warblers require much larger tracts, and their frequency steadily increases with patch size, even above 2,500 acres (1,000 ha) **(Fig. 10–62)**. Therefore, preserving the complete community of woodland bird species requires preserving tracts of considerable size.

Grassland birds have the same tendency to go extinct in smaller habitat patches. As humans convert grasslands into farm fields, remnant patches of native grassland form "island archipelagos" amid

Figure 10–63. Upland Sandpiper at its Nest: With a soft, mournful whistle as distinctive as the calls of wolves or loons, the Upland Sandpiper relies entirely on grasslands for both breeding and nonbreeding activities. Ideal habitat is tallgrass and mixed grass prairie, and the species' range generally reflects the areas where such habitat remains. Upland Sandpipers breed from central Canada south to the central United States, and from the Rocky Mountains east to the Appalachian Mountains, with scattered populations in the high altitude meadows of Alaska, the Yukon, and the Northwest Territories. As North America's native grasslands are carved up to make way for suburbs, shopping malls, and agricultural land, the Upland Sandpiper is disappearing from all but the largest grassland patches. In the eastern United States the majority of nesting pairs are found in large grassy areas at airports. In Ohio, for instance, 74 percent of all observed Upland Sandpipers were at airports (Osborne and Peterson 1984). The species is listed as threatened or endangered in most states where it occurs. Photo courtesy of Michael Hopiak/CLO.

oceans of wheat, alfalfa, or corn. Grassland species such as Horned Larks, Grasshopper Sparrows, Chestnut-collared Longspurs, and Sprague's Pipits continue to occupy these islands, but the smallest islands gradually lose their species. Birds having the smallest population densities under native conditions, such as Sharp-tailed Grouse, Upland Sandpiper **(Fig. 10–63)**, and Sprague's Pipit, fail to persist in any but the largest prairie islands. Preserving or restoring large expanses of native grassland is essential for area-sensitive grassland birds, and this has become one of the most urgent needs in bird conservation the world over.

Adding to the problems facing birds undergoing habitat fragmentation are **edge effects**, or factors causing habitat near the edge of the patch to be less suitable for survival or reproduction than habitat in the middle. The most important edge effects are: **(1) changes in microclimate** (sunlight, temperature, and humidity) near a forest edge, leading to dramatic changes in plant composition and structure; **(2) introduced plants and animals**, which often penetrate habitat fragments and alter their characteristics near the edge; **(3) increased frequency of habitat disturbance** such as fire or wind damage; and **(4) predators** (such as squirrels, raccoons, and Blue Jays) **and brood parasites** (for example, the Brown-headed Cowbird), which often cruise along the edges of habitat patches and reduce survival or reproductive success. In forest fragments of southern Illinois, from 50 to 100 percent of songbird nests are parasitized by cowbirds **(Fig. 10–64)**, and 70 to 99 percent are destroyed by predators before

Figure 10–64. Veery Nest Parasitized by Brown-headed Cowbird: In this nest of a Veery—a species of forest-dwelling thrush—the speckled egg (upper left) of a Brown-headed Cowbird stands out clearly from the three uniformly blue Veery eggs. A study conducted in Canada found that 87 percent of Veery nests in Alberta and Manitoba were parasitized by cowbirds, and thus Veeries were raising fewer of their own young than necessary to sustain their population (Friedmann et al. 1977). Because cowbirds live near forest edges and often search visually for nests to parasitize, they cause particular problems for Veeries and other forest songbirds that build open-cup nests and live where forests are subject to fragmentation. For more information about cowbird nest parasitism, see Ch. 8, The Brown-headed Cowbird: History and Conservation. Photo by M. Patrikeev/VIREO.

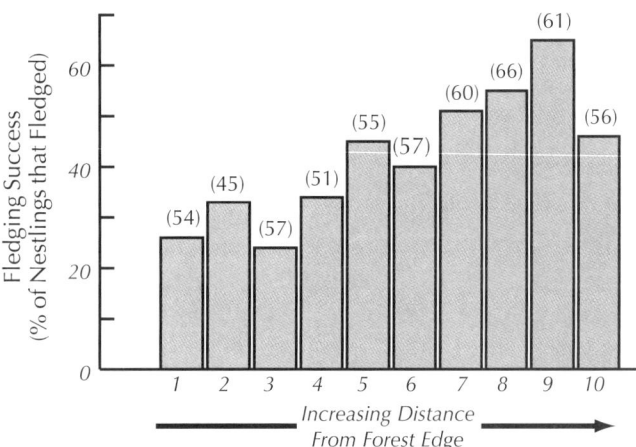

Approximate Distance Categories in Feet (Meters)		
(1)	0-3	(0–1)
(2)	3–7	(1–2)
(3)	7–14	(2–4)
(4)	14–23	(4–7)
(5)	23–33	(7–10)
(6)	33–47	(10–14)
(7)	47–88	(14–27)
(8)	88–152	(27–46)
(9)	152–215	(46–66)
(10)	215–403	(66–123)

Figure 10–65. Nesting Success of Song-birds as a Function of Distance from the Forest Edge: When continuous habitat such as a large forest is divided into smaller patches of woodland, isolated from each other by a new habitat such as agricultural fields, the small woodland patches have much more edge relative to their interior area than does the original unfragmented forest. Edges provide easy hunting routes for many nest predators, such as jays, crows, grackles, chipmunks, weasels, and raccoons. Edges also attract the infamous brood parasite, the Brown-headed Cowbird. As a result, when certain forest interior birds are forced to nest in small woodlots, their nesting attempts are more likely to fail because the nests are close to an edge—a phenomenon known as "edge effect." This famous graph shows that fledging success (the percentage of nestlings that survived to fledge) for several species of open-cup-nesting songbirds increases with greater distance from the edge of a Michigan forest patch. For each distance category, distance range (rounded to the nearest foot [meter]) is presented on the right. Sample size is given in parentheses above its respective bar on the graph. Graph from Wilcove et al. (1986), after Gates and Gysel (1978).

hatching (Brawn and Robinson 1996). As a result of all these factors, bird species that typically nest in the middle of the forest show lower success when they nest near a forest edge **(Fig. 10–65)**.

Isolated populations are vulnerable to unpredictable episodes of catastrophe, a fact demonstrated most dramatically when Hurricane Hugo pounded into South Carolina in 1989. The full brunt of this fierce autumn storm virtually leveled the Francis Marion National Forest, which until then had harbored the largest continuous population of the endangered Red-cockaded Woodpecker. Recall that this species drills its nest cavities in very old but living pine trunks, and the storm destroyed more than 80 percent of the cavity trees. The woodpecker population plummeted, and remains low to this day. Emergency installation of artificial nest cavities is helping to save this vital population from outright extinction during the long recovery period required to grow trees of sufficient size to accommodate natural cavities (Watson et al. 1995).

Conservation Genetics

When populations exist at very small sizes, they face risks not only from environmental fluctuations and chance perturbations, but also from genetic effects. Recall that genes are located on long, paired DNA molecules called chromosomes, and that each pair of chromosomes consists of one from the individual's mother and one from the father (see Fig. 4–114). A gene is a specific segment, or locus, along the DNA molecule where certain information about structure or maintenance of the living organism is encoded. Alternate forms of genes are called alleles. (For example, eye color in humans is mainly determined by blue-eyed versus brown-eyed alleles at the eye-color locus.) Individuals bearing two different alleles for a given gene are referred to as **heterozygous** at that locus, and those bearing the same allele on both DNA strands are **homozygous** at that locus. For a variety of reasons, many alleles supply important functions when they occur singly, but can be harmful or even deadly if they are carried in duplicate. Individuals that are heterozygous at many gene loci usually are better able to survive, reproduce, and resist disease than those that are more homozygous. Because close inbreeding increases the chances that offspring will be homozygous for harmful alleles, the **incest taboo** is exceptionally strong among all birds. Through this behavioral mechanism, individuals typically avoid pairing and breeding with members of their immediate family. Because of this taboo, and through processes of genetic mutation, recombination, and sexual reproduction, all large populations of birds maintain vast pools of genetic diversity. This keeps the average level of heterozygosity high—in most bird populations the average individual is heterozygous at 30 percent or more of its genetic loci.

When a population becomes very small, three things begin to happen that cause a **loss of genetic variation**. (1) Simply by chance, some alleles—beginning with the rarest ones—disappear altogether. (2) The rate at which new alleles crop up in the population as a whole declines, because there are fewer individuals in which new mutations can occur. (3) Opportunities to pair with genetically dissimilar individuals become reduced, so that close inbreeding begins to occur more frequently. The resulting loss of genetic variation then leads to both increased mortality and decreased birth rates within the population as a whole. Thus, extremely small populations are subject to an ever-worsening spiral: as each succeeding generation fails to replace itself fully, more alleles are lost and the population becomes ever more inbred, or genetically homogeneous. Such populations are said to have experienced a **genetic bottleneck (Fig. 10–66)**.

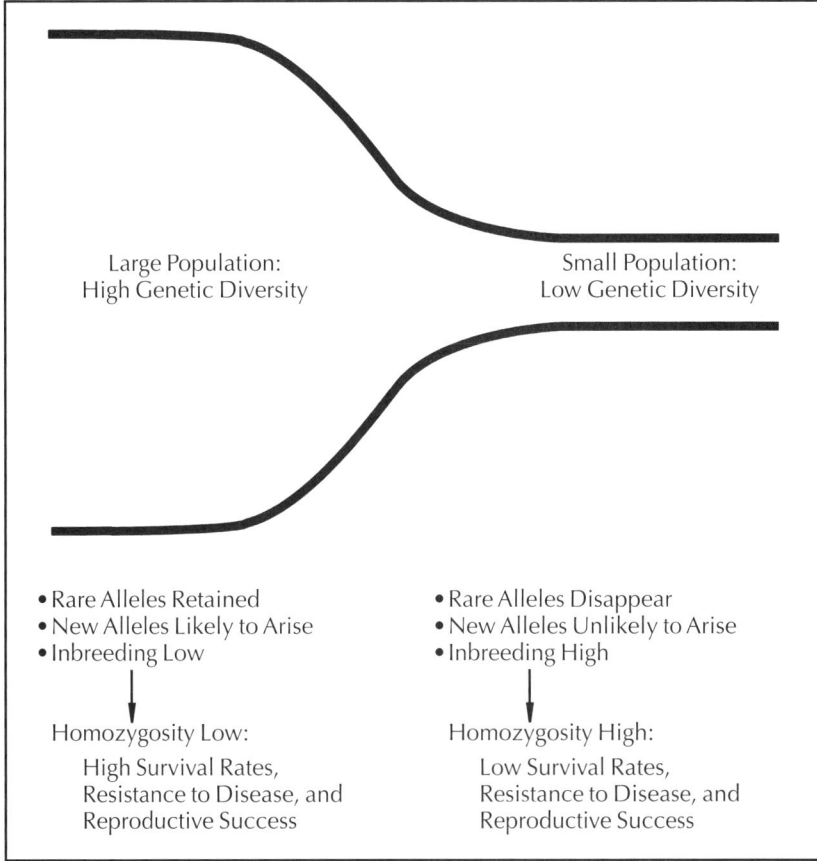

Figure 10–66. Development of a Genetic Bottleneck: Very small populations usually become inbred and lose much of their genetic diversity. This situation, known as a **genetic bottleneck**, results from three factors. First, in small populations some alleles (alternative versions of genes), especially the rarest ones, disappear by chance. Second, the rate of appearance of new alleles decreases, because fewer individuals exist in which new, potentially beneficial mutations can arise. Finally, fewer genetically dissimilar individuals are available as potential reproductive partners. When populations become inbred they face a higher chance that offspring will be homozygous for deleterious alleles, with the result that they suffer lower survival rates, increased susceptibility to disease, or reduced reproductive success.

Conservation Solutions: Tools and Prescriptions for Stabilizing Populations

■ To accomplish bird conservation in real world settings, our first job is to understand the biological underpinnings of a problem, such as a population decline in a species of concern. Once we understand the problem, we must bring together basic principles of genetics, evolutionary biology, life history theory, population biology, community ecology, animal behavior, and mathematical modeling to "prescribe" steps that will result in a solution. In this section I explore some tools for applying the principles of conservation biology in the effort to stabilize bird populations around the world.

DNA Fingerprinting and Genetic Augmentation

Close inbreeding and genetic bottlenecking probably explain why some of the most endangered bird species begin failing to breed successfully as they reach critically low numbers. In the Hawaiian Crow, for example, the last few pairs of wild individuals failed to fledge any young between 1991 and 2001. A number of the eggs they produced appeared to be infertile when examined after the breeding season, suggesting that inbreeding might be playing a role. Analysis of the DNA of the remaining individuals—via the same kind of "genetic fingerprinting" used today as evidence in court cases and forensics—revealed them to be unusually similar to one another genetically. Successful reproduction within the captive Hawaiian Crow flock therefore demanded efforts to pair individual crows with mates as genetically dissimilar as possible. To accomplish this, the project managers maintain a **studbook**, a technique adapted from standard practices used in zoos. DNA fingerprinting, and keeping careful records of individual breeders and their offspring, allow biologists to select the genetically least similar individuals for pairing with one another, and especially to avoid matings between siblings or parents and their offspring. Fingerprinting and studbooks have become standard tools in captive-rearing programs involving rare birds, and they are essential for maximizing the success of a released population. For example, when captive-reared Guam Rails were reintroduced to the wild on the island of Rota in 1989, individuals were chosen for release on the basis of their ability to contribute to the overall genetic diversity of any newly established population (Haig et al. 1990).

Human intervention also can prevent tiny, wild populations from experiencing a genetic bottleneck and the associated extinction spiral caused by loss of genetic variability. A good demonstration of successful intervention took place in Illinois, where a small and isolated population of Greater Prairie-Chickens began to decline precipitously in the 1980s. By 1993, with fewer than 50 individuals remaining, it was determined that this population possessed only about two-thirds of the allelic diversity found in larger populations in Nebraska, Kansas, and Minnesota. As is typical of inbred populations, only about half of the eggs laid by females in this tiny population were hatching, compared with 93 percent hatchability elsewhere (Westemeier et al. 1991). When birds from other populations were experimentally introduced into the declining population, hatching rate increased to pre-bottleneck levels and the population was stabilized (Bouzat et al. 1998).

Population Viability Analysis and Metapopulations

Small, localized populations of virtually any species are likely to go extinct if cut off from contact with the rest of their kind. Random population fluctuations, depressed reproduction and survival from edge effects, and the possibility of catastrophic disease or weather events make small populations likely to disappear altogether. To allow some precision in measuring their vulnerability, conservation biologists in the 1980s developed a tool for predicting the likelihood of local extinction for populations of different starting sizes. The tool—**population viability analysis (or PVA)**—uses computers to simulate the progression of real world populations through the annual cycles of reproduction, dispersal, and mortality. The simplest PVAs are **single-population models**, which simulate population numbers through time based on an assumed initial population size plus demographic variables such as birth rates, death rates, and amounts of annual variation. Typically, such a model is run until the "virtual population" goes extinct, then the process is repeated a number of times (say, 100 or 1,000 replications) for the same starting population size. The process is also repeated for different starting population sizes, ultimately yielding a graph that projects the probability of extinction as a function of time for many starting population sizes **(Fig. 10-67)**. For most endangered species in the real world, the "starting" population size closely depends on the size of its habitat patch. Therefore, PVAs can be used to set goals for how big a habitat preserve needs to be to establish a low probability that a target species within it will go extinct in, say, 100 or 500 years.

If the bad news is that local extinctions of small populations are inevitable in the real world, then the good news is that such extinctions can be reversed through **recolonization**. Because animals disperse, separate populations of a given species often exchange individuals and genetic information with one another. Dispersal is vitally important in conservation, for two reasons: (1) it helps maintain genetic variability across large areas and among separate populations, and (2) it provides opportunity for re-establishment of a local population that has gone extinct. The idea that local populations distributed across a landscape can "wink" out, and then "wink" back on, has achieved landmark status in conservation biology, and is known as **metapopulation dynamics (Fig. 10–68)**. If we know (or if we can guess) the rate at which individuals from one population disperse across habitat gaps to colonize other populations, then we can conduct a PVA

10

Figure 10–67. Population Viability Analysis of the Florida Scrub-Jay: In order to set targets for management, conservation biologists often must predict the chances of a population going extinct over a projected time frame, or must determine the minimum viable size of a threatened population. To do this they developed **population viability analysis (PVA)**, a tool that allows prediction of the probability of local extinction for populations of given starting sizes. Building equations that incorporate numerous demographic variables (such as birth rates and death rates at various ages), biologists use computers to simulate (or "model") the progression of populations over time, until extinction of these "virtual populations" occurs. By running the same model many times for each population size, graphs emerge that show the proportion of simulations in which the population went to zero during a given time interval. Shown here are the results of a PVA of the Florida Scrub-Jay, a cooperatively breeding, threatened species that is strictly limited to remnant patches of stunted oak scrub in central Florida. Each line is the cumulative extinction probability versus time for a Florida Scrub-Jay population of a given starting size (the number in the circle associated with the line), with size measured as the maximum number of territories that can exist in the habitat patch. Each territory contains only one breeding pair of scrub-jays plus from zero to three helpers (see Ch. 1, Sidebar 2, Fig. C). The x-axis shows time after the population was isolated (the simulation assumes no immigration into the population). The y-axis shows the cumulative probability of extinction. Each small point on each line is a five-year interval. From this graph we can estimate, for instance, that a population containing only one territory has a 60 percent chance of extinction in just 10 years from isolation, and certainly will be extinct within 45 years after isolation. Although populations containing 10 territories are more persistent, even these have a 50 percent chance of being extinct in 100 years. Only very large populations, such as those containing 40 to 100 territories, have good chances of persisting as long as 400 years. Species recovery plans often aim at target probabilities and time frames (for example, less than a 10 percent chance of extinction in 100 years) using PVAs such as this. Graph from Fitzpatrick et al. (1991).

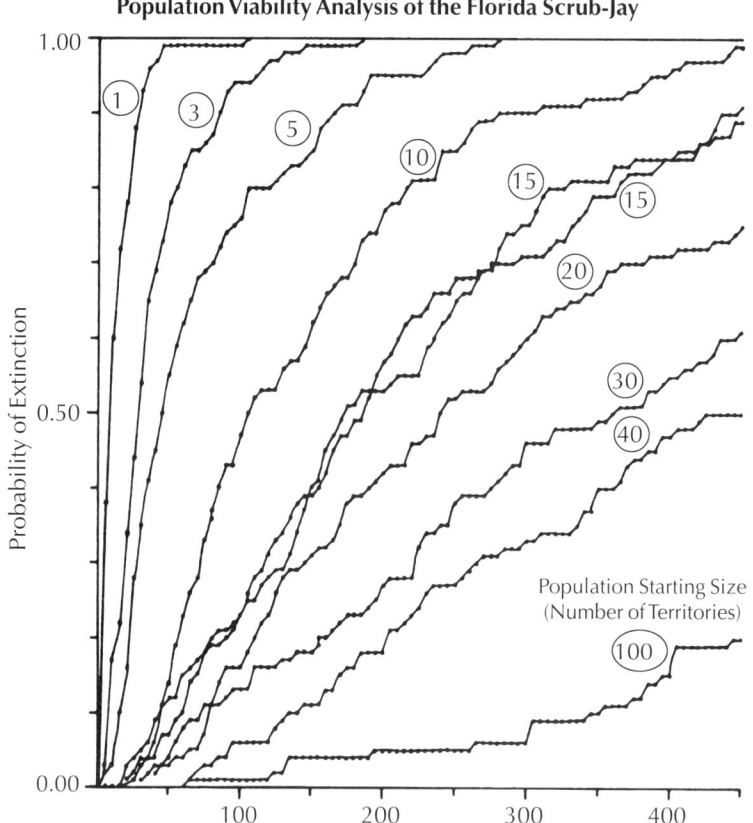

Population Viability Analysis of the Florida Scrub-Jay

Probability of Extinction

Years Since Theoretical Population was Isolated

Population Starting Size
(Number of Territories)

using metapopulation models. These two-dimensional models based on many populations distributed across a landscape usually present a more realistic projection of overall extinction probability than do single-population models. However, they also require much more detailed information about the biology and movement pattern of a target species, including its patterns of behavior and mortality while moving across inhospitable habitats. For this reason most modeling efforts are subject to considerable error, and must be interpreted with caution. Nevertheless, beginning in about 1995, metapopulation models helped to shape recovery efforts for numerous endangered species that are highly susceptible to local extinction because they breed in patchy habitats (for instance, Spotted Owl, California Gnatcatcher, Black-capped Vireo, and Florida Scrub-Jay).

Preserve Design

Because of all the problems that face populations when their habitat shrinks and is fragmented into isolated islands, few species can persist if their habitat is reduced to a single island, especially if that island is small. The key to conserving rare species and vanishing ecological communities is to secure **habitat preserve networks** that are designed to accomplish long-term protection. We now understand a great deal about how to design these networks so as to minimize the risks rare species face. Single-population modeling helps us project

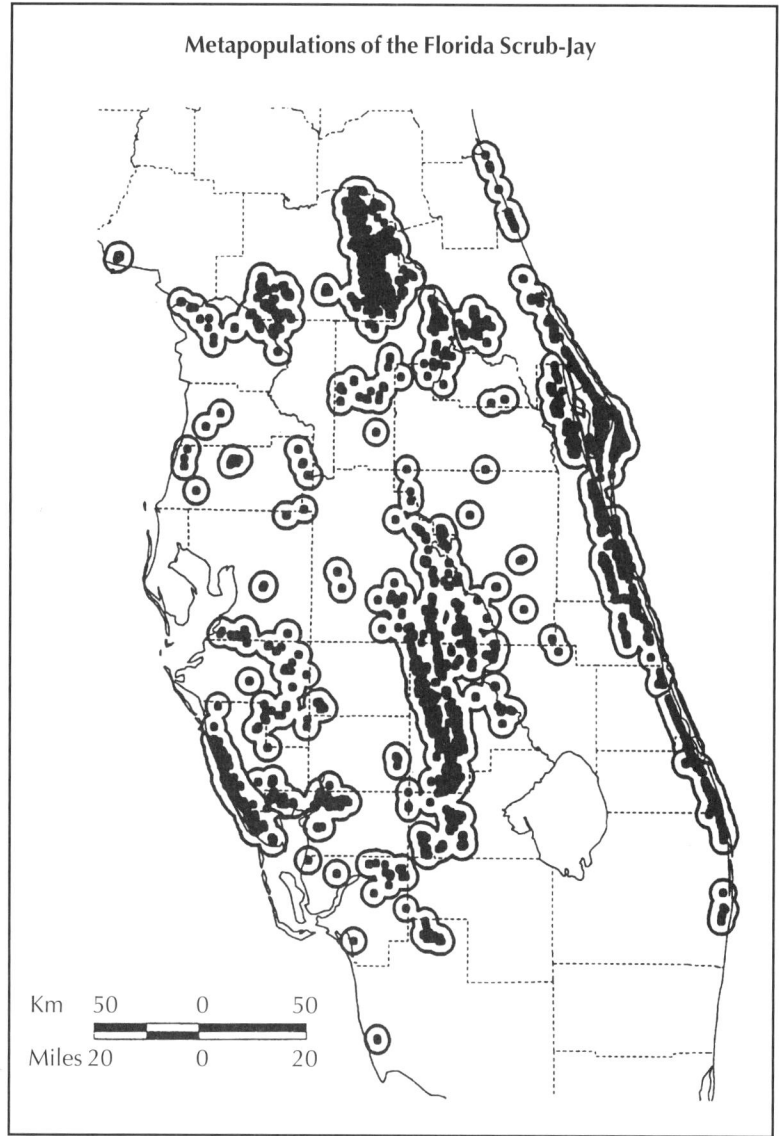

Metapopulations of the Florida Scrub-Jay

Km 50   0   50
Miles 20   0   20

Figure 10–68. Metapopulations of the Florida Scrub-Jay: This map of the Florida peninsula (dashed lines are county boundaries) shows the current geographic range of the Florida Scrub-Jay, a federally threatened species. Each black dot represents one or more territorial family groups, of which about 3,000 still exist. Solid lines enclosing clusters of territories identify 42 separate **metapopulations**, based on detailed information about Florida Scrub-Jay dispersal ability. Metapopulations consist of subpopulations close enough together to allow dispersing individuals from occupied patches to recolonize patches where the species has gone extinct. Protecting a patchily distributed species such as the Florida Scrub-Jay requires understanding its dispersal capacities in order to identify meta-populations, and protecting and properly managing multiple habitat patches within each metapopulation. From Metapopulations and Wildlife Conservation, Dale R. McCullough, ed. Copyright 1996 by Island Press. Reproduced by permission of Island Press, Washington, D.C. and Covelo, CA. All rights reserved.

the minimum size required to reduce local extinctions to acceptably low levels, and metapopulation modeling can help estimate the extinction probabilities for alternative network configurations. These modeling exercises have resulted in a series of "rules" guiding the design of preserves and preserve networks **(Fig. 10–69)**. The most important of these rules are:

1. **Bigger is usually better**, because large preserves contain larger total population sizes, reducing vulnerability to stochastic variation.

2. **Round or square is better than long and skinny**, because round preserves have the least amount of edge per area.

3. **Create multiple preserves whenever possible,** as this minimizes the chance that a catastrophe will wipe out an entire population.

4. **Ensure at least one large preserve in the network**, as this provides a reliable source of dispersers to recolonize smaller patches following local extinctions.

5. **Minimize distances among preserves,** to facilitate local movements between patches and reduce genetic isolation among them.

6. **Add "stepping stones" and "corridors" between preserves,** as these small pieces of habitat greatly facilitate movement among preserves, thereby allowing a fragmented preserve system to act more as if it were larger and more continuous.

7. **A two-dimensional landscape configuration is better than linear,** because it promotes opportunities for recolonization among all habitat patches.

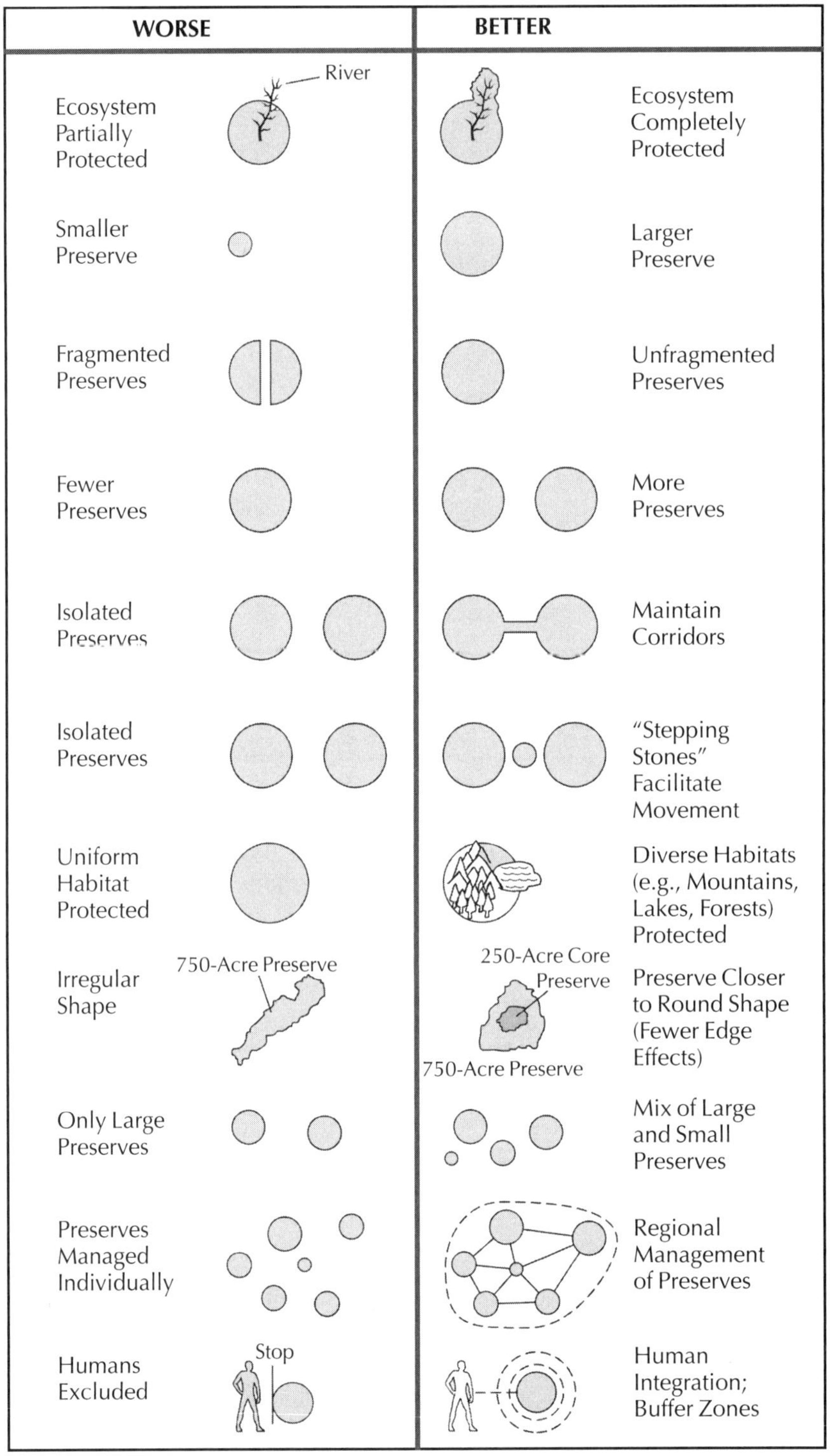

Figure 10–69. Nature Preserve Design: Conservation biologists have developed guidelines for preserve design that take into consideration such factors as the size and shape of individual preserves and how they are grouped into networks across the landscape. In general, the designs on the right are considered preferable to those on the left. See text for details. Adapted from Shafer (1997).

WORSE	BETTER
Ecosystem Partially Protected (River)	Ecosystem Completely Protected
Smaller Preserve	Larger Preserve
Fragmented Preserves	Unfragmented Preserves
Fewer Preserves	More Preserves
Isolated Preserves	Maintain Corridors
Isolated Preserves	"Stepping Stones" Facilitate Movement
Uniform Habitat Protected	Diverse Habitats (e.g., Mountains, Lakes, Forests) Protected
Irregular Shape (750-Acre Preserve)	Preserve Closer to Round Shape (Fewer Edge Effects) (250-Acre Core Preserve, 750-Acre Preserve)
Only Large Preserves	Mix of Large and Small Preserves
Preserves Managed Individually	Regional Management of Preserves
Humans Excluded (Stop)	Human Integration; Buffer Zones

Protecting a series of habitat preserves by properly managing "islands" of habitat has become the single most important means for protecting the rarest endangered birds. For example, two endangered subspecies in southern California, the Light-footed Clapper Rail and the Belding's Savannah Sparrow **(Fig. 10–70)** are protected in a string of jealously guarded salt marshes, even though each marsh is surrounded by cities, suburbs, beach houses, and yacht clubs. Likewise, Florida Scrub-Jays and California Gnatcatchers are protected in carefully managed scrub preserves, where suitable habitat patches are surrounded by orange groves, shopping malls, and retirement communities. Piping Plovers breed on isolated sand beaches of the Great Lakes and the Atlantic Coast, where their fragile sand-dune habitats are surrounded by beach houses and off-road vehicles. In many respects, each of these separate populations is exposed to the same dangers of island existence as populations living on small islands in the remote Pacific. The difference is that for species that exist only on one oceanic island, extinction is forever. On continental islands, each habitat patch retains some opportunity for recolonization as long as we can ensure that multiple sources of colonists continue to exist in separate patches.

Habitat Management

As discussed earlier in this chapter, most species that are endangered today were vulnerable early on, because they have narrowly specialized habitat tolerances. And as discussed in the preceding paragraph, our principal means for maintaining such species is to protect many separate populations by preserving patches of their required habitat. However, simply setting a habitat patch aside rarely

Figure 10–70. Belding's Savannah Sparrow: A male Belding's Savannah Sparrow (inset photo) sings to attract a mate in early spring. This endangered subspecies is restricted to dwindling salt marshes along the coast of southern California, an area undergoing rapid human development. One of the sparrow's strongholds is Bolsa Chica Ecological Reserve (main photo), a salt marsh sanctuary surrounded by the beachfront hotels, suburban homes, and shopping malls of Huntington Beach, California, south of Los Angeles. Still other threats come from encroaching oilfields nearby. One survey estimates that 163 pairs of Belding's Savannah Sparrows live on the reserve, where they nest exclusively in pickleweed, a salt marsh plant. Photos by Marie Read.

suffices to ensure its long-term preservation. Left unattended, most habitats remain subject to continuing influences from humans, either directly (such as continued introduction of predators) or indirectly (for instance, artificially altered hydroperiods). Therefore, most natural preserves require **habitat management** to remain suitable for the species and ecological systems they are designed to protect. For example, wetland habitats often require restoration and artificial maintenance of their natural water flows to maintain the annual cycles to which herons, shorebirds, and waterfowl are adapted. Invasive plant species must be held in check by active removal programs. Introduced predators must be trapped and removed, especially on islands, so that sensitive native birds can reproduce successfully.

Ironically, mimicking natural disturbances can be the most important and difficult responsibility in modern habitat management. For example, many western grasslands support higher species diversity when they are subject to periodic disturbances that mimic the temporary passage of a bison herd. Thus, in the absence of bison, carefully managed "short-rotation" grazing by cattle can improve some western grasslands as bird habitat. On the other hand, cattle tend to loaf and forage near streams, causing erosion of stream banks and degradation of the riparian thickets that are so important for breeding bird communities throughout the West. Proper management of cattle grazing, therefore, requires construction and maintenance of fencing to permit control of stocking rates and exposure periods within each management unit, and to limit the access of cattle to riparian habitats **(Fig. 10–71)**. The financial burden of such management practices has made cattle grazing and ranchland management one of the most hotly debated issues in bird conservation across the United States.

Beginning in Aldo Leopold's day, ecologists and wildlife managers increasingly have become aware that species diversity in many habitats around the world (such as native grasslands, savannas, scrubs,

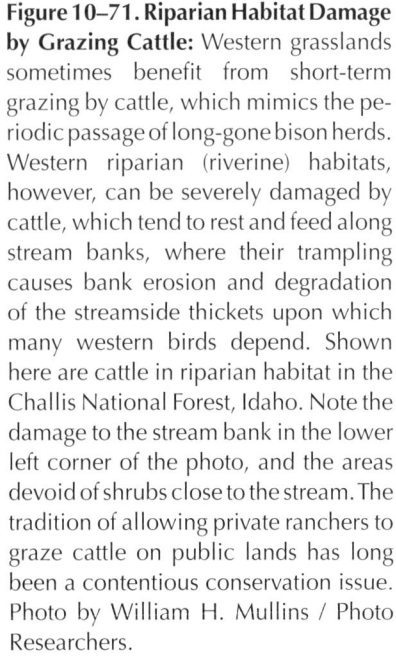

Figure 10–71. Riparian Habitat Damage by Grazing Cattle: Western grasslands sometimes benefit from short-term grazing by cattle, which mimics the periodic passage of long-gone bison herds. Western riparian (riverine) habitats, however, can be severely damaged by cattle, which tend to rest and feed along stream banks, where their trampling causes bank erosion and degradation of the streamside thickets upon which many western birds depend. Shown here are cattle in riparian habitat in the Challis National Forest, Idaho. Note the damage to the stream bank in the lower left corner of the photo, and the areas devoid of shrubs close to the stream. The tradition of allowing private ranchers to graze cattle on public lands has long been a contentious conservation issue. Photo by William H. Mullins / Photo Researchers.

and numerous types of forests) is maintained by periodic wildfire. As early as the 1930s, a game manager named Herbert Stoddard at Tall Timbers Research Station near Tallahassee, Florida showed that Northern Bobwhite populations responded immediately and favorably to regular burning of the wiregrass understory beneath mature longleaf pines. (Several decades later, ecologists realized that the Red-cockaded Woodpeckers sharing this habitat were even more fire-dependent than the quail.) In fact, most habitats around the world where lightning-caused fires are frequent contain a host of species

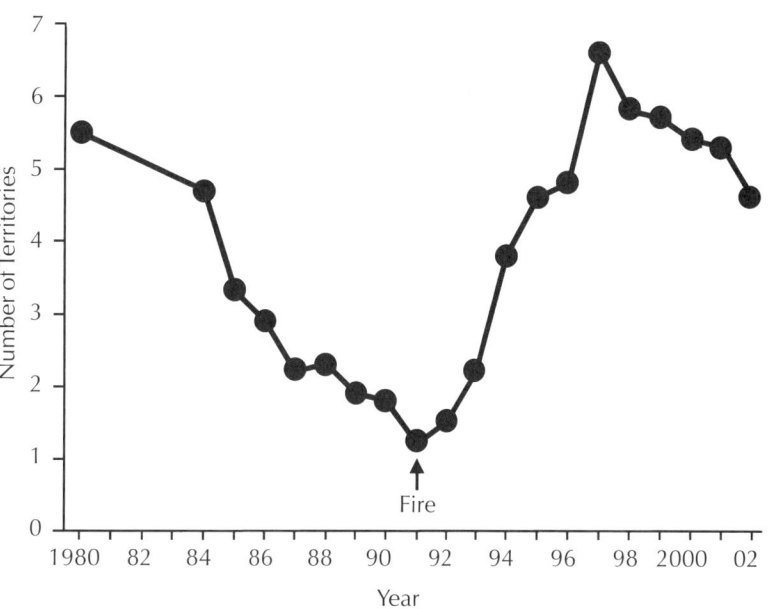

Effect of Fire on the Number of Territories in a Florida Scrub-Jay Population

that eventually will go extinct if fire is suppressed or excluded. Many of these "post-fire specialists" may be more endangered today by fire suppression than by outright habitat loss. In these cases, proper habitat management requires **prescribed burning** to maintain the ecological conditions under which the species evolved. Florida Scrub-Jays, for example, gradually decline in intact oak scrub where fire has been suppressed, because they are increasingly exposed to competition from Blue Jays and to predation by hawks, snakes, and various mammals. However, Florida Scrub-Jay numbers can bounce back rapidly once fire is reintroduced to their habitat **(Fig. 10–72)**.

Ecosystem Management

Over the past half century, ecologists have begun to appreciate the complexity of relationships among species and communities across large landscapes. Because these relationships occur at many scales, protection of a single target species (for example, a declining bird) actually requires protecting a host of ecological processes such as fire, periodic flooding, nutrient cycling, plant succession, and seasonal migrations. Such processes may occur unpredictably (for example, damaging storms), or in cycles that may be daily (such as the activity patterns of insectivorous birds and their prey); annual (for instance, seasonal flooding of vernal pools); regular multi-year (predator-prey cycles, for example); or irregular multi-year (such as floods or fires). Thus, the most important management objective for preserving native species becomes ensuring that these processes continue to exist as they did prior to human impact. By the beginning of the 21st century, conservationists began referring to the understanding and maintenance of the whole range of natural processes as **ecosystem management**.

Figure 10–72. Effects of Fire on a Florida Scrub-Jay Population: Many habitats around the world are subjected to frequent fires naturally caused by lightning. To maintain the health of these habitats today, land managers must engage in **prescribed burning**—purposefully starting fires under controlled conditions, and at intervals that mimic the natural fire cycle of the landscape. These fires allow the habitat to "reset itself," keeping many species from disappearing. From 1980 to 1990, after a long period of fire suppression in an otherwise undisturbed 136-acre (55-ha) area of oak scrub in central Florida, the population of resident Florida Scrub-Jays gradually declined, as can be seen from this graph. After a prescribed burn was conducted in 1990, Florida Scrub-Jay numbers immediately rebounded. During recent years, the population has begun to decline again. Biological information such as this, now available for many species of animals and plants, suggests that the natural fire cycle in this oak scrub habitat was about once every 8 to 15 years. This graph shows the number of scrub-jay territories in the population; each territory may be defended by a group of up to eight birds. Graph courtesy of John Fitzpatrick and Glen Woolfenden.

Figure 10–73. Wood Stork: The endangered Wood Stork is the largest wading bird breeding in the United States. It stalks shallow wetlands of the Southeast, feeling for food in muddy water with its highly sensitive bill. In the 1930s huge breeding colonies existed in the Everglades and Big Cypress basin, with a total population of 10,000 nesting pairs. However, starting in the 1960s, water management practices in south Florida—in which natural flow patterns were dramatically altered (see Fig. 10–74)—and wetland degradation, for instance by contamination with pesticides and chemical fertilizers, caused the Wood Stork's habitat to decline in quality and its numbers to decrease dramatically. The species was listed as endangered in 1984, and the once large breeding colonies of south Florida have been reduced to about 10 percent of their former numbers. The species has become more numerous farther north, with nesting colonies appearing in Georgia and South Carolina. Unfortunately the bulk of these populations migrate to southern Florida each winter, where they face the same problems that forced them farther north. Considered a bio-indicator of the overall health of southeastern wetlands, the Wood Stork and its ecosystem are now the focus of extensive conservation efforts. Photo by Marie Read.

Under ecosystem management, restoring stable populations of endangered Wood Storks (**Fig. 10–73**) in the Everglades requires re-establishing the seasonal flow of water through the entire southern Florida ecosystem. Before humans ditched and drained southern Florida during the 1950s and 1960s, heavy summer rains enriched the Everglades by producing vast flows of fresh water through marshlands originating hundreds of miles to the north (where Orlando lies today). The spring dry season then caused wet areas to contract, concentrating fish, frogs, snakes, and sirens (large, legless amphibians) into small pools that provided enormous concentrations of prey and supported huge nesting colonies of wading birds. To restore native biological functions to this system, the multibillion dollar Kissimmee River Restoration Project (**Fig. 10–74**) and Everglades Restoration Project were initiated in the late 1990s. These colossal—and controversial—projects constitute the most ambitious examples of ecosystem restoration ever undertaken by humankind, and are expected to require two decades to complete.

Adaptive Management

Many of our biggest challenges in conserving birds and their habitats occur because we almost always make management decisions without knowing everything we'd like to know about a system. Therefore, effective ecosystem management requires that before we begin we define our target and recognize the gaps in our knowledge. We then proceed using a three-step process: **(1) set goals** in terms of desired outcomes of management actions (for example, strive to sustain a certain population size or density of a target bird species, or a specified range of abundance values for different species in a habitat); **(2) keep learning** new details about how the ecosystem works, through experimentation and monitoring, so that we know how our target species respond to different management actions; and **(3) modify our management plans** so that they better accomplish the goals. As time goes on, therefore, we adapt management techniques to incorporate our new understanding about the system. Beginning in the 1980s this concept was formalized into a new approach to hands-on habitat management called **adaptive management**. The approach stresses the importance of learning as we go by incorporating landscape-scale experiments (such as different burn regimes, grazing frequencies, or water depths) accompanied by question-driven monitoring of key species and processes (bird censuses or measurements of nest success in different habitats, for example). Adaptive management also calls for being ready, willing, and able to modify management techniques as we learn new information. This

a. Kissimmee River Before Channelization, Around 1961

b. Kissimmee River After Channelization

approach can be challenging to carry out, especially by public land management bureaucracies such as park services, wildlife agencies, forestry divisions, and county governments that are accustomed to operating with fixed and long-standing management formulas, and that lack a scientific infrastructure for conducting valid experiments and monitoring. Therefore, adaptive management represents an important opportunity for public agencies to partner with private conservation groups and research institutions to accomplish the most effective long-term management of the land **(Sidebar 4: Conservation Planning at Ecoregional Scales)**.

(Continued on p. 10·89)

Figure 10–74. Kissimmee River Restoration Project: Restoring declining species often requires **ecosystem management**. For example, to return wading bird populations to their historical levels in the Florida Everglades, attention must be given to the entire southern Florida ecosystem, especially its hydrology (the science of water, its flow patterns, and its distribution). **a. Before Channelization:** The Kissimmee River historically meandered southward from Lake Kissimmee in central Florida to Lake Okeechobee, whose waters eventually reach the Everglades at the southernmost tip of the Florida peninsula. The original floodplain supported extensive wetlands including wet prairies, marshes, willow and cypress swamps, and hammocks (clumps of trees surrounded by wet areas). These diverse habitats were year-round home to thousands of wading birds, and served as wintering areas for tens of thousands of ducks. After World War II, human habitation and subsequent concerns about flooding began to increase in the region. **b. After Channelization:** Beginning in 1962, the Kissimmee River was channelized into a 30-foot deep, 56-mile long canal (shown here on the right side of the photo). Confined by levees and compartmentalized by dams and other water-control structures, the river became virtually stagnant. Although channelization provided flood control for the region's burgeoning human population, it had profound and unforeseen effects on local wildlife. Because an estimated 35,000 acres (14,000 ha) of floodplain wetlands were lost, numbers of wading birds and waterfowl that used the area were tremendously reduced. In addition, game fish populations began a long-term decline. In response to this loss of biological diversity, a project to restore the river's natural course and floodplain began in 1992. The hope is to re-establish natural flow patterns, with seasonal highs and lows, by filling in 22 contiguous miles of the flood-control canal and by removing many of the water control structures. Eventually, 40 square miles (10,300 ha) of river and adjacent floodplain will be restored to benefit fish and wildlife, including endangered species such as the Wood Stork, Bald Eagle, and Snail Kite. Ecological mistakes are costly to remedy—a 1997 estimate put the total cost of the Kissimmee River Restoration Project at $414 million, shared equally by the state of Florida and the federal government. The first phase of filling began in spring 1999, and the project is scheduled for completion in 2010. (For updates visit the South Florida Water Management District's web site <**www.sfwmd.gov**>). Photos courtesy of South Florida Water Management District.

10

Sidebar 4: CONSERVATION PLANNING AT ECOREGIONAL SCALES
John W. Fitzpatrick

All efforts to conserve birds and their ecosystems in the face of pressure from a spreading human population come face to face with an icy reality: we cannot possibly save it all. Every natural area and system on our continent today faces a growing list of threats, from suburban sprawl and agricultural runoff to alien species and global climate change. To safeguard ecosystems against these threats over the long term, we must engage in careful, biologically driven planning at the scales most appropriate to each landscape and system.

Because the natural world rarely conforms to political units such as counties, states, or countries, effective conservation planning organizes the world by ecological, not political, boundaries. To do this we divide the world into biologically distinct regions, called **ecoregions** or **physiographic regions**, that are based on climate conditions, topography, soil types, and plant communities. Careful, biologically driven planning at the level of ecoregions is called **ecoregional planning**.

Ecoregional planning in one form or another is practiced by many of the largest and most important conservation organizations in the world. In 1995, for example, The Nature Conservancy defined a science-based planning system called "Conservation by Design" by which their entire conservation enterprise and management structure are organized **(Fig. A)**. And, in 1990, to aid in long-term ecoregional planning for North American birds, a broad group of public agencies, private conservation groups, and industry representatives formed the consortium Partners in Flight.

Ecoregional planning follows a series of steps to set goals and plans for long-term conservation.

• **Map ecoregions:** For North America, many recent ecoregional maps are adapted from an outstanding example created in 1978 by Robert Bailey of the U. S. Forest Service (Bailey et al. 1994). Bailey divided the continental United States into ecologically distinct "provinces," then divided each province into "sections." Partners In Flight modified Bailey's map slightly and added Canada and Mexico to create their map of physiographic regions within which to conduct ecological planning for bird conservation **(Fig. B)**.

• **Catalog ecosystems:** Within each ecoregion we identify its important plant communities (for instance, mixed tallgrass prairie), along with key indicator species (for example, big bluestem grass and Greater Prairie-Chicken) and dominant ecological forces (such as frequent summertime wildfire).

• **Map sites:** To be specific about where our actions should be directed, we must generate detailed maps of the actual sites and habitat tracts where the best examples of each ecological community remain, or where habitat restoration could re-create them. Today

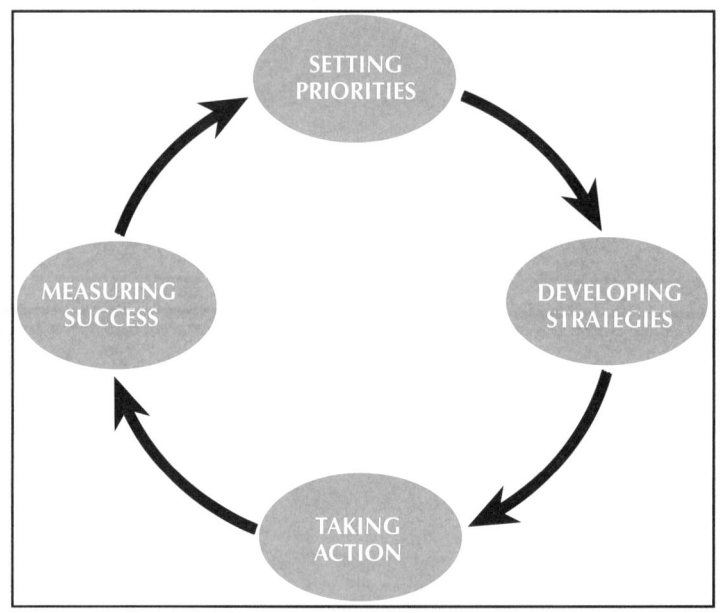

Figure A. Conservation by Design: The mission of The Nature Conservancy (TNC) is to preserve the plants, animals, and natural communities that represent the diversity of life on earth by protecting the lands and waters they need to survive. TNC's strategic, science-based planning process, called Conservation by Design, helps identify the highest-priority places—landscapes and seascapes—that, if conserved, promise to ensure biodiversity over the long term. TNC calls these places **functional conservation areas**. A set of such areas is a **portfolio**. TNC plans to protect portfolios within and across large areas of land or water defined by their distinct climate, geology, and native species. These large areas are called **ecoregions**. Using a collaborative, science-based approach to conservation, TNC and its partners identify the areas that need to be protected to ensure the survival of each ecoregion's biological diversity. The four fundamental, related parts to TNC's conservation approach are shown in the figure: setting priorities, developing strategies, taking action, and measuring success. For more information about TNC's Conservation by Design program visit their web site: <**nature.org/aboutus/howwework/**>.

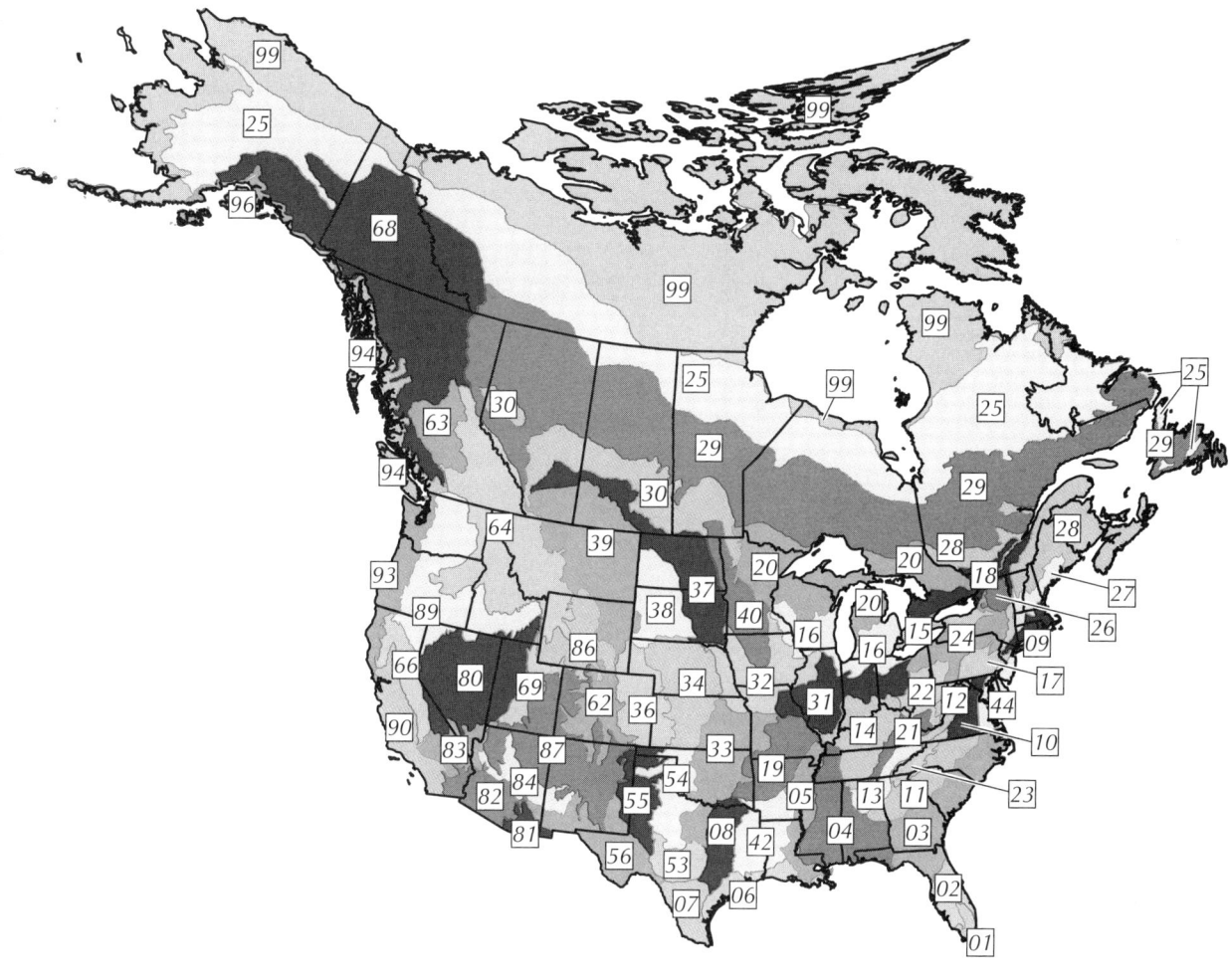

KEY

01	Subtropical Florida	23	Southern Blue Ridge
02	Peninsular Florida	24	Allegheny Plateau
03	South Atlantic Coastal Plain	25	Open Boreal Forest
04	East Gulf Coastal Plain	26	Adirondack Mountains
05	Mississippi Alluvial Valley	27	Northern New England
06	Coastal Prairies	28	Spruce-Hardwood Forest
07	South Texas Brushlands	29	Closed Boreal Forest
08	Oaks and Prairies	30	Aspen Parklands
09	Southern New England	31	Prairie Peninsula
10	Mid-Atlantic Piedmont	32	Dissected Till Plains
11	Southern Piedmont	33	Osage Plains
12	Mid-Atlantic Ridge and Valley	34	Central Mixed Grass Prairie
13	Southern Ridge and Valley	36	Central Shortgrass Prairie
14	Interior Low Plateaus	37	Northern Mixed Grass Prairie
15	Lower Great Lakes Plain	38	West River
16	Upper Great Lakes Plain	39	Northern Shortgrass Prairie
17	Northern Ridge and Valley	40	Northern Tallgrass Prairie
18	St. Lawrence Plain	42	West Gulf Coastal Plain
19	Ozark-Ouachita Plateau	44	Mid-Atlantic Coastal Plain
20	Boreal Hardwood Transition	53	Edwards Plateau
21	Northern Cumberland Plateau	54	Rolling Red Plains
22	Ohio Hills	55	Pecos and Staked Plains

56	Chihuahuan Desert
62	Southern Rocky Mountains
64	Central Rocky Mountains
66	Sierra Nevada
68	Northern Rocky Mountains
69	Utah Mountains
80	Basin and Range
81	Mexican Highlands
82	Sonoran Desert
83	Mohave Desert
84	Mogollon Rim
86	Wyoming Basin
87	Colorado Plateau
89	Columbia Plateau
90	Central and Southern California Coasts and Valleys
93	Southern Pacific Rain Forests
94	Northern Pacific Rain Forests
96	Southern Coastal Alaska
99	Tundra

Figure B. North American Physiographic Regions: Shown here are the **physiographic regions**, also known as **ecoregions**, of North America, north of Mexico—part of the map used in Partners In Flight bird conservation programs. Numbers on the map refer to physiographic regions identified in the key below the map. Although all physiographic regions in North America are listed here, note that some numbers are missing from the sequence. This has developed as Partners in Flight scientists continue to revise the map based on new information—as they restructure, combine, or split various physiographic regions, the sequence is altered.

10

this process typically uses aerial photographs and satellite imagery, together with sophisticated computer mapping technology (called "geographic information systems" or "GIS") to archive, display, and modify these all-important maps.

- **Prioritize species and communities:** Because certain species and ecological communities are more rare, more threatened, or more rapidly declining than others, it is imperative that we identify those that are in most immediate need of protective action. Often, these elements will be the most sensitive indicators of the health of target ecosystems, so monitoring them becomes a vital component of any long-term conservation strategy.

- **Prioritize sites:** Money is limited, and human population pressures are inexorable. Therefore, it is vital that we analyze each site for its contribution—alone or in concert with other sites—to maintaining the overall health and integrity of a regional ecosystem. Although the process can be difficult, we must rank sites according to their contributions, and we must dedicate ourselves to the protection and long-term management of the highest priority sites first. The Nature Conservancy, for example, identifies a "portfolio" of highest-priority sites for each ecoregion, and dedicates the majority of its human and financial resources to these sites (see Fig. A). Establishing criteria for ranking sites can be difficult because it often requires choosing among numerous and conflicting variables (such as overall species diversity, presence or population density of rare species, geographic role as a corridor or stepping-stone, or capacity for long-term management).

- **Set goals:** Only by establishing biological goals in advance can we expect to judge our success in conserving ecosystems as decades and generations go by. For each ecoregion our targets must be as specific as possible, but they may take many forms. For example, in the Partners In Flight handbook called Conservation of the Land Birds of the United States (Pashley et al. 2000), targets are expressed as minimum acceptable population counts for critical species (for example, 500 pairs of Henslow's Sparrows in the Lower Great Lakes Plain); as total area for a critically threatened vegetation type and its birds (for example, 67,000 acres [25,000 ha]) of long-rotation or old-growth hemlock-white pine forest in the Northern Cumberland Plateau); as minimum sizes of continuous habitat blocks (for instance, 10,000-acre [4,000-ha] forest blocks to support Swainson's Warblers); and as long-term population trajectories (stable or increasing numbers of Lesser Prairie-Chickens in the Pecos and Staked Plains region of western Texas, for instance).

- **Identify threats:** Most conservation sites face a barrage of threats, and some threats are so destructive that their elimination is mandatory in order to accomplish long-term conservation (for example, eliminating or redirecting off-road vehicle traffic in coastal sand dune communities is essential for the successful breeding of Least Terns and Piping Plovers). Conservation plans for individual sites must include specific actions to reduce the impacts of these threats.

- **Outline plans and partnerships:** Conservation action is "where the rubber meets the road," but usually a wide array of activities and participants must be identified in order to accomplish the long-term goals within an ecoregion. Activities include outright land protection (for instance, through acquisition, land trusts, easements, voluntary agreements, mitigation banks, and so on), land management (for example, by private owners, conservation organizations, or government agencies), community or political action (such as local ordinances, conservation bond referenda, or national and international policy-setting), and community education (for example, through interpretive centers, Important Bird Areas [see Fig. 10–96], or school and after-school programs). Clearly, no single organization can accomplish all of these actions; thus conservation planning at the ecoregional scale requires identifying and establishing the partnerships required to make conservation efforts sustainable over the long term.

- **Monitor results:** The best conservation planning acknowledges its own imperfection. Having properly identified the indicator species and communities representing each target system, we must develop long-term monitoring protocols in order to measure the success of our efforts. Because many changes happen at long time scales and at unpredictable intervals, it is vital to have monitoring schemes that can be sustained over the long term. Sustained monitoring requires participation by organizations and agencies that can commit to the process for the long term, that are realistic about the work load and demands, and that can accomplish long-term archiving and analysis. Monitoring activities at real places on the ground provides some of the most important ways for interested and committed volunteer citizens to participate in long-term conservation. ∎

Translocation

In New Zealand, almost all of the lowland habitat has been converted into grazing land, and introduced birds dominate all but the most remote areas. Moreover, the main islands are teeming with introduced rats, cats, ferrets, and stoats. All but a few of the remaining native land bird species are threatened or endangered. The good news is that New Zealand's Department of Conservation has become an international leader in developing modern techniques for habitat restoration, predator removal, and species reintroduction. These efforts have produced some dramatic conservation success stories, including last-ditch rescues of some truly spectacular birds.

The Kakapo **(Fig. 10–75)**—a large, nocturnal, parrot that roosts and breeds in burrows—was once widespread but is now entirely gone from all three of New Zealand's main islands. To provide the last few individuals with a predator-free environment, scientists transported them onto several small offshore islands (including Little Barrier Island) where all introduced mammals were (and still are) systematically destroyed. A similar strategy paid off for the Chatham Robin, which at its low point was reduced to just two breeding females and three males, all translocated to Mangere Island. In this case, wildlife professionals more than doubled the species' reproductive output (typically only two eggs and one fledgling per year) by placing the first-laid clutch into

a. Kakapo

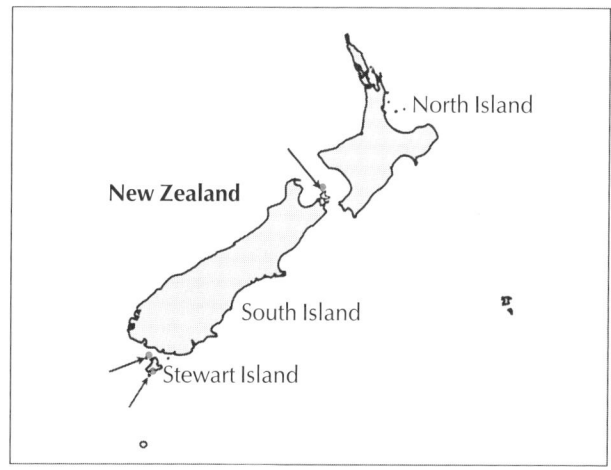

b. Current Range of the Kakapo

Figure 10–75. Kakapo: Nocturnal and flightless, the Kakapo (formerly known as the Owl Parrot, but now better known by its Maori name) **(a)**, is endemic to forested regions of New Zealand. During the breeding season males congregate at leks, giving repetitive booming calls at night to attract females, which then nest and raise young in burrows. Critically endangered, the Kakapo once occurred throughout North, South, and Stewart Islands. However, soon after New Zealand was settled by Europeans, Kakapo populations declined precipitously owing to predation on both adults and nests by introduced stoats, rats, and domestic cats. By 1976, only 18 birds (all males) were known to exist, all located in the Fiordland region in the southwest of South Island. Fortunately, in 1977 another population of around 150 birds was discovered on Stewart Island. Because this population also was declining rapidly, between 1980 and 1992 researchers transferred all remaining individuals to several small, predator-free offshore islands, indicated by the arrows on the map **(b)**. Compounding the Kakapo's predicament is its low reproductive rate: individuals often do not breed until they are six to nine years old, and even then they may breed only once every three to five years, depending on the abundance of certain fruits and seeds important in their diet. Currently safe from predators, and helped by intensive management practices such as tracking of radio-tagged adults, constant infrared video monitoring of natural nests, and supplemental hand rearing, the island populations have gradually increased. In 1999, 26 females and 36 males existed. Photo by D. Merton/VIREO.

10

the nest of another bird (especially the Chatham Gerygone), thereby inducing a second clutch from the Chatham Robin females. As another example, Little Spotted Kiwis were transferred from South Island (where introduced predators were driving the species to extinction) to Kapiti and Maud islands. Finally, the bizarre Takahe **(Fig. 10–76)**—a giant, flightless gallinule that was long thought extinct but was rediscovered in 1948—also has been transferred to several small islands including Maud Island. A few Takahes also remain within a grassland habitat preserve in the Murchison Mountains of New Zealand's South Island, where they face competition for wintertime food from introduced red deer. Year-round vigilance to control the deer population is under way in a concerted effort to save the Takahe within its original habitat.

Legal Protection

The long history of human culture and its effects on the rest of the world's species and ecosystems bear witness that conservation of species and ecosystems does not come naturally to us. The world over, individuals and cultures alike tend to place human wants and needs over those of nonhuman species. Although most people agree that saving natural species and ecosystems is beneficial in many ways, relatively few willingly sacrifice their own interests or directly reduce their own economic prospects in order to protect nature. Thus, natural systems represent a classic example of the "tragedy of the commons": even when a resource benefits everyone collectively, it will tend to be depleted through individual selfishness unless it is collectively protected.

a. Takahe

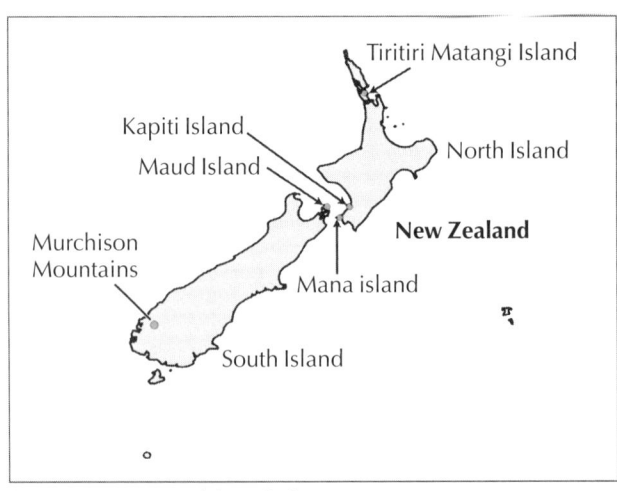

b. Current Range of the Takahe

Figure 10–76. Takahe: Another bizarre and endangered New Zealand endemic is the Takahe **(a)**, a large, flightless, blue-and-green rail with strong red legs and a huge red bill and frontal shield. Historically the Takahe was probably widespread throughout the North and South Islands, but by 1948 the population consisted of only 250 to 300 birds restricted to the alpine grasslands of the Murchison Mountains in the Fiordland region of South Island, shown in red on the map **(b)**. After declining further, probably because of competition with introduced red deer for the grasses that form the bird's diet, this population fluctuated between 100 and 160 individuals for 20 years. To boost the threatened South Island population, some of the Takahes were introduced to several small, predator-free, offshore islands (shown by arrows on the map). Intensive management practices, such as supplementary feeding to aid reproduction, inter-island transfers to avoid inbreeding, and removal of infertile eggs to encourage renesting, have helped the number of Takahes to slowly increase, and the current population numbers about 220. Photo by R. L. Pitman/VIREO.

Societies construct laws to accomplish collective aims, and conservation laws are no exception. By the year 2000, most countries in the world were contemplating or establishing conservation laws and enforcement practices aimed at reducing the rate at which species and ecosystems vanish from the earth. In the United States, the two most far-reaching examples are the **Endangered Species Act (ESA)**, discussed earlier, and the **Clean Water Act**, which contains a large and diverse set of wetland laws—local, regional, and federal—aimed at prohibiting the net loss of any natural wetland larger than about a quarter of an acre (0.1 ha). Internationally, the most important treaty affecting the conservation of endangered species is the **Convention on International Trade in Endangered Species (CITES)**. These three examples are described in detail below.

Endangered Species Act

The Endangered Species Act was signed in 1973 by President Richard Nixon, and was amended in 1982. This act commits the U. S. government to a variety of steps to identify and protect those animals and plants most in danger of extinction. The body of the ESA consists of 11 numbered sections, many of which are cited often in discussions about the act's biological and economic implications. The ESA contains truly historic language specifying both the biological and legal basis for species protection in the United States:

ESA Section 1. Introduction

Empowers the Secretary of the Interior to oversee the process and the budget, administered principally through the U. S. Fish and Wildlife Service (USFWS).

ESA Section 2. Findings and Purpose

Lays out the biological rationale for legally protecting species, and defines the principal purpose as "to provide a means whereby the ecosystems upon which endangered species and threatened species depend may be conserved."

ESA Section 3. Definitions

Defines 20 terms, including "endangered" and "threatened"; defines "species" for vertebrates to mean any natural unit that may receive legal protection, including the whole species, any named subspecies, and even a "distinct population segment"; and defines "critical habitat." Among the ESA's most famous words are those in this section defining the prohibited act "take," meaning "to harass, harm, pursue, hunt, shoot, wound, kill, trap, capture, or collect, or attempt to engage in any such conduct." The courts continue to uphold interpretation of this phrase to include destruction of habitat.

ESA Section 4. Determination of Status

Specifies the process by which a species is nominated and placed on endangered and threatened species lists; mandates that species shall be listed "solely on the basis of the best scientific and commercial data available"; provides for development of Recovery Plans and special Recovery Teams; outlines procedures for down-listing and de-listing.

Figure 10–77. Northern Spotted Owl: In many respects, the "poster child" for conflicts over the Endangered Species Act (ESA) is the Northern Spotted Owl, a threatened subspecies that breeds primarily in old-growth coniferous forests of the Pacific Northwest from California north to British Columbia. Section 7 of the ESA specifies that no federal agency can release grants or contracts that could jeopardize a species listed as threatened or endangered. Therefore, during the 1990s the U. S. Forest Service was required by the courts to conduct a thorough review of its timber harvest concessions, and as a result to declare significant areas of old-growth forest in Oregon and Washington off limits to further logging. The logging restrictions were viewed as too lenient by many environmental groups, but as too severe by many local loggers, who perceived a serious threat to their jobs and the local economy. Protection of Spotted Owls by protecting the old-growth ecosystem upon which they depend remains a contentious issue to this day, along with many other such conflicts between long-established land-use practices and modern efforts to protect endangered species. Photo by K. Schafer/VIREO.

ESA Section 5. Land Acquisition

Empowers the Department of the Interior to take necessary steps to acquire habitat deemed critical for the survival and recovery of listed species.

ESA Section 6. Cooperation with States

Specifies that the USFWS establish cooperative agreements with individual states for administering federal funds designated for endangered species protection or management.

ESA Section 7. Interagency Cooperation

This critically important section holds every federal agency responsible for protecting its endangered species, and requires that biological assessments be carried out before any federal funds, grants, or contracts may be used that could jeopardize an endangered species **(Fig. 10–77)**. It also specifies that judicial review of potential endangered species violations may be launched by any private citizen of the United States.

ESA Section 8. International Cooperation

Specifies that the United States shall abide by international treaties governing trade in endangered species, and empowers the USFWS as the federal agency responsible for promoting and enforcing this provision.

ESA Section 9. Prohibited Acts

Details what actions involving listed species are prohibited under the ESA, notably "take," "possession," and all manner of commercial trade.

ESA Section 10. Exceptions

This crucial section contains the 1982 amendment specifying that the USFWS may grant permission to "take" endangered species "if such taking is incidental to, and not the purpose of, the carrying out of an otherwise lawful activity"; applicants for such a permit must submit a "conservation plan" (known widely today as a "habitat conservation plan," or HCP) specifying the steps by which the applicant intends to minimize take, and must assess the impact of the proposed action on the species as a whole and explain why other alternatives are impossible. This section also outlines the so-called "safe harbor" provision, exempting landowners from responsibility to protect a population deemed to be "experimental" (for instance, California Condors reintroduced into Arizona).

ESA Section 11. Penalties and Enforcement

Provides that ESA violations can result in both civil and criminal penalties, under the jurisdiction of district courts, and that "any person may commence a civil suit on his own behalf" if no action is undertaken to correct or punish the violation after 60 days' notice of the violation.

Clean Water Act, Section 404

Swamps, marshes, bogs, river beds, floodplains, springs, oases, temporary sloughs, seasonal ponds, and other wet places provide vital habitat for both breeding and wintering populations of hundreds of bird species. Unfortunately, more than half of all these wetland habitats in the United States have been destroyed during the past two centuries (Dahl 1990). Loss of wet habitats has had devastating impacts on bird numbers, and significant laws protecting wetlands began to proliferate in the 1960s. Still, by the time the first Clean Water Act was passed in the United States in 1972, wetlands were being destroyed at the rate of 460,000 acres (186,300 ha) annually. Today a bewildering variety of local, state, and federal regulations protect water resources and wetland habitats across North America. Although losses continue to occur, the rate of loss has been cut to below 10 percent of its 1972 level.

Of all wetland laws on the books, none is more widely known— nor, in some circles, more adamantly cursed —than Section 404 of the Federal Clean Water Act, last modified in 1994. This provision regulates all filling of, or discharge of dredged material into, the fresh waters of the United States, including wetlands. The law covers activities such as filling for residential and commercial development; construction of dams, levees, roads, and airports; and conversion of wetlands to uplands for farming or forestry. Its basic premise is that no dredging or filling of a wetland can be permitted if an alternative exists that is less damaging to the aquatic environment, or if waters would be degraded significantly. Permit applicants must (1) take steps to avoid wetland impacts where practicable, (2) show how they are minimizing any wetland impact, and (3) provide compensation for all wetland impacts deemed unavoidable, so that a biological replacement can be created. Compensation, commonly known today as **wetland mitigation**, typically is accomplished either through direct wetland restoration or re-creation—on at least an acre-for-acre basis—or through cash payment into a **mitigation bank** managed by a public agency or private conservation group. Mitigation banking allows for well-planned, landscape-scale wetland restoration projects using funds accumulated from a large number of small-impact projects.

CITES

International commerce in pet birds is directly responsible for the plight of a large number of species currently in danger of extinction, and for several, such as Spix's Macaw, that appear to have gone extinct already. In 1963, members of the **World Conservation Union (IUCN)** drafted the original text of a historic document finally ratified in 1975 as the **Convention on International Trade in Endangered Species (CITES)**. This international agreement, to which 150 countries now voluntarily subscribe, binds participating parties to monitor, regulate, or prohibit the import and export of species that the group deems worthy of global protection. The legal framework for enforcement within each country is provided by domestic laws, such as our own ESA Section 8.

The heart of CITES lies in three appendices constituting the list of species for which international trade is restricted. Appendix I includes the most endangered species (including all macaws, many other parrots, and a few falcons), for which all commercial trade is strictly prohibited and special import/export permits are required for scientific transport. Appendix II lists species not in immediate danger of extinction, but which could become so if trade is not regulated (for example, all parrots, hawks, and falcons not in Appendix I). Species may be added to or removed from Appendices I and II, or moved between them, only via discussion and vote at periodic CITES conferences. Appendix III lists species added by individual countries that, for any reason, request international assistance in regulating their trade. CITES has certainly helped to stem the tide of extinction. Until the disappearance of Spix's Macaw in Brazil in 2000, not a single species regulated by CITES had become extinct as a result of international trade.

Bringing Birds Back from the Brink

Around the world, scores of endangered birds continue to be signals of ecosystems that are on the brink of collapse. In Africa, for example, the IUCN—in its Red Data Book (Collar and Stuart 1985) and "Redlist" web site—lists 43 bird species as endangered or vulnerable, with another 109 sufficiently rare to warrant concern. In the Western Hemisphere, the IUCN lists no fewer than 92 endangered bird species (Collar et al. 1992). Of these, 23 are viewed as "terminal" (possibly extinct, action is urgently needed if populations are rediscovered) and another 23 are deemed "critical" (action is urgently needed now). For many of the most endangered birds, conservation measures are under way. Efforts include cracking down on illegal trade in popular pets such as parrots; establishing public and private reserves to protect native forests and watersheds; trapping, poisoning, snaring, and shooting introduced mammals; promoting nondestructive harvesting of tropical forest resources; and developing ecotourism in and around protected areas as an incentive for local and indigenous people to maintain these habitats in their natural state.

It is also true, however, that we have learned from our mistakes. The sad declines of Passenger Pigeons and Carolina Parakeets cannot be reversed, but they shall never be forgotten. North Americans instituted sweeping changes in behavior and laws as a direct consequence of 19th century excesses, demonstrating that even century-long declines in bird populations could be reversed. Snowy Egret and Great Egret numbers began growing immediately after the ban on plume hunting. Ducks that were rare at the turn of the century occur by the millions across North America today. Conservation success stories are now evident all over the world, attesting to the resilience that bird populations can exhibit when the agents causing their declines are removed, and when birth and death rates come back into balance. The following success stories illustrate how real-world actions can reverse even the most desperate situations.

Wood Duck—Regulated Hunting and Adaptive Management

Perhaps the most spectacular recovery of a North American bird is that of the Wood Duck **(Fig. 10–78)**, hunted so extensively across eastern North America that by 1900 ornithologists generally predicted it would follow the Passenger Pigeon into oblivion. Ornithologist Joseph Grinnell—editor of Forest and Stream magazine—wrote, "Being shot at all seasons of the year [Wood Ducks] are becoming very scarce and are likely to be exterminated before long." The lovely Wood Duck, however, would become the first clear beneficiary of laws passed to prevent such a fate. Complete closing of the hunting season for Wood Ducks in 1918 allowed the species to begin staging one of the most complete population recoveries ever witnessed in North America (Bellrose 1976). By the 1930s the Wood Duck had recovered over much of its former range, and by 1941 sufficient numbers existed that hunting was permitted again in 14 states. By the 1960s population estimates ranged between one and two million, and during the 1970s and 1980s the Wood Duck began spreading into previously unoccupied areas of the Great Plains, northwestern Canada, and northern Mexico (Hepp and Bellrose 1995).

The Wood Duck was able to recover so rapidly over a 70-year period because of its extraordinarily high reproductive rate. Typical nests **(Fig. 10–79)** contain 10 to14 eggs from a single female, and females often "dump" partial or even whole clutches into the nests of other Wood Ducks before laying their own. High natural birth rates were augmented in many areas by the placement of artificial nest boxes by humans, a strategy that appears to increase Wood Duck densities at least locally (Nichols and Johnson 1990). By far the most important factor in early phases of the Wood Duck's recovery, however, was the strict removal of hunting as an extra source of mortality for several decades. This was followed by careful population monitoring by wildlife personnel, and strict management of hunting limits throughout the range of the species. What would North America have been like today if the same management strategies had been imposed for Passenger Pigeons just a few years earlier?

For many decades certain ducks did not fare as well as the Wood Duck. State and federal regulation of waterfowl hunting were strengthened steadily throughout the 20th century in the face of huge

Figure 10–78. Adult Male Wood Duck: When populations of the handsome Wood Duck began to plummet in the late 19th century from overhunting and loss of wetland habitat, many North American ornithologists predicted that the species would soon become extinct. However, several decades of closed hunting seasons for Wood Ducks allowed the species to recover dramatically, and today it occurs over a wider area of North America than ever. Photo courtesy of Arthur A. Allen/CLO.

Figure 10–79. Nestling Wood Duck: A young Wood Duck contemplates taking the plunge from the safety of its nest box to the water below. All individuals of a brood leave the nest within a day of hatching, by jumping from their nest cavity while their mother encourages them with soft clucking calls from below. One factor that has aided the Wood Duck's recovery to historically high population levels is its high reproductive rate. Clutch size is typically 10 to14 eggs, and some females may deposit eggs in other Wood Duck nests before laying full clutches of their own. In many locations artificial nest boxes have helped the Wood Duck's recovery by providing nest sites where tree cavities, the bird's natural choice, are unavailable. Photo courtesy of Warren Greene/CLO.

10

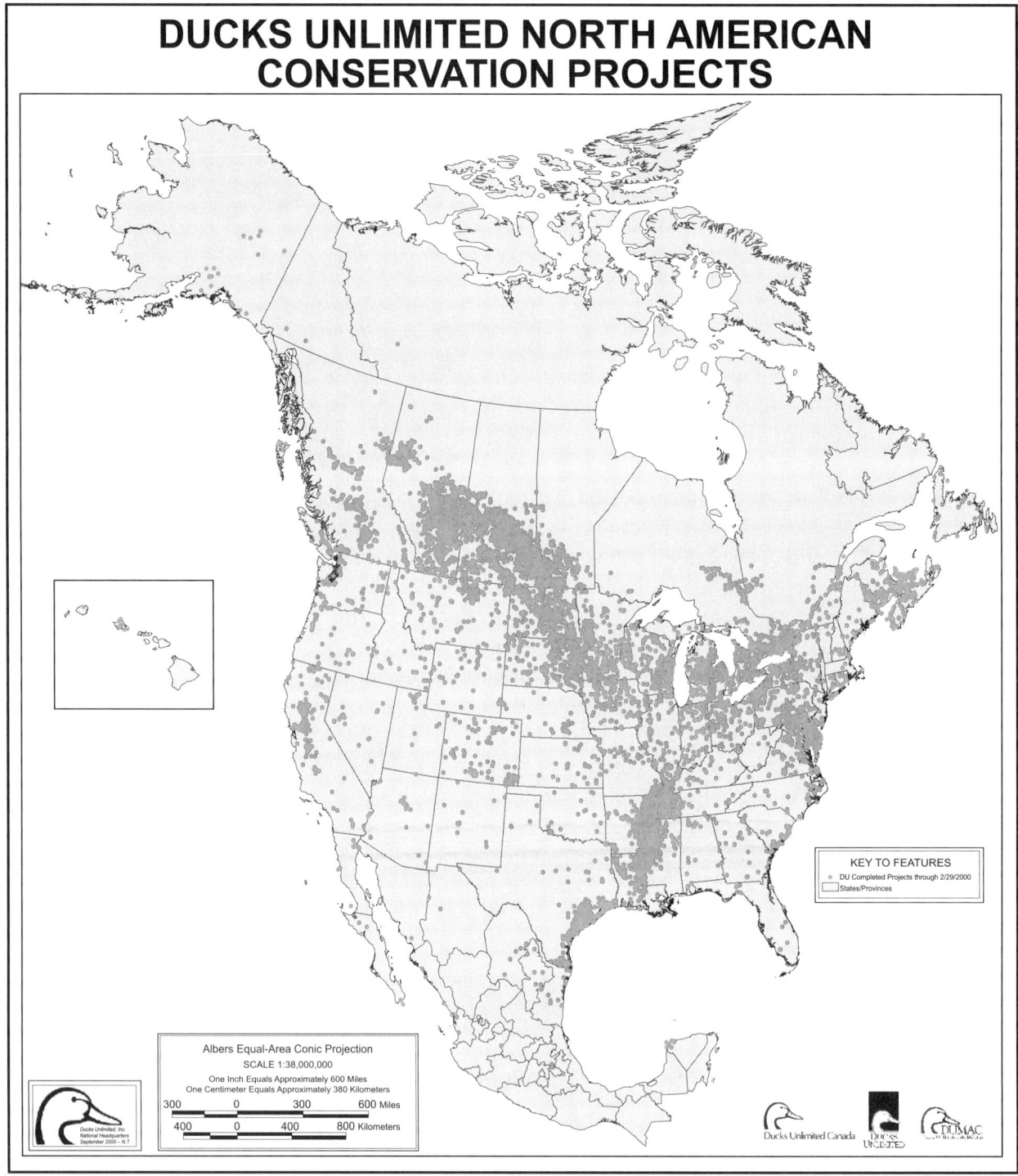

DUCKS UNLIMITED NORTH AMERICAN CONSERVATION PROJECTS

Figure 10–80. Ducks Unlimited's North American Conservation Projects: With more than 900,000 members in 2001, Ducks Unlimited (DU) is the leading nonprofit waterfowl and wetlands conservation organization in the world. According to the organization, "The idea for Ducks Unlimited was born by the winds that created the Dust Bowl. Untold acres of wetland habitat vanished in the 1930s, taking with them the promise of generations of waterfowl. In 1937 a small group of conservationists who realized that the majority of North America's waterfowl breed in the Canadian prairies, organized to raise money in the United States for waterfowl conservation in Canada." Although its membership consists primarily of waterfowl hunting enthusiasts, the organization also has many nonhunting members. As of October 2001, DU had conserved 9.4 million acres (3.8 million ha) of waterfowl habitat in North America. Each colored dot on the map represents one DU project completed as of February 2000. Note the concentration of projects in the prairie pothole region of North/South Dakota, United States and Saskatchewan/Manitoba, Canada—the most important waterfowl breeding area—and in the Mississippi and Atlantic flyways, which are important waterfowl migration routes and wintering areas. Map copyright Ducks Unlimited, Inc.

reductions in habitat for breeding, migration stopover, and wintering, which caused steady declines in duck populations. Conservation efforts increasingly turned from hunting regulations to habitat protection, both on the breeding grounds (especially the prairie pothole region of the north central United States and Canada) and on wintering grounds in states bordering the Gulf of Mexico. Habitat protection by the U. S. Fish and Wildlife Service (through its system of National Wildlife Refuges) was augmented by private efforts, especially those of Ducks Unlimited **(Fig. 10–80)**.

One of the first and best examples of adaptive management was introduced in 1985, when the U. S. Fish and Wildlife Service adopted a 10-year Waterfowl Habitat Acquisition Plan. A year later, the U. S. Secretary of the Interior and the Canadian Minister of the Environment signed the historic **North American Waterfowl Management Plan** outlining far-reaching cooperative efforts to protect habitat, restore declining species, and conduct population research. Detailed population estimates are made every year, and are combined with rigorous statistical models to produce the following year's "bag limits" for hunters. Biologists then study their success at predicting subsequent year's numbers, in an effort to improve their ability to predict the consequences of alternative harvest scenarios. As a result, waterfowl numbers during 1999 reached their highest point in 50 years, aided by several consecutive years of good rainfall on the breeding grounds. Today, fluctuations in most waterfowl densities result mainly from annual variability in weather conditions.

Whooping Crane—Protected Habitat and Captive Rearing

No bird more dramatically symbolizes species-level conservation efforts of the late 20th century than the Whooping Crane **(Fig. 10–81)**. A "charter member" of the endangered species lists of the United States and Canada, this species reached its historic low point when only 16 individuals were counted at its traditional wintering marshes along the Texas coast in 1941 (Lewis 1995). Although most prominent in the northern Great Plains, this magnificent species once occupied most of eastern North America south to the Gulf Coasts of Louisiana, Florida, and Georgia **(Fig. 10–82)**. Its decline began in the 1800s as a consequence of sport shooting, but was dramatically hastened through the early and middle 1900s by the draining and plowing of vast tallgrass prairie marshes of the Great Plains for agriculture. In contrast to Wood Ducks, this huge crane has an extremely low reproductive rate (typically, only one young is fledged each year, and breeding does not begin until five years of age). Consequently, even

Figure 10–81. Whooping Crane: No bird better symbolizes national and international conservation efforts than the Whooping Crane. Probably never common, the species suffered from sport hunting and loss of prairie marsh habitat. Its population reached its lowest level in the 1940s, when only 16 wild individuals returned to their traditional wintering area on the Texas coast. Intensive conservation efforts—including protection of breeding, migration, and wintering areas, as well as captive-rearing and release programs—have raised the Whooping Crane's numbers to more secure levels. Here a recently released, captive-raised Whooping Crane fitted with leg bands and radio transmitter lands in a Florida lake. Photo by Arthur Morris/Birds As Art.

Whooping Crane Distribution

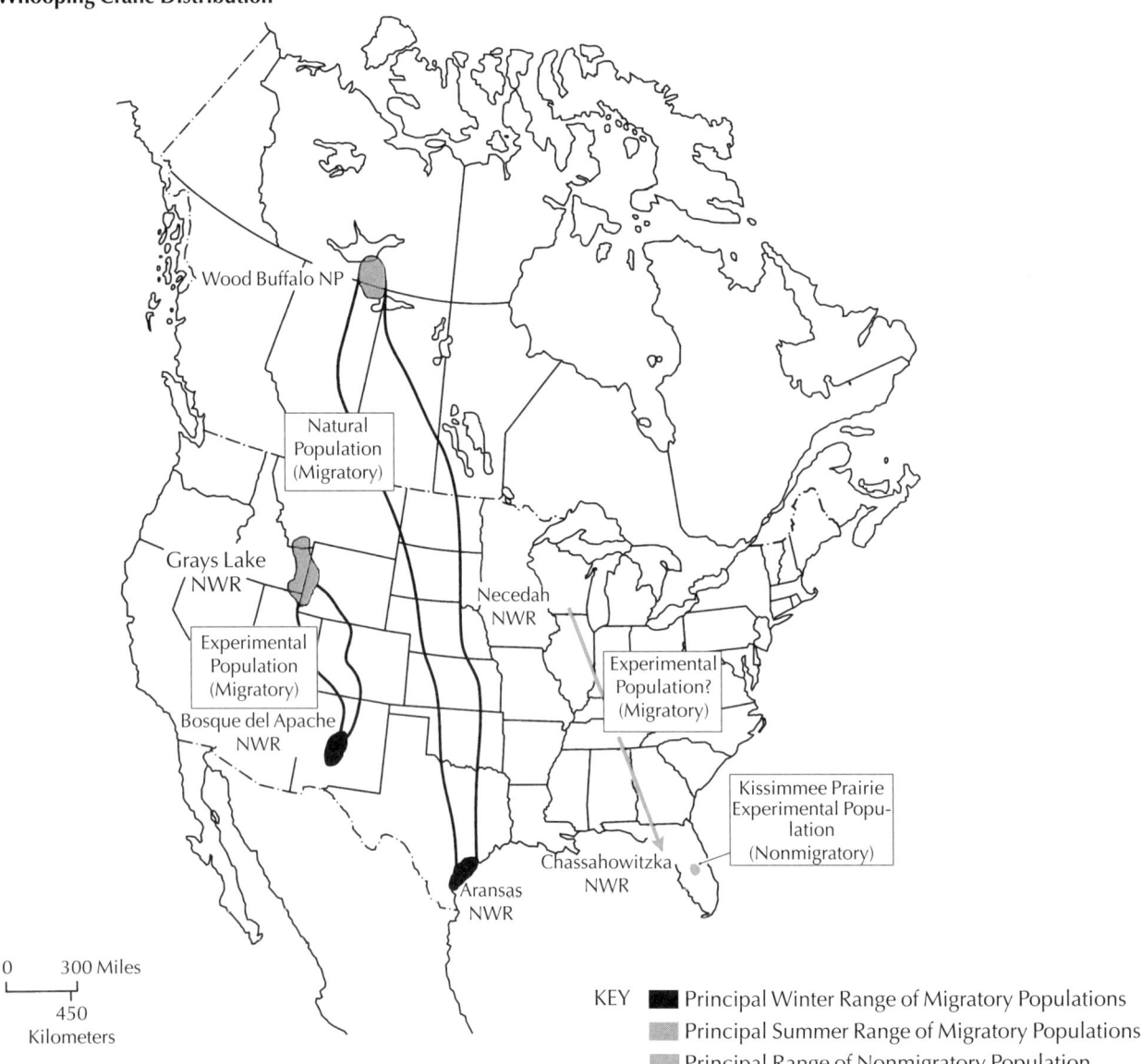

KEY
■ Principal Winter Range of Migratory Populations
▨ Principal Summer Range of Migratory Populations
▨ Principal Range of Nonmigratory Population

Figure 10–82. Current Distribution of the Whooping Crane: This map shows the distribution of wild and introduced Whooping Crane populations. Gray areas denote breeding grounds of wild populations, black areas are wintering grounds, and dark lines enclose migration corridors between the two. As of 2001, the only self-sustaining wild population was nesting in Wood Buffalo National Park in Canada's Northwest Territories, and wintering at the Aransas National Wildlife Refuge on the Texas coast. In the 1970s, efforts were made to develop an experimental wild Whooping Crane population in the Rocky Mountains of Idaho, Wyoming, and Montana. First, several Whooping Crane eggs were hatched. Then the young whoopers were placed in the nests of Sandhill Cranes at Grays Lake National Wildlife Refuge, Idaho, who "cross fostered" the young and then led them to winter at Bosque del Apache National Wildlife Refuge in New Mexico. Also in the Rockies, other young whoopers were trained to migrate by following an ultralight aircraft. Unfortunately, the Rocky Mountain population has dwindled as a result of high mortality during migration and because the fostered whoopers—imprinted on

Sandhill Crane foster parents—have neither paired with each other nor bred. During the 1990s a nonmigratory Whooping Crane flock was established in the Kissimmee Prairie, Florida (red area on the map), using birds from captive rearing programs at the International Crane Foundation, the Patuxent Environmental Science Center, and the Calgary Zoo. Numbering 91 individuals by February 2000, some of these birds have started to breed, although their survival is not without problems (see text). In October 2001, an effort was launched to introduce a migratory Whooping Crane flock to Florida, as eight Whooping Crane juveniles, trained to follow an ultralight aircraft, set off from Wisconsin's Necedah National Wildlife Refuge en route to Chassahowitzka National Wildlife Refuge, 1,250 miles (2,000 km) to the south on Florida's Gulf coast (red arrow on the map). Biologists hoped that the young cranes would make the return journey northward in spring alone. Similar ambitious plans exist for migratory populations in the prairie provinces of Canada. Thanks to these and other conservation efforts, the Whooping Crane's once-tenuous future seems more assured. Map adapted from Gill (1995).

in fully restored habitat, the species would take hundreds of years to recover on its own. Clearly, as the population dwindled to just a handful of birds, dramatic steps were required to prevent its final demise.

In 1937 the Whooping Crane's entire wintering habitat was protected as the Aransas National Wildlife Refuge. Fortunately, its breeding grounds in Canada had been similarly protected in 1922 as Wood Buffalo National Park, although the fact that all remaining Whooping Cranes bred there was not known until the 1950s! Beginning in 1967, through a cooperative project between the Canadian Wildlife Service and the U. S. Fish and Wildlife Service, eggs were removed from wild nests and the young reared in captivity to produce several large captive flocks. Despite such "intrusion" by humans, replacement laying by wild individuals and continued protection of breeding, migratory, and wintering habitat resulted in steady population growth, from nine wild pairs in 1967 to 45 pairs in 1993. In captivity, cranes began to pair and breed as well, aided greatly by a young researcher at Cornell University in the early 1970s. George Archibald personally helped induce the first courtship displays by captive Whooping Cranes by dressing as a crane and dancing himself **(Fig. 10–83)**! Archibald established the International Crane Foundation in Baraboo, Wisconsin, where all 16 species of the world's cranes are reared and studied for conservation.

By the 1990s, captive flocks at three locations (Patuxent, Maryland; Baraboo, Wisconsin; and Calgary, Alberta) had become large enough to provide birds for reintroduction into the wild. Releases were begun in 1995 to attempt establishment of a "nonmigratory" population of Whooping Cranes in the Kissimmee Prairie of central Florida. This experiment has experienced significant problems with local predators (bobcats and coyotes), and it remains far from certain that the new population will become viable. However, an unexpected success occurred in 1999, when to the surprise of the entire Recovery

Figure 10–83. George Archibald Dancing with Captive Whooping Crane: In the early 1970s, George Archibald, together with Ron Sauey, founded the International Crane Foundation (ICF) in Baraboo, Wisconsin. ICF has made monumental strides to save the world's endangered species of cranes and their associated wetland ecosystems. An early project was to induce Whooping Cranes to breed in captivity. One captive female whooper had been sexually imprinted upon humans, so to stimulate her into breeding condition, Archibald set up housekeeping with the bird— he accompanied her constantly and performed courtship displays with her, as shown here. The project met with success: the imprinted female produced an egg after she was artificially inseminated with sperm from a captive male crane. Later, Archibald and his co-workers began dressing as cranes while interacting with their charges, to avoid the problematic imprinting that might otherwise prevent captive-reared whoopers from breeding in the wild. Photo courtesy of International Crane Foundation/David Thompson.

Figure 10–84. Peregrine Falcon in Flight: Three decades after being placed on the endangered species list, the American Peregrine Falcon has made a remarkable comeback. A victim of reproductive failure owing to the eggshell-thinning effects of DDT contamination (see Fig. 10–41), by the 1960s the bird's numbers had dwindled to the point that no nesting pairs could be found in the eastern United States, and it was fast diminishing elsewhere on the continent. Fortunately, the banning of DDT in 1972 along with intensive conservation efforts led to the species' successful recovery. In 1999 the Peregrine Falcon was removed from the endangered species list ("de-listed"), although it will continue to be monitored for several years, as required by law. This juvenile Peregrine Falcon was photographed by Tom Vezo.

Team, two of these Florida Whooping Cranes—which had been missing and believed dead—were found nesting in a farmer's field in the Upper Peninsula of Michigan. Plans call for the establishment of one or two additional, migratory populations in the prairie provinces of Canada. Thanks to heroic, cooperative efforts of both Canadian and U. S. wildlife agencies—and the International Crane Foundation—the sight of migrating Whooping Cranes may grace the skies of eastern North America once again.

Peregrine Falcon—Pesticide Regulation and "Soft Release" Reintroduction

No sight is more thrilling than that of a Peregrine Falcon **(Fig. 10–84)** beating a low, powerful trajectory along the mud flats during migration, scaring up thousands of shorebirds and ducks before it. The peregrine is the most widespread of the large falcons, and in North America it once bred in all regions except the southeast. As detailed earlier, the Peregrine Falcon became the "poster child" for the lethal effects of pesticides, especially DDT and dieldrin. By the time DDT was banned in 1972, peregrines had completely disappeared as breeding birds essentially everywhere except in Alaska and far northern Canada.

Again, a researcher at Cornell University pioneered a dramatic recovery plan. Dr. Tom J. Cade **(Fig. 10–85)** organized a program of captive rearing, initially using eggs from Peregrine Falcons belonging to the tundra-breeding subspecies still found commonly in Alaska. He began releasing ("hacking") young falcons from wooden boxes that served as artificial nests, constructed to provide human access from the rear for feeding. Like many raptors, Peregrine Falcons naturally nest high on cliffs, away from dangerous predators (especially Great Horned Owls) and high enough to provide fledging young with plenty of air during their initial, awkward first flights. For Cade's group, the most convenient and predator-free cliffs were those atop the urban skyscrapers of eastern North America **(Fig. 10–86)**. During the 1970s and 1980s, news stories flashed through Baltimore, New York, Chicago, Toronto, and Minneapolis as brood after brood of falcons were hacked in these cities. Eventually, released birds began returning and pairing

Figure 10–85. Dr. Tom J. Cade with Peregrine Falcon: Cornell Lab of Ornithology researcher and lifelong falconer Tom J. Cade (shown here with a juvenile Peregrine Falcon) founded The Peregrine Fund in 1970, beginning an ambitious program to rear Peregrine Falcons in captivity and reintroduce them to the wild. Using a variety of techniques—artificial insemination, artificial incubation of fragile eggs rescued from wild nests, fostering of nestlings by other raptor species, and eventually releasing young peregrines at natural or artificial nest sites—Cade's team and their collaborators around the country gradually boosted peregrine numbers such that natural breeding started to occur, allowing the population to recover. In 1984 The Peregrine Fund relocated to help establish the World Center for Birds of Prey in Boise, Idaho, where its researchers are now helping to save threatened raptors and other birds worldwide, using captive-rearing and reintroduction methods similar to those developed for the Peregrine Falcon. Photo courtesy of The Peregrine Fund.

Figure 10–86: Peregrine Falcon "Scarlet" in Baltimore: Ledges atop the urban skyscrapers of eastern North America proved to be ideal hack (release) sites for captive-reared Peregrine Falcons. This adult female, known as "Scarlet," made her home atop a building in Baltimore, Maryland. Cities provide ample prey, in the form of feral pigeons and starlings, and compared to rural sites they have fewer potential predators, such as Great Horned Owls. Also, ledges of tall buildings provide the same height and protection as cliffs upon which peregrines nested in their traditional habitat. Today more than 100 pairs of peregrines nest in urban or industrial areas in North America, allowing many people to enjoy views of this once-endangered bird. Photo courtesy of Tom Maechtle/The Peregrine Fund.

10

with one another, establishing new territories and new nest sites. The magnificent sight of Peregrine Falcons once again became commonplace, especially in the midst of large cities. In 1999, the Peregrine Falcon was removed from the United States endangered species list, culminating one of the most dramatically successful hands-on conservation stories in history.

In 1988 Cade moved his growing raptor recovery program (known since 1973 as The Peregrine Fund) to a new site provided by the University of Idaho near Boise, Idaho. That site became the World Center for Birds of Prey, and to this day is the world's leading center for experimentation with new techniques for rearing and releasing endangered raptors around the world. Following its success with Peregrine Falcons, The Peregrine Fund used similar procedures to rescue the critically endangered Mauritius Kestrel, on the same island where the Dodo went extinct two centuries ago.

California Condor—Wild Capture, Captive-rearing, and Study of "Surrogates"

Today, The Peregrine Fund plays a central role in perhaps its biggest challenge ever—the rearing and releasing of California Condors. Following the species' century-long population collapse (see Ch. 10, Chemical Toxins), all 21 remaining wild individuals of this magnificent bird were captured during the mid-1980s. The goals of the capture were twofold. First, it had become clear that these final survivors would remain alive only if they were medically treated for lead poisoning and sheltered from additional lead exposure. Second, with some luck and considerable effort, they could become a nucleus for a captive-breeding program designed to bolster the total population size. Successful programs were launched at the Los Angeles Zoo and the San Diego Wild Animal Park (where the first successful breeding by captive individuals occurred in 1988 [see Fig. 8–86]). With condor numbers beginning to increase, these two programs were augmented by the World Center for Birds of Prey in 1994. Today the total population of California Condors exceeds 150, and all three institutions cooperatively maintain a joint studbook to ensure the maximum amount of genetic cross-breeding among their birds. The first releases of California Condors back into the wild were preceded by experiments using Andean Condors as **surrogates**, allowing biologists to perfect their techniques for tracking and monitoring of released birds by practicing with a biologically similar, but much less endangered, species. (The Andean Condors were prohibited from breeding by the fact that only females were released into the wild. They were later recaptured and transported back to South America.)

Two young California Condors were released in 1992, but one was found dead less than a year later after consuming ethylene glycol, a component of antifreeze. Other releases since that time demonstrate that the problems plaguing the species before the mid-1980s continue. Deaths still occur from power-line collisions, illegal shooting, and environmental poisoning, especially—once again—from lead.

a. Maui Parrotbill

b. Akohekohe

To this day, although a wild population of up to several dozen condors has been created in southern California and northern Arizona, the population suffers far too high a mortality rate to sustain itself without regular augmentation from the captive-rearing programs. The long-term challenges will be to concentrate release programs in the biggest wilderness areas available within the birds' former range, and to train young condors to avoid the most human-modified habitats.

Maui Parrotbill and Akohekohe—Protected Habitat and Feral Mammal Control

High on the windward slopes of the Haleakala Volcano on the island of Maui, one of the world's wettest rain forests struggles to become a success story. As the Hawaiian bird fauna disappeared through the 19th and 20th centuries, populations of two peculiar and spectacular Hawaiian honeycreepers—both feared extinct—were rediscovered. About 500 Maui Parrotbills and perhaps twice as many Akohekohes **(Fig. 10–87)** were found to persist in a swath of drenched forest above the main elevation for mosquitos, where stunted ohia trees are nearly always shrouded with clouds. Then, scientists during the 1960s made an even more startling discovery in these same forests. They discovered a previously unknown and critically endangered new bird, which they called Poo-uli. These rare birds were just the tip of an iceberg that was melting fast. Amid the billowy mosses covering every gnarled branch and trunk grew more than a dozen critically endangered plants, including some spectacular lobelias that had not been seen flowering in decades. The problem: an abundance of feral pigs, introduced centuries earlier by Polynesian colonists and hunted to this day in a tradition deeply rooted among Hawaiians. Pigs root the soil and eat young shoots of emerging plants, effectively "mowing the lawn" of rich understory vegetation. Overall biological diversity of this spectacular forest was rapidly dwindling.

The rare birds and plants persisting on the slopes of east Maui sparked a concerted effort spearheaded by the Hawaiian chapter of The Nature Conservancy to secure the future of this forest. Two steps were

Figure 10–87. Endangered Hawaiian Honeycreepers: In 1950, surveys of the misty mountains of eastern Maui, in the Hawaiian Islands, resulted in rediscovery of two bird species previously thought to be extinct. The Maui Parrotbill **(a)** is an olive-green bird with a large, parrot-like bill used to dig into branches and stems, to lift bark or lichens in search of hidden invertebrates, and to bite open fruit. The Akohekohe **(b)** is a colorful, nectar-eating bird whose prominent, white bushy crest—curved forward over the bill—provided its English name of Crested Honeycreeper. After these birds were rediscovered, the Hawaii chapter of The Nature Conservancy successfully coordinated an effort to protect their isolated forest refuge, which also is home to many endangered plants. As part of the Hawaiian Forest Bird Recovery Project, captive propagation and release programs are under way for both species. Photos by Jack Jeffrey.

Figure 10–88. Conservation in Action in Hawaii: Nature Conservancy staff erect a fence at Kamakou Preserve on the island of Molokai, Hawaii to keep feral pigs out of the upper-elevation forest. In addition, snaring, although controversial, is used to remove feral pigs, which severely damage native vegetation. Despite the difficulty and high cost of these efforts, similar projects have been carried out successfully at The Nature Conservancy's Waikamoi Preserve on Maui, whose mountain rain forests are the last stronghold of the Maui Parrotbill, the Akohekohe, and other declining Hawaiian birds. Photo courtesy of The Nature Conservancy.

key to the process. First, numerous private and public landowners had to be convinced of the importance of protecting and restoring the forest. Global significance for a few rare species would not alone suffice as a rationale. What made the difference was the forest's importance in protecting and purifying the watershed for the burgeoning human population occupying the Maui coastline. A consortium of owners was brought together to form the East Maui Watershed Preserve. The second step was to restore biological integrity inside a badly degraded forest, and this would require making some tough and unpopular decisions. Most controversial of these was to fence the entire lower boundary of the watershed preserve **(Fig. 10–88)** and to eliminate feral pigs uphill of the fence by various means, including lethal snaring. Although initially met with disapproval from both local and international circles, this difficult and expensive process almost instantly began producing its desired effect. Today the forest amid the clouds of Haleakala is teeming with new life, the lobelias are flowering once again, and Maui Parrotbills and Akohekohes thrive above them. Whether the critically rare Poo-uli can be brought back from the brink remains to be seen.

Why Protect Birds?

■ Why should we bother trying to protect birds, especially in the face of such huge odds against us? This question could be phrased in market terms: what is the worth of birds to society, and what are the values we gain by choosing to invest in their protection? After all, protecting birds from extinction sometimes requires considerable expenditure of energy and resources, both of which could be used for other things that are not "for the birds." For example, as we've seen above, the cornerstone of almost all conservation efforts is habitat protection, but setting wild land aside to protect plants and animals often reduces local opportunity to use the land for scenic homes, recreational pastimes, or profit-making ventures. Is such sacrifice worth the cost?

Certainly for many humans, the answer is yes, because as with all of biodiversity, birds do have considerable tangible and intangible values in our lives. Birds fill material needs for food, clothing, tools, and enjoyment (direct benefits). They also perform services useful to humans, without which the world would be less healthy or satisfying for us (indirect benefits). Many of these benefits are widely recognized already, but others certainly remain to be identified by future generations. Therefore, protecting birds and their habitats today ensures that we will not sacrifice future benefits before they can be realized.

Direct Benefits

Food

Humans the world over use birds or their eggs for food. Game birds in North America include dozens of species of ducks, grouse, quail, and shorebirds. Elsewhere around the world, humans consume wild birds or their eggs belonging to a wide variety of other groups. In Asia, even the candylike nest of the Edible-nest Swiftlet (made of hardened saliva [see Fig. 4–93] is eaten in soups as a delicacy. Where harvesting practices have become strictly controlled to ensure long-term sustainability, the practice of consuming birds or their by-products is fully compatible with conservation. Many birds once eaten bountifully from wild populations have been domesticated to increase production and to ease the cost of harvesting. Even Emu and Ostrich are now served in American restaurants, along with duck, quail, pheasant, chicken, and turkey. (However, as discussed earlier in this chapter, unregulated killing of wild birds for food remains a serious problem in many parts of the world.)

Clothing

Besides providing steaks for the table, Ostriches and Emus provide leather for fine clothing, especially boots and belts. Native Americans continue to use eagle feathers for ceremonial clothing and adornments, but most other uses of birds for clothing have been banned in North America following the excesses of the 19th century.

Other Utilitarian Uses

Down feathers from wild geese and ducks have been used for centuries as insulation and cushioning in pillows and blankets. Nitrogen-rich guano from cormorants and boobies is harvested from offshore islands of South America to produce agricultural fertilizer.

Recreational Hunting

Regulated hunting of wild birds is an important industry and pastime throughout the world. In the United States and Canada each year, hunters harvest tens of millions of ducks, geese, and upland game birds, spending hundreds of millions of dollars in the process. Important conservation achievements have been spawned by recreational hunters. For example, a huge amount of natural habitat across

Figure 10–89. Birders Scan the Sky for Migrating Hawks at Hawk Mountain, Pennsylvania: Birding is North America's fastest growing outdoor pastime. The money that birders spend on optical and other birding equipment, field guides, outdoor apparel, ecotourism, and other birding-related travel contributes significantly to local economies. As an example, more than 80,000 people visit Pennsylvania's Hawk Mountain each year to watch hawks migrating south. Businesses in such birding hotspots make major efforts to attract bird watchers. A U.S. Department of the Interior report estimated that in 1991, 24 million Americans took trips for the express purpose of watching wild birds (as compared to 14 million hunters and 35 million anglers). Another study (Wiedner and Kerlinger 1990) estimated that the average birder spends $350 per year on birding-related purchases, whereas serious birders may spend up to $2,000 annually, with half the amount being spent on travel. Photo courtesy of Tim Gallagher.

North America has been protected specifically because of its value as breeding habitat for game birds. The three organizations leading this effort (Ducks Unlimited, the U. S. Fish and Wildlife Service, and the Canadian Wildlife Service) have protected more than 100 million acres (40.5 million ha) of wetland and adjacent upland habitat for the benefit of game birds, especially ducks and geese. Such protection efforts also provide habitat for nongame bird species, as well as thousands of other species of plants and animals.

Bird Watching

Survey after survey has revealed that bird watching is now the most rapidly growing outdoor pastime in North America **(Fig. 10–89)**. A survey in 2000–2001 estimated that around 70 million people participated in some form of bird watching. A 1996 report by the U. S. Fish and Wildlife Service estimated that the U. S. public directly spent about $29 billion watching wildlife, and that the pastime produced an overall economic impact of $85 billion and more than one million jobs. Therefore, bird conservation should be viewed as essential in maintaining an important and rapidly expanding economic influence in the world.

Indirect Benefits

Ecological and Evolutionary Roles

As described in Chapter 9, birds play important roles in ecological systems throughout the world. For example, by eating or caching fruits, seeds, and nuts, birds disperse thousands of species of plants, especially in the tropics. In fact, certain species of trees (such as pinyon and whitebark pines in western North America) have co-evolved with birds (Pinyon Jay and Clark's Nutcracker, for instance). The trees produce large and nutritious seeds, yielding a staple food for the birds throughout winter. In turn, the birds disperse the pines by gathering and burying their seeds. Birds pollinate a wide array of plants in the tropics of both the New World (especially hummingbirds **[Fig. 10–90]**) and Old World (sunbirds and honeyeaters). By creating nest holes in dead trees, cacti, river banks, and even flat prairie ground, birds such as woodpeckers, swallows, kingfishers, and Burrowing Owls create cavities that last for many years and provide shelter and breeding places for whole complexes of insects, mammals, reptiles, and other birds. Because nearly all birds eat some animal prey, they play active roles as mid-level and top-level predators in food chains of almost every ecosystem in the world. As a result, birds have influenced the evolution of such features as mimicry in butterflies, cryptic coloration in insects, schooling behavior in fishes, and sentinel behavior in other birds.

Figure 10–90. Anna's Hummingbird: Birds play significant roles in ecological systems. The pollen on the forehead and upper beak of this Anna's Hummingbird shows that it is acting in the important role of pollinator, thereby helping wildflower populations through enhanced reproduction and cross-fertilization. Photo by Hugh P. Smith/VIREO.

Environmental Services

Environmental services performed by birds are no figment of lore and legend. In the middle of Temple Square in Salt Lake City, Utah, a gold statue of a California Gull celebrates the fact that humans the world over depend on birds in some unexpected ways. Ravaged by occasional plagues of swarming locusts, the crops of early Mormon communities were saved by California Gulls, which breed in huge colonies and feast on insect outbreaks. Similarly, the health and condition of hardwood forests in the mountains of New Hampshire are vitally dependent on the voracious consumption of caterpillars by migrant songbirds (Holmes et al. 1979) **(Fig 10–91)**. Around the world,

Figure 10–91. Bird Predation on Caterpillars: An experiment in a northern hardwood forest at Hubbard Brook, New Hampshire provides evidence that predation by birds significantly reduces the density of caterpillars feeding on shrubbery in the forest understory. Using crop-protection netting, researchers excluded birds from 10 patches of understory shrubs. Ten similar plots nearby, where birds were not excluded, acted as a control. Each week during the summer, the researchers counted the number of caterpillars found on 400 leaves within each of the 10 exclosures and on 400 leaves within each of the ten control plots (a total of 4,000 leaves in each plot). The solid line on the graph presents the results of counts taken within the exclosures (that is, where birds were prevented from entering), whereas the dashed line shows the results from the control plots. Throughout the season the number of caterpillars per 4,000 leaves was higher within the exclosures than in the unprotected control plots. Two asterisks show where these differences were statistically significant (p< 0.05): at the very end of June and in the third week of July, coinciding with peak nestling and fledgling periods of insectivorous birds—warblers, thrushes, and vireos—in this forest. By helping to control populations of caterpillars and other foliage-eating insects, birds contribute to forest health. Adapted from Holmes et al. (1979).

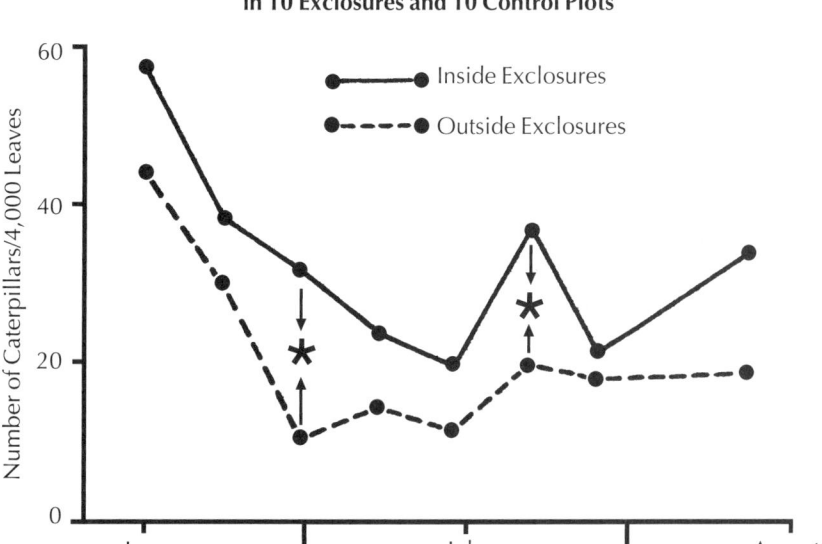

Densities of Caterpillars on Leaves of Striped Maple in 10 Exclosures and 10 Control Plots

● Inside Exclosures
● Outside Exclosures

Number of Caterpillars/4,000 Leaves

June   July   August

we can thank birds for promoting the evolution of hundreds of species of brightly colored, sugar-filled fruits and berries. These plants use birds as dispersal agents, packaging their precious seeds inside sugary, nutritious coatings in order to send them far and wide inside the guts of birds. By providing this "environmental courier service" for hundreds of species of trees and shrubs, birds constitute essential elements in the maintenance and restoration of forest systems all over the world. Finally, consider what our yards and roadsides would be like without the clean-up services provided by vultures and crows **(Fig. 10–92)**.

Biological Indicators

The expression "canary in the coal mine" sings out one of the most important ways in which humans benefit indirectly from services provided by birds. For the hundreds of years that preceded more modern detection devices, coal miners around the world descended into gas-laden coal deposits armed with carbon lights, pick-axes, and caged canaries **(Fig. 10–93)**. Because deadly gases often are odorless, the miners paid strict attention to their canaries, and if a bird gasped or succumbed, the miners fled the mine immediately before their own more robust bodies met a similar fate.

Birds announce environmental changes and dangers today in far more diverse ways than sensitivity to poisonous gas (see Chemical Toxins, above). Through their position atop natural food chains, for example, hawks and eagles focus our attention on dangerous environmental poisons that would otherwise go overlooked. Cadmium poisoning of White-tailed Ptarmigan in the Rocky Mountains alerts us to the lethal effects of heavy metals in mining areas worldwide (Larison et al. 2000). Population declines of many songbirds in eastern North America provide vital clues about the devastating ecological impacts of forest fragmentation. Finally, because many birds are narrow habitat specialists, we can monitor their populations—even in areas already under protection—to detect subtle habitat changes resulting from mistakes in long-term management. For example, as discussed above, understanding the delicate ecological needs of Northern Bobwhite in

Figure 10–92. Black Vultures on Road-killed Deer: Many birds and mammals are the unfortunate victims of traffic collisions. North Americans would sorely miss the roadside clean-up crews of crows and vultures, just one example of the many environmental services that birds provide. Photo by Marie Read.

the 1930s and Red-cockaded Woodpeckers in the 1980s provided the first key evidence that fire suppression was harming the southeastern pine forest ecosystems of North America.

Genetic Information

DNA molecules constituting the genetic makeup of one individual bird contain as much information as do all 17 printed editions of the Encyclopedia Britannica. A major reason why humans strive to protect all the world's species of animals and plants is that we have only begun to glance inside the cover of these myriad "genetic encyclopedias." The information contained in the DNA of any wild organism has survived millions of years of evolutionary challenges, and much of this information is of potential use to humans. We possess only rudimentary understanding of the genetic bases for disease resistance, protein synthesis, detoxification of natural compounds, complex behavior patterns, neural pathways, and many other natural functions of organisms. Every time we allow a species to go extinct we lose incalculable amounts of information about the earth's evolutionary history, its evolutionary potential, and its capacity for helping us lead our own lives more fully.

Scientific Study

Perhaps most important of all, birds continue to teach us how nature works. They are mostly diurnal and relatively easy to study. They occur in virtually every habitat on land and sea. They are extremely diverse, both socially and structurally. They migrate from pole to pole and from mountaintop to valley bottom. They exhibit remarkable feats of both short-term problem-solving and long-term memory. They are environmentally sensitive, and their birth rates and death rates reveal much about the world around them. For all these reasons and more, birds provide raw materials for never-ending scientific inquiry into all aspects of the living world.

Charles Darwin and Alfred Russell Wallace used birds to derive the theory of evolution by natural selection. Ernst Mayr used birds to deduce the genetic basis of the creation of new species. Robert MacArthur and Edward O. Wilson—through their studies of birds— led a host of other pioneering scientists in deriving profoundly important principles of ecology, island biogeography, population biology, and community structure. Konrad Lorenz and Niko Tinbergen won Nobel prizes for their studies of bird behavior. The most successful game management theories in the world were derived from studies of birds. Modern studies of environmental toxicology and wildlife disease began by focusing on birds. The biology of migration, navigation, and homing behavior, and the neurophysiology of memory, are all illuminated through studies of birds both in and out of the laboratory. Today, hundreds of scientific journals around the world focus on bird biology, and probably a thousand journals regularly publish ornithological research.

Figure 10–93. Canary in a Coal Mine: For decades, British coal miners used caged canaries to detect the presence of dangerous gases, particularly carbon monoxide, an odorless but deadly substance. Canaries react to carbon monoxide more rapidly than do humans, so if their birds showed signs of distress, the miners knew that the poisonous gas was present and that they needed to escape. Not until 1992 did engineers develop electronic equipment as reliable as the birds. Photo courtesy of British Coal Corporation.

Aesthetics and Spiritual Values

Can we put a dollar figure on the collective value derived across the land from seeing the first robin in the spring? From hearing Wood Thrushes in our parks? From watching gulls wheel at the seashore? From admiring an impossibly red cardinal at a feeder, or an impossibly nimble hummingbird at a flower? We cannot fathom the scale of enjoyment that birds bring to people all over the world, nor could we estimate the true monetary value of such pleasure, but we do know that birds supply it. Birds contribute to the total pleasure that humans derive from the outdoors. (This pleasure has remarkable healing power: hospital patients provided with good outdoor views heal significantly faster than patients lacking such a view.) Birds represent beauty in nature, providing spiritual and even religious experiences for millions of humans. It is no accident that Rachel Carson chose the title Silent Spring for her landmark book, as she intended the title alone to shock a public into contemplating a landscape so impoverished as to lack bird songs in the spring.

Finally, many thoughtful people believe that, given our phenomenal power to destroy life, humans have a philosophical or spiritual obligation to protect the natural and evolutionary heritage of the earth. By this view, nonhuman life forms such as birds have "rights to exist" that are fully equal to those of humans. The entirety of nature—created over three billion years of evolutionary history—represents a work of art at least as exquisitely complex, as mysterious, and as beautiful as the cathedrals, paintings, literature, and poetry of humankind. For anyone holding such a view of nature, it is especially easy to answer the question "why should we save it?"

What Can Each of Us Do?

■ Although we often count on institutions such as governments or conservation groups to accomplish large-scale bird conservation, individual citizens can accomplish a great deal to help keep common birds common and to help keep rare birds from going extinct. Biological diversity exists everywhere, and its long-term protection depends just as much on managing millions of small places across the landscape as it does on establishing and managing large parks and preserves. Here is a brief checklist of some ways in which every individual can help conserve birds around the world.

Backyard Conservation

From a few square feet in the midst of a city, to a backyard in the suburbs, or to a sweeping landscape of rural farmlands, every owner of outdoor space can make a difference in the effort to conserve birds and their habitats. Most important is to provide in ample quantity the elements that birds need to survive—shelter, nesting habitat, food, and water. These can be provided in imaginative and biologically useful ways in all kinds of backyards and schoolyards, both large and small.

Figure 10–94. Bird-friendly Backyard: The backyard can be excellent bird habitat if it caters to birds' survival needs by providing food, water, shelter, and nesting habitat. A varied habitat structure is best—intersperse flower gardens within patches of lawn, and plant shrubs and trees of various species and heights. The yard in this photo demonstrates some of these characteristics. Dense coniferous shrubs and tall, deciduous trees provide shelter and nest sites; flowers such as Black-eyed Susans are left to go to seed, providing winter food for finches and sparrows; a hanging feeder attracts seed-eaters; and a nectar feeder provides food for hummingbirds and orioles. Adding native shrubs that produce fruit in fall, leaving dead trees for cavity-nesting species, and providing a water source would make the yard even more attractive to birds. For other tips see one of the many books on the topic of gardening for birds and other wildlife, as well as Ch. 2, Sidebar 1: Attracting Birds to Your Yard. Photo by Marie Read.

Most important for providing habitat useful to birds is proper use of plantings, for which numerous specific suggestions are provided in bird-gardening books (for instance, Kress 1995). Just as in natural ecosystems, more species of birds can live in yards having varied habitat structure—lawn or native prairie grass; edged by open gardens; interspersed with boulders, ferns, and hedges; and overshadowed by trees of different species and heights—than in yards lacking such variation. Crucial to any bird-friendly yard is dense shrubbery, preferably representing plant species native to the region, as these provide vital nesting habitat and protection from predators for a wide variety of songbirds **(Fig. 10–94)**. An overstory of trees attracts both year-round residents and migratory songbirds. Birdbaths can be maintained year round, and are remarkably successful at attracting and keeping birds in the yard.

Healthy, native woodlands and forests are filled with dead or dying trunks and limbs, but humans have an aversion to dead trees. Fight this aversion. Allowing dead limbs and trees to remain in place makes a yard much more attractive to a large number of bird species. It is speculated, for example, that the widespread practice of removing dead trees from forests, parks, and yards may be contributing to the rapid decline of Northern Flickers from eastern North America. In addition to being essential for woodpeckers, however, dead limbs and snags also provide highly favorable perch sites for raptors, flycatchers, bluebirds, shrikes, waxwings, and countless other territorial songbirds.

Keep your pet cats indoors at all times. This habit is especially important during spring and summer when recently fledged juveniles are extremely vulnerable to predation by house cats **(Fig. 10–95)**. (This is so important to successful reproduction by birds that many conservation-minded homeowners fence their yards in order to keep neighboring house cats out!) Avoiding pesticides and lawn chemicals eliminates any chance of accidental ingestion by birds, and also promotes more natural insect populations throughout any backyard.

10

Figure 10–95. The Backyard Predator: According to the American Bird Conservancy, 66 million pet cats currently reside in the United States, and only about one-third of them are kept exclusively indoors. Added to their numbers are the many stray and feral cats that roam our cities, suburbs, farmlands, and natural areas, such as this feral cat that has captured a Steller's Jay. The number of these homeless cats is unknown, but estimates range from 60 to 100 million. Nationwide, free-roaming domestic cats are thought to be responsible for the deaths of hundreds of millions of birds and more than a billion small mammals (the natural food of hawks and owls) each year. A study in Great Britain estimated that each year pet cats kill 300 million birds and small mammals in that small country alone. Ground-nesting and ground-feeding birds, as well as nestlings and fledglings of all species, are particularly vulnerable. Despite myths to the contrary, well-fed cats and even those outfitted with bells on their collars still pose a threat to birds, and once a cat has captured and damaged a bird, rescuing it rarely saves its life. Through its Cats Indoors! campaign, the American Bird Conservancy encourages cat owners to help protect birds and other wildlife from domestic cat predation by keeping their cats indoors, on a leash, or in outdoor enclosures or cat runs. Photo by Steven Holt / Aigrette Photography.

Be a Citizen Scientist

Modern conservation practices depend in many ways on population monitoring, which permits detection of population declines, helps set priorities for conservation action, and aids in measuring success at sustaining a healthy habitat. However, resource limitations prohibit professional conservation managers from monitoring all possible nature preserves, let alone "nature at large" across the continent. We now know that we can resolve this paradox—the need for monitoring, but the seemingly impossible scale of the task—by taking advantage of the growing population of "citizen scientists." Bird monitoring programs organized by professionals and carried out by ordinary citizens (Breeding Bird Survey, Christmas Bird Count, Colonial Waterbird Registry, Project FeederWatch, and so on) now supply some of the most critical information on bird populations across North America. Anyone with an interest in birds can take part, for example, by spending a few minutes at the window (Great Backyard Bird Count), a whole day in the field (Christmas Bird Count), or a series of morning-long visits to a sample of forest tracts (Birds in Forested Landscapes). Participation in citizen science provides fun and enriching experiences to the participant, and helps accumulate crucial information for science and conservation.

Adopt a Place

Sustaining bird populations and biodiversity so that future generations can enjoy them requires that we recognize a key fact about today's world: directly or indirectly, humans now manage essentially every square foot of the earth. Whether we like it or not, our global job now includes being stewards of the natural systems of the earth. To do this job well requires adoption of a "culture of conservation" throughout the world, so that Aldo Leopold's "land ethic" becomes comparable in weight to our system of ethics for human behavior and spirituality. The consequence of such a shift will be that each of us strives to preserve natural systems and species side by side with human systems.

A powerful way for each of us to affect the long-term result is to choose a favorite place and to invest in its future as a functioning piece of the natural system. Our places need not be spectacular, nor must they harbor endangered species. After all, "keeping common birds common" may become just as difficult as rescuing endangered ones. Besides, many of the big and spectacular places are increasingly well equipped to accomplish long-term conservation without our individual attention. As private citizens, each of us can make a difference for conservation simply by making a personal commitment to a local park or recreation area, a commercial woodlot, a beachfront dune, or the grounds of a corporate headquarters. We might count its birds at key times of the year and submit our counts into citizen-science databases. We might study the management challenges facing our place, and perhaps even work with its owners to develop management alternatives (for instance, modify the landscaping plan to accommodate migratory birds; extend the timbering or mowing cycles to provide an unbroken reproductive season; alter water-flow regimes to restore seasonal flooding; or remove invasive alien plants to promote species diversity and reproduction by native species). We might even contribute financially to a management endowment for our place, so that future generations can continue the vigilance.

Local Vigilance and Grassroots Activism

Promoting personal commitment to a place is a central tenet of the National Audubon Society, and nowhere is this concept better exemplified than in Audubon's **Important Bird Areas** (IBA) program. Initially developed as an international effort by BirdLife International, Audubon's program in the United States is organized on a state-by-state basis **(Fig. 10–96)**. Through input from citizens and scientists, specialists identify several hundred places in each state having special significance for bird populations, and Audubon staff encourage both private and public investment in their long-term care. An IBA may be a pristine woodland, a stream system supporting intact or restorable riparian habitat, a local wetland where waterfowl breed or shorebirds stage during migration, a cliff system supplying nesting habitat for raptors, or even a local airport or cemetery harboring native grassland around its edges. The key feature of an IBA is its capacity to motivate people to invest in its future through energy, vigilance, and grassroots activism.

Anyone can participate in grassroots conservation simply by promoting the protection, enhancement, and management of local open spaces and by promoting regional land-use planning. Active conservation may entail vigilance (noticing and reporting violations of wetlands or endangered species laws, town ordinances, and so on), persuasion (for instance, convincing county commissioners to adopt open space ordinances), and involvement in the electoral process by electing officials sympathetic with conservation of open space and biodiversity. Often, the most important activism stems from searching out and acting upon opportunities to promote long-term protection of key parcels of land through nonregulatory processes (for example, through purchases from willing sellers, voluntary easements, or gifts to local land trusts).

Environmental Education

Participating directly or indirectly in local environmental education programs provides one of the most important ways for individuals to accomplish conservation. A tried and true adage holds that "humans will protect what they love, and they will love what they

Figure 10–96. The Important Bird Areas Program: Originally developed as an international bird conservation effort by BirdLife International, the North American Important Bird Areas Program (IBA) is organized on a state-by-state basis by the National Audubon Society. Local citizens, bird watchers, and professional ornithologists nominate potential IBA sites based on the importance of each to local bird populations as breeding, wintering, or migration habitat. Technical committees composed of bird experts and conservationists review each nomination. After a nomination has been approved, IBA staff members advocate public and private stewardship of the area. By 2001, several hundred locations across the country had been recognized formally as IBAs. Shown here is a 600-acre (243-ha) IBA near Ithaca, New York where mature cottonwoods, sycamores, and walnuts in a rare undeveloped stretch of riparian habitat support a major breeding concentration of Cerulean Warblers. Forty-six pairs of this declining species—whose range has been severely fragmented owing to habitat loss—nested here in 1997. Several private landowners as well as the local Finger Lakes Land Trust share stewardship of this IBA. For more information about IBAs visit <**www.audubon. org/bird/iba**>. Photo courtesy of Marie Read/ CLO.

understand." In the industrially developed countries today, fewer individuals than ever grow up directly in contact with nature. Any contact with the environment helps fill the void, from books and media to museums, nature centers, and classroom curricula. Birds supply an unusually effective "hook" by which many people first become environmentally engaged.

Every individual who appreciates and strives to protect nature can help teach others somewhere in the community. Nature centers constantly seek volunteers to help provide interpretation and hands-on experiences for their visitors. Classroom teachers who incorporate environmental materials often seek local experts to assist in teaching, both in the classroom and on field trips. Outstanding supplementary materials in environmental education are now available for elementary and middle school teachers, but these may be either unknown to, or unaffordable by, the teachers who could use them. Donating such materials to schools is a vital way in which individuals can help teach the next generation about saving birds, wildlife, and habitats.

Take a Child Birding

I have never forgotten the ecstatic shout and wide eyes of a young boy that I was with as he first peered into a telescope fixed on an American Oystercatcher probing in a tidal flat. Indeed, many people who enjoy birds through their lifetime can identify a person or an event that first sparked their interest. Very often, these experiences occurred during childhood. One of the best ways to help tomorrow's generation accept responsible land stewardship is to share one's own knowledge and passion with the children of today. Kids have an enormous capacity to enjoy nature in the company of adults or other kids who love it. Lifelong sympathy and curiosity often can be sparked by the smallest events, and among the most memorable is the singular thrill of seeing a beautiful bird through binoculars or a telescope for the first time **(Fig. 10–97)**. Seeing birds well, in real life, and being able to look back and forth between the bird and the pictures in a field guide, can open a child's mind to a lifelong appreciation of diversity and beauty in nature. This is the first and most important step that any individual can take toward recognizing the need to conserve such wonderful things.

Contribute to Conservation Organizations

Not-for-profit organizations (also known as nongovernmental organizations, or NGOs) play profoundly important roles in the conservation of birds and biodiversity, both in the United States and around the world. These charitable institutions could not survive without significant financial contributions and bequests from the public. Virtually all in the United States are tax exempt (classified 501(c)(3) by the U. S. Internal Revenue Service), which allows most membership dues and gifts to be tax deductible for the donor. Most conservation organizations are managed extremely well, staffed by committed and hard-working individuals earning only modest salaries, and overseen by independent governing boards who themselves contribute vitally through a combination of "wealth, work, or wisdom." As outlined each year in their annual reports, most of these organizations operate on three main sources of revenue: (1) gifts and membership dues from the public, (2) foundation grants and government contracts to carry out specific projects, and (3) proceeds from permanently invested endowments. A standard problem faces every one of these organizations: money is insufficient to accomplish the programmatic tasks, yet significant resources are always required to support the essential operating costs (such as salaries for management, development, and clerical staff; building or rental costs; utilities; travel to meetings; office furnishings; computer hardware and software; and web site construction and maintenance). Because virtually no foundations provide unrestricted operating support for NGOs, the chief source of these dollars is direct contributions from the public.

There are hundreds, perhaps thousands, of conservation-oriented NGOs, and their roles vary widely. Each typically seeks to fill a niche that is both essential for conservation and distinctive enough to provide "brand identity." Local land trusts and a few large, international organizations specialize in land acquisition and land management, often leveraging their own resources by forming partnerships with governmental agencies on public land. Most NGOs have a strong educational component, and a few concentrate in this activity through publications, web sites, citizen-science projects, and interpretive nature centers. Certain conservation groups invest strategically in lobbying state and federal legislators to bring about public policy, public investment, and regulations designed to improve environmental protection. Others specialize in environmental law, devoting resources to legal research and environmental litigation. Finally, a host of conservation organizations invest in scientific research to further our understanding

Figure 10–97. Adult and Child Birding: Early experiences leave powerful memories! Sharing the excitement of birding with a young child is one of the best ways to ensure that the next generation will care enough about birds and the natural world to protect them. Main photo courtesy of Terry Mingle. Photo of American Oystercatcher by Marie Read.

of conservation biology. Many of these develop strong partnerships with research universities or natural history museums. All of these roles are vital, because the challenges facing the natural world are far too great to be met by government, or by any organization acting in isolation. Therefore, making a financial contribution to any of these groups constitutes one of the most tangible ways in which an individual can participate in the process.

Never Give Up

Some individuals reading the Handbook of Bird Biology will be new to the idea of personal investment in protecting birds and conserving nature. Others have been passionately involved since childhood. For both groups, the message of this final section is the same: never give up. Do not ever resign yourself to the idea that the battle is lost, or that it cannot be won.

We should not sugarcoat the tragedies: we have lost some important skirmishes for good. We cannot bring back the Dodo, Reunion Solitaire, elephantbird, or moa. Sightings of Labrador Ducks and Kona Finches ended forever more than a century ago. Despite reaching abundances that defy imagination, Passenger Pigeons and Eskimo Curlews contribute today only as history lessons. They teach us about the immortality of the mistake we commit when we allow a species to disappear.

We do, however, still have enormous powers at our fingertips. We have power to repair ecosystems that have been damaged for centuries by cultures less aware than our own about the values that such places provide. We have power to measure the trajectory of species and ecosystems, and to warn ourselves about the systems that are faltering before it's too late. We have power to predict ecological consequences of our actions, at scales both large and small, before we commit to them. We have power to influence new generations of humans—the next managers of the earth—about how to avoid committing the mistakes that their forefathers made. In short, we still have immense power to influence the course of earth's natural history simply by deciding that we value other species, that we can protect natural systems, and that we can improve on our own behavior.

Although we have used our powers to damage ecosystems and destroy birds for centuries, within the last few decades we have also begun to master our powers to heal and protect. We are getting better at this every year, and the number of individuals committed to success is growing. College students today study conservation in unprecedented droves, and with passion to become involved. It is by no means too late to imagine that human society can peacefully and gainfully coexist with natural systems, but such a day will not come without effort. It will come only because we put our collective minds to the task. Fortunately, if we enjoy birds we can always have great fun along the way, because birds will never be far from the center of our focus.

Species Table

A

Abert's Towhee	*Pipilo aberti*
Acadian Flycatcher	*Empidonax virescens*
Acorn Woodpecker	*Melanerpes formicivorus*
Adelie Penguin	*Pygoscelis adeliae*
African Harrier-Hawk	*Polyboroides typus*
African Jacana	*Actophilornis africanus*
African Masked-Weaver	*Ploceus velatus*
African Palm-Swift	*Cypsiurus parvus*
Akialoa	
Lesser Akialoa	*Hemignathus obscurus*
Akiapolaau	*Hemignathus munroi*
Akikiki	*Oreomystis bairdi*
Akohekohe	*Palmeria dolei*
Alaotra Grebe	*Tachybaptus rufolavatus*
Albatross	
Laysan Albatross	*Phoebastria immutabilis*
Royal Albatross	*Diomedea epomophora*
Wandering Albatross	*Diomedea exulans*
Alder Flycatcher	*Empidonax alnorum*
Allen's Hummingbird	*Selasphorus sasin*
Alpine Swift	*Tachymarptis melba*
Altamira Oriole	*Icterus gularis*
Amakihi	
Hawaii Amakihi	*Hemignathus virens*
American Avocet	*Recurvirostra americana*
American Bittern	*Botaurus lentiginosus*
American Black Duck	*Anas rubripes*
American Coot	*Fulica americana*
American Crow	*Corvus brachyrhynchos*
American Dipper	*Cinclus mexicanus*
American Golden-Plover	*Pluvialis dominica*
American Goldfinch	*Carduelis tristis*
American Kestrel	*Falco sparverius*
American Oystercatcher	*Haematopus palliatus*
American Redstart	*Setophaga ruticilla*
American Robin	*Turdus migratorius*
American Tree Sparrow	*Spizella arborea*
American White Pelican	*Pelecanus erythrorhynchos*
American Woodcock	*Scolopax minor*
Ancient Murrelet	*Synthliboramphus antiquus*
Andean Cock-of-the-rock	*Rupicola peruviana*
Andean Condor	*Vultur gryphus*
Anhinga	*Anhinga anhinga*
Ani	
Greater Ani	*Crotophaga major*
Groove-billed Ani	*Crotophaga sulcirostris*
Anna's Hummingbird	*Calypte anna*
Antbird	
Bicolored Antbird	*Gymnopithys leucaspis*
Blackish Antbird	*Cercomacra nigrescens*
Dusky Antbird	*Cercomacra tyrannina*
Ocellated Antbird	*Phaenostictus mcleannani*
Spotted Antbird	*Hylophylax naevioides*
White-bellied Antbird	*Myrmeciza longipes*
White-plumed Antbird	*Pithys albifrons*
Antpipit	
Ringed Antpipit	*Corythopis torquata*
Antpitta	
Black-crowned Antpitta	*Pittasoma michleri*
Antshrike	
Barred Antshrike	*Thamnophilus doliatus*
Antthrush	
Black-faced Antthrush	*Formicarius analis*

Apapane	*Himatione sanguinea*
Apostlebird	*Struthidea cinerea*
Aracari	
Green Aracari	*Pteroglossus viridis*
Lettered Aracari	*Pteroglossus inscriptus*
Archaeopteryx	*Archaeopteryx lithographica*
Arctic Loon	*Gavia arctica*
Arctic Tern	*Sterna paradisaea*
Ash-throated Flycatcher	*Myiarchus cinerascens*
Asian Fairy-bluebird	*Irena puella*
Atlantic Puffin	*Fratercula arctica*
Audubon's Shearwater	*Puffinus lherminieri*
Auk	
Great Auk	*Pinguinus impennis*
Australasian Magpie	*Gymnorhina tibicen*
Australian Brush-turkey	*Alectura lathami*
Australian Raven	*Corvus coronoides*
Avocet	
American Avocet	*Recurvirostra americana*

B

Bachman's Warbler	*Vermivora bachmanii*
Baird's Sparrow	*Ammodramus bairdii*
Bald Eagle	*Haliaeetus leucocephalus*
Bali Myna	*Leucopsar rothschildi*
Baltimore Oriole	*Icterus galbula*
Bananaquit	*Coereba flaveola*
Banded Martin	*Riparia cincta*
Bank Swallow	*Riparia riparia*
Bar-headed Goose	*Anser indicus*
Barbet	
Golden-throated Barbet	*Megalaima oorti*
Scarlet-hooded Barbet	*Eubucco tucinkae*
Bare-eye	
Black-spotted Bare-eye	*Phlegopsis nigromaculata*
Barn Owl	*Tyto alba*
Barn Swallow	*Hirundo rustica*
Barred Antshrike	*Thamnophilus doliatus*
Barred Becard	*Pachyramphus versicolor*
Barred Owl	*Strix varia*
Barred-Woodcreeper	
Northern Barred-Woodcreeper	*Dendrocolaptes sanctithomae*
Barrow's Goldeneye	*Bucephala islandica*
Bat Falcon	*Falco rufigularis*
Bay Wren	*Thryothorus nigricapillus*
Bay-breasted Warbler	*Dendroica castanea*
Baya Weaver	*Ploceus philippinus*
Bearded Bellbird	*Procnias averano*
Becard	
Barred Becard	*Pachyramphus versicolor*
Bee Hummingbird	*Mellisuga helenae*
Bee-eater	
European Bee-eater	*Merops apiaster*
Purple-bearded Bee-eater	*Meropogon forsteni*
White-fronted Bee-eater	*Merops bullockoides*
Belding's Savannah Sparrow	*Passerculus sandwichensis beldingi*
Bellbird	
Bearded Bellbird	*Procnias averano*
Bell's Vireo	*Vireo bellii*
Belted Kingfisher	*Ceryle alcyon*
Bendire's Thrasher	*Toxostoma bendirei*

Berrypecker
   Crested Berrypecker *Paramythia montium*
   Tit Berrypecker *Oreocharis arfaki*
Bewick's Wren *Thryomanes bewickii*
Bicknell's Thrush *Catharus bicknelli*
Bicolored Antbird *Gymnopithys leucaspis*

Bird-of-paradise
   Blue Bird-of-paradise *Paradisaea rudolphi*
   Greater Bird-of-paradise *Paradisaea apoda*
   King Bird-of-paradise *Cicinnurus regius*
   King-of-Saxony Bird-of-paradise *Pteridophora alberti*
   Magnificent Bird-of-paradise *Cicinnurus magnificus*
   Superb Bird-of-paradise *Lophorina superba*

Bittern
   American Bittern *Botaurus lentiginosus*
   Least Bittern *Ixobrychus exilis*
Black Crowned-Crane *Balearica pavonina*
Black Grouse *Tetrao tetrix*
Black Guillemot *Cepphus grylle*
Black Kite *Milvus migrans*
Black Mamo *Drepanis funerea*
Black Sawwing *Psalidoprocne holomelas*
Black Scoter *Melanitta nigra*
Black Skimmer *Rynchops niger*
Black Tern *Chlidonias niger*
Black Turnstone *Arenaria melanocephala*
Black Vulture *Coragyps atratus*
Black Woodpecker *Dryocopus martius*
Black-and-red Broadbill *Cymbirhynchus macrorhynchos*
Black-and-white Warbler *Mniotilta varia*
Black-backed Woodpecker *Picoides arcticus*
Black-billed Cuckoo *Coccyzus erythropthalmus*
Black-billed Magpie *Pica pica*
Black-billed Scythebill *Campylorhamphus falcularius*
Black-capped Chickadee *Poecile atricapillus*
Black-capped Vireo *Vireo atricapillus*
Black-chinned Hummingbird *Archilochus alexandri*
Black-crowned Antpitta *Pittasoma michleri*
Black-crowned Night-Heron *Nycticorax nycticorax*
Black-faced Antthrush *Formicarius analis*
Black-faced Woodswallow *Artamus cinereus*
Black-fronted Nunbird *Monasa nigrifrons*
Black-headed Duck *Heteronetta atricapilla*
Black-headed Grosbeak *Pheucticus melanocephalus*
Black-headed Gull *Larus ridibundus*
Black-headed Trogon *Trogon melanocephalus*
Black-legged Kittiwake *Rissa tridactyla*
Black-necked Stilt *Himantopus mexicanus*
Black-spotted Bare-eye *Phlegopsis nigromaculata*
Black-tailed Gnatcatcher *Polioptila melanura*
Black-tailed Trogon *Trogon melanurus*
Black-throated Blue Warbler *Dendroica caerulescens*
Black-throated Gray Warbler *Dendroica nigrescens*
Black-throated Green Warbler *Dendroica virens*
Black-throated Sparrow *Amphispiza bilineata*

Blackbird
   Brewer's Blackbird *Euphagus cyanocephalus*
   Eurasian Blackbird *Turdus merula*
   Red-winged Blackbird *Agelaius phoeniceus*
   Tricolored Blackbird *Agelaius tricolor*
   Yellow-headed Blackbird *Xanthocephalus xanthocephalus*
Blackburnian Warbler *Dendroica fusca*
Blackcap *Sylvia atricapilla*
Blackish Antbird *Cercomacra nigrescens*
Blackpoll Warbler *Dendroica striata*
Blacksmith Plover *Vanellus armatus*
Blue Bird-of-paradise *Paradisaea rudolphi*
Blue Grosbeak *Guiraca caerulea*
Blue Grouse *Dendragapus obscurus*

Blue Jay *Cyanocitta cristata*
Blue Manakin *Chiroxiphia caudata*
Blue Tit *Cyanistes caeruleus*
Blue Whistling-Thrush *Myophonus caeruleus*
Blue-and-white Swallow *Pygochelidon cyanoleuca*
Blue-and-yellow Macaw *Ara ararauna*
Blue-capped Fruit-Dove *Ptilinopus monacha*
Blue-footed Booby *Sula nebouxii*
Blue-fronted Parrot *Amazona aestiva*
Blue-gray Gnatcatcher *Polioptila caerulea*
Blue-gray Tanager *Thraupis episcopus*
Blue-headed Vireo *Vireo solitarius*
Blue-throated Hummingbird *Lampornis clemenciae*
Blue-winged Teal *Anas discors*
Blue-winged Warbler *Vermivora pinus*

Bluebird
   Eastern Bluebird *Sialia sialis*
   Mountain Bluebird *Sialia currucoides*
Bluethroat *Luscinia svecica*
Boat-billed Flycatcher *Megarynchus pitangua*
Boat-tailed Grackle *Quiscalus major*
Bobolink *Dolichonyx oryzivorus*

Bobwhite
   Northern Bobwhite *Colinus virginianus*
Bonaparte's Gull *Larus philadelphia*
Bonin Petrel *Pterodroma hypoleuca*

Booby
   Blue-footed Booby *Sula nebouxii*
   Masked Booby *Sula dactylatra*
   Peruvian Booby *Sula variegata*
   Red-footed Booby *Sula sula*
Boreal Chickadee *Poecile hudsonicus*
Boreal Owl *Aegolius funereus*

Bowerbird
   Fire-maned Bowerbird *Sericulus bakeri*
   Macgregor's Bowerbird *Amblyornis macgregoriae*
   Satin Bowerbird *Ptilonorhynchus violaceus*
   Vogelkop Bowerbird *Amblyornis inornatus*
Brant *Branta bernicla*
Brewer's Blackbird *Euphagus cyanocephalus*
Brewer's Sparrow *Spizella breweri*
Bridled Titmouse *Baeolophus wollweberi*

Brilliant
   Pink-throated Brilliant *Heliodoxa gularis*
Broad-tailed Hummingbird *Selasphorus platycercus*
Broad-winged Hawk *Buteo platypterus*

Broadbill
   Black-and-red Broadbill *Cymbirhynchus macrorhynchos*
   Wattled Broadbill *Eurylaimus steerii*
Brown Creeper *Certhia americana*
Brown Kiwi *Apteryx australis*
Brown Noddy *Anous solidus*
Brown Pelican *Pelecanus occidentalis*
Brown Skua *Catharacta antarctica*
Brown Thrasher *Toxostoma rufum*
Brown-chested Martin *Progne tapera*
Brown-crested Flycatcher *Myiarchus tyrannulus*
Brown-headed Cowbird *Molothrus ater*

Brush-turkey
   Australian Brush-turkey *Alectura lathami*

Brush-Warbler
   Seychelles Brush-Warbler *Acrocephalus sechellensis*
Budgerigar *Melopsittacus undulatus*
Buff-breasted Sandpiper *Tryngites subruficollis*
Buff-breasted Wren *Thryothorus leucotis*
Buff-throated Purpletuft *Iodopleura pipra*
Bufflehead *Bucephala albeola*

Bulbul
   Yellow-wattled Bulbul *Pycnonotus urostictus*
Bullock's Oriole *Icterus bullockii*

Bunting
Indigo Bunting	*Passerina cyanea*
Lark Bunting	*Calamospiza melanocorys*
Lazuli Bunting	*Passerina amoena*
Painted Bunting	*Passerina ciris*
Snow Bunting	*Plectrophenax nivalis*
Burrowing Owl	*Athene cunicularia*
Bushtit	*Psaltriparus minimus*

Bustard
Great Bustard	*Otis tarda*
Kori Bustard	*Ardeotis kori*

Buzzard
Eurasian Buzzard	*Buteo buteo*

C

Cactus-Finch
Common Cactus-Finch	*Geospiza scandens*
Large Cactus-Finch	*Geospiza conirostris*
Cactus Wren	*Campylorhynchus brunneicapillus*
California Condor	*Gymnogyps californianus*
California Gnatcatcher	*Polioptila californica*
California Gull	*Larus californicus*
California Quail	*Callipepla californica*
California Thrasher	*Toxostoma redivivum*
California Towhee	*Pipilo crissalis*
Canada Goose	*Branta canadensis*
Canada Warbler	*Wilsonia canadensis*

Canary
Island Canary	*Serinus canaria*
Canvasback	*Aythya valisineria*
Canyon Towhee	*Pipilo fuscus*
Canyon Wren	*Catherpes mexicanus*
Cape May Warbler	*Dendroica tigrina*
Cape Petrel	*Daption capense*

Capercaillie
Eurasian Capercaillie	*Tetrao urogallus*

Caracara
Chimango Caracara	*Milvago chimango*
Crested Caracara	*Caracara plancus*

Cardinal
Northern Cardinal	*Cardinalis cardinalis*
Carolina Chickadee	*Poecile carolinensis*
Carolina Parakeet	*Conuropsis carolinensis*
Carolina Wren	*Thryothorus ludovicianus*
Carrion Crow	*Corvus corone*
Caspian Tern	*Sterna caspia*
Cassin's Finch	*Carpodacus cassinii*

Cassowary
Northern Cassowary	*Casuarius unappendiculatus*
Southern Cassowary	*Casuarius casuarius*

Catbird
Gray Catbird	*Dumetella carolinensis*
Cattle Egret	*Bubulcus ibis*
Cave Swallow	*Petrochelidon fulva*
Cedar Waxwing	*Bombycilla cedrorum*
Cerulean Warbler	*Dendroica cerulea*
Chaffinch	*Fringilla coelebs*
Channel-billed Toucan	*Ramphastos vitellinus*

Chat
Yellow-breasted Chat	*Icteria virens*
Chatham Gerygone	*Gerygone albofrontata*
Chatham Robin	*Petroica traversi*
Chestnut-collared Longspur	*Calcarius ornatus*
Chestnut-headed Oropendola	*Psarocolius wagleri*
Chestnut-sided Warbler	*Dendroica pensylvanica*

Chickadee
Black-capped Chickadee	*Poecile atricapillus*
Boreal Chickadee	*Poecile hudsonicus*
Carolina Chickadee	*Poecile carolinensis*
Mexican Chickadee	*Poecile sclateri*
Mountain Chickadee	*Poecile gambeli*
Chilean Tinamou	*Nothoprocta perdicaria*
Chimango Caracara	*Milvago chimango*
Chimney Swift	*Chaetura pelagica*
Chinstrap Penguin	*Pygoscelis antarctica*
Chipping Sparrow	*Spizella passerina*
Choco Toucan	*Ramphastos brevis*

Chough
White-winged Chough	*Corcorax melanorhamphos*
Yellow-billed Chough	*Pyrrhocorax graculus*
Chuck-will's-widow	*Caprimulgus carolinensis*
Cinnamon Teal	*Anas cyanoptera*
Cinnamon-headed Pigeon	*Treron fulvicollis*

Cisticola
Zitting Cisticola	*Cisticola juncidis*
Citron-headed Yellow-Finch	*Sicalis luteocephala*
Clapper Rail	*Rallus longirostris*
Claret-breasted Fruit-Dove	*Ptilinopus viridis*
Clark's Grebe	*Aechmophorus clarkii*
Clark's Nutcracker	*Nucifraga columbiana*
Cliff Flycatcher	*Hirundinea ferruginea*
Cliff Swallow	*Petrochelidon pyrrhonota*
Cobalt-winged Parakeet	*Brotogeris cyanoptera*

Cock-of-the-rock
Andean Cock-of-the-rock	*Rupicola peruviana*
Guianan Cock-of-the-rock	*Rupicola rupicola*
Cock-tailed Tyrant	*Alectrurus tricolor*

Cockatoo
Sulphur-crested Cockatoo	*Cacatua galerita*
Cocos Island Finch	*Pinaroloxias inornata*
Colima Warbler	*Vermivora crissalis*
Common Cactus-Finch	*Geospiza scandens*
Common Crane	*Grus grus*
Common Cuckoo	*Cuculus canorus*
Common Diving-Petrel	*Pelecanoides urinatrix*
Common Eider	*Somateria mollissima*
Common Goldeneye	*Bucephala clangula*
Common Grackle	*Quiscalus quiscula*
Common Kingfisher	*Alcedo atthis*
Common Loon	*Gavia immer*
Common Merganser	*Mergus merganser*
Common Moorhen	*Gallinula chloropus*
Common Murre	*Uria aalge*
Common Nighthawk	*Chordeiles minor*
Common Nightingale	*Luscinia megarhynchos*
Common Poorwill	*Phalaenoptilus nuttallii*
Common Raven	*Corvus corax*
Common Redpoll	*Carduelis flammea*
Common Redshank	*Tringa totanus*
Common Redstart	*Phoenicurus phoenicurus*
Common Shelduck	*Tadorna tadorna*
Common Snipe	*Gallinago gallinago*
Common Swift	*Apus apus*
Common Tailorbird	*Orthotomus sutorius*
Common Tern	*Sterna hirundo*
Common Wood-Pigeon	*Columba palumbus*
Common Yellowthroat	*Geothlypis trichas*

Condor
Andean Condor	*Vultur gryphus*
California Condor	*Gymnogyps californianus*
Cooper's Hawk	*Accipter cooperii*

Coot
American Coot	*Fulica americana*
Horned Coot	*Fulica cornuta*
Red-fronted Coot	*Fulica rufifrons*
Red-gartered Coot	*Fulica armillata*
Cordilleran Flycatcher	*Empidonax occidentalis*

Cormorant

Double-crested Cormorant	*Phalacrocorax auritus*
Great Cormorant	*Phalacrocorax carbo*
Guanay Cormorant	*Phalacrocorax bougainvillii*
Corn Crake	*Crex crex*
Costa's Hummingbird	*Calypte costae*
Cotinga	
Pompadour Cotinga	*Xipholena punicea*
Cowbird	
Brown-headed Cowbird	*Molothrus ater*
Giant Cowbird	*Scaphidura oryzivora*
Screaming Cowbird	*Molothrus rufoaxillaris*
Shiny Cowbird	*Molothrus bonariensis*
Crab Plover	*Dromas ardeola*
Crake	
Corn Crake	*Crex crex*
Crowned-Crane	
Black Crowned-Crane	*Balearica pavonina*
Crane	
Common Crane	*Grus grus*
Sandhill Crane	*Grus canadensis*
Whooping Crane	*Grus americana*
Craveri's Murrelet	*Synthliboramphus craveri*
Creeper	
Brown Creeper	*Certhia americana*
Crested Berrypecker	*Paramythia montium*
Crested Caracara	*Caracara plancus*
Crimson Rosella	*Platycercus elegans*
Crissal Thrasher	*Toxostoma crissale*
Crossbill	
Red Crossbill	*Loxia curvirostra*
White-winged Crossbill	*Loxia leucoptera*
Crow	
American Crow	*Corvus brachyrhynchos*
Carrion Crow	*Corvus corone*
Hawaiian Crow	*Corvus hawaiiensis*
New Caledonian Crow	*Corvus moneduloides*
Northwestern Crow	*Corvus caurinus*
Cuckoo	
Black-billed Cuckoo	*Coccyzus erythropthalmus*
Common Cuckoo	*Cuculus canorus*
Great Spotted Cuckoo	*Clamator glandarius*
Guira Cuckoo	*Guira guira*
Mangrove Cuckoo	*Coccyzus minor*
Pavonine Cuckoo	*Dromococcyx pavoninus*
Pearly-breasted Cuckoo	*Coccyzus euleri*
Pheasant Cuckoo	*Dromococcyx phasianellus*
Striped Cuckoo	*Tapera naevia*
Yellow-billed Cuckoo	*Coccyzus americanus*
Cuckoo-Roller	*Leptosomus discolor*
Curassow	
Great Curassow	*Crax rubra*
Curlew	
Eskimo Curlew	*Numenius borealis*
Long-billed Curlew	*Numenius americanus*
Currawong	
Pied Currawong	*Strepera graculina*

D

Dalmatian Pelican	*Pelecanus crispus*
Dark-eyed Junco	*Junco hyemalis*
Dickcissel	
Dickcissel	*Spiza americana*
Dipper	
American Dipper	*Cinclus mexicanus*
Diving-Petrel	
Common Diving-Petrel	*Pelecanoides urinatrix*
Dodo	*Raphus cucullatus*
Double-crested Cormorant	*Phalacrocorax auritus*
Double-toothed Kite	*Harpagus bidentatus*

Dove	
Mourning Dove	*Zenaida macroura*
Rock Dove	*Columba livia*
Socorro Dove	*Zenaida graysoni*
Dovekie	*Alle alle*
Downy Woodpecker	*Picoides pubescens*
Duck	
American Black Duck	*Anas rubripes*
Black-headed Duck	*Heteronetta atricapilla*
Harlequin Duck	*Histrionicus histrionicus*
Labrador Duck	*Camptorhynchus labradorius*
Long-tailed Duck	*Clangula hyemalis*
Mandarin Duck	*Aix galericulata*
Muscovy Duck	*Cairina moschata*
Ruddy Duck	*Oxyura jamaicensis*
Wood Duck	*Aix sponsa*
Dunlin	*Calidris alpina*
Dunnock	*Prunella modularis*
Dusky Antbird	*Cercomacra tyrannina*
Dusky Seaside Sparrow	*Ammodramus maritimus nigrescens*
Dusky-capped Flycatcher	*Myiarchus tuberculifer*

E

Eagle	
Bald Eagle	*Haliaeetus leucocephalus*
Golden Eagle	*Aquila chrysaetos*
Eared Grebe	*Podiceps nigricollis*
Eastern Bluebird	*Sialia sialis*
Eastern Kingbird	*Tyrannus tyrannus*
Eastern Meadowlark	*Sturnella magna*
Eastern Phoebe	*Sayornis phoebe*
Eastern Screech-Owl	*Otus asio*
Eastern Towhee	*Pipilo erythrophthalmus*
Eastern Wood-Pewee	*Contopus virens*
Edible-nest Swiftlet	*Aerodramus fuciphagus*
Egret	
Cattle Egret	*Bubulcus ibis*
Great Egret	*Ardea alba*
Reddish Egret	*Egretta rufescens*
Snowy Egret	*Egretta thula*
Egyptian Plover	*Pluvianus aegyptius*
Eider	
Common Eider	*Somateria mollissima*
King Eider	*Somateria spectabilis*
Steller's Eider	*Polysticta stelleri*
Elegant Trogon	*Trogon elegans*
Eleonora's Falcon	*Falco eleonorae*
Elepaio	*Chasiempis sandwichensis*
Elephantbird	
Giant Elephantbird	*Aepyornis maximus*
Elf Owl	*Micrathene whitneyi*
Emperor Penguin	*Aptenodytes forsteri*
Emu	*Dromaius novaehollandiae*
Eskimo Curlew	*Numenius borealis*
Euphonia	
Thick-billed Euphonia	*Euphonia laniirostris*
Eurasian Blackbird	*Turdus merula*
Eurasian Buzzard	*Buteo buteo*
Eurasian Capercaillie	*Tetrao urogallus*
Eurasian Golden Oriole	*Oriolus oriolus*
Eurasian Golden-Plover	*Pluvialis apricaria*
Eurasian Kestrel	*Falco tinnunculus*
Eurasian Nightjar	*Caprimulgus europaeus*
Eurasian Penduline-Tit	*Remiz pendulinus*
Eurasian Reed-Warbler	*Pipilo fuscus*
Eurasian Sparrowhawk	*Accipiter nisus*
Eurasian Tree Sparrow	*Passer montanus*
Eurasian Wigeon	*Anas penelope*

European Bee-eater	*Merops apiaster*
European Goldfinch	*Carduelis carduelis*
European Greenfinch	*Carduelis chloris*
European Pied Flycatcher	*Ficedula hypoleuca*
European Robin	*Erithacus rubecula*
European Starling	*Sturnus vulgaris*
Evening Grosbeak	*Coccothraustes vespertinus*

F

Fairy-bluebird
Asian Fairy-bluebird	*Irena puella*

Fairywren
Superb Fairywren	*Malurus cyaneus*

Falcon
Bat Falcon	*Falco rufigularis*
Eleonora's Falcon	*Falco eleonorae*
Peregrine Falcon	*Falco peregrinus*
Prairie Falcon	*Falco mexicanus*
Fan-tailed Warbler	*Euthlypis lachrymosa*
Ferruginous Hawk	*Buteo regalis*
Ferruginous Pygmy-Owl	*Glaucidium brasilianum*

Field-Tyrant
Short-tailed Field-Tyrant	*Muscigralla brevicauda*
Field Sparrow	*Spizella pusilla*
Fiery-throated Hummingbird	*Panterpe insignis*

Finch
Cassin's Finch	*Carpodacus cassinii*
Cocos Island Finch	*Pinaroloxias inornata*
House Finch	*Carpodacus mexicanus*
Laysan Finch	*Telespiza cantans*
Mangrove Finch	*Camarhynchus heliobates*
Plush-capped Finch	*Catamblyrhynchus diadema*
Purple Finch	*Carpodacus purpureus*
Vegetarian Finch	*Camarhynchu crassirostris*
Warbler Finch	*Certhidea olivacea*
Woodpecker Finch	*Camarhynchus pallidus*
Zebra Finch	*Taeniopygia guttata*

Fire-eye
White-backed Fire-eye	*Pyriglena leuconota*
Fire-maned Bowerbird	*Sericulus bakeri*

Firefinch
Jameson's Firefinch	*Lagonosticta rhodopareia*
Red-billed Firefinch	*Lagonosticta senegala*

Flamingo
Greater Flamingo	*Phoenicopterus ruber*
Lesser Flamingo	*Phoenicopterus minor*

Flicker
Gilded Flicker	*Colaptes chrysoides*
Northern Flicker	*Colaptes auratus*
Florida Scrub-Jay	*Aphelocoma coerulescens*

Flycatcher
Acadian Flycatcher	*Empidonax virescens*
Alder Flycatcher	*Empidonax alnorum*
Ash-throated Flycatcher	*Myiarchus cinerascens*
Boat-billed Flycatcher	*Megarynchus pitangua*
Brown-crested Flycatcher	*Myiarchus tyrannulus*
Cliff Flycatcher	*Hirundinea ferruginea*
Cordilleran Flycatcher	*Empidonax occidentalis*
Dusky-capped Flycatcher	*Myiarchus tuberculifer*
European Pied Flycatcher	*Ficedula hypoleuca*
Gray Flycatcher	*Empidonax wrightii*
Great Crested Flycatcher	*Myiarchus crinitus*
Hammond's Flycatcher	*Empidonax hammondii*
Least Flycatcher	*Empidonax minimus*
Olive-sided Flycatcher	*Contopus cooperi*
Pacific-slope Flycatcher	*Empidonax difficilis*
Royal Flycatcher	*Onychorhynchus coronatus*
Scissor-tailed Flycatcher	*Tyrannus forficatus*
Swainson's Flycatcher	*Myiarchus swainsoni*

Vermilion Flycatcher	*Pyrocephalus rubinus*
Willow Flycatcher	*Empidonax traillii*
Yellow-bellied Flycatcher	*Empidonax flaviventris*

Fody
Red Fody	*Foudia madagascariensis*

Forktail
Spotted Forktail	*Enicurus maculatus*
Formosan Whistling-Thrush	*Myophonus insularis*
Forster's Tern	*Sterna forsteri*
Fox Sparrow	*Passerella iliaca*

Frigatebird
Magnificent Frigatebird	*Fregata magnificens*

Frogmouth
Papuan Frogmouth	*Podargus papuensis*

Fruit-Dove
Blue-capped Fruit-Dove	*Ptilinopus monacha*
Claret-breasted Fruit-Dove	*Ptilinopus viridis*

Fulmar
Northern Fulmar	*Fulmarus glacialis*

G

Gabar Goshawk	*Micronisus gabar*
Gadwall	*Anas strepera*
Galah	*Eolophus roseicapillus*

Gallinule
Purple Gallinule	*Porphyrula martinica*
Gambel's Quail	*Callipepla gambelii*

Gannet
Northern Gannet	*Morus bassanus*
Garden Warbler	*Sylvia borin*
Gentoo Penguin	*Pygoscelis papua*

Gerygone
Chatham Gerygone	*Gerygone albofrontata*
Giant Cowbird	*Scaphidura oryzivora*
Giant Elephantbird	*Aepyornis maximus*
Gila Woodpecker	*Melanerpes uropygialis*
Gilded Flicker	*Colaptes chrysoides*
Gilt-edged Tanager	*Tangara cyanoventris*
Glaucous Gull	*Larus hyperboreus*
Glaucous-winged Gull	*Larus glaucescens*
Glistening-green Tanager	*Chlorochrysa phoenicotis*
Glossy Ibis	*Plegadis falcinellus*

Gnatcatcher
Black-tailed Gnatcatcher	*Polioptila melanura*
Blue-gray Gnatcatcher	*Polioptila caerulea*
California Gnatcatcher	*Polioptila californica*
Tropical Gnatcatcher	*Polioptila plumbea*

Godwit
Hudsonian Godwit	*Limosa haemastica*
Golden Eagle	*Aquila chrysaetos*
Golden-cheeked Warbler	*Dendroica chrysoparia*
Golden-crowned Kinglet	*Regulus satrapa*

Golden-Plover
American Golden-Plover	*Pluvialis dominica*
Eurasian Golden-Plover	*Pluvialis apricaria*
Golden-throated Barbet	*Megalaima oorti*
Golden-winged Warbler	*Vermivora chrysoptera*

Goldeneye
Barrow's Goldeneye	*Bucephala islandica*
Common Goldeneye	*Bucephala clangula*

Goldfinch
American Goldfinch	*Carduelis tristis*
European Goldfinch	*Carduelis carduelis*
Lesser Goldfinch	*Carduelis psaltria*

Goose
Bar-headed Goose	*Anser indicus*
Canada Goose	*Branta canadensis*
Greater White-fronted Goose	*Anser albifrons*
Greylag Goose	*Anser anser*

Lesser White-fronted Goose	*Anser erythropus*
Ross's Goose	*Chen rossii*
Snow Goose	*Chen caerulescens*
Goshawk	
Gabar Goshawk	*Micronisus gabar*
Northern Goshawk	*Accipiter gentilis*
Grackle	
Boat-tailed Grackle	*Quiscalus major*
Common Grackle	*Quiscalus quiscula*
Grasshopper Sparrow	*Ammodramus savannarum*
Grasshopper Warbler	*Locustella naevia*
Gray Catbird	*Dumetella carolinensis*
Gray Flycatcher	*Empidonax wrightii*
Gray Jay	*Perisoreus canadensis*
Gray Partridge	*Perdix perdix*
Gray Vireo	*Vireo vicinior*
Gray-cheeked Thrush	*Catharus minimus*
Gray-headed Tanager	*Eucometis penicillata*
Gray-headed Woodpecker	*Dendropicos spodocephalus*
Gray-rumped Swallow	*Hirundo griseopyga*
Great Auk	*Pinguinus impennis*
Great Black-backed Gull	*Larus marinus*
Great Blue Heron	*Ardea herodias*
Great Bustard	*Otis tarda*
Great Cormorant	*Phalacrocorax carbo*
Great Crested Flycatcher	*Myiarchus crinitus*
Great Crested Grebe	*Podiceps cristatus*
Great Curassow	*Crax rubra*
Great Egret	*Ardea alba*
Great Horned Owl	*Bubo virginianus*
Great Jacamar	*Jacamerops aurea*
Great Kiskadee	*Pitangus sulphuratus*
Great Potoo	*Nyctibius grandis*
Great Reed-Warbler	*Acrocephalus arundinaceus*
Great Skua	*Catharacta skua*
Great Snipe	*Gallinago media*
Great Spotted Cuckoo	*Clamator glandarius*
Great Spotted Kiwi	*Apteryx haastii*
Great Spotted Woodpecker	*Dendrocopos major*
Great Tit	*Parus major*
Great White Pelican	*Pelecanus onocrotalus*
Greater Ani	*Crotophaga major*
Greater Bird-of-paradise	*Paradisaea apoda*
Greater Flamingo	*Phoenicopterus ruber*
Greater Pewee	*Contopus pertinax*
Greater Prairie-Chicken	*Tympanuchus cupido*
Greater Rhea	*Rhea americana*
Greater Roadrunner	*Geococcyx californianus*
Greater Scaup	*Aythya marila*
Greater Striped-Swallow	*Hirundo cucullata*
Greater White-fronted Goose	*Anser albifrons*
Greater Yellowlegs	*Tringa melanoleuca*
Grebe	
Alaotra Grebe	*Tachybaptus rufolavatus*
Clark's Grebe	*Aechmophorus clarkii*
Eared Grebe	*Podiceps nigricollis*
Great Crested Grebe	*Podiceps cristatus*
Horned Grebe	*Podiceps auritus*
Little Grebe	*Tachybaptus ruficollis*
Pied-billed Grebe	*Podilymbus podiceps*
Red-necked Grebe	*Podiceps grisegena*
Western Grebe	*Aechmophorus occidentalis*
Green Aracari	*Pteroglossus viridis*
Green Heron	*Butorides virescens*
Green Jay	*Cyanocorax yncas*
Green Kingfisher	*Chloroceryle americana*
Green Violet-ear	*Colibri thalassinus*
Green Woodpecker	*Picus viridis*
Green-rumped Parrotlet	*Forpus passerinus*
Green-tailed Towhee	*Pipilo chlorurus*

Green-winged Teal	*Anas crecca*
Greenfinch	
European Greenfinch	*Carduelis chloris*
Greenlet	
Lesser Greenlet	*Hylophilus decurtatus*
Grey Parrot	*Psittacus erithacus*
Greylag Goose	*Anser anser*
Griffon	
Rueppell's Griffon	*Gyps rueppellii*
Groove-billed Ani	*Crotophaga sulcirostris*
Grosbeak	
Black-headed Grosbeak	*Pheucticus melanocephalus*
Blue Grosbeak	*Guiraca caerulea*
Evening Grosbeak	*Coccothraustes vespertinus*
Kona Grosbeak	*Chloridops kona*
Pine Grosbeak	*Pinicola enucleator*
Rose-breasted Grosbeak	*Pheucticus ludovicianus*
Grosbeak Weaver	*Amblyospiza albifrons*
Ground-Cuckoo	
Rufous-vented Ground-Cuckoo	*Neomorphus geoffroyi*
Ground-Finch	
Large Ground-Finch	*Geospiza magnirostris*
Medium Ground-Finch	*Geospiza fortis*
Sharp-beaked Ground-Finch	*Geospiza difficilis*
Small Ground-Finch	*Geospiza fuliginosa*
Ground-Hornbill	
Southern Ground-Hornbill	*Bucorvus leadbeateri*
Grouse	
Black Grouse	*Tetrao tetrix*
Blue Grouse	*Dendragapus obscurus*
Ruffed Grouse	*Bonasa umbellus*
Sage Grouse	*Centrocercus urophasianus*
Sharp-tailed Grouse	*Tympanuchus phasianellus*
Spruce Grouse	*Falcipennis canadensis*
Guadalcanal Honeyeater	*Guadalcanaria inexpectata*
Guam Micronesian Kingfisher	*Todirhamphus cinnamominus cinnamominus*
Guam Rail	*Gallirallus owstoni*
Guanay Cormorant	*Phalacrocorax bougainvillii*
Guianan Cock-of-the-rock	*Rupicola rupicola*
Guillemot	
Black Guillemot	*Cepphus grylle*
Pigeon Guillemot	*Cepphus columba*
Guineafowl	
Plumed Guineafowl	*Guttera plumifera*
Vulturine Guineafowl	*Acryllium vulturinum*
Guira Cuckoo	*Guira guira*
Gull	
Black-headed Gull	*Larus ridibundus*
Bonaparte's Gull	*Larus philadelphia*
California Gull	*Larus californicus*
Glaucous Gull	*Larus hyperboreus*
Glaucous-winged Gull	*Larus glaucescens*
Great Black-backed Gull	*Larus marinus*
Herring Gull	*Larus argentatus*
Laughing Gull	*Larus atricilla*
Lesser Black-backed Gull	*Larus fuscus*
Mew Gull	*Larus canus*
Ring-billed Gull	*Larus delawarensis*
Ross's Gull	*Rhodostethia rosea*
Sabine's Gull	*Xema sabini*
Western Gull	*Larus occidentalis*
Gull-billed Tern	*Sterna nilotica*
Gyrfalcon	*Falco rusticolus*

H

Hairy Woodpecker	*Picoides villosus*
Hamerkop	*Scopus umbretta*
Hammond's Flycatcher	*Empidonax hammondii*

Harlequin Duck	*Histrionicus histrionicus*
Harrier	
Northern Harrier	*Circus cyaneus*
Harrier-Hawk	
African Harrier-Hawk	*Polyboroides typus*
Harris's Sparrow	*Zonotrichia querula*
Hawaii Amakihi	*Hemignathus virens*
Hawaii Mamo	*Drepanis pacifica*
Hawaii Oo	*Moho nobilis*
Hawaiian Crow	*Corvus hawaiiensis*
Hawk	
Broad-winged Hawk	*Buteo platypterus*
Cooper's Hawk	*Accipter cooperii*
Ferruginous Hawk	*Buteo regalis*
Red-shouldered Hawk	*Buteo lineatus*
Red-tailed Hawk	*Buteo jamaicensis*
Rough-legged Hawk	*Buteo lagopus*
Sharp-shinned Hawk	*Accipter striatus*
Short-tailed Hawk	*Buteo brachyurus*
Swainson's Hawk	*Buteo swainsoni*
Hen	
Heath Hen	*Tympanuchus cupido cupido*
Henslow's Sparrow	*Ammodramus henslowii*
Hermit	
Little Hermit	*Phaethornis longuemareus*
Long-tailed Hermit	*Phaethornis superciliosus*
Planalto Hermit	*Phaethornis pretrei*
Sooty-capped Hermit	*Phaethornis augusti*
Hermit Thrush	*Catharus guttatus*
Hermit Warbler	*Dendroica occidentalis*
Heron	
Great Blue Heron	*Ardea herodias*
Green Heron	*Butorides virescens*
Little Blue Heron	*Egretta caerulea*
Tricolored Heron	*Egretta tricolor*
Herring Gull	*Larus argentatus*
Hoatzin	*Opisthocomus hoazin*
Honeycreeper	
Shining Honeycreeper	*Cyanerpes lucidus*
Honeyeater	
Guadalcanal Honeyeater	*Guadalcanaria inexpectata*
Hooded Merganser	*Lophodytes cucullatus*
Hooded Oriole	*Icterus cucullatus*
Hooded Warbler	*Wilsonia citrina*
Hook-billed Kite	*Chondrohierax uncinatus*
Hornbill	
Red-billed Hornbill	*Tockus erythrorhynchus*
Rhinoceros Hornbill	*Buceros rhinoceros*
Horned Coot	*Fulica cornuta*
Horned Grebe	*Podiceps auritus*
Horned Lark	*Eremophila alpestris*
Horned Puffin	*Fratercula corniculata*
Hornero	
Rufous Hornero	*Furnarius rufus*
House Finch	*Carpodacus mexicanus*
House Martin	*Delichon urbica*
House Sparrow	*Passer domesticus*
House Wren	*Troglodytes aedon*
Hudsonian Godwit	*Limosa haemastica*
Huia	*Heteralocha acutirostris*
Humboldt Penguin	*Spheniscus humboldti*
Hummingbird	
Allen's Hummingbird	*Selasphorus sasin*
Anna's Hummingbird	*Calypte anna*
Bee Hummingbird	*Mellisuga helenae*
Black-chinned Hummingbird	*Archilochus alexandri*
Blue-throated Hummingbird	*Lampornis clemenciae*
Broad-tailed Hummingbird	*Selasphorus platycercus*
Costa's Hummingbird	*Calypte costae*
Fiery-throated Hummingbird	*Panterpe insignis*

Magnificent Hummingbird	*Eugenes fulgens*
Ruby-throated Hummingbird	*Archilochus colubris*
Ruby-topaz Hummingbird	*Chrysolampis mosquitus*
Rufous Hummingbird	*Selasphorus rufus*
Volcano Hummingbird	*Selasphorus flammula*
Hutton's Vireo	*Vireo huttoni*
Hyacinth Macaw	*Anodorhynchus hyacinthinus*
Hypocolius	*Hypocolius ampelinus*

I

Ibis	
Glossy Ibis	*Plegadis falcinellus*
Scarlet Ibis	*Eudocimus ruber*
White Ibis	*Eudocimus albus*
White-faced Ibis	*Plegadis chihi*
Ibisbill	*Ibidorhyncha struthersii*
Iiwi *Vestiaria coccinea*	
Imperial-Pigeon	
Torresian Imperial-Pigeon	*Ducula spilorrhoa*
Imperial Parrot	*Amazona imperialis*
Imperial Woodpecker	*Campephilus imperialis*
Indian Peafowl	*Pavo cristatus*
Indigo Bunting	*Passerina cyanea*
Indigobird	
Purple Indigobird	*Vidua purpurascens*
Village Indigobird	*Vidua chalybeata*
Island Canary	*Serinus canaria*
Ivory-billed Woodpecker	*Campephilus principalis*

J

Jacamar	
Great Jacamar	*Jacamerops aurea*
Jacana	
African Jacana	*Actophilornis africanus*
Northern Jacana	*Jacana spinosa*
Wattled Jacana	*Jacana jacana*
Jackass Penguin	*Spheniscus demersus*
Jaeger	
Long-tailed Jaeger	*Stercorarius longicaudus*
Parasitic Jaeger	*Stercorarius parasiticus*
Pomarine Jaeger	*Stercorarius pomarinus*
Jameson's Firefinch	*Lagonosticta rhodopareia*
Japanese Quail	*Coturnix japonica*
Jay	
Blue Jay	*Cyanocitta cristata*
Gray Jay	*Perisoreus canadensis*
Green Jay	*Cyanocorax yncas*
Mexican Jay	*Aphelocoma ultramarina*
Pinyon Jay	*Gymnorhinus cyanocephalus*
Steller's Jay	*Cyanocitta stelleri*
Junco	
Dark-eyed Junco	*Junco hyemalis*
Yellow-eyed Junco	*Junco phaeonotus*
Junglefowl	
Red Junglefowl	*Gallus gallus*
Juniper Titmouse	*Baeolophus griseus*

K

Kagu	*Rhynochetos jubatus*
Kakapo	*Strigops habroptilus*
Kauai Oo	*Moho braccatus*
Kea *Nestor notabilis*	
Keel-billed Toucan	*Ramphastos sulfuratus*
Kentucky Warbler	*Oporornis formosus*

Kestrel
American Kestrel	*Falco sparverius*
Eurasian Kestrel	*Falco tinnunculus*
Mauritius Kestrel	*Falco punctatus*
Killdeer	*Charadrius vociferus*
King Bird-of-paradise	*Cicinnurus regius*
King Eider	*Somateria spectabilis*
King Penguin	*Aptenodytes patagonicus*
King Rail	*Rallus elegans*
King Vulture	*Sarcoramphus papa*
King-of-Saxony Bird-of-paradise	*Pteridophora alberti*

Kingbird
Eastern Kingbird	*Tyrannus tyrannus*
Western Kingbird	*Tyrannus verticalis*

Kingfisher
Belted Kingfisher	*Ceryle alcyon*
Common Kingfisher	*Alcedo atthis*
Green Kingfisher	*Chloroceryle americana*
Guam Micronesian Kingfisher	*Todirhamphus cinnamominus cinnamominus*
Pied Kingfisher	*Ceryle rudis*
Ringed Kingfisher	*Ceryle torquata*

Kinglet
Golden-crowned Kinglet	*Regulus satrapa*
Ruby-crowned Kinglet	*Regulus calendula*
Kirtland's Warbler	*Dendroica kirtlandii*

Kiskadee
Great Kiskadee	*Pitangus sulphuratus*

Kite
Black Kite	*Milvus migrans*
Double-toothed Kite	*Harpagus bidentatus*
Hook-billed Kite	*Chondrohierax uncinatus*
Snail Kite	*Rostrhamus sociabilis*
Swallow-tailed Kite	*Elanoides forficatus*
White-tailed Kite	*Elanus leucurus*

Kittiwake
Black-legged Kittiwake	*Rissa tridactyla*

Kiwi
Brown Kiwi	*Apteryx australis*
Great Spotted Kiwi	*Apteryx haastii*
Little Spotted Kiwi	*Apteryx owenii*

Knot
Red Knot	*Calidris canutus*

Koel
Long-tailed Koel	*Eudynamys taitensis*
Kona Grosbeak	*Chloridops kona*
Kori Bustard	*Ardeotis kori*

L

Labrador Duck	*Camptorhynchus labradorius*
Ladder-backed Woodpecker	*Picoides scalaris*
Lammergeier	*Gypaetus barbatus*
Lapland Longspur	*Calcarius lapponicus*
Lappet-faced Vulture	*Torgos tracheliotus*

Lapwing
Northern Lapwing	*Vanellus vanellus*
Large Cactus-Finch	*Geospiza conirostris*
Large Ground-Finch	*Geospiza magnirostris*
Large Tree-Finch	*Camarhynchus psittacula*

Lark
Horned Lark	*Eremophila alpestris*
Sky Lark	*Alauda arvensis*
Lark Bunting	*Calamospiza melanocorys*
Lark Sparrow	*Chondestes grammacus*
Laughing Gull	*Larus atricilla*
Lawrence's Thrush	*Turdus lawrencii*
Laysan Albatross	*Phoebastria immutabilis*
Laysan Finch	*Telespiza cantans*

Lazuli Bunting	*Passerina amoena*
Le Conte's Thrasher	*Toxostoma lecontei*
Leach's Storm-Petrel	*Oceanodroma leucorhoa*
Least Bittern	*Ixobrychus exilis*
Least Flycatcher	*Empidonax minimus*
Least Sandpiper	*Calidris minutilla*
Least Tern	*Sterna antillarum*
Lemon-spectacled Tanager	*Chlorothraupis olivacea*
Lesser Akialoa	*Hemignathus obscurus*
Lesser Black-backed Gull	*Larus fuscus*
Lesser Flamingo	*Phoenicopterus minor*
Lesser Goldfinch	*Carduelis psaltria*
Lesser Greenlet	*Hylophilus decurtatus*
Lesser Nighthawk	*Chordeiles acutipennis*
Lesser Prairie-Chicken	*Tympanuchus pallidicinctus*
Lesser Roadrunner	*Geococcyx velox*
Lesser Scaup	*Aythya affinis*
Lesser Spotted Woodpecker	*Dendrocopos minor*
Lesser Striped-Swallow	*Hirundo abyssinica*
Lesser White-fronted Goose	*Anser erythropus*
Lesser Yellowlegs	*Tringa flavipes*
Lettered Aracari	*Pteroglossus inscriptus*
Lewis's Woodpecker	*Melanerpes lewis*
Light-footed Clapper Rail	*Rallus longirostris levipes*
Lilac-tailed Parrotlet	*Touit batavica*
Limpkin	*Aramus guarauna*
Lincoln's Sparrow	*Melospiza lincolnii*
Little Blue Heron	*Egretta caerulea*
Little Grebe	*Tachybaptus ruficollis*
Little Hermit	*Phaethornis longuemareus*
Little Owl	*Athene noctua*
Little Penguin	*Eudyptula minor*
Little Spotted Kiwi	*Apteryx owenii*
Little Swift	*Apus affinis*
Loggerhead Shrike	*Lanius ludovicianus*
Long-billed Curlew	*Numenius americanus*
Long-eared Owl	*Asio otus*
Long-tailed Duck	*Clangula hyemalis*
Long-tailed Hermit	*Phaethornis superciliosus*
Long-tailed Jaeger	*Stercorarius longicaudus*
Long-tailed Koel	*Eudynamys taitensis*
Long-tailed Manakin	*Chiroxiphia linearis*
Long-tailed Tit	*Aegithalos caudatus*
Long-tailed Tyrant	*Colonia colonus*
Long-wattled Umbrellabird	*Cephalopterus penduliger*

Longclaw
Yellow-throated Longclaw	*Macronyx croceus*

Longspur
Chestnut-collared Longspur	*Calcarius ornatus*
Lapland Longspur	*Calcarius lapponicus*
McCown's Longspur	*Calcarius mccownii*
Smith's Longspur	*Calcarius pictus*

Loon
Arctic Loon	*Gavia arctica*
Common Loon	*Gavia immer*
Red-throated Loon	*Gavia stellata*

Lory
Violet-necked Lory	*Eos squamata*
Louisiana Waterthrush	*Seiurus motacilla*
Lucy's Warbler	*Vermivora luciae*

Lyrebird
Superb Lyrebird	*Menura novaehollandiae*

M

Macaroni Penguin	*Eudyptes chrysolophus*

Macaw
Blue-and-yellow Macaw	*Ara ararauna*
Hyacinth Macaw	*Anodorhynchus hyacinthinus*
Red-and-green Macaw	*Ara chloroptera*

Scarlet Macaw	*Ara macao*
Spix's Macaw	*Cyanopsitta spixii*
MacGillivray's Warbler	*Oporornis tolmiei*
Macgregor's Bowerbird	*Amblyornis macgregoriae*
Madagascar Pochard	*Aythya innotata*
Madagascar Serpent-Eagle	*Eutriorchis astur*
Magellanic Plover	*Pluvianellus socialis*
Magnificent Bird-of-paradise	*Cicinnurus magnificus*
Magnificent Frigatebird	*Fregata magnificens*
Magnificent Hummingbird	*Eugenes fulgens*
Magnolia Warbler	*Dendroica magnolia*
Magpie	
Australasian Magpie	*Gymnorhina tibicen*
Black-billed Magpie	*Pica pica*
Magpie-lark	*Grallina cyanoleuca*
Malachite Sunbird	*Nectarinia famosa*
Maleo	*Macrocephalon maleo*
Malimbe	
Red-vented Malimbe	*Malimbus scutatus*
Mallard	*Anas platyrhynchos*
Malleefowl	*Leipoa ocellata*
Mamo	
Black Mamo	*Drepanis funerea*
Hawaii Mamo	*Drepanis pacifica*
Manakin	
Blue Manakin	*Chiroxiphia caudata*
Long-tailed Manakin	*Chiroxiphia linearis*
Opal-crowned Manakin	*Pipra iris*
White-bearded Manakin	*Manacus manacus*
Mandarin Duck	*Aix galericulata*
Mangrove Cuckoo	*Coccyzus minor*
Mangrove Finch	*Camarhynchus heliobates*
Manx Shearwater	*Puffinus puffinus*
Many-colored Rush-Tyrant	*Tachuris rubrigastra*
Marsh Warbler	*Acrocephalus palustris*
Marsh Wren	*Cistothorus palustris*
Martin	
Banded Martin	*Riparia cincta*
Brown-chested Martin	*Progne tapera*
House Martin	*Delichon urbica*
Purple Martin	*Progne subis*
Rock Martin	*Hirundo fuligula*
Marvelous Spatuletail	*Loddigesia mirabilis*
Masked Booby	*Sula dactylatra*
Masked-Weaver	
African Masked-Weaver	*Ploceus velatus*
Northern Masked-Weaver	*Ploceus taeniopterus*
Maui Parrotbill	*Pseudonestor xanthophrys*
Mauritius Kestrel	*Falco punctatus*
Mauritius Parakeet	*Psittacula echo*
McCown's Longspur	*Calcarius mccownii*
Meadowlark	
Eastern Meadowlark	*Sturnella magna*
Western Meadowlark	*Sturnella neglecta*
Mealy Parrot	*Amazona farinosa*
Medium Ground-Finch	*Geospiza fortis*
Medium Tree-Finch	*Camarhynchus pauper*
Merganser	
Common Merganser	*Mergus merganser*
Hooded Merganser	*Lophodytes cucullatus*
Red-breasted Merganser	*Mergus serrator*
Merlin	*Falco columbarius*
Mew Gull	*Larus canus*
Mexican Chickadee	*Poecile sclateri*
Mexican Jay	*Aphelocoma ultramarina*
Mockingbird	
Northern Mockingbird	*Mimus polyglottos*
Socorro Mockingbird	*Mimodes graysoni*
Monarch	
Rarotonga Monarch	*Pomarea dimidiata*

Monk Parakeet	*Myiopsitta monachus*
Montezuma Oropendola	*Gymnostinops montezuma*
Moorhen	
Common Moorhen	*Gallinula chloropus*
Motmot	
Turquoise-browed Motmot	*Eumomota superciliosa*
Mountain Bluebird	*Sialia currucoides*
Mountain Chickadee	*Poecile gambeli*
Mourning Dove	*Zenaida macroura*
Mousebird	
Speckled Mousebird	*Colius striatus*
Murre	
Common Murre	*Uria aalge*
Thick-billed Murre	*Uria lomvia*
Murrelet	
Ancient Murrelet	*Synthliboramphus antiquus*
Craveri's Murrelet	*Synthliboramphus craveri*
Muscovy Duck	*Cairina moschata*
Musician Wren	*Cyphorhinus aradus*
Mute Swan	*Cygnus olor*
Myna	
Bali Myna	*Leucopsar rothschildi*

N

Namaqua Sandgrouse	*Pterocles namaqua*
Nashville Warbler	*Vermivora ruficapilla*
New Caledonian Crow	*Corvus moneduloides*
Nicobar Pigeon	*Caloenas nicobarica*
Night-Heron	
Black-crowned Night-Heron	*Nycticorax nycticorax*
Yellow-crowned Night-Heron	*Nyctanassa violacea*
Nighthawk	
Common Nighthawk	*Chordeiles minor*
Lesser Nighthawk	*Chordeiles acutipennis*
Nightingale	
Common Nightingale	*Luscinia megarhynchos*
Nightjar	
Eurasian Nightjar	*Caprimulgus europaeus*
Standard-winged Nightjar	*Macrodipteryx longipennis*
Noddy	
Brown Noddy	*Anous solidus*
Northern Barred-Woodcreeper	*Dendrocolaptes sanctithomae*
Northern Bobwhite	*Colinus virginianus*
Northern Cardinal	*Cardinalis cardinalis*
Northern Cassowary	*Casuarius unappendiculatus*
Northern Flicker	*Colaptes auratus*
Northern Fulmar	*Fulmarus glacialis*
Northern Gannet	*Morus bassanus*
Northern Goshawk	*Accipiter gentilis*
Northern Harrier	*Circus cyaneus*
Northern Jacana	*Jacana spinosa*
Northern Lapwing	*Vanellus vanellus*
Northern Masked-Weaver	*Ploceus taeniopterus*
Northern Mockingbird	*Mimus polyglottos*
Northern Parula	*Parula americana*
Northern Pintail	*Anas acuta*
Northern Pygmy-Owl	*Glaucidium gnoma*
Northern Rough-winged Swallow	*Stelgidopteryx serripennis*
Northern Screamer	*Chauna chavaria*
Northern Shoveler	*Anas clypeata*
Northern Shrike	*Lanius excubitor*
Northern Spotted Owl	*Strix occidentalis caurina*
Northern Waterthrush	*Seiurus noveboracensis*
Northwestern Crow	*Corvus caurinus*
Nukupuu	*Hemignathus lucidus*
Nunbird	
Black-fronted Nunbird	*Monasa nigrifrons*
Nutcracker	
Clark's Nutcracker	*Nucifraga columbiana*

Nuthatch
  Pygmy Nuthatch   *Sitta pygmaea*
  Red-breasted Nuthatch   *Sitta canadensis*
  Rock Nuthatch   *Sitta neumayer*
  White-breasted Nuthatch   *Sitta carolinensis*
Nuttall's Woodpecker   *Picoides nuttallii*

O

Oak Titmouse   *Baeolophus inornatus*
Ocellated Antbird   *Phaenostictus mcleannani*
Oilbird   *Steatornis caripensis*
Oldsquaw   *Clangula hyemalis*
Olive Warbler   *Peucedramus taeniatus*
Olive-sided Flycatcher   *Contopus cooperi*
Oo
  Hawaii Oo   *Moho nobilis*
  Kauai Oo   *Moho braccatus*
Opal-crowned Manakin   *Pipra iris*
Orange-crowned Warbler   *Vermivora celata*
Orchard Oriole   *Icterus spurius*
Oriole
  Altamira Oriole   *Icterus gularis*
  Baltimore Oriole   *Icterus galbula*
  Bullock's Oriole   *Icterus bullockii*
  Eurasian Golden Oriole   *Oriolus oriolus*
  Hooded Oriole   *Icterus cucullatus*
  Orchard Oriole   *Icterus spurius*
Oropendola
  Chestnut-headed Oropendola   *Psarocolius wagleri*
  Montezuma Oropendola   *Gymnostinops montezuma*
Osprey   *Pandion haliaetus*
Ostrich   *Struthio camelus*
Ou   *Psittirostra psittacea*
Ovenbird   *Seiurus aurocapillus*
Owl
  Barn Owl   *Tyto alba*
  Barred Owl   *Strix varia*
  Boreal Owl   *Aegolius funereus*
  Burrowing Owl   *Athene cunicularia*
  Elf Owl   *Micrathene whitneyi*
  Great Horned Owl   *Bubo virginianus*
  Little Owl   *Athene noctua*
  Long-eared Owl   *Asio otus*
  Northern Spotted Owl   *Strix occidentalis caurina*
  Short-eared Owl   *Asio flammeus*
  Snowy Owl   *Nyctea scandiaca*
  Spotted Owl   *Strix occidentalis*
  Tawny Owl   *Strix aluco*
Oystercatcher
  American Oystercatcher   *Haematopus palliatus*

P

Pacific-slope Flycatcher   *Empidonax difficilis*
Painted Bunting   *Passerina ciris*
Painted Redstart   *Myioborus pictus*
Palm Warbler   *Dendroica palmarum*
Palm-Swift
  African Palm-Swift   *Cypsiurus parvus*
Palmchat   *Dulus dominicus*
Papuan Frogmouth   *Podargus papuensis*
Paradise Tanager   *Tangara chilensis*
Parakeet
  Carolina Parakeet   *Conuropsis carolinensis*
  Cobalt-winged Parakeet   *Brotogeris cyanoptera*
  Mauritius Parakeet   *Psittacula echo*
  Monk Parakeet   *Myiopsitta monachus*
Parasitic Jaeger   *Stercorarius parasiticus*

Parasitic Weaver   *Anomalospiza imberbis*
Parrot
  Blue-fronted Parrot   *Amazona aestiva*
  Grey Parrot   *Psittacus erithacus*
  Imperial Parrot   *Amazona imperialis*
  Mealy Parrot   *Amazona farinosa*
  White-fronted Parrot   *Amazona albifrons*
  Yellow-headed Parrot   *Amazona oratrix*
Parrotbill
  Maui Parrotbill   *Pseudonestor xanthophrys*
Parrotlet
  Green-rumped Parrotlet   *Forpus passerinus*
  Lilac-tailed Parrotlet   *Touit batavica*
  Sapphire-rumped Parrotlet   *Touit purpurata*
  Spectacled Parrotlet   *Forpus conspicillatus*
Partridge
  Gray Partridge   *Perdix perdix*
  Rock Partridge   *Alectoris graeca*
Parula
  Northern Parula   *Parula americana*
Passenger Pigeon   *Ectopistes migratorius*
Pavonine Cuckoo   *Dromococcyx pavoninus*
Peafowl
  Indian Peafowl   *Pavo cristatus*
Pearly-breasted Cuckoo   *Coccyzus euleri*
Pelican
  American White Pelican   *Pelecanus erythrorhynchos*
  Brown Pelican   *Pelecanus occidentalis*
  Dalmatian Pelican   *Pelecanus crispus*
  Great White Pelican   *Pelecanus onocrotalus*
  Peruvian Pelican   *Pelecanus thagus*
Penduline-Tit
  Eurasian Penduline-Tit   *Remiz pendulinus*
  Southern Penduline-Tit   *Anthoscopus minutus*
Penguin
  Adelie Penguin   *Pygoscelis adeliae*
  Chinstrap Penguin   *Pygoscelis antarctica*
  Emperor Penguin   *Aptenodytes forsteri*
  Gentoo Penguin   *Pygoscelis papua*
  Humboldt Penguin   *Spheniscus humboldti*
  Jackass Penguin   *Spheniscus demersus*
  King Penguin   *Aptenodytes patagonicus*
  Little Penguin   *Eudyptula minor*
  Macaroni Penguin   *Eudyptes chrysolophus*
  Yellow-eyed Penguin   *Megadyptes antipodes*
Peppershrike
  Rufous-browed Peppershrike   *Cyclarhis gujanensis*
Peregrine Falcon   *Falco peregrinus*
Peruvian Booby   *Sula variegata*
Peruvian Pelican   *Pelecanus thagus*
Petrel
  Bonin Petrel   *Pterodroma hypoleuca*
  Cape Petrel   *Daption capense*
Pewee
  Greater Pewee   *Contopus pertinax*
Phainopepla   *Phainopepla nitens*
Phalarope
  Red-necked Phalarope   *Phalaropus lobatus*
  Wilson's Phalarope   *Phalaropus tricolor*
Pheasant
  Ring-necked Pheasant   *Phasianus colchicus*
Pheasant Cuckoo   *Dromococcyx phasianellus*
Philadelphia Vireo   *Vireo philadelphicus*
Phoebe
  Eastern Phoebe   *Sayornis phoebe*
  Say's Phoebe   *Sayornis saya*
Pied Currawong   *Strepera graculina*
Pied Kingfisher   *Ceryle rudis*
Pied-billed Grebe   *Podilymbus podiceps*
Pigeon

Cinnamon-headed Pigeon	*Treron fulvicollis*
Nicobar Pigeon	*Caloenas nicobarica*
Passenger Pigeon	*Ectopistes migratorius*
Pigeon Guillemot	*Cepphus columba*
Pileated Woodpecker	*Dryocopus pileatus*
Pin-tailed Whydah	*Vidua macroura*
Pine Grosbeak	*Pinicola enucleator*
Pine Siskin	*Carduelis pinus*
Pine Warbler	*Dendroica pinus*
Pink-throated Brilliant	*Heliodoxa gularis*
Pintail	
Northern Pintail	*Anas acuta*
Pinyon Jay	*Gymnorhinus cyanocephalus*
Piping Plover	*Charadrius melodus*
Pipit	
Sprague's Pipit	*Anthus spragueii*
Water Pipit	*Anthus spinoletta*
Plain Wren	*Thryothorus modestus*
Plain-brown Woodcreeper	*Dendrocincla fuliginosa*
Plains-wanderer	*Pedionomus torquatus*
Planalto Hermit	*Phaethornis pretrei*
Plover	
Blacksmith Plover	*Vanellus armatus*
Crab Plover	*Dromas ardeola*
Egyptian Plover	*Pluvianus aegyptius*
Magellanic Plover	*Pluvianellus socialis*
Piping Plover	*Charadrius melodus*
Snowy Plover	*Charadrius alexandrinus*
Wilson's Plover	*Charadrius wilsonia*
Plumed Guineafowl	*Guttera plumifera*
Plush-capped Finch	*Catamblyrhynchus diadema*
Pochard	
Madagascar Pochard	*Aythya innotata*
Pomarine Jaeger	*Stercorarius pomarinus*
Pompadour Cotinga	*Xipholena punicea*
Poo-uli	*Melamprosops phaeosoma*
Poorwill	
Common Poorwill	*Phalaenoptilus nuttallii*
Potoo	
Great Potoo	*Nyctibius grandis*
Prairie Falcon	*Falco mexicanus*
Prairie Warbler	*Dendroica discolor*
Prairie-Chicken	
Greater Prairie-Chicken	*Tympanuchus cupido*
Lesser Prairie-Chicken	*Tympanuchus pallidicinctus*
Prothonotary Warbler	*Protonotaria citrea*
Ptarmigan	
Rock Ptarmigan	*Lagopus mutus*
White-tailed Ptarmigan	*Lagopus leucurus*
Willow Ptarmigan	*Lagopus lagopus*
Puffin	
Atlantic Puffin	*Fratercula arctica*
Horned Puffin	*Fratercula corniculata*
Tufted Puffin	*Fratercula cirrhata*
Purple Finch	*Carpodacus purpureus*
Purple Gallinule	*Porphyrula martinica*
Purple Indigobird	*Vidua purpurascens*
Purple Martin	*Progne subis*
Purple-backed Sunbeam	*Aglaeactics alciae*
Purple-bearded Bee-eater	*Meropogon forsteni*
Purpletuft	
Buff-throated Purpletuft	*Iodopleura pipra*
Pygmy Nuthatch	*Sitta pygmaea*
Pygmy-Owl	
Ferruginous Pygmy-Owl	*Glaucidium brasilianum*
Northern Pygmy-Owl	*Glaucidium gnoma*
Pygmy-Tyrant	
Short-tailed Pygmy-Tyrant	*Myiornis ecaudatus*

Q

Quail	
California Quail	*Callipepla californica*
Gambel's Quail	*Callipepla gambelii*
Japanese Quail	*Coturnix japonica*
Quelea	
Red-billed Quelea	*Quelea quelea*
Quetzal	
Resplendent Quetzal	*Pharomachrus mocinno*

R

Rail	
Clapper Rail	*Rallus longirostris*
Guam Rail	*Gallirallus owstoni*
King Rail	*Rallus elegans*
Light-footed Clapper Rail	*Rallus longirostris levipes*
Virginia Rail	*Rallus limicola*
Rarotonga Monarch	*Pomarea dimidiata*
Raven	
Australian Raven	*Corvus coronoides*
Common Raven	*Corvus corax*
White-necked Raven	*Corvus albicollis*
Razorbill	*Alca torda*
Red Crossbill	*Loxia curvirostra*
Red Fody	*Foudia madagascariensis*
Red Junglefowl	*Gallus gallus*
Red Knot	*Calidris canutus*
Red-and-green Macaw	*Ara chloroptera*
Red-bellied Woodpecker	*Melanerpes carolinus*
Red-billed Firefinch	*Lagonosticta senegala*
Red-billed Hornbill	*Tockus erythrorhynchus*
Red-billed Quelea	*Quelea quelea*
Red-billed Tropicbird	*Phaethon aethereus*
Red-breasted Merganser	*Mergus serrator*
Red-breasted Nuthatch	*Sitta canadensis*
Red-cockaded Woodpecker	*Picoides borealis*
Red-eyed Vireo	*Vireo olivaceus*
Red-faced Warbler	*Cardellina rubrifrons*
Red-footed Booby	*Sula sula*
Red-fronted Coot	*Fulica rufifrons*
Red-gartered Coot	*Fulica armillata*
Red-headed Woodpecker	*Melanerpes erythrocephalus*
Red-naped Sapsucker	*Sphyrapicus nuchalis*
Red-necked Grebe	*Podiceps grisegena*
Red-necked Phalarope	*Phalaropus lobatus*
Red-shouldered Hawk	*Buteo lineatus*
Red-tailed Hawk	*Buteo jamaicensis*
Red-throated Loon	*Gavia stellata*
Red-vented Malimbe	*Malimbus scutatus*
Red-winged Blackbird	*Agelaius phoeniceus*
Red-winged Tinamou	*Rhynchotus rufescens*
Reddish Egret	*Egretta rufescens*
Redhead	*Aythya americana*
Redpoll	
Common Redpoll	*Carduelis flammea*
Redshank	
Common Redshank	*Tringa totanus*
Redstart	
American Redstart	*Setophaga ruticilla*
Common Redstart	*Phoenicurus phoenicurus*
Painted Redstart	*Myioborus pictus*
Saffron-breasted Redstart	*Myioborus cardonai*
Reed-Warbler	
Eurasian Reed-Warbler	*Acrocephalus scirpaceus*
Great Reed-Warbler	*Acrocephalus arundinaceus*
Resplendent Quetzal	*Pharomachrus mocinno*
Reunion Solitaire	*Raphus solitarius*

Rhea
    Greater Rhea — *Rhea americana*
Rhinoceros Hornbill — *Buceros rhinoceros*
Rifleman — *Acanthisitta chloris*
Ring-billed Gull — *Larus delawarensis*
Ring-necked Pheasant — *Phasianus colchicus*
Ringed Antpipit — *Corythopis torquata*
Ringed Kingfisher — *Ceryle torquata*
Riverside Wren — *Thryothorus semibadius*
Roadrunner
    Greater Roadrunner — *Geococcyx californianus*
    Lesser Roadrunner — *Geococcyx velox*
Robin
    American Robin — *Turdus migratorius*
    Chatham Robin — *Petroica traversi*
    European Robin — *Erithacus rubecula*
Rock Dove — *Columba livia*
Rock Martin — *Hirundo fuligula*
Rock Nuthatch — *Sitta neumayer*
Rock Partridge — *Alectoris graeca*
Rock Ptarmigan — *Lagopus mutus*
Rock Wren — *Salpinctes obsoletus*
Rodrigues Solitaire — *Pezophaps solitaria*
Rose-breasted Grosbeak — *Pheucticus ludovicianus*
Roseate Spoonbill — *Ajaia ajaja*
Roseate Tern — *Sterna dougallii*
Rosella
    Crimson Rosella — *Platycercus elegans*
Ross's Goose — *Chen rossii*
Ross's Gull — *Rhodostethia rosea*
Rough-legged Hawk — *Buteo lagopus*
Royal Albatross — *Diomedea epomophora*
Royal Flycatcher — *Onychorhynchus coronatus*
Royal Tern — *Sterna maxima*
Ruby-crowned Kinglet — *Regulus calendula*
Ruby-throated Hummingbird — *Archilochus colubris*
Ruby-topaz Hummingbird — *Chrysolampis mosquitus*
Ruddy Duck — *Oxyura jamaicensis*
Ruddy Turnstone — *Arenaria interpres*
Rueppell's Griffon — *Gyps rueppellii*
Ruff — *Philomachus pugnax*
Ruffed Grouse — *Bonasa umbellus*
Rufous Hornero — *Furnarius rufus*
Rufous Hummingbird — *Selasphorus rufus*
Rufous-and-white Wren — *Thryothorus rufalbus*
Rufous-browed Peppershrike — *Cyclarhis gujanensis*
Rufous-chested Swallow — *Hirundo semirufa*
Rufous-collared Sparrow — *Zonotrichia capensis*
Rufous-naped Wren — *Campylorhynchus rufinucha*
Rufous-vented Ground-Cuckoo — *Neomorphus geoffroyi*
Rush-Tyrant
    Many-colored Rush-Tyrant — *Tachuris rubrigastra*

S

Sabine's Gull — *Xema sabini*
Sabrewing
    Violet Sabrewing — *Campylopterus hemileucurus*
Saffron-breasted Redstart — *Myioborus cardonai*
Sage Grouse — *Centrocercus urophasianus*
Sage Sparrow — *Amphispiza belli*
Sage Thrasher — *Oreoscoptes montanus*
Saltmarsh Sharp-tailed Sparrow — *Ammodramus caudacutus*
Sanderling — *Calidris alba*
Sandgrouse
    Namaqua Sandgrouse — *Pterocles namaqua*
Sandhill Crane — *Grus canadensis*
Sandpiper
    Buff-breasted Sandpiper — *Tryngites subruficollis*
    Least Sandpiper — *Calidris minutilla*
    Semipalmated Sandpiper — *Calidris pusilla*
    Solitary Sandpiper — *Tringa solitaria*
    Spotted Sandpiper — *Actitis macularia*
    Upland Sandpiper — *Bartramia longicauda*
    Western Sandpiper — *Calidris mauri*
Sandwich Tern — *Sterna sandvicensis*
Sapphire-rumped Parrotlet — *Touit purpurata*
Sapsucker
    Red-naped Sapsucker — *Sphyrapicus nuchalis*
    Williamson's Sapsucker — *Sphyrapicus thyroideus*
    Yellow-bellied Sapsucker — *Sphyrapicus varius*
Satin Bowerbird — *Ptilonorhynchus violaceus*
Savannah Sparrow — *Passerculus sandwichensis*
Sawwing
    Black Sawwing — *Psalidoprocne holomelas*
Say's Phoebe — *Sayornis saya*
Scarlet Ibis — *Eudocimus ruber*
Scarlet Macaw — *Ara macao*
Scarlet Tanager — *Piranga olivacea*
Scarlet-chested Sunbird — *Chalcomitra senegalensis*
Scarlet-hooded Barbet — *Eubucco tucinkae*
Scaup
    Greater Scaup — *Aythya marila*
    Lesser Scaup — *Aythya affinis*
Schalow's Turaco — *Tauraco schalowi*
Scissor-tailed Flycatcher — *Tyrannus forficatus*
Scoter
    Black Scoter — *Melanitta nigra*
    Surf Scoter — *Melanitta perspicillata*
    White-winged Scoter — *Melanitta fusca*
Screamer
    Northern Screamer — *Chauna chavaria*
Screaming Cowbird — *Molothrus rufoaxillaris*
Screech-Owl
    Eastern Screech-Owl — *Otus asio*
    Western Screech-Owl — *Otus kennicottii*
Scrub-Jay
    Florida Scrub-Jay — *Aphelocoma coerulescens*
    Western Scrub-Jay — *Aphelocoma californica*
Scrubwren
    White-browed Scrubwren — *Sericornis frontalis*
Scythebill
    Black-billed Scythebill — *Campylorhamphus falcularius*
Seaside Sparrow — *Ammodramus maritimus*
Secretary-bird — *Sagittarius serpentarius*
Sedge Wren — *Cistothorus platensis*
Semipalmated Sandpiper — *Calidris pusilla*
Serpent-Eagle
    Madagascar Serpent-Eagle — *Eutriorchis astur*
Seychelles Brush-Warbler — *Acrocephalus sechellensis*
Sharp-beaked Ground-Finch — *Geospiza difficilis*
Sharp-shinned Hawk — *Accipter striatus*
Sharp-tailed Grouse — *Tympanuchus phasianellus*
Sharp-tailed Tyrant — *Culicivora caudacuta*
Sharpbill — *Oxyruncus cristatus*
Shearwater
    Audubon's Shearwater — *Puffinus lherminieri*
    Manx Shearwater — *Puffinus puffinus*
    Short-tailed Shearwater — *Puffinus tenuirostris*
    Sooty Shearwater — *Puffinus griseus*
Sheathbill
    Snowy Sheathbill — *Chionis alba*
Shelduck
    Common Shelduck — *Tadorna tadorna*
Shining Honeycreeper — *Cyanerpes lucidus*
Shiny Cowbird — *Molothrus bonariensis*
Shoebill — *Balaeniceps rex*
Short-eared Owl — *Asio flammeus*
Short-tailed Field-Tyrant — *Muscigralla brevicauda*
Short-tailed Hawk — *Buteo brachyurus*

Short-tailed Pygmy-Tyrant	*Myiornis ecaudatus*
Short-tailed Shearwater	*Puffinus tenuirostris*
Shoveler	
Northern Shoveler	*Anas clypeata*
Shrike	
Loggerhead Shrike	*Lanius ludovicianus*
Northern Shrike	*Lanius excubitor*
Shrike-Tanager	
White-winged Shrike-Tanager	*Lanio versicolor*
Sicklebill	
White-tipped Sicklebill	*Eutoxeres aquila*
Silvery-throated Spinetail	*Synallaxis subpudica*
Siskin	
Pine Siskin	*Carduelis pinus*
Sittella	
Varied Sittella	*Neositta chrysoptera*
Skimmer	
Black Skimmer	*Rynchops niger*
Skua	
Brown Skua	*Catharacta antarctica*
Great Skua	*Catharacta skua*
South Polar Skua	*Catharacta maccormicki*
Sky Lark	*Alauda arvensis*
Slaty-tailed Trogon	*Trogon massena*
Small Ground-Finch	*Geospiza fuliginosa*
Small Tree-Finch	*Camarhynchus parvulus*
Smith's Longspur	*Calcarius pictus*
Snail Kite	*Rostrhamus sociabilis*
Snipe	
Common Snipe	*Gallinago gallinago*
Great Snipe	*Gallinago media*
Snow Bunting	*Plectrophenax nivalis*
Snow Goose	*Chen caerulescens*
Snowy Egret	*Egretta thula*
Snowy Owl	*Nyctea scandiaca*
Snowy Plover	*Charadrius alexandrinus*
Snowy Sheathbill	*Chionis alba*
Social Weaver	*Philetairus socius*
Socorro Dove	*Zenaida graysoni*
Socorro Mockingbird	*Mimodes graysoni*
Socorro Wren	*Thryomanes sissonii*
Solitaire	
Reunion Solitaire	*Raphus solitarius*
Rodrigues Solitaire	*Pezophaps solitaria*
Solitary Sandpiper	*Tringa solitaria*
Song Sparrow	*Melospiza melodia*
Song Thrush	*Turdus philomelos*
Sooty Shearwater	*Puffinus griseus*
Sooty Tern	*Sterna fuscata*
Sooty-capped Hermit	*Phaethornis augusti*
Sora *Porzana carolina*	
South African Swallow	*Hirundo spilodera*
South Polar Skua	*Catharacta maccormicki*
Southern Cassowary	*Casuarius casuarius*
Southern Ground-Hornbill	*Bucorvus leadbeateri*
Southern Penduline-Tit	*Anthoscopus minutus*
Spadebill	
White-crested Spadebill	*Platyrinchus platyrhynchos*
Sparrow	
American Tree Sparrow	*Spizella arborea*
Baird's Sparrow	*Ammodramus bairdii*
Belding's Savannah Sparrow	*Passerculus sandwichensis beldingi*
Black-throated Sparrow	*Amphispiza bilineata*
Brewer's Sparrow	*Spizella breweri*
Chipping Sparrow	*Spizella passerina*
Dusky Seaside Sparrow	*Ammodramus maritimus nigrescens*
Eurasian Tree Sparrow	*Passer montanus*
Field Sparrow	*Spizella pusilla*

Fox Sparrow	*Passerella iliaca*
Grasshopper Sparrow	*Ammodramus savannarum*
Harris's Sparrow	*Zonotrichia querula*
Henslow's Sparrow	*Ammodramus henslowii*
House Sparrow	*Passer domesticus*
Lark Sparrow	*Chondestes grammacus*
Lincoln's Sparrow	*Melospiza lincolnii*
Rufous-collared Sparrow	*Zonotrichia capensis*
Sage Sparrow	*Amphispiza belli*
Saltmarsh Sharp-tailed Sparrow	*Ammodramus caudacutus*
Savannah Sparrow	*Passerculus sandwichensis*
Seaside Sparrow	*Ammodramus maritimus*
Song Sparrow	*Melospiza melodia*
Swamp Sparrow	*Melospiza georgiana*
Vesper Sparrow	*Pooecetes gramineus*
White-crowned Sparrow	*Zonotrichia leucophrys*
White-throated Sparrow	*Zonotrichia albicollis*
Sparrowhawk	
Eurasian Sparrowhawk	*Accipiter nisus*
Spatuletail	
Marvelous Spatuletail	*Loddigesia mirabilis*
Speckled Mousebird	*Colius striatus*
Spectacled Parrotlet	*Forpus conspicillatus*
Spectacled Tyrant	*Hymenops perspicillatus*
Spinetail	
Silvery-throated Spinetail	*Synallaxis subpudica*
Spix's Macaw	*Cyanopsitta spixii*
Spoonbill	
Roseate Spoonbill	*Ajaia ajaja*
Spotted Antbird	*Hylophylax naevioides*
Spotted Forktail	*Enicurus maculatus*
Spotted Owl	*Strix occidentalis*
Spotted Sandpiper	*Actitis macularia*
Spotted Towhee	*Pipilo maculatus*
Sprague's Pipit	*Anthus spragueii*
Spruce Grouse	*Falcipennis canadensis*
Standard-winged Nightjar	*Macrodipteryx longipennis*
Standardwing	
Wallace's Standardwing	*Semioptera wallacii*
Starling	
European Starling	*Sturnus vulgaris*
Steller's Eider	*Polysticta stelleri*
Steller's Jay	*Cyanocitta stelleri*
Stephens Island Wren	*Xenicus lyalli*
Stilt	
Black-necked Stilt	*Himantopus mexicanus*
Stork	
White Stork	*Ciconia ciconia*
Wood Stork	*Mycteria americana*
Storm-Petrel	
Leach's Storm-Petrel	*Oceanodroma leucorhoa*
Wilson's Storm-Petrel	*Oceanites oceanicus*
Strange-tailed Tyrant	*Alectrurus risora*
Strickland's Woodpecker	*Picoides stricklandi*
Stripe-backed Wren	*Campylorhynchus nuchalis*
Striped Cuckoo	*Tapera naevia*
Striped-Swallow	
Greater Striped-Swallow	*Hirundo cucullata*
Lesser Striped-Swallow	*Hirundo abyssinica*
Sulphur-crested Cockatoo	*Cacatua galerita*
Summer Tanager	*Piranga rubra*
Sunbeam	
Purple-backed Sunbeam	*Aglaeactis alciae*
Sunbird	
Malachite Sunbird	*Nectarinia famosa*
Scarlet-chested Sunbird	*Chalcomitra senegalensis*
Sunbittern	*Eurypyga helias*
Sungrebe	*Heliornis fulica*
Superb Bird-of-paradise	*Lophorina superba*
Superb Fairywren	*Malurus cyaneus*

Superb Lyrebird	*Menura novaehollandiae*
Surf Scoter	*Melanitta perspicillata*
Swainson's Flycatcher	*Myiarchus swainsoni*
Swainson's Hawk	*Buteo swainsoni*
Swainson's Thrush	*Catharus ustulatus*
Swallow	
Bank Swallow	*Riparia riparia*
Barn Swallow	*Hirundo rustica*
Blue-and-white Swallow	*Pygochelidon cyanoleuca*
Cave Swallow	*Petrochelidon fulva*
Cliff Swallow	*Petrochelidon pyrrhonota*
Gray-rumped Swallow	*Hirundo griseopyga*
Northern Rough-winged Swallow	*Stelgidopteryx serripennis*
Rufous-chested Swallow	*Hirundo semirufa*
South African Swallow	*Hirundo spilodera*
Tree Swallow	*Tachycineta bicolor*
Violet-green Swallow	*Tachycineta thalassina*
Swallow-tailed Kite	*Elanoides forficatus*
Swamp Sparrow	*Melospiza georgiana*
Swan	
Mute Swan	*Cygnus olor*
Trumpeter Swan	*Cygnus buccinator*
Tundra Swan	*Cygnus columbianus*
Swift	
Alpine Swift	*Tachymarptis melba*
Chimney Swift	*Chaetura pelagica*
Common Swift	*Apus apus*
Little Swift	*Apus affinis*
Swiftlet	
Edible-nest Swiftlet	*Aerodramus fuciphagus*

T

Tailorbird	
Common Tailorbird	*Orthotomus sutorius*
Takahe	*Porphyrio mantelli*
Tanager	
Blue-gray Tanager	*Thraupis episcopus*
Gilt-edged Tanager	*Tangara cyanoventris*
Glistening-green Tanager	*Chlorochrysa phoenicotis*
Gray-headed Tanager	*Eucometis penicillata*
Lemon-spectacled Tanager	*Chlorothraupis olivacea*
Paradise Tanager	*Tangara chilensis*
Scarlet Tanager	*Piranga olivacea*
Summer Tanager	*Piranga rubra*
Western Tanager	*Piranga ludoviciana*
Yellow-crested Tanager	*Tachyphonus rufiventer*
Tawny Owl	*Strix aluco*
Teal	
Blue-winged Teal	*Anas discors*
Cinnamon Teal	*Anas cyanoptera*
Green-winged Teal	*Anas crecca*
Tennessee Warbler	*Vermivora peregrina*
Tern	
Arctic Tern	*Sterna paradisaea*
Black Tern	*Chlidonias niger*
Caspian Tern	*Sterna caspia*
Common Tern	*Sterna hirundo*
Forster's Tern	*Sterna forsteri*
Gull-billed Tern	*Sterna nilotica*
Least Tern	*Sterna antillarum*
Roseate Tern	*Sterna dougallii*
Royal Tern	*Sterna maxima*
Sandwich Tern	*Sterna sandvicensis*
Sooty Tern	*Sterna fuscata*
White Tern	*Gygis alba*
Thick-billed Euphonia	*Euphonia laniirostris*
Thick-billed Murre	*Uria lomvia*
Thick-knee	
Water Thick-knee	*Burhinus vermiculatus*

Thrasher	
Bendire's Thrasher	*Toxostoma bendirei*
Brown Thrasher	*Toxostoma rufum*
California Thrasher	*Toxostoma redivivum*
Crissal Thrasher	*Toxostoma crissale*
Le Conte's Thrasher	*Toxostoma lecontei*
Sage Thrasher	*Oreoscoptes montanus*
Three-toed Woodpecker	*Picoides tridactylus*
Thrush	
Bicknell's Thrush	*Catharus bicknelli*
Gray-cheeked Thrush	*Catharus minimus*
Hermit Thrush	*Catharus guttatus*
Lawrence's Thrush	*Turdus lawrencii*
Song Thrush	*Turdus philomelos*
Swainson's Thrush	*Catharus ustulatus*
Varied Thrush	*Ixoreus naevius*
Wood Thrush	*Hylocichla mustelina*
Tinamou	
Chilean Tinamou	*Nothoprocta perdicaria*
Red-winged Tinamou	*Rhynchotus rufescens*
Tit	
Blue Tit	*Cyanistes caeruleus*
Great Tit	*Parus major*
Long-tailed Tit	*Aegithalos caudatus*
Tit Berrypecker	*Oreocharis arfaki*
Titmouse	
Bridled Titmouse	*Baeolophus wollweberi*
Juniper Titmouse	*Baeolophus griseus*
Oak Titmouse	*Baeolophus inornatus*
Tufted Titmouse	*Baeolophus bicolor*
Torresian Imperial-Pigeon	*Ducula spilorrhoa*
Toucan	
Channel-billed Toucan	*Ramphastos vitellinus*
Choco Toucan	*Ramphastos brevis*
Keel-billed Toucan	*Ramphastos sulfuratus*
Towhee	
Abert's Towhee	*Pipilo aberti*
California Towhee	*Pipilo crissalis*
Canyon Towhee	*Pipilo fuscus*
Eastern Towhee	*Pipilo erythrophthalmus*
Green-tailed Towhee	*Pipilo chlorurus*
Spotted Towhee	*Pipilo maculatus*
Townsend's Warbler	*Dendroica townsendi*
Tree-Finch	
Large Tree-Finch	*Camarhynchus psittacula*
Medium Tree-Finch	*Camarhynchus pauper*
Small Tree-Finch	*Camarhynchus parvulus*
Tree Swallow	*Tachycineta bicolor*
Tricolored Blackbird	*Agelaius tricolor*
Tricolored Heron	*Egretta tricolor*
Trogon	
Black-headed Trogon	*Trogon melanocephalus*
Black-tailed Trogon	*Trogon melanurus*
Elegant Trogon	*Trogon elegans*
Slaty-tailed Trogon	*Trogon massena*
Violaceous Trogon	*Trogon violaceus*
Tropical Gnatcatcher	*Polioptila plumbea*
Tropicbird	
Red-billed Tropicbird	*Phaethon aethereus*
Trumpeter Swan	*Cygnus buccinator*
Tufted Puffin	*Fratercula cirrhata*
Tufted Titmouse	*Baeolophus bicolor*
Tui	*Prosthemadera novaeseelandiae*
Tundra Swan	*Cygnus columbianus*
Turaco	
Schalow's Turaco	*Tauraco schalowi*
Turkey	
Wild Turkey	*Meleagris gallopavo*
Turkey Vulture	*Cathartes aura*

Turnstone

Black Turnstone	*Arenaria melanocephala*
Ruddy Turnstone	*Arenaria interpres*
Turquoise-browed Motmot	*Eumomota superciliosa*

Tyrant

Cock-tailed Tyrant	*Alectrurus tricolor*
Long-tailed Tyrant	*Colonia colonus*
Sharp-tailed Tyrant	*Culicivora caudacuta*
Spectacled Tyrant	*Hymenops perspicillatus*
Strange-tailed Tyrant	*Alectrurus risora*

U

Ula-ai-hawane	*Ciridops anna*

Umbrellabird

Long-wattled Umbrellabird	*Cephalopterus penduliger*
Upland Sandpiper	*Bartramia longicauda*

V

Varied Sittella	*Neositta chrysoptera*
Varied Thrush	*Ixoreus naevius*
Veery	*Catharus fuscescens*
Vegetarian Finch	*Camarhynchus crassirostris*
Verdin	*Auriparus flaviceps*
Vermilion Flycatcher	*Pyrocephalus rubinus*
Vesper Sparrow	*Pooecetes gramineus*
Village Indigobird	*Vidua chalybeata*
Village Weaver	*Ploceus cucullatus*
Violaceous Trogon	*Trogon violaceus*
Violet Sabrewing	*Campylopterus hemileucurus*

Violet-ear

Green Violet-ear	*Colibri thalassinus*
Violet-green Swallow	*Tachycineta thalassina*
Violet-necked Lory	*Eos squamata*

Vireo

Bell's Vireo	*Vireo bellii*
Black-capped Vireo	*Vireo atricapillus*
Blue-headed Vireo	*Vireo solitarius*
Gray Vireo	*Vireo vicinior*
Hutton's Vireo	*Vireo huttoni*
Philadelphia Vireo	*Vireo philadelphicus*
Red-eyed Vireo	*Vireo olivaceus*
Warbling Vireo	*Vireo gilvus*
White-eyed Vireo	*Vireo griseus*
Yellow-throated Vireo	*Vireo flavifrons*
Virginia Rail	*Rallus limicola*
Virginia's Warbler	*Vermivora virginiae*
Vogelkop Bowerbird	*Amblyornis inornatus*
Volcano Hummingbird	*Selasphorus flammula*

Vulture

Black Vulture	*Coragyps atratus*
King Vulture	*Sarcoramphus papa*
Lappet-faced Vulture	*Torgos tracheliotus*
Turkey Vulture	*Cathartes aura*
White-backed Vulture	*Gyps africanus*
Vulturine Guineafowl	*Acryllium vulturinum*

W

Wagtail

White Wagtail	*Motacilla alba*
Wallace's Standardwing	*Semioptera wallacii*
Wandering Albatross	*Diomedea exulans*

Warbler

Bachman's Warbler	*Vermivora bachmanii*
Bay-breasted Warbler	*Dendroica castanea*
Black-and-white Warbler	*Mniotilta varia*
Black-throated Blue Warbler	*Dendroica caerulescens*
Black-throated Gray Warbler	*Dendroica nigrescens*
Black-throated Green Warbler	*Dendroica virens*
Blackburnian Warbler	*Dendroica fusca*
Blackpoll Warbler	*Dendroica striata*
Blue-winged Warbler	*Vermivora pinus*
Canada Warbler	*Wilsonia canadensis*
Cape May Warbler	*Dendroica tigrina*
Cerulean Warbler	*Dendroica cerulea*
Chestnut-sided Warbler	*Dendroica pensylvanica*
Colima Warbler	*Vermivora crissalis*
Fan-tailed Warbler	*Euthlypis lachrymosa*
Garden Warbler	*Sylvia borin*
Golden-cheeked Warbler	*Dendroica chrysoparia*
Golden-winged Warbler	*Vermivora chrysoptera*
Grasshopper Warbler	*Locustella naevia*
Hermit Warbler	*Dendroica occidentalis*
Hooded Warbler	*Wilsonia citrina*
Kentucky Warbler	*Oporornis formosus*
Kirtland's Warbler	*Dendroica kirtlandii*
Lucy's Warbler	*Vermivora luciae*
MacGillivray's Warbler	*Oporornis tolmiei*
Magnolia Warbler	*Dendroica magnolia*
Marsh Warbler	*Acrocephalus palustris*
Nashville Warbler	*Vermivora ruficapilla*
Olive Warbler	*Peucedramus taeniatus*
Orange-crowned Warbler	*Vermivora celata*
Palm Warbler	*Dendroica palmarum*
Pine Warbler	*Dendroica pinus*
Prairie Warbler	*Dendroica discolor*
Prothonotary Warbler	*Protonotaria citrea*
Red-faced Warbler	*Cardellina rubrifrons*
Tennessee Warbler	*Vermivora peregrina*
Townsend's Warbler	*Dendroica townsendi*
Virginia's Warbler	*Vermivora virginiae*
Worm-eating Warbler	*Helmitheros vermivorus*
Yellow Warbler	*Dendroica petechia*
Yellow-rumped Warbler	*Dendroica coronata*
Yellow-throated Warbler	*Dendroica dominica*
Warbler Finch	*Certhidea olivacea*
Warbling Vireo	*Vireo gilvus*
Water Pipit	*Anthus spinoletta*
Water Thick-knee	*Burhinus vermiculatus*

Waterthrush

Louisiana Waterthrush	*Seiurus motacilla*
Northern Waterthrush	*Seiurus noveboracensis*
Wattled Broadbill	*Eurylaimus steerii*
Wattled Jacana	*Jacana jacana*

Waxwing

Cedar Waxwing	*Bombycilla cedrorum*

Weaver

Baya Weaver	*Ploceus philippinus*
Grosbeak Weaver	*Amblyospiza albifrons*
Parasitic Weaver	*Anomalospiza imberbis*
Social Weaver	*Philetairus socius*
Village Weaver	*Ploceus cucullatus*
Weka	*Gallirallus australis*
Western Grebe	*Aechmophorus occidentalis*
Western Gull	*Larus occidentalis*
Western Kingbird	*Tyrannus verticalis*
Western Meadowlark	*Sturnella neglecta*
Western Sandpiper	*Calidris mauri*
Western Screech-Owl	*Otus kennicottii*
Western Scrub-Jay	*Aphelocoma californica*
Western Tanager	*Piranga ludoviciana*
Western Wood-Pewee	*Contopus sordidulus*
Whip-poor-will	*Caprimulgus vociferus*

Whistling-Thrush

Blue Whistling-Thrush	*Myophonus caeruleus*
Formosan Whistling-Thrush	*Myophonus insularis*

White Ibis	*Eudocimus albus*	Nuttall's Woodpecker	*Picoides nuttallii*
White Stork	*Ciconia ciconia*	Pileated Woodpecker	*Dryocopus pileatus*
White Tern	*Gygis alba*	Red-bellied Woodpecker	*Melanerpes carolinus*
White Wagtail	*Motacilla alba*	Red-cockaded Woodpecker	*Picoides borealis*
White-backed Fire-eye	*Pyriglena leuconota*	Red-headed Woodpecker	*Melanerpes erythrocephalus*
White-backed Vulture	*Gyps africanus*	Strickland's Woodpecker	*Picoides stricklandi*
White-backed Woodpecker	*Dendrocopos leucotos*	Three-toed Woodpecker	*Picoides tridactylus*
White-bearded Manakin	*Manacus manacus*	White-backed Woodpecker	*Dendrocopos leucotos*
White-bellied Antbird	*Myrmeciza longipes*	White-headed Woodpecker	*Picoides albolarvatus*
White-breasted Nuthatch	*Sitta carolinensis*	Woodpecker Finch	*Camarhynchus pallidus*
White-browed Scrubwren	*Sericornis frontalis*	**Woodswallow**	
White-crested Spadebill	*Platyrinchus platyrhynchos*	Black-faced Woodswallow	*Artamus cinereus*
White-crowned Sparrow	*Zonotrichia leucophrys*	Worm-eating Warbler	*Helmitheros vermivorus*
White-eyed Vireo	*Vireo griseus*	**Wren**	
White-faced Ibis	*Plegadis chihi*	Bay Wren	*Thryothorus nigricapillus*
White-fronted Bee-eater	*Merops bullockoides*	Bewick's Wren	*Thryomanes bewickii*
White-fronted Parrot	*Amazona albifrons*	Buff-breasted Wren	*Thryothorus leucotis*
White-headed Woodpecker	*Picoides albolarvatus*	Cactus Wren	*Campylorhynchus brunneicapillus*
White-necked Raven	*Corvus albicollis*		
White-plumed Antbird	*Pithys albifrons*	Canyon Wren	*Catherpes mexicanus*
White-tailed Kite	*Elanus leucurus*	Carolina Wren	*Thryothorus ludovicianus*
White-tailed Ptarmigan	*Lagopus leucurus*	House Wren	*Troglodytes aedon*
White-throated Sparrow	*Zonotrichia albicollis*	Marsh Wren	*Cistothorus palustris*
White-tipped Sicklebill	*Eutoxeres aquila*	Musician Wren	*Cyphorhinus aradus*
White-winged Chough	*Corcorax melanorhamphos*	Plain Wren	*Thryothorus modestus*
White-winged Crossbill	*Loxia leucoptera*	Riverside Wren	*Thryothorus semibadius*
White-winged Scoter	*Melanitta fusca*	Rock Wren	*Salpinctes obsoletus*
White-winged Shrike-Tanager	*Lanio versicolor*	Rufous-and-white Wren	*Thryothorus rufalbus*
Whooping Crane	*Grus americana*	Rufous-naped Wren	*Campylorhynchus rufinucha*
Whydah			
Pin-tailed Whydah	*Vidua macroura*	Sedge Wren	*Cistothorus platensis*
Wigeon		Socorro Wren	*Thryomanes sissonii*
Eurasian Wigeon	*Anas penelope*	Stephens Island Wren	*Xenicus lyalli*
Wild Turkey	*Meleagris gallopavo*	Stripe-backed Wren	*Campylorhynchus nuchalis*
Willet	*Catoptrophorus semipalmatus*	Winter Wren	*Troglodytes troglodytes*
Williamson's Sapsucker	*Sphyrapicus thyroideus*	Wrentit	*Chamaea fasciata*
Willie-wagtail	*Rhipidura leucophrys*	Wrybill	*Anarhynchus frontalis*
Willow Flycatcher	*Empidonax traillii*		
Willow Ptarmigan	*Lagopus lagopus*		
Wilson's Phalarope	*Phalaropus tricolor*		
Wilson's Plover	*Charadrius wilsonia*	**Y**	
Wilson's Storm-Petrel	*Oceanites oceanicus*	Yellow Warbler	*Dendroica petechia*
Winter Wren	*Troglodytes troglodytes*	Yellow-bellied Flycatcher	*Empidonax flaviventris*
Wood Duck	*Aix sponsa*	Yellow-bellied Sapsucker	*Sphyrapicus varius*
Wood Stork	*Mycteria americana*	Yellow-billed Chough	*Pyrrhocorax graculus*
Wood Thrush	*Hylocichla mustelina*	Yellow-billed Cuckoo	*Coccyzus americanus*
Wood-Pewee		Yellow-breasted Chat	*Icteria virens*
Eastern Wood-Pewee	*Contopus virens*	Yellow-crested Tanager	*Tachyphonus rufiventer*
Western Wood-Pewee	*Contopus sordidulus*	Yellow-crowned Night-Heron	*Nyctanassa violacea*
Wood-Pigeon		Yellow-eyed Junco	*Junco phaeonotus*
Common Wood-Pigeon	*Columba palumbus*	Yellow-eyed Penguin	*Megadyptes antipodes*
Woodcock		**Yellow-Finch**	
American Woodcock	*Scolopax minor*	Citron-headed Yellow-Finch	*Sicalis luteocephala*
Woodcreeper		Yellow-headed Blackbird	*Xanthocephalus xanthocephalus*
Plain-brown Woodcreeper	*Dendrocincla fuliginosa*		
Woodpecker		Yellow-headed Parrot	*Amazona oratrix*
Acorn Woodpecker	*Melanerpes formicivorus*	Yellow-rumped Warbler	*Dendroica coronata*
Black Woodpecker	*Dryocopus martius*	Yellow-throated Longclaw	*Macronyx croceus*
Black-backed Woodpecker	*Picoides arcticus*	Yellow-throated Vireo	*Vireo flavifrons*
Downy Woodpecker	*Picoides pubescens*	Yellow-throated Warbler	*Dendroica dominica*
Gila Woodpecker	*Melanerpes uropygialis*	Yellow-wattled Bulbul	*Pycnonotus urostictus*
Gray-headed Woodpecker	*Dendropicos podocephalus*	**Yellowlegs**	
Great Spotted Woodpecker	*Dendrocopos major*	Greater Yellowlegs	*Tringa melanoleuca*
Green Woodpecker	*Picus viridis*	Lesser Yellowlegs	*Tringa flavipes*
Hairy Woodpecker	*Picoides villosus*	**Yellowthroat**	
Imperial Woodpecker	*Campephilus imperialis*	Common Yellowthroat	*Geothlypis trichas*
Ivory-billed Woodpecker	*Campephilus principalis*		
Ladder-backed Woodpecker	*Picoides scalaris*	**Z**	
Lesser Spotted Woodpecker	*Dendrocopos minor*	Zebra Finch	*Taeniopygia guttata*
Lewis's Woodpecker	*Melanerpes lewis*	Zitting Cisticola	*Cisticola juncidis*

Glossary

Note: In general, terms have been defined as they apply to birds. Nevertheless, many terms (especially those naming basic anatomical structures or biological principles) apply to a range of living things beyond birds. In most cases, terms that apply only to birds are noted as such. Most terms that are bolded in the text of the *Handbook of Bird Biology* appear here. Numbers in brackets following each entry give the primary pages on which the term is defined.

Please note that this glossary is also available on the Internet at <www.birds.cornell.edu/homestudy>.

A

abdominal air sacs: A pair of air sacs in the abdominal region of birds that may have connections into the bones of the pelvis and femur; their position within the abdominal cavity may shift during the day to maintain the bird's streamlined shape during digestion and egg laying. [4·101]

abducent nerve: The sixth cranial nerve; it stimulates a muscle of the eyeball and two skeletal muscles that move the nictitating membrane across the eyeball. [4·41]

abiotic: Nonliving; includes both things that are dead (such as dead leaves) and those that have never been alive (for example, rocks). [9·7]

accessory nerve: The eleventh cranial nerve; it carries motor output to constrict the neck muscles. [4·42]

accommodation: The changes in the curvature of the lens (and cornea, in birds) of the eye brought about by the action of the ciliary muscles. These changes allow the eye to focus on objects at different distances. [4·50]

acetabulum: At the hip joint, the hollow on the pelvic girdle into which the head of the femur fits. [4·24]

acoustic nerve: See **vestibulocochlear nerve**. [4·42]

adaptation: A genetically controlled trait that increases an individual's fitness relative to that of other individuals. [1·35]

adaptive: Describes a trait that better promotes an individual's fitness than does some alternative form of that characteristic. [6·42]

adaptive management: A type of **ecosystem management** (see separate entry) in which managers continue to learn more about the ecosystem as they proceed, and continually modify their management techniques to incorporate the new information. [10·84]

adaptive radiation: The evolution, from a common ancestor, of a variety of different species adapted to different niches; the species usually have different morphologies and behaviors. [1·59]

adherent cup nest: A cup nest made of mud or saliva that relies on chemical forces to hold it to a vertical surface; built by many swifts, including the Edible-nest Swiftlets of Southeast Asia, whose nests are used in the Asian delicacy bird's-nest soup. [8·32]

adoption: In avian biology, the peaceful acquisition of a lone chick or chicks by a pair of adults other than the biological parents. [8·128]

adrenal glands: Small yellow or orange endocrine glands at the cranial end of each kidney; they produce a variety of hormones (including adrenaline, steroids, and the sex hormones) that are involved with circulation, digestion, and reproduction. [4·74]

advertising displays: Displays performed by one sex (usually the male) to attract a mate of the opposite sex; also called **mate attraction displays**. [6·37]

aerodynamic valve: A vortex-like movement of air within the air tubes of each avian lung, at the junction between the mesobronchus and the first secondary bronchus; it prevents the backflow of air into the mesobronchus by forcing the incoming air along the mesobronchus and into the posterior air sacs. [4·102]

African barbets: A family (Lybiidae, 42 species) of small, colorful, stocky African birds with large, sometimes serrated, beaks; they dig their nest cavities in trees, earthen banks, or termite nests. [1·85]

Afrotropical region: Zoogeographic region including Madagascar, southern Arabia, and all of Africa south of the Sahara Desert. Sometimes called the Ethiopian Region. [1·70, 1·81]

afterfeather: A small feather that grows from the lower shaft of a contour feather and resembles the main feather but in miniature. [3·13]

age-specific fecundity: The average birth rate for females in a particular age group in a population. [9·64]

age-specific survival rate: The proportion of individuals in a particular age group in a population that survive a particular interval of time—usually a year. [9·64]

age structure: The relative proportions of individuals of different ages—usually noted for a given population. [9·80]

airfoil: Any structure designed to help lift or control a flying object by using the air currents through which it moves. A typical airfoil, such as the wing of a bird or airplane, is rounded on top and curved inward below. [5·10]

air sacs: Thin-walled, transparent sacs extending from the mesobronchi or the lungs to different regions of the body; they act as bellows to bring air into the body and store it until expiration. They are found only in birds. [4·100]

air speed: A flying individual's speed relative to the air through which it is moving. It does not include being carried along or slowed down by the wind, so it may or may not reflect a bird's speed relative to the ground. [5·45]

albino: An individual that lacks the pigment melanin all over its body. An individual that lacks all types of pigments is called a **complete albino**. [3·52]

albumen: Egg white; albumen is composed primarily of water and protein. [8·63]

alimentary canal: The tube for the passage, digestion, and absorption of food; in most birds, it includes the esophagus, crop, two-part stomach, small intestine, ceca, and the large intestine. It is also called the **gastrointestinal tract**, **digestive tract**, or **gut**. [4·103]

allantois: In avian biology, the extra-embryonic membrane inside the egg that forms a sac into which the developing embryo shunts all metabolic wastes that cannot evaporate through the shell, such as uric acid crystals. [8·65, 8·69]

Allee effect: The response, shown in some species when population density falls below some threshold level, by which reproductive behavior and/or social structure become disrupted in various ways. In some circumstances, this effect may cause species to be unusually prone to extinction. [10·38]

alleles: Alternate forms of genes. Most animals have two alleles for each trait: one allele is on the chromosome they received from their mother, and the other is on the chromosome they received from their father. As an example consider eye color in humans, which is determined primarily by whether someone has two alleles for blue eyes, two alleles for brown eyes, or one of each (in which case, the brown dominates and the person has brown eyes). [10·74]

allopreening: Mutual preening during which two birds preen each other, usually around the head and neck. In many species allopreening not only keeps the plumage clean and orderly, but also helps to establish social bonds between individuals. [3·19]

alpine tundra: Ecosystem found above the tree line on mountains; it consists of rugged, well-drained terrain interspersed with meadows with low-growing vegetation and a profusion of summer-blooming wildflowers. Very few birds breed in this harsh environment. [9·114]

alternate plumage: In the Humphrey-Parkes system of nomenclature, alternate plumage is the plumage worn by an adult bird during the breeding season, if that plumage is produced by a partial molt before breeding. If a bird does not molt before breeding, it continues to wear its basic plumage during breeding. In the traditional system, the alternate plumage was known as the **nuptial plumage** or the **breeding plumage**. [3·33]

altricial: Describes young birds that hatch undeveloped and in many cases naked or with sparse down; such helpless young require complete parental care. [8·106]

alula: A group of two to six feathers projecting from the phalanx of the bird's first finger (its thumb) at the bend of the wing. It reduces turbulence by allowing fine control of airflow over the wing. [1·11, 5·15]

American Ornithologists' Union (AOU): The largest organization of professional ornithologists in North America; it publishes the research journal *The Auk* and the *Check-list of North American Birds* (commonly called the **AOU Check-list**; see separate entry). [1·39]

amnion: In avian biology, the extra-embryonic membrane inside the egg that becomes filled with fluid and surrounds the developing embryo, allowing it to move and stay moist, and preventing its various growing parts from sticking to or blocking one another. [8·65]

ampulla: A membranous chamber at the base of each of the three semicircular ducts in the inner ear; it contains sensory hair cells embedded in a gelatinous material and surrounded by endolymph, and it senses changes in the animal's speed or direction in a particular plane of space. Information from all three ampullae, one in each plane, is combined by the brain to determine the animal's motion and thus to aid balance. [4·58]

angle of attack: In a flying bird, the angle between the cranial-caudal axis of the wing and the oncoming airstream. [5·13]

anisodactyl feet: Foot arrangement in which the hallux points backward and the other three toes point forward. Found in most passerines. [1·21]

annual survival rate: See **survival rate**. [8·3]

ant-acacias: Various species of tropical and subtropical trees in the genus *Acacia* that harbor ants inside their hollow thorns. The ants receive shelter and extra nutrition from special substances produced by the trees exclusively for the ants, and in turn keep away insects, mammals, and other herbivores that might feed on the trees by attacking them. The ants also prevent other vegetation from growing nearby by biting off shoots as they emerge from the ground. [8·19]

Antarctic Convergence: Region of the oceans between about 50 and 60 degrees south latitude, where cold, north-flowing currents meet warmer, south-flowing currents, resulting in large-scale upwelling of nutrient-rich water. The nutrients support abundant plankton, which attract a great diversity and abundance of seabirds. [1·104]

antbirds: A Neotropical suboscine family (Thamnophilidae, 197 species) of small, insectivorous, forest birds; some species specialize in the technique of following columns of army ants to prey on the insects and other arthropods stirred up by the numerous moving ants. [1·79]

antebrachium: The middle portion of the forelimb, consisting of the radius and ulna. The secondary feathers attach to the ulna. [1·9]

anterior: Toward the front of an organism, using the earth as a frame of reference. With birds, technically used only within the eye and inner ear. But in practice, often used interchangeably with the term **cranial**. [1·4]

anterior air sacs: General term referring to the air sacs nearest the bird's front end—the cervical, clavicular, and anterior thoracic air sacs. [4·102]

anterior chamber: Space within the eye between the iris and the cornea; it is filled with aqueous fluid, which nourishes the eye and removes wastes. [4·48]

anterior lobe of the pituitary gland: Portion of the pituitary gland that receives instructions from the nervous system in the form of neurohormones from the hypothalamus. As a result, the anterior lobe secretes various hormones into the blood that may act directly on organs or on other endocrine glands (such as the gonads, adrenals, and thyroids); because of this central controlling role in the endocrine system, the anterior lobe is nicknamed "the master gland." [4·72]

anting, active: Picking up an ant or other chemically potent object, such as a millipede, and deliberately rubbing it in the feathers—presumably to deter ectoparasites. [3·22]

anting, passive: Positioning oneself among a swarm of ants, permitting them to run all over the body and to move in and out among the feathers, presumably to deter ectoparasites. [3·22]

antiphonal: Describes a singing interaction in which two individuals alternate their contributions. Often used to describe duets between a male and female bird. [7·78]

antorbital fenestra: An opening on each side of the skull in front of the eye socket; found in all archosaurs. [E·8]

antpittas: Together with antthrushes, antpittas form the suboscine family Formicariidae (60 species). They are small, drab Neotropical birds with loud, ringing songs; many haunt the rain forest floor and may follow army ant swarms. [1·79]

antthrushes: Together with antpittas, antthrushes form the suboscine family Formicariidae (60 species). They are small, drab Neotropical birds with loud, ringing songs; many haunt the rain forest floor and may follow army ant swarms. [1·79]

aorta: The main artery exiting the heart; its branches distribute oxygenated blood to all parts of the body. [4·78]

aortic arch: The curve of the aorta, just after it exits the left ventricle of the heart. In birds, the aorta curves to the bird's right as it passes dorsal to the heart and toward the backbone, but in mammals, the aorta curves to the left. [4·83]

AOU: See **American Ornithologists' Union**. [1·39]

AOU Check-list: Bird checklist produced by the Committee on Classification and Nomenclature of the American Ornithologists' Union; it contains common and scientific names of all birds that occur in North America north of Mexico, or near North American coasts, including Hawaii, and is the generally accepted reference for common names of North American birds. [1·64, 1·111]

aponeuroses: Shiny, broad sheets of connective tissue that bind muscle fibers together to form muscles. [4·27]

aposematic: Having bright, bold colors and patterns (often reds and oranges) that advertise to potential predators that an individual is bad tasting or poisonous. A few species that are perfectly palatable also have evolved these warning colors and patterns. The strategy of sporting aposematic coloration is called **aposematism**. [9·26]

appeasement displays: Displays given to decrease the aggression of another individual; they usually consist of stereotyped postures that de-emphasize the performer's weapons or size and expose vulnerable parts of its body. For example, a submissive bird may point its beak down or away, fold its wings, lower or turn away its head, point its tail down, or adopt some combination of these postures. [6·28]

appendicular skeleton: Portion of the skeleton consisting of the sternum (breast bone), the pectoral girdle including the front limbs (wings), and the pelvic girdle including the hind limbs (legs). [4·6, 4·18]

apteria (singular, **apterium**): Regions of bare or less-feathered skin between the feather tracts of birds. [3·2]

aquatic birds: All birds with webbed feet that commonly swim, including the Anatidae; also, all deep-water waders belonging to the order Ciconiiformes, such as herons and storks. Also called **water birds**. [1·65]

aqueous fluid: A cell-free fluid, similar in structure to blood plasma; it fills the anterior and posterior chambers of the eye (between the lens and the cornea), providing nourishment and waste removal. [4·47]

arachnoid: The middle of the three vascularized membranes, called meninges, surrounding the brain and spinal cord. The meninges provide sustenance and waste removal for the cells of the brain and spinal cord, which are not served by the circulatory system. [4·36]

arboreal: Living in trees. [E·18]

arboreal theory (of the origin of avian flight): Theory suggesting that flight originated in small, arboreal (tree-dwelling), reptile-like birds that jumped or glided among tree branches. Proponents suggest that feathers first evolved to keep the animals warmer in the cooler arboreal environment, and then were used to extend jumps or glides. First proposed in 1880 by O. C. Marsh. Also called the **trees-down theory**. [E·18]

Archaeopteryx: A feathered reptile from 150-million-year-old Jurassic limestone deposits. Because it had a mosaic of bills and gaudy, clashing colors. They are endemic to the Oriental zoogeographic region. [1·89]

Asities: Members (with false sunbirds) of the family Philepittidae, endemic to Madagascar. These two species of suboscine passerines feed on fruits. [1·88]

aspect ratio: The ratio of an object's length to its width. It is used to refer to the shape of a bird's wing: the larger the aspect ratio, the more elongated the wing. [5·37]

asynchronous hatching: Pattern of hatching in which the eggs of a single clutch hatch over a period of several days, resulting in a brood of young of different ages. This pattern occurs when incubation begins at the time the first egg is laid. Because eggs are laid one per day, at one- to two-day intervals, the embryos of the earliest-laid eggs have already started to develop by the time the later eggs are laid, and they hatch sooner. [6·88]

atlas: The first cervical vertebra; in birds, it articulates with the single occipital condyle on the base of the skull. (Mammals have two occipital condyles.) [4·16]

atria (singular, **atrium**): The two thin-walled, anterior chambers of the heart; they receive blood returning to the heart from the lungs (**left atrium**) or body (**right atrium**). [4·77]

atrioventricular valve: The valve between each atrium and its corresponding ventricle; it prevents the backflow of blood into the atrium as the ventricle contracts. [4·78]

atrophy: To shrivel or die back; may be pathological or part of the normal course of development. [4·27]

attentive periods: In avian biology, periods spent on the nest during incubation. [8·99]

auditory nerve: See **vestibulocochlear nerve**. [4·42]

auditory tube: Air-filled tube leading from the middle ear to the throat; it helps to equalize the air pressure on the two sides of the eardrum. In birds, the right and left tubes join and enter the roof of the mouth caudally through just one opening. In mammals, the right and left tubes enter the mouth separately. Also called **eustachian tube**. [4·57]

auricular feathers: A patch of feathers covering the external ear opening. Their open texture protects the ear from debris and wind noise, yet helps to channel sounds into the ear. [1·8, 4·55]

Australasian region: One of the major zoogeographic regions of the world, stretching from a line termed "Wallace's Line" east of the islands of Timor and Sulawesi in Indonesia, southeast to New Zealand, and including New Guinea, bird and reptile characteristics, its relationship to modern birds and other reptiles has been highly controversial since its discovery in the early 1860s. [E·2]

archosaurs: All reptiles in the Archosauromorpha, the large group of diapsid reptiles that includes all thecodonts and their descendants (including birds and crocodiles), but does not include lizards, snakes, or turtles. All archosaurs have an opening on each side of the skull, in front of the eye socket, called the antorbital fenestra. [E·8]

arctic tundra: Ecosystem found around the world in a belt extending north from the limit of trees; it consists of swampy flatland covered with low-growing vegetation, interrupted by countless shallow lakes. Very few birds live in the arctic tundra year round, but many migrants breed there, taking advantage of the long hours of daylight and numerous spring insects for feeding their young. [9·114]

arctic tundra/coniferous forest ecotone: Transitional zone between the arctic tundra and coniferous forest ecosystems, sometimes called the **northern timberline**. It consists of fingers of open, stunted spruce forest extending north between fingers of low, shrubby tundra extending south. [9·123]

area-sensitive species: Species that require large tracts of habitat in order to persist and breed successfully. [10·72]

army ants: A subfamily of primarily tropical ants that are highly social and nomadic: colonies make daily raids in long streams of thousands of individuals moving across the forest floor, looking for insects and other arthropods and flushing out many other small creatures as they move. Numerous birds (see **antbirds**) and some primates follow moving army ant swarms to take advantage of the prey they flush. [9·88]

arterioles: Small blood vessels branching from arteries. They carry blood from the arteries to the capillaries. [4·81, 4·84]

artery: A vessel conducting blood away from the heart. All arteries except the pulmonary artery carry oxygen-rich blood. [4·81, 4·82]

articular bone: Bone on the upper surface of each side of the lower jaw of many vertebrates, near the caudal end; in birds, it links with the quadrate bone of the upper jaw, forming the joint between the jaws. [4·12]

arytenoid cartilages: Two cartilages of the larynx; they stiffen and hold the shape of the fleshy folds surrounding the glottis—the opening of the larynx. [4·91]

Asian barbets: A family (Megalaimidae, 26 species) of chunky birds slightly smaller than a Belted Kingfisher, with thick

Australia, Hawaii, and other islands of the mid-Pacific Ocean. Sometimes called the Australian region. [1·70, 1·91]

Australasian robins: A family (Eopsaltriidae, 44 species) of songbirds endemic to the Australasian region. Reminiscent of both New and Old World robins in coloration, they actually are more like flycatchers, although they usually snatch food from the ground. [1·95]

autonomic nervous system: A set of nerves considered as a group because of their similar functions. Acting primarily unconsciously, they innervate the smooth muscle of the viscera, glands, and blood vessels, thus controlling the automatic function of the internal organs. They are under direct chemical control from substances circulating in the blood. [4·43, 4·44]

avifauna: The set of bird species living in a region. [1·69]

axial skeleton: The portion of the skeleton consisting of the skull, hyoid apparatus, and the vertebral column of the neck, trunk, and tail. [4·6]

axillaries: A cluster of feathers in the bird's "armpit"; they are recognizably longer than those lining the wing. [1·12, 1·13]

axis: The second cervical vertebra. [4·16]

axon: The cable-like, impulse-conducting, main axis of a neuron. [4·32]

B

babblers: A diverse family (Timaliidae, 267 species) of gregarious, insectivorous birds, many of which have complex social systems and breed cooperatively. They are found in the Afrotropical, Oriental, and Australasian regions. [1·90]

back: In birds, the dorsal side of the body, between the neck and the rump. [1·6, 1·9]

barbs: The parallel branches extending from each side of the rachis of the feather shaft; collectively they form the vanes. [3·3]

barbules: Branchlets coming off both sides of the barbs of a feather, at right angles to the barbs and in the same plane. Adjacent barbules hook together, holding the vane intact. [3·3]

basal archosaurs: See **thecodonts**. [E·8, E·33]

basal metabolism: The number of calories an organism uses when completely at rest, which indicates the amount of energy needed to maintain minimal body functions. [4·144]

basic plumage: In the Humphrey-Parkes system of nomenclature, the plumage

worn by an adult bird for the longest time each year; it usually is produced by a complete molt. In the traditional system, this plumage was known as the **nonbreeding** or **winter plumage**. If a bird does not molt before breeding, it continues to wear its basic plumage during breeding. [3·33]

basilar papilla: The lower membrane of the cochlear duct in the inner ear of birds; it is coated with a layer of sensory hair cells. Sound waves set the basilar papilla into motion, causing the hair cells to push against the tectorial membrane, triggering nerve impulses in the hair cells that are sent to the brain in the process of sound perception. In mammals, the corresponding structure is called the "basilar membrane." [4·59]

BBS: See **Breeding Bird Survey**. [9·62]

B cells: Special white blood cells found in the lymphatic tissues; they are important in the immune response because they produce antibodies. B cells are produced by the cloacal bursa and are also important in understanding the development of AIDS in humans. [4·123]

beak: A bird's upper and lower jaws, including the external covering; also called the **bill**. [1·6]

bee-eaters: An Old World family (Meropidae, 26 species) of brightly colored birds with long, slender beaks. They catch stinging insects in a manner similar to that of flycatchers, and then beat them to remove the stingers before eating them. [1·85]

belly: In birds, part of the lower (ventral) surface of the body between the breast and the vent. [1·6, 1·9]

bend of the wing: The prominent angle at the wrist, where the bird's wing bends noticeably. [1·11]

benefits of philopatry hypotheses: A set of possible explanations for why certain individual birds might forego their own breeding in a particular breeding season and act as helpers at the nest of other breeding pairs (usually their parents or other close relatives); the benefits of philopatry hypotheses focus on the possible benefits to young adults of remaining with their parents. Examples of these hypotheses include (1) the survival of young adults may be improved when they remain in a group, (2) by helping, young adults may improve the survival of close relatives (and thus increase their own indirect fitness), and (3) by staying with their parents, young adults may increase their own chance of acquiring a superior territory, either by monitoring vacancies in neighboring sites or by inheriting their natal territory. [6·89]

Bergmann's rule: Rule describing the pattern of body sizes found within most

bird and mammal species, in that individuals living in colder regions tend to be larger than those living in warmer areas. [9·16]

Bernoulli's law: Physical law stating that, in any system of airflow, static and dynamic pressure must always add up to a constant. Because the airflow over a moving airfoil (such as a bird's wing) is faster above than below, the dynamic pressure is higher and thus the static pressure is lower. Because the static pressure below the wing is higher than that above, lift is created and the bird can remain aloft. [5·13]

biconical: Describes an egg that is slightly longer than subelliptical; also called **fusiform** or **long subelliptical**. [8·73]

bile: Substance, produced by the liver, that emulsifies fats to facilitate their digestion. In birds with no gall bladder, bile is released directly into the small intestine through the hepatoenteric ducts; in birds with a gall bladder (and mammals), bile is stored in the gall bladder and released through the bile ducts. [4·124]

bill: A bird's upper and lower jaws, including the external covering; also called the **beak**. [1·6]

bill tip organ: An aggregation of sensory cells at the tip of both the upper and lower beak, best developed in ducks, geese, sandpipers, and snipe; it is thought to sense tactile stimuli during feeding. [4·57]

bill-wiping: A maintenance behavior in which a bird swipes its bill sideways on tree branches, the ground, or other surfaces, especially after eating messy foods such as oily insects or suet. [3·40]

binocular vision: A type of vision that produces three-dimensional images, in contrast to monocular vision, which produces flat images. Binocular vision results when the eyes are positioned toward the front of the head, so that objects are detected by both eyes simultaneously. [4·51]

binomial nomenclature: The currently accepted system of naming organisms, devised by Linnaeus, in which each species is designated by two words: the genus and the species names. [1·47]

bioaccumulation: The increase in the concentration of toxic or other foreign substances in organisms' bodies as a result of taking up the substances from the environment (through plant roots, or in ingested food or water) at a rate higher than that at which they are excreted from the body. Many toxic substances that do not occur naturally, such as DDT and PCBs, are readily deposited in body tissues but excreted at very slow rates, because organisms have not evolved mechanisms to metabolize them effectively. [9·126, 9·127]

bioconcentration: The stepwise increase, found at each higher level of the food chain, in the concentration of certain chemicals in the bodies of organisms. Chemicals that usually bioconcentrate in the food chain are those toxins, such as DDT, that tend to accumulate (see **bioaccumulation**) in organisms because they are taken up from the environment faster than they are excreted. Also called **biological magnification**. [9·125, 9·127]

biodiversity: The great wealth of living organisms that occur on earth. [1·106]

biogeography: The study of the distribution patterns of living things. [9·50]

biological indicators: See **indicator species**. [10·108]

biological magnification: See **bioconcentration**. [9·126, 9·127]

biomass: The total mass of all the living organisms in a particular population, community, or area at a given point in time. [9·87]

biome: A major terrestrial ecosystem; in North America north of Mexico, the nine biomes are tundra, coniferous forest, deciduous forest, grassland, southwestern oak woodland, pinyon-juniper woodland, chaparral, sagebrush, and scrub desert. Biomes are named for the dominant vegetation of the region's climax community, but they include all the living organisms in that community, the physical surroundings, and all the ecological processes that occur there. [9·109]

biotic: Living. [9·7]

bipedal: Walking (or running) on two legs. [E·16]

bird community: See **community**. [9·82]

birds-of-paradise: A family (Paradisaeidae, 46 species) of forest-dwelling songbirds found primarily in New Guinea; they are famous for the spectacular, colorful plumages and displays of the males. [1·94]

birth rate: The number of young born to an individual or set of individuals, or born into a population or species, in a given period of time (often a year). Usually (especially for humans) expressed as the number of births per 1,000 individuals per year. Also called **fecundity**. [9·49]

blastoderm: In avian biology, the flattened disc of dividing cells that lies on the upper surface of the yolk and is the first stage in the development of the embryo. [8·63]

blind: A structure used to conceal a person so that he or she may observe birds or other wildlife; known as a "hide" in Great Britain. [2·29]

blind spot: Site on the retina where the optic nerve penetrates the retina and leaves the eye; because no rods or cones are present at this spot to capture incoming light, an object whose image falls on the spot is not perceived. [4·49]

blocking: A method of estimating the size of large flocks of birds by counting the birds in a block of typical density, beginning at the trailing end of the flock (so that birds are not flying into the area you are counting), and then visually superimposing the block onto the rest of the flock to see how many times it will fit. [2·56]

blood/brain barrier: The specialized arrangement of capillaries in the brain that prevents blood from reaching the nerve cells, thus protecting the brain from potentially toxic substances that circulate in the blood. Capillaries penetrate the meninges surrounding the brain, but do not reach the actual neurons of the brain. [4·36]

blood plasma: See **plasma**. [4·86]

blood vascular system: Alternate name for the circulatory system. [4·76]

body downs: The down feathers of adult birds, found under the contour feathers. Body downs are most common in water birds and hawks, as they provide extra insulation. [3·16]

bone: Tissue composed of living cells in a mineralized matrix; the bones provide support for the body and attachment sites for the muscles. [4·3]

bony labyrinth: The bony, outer system of fluid-filled canals (containing perilymph) that make up the inner ear. It forms the cochlea, semicircular canals, and vestibule, and encloses the membranous labyrinth. [4·57]

booming sacs: Brightly colored outpocketings of the esophagus of some North American grouse that appear at the sides of the neck; they fill with air and act as resonators to produce loud sounds during displays. [4·114]

booted podotheca: A smooth podotheca, divided into long, continuous, nonoverlapping scales. [1·20]

boreal forest: Coniferous forest ecosystem dominated by spruce and fir trees and found around the world, generally in a belt north of temperate zone deciduous forests and south of the arctic tundra. Many birds migrate to the boreal forests to breed, taking advantage of the long daylight hours and abundant insects for feeding their young. Some bird species are year-round residents. Also called **taiga**. [9·116]

bounding: A flight pattern in which a bird alternates flapping (during which it rises slightly) with glides on closed wings (during which it descends slightly). [5·26]

bower: A complex mating structure built by male bowerbirds—members of the passerine family Ptilonorhynchidae. See **bowerbirds**. [6·76]

bowerbirds: A passerine (oscine) family (Ptilonorhynchidae, 20 species) of New Guinea and Australia whose polygynous males attract females by building and decorating remarkably complex "bowers" out of twigs and other objects. See **bower**. [6·76]

brachial plexus: Plexus along the spinal cord of birds at the level of the wing; it is associated with a cervical enlargement of the spinal cord. [4·39]

brachium: The upper (proximal) portion of the forelimb (wing); it contains the humerus. [1·9]

brain stem: See **medulla oblongata**. [4·37]

breast: In birds, part of the lower (ventral) surface of the body, between the throat and belly. [1·6]

Breeding Bird Survey (BBS): A count of the breeding birds of North America conducted each summer since 1967 and coordinated by the U. S. Fish and Wildlife Service. Observers count birds seen or heard during three-minute periods at half-mile (0.8-km) intervals along a 25-mile (40-km) stretch of road. Because the same routes and stops are sampled each year, BBS data can be used to track population trends, and the results can be correlated with habitat. [9·62]

breeding plumage: See **alternate plumage**. [3·33]

breeding season: The period of time during the year when a particular species may breed. [8·11]

bristles: Highly specialized contour feathers in which the rachis is stiffened and lacks barbs along its outermost parts. [3·13]

broadbills: A family (Eurylaimidae, 15 species) of stocky, brightly colored, suboscine birds found throughout forests and scrublands of the Old World tropics, particularly in the Oriental region. Broadbills use their wide, flat, colorful bills to snatch up large insects. [1·81]

bronchi (singular, **bronchus**)**:** The two major air tubes branching off the lower end of the trachea; one goes into each lung. [4·92]

brooding: Sitting on hatched young, or sheltering them under the wings, primarily to keep them warm, but also to protect them from sun, rain, or predators. Occurs either in the nest, or outside the nest in those species whose young leave the nest shortly after hatching. [8·95]

brood parasite: A bird that lays eggs in the nests of other species, leaving the resulting young to be raised entirely by the host parents. Some species parasitize others only occasionally; others, called **obligate brood parasites**, never build their own nests, and lay all their eggs in other species' nests. [8·139]

brood patch: A patch of skin on the breast and belly of birds that has lost feathers and become swollen through both the retention of large amounts of water in the tissues and the expansion of blood vessels feeding the skin. It develops a few days before egg laying in most individual birds (either male or female) that incubate their eggs by sitting on them, and increases the efficiency of heat transfer to the eggs. One large patch or several smaller ones may develop, depending on the species. [7·94]

brood reduction: Practice carried out by some parent birds whose young hatch asynchronously and thus vary in size, in which the parents first feed the most vigorously begging offspring until it can swallow no more, and then move on to another. Thus, in years with low food supplies only the largest and strongest young will survive, but in years with abundant food smaller young will survive as well. Brood reduction ensures that at least some young will survive in years when food is scarce. [8·98]

Brown-headed Cowbird: See **cowbird**. [8·142, 8·148]

bursa of Fabricius: Former name for the **cloacal bursa**. [4·123]

bustards: A family (Otididae, 25 species) of large, heavy-bodied, flat-headed birds with long legs and necks. These ground-dwelling birds are strong runners, and frequent open areas in the Oriental, Australasian, and southern Palearctic regions, although most species live in Africa. [1·83]

C

calamus: The hollow lower portion of the feather shaft, part of which lies beneath the skin; it has no vanes. [3·3]

call notes: Bird sounds that are generally shorter and simpler than songs. Many seem to convey a specific message, such as begging calls (hunger), alarm calls (danger), and contact calls (the caller's location). [7·14, 7·72]

CAM: See **chorioallantoic membrane**. [8·69]

camouflage: Coloration of an organism or structure serving to conceal it from predators, other enemies, or prey. [3·61]

canopy: The upper, continuous level of vegetation in a forest; it contains the branches of tall, mature trees and the epiphytes that grow on these branches. In dense forests, the canopy receives most of the sunlight and thus has the greatest

productivity. The tallest trees that stick above the main canopy, somewhat like "lollipops," are not considered part of the canopy; they are called the **emergent layer**. [9·93]

capillaries: The smallest blood vessels. Within all body tissues except the epidermis and those in the central nervous system, capillaries form networks called **capillary beds**, which connect the arterioles and venules. The thin walls of the capillaries allow materials to be exchanged between the blood and the body cells—the body cells absorb oxygen from the blood, and the blood takes up wastes from the body cells. [4·81]

capillary bed: See **capillaries**. [4·81]

cardiac muscle: A special type of smooth muscle that forms the bulk of the heart. Cardiac muscle can contract without being stimulated by the nervous system. [4·31]

carina: See **keel**. [4·23]

carinates: Birds that have a keel (carina) on their sternum; includes all flying birds and many smaller flightless ones. [4·23]

carnivores: Meat-eaters. [9·123]

carotenoids: Pigments producing bright yellow, orange, or red colors. They are synthesized only by plants, so birds must obtain them via their diets. [3·51]

carotid artery: Main artery supplying the head and neck region; among birds it is highly variable, with different species having none, one, two equal in size, or one large and one small. [4·84]

carpals: The bones of the wrist. In birds, they are fused and reduced to just two bones—the radiale and ulnare. [1·9]

carpometacarpus: The largest bone of the manus of birds, formed by the fusion of some of the carpals (wrist bones) with the metacarpals (palm bones). [1·9]

carrion: The flesh of dead animals. [9·28]

carrying capacity: The maximum population size or density that a particular area can support over the long term, without any degradation in the quality of the area or its resources. [9·70]

cartilage: A tissue with living cells embedded in a nonmineralized matrix; cartilage is found in the flexible joints and is capable of growth or resorption as well as transformation into bone. [4·6]

casques: Enlargements on the top of the bill or the front of the head, usually involving the underlying bone. Found in cassowaries and hornbills, among others. [3·48]

cassowaries: A family (Casuariidae, 3 species) of large, flightless ratites inhabiting New Guinea and Australia. They have a distinctive bony casque on top of their head. [3·49]

cast: See **pellet**. [4·119, 4·120]

caudal: 1. Toward the tail or posterior part of the body. [1·4] 2. Pertaining to the tail. [4·15]

caudal mesenteric vein: Vein that in birds brings blood from the lower portion of the digestive tract to the renal portal veins that form the venous ring connecting the lobes of the two kidneys; it is part of the avian renal portal system. [4·85]

caudal vena cava: Single large vein that gathers blood from veins coming from the legs, tail, kidneys, and caudal regions of the body and delivers the blood to the right atrium of the heart. [4·83, 4·84]

caudal vertebrae: The vertebrae of the tail. In birds, the anterior caudal vertebrae fuse with the sacral and lumbar vertebrae and some of the thoracic vertebrae to form the synsacrum; the next four to nine vertebrae articulate freely; and the posterior ones fuse to form the pygostyle (tail bone). [4·15, 4·18]

caval veins: General term for the right and left **cranial vena cavae** and the **caudal vena cava**; these large veins gather deoxygenated blood from the body and return it to the right atrium of the heart. [4·77, 4·84]

cavity adopters: Birds that nest in cavities but do not excavate their own, instead obtaining cavities that were created by physical forces (such as decay or erosion) or by other species. Bluebirds, Great Crested Flycatchers, Tree Swallows, and House Wrens are all cavity adopters. Also called **secondary cavity nesters**. [8·39]

CBC: See **Christmas Bird Count**. [9·60]

cell: Membrane-enclosed unit capable of metabolism and reproduction; the basic structural and functional unit of life. [4·2]

cell body: The part of a neuron containing the nucleus and surrounding cytoplasm; also called a **soma**. [4·32]

center of origin (of a species): The geographic area where a species evolved. [9·50]

central fovea: Area in the central part of the retina of most birds and mammals where the cones are most concentrated and the neural layer (the nerves from the rods and cones, which overlay the rods and cones and block some light) is the thinnest, and thus vision is the sharpest. The central foveae provide sharp monocular views of the areas to the sides of the bird. [4·49, 4·53]

central nervous system (CNS): The brain and spinal cord. [4·35]

centrum: The main body, or central axis, of a vertebra; the anterior end of the centrum connects to the preceding vertebra, and the posterior end connects to the following vertebra. [4·15]

cere: A leathery band of skin covering the base of the bill, into which the nostrils open; presumably the cere protects the nostrils. Present only in certain birds, such as hawks, pigeons, and some parrots. [3·40]

cerebellum: A large, deeply folded structure on the dorsal surface of the hindbrain, attached to the brain stem by two pairs of stout neural tracts; it controls muscular coordination and plays an important role in balance, posture, and proprioception. The cerebellum of birds is particularly large, due to the demands of flight. [4·37]

cerebral hemispheres: The two large, smooth lobes on the dorsal anterior region of the forebrain; together they form the **cerebrum**, which coordinates and controls complex behaviors, including memory and learning. In mammals, these lobes have folds and grooves. [4·36]

cerebrum: See **cerebral hemispheres**. [4·36]

cervical air sacs: A pair of air sacs—one sac on each side of the body—located in the neck region of birds; one sac usually extends from each lung, but sometimes a series of cervical sacs are located along the neck, as in geese. [4·101]

cervical enlargement: A swelling along the spinal cord at the level of the wings, associated with the brachial plexus. [4·40]

cervical vertebrae: The vertebrae of the neck region; the cervical vertebrae of birds have uniquely shaped centrum ends (see **heterocoelous centrum ends**). [4·15]

chalazae: In avian biology, the gelatinous, usually milky white, stringy coils of albumen (egg white) that surround and protect the egg yolk, and are visible at either end of the yolk as twisted cords. The chalazae attach to the far ends of the eggshell and form a suspension system for the yolk that allows it to rotate throughout embryonic development. [8·62]

channelization: The process by which humans deepen and straighten natural streams, converting them into water-filled ditches. Theoretically channelization controls flooding along the stream, but it destroys the stream ecosystem and often increases flooding downstream. [10·85]

chaparral: North American ecosystem found on the low hillsides of southwestern California; it consists of dense stands of broad-leaved evergreen shrubs, dominated by chamise and manzanita. This ecosystem is dry, with long hot summers with frequent fires; rain falls only in winter. A moderate number of birds breed in the chaparral. [9·120]

checklist: A printed list of the birds found in a particular area. [2·46]

cheek: See **malar region**. [1·7, 1·8]

chin: A small area under the lower beak of birds. [1·7, 1·8]

choana: A single slit in the roof of the mouth, running in an anterior-posterior direction, through which the two nasal cavities open to the mouth. [4·90]

chorioallantoic membrane (CAM): Membrane covering the entire inner surface of the avian eggshell, inside the inner shell membrane; it develops partway through embryonic development from a fusion of two extra-embryonic membranes, the chorion and allantois. The CAM is richly invested with blood capillaries, and together with the pores in the eggshell, it allows the embryo to carry out gas exchange as it receives oxygen from outside the egg and expels carbon dioxide. [8·66, 8·69]

chorion: In avian biology, the extra-embryonic membrane inside the egg that surrounds the entire avian embryo and the other three extra-embryonic membranes (the allantois, amnion, and yolk sac). It is homologous (evolutionarily related) to the mammalian membrane, also called the chorion, which forms much of the placenta in most mammals. [8·65, 8·69]

choroid: The middle layer of the three main layers of the eye; it lies just inside the sclera. It is pigmented and forms the iris and ciliary processes. [4·48]

Christmas Bird Count (CBC): A count of the wintering birds of North America conducted each year since 1900 and coordinated by the National Audubon Society. Observers count as many individual birds as possible within one of the circular count areas 15 miles (24 km) in diameter that are scattered across North America and beyond. Observers record their time spent and distance covered, so the numbers of birds seen can be adjusted for observer effort. The data can be used to track winter bird distribution and abundance, as well as long-term population trends. [9·60]

chromosomes: Long strands of DNA found in the nuclei of most cells. Each section of the chromosome (a specific sequence of nucleotides) that codes for a specific protein is called a gene. [1·44, 4·134]

ciliary muscles: Muscles that, in birds, attach to the ciliary processes, which attach to the lens of the eye. When the ciliary muscles contract, they move the ciliary processes, which squeeze the lens and make it become more round. In mammals, ciliary muscle contraction relaxes the lens, allowing it to become round by elastic rebound. [4·48]

ciliary processes: Structures of the choroid of the eye that attach to the lens and hold it in place. Ciliary muscles move the ciliary processes, which in turn move the lens, changing its shape. Only in birds do the processes attach directly to the lens. [4·47]

circadian rhythms: Daily cycles of behavioral and physiological events exhibited by organisms; they are regulated by an internal biological clock and persist even when organisms are kept under constant environmental conditions. Examples include daily patterns of activity, body temperature, or nectar production. [5·61]

circannual rhythms: Cycles of behavior, growth, or other physiological activities that occur on approximately a yearly basis; like circadian rhythms, they are regulated by an internal biological clock and persist even when organisms are kept under constant environmental conditions. Examples include yearly patterns of migration, shedding and regrowth of antlers, and hibernation. [5·62]

circumpolar constellations: The stars near to and surrounding the North Star (Polaris). [5·87]

cisticolas: A family (Cisticolidae, 117 species) of small, drab, insectivorous warbler-like birds of open, grassy areas of the Old World, primarily Africa. [8·7]

CITES: See **Convention on International Trade in Endangered Species**. [10·93]

class: Level of classification of organisms above "order" and below "phylum"; similar orders are placed within the same class. All birds are in the class Aves. [1·52]

clavicle: The collar bone; in nearly all birds, the left and right clavicles fuse with a small interclavicle bone to form the V-shaped furcula (wishbone). [4·19]

clavicular air sac: A single, median air sac between the clavicles and surrounding the bifurcation of the trachea of birds. [4·101]

claw arc: The angle between the tip and base of a claw, considering the claw as a section (arc) of a circle; used by researchers to compare the curvatures of different claws. It is significant because claw curvature is related to a bird's habits: ground dwellers have flatter, straighter claws (smaller claw arcs), perching birds have moderately curved claws, and trunk climbers have the most highly curved claws (largest claw arcs). [E·6]

clay lick: Clay banks rich in minerals, such as calcium, that are visited by frugivorous or seed-eating birds, such as parrots and macaws, who eat the clay to obtain minerals that are otherwise lacking in their diet. Clay licks are one type of **mineral lick**. [9·28]

Clean Water Act: Federal law passed in the United States in 1972 that attempted to improve the quality of surface waters by controlling pollution and requiring sewage treatment. **Section 404** (see separate entry) contains a set of protections for wetlands. [10·93]

climax community: The association of plants and animals (and other organisms) that can perpetuate itself in a given area in the absence of large-scale climatic change, disease, or human disturbance. The climax community is the final stage in the process of ecological succession in any given area. [9·109, 9·110]

cline: A gradual change in certain characteristics of individuals of the same species, which is evident in a geographic progression from one population to the next. [1·55]

cloaca: Common opening at the lower end of the avian digestive tract for the digestive, excretory, and reproductive systems; it receives feces from the large intestine, uric acid from the kidneys, and eggs or sperm from the gonads, and releases these materials through the vent. [4·123]

cloacal bursa: A lymphoid organ that opens into the roof of the cloaca in young birds and atrophies in later life; it produces special white blood cells called B cells, which are important in immune function and are of interest to researchers for their role in the development of AIDS in humans. Previously known as the **bursa of Fabricius**. [4·123]

cloacal phallus: The copulatory organ of male ratites and waterfowl; it is an elongate, spiral, ridged structure that erects by lymphatic pressure during copulation. Sperm travel along its surface to reach the cloaca of the female. [4·129]

cloacal protuberance: A swelling at the caudal end of the deferent ducts, visible externally in hand-held birds with the feathers parted; it is present only in breeding male passerines, and is used by researchers to determine sex and breeding condition. The swelling is caused by the enlargement of structures in the terminal regions of each deferent duct during the breeding season, and may help to keep sperm slightly cooler than the body's core temperature (as does the scrotum of mammals). [4·129]

clutch: A complete set of eggs; those laid in an uninterrupted series, for a single nesting, by one female. [8·78]

clutch size: The number of eggs in a given clutch. [8·78]

CNS: Central nervous system; the brain and spinal cord. [4·35]

coastal: Of the coast; coastal bird species primarily occupy the shallower waters around oceanic islands or above the

continental shelf, feeding mainly on fish, crustacea, and mollusks, which they find on or near beaches and other shorelines. They visit land frequently, during both the breeding and the nonbreeding season. [1·98]

cochlea: Elongate, bony, structure of the inner ear that is concerned with hearing; it is part of the bony labyrinth and thus is filled with perilymph. It contains the perilymph-filled vestibular and tympanic canals, which are connected to one another at one end, and the endolymph-filled cochlear duct, which lies between them. In mammals, but not in birds, the cochlea is coiled like a snail's shell. [4·59]

cochlear duct: Membranous canal inside the cochlea, bounded above by the tectorial membrane and below by a membrane that in birds is called the basilar papilla. The cochlear duct is part of the membranous labyrinth and thus is filled with endolymph. [4·59]

cochlear window: A soft spot at the end of the bony cochlea farthest from the vestibular window; it acts both as a pressure-release valve and as a damper for the waves in the cochlea. Formerly called the **round window**. [4·59]

coevolution: An evolutionary interaction between two or more species in which one evolves an adaptation that affects another, and then the other evolves an adaptation in response, and then the first evolves another adaptation in response to the response, and so on. Coevolution sometimes results in a kind of "battle" or "arms race" between two species. Notable examples of coevolution include interactions between predators and prey, plants and pollinators, and brood parasites and hosts. [8·146]

cohort: A group of individuals born during the same time period (often a year). [8·5]

cold-blooded: See **ectothermic**. [1·1]

cold front: The interface between a mass of cold air and the warm air mass it is overtaking; the dense, cold air tends to wedge under the warm air, forcing the warm air up and cooling it abruptly, forming precipitation that is often accompanied by strong winds and lightning. [5·69]

colic ceca (singular, **cecum**): Pouches extending from the junction between the small and large intestines that hold partly digested food long enough for bacterial action to further break it down; digested material is released to the large intestine, where any released nutrients are absorbed. Birds may have none, one, or one or two pairs of colic ceca, and the size is highly variable among species. Also spelled **caeca** or **caecum**. [4·123]

colies: See **mousebirds**. [1·85]

color phases: Polymorphisms in which the

morphs differ in color; for example, the red and gray phases of Eastern Screech-Owls and Ruffed Grouse, and the blue and white phases of Snow Geese. [9·81]

columella: Small, thin bone extending across the middle ear of birds, attached at one end to the inner surface of the eardrum and at the other end to the vestibular window of the inner ear; it transmits sound waves from the eardrum to the fluid-filled cochlea. [4·56]

comb: Fleshy, erect structure positioned longitudinally on top of the head of a bird, often with a serrated margin (like a hair comb), as in domestic chickens. [3·48]

commensal: Describes a relationship between two species or individuals in which one benefits and the other neither benefits nor is harmed. For example, when Cattle Egrets follow large herbivorous mammals to eat the insects they stir up as they move, the egrets benefit but the mammals appear unaffected in any way. [9·91]

communal roost: A group of birds gathered to spend the night together, sleeping; may consist of just one species or a number of different species. Birds that form particularly large and noisy communal roosts include vultures, ravens, crows, starlings, herons, egrets, ibis, grackles, blackbirds, cowbirds, robins, and the extinct Passenger Pigeon. [6·65]

community: All the populations of species living and interacting with one another in the same place. Communities also can be defined as containing only certain types of species in one location, such as a forest bird community or a stream insect community. [9·3]

competitive exclusion principal: Rule stating that no two species can occupy exactly the same ecological niche in a community; if they did, then eventually one would outcompete the other and cause it to go extinct. [9·102]

complete albino: An individual lacking all types of pigments in the plumage, eyes, and skin. [3·52]

complete molt: A type of molt in which the entire feather coat is replaced. [3·28]

conchae: Two thin, scroll-like structures (one median, one anterior) extending from the lateral wall of each nasal cavity; they are covered with a mucus-secreting membrane that contains the nerve endings of the olfactory nerves, which sense odors. The mucus traps dust, and blood vessels in the membrane warm the inhaled air. [4·90]

cones: One of the two kinds of light-sensitive cells lining the retina of the eye; they are responsible for visual acuity (due to their tight packing) and for sensing color information (via four or five pigments and specialized oil droplets in birds), but are not very sensitive to low light levels. When light energy stimulates a cone cell, it sends a nerve impulse to the brain via the optic tract. [4·48]

conical: See **pyriform**. [8·73]

coniferous forest: Ecosystem of colder parts of the temperate zone, where there is sufficient moisture to support a forest; cone-bearing trees, especially spruce and fir, dominate the vegetation. North America contains three main types of coniferous forests: **western montane forest**, **boreal forest (taiga)**, and **eastern montane forest**. See each forest type for more information. [9·116]

coniferous forest/deciduous forest ecotone: Transitional zone between the coniferous forest and deciduous forest ecosystems; it has a mixture of the two types of trees and, in North America, has more species of breeding birds than any other region. [9·123]

conservation biology: An applied science that combines information gained through the biological fields of ecology, population biology, animal behavior, and genetics to attempt to reverse the widespread declines and extinctions of species occurring throughout the world today. [10·39]

conservation plan: See **habitat conservation plan**. [10·92]

Conservation Reserve Program (CRP): Part of the 1985 Food Security Act that authorized the USDA to lease millions of areas of marginal croplands from farmers each year, paying them to keep the land in perennial vegetation to reduce soil erosion and crop surpluses, and to restore wildlife habitat. The program has helped prairie songbirds by providing millions of acres of grassland habitat. [10·35]

conspecifics: Members of the same species. [3·66]

contact call: A sound produced by a bird that appears to inform a nearby bird (usually a family member) of the caller's location. Often uttered by a mated male and female as they forage relatively close to one another. [7·73]

contour feathers: Feathers that make up the exterior surface of a bird, including the wings and tail; they streamline and shape the bird, and usually have well-developed barbules and hooklets. [3·5]

Convention on International Trade in Endangered Species (CITES): International agreement to which 150 nations voluntarily subscribe that binds participating parties to monitor, regulate, or prohibit the import and export of species that the group has deemed worthy of global pro-

tection. The species are listed in three Appendices: Appendix I lists the most endangered species, for which all commercial trade is prohibited, Appendix II lists species that would be in immediate danger if trade were not regulated, and Appendix III lists species added by individual countries that are requesting international help in regulating their trade. [10·93]

convergent evolution: The process by which organisms evolve similar forms, behaviors, or ecological characteristics not because they are related, but to meet similar environmental challenges. [1·40, 9·106]

cooperative breeding: Breeding system in which adults other than the breeding pair help the breeding pair to rear their offspring. In birds, if the helpers assist with incubation or care of nestlings, they are called "**helpers at the nest**" (see separate entry). [6·88]

cooperative foraging: A technique in which a group of individuals work together to obtain prey. For example, American White Pelicans may swim in a line or semicircle and beat their wings to drive schools of fish into shallow water. Other pelicans, cormorants, and mergansers, among others, forage cooperatively some of the time. Also called **social foraging**. [6·64]

coracoids: Strong, stout, paired bones of a bird's pectoral girdle; they are not present in mammals. During flight the coracoids function as a powerful brace holding the shoulder joint, and thus the wing, away from the body while the pectoral muscles pull on the wing in the opposite direction. [4·19]

corcoracids: The two members of the passerine (oscine) family Corcoracidae—the blackbird-like White-winged Chough and the smaller, seed-eating Apostlebird. These large, Australian cooperative breeders range over agricultural fields in huge flocks when not breeding. [1·95]

core area: The portion of a large, continuous habitat, such as a forest, that is far from the edges and thus is suitable to host species that would be adversely affected if forced to live near the edges. [9·100]

cornea: The transparent anterior surface of the sclera; it allows light to enter the eye. [4·48]

coronary arteries: Arteries (usually two) arising from the first part of the aorta; they carry blood that nourishes and supplies oxygen to the heart muscle itself. [4·79]

coronary veins: Veins that carry deoxygenated blood and wastes from the muscle tissue forming the heart back into the right atrium of the heart. [4·79]

corpuscles: An alternate name for blood cells. [4·86]

corridors: In conservation biology, long, narrow areas of wildlife habitat that connect larger areas, thus allowing individuals to move between the larger areas. [10·80]

countercurrent exchange system: A system in which two fluids (liquids or gases) flow adjacent to one another, but in opposite directions, while heat energy or materials move passively from one to the other, from higher to lower temperature or from higher to lower concentration. An example is the countercurrent heat-exchange system located in the legs of many gulls and waterfowl: warm blood on its way to the feet flows through a network of small vessels, which intertwines with another network of small vessels carrying cold blood flowing from the feet back to the body core. As a result, the blood returning from the feet is warmed, conserving body heat. [4·151] Another example is the countercurrent exchange system that efficiently accomplishes gas exchange in a bird's lungs: blood in the capillaries around the parabronchi moves in the opposite direction to the air in the air spaces. [4·100]

countershading: A type of coloration in which an organism is darker on top than below; countershading provides camouflage by reducing the contrast between the top and shadowed underside of an organism so that it appears less three-dimensional. [3·63]

countersinging: Interaction in which two birds sing back and forth to each other, alternating their songs. [7·36]

coursers: Together with pratincoles, form the family Glareolidae (17 species). These slender, plover-like ground nesters live in open areas in tropical parts of the Old World, and sometimes cool their eggs or young by partially burying them in sand or by bringing water to the nest in their breast feathers. [1·83]

courting nests: Nests, usually unlined, produced by male Winter Wrens and some male Marsh Wrens to attract females. A male may build a number of courting nests on his territory. In Marsh Wrens, females may add a lining and lay eggs in the courting nest of their chosen mate. [8·24]

courtship displays: Displays performed for the opposite sex to acquire a mate of the same species, maintain a pair bond, and/or stimulate and synchronize breeding behavior. [6·34]

courtship feeding: Feeding of a female by her mate that occurs during courtship and incubation, often in response to ritualized begging calls and postures that resemble those of the young. The function of courtship feeding is unknown, but hypotheses (which are not mutually exclusive) include (1) it strengthens the pair bond, (2) the food improves the female's condition, and (3) it allows the female to assess the male's ability to provide food for the young. [6·6, 8·104]

coverts: The smaller feathers that partly overlie the flight feathers of the wing and tail at their bases, like evenly spaced shingles on a roof. For more information, see specific types. [1·11, 1·13]

cowbirds: Six species of dark, slender, medium-sized birds that forage on seeds and insects; they are members of the New World family Icteridae (blackbirds and New World orioles). All species except the Bay-winged Cowbird are obligate brood parasites: they lay their eggs only in the nests of other species. The Brown-headed Cowbird, found throughout the United States and much of Canada, successfully parasitizes more than 140 host species and is thought to have caused severe population declines in some. It is the only brood parasite common in North America north of Mexico. [8·142, 8·148]

cracticids: Members of a distinctive family (Cracticidae, 10 species) of songbirds endemic to Australasia. They have stout, straight beaks; loud, melodic calls; and a generalist, crow-like diet of small vertebrates, eggs, insects, and fruits. Cracticids include butcherbirds, currawongs, and the Australasian Magpie. [1·95]

cranial: 1. Toward the head. In practice, may be used interchangeably with **anterior** (but see separate entry). [1·4] 2. Pertaining to the head, brain, or skull. [4·13]

cranial kinesis: The ability of the bird's upper jaw (upper beak) to move upward at the same time that the lower jaw (lower beak) is depressed, an action permitted by the highly flexible craniofacial hinge. [4·13]

cranial nerves: Twelve sets of paired nerves, each with a specific function, serving the head, neck, and thorax region. Most exit from the medulla oblongata of the brain. [4·37]

cranial vena cavae: Two large veins (**right** and **left**) that gather deoxygenated blood from large veins coming from the wings, head, and neck and deliver it to the right atrium of the heart. [4·83, 4·84]

craniofacial hinge: Flexible joint where the upper beak connects to the rest of the skull. Also called the **nasal-frontal hinge**. [4·11]

cranium: The part of the skull enclosing the brain; the braincase. [4·11]

creche: An assemblage of the still-dependent young of two or more (usually many) breeding females, attended by one or more adults. Bird species whose young form creches include some pelicans, fla-

mingos, geese, penguins, parrots, jays, and terns; creches of King Penguins and Emperor Penguins may contain several thousand young. Creche formation allows some birds (notably penguins) to conserve energy. It also may decrease predation on the young, and may free parents to spend time foraging in distant areas, thereby allowing them to bring more food back for the young and to prepare themselves for the next breeding. [8·126]

crepuscular: Active at dawn and dusk. [9·31]

crest: Tuft of feathers on the peak of the head that either stick up or can be raised. [2·10]

cricoid cartilages: Two major cartilages that make up the sides and floor of the larynx. Also called the **laryngeal cartilages**. [4·91]

crop: A dilation of the lower esophagus that stores food; it is found in many birds that eat dry seeds or fruit containing seeds. [4·115]

crop milk: Milk-like substance produced by pigeons and doves; it is composed of fluid-filled cells sloughed from the lining of the crop and is regurgitated to feed to nestlings. Crop milk is high in lipids and vitamins A and B, and has a greater protein and fat content than human or cow milk. Also called **pigeon's milk**. [4·116]

cross-fostering experiments: Studies in which young of one species are placed with host "parents" of another species, who then rear their "adopted" young. [9·11]

crown: The top of the head. [1·7]

CRP: See **Conservation Reserve Program**. [10·35]

crus: The lower leg; in birds, supported by the tibiotarsus bone. [1·14]

cuckoo: General term for birds in the large family Cuculidae. The nearly 50 species of Old World cuckoos are in subfamily Cuculinae, and are all obligate brood parasites. The familiar two-note chime of the cuckoo-clock mimics the song of one of these species—the Common Cuckoo of Eurasia. In the New World, cuckoos are in two subfamilies, Coccyzinae, whose members, all nonparasitic, are termed "New World Cuckoos" and include the Yellow-billed Cuckoo, Black-billed Cuckoo, and Mangrove Cuckoo of North America; and Neomorphinae (the neomorphine cuckoos), whose members are all Neotropical and include three species that are obligate brood parasites. [8·141]

Cuckoo-Roller: A crow-sized, arboreal bird with a stout, broad bill; it is the only member of its family (Leptosomatidae) and is found only in Madagascar. [1·88]

cuckoo-shrikes: A diverse group of arboreal songbirds, many of which have a shrike-like bill and are slender and (in some cases) barred like cuckoos; however, they are related to neither group. Together with minivets and trillers, they make up the family Campephagidae, found in the warmer parts of the Old World. [1·96]

cup nest: A nest in the shape of a cup, usually constructed of mud, or small twigs and dried grass, with a depression in the center to hold the eggs; built by the majority of bird species. [8·30]

cursorial: Adapted for running. [E·14]

cursorial theory (of the origin of avian flight): A theory suggesting that the ancestors of birds evolved to fly by first running along the ground, and then by jumping and leaping, which was augmented by the evolution of wings and feathers, which eventually led to full flight. First proposed by Samuel Williston in 1879. Also called the "Ground-Up Theory." [E·14]

cytoplasm: The contents of a cell outside the nucleus but within the cell membrane. [1·44]

D

dabbling: A foraging technique in which a bird moves the beak rapidly on the surface of shallow water to pick up small aquatic animals and plant materials; it is used by "dabbling ducks" and a few other species. [6·47]

dabbling ducks: Ducks (such as Mallards, teal, wigeon, and pintails) that feed by dabbling on the water's surface, in contrast to diving ducks, which dive under water to search for plant material or aquatic organisms. Dabbling ducks also may tip "bottom up" and reach down under the water to obtain submerged food. Also called **puddle ducks**. [6·47]

dark meat: See **red fibers**. [5·7]

Darwin's Finches: See **Galapagos Finches**. [1·60]

dawn chorus: The great amount of bird song heard around dawn. At this time the largest number of bird species are singing, and they sing more frequently, and often more energetically and with more variety than at other times of the day. Why birds sing most at dawn, and in different ways than they sing during the day, is not known. [7·75]

dawn song: Bird song usually given only during the early morning hours; it differs from a species' normal daytime song. [7·75]

DDE: A stable, persistent, toxic organic compound (1,1-dichloro-2,2-bis(p-chlorophenyl)eth-ylene) formed in the body by the metabolism of the organic pesticide **DDT** (see separate entry). DDE accumulates in fatty tissues and is excreted very slowly, and when concentrations become high it can cause death or other toxic effects such as reproductive failure resulting from eggshell thinning (due to the disruption of calcium metabolism). Thin eggshells severely decreased reproductive success in North American raptors in the 1950s and 1960s, causing populations of most raptor species to plummet. [10·52]

DDT: An organic pesticide (dichloro-diphenyl-trichloroethane) used commonly in the United States from the mid-1940s to the early 1970s to control Mexican boll weevils, gypsy moths, mosquitoes, and other insect pests. DDT is highly persistent in the environment and is taken in by organisms and converted to DDE, a toxic compound that accumulates in fatty tissues and is excreted very slowly. DDT was banned in the United States in 1972, but is still used in other countries, including Argentina, Belize, Ecuador, Guyana, Peru, and Mexico. See **DDE** for more information. [9·125, 10·52, 10·53]

death rate: The number of individuals in a species or population that die in a given period of time; also called the **mortality rate**. [9·49]

deciduous: Describing trees or shrubs that lose their leaves during part of the year. They generally drop their leaves during the drier season (which in the temperate zone is winter) to conserve moisture, which might otherwise be lost through evaporation. [9·117]

decomposers: Level in a food chain or web that consists of organisms (such as earthworms, fungi, and bacteria) that eat dead organisms and waste products, breaking them down into basic nutrients (such as oxygen, nitrogen, and phosphorus) and returning them to the soil for plants to use. [9·123]

decurved: Curved downward. Used to describe beaks such as that of the White Ibis. [1·17]

deferent ducts: Highly convoluted tubes (also called **vasa deferentia**) that carry sperm from the testes to (in birds) the cloaca; in birds, the lower portion of each is enlarged to form a temporary storage receptacle for sperm. [4·128]

definitive plumages: Any of the plumages of a fully mature bird; they may change seasonally, but do not change from year to year as the bird ages. Some species, such as gulls, large raptors, and pelagic seabirds, take several years to reach their definitive plumage. [3·30]

deflective coloration: Conspicuous markings found on otherwise cryptically colored organisms; on birds, these mark-ings show only in flight or when they are flashed. They may have antipredator or social functions. [3·64]

delayed plumage maturation: A situation found in some species, in which one sex remains in subadult plumage longer than the other sex. [3·32]

dendrites: Rootlet-like extensions from the cell body of a neuron; they usually receive nerve impulses across a synapse from other nerve cells, but may transmit them as well. [4·32]

dens: The upwardly projecting knob or peg of the axis (second cervical vertebra); it fits into a hole in the ventral surface of the atlas (first cervical vertebra). [4·16]

density: See **population density**. [9·58]

density-dependent factor: A factor that regulates population density in such a way that the magnitude of its effect is determined by the population density. For example, population density is decreased by disease, which has a greater and greater effect as population density increases and thus rates of disease transmission go up. Other density-dependent factors include predation, parasite levels, and competition for resources. [9·69]

density-independent factor: A factor that affects population density in such a way that the magnitude of its effect does not depend on the population density. Examples include severe weather, natural disasters, and the failure of food supplies. [9·75]

dentary: In birds, the bone forming each side of the lower beak. [4·11]

deoxygenated blood: Blood whose red blood cells carry very little oxygen. Also called **oxygen-poor blood**, it is found in all veins except the pulmonary veins and venules. [4·81]

depolarizing material: A substance that can be used to take a polarized light beam and vibrate it in all directions, creating waves of all orientations, to form an unpolarized beam. [5·94]

dermal papilla: The portion of a feather papilla formed from dermal tissue; blood vessels extend from the dermal papilla into the developing feather, providing nourishment. This core of tissue remains in a feather follicle throughout a bird's life, ready to aid each round of feather development after a feather is lost. [3·26]

dermis: The inner layer of the skin; it lies just beneath the epidermis and contains blood vessels, muscles, and nerves. [3·26]

descending aorta: The continuation of the aorta (after it exits the heart and curves) down through the body and toward the tail. [4·82]

determinate layers: Bird species that will not lay additional (replacement) eggs if one or more are removed from a clutch during laying. Most determinate layers will lay a new clutch if the entire clutch is destroyed, however. For comparison, see **indeterminate layers**. [8·90]

deterministic: In biology, describes events whose occurrence and/or outcomes are inevitable, based on a certain set of starting conditions. For comparison, see **stochastic**. [9·76]

dialect: A geographic cluster of similar vocalizations (bird song, human speech, or the sounds of other animals) that is a consequence of those vocalizations being learned. Dialects may exist over very small or very large areas, depending on the details of dispersal and learning. [7·53]

diapsid: Describes a condition in which the skull has two openings on each side in the temporal region, posterior to the eye socket. **Diapsid skulls** have this arrangement, and **diapsid reptiles** are those reptiles that have diapsid skulls—including thecodonts and their descendants (including birds and crocodiles), snakes, and lizards. Turtles are the only living group of reptiles that are not diapsids. [E·7]

differential exploitation: A situation in which direct interactions among two or more individuals that seek the same resources are avoided because the individuals use the resources in slightly different ways or use slightly different resources. Differential exploitation may result from the evolution of either morphological or behavioral differences among individuals. [9·39]

dilution effect hypothesis: The idea that colonial breeding may benefit individuals because their large numbers can overwhelm the consumption capacity of local predators. [6·59]

dinosaur theory (of bird evolution): The theory that birds evolved from theropods such as *Compsognathus* approximately 150 million years ago (proposed by Thomas Huxley in 1868) or from Dromaeosaurs such as *Deinonychus* approximately 110 to 120 million years ago (proposed by John Ostrom in 1973). [E·8, E·10]

diopter adjustment ring: The ring on binoculars, usually on one of the eyepieces but sometimes on the hinge post, that allows the eyepieces to be focused independently to make up for the differences in visual acuity between an individual's two eyes. [2·30]

dip angle: The angle at which the magnetic field lines around the earth contact the earth. The dip angle is 0 degrees at the magnetic equator, and approaches 90 degrees near the magnetic poles. [5·90]

direct benefits hypothesis: One possible explanation for why females of some species choose the males with the most elaborate ornaments (such as ornate plumage) to copulate with. The explanation applies primarily to females choosing nonpaternal sexual partners, either (1) males for extrapair copulations, or (2) mates in species in which males do not provide parental care or other resources for their offspring (such as territories with food or nesting sites). The direct benefits hypothesis suggests that females may choose the most ornamented males because they are least likely to infect the females with mites, a disease, or some other affliction. Thus the female gains reproductive advantages because her health is not diminished by her mate choice. [6·84]

direct fitness: An individual's direct fitness is the portion of its genes that is transferred into the next generation (and eventually beyond) through the production of its own offspring. This contrasts with **indirect fitness**, which is the portion transferred as a result of an individual's blood relatives producing offspring. [6·86]

disjunct range: A range of a species or population that is not continuous, but rather is divided into geographically separate areas. [9·50]

dispersal: In biology, usually the movement of individuals away from the area where they were born, or away from areas containing concentrations of individuals. [9·62]

displacement activities: Behaviors or actions that seem irrelevant or inappropriate to the current situation. For example, a Herring Gull may stop to preen in the middle of a territorial conflict. Ethologists hypothesize that displacement activities occur because of conflicting motivations or indecision. Some displacement activities have become exaggerated and ritualized into displays. [6·32]

disruptive coloration: A type of cryptic coloration with patches, streaks, or other bold patterns of color that break up the shape of the organism, catching the eye and distracting the observer from recognizing the whole organism. [3·62]

distal: Away from the center of the body (fingers are distal to the elbow) or from the origin of the structure (the tip of a feather is distal to its base—where it is attached). [1·4]

distraction displays: Displays in which a bird or other animal feigns injury or in some other way creates a highly noticeable fuss or disturbance, in order to shift a potential predator's attention away from the bird's nest or young. For example, Killdeer give a broken-wing distraction display by dragging and flapping one wing and uttering distress calls, while slowly fluttering along the ground away from the nest or young. [6·53]

diurnal: Active during daylight. [9·31]

divergent evolution: A type of evolution in which different populations of the same species become increasingly distinct from one another over many generations (due to exposure to different ecological factors), eventually diverging into two or more new species. [1·57]

diving ducks: Ducks (such as Canvasbacks, Redheads, scaup, and goldeneyes) that feed by diving under the water's surface to obtain aquatic plants or animals, in contrast to dabbling ducks, which remain at the water's surface. [6·45]

DNA (deoxyribonucleic acid): The genetic material of all cellular organisms and some viruses, forming the chromosomes of these organisms. A molecule of DNA consists of two long strands of nucleotides, held together by bonds between nitrogenous bases on the two strands, with the whole structure twisted to form a double helix. The sequence of the nucleotides is the "genetic code," which contains the instructions for making proteins, which in turn determine the characteristics of an organism. [1·42]

DNA-DNA hybridization: A technique used to determine the degree of similarity (in nucleotide sequence) between two different samples of DNA. It is often used to compare the DNA of two different species, to estimate how closely related they are and to hypothesize their evolutionary relationships. [1·46]

DNA fingerprinting: A technique by which the nucleotide sequences of selected portions of the DNA of individuals of the same species are analyzed and compared to determine how closely related the individuals are likely to be. Used in criminology to determine if the DNA of clues such as hair or semen match the DNA of a suspected criminal; also used in biological research to determine the relatedness among individuals whose family histories are not known—sometimes to avoid the breeding of closely related individuals of endangered species. [10·76]

domed nest: A cup nest with a woven overhead dome that likely helps to conceal the eggs or nestlings; built by meadowlarks, snipe, Ovenbirds, and others. [8·34]

dominance hierarchy: A ranking system of social status among members of a group; often established and maintained through displays and various aggressive-submissive behaviors, including, on occasion, physical combat. Many dominance hierarchies are linear—A dominates B, B dominates C, and C dominates D—but other arrangements exist. [6·24]

dorsal: Toward the back (the vertebral column) of an organism. [1·4]

double-clutching: A technique in which biologists remove one or several eggs from the nests of indeterminate layers and rear those eggs in the lab while the birds lay replacement eggs and rear the resulting young in the wild. Used by conservationists to increase the population size in declining species such as California Condors and Peregrine Falcons. [8·91]

down feathers: Soft, fluffy feathers, typically lacking a rachis. Because the barbules lack hooks, the barbs do not cling together, so they trap more air and thus provide extra insulation. Some adult birds have **body downs** under their contour feathers, and young birds have **natal down** before molting into their juvenal plumage. [3·16]

drafting: Driving closely behind another vehicle and being pulled forward by air currents moving back over the top of the first vehicle and swirling down and forward, as well as by air currents swirling up and forward from below the first vehicle. Used by race car drivers or cars following trucks on a highway; the energetic advantages are similar to those experienced by a bird flying close to and straight behind another bird, as in a straight-line formation. [5·47]

drag: A force on a moving object, resulting from the friction between the object and the fluid (such as air) that it is moving through. [5·16]

dromornithids: A group of giant, flightless birds that lived in Australia beginning in the Tertiary period, but that became extinct about 10,000 years ago during the Ice Age. They are thought to be neognathous birds most closely related to the Anseriformes (ducks, geese, and swans). [E·23]

drumming: 1. Nonvocal sounds produced by woodpeckers banging on dead trees or other resonant objects with their beaks; most sound like short, emphatic drum rolls, and are given by both sexes to proclaim territory and attract mates. [7·15] 2. A series of accelerating, nonvocal, muffled "thumps" at a very low frequency, produced by male Ruffed Grouse to proclaim territory and attract mates. Produced by repeatedly bringing the wings forward and upward so rapidly that with each stroke they compress a parcel of air between the chest and wings, creating a sound wave without the wings and chest ever touching. Usually performed while standing on a hollow log, which acts as a resonating chamber to amplify the sound. [7·15]

duet: Singing performance by two individuals. Avian duets usually are given by a paired male and female who may sing in synchrony, overlap one another, or alternate their songs, depending on the species. Most duetting species are tropical, and include some parrots, woodpeckers, antbirds, flycatchers, shrikes, and wrens. [7·78]

dump laying: Laying an egg or eggs in the nest of a conspecific (or sometimes, a similar species). Females that dump lay usually also build their own nests and incubate their own eggs. Wood Ducks, Hooded Mergansers, and some other cavity-nesting waterfowl commonly dump lay, sometimes resulting in large numbers of eggs in one nest. Also called **dump nesting** or **egg dumping**. [8·76, 9·95]

duodenum: The U-shaped first loop of the small intestine, running from the stomach to the jejunum. [4·122]

dura: The outermost, toughest, and most fibrous vascularized membrane surrounding the brain and spinal cord. Along with the other meninges, it provides sustenance and waste removal for the cells of the brain and spinal cord, which are not served by the circulatory system. [4·36]

dust-bathing: Driving fine particles through the feathers by rolling the body, fluffing the feathers, and wiping the head and bill in a dusty area or by picking up dust and throwing it over the body, after which dust is shaken or preened out; important for feather maintenance and removal of ectoparasites. [3·21]

dynamic pressure: The pressure of a flowing fluid, or of movement through a fluid such as air or water. You feel dynamic pressure when wind blows against your face. [5·12]

dynamic soaring: A type of flight in which birds use the gradient in wind speed that exists over the surface of the ocean to travel for long distances without spending much of their own energy: the bird glides down the gradient at an angle, then turns and abruptly rises into the wind, using its momentum to gain height quickly, then turns and glides down again, crossing the ocean in large zig-zags. Dynamic soaring is used most by albatrosses and other large pelagic birds with high-aspect-ratio wings. [5·44]

E

eardrum: See **tympanic membrane**. [4·56]

eastern montane forest: North American coniferous forest ecosystem dominated by spruce and fir trees and found below alpine tundra but on the higher summits, ridges, and slopes of the Appala-chians from northern New England south through Georgia and Alabama; it hosts relatively few birds. [9·116]

eclipse plumage: The set of dull-colored feathers worn briefly after the breeding season by some adult birds, such as ducks. In eclipse plumage, male ducks look like females, which do not change much in appearance. Eclipse plumage is acquired by a complete molt after the breeding season, and is soon replaced through a partial molt that produces the brighter colors of the breeding plumage. [3·34]

ecological constraints hypotheses: A set of hypotheses that each give an explanation for why certain individual birds might forego their own breeding in a particular breeding season, instead acting as helpers at the nest of other breeding pairs (usually their parents or other close relatives). The ecological constraints hypotheses focus on the possible costs to young birds of dispersing from their natal territory; examples of these hypotheses include (1) few vacant territories of good quality may be available, (2) few suitable breeding partners may be available, and (3) the birds may have little chance of reproducing successfully until they gain "parenting" experience. [6·89]

ecological isolating mechanisms: Structural, physiological, or behavioral adaptations that have evolved in species that have very similar niches, and that allow the species to divide up the resources in various ways (such as using slightly different resources or foraging in different parts of the habitat), and thus to coexist. For example, certain wood-warbler species can coexist in spruce forests of northeastern North America even though they all eat similar small insects, because they forage in different parts of trees. [9·103]

ecological niche: See **niche**. [9·102]

ecological release: An expansion of the niche of certain populations of a species, such that a greater breadth of resources such as habitat and food are used, in areas where interspecific competition is lower, as on islands. (In these situations, the niche is "released.") [9·104]

ecological succession: The process by which one association of plants and animals is replaced by another, then that one is replaced by another, and so on until the climax community for that area is reached. The types of communities, and the order in which each succeeds the previous one, is fairly predictable for a given habitat and geographical region. **Primary succession** begins on a substrate, such as rock, sand, or lava, that has never before supported a community. The process by which a lake gradually fills in to form a bog community also is considered a form of primary succession. **Secondary** succession begins with bare soil or an existing community. [9·109, 9·110]

ecology: The study of the relationships between organisms and their environment. [9·1]

ecoregional planning: Careful, biologically driven planning at the level of ecoregions, with the ultimate goal of preserving the species and ecosystem processes (for example, fire and pollination) that occur within each ecoregion. Ecoregional planning involves geographically delineating ecoregions; cataloging the ecosystems within each ecoregion; identifying and mapping the most important species, communities, and habitats; determining potential threats to species and sites; and then prioritizing species, communities, and the key habitats as to which require the most urgent conservation action. Ecoregional planners then set conservation goals, outline plans, and monitor the results. [10·86]

ecoregions: Regions of the world that are biologically distinct in terms of climate conditions, topography, soil types, and plant communities; also called **physiographic regions**. [10·86, 10·87]

ecosystem: Both the living and nonliving components of a particular area (including the physical surroundings), as well as the ecological processes that bind them all together (such as decomposition, soil erosion, and water and nutrient cycling). Ecosystems may be small (for example, the ecosystem of a rotting log) or large (a deciduous forest ecosystem). [9·109]

ecosystem management: Understanding and maintaining entire **ecosystems** (see separate entry), instead of focusing on particular species or habitat types. [10·83]

ecotone: A zone of transition between two ecosystems, such as the oak savanna ecotone between the eastern deciduous forests and prairie grasslands of North America. Ecotones host a greater number of species than either adjacent ecosystem, because some species in each of the two adjacent ecosystems frequent the transition zone, and because some species prefer the greater variety of resources found at habitat edges and thus live specifically in those areas. [9·93]

ectoparasites: Parasites, such as flies, ticks, fleas, lice, and mites, inhabiting the exterior of a host's body. [3·19, 3·23]

ectothermic: Describes organisms that must rely on sources of heat outside their own bodies to keep warm; also called **cold-blooded**. [1·1]

edge effect: The tendency for areas near the edge of a habitat patch to differ from areas near the center in a number of different ways. For example, areas where two habitats meet fairly abruptly (such as a forest/field boundary) often host a higher number of species than do either of the two adjacent habitats. Other ways in which the edge of a habitat patch may differ from the center include changes in microclimate, such as amount of sunlight or humidity; increased habitat disturbances, such as fire and wind damage; and higher numbers of introduced plants or animals (because these species generally invade from the edge). Forest birds breeding near edges may experience higher rates of predation and/or brood parasitism than birds breeding near the center, because predators such as raccoons, squirrels, and Blue Jays and parasites such as Brown-headed Cowbirds are more common near edges. [9·93, 10·73]

egg: 1. The ovum; the female reproductive cell sometimes called the egg cell, both before and just after it is fertilized by a sperm cell. [4·130] 2. The hard-shelled structure laid by birds, containing the embryo, yolk, and white. [4·130]

egg tooth: A short, pointed, calcareous structure on the tip of the upper beak (and sometimes the lower beak as well) that develops in bird embryos shortly before hatching; the embryo rubs and pounds the egg tooth against the inner wall of the eggshell to break it open and hatch. The egg tooth sloughs off or is resorbed by the growing chick within a few days after hatching. [8·104]

electrophoresis: A method of separating large, charged molecules of different lengths or charges (DNA fragments or proteins treated to carry a charge) by their rate of movement through a thin slab of gel in an electric field. [1·43]

elephantbirds: A family (Aepyornithidae) of huge, flightless ratites that lived on the island of Madagascar beginning 10 to 20 million years ago, but were exterminated about 2,000 years ago by human activity. The tallest stood about 10 feet (3 m) and weighed about half a ton. [E·23]

elliptical wings: Short, broad wings having a low aspect ratio; they allow great maneuverability, but do not promote efficient or rapid flight. Elliptical wings are common in birds that live in forests, woodlands, or shrubby areas, such as crows, grouse, quail, and most songbirds. [5·37]

emarginate tail: Tail shape in which the rectrices become slightly longer from the inside out. Also called **notched**. [1·19]

embryo: A developing young animal that is still inside its egg or mother; in some animals, especially mammals, refers only to the earlier developmental stages. [8·64]

emigration: The movement of individuals out of a population. [9·49]

Emu: A large, flightless ratite of Australia; it is the only member of its family (Dromiceidae) and is the second largest living bird, next to the Ostrich. [1·93]

enantiornithine birds: See **opposite birds**. [E·20]

Endangered Species Act (ESA): Federal law passed in the United States in 1973 that commits the government to take action to prevent the extinction of native species and to protect their habitat. It also establishes a procedure to develop a list of threatened and endangered species, identify their critical habitat, and develop and carry out Recovery Plans. [10·23, 10·91]

endemic: Found only in a particular region; describes a species or other taxonomic group. For example, kiwis are *endemic* to New Zealand. [1·70]

endocrine glands: Structures that secrete hormones directly into the blood; the hormones are carried to other parts of the body, where they stimulate or regulate the activities of other glands or organs. The major endocrine glands of birds are the pituitary, thyroids, parathyroids, ultimobranchials, adrenals, gonads, pancreas, pineal, thymus, and cloacal bursa. [4·71]

endocrine system: Organ system that acts with the nervous system to initiate, coordinate, and regulate body functions, including reproduction and development. It consists of the endocrine glands and their secretions, called hormones. [4·69]

endolymph: The fluid that fills the structures forming the membranous inner labyrinth of the inner ear: the cochlear duct, the semicircular ducts, and the utriculus and sacculus. [4·57]

endothermic: Having the ability to generate one's own body heat through metabolic processes; also called **warm-blooded**. Only birds and mammals are endothermic. [1·1, 9·13]

entoglossal: The bone (part of the hyoid apparatus) that supports the tongue; also called the **tongue bone**. [4·14]

environment: The surroundings of an organism, including physical features, chemical and energetic factors, and other living organisms. [9·7]

environmental stochasticity: The tendency of nearly all environments to experience many random (or at least unpredictable) events, such as natural disasters or severe weather. If a population is small and "unlucky" enough to be struck by one of these events, it may be entirely wiped out. [9·77]

enzyme: A protein that catalyzes (assists) a biochemical reaction without being consumed in the reaction. [1·41]

epaulettes: Patches of colored feathers on a bird's shoulders, such as the red feathers on a Red-winged Blackbird. [3·5, 6·73]

EPC: See **extrapair copulations**. [6·79]

epidermal collar: Structure formed during feather development as a feather papilla elongates: the outer cells harden, fuse, and form a ring or "collar" of epidermal tissue surrounding the dermal portion of the original papilla. The collar cells multiply to produce most structures in the developing feather. [3·26]

epidermis: The outer layer of the skin; it protects the inner layer—the dermis—and does not contain blood vessels. In birds, it gives rise to the feathers and to the horny sheath covering the bill, legs, and feet, including the claws. [3·26]

epiphytes: Plants that grow on other plants but, in contrast to parasitic plants, use their roots for attachment rather than to obtain nutrients from their support plant; examples include many bromeliads ("air plants") and orchids, and a few cacti. [9·88]

epizootic: 1. A disease that spreads quickly among crowded animals, such as viral enteritis (duck plague), which may kill hundreds or thousands of waterfowl. [9·69] 2. Term used to describe a disease that spreads quickly among animals when they are highly crowded. [9·69]

epoch: A unit of geological time. Successive epochs make up a period. [1·113]

equilibrium theory: Theory proposing that the number of species on an island at any one time represents a balance between the number of new species colonizing the island (immigration) and the number of species becoming extinct. A consequence of this relationship is that smaller islands tend to have fewer species because (1) they tend to have lower immigration rates (they are less likely to be "discovered" by colonists), (2) they have higher extinction rates (the populations are smaller and thus more likely to go extinct due to stochastic factors, and the fewer resources and reduced habitat diversity are more likely to lead to competitive exclusion), and (3) they typically contain a lower diversity of habitat types than larger islands, thereby providing niches for fewer kinds of species. This theory applies only in situations in which speciation is not a major source of new species, for example over short periods of time, or on islands too small or too far from other islands to permit speciation through geographic isolation. [9·95]

era: A unit of geological time. Eras are divided into periods. [1·113]

erythrocytes: See **red blood cells**. [4·88]

ESA: See **Endangered Species Act**. [10·23, 10·91]

esophagus: Thin, straight muscular tube carrying food from the pharynx to the stomach. It is lined with mucus-secreting glands that moisten food to ease its passage, but contains no digestive glands. [4·113]

ethology: The study of animal behavior, primarily from a proximate approach, by comparing similar behaviors in related species to understand how certain behaviors evolved, and by investigating releasers and instinctive behaviors and the underlying physiological processes. Some of the earliest animal behaviorists (primarily Europeans) were concerned mainly with ethology, and the two pioneers of the field, Konrad Lorenz and Niko Tinbergen, laid the groundwork for animal behavior studies carried out today. [6·7]

eustachian tube: See **auditory tube**. [4·57]

eutrophication: The changes in a lake, estuary, or slow-moving stream when it receives excess plant nutrients, especially nitrates and phosphates. Some nutrient input occurs naturally, through the erosion of soil and run-off from adjacent land. In many cases, however, human activity greatly increases the rate of nutrient input or adds new sources, such as discharges from industries and sewage treatment plants; when human activity is involved, the changes are sometimes termed **cultural eutrophication**. One important change is the dramatic increase in the growth of plants and (especially) algae, which can choke out native plants. Then, as the masses of plants and algae die and sink to the bottom they decompose, depleting the water of oxygen, and thus few aquatic organisms can continue to survive. In addition, some algae may produce toxins, dyes, or odors that decrease water quality. [10·56]

eventual variety: Pattern of singing in which a bird repeats one song type many times before switching to a different type, which it then repeats many times, and then switches to another type, and so on. [7·86]

evolution: A change over time. The evolution of living things is the set of cumulative changes in the characteristics of a species or population over successive generations that result from natural selection acting on the genetic variation among individuals. [1·34]

excavators: In avian biology, bird species that dig their own nest cavities or tunnels, either in sandy soil (such as Belted Kingfishers, Bank Swallows, bee-eaters, mot-mots, and Crab Plovers) or in wood (woodpeckers). [8·39]

excretory system: Organ system consisting of the kidneys and their ducts, the ureters (in the rheas of South America,

but in no other birds, a urinary bladder is also present); it removes toxic nitrogenous wastes from the blood by producing, storing, and excreting urine. In birds, the urine is composed of uric acid. Also called the **urinary system**; it is often considered together with the reproductive system as the urogenital system. [4·125]

exit pupil: The space through which the light beam exiting the eyepiece of binoculars or a telescope passes; this is, in effect, the hole through which the observer looks. The diameter of the exit pupil is calculated by dividing the size of the objective lens by the magnification; the larger the exit pupil, the brighter the image. [2·35]

extensor: A muscle that pulls one bone away from another bone. [4·27]

external ear: The portion of the ear containing the external ear canal and the eardrum (tympanic membrane. [4·56]

external ear canal: The channel through which sounds enter the ear; it leads from outside the body to the eardrum. [4·56]

extinction events: Large-scale extinctions of many species; they have occurred periodically throughout Earth's history. [1·113]

extra-embryonic membranes: Membranes that protect and nourish the growing embryo, but do not become part of the adult body. The four main ones in the avian egg are the yolk sac, amnion, allantois, and chorion. [8·65]

extrapair copulations (EPCs): Copulations with birds other than one's mate. [6·79]

eyeball: The eye; in birds it is flat to tubular, in contrast to the spherical eyeball of mammals. Its outer layer is the sclera, and the entire eyeball sits in the socket and is protected on its exposed side by the eyelids and nictitating membrane. [4·47]

eyebrow stripe: A distinctively colored line running from the upper beak toward the back of the head, located ventral to the boundary of the forehead and crown; also called the **superciliary line**. [1·7, 2·10]

eyeline: A distinctively colored line that passes through the eye. [2·10]

eye ring: A circle of distinctively colored feathers or skin surrounding the eye. [2·10]

F

facial disc: Flat, relatively round, forward-facing part of the head of owls; it probably funnels sounds into the bird's ear openings. Also spelled **facial disk**. [4·55]

facial nerve: The seventh cranial nerve; it carries motor signals to muscles that

protrude the tongue, lower the lower beak, constrict the neck, and (in birds) tense the columella ear bone. It also may carry some taste sensory input from the tongue. [4·41]

facultative partial migration: See **partial migration**. [5·56]

fairy-bluebirds: Two species of arboreal songbirds endemic to the Oriental zoogeographic region; they are named for the brilliant blue and black plumage of the males, and feed primarily on figs and other fruits. Together with leafbirds, they make up the family Irenidae. [1·89]

fairywrens: An Australasian family (Meluridae, 26 species) of cooperatively breeding, wrenlike birds with long, cocked tails. [1·95]

false sunbirds: Members (with asities) of the family Philepittidae, endemic to Madagascar. These two species of suboscine passerines feed on insects and nectar. [1·88]

family: Level of classification of organisms above "genus" and below "order"; similar genera are placed in the same family. The scientific names of bird families end in "idae" (for example, Corvidae). [1·52]

FAP: See **fixed action pattern**. [6·8]

fascia: Connective tissue binding together hundreds or thousands of muscle fibers to form a skeletal muscle; it may be in the form of bandlike tendons or broad, shiny sheets called aponeuroses. [4·27]

fat bodies: In birds, yellowish fat deposits laid down just under the skin, usually in individuals storing fat in preparation for migration; the most conspicuous fat bodies lie over the abdomen and in the depression formed anterior to the breast muscles where the clavicles fuse to form the wishbone, and are visible in a handheld bird with the feathers parted. Researchers often use the degree of fat accumulation in a migrant as an indication of its energetic condition and potential to continue migration. [5·64]

feather comb: See **pectinate claw**. [3·46]

feather papillae: Small bumps covering the surface of the skin of birds during embryonic development. They consist of a core of dermis and a covering of epidermis, and each will eventually form an embryonic feather. Also simply called **papillae**. [3·26]

feather sheath: A thin, cylindrical tube of keratin surrounding and protecting a developing feather. It eventually breaks open to let the mature feather unfurl. [3·26]

feather tracts: Areas of a bird's skin where feathers are attached; also called **pterylae**. [3·2]

fecal sac: A tough, flexible bag enclosing the feces of most passerine nestlings; it allows the parents to remove and dispose of the feces more easily—parents sometimes grab the fecal sacs as they emerge from a nestling's cloaca. Many parents carry the sacs some distance from the nest and drop them, but others consume them. [8·136]

fecundity: See **birth rate**. [9·49]

female-defense polygyny: Mating system in which males compete fiercely for control of clusters of nesting females. In some species, such as Montezuma Oropendolas of Central America, a male dominance hierarchy results with the top few males securing most of the matings. Because males do not help to rear the young, this system may evolve when male parental care is less important to the survival of the young than are safe nesting sites or rich food supplies. [6·74]

femur: Bone that supports the upper hind limb (thigh) of many vertebrates, including birds and humans. [1·14]

fibula: The thinner of the two lower hind limb bones in many vertebrates, including humans; in birds, the fibula is reduced and present only as a thin, needlelike bone running two-thirds of the way down the side of the tibiotarsus. [1·14, 4·25]

field of view: 1. The width of the area visible (usually at 1,000 yards from the observer) through binoculars or a telescope. If the field of view is labeled in degrees, multiply degrees by 52.5 to get the width in feet. 2. The view attained by a particular species, due to the placement of its eyes; also called an organism's **visual field**. [2·37, 4·51]

filoplumes: Hairlike but relatively stiff feathers having a rachis but few or no barbs (any barbs are present only at the tip). In the skin next to their follicles they have sensory receptors that allow them to monitor movement within the feather coat. [3·17]

fitness: An individual's degree of success at contributing its genes to the next generation; often measured as the number of its offspring that survive to reproduce. [1·35, 9·7]

fixed action pattern (FAP): A behavior that occurs in complete form each time the animal encounters the releasing stimulus—even upon the animal's first exposure to that stimulus. An FAP may be a simple or complex behavior, or a series of behaviors, but once begun, it is played out to the end regardless of any response that occurs or any intervening stimuli. [6·8]

flanges: See **oral flanges**. [3·43, 8·107]

flank: The side of a bird, dorsal and caudal to the leg. [1·6]

fledging: Term commonly used to describe the time at which nestlings that are reared in the nest leave the nest, even though their flight abilities may not yet be well developed. But, the term is sometimes used to describe the time at which a young bird has finished acquiring its first complete set of flight feathers—generally the time at which it is capable of flight. The term is used less often in precocial species that leave the nest shortly after hatching, but sometimes it refers to the time at which they begin to fly. "Fledging" may also be used to refer to the process of reaching the moment of fledging. [8·115]

fledging period: The period of time from hatching to the moment of **fledging** (see separate entry). [8·116]

fledgling: A young bird that has recently fledged (see separate entry for **fledging**). [8·115]

flexor: A muscle that pulls one bone toward another bone. [4·27]

flight feathers: The remiges of the wings and rectrices of the tail. [1·11, 1·12, 1·13]

flight songs: Songs given by birds during flight; they are particularly common among birds of open areas, such as grasslands and the tundra, where few perches are available. Singing from higher up generally increases the distance over which a song can be heard. Nevertheless, some species that do have ample perches, such as the forest-dwelling Ovenbird, also give flight songs; these usually begin with a jumble of notes that appear to draw attention to the singer, and then proceed with the bird's normal song. Their function is unknown. [7·84]

floaters: Animals, generally males, that do not hold territories or form pair bonds, but cruise around areas containing territorial individuals, waiting for a chance to take over a territory or sneak a copulation with a paired bird. [9·70]

flowerpeckers: A family (Dicaeidae, 43 species) of small, busy, noisy songbirds, primarily of the Oriental zoogeographic region, that forage high in trees on berries, nectar, and insects. [1·90]

follicle: A small, epidermis-lined pit in the skin of a bird, from which a feather emerges and to which it is attached. [3·2, 3·26]

food chain: The sequence in which organisms in an ecosystem feed upon other organisms. Most food chains consist of producers, consumers, and decomposers. A food chain is a fairly linear and simplistic model of feeding relationships in communities, which are more realistically represented as **food webs** because of the numerous interconnections among species and levels. [9·123]

food web: See **food chain**. [9·123]

foot: In birds, refers to the portion of the leg distal to the tibiotarsus bone, and has two sections. The upper section is supported by the tarsometatarsus bone, which does not touch the ground when the bird walks; the lower section consists of the phalanges of the toes, upon which the bird walks. [1·14]

foramen triosseum: An opening at a bird's shoulder joint, formed by the junction of the scapula, coracoid, and clavicle bones. This hole acts as part of a pulley system that allows the force of the supracoracoideus muscle to be redirected: because the tendon of the supracoracoideus passes through this hole, the muscle can be located ventrally yet still raise the wing. Also called the **triosseal canal** or **supracoracoid foramen**. [4·20, 5·6]

forebrain: The anterior portion of the brain; it consists of the two cerebral hemispheres with the olfactory lobes at their anterior ends. [4·37]

forehead: The front of the head, from the crown to the base of the bill. [1·7]

forked tail: Tail shape in which the rectrices become abruptly longer from the inside out. [1·19]

formation: An ordered arrangement of a group of birds in flight, such as V-shaped flocks of geese or single lines of Brown Pelicans and cormorants. [5·47]

frequency: Also known as **pitch**, the frequency is the rate at which a sound causes the air through which it is moving to compress and thin (one compressing and thinning is called one cycle), and is measured in cycles per second, or Hertz (Hz). The more cycles per second, the higher the frequency, and the higher the pitch. [7·4]

fright molt: See **shock molt**. [3·38]

frogmouths: A family (Podargidae, 14 species) of nocturnal forest birds of the Oriental and Australasian regions that resemble their smaller nightjar relatives in both cryptic appearance and behavior. [1·91]

frontal plane: A plane, usually horizontal, through an organism, dividing the body into dorsal and ventral portions. [1·4]

frugivorous: Feeding mainly or exclusively on fruits. [1·81]

functional response: A change in the amount of a certain type of prey taken by a predator, as a result of a change in the population density of that prey. For example, as insect populations in an area increase during spring, American Kestrels may begin to eat more insects and fewer voles. [9·25]

furcula: V-shaped bone of the pectoral girdle of birds, formed by the fusion of the right and left clavicles with a small interclavicle bone. The furcula is also called the **wishbone**. [4·19]

fusiform: Describes an egg that is slightly longer than subelliptical; also called **biconical** or **long subelliptical**. [8·73]

G

Galapagos Finches: A group of 15 finch species in the family Emberizidae living in and near the Galapagos Islands. They are a classic example of adaptive radiation, as they have a wide array of beak sizes and shapes and are all thought to have evolved from a common ancestor. Also called **Darwin's Finches**. [1·60]

gall bladder: Small organ for storing bile; it is located under the liver and is not present in all birds. [4·124]

gallinaceous birds: Grouse, quails, turkeys, pheasants, and all other birds in the order Galliformes; includes domestic chickens. [1·66]

gametes: The ova (egg cells) and sperm cells. [4·134]

ganglia (singular, **ganglion**)**:** Aggregations of nerve cell bodies; they form nerve centers outside the central nervous system (in the peripheral nervous system). [4·36, 4·44]

gaping: In avian biology, begging behavior of young birds that begins shortly after hatching in which they open the mouth widely; may be accompanied by a begging call. Given by altricial young and those precocial young whose parents feed them. [8·107]

gas exchange: The movement of gases between an organism and the environment; for example, in the lungs of many organisms including birds, the blood takes up oxygen from the air and discharges carbon dioxide and water. [4·89]

gastric cuticle: Leathery or sandpaper-like material that forms the lining of the gizzard; it is a combination of carbohydrate and protein secreted by glands in the wall of the gizzard. Also called **koilin**. [4·119]

gastrointestinal tract: See **alimentary canal**. [4·103]

gene: The sequence of base pairs within a molecule of DNA that codes for one specific protein. [1·41, 4·134]

gene flow: The movement of genetic material between populations. In mobile animals, gene flow generally occurs as individuals emigrate, immigrate, or breed with individuals from other populations. In organisms such as plants and fungi, gene flow occurs as spores, pollen, or seeds are carried by water, wind, or animals. [1·53]

gene pool: All the genes existing in a population at a given time. [6·22]

generalist: In biology, an organism that is able to use a wide range of some type of resource; for example, animals with generalist diets eat many different types of foods. [9·5]

genetic bottleneck: The loss of genetic diversity experienced by most populations as they become very small. Such loss occurs for three main reasons: (1) some alleles are lost simply by chance, when the only individuals that possess them die, (2) the rate at which new alleles arise in the population declines, because there are fewer individuals in which new mutations can occur, and (3) inbreeding occurs more frequently because fewer genetically dissimilar individuals are available as potential reproductive partners. [10·75]

genetic structure: The relative proportions of individuals with different genetic types—usually noted for a given population. [9·80]

genital system: The reproductive system. [4·127]

genus (plural, **genera**)**:** Level of classification of organisms above "species" and below "family." Genus is always capitalized, and is underlined or printed in italics. [1·47]

geographic range: The geographic area within which a species or population generally remains at a particular time of year; a species may have different breeding and nonbreeding ranges. Also called the **range**. [2·19, 9·49]

germinal spot: In avian biology, the light-colored site on the egg yolk where the embryo will eventually develop. The germinal spot sits atop a cylinder of light-colored yolk that stretches from the yolk's core to its surface. [8·63]

gizzard: The lower part of the bird's two-part stomach; it is rounded and has a tough lining and thick, muscular walls, often with internal ridges. It grinds and softens foods, and in birds that eat seeds, the gizzard has more muscular walls than in birds that eat meat. Seed-eating birds may eat grit or small stones, which reside in the gizzard to aid in grinding. In birds that eat fish or other meat, the gizzard molds indigestible material, such as bones and feathers, into compact balls (pellets) that are then ejected through the mouth. [4·118]

gleaning: A foraging technique in which a bird takes insects and other small invertebrates from the surface of vegetation or other substrates. In **perch gleaning**, practiced by many wood-warblers and other species, a bird grabs prey without flying from its perch. In **sally gleaning**, practiced by birds such as Red-eyed Vireos, chickadees, titmice, and some small flycatchers, the bird sits still and watches the surrounding vegetation until it sees an insect move, then flies out and grabs it from the surface. In **hover gleaning**, practiced by kinglets, phoebes, and Great Crested Flycatchers, among others, the bird hovers while taking food from the surface of vegetation. [6·44]

glenoid fossa: In birds, a cup-shaped depression formed where the coracoid and scapula meet; it receives the rounded end of the humerus, forming a ball-and-socket joint that enables the humerus to rotate freely around the shoulder joint. [4·20]

glial cell: See **neuroglia**. [4·36]

gliding: Unpowered flight (no thrust is provided) in which the flying object loses altitude. In a bird or other animal, flying without flapping the wings or limbs, while losing altitude (as compared to soaring, in which the animal rises). [5·36]

globular nest: A spherical dome nest with a top that completely encloses the nest; usually entered through a hole on the side. Examples include the nests of Cactus Wrens, Black-billed Magpies, and Southern Penduline-Tits. [8·36]

glossopharyngeal nerve: The ninth cranial nerve; it carries sensory and motor information between the brain and the tongue, pharynx, esophagus, and throat. It also carries motor output to the salivary glands. [4·42]

glottis: Small, slit-like opening to the larynx; it is surrounded by fleshy folds whose muscles regulate the passage of air into the respiratory system. [4·91]

glycogen body: A gelatinous mass of neuroglial cells rich in the nutritive sugar glycogen; it is located in the rhomboid sinus, and its function is unknown. The rhomboid sinus and glycogen body are unique to birds. [4·39]

gonads: The primary sexual organs, the testes and ovary, which produce, respectively, sperm and eggs. They also are endocrine glands, secreting the sex hormones testosterone, estrogen, and, from the ovary only, proges-terone. [4·75]

Gondwanaland: The southern land mass formed 200 million years ago when Pangea split into two large land masses. It consists of present-day South America, Africa, Madagascar, India, Australia, New Zealand, and Antarctica. [1·68]

good genes hypothesis: One possible explanation for why females of some species choose the males with the most elaborate ornaments (such as ornate plumage) to copulate with. The explanation applies primarily to females choosing nonpaternal sexual partners, either (1) males for extrapair copulations, or (2) mates in species in which males do not provide parental care or other resources for their offspring (such as territories with food or nesting sites). The good genes hypothesis suggests that the most ornamented males have genes that increase their own survival in some way (for example, they may have greater skill at foraging, avoiding predators, or obtaining good territories), and that females choose them because then their own male and female offspring may inherit those traits and have an increased chance of surviving and producing offspring of their own. [6·84]

graduated tail: Tail shape in which the rectrices become abruptly longer from the outside in. [1·19]

granivorous: Seed-eating. [9·27]

gravity: The attractive force between two masses of matter; this force, for example, tends to draw objects toward the center of the earth. [5·10]

gray matter: Darker-colored tissue (compared to white matter) that makes up much of the brain and spinal cord. It consists of numerous nuclei, which are collections of nerve cell bodies, and is found at the core of the spinal cord and in the outer areas of the brain. [4·39]

grazing: A foraging technique in which an animal bites off clumps of grass or other vegetation; used by geese, antelope, and others. [6·46]

greater coverts: The feathers partly overlying each remex on the upper surface of the wing. A **greater primary covert** overlies each primary feather, and a **greater secondary covert** overlies each secondary feather. [1·12, 1·13]

gross primary productivity: See **primary productivity**. [9·87]

ground-rollers: A family (Brachypteraciidae, 5 species) of solitary, terrestrial insect-eaters with stout bills, short wings, and moderately long legs and tails. They are endemic to Madagascar and are declining in number dramatically. [1·88]

ground speed: The speed of a moving object (such as a flying bird) in relation to the ground; it is equal to the bird's speed with respect to the air (the air speed) plus or minus the wind speed, depending on the direction the bird is flying with respect to the wind. [5·68]

ground-up theory (of the origin of avian flight): See **cursorial theory**. [E·14]

group selection: Theory proposed by V. C. Wynne-Edwards in 1962 suggesting that different groups of individuals might experience differential survival based on variations among the groups, and that this might be one way in which natural selection worked—passing down through the generations a greater proportion of the genes of the groups with higher survival rates. Theorists have basically disproved the idea as originally presented, but are still exploring whether some other form of group selection might be plausible.

[6·28]

guano: The accumulation of seabird droppings, particularly at a breeding colony, dried to a hard, crusty state. Guano is often mined and used as fertilizer. [1·104]

guild: In avian biology, a group of ecologically similar (but not necessarily related) birds that use the same resources in similar ways. For example, Henslow's Sparrows, Grasshopper Sparrows, Eastern Meadowlarks, and Bobolinks all nest on the ground in extensive grasslands, and thus are considered to be in the same "grassland nesting guild." [9·102]

guineafowl: Gregarious, chicken-like birds with distinctively spotted and striped plumage; they are often domesticated or found in zoos. The six species are endemic to Africa and form the subfamily Numidinae of the family Phasianidae. [1·85]

gular fluttering: Opening the bill wide and vibrating the thin, expansive gular membranes of the throat, in order to dissipate heat. This cooling method is used by pelicans, cormorants, herons, owls, and nighthawks. [4·153]

gular region: Upper part of the throat, just below the chin. [1·7, 1·8]

gut: See **alimentary canal**. [4·103]

H

habitat: The physical surroundings in which an organism lives. It consists of physical factors, such as light, temperature, and moisture, as well as living organisms, such as plants and animals. Habitats are often characterized by a dominant plant type or physical feature, such as a grassland habitat or stream habitat. [9·8]

habitat conservation plan (HCP): A plan that must be submitted to the U. S. Fish and Wildlife Service by anyone who applies for a permit to destroy endangered species or their habitats (as allowed under a 1982 amendment to the Endangered Species Act). The plan must specify the steps that the applicant will take to minimize the number of individuals killed and to minimize the impact on the species as a whole, and also must explain why other alternatives are not feasible. [10·92]

habitat fragmentation: The process by which a large, continuous habitat is broken into a number of small, isolated patches by activities such as development, logging, or farming. [9·97]

habitat generalists: Species that can live and breed successfully in a wide range of different habitats. If all else is equal, these species are less likely to go extinct than species that can live and breed only

in one or a few specific types of habitats (**habitat specialists**). [10·59]

habitat imprinting: The process by which a young animal, especially a bird, learns the characteristics of appropriate habitat by observing its surroundings while it is still living within its parents' territory. When adult birds eventually settle on their own territories, most choose to live and/or breed in areas similar to those in which they were raised. [9·11]

habitat richness theory: The idea that smaller islands generally hold fewer species than larger islands because they tend to have fewer different habitats, and thus species immigrating to smaller islands are less likely to find suitable habitats than those colonizing larger islands, and are more likely to perish (or leave). [9·96]

habitat specialists: Species that live and breed successfully in only one or a few specific types of habitats. If all else is equal, these species are more likely to go extinct than species that can live and breed successfully in a wide range of habitats (**habitat generalists**). [10·59]

habitat specificity: How wide a range of different habitat types in which a given species can live and breed successfully. [10·62]

habituation: The permanent loss of a response as a result of repeated stimulation without reward or punishment; learning to ignore unimportant stimuli. [6·9]

hacking: The technique of introducing young, captive-raised birds of prey, especially falcons, to appropriate habitat by releasing them from an enclosure that serves as an artificial nest and in which biologists continue to place food until the bird has learned to hunt on its own. [10·100]

hallux: The first toe, composed of two phalanges. In nearly all birds, it points backward. The hallux is well developed in perching birds and is reduced or absent in many running birds. [4·25, 4·26]

Hamerkop: A stork-like water bird whose shaggy, crested nape and stout, tapering bill make the head appear hammershaped. It builds an enormous mound nest. It is endemic to Africa, and is the only member of its family, Scopidae. [1·85]

hand: See **manus**. [1·9]

hatching: Emerging from the egg. A clutch may hatch synchronously (all at about the same time—see **synchronous hatching**) or asynchronously (over a period of several days—see **asynchronous hatching**). [6·88]

hatchling: A newly hatched animal. [8·61]

Hawaiian Honeycreepers: A group of 32 species, many of them extinct, in the subfamily Drepanidinae, within the family Fringillidae. These small, colorful birds have a wide array of beak shapes and are a dramatic example of adaptive radiation, as they all are thought to have evolved from a common ancestor. [1·62]

hawking: A foraging technique in which a bird sits very still on a high or exposed perch, and when it sees an insect, flies out and snatches it in midair, returning to the same or a nearby perch; used by many flycatchers, kingbirds, bee-eaters, waxwings (sometimes), and some woodpeckers, among others. [6·44]

HCP: See **habitat conservation plan**. [10·92]

heading: The compass direction in which a bird is pointing its beak and propelling itself through the air. Because of crosswinds, the heading may not be the actual direction that the bird is progressing with respect to the ground. Also applies to other flying animals and aircraft. [5·68]

heel pad: Calloused enlargement of the upper end of the tarsus (at the heel), found in the nestlings of many cavity nesters, such as woodpeckers and trogons. It is thought to reduce abrasion of the tarsus from the rough lining of the nest cavity. [3·44]

helpers at the nest: Adult birds that are not currently breeding themselves, but assist other breeding pairs (usually, but not always, their relatives—especially their parents) in rearing their offspring. Some helpers are birds whose own breeding attempts have failed, whereas others are unpaired birds or those without territories; helpers often breed on their own in subsequent years. Bird species in which helpers are common are called cooperative breeders. [6·88]

hepatic portal system: Pattern of circulation in which blood from the capillary beds of the small intestine is brought by the hepatic portal vein to the capillary beds of the liver, for further processing. [4·85]

hepatic portal vein: Large vein carrying blood from the upper part of the small intestine, where the blood has absorbed digested nutrients, to the liver, where it undergoes further chemical processing before returning to the heart. "Portal" veins carry blood between two capillary beds—in this case, the capillary beds in the small intestine and the liver—instead of between a capillary bed and the heart. [4·85]

hepatic veins: Two large veins (left and right) that carry blood from the capillary beds of the liver to the caudal vena cava near its entry into the right atrium of the heart. [4·85]

hepatoenteric ducts: In birds that lack a gall bladder, ducts that transport bile from the liver to the small intestine. [4·121]

herbivores: Organisms that eat primarily plants. [9·123]

Hertz (Hz): A measure of the frequency of a sound, in cycles (of compression and thinning of the air) per second. [7·4]

heterocoelous centrum ends: The saddle-shaped, interlocking ends of the centrum (main body) of avian cervical vertebrae. Because the anterior end is concave in a lateral direction and the posterior end is concave in a dorso-ventral direction, articulating vertebrae can rotate freely against one another, allowing the neck to be highly flexible. [4·15]

heterodactyl foot: Foot arrangement in which the third and fourth toes point forward and the hallux and second toe point backward; found in trogons. [1·21]

heterogametic: Describes individuals whose two sex chromosomes are different; in birds, the chromosomes are called ZW and heterogametic individuals are females, but in mammals, the chromosomes are called XY and heterogametic individuals are males. Heterogametic individuals are capable of producing two different types of gametes (eggs or sperm)—one with one type of sex chromosome and one with the other. For comparison, see **homogametic**. [4·136]

heterozygous: Possessing two alleles that are different for a given gene; for example, having one allele for blue eyes and one allele for brown eyes. (Each allele comes from a different parent.) For comparison, see **homozygous**. [10·74]

high-aspect-ratio wings: Wings that are long, narrow, and unslotted; the length derives primarily from the lengthened inner wing (as compared to high-speed wings, in which the length results primarily from the long outer wing). High-aspect-ratio wings are highly efficient at producing lift at relatively high flight speeds, but they are difficult to maneuver, especially during take-offs. They are found in a few seabirds that are highly specialized for dynamic soaring over the ocean, such as albatrosses, shearwaters, petrels, and some gulls. [5·42]

high-speed wings: Wings that are tapered, pointed, and in many cases sweptback, with unslotted primaries and a high aspect ratio; found in birds such as falcons, swifts, swallows, terns, ducks, and many shorebirds. High-speed wings allow good control and high speeds, but are energetically expensive to use because the bird must flap constantly. [5·38]

hindbrain: The posterior portion of the brain; it includes the cerebellum and medulla oblongata (brain stem). [4·36]

hinge post: Central rod of binoculars, around which the two barrels pivot. [2·30]

Hoatzin: An odd-looking, leaf-eating bird that nests in branches overhanging lakes and slow-moving streams in the Neotropics; it is the only member of its order and family (Opisthocomidae). To avoid predators, nestlings may temporarily leave the nest and clamber about in trees, aided by small claws on their wings. [1·77]

Holarctic: A combination of the Nearctic and Palearctic zoogeographic regions; the Holarctic encompasses the Northern Hemisphere north of the tropics. [1·70]

homeostatic mechanisms: Mechanisms that act in various ways to preserve balance. For example, various physiological factors keep the body temperatures of birds and mammals relatively constant. And, various ecological factors keep many populations at a density that fluctuates around a fairly stable level. [9·67]

homeothermic: Describes animals that are able to keep their internal body temperature constant even when the outside temperature varies. In birds and mammals, the hypothalamus monitors body temperature and, if required, triggers responses that warm or cool the individual to bring it back to normal body temperature. [9·13]

homing ability: The ability to return to a specific place. For example, homing pigeons can return to their lofts when released from great distances away. [5·80]

homogametic: Describes individuals whose two sex chromosomes are the same; in birds, the chromosomes are called ZZ and homogametic individuals are males, but in mammals, the chromosomes are called XX and the homogametic individuals are females. Homogametic individuals are capable of producing only one type of gamete (eggs or sperm)—one that contains the same type of sex chromosome as the parent. For comparison, see **heterogametic**. [4·136]

homozygous: Possessing two identical alleles for a given gene; for example, having two alleles for blue eyes. For comparison, see **heterozygous**. [10·74]

honeyeaters: A huge, diverse Australasian family (Meliphagidae, 181 species) of dull-colored, arboreal songbirds with medium-length, curved bills. Honeyeaters have a distinctive brush-tipped tongue for gathering nectar and are important pollinators, but they also eat insects and fruits. They feed busily, and often congregate at flowering trees. [1·95]

honeyguides: A nonpasserine family (Indicatoridae, 17 species) of the warmer parts of the Old World, whose members are peculiar in their ability to digest wax, especially beeswax, in addition to their insect prey. Some honeyguides lead humans or other mammals to bee nests, and then eat the wax that is exposed as the mammal breaks open the nest. [1·87]

hooklets: Tiny hooks found on each of the barbules that branch from the distal side of each barb of a contour feather; hooklets catch onto the barbules branching from the proximal side of the next barb (toward the feather tip), lightly holding the barbs together to form a smooth, continuous vane. [3·4]

hormone: Chemical substance secreted into the blood and thus carried to other parts of the body, where it may stimulate or regulate the activities of glands or organs. Hormones are the messengers of the endocrine system. [4·69]

hornbills: A family (Bucerotidae, 55 species) of large, toucan-like birds of the Oriental and Afrotropical regions; hornbills have long tails and huge, down-curved bills topped by a distinctive casque. They are famous for the female's nesting behavior—she seals herself inside the nest cavity with the eggs and young. [1·91]

horns of the hyoid: Part of the hyoid apparatus; the two bones on each side of the hyoid that extend backward (caudally) from the tongue bone, running beneath the skull and then curving upward around the back of the head. [4·13]

House Finch eye disease: See **mycoplasmal conjunctivitis**. [9·71]

hover gleaning: See gleaning. [6·44]

hovering: A type of flight in which an aircraft or flapping animal remains suspended in air in one place. In birds, hovering usually is achieved by positioning the body nearly vertically and beating the wings more or less horizontally, producing just enough forward thrust to balance wind speed, and just enough lift to compensate for gravity. Hovering is an energetically expensive undertaking. [5·26]

humeral patagium: A flap of skin extending from the brachium to the trunk of the bird. [1·11]

humerus: The bone supporting the brachium (upper arm or wing). [1·10]

Humphrey-Parkes Nomenclature: The system of naming plumages and molts most commonly used by ornithologists today. In this system, the plumage worn for the longest time each year (the non-breeding plumage), usually produced by a complete molt, is called the bird's basic plumage, and other plumages are called alternate and supplemental. In addition, molts are named for the type of plumage they produce, not the feathers they shed. In the traditional system, a bird's breeding plumage is considered to be the main plumage. [3·33, 3·35]

hybridization: Breeding that occurs between two individuals of different species. [3·66]

hydrology: The study of water: its distribution, properties, and patterns of flow on the earth and in the atmosphere. [10·85]

hyoid apparatus: A V-shaped unit composed of bones and cartilage, located between the two halves of the lower jaw. The hyoid apparatus consists of the tongue bone (entoglossal) and the horns of the hyoid, which, respectively, support the tongue and the muscles that control tongue movement. [4·13]

hyperphagia: The dramatic increase in the amount of food that birds consume as they prepare to migrate; because these birds eat more than their body requires in the short term, they store body fat to be metabolized for energy during migration. [5·63, 9·24]

hyperthermia: Condition in which the body temperature rises a few degrees above normal; if the animal cannot bring its temperature down, it soon dies. [9·18]

hypoglossal nerve: The twelfth cranial nerve; it combines with the vagus (tenth) and glossopharyngeal (eleventh) cranial nerves to form a combined trunk that controls movement of the tongue, larynx, trachea, and syrinx. [4·42]

hypothalamus: The ventral portion of the brain region between the forebrain and midbrain. The hypothalamus plays a major role in hormonal control of body processes by using special neurosecretory neurons to control the pituitary gland, which extends from a stalk below the hypothalamus. [4·36, 4·72]

hypothermia: Condition in which the body temperature drops below normal. If the animal cannot bring its temperature back to near normal, it soon dies. [9·16]

hypothesis: A tentative explanation for an observation or phenomenon; hypotheses are usually stated in a way such that they can be tested via the scientific method. [6·5]

Hz: See **Hertz**. [7·4]

I

IBA: Important Bird Area. See **Important Bird Areas Program**. [10·113]

ileum: The short, final portion of the small intestine, running from the jejunum to the large intestine. [4·122]

ilium (plural, **ilia**): One of the three paired bones fused to form the pelvic girdle; the ilium forms the cranial and lateral portion, and in birds is completely fused with the ischium and the synsacrum. It has a cup-shaped depression for the attachment of the femur (thigh bone). [4·24]

immediate variety: Pattern of singing in which a bird sings a song type once, then moves on to a different song type and sings it once, then goes on to yet another type, and so on, such that the bird sings many different songs in its repertoire before ever repeating one. [7·86]

immigration: The movement of individuals into a population. [9·49]

imperforate septum: A condition in which the nasal septum (tissue separating the left and right nasal cavities) has no opening. In contrast with a **perforate septum** (see separate entry), an imperforate septum appears to decrease an animal's sensitivity in detecting odors. However, it increases the ability to locate an odor's source, because odors entering one nostril do not mix with those from the other. [4·90]

Important Bird Areas (IBA) Program: A program to identify sites (called IBAs) in each state that are particularly important to local populations of breeding, wintering, or migrating birds. The program gathers input from bird watchers and professional ornithologists, designates the sites it feels are most important, and then encourages public and private stewardship of the sites. Developed as an international effort by BirdLife International, the program is conducted in North America by the National Audubon Society. [10·113]

imprinting: A type of early learning in which a young animal quickly acquires specific information for certain experiences. For example, after hatching, Greylag Goose goslings rapidly learn to follow the first moving object they encounter, which is usually their parent. [6·12]

inattentive periods: In avian biology, blocks of time spent off the nest during incubation. [8·99]

inbreeding: The mating of closely related individuals within a population. The greater the degree of inbreeding, the greater the chance that deleterious genetic traits will occur in the population. See **inbreeding depression**. [9·66]

inbreeding depression: The reduction in the average fitness of offspring born to parents that are closely related to each other, compared to the fitness of offspring born to unrelated parents. Inbreeding depression occurs because closely related parents share more genes, and thus their offspring are more likely to receive two copies (one from each parent) of alleles that cause deleterious traits or genetic diseases. For example, in humans, the genetic disease hemophilia was once common among some of the inbred royal families of Eurasia. [9·66]

incest: A mating between close relatives. [6·94]

incest taboo: The instinctive avoidance of breeding with close relatives; it has undoubtedly evolved in many organisms because it reduces **inbreeding depression** (see separate entry). Because determining relatedness can be difficult, many species seem to avoid incest by not breeding with any individual with which they have been in close contact during their developmental period (which is usually a sibling or parent). [10·75]

incubation: The process by which animals that lay external eggs keep those eggs at the proper temperature for embryonic development until they hatch (or the nest fails). Only birds, crocodiles, pythons, and monotremes (egg-laying mammals) incubate their eggs. In most cases, birds sit on their eggs to keep them warm, but many megapodes bury them—in piles of decaying vegetation, in long tunnels or broad pits where the earth is warmed from nearby hot streams or volcanic cinder fields, or in pits or burrows where bare sand or soil is heated by the sun. In very hot environments incubation may require cooling the eggs by shading them, burying them in sand, or keeping them moist. [8·93]

incubation patch: See **brood patch.** [8·94]

incubation period: The time from the start of regular, uninterrupted incubation to hatching. [8·96]

incubation pouch: A type of brood patch, found on the breast of some albatrosses, that is a featherless cavity surrounded by thick feathers into which the single egg fits so snugly that it may remain inside even when the bird stands. [8·96]

indeterminate layers: Bird species that will lay additional (replacement) eggs if one or more eggs are removed from the clutch during laying. Once they have begun incubation on a full clutch, however, they cease to lay replacement eggs if more are removed. For comparison, see **determinate layers.** [8·90]

index: A numerical sequence or other representation that indicates the relative level, degree, or amount of something or some property. For example, an index of population density, or an index of health. [9·61]

indicator species: A species that acts as a "canary in a coal mine," because changes in its population density or distribution provide an early warning that its habitat is changing in some way—often through degradation by human activity. Birds are good indicators because they are relatively noticeable and easy to survey, because many bird species are high in the food chain and thus sensitive to the **bioconcentration** (see separate entry) of toxins, and because many species have narrow habitat requirements. Also called **biological indicators.** [10·108]

indigobirds: Members, along with whydahs, of the African genus *Vidua* (members of this genus are called **viduines**). Viduines are colorful seed-eaters that are brood parasites on other members of their family, Estrildidae; they are known for the intricate patterns inside the mouths of their nestlings, which strikingly resemble those of the nestlings of their hosts. [8·142]

indirect fitness: An individual's indirect fitness is the portion of its genes that is transferred into the next generation (and eventually beyond) as a result of that individual's blood relatives producing offspring. Some of an individual's genes are propagated when a relative breeds because the individual shares a percentage of its genes with that relative (the percentage depends on the relatedness), and because the relative passes 50 percent of its genes (both those that it shares and those that it doesn't share) to its offspring. This contrasts with **direct fitness**, which is the portion of genes transferred through the production of an individual's own offspring. [6·86]

individual recognition: The ability of an animal to identify other specific individuals. [7·42]

inertial navigation: Finding your way by keeping track of all the turns and accelerations you have taken. For example, logging in your brain the turns and accelerations of an outward trip and then integrating them to compute a direct route home. Inertial navigation requires no outside reference points. [5·82]

information center hypothesis: The idea that one benefit of colonial breeding is that individuals who have been unsuccessful at finding food might find better feeding grounds by watching at the colony for successful foragers (those who return with food), and then following them to good hunting spots. [6·59]

infundibulum: The flattened, funnel-shaped opening of the oviduct. In birds, when the ovary releases an egg, the infundibulum moves up to the ovary and opens, "swallowing" the egg much like a snake swallows a rat. [4·130]

innate behavior: See **instinctive behavior.** [6·5]

innate rhythmicity: The ability to contract without being stimulated by the nervous system, as found in cardiac muscle cells. [4·31]

inner ear: Complex structure inside the ear in the form of a membranous, fluid-filled sac (the membranous labyrinth) floating inside a bony, fluid-filled sac (the bony labyrinth). The inner ear consists of the semicircular canals enclosing their ducts and the vestibule enclosing the utriculus and the sacculus—all of which are concerned with balance— and the cochlea and its enclosed cochlear duct, which are concerned with hearing. [4·57]

inner shell membrane: In avian biology, the membrane just inside an egg's outer shell membrane; it is thinner and less coarse than the outer membrane. The inner shell membrane contains the albumen and adheres tightly to the white of a hard-boiled egg, making it difficult to peel. [8·62, 8·69]

inner vane: The vane located on the medial side of a wing or tail feather. On a wing feather, the inner vane is on the edge of the wing that trails in flight. In flying birds, the inner vane is wider than the outer vane, producing an asymmetry that aids in flight. [3·3]

inner wing: The portion of the wing from the wrist to the shoulder; the secondary feathers are located on one section of the inner wing. [5·23]

innominate arteries: Two large arteries (left and right) that branch from the aortic arch and then soon branch again into the carotid arteries to the head and neck and the subclavian arteries to the front limbs (wings). [4·82, 4·84]

insect-net theory: A variation on the cursorial theory of the origin of avian flight, first proposed by John Ostrom in 1976. The insect-net theory suggests that dinosaurian bird ancestors first evolved feathers as thermal insulation; and then, as running dinosaurs clapped the feathered forelimbs together to catch insects, eventually the motion became flapping flight. [E·16]

insemination: The transfer of sperm from the male into the genital tract of the female; in birds, sperm enter the cloaca. [4·133]

insertion: On a skeletal muscle, the end (attachment site) that moves the most during contraction. [4·27]

inshore feeders: Birds (or other animals) that forage close to shore, fairly near their nesting areas; includes birds such as terns and many gulls. [6·68]

insight learning: A modification of behavior that occurs by evaluation of a situation rather than as a result of previous experience with a particular problem. [6·14]

instinctive behavior: A behavior that is triggered in full form, without any learning, the first time an individual responds to the releaser; also called **innate behavior** or **instinct.** [6·5]

integumentary system: The skin and structures that are produced by the skin, such as (in birds) feathers, color pigments, scales, claws, beak, wattles, and comb. [4·3]

intention movements: Movements that are either incomplete (such as the initial stages of a behavior) or that indicate what the performer is about to do. For example, a bird about to attack may crouch down and tense its muscles. Many intention movements have become exaggerated and incorporated into displays. [6·30]

interactive playback experiments: Computer-controlled playback experiments in which researchers can change various aspects of songs played to territorial birds in the middle of a singing encounter, based on the subject bird's singing behavior. For example, researchers might choose to play longer or shorter songs, or songs that match those of the subject, or they might respond more slowly or more quickly, or overlap the subject's singing. [7·49]

interference: In ecology, a situation in which one individual actively prevents other individuals from obtaining a limited resource. It may do so through various behaviors, such as aggressive interactions that set up dominance hierarchies or territories. [9·42]

International Union for the Conservation of Nature (IUCN): International nongovernmental organization based in Switzerland and devoted to the conservation of species. In 1963, the IUCN drafted the original text of the **Convention on International Trade in Endangered Species**, also known as **CITES** (see separate entry), which was finally ratified in 1975. [10·93]

interpupillary distance: The distance between the centers of an individual's two pupils; important in fitting binoculars. [2·30]

interspecific competition: Direct competition for limited resources among individuals of different species. [9·56]

intestinal lymph trunk: Lymph vessel that carries the products of fat digestion from lymph vessels coming from the small intestine to the thoracic duct. These digestive products are first picked up at the small intestine by lymph capillaries, which carry them to progressively larger lymph vessels running along the surface of blood vessels in the intestinal wall. These vessels eventually join to form the intestinal lymph trunk, which carries the products to one of the two thoracic ducts, which eventually enter the cranial vena cava. The reason for this circulatory pattern, which occurs in (at least) all birds and mammals, is not understood. [4·89]

intraspecific competition: Competition (for food, territories, mates, and so on) that occurs among members of the same species. [9·39]

ioras: Small, arboreal songbirds that search leaves, often in dense foliage, for insects. The four species form the family Aegithinidae, endemic to the Oriental zoogeographic region. [1·89]

iridescent colors: Structural colors, sometimes brilliant, that shimmer and glitter because they change in brightness as the angle of view changes. The colors are produced when light waves reflected off thin films (in birds, structural layers within flattened feather barbules) interfere with one another, as can be seen in soap bubbles and most hummingbirds. [3·54]

iris: The colored part of the eye surrounding the pupil; it is part of the eye's choroid (middle) layer. The iris contains muscle fibers and controls the diameter of the pupil and thus the amount of light that enters the eye. [4·4]

irruptive migration: Migratory movements that are irregular in time and space, depending upon factors other than a change of seasons, such as food availability. For example, the seeds and buds eaten by finches such as Pine Siskins and redpolls fluctuate in abundance not only seasonally but from year to year and from region to region, so in some years large numbers of the birds move out of northern forests to breed, yet in others they stay put. [5·55]

ischium (plural, **ischia**): One of the three paired bones fused to form the pelvic girdle; the ischium forms the caudal and lateral portion. [4·24]

island endemics: Species that evolved on an island and are found only upon that island. [9·55]

isthmus: The region of the avian oviduct after the magnum and before the shell gland; the isthmus secretes the egg and shell membranes and fluid albumen around the fertilized ovum as the ovum travels down the oviduct. [4·130]

IUCN: See **International Union for the Conservation of Nature**. [10·93]

jejunum: The long middle portion of the small intestine, running from the duodenum to the ileum. [4·122]

jizz: Birding term for a quick impression of a bird's major features. Jizz harkens back to the "general impression of size and shape" (G. I. S. S.) that British observers used during World War II to distinguish between enemy and friendly aircraft. [1·24]

journal: A written record of field observations of birds, other animals, or natural phenomena. [2·47]

jugal arch: Bony rod on each side of the upper jaw below the palatine (another set of bony rods). In birds, as the lower jaw opens it moves the quadrate, which pushes the palatine and jugal arch forward, pushing on the premaxillary bones to raise the upper jaw. [4·12]

jugular veins: Paired veins (left and right) carrying blood from the head and neck region; they merge on each side with the subclavian and pectoral veins to form the cranial vena cavae, which return the blood to the heart. [4·83]

jugulum: The lower part of the throat of birds, just below the gular region. [1·7, 1·8]

juvenal plumage: Feather coat worn by juvenile birds after they have molted their natal down; it consists of the first true contour feathers. [3·29]

juvenile: A young bird. [3·29]

K

keel: A midventral ridge of bone that projects outward from the sternum and provides a site for the attachment of the large pectoral flight muscles. Large, flightless birds lack a keel, as do the tinamous. Also called a **carina**. [4·23]

keratin: A hard protein that forms scales and claws and is the primary structural component of mature feathers. Avian keratin differs from the keratin of all other animals in its amino acid sequence. [3·28]

kettle: A large aggregation of birds, usually hawks, that are spiraling upward in a thermal. [5·39]

keystone species: Species that affect many other species in their community, and whose removal would precipitate a reduction in species diversity or would produce other significant changes in community structure. [9·126, 9·128]

kHz: See **kilohertz**. [7·4]

kidnapping: In avian biology, the aggressive takeover of a brood of young by adults that are not the parents. [8·128]

kidneys: Paired organs of the excretory system that are irregular in shape and, in birds, are each composed of three interconnecting lobes; they remove waste products from the blood, especially nitrogenous wastes, and form a highly concentrated urine that, in birds, is passed to the cloaca via the ureters. Kidneys also maintain a balance of salt ions in the blood. [4·125]

kilohertz (kHz): A measure of the frequency of a sound, in thousands of cycles (of compression and thinning of the air) per second. One kilohertz equals 1,000 Hertz. Also written **kiloHertz**. [7·4]

kingdom: Level of classification of organisms above "phylum"; similar phyla are placed within the same kingdom. All birds are in the kingdom Animalia. [1·52]

kin selection theory: A line of reasoning suggesting that the closer the kinship (degree of relatedness) between two animals of the same species, the greater will be their tendency to cooperate with one another in various ways. Such cooperation results from natural selection because if an animal enhances the survival of a relative, it also enhances its own **indirect fitness** (see separate entry); the closer the relative, the greater the degree to which its own indirect fitness is enhanced. [6·93]

kiwi: A family (Apterygidae) of three species of grouse-sized, flightless, nocturnal ratites endemic to New Zealand; kiwis probe soil with their long beaks, using their keen sense of smell to locate earthworms. [1·96]

koilin: See **gastric cuticle**. [4·119]

K-selected species: Species in which individuals have a life history strategy that relies on longevity, rather than a high reproductive rate, to maximize the number of offspring produced during their lifetimes. Such species tend to be large, develop slowly, begin breeding at a relatively old age, have few young per cycle (small clutches or litters), breed infrequently, take care of their young for extended periods, and have long life spans. [9·45]

K·T event: The massive, worldwide extinctions 65 million years ago at the end of the Cretaceous Period and the beginning of the Tertiary Period, which wiped out dinosaurs as well as many other plants and animals. The extinctions are thought to have resulted from worldwide climatic disturbance caused by a large meteor colliding with Earth. [E·25]

L

lacrimal gland: One of the two tear glands on each side of the eye; the lacrimal gland lies in the lower part of the orbit of the eye and has many ducts entering the space between the lower lid and the cornea. Its secretions moisten the eye and nourish the cornea. [4·46]

laminar flow: A smooth flow of air over an airfoil. [5·13]

large intestine: Short, straight tube extending from the small intestine to (in birds) the cloaca; it holds the intestinal contents while water is being reabsorbed, and passes the remainder to the cloaca. [4·123]

laryngeal cartilages: Two major cartilages that make up the sides and floor of the larynx. Also called the **cricoid cartilages**. [4·91]

laryngeal folds: Fleshy folds in the lower surface of the pharynx that surround the glottis, the opening to the larynx. [4·112]

larynx: Structure at the upper end of the trachea that consists of several cartilages;

it acts as a valve, regulating the flow of air into the trachea. In mammals the larynx is the voice box, but in birds it produces no sound, leaving sound production to the syrinx. [4·91]

lateral: Toward the side of the body; away from the midline. [1·4]

lateral labia: Important sound-producing membranes in the lateral wall of each half of the syrinx. During sound production muscles move the bronchi upward into the trachea, twisting the third bronchial cartilages so that the lateral labia and medial labia (in the medial walls) move toward (and close to) each other and into the path of air flowing out of the respiratory system. Sound is produced as air rushes between the lateral and medial labia, causing these soft tissues to vibrate. [4·96]

Laurasia: The northern land mass formed 200 million years ago when Pangea split into two large land masses. It consisted of present-day North America, Europe, and Asia. [1·68]

leafbirds: Eight species of oriole-sized arboreal songbirds endemic to the Oriental zoogeographic region; leafbirds are mostly green and yellow, and feed mainly on insects and fruit. Together with fairy-bluebirds, they form the family Irenidae. [1·88]

leaf tossing: A foraging technique in which a bird tosses aside the leaf litter (with its beak or feet) to search for food; used by towhees, turkeys, thrashers, and many sparrows. [6·47]

learned behavior: A behavior that requires some amount of previous experience—such as exposure to a stimulus—as well as a memory of that experience, before it is carried out fully. [6·9]

learning: The modification of a behavior as a result of experience. [6·8]

leg spurs: Bony outgrowths near the distal end of the tarsometatarsus, covered by a pointed horny sheath; used as weapons by male chickens, peafowl, and many other pheasant relatives. [4·26]

lek: 1. A traditional courtship area where many males of the same species gather to attract females for mating; each male spends a large amount of time defending a small site at which he displays to compete with other males and, in particular, to earn copulations with sexually receptive females who visit the lek to choose among the males. In most species, the top few males secure most of the matings. The lek contains no nest sites, food, or other resources useful to nesting females. [6·75] 2. The group of males gathered at the traditional courtship site described in definition 1. [6·75]

lek polygyny: Mating system in which males gather in leks and display to earn

copulations with visiting females. In most lekking species, the top few males secure most of the matings, resulting in a high degree of polygyny. [6·75]

lens: Spherical or ovoid structure near the front of the eye; it changes its curvature to sharply focus images from varying distances on the retina. The lens is crystalline-like and composed of regularly oriented layers of collagen fibers. [4·47]

Lepidosauromorpha: One of the two major groups of diapsid reptiles; it contains all snakes and lizards as well as the ancient ichthyosaurs and plesiosaurs. [E·8]

lesser coverts: The feathers on the upper surface of the wing that partly overlie the median coverts and extend to the marginal coverts. [1·12, 1·13]

leukocytes: See **white blood cells**. Also spelled **leucocytes**. [4·88]

liana: A woody vine. [9·88]

life history theory: A set of ideas that attempt to explain the diversity in the breeding strategies of living things by looking at why various characteristics relating to an organism's birth, death, and reproduction have evolved. These characteristics, called **life history traits** (see separate entry), include such things as number of offspring, age at first breeding, and length of the developmental period. [8·3]

life history traits: Characteristics of living things that are related to birth, death, and reproduction. Examples include the number of offspring, the age at first breeding, the interval between breeding cycles, and the chance of surviving to various ages. Collectively, these characteristics make up an organism's **life history** or **life history strategy**. [8·3]

life list: Record of every species seen by a particular person, noting the date and location of the first sighting. People may keep life lists of different groups of organisms, such as birds, butterflies, or reptiles. [2·54]

lift: Force acting on a moving airfoil (such as a bird's wing) perpendicular to the direction of airflow; in a bird that is gliding or flying horizontally, lift acts upward in opposition to gravity. [5·10, 5·13]

lift-to-drag ratio: The lift produced by a wing divided by the drag it experiences while flying. Long, narrow wings have the greatest lift-to-drag ratios and are the most efficient. [5·35]

ligament: Fibrous connective tissue that connects one bone to another across a joint. [4·6]

limiting factor: Something that is present in (or missing from) a particular environment and as a result prevents a particular species from living or breeding in that place. Examples of limiting factors include a large number of predators and a low level of a critical resource, such as food. [9·10]

liver: The largest internal organ in the body; in birds the liver has two lobes and performs a variety of functions, including secreting bile to help emulsify fats for digestion, storing sugars and fats, forming uric acid, and removing foreign substances from the blood. [4·124]

lithornithids: Extinct, flying, chicken-like palaeognathous birds thought to have been common in North America and Europe during the early Tertiary; they may be the ancestral stock that gave rise to ratites all over the world. [E·23]

local population: A population confined to a small area; for example, all the Blue Jays in one neighborhood. [9·49]

lore: Small area between the eye and the base of the upper beak. [1·7, 1·8]

lower beak: The lower half of the beak; sometimes called the **mandible** or **lower jaw**. [1·6]

lower critical temperature: The environmental temperature below which the body of a bird or mammal increases its metabolic rate and employs physiological responses to warm itself, assuming behavioral responses are no longer adequate. [9·16]

lower jaw: See **lower beak**. [1·6]

lower lethal temperature: The environmental temperature below which a bird or mammal cannot keep its body warm enough to survive. [9·16]

low pressure system: A weather system consisting of cold and warm air masses circulating around an area of low barometric pressure. In the Northern Hemisphere the masses circulate counterclockwise; in the Southern Hemisphere they move clockwise. [5·69]

lumbar vertebrae: The vertebrae of the lower back; in birds they are all fused with the sacral vertebrae and some of the thoracic and caudal vertebrae to form the synsacrum. [4·15, 4·18]

lumbosacral enlargement: Swelling along the spinal cord at the level of the legs; it is associated with the lumbosacral plexus. [4·40]

lumbosacral plexus: Plexus (see separate entry) along the spinal cord at the level of the hind limbs; it is associated with the lumbosacral enlargement of the spinal cord. [4·39]

lymph: Fluid carried in lymph vessels; also see **lymphatic capillaries** and **lymphatic system**. [4·87]

lymphatic capillaries: Tiny vessels that form a network intertwining with the capillary beds of the arterial and venous systems throughout most body tissues. Some of the tissue fluid that has lost its

oxygen and nutrients to the body (tissue) cells and picked up carbon dioxide and other wastes diffuses into the lymphatic capillaries, and is then termed **lymph**. The lymphatic capillaries lead to progressively larger lymphatic vessels, which eventually dump their contents into the venous system. [4·88]

lymphatic system: Organ system—composed of lymphatic vessels, ducts, and nodes—that gathers tissue fluid that has leaked from the blood capillaries, filters foreign substances and old or damaged cells from the fluid, and then returns the fluid to the general blood circulation. The lymphatic system also releases antibodies and transports the products of fat digestion from the intestines to the venous system, bypassing the liver. [4·88]

lyrebirds: Two pheasant-sized songbirds (forming the family Menuridae) of Australian rain forests, named for the elaborate, harp-shaped tail of the Superb Lyrebird. They have loud songs and calls and are fantastic mimics, sometimes clearly reproducing mechanical sounds, such as logging trucks and chain saws, along with the songs of other birds. [1·94]

M

macroevolution: The evolution of new species over long periods of time, such as thousands of years (see **microevolution**). [1·37]

magnetic anomaly: A place where the earth's magnetic field is disturbed, usually by large deposits of iron near the surface. [5·98]

magnetic compass: A mechanism by which birds and some other animals can use the earth's magnetic field to determine compass direction. [5·89]

magnetic field of the earth: The region around the earth in which objects experience magnetic force (which is a vector—a quantity with both strength and direction). The field is created by currents generated by the constant motion of molten iron in the earth's core. In effect, the earth is a large magnet with two magnetic poles (near the North and South Pole). The field is stronger closer to the magnetic poles and weaker toward the equator. [5·90]

magnetic map: A mechanism by which birds (and some other animals) may be able to use the earth's magnetic field as a map. For example, because the magnetic field is strongest toward the poles, a bird might use the strength of the field at any given point to estimate latitude. [5·97]

magnum: The first region of the avian oviduct, after the infundibulum; the magnum is glandular and secretes the first of the albumen or "white" of the egg around

the fertilized ovum as it travels down the oviduct. [4·130]

malar region: Small area caudal to the base of the lower beak; also called the **cheek**. [1·7, 1·8]

malar stripe: A distinctively colored stripe in the malar region of birds; also called a **mustache stripe** or **whisker stripe**. [1·8, 2·10]

manakins: A Neotropical suboscine family (Pipridae, 44 species) of small, frugivorous birds noted for the stereotyped courtship displays of the colorful males. Many manakin species form leks, and most displays involve intricate maneuvers that may be enhanced by mechanical and/or vocal sounds and plumage displays. [6·78]

mandible: In birds, usually refers to the lower half of the beak; called the **lower beak** or **lower jaw** in the *Handbook of Bird Biology*. [4·11]

mandibular nerve: A division of the trigeminal nerve (the fifth cranial nerve) after it leaves the brain; the mandibular nerve carries sensory input from the lower beak and corner of the mouth, and motor output from the brain to the muscles of the lower beak. [4·41]

manus: The portion of the forelimb distal to the wrist; also called the **hand**. The primary feathers attach to the manus. [1·9]

map and compass model: The theory (proposed by Gustav Kramer in 1953) that two things are required to navigate to a particular destination from an unfamiliar location: (1) a map indicating current location with respect to destination, and (2) a compass indicating the desired direction of travel. [5·84]

marginal coverts: The feathers covering the upper surface of the wing, from the leading edge back to the greater coverts, which they partly overlie. [1·12, 1·13]

market hunting: Intensive hunting of species to obtain meat, feathers, hides, or other body parts highly prized by commercial consumers. [10·11]

marine birds: See **seabirds**. [1·65]

matched countersinging: Interaction in which two birds sing back and forth to one another, each choosing songs that are identical or similar to those of the other, or that contain similar phrases. Not well understood, but thought to be one way in which birds "duel" through song. [7·36]

matrilineal societies: Groups of animals in which the females remain in their natal colonies but the males disperse when they become adults. Thus, within a colony, the females share more genes than do the males, and are more likely than males to engage in acts that appear altruistic toward one another. [6·56]

maxillary nerve: A division of the trigeminal nerve (the fifth cranial nerve) as it leaves the brain. It carries sensory input from the skin of the face, upper jaw, upper eyelid, and conjunctiva (tissue covering the eye). [4·41]

medial: Toward the midline of the body. [1·4]

medial labia: Important sound-producing membranes in the medial wall of each half of the syrinx. During sound production muscles move the bronchi upward into the trachea, pushing the medial labia and lateral labia (in the lateral walls) toward (and close to) each other and into the path of air flowing out of the respiratory system. Sound is produced as air rushes between the lateral and medial labia, causing these soft tissues to vibrate. [4·96]

median: On the midline of the body. [1·4]

median coverts: The covert feathers lying between the lesser and greater coverts on the upper surface of the wing. [1·12, 1·13]

medulla oblongata: The most posterior portion of the brain, also called the **brain stem**, where the nuclei of most of the cranial nerves are located. The medulla oblongata extends caudally through the foramen magnum to become the spinal cord. [4·37]

megapodes: A family (Megapodiidae, 21 species) of chicken-like terrestrial birds of the Australasian and Oriental regions. Megapodes do not use their own body heat to incubate their eggs; instead, in many species the males tend the eggs of several females in huge, warm mounds of decaying vegetation. Other species use geothermal heat to warm their eggs. [1·93, 6·70]

melanin: Pigment, usually present as tiny granules, that produces a range of earthy colors from dark black, brown, and red-brown to gray, yellow-brown and pale yellow. Birds can synthesize their own melanin by oxidizing the amino acid tyrosine. [3·50]

membranous labyrinth: A system of interconnected, fluid-filled canals (containing endolymph) floating in the bony labyrinth of the inner ear. The labyrinth forms the cochlear duct, the semicircular ducts, and the utriculus and sacculus. [4·57]

meninges: General term for the vascularized membranes surrounding the brain and spinal cord. Specifically, these are the outer fibrous dura and the inner arachnoid and pia layers. Meninges provide sustenance and waste removal for the cells of the brain and spinal cord, which are not served by the circulatory system. [4·36]

mesencephalon: See **midbrain**. [4·38]

mesites: A family (Mesitornithidae, 3 species) of rail-like, ground-dwelling birds endemic to Madagascar; mesites are about the size of a Mourning Dove. [1·87]

mesobronchus: The continuation of each bronchus (right and left) as it enters the lung and loses its cartilaginous half-rings; also called the **primary bronchus**. The mesobronchi gradually decrease in diameter and branch into secondary bronchi. Only in birds is the term "mesobronchus" used as an alternate name for the primary bronchus. [4·98]

metabolic water: Water obtained by an organism's body as a byproduct of the chemical breakdown of fats, carbohydrates, and proteins within the body. [9·20]

metabolism: All the chemical processes that take place in the cells and tissues of the body. [4·144]

metacarpals: Palm bones. In humans they remain distinct, but in birds, they are fused with some of the carpals (wrist bones) to form the large carpometacarpus. [4·21]

metapopulation: All the individuals of a species living in an area that contains a set of subpopulations (local populations) close enough together so that individuals from occupied habitat patches can disperse and recolonize patches where the species has gone extinct. [10·79]

metapopulation dynamics: The changes in local populations found within a metapopulation over time, in which the local populations (in a habitat patch) may go extinct and then individuals from within the metapopulation may recolonize the patch at a later date, producing a long-term situation in which local populations "wink" out and then "wink" back on. [10·77]

metapopulation models: See **population viability analysis**. [10·77]

metatarsals: Instep bones. In humans they remain distinct, but in birds they fuse with the distal tarsals (ankle bones) to form the tarsometatarsus, the long bone supporting the upper section of the foot. [1·14, 4·25]

metatarsus: See **tarsometatarsus**. [1·14]

microevolution: A change in the frequency of a genetically controlled characteristic in a population over a relatively short period of time (see **macroevolution**). [1·37]

midbrain: The middle region of the vertebrate brain; it contains the optic lobes (which in birds are huge and dominant), as well as regions for the input and processing of information on hearing and balance. Also called the **mesencephalon**.

[4·38]

middle ear: Air-filled chamber between the eardrum and the cochlea (inner ear); in birds it houses the columella. [4·57]

mid-story vegetation: Layer of vegetation in a forest that consists of trees whose crowns are below the level of the main canopy and above the **understory vegetation** (see separate entry). [9·93]

migration: The regular movement of all or part of a population to and from an area; usually refers to seasonal journeys to and from breeding grounds or feeding areas. [5·52]

migratory program: The genetically programmed information that guides an inexperienced bird on its first migration. The timing of that migration is controlled by an internal biological clock that controls circannual rhythms, and the bird chooses the approximate direction and distance through its built-in ability to carry out vector navigation. [5·79]

migratory restlessness: Nocturnal hopping and fluttering during the normal migratory period, performed by caged birds that normally are inactive at night. Researchers often use the degree of this "unrest" as an indication of a bird's desire to migrate, and the orientation of the hopping and fluttering to indicate the compass direction in which the bird would normally be migrating. First called **Zugunruhe** by the German scientists who discovered it. [5·60]

mimicry: A situation in which one individual or species has evolved or learned to be similar to another in appearance, behavior, or sound. For example, Brown Thrashers and Northern Mockingbirds imitate the songs of other bird species. [7·81]

mineral lick: General term for a spot that animals visit to obtain minerals, either naturally occurring in the soil or provided by humans. A **clay lick** (see separate entry) is one type of mineral lick. [9·28]

mitigation bank: See **wetland mitigation**. [10·93]

mitochondria: Membrane-bounded units in all cells except bacteria; they generate most of the cell's energy. Mitochondria contain DNA in the form of a ring. [1·44]

mnemonic device: A memory aid. In birding, usually refers to the association of phrases from human speech with the songs of particular birds, to help people remember the songs more easily. [2·16]

moas: A family (Dinornithidae) of at least 22 species of enormous, flightless ratites that roamed the open foothills and tussock lands of interior New Zealand. Moas evolved in the late Tertiary, but were extirpated by the Polynesians approximately

400 years ago. The largest moa stood 14 feet tall. [1·96, E·24]

mobbing: Behavior in which a number of birds (often different species) swoop and dash at a potential predator; they usually give broad-band, raspy calls (**mobbing calls**) that are easy to locate and thus attract additional birds. [2·26, 6·51]

model song: A song that a bird listens to and attempts to duplicate during the process of song learning. [7·26]

molting: The process of shedding all or part of the feather coat and replacing it with new growth. [3·28] **Note:** For specific names of molts, see p. 3·35.

monocular vision: Type of vision that produces flat, two-dimensional images, in contrast to binocular vision, which produces three-dimensional images. Monocular vision results when the eyes are positioned on the sides of the head such that an object can be seen by one eye or the other, but not by both eyes at the same time. [4·51]

monogamy: Mating system in which one male pairs with one female, at least for a given breeding season. [6·68]

morph: A set of individuals within a species that are similar to one another in some genetically determined morphological characteristic, but are distinctly different from other sets of individuals within that species. Morphs may differ in characteristics such as color, body size, or bill length or shape, but not in characteristics that are related to sex, age, locality, or season. [9·42]

mortality rate: See **death rate**. [9·49]

motor neurons: Nerve cells that convey impulses from the brain and spinal cord to stimulate a muscle to contract or permit it to relax, or to cause a gland to secrete. [4·33]

mound nest: A nest composed of a pile of material with an egg chamber in the middle; in some species the chamber is entered through a tunnel from the outside. Mound nests are built by many megapodes, which bury their eggs in a pile of decomposing organic material; the Hamerkop of Africa, which builds a massive mound of sticks; and Monk Parakeets and Palmchats, which build large, colonial mound nests. [8·39]

mousebirds: Long-tailed African birds (family Coliidae, 6 species), about the size of a Mourning Dove, that climb through vegetation using their stout, hooked beaks; frequently mousebirds perch in a "chin-up position," hanging vertically from a branch. Also known as **colies**. [1·85]

mudnest builders: A family (Grallinidae) of two striking, black-and-white, robin-sized birds named for their large, cup-shaped, mud nests. The Magpie-lark is widespread, abundant, and well-known throughout open areas in much of Australia; the Torrent-lark inhabits fast-flowing streams in the mountainous areas of New Guinea. [1·96]

muscle fibers: Bundles of long, cylindrical, contractile cells (those that can contract) that are the functional units of muscle tissue; they shorten when stimulated by a nerve impulse. [4·26]

mustache stripe: A distinctively colored stripe in the malar region of birds; also called a **malar stripe** or **whisker stripe**. [1·8, 2·10]

mutation: A change in the sequence of nitrogenous bases in DNA. Most mutations have little or no effect on an organism, but a few result in changes in the structure or type of proteins produced, thus creating the genetic diversity upon which natural selection can act. [1·36, 1·41]

mutual displays: Intricate, synchronized displays or dances performed by members of a mated pair; the displays appear to stimulate and coordinate breeding behavior between pair members and to reaffirm the pair bond. Given by many long-lived birds that mate for life, such as albatrosses, gannets, grebes, and penguins. [6·38]

mutualism: An association between two (or more) organisms in which both (or all) organisms benefit. [9·7]

mycoplasmal conjunctivitis: A highly contagious infection of the conjunctiva of the eyes that has spread rapidly through House Finch populations in eastern North America since its discovery in Maryland in 1994. The disease also affects Purple Finches, American Goldfinches, Evening Grosbeaks, and Pine Grosbeaks, although to a far lesser extent. Many birds that contract the disease eventually die—usually from starvation, predation, exposure to the elements, or some other factor that results from having impaired eyesight. The disease is caused by a previously unknown strain of the bacterium *Mycoplasma gallisepticum*; other strains had long been known in poultry. Also called **House Finch eye disease**. [9·71]

myelin: Fatty material forming the nerve sheath around many axons. The process by which myelin is deposited in a developing embryo and young animal is called **myelination**. [4·33]

myelination: See **myelin**. [4·33]

myelin sheath: The pale, fatty wrapping that surrounds the axons of many neurons; it insulates the axon, functioning, in part, like the insulation around household wiring. [4·33]

myoglobin: A red-pigmented protein found in muscle cells. Similar in both form and function to the hemoglobin in blood cells, myoglobin binds with oxygen, storing it until the muscle cell needs it to release energy. [5·7]

myopia: Visual condition of an individual in which distant objects blur because images are focussed in front of, rather than on, the retina; also called **nearsightedness**. Some types of birds, such as penguins, are myopic on land because their eye is designed for vision under water. [4·50]

N

nape: The back of the neck. [1·7]

naris (plural, **nares**): The openings of the nasal cavity; they are located in the upper beak, usually near its base. Also called **nostrils**. [3·40]

nasal-frontal hinge: See **craniofacial hinge**. [4·11]

nasal region: Portion of the skull containing the nostrils. [4·11]

natal dispersal: The movement of a young animal away from the area in which it was born and raised. In birds, natal dispersal usually occurs between fledging and the first breeding season. [9·65]

natal down: The soft down feathers covering young birds before they molt into juvenal plumage. [3·16]

natural selection: The process by which evolution occurs: as individuals with traits that allow them to compete better for essential resources survive and reproduce better than other members of their population, they contribute more of their genes to the next generation, and these favorable (**adaptive**) traits increase in the population. [1·34]

Nearctic region: Zoogeographic region including arctic, temperate, and subtropical North America, reaching south to the northern border of tropical rain forest in Mexico. In the *Handbook of Bird Biology* Greenland is included in the Nearctic, but some sources place it in the Palearctic. [1·70, 1·71]

neighbor recognition: In avian biology, the ability of territorial birds to identify their neighbors—usually on the basis of song alone. [7·43]

Neognathae: One of the two Superorders of ornithurine birds; it contains all birds with a **neognathous** palate—one in which the vomer and basipterygoid process are reduced and a flexible joint exists between the pterygoid and palatine bones (as compared to a palaeognathous palate, which is formed by larger, more rigid bones). Neognathae includes all modern birds except ratites, as well as the extinct *Diatryma* and dromornithids. [E·23]

neognathous: See **Neognathae**. [E·23]

neomorphine cuckoo: See **cuckoo**. [8·141]

neoteny: An evolutionary phenomenon in which juvenile traits are retained into adulthood. [5·51]

Neotropical migrants: Birds that winter in the Neotropics but migrate to the Nearctic region to breed. Examples include many wood-warblers, tanagers, and orioles. [1·71]

Neotropical region: Zoogeographic region including the West Indies, South America, and Central America north to the northern edge of the tropical forests in Mexico. Also called the **Neotropics**. [1·70, 1·72]

nerve: A bundle of many nerve cell fibers, surrounded and bound together by connective tissue; one nerve is large enough to be seen with the naked eye. [4·32]

nerve cell: See **neuron**. [4·32]

nest: In avian biology, a structure built, excavated, or taken over by a bird, in which the eggs are laid and remain until they hatch. In many species, the young remain in the nest until they are able to fly. In some species, the "nest" is simply a scrape or depression on the ground. See specific nest types, such as **cup nest**, for more information. [8·18]

nest appropriation: Nesting in a nest that previously was used by another species or another member of the same species—usually after the previous breeding attempt has ended. Nest appropriators include cavity adopters as well as species that take over open nests, such as Solitary Sandpipers, Bonaparte's Gulls, House Sparrows, Great Horned Owls, Little Swifts, and many others. [8·58]

net primary productivity: See **primary productivity**. [9·87]

neuroglia: Cells that form a supporting, protective, felt-like bed for neurons and also provide electrical insulation. Not nerve cells themselves, they also are called **glia** or **glial cells**. [4·36]

neuron: A **nerve** cell. The basic unit of the nervous system, it consists of a cell body, axon, and one or more dendrites. It is capable of generating, conducting, and receiving nerve impulses. [4·32]

New World cuckoo: See **cuckoo**. [8·141]

New World warblers: A family (Parulidae) of 115 species of small, insectivorous birds, many of which are colorful, found in the New World. Many are Neotropical migrants. Also called **wood-warblers**. [1·71]

New Zealand Wrens: A family (Acanthisittidae) of tiny (smaller than a kinglet), nearly tailless, primarily insectivorous suboscine birds endemic to New Zealand. One of the four known species, the Stephens Island Wren, became extinct in the late 1800s. [1·96]

NGO: See **nongovernmental organization**. [10·115]

niche: The role played by a particular species in its environment. Niche includes the many ways in which a species interacts with its physical and biotic environment, such as what and how it eats, what temperatures it requires, where it spends time, when it is active, whether it disperses seeds or pollinates plants (if an animal), what animals pollinate it (if a plant), what preys on it, and so on. Also called **ecological niche**. [9·102]

niche shifting: Describes a situation in which a species occupies a different niche in different communities, depending on which competitors are found in each. [9·103]

nictitating membrane: A thin, translucent fold of skin that sweeps sideways across the eye from front to back, moistening and cleaning the eye and protecting its surface. Found in many vertebrates, including birds, but not in humans. [1·7]

night flight calls: Calls given by migrating birds as they fly at night. [7·44]

nocturnal: Active at night. [9·31]

node: A clump or mass of tissue, for example lymph nodes [4·88] or the nodes in the heart muscle that stimulate the heartbeat. [4·79]

nomadic: Pattern of movement in which individuals are constantly on the move, showing no tendency to return to previously occupied places. Crossbills and perhaps Budgerigars may be considered nomadic. [5·56]

nonbreeding plumage: See **basic plumage**. [3·33]

nongovernmental organization (NGO): A nonprofit organization that is independent from the government. [10·115]

noniridescent structural colors: Colors produced in birds when tiny vacuoles (pockets) of air within cells in the feather barbs scatter incoming light. All blues and whites of birds are noniridescent structural colors, as are many greens. [3·55]

North American Waterfowl Management Plan: A 1986 agreement between the United States and Canada to cooperatively protect waterfowl by jointly protecting habitat, restoring declining species, and conducting population research. Each year biologists make detailed population estimates, and then use them to help regulate the number of individuals of each waterfowl species harvested during the waterfowl hunting season. [10·97]

northeaster: Winter storm that sweeps into New England across the Atlantic Ocean from the northeast, the high winds circulating around a low pressure area. Often northeasters push pelagic birds from the open ocean near or over land; in severe storms, many birds die. Also called **nor'easter**. [5·72]

Northern Marine Region: The major faunal region of the seas that includes the frigid waters of the Arctic south to about 35 degrees north latitude. [1·99]

northern timberline: See **arctic tundra/ coniferous forest ecotone**. [9·123]

notched tail: See **emarginate**. [1·19]

nucleus (plural, **nuclei**): 1. The membrane-bounded command center of all cells except bacteria; usually the nucleus contains the chromosomes. [1·44] 2. Clusters of nerve cell bodies within the central nervous system. Collectively nuclei form the gray matter, found at the core of the spinal cord and in the outer areas of the brain. [4·36]

nuptial plumage: See **alternate plumage**. [3·33]

objective lens: The large lens in a pair of binoculars or a telescope that is farthest from the eye; it receives the image viewed in the eyepiece. [2·35]

obligate brood parasites: Bird species that always lay their eggs in the nests of other species, leaving the resulting young to be raised entirely by the host parents. For comparison, see **brood parasite**. [8·139]

obligate partial migrant: See **partial migration**. [5·56]

obligatory annual migration: A type of migration in which all individuals of a species migrate to and from a particular area each year. Usually obligatory migration occurs in species whose breeding-area resources vary greatly from season to season in a predictable way—for example, in birds that eat insects or nectar and that nest at high latitudes where both insects and nectar become scarce or absent during winter. [5·54]

occipital condyle: A prominent "bump" or peg on the base of the skull, with which the atlas (the first cervical vertebra) articulates. Birds have one occipital condyle, but mammals have two. [4·16]

oculomotor nerve: The third cranial nerve. It controls eye movement by carrying motor output to the eye muscles; it also carries motor signals to the eyelid muscles and the tear gland of the nictitating membrane. [4·41]

offshore feeders: Birds (or other animals) that hunt schooling fish far out to sea and far from their nests; includes most pelagic species such as albatrosses and gannets. Avian offshore feeders are also called **pelagic feeders**. [6·68]

Oilbird: A large, nocturnal frugivore of South America, related to nighthawks; the Oilbird is the only member of its family, Steatornithidae. Oilbirds use echolocation to reach their nests, which are located deep within caves. [4·63]

oil gland: A gland, located at the base of the tail on the dorsal side of the bird's body, that secretes oils that birds spread over their feathers during preening. The oils keep the skin supple and the feathers and scales from becoming brittle, but they do not appear to waterproof the feathers. Also called the **uropygial gland** or **preen gland**. [3·20]

old-growth forests: Virgin (uncut) forests or forests that have remained uncut for a very long time and thus contain trees ranging from hundreds to thousands of years in age. [10·22]

Old World cuckoo: See **cuckoo**. [8·141]

olfaction: The sense of smell. [4·62]

olfactory: Relating to the sense of smell. [4·62]

olfactory epithelium: The lining or surface tissue of the nasal cavities; it contains the sensory endings of the olfactory nerves, which carry input about odors to the brain. [4·62]

olfactory lobes: Two lobes at the anterior ends of the cerebral hemispheres of the brain that are concerned with the sense of smell. They are relatively small in birds. [4·37]

olfactory map hypothesis: The idea that homing pigeons may learn a gradient odor map of the vicinity of their home loft by associating airborne odors with the directions from which winds carry them past the loft. This gradient odor map would be based on small but systematic changes in the intensity or composition of odors over a large area; as a bird moves in a given direction, particular odors become steadily stronger or weaker. [5·93]

olfactory nerve: The first cranial nerve; it carries the sensations of smell from the lining of the nasal cavity to the olfactory bulb of the brain. [4·40]

omnivores: Organisms that eat both plants and animals. [9·123]

oölogy: The study of birds' eggs. [8·80]

operational sex ratio: The ratio of fertilizable females to sexually active males in a given population. Thus, the ratio measures not just the breeders, but all individuals physiologically and behaviorally capable of breeding. [9·80]

operculum: A flap partially covering the nares; it may help to keep out debris.

Found in some ground-feeding birds such as starlings, pigeons, and domestic chickens. [3·40]

ophthalmic nerve: A division of the trigeminal nerve (the fifth cranial nerve) as it leaves the brain; the ophthalmic carries sensory input from the nasal cavity (for nonolfactory nasal sensations), eyeball (for nonvisual eye sensations), upper eyelid, forehead, and upper beak. In ducks and geese, it carries sensory input from the bill tip organ. [4·41]

opposite birds: A group of small to medium birds that lived in the Cretaceous period between 65 and 140 million years ago. They are called "opposite" birds because their metatarsals (the instep bones of humans) fuse to form part of the tarsometatarsus from the proximal end to the distal end, a direction *opposite* to that of modern birds. Also called **enantiornithines**. [E·20]

optic chiasma: Site at which the optic tract coming from the left eye crosses the tract coming from the right eye just before entering the optic lobes of the brain. [4·41]

optic lobes: Two large lobes of the midbrain; in birds, they dominate the midbrain and are proportionally larger than those of mammals. The optic lobes receive the optic tracts from the eyes and are the sites of much initial processing of visual information. [4·37]

optic nerve: See **optic tract**. [4·41]

optic tract: Bundle of sensory nerves carrying visual sensations from the retina of the eye to the optic lobe on the opposite side of the brain. This nerve cable is considered the second cranial nerve, although it is really a tract. [4·41]

optimal foraging: A feeding strategy that maximizes caloric intake while minimizing the costs of obtaining food, such as those associated with searching for, capturing, and handling prey. [9·35]

oral flanges: Brightly colored enlargements around the base of the bill in nestlings of many species in which the parents feed the young. The flanges extend from the corner of the mouth and taper toward the tip of the bill, and are well-supplied with tactile nerve endings. Touching a flange causes the mouth to spring open, and the colors may help parents to place the food properly. [3·43, 8·107]

orbit: Cavity in the skull that houses the eye. The eyes of most birds are so large that the left and right orbits nearly meet at the midline of the skull. [4·11, 4·13]

order: Level of classification of organisms above "family" and below "class"; similar families are placed in the same order. The scientific names of bird orders end in "iformes" (for example, Passeriformes). [1·52]

organ: A group of tissues, often of distinctive character and function, that aggregate to form a discrete structure with a particular function, such as the heart, stomach, or lung. [4·2]

organ system: A group of organs whose various functions are coordinated to accomplish one or more of the basic functions of life; examples include the digestive system and the respiratory system. [4·2]

Oriental region: Zoogeographic region including all of Asia south and east of the Himalayan Mountains (India and Southeast Asia), as well as southern China and the islands of Indonesia and the Philippines, east to include the islands of Timor and Sulawesi. [1·70, 1·88]

origin: The end (attachment site) of a skeletal muscle that moves the least during contraction. [4·27]

ornithischian dinosaurs: One of the two major groups of dinosaurs, called *ornithischian* (bird-hipped) dinosaurs because their hips superficially resembled those of modern birds. They were highly specialized herbivores. [E·8, E·32]

ornithopters: Early flying machines whose crude wings were lifted by humans flapping their arms; they never got off the ground. [5·8]

Ornithurae: One of the two major subclasses of birds; it includes all modern birds as well as the Lithornithids, Ambiortiformes, Hesperornithiformes, and Ichthyornithiformes. [E·20]

ornithurine: Describes birds in the Subclass Ornithurae. [E·20]

oscillogram: A visual representation of a sound, plotted on a graph as relative loudness (vertical axis) versus time (horizontal axis). The vertical axis actually is a measure of the increase or decrease in air pressure (measured in micropascals) associated with the sound wave, which determines the loudness. The oscillogram does not show sound frequency. [7·7]

oscines: Members of Suborder Passeri, which is one of the two large suborders of Order Passeriformes (perching birds); oscines also are known as **songbirds** or **true songbirds**. Oscines have particularly complex voice boxes, which allow them to sing more complex songs than other birds. [7·25]

ossification: Hardening or calcification of soft tissue (such as cartilage or tendon) into bone or a bone-like material. [4·6]

Ostrich: The largest living bird and the only member of its order and family, Struthionidae. This familiar, flightless ratite of African deserts and savannas is almost entirely herbivorous. Ostriches breed communally, with several females laying eggs in the same nest. [1·83]

otic: Of or relating to the ear. [4·11]

outbreeding depression: The reduction in the degree to which local populations are adapted to their local environments as a result of individuals dispersing and mating with individuals in other populations (outbreeding). The reduction occurs because outbreeding individuals bring alleles adapted to one local environment into another, where they may be less advantageous. [9·66]

outer shell membrane: A loose, fibrous membrane that lines the inner surface of the avian eggshell; it sticks tightly to the eggshell, helping to hold it together. [8·62, 8·69]

outer vane: The vane located on the side of a wing or tail feather that is away from the midline of the bird. On a wing feather, the outer vane is on the edge of the wing that leads in flight. In flying birds, the outer vane is narrower than the inner vane, producing an asymmetry that aids in flight. [3·3]

outer wing: The portion of the wing from the wrist to the wing tip; the primary feathers are located on the outer wing. [5·23]

ova (singular, **ovum**): Female reproductive cells, also called **egg cells** or **eggs**, both before and just after they are fertilized by a sperm cell. [4·128, 4·130]

oval: Describes an egg that is shaped like a chicken egg; also called **ovate**. [8·73]

oval window: Former name for the **vestibular window**. [4·57]

ovary: The female gonad; it matures and releases egg cells (ova) periodically throughout the breeding season in a process called ovulation, and also produces the sex hormones estrogen, progesterone, and testosterone. In most birds only the left ovary is functional, and it enlarges greatly during the breeding season. [4·128]

ovate: See **oval**. [8·73]

Ovenbird: A wood-warbler (family Parulidae) that breeds in mature deciduous forests of the eastern and central United States and migrates to the Neotropics for the winter. [8·35]

ovenbirds: A diverse Neotropical suboscine family (Furnariidae, 240 species) whose members are especially numerous in temperate South America; ovenbirds are named for the oven-shaped, clay nests built by some species. [1·79]

overstory vegetation: The mature trees that make up the canopy of a forest. [9·93]

oviduct: The tube that transports the egg from the ovary to (in birds) the cloaca; it is suspended from the dorsal body wall by a curtain-like membrane. After the egg enters the oviduct as an egg cell (ovum) and is fertilized, different sections of the ovary add different substances to the ovum to produce the hard-shelled egg. In most birds, only the left oviduct is functional. [4·131]

oxidation: The process by which a substance is chemically combined with oxygen; also called "burning." When body cells combine oxygen with the products of food digestion, such as carbohydrates, the reaction releases energy and two waste products—water and carbon dioxide. [4·86]

oxpeckers: Two bluebird-sized species in the starling family (Sturnidae) that climb upon large grazing mammals in the African savannas, removing ticks, insects, and the scabs of skin wounds—a behavior that benefits both the birds and their hosts. Also called **tickbirds**. [1·87]

oxygenated blood: Blood whose red blood cells contain high levels of oxygen. Also called **oxygen-rich blood**, it is found in all arteries except the pulmonary artery and its branches. [4·81]

oxygen-poor blood: See **deoxygenated blood**. [4·81]

oxygen-rich blood: See **oxygenated blood**. [4·81]

Palaeognathae: One of the two Superorders of ornithurine birds; it contains all birds with a **palaeognathous** palate—one formed by large, rigid bones (compared to a neognathous palate in which the vomer and basipterygoid process are reduced and a flexible joint exists between the pterygoid and palatine bones). Palaeognathae includes all living ratites as well as the extinct elephantbirds and moas. [E·21, E·23]

palaeognathous: See **Palaeognathae**. [E·21, E·23]

palate: The roof of the mouth. [4·91, 4·112]

palatine: Bony rod on each side of the upper jaw above the jugal arch (another set of bony rods). In birds, as the lower jaw opens it moves the quadrate, which pushes the palatine and jugal arch forward, pushing on the premaxillary bones to raise the upper jaw. [4·12]

Palearctic region: Zoogeographic region that consists of most of the large landmass of Eurasia, as well as northern Africa and most of the Sahara Desert. In some systems, but not the one used by the *Handbook of Bird Biology*, Greenland is included in the Palearctic. [1·69, 1·70]

pamprodactyl feet: Foot arrangement in which all four toes, including the hallux, point forward. Found in some swifts. [1·21]

pancreas: Digestive organ and endocrine gland located in the uppermost loop of the small intestine; the pancreas secretes digestive juices into the small intestine and produces hormones (insulin, glucagon, and somatostatin) that regulate carbohydrate metabolism and blood sugar levels. [4·75, 4·122]

Pangea: The single land mass that formed 245 million years ago from all the existing continents and persisted throughout the Triassic period. [1·68]

Pantanal: An enormous expanse of grasslands extending across parts of Brazil, Paraguay, and Bolivia that each year is transformed into a huge marsh by torrential seasonal rains. It hosts one of the greatest concentrations of wildlife on the continent of South America, including numerous waders and other water birds, as well as the few remaining Hyacinth Macaws. [1·73]

pantropical: Distributed throughout the tropics of the world. [1·81]

papilla (plural, **papillae**): A small bump. 1. The bump at the end of each of a bird's deferent ducts where it opens into the cloaca; also called the **papilla of the deferent duct**. When the cloaca is everted during copulation, the papillae may slightly enter the female's oviduct. [4·128] 2. One of many small bumps covering the surface of the skin of birds during embryonic development. Each papilla consists of a core of dermis and a covering of epidermis and will eventually form an embryonic feather. Also called a **feather papilla**. [3·26] 3. One of many spiny-tipped bumps on the undersides of the feet of some birds, such as Osprey, to aid them in grasping slippery prey. [3·45] 4. One of many tiny bumps in the lining of the shell gland of the oviduct; they secrete albumen and the hard, calcium-rich outer shell around the fertilized ovum. [4·131]

parabronchi: Tiny (microscopic) air tubes formed by the branching of secondary bronchi within the avian lung; they form a network within the lung tissue. Air carrying oxygen passes out of openings in the thick walls of the parabronchi and into a network of air spaces surrounded by a network of blood capillaries. Here, gas exchange occurs—oxygen dissolves into the blood and carbon dioxide and water vapor move from the blood into the air and back into the parabronchi, eventually to be exhaled. The avian system of airflow is much more efficient than that of mammals (in which air exchange occurs in dead-end pockets called alveoli), because in birds air flows continuously across the surface of the capillary bed. [4·98]

paramo: Humid grassland area of the high Andes in South America above the tree line. It contains some shrubs and is dotted with lakes and bogs. [1·75]

parasitic: Describes a relationship between two species or individuals in which one benefits as a result of some cost to the other. [9·91]

parasitoids: Animals (usually insects) that feed in or on a host animal for a long period of time, consuming most of its tissues and eventually killing it. Some parasitoids (such as certain small wasps) live in the nests of cavity-nesting birds, feeding on avian ectoparasites and actually benefiting the birds. [8·58]

parasympathetic system: The part of the autonomic nervous system that acts on smooth muscle to reduce the heart rate and to promote feeding, egg laying, and other "peaceful" activities, such as digestion. Its nerves originate in the cranial and sacral region. The other part of the autonomic nervous system, the sympathetic system, functions under conditions of stress. [4·35]

parathyroid glands: Several small endocrine glands located on or caudal to the thyroid glands; they secrete parathormone, a protein that causes calcium resorption from the bones. [4·74]

partial migration: Migratory pattern in which some individuals in a population migrate while others remain as year-round residents. When the specific individuals that migrate are determined genetically, a situation that occurs in environments where the resources always are sufficient to enable some, but not all, individuals to overwinter, the migrating birds are called **obligate partial migrants**; examples include the European Robin and populations of the Blackcap in southern Europe. When the number of birds the environment can support varies from year to year, **facultative partial migration** evolves, in which the number and identity of the individuals migrating change from year to year in response to the availability of resources, usually food. In this case, the individuals that migrate are not determined genetically. Facultative partial migration occurs in the Blue Tit and some North American chickadees. [5·56]

partial molt: A type of molt in which only some of the feathers are replaced. [3·28]

Passeri: One of the two large suborders of Order Passeriformes (perching birds); members are also known as **oscines**, **songbirds**, or **true songbirds**. Songbirds have particularly complex voice boxes, which allow them to sing more complex songs than other birds. [7·25]

Passeriformes: See **passerines**. [1·66, 7·25]

passerines: All birds in the Order Passeriformes; also called **perching birds**. The order contains approximately 4,600 species, nearly one-half the world's bird species, all of which have a foot adapted for perching on branches or stems. [1·66, 7·25]

patagium: A fold of tough skin that extends from the brachium to the antebrachium; the patagium connects the shoulder to the wrist and forms the leading edge of the inner wing in flight. [1·11]

patella: The kneecap; an ossification within a tendon at the lower end of the femur (thigh bone). The patella glides in a deep groove and adds stability to the knee joint. [4·24]

pecten: Highly vascularized structure of the choroid layer of the eyes of all birds and some reptiles; it projects into the vitreous body where the optic nerve exits the eyeball. The pecten is believed to nourish the retina and to control the pH of the vitreous body. [4·47]

pectinate claw: A modification of the side of the middle toe claw into a comblike, serrated edge thought to be used as a preening tool. Found in only a few birds, such as Barn Owls, nightjars, bitterns, and herons. Also called a **feather comb**. [3·46]

pectoral girdle: The three bones (clavicle, coracoid, and scapula) on each side of the avian body that form a "free-floating" support for the wings. [4·19]

pectoral veins: Paired veins (left and right) that carry blood from the pectoral region; they merge with the subclavian and jugular veins on each side to form the cranial vena cavae leading to the heart. [4·83]

pectoralis: Large, powerful flight muscle of birds that attaches to the sternum. The pectoralis has two portions: the larger part pulls the wing down, slows it down at the end of the upstroke, and pulls the wing forward; the smaller part pulls the wing down and back. [5·6, 5·7]

peents: Short, nasal vocalizations given from the ground at twilight by a male American Woodcock as part of his courtship display. They are similar to the buzzy call of a Common Nighthawk. [7·16]

pelagic: Of the ocean; pelagic birds spend most of their life on the open sea, feeding at the surface or just below it and coming to land only to nest. Pelagic birds include species in the order Procellariiformes, as well as tropicbirds, some penguins, the boobies and gannets, most alcids, the skuas and jaegers, the noddies, kittiwakes, Sabine's Gull, and some terns. [1·98]

pelagic feeders: See **offshore feeders**. [6·68]

pellet: A compact ball of indigestible food, such as bones, fur, feathers, and insect exoskeletons, that is formed by the gizzard of birds that eat meat or fish—such as owls, hawks, and kingfishers—and is regurgitated through the mouth; also called a **cast**. [4·119, 4·120]

pelvic girdle: The three bones (ilium, ischium, and pubis) on each side of the body that are partly fused with one another and with the synsacrum in birds to form a strong but lightweight attachment site for the muscles of the legs, tail, and abdomen. They also provide protection for the abdominal organs. Also called the **pelvis**. [4·24]

pelvis: See **pelvic girdle**. [4·24]

pendulous cup nest: A cup nest whose rim is supported but whose unsupported belly (nest chamber) hangs from around 4 inches (10 cm) to over 1 yard (1 m) below the supports. It is generally entered from the top. Pendulous cup nests are built by New World orioles, oropendolas, and caciques, as well as some Old World weavers, and usually are woven from plant strips or fibers. Shallow pendulous cup nests grade into **pensile cup nests** (see separate entry). [8·32]

pensile cup nest: A cup nest whose rim is supported but whose unsupported belly (nest chamber) hangs below. Pensile cups are built by New World blackbirds, vireos, and kinglets; by many songbirds in Australia and Asia; and by a host of orioles and flycatchers and their relatives in the Old World. Deep pensile cup nests grade into **pendulous cup nests** (see separate entry). [8·30]

perch gleaning: See gleaning. [6·44]

perching birds: See **passerines**. [1·66, 7·25]

perforate septum: A condition in which the nasal septum (tissue separating the left and right nasal cavities) has an opening or is absent. In contrast with having an **imperforate septum** (see separate entry), having a perforate septum appears to increase an animal's sensitivity in detecting odors. However, it decreases the ability to locate an odor's source, because odors entering one nostril mix with those from the other. [4·90]

pericardium: Thin membrane surrounding the heart; its inner lining secretes pericardial fluid, which reduces the friction of the beating heart against adjacent tissues. [4·80]

perilymph: The fluid that fills the structures that make up the bony outer labyrinth of the inner ear (the cochlea, semicircular canals, and vestibule), in which floats the membranous labyrinth. [4·57]

period: A unit of geological time. Periods are divided into epochs, and successive periods make up an era. [1·113]

peripheral nervous system (PNS): The portion of the nervous system outside the brain and spinal cord; it consists of the cranial nerves as well as the spinal nerves and their associated ganglia. [4·40]

pessulus: Cartilaginous structure found only in songbirds; it sticks up into the syrinx at the bifurcation of the bronchi. [4·93]

phalanges (singular, **phalanx**)**:** The bones of the fingers and toes. [1·9, 4·25]

phallus: See **cloacal phallus**. [4·129]

pharynx: Area, also known as the **throat**, that begins at the back of the tongue and connects the mouth and esophagus, and where the digestive and respiratory pathways cross one another. The nasal cavities and auditory tube open into the pharynx. [4·91, 4·112]

Phasianidae: The pheasant family; it includes pheasants, partridges, ptarmigan, grouse, guineafowl, turkeys, and Old World quail—as well as the Red Junglefowl (the wild relative of domestic chickens) and the spectacular peafowl. Its 177 species are found throughout most of the world, but few species inhabit the Neotropics. [1·89, 1·90]

philopatry: The tendency of individuals of certain species to eventually take up residence (and breed) near the area where they were born. [9·65]

photoperiod: The amount of time during each 24-hour period that an organism is exposed to light. [5·63]

photoperiodism: A behavioral or physiological response to day length. [4·140]

phylogeny: The evolutionary history of an organism. [1·32]

phylum (plural, **phyla**)**:** Level of classification of organisms above "class" and below "kingdom"; similar classes are placed within the same phylum. All birds are in the phylum Chordata. [1·52]

physiographic regions: Regions of the world that are biologically distinct based on climate conditions, topography, soil types, and plant communities; also called **ecoregions**. [10·86, 10·87]

phytoplankton: Microscopic plants and algae that float freely in aquatic environments; phytoplankton make up the "plant" portion of plankton. [1·103]

pia: The innermost of the three vascularized membranes called meninges that surround the brain and spinal cord. The meninges provide sustenance and waste removal for the cells of the brain and spinal cord, which are not served by the circulatory system. [4·36]

pigeon's milk: See **crop milk**. [4·116]

pigments: Colored substances that give color to the structures, such as skin and feathers, in which they occur. In theory, pigments can be extracted from these structures. [3·50]

pineal gland: Endocrine gland located in the dorsal midbrain; it secretes the hormone melatonin, which plays a role in regulating daily activity cycles (circadian rhythms). [4·71]

pin feathers: Developing feathers that are still surrounded by a feather sheath. [3·26]

pinyon-juniper woodland: Ecosystem of southwestern North America consisting of park-like stands of small pinyon pines and junipers growing on hills and mountain slopes below the montane forests. Five species of breeding birds are particularly characteristic to this dry ecosystem: Gray Flycatcher, Pinyon Jay, Juniper Titmouse, Bushtit, and Bewick's Wren. [9·121]

pipped: Describes an egg about to hatch, in which the embryo within has punctured a small hole. [8·105]

piracy: A foraging technique in which a bird steals food from another bird; used by Parasitic Jaegers, frigatebirds, and some gulls, among others. [6·48]

pitch: See **frequency**. [7·4]

pittas: An Old World family (Pittidae, 31 species) of secretive, stocky suboscines with long legs and a short tail; they live on the tropical forest floor, and many are brightly colored below and cryptic above. Pittas use their heavy bills to catch a variety of insects and other small animals, especially snails. [1·90]

pituitary gland: An endocrine gland attached to the ventral surface of the hypothalamus; it consists of an anterior and a posterior lobe. The pituitary synthesizes, stores, and releases a wide variety of hormones. [4·72]

plankton: Microscopic plants (and algae) termed **phytoplankton** and microscopic animals (and protozoa) termed **zooplankton** that float freely in aquatic environments. [1·101]

plasma: The fluid portion of the blood; it contains sugars, inorganic salts, and certain proteins (plasma proteins) found only in the plasma, and carries numerous types of dissolved substances. Avian plasma has a higher sugar, fat, and uric acid content than the plasma of most mammals. [4·86]

platform nest: Nest consisting of a very shallow depression in the top of a mound of nest material. Varies in complexity from the flimsy platform of twigs built by Mourning Doves to the enormous stick piles amassed by storks or large birds of prey such as Osprey and eagles. Some platform nests, such as those of some terns and rails, float. [8·28]

playback experiments: Experiments in which researchers play songs (through a speaker) to territorial birds, and then note their responses. [7·48]

plexuses: Complex web- or net-like structures along the spinal cord at the level of the front limbs or wings (brachial plexus) and hind limbs (lumbosacral plexus); they are formed by a merging of the large spinal nerves within each region. [4·39]

plumage: 1. A bird's entire feather coat. [3·33] 2. The set of feathers produced by a particular molt. With this usage, a bird wears parts of two different plumages after a partial molt. [3·33]

plunge diving: A foraging technique in which an airborne bird dives under water to pursue aquatic prey; performed by birds such as Brown Pelicans, auks, gannets, Osprey, and kingfishers. [6·45]

pneumatic: Describes bones that are filled with air spaces and may contain air sacs; nearly all the bones of most birds are pneumatic. [4·5]

pneumatic foramen: Opening in the avian humerus (upper wing bone) leading to the air space within the bone; extensions of the clavicular air sac lead into the pneumatic foramen and thus connect with the air space. [4·100]

PNS: See **peripheral nervous system**. [4·40]

podotheca: The tough skin covering the tarsus. [1·20]

polarized light: Light in which the waves are all oriented in the same plane, or in a reduced number of planes, compared to unpolarized light. [5·94]

polarizing lens or **filter:** Something that selectively transmits only those light waves with certain orientations, creating a polarized beam of light. [5·94]

polyandry: Mating system in which one female mates with several males within the same breeding season. [6·77]

polygamy: General term for a mating system in which individuals of one sex (either the males or females) mate with more than one partner during the same breeding season. Both polygyny and polyandry are forms of polygamy. [6·79]

polygyny: Mating system in which one male mates with several females within the same breeding season. Polyandrous species include Spotted Sandpipers, most jacanas, and certain phalaropes. [6·68]

polymorphism: A situation in which one species contains two or more distinct morphological types (**morphs**) of individuals, which are determined genetically. Morphs may differ in physical characteristics such as color, body size, or bill length or shape, but not in characteristics that are related to sex, age, locality, or season. [9·42]

population: All the individuals of a species that live in the same area. [1·45, 9·3]

population density: The number of individuals per unit area in a given population. [9·58]

population dynamics: 1. Changes in population size, density, dispersion, and structure over time. [9·49] 2. The study of how populations change over time. [9·49]

population ecology: The study of how animal populations are related to, and respond to, their environments. It involves monitoring and studying reproductive rates, survival rates, movements of individuals and populations, and changes in population densities over time and from one area to another. [9·49]

population structure: The relative proportions of males and females, individuals of different ages, and individuals with other definable differences in a population. [9·49]

population viability analysis (PVA): A computer technique developed by conservation biologists in the 1980s that simulates the growth of populations over time. Given numerous demographic variables, such as birth and death rates at various ages and the annual variation in these rates, PVA can predict the probability that local populations of different sizes will go extinct over a designated period of time. The simplest PVAs are **single-population models**, which focus on just one population (of different starting sizes and over different time periods) and ignore any dispersal of individuals between populations. More complex PVAs focus at the **metapopulation** level (see separate entry); these use more detailed information, such as dispersal rates and patterns of behavior and mortality while moving across inhospitable habitats, to more realistically predict the probability of extinction. [10·77]

porphyrins: Pigments producing red, brown, green, or pink colors in a number of avian orders; they are especially well known for producing the bright reds and greens of the turacos. Porphyrins are complex, nitrogen-containing molecules related to hemoglobin, that birds and other organisms synthesize by modifying amino acids. [3·51]

Porro prism binoculars: Binoculars in which the eyepieces are closer together than are the objective lenses. [2·38]

portal system: A blood circulatory pattern involving **portal veins**. [4·85]

portal veins: Veins (such as the hepatic portal vein and the renal portal vein) that carry blood between two capillary beds, rather than connecting a capillary bed and the heart, as do most veins. Portal veins allow a second organ to process the blood before it is returned to the heart and thus the general circulation. [4·85]

posterior: Toward the back of an organism, using the earth as a frame of reference.

With birds, technically used only within the eye and inner ear. But in practice, often used interchangeably with **caudal**. [1·4]

posterior air sacs: General term referring to the air sacs nearest the bird's caudal end—the posterior thoracic and abdominal air sacs. [4·102]

posterior chamber: Space between the iris and lens of the eye; it is filled with aqueous fluid, which nourishes the eye and removes wastes. [4·48]

posterior lobe of the pituitary gland: Part of the pituitary gland that does not manufacture hormones itself, but instead stores and releases neurohormones such as mesotocin and arginine vasotocin. [4·73]

post-hatch brood amalgamation: A broad term describing a variety of methods by which young from several broods may combine into larger groups. [8·128]

postpatagium: A tough band of tendinous tissue running along the edge of the antebrachium and manus that trails in flight. The postpatagium holds the primary and secondary feathers firmly in place and supports each quill. [1·11]

powder downs: Down feathers that are never molted but grow continuously, disintegrating at their tips to produce a fine powder resembling talcum powder, which may help to waterproof the feathers. [3·18]

powered flight: Type of flight in which the flying object produces thrust. In a bird, refers to flight powered by flapping, which produces thrust to propel the bird forward. [5·36]

precocial: Describes young birds that hatch in a relatively developed state—downy and with their eyes open. Many are soon able to walk or swim and even eat on their own. [8·106]

precopulatory displays: Displays given by either sex immediately before copulation that appear to invite or solicit copulation. Female birds of many species perform a stereotyped precopulatory display in which they raise the tail and hold the wings down or out, quivering them. [6·40, 7·71]

preen gland: See **oil gland**. [3·20]

preening: Feather maintenance behavior in which a bird grasps a feather near its base, then nibbles along the shaft toward the tip with a quivering motion; this cleans and smooths the feather. Many birds gather on their bill oily secretions from the oil gland, and then spread them on their feathers as they preen. [3·18]

premaxilla: Main bone forming each side of the upper beak of a bird. The two halves together are called the **premaxillary bones**. [4·11]

prescribed burning: Purposefully starting fires under controlled conditions, at in-

tervals that mimic the natural fire cycle of the landscape, to maintain certain types of habitats and the species that depend on them. [10·83]

primary bronchus: See **mesobronchus**. [4·98]

primary consumers: Level in a food chain or web that consists of organisms—usually animals—that directly eat plants or other producers; most primary consumers are herbivores. [9·123]

primary feathers: The flight feathers of the outer wing; they are attached to the manus. [1·11, 1·10]

primary productivity: **Gross primary productivity** is the rate at which green plants and the few other producers (such as green algae) in an ecosystem produce biomass, including the biomass that they use themselves during respiration. **Net primary productivity** is the gross primary productivity *minus* the biomass that the producers use themselves; in other words, it is the rate at which the producers create biomass that is available as food for other organisms. [9·87]

primary succession: Ecological succession that begins on a substrate, such as rock, sand, or lava, that has never before supported a community. Also refers to the process by which a lake gradually fills in to form a bog community. See **ecological succession**. [9·110]

primitive streak: In avian biology, very early embryonic stage consisting of a long trough (the "streak") with raised sides that establishes the cranial-caudal, left-right, and dorsal-ventral body axes. The primitive streak is formed in the flattened mass of cells that lies on the upper surface of the yolk. [8·64]

producers: Level in a food chain or web that consists of green plants and a few other organisms (such as green algae) that capture energy from the sun and then introduce the energy into the food chain by incorporating it into organic compounds, which the producers synthesize from water and inorganic substances through the process of photosynthesis. [9·123]

productivity: In biology, the total number of grams of living material produced per square meter per year ($g/m^2/yr$) in a particular area or habitat. Productivity includes the increase in size of all living things, as well as new individuals added through birth or germination. Productivity may be measured as **gross productivity** or **net productivity** (see similar definitions under **primary productivity**). [9·92]

projectile vomiting: Vomiting that is sufficiently forceful to propel the regurgitated substance a short distance away. Birds such as vultures, herons, gulls, and tubenoses (Fulmars, albatrosses, petrels, storm-petrels, and diving petrels) use this technique to ward off predators; Fulmars can spew an oily, foul-smelling mixture of flesh and fluid several feet. [6·54]

proprioception: A "body-parts-position sense": the subconscious perception of impulses from the muscles, tendons, and joints that allows the body to know the position of its parts, which is crucial for balance and movement. [4·33]

proteins: Complex molecules composed of strings of amino acids; proteins are the main building blocks of all living organisms and also act as enzymes, assisting chemical reactions. [1·41]

protoplasm: The living material composing all organisms. [4·144]

proventriculus: The upper part of the bird's two-part stomach; it is elongate and glandular, and secretes mucus, hydrochloric acid, and an inactive precursor to pepsin, an enzyme that digests protein. [4·118]

proximal: Toward the center of the body (the elbow is proximal to the fingers) or toward the origin of a structure (the base of a feather—where it is attached—is proximal to its tip). [1·4]

proximate: In biology, pertaining to immediate, internal causes; mechanistic. **Proximate questions** about a behavior, for example, look into the internal causes of the behavior—*how* it is carried out. For comparison, see **ultimate**. [6·2]

proximate factors: In biology, the actual cues that trigger the body to carry out a particular process or behavior. For example, day length is a proximate factor that affects the timing of breeding in birds. For comparison, see **ultimate factors**. [8·11]

Prunellidae: A family of 13 sparrow-like birds with slender, pointed bills; it includes the accentors and Dunnock. Prunellidae is the only bird family endemic to the Palearctic region. [1·70]

pruning: A foraging technique in which a bird bites off and eats plant buds; used by Ruffed Grouse and other birds. [6·46]

pseudo-babblers: A family (Pomatostomidae, 5 species) of noisy, busy, social ground-feeding songbirds with long, curved beaks and long, towhee-like tails; they are endemic to the Australasian region. [1·95]

pseudosuchian thecodont: Term used primarily by Robert Broom in his "pseudosuchian thecodont hypothesis" to refer to the **thecodonts** (see separate entry). [E·33]

pseudosuchian thecodont hypothesis (of bird evolution): Theory that birds evolved approximately 230 million years ago from small, arboreal thecodonts; first proposed by Robert Broom in 1913. [E·8]

pterosaurs: Flying reptiles from the Triassic that radiated and diversified in the Jurassic and Cretaceous. They had fea-

tures convergent with those of birds, such as hollow bones and a slight keel on the sternum, but their large, membranous wings were structurally unique: they were supported by a single, greatly elongated fourth finger and attached to the side of the body and possibly also to the hind limb. [E·33]

pterylae (singular, **pteryla**): Areas of a bird's skin where feathers are attached; also called **feather tracts**. [3·2]

pterylosis: The arrangement of feather tracts and bare patches on a bird. [3·2]

pubis: One of the three paired bones that fuse to form the pelvic girdle; the long, thin pubis runs caudally along the ventral side of the ischium, and is only partly fused with it. [4·24]

puddle ducks: See **dabbling ducks**. [6·47]

pulmonary arteries: Two large arteries (right and left) carrying oxygen-poor blood from the pulmonary trunk (coming from the right ventricle of the heart) to the lungs; the pulmonary arteries and their branches are the only arteries that carry deoxygenated blood. [4·78]

pulmonary arterioles: Tiny arteries within the lung that supply deoxygenated blood to the capillaries surrounding the network of air spaces adjacent to the parabronchi, where air exchange occurs in birds. [4·99]

pulmonary circulation: Portion of the circulatory system carrying blood from the heart to the lungs and back to the heart. [4·77]

pulmonary trunk: A large artery that carries oxygen-poor blood from the right ventricle of the heart and then branches to the right and left lungs as the right and left pulmonary arteries. [4·78]

pulmonary veins: Two large veins (left and right) carrying oxygen-rich blood from the lungs to the left atrium of the heart; the only veins (along with the pulmonary venules) that carry oxygenated blood. [4·78]

pulmonary venules: Tiny veins within the lung that carry oxygenated blood away from the capillaries surrounding the network of air spaces adjacent to the parabronchi, where air exchange occurs. [4·99]

pupil: The opening in the center of the iris; it controls the amount of light entering the eye. [4·47]

pus: The yellowish fluid seen in infections; it is a mixture of bacteria, dead white blood cells, and fluid. [4·88]

PVA: See **population viability analysis**. [10·77]

pygostyle: The **tail bone** of birds; it is formed by the fusion of the final few caudal

vertebrae. The pygostyle is shaped like a plowshare and provides an attachment site for the flight feathers of the tail. [4·18]

pylorus: A muscular sphincter (circular band of muscle) at the lower end of the stomach; it regulates the passage of food from the stomach into the small intestine. [4·120]

pyriform: Pear-shaped. Describes an egg that is rounded at one end and relatively pointed at the other; also called **conical**. [8·73]

Q

quadrate bone: Bone on each side of the skull, at the caudal and lower end of the upper jaw; in birds, it links with the articular bone on each side of the lower jaw. This linkage forms the joint between the upper and lower jaws, allowing a bird to open its mouth widely. [4·12]

quill: Feather. [1·11]

R

race: A population of a species, usually in a particular geographic area, that is morphologically distinct from other populations of the same species, but whose members remain capable of interbreeding with members of those other populations. Also called a **subspecies**. [1·55]

rachis: The portion of the feather shaft above the calamus; it is solid and has a vane on each side. [3·3]

radiale: The larger of the two wrist bones of birds; it consists of fused carpals. [1·9]

radius: The smaller of the two bones supporting the antebrachium (forearm). [1·9]

range: The geographic area within which a species or population generally remains at a particular time of the year; a species may have different breeding and nonbreeding ranges. Also called the **geographic range**. [2·19, 9·49]

raptors: Members of the orders Falconiformes and Strigiformes, which contain all the diurnal and nocturnal birds of prey. [1·66]

ratites: All birds lacking a keel on the sternum. Includes flightless birds—the Ostrich, rheas, Emu, cassowaries, and kiwis—as well as the tinamous, which are fully capable of flight. [E·22]

rattle: A harsh trill. [2·16]

recolonization: The establishment of a population of a particular species in an area where that species formerly occurred, generally through natural dispersal. [10·77]

rectrices (singular, **rectrix**): The long, stiff flight feathers of the tail. [1·12, 1·13]

recurrent bronchi: Secondary bronchi (branches off the main air tube, the mesobronchus) within the avian lung that connect to the air sacs outside the lung. [4·98]

recurved: Curved upward. Describes beaks such as that of the American Avocet. [1·17]

red blood cells: Flattened, elliptical, red-colored cells found in the blood; they house the iron-containing pigment hemoglobin, which carries large amounts of oxygen and gives the blood its red color. The red blood cells of birds have a nucleus, unlike those of mammals. Also called **erythrocytes**. [4·86]

red fibers: A type of muscle fiber that appears red because it contains a large amount of myoglobin (which carries oxygen) and is permeated by massive capillary beds (containing many red blood cells, which also carry oxygen). Muscles with many red fibers produce energy aerobically (using oxygen), without building up the quantities of lactic acid that cause fatigue, and thus can sustain actions for long periods of time. Therefore, long-distance fliers often have a preponderance of red fibers. Muscles with many red fibers often are called "**dark meat**" (as in the drumstick of a turkey). [5·7]

redirected activities: Actions that are appropriate for a given situation, but are directed at an inappropriate subject; often they appear at times of stress or conflicting motivations. Some have been exaggerated and incorporated into displays. [6·33]

reflex: A relatively simple and stereotyped response to a stimulus; it may be either automatic or learned. [4·34]

refugia: Areas remaining stable when surrounding areas are undergoing change (such as that caused by glaciation); often used to describe isolated patches of rain forest separated by grasslands during glacial periods. [1·77]

regional population: A population that inhabits a fairly large area; for example, all the Canada Geese in the central migratory flyway. [9·49]

reintroduction: The establishment of individuals of a species (through human effort) in an area where that species used to live—using individuals from a different area (**translocation**) or from a captive breeding program. [10·89]

relative abundance (of species): The number of individuals of each species living in a particular community or area compared to the numbers of individuals of other species in that area. Relative abundance measures how evenly individuals are distributed among species. [9·82]

releasers: Specific objects, physical features, or behaviors that activate (or "release") a specific response in an individual. [6·5]

relict population: A small, isolated population that is a remnant of a much larger population that once existed over a wider geographic area. [9·57]

remiges (singular, **remex**): The longest wing feathers, also called the **flight feathers of the wings**; they include the primary and secondary feathers. [1·11, 1·13]

renal efferent veins: Veins that collect blood from the capillary beds of the kidneys and carry it to the caudal vena cava, which returns it to the heart. [4·85]

renal portal system: System of blood flow, found in birds and reptiles but not in mammals, that collects venous blood from the lower portion of the digestive tract via the caudal mesenteric vein, and conveys it to the venous ring at the kidneys, where some of it passes through a capillary bed before being conveyed back to the heart. The reason for this pattern of blood flow is unknown. [4·85]

renal portal veins: Two veins (left and right) carrying blood from the caudal mesenteric vein to the capillary beds of the two kidneys; part of the **renal portal system** of birds. [4·83]

repertoire: See **song repertoire** [7·14, 7·19, 7·85] or **vocal repertoire**. [7·10]

residents: Individuals that live year round in a particular area. [5·54]

resource-defense polygyny: Mating system in which one male may have several mates, and the females choose their male partners by evaluating the quality of their breeding territories; males with better territories attract more mates. This system may evolve when safe nesting sites or rich food supplies are more important to the survival of the young than is male parental care. Found in Red-winged Blackbirds and a few other species. [6·73]

reticulate podotheca: A podotheca that is divided into a network of small, irregular, nonoverlapping plates. [1·20]

retina: The innermost layer of the eyeball, upon which images are focused; it is pigmented and contains the light-sensitive rod cells and cone cells. [4·49]

retort nest: A globular nest with an entrance tunnel; built out of mud by many mud-nesting swallows such as Cliff Swallows, and constructed from grass by many African weavers and New World swifts in the genus *Panyptila*. [8·36]

reverse Porro prism binoculars: Binoculars in which the objective lenses are closer together than the eyepieces; characteristic of some compact binoculars. [2·38]

reverse sexual size dimorphism: A situation in which females are significantly larger than males of the same species; found in many predatory birds such as raptors, jaegers, skuas, frigatebirds, and boobies. Because the sexes are different sizes, they can take different sizes and thus different types of prey. This may reduce competition between mates, but why females of these species are larger than males (rather than the more typical situation, in which females are smaller) remains a matter of much speculation. [9·40]

rhamphotheca: Outer covering (sheath) of the beak; it grows throughout a bird's life. In most birds it is hard and horny, but in waterfowl, pigeons, and some shorebirds it is soft and leathery. [3·39]

rheas: Two species (family Rheidae) of large, flightless ratites of temperate open country in South America. [1·74]

rhomboid sinus: Opening on the dorsal midline of the lumbosacral enlargement of the spinal cord; it contains a gelatinous mass of neuroglial cells rich in glycogen, known as the glycogen body, whose function is unknown. The rhomboid sinus and glycogen body are unique to birds. [4·39]

rictal bristles: Stiff, hairlike feathers projecting from the base of the beak in birds that catch insects. They may protect the face and eyes of some birds that capture large, scaly insects; they also may help birds detect movements of prey held in the beak. [3·13]

right heart: The right atrium and right ventricle of the heart; these chambers contain deoxygenated blood, which does not mix with the oxygenated blood of the left heart (left atrium and ventricle). [4·78]

ritualization: The evolutionary process by which certain everyday motions become exaggerated, repeated, and stereotyped into displays presenting a clear message. [6·32]

rockfowl: Dull-colored African songbirds (family Picathartidae, 2 species) with colorful, bare heads; they hop along the rain forest floor in rocky areas and nest colonially in caves. [1·85]

rods: One of the two kinds of light-sensitive cells lining the retina of the eye; they are particularly good at detecting low light levels (but not in differentiating colors, which is the function of the cones). When a rod cell is stimulated by light energy, it sends a nerve impulse to the brain via the optic tract. [4·48]

rollers: Old World nonpasserines (family Coraciidae, 12 species) colored in shades of blue, pink, olive, or chestnut; they are named for their rolling or rocking dives during courtship flights. They have long, slender beaks and forage by perching and then flying to the ground to catch large arthropods. [1·87]

roof prism binoculars: Binoculars with straight barrels, in which the eyepieces and objective lenses are directly in line. [2·38]

rostral: Toward the beak; used for positions on the head and neck. [1·4]

round window: Former name of the **cochlear window**. [4·59]

r-selected species: Species in which individuals have a life history strategy that emphasizes high rates of reproduction instead of (and probably at the expense of) long-term survival. These species are generally small and tend to develop rapidly, breed at an early age, have many young per cycle (large litters or clutches), breed frequently, care for their young only briefly, and have short life spans. [9·45]

rump: The lower back of the bird, above the tail coverts; conspicuously colored in some birds. [1·6]

runaway selection: One possible explanation for why females of some species choose the males with the most elaborate ornaments (such as ornate plumage) to copulate with. The explanation applies primarily to females choosing nonpaternal sexual partners, either (1) males for extrapair copulations, or (2) mates in species in which males do not provide parental care or other resources for their offspring (such as territories with food or nesting sites). The runaway selection hypothesis suggests that females may choose males with elaborate ornaments simply because their sons will then inherit these attributes and be especially attractive to females, and thus produce more offspring (of both sexes); in this way, both the genes for possessing elaborate ornaments and the genes for choosing male partners with those ornaments will increase in frequency in future generations. As each generation of females chooses the most ornate males available, the ornaments evolve to be more and more elaborate, and selection is said to "run away" with the trait—even to the extent that having an elaborate ornament may reduce a male's longevity. For runaway selection to get started, females must gain some other advantage by choosing males with elaborate ornaments than simply having attractive sons. Once the trend is begun, however, the advantage need no longer exist for runaway selection to proceed. [6·84]

S

sacculus: One of the two membranous chambers inside the bony vestibule of the inner ear (the other is the utriculus); these chambers contain hair cells and dense crystals called statoconia, both of which are embedded in a gelatinous material that is surrounded by endolymph. The statoconia and hair cells perceive the position of the head with respect to gravity (see entry for **statoconia**). [4·58]

sacral vertebrae: The vertebrae of the pelvic region; in birds, they are all fused with the lumbar vertebrae and with some of the thoracic and caudal vertebrae to form the synsacrum. [4·15, 4·18]

sagebrush ecosystem: Fairly dry ecosystem found above the lowest elevations of the Great Basin of North America; it is a shrubland dominated by sagebrush, shadscale, and other woody growth 2 to 5 feet (0.6 to 1.5 m) high. Of the few breeding birds, four species are particularly characteristic: Sage Grouse, Sage Thrasher, Sage Sparrow, and Brewer's Sparrow. [9·121]

sagittal plane: A vertical plane through the long axis of an organism, extending from head to tail. It divides the body into left and right portions. [1·4]

salivary glands: Glands that secrete saliva, which primarily acts to moisten food in the mouth. Some birds that obtain their food from the water, such as Anhingas, have no salivary glands. The salivary glands of others, such as certain swifts and swiftlets, secrete an adhesive substance used in nest building. [4·111]

sally gleaning: See **gleaning**. [6·44]

sandgrouse: Sixteen bird species that form the family Pteroclidae and that inhabit the deserts and dry savannas of Africa and Eurasia; they are similar in size and shape to pigeons, but taxonomists are not sure of their relationships to other families, and thus their order is not determined. Sandgrouse feed primarily on dry seeds and often fly long distances to water holes to obtain water. They are well known for their specialized belly feathers, which are structurally adapted to absorb and retain water. Parents (usually the male) soak the belly feathers at a water hole, and then carry the water back to their nestlings. [8·133, 9·21]

sap well: A rectangular hole up to a few inches long, drilled through the bark of trees and shrubs by a sapsucker. The bird may return to the well many times to feed, both on the sap that flows from the wound and on insects that are attracted to the flowing sap. Other birds and mammals also may feast at the sapsucker's well. [9·128]

saurischian dinosaurs: One of the two major groups of dinosaurs, the saurischian dinosaurs are also known as "reptile-hipped" dinosaurs because of the structure of their hip joint. They are composed of two groups, the herbivorous sauropods and the carnivorous theropods. [E·8, E·33]

Sauriurae: One of the two major subclasses of birds; it contains primitive birds such as *Archaeopteryx* and the enantiornithines (opposite birds), but no living bird groups. [E·20]

sauriurine: Describes birds in the Subclass Sauriurae. [E·20]

scapula: The shoulder blade; one of the three sets of paired bones that make up the pectoral girdle. [4·19]

scapulars: A group of feathers in the shoulder region. [1·12, 1·13]

scavengers: Organisms that eat dead organisms and wastes. [6·63]

scientific method: A procedure that scientists use to investigate how the world works. The specific steps typically followed by scientists to investigate aspects of the world vary among the different scientific disciplines, but in the biological sciences the scientific method usually involves the following: asking a question about the world, formulating the question into a testable hypothesis, designing a study and collecting unbiased data, analyzing the data to see if the hypothesis is supported or rejected (this often involves **statistical testing**—see separate entry), and then drawing conclusions. [6·97]

sclera: Tough, whitish, outer layer of connective tissue surrounding the eyeball. [4·48]

scleral ossicles: Bony rings supporting the eyes of all birds, lizards, turtles, and fishes; they are not present in mammals. [4·47]

scrape: A shallow depression created by certain types of birds for use as a nest. It usually contains little or no nest material; in some species it may be lined with a few flat pebbles. Scrapes are built by many plovers, terns, skimmers, and penguins. [8·26]

screamers: A South American family (Anhimidae, 3 species) of heavy-bodied, goose-like birds with far-reaching calls. [1·74, 4·22]

scrub desert: Hot, dry ecosystem of the lowlands and valley floors of southwestern North America; it is composed of bare ground with widely spaced plants such as creosote bush, ocotillo, and bur yucca. The scrub desert ecosystem hosts a moderate number of breeding birds, which tend to synchronize their nesting with periods of rainfall. [9·122]

scutellate podotheca: A podotheca that is broken up into overlapping scales. [1·20]

seabirds: All birds directly associated with the open seas and consistently dependent on the seas for food; also called **marine birds**. [1·65]

search image: A general idea or internal representation of an object being sought that allows an animal to find it easily. If an animal has a search image for a particular type of prey, it is able to find the prey more efficiently than it can find other prey for which it has no search image. [9·25]

secondary bronchi: Branches off the mesobronchus (main air tube) of each avian lung; some connect to the air sacs and are called recurrent bronchi, and others divide to form the tiny parabronchi, the major respiratory units of the avian lung. [4·98]

secondary cavity nesters: See **cavity adopters**. [8·39]

secondary consumers: Level in a food chain or web that consists of organisms—usually animals—that directly eat primary consumers. [9·123]

secondary feathers: The flight feathers of the inner wing; they are attached to the ulna of the antebrachium. [1·11, 1·13]

secondary sex characters: Anatomical features that distinguish the sexes, but are not directly involved in the production of eggs or sperm. [4·137]

secondary succession: Ecological succession (see separate entry) that begins with bare soil or an existing community. [9·110]

Secretary-bird: A long-legged bird of prey that stalks the African savannas on foot for snakes, other small vertebrates, and large insects. It is the only member of its family, Sagittariidae. [1·84]

Section 404, Clean Water Act: Part of the **Clean Water Act** (see separate entry) passed in the United States in 1972; Section 404 strives to protect wetlands by regulating the filling of, or discharge of dredged material into, all fresh waters of the United States, including wetlands. It allows wetland filling by permit if applicants show that less damaging alternatives are not feasible, and that the water will not be degraded significantly. Applicants must avoid wetland impacts when possible, minimize wetland damage, and provide compensation (**wetland mitigation**—see separate entry) for all wetland impacts deemed unavoidable. [10·93]

seedsnipes: A family (Thinocoridae, 4 species) of ptarmigan-like birds that nest both in the southern lowlands and the high mountains of South America. [1·74]

semicircular canals: Three bony, ring-like canals of the inner ear, arranged at right angles to one another (one in each plane of space); they are part of the bony labyrinth, and thus are filled with perilymph. They enclose the semicircular ducts, which contain the hair cells that

sense changes in the organism's speed or direction, thus guiding balance. [4·58]

semicircular ducts: Three ducts floating in the perilymph of the semicircular canals of the inner ear, one in each plane of space; they contain the hair cells that guide balance by sensing changes in the organism's speed or direction. They are part of the membranous labyrinth of the inner ear and thus are filled with endolymph. [4·60]

semilunar membrane: Membrane that extends from the pessulus into the cavity of the syrinx; it is found only in songbirds. [4·93]

semilunar valves: Valves, each composed of three cusps (flaps), that are located at the beginning of the pulmonary trunk and the aorta; the semilunar valves prevent the backflow of blood into the ventricles as the ventricles relax after each heartbeat. [4·79]

semiplumes: Feathers that occur in a continuum of forms between down and contour feathers. Located at the edges of the contour feather tracts, semiplumes provide insulation and help to maintain the bird's streamlined shape. [3·17]

sensitive period: A relatively brief span of time early in life during which songbirds are best able to memorize the details of songs they hear. [7·26]

sensory neuron: Nerve cell that conveys impulses to the spinal cord and brain; these impulses are interpreted as sensations—either conscious or subconscious. [4·33]

sentinel system: A system in which group members take turns watching for danger. Birds using sentinel systems include Florida Scrub-Jays and some American Crows. [6·56]

septum: Tissue that extends ventrally from the upper wall of the nasal cavities and separates the left and right nasal cavities from one another; the septum may be **perforate** or **imperforate** (see separate entries). [4·90]

seriemas: A family (Cariamidae, 2 species) of fast, long-legged, grassland birds of South America; seriemas chase down their reptile prey on foot. [1·75]

sex hormones: Hormones produced by the anterior lobe of the pituitary gland, the adrenal glands, and (primarily) the gonads; they regulate the anatomical structures, physiological processes, and behaviors that are essential for reproduction. [4·137]

sex ratio: The ratio of males to females (or females to males)—usually noted for a given population. [9·80]

sexual dimorphism: A situation in which males and females of the same species differ from each other in size or form. [9·39]

sexual selection: Form of natural selection that occurs when individuals differ in their ability to acquire mates. In sexual selection the characters that enhance mating success (either by increasing an individual's ability to compete with other members of the same sex and thus gain mates, or by directly increasing an individual's attractiveness to the opposite sex) are selected for, in some cases even if they decrease individual survival (see **runaway selection**). [3·68, 6·82]

shaft: The stiff, central rod of a feather, from which the vanes extend. [3·3]

shell gland: Region of the avian oviduct after the isthmus and before the vagina; it is lined with papillae that secrete fluid albumen and the calcium-rich shell around the fertilized ovum. Shell pigments, if any, also are added here; the patterns of pigmentation reflecting the speed of the egg's passage and rotation through the region. [4·131]

shivering: Uncoordinated muscle fiber contraction; a way to produce heat by muscle contraction without directed movement. [4·27]

shock molt: A rare situation in which a bird sheds many feathers all at once; it usually is the result of stress, such as when a bird is grabbed by a predator (tail feathers may be shed if grabbed by the tail), handled by humans, or exposed to violent natural events such as earthquakes and tornados. Also called **fright molt**. [3·38]

Shoebill: A large, stork-like water bird endemic to Africa; it has a huge, shoe-shaped, hooked bill for seizing large fish. It is the only member of its family, Balaenicipitidae. [1·84]

shorebirds: Oystercatchers, plovers, snipes, sandpipers, curlews, phalaropes, and sheathbills. Ornithologists in Britain and the British Commonwealth, except Canada, speak of shorebirds as "waders." [1·65]

shore feeders: Birds (or other animals) that hunt on shore in the intertidal zone (between the low and high tide marks), relatively close to their nests; shore feeders include marine waders, such as the American Oystercatcher, and many other shorebirds. [6·68]

siblicide: The killing of one's sibling; in birds whose offspring practice siblicide, usually the stronger, older nestlings kill the younger, weaker nestlings, either by direct attack or by pushing them out of the nest. [6·87]

sifting: A foraging technique in which a bird sweeps its partly opened bill from side to side in water or mud, straining out small animals and plant material with the specialized edges of the bill as the water (or mud) drains out. Sifting is used by

Roseate Spoonbills, Northern Shovelers, and others. [6·46]

single-population models (in PVA): See **population viability analysis**. [10·77]

site fidelity: Loyalty shown by birds or other organisms to places they previously occupied; the places may be breeding locations, nonbreeding locations, or stopover points between the two. Also called **site tenacity**. [5·75, 9·13]

site tenacity: See **site fidelity**. [5·75]

skeletal muscles: Muscles that move the bones; they constitute the "meat" of an animal and are under conscious control. Also called **voluntary muscles**. [4·26]

skulling: A technique by which field workers use the degree of skull ossification and pneumatization to determine the age of live passerines captured in the fall. [4·10]

slope soaring: A type of static soaring in which a bird derives lift from air deflected upward when wind strikes a hill, ridge, or cliff; or from rising eddies created when wind spills over a cliff. Slope soaring is particularly common along seacoasts among gulls, terns, fulmars, and gannets; it also is common along mountain ridges, where ravens, crows, and migrating hawks may soar. [5·42]

slots: Gaps between the feathers of the wing tips, created when a bird having narrow-tipped primaries spreads them during flight. Slots are most common in large, soaring birds, such as vultures, condors, eagles, and certain hawks, because they allow these birds, which have broad, rounded wing tips, to increase their lift-to-drag ratio. Slots reduce tip vortex by turning each primary feather into an individual, narrow, pointed "wing tip"; they also increase lift because each separated primary feather acts as an individual airfoil. [5·35]

slotted high-lift wings: Broad wings with a deep camber and very prominent slotting; the broadness reduces wing loading and allows low flight speeds, and the slotting increases lift, allowing the birds to carry heavy prey. Birds such as vultures, eagles, condors, storks, and some owls and hawks have this type of wing; they use it for static soaring. [5·39]

small intestine: The longest part of the digestive tract, running from the stomach to the large intestine. The final processes of digestion take place inside the small intestine, as proteins are broken down into amino acids, carbohydrates are converted to simple sugars, and fats are reduced to glycerol and fatty acids. The small intestine is longer in birds that feed on foliage or grain than in those that feed on fruit or meat, reflecting the difficulty of digesting the cellulose in plant material. [4·120]

smooth muscle: Type of muscle found in the walls of the hollow organs (such as the stomach or intestines) and in the walls of blood vessels larger than capillaries, especially arteries and arterioles. The contraction of smooth muscle is not controlled consciously. [4·28]

snood: The limp, red, fingerlike, fleshy structure that projects over the bill from the forehead of a Wild Turkey. [3·48]

soaring: In a bird or other animal, flying without flapping the wings or limbs, while gaining altitude or remaining horizontal (as compared to gliding, in which the animal loses altitude). [5·36]

social foraging: See **cooperative foraging**. [6·64]

Solnhofen limestone: Fine-grained limestone deposits near the village of Solnhofen in Bavaria that, due to meticulous mining, have produced an array of amazingly well-preserved fossils of plants, invertebrates, fishes, and reptiles, as well as several specimens of *Archaeopteryx*. [E·2]

soma: See **cell body**. [4·32]

somites: In avian biology, paired segments that appear along the middle region of the cranial-caudal axis of the very early embryo; eventually they will become vertebrae and muscle. [8·65]

sonagram: A visual representation of sounds, such as bird song, plotted as a graph of frequency or pitch (vertical axis) versus time (horizontal axis). The darkness of the images gives a rough indication of the relative loudness of the different sounds. Also spelled **sonogram**. [7·7]

songbirds: Members of Suborder Passeri, which is one of the two large suborders of Order Passeriformes (perching birds); songbirds also are known as **oscines** or **true songbirds**. Songbirds have particularly complex voice boxes, which allow them to sing more complex songs than other birds. [7·25]

song control centers: Groups of nerve cells within the brains of songbirds that are involved with both the hearing and production of songs; they are interconnected with one another in a series of complex pathways. Individuals that sing frequently or that have large song repertoires have large song control centers. The centers shrink in the nonbreeding season—a time when most birds sing less often. [7·39]

song repertoire: All the songs (different song types) sung by an individual bird; for comparison, see **vocal repertoire**. [7·14, 7·19, 7·85]

songs: Loud vocalizations, often delivered from an exposed perch, that are presumed to attract mates or repel territorial

intruders. Strictly speaking, songs are given only by songbirds—passerines in the suborder Passeri (oscines). In practice, however, similar vocalizations made by nonsongbirds (and even other animals) often are called songs. [7·14]

Southern Marine Region: The major faunal region of the seas that includes the frigid waters around Antarctica north to about 35 degrees south latitude. [1·99, 1·100]

southwestern oak woodland: North American ecosystem that is scattered on hills and mountain slopes and that contains open woodlands consisting primarily of oaks and occasional large ponderosa pines; southwestern oak woodland is found in Utah, Nevada, California, New Mexico, Arizona, and parts of Colorado. Rainfall is low to moderate. [9·120]

spatial learning: Learning the location of objects. [6·12]

speaker replacement experiments: Experiments in which researchers remove animals (usually male birds) from their territories, and then replace them with tape recorders and loudspeakers broadcasting various sounds (the male's song, the song of a different species, background noise with no song, and so on), depending on the specific experiment. Then the researchers watch the reaction of other birds in the area. [7·69]

speciation: The formation (evolution) of new species through natural selection. [1·57]

species: 1. Level of classification below "genus"; similar species are placed within the same genus. A species usually is defined as a group of potentially interbreeding individuals that share distinctive characteristics and are unlikely to breed with other such groups of individuals. Distinguishing which groups of organisms constitute species is not always easy, however, and slightly different criteria may be used by different scientists. [1·53] 2. The two-word designation constituting the Latin or scientific name of a "species" (see definition 1); the first word, capitalized, is the "genus" and the second word, lowercase, is the "species" name (see definition 3). Both words are underlined or printed in italics. [1·47] 3. The second word in the two-word scientific name of a species (see definition 2). It is always lowercase, and is underlined or printed in italics. [1·47]

species account: A set of written field notes on a particular species. [2·52]

species-area relationship: Typical relationship between the size of an area and the number of species it supports: as area size increases, so does the number of species. This relationship holds (1) for patches of the same type of habitat either in the same or different areas (the larger the habitat patch, the greater the number of species it supports), and (2) for areas of increasing size centered on the same point (as the boundaries of the area being considered are expanded outward, the area contains more species—both because the size of a patch of one type of habitat may increase, and because a larger area may contain more different types of habitats). The species-area relationship also holds for islands surrounded by water. [9·95, 10·72]

species population: A population of all the living individuals of a particular species; also called simply a **species**. [9·49]

species richness: The number of species living in a community (or in a particular area). The more species, the greater the species richness. [9·82]

sperm: See **spermatozoa**. [4·127]

spermatozoa (singular, **spermatozoan**): Sperm cells or sperm; each spermatozoan is a single cell composed of a DNA-containing head and a propulsive tail. Sperm are short and simple in nonpasserines, and longer and spiral-shaped in passerines. [4·127]

spinal cord: A cable of neurons conducting nerve impulses to and from the brain; it runs inside the vertebral canal of the vertebral column, extending from the medulla oblongata of the brain to the tail. [4·38]

spinal nerves: Nerves that arise from the spinal cord in pairs—one on each side of the cord—all along the spinal cord; each pair carries sensory input and motor output, and serves a very specific region of the skin, or very specific muscles or other organs. [4·42]

spinous process: A ridge of bone projecting from the dorsal surface of each vertebra; it is particularly well developed in the thoracic vertebrae. [4·16]

spotting scope: A medium-range telescope commonly used for bird watching, usually with magnification power between 15x and 60x. [2·42]

spread-wing posture: A stance in which a bird stands motionless with the wings extended to the side, either to dry the wings or to absorb sunlight. Adopted by many large birds, such as cormorants, anhingas, pelicans, storks, and New World vultures. [3·22]

spurs: Bony outgrowths from any part of the skeleton, such as the wing (see **wing spurs**). [4·22]

square tail: Tail shape in which the rectrices are all about the same length. [1·19]

stabilizing selection: Natural selection that acts against any change in a current characteristic. Stabilizing selection occurs when an organism is already very well adapted to current conditions, and it tends to keep the species at an optimal middle point. [1·37]

stable age distribution: Property of some populations such that the relative proportions of individuals of different ages are the same each year, when the same time of year (such as the start of the breeding season) is compared. Note, however, that in a stable age distribution these proportions still change predictably throughout each year: in most birds, for example, the proportion of very young individuals increases in late spring as eggs hatch, and then gradually decreases as many juveniles die before reaching one year of age. [9·80]

star compass: The mechanism by which nocturnally migrating birds are able to use the star patterns surrounding the North Star (in the Northern Hemisphere) to determine which way is north (and thus the other compass directions as well). A young bird initially learns how to use the star patterns by observing their rotation and using the point at the center of their rotation (the North Star, Polaris) as the direction to North; but once a bird learns the star patterns, it can find north even in a stationary planetarium. [5·86]

statant cup nest: Cup nest constructed on top of a hard physical support or supports, including the ground, building ledges, or tree branches; built by American Robins, Horned Larks, and many other birds. [8·30]

static pressure: The force, produced by random motion of molecules, that air exerts uniformly in all directions. When you squeeze a balloon, you feel the static pressure from the air inside. [5·11]

static soaring: A type of soaring (flying without flapping or losing altitude) in which birds take advantage of the energy in rising air masses (either in thermals, called **thermal soaring**, or along hills or ridges, called **slope soaring**) to obtain lift with little or no energy expenditure on their part. Performed by birds with slotted high-lift wings. [5·39]

statistical testing: Mathematical procedures used by scientists to determine whether a set of observations or the results of an experiment either (1) support their hypothesis, or (2) could result from chance alone. Most statistical tests generate a number that indicates the probability that the observations or results could have occurred from chance alone. The lower the probability, the greater the statistical significance of the data. Many scientists choose a 1 in 20 probability (.05) as the cut-off for significance: if the probability is greater than 1 in 20—for example, 2 in 20 (.10)—then the hypothesis is rejected, but if the probability is equal to or smaller than
1 in 20—for example, 1 in 100 (.01)—the hypothesis is accepted. In the biomedical sciences, where the risk of making a mistake may carry serious consequences, acceptable levels of significance often are set much lower. [6·97]

statoacoustic nerve: See **vestibulocochlear nerve**. [4·42]

statoconia: Dense crystals of calcium carbonate embedded, along with sensory hair cells, in a gelatinous material inside the utriculus and sacculus of the inner ear. As the head's orientation changes with respect to gravity, the crystals settle against different hair cells, stimulating them; the brain determines the head's position by detecting which hair cells are stimulated. [4·61]

stepping stones: In conservation biology, small pieces of habitat, suitable for a particular species, that are located between larger habitat preserves; they allow individuals to move between the larger preserves. [10·80]

sternal rib: The ventral portion of each thoracic rib; it articulates with the sternum ventrally, and with the vertebral rib dorsally. [4·18]

sternum: The breastbone; it provides an attachment site for the ribs and pectoral muscles. In all flying birds except tinamous, the sternum has a keel (carina). [4·23]

stochastic: In biology, describes events whose occurrence and/or outcomes are the result of random (or at least unpredictable) factors. For comparison, see **deterministic**. [9·77, 10·67]

stomach: Saclike, expandable digestive organ between the esophagus and the small intestine; in birds, the stomach consists of two parts: the proventriculus (upper) and gizzard (lower). [4·118]

stooping: A foraging technique used by falcons in which a bird drops through the air at great speed in pursuit of a flying bird or insect. [6·45]

structural colors: Colors not produced by pigments, but instead by the reflection of certain wavelengths of light off the actual physical structure of the object (such as a feather of a bird or a scale of a butterfly wing). [3·50]

studbook: A collection of information on the individuals of a particular species. Key to a studbook is data on the relatedness among individuals—for example, which individuals are the parents, siblings, and offspring of a given individual. But studbooks may contain a variety of other information as well, such as data on each individual's disease history, birth location and date, and location in the wild or in captivity. Studbooks are used primarily by researchers who are breeding endangered or declining species; the studbook information helps the researchers to pair individuals with the partners to whom they are least related, thus promoting and preserving as much genetic diversity as possible. [10·76]

subadult: A bird that has not yet reached maturity; immature. [3·30]

subadult plumage: Any of the plumages worn by young birds before they reach their definitive plumages (those of a mature bird). [3·30]

subclavian arteries: Two arteries (left and right) that carry blood to the front limbs (wings). [4·82, 4·84]

subclavian veins: Paired veins (left and right) that carry blood from the front limbs (wings); on each side, they merge with the jugular and pectoral veins to form the cranial vena cava, which returns blood to the heart. [4·83]

subelliptical: Describes an egg that is longer than an elliptical egg, with a slight bulge to the sides. [8·73]

suboscines: Members of Suborder Tyranni, one of the two large suborders of Order Passeriformes (perching birds). The suboscines have less complex voice boxes than members of the other passerine order, Passeri, and thus must sing simpler songs. Includes birds such as tyrant flycatchers, antbirds, manakins, and cotingas. [7·25]

subsong: Practice singing (somewhat like human infant babbling) that begins shortly after songbirds leave the nest. Subsong usually is quiet, garbled, and rambling compared to adult song; it contains some of the same elements, but they are strung together in odd ways. It gradually comes to resemble adult song. [7·27]

subspecies: A subset of a species, usually in a particular geographic area, that contains individuals that are morphologically distinct from other individuals of the same species, but are still capable of interbreeding with those other individuals. Also called a **race**. [1·55]

sugarbirds: A family (Promeropidae, 2 species) of long-billed, long-tailed, nectar feeders that specialize on *Protea* plants on the mountainous slopes of South Africa. [1·85]

sunbirds: A family (Nectariniidae, 124 species) of tiny birds with long, slender, down-curved bills; they feed on nectar and insects and serve as pollinators of flowers. Sunbirds are similar to hummingbirds, but do not hover while feeding. Many males are brilliantly colored. Most species inhabit Africa, but some are found in tropical parts of the other Old World regions. [1·86]

sun compass: A mechanism by which birds and some other animals use the position of the sun in the sky to indicate compass direction. To do this, they must be able to compensate for the changing position of the sun during the day. [5·84]

sunning: Exposing the plumage to the sun, usually by spreading the wings and/or tail (often against the ground), fluffing the feathers, and remaining motionless; sunning may aid Vitamin D production and remove ectoparasites, as well as warm the bird. [3·22]

superciliary line: See **eyebrow stripe**. [1·7, 2·10]

supplemental plumage: In the Humphrey-Parkes system of nomenclature, a plumage that may occur in addition to the basic and alternate plumages; it is found in just a few species, such as ptarmigans and certain buntings. [3·33]

supracoracoideus: The powerful flight muscle (also called the **supracoracoid muscle**) that raises a bird's wing; it also slows down the wing at the end of the downstroke and accelerates it at the beginning of the upstroke. [5·6]

supracoracoid foramen: See **foramen triosseum**. [4·20, 5·6]

surface diving: Diving under water from a swimming position on the water's surface; performed by birds such as diving ducks, grebes, cormorants, and loons. [6·45]

surface-to-volume ratio: An individual's surface area (through which heat is lost) divided by its volume (which generates heat); the larger the ratio, the faster an individual will lose heat. Smaller birds have larger ratios than larger birds, and thus to compensate they must have higher metabolic rates and must consume relatively more food. Also called **surface area-to-volume ratio**. [3·11, 4·146]

surrogate: In conservation biology, a stand-in for another individual or species. For example, researchers practiced releasing techniques for the highly endangered California Condor by first using the biologically similar but less endangered Andean Condor as a surrogate. [10·102]

survival rate: 1. The proportion of individuals in a population that survive for a particular interval of time—usually a year. [9·64] 2. The chance that a particular individual will survive a given period of time, usually one year. For example, an adult Royal Albatross has an **annual survival rate** of 95 percent. [8·3]

survivorship: The proportion of individuals from a cohort (a group containing individuals born during the same period of time) that survive to a given age. [8·5]

survivorship curve: A graph of the proportions of individuals in a cohort (a group containing individuals born during the same period of time) that remain alive at different ages. The survivorship curve shows the chance of mortality at different stages in an organism's lifetime; the basic shape of the curve for a particular species or group of species illustrates the general pattern of mortality. [8·5]

sutures: Boundary lines along the junction between two bones; sutures are visible in the skull of a young bird. [4·10]

sweeping: A foraging technique in which a bird pursues insects in flight, capturing them in midair in its mouth; used by swifts, nighthawks, and many swallows, among others. [6·44]

syllable: In a bird song, a note or cluster of notes that forms a unit that is repeated. [7·9]

symbiosis: An ecological relationship between organisms of two different species that live in close association with one another; types of symbioses include **mutualism**, **commensalism**, and **parasitism** (see separate entries for each). [9·132]

sympathetic ganglion: A clump of cell bodies of sympathetic neurons that arise in the spinal cord. A chain of sympathetic ganglia runs the length of the spinal cord near its ventral surface. [4·44]

sympathetic system: The part of the autonomic nervous system that functions under conditions of stress; the sympathetic system prepares the body for "fight or flight" by increasing the heart rate, breathing rate, and blood pressure. It consists of nerves that leave the spinal cord from the thoracic and lumbar regions. [4·44]

synapse: A small gap between nerve cells, across which nerve impulses travel. [4·32]

synchronous hatching: Pattern of hatching in which all the eggs of a single clutch hatch at about the same time (on the same day), resulting in a brood of young all the same size and age. This hatching pattern occurs when incubation is delayed until the last egg is laid. Because development of laid eggs does not begin until they are warmed, all the embryos begin to develop at the same time and are ready to hatch at the same time. [6·88]

syndactyl feet: Foot arrangement in which the hallux points backward and toes two, three, and four point forward, with toes two and three (the inner and middle toes) fused for much of their length. Found in many kingfishers and hornbills. [1·21]

synsacrum: The segment of the vertebral column of birds that is formed by the fusion of some thoracic vertebrae with all of the lumbar, all of the sacral, and the first few caudal vertebrae; the synsacrum is in turn fused on either side with the ilium bones of the pelvis. [4·18, 4·24]

syringeal muscles: The muscles of the syrinx; they allow the syrinx to change shape and thus to produce different types of sounds. [4·93]

syrinx: The sound-producing organ of birds; located at the point where the trachea divides to form the two bronchi. [4·93, 7·38]

systemic circulation: Portion of the circulatory system that carries blood from the heart to the body tissues (excluding the lungs) and back to the heart. [4·77]

T

taiga: See **boreal forest**. [9·116]

tail bone: See **pygostyle**. [4·18]

tapaculos: A Neotropical suboscine family (Rhino-cryptidae, 52 species) related to antbirds; tapaculos have cocked tails and frequent more southerly, open, and dry areas than do antbirds. [1·79]

tarsals: Ankle bones. In humans, they remain distinct; in birds, the proximal tarsals fuse with the tibia to form the tibiotarsus, and the distal tarsals fuse with the metatarsals (instep bones) to form the tarsometatarsus. [1·14, 4·25]

tarsometatarsus: The long bone supporting the upper section of the bird foot; the tarsometatarsus is formed by a fusion of the distal tarsals (ankle bones) and the metatarsals (instep bones). Also called simply the **metatarsus.** [1·14]

tarsus: The upper section of the avian foot, between the heel and the toes. [1·15]

taste buds: Simple structures, usually embedded in the epithelium of the oral cavity and the tongue, that are the receptors for taste sensations in all vertebrates. Humans have numerous taste buds located on the tongue, but birds have few taste buds, which are located primarily on the roof of the mouth or deep in the oral cavity, with none on the tongue. [4·65]

taxonomy: The classification of organisms—assigning names and relationships. [1·32]

tectorial membrane: The upper membrane of the cochlear duct in the inner ear. Sound waves set the lower membrane—called the basilar papilla in birds—into motion, moving its hair cells against the tectorial membrane and triggering nerve impulses in the hair cells, which are sent to the brain in the process of sound perception. [4·59]

telencephalon: The anterior portion of the forebrain; it contains the olfactory lobes. [4·36]

temperate: Moderate; describes a climate free from extreme heat and cold, but experiencing some of both; generally found in the middle latitudes. [1·69]

temporal fovea: Area—in the posterior quadrant of the retina of hawks and other fast-flying diurnal avian predators—where the cones are most concentrated and the neural layer (the nerves from the rods and cones, which overlie the rods and cones and block some light) is the thinnest, and thus vision is the sharpest. The temporal foveae provide sharp binocular views of the area in front of the bird. [4·49, 4·53]

tendon: A band of fibrous connective tissue that attaches muscles to bones, and/or binds many muscle fibers together to form a skeletal muscle. [4·6]

teratorns: Giant, soaring, scavenging birds of the New World (subfamily Teratornithinae) whose fossils date from the late Tertiary period; teratorns became extinct around 10,000 years ago when many large mammals were wiped out. The largest species had a wingspan approaching 19.4 feet (5.9 m). [10·8]

termitary: An active termite nest. [8·19]

territory: A defended area. [6·22]

tertiary consumers: Level in a food chain or web that consists of organisms—usually animals—that directly eat secondary consumers. [9·123]

testes (singular, **testis**): The male gonads; one testis lies at the cranial end of each kidney. The testes produce sperm as well as the male sex hormone, testosterone, and in some species, a female sex hormone, estrogen. [4·127]

thecodonts: A diverse group of reptiles from the early Mesozoic, all of which have a diapsid skull, teeth set in sockets, and an antorbital fenestra (an opening on each side of the skull in front of the eye socket). Also known as **basal archosaurs** or **pseudosuchian thecodonts**. [E·8, E·33]

thermal: A rising column or large bubble of warm air that results from differential heating of land surfaces by the sun; thermals develop over areas that are darker than surrounding areas, and over south-facing slopes. [5·39]

thermal soaring: A type of **static soaring** (see separate entry) in which birds use the rising air in thermals (rising columns of warm air) to propel themselves upward, circling higher and higher with little energy expenditure of their own. Once high in the air, they can glide out of the thermal and across the countryside in whatever direction they wish to travel. [5·39]

thigh: The upper leg, supported by the femur. [1·14]

thoracic air sacs: Two pairs of air sacs in the chest region of birds, one pair (a cranial and caudal sac) located on each side of the body. [4·101]

thoracic cage: The rib cage; it consists of the ribs connected to the thoracic vertebrae above and to the sternum (breastbone) below. It forms a flexible but strong enclosure for the heart, liver, and lungs, as well as for the thoracic air sacs of birds. [4·19]

thoracic ducts: Paired lymph ducts that collect the products of fat digestion from the **intestinal lymph trunk** (coming from the small intestine; see separate entry); the thoracic ducts run along the surface of the aorta and eventually deliver their contents to the venous system at the cranial vena cavae just before these large veins enter the heart. [4·89]

thoracic vertebrae: The vertebrae of the thorax (chest) region; they articulate with the ribs and thus form part of the rib cage. In birds, some thoracic vertebrae are fused with lumbar, sacral, and some caudal vertebrae to form the synsacrum. [4·15, 4·18]

thrombocytes: Nucleated blood cells that resemble red blood cells, but are more dense and complex and are highly specialized to carry out blood clotting. They are not present in mammals, which instead have platelets (not found in birds) to carry out blood clotting. [4·88]

thrust: In a bird, the portion of the force generated by the flapping wings that propels the bird forward (in the direction the bird is moving, which is always the direction opposite to drag). [5·18]

thyroid glands: Paired endocrine glands at the base of the neck. Under the control of the anterior pituitary, the thyroid glands secrete the hormones thyroxin and triiodothyronine, which regulate the annual increase in gonad size, sperm and egg production, the growth and pigmentation of feathers, and molting. [4·74]

tibia: In humans, the larger of the two lower leg bones; in birds, the tibia fuses with several tarsal (ankle) bones to form the tibiotarsus, which supports the crus (lower leg). [1·14]

tibiotarsus: Bone supporting the lower leg (crus) of birds; the tibiotarsus is formed by the fusion of the tibia with the proximal tarsal (ankle) bones. [1·14, 4·25]

tickbirds: See **oxpeckers**. [1·87]

time compensation: In the course of orientation or navigation, making allowances for the changes in the position of the sun (or other celestial bodies) in the sky throughout the day. [5·84]

tinamous: Primitive, grouse-like birds (family Tinamidae, 46 species) whose eerie calls haunt both forests and pampas of South and Central America; tinamous are the only ratites capable of flight. [7·24]

tip vortex: A type of turbulence created at the tip of a wing; it consists of air currents spiraling off the wing tip behind a bird or other flying object. These eddies sometimes may be seen as trails of white behind the wing tips of large airplanes as they take off or land. [5·34]

tissue: An aggregation of cells with related and often very similar characteristics; examples include muscle tissue and bone. [4·2]

tissue fluid: Blood plasma that has moved out of blood cells at the capillary beds, and is in the minute spaces between the body cells. Oxygen and nutrients diffuse out of it to the body cells, and carbon dioxide and other wastes diffuse in. Tissue fluid eventually returns to the blood system, either by diffusing directly back into a blood capillary near the venous side, or by diffusing into a lymphatic capillary (in which it is called lymph), which carries it to larger lymph vessels and eventually into the venous system. [4·87]

tongue bone: See **entoglossal**. [4·14]

torpor: A profound state of sleep in which the body temperature drops and consequently all metabolic processes and stimulus-reaction processes slow down. Used by swifts, hummingbirds, nighthawks, and some other birds to conserve energy when food resources are unavailable. Birds may enter torpor on a nightly basis, or for extended periods up to an entire season of cold weather. Also called **hibernation**. [4·154]

trachea: The windpipe; the tube conducting air from the larynx to the lungs. [4·92]

tracheal bulla: An expanded sac on one side of the lower end of the trachea in males of many duck species; it is thought to modify the sounds produced by the syrinx. [4·92]

track: A flying bird's actual direction of movement with respect to the ground; due to crosswinds, a bird's track may not be the same as its heading. Also applies to aircraft and other flying organisms. [5·68]

tracts: Bundles of axons and their myelin sheaths within the central nervous system; collectively, tracts form the white matter, found in the outer portions of the spinal cord and the inner areas of the brain. [4·36]

transequatorial migrants: Birds that migrate back and forth across the equator to take advantage of spring and summer in both hemispheres. [1·103]

transitive inference: The ability to understand and infer linear relationships among individuals. For example, the ability to understand that "if A is dominant to B, and B is dominant to me, then A is dominant to me." [6·18]

translocation: Establishing individuals of a species in an area in which that species formerly lived by importing individuals from a different area. [10·89]

transverse plane: A vertical plane through an organism, dividing the body into cranial and caudal portions. [1·4]

trees-down theory (of the origin of avian flight): See **arboreal theory**. [E·18]

trial-and-error learning: Learning to associate one's own behaviors with either a reward or punishment; behaviors that result in reward are repeated, and those that result in punishment are abandoned. [6·10]

trigeminal nerve: The fifth cranial nerve; it divides into the **ophthalmic**, **maxillary**, and **mandibular nerves** (see separate entries) after exiting the brain. [4·41]

trill: A phrase or song that consists of a single note or cluster of notes (a syllable) that is repeated many times in rapid succession. [2·16]

triosseal canal: See **foramen triosseum**. [4·20, 5·6]

trochlear nerve: The fourth cranial nerve; it carries motor output from the brain to a single eye muscle. [4·41]

trophic: Pertaining to feeding. [9·123]

trophic levels: The different levels of food production or consumption within a food chain or web; for example, producers, primary consumers, secondary consumers, and decomposers. [9·123]

trophic structure: Organizational scheme of a community that is based on the feeding relationships between organisms: species are assigned to different trophic levels according to what they eat and what eats them (see **trophic levels**). [9·123]

Tropical Marine region: The major faunal region of the seas that includes the warm equatorial waters between the Subtropical Convergences of both hemispheres. [1·99, 1·101]

true navigation: The ability to take the proper course toward a specific goal, even from completely unfamiliar sites at vast distances from known areas. [5·83]

trumpeters: A family (Psophiidae, 3 species) of large, hump-backed, chicken-like birds that inhabit the rain forests and floodplains of Amazonia. Trumpeters forage on the ground in flocks, but roost and nest colonially in trees. They eat both plants and animals (including reptiles and amphibians, and even carrion), and sometimes follow army ants for the arthropod prey they disturb. [1·74]

tubenoses: Birds in the order Procellariiformes, including albatrosses, shearwaters, fulmars, petrels, and storm-petrels, all of which have **tubular nares** (see separate entry). [3·40]

tubular nares: Nares (nostrils) that open into a horny, tubular structure that sits on top of the bill; in some species the tube for

each nostril is separate, but in others the two adjacent tubes fuse to form a double tube—within which the tubes coming from each nostril may remain separate or may be partially fused. Tubular nares are found in the Procellariiformes, an order of pelagic seabirds that excrete large amounts of salty fluid through their nostrils. The tube is thought to reduce heat and airflow at the nostril (where the nasal cavity opens into the upper beak), allowing the salty fluid to flow down the beak and away from the nostril before evaporating. Thus the salts remaining after evaporation are less likely to clog the nostril and salt gland than if evaporation occurred right at the nostril. [3·39, 3·40]

tufa: A porous, white limestone formed under water when springs or rainwater carry dissolved calcium into lakes that have a high concentration of bicarbonate ions (found in baking soda). When the two solutions mix they form calcium carbonate, which precipitates (comes out of solution), creating tufa—which often accumulates into underwater towers or crusts covering the lake bottom. Mono Lake, in California, is well known for its tufa formations, which became exposed when the lake's water level dropped dramatically after water was diverted for use in the Los Angeles area. [8·60]

tundra: See **arctic tundra** or **alpine tundra**. [9·114]

turacos: African relatives of cuckoos, turacos are noisy and arboreal, and tend to run squirrel-like along branches. The plumage of turacos is soft green (one of the few green pigments known from birds), blue, or gray, often with bright red (or in a few species, purple) on the wings. The 18 species of turacos, together with plantain-eaters and go-away-birds, make up the family Musophagidae. [1·86]

turbulence: A disorderly flow of air; turbulence may interfere with smooth airflow over a bird's wings and thus may disrupt flight. [5·13]

tutor song: A song played to a bird in the laboratory during experiments on song learning. [7·26]

tympanic canal: The lower perilymph-filled canal inside the bony cochlea of the inner ear; the tympanic canal is connected to the (upper) vestibular canal at one end. Pressure waves in the perilymph of the two canals are transmitted to the endolymph of the cochlear duct between them, setting the (in birds) basilar papilla in motion and stimulating the sensory hair cells. [4·57]

tympanic membrane: A membrane that stretches tightly across the external ear canal; the tympanic membrane vibrates when struck by the pressure waves of a sound, and transmits those vibrations to the (in birds) columella. Also called the **eardrum**. [4·56]

tympaniform membranes: Flexible membranes that stretch between successive cartilaginous rings of the syrinx, allowing the syrinx to change shape (in response to contractions of the syringeal muscles) and thus to produce different types of sounds. [4·93]

Tyranni: One of the two large suborders of Order Passeriformes (perching birds). The Tyranni (also called **suboscines**) have less complex voice boxes and thus sing simpler songs than members of the other passerine order, Passeri (also called **oscines**). Suborder Tyranni includes birds such as tyrant flycatchers, antbirds, woodcreepers, ovenbirds, tapaculos, manakins, and cotingas. [7·25]

Tyrannidae: Extremely diverse New World suboscine family containing nearly 400 species; members of the family Tyrannidae are commonly known as **tyrant flycatchers**. No other bird family has more species. [1·80]

tyrant flycatchers: The nearly 400 species of the diverse New World suboscine family Tyrannidae. [1·80]

U

ulna: The larger and thicker of the two bones that make up the antebrachium. [1·9]

ulnare: The larger of the two wrist bones of birds; the ulnare consists of several fused carpals. [1·9]

ultimate: In biology, pertaining to long-term causes; evolutionary. **Ultimate questions** about a behavior, for example, look into the factors that might have caused the particular behavior to evolve—*why* it evolved. For comparison, see **proximate**. [6·3]

ultimate factors: In biology, the variables affecting the survival and reproduction of a population and its ancestors, and therefore influencing the evolution of certain traits, behaviors, or physiological processes, or the timing of such things. For example, the food supply at different times of the year is a major ultimate factor that affects the timing of breeding in birds. For comparison, see **proximate factors**. [8·11]

ultimobranchial glands: Two small, light-colored glands located near the parathyroid glands, usually on an artery; the ultimobranchial glands secrete the protein calcitonin, which lowers the blood calcium concentration. [4·74]

uncinate process: A flattened, hook-shaped extension of bone that projects caudally from the vertebral segment of

each rib. Because each uncinate process overlaps with the rib behind it, the processes help to strengthen the rib cage. Found only in birds. [4·18]

understory vegetation: The shrubs and shorter trees in a forest; the understory vegetation grows below the level of the main canopy and **mid-story vegetation** (see separate entry), and above the ground vegetation. [9·93]

undertail coverts: The short contour feathers that cover the bases of the rectrices on the ventral side of the tail. [1·6, 1·12]

underwing coverts: The feathers lining the underside of the wing, from the leading edge of the wing to the flight feathers. [1·12, 1·13]

upper beak: The upper half of the beak; also called the **upper jaw**. [1·6]

upper jaw: See **upper beak**. [1·6]

uppertail coverts: The short contour feathers that cover the bases of the rectrices on the dorsal side of the tail. [1·12, 1·13]

upwelling: A process by which cold, nutrient-rich water from the lower ocean depths rises to the surface. Upwelling may occur where certain ocean currents meet, or where winds blowing away from the coast move the surface water away from shore, causing water from below to rise to replace it. [1·103, 1·105]

ureters: Two tubes, one leading from each kidney, that carry urine from the kidneys to the (in birds) cloaca. [4·125]

uric acid: A complex but less toxic molecule to which the liver of birds converts the highly toxic nitrogenous waste products of protein metabolism. Compared to urea, the nitrogenous waste product of mammals, uric acid requires more energy to produce, but it is less soluble in water and more highly concentrated, allowing birds to excrete more toxic waste per molecule excreted, while minimizing water loss. [4·125]

urinary bladder: Sac that stores urine before it is excreted; among birds, the urinary bladder is pre-sent only in the South American rheas. [4·125]

urinary system: See **excretory system**. [4·125]

urogenital system: The combination of two organ systems: the urinary (excretory) system, which removes toxic nitrogenous wastes from the blood, and the genital (reproductive) system. The two systems often are considered together because they are located close together in the body, and because they may share some structures (such as the cloaca, in birds). [4·124]

uropygial gland: See **oil gland**. [3·20]

urvogel: General term (used primarily by the Germans) that refers to early, primi-

tive birds such as *Archaeopteryx*. [E·4]

utriculus: One of the two membranous chambers inside the bony vestibule of the inner ear (the other is the sacculus); these chambers contain hair cells and dense crystals called statoconia, both of which are embedded in a gelatinous material that is surrounded by endolymph. The statoconia and hair cells perceive the position of the head with respect to gravity (see entry for **statoconia**). [4·58]

V

vagina: In birds, the short, lower section of the oviduct, after the shell gland; the vagina opens into the left side of the cloaca. [4·130]

vagus nerve: The tenth cranial nerve; it carries sensory and motor information between the brain and the pharynx, larynx, heart, lungs, gizzard, liver, and intestines. [4·42]

vane: The broad, flat surface on each side of the shaft of a contour feather; the vanes are formed by the interlocking of adjacent barbs. [3·3]

vangas: A diverse songbird family (Vangidae, 14 species) of shrike-like birds. Most vanga species are gregarious and noisy, gleaning insects and other small animals as they move through the trees in groups. [1·88]

vasa deferentia: See **deferent ducts**. [4·128]

vector: 1. A quantity with both magnitude (size) and direction; for example, a bird's flight at 25 miles per hour to the south is considered a vector. A vector is often shown as an arrow whose length represents magnitude and whose orientation indicates compass direction. [5·68] 2. An organism that acts as a carrier, transmitting pathogens (viruses, bacteria, and so on) from one individual to another. For example, certain mosquitoes are vectors because they carry malaria from one person to another. [9·70]

vector navigation: Finding one's way by adding vectors composed of compass direction and distance in order to determine the straight-line path to a desired destination. [5·79]

vein: A vessel that carries blood toward the heart. All veins except the pulmonary vein and its tributaries carry oxygen-poor blood. [4·81, 4·84]

vena cava: See **caval veins** or **cranial vena cavae** or **caudal vena cava**. [4·84]

venous ring: A ring of veins formed by the left and right renal portal veins of birds; the venous ring connects the various lobes of the two kidneys, and is part of the avian renal portal system. [4·83]

vent: The opening of the cloaca to the exterior of the body. In birds, the vent is the only posterior opening, and thus it releases feces from the digestive system, uric acid from the excretory system, and sperm or eggs from the reproductive system. [4·123]

ventral: Toward the belly of an organism. [1·4]

ventricles: The two thick-walled, muscular, posterior chambers of the heart; the ventricles receive blood from the atria and then pump it to other parts of the body: the **right ventricle** pumps blood to the lungs and the **left ventricle** pumps blood to the body arterial system. [4·77]

venules: Small veins that carry blood from the capillaries to larger veins. [4·81, 4·84]

vertebrae (singular, **vertebra**): The individual bones that form the vertebral column (backbone). [4·14]

vertebral canal: A long tube that runs the length of the back, inside the backbone; the vertebral canal is formed by the vertebral arches of successive vertebrae. The vertebral canal protects the spinal cord, which runs inside it from the brain to the tail. [4·38]

vertebral column: Commonly called the backbone or spine, the vertebral column is a series of complex, uniquely articulating or rigidly fused vertebrae that provide support for the head, neck, ribs, back, and tail. [4·14]

vertebral rib: The dorsal portion of each thoracic rib; the vertebral rib articulates with the thoracic vertebra dorsally and with the sternal rib ventrally. [4·18]

vertebrates: Animals that have a backbone: birds, fish, amphibians, reptiles, and mammals. [1·1]

vertical migration: Type of migration in which animals move seasonally up and down mountainsides. [5·53]

vestibular canal: The upper perilymph-filled canal inside the bony cochlea of the inner ear; the vestibular canal is connected to the lower perilymph-filled canal, the tympanic canal, at one end. Pressure waves in the perilymph of the two canals are transmitted to the endolymph of the cochlear duct between them, setting the (in birds) basilar papilla in motion and stimulating the sensory hair cells. [4·57]

vestibular window: A soft, oval region on the bony cochlea, to which (in birds) the columella of the middle ear is attached. Movement of the columella moves the vestibular window, sending pressure waves through the fluid (perilymph) inside the cochlea. Formerly called the **oval window**. [4·57]

vestibule: Bony structure of the inner ear

containing two membranous chambers, the utriculus and sacculus, which contain hair cells that perceive the position of the head with respect to gravity. The vestibule is part of the bony labyrinth of the inner ear and thus contains perilymph. [4·48]

vestibulocochlear nerve: The eighth cranial nerve; the vestibulocochlear nerve carries sensory input for balance and hearing. Its vestibular (balance) portion carries sensory input from the semicircular canals and vestibule of the inner ear, and its cochlear (hearing) portion carries sensations from the hair cells of the inner ear. Formerly called the **auditory**, **acoustic**, or **statoacoustic nerve**. [4·42]

vestigial structure: A structure that has little or no apparent function and is thought to be present in an organism only because it has been inherited from an ancestor. Many vestigial structures are reduced in size from the ancestral (and functional) condition, and presumably are in the process of being eliminated through natural selection. A vestigial structure in humans is the remains of a **nictitating membrane** in the inner corner of each eye—in reptiles, birds, and some other mammals the nictitating membrane is functional. [3·46]

viduines: Refers to members of the African genus *Vidua*, which consists of whydahs and indigobirds. Viduines are colorful seed-eaters that are brood parasites on other members of their family, Estrildidae; they are known for the intricate patterns inside the mouths of their nestlings, which strikingly resemble those of their host's nestlings. [8·142]

villi (singular, **villus**): Minute fingerlike projections. Villi often are produced by folds in a tissue lining an organ—as in the inner surface of the small intestine of vertebrates, where the presence of villi greatly increases the surface area available for absorbing nutrients. [4·121]

visual field: See **field of view**. [4·51]

vitelline diverticulum: Small pouch of tissue at the junction of the jejunum and ileum of the small intestine in many birds; the vitelline diverticulum is a remnant of the yolk sac of the embryo. [4·122]

vitelline membrane: In avian biology, the transparent membrane surrounding and holding together the yolk of an egg. [8·62]

vitreous body: Clear, gelatinous material filling the **vitreous chamber** (see separate entry) inside the eyeball. The vitreous body gives rigidity to the eyeball and helps to maintain its shape. [4·48]

vitreous chamber: The largest chamber inside the eyeball; the vitreous chamber is located on the posterior side of the eyeball and is filled with a clear, gelatinous material called the **vitreous body** (see

separate entry), which gives rigidity to the eyeball. [4·48]

vocal repertoire: All the different types of vocalizations that are produced by an individual bird, including both songs and calls; for comparison, see **song repertoire**. [7·10]

vocal signals: Vocalizations that have evolved for a specific function. [7·11]

voluntary muscles: See **skeletal muscles**. [4·26]

W

warm-blooded: See **endothermic**. [1·1]

warm front: The interface between a warm air mass and the cold air mass it is overtaking; the warm air tends to push up over the denser cold air and become cooler, which results in the formation of clouds and eventually precipitation. [5·69]

water birds: See **aquatic birds**. [1·65]

waterfowl: Ducks, geese, and swans; family Anatidae. [1·65]

wattlebirds: Common name for members of the family Callaeidae, which consists of two living species (the Saddleback and Kokako) and one extinct species (the Huia). Wattlebirds are forest-dwelling songbirds endemic to New Zealand; they are named for the pair of colorful, fleshy wattles at the corners of their mouths. [1·97]

wattles: Ornamental flaps or folds of skin that dangle from the head or neck of some birds. Wattles are found in turkeys, pheasants, cotingas, wattlebirds, and many others. [3·48]

waxbills: Together with whydahs, indigobirds, and numerous related species, these colorful seed-eaters of the Old World form the large family Estrildidae (159 species). The Zebra Finch, a well-known cage bird, is a waxbill. [1·83]

weavers: A diverse Old World family (Ploceidae, 117 species) of colorful seed-eaters; weavers are found primarily in open tropical areas and are named for the ability of many species to weave large, complex nests. [1·83, 8·53]

well: See **sap well**. [9·128]

western montane forest: North American coniferous forest ecosystem that is dominated by spruce and fir trees and is found along the North Pacific slope and in the Cascades, Sierra Nevada, and Rocky Mountains; western montane forest hosts few breeding birds. [9·116]

wetland mitigation: Compensation that the recipient of a permit to fill or otherwise destroy wetland areas is legally required to provide, to make up for the wetland area lost. Compensation may consist of restoring or re-creating other wetland areas of equal or greater size compared to the wetland area destroyed, or may be a cash payment into a **mitigation bank**—a fund managed by a public agency or private conservation group that carries out large-scale wetland restoration projects using money accumulated from a large number of small-impact projects. [10·93]

whisker stripe: A distinctively colored stripe in the malar region of birds; also called a **mustache stripe** or **malar stripe**. [1·8, 2·10]

whistle: A song or phrase in which clear tones are produced one pitch at a time. [2·16]

whistlers: Large, round-headed songbirds with slightly hooked bills; whistlers are known for their explosive, often-beautiful, songs. The nearly 40 species of whistlers, along with shrike-thrushes, pitohuis, and a few related species, form the family Pachycephalidae, which is found primarily in the Australasian region. [1·95]

white blood cells: Unpigmented blood cells of several different types, all of which contain a nucleus and can leave the capillaries and move about in the spaces between the body cells. Many types of white blood cells are important in fighting infections. Also called **leukocytes**. [4·88]

white fibers: Muscle fibers that appear lighter in color than red fibers because they have fewer capillaries (and thus fewer red blood cells) and less of the reddish oxygen-transport molecule myoglobin; white fibers also are larger in diameter than red fibers. Muscles with many white fibers are used for quick bursts of action, but they cannot carry out sustained activity because they produce energy anaerobically (without oxygen)—so lactic acid builds up quickly and causes fatigue. Muscles with many white fibers often are called "**white meat**" (as in the breast of a turkey). [5·7]

white matter: Lighter-colored tissue (compared to gray matter) that makes up much of the brain and spinal cord. White matter consists of numerous tracts, which are bundles of nerve cell axons and their myelin sheaths, and is found in the outer portions of the spinal cord and the inner areas of the brain. [4·39]

white meat: See **white fibers**. [5·7]

whydahs: Members, along with indigobirds, of the African genus *Vidua* (members of this genus are called **vidu-**

ines). Viduines are colorful seed-eaters that are brood parasites on other members of their family, Estrildidae; they are known for the intricate patterns inside the mouths of their nestlings, which strikingly resemble those of their host's nestlings. [8·142]

wick: In avian biology, a beakful of vegetation that is dipped in mud and then used to transport the mud to a nest that is under construction; usually the entire wick is added to the nest. [8·54]

wing-flashing: A foraging technique used by some herons, egrets, and storks in which a bird wading through the water quickly raises or brings forward one or both wings, apparently to frighten prey out of hiding or to provide a shady place where unsuspecting prey may try to hide, thus bringing the prey within the bird's reach. [6·46]

wing loading: The ratio of body weight to wing area; wing loading is a measure of how much "load" each unit area of wing must carry. [5·31]

wing pouch: A pocket under each wing of male Sungrebes (Neotropical inhabitants of wooded streams); the wing pouch is formed from a pleat of skin and is used to carry the young while the male flies or swims. [8·137]

wing spurs: Bony outgrowths of the carpometacarpus (the main bone of the avian manus); wing spurs are found in various birds such as cassowaries, plovers, sheathbills, screamers, and jacanas. [4·22]

winnowing: Part of a twilight territorial and courtship display given by both male and female Common Snipe. The birds circle high in the air and produce a series of rapid, pulsating, whistle-hums (called **winnows**) by spreading the tail and diving at high speeds—causing the stiff outer tail feathers to vibrate. European Snipe also winnow. [7·17]

winter plumage: See **basic plumage**. [3·33]

wishbone: See **furcula**. [4·19]

woodcreepers: A Neotropical family (Dendrocolaptidae, 51 species) of similar-looking, rust-colored, suboscines with a wide array of beak shapes; woodcreepers typically forage in tree bark much like the unrelated Brown Creeper of North America. [1·78]

woodhoopoes: A family (Phoeniculidae, 8 species) of sociable African birds with glossy, dark plumage and long tails; woodhoopoes nest in tree cavities and breed cooperatively—with additional

adult birds helping the parents to tend the nest. [1·85]

woodswallows: A family (Artamidae, 14 species) of small, chunky songbirds with a graceful, swallow-like flight; woodswallows inhabit the Australasian and Oriental regions and are known for huddling together on branches or in tree cavities in groups of up to 50 or more individuals. [1·96]

wood-warblers: A family (Parulidae, 115 species) of small, insectivorous birds, many of which are colorful, found in the New World. Many wood-warblers are Neotropical migrants. Also called **New World warblers**. [1·71]

Y

yolk: In avian biology, the familiar, yellow portion of an egg; the yolk contains nearly all of the lipids (fat) and most of the protein needed by the developing embryo. The yolk is surrounded and held together by the transparent vitelline membrane. [8·62]

yolking up: The deposition of yolk within the vitelline membrane in alternating bands of darker and lighter yolk; yolking up is part of the egg-formation process in birds and occurs before egg laying. [8·63]

yolk sac: In avian biology, the extra-embryonic membrane inside the egg that surrounds the yolk of the developing embryo. [8·65]

Z

zooplankton: Microscopic animals and protozoa that float freely in aquatic environments; zoo-plankton make up the "animal" portion of plankton. [1·103]

Zugunruhe: See **migratory restlessness**. [5·60]

zygodactyl feet: Foot arrangement in which toes two and three point forward, and the hallux and toe four point backward. Found in woodpeckers, cuckoos, toucans, owls, Osprey, turacos, most parrots, and some other birds. [1·21]

zygote: The fertilized egg; the single-celled product resulting from the union of (1) the nucleus of the sperm cell from the male and (2) the nucleus of the ovum (egg cell) from the female. In birds, fertilization occurs in the infundibulum of the oviduct. [4·133]

About the Authors

To ensure scientific accuracy and a high level of expertise, one or more professional ornithologists with specific knowledge and research experience have authored each chapter of this book. To achieve maximum clarity and readability, the twelve chapters were carefully edited by Lab of Ornithology Education staff, resulting in better consistency and flow among the primary subjects. However, because each author has a unique writing style, students may notice some differences in tone and presentation among the chapters.

Birds and Humans: A Historical Perspective

Sandra G. Podulka, Marie Z. Eckhardt, and Daniel R. Otis

Sandra G. Podulka received a B.S. in Wildlife Biology from Cornell University and an M.S. in Zoology (Animal Behavior) from the University of Maryland, where she studied the function of song repertoires in Song Sparrows. After graduate school, Sandy was a Research Technician at Cornell for Dr. Stephen T. Emlen, analyzing the social behavior of White-fronted Bee-eaters, but discovered she wanted to spend more time sharing her love of nature with others, so turned toward environmental education. She worked at the Cayuga Nature Center for several years, and as an Adjunct Professor of Biology at Tompkins Cortland Community College from 1988 to 1996, teaching courses in biology and conservation. Since 1986, she has worked at the Cornell Lab of Ornithology, in research, writing, public education, and editing—most recently as one of the editors of the Handbook of Bird Biology and the Home Study Course in Bird Biology.

Sandy spent her childhood knee-deep in muddy ponds trying to catch tadpoles and frogs, and roaming fields collecting butterflies—her first real love. She has enjoyed birds as long as she can remember, but they did not take center stage until she took a summer ethology course at Cornell University from Dr. Bill Dilger. He hauled his students out before dawn every morning and taught them to recognize birds by their songs, and Sandy has been listening to them ever since. She has participated in Christmas Bird Counts and Breeding Bird Atlases, and has traveled to Costa Rica, Belize, Trinidad, and Peru to watch birds, but her favorite bird-watching site is her yard—which overlooks a beaver pond with an ever-changing cast of avian actors.

Marie Z. Eckhardt was a biologist in the Education Program at the Cornell Lab of Ornithology. Her academic background in vertebrate zoology and experience as a museum consultant and a collections manager at the New York State Museum provided a strong foundation for her varied contributions at the Lab. Marie has been interested in animals and the natural world for as long as she can remember—an interest she credits to early and regular museum visits.

Daniel R. Otis grew up on a farm in Upstate New York. Since 1988 he has worked as a freelance editor, proofreader, and writer for Living Bird and other Lab of Ornithology projects. Currently a Ph.D. student in Horticulture at Cornell, his research focuses on conservation of the world's maple species and on assessing the extent and effects of Norway maple invasiveness on northeastern forests.

Although plants rather than birds became his main interest, his appreciation of nature derives partly from memorable experiences with birds: "Once when I was a teenager, on a stormy November day, as I walked along a hedgerow, I was astonished to come across an isolated little tree crowded with silent birds I'd never seen before—Cedar Waxwings. Too exhausted to fly, they merely shuffled down their perches a bit when I came close. And in March of some years, the air was alive day and night with the sound of immense flocks of Canada Geese flying between the lake and the muddy cornfields around our house." Experiences such as these, he notes, quicken one's appreciation of the vivacity and fascination of the natural world, and help us to understand what conservationists are fighting for.

Introduction: The World of Birds

Dr. Kevin J. McGowan

Kevin J. McGowan is a research associate at the Cornell Lab of Ornithology. He received his B.S. and M.S. degrees in Zoology from Ohio State University, and a Ph.D. in Biology from the University of South Florida, where he studied the social development of Florida Scrub-Jays. In 1988, after working as a non-game biologist for the Florida Game and Freshwater Fish Commission, he came to Cornell University, taking a position as Curator of the Ornithology and Mammalogy Collections in the Department of Ecology & Evolutionary Biology. In addition to caring for the collections, he conducted research on crows and taught classes in specimen preparation, field collecting, the relationships among birds, and Neotropical canopy biology. He moved to his present position at the Lab in 2001.

Kevin's primary research focuses on the behavioral ecology of birds. Currently he is studying the reproductive and social behavior of two crow species in central New York State and investigating the impact of West Nile virus on crow populations. He is an Elected Member of the American Ornithologists' Union, Webmaster and a Director of the Federation of New York State Bird Clubs, and a member of the New York State Avian Records Committee. Formerly, he was Secretary of the Ornithological Societies of North America (OSNA), and editor of the Ornithological Newsletter, a bi-monthly OSNA publication.

An avocational birder since childhood, Kevin has traveled throughout North America and to Europe, Africa, and Central and South America to watch and study birds.

A Guide to Bird Watching

Dr. Stephen W. Kress

Stephen W. Kress is vice-president for bird conservation for the National Audubon Society and Manager of the Society's Maine Coast Seabird Sanctuaries. He also is ornithology Program Director for the Audubon Camp in Maine, an adjunct faculty member in the Wildlife Department at the University of Maine, Orono, and a Visiting Fellow at the Cornell Lab of Ornithology, where each year he teaches the popular course Spring Field Ornithology.

As Director of Audubon's Seabird Restoration Program, Steve advises and manages the development of techniques for re-establishing in Maine colonies of various seabirds, such as Atlantic Puffins; Leach's Storm-Petrels; and Arctic, Common, and Roseate terns. In the Pacific region, he has studied the role of vocalizations in attracting endangered Dark-rumped Petrels to artificial burrows in the Galápagos Islands, and Short-tailed Albatross to decoys on Midway Island. He is author of many books, including The National Audubon Society's Birder's Handbook, The Bird Garden, and Project Puffin, as well as the Golden Guide to Bird Life. He also has written numerous scientific papers on seabird biology and conservation.

During most of the year Steve lives on 33 acres of woods and meadows near Ithaca, New York, where he manages his land for songbirds and works on methods for restoring populations of Northern Bobwhite. He spends summers on the Maine coast, continuing his lifelong interest in restoring nesting seabird colonies.

Form and Function: The External Bird

Dr. George A. Clark, Jr.

George A. Clark, Jr. is Professor Emeritus of Ecology and Evolutionary Biology at the University of Connecticut in Storrs. He received his Ph.D. in Biology from Yale University, where he specialized in Ornithology. After spending two years at the University of Washington in Seattle, he moved to the University of Connecticut, where he spent 32 years as a faculty member.

His 200 publications reflect his research interests in the structure, behavior, distribution, and evolution of birds. He is past president of the Association of Field Ornithologists, served as co-editor of the book Perspectives in Ornithology/ Essays Presented for the Centennial of the American Ornithologists' Union, and has led educational field trips for groups to observe birds in North and South America, Europe, and Africa.

A birder since his high school days in Pennsylvania, George now resides in Vermont and enjoys seeking birds by walking or snowshoeing in the northern New England hills.

What's Inside: Anatomy and Physiology

Dr. Howard E. Evans and Dr. J. B. Heiser

Howard E. Evans received both his B.S. and Ph.D. in Comparative Anatomy from Cornell University, where he became a faculty member in the Veterinary College in 1950. There, he taught gross anatomy of the horse and cow for seven years, and anatomy of the dog, bird, and fish for 36 years. He was Secretary of the college for 12 years and served as Chairman of Anatomy from 1976 to 1986, when he retired. He continues to teach a course on the literature and materials of natural history, and to lecture in several other courses.

Howie's research concerned the anatomy of reptiles and birds, the replacement of teeth in fishes, the plant-induced cyclopis in sheep, fetal development of the dog, and anatomy of tropical fishes. His most recent works include the third edition of Miller's Anatomy of the Dog (1993), the fifth edition of Guide to the Dissection of the Dog (2000, with Dr. Alexander deLahunta), and the third edition of Anatomy of the Budgerigar and Other Birds (1996). He is co-editor of the Handbook of Avian Anatomy, published by the Nuttall Ornithological Club of Harvard University. He has served as President of both the American and the World Association of Veterinary Anatomists, and has been an associate editor of the American Journal of Anatomy and the Journal of Morphology.

A native of New York City, Howie and his wife Erica have led natural history trips for Cornell Adult University to the Virgin Islands; Hawaii; Sapelo Island, Georgia; East and South Africa; Papua, New Guinea; and Antarctica. He has lectured in China, Russia, Taiwan, Hungary, Switzerland, Germany, England, Thailand, Brazil, Mexico, and Japan.

J. B. Heiser is a Senior Lecturer in the Department of Ecology and Evolutionary Biology at Cornell University. A vertebrate evolutionary ecologist, J. B. received his B.S. from Purdue University and his Ph. D. from Cornell University. For the past 35 years, he has taught a variety of courses in vertebrate comparative anatomy and ecology.

J. B. began teaching on the Ithaca campus, but soon was teaching field courses at the Shoals Marine Laboratory, an isolated island facility in the Gulf of Maine that is cooperatively run by Cornell University and the University of New Hampshire. Eventually he became Director of the Shoals program, a position that he held for 15 years. He won the Clark Award for distinguished teaching and has traveled to every continent and ocean to teach natural history "on location."

Although J. B. was trained as a marine biologist and his research focuses on the evolution and interrelationships of coral reef fish, he has considerable natural history experience in tropical forests worldwide. In his travels he has watched birds (and fish) in every major biome and biogeographic region that the planet has to offer, but he still gets a thrill out of backyard birding in Upstate New York.

Birds on the Move: Flight and Migration

Dr. Kenneth P. Able

Kenneth P. Able is Professor Emeritus of Biology in the Department of Biological Sciences at the State University of New York in Albany, where he had been a member of the faculty since 1971. He received his B.S. and M.S. degrees from the University of Louisville, and his Ph.D. from the University of Georgia. Ken's research focuses on bird migration, particularly the mechanisms of orientation and navigation.

Ken has been passionately interested in birds since childhood. He has birded extensively across North America and in Mexico, Costa Rica, Peru, Spain, France, Germany, Italy, Switzerland, the United Kingdom, Australia, New Zealand, and South Africa. His favorite destination is Australia, because the avifauna is unique and the species fantastic.

Evolution of Birds and Avian Flight

Dr. Alan Feduccia

Alan Feduccia is S. K. Heninger Professor of Biology at the University of North Carolina at Chapel Hill, where he has been for 30 years. He received his Ph.D. at the University of Michigan, where his thesis focused on the evolution of woodhewers and ovenbirds.

Alan's career has focused on vertebrate evolution, the evolution of birds, and the tempo and mode of the evolution of modern groups of birds. His interest in the origin of birds blossomed in the late 1970s when he wrote The Age of Birds for Harvard University Press. In 1973 he wrote a rebuttal to the theory of hot-blooded dinosaurs in Evolution, and since that time has been involved in the debate on bird origins. His latest book, The Origin and Evolution of Birds (Yale University Press, 1999), addresses many of the main issues in the bird evolution controversy, and takes what he calls the "ornithological" position—that birds were already arboreal when they evolved to fly, and that they evolved from a common ancestor with dinosaurs, but not directly from them.

Alan's interest in birds began as a teenager. Later, as an undergraduate student at Louisiana State University, he had the good fortune to participate in "bird" expeditions to Honduras, El Salvador, and Peru. He has maintained an interest in Neotropical birds ever since.

Understanding Bird Behavior

Dr. John Alcock

John Alcock is Regents' Professor of Biology at Arizona State University, Tempe. John is currently researching the evolution of insect mating systems, with a special emphasis on the diversity of male mating tactics among bee species native to the Sonoran Desert. Earlier research has focused on birds, however, and his textbook, Animal Behavior, An Evolutionary Approach, uses many bird examples to illustrate all aspects of the modern study of behavior.

John began bird watching at age five (bird number one was the Mallard), and seeing a good bird still boosts his heart rate. In attempts to keep his heart rate up, he has visited Costa Rica, Ecuador, Argentina, Australia, and several countries in Europe.

Vocal Behavior

Dr. Donald E. Kroodsma

Birds enter our lives in different ways. Some of us have been bird-crazy as long as we can remember, but others discovered birds later. I was a late-comer, as birds grabbed me from the chemistry lab during my last year of college. How grateful I am that they have never let go. Immediately after college I took two summer bird courses from the famed Olin Sewall Pettingill, who was then director of the Cornell Lab of Ornithology but taught at the University of Michigan field station in Pellston, Michigan. He put a tape recorder, headphones, and parabolic microphone in my hands and told me to "go out and tape record some birds." In doing so, he changed my life. I went to graduate school at Oregon State to study how young wrens learn their songs; and then for eight years I had the good fortune to work with Peter Marler at Rockefeller University in New York, on all aspects of bird song. I am currently Professor of Biology at the University of Massachusetts, Amherst. My time since college has been spent reveling in the who, what, when, where, how, and why of bird song.

Nests, Eggs, and Young: Breeding Biology of Birds

Dr. David W. Winkler

David W. Winkler was born and raised in Sacramento, California. Unlike the rest of his family, he was a naturalist from about age four. After a progression of enthusiastic interest in butterflies, wildflowers, and herps, he finally settled on birds in his early teens. He learned and studied local birds alone for two years, when, at his first Christmas Bird Count, he ran into Rich Stallcup. David then spent the last two years of high school learning a great deal from Rich about the birds of California. While in high school, an American Birds article introduced David to The Herring Gull's World by Niko Tinbergen, and he carried out a senior English project on gull taxonomy within a couple of years. By his freshman year at U. C. Davis, David was dreaming about all the species he might someday study, as opposed to just see, for his list.

While attending U. C. Davis, David and friends secured National Science Foundation funding for an ecological study of Mono Lake. This cemented his attachment to the Mono Basin and its birdlife, and his friendship with David Gaines, with whom he co-founded the Mono Lake Committee in 1978. His dissertation on the clutch sizes of California Gulls (with Frank Pitelka at Berkeley) at Mono and Great Salt Lakes was followed by post-docs at the University of Gothenburg, Sweden (with

Malte Andersson); Oxford University (with John Krebs); and Cornell University (with Paul Sherman). David joined the Cornell faculty in 1988 upon the retirement of Tom Cade, and his research on swallows worldwide, and Tree Swallows in Ithaca in particular, has thrived ever since.

Individuals, Populations, and Communities: The Ecology of Birds

Dr. Stanley A. Temple

Stanley A. Temple is the Beers-Bascom Professor of Conservation in the Department of Wildlife Ecology at the University of Wisconsin in Madison. He is also the Chair of the graduate program in Conservation Biology and Sustainable Development in the Institute for Environmental Studies at Madison. A quintessential Cornellian, Stan earned his B.S., M.S., and Ph.D. all at Cornell, and for years has served as a member of the Lab of Ornithology's Administrative Board.

Stan's professional activities focus on avian ecology and bird conservation, with a special emphasis on endangered species. He and his students have worked with some of the world's most endangered birds, including Peregrine Falcons, California Condors, Whooping Cranes, and dozens of endangered species endemic to islands around the world. To date, none of those species has become extinct, and most are doing significantly better as a result of his work.

Stan has been interested in birds as long as he can remember. Birds of prey have always been among his favorites, and he has been a falconer for 45 years. He feels fortunate to have incorporated all of his ornithological pleasures into his professional life.

Bird Conservation

Dr. John W. Fitzpatrick

John W. Fitzpatrick (Ph.D., Princeton University, 1978) is Director of the Cornell Lab of Ornithology, where he arrived in September 1995. He was Executive Director of Archbold Biological Station, a private ecological research foundation in central Florida, from 1988 through August 1995, and was Curator of Birds at the Field Museum of Natural History (Chicago) from 1978 to 1989. He is a Fellow of the American Ornithologists' Union (AOU). In 1985 the AOU awarded him its highest research honor, the Brewster Award, for his book and numerous research articles (co-authored with Glen Woolfenden) on demography, social behavior, and conservation of the endangered Florida Scrub-Jay. Fitz has led many expeditions to remote areas of South America, especially the western Amazonian basin and the Andean foothills. He has published numerous papers on Neotropical birds, including descriptions of seven bird species new to science. Co-author of the book Neotropical Birds: Ecology and Conservation (University of Chicago Press, 1996), he has been engaged in applying science to real-world conservation issues throughout his career. Most recently, he helped design and implement a major network of ecological preserves in central Florida by convening panels of scientific experts and by engaging county, state, and federal agencies; nongovernmental organizations; and private industry in the process. He serves on the national governing boards of The Nature Conservancy, the National Audubon Society, and the Center for Biodiversity and Conservation at the American Museum of Natural History. He is on two Endangered Species Recovery Teams, including that of the world's rarest bird, the Hawaiian Crow. He enjoys watercolor painting, and has been a bird watcher since kindergarten.

Illustrations

N. John Schmitt

N. John Schmitt is a wildlife illustrator with a lifelong interest in birds. Since leaving the United States Army in 1973, John has devoted his life to a variety of ornithology-related endeavors that have taken him to many countries in Latin America, Asia, and Europe. He has worked as a field biologist for both the Peregrine Falcon and California Condor recovery programs.

John's work as a field biologist led to more serious devotion to illustrating birds. He has illustrated several books, including the National Geographic Society's Third Edition of Birds of North America; Clark's A Field Guide to the Raptors of Europe, the Middle East, and North Africa; and Skutch's Birds Asleep. Currently, he is working on field guides to the birds of Peru and India. John is a self-taught taxidermist/museum preparator whose work is on display in several California museums. He also co-leads ecotours in the United States and abroad.

A native of California, John is an avid bird watcher. His time in the field provides valuable inspiration and is integral to maintaining his enthusiasm. He considers his notebooks, which he fills with written and sketch notes, as the source of his most valuable references and inspirations. John encourages everyone interested in natural history to keep a notebook.

References

Able, K.P. 1989. Skylight polarization patterns and the orientation of migratory birds. *Journal of Experimental Biology* 141:241–256.

Able, K.P. 1994. Magnetic orientation and magnetoreception in birds. *Progress in Neurobiology* 42:449–473.

Able, K.P. 1995. Orientation and navigation: a perspective on fifty years of research. *Condor* 97:592–604.

Able, K.P. 1996. The debate over olfactory navigation by homing pigeons. *Journal of Experimental Biology* 199:121–124.

Able, K.P., ed. 1999. *Gatherings of Angels: Migrating Birds and Their Ecology.* Comstock Books, Ithaca, NY. 193 pp.

Alexander, R.D. 1974. The evolution of social behaviour. *Annual Review of Ecology and Systematics* 5:325–383.

Allen, A.A. 1914. The Red-winged Blackbird: a study in the ecology of the cattail marsh. *Proceedings of the Linnaean Society of New York,* No. 24–25, pp. 43–128.

Allen, A.A. 1961. *The Book of Bird Life, 2nd Edition.* D. Van Nostrand Company, New York.

Allen, R.P. 1947. *The Flame Birds.* Dodd, Mead, and Co., New York.

Alvarez del Toro, M. 1971. On the breeding biology of the American Finfoot in southern Mexico. *Living Bird* 10:79–88.

American Ornithologists' Union. 1998. *Check-list of North American Birds, 7th Edition.* American Ornithologists' Union, Washington, D.C. 829 pp.

American Ornithologists' Union. 2000. 42nd Supplement to the AOU Check-list of North American Birds. *Auk* 117:847–858.

Ames, P.L. 1971. *The Morphology of the Syrinx in Passerine Birds.* Peabody Museum of Natural History, Bulletin 37. Yale University, New Haven.

Ammann, G.A. 1937. Number of contour feathers in *Cygnus* and *Xanthocephalus.* *Auk* 54:201–202.

Anderson, M.G. 1985. Variations on monogamy in Canvasbacks (*Aythya valisineria*). In *Avian Monogamy,* pp. 57–67. (Edited by P.A. Gowaty and D.W. Mock). Ornithological Monographs, No. 37. American Ornithologists' Union, Washington, D.C.

Andersson, S. 1996. Bright ultraviolet colouration in the Asian whistling-thrushes (*Myiophonus spp.*). *Proceedings of the Royal Society of London B* 263:843–848.

Andersson, S. and T. Amundsen. 1997. Ultraviolet colour vision and ornamentation in Bluethroats. *Proceedings of the Royal Society of London B* 264(1388):1587–1591.

Andersson, S., J. Ornborg, and M. Andersson. 1998. Ultraviolet sexual dimorphism and assortative mating in Blue Tits. *Proceedings of the Royal Society of London B* 265(1395):445–450.

Ankney, C.D. and C.D. MacInnes. 1978. Nutrient reserves and reproductive performance of female Lesser Snow Geese. *Auk* 95(3):459–471.

Arad, Z., U. Midtgard, and M.H. Bernstein. 1989. Thermoregulation in Turkey Vultures: vascular anatomy, arteriovenous heat exchange, and behavior. *Condor* 91(3):505–514.

Arcese, P.A. 1989. Territory acquisition and loss in male Song Sparrows. *Animal Behaviour* 37:45–55.

Arias-de-Reyna, L. 1998. Coevolution of the Great Spotted Cuckoo and its hosts. In *Parasitic Birds and their Hosts,* pp. 129–142. (Edited by S.I. Rothstein, and S.K. Robinson). Oxford University Press, Oxford.

Armstrong, E.A. 1975. *Life and Lore of the Bird in Nature, Art, Myth, and Literature.* Crown Publishers, New York. 250 pp.

Armstrong, T.A. 1994. *Female Red-winged Blackbird* (Agelaius phoeniceus) *Vocalizations: Variations and Ontogeny.* Ph.D. Dissertation, University of Massachusetts, Amherst, MA.

Ashmole, N.P. 1963. The regulation of numbers of tropical oceanic birds. *Ibis* 103:458–473.

Askins, R.A., J.F. Lynch, and R. Greenberg. 1990. Population declines in migratory birds in eastern North America. In *Current Ornithology,* Vol. 7, pp. 1–57. (Edited by D.M. Power). Plenum Press, New York and London.

Atwood, J.L., C.C. Rimmer, K.P. McFarland, S.H. Tsai, and L.H. Nagy. 1996. Distribution of Bicknell's Thrush in New England and New York. *Wilson Bulletin* 108(4):650–661.

Audubon, J.J. 1827–1838. *The Birds of America.* 4 Volumes. Published by the author, London.

Azanza, M. and A. Del Moral. 1994. Cell membrane biochemistry and neurobiological approach to biomagnetism. *Progress in Neurobiology* 44:517–601.

Badyaev, A.V. 1997. Avian life history variation along altitudinal gradients: an example with carduelline finches. *Oecologia* 111(3):365–374.

Bailey, A.M. and R.J. Niedrach. 1936. Community nesting of western robins and House Finches. *Condor* 38:214.

Baird, S.F. 1858. *Reports of Explorations and Surveys to Ascertain the Most Practical and Economic Route for a Railroad from the Mississippi River to the Pacific Ocean.* Vol. 9, Part 2. A.O.P. Nicholson, Printer, Washington, D.C.

Baird, S.F. 1860. *The Birds of North America.* J.B. Lippincott & Co., Philadelphia, PA.

Baker, J.R. 1938. The Evolution of Breeding Seasons. In *Evolution: Essays on Aspects of Evolutionary Biology,* pp. 161–177. (Edited by G.R. deBeer). The Clarendon Press, Oxford.

Baker, M.C. and M.A. Cunningham. 1985. The biology of bird-song dialects. *The Behavioral and Brain Sciences* 8:85–133.

Baker, R. 1984. *Bird Navigation: The Solution of a Mystery?* Holmes and Meier, New York. 256 pp.

Balda, R.P. and J.H. Balda. 1978. The care of young Piñon Jays (*Gymnorhinus cyanocephalus*) and their integration into the flock. *Journal für Ornithologie* 119(2):146–171.

Balda, R.P. and A.C. Kamil. 1992. Long-term spatial memory in Clark's Nutcracker *Nucifraga columbiana.* *Animal Behaviour* 44:761–769.

Ball, G.F. 1983. *Evolutionary and Ecological Aspects of the Sexual Division of Parental Care in Barn Swallows.* Ph.D. Dissertation, Rutgers University, Newark, NJ.

Ball, S.C. 1952. *Fall bird migration of the Gaspé Peninsula.* Peabody Museum of Natural History, Bulletin 7. Yale University, New Haven. 211 pp.

Baptista, L.F. 1996. Nature and its nurture in avian vocal development. In *Ecology and Evolution of Acoustic Communication in Birds,* pp. 39–60. (Edited by D.E. Kroodsma and E.H. Miller). Cornell University Press, Ithaca, NY.

Baptista, L.F. and K.L. Schuchmann. 1990. Song learning in the Anna Hummingbird (*Calypte anna*). *Ethology* 84:15–26.

Barrow, M.V., Jr. 1998. *A Passion for Birds: American Ornithology after Audubon.* Princeton University Press, Princeton, NJ. 326 pp.

Bartram, W. 1791. *Travels Through North and South Carolina.* James & Johnson, Philadelphia, PA.

Baumel, J.J., S.S. King, J.E. Breazile, H.E. Evans, and J.C. VandenBerge. 1993. *Handbook of Avian Anatomy: Nomina Anatomica Avium, 2nd Edition.* Nuttall Ornithological Club, Harvard University, Cambridge, MA.

Baylis, J.R. 1982. Avian vocal mimicry: its function and evolution. In *Acoustic Communication in Birds,* Vol. 2, pp. 51–83. (Edited by D.E. Kroodsma and E.H. Miller). Academic Press, New York.

Beebe, W. 1924. The rarest of nests on the tallest of grass stems. *Bulletin of the New York Zoological Society* 27(5):114–118.

Beecher, M.D. 1996. Birdsong learning in the laboratory and field. In *Ecology and Evolution of Acoustic Communication in Birds,* pp. 61–78. (Edited by D.E. Kroodsma and E.H. Miller). Cornell University Press, Ithaca, NY.

Beecher, M.D. and P.K. Stoddard. 1990. The role of bird song and calls in individual recognition: contrasting field and laboratory perspectives. In *Comparative Perception,* Vol. 2, pp. 375–408. (Edited by W.C. Stebbins and M.A. Berkeley). Wiley, New York.

Beer, C.G. 1969. Laughing Gull chicks: recognition of their parents' voices. *Science* 166:1030–1032.

Beissinger, S.R., S. Tygielski, and B. Elderd. 1998. Social constraints on the onset of incubation in a Neotropical parrot: a nest-box addition experiment. *Animal Behaviour* 55(1):21–32.

Beletsky, L.D. 1983. Aggressive and pair-bond maintenance songs of female Red-winged Blackbirds. *Zeitschrift für Tierpsychologie* 62:47–54.

Beletsky, L.D., B.J. Higgins, and G. Orians. 1986. Communication by changing signals: call switching in Red-winged Blackbirds. *Behavioral Ecology and Sociobiology* 18:221–229.

Beletsky, L.D. and G.H. Orians. 1985. Nest-associated vocalizations of female Red-winged Blackbirds. *Zeitschrift für Tierpsychologie* 69:329–339.

Bellrose, F.C. 1976. The comeback of the Wood Duck. *Wildlife Society Bulletin* 4:107–110.

Bennett, A.T.D. and I. Cuthill. 1994. Ultraviolet vision in birds: what is its function? *Vision Research* 14(11):1471–1478.

Bennett, A.T.D., I.C. Cuthill, and K.J. Norris. 1994. Sexual selection and the mismeasurement of color. *American Naturalist* 144:848–860.

Bennett, A.T.D., I. Cuthill, J. Partridge, and K. Lunau. 1997. Ultraviolet plumage colors predict mate preferences in starlings. *Proceedings of the National Academy of Sciences of the United States of America* 94(16):8618–8621.

Bennett, A.T.D., I. Cuthill, J. Partridge, and E. Maier. 1996. Ultraviolet vision and mate choice in Zebra Finches. *Nature* 380:433–435.

Bent, A.C. 1937. *Life Histories of North American Birds of Prey, Part 1.* U.S. National Museum Bulletin, No. 167. United States Government Printing Office, Washington, D.C.

Bent, A.C. 1938. *Life Histories of North American Birds of Prey, Part 2.* U.S. National Museum Bulletin, No. 170. United States Government Printing Office, Washington, D.C.

Bent, A.C. 1940. *Life Histories of North American Cuckoos, Goatsuckers, Hummingbirds, and their Allies.* United States National Museum Bulletin, No. 176. United States Government Printing Office, Washington, D.C.

Bent, A.C. 1953. *Life Histories of North American Wood Warblers.* U.S. National Museum Bulletin, No. 203. United States Government Printing Office, Washington, D.C.

Berthold, P. 1993. *Bird Migration: A General Survey.* Oxford University Press, Oxford. 239 pp.

Berthold, P. 1996. *Control of Bird Migration.* Chapman and Hall, London. 355 pp.

Berthold, P. and U. Querner. 1981. Genetic Basis of Migratory Behavior in European Warblers. *Science* 212:77–79.

Bertram, B.C.R. 1980. Vigilance and group size in Ostriches. *Animal Behaviour* 28:278–286.

Best, L.B. 1978. Field Sparrow reproductive success and nesting ecology. *Auk* 95:9–22.

Bewick, T. 1809. *A History of British Birds.* Printed by Edward Walker, Newcastle.

Bildstein, K. 1992. Causes and consequences of reversed sexual size dimorphism in raptors: the head start hypothesis. *Journal of Raptor Research* 26:115–123.

Birdlife International. 2000. *Threatened Birds of the World.* Lynx Edicions, Barcelona and BirdLife International, Cambridge, UK. 852 pp.

Birkhead, T.R. and A.P. Møller. 1992. *Sperm Competition in Birds: Evolutionary Causes and Consequences.* Academic Press, London.

Blackburn, D.G. and H.E. Evans. 1986. Why are there no viviparous birds? *American Naturalist* 128(2):165–190.

Blake, J.G. and J.R. Karr. 1984. Species composition of bird communities and the conservation benefit of large versus small forests. *Biological Conservation* 30:173–187.

Blanchan, N. 1902. *How to Attract the Birds: And other Talks about Bird Neighbors.* Doubleday Page, New York. 224 pp.

Blondel, J. 1985. Breeding strategies of the Blue Tit and Coal Tit (*Parus*) in mainland and island Mediterranean habitats: a comparison. *Journal of Animal Ecology* 54(2):531–556.

Bock, W.J. 1985. The arboreal theory for the origin of birds. In *The Beginnings of Birds*, pp. 199–207. (Edited by M.K. Hecht, et al.). Freunde des Jura-Museum, Eichstätt.

Bock, W.J. 1986. The arboreal origin of avian flight. *Memoires of the California Academy of Sciences* 8:57–72.

Bock, W.J. 1999. Review of *The Mistaken Extinction. Dinosaur Evolution and the Origin of Birds*, by Lowell Dingus and Timothy Rowe. *Auk* 116(2):566–568.

Bol, M.C., ed. 1998. *Stars Above, Earth Below: American Indians and Nature.* Roberts Rinehart Publishers for Carnegie Museum of Natural History, Niwot, CO. 272 pp.

Bolles, F. 1893. *At the North of Bearcamp Water.* Houghton, Mifflin and Company, Boston, MA.

Bollinger, E.K. and T.A. Gavin. 1992. Eastern Bobolink populations: ecology and conservation in an agricultural landscape. In *Ecology and Conservation of Neotropical Migrant Landbirds*, pp. 497–506. (Edited by J.M. Hagan, III and D.W. Johnston). Smithsonian Institution Press, Washington, D.C. and London.

Bonney, R.E., Jr. 1991. Write on. *The Living Bird Quarterly* 10(2):8–9.

Borgia, G. 1985. Bower quality, number of decorations, and mating success of male Satin Bowerbirds (*Ptilonorhynchus violaceus*): an experimental analysis. *Animal Behaviour* 33(1):266–271.

Borgia, G. 1986. Sexual selection in bowerbirds. *Scientific American* 254(June):92–100.

Borror, D.L. 1959. Songs of the Chipping Sparrow. *Ohio Journal of Science* 59:347–356.

Bosque, C. and O. DeParra. 1992. Digestive efficiency and rate of food passage in Oilbird nestlings. *Condor* 94(3):557–571.

Bouzat, J.L., H.H. Cheng, H.A. Lewin, R.L. Westemeier, J.D. Brawn, and K.N. Paige. 1998. Genetic evaluation of a demographic bottleneck in the Greater Prairie Chicken (*Tympanuchus cupido*). *Conservation Biology* 12:836–843.

Boxall, P.C. and M.R. Lein. 1982. Territoriality and habitat selection of female Snowy Owls (*Nyctea scandiaca*) in winter. *Canadian Journal of Zoology* 60:2344–2350.

Bradley, C.C. 1978. Play behaviour in Northern Ravens. *The Passenger Pigeon* 40:493–495.

Brawn, J.D. and S.K. Robinson. 1996. Source-sink population dynamics may complicate the interpretation of long-term census data. *Ecology* 77(1):3–12.

Brenowitz, E.A. and D.E. Kroodsma. 1996. The neuroethology of birdsong. In *Ecology and Evolution of Acoustic Communication in Birds*, pp. 285–304. (Edited by D.E. Kroodsma and E.H. Miller). Cornell University Press, Ithaca, NY.

Briggs, K. 1984. The breeding ecology of coastal and inland oystercatchers *Haematopus ostralegus* in North Lancashire, UK. *Bird Study* 31(2):141–147.

Brightsmith, D.J. 2000. Use of arboreal termitaria by nesting birds in the Peruvian Amazon. *Condor* 102(3):529–538.

Brittingham, M.C. and S.A. Temple. 1983. Have cowbirds caused forest songbirds to decline? *Bioscience* 33(1):31–35.

Broley, C.L. 1947. Migration and nesting of Florida Bald Eagles. *Wilson Bulletin* 59(1):3–20.

Brooke, M. and T. Birkhead, eds. 1991. *The Cambridge Encyclopedia of Ornithology.* Cambridge University Press, Cambridge and New York. 362 pp.

Broom, R. 1913. On the South African pseudosuchian *Euparkeria* and allied genera. *Proceedings of the Zoological Society of London* 1913:619–633.

Brower, L.P. 1969. Ecological chemistry. *Scientific American* 220 (February):22–29.

Brown, C.R. and M.B. Brown. 1988. The costs and benefits of egg destruction by conspecifics in colonial Cliff Swallows. *Auk* 105(4):737–748.

Brown, R.E. and M.R. Fedde. 1993. Airflow sensors in the avian wing. *Journal of Experimental Biology* 179:13–30.

Brush, A.H. 1990. Metabolism of carotenoid pigments in birds. *FASEB Journal* 4:2969–2977.

Brush, A.H. 1993. The origin of feathers: a novel approach. In *Avian Biology*, Vol. 9, pp. 121–162. (Edited by D.S. Farner, J.R. King, and K.C. Parkes). Academic Press, London.

Bull, J. 1985. *Birds of New York State, including the 1976 Supplement.* Comstock Publishing Associates, Ithaca, NY. 655 pp.

Burkhardt, D. 1989. UV vision: a bird's eye view of feathers. *Journal of Comparative Physiology A* 164:787–796.

Burroughs, J. 1871. *Wake-robin.* Hurd and Houghton, New York and Riverside Press, Cambridge. 231 pp.

Burton, R. 1990. *Bird Flight: An Illustrated Study of Birds' Aerial Mastery.* Facts on File, New York. 160 pp.

Burtt, E.H., Jr. 1984. Colour of the upper mandible: an adaptation to reduce reflectance. *Animal Behaviour* 32:652–658.

Burtt, E.H., Jr. 1986. *An Analysis of Physical, Physiological, and Optical Aspects of Avian Coloration with Emphasis on Wood-Warblers.* Ornithological Monographs, No. 38. American Ornithologists' Union, Washington, D.C.

Butcher, G.S. and S. Rohwer. 1988. Winter versus summer explanations of delayed plumage maturation in temperate passerine birds.

American Naturalist 131(4):556–572.

Butcher, G.S. and S. Rohwer. 1989. The evolution of conspicuous and distinctive coloration for communication in birds. In *Current Ornithology*, Vol. 6, pp. 51–108. (Edited by D.M. Power). Plenum Press, New York.

Cade, T.J. and J.A. Dybas, Jr. 1962. Water Economy in the Budgerigar. *Auk* 79:345–364.

Cade, T.J., J.H. Enderson, C.G. Thelander, and C.M. White, eds. 1988. *Peregrine Falcon Populations: Their Management and Recovery.* The Peregrine Fund, Inc., Boise, ID.

Cade, T.J. and G.L. Maclean. 1967. Transport of water by adult sandgrouse to their young. *Condor* 69:323–343.

Calder, W.A. 1973. Microhabitat selection during nesting of hummingbirds in the Rocky Mountains. *Ecology* 54:127–134.

Campbell, B. and E. Lack, eds. 1985. *A Dictionary of Birds.* Published for the British Ornithologists' Union, Buteo Books, Vermillion, SD. 670 pp.

Campbell, N.A. 1990. *Biology, 2nd Edition.* Benjamin/Cummings Publishing Co., Redwood City, CA. 1165 pp.

Carey, M. and V. Nolan, Jr. 1979. Population dynamics of Indigo Buntings and the evolution of avian polygyny. *Evolution* 33:1180–1192.

Carson, R. 1962. *Silent Spring.* Houghton Mifflin, Boston, MA. 368 pp.

Cassin, J. 1856. *Illustrations of the Birds of California, Texas, Oregon, British and Russian America: Intended to contain descriptions and figures of all North American Birds not given by former American Authors, and a General Synopsis of North American Ornithology. 1853–1856.* J.B. Lippincott & Co., Philadelphia, PA.

Catesby, M. 1731–1743. *The Natural History of Carolina, Florida, and the Bahama Islands.* Printed at the expense of the author, London.

Chace, J.F. and R.C. Tweit. 1999. Greater Pewee (*Contopus pertinax*). In *The Birds of North America*, No. 450. (Edited by A. Poole and F. Gill). The Birds of North America, Inc., Philadelphia, PA.

Chapman, F. 1895. *Handbook of Birds of Eastern North America.* D. Appleton and Company, New York.

Chatterjee, S. 1991. Cranial anatomy and relationships of a new Triassic bird from Texas. *Philosophical Transactions of the Royal Society of London (Bio)* 332:277–342.

Chatterjee, S. 1997. *The Rise of Birds: 225 Million Years of Evolution.* The Johns Hopkins University Press, Baltimore and London. 312 pp.

Clark, G.A., Jr. 1969. Oral flanges of juvenile birds. *Wilson Bulletin* 81(3):270–279.

Clark, G.A., Jr. 1972. Passerine foot-scutes. *Auk* 89:549–558.

Clark, G.A., Jr. 1977. Foot scutes in North American oscines. *Bird-Banding* 48:301–308.

Clark, K.L. and R.J. Robertson. 1981. Cowbird parasitism and evolution of anti-parasite strategies in the Yellow Warbler. *Wilson Bulletin* 93:249–258.

Clark, L. and J.R. Mason. 1985. Use of nest material as insecticidal and anti-pathogenic agents by the European Starling. *Oecologia* 67(2):169–176.

Clark, L. and J.R. Mason. 1988. Effect of biologically active plants used as nest material and the derived benefit to starling nestlings. *Oecologia* 77(2):174–180.

Clayton, D.H. 1991. Coevolution of avian grooming and ectoparasite avoidance. In *Bird–parasite Interactions*, pp. 258–289. (Edited by J.E. Loye and M. Zuk). Oxford University Press, Oxford.

Clements, J.F. 1993. *Supplement to Birds of the World: A Check List.* Ibis Publishing Co., Vista, CA.

Clements, J.F. 2000. *Birds of the World: A Checklist, 5th Edition.* Ibis Publishing Co., Vista, CA. 867 pp.

Clode, D. 1993. Colonially breeding seabirds: predators or prey? *Trends in Ecology and Evolution* 8(9):336–338.

Cochran, W.W. 1987. Orientation and other migratory behavior of a Swainson's Thrush followed for 1500 km. *Animal Behaviour* 35:927–929.

Cody, M.L. 1985. Habitat selection in grassland and open-country birds. In *Habitat Selection in Birds*, pp. 191–226. (Edited by M.L. Cody). Academic Press, Orlando, FL.

Colbert, E.H. 1955. *Evolution of Vertebrates.* John Wiley and Sons, New York.

Collar, N.J., L.P. Gonzaga, N. Krabbe, A. Madroño Nieto, L.G. Naranjo, T.A. Parker, and D.C. Wege. 1992. *Threatened Birds of the Americas: The ICBP/IUCN Red Data Book.* International Council for Bird Preservation, Cambridge, UK.

Collar, N.J. and S.N. Stuart. 1985. *Threatened Birds of Africa and Related Islands: The ICBP/IUCN Red Data Book.* International Council for Bird Preservation and International Union for Conservation of Nature and Natural Resources, Cambridge, UK.

Collias, N.E. and E.C. Collias. 1984. *Nest Building and Bird Behavior.* Princeton University Press, Princeton, NJ. 336 pp.

Conover, M.R. and D.E. Miller. 1980. Rictal bristle function in Willow Flycatcher. *Condor* 82:469–471.

Conway, C.J. and T.E. Martin. 2000. Evolution of passerine incubation behavior: influence of food, temperature, and nest predation. *Evolution* 54(2):670–685.

Coues, E. 1872. *Key to North American Birds.* Dodd and Mead, Salem Naturalists' Agency, New York. 361 pp.

Coues, E. and D.W. Prentiss. 1862. List of birds ascertained to inhabit the District of Columbia, with the times of arrival and departure of such as are non-residents, and brief notes of habits, etc. In *Annual Report of the Board of Regents of the Smithsonian Institution for 1861*, pp. 399–421. Miscellaneous Document 77. 37th Congress 2nd Session, House of Representatives.

Craig, W. 1943. *The Song of the Wood Pewee, Myiochanes virens, Linnaeus: A Study of Bird Music.* New York State Museum Bulletin, No. 334. The University of the State of New York, Albany, NY. 186 pp.

Crane, W. 1887. 7 Routledge & 56 pp.

Crick, H.Q.P., [1993. Seasona. in British birds. *Journa.* 63:263–273.

Crook, J.H. 1960. Nest form and construc. in certain West African weaver birds. *Ibis* 102:1–25.

Cuisin, M. 1983. Note sur certaines adaptations du Pic Noir (*Dryocopus martius* [L.]) et sa niche ecologique dans deux biocenoses. *Oiseau Revue Francaise Ornithologie* 53:63–77.

Curio, E. 1978. The adaptive significance of avian mobbing. I. Teleonomic hypotheses and predictions. *Zeitschrift für Tierpsychologie* 48:175–183.

Curtis, T.J. 1956. The modification of mid-latitude grasslands and forests by man. In *Man's Role in Changing the Face of the Earth*, pp. 721–736. (Edited by W.L. Thomas). Chicago University Press, Chicago, IL.

Daan, S., C. Dijkstra, R. Drent, and T. Meijer. 1988. Food supply and the annual timing of avian reproduction. In *Acta XIX Congressus Internationalis Ornithologici*, Vol. 1, pp. 392–407. (Edited by H. Ouellet). Published for the National Museum of Natural Sciences by University of Ottawa Press, Ottawa.

Daan, S. and J. Tinbergen. 1979. Young Guillemots (*Uria lomvia*) leaving their arctic breeding cliffs: a daily rhythm in numbers and risk. *Ardea* 67:96–100.

Dabelsteen, T. and P.K. McGregor. 1996. Dynamic acoustic communication and interactive playback. In *Ecology and Evolution of Acoustic Communication in Birds*, pp. 398–425. (Edited by D.E. Kroodsma and E.H. Miller). Cornell University Press, Ithaca, NY.

Dahl, T.E. 1990. *Wetland Losses in the United States, 1780's to 1980's.* U.S. Fish and Wildlife Service, Washington, D.C.

Dare, P. 1961. *Ecological Observations on a Breeding Population of the Common Buzzard* (Buteo buteo). Ph.D. Dissertation, Exeter University.

Darwin, C. 1859. *On the Origin of Species by Means of Natural Selection, or the Preservation of Favoured Races in the Struggle for Life, 2nd Edition.* John Murray, London.

Darwin, C. 1874. *The Descent of Man and Selection in Relation to Sex.* Hurst and Co., New York. 705 pp.

Davies, N.B. 1976. Food, flocking, and territorial behavior of the Pied Wagtail (*Motacilla alba yarrelli* Gould) in winter. *Journal of Animal Ecology* 45:235–253.

Davies, N.B. and M.D.L. Brooke. 1988. Cuckoos versus reed warblers: adaptations and counteradaptations. *Animal Behaviour* 36(1):262–284.

Davies, N.B. and M.D.L. Brooke. 1998. Cuckoos versus hosts. In *Parasitic Birds and their Hosts*, pp. 59–79. (Edited by S.I. Rothstein and S.K. Robinson). Oxford University Press, Oxford.

Davis, L.S. 1982. Creching behaviour of Adelie Penguin chicks (*Pygoscelis adeliae*). *New Zealand Journal of Zoology* 9:279–286.

, T.A. and R.A. Ackerman. 1985. Adaptations of Black Tern *(Chlidonias niger)* eggs for water loss in a moist nest. *Auk* 102:640–643.

Davis, W.H., P.J. Kalisz, and R.J. Wells. 1994. Eastern Bluebirds prefer boxes containing old nests. *Journal of Field Ornithology* 65(2):250–253.

Dawkins, R. 1987. *The Blind Watchmaker.* W.W. Norton, New York. 332 pp.

Dawkins, R. and J.R. Krebs. 1978. Animal signals: information or manipulation? In *Behavioural Ecology, An Evolutionary Approach, 1st Edition*, pp. 282–309. (Edited by J.R. Krebs and N.B. Davies). Blackwell, Oxford.

Delius, J.D. 1988. Preening and associated comfort behaviour in birds. In *Neural Mechanisms and Biological Significance of Grooming Behavior*, pp. 40–55. (Edited by D.L. Colbern and W.H. Gispen). New York Academy of Sciences, New York.

DeVoogd, T.J. and C.H.A. Lauay. 2001. Emerging psychobiology of the avian song system. In *Handbook of Behavioral Neurobiology*, pp. 357–392. (Edited by E. Blass). Kluwer/Plenum, New York.

Dhondt, A.A., W.M. Hochachka, S.M. Altizer, and B.K. Hartup. 2001. The House Finch hot zone. *Living Bird* 20(4):24–30.

Diamond, J.M., K.D. Bishop, and S. Van Balen. 1987. Bird survival in an isolated Javan Indonesian woodland island or mirror. *Conservation Biology* 1(2):132–142.

Dobzhansky, T. 1950. Evolution in the tropics. *American Scientist* 38:209–221.

Dooling, R.J. 1982. Auditory perception in birds. In *Acoustic Communication in Birds*, Vol. 1, pp. 95–130. (Edited by D.E. Kroodsma and E.H. Miller). Academic Press, New York.

Drent, R.H. 1975. Incubation. In *Avian Biology*, Vol. 5, pp. 333–419. (Edited by D.S. Farner, and J.R. King). Academic Press, New York.

Dunning, J.B., Jr. 1984. *Body Weights of 686 Species of North American Birds.* Western Bird Banding Association, Monograph No. 1. Western Bird Banding Association, Iverness, CA.

Eadie, J.M., F.P. Kehoe, and T.D. Nudds. 1988. Pre-hatch and post-hatch brood amalgamation in North American Anatidae: a review of hypotheses. *Canadian Journal of Zoology* 66:1709–1721.

Eadie, J.M., P.W. Sherman, and B. Semel. 1998. Conspecific brood parasitism, population dynamics, and the conservation of cavity-nesting birds. In *Behavioral Ecology and Conservation Biology*, pp. 306–340. (Edited by T. Caro). Oxford University Press, Oxford.

Ede, D.A. 1964. *Bird Structure.* Hutchinson Educational Ltd., London. 120 pp.

Edwards, G. 1743–1751. *A Natural History of Birds.* Printed for the author at the Royal College of Physicians, London.

Ehrlich, P.R., D.S. Dobkin, and D. Wheye. 1988. *The Birder's Handbook.* Simon and Schuster, New York.

Ekman, J. 1986. Reproductive cost, age-specific survival, and a comparison of the reproductive strategy in two European tits (genus *Parus*). *Evolution* 40(1):159–168.

Elgar, M.A. 1986. House Sparrows *(Passer domesticus)* establish foraging flocks by giving chirrup calls if the resources are divisible. *Animal Behaviour* 34(1):169–174.

Ely, C.A., P.J. Latas, and R.R. Lohoefener. 1977. Additional returns and recoveries of North American birds banded in southern Mexico. *Bird–Banding* 48:275–276.

Emlen, J.T., Jr. 1940. Sex and age ratios in survival of the California Quail. *Journal of Wildlife Management* 4:92–99.

Emlen, J.T., Jr. 1963. Determinants of cliff edge and escape responses in Herring Gull chicks in nature. *Behaviour* 22:1–15.

Emlen, S.T. 1967a. Migratory orientation in the Indigo Bunting, *Passerina cyanea*. Part I. Evidence for use of celestial cues. *Auk* 84:309–342.

Emlen, S.T. 1967b. Migratory orientation in the Indigo Bunting, *Passerina cyanea*. Part II. Mechanism of celestial orientation. *Auk* 84:463–489.

Emlen, S.T. 1975. The stellar-orientation system of a migratory bird. *Scientific American* 233(2):102–111.

Emlen, S.T. 1982. The evolution of helping 1. An ecological constraints model. *American Naturalist* 119(1):29–39.

Emlen, S.T. 1994. Benefits, constraints, and the evolution of the family. *Trends in Ecology and Evolution* 9(8):282–285.

Emlen, S.T. and L.W. Oring. 1977. Ecology, sexual selection, and the evolution of mating systems. *Science* 197:215–223.

Emlen, S.T., P.H. Wrege, and N.J. Demong. 1995. Making decisions in the family—an evolutionary perspective. *American Scientist* 83(2):148–157.

Evans, H.E. 1996. Anatomy of the Budgerigar and other birds. In *Diseases of Cage and Aviary Birds, 3rd Edition*, pp. 79–162. (Edited by W.J. Rosskopf and R.W. Woerpel). Williams and Wilkins, Baltimore.

Evans, M.R. and J.L. Burn. 1996. An experimental analysis of mate choice in the wren: a monomorphic, polygynous passerine. *Behavioral Ecology* 7(1):101–108.

Evans, R.M. 1984. Some causal and functional correlates of creching in young white pelicans. *Canadian Journal of Zoology* 62(5):814–819.

Evans Ogden, L.J. and B.J. Stutchbury. 1994. Hooded Warbler *(Wilsonia citrina)*. In *The Birds of North America*, No. 110. (Edited by A. Poole and F. Gill). The Academy of Natural Sciences, Philadelphia, PA and The American Ornithologists' Union, Washington, D.C.

Ewert, D.N. and D.E. Kroodsma. 1994. Song sharing and repertoires among migratory and resident Rufous-sided Towhees. *Condor* 96(1):190–196.

Faaborg, J. and S.B. Chaplin. 1988. *Ornithology: An Ecological Approach.* Prentice-Hall, Inc., Englewood Cliffs, NJ. 470 pp.

Falls, J.B. 1988. Does song deter territorial intrusion in White-throated Sparrows *(Zonotrichia albicollis)*? *Canadian Journal of Zoology* 66:206–211.

Falls, J.B. and J.G. Kopachena. 1994. White-throated Sparrow *(Zonotrichia albicollis)*. In *The Birds of North America*, No. 128. (Edited by A. Poole and F. Gill). The Academy of Natural Sciences, Philadelphia, PA and The American Ornithologists' Union, Washington, D.C.

Farabaugh, S.M. 1982. The ecological and social significance of duetting. In *Acoustic Communication in Birds*, Vol. 2, pp. 85–124. (Edited by D.E. Kroodsma and E.H. Miller). Academic Press, New York.

Feduccia, A. 1993. Evidence from claw geometry indicating arboreal habits of *Archaeopteryx*. *Science* 259:790–793.

Feduccia, A. 1996. *The Origin and Evolution of Birds.* Yale University Press, New Haven and London. 420 pp.

Feduccia, A. and H.B. Tordoff. 1979. Feathers of *Archaeopteryx*: asymmetric vanes indicate aerodynamic function. *Science* 203:1021–1022.

Fee, M.S., B. Shraiman, B. Pesaran, and P. Mitra. 1998. The role of nonlinear dynamics of the syrinx in the vocalizations of a songbird. *Nature* 395:67–71.

Feld, S. 1990. *Sound and Sentiment: Birds, Weeping, Poetics, and Song in Kaluli Expression, 2nd Edition.* University of Pennsylvania Press, Philadelphia, PA. 297 pp.

Ficken, M.S. 1977. Avian play. *Auk* 94:573–582.

Ficken, M.S. and R.W. Ficken. 1968. Reproductive isolating mechanisms in the Blue-winged Warbler—Golden-winged Warbler complex. *Evolution* 22:166–179.

Firstencel, H. 1987. *The Black Tern* (Chlidonias niger *Linn.): Breeding Ecology in Upstate New York and Results of Pesticide Residue Analyses.* M.S. Thesis, State University of New York, Rockport.

Fisher, A.K. 1893. *The Hawks and Owls of the United States and their Relation to Agriculture.* Government Printing Office, Washington, D.C. 210 pp.

Fitzpatrick, J.W., G.E. Woolfenden, and M.T. Kopeny. 1991. *Ecology and Development-related Habitat Requirements of the Florida Scrub-Jay* (Aphelocoma coerulescens coerulescens). Nongame Wildlife Program Technical Report No. 8, Florida Game and Fresh Water Fish Commission, Tallahassee, FL. 49 pp.

Forbush, T.R. 1925–1929. *Birds of Massachusetts and Other New England States, 3 Volumes.* Printed by Berwick and Smith Company, Norwood, MA.

Foster, M.S. 1978. Total frugivory in tropical passerines: a reappraisal. *Tropical Ecology* 19(2):131–154.

Fraga, R.M. 1998. Interactions of the parasitic Screaming and Shiny cowbirds (*Molothrus rufoaxillaris* and *M. bonariensis*) with a shared host, the Bay-winged Cowbird (*M. badius*). In *Parasitic Birds and their Hosts*, pp. 173–193. (Edited by S.I. Rothstein and S.K. Robinson). Oxford University Press, New York.

Freemark, K.E. and B. Collins. 1992. Landscape ecology of birds breeding in temperate forest fragments. In *Ecology and Conservation of Neotropical Migrant Landbirds*, pp. 443–454. (Edited by J.M. Hagan and D.W. Johnston). Smithsonian Institution Press, Washington, D.C. 609 pp.

Freemark, K.E., J.B. Dunning, S.J. Heijl, and J.R. Probst. 1995. A landscape ecology perspective for research, conservation, and management. In *Ecology and Management of Neotropical Migratory Birds*, pp. 381–427. (Edited by T.E. Martin and D.M. Finch). Oxford University Press, New York.

Friedmann, H., L.F. Kiff, and S.J. Rothstein. 1977. *A Further Contribution to Knowledge of the Host Relations of Parasitic Cowbirds*. Smithsonian Contributions to Zoology, No. 235. Smithsonian Institution Press, Washington, D.C. 75 pp.

Fry, C.H., S. Keith, and E. Urban, eds. 1988. *The Birds of Africa, Vol. 3*. Academic Press Ltd., London.

Gahr, M., H.R. Güttinger, and D.E. Kroodsma. 1993. Estrogen receptors in the avian brain: survey reveals general distribution and forebrain areas unique to songbirds. *Journal of Comparative Neurology* 327:112–122.

Gates, J.E. and L.W. Gysel. 1978. Avian nest dispersion and fledging success in field-forest ecotones. *Ecology* 59(5):871–883.

Gaunt, A.S. and S. Nowicki. 1997. Sound production in birds: acoustics and physiology revisited. In *Animal Acoustic Communication: Sound Analysis and Research Methods*, pp. 291–321. (Edited by S.L. Hopp, M.J. Owren, and C.S. Evans). Springer-Verlag, New York and Berlin.

Gill, F.B. 1988. Trapline foraging by hermit hummingbirds: competition for an undefended, renewable resource. *Ecology* 69(6):1933–1942.

Gill, F.B. 1990. *Ornithology*. W.H. Freeman and Co., New York. 660 pp.

Gill, F.B. 1995. *Ornithology, 2nd Edition*. W.H. Freeman and Co., New York. 766 pp.

Gill, R.E., Jr., P. Canevari, and E.H. Iversen. 1998. Eskimo Curlew *(Numenius borealis)*. In *The Birds of North America*, No. 235. (Edited by A. Poole and F. Gill). The Birds of North America, Inc., Philadelphia, PA.

Goguen, C.B. and N.E. Mathews. 1998. Songbird community composition and nesting success in grazed and ungrazed pinyon-juniper woodlands. *Journal of Wildlife Management* 62:474–484.

Goller, F. 1998. Vocal gymnastics and the bird brain. *Nature* 395:11–12.

Goller, F. and O.N. Larsen. 1997. A new mechanism of sound generation in songbirds. *Proceedings of the National Academy of Sciences of the United States of America* 94:14787–14791.

Goodenough, J., B. McGuire, and R. Wallace. 1993. *Perspectives on Animal Behavior*. John Wiley and Sons, Inc., New York.

Goodfellow, P. 1977. *Birds as Builders*. Arco Publishing Company, Inc., New York. 168 pp.

Goossen, J.P. and S.G. Sealy. 1982. Production of young in a dense nesting population of Yellow Warblers, *Dendroica petechia*, in Manitoba. *Canadian Field-Naturalist* 96:189–199.

Gorman, M.L. and H. Milne. 1972. Creche behaviour in the Common Eider *Somateria m. mollissima* L. *Ornis Scandinavica* 3(1):21–25.

Goss-Custard, J.D. 1976. Variation in the dispersion of Redshank *(Tringa totanus)* on their winter feeding grounds. *Ibis* 118:257–263.

Goss-Custard, J.D. 1977. Optimal foraging and the size selection of worms by Redshank, *Tringa totanus*, in the field. *Animal Behaviour* 25:10–29.

Götmark, F. 1987. White underparts in gulls function in hunting camouflage. *Animal Behaviour* 35:1786–1792.

Gottschaldt, K.M. and S. Lausmann. 1974. The peripheral morphological basis of tactile sensibility in the beak of geese. *Cell and Tissue Research* 153:477–496.

Gould, J. 1850–1883. *The Birds of Asia*. Printed for the author by Taylor and Francis, London.

Gould, J. 1849–1861. *A Monograph of the Trochilidae, or Family of Humming-birds*. Printed for the author by Taylor and Francis, London.

Goutis, C. and D.W. Winkler. 1992. Hungry chicks and mortal parents: a state-variable approach to the breeding seasons of birds. *Bulletin of Mathematical Biology* 54:379–400.

Graber, R.R. 1965. Night flight with a thrush. *Audubon* 67(6):368–374.

Graber, R.R. 1968. Nocturnal migration in Illinois—Different points of view. *Wilson Bulletin* 80:36–71.

Graber, R.R. and W.W. Cochran. 1959. An audio technique for the study of nocturnal migration of birds. *Wilson Bulletin* 71:220–236.

Graber, R.R. and W.W. Cochran. 1960. Evaluation of an aural record of nocturnal migration. *Wilson Bulletin* 72:252–273.

Graham, F., Jr. 1989. Oilspeak, common sense, and soft science. *Audubon* 91(5):102–111.

Grahame, J. 1987. *Plankton and Fisheries*. Edward Arnold Publishers, Ltd., London. 140 pp.

Grant, G.S. 1982. *Avian Incubation: Egg Temperature, Nest Humidity, and Behavioral Thermoregulation in a Hot Environment*. Ornithological Monographs, No. 30. American Ornithologists' Union, Washington, D.C. 75 pp.

Grau, C.R. 1976. Ring structure of avian egg yolk. *Poultry Science* 55:1418–1422.

Greenewalt, C.H. 1960. *Hummingbirds*. Published for the American Museum of Natural History by Doubleday and Co., Inc., Garden City, NY. 250 pp.

Greenewalt, C.H., W. Brandt, and D.D. Friel. 1960. Iridescent colors of hummingbird feathers. *Journal of the Optical Society of America* 50:1005–1013.

Griffin, D.R. 1974. *Bird Migration*. Dover Publications, New York. 180 pp.

Griscom, L. 1923. *Birds of the New York City Region*. The Museum, New York. 400 pp.

Grubb, T.C., Jr. 1995. Ptilochronology: a review and prospectus. In *Current Ornithology*, Vol. 12, pp. 85–114. (Edited by D.M. Power). Plenum Press, New York.

Gwinner, E. 1966. Über einige Bewegungsspiele des Kolkraben. *Zeitschrift für Tierpsychologie* 23:28–36.

Gwinner, E. 1986. Circannual rhythms in the control of avian migrations. *Advances in the Study of Behaviour* 16:191–228.

Gwinner, E. and W. Wiltschko. 1978. Endogenously controlled changes in migratory direction of the Garden Warbler, *Sylvia borin. Journal of Comparative Physiology* 125:267–273.

Gwinner, E. and W. Wiltschko. 1980. Circannual changes in migratory orientation of the Garden Warbler, *Sylvia borin. Behavioral Ecology and Sociobiology* 7:73–78.

Haffer, J. 1974. *Avian Speciation in Tropical South America*. Publications of the Nuttall Ornithological Club, No. 14. Nuttall Ornithological Club, Cambridge, MA. 390 pp.

Haggerty, T.M. and E.S. Morton. 1995. Carolina Wren *(Thryothorus ludovicianus)*. In *The Birds of North America*, No. 188. (Edited by A. Poole and F. Gill). The Academy of Natural Sciences, Philadelphia, PA and The American Ornithologists' Union, Washington, D.C.

Haig, S.M., J.D. Ballou, and S.R. Derrickson. 1990. Management options for preserving genetic diversity: reintroduction of Guam Rails to the wild. *Conservation Biology* 4:290–300.

Hailman, J.P. 1989. The organization of major vocalizations in the Paridae. *Wilson Bulletin* 101:305–343.

Hailman, J.P. and M.S. Ficken. 1996. Comparative analysis of vocal repertoires, with reference to chickadees. In *Ecology and Evolution of Acoustic Communication in Birds*, pp. 136–159. (Edited by D.E. Kroodsma and E.H. Miller). Cornell University Press, Ithaca, NY.

Halliday, T. 1978. *Vanishing Birds: Their Natural History and Conservation*. Sidgwick & Jackson, London. 296 pp.

Hamas, M.J. 1994. Belted Kingfisher *(Ceryle alcyon)*. In *The Birds of North America*, No. 84. (Edited by A. Poole and F. Gill). The Academy of Natural Sciences, Philadelphia, PA and The American Ornithologists' Union, Washington, D.C.

Hamel, P.B. 1995. Bachman's Warbler *(Vermivora bachmanii)*. In *The Birds of North America*, No. 150. (Edited by A. Poole and F. Gill). The Academy of Natural Sciences, Philadelphia, PA and The American Ornithologists' Union, Washington, D.C.

Hamilton, W.D. and M. Zuk. 1982. Heritable true fitness and bright birds: a role for parasites? *Science* 218:384–387.

Hann, H.W. 1937. Life history of the Ovenbird in southern Michigan. *Wilson Bulletin* 49:145–237.

Hansell, M. 2001. *Bird Nests and Construction Behaviour*. Cambridge University Press, Cambridge, England. 288 pp.

Harrison, H.H. 1975. *A Field Guide to Birds' Nests of the Eastern United States*. Houghton Mifflin Company, Boston, MA. 257 pp.

Hartshorne, C. 1973. *Born to Sing. An Interpretation and World Survey of Bird Song*. Indiana University Press, Bloomington.

Hartzler, J.E. 1970. Winter dominance relationships in Black-capped Chickadees. *Wilson Bulletin* 82(4):427–434.

Hasselquist, D. 1994. *Male Attractiveness, Mating Tactics and Realized Fitness in the Polygynous Great Reed Warbler*. Ph.D. Dissertation, Lund University, Lund, Sweden.

Hasselquist, D., S. Bensch, and T. von Schantz.

1996. Correlation between male song repertoire, extra-pair paternity and offspring survival in the Great Reed Warbler. *Nature* 381:229–232.

Hayden, T.J. 1985. *Minimum Area Requirements of Some Breeding Bird Species in Fragmented Habitats in Missouri*. M.S. Thesis, University of Missouri, Columbia.

Hayden, T.J., J. Faaborg, and R.L. Clawson. 1985. Estimates of minimum area requirements of Missouri forest birds. *Transactions of the Missouri Academy of Science* 19:11–22.

Heilmann, G. 1926. *The Origin of Birds*. Witherby, London.

Heinrich, B. 1995. An experimental investigation of insight in Common Ravens *(Corvus corax)*. *Auk* 112:994–1003.

Heinroth, O. 1938. *Aus dem Leben der Vögel*. Julius Springer, Berlin.

Helbig, A.J. 1989. *Angeborene Zugrichtungen nachts ziehender Singvögel: Orientierungsmechanismen, geographische Variation und Vererbung*. Ph.D. Dissertation, University of Frankfurt/Main.

Hentschel, E. 1933. Allgemeine Biologie des südatlantischen Ozeans, Part 1: Das Pelagial der obersten Wasserschicht. *Wissenschaftliche Ergebnisse der Deutschen Atlantischen Expedition auf dem Forschungsund Vermessungsschiff "Meteor", 1925–1927.* 11:1–168.

Hepp, G.R. and F.C. Bellrose. 1995. Wood Duck *(Aix sponsa)*. In *The Birds of North America*, No. 169. (Edited by A. Poole and F. Gill). The Academy of Natural Sciences, Philadelphia, PA and The American Ornithologists' Union, Washington, D.C.

Herkert, J.R. 1991. *An Ecological Study of the Breeding Birds of Grassland Habitats within Illinois*. Ph.D. Dissertation, University of Illinois, Urbana.

Herkert, J.R., R.E. Szafoni, V.M. Kleen, and J.E. Schwegman. 1993. *Habitat Establishment, Enhancement and Management for Forest and Grassland Birds in Illinois*. National Heritage Publication No. 1. Division of Natural Heritage, Illinois Department of Conservation, Springfield, IL.

Herman, S.G. 1986. *The Naturalist's Field Journal: A Manual of Instruction Based on a System Established by Joseph Grinnell*. Buteo Books, Vermillion, SD. 200 pp.

Herrick, F.H. 1934. *The American Eagle*. D. Appleton-Century Company, Inc., New York.

Hickey, J.J., ed. 1969. *Peregrine Falcon Populations: Their Biology and Decline*. University of Wisconsin Press, Madison, WI. 596 pp.

Hickey, J.J. and D.W. Anderson. 1968. Chlorinated hydrocarbons and eggshell changes in raptorial and fish-eating birds. *Science* 162:271–273.

Highland, H.J. 1963. *The How and Why Wonder Book of Light and Color*. Wonder Books, Inc., New York. 48 pp.

Higuchi, H. 1987. Cast master. *Natural History* 96:40–43.

Hill, G.E. and R. Montgomerie. 1994. Plumage colour signals nutritional condition in the House Finch. *Proceedings of the Royal Society of London B* 258(1351):47–52.

Hill, R.A. 1976. Host-parasite relationships of the Brown-headed Cowbird in a prairie habitat of west-central Kansas. *Wilson Bulletin* 88:555–565.

Hill, S.R. and J.E. Gates. 1988. Nesting ecology and microhabitat of the Eastern Phoebe in the central Appalachians. *American Midland Naturalist* 120:313–324.

Högstedt, G. 1980. Evolution of clutch size in birds: adaptive variation in relation to territory quality. *Science* 210:1148–1150.

Holmes, R.T., J.C. Schultz, and P.J. Nothnagle. 1979. Bird predation on forest insects: an exclosure experiment. *Science* 206:462–463.

Hoogland, J.L. and C.R. Brown. 1986. Risk in mobbing for solitary and colonial swallows. *Animal Behaviour* 34(5):1319–1323.

Horn, A.G., M.L. Leonard, L. Ratcliffe, S.A. Shackleton, and R.G. Weisman. 1992. Frequency variation in songs of Black-capped Chickadees *(Parus atricapillus)*. *Auk* 109:847–852.

Hovis, J.C., D.M. Evans, J.C. Bednarz, and C.R. Davis. 1997. *The Effects of Forest Landscape Modification and Management on Neotropical Migratory Songbird Populations in West-Central Idaho*. 1996 Report. Unpublished Report of the Department of Biological Sciences, Arkansas State University, State University, AR.

Howell, J.C. 1973. *Communicative Behavior in the Cedar Waxwing (Bombycilla cedrorum) and the Bohemian Waxwing (Bombycilla garrulus)*. Ph.D. Dissertation, University of Michigan, Ann Arbor, Michigan.

Howell, T.R., B. Araya, and W.R. Millie. 1974. *Breeding Biology of the Gray Gull*, Larus modestus. University of California Publications in Zoology, Vol. 104. University of California Press, Berkeley, CA. 57 pp.

Howes, J.D. 1985. Nesting habits and activity patterns of Warbling Vireos (Vireo gilvus) in Southern Ontario, Canada. *Canadian Field-Naturalist* 99(4):484–489.

Hudon, J. 1991. Unusual carotenoid use by the Western Tanager *(Piranga ludovicianus)* and its evolutionary implications. *Canadian Journal of Zoology* 69:2311–2320.

Hudon, J. 1999. Western Tanager *(Piranga ludoviciana)*. In *The Birds of North America*, No. 432. (Edited by A. Poole and F. Gill). The Birds of North America, Inc., Philadelphia, PA.

Hunt, G.R. 1996. Manufacture and use of hook-tools by New Caledonian Crows. *Nature* 379:249–251.

Hunt, S., I. Cuthill, J. Swaddle, and A. Bennett. 1997. Ultraviolet vision and band colour preferences in female Zebra Finches, *Taeniopygia guttata*. *Animal Behaviour* 54(6):1383–1392.

Hurd, P.L., P.J. Weatherhead, and S.B. McRae. 1991. Parental consumption of nestling feces: good food or sound economics? *Behavioral Ecology* 2(1):69–76.

Huxley, T.H. 1868. On the animals which are most nearly intermediate between the birds and reptiles. *Annals and Magazine of Natural History* 2:66–75.

Ignatiuk, J.B. and R.G. Clark. 1991. Breeding biology of American Crows in Saskatchewan parkland habitat. *Canadian Journal of Zoology* 69:168–175.

Ioale, P., M. Nozzolini, and F. Papi. 1990. Homing pigeons do extract directional information from olfactory stimuli. *Behavioral Ecology and Sociobiology* 26(5):301–306.

Iredale, T. 1956. *Birds of New Guinea*. Georgian House, Melbourne, Australia.

Jackson, W.M. 1992. Estimating conspecific nest parasitism in the Northern Masked-Weaver based on within-female variability in egg appearance. *Auk* 109(3):435–443.

Jackson, W.M. 1993. Causes of conspecific nest parasitism in the Northern Masked-Weaver. *Behavioral Ecology and Sociobiology* 32(2):119–126.

Jacob, J. and V. Ziswiler. 1982. The uropygial gland. In *Avian Biology*, Vol. 6, pp. 199–324. (Edited by D.S. Farner, J.R. King, and K.C. Parkes). Academic Press, New York.

Jaeger, E.C. 1955. *A Source-Book of Biological Names and Terms, 3rd Edition*. Charles C Thomas, Springfield, IL.

Jashemski, W.F.A. 1993. *The Gardens of Pompeii*. Arastide C. Cavatzas, New Rochelle, NY.

Jenkins, F.A., K.P. Dial, and G.E. Goslow, Jr. 1988. A cineradiographic analysis of bird flight: the wishbone in starlings is a spring. *Science* 241:1495–1498.

Jepsen, G.L. 1963. Terrible lizards revisited. *Princeton Alumni Weekly*, Nov. 26, 1963.

Johnsen, A., S. Andersson, J. Ornborg, and J. Lifjeld. 1998. Ultraviolet plumage ornamentation affects social mate choice and sperm competition in Bluethroats (Aves: *Luscinia s. svecica*): a field experiment. *Proceedings of the Royal Society of London B* 265(1403):1313–1318.

Johnson, N.K. and J.A. Marten. 1988. Evolutionary genetics of flycatchers. II. Differentiation in the *Empidonax difficilis* complex. *Auk* 105:177–191.

Johnstone, R.A. 1997. The evolution of animal signals. In *Behavioral Ecology, An Evolutionary Approach, 4th Edition*, pp. 155–178. (Edited by J.R. Krebs and N.B. Davies). Blackwell, Oxford.

Jones, D.N., R.W.R.J. Dekker, and C.S. Roselaar. 1995. *The Megapodes: Megapodiidae*. Oxford University Press, Oxford. 262 pp.

Joyce, F.J. 1993. Nesting success of Rufous-naped Wrens *(Campylorhynchus rufinucha)* is greater near wasp nests. *Behavioral Ecology and Sociobiology* 32:71–77.

Kahl, M.P. 1967. Observations on the behaviour of the Hamerkop *Scopus umbretta* in Uganda. *Ibis* 109:25–32.

Kaufman, K. 1996. *Lives of North American Birds*. Houghton Mifflin, Boston. 675 pp.

Keeton, W.T. 1980. *Biological Science, 3rd Edition*. W.W. Norton and Co., New York and London.

Kehoe, F.P. 1989. The adaptive significance of creching behavior in the White-winged Scoter *(Melanitta fusca deglandi)*. *Canadian Journal of Zoology* 67(2):406–411.

Kendeigh, S.C. 1952. *Parental Care and its Evolution in Birds*. Illinois Biological Monographs, Vol. 22, Nos. 1–3. University of Illinois Press, Urbana, IL. 356 pp.

Kenward, R.E. 1978. Hawks and doves: factors affecting success and selection in goshawk attacks on wood pigeons. *Journal of Animal Ecology* 47:449–460.

Kerlinger, P. 1998. Showdown at Delaware Bay. *Natural History* 107(4):56–58.

Kerlinger, P. and F.R. Moore. 1989. Atmospheric structure and avian migration. In *Current Ornithology*, Vol. 6, pp. 109–136. (Edited by D.M. Power). Plenum Press, New York.

Kessel, B. 1976. Winter activity patterns in Black-capped Chickadees in interior Alaska. *Wilson Bulletin* 88:36–61.

Kilgore, D.L., Jr. and K.L. Knudsen. 1977. Analysis of materials in Cliff Swallow and Barn Swallow nests: relationships between mud selection and nest architecture. *Wilson Bulletin* 89:562–571.

Kilham, L. 1974. Biology of young Belted Kingfishers. *American Midland Naturalist* 92:245–247.

King, A.S. and J. McLelland. 1975. *Outlines of Avian Anatomy*. Bailliere Tindall, London. 154 pp.

King, A.S. and J. McLelland, eds. 1981. *Form and Function in Birds*, Vols. 1–4. Academic Press, London and New York.

King, B. 1969. Hooded Crows dropping and transferring objects from bill to foot in flight. *British Birds* 62:201.

Klaas, E.E. 1975. Cowbird parasitism and nesting success in the Eastern Phoebe. *Occasional Papers of the Museum of Natural History, University of Kansas* 41:1–18.

Klomp, H. 1970. The determination of clutch-size in birds, a review. *Ardea* 58:1–124.

Knight, R.L., D.J. Grout, and S.A. Temple. 1987. Nest-defense behavior of the American Crow in urban and rural areas. *Condor* 89:175–177.

Koenig, W.D. 1984. Geographic variation in clutch size in the Northern Flicker *(Colaptes auratus)*: support for Ashmole's hypothesis. *Auk* 101:698–706.

Komdeur, J. 1992. Importance of habitat saturation and territory quality for evolution of cooperative breeding in the Seychelles Warbler. *Nature* 358(6386):493–495.

Korpimäki, E. 1987. Timing of breeding of Tengmalm's Owl *Aegolius funereus* in relation to vole dynamics in western Finland. *Ibis* 129(1):58–68.

Korpimäki, E. 1988. Costs of reproduction and success of manipulated broods under varying food conditions in Tengmalm's Owls. *Journal of Animal Ecology* 57(3):1027–1040.

Korpimäki, E. and H. Hakkarainen. 1991. Fluctuating food supply affects the clutch size of Tengmalm's Owl independent of laying date. *Oecologia* 85(4):543–552.

Kramer, G. 1951. Eine neue Methode zur Erforschung der Zugorientierung und edie bisher damit erzielten Ergebnisse. Proceedings of the 10th Ornithological Congress, Uppsala. pp. 269–280.

Kramer, G. 1953. Wird die Sonnenhöhe bei der Heimfindorientierung verwertet? *Journal für Ornithologie* 94:201–219.

Krebs, E.A. 1998. Breeding biology of Crimson Rosellas *(Platycercus elegans)* on Black Mountain, Australian Capital Territory. *Australian Journal of Zoology* 46(2):119–136.

Krebs, J.R. 1971. Territory and breeding density in the Great Tit, *Parus major* L. *Ecology* 52:2–22.

Krebs, J.R., R. Ashcroft, and M. Webber. 1978. Song repertoires and territory defence in the Great Tit. *Nature* 271:539–542.

Krebs, J.R. and N.B. Davies. 1981. *An Introduction to Behavioural Ecology*. Sinauer Associates Inc., Sunderland, MA. 292 pp.

Krebs, J.R. and D.E. Kroodsma. 1980. Repertoires and geographical variation in bird song. In *Advances in the Study of Behavior*, Vol. 11, pp. 143–177. (Edited by J.S. Rosenblatt, R.A. Hinde, C. Beer, and M.C. Busnel). Academic Press, New York.

Kress, S.W. 1981. *The Audubon Society Handbook for Birders*. Charles Scribner's Sons, New York. 322 pp.

Kress, S.W. 1988. *The McGraw-Hill Wildlife Birding Series*. McGraw-Hill, Inc., Washington, D.C.

Kress, S.W. 1995. *National Audubon Society's The Bird Garden*. Dorling Kindersley Limited, London. 176 pp.

Kress, S.W. 2000. The winter banquet. *Audubon* 102(1):80–83.

Kroodsma, D.E. 1981. Winter Wren singing behavior: a pinnacle of song complexity. *Condor* 82:357–365.

Kroodsma, D.E. 1983. Marsh wrenditions. *Natural History* 92(9):43–46.

Kroodsma, D.E. 1986. The spice of bird song. *The Living Bird Quarterly* 5:12–16.

Kroodsma, D.E. 1988. Contrasting styles of song development and their consequences among the Passeriformes. In *Evolution and Learning*, pp. 157–184. (Edited by R.C. Bolles and M.D. Beecher). Erlbaum Assoc., Inc., Hillsdale, NJ.

Kroodsma, D.E. 1989. What, when, where, and why warblers warble. *Natural History* 98(5):50–59.

Kroodsma, D.E. 1993. Cousin songbird. *Birder's World* 7(5):17–21.

Kroodsma, D.E. 1996a. Ecology of passerine song development. In *Ecology and Evolution of Acoustic Communication in Birds*, pp. 3–19. (Edited by D.E. Kroodsma and E.H. Miller). Cornell University Press, Ithaca, NY.

Kroodsma, D.E. 1996b. A song of their own. *Living Bird* 15(3):10–17.

Kroodsma, D.E., D.J. Albano, P.W. Houlihan, and J.A. Wells. 1995. Song development by Black-capped Chickadees *(Parus atricapillus)* and Carolina Chickadees *(P. carolinensis)*. *Auk* 112:29–43.

Kroodsma, D.E., B.E. Byers, S.L. Halkin, C. Hill, D. Minis, J.R. Bolsinger, J.-A. Dawson, E. Donelan, J. Farrington, F. Gill, P. Houlihan, D. Innes, G. Keller, L. Macaulay, C.A. Marants, J. Ortiz, P.K. Stoddard, and K. Wilda. 1999. Geographic variation in Black-capped Chickadee songs and singing behavior. *Auk* 116:387–402.

Kroodsma, D.E. and F.C. James. 1994. Song variation among populations of the Red-winged Blackbird. *Wilson Bulletin* 106:156–162.

Kroodsma, D.E., W.–C. Liu, E. Goodwin, and P.A. Bedell. 1999. The ecology of song improvisation as illustrated by North American Sedge Wrens. *Auk* 116:373–386.

Kroodsma, D.E. and H. Momose. 1991. Songs of the Japanese population of the Winter Wren *(Troglodytes troglodytes)*. *Condor* 93:424–432.

Kroodsma, D.E. and J. Verner. 1978. Complex singing behaviors among *Cistothorus* wrens. *Auk* 95:703–716.

Kroodsma, D.E., J.M.E. Vielliard, and F.G. Stiles. 1996. Study of bird sounds in the Neotropics: urgency and opportunity. In *Ecology and Evolution of Acoustic Communication in Birds*, pp. 269–281. (Edited by D.E. Kroodsma and E.H. Miller). Cornell University Press, Ithaca, NY.

Krueper, D. 2000. Conservation priorities in naturally fragmented and human-altered riparian habitats of the arid west. In *Strategies for Bird Conservation: The Partners in Flight Planning Process; Proceedings of the 3rd Partners in Flight Workshop; 1995 October 1–5; Cape May, NJ*, pp. 88–90. (Edited by R. Bonney, D.N. Pashley, R.J. Cooper, and L. Niles). Proceedings RMRS–P–16. U.S. Department of Agriculture, Forest Service, Rocky Mountain Research Station, Ogden, UT.

Kruuk, H. 1964. Predators and antipredator behaviour of the Black-headed Gull *Larus ridibundus*. *Behaviour Supplement* 11:1–129.

Kurlansky, M. 1997. *Cod: A Biography of the Fish that Changed the World*. Walker and Co., New York. 294 pp.

Lack, D. 1947. The significance of clutch size, Part 1. *Ibis* 89:302–352.

Lack, D. 1948. The significance of clutch size, Part 2. *Ibis* 90:25–45.

Lack, D. 1954. *The Natural Regulation of Animal Numbers*. Oxford University Press, Oxford. 343 pp.

Lack, D. 1966. *Population Studies of Birds*. Oxford University Press, Oxford. 341 pp.

Lack, D. 1968. *Ecological Adaptations for Breeding in Birds*. Methuen and Company, London. 409 pp.

Lack, D. 1973. *Swifts in a Tower*. Chapman and Hall, London. 239 pp.

Lammertink, J.M., J.A. Rojas-Tomé, F.M. Casillas-Orona, and R.L. Otto. 1996. *Status and Conservation of Old-growth Forests and Endemic Birds in the Pine-oak Zone of the Sierra Madre Occidental, Mexico*. Verslagen en Technische Gegevens No. 69. Institute for Systematics and Population Biology (Zoological Museum), University of Amsterdam, The Netherlands.

Lanyon, S.M. 1992. Interspecific brood parasitism in blackbirds (Icterinae): a phylogenetic perspective. *Science* 225:77–79.

Lanyon, W.E. 1969. Vocal characters and avian systematics. In *Bird Vocalizations: Their Relation to Current Problems in Biology and Psychology*, pp. 291–310. (Edited by R.A. Hinde). Cambridge University Press, Cambridge, England.

Lanyon, W.E. 1978. Revision of the *Myiarchus* flycatchers of South America. *Bulletin of the American Museum of Natural History* 161:427–628.

Larison, J.R., G.E. Likens, J.W. Fitzpatrick, and J.G. Crock. 2000. Cadmium toxicity among wildlife in the Colorado Rocky Mountains. *Nature* 406:181–183.

Latham, J. 1821–1828. *A General History of Birds.* Printed by Jacob and Johnson for the author, Winchester.

Laubin, R. and G. Laubin. 1977. *Indian Dances of North America.* University of Oklahoma Press, Norman, OK. 538 pp.

Le Maho, Y., P. Delclitte, and J. Chatonnet. 1976. Thermoregulation in fasting Emperor Penguins under natural conditions. *American Journal of Physiology* 231:913–922.

Leader, N. and Y. Yom-Tov. 1998. The possible function of stone ramparts at the nest entrance of the Blackstart. *Animal Behaviour* 56:207–217.

Lefevre, K., R. Montgomerie, and A.J. Gaston. 1998. Parent-offspring recognition in Thick-billed Murres (Aves: Alcidae). *Animal Behaviour* 55(4):925–938.

Leonard, M.L., A.G. Horn, C.R. Brown, and N.J. Fernandez. 1997. Parent-offspring recognition in Tree Swallows, *Tachycineta bicolor.* *Animal Behaviour* 54(5):1107–1116.

Leonard, M.L. and J. Picman. 1987. The adaptive significance of multiple nest building by male Marsh Wrens. *Animal Behaviour* 35:271–277.

Leopold, A. 1933. *Game Management.* C. Scribner's Sons, New York. 481 pp.

Leopold, A. 1949. *A Sand County Almanac.* Oxford University Press, New York. 226 pp.

Lewis, J.C. 1995. Whooping Crane *(Grus americana).* In *The Birds of North America,* No. 153. (Edited by A. Poole and F. Gill). The Academy of Natural Sciences, Philadelphia, PA and The American Ornithologists' Union, Washington, D.C.

Lickliter, R. and T.B. Hellewell. 1992. Contextual determinants of auditory learning in bobwhite quail embryos and hatchlings. *Developmental Psychobiology* 25(1):17–31.

Lill, A. 1974. Sexual behavior of the lek-forming White-bearded Manakin *(Manacus manacus trinitatis* Hartert). *Zeitschrift für Tierpsychologie* 36:1–36.

Lima, S.L. 1985. Maximizing feeding efficiency and minimizing time exposed to predators: a trade-off in the Black-capped Chickadee *(Parus atricapillus).* *Oecologia* 66(1):60–67.

Limpert, R.J. and S.L. Earnst. 1994. Tundra Swan *(Cygnus columbianus).* In *The Birds of North America,* No. 89. (Edited by A. Poole and F. Gill). The Academy of Natural Sciences, Philadelphia, PA and The American Ornithologists' Union, Washington, D.C.

Lindell, C. 1996. Benefits and costs to Plain-fronted Thornbirds *(Phacellodomus rufifrons)* of interactions with avian nest associates. *Auk* 113(3):565–577.

Linnaeus, C. 1758. *Systema Naturae.* Impensis L. Salvii, Holmiae.

Lipske, M. 1982. To down detective Roxie Laybourne, a feather's the clue. *Smithsonian* 12(12):98–105.

Liversidge, R. 1971. The biology of the Jacobin Cuckoo *Clamator jacobinus.* In *Proceedings of the 3rd Pan-African Ornithological Conference,* 1969, pp. 117–137. (Edited by G.L. Maclean). Ostrich Supplement, No. 8.

Lockwood, J.A. and L.D. DeBrey. 1990. A solution for the sudden and unexplained extinction of the Rocky Mountain grasshopper (Orthoptera: Acrididae). *Environmental Entomology* 19:1194–1205.

Longfellow, H.W. 1893. The Birds of Killingworth, from "The Poet's Tale". In *Tales of a Wayside Inn, The Complete Poetical Works of Longfellow,* p. 242. (Edited by H.E. Scudder). Houghton Mifflin, Boston.

Lorant, S. 1946. *The New World; The First Pictures of America, made by John White and Jacques LeMoyne and engraved by Theodore DeBry, with Contemporary Narratives of the Huguenot Settlement in Florida, 1562–1565, and the Virginia Colony, 1585–1590.* Duell, Sloan & Pearce, New York. 292 pp.

Lorenz, K.Z. 1952. *King Solomon's Ring: New Light on Animal Ways.* Thomas Y. Crowell Co., New York.

Lorenz, K.Z. 1970. Companions as factors in the bird's environment. In *Studies on Animal and Human Behavior,* Vol. 1, pp. 101–258. Harvard University Press, Cambridge, MA. Translated by Robert Martin.

Lotem, A. 1992. Rejection of cuckoo eggs in relation to host age: a possible evolutionary equilibrium. *Behavioral Ecology* 3(2):128–132.

Loud, L.L. and M.R. Harrington. 1929. *Lovelock Cave.* University of California Press, Berkeley, CA. 183 pp.

Lowther, P.E. 1993. Brown-headed Cowbird *(Molothrus ater).* In *The Birds of North America,* No. 47. (Edited by A. Poole, and F. Gill). The Academy of Natural Sciences, Philadelphia, PA and The American Ornithologists' Union, Washington, D.C.

Lucas, A.M. and P.R. Stettenheim. 1972. *Avian Anatomy: Integument, Parts I and II.* Agriculture Handbook 362. U.S. Government Printing Office, Washington, D.C. 750 pp.

Lutgens, F.K. and E.J. Tarbuck. 1998. *The Atmosphere: An Introduction to Meteorology,* 7th Edition. Prentice Hall, Upper Saddle River, NJ.

Lynch, A. 1996. The population memetics of bird song. In *Ecology and Evolution of Acoustic Communication in Birds,* pp. 181–197. (Edited by D.E. Kroodsma and E.H. Miller). Cornell University Press, Ithaca, NY.

Lyon, B.E., R.D. Montgomerie, and L.D. Hamilton. 1987. Male parental care and monogamy in Snow Buntings. *Behavioral Ecology and Sociobiology* 20:372–382.

MacArthur, R.H. 1958. Population ecology of some warblers of northeastern forests. *Ecology* 39:599–619.

MacArthur, R.H. and E.O. Wilson. 1967. *The Theory of Island Biogeography.* Princeton University Press, Princeton, NJ. 203 pp.

Maclean, G.L. 1983. Water transport in sandgrouse. *Bioscience* 33(6):365–369.

MacPhee, R.D.E. and P.A. Marx. 1997. The 40,000-year plague: humans, hyperdiseases, and first-contact extinction. In *Natural Change and Human Impact in Madagascar,* pp. 169–217. (Edited by S.M. Goodman and B.D. Patterson). Smithsonian Institution Press, Washington, D.C.

Mader, S. 1988. *Inquiry into Life, 5th Edition.* William C. Brown Publishers, Dubuque, IA. 802 pp.

Mark, D. and B.J. Stutchbury. 1994. Response of a forest-interior songbird to the threat of cowbird parasitism. *Animal Behaviour* 47(2):275–280.

Marler, P. 1955. Characteristics of some alarm calls. *Nature* 176:6–8.

Marler, P., P. Mundinger, M.S. Waser, and A. Lutjen. 1972. Effects of acoustical stimulation and deprivation on song development in Red-winged Blackbirds *(Agelaius phoeniceus).* *Animal Behaviour* 20:586–606.

Marsh, O.C. 1880. *Odontornithes: A Monograph on the Extinct Toothed Birds of North America.* Report of the U.S. Geological Exploration of the Fortieth Parallel, No. 7. Washington, D.C.

Marsh, O.C. 1896. *The Dinosaurs of North America.* Sixteenth Annual Report of the U.S. Geological Survey. Washington, D.C. 414 pp.

Marshall, A.J., ed. 1961. *Biology and Comparative Physiology of Birds, Vol. 2.* Academic Press, New York.

Martens, J. 1996. Vocalizations and speciation of Palearctic birds. In *Ecology and Evolution of Acoustic Communication in Birds,* pp. 221–240. (Edited by D.E. Kroodsma and E.H. Miller). Cornell University Press, Ithaca, NY.

Martin, L.D. and A.M. Neuner. 1978. The end of the Pleistocene in North America. *Transactions of the Nebraska Academy of Science* 6:117–126.

Martin, P.S. and D.W. Steadman. 1999. Prehistoric extinctions on islands and continents. In *Extinction in Near Time: Causes, Contexts, and Consequences,* pp. 17–55. (Edited by R.D.E. MacPhee). Kluwer Academic/Plenum Publishers, New York.

Martin, T.E. 1987. Food as a limit on breeding birds: a life history perspective. *Annual Review of Ecology and Systematics* 18:453–487.

Martin, T.E. 1993a. Nest predation among vegetation layers and habitat types: revising the dogmas. *American Naturalist* 141:897–913.

Martin, T.E. 1993b. Evolutionary determinants of clutch size in cavity-nesting birds: nest predation or limited breeding opportunities? *American Naturalist* 142:937–946.

Mayfield, H. 1960. *The Kirtland's Warbler.* Cranbrook Institute of Science Bulletin No. 40. Cranbrook Institute of Science, Bloomfield Hills, MI. 242 pp.

Mayr, E. and J. Bond. 1943. Notes on the generic classification of the swallows. *Ibis* 85:334–341.

McCarty, J.P. 1995. *Effects of Short-term Changes in Environmental Conditions on the Foraging Ecology and Reproductive Success of Tree Swallows.* Ph.D. Dissertation, Cornell University, Ithaca, NY.

McCarty, J.P. and D.W. Winkler. 1999. Relative importance of environmental variables in determining the growth of nestling Tree Swallows *Tachycineta bicolor. Ibis* 141:286–296.

McDonald, D.B. 1989. Cooperation under sexual selection: age-graded changes in a lekking bird. *American Naturalist* 134(5):709–730.

McDonald, D.B. and W.K. Potts. 1994. Cooperative display and relatedness among males in a lek-mating bird. *Science* 266(5187):1030–1032.

McGowan, K.J. 2001. Demographic and behavioral comparisons of suburban and rural American Crows. In *Avian Ecology and Conservation in an Urbanizing World*, pp. 365–381. (Edited by J.M. Marzluff, R. Bowman, and R. Donnelly). Kluwer Academic Press, Norwell, MA.

McGowan, K.J. and G.E. Woolfenden. 1989. A sentinel system in the Florida Scrub Jay. *Animal Behaviour* 37(6):1000–1006.

McGregor, P.K. and T. Dabelsteen. 1996. Communication networks. In *Ecology and Evolution of Acoustic Communication in Birds*, pp. 409–425. (Edited by D.E. Kroodsma and E.H. Miller). Cornell University Press, Ithaca, NY.

McGregor, P.K., J.R. Krebs, and C.M. Perrins. 1981. Song repertoires and lifetime reproductive success in the Great Tit *(Parus major)*. *American Naturalist* 118:149–159.

Medvin, M.B. and M.D. Beecher. 1986. Parent-offspring recognition in the Barn Swallow *(Hirundo rustica)*. *Animal Behaviour* 34(6):1627–1639.

Merriam, F.A. 1889. *Birds Through an Opera Glass*. Houghton, Mifflin and Company, Boston and New York. 223 pp.

Merriam, F.A. 1898. *Birds of Village and Field: A Bird Book for Beginners*. Houghton, Mifflin and Company, Boston and New York. 406 pp.

Middleton, A.L. 1993. American Goldfinch *(Carduelis tristis)*. In *The Birds of North America*, No. 80. (Edited by A. Poole and F. Gill). The Academy of Natural Sciences, Philadelphia, PA and The American Ornithologists' Union, Washington, D.C.

Middleton, A.L. 1998. Chipping Sparrow *(Spizella passerina)*. In *The Birds of North America*, No. 334. (Edited by A. Poole and F. Gill). The Birds of North America, Inc., Philadelphia, PA.

Miller, O.T. 1899. *The First Book of Birds*. Houghton, Mifflin and Company, Boston and New York. 149 pp.

Mock, D.W. 1976. Pair formation displays of the Great Blue Heron. *Wilson Bulletin* 88(2):185–230.

Mock, D.W., H. Drummond, and C.H. Stinson. 1990. Avian siblicide. *American Scientist* 78:438–449.

Mock, D.W. and B.J. Ploger. 1987. Parental manipulation of optimal hatch asynchrony in Cattle Egrets: an experimental study. *Animal Behaviour* 35:150–160.

Møller, A.P. 1992. Female swallow sexual preference for symmetrical male sexual ornaments. *Nature* 357(6375):238–240.

Møller, A.P. 1994. *Sexual Selection and the Barn Swallow*. Oxford University Press, Oxford, England.

Morse, D.H. 1980. Foraging and coexistence of spruce-woods warblers. *Living Bird* 18:7–25.

Morse, D.H. 1989. *American Warblers: An Ecological and Behavioral Perspective*. Harvard University Press, Cambridge, MA. 406 pp.

Morton, E.S. 1976. Vocal mimicry in the Thick-billed Euphonia. *Wilson Bulletin* 88:485–487.

Morton, E.S. 1982. Grading, discreteness, redundancy, and motivation-structural rules. In *Acoustic Communication in Birds*, Vol. 1, pp. 183–212. (Edited by D.E. Kroodsma and R.H. Miller). Academic Press, New York.

Morton, E.S. 1996. A comparison of vocal behavior among tropical and temperate passerine birds. In *Ecology and Evolution of Acoustic Communication in Birds*, pp. 258–268. (Edited by D.E. Kroodsma and E.H. Miller). Cornell University Press, Ithaca, NY.

Mountjoy, J. and R.J. Robertson. 1988. Why are waxwings "waxy"? Delayed plumage maturation in the Cedar Waxwing. *Auk* 105:61–69.

Muir, J. 1961. *The Mountains of California*. Doubleday, Garden City, NY. 300 pp.

Mulder, R.A. 1994. Faithful philanderers. *Natural History* 103(11):56–63.

Mundinger, P.C. 1995. Behaviour-genetic analysis of canary song: inter-strain differences in sensory learning, and epigenetic rules. *Animal Behaviour* 50:1491–1511.

Munn, C.A. 1986. Birds that "cry wolf." *Nature* 319:143–145.

Munro, J. and J. Bedard. 1977. Gull predation and creching behaviour in the Common Eider. *Journal of Animal Ecology* 46:799–810.

Murphy, M.E., B.T. Miller, and J.R. King. 1989. A structural comparison of fault bars with feather defects known to be nutritionally induced. *Canadian Journal of Zoology* 67:1311–1317.

Murray, B.G., Jr. and V. Nolan, Jr. 1989. The evolution of clutch size. I. An equation for predicting clutch size. *Evolution* 43:1699–1705.

Murray, P.F. and D. Megirian. 1998. The skull of Dromornithid birds: anatomical evidence for their relationship to Anseriformes. *Records of the South Australian Museum* 31(1):51–97.

Necker, R. 1985. Receptors in the skin of the wing of the pigeon and their possible role in bird flight. In *Bird Flight*, pp. 433–444. (Edited by W. Nachtigall). Gustav Fischer Verlag, Stuttgart.

Neuchterlein, G.L. and R.W. Storer. 1982. The pair-formation displays of the Western Grebe (Weed Ceremony of the Grebe). *Condor* 84:351–369.

Nevitt, G. 1999. Foraging by seabirds on an olfactory landscape. *American Scientist* 87:46–53.

Newman, G.A. 1970. Cowbird parasitism and nesting success of Lark Sparrows in southern Oklahoma. *Wilson Bulletin* 82:304–309.

Newton, I. 1979. *Population Ecology of Raptors*. T. & A.D. Poyser, Berkhamsted, England.

Newton, I. 1998. *Population Limitation in Birds*. Academic Press, San Diego, CA. 597 pp.

Nichols, J.D. and F.A. Johnson. 1990. Wood Duck population dynamics: a review. In *Proceedings of the 1988 North American Wood Duck Symposium*. (Edited by L.H. Fredrickson, G. Burger, S. Havena, D. Graber, R. Kirby, and T. Taylor). St. Louis, MO.

Nice, M.M. 1937. Studies in the Life History of the Song Sparrow. I. A Population Study of the Song Sparrow. *Transactions of the Linnaean Society of New York* 4:1–127.

Nice, M.M. 1943. Studies in the Life History of the Song Sparrow. II. The Behavior of the Song Sparrow and Other Passerines. *Transactions of the Linnaean Society of New York* 6:1–238.

Nielsen, B.M.B. and S.L. Vehrencamp. 1995. Responses of Song Sparrows to song-type matching via interactive playback. *Behavioral Ecology and Sociobiology* 37:109–117.

Nisbet, I.C.T. 1977. Courtship, feeding, and clutch size in Common Terns, *Sterna hirundo*. In *Evolutionary Ecology*, pp. 101–109. (Edited by B. Stonehouse and C. Perrins). University Park Press, Baltimore, MD.

Noble, G.K. 1936. Courtship and sexual selection of the flicker *(Colaptes auratus luteus)*. *Auk* 53:269–282.

Noon, B.R. 1981. The distribution of an avian guild along a temperate elevational gradient: the importance and expression of competition. *Ecological Monographs* 51(1):105–124.

Norberg, R.A. 1978. Skull asymmetry, ear structure and function, and auditory localization in Tengmalm's Owl, *Aegolius funereus* (Linne). *Philosophical Transactions of the Royal Society of London (Bio)* 282:325–410.

Norberg, U.M. 1990. *Vertebrate Flight*. Springer-Verlag, Berlin.

Nottebohm, F. 1987. Plasticity in adult avian central nervous system: possible relation between hormones, learning, and brain repair. In *Handbook of Physiology, Section 1*. (Edited by F. Plum). Williams and Wilkins, Baltimore, MD.

Nowicki, S. 1989. Vocal plasticity in captive Black-capped Chickadees: the acoustic basis and rate of call convergence. *Animal Behaviour* 37:64–73.

Nowicki, S. and P. Marler. 1988. How do birds sing? *Music Perception* 5(3):391–426.

Ohmart, R.D. 1994. The effects of human-induced changes on the avifauna of western riparian habitats. In *Studies in Avian Biology*, No. 15, pp. 273–285. (Edited by J.R. Jehl, Jr. and N.K. Johnson). Cooper Ornithological Society, Camarillo, CA.

Orians, G.H. 1969. On the evolution of mating systems in birds and mammals. *American Naturalist* 103:589–603.

Orians, G.H. and G.M. Christman. 1968. *A Comparative Study of the Behavior of Red-winged, Tricolored, and Yellow-headed Blackbirds*. University of California Publications in Zoology, Vol. 84. University of California Press, Berkeley, CA.

Oring, L.W. 1985. Avian polyandry. In *Current Ornithology*, Vol. 3, pp. 309–351. (Edited by R.F. Johnston). Plenum Press, New York.

Ortega, J.C. and M. Bekoff. 1987. Avian play: comparative evolutionary and developmental trends. *Auk* 104:338–341.

Osborne, D.R. and A.T. Peterson. 1984. Decline of the Upland Sandpiper (*Bartramia longicauda*) in Ohio: an endangered species. *Ohio Journal of Science* 84:8–10.

Ostrom, J.H. 1973. The ancestry of birds. *Nature* 242:136.

Ostrom, J.H. 1976. Some hypothetical anatomical stages in the evolution of avian flight. *Smithsonian Contributions to Paleobiology* 27:1–21.

Ostrom, J.H. 1978. The osteology of *Compsognathus longipes* Wagner. *Zitteliana* 4:73–118.

Ottoson, U., J. Bäckman, and H.G. Smith. 1997. Begging affects parental effort in the Pied Flycatcher, *Ficedula hypoleuca*. *Behaviorial Ecology and Sociobiology* 41:381–384.

Owen-Smith, N. 1987. Pleistocene extinctions: the pivotal role of megaherbivores. *Paleobiology* 13:351–362.

Owens, O.S. and D.D. Chiras. 1990. *Resource Conservation: An Ecological Approach, 5th Edition*. Macmillan Publishing Co., New York.

Palmer, J.D. 1966. How a bird tells the time of day. *Natural History* 75:48–53.

Papi, F., P. Ioale, P. Dall'Antonia, and S. Benvenuti. 1991. Homing strategies of pigeons investigated by clock shift and flight path reconstruction. *Naturwissenschaften* 78(8):370–373.

Papi, F. and H.G. Wallraff. 1992. Birds. In *Animal Homing*, pp. 263–319. (Edited by F. Papi). Chapman and Hall, London.

Parmelee, D. 1992. Snowy Owl (*Nyctea scandiaca*). In *The Birds of North America*, No. 10. (Edited by A. Poole, P. Stettenheim, and F. Gill). The Academy of Natural Sciences, Philadelphia, PA and The American Ornithologists' Union, Washington, D.C.

Patten, B.M. 1971. *Early Embryology of the Chick, 5th Edition*. McGraw-Hill, Inc., New York.

Payne, R.B. 1973. *Behavior, Mimetic Songs and Song Dialects, and Relationships of the Parasitic Indigobirds (Vidua) of Africa*. Ornithological Monographs, No. 11. American Ornithologists' Union, Anchorage, KY. 333 pp.

Payne, R.B. 1977. The ecology of brood parasitism in birds. *Annual Review of Ecology and Systematics* 8:1–28.

Payne, R.B. 1989. Indigo Bunting. In *Lifetime Reproduction in Birds*, pp. 153–172. (Edited by I. Newton). Academic Press, London.

Payne, R.B. 1996. Song traditions in Indigo Buntings: origin, improvisation, dispersal, and extinction. In *Ecology and Evolution of Acoustic Communication in Birds*, pp. 198–220. (Edited by D.E. Kroodsma and E.H. Miller). Cornell University Press, Ithaca, NY.

Paynter, R.A., Jr. 1966. A new attempt to construct life tables for Kent Island Herring Gulls. *Bulletin of the Museum of Comparative Zoology* 133:489–528.

Peakall, D.B. and L.F. Kiff. 1988. DDE contamination in peregrines and American Kestrels and its effects on reproduction. In *Peregrine Falcon Populations: Their Management and Recovery*, pp. 337–350. (Edited by T.J. Cade, J.H. Enderson, C.F. Thelander, and C.M. White). The Peregrine Fund, Boise, ID.

Pearson, O.P. 1953. The metabolism of hummingbirds. *Scientific American* 188(1):69–72.

Pennant, T. 1784–1785. *Arctic Zoology*. Printed by H. Hughes, London.

Pepperberg, I.M. 1990. Some cognitive capacities of an African Grey Parrot. In *Advances in the Study of Behavior*, Vol. 19, pp. 357–409. (Edited by P.J.B. Slater, J.S. Rosenblatt, and C. Beer). Academic Press, Inc., Troy, MD.

Perdeck, A.C. 1958. Two types of orientation in migrating starlings, *Sturnus vulgaris* L., and Chaffinches, *Fringilla coelebs* L., as revealed by displacement experiments. *Ardea* 46:1–37.

Perrins, C.M. 1970. The timing of birds' breeding seasons. *Ibis* 112(2):242–255.

Peterjohn, B.G., J.R. Sauer, and C.S. Robbins. 1995. Population trends from the North American Breeding Bird Survey. In *Ecology and Management of Neotropical Migratory Birds: A Synthesis of the Critical Issues*, pp. 3–39. (Edited by T.E. Martin and D.M. Finch). Oxford University Press, New York.

Peterson, R.T. 1934. *A Field Guide to the Birds: Giving Field Marks of all Species Found in Eastern North America*. Houghton, Mifflin and Company, Boston and New York. 167 pp.

Peterson, R.T. 1939. *A Field Guide to the Birds: Giving Field Marks of all Species Found in Eastern North America*. Houghton, Mifflin and Company, Boston and New York. 180 pp.

Peterson, R.T. 1980. *A Field Guide to the Birds: A Completely New Guide to all the Birds of Eastern and Central North America, 4th Edition*. Houghton Mifflin, Boston, MA. 384 pp.

Petrie, M. and T. Halliday. 1994. Experimental and natural changes in the peacock's (*Pavo cristatus*) train can affect mating success. *Behavioral Ecology and Sociobiology* 35:213–217.

Pettingill, O.S., Jr. 1972. *Seminars in Ornithology: A Home Study Course in Bird Biology*. Cornell Laboratory of Ornithology, Cornell University, Ithaca, NY.

Poole, E.L. 1938. Weights and wing areas in North American birds. *Auk* 55:511–517.

Post, W. and J.W. Wiley. 1992. The head-down display in Shiny Cowbirds and its relation to dominance behavior. *Condor* 94:999–1002.

Potts, W.K. 1984. The chorus line hypothesis of maneuver coordination in avian flocks. *Nature* 309(5966):344–345.

Pough, F.H., J.B. Heiser, and W.N. McFarland. 1996. *Vertebrate Life, 4th Edition*. Prentice-Hall, Inc., Upper Saddle River, NJ. 798 pp.

Pough, F.H., C.M. Janis, and J.B. Heiser. 1999. *Vertebrate Life, 5th Edition*. Prentice-Hall, Inc., Upper Saddle River, NJ. 773 pp.

Pregill, G.K., D.W. Steadman, and D.R. Watters. 1994. *Late Quaternary Vertebrate Faunas of the Lesser Antilles: Historical Components of Caribbean Biogeography*. Bulletin of the Carnegie Museum of Natural History, No. 30. Carnegie Museum of Natural History, Pittsburgh, PA. 51 pp.

Preston, F.W. 1953. Shapes of bird eggs. *Auk* 70(2):160–182.

Price, T., M. Kirkpatrick, and S.J. Arnold. 1988. Directional selection and the evolution of breeding date in birds. *Science* 248(4853):798–799.

Proctor, N.S. and P.J. Lynch. 1993. *Manual of Ornithology: Avian Structure and Function*. Yale University Press, New Haven and London. 340 pp.

Pugesek, B.H. 1981. Increased reproductive effort with age in the California Gull (*Larus californicus*). *Science* 212:822–823.

Pugesek, B.H. and P. Wood. 1992. Alternate reproductive strategies in the California Gull. *Evolutionary Ecology* 6(4):279–295.

Pycraft, W.P. 1910. *A History of Birds*. Methuen and Co., London. 458 pp.

Pye, J.D. and W.R. Langbauer, Jr. 1998. Ultrasound and infrasound. In *Animal Acoustic Communication: Sound Analysis and Research Methods*, pp. 221–250. (Edited by S.L. Hopp, M.J. Owren, and C.S. Evans). Springer-Verlag, New York and Berlin.

Pyle, P., S.N.G. Howell, R.P. Yunick, and D.F. DeSante. 1987. *Identification Guide to North American Passerines*. Slate Creek Press, Bolinas, CA. 273 pp.

Rabinowitz, D., S. Cairns, and T. Dillon. 1986. Seven forms of rarity and their frequency in the flora of the British Isles. In *Conservation Biology: The Science of Scarcity and Diversity*, pp. 182–204. (Edited by M.E. Soulé). Sinauer Associates, Inc., Sunderland, MA.

Rahn, H. and A. Ar. 1974. The avian egg: incubation time and water loss. *Condor* 76:147–152.

Rappole, J.H. 1995. *The Ecology of Migrant Birds: A Neotropical Perspective*. Smithsonian Institution Press, Washington and London. 269 pp.

Ratcliffe, D.A. 1967. Decrease in eggshell weight in certain birds of prey. *Nature* 215:208–210.

Raveling, D.G. 1970. Dominance relationships and agonistic behavior of Canada Geese in winter. *Behaviour* 37(3–4):291–319.

Rayner, J.M.V. 1985. Mechanical and ecological constraints on flight evolution. In *The Beginnings of Birds*, pp. 279–288. (Edited by M.K. Hecht, J.H. Ostrom, G. Viohl, and P. Wellnhofer). Freunde des Jura-Museum, Eichstätt.

Rensch, B. 1947. *Neure Probleme der Abstammungslehne*. Ferdinand Enke Verlag, Stuttgart.

Reyer, H.-U. 1984. Investment and relatedness: a cost/benefit analysis of breeding and helping in the Pied Kingfisher. *Animal Behaviour* 32:1163–1178.

Reynolds, J.D. and M.R. Gross. 1990. Costs and benefits of female mate choice: is there a lek paradox? *American Naturalist* 136:230–243.

Richardson, W.J. 1976. Autumn migration over Puerto Rico and the western Atlantic: a radar study. *Ibis* 118(3):309–332.

Ricklefs, R.E. 1973a. *Ecology*. Chiron Press, Newton, MA. 861 pp.

Ricklefs, R.E. 1973b. Fecundity, mortality and avian demography. In *Breeding Biology of Birds*, pp. 366–435. (Edited by D.S. Farner). National Academy of Sciences, Washington, D.C.

Ricklefs, R.E. 1979. *Ecology, 2nd Edition*. Chiron Press, New York. 966 pp.

Ricklefs, R.E. 1980a. Geographical variation

in clutch size among passerine birds: Ashmole's hypothesis. *Auk* 97:38–49.

Ricklefs, R.E. 1980b. Postnatal development of Leach's Storm-Petrel *Oceanodroma leucorhoa. Auk* 97(4):768–781.

Ridgely, R.S. and J.A. Gwynne. 1989. *A Guide to the Birds of Panama*. Princeton University Press, New Jersey.

Ridgway, R. 1887. *A Manual of North American Birds*. J.B. Lippincott & Co., Philadelphia, PA.

Ridgway, R. 1901–1919. *Birds of Middle and North America*. Government Printing Office, Washington, D.C.

Rising, J.D. 1983. The Great Plains hybrid zones. In *Current Ornithology*, Vol. 1, pp. 131–157. (Edited by R.F. Johnston). Plenum Press, New York and London.

Robbins, C.S., B. Bruun, and H.S. Zim. 1966. *Birds of North America, A Guide to Field Identification*. Golden Press, New York. 340 pp.

Robbins, C.S., D. Bystrak, and P.H. Geissler. 1986. *The Breeding Bird Survey: Its First Fifteen Years, 1965–1979*. U.S. Fish and Wildlife Service Resource Publication 157. U.S. Fish and Wildlife Service, Washington, D.C.

Robbins, C.S., D.K. Dawson, and B.A. Dowell. 1989. Habitat area requirements of breeding forest birds of the middle Atlantic states. *Wildlife Monographs* 103:1–34.

Robbins, C.S., J.R. Sauer, R.W. Greenberg, and S. Droege. 1989. Population declines in North American birds that migrate to the Neotropics. *Proceedings of the National Academy of Sciences* 86:7658–7662.

Robbins, S.D., Jr. 1991. *Wisconsin Birdlife: Population and Distribution Past and Present*. University of Wisconsin Press, Madison, WI. 702 pp.

Roberts, B. 1934. Notes on the birds of central and south-east Iceland with special reference to food habits. *Ibis* 13:239–264.

Robidoux, Y.P. and A. Cyr. 1989. Sélection granulométrique pour la construction du nid chez l'Hirondelle à front blanc, *Hirundo pyrrhonota. Canadian Field-Naturalist* 103(4):577–583.

Robinson, S.K. 1992. Population dynamics of breeding Neotropical migrants in a fragmented Illinois landscape. In *Ecology and Conservation of Neotropical Migrant Landbirds*, pp. 408–418. (Edited by J.M. Hagan III and D.W. Johnston). Smithsonian Institution Press, Washington, D.C.

Robinson, S.K., F.R. Thompson, III, T.M. Donovan, D.R. Whitehead, and J. Faaborg. 1995. Regional forest fragmentation and the nesting success of migratory birds. *Science* 267:1987–1990.

Robinson, S.K. and D.S. Wilcove. 1994. Forest fragmentation in the temperate zone and its effects on migratory songbirds. *Bird Conservation International* 4(2–3):233–249.

Rockwell, R.F., E.G. Cooch, and S. Brault. 1997. Dynamics of the mid-continent population of Lesser Snow Geese: projected impacts of reductions in survival and fertility on population growth rates. In *Arctic Ecosystems in Peril: Report of the Arctic Goose Habitat Working Group*, pp. 73–100. (Edited by B.D.J. Batt). Arctic Goose Joint Venture Special Publication, U.S. Fish and Wildlife Service, Washington, D.C., and Canadian Wildlife Service, Ottawa, Ontario.

Rodewald, P.G. and R.D. James. 1996. Yellow-throated Vireo *(Vireo flavifrons).* In *The Birds of North America*, No. 247. (Edited by A. Poole and F. Gill). The Academy of Natural Sciences, Philadelphia, PA and The American Ornithologists' Union, Washington, D.C.

Rohwer, F.C. and S. Freeman. 1989. The distribution of conspecific nest parasitism in birds. *Canadian Journal of Zoology* 67:239–253.

Rohwer, S., S.D. Fretwell, and D.M. Niles. 1980. Delayed maturation in passerine plumages and the deceptive acquisition of resources. *American Naturalist* 115:400–437.

Romanoff, A.L. and A.J. Romanoff. 1949. *The Avian Egg*. John Wiley and Sons, New York. 918 pp.

Romer, A.S. 1966. *Vertebrate Paleontology, 3rd Edition*. University of Chicago Press, Chicago.

Root, T. 1988. Energy constraints on avian distributions and abundances. *Ecology* 69(2):330–339.

Roseman, M. 1991. *Healing Sounds from the Malaysian Rainforest: Temiar Music and Medicine*. University of California Press, Berkeley and Los Angeles, CA. 233 pp.

Rosenberg, K.V., R.W. Rohrbaugh, Jr., S.E. Barker, J.D. Lowe, R.S. Hames, and A.A. Dhondt. 1999. *A Land Manager's Guide to Improving Habitat for Scarlet Tanagers and other Forest-interior Birds*. The Cornell Lab of Ornithology, Ithaca, NY.

Rosenfield, R.N., J. Bielefeldt, and K.R. Nolte. 1992. Eye color of Cooper's Hawks breeding in Wisconsin. *Journal of Raptor Research* 26(3):189–191.

Røskaft, E. 1985. The effect of enlarged brood size on the future reproductive potential of the Rook *Corvus frugilegus. Journal of Animal Ecology* 54(1):255–260.

Rothstein, S.I., M.A. Patten, and R.C. Fleischer. 2002. Phylogeny, specialization, and brood parasite-host coevolution: some possible pitfalls of parsimony. *Behavioral Ecology* 13:1–10.

Rothstein, S.I. and S.K. Robinson. 1998. *Parasitic Birds and their Hosts*. Oxford University Press, New York. 444 pp.

Rowley, I. 1971. The use of mud in nest-building—a review of the incidence and taxonomic importance. In *Proceedings of the 3rd Pan-African Ornithological Conference, 1969*, pp. 139–148. (Edited by G.L. Maclean). Ostrich Supplement, No. 8.

Rowley, I. 1980. Parent offspring recognition in a cockatoo, the Galah *(Cacatua roseicapilla). Australian Journal of Zoology* 28(3):445–456.

Royama, T. 1966. Factors governing feeding rate, food requirement and brood size of nestling Great Tits, *Parus major. Ibis* 108:313–347.

Ruppell, G. 1975. *Bird Flight*. Van Nostrand Reinhold Co., New York. 191 pp.

Russell, D.A. 1972. Ostrich dinosaurs from the late Cretaceous of western Canada. *Canadian Journal of Earth Sciences* 9:375–402.

Ryan, M.J. 1997. Sexual selection and mate choice. In *Behavioral Ecology, An Evolutionary Approach, 4th Edition*, pp. 179–202. (Edited by J.R. Krebs and N.B. Davies). Blackwell, Oxford.

Sallabanks, R. and F.C. James. 1999. American Robin *(Turdus migratorius).* In *The Birds of North America*, No. 462. (Edited by A. Poole and F. Gill). The Birds of North America, Inc., Philadelphia, PA.

Sargent, S. 1993. Nesting biology of the Yellow-throated Euphonia: large clutch size in a Neotropical frugivore. *Wilson Bulletin* 105(2):285–300.

Sauer, E.G.F. 1957. Die Sternenorientieurung nächtlich ziehender Grasmücken, *Sylvia atricapilla, borin* und *curruca. Zeitschrift für Tierpsychologie* 14:29–70.

Savalli, U.M. 1995. The evolution of bird coloration and plumage elaboration: a review of hypotheses. In *Current Ornithology*, Vol. 12, pp. 141–190. (Edited by D.M. Power). Plenum Press, New York.

Savard, J.-P.L. 1987. Causes and functions of brood amalgamation in Barrow's Goldeneye and Bufflehead. *Canadian Journal of Zoology* 65:1548–1553.

Schenkeveld, L.E. and R.C. Ydenberg. 1985. Synchronous diving by Surf Scoter *(Melanitta perspicillata)* flocks. *Canadian Journal of Zoology* 63(11):2516–2519.

Schermuly, L. and R. Klinke. 1990. Infrasound sensitive neurons in the pigeon cochlear ganglion. *Journal of Comparative Physiology A* 166:355–363.

Schmidt-Koenig, K. 1960. Internal clocks and homing. *Cold Spring Harbor Symposia on Quantitative Biology* 25:389–393.

Schmidt-Koenig, K. and H.J. Schlichte. 1972. Homing in pigeons with impaired vision. *Proceedings of the National Academy of Sciences of the United States of America* 69:2446–2447.

Schmidt-Nielsen, K. 1959. Salt Glands. *Scientific American* 200(1):109–116.

Schorger, A.W. 1955. *The Passenger Pigeon: Its Natural History and Extinction*. University of Wisconsin Press, Madison, WI. 424 pp.

Schultz, H.-P. and L. Trueb, eds. 1991. *Origins of the Higher Groups of Tetrapods: Controversy and Consensus*. Cornell University Press, Ithaca, NY.

Scott, D.M. and C.D. Ankney. 1980. Fecundity of the Brown-headed Cowbird in southern Ontario. *Auk* 97:677–683.

Sealy, S.G. 1995. Burial of cowbird eggs by parasitized Yellow Warblers—an empirical and experimental study. *Animal Behaviour* 49:877–889.

Searcy, W.A. and K. Yasukawa. 1983. Sexual selection and Red-winged Blackbirds. *American Scientist* 71:166–174.

Searcy, W.A. and K. Yasukawa. 1996. Song and female choice. In *Ecology and Evolution of Acoustic Communication in Birds*, pp. 454–473. (Edited by D.E. Kroodsma and E.H. Miller). Cornell University Press, Ithaca, NY.

Seel, D.C. 1973. Egg-laying by the cuckoo. *British Birds* 66:528–535.

Semel, B. and P.W. Sherman. 2001. Intraspecific parasitism and nest-site competition in Wood Ducks. *Animal Behaviour* 61(4):787–803.

Serventy, D.L. 1967. Aspects of the population ecology of the Short-tailed Shearwater, *Puffinus tenuirostris.* Proceedings of the XIVth International Ornithological Congress, Blackwell Scientific Publications, Oxford. pp. 165–190.

Shackelford, C. E., R. E. Brown, and R. N. Conner. 2000. Red-bellied Woodpecker (*Melanerpes carolinus*). In *The Birds of North America*, No. 500. (Edited by A. Poole and F. Gill). The Birds of North America, Inc., Philadelphia, PA.

Shafer, C.L. 1997. Terrestrial nature reserve design at the urban/rural interface. In *Conservation in Highly Fragmented Landscapes*, pp. 345–378. (Edited by M.W. Schwartz). Chapman and Hall, Inc., New York.

Sheldon, F.H., L.A. Whittingham, and D.W. Winkler. 1999. A comparison of cytochrome b and DNA hybridization data bearing on the phylogeny of swallows (Aves: Hirundinidae). *Molecular Phylogenetics and Evolution* 11(2):320–331.

Shelford, V.E. 1945. The relation of Snowy Owl migration to the abundance of the collared lemming. *Auk* 62:592–596.

Sherman, P.W. 1977. Nepotism and the evolution of alarm calls. *Science* 197:1246–1253.

Sherry, T.W. and R.T. Holmes. 1992. Population fluctuations in a long-distance Neotropical migrant: demographic evidence for the importance of breeding season events in the American Redstart. In *Ecology and Conservation of Neotropical Migrant Landbirds*, pp. 431–442. (Edited by J.M. Hagan III and D.W. Johnston). Smithsonian Institution Press, Washington, D.C.

Sherwood, G.A. 1967. Behavior of family groups of Canada Geese. *Transactions of the North American Wildlife and Natural Resources Conference* 32:340–355.

Shields, W.M. 1984. Barn Swallow (*Hirundo rustica*) mobbing: self defense, collateral kin defense, group defense, or parental care? *Animal Behaviour* 32(1):132–148.

Shuz, E. 1940. Bericht der Vogelwarte Rossitten der Kaiser Wilhelm-Gessellschaft zur Forderung der Wissenchafter (April 1938 bis June 1940). *Vogelzug* 11:109–120.

Sibley, C.G. and J.E. Ahlquist. 1985. The phylogeny and classification of the Australo-Papuan passerine birds. *Emu* 85:1–14.

Sibley, C.G. and J.E. Ahlquist. 1990. *Phylogeny and Classification of Birds: A Study in Molecular Evolution.* Yale University Press, New Haven.

Sibley, C.G. and B.L. Monroe. 1990. *Distribution and Taxonomy of Birds of the World.* Yale University Press, New Haven.

Simmons, K.E.L. 1986. *The Sunning Behaviour of Birds: A Guide for Ornithologists.* Bristol Ornithological Club, Bristol, England.

Simon, H. 1971. *The Splendor of Iridescence: Structural Colors in the Animal World.* Dodd, Mead, and Co., New York. 268 pp.

Skeate, S.T. 1985. Social play behaviour in captive White-fronted Amazon Parrots. *Bird Behaviour* 6:46–48.

Skutch, A.F. 1976. *Parents and their Young.* University of Texas Press, Austin, TX and London.

Slagsvold, T. 1982. Clutch size variation in passerine birds: the nest predation hypothesis. *Oecologia* 54:159–169.

Slagsvold, T. and J.T. Lifjeld. 1985. Variation in plumage colour of the Great Tit *Parus major* in relation to habitat, season and food. *Journal of Zoology (London)* 206:321–328.

Slinger, J. 1996. *Down & Dirty Birding.* Key Porter Books, Toronto. 240 pp.

Slip, D.J. and R. Shine. 1988. Reptilian endothermy: a field study of thermoregulation by brooding diamond pythons. *Journal of Zoology* 216:367–378.

Small, M.F. and M.L. Hunter. 1988. Forest fragmentation and avian nest predation in forested landscapes. *Oecologia* 76:62–64.

Smith, J.M., E. Stauber, and M.J. Bechard. 1993. Identification of Peregrine Falcons using a computerized classification system of toe-scale pattern analysis. *Journal of Raptor Research* 27:191–195.

Smith, N.G. 1968. The advantage of being parasitized. *Nature* 219:690–694.

Smith, R.L. 1974. *Ecology and Field Biology, 2nd Edition.* Harper and Row Publishers, New York. 850 pp.

Smith, S.M. 1975. Innate recognition of coral snake pattern by a possible avian predator. *Science* 187:759–760.

Smith, S.M. 1988. Extrapair copulations in Black-capped Chickadees: the role of the female. *Behaviour* 107:15–23.

Smith, S.M. 1991. *The Black-capped Chickadee.* Cornell University Press, Ithaca, NY.

Smith, T.B. and S.A. Temple. 1982. Feeding habits and bill polymorphism in Hook-billed Kites. *Auk* 99:197–207.

Smith, W.J. 1969. Displays of *Sayornis phoebe* (Aves, Tyrannidae). *Behaviour* 33:283–322.

Smith, W.J. 1977. *The Behavior of Communicating: An Ethological Approach.* Harvard University Press, Cambridge, MA. 545 pp.

Smith, W.J. 1996. Using interactive playback to study how songs and singing contribute to communication and behavior. In *Ecology and Evolution of Acoustic Communication in Birds*, pp. 377–397. (Edited by D.E. Kroodsma and E.H. Miller). Cornell University Press, Ithaca, NY.

Snow, D.W. 1976. *The Web of Adaptation: Bird Studies in the American Tropics.* Cornell University Press, Ithaca, NY. 176 pp.

Sorenson, M.D. 1998. Patterns of parasitic egg laying and typical nesting in Redhead and Canvasback ducks. In *Parasitic Birds and their Hosts*, pp. 357–375. (Edited by S.I. Rothstein, and S.K. Robinson). Oxford University Press, New York.

Southern, W.E. 1958. Nesting of the Red-eyed Vireo in the Douglas Lake region, Michigan, Pt. 2. *Jack-Pine Warbler* 36:185–207.

Spector, D.A. 1992. Wood-warbler song systems. A review of paruline singing behaviors. In *Current Ornithology*, Vol. 9, pp. 199–238. (Edited by D.M. Power). Plenum Press, New York.

Spector, D.A. 1994. Definition in biology: the case of "bird song." *Journal of Theoretical Biology* 168:373–381.

Speiss, F. 1928. *Die 'Meteor'-Fahrt.* Berlin.

Staebler, A.E. 1941. The number of feathers in the English Sparrow. *Wilson Bulletin* 53:126–127.

Staicer, C.A., D.A. Spector, and A.G. Horn. 1996. The dawn chorus and other diel patterns in acoustic signaling. In *Ecology and Evolution of Acoustic Communication in Birds*, pp. 426–453. (Edited by D.E. Kroodsma and E.H. Miller). Cornell University Press, Ithaca, NY.

Starck, J.M. and R.E. Ricklefs. 1998. Patterns of development: the altricial-precocial spectrum. In *Avian Growth and Development: Evolution within the Altricial-Precocial Spectrum*, pp. 3–30. (Edited by J.M. Starck, and R.E. Ricklefs). Oxford University Press, Oxford.

Steadman, D.W. 1995. Prehistoric extinctions of Pacific Island birds: biodiversity meets zooarchaeology. *Science* 267:1123–1131.

Stein, B.A., L.S. Kutner, G.A. Hammerson, L.L. Master, and L.E. Morse. 2000. State of the states: geographic patterns of diversity, rarity, and endemism. In *Precious Heritage: The Status of Biodiversity in the United States*, pp. 119–157. (Edited by B.A. Stein, L.S. Kutner, and J.S. Adams). Oxford University Press, Oxford.

Stettenheim, P. 1974. The bristles of birds. *Living Bird* 12:201–234.

Stewart, R.E. 1953. A life history study of the Yellowthroat. *Wilson Bulletin* 65:99–115.

Stith, B.M., J.W. Fitzpatrick, G.E. Woolfendon, and B. Pranty. 1996. Classification and conservation of metapopulations: a case study of the Florida Scrub-Jay. In *Metapopulations and Wildlife Conservation*, pp. 187–215. (Edited by D.R. McCullough). Island Press, Washington, D.C.

Stoddard, P.K. 1996. Vocal recognition of neighbors by territorial passerines. In *Ecology and Evolution of Acoustic Communication in Birds*, pp. 356–374. (Edited by D.E. Kroodsma and E.H. Miller). Cornell University Press, Ithaca, NY.

Stokes, D.W. and L.Q. Stokes. 1989. *A Guide to Bird Behavior, Vol. III.* Stokes Nature Guides. Little, Brown and Company, Boston, MA. 397 pp.

Stoleson, S.H. and S.R. Beissinger. 1995. Hatching asynchrony and the onset of incubation in birds, revisited: when is the critical period? In *Current Ornithology*, Vol. 12, pp. 191–270. (Edited by D.M. Power). Plenum Press, New York.

Streseman, E. 1975. *Ornithology, From Aristotle to the Present.* Harvard University Press, Cambridge, MA. 432 pp.

Stresemann, E. 1927–1934. *Kükenthal Handbuch der Zoologie. Sauropsida: Aves. Band VII, Part 2.* Edited by W. Krumbach and T. Krumbach. W. de Gruyter and Co., Berlin and Leipzig. 899 pp.

Suthers, R.A. and F. Goller. 1997. Motor Correlates of Vocal Diversity in Songbirds. In *Current Ornithology*, Vol. 14, pp. 235–288. (Edited by J. V. Nolan, E.D. Ketterson, and C.F. Thompson). Plenum Press, New York.

Tarsitano, S. 1985. The morphological and

aerodynamic constraints on the origin of avian flight. In *The Beginnings of Birds*, pp. 319–332. (Edited by M.K. Hecht, J. Ostrom, G. Viohl, and P. Wellnhofer). Freunde des Jura-Museum, Eichstätt.

Temple, S.A. 1986. Predicting impacts of habitat fragmentation on forest birds: a comparison of two models. In *Wildlife 2000; Modeling Habitat Relationships of Terrestrial Vertebrates*, pp. 301–304. (Edited by J. Verner, M.L. Morrison, and C.J. Ralph). University of Wisconsin Press, Madison, WI.

Temple, S.A. 1990a. Patterns of abundance and diversity of birds on the Apostle Islands. *The Passenger Pigeon* 52(3):219–223.

Temple, S.A. 1990b. How long do birds live? *The Passenger Pigeon* 52(3):255–257.

Temple, S.A. and M.P. Wallace. 1989. Survivorship patterns in a population of Andean Condors *Vultur gryphus*. In *Raptors of the Modern World, Proceedings of the III World Conference on Birds of Prey and Owls, Eilat, Israel, 22–27 March 1987*, pp. 247–251. World Working Group on Birds of Prey and Owls; Berlin, London, and Paris.

Terborgh, J. 1989. *Where Have All the Birds Gone?* Princeton University Press, Princeton, NJ.

Terborgh, J., S.K. Robinson, T.A. Parker, III, C.A. Munn, and N. Pierpont. 1990. Structure and organization of an Amazonian forest bird community. *Ecological Monographs* 60:213–238.

Terres, J.K. 1980. *The Audubon Society Encyclopedia of North American Birds*. Alfred A. Knopf, New York. 1109 pp.

Thayer, G.H. 1909. *Concealing-coloration in the Animal Kingdom*. Macmillan, New York.

Thomas, A.L.R. and A. Balmford. 1995. How natural selection shapes birds' tails. *American Naturalist* 146:848–868.

Thorpe, W.H. 1966a. Ritualization in ontogeny. I. Animal play. *Philosophical Transactions of the Royal Society of London (Bio)* 251:311–319.

Thorpe, W.H. 1966b. Ritualization in ontogeny. II. Ritualization in the development of bird song. *Philosophical Transactions of the Royal Society of London (Bio)* 251:351–358.

Tilghman, N.G. 1987. Characteristics of urban woodlands affecting breeding bird diversity and abundance. *Landscape and Urban Planning* 14:481–495.

Tinbergen, J.M. 1987. Costs of reproduction in the Great Tit: intraseasonal costs associated with brood size. *Ardea* 75:111–122.

Tinbergen, N. 1951. *The Study of Instinct*. Oxford University Press, Oxford.

Todt, D. and H. Hultsch. 1996. Acquisition and performance of song repertoires: ways of coping with diversity and versatility. In *Ecology and Evolution of Acoustic Communication in Birds*, pp. 79–96. (Edited by D.E. Kroodsma and E.H. Miller). Cornell University Press, Ithaca, NY.

Torrey, B. 1885. *Birds in the Bush*. Houghton, Mifflin and Company, Boston, MA. 300 pp.

Torrey, B. 1892. *The Foot-Path Way*. Houghton, Mifflin and Company, Boston, MA.

Traylor, M.A. and J.W. Fitzpatrick. 1982. A survey of tyrant flycatchers. *Living Bird, Nineteenth Annual* 1980–1981:7–50.

Tucker, V.A. 1995. Drag reduction by wing tip slots in a gliding Harris' Hawk, *Parabuteo unicinctus*. *Journal of Experimental Biology* 198:775–781.

Turk, J. and A. Turk. 1984. *Environmental Science, 3rd Edition*. Saunders College Publishing, Philadelphia, PA. 544 pp.

Turner, G.E.S. 1985. Birds in folklore. In *A Dictionary of Birds*, pp. 233–234. (Edited by B. Campbell and E. Lack). Buteo Books, Vermillion, SD.

Turner, S. 1997. On the thermal capacity of a bird's egg warmed by a brood patch. *Physiological Zoology* 70(4):470–480.

Tuttle, E.M. 1993. *Mate Choice and Stable Polymorphism in the White-throated Sparrow*. Ph.D. Dissertation, State University of New York at Albany.

Tyler, W.N. 1916. The call-notes of some nocturnal migrating birds. *Auk* 33:132–141.

Ueda, K. 1985. Juvenile female breeding of the Fan-tailed Warbler *Cisticola juncidis*: occurrence of two generations in the year. *Ibis* 127:111–116.

Valera, F., H. Hoi, and B. Schleicher. 1997. Egg burial in penduline tits, *Remiz pendulinus*: its role in mate desertion and female polyandry. *Behavioral Ecology* 8:20–27.

Valli, E. and D. Summers. 1990. Nest gatherers of Tiger Cove. *National Geographic* 177(1):106–133.

Van Tyne, J. and A.J. Berger. 1959. *Fundamentals of Ornithology*. Wiley, New York. 624 pp.

Verbeek, N.A.M. 1973. *The Exploitation System of the Yellow-billed Magpie*. University of California Publications in Zoology, Vol. 99. University of California Press, Berkeley, CA. 58 pp.

Verner, J. 1976. Complex song repertoire of male Long-billed Marsh Wrens in eastern Washington. *The Living Bird Quarterly* 14:263–300.

Verner, J. and G.H. Engelsen. 1970. Territories, multiple nest building, and polygyny in the Long-billed Marsh Wren. *Auk* 87:557–567.

Verner, J. and M.F. Willson. 1966. The influence of habitats on mating systems of North American passerine birds. *Ecology* 47:143–147.

Vieillot, L. 1807. *L'Histoire Naturelle des Oiseaux de L'Amerique Septentrionale*. Chez Desray, Paris.

Vitala, J., E. Korpimaki, P. Palokangas, and M. Kolvula. 1995. Attraction of kestrels to vole scent marks visible in ultraviolet light. *Nature* 373:425–427.

von Haartman, L. 1953. Was reizt den Trauerfliegenschnäpper (*Muscicapa hypoleuca*) zu füttern? *Vogelwarte* 16:157–164.

von Haartman, L. 1971. Population Dynamics. In *Avian Biology*, Vol. 1, pp. 391–459. (Edited by D.S. Farner, and J.R. King). Academic Press, New York.

von Haartman, L. 1982. Two modes of clutch size determination in passerine birds. *Journal of the Yamashina Institute for Ornithology* 14:214–219.

Vuilleumier, B.S. 1971. Pleistocene changes in the fauna and flora of South America. *Science* 173(3999):771–780.

Walcott, C. 1996. Pigeon homing: observations, experiments and confusion. *Journal of Experimental Biology* 199:21–27.

Waldvogel, J. 1983. Olfactory orientation by birds. In *Current Ornithology*, Vol. 6, pp. 269–321. (Edited by D.M. Power). Plenum Press, New York.

Walker, M.M., C.E. Diebel, C.V. Haugh, P.M. Pankhurst, J.C. Montgomery, and C.R. Green. 1997. Structure and function of the vertebrate magnetic sense. *Nature* 390:371–376.

Walkinshaw, L.H. 1961. The effect of parasitism by the Brown-headed Cowbird on *Empidonax* flycatchers in Michigan. *Auk* 78:266–268.

Wallraff, H.G. 1980. Does pigeon homing depend on stimuli perceived during displacement? I. Experiments in Germany. *Journal of Comparative Physiology* 139:193–201.

Wallraff, H.G. 1996. Seven theses on pigeon homing deduced from empirical findings. *Journal of Experimental Biology* 199:105–111.

Walsberg, G.E. 1977. *Ecology and Energetics of Contrasting Social Systems in* Phainopepla nitens *(Aves: Ptilogonatidae)*. University of California Publications in Zoology, Vol. 108. University of California Press, Berkeley, CA. 63 pp.

Walter, H. 1979. *Eleonora's Falcon: Adaptations to Prey and Habitat in a Social Raptor*. University of Chicago Press, Chicago.

Wanker, R., B.L. Cruz, and D. Franck. 1996. Socialization of Spectacled Parrotlets (*Forpus conspicillatus*): the role of parents, creches, and sibling groups in nature. *Journal für Ornithologie* 137(4):447–461.

Warner, R.E. 1994. Agricultural land use and grassland habitat in Illinois: future shock for midwestern birds? *Conservation Biology* 8:147–156.

Waterman, T.H. 1989. *Animal Navigation*. Scientific American Library, W.H. Freeman, New York. 243 pp.

Watson, D.M.S. 1951. *Paleontology and Modern Biology*. Yale University Press, New Haven, CT.

Watson, J.C., D.L. Carlson, W.E. Taylor, and T.E. Milling. 1995. Restoration of the Red-cockaded Woodpecker population in the Francis Marion National Forest: three years post Hugo. In *Red-cockaded Woodpecker: Recovery, Ecology and Management*, pp. 172–182. (Edited by D.L. Kulhavy, R.G. Hooper, and R. Costa). Stephen F. Austin State University, Nacogdoches, TX.

Weathers, W.W., R.S. Seymour, and R.V. Baudinette. 1993. Energetics of mound-tending behaviour in the Malleefowl, *Leipoa ocellata* (Megapodiidae). *Animal Behaviour* 45:333–341.

Weathers, W.W. and R.B. Siegel. 1995. Body size establishes the scaling of avian postnatal metabolic rate: an interspecific analysis using phylogenetically independent contrasts. *Ibis* 137:532–542.

Webster, M. 1994. Female-defense polygyny in a Neotropical bird, the Montezuma Oropendola. *Animal Behaviour* 48(4):779–794.

Weeden, J.S. and J.B. Falls. 1959. Differential

responses of male Ovenbirds to recorded songs of neighboring and more distant individuals. *Auk* 76:343–351.

Weeks, H.P., Jr. 1979. Nesting ecology of the Eastern Phoebe in southern Indiana. *Wilson Bulletin* 91:441–454.

Weid, M.P. 1833. *Beitrage zur Naturgeschichte von Brasilien.* Vol. 4, Part 2:827–828.

Wells, J.V., K.V. Rosenberg, D.L. Tessaglia, and A.A. Dhondt. 1996. Population cycles in the Varied Thrush (*Ixoreus naevius*). *Canadian Journal of Zoology* 74:2062–2069.

Welty, J.C. and L. Baptista. 1988. *The Life of Birds, 4th Edition.* Harcourt Brace Jovanovich College Publishers, New York. 581 pp.

West, M. and A. King. 1996. Eco-gen-actics: a systems approach to the ontogeny of avian communication. In *Ecology and Evolution of Acoustic Communication in Birds,* pp. 20–38. (Edited by D.E. Kroodsma and E.H. Miller). Cornell University Press, Ithaca, NY.

Westemeier, R.L., S.A. Simpson, and D.A. Cooper. 1991. Successful exchange of prairie-chicken eggs between nests in two remnant populations. *Wilson Bulletin* 103:717–720.

Wetmore, A. 1936. The number of contour feathers in passeriform and related birds. *Auk* 53:159–169.

Whitcomb, R.F., C.S. Robbins, J.F. Lynch, B.L. Whitcomb, M.K. Klimkiewicz, and D. Bystrak. 1981. Effects of forest fragmentation on avifauna of the eastern deciduous forest. In *Forest Island Dynamics in Man-dominated Landscapes,* pp. 125–206. (Edited by R.L. Burgess and B.M. Sharpe). Springer-Verlag, New York.

White, F.N., G.A. Bartholomew, and T.R. Howell. 1975. The thermal significance of the nest of the Sociable Weaver *Philetairus socius:* winter observations. *Ibis* 117:171–179.

White, H.C. 1939. Change in gastric digestion of kingfishers with development. *American Naturalist* 73:188–190.

Whitfield, D.P. 1986. Plumage variability and territoriality in breeding turnstone *Arenaria interpres:* status signalling or individual recognition? *Animal Behaviour* 34:1471–1482.

Whitney, G.G. 1994. *From Coastal Wilderness to Fruited Plain: A History of Environmental Change in Temperate North America, 1500 to the Present.* Cambridge University Press, Cambridge, England and New York, NY. 451 pp.

Whittick, A. 1971. *Symbols: Signs and Their Meaning and Uses in Design, 2nd Edition.* L. Hill, London. 383 pp.

Whittow, G.C. 1993. Laysan Albatross (*Diomedea immutabilis*). In *The Birds of North America,* No. 66. (Edited by A. Poole, and F. Gill). The Academy of Natural Sciences, Philadelphia, PA and The American Ornithologists' Union, Washington, D.C.

Wilcove, D.S., C.H. McClellan, and A.P. Dobson. 1986. Habitat fragmentation in the temperate zone. In *Conservation Biology: The Science of Scarcity and Diversity,* pp. 237–256. (Edited by M.E. Soulé). Sinauer Associates, Inc., Sunderland, MA.

Wiley, R.H. and D.G. Richards. 1982. Adapta-

tions for acoustic communication in birds: sound transmission and signal detection. In *Acoustic Communication in Birds,* Vol. 1, pp. 131–181. (Edited by D.E. Kroodsma and E.H. Miller). Cornell University Press, Ithaca, NY.

Williams, G.C. 1966. *Adaptation and Natural Selection; A Critique of Some Current Evolutionary Thought.* Princeton University Press, Princeton, NJ. 307 pp.

Williams, T.C., J.M. Williams, L.C. Ireland, and J.M. Teal. 1977. Autumnal bird migration over the western North Atlantic Ocean. *American Birds* 31:251–267.

Williams, T.D. 1994. Adoption in a precocial species, the Lesser Snow Goose: intergenerational conflict, altruism or a mutually beneficial strategy? *Animal Behaviour* 47:101–107.

Williams, T.D. 1995. *The Penguins.* Oxford University Press, Oxford. 295 pp.

Willis, E.O. and Y. Oniki. 1978. Birds and army ants. *Annual Review of Ecology and Systematics* 9:243–263.

Williston, S.W. 1879. Are birds derived from dinosaurs? *Kansas City Review of Science* 3:457–460.

Willughby, F. 1678. *The Ornithology of Francis Willughby.* Printed by A.C. for John Martyn, London.

Wilmore, S.B. 1979. *Swans of the World.* Taplinger Publishing Company, Inc., New York. 229 pp.

Wilson, A. 1808–1814. *American Ornithology; or, The Natural History of the Birds of the United States.* 9 Volumes. Published by Bradford and Inskeep, printed by Robert Carr, Philadelphia, PA.

Wilson, B.W. 1980. *Birds: Readings from Scientific American.* W.H. Freeman, New York. 276 pp.

Wiltschko, R. 1996. The function of olfactory input in pigeon orientation: does it provide navigational information or play another role? *Journal of Experimental Biology* 199:113–119.

Wiltschko, R., M. Schöps, and U. Kowalski. 1989. Pigeon homing: wind exposition determines the importance of olfactory input. *Naturwissenschaften* 76:229–231.

Wiltschko, R. and W. Wiltschko. 1995. *Magnetic Orientation in Animals.* Springer-Verlag, Berlin. 297 pp.

Wiltschko, W. 1968. Über den einfluss statischer Magnetfelder auf die Zugorientierung der Rotkehlchen (*Erithacus rubecula*). *Zeitschrift für Tierpsychologie* 25:537–558.

Wiltschko, W. and R. Wiltschko. 1972. Magnetic compass of European Robins. *Science* 176:62–64.

Wiltschko, W. and R. Wiltschko. 1991. Magnetic orientation and celestial cues in migratory orientation. In *Orientation in Birds,* pp. 16–37. (Edited by P. Berthold). Springer, Berlin.

Wiltschko, W. and R. Wiltschko. 1996. Magnetic orientation in birds. *Journal of Experimental Biology* 199:29–38.

Wimberger, P.H. 1984. The use of green plant material in bird nests to avoid ectoparasites.

Auk 101(3):615–618.

Winkler, D.W. 1985. Factors determining a clutch size reduction in California Gulls *Larus californicus:* a multi-hypothesis approach. *Evolution* 39(3):667–677.

Winkler, D.W. 1991. Parental investment decision rules in Tree Swallows: parental defense, abandonment, and the so-called Concorde Fallacy. *Behavioral Ecology* 2(2):133–142.

Winkler, D.W. 1992. Causes and consequences of variation in parental defense behavior by Tree Swallows. *Condor* 94(2):502–520.

Winkler, D.W. 1993. Use and importance of feathers as nest lining in Tree Swallows (*Tachycineta bicolor*). *Auk* 110(1):29–36.

Winkler, D.W. and P.E. Allen. 1996. The seasonal decline in Tree Swallow clutch size: physiological constraint or strategic adjustment? *Ecology* 77:922–932.

Winkler, D.W. and F.H. Sheldon. 1993. Evolution of nest construction in swallows (Hirundinidae): a molecular phylogenetic perspective. *Proceedings of the National Academy of Sciences* 90(12):5705–5707.

Winkler, D.W. and J.R. Walters. 1983. The determination of clutch size in precocial birds. In *Current Ornithology,* Vol. 1, pp. 33–68. (Edited by R.F. Johnston). Plenum Press, New York and London.

Winkler, D.W. and G.W. Wilkinson. 1988. Parental effort in birds and mammals: theory and measurement. In *Oxford Surveys in Evolutionary Biology,* Vol. 5, pp. 185–214. (Edited by P. Harvey and L. Partridge). Oxford University Press, Oxford and New York.

Winnett-Murray, K. 1979. The influence of cover on offspring survival in Western Gulls. *Proceedings of the Colonial Waterbird Group* 3:33–43.

Withers, P.C. 1979. Aerodynamics and hydrodynamics of the "hovering" flight of Wilson's Storm-Petrel. *Journal of Experimental Biology* 80:83–91.

Witmer, L.M. 1991. Perspectives on avian origins. In *Origins of the Higher Groups of Tetrapods: Controversy and Consensus,* pp. 427–446. (Edited by H.P. Schultz, and L. Trueb). Cornell University Press, Ithaca, NY.

Witmer, L.M. and K.D. Rose. 1991. Diatryma: implications for diet and mode of life. *Paleobiology* 17:95–120.

Witmer, M.C., D.J. Mountjoy, and L. Elliott. 1997. Cedar Waxwing (*Bombycilla cedrorum*). In *The Birds of North America,* No. 309. (Edited by A. Poole and F. Gill). The Academy of Natural Sciences, Philadelphia, PA and The American Ornithologists' Union, Washington, DC.

Wolf, B.O. and G.E. Walsberg. 2000. The role of the plumage in heat transfer processes of birds. *American Zoologist* 40:575–584.

Wolf, L.L., F.G. Stiles, and F.R. Hainsworth. 1976. Ecological organization of a tropical, highland hummingbird community. *Journal of Animal Ecology* 45:349–379.

Woolfenden, G.E. and J.W. Fitzpatrick. 1984. *The Florida Scrub Jay: Demography of a Cooperative-Breeding Bird.* Princeton University Press, Princeton, NJ. 406 pp.

Wootton, J.T., B.E. Young, and D.W. Winkler. 1991. Ecological versus evolutionary hypotheses: demographic stasis and the Murray-Nolan clutch size equation. *Evolution* 45:1947–1950.

Wright, M.O. 1897. *Citizen Bird: Scenes from Bird-life in Plain English for Beginners.* The Macmillan Company, New York. 430 pp.

Wynne-Edwards, V.C. 1962. *Animal Dispersion in Relation to Social Behaviour.* Oliver and Boyd, Edinburgh. 653 pp.

Yalden, D.W. 1985. Forelimb function in *Archaeopteryx.* In *The Beginnings of Birds*, pp. 91–97. (Edited by M.K. Hecht, J. Ostrom, G. Viohl, and P. Wellnhofer). Freunde des Jura-Museum, Eichstätt.

Yasukawa, K. and W.A. Searcy. 1995. Red-winged Blackbird *(Agelaius phoeniceus).* In *The Birds of North America*, No. 184. (Edited by A. Poole and F. Gill). The Academy of Natural Sciences, Philadelphia, PA and The American Ornithologists' Union, Washington, D.C.

Yom-Tov, Y. and A. Ar. 1980 (1981). Breeding ecology of the Dead Sea Israel Sparrow *Passer moabiticus. Israel Journal of Zoology* 29(4):171–187.

Young, B.E. 1994. Geographic and seasonal patterns of clutch-size variation in House Wrens. *Auk* 111:545–555.

Young, B.E., M. Kaspari, and T.E. Martin. 1990. Species-specific nest site selection by birds in ant-acacia trees. *Biotropica* 22:310–315.

Zach, R. 1979. Shell dropping: decision-making and optimal foraging in Northwestern Crows. *Behaviour* 68(1–2):106–117.

Index

Note: Page numbers in *italics* indicate figures; those in **bold** indicate definitions; those followed by a "t" indicate tables; those followed by an "s" indicate sonagrams. "Track" refers to the Track on the CD that accompanies Chapter 7. "H" indicates pages in the section *Birds and Humans;* "E" indicates pages in the section *Evolution.* When a topic is mentioned both in a figure and in the text on the same page, generally only the more extensive entry is indexed.

Please note that this index is also available on the Internet at <www.birds.cornell.edu/homestudy>.

R

T